Atomic Numbers and Atomic Masses of the Elements

Based on $^{12}_{6}C$. Numbers in parentheses are the mass numbers of the most stable isotopes of radioactive elements.

Element	*Symbol*	*Atomic Number*	*Atomic Weight*
Actinium	Ac	89	(227)
Aluminum	Al	13	26.98
Americium	Am	95	(243)
Antimony	Sb	51	121.76
Argon	Ar	18	39.95
Arsenic	As	33	74.92
Astatine	At	85	(210)
Barium	Ba	56	137.33
Berkelium	Bk	97	(247)
Beryllium	Be	4	9.01
Bismuth	Bi	83	208.98
Bohrium	Bh	107	(264)
Boron	B	5	10.81
Bromine	Br	35	79.90
Cadmium	Cd	48	112.41
Calcium	Ca	20	40.08
Californium	Cf	98	(251)
Carbon	C	6	12.01
Cerium	Ce	58	140.12
Cesium	Cs	55	132.91
Chlorine	Cl	17	35.45
Chromium	Cr	24	52.00
Cobalt	Co	27	58.93
Copper	Cu	29	63.55
Curium	Cm	96	(248)
Dubnium	Db	105	(262)
Dysprosium	Dy	66	162.50
Einstenium	Es	99	(252)
Erbium	Er	68	167.26
Europium	Eu	63	151.96
Fermium	Fm	100	(257)
Fluorine	F	9	19.00
Francium	Fr	87	(223)
Gadolinium	Gd	64	157.25
Gallium	Ga	31	69.72
Germanium	Ge	32	72.59
Gold	Au	79	196.97
Hafnium	Hf	72	178.49
Hassium	Hs	108	(269)
Helium	He	2	4.00
Holmium	Ho	67	164.93
Hydrogen	H	1	1.01
Indium	In	49	114.82
Iodine	I	53	126.90
Iridium	Ir	77	192.22
Iron	Fe	26	55.85
Krypton	Kr	36	83.80
Lanthanum	La	57	138.91
Lawrencium	Lr	103	(262)
Lead	Pb	82	207.2
Lithium	Li	3	6.94
Lutetium	Lu	71	174.97
Magnesium	Mg	12	24.30
Manganese	Mn	25	54.94
Meitnerium	Mt	109	(268)
Mendelevium	Md	101	(260)
Mercury	Hg	80	200.59
Molybdenum	Mo	42	95.94
Neodymium	Nd	60	144.24
Neon	Ne	10	20.18
Neptunium	Np	93	(237)
Nickel	Ni	28	58.69
Niobium	Nb	41	92.91
Nitrogen	N	7	14.01
Nobelium	No	102	(259)
Osmium	Os	76	190.23
Oxygen	O	8	16.00
Palladium	Pd	46	106.42
Phosphorus	P	15	30.97
Platinum	Pt	78	195.08
Plutonium	Pu	94	(242)
Polonium	Po	84	(209)
Potassium	K	19	39.10
Praseodymium	Pr	59	140.91
Promethium	Pm	61	(145)
Protactinium	Pa	91	(231)
Radium	Ra	88	(226)
Radon	Rn	86	(222)
Rhenium	Re	75	186.21
Rhodium	Rh	45	102.91
Rubidium	Rb	37	85.47
Ruthenium	Ru	44	101.07
Rutherfordium	Rf	104	(261)
Samarium	Sm	62	150.36
Scandium	Sc	21	44.96
Seaborgium	Sg	106	(266)
Selenium	Se	34	78.96
Silicon	Si	14	28.09
Silver	Ag	47	107.87
Sodium	Na	11	22.99
Strontium	Sr	38	87.62
Sulfur	S	16	32.07
Tantalum	Ta	73	180.95
Technetium	Tc	43	(98)
Tellurium	Te	52	127.60
Terbium	Tb	65	158.93
Thallium	Tl	81	204.38
Thorium	Th	90	(232)
Thulium	Tm	69	168.93
Tin	Sn	50	118.71
Titanium	Ti	22	47.87
Tungsten	W	74	183.84
Uranium	U	92	(238)
Vanadium	V	23	50.94
Xenon	Xe	54	131.29
Ytterbium	Yb	70	173.04
Yttrium	T	39	88.91
Zinc	Zn	30	65.38
Zirconium	Zr	40	91.22

SECOND EDITION

General, Organic, and Biological Chemistry

H. Stephen Stoker
Weber State University

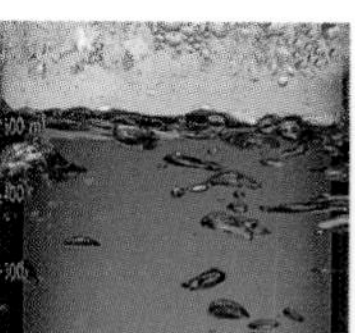

Sharon K. Stoffels, RNC, MSN
Project Consultant
Boise State University

Editor-in-Chief: Kathi Prancan
Senior Sponsoring Editor: Richard Stratton
Senior Development Editor: Karla Paschkis
Editorial Associate: Sarah Gessner
Assistant Editor: Sara Wise
Internet Producer: Pamela Bachorz
Senior Project Editor: Nancy Blodget
Editorial Assistant: Elisabeth Kehrer
Senior Production/Design Coordinator: Jill Haber
Senior Manufacturing Coordinator: Sally Culler
Executive Marketing Manager: Andy Fisher

For permission to use photographs, grateful acknowledgment is made to the copyright holders listed under Photo Credits on page A-46, hereby considered an extension of this copyright page.

Cover design: Diana Coe / ko Design Studio
Cover photograph: © Mark A. Johnson/The Stock Market

Printed in the U.S.A.

Library of Congress Catalog Card Numbers:
General, Organic, and Biological Chemistry: 00-104439
Organic and Biological Chemistry: 00-104440

International Standard Book Numbers:
General, Organic, and Biological Chemistry: 0-618-05206-2
Organic and Biological Chemistry: 0-618-05207-0

2 3 4 5 6 7 8 9—VH—04 03 02 01

Brief Contents

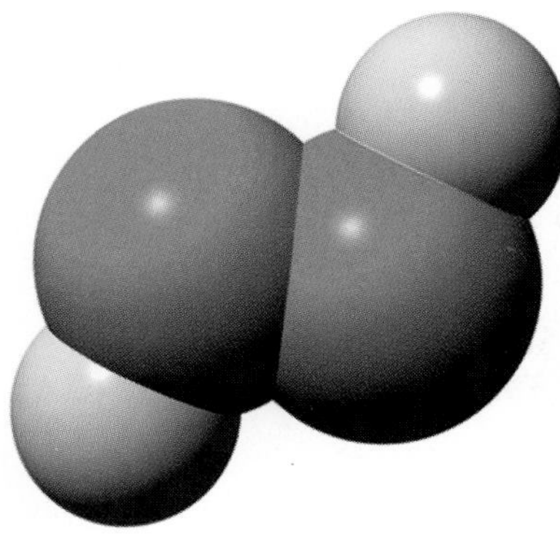

PART I General Chemistry

1 Basic Concepts About Matter 1

2 Measurements in Chemistry 21

3 Atomic Structure and the Periodic Table 48

4 Chemical Bonding: The Ionic Bond Model 79

5 Chemical Bonding: The Covalent Bond Model 102

6 Chemical Calculations: Formula Masses, Moles, and Chemical Equations 127

7 Gases, Liquids, and Solids 150

8 Solutions 179

9 Chemical Reactions 208

10 Acids, Bases, and Salts 237

11 Nuclear Chemistry 271

PART II Organic Chemistry

12 Saturated Hydrocarbons 298

13 Unsaturated Hydrocarbons 332

14 Alcohols, Phenols, and Ethers 371

15 Aldehydes and Ketones 406

16 Carboxylic Acids and Esters 436

17 Amines and Amides 471

PART III Biological Chemistry

18 Carbohydrates 506

19 Lipids 554

20 Proteins 587

21 Enzymes, Vitamins, and Minerals 620

22 Nucleic Acids 652

23 Biochemical Energy Production 688

24 Carbohydrate Metabolism 715

25 Lipid Metabolism 743

26 Protein Metabolism 769

Answers to Selected Exercises A-1

Glossary/Index A-24

Photo Credits A-46

Contents

Preface xiii

PART I General Chemistry

Chapter 1 Basic Concepts About Matter 1

1.1 Chemistry—The Study of Matter 1
1.2 Physical States of Matter 2
1.3 Properties of Matter 3
1.4 Changes in Matter 4
• **Chemistry at a Glance:**
Use of the Terms Physical *and* Chemical *5*
1.5 Pure Substances and Mixtures 6
1.6 Elements and Compounds 7
• **Chemistry at a Glance:**
Classes and Properties of Matter 9
1.7 Discovery and Abundance of the Elements 9
1.8 Names and Symbols of the Elements 11
1.9 Atoms and Molecules 13
1.10 Chemical Formulas 14
• **Chemical Connections**
1.1 "Good" versus "Bad" Properties for a Chemical Substance 4
1.2 Elements Necessary for Human Life 11
Concepts to Remember 15
Key Terms 16
Exercises and Problems 16
Additional Problems 19
Grid Problems 19

Chapter 2 Measurements in Chemistry 21

2.1 Measurement Systems 21
2.2 Metric System Units 22
2.3 Uncertainty in Measurement and Significant Figures 24
2.4 Significant Figures and Mathematical Operations 26
2.5 Scientific Notation 28
• **Chemistry at a Glance:**
Important Concepts Associated with Measurement 29
2.6 Conversion Factors and Dimensional Analysis 32
• **Chemistry at a Glance:**
Conversion Factors 36
2.7 Density 36
2.8 Temperature and Heat Energy 39
• **Chemical Connections**
2.1 Body Density and Percent Body Fat 38
2.2 Normal Human Body Temperature 42
Concepts to Remember 43
Key Reactions and Equations 43
Key Terms 44
Exercises and Problems 44
Additional Problems 46
Grid Problems 46

Chapter 3 Atomic Structure and the Periodic Table 48

3.1 Internal Structure of the Atom 48
3.2 Experimental Evidence for the Existence of Subatomic Particles 50
3.3 Atomic Number and Mass Number 52
• **Chemistry at a Glance:**
Atomic Structure 56
3.4 The Periodic Law and the Periodic Table 57
3.5 Metals and Nonmetals 59
3.6 Electron Arrangements Within Atoms 61
3.7 Specification of Electronic Structure 65
3.8 The Electronic Basis for the Periodic Law and the Periodic Table 69
3.9 Classification of the Elements 71
• **Chemistry at a Glance:**
Element Classification Schemes and the Periodic Table 72
• **Chemical Connections**
3.1 Protium, Deuterium, and Tritium—The Three Isotopes of Hydrogen 54
3.2 Calcium, Your Body, and the Periodic Law 60
3.3 Importance of Metallic and Nonmetallic Trace Elements for Human Health 62
3.4 Electrons in Excited States 67
Concepts to Remember 73
Key Reactions and Equations 73
Key Terms 74
Exercises and Problems 74
Additional Problems 77
Grid Problems 77

Chapter 4 Chemical Bonding: The Ionic Bond Model 79

4.1 Chemical Bonds 79
4.2 Valence Electrons and Lewis Structures 80
4.3 The Octet Rule 82
4.4 The Ionic Bond Model 83

4.5 Determination of Ionic Charge Magnitude 84
4.6 Ionic Compound Formation 86
4.7 Formulas for Ionic Compounds 87
4.8 The Structure of Ionic Compounds 88
4.9 Naming Binary Ionic Compounds 90
• **Chemistry at a Glance:**
Ionic Bonds and Ionic Compounds 93
4.10 Polyatomic Ions 94
4.11 Formulas and Names for Ionic Compounds Containing Polyatomic Ions 95
• **Chemistry at a Glance:**
Nomenclature of Ionic Compounds 98
• **Chemical Connections**
4.1 Fresh Water, Seawater, Hard Water, and Soft Water: A Matter of Ions 85
4.2 Sodium Chloride: The World's Most Common Food Additive 89
Concepts to Remember 97
Key Reactions and Equations 98
Key Terms 98
Exercises and Problems 98
Additional Problems 100
Grid Problems 100

Chapter 5 Chemical Bonding: The Covalent Bond Model 102

5.1 The Covalent Bond Model 102
5.2 Lewis Structures for Molecular Compounds 103
5.3 Single, Double, and Triple Covalent Bonds 104
5.4 Valence Electrons and Number of Covalent Bonds Formed 105
5.5 Coordinate Covalent Bonds 106
5.6 Systematic Procedures for Drawing Lewis Structures 106
5.7 Bonding in Compounds with Polyatomic Ions Present 110
5.8 The Shapes of Molecules 111
5.9 Electronegativity 114
• **Chemistry at a Glance:**
The Shape (Geometry) of Molecules 115
5.10 Bond Polarity 116
• **Chemistry at a Glance:**
Covalent Bonds and Molecular Compounds 118
5.11 Molecular Polarity 119
5.12 Naming Binary Molecular Compounds 120
• **Chemical Connections**
5.1 Nitric Oxide: A Molecule Whose Bonding Does Not Follow "The Rules" 109
5.2 Molecular Shape and Odor 113
Concepts to Remember 121
Key Reactions and Equations 122
Key Terms 122
Exercises and Problems 122
Additional Problems 125
Grid Problems 125

Chapter 6 Chemical Calculations: Formula Masses, Moles, and Chemical Equations 127

6.1 Formula Masses 127
6.2 The Mole: A Counting Unit for Chemists 129
6.3 The Mass of a Mole 130
6.4 The Mole and Chemical Formulas 132
6.5 The Mole and Chemical Calculations 133
6.6 Writing and Balancing Chemical Equations 136
6.7 Chemical Equations and the Mole Concept 140
6.8 Chemical Calculations Using Chemical Equations 140
• **Chemistry at a Glance:**
Relationships Involving the Mole Concept 141
• **Chemical Connections**
6.1 "Laboratory-Sized Amounts" versus "Industrial Chemistry-Sized Amounts" 142
Concepts to Remember 145
Key Reactions and Equations 146
Key Terms 146
Exercises and Problems 146
Additional Problems 148
Grid Problems 149

Chapter 7 Gases, Liquids, and Solids 150

7.1 The Kinetic Molecular Theory of Matter 150
7.2 Kinetic Molecular Theory and Physical States 152
7.3 Gas Law Variables 155
7.4 Boyle's Law: A Pressure–Volume Relationship 155
7.5 Charles's Law: A Temperature–Volume Relationship 157
7.6 The Combined Gas Law 159

7.7 The Ideal Gas Law 160
7.8 Dalton's Law of Partial Pressures 161
• **Chemistry at a Glance:**
The Gas Laws 163
7.9 Changes of State 164
7.10 Evaporation of Liquids 164
7.11 Vapor Pressure of Liquids 165
7.12 Boiling and Boiling Point 166
7.13 Intermolecular Forces in Liquids 168
• **Chemistry at a Glance:**
Intermolecular Forces 172
• **Chemical Connections**
7.1 The Importance of Gas Densities 154
7.2 The Exchange of the Respiratory Gases, Oxygen and Carbon Dioxide 162
Concepts to Remember 173
Key Reactions and Equations 174
Key Terms 174
Exercises and Problems 175
Additional Problems 177
Grid Problems 177

Chapter 8 Solutions 179

8.1 Characteristics of Solutions 179
8.2 Solubility and Saturated Solutions 181
8.3 Solution Formation 183
8.4 Solubility Rules 184
8.5 Solution Concentration Units 186
• **Chemistry at a Glance:**
Solutions and Solution Concentrations 193
8.6 Colloidal Dispersions 194
8.7 Colligative Properties of Solutions 195
8.8 Osmosis and Osmotic Pressure 196
• **Chemistry at a Glance:**
Colligative Properties of Solutions 199
8.9 Dialysis 201
• **Chemical Connections**
8.1 Factors Affecting Gas Solubility 182
8.2 Solubility of Vitamins 186
8.3 Controlled-Release Drugs: Regulating Concentration, Rate, and Location of Release 192
8.4 The Artificial Kidney: A Hemodialysis Machine 202
Concepts to Remember 203
Key Reactions and Equations 203
Key Terms 203
Exercises and Problems 204
Additional Problems 206
Grid Problems 207

Chapter 9 Chemical Reactions 208

9.1 Types of Chemical Reactions 208
9.2 Redox and Nonredox Reactions 211
• **Chemistry at a Glance:**
Types of Chemical Reactions 212
9.3 Terminology Associated with Redox Processes 214
9.4 Collision Theory and Chemical Reactions 216
9.5 Exothermic and Endothermic Reactions 218
9.6 Factors That Influence Reaction Rates 218
• **Chemistry at a Glance:**
Factors That Influence Reaction Rates 221
9.7 Chemical Equilibrium 221
9.8 Equilibrium Constants 224
9.9 Altering Equilibrium Conditions: Le Châtelier's Principle 227
• **Chemical Connections**
9.1 Smog Formation: A Set of Simple Reactions 210
9.2 Stratospheric Ozone: An Equilibrium Situation 223
9.3 Oxygen, Hemoglobin, Equilibrium, and Le Châtelier's Principle 230
Concepts to Remember 231
Key Reactions and Equations 232
Key Terms 232
Exercises and Problems 232
Additional Problems 235
Grid Problems 236

Chapter 10 Acids, Bases, and Salts 237

10.1 Arrhenius Acid–Base Theory 237
10.2 Brønsted–Lowry Acid–Base Theory 238
10.3 Mono-, Di-, and Triprotic Acids 241
10.4 Strengths of Acids and Bases 242
10.5 Ionization Constants for Acids and Bases 243
10.6 Salts 244
10.7 Acid–Base Neutralization Reactions 245
10.8 Self-Ionization of Water 246
10.9 The pH Concept 248
10.10 pK_a Method for Expressing Acid Strength 251
• **Chemistry at a Glance:**
Acids and Acidic Solutions 252
10.11 The pH of Aqueous Salt Solutions 252
10.12 Buffers 256
• **Chemistry at a Glance:**
Buffer Systems 262
10.13 The Henderson–Hasselbalch Equation 259
10.14 Electrolytes 262
10.15 Acid–Base Titrations 264
• **Chemical Connections**
10.1 Acid Rain: Excess Acidity 254
10.2 Blood Plasma pH and Hydrolysis 256
10.3 Buffering Action in Human Blood 260
10.4 Electrolytes and Body Fluids 263
Concepts to Remember 265
Key Reactions and Equations 266
Key Terms 266

Exercises and Problems 267
Additional Problems 270
Grid Problems 270

Chapter 11 Nuclear Chemistry 271

11.1 Stable and Unstable Nuclides 271
11.2 The Nature of Radioactivity 272
11.3 Radioactive Decay 273
11.4 Rate of Radioactive Decay 276
11.5 Transmutation and Bombardment Reactions 279
11.6 Radioactive Decay Series 280
11.7 Ionizing Effects of Radiation 282
• **Chemistry at a Glance:**
Terminology Associated with Nuclear Reactions 283
11.8 Biological Effects of Radiation 283
• **Chemistry at a Glance:**
Ionizing Radiation 285
11.9 Detection of Radiation 285
11.10 Sources of Radiation Exposure 286
11.11 Nuclear Medicine 287
11.12 Nuclear Fission and Nuclear Fusion 290
11.13 Nuclear and Chemical Reactions Compared 293
• **Chemical Connections**
11.1 Tobacco Radioactivity and the Uranium-238 Decay Series 281
11.2 The PET Scan: Seeing the Brain in Action 289
Concepts to Remember 293
Key Reactions and Equations 294
Key Terms 294
Exercises and Problems 294
Additional Problems 296
Grid Problems 296

PART II Organic Chemistry

Chapter 12 Saturated Hydrocarbons 298

12.1 Organic and Inorganic Compounds 298
12.2 Bonding Characteristics of the Carbon Atom 299
12.3 Hydrocarbons and Hydrocarbon Derivatives 300
12.4 Structural Characteristics of Simple Alkanes 300
12.5 Structural Formulas 300
12.6 Structural Isomerism 302
12.7 Conformations of Alkanes 304
12.8 IUPAC Nomenclature for Alkanes 306
12.9 Classification of Carbon Atoms 311
12.10 Branched-Chain Alkyl Groups 311
12.11 Cycloalkanes 312
12.12 IUPAC Nomenclature for Cycloalkanes 313
12.13 Isomerism in Cycloalkanes 314
12.14 Sources of Alkanes and Cycloalkanes 316
12.15 Physical Properties of Alkanes and Cycloalkanes 317
12.16 Chemical Properties of Alkanes and Cycloalkanes 320
12.17 Nomenclature and Properties of Halogenated Alkanes 322
• **Chemistry at a Glance:**
Properties of Alkanes and Cycloalkanes 323
• **Chemical Connections**
12.1 The Occurrence of Methane 302
12.2 The Physiological Effects of Alkanes 319
12.3 Chlorofluorocarbons and the Ozone Layer 324
Concepts to Remember 324
Key Reactions and Equations 325
Key Terms 325
Exercises and Problems 325
Additional Problems 330
Grid Problems 331

Chapter 13 Unsaturated Hydrocarbons 332

13.1 Unsaturated Hydrocarbons 332
13.2 Characteristics of Alkenes and Cycloalkenes 333
13.3 Names for Alkenes and Cycloalkenes 334
13.4 The Nature of Carbon–Carbon Multiple Bonds 337
13.5 Isomerism in Alkenes 339
13.6 Naturally Occurring Alkenes 343
13.7 Physical and Chemical Properties of Alkenes 344
13.8 Polymerization of Alkenes: Addition Polymers 350
13.9 Alkynes 352
• **Chemistry at a Glance:**
Chemical Reactions of Alkenes 353

13.10 Aromatic Hydrocarbons 354
13.11 Names for Aromatic Hydrocarbons 356
13.12 Aromatic Hydrocarbons: Physical Properties and Sources 359
13.13 Chemical Reactions of Aromatic Hydrocarbons 360
• **Chemistry at a Glance:**
Benzene Substitution Reactions 361
13.14 Fused-Ring Aromatic Compounds 361
• **Chemical Connections**
13.1 Ethene: A Plant Hormone and High-Volume Industrial Chemical 337
13.2 Cis–Trans *Isomerism and Vision 342*
13.3 Carotenoids: A Source of Color 345
13.4 Fused-Ring Aromatic Hydrocarbons and Cancer 362
Concepts to Remember 362
Key Reactions and Equations 363
Key Terms 363
Exercises and Problems 364
Additional Problems 369
Grid Problems 369

Chapter 14 Alcohols, Phenols, and Ethers 371
14.1 Bonding Characteristics of Oxygen Atoms in Organic Compounds 371
14.2 Structural Features of Alcohols, Phenols, and Ethers 372
14.3 Nomenclature of Alcohols and Phenols 373
14.4 Important Commonly Encountered Alcohols 376
14.5 Physical Properties of Alcohols 380
14.6 Preparation of Alcohols 383
14.7 Reactions of Alcohols 383
14.8 Polymeric Alcohols 388
• **Chemistry at a Glance:**
Summary of Reactions Involving Alcohols 389
14.9 Properties and Uses of Phenols 389
14.10 Nomenclature for Ethers 392
14.11 Physical and Chemical Properties of Ethers 394
14.12 Cyclic Ethers 395
14.13 Sulfur Analogs of Alcohols and Ethers 396
• **Chemical Connections**
14.1 Menthol: A Useful Naturally Occurring Terpene Alcohol 380
14.2 Marijuana: The Most Commonly Used Illicit Drug 391
14.3 Ethers as General Anesthetics 393
Concepts to Remember 398
Key Reactions and Equations 398
Key Terms 398
Exercises and Problems 399
Additional Problems 404
Grid Problems 405

Chapter 15 Aldehydes and Ketones 406
15.1 The Carbonyl Group 406
15.2 Structure of Aldehydes and Ketones 407
15.3 Nomenclature for Aldehydes 408
15.4 Nomenclature for Ketones 410
15.5 Selected Common Aldehydes and Ketones 413
15.6 Physical Properties of Aldehydes and Ketones 414
15.7 Preparation of Aldehydes and Ketones 417
15.8 Oxidation and Reduction of Aldehydes and Ketones 418
15.9 Reaction of Aldehydes and Ketones with Alcohols 420
• **Chemistry at a Glance:**
Reactions Involving Aldehydes and Ketones 425
15.10 Formaldehyde-Based Polymers 425
15.11 Sulfur-Containing Carbonyl Groups 426
• **Chemical Connections**
15.1 Lachrymatory Aldehydes and Ketones 412
15.2 Suntan, Sunburn, and Ketones 415
15.3 Diabetes, Aldehyde Oxidation, and Glucose Testing 418
Concepts to Remember 428
Key Reactions and Equations 428
Key Terms 428
Exercises and Problems 429
Additional Problems 434
Grid Problems 435

Chapter 16 Carboxylic Acids and Esters 436
16.1 Structure of Carboxylic Acids 436
16.2 IUPAC Nomenclature for Carboxylic Acids 437
16.3 Common Names for Carboxylic Acids 440
16.4 Polyfunctional Carboxylic Acids 443
16.5 Physical Properties of Carboxylic Acids 445
16.6 Preparation of Carboxylic Acids 446
16.7 Acidity of Carboxylic Acids 447
16.8 Carboxylic Acid Salts 448
16.9 Structures of Esters 450
16.10 Nomenclature for Esters 451
16.11 Selected Common Esters 453
16.12 Physical Properties of Esters 455
16.13 Chemical Reactions of Esters 456
16.14 Sulfur Analogs of Esters 458
• **Chemistry at a Glance:**
Summary of Reactions Involving Carboxylic Acids and Esters 459
16.15 Polyesters 459
16.16 Esters of Inorganic Acids 461

• **Chemical Connections**

16.1 Nonprescription Pain Relievers Derived from Propanoic Acid 442

16.2 Aspirin 455

16.3 Nitroglycerin: An Inorganic Triester 462

Concepts to Remember 463

Key Reactions and Equations 463

Key Terms 464

Exercises and Problems 464

Additional Problems 469

Grid Problems 470

Chapter 17 Amines and Amides 471

17.1 Organic Nitrogen-Containing Compounds 471

17.2 Structure and Classification of Amines 472

17.3 Nomenclature for Amines 473

17.4 Physical Properties of Amines 475

17.5 Basicity of Amines 476

17.6 Amine Salts 478

17.7 Preparation of Amines 480

17.8 Heterocyclic Amines 481

17.9 Selected Biologically Important Amines 482

17.10 Alkaloids 485

17.11 Structure of and Nomenclature for Amides 486

17.12 Selected Amides and Their Uses 488

17.13 Properties of Amides 488

17.14 Preparation of Amides 490

17.15 Hydrolysis of Amides 492

• **Chemistry at a Glance:**

Summary of Reactions Involving Amines and Amides 494

17.16 Polyamides and Polyurethanes 494

• **Chemical Connections**

17.1 Caffeine: The Most Widely Used Central Nervous System Stimulant 482

17.2 Nicotine Addiction: A Widespread Example of Drug Dependence 483

17.3 Acetaminophen: A Substituted Amide 489

Concepts to Remember 496

Key Terms 496

Key Reactions and Equations 497

Exercises and Problems 497

Additional Problems 504

Grid Problems 505

PART III Biological Chemistry

Chapter 18 Carbohydrates 506

18.1 Biochemistry—An Overview 506

18.2 Occurrence and Functions of Carbohydrates 507

18.3 Carbohydrate Classifications 508

18.4 Chirality: Handedness in Molecules 509

18.5 Stereoisomerism: Enantiomers and Diastereomers 511

18.6 Fischer Projections 513

18.7 Properties of Enantiomers 517

• **Chemistry at a Glance:**

Isomers 519

18.8 Classification of Monosaccharides 520

18.9 Biologically Important Monosaccharides 523

18.10 Cyclic Forms of Monosaccharides 525

18.11 Haworth Projection Formulas 527

18.12 Reactions of Monosaccharides 528

18.13 Disaccharides 532

• **Chemistry at a Glance:**

Reactions of Monosaccharides 539

18.14 Polysaccharides 539

18.15 Mucopolysaccharides 543

18.16 Glycolipids and Glycoproteins 544

• **Chemical Connections**

18.1 Blood Types and Monosaccharides 530

18.2 Lactose Intolerance and Galactosemia 535

18.3 Artificial Sweeteners 537

18.4 Sucrose Derivatives and Flatulence 538

Concepts to Remember 545

Key Reactions and Equations 545

Key Terms 546

Exercises and Problems 546

Additional Problems 551

Grid Problems 552

Chapter 19 Lipids 554

19.1 Characteristics of Lipids 554

19.2 Fatty Acids: Saponifiable Lipid Building Blocks 555

19.3 Physical Properties of Fatty Acids 558

• **Chemistry at a Glance:**

Classification Schemes for Fatty Acids 559

19.4 Fats and Oils 559

19.5 Chemical Reactions of Triacylglycerols 563

19.6 Phosphoacylglycerols 567

19.7 Waxes 569

19.8 Sphingolipids 570

• **Chemistry at a Glance:**

Saponifiable Lipids 572

19.9 Nonsaponifiable Lipids 572

19.10 Steroids 573

19.11 Eicosanoids 577

19.12 Plasma Membranes 578

19.13 Transport Across Cell Membranes 580

• **Chemical Connections**

19.1 Artificial Fat Substitutes 563

19.2 The Cleansing Action of Soap 565

19.3 Trans *Fatty Acids and Blood Cholesterol Levels 566*
19.4 Steroid Drugs in Sports 576
Concepts to Remember 581
Key Reactions and Equations 581
Key Terms 582
Exercises and Problems 582
Additional Problems 585
Grid Problems 586

Chapter 20 Proteins 587

20.1 Characteristics of Proteins 587
20.2 Amino Acids: The Building Blocks for Proteins 588
20.3 Chirality and Amino Acids 590
20.4 Acid–Base Properties of Amino Acids 591
20.5 Peptide Formation 595
20.6 Levels of Protein Structure 598
20.7 Primary Structure of Proteins 598
20.8 Secondary Structure of Proteins 600
20.9 Tertiary Structure of Proteins 602
20.10 Quaternary Structure of Proteins 604
20.11 Globular and Fibrous Proteins 605
• **Chemistry at a Glance:**
Protein Structure 606
20.12 Simple and Conjugated Proteins 607
20.13 Protein Hydrolysis 608
20.14 Protein Denaturation 609
20.15 Glycoproteins 610
20.16 Lipoproteins 615
• **Chemical Connections**
20.1 The Essential Amino Acids 590
20.2 Substitutes for Human Insulin 599
20.3 Denaturation and Human Hair 610
20.4 Cyclosporine: An Antirejection Drug 613
20.5 Lipoproteins and Heart Attack Risk 614
Concepts to Remember 615
Key Reactions and Equations 615
Key Terms 616
Exercises and Problems 616
Additional Problems 619
Grid Problems 619

Chapter 21 Enzymes, Vitamins, and Minerals 620

21.1 General Characteristics of Enzymes 621
21.2 Enzyme Nomenclature 621
21.3 Enzyme Structure 622
21.4 Models of Enzyme Action 622
21.5 Enzyme Specificity 625
21.6 Factors That Affect Enzyme Activity 626
• **Chemistry at a Glance:**
Enzyme Activity 629
21.7 Enzyme Inhibition 630
• **Chemistry at a Glance:**
Enzyme Inhibition 632
21.8 Regulation of Enzyme Activity: Allosteric Enzymes 632
21.9 Regulation of Enzyme Activity: Zymogens 633
21.10 Antibiotics That Inhibit Enzyme Activity 634
21.11 Medical Uses of Enzymes 636
21.12 Vitamins 636
21.13 Water-Soluble Vitamins 638
21.14 Fat-Soluble Vitamins 639
21.15 Minerals 642
• **Chemical Connections**
21.1 Enzymatic Browning: Discoloration of Fruits and Vegetables 627
21.2 Heart Attacks and Enzyme Analysis 637
21.3 Iron: The Most Abundant Trace Mineral in the Human Body 645
Concepts to Remember 646
Key Reactions and Equations 647
Key Terms 647
Exercises and Problems 647
Additional Problems 650
Grid Problems 651

Chapter 22 Nucleic Acids 652

22.1 Types of Nucleic Acids 652
22.2 Nucleotides: Building Blocks of Nucleic Acids 653
22.3 Primary Structure of Nucleic Acids 656
22.4 The DNA Double Helix 658
22.5 Replication of DNA Molecules 660
• **Chemistry at a Glance:**
DNA Replication 665
22.6 Overview of Protein Synthesis 665

22.7 Ribonucleic Acids 666
22.8 Transcription: RNA Synthesis 667
22.9 The Genetic Code 669
22.10 Anticodons and tRNA Molecules 671
22.11 Translation: Protein Synthesis 672
22.12 Mutations 676
• **Chemistry at a Glance:**
Protein Synthesis 677
22.13 Nucleic Acids and Viruses 678
22.14 Recombinant DNA and Genetic Engineering 678
• **Chemical Connections**
22.1 Use of Synthetic Nucleic Acid Bases in Medicine 662
22.2 Chromosomes in Cell Division and Reproduction 664
22.3 Antibiotics That Inhibit Bacterial Protein Synthesis 676
22.4 Polymerase Chain Reaction and DNA Sequencing 680
Concepts to Remember 682
Key Reactions and Equations 683
Key Terms 683
Exercises and Problems 683
Additional Problems 686
Grid Problems 687

Chapter 23 Biochemical Energy Production 688
23.1 Metabolism 688
23.2 Metabolism and Cell Structure 689
23.3 Important Intermediate Compounds in Metabolic Pathways 691
23.4 High-Energy Phosphate Compounds 695
23.5 An Overview of Biochemical Energy Production 696
• **Chemistry at a Glance:**
Catabolic Pathways 697
23.6 The Citric Acid Cycle 698
23.7 The Electron Transport Chain 702
23.8 Oxidative Phosphorylation 706
23.9 ATP Production for the Common Metabolic Pathway 707
23.10 The Importance of ATP 708
• **Chemical Connections**
23.1 Cyanide Poisoning 706
23.2 Brown Fat, Newborn Babies, and Hibernating Animals 709
Concepts to Remember 710
Key Reactions and Equations 710
Key Terms 710
Exercises and Problems 710
Additional Problems 713
Grid Problems 713

Chapter 24 Carbohydrate Metabolism 715
24.1 Digestion and Absorption of Carbohydrates 715
24.2 Glycolysis 717
24.3 Fates of Pyruvate 724
24.4 ATP Production for the Complete Oxidation of Glucose 727
24.5 Glycogen Synthesis and Degradation 728
24.6 Gluconeogenesis 730
24.7 Terminology for Glucose Metabolic Pathways 733
24.8 The Pentose Phosphate Pathway 733
• **Chemistry at a Glance:**
Glucose Metabolism 735
24.9 Hormonal Control of Carbohydrate Metabolism 736
• **Chemical Connections**
24.1 Lactate Accumulation 726
24.2 Diabetes Mellitus 737
Concepts to Remember 738
Key Reactions and Equations 738
Key Terms 739
Exercises and Problems 739
Additional Problems 741

Chapter 25 Lipid Metabolism 743
25.1 Digestion and Absorption of Lipids 743
25.2 Triacylglycerol Storage and Mobilization 744
25.3 Glycerol Metabolism 746
25.4 Oxidation of Fatty Acids 747
25.5 ATP Production from Fatty Acid Oxidation 751
25.6 Ketone Bodies 753
25.7 Biosynthesis of Fatty Acids: Lipogenesis 756
25.8 Biosynthesis of Cholesterol 759
25.9 Relationships Between Lipid and Carbohydrate Metabolism 762
• **Chemistry at a Glance:**
Interrelationships Between Carbohydrate and Lipid Metabolism 763
• **Chemical Connections**
25.1 High-Intensity Versus Low-Intensity Workouts 752
25.2 Statins: Drugs That Lower Plasma Levels of Cholesterol 761
Concepts to Remember 764
Key Reactions and Equations 764
Key Terms 764
Exercises and Problems 764
Additional Problems 767
Grid Problems 767

Chapter 26 Protein Metabolism 769
26.1 Protein Digestion and Absorption 769
26.2 Amino Acid Utilization 770

26.3 Transamination and Oxidative Deamination 772
26.4 The Urea Cycle 776
26.5 Amino Acid Carbon Skeletons 781
26.6 Amino Acid Biosynthesis 782
26.7 Hemoglobin Catabolism 783
• **Chemistry at a Glance:**
Interrelationships Among Lipid, Carbohydrate, and Protein Metabolism 787
26.8 Interrelationships Among Metabolic Pathways 789
• **Chemical Connections**
26.1 The Chemical Composition of Urine 780
26.2 Arginine, Citrulline, and the Chemical Messenger Nitric Oxide 781
Concepts to Remember 789
Key Reactions and Equations 789
Key Terms 789
Exercises and Problems 790
Additional Problems 792
Grid Problems 792

Answers to Selected Exercises A-1
Glossary/Index A-24
Photo Credits A-46

Preface

Writing an introductory college text, particularly one that encompasses as wide a range of topics as *General, Organic, and Biological Chemistry* does, is a huge undertaking. When the first edition of the text was published three years ago, my hopes were high. Thus, the positive responses of instructors and students who used the first edition and found it a notably accessible teaching and learning tool have been gratifying—and have led to the new edition you hold in your hands. This second edition represents a renewed commitment to the goals I initially set out to meet.

I wrote the text to fit the needs of the many students in the fields of nursing, allied health, biological sciences, agricultural sciences, food sciences, and public health, who are required to take such a course. The students who will use this text often have little or no background in chemistry and hence, approach the course with a good deal of trepidation. Thus, this text's development of chemical topics always starts at *ground level.* Though clearly some chemical principles cannot be divorced entirely from mathematics, the amount and level of mathematics used is purposefully minimized. Early chapters of the text focus on fundamental chemical principles. The later chapters, built on the foundation of these principles, develop the concepts and applications central to the fields of organic chemistry and biochemistry.

Focus on Biochemistry. Most students taking this course have a greater interest in the biochemistry portion of the course than the preceding two parts. But biochemistry, of course, cannot be understood without a knowledge of the fundamentals of organic chemistry, and understanding organic chemistry in turn depends on knowing the key concepts of general chemistry. Thus, in writing this text, I essentially started from the back and worked forward. I began by determining what topics would be considered in the biochemistry chapters and then tailored the organic and then the general sections to support that presentation. Users of the first edition confirm that this approach ensures an efficient but thorough coverage of the principles needed to understand biochemistry.

Emphasis on Visual Support. I believe strongly in visual reinforcement of key concepts in a textbook; thus, this book uses art and photos wherever possible to teach key concepts. The book uses artwork to make connections and highlight what's important for the student to know. The use of color in reaction equations to emphasize the portions of a molecule that undergo change has been greatly expanded in this new edition. Colors are likewise assigned to things like valence shells and classes of compounds, to help students follow trends. Computer-generated, three-dimensional molecular models now accompany many discussions in the organic and biochemistry sections of the text. Color photographs show applications of chemistry to help make concepts real and more readily remembered.

Visual summary features, called ***Chemistry at a Glance,*** pull together material from several sections of a chapter to help students see the larger picture. See, for example, the *Chemistry at a Glance* on page 172 that summarizes intermolecular forces; the one on page 262 summarizing buffer solutions; the one on page 323 that summarizes the physical and chemical properties of alkanes, the simplest type of organic compound; the one on page 519 that summarizes the thought processes involved in classifying molecules as enantiomers or diastereomers; or the one on page 663 on DNA replication. The *Chemistry at a Glance* features serve both as overviews for the student reading the material for the first time and as review tools for the student preparing for exams. Given the popularity of the *Chemistry at a Glance* summaries in the first edition, this feature has been expanded to all chapters; nearly two thirds of the book's forty *Chemistry at a Glance* features are new or extensively modified from the first edition.

Commitment to Student Learning. In addition to the study help *Chemistry at a Glance* offers, the text is built on a strong foundation of learning aids designed to help students master the course material.

- **Problem-solving pedagogy:** Because problem solving is often difficult for students in this course to master, I have taken special care to provide support to help students build their skills. Within the chapters, worked ***Examples*** follow the explanation of many concepts. These examples walk students through the thought processes involved in problem-solving, carefully outlining all the steps involved. Each example is immediately followed by a ***Practice Exercise,*** to reinforce the information just presented.
- ***Chemical Connections*** in every chapter show chemistry as it appears in everyday life. These boxes focus on topics that are relevant to students' future careers in the health and environmental fields and on those that are important for informed citizens to understand. Many of the health-related *Chemical Connections* have been updated to include the latest research findings and several new boxes on environmental and social issues have been added. Nearly a quarter of the *Chemical Connections* are modified or new to this edition.
- **Margin Notes** distributed liberally throughout the text provide tips for remembering and distinguishing between concepts, highlight links across chapters, and describe interesting historical background information.
- **Defined Terms** are highlighted in the text when they are first presented, using boldface and italic type. Each definition appears in a complete sentence; students are never forced to deduce a definition from context. In addition, the definitions of all terms appear in the ***combined Glossary/Index*** found at the end of the text.
- **Review Aids** appear at the end of each chapter. ***Concepts to Remember*** and ***Key Reactions and Equations*** provide concise review of the material presented in the chapter. A new ***Key Terms Review*** lists all the key terms in the chapter alphabetically and cross-references the section of the chapter in which they appear. These aids help students prepare for exams.
- **End-of-chapter problems** complement the worked examples within the chapters. Each end-of-chapter problem set is divided into three sections: Exercises and Problems, Additional Problems, and Grid Problems. The ***Exercises and Problems*** are organized by topic and paired, with each pair testing similar material. These problems always involve only a single concept. The answers to the odd-numbered problems are at the back of the book. The ***Additional Problems*** involve more than one concept and are more difficult than the *Exercises and Problems;* all *Additional Problems* are new to this edition, and every chapter now includes eight such problems. Each chapter ends with four ***Grid Problems***. Here, students apply the principles from the chapter in a setting where multiple concepts are examined at once. Answers are selected from a grid of choices, and most often multiple answers are correct. Answers to all *Additional Problems* and *Grid Problems* are provided at the back of the book.

Content Changes. New textual materials have been added to this edition at the request of reviewers, including:

- Discussion of experimental evidence for the subatomic makeup of the atom in Chapter 3
- The pK_a method for expressing acid strength and the Henderson–Hasselbalch equation in Chapter 10
- A brief section on sulfur analogs of aldehydes and ketones in Chapter 15
- Mucopolysaccharides in Chapter 18
- Transport processes by which molecules cross membranes in Chapter 19
- Globular and fibrous proteins in Chapter 20

In addition, several existent sections have been heavily modified in response to reviewer recommendations. These include:

- Chapter 7's treatment of gas laws, which has been simplified. Avogadro's law has been deleted, while the other simple gas laws, the ideal gas law, and Dalton's law of partial pressure are still covered.
- Chapter 14's discussion of intra- and intermolecular alcohol dehydration reactions, which has been expanded for greater clarity
- Chapter 23's more detailed discussion of NAD^+ and FAD
- The expanded discussion of transamination and deamination reactions in Chapter 26

Three new margin note features have also been incorporated into all organic chemistry chapters. The first summarizes physical state information at room temperature (solid, liquid, gas) for the simplest members of each family of organic compounds. The second contrasts common and IUPAC naming systems for the simple members of each family of compounds. The third introduces line-angle drawing representations of the structures of the simplest members of each family of organic compounds to prepare the student for the further use of such line-angle notation in the structures of biochemical molecules.

• The Package

Alternate Edition. For instructors who prefer to use only the organic and biochemistry portions of a text, an alternate edition of this book, *Organic and Biological Chemistry,* is available.

Study Help for Students

Student Website (accessible through http://college.hmco.com/chemistry). Developed with input from project consultant Sharon K. Stoffels, RNC, MSN, Associate Professor of Nursing, Boise State University and available free of charge, this dedicated website offers a wealth of resources to help students succeed, including:

- Chapter overviews
- Self-quizzing using Houghton Mifflin's ACE system
- Glossary terms, key reactions, and key concepts in flashcard format
- Additional *Chemical Connections* with links to related sites
- Additional real-life applications with relevant links
- Career preparation information, including practice questions for the NCLEX exam, descriptions of common nursing specialties, frequently asked questions about entering the nursing field (and their answers), and web links to nursing-related job information.

Study Guide with Answers to Selected Problems by Danny V. White of American River College and Joanne A. White includes, for each chapter, a brief overview, activities and practice problems to reinforce skills, and a practice test. The answer section includes answers for all odd-numbered end-of-chapter exercises.

Experiments in General, Organic, and Biological Chemistry by Robert J. Brenstein and Conrad C. Hinkley, both of the University of Illinois–Carbondale, is a comprehensive collection of 56 detailed experiments that have been thoroughly class tested and include extensive safety and waste disposal instructions. These experiments are available only through **Bibliobase customizable lab manuals** (www.bibliobase.com), which allows instructors to create a customized lab manual, using selections from these experiments.

Experiments for General, Organic, and Biological Chemistry by Michael S. Matta of Southern Illinois University, is a printed lab manual containing 45 experiments. Suggestions for demonstrations and structural studies are also included.

Math Review CD-ROM. This CD, included at no additional charge with all new texts, offers brief tutorials on basic mathematical concepts that appear in the text, including

solving simple algebraic equations, scientific notation, conversions, reading a graph, and ratio and proportion.

Course Support for Instructors

Instructor Website (accessible through http://college.hmco.com/chemistry) allows access to all student website resources (above) as well as additional instructor course/classroom resources such as downloadable PowerPoint slides and useful links.

Instructor's Resource Manual with Test Bank by H. Stephen Stoker includes answers to all end-of-chapter exercises and a printed test bank of over 1,500 multiple choice and matching problems. The **Computerized Test Bank**, available for Windows and Macintosh, is a customizable electronic version of the printed test bank. Chapter tests, midterms, and final exams are created easily and quickly with **Brownstone's Diploma Computerized Testing System**, a flexible test-creation program with a comprehensive gradebook for easy administration and tracking of paper quizzes, network-based tests, and internet exams. The *Diploma* Exam database contains a wealth of questions, including algorithm-based questions that can automatically regenerate similar questions with different values. The powerful, easy-to-use program can produce multiple-choice, true/false, short-answer, matching, fill-in-the-blank, and essay questions. Professors can select questions based upon chapter, question format, type, difficulty, subject, and topic. At the professor's discretion, students can take these tests and quizzes at a computer and obtain immediate feedback. The Gradebook sorts students, curves tests, and generates reports. It can also separate tests, homework, lab work, etc. into different grade categories, facilitating averaging. *Diploma* runs on Windows 95, 98, and NT operating systems, and on Macs with at least 4 MB of free space on their hard drives.

Instructor's Guide for Experiments in General, Organic, and Biological Chemistry offers reagent/solvent lists, equipment lists, waste disposal instructions, mixing instructions, preparatory notes, planning notes, teaching notes, quiz questions, expected results, and answers to questions and problems for each experiment in the Bibliobase customizable lab manuals.

PowerPoint Presentation Slides CD includes all of the line art from the text and a selection of *Chemistry at a Glance* visual summaries that can be used as the basis for creating classroom presentations. This material is also available in downloadable files at the Instructor Website.

Overhead Transparencies provide over 140 full-color acetates of important figures, tables, and images from the text.

• Acknowledgments

I gratefully acknowledge the valuable comments and suggestions of the following reviewers, whose thoughtful pre-revision critiques of all or parts of the manuscript guided by revision efforts:

Vicky L. H. Bevilacqua
Kennesaw State University

David R. Bjorkman
East Carolina University

Frank B. Day
North Shore Community College

Alison J. Dobson
Georgia Southern University

Naomi Eliezer
Oakland University

Wes Fritz
College of DuPage

Caroline Gil
Lexington Community College

Robert Gooden
Southern University–Baton Rouge

Ellen Kime-Hunt
Riverside Community College

Peter Kreiger
Palm Beach Community College

Cathy MacGowan
Armstrong Atlantic State University
Lawrence L. Mack
Bloomsburg University
Charmaine B. Mamantov
University of Tennessee, Knoxville
Joanne S. Monko
Kutztown University
Elva Mae Nicholson
Eastern Michigan University
Michael Shanklin
Palo Alto College

These instructors, who reviewed selected chapters of revised manuscript, likewise provided valuable guidance:

Hugh Akers
Lamar University
Eric Holmberg
University of Alaska
Marvin Jaffe
Borough of Manhattan Community College
David Johnson
Biola University
Fred Johnson
Brevard Community College
Daniel Jones
University of North Carolina–Charlotte
Peter Krieger
Palm Beach Community College
Da-hong Lu
Fitchburg State College
Cynthia Martin
Des Moines Area Community College
Elva Mae Nicholson
Eastern Michigan University
Mary Palaszek
Grand Valley State University
Diane Payne
Villa Julie College
Janet Rogers
Edinboro University of Pennsylvania
Jackie Scholars
Bellevue University
Michelle Sulikowski
Texas A&M University
Joanne Tscherne
Bergen Community College
James Yuan
Old Dominion University

I would also like to thank reviewers of the first edition, whose influence continues to be felt: Hugh Akers, *Lamar University;* Steven Albrecht, *Oregon State University;* Margaret Asirvatham, *University of Colorado;* George Bandik, *University of Pittsburgh;* Gerald Berkowitz, *Erie Community College;* Robert Bogess, *Radford University;* Christine Brzezowski, *University of Utah;* Harry Conley, *Murray State;* Karen Eichstadt, *Ohio University;* William Euler, *University of Rhode Island;* Arthur Glasfeld, *Reed College;* Fabian Fang, *California State University–Long Beach;* John Fulkrod, *University of Minnesota–Duluth;* Marvin Hackert, *University of Texas at Austin;* Henry Harris, *Armstrong State College;* Leland Harris, *University of Arizona;* Larry Jackson, *Montana State University;* James Jacob, *University of Rhode Island;* James Johnson, *Sinclair Community College;* Eugene Klein, *Tennessee Technological University;* Norman Kulevsky, *University of North Dakota;* James W. Long, *University of Oregon;* Ralph Martinez, *Humboldt State University;* Scott Mohr, *Boston University;* Melvyn Mosher, *Missouri Southern University;* Elva Mae Nicholson, *Eastern Michigan University;* Frasier Nyasulu, *University of Washington;* John Ohlsson, *University of Colorado;* Roger Penn, *Sinclair Community College;* Helen Place, *Washington State University;* John Reasoner, *Western Kentucky University;* Norman Rose, *Portland State University;* Michael Ryan, *Marquette University;* John Searle, *College of San Mateo;* Dan Sullivan, *University of Nebraska at Omaha;* Emanuel Terezakis, *Community College of Rhode Island;* Ruiess Van Fossen Bravo, *Indiana University of Pennsylvania;* Donald Williams, *University of Louisville;* Les Wynston, *California State University–Long Beach.*

Special thanks go to Tim Champion, Johnson C. Smith University, for his help in ensuring this book's accuracy by reviewing manuscript, proofs, and artwork.

I also give special thanks to the people at Houghton Mifflin who guided the revision through various stages of development and production: Richard Stratton, Senior Sponsoring Editor, Chemistry, Karla Paschkis, Senior Development Editor, and Andy Fisher, Executive Marketing Manager, Science, Math, and Market Development and Research, made especially significant suggestions for refining the features and text. Nancy Blodget, Senior Project Editor, Elisabeth Kehrer, Editorial Assistant, Jill Haber, Production/Design Coordinator, and Sally Culler, Senior Manufacturing Coordinator, adroitly marshalled and managed the myriad details that transformed the manuscript into this beautiful bound book. I would also like to thank Sarah Gessner, Associate Editor, for her skillful management of the ancillary package; Sara Wise, Assistant Editor, for her development of the website and Math Review CD; Pamela Bachorz, Internet Producer, for her management of the website; Charlotte Miller, Art Editor, for her thoughtful and creative contributions to the illustration program; Naomi Kornhauser and Jessyca Broekman, Photo Editors, for helping to enrich the text with photographs and molecular models; and Janet Theurer, Designer, for the complementary design.

H. Stephen Stoker, Weber State University

Exciting Photo Program

Throughout the text, photos help students see the everyday applications of the chemistry they are learning.

Acids, Bases, and Salts

CHAPTER OUTLINE

10.1 Arrhenius Acid–Base Theory 237
10.2 Brønsted–Lowry Acid–Base Theory 238
10.3 Mono-, Di-, and Triprotic Acids 241
10.4 Strengths of Acids and Bases 242
10.5 Ionization Constants for Acids and Bases 243
10.6 Salts 244
10.7 Acid–Base Neutralization Reactions 245
10.8 Self-Ionization of Water 246
10.9 The pH Concept 248
10.10 pK_a Method for Expressing Acid Strength 251

Chemistry at a Glance: **Acids and Acidic Solutions** 252
10.11 The pH of Aqueous Salt Solutions 252
10.12 Buffers 256

Chemistry at a Glance: **Buffer Systems** 262
10.13 The Henderson–Hasselbalch Equation 261
10.14 Electrolytes 262
10.15 Acid–Base Titrations 264

Chemical Connections
10.1 Acid Rain: Excess Acidity 254
10.2 Blood Plasma pH and Hydrolysis 256
10.3 Buffering Action in Human Blood 260
10.4 Electrolytes and Body Fluids 263

Fish are very sensitive to the acidity of the water present in an aquarium.

Acids, bases, and salts are among the most common and important compounds known. In the form of aqueous solutions, these compounds are key materials in both biological systems and the chemical industry. A major ingredient of gastric juice in the stomach is hydrochloric acid. Quantities of lactic acid are produced when the human body is subjected to strenuous exercise. The lye used in making soap contains the base sodium hydroxide. Bases are ingredients in many stomach antacid formulations. The white crystals you sprinkle on your food to make it taste better represent ... s that exist.

...ase Theory

...nte August Arrhenius (1859–1927) proposed that ...s of the chemical species they form when they dis- ... *is a hydrogen-containing compound that, in water,* ...e acidic species in Arrhenius theory is thus the ... *is a hydroxide-containing compound that, in water,*

237

Chapter Outlines give students a road map for where they are going.

16.12 Physical Properties of Esters 455

Chemical CONNECTIONS

16.2 Aspirin

Aspirin, an ester of salicylic acid (Section 16.11), is a drug that has the ability to decrease pain (analgesic properties), to lower body temperature (antipyretic properties), and to reduce inflammation (anti-inflammatory properties). It is most frequently taken in tablet form, and the tablet usually contains 325 mg of aspirin held together with an inert starch binder.

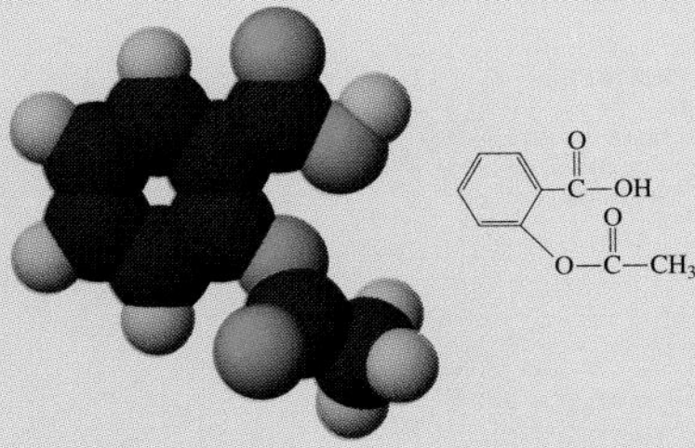

After ingestion, aspirin undergoes hydrolysis to produce salicylic acid and acetic acid. Salicylic acid is the active ingredient of aspirin—the substance that has analgesic, antipyretic, and anti-inflammatory effects.

Salicylic acid is capable of irritating the lining of the stomach, inducing a small amount of bleeding. Breaking (or chewing) an aspirin tablet, rather than taking it whole, reduces the chance of bleeding by eliminating drug concentration on one part of the stomach lining. Buffered aspirin products contain alkaline chemicals (such as aluminum glycinate or aluminum hydroxide) to neutralize the acidity of the aspirin when it contacts the stomach lining.

Aspirin—that is, salicylic acid—inhibits the synthesis of a class of hormones called prostaglandins (Section 19.11), molecules that cause pain, fever, and inflammation when present in the bloodstream in higher-than-normal levels. Salicylic acid's mode of action is irreversible inhibition (Section 21.7) of *cyclooxygenase,* an enzyme necessary for the production of prostaglandins.

Recent studies show that aspirin also increases the time it takes blood to coagulate (clot). For blood to coagulate, platelets must first be able to aggregate, and prostaglandins (which aspirin inhibits) appear to be necessary for platelet aggregation to occur. One study suggests that healthy men can cut their risk of heart attacks nearly in half by taking one baby aspirin per day (85 mg compared to the 325 mg in a regular tablet). Aspirin acts by making the blood less likely to clot. Heart attacks usually occur when clots form in the coronary arteries, cutting off blood supply to the heart.

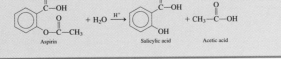

Aspirin $+ H_2O \xrightarrow{H^+}$ Salicylic acid $+ CH_3—C(=O)—OH$ (Acetic acid)

Oil of wintergreen, also called methyl salicylate, is used in skin rubs and liniments to help decrease the pain of sore muscles. It is absorbed through the skin, where it is hydrolyzed to produce salicylic acid. Salicylic acid is the actual pain reliever.

Chemical Connections boxes show chemistry as it appears in everyday life. Topics are relevant to students' future careers in the health and environmental fields and are important for informed citizens to understand. See page 455, for example.

Chemistry at a Glance, a visual summary feature, pulls together material from a group of sections or a whole chapter to help students see the larger picture. Many of the Chemistry at a Glance features in the second edition are new or substantially revised.

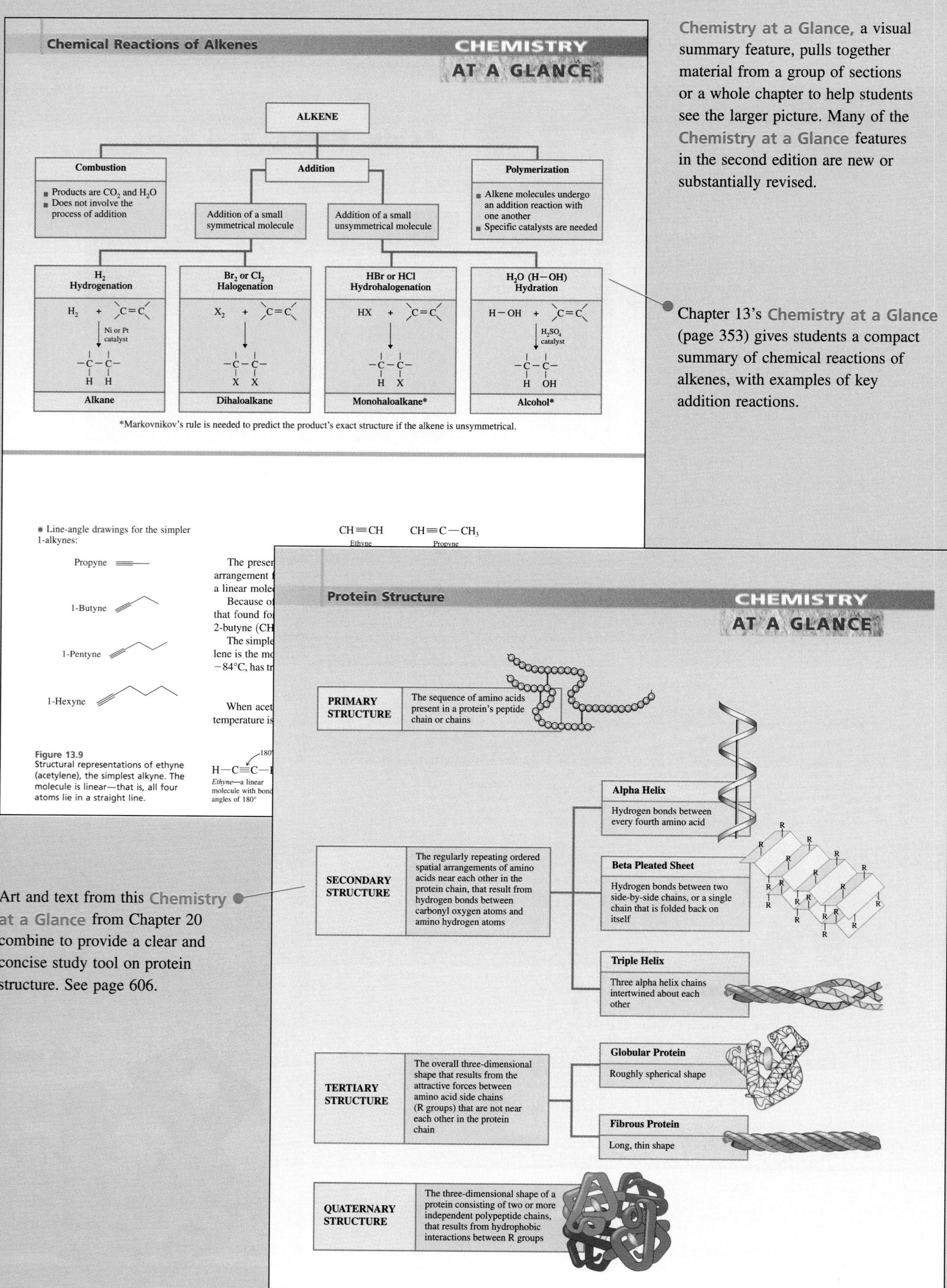

Chapter 13's Chemistry at a Glance (page 353) gives students a compact summary of chemical reactions of alkenes, with examples of key addition reactions.

Art and text from this Chemistry at a Glance from Chapter 20 combine to provide a clear and concise study tool on protein structure. See page 606.

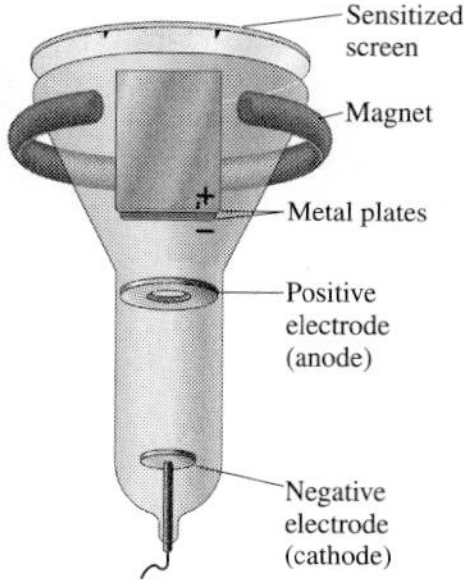

Figure 3.2
A simplified version of a gas discharge tube.

are mostly empty space), the coin would weigh 190,000,000 tons! Nuclei are indeed very dense matter.

Despite the existence of subatomic particles, we will continue to use the concept of atoms as the fundamental building blocks for all types of matter. Subatomic particles do not lead an independent existence for any appreciable length of time; they gain stability by joining together to form atoms.

3.2 Experimental Evidence for the Existence of Subatomic Particles

A large amount of experimental evidence exists that is consistent with and supports the existence, nature, and arrangement of subatomic particles within an atom (Section 3.1). Two landmark types of experiments performed in the course of developing our present concepts of the atom illustrate some of the sources of this evidence. *Discharge tube experiments* originally suggested that the atom contained negatively and positively charged particles. *Metal foil experiments* provided evidence for the existence of a nucleus within the atom.

Marginal Notes, such as these on pages 50 and 110, summarize key information, give tips for remembering or distinguishing between similar ideas, and provide additional details and links between concepts.

• Discharge Tube Experiments

• Neon signs, fluorescent lights, television tubes, and computer monitors are all basic components of our modern technological society. The forerunner for all four of these developments was the gas discharge tube.

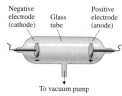

Experiments involving gas discharge tubes provided the first evidence that *electrons* and *protons* are present within an atom. A simplified diagram of a gas discharge tube is shown in Figure 3.2. The apparatus consists of a sealed glass tube containing two metal disks called *electrodes*. The glass tube also has a side arm for attachment to a vacuum pump. During operation, the electrodes are connected to a source of electrical power. (The electrode attached to the positive side of the electrical power source is called the *anode;* the one attached to the negative side is known as the *cathode*.) Use of the vacuum pump allows the amount of gas within the tube to be varied. The smaller the amount of gas present, the lower the pressure within the tube.

Early studies with gas discharge tubes, conducted in the mid-1800s, showed that when the tube was almost ev
other, and the residual
different colors to the
els (very little gas rem
trical conductance co
walls. This glow was t
discovery marked the
ultimately led to the c

The term *cathode r*
tween the negative ele
shape of the obstacle i
cathode.

Further studies sho
lerite (zinc sulfide) to
high magnification wh
sphalerite coating con
cathode rays were in r

Joseph John Thoms
the nature of cathode r
ode ray production is
ray tube (see Figure 3
plates or a magnetic fi
attracted to the south
charged. In 1897, Thor
particles, which today
conclusions were corr

Sensitized screen
Magnet
Metal plates
Positive electrode (anode)
Negative electrode (cathode)

Figure 3.3
A cathode-ray tube with perpendicular magnetic and electric fields was used in characterizing cathode rays (electrons).

5.7 Bonding in Compounds with Polyatomic Ions Present

Ionic compounds containing polyatomic ions (Section 4.10) present an interesting combination of both ionic and covalent bonds: covalent bonding *within* the polyatomic ion and ionic bonding *between* it and ions of opposite charge.

Polyatomic ion Lewis structures, which show the covalent bonding within such ions, are drawn using the same procedures as for molecular compounds (Section 5.6), with the accommodation that the total number of electrons used in the structure must be adjusted (increased or decreased) to take into account ion charge. The number of electrons is increased in the case of negatively charged ions and decreased in the case of positively charged ions.

• Students often erroneously assume that the charge associated with a polyatomic ion is assigned to a particular atom within the ion. Polyatomic ion charge is not localized on a particular atom but rather is associated with the ion as a whole.

In the Lewis structure for an *ionic compound* that contains a polyatomic ion, the positive and negative ions are treated separately to show that they are individual ions not linked by covalent bonds. The Lewis structure of potassium sulfate, K_2SO_4, is written as

• When we write the Lewis structure of an ion (monatomic or polyatomic), it is customary to use brackets and to show ionic charge outside the brackets.

$[K]^+ [K]^+ [:\ddot{O}-S(:\ddot{O}:)(:\ddot{O}:)-\ddot{O}:]^{2-}$ and not as K—O—S—O—K

Correct structure   Incorrect structure

Example 5.2

Drawing Lewis Structures for Polyatomic Ions

Draw a Lewis structure for SO_4^{2-}, a polyatomic ion in which S is the central atom and all O atoms are bonded to the S atom.

Solution

Step 1: Both S and O are Group VIA elements. Thus each of the atoms has 6 valence electrons. Two extra electrons are also present, which accounts for the −2 charge on the ion. The total electron count is 6 + 4(6) + 2 = 32.

Step 2: Drawing the molecular skeleton with single covalent bonds between bonded atoms gives

$[O:S:O]^{2-}$ (with O above and below S)

Step 3: Adding nonbonding electron pairs to give each oxygen atom an octet of electrons yields

$[:\ddot{O}:\ddot{S}:\ddot{O}:]^{2-}$ (with :Ö: above and below S)

Step 4: The Step 3 structure has 32 electrons, the total number available. No more electrons can be added to the structure, and indeed, none need to be added because the central S atom has an octet of electrons. There is no need to proceed to Step 5.

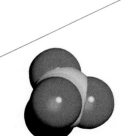

The sulfate ion (SO_4^{2-}). A computer-generated model.

Practice Exercise 5.2

Draw a Lewis structure for BrO_3^-, a polyatomic ion in which Br is the central atom and all O atoms are bonded to it.

• *Answer.* $[:\ddot{O}:\ddot{Br}:\ddot{O}:]^-$ (with :Ö: below Br)

Within the chapters, worked-out Examples follow the explanation of many concepts. These examples walk students through the thought processes involved in problem-solving, carefully outlining all the steps involved. They are immediately followed by a Practice Exercise, to reinforce the information just presented. See, for example, page 110.

convert ADP back to ATP, because ATP synthesis requires an energy input equal to or greater than the hydrolysis energy, and such an unusually high amount of energy would not be available.

Concepts to Remember, Key Reactions and Equations, and **Key Terms** provide concise review of the material presented in the chapter, helping students prepare for exams.

Concepts to Remember

Metabolism. Metabolism is the sum total of all the chemical reactions that take place in a living organism. Metabolism consists of catabolism and anabolism. Catabolic reactions involve the breakdown of large molecules into smaller fragments. Anabolic reactions synthesize large molecules from smaller ones.

Mitochondria. Mitochondria are membrane-enclosed subcellular structures that are the site of energy production in the form of ATP molecules. Enzymes for both the citric acid cycle and the electron transport chain are housed in the mitochondria.

Important coenzymes. Three very important coenzymes involved in catabolism are NAD^+, FAD, and CoA. NAD^+ and FAD are oxidizing agents that participate in the oxidation reactions of the citric acid cycle. They transport hydrogen atoms and electrons from the citric acid cycle to the electron transport chain. CoA interacts with acetyl groups produced from food degradation to form acetyl CoA. Acetyl CoA is the "fuel" for the citric acid cycle.

High-energy compounds. A high-energy compound liberates a larger-than-normal amount of free energy upon hydrolysis, because structural features in the molecule contribute to repulsive strain in one or more bonds. Most high-energy biochemical molecules contain phosphate groups.

Common catabolic pathway. The common catabolic pathway includes the reactions of the citric acid cycle and those of the electron transport chain and oxidative phosphorylation. The degradation products from all types of foods (carbohydrates, fats, and proteins) participate in these reactions.

Citric acid cycle. The citric acid cycle is a cyclic series of eight reactions that oxidize the acetyl portion of acetyl CoA, resulting in the production of two molecules of CO_2. The complete oxidation of one acetyl group produces three molecules of NADH, one of $FADH_2$, and one of GTP.

Electron transport chain. The electron transport chain is a series of reactions that passes electrons from NADH and $FADH_2$ to molecular oxygen. Each electron carrier that participates in the chain has an increasing affinity for electrons. Upon accepting the electrons and hydrogen ions, the O_2 is reduced to H_2O.

Oxidative phosphorylation. In the electron transport chain, released energy is used to convert ADP to ATP. One molecule of NADH produces three molecules of ATP. $FADH_2$, which enters the chain later than NADH, produces only two molecules of ATP.

Chemiosmotic coupling. Chemiosmotic coupling explains how the energy needed for ATP synthesis is obtained. Synthesis takes place because of a flow of protons across the inner mitochondrial membrane.

Importance of ATP. ATP is the link between energy production and energy use in cells. The conversion of ATP to ADP powers life processes, and the conversion of ADP back to ATP regenerates the energy expended in cell operation.

Extensive and varied **Exercises and Problems** at the end of each chapter are organized by topic and paired, with answers to the odd-numbered problems at the back of the book. These problems always involve only a single concept.

Key Reactions and Equations

1. Oxidation by NAD^+ (Section 23.3)

$$NAD^+ + 2H^+ + 2e^- \longrightarrow NADH + H^+$$

2. Oxidation by FAD (Section 23.3)

$$FAD + 2H^+ + 2e^- \longrightarrow FADH_2$$

3. The citric acid cycle (Section 23.6)

$$\text{Acetyl CoA} + 3NAD^+ + FAD + GDP + P_i + 2H_2O \longrightarrow$$
$$2CO_2 + CoA + 3NADH + 2H^+ + FADH_2 + GT$$

Key Terms

Anabolism (23.1)
Catabolism (23.1)
Chemiosmotic coupling (23.8)
Citric acid cycle (23.6)
Common metabolic pathway (23.5)
Coupled reactions
Cytochromes (23.7
Electron transport
High-energy comp
Metabolic pathway

Exercises and Problems

The members of each pair of problems in this section test similar ma

Metabolism (Section 23.1)

23.1 Classify anabolism and catabolism as synthetic or degradative processes.

24.79 What compound contains the carbon atom lost from glucose (a hexose) in its conversion to ribose (a pentose)?

24.80 How many molecules of NADPH are produced per glucose 6-phosphate in the pentose phosphate pathway?

Control of Carbohydrate Metabolism (Section 24.9)

24.81 What effect does insulin have on glycogen metabolism?

24.82 What effect does insulin have on blood glucose levels?

24.83 What effect does glucagon have on blood glucose levels?

24.84 What effect does glucagon have on glycogen metabolism?

24.85 What organ is the source of insulin?

24.86 What organ is the source of glucagon?

24.87 The hormone epinephrine generates a "second messenger." Explain.

24.88 What is the relationship between cAMP and the hormone epinephrine?

24.89 Compare the target tissues for glucagon and epinephrine.

24.90 Compare the biological functions of glucagon and epinephrine.

Additional Problems

24.91 Indicate in which of the four processes *glycolysis, glycogenesis, glycogenolysis,* and *gluconeogenesis* each of the following compounds is encountered. There may be more than one correct answer for a given compound.
a. Glucose 6-phosphate
b. Glucose 1-phosphate
c. Dihydroxyacetone phosphate
d. Oxaloacetate

24.92 Indicate in which of the four processes *glycolysis, glycogenesis, glycogenolysis,* and *gluconeogenesis* each of the following situations is encountered. There may be more than one correct answer for a given situation.
a. NAD^+ is consumed
b. ATP is produced
c. ATP is consumed
d. UDP is involved

24.93 Indicate in which of the four processes *glycolysis, glycogenesis, glycogenolysis,* and *glyconeogenesis* each of the following characterizations applies.
a. Glucose is converted to two pyruvates.
b. Glycogen is synthesized from glucose.
c. Glycogen is broken down into free glucose units.
d. Glucose is synthesized from pyruvate.

24.94 What is the ATP yield *per glucose molecule* in each of the following processes?
a. Glycolysis
b. Glycolysis, acetyl CoA formation, and the common metabolic pathway
c. Glycolysis plus oxidation of pyruvate to acetyl CoA
d. Glycolysis plus reduction of pyruvate to lactate

24.95 Which one of these characterizations, (1) Cori cycle, (2) an anaerobic process, (3) oxidative stage of pentose phosphate pathway, or (4) nonoxidative stage of pentose phosphate pathway, applies to each of the following chemical changes?
a. Pyruvate to lactate
b. Pyruvate to ethanol
c. Glucose 6-phosphate to ribulose 5-phosphate
d. Ribulose 5-phosphate to ribose 5-phosphate

24.96 What condition or conditions determine that pyruvate is involved in each of the following?
a. Gluconeogenesis
b. Converted to lactate
c. Citric acid cycle
d. Converted to ethanol

24.97 In the complete metabolism of 1 mole of sucrose, how many moles of each of the following are produced?
a. CO_2
b. Pyruvate
c. Acetyl CoA
d. ATP

24.98 Under what conditions does glucose 6-phosphate enter each of the following pathways?
a. Glycogenesis
b. Glycolysis
c. Pentose phosphate pathway
d. Hydrolysis to free glucose

Grid Problems

24.99

1. pyruvate	2. lactate	3. acetyl part of acetyl CoA
4. dihydroxyacetone	5. glycerate	6. glyceraldehyde

Select from the grid *all* correct responses for each of the following situations.
a. Substances that are C_3 species
b. Substances for which one or more phosphoderivatives are intermediates in glycolysis
c. Substances for which one or more bisphosphoderivatives are intermediates in glycolysis
d. Substances that are not part of the glycolysis process

24.100

1. glucose 6-phosphate	2. dihydroxyacetone phosphate	3. 3-phosphoglycerate
4. glyceraldehyde 3-phosphate	5. fructose 6-phosphate	6. phosphoenolpyruvate

Select from the grid *all* correct responses for each of the following situations.

Additional Problems involve more than one concept and are more difficult than the Exercises and Problems. All additional problem sets have been expanded in the second edition and a majority of the problems are new.

Grid Problems examine multiple concepts at once. Answers are selected from a grid of choices, and most often multiple answers are correct.

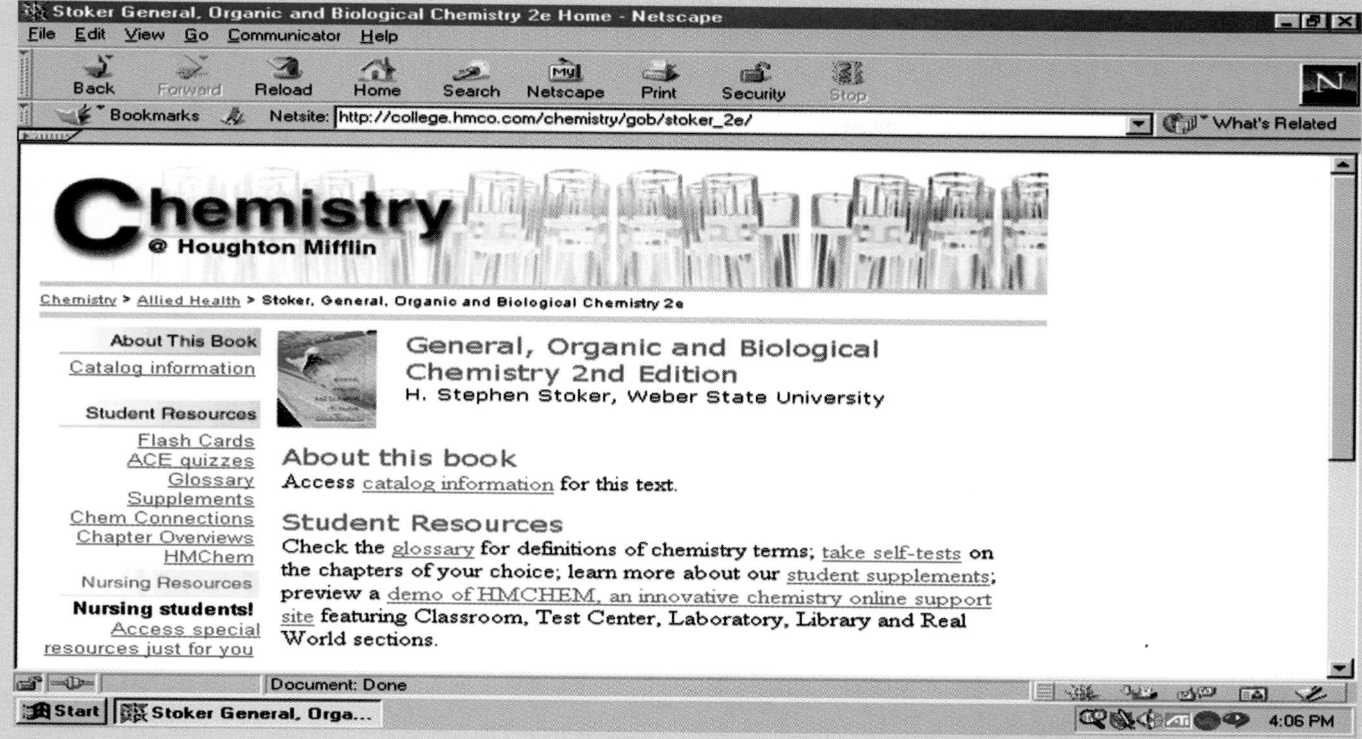

A password-protected **Student Website** is accessible through http://college.hmco.com (select "chemistry"). It includes a wealth of resources to help students in the course, including:

- Chapter overviews
- Self-quizzes with immediate scoring using Houghton Mifflin's ACE system
- Electronic flashcards of key terms, reactions, and concepts
- Additional *Chemical Connections* features
- Additional real-life applications with relevant links
- Career-related information

The **Instructor Website,** accessible through the address above, allows access to all student resources, plus instructor/classroom resources such as downloadable PowerPoint slides and useful links.

PART I

Basic Concepts About Matter

CHAPTER OUTLINE

1.1 Chemistry: The Study of Matter 1
1.2 Physical States of Matter 2
1.3 Properties of Matter 3
1.4 Changes in Matter 4

Chemistry at a Glance: **Use of the Terms *Physical* and *Chemical*** 5

1.5 Pure Substances and Mixtures 6
1.6 Elements and Compounds 7
1.7 Discovery and Abundance of the Elements 9

Chemistry at a Glance: **Classes of Matter** 9

1.8 Names and Symbols of the Elements 11
1.9 Atoms and Molecules 13
1.10 Chemical Formulas 14

Chemical Connections

1.1 "Good" versus "Bad" Properties for a Chemical Substance 4
1.2 Elements Necessary for Human Life 11

Heat and light generation, as well as numerous changes in matter, occur as wood is burned in a bonfire.

In this chapter we address the question "What exactly is chemistry about?" In addition, we consider common terminology associated with the field of chemistry. Much of this terminology is introduced in the context of the ways in which matter is classified. Like all other sciences, chemistry has its own specific language. It is necessary to restrict the meanings of some words so that all chemists (and those who study chemistry) can understand a given description of a chemical phenomenon in the same way.

1.1 Chemistry: The Study of Matter

Chemistry *is the field of study concerned with the characteristics, composition, and transformations of matter.* What is matter? **Matter** *is anything that has mass and occupies space.* The term *mass* refers to the amount of matter present in a sample.

Matter includes all things—both living and nonliving—that can be seen (such as plants, soil, and rocks) as well as things that cannot be seen (such as air and bacteria). Not considered to be matter are the various forms of energy, such as heat, light, and

Everyday activities such as making a batch of muffins, involve chemistry.

• The universe is composed entirely of matter and energy.

electricity. However, chemists must be concerned with energy as well as matter, because almost all changes that matter undergoes involve the release or absorption of energy.

• The term *chemistry* is derived from the word *alchemy,* which denotes practices carried out during the Middle Ages in an attempt to transform something common into something precious (in particular, lead into gold). Alchemy originated in Alexandrian Egypt, and the term *alchemy* is derived from the Greek *al* ("the") and *khemia* (a native name for Egypt).

The scope of chemistry is extremely broad, and it touches every aspect of our lives. An iron gate rusting, a chocolate cake baking, the diagnosis and treatment of a heart attack, the propulsion of a jet airliner, and the digesting of food all fall within the realm of chemistry. The key to understanding such diverse processes is an understanding of the fundamental nature of matter, which is what we are going to talk about for the rest of this chapter.

1.2 Physical States of Matter

Three physical states exist for matter: solid, liquid, and gas. The classification of a given matter sample in terms of physical state is based on whether its shape and volume are definite or indefinite.

• The *volume* of a sample of matter is a measure of the amount of space occupied by the sample.

A **solid** *has a definite shape and a definite volume.* A silver dollar has the same shape and volume whether it is placed in a large container or on a table top (Figure 1.1a). For solids in powdered or granulated forms, such as sugar or salt, a quantity of the solid takes the shape of the portion of the container it occupies, but each individual particle has a definite shape and volume. A **liquid** *has an indefinite shape and a definite volume.* A liquid always takes the shape of its container to the extent that it fills the container (Figure

(a) (b) (c)

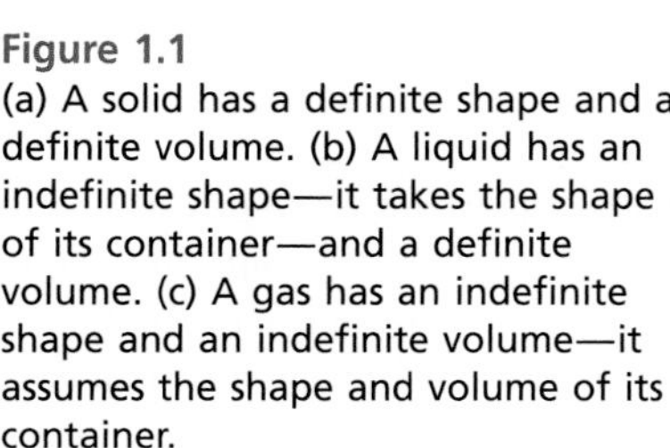

Figure 1.1
(a) A solid has a definite shape and a definite volume. (b) A liquid has an indefinite shape—it takes the shape of its container—and a definite volume. (c) A gas has an indefinite shape and an indefinite volume—it assumes the shape and volume of its container.

Figure 1.2
Water can be found in the solid, liquid, and gaseous forms simultaneously, as shown here at Yellowstone National Park.

Figure 1.3
The green color of the Statue of Liberty (present before it was restored) results from the reaction of the copper skin of the statue with the components of air. The fact that copper will react with the components of air is a chemical property of copper.

• Chemical properties describe the ability of a substance to form new substances, either by reaction with other substances or by decomposition. Physical properties are properties associated with a substance's physical existence. They can be determined without reference to any other substance, and their determination causes no change in the identity of the substance.

1.1b). A **gas** *has an indefinite shape and an indefinite volume.* A gas always completely fills its container, adopting both its volume and its shape (Figure 1.1c).

The state of matter observed for a particular substance depends on its temperature, the surrounding pressure, and the strength of the forces holding its structural particles together. At the temperatures and pressures normally encountered on Earth, water is one of the few substances found in all three of its physical states: solid ice, liquid water, and gaseous steam (Figure 1.2). Under laboratory conditions, states other than those commonly observed can be attained for almost all substances. Oxygen, which is nearly always thought of as a gas, becomes a liquid at −183°C and a solid at −218°C. The metal iron is a gas at extremely high temperatures (above 3000°C).

1.3 Properties of Matter

Various kinds of matter are distinguished from each other by their properties. **Properties** *are the distinguishing characteristics of a substance that are used in its identification and description.* Each substance has a unique set of properties that distinguishes it from all other substances. Properties of matter are of two general types: physical and chemical.

A **physical property** *is a characteristic of a substance that we can observe without changing the basic identity of the substance.* Common physical properties include color, odor, physical state (solid, liquid, or gas), melting point, boiling point, and hardness.

During the process of determining a physical property, the physical appearance of a substance may change, but the substance's identity does not. For example, it is impossible to measure the melting point of a solid without changing the solid into a liquid. Although the liquid's appearance is much different from that of the solid, the substance is still the same; its chemical identity has not changed. Hence melting point is a physical property.

A **chemical property** *is a characteristic of a substance that describes the way the substance undergoes or resists change to form a new substance.* For example, copper objects turn green when exposed to moist air for long periods of time (Figure 1.3); this is a chemical property of copper. The green coating formed on the copper is a new substance that results from the copper's reaction with oxygen, carbon dioxide, and water present in air. The properties of this new substance (the green coating) are very different from those of

Chemical CONNECTIONS

1.1 "Good" versus "Bad" Properties for a Chemical Substance

It is important not to judge the significance or usefulness of a chemical substance on the basis of just one or two of the many chemical and physical properties it exhibits. Possession of a "bad" property, such as toxicity or a strong noxious odor, does not mean that a chemical substance has nothing to contribute to the betterment of human society.

A case in point is the substance carbon monoxide. Everyone knows that it is a gaseous air pollutant present in automobile exhaust and cigarette smoke and that it is toxic to human beings. For this reason, some people automatically label carbon monoxide a "bad" substance, a substance we do not need or want.

Indeed, carbon monoxide is toxic to human beings. It impairs human health by reducing the oxygen-carrying capacity of the blood. Carbon monoxide does this by interacting with the hemoglobin in red blood cells in a way that prevents the hemoglobin from distributing oxygen throughout the body. Someone who dies from carbon monoxide poisoning actually dies from lack of oxygen.

The fact that carbon monoxide is colorless, odorless, and tasteless is very significant. Because of these properties, carbon monoxide gives no warning of its initial presence. There are several other common air pollutants that are more toxic than carbon monoxide. However, they have properties that give warning of their presence and hence are not considered as "dangerous" as carbon monoxide.

Despite its toxicity, carbon monoxide plays an important role in the maintenance of the high standard of living we now enjoy. Its contribution lies in the field of iron metallurgy and the production of steel. The isolation of iron from iron ores, necessary for the production of steel, involves a series of high-temperature reactions, carried out in a blast furnace, in which the iron content of molten iron ores reacts with carbon monoxide. These reactions release the iron from its ores. The carbon monoxide needed in steel making is obtained by reacting coke (a product derived by heating coal to a high temperature without air being present) with oxygen.

The industrial consumption of the metal iron, both in the United States and worldwide, is approximately 10 times greater than that of all other metals combined. Steel production accounts for nearly all of this demand for iron. Without steel, our standard of living would drop dramatically, and carbon monoxide is necessary for the production of steel.

Is carbon monoxide a "good" or a "bad" chemical substance? The answer to this question depends on the context in which the carbon monoxide is encountered. In terms of air pollution, it is a "bad" substance. In terms of steel making, it is a "good" substance. A similar "good–bad" dichotomy exists for almost every chemical substance.

metallic copper. On the other hand, gold objects resist change when exposed to air for long periods of time. The lack of reactivity of gold with air is a chemical property of gold.

Most often the changes associated with chemical properties result from the interaction (reaction) of a substance with one or more other substances. However, the presence of a second substance is not an absolute requirement. Sometimes the presence of energy (usually heat or light) can trigger the change called decomposition. The fact that hydrogen peroxide, in the presence of either heat or light, decomposes into the substances water and oxygen is a chemical property of hydrogen peroxide.

The melting of ice cream is a physical change.

When we specify chemical properties, we usually give conditions such as temperature and pressure because they influence the interactions between substances. For example, the gases oxygen and hydrogen are unreactive toward each other at room temperature, but they interact explosively at a temperature of several hundred degrees.

1.4 Changes in Matter

Changes in matter are common and familiar occurrences. Changes take place when food is digested, paper is burned, and a pencil is sharpened. Like properties of matter, changes in matter are classified into two categories: physical and chemical.

A **physical change** *is a process that does not alter the basic nature (chemical composition) of the substance undergoing change.* A new substance is never formed as a result of a physical change.

A change in physical state is the most common type of physical change. Melting, freezing, evaporation, and condensation are all changes of state. In any of these processes, the

Figure 1.4
As a result of chemical change, bright steel girders become rusty when exposed to moist air.

composition of the substance undergoing change remains the same even though its physical state and appearance change. The melting of ice does not produce a new substance; the substance is water both before and after the change. Similarly, the steam produced from boiling water is still water.

A **chemical change** *is a process that involves a change in the basic nature (chemical composition) of the substance.* Chemical changes always involve conversion of the material or materials under consideration into one or more new substances, each of which has distinctly different properties and composition from the original materials. Consider, for example, the rusting of iron objects left exposed to moist air (Figure 1.4). The reddish brown substance (the rust) that forms is a new substance with chemical properties that are obviously different from those of the original iron.

Chemists study the nature of changes in matter to learn how to bring about favorable changes and prevent undesirable ones. The control of chemical change has been a major factor in attainment of the modern standard of living now enjoyed by most people in the developed world. The many plastics, synthetic fibers, and prescription drugs now in common use are the result of controlled chemical change.

Chemistry at a Glance reviews the ways in which the terms *physical* and *chemical* are used to describe the properties of substances and the changes that substances undergo. Note that the term *physical*, used as a modifier, always conveys the idea that the composition (chemical identity) of a substance did not change, and that the term *chemical*, used as a modifier, always conveys the idea that the composition of a substance did change.

• Physical changes need not involve a change of state. Pulverizing an aspirin tablet into a powder and cutting a piece of adhesive tape into small pieces are physical changes that involve only the solid state.

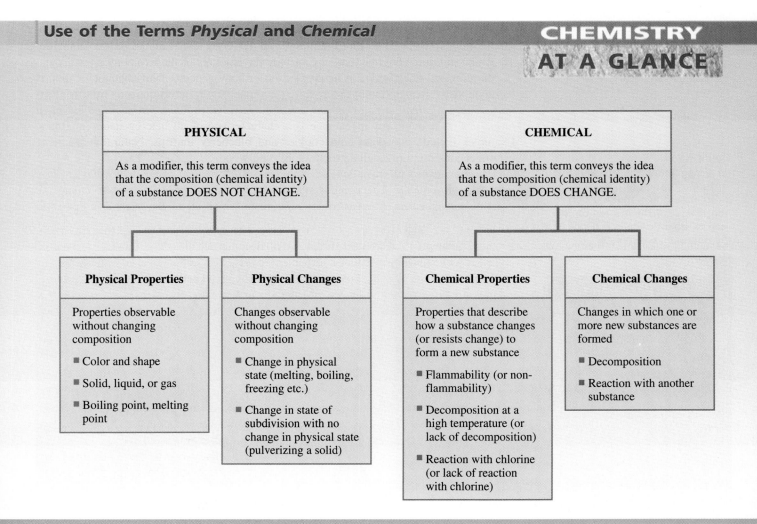

Use of the Terms *Physical* and *Chemical*

CHEMISTRY AT A GLANCE

PHYSICAL

As a modifier, this term conveys the idea that the composition (chemical identity) of a substance DOES NOT CHANGE.

Physical Properties

Properties observable without changing composition

- Color and shape
- Solid, liquid, or gas
- Boiling point, melting point

Physical Changes

Changes observable without changing composition

- Change in physical state (melting, boiling, freezing etc.)
- Change in state of subdivision with no change in physical state (pulverizing a solid)

CHEMICAL

As a modifier, this term conveys the idea that the composition (chemical identity) of a substance DOES CHANGE.

Chemical Properties

Properties that describe how a substance changes (or resists change) to form a new substance

- Flammability (or non-flammability)
- Decomposition at a high temperature (or lack of decomposition)
- Reaction with chlorine (or lack of reaction with chlorine)

Chemical Changes

Changes in which one or more new substances are formed

- Decomposition
- Reaction with another substance

1.5 Pure Substances and Mixtures

In addition to its classification by physical state (Section 1.2), matter can also be classified in terms of its chemical composition as a pure substance or as a mixture. A **pure substance** *is a single kind of matter that cannot be separated into other kinds of matter by any physical means.* All samples of a pure substance contain only that substance and nothing else. Pure water is water and nothing else. Pure sucrose (table sugar) contains only that substance and nothing else.

• *Substance* is a general term used to denote any variety of matter. *Pure substance* is a specific term that applies only to matter that contains a single substance.

A pure substance always has a definite and constant composition. This invariant composition dictates that the properties of a pure substance are always the same under a given set of conditions. Collectively, these definite and constant physical and chemical properties constitute the means by which we identify the pure substance.

• All samples of a pure substance, no matter what their source, have the same properties under the same conditions.

A **mixture** *is a physical combination of two or more pure substances in which each substance retains its own chemical identity.* Components of the mixture retain their identity because they are physically mixed rather than chemically combined. Consider a mixture of small rock salt crystals and ordinary sand. Mixing these two substances changes neither the salt nor the sand in any way. The larger, colorless salt particles are easily distinguished from the smaller, light-gray sand granules.

One characteristic of any mixture is that its components can be separated by using physical means. In our salt–sand mixture, the larger salt crystals could be—though very tediously—"picked out" from the sand. A somewhat easier separation method would be to dissolve the salt in water, which would leave the undissolved sand behind. The salt could then be recovered by evaporation of the water. Figure 1.5 shows a heterogeneous mixture of potassium dichromate (orange crystals) and iron filings. A magnet can be used to separate the components of this mixture.

• Most naturally occurring samples of matter are mixtures. Gold and diamond are two of the few naturally occurring pure substances. Despite their scarcity in nature, numerous pure substances are known. They are obtained from natural mixtures by using various types of separation techniques or are synthesized in the laboratory from naturally occurring materials.

Another characteristic of a mixture is variable composition. Numerous different salt–sand mixtures, with compositions ranging from a slightly salty sand mixture to a slightly sandy salt mixture, could be made by varying the amounts of the two components.

Mixtures are subclassified as heterogeneous or homogeneous. This subclassification is based on visual recognition of the mixture's components. A **heterogeneous mixture** *contains visibly different parts, or phases, each of which has different physical and chemical properties.* A nonuniform appearance is a characteristic of all heterogeneous mixtures. Examples include chocolate chip cookies and blueberry muffins. Naturally occurring heterogeneous mixtures include rocks, soils, and wood.

A **homogeneous mixture** *contains only one visibly distinct phase, which has uniform properties throughout.* The components present in a homogeneous mixture cannot be visually distinguished. A sugar–water mixture in which all of the sugar has dissolved

(a)

(b)

Figure 1.5
(a) A magnet (on the left) and a mixture consisting of potassium dichromate (the orange crystals) and iron filings. (b) The magnet can be used to separate the iron filings from the potassium dichromate.

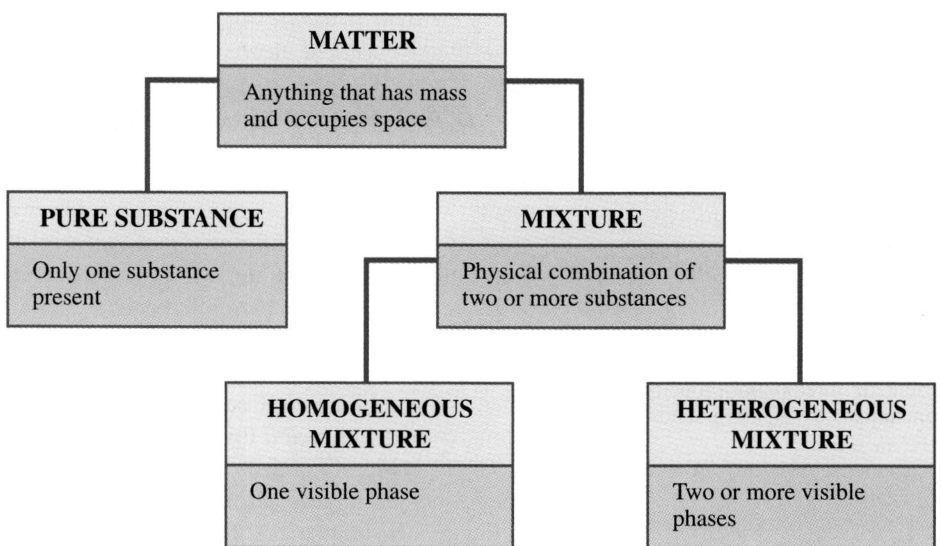

Figure 1.6 Matter falls into two basic classes: pure substances and mixtures. Mixtures, in turn, may be homogeneous or heterogeneous.

has an appearance similar to that of pure water. Air is a homogeneous mixture of gases; motor oil and gasoline are multicomponent homogeneous mixtures of liquids; and metal alloys such as 14-karat gold (a mixture of copper and gold) are examples of homogeneous mixtures of solids.

Figure 1.6 summarizes what we have learned thus far about various classifications of matter.

1.6 Elements and Compounds

• Both elements and compounds are pure substances.

Chemists have isolated and characterized an estimated 8.5 million pure substances. A very small number of these pure substances, 115 to be exact, are different from all of the others. They are elements. All of the rest, the remaining millions, are compounds. What distinguishes an element from a compound?

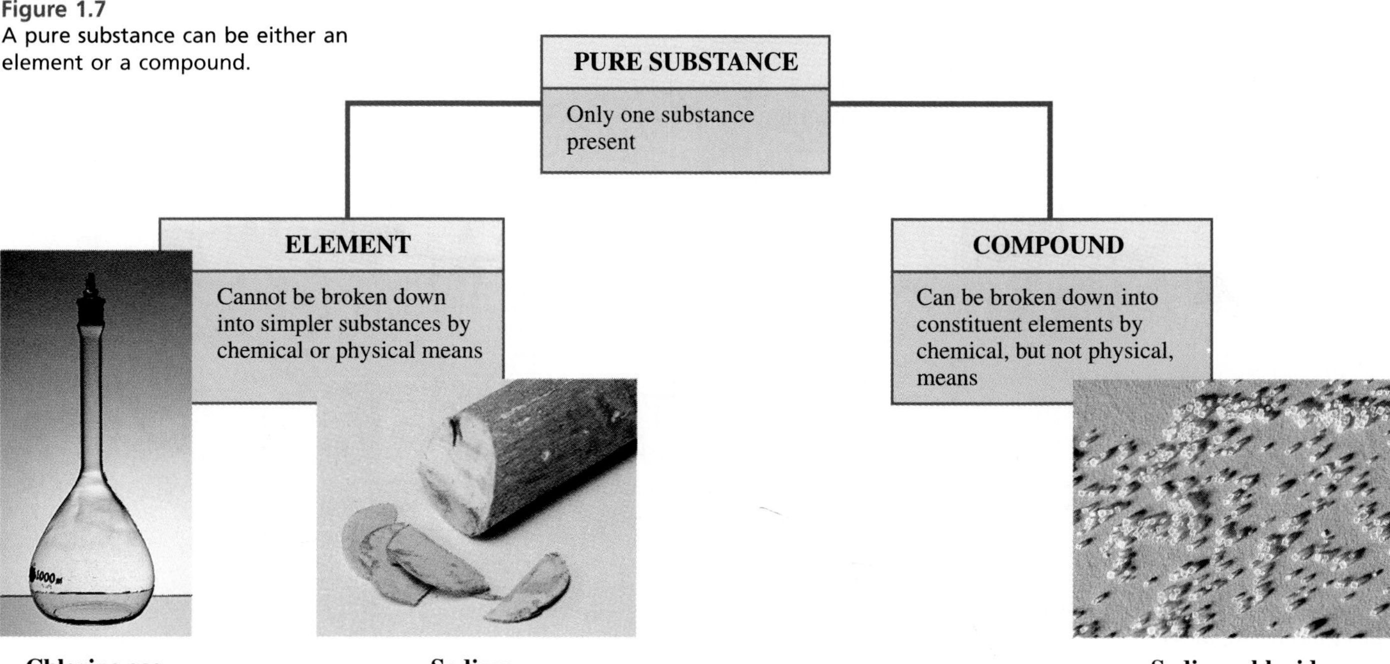

Figure 1.7 A pure substance can be either an element or a compound.

• The definition for the term *element* that is given here will do for now. After considering the concept of atomic number (Section 3.3), we will give a more precise definition.

• Every known compound is made up of some combination of the 115 known elements. In any given compound, the elements are combined chemically in fixed proportions by mass.

• There are three major property distinctions between compounds and mixtures.

1. Compounds have properties distinctly different from those of the substances that combined to form the compound. The components of mixtures retain their individual properties.
2. Compounds have a definite composition. Mixtures have a variable composition.
3. Physical methods are sufficient to separate the components of a mixture. The components of a compound cannot be separated by physical methods; chemical methods are required.

An **element** *is a pure substance that cannot be broken down into simpler substances by ordinary chemical means such as a reaction, an electric current, heat, or a beam of light.* The metals gold, silver, and copper are all elements.

A **compound** *is a pure substance that can be broken down into two or more simpler substances by chemical means.* Water is a compound. By means of an electric current, water can be broken down into the gases hydrogen and oxygen, both of which are elements. The ultimate breakdown products for any compound are elements. A compound's properties are always different from those of its component elements, because the elements are chemically rather than physically combined in the compound (Figure 1.7).

Even though two or more elements are obtained from decomposition of compounds, compounds are not mixtures. Why is this so? Remember, substances can be combined either physically or chemically. Physical combination of substances produces a mixture. Chemical combination of substances produces a compound, a substance in which combining entities are *bound* together. No such binding occurs during physical combination.

The following Chemistry at a Glance summarizes what we have learned thus far about matter, including pure substances, elements, compounds, and mixtures.

Figure 1.8 summarizes the thought processes that a chemist goes through in classifying a sample of matter as a heterogeneous mixture, homogeneous mixture, element, or compound. This figure is based on the following three questions about a sample of matter:

1. Does the sample of matter have the same properties throughout?
2. Are two or more different substances present?
3. Can the pure substance be broken down into simpler substances?

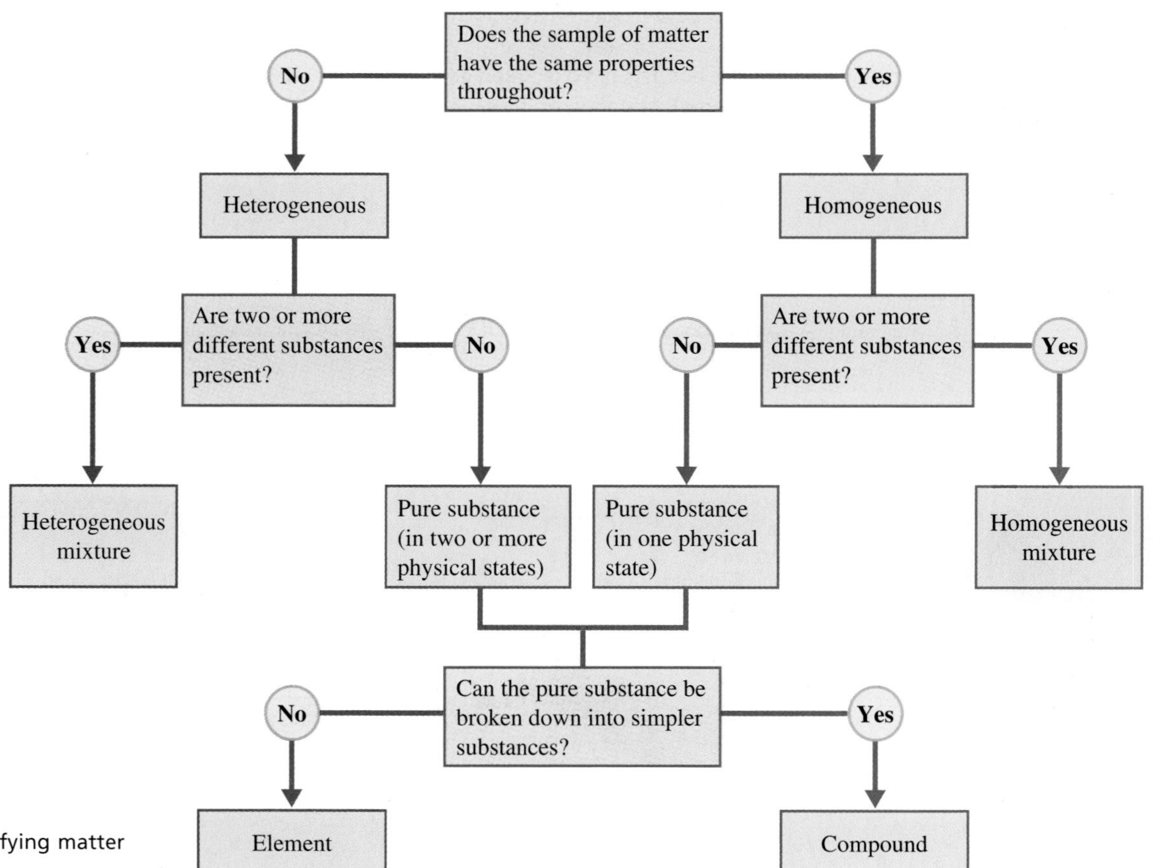

Figure 1.8
Questions used in classifying matter into various categories.

CHEMISTRY AT A GLANCE

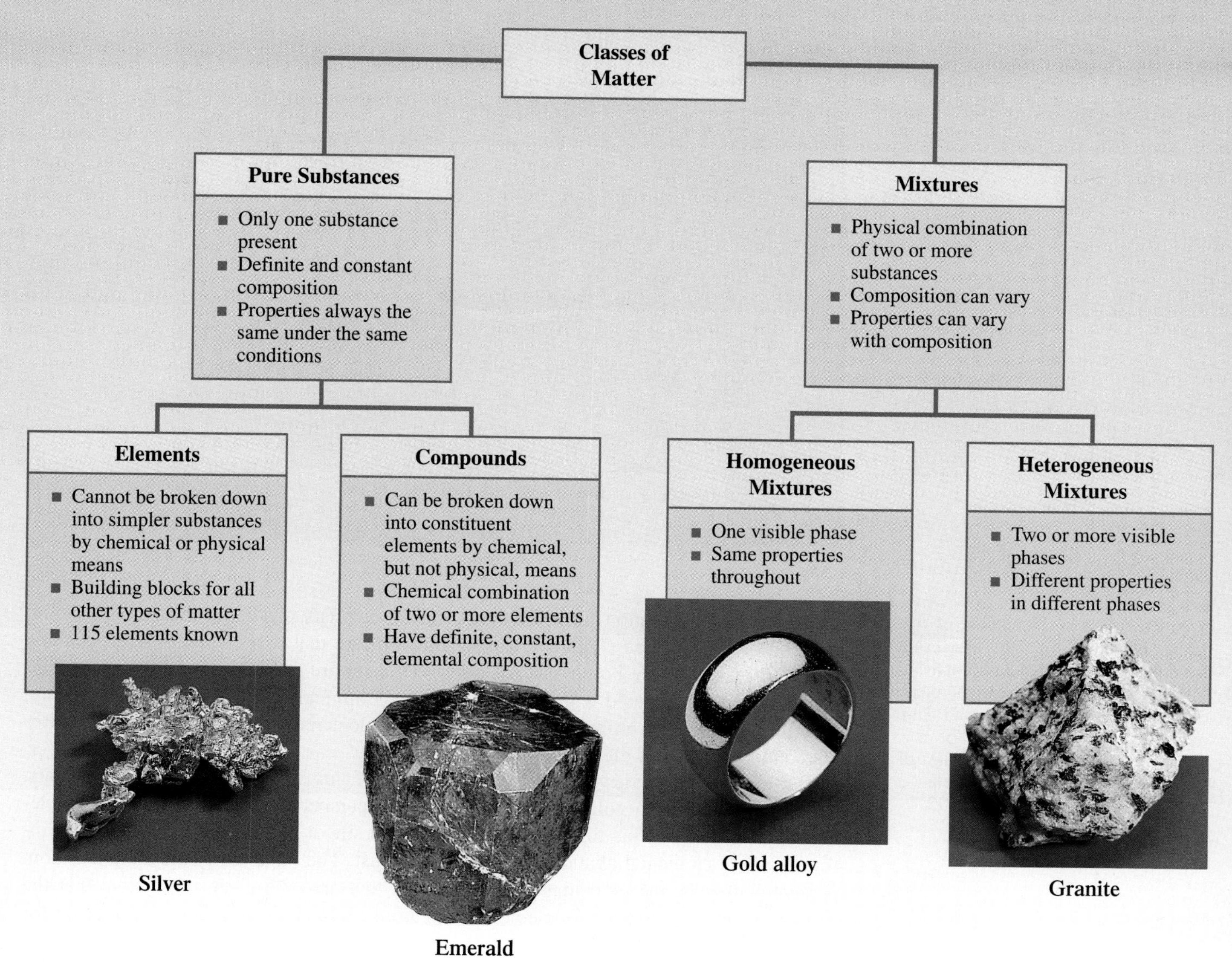

Classes of Matter
Pure Substances
■ Only one substance present
■ Definite and constant composition
■ Properties always the same under the same conditions
Mixtures
■ Physical combination of two or more substances
■ Composition can vary
■ Properties can vary with composition
Elements
■ Cannot be broken down into simpler substances by chemical or physical means
■ Building blocks for all other types of matter
■ 115 elements known
Compounds
■ Can be broken down into constituent elements by chemical, but not physical, means
■ Chemical combination of two or more elements
■ Have definite, constant, elemental composition
Homogeneous Mixtures
■ One visible phase
■ Same properties throughout
Heterogeneous Mixtures
■ Two or more visible phases
■ Different properties in different phases
Silver
Emerald
Gold alloy
Granite

1.7 Discovery and Abundance of the Elements

• A student who attended a university in the year 1700 would have been taught that 13 elements existed. In 1750 he or she would have learned about 16 elements, in 1800 about 34, in 1850 about 59, in 1900 about 82, and in 1950 about 98. Today's total of 115 elements was reached in 1999.

The discovery and isolation of the 115 known elements, the building blocks for all matter, have taken place over a period of several centuries. Most of the discoveries have occurred since 1700, the 1800s being the most active period.

Eighty-eight of the 115 elements occur naturally, and 27 have been synthesized in the laboratory by bombarding samples of naturally occurring elements with small particles. Figure 1.9 shows samples of selected naturally occurring elements. The synthetic (laboratory-produced) elements are all unstable (radioactive) and usually revert quickly to the naturally occurring elements (see Section 11.5).

The naturally occurring elements are not evenly distributed on Earth and in the universe. What is startling is the nonuniformity of the distribution. A small number of elements account for the majority of elemental particles (atoms). (An atom is the smallest particle of an element that can exist. See Section 1.9.)

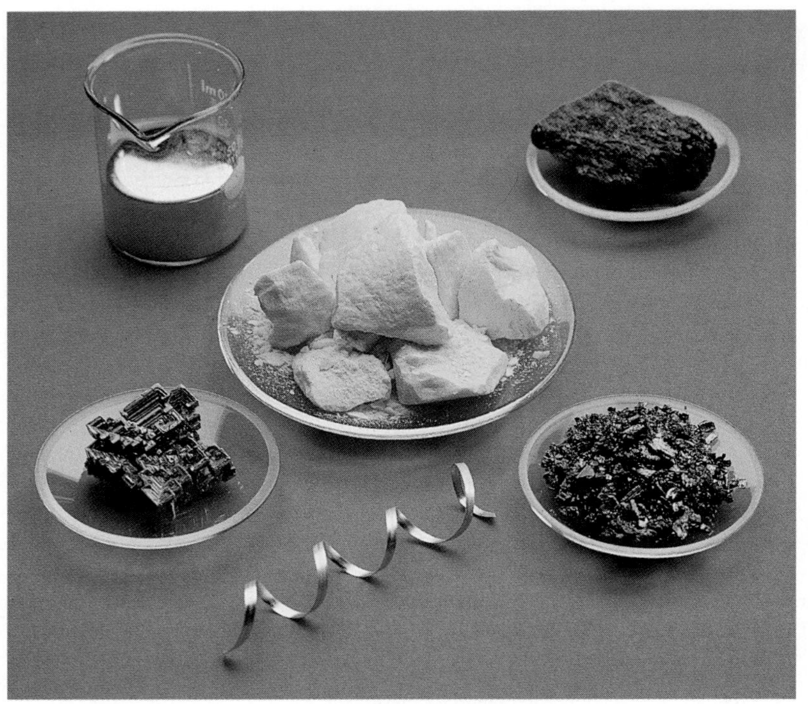

Figure 1.9
Outward physical appearance of selected naturally occurring elements. *Center:* Sulfur. *From upper right, clockwise:* Arsenic, iodine, magnesium, bismuth, and mercury.

• Any increase in the number of known elements from 115 will result from the production of additional synthetic elements. Current chemical theory strongly suggests that all naturally occurring elements have been identified. The isolation of the last of the known naturally occurring elements, rhenium, occurred in 1925.

Studies of the radiation emitted by stars enable scientists to estimate the elemental composition of the universe (Figure 1.10). Results indicate that two elements, hydrogen and helium, are absolutely dominant. All other elements are mere "impurities" when their abundances are compared with those of these two dominant elements. In this big picture, in which Earth is but a tiny microdot, 91% of all elemental particles (atoms) are hydrogen, and almost all of the remaining 9% are helium.

If we narrow our view to the chemical world of humans—Earth's crust (its waters, atmosphere, and outer solid surface)—a different perspective emerges. Again, two elements dominate, but this time they are oxygen and silicon. Figure 1.10 also provides information on elemental abundances for Earth's crust. The numbers in the table are atom percents—that is, the percentage of total atoms that are of a given type. Note that the eight elements listed (the only elements with atom percents greater than 1%) account for

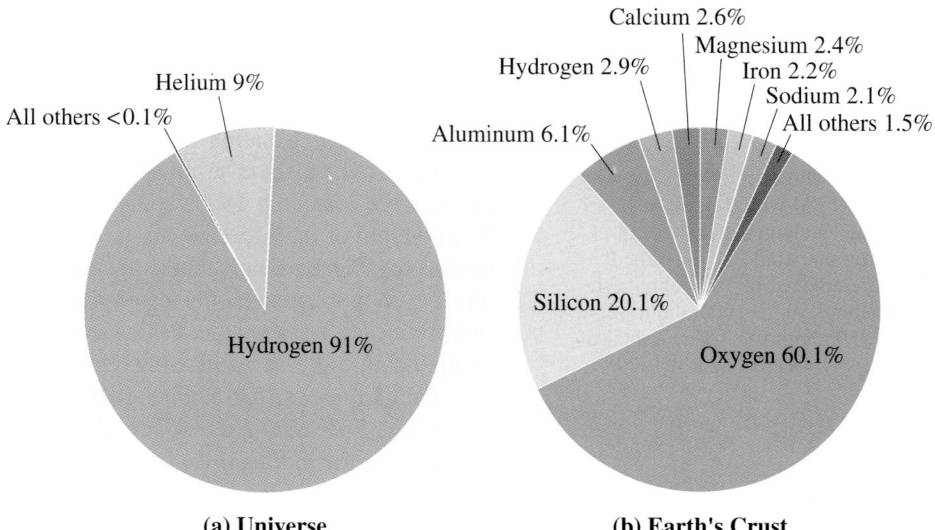

Figure 1.10
Abundance of elements in the universe and in Earth's crust (in atom percent).

Chemical CONNECTIONS

1.2 Elements Necessary for Human Life

The distribution of elements in the human body and other living systems is very different from that found in Earth's crust. This distribution is the result of living systems *selectively* taking up matter from their external environment rather than simply accumulating matter representative of their surroundings.

Eleven elements are found in the human body in atom percent levels of 0.01 or greater, as shown in the accompanying table.

Element	Percent of total number of atoms in the human body
hydrogen	63
oxygen	25.5
carbon	9.5
nitrogen	1.4
calcium	0.31
phosphorus	0.22
potassium	0.06
sulfur	0.05
chlorine	0.03
sodium	0.03
magnesium	0.01

The high abundances of hydrogen and oxygen in the body reflect its high water content. Hydrogen is over twice as abundant as oxygen, largely because water contains hydrogen and oxygen in a 2-to-1 atom ratio.

The compositions of the four major classes of compounds present in the human body (carbohydrates, fats, proteins, and nucleic acids) involve the elements carbon, nitrogen, sulfur, and phosphorus, together with hydrogen and oxygen. All four of these types of compounds contain hydrogen, oxygen, and carbon. Proteins also contain nitrogen and sulfur, and nucleic acids (DNA, RNA) also contain nitrogen and phosphorus.

The following table offers additional information about the functions of the 11 most abundant elements in the human body. Several other elements, including iron, copper, and iodine, are needed in trace amounts. These trace elements are considered in Chemical Connections 3.2.

Elements Found in Major Types of Biochemical Molecules	
hydrogen	present in water, carbohydrates, fats, proteins, and nucleic acids
oxygen	present in water, carbohydrates, fats, proteins, and nucleic acids
carbon	present in carbohydrates, fats, proteins, and nucleic acids
nitrogen	present in proteins and nucleic acids
sulfur	present in proteins
phosphorus	present in nucleic acids

Additional Elements Needed by the Human Body	
calcium	present in bones and teeth; needed for proper blood clotting
potassium	helps regulate movement of body fluids
chlorine	helps regulate movement of body fluids
sodium	helps regulate movement of body fluids
magnesium	enhances nerve action and muscle tissue action

over 98% of total atoms in Earth's crust. Note also the dominance of oxygen and silicon; these two elements account for 80% of the atoms that make up the chemical world of humans.

1.8 Names and Symbols of the Elements

Each element has a unique name that, in most cases, was selected by its discoverer. A wide variety of rationales for choosing a name can be found when studying name origins. Some elements bear geographical names; germanium is named after the native country of its German discoverer, and the elements francium and polonium are named after France and Poland. The elements mercury, uranium, neptunium, and plutonium are all named for planets. Helium gets its name from the Greek word *helios,* for "sun," because it was first observed spectroscopically in the sun's corona during an eclipse. Some elements carry names that reflect specific properties of the element or of the compounds that contain it. Chlorine's name is derived from the Greek *chloros,* denoting "greenish-yellow," the color of chlorine gas. Iridium gets its name from the Greek *iris,* meaning "rainbow"; this alludes to the varying colors of the compounds from which it was isolated.

• Learning the symbols of the more common elements is an important key to success in studying chemistry. Knowledge of chemical symbols is essential for writing chemical formulas (Section 1.10) and chemical equations (Section 6.6).

Abbreviations called chemical symbols exist for the names of the elements. A **chemical symbol** *is a one- or two-letter notation used to represent the name of an element.* These symbols are used more frequently than the elements' names. They can be written more quickly than the names, and they occupy less space. A complete list of the known elements and their symbols is given in Table 1.1. The symbols and names of the more frequently encountered elements are shown in color in this table.

Note that the first letter of a chemical symbol is always capitalized and that the second is not. Two-letter symbols are often, but not always, the first two letters of the element's name.

Eleven elements have symbols that bear no relationship to the element's English-language name. In ten of these cases, the symbol is derived from the Latin name of the element; in the case of the element tungsten, a German name is the symbol's source.

Table 1.1
The Chemical Symbols for the Elements[a]

Ac	actinium	Ge	germanium	Pr	praseodymium
Ag	silver*	H	hydrogen	Pt	platinum
Al	aluminum	He	helium	Pu	plutonium
Am	americium	Hf	hafnium	Ra	radium
Ar	argon	Hg	mercury*	Rb	rubidium
As	arsenic	Ho	holmium	Re	rhenium
At	astatine	Hs	hassium	Rf	rutherfordium
Au	gold*	I	iodine	Rh	rhodium
B	boron	In	indium	Rn	radon
Ba	barium	Ir	iridium	Ru	ruthenium
Be	beryllium	K	potassium*	S	sulfur
Bh	bohrium	Kr	krypton	Sb	antimony*
Bi	bismuth	La	lanthanum	Sc	scandium
Bk	berkelium	Li	lithium	Se	selenium
Br	bromine	Lu	lutetium	Sg	seaborgium
C	carbon	Lr	lawrencium	Si	silicon
Ca	calcium	Md	mendelevium	Sm	samarium
Cd	cadmium	Mg	magnesium	Sn	tin*
Ce	cerium	Mn	manganese	Sr	strontium
Cf	californium	Mo	molybdenum	Ta	tantalum
Cl	chlorine	Mt	meitnerium	Tb	terbium
Cm	curium	N	nitrogen	Tc	technetium
Co	cobalt	Na	sodium*	Te	tellurium
Cr	chromium	Nb	niobium	Th	thorium
Cs	cesium	Nd	neodymium	Ti	titanium
Cu	copper*	Ne	neon	Tl	thallium
Db	dubnium	Ni	nickel	Tm	thulium
Dy	dysprosium	No	nobelium	U	uranium
Er	erbium	Np	neptunium	V	vanadium
Es	einsteinium	O	oxygen	W	tungsten*
Eu	europium	Os	osmium	Xe	xenon
F	fluorine	P	phosphorus	Y	yttrium
Fe	iron*	Pa	protactinium	Yb	ytterbium
Fm	fermium	Pb	lead*	Zn	zinc
Fr	francium	Pd	palladium	Zr	zirconium
Ga	gallium	Pm	promethium		
Gd	gadolinium	Po	polonium		

[a]Only 109 elements are listed in this table. Elements 110–115, discovered (synthesized) in the period 1994–1999, are yet to be named.

*These elements have symbols that were derived from non-English names.

Most of these elements have been known for hundreds of years and date back to the time when Latin was the language of scientists. Elements whose symbols are derived from non-English names are marked with an asterisk in Table 1.1.

Figure 1.11
A computer reconstruction of the surface of a sample of graphite (carbon) as observed with a scanning tunneling microscope. The image reveals the regular pattern of individual carbon atoms. The color was added to the image by computer.

1.9 Atoms and Molecules

Consider the process of subdividing a sample of the element gold (or any other element) into smaller and smaller pieces. It seems reasonable that eventually a "smallest possible piece" of gold would be reached that could not be divided further and still be the element gold. This smallest possible unit of gold is called a gold atom. An **atom** *is the smallest particle of an element that can exist and still have the properties of the element.*

A sample of any element is composed of atoms of a single type, those of that element. In contrast, a compound must have two or more types of atoms present, because by definition at least two elements must be present (Section 1.6).

No one ever has seen or ever will see an atom with the naked eye; they are simply too small for such observation. However, sophisticated electron microscopes, with magnification factors in the millions, have made it possible to photograph "images" of individual atoms (Figure 1.11).

Atoms are incredibly small particles. Atomic dimensions, although not directly measurable, can be calculated from measurements made on large-size samples of elements. The diameter of an atom is approximately four-billionths of an inch. If atoms of such diameter were arranged in a straight line, it would take 254 million of them to extend a distance of 1 inch.

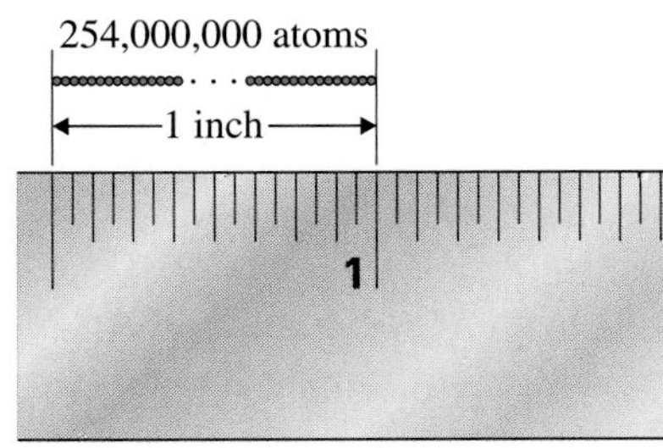

For some elements, such as helium, neon, and argon, the atoms exist as *separate individual* entities. For other elements, such as hydrogen, nitrogen, and oxygen, the atoms exist in groupings called *molecules.* A **molecule** *is a group of two or more atoms that function as a unit because the atoms are tightly bound together.* In the case of the elements hydrogen, oxygen, chlorine, and nitrogen, two atoms are present in a molecule (Figure 1.12). There are four atoms present in a phosphorus molecule and eight atoms present in a sulfur molecule. On the basis of the *number of atoms* present, molecules may be classified as *diatomic* (two atoms), *triatomic* (three atoms), *tetraatomic* (four atoms), and so on.

Either all of the atoms in a molecule are the same or two or more kinds of atoms are present. This leads to two further molecular categories. A **homoatomic molecule** *is a molecule in which all atoms present are of the same kind.* A substance containing homoatomic molecules must be an element. A **heteroatomic molecule** *is a molecule in which two or more kinds of atoms are present.* Substances that contain heteroatomic molecules must be compounds, because the presence of two or more kinds of atoms reflects the presence of two or more kinds of elements.

• The Latin word *mole* means "a mass." The word *molecule* denotes a "little mass."

The number of atoms present in the heteroatomic molecule of a compound varies with the compound's identity. A water molecule contains 3 atoms: 2 hydrogen atoms and 1 oxygen atom. The compound sucrose (table sugar) has a much larger molecule: 45 atoms are present, of which 12 are carbon atoms, 22 are hydrogen atoms, and 11 are oxygen

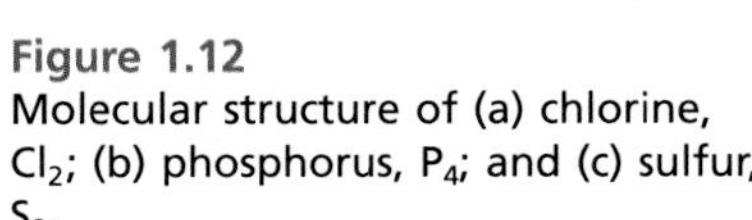

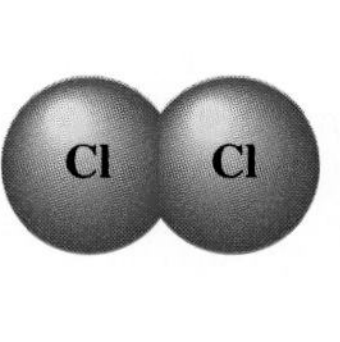

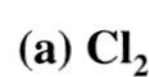

(a) Cl_2

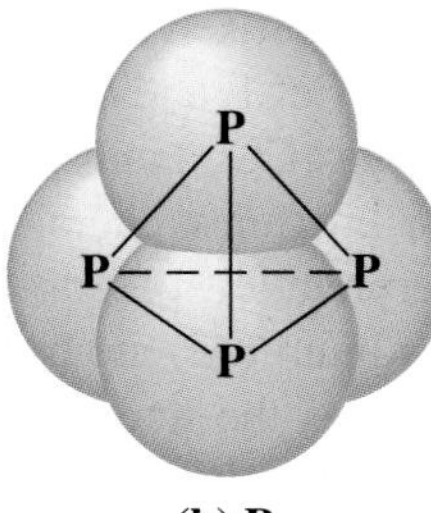

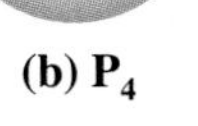

(b) P_4

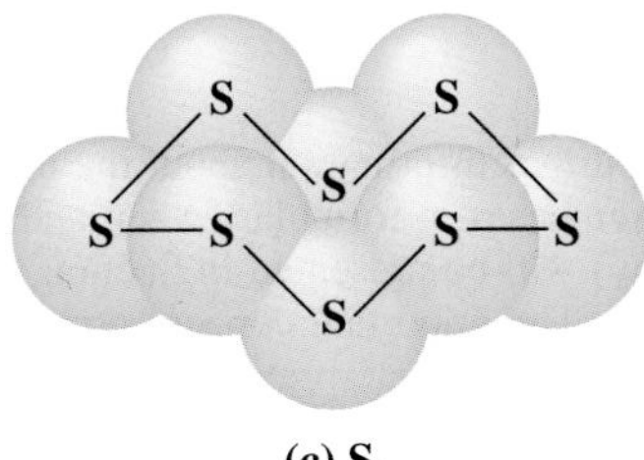

(c) S_8

Figure 1.12
Molecular structure of (a) chlorine, Cl_2; (b) phosphorus, P_4; and (c) sulfur, S_8.

Figure 1.13
Depictions of various simple heteroatomic molecules using models. Spheres of different sizes and colors represent different kinds of atoms.

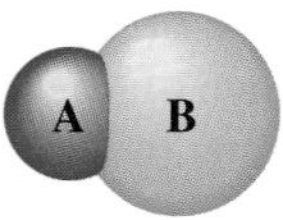

(a) A diatomic molecule containing one atom of A and one atom of B

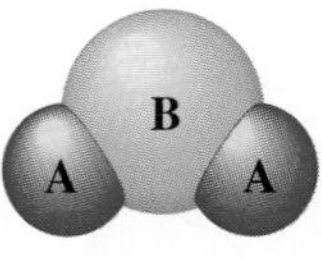

(b) A triatomic molecule containing two atoms of A and one atom of B

(c) A tetraatomic molecule containing two atoms of A and two atoms of B

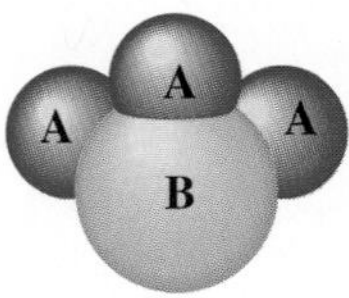

(d) A tetraatomic molecule containing three atoms of A and one atom of B

atoms. Figure 1.13 shows general models for four simple types of heteroatomic molecules. Comparison of parts (c) and (d) of this figure shows that molecules with the same number of atoms need not have the same arrangement of atoms.

• The concept that heteroatomic molecules are the building blocks for all compounds will have to be modified when certain solids, called ionic solids, are considered in Section 4.6.

A molecule is the smallest particle of a compound capable of a stable independent existence. Continued subdivision of a quantity of table sugar to yield smaller and smaller amounts would ultimately lead to the isolation of one single "unit" of table sugar: a molecule of table sugar. This table sugar molecule could not be broken down any further and still exhibit the physical and chemical properties of table sugar. The table sugar molecule could be broken down further by chemical (not physical) means to produce atoms, but if that occurred, we would no longer have table sugar. The *molecule* is the limit of *physical* subdivision. The *atom* is the limit of *chemical* subdivision.

1.10 Chemical Formulas

Information about compound composition can be presented in a concise way by using a chemical formula. A **chemical formula** *is a notation made up of the symbols of the elements present in a compound and the numerical subscripts (located to the right of each symbol) that indicate the number of atoms of each element present in a molecule of the compound.*

The chemical formula for the compound aspirin is $C_9H_8O_4$. This chemical formula conveys the information that an aspirin molecule contains three different elements—carbon (C), hydrogen (H), and oxygen (O)—and 21 atoms—9 carbon atoms, 8 hydrogen atoms, and 4 oxygen atoms.

When only one atom of a particular element is present in a molecule of a compound, that element's symbol is written without a numerical subscript in the formula for the compound. The formula for rubbing alcohol, C_3H_6O, reflects this practice for the element oxygen.

In order to write formulas correctly, one must follow the capitalization rules for elemental symbols (Section 1.8). Making the error of capitalizing the second letter of an element's symbol can dramatically alter the meaning of a chemical formula. The formulas $CoCl_2$ and $COCl_2$ illustrate this point; the symbol Co stands for the element cobalt, whereas CO stands for one atom of carbon and one atom of oxygen.

• Further information about the use of parentheses in formulas (when and why) will be presented in Section 4.11. The important concern now is being able to interpret formulas that contain parentheses in terms of total atoms present.

Sometimes chemical formulas contain parentheses; an example is $Al_2(SO_4)_3$. The interpretation of this formula is straightforward; in a formula unit, there are present 2 aluminum (Al) atoms and 3 SO_4 groups. The subscript following the parentheses always indicates the number of units in the formula of the polyatomic entity inside the parentheses. In terms of atoms, the formula $Al_2(SO_4)_3$ denotes 2 aluminum (Al) atoms, $3 \times 1 = 3$ sulfur (S) atoms, and $3 \times 4 = 12$ oxygen (O) atoms. Example 1.1 contains further comments about chemical formulas that contain parentheses.

Example 1.1

Interpreting Chemical Formulas

For each of the following chemical formulas, state how many atoms of each element are present in one molecule of the substance.

a. HCN—hydrogen cyanide, a poisonous gas
b. $C_{18}H_{21}NO_3$—codeine, a pain-killing drug
c. $Ca_{10}(PO_4)_6(OH)_2$—hydroxyapatite, present in tooth enamel

Solution

a. One atom each of the elements hydrogen, carbon, and nitrogen is present. Remember that the subscript 1 is implied when no subscript is written.
b. This formula indicates that 18 carbon atoms, 21 hydrogen atoms, 1 nitrogen atom, and 3 oxygen atoms are present in one molecule of the compound.
c. There are 10 calcium atoms. The amounts of phosphorus, hydrogen, and oxygen are affected by the subscripts outside the parentheses. There are 6 phosphorus atoms and 2 hydrogen atoms present. Oxygen atoms are present in two locations in the formula. There are a total of 26 oxygen atoms: 24 from the PO_4 subunits (6×4) and 2 from the OH subunits (2×1).

Practice Exercise 1.1

For each of the following chemical formulas, state how many atoms of each element are present in one molecule of the substance.

a. H_2SO_4—sulfuric acid, an industrial acid
b. $C_{17}H_{20}N_4O_6$—riboflavin, a B vitamin
c. $Ca(NO_3)_2$—calcium nitrate, used in fireworks to give a reddish color

• *Answers* **a.** 2 hydrogen, 1 sulfur, and 4 oxygen atoms; **b.** 17 carbon, 20 hydrogen, 4 nitrogen, and 6 oxygen atoms; **c.** 1 calcium, 2 nitrogen, and 6 oxygen atoms

Concepts to Remember

Chemistry. Chemistry is the field of study that is concerned with the characterization, composition, and transformations of matter.

Matter. Matter, the substances of the physical universe, is anything that has mass and occupies space. Matter exists in three physical states: solid, liquid, and gas.

Properties of matter. Properties, the distinguishing characteristics of a substance that are used in its identification and description, are of two types: physical and chemical. Physical properties are properties that we can observe without changing a substance into another substance. Chemical properties are properties that matter exhibits as it undergoes or resists changes in chemical composition. The failure of a substance to undergo change in the presence of another substance is considered a chemical property.

Changes in matter. Changes that can occur in matter are classified into two types: physical and chemical. A physical change is a process that does not alter the basic nature (chemical composition) of the substance under consideration. No new substances are ever formed as a result of a physical change. A chemical change is a process that involves a change in the basic nature (chemical composition) of the substance. Such changes always involve conversion of the material or materials under consideration into one or more new substances that have properties and composition distinctly different from those of the original materials.

Pure substances and mixtures. All specimens of matter are either pure substances or mixtures. A pure substance is a form of matter that always has a definite and constant composition. A mixture is a physical combination of two or more pure substances in which the pure substances retain their identity.

Types of mixtures. Mixtures can be classified as heterogeneous or homogeneous on the basis of the visual recognizability of the components present. A heterogeneous mixture contains visibly different parts or phases, each of which has different properties. A homogeneous mixture contains only one phase, which has uniform properties throughout it.

Types of pure substances. A pure substance can be classified as either an element or a compound on the basis of whether it can be broken down into two or more simpler substances by ordinary chemical means. Elements cannot be broken down into simpler substances. Compounds yield two or more simpler substances when broken down. There are 115 pure substances that qualify as elements. There are millions of compounds.

Atoms and molecules. An atom is the smallest particle of an element that can exist and still have the properties of the element. Free isolated atoms are rarely encountered in nature. Instead, atoms are almost always found together in aggregates or clusters. A molecule is a group of two or more atoms that functions as a unit because the atoms are tightly bound together.

Types of molecules. Molecules are of two types: homoatomic and heteroatomic. Homoatomic molecules are molecules in which all atoms present are of the same kind. A pure substance containing homoatomic molecules is an element. Heteroatomic molecules are molecules in which two or more different kinds of atoms are present. Pure substances that contain heteroatomic molecules must be compounds.

Chemical symbols. Chemical symbols are a shorthand notation for the names of the elements. Most consist of two letters; a few involve a single letter. The first letter of a chemical symbol is always capitalized, and the second letter is always lower-case.

Chemical formulas. Chemical formulas are used to specify compound composition in a concise manner. They consist of the symbols of the elements present in the compound and numerical subscripts (located to the right of each symbol) that indicate the number of atoms of each element present in a molecule of the compound.

Key Terms

Atom (1.9)
Chemical change (1.4)
Chemical formula (1.10)
Chemical property (1.3)
Chemical symbol (1.8)
Chemistry (1.1)
Compound (1.6)
Element (1.6)
Gas (1.2)
Heteroatomic molecule (1.9)
Heterogeneous mixture (1.5)
Homoatomic molecule (1.9)
Homogeneous mixture (1.5)
Liquid (1.2)
Matter (1.1)
Mixture (1.5)
Molecule (1.9)
Physical change (1.4)
Physical property (1.3)
Properties (1.3)
Pure substance (1.5)
Solid (1.2)

Exercises and Problems

The members of each pair of problems in this section test similar material.

Chemistry: The Study of Matter (Section 1.1)

1.1 Classify each of the following as matter or energy (nonmatter).

a. Air b. Pizza c. Sound
d. Light e. Gold f. Virus

1.2 Classify each of the following as matter or energy (nonmatter).

a. Electricity b. Bacteria c. Silver
d. Cake e. Water f. Magnetism

Physical States of Matter (Section 1.2)

1.3 Give a characteristic that distinguishes

a. liquids from solids b. gases from liquids

1.4 Give a characteristic that is the same for

a. liquids and solids b. gases and liquids

1.5 Indicate whether each of the following would take the shape of its container and also have a definite volume.

a. Copper wire b. Oxygen gas
c. Granulated sugar d. Liquid water

1.6 Indicate whether each of the following would take the shape of its container and also have an indefinite volume.

a. Aluminum powder b. Carbon dioxide gas
c. Clean air d. Gasoline

Properties of Matter (Section 1.3)

1.7 The following are properties of the substance magnesium. Classify each property as physical or chemical.

a. Solid at room temperature
b. Ignites upon heating in air
c. Hydrogen gas is produced when it is dissolved in acids
d. Has a density of 1.738 g/cm^3 at 20°C

1.8 The following are properties of the substance magnesium. Classify each property as physical or chemical.

a. Silvery-white in color
b. Does not react with cold water
c. Melts at 651°C
d. Finely divided form burns in oxygen with a dazzling white flame

1.9 Indicate whether each of the following statements describes a physical or a chemical property.

a. Silver salts discolor the skin by reacting with skin protein.
b. Hemoglobin molecules have a red color.
c. Beryllium metal vapor is extremely toxic to humans.
d. Aspirin tablets can be pulverized with a hammer.

1.10 Indicate whether each of the following statements describes a physical or a chemical property.

a. Diamonds are very hard substances.
b. Gold metal does not react with nitric acid.
c. Lithium metal is light enough to float on water.
d. Mercury is a liquid at room temperature.

Changes in Matter (Section 1.4)

1.11 Does each of the following statements describe a physical change or a chemical change?

a. An Alka-Seltzer tablet is dropped into water.
b. A table leg is fashioned from a piece of wood.
c. Water is made into ice cubes.
d. Leaves turn red in autumn.

1.12 Does each of the following statements describe a physical change or a chemical change?

a. A board is sawed into two pieces.
b. Frozen orange juice is reconstituted by adding water to it.

c. A container of grape juice left for an extended period of time becomes fermented.

d. Ski goggles get fogged up.

1.13 Correctly complete each of the following sentences by placing the word *chemical* or *physical* in the blank.

a. The freezing over of a pond's surface is a ____________ process.

b. The crushing of some ice to make ice chips is a ____________ procedure.

c. The destruction of a newspaper through burning it is a ____________ process.

d. Pulverizing a hard sugar cube using a mallet is a ____________ procedure.

1.14 Correctly complete each of the following sentences by placing the word *chemical* or *physical* in the blank.

a. The reflection of light by a shiny metallic object is a ____________ process.

b. The heating of a blue powdered material to produce a white glassy-type substance and a gas is a ____________ procedure.

c. A burning candle produces light by ____________ means.

d. The grating of a piece of cheese is a ____________ technique.

Pure Substances and Mixtures (Section 1.5)

1.15 Classify each of the following statements as true or false.

a. All heterogeneous mixtures must contain three or more substances.

b. Pure substances cannot have a variable composition.

c. Substances maintain their identity in a heterogeneous mixture but not in a homogeneous mixture.

d. Pure substances are seldom encountered in the "everyday" world.

1.16 Classify each of the following statements as true or false.

a. All homogeneous mixtures must contain at least two substances.

b. Heterogeneous mixtures, but not homogeneous mixtures, can have a variable composition.

c. Pure substances cannot be separated into other kinds of matter by physical means.

d. The number of known pure substances is less than 100,000.

1.17 Assign each of the following descriptions of matter to one of the following categories: *heterogeneous mixture, homogeneous mixture,* or *pure substance.*

a. Two substances present, two phases present

b. Two substances present, one phase present

c. One substance present, two phases present

d. Three substances present, three phases present

1.18 Assign each of the following descriptions of matter to one of the following categories: *heterogeneous mixture, homogeneous mixture,* or *pure substance.*

a. Three substances present, one phase present

b. One substance present, three phases present

c. One substance present, one phase present

d. Two substances present, three phases present

1.19 Classify each of the following as a *heterogeneous mixture,* a *homogeneous mixture,* or a *pure substance.* Also indicate how many phases are present, assuming all components are present in the same container.

a. Water and dissolved salt

b. Water and sand

c. Water, ice, and oil

d. Carbonated water (soda water) and ice

1.20 Classify each of the following as a *heterogeneous mixture,* a *homogeneous mixture,* or a *pure substance.* Also indicate how many phases are present, assuming all components are present in the same container.

a. Water and dissolved sugar

b. Water and oil

c. Water, wax, and pieces of copper metal

d. Salt water and sugar water

Elements and Compounds (Section 1.6)

1.21 From the information given, classify each of the pure substances A through D as elements or compounds, or indicate that no such classification is possible because of insufficient information.

a. Analysis with an elaborate instrument indicates that substance A contains two elements.

b. Substance B decomposes upon heating.

c. Heating substance C to 1000°C causes no change in it.

d. Heating substance D to 500°C causes it to change from a solid to a liquid.

1.22 From the information given, classify each of the pure substances A through D as elements or compounds, or indicate that no such classification is possible because of insufficient information.

a. Substance A cannot be broken down into simpler substances by chemical means.

b. Substance B cannot be broken down into simpler substances by physical means.

c. Substance C readily dissolves in water.

d. Substance D readily reacts with the element chlorine.

1.23 From the information given in the following equations, classify each of the pure substances A through G as elements or compounds, or indicate that no such classification is possible because of insufficient information.

a. $A + B \rightarrow C$ b. $D \rightarrow E + F + G$

1.24 From the information given in the following equations, classify each of the pure substances A through G as elements or compounds, or indicate that no such classification is possible because of insufficient information.

a. $A \rightarrow B + C$ b. $D + E \rightarrow F + G$

1.25 Indicate whether each of the following statements is true or false.

a. Both elements and compounds are pure substances.

b. A compound results from the physical combination of two or more elements.

c. In order for matter to be heterogeneous, at least two compounds must be present.

d. Compounds, but not elements, can have a variable composition.

1.26 Indicate whether each of the following statements is true or false.

a. Compounds can be separated into their constituent elements by chemical means.

b. Elements can be separated into their constituent compounds by physical means.

c. A compound must contain at least two elements.

d. A compound is a physical mixture of different elements.

Discovery and Abundance of the Elements (Section 1.7)

1.27 Indicate whether each of the following statements about elements is true or false.

a. All except two of the elements are naturally occurring.

b. New elements have been identified within the last 10 years.

c. Oxygen is the most abundant element in the universe as a whole.

d. Two elements account for over 75% of the atoms in Earth's crust.

1.28 Indicate whether each of the following statements about elements is true or false.

a. The majority of the known elements have been discovered since 1900.

b. At present, 113 elements are known.

c. Silicon is the second most abundant element in Earth's crust.

d. Elements that do not occur in nature can be produced in a laboratory setting.

Names and Symbols of the Elements (Section 1.8)

1.29 Name the element that each of the following symbols represents.

a. Ag b. Au c. Ca
d. Na e. P f. S

1.30 Name the element that each of the following symbols represents.

a. Cl b. Fe c. He
d. Hg e. Ni f. Pb

1.31 What are the chemical symbols of the following elements?

a. Tin b. Copper c. Aluminum
d. Boron e. Barium f. Argon

1.32 What are the chemical symbols of the following elements?

a. Zinc b. Silicon c. Beryllium
d. Magnesium e. Cobalt f. Fluorine

1.33 In which of the following sequences of elements do all of the elements have two-letter symbols?

a. Magnesium, nitrogen, phosphorus

b. Bromine, iron, calcium

c. Aluminum, copper, chlorine

d. Boron, barium, beryllium

1.34 In which of the following sequences of elements do all of the elements have symbols that start with a letter that is *not* the first letter of the element's English name?

a. Silver, gold, mercury b. Copper, helium, neon
c. Cobalt, chromium, sodium d. Potassium, iron, lead

Atoms and Molecules (Section 1.9)

1.35 Indicate whether each of the following statements is true or false. If a statement is false, change it to make it true. (Such a rewriting should involve more than merely converting the statement to the negative of itself.)

a. The atom is the limit of chemical subdivision for both elements and compounds.

b. Triatomic molecules must contain at least two kinds of atoms.

c. A molecule of a compound must be heteroatomic.

d. Only heteroatomic molecules may contain three or more atoms.

1.36 Indicate whether each of the following statements is true or false. If a statement is false, change it to make it true. (Such a rewriting should involve more than merely converting the statement to the negative of itself.)

a. A molecule of an element may be homoatomic or heteroatomic, depending on which element is involved.

b. The limit of chemical subdivision for a compound is a molecule.

c. Heteroatomic molecules do not maintain the properties of their constituent elements.

d. Only one kind of atom may be present in a homoatomic molecule.

1.37 Draw a diagram of each of the following molecules, using circular symbols of your choice to represent atoms.

a. A diatomic molecule of a compound

b. A molecule that is triatomic and homoatomic

c. A molecule that is tetraatomic and contains three kinds of atoms

d. A molecule that is triatomic, is symmetrical, and contains two elements

1.38 Draw a diagram of each of the following molecules, using circular symbols of your choice to represent atoms.

a. A triatomic molecule of an element

b. A molecule that is diatomic and heteroatomic

c. A molecule that is triatomic and contains three elements

d. A molecule that is triatomic, is not symmetrical, and contains two kinds of atoms

Chemical Formulas (Section 1.10)

1.39 On the basis of its formula, classify each of the following substances as an element or a compound.

a. $LiClO_3$ b. CO c. Co d. $CoCl_2$
e. $COCl_2$ f. BN g. S_8 h. Sn

1.40 On the basis of its formula, classify each of the following substances as an element or a compound.

a. BaO b. Ba c. BaO_2 d. OF_2
e. O_3 f. O_2 g. Pm h. $NaNO_3$

1.41 Write a formula for each of the following substances by using the information given about a molecule of the substance.

a. A molecule of caffeine contains 8 atoms of carbon, 10 atoms of hydrogen, 4 atoms of nitrogen, and 2 atoms of oxygen.

b. A molecule of table sugar contains 12 atoms of carbon, 22 atoms of hydrogen, and 11 atoms of oxygen.

c. A molecule of hydrogen cyanide is triatomic and contains the elements hydrogen, carbon, and nitrogen.

d. A molecule of sulfuric acid contains 2 atoms of hydrogen, 1 atom of sulfur, and the element oxygen and is heptaatomic.

1.42 Write a formula for each of the following substances by using the information given about a molecule of the substance.

a. A molecule of nicotine contains 10 atoms of carbon, 14 atoms of hydrogen, and 2 atoms of nitrogen.

b. A molecule of vitamin C contains 6 atoms of carbon, 8 atoms of hydrogen, and 6 atoms of oxygen.

c. A molecule of nitrous oxide contains twice as many atoms of nitrogen as of oxygen and is triatomic.

d. A molecule of nitric acid contains 3 atoms of oxygen and the elements hydrogen and nitrogen and is pentaatomic.

1.43 Determine the number of elements and the number of each type of atom present in molecules represented by the following formulas.

a. H_2CO_3 b. NH_4ClO_4 c. $CaSO_4$ d. C_4H_{10}

1.44 Determine the number of elements and the number of each type of atom present in molecules represented by the following formulas.

a. $KHCO_3$ b. NaSCN c. $Na_3P_3O_{10}$ d. NH_4ClO

Additional Problems

1.45 A hard sugar cube is pulverized, and the resulting granules are heated in air until they discolor and then finally burst into flame and burn.

a. List all *physical* changes to substances mentioned in the preceding narrative.

b. List all *chemical* changes to substances mentioned in the preceding narrative.

1.46 Assign each of the following descriptions of matter to one of the following categories: *element, compound,* or *mixture.*

a. One substance present, one phase present, substance cannot be decomposed by chemical means

b. One substance present, three elements present

c. Two substances present, two phases present

d. Two elements present, composition is definite and constant

1.47 Indicate whether each of the following samples of matter is a *heterogeneous mixture,* a *homogeneous mixture,* a *compound,* or an *element.*

a. A colorless gas, only part of which reacts with hot iron

b. A "cloudy" liquid that separates into two layers upon standing for 2 hours

c. A green solid, all of which melts at the same temperature to produce a liquid that decomposes upon further heating

d. A colorless gas that cannot be separated into simpler substances using physical means and that reacts with copper to produce both a copper–nitrogen and a copper–oxygen compound

1.48 Assign each of the following descriptions of matter to one of the following categories: *element, compound,* or *mixture.*

a. One substance present, one phase present, one kind of homoatomic molecule present

b. Two substances present, two phases present, all molecules are heteroatomic

c. One phase present, two kinds of homoatomic molecules present

d. One phase present, all molecules are triatomic, all molecules are heteroatomic, all molecules are identical

1.49 Certain words can be viewed whimsically as sequential combinations of symbols of elements. For example, the given name Stephen is made up of the following sequence of chemical symbols: S-Te-P-He-N. Analyze each of the following given names in a similar manner.

a. Barbara

b. Eugene

c. Heather

d. Allan

1.50 In each of the following pairs of formulas, indicate whether the first formula listed denotes *more total atoms, the same number of total atoms,* or *fewer total atoms* than the second formula listed.

a. HN_3 and NH_3

b. $CaSO_4$ and $Mg(OH)_2$

c. $NaClO_3$ and $Be(CN)_2$

d. $Be_3(PO_4)_2$ and $Mg(C_2H_3O_2)_2$

1.51 On the basis of the given information, determine the numerical value of the subscript x in each of the following chemical formulas.

a. BaS_2O_x; formula unit contains 6 atoms

b. $Al_2(SO_x)_3$; formula unit contains 17 atoms

c. SO_xCl_x; formula unit contains 5 atoms

d. $C_xH_{2x}Cl_x$; formula unit contains 8 atoms

1.52 A mixture contains the following five pure substances: N_2, N_2H_4, NH_3, CH_4, and CH_3Cl.

a. How many different kinds of molecules that contain four or fewer atoms are present in the mixture?

b. How many different kinds of atoms are present in the mixture?

c. How many total atoms are present in a mixture sample containing five molecules of each component?

d. How many total hydrogen atoms are present in a mixture sample containing four molecules of each component?

Grid Problems

1.53

1. chemical change	2. physical change	3. change of state
4. change in composition	5. new substance produced	6. chemical reaction occurs

Select from the grid *all* correct responses for each of the following situations.

a. Gold metal is hammered into a thin sheet.

b. A piece of silverware tarnishes.

c. An ice cube melts when heated.

d. Milk curdles when left unrefrigerated.

1.54

1. homogeneous mixture	2. element	3. pure substance
4. one phase	5. two phases	6. three or more phases

Select from the grid *all* correct responses for each of the following situations.

a. Sample of copper wire
b. Beef noodle soup
c. Ice cubes in pure water
d. Sample of human blood

1.55

1. phosphorus aluminum	2. sodium potassium	3. chlorine fluorine
4. magnesium beryllium	5. iron copper	6. nitrogen sulfur

Select from the grid *all* correct responses for each of the following situations.

a. Both elements have one-letter chemical symbols.
b. Both elements have two-letter chemical symbols.
c. Both elements have symbols that come from non-English names for the elements.
d. Both elements have symbols in which the first letter is capitalized.

1.56

1. HF	2. Hf	3. NO_2
4. KNO_3	5. Cl_2O_3	6. H_2Se

Select from the grid *all* correct responses for each of the following situations.

a. Formulas that represent elements
b. Formulas that involve pentaatomic units
c. Formulas in which there are equal numbers of two kinds of atoms (other atoms may also be present)
d. Pairs of formulas in which there are the same number of atoms and also the same number of elements

Measurements in Chemistry

CHAPTER OUTLINE

2.1 Measurement Systems 21
2.2 Metric System Units 22
2.3 Uncertainty in Measurement and Significant Figures 24
2.4 Significant Figures and Mathematical Operations 26
2.5 Scientific Notation 28
Chemistry at a Glance: **Significant Figures** 29
2.6 Conversion Factors and Dimensional Analysis 32
Chemistry at a Glance: **Conversion Factors** 36
2.7 Density 36
2.8 Temperature and Heat Energy 39
Chemical Connections
2.1 Body Density and Percent Body Fat 38
2.2 Normal Human Body Temperature 42

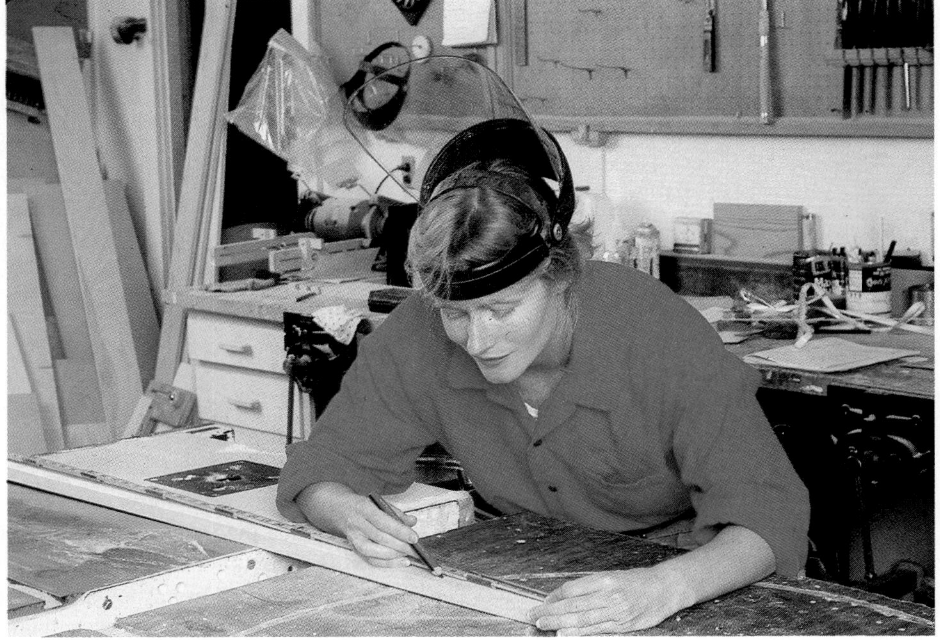

Measurements can never be exact; there is always some degree of uncertainty.

It would be extremely difficult for a carpenter to build cabinets without being able to use hammers, saws, and drills. They are the tools of a carpenter's trade. Chemists also have "tools of the trade." The tool they use most is called *measurement.* Understanding measurement is indispensable in the study of chemistry. Questions such as "How much . . . ?," "How long . . . ?," and "How many . . . ?" simply cannot be answered without resorting to measurements. This chapter will help you learn what you need to know to deal properly with measurement. Much of the material in the chapter is mathematical. This is necessary; measurements require the use of numbers.

2.1 Measurement Systems

We all make measurements on a routine basis. For example, measurements are involved in following a recipe for making brownies, in determining our height and weight, and in fueling a car with gasoline. **Measurement** *is the determination of the dimensions, capacity, quantity, or extent of something.* In chemical laboratories, the most common types of measurements are those of mass, volume, length, time, temperature, pressure, and concentration.

Figure 2.1
Metric system units are becoming increasingly evident on highway signs and consumer products.

• The word *metric* is derived from the Greek word *metron,* which means "measure.

Two systems of measurement are in use in the United States at present: (1) the English system of units, and (2) the metric system of units. Common measurements of commerce, such as those used in a grocery store, are made in the *English system.* The units of this system include the inch, foot, pound, quart, and gallon. The *metric system* is used in scientific work. The units of this system include the gram, meter, and liter.

The United States is in the process of voluntary conversion to the metric system for measurements of commerce. Metric system units now appear on numerous consumer products (Figure 2.1). Soft drinks now come in 1-, 2- and 3-liter containers. Road signs in some states display distances in both miles and kilometers. Canned and packaged goods such as cereals and mixes on grocery store shelves now have the masses of their contents listed in grams as well as in pounds and ounces.

Interrelationships between units of the same type, such as volume or length, are less complicated in the metric system than in the English system. Within the metric system, conversion from one unit size to another can be accomplished simply by multiplying or dividing by units of 10, because the metric system is a decimal unit system—that is, it is based on multiples of 10. The metric system is simply more convenient to use.

2.2 Metric System Units

• The modern version of the metric system is called the International System, or SI (the abbreviation is taken from the French name, *le Système International*).

• The use of numerical prefixes should not be new to you. Consider the use of the prefix *tri-* in the words *tri*angle, *tri*cycle, *trio, tri*nity, and *tri*ple. Each of these words conveys the idea of three of something. The metric system prefixes are used in the same way.

In the metric system, there is one base unit for each type of measurement (length, mass, volume, and so on). The names of fractional parts of the base unit and multiples of the base unit are constructed by adding prefixes to the base unit. These prefixes indicate the size of the unit relative to the base unit. Table 2.1 lists common metric system prefixes, along with their symbols or abbreviations and mathematical meanings. The prefixes in color are the ones most frequently used.

The meaning of a metric system prefix is independent of the base unit it modifies and always remains constant. For example, the prefix *kilo-* always means 1000; a *kilo*second is 1000 seconds, a *kilo*watt is 1000 watts, and a *kilo*calorie is 1000 calories. Similarly, the prefix *nano-* always means one-billionth; a *nano*meter is one-billionth of a meter, a *nano*gram is one-billionth of a gram, and a *nano*liter is one-billionth of a liter.

Table 2.1
Common Metric System Prefixes with Their Symbols and Mathematical Meanings

	Prefix[a]	Symbol	Mathematical meaning[b]
Multiples	giga-	G	1,000,000,000 (10^9, billion)
	mega-	M	1,000,000 (10^6, million)
	kilo-	k	1000 (10^3, thousand)
Fractional parts	deci-	d	0.1 (10^{-1}, one-tenth)
	centi-	c	0.01 (10^{-2}, one-hundredth)
	milli-	m	0.001 (10^{-3}, one-thousandth)
	micro-	μ (Greek mu)	0.000001 (10^{-6}, one-millionth)
	nano-	n	0.000000001 (10^{-9}, one-billionth)
	pico-	p	0.000000000001 (10^{-12}, one-trillionth)

[a]Other prefixes also are available but are less commonly used.
[b]The power-of-10 notation for denoting numbers is considered in Section 2.5.

• Metric Length Units

• *Length* is measured by determining the distance between two points.

The **meter** (m) *is the base unit of length in the metric system.* It is about the same size as the English yard; 1 meter equals 1.09 yards (Figure 2.2a). The prefixes listed in Table 2.1 enable us to derive other units of length from the meter. The kilometer (km) is 1000 times larger than the meter; the centimeter (cm) and millimeter (mm) are, respectively, one-hundredth and one-thousandth of a meter. Most laboratory length measurements are made in centimeters rather than meters because of the meter's relatively large size.

• Metric Mass Units

• *Mass* is measured by determining the amount of matter in an object.

The **gram** (g) *is the base unit of mass in the metric system.* It is a very small unit compared with the English ounce and pound (Figure 2.2b). It takes approximately 28 grams to equal 1 ounce and nearly 454 grams to equal 1 pound. Both grams and milligrams (mg) are commonly used in the laboratory, where the kilogram (kg) is generally too large.

The terms *mass* and *weight* are often used interchangeably in measurement discussions; technically, however, they have different meanings. **Mass** *is a measure of the total*

(a) Length	(b) Mass	(c) Volume
A meter is slightly larger than a yard. 1 meter = 1.09 yards. A baseball bat is about 1 meter long.	A gram is a small unit compared to a pound. 1 gram = 1/454 pound. Two pennies, five paper-clips, and a marble have masses of about 5, 2, and 5 grams, respectively.	A liter is slightly larger than a quart. 1 liter = 1.06 quarts. Most beverages are now sold by the liter rather than by the quart.

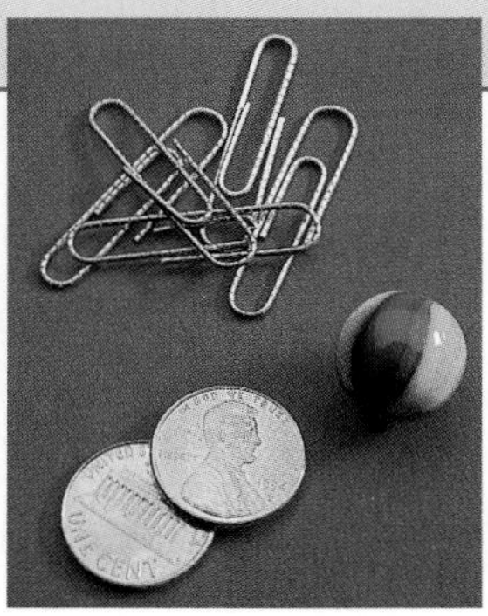

Figure 2.2
Comparisons of the base metric system units of length (meter), mass (gram), and volume (liter) with common objects.

quantity of matter in an object. **Weight** *is a measure of the force exerted on an object by the pull of gravity.*

The mass of a substance is a constant; the weight of an object varies with the object's geographical location. For example, matter at the equator weighs less than it would at the North Pole because the pull of gravity is less at the equator. Because Earth is not a perfect sphere, but bulges at the equator, the magnitude of gravitational attraction is less at the equator. An object would weigh less on the moon than on Earth because of the smaller size of the moon and the correspondingly lower gravitational attraction. Quantitatively, a 22.0-lb mass weighing 22.0 lb at Earth's North Pole would weigh 21.9 lb at Earth's equator and only 3.7 lb on the moon. In outer space, an astronaut may be weightless but never massless. In fact, he or she has the same mass in space as on Earth.

• Students often erroneously think that the terms *mass* and *weight* have the same meaning. *Mass* is a measure of the amount of material present in a sample. *Weight* is a measure of the force exerted on an object by the pull of gravity.

• Metric Volume Units

The **liter** (L) *is the base unit of volume in the metric system.* The abbreviation for liter is a capital L rather than a lower-case l because a lower-case l is easily confused with the number 1. A liter is a volume equal to that occupied by a cube that is 10 centimeters on each side. Because the volume of a cube is calculated by multiplying length times width times height (which are all the same for a cube), we have

• Measuring volume involves determining the amount of space occupied by a three-dimensional object.

$$\begin{aligned} 1 \text{ liter} &= \text{volume of a cube with edge 10 cm} \\ &= 10 \text{ cm} \times 10 \text{ cm} \times 10 \text{ cm} \\ &= 1000 \text{ cm}^3 \end{aligned}$$

A liter is also equal to 1000 milliliters; the prefix *milli-* means one-thousandth. Therefore,

$$1000 \text{ mL} = 1000 \text{ cm}^3$$

Dividing both sides of this equation by 1000 reveals that

$$1 \text{ mL} = 1 \text{ cm}^3$$

Consequently, the units mL and cm^3 are the same. In practice, mL is used for volumes of liquids and gases, cm^3 for volumes of solids. Figure 2.3 shows the relationship between 1 mL (1 cm^3) and its parent unit, the liter, in terms of cubic measurements.

• Another abbreviation for the unit cubic centimeter, used in medical situations, is cc.

$$1 \text{ cm}^3 = 1 \text{ cc}$$

A liter and a quart have approximately the same volume; 1 liter equals 1.06 quarts (Figure 2.2c). The milliliter and deciliter (dL) are commonly used in the laboratory. Deciliter units are routinely encountered in clinical laboratory reports detailing the composition of body fluids (Figure 2.4). A deciliter is equal to 100 mL (0.100 L).

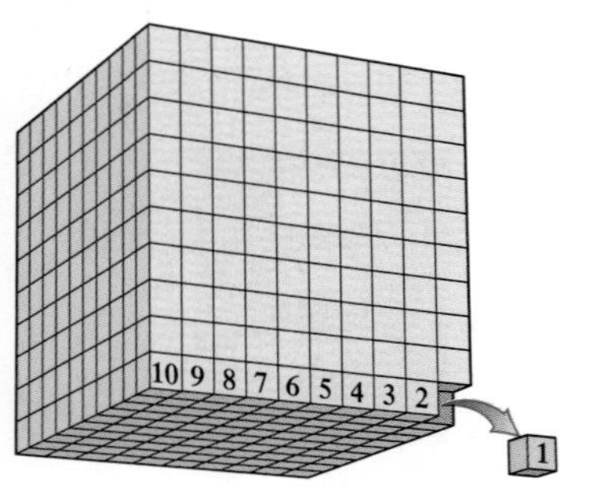

Figure 2.3
A cube 10 cm on a side has a volume of 1000 cm^3, which is equal to 1 L. A cube 1 cm on a side has a volume of 1 cm^3, which is equal to 1 mL.

2.3 Uncertainty in Measurement and Significant Figures

Because of the limitations of the measuring device and the limited powers of observation of the individual making the measurement, every measurement carries a degree of uncertainty or error. Even when very elaborate measuring devices are used, some degree of uncertainty is always present.

• Origin of Measurement Uncertainty

To illustrate how measurement uncertainty arises, let us consider how two different rulers, shown in Figure 2.5, are used to measure a given length. Using ruler A, we can say with certainty that the length of the rod is between 3 and 4 centimeters. We can further say that the actual length is closer to 4 centimeters and estimate it to be 3.7 centimeters. Ruler B has more subdivisions on its scale than ruler A. It is marked off in tenths of a centimeter instead of in centimeters. Using ruler B, we can definitely say that the length of the rod is between 3.7 and 3.8 centimeters and can estimate it to be 3.74 centimeters.

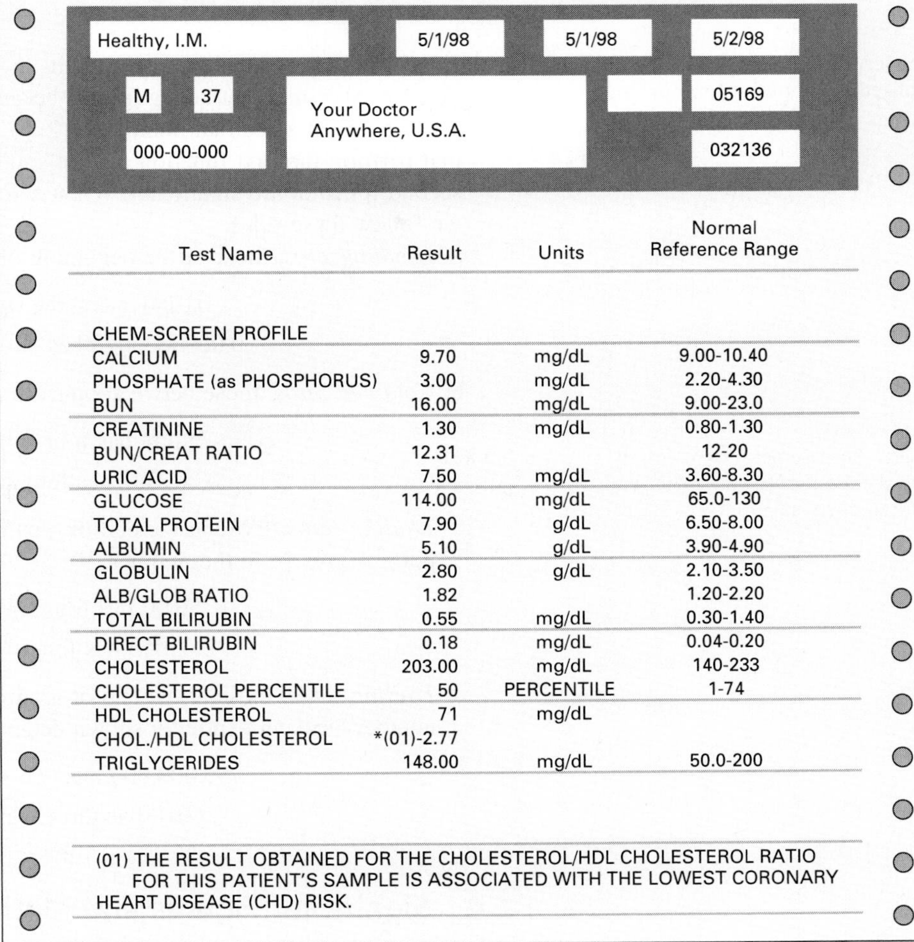

Healthy, I.M. | 5/1/98 | 5/1/98 | 5/2/98

M | 37 | Your Doctor Anywhere, U.S.A. | 05169

000-00-000 | 032136

Test Name	Result	Units	Normal Reference Range
CHEM-SCREEN PROFILE			
CALCIUM	9.70	mg/dL	9.00-10.40
PHOSPHATE (as PHOSPHORUS)	3.00	mg/dL	2.20-4.30
BUN	16.00	mg/dL	9.00-23.0
CREATININE	1.30	mg/dL	0.80-1.30
BUN/CREAT RATIO	12.31		12-20
URIC ACID	7.50	mg/dL	3.60-8.30
GLUCOSE	114.00	mg/dL	65.0-130
TOTAL PROTEIN	7.90	g/dL	6.50-8.00
ALBUMIN	5.10	g/dL	3.90-4.90
GLOBULIN	2.80	g/dL	2.10-3.50
ALB/GLOB RATIO	1.82		1.20-2.20
TOTAL BILIRUBIN	0.55	mg/dL	0.30-1.40
DIRECT BILIRUBIN	0.18	mg/dL	0.04-0.20
CHOLESTEROL	203.00	mg/dL	140-233
CHOLESTEROL PERCENTILE	50	PERCENTILE	1-74
HDL CHOLESTEROL	71	mg/dL	
CHOL./HDL CHOLESTEROL	*(01)-2.77		
TRIGLYCERIDES	148.00	mg/dL	50.0-200

(01) THE RESULT OBTAINED FOR THE CHOLESTEROL/HDL CHOLESTEROL RATIO FOR THIS PATIENT'S SAMPLE IS ASSOCIATED WITH THE LOWEST CORONARY HEART DISEASE (CHD) RISK.

Figure 2.4
The use of the concentration unit mg/dL is common in clinical laboratory reports dealing with the composition of human body fluids.

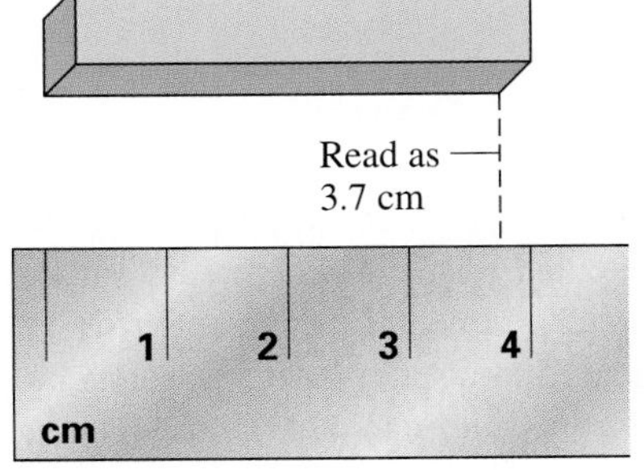

Ruler A

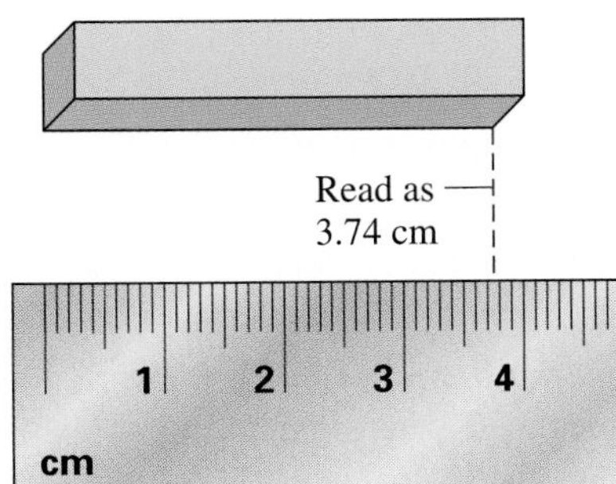

Ruler B

Figure 2.5
The scale on a measuring device determines the magnitude of the uncertainty for the recorded measurement. Measurements made with ruler A will have greater uncertainty than those made with ruler B.

Note how both length measurements (ruler A and ruler B) contain some digits (all those except the last one) that are exactly known and one digit (the last one) that is estimated. It is this last digit, the estimated one, that produces uncertainty in a measurement. Note also that the uncertainty in the second length measurement is less than that in the first one—an uncertainty in the hundredths place compared with an uncertainty in the tenths place. We say that the second measurement is more *precise* than the first one; that is, it has less uncertainty than the first measurement.

Only one estimated digit is ever recorded as part of a measurement. It would be incorrect for a scientist to report that the length of the metal rod in Figure 2.5 is 3.745 centimeters as read by using ruler B. The value 3.745 contains two estimated digits, the 4 and the 5, and indicates a measurement of precision greater than what is actually obtainable with that particular measuring device. Again, only one estimated digit is ever recorded as part of a measurement.

Because measurements are never exact, two types of information must be conveyed whenever a numerical value for a measurement is recorded: (1) the magnitude of the measurement and (2) the precision or uncertainty of the measurement. The magnitude is indicated by the digit values. Precision is indicated by the number of significant figures recorded. **Significant figures** *are the digits in any measurement that are known with certainty plus one digit that is uncertain.*

• Guidelines for Determining Significant Figures

Recognizing the number of significant figures in a measured quantity is easy for measurements we make ourselves, because we know the type of instrument we are using and its limitations. However, when someone else makes the measurement, such information

• The term *significant figures* is often verbalized in shortened form as "sig figs."

is often not available. In such cases, we follow a set of guidelines for determining the number of significant figures in a measured quantity.

1. In any measurement, all nonzero digits are significant.
2. *Zeros* may or may not be significant because zeros can be used in two ways: (1) to position a decimal point, and (2) to indicate a measured value. Zeros that perform the first function are not significant, and zeros that perform the second function are significant. When zeros are present in a measured number, we follow these rules:
   a. *Leading zeros,* those at the beginning of a number, are never significant.

      0.0141 has three significant figures.

      0.0000000048 has two significant figures.

   b. *Confined zeros,* those between nonzero digits, are always significant.

      3.063 has four significant figures.

      0.001004 has four significant figures.

   c. *Trailing zeros,* those at the end of a number, are significant if a decimal point is present in the number.

      56.00 has four significant figures.

      0.05050 has four significant figures.

   d. *Trailing zeros,* those at the end of a number, are not significant if the number lacks an explicitly shown decimal point.

      59,000,000 has two significant figures.

      6010 has three significant figures.

2.4 Significant Figures and Mathematical Operations

When measurements are added, subtracted, multiplied, or divided, consideration must be given to the number of significant figures in the computed result. Mathematical operations should not increase (or decrease) the precision of experimental measurements.

Hand-held electronic calculators generally "complicate" preciseness considerations because they are not programmed to take significant figures into account. Consequently, the digital readouts display more digits than are warranted (Figure 2.6). It is a mistake to record these extra digits, because they are not significant figures and hence are meaningless.

Figure 2.6
The digital readout on an electronic calculator usually shows more digits than are needed—and more than are acceptable. Calculators are not programmed to account for significant figures.

• Rounding Off Numbers

When we obtain calculator answers that contain too many digits, it is necessary to drop the nonsignificant digits, a process that is called rounding off. **Rounding off** *is the process of deleting unwanted (nonsignificant) digits from calculated numbers.* There are two rules for rounding off numbers.

1. *If the first digit to be deleted is 4 or less, simply drop it and all the following digits.* For example, the number 3.724567 becomes 3.72 when rounded to three significant figures.
2. *If the first digit to be deleted is 5 or greater, that digit and all that follow are dropped and the last retained digit is increased by one.* The number 5.00673 becomes 5.01 when rounded to three significant figures.

• Operational Rules

Significant-figure considerations in mathematical operations that involve measured numbers are governed by two rules, one for multiplication and division and one for addition and subtraction.

1. *In multiplication and division, the number of significant figures in the answer is the same as the number of significant figures in the measurement that contains the fewest significant figures.* For example,

Four significant figures (6.038) ↓ Three significant figures (2.57) ↓

$$6.038 \times 2.57 = 15.51766 \quad \text{(calculator answer)}$$

$$= 15.5 \quad \text{(correct answer)}$$

↑ Three significant figures

The calculator answer is rounded to three significant figures because the measurement with the fewest significant figures (2.57) contains only three significant figures.

2. *In addition and subtraction, the answer has no more digits to the right of the decimal point than are found in the measurement with the fewest digits to the right of the decimal point.* For example,

9.333	← Uncertain digit (thousandths)
+ 1.4	← Uncertain digit (tenths)
10.733	(calculator answer)
10.7	(correct answer)
↑ Uncertain digit (tenths)	

The calculator answer is rounded to the tenths place because the uncertainty in the number 1.4 is in the tenths place.

• Concisely stated, the significant-figure operational rules are

$\times$ or $\div$: Keep smallest number of significant figures in answer.

$+$ or $-$: Keep smallest number of decimal places in answer.

Note the contrast between the rule for multiplication and division and the rule for addition and subtraction. In multiplication and division, significant figures are counted; in addition and subtraction, decimal places are counted. It is possible to gain or lose significant figures during addition or subtraction, but *never* during multiplication or division. In our previous sample addition problem, one of the input numbers (1.4) has two significant figures and the correct answer (10.7) has three significant figures. This is allowable in addition (and subtraction) because we are counting decimal places, not significant figures.

Example 2.1

Expressing Answers to the Proper Number of Significant Figures

Perform the following computations, expressing your answers to the proper number of significant figures.

a. 6.7321×0.0021 **b.** $\dfrac{16{,}340}{23.42}$

c. $8.3 + 1.2 + 1.7$ **d.** $3.07 \times (17.6 - 13.73)$

Solution

a. The calculator answer to this problem is

$$6.7321 \times 0.0021 = 0.01413741$$

The input number with the least number of significant figures is 0.0021.

$$6.7321 \times 0.0021$$

6.7321: Five significant figures ↗ ↖ 0.0021: Two significant figures

Thus the calculator answer must be rounded to two significant figures.

0.01413741	becomes	0.014
Calculator answer		Correct answer

(continued)

b. The calculator answer to this problem is

$$\frac{16{,}340}{23.42} = 697.69427$$

Both input numbers contain four significant figures. Thus the correct answer will also contain four significant figures.

697.69427	becomes	697.7
Calculator answer		Correct answer

c. The calculator answer to this problem is

$$8.3 + 1.2 + 1.7 = 11.2$$

All three input numbers have uncertainty in the tenths place. Thus the last retained digit in the correct answer will be that of tenths. (In this particular problem, the calculator answer and the correct answer are the same, a situation that does not occur very often.)

d. This problem involves the use of both multiplication and subtraction significant-figure rules. We do the subtraction first.

$$17.6 - 13.73 = 3.87 \quad \text{(calculator answer)}$$
$$= 3.9 \quad \text{(correct answer)}$$

This answer must be rounded to tenths because the input number 17.6 involves only tenths. We now do the multiplication.

$$3.07 \times 3.9 = 11.973 \quad \text{(calculator answer)}$$
$$= 12 \quad \text{(correct answer)}$$

The number 3.9 limits the answer to two significant figures.

Practice Exercise 2.1

Perform the following computations, expressing your answers to the proper number of significant figures. Assume that all numbers are measured numbers.

a. 5.4430×1.203 **b.** $\frac{17.4}{0.0031}$

c. $7.4 + 20.74 + 3.03$ **d.** $4.73 \times (2.2 + 8.9)$

• *Answer:* **a.** 6.548; **b.** 5600; **c.** 31.2; **d.** 52.5

• Measurements can never be exact; there is always some degree of uncertainty.

Some numbers used in computations are *exact numbers* rather than measured numbers. Exact numbers occur in definitions (for example, there are exactly 12 objects in a dozen, not 12.01 or 12.02); in counting (for example, there can be 7 people in a room, but never 6.99 or 7.02); and in simple fractions ($\frac{1}{3}$, $\frac{3}{5}$, or $\frac{5}{9}$). Because exact numbers have no uncertainty associated with them, they possess an unlimited number of significant figures. Therefore, such numbers never limit the number of significant figures in a computational answer.

Chemistry at a Glance on the next page reviews the rules that govern which digits in a measurement are significant.

2.5 Scientific Notation

Up to this point in the chapter, we have expressed all numbers in decimal notation, the everyday method for expressing numbers. Such notation becomes cumbersome for very large and very small numbers (which occur frequently in scientific work). For example, in one drop of blood, which is 92% water by mass, there are approximately

CHEMISTRY AT A GLANCE

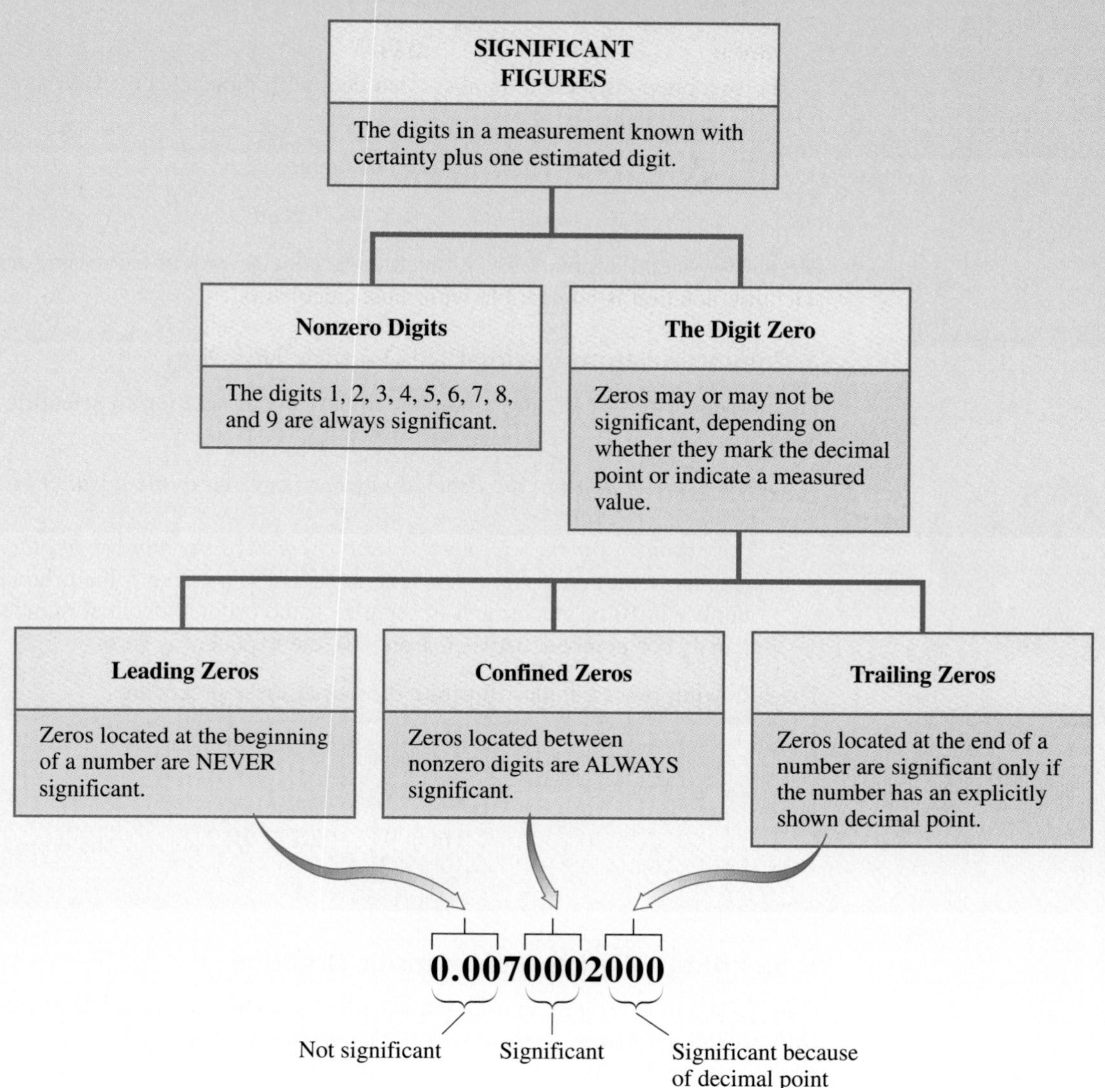

1,600,000,000,000,000,000,000 molecules

of water, each of which has a mass of

0.000000000000000000000030 gram

Recording such large and small numbers is not only time-consuming but also open to error; often, too many or too few zeros are recorded. Also, it is impossible to multiply or divide such numbers with most calculators because they can't accept that many digits. (Most calculators accept either 8 or 10 digits.)

• Scientific notation is also called exponential notation.

A method called *scientific notation* exists for expressing multidigit numbers involving many zeros in compact form. **Scientific notation** *is a system in which an ordinary decimal number is expressed as the product of a number between 1 and 10 times 10 raised to a power.* The ordinary decimal number is called a *coefficient* and is written first. The number 10 raised to a power is called an *exponential term.* The coefficient is always multiplied by the exponential term, as in the example at the top of the next page.

Coefficient — Exponent

$$1.07 \times 10^4$$

Multiplication sign — Exponential term

The two previously cited numbers that deal with molecules of water are expressed in scientific notation as

$$1.6 \times 10^{21} \text{ molecules}$$

and

$$3.0 \times 10^{-22} \text{ gram}$$

Obviously, scientific notation is a much more concise way of expressing numbers. Such scientific notation is compatible with most calculators.

• Converting from Decimal to Scientific Notation

The procedure for converting a number from decimal notation to scientific notation has two parts.

1. *The decimal point in the decimal number is moved to the position behind the first nonzero digit.*
2. *The exponent for the exponential term is equal to the number of places the decimal point has been moved.* The exponent is positive if the original decimal number is 10 or greater and is negative if the original decimal number is less than 1. For numbers between 1 and 10, the exponent is zero.

The following two examples illustrate the use of these procedures:

$$93{,}000{,}000 = 9.3 \times 10^7$$

Decimal point is moved 7 places

$$0.0000037 = 3.7 \times 10^{-6}$$

Decimal point is moved 6 places

• Significant Figures and Scientific Notation

How do significant-figure considerations affect scientific notation? The answer is simple. *Only significant figures become part of the coefficient.* The numbers 63, 63.0, and 63.00, which respectively have two, three, and four significant figures, when converted to scientific notation become, respectively,

• The decimal and scientific notation forms of a number *always* contain the same number of significant figures.

6.3×10^1 (two significant figures)

6.30×10^1 (three significant figures)

6.300×10^1 (four significant figures)

Multiplication and division of numbers expressed in scientific notation are common procedures. For these two types of operations, the coefficients, which are decimal numbers, are combined in the usual way. The rules for handling the exponential terms are

1. To multiply exponential terms, *add* the exponents.
2. To divide exponential terms, *subtract* the exponents.

Example 2.2

Multiplication and Division in Scientific Notation

Carry out the following multiplications and divisions in scientific notation.

a. $(2.33 \times 10^3) \times (1.55 \times 10^4)$ **b.**

c. $\dfrac{8.42 \times 10^6}{3.02 \times 10^4}$ **d.** $\dfrac{4.20 \times 10^{-3}}{1.1 \times 10^{-7}}$

Solution

a. Multiplying the two coefficients gives

$$2.33 \times 1.55 = 3.6115 \quad \text{(calculator answer)}$$
$$= 3.61 \quad \text{(correct answer)}$$

Remember that the coefficient obtained by multiplication can have only three significant figures in this case, the same number as in both input numbers for the multiplication.

Multiplication of the two powers of 10 to give the exponential term requires that we add the exponents.

$$10^3 \times 10^4 = 10^{3+4} = 10^7$$

Combining the new coefficient with the new exponential term gives the answer.

$$3.61 \times 10^7$$

b. Multiplying the two coefficients gives

$$1.130 \times 1.33 = 1.5029 \quad \text{(calculator answer)}$$
$$= 1.50 \quad \text{(correct answer)}$$

The input number 1.33 limits the answer to three significant figures.

In multiplying the powers of 10, we add exponents of differing signs.

$$10^{-3} \times 10^4 = 10^{(-3)+(4)} = 10^1$$

Combining the coefficient and the exponential term gives

$$1.50 \times 10^1$$

Writing the final answer as 1.5×10^1 instead of 1.50×10^1 is incorrect. The former contains two significant figures, and the latter three significant figures. We need three significant figures because the least number of significant figures in the input numbers is three.

c. Performing the indicated division of the coefficients gives

$$\frac{8.42}{3.02} = 2.7880794 \quad \text{(calculator answer)}$$
$$= 2.79 \quad \text{(correct answer)}$$

Because both input numbers have three significant figures, the answer also has three significant figures.

The division of exponential terms requires that we subtract the exponents.

$$\frac{10^6}{10^4} = 10^{(+6)-(+4)} = 10^2$$

Combining the coefficient and the exponential term gives

$$2.79 \times 10^2$$

d. The new coefficient is obtained by dividing 4.20 by 1.1.

$$\frac{4.20}{1.1} = 3.8181818 \quad \text{(calculator answer)}$$
$$= 3.8 \quad \text{(correct answer)}$$

The input number 1.1 limits the answer to two significant figures.

When dividing powers of 10, we obtain the exponent by subtracting.

$$\frac{10^{-3}}{10^{-7}} = 10^{(-3)-(-7)} = 10^4$$

In performing this subtraction, remember that a minus times a minus is a plus; that is, $-(-7) = +7$.

Combining the coefficient and the exponential term gives

$$3.8 \times 10^4$$

Practice Exercise 2.2

Carry out the following mathematical operations in scientific notation.

a. $(4.057 \times 10^3) \times (2.001 \times 10^7)$ **b.** $(3.40 \times 10^{-3}) \times (7.500 \times 10^8)$

c. $\dfrac{4.1 \times 10^{-10}}{3.112 \times 10^{-7}}$ **d.** $\dfrac{5.723 \times 10^{-1}}{3.00 \times 10^{-4}}$

• *Answer.* **a.** 8.118×10^{10}; **b.** 2.55×10^6; **c.** 1.3×10^{-3}; **d.** 1.91×10^3

2.6 Conversion Factors and Dimensional Analysis

With both the English unit and metric unit systems in common use in the United States, we often must change measurements from one system to their equivalent in the other system. The mathematical tool we use to accomplish this task is a general method of problem solving called *dimensional analysis.* Central to the use of dimensional analysis is the concept of conversion factors. A **conversion factor** *is a ratio that specifies how one unit of measurement is related to another.*

Conversion factors are derived from equations (equalities) that relate units. Consider the quantities "1 minute" and "60 seconds," both of which describe the same amount of time. We may write an equation describing this fact.

$$1 \text{ min} = 60 \text{ sec}$$

This fixed relationship is the basis for the construction of a pair of conversion factors that relate seconds and minutes.

$$\frac{1 \text{ min}}{60 \text{ sec}} \quad \text{and} \quad \frac{60 \text{ sec}}{1 \text{ min}}$$

These two quantities are the same.

Note that conversion factors always come in pairs, one member of the pair being the reciprocal of the other. Also note that the numerator and the denominator of a conversion factor always describe the same amount of whatever we are considering. One minute and 60 seconds denote the same amount of time.

• Conversion Factors Within a System of Units

Most students are familiar with and have memorized numerous conversion factors within the English system of measurement (English-to-English conversion factors). Some of these factors, with only one member of a conversion factor pair being listed, are

• In order to avoid confusion with the word *in*, the abbreviation for inches, in., includes a period. This is the only unit abbreviation in which a period appears.

$$\frac{12 \text{ in.}}{1 \text{ ft}} \quad \frac{3 \text{ ft}}{1 \text{ yd}} \quad \frac{4 \text{ qt}}{1 \text{ gal}} \quad \frac{16 \text{ oz}}{1 \text{ lb}}$$

Such conversion factors contain an unlimited number of significant figures because the numbers within them arise from definitions.

Metric-to-metric conversion factors are similar to English-to-English conversion factors in that they arise from definitions. Individual conversion factors are derived from the meanings of the metric system prefixes (Section 2.2). For example, the set of conversion factors involving kilometer and meter come from the equality

$$1 \text{ kilometer} = 10^3 \text{ meters}$$

and those relating microgram and gram come from the equality

$$1 \text{ microgram} = 10^{-6} \text{ gram}$$

• In order to obtain metric-to-metric conversion factors, you need to know the meaning of the metric system prefixes in terms of powers of 10 (see Table 2.1).

The two pairs of conversion factors are

$$\frac{10^3 \text{ m}}{1 \text{ km}} \quad \text{and} \quad \frac{1 \text{ km}}{10^3 \text{ m}} \qquad \frac{1\ \mu\text{g}}{10^{-6} \text{ g}} \quad \text{and} \quad \frac{10^{-6} \text{ g}}{1\ \mu\text{g}}$$

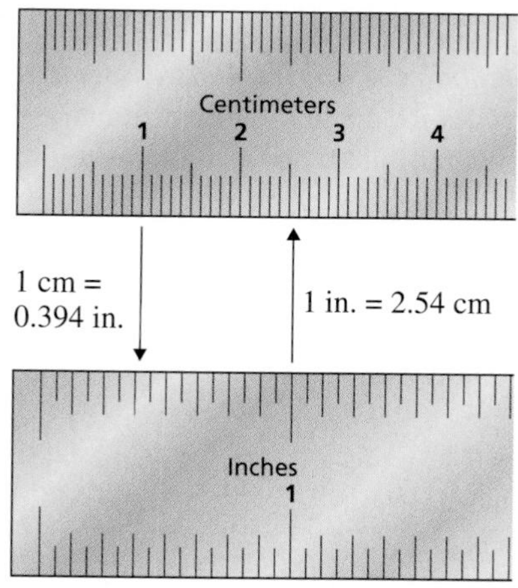

One inch equals 2.54 centimeters or one centimeter equals 0.394 inch.

Note that the numerical equivalent of the prefix is always associated with the base (unprefixed) unit in a metric-to-metric conversion factor.

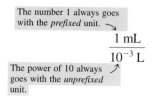

The number 1 always goes with the *prefixed* unit.

$$\frac{1 \text{ mL}}{10^{-3} \text{ L}}$$

The power of 10 always goes with the *unprefixed* unit.

• Conversion Factors Between Systems of Units

Conversion factors that relate metric units to English units and vice versa are not exact defined quantities because they involve two different systems of measurement. The numbers associated with these conversion factors must be determined experimentally. Table 2.2 lists commonly encountered relationships between metric system and English system units. These few conversion factors are sufficient to solve most of the problems that we will encounter.

Metric-to-English conversion factors can be specified to differing numbers of significant figures. For example,

$$1.00 \text{ lb} = 454 \text{ g}$$
$$1.000 \text{ lb} = 453.6 \text{ g}$$
$$1.0000 \text{ lb} = 453.59 \text{ g}$$

In a problem-solving context, which "version" of a conversion factor is used depends on how many significant figures there are in the other numbers of the problem. Conversion factors should never limit the number of significant figures in the answer to a problem. The conversion factors in Table 2.2 are given to three significant figures, which is sufficient for the applications we will make of them.

Table 2.2
Equalities and Conversion Factors That Relate the English and Metric Systems of Measurement

	Metric to English	English to Metric
Length		
1.00 inch = 2.54 centimeters	$\frac{1.00 \text{ in.}}{2.54 \text{ cm}}$	$\frac{2.54 \text{ cm}}{1.00 \text{ in.}}$
1.00 meter = 39.4 inches	$\frac{39.4 \text{ in.}}{1.00 \text{ m}}$	$\frac{1.00 \text{ m}}{39.4 \text{ in.}}$
1.00 kilometer = 0.621 mile	$\frac{0.621 \text{ mi}}{1.00 \text{ km}}$	$\frac{1.00 \text{ km}}{0.621 \text{ mi}}$
Mass		
1.00 pound = 454 grams	$\frac{1.00 \text{ lb}}{454 \text{ g}}$	$\frac{454 \text{ g}}{1.00 \text{ lb}}$
1.00 kilogram = 2.20 pounds	$\frac{2.20 \text{ lb}}{1.00 \text{ kg}}$	$\frac{1.00 \text{ kg}}{2.20 \text{ lb}}$
1.00 ounce = 28.3 grams	$\frac{1.00 \text{ oz}}{28.3 \text{ g}}$	$\frac{28.3 \text{ g}}{1.00 \text{ oz}}$
Volume		
1.00 quart = 0.946 liter	$\frac{1.00 \text{ qt}}{0.946 \text{ L}}$	$\frac{0.946 \text{ L}}{1.00 \text{ qt}}$
1.00 liter = 0.265 gallon	$\frac{0.265 \text{ gal}}{1.00 \text{ L}}$	$\frac{1.00 \text{ L}}{0.265 \text{ gal}}$
1.00 milliliter = 0.034 fluid ounce	$\frac{0.034 \text{ fl oz}}{1.00 \text{ mL}}$	$\frac{1.00 \text{ mL}}{0.034 \text{ fl oz}}$

• Dimensional Analysis

Dimensional analysis *is a general problem-solving method in which the units associated with numbers are used as a guide in setting up calculations.* In this method, units are treated in the same way as numbers; that is, they can be multiplied, divided, or canceled. For example, just as

$$5 \times 5 = 5^2 \qquad \text{(5 squared)}$$

we have

$$\text{cm} \times \text{cm} = \text{cm}^2 \qquad \text{(cm squared)}$$

Also, just as the 3s cancel in the expression

$$\frac{\cancel{3} \times 5 \times 7}{\cancel{3} \times 2}$$

the centimeters cancel in the expression

$$\frac{\cancel{(\text{cm})} \times (\text{in.})}{\cancel{(\text{cm})}}$$

"Like units" found in the numerator and denominator of a fraction will always cancel, just as like numbers do.

The following steps show how to set up a problem using dimensional analysis.

Step 1: *Identify the known or given quantity (both numerical value and units) and the units of the new quantity to be determined.*

This information will always be found in the statement of the problem. Write an equation with the given quantity on the left and the units of the desired quantity on the right.

Step 2: *Multiply the given quantity by one or more conversion factors in such a manner that the unwanted (original) units are canceled, leaving only the desired units.*

The general format for the multiplication is

$$\text{(Information given)} \times \text{(conversion factors)} = \text{(information sought)}$$

The number of conversion factors depends on the individual problem.

Step 3: *Perform the mathematical operations indicated by the conversion factor setup.*

When performing the calculation, double-check to make sure that all units except the desired set have canceled.

Example 2.3

Unit Conversions Within the Metric System

A standard aspirin tablet contains 324 mg of aspirin. How many grams of aspirin are in a standard aspirin tablet?

Solution

Step 1: The given quantity is 324 mg, the mass of aspirin in the tablet. The unit of the desired quantity is grams.

$$324 \text{ mg} = ? \text{ g}$$

Step 2: Only one conversion factor will be needed to convert from milligrams to grams, one that relates milligrams to grams. The two forms of this conversion factor are

$$\frac{1 \text{ mg}}{10^{-3} \text{ g}} \quad \text{and} \quad \frac{10^{-3} \text{ g}}{1 \text{ mg}}$$

The second factor is used because it allows for cancellation of the milligram units, leaving us with grams as the new units.

$$324\ \cancel{\text{mg}} \times \left(\frac{10^{-3}\ \text{g}}{1\ \cancel{\text{mg}}}\right) = ?\ \text{g}$$

Step 3: Combining numerical terms as indicated generates the final answer.

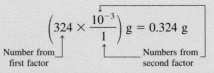

$$\left(324 \times \frac{10^{-3}}{1}\right)\ \text{g} = 0.324\ \text{g}$$

Number from first factor; Numbers from second factor

The answer is given to three significant figures because the given quantity in the problem, 324 mg, has three significant figures. The conversion factor used arises from a definition and thus does not limit significant figures in any way.

Practice Exercise 2.3

Analysis shows the presence of 203 μg of cholesterol in a sample of blood. How many grams of cholesterol are present in this blood sample?

- *Answer.* 2.03×10^{-4} g

Example 2.4

Unit Conversions Between the Metric and English Systems

Capillaries, the microscopic vessels that carry blood from small arteries to small veins, are on the average only 1 mm long. What is the average length of a capillary in inches?

Solution

Step 1: The given quantity is 1 mm, and the units of the desired quantity are inches.

$$1\ \text{mm} = ?\ \text{in.}$$

Step 2: The conversion factor needed for a one-step solution, millimeters to inches, is not given in Table 2.2. However, a related conversion factor, meters to inches, is given. Therefore, we first convert millimeters to meters and then use the meters-to-inches conversion factor in Table 2.2.

$$\text{mm} \longrightarrow \text{m} \longrightarrow \text{in.}$$

The correct conversion factor setup is

$$1\ \cancel{\text{mm}} \times \left(\frac{10^{-3}\ \cancel{\text{m}}}{1\ \cancel{\text{mm}}}\right) \times \left(\frac{39.4\ \text{in.}}{1.00\ \cancel{\text{m}}}\right) = ?\ \text{in.}$$

All of the units except for inches cancel, which is what is needed. The information for the middle conversion factor was obtained from the meaning of the prefix *milli-*.

This setup illustrates the fact that sometimes the given units must be changed to intermediate units before common conversion factors, such as those found in Table 2.2, are applicable.

Step 3: Collecting the numerical factors and performing the indicated math gives

$$\left(\frac{1 \times 10^{-3} \times 39.4}{1 \times 1.00}\right)\ \text{in.} = 0.0394\ \text{in.} \qquad \text{(calculator answer)}$$

$$= 0.04\ \text{in.} \qquad \text{(correct answer)}$$

The calculator answer must be rounded to one significant figure because 1 mm, the given quantity, contains only one significant figure.

Practice Exercise 2.4

Blood analysis reports often give the amounts of various substances present in the blood in terms of milligrams per deciliter. What is the measure, in quarts, of 1.00 deciliter?

- *Answer.* 0.106 qt

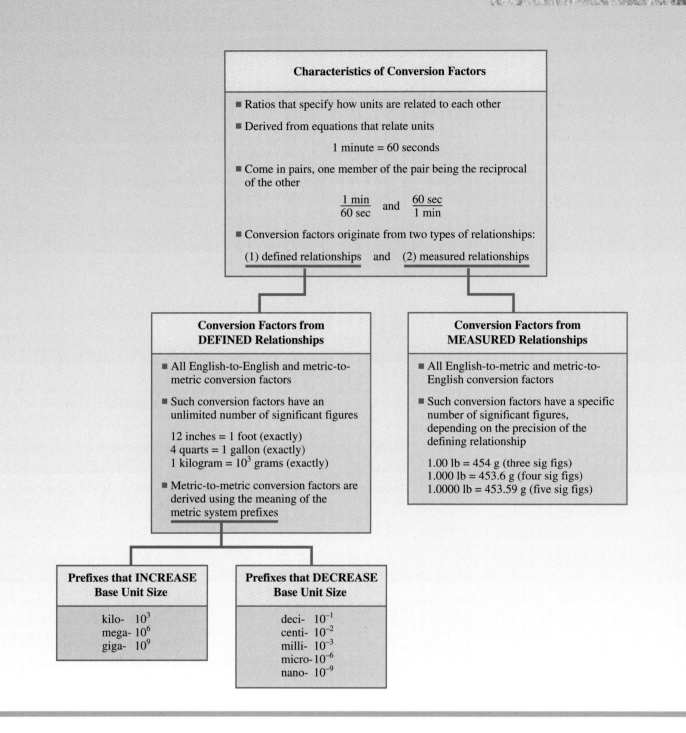

CHEMISTRY AT A GLANCE

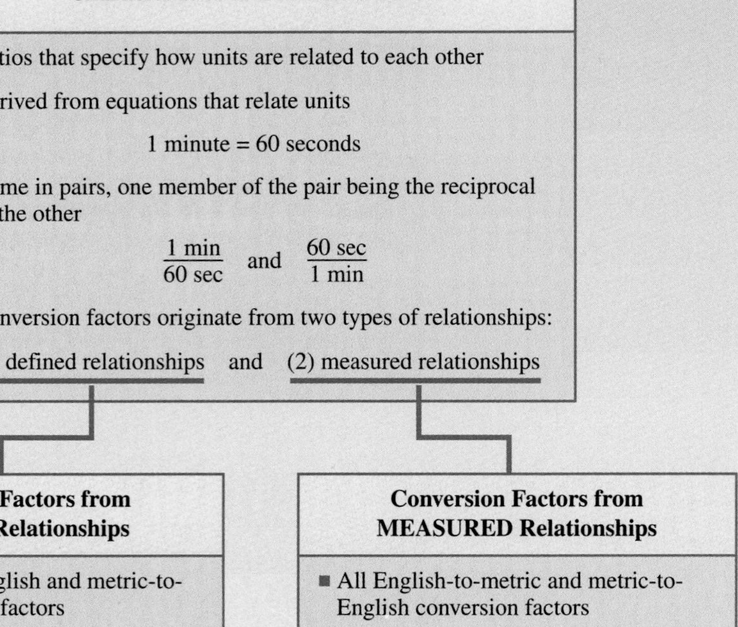

Characteristics of Conversion Factors

- Ratios that specify how units are related to each other
- Derived from equations that relate units

  1 minute = 60 seconds
- Come in pairs, one member of the pair being the reciprocal of the other

  $$\frac{1\text{ min}}{60\text{ sec}} \quad \text{and} \quad \frac{60\text{ sec}}{1\text{ min}}$$
- Conversion factors originate from two types of relationships:

  (1) defined relationships and (2) measured relationships

Conversion Factors from DEFINED Relationships

- All English-to-English and metric-to-metric conversion factors
- Such conversion factors have an unlimited number of significant figures

  12 inches = 1 foot (exactly)
  4 quarts = 1 gallon (exactly)
  1 kilogram = 10^3 grams (exactly)
- Metric-to-metric conversion factors are derived using the meaning of the metric system prefixes

Conversion Factors from MEASURED Relationships

- All English-to-metric and metric-to-English conversion factors
- Such conversion factors have a specific number of significant figures, depending on the precision of the defining relationship

  1.00 lb = 454 g (three sig figs)
  1.000 lb = 453.6 g (four sig figs)
  1.0000 lb = 453.59 g (five sig figs)

Prefixes that INCREASE Base Unit Size

kilo- 10^3
mega- 10^6
giga- 10^9

Prefixes that DECREASE Base Unit Size

deci- 10^{-1}
centi- 10^{-2}
milli- 10^{-3}
micro- 10^{-6}
nano- 10^{-9}

Chemistry at a Glance reviews what we have said about conversion factors.

2.7 Density

Density *is the ratio of the mass of an object to the volume occupied by that object.*

$$\text{Density} = \frac{\text{mass}}{\text{volume}}$$

(a)

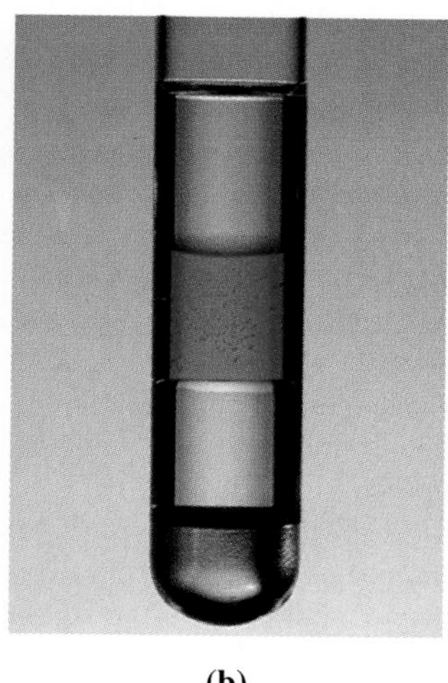

(b)

Figure 2.7
(a) The penny is less dense than the mercury it floats on. (b) Liquids that do not dissolve in one another and that have different densities float on one another, forming layers. The top layer is gasoline, with a density of about 0.8 g/mL. Next is water (plus food coloring), with a density of 1.0 g/mL. The next layer is carbon tetrachloride, with a density of 1.6 g/mL. The bottom layer is mercury, with a density of 13.6 g/mL.

People often speak of a substance as being heavier or lighter than another substance. What they actually mean is that the two substances have different densities; a specific volume of one substance is heavier or lighter than the same volume of the second substance (Figure 2.7).

A correct density expression includes a number, a mass unit, and a volume unit. Although any mass and volume units can be used, densities are usually expressed in grams per cubic centimeter (g/cm^3) for solids, grams per milliliter (g/mL) for liquids, and grams per liter (g/L) for gases. Table 2.3 gives density values for a number of substances. Note

Table 2.3
Densities of Selected Substances

Solids (25°C)			
gold	19.3 g/cm^3	table salt	2.16 g/cm^3
lead	11.3 g/cm^3	bone	1.7–2.0 g/cm^3
copper	8.93 g/cm^3	table sugar	1.59 g/cm^3
aluminum	2.70 g/cm^3	wood, pine	0.30–0.50 g/cm^3
Liquids (25°C)			
mercury	13.55 g/mL	water	0.997 g/mL
milk	1.028–1.035 g/mL	olive oil	0.92 g/mL
blood plasma	1.027 g/mL	ethyl alcohol	0.79 g/mL
urine	1.003–1.030 g/mL	gasoline	0.56 g/mL
Gases (25°C and 1 atmosphere pressure)			
chlorine	3.17 g/L	nitrogen	1.25 g/L
carbon dioxide	1.96 g/L	methane	0.66 g/L
oxygen	1.42 g/L	hydrogen	0.08 g/L
air (dry)	1.29 g/L		

Chemical CONNECTIONS

2.1 Body Density and Percent Body Fat

More than half the adult population of the United States is overweight. But what does "overweight" mean? In years past, people were considered overweight if they weighed more for their height than called for in standard height/mass charts. Such charts are now considered outdated. Today, we realize that body composition is more important than total body mass. The proportion of fat to total body mass—that is, the percent of body fat—is the key to defining *overweight.* A very muscular person, for example, can be overweight according to height/mass charts although he or she has very little body fat. Some athletes fall into this category. Body composition ratings, tied to percent body fat, are listed here.

Body composition rating	Percent body fat	
	Men	**Women**[a]
excellent	less than 13	less than 18
good	13–17	18–22
average	18–21	23–26
fair	22–30	27–35
poor	greater than 30	greater than 35

[a]Women are genetically predisposed to maintain a higher percentage of body fat.

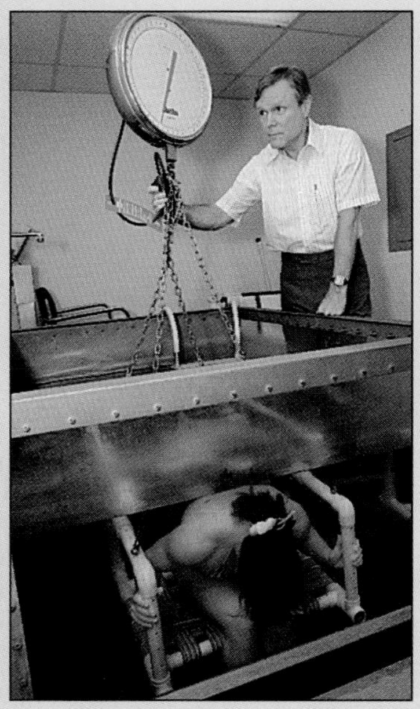

The percentage of fat in a person's body can be determined by hydrostatic (underwater) weighing. Fat cells, unlike most other human body cells and fluids, are less dense than water. Consequently, a person with a high percentage of body fat is buoyed up by water more than is a lean person. The hydrostatic-weighing technique for determining body fat is based on this difference in density. A person is first weighed in air and then weighed again submerged in water. The difference between these two masses (with a correction for residual air in the lungs and for the temperature of the water) is used to calculate body density. The higher the density of the body, the lower the percent of body fat. Sample values relating body density and percent body fat are given here.

Body density (g/mL)	Percent body fat
1.070	12.22
1.062	15.25
1.052	19.29
1.036	25.35
1.027	29.39

that temperature must be specified with density values, because substances expand and contract with changes in temperature. For the same reason, the pressure of gases is also given with their density values.

• Density may be used as a conversion factor to convert from mass to volume or vice versa.

Density can be used as a conversion factor that relates the volume of a substance to its mass. This use of density enables us to calculate the volume of a substance if we know its mass. Conversely, the mass can be calculated if the volume is known.

Density conversion factors, like all other conversion factors, have two reciprocal forms. For a density of 1.03 g/mL, the two conversion factor forms are

$$\frac{1.03\text{ g}}{1\text{ mL}} \quad \text{and} \quad \frac{1\text{ mL}}{1.03\text{ g}}$$

Example 2.5

Converting from Mass to Volume by Using Density as a Conversion Factor

Blood plasma has a density of 1.027 g/mL at 25°C. What volume, in milliliters, does 125 g of plasma occupy?

Both of these items have a mass of 23 grams, but very different volumes; therefore, their densities are different as well.

Solution

Step 1: The given quantity is 125 g of blood plasma. The units of the desired quantity are milliliters. Thus our starting point is

$$125\ \text{g} = ?\ \text{mL}$$

Step 2: The conversion from grams to milliliters can be accomplished in one step because the given density, used as a conversion factor, directly relates grams to milliliters. Of the two conversion factor forms

$$\frac{1.027\ \text{g}}{1\ \text{mL}} \quad \text{and} \quad \frac{1\ \text{mL}}{1.027\ \text{g}}$$

we will use the latter, because it allows for cancellation of gram units, leaving milliliters.

$$125\ \cancel{\text{g}} \times \left(\frac{1\ \text{mL}}{1.027\ \cancel{\text{g}}}\right) = ?\ \text{mL}$$

Step 3: Doing the necessary arithmetic gives us our answer:

$$\left(\frac{125 \times 1}{1.027}\right)\text{mL} = 121.71372\ \text{mL} \qquad \text{(calculator answer)}$$

$$= 122\ \text{mL} \qquad \text{(correct answer)}$$

Even though the given density contained four significant figures, the correct answer is limited to three significant figures. This is because the other given number, the mass of blood plasma, had only three significant figures.

Practice Exercise 2.5

If your blood has a density of 1.05 g/mL at 25°C, how many grams of blood would you lose if you made a blood bank donation of 1.00 pint (473 mL) of blood?

• *Answer:* 497 g

2.8 Temperature and Heat Energy

Heat is a form of energy. Temperature is an indicator of the tendency of heat energy to be transferred. Heat energy flows from objects of higher temperature to objects of lower temperature.

Three different temperature scales are in common use: Celsius, Kelvin, and Fahrenheit (Figure 2.8). Both the Celsius and the Kelvin scales are part of the metric measurement system; the Fahrenheit scale belongs to the English measurement system. Degrees of different size and different reference points are what produce the various temperature scales.

The *Celsius scale* is the scale most commonly encountered in scientific work. The normal boiling and freezing points of water serve as reference points on this scale, the former having a value of 100° and the latter 0°. Thus there are 100 "degree intervals" between the two reference points.

The *Kelvin scale* is a close relative of the Celsius scale. Both have the same size of degree, and the number of degrees between the freezing and boiling points of water is the same. The two scales differ only in the numbers assigned to the reference points. On the Kelvin scale, the boiling point of water is 373 kelvins (K) and the freezing point of water is 273 K. The choice of these reference points makes all temperature readings on the Kelvin scale positive values. Note that the degree sign (°) is not used with the Kelvin scale. For example, we say that an object has a temperature of 350 K (*not* 350° K).

• Zero on the Kelvin scale is known as *absolute zero.* It corresponds to the lowest temperature allowed by nature. How fast particles (molecules) move depends on temperature. The colder it gets, the more slowly they move. At absolute zero, movement stops. Scientists in laboratories have been able to attain temperatures as low as 0.0001 K, but a temperature of 0 K is impossible.

The *Fahrenheit scale* has a smaller degree size than the other two temperature scales. On this scale, there are 180 degrees between the freezing and boiling points of water as contrasted to 100 degrees on the other two scales. Thus the Celsius (and Kelvin) degree size is almost two times ($\frac{9}{5}$) larger than the Fahrenheit degree. Reference points on the

Figure 2.8
The relationships among the Celsius, Kelvin, and Fahrenheit temperature scales are determined by the degree sizes and the reference point values.

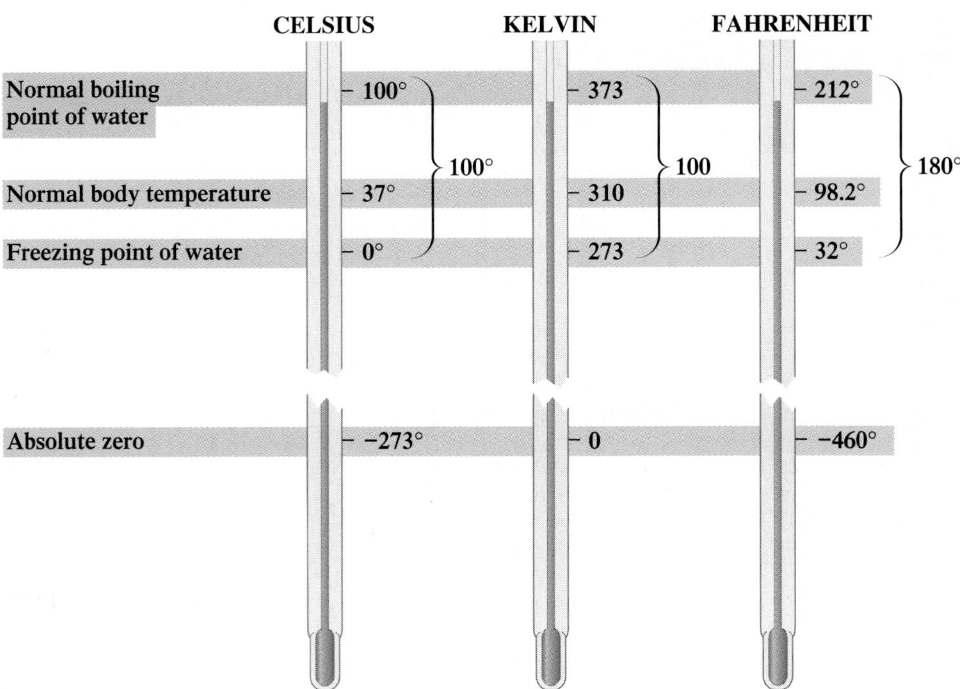

Fahrenheit scale are 32° for the freezing point of water and 212° for the normal boiling point of water.

• Conversions Between Temperature Scales

Because the size of the degree is the same, the relationship between the Kelvin and Celsius scales is very simple. No conversion factors are needed; all that is required is an adjustment for the differing numerical scale values. The adjustment factor is 273, the number of degrees by which the two scales are offset from one another.

$$K = {}^\circ C + 273$$
$${}^\circ C = K - 273$$

The relationship between the Fahrenheit and Celsius scales can also be stated in an equation format.

$${}^\circ F = \frac{9}{5}({}^\circ C) + 32 \qquad \text{or} \qquad {}^\circ C = \frac{5}{9}({}^\circ F - 32)$$

Example 2.6

Converting from One Temperature Scale to Another

Body temperature for a person with a high fever is found to be 104°F. To what is this temperature equivalent on the following scales?

a. Celsius scale **b.** Kelvin scale

Solution

a. We substitute 104° for °F in the equation

$${}^\circ C = \frac{5}{9}({}^\circ F - 32)$$

Then solving for °C gives

$${}^\circ C = \frac{5}{9}(104 - 32) = \frac{5}{9}(72) = 40^\circ$$

b. Using the answer from part **a** and the equation

$$K = °C + 273$$

we get, by substitution,

$$K = 40° + 273 = 313$$

Practice Exercise 2.6

In the human body, heat stroke occurs at a temperature of 41°C. To what is this temperature equivalent on the following scales?

a. Fahrenheit scale **b.** Kelvin scale

• *Answer.* **a.** 106°F; **b.** 314 K

Table 2.4
Specific Heats of Selected Common Substances

Substance	Specific heat (cal/g · °C)[a]
water, liquid	1.00
ethyl alcohol	0.58
olive oil	0.47
wood	0.42
aluminum	0.21
glass	0.12
silver	0.057
gold	0.031

[a]The unit notation cal/g · °C means calories per gram per degree Celsius.

• Heat Energy and Specific Heat

The form of energy most often required for or released by chemical reactions and physical changes is *heat energy.* A commonly used unit for the measurement of heat energy is the calorie. A **calorie** (cal) *is the amount of heat energy needed to raise the temperature of 1 gram of water by 1 degree Celsius.* For large amounts of heat energy, the measurement is usually expressed in kilocalories.

$$1 \text{ kilocalorie} = 1000 \text{ calories}$$

• In discussions involving nutrition, the energy content of foods, and dietary tables, the term *Calorie* (spelled with a capital C) is used. The dietetic Calorie is actually 1 kilocalorie (1000 calories). The statement that an oatmeal raisin cookie contains 60 Calories means that 60 kcal (60,000 cal) of energy is released when the cookie is metabolized (undergoes chemical change) within the body.

Another unit for heat energy that is used with increasing frequency is the joule (J). The relationship between the joule (which rhymes with *pool*) and the calorie is

$$1 \text{ calorie} = 4.184 \text{ joules}$$

Heat energy values in calories can be converted to joules by using the conversion factor

$$\frac{4.184 \text{ J}}{1 \text{ cal}}$$

The **specific heat** *of a substance is the quantity of heat energy, in calories, that is necessary to raise the temperature of 1 gram of the substance by 1 degree Celsius.* Specific heats for a number of substances in various states are given in Table 2.4.

The higher the specific heat of a substance, the less its temperature will change as it absorbs a given amount of heat. Water has a relatively high specific heat; it is thus a very

Water's high specific heat causes it to have a moderating effect in climates like that of Hawaii. The desert, lacking water, experiences extreme temperatures.

• Water has the highest specific heat of all common substances.

• The property of specific heat varies *slightly* with temperature and pressure. We will ignore such variations in this text.

effective coolant. The moderate climates of geographical areas near large bodies of water—the Hawaiian Islands, for example—are related to the ability of water to absorb large amounts of heat without undergoing drastic temperature changes. Desert areas (obviously lacking in water) experience low- and high-temperature extremes.

Specific heat is an important quantity because it can be used to calculate the number of calories required to heat a known mass of a substance from one temperature to another. It can also be used to calculate how much the temperature of a substance increases when it absorbs a known number of calories of heat. The equation used for such calculations is

$$\text{Heat absorbed} = \text{specific heat} \times \text{mass} \times \text{temperature change}$$

If any three of the four quantities in this equation are known, the fourth quantity can be calculated. If the units for specific heat are cal/g · °C, the units for mass are grams, and the units for temperature change are °C, then the heat absorbed has units of calories.

Example 2.7

Calculating the Amount of Heat Released as the Result of a Temperature Decrease

If a hot-water bottle contains 1200 g of water at 65°C, how much heat, in calories, will it have supplied to a person's "aching back" by the time it has cooled to 37°C (assuming all of the heat energy goes into the person's back)?

(continued)

Chemical CONNECTIONS

2.2 Normal Human Body Temperature

Studies show that "normal" human body temperature varies from individual to individual. For oral temperature measurements, this individual variance spans the range from 96°F to 101°F.

Furthermore, individual body temperatures vary with exercise and with the temperature of the surroundings. When excessive heat is produced in the body by strenuous exercise, oral temperature can rise as high as 103°F. On the other hand, when the body is exposed to cold, oral temperature can fall to values considerably below 96°F. A rapid fall in temperature of 2°F to 3°F produces uncontrollable shivering.

Each individual also has a characteristic pattern of temperature variation during the day, with differences of as much as 1°F to 3°F between high and low points. Body temperature is typically lowest in the very early morning, after several hours of sleep, when one is inactive and not digesting food. During the day, body temperature rises to a peak and begins to fall again. "Morning people"—people who are most productive early in the day—have a body temperature peak at midmorning or midday. "Night people"—people who feel as though they are just getting started as evening approaches and who work best late at night—have a body temperature peak in the evening.

What, then, is the average (normal) human body temperature? Reference books list the value 98.6°F (37.0°C) as the answer to this question. The source for this value is a study involving over 1 million human body temperature readings that was published in 1868, over 130 years ago.

A 1992 study, published in the *Journal of the American Medical Association,* questions the validity of this average value (98.6°F). This new study notes that the 1868 study was carried out using thermometers that were more difficult to get accurate readings from than modern thermometers. The 1992 study is based on oral temperature readings obtained using electronic thermometers. Findings of this new study include the following:

1. The range of temperatures was 96.0°F to 100.8°F.
2. The mean (average) temperature was 98.2°F (36.8°C).
3. At 6 A.M., the temperature 98.9°F is the upper limit of the normal temperature range.
4. In late afternoon (4 P.M.), the temperature 99.9°F is the upper limit of the normal temperature range.
5. Women have a slightly higher average temperature than men (98.4°F versus 98.1°F).
6. Over the temperature range 96°F to 101°F, there is an average increase in heart rate of 2.44 beats per minute for each 1°F rise in temperature.

As a result of this study, future reference books will probably use 98.2°F rather than 98.6°F as the value for average (normal) human body temperature.

The water in sweat effectively transfers excess heat out of the body

Solution

We will substitute known quantities into the equation

$$\text{Heat released} = \text{specific heat} \times \text{mass} \times \text{temperature change}$$

Table 2.4 shows that the specific heat of liquid water is 1.00 cal/g · °C. The mass of the water is given as 1200 g. The temperature change in going from 65°C to 37°C is 28°C. Substituting these values into the preceding equation gives

$$\text{Heat released} = \left(\frac{1.00\ \text{cal}}{\cancel{\text{g}} \cdot \cancel{°\text{C}}}\right) \times (1200\ \cancel{\text{g}}) \times (28\cancel{°\text{C}})$$

$$= 33{,}600\ \text{cal} \qquad \text{(calculator answer)}$$

$$= 34{,}000\ \text{cal} \qquad \text{(correct answer)}$$

The given quantity of 1200 g and the temperature difference of 28°C, both of which have only two significant figures, limit the answer to two significant figures.

Practice Exercise 2.7

How much heat energy, in calories, must be absorbed by 125.0 g of water to raise its temperature by 12°C?

- *Answer.* 1500 cal

Concepts to Remember

The metric system. The metric system, the measurement system preferred by scientists, is a decimal system in which larger and smaller units of a quantity are related by factors of 10. Prefixes are used to designate relationships between the basic unit and larger or smaller units of a quantity. Units in the metric system include the gram (mass), liter (volume), and meter (length).

Significant figures. Significant figures in a measurement are those digits that are certain, plus a last digit that has been estimated. The maximum number of significant figures possible in a measurement is determined by the design of the measuring device.

Calculations and significant figures. Calculations should never improve (or decrease) the precision of experimental measurements. In multiplication and division, the number of significant figures in the answer is the same as that in the measurement containing the fewest significant figures. In addition and subtraction, the answer has no more digits to the right of the decimal point than are found in the measurement with the fewest digits to the right of the decimal point.

Scientific notation. Scientific notation is a system for writing decimal numbers in a more compact form that greatly simplifies the mathematical operations of multiplication and division. In this system, numbers are expressed as the product of a number between 1 and 10 and 10 raised to a power.

Dimensional analysis. Dimensional analysis is a general problem-solving method in which the units associated with numbers are used as a guide in setting up calculations. A given quantity is multiplied by one or more conversion factors in such a manner that the unwanted (original) units are canceled, leaving only the desired units.

Density. Density is the ratio of the mass of an object to the volume occupied by that object. A correct density expression includes a number, a mass unit, and a volume unit.

Temperature scales. The three major temperature scales are the Celsius, Kelvin, and Fahrenheit scales. The size of the degree for the Celsius and Kelvin scales is the same. They differ only in the numerical values assigned to the reference points. The Fahrenheit scale has a smaller degree size than the other two temperature scales.

Heat energy and specific heat. The most commonly used unit of measurement for heat energy is the calorie. A calorie is the amount of heat energy needed to raise the temperature of 1 gram of water by 1 degree Celsius. The specific heat of a substance is the quantity of heat energy, in calories, that is necessary to raise the temperature of 1 gram of the substance by 1 degree Celsius.

Key Reactions and Equations

1. Density of a substance (Section 2.7)

$$\text{Density} = \frac{\text{mass}}{\text{volume}}$$

2. Conversion of temperature readings from one scale to another (Section 2.8)

$$\text{K} = °\text{C} + 273 \qquad °\text{C} = \text{K} - 273$$

$$°\text{F} = \frac{9}{5}(°\text{C}) + 32 \qquad °\text{C} = \frac{5}{9}(°\text{F} - 32)$$

3. Heat energy absorbed by a substance (Section 2.8)

$$\begin{matrix}\text{Heat energy} \\ \text{absorbed}\end{matrix} = \begin{matrix}\text{specific} \\ \text{heat}\end{matrix} \times \text{mass} \times \begin{matrix}\text{temperature} \\ \text{change}\end{matrix}$$

Key Terms

Calorie (2.8)
Conversion factor (2.6)
Density (2.7)
Dimensional analysis (2.6)
Gram (2.1)
Liter (2.2)
Mass (2.2)
Measurement (2.1)
Meter (2.2)
Rounding off (2.4)
Scientific notation (2.5)
Specific heat (2.8)
Significant figures (2.3)
Weight (2.2)

Exercises and Problems

The members of each pair of problems in this section test similar material.

Metric System Units (Section 2.2)

2.1 Write the name of the metric system prefix associated with each of the following mathematical meanings.
a. 10^3 b. 10^{-3} c. 10^{-6} d. 1/10

2.2 Write the name of the metric system prefix associated with each of the following mathematical meanings.
a. 10^{-2} b. 10^{-9} c. 10^6 d. 1/1000

2.3 Write out the names of the metric system units that have the following abbreviations.
a. cm b. kL c. μL d. ng

2.4 Write out the names of the metric system units that have the following abbreviations.
a. mg b. pg c. Mm d. dL

2.5 Arrange each of the following from smallest to largest.
a. milligram, centigram, nanogram
b. gigameter, megameter, kilometer
c. microliter, deciliter, picoliter
d. milligram, kilogram, microgram

2.6 Arrange each of the following from smallest to largest.
a. milliliter, gigaliter, microliter
b. centigram, megagram, decigram
c. micrometer, picometer, kilometer
d. nanoliter, milliliter, centiliter

Uncertainty in Measurement and Significant Figures (Section 2.3)

2.7 Indicate to what decimal position readings should be recorded (nearest 0.1, 0.01, etc.) for measurements made with the following devices.
a. A thermometer with a smallest scale marking of 1°C
b. A graduated cylinder with a smallest scale marking of 0.1 mL
c. A volumetric device with a smallest scale marking of 10 mL
d. A ruler with a smallest scale marking of 1 mm

2.8 Indicate to what decimal position readings should be recorded (nearest 0.1, 0.01, etc.) for measurements made with the following devices.
a. A ruler with a smallest scale marking of 1 cm
b. A device for measuring angles with a smallest scale marking of 1°
c. A thermometer with a smallest scale marking of 0.1°F
d. A graduated cylinder with a smallest scale marking of 10 mL

2.9 Determine the number of significant figures in each of the following measured values.
a. 6.000 b. 0.0032 c. 0.01001
d. 65,400 e. 76.010 f. 0.03050

2.10 Determine the number of significant figures in each of the following measured values.
a. 23,009 b. 0.00231 c. 0.3330
d. 73,000 e. 73.000 f. 0.40040

2.11 In which of the following pairs of numbers do both members of the pair contain the same number of significant figures?
a. 11.01 and 11.00 b. 2002 and 2020
c. 0.000066 and 660,000 d. 0.05700 and 0.05070

2.12 In which of the following pairs of numbers do both members of the pair contain the same number of significant figures?
a. 345,000 and 340,500 b. 2302 and 2320
c. 0.6600 and 0.66 d. 936 and 936,000

Significant Figures and Mathematical Operations (Section 2.4)

2.13 Round off each of the following numbers to the number of significant figures indicated in parentheses.
a. 0.350763 (three) b. 653,899 (four)
c. 22.55555 (five) d. 0.277654 (four)

2.14 Round off each of the following numbers to the number of significant figures indicated in parentheses.
a. 3883 (two) b. 0.0003011 (two)
c. 4.4050 (three) d. 2.1000 (three)

2.15 Without actually solving, indicate the number of significant figures that should be present in the answers to the following multiplication and division problems.
a. $10.300 \times 0.30 \times 0.300$ b. $3300 \times 3330 \times 333.0$
c. $\frac{6.0}{33.0}$ d. $\frac{6.000}{33}$

2.16 Without actually solving, indicate the number of significant figures that should be present in the answers to the following multiplication and division problems.
a. $3.00 \times 0.0003 \times 30.00$ b. $0.3 \times 0.30 \times 3.0$
c. $\frac{6.00}{33{,}000}$ d. $\frac{6.00000}{3}$

2.17 Carry out the following multiplications and divisions, expressing your answer to the correct number of significant figures. Assume that all numbers are measured numbers.
a. $2.0000 \times 2.00 \times 0.0020$ b. 4.1567×0.00345

c. $0.0037 \times 3700 \times 1.001$ d. $\frac{6.00}{33.0}$

e. $\frac{530{,}000}{465{,}300}$ f. $\frac{4.670 \times 3.00}{2.450}$

2.18 Carry out the following multiplications and divisions, expressing your answer to the correct number of significant figures. Assume that all numbers are measured numbers.

a. $2.000 \times 0.200 \times 0.20$ b. 3.6750×0.04503

c. $0.0030 \times 0.400 \times 4.00$ d. $\frac{6.0000}{33.00}$

e. $\frac{45{,}000}{1.2345}$ f. $\frac{3.000 \times 6.53}{13.567}$

2.19 Carry out the following additions and subtractions, expressing your answer to the correct number of significant figures. Assume that all numbers are measured numbers.

a. $12 + 23 + 127$ b. $3.111 + 3.11 + 3.1$

c. $1237.6 + 23 + 0.12$ d. $43.65 - 23.7$

2.20 Carry out the following additions and subtractions, expressing your answer to the correct number of significant figures. Assume that all numbers are measured numbers.

a. $237 + 37.0 + 7.0$ b. $4.000 + 4.002 + 4.20$

c. $235.45 + 37 + 36.4$ d. $3.111 - 2.07$

Scientific Notation (Section 2.5)

2.21 Express the following numbers in scientific notation.

a. 120.7 b. 0.0034 c. 231.00

d. 23,000 e. 0.200 f. 0.1011

2.22 Express the following numbers in scientific notation.

a. 37.06 b. 0.00571 c. 437.0

d. 4370 e. 0.20340 f. 230,000

2.23 Carry out the following multiplications and divisions of exponential terms.

a. $10^5 \times 10^3$ b. $10^5 \times 10^{-3}$ c. $10^{-5} \times 10^{-3}$

d. $\frac{10^5}{10^{-3}}$ e. $\frac{10^{-5}}{10^3}$ f. $\frac{10^{-5}}{10^{-3}}$

2.24 Carry out the following multiplications and divisions of exponential terms.

a. $10^4 \times 10^2$ b. $10^{-4} \times 10^2$ c. $10^{-4} \times 10^{-2}$

d. $\frac{10^4}{10^{-2}}$ e. $\frac{10^{-4}}{10^2}$ f. $\frac{10^{-4}}{10^{-2}}$

2.25 Carry out the following multiplications and divisions, expressing your answer in scientific notation to the correct number of significant figures.

a. $(3.20 \times 10^7) \times (1.720 \times 10^5)$

b. $(3.71 \times 10^{-4}) \times (1.117 \times 10^2)$

c. $(1.00 \times 10^3) \times (5.00 \times 10^3) \times (3.0 \times 10^{-3})$

d. $\frac{3.0 \times 10^{-5}}{1.5 \times 10^2}$ e. $\frac{4.56 \times 10^7}{3.0 \times 10^{-4}}$

f. $\frac{(2.2 \times 10^6) \times (2.3 \times 10^{-6})}{(1.2 \times 10^{-3}) \times (3.5 \times 10^{-3})}$

2.26 Carry out the following multiplications and divisions, expressing your answer in scientific notation to the correct number of significant figures.

a. $(4.0 \times 10^4) \times (1.32 \times 10^8)$

b. $(2.23 \times 10^{-6}) \times (1.230 \times 10^{-2})$

c. $(3.200 \times 10^7) \times (1.10 \times 10^{-2}) \times (2.3 \times 10^{-7})$

d. $\frac{6.0 \times 10^{-5}}{3.0 \times 10^3}$ e. $\frac{5.132 \times 10^7}{1.12 \times 10^3}$

f. $\frac{(3.2 \times 10^2) \times (3.31 \times 10^6)}{(4.00 \times 10^{-3}) \times (2.0 \times 10^6)}$

Conversion Factors and Dimensional Analysis (Section 2.6)

2.27 Give both forms of the conversion factor that you would use to relate the following sets of units to each other.

a. gram and kilogram b. meter and nanometer

c. liter and milliliter d. gram and pound

e. kilometer and mile f. gallon and liter

2.28 Give both forms of the conversion factor that you would use to relate the following sets of units to each other.

a. liter and milliliter b. meter and micrometer

c. gram and centigram d. inch and meter

e. kilogram and pound f. liter and quart

2.29 Convert each of the following measurements to meters.

a. 1.6×10^3 dm b. 24 nm

c. 0.003 km d. 3.0×10^8 mm

2.30 Convert each of the following measurements to meters.

a. 2.7×10^3 mm b. 24 μm

c. 0.003 pm d. 4.0×10^5 cm

2.31 The human stomach produces approximately 2500 mL of gastric juice per day. What is the volume, in liters, of gastric juice produced?

2.32 A typical normal loss of water through sweating per day for a human is 450 mL. What is the volume, in liters, of sweat produced per day?

2.33 The mass of premature babies is customarily determined in grams. If a premature baby weighs 1550 g, what is its mass in pounds?

2.34 The smallest bone in the human body, which is in the ear, has a mass of 0.0030 g. What is the mass of this bone in pounds?

2.35 What volume of water, in gallons, would be required to fill a 25-mL container?

2.36 What volume of gasoline, in milliliters, would be required to fill a 17.0-gal gasoline tank?

2.37 An individual weighs 83.2 kg and is 1.92 m tall. What are the person's equivalent measurements in pounds and feet?

2.38 An individual weighs 135 lb and is 5 ft 4 in. tall. What are the person's equivalent measurements in kilograms and meters?

Density (Section 2.7)

2.39 A sample of mercury is found to have a mass of 524.5 g and a volume of 38.72 cm^3. What is its density in grams per cubic centimeter?

2.40 A sample of sand is found to have a mass of 12.0 g and a volume of 2.69 cm^3. What is its density in grams per cubic centimeter?

2.41 Acetone, the solvent in nail polish remover, has a density of 0.791 g/mL. What is the volume, in milliliters, of 20.0 g of acetone?

2.42 Silver metal has a density of 10.40 g/cm^3. What is the volume, in cubic centimeters, of a 100.0-g bar of silver metal?

2.43 The density of homogenized milk is 1.03 g/mL. How much does 1 cup (236 mL) of homogenized milk weigh in grams?

2.44 Nickel metal has a density of 8.90 g/cm^3. How much does 15 cm^3 of nickel metal weigh in grams?

Temperature and Heat Energy (Section 2.8)

2.45 An oven for baking pizza operates at approximately 525°F. What is this temperature in degrees Celsius?

2.46 A comfortable temperature for bathtub water is 95°F. What temperature is this in degrees Celsius?

2.47 Mercury freezes at −38.9°C. What is the coldest temperature, in degrees Fahrenheit, that can be measured using a mercury thermometer?

2.48 The body temperature for a hypothermia victim is found to have dropped to 29.1°C. What is this temperature in degrees Fahrenheit?

2.49 Which is the higher temperature, −10°C or 10°F?

2.50 Which is the higher temperature, −15°C or 4°F?

2.51 A substance has a specific heat of 0.63 cal/g · °C. What is its specific heat in J/g · °C?

2.52 A substance has a specific heat of 0.24 cal/g · °C. What is its specific heat in J/g · °C?

2.53 If it takes 18.6 cal of heat to raise the temperature of 12.0 g of a substance by 10.0°C, what is the specific heat of the substance?

2.54 If it takes 35.0 cal of heat to raise the temperature of 25.0 g of a substance by 12.0°C, what is the specific heat of the substance?

2.55 How many calories of heat energy are required to raise the temperature of 42.0 g of each of the following substances from 20.0°C to 40.0°C?

a. Silver b. Liquid water c. Aluminum

2.56 How many calories of heat energy are required to raise the temperature of 20.0 g of each of the following substances from 25°C to 55°C?

a. Gold b. Ethyl alcohol c. Olive oil

Additional Problems

2.57 Round off the number 4.7205059 to the indicated number of significant figures.

a. six b. five c. four d. two

2.58 Write each of the following numbers in scientific notation to the number of significant figures indicated in parentheses.

a. 0.00300300 (three) b. 936,000 (two)
c. 23.5003 (three) d. 450,000,001 (six)

2.59 For each of the pairs of units listed, indicate whether the first unit is larger or smaller than the second unit and then indicate how many times larger or smaller it is.

a. milliliter, liter b. kiloliter, microliter
c. nanoliter, deciliter d. centiliter, megaliter

2.60 Indicate how each of the following conversion factors should be interpreted in terms of significant figures present.

a. $\frac{2.540 \text{ cm}}{1.000 \text{ in.}}$ b. $\frac{453.6 \text{ g}}{1.000 \text{ lb}}$ c. $\frac{2.113 \text{ pt}}{1.00 \text{ L}}$ d. $\frac{10^{-9} \text{ m}}{1 \text{ nm}}$

2.61 A one-gram sample of a powdery white solid is found to have a volume of two cubic centimeters. Calculate the solid's density using the following precision specifications and express your answers in scientific notation.

a. 1.0 g and 2.0 cm^3
b. 1.000 g and 2.00 cm^3
c. 1.0000 g and 2.0000 cm^3
d. 1.000 g and 2.0000 cm^3

2.62 Calculate the volume, in milliliters, for each of the following.

a. 75.0 g of gasoline (density = 0.56 g/mL)
b. 75.0 g of sodium metal (density = 0.93 g/cm^3)
c. 75.0 g of ammonia gas (density = 0.759 g/L)
d. 75.0 g of mercury (density = 13.6 g/mL)

2.63 The concentration of salt in a salt solution is found to be 4.5 mg/mL. What is the salt concentration in each of the following units?

a. mg/L b. pg/mL c. g/L d. kg/m^3

2.64 If one U.S. dollar is equal to 5.94 French francs and one franc is equal to 0.102 British pound, then what is the value in dollars of seven pounds?

Grid Problems

2.65

1. milli-	2. centi-	3. mega-
4. micro-	5. kilo-	6. nano-

Select from the grid *all* correct responses for each of the following situations.

a. Metric system prefixes that increase the size of the base unit
b. Metric system prefixes whose power-of-10 value is 10^{-3} or less
c. Pairs of metric system prefixes whose magnitudes are the reciprocal of one another (for example, 100 and 1/100)
d. Pairs of metric system prefixes whose magnitudes differ by a factor of 10^3

2.66

1. 0.0216	2. 1.670	3. 16,003
4. 244,000	5. 102.0	6. 0.300

Select from the grid *all* correct responses for each of the following situations.

a. Numbers that contain three significant figures
b. Numbers in which all zeros are significant
c. Numbers in which none of the zeros are significant
d. Numbers in which all nonzero digits are significant

2.67

1. 3.3×10^{-1}	2. 3.30×10^{-1}	3. 3.300×10^{-1}
4. 3.3×10^{1}	5. 3.30×10^{1}	6. 3.300×10^{1}

Select from the grid *all* correct responses for each of the following situations.

a. Numbers that contain three significant figures
b. Numbers that have a magnitude of less than 1
c. Numbers that, when multiplied by 1.000×10^{3}, would give an answer that has three significant figures
d. Numbers that, when multiplied by 1.000×10^{2}, would give an answer with a magnitude greater than 1

2.68

1. $\frac{1\text{ in.}}{2.54\text{ cm}}$	2. $\frac{453.6\text{ g}}{1\text{ lb}}$	3. $\frac{24\text{ hr}}{1\text{ day}}$
4. $\frac{10^{3}\text{ g}}{1\text{ kg}}$	5. $\frac{1\text{ mm}}{10^{-3}\text{ m}}$	6. $\frac{10^{-2}\text{ L}}{1\text{ cL}}$

Select from the grid *all* correct responses for each of the following situations.

a. Conversion factors that would limit a calculation to three significant figures
b. Conversion factors that, when used as written, would bring about a change from English to metric units
c. Conversion factors that, when used as written, would decrease unit size
d. Conversion factors that, when used as written, would change unit size by a factor of 1000

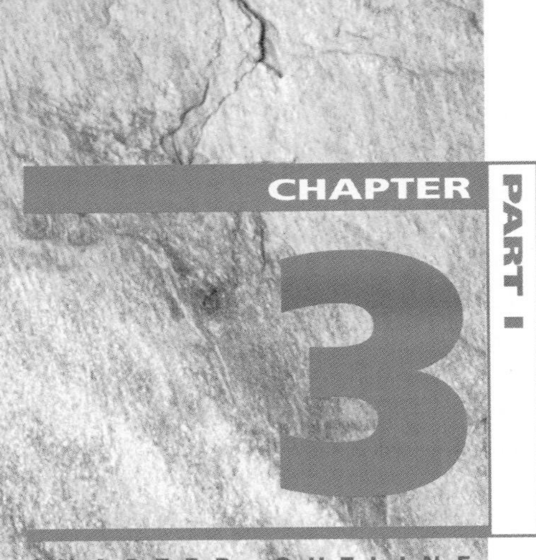

CHAPTER 3

PART I

Atomic Structure and the Periodic Table

CHAPTER OUTLINE

3.1 Internal Structure of an Atom 48
3.2 Experimental Evidence for the Existence of Subatomic Particles 50
3.3 Atomic Number and Mass Number 52

Chemistry at a Glance: **Atomic Structure** 56

3.4 The Periodic Law and the Periodic Table 57
3.5 Metals and Nonmetals 59
3.6 Electron Arrangements Within Atoms 61
3.7 Specification of Electronic Structure 65
3.8 The Electronic Basis for the Periodic Law and the Periodic Table 69
3.9 Classification of the Elements 71

Chemistry at a Glance: **Element Classification Schemes and the Periodic Table** 72

Chemical Connections

3.1 Protium, Deuterium, and Tritium: The Three Isotopes of Hydrogen 54
3.2 Calcium, Your Body, and the Periodic Law 60
3.3 Importance of Metallic and Nonmetallic Trace Elements for Human Health 62
3.4 Electrons in Excited States 67

Music consists of a series of tones that build octave after octave. Similarly, elements have properties that recur period after period.

In Chapter 1 we learned that all matter is made up of small particles called atoms and that 115 different types of atoms are known, each type of atom corresponding to a different element. Furthermore, we found that compounds result from the chemical combination of different types of atoms in various ratios and arrangements.

Until the last two decades of the nineteenth century, scientists believed that atoms were solid, indivisible spheres without an internal structure. Today, this model of the atom is known to be incorrect. Evidence from a variety of sources indicates that atoms are made up of even smaller particles called *subatomic particles.* In this chapter we consider the fundamental types of subatomic particles, how they arrange themselves within an atom, and the relationship between an atom's subatomic makeup and its chemical identity.

3.1 Internal Structure of an Atom

Atoms possess internal structure; that is, they are made up of even smaller particles, which are called subatomic particles. **Subatomic particles** *are very small particles that are the building blocks from which atoms are made.* Three types of subatomic parti-

Table 3.1
Charge and Mass Characteristics of Electrons, Protons, and Neutrons

	Electron	Proton	Neutron
Charge	−1	+1	0
Actual mass (g)	9.109×10^{-28}	1.673×10^{-24}	1.675×10^{-24}
Relative mass (based on the electron being 1 unit)	1	1837	1839

• Atoms of all 115 elements contain the same three types of subatomic particles. Different elements differ only in the numbers of the various subatomic particles they contain.

cles are found within atoms: electrons, protons, and neutrons. Key properties of these three types of particles are summarized in Table 3.1. **Electrons** *are subatomic particles that possess a negative (−) electrical charge.* They are the smallest, in terms of mass, of the three types of subatomic particles. **Protons** *are subatomic particles that possess a positive (+) electrical charge.* Protons and electrons carry the *same amount* of charge; the charges, however, are opposite (positive versus negative). **Neutrons** *are subatomic particles that have no charge associated with them; that is, they are neutral.* Both protons and neutrons are massive particles compared to electrons; they are almost 2000 times heavier.

• Arrangement of Subatomic Particles Within an Atom

The arrangement of subatomic particles within an atom is not haphazard. *All* protons and *all* neutrons present are found at the center of an atom in a very tiny volume called the *nucleus* (Figure 3.1). The **nucleus** *is the very small, dense, positively charged center of an atom.* A nucleus is always positively charged because it contains positively charged protons. Because the nucleus houses the heavy subatomic particles (protons and neutrons), almost all (over 99.9%) of the mass of an atom is concentrated in its nucleus. The small size of the nucleus, coupled with its large amount of mass, causes nuclear material to be extremely dense.

• Because they are found in the nucleus of an atom, protons and neutrons are often collectively referred to as *nucleons.*

The outer (extranuclear) region of an atom contains all of the electrons. In this region, which accounts for most of the volume of an atom, the electrons move rapidly about the nucleus. The electrons are attracted to the positively charged protons of the nucleus by the forces that exist between particles of opposite charge. The motion of the electrons in the extranuclear region determines the volume (size) of the atom in the same way that the blade of a fan determines a volume by its circular motion. The volume occupied by the electrons is sometimes referred to as the *electron cloud.* Because electrons are negatively charged, the electron cloud is also negatively charged. Figure 3.1 illustrates the nuclear and extranuclear regions of an atom.

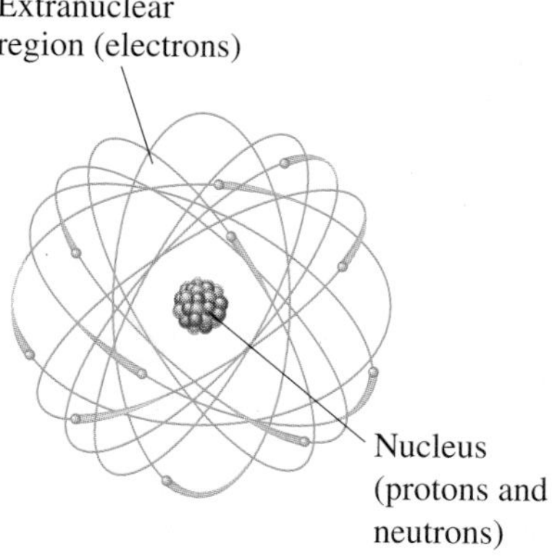

Figure 3.1
The protons and neutrons of an atom are found in the central nuclear region, or nucleus, and the electrons are found in an electron cloud outside the nucleus. Note that this figure is not drawn to scale; the correct scale would be comparable to a penny (the nucleus) in the center of a baseball field (the atom).

• Charge Neutrality of an Atom

An atom as a whole is electrically neutral; that is, it has no *net* electrical charge. For this to be the case, the same amount of positive and negative charge must be present in the atom. Equal amounts of positive and negative charges cancel one another. Thus equal numbers of protons and electrons are present in an atom.

$$\text{Number of protons} = \text{number of electrons}$$

• Size Relationships Within an Atom

To help you visualize the size relationships among the parts of an atom, imagine enlarging (magnifying) the nucleus until it is the size of a baseball (about 2.9 inches in diameter). If the nucleus were this large, the whole atom would have a diameter of approximately 2.5 miles. The electrons would still be smaller than the periods used to end sentences in this text, and they would move about at random within that 2.5-mile region.

The concentration of nearly all of the mass of an atom in the nucleus can also be illustrated by using our imagination. If a coin the same size as a copper penny contained copper nuclei (copper atoms stripped of their electrons) rather than copper atoms (which

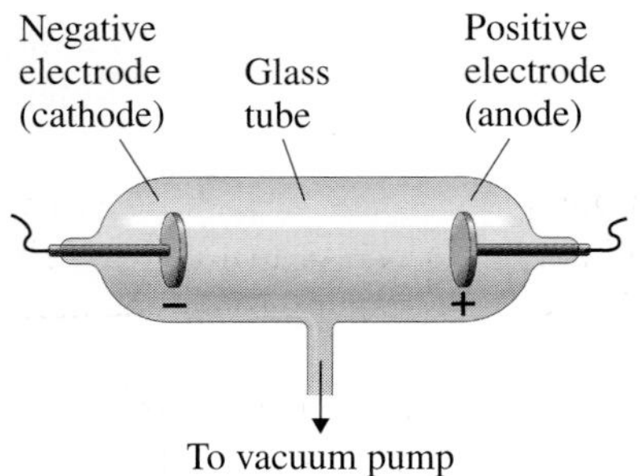

Figure 3.2
A simplified version of a gas discharge tube.

are mostly empty space), the coin would weigh 190,000,000 tons! Nuclei are indeed very dense matter.

Despite the existence of subatomic particles, we will continue to use the concept of atoms as the fundamental building blocks for all types of matter. Subatomic particles do not lead an independent existence for any appreciable length of time; they gain stability by joining together to form atoms.

3.2 Experimental Evidence for the Existence of Subatomic Particles

A large amount of experimental evidence exists that is consistent with and supports the existence, nature, and arrangement of subatomic particles within an atom (Section 3.1). Two landmark types of experiments performed in the course of developing our present concepts of the atom illustrate some of the sources of this evidence. *Discharge tube experiments* originally suggested that the atom contained negatively and positively charged particles. *Metal foil experiments* provided evidence for the existence of a nucleus within the atom.

• Discharge Tube Experiments

Experiments involving gas discharge tubes provided the first evidence that *electrons* and *protons* are present within an atom. A simplified diagram of a gas discharge tube is shown in Figure 3.2. The apparatus consists of a sealed glass tube containing two metal disks called *electrodes*. The glass tube also has a side arm for attachment to a vacuum pump. During operation, the electrodes are connected to a source of electrical power. (The electrode attached to the positive side of the electrical power source is called the *anode;* the one attached to the negative side is known as the *cathode.*) Use of the vacuum pump allows the amount of gas within the tube to be varied. The smaller the amount of gas present, the lower the pressure within the tube.

• Neon signs, fluorescent lights, television tubes, and computer monitors are all basic components of our modern technological society. The forerunner for all four of these developments was the gas discharge tube.

Early studies with gas discharge tubes, conducted in the mid-1800s, showed that when the tube was almost evacuated (low pressure), electricity flowed from one electrode to the other, and the residual gas became luminous (it glowed). Different gases in the tube gave different colors to the glow. After the pressure in the tube was reduced to still lower levels (very little gas remaining), it was found that the luminosity disappeared but the electrical conductance continued, as shown by a greenish glow given off by the tube's glass walls. This glow was the initial indication of what became known as *cathode rays.* Their discovery marked the beginning of nearly 40 years of discharge tube experimentation that ultimately led to the characterization of both the electron and the proton.

The term *cathode rays* comes from the observation that when an obstacle is placed between the negative electrode (cathode) and the opposite glass wall, a sharp shadow the shape of the obstacle is cast on that wall. This indicates that the rays are coming from the cathode.

Further studies showed that these cathode rays caused certain minerals such as sphalerite (zinc sulfide) to glow. Glass plates were coated with sphalerite and observed under high magnification while being bombarded with cathode rays. The light emitted by the sphalerite coating consisted of many pinpoint flashes. This observation suggested that cathode rays were in reality a stream of extremely small particles.

Joseph John Thomson (1856–1940), an English physicist, discovered a great deal about the nature of cathode rays. Using a variety of materials as cathodes, he showed that cathode ray production is a general property of matter. By using a specially designed cathode ray tube (see Figure 3.3), he also found that cathode rays can be deflected by charged plates or a magnetic field. The rays were repelled by the north pole or negative plate and attracted to the south pole or positive plate, which indicated that they were negatively charged. In 1897, Thomson concluded that cathode rays were streams of negatively charged particles, which today we call *electrons.* Further experiments by others proved that his conclusions were correct.

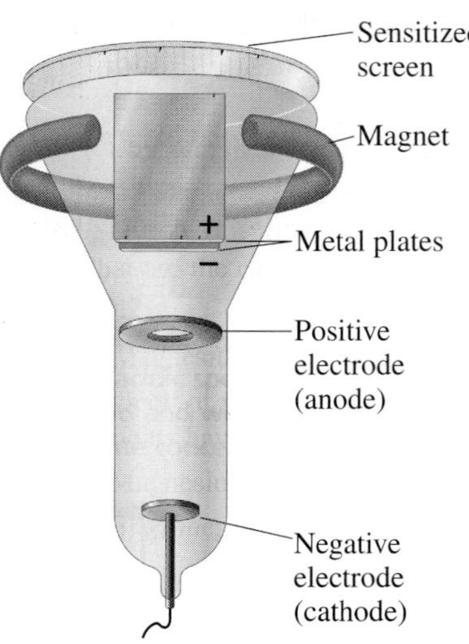

Figure 3.3
A cathode-ray tube with perpendicular magnetic and electric fields was used in characterizing cathode rays (electrons).

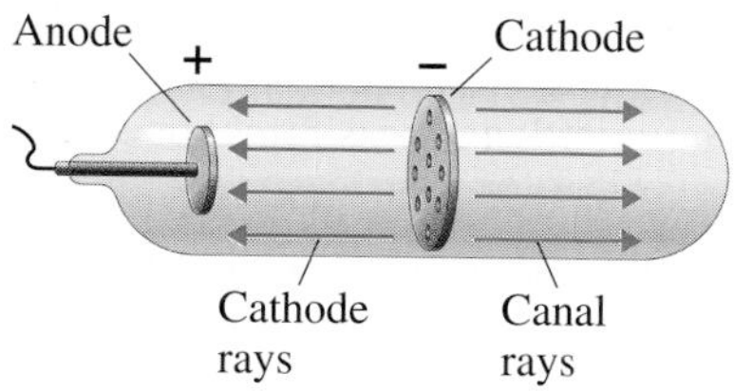

Figure 3.4
A gas discharge tube modified to detect canal rays.

In 1886, the German physicist Eugene Goldstein (1850–1930) showed that positive particles were also present in discharge tubes. He used a discharge tube in which the cathode was a metal plate with a large number of holes drilled in it. The usual cathode rays were observed to stream from cathode to anode. In addition, rays of light appeared to stream from each of the holes in the cathode in a direction opposite to that of the cathode rays (see Figure 3.4). Because these rays were observed streaming through the holes or channels in the cathode, Goldstein called them *canal rays.*

Further research showed that canal rays are of many different types, in contrast to cathode rays, which are of only one type, and that the particles making up canal rays were much heavier than those of cathode rays. The type of canal ray produced depended on the gas in the tube. The simplest canal rays were eventually identified as the particles now called *protons*.

Canal rays are now known to be gas atoms that have lost one or more electrons. Their origin and behavior in a discharge tube can be understood as follows. Electrons (cathode rays) emitted from the cathode collide with residual gas molecules (air) on the way to the anode. Some of these electrons have enough energy to knock electrons away from the gas molecules, leaving behind a positive particle (the remainder of the gas molecule). These positive particles are attracted to the cathode, and some of them pass through the holes or channels. The fact that atoms can lose electrons under certain conditions will be discussed further in Chapter 4.

On the basis of discharge tube experiments, Thomson proposed in 1898 that the atom was composed of a sphere of positive electricity, which contained most of the mass, and that small negative electrons were attached to the surface of the positive sphere. He postulated that a high voltage could pull off surface electrons to produce cathode rays. Thomson's model of the atom, sometimes referred to as the "raisin muffin" or "plum pudding" model—with the electrons as the raisins—is known to be incorrect. Its significance is that it set the stage for an experiment, commonly called the gold foil experiment, that led to the currently accepted concept of the arrangement of protons and electrons in the atom.

• Metal Foil Experiments

In 1911, Ernest Rutherford (1871–1937) designed an experiment to test the Thomson model of the atom. In this experiment, thin sheets of metal were bombarded with alpha particles from a radioactive source. Alpha particles, which are positively charged, are ejected at high speeds from some radioactive materials (see Section 11.3). The phenomenon of radioactivity had been discovered in 1896 and gave further evidence that electrical charges exist in atoms. Gold was chosen as the target metal because it is easily hammered into very thin sheets. The experimental setup for Rutherford's experiment is shown in Figure 3.5. Alpha particles do not appreciably penetrate lead, so a lead plate with a slit was used to produce a narrow alpha particle beam.

Rutherford expected that all the alpha particles, because they were so energetic, would pass straight through the thin gold foil, hit the fluorescent screen, and produce a flash of light. His reasoning was based on the Thomson model, in which the mass and positive charge of the gold atoms were distributed uniformly through each atom. As each positive

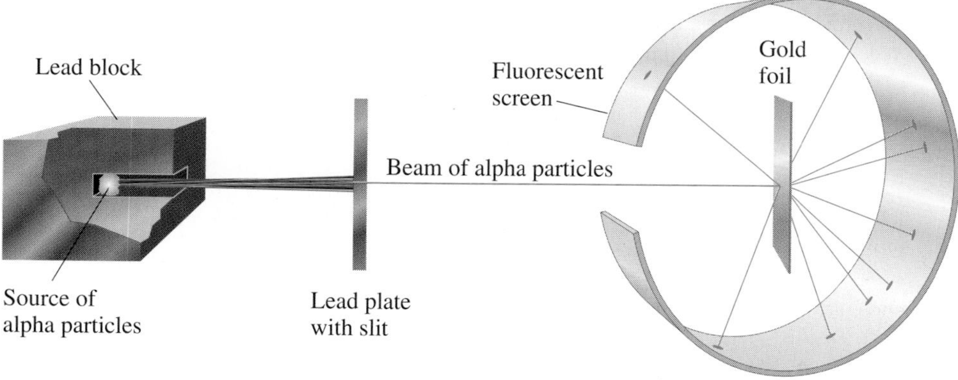

Figure 3.5
Rutherford's gold foil–alpha particle experiment. Most of the alpha particles went straight through the foil, but a few were deflected at large angles.

alpha particle neared the foil, Rutherford assumed that it would be confronted by a uniform positive charge. All particles would be affected the same way (no deflection), which would support the Thomson model of the atom.

• At a later date, Rutherford described the unexpected results of his gold foil experiment as follows: "It is about as credible as if you had fired a 15-inch shell at a piece of tissue paper and it came back and hit you."

The results from the experiment were very surprising. Most of the particles—more than 99%—went straight through as expected. A few, however, were appreciably deflected by something that had to be much heavier than the alpha particles themselves. A very few particles were deflected almost directly back toward the alpha particle source. Similar results were obtained when elements other than gold were used as targets.

Extensive study of his results led Rutherford to propose the following explanation.

1. A very dense, small nucleus exists in the center of the atom. This nucleus contains most of the mass of the atom and all of the positive charge.
2. Electrons occupy most of the total volume of the atoms and are located outside the nucleus.
3. When an alpha particle scores a direct hit on a nucleus, it is deflected back along the incoming path.
4. A near miss of a nucleus by an alpha particle results in repulsion and deflection.
5. Most of the alpha particles pass through without any interference, because most of the atomic volume is empty space.
6. Electrons have so little mass that they do not deflect the much larger alpha particles (an alpha particle is almost 8000 times heavier than an electron).

Many other experiments have since verified Rutherford's conclusions that at the center of an atom there is a nucleus that is very small and very dense.

3.3 Atomic Number and Mass Number

• Atomic number and mass number are always *whole* numbers because they are obtained by counting whole objects (protons, neutrons, and electrons).

The **atomic number** *of an atom is the number of protons in its nucleus.* Because an atom has the same number of electrons as protons (Section 3.1), the atomic number also specifies the number of electrons present.

$$\text{Atomic number} = \text{number of protons} = \text{number of electrons}$$

The **mass number** *of an atom is the sum of the numbers of protons and neutrons in its nucleus.* Thus the mass number gives the number of subatomic particles present in the nucleus.

$$\text{Mass number} = \text{number of protons} + \text{number of neutrons}$$

The mass of an atom is almost totally accounted for by the protons and neutrons present—hence the term *mass number.*

The number and identity of subatomic particles present in an atom can be calculated from its atomic and mass numbers in the following manner.

• The *sum* of the mass number and the atomic number for an atom corresponds to the total number of subatomic particles present in the atom (protons, neutrons, and electrons).

$$\text{Number of protons} = \text{atomic number}$$
$$\text{Number of electrons} = \text{atomic number}$$
$$\text{Number of neutrons} = \text{mass number} - \text{atomic number}$$

Note that neutron count is obtained by subtracting atomic number from mass number.

Example 3.1

Determining the Subatomic Particle Makeup of an Atom Given Its Atomic Number and Mass Number

An atom has an atomic number of 9 and a mass number of 19.

a. Determine the number of protons present.
b. Determine the number of neutrons present.
c. Determine the number of electrons present.

Solution

a. There are 9 protons because the atomic number is always equal to the number of protons present.

b. There are 10 neutrons because the number of neutrons is always obtained by subtracting the atomic number from the mass number.

$$\underbrace{\text{(Protons + neutrons)}}_{\text{Mass number}} - \underbrace{\text{protons}}_{\text{Atomic number}} = \text{neutrons}$$

c. There are 9 electrons because the number of protons and the number of electrons are always the same in an atom.

Practice Exercise 3.1

An atom has an atomic number of 11 and a mass number of 23.

a. Determine the number of protons present.
b. Determine the number of neutrons present.
c. Determine the number of electrons present.

• *Answer.* **a.** 11; **b.** 12; **c.** 11

• Electrons and Chemical Properties

The chemical properties of an atom, which are the basis for its identification, are determined by the number and arrangement of the electrons about the nucleus. When two atoms interact, the outer part (electrons) of one interacts with the outer part (electrons) of the other. The small nuclear centers never come in contact with each other in a chemical reaction. The number of electrons about a nucleus may be considered to be determined by the number of protons in the nucleus; charge balance requires an equal number of the two (Section 3.1). Hence the number of protons (which is the atomic number) characterizes an atom. All atoms with the same atomic number have the same chemical properties and are atoms of the same element.

In Section 1.6, an element was defined as a pure substance that cannot be broken down into simpler substances by ordinary chemical means. Although this is a good historical definition for an element, we can now give a more rigorous definition by using the concept of atomic number. An **element** *is a pure substance in which all atoms present have the same atomic number.*

An alphabetical listing of the 115 known elements, with their atomic numbers as well as other information, is found on the inside front cover of this text. If you check the atomic number column in this tabulation, you will find an entry for each of the numbers in the sequence 1 to 112 plus 114, 116, and 118. The highest-atomic-numbered element that occurs naturally is uranium (element 92); elements 93 to 115 have been made in the laboratory but are not found in nature (Section 1.7). The fact that there are no gaps in the numerical sequence 1 to 92 is interpreted by scientists to mean that there are no "missing elements" yet to be discovered in nature.

• Isotopes

Charge neutrality (Section 3.1) requires the presence in an atom of an equal number of protons and electrons. However, because neutrons have no electrical charge, their numbers in atoms do not have to be the same as the number of protons or electrons. Most atoms contain more neutrons than either protons or electrons.

Studies of atoms of various elements also show that the number of neutrons present in atoms of an element is not constant; it varies over a small range. This means that not all atoms of an element have to be identical. They must have the same number of protons and electrons, but they can differ in the number of neutrons.

Atoms of an element that differ in neutron count are called isotopes. **Isotopes** *are atoms of an element that have the same number of protons and electrons but different numbers*

• The word *isotope* comes from the Greek *iso,* meaning "equal," and *topos,* meaning "place." Isotopes occupy an equal place (location) in listings of elements because all isotopes of an element have the same atomic number.

of neutrons. Different isotopes always have the same atomic number and different mass numbers.

Most elements found in nature exist in isotopic forms, with the number of naturally occurring isotopes ranging from two to ten. For example, all silicon atoms have 14 protons and 14 electrons. Most silicon atoms also contain 14 neutrons. However, some silicon atoms contain 15 neutrons and others contain 16 neutrons. Thus three different kinds of silicon atoms exist.

Isotopes of an element have the same chemical properties, but their physical properties are often slightly different. Isotopes of an element have the same chemical properties because they have the same number of electrons. They have slightly different physical properties because they have different numbers of neutrons and therefore different masses.

When it is necessary to distinguish between isotopes of an element, the following notation is used:

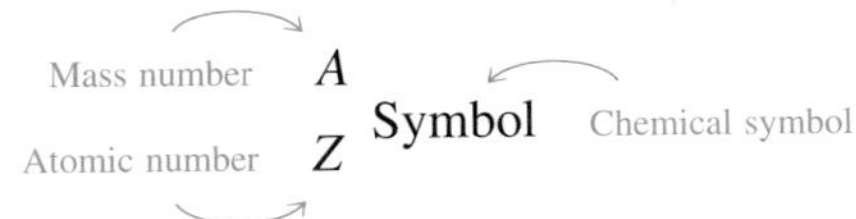

• There are a few elements for which all naturally occurring atoms have the same number of neutrons—that is, for which all atoms are identical. They include the elements Be, F, Na, Al, P, and Au.

The atomic number is written as a *subscript* to the left of the elemental symbol for the atom. The mass number is written as a *superscript* to the left of the elemental symbol. Thus the three silicon isotopes are designated, respectively, as

$$^{28}_{14}\text{Si}, \quad ^{29}_{14}\text{Si}, \quad \text{and} \quad ^{30}_{14}\text{Si}$$

Names for isotopes include the mass number. $^{28}_{14}\text{Si}$ is called silicon-28, and $^{29}_{14}\text{Si}$ is called silicon-29. The atomic number is not included in the name because it is the same for all isotopes of an element.

Chemical CONNECTIONS

3.1 Protium, Deuterium, and Tritium: The Three Isotopes of Hydrogen

Measurable differences in physical properties are found among isotopes for elements with low atomic numbers. This results from differences in mass among isotopes being relatively large compared to the masses of the isotopes themselves. The situation is greatest for the element hydrogen, the element with the lowest atomic number.

Three isotopes of hydrogen exist: ^{1}H, ^{2}H, and ^{3}H. Hydrogen-1, with a single proton and no neutrons in its nucleus, is by far the most abundant isotope (99.985%). Hydrogen-2, with a neutron in addition to a proton in its nucleus, has an abundance of 0.015%. The presence of the additional neutron in ^{2}H doubles its mass compared to that of ^{1}H. Hydrogen-3 has two neutrons and a proton in its nucleus and has a mass triple that of ^{1}H. Only minute amounts of ^{3}H, which is radioactive (unstable: see Section 11.1), occur naturally.

In discussions involving hydrogen isotopes, special names and symbols are given to the isotopes—something that does not occur for any other element. Hydrogen-1 is usually called hydrogen but is occasionally called *protium.* Hydrogen-2 has the name *deuterium* (symbol D), and hydrogen-3 is called *tritium* (symbol T). The following table contrasts the properties of H_2 and D_2.

Isotope	Melting point	Boiling point	Density (at 0°C and 1 atmosphere pressure)
H_2	−259°C	−253°C	0.090 g/L
D_2	−253°C	−250°C	0.18 g/L

Water in which both hydrogen atoms are deuterium (D_2O) is called "heavy water." The properties of heavy water are measurably different from those of "ordinary" (H_2O).

Compound	Melting point	Boiling point	Density (at 0°C and 1 atmosphere pressure)
H_2O	0.0°C	100.0°C	0.99987 g/mL
D_2O	3.82°C	101.4°C	1.1047 g/mL

Heavy water (D_2O) can be obtained from natural water by distilling a sample of natural water, because the D_2O has a slightly higher boiling point than H_2O. Pure deuterium (D_2) is produced by decomposing the D_2O. Heavy water is used in the operation of nuclear power plants (to slow down free neutrons present in the reactor core).

Tritium, the heaviest hydrogen isotope, is used in nuclear weapons. Because of the minute amount of naturally occurring tritium, it must be synthesized in the laboratory using bombardment reactions (Section 11.5 describes these in detail).

• A mass number, in contrast to an atomic number, lacks uniqueness. Atoms of different elements can have the same mass number. For example, carbon-14 and nitrogen-14 have the same mass numbers. Atoms of different elements, however, cannot have the same atomic number.

• An analogy involving isotopes and identical twins may be helpful: Identical twins need not weigh the same, even though they have identical "gene packages." Likewise, isotopes, even though they have different masses, have the same number of protons.

• The terms *atomic mass* and *atomic weight* are often used interchangeably. Atomic mass, however, is the correct term.

The various isotopes of a given element are of varying abundance; usually one isotope is predominant. Silicon is typical of this situation. The percentage abundances for its three isotopes are 92.21% ($^{28}_{14}Si$), 4.70% ($^{29}_{14}Si$), and 3.09% ($^{30}_{14}Si$). Percentage abundances are number percentages (numbers of atoms) rather than mass percentages. A sample of 10,000 silicon atoms contains 9221 $^{28}_{14}Si$ atoms, 470 $^{29}_{14}Si$ atoms, and 309 $^{30}_{14}Si$ atoms.

There are 286 isotopes that occur naturally. In addition, over 2000 more have been synthesized in the laboratory via nuclear rather than chemical reactions (Section 11.5). All these synthetic isotopes are unstable (radioactive). Despite their instability, many are used in chemical and biological research, as well as in medicine.

• Atomic Masses

The existence of isotopes means that atoms of an element can have several different masses. For example, silicon atoms can have any one of three masses because there are three silicon isotopes. Which of these three silicon isotopic masses is used in situations in which the mass of the element silicon needs to be specified? The answer is none of them. Instead we use a *weighted-average mass* that takes into account the existence of isotopes and their relative abundances.

The *weighted-average mass* of the isotopes of an element is known as the element's atomic mass. An **atomic mass** *is the calculated average mass for the isotopes of an element, expressed on a scale using atoms of* $^{12}_{6}C$ *as the reference.* What we need to calculate an atomic mass are the masses of the various isotopes on the $^{12}_{6}C$ reference scale and the percentage abundance of each isotope.

The $^{12}_{6}C$ reference scale mentioned in the definition of *atomic mass* is a scale scientists have set up for comparing the masses of atoms. On this scale, the mass of a $^{12}_{6}C$ atom is defined to be exactly 12 atomic mass units (amu). The masses of all other atoms are then determined relative to that of $^{12}_{6}C$. For example, if an atom is twice as heavy as $^{12}_{6}C$, its mass is 24 amu, and if an atom weighs half as much as an atom of $^{12}_{6}C$, its mass is 6 amu.

Example 3.2 shows how an atomic mass is calculated by using the amu ($^{12}_{6}C$) scale, the percentage abundances of isotopes, and the number of isotopes of an element.

Example 3.2

Calculation of an Element's Atomic Mass

Naturally occurring chlorine exists in two isotopic forms, $^{35}_{17}Cl$ and $^{37}_{17}Cl$. The relative mass of $^{35}_{17}Cl$ is 34.97 amu, and its abundance is 75.53%; the relative mass of $^{37}_{17}Cl$ is 36.97 amu, and its abundance is 24.47%. What is the atomic mass of chlorine?

Solution

An element's atomic mass is calculated by multiplying the relative mass of each isotope by its fractional abundance and then totaling the products. The fractional abundance for an isotope is its percentage abundance converted to decimal form (divided by 100).

$$^{35}_{17}Cl: \quad \left(\frac{75.53}{100}\right) \times 34.97\,\text{amu} = (0.7553) \times 34.97\,\text{amu} = 26.41\,\text{amu}$$

$$^{37}_{17}Cl: \quad \left(\frac{24.47}{100}\right) \times 36.97\,\text{amu} = (0.2447) \times 36.97\,\text{amu} = 9.047\,\text{amu}$$

$$\text{Atomic mass of Cl} = (26.41 + 9.047)\,\text{amu}$$
$$= 35.46\,\text{amu}$$

This calculation involved an element containing just two isotopes. A similar calculation for an element having three isotopes would be carried out the same way, but it would have three terms in the final sum; an element possessing four isotopes would have four terms in the final sum.

CHEMISTRY AT A GLANCE

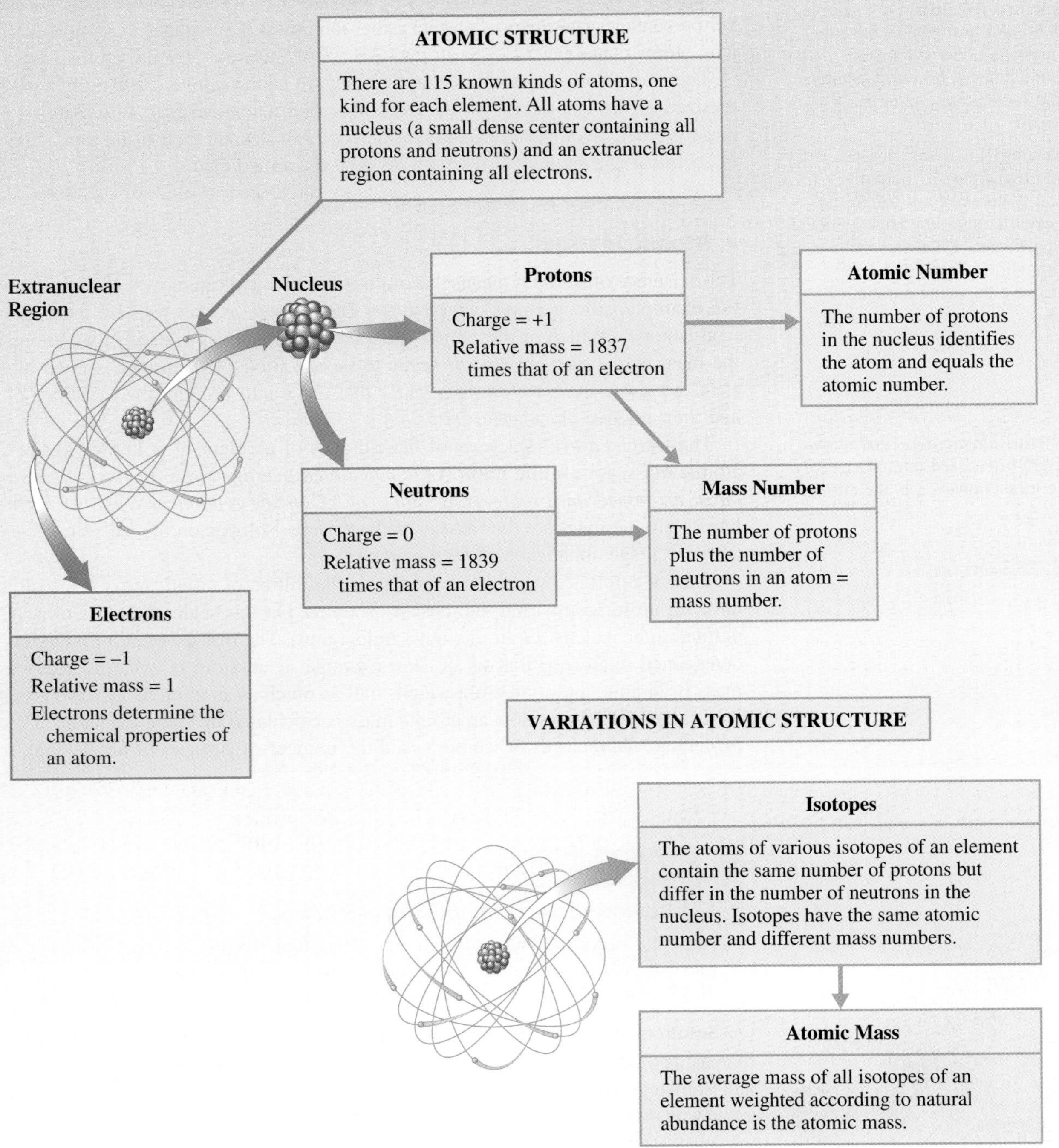

ATOMIC STRUCTURE
There are 115 known kinds of atoms, one kind for each element. All atoms have a nucleus (a small dense center containing all protons and neutrons) and an extranuclear region containing all electrons.
Extranuclear Region
Nucleus
Protons
Charge = +1
Relative mass = 1837 times that of an electron
Atomic Number
The number of protons in the nucleus identifies the atom and equals the atomic number.
Neutrons
Charge = 0
Relative mass = 1839 times that of an electron
Mass Number
The number of protons plus the number of neutrons in an atom = mass number.
Electrons
Charge = –1
Relative mass = 1
Electrons determine the chemical properties of an atom.
VARIATIONS IN ATOMIC STRUCTURE
Isotopes
The atoms of various isotopes of an element contain the same number of protons but differ in the number of neutrons in the nucleus. Isotopes have the same atomic number and different mass numbers.
Atomic Mass
The average mass of all isotopes of an element weighted according to natural abundance is the atomic mass.

Practice Exercise 3.2

Naturally occurring copper exists in two isotopic forms, $^{63}_{29}Cu$ and $^{65}_{29}Cu$. The relative mass of $^{63}_{29}Cu$ is 62.93 amu, and its abundance is 69.09%; the relative mass of $^{65}_{29}Cu$ is 64.93 amu, and its abundance is 30.91%. What is the atomic mass of copper?

- *Answer.* 63.55 amu

1 Hydrogen	2 Helium	3 Lithium
$^{1}_{1}H$ 1.008 amu 99.985% $^{2}_{1}H$ 2.014 amu 0.015% $^{3}_{1}H$ 3.016 amu trace	$^{3}_{2}He$ 3.016 amu trace $^{4}_{2}He$ 4.003 amu 100%	$^{6}_{3}Li$ 6.015 amu 7.42% $^{7}_{3}Li$ 7.016 amu 92.58%
4 Beryllium	**5 Boron**	**6 Carbon**
$^{9}_{4}Be$ 9.012 amu 100%	$^{10}_{5}B$ 10.013 amu 19.6% $^{11}_{5}B$ 11.009 amu 80.4%	$^{12}_{6}C$ 12.000 amu 98.89% $^{13}_{6}C$ 13.003 amu 1.11% $^{14}_{6}C$ 14.003 amu trace
7 Nitrogen	**8 Oxygen**	**9 Fluorine**
$^{14}_{7}N$ 14.003 amu 99.63% $^{15}_{7}N$ 15.000 amu 0.37%	$^{16}_{8}O$ 15.995 amu 99.759% $^{17}_{8}O$ 16.999 amu 0.037% $^{18}_{8}O$ 17.999 amu 0.204%	$^{19}_{9}F$ 18.998 amu 100%
10 Neon	**11 Sodium**	**12 Magnesium**
$^{20}_{10}Ne$ 19.992 amu 90.92% $^{21}_{10}Ne$ 20.994 amu 0.26% $^{22}_{10}Ne$ 21.991 amu 8.82%	$^{23}_{11}Na$ 22.990 amu 100%	$^{24}_{12}Mg$ 23.985 amu 78.70% $^{25}_{12}Mg$ 24.986 amu 10.13% $^{26}_{12}Mg$ 25.983 amu 11.17%

Table 3.2
Isotopic Data for Elements with Atomic Numbers 1 Through 12. Information given for each isotope includes mass number, isotopic mass in terms of amu, and percentage abundance.

The alphabetical listing of the known elements printed inside the front cover of this text gives the calculated atomic mass for each of the elements; it is the last column of numbers. Table 3.2 gives isotopic data for the elements with atomic numbers 1 through 12.

The accompanying Chemistry at a Glance summarizes all that we have said about atoms thus far.

Dmitri Ivanovich Mendeleev (1834–1907). Mendeleev constructed a periodic table as part of his effort to systematize chemistry. He received many international honors for his work, but his reception at home in czarist Russia was mixed. Element 101 carries his name.

3.4 The Periodic Law and the Periodic Table

During the early part of the nineteenth century, scientists began to look for order in the increasing amount of chemical information that had become available. They knew that certain elements had properties that were very similar to those of other elements, and they sought reasons for these similarities in the hope that these similarities would suggest a method for arranging or classifying the elements.

In 1869, these efforts culminated in the discovery of what is now called the *periodic law,* proposed independently by the Russian chemist Dmitri Mendeleev and the German chemist Julius Lothar Meyer. Given in its modern form, the **periodic law** *states that when elements are arranged in order of increasing atomic number, elements with similar properties occur at periodic (regularly recurring) intervals.*

A periodic table represents a compact graphical method for representing the behavior described by the periodic law. A **periodic table** *is a graphical display of the elements in order of increasing atomic number in which elements with similar properties fall in the same column of the display.* The most commonly used form of the periodic table is shown in Figure 3.6 (see also the inside front cover of the text). Within the table, each element is represented by a rectangular box, which contains the symbol, atomic number, and atomic mass of the element.

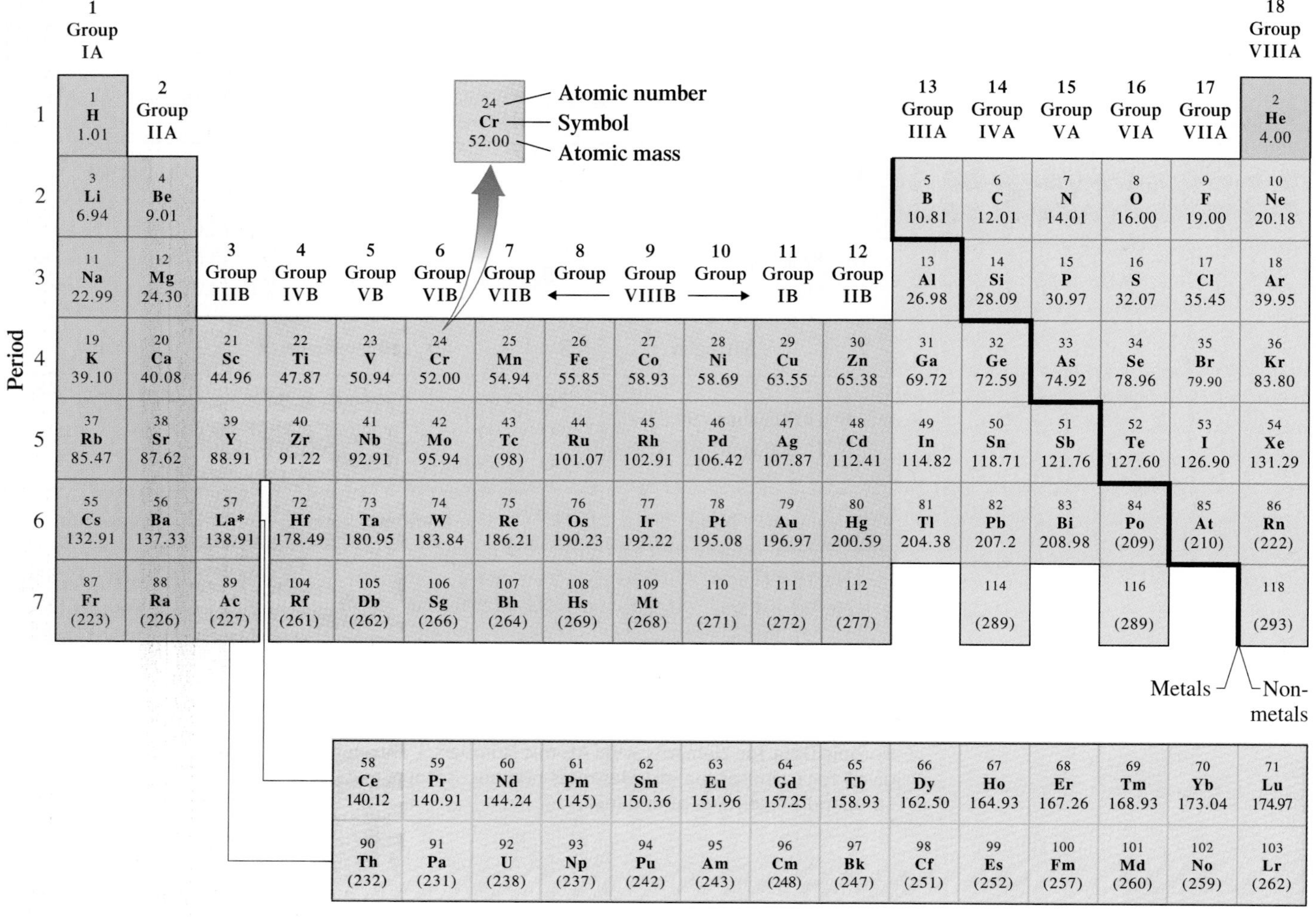

1 Group IA
2 Group IIA
3 Group IIIB
4 Group IVB
5 Group VB
6 Group VIB
7 Group VIIB
8 Group
9 Group VIIIB
10 Group
11 Group IB
12 Group IIB
13 Group IIIA
14 Group IVA
15 Group VA
16 Group VIA
17 Group VIIA
18 Group VIIIA
Period
1
2
3
4
5
6
7
24 Cr 52.00
Atomic number
Symbol
Atomic mass
1 H 1.01
2 He 4.00
3 Li 6.94
4 Be 9.01
5 B 10.81
6 C 12.01
7 N 14.01
8 O 16.00
9 F 19.00
10 Ne 20.18
11 Na 22.99
12 Mg 24.30
13 Al 26.98
14 Si 28.09
15 P 30.97
16 S 32.07
17 Cl 35.45
18 Ar 39.95
19 K 39.10
20 Ca 40.08
21 Sc 44.96
22 Ti 47.87
23 V 50.94
24 Cr 52.00
25 Mn 54.94
26 Fe 55.85
27 Co 58.93
28 Ni 58.69
29 Cu 63.55
30 Zn 65.38
31 Ga 69.72
32 Ge 72.59
33 As 74.92
34 Se 78.96
35 Br 79.90
36 Kr 83.80
37 Rb 85.47
38 Sr 87.62
39 Y 88.91
40 Zr 91.22
41 Nb 92.91
42 Mo 95.94
43 Tc (98)
44 Ru 101.07
45 Rh 102.91
46 Pd 106.42
47 Ag 107.87
48 Cd 112.41
49 In 114.82
50 Sn 118.71
51 Sb 121.76
52 Te 127.60
53 I 126.90
54 Xe 131.29
55 Cs 132.91
56 Ba 137.33
57 La* 138.91
72 Hf 178.49
73 Ta 180.95
74 W 183.84
75 Re 186.21
76 Os 190.23
77 Ir 192.22
78 Pt 195.08
79 Au 196.97
80 Hg 200.59
81 Tl 204.38
82 Pb 207.2
83 Bi 208.98
84 Po (209)
85 At (210)
86 Rn (222)
87 Fr (223)
88 Ra (226)
89 Ac (227)
104 Rf (261)
105 Db (262)
106 Sg (266)
107 Bh (264)
108 Hs (269)
109 Mt (268)
110 (271)
111 (272)
112 (277)
114 (289)
116 (289)
118 (293)
Metals
Nonmetals
58 Ce 140.12
59 Pr 140.91
60 Nd 144.24
61 Pm (145)
62 Sm 150.36
63 Eu 151.96
64 Gd 157.25
65 Tb 158.93
66 Dy 162.50
67 Ho 164.93
68 Er 167.26
69 Tm 168.93
70 Yb 173.04
71 Lu 174.97
90 Th (232)
91 Pa (231)
92 U (238)
93 Np (237)
94 Pu (242)
95 Am (243)
96 Cm (248)
97 Bk (247)
98 Cf (251)
99 Es (252)
100 Fm (257)
101 Md (260)
102 No (259)
103 Lr (262)

Figure 3.6
The periodic table of the elements is a graphical way to show relationships among the elements. Elements with similar chemical properties fall in the same vertical column.

• Using the information on a periodic table, you can quickly determine the number of protons and electrons for atoms of an element. However, no information concerning neutrons is available from a periodic table; mass numbers are not part of the information given, because they are not unique to an element.

• The elements within a given periodic-table group show numerous similarities in properties, the degree of similarity varying from group to group. In no case are the group members "clones" of one another. Each element has some individual characteristics not found in other elements of the group. By analogy, the members of a human family often bear many resemblances to each other, but each member also has some (and often much) individuality.

Groups and Periods of Elements

The location of an element within the periodic table is specified by giving its group number and period number.

A **group** *in the periodic table is a vertical column of elements.* There are two notations in use for designating individual periodic-table groups. In the first notation, which has been in use for many years, groups are designated by using Roman numerals and the letters A and B. In the second notation, which has recently been recommended for use by an international scientific commission, the Arabic numbers 1 through 18 are used. Note that in Figure 3.6 both group notations are given at the top of each group. The elements with atomic numbers 8, 16, 34, 52, and 84 (O, S, Se, Te, and Po) constitute Group VIA (old notation) or Group 16 (new notation).

Several groups of elements have common (non-numerical) names that are used so frequently that they should be learned. The Group IA elements, except for hydrogen, are called the *alkali metals*, and the Group IIA elements the *alkaline earth metals.* On the opposite side of the periodic table from the IA and IIA elements are found the *halogens* (the Group VIIA elements) and the *noble gases* (Group VIIIA).

A **period** *in the periodic table is a horizontal row of elements.* For identification purposes, the periods are numbered sequentially with Arabic numbers, starting at the top of the periodic table. In Figure 3.7, period numbers are found on the left side of the table. The elements Na, Mg, Al, Si, P, S, Cl, and Ar are all members of Period 3, the third row of elements. Period 4 is the fourth row of elements, and so on. There are only two elements in Period 1, H and He.

The location of any element in the periodic table is specified by giving its group number and its period number. The element gold, with an atomic number of 79, belongs to

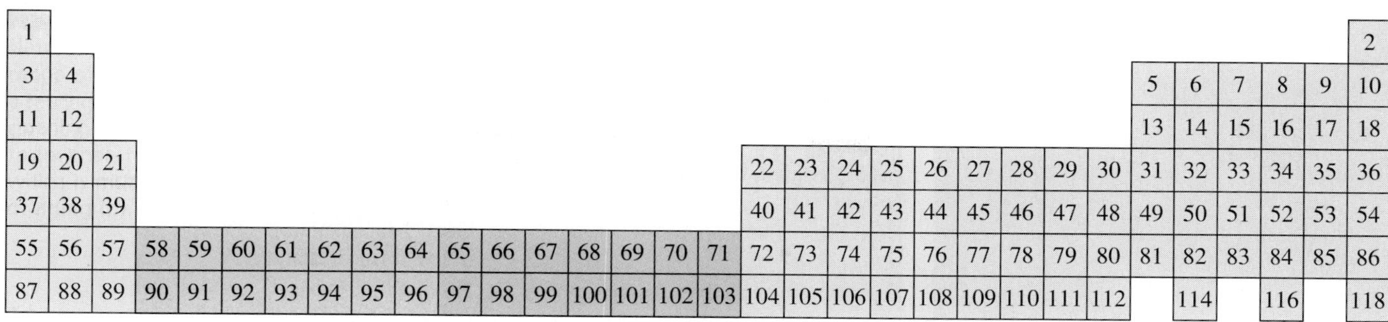

Figure 3.7
In this periodic table, elements 58 through 71 and 90 through 103 (in color) are shown in their proper positions.

Group IB (or 11) and is in Period 6. The element nitrogen, with an atomic number of 7, belongs to Group VA (or 15) and is in Period 2.

• When the statement "the first ten elements" is used, it means the first ten elements in the periodic table, the elements with atomic numbers 1 through 10.

• The Shape of the Periodic Table

Within the periodic table of Figure 3.6, the practice of arranging the elements according to increasing atomic number is violated in Groups IIIB and IVB. Element 72 follows element 57, and element 104 follows element 89. The missing elements, elements 58 through 71 and 90 through 103 are located in two rows at the bottom of the periodic table. Technically, the elements at the bottom of the table should be included in the body of the table, as shown in Figure 3.7. However, in order to have a more compact table, we place them at the bottom of the table as shown in Figure 3.6.

3.5 Metals and Nonmetals

In the previous section, we noted that the Group IA and IIA elements are known, respectively, as the alkali metals and the alkaline earth metals. Both of these designations contain the word *metal.* But what is a metal?

On the basis of selected physical properties, elements are classified into the categories metal and nonmetal (Figure 3.8). A **metal** *is an element that has the characteristic properties of luster, thermal conductivity, electrical conductivity, and malleability.* With the exception of mercury, all metals are solids at room temperature (25°C). Metals are good conductors of heat and electricity. Most metals are ductile (can be drawn into wires) and malleable (can be rolled into sheets). Most metals have high luster (shine), high density, and high melting points. Among the more familiar metals are the elements iron, aluminum, copper, silver, gold, lead, tin, and zinc (see Figure 3.8b).

Figure 3.8
(a) This portion of the periodic table shows the dividing line between metals and nonmetals. All elements that are not shown are metals. (b) Some familiar metals are aluminum, lead, tin, and zinc. (c) Some familiar nonmetals are sulfur (yellow), phosphorus (dark red), and bromine (reddish-brown).

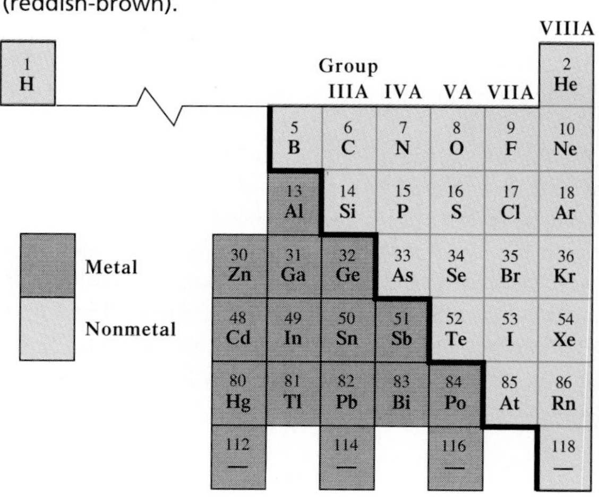

(a)

(b)

(c)

Table 3.3
Selected Physical Properties of Metals and Nonmetals

Metals	Nonmetals
1. High electrical conductivity that decreases with increasing temperature	1. Poor electrical conductivity (except carbon in the form of graphite)
2. High thermal conductivity	2. Good heat insulators (except carbon in the form of diamond)
3. Metallic gray or silver luster[a]	3. No metallic luster
4. Almost all are solids[b]	4. Solids, liquids, or gases
5. Malleable (can be hammered into sheets)	5. Brittle in solid state
6. Ductile (can be drawn into wires)	6. Nonductile

[a]Except copper and gold.
[b]Except mercury; cesium and gallium melt on a hot summer day (85°F) or when held in a person's hand.

• *Metals* generally are malleable, ductile, and lustrous and are good thermal and electrical conductors. *Nonmetals* tend to lack these properties. In many ways, the general properties of metals and nonmetals are opposites.

A **nonmetal** *is an element characterized by the absence of the properties of luster, thermal conductivity, electrical conductivity, and malleability.* Many of the nonmetals, such as hydrogen, oxygen, nitrogen, and the noble gases, are gases. The only nonmetal found as a liquid at room temperature is bromine. Solid nonmetals include carbon, iodine, sulfur, and phosphorus (Figure 3.8c). In general, the nonmetals have lower densities and lower melting points than metals. Table 3.3 contrasts selected physical properties of metals and nonmetals.

Chemical CONNECTIONS

3.2 Calcium, Your Body, and the Periodic Law

The element calcium is essential to bone formation. As bones begin to form, crystals of calcium-containing compounds deposit on a foundation material of the protein collagen. The crystals invade the collagen matrix and gradually give more and more rigidity to the maturing bones. The formation of teeth follows a similar pattern.

Most of the body's calcium (about 99%) is found in bones and teeth. The remaining 1% is found in the blood. Here it is involved in regulating the transport of other chemical substances across cell membranes. It also helps maintain normal blood pressure and is essential for muscle contraction and therefore for maintaining the heartbeat.

Calcium is a member of Group IIA. Heavier members of this family of elements include strontium, barium, and radium. As predicted by the periodic law, these elements and calcium have similar chemical properties. The similarities are such that these elements can also be incorporated into bone structure, just as calcium is. This would normally be of little significance, because we do not ingest significant amounts of any of these elements. It is, however, a very important observation because of the world's use of nuclear energy and nuclear weapons.

Radioactive isotopes of strontium, barium, and radium are produced in nuclear weapon explosions and in the operation of nuclear power plant reactors. Release of these substances into the environment by nuclear weapons testing (open-air testing is now banned) or nuclear power plant accidents (such as that in Chernobyl in 1986) can result in their reaching human food chains. For example, consumption of contaminated plant forage by cows results in cows' milk containing radioactive strontium, barium, and radium. Drinking such milk subjects the individual ingesting the milk to radiation exposure.

If these calcium-like elements could not be used by the body, they would be excreted in a matter of days and their ingestion would be of little consequence. However, because of their periodic-law relationship to calcium, the body incorporates them into teeth and bones as it would calcium. Thus they remain within the body for an indefinite time. The release of radiation by radioactive isotopes of these calcium-like elements can adversely affect an individual and, with time, may cause diseases such as leukemia.

Metals are malleable and ductile and thus lend themselves to uses like this decorative ironwork.

Periodic Table Locations for Metals and Nonmetals

The majority of the elements are metals. Only 23 elements are nonmetals. It is not necessary to memorize which elements are nonmetals and which are metals; this information is obtainable from a periodic table (Figure 3.8a). The steplike heavy line that runs through the right third of the periodic table separates the metals on the left from the nonmetals on the right. Note also that the element hydrogen is a nonmetal.

The fact that the vast majority of elements are metals in no way indicates that metals are more important than nonmetals. Most nonmetals are relatively common and are found in many important compounds. For example, water (H_2O) is a compound involving two nonmetals.

An analysis of the abundance of the elements in Earth's crust (Figure 1.10) in terms of metals and nonmetals shows that the two most abundant elements, which account for 80.2% of all atoms, are nonmetals—oxygen and silicon. The four most abundant elements in the human body (Chemical Connections 1.2), which comprise over 99% of all atoms in the body, are nonmetals—hydrogen, oxygen, carbon, and nitrogen.

3.6 Electron Arrangements Within Atoms

Current chemical theory indicates that as electrons move about an atom's nucleus, they are restricted to specific regions within the extranuclear portion of the atom. Such restrictions are determined by the amount of energy the electrons possess. Furthermore, chemical theory indicates that electron energies are limited to certain values and that a specific "behavior" is associated with each allowed energy value.

The space in which electrons move rapidly about a nucleus is divided into subspaces called *shells, subshells,* and *orbitals.*

Electron Shells

Electrons within an atom are grouped into main energy levels called electron shells. An **electron shell** *is a region of space about a nucleus that contains electrons that have approximately the same energy and that spend most of their time approximately the same distance from the nucleus.*

• Electrons that occupy the first electron shell are closer to the nucleus and have a lower energy than electrons in the second electron shell.

Electron shells are numbered 1, 2, 3, and so on, outward from the nucleus. Electron energy increases as the distance of the electron shell from the nucleus increases. An electron in shell 1 has the minimum amount of energy that an electron can have.

Electron Subshells

Within each electron shell, electrons are further grouped into electron subshells. An **electron subshell** *is a region of space within an electron shell that contains electrons that have the same energy.* We can draw an analogy between the relationship of shells and subshells and the physical layout of a high-rise apartment complex. The shells are analogous to the floors of the apartment complex, and the subshells are the counterparts of the various apartments on each floor.

The number of subshells within a shell is the same as the shell number. Shell 1 contains one subshell, shell 2 contains two subshells, shell 3 contains three subshells, and so on.

• The letters used to label the different types of subshells come from old spectroscopic terminology associated with the lines in the spectrum of the element hydrogen. These lines were denoted as *s*harp, *p*rincipal, *d*iffuse, and *f*undamental. Relationships exist between such lines and the arrangement of electrons in an atom.

Subshells within a shell differ in size (that is, the maximum number of electrons they can accommodate) and energy. The higher the energy of the contained electrons, the larger the subshell.

Subshell size (type) is designated using the letters *s, p, d,* and *f.* Listed in this order, these letters denote subshells of increasing energy and size. The lowest-energy subshell within a shell is always the *s* subshell, the next highest is the *p* subshell, then the *d* subshell, and finally the *f* subshell. An *s* subshell can accommodate 2 electrons, a *p* subshell 6 electrons, a *d* subshell 10 electrons, and an *f* subshell 14 electrons.

Chemical CONNECTIONS

3.3 Importance of Metallic and Nonmetallic Trace Elements for Human Health

Within the past three decades, biochemists have developed techniques that can detect substances in smaller and smaller quantities in living cells. As a result, we now know that living organisms need minute amounts of certain elements—called trace elements—to function properly. These trace elements now number 15, and more may be discovered. Ten of the 15 known trace elements are metals, and 5 are nonmetals. The identity and functions of these trace elements are listed in the accompanying table.

The trace elements are present in milligram quantities in the human body. If you could collect them all together, you would not have enough material to fill a teaspoon. Each, however, plays a vital role in the operation of living cells, and each is responsible for a function for which there is no substitute.

Knowledge about the functions of trace elements is difficult to obtain, because it is so difficult to provide an experimental diet that lacks the one element under study. Infinitesimal amounts of some of the elements are all that is needed to invalidate the experiment. Thus research in this area consists primarily of animal studies involving highly refined, purified diets in environments that are free of all contamination.

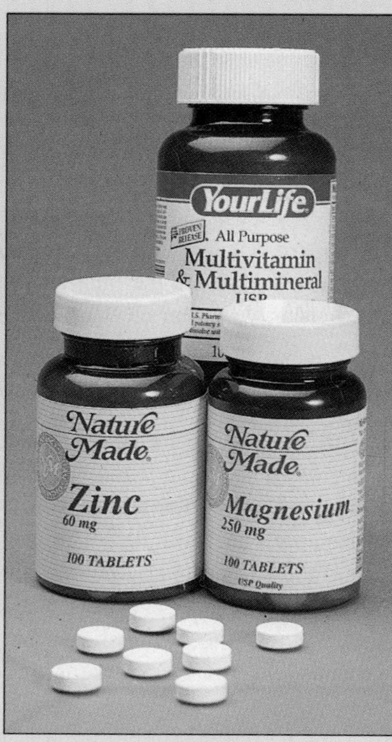

Many food supplements incorporate trace elements required by humans.

Need in Humans Established and Quantified

Metals

Iron: forms part of hemoglobin (the oxygen-carrying protein of red blood cells) and myoglobin (the oxygen-holding protein in muscle cells)

Zinc: occurs in more than 70 enzymes that perform specific tasks in the eyes, liver, kidneys, muscles, skin, bones, and male reproductive organs

Nonmetals

Iodine: occurs in three thyroid gland hormones that regulate metabolic rate

Selenium: part of an enzyme that acts as an antioxidant for polyunsaturated fatty acids

Need in Humans Established but not Completely Quantified

Metals

Copper: necessary for the absorption and use of iron in the formation of hemoglobin; also a factor in the formation of the protective covering of nerves

Manganese: facilitator, with enzymes, of many different metabolic processes

Cobalt: part of vitamin B_{12}; necessary for nerve cell function and blood formation

Molybdenum: facilitator, with enzymes, of numerous cell processes

Chromium: associated with insulin and required for the release of energy from glucose

Nonmetals

Fluorine: involved in the formation of bones and teeth; helps make teeth resistant to decay

Need Established in Animals but not Yet in Humans

Metals

Nickel: deficiencies harm the liver and other organs

Tin: necessary for growth

Vanadium: necessary for growth, bone development, and normal reproduction

Nonmetals

Silicon: involved in bone calcification

Boron: involved in bone development and minimization of demineralization in osteoporosis

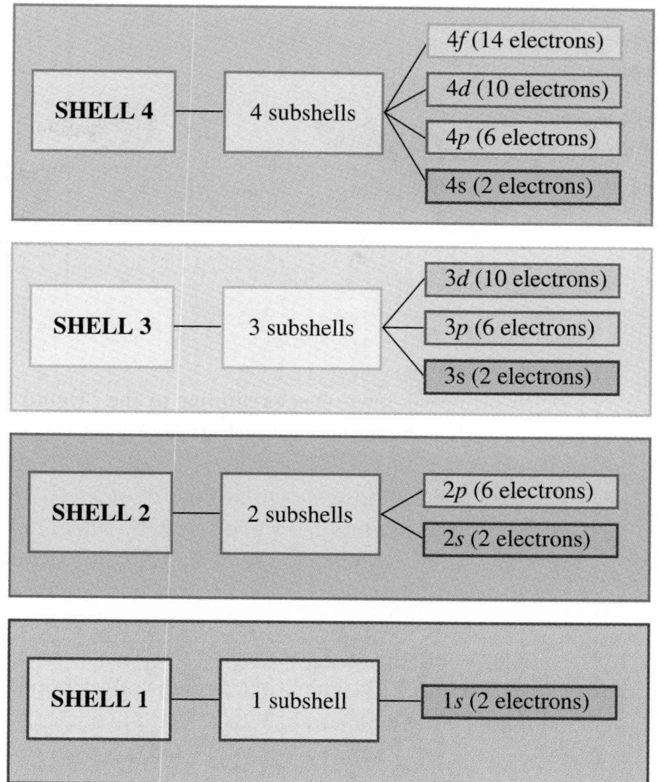

Figure 3.9
The number of subshells within a shell is equal to the shell number, as shown here for the first four shells. Each individual subshell is denoted with both a number (its shell) and a letter (the type of subshell it is in).

Figure 3.9 summarizes the relationships between electron shells and electron subshells for the first four shells.

• Electron Orbitals

Electron subshells have within them a certain, definite number of locations (regions of space), called electron orbitals, where electrons may be found. In our apartment complex analogy, if shells are the counterparts of floor levels and subshells are the apartments, then electron orbitals are the rooms of the apartments. An **electron orbital** *is a region of space within an electron subshell where an electron with a specific energy is most likely to be found.*

• An electron orbital is also often called an atomic orbital.

An electron orbital, independent of all other considerations, can accommodate a maximum of 2 electrons. Thus an *s* subshell (2 electrons) contains one orbital, a *p* subshell (6 electrons) contains three orbitals, a *d* subshell (10 electrons) contains five orbitals, and an *f* subshell (14 electrons) contains seven orbitals.

Orbitals have distinct shapes that are related to the type of subshell in which they are found. Note that we are talking not about the shape of an electron but, rather, about the shape of the region in which the electron is found. An orbital in an *s* subshell, which is called an *s* orbital, has a spherical shape (Figure 3.10a). Orbitals found in *p* subshells—

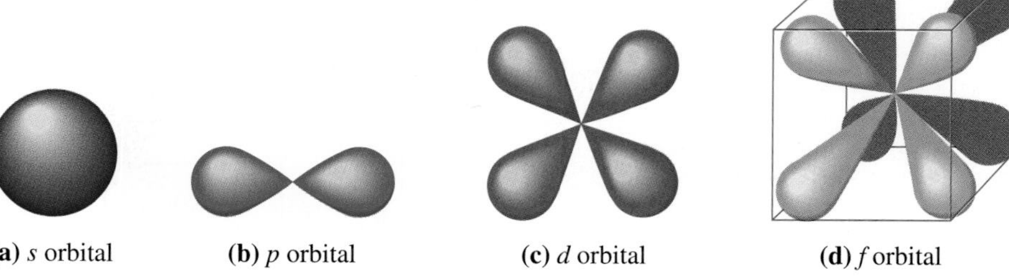

Figure 3.10
An *s* orbital has a spherical shape, a *p* orbital has two lobes, a *d* orbital has four lobes, and an *f* orbital has eight lobes. The *f* orbital is shown within a cube to illustrate that its lobes are directed toward the corners of a cube. Some *d* and *f* orbitals have shapes related to, but not identical to, those shown.

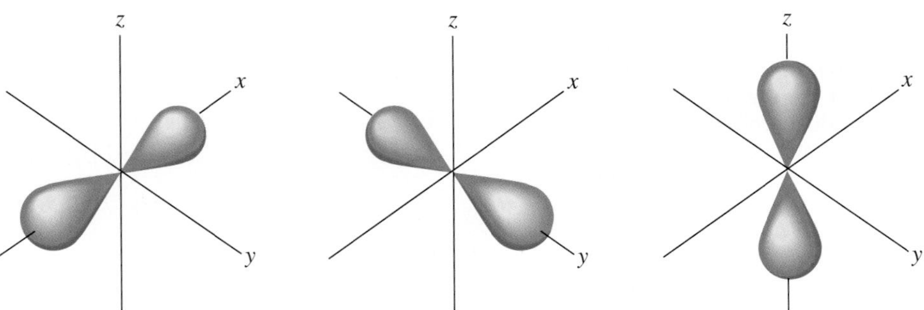

Figure 3.11
Orbitals within a subshell differ mainly in orientation. For example, the three *p* orbitals within a *p* subshell lie along the *x, y,* and *z* axes of a Cartesian coordinate system.

p orbitals—have shapes similar to the "figure 8" of an ice skater (Figure 3.10b). More complex shapes involving four and eight lobes, respectively, are associated with *d* and *f* orbitals (Figures 3.10c and 3.10d). Some *d* and *f* orbitals have shapes related to, but not identical to, those shown in Figure 3.10.

Orbitals within the same subshell, which have the same shape, differ mainly in orientation. For example, the three 2*p* orbitals extend out from the nucleus at 90° angles to one another (along the *x, y,* and *z* axes in a Cartesian coordinate system), as is shown in Figure 3.11. Such orientation of the *p* orbitals is of particular importance to the structural characteristics of hydrocarbon molecules (Section 13.4).

Figure 3.12, which is an extension of Figure 3.9, summarizes the important relationships among electron shells, electron subshells, and electron orbitals.

Electron Spin

Experimental studies indicate that as an electron "moves about" within an orbital, it spins on its own axis in either a clockwise or a counterclockwise direction. Furthermore, when two electrons are present in an orbital, they always have opposite spins; that is, one is spinning clockwise and the other counterclockwise. This situation of opposite spins is en-

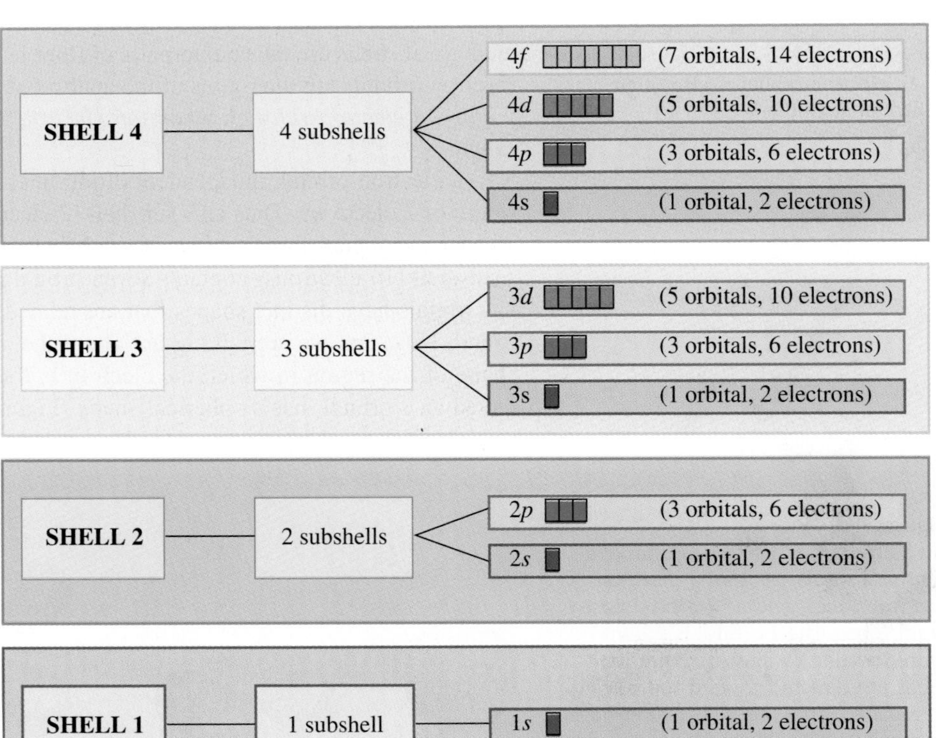

Figure 3.12
A summary of the interrelationships among electron shells, electron subshells, and electron orbitals for the first four shells. Similar relationship patterns exist for high-numbered shells.

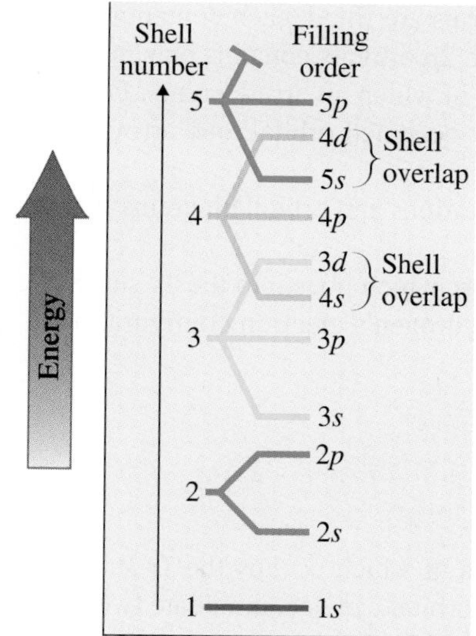

Figure 3.13
The order of filling of various electron subshells is shown on the right-hand side of this diagram. Above the 3*p* subshell, subshells of different shells "overlap."

ergetically the most favorable state for two electrons in the same orbital. We will have more to say about electron spin when we discuss orbital diagrams in Section 3.7.

3.7 Specification of Electronic Structure

Electron shells, subshells, and orbitals describe "permissible" locations for electrons—that is, where electrons *can* be found. We are now ready to discuss *actual* locations of the electrons in specific atoms.

There are many orbitals about the nucleus of an atom. Electrons do not occupy these orbitals in a random, haphazard fashion; a very predictable pattern exists for electron orbital occupancy. There are three rules, all quite simple, for assigning electrons to various shells, subshells, and orbitals.

1. *Electron subshells are filled in order of increasing energy.*
2. *Electrons occupy the orbitals of a subshell such that each orbital acquires one electron before any orbital acquires a second electron.* All electrons in such singly occupied orbitals must have the same spin.
3. *No more than two electrons may exist in a given orbital—and then only if they have opposite spins.*

• Subshell Energy Order

The ordering of electron subshells in terms of increasing energy, which is experimentally determined, is more complex than might be expected. This is because the energies of subshells in different shells often "overlap," as shown in Figure 3.13. This diagram shows, for example, that the 4*s* subshell has lower energy than the 3*d* subshell.

• All electrons in a given subshell have the same energy because all orbitals within a subshell have the same energy.

A useful mnemonic (memory) device for remembering subshell filling order, which incorporates "overlap" situations such as those in Figure 3.13, is given in Figure 3.14. This diagram, which lists all subshells needed to specify the electron arrangements for all 115 elements, is constructed by locating all *s* subshells in column 1, all *p* subshells in column 2, and so on. Subshells that belong to the same shell are found in the same row. The order of subshell filling is given by following the diagonal arrows, starting at the top. The 1*s* subshell fills first. The second arrow points to (goes through) the 2*s* subshell, which fills next. The third arrow points to both the 2*p* and the 3*s* subshells. The 2*p* fills first, followed by the 3*s*. Any time a single arrow points to more than one subshell, we start at the tail of the arrow and work to its tip to determine the proper filling sequence.

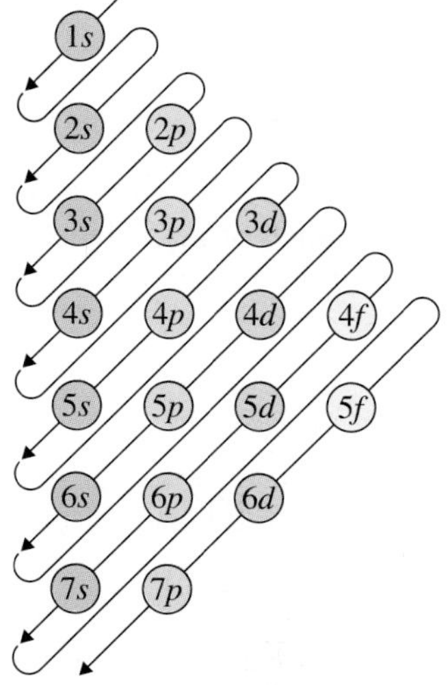

Figure 3.14
The order for filling electron subshells with electrons follows the order given by the arrows in this diagram. Start with the arrow at the top of the diagram and work toward the bottom of the diagram, moving from the bottom of one arrow to the top of the next-lower arrow.

• Electron Configuration and Orbital Diagrams

An **electron configuration** *is a statement of how many electrons an atom has in each of its subshells.* Because subshells group electrons according to energy, electron configurations indicate how many electrons of various energies an atom has.

Electron configurations are not written out in words; rather, a shorthand system with symbols is used. Subshells containing electrons, listed in order of increasing energy, are designated by using number–letter combinations (1*s*, 2*s*, and 2*p*). A superscript following each subshell designation indicates the number of electrons in that subshell. The electron configuration for nitrogen in this shorthand notation is

$$1s^22s^22p^3$$

Thus, a nitrogen atom has an electron arrangement of two electrons in the 1*s* subshell, two electrons in the 2*s* subshell, and three electrons in the 2*p* subshell.

An **orbital diagram** *is a statement of how many electrons an atom has in each of its orbitals.* Note that electron configurations deal with *subshell* occupancy and orbital diagrams deal with *orbital* occupancy. The orbital diagram for the element nitrogen is

1*s*	2*s*	2*p*
↑↓	↑↓	↑ ↑ ↑

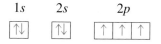

This diagram indicates that both the 1*s* and the 2*s* orbitals are filled, each containing two electrons of opposite spin. In addition, each of the three 2*p* orbitals contains one electron. Electron spin is denoted by the direction (up or down) in which an arrow points. For two electrons of opposite spin, which is the case in a fully occupied orbital, one arrow must point up and the other down.

Let us now systematically consider electron configurations and orbital diagrams for the first few elements in the periodic table.

Hydrogen (atomic number = 1) has only one electron, which goes into the 1*s* subshell; this subshell has the lowest energy of all subshells. Hydrogen's electron configuration is written as $1s^1$, and its orbital diagram is

	1*s*
H:	↑

Helium (atomic number = 2) has two electrons, both of which occupy the 1*s* subshell. (Remember, an *s* subshell contains one orbital, and an orbital can accommodate two electrons.) Helium's electron configuration is $1s^2$, and its orbital diagram is

	1*s*
He:	↑↓

The two electrons present are of opposite spin.

Lithium (atomic number = 3) has three electrons, and the third electron cannot enter the 1*s* subshell because its maximum capacity is two electrons. (All *s* subshells are completely filled with two electrons.) The third electron is placed in the next-highest-energy subshell, the 2*s*. The electron configuration for lithium is $1s^22s^1$, and its orbital diagram is

	1*s*	2*s*
Li:	↑↓	↑

For *beryllium* (atomic number = 4), the additional electron is placed in the 2*s* subshell, which is now completely filled, giving beryllium the electron configuration $1s^22s^2$. The orbital diagram for beryllium is

	1*s*	2*s*
Be:	↑↓	↑↓

For *boron* (atomic number = 5), the 2*p* subshell, which is the subshell of next highest energy (Figures 3.13 and 3.14), becomes occupied for the first time. Boron's electron configuration is $1s^22s^22p^1$, and its orbital diagram is

	1*s*	2*s*	2*p*		
B:	↑↓	↑↓	↑		

The 2*p* subshell contains three orbitals of equal energy. It does not matter which of the 2*p* orbitals is occupied, because they are of equivalent energy.

With the next element, *carbon* (atomic number = 6), we come to a new situation. We know that the sixth electron must go into a 2*p* orbital. However, does this new electron go into the 2*p* orbital that already has one electron or into one of the others? Rule 2, at the start of this section, covers this situation. Electrons will occupy equal-energy orbitals singly to the maximum extent possible before any orbital acquires a second electron. Thus, for carbon, we have the electron configuration $1s^22s^22p^2$ and the orbital diagram

	1*s*	2*s*	2*p*		
C:	↑↓	↑↓	↑	↑	

A *p* subshell can accommodate six electrons because there are three orbitals within it. The 2*p* subshell can thus accommodate the additional electrons found in the elements with atomic numbers 7 through 10: *nitrogen* (N), *oxygen* (O), *fluorine* (F), and *neon* (Ne). The electron configurations and orbital diagrams for these elements are

• The symbols $1s^2$, $2s^2$, and $2p^3$ are read as "one *s* two," "two s two," and "two p three," not as "one s squared," "two s squared," and "two p cubed."

			1s	2s	2p
N:	$1s^2 2s^2 2p^3$	N:	↑↓	↑↓	↑ | ↑ | ↑
O:	$1s^2 2s^2 2p^4$	O:	↑↓	↑↓	↑↓ | ↑ | ↑
F:	$1s^2 2s^2 2p^5$	F:	↑↓	↑↓	↑↓ | ↑↓ | ↑
Ne:	$1s^2 2s^2 2p^6$	Ne:	↑↓	↑↓	↑↓ | ↑↓ | ↑↓

Chemical CONNECTIONS

3.4 Electrons in Excited States

When an atom has its electrons positioned in the lowest-energy orbitals available, it is said to be in its *ground state.* An atom's ground state is the normal (most stable) state for the atom.

It is possible to elevate electrons in an atom to higher-energy unoccupied orbitals by subjecting the atom to a beam of light energy, an electrical discharge, or an influx of heat energy. With electrons in higher, normally unoccupied orbitals, the atom is said to be in an *excited state.* An excited state is an unstable state that has a short life span. Quickly, the excited electrons drop back down to their previous positions (the ground state). Accompanying the transition from excited state to ground state is a release of energy. Often this release of energy is in the form of visible light.

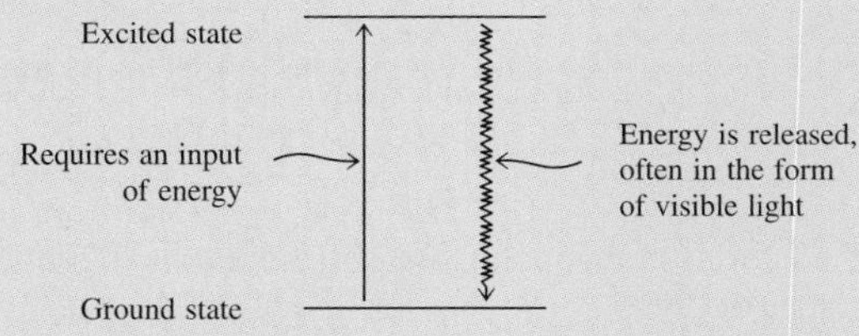

The principle of electron excitation through input of energy has been found to have useful applications in many areas.

1. *"Neon" Advertising Signs.* In such signs, gaseous atoms are excited by an electric discharge to produce a variety of colors; the color depends on the identity of the gas. Neon gas produces an orange-red light, argon gas a blue-purple light, and krypton gas a white light.
2. *Street and Highway Lights.* Such lights involve energy emitted by electrically excited metal atoms. Mercury vapor lamps produce a yellow light that can penetrate fog farther than does light from a sodium vapor lamp. On the other hand, sodium vapor lamps are more energy-efficient in their operation.
3. *Fireworks.* Metal atoms excited by heat are responsible for the color of fireworks. Strontium (red color), barium (green color), copper (blue color), and aluminum (white color) are some of the metals involved. The metals are present in the fireworks in the form of metal-containing compounds rather than as pure metals.
4. *Identification of Elements.* Each of the known elements, when electronically excited in the gaseous state, produces a unique pattern of wavelengths of emitted light (radiation). This emission pattern, which is called an atomic spectrum, serves as a "fingerprint" for the element and can be used to distinguish the element from any other.
5. *Analysis of Human Body Fluids.* Instruments called atomic spectrometers are now used to analyze body fluids for the presence of particular elements (in free or combined form). It is possible to determine concentrations of species with such instruments, which are now found in almost all clinical chemistry laboratories. The concentration of sodium and potassium in a particular fluid can be obtained from the intensity of the light emitted by excited atoms of these elements. Atomic spectroscopy makes it possible to measure the amount of lead in a patient's blood or urine (in cases of lead poisoning) by using a sample as small as 0.01 cm^3.

The different colors of fireworks result when heat excites the electrons of different kinds of metal atoms present.

With *sodium* (atomic number = 11), the 3*s* subshell acquires an electron for the first time. Sodium's electron configuration is

$$1s^22s^22p^63s^1$$

Note the pattern that is developing in the electron configurations we have written so far. Each element has an electron configuration that is the same as the one just before it except for the addition of one electron.

• An *electron configuration* is a shorthand notation designating the subshells in an atom that are occupied by electrons. The sum of the superscripts in an electron configuration equals the total number of electrons present and hence must equal the atomic number of the element.

Electron configurations for other elements are obtained by simply extending the principles we have just illustrated. A subshell of lower energy is always filled before electrons are added to the next highest subshell; this continues until the correct number of electrons have been accommodated.

For a few elements in the middle of the periodic table, the actual distribution of electrons within subshells differs slightly from that obtained by using the procedures outlined in this section. These exceptions are caused by very small energy differences between some subshells and are not important in the uses we shall make of electron configurations.

Example 3.3

Writing an Electron Configuration

Write the electron configurations for the following elements.

a. Strontium (atomic number = 38) **b.** Lead (atomic number = 82)

Solution

a. The number of electrons in a strontium atom is 38. Remember that the atomic number gives the number of electrons (Section 3.3). We will need to fill subshells, in order of increasing energy, until 38 electrons have been accommodated.

The 1*s*, 2*s*, and 2*p* subshells fill first, accommodating a total of 10 electrons among them.

$$1s^22s^22p^6\ldots$$

Next, according to Figures 3.13 and 3.14, the 3*s* subshell fills and then the 3*p* subshell.

$$1s^22s^22p^6\boxed{3s^23p^6}\ldots$$

We have accommodated 18 electrons at this point. We still need to add 20 more electrons to get our desired number of 38.

The 4*s* subshell fills next, followed by the 3*d* subshell, giving us 30 electrons at this point.

$$1s^22s^22p^63s^23p^6\boxed{4s^23d^{10}}\ldots$$

Note that the maximum electron population for *d* subshells is 10 electrons.

Eight more electrons are needed, which are added to the next two higher subshells, the 4*p* and the 5*s*. The 4*p* subshell can accommodate 6 electrons, and the 5*s* can accommodate 2 electrons.

$$1s^22s^22p^63s^23p^64s^23d^{10}\boxed{4p^65s^2}$$

To double-check that we have the correct number of electrons, 38, we add the superscripts in our final electron configuration.

$$2 + 2 + 6 + 2 + 6 + 2 + 10 + 6 + 2 = 38$$

The sum of the superscripts in any electron configuration should add up to the atomic number if the configuration is for a neutral atom.

b. To write this configuration, we continue along the same lines as in part **a**, remembering that the maximum electron subshell populations are $s = 2$, $p = 6$, $d = 10$, and $f = 14$.

Lead, with an atomic number of 82, contains 82 electrons, which are added to subshells in the following order. (The line of numbers beneath the electron configuration is a running total of added electrons and is obtained by adding the superscripts up to that point. We stop when we have 82 electrons.)

$$1s^2 2s^2 2p^6 3s^2 3p^6 4s^2 3d^{10} 4p^6 5s^2 4d^{10} 5p^6 6s^2 4f^{14} 5d^{10} 6p^2$$

2 4 10 12 18 20 30 36 38 48 54 56 70 80 82

Running total of electrons added

Note in this electron configuration that the $6p$ subshell contains only 2 electrons, even though it can hold a maximum of 6. We put only 2 electrons in this subshell because that is sufficient to give 82 total electrons. If we had completely filled this subshell, we would have had 86 total electrons, which is too many.

Practice Exercise 3.3

Write the electron configurations for the following elements.

a. Manganese (atomic number = 25) **b.** Xenon (atomic number = 54)

• *Answer.* **a.** $1s^2 2s^2 2p^6 3s^2 3p^6 4s^2 3d^5$ **b.** $1s^2 2s^2 2p^6 3s^2 3p^6 4s^2 3d^{10} 4p^6 5s^2 4d^{10} 5p^6$

3.8 The Electronic Basis for the Periodic Law and the Periodic Table

For many years, there was no explanation available for either the periodic law or why the periodic table has the shape that it has. We now know that the theoretical basis for both the periodic law and the periodic table is found in electronic theory. As we saw earlier in the chapter (Section 3.3), when two atoms interact, it is their electrons that interact. Thus the number and arrangement of electrons determine how an atom reacts with other atoms—that is, what its chemical properties are. The properties of the elements repeat themselves in a periodic manner because the arrangement of electrons about the nucleus of an atom follows a periodic pattern, as we saw in Section 3.7.

• Electron Configurations and the Periodic Law

The periodic law (Section 3.4) points out that the properties of the elements repeat themselves in a regular manner when the elements are arranged in order of increasing atomic number. The elements that have similar chemical properties are placed under one another in vertical columns (groups) in the periodic table.

Groups of elements have similar chemical properties because of similarities in their electron configuration. *Chemical properties repeat themselves in a regular manner among the elements because electron configurations repeat themselves in a regular manner among the elements.*

To illustrate this correlation between similar chemical properties and similar electron configurations, let us look at the electron configurations of two groups of elements known to have similar chemical properties.

We begin with the elements lithium, sodium, potassium, and rubidium, all members of Group IA of the periodic table. The electron configurations for these elements are

• The electron arrangement in the outermost shell is the same for elements in the same group. This is why elements in the same group have similar chemical properties.

$$_{3}\text{Li: } 1s^2 \boxed{2s^1}$$
$$_{11}\text{Na: } 1s^2 2s^2 2p^6 \boxed{3s^1}$$
$$_{19}\text{K: } 1s^2 2s^2 2p^6 3s^2 3p^6 \boxed{4s^1}$$
$$_{37}\text{Rb: } 1s^2 2s^2 2p^6 3s^2 3p^6 4s^2 3d^{10} 4p^6 \boxed{5s^1}$$

Note that each of these elements has one electron in its outermost shell. (The outermost shell is the shell with the highest number.) This similarity in outer-shell electron arrangements causes these elements to have similar chemical properties. In general, elements with similar outer-shell electron configurations have similar chemical properties.

Let us consider another group of elements known to have similar chemical properties: fluorine, chlorine, bromine, and iodine of Group VIIA of the periodic table. The electron configurations for these four elements are

$_{9}F$: $1s^2(2s^22p^5)$
$_{17}Cl$: $1s^22s^22p^6(3s^23p^5)$
$_{35}Br$: $1s^22s^22p^63s^23p^6(4s^2)\,3d^{10}(4p^5)$
$_{53}I$: $1s^22s^22p^63s^23p^64s^23d^{10}4p^6(5s^2)\,4d^{10}(5p^5)$

Once again, similarities in electron configuration are readily apparent. This time, the repeating pattern involves an outermost *s* and *p* subshell containing seven electrons (shown in color). Remember that for Br and I, shell numbers 4 and 5 designate, respectively, electrons in the outermost shells.

Electron Configurations and the Periodic Table

One of the strongest pieces of supporting evidence for the assignment of electrons to shells, subshells, and orbitals is the periodic table itself. The basic shape and structure of this table, which was determined many years before electrons were even discovered, is consistent with and can be explained by electron configurations. Indeed, the specific location of an element in the periodic table can be used to obtain information about its electron configuration.

As the first step in linking electron configurations to the periodic table, let us analyze the general shape of the periodic table in terms of columns of elements. As shown in Figure 3.15, on the extreme left of the table, there are 2 columns of elements; in the center, there is a region containing 10 columns of elements; to the right there is a block of 6 columns of elements; and in the two rows at the bottom of the table, there are 14 columns of elements.

The number of columns of elements in the various regions of the periodic table—2, 6, 10, and 14—is the same as the maximum number of electrons that the various types of subshells can accommodate. We will see shortly that this is a very significant observation; the number matchup is no coincidence. The various columnar regions of the periodic table are called the *s* area (2 columns), the *p* area (6 columns), the *d* area (10 columns), and the *f* area (14 columns), as shown in Figure 3.15.

The concept of *distinguishing electrons* is the key to obtaining electron configuration information from the periodic table. The **distinguishing electron** *for an element is the last electron that is added to its electron configuration when subshells are filled in order*

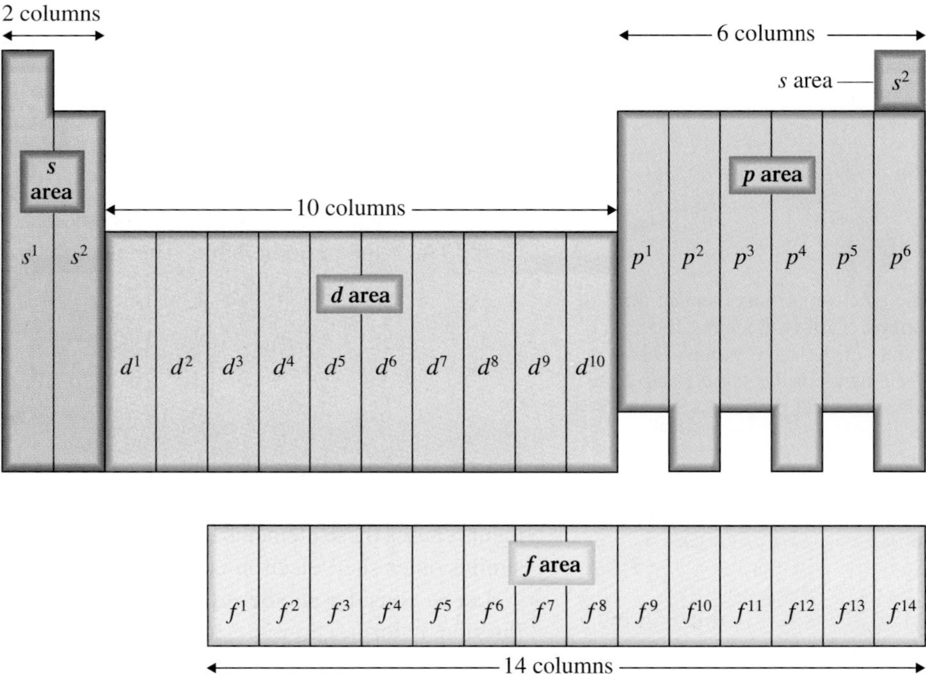

Figure 3.15 Electron configurations and the positions of elements in the periodic table. The periodic table can be divided into four areas that are 2, 6, 10, and 14 columns wide. The four areas contain elements whose distinguishing electron is located, respectively, in *s, p, d,* and *f* subshells. The extent of filling of the subshell that contains an element's distinguishing electron can be determined from the element's position in the periodic table.

of increasing energy. This last electron is the one that causes an element's electron configuration to differ from that of the element immediately preceding it in the periodic table.

For all elements that are located in the *s* area of the periodic table, the distinguishing electron is always found in an *s* subshell. All *p* area elements have distinguishing electrons in *p* subshells. Similarly, elements in the *d* and *f* areas of the periodic table have distinguishing electrons located in *d* and *f* subshells, respectively. Thus the area location of an element in the periodic table can be used to determine the type of subshell that contains the distinguishing electron. Note that the element helium belongs to the *s* rather than the *p* area of the periodic table, even though its table position is on the right-hand side. (The reason for this placement of helium will be explained in Section 4.3).

The extent to which the subshell containing an element's distinguishing electron is filled can also be determined from the element's position in the periodic table. All elements in the first column of a specific area contain only one electron in the subshell; all elements in the second column contain two electrons in the subshell; and so on. Thus, all elements in the first column of the *p* area (Group IIIA) have an electron configuration ending in p^1. Elements in the second column of the *p* area (Group IVA) have electron configurations ending in p^2; and so on. Similar relationships hold in other areas of the table, as shown in Figure 3.15.

3.9 Classification of the Elements

The elements can be classified in several ways. The two most common classification systems are

1. A system based on selected physical properties of the elements, in which they are described as metals or nonmetals. This classification scheme was discussed in Section 3.5.
2. A system based on the electron configurations of the elements, in which elements are described as *noble gas, representative, transition,* or *inner transition elements.*

The classification scheme based on electron configurations of the elements, along with a summary of what we have said about the periodic table, is depicted in Figure 3.16 and in the accompanying Chemistry at a Glance.

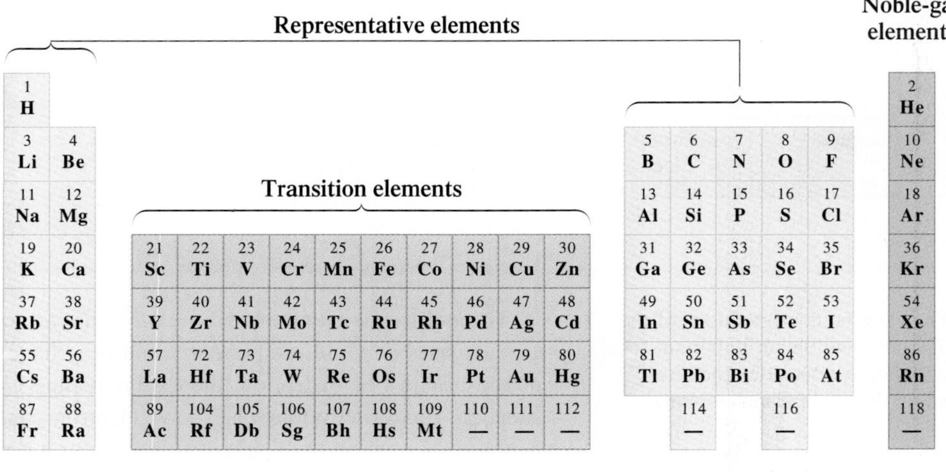

Figure 3.16
A classification scheme for the elements based on their electron configurations. Representative elements occupy the *s* area and most of the *p* area shown in Figure 3.15. The noble-gas elements occupy the last column of the *p* area. The transition elements are found in the *d* area, and the inner transition elements are found in the *f* area.

Element Classification Schemes and the Periodic Table

CHEMISTRY AT A GLANCE

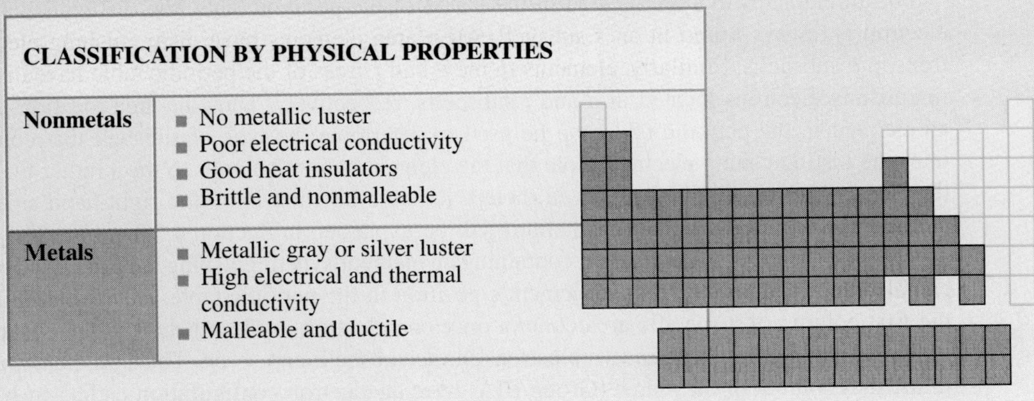

CLASSIFICATION BY PHYSICAL PROPERTIES	
Nonmetals	■ No metallic luster ■ Poor electrical conductivity ■ Good heat insulators ■ Brittle and nonmalleable
Metals	■ Metallic gray or silver luster ■ High electrical and thermal conductivity ■ Malleable and ductile

CLASSIFICATION BY ELECTRONIC PROPERTIES	
Representative elements	■ Found in *s* area and first five columns of the *p* area ■ Some are metals, some nonmetals
Noble-gas elements	■ Found in last column of *p* area plus He (*s* area) ■ All are nonmetals
Transition elements	■ Found in *d* area ■ All are metals
Inner transition elements	■ Found in *f* area ■ All are metals

PERIODIC TABLE GROUPS WITH SPECIAL NAMES	
Alkali metals	■ Group IA elements (except for H, a nonmetal) ■ Electron configurations end in s^1
Alkaline earth metals	■ Group IIA elements ■ Electron configurations end in s^2
Halogens	■ Group VIIA ■ Electron configurations end in p^5
Noble gases	■ Group VIIIA elements ■ Electron configurations end in p^6, except for He, which ends in s^2

IA IIA VIIA VIIIA

• The electron configurations of the noble gases will be an important focal point when we consider chemical bonding theory in Chapters 4 and 5.

The **noble-gas elements** *are found in the far right column of the periodic table.* They are all gases at room temperature, and they have little tendency to form chemical compounds. With one exception, the distinguishing electron for a noble gas completes the *p* subshell; therefore, noble gases have electron configurations ending in p^6. The exception is helium, in which the distinguishing electron completes the first shell—a shell that has only two electrons. Helium's electron configuration is $1s^2$.

The **representative elements** *are all the elements of the* s *and* p *areas of the periodic table, with the exception of the noble gases.* The distinguishing electron in these elements partially or completely fills an *s* subshell or partially fills a *p* subshell. The representative elements include most of the more common elements.

The **transition elements** *are all the elements of the* d *area of the periodic table.* Each has its distinguishing electron in a *d* subshell.

The **inner transition elements** *are all the elements of the* f *area of the periodic table.* Each has its distinguishing electron in an *f* subshell. There is very little variance in the properties of either the 4*f* or the 5*f* series of inner transition elements.

Concepts to Remember

Subatomic particles. Subatomic particles, the very small building blocks from which atoms are made, are of three major types: electrons, protons, and neutrons. Electrons are negatively charged, protons are positively charged, and neutrons have no charge. All neutrons and protons are found at the center of the atom in the nucleus. The electrons occupy the region about the nucleus. Protons and neutrons have much larger masses than the electron.

Atomic number and mass number. Each atom has a characteristic atomic number and mass number. The atomic number is equal to the number of protons in the nucleus of the atom. The mass number is equal to the total number of protons and neutrons in the nucleus.

Isotopes. Isotopes are atoms that have the same number of protons and electrons but have different numbers of neutrons. The isotopes of an element always have the same atomic number and different mass numbers. Isotopes of an element have the same chemical properties.

Atomic mass. The atomic mass of an element is a calculated average mass. It depends on the percentage abundances and masses of the naturally occurring isotopes of the element.

Periodic law and periodic table. The periodic law states that when elements are arranged in order of increasing atomic number, elements with similar chemical properties occur at periodic (regularly recurring) intervals. The periodic table is a graphical representation of the behavior described by the periodic law. In a modern periodic table, vertical columns contain elements with similar chemical properties. A group in the periodic table is a vertical column of elements. A period in the periodic table is a horizontal row of elements.

Metals and nonmetals. Metals exhibit luster, thermal conductivity, electrical conductivity, and malleability. Nonmetals are characterized by the absence of the properties associated with metals. The majority of the elements are metals. The steplike heavy line that runs through the right third of the periodic table separates the metals on the left from the nonmetals on the right.

Electron shell. A shell contains electrons that have approximately the same energy and spend most of their time approximately the same distance from the nucleus.

Electron subshell. A subshell contains electrons that all have the same energy. The number of subshells in a particular shell is equal to the shell number. Each subshell can hold a specific maximum number of electrons. These values are 2, 6, 10, and 14 for *s, p, d,* and *f* subshells, respectively.

Electron orbital. An orbital is a region of space about a nucleus where an electron with a specific energy is most likely to be found. Each subshell consists of one or more orbitals. For *s, p, d,* and *f* subshells there are 1, 3, 5, and 7 orbitals, respectively. No more than two electrons may occupy any orbital.

Electron configuration. An electron configuration is a statement of how many electrons an atom has in each of its subshells. The principle that electrons normally occupy the lowest-energy subshell available is used to write electron configurations.

Orbital diagram. An orbital diagram is a statement of how many electrons an atom has in each of its orbitals. Electrons occupy the orbitals of a subshell such that each orbital within the subshell acquires one electron before any orbital acquires a second electron. All electrons in such singly occupied orbitals must have the same spin.

Electron configurations and the periodic law. Chemical properties repeat themselves in a regular manner among the elements because electron configurations repeat themselves in a regular manner among the elements.

Electron configurations and the periodic table. The groups of the periodic table consist of elements with similar electron configurations. Thus the location of an element in the periodic table can be used to obtain information about its electron configuration.

Classification system for the elements. On the basis of electron configuration, elements can be classified into four categories: noble gases (far right column of the periodic table); representative elements (*s* and *p* areas of the periodic table, with the exception of the noble gases); transition elements (*d* area of the periodic table); and inner transition elements (*f* area of the periodic table).

Key Reactions and Equations

1. Relationships involving atomic number and mass number for a neutral atom (Section 3.3)

Atomic number = number of protons = number of electrons

Mass number = number of protons + number of neutrons

Mass number = total number of subatomic particles in the nucleus

Mass number − atomic number = number of neutrons

Mass number + atomic number = total number of subatomic particles

2. Relationships involving electron shells, electron subshells, and electron orbitals (Section 3.6)

Number of subshells in a shell = shell number

Maximum number of electrons in an *s* subshell = 2

Maximum number of electrons in a *p* subshell = 6

Maximum number of electrons in a *d* subshell = 10

Maximum number of electrons in an *f* subshell = 14

Maximum number of electrons in an orbital = 2

3. Order of filling of subshells in terms of increasing energy (Section 3.7)

1*s*, 2*s*, 2*p*, 3*s*, 3*p*, 4*s*, 3*d*, 4*p*, 5*s*, 4*d*, 5*p*, 6*s*, 4*f*, 5*d*, 6*p*, 7*s*, 5*f*, 6*d*, 7*p*

Key Terms

Atomic mass (3.2)
Atomic number (3.2)
Distinguishing electron (3.7)
Electron configuration (3.6)
Electron orbital (3.5)
Electron shell (3.5)
Electron subshell (3.5)
Electrons (3.1)
Element (3.2)
Group (3.3)
Inner transition elements (3.8)
Isotopes (3.2)
Mass number (3.2)
Metal (3.4)
Noble-gas elements (3.8)
Nucleus (3.1)
Neutrons (3.1)
Nonmetal (3.4)
Orbital diagram (3.6)
Period (3.3)
Periodic law (3.3)
Periodic table (3.3)
Protons (3.1)
Representative elements (3.8)
Subatomic particles (3.1)
Transition elements (3.8)

Exercises and Problems

The members of each pair of problems in this section test similar material.

Internal Structure of the Atom (Sections 3.1 and 3.2)

3.1 Indicate which subatomic particle (proton, neutron, or electron) correctly matches each of the following statements. More than one particle can be used as an answer.

a. Possesses a negative charge
b. Has no charge
c. Has a mass slightly less than that of a neutron
d. Has a charge equal to, but opposite in sign from, that of an electron

3.2 Indicate which subatomic particle (proton, neutron, or electron) correctly matches each of the following statements. More than one particle can be used as an answer.

a. Is not found in the nucleus
b. Has a positive charge
c. Can be called a nucleon
d. Has a relative mass of 1837 if the relative mass of an electron is 1

3.3 Indicate whether each of the following statements about the nucleus of an atom is true or false.

a. The nucleus of an atom is neutral.
b. The nucleus of an atom contains only neutrons.
c. The number of nucleons present in the nucleus is equal to the number of electrons present outside the nucleus.
d. The nucleus accounts for almost all the mass of an atom.

3.4 Indicate whether each of the following statements about the nucleus of an atom is true or false.

a. The nucleus of an atom contains all of the "heavy" subatomic particles.
b. The nucleus of an atom accounts for almost all of the volume of the atom.
c. The nucleus of an atom has an extremely low density compared to that of the atom as a whole.
d. The nucleus of an atom can be positively or negatively charged, depending on the identity of the atom.

3.5 Indicate whether each of the following statements about discharge tube experiments and metal foil experiments is *true* or *false*.

a. Information obtained from discharge tube experiments led to the characterization of both protons and electrons.
b. Canal rays are negatively charged particles produced in a discharge tube.
c. Metal foil experiments led to the discovery of neutrons.
d. Many different types of cathode rays are known.

3.6 Indicate whether each of the following statements about discharge tube experiments and metal foil experiments is *true* or *false*.

a. In metal foil experiments, nearly all of the bombarding particles were stopped by the metal foil.
b. Metal foil experiments led to the concept that an atom has a nucleus.
c. Cathode rays and canal rays move in opposite directions in a discharge tube.
d. Many different types of canal rays have been observed, but only one type of cathode ray is known.

Atomic Number and Mass Number (Section 3.3)

3.7 Determine the atomic number and mass number for atoms with the following subatomic makeups.

a. 2 protons, 2 neutrons, and 2 electrons
b. 4 protons, 5 neutrons, and 4 electrons
c. 5 protons, 4 neutrons, and 5 electrons
d. 28 protons, 30 neutrons, and 28 electrons

3.8 Determine the atomic number and mass number for atoms with the following subatomic makeups.

a. 1 proton, 1 neutron, and 1 electron
b. 10 protons, 12 neutrons, and 10 electrons
c. 12 protons, 10 neutrons, and 12 electrons
d. 50 protons, 69 neutrons, and 50 electrons

3.9 Determine the number of protons, neutrons, and electrons present in atoms with the following characteristics.

a. Atomic number = 8 and mass number = 16
b. Mass number = 18 and $Z = 8$
c. Atomic number = 20 and $A = 44$
d. $A = 257$ and $Z = 100$

3.10 Determine the number of protons, neutrons, and electrons present in atoms with the following characteristics.

a. Atomic number = 10 and mass number = 20
b. Mass number = 110 and $Z = 48$
c. $A = 11$ and atomic number = 5
d. $Z = 92$ and $A = 238$

3.11 Arrange the following atoms in the orders specified.

$^{32}_{16}S$ $^{40}_{18}Ar$ $^{35}_{17}Cl$ $^{37}_{19}K$

a. Order of increasing atomic number
b. Order of decreasing mass number
c. Order of increasing number of electrons
d. Order of increasing number of neutrons

3.12 Arrange the following atoms in the orders specified.

$^{14}_{6}C$ $^{17}_{8}O$ $^{13}_{7}N$ $^{19}_{9}F$

a. Order of decreasing atomic number
b. Order of increasing mass number
c. Order of decreasing number of neutrons
d. Order of increasing number of nucleons

3.13 Determine the number of protons, neutrons, electrons, nucleons, and total subatomic particles for each of the following atoms.

a. $^{53}_{24}Cr$ b. $^{256}_{101}Md$ c. $^{67}_{30}Zn$ d. $^{40}_{20}Ca$

3.14 Determine the number of protons, neutrons, electrons, nucleons, and total subatomic particles for each of the following atoms.

a. $^{103}_{44}Ru$ b. $^{34}_{16}S$ c. $^{9}_{4}Be$ d. $^{4}_{2}He$

Isotopes (Section 3.3)

3.15 With the help of the information listed on the inside front cover, write complete symbols for the five naturally occurring isotopes of zirconium, given that the heaviest isotope has a mass number of 96 and that the other isotopes have, respectively, 2, 4, 5, and 6 fewer neutrons.

3.16 With the help of the information listed on the inside front cover, write complete symbols for the four naturally occurring isotopes of strontium, given that the lightest isotope has a mass number of 84 and that the other isotopes have, respectively, 2, 4, and 5 more neutrons.

3.17 Indicate whether each of the following statements about sodium isotopes is true or false.

a. $^{23}_{11}Na$ has one more electron than $^{24}_{11}Na$.
b. $^{23}_{11}Na$ and $^{24}_{11}Na$ contain the same number of neutrons.
c. $^{23}_{11}Na$ has one less subatomic particle than $^{24}_{11}Na$.
d. $^{23}_{11}Na$ and $^{24}_{11}Na$ have the same atomic number.

3.18 Indicate whether each of the following statements about magnesium isotopes is true or false.

a. $^{24}_{12}Mg$ has one more proton than $^{25}_{12}Mg$.
b. $^{24}_{12}Mg$ and $^{25}_{12}Mg$ contain the same number of subatomic particles.
c. $^{24}_{12}Mg$ has one less neutron than $^{25}_{12}Mg$.
d. $^{24}_{12}Mg$ and $^{25}_{12}Mg$ have different mass numbers.

3.19 The following are selected properties for the most abundant isotope of a particular element. Which of these properties would also be the same for the second-most-abundant isotope of the element?

a. Mass number is 70
b. 31 electrons are present
c. Isotopic mass is 69.92 amu
d. Isotope reacts with chlorine to give a green compound

3.20 The following are selected properties for the most abundant isotope of a particular element. Which of these properties would also be the same for the second-most-abundant isotope of the element?

a. Atomic number is 31
b. Does not react with the element gold
c. 40 neutrons are present
d. Density is 1.03 g/mL

Atomic Mass (Section 3.3)

3.21 A certain isotope of silver is 8.91 times heavier than $^{12}_{6}C$. What is the mass of this silver isotope on the amu scale?

3.22 A certain isotope of silicon is 2.42 times heavier than $^{12}_{6}C$. What is the mass of this silicon isotope on the amu scale?

3.23 Calculate the atomic mass of each of the following elements using the given data for the percentage abundance and mass of each isotope.

a. Lithium: 7.42% ^{6}Li (6.01 amu) and 92.58% ^{7}Li (7.02 amu)
b. Magnesium: 78.99% ^{24}Mg (23.99 amu), 10.00% ^{25}Mg (24.99 amu), and 11.01% ^{26}Mg (25.98 amu)

3.24 Calculate the atomic mass of each of the following elements using the given data for the percentage abundance and mass of each isotope.

a. Silver: 51.82% ^{105}Ag (106.9 amu) and 48.18% ^{107}Ag (108.9 amu)
b. Silicon: 92.21% ^{28}Si (27.98 amu), 4.70% ^{29}Si (28.98 amu), and 3.09% ^{30}Si (29.97 amu)

3.25 The arbitrary standard for the atomic mass scale is the exact number 12 for the mass of $^{12}_{6}C$. Why, then, is the atomic mass of carbon listed as 12.011?

3.26 The atomic mass of fluorine is 18.998 amu, and the atomic mass of iron is 55.847 amu. All fluorine atoms have a mass of 18.998 amu, and not a single iron atom has a mass of 55.847 amu. Explain.

The Periodic Law and the Periodic Table (Section 3.4)

3.27 Give the symbol of the element that occupies each of the following positions in the periodic table.

a. Period 4, Group IIA
b. Period 5, Group VIB
c. Group IA, Period 2
d. Group IVA, Period 5

3.28 Give the symbol of the element that occupies each of the following positions in the periodic table.

a. Period 1, Group IA
b. Period 6, Group IB
c. Group IIIB, Period 4
d. Group VIIA, Period 3

3.29 For each of the following sets of elements, choose the two that would be expected to have similar chemical properties.

a. $_{19}K$, $_{29}Cu$, $_{37}Rb$, $_{41}Nb$
b. $_{13}Al$, $_{14}Si$, $_{15}P$, $_{33}As$
c. $_{9}F$, $_{40}Zr$, $_{50}Sn$, $_{53}I$
d. $_{11}Na$, $_{12}Mg$, $_{54}Xe$, $_{55}Cs$

3.30 For each of the following sets of elements, choose the two that would be expected to have similar chemical properties.

a. $_{11}Na$, $_{14}Si$, $_{23}V$, $_{55}Cs$
b. $_{13}Al$, $_{19}K$, $_{32}Ge$, $_{50}Sn$
c. $_{37}Rb$, $_{38}Sr$, $_{54}Xe$, $_{56}Ba$
d. $_{2}He$, $_{6}C$, $_{8}O$, $_{10}Ne$

3.31 The following statements either define or are closely related to the terms *periodic law, period,* and *group.* Match each statement with the appropriate term.

a. This is a vertical arrangement of elements in the periodic table.
b. The properties of the elements repeat in a regular way as atomic numbers increase.

c. The chemical properties of elements 12, 20, and 38 demonstrate this principle.

d. Carbon is the first member of this arrangement.

3.32 The following statements either define or are closely related to the terms *periodic law, period,* and *group*. Match each statement with the appropriate term.

a. This is a horizontal arrangement of elements in the periodic table.

b. Element 19 begins this arrangement in the periodic table.

c. Elements 24 and 33 belong to this arrangement.

d. Elements 10, 18, and 36 belong to this arrangement.

3.33 Classify each of the following elements as a *halogen, noble gas, alkali metal,* or *alkaline earth metal.*

a. Cs b. K c. Ne d. Mg
e. F f. Ar g. Ca h. I

3.34 Classify each of the following elements as a *halogen, noble gas, alkali metal,* or *alkaline earth metal.*

a. Br b. Na c. Li d. Xe
e. Be f. Ba g. Kr h. Cl

Metals and Nonmetals (Section 3.5)

3.35 In which of the following pairs of elements are both members of the pair metals?

a. ${}_{17}Cl$ and ${}_{35}Br$ b. ${}_{13}Al$ and ${}_{14}Si$
c. ${}_{29}Cu$ and ${}_{42}Mo$ d. ${}_{30}Zn$ and ${}_{83}Bi$

3.36 In which of the following pairs of elements are both members of the pair metals?

a. ${}_{7}N$ and ${}_{34}Se$ b. ${}_{16}S$ and ${}_{48}Cd$
c. ${}_{3}Li$ and ${}_{26}Fe$ d. ${}_{50}Sn$ and ${}_{53}I$

3.37 Identify the nonmetal in each of the following sets of elements.

a. S, Na, K b. Cu, Li, P
c. Be, I, Ca d. Fe, Cl, Ga

3.38 Identify the nonmetal in each of the following sets of elements.

a. Al, H, Mg b. C, Sn, Sb
c. Ti, V, F d. Sr, Se, Sm

Electron Arrangements Within Atoms (Section 3.6)

3.39 The following statements define or are closely related to the terms *shell, subshell,* and *orbital*. Match each statement with the appropriate term.

a. In terms of electron capacity, this unit is the smallest of the three.

b. This unit can contain a maximum of two electrons.

c. This unit is designated just by a number.

d. The term *energy level* is closely associated with this unit.

3.40 The following statements define or are closely related to the terms *shell, subshell,* and *orbital*. Match each statement with the appropriate term.

a. This unit can contain as many electrons as, or more electrons than, either of the other two.

b. The term *energy sublevel* is closely associated with this unit.

c. Electrons that occupy this unit do not need to have identical energies.

d. The unit is designated in the same way as the orbitals contained within it.

3.41 Indicate whether each of the following statements is true or false.

a. An orbital has a definite size and shape, which are related to the energy of the electrons it could contain.

b. All the orbitals in a subshell have the same energy.

c. All subshells accommodate the same number of electrons.

d. A 2*p* subshell and a 3*p* subshell contain the same number of orbitals.

3.42 Indicate whether each of the following statements is true or false.

a. All the subshells in a shell have the same energy.

b. An *s* orbital has a shape that resembles a four-leaf clover.

c. The third shell can accommodate a maximum number of 18 electrons.

d. All orbitals accommodate the same number of electrons.

3.43 Give the maximum number of electrons that can occupy each of the following electron-accommodating units.

a. One of the orbitals in the 2*p* subshell

b. One of the orbitals in the 3*d* subshell

c. The 4*p* subshell

d. The third shell

3.44 Give the maximum number of electrons that can occupy each of the following electron-accommodating units.

a. One of the orbitals in the 4*d* subshell

b. One of the orbitals in the 5*f* subshell

c. The 3*d* subshell

d. The second shell

Specification of Electronic Structure (Section 3.7)

3.45 Write complete electron configurations for atoms of each of the following elements.

a. ${}_{6}C$ b. ${}_{11}Na$ c. ${}_{16}S$ d. ${}_{18}Ar$

3.46 Write complete electron configurations for atoms of each of the following elements.

a. ${}_{10}Ne$ b. ${}_{13}Al$ c. ${}_{19}K$ d. ${}_{22}Ti$

3.47 Write *complete* electron configurations for atoms whose electron configurations *end* as follows.

a. $3p^5$ b. $4d^7$ c. $4s^2$ d. $3d^1$

3.48 Write *complete* electron configurations for atoms whose electron configurations *end* as follows.

a. $4p^2$ b. $3d^{10}$ c. $5s^1$ d. $4p^6$

3.49 Draw the orbital diagram associated with each of the following electron configurations.

a. $1s^22s^22p^2$ b. $1s^22s^22p^63s^2$
c. $1s^22s^22p^63s^23p^3$ d. $1s^22s^22p^63s^23p^64s^23d^7$

3.50 Draw the orbital diagram associated with each of the following electron configurations.

a. $1s^22s^22p^5$ b. $1s^22s^22p^63s^1$
c. $1s^22s^22p^63s^23p^1$ d. $1s^22s^22p^63s^23p^64s^23d^5$

3.51 How many unpaired electrons are present in the orbital diagram for each of the following elements?

a. ${}_{7}N$ b. ${}_{12}Mg$ c. ${}_{17}Cl$ d. ${}_{25}Mn$

3.52 How many unpaired electrons are present in the orbital diagram for each of the following elements?

a. ${}_{9}F$ b. ${}_{16}S$ c. ${}_{20}Ca$ d. ${}_{30}Zn$

Electron Configurations and the Periodic Law (Section 3.8)

3.53 Indicate whether the elements represented by the given pairs of electron configurations have similar chemical properties.

a. $1s^22s^1$ and $1s^22s^2$

b. $1s^22s^22p^6$ and $1s^22s^22p^63s^23p^6$

c. $1s^22s^22p^3$ and $1s^22s^22p^63s^23p^64s^23d^3$

d. $1s^22s^22p^63s^23p^4$ and $1s^22s^22p^63s^23p^64s^23d^{10}4p^4$

3.54 Indicate whether the elements represented by the given pairs of electron configurations have similar chemical properties.

a. $1s^22s^22p^4$ and $1s^22s^22p^5$

b. $1s^22s^2$ and $1s^22s^22p^2$

c. $1s^22s^1$ and $1s^22s^22p^63s^23p^64s^1$

d. $1s^22s^22p^6$ and $1s^22s^22p^63s^23p^64s^23d^6$

Electron Configurations and the Periodic Table (Section 3.8)

3.55 For each of the following elements, specify the extent to which the subshell containing the distinguishing electron is filled (s^2, p^3, p^5, d^4, etc.).

a. ${}_{13}Al$ b. ${}_{23}V$ c. ${}_{20}Ca$ d. ${}_{36}Kr$

3.56 For each of the following elements, specify the extent to which the subshell containing the distinguishing electron is filled (s^2, p^3, p^5, d^4, etc.).

a. ${}_{10}Ne$ b. ${}_{19}K$ c. ${}_{33}As$ d. ${}_{30}Zn$

Classification of the Elements (Section 3.8)

3.57 Classify each of the following elements as a noble gas, representative element, transition element, or inner transition element.

a. ${}_{15}P$ b. ${}_{18}Ar$ c. ${}_{79}Au$ d. ${}_{92}U$

3.58 Classify each of the following elements as a noble gas, representative element, transition element, or inner transition element.

a. ${}_{1}H$ b. ${}_{44}Ru$ c. ${}_{51}Sb$ d. ${}_{86}Rn$

3.59 Classify the element with each of the following electron configurations as a representative element, transition element, noble gas, or inner transition element.

a. $1s^22s^22p^6$ b. $1s^22s^22p^63s^23p^4$

c. $1s^22s^22p^63s^23p^64s^23d^1$ d. $1s^22s^22p^63s^23p^64s^2$

3.60 Classify the element with each of the following electron configurations as a representative element, transition element, noble gas, or inner transition element.

a. $1s^22s^22p^63s^1$ b. $1s^22s^22p^63s^23p^6$

c. $1s^22s^22p^63s^23p^64s^23d^7$ d. $1s^22s^22p^63s^23p^64s^23d^{10}4p^5$

Additional Problems

3.61 Write complete symbols (${}^{A}_{Z}E$), with the help of a periodic table, for atoms with the following characteristics.

a. Contains 20 electrons and 24 neutrons

b. Radon atom with a mass number of 211

c. Silver atom that contains 157 subatomic particles

d. Beryllium atom that contains 9 nucleons

3.62 Characterize each of the following pairs of atoms as containing (1) the same number of neutrons, (2) the same number of electrons, or (3) the same total number of subatomic particles.

a. ${}^{13}_{6}C$ and ${}^{14}_{7}N$ b. ${}^{18}_{8}O$ and ${}^{19}_{9}F$

c. ${}^{37}_{17}Cl$ and ${}^{36}_{18}Ar$ d. ${}^{35}_{17}Cl$ and ${}^{37}_{17}Cl$

3.63 Write the complete symbol (${}^{A}_{Z}E$) for the isotope of chromium with each of the following characteristics.

a. Two more neutrons than ${}^{55}_{24}Cr$

b. Two fewer subatomic particles than ${}^{52}_{24}Cr$

c. The same number of neutrons as ${}^{60}_{29}Cu$

d. The same number of subatomic particles as ${}^{60}_{29}Cu$

3.64 How many electrons are present in nine molecules of the compound $C_{12}H_{22}O_{11}$ (table sugar)?

3.65 Which of the six elements nitrogen, beryllium, argon, aluminum, silver, and gold belong(s) in each of the following classifications?

a. Period and Roman numeral group numbers are numerically equal

b. Readily conducts electricity and heat

c. Has an atomic mass greater than its atomic number

d. All atoms have a nuclear charge greater than +20

3.66 The electron configuration of the isotope ${}^{16}_{8}O$ is $1s^22s^22p^4$. What is the electron configuration for the isotope ${}^{18}_{8}O$?

3.67 Write electron configurations for the following elements.

a. The Group IIIA element in the same period as ${}_{4}Be$

b. The Period 3 element in the same group as ${}_{5}B$

c. The lowest-atomic-numbered metal in Group IA

d. The Period 3 element that has three unpaired electrons

3.68 Referring only to the periodic table, determine the element of lowest atomic number whose electron configuration contains each of the following.

a. Three completely filled orbitals

b. Three completely filled subshells

c. Three completely filled shells

d. Three completely filled *s* subshells

Grid Problems

3.69

1. ${}^{28}_{14}Si$	2. ${}^{31}_{15}P$	3. ${}^{35}_{17}Cl$
4. ${}^{36}_{18}Ar$	5. ${}^{38}_{18}Ar$	6. ${}^{39}_{19}K$

Select from the grid *all* correct responses for each of the following situations.

a. Atoms that contain an equal number of protons and electrons

b. Atoms that contain more neutrons than electrons

c. Atoms that contain equal numbers of all three kinds of subatomic particles

d. Pairs of atoms that contain the same number of neutrons

3.70

1. beryllium	2. aluminum	3. scandium
4. arsenic	5. silver	6. tellurium

Select from the grid *all* correct responses for each of the following situations.

a. Elements whose Roman numeral group number and period number are numerically equal
b. Transition elements
c. Metallic representative elements
d. Pairs of elements that are in the same period of the periodic table

3.71

1. N	2. F	3. Al
4. P	5. K	6. Ca

Select from the grid *all* correct responses for each of the following situations.

a. Is chemically similar to the element Cl
b. Has an electron configuration that ends in p^3
c. Has an electron configuration that contains the notation $2p^6$
d. Has an orbital diagram that shows unpaired electrons present

3.72

1. $1s^22s^2$	2. $1s^22s^22p^2$	3. $1s^22s^22p^4$
4. $1s^22s^22p^6$	5. $1s^22s^22p^63s^1$	6. $1s^22s^22p^63s^2$

Select from the grid *all* correct responses for each of the following situations.

a. Electron configuration in which all occupied subshells are completely filled
b. Electron configuration in which electrons are present in three different shells
c. Electron configuration in which four different subshells contain electrons
d. Electron configuration in which four different orbitals contain electrons

CHAPTER 4

PART I

Chemical Bonding: The Ionic Bond Model

CHAPTER OUTLINE

4.1 Chemical Bonds 79
4.2 Valence Electrons and Lewis Structures 80
4.3 The Octet Rule 82
4.4 The Ionic Bond Model 83
4.5 Determination of Ionic Charge Magnitude 84
4.6 Ionic Compound Formation 86
4.7 Formulas for Ionic Compounds 87
4.8 The Structure of Ionic Compounds 88
4.9 Naming Binary Ionic Compounds 90

Chemistry at a Glance: **Ionic Bonds and Ionic Compounds** 93

4.10 Polyatomic Ions 94
4.11 Formulas and Names for Ionic Compounds Containing Polyatomic Ions 95

Chemistry at a Glance: **Nomenclature of Ionic Compounds** 96

Chemical Connections
4.1 Fresh Water, Seawater, Hard Water, and Soft Water: A Matter of Ions 85
4.2 Sodium Chloride: The World's Most Common Food Additive 89

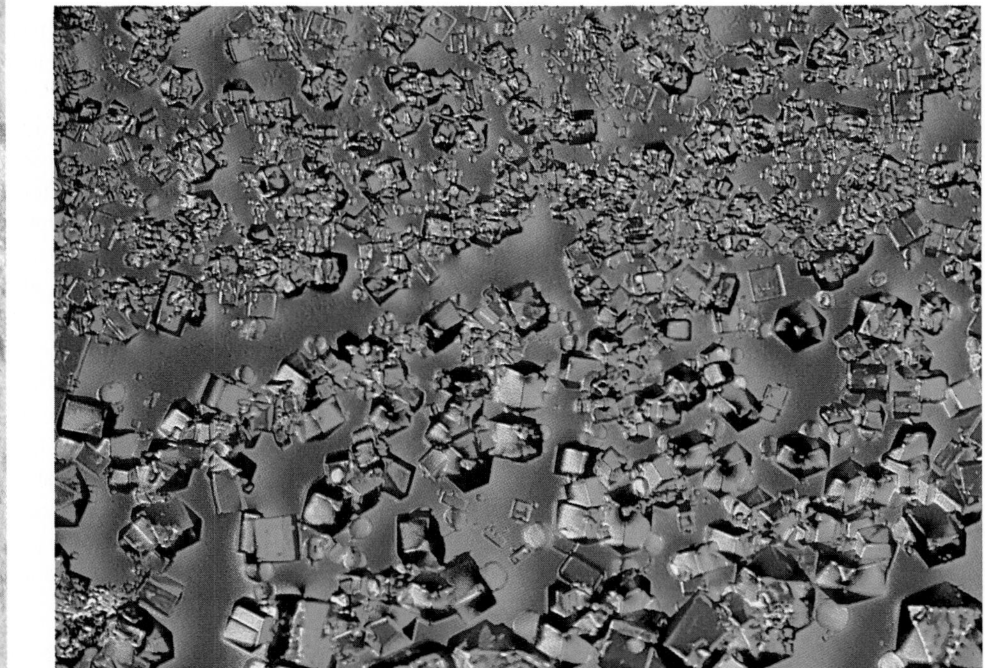

Magnification of crystals of sodium chloride (table salt), one of the most commonly encountered ionic compounds.

As scientists study living organisms and the world in which we live, they rarely encounter free isolated atoms. Instead, under normal conditions of temperature and pressure, they nearly always find atoms associated in aggregates or clusters ranging in size from two atoms to numbers too large to count. In this chapter, we will explain why atoms tend to join together in larger units, and we will discuss the binding forces (chemical bonds) that hold them together.

As we examine the nature of attractive forces between atoms, we will discover that both the tendency and the capacity of an atom to be attracted to other atoms are dictated by its electron configuration.

4.1 Chemical Bonds

Chemical compounds are conveniently divided into two broad classes called *ionic compounds* and *molecular compounds.* Ionic and molecular compounds can be distinguished from each other on the basis of general physical properties. Ionic compounds tend to have high melting points (500°C − 2000°C) and are good conductors of electricity when they are in a molten (liquid) state or in solution. Molecular compounds, on the other hand, generally have much lower melting points and tend to be gases, liquids, or low-melting solids. They do not conduct electricity in the molten state. Ionic compounds, unlike molecular compounds, do not have molecules as their basic structural unit. Instead, an extended array of positively and negatively charged particles called *ions* is present (Section 4.8).

Some combinations of elements produce ionic compounds, whereas other combinations of elements form molecular compounds. What determines whether the interaction of two elements produces ions (an ionic compound) or molecules (a molecular compound)? To answer this question, we need to learn about chemical bonds. **Chemical bonds** *are the attractive forces that hold atoms together in more complex units.* Chemical bonds form as a result of interactions between electrons found in the combining atoms. Thus the nature of chemical bonds is closely linked to electron configurations (Section 3.7).

• Another designation for *molecular compound* is *covalent compound.* The two designations are used interchangeably. The modifier *molecular* draws attention to the basic structural unit present (the molecule), and the modifier *covalent* focuses on the mode of bond formation (electron sharing).

Corresponding to the two broad categories of chemical compounds are two types of chemical attractive forces (chemical bonds): ionic bonds and covalent bonds. An **ionic bond** *results from the transfer of one or more electrons from one atom or group of atoms to another.* As its name suggests, the ionic bond model (electron transfer) is used in describing the attractive forces in ionic compounds. A **covalent bond** *results from the sharing of one or more pairs of electrons between atoms.* The covalent bond model (electron sharing) is used in describing the attractions between atoms in molecular compounds.

• *Purely* ionic bonds involve a complete transfer of electrons from one atom to another. *Purely* covalent bonds involve equal sharing of electrons. Experimentally, it is found that most actual bonds have some degree of both ionic and covalent character. The exceptions are bonds between identical atoms; here, the bonding is purely covalent.

Even before we consider the details of these two bond models, it is important to emphasize that the concepts of ionic and covalent bonds are actually "convenience concepts." Most bonds are not 100% ionic or 100% covalent. Instead, most bonds have some degree of both ionic and covalent character—that is, some degree of both the transfer and the sharing of electrons. However, it is easiest to understand these intermediate bonds (the real bonds) by relating them to the pure or ideal bond types called ionic and covalent.

Two fundamental concepts are necessary for understanding both the ionic and the covalent bonding models.

1. Not all electrons in an atom are available for bonding. Those that are available are called *valence electrons.*
2. Certain arrangements of electrons are more stable than others, as is explained by the *octet rule.*

Section 4.2 addresses the concept of valence electrons, and Section 4.3 discusses the octet rule.

4.2 Valence Electrons and Lewis Structures

Certain electrons, called valence electrons, are particularly important in determining the bonding characteristics of a given atom. For representative and noble-gas elements, **valence electrons** *are the electrons in the outermost electron shell, which is the shell with the highest shell number (n).* Valence electrons are always found in either *s* or *p* subshells. Note the restriction on the use of this definition; it applies only to representative and noble-gas elements. Because most of the common elements are representative elements, this definition is quite useful. (We will not consider in this text the more complicated valence electron definitions for transition or inner transition elements; here, the presence of incompletely filled *inner d* or *f* subshells is a complicating factor.)

• The term *valence* is derived from the Latin word *valentia,* which means "capacity" (to form bonds).

The number of valence electrons in an atom of a representative element can be determined from the atom's electron configuration, as is illustrated in Example 4.1.

Scientists have developed a shorthand system for designating the number of valence electrons present in atoms of an element. This system involves the use of Lewis structures. A **Lewis structure** *consists of an element's symbol with one dot for each valence*

Example 4.1

Determining the Number of Valence Electrons in an Atom

Determine the number of valence electrons in atoms of each of the following elements.

a. ${}_{12}Mg$ **b.** ${}_{14}Si$ **c.** ${}_{33}As$

Solution

a. The element magnesium has two valence electrons, as can be seen by examining its electron configuration.

$$1s^2 2s^2 2p^6 \textcircled{3}s^{\textcircled{2}}$$

Number of valence electrons (points to the circled 2); Highest value of the electron shell number (points to the circled 3)

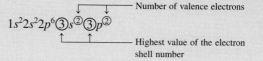

The highest value of the electron shell number is $n = 3$. Only two electrons are found in shell 3: the two electrons in the 3s subshell.

b. The element silicon has four valence electrons

$$1s^2 2s^2 2p^6 \textcircled{3}s^{\textcircled{2}} \textcircled{3}p^{\textcircled{2}}$$

Number of valence electrons; Highest value of the electron shell number

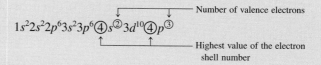

Electrons in two different subshells can simultaneously be valence electrons. The highest shell number is 3, and both the 3s and the 3p subshells belong to this shell. Hence all of the electrons in both of these subshells are valence electrons.

c. The element arsenic has five valence electrons.

$$1s^2 2s^2 2p^6 3s^2 3p^6 \textcircled{4}s^{\textcircled{2}} 3d^{10} \textcircled{4}p^{\textcircled{3}}$$

Number of valence electrons; Highest value of the electron shell number

The 3d electrons are not counted as valence electrons because the 3d subshell is in shell 3, and this shell does not have maximum n value. Shell 4 is the outermost shell and has maximum n value.

Practice Exercise 4.1

Determine the number of valence electrons in atoms of each of the following elements.

a. ${}_{11}Na$ **b.** ${}_{16}S$ **c.** ${}_{35}Br$

- *Answers.* **a.** 1; **b.** 6; **c.** 7

electron placed around the elemental symbol. Lewis structures for the first 20 elements (all representative or noble-gas elements), arranged as in the periodic table, are given in Figure 4.1. Lewis structures, named in honor of the American chemist Gilbert N. Lewis (1875–1946), who first introduced them, are also frequently called *electron-dot structures.*

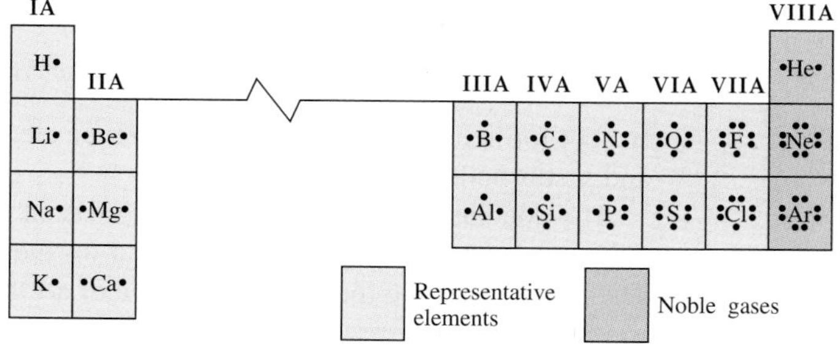

Figure 4.1
Lewis structures for selected representative and noble-gas elements.

Gilbert Newton Lewis (1875–1946), one of the foremost chemists of the twentieth century, made significant contributions in other areas of chemistry besides his pioneering work in describing chemical bonding. He formulated a generalized theory for describing acids and bases and was the first to isolate deuterium (heavy hydrogen).

In positioning electrons in a Lewis structure, always place one electron on all four sides of the elemental symbol before putting a second electron on any side. Note that which side of the symbol the dot is placed on is not critical. The following notations all have the same meaning:

$$\dot{\text{Ca}}\cdot \qquad \underset{\cdot}{\text{Ca}}\cdot \qquad \cdot\dot{\text{Ca}} \qquad \cdot\text{Ca}\cdot$$

Three important generalizations about valence electrons can be drawn from a study of the structures shown in Figure 4.1.

1. *Representative elements in the same group of the periodic table have the same number of valence electrons.* This should not be surprising. Elements in the same group in the periodic table have similar chemical properties as a result of their similar outer-shell electron configurations (Section 3.8). The electrons in the outermost shell are the valence electrons.
2. *The number of valence electrons for representative elements in a group is the same as the Roman numeral periodic-table group number.* For example, the Lewis structures for oxygen and sulfur, which are both members of Group VIA, show six dots. Similarly, the Lewis structures of hydrogen, lithium, sodium, and potassium, which are all members of Group IA, show one dot.
3. *The maximum number of valence electrons for any element is eight.* Only the noble gases (Section 3.9), beginning with neon, have the maximum number of eight electrons. Helium, which has only two valence electrons, is the exception in the noble-gas family. Obviously, an element with a total of two electrons cannot have eight valence electrons. Although shells with n greater than 2 are capable of holding more than eight electrons, they do so only when they are no longer the outermost shell and are thus not the valence shell. For example, arsenic has 18 electrons in its third shell; however, shell 4 is the valence shell for arsenic.

4.3 The Octet Rule

A key concept in modern elementary bonding theory is that certain arrangements of valence electrons are more stable than others. The term *stable* as used here refers to the idea that a system, which in this case is an arrangement of electrons, does not easily undergo spontaneous change.

The valence electron configurations of the noble gases (helium, neon, argon, krypton, xenon, and radon) are considered the *most stable of all valence electron configurations.* All of the noble gases except helium possess eight valence electrons, which is the maximum number possible. Helium's valence electron configuration is $1s^2$. All of the other noble gases possess ns^2np^6 valence electron configurations, where n has the maximum value found in the atom.

He: $1s^2$
Ne: $1s^2 2s^2 2p^6$
Ar: $1s^2 2s^2 2p^6 3s^2 3p^6$
Kr: $1s^2 2s^2 2p^6 3s^2 3p^6 4s^2 3d^{10} 4p^6$
Xe: $1s^2 2s^2 2p^6 3s^2 3p^6 4s^2 3d^{10} 4p^6 5s^2 4d^{10} 5p^6$
Rn: $1s^2 2s^2 2p^6 3s^2 3p^6 4s^2 3d^{10} 4p^6 5s^2 4d^{10} 5p^6 6s^2 4f^{14} 5d^{10} 6p^6$

• The *outermost* electron shell of an atom is also called the *valence* electron shell.

Except for helium, all the noble-gas valence electron configurations have the outermost s and p subshells *completely filled.*

The conclusion that an ns^2np^6 configuration ($1s^2$ for helium) is the most stable of all valence electron configurations is based on the chemical properties of the noble gases. The noble gases are the *most unreactive* of all the elements. They are the only elemental gases found in nature in the form of individual uncombined atoms. There are no known compounds of helium, neon, and argon, and only a very few compounds of krypton, xenon, and radon are known. The noble gases have little or no tendency to form bonds to other atoms.

• Some compounds exist whose formulation is not consistent with the octet rule, but the vast majority of simple compounds have formulas that are consistent with its precepts.

Atoms of many elements that lack this very stable noble-gas valence electron configuration tend to acquire it through chemical reactions that result in compound formation. This observation is known as the **octet rule:** *In compound formation, atoms of elements lose, gain, or share electrons in such a way that their electron configurations become identical to that of the noble gas nearest them in the periodic table.*

4.4 The Ionic Bond Model

• The word *ion* is pronounced "eye-on."

Electron transfer between two or more atoms is central to the ionic bond model. This electron transfer process produces charged particles called ions. An **ion** *is an atom (or group of atoms) that is electrically charged as a result of the loss or gain of electrons.* An atom is neutral when the number of protons (positive charges) is equal to the number of electrons (negative charges). Loss or gain of electrons destroys this proton–electron balance and leaves a net charge on the atom.

• An atom's nucleus *never* changes during the process of ion formation. The number of neutrons and protons remains constant.

If an atom *gains* one or more electrons, it becomes a *negatively* charged ion; excess negative charge is present because electrons outnumber protons. If an atom *loses* one or more electrons, it becomes a *positively* charged ion; more protons are present than electrons. There is excess positive charge (Figure 4.2). Note that the excess positive charge associated with a positive ion is never caused by proton gain but always by electron loss. If the number of protons remains constant and the number of electrons decreases, the result is net positive charge. The number of protons, which determines the identity of an element, never changes during ion formation.

• A loss of electrons by an atom always produces a positive ion. A gain of electrons by an atom always produces a negative ion.

The charge on an ion depends on the number of electrons that are lost or gained. Loss of one, two, or three electrons gives ions with +1, +2, and +3 charges, respectively. A gain of one, two, or three electrons gives ions with −1, −2, and −3 charges, respectively. (Ions that have lost or gained more than three electrons are very seldom encountered.)

The notation for charges on ions is a superscript placed to the right of the elemental symbol. Some examples of ion symbols are:

Positive ions: Na^+, K^+, Ca^{2+}, Mg^{2+}, Al^{3+}

Negative ions: Cl^-, Br^-, O^{2-}, S^{2-}, N^{3-}

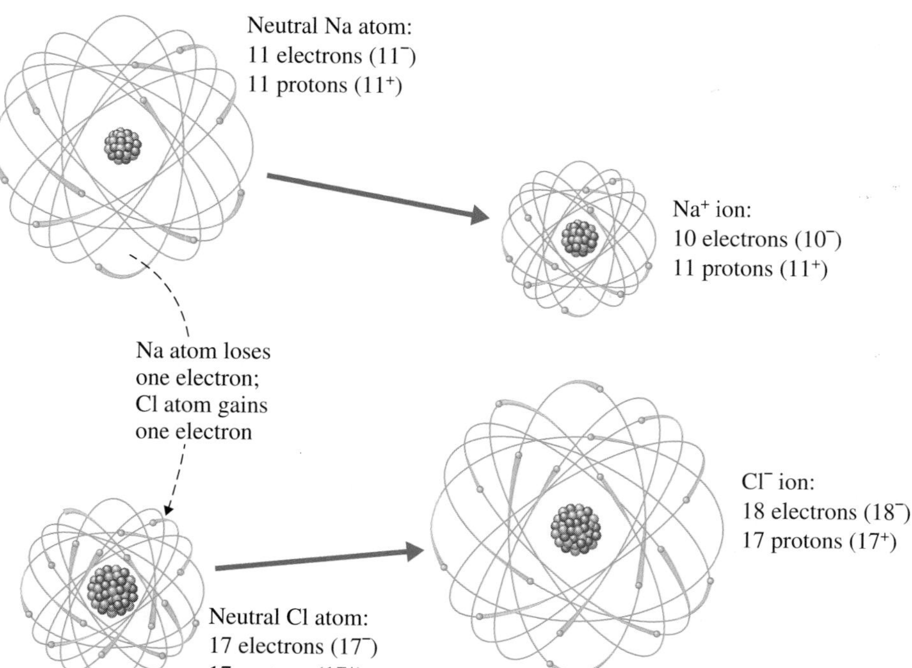

Figure 4.2
Loss of an electron from a sodium atom leaves it with one more proton than electrons, so it has a net electrical charge of +1. When chlorine gains an electron, it has one more electron than protons, so it has a net electrical charge of −1.

Note that we use a single plus or minus sign to denote a charge of 1, instead of using the notation $^{1+}$ or $^{1-}$. Also note that in multicharged ions, the number precedes the charge sign; that is, the notation for a charge of plus two is $^{2+}$ rather than $^{+2}$.

Example 4.2

Writing Symbols for Ions

Give the symbol for each of the following ions.

a. The ion formed when a barium atom loses two electrons.
b. The ion formed when a phosphorus atom gains three electrons.

Solution

a. A neutral barium atom contains 56 protons and 56 electrons because barium has an atomic number of 56. The barium ion formed by the loss of 2 electrons would still contain 56 protons but would have only 54 electrons because 2 electrons were lost.

$$\begin{array}{r} \text{56 protons} = \text{56 + charges} \\ \underline{\text{54 electrons} = \text{54 − charges}} \\ \text{Net charge} = 2+ \end{array}$$

The symbol of the barium ion is thus Ba^{2+}.

b. The atomic number of phosphorus is 15. Thus 15 protons and 15 electrons are present in a neutral phosphorus atom. A gain of 3 electrons raises the electron count to 18.

$$\begin{array}{r} \text{15 protons} = \text{15 + charges} \\ \underline{\text{18 electrons} = \text{18 − charges}} \\ \text{Net charge} = 3- \end{array}$$

The symbol for the ion is P^{3-}.

Practice Exercise 4.2

Give the symbol for each of the following ions.

a. The ion formed when cesium loses one electron.
b. The ion formed when selenium gains two electrons.

• *Answers.* **a.** Cs^+; **b.** Se^{2-}

The chemical properties of a particle (atom or ion) depend on its electron arrangement. Because an ion has a different electron configuration (fewer or more electrons) from the atom from which it was formed, it has different chemical properties as well. For example, the drug many people call lithium, which is used to treat mental illness (manic-depressive symptoms), does not involve lithium (Li, the element) but rather lithium ions (Li^+). The element lithium, if ingested, would be poisonous and possibly fatal. The lithium ion, ingested in the form of lithium carbonate, has entirely different effects on the human body.

4.5 Determination of Ionic Charge Magnitude

The octet rule provides a very simple and straightforward explanation for the charge magnitude associated with ions of the representative elements. *Atoms tend to gain or lose electrons until they have obtained an electron configuration that is the same as that of a noble gas.* The element sodium has the electron configuration

$$1s^22s^22p^63s^1$$

One valence electron is present. Sodium can attain a noble-gas electron configuration by losing this valence electron (to give it the electron configuration of neon) or by gaining seven electrons (to give it the electron configuration of argon).

$$Na\ (1s^22s^22p^63s^1) \xrightarrow{\text{Loss of 1 e}^-} Na^+\ (1s^22s^22p^6)$$

Electron configuration of neon

$$Na\ (1s^22s^22p^63s^1) \xrightarrow{\text{Gain of 7 e}^-} Na^{7-}\ (1s^22s^22p^63s^23p^6)$$

Electron configuration of argon

The electron loss or gain that involves the fewest electrons will always be the more favorable process, from an energy standpoint, and will be the process that occurs. Thus for sodium, the loss of one electron to form the Na^+ ion is the process that occurs.

The element chlorine has the electron configuration

$$1s^22s^22p^63s^23p^5$$

Seven valence electrons are present. Chlorine can attain a noble-gas electron configuration by losing seven electrons (to give it the electron configuration of neon) or by gaining one electron (to give it the electron configuration of argon). The latter occurs for the reason we have cited.

$$Cl\ (1s^22s^22p^63s^23p^5) \xrightarrow{\text{Loss of 7 e}^-} Cl^{7+}\ (1s^22s^22p^6)$$

Electron configuration of neon

$$Cl\ (1s^22s^22p^63s^23p^5) \xrightarrow{\text{Gain of 1 e}^-} Cl^-\ (1s^22s^22p^63s^23p^6)$$

Electron configuration of argon

The considerations we have just applied to sodium and chlorine lead to the following generalizations:

1. Metal atoms containing one, two, or three valence electrons (the metals in Groups IA, IIA, and IIIA of the periodic table) tend to lose electrons to

Chemical CONNECTIONS

4.1 Fresh Water, Seawater, Hard Water, and Soft Water: A Matter of Ions

Water is the most abundant compound on the face of the earth. We encounter it everywhere we go: as water vapor in the air, as a liquid in rivers, lakes, and oceans, and as a solid (ice and snow) both on land and in the oceans.

All water as it occurs in nature is impure in a chemical sense. The impurities present include suspended matter, microbiological organisms, dissolved gases, and dissolved minerals. Minerals dissolved in water produce ions. For example, rock salt (NaCl) dissolves in water to produce Na^+ and Cl^- ions.

The major distinction between *fresh water* and *seawater* (salt water) is the number of ions present. On a relative scale, where the total concentration of ions in fresh water is assigned a value of 1, seawater has a value of approximately 500; that is, seawater has a concentration of dissolved ions 500 times greater than that of fresh water.

The dominant ions in fresh water and seawater are not the same. In seawater, Na^+ ion is the dominant positive ion and Cl^- ion is the dominant negative ion. This contrasts with fresh water, where Ca^{2+} and Mg^{2+} ions are the most abundant positive ions and HCO_3^- (a polyatomic ion; Section 4.10) is the most abundant negative ion.

When fresh water is purified for drinking purposes, suspended particles, disease-causing agents, and objectionable odors are removed. Dissolved ions are not removed. At the concentrations at which they are normally present in fresh water, dissolved ions are not harmful to health. Indeed, some of the taste of water is caused by the ions present; water without any ions present would taste "unpleasant" to most people.

Hard water is water that contains Ca^{2+}, Mg^{2+}, and Fe^{2+} ions. The presence of these ions does not affect the drinkability of water, but it does affect other uses for the water. The hard-water ions form insoluble compounds with soap (producing scum) and lead to the production of deposits of scale in steam boilers, tea kettles, and hot water pipes.

The most popular method for obtaining *soft water* from hard water involves the process of "ion exchange." In this process, the offending hard-water ions are exchanged for Na^+ ions. Sodium ions do not form insoluble soap compounds or scale. People with high blood pressure or kidney problems are often advised to avoid drinking soft water because of its high sodium content.

acquire a noble-gas electron configuration. The noble gas involved is the one preceding the metal in the periodic table.

Group IA metals form 1^+ ions.
Group IIA metals form 2^+ ions.
Group IIIA metals form 3^+ ions.

2. Nonmetal atoms containing five, six, or seven valence electrons (the nonmetals in Groups VA, VIA, and VIIA of the periodic table) tend to gain electrons to acquire a noble-gas electron configuration. The noble gas involved is the one following the nonmetal in the periodic table.

Group VIIA nonmetals form 1^- ions.
Group VIA nonmetals form 2^- ions.
Group VA nonmetals form 3^- ions.

• The positive charge on metal ions from Groups IA, IIA, and IIIA has a magnitude equal to the metal's periodic-table group number.

• Nonmetals from Groups VA, VIA, and VIIA form negative ions whose charge is equal to the group number minus 8. For example, S, in Group VIA, forms S^{2-} ions $(6 - 8 = -2)$.

Elements in Group IVA occupy unique positions relative to the noble gases. They would have to gain or lose four electrons to attain a noble-gas structure. Theoretically, ions with charges of +4 or −4 could be formed by elements in this group, but in most cases, these elements form covalent bonds instead, which are discussed in Chapter 5.

4.6 Ionic Compound Formation

Ion formation, through the loss or gain of electrons by atoms, is not an isolated, singular process. In reality, electron loss and electron gain are always partner processes; if one occurs, the other also occurs. Ion formation requires the presence of two elements: a metal that can donate electrons and a nonmetal that can accept electrons. The electrons lost by the metal are the same ones gained by the nonmetal. The positive and negative ions simultaneously formed from such *electron transfer* attract one another. The result is the formation of an ionic compound.

• No atom can lose electrons unless another atom is available to accept them.

Lewis structures (Section 4.2) are helpful in visualizing the formation of simple ionic compounds. The reaction between the element sodium (with one valence electron) and chlorine (with seven valence electrons) is represented as follows with such structures:

$$\text{Na}\cdot + \cdot\ddot{\underset{..}{\text{Cl}}}: \longrightarrow [\text{Na}]^+ \left[:\ddot{\underset{..}{\text{Cl}}}:\right]^- \longrightarrow \text{NaCl}$$

The loss of an electron by sodium empties its valence shell. The next inner shell, which contains eight electrons (a noble-gas configuration), then becomes the valence shell. After the valence shell of chlorine gains one electron, it has the needed eight valence electrons.

When sodium, which has one valence electron, combines with oxygen, which has six valence electrons, the oxygen atom requires the presence of two sodium atoms to acquire two additional electrons.

$$\begin{matrix}\text{Na}\cdot \\ \text{Na}\cdot\end{matrix} + \cdot\ddot{\underset{..}{\text{O}}}: \longrightarrow \begin{matrix}[\text{Na}]^+ \\ [\text{Na}]^+\end{matrix} \left[:\ddot{\underset{..}{\text{O}}}:\right]^{2-} \longrightarrow \text{Na}_2\text{O}$$

Note that because oxygen has room for two additional electrons, two sodium atoms are required per oxygen atom—hence the formula Na_2O.

An opposite situation occurs in the reaction between calcium, which has two valence electrons, and chlorine, which has seven valence electrons. Here, two chlorine atoms are required to accommodate electrons transferred from one calcium atom, because a chlorine atom can accept only one electron. (It has seven valence electrons and needs only one more.)

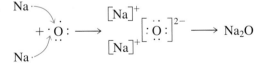

$$\text{Ca}: + \begin{matrix}\cdot\ddot{\underset{..}{\text{Cl}}}: \\ \cdot\ddot{\underset{..}{\text{Cl}}}:\end{matrix} \longrightarrow [\text{Ca}]^{2+} \begin{matrix}\left[:\ddot{\underset{..}{\text{Cl}}}:\right]^- \\ \left[:\ddot{\underset{..}{\text{Cl}}}:\right]^-\end{matrix} \longrightarrow \text{CaCl}_2$$

Many processed foods are high in compounds containing Na^+ ions, such as NaCl (table salt).

Example 4.3

Using Lewis Structures to Depict Ionic Compound Formation

Show the formation of the following ionic compounds using Lewis structures.

a. Na_3N **b.** MgO **c.** Al_2S_3

Solution

a. Sodium (a Group IA element) has one valence electron, which it would "like" to lose. Nitrogen (a Group VA element) has five valence electrons and would thus "like" to acquire three more. Three sodium atoms are needed to supply enough electrons for one nitrogen atom.

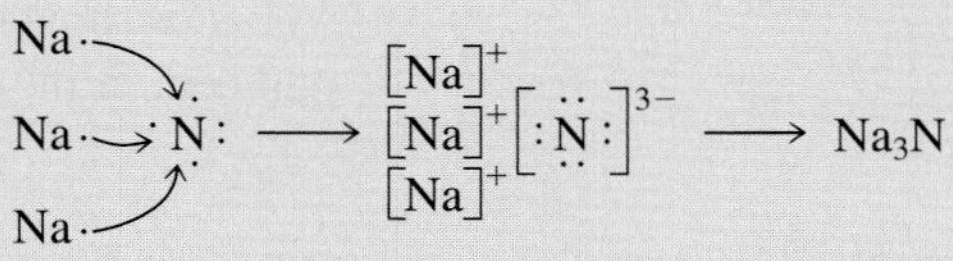

$$3\,Na\cdot + \cdot\ddot{N}: \longrightarrow [Na]^+[Na]^+[Na]^+[:\ddot{N}:]^{3-} \longrightarrow Na_3N$$

b. Magnesium (a Group IIA element) has two valence electrons, and oxygen (a Group VIA element) has six valence electrons. The transfer of the two magnesium valence electrons to an oxygen atom results in each atom having a noble-gas electron configuration. Thus these two elements combine in a one-to-one ratio.

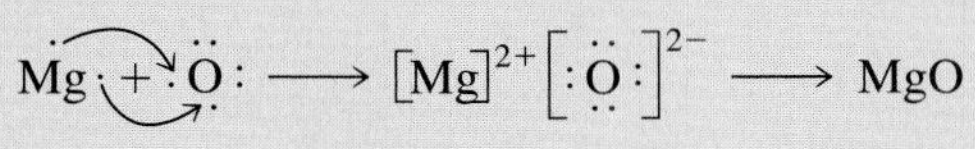

$$Mg: + \cdot\ddot{O}: \longrightarrow [Mg]^{2+}[:\ddot{O}:]^{2-} \longrightarrow MgO$$

c. Aluminum (a Group IIIA element) has three valence electrons, all of which need to be lost through electron transfer. Sulfur (a Group VIA element) has six valence electrons and thus needs to acquire two more. Three sulfur atoms are needed to accommodate the electrons given up by two aluminum atoms.

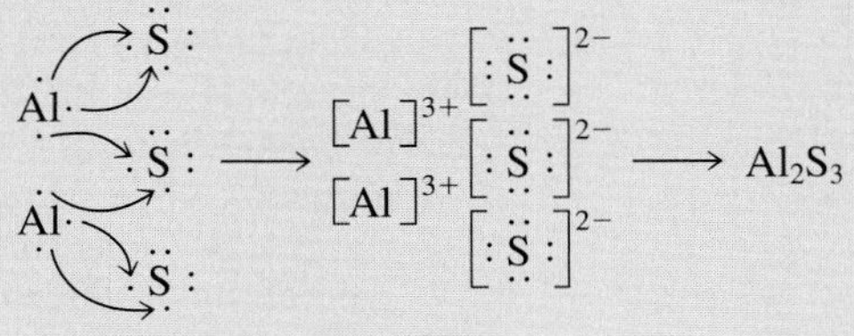

$$2\,Al\cdot + 3\,\cdot\ddot{S}: \longrightarrow [Al]^{3+}[Al]^{3+}[:\ddot{S}:]^{2-}[:\ddot{S}:]^{2-}[:\ddot{S}:]^{2-} \longrightarrow Al_2S_3$$

Practice Exercise 4.3

Show the formation of the following ionic compounds using Lewis structures.

a. KF **b.** Li_2O **c.** Ca_3P_2

• *Answers.* **a.** $K\cdot + \cdot\ddot{F}:$ **b.** $Li\cdot$, $Li\cdot$ → $\cdot\ddot{O}:$ **c.** $Ca\cdot$, $Ca\cdot$, $Ca\cdot$ → $\cdot\dot{P}:$, $\cdot\dot{P}:$

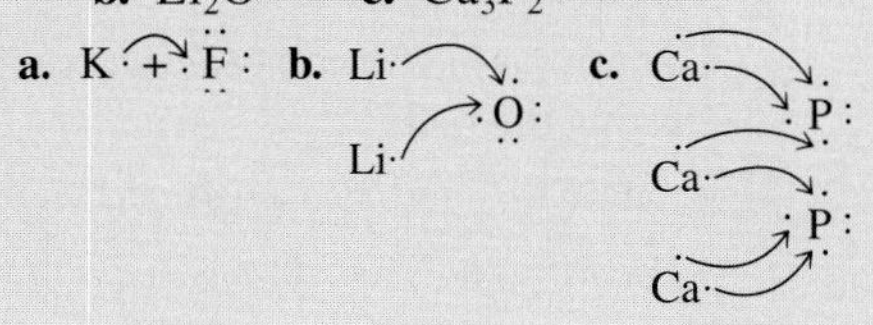

4.7 Formulas for Ionic Compounds

Electron loss always equals electron gain in an electron transfer process. Consequently, ionic compounds are always neutral; no net charge is present. The total positive charge present on the ions that have lost electrons always is exactly counterbalanced by the total negative charge on the ions that have gained electrons. Thus *the ratio in which positive and negative ions combine is the ratio that achieves charge neutrality for the resulting compound.* This generalization can be used, instead of Lewis structures, to determine ionic compound formulas. Ions are combined in the ratio that causes the positive and negative charges to add to zero.

The correct combining ratio when K^+ ions and S^{2-} ions combine is two to one. Two K^+ ions (each of +1 charge) will be required to balance the charge on a single S^{2-} ion.

$$\begin{array}{rr} 2(K^+): & (2 \text{ ions}) \times (\text{charge of } +1) = +2 \\ S^{2-}: & (1 \text{ ion}) \times (\text{charge of } -2) = -2 \\ \hline & \text{Net charge} = 0 \end{array}$$

The formula of the compound formed is thus K_2S.

There are three rules to remember when writing formulas for all ionic compounds.

1. The symbol for the positive ions is always written first.
2. The charges on the ions that are present are *not* shown in the formula. You need to know the charges to determine the formula; however, the charges are not explicitly shown in the formula.
3. The numbers in the formula (the subscripts) give the combining ratio for the ions.

Example 4.4

Using Ionic Charges to Determine the Formula of an Ionic Compound

Determine the formula for the compound that is formed when each of the following pairs of ions interact.

a. Na^+ and P^{3-} **b.** Be^{2+} and P^{3-}

Solution

a. The Na^+ and P^{3-} ions combine in a three-to-one ratio because this combination causes the charges to add to zero. Three Na^+ ions give a total positive charge of 3. One P^{3-} ion results in a total negative charge of 3. Thus the formula for the compound is Na_3P.

b. The numbers in the charges for these ions are 2 and 3. The lowest common multiple of 2 and 3 is 6 ($2 \times 3 = 6$). Thus we need 6 units of positive charge and 6 units of negative charge. Three Be^{2+} ions are needed to give the 6 units of positive charge, and two P^{3-} ions are needed to give the 6 units of negative charge. The combining ratio of ions is three to two, and the formula is Be_3P_2.

The strategy of finding the lowest common multiple of the numbers in the charges of the ions always works, and it saves you the inconvenience of drawing the Lewis structures.

Practice Exercise 4.4

Determine the formula for the compound that is formed when each of the following pairs of ions interact.

a. Ca^{2+} and F^- **b.** Al^{3+} and O^{2-}

• *Answers.* **a.** CaF_2; **b.** Al_2O_3

4.8 The Structure of Ionic Compounds

An ionic compound, in the solid state, consists of positive and negative ions arranged in such a way that each ion is surrounded by nearest neighbors of the opposite charge. Any given ion is bonded by electrostatic (positive – negative) attractions to all the other ions of opposite charge immediately surrounding it. Figure 4.3 shows a two-dimensional cross section and a three-dimensional view of the arrangement of ions in the ionic compound sodium chloride (NaCl). Note in these structural representations that no given ion has a single partner. A given sodium ion has six immediate neighbors (chloride ions) that are equidistant from it. A chloride ion in turn has six immediate sodium ion neighbors.

The alternating array of positive and negative ions present in an ionic compound means that discrete molecules do not exist in such compounds. Therefore, the formulas of ionic

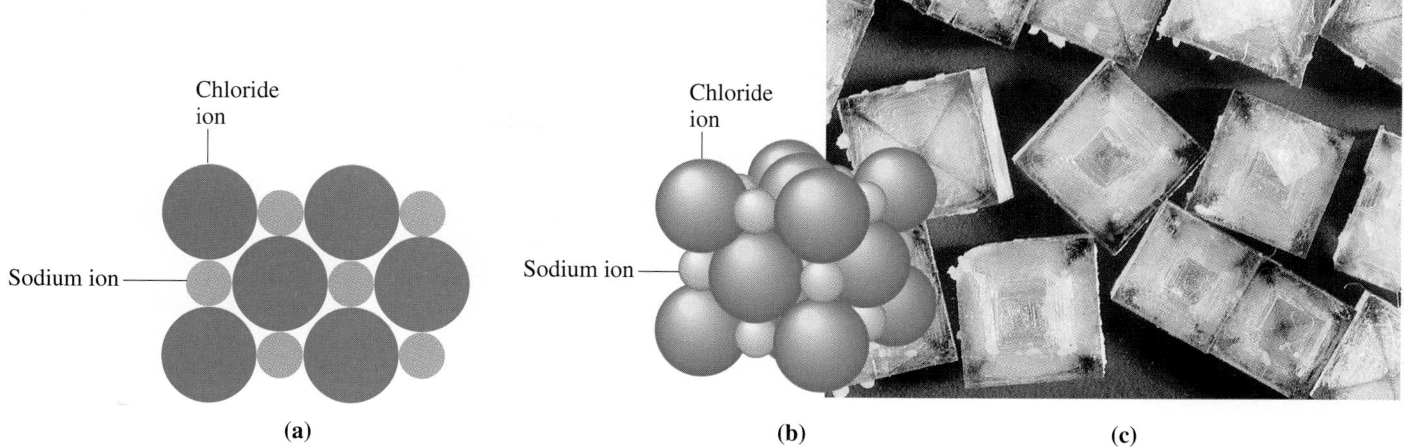

Figure 4.3
(a, b) A two-dimensional cross section and a three-dimensional view of sodium chloride (NaCl), an ionic solid. Both views show an alternating array of positive and negative ions.
(c) Sodium chloride crystals.

• In Section 1.9 the molecule was described as the smallest unit of a pure substance that is capable of a stable, independent existence. Ionic compounds, with their formula units, are exceptions to this generalization.

compounds cannot represent the composition of molecules of these substances. Instead, such formulas represent the simplest combining ratio for the ions present. The formula for sodium chloride, NaCl, indicates that sodium and chloride ions are present in a one-to-one ratio in this compound. Chemists use the term *formula unit,* rather than molecule, to refer to the smallest unit of an ionic compound. Ionic compound formulas, although they represent only ratios, can still be used in chemical equations and chemical calculations in the same manner as formulas for molecules.

The ions present in an ionic solid adopt an arrangement that maximizes attractions between ions of opposite charge and minimizes repulsions between ions of like charge. The specific arrangement that is adopted depends on ion sizes and on the ratio between

Chemical CONNECTIONS

4.2 Sodium Chloride: The World's Most Common Food Additive

The *ionic compound* sodium chloride (table salt), which contains Na^+ and Cl^- ions, is a white crystalline solid that occurs naturally in large underground deposits (rock salt). It also can be obtained by solar evaporation of seawater.

Sodium chloride is the world's most common food additive. Most people find its taste innately appealing. Salt use tends to enhance other flavors, probably by suppressing the bitter flavors. In general, processed foods contain the most sodium chloride, and unprocessed foods, such as fresh fruits and vegetables, contain the least. Studies on the sodium content of foods show that as much as 75% of it is added during processing and manufacturing, 15% comes from salt added during cooking and at the table, and only 10% is naturally present in the food.

Physiologically, sodium chloride is an extremely important substance. It is the major source for both Na^+ and Cl^- ions, the two most abundant ions in blood plasma and in interstitial fluid (the fluid outside cells). Sodium ions play a major role in determining interstitial fluid volume and are essential to nerve transmission and muscle contraction. Chloride ions are essential to acid–base balance within the body and help regulate fluid flow.

Most people are aware that numerous studies on health and diet have shown a relationship between high sodium (sodium chloride) intake and hypertension (high blood pressure). Too much sodium commonly aggravates hypertension.

Many factors, including genetics, contribute to the regulation of blood pressure in the human body. Recent research on the role of sodium ions in hypertension indicates that their contribution to hypertension may be more complex than was originally thought. Sodium chloride has a greater effect on blood pressure than either sodium or chloride ions alone or in combination with other ions. The question now is whether it is the sodium ion (from any source) or just salt (sodium chloride) that is responsible for the effects seen. Much of the early research that implicated sodium ion as a cause of hypertension may actually have reflected the effect of sodium's "silent partner," the chloride ion. Research continues.

How sodium ion (or sodium chloride) contributes to high blood pressure is not completely known, but there are indications that elevated sodium ion levels in body fluids lead to a contraction of small arteries, increasing resistance to blood flow.

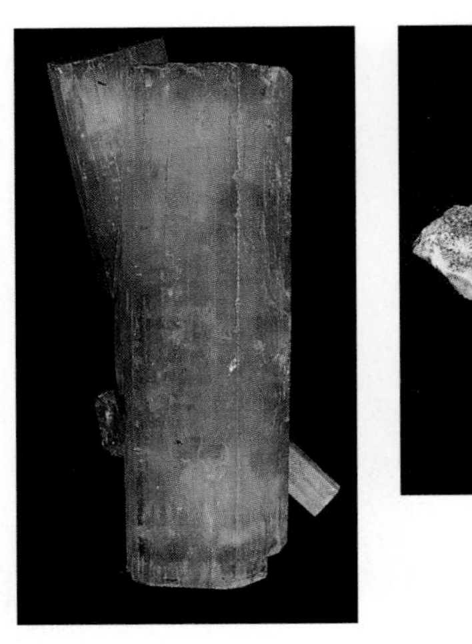

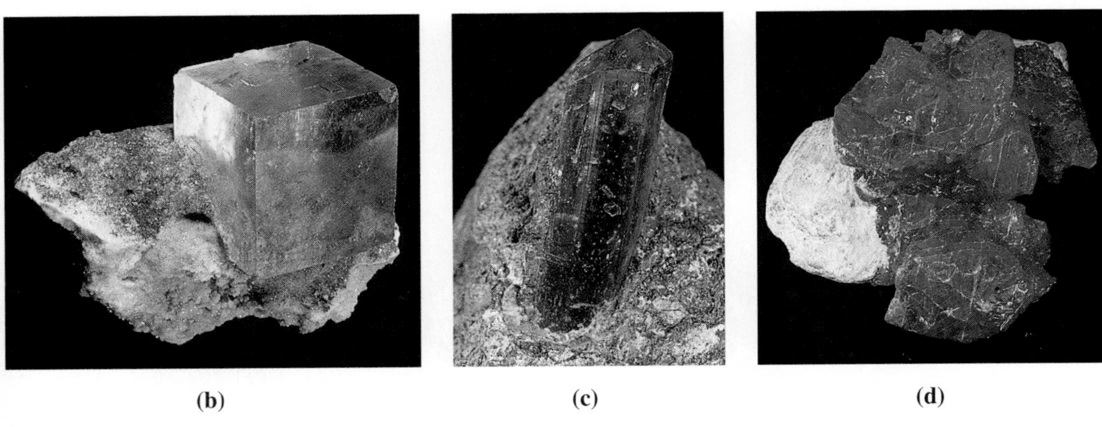

(b) (c) (d)

(a)

Figure 4.4
Ionic compounds usually have crystalline forms in the solid state, such as those associated with (a) emerald, (b) fluorite, (c) topaz, and (d) ruby.

positive and negative ions. Arrangements are usually very symmetrical and result in crystalline solids—that is, solids with highly regular shapes. Crystalline solids usually have flat surfaces or faces that make definite angles with one another. Figure 4.4 shows crystals of various minerals.

4.9 Naming Binary Ionic Compounds

A **binary ionic compound** *is an ionic compound in which only two elements are present.* Such compounds represent the simplest type of ionic compound and are the first type of ionic compound we will learn how to name. In all binary ionic compounds, one element is a metal and the other a nonmetal. The metal is always present as the positive ion, and the nonmetal is always present as the negative ion.

Binary ionic compounds are named using the following rule: *The full name of the metallic element is given first, followed by a separate word containing the stem of the nonmetallic element name and the suffix* -ide. Thus, in order to name the compound NaF, we start with the name of the metal (sodium), follow it with the stem of the name of the nonmetal (fluor-), and then add the suffix *-ide.* The name becomes *sodium fluoride.*

The stem of the name of the nonmetal is the name of the nonmetal with its ending chopped off. Table 4.1 gives the stem part of the name for each of the most common nonmetallic elements. The name of the metal ion is always exactly the same as the name of the metal itself; the metal's name is never shortened. Example 4.5 illustrates the use of the rule for naming binary ionic compounds.

Table 4.1
Names of Selected Common Nonmetallic Ions

Element	Stem	Name of ion	Formula of ion
bromine	brom-	bromide	Br^-
carbon	carb-	carbide	C^{4-}
chlorine	chlor-	chloride	Cl^-
fluorine	fluor-	fluoride	F^-
hydrogen	hydr-	hydride	H^-
iodine	iod-	iodide	I^-
nitrogen	nitr-	nitride	N^{3-}
oxygen	ox-	oxide	O^{2-}
phosphorus	phosph-	phosphide	P^{3-}
sulfur	sulf-	sulfide	S^{2-}

Example 4.5

Naming Binary Ionic Compounds

Name the following binary ionic compounds.

a. MgO **b.** Al_2S_3 **c.** K_3N **d.** $CaCl_2$

Solution

The general pattern for naming binary ionic compounds is

Name of metal + stem of name of nonmetal + *-ide*

a. The metal is magnesium and the nonmetal is oxygen. Thus the compound's name is *magnesium oxide.*

b. The metal is aluminum and the nonmetal is sulfur; the compound's name is *aluminum sulfide.* Note that no mention is made of the subscripts present in the formula—the 2 and the 3. The name of an ionic compound never contains any reference to formula subscript numbers. There is only one ratio in which aluminum and sulfur atoms combine. Thus, just telling the names of the elements present in the compound is adequate nomenclature.

c. Potassium (K) and nitrogen (N) are present in the compound, and its name is *potassium nitride.*

d. The compound's name is *calcium chloride.*

Practice Exercise 4.5

Name the following binary ionic compounds.

a. Na_2S **b.** BeO **c.** Li_3P **d.** BaI_2

• *Answers.* **a.** sodium sulfide; **b.** beryllium oxide; **c.** lithium phosphide; **d.** barium iodide

• All the inner transition elements (*f* area of the periodic table), most of the transition elements (*d* area), and a few representative metals (*p* area) exhibit variable ionic charge behavior.

• An older method for indicating the charge on metal ions uses the suffixes *-ic* and *-ous* rather than the Roman numeral system. It is mentioned here because it is still sometimes encountered. In this system, when a metal has two common ionic charges, the suffix *-ous* is used for the ion of lower charge and the suffix *-ic* for the ion of higher charge. The metal's Latin name is also used. In this older system, iron(II) ion is called ferrous ion, and iron(III) ion is called ferric ion.

Thus far in our discussion of ionic compounds, it has been assumed that the only behavior allowable for an element is predicted by the octet rule. This is a good assumption for nonmetals and for most representative element metals. However, there are many other metals that exhibit a less predictable behavior because they are able to form more than one type of ion. For example, iron forms both Fe^{2+} ions and Fe^{3+} ions, depending on chemical circumstances.

When we name compounds that contain metals with variable ionic charges, the charge on the metal ion must be incorporated into the name. This is done by using Roman numerals. For example, the chlorides of Fe^{2+} and Fe^{3+} ($FeCl_2$ and $FeCl_3$, respectively) are named iron(II) chloride and iron(III) chloride (Figure 4.5). Likewise, CuO is named copper(II) oxide. If you are uncertain about the charge on the metal ion in an ionic

(a) (b)

Figure 4.5
(a) Copper(II) oxide (CuO) is black, whereas copper(I) oxide (Cu_2O) is reddish brown. (b) Iron(II) chloride ($FeCl_2$) is green, whereas iron(III) chloride ($FeCl_3$) is bright yellow.

compound, use the charge on the nonmetal ion (which does not vary) to calculate it. For example, in order to determine the charge on the copper ion in CuO, you can note that the oxide ion carries a −2 charge because oxygen is in Group VIA. This means that the copper ion must have a +2 charge to counterbalance the −2 charge.

Example 4.6

Using Roman Numerals in the Naming of Binary Ionic Compounds

Name the following binary ionic compounds, each of which contains a metal whose ionic charge can vary.

a. AuCl **b.** Fe_2O_3

Solution

We will need to indicate the magnitude of the charge on the metal ion in the name of each of these compounds by means of a Roman numeral.

a. To calculate the metal ion charge, use the fact that total ionic charge (both positive and negative) must add to zero.

$$(\text{Gold charge}) + (\text{chlorine charge}) = 0$$

The chloride ion has a −1 charge (Section 4.5). Therefore,

$$(\text{Gold charge}) + (-1) = 0$$

Thus,

$$\text{Gold charge} = +1$$

Therefore, the gold ion present is Au^+, and the name of the compound is *gold(I) chloride.*

b. For charge balance in this compound we have the equation

$$2(\text{iron charge}) + 3(\text{oxygen charge}) = 0$$

Note that we have to take into account the number of each kind of ion present (2 and 3 in this case). Oxide ions carry a −2 charge (Section 4.5). Therefore,

$$2(\text{iron charge}) + 3(-2) = 0$$
$$2(\text{iron charge}) = +6$$
$$\text{Iron charge} = +3$$

Here, we are interested in the charge on a single iron ion (+3) and not in the total positive charge present (+6). The compound is named *iron(III) oxide* because Fe^{3+} ions are present. As is the case for all ionic compounds, the name does not contain any reference to the numerical subscripts in the compound's formula.

Practice Exercise 4.6

Name the following binary ionic compounds, each of which contains a metal whose ionic charge can vary.

a. PbO_2 **b.** Cu_2O

• *Answers.* **a.** lead(IV) oxide; **b.** copper(I) oxide

• The fixed-charge metals are those in Group IA (+1 ionic charge), those in Group IIA (+2 ionic charge), and five others (Al^{3+}, Ga^{3+}, Zn^{2+}, Cd^{2+}, and Ag^+).

In order to know when to use Roman numerals in binary ionic compound names, you must know which metals exhibit variable ionic charge and which have a fixed ionic charge. There are many more of the former (Roman numeral required) than of the latter (no Roman numeral required). Thus you should learn the identity of the metals that have a fixed ionic charge (the short list); any metal not on the short list must exhibit variable charge. Figure 4.6 shows the metals that always form a single type of ion in ionic compound formation. Ionic compounds that contain these metals are the only ones without Roman numerals in their names.

Chemistry at a Glance reviews the formation of ionic bonds and ionic compounds.

Ionic Bonds and Ionic Compounds

CHEMISTRY AT A GLANCE

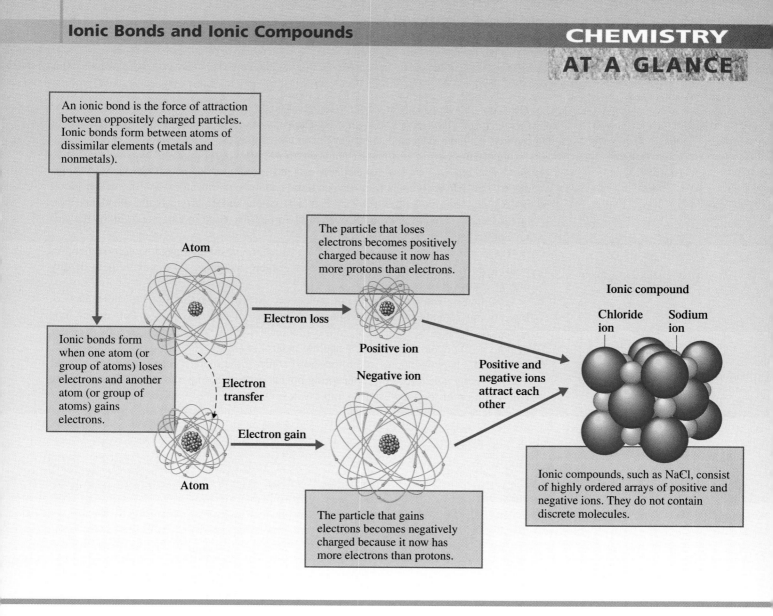

An ionic bond is the force of attraction between oppositely charged particles. Ionic bonds form between atoms of dissimilar elements (metals and nonmetals).

Ionic bonds form when one atom (or group of atoms) loses electrons and another atom (or group of atoms) gains electrons.

Atom

Electron loss

Positive ion

The particle that loses electrons becomes positively charged because it now has more protons than electrons.

Electron transfer

Atom

Electron gain

Negative ion

The particle that gains electrons becomes negatively charged because it now has more electrons than protons.

Positive and negative ions attract each other

Ionic compound

Chloride ion

Sodium ion

Ionic compounds, such as NaCl, consist of highly ordered arrays of positive and negative ions. They do not contain discrete molecules.

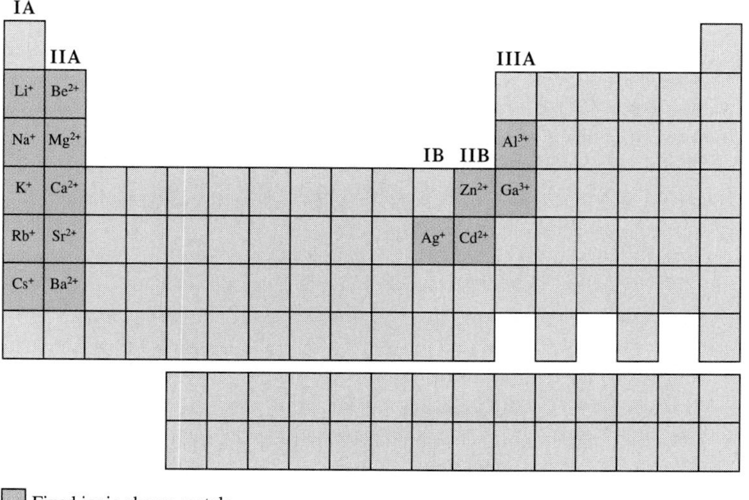

Figure 4.6
A periodic table in which the metallic elements that exhibit a fixed ionic charge are highlighted.

(a)

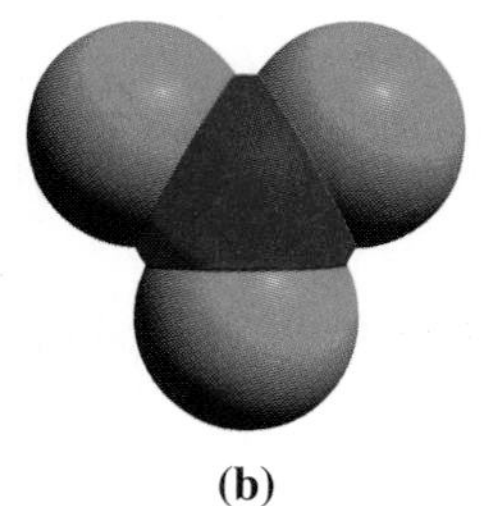

(b)

Figure 4.7
Models of (a) a sulfate ion (SO_4^{2-}) and (b) a nitrate ion (NO_3^-).

4.10 Polyatomic Ions

There are two categories of ions: monatomic and polyatomic. A **monatomic ion** *is formed from a single atom that has lost or gained electrons and thus acquired a charge.* All of the ions we have discussed so far have been monatomic (Cl^-, Na^+, Ca^{2+}, N^{3-}, and so on).

A **polyatomic ion** *is formed from a group of atoms, held together by covalent bonds, that has acquired a charge.* An example of a polyatomic ion is the sulfate ion, SO_4^{2-} (Figure 4.7). This ion contains four oxygen atoms and one sulfur atom, and the whole group of five atoms has acquired a −2 charge. The whole sulfate group is the ion rather than any one atom within the group. Covalent bonding, discussed in Chapter 5, holds the sulfur and oxygen atoms together.

There are numerous ionic compounds in which the positive or negative ion (sometimes both) is polyatomic. Polyatomic ions are very stable and generally maintain their identity during chemical reactions.

Note that polyatomic ions are not molecules. They never occur alone as molecules do. Instead, they are always found associated with ions of opposite charge. Polyatomic ions are *charged pieces* of compounds, not compounds. Ionic compounds require the presence of both positive and negative ions and are neutral overall.

Table 4.2 lists the names and formulas of some of the more common polyatomic ions. The following generalizations concerning polyatomic ion names and charges emerge from consideration of the ions listed in Table 4.2.

Table 4.2
Formulas and Names of Some Common Polyatomic Ions

Key element present	Formula	Name of ion
nitrogen	NO_3^-	nitrate
	NO_2^-	nitrite
	NH_4^+	ammonium
	N_3^-	azide
sulfur	SO_4^{2-}	sulfate
	HSO_4^-	bisulfate or hydrogen sulfate
	SO_3^{2-}	sulfite
	HSO_3^-	bisulfite or hydrogen sulfite
	$S_2O_3^{2-}$	thiosulfate
phosphorus	PO_4^{3-}	phosphate
	HPO_4^{2-}	hydrogen phosphate
	$H_2PO_4^-$	dihydrogen phosphate
	PO_3^{3-}	phosphite
carbon	CO_3^{2-}	carbonate
	HCO_3^-	bicarbonate or hydrogen carbonate
	$C_2O_4^{2-}$	oxalate
	$C_2H_3O_2^-$	acetate
	CN^-	cyanide
chlorine	ClO_4^-	perchlorate
	ClO_3^-	chlorate
	ClO_2^-	chlorite
	ClO^-	hypochlorite
hydrogen	H_3O^+	hydronium
	OH^-	hydroxide
metals	MnO_4^-	permanganate
	CrO_4^{2-}	chromate
	$Cr_2O_7^{2-}$	dichromate

1. Most of the polyatomic ions have a negative charge, which can vary from −1 to −3. Only two positive ions are listed in the table: NH_4^+ (ammonium) and H_3O^+ (hydronium).
2. Two of the negatively charged polyatomic ions, OH^- (hydroxide) and CN^- (cyanide), have names ending in *-ide*, and the rest of them have names ending in either *-ate* or *-ite*.
3. A number of *-ate*, *-ite* pairs of ions exist, as in SO_4^{2-} (sulfate) and SO_3^{2-} (sulfite). The *-ate* ion always has one more oxygen atom than the *-ite* ion. Both the *-ate* and *-ite* ions of a pair carry the same charge.
4. A number of pairs of ions exist wherein one member of the pair differs from the other by having a hydrogen atom present, as in CO_3^{2-} (carbonate) and HCO_3^- (hydrogen carbonate or bicarbonate). In such pairs, the charge on the ion that contains hydrogen is always 1 less than that on the other ion.

• Learning the names of the common polyatomic ions is a memorization project. There is no shortcut. The charges and formulas for the various polyatomic ions cannot be easily related to the periodic table, as was the case for many of the monatomic ions.

4.11 Formulas and Names for Ionic Compounds Containing Polyatomic Ions

Formulas for ionic compounds that contain polyatomic ions are determined in the same way as those for ionic compounds that contain monatomic ions (Section 4.7). The positive and negative charges present must add to zero.

Two conventions not encountered previously in formula writing often arise when we write formulas containing polyatomic ions.

1. When more than one polyatomic ion of a given kind is required in a formula, the polyatomic ion is enclosed in parentheses, and a subscript, placed outside the parentheses, is used to indicate the number of polyatomic ions needed. An example is $Fe(OH)_3$.
2. So that the identity of polyatomic ions is preserved, the same elemental symbol may be used more than once in a formula. An example is the formula NH_4NO_3, where *N* appears in two locations.

Example 4.7 illustrates the use of both of these new conventions.

Example 4.7

Writing Formulas for Ionic Compounds Containing Polyatomic Ions

Determine the formulas for the ionic compounds that contain these pairs of ions.

a. Na^+ and SO_4^{2-} **b.** Mg^{2+} and NO_3^- **c.** NH_4^+ and CN^-

Solution

a. In order to equalize the total positive and negative charge, we need two sodium ions (+1 charge) for each sulfate ion (−2 charge). We indicate the presence of two Na^+ ions with the subscript 2 following the symbol of this ion. The formula of the compound is Na_2SO_4. The convention that the positive ion is always written first in the formula still holds when polyatomic ions are present.

b. Two nitrate ions (−1 charge) are required to balance the charge on one magnesium ion (+2 charge). Because more than one polyatomic ion is needed, the formula contains parentheses, $Mg(NO_3)_2$. The subscript 2 outside the parentheses indicates two of what is inside the parentheses. If parentheses were not used, the formula would appear to be $MgNO_{32}$, which is not intended and conveys false information.

c. In this compound, both ions are polyatomic, which is a perfectly legal situation. Because the ions have equal but opposite charges, they combine in a one-to-one ratio. Thus the formula is NH_4CN. No parentheses are necessary because we need only one polyatomic ion of each type in a formula unit. The appearance of the symbol for the

(*continued*)

Nomenclature of Ionic Compounds

CHEMISTRY AT A GLANCE

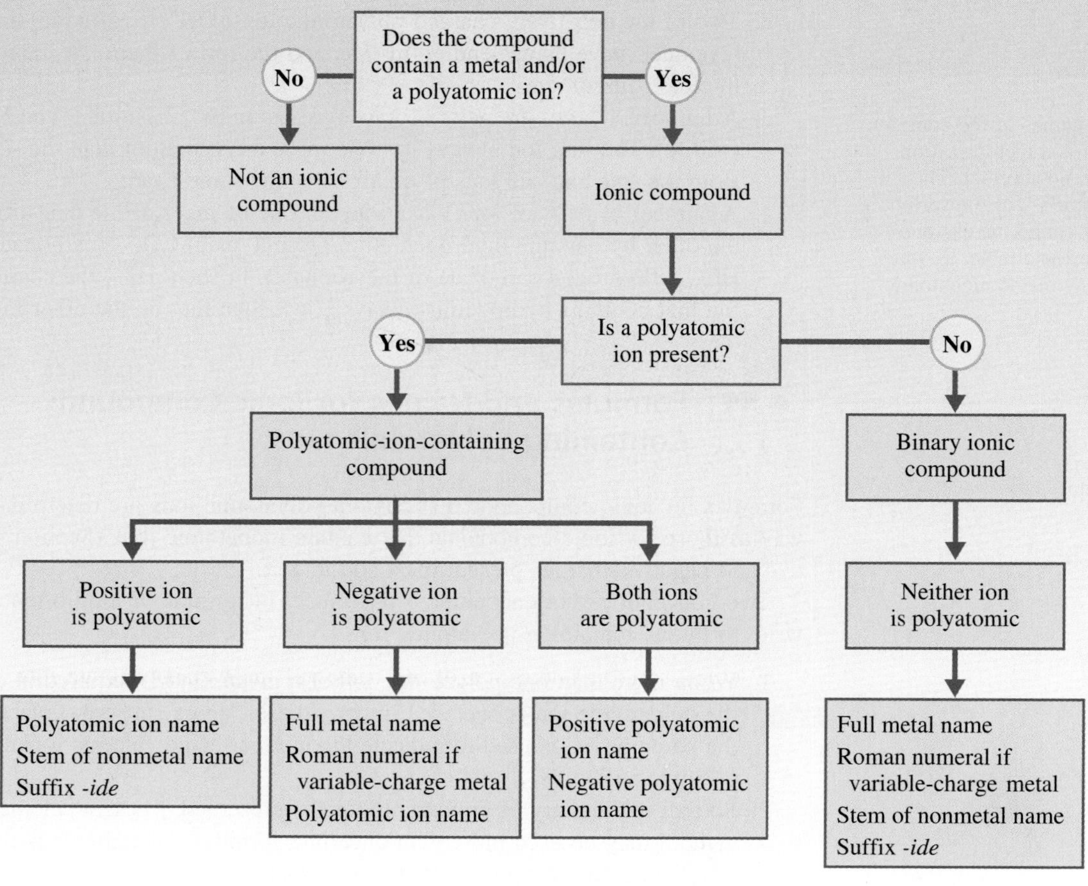

element nitrogen (N) at two locations in the formula could be prevented by combining the two nitrogens, resulting in N_2H_4C. But the formula N_2H_4C does not convey the message that NH_4^+ and CN^- ions are present. Thus, when writing formulas that contain polyatomic ions, we always maintain the identities of these ions, even if it means having the same elemental symbol at more than one location in the formula.

Practice Exercise 4.7

Determine the formulas for the ionic compounds that contain the following pairs of ions.

a. K^+ and CO_3^{2-} **b.** Ca^{2+} and OH^- **c.** NH_4^+ and HPO_4^{2-}

• *Answers.* **a.** K_2CO_3; **b.** $Ca(OH)_2$; **c.** $(NH_4)_2HPO_4$

The names of ionic compounds containing polyatomic ions are derived in a manner similar to that for binary ionic compounds (Section 4.9). The rule for naming binary ionic compounds is as follows: Give the name of the metallic element first (including, when needed, a Roman numeral indicating ion charge), and then give a separate word containing the stem of the nonmetallic name and the suffix *-ide.*

For our present situation, *if the polyatomic ion is positive, its name is substituted for that of the metal. If the polyatomic ion is negative, its name is substituted for the nonmetal stem plus* -ide. Where both positive and negative ions are polyatomic, dual substitution occurs, and the resulting name includes just the names of the polyatomic ions.

Example 4.8

Naming Ionic Compounds in Which Polyatomic Ions Are Present

Name the following compounds, which contain one or more polyatomic ions.

a. $Ca_3(PO_4)_2$ **b.** $Fe_2(SO_4)_3$ **c.** $(NH_4)_2CO_3$

Solution

a. The positive ion present is the calcium ion (Ca^{2+}). We will not need a Roman numeral to specify the charge on a Ca^{2+} ion because it is always +2. The negative ion is the polyatomic phosphate ion (PO_4^{3-}). The name of the compound is *calcium phosphate*. As in naming binary ionic compounds, subscripts in the formula are not incorporated into the name.

b. The positive ion present is iron(III). The negative ion is the polyatomic sulfate ion (SO_4^{2-}). The name of the compound is *iron(III) sulfate*. The determination that iron is present as iron(III) involves the following calculation dealing with charge balance:

$$2(\text{iron charge}) + 3(\text{sulfate charge}) = 0$$

The sulfate charge is −2. (You had to memorize that.) Therefore,

$$2(\text{iron charge}) + 3(-2) = 0$$
$$2(\text{iron charge}) = +6$$
$$\text{Iron charge} = +3$$

c. Both the positive and the negative ions in this compound are polyatomic—the ammonium ion (NH_4^+) and the carbonate ion (CO_3^{2-}). The name of the compound is simply the combination of the names of the two polyatomic ions: *ammonium carbonate.*

Practice Exercise 4.8

Name the following compounds, which contain one or more polyatomic ions.

a. $Ba(NO_3)_2$ **b.** Cu_3PO_4 **c.** $(NH_4)_2SO_4$

- *Answers.* **a.** barium nitrate; **b.** copper(I) phosphate; **c.** ammonium sulfate

Concepts to Remember

Chemical bonds. Chemical bonds are the attractive forces that hold atoms together in more complex units. Chemical bonds result from the transfer of valence electrons between atoms (ionic bond) or from the sharing of electrons between atoms (covalent bond).

Valence electrons. Valence electrons, for representative elements, are the electrons in the outermost electron shell, which is the shell with the highest shell number. These electrons are particularly important in determining the bonding characteristics of a given atom.

Octet rule. In compound formation, atoms of representative elements lose, gain, or share electrons in such a way that their electron configurations become identical to those of the noble gas nearest them in the periodic table.

Ionic compounds. Ionic compounds usually involve a metal atom and a nonmetal atom. Metal atoms lose one or more electrons, producing positive ions. Nonmetal atoms acquire the electrons lost by the metal atoms, producing negative ions. The oppositely charged ions attract one another, creating ionic bonds.

Charge magnitude for ions. Metal atoms containing one, two, or three valence electrons tend to lose such electrons, producing ions of +1, +2, and +3 charge, respectively. Nonmetal atoms containing five, six, or seven valence electrons tend to gain electrons, producing ions of −3, −2, and −1 charge, respectively.

Formulas for ionic compounds. The ratio in which positive and negative ions combine is the ratio that causes the total amount of positive and negative charges to add up to zero.

Structure of ionic compounds. Ionic solids consist of positive and negative ions arranged in such a way that each ion is surrounded by ions of the opposite charge.

Binary ionic compound nomenclature. Binary ionic compounds are named by giving the full name of the metallic element first, followed by a separate word containing the stem of the nonmetallic element name and the suffix *-ide*. A Roman numeral specifying ionic charge is appended to the name of the metallic element if it is a metal that exhibits variable ionic charge.

Polyatomic ions. A polyatomic ion is a group of covalently bonded atoms that has acquired a charge through the loss or gain of electrons. Polyatomic ions are very stable entities that generally maintain their identity during chemical reactions.

Key Reactions and Equations

1. Number of valence electrons for representative elements (Section 4.2)

   Number of valence electrons = periodic-table group number

2. Charges on metallic monatomic ions (Section 4.5)

   Group IA metals form 1^+ ions.
   Group IIA metals form 2^+ ions.
   Group IIIA metals form 3^+ ions.

3. Charges on nonmetallic monatomic ions (Section 4.5)

   Group VA nonmetals form 3^- ions.
   Group VIA nonmetals form 2^- ions.
   Group VIIA nonmetals form 1^- ions.

Key Terms

Binary ionic compound (4.9)
Chemical bonds (4.1)
Covalent bond (4.1)
Ion (4.4)
Ionic bond (4.1)
Lewis structure (4.2)
Monatomic ion (4.10)
Octet rule (4.3)
Polyatomic ion (4.10)
Valence electrons (4.2)

Exercises and Problems

The members of each pair of problems in this section test similar material.

Valence Electrons (Section 4.2)

4.1 How many valence electrons do atoms with the following electron configurations have?

a. $1s^22s^2$ b. $1s^22s^22p^63s^2$
c. $1s^22s^22p^63s^23p^1$ d. $1s^22s^22p^63s^23p^64s^23d^{10}4p^2$

4.2 How many valence electrons do atoms with the following electron configurations have?

a. $1s^22s^22p^6$ b. $1s^22s^22p^63s^23p^1$
c. $1s^22s^22p^63s^1$ d. $1s^22s^22p^63s^23p^64s^23d^{10}4p^5$

4.3 How many valence electrons do atoms of each of the following elements have?

a. $_{3}$Li b. $_{10}$Ne c. $_{20}$Ca d. $_{53}$I

4.4 How many valence electrons do atoms of each of the following elements have?

a. $_{8}$O b. $_{13}$Al c. $_{34}$Se d. $_{55}$Cs

4.5 Write the complete electron configuration for each of the following representative elements.

a. Period 2 element with four valence electrons
b. Period 2 element with seven valence electrons
c. Period 3 element with two valence electrons
d. Period 3 element with five valence electrons

4.6 Write the complete electron configuration for each of the following representative elements.

a. Period 2 element with one valence electron
b. Period 2 element with six valence electrons
c. Period 3 element with seven valence electrons
d. Period 3 element with three valence electrons

Lewis Structures for Atoms (Section 4.2)

4.7 Draw Lewis structures for atoms of each of the following elements.

a. $_{12}$Mg b. $_{19}$K c. $_{15}$P d. $_{36}$Kr

4.8 Draw Lewis structures for atoms of each of the following elements.

a. $_{13}$Al b. $_{20}$Ca c. $_{17}$Cl d. $_{4}$Be

4.9 Each of the following Lewis structures represents a Period 2 element. Determine each element's identity.

a. $X\cdot$ b. $:\underset{\cdot}{\ddot{X}}:$ c. $\dot{X}\cdot$ d. $\cdot\underset{\cdot}{\ddot{X}}\cdot$

4.10 Each of the following Lewis structures represents a Period 3 element. Determine each element's identity

a. $\cdot\dot{X}\cdot$ b. $\cdot\underset{\cdot}{\dot{X}}\cdot$ c. $\cdot\underset{\cdot}{\ddot{X}}:$ d. $:\underset{\cdot\cdot}{\ddot{X}}:$

Notation for Ions (Section 4.4)

4.11 Give the symbol for each of the following ions.

a. An oxygen atom that has gained two electrons
b. A magnesium atom that has lost two electrons
c. A fluorine atom that has gained one electron
d. An aluminum atom that has lost three electrons

4.12 Give the symbol for each of the following ions.

a. A chlorine atom that has gained one electron
b. A sulfur atom that has gained two electrons
c. A potassium atom that has lost one electron
d. A beryllium atom that has lost two electrons

4.13 What would be the symbol for an ion with each of the following numbers of protons and electrons?

a. 20 protons and 18 electrons
b. 8 protons and 10 electrons
c. 11 protons and 10 electrons
d. 13 protons and 10 electrons

4.14 What would be the symbol for an ion with each of the following numbers of protons and electrons?

a. 15 protons and 18 electrons
b. 17 protons and 18 electrons
c. 12 protons and 10 electrons
d. 19 protons and 18 electrons

4.15 Calculate the number of protons and electrons in each of the following ions.

a. P^{3-} b. N^{3-} c. Mg^{2+} d. Li^+

4.16 Calculate the number of protons and electrons in each of the following ions.

a. S^{2-} b. F^- c. K^+ d. H^+

Ionic Charge Magnitude (Section 4.5)

4.17 What is the charge on the monatomic ion formed by each of the following elements?

a. $_{12}Mg$ b. $_7N$ c. $_{19}K$ d. $_9F$

4.18 What is the charge on the monatomic ion formed by each of the following elements?

a. $_3Li$ b. $_{15}P$ c. $_{16}S$ d. $_{13}Al$

4.19 Indicate the number of electrons lost or gained when each of the following atoms forms an ion.

a. $_4Be$ b. $_{35}Br$ c. $_{38}Sr$ d. $_{34}Se$

4.20 Indicate the number of electrons lost or gained when each of the following atoms forms an ion.

a. $_{37}Rb$ b. $_{53}I$ c. $_8O$ d. $_{11}Na$

4.21 Write the electron configuration of the following.

a. An aluminum atom

b. An aluminum ion

4.22 Write the electron configuration of the following.

a. An oxygen atom

b. An oxygen ion

Ionic Compound Formation (Section 4.6)

4.23 Using Lewis structures, show how ionic compounds are formed by atoms of

a. Be and O b. Mg and S

c. K and N d. F and Ca

4.24 Using Lewis structures, show how ionic compounds are formed by atoms of

a. Na and F b. Li and S

c. Be and S d. P and K

Formulas for Ionic Compounds (Section 4.7)

4.25 Write the formula for an ionic compound formed from Ba^{2+} ions and each of the following ions.

a. Cl^- b. Br^- c. N^{3-} d. O^{2-}

4.26 Write the formula for an ionic compound formed from K^+ ions and each of the following ions.

a. Cl^- b. Br^- c. N^{3-} d. O^{2-}

4.27 Write the formula for an ionic compound formed from F^- ions and each of the following ions.

a. Mg^{2+} b. Be^{2+} c. Li^+ d. Al^{3+}

4.28 Write the formula for an ionic compound formed from S^{2-} ions and each of the following ions.

a. Mg^{2+} b. Be^{2+} c. Li^+ d. Al^{3+}

4.29 Write the formulas for the compounds formed from the following ions.

a. Na^+ and S^{2-} b. Ca^{2+} and I^-

c. Li^+ and N^{3-} d. Al^{3+} and Br^-

4.30 Write the formulas for the compounds formed from the following ions.

a. Li^+ and O^{2-} b. Al^{3+} and N^{3-}

c. K^+ and Cl^- d. Mg^{2+} and I^-

Binary Ionic Compound Nomenclature (Section 4.9)

4.31 Name the following binary ionic compounds, each of which contains a fixed-charge metal.

a. KI b. BeO c. AlF_3 d. Na_3P

4.32 Name the following binary ionic compounds, each of which contains a fixed-charge metal.

a. $CaCl_2$ b. Ca_2C c. Be_3N_2 d. K_2S

4.33 Calculate the charge on the metal ion in the following binary ionic compounds, each of which contains a variable-charge metal.

a. Au_2O b. CuO c. SnO_2 d. SnO

4.34 Calculate the charge on the metal ion in the following binary ionic compounds, each of which contains a variable-charge metal.

a. Fe_2O_3 b. FeO c. $SnCl_4$ d. Cu_2S

4.35 Name the following binary ionic compounds, each of which contains a variable-charge metal.

a. FeO b. Au_2O_3 c. CuS d. $CoBr_2$

4.36 Name the following binary ionic compounds, each of which contains a variable-charge metal.

a. PbO b. $FeCl_3$ c. SnO_2 d. NiI_2

4.37 Name each of the following binary ionic compounds.

a. AuCl b. KCl c. AgCl d. $CuCl_2$

4.38 Name each of the following binary ionic compounds.

a. NiO b. FeN c. AlN d. BeO

4.39 Write formulas for the following binary ionic compounds.

a. Potassium bromide b. Silver oxide

c. Beryllium fluoride d. Barium phosphide

4.40 Write formulas for the following binary ionic compounds.

a. Gallium nitride b. Zinc chloride

c. Magnesium sulfide d. Aluminum nitride

4.41 Write formulas for the following binary ionic compounds.

a. Cobalt(II) sulfide b. Cobalt(III) sulfide

c. Tin(IV) iodide d. Lead(II) nitride

4.42 Write formulas for the following binary ionic compounds.

a. Iron(III) oxide b. Iron(II) oxide

c. Nickel(III) sulfide d. Copper(I) bromide

Compounds Containing Polyatomic Ions (Sections 4.10 and 4.11)

4.43 Write formulas (including charge) for each of the following polyatomic ions.

a. Sulfate b. Chlorate

c. Hydroxide d. Cyanide

4.44 Write formulas (including charge) for each of the following polyatomic ions.

a. Ammonium b. Nitrate

c. Perchlorate d. Phosphate

4.45 Write formulas (including charge) for each of the following pairs of polyatomic ions.
a. Phosphate and hydrogen phosphate
b. Nitrate and nitrite
c. Hydronium and hydroxide
d. Chromate and dichromate

4.46 Write formulas (including charge) for each of the following pairs of polyatomic ions.
a. Chlorate and perchlorate
b. Hydrogen phosphate and dihydrogen phosphate
c. Carbonate and bicarbonate
d. Sulfate and hydrogen sulfate

4.47 Write formulas for the compounds formed between the following positive and negative ions.
a. Na^+ and ClO_4^- b. Fe^{3+} and OH^-
c. Ba^{2+} and NO_3^- d. Al^{3+} and CO_3^{2-}

4.48 Write formulas for the compounds formed between the following positive and negative ions.
a. K^+ and CN^- b. NH_4^+ and SO_4^{2-}
c. Co^{2+} and $H_2PO_4^-$ d. Ca^{2+} and PO_4^{3-}

4.49 Name the following compounds, all of which contain polyatomic ions and fixed-charge metals.
a. $MgCO_3$ b. $ZnSO_4$
c. $Be(NO_3)_2$ d. Ag_3PO_4

4.50 Name the following compounds, all of which contain polyatomic ions and fixed-charge metals.
a. LiOH b. $Al(CN)_3$
c. $Ba(ClO_3)_2$ d. $NaNO_3$

4.51 Name the following compounds, all of which contain polyatomic ions and variable-charge metals.
a. $Fe(OH)_2$ b. $CuCO_3$
c. AuCN d. $Mn_3(PO_4)_2$

4.52 Name the following compounds, all of which contain polyatomic ions and variable-charge metals.
a. $Fe(NO_3)_3$ b. $Co_2(CO_3)_3$
c. Cu_3PO_4 d. $Pb(SO_4)_2$

4.53 Write formulas for the following compounds, all of which contain polyatomic ions.
a. Potassium bicarbonate b. Gold(III) sulfate
c. Silver nitrate d. Copper(II) phosphate

4.54 Write formulas for the following compounds, all of which contain polyatomic ions.
a. Aluminum nitrate b. Iron(III) sulfate
c. Calcium cyanide d. Lead(IV) hydroxide

Additional Problems

4.55 What would be the symbol for an ion with each of the following characteristics?
a. A sodium ion with ten electrons
b. A fluorine ion with ten electrons
c. A sulfur ion with two fewer protons than electrons
d. A calcium ion with two more protons than electrons

4.56 Write the formula of the ionic compound that could form from the elements X and Z if
a. X has two valence electrons and Z has seven valence electrons
b. X has one valence electron and Z has six valence electrons
c. X has three valence electrons and Z has five valence electrons
d. X has six valence electrons and Z has two valence electrons

4.57 Identify the Period 3 element that most commonly produces each of the following ions.
a. X^{2-} b. X^{2+} c. X^{3-} d. X^{3+}

4.58 Write formulas (symbol and charge) for both kinds of ions present in each of the following compounds.
a. KCl b. CaS c. BeF_2 d. Al_2S_3

4.59 Name each compound in the following pairs of binary ionic compounds.
a. $SnCl_4$ and $SnCl_2$ b. FeS and Fe_2S_3
c. Cu_3N and Cu_3N_2 d. NiI_2 and NiI_3

4.60 In which of the following pairs of binary ionic compounds do both members of the pair contain positive ions with the same charge?
a. Co_2O_3 and $CoCl_3$ b. Cu_2O and CuO
c. K_2O and Al_2O_3 d. MgS and NaI

4.61 Name each compound in the following pairs of polyatomic-ion-containing compounds.
a. $CuNO_3$ and $Cu(NO_3)_2$
b. $Pb_3(PO_4)_2$ and $Pb_3(PO_4)_4$
c. $Mn(CN)_3$ and $Mn(CN)_2$
d. $Co(ClO_3)_2$ and $Co(ClO_3)_3$

4.62 Write formulas for the following compounds.
a. Sodium sulfide b. Sodium sulfate
c. Sodium sulfite d. Sodium thiosulfate

Grid Problems

4.63

1. beryllium	2. magnesium	3. sulfur
4. fluorine	5. phosphorus	6. aluminum

Select from the grid *all* correct responses for each of the following situations.
a. Elements for which monatomic ion formation involves the loss of two electrons
b. Elements that form monatomic ions that have a charge of −1
c. Pairs of elements that form an ionic compound in which there is a one-to-one ratio between positive and negative ions
d. Pairs of elements that form an ionic compound whose formula unit contains five ions

4.64

1. $CaCl_2$	2. MgO	3. NaCl
4. K_2S	5. Al_2O_3	6. Na_3N

Select from the grid *all* correct responses for each of the following situations.

a. Compounds in which 2^+ ions are present
b. Compounds in which 1^- ions are present
c. Compounds in which the number of electrons transferred per formula unit is two
d. Compounds in which both types of ions present have the same electron configuration as the noble gas neon

4.65

1. sulfate phosphate	2. nitrate carbonate	3. ammonium cyanide
4. hydroxide chlorate	5. sulfate nitrate	6. hydroxide cyanide

Select from the grid *all* correct responses for each of the following situations.

a. Paired ions that have the same ionic charge
b. Paired ions that have the same number of atoms
c. Paired ions that contain no oxygen atoms
d. Paired ions that could combine with one another to form an ionic compound

4.66

1. NH_4Cl	2. K_2O	3. $Cu(NO_3)_2$
4. SnF_2	5. KOH	6. K_2SO_4

Select from the grid *all* correct responses for each of the following situations.

a. Compounds whose names end in *-ide*
b. Compounds whose names contain a Roman numeral
c. Compounds in which polyatomic ions are present
d. Compounds in which positive and negative ions are present in a one-to-one ratio

CHAPTER 5

PART I

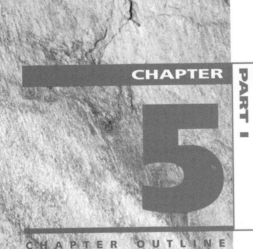

Chemical Bonding: The Covalent Bond Model

CHAPTER OUTLINE

5.1 The Covalent Bond Model 102
5.2 Lewis Structures for Molecular Compounds 103
5.3 Single, Double, and Triple Covalent Bonds 104
5.4 Valence Electrons and Number of Covalent Bonds Formed 105
5.5 Coordinate Covalent Bonds 106
5.6 Systematic Procedures for Drawing Lewis Structures 106
5.7 Bonding in Compounds with Polyatomic Ions Present 110
5.8 The Shapes of Molecules 111

Chemistry at a Glance: **The Shape (Geometry) of Molecules** 114
5.9 Electronegativity 115
5.10 Bond Polarity 116

Chemistry at a Glance: **Covalent Bonds and Molecular Compounds** 118
5.11 Molecular Polarity 119
5.12 Naming Binary Molecular Compounds 120

Chemical Connections
5.1 Nitric Oxide: A Molecule Whose Bonding Does Not Follow "The Rules" 109
5.2 Molecular Shape and Odor 113

Propellant systems used to launch space vehicles contain molecular compounds, compounds in which covalent bonds are present.

The forces that hold atoms in compounds together as a unit are of two general types: (1) ionic bonds (which involve electron transfer) and (2) covalent bonds (which involve electron sharing). The ionic bond model was the subject of Chapter 4. We now consider the covalent bond model.

5.1 The Covalent Bond Model

We begin our discussion of covalent bonding and the molecular compounds that result from such bonding by listing three key differences between ionic and covalent bonding.

1. Ionic bonds form between atoms of dissimilar elements (a metal and a nonmetal). Covalent bond formation occurs between *similar* or even *identical* atoms. Most often two nonmetals are involved.
2. Electron transfer is the mechanism by which ionic bond formation occurs. Covalent bond formation involves *electron sharing.*

3. Ionic compounds do not contain discrete molecules. Instead, such compounds consist of an extended array of alternating positive and negative ions. In covalently bonded compounds, the basic structural unit is a molecule. Indeed, such compounds are called molecular compounds.

• Among the millions of compounds that are known, those that have covalent bonds are dominant. Almost all compounds encountered in the fields of organic chemistry and biochemistry contain covalent bonds.

Consideration of the hydrogen molecule (H_2), the simplest of all molecules, provides initial insights into the nature of the covalent bond and its formation. When two hydrogen atoms, each with a single electron, are brought together, the orbitals that contain the valence electrons *overlap* to create an orbital common to both atoms. This overlapping is shown in Figure 5.1. The two electrons, one from each H atom, now move throughout this new orbital and are said to be *shared* by the two nuclei.

• Covalent bonds result from a common attraction of two nuclei for one or more shared pairs of electrons.

Once two orbitals overlap, the most favorable location for the shared electrons is the area directly between the two nuclei. Here the two electrons can simultaneously interact with (be attracted to) both nuclei, a situation that produces increased stability. This concept of increased stability can be explained by using an analogy. Consider the nuclei of the two hydrogen atoms in H_2 to be "old potbellied stoves" and the two electrons to be running around each of the stoves trying to keep warm. When the two nuclei are together (an H_2 molecule) the electrons have two sources of heat. In particular, in the region between the nuclei (the overlap region) the electrons can keep both front and back warm at the same time. This is a better situation than when each electron has only one "stove" (nucleus) as a source of heat.

In terms of Lewis notation, this sharing of electrons by the two hydrogen atoms is diagrammed as follows:

Shared electron pair

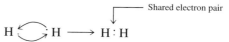

H· ·H ⟶ H:H

The two shared electrons do double duty, helping each hydrogen atom achieve a helium noble-gas configuration.

5.2 Lewis Structures for Molecular Compounds

Using the octet rule, which applies to both electron transfer and electron sharing (Section 4.3), and Lewis structures (Section 4.2), let us now consider the formation of selected simple covalently bonded molecules that contain the element fluorine. Fluorine, located in Group VIIA of the periodic table, has seven valence electrons. Its Lewis structure is

·F:

Fluorine needs only one electron to achieve the octet of electrons that enables it to have a noble-gas electron configuration. When fluorine bonds to other nonmetals, the octet of electrons is completed by means of electron sharing. The molecules HF, F_2, and BrF, whose electron-dot structures follow, are representative of this situation.

H· ·F: ⟶ H:F:

:F· ·F: ⟶ :F:F:

:Br· ·F: ⟶ :Br:F:

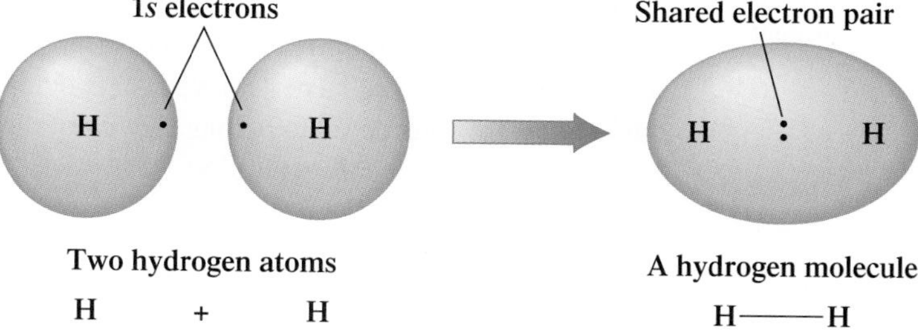

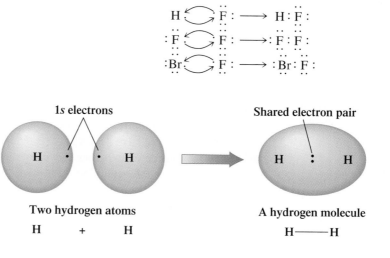

Figure 5.1
Electron sharing can occur only when electron orbitals from two different atoms overlap.

The HF and BrF molecules illustrate the point that the two atoms involved in a covalent bond need not be identical (as is the case with H_2 and F_2).

A common practice in writing Lewis structures for covalently bonded molecules is to represent the *shared* electron pairs with dashes. Using this notation, the H_2, HF, F_2, and BrF molecules are written as

H—H H—F: :F—F: :Br—F:

The atoms in covalently bonded molecules often possess both *bonding* and *nonbonding* electrons. **Bonding electrons** *are pairs of valence electrons that are shared between atoms in a covalent bond.* Each of the fluorine atoms in the molecules HF, F_2, and BrF possesses one pair of bonding electrons. **Nonbonding electrons** *are pairs of valence electrons that are not involved in electron sharing.* Each of the fluorine atoms in HF, F_2, and BrF possesses three pairs of nonbonding electrons, as does the bromine atom in BrF.

• Nonbonding electron pairs are often also referred to as *unshared electron pairs* or *lone electron pairs* (or simply *lone pairs*).

• In Section 5.7 we will learn that nonbonding electrons play an important role in determining the shape (geometry) of molecules when three or more atoms are present.

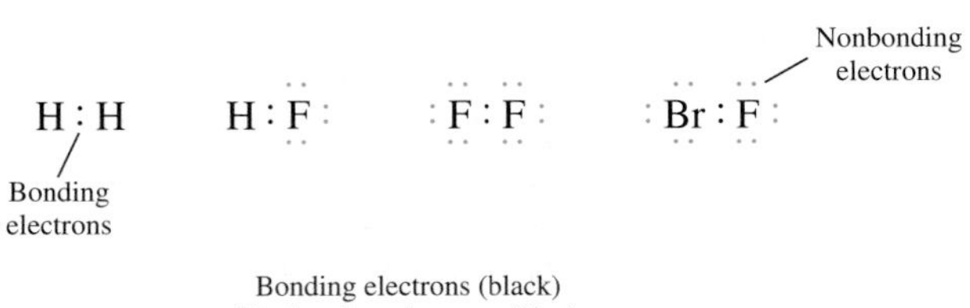

5.3 Single, Double, and Triple Covalent Bonds

A **single covalent bond** *is a bond in which two atoms share one pair of electrons.* The H_2, HF, F_2, and BrF molecules of the previous section all contain *single* covalent bonds.

Single covalent bonds are not adequate to explain covalent bonding in all molecules. Sometimes two atoms must share two or three pairs of electrons in order to provide a complete octet of electrons for each atom involved in the bonding. Such bonds are called *double* covalent bonds and *triple* covalent bonds. A **double covalent bond** *is a bond in which two atoms share two pairs of electrons.* A double covalent bond between two atoms is approximately twice as strong as a single covalent bond between the same two atoms; that is, it takes approximately twice as much energy to break the double bond as it does the single bond. A **triple covalent bond** *is a bond in which two atoms share three pairs of electrons.* A triple covalent bond is approximately three times as strong as a single covalent bond between the same two atoms. The term *multiple covalent bond* is a designation that applies to both double and triple covalent bonds.

One of the simplest molecules possessing a multiple covalent bond is the N_2 molecule, which has a triple covalent bond. A nitrogen atom has five valence electrons and needs three additional electrons to complete its octet.

·N̈·

In order to acquire a noble-gas electron configuration, each nitrogen atom must share three of its electrons with the other nitrogen atom.

:N· + ·N: ⟶ :N:::N: or :N≡N:

• A single line (dash) is used to denote a single covalent bond, two lines to denote a double covalent bond, and three lines to denote a triple covalent bond.

Note that all three shared electron pairs are placed in the area between the two nitrogen atoms in this bonding diagram. Just as one line is used to denote a single covalent bond, three lines are used to denote a triple covalent bond.

When you are "counting" electrons in an electron-dot structure to make sure that all atoms in the molecule have achieved their octet of electrons, *all* electrons in a double or triple bond are considered to belong to *both* of the atoms involved in that bond. The "counting" for the N_2 molecule would be

:N:::N:

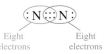

Eight electrons Eight electrons

Each of the circles around a nitrogen atom contains eight valence electrons. Circles are never drawn to include just some of the electrons in a double or triple bond.

A slightly more complicated molecule containing a triple covalent bond is the molecule C_2H_2 (acetylene). A carbon–carbon triple covalent bond is present as well as two carbon–hydrogen single bonds. The arrangement of valence electrons in C_2H_2 is as follows:

$$\mathrm{H\cdot\dot{\underset{\cdot}{C}}\cdot} \longrightarrow \longleftarrow \mathrm{\cdot\dot{\underset{\cdot}{C}}\cdot H} \longrightarrow \mathrm{H:C:::C:H} \quad \text{or} \quad \mathrm{H{-}C{\equiv}C{-}H}$$

The two atoms in a triple covalent bond are commonly the same element. However, they do not have to be. The molecule HCN contains a heteroatomic triple covalent bond.

$$\mathrm{H:C:::N:} \quad \text{or} \quad \mathrm{H{-}C{\equiv}N:}$$

A common molecule that contains a double covalent bond is carbon dioxide (CO_2). In fact, there are two carbon–oxygen double covalent bonds present in CO_2.

$$\mathrm{:\ddot{O}\cdot\ \cdot\dot{\underset{\cdot}{C}}\cdot\ \cdot\ddot{O}:} \longrightarrow \mathrm{:\underset{\cdot\cdot}{O}::C::\ddot{O}:} \quad \text{or} \quad \mathrm{:\underset{\cdot\cdot}{O}{=}C{=}\ddot{O}:}$$

Note in the following diagram how the circles are drawn for the octet of electrons about each of the atoms in carbon dioxide.

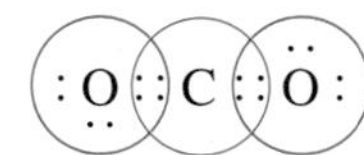

5.4 Valence Electrons and Number of Covalent Bonds Formed

Not all elements can form double or triple covalent bonds. There must be at least two vacancies in an atom's valence electron shell prior to bond formation if it is to participate in a double bond, and at least three vacancies are necessary for triple-bond formation. This requirement eliminates Group VIIA elements (fluorine, chlorine, bromine, iodine) and hydrogen from participating in such bonds. The Group VIIA elements have seven valence electrons and one vacancy, and hydrogen has one valence electron and one vacancy. All covalent bonds formed by these elements are single covalent bonds.

O, with six valence electrons, forms two covalent bonds.

Water, H_2O

Double bonding becomes possible for elements that need two electrons to complete their octet, and triple bonding becomes possible when three or more electrons are needed to complete an octet. Note that the word *possible* was used twice in the previous sentence. Multiple bonding does not have to occur when an element has two, three, or four vacancies in its octet; single covalent bonds can be formed instead. When more than one behavior is possible, the "bonding behavior" of an element is determined by the element or elements to which it is bonded.

N, with five valence electrons, forms three covalent bonds.

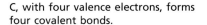

Ammonia, NH_3

Let us consider the possible "bonding behaviors" for O (six valence electrons, two octet vacancies), N (five valence electrons, three octet vacancies), and C (four valence electrons, four octet vacancies).

To complete its octet by electron sharing, an oxygen atom can form either two single bonds or one double bond.

$$\mathrm{:\underset{\cdot\cdot}{\overset{|}{O}}{-}} \qquad \mathrm{:\underset{\cdot\cdot}{O}{=}}$$

Two single bonds — One double bond

C, with four valence electrons, forms four covalent bonds.

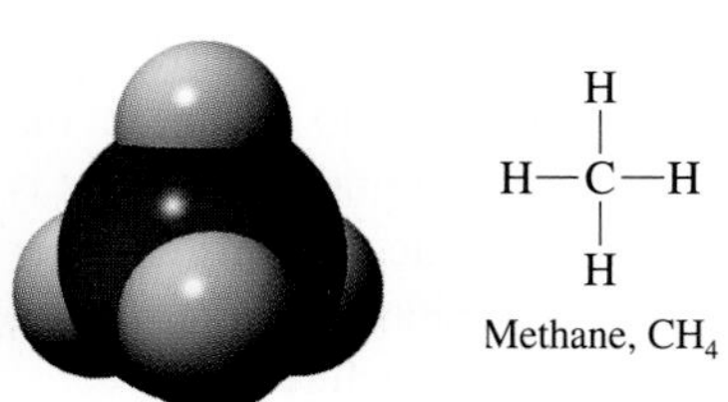

Methane, CH_4

Nitrogen is a very versatile element with respect to bonding. It can form single, double, or triple covalent bonds as dictated by the other atoms present in a molecule.

$$\mathrm{{-}\underset{|}{\ddot{N}}{-}} \qquad \mathrm{{-}\ddot{N}{=}} \qquad \mathrm{:N{\equiv}}$$

Three single bonds — One single and one double bond — One triple bond

Note that the nitrogen atom forms three bonds in each of these bonding situations. A double bond counts as two bonds, a triple bond as three. Because nitrogen has only five valence electrons, it must form three covalent bonds to complete its octet.

Carbon is an even more versatile element than nitrogen with respect to variety of types of bonding, as illustrated by the following possibilities. In each case, carbon forms four bonds.

• There is a strong tendency for atoms of nonmetallic elements to form a specific number of covalent bonds. The number of bonds formed is equal to the number of electrons the nonmetallic atom must share to obtain an octet of electrons.

—C— (with bonds above and below)	—C= (with bond above)	=C=	—C≡
Four single bonds	Two single bonds and one double bond	Two double bonds	One single bond and one triple bond

5.5 Coordinate Covalent Bonds

In the covalent bonds we have considered so far (single, double, and triple), all of the participating atoms in the bond contributed the same number of electrons to the bond. There is another, *less common* way in which a covalent bond can form. It is possible for one atom to supply two electrons and the other atom none to a shared electron pair. A **coordinate covalent bond** *is a bond in which both electrons of a shared pair come from one of the two atoms involved in the bond.* Coordinate covalent bonding enables an atom that has two or more vacancies in its valence shell to share a pair of nonbonding electrons that are located on another atom.

• An "ordinary" covalent bond can be thought of as a "Dutch-treat" bond; each atom "pays" its part of the bill. A coordinate covalent bond can be thought of as a "you-treat" bond; one atom pays the whole bill.

The element oxygen, with two vacancies in its valence octet, quite often forms coordinate covalent bonds. Consider the Lewis structures of the molecules HOCl (hypochlorous acid) and $HClO_2$ (chlorous acid).

$$\mathrm{H:\ddot{\underset{..}{O}}:\ddot{\underset{..}{Cl}}:} \qquad \mathrm{H:\ddot{\underset{..}{O}}:\ddot{\underset{..}{Cl}}:\overset{xx}{\underset{xx}{O}}{}^{x}_{x}}$$

Hypochlorous acid   Chlorous acid

In hypochlorous acid, all the bonds are "ordinary" covalent bonds. In chlorous acid, which differs from hypochlorous acid in that a second oxygen atom is present, the "new" chlorine–oxygen bond is a coordinate covalent bond. The second oxygen atom with six valence electrons (denoted by x's) needs two more for an octet. It shares one of the nonbonding electron pairs present on the chlorine atom. (The chlorine atom does not need any of the oxygen's electrons because it already has an octet.)

Atoms participating in coordinate covalent bonds generally deviate from the common bonding pattern (Section 5.5) expected for that type of atom. For example, oxygen normally forms two bonds; yet in the molecules N_2O and CO, which contain coordinate covalent bonds, oxygen forms one and three bonds, respectively.

• Atoms participating in coordinate covalent bonds generally do not form their normal number of covalent bonds.

$$:\mathrm{N{\equiv}N{-}\ddot{\underset{..}{O}}}: \qquad :\mathrm{C{\equiv}O}:$$

Once a coordinate covalent bond is formed, there is no way to distinguish it from any of the other covalent bonds in a molecule; all electrons are identical regardless of their source. The main use of the concept of coordinate covalency is to help rationalize the existence of certain molecules and ions whose electron-bonding arrangement would otherwise present problems. Figure 5.2 contrasts the formation of a "regular" covalent bond with that of a coordinate covalent bond.

• Once a coordinate covalent bond forms, it is indistinguishable from other covalent bonds in a molecule.

5.6 Systematic Procedures for Drawing Lewis Structures

Could you draw the Lewis structures of HOCl, $HClO_2$, N_2O, and CO given in the preceding section without any help? Drawing Lewis structures for diatomic molecules is usually straightforward and uncomplicated. However, with triatomic and even larger molecules, students often have trouble. Here is a stepwise procedure for distributing valence electrons as bonding and nonbonding pairs within a Lewis structure.

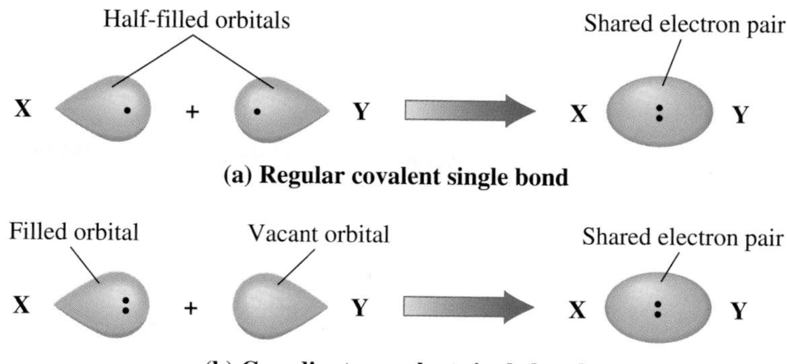

Figure 5.2
(a) A "regular" covalent single bond is the result of overlap of two half-filled orbitals. (b) A coordinate covalent single bond is the result of overlap of a filled and a vacant orbital.

The sulfur dioxide (SO_2) molecule. A computer-generated model.

Let us apply this stepwise procedure to the molecule SO_2, a molecule in which two oxygen atoms are bonded to a central sulfur atom.

Step 1: *Calculate the total number of valence electrons available in the molecule by adding together the valence electron counts for all atoms in the molecule.* The periodic table is a useful guide for determining this number.

An SO_2 molecule has 18 valence electrons available for bonding. Sulfur (Group VIA) has 6 valence electrons, and each oxygen (also Group VIA) has 6 valence electrons. The total number is therefore $6 + 2(6) = 18$.

Step 2: *Write the symbols of the atoms in the molecule in the order in which they are bonded to one another, and then place a single covalent bond, involving two electrons, between each pair of bonded atoms.* For SO_2, the S atom is the central atom. Thus we have

O : S : O

Determining which atom is the *central atom*—that is, which atom has the most other atoms bonded to it—is the key to determining the arrangement of atoms in a molecule or ion. Most other atoms present will be bonded to the central atom. For common binary molecular compounds, the molecular formula can help us determine the identity of the central atom. The central atom is the atom that appears only once in the formula; for example, S is the central atom in SO_3, O is the central atom in H_2O, and P is the central atom in PF_3. In molecular compounds containing hydrogen, oxygen, and an additional element, that additional element is the central atom; for example, N is the central atom in HNO_3, and S is the central atom in H_2SO_4. In compounds of this type, the oxygen atoms are bonded to the central atom, and the hydrogen atoms are bonded to the oxygens. Carbon is the central atom in nearly all carbon-containing compounds. Neither hydrogen nor fluorine is ever the central atom.

Step 3: *Add nonbonding electron pairs to the structure such that each atom bonded to the central atom has an octet of electrons. Remember that for hydrogen, an "octet" is only 2 electrons.*

For SO_2, addition of the nonbonding electrons gives

$$:\ddot{\underset{\cdot\cdot}{O}}:S:\ddot{\underset{\cdot\cdot}{O}}:$$

At this point, 16 of the 18 available electrons have been used.

Step 4: *Place any remaining electrons on the central atom of the structure.*

Placing the two remaining electrons on the S atom gives

$$:\ddot{\underset{\cdot\cdot}{O}}:\ddot{S}:\ddot{\underset{\cdot\cdot}{O}}:$$

Step 5: *If there are not enough electrons to give the central atom an octet, then use one or more pairs of nonbonding electrons on the atoms bonded to the central atom to form double or triple bonds.*

The S atom has only 6 electrons. Thus a nonbonding electron pair from an O atom is used to form a sulfur–oxygen double bond.

: O : S : O : ⟶ : O : S :: O :

This structure now obeys the octet rule.

Step 6: *Count the total number of electrons in the completed Lewis structure to make sure it is equal to the total number of valence electrons available for bonding, as calculated in Step 1.* This step serves as a "double-check" on the correctness of the Lewis structure.

For SO_2, there are 18 valence electrons in the Lewis structure of Step 5, the same number we calculated in Step 1.

Example 5.1

Drawing a Lewis Structure Using Systematic Procedures

Draw Lewis structures for the following molecules.

a. PF_3, a molecule in which P is the central atom and all F atoms are bonded to it
b. HCN, a molecule in which C is the central atom

Solution

Part a.

Step 1: Phosphorus (Group VA) has 5 valence electrons, and each of the fluorine atoms (Group VIIA) has 7 valence electrons. The total electron count is 5 + 3(7) = 26.

Step 2: Drawing the molecular skeleton with single covalent bonds (two electrons) placed between all bonded atoms gives

F : P : F
F

Step 3: Adding nonbonding electrons to the structure to complete the octets of all atoms bonded to the central atom gives

: F : P : F :
: F :

At this point, we have used 24 of the 26 available electrons.

Step 4: The central P atom has only 6 electrons; it needs 2 more. The 2 remaining available electrons are placed on the P atom, completing its octet. All atoms now have an octet of electrons.

: F : P : F :
: F :

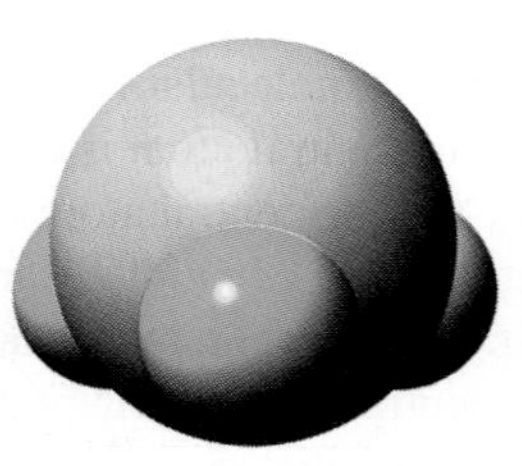

The phosphorus trifluoride (PF_3) molecule. A computer-generated model.

Step 5: This step is not needed; the central atom already has an octet of electrons.

Step 6: There are 26 electrons in the Lewis structure, the same number of electrons we calculated in Step 1.

Part b.

Step 1: Hydrogen (Group IA) has 1 valence electron, carbon (Group IVA) has 4 valence electrons, and nitrogen (Group VA) has 5 valence electrons. The total number of electrons is 10.

Step 2: Drawing the molecular skeleton with single covalent bonds between bonded atoms gives

H : C : N

The hydrogen cyanide (HCN) molecule. A computer-generated model.

Step 3: Adding nonbonding electron pairs to the structure such that the atoms bonded to the central atom have "octets" gives

$$\mathrm{H:C:\overset{..}{\underset{..}{N}}:}$$

Remember that hydrogen needs only 2 electrons.

Step 4: The structure in Step 3 has 10 valence electrons, the total number available. Thus there are no additional electrons available to place on the carbon atom to give it an octet of electrons.

Step 5: To give the central carbon atom its octet, 2 nonbonding electron pairs on the nitrogen atom are used to form a carbon–nitrogen triple bond.

$$\mathrm{H:C:\overset{..}{\underset{..}{N}}:} \longrightarrow \mathrm{H:C:::N:}$$

Step 6: The Lewis structure has 10 electrons, as calculated in Step 1.

Practice Exercise 5.1

Draw Lewis structures for the following molecules.

a. $SiCl_4$, a molecule in which Si is the central atom and all Cl atoms are bonded to it

b. H_2CO, a molecule in which C is the central atom and to which all other atoms are bonded.

• *Answer.* **a.** $\mathrm{:\overset{..}{\underset{..}{Cl}}:\overset{:\overset{..}{Cl}:}{\underset{:\underset{..}{Cl}:}{Si}}:\overset{..}{\underset{..}{Cl}}:}$ **b.** $\mathrm{H:\overset{\overset{..}{O}:}{\overset{::}{C}}:H}$

Chemical CONNECTIONS

5.1 Nitric Oxide: A Molecule Whose Bonding Does Not Follow "The Rules"

The bonding in most, *but not all,* simple molecules is easily explained using the systematic procedures for drawing Lewis structures described in Section 5.6. The molecule NO (nitric oxide) is an example of a simple molecule whose bonding does not conform to the standard rules for bonding. The presence of an odd number (11) of valence electrons in nitric oxide (5 from nitrogen and 6 from oxygen) makes it impossible to write a Lewis structure in which all electrons are paired as required by the octet rule. Thus an unpaired electron is present in the Lewis structure of NO.

Unpaired electron $\mathrm{:\dot{N}{=}\ddot{O}:}$

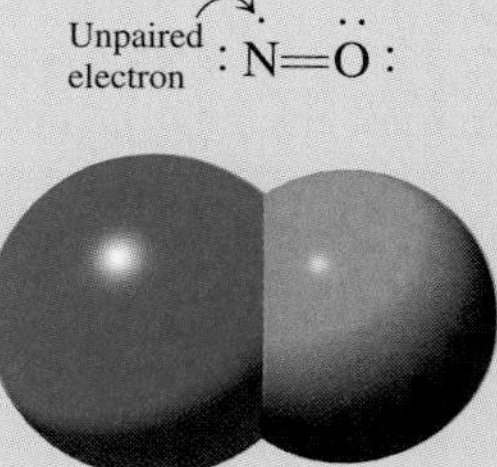

Despite the "nonconforming" nature of the bonding in nitric oxide, it is an abundant and important molecule within our environment. This colorless, odorless, nonflammable gas is generated by numerous natural and human-caused processes, including (1) lightning passing through air, which causes the N_2 and O_2 of air to react (to a small extent) with each other to produce NO, (2) automobile engines, within which the hot walls of the cylinders again cause N_2 and O_2 of air to become slightly reactive toward each other, and (3) a burning cigarette.

The fact that NO is produced in the preceding ways has been known for many years. The environmental effects of such NO have also been well documented. The NO serves as a precursor for the formation of both acid rain and smog.

During the early 1990s, it was found that NO is also an important biochemical that is naturally present in the human body. The body generates its own NO, usually from amino acids, and once formed, the NO has a life of about 10 seconds or less. Its biochemical functions within the human body include (1) helping maintain blood pressure by dilating blood vessels, (2) helping kill foreign invading molecules as part of the body's immune system response, and (3) serving as a biochemical messenger in the brain for processes associated with long-term memory.

5.7 Bonding in Compounds with Polyatomic Ions Present

Ionic compounds containing polyatomic ions (Section 4.10) present an interesting combination of both ionic and covalent bonds: covalent bonding *within* the polyatomic ion and ionic bonding *between* it and ions of opposite charge.

Polyatomic ion Lewis structures, which show the covalent bonding within such ions, are drawn using the same procedures as for molecular compounds (Section 5.6), with the accommodation that the total number of electrons used in the structure must be adjusted (increased or decreased) to take into account ion charge. The number of electrons is increased in the case of negatively charged ions and decreased in the case of positively charged ions.

• Students often erroneously assume that the charge associated with a polyatomic ion is assigned to a particular atom within the ion. Polyatomic ion charge is not localized on a particular atom but rather is associated with the ion as a whole.

In the Lewis structure for an *ionic compound* that contains a polyatomic ion, the positive and negative ions are treated separately to show that they are individual ions not linked by covalent bonds. The Lewis structure of potassium sulfate, K_2SO_4, is written as

• When we write the Lewis structure of an ion (monatomic or polyatomic), it is customary to use brackets and to show ionic charge outside the brackets.

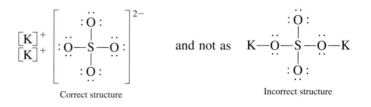

```
            ┌     ..     ┐2−                          ..
            │    :O:     │                           :O:
[K]+        │ ..  |   .. │                        ..  |  ..
[K]+        │:O—S—O:│       and not as    K—O—S—O—K
            │ ..  |   .. │                        ..  |  ..
            │    :O:     │                           :O:
            └     ..     ┘                            ..
          Correct structure                   Incorrect structure
```

Example 5.2

Drawing Lewis Structures for Polyatomic Ions

Draw a Lewis structure for SO_4^{2-}, a polyatomic ion in which S is the central atom and all O atoms are bonded to the S atom.

Solution

Step 1: Both S and O are Group VIA elements. Thus each of the atoms has 6 valence electrons. Two extra electrons are also present, which accounts for the −2 charge on the ion. The total electron count is 6 + 4(6) + 2 = 32.

Step 2: Drawing the molecular skeleton with single covalent bonds between bonded atoms gives

```
┌    O    ┐2−
│    ..   │
│ O : S : O │
│    ..   │
└    O    ┘
```

Step 3: Adding nonbonding electron pairs to give each oxygen atom an octet of electrons yields

```
┌     ..     ┐2−
│    :O:     │
│ ..  ..  .. │
│:O : S : O:│
│ ..  ..  .. │
│    :O:     │
└     ..     ┘
```

Step 4: The Step 3 structure has 32 electrons, the total number available. No more electrons can be added to the structure, and indeed, none need to be added because the central S atom has an octet of electrons. There is no need to proceed to Step 5.

The sulfate ion (SO_4^{2-}). A computer-generated model.

Practice Exercise 5.2

Draw a Lewis structure for BrO_3^-, a polyatomic ion in which Br is the central atom and all O atoms are bonded to it.

• *Answer:*

```
┌ ..  ..  .. ┐−
│:O :Br: O:│
│ ..  ..  .. │
│    :O:     │
└     ..     ┘
```

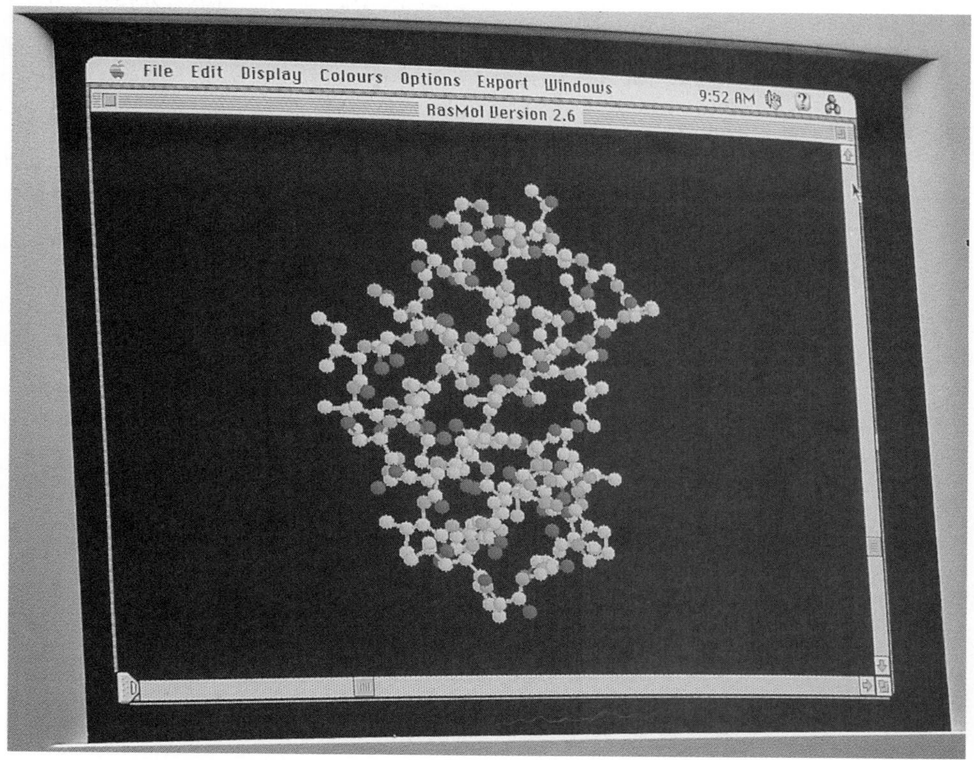

Figure 5.3
Computer-generated models that can be rotated on screen make it easy to visualize the three-dimensional nature of a given molecular structure.

5.8 The Shapes of Molecules

Lewis structures show the numbers and types of bonds present in molecules. They do not, however, convey any information about *molecular shape*—that is, about the actual three-dimensional arrangement of atoms in a molecule. Indeed, Lewis structures falsely imply that all molecules have flat, two-dimensional shapes.

Molecular shape is an important factor in determining the physical and chemical properties of a substance (Figure 5.3). Dramatic relationships between shape and properties are often observed in research associated with the development of prescription drugs. A small change in overall molecular shape, caused by the addition or removal of atoms, can enhance drug effectiveness and/or decrease drug side effects. Studies also show that the human senses of taste and smell depend in part on the shapes of molecules.

For molecules that contain only a few atoms, molecular shape can be predicted by using the information present in a molecule's Lewis structure and a procedure called valence shell electron pair repulsion (VSEPR) theory. **VSEPR theory** *is a set of procedures for predicting the three-dimensional shape of a molecule using the information contained in the molecule's Lewis structure.*

The central concept of VSEPR theory is that electron pairs in the valence shell of an atom adopt an arrangement in space that minimizes the repulsions between the like-charged (all negative) electron pairs. The specific arrangement adopted by the electron pairs depends on the number of electron pairs present. The electron pair arrangements about a *central atom* in the cases of two, three, and four electron pairs are as follows:

1. Two electron pairs, to be as far apart as possible from one another, are found on opposite sides of a nucleus—that is, at 180° angles to one another (Figure 5.4a). Such an electron pair arrangement is said to be *linear.*
2. Three electron pairs are as far apart as possible when they are found at the corners of an equilateral triangle. In such an arrangement, they are separated by 120° angles, giving a *trigonal planar* arrangement of electron pairs (Figure 5.4b).

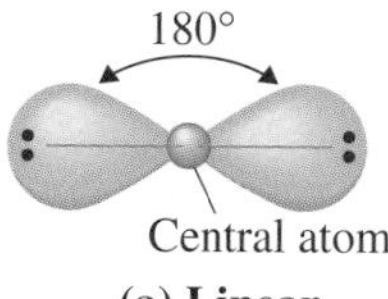

(a) Linear

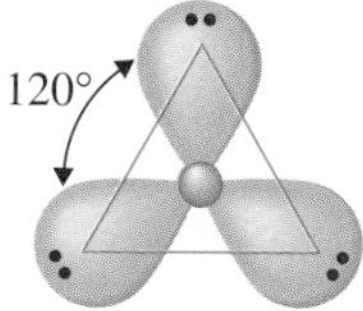

(b) Trigonal planar

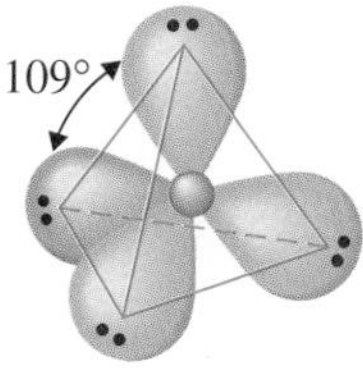

(c) Tetrahedral

Figure 5.4
Arrangements of valence electron pairs about a central atom that minimize repulsions between the pairs.

• The preferred arrangement of a given number of valence electron pairs about a central atom is the one that maximizes the separation among them. Such an arrangement minimizes repulsions between electron pairs.

3. A *tetrahedral* arrangement of electron pairs minimizes repulsions among four sets of electron pairs (Figure 5.4c). A tetrahedron is a four-sided solid in which all four sides are identical equilateral triangles. The angle between any two electron pairs is 109°.

For actual molecules, the number of valence electron pairs about a *central* atom is determined by using the following VSEPR theory conventions:

1. No distinction is made between bonding and nonbonding electron pairs. Both are counted.
2. Single, double, and triple bonds are all counted equally as "one pair," because each takes up only one region of space about the central atom.

• The acronym VSEPR is pronounced "vesper."

Let us now apply VSEPR theory to molecules in which two, three, and four VSEPR electron pairs are present about a central atom. Our operational rules will be

1. Draw a Lewis structure for the molecule and identify the specific atom for which geometrical information is desired. (This atom will usually be the central atom in the molecule.)
2. Count the number of VSEPR electron pairs present about the atom of interest.
3. Predict the VSEPR electron pair arrangement about the atom by assuming that the electron pairs orient themselves in a manner that minimizes repulsions (see Figure 5.4).

• Molecules with Two VSEPR Electron Pairs

All molecules with two VSEPR electron pairs are *linear.* Two common molecules with two VSEPR electron pairs are carbon dioxide (CO_2) and hydrogen cyanide (HCN), whose Lewis structures are

$$:\ddot{O}{=}C{=}\ddot{O}: \qquad H{-}C{\equiv}N:$$

In CO_2, the central carbon atom's two VSEPR pairs are the two double bonds. In HCN, the central carbon atom's two VSEPR pairs are a single bond and a triple bond. In both molecules, the VSEPR electron pairs arrange themselves on opposite sides of the carbon atom, which produces a linear molecule.

• Molecules with Three VSEPR Electron Pairs

Molecules with three VSEPR electron pairs have two possible molecular structures: *trigonal planar* and *angular.* The former occurs when all three VSEPR pairs are bonding and the latter when one of the three VSEPR pairs is nonbonding. The molecules H_2CO (formaldehyde) and SO_2 (sulfur dioxide) illustrate these two possibilities. Their Lewis structures are

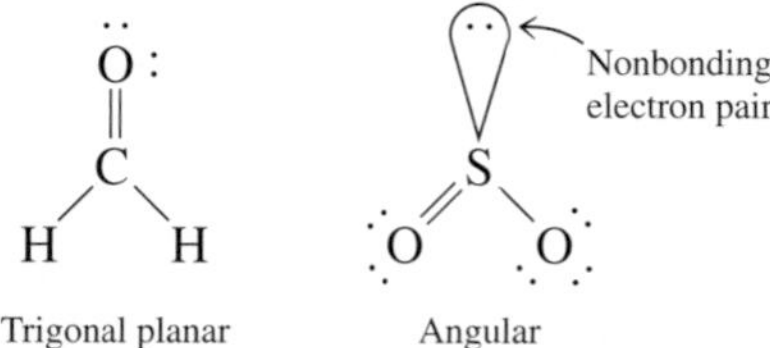

In both molecules, the VSEPR electron pairs are found at the corners of an equilateral triangle.

• VSEPR electron pair arrangement and molecular shape are not the same when a central atom possesses nonbonding electron pairs. The word used to describe the shape in such cases does not include the positions of the nonbonding electron pairs.

The shape of the SO_2 molecule is described as *angular* rather than *trigonal planar,* because molecular shape describes only *atom positions.* The positions of nonbonding electron pairs are not taken into account in describing molecular shape. Do not interpret this to mean that nonbonding electron pairs are unimportant in molecular shape determinations; indeed, in the case of SO_2, it is the presence of the nonbonding electron pair that makes the molecule angular rather than linear.

• "Dotted line" and "wedge" bonds can be used to indicate the directionality of bonds, as shown below.

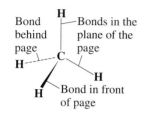

Bond behind page — Bonds in the plane of the page — Bond in front of page

• Molecules with Four VSEPR Electron Pairs

Molecules with four VSEPR electron pairs have three possible molecular shapes: *tetrahedral* (no nonbonding electron pairs present), *trigonal pyramidal* (one nonbonding electron pair present), and *angular* (two nonbonding electron pairs present). The molecules CH_4 (methane), NH_3 (ammonia), and H_2O (water) illustrate this sequence of molecular shapes.

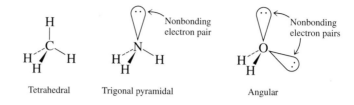

Nonbonding electron pair — Nonbonding electron pairs

Tetrahedral — Trigonal pyramidal — Angular

In all three molecules, the VSEPR electron pairs arrange themselves at the corners of a tetrahedron. Again, note that the word used to describe the shape of the molecule does not take into account the positioning of nonbonding electron pairs.

Chemical CONNECTIONS

5.2 Molecular Shape and Odor

Whether a substance has an odor depends on whether it can excite the olfactory nerve endings in the nose. Factors that determine whether excitation of nerve endings in the nose occurs include:

1. *Molecular volatility.* Odoriferous substances must be either gases or easily vaporized liquids or solids at room temperature; otherwise, the molecules of such substances would never reach the nose.
2. *Molecular solubility.* Olfactory nerve endings are covered with *mucus,* an aqueous solution that contains dissolved proteins and carbohydrates. Odoriferous molecules must be at least slightly soluble in this mucus.
3. *Molecular shape and polarity.* Olfactory nerve endings have receptor sites that accommodate molecules only of particular shapes and polarity. The interaction that causes excitation is similar to that between a key and a lock; the key must have a particular shape in order to open the lock.

One current theory for olfaction suggests that every known odor can be made from a combination of seven primary odors (see the accompanying table) in much the same way that all colors can be made from the three primary colors of red, yellow, and blue. Associated with each of the primary odors is a particular molecular polarity and/or molecular shape.

Primary odor	Familiar substance with this odor	Molecular characteristics
ethereal	dry cleaning fluid	rodlike shape
peppermint	mint candy	wedgelike shape
musk	some perfumes	disclike shape
camphoraceous	moth balls	spherical shape
floral	rose	disc with tail (kite) shape
pungent[a]	vinegar	negative polarity interaction
putrid[a]	skunk odor	positive polarity interaction

[a]These odors are less specific than are the other five odors, which indicates that they involve different interactions.

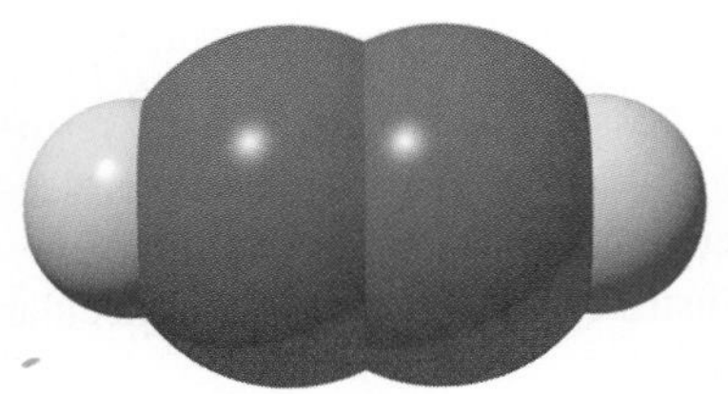

The acetylene (C_2H_2) molecule. A computer-generated model.

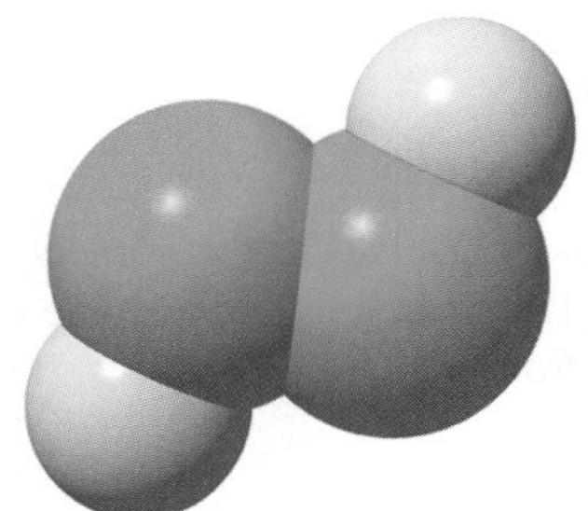

The hydrogen peroxide (H_2O_2) molecule. A computer-generated model.

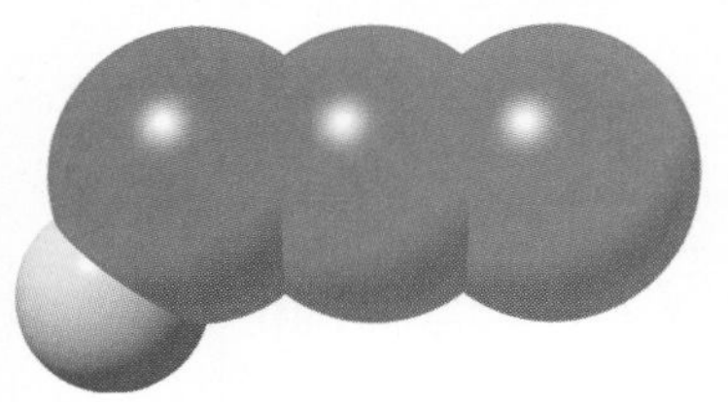

The hydrogen azide (HN_3) molecule. A computer-generated model.

• Molecules with More Than One Central Atom

The molecular shape of molecules that contain more than one central atom can be obtained by considering each central atom separately and then combining the results. Let us apply this principle to the molecules C_2H_2 (acetylene), H_2O_2 (hydrogen peroxide), and HN_3 (hydrogen azide), all of which have a four-atom "chain" structure. Their Lewis structures and VSEPR electron pair counts are as follows:

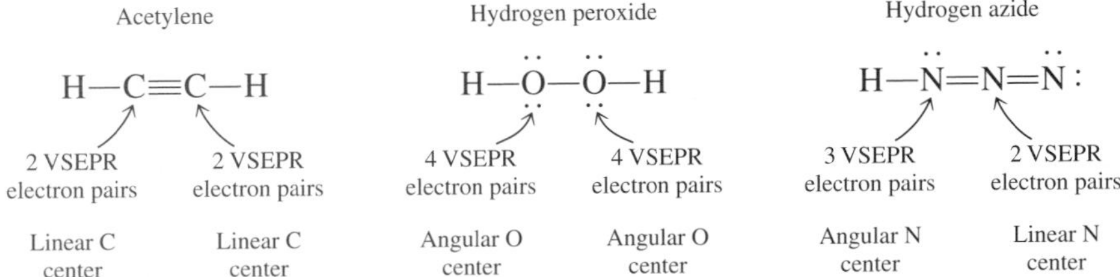

The molecular shapes for these three molecules are as follows:

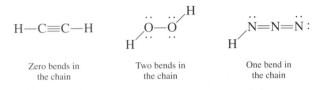

H—C≡C—H — Zero bends in the chain

H—O—O—H (bent) — Two bends in the chain

H—N=N=N: (bent at first N) — One bend in the chain

What we have said about molecular geometry is reviewed in Chemistry at a Glance on the following page.

5.9 Electronegativity

The ionic and covalent bonding models seem to represent two very distinct forms of bonding. Actually, the two models are closely related; they are the extremes of a broad continuum of bonding patterns. The close relationship between the two bonding models becomes apparent when the concepts of *electronegativity* (discussed in this section) and *bond polarity* (discussed in the next section) are considered.

The electronegativity concept has its origins in the fact that the nuclei of various elements have differing abilities to attract shared electrons (in a bond) to themselves. Some elements are better electron attractors than other elements. **Electronegativity** *is a measure of the relative attraction that an atom has for the shared electrons in a bond.*

Linus Carl Pauling (1901–1994). Pauling received the Nobel Prize in chemistry in 1954 for his work on the nature of the chemical bond. In 1962 he received the Nobel Peace Prize in recognition of his efforts to end nuclear weapons testing.

Linus Pauling, whose contributions to chemical bonding theory earned him a Nobel Prize in chemistry, was the first chemist to develop a *numerical* scale of electronegativity. Figure 5.5 gives Pauling electronegativity values for the more frequently encountered representative elements. The higher the electronegativity value for an element, the greater the attraction of atoms of that element for the shared electrons in bonds. The element fluorine, whose Pauling electronegativity value is 4.0, is the most electronegative of all elements; that is, it possesses the greatest electron-attracting ability for electrons in a bond.

As Figure 5.5 shows, electronegativity values increase from left to right across periods and from bottom to top within groups of the periodic table. These two trends result in nonmetals generally having higher electronegativities than metals. This fact is consistent with our previous generalization (Section 4.5) that metals tend to lose electrons and nonmetals tend to gain electrons when an ionic bond is formed. Metals (low electronegativities, poor electron attractors) give up electrons to nonmetals (high electronegativities, good electron attractors).

The Shape (Geometry) of Molecules

CHEMISTRY AT A GLANCE

PREDICTING MOLECULAR GEOMETRY USING VSEPR THEORY	**Operational rules** 1. Draw a Lewis structure for the molecule. 2. Count the number of VSEPR electron pairs about the central atom in the Lewis structure. 3. Assign a geometry based on minimizing repulsions between electron pairs.

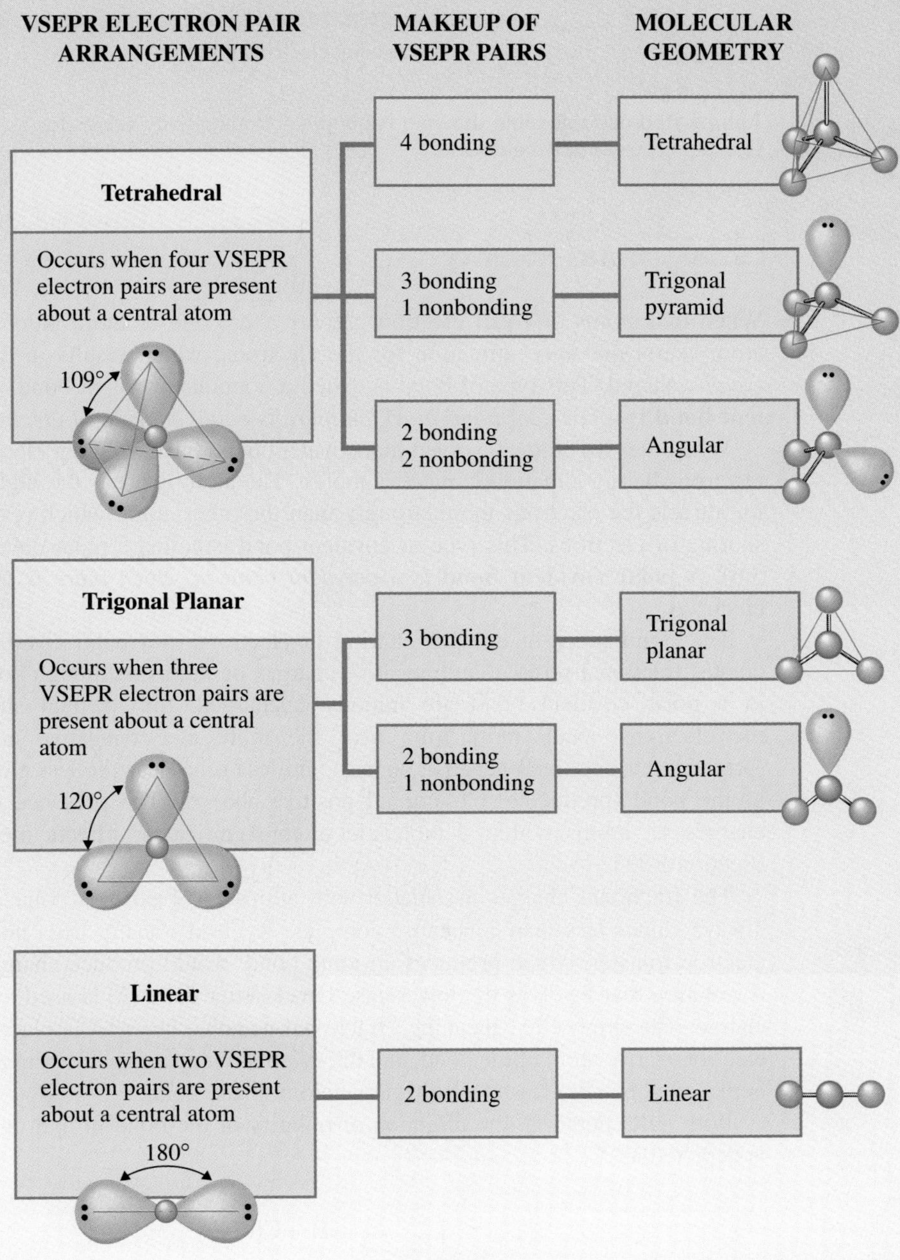

Note that the electronegativity for an element is not a directly measurable quantity. Rather, electronegativity values are calculated from bond energy information and other related experimental data. Values differ from element to element because of differences in atom size, nuclear charge, and number of inner-shell (nonvalence) electrons.

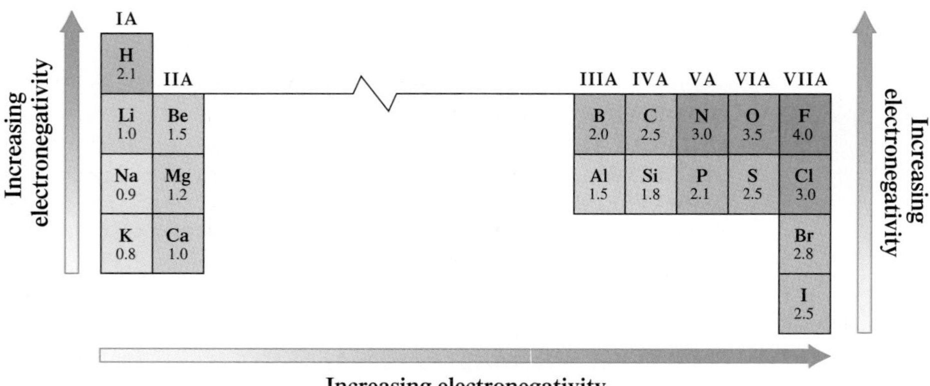

Figure 5.5
Abbreviated periodic table showing Pauling electronegativity values for selected representative elements.

5.10 Bond Polarity

When two atoms of equal electronegativity share one or more pairs of electrons, each atom exerts the same attraction for the electrons, which results in the electrons being *equally* shared. This type of bond is called a nonpolar covalent bond. A **nonpolar covalent bond** *is a covalent bond in which there is equal sharing of electrons.*

When the two atoms involved in a covalent bond have different electronegativities, the electron-sharing situation is more complex. The atom that has the higher electronegativity attracts the electrons more strongly than the other atom, which results in an *unequal* sharing of electrons. This type of covalent bond is called a polar covalent bond (Figure 5.6). A **polar covalent bond** *is a covalent bond in which there is unequal sharing of electrons.*

The significance of unequal sharing of electrons in a polar covalent bond is that it creates fractional positive and negative charges on atoms. Although both atoms involved in a polar covalent bond are initially uncharged, the unequal sharing means that the electrons spend more time near the more electronegative atom of the bond (producing a fractional negative charge) and less time near the less electronegative atom of the bond (producing a fractional positive charge). The presence of such fractional charges on atoms within a molecule often significantly affects molecular properties (Section 5.11).

• The δ^+ and δ^- symbols are pronounced "delta plus" and "delta minus." Whatever the magnitude of δ^+, it must be the same as that of δ^- because the sum of δ^+ and δ^- must be zero.

The fractional charges associated with atoms involved in a polar covalent bond are always values less than 1 because complete electron transfer does not occur. Complete electron transfer, which produces an ionic bond, would produce charges of $+1$ and -1. A notation that involves the lower-case Greek letter delta (δ) is used to denote fractional charge. The symbol δ^-, meaning "fractional negative charge," is placed above the more electronegative atom of the bond, and the symbol δ^+, meaning "fractional positive charge," is placed above the less electronegative atom of the bond.

With delta notation, the direction of polarity of the bond in hydrogen chloride (HCl) is depicted as

$$\overset{\delta^+}{\mathrm{H}}-\overset{\delta^-}{\ddot{\underset{\cdot\cdot}{\mathrm{Cl}}}}:$$

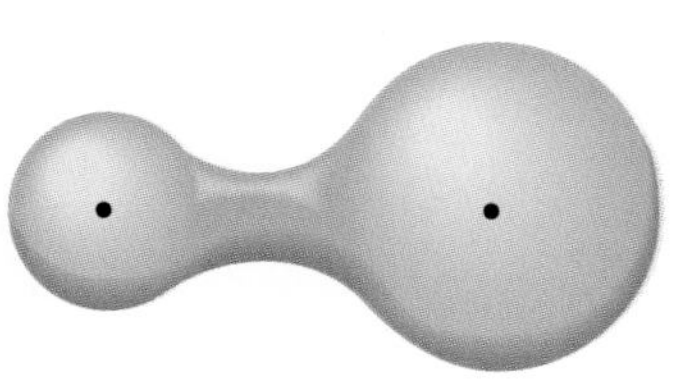

Figure 5.6
A polar covalent bond—a bond in which there is unequal sharing of electrons.

Chlorine is the more electronegative of the two elements; it dominates the electron-sharing process and draws the shared electrons closer to itself. Hence the chlorine end of the bond has the δ^- designation (the more electronegative element always has the δ^- designation).

The direction of polarity of a polar covalent bond can also be designated by using an arrow with a cross at one end ($+\!\!\longrightarrow$). The cross is near the end of the bond that is "positive," and the arrowhead is near the "negative" end of the bond. Using this notation, we would denote the bond in the molecule HCl as

$$\overset{+\!\!\longrightarrow}{\text{H—}\ddot{\underset{..}{\text{Cl}}}\text{:}}$$

An extension of the reasoning used in characterizing the covalent bond in the HCl molecule as polar leads to the generalization that most chemical bonds are not 100% covalent (equal sharing) or 100% ionic (no sharing). Instead, most bonds are somewhere in between (unequal sharing).

• Prediction of bond type, on the basis of electronegativity differences, is as follows:

Nonpolar covalent: zero difference
Polar covalent: greater than 0 but less than 2.0
Ionic: 2.0 or greater

The numerical value of the electronegativity difference between two bonded atoms gives an approximate measure of the polarity of the bond. The greater the numerical difference, the greater the inequality of electron sharing and the greater the polarity of the bond. For electronegativity differences of 2.0 or greater, inequality in electron sharing is sufficiently large that the bond is considered ionic (electron transfer).

The Chemistry at a Glance that follows summarizes much that we have said about chemical bonds in this chapter.

Example 5.3

Using Electronegativity Difference to Predict Bond Polarity and Bond Type

Consider the following bonds

N—Cl Ca—F C—O B—H N—O

a. Rank the bonds in order of increasing polarity.
b. Determine the direction of polarity for each bond.
c. Classify each bond as nonpolar covalent, polar covalent, or ionic.

Solution

Let us first calculate the electronegativity difference for each of the bonds by using the electronegativity values in Figure 5.5.

N—Cl: $3.0 - 3.0 = 0.0$
Ca—F: $4.0 - 1.0 = 3.0$
C—O: $3.5 - 2.5 = 1.0$
B—H: $2.1 - 2.0 = 0.1$
N—O: $3.5 - 3.0 = 0.5$

a. Bond polarity increases as electronegativity difference increases. Using the mathematical symbol $<$, which means "is less than," we can rank the bonds in terms of increasing bond polarity as follows:

$$\underset{0.0}{\text{N—Cl}} < \underset{0.1}{\text{B—H}} < \underset{0.5}{\text{N—O}} < \underset{1.0}{\text{C—O}} < \underset{3.0}{\text{Ca—F}}$$

b. The direction of bond polarity is from the least electronegative atom to the most electronegative atom. The more electronegative atom bears the partial negative charge (δ^-).

$$\text{N—Cl} \quad \overset{+\!\!\longrightarrow}{\text{B—H}} \quad \overset{+\!\!\longrightarrow}{\text{N—O}} \quad \overset{+\!\!\longrightarrow}{\text{C—O}} \quad \overset{+\!\!\longrightarrow}{\text{Ca—F}}$$

c. Nonpolar covalent bonds require zero difference in electronegativity, and ionic bonds require an electronegativity difference of 2.0 or greater. The in-between region characterizes polar covalent bonds.

Nonpolar covalent: N—Cl
Polar covalent: B—H, N—O, and C—O
Ionic: Ca—F

Covalent Bonds and Molecular Compounds — CHEMISTRY AT A GLANCE

COVALENT BONDS AND MOLECULAR COMPOUNDS

- A covalent bond results from the sharing of one or more pairs of electrons between atoms.
- A molecule is the basic structural unit in a covalently bonded compound.
- Covalent bonds form between similar or identical atoms—most often between nonmetals.
- Covalent bonds form by the sharing of electrons through an overlap of electron orbitals.

H• + •H ⟶ H:H

Hydrogen atoms — Shared electrons of covalent bond

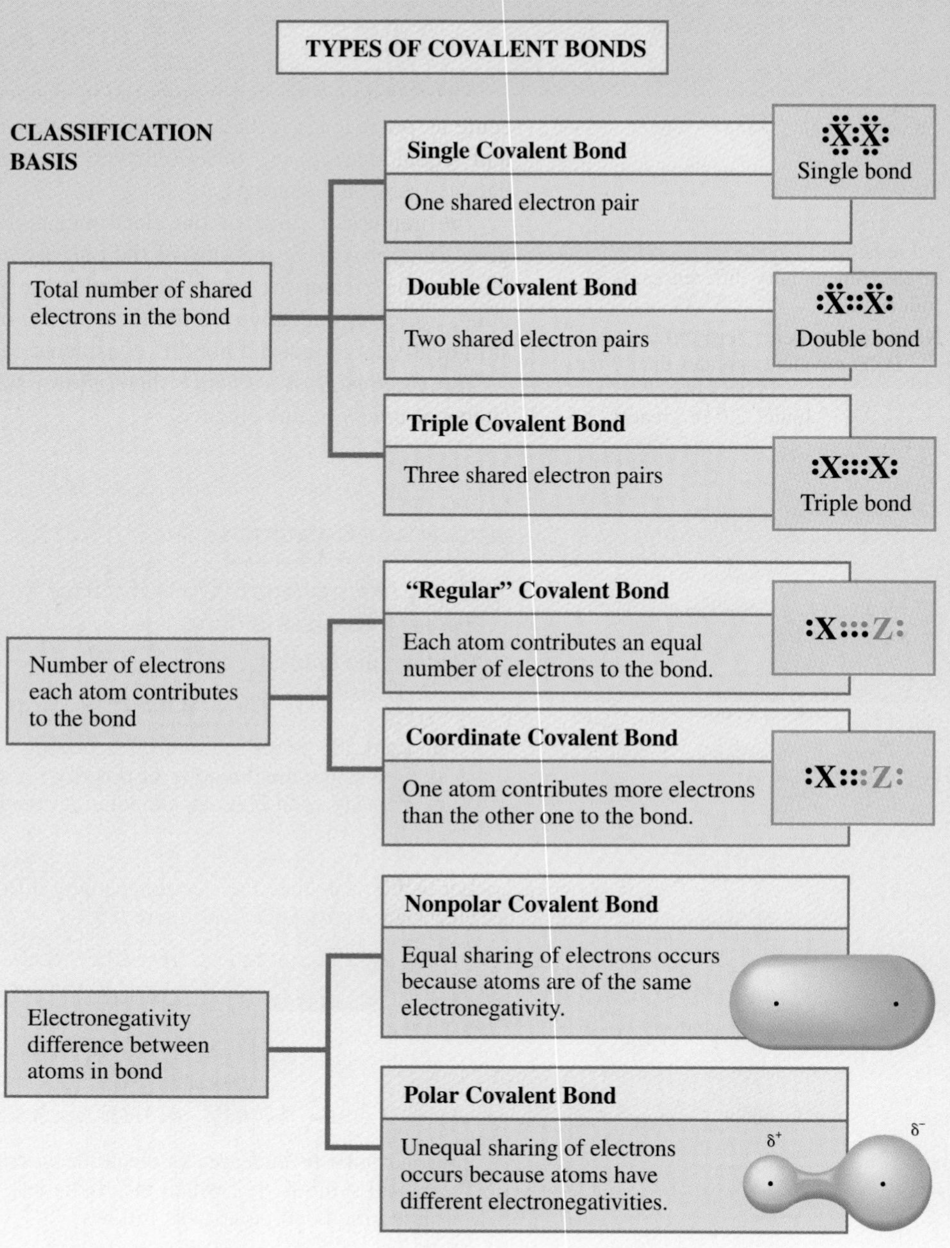

TYPES OF COVALENT BONDS
CLASSIFICATION BASIS
Total number of shared electrons in the bond
Single Covalent Bond
One shared electron pair
:X:X:
Single bond
Double Covalent Bond
Two shared electron pairs
:X::X:
Double bond
Triple Covalent Bond
Three shared electron pairs
:X:::X:
Triple bond
Number of electrons each atom contributes to the bond
"Regular" Covalent Bond
Each atom contributes an equal number of electrons to the bond.
:X:::Z:
Coordinate Covalent Bond
One atom contributes more electrons than the other one to the bond.
:X:::Z:
Electronegativity difference between atoms in bond
Nonpolar Covalent Bond
Equal sharing of electrons occurs because atoms are of the same electronegativity.
Polar Covalent Bond
Unequal sharing of electrons occurs because atoms have different electronegativities.
δ+
δ−

Practice Exercise 5.3

Consider the following bonds:

N—S H—H Na—F K—Cl F—Cl

a. Rank the bonds in order of increasing polarity.
b. Determine the direction of polarity for each bond.
c. Classify each bond as nonpolar covalent, polar covalent, or ionic.

• *Answers.* **a.** H—H < N—S < F—Cl < K—Cl < Na—F;

b. H—H N—S (⟵+) F—Cl (⟵+) K—Cl (+⟶) Na—F (+⟶);

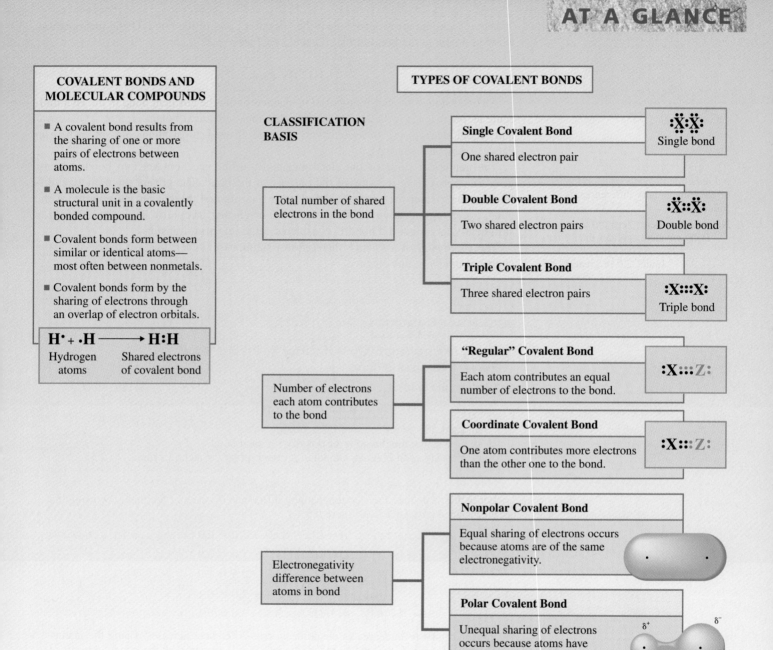

c. Nonpolar covalent: H—H
Polar covalent: N—S, F—Cl
Ionic: K—Cl, Na—F

5.11 Molecular Polarity

• A prerequisite for determining molecular polarity is a knowledge of molecular shape.

Molecules, as well as bonds, can have polarity. A molecule, as a whole, is polar if electrons are more attracted to one part of the molecule than to other parts. Molecular polarity depends on two factors: (1) bond polarities and (2) molecular shape (Section 5.7). In molecules that are symmetrical, the effects of polar bonds may cancel each other, resulting in the molecule as a whole having no polarity.

Determining the molecular polarity of a diatomic molecule is simple because only one bond is present. If that bond is nonpolar, then the molecule is nonpolar; if the bond is polar, then the molecule is polar.

Determining molecular polarity for triatomic molecules is more complicated. Two different molecular geometries are possible: linear and angular. In addition, the symmetrical nature of the molecule must be considered. Let us consider the polarities of three specific triatomic molecules: CO_2 (linear), H_2O (angular), and HCN (linear).

• Molecules in which all bonds are polar can be nonpolar if the bonds are so oriented in space that the polarity effects cancel each other.

In the linear CO_2 molecule, both bonds are polar (oxygen is more electronegative than carbon). Despite the presence of these polar bonds, CO_2 molecules are *nonpolar*. The effects of the two polar bonds are canceled as a result of the oxygen atoms being arranged symmetrically around the carbon atom. The shift of electronic charge toward one oxygen atom is exactly compensated for by the shift of electronic charge toward the other oxygen atom. Thus one end of the molecule is not negatively charged relative to the other end (a requirement for polarity), and the molecule is nonpolar. This cancellation of individual bond polarities, with crossed arrows used to denote the polarities, is diagrammed as follows:

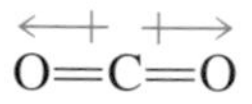

$$\overset{\longleftarrow\!+}{\text{O}{=}}\overset{+\!\longrightarrow}{\text{C}{=}\text{O}}$$

The nonlinear (angular) triatomic H_2O molecule is polar. The bond polarities associated with the two hydrogen–oxygen bonds do not cancel one another because of the nonlinearity of the molecule.

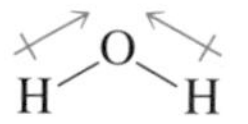

$$\text{H}\diagup^{\text{O}}\diagdown\text{H}$$

As a result of their orientation, both bonds contribute to an accumulation of negative charge on the oxygen atom. The two bond polarities are equal in magnitude but are not opposite in direction.

The generalization that linear triatomic molecules are nonpolar and nonlinear triatomic molecules are polar, which you might be tempted to make on the basis of our discussion of CO_2 and H_2O molecular polarities, is not valid. The linear molecule HCN, which is polar, invalidates this statement. Both bond polarities contribute to nitrogen's acquiring a partial negative charge relative to hydrogen in HCN.

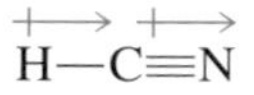

$$\overset{+\!\longrightarrow}{\text{H}{-}}\overset{+\!\longrightarrow}{\text{C}{\equiv}\text{N}}$$

(The two polarity arrows point in the same direction because nitrogen is more electronegative than carbon, and carbon is more electronegative than hydrogen.)

Molecules that contain four and five atoms commonly have trigonal planar and tetrahedral geometries, respectively. Such molecules in which all of the atoms attached to the central atom are identical, such as SO_3 (trigonal planar) and CH_4 (tetrahedral), are *nonpolar*. The individual bond polarities cancel as a result of the highly symmetrical arrangement of atoms around the central atom.

If two or more kinds of atoms are attached to the central atom in a trigonal planar or tetrahedral molecule, the molecule is polar. The high degree of symmetry required for cancellation of the individual bond polarities is no longer present. For example, if one of the hydrogen atoms in CH_4 (a nonpolar molecule) is replaced by a chlorine atom, then a polar molecule results, even though the resulting CH_3Cl is still a tetrahedral molecule. A carbon–chlorine bond has a greater polarity than a carbon–hydrogen bond; chlorine has an electronegativity of 3.0 and hydrogen has an electronegativity of only 2.1. Figure 5.7

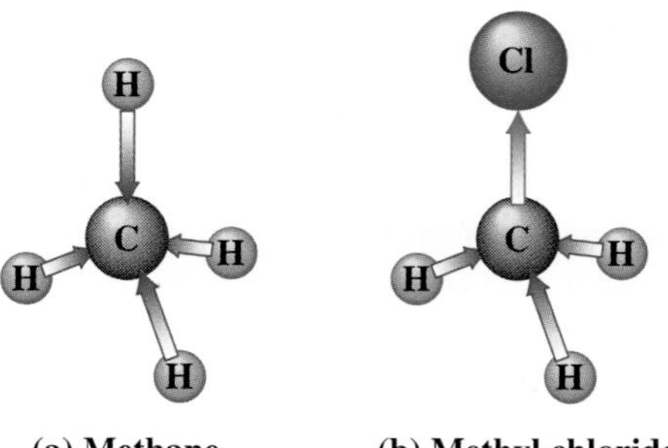

Figure 5.7
(a) Methane (CH_4) is a nonpolar tetrahedral molecule. (b) Methyl chloride (CH_3Cl) is a polar tetrahedral molecule. Bond polarities cancel in the first case, but not in the second.

contrasts the polar CH_3Cl and nonpolar CH_4 molecules. Note that the direction of polarity of the carbon–chlorine bond is opposite to that of the carbon–hydrogen bonds.

5.12 Naming Binary Molecular Compounds

• Numerical prefixes are used in naming binary molecular compounds. They are *never* used, however, in naming binary ionic compounds.

The names of binary molecular compounds are derived by using a rule very similar to that used for naming binary ionic compounds (Section 4.9). However, one major difference exists. Names for binary molecular compounds always contain numerical prefixes that give the number of each type of atom present in addition to the names of the elements present. This is in direct contrast to binary ionic compound nomenclature, where formula subscripts are never mentioned in the names.

Here is the basic rule to use when constructing the name of a binary molecular compound: *The full name of the nonmetal of lower electronegativity is given first, followed by a separate word containing the stem of the name of the more electronegative nonmetal and the suffix* -ide. *Numerical prefixes, giving numbers of atoms, precede the names of both nonmetals.* Thus the compounds N_2O, N_2O_3, and N_2O_4 are dinitrogen monoxide, dinitrogen trioxide, and dinitrogen tetroxide, respectively.

• When an element name begins with a vowel, an *a* or *o* at the end of the Greek prefix is dropped for phonetic reasons, as in pentoxide instead of pentaoxide.

Prefixes are necessary because several different compounds exist for most pairs of nonmetals. For example, all of the following nitrogen–oxygen compounds exist: NO, NO_2, N_2O, N_2O_3, N_2O_4, and N_2O_5. Such diverse behavior between two elements is related to the fact that single, double, and triple covalent bonds exist. The prefixes used are the standard numerical prefixes, which are given for the numbers 1 through 10 in Table 5.1. Example 5.4 shows how these prefixes are used in nomenclature for binary covalent compounds.

Table 5.1
Numerical Prefixes for the Numbers 1 Through 10

Prefix	Number
mono-	1
di-	2
tri-	3
tetra-	4
penta-	5
hexa-	6
hepta-	7
octa-	8
ennea-	9
deca-	10

Table 5.2
Selected Binary Molecular Compounds That Have Common Names

Compound formula	Accepted common name
H_2O	water
H_2O_2	hydrogen peroxide
NH_3	ammonia
N_2H_4	hydrazine
CH_4	methane
C_2H_6	ethane
PH_3	phosphine
AsH_3	arsine

The tetraphosphorus decoxide (P_4O_{10}) molecule. A computer-generated molecular model.

Example 5.4

Naming Binary Molecular Compounds

Name the following binary molecular compounds.

a. S_2Cl_2 **b.** CS_2 **c.** P_4O_{10} **d.** CBr_4

Solution

The names of each of these compounds will consist of two words. These words will have the following general formats:

$$\text{First word:} \quad (\text{prefix}) + \begin{pmatrix} \text{full name of least} \\ \text{electronegative nonmetal} \end{pmatrix}$$

$$\text{Second word:} \quad (\text{prefix}) + \begin{pmatrix} \text{stem of name of more} \\ \text{electronegative nonmetal} \end{pmatrix} + (\text{ide})$$

a. The elements present are sulfur and chlorine. The two portions of the name (including prefixes) are *disulfur and dichloride*, which are combined to give the name *disulfur dichloride.*

b. When only one atom of the first nonmetal is present, it is customary to omit the initial prefix *mono-*. Thus the name of this compound is *carbon disulfide.*

c. The prefix for four atoms is *tetra-* and for ten atoms is *deca-*. This compound has the name *tetraphosphorus decoxide.*

d. Omitting the initial *mono-* (see part **b**), we name this compound *carbon tetrabromide.*

Practice Exercise 5.4

Name the following binary molecular compounds.

a. PF_3 **b.** SO_2 **c.** P_4S_{10} **d.** $SiCl_4$

• *Answers.* **a.** phosphorus trifluoride; **b.** sulfur dioxide; **c.** tetraphosphorus decasulfide; **d.** silicon tetrachloride

• In Section 10.3, we will learn that placing hydrogen first in a formula conveys the message that the compound behaves as an acid in aqueous solution.

• Classification of a compound as ionic or molecular determines which set of nomenclature rules is used. For *nomenclature purposes*, binary compounds in which a metal and a nonmetal are present are considered ionic, and binary compounds that contain two nonmetals are considered covalent. Electronegativity differences are *not* used in classifying a compound as ionic or molecular for nomenclature purposes.

There is one standard exception to the use of numerical prefixes when naming binary molecular compounds. Compounds in which hydrogen is the first listed element in the formula are named without numerical prefixes. Thus the compounds H_2S and HCl are hydrogen sulfide and hydrogen chloride, respectively.

A few binary molecular compounds have names that are completely unrelated to the rules we have been discussing. They have common names that were coined prior to the development of systematic rules. At one time, in the early history of chemistry, all compounds had common names. With the advent of systematic nomenclature, most common names were discontinued. A few, however, have persisted and are now officially accepted. The most "famous" example is the compound H_2O, which has the systematic name hydrogen oxide, a name that is never used. The compound H_2O is *water,* a name that will never change. Table 5.2 lists other compounds for which common names are used in preference to systematic names.

Concepts to Remember

Covalent bonds. Covalent bond formation occurs through electron sharing between two nonmetal atoms. The covalent bond results from the common attraction of the two nuclei for the shared electrons.

Bonding and nonbonding electron pairs. Bonding electrons are pairs of valence electrons that are shared between atoms in a covalent bond. Nonbonding electrons are pairs of valence electrons about an atom that are not involved in electron sharing.

Types of covalent bonds. One shared pair of electrons constitutes a single covalent bond. Two or three pairs of electrons may be shared between atoms to give double and triple covalent bonds. Most often, both atoms of the bond contribute an equal number of electrons to the bond. In a few cases, however, both electrons of a shared pair come from the same atom; this is a coordinate covalent bond.

Number of covalent bonds formed. There is a strong tendency for nonmetals to form a particular number of covalent bonds. The number of valence electrons the nonmetal has and the number of covalent bonds it forms give a sum of eight.

Molecular shape. Molecular shape describes the way atoms in a molecule are arranged in space relative to one another. VSEPR theory is a set of procedures used to predict molecular shape from a compound's Lewis structure. VSEPR theory is based on the concept that valence shell electron pairs about an atom (bonding or nonbonding) orient themselves as far away from one another as possible (to minimize repulsions).

Electronegativity. Electronegativity is a measure of the relative attraction that an atom has for the shared electrons in a bond. Electronegativity values are useful in predicting the type of bond that forms (ionic or covalent).

Bond polarity. When atoms of like electronegativity participate in a bond, the bonding electrons are equally shared and the bond is nonpolar. When atoms of differing electronegativity participate in a bond, the bonding electrons are unequally shared and the bond is polar. In a polar bond, the more electronegative atom dominates the sharing process. The greater the electronegativity difference between two bonded atoms, the greater the polarity of the bond.

Molecular polarity. Molecules as a whole can have polarity. If individual bond polarities do not cancel because of the symmetrical nature of a molecule, then the molecule as a whole is polar.

Key Reactions and Equations

1. Molecular shape and central atom VSEPR electron pair count (Section 5.7)

Four VSEPR electron pairs none of which is nonbonding	= tetrahedral shape
Four VSEPR electron pairs one of which is nonbonding	= trigonal pyramidal shape
Four VSEPR electron pairs two of which are nonbonding	= angular shape
Three VSEPR electron pairs none of which is nonbonding	= trigonal planar shape
Three VSEPR electron pairs one of which is nonbonding	= angular shape
Two VSEPR electron pairs none of which is nonbonding	= linear shape

2. Bond characterization and electronegativity difference (Section 5.10)

Electronegativity difference of 2.0 or greater	= ionic bond
Electronegativity difference greater than 0 but less than 2.0	= polar covalent bond
Electronegativity difference of 0	= nonpolar covalent bond

Key Terms

Bonding electrons (5.2)
Coordinate covalent bond (5.5)
Double covalent bond (5.3)
Electronegativity (5.9)
Nonbonding electrons (5.2)
Nonpolar covalent bond (5.10)
Polar covalent bond (5.10)
Single covalent bond (5.3)
Triple covalent bond (5.3)
VSEPR theory (5.8)

Exercises and Problems

The members of each pair of problems in this section test similar material.

The Covalent Bond (Sections 5.1 through 5.5)

5.1 Draw Lewis structures to illustrate the covalent bonding in the following diatomic molecules.

a. Br_2 b. HI c. IBr d. BrF

5.2 Draw Lewis structures to illustrate the covalent bonding in the following diatomic molecules.

a. I_2 b. HF c. IF d. F_2

5.3 How many nonbonding electron pairs are present in each of the following Lewis structures?

a. $:N:::N:$ b. $H:\underset{\cdot\cdot}{\ddot{O}}:H$

c. $H:H$ d. $:\ddot{O}::\ddot{O}:\underset{\cdot\cdot}{\ddot{O}}:$

5.4 How many nonbonding electron pairs are present in each of the following Lewis structures?

a. $:C:::O:$ b. $H:C:::N:$

c. $:\ddot{O}::C::\ddot{O}:$ d. $:\underset{\cdot\cdot}{\ddot{Cl}}:\underset{\cdot\cdot}{\ddot{Cl}}:$

5.5 Convert each of the following Lewis structures into the form in which lines are used to denote shared electron pairs. Include nonbonding electron pairs in the rewritten structures.

a. $:N:::N:$ b. $H:\underset{\cdot\cdot}{\ddot{O}}:\underset{\cdot\cdot}{\ddot{O}}:H$

c. $\begin{array}{c} H:C:H \\ :: \\ :\underset{\cdot\cdot}{O} \end{array}$ d. $\begin{array}{c} H:C::C:H \\ \cdot\cdot \quad \cdot\cdot \\ H \quad H \end{array}$

5.6 Convert each of the following Lewis structures into the form in which lines are used to denote shared electron pairs. Include nonbonding electron pairs in the rewritten structures.

a. :C:::O:
b. H:N:N:H (each N has a nonbonding pair above and an H below)
c. :O:S:O: with :O bonded below S (all O atoms and S with nonbonding pairs)
d. H:C:::C:H

5.7 What would be the predicted formula for the simplest molecular compound formed between the following pairs of elements?

a. Nitrogen and fluorine
b. Chlorine and oxygen
c. Hydrogen and sulfur
d. Carbon and hydrogen

5.8 What would be the predicted formula for the simplest molecular compound formed between the following pairs of elements?

a. Nitrogen and hydrogen
b. Oxygen and fluorine
c. Sulfur and bromine
d. Carbon and chlorine

5.9 Identify the Period 2 nonmetal that would normally be expected to exhibit each of the following bonding capabilities.

a. Forms three single bonds
b. Forms two double bonds
c. Forms one single bond and one double bond
d. Forms two single bonds and one double bond

5.10 Identify the Period 3 nonmetal that would normally be expected to exhibit each of the following bonding capabilities.

a. Forms one triple bond
b. Forms one single bond and one triple bond
c. Forms four single bonds
d. Forms one double bond

5.11 What aspect of the following Lewis structure gives you a "hint" that the concept of coordinate covalency is needed to explain the bonding in the molecule?

:C:::O:

5.12 What aspect of the following Lewis structure gives you a "hint" that the concept of coordinate covalency is needed to explain the bonding in the molecule?

:N:::N:Ö: (O with nonbonding pairs above and below)

Drawing Lewis Structures (Section 5.6)

5.13 Draw Lewis structures to illustrate the covalent bonding in the following polyatomic molecules. The first atom in each formula is the central atom to which all other atoms are bonded.

a. PH_3 b. PCl_3 c. $SiBr_4$ d. OF_2

5.14 Draw Lewis structures to illustrate the covalent bonding in the following polyatomic molecules. The first atom in each formula is the central atom to which all other atoms are bonded.

a. AsH_3 b. $AsCl_3$ c. CBr_4 d. SCl_2

5.15 Draw Lewis structures for the simplest molecular compound likely to form between these pairs of elements.

a. Sulfur and fluorine
b. Carbon and iodine
c. Nitrogen and bromine
d. Selenium and hydrogen

5.16 Draw Lewis structures for the simplest molecular compound likely to form between these pairs of elements.

a. Nitrogen and chlorine
b. Bromine and hydrogen
c. Phosphorus and fluorine
d. Selenium and bromine

5.17 Draw Lewis structures to illustrate the bonding in the following molecules. In each case, there will be at least one multiple bond present in a molecule.

a. C_3H_4: A central carbon atom has two other carbon atoms bonded to it. Each of the noncentral carbon atoms also has two hydrogen atoms bonded to it.
b. N_2F_2: The two nitrogen atoms are bonded to one another, and each nitrogen atom also has a fluorine atom bonded to it.
c. C_2H_3N: The two carbon atoms are bonded to each other. One of the carbon atoms has a nitrogen atom bonded to it, and the other carbon atom has three hydrogen atoms bonded to it.
d. C_3H_4: A central carbon atom has two other carbon atoms bonded to it. One of the noncentral carbon atoms also has one hydrogen atom bonded to it, and the other one has three hydrogen atoms bonded to it.

5.18 Draw Lewis structures to illustrate the bonding in the following molecules. In each case, there will be at least one multiple bond present in a molecule.

a. $COCl_2$: Both chlorine atoms and the oxygen atom are bonded to the carbon atom.
b. $C_2H_2Br_2$: The two carbon atoms are bonded to one another. Each carbon atom also has a bromine atom and a hydrogen atom bonded to it.
c. C_2N_2: The two carbon atoms are bonded to one another, and each carbon atom also has a nitrogen bonded to it.
d. CH_2N_2: A central carbon atom has both nitrogen atoms bonded to it. Both hydrogen atoms are bonded to one of the two nitrogen atoms.

Lewis Structures for Polyatomic Ions (Section 5.7)

5.19 Draw Lewis structures for the following polyatomic ions.

a. OH^- b. BeH_4^{2-}
c. $AlCl_4^-$ d. NO_3^-

5.20 Draw Lewis structures for the following polyatomic ions.

a. CN^- b. PF_4^+
c. BH_4^- d. ClO_3^-

5.21 Draw Lewis structures for the following compounds that contain polyatomic ions.

a. NaCN b. K_3PO_4

5.22 Draw Lewis structures for the following compounds that contain polyatomic ions.

a. KOH b. NH_4Br

The Shapes of Molecules (VSEPR Theory) (Section 5.8)

5.23 Using VSEPR theory, predict whether each of the following triatomic molecules is linear or angular (bent).

a. H:S̈:H
b. H:Ö:C̈l:
c. :Ö::Ö:Ö:
d. :N::N::Ö:

5.24 Using VSEPR theory, predict whether each of the following triatomic molecules is linear or angular (bent).

a. :H:C:::N: b. :N̈::S̈:F̤̈:

c. :F̤̈:S̤̈:F̤̈: d. :C̤̈l:Ö̤:C̤̈l:

5.25 Using VSEPR theory, predict the shape of the following molecules.

a. :F̤̈:N̈:F̤̈: with :F̤: below N b. :C̤̈l:C̈:C̤̈l: with :Ö̤ below C

c. :Ö: above P; :C̤̈l:P:C̤̈l: with :C̤̈l: below P d. H above C; :C̤̈l:C:C̤̈l: with :C̤̈l: below C

5.26 Using VSEPR theory, predict the shape of the following molecules.

a. H:P̈:H with :C̤̈l: below P b. :C̤̈l:C::Ö: with H below C

c. H above C; :C̤̈l:C:C̤̈l: with H below C d. :F̤̈: above Si; H:Si:H with :C̤̈l: below Si

5.27 Using VSEPR theory, predict the shape of the following molecules.

a. NCl_3 b. $SiCl_4$ c. H_2Se d. SBr_2

5.28 Using VSEPR theory, predict the shape of the following molecules.

a. HOBr b. H_2Te c. NBr_3 d. SiF_4

5.29 Using VSEPR theory, predict the shape of the following molecules.

a. H:C::C:H with H H below the two C atoms b. H above C; H:C:Ö̤:H with H below C

5.30 Using VSEPR theory, predict the shape of the following molecules.

a. :Ö: above N; :O::N:Ö̤:H b. H above first C, :Ö̤: above second C; H:C:C:H with H below first C

Electronegativity (Section 5.9)

5.31 Using a periodic table, but not a table of electronegativity values, arrange each of the following sets of atoms in order of increasing electronegativity.

a. Na, Al, P, Mg b. Cl, Br, I, F

c. S, P, O, Al d. Ca, Mg, O, C

5.32 Using a periodic table, but not a table of electronegativity values, arrange each of the following sets of atoms in order of increasing electronegativity.

a. Be, N, O, B b. Li, C, B, K

c. S, Te, Cl, Se d. S, Mg, K, Ca

5.33 Use the information in Figure 5.5 as a basis for answering the following questions.

a. Which elements have electronegativity values that exceed that of the element carbon?

b. Which elements have electronegativity values of 1.0 or less?

c. What are the four most electronegative elements listed in Figure 5.5?

d. By what constant amount do the electronegativity values for sequential Period 2 elements differ?

5.34 Use the information in Figure 5.5 as a basis for answering the following questions.

a. Which elements have electronegativity values that exceed that of the element sulfur?

b. What are the four least electronegative elements listed in Figure 5.5?

c. Which three elements in Figure 5.5 have numerically equal electronegativities?

d. How does the electronegativity of the element hydrogen compare to that of the Period 2 elements?

Bond Polarity (Section 5.10)

5.35 Place δ^+ above the atom that is relatively positive and δ^- above the atom that is relatively negative in each of the following bonds. Try to answer this question without referring to Figure 5.5.

a. B—N b. Cl—F c. N—C d. F—O

5.36 Place δ^+ above the atom that is relatively positive and δ^- above the atom that is relatively negative in each of the following bonds. Try to answer this question without referring to Figure 5.5.

a. Cl—Br b. Al—S c. Br—S d. O—N

5.37 Rank the following bonds in order of increasing polarity.

a. H—Cl, H—O, H—Br b. O—F, P—O, Al—O

c. H—Cl, Br—Br, B—N d. P—N, S—O, Br—F

5.38 Rank the following bonds in order of increasing polarity.

a. H—Br, H—Cl, H—S b. N—O, Be—N, N—F

c. N—P, P—P, P—S d. B—Si, Br—I, C—H

5.39 Classify each of the following bonds as nonpolar covalent, polar covalent, or ionic on the basis of electronegativity differences.

a. C—O b. Na—Cl c. C—I d. Ca—S

5.40 Classify each of the following bonds as nonpolar covalent, polar covalent, or ionic on the basis of electronegativity differences.

a. Cl—F b. P—H c. C—H d. Ca—O

Molecular Polarity (Section 5.11)

5.41 Indicate whether each of the following hypothetical triatomic molecules is polar or nonpolar. Assume that A, X, and Y have different electronegativities.

a. A linear X—A—X molecule

b. A linear X—X—A molecule

c. An angular A—X—Y molecule

d. An angular X—A—Y molecule

5.42 Indicate whether each of the following hypothetical triatomic molecules is polar or nonpolar. Assume that A, X, and Y have different electronegativities.

a. A linear X—A—Y molecule

b. A linear A—Y—A molecule

c. An angular X—A—X molecule

d. An angular X—X—X molecule

5.43 Indicate whether each of the following molecules is polar or nonpolar. The molecular shape is given in parentheses.

a. NCl_3 (trigonal pyramid with N at the apex)

b. H_2Se (angular with Se in the center position)

c. CS_2 (linear with C in the center position)

d. $CHCl_3$ (tetrahedral with C in the center position)

5.44 Indicate whether each of the following molecules is polar or nonpolar. The molecular shape is given in parentheses.

a. PH_2Cl (trigonal pyramid with P at the apex)

b. SO_3 (trigonal planar with S in the center position)

c. CH_2Cl_2 (tetrahedral with C in the center position)

d. CCl_4 (tetrahedral with C in the center position)

Naming Binary Molecular Compounds (Section 5.12)

5.45 Name the following binary molecular compounds.

a. SF_4 b. P_4O_6 c. ClO_2 d. H_2S

5.46 Name the following binary molecular compounds.

a. Cl_2O b. CO c. PI_3 d. HI

5.47 Write formulas for the following binary molecular compounds.

a. Iodine monochloride b. Dinitrogen monoxide

c. Nitrogen trichloride d. Hydrogen bromide

5.48 Write formulas for the following binary molecular compounds.

a. Bromine monochloride b. Tetrasulfur dinitride

c. Sulfur trioxide d. Dioxygen difluoride

5.49 Write formulas for the following binary molecular compounds.

a. Hydrogen peroxide b. Methane

c. Ammonia d. Phosphine

5.50 Write formulas for the following binary molecular compounds.

a. Ethane b. Water

c. Hydrazine d. Arsine

Additional Problems

5.51 How many electron dots should appear in the Lewis structure for each of the following molecules or ions?

a. O_2F_2 b. $C_2H_2Br_2$

c. S_2^{2-} d. NH_4^+

5.52 In which of the following pairs of diatomic species do both members of the pair have bonds of the same multiplicity (single, double, triple)?

a. HCl and HF b. S_2 and Cl_2

c. CO and NO^+ d. OH^- and HS^-

5.53 Specify the reason why each of the following Lewis structures is incorrect using the following choices: (1) not enough electron dots, (2) too many electron dots, or (3) improper placement of a correct number of electron dots.

a. $:O{\equiv}O:$ b. $H:\ddot{\underset{..}{O}}:Cl$

c. $H:\ddot{O}:::\ddot{O}:H$ d. $\left[:\ddot{N}::\ddot{O}:\right]^+$

5.54 Specify both the electron pair geometry about the central atom and the molecular geometry for each of the following species.

a. SiH_4 b. NH_4^+

c. ClNO d. NO_3^-

5.55 Indicate which molecule in each of the following pairs of molecules is *more* polar.

a. BrCl and BrI b. CO_2 and SO_2

c. SO_3 and NF_3 d. H_3CF and Cl_3CF

5.56 Four hypothetical elements, A, B, C, and D, have electronegativities A = 3.8, B = 3.3, C = 2.8, and D = 1.3. These elements form the compounds BA, DA, DB, and CA. Arrange these compounds in order of increasing *ionic* bond character.

5.57 Successive substitution of F atoms for H atoms in the molecule CH_4 produces the molecules CH_3F, CH_2F_2, CHF_3, and CF_4.

a. Draw Lewis structures for each of the five molecules.

b. Using VSEPR theory, predict the geometric shape of each of the five molecules.

c. Give the polarity (polar or nonpolar) of each of the five molecules

5.58 Name each of the following binary compounds. (*Caution:* At least one of the compounds is ionic.)

a. NaCl b. BrCl

c. K_2S d. Cl_2O

Grid Problems

5.59

1. chlorine	2. hydrogen	3. fluorine
4. nitrogen	5. carbon	6. oxygen

Select from the grid *all* correct responses for each of the following situations.

a. Elements that would normally form three single covalent bonds

b. Elements that would normally form two double covalent bonds

c. Pairs of elements whose simplest compound would fit the formula XY_3

d. Pairs of elements whose simplest compound would be a diatomic molecule

5.60

1.	2.	3.
HCl	N_2	F_2
4. CO	5. CN^-	6. OH^-

Select from the grid *all* correct responses for each of the following situations.

a. Molecules or ions in which no multiple bonds are present
b. Molecules or ions that are polar
c. Molecules or ions that possess more bonding electrons than nonbonding electrons
d. Pairs of molecules or ions that have the same number of bonding electrons and the same number of nonbonding electrons

5.61

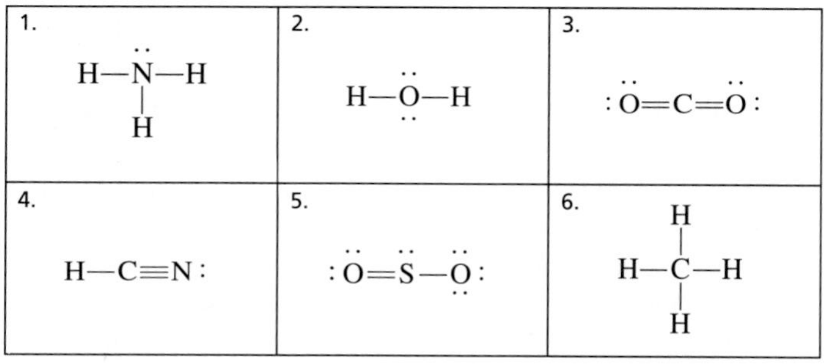

1.	2.	3.
H—N̈—H with H below N (N has one lone pair)	H—Ö—H (O has two lone pairs)	:Ö=C=Ö:
4. H—C≡N:	5. :Ö=S̈—Ö: (S has one lone pair; single-bonded O has three lone pairs)	6. H—C—H with H above and below C

Select from the grid *all* correct responses for each of the following situations.

a. Molecules that have an angular molecular shape
b. Molecules that have a tetrahedral arrangement of VSEPR electron pairs about the central atom
c. Molecules that have a trigonal planar arrangement of VSEPR electron pairs about the central atom
d. Molecules that are polar

5.62

1.	2.	3.
H_2S	CO_2	C_2H_6
4. N_2O	5. SO_3	6. NH_3

Select from the grid *all* correct responses for each of the following situations.

a. Molecules whose name includes the prefix *di-*
b. Molecules whose name includes the prefix *tri-*
c. Molecules whose name contains two numerical prefixes
d. Molecules whose name ends in *-ide*

CHAPTER 6

PART I

Chemical Calculations: Formula Masses, Moles, and Chemical Equations

CHAPTER OUTLINE

6.1 Formula Masses 127
6.2 The Mole: A Counting Unit for Chemists 129
6.3 The Mass of a Mole 130
6.4 The Mole and Chemical Formulas 132
6.5 The Mole and Chemical Calculations 133
6.6 Writing and Balancing Chemical Equations 136
6.7 Chemical Equations and the Mole Concept 140
6.8 Chemical Calculations Using Chemical Equations 140

Chemistry at a Glance:
Relationships Involving the Mole Concept 141

Chemical Connections
6.1 "Laboratory-Sized Amounts" versus "Industrial Chemistry-Sized Amounts" 142

The energy associated with a lightning discharge causes many different chemical reactions to occur within the atmosphere.

In this chapter we discuss "chemical arithmetic," the quantitative relationships between elements and compounds. Anyone who deals with chemical processes needs to understand at least the simpler aspects of this topic. All chemical processes, regardless of where they occur—in the human body, at a steel mill, on top of the kitchen stove, or in a clinical laboratory setting—are governed by the same mathematical rules.

We have already presented some information about chemical formulas (Section 1.10). In this chapter we discuss formulas again, and here we look beyond describing the composition of compounds in terms of constituent atoms. A new unit, the mole, will be introduced and its usefulness discussed. Chemical equations will be considered for the first time. We will learn how to represent chemical reactions by using chemical equations and how to derive quantitative relationships from these equations.

6.1 Formula Masses

Our entry point into the realm of "chemical arithmetic" is a discussion of the quantity called formula mass. The **formula mass** *of a substance is the sum of the atomic masses*

• Many chemists use the term *molecular mass* interchangeably with *formula mass* when dealing with substances that contain discrete molecules. It is incorrect, however, to use the term *molecular mass* when dealing with ionic compounds, because such compounds do not have molecules as their basic structural unit (Section 4.8).

of the atoms in its formula. Formula masses, like the atomic masses from which they are calculated, are relative masses based on the ${}^{12}_{6}C$ relative-mass scale (Section 3.3). Example 6.1 illustrates how formula masses are calculated.

Example 6.1

Using a Compound's Formula and Atomic Masses to Calculate Formula Mass

Calculate the formula mass of each of the following substances.

a. SnF_2 (tin(II) fluoride, a toothpaste additive)
b. $Al(OH)_3$ (aluminum hydroxide, a water purification chemical)

Solution

Formula masses are obtained simply by adding the atomic masses of the constituent elements, counting each atomic mass as many times as the symbol for the element occurs in the formula.

a. A formula unit of SnF_2 contains three atoms: one atom of Sn and two atoms of F. The formula mass, the collective mass of these three atoms, is calculated as follows:

$$1 \text{ atom Sn} \times \left(\frac{118.71 \text{ amu}}{1 \text{ atom Sn}}\right) = 118.71 \text{ amu}$$

$$2 \text{ atoms F} \times \left(\frac{19.00 \text{ amu}}{1 \text{ atom F}}\right) = 38.00 \text{ amu}$$

$$\text{Formula mass} = 156.71 \text{ amu}$$

We derive the conversion factors in the calculation from the atomic masses listed on the inside front cover of the text. Our rules for the use of conversion factors are the same as those discussed in Section 2.6.

Conversion factors are usually not explicitly shown in a formula mass calculation, as they are in our example; the calculation is simplified as follows:

$$\begin{aligned} \text{Sn:} \quad & 1 \times 118.71 \text{ amu} = 118.71 \text{ amu} \\ \text{F:} \quad & 2 \times 19.00 \text{ amu} = 38.00 \text{ amu} \\ & \text{Formula mass} = 156.71 \text{ amu} \end{aligned}$$

b. The formula for this compound contains parentheses. Improper interpretation of parentheses (see Section 4.11) is a common error made by students doing formula mass calculations. In the formula $Al(OH)_3$, the subscript 3 outside the parentheses affects only the symbols inside the parentheses. Thus we have

$$\begin{aligned} \text{Al:} \quad & 1 \times 26.98 \text{ amu} = 26.98 \text{ amu} \\ \text{O:} \quad & 3 \times 16.00 \text{ amu} = 48.00 \text{ amu} \\ \text{H:} \quad & 3 \times 1.01 \text{ amu} = 3.03 \text{ amu} \\ & \text{Formula mass} = 78.01 \text{ amu} \end{aligned}$$

In this text, we will always use atomic masses rounded to the hundredths place, as we have done in this example. This rule allows us to use, without rounding, the atomic masses given inside the front cover of the text. A benefit of this approach is that we always use the same atomic mass for a given element and thus become familiar with the atomic masses of the common elements.

Practice Exercise 6.1

Calculate the formula mass of each of the following substances.

a. $Na_2S_2O_3$ (sodium thiosulfate, a photographic chemical)
b. $(NH_2)_2CO$ (urea, a chemical fertilizer for crops)

• *Answers:* **a.** 158.12 amu; **b.** 60.07 amu

Figure 6.1
Oranges may be bought in units of mass (4-lb bag) or units of amount (3 oranges).

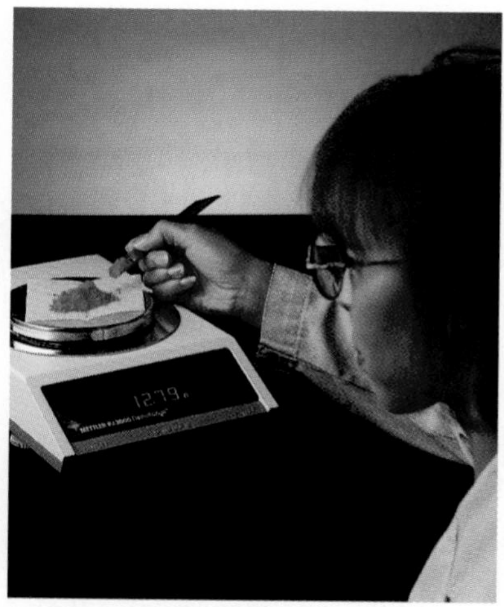

A basic process in chemical laboratory work is determining the mass of a substance.

6.2 The Mole: A Counting Unit for Chemists

The quantity of material in a sample of a substance can be specified either in terms of units of mass or in terms of units of *amount.* Mass is specified in terms of units such as grams, kilograms, and pounds. The amount of a substance is specified by indicating the number of objects present—3, 17, or 437, for instance.

We all use both units of mass and units of amount on a daily basis. For example, when buying oranges at the grocery store, we can decide on quantity in either mass units (4-lb bag or 10-lb bag) or amount units (3 oranges or 8 oranges) (Figure 6.1). In chemistry, as in everyday life, both mass and amount methods of specifying quantity are used. In laboratory work, practicality dictates working with quantities of known mass. Counting out a given number of atoms for a laboratory experiment is impossible because we cannot see individual atoms.

When we perform chemical calculations after the laboratory work has been done, it is often useful and even necessary to think of the quantities of substances present in terms of numbers of atoms or molecules instead of mass. When this is done, very large numbers are always encountered. Any macroscopic-sized sample of a chemical substance contains many trillions of atoms or molecules.

In order to cope with this large-number problem, chemists have found it convenient to use a special unit when counting atoms and molecules. Specialized counting units are used in many areas—for example, a *dozen* eggs or a *ream* (500 sheets) of paper.

• How large is the number 6.02×10^{23}? It would take an ultramodern computer that can count 100 million times a second 190 million years to count 6.02×10^{23} times. If each of the 6 billion people on Earth were made a millionaire (receiving 1 million dollar bills), we would still need 100 million other worlds, each inhabited with the same number of millionaires, in order to have 6.02×10^{23} dollar bills in circulation.

The chemist's counting unit is the *mole.* What is unusual about the mole is its magnitude. A **mole** *is* 6.02×10^{23} *objects.* The extremely large size of the mole unit is necessitated by the extremely small size of atoms and molecules. To the chemist, *one mole* always means 6.02×10^{23} objects, just as *one dozen* always means 12 objects. Two moles of objects is two times 6.02×10^{23} objects, and five moles of objects is five times 6.02×10^{23} objects.

Avogadro's number *is the name given to the numerical value* 6.02×10^{23}. This designation honors Amedeo Avogadro (1776–1856), an Italian physicist whose pioneering work on gases later proved valuable in determining the number of particles present in given volumes of substances. When we solve problems dealing with the number of objects (atoms or molecules) present in a given number of moles of a substance, Avogadro's number becomes part of the conversion factor used to relate the number of objects present to the number of moles present.

• Why the number 6.02×10^{23}, rather than some other number, was chosen as the counting unit of chemists is discussed in Section 6.3. A more formal definition of the mole will also be presented in that section.

From the definition

$$1 \text{ mole} = 6.02 \times 10^{23} \text{ objects}$$

two conversion factors can be derived:

$$\frac{6.02 \times 10^{23} \text{ objects}}{1 \text{ mole}} \quad \text{and} \quad \frac{1 \text{ mole}}{6.02 \times 10^{23} \text{ objects}}$$

Example 6.2 illustrates the use of these conversion factors in solving problems.

Everyday counting units—a dozen, a pair, and a ream.

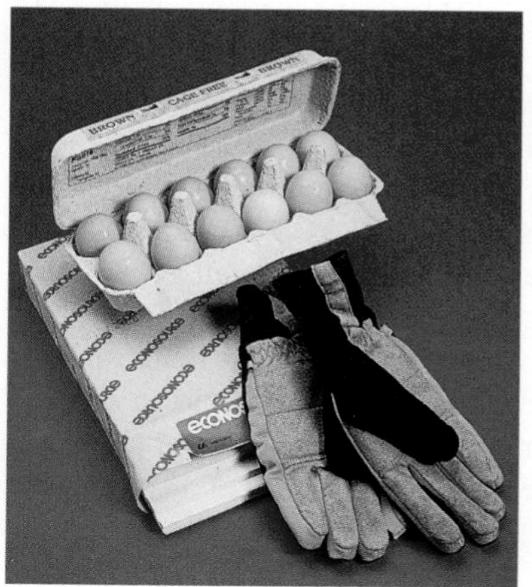

Example 6.2

Calculating the Number of Objects in a Molar Quantity

How many objects are there in each of the following quantities?

a. 0.23 mole of aspirin molecules **b.** 1.6 moles of oxygen atoms

Solution

Dimensional analysis (Section 2.9) will be used to solve each of these problems. Both of the problems are similar in that we are given a certain number of moles of substance and want to find the number of objects present in the given number of moles. We will need Avogadro's number to solve each of these moles-to-particles problems.

Amedeo Avogadro (1776–1856) was the first scientist to distinguish between atoms and molecules. His name is associated with the number 6.02×10^{23}, the number of particles (atoms or molecules) in a mole.

Moles of Substance → (Conversion factor involving Avogadro's number) → Particles of Substance

a. The objects of concern are molecules of aspirin. The given quantity is 0.23 mole of aspirin molecules, and the desired quantity is the number of aspirin molecules.

$$0.23 \text{ mole aspirin molecules} = \text{? aspirin molecules}$$

Applying dimensional analysis here involves the use of a single conversion factor, one that relates moles and molecules.

$$0.23 \cancel{\text{mole aspirin molecules}} \times \left(\frac{6.02 \times 10^{23} \text{ aspirin molecules}}{1 \cancel{\text{mole aspirin molecules}}}\right)$$
$$= 1.4 \times 10^{23} \text{ aspirin molecules}$$

b. This time we are dealing with atoms instead of molecules. This switch does not change the way we work the problem. We will need the same conversion factor.

The given quantity is 1.6 moles of oxygen atoms, and the desired quantity is the actual number of oxygen atoms present.

$$1.6 \text{ moles oxygen atoms} = \text{? oxygen atoms}$$

The setup is

$$1.6 \cancel{\text{moles oxygen atoms}} \times \left(\frac{6.02 \times 10^{23} \text{ oxygen atoms}}{1 \cancel{\text{mole oxygen atoms}}}\right)$$
$$= 9.6 \times 10^{23} \text{ oxygen atoms}$$

Practice Exercise 6.2

How many objects are there in each of the following quantities?

a. 0.46 mole of vitamin C molecules **b.** 1.27 moles of copper atoms

• *Answers:* **a.** 2.8×10^{23} vitamin C molecules; **b.** 7.65×10^{23} copper atoms

6.3 The Mass of a Mole

How much does a mole weigh? Are you uncertain about the answer to that question? Let us consider a similar but more familiar question first: "How much does a dozen weigh?" Your response is now immediate: "A dozen what?" The mass of a dozen identical objects obviously depends on the identity of the object. For example, the mass of a dozen elephants is greater than the mass of a dozen peanuts. The mass of a mole, like the mass of a dozen, depends on the identity of the object. Thus the mass of a mole, or *molar mass,* is not a set number; it varies and is different for each chemical substance (see Figure 6.2). This is in direct contrast to the *molar number,* Avogadro's number, which is the same for all chemical substances.

The **molar mass** *of a substance is the mass in grams that is numerically equal to the substance's formula mass.* For example, the formula mass (atomic mass) of the element sodium is 22.99 amu; therefore, 1 mole of sodium weighs 22.99 g. In Example 6.1, we calculated that the formula mass of tin(II) fluoride is 156.71 amu; therefore, 1 mole of tin(II) fluoride weighs 156.71 g. We can obtain the actual mass in grams of 1 mole of any substance by computing its formula mass (atomic mass for elements) and writing "grams" after it. Thus, when we add atomic masses to get the formula mass (in amu's) of a compound, we are simultaneously finding the mass of 1 mole of that compound (in grams).

• The mass value below each symbol in the periodic table is both an atomic mass in atomic mass units and a molar mass in grams. For example, the mass of one nitrogen atom is 14.01 amu, and the mass of 1 mole of nitrogen atoms is 14.01 g.

It is not a coincidence that the molar mass of a substance and its formula mass or atomic mass match numerically. Avogadro's number has the value that it has in order to cause this relationship to exist. The numerical match between molar mass and atomic or formula mass makes calculating the mass of any given number of moles of a substance a very simple procedure. When you solve problems of this type, the numerical value of the molar mass becomes part of the conversion factor used to convert from moles to grams.

Figure 6.2
The mass of a mole is not a set number of grams; it depends on the substance. For the substances shown, the mass of 1 mole (clockwise from sulfur, the yellow solid) is as follows: sulfur, 32.07 g; zinc, 65.38 g; carbon, 12.0 g; magnesium, 24.30 g; lead, 207.2 g; silicon, 28.09 g; copper, 63.55 g; and in the center, mercury, 200.6 g.

For example, for the compound CO_2, which has a formula mass of 44.01 amu, we can write the equality

$$44.01 \text{ g } CO_2 = 1 \text{ mole } CO_2$$

From this statement (equality), two conversion factors can be written:

$$\frac{44.01 \text{ g } CO_2}{1 \text{ mole } CO_2} \quad \text{and} \quad \frac{1 \text{ mole } CO_2}{44.01 \text{ g } CO_2}$$

Example 6.3 illustrates the use of gram-to-mole conversion factors like these in solving problems.

Example 6.3

Calculating the Mass of a Molar Quantity of Compound

Acetaminophen, the pain-killing ingredient in Tylenol formulations, has the formula $C_8H_9O_2N$. Calculate the mass, in grams, of a 0.30-mole sample of this pain reliever.

Solution

We will use dimensional analysis to solve this problem. The relationship between molar mass and formula mass will serve as a conversion factor in the setup of this problem.

Moles of Substance → (Conversion factor involving molar mass) → Grams of Substance

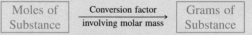

The given quantity is 0.30 mole of $C_8H_9O_2N$, and the desired quantity is grams of this same substance.

$$0.30 \text{ mole } C_8H_9O_2N = ? \text{ grams } C_8H_9O_2N$$

The calculated formula mass of $C_8H_9O_2N$ is 151.18 amu. Thus,

$$151.18 \text{ grams } C_8H_9O_2N = 1 \text{ mole } C_8H_9O_2N$$

With this relationship in the form of a conversion factor, the setup for the problem becomes

$$0.30 \text{ mole } C_8H_9O_2N \times \left(\frac{151.18 \text{ g } C_8H_9O_2N}{1 \text{ mole } C_8H_9O_2N}\right) = 45 \text{ g } C_8H_9O_2N$$

Practice Exercise 6.3

Carbon monoxide (CO) is an air pollutant that enters the atmosphere primarily in automobile exhaust. Calculate the mass in grams of a 2.61-mole sample of this air pollutant.

• *Answer:* 73.1 g CO

The atomic mass unit (amu) and the grams (g) unit are related to one another through Avogadro's number.

$$6.02 \times 10^{23} \text{ amu} = 1.00 \text{ g}$$

That this is the case can be deduced from the following equalities:

$$\text{Atomic mass of N} = \text{mass of 1 N atom} = 14.01 \text{ amu}$$
$$\text{Molar mass of N} = \text{mass of } 6.02 \times 10^{23} \text{ N atoms} = 14.01 \text{ g}$$

Because the second line of the equalities involves 6.02×10^{23} times as many atoms and the masses come out numerically equal, the gram unit must be 6.02×10^{23} times larger than the amu unit.

In Section 6.2 we defined the mole simply as

$$1 \text{ mole} = 6.02 \times 10^{23} \text{ objects}$$

• The numerical relationship between the amu unit and the gram unit can be expressed as

$$6.02 \times 10^{23} \text{ amu} = 1.00 \text{ g}$$

or as

$$1 \text{ amu} = 1.66 \times 10^{-24} \text{ g}$$

This second equality is obtained from the first by dividing each side of the first equality by 6.02×10^{23}.

Although this statement conveys correct information (the value of Avogadro's number to three significant figures is 6.02×10^{23}), it is not the officially accepted definition for Avogadro's number. The official definition, which is based on mass, is as follows: The **mole** *is the amount of substance in a system that contains as many elementary particles (atoms, molecules, or formula units) as there are* $^{12}_{6}C$ *atoms in exactly 12 grams of* $^{12}_{6}C$. The value of Avogadro's number is an experimentally determined quantity (the number of atoms in exactly 12 g of $^{12}_{6}C$ atoms) rather than a defined quantity. Its value is not even mentioned in the definition. The most up-to-date experimental value for Avogadro's number is 6.022137×10^{23}, which is consistent with our previous definition (Section 6.2).

6.4 The Mole and Chemical Formulas

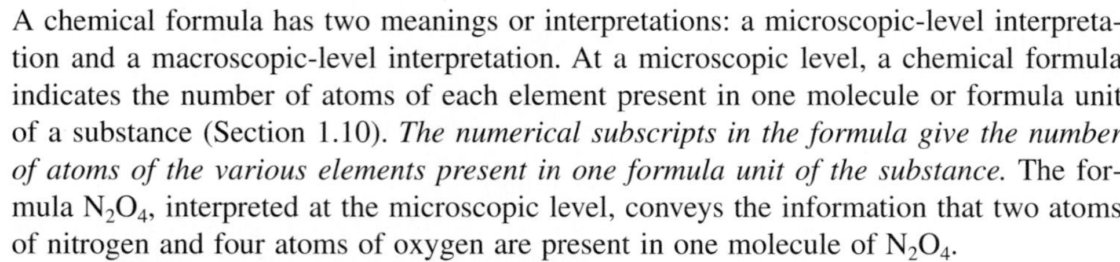

A chemical formula has two meanings or interpretations: a microscopic-level interpretation and a macroscopic-level interpretation. At a microscopic level, a chemical formula indicates the number of atoms of each element present in one molecule or formula unit of a substance (Section 1.10). *The numerical subscripts in the formula give the number of atoms of the various elements present in one formula unit of the substance.* The formula N_2O_4, interpreted at the microscopic level, conveys the information that two atoms of nitrogen and four atoms of oxygen are present in one molecule of N_2O_4.

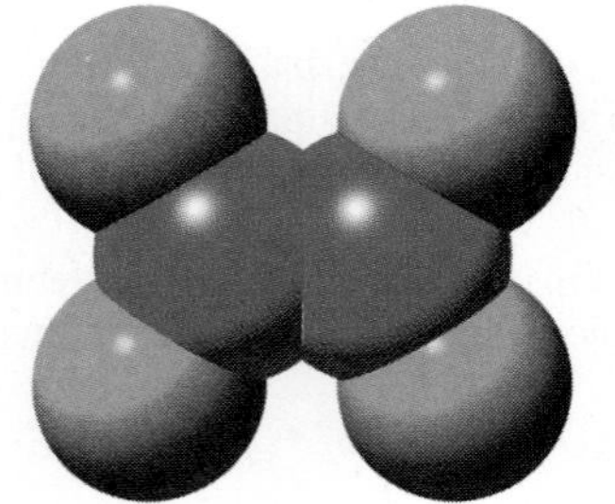

A computer-generated model of the molecular structure of the compound N_2O_4.

Now that the mole concept has been introduced, a macroscopic interpretation of chemical formulas is possible. At a macroscopic level, a chemical formula indicates the number of moles of atoms of each element present in one mole of a substance. *The numerical subscripts in the formula give the number of moles of atoms of the various elements present in 1 mole of the substance.* The designation *macroscopic* is given to this molar interpretation because moles are laboratory-sized quantities of atoms. The formula N_2O_4, interpreted at the macroscopic level, conveys the information that 2 moles of nitrogen atoms and 4 moles of oxygen atoms are present in 1 mole of N_2O_4 molecules. Thus the subscripts in a formula always carry a dual meaning: atoms at the microscopic level and moles of atoms at the macroscopic level.

• The molar (macroscopic-level) interpretation of a chemical formula is used in calculations where information about a *particular element within a compound* is needed.

When it is necessary to know the number of moles of a particular element *within* a compound, the subscript of that element's symbol in the chemical formula becomes part of the conversion factor used to convert from moles of compound to moles of element *within* the compound. Using N_2O_4 as our chemical formula, we can write the following conversion factors:

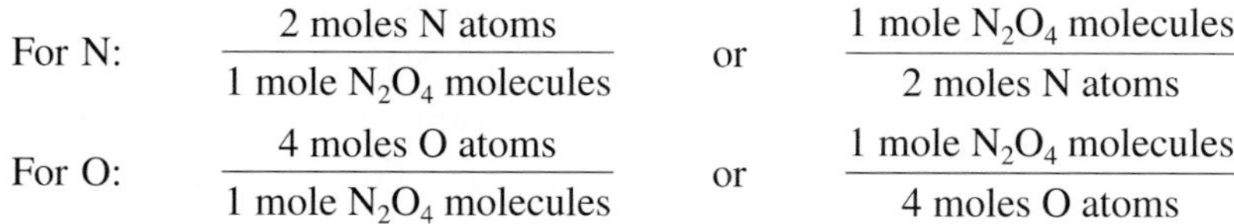

$$\text{For N:} \quad \frac{2 \text{ moles N atoms}}{1 \text{ mole } N_2O_4 \text{ molecules}} \quad \text{or} \quad \frac{1 \text{ mole } N_2O_4 \text{ molecules}}{2 \text{ moles N atoms}}$$

$$\text{For O:} \quad \frac{4 \text{ moles O atoms}}{1 \text{ mole } N_2O_4 \text{ molecules}} \quad \text{or} \quad \frac{1 \text{ mole } N_2O_4 \text{ molecules}}{4 \text{ moles O atoms}}$$

Example 6.4 illustrates the use of this type of conversion factor in problem solving.

• Conversion factors that relate a component of a substance to the substance as a whole are dependent on the formula of the substance. By analogy, the relationship of body parts of an animal to the animal as a whole is dependent on the animal's identity. For example, in 1 mole of elephants there would be 4 moles of elephant legs, 2 moles of elephant ears, 1 mole of elephant tails, and 1 mole of elephant trunks.

Example 6.4

Calculating Molar Quantities of Compound Components

Lactic acid, the substance that builds up in muscles and causes them to hurt when they are worked hard, has the formula $C_3H_6O_3$. How many moles of carbon atoms, hydrogen atoms, and oxygen atoms are present in a 1.2-mole sample of lactic acid?

Solution

One mole of $C_3H_6O_3$ contains 3 moles of carbon atoms, 6 moles of hydrogen atoms, and 3 moles of oxygen atoms. We obtain the following conversion factors from this statement:

$$\left(\frac{3 \text{ moles C atoms}}{1 \text{ mole } C_3H_6O_3}\right) \quad \left(\frac{6 \text{ moles H atoms}}{1 \text{ mole } C_3H_6O_3}\right) \quad \left(\frac{3 \text{ moles O atoms}}{1 \text{ mole } C_3H_6O_3}\right)$$

Using the first conversion factor, the moles of carbon atoms present are calculated as follows:

$$1.2 \text{ moles } C_3H_6O_3 \times \left(\frac{3 \text{ moles C atoms}}{1 \text{ mole } C_3H_6O_3}\right) = 3.6 \text{ moles C atoms}$$

Similarly, from the second and third conversion factors, the moles of hydrogen and oxygen atoms present are calculated as follows:

$$1.2 \text{ moles } C_3H_6O_3 \times \left(\frac{6 \text{ moles H atoms}}{1 \text{ mole } C_3H_6O_3}\right) = 7.2 \text{ moles H atoms}$$

$$1.2 \text{ moles } C_3H_6O_3 \times \left(\frac{3 \text{ moles O atoms}}{1 \text{ mole } C_3H_6O_3}\right) = 3.6 \text{ moles O atoms}$$

Practice Exercise 6.4

The compound deoxyribose, whose chemical formula is $C_5H_{10}O_5$, is an important component of DNA molecules, the molecules responsible for the transfer of genetic information from one generation to the next in living organisms. How many moles of carbon atoms, hydrogen atoms, and oxygen atoms are present in a 0.456-mole sample of deoxyribose?

- *Answer:* 2.28 moles C, 4.56 moles H, and 2.28 moles O

6.5 The Mole and Chemical Calculations

In this section, we will combine the major points we have learned about moles to produce a general approach to problem solving that is applicable to a variety of chemical situations. In Section 6.2, we learned that *Avogadro's number* provides a relationship between the number of particles of a substance and the number of moles of that same substance:

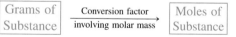

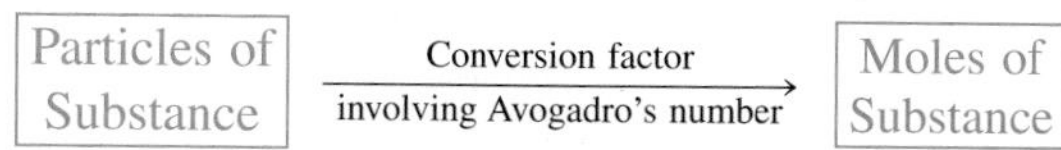

In Section 6.3, we learned that *molar mass* provides a relationship between the number of grams of a substance and the number of moles of that substance:

Grams of Substance → (Conversion factor involving molar mass) → Moles of Substance

In Section 6.4, we learned that the *molar interpretation of chemical formula subscripts* provides a relationship between the number of moles of a substance and the number of moles of its components:

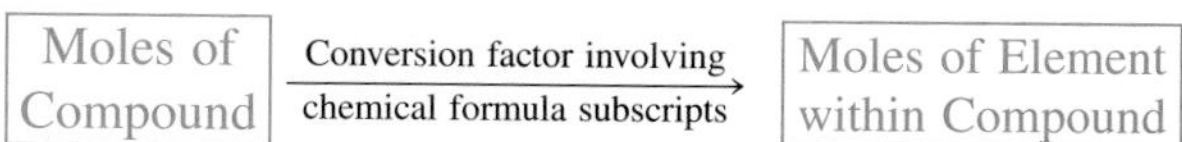

The preceding three concepts can be combined into a single diagram that is very useful in problem solving. This diagram, Figure 6.3, can be viewed as a road map from which conversion factor sequences (pathways) may be obtained. It gives all of the needed relationships for solving two general types of problems:

1. Calculations where information (moles, particles, or grams) is given about a particular substance, and additional information (moles, particles, or grams) is needed concerning the *same* substance.
2. Calculations where information (moles, particles, or grams) is given about a particular substance, and information is needed concerning a *component* of that same substance.

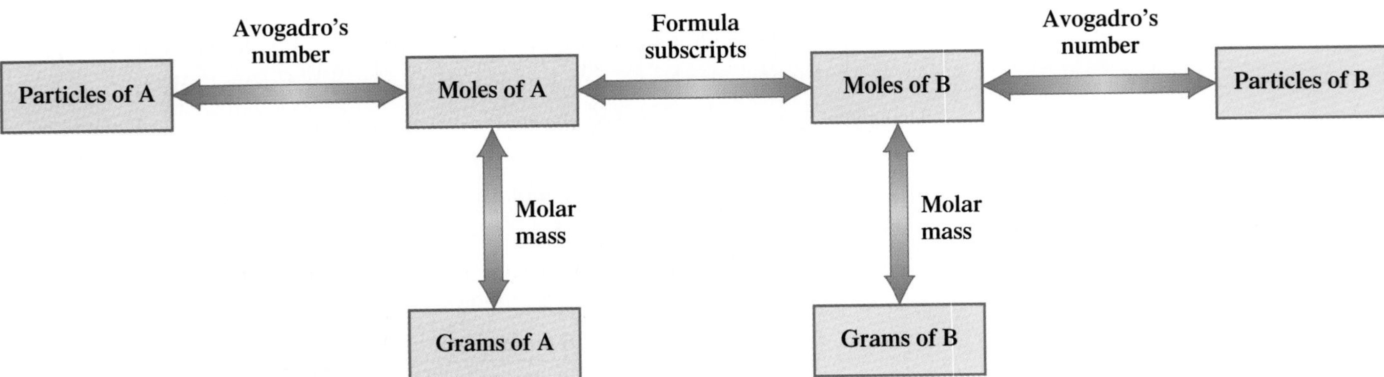

Figure 6.3
In solving chemical formula–based problems, the only "transitions" allowed are those between quantities (boxes) connected by arrows. Associated with each arrow is the concept on which the required conversion factor is based.

For the first type of problem, only the left side of Figure 6.3 (the "A" boxes) is needed. For problems of the second type, both sides of the diagram (both "A" and "B" boxes) are used.

The thinking pattern needed to use Figure 6.3 is very simple.

1. Determine which box in the diagram represents the *given* quantity in the problem.
2. Locate the box that represents the *desired* quantity.
3. Follow the indicated pathway that takes you from the given quantity to the desired quantity. This involves simply following the arrows. There will always be only one pathway possible for the needed transition.

Examples 6.5 and 6.6 illustrate some of the types of problems that can be solved by using the relationships shown in Figure 6.3.

Example 6.5

Calculating the Number of Particles in a Given Mass of Substance

Vitamin C has the formula $C_6H_8O_6$. Calculate the number of vitamin C molecules present in a 0.250-g tablet of vitamin C.

Solution

We will solve this problem by using the three steps of dimensional analysis (Section 2.6) and Figure 6.3.

Step 1: The given quantity is 0.250 g of $C_6H_8O_6$, and the desired quantity is molecules of $C_6H_8O_6$.

$$0.250 \text{ g } C_6H_8O_6 = \text{? molecules } C_6H_8O_6$$

In terms of Figure 6.3, this is a "grams of A" to "particles of A" problem. We are given grams of a substance, A, and desire to find molecules (particles) of that same substance.

Step 2: Figure 6.3 gives us the pathway we need to solve this problem. Starting with "grams of A," we convert to "moles of A" and finally reach "particles of A." The arrows between the boxes along our path give the type of conversion factor needed for each step.

$$\boxed{\text{Grams of A}} \xrightarrow[\text{mass}]{\text{molar}} \boxed{\text{Moles of A}} \xrightarrow[\text{number}]{\text{Avogadro's}} \boxed{\text{Particles of A}}$$

From dimensional analysis, the setup for this sequence of conversion factors is

$$0.250 \text{ g } \cancel{C_6H_8O_6} \times \left(\frac{1 \text{ mole } \cancel{C_6H_8O_6}}{176.14 \text{ g } \cancel{C_6H_8O_6}}\right) \times \left(\frac{6.02 \times 10^{23} \text{ molecules } C_6H_8O_6}{1 \text{ mole } \cancel{C_6H_8O_6}}\right)$$

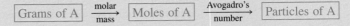

$$\text{g } C_6H_8O_6 \longrightarrow \text{moles } C_6H_8O_6 \longrightarrow \text{molecules } C_6H_8O_6$$

The number 176.14 that is used in the first conversion factor is the formula mass of $C_6H_8O_6$. It was not given in the problem but had to be calculated by using atomic masses and the method for calculating formula masses shown in Example 6.1.

Step 3: The solution to the problem, obtained by doing the arithmetic, is

$$\frac{0.250 \times 1 \times 6.02 \times 10^{23}}{176.14 \times 1} \text{ molecules } C_6H_8O_6$$
$$= 8.54 \times 10^{20} \text{ molecules } C_6H_8O_6$$

Practice Exercise 6.5

The compound lithium carbonate, used to treat manic depression, has the formula Li_2CO_3. Calculate the number of formula units of lithium carbonate present in a 0.500-g sample of lithium carbonate.

- *Answer:* 4.07×10^{21} formula units Li_2CO_3

Example 6.6

Calculating the Mass of an Element Present in a Given Mass of Compound

How many grams of nitrogen are present in a 0.10-g sample of caffeine, the stimulant in coffee and tea? The formula of caffeine is $C_8H_{10}N_4O_2$.

Solution

Step 1: There is an important difference between this problem and the preceding one; here we are dealing with not one but two substances, caffeine and nitrogen. The given quantity is grams of caffeine (substance A), and we are asked to find the grams of nitrogen (substance B). This is a "grams of A" to "grams of B" problem.

$$0.10 \text{ g } C_8H_{10}N_4O_2 = ? \text{ g N}$$

Step 2: The appropriate set of conversions for a "grams of A" to "grams of B" problem, from Figure 6.3, is

$$\boxed{\text{Grams of A}} \xrightarrow[\text{mass}]{\text{Molar}} \boxed{\text{Moles of A}} \xrightarrow[\text{subscripts}]{\text{Formula}} \boxed{\text{Moles of B}} \xrightarrow[\text{mass}]{\text{Molar}} \boxed{\text{Grams of B}}$$

The conversion factor setup is

$$0.10 \text{ g } \cancel{C_8H_{10}N_4O_2} \times \left(\frac{1 \text{ } \cancel{\text{mole } C_8H_{10}N_4O_2}}{194.26 \text{ } \cancel{\text{g } C_8H_{10}N_4O_2}}\right) \times \left(\frac{4 \text{ } \cancel{\text{moles N}}}{1 \text{ } \cancel{\text{mole } C_8H_{10}N_4O_2}}\right) \times \left(\frac{14.01 \text{ g N}}{1 \text{ } \cancel{\text{mole N}}}\right)$$

The number 194.26 that is used in the first conversion factor is the formula mass for caffeine. The conversion from "moles of A" to "moles of B" (the second conversion factor) is made by using the information contained in the formula $C_8H_{10}N_4O_2$. One mole of caffeine contains 4 moles of nitrogen. The number 14.01 in the final conversion factor is the molar mass of nitrogen.

Step 3: Collecting the numbers from the various conversion factors and doing the arithmetic give us our answer.

$$\left(\frac{0.10 \times 1 \times 4 \times 14.01}{194.26 \times 1 \times 1}\right) \text{ g N} = 0.029 \text{ g N}$$

Practice Exercise 6.6

How many grams of oxygen are present in a 0.10-g sample of adrenaline, a hormone secreted into the bloodstream in times of stress? The formula of adrenaline is $C_9H_{13}NO_3$.

- *Answer:* 0.026 g O

6.6 Writing and Balancing Chemical Equations

A **chemical equation** *is a written statement that uses symbols and formulas instead of words to describe the changes that occur in a chemical reaction.* The following example shows the contrast between a word description of a chemical reaction and a chemical equation for the same reaction.

Word description: Calcium sulfide reacts with water to produce calcium oxide and hydrogen sulfide.

Chemical equation: $CaS + H_2O \longrightarrow CaO + H_2S$

In the same way that chemical symbols are considered the *letters* of chemical language, and formulas are considered the *words* of the language, chemical equations can be considered the *sentences* of chemical language.

• Conventions Used in Writing Chemical Equations

Four conventions are used to write chemical equations.

• In a chemical equation, the *reactants* (starting materials in a chemical reaction) are always written on the left side of the equation, and the *products* (substances produced in a chemical reaction) are always written on the right side of the equation.

1. *The correct formulas of the* ***reactants*** *(starting materials) are always written on the* ***left*** *side of the equation.*

$$\text{(CaS)} + \text{(H}_2\text{O)} \longrightarrow CaO + H_2S$$

2. *The correct formulas of the* ***products*** *(substances produced) are always written on the* ***right*** *side of the equation.*

$$CaS + H_2O \longrightarrow \text{(CaO)} + \text{(H}_2\text{S)}$$

3. *The reactants and products are separated by an arrow pointing toward the products.*

$$CaS + H_2O \,(\longrightarrow)\, CaO + H_2S$$

This arrow means "to produce."

4. *Plus signs are used to separate different reactants or different products.*

$$CaS \,(+)\, H_2O \longrightarrow CaO \,(+)\, H_2S$$

Plus signs on the reactant side of the equation mean "reacts with," and plus signs on the product side mean "and."

A *valid* chemical equation must satisfy two conditions:

• The diatomic elemental gases are the elements whose names end in *-gen* (hydrogen, oxygen, and nitrogen) or *-ine* (fluorine, chlorine, bromine, and iodine).

• Atoms are neither created nor destroyed in an ordinary chemical reaction. The production of new substances in a reaction results from the rearrangement of the existent groupings of atoms into new groupings. Because only rearrangement occurs, the products always contain the same number of atoms of each kind as do the reactants. This generalization is often referred to as the *law of conservation of mass.* The mass of the reactants and the mass of the products are the same, because both contain exactly the same number of atoms of each kind present.

1. *It must be consistent with experimental facts.* Only the reactants and products that are actually involved in a reaction are shown in an equation. An accurate formula must be used for each of these substances. Elements in solid and liquid states are represented in equations by the chemical symbol for the element. Elements that are gases at room temperature are represented by the molecular formula denoting the form in which they actually occur in nature. The following monatomic, diatomic, and tetratomic elemental gases are known.

Monatomic:	He, Ne, Ar, Kr, Xe
Diatomic:	H_2, O_2, N_2, F_2, Cl_2, Br_2 (vapor), I_2 (vapor)
Tetratomic:	P_4 (vapor), As_4 (vapor)*

2. *There must be the same number of atoms of each kind on both sides of the equation.* Equations that satisfy this condition are said to be balanced. A **balanced chemical equation** *has the same number of atoms of each element involved in the reaction on each side of the equation.* Because the conventions previously listed for writing equations do not guarantee that an equation will be balanced, we now consider procedures for balancing equations.

*The four elements listed as vapors are not gases at room temperature but vaporize at slightly higher temperatures. The resultant vapors contain molecules with the formulas indicated.

• Guidelines for Balancing Chemical Equations

An unbalanced equation is brought into balance by adding *coefficients* to the equation to adjust the number of reactant or product molecules present. A **coefficient** *is a number that is placed to the left of the formula of a substance and that changes the amount, but not the identity, of the substance.* In the notation $2H_2O$, the 2 on the left is a coefficient; $2H_2O$ means two molecules of H_2O, and $3H_2O$ means three molecules of H_2O. Thus coefficients tell how many formula units of a given substance are present.

• The coefficients of a balanced equation represent numbers of molecules or formula units of various species involved in the chemical reaction.

The following is a balanced chemical equation, with the coefficients shown in color.

$$4NH_3 + 3O_2 \longrightarrow 2N_2 + 6H_2O$$

This balanced equation tells us that four NH_3 molecules react with three O_2 molecules to produce two N_2 molecules and six H_2O molecules.

A coefficient of 1 in a balanced equation is not explicitly written; it is considered to be understood. Both Na_2SO_4 and Na_2S have "understood coefficients" of 1 in the following balanced equation:

$$Na_2SO_4 + 2C \longrightarrow Na_2S + 2CO_2$$

• In balancing a chemical equation, formula subscripts are *never changed.* You must use the formulas just as they are given. *The only thing you can do is add coefficients.*

A coefficient placed in front of a formula applies to the whole formula. By contrast, subscripts, which are also present in formulas, affect only parts of a formula.

Coefficient (affects both H and O)

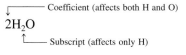

$$2H_2O$$

Subscript (affects only H)

The preceding notation denotes two molecules of H_2O; it also denotes a total of four H atoms and two O atoms.

Let's look at the mechanics involved in determining the coefficients needed to balance an equation. Suppose we want to balance the chemical equation

$$FeI_2 + Cl_2 \longrightarrow FeCl_3 + I_2$$

Step 1: *Examine the equation and pick one element to balance first.* It is often convenient to start with the compound that contains the greatest number of atoms, whether a reactant or a product, and key in on the element in that compound that has the greatest number of atoms. Using this guideline, we select $FeCl_3$ and the element chlorine within it.

We note that there are three chlorine atoms on the right side of the equation and two atoms of chlorine on the left (in Cl_2). In order for the chlorine atoms to balance, we will need six on each side; 6 is the lowest number that both 3 and 2 will divide into evenly. In order to obtain six atoms on each side of the equation, we place the coefficient 3 in front of Cl_2 and the coefficient 2 in front of $FeCl_3$.

$$FeI_2 + ③Cl_2 \longrightarrow ②FeCl_3 + I_2$$

We now have six chlorine atoms on each side of the equation.

$$3Cl_2: \quad 3 \times 2 = 6$$

$$2FeCl_3: \quad 2 \times 3 = 6$$

Step 2: *Now pick a second element to balance.* We will balance the iron next. The number of iron atoms on the right side has already been set at 2 by the coefficient previously placed in front of $FeCl_3$. We will need two iron atoms on the reactant side of the equation instead of the one iron atom now present. This is accomplished by placing the coefficient 2 in front of FeI_2.

$$②FeI_2 + 3Cl_2 \longrightarrow 2FeCl_3 + I_2$$

It is always wise to pick, as the second element to balance, one whose amount is already set on one side of the equation by a previously determined coefficient. If we had chosen iodine as the second element to balance instead of iron,

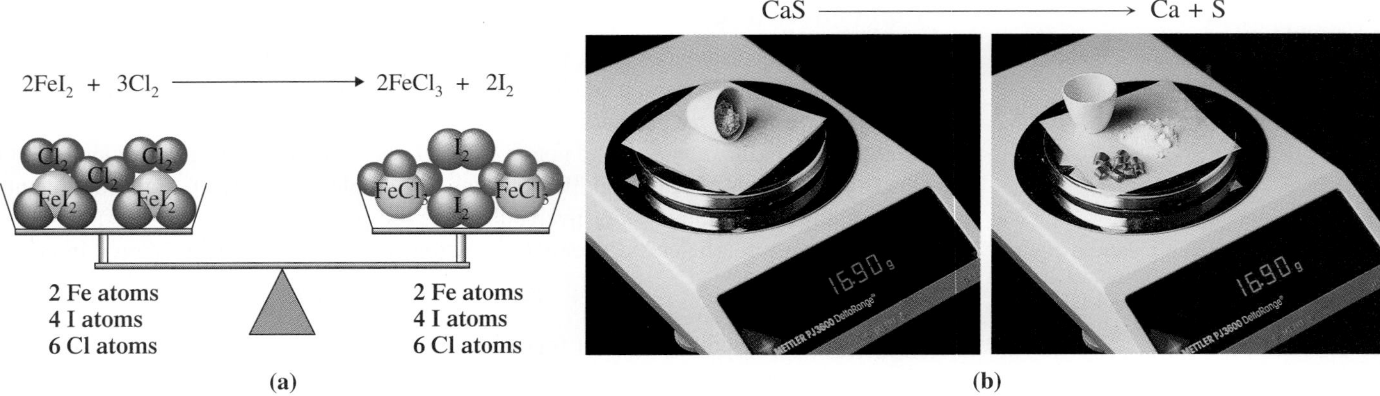

Figure 6.4
(a) Atoms are neither created nor destroyed in a chemical reaction. They merely change partners to give new groupings of atoms. Thus the masses of reactants and products in a chemical reaction are the same, because reactants and products contain the same number of atoms of each kind. Balancing a chemical equation makes the equation consistent with these principles.
(b) The mass of CaS, the product, equals the combined mass of Ca and S, the reactants.

we would have run into problems. Because the coefficient for neither FeI_2 nor I_2 had been determined, we would have had no guidelines for deciding on the amount of iodine needed.

Step 3: *Now pick a third element to balance.* Only one element is left to balance—iodine. The number of iodine atoms on the left side of the equation is already set at four ($2FeI_2$). In order to obtain four iodine atoms on the right side of the equation, we place the coefficient 2 in front of I_2.

$$2FeI_2 + 3Cl_2 \longrightarrow 2FeCl_3 + \textcircled{2}I_2$$

The addition of the coefficient 2 in front of I_2 completes the balancing process; all the coefficients have been determined.

Step 4: *As a final check on the correctness of the balancing procedure, count atoms on each side of the equation.* The following table can be constructed from our balanced equation.

$$2FeI_2 + 3Cl_2 \longrightarrow 2FeCl_3 + 2I_2$$

Atom	Left side	Right side
Fe	$2 \times 1 = 2$	$2 \times 1 = 2$
I	$2 \times 2 = 4$	$2 \times 2 = 4$
Cl	$3 \times 2 = 6$	$2 \times 3 = 6$

All elements are in balance: two iron atoms on each side, four iodine atoms on each side, and six chlorine atoms on each side (see Figure 6.4).

Notice that the elements chlorine and iodine in the equation are written in the form of diatomic molecules (Cl_2 and I_2). This is in accordance with the guideline given at the start of this section on the use of molecular formulas for elements that are gases at room temperature.

In Example 6.7 we will balance another chemical equation.

Example 6.7

Balancing a Chemical Equation

Balance the following chemical equation.

$$C_2H_6O + O_2 \longrightarrow CO_2 + H_2O$$

Solution

The element oxygen appears in four different places in this equation. This means we do not want to start the balancing process with the element oxygen. Always start the balancing process with an element that appears only once on both the reactant and product sides of the equation.

Step 1: *Balancing of H atoms.* There are six H atoms on the left and two H atoms on the right. Placing the coefficient 3 in front of H_2O balances the H atoms at six on each side.

$$1C_2H_6O + O_2 \longrightarrow CO_2 + 3H_2O$$

Step 2: *Balancing of C atoms.* An effect of balancing the H atoms at six (Step 1) is the setting of the C atoms on the left side at two; the coefficient in front of C_2H_6O is 1. Placing the coefficient 2 in front of CO_2 causes the carbon atoms to balance at two on each side of the equation.

$$1C_2H_6O + O_2 \longrightarrow 2CO_2 + 3H_2O$$

Step 3: *Balancing of O atoms.* The oxygen content of the right side of the equation is set at seven atoms: four oxygen atoms from $2CO_2$ and three oxygen atoms from $3H_2O$. To obtain seven oxygen atoms on the left side of the equation, we place the coefficient 3 in front of O_2; $3O_2$ gives six oxygen atoms, and there is an additional O in $1C_2H_6O$. The element oxygen is present in all four formulas in the equation.

$$1C_2H_6O + 3O_2 \longrightarrow 2CO_2 + 3H_2O$$

Step 4: *Final check.* The equation is balanced. There are two carbon atoms, six hydrogen atoms, and seven oxygen atoms on each side of the equation.

$$C_2H_6O + 3O_2 \longrightarrow 2CO_2 + 3H_2O$$

Practice Exercise 6.7

Balance the following chemical equation.

$$C_4H_{10}O + O_2 \longrightarrow CO_2 + H_2O$$

• *Answer:* $C_4H_{10}O + 6O_2 \longrightarrow 4CO_2 + 5H_2O$

Some additional comments and guidelines concerning equations in general, and the process of balancing in particular, are given here.

1. The coefficients in a balanced equation are always the *smallest set of whole numbers* that will balance the equation. We mention this because more than one set of coefficients will balance an equation. Consider the following three equations:

$$2H_2 + O_2 \longrightarrow 2H_2O$$
$$4H_2 + 2O_2 \longrightarrow 4H_2O$$
$$8H_2 + 4O_2 \longrightarrow 8H_2O$$

   All three of these equations are mathematically correct; there are equal numbers of hydrogen and oxygen atoms on both sides of the equation. However, the first equation is considered the correct form because the coefficients used there are the smallest set of whole numbers that will balance the equation. The coefficients in the second equation are two times those in the first equation, and the third equation has coefficients that are four times those of the first equation.
2. At this point, you are not expected to be able to write down the products for a chemical reaction when given the reactants. After learning how to balance equations, students sometimes get the mistaken idea that they ought to be able to write down equations from scratch. This is not so. You will need more chemical knowledge before attempting this task. At this stage, you should be able to balance simple equations, given *all* of the reactants and *all* of the products.
3. It is often useful to know the physical state of the substances involved in a chemical reaction. We specify physical state by using the symbols (*s*) for solid,

(*l*) for liquid, (*g*) for gas, and (*aq*) for aqueous solution (a substance dissolved in water). Two examples of such symbol use in chemical equations are

$$2Fe_2O_3(s) + 3C(s) \longrightarrow 4Fe(s) + 3CO_2(g)$$
$$2HNO_3(aq) + 3H_2S(aq) \longrightarrow 2NO(g) + 3S(s) + 4H_2O(l)$$

6.7 Chemical Equations and the Mole Concept

The coefficients in a balanced chemical equation, like the subscripts in a chemical formula (Section 6.5), have two levels of interpretation—a microscopic level of meaning and a macroscopic level of meaning. The microscopic level of interpretation was used in the previous two sections. *The coefficients in a balanced equation give the numerical relationships among formula units consumed (used up) or produced in the chemical reaction.* Interpreted at the microscopic level, the equation

$$N_2 + 3H_2 \longrightarrow 2NH_3$$

conveys the information that one molecule of N_2 reacts with three molecules of H_2 to produce two molecules of NH_3.

At the macroscopic level of interpretation, chemical equations are used to relate mole-sized quantities of reactants and products to each other. At this level, *the coefficients in the equation give the fixed molar ratios between substances consumed or produced in the chemical reaction.* Interpreted at the macroscopic level, the equation

$$N_2 + 3H_2 \longrightarrow 2NH_3$$

conveys the information that 1 mole of N_2 reacts with 3 moles of H_2 to produce 2 moles of NH_3.

The coefficients in an equation can be used to generate conversion factors to be used in solving problems. Numerous conversion factors are obtainable from a single balanced equation. Consider the following balanced equation:

$$4Fe + 3O_2 \longrightarrow 2Fe_2O_3$$

Three mole-to-mole relationships are obtainable from this equation:

• Conversion factors that relate two different substances to one another are valid only for systems governed by the chemical equation from which they were obtained.

4 moles of Fe produces 2 moles of Fe_2O_3.

3 moles of O_2 produces 2 moles of Fe_2O_3.

4 moles of Fe reacts with 3 moles of O_2.

From each of these macroscopic-level relationships, two conversion factors can be written. The conversion factors for the first relationship are

$$\left(\frac{4 \text{ moles Fe}}{2 \text{ moles } Fe_2O_3}\right) \quad \text{and} \quad \left(\frac{2 \text{ moles } Fe_2O_3}{4 \text{ moles Fe}}\right)$$

All balanced chemical equations are the source of numerous conversion factors. The more reactants and products there are in the equation, the greater the number of derivable conversion factors. The next section details how conversion factors such as those in the preceding illustration are used in solving problems.

Chemistry at a Glance reviews the relationships that involve the mole.

6.8 Chemical Calculations Using Chemical Equations

When the information contained in a chemical equation is combined with the concepts of molar mass (Section 6.3) and Avogadro's number (Section 6.2), many useful types of chemical calculations can be carried out. A typical chemical equation–based calculation gives information about one reactant or product of a reaction (number of grams, moles,

Relationships Involving the Mole Concept

CHEMISTRY AT A GLANCE

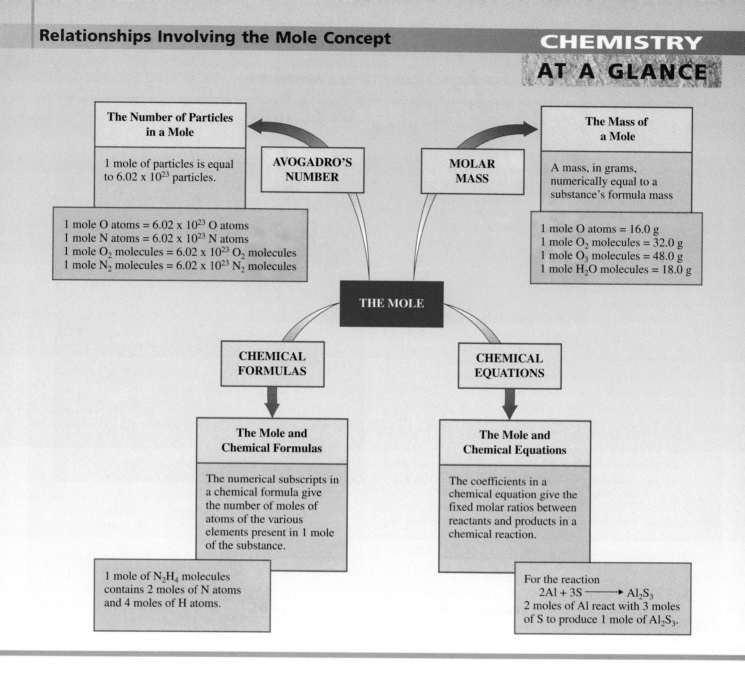

The Number of Particles in a Mole

1 mole of particles is equal to 6.02×10^{23} particles.

1 mole O atoms = 6.02×10^{23} O atoms
1 mole N atoms = 6.02×10^{23} N atoms
1 mole O_2 molecules = 6.02×10^{23} O_2 molecules
1 mole N_2 molecules = 6.02×10^{23} N_2 molecules

AVOGADRO'S NUMBER

MOLAR MASS

The Mass of a Mole

A mass, in grams, numerically equal to a substance's formula mass

1 mole O atoms = 16.0 g
1 mole O_2 molecules = 32.0 g
1 mole O_3 molecules = 48.0 g
1 mole H_2O molecules = 18.0 g

THE MOLE

CHEMICAL FORMULAS

The Mole and Chemical Formulas

The numerical subscripts in a chemical formula give the number of moles of atoms of the various elements present in 1 mole of the substance.

1 mole of N_2H_4 molecules contains 2 moles of N atoms and 4 moles of H atoms.

CHEMICAL EQUATIONS

The Mole and Chemical Equations

The coefficients in a chemical equation give the fixed molar ratios between reactants and products in a chemical reaction.

For the reaction
$2Al + 3S \longrightarrow Al_2S_3$
2 moles of Al react with 3 moles of S to produce 1 mole of Al_2S_3.

or particles) and requests information about another reactant or product of the same reaction. The substances involved in such a calculation may both be reactants or products or may be a reactant and a product.

The conversion factor relationships needed to solve problems of this general type are given in Figure 6.5. This diagram should seem very familiar to you; it is almost identical

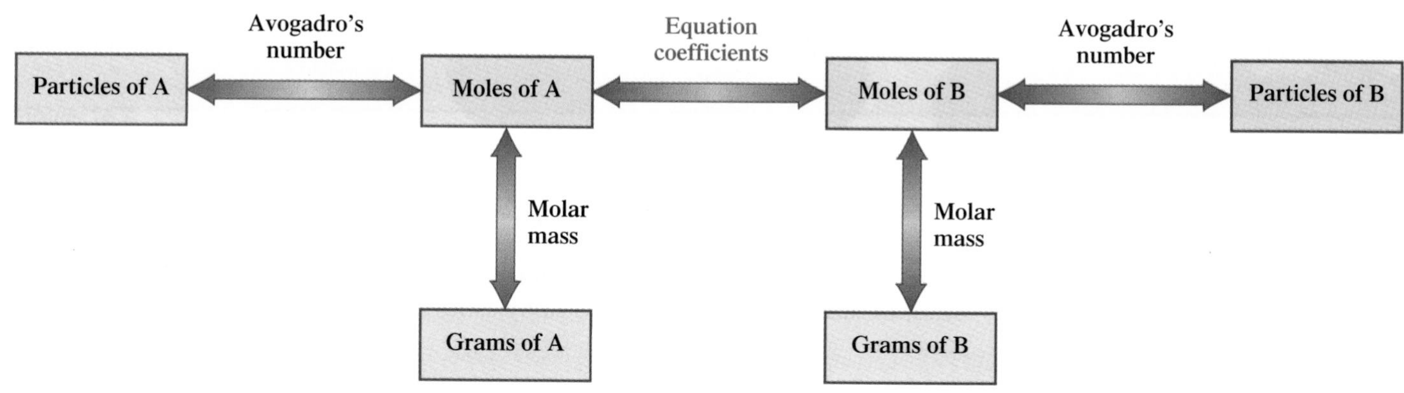

Figure 6.5 In solving chemical equation–based problems, the only "transitions" allowed are those between quantities (boxes) connected by arrows. Associated with each arrow is the concept on which the required conversion factor is based.

Chemical CONNECTIONS

6.1 "Laboratory-Sized Amounts" versus "Industrial Chemistry-Sized Amounts"

The various calculations in this chapter can be considered to be "laboratory-based" calculations. Chemical substance amounts are always specified in grams, the common laboratory unit for mass. These gram-sized laboratory amounts are very small, almost "infinitesimal," when compared with industrial production figures for various "high-volume" chemicals, which are specified in terms of billions of pounds per year. About 50 of the millions of compounds known are produced in amounts exceeding 1 billion pounds per year in the United States.

The number-one chemical in the United States, in terms of production amount, is sulfuric acid (H_2SO_4), with an annual production approaching 100 billion pounds. Its production amount is almost twice that of any other chemical. So important is sulfuric acid production in the United States (and the world) that some economists use sulfuric acid production as a measure of a nation's industrial strength.

Why is so much sulfuric acid produced in the United States? What are its uses? What are its properties? Where do we encounter it in our everyday life?

Pure sulfuric acid is a colorless, corrosive, oily liquid. It is usually marketed as a concentrated (96% by mass) aqueous solution. People rarely have direct contact with this strong acid because it is seldom part of finished consumer products. The closest encounter most people have with the acid (other than in a chemical laboratory) is involvement with automobile batteries. The acid in a standard automobile battery is a 38%-by-mass aqueous solution of sulfuric acid. However, less than 1% of annual sulfuric acid production ends up in car batteries.

Approximately two-thirds of sulfuric acid production is used in the manufacture of chemical fertilizers. These fertilizer compounds are an absolute necessity if the food needs of an ever-increasing population are to be met. The connection between sulfuric acid and fertilizer revolves around the element phosphorus, which is necessary for plant growth. The starting material for phosphate fertilizer production is phosphate rock, a highly insoluble material containing calcium phosphate, $Ca_3(PO_4)_2$. The treatment of phosphate rock with H_2SO_4 results in the formation of phosphoric acid, H_3PO_4.

$$Ca_3(PO_4)_2 + 3H_2SO_4 \longrightarrow 3CaSO_4 + 2H_3PO_4$$

The phosphoric acid so produced is then used to produce soluble phosphate compounds that plants can use as a source

of phosphorus. The major phosphoric acid fertilizer derivative is ammonium hydrogen phosphate—$(NH_4)_2HPO_4$.

The raw materials needed for sulfuric acid production are simple: sulfur, air, and water. In the first step of production, elemental sulfur is burned to give sulfur dioxide gas.

$$S + O_2 \longrightarrow SO_2$$

Some SO_2 is also obtainable as a by-product of metallurgical operations associated with zinc and copper production. Next, the SO_2 gas is combined with additional O_2 (air) to produce sulfur trioxide gas.

$$2SO_2 + O_2 \longrightarrow 2SO_3$$

The SO_3 is then dissolved in water, which yields sulfuric acid as the product.

$$SO_3 + H_2O \longrightarrow H_2SO_4$$

Reactions similar to the last two steps in commercial H_2SO_4 production can also occur naturally in the atmosphere. The H_2SO_4 so produced is a major contributor to the phenomenon called acid rain (see Chemical Connection 10.1).

to Figure 6.3, which you used in solving problems based on chemical formulas. There is only one difference between the two diagrams. In Figure 6.3, the subscripts in a chemical formula are listed as the basis for relating "moles of A" to "moles of B." In Figure 6.5, the same two quantities are related by using the coefficients of a balanced chemical equation.

• The quantitative study of the relationships among reactants and products in a chemical reaction is called *chemical stoichiometry.* The word *stoichiometry,* pronounced stoy-key-om-eh-tree, is derived from the Greek *stoicheion* ("element") and *metron* ("measure"). The stoichiometry of a chemical reaction always involves the *molar relationships* between reactants and products and thus is given by the coefficients in the balanced equation for the chemical reaction.

The most common type of chemical equation–based calculation is a "grams of A" to "grams of B" problem. In this type of problem, the mass of one substance involved in a chemical reaction (either reactant or product) is given, and information is requested about the mass of another substance involved in the reaction (either reactant or product). This type of problem is frequently encountered in laboratory settings. For example, a chemist may have a certain number of grams of a chemical available and may want to know how many grams of another substance can be produced from it or how many grams of a third substance are needed to react with it. Examples 6.8 and 6.9 illustrate this type of problem.

Example 6.8

Calculating the Mass of a Product in a Chemical Reaction

The human body converts the glucose, $C_6H_{12}O_6$, contained in foods to carbon dioxide, CO_2, and water, H_2O. The equation for the chemical reaction is

$$C_6H_{12}O_6 + 6O_2 \longrightarrow 6CO_2 + 6H_2O$$

Assume a person eats a candy bar containing 14.2 g (1/2 oz) of glucose. How many grams of water will the body produce from the ingested glucose, assuming all of the glucose undergoes reaction?

Solution

Step 1: The given quantity is 14.2 g of glucose. The desired quantity is grams of water.

$$14.2 \text{ g } C_6H_{12}O_6 = ? \text{ g } H_2O$$

In terms of Figure 6.5, this is a "grams of A" to "grams of B" problem.

Step 2: Using Figure 6.5 as a road map, we determine that the pathway for this problem is

$$\boxed{\text{Grams of A}} \xrightarrow[\text{mass}]{\text{Molar}} \boxed{\text{Moles of A}} \xrightarrow[\text{coefficients}]{\text{Equation}} \boxed{\text{Moles of B}} \xrightarrow[\text{mass}]{\text{Molar}} \boxed{\text{Grams of B}}$$

The mathematical setup for this problem is

$$14.2 \cancel{\text{ g } C_6H_{12}O_6} \times \left(\frac{1 \cancel{\text{ mole } C_6H_{12}O_6}}{180.18 \cancel{\text{ g } C_6H_{12}O_6}}\right) \times \left(\frac{6 \cancel{\text{ moles } H_2O}}{1 \cancel{\text{ mole } C_6H_{12}O_6}}\right) \times \left(\frac{18.02 \text{ g } H_2O}{1 \cancel{\text{ mole } H_2O}}\right)$$

$$\text{g } C_6H_{12}O_6 \longrightarrow \text{moles } C_6H_{12}O_6 \longrightarrow \text{moles } H_2O \longrightarrow \text{g } H_2O$$

The 180.18 g in the first conversion factor is the molar mass of glucose, the 6 and 1 in the second conversion factor are the coefficients, respectively, of H_2O and $C_6H_{12}O_6$ in the balanced chemical equation, and the 18.02 g in the third conversion factor is the molar mass of H_2O.

Step 3: The solution to the problem, obtained by doing the arithmetic after all the numerical factors have been collected, is

$$\left(\frac{14.2 \times 1 \times 6 \times 18.02}{180.18 \times 1 \times 1}\right) \text{g } H_2O = 8.52 \text{ g } H_2O$$

Practice Exercise 6.8

Silicon carbide, SiC, which is used as an abrasive on sandpaper, is prepared using the chemical reaction

$$SiO_2 + 3C \longrightarrow SiC + 2CO$$

How many grams of SiC can be produced from 15.0 g of C?

• *Answer:* 16.7 g SiC

Example 6.9

Calculating the Mass of a Substance Taking Part in a Chemical Reaction

The active ingredient in many commercial antacids is magnesium hydroxide, $Mg(OH)_2$, which reacts with stomach acid (HCl) to produce magnesium chloride ($MgCl_2$) and water. The equation for the reaction is

$$Mg(OH)_2 + 2HCl \rightarrow MgCl_2 + 2H_2O$$

How many grams of $Mg(OH)_2$ are needed to react with 0.30 g of HCl?

Solution

Step 1: This problem, like Example 6.8, is a "grams of A" to "grams of B" problem. It differs from the previous problem in that both the given and the desired quantities involve reactants.

$$0.30 \text{ g HCl} \rightarrow ? \text{ g } Mg(OH)_2$$

Step 2: The pathway used to solve it will be the same as in Example 6.8.

Grams of A $\xrightarrow[\text{mass}]{\text{molar}}$ Moles of A $\xrightarrow[\text{coefficients}]{\text{equation}}$ Moles of B $\xrightarrow[\text{mass}]{\text{molar}}$ Grams of B

The dimensional-analysis setup is

$$0.30 \text{ g HCl} \times \left(\frac{1 \text{ mole HCl}}{36.46 \text{ g HCl}}\right) \times \left(\frac{1 \text{ mole } Mg(OH)_2}{2 \text{ moles HCl}}\right) \times \left(\frac{58.32 \text{ g } Mg(OH)_2}{1 \text{ mole } Mg(OH)_2}\right)$$

$$\text{g HCl} \longrightarrow \text{moles HCl} \longrightarrow \text{moles } Mg(OH)_2 \longrightarrow \text{g } Mg(OH)_2$$

The balanced chemical equation for the reaction is used as the bridge that enables us to go from HCl to $Mg(OH)_2$. The numbers in the second conversion factor are coefficients from this equation.

Step 3: The solution obtained by combining all of the numbers in the manner indicated in the setup is

$$\left(\frac{0.30 \times 1 \times 1 \times 58.32}{36.46 \times 2 \times 1}\right) \text{g } Mg(OH)_2 = 0.24 \text{ g } Mg(OH)_2$$

To put our answer in perspective, we note that a common brand of antacid tablets has tablets containing 0.10 g of $Mg(OH)_2$.

Practice Exercise 6.9

The chemical equation for the photosynthesis reaction in plants is

$$6CO_2 + 6H_2O \longrightarrow C_6H_{12}O_6 + 6O_2$$

How many grams of H_2O are consumed at the same time that 20.0 g of CO_2 is consumed?

• *Answer:* 8.19 g H_2O

"Grams of A" to "grams of B" problems (Examples 6.8 and 6.9) are not the only type of problem for which the coefficients in a balanced equation can be used to relate the quantities of two substances. As a further example of the use of equation coefficients in problem solving, consider Example 6.10 (a "particles of A" to "moles of B" problem).

Example 6.10

Calculating the Amount of a Substance Taking Part in a Chemical Reaction

Automotive airbags inflate when sodium azide, NaN_3, rapidly decomposes to its constituent elements. The equation for the chemical reaction is

$$2NaN_3(s) \longrightarrow 2Na(s) + 3N_2(g)$$

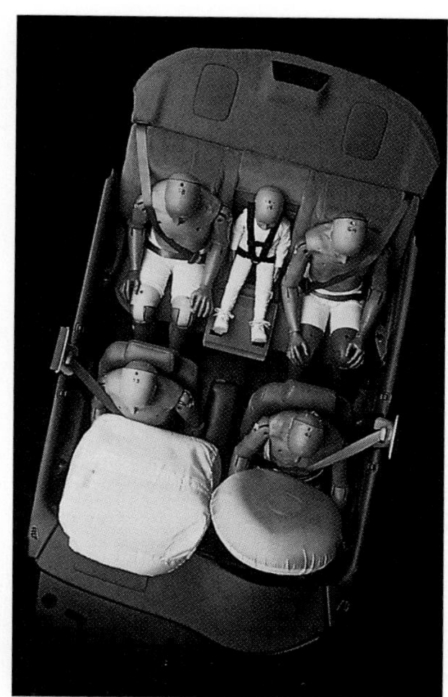

Testing apparatus for measuring the effects of air bag deployment

The gaseous N_2 so generated inflates the airbag. How many moles of NaN_3 would have to decompose in order to generate 253 million (2.53×10^8) molecules of N_2?

Solution

Although a calculation of this type does not have a lot of practical significance, it tests your understanding of the problem-solving relationships discussed in this section of the text.

Step 1: The given quantity is 2.53×10^8 molecules of N_2, and the desired quantity is moles of NaN_3.

$$2.53 \times 10^8 \text{ molecules } N_2 = ? \text{ moles } NaN_3$$

In terms of Figure 6.5, this is a "particles of A" to "moles of B" problem.

Step 2: Using Figure 6.5 as a road map, we determine that the pathway for this problem is

$$\boxed{\text{Particles of A}} \xrightarrow[\text{number}]{\text{Avogadro's}} \boxed{\text{Moles of A}} \xrightarrow[\text{coefficients}]{\text{equation}} \boxed{\text{Moles of B}}$$

The mathematical setup is

$$2.53 \times 10^8 \text{ molecules } N_2 \times \left(\frac{1 \text{ mole } N_2}{6.02 \times 10^{23} \text{ molecules } N_2}\right) \times \left(\frac{2 \text{ moles } NaN_3}{3 \text{ moles } N_2}\right)$$

Avogadro's number is found in the first conversion factor. The 2 and 3 in the second conversion factor are the coefficients, respectively, of NaN_3 and N_2 in the balanced chemical equation.

Step 3: The solution to the problem, obtained by doing the arithmetic after all the numerical factors have been collected, is

$$\left(\frac{2.53 \times 10^8 \times 1 \times 2}{6.02 \times 10^{23} \times 3}\right) \text{mole } NaN_3 = 2.80 \times 10^{-16} \text{ mole } NaN_3$$

Practice Exercise 6.10

Decomposition of $KClO_3$ serves as a convenient laboratory source of small amounts of oxygen gas. The reaction is

$$2KClO_3 \longrightarrow 2KCl + 3O_2$$

How many moles of $KClO_3$ must be decomposed to produce 64 billion (6.4×10^{10}) O_2 molecules?

- *Answer:* 7.1×10^{-14} mole $KClO_3$

Concepts to Remember

Formula mass. The formula mass of a substance is the sum of the atomic masses of the atoms in its formula.

The mole concept. The mole is the chemist's counting unit. One mole of any substance—element or compound—consists of 6.02×10^{23} formula units of the substance. Avogadro's number is the name given to the numerical value 6.02×10^{23}.

Molar mass. The molar mass of a substance is the mass in grams that is numerically equal to the substance's formula mass. Molar mass is not a set number; it varies and is different for each chemical substance.

The mole and chemical formulas. The numerical subscripts in a chemical formula give the number of moles of atoms of the various elements present in 1 mole of the substance.

Chemical equation. A chemical equation is a written statement that uses symbols and formulas instead of words to represent how reactants undergo transformation into products in a chemical reaction.

Balanced chemical equation. A balanced chemical equation has the same number of atoms of each element involved in the reaction on each side of the equation. An unbalanced equation is brought into balance through the use of coefficients. A coefficient is a number that is placed to the left of the formula of a substance and that changes the amount, but not the identity, of the substance.

The mole and chemical equations. The coefficients in a balanced chemical equation give the molar ratios between substances consumed or produced in the chemical reaction described by the equation.

Key Reactions and Equations

1. Calculation of formula mass (Section 6.1)

   Formula mass = sum of atomic masses of all components

2. The mole (Section 6.3)

   1 mole = 6.02×10^{23} objects

3. Avogadro's number (Section 6.3)

   Avogadro's number = 6.02×10^{23}

4. Mass of a mole (Section 6.4)

   Molar mass = mass, in grams, numerically equal to a substance's formula mass

5. Balanced chemical equation (Section 6.7)

   Balanced chemical equation = same number of atoms of each kind on each side of the equation

Key Terms

Avogadro's number (6.2)
Balanced chemical equation (6.6)
Chemical equation (6.6)
Coefficient (6.6)
Formula mass (6.1)
Molar mass (6.3)
Mole (6.2 and 6.3)

Exercises and Problems

The members of each pair of problems in this section test similar material.

Formula Masses (Section 6.1)

6.1 Calculate the formula mass, to two decimal places, of each of the following substances. Obtain the needed atomic masses from the inside front cover of the text.

a. $C_{12}H_{22}O_{11}$ (sucrose, table sugar)
b. C_7H_{16} (heptane, a component of gasoline)
c. $C_7H_5NO_3S$ (saccharin, an artificial sweetener)
d. $(NH_4)_2SO_4$ (ammonium sulfate, a lawn fertilizer)

6.2 Calculate the formula mass, to two decimal places, of each of the following substances. Obtain the needed atomic masses from the inside front cover of the text.

a. $C_{20}H_{30}O$ (vitamin A)
b. $C_{14}H_9Cl_5$ (DDT, formerly used as an insecticide)
c. $C_8H_{10}N_4O_2$ (caffeine, a central nervous system stimulant)
d. $Ca(NO_3)_2$ (calcium nitrate, gives fireworks their red color)

The Mole as a Counting Unit (Section 6.2)

6.3 How many molecules are present in each of the following amounts of substance?

a. 2.00 moles H_2O molecules
b. 3.25 moles CO_2 molecules
c. 0.356 mole CO molecules
d. Avogadro's number of CO molecules

6.4 How many molecules are present in each of the following amounts of substance?

a. 3.00 moles NH_3 molecules
b. 4.52 moles SO_2 molecules
c. 0.275 mole SO_3 molecules
d. Avogadro's number of SO_3 molecules

6.5 You are given a sample containing 0.542 mole of a substance.

a. How many atoms are present if the substance is copper metal?
b. How many atoms are present if the substance is iron metal?
c. How many molecules are present if the sample is nitric acid, HNO_3?
d. How many molecules are present if the sample is aspirin, $C_9H_8O_4$?

6.6 You are given a sample containing 1.43 moles of a substance.

a. How many atoms are present if the substance is silver metal?
b. How many atoms are present if the substance is chromium metal?
c. How many molecules are present if the sample is sulfuric acid, H_2SO_4?
d. How many molecules are present if the sample is glucose, $C_6H_{12}O_6$?

Molar Mass (Section 6.3)

6.7 How much, in grams, does 1.00 mole of each of the following substances weigh?

a. CO (carbon monoxide)
b. CO_2 (carbon dioxide)
c. NaCl (table salt)
d. $C_{12}H_{22}O_{11}$ (table sugar)

6.8 How much, in grams, does 1.00 mole of each of the following substances weigh?

a. H_2O (water)
b. H_2O_2 (hydrogen peroxide)
c. NaCN (sodium cyanide)
d. KCN (potassium cyanide)

6.9 What is the mass, in grams, of each of the following quantities of matter?

a. 0.034 mole of gold atoms
b. 0.034 mole of silver atoms
c. 3.00 moles of oxygen atoms
d. 3.00 moles of oxygen molecules (O_2)

6.10 What is the mass, in grams, of each of the following quantities of matter?

a. 0.85 mole of copper atoms

b. 0.85 mole of nickel atoms

c. 2.50 moles of nitrogen atoms

d. 2.50 moles of nitrogen molecules (N_2)

6.11 How many moles are present in a sample of each of the following substances if each sample weighs 5.00 g?

a. CO molecules b. CO_2 molecules

c. B_4H_{10} molecules d. U atoms

6.12 How many moles are present in a sample of each of the following substances if each sample weighs 7.00 g?

a. N_2O molecules b. NO_2 molecules

c. P_4O_{10} molecules d. V atoms

Molar Interpretation of Chemical Formulas (Section 6.4)

6.13 Write the six mole-to-mole conversion factors that can be derived from each of the following chemical formulas.

a. H_2SO_4 b. $POCl_3$

6.14 Write the six mole-to-mole conversion factors that can be derived from each of the following chemical formulas.

a. HNO_3 b. $C_2H_4Br_2$

6.15 List the number of moles of each type of atom that are present in each of the following molar quantities.

a. 2.00 moles SO_2 molecules

b. 2.00 moles SO_3 molecules

c. 3.00 moles NH_3 molecules

d. 3.00 moles N_2H_4 molecules

6.16 List the number of moles of each type of atom that are present in each of the following molar quantities.

a. 4.00 moles NO_2 molecules

b. 4.00 moles N_2O molecules

c. 7.00 moles H_2O molecules

d. 7.00 moles H_2O_2 molecules

Calculations Based on Chemical Formulas (Section 6.5)

6.17 Determine the number of atoms in each of the following quantities of an element.

a. 10.0 g B b. 32.0 g Ca

c. 2.0 g Ne d. 7.0 g N

6.18 Determine the number of atoms in each of the following quantities of an element.

a. 10.0 g S b. 39.1 g K

c. 3.2 g U d. 7.0 g Be

6.19 Determine the mass, in grams, of each of the following quantities of substance.

a. 6.02×10^{23} copper atoms

b. 3.01×10^{23} copper atoms

c. 557 copper atoms

d. 1 copper atom

6.20 Determine the mass, in grams, of each of the following quantities of substance.

a. 6.02×10^{23} silver atoms b. 3.01×10^{23} silver atoms

c. 1.00×10^{6} silver atoms d. 1 silver atom

6.21 Determine the number of moles of substance present in each of the following quantities.

a. 10.0 g He b. 10.0 g N_2O

c. 4.0×10^{10} atoms P d. 4.0×10^{10} atoms Be

6.22 Determine the number of moles of substance present in each of the following quantities.

a. 25.0 g N

b. 25.0 g Li

c. 8.50×10^{15} atoms S

d. 8.50×10^{15} atoms Cl

6.23 Determine the number of atoms of sulfur present in each of the following quantities.

a. 10.0 g H_2SO_4 b. 20.0 g SO_3

c. 30.0 g Al_2S_3 d. 2.00 moles S_2O

6.24 Determine the number of atoms of nitrogen present in each of the following quantities.

a. 10.0 g N_2H_4 b. 20.0 g HN_3

c. 30.0 g $LiNO_3$ d. 4.00 moles N_2O_5

6.25 Determine the number of grams of sulfur present in each of the following quantities.

a. 3.01×10^{23} S_2O molecules

b. 3 S_4N_4 molecules

c. 2.00 moles SO_2 molecules

d. 4.50 moles S_8 molecules

6.26 Determine the number of grams of oxygen present in each of the following quantities.

a. 4.50×10^{22} SO_3 molecules

b. 7 P_4O_{10} molecules

c. 3.00 moles H_2SO_4 molecules

d. 1.50 moles O_3 molecules

Writing and Balancing Chemical Equations (Section 6.6)

6.27 Balance the following equations.

a. $H_2 + O_2 \rightarrow H_2O$ b. $NO + O_2 \rightarrow NO_2$

c. $Fe_2O_3 \rightarrow Fe + O_2$ d. $NH_4NO_2 \rightarrow N_2 + H_2O$

6.28 Balance the following equations.

a. $Al + O_2 \rightarrow Al_2O_3$ b. $SO_2 + O_2 \rightarrow SO_3$

c. $KClO_3 \rightarrow KCl + O_2$ d. $NH_3 \rightarrow N_2 + H_2$

6.29 Balance the following equations.

a. $Na + H_2O \rightarrow NaOH + H_2$

b. $Na + ZnSO_4 \rightarrow Na_2SO_4 + Zn$

c. $NaBr + Cl_2 \rightarrow NaCl + Br_2$

d. $ZnS + O_2 \rightarrow ZnO + SO_2$

6.30 Balance the following equations.

a. $H_2S + O_2 \rightarrow SO_2 + H_2O$

b. $Ni + HCl \rightarrow NiCl_2 + H_2$

c. $IBr + NH_3 \rightarrow NH_4Br + NI_3$

d. $C_2H_6 + O_2 \rightarrow CO_2 + H_2O$

6.31 Balance the following equations.

a. $CH_4 + O_2 \rightarrow CO_2 + H_2O$

b. $C_6H_6 + O_2 \rightarrow CO_2 + H_2O$

c. $C_4H_8O_2 + O_2 \rightarrow CO_2 + H_2O$

d. $C_5H_{10}O + O_2 \rightarrow CO_2 + H_2O$

6.32 Balance the following equations.

a. $C_2H_4 + O_2 \rightarrow CO_2 + H_2O$

b. $C_6H_{12} + O_2 \rightarrow CO_2 + H_2O$

c. $C_3H_6O + O_2 \rightarrow CO_2 + H_2O$

d. $C_5H_{10}O_2 + O_2 \rightarrow CO_2 + H_2O$

6.33 Balance the following equations.

a. $PbO + NH_3 \rightarrow Pb + N_2 + H_2O$

b. $Fe(OH)_3 + H_2SO_4 \rightarrow Fe_2(SO_4)_3 + H_2O$

6.34 Balance the following equations.

a. $SO_2Cl_2 + HI \rightarrow H_2S + H_2O + HCl + I_2$

b. $Na_2CO_3 + Mg(NO_3)_2 \rightarrow MgCO_3 + NaNO_3$

Chemical Equations and the Mole Concept (Section 6.7)

6.35 Write the 12 mole-to-mole conversion factors that can be derived from the following balanced equation.

$$2Ag_2CO_3 \rightarrow 4Ag + 2CO_2 + O_2$$

6.36 Write the 12 mole-to-mole conversion factors that can be derived from the following balanced equation.

$$N_2H_4 + 2H_2O_2 \rightarrow N_2 + 4H_2O$$

6.37 Using each of the following equations, calculate the number of moles of CO_2 that can be obtained from 2.00 moles of the first listed reactant with an excess of the other reactant.

a. $C_7H_{16} + 11O_2 \rightarrow 7CO_2 + 8H_2O$

b. $2HCl + CaCO_3 \rightarrow CaCl_2 + CO_2 + H_2O$

c. $Na_2SO_4 + 2C \rightarrow Na_2S + 2CO_2$

d. $Fe_3O_4 + CO \rightarrow 3FeO + CO_2$

6.38 Using each of the following equations, calculate the number of moles of CO_2 that can be obtained from 3.50 moles of the first listed reactant with an excess of the other reactant.

a. $FeO + CO \rightarrow Fe + CO_2$

b. $3O_2 + CS_2 \rightarrow CO_2 + 2SO_2$

c. $2C_8H_{18} + 25O_2 \rightarrow 16CO_2 + 18H_2O$

d. $C_6H_{12}O_6 + 6O_2 \rightarrow 6H_2O + 6CO_2$

Calculations Based on Chemical Equations (Section 6.8)

6.39 How many grams of the first reactant in each of the following equations would be needed to produce 20.0 g of N_2 gas?

a. $4NH_3 + 3O_2 \rightarrow 2N_2 + 6H_2O$

b. $(NH_4)_2Cr_2O_7 \rightarrow N_2 + 4H_2O + Cr_2O_3$

c. $N_2H_4 + 2H_2O_2 \rightarrow N_2 + 4H_2O$

d. $2NH_3 \rightarrow N_2 + 3H_2$

6.40 How many grams of the first reactant in each of the following equations would be needed to produce 20.0 g of H_2O?

a. $N_2H_4 + 2H_2O_2 \rightarrow N_2 + 4H_2O$

b. $H_2O_2 + H_2S \rightarrow 2H_2O + S$

c. $2HNO_3 + NO \rightarrow 3NO_2 + H_2O$

d. $3H_2 + WO_3 \rightarrow W + 3H_2O$

6.41 The principal constituent of natural gas is methane, which burns in air according to the reaction

$$CH_4 + 2O_2 \rightarrow CO_2 + 2H_2O$$

How many grams of O_2 are needed to produce 3.50 g of CO_2?

6.42 Tungsten (W) metal, which is used to make incandescent bulb filaments, is produced by the reaction

$$WO_3 + 3H_2 \rightarrow 3H_2O + W$$

How many grams of H_2 are needed to produce 1.00 g of W?

6.43 The catalytic converter that is now standard equipment on American automobiles converts carbon monoxide (CO) to carbon dioxide (CO_2) by the reaction

$$2CO + O_2 \rightarrow 2CO_2$$

What mass of O_2, in grams, is needed to react completely with 25.0 g of CO?

6.44 A mixture of hydrazine (N_2H_4) and hydrogen peroxide (H_2O_2) is used as a fuel for rocket engines. These two substances react as shown by the equation

$$N_2H_4 + 2H_2O_2 \rightarrow N_2 + 4H_2O$$

What mass of N_2H_4, in grams, is needed to react completely with 35.0 g of H_2O_2?

6.45 Both water and sulfur dioxide are products from the reaction of sulfuric acid (H_2SO_4) with copper metal, as shown by the equation

$$2H_2SO_4 + Cu \rightarrow SO_2 + 2H_2O + CuSO_4$$

How many grams of H_2O will be produced at the same time that 10.0 g of SO_2 is produced?

6.46 Potassium thiosulfate ($K_2S_2O_3$) is used to remove any excess chlorine from fibers and fabrics that have been bleached with that gas. The reaction is

$$K_2S_2O_3 + 4Cl_2 + 5H_2O \rightarrow 2KHSO_4 + 8HCl$$

How many grams of HCl will be produced at the same time that 25.0 g of $KHSO_4$ is produced?

Additional Problems

6.47 The compound 1-propanethiol, which is the eye irritant that is released when fresh onions are chopped up, has a formula mass of 76.18 amu and the formula C_3H_yS. What number does y stand for in the formula?

6.48 Select the quantity that has the greater number of atoms in each of the following pairs of quantities. Make your selection using the periodic table but without performing an actual calculation.

a. 1.00 mole S or 1.00 mole S_8

b. 28.0 g Al or 1.00 mole Al

c. 28.1 g Si or 30.0 g Mg

d. 2.00 g Na or 6.02×10^{23} atoms He

6.49 What amount or mass of each of the following substances would be needed to obtain 1.000 g of Si?

a. moles of SiH_4

b. grams of SiO_2

c. molecules of $(CH_3)_3SiCl$

d. atoms of Si

6.50 How many grams of Si would contain the same number of atoms as there are in 2.10 moles of Ar?

6.51 After the following equation was balanced, the name of one of the reactants was substituted for its formula.

$$2 \text{ butyne} + 11O_2 \longrightarrow 8CO_2 + 6H_2O$$

Using only the information found within the equation, determine the molecular formula of butyne.

6.52 Ammonium dichromate decomposes according to the following reaction.

$$(NH_4)_2Cr_2O_7 \longrightarrow N_2 + 4H_2O + Cr_2O_3$$

How many grams of each of the products can be formed from the decomposition of 75.0 g of ammonium dichromate?

6.53 Black silver sulfide can be produced from the reaction of silver metal with sulfur.

$$2Ag + S \longrightarrow Ag_2S$$

How many grams of Ag and how many grams of S are needed to produce 125 g of Ag_2S?

6.54 How many grams of beryllium (Be) are needed to react completely with 45.0 g of nitrogen (N_2) in the synthesis of Be_3N_2?

Grid Problems

6.55

1. N_2H_4	2. C_2H_6	3. NO
4. C_3H_6	5. CO_2	6. N_2O

Select from the grid *all* correct responses for each of the following situations.

a. Pairs of substances that have the same molecular mass (to two significant figures)
b. Pairs of substances that have the same molar mass (to two significant figures)
c. Pairs of substances in which there is the same number of atoms per molecule
d. Pairs of substances in which there is the same number of atoms per mole

6.56

1. $Na + N_2 \rightarrow NaN_3$	2. $C + H_2 \rightarrow CH_4$	3. $NO + O_2 \rightarrow NO_2$
4. $Al + S \rightarrow Al_2S_3$	5. $N_2 + H_2 \rightarrow NH_3$	6. $H_2 + Cl_2 \rightarrow HCl$

Select from the grid *all* correct responses for each of the following situations.

a. Balanced equations where the coefficient of the product is 2
b. Balanced equations where the sum of the coefficients is 4
c. Reactions where 5 moles of reactants produce 2 moles of product
d. Reactions where the moles of reactants consumed and the moles of products produced are equal

6.57

1. NO	2. N_2O	3. NO_2
4. NH_3	5. HN_3	6. N_2H_4

Select from the grid *all* correct responses for each of the following situations.

a. Compounds for which the following conversion factor is valid:

$$\frac{1 \text{ mole N}}{1 \text{ mole compound}}$$

b. Compounds for which the following conversion factor is valid:

$$\frac{2 \text{ moles N}}{1 \text{ mole compound}}$$

c. Compounds where the ratio of moles of atoms to moles of molecules is 3 or greater
d. Compounds where the molar mass is less than 40 g

6.58

1. 1.0 mole O_2	2. 2.0 moles O_3	3. 2.0 moles NO_2
4. 3.0 moles N_2O	5. 1.0 mole CO_2	6. 3.0 moles CO

Select from the grid *all* correct responses for each of the following situations.

a. Quantities in which 2.0 moles of molecules are present
b. Quantities in which 6.0 moles of atoms are present
c. Quantities that have a mass exceeding 75.0 g
d. Quantities in which more than 50.0 g of oxygen is present

Gases, Liquids, and Solids

CHAPTER OUTLINE

7.1 The Kinetic Molecular Theory of Matter 150
7.2 Kinetic Molecular Theory and Physical States 152
7.3 Gas Law Variables 155
7.4 Boyle's Law: A Pressure–Volume Relationship 155
7.5 Charles's Law: A Temperature–Volume Relationship 157
7.6 The Combined Gas Law 159
7.7 The Ideal Gas Law 160
7.8 Dalton's Law of Partial Pressures 161
7.9 Changes of State 164
Chemistry at a Glance: **The Gas Laws** 163
7.10 Evaporation of Liquids 164
7.11 Vapor Pressure of Liquids 165
7.12 Boiling and Boiling Point 166
7.13 Intermolecular Forces in Liquids 168
Chemistry at a Glance: **Intermolecular Forces** 172
Chemical Connections
7.1 The Importance of Gas Densities 154
7.2 The Exchange of the Respiratory Gases, Oxygen and Carbon Dioxide 162

A high-rise building being demolished through the use of explosives. Changes in volume that occur as the explosive (a solid) is converted into gases via a chemical reaction are the basis for the destructive effects of explosives.

In Chapters 3, 4, and 5, we considered the structure of matter from a submicroscopic point of view—in terms of molecules, atoms, protons, neutrons, and electrons. In this chapter, we are concerned with the macroscopic characteristics of matter as represented by the physical states—solid, liquid, and gas. Of particular concern are the properties exhibited by matter in the various physical states and a theory that correlates these properties with molecular behavior.

7.1 The Kinetic Molecular Theory of Matter

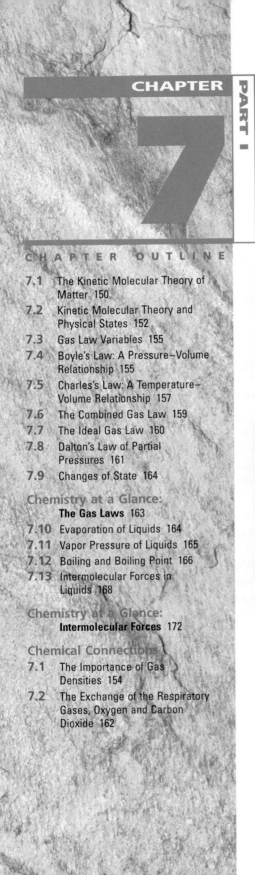

Solids, liquids, and gases (Section 1.2) are easily distinguished by using four common physical properties of matter (Table 7.1): (1) volume and shape, (2) density, (3) compressibility, and (4) thermal expansion. We discussed the property of density in Section 2.7. *Compressibility* involves the change in volume that results from a pressure

Figure 7.1
The water in the lake behind the dam has potential energy as a result of its position. When the water flows over the dam, its potential energy becomes kinetic energy that can be used to turn the turbines of a hydroelectric plant.

change. *Thermal expansion* involves the volume change that results from a temperature change. The physical characteristics of the solid, liquid, and gaseous states listed in Table 7.1 can be explained by kinetic molecular theory, which is one of the fundamental theories of chemistry.

The **kinetic molecular theory of matter** *is a set of five statements that are used to explain the physical behavior of the three states of matter (solids, liquids, and gases).* The basic idea of this theory is that the particles (atoms, molecules, or ions) present in a substance, independent of the physical state of the substance, are always in motion.

The five statements of the kinetic molecular theory of matter are as follows:

Statement 1: *Matter is ultimately composed of tiny particles (atoms, molecules, or ions) that have definite and characteristic sizes that do not change.*

Statement 2: *The particles are in constant random motion and therefore possess kinetic energy.*

Kinetic energy *is energy that matter possesses because of its motion.* An object that is in motion has the ability to transfer its kinetic energy to another object upon collision with that object.

Statement 3: *The particles interact with one another through attractions and repulsions and therefore possess potential energy.*

Potential energy *is stored energy that matter possesses as a result of its position, condition, and/or composition* (Figure 7.1). The potential energy of greatest importance to the differences among the three states of matter is that which originates from electrostatic interactions between particles. **Electrostatic interactions** *are attractions and repulsions that occur between particles.* Particles of opposite charge (one positive and the other negative) attract one another, and particles of like charge (both positive or both negative) repel one another.

Statement 4: *The velocity of the particles increases as the temperature is increased.*

The *average* velocity (kinetic energy) of all particles in a system depends on the temperature; velocity increases as temperature increases.

Statement 5: *The particles in a system transfer energy to each other through elastic collisions.*

In an elastic collision, the total kinetic energy remains constant; no kinetic energy is lost. The difference between an *elastic* and an *inelastic* collision is illustrated by comparing the collision of two hard steel spheres with the collision of two masses of putty. The collision of spheres approximates an elastic collision (the spheres bounce off one another and continue moving, as in Figure 7.2); the putty collision has none of these characteristics (the masses "glob" together with no resulting movement).

• The word *kinetic* comes from the Greek *kinesis,* which means "movement." The kinetic molecular theory deals with the movement of particles.

• The energy released when gasoline is burned represents potential energy associated with chemical bonds.

• For gases, the attractions between particles (statement 3) are minimal and as a first approximation are considered to be zero (see Section 7.2).

• Two consequences of the elasticity of particle collisions (statement 5) are that (1) the energy of any given particle is continually changing, and (2) particle energies for a system are not all the same; a range of particle energies is always encountered.

Table 7.1
Distinguishing Properties of Solids, Liquids, and Gases

Property	Solid state	Liquid state	Gaseous state
volume and shape	definite volume and definite shape	definite volume and indefinite shape; takes the shape of container to the extent that it is filled	indefinite volume and indefinite shape; takes the volume and shape of container it fills
density	high	high, but usually lower than corresponding solid	low
compressibility	small	small, but usually greater than corresponding solid	large
thermal expansion	very small: about 0.01% per °C	small: about 0.10% per °C	moderate: about 0.30% per °C

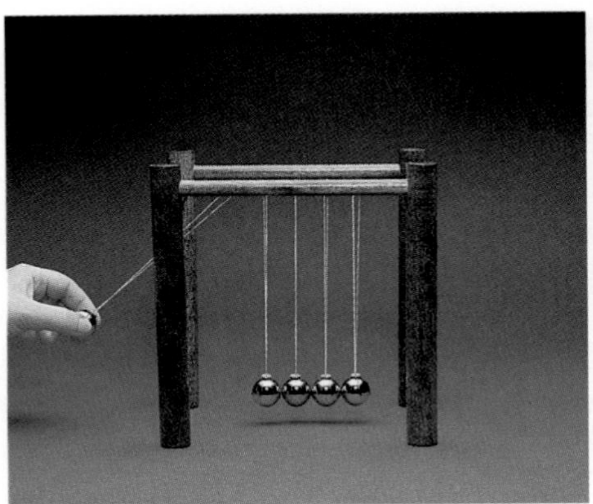
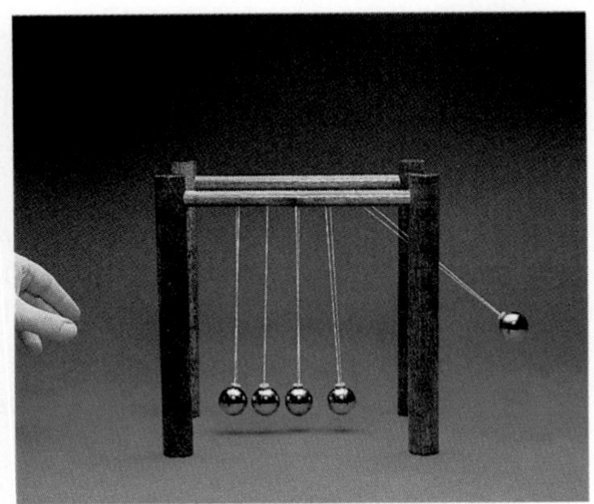

Figure 7.2
Upon release, the steel ball on the left transmits its kinetic energy through a series of elastic collisions to the ball on the right.

The differences among the solid, liquid, and gaseous states of matter can be explained by the relative magnitudes of kinetic energy and potential energy (in this case, electrostatic attractions) associated with the physical state. Kinetic energy can be considered a *disruptive force* that tends to make the particles of a system increasingly independent of one another. This is because the particles tend to move away from one another as a result of the energy of motion. Potential energy of attraction can be considered a *cohesive force* that tends to cause order and stability among the particles of a system.

How much kinetic energy a chemical system has depends on its temperature. Kinetic energy increases as temperature increases (statement 4 of the kinetic molecular theory of matter). Thus the higher the temperature, the greater the magnitude of disruptive influences within a chemical system. Potential energy magnitude, or cohesive force magnitude, is essentially independent of temperature. The fact that one of the types of forces depends on temperature (disruptive forces) and the other does not (cohesive forces) causes temperature to be the factor that determines in which of the three physical states a given sample is found. We will discuss the reason for this in Section 7.2.

7.2 Kinetic Molecular Theory and Physical States

The **solid state** *is characterized by a dominance of potential energy (cohesive forces) over kinetic energy (disruptive forces).* The particles in a solid are drawn close together in a regular pattern by the strong cohesive forces present (Figure 7.3a). Each particle occupies a fixed position, about which it vibrates because of disruptive kinetic energy. With this model, the characteristic properties of solids (Table 7.1) can be explained as follows:

1. *Definite volume and definite shape.* The strong, cohesive forces hold the particles in essentially fixed positions, resulting in definite volume and definite shape.

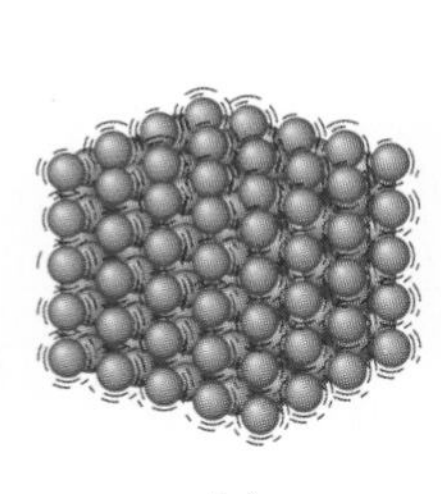

(a)

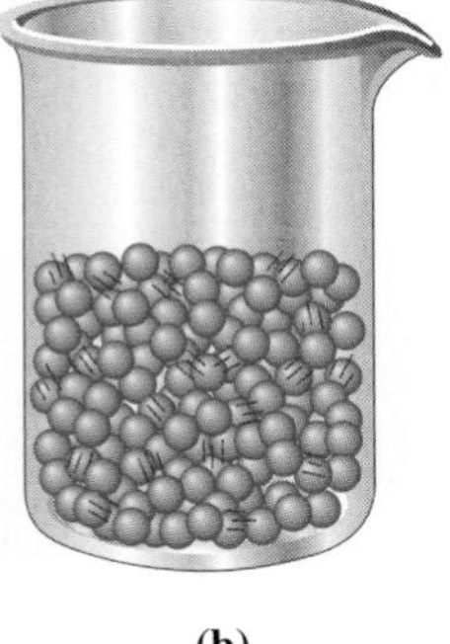

(b)

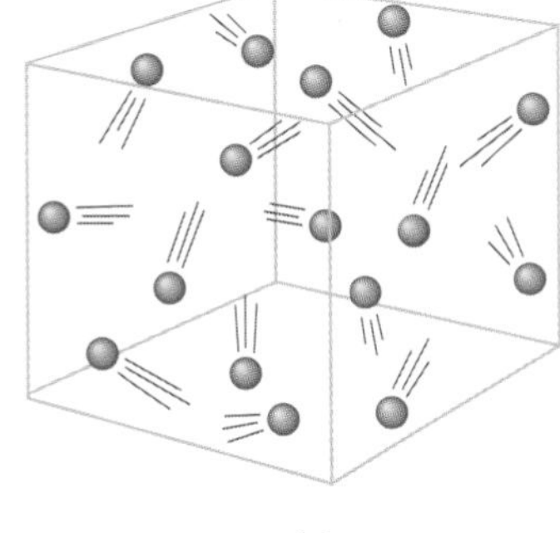

(c)

Figure 7.3
(a) In a solid, the particles (atoms, molecules, or ions) are close together and vibrate about fixed sites. (b) The particles in a liquid, though still close together, freely slide over one another. (c) In a gas, the particles are in constant random motion, each particle being independent of the others present.

2. *High density.* The constituent particles of solids are located as close together as possible (essentially touching each other). Therefore, a given volume contains large numbers of particles, resulting in a high density.
3. *Small compressibility.* Because there is very little space between particles, increased pressure cannot push the particles any closer together; therefore, it has little effect on the solid's volume.
4. *Very small thermal expansion.* An increased temperature increases the kinetic energy (disruptive forces), thereby causing more vibrational motion of the particles. Each particle occupies a slightly larger volume, and the result is a slight expansion of the solid. The strong, cohesive forces prevent this effect from becoming very large.

The **liquid state** *is characterized by potential energy (cohesive forces) and kinetic energy (disruptive forces) of about the same magnitude.* The liquid state consists of particles that are randomly packed but relatively near one another (Figure 7.3b). The molecules are in constant, random motion; they slide freely over one another but do not move with enough energy to separate. The fact that the particles freely slide over each other indicates the influence of disruptive forces; however, the fact that the particles do not separate indicates fairly strong cohesive forces. With this model, the characteristic properties of liquids (Table 7.1) can be explained as follows:

1. *Definite volume and indefinite shape.* The attractive forces are strong enough to restrict particles to movement within a definite volume. They are not strong enough, however, to prevent the particles from moving over each other in a random manner that is limited only by the container walls. Thus liquids have no definite shape, except that they maintain a horizontal upper surface in containers that are not completely filled.
2. *High density.* The particles in a liquid are not widely separated; they are still touching one another. Therefore, there will be a large number of particles in a given volume—a high density.
3. *Small compressibility.* Because the particles in a liquid are still touching each other, there is very little empty space. Therefore, an increase in pressure cannot squeeze the particles much closer together.
4. *Small thermal expansion.* Most of the particle movement in a liquid is vibrational because a particle can move only a short distance before colliding with a neighbor. The increased particle velocity that accompanies a temperature increase results only in increased vibrational amplitudes. The net effect is an increase in the effective volume a particle occupies, which causes a slight volume increase in the liquid.

The **gaseous state** *is characterized by a complete dominance of kinetic energy (disruptive forces) over potential energy (cohesive forces).* As a result, the particles of a gas move essentially independently of one another, in a totally random manner (Figure 7.3c). Under ordinary pressure, the particles are relatively far apart, except when they collide with one another. In between collisions with one another or the container walls, gas particles travel in straight lines (Figure 7.4).

Figure 7.4
Gas molecules can be compared to billiard balls in random motion, bouncing off one another and off the sides of the pool table.

Chemical CONNECTIONS

7.1 The Importance of Gas Densities

In the gaseous state, particles are approximately 10 times farther apart than in the solid or liquid state at a given temperature and pressure. Consequently, gases have densities much lower than those of solids and liquids. The fact that gases have low densities is a major factor in explaining many commonly encountered phenomena.

Popcorn pops because of the difference in density between liquid and gaseous water (1.0 g/mL versus 0.001 g/mL). As the corn kernels are heated, water within the kernels is converted to steam. The steam's volume, approximately 1000 times greater than that of the water from which it was generated, causes the kernels of corn to "blow up."

Changes in density that occur as a solid is converted to gases via a chemical reaction are the basis for the operation of automobile air bags and the effects of explosives. Automobile air bags are designed to inflate rapidly (in a fraction of a second) in the event of a crash and then to deflate immediately. Their activation involves mechanical shock causing a steel ball to compress a spring that eletronically ignites a detonator cap, which in turn causes solid sodium azide (NaN_3) to decompose. The decomposition reaction is

$$2NaN_3(s) \longrightarrow 2Na(l) + 3N_2(g)$$

The nitrogen gas so generated inflates the air bag. A small amount of NaN_3 (high density) will generate over 50 L of N_2 gas at 25°C. Because the air bag is porous, it goes limp quickly as the generated N_2 gas escapes.

Millions of hours of hard manual labor are saved annually by the use of industrial explosives in quarrying rock, constructing tunnels, and mining coal and metal ores. The active ingredient in dynamite, a heavily used industrial explosive, is nitroglycerin, whose destructive power comes from the generation of large volumes of gases at high temperatures. The reaction is

$$\underset{\text{Nitroglycerin}}{4C_3H_5O_3(NO_2)_3(s)} \longrightarrow 12CO_2(g) + 10H_2O(g) + 6N_2(g) + O_2(g)$$

At the temperature of the explosion, about 5000°C, there is an approximately 20,000-fold increase in volume as the result of density changes. No wonder such explosives can blow materials to pieces!

The density difference associated with temperature change is the basis for the operation of hot air balloons. Hot air, which is less dense than cold air, rises.

Weather balloons and blimps are filled with helium, a gas less dense than air. Thus, such objects rise in air.

Water vapor is less dense than air. Thus, moist air is less dense than dry air. Decreasing barometric pressure (from lower-density moist air) is an indication that a storm front is approaching.

Figure 7.5
When a gas is compressed, the amount of empty space in the container is decreased. The size of the molecules does not change; they simply move closer together.

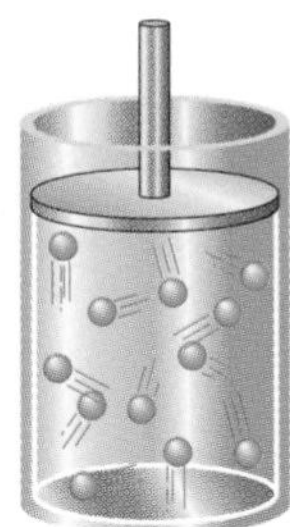

Gas at low pressure

Gas at higher pressure

The kinetic theory explanation of the properties of gases follows the same pattern that we saw earlier for solids and liquids.

1. *Indefinite volume and indefinite shape.* The attractive (cohesive) forces between particles have been overcome by high kinetic energy, and the particles are free to travel in all directions. Therefore, gas particles completely fill their container, and the shape of the gas is that of the container.
2. *Low density.* The particles of a gas are widely separated. There are relatively few particles in a given volume (compared with liquids and solids), which means little mass per volume (a low density).
3. *Large compressibility.* Particles in a gas are widely separated; essentially, a gas is mostly empty space. When pressure is applied, the particles are easily pushed closer together, decreasing the amount of empty space and the volume of the gas (see Figure 7.5).
4. *Moderate thermal expansion.* An increase in temperature means an increase in particle velocity. The increased kinetic energy of the particles enables them to push back whatever barrier is confining them into a given volume, and the

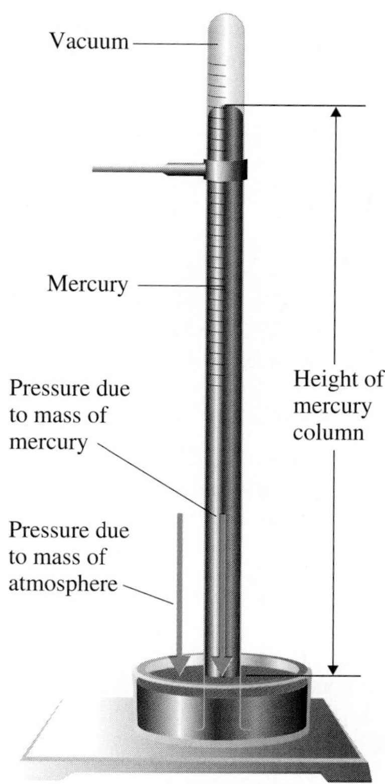

Figure 7.6
The essential components of a mercury barometer are a graduated glass tube, a glass dish, and liquid mercury.

• Millimeters of mercury is the pressure unit most often encountered in clinical work in allied health fields. For example, oxygen and carbon dioxide pressures in respiration are almost always specified in millimeters of mercury.

• Blood pressure is measured with the aid of an apparatus known as a sphygmomanometer, which is essentially a barometer tube connected to an inflatable cuff by a hollow tube. A typical blood pressure is 120/80; this ratio means a systolic pressure of 120 mm Hg above atmospheric pressure and a diastolic pressure of 80 mm Hg above atmospheric pressure.

volume increases. You will note that the size of the particles is not changed during expansion or compression of gases, solids, or liquids; they merely move either farther apart or closer together. It is the space between the particles that changes.

7.3 Gas Law Variables

Behavior of matter in the gaseous state can be described by some very simple quantitative relationships called gas laws. **Gas laws** *are generalizations that describe in mathematical terms the relationships among the pressure, temperature, and volume of a specific quantity of gas.*

Gas laws involve three variables: volume, temperature, and pressure. Two of these variables, volume and temperature, were discussed in Sections 2.2 and 2.8. **Pressure** *is the force applied per unit area—that is, the total force on a surface divided by the area of that surface.* The mathematical equation for pressure is

$$P\text{ (pressure)} = \frac{F\text{ (force)}}{A\text{ (area)}}$$

For a gas, the force that creates pressure is that which is exerted by the gas molecules or atoms as they constantly collide with the walls of their container. Barometers, manometers, and gauges are the instruments most commonly used to measure gas pressures.

The air that surrounds Earth exerts pressure on every object it touches. A **barometer** *is a device used to measure atmospheric pressure.* The essential components of a simple barometer are shown in Figure 7.6. Atmospheric pressure is expressed in terms of the height of the barometer's mercury column, usually in millimeters of mercury (mm Hg). Another name for millimeters of mercury is *torr,* used in honor of Evangelista Torricelli, the Italian physicist who invented the barometer.

$$1\text{ mm Hg} = 1\text{ torr}$$

Atmospheric pressure varies with the weather and the altitude. It averages about 760 mm Hg at sea level, and it decreases by approximately 25 mm Hg for every 1000-ft increase in altitude. The pressure unit *atmosphere* (atm) is defined in terms of this average pressure at sea level. By definition,

$$1\text{ atm} = 760\text{ mm Hg} = 760\text{ torr}$$

Another commonly used pressure unit is *pounds per square inch* (psi or lb/in^2). One atmosphere is equal to 14.7 psi.

$$1\text{ atm} = 14.7\text{ psi}$$

7.4 Boyle's Law: A Pressure–Volume Relationship

Of the several relationships that exist among gas law variables, the first to be discovered relates gas pressure to gas volume. It was formulated over 300 years ago, in 1662, by the British chemist and physicist Robert Boyle. **Boyle's law** states that *the volume of a sample of a gas is* inversely proportional *to the pressure applied to the gas if the temperature is kept constant.* This means that if the pressure on the gas increases, the volume decreases proportionally; conversely, if the pressure decreases, the volume increases. Doubling the pressure cuts the volume in half; tripling the pressure cuts the volume to one-third its original value; quadrupling the pressure cuts the volume to one-fourth; and so on. Figure 7.7 illustrates Boyle's law.

The mathematical equation for Boyle's law is

$$P_1 \times V_1 = P_2 \times V_2$$

Robert Boyle (1627–1691), like most men of the seventeenth century who devoted themselves to science, was self-taught. It was through his efforts that the true value of experimental investigation was first recognized.

• When we know any three of the four quantities in the Boyle's law equation, we can calculate the fourth, which is usually the final pressure, P_2, or the final volume, V_2. The Boyle's law equation is valid only if the temperature of the gas remains constant.

where P_1 and V_1 are the pressure and volume of a gas at an initial set of conditions, and P_2 and V_2 are the pressure and volume of the same sample of gas under a new set of conditions, with the temperature remaining constant.

Example 7.1

Using Boyle's Law to Calculate the New Volume of a Gas

A sample of O_2 gas occupies a volume of 1.50 L at a pressure of 735 mm Hg and a temperature of 25°C. What volume will it occupy, in liters, if the pressure is increased to 770 mm Hg with no change in temperature?

Solution

A suggested first step in working gas law problems that involve two sets of conditions is to analyze the given data in terms of initial and final conditions.

$$P_1 = 735 \text{ mm Hg} \qquad P_2 = 770 \text{ mm Hg}$$

$$V_1 = 1.50 \text{ L} \qquad V_2 = ?\text{ L}$$

We know three of the four variables in the Boyle's law equation, so we can calculate the fourth, V_2. We will rearrange Boyle's law to isolate V_2 (the quantity to be calculated) on one side of the equation. This is accomplished by dividing both sides of the Boyle's law equation by P_2.

$$P_1V_1 = P_2V_2 \qquad \text{(Boyle's law)}$$

$$\frac{P_1V_1}{P_2} = \frac{\cancel{P_2}V_2}{\cancel{P_2}} \qquad \text{(Divide each side of the equation by } P_2\text{.)}$$

$$V_2 = V_1 \times \frac{P_1}{P_2}$$

Substituting the given data into the rearranged equation and doing the arithmetic give

$$V_2 = 1.50 \text{ L} \times \left(\frac{735 \cancel{\text{ mm Hg}}}{770 \cancel{\text{ mm Hg}}}\right) = 1.43 \text{ L}$$

Practice Exercise 7.1

A sample of H_2 gas occupies a volume of 2.25 L at a pressure of 628 mm Hg and a temperature of 35°C. What volume will it occupy, in liters, if the pressure is decreased to 428 mm Hg with no change in temperature?

• *Answer:* 3.30 L

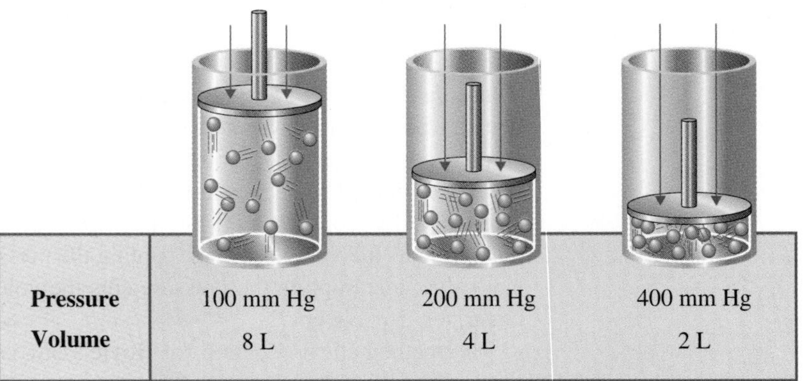

Pressure	100 mm Hg	200 mm Hg	400 mm Hg
Volume	8 L	4 L	2 L

Figure 7.7
Data illustrating the inverse proportionality associated with Boyle's law.

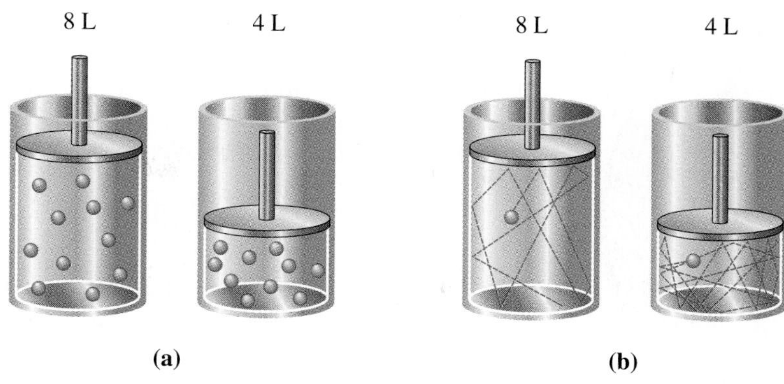

Figure 7.8
When the volume of a gas, at constant temperature, decreases by half (a), the average number of times a molecule hits the container walls is doubled (b).

• Boyle's law explains the process of breathing. Breathing in occurs when the diaphragm flattens out (contracts). This contraction causes the volume of the thoracic cavity to increase and the pressure within the cavity to drop (Boyle's law) below atmospheric pressure. Air flows into the lungs and expands them, because the pressure is greater outside the lungs than within them. Breathing out occurs when the diaphragm relaxes (moves up), decreasing the volume of the thoracic cavity and increasing the pressure (Boyle's law) within the cavity to a value greater than the external pressure. Air flows out of the lungs. The air flow direction is always from a high-pressure region to a low-pressure region.

Boyle's law is consistent with kinetic molecular theory. The pressure that a gas exerts results from collisions of the gas molecules with the sides of the container. If the volume of a container holding a specific number of gas molecules is increased, the total wall area of the container will also increase, and the number of collisions in a given area (the pressure) will decrease because of the greater wall area. Conversely, if the volume of the container is decreased, the wall area will be smaller and there will be more collisions within a given wall area. Figure 7.8 illustrates this concept.

Filling a medical syringe with a liquid demonstrates Boyle's law. As the plunger is drawn out of the syringe (see Figure 7.9), the increase in volume inside the syringe chamber results in decreased pressure there. The liquid, which is at atmospheric pressure, flows into this reduced-pressure area. This liquid is then expelled from the chamber by pushing the plunger back in. This ejection of the liquid does not involve Boyle's law; a liquid is incompressible and mechanical force pushes it out.

Figure 7.9
Filling a syringe with a liquid is an application of Boyle's law.

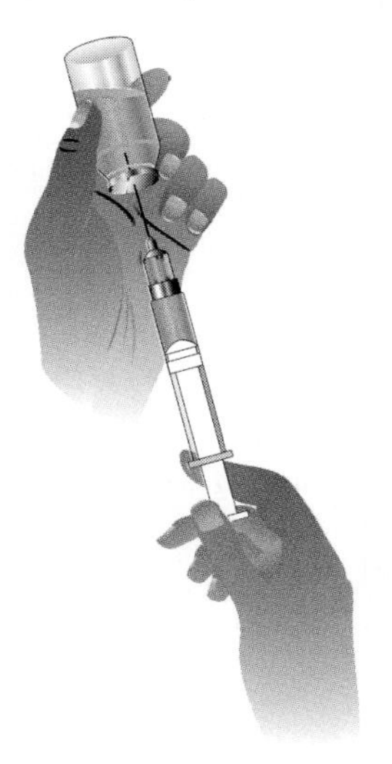

7.5 Charles's Law: A Temperature–Volume Relationship

The relationship between the temperature and the volume of a gas at constant pressure is called *Charles's law* after the French scientist Jacques Charles. This law was discovered in 1787, over 100 years after the discovery of Boyle's law. **Charles's law** states that *the volume of a sample of gas is* directly proportional *to its Kelvin temperature if the pressure is kept constant* (Figure 7.10). Whenever a *direct* proportion exists between two quantities, one increases when the other increases and one decreases when the other decreases. The direct proportion relationship of Charles's law means that if the temperature increases, the volume will also increase and that if the temperature decreases, the volume will also decrease.

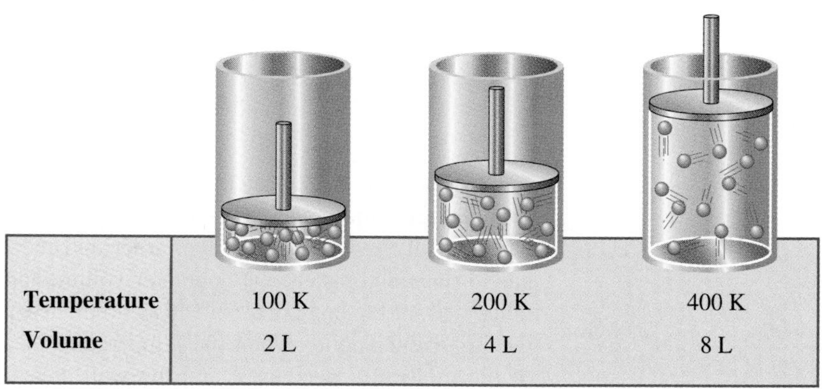

Temperature	100 K	200 K	400 K
Volume	2 L	4 L	8 L

Figure 7.10
Data illustrating the direct proportionality associated with Charles's law.

Jacques Charles (1746–1823), a French physicist, in the process of working with hot air balloons, made the observations that ultimately led to the formulation of what is now known as Charles's law.

A balloon filled with air illustrates Charles's law. If the balloon is placed near a heat source such as a light bulb that has been on for some time, the heat will cause the balloon to increase visibly in size (volume). Putting the same balloon in the refrigerator will cause it to shrink.

Charles's law, stated mathematically, is

$$\frac{V_1}{T_1} = \frac{V_2}{T_2}$$

where V_1 is the volume of a gas at a given pressure, T_1 is the Kelvin temperature of the gas, and V_2 and T_2 are the volume and Kelvin temperature of the gas under a new set of conditions, with the pressure remaining constant.

• When you use the mathematical form of Charles's law, the temperatures used *must be* Kelvin scale temperatures.

• Charles's law predicts that gas volume will become smaller and smaller as temperature is reduced, until eventually a temperature is reached at which gas volume becomes zero. This "zero-volume" temperature is calculated to be −273°C and is known as *absolute zero* (see Section 2.8). Absolute zero is the basis for the Kelvin temperature scale. In reality, gas volume never vanishes. As temperature is lowered, at some point before absolute zero, the gas condenses to a liquid, at which point Charles's law is no longer valid.

Example 7.2

Using Charles's Law to Calculate the New Volume of a Gas

A sample of the gaseous anesthetic cyclopropane, with a volume of 425 mL at a temperature of 27°C, is cooled at constant pressure to 20°C. What is the new volume, in milliliters, of the sample?

Solution

First, we will analyze the data in terms of initial and final conditions.

$$V_1 = 425 \text{ mL} \qquad V_2 = ? \text{ mL}$$

$$T_1 = 27°\text{C} + 273 = 300 \text{ K} \qquad T_2 = 20°\text{C} + 273 = 293 \text{ K}$$

Note that both of the given temperatures have been converted to Kelvin scale readings. This change is accomplished by simply adding 273 to the Celsius scale value (Section 2.8).

We know three of the four variables in the Charles's law equation, so we can calculate the fourth, V_2. We will rearrange Charles's law to isolate V_2 (the quantity desired) by multiplying each side of the equation by T_2.

$$\frac{V_1}{T_1} = \frac{V_2}{T_2} \qquad \text{(Charles's law)}$$

$$\frac{V_1 T_2}{T_1} = \frac{V_2 \cancel{T_2}}{\cancel{T_2}} \qquad \text{(Multiply each side by } T_2\text{.)}$$

$$V_2 = V_1 \times \frac{T_2}{T_1}$$

Substituting the given data into the equation and doing the arithmetic give

$$V_2 = 425 \text{ mL} \times \left(\frac{293 \cancel{\text{K}}}{300 \cancel{\text{K}}}\right) = 415 \text{ mL}$$

Practice Exercise 7.2

A sample of dry air, with a volume of 125 mL at a temperature of 53°C, is heated at constant pressure to 95°C. What is the new volume, in milliliters, of the sample?

• *Answer:* 141 mL

Charles's law is consistent with kinetic molecular theory. When the temperature of a gas increases, the velocity (kinetic energy) of the gas molecules increases. The speedier particles hit the container walls harder and more often. In order for the pressure of the gas to remain constant, the container volume must increase. In a larger volume, the particles will hit the container walls less often, and the pressure can remain the same. A similar argument applies when the temperature of a gas is lowered. This time the velocity of the molecules decreases, and the wall area (volume) must also decrease in order to increase the number of collisions in a given area in a given time.

The effects of immersing a balloon in liquid nitrogen, which has a temperature of 77 K (−196°C)

Charles's law is the principle used in the operation of a convection heater. When air comes in contact with the heating element, it expands (its density becomes less). The hot, less dense air rises, causing continuous circulation of warm air. This same principle has ramifications in closed rooms that lack effective air circulation. The warmer and less dense air stays near the top of the room. This is desirable in the summer but not in the winter.

7.6 The Combined Gas Law

The **combined gas law** *is an expression obtained by mathematically combining Boyle's and Charles's laws.* The mathematical equation for the combined gas law is

$$\frac{P_1V_1}{T_1} = \frac{P_2V_2}{T_2}$$

• Any time a gas law contains temperature terms, as is the case for both Charles's law and the combined gas law, these temperatures must be specified on the Kelvin temperature scale.

This combined gas law is a much more versatile equation than either of the laws from which it is derived. Using this equation, we can calculate the change in pressure, temperature, or volume that is brought about by changes in the other two variables.

Example 7.3

Using the Combined Gas Law to Calculate the New Volume of a Gas

A sample of O_2 gas occupies a volume of 1.62 L at 755 mm Hg pressure and has a temperature of 0°C. What volume, in liters, will this gas sample occupy at 725 mm Hg pressure and 50°C?

Solution

First, we analyze the data in terms of initial and final conditions.

$P_1 = 755 \text{ mm Hg}$ $\quad$ $P_2 = 725 \text{ mm Hg}$

$V_1 = 1.62 \text{ L}$ $\quad$ $V_2 = ? \text{ L}$

$T_1 = 0°\text{C} + 273 = 273 \text{ K}$ $\quad$ $T_2 = 50°\text{C} + 273 = 323 \text{ K}$

We are given five of the six variables in the combined gas law, so we can calculate the sixth one, V_2. Rearranging the combined gas law to isolate the variable V_2 on a side by itself gives

$$V_2 = \frac{V_1P_1T_2}{P_2T_1}$$

Substituting numerical values into this "version" of the combined gas law gives

$$V_2 = 1.62 \text{ L} \times \frac{755 \cancel{\text{mm Hg}}}{725 \cancel{\text{mm Hg}}} \times \frac{323 \cancel{\text{K}}}{273 \cancel{\text{K}}} = 2.00 \text{ L}$$

Practice Exercise 7.3

A helium-filled weather balloon, when released, has a volume of 10.0 L at 27°C and a pressure of 663 mm Hg. What volume, in liters, will the balloon occupy at an altitude where the pressure is 96 mm Hg and the temperature is −30.0°C?

- *Answer:* 56 L

7.7 The Ideal Gas Law

The **ideal gas law** *is an equation that includes the quantity of gas in a sample as well as the temperature, pressure, and volume of the sample.* Mathematically, the ideal gas law has the form

$$PV = nRT$$

In this equation, pressure, temperature, and volume are defined in the same manner as in the gas laws we have already discussed. The symbol n stands for the *number of moles* of gas present in the sample. The symbol R represents the *ideal gas constant,* the proportionality constant that makes the equation valid.

• The ideal gas law is used in calculations when *one* set of conditions is given with one missing variable. The combined gas law (Section 7.6) is used when *two* sets of conditions are given with one missing variable.

The value of the ideal gas constant (R) varies with the units chosen for pressure and volume. With pressure in atmospheres and volume in liters, R has the value

$$R = \frac{PV}{nT} = 0.0821 \frac{\text{atm} \cdot \text{L}}{\text{mole} \cdot \text{K}}$$

The value of R is the same for all gases under normally encountered conditions of temperature, pressure, and volume.

If three of the four variables in the ideal gas law equation are known, then the fourth can be calculated using the equation. Example 7.4 illustrates the use of the ideal gas law.

Example 7.4

Using the Ideal Gas Law to Calculate the Volume of a Gas

The colorless, odorless, tasteless gas carbon monoxide, CO, is a by-product of incomplete combustion of any material that contains the element carbon. Calculate the volume, in liters, occupied by 1.52 moles of this gas at 0.992 atm pressure and a temperature of 65°C.

Solution

This problem deals with only one set of conditions, so the ideal gas equation is applicable. Three of the four variables in the ideal gas equation (P, n, and T) are given, and the fourth (V) is to be calculated.

$$P = 0.992 \text{ atm} \qquad n = 1.52 \text{ moles}$$

$$V = ?\text{ L} \qquad T = 65°\text{C} = 338 \text{ K}$$

Rearranging the ideal gas equation to isolate V on the left side of the equation gives

$$V = \frac{nRT}{P}$$

Because the pressure is given in atmospheres and the volume unit is liters, the R value 0.0821 is valid. Substituting known numerical values into the equation gives

$$V = \frac{(1.52 \text{ moles}) \times \left(0.0821 \frac{\text{atm} \cdot \text{L}}{\text{mole} \cdot \text{K}}\right)(338 \text{ K})}{0.992 \text{ atm}}$$

Note that all the parts of the ideal gas constant unit cancel except for one, the volume part.

Doing the arithmetic yields the volume of CO.

$$V = \left(\frac{1.52 \times 0.0821 \times 338}{0.992}\right) \text{L} = 42.5 \text{ L}$$

Practice Exercise 7.4

Calculate the volume, in liters, occupied by 3.25 moles of Cl_2 gas at 1.54 atm pressure and a temperature of 213°C.

- *Answer:* 84.2 L Cl_2

John Dalton (1766–1844) throughout his life had a particular interest in the study of weather. From "weather" he turned his attention to the nature of the atmosphere and then to the study of gases in general.

7.8 Dalton's Law of Partial Pressures

In a mixture of gases that do not react with one another, each type of molecule moves around in the container as though the other kinds were not there. This type of behavior is possible because a gas is mostly empty space, and attractions between molecules in the gaseous state are negligible at most temperatures and pressures. Each gas in the mixture occupies the entire volume of the container; that is, it distributes itself uniformly throughout the container. The molecules of each type strike the walls of the container as frequently and with the same energy as if they were the only gas in the mixture. Consequently, the pressure exerted by each gas in a mixture is the same as it would be if the gas were alone in the same container under the same conditions.

The English scientist John Dalton was the first to notice the independent behavior of gases in mixtures. In 1803, he published a summary statement concerning this behavior that is now known as Dalton's law of partial pressures. **Dalton's law of partial pressures** states that *the total pressure exerted by a mixture of gases is the sum of the partial pressures of the individual gases.* A **partial pressure** *is the pressure that a gas in a mixture would exert if it were present alone under the same conditions.*

- A sample of clean air is the most common example of a mixture of gases that do not react with one another.

Expressed mathematically, Dalton's law states that

$$P_{\text{Total}} = P_1 + P_2 + P_3 + \cdots$$

where P_{Total} is the total pressure of a gaseous mixture and P_1, P_2, P_3, and so on are the partial pressures of the individual gaseous components of the mixture.

As an illustration of Dalton's law, consider the four identical gas containers shown in Figure 7.11. Suppose we place amounts of three different gases (represented by *A*, *B*, and *C*) into three of the containers and measure the pressure exerted by each sample. We then

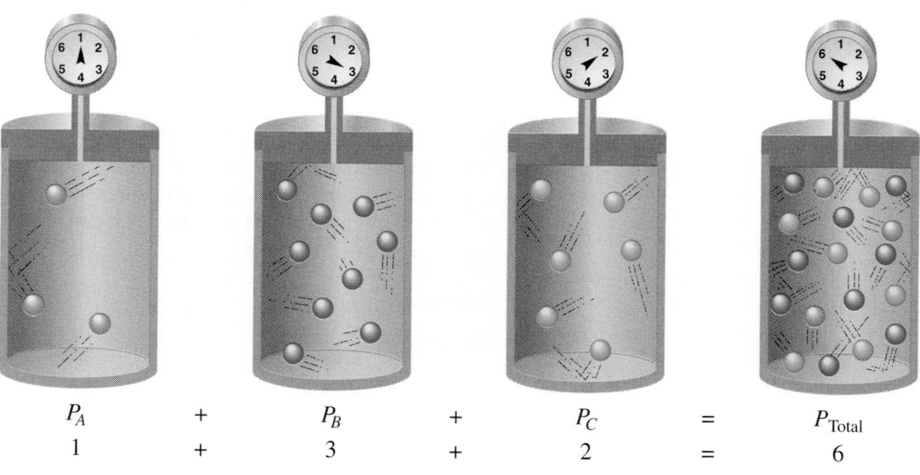

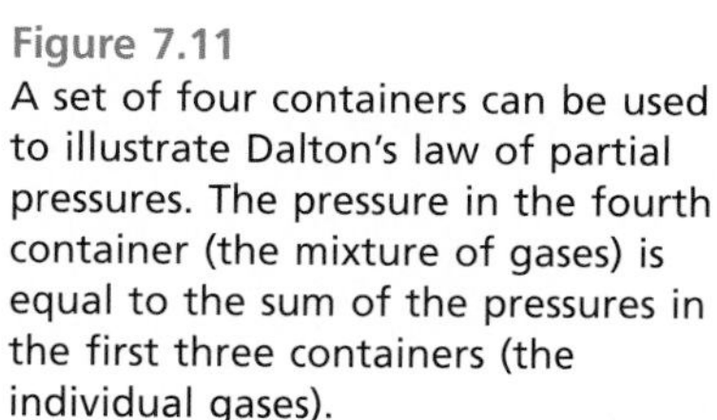
Figure 7.11
A set of four containers can be used to illustrate Dalton's law of partial pressures. The pressure in the fourth container (the mixture of gases) is equal to the sum of the pressures in the first three containers (the individual gases).

Chemical CONNECTIONS

7.2 The Exchange of the Respiratory Gases, Oxygen and Carbon Dioxide

The concept of partial pressures helps explain the process of respiration in the human body. In general, inhaling replenishes oxygen in the blood and exhaling removes carbon dioxide from the blood. The oxygen is consumed by our cells in various cellular processes, and the carbon dioxide is the waste product from these cellular processes. Pressure gradients resulting from partial-pressure differences govern the movement of oxygen and carbon dioxide throughout the human body. Both gases always move from areas of higher partial pressure to areas of lower partial pressure. The overall exchange system is shown in the accompanying diagram.

As shown on the left side of the diagram, venous blood—blood returning to the lungs from the tissues—has been depleted of its oxygen supply (low oxygen partial pressure) and is carrying waste product carbon dioxide (high carbon dioxide partial pressure). Within the lungs (left side of diagram), carbon dioxide leaves the blood and oxygen enters the blood. The partial pressure of carbon dioxide is greater in the blood than within the lungs; for oxygen, the opposite is true.

As shown at the top of the diagram, arterial blood—blood that circulates from the lungs to the tissues—has been replenished in oxygen (high oxygen partial pressure). Oxygen from this blood diffuses into the tissues (low oxygen partial pressure). At the same time, carbon dioxide produced in the cells (high carbon dioxide partial pressure) diffuses into the bloodstream (low carbon dioxide partial pressure).

Note the actual values given in the diagram of the partial pressures for oxygen and carbon dioxide in the body. For carbon dioxide, the partial pressure varies between 40 mm Hg and 46 mm Hg; this is a small pressure gradient. For oxygen, a much larger pressure gradient exists; the gradient between arterial and venous blood is 65 mm Hg.

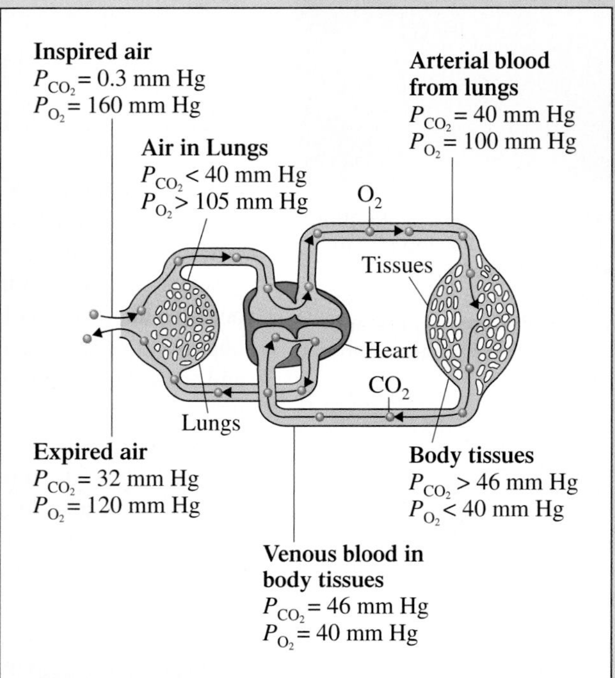

place all three samples in the fourth container and measure the pressure exerted by this mixture of gases. We find that

$$P_{\text{Total}} = P_A + P_B + P_C$$

Using the actual gauge pressure values given in Figure 7.11, we see that

$$P_{\text{Total}} = 1 + 3 + 2 = 6$$

Example 7.5

Using Dalton's Law to Calculate Partial Pressure

The total pressure exerted by a mixture of the three gases oxygen, nitrogen, and water vapor is 742 mm Hg. The partial pressures of the nitrogen and oxygen in the sample are 581 mm Hg and 143 mm Hg, respectively. What is the partial pressure of the water vapor present in the mixture?

Solution

Dalton's law says that

$$P_{\text{Total}} = P_{N_2} + P_{O_2} + P_{H_2O}$$

The known values for variables in this equation are

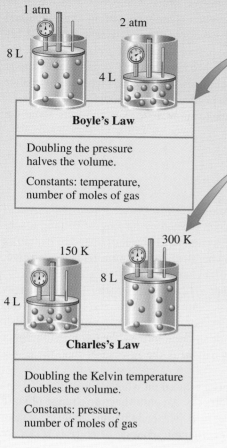

CHEMISTRY AT A GLANCE

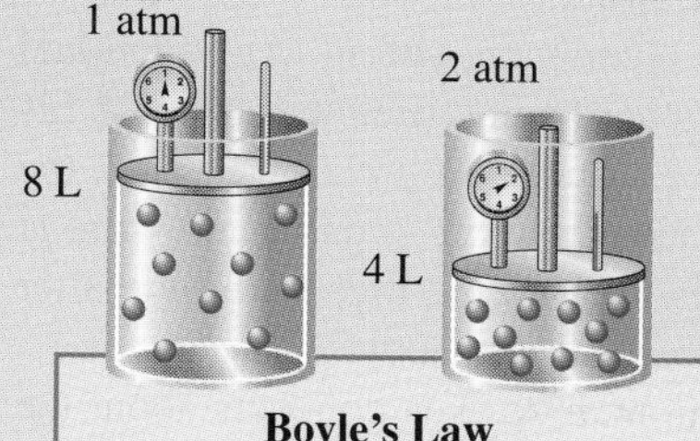

Boyle's Law

Doubling the pressure halves the volume.

Constants: temperature, number of moles of gas

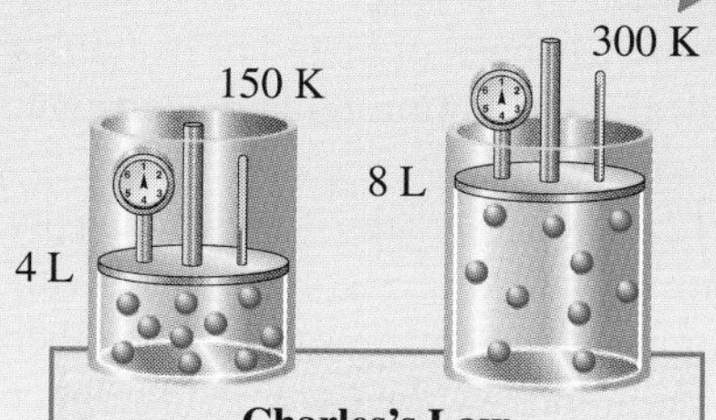

Charles's Law

Doubling the Kelvin temperature doubles the volume.

Constants: pressure, number of moles of gas

GAS LAW	SYNOPSIS	CONSTANTS	VARIABLES
Boyle's Law $P_1V_1 = P_2V_2$	At constant temperature, the volume of a given gas sample is *inversely proportional* to the pressure applied to it.	temperature, number of moles of gas	pressure, volume
Charles's Law $\frac{V_1}{T_1} = \frac{V_2}{T_2}$	At constant pressure, the volume of a given gas sample is *directly proportional* to its Kelvin temperature.	pressure, number of moles of gas	volume, temperature
Combined Gas Law $\frac{P_1V_1}{T_1} = \frac{P_2V_2}{T_2}$	The product of the pressure and the volume of a given gas sample is *directly proportional* to its Kelvin temperature.	number of moles of gas	pressure, temperature, volume
Ideal Gas Law $PV = nRT$	Relates volume, pressure, temperature, and molar amount of a gas sample under one set of conditions. If three of the four variables are known, the fourth can be calculated from the equation.	$R = 0.0821 \frac{\text{L} \cdot \text{atm}}{\text{mole} \cdot \text{K}}$	pressure, volume, temperature, number of moles
Dalton's Law $P_{Total} = P_1 + P_2 + P_3$	The total pressure exerted by a sample that consists of a mixture of gases is equal to the sum of the partial pressures of the individual gases.		

Dalton's Law	P_1	+	P_2	+	P_3	=	P_{Total}
	1	+	3	+	2	=	6

$$P_{Total} = 742 \text{ mm Hg}$$
$$P_{N_2} = 581 \text{ mm Hg}$$
$$P_{O_2} = 143 \text{ mm Hg}$$

Rearranging Dalton's law to isolate P_{H_2O} on the left side of the equation gives

$$P_{H_2O} = P_{Total} - P_{N_2} - P_{O_2}$$

Substituting the known numerical values into this equation and doing the arithmetic give

$$P_{H_2O} = 742 \text{ mm Hg} - 581 \text{ mm Hg} - 143 \text{ mm Hg} = 18 \text{ mm Hg}$$

Practice Exercise 7.5

A gaseous mixture contains the three noble gases He, Ar, and Kr. The total pressure exerted by the mixture is 1.57 atm, and the partial pressures of the He and Ar are 0.33 atm and 0.39 atm, respectively. What is the partial pressure of the Kr in the mixture?

- *Answer:* 0.85 atm

Dalton's law of partial pressures is important when we consider the air of our atmosphere, which is a mixture of numerous gases. At higher altitudes, the total pressure of air decreases, as do the partial pressures of the individual components of air. An individual going from sea level to a higher altitude usually experiences some tiredness because his or her body is not functioning as efficiently at the higher altitude. At higher elevation, the red blood cells absorb a smaller amount of oxygen because the oxygen partial pressure at the higher altitude is lower. A person's body acclimates itself to the higher altitude after a period of time as additional red blood cells are produced by the body.

The Chemistry at a Glance on the preceding page summarizes key concepts about the gas laws we have considered in this chapter.

7.9 Changes of State

A **change of state** *is a process in which a substance is transformed from one physical state to another.* Changes of state are usually accomplished by heating or cooling a substance. Pressure change is also a factor in some systems. Changes of state are examples of physical changes—that is, changes in which chemical composition remains constant. No new substances are ever formed as a result of a change of state.

There are six possible changes of state. Figure 7.12 identifies each of these changes and gives the terminology used to describe them. Four of the six terms used in describing state changes are familiar: freezing, melting, evaporation, and condensation. The other two terms—sublimation and deposition—are not so common. *Sublimation* is the direct change from the solid to the gaseous state; *deposition* is the reverse of this, the direct change from the gaseous to the solid state.

• Although the processes of sublimation and deposition are not common, they are encountered in everyday life. Dry ice sublimes, as do mothballs placed in a clothing storage area. It is because of sublimation that ice cubes left in a freezer get smaller as time passes. Ice or snow forming in clouds (from water vapor) during the winter season is an example of deposition.

Changes of state are classified into two categories based on whether heat (thermal energy) is given up or absorbed during the change process. An **endothermic change of state** *is a change that requires the input (absorption) of heat.* The endothermic changes of state are melting, sublimation, and evaporation. An **exothermic change of state** *is a change that requires heat to be given up (released).* Exothermic changes of state are the reverse of endothermic changes of state; they are freezing, condensation, and deposition.

7.10 Evaporation of Liquids

Evaporation *is the process by which molecules escape from the liquid phase to the gas phase.* We are all aware that water left in an open container at room temperature slowly disappears by evaporation.

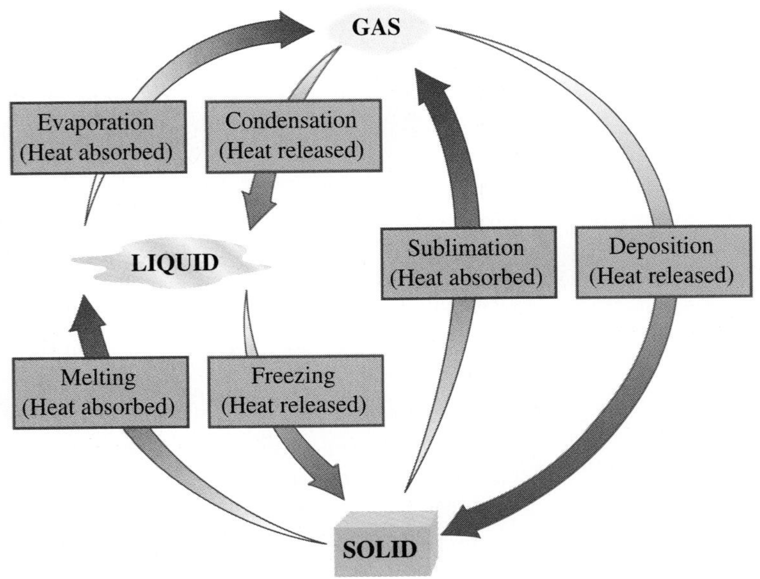

Figure 7.12
There are six changes of state possible for substances. The three endothermic changes, which require the input of heat, are melting, evaporation, and sublimation. The three exothermic changes, which release heat, are freezing, condensation, and deposition.

(a)

(b)

Sublimation of iodine. (a) The beaker contains iodine crystals, I_2; a dish of ice rests on top of the beaker. (b) Iodine has an appreciable vapor pressure even below its melting point (114°C); thus, when heated carefully, the solid sublimes without melting. The vapor deposits crystals on the cool underside of the dish.

Evaporation can be explained using kinetic molecular theory. Statement 5 of this theory (Section 7.1) indicates that the molecules in a liquid (or solid or gas) do not all possess the same kinetic energy. At any given instant, some molecules will have above-average kinetic energies and others will have below-average kinetic energies as a result of collisions between molecules. A given molecule's energy constantly changes as a result of collisions with neighboring molecules. When molecules that happen to be considerably above average in kinetic energy at a given moment are on the liquid surface and are moving in a favorable direction relative to the surface, they can overcome the attractive forces (potential energy) holding them in the liquid and escape.

• For a liquid to evaporate, its molecules must gain enough energy to overcome the attractive forces between them.

Evaporation is a surface phenomenon. Surface molecules are subject to fewer attractive forces because they are not completely surrounded by other molecules; thus escape is much more probable. Liquid surface area is an important factor to consider when determining the rate at which evaporation occurs. Increased surface area results in an increased evaporation rate, because a greater fraction of the total molecules are on the surface.

• Rate of Evaporation and Temperature

Water evaporates faster from a glass of hot water than from a glass of cold water, because a certain minimum kinetic energy is required for molecules to escape from the attractions of neighboring molecules. As the temperature of a liquid increases, a larger fraction of the molecules present acquire this minimum kinetic energy. Consequently, the rate of evaporation always increases as liquid temperature increases.

The escape of high-energy molecules from a liquid during evaporation affects the liquid in two ways: The amount of liquid decreases, and the liquid temperature is lowered. The lower temperature reflects the loss of the most energetic molecules. (Analogously, when all the tall people are removed from a classroom of students, the average height of the remaining students decreases.) A lower average kinetic energy corresponds to a lower temperature (statement 4 of the kinetic molecular theory); hence a cooling effect is produced.

• Evaporative cooling is important in many processes. Our own bodies use evaporation to maintain a constant temperature. We perspire in hot weather, and evaporation of the perspiration cools our skin. The cooling effect of evaporation is quite noticeable when one first comes out of an outdoor swimming pool on a hot, breezy day.

The molecules that escape from an evaporating liquid are often collectively referred to as vapor, rather than gas. The term **vapor** *describes gaseous molecules of a substance at a temperature and pressure at which we ordinarily would think of the substance as a liquid or solid.* For example, at room temperature and atmospheric pressure, the normal state for water is the liquid state. Molecules that escape (evaporate) from liquid water at these conditions are frequently called *water vapor.*

7.11 Vapor Pressure of Liquids

The evaporative behavior of a liquid in a closed container is quite different from its behavior in an open container. Some liquid evaporation occurs in a closed container; this is indicated by a drop in liquid level. However, unlike the liquid level in an open-container

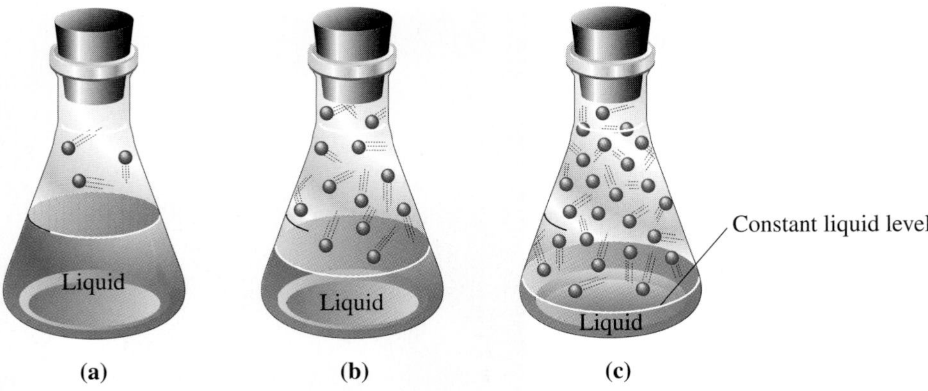

Figure 7.13
In the evaporation of a liquid in a *closed* container (a), the liquid level drops for a time (b) and then becomes constant (ceases to drop). At that point a state of equilibrium has been reached in which the rate of evaporation equals the rate of condensation (c).

system, the liquid level in a closed-container system eventually ceases to drop (becomes constant).

Kinetic molecular theory explains these observations in the following way. The molecules that evaporate in a closed container do not leave the system, as they do in an open container. They find themselves confined in a fixed space immediately above the liquid (see Figure 7.13a). These trapped vapor molecules undergo many random collisions with the container walls, other vapor molecules, and the liquid surface. Molecules that collide with the liquid surface are recaptured by the liquid. Thus two processes, evaporation (escape) and condensation (recapture), take place in a closed container (see Fig. 7.13b).

For a short time, the rate of evaporation in a closed container exceeds the rate of condensation, and the liquid level drops. However, as more of the liquid evaporates, the number of vapor molecules increases; the chance of their recapture through striking the liquid surface also increases. Eventually, the rate of condensation becomes equal to the rate of evaporation, and the liquid level stops dropping (see Figure 7.13c). At this point, the number of molecules that escape in a given time is the same as the number recaptured; a steady-state situation has been reached. The amounts of liquid and vapor in the container do not change, even though both evaporation and condensation are still occurring.

• Remember that for a system at equilibrium, change at the molecular level is still occurring even though you cannot see it.

This steady-state situation, which will continue as long as the temperature of the system remains constant, is an example of a state of equilibrium. A **state of equilibrium** *is a situation in which two opposite processes take place at equal rates.* For systems in a state of equilibrium, no net macroscopic changes can be detected. However, the system is dynamic; the forward and reverse processes are occurring at equal rates.

When there is a liquid–vapor equilibrium in a closed container, the vapor in the fixed space immediately above the liquid exerts a constant pressure on both the liquid surface and the walls of the container. This pressure is called the *vapor pressure* of the liquid. **Vapor pressure** *is the pressure exerted by a vapor above a liquid when the liquid and vapor are in equilibrium.*

The magnitude of a vapor pressure depends on the nature and temperature of the liquid. Liquids that have strong attractive forces between molecules have lower vapor pressures than liquids that have weak attractive forces between particles. Substances that have high vapor pressures (weak attractive forces) evaporate readily—that is, they are *volatile.* A **volatile substance** *is a substance that readily evaporates at room temperature because of a high vapor pressure.* Gasoline is a substance whose components are very volatile.

The vapor pressure of all liquids increases with temperature because an increase in temperature results in more molecules having the minimum kinetic energy required for evaporation. Table 7.2 shows the variation in vapor pressure, as temperature increases, of water.

7.12 Boiling and Boiling Point

In order for a molecule to escape from the liquid state, it usually must be on the surface of the liquid. **Boiling** *is a special form of evaporation where conversion from the liquid*

Table 7.2
Vapor Pressure of Water at Various Temperatures

Temperature (°C)	Vapor pressure (mm Hg)	Temperature (°C)	Vapor pressure (mm Hg)
0	4.6	50	92.5
10	9.2	60	149.4
20	17.5	70	233.7
25[a]	23.8	80	355.1
30	31.8	90	525.8
37[b]	37.1	100	760.0
40	55.3		

[a]Room temperature
[b]Body temperature

state to the vapor state occurs within the body of a liquid through bubble formation. This phenomenon begins to occur when the vapor pressure of a liquid, which steadily increases as the liquid is heated, reaches a value equal to that of the prevailing external pressure on the liquid; for liquids in open containers, this value is atmospheric pressure. When these two pressures become equal, bubbles of vapor form around any speck of dust or around any irregularity associated with the container surface. These vapor bubbles quickly rise to the surface and escape because they are less dense than the liquid itself. We say the liquid is boiling.

Bubbles of vapor form within a liquid when the temperature of the liquid reaches the liquid's boiling point.

The **boiling point** *of a liquid is the temperature at which the vapor pressure of a liquid becomes equal to the external (atmospheric) pressure exerted on the liquid.* Because the atmospheric pressure fluctuates from day to day, the boiling point of a liquid does also. Thus, in order for us to compare the boiling points of different liquids, the external pressure must be the same. The boiling point of a liquid that is most often used for comparison and tabulation purposes is called the normal boiling point. A liquid's **normal boiling point** *is the temperature at which a liquid boils under a pressure of 760 mm Hg.*

• Conditions That Affect Boiling Point

At any given location, the changes in the boiling point of a liquid caused by *natural* variations in atmospheric pressure seldom exceed a few degrees; in the case of water, the maximum is about 2°C. However, variations in boiling points *between* locations at different elevations can be quite striking, as shown in Table 7.3.

The boiling point of a liquid can be increased by increasing the external pressure. This principle is used in the operation of a pressure cooker. Foods cook faster in pressure cookers because the elevated pressure causes water to boil above 100°C. An increase in temperature of only 10°C will cause food to cook in approximately half the normal time.

Table 7.3
Boiling Point of Water at Various Locations That Differ in Elevation

Location	Feet above sea level	P_{atm} (mm Hg)	Boiling point (°C)
top of Mt. Everest, Tibet	29,028	240	70
top of Mt. McKinley, Alaska	20,320	340	79
Leadville, Colorado	10,150	430	89
Salt Lake City, Utah	4,390	650	96
Madison, Wisconsin	900	730	99
New York, New York	10	760	100
Death Valley, California	−282	770	100.4

Table 7.4
Boiling Point of Water at Various Pressure Cooker Settings When Atmospheric Pressure is 1 Atmosphere

Pressure cooker setting (additional pressure beyond atmospheric, lb/in²)	Internal pressure in cooker (atm)	Boiling point of water (°C)
5	1.34	108
10	1.68	116
15	2.02	121

Table 7.4 gives the boiling temperatures reached by water under normal household pressure cooker conditions. Hospitals use this same principle to sterilize instruments and laundry in autoclaves; there, sufficiently high temperatures are reached to destroy bacteria.

Liquids that have high normal boiling points or that undergo undesirable chemical reactions at elevated temperatures can be made to boil at low temperatures by reducing the external pressure. This principle is used in the preparation of numerous food products, including frozen fruit juice concentrates. Some of the water in a fruit juice is boiled away at a reduced pressure, thus concentrating the juice without heating it to a high temperature (which spoils the taste of the juice and reduces its nutritional value).

The converse of the pressure cooker "phenomenon" is that food cooks more slowly at reduced pressures. The pressure reduction associated with higher altitudes, and the accompanying reduction in boiling points of liquids, mean that food cooked over a campfire in the mountains requires longer cooking times.

7.13 Intermolecular Forces in Liquids

Boiling points vary greatly among substances. The boiling points of some substances are well below zero; for example, oxygen has a boiling point of −183°C. Numerous other substances do not boil until the temperature is much higher. An explanation of this variation in boiling points involves a consideration of the nature of the intermolecular forces that must be overcome in order for molecules to escape from the liquid state into the vapor state. **Intermolecular forces** *are forces that act* between *a molecule and another molecule.*

Intermolecular forces are similar in one way to the previously discussed *intramolecular* forces (forces *within* molecules) that are involved in covalent bonding (Sections 5.3 and 5.4); they are electrostatic in origin. A major difference between inter- and intramolecular forces is their strength. Intermolecular forces are weak compared to intramolecular forces (true chemical bonds). Generally, their strength is less than one-tenth that of a single covalent bond. However, intermolecular forces are strong enough to influence the behavior of liquids, and they often do so in very dramatic ways.

There are three main types of intermolecular forces: dipole–dipole interactions, hydrogen bonds, and London forces.

• Dipole–Dipole Interactions

Dipole–dipole interactions *are electrostatic attractions between polar molecules.* A polar molecule (Section 5.10) has a negative end and a positive end; that is, it has a *dipole* (two poles resulting from opposite charges being separated from one another). As a consequence, the positive end of one molecule attracts the negative end of another molecule, and vice versa. This attraction constitutes a dipole–dipole interaction. The greater the polarity of the molecules, the greater the strength of the dipole–dipole interactions. And the greater the strength of the dipole–dipole interactions, the higher the boiling point of the liquid. Figure 7.14 shows the many dipole–dipole interactions that are possible for a random arrangement of polar chlorine monofluoride (ClF) molecules.

• Hydrogen Bonds

Unusually strong dipole–dipole interactions are observed among hydrogen-containing molecules in which hydrogen is covalently bonded to a highly electronegative element of small atomic size (fluorine, oxygen, and nitrogen). Two factors account for the extra strength of these dipole–dipole interactions.

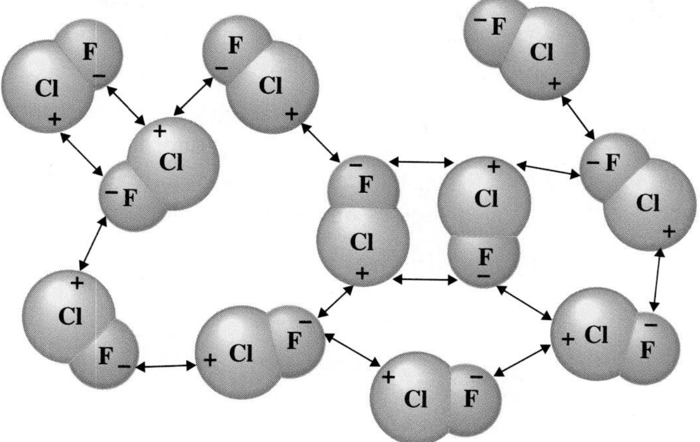

Figure 7.14
There are many dipole–dipole interactions possible between randomly arranged ClF molecules. In each interaction, the positive end of one molecule is attracted to the negative end of a neighboring ClF molecule.

1. The highly electronegative element to which hydrogen is covalently bonded attracts the bonding electrons to such a degree that the hydrogen atom is left with a significant δ^+ charge.

$$\overset{\delta^+}{\text{H}}\text{—}\overset{\delta^-}{\text{F}} \qquad \overset{\delta^+}{\text{H}}\text{—}\overset{\delta^-}{\text{O}} \qquad \overset{\delta^+}{\text{H}}\text{—}\overset{\delta^-}{\text{N}}$$

   Indeed, the hydrogen atom is essentially a "bare" nucleus, because it has no electrons besides the one attracted to the electronegative element—a unique property of hydrogen.
2. The small size of the hydrogen atom allows the "bare" nucleus to approach closely, and be strongly attracted to, a lone pair of electrons on the electronegative atom of another molecule.

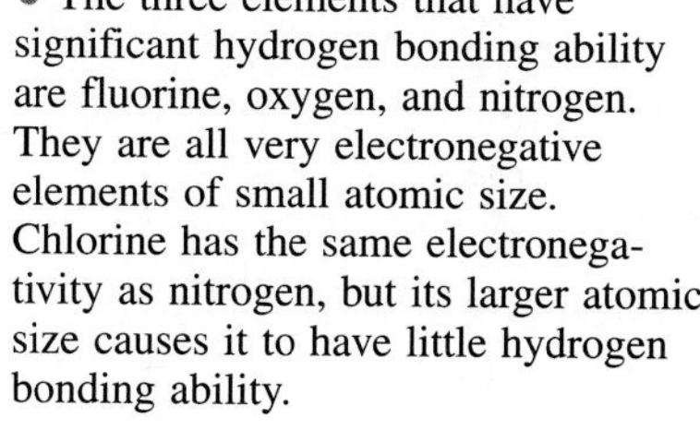
• The three elements that have significant hydrogen bonding ability are fluorine, oxygen, and nitrogen. They are all very electronegative elements of small atomic size. Chlorine has the same electronegativity as nitrogen, but its larger atomic size causes it to have little hydrogen bonding ability.

Dipole–dipole interactions of this type are given a special name, hydrogen bonds. A **hydrogen bond** *is an extra-strong dipole–dipole interaction between a hydrogen atom covalently bonded to a small, very electronegative atom (F, O, or N) and a lone pair of electrons on another small, very electronegative atom (F, O, or N).*

Water (H_2O) is the most commonly encountered substance wherein hydrogen bonding is significant. Figure 7.15 depicts the process of hydrogen bonding among water molecules. Note that each oxygen atom in water can participate in two hydrogen bonds—one involving each of its nonbonding electron pairs.

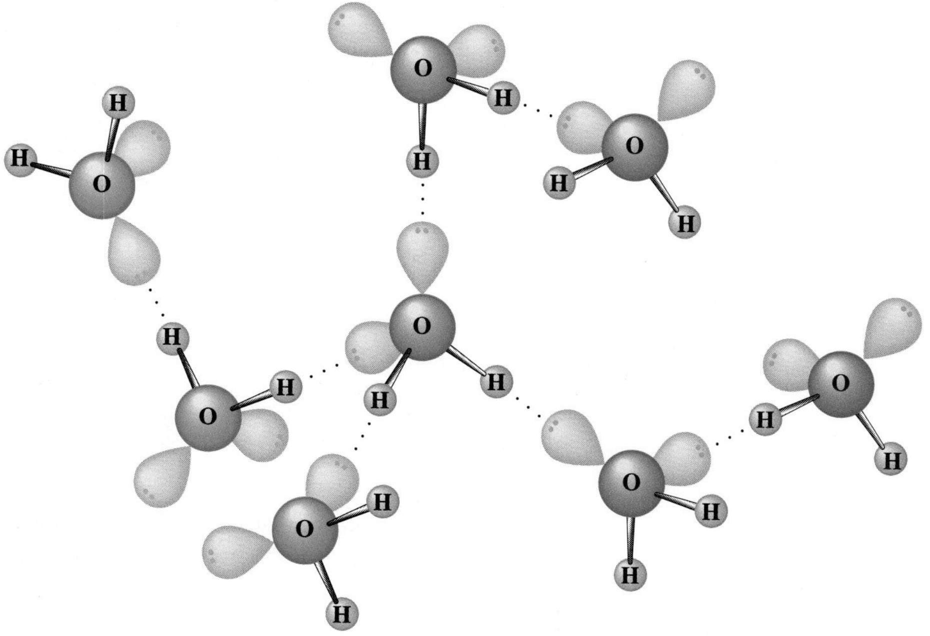

Figure 7.15
Depiction of hydrogen bonding among water molecules. The dotted lines are the hydrogen bonds.

The two molecules that participate in a hydrogen bond need not be identical. Hydrogen bond formation is possible whenever two molecules, the same or different, have the following characteristics.

• A series of dots is used to represent a hydrogen bond, as in the notation

—X—H···Y—

X and Y represent small, highly electronegative elements (fluorine, oxygen, or nitrogen).

1. One molecule has a hydrogen atom attached by a covalent bond to an atom of nitrogen, oxygen, or fluorine.
2. The other molecule has a nitrogen, oxygen, or fluorine atom present that possesses one or more nonbonding electron pairs.

Figure 7.16 gives additional examples of hydrogen bonding involving simple molecules.

Example 7.6

Predicting Whether Hydrogen Bonding Will Occur Between Molecules

Indicate whether hydrogen bonding should occur between two molecules of each of the following substances.

a. Ethyl amine

```
    H  H  H
    |  |  |
 H—C—C—N—H
    |  |  ··
    H  H
```

b. Methyl alcohol

```
    H
    |  ··
 H—C—O—H
    |  ··
    H
```

c. Diethyl ether

```
    H  H      H  H
    |  |  ··  |  |
 H—C—C—O—C—C—H
    |  |  ··  |  |
    H  H      H  H
```

Solution

a. Hydrogen bonding should occur, because we have an N—H bond and a nitrogen atom with a nonbonding electron pair.
b. Hydrogen bonding should occur, because we have an O—H bond and an oxygen atom with nonbonding electron pairs.
c. Hydrogen bonding should not occur. We have an oxygen atom with nonbonding electron pairs, but no N—H, O—H, or F—H bond is present.

Practice Exercise 7.6

Would you expect hydrogen bonding to occur between two molecules of each of the following substances?

a. Nitrogen trifluoride

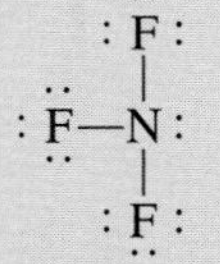

b. Ethyl alcohol

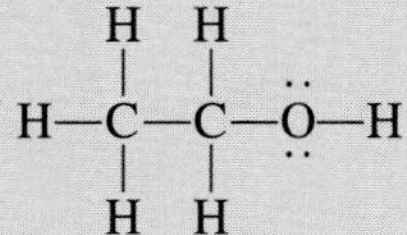

c. Formaldehyde

```
     :O:
      ‖
  H—C—H
```

• *Answer:* **a.** no; **b.** yes; **c.** no

The vapor pressures (Section 7.11) of liquids that have significant hydrogen bonding are much lower than those of similar liquids wherein little or no hydrogen bonding occurs. This is because the presence of hydrogen bonds makes it more difficult for molecules to escape from the condensed state; additional energy is needed to overcome the hy-

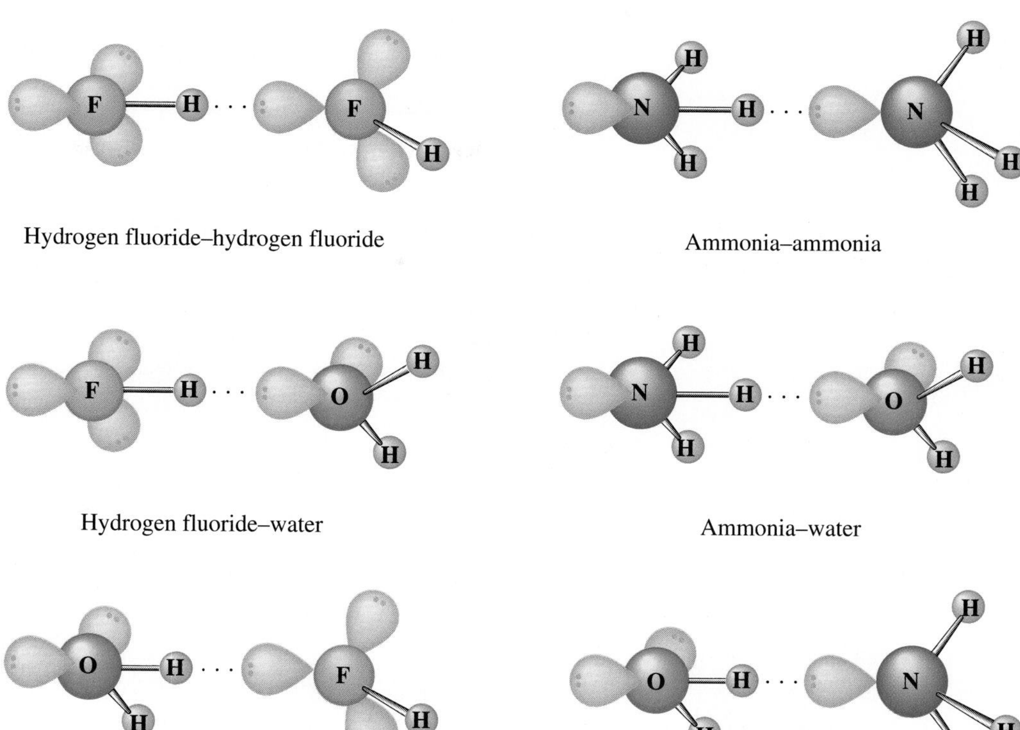

Figure 7.16 Diagrams of hydrogen bonding involving selected simple molecules. The solid lines represent covalent bonds; the dotted lines represent hydrogen bonds.

drogen bonds. For this reason, boiling points are much higher for liquids in which hydrogen bonding occurs. The effect that hydrogen bonding has on boiling point can be seen by comparing water's boiling point with those of other hydrogen compounds of Group VIA elements—H_2S, H_2Se, and H_2Te (see Figure 7.17). Water is the only compound in this series where significant hydrogen bonding occurs.

• London Forces

The third type of intermolecular force, and the weakest, is the London force, named after the German physicist Fritz London (1900–1954), who first postulated its existence. **London forces** *are instantaneous dipole–dipole interactions that exist between all atoms and molecules, polar as well as nonpolar.* The origin of London forces is more difficult to visualize than that of dipole–dipole interactions.

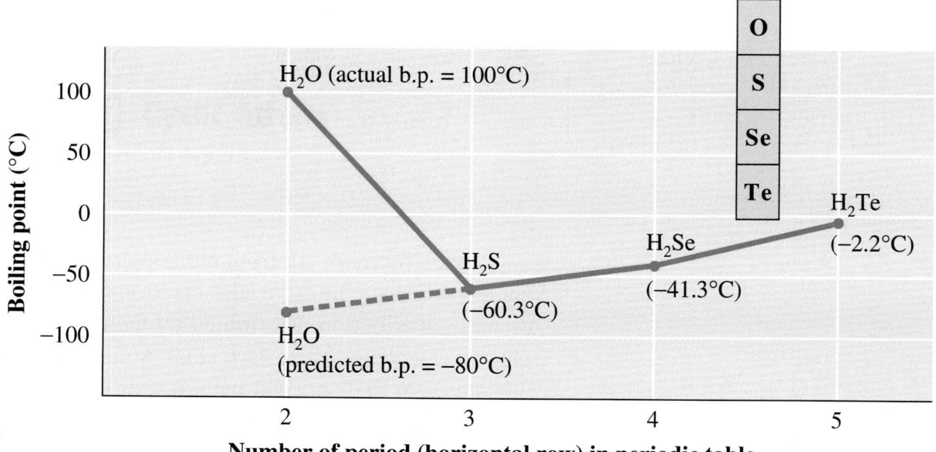

Figure 7.17 If there were no hydrogen bonding between water molecules, the boiling point of water would be approximately −80°C; this value is obtained by extrapolation (extension of the line connecting the three heavier compounds). Because of hydrogen bonding, the actual boiling point of water, 100°C, is nearly 200°C higher than predicted. Indeed, in the absence of hydrogen bonding, water would be a gas at room temperature, and life as we know it on Earth would not be possible.

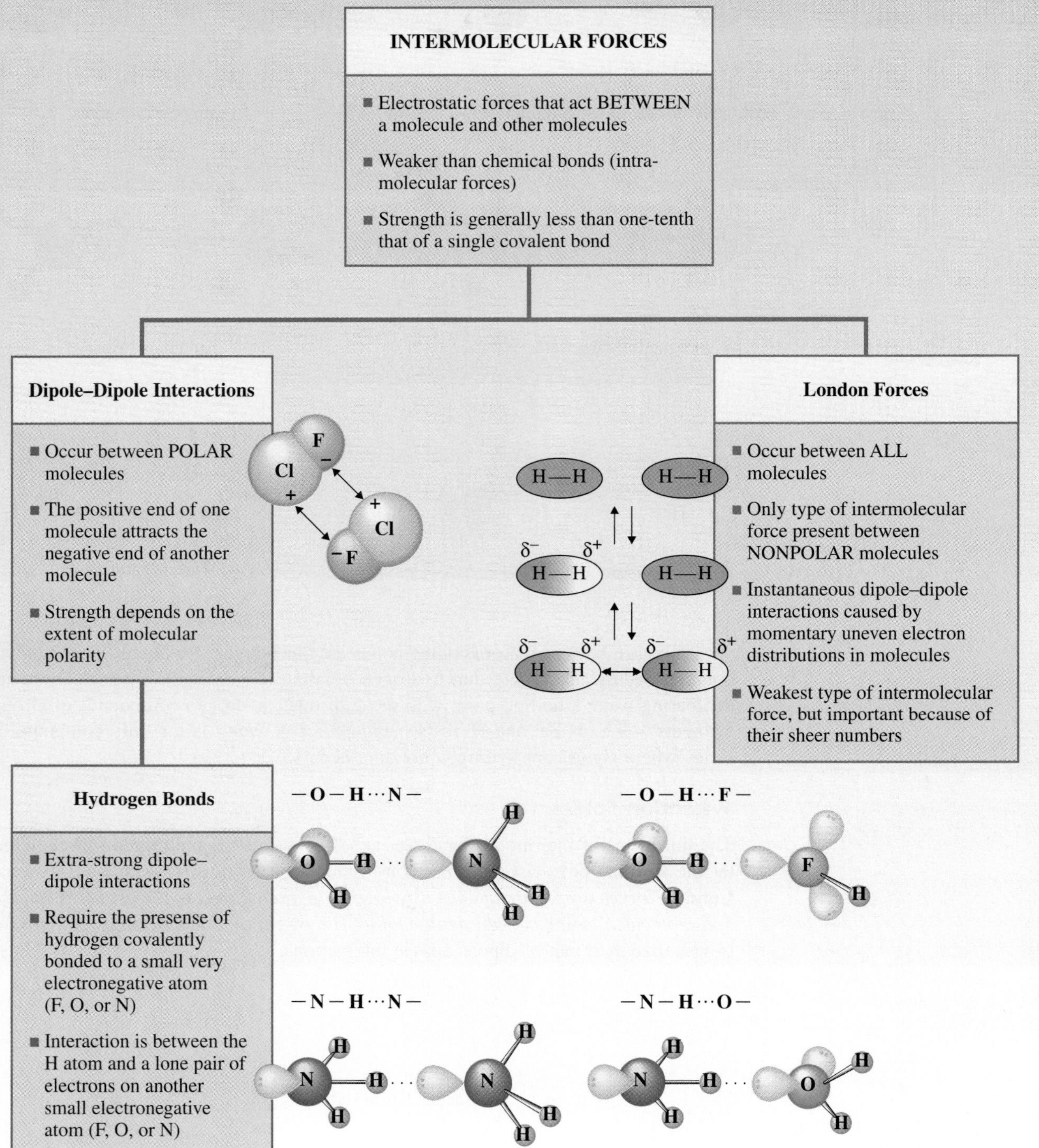
INTERMOLECULAR FORCES
■ Electrostatic forces that act BETWEEN a molecule and other molecules
■ Weaker than chemical bonds (intramolecular forces)
■ Strength is generally less than one-tenth that of a single covalent bond
Dipole–Dipole Interactions
■ Occur between POLAR molecules
■ The positive end of one molecule attracts the negative end of another molecule
■ Strength depends on the extent of molecular polarity
Cl
F
London Forces
■ Occur between ALL molecules
■ Only type of intermolecular force present between NONPOLAR molecules
■ Instantaneous dipole–dipole interactions caused by momentary uneven electron distributions in molecules
■ Weakest type of intermolecular force, but important because of their sheer numbers
H—H
δ−
δ+
Hydrogen Bonds
■ Extra-strong dipole–dipole interactions
■ Require the presense of hydrogen covalently bonded to a small very electronegative atom (F, O, or N)
■ Interaction is between the H atom and a lone pair of electrons on another small electronegative atom (F, O, or N)
—O—H···N—
—O—H···F—
—N—H···N—
—N—H···O—

London forces result from momentary (temporary) uneven electron distributions in molecules. Most of the time, the electrons in a molecule can be considered to have a predictable distribution determined by their energies and the electronegativities of the atoms present. However, there is a small statistical chance (probability) that the electrons will deviate from their normal pattern. For example, in the case of a nonpolar diatomic molecule, more electron density may temporarily be located on one side of the molecule than

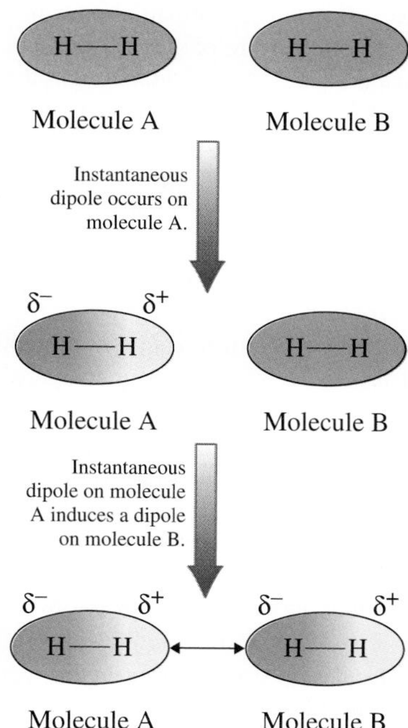

Figure 7.18
Nonpolar molecules such as H_2 can develop instantaneous dipoles and induced dipoles. The attractions between such dipoles, even though they are transitory, create London forces.

Table 7.5
Boiling Point Trends for Related Series of Nonpolar Molecules: (a) noble gases, (b) halogens

(a) Noble Gases			(b) Halogens (Group VIIA Elements)		
Substance	Molecular mass (amu)	Boiling point (°C)	Substance	Molecular mass (amu)	Boiling point (°C)
He	4.0	−269	F_2	39.0	−187
Ne	20.2	−246	Cl_2	70.9	−35
Ar	39.9	−186	Br_2	159.8	+59
Kr	83.8	−153			
Xe	131.3	−107			
Rn	222.0	−62			

on the other. This condition causes the molecule to become polar for an instant. The negative side of this *instantaneously* polar molecule tends to repel electrons of adjoining molecules and causes these molecules also to become polar (*induced polarity*). The original polar molecule and all of the molecules with induced polarity are then attracted to one another. This happens many, many times per second throughout the liquid, resulting in a net attractive force. Figure 7.18 depicts the situation that prevails when London forces exist.

As an analogy for London forces, consider what happens when a bucket filled with water is moved. The water will "slosh" from side to side. This is similar to the movement of electrons. The "sloshing" from side to side is instantaneous; a given "slosh" quickly disappears. "Uneven" electron distribution is likewise a temporary situation.

The strength of London forces depends on the ease with which an electron distribution in a molecule can be distorted (polarized) by the polarity present in another molecule. In large molecules, the outermost electrons are necessarily located farther from the nucleus than are the outermost electrons in small molecules. The farther electrons are from the nucleus, the weaker the attractive forces on them, the more freedom they have, and the more susceptible they are to polarization. This leads to the observation that for *related* molecules, boiling points increase with molecular mass, which usually parallels size. This trend is reflected in the boiling points given in Table 7.5 for two series of related substances: the noble gases and the halogens (Group VIIA).

• The boiling points of substances with similar molar masses increase in this order: nonpolar molecules < polar molecules with no hydrogen bonding < polar molecules with hydrogen bonding.

Chemistry at a Glance provides a summary of what we have learned about intermolecular forces.

Concepts to Remember

Kinetic molecular theory. The kinetic molecular theory of matter is a set of five statements that explain the physical behavior of the three states of matter (solids, liquids, and gases). The basic idea of this theory is that the particles (atoms, molecules, or ions) present in a substance are in constant motion and are attracted or repelled by each other.

The solid state. The solid state is characterized by a dominance of potential energy (cohesive forces) over kinetic energy (disruptive forces). As a result, the particles of solids are held in rigid three-dimensional lattices in which the particle's kinetic energy takes the form of vibrations about each lattice site.

The liquid state. The liquid state is one in which neither potential energy (cohesive forces) nor kinetic energy (disruptive forces) dominates. As a result, particles of liquids are randomly arranged but are relatively close to each other and are in constant random motion, sliding freely over each other but without enough kinetic energy to become separated.

The gaseous state. The gaseous state is characterized by a complete dominance of kinetic energy (disruptive forces) over potential energy (cohesive forces). As a result, particles move randomly, essentially independently of each other. Under ordinary pressure, the particles are separated from each other by relatively large distances, except when they collide.

Gas laws. Gas laws are generalizations that describe, in mathematical terms, the relationships among the pressure, temperature, and volume of a specific quantity of gas. When these relationships are used, it is necessary to express the temperature on the Kelvin scale. Pressure is usually expressed in atm, mm Hg, or torr.

Boyle's law. Boyle's law, the pressure–volume law, states that the volume of a sample of a gas is inversely proportional to the pressure applied to the gas if the temperature is kept constant. This means that when the pressure on the gas increases, the volume decreases proportionally; conversely, when the volume decreases, the pressure increases.

Charles's law. Charles's law, the volume–temperature law, states that the volume of a sample of gas is directly proportional to its Kelvin temperature if the pressure is kept constant. This means that when the temperature increases, the volume also increases and that when the temperature decreases, the volume also decreases.

The combined gas law. The combined gas law is an expression obtained by mathematically combining Boyle's and Charles's laws. A change in pressure, temperature, or volume that is brought about by changes in the other two variables can be calculated by using this law.

Ideal gas law. The ideal gas law has the form $PV = nRT$, where R is the ideal gas constant (0.0821 atm · L/mole · K). This equation enables us to calculate any one of the characteristic gas properties (P, V, T, or n), given the other three.

Dalton's law of partial pressures. Dalton's law of partial pressures states that the total pressure exerted by a mixture of gases is the sum of the partial pressures of the individual gases. A partial pressure is the pressure that a gas in a mixture would exert if it were present alone under the same conditions.

Changes of state. Most matter can be changed from one physical state to another by heating, cooling, or changing pressure. The state changes that release heat are called exothermic (condensation, deposition, and freezing), and those that absorb heat are called endothermic (melting, evaporation, and sublimation).

Vapor pressure. The pressure exerted by vapor in equilibrium with its liquid is the vapor pressure of the liquid. Vapor pressure increases as liquid temperature increases.

Boiling and boiling point. Boiling is a special form of evaporation in which bubbles of vapor form within the liquid and rise to the surface. The boiling point of a liquid is the temperature at which the vapor pressure of the liquid becomes equal to the external (atmospheric) pressure exerted on the liquid. The boiling point of a liquid increases or decreases as the prevailing atmospheric pressure increases or decreases.

Intermolecular forces. Intermolecular forces are forces that act between a molecule and another molecule. The three principal types of intermolecular forces in liquids are dipole–dipole interactions, hydrogen bonds, and London forces.

Hydrogen bonds. A hydrogen bond is an extra-strong dipole–dipole interaction between a hydrogen atom covalently bonded to a very electronegative atom (F, O, or N) and a lone pair of electrons on another small, very electronegative atom (F, O, or N).

Key Reactions and Equations

1. Boyle's law (Section 7.4)

$$P_1V_1 = P_2V_2 \qquad (n, T \text{ constant})$$

2. Charles's law (Section 7.5)

$$\frac{V_1}{T_1} = \frac{V_2}{T_2} \qquad (n, P \text{ constant})$$

3. Combined gas law (Section 7.6)

$$\frac{P_1V_1}{T_1} = \frac{P_2V_2}{T_2} \qquad (n \text{ constant})$$

4. Ideal gas law (Section 7.7)

$$PV = nRT$$

5. Ideal gas constant (Section 7.7)

$$R = 0.0821 \text{ atm} \cdot \text{L/mole} \cdot \text{K}$$

6. Dalton's law of partial pressures (Section 7.8)

$$P_{\text{Total}} = P_1 + P_2 + P_3 + \cdots$$

Key Terms

Barometer (7.3)
Boiling (7.12)
Boiling point (7.12)
Boyle's law (7.4)
Change of state (7.9)
Charles's Law (7.5)
Combined gas law (7.6)
Dalton's law of partial pressures (7.8)
Dipole–dipole interactions (7.13)
Electrostatic interactions (7.1)
Endothermic change of state (7.9)
Evaporation (7.10)
Exothermic change of state (7.9)
Gas laws (7.3)
Gaseous state (7.2)
Hydrogen bond (7.13)
Ideal gas law (7.7)
Intermolecular forces (7.13)
Kinetic energy (7.1)
Kinetic molecular theory of matter (7.1)
Liquid state (7.2)
London forces (7.13)
Normal boiling point (7.12)
Partial pressure (7.8)
Potential energy (7.1)
Pressure (7.3)
Solid state (7.2)
State of equilibrium (7.11)
Vapor (7.11)
Vapor pressure (7.11)
Volatile substance (7.11)

Exercises and Problems

The members of each pair of problems in this section test similar material.

Kinetic Molecular Theory (Sections 7.1 and 7.2)

7.1 Using kinetic molecular theory concepts, answer the following questions.

a. What is the relationship between temperature and the average velocity with which particles move?

b. What type of energy is related to cohesive forces?

c. What effect does temperature have on the magnitude of disruptive forces?

d. In which of the three states of matter are disruptive forces greater than cohesive forces?

7.2 Using kinetic molecular theory concepts, answer the following questions.

a. How do molecules transfer energy from one to another?

b. What type of energy is related to disruptive forces?

c. What are the effects of cohesive forces on a system of particles?

d. In which of the three states of matter are disruptive forces and cohesive forces of about the same magnitude?

7.3 Explain each of the following observations using kinetic molecular theory.

a. Liquids show little change in volume with changes in temperature.

b. Gases have a low density.

7.4 Explain each of the following observations using kinetic molecular theory.

a. Both liquids and solids are practically incompressible.

b. A container can be half full of a liquid but not half full of a gas.

Gas Law Variables (Section 7.3)

7.5 Carry out the following pressure unit conversions using the dimensional-analysis method of problem solving.

a. 735 mm Hg to atmospheres

b. 0.530 atm to millimeters of mercury

c. 0.530 atm to torr

d. 12.0 psi to atmospheres

7.6 Carry out the following pressure unit conversions using the dimensional-analysis method of problem solving.

a. 73.5 mm Hg to atmospheres

b. 1.75 atm to millimeters of mercury

c. 735 torr to atmospheres

d. 1.61 atm to pounds per square inch

Boyle's Law (Section 7.4)

7.7 At constant temperature, a sample of 6.0 L of O_2 at 3.0 atm pressure is compressed until the volume decreases to 2.5 L. What is the new pressure, in atmospheres?

7.8 At constant temperature a sample of 6.0 L of N_2 at 2.0 atm pressure is allowed to expand until the volume reaches 9.5 L. What is the new pressure, in atmospheres?

7.9 A sample of ammonia (NH_3), a colorless gas with a pungent odor, occupies a volume of 3.00 L at a pressure of 655 mm Hg and a temperature of 25°C. What volume, in liters, will this NH_3 sample occupy, at the same temperature, if the pressure is increased to 725 mm Hg?

7.10 A sample of nitrogen dioxide (NO_2), a toxic gas with a reddish-brown color, occupies a volume of 4.00 L at a pressure of 725 mm Hg and a temperature of 35°C. What volume, in liters, will this NO_2 sample occupy, at the same temperature, if the pressure is decreased to 125 mm Hg?

Charles's Law (Section 7.5)

7.11 At atmospheric pressure, a sample of H_2 gas has a volume of 2.73 L at 27°C. What volume, in liters, will the H_2 gas occupy if the temperature is increased to 127°C and the pressure is held constant?

7.12 At atmospheric pressure, a sample of O_2 gas has a volume of 55 mL at 27°C. What volume, in milliliters, will the O_2 gas occupy if the temperature is decreased to 0°C and the pressure is held constant?

7.13 A sample of N_2 gas occupies a volume of 375 mL at 25°C and a pressure of 2.0 atm. Determine the temperature, in degrees Celsius, at which the volume of the gas would be 525 mL at the same pressure.

7.14 A sample of Ar gas occupies a volume of 1.2 L at 125°C and a pressure of 1.0 atm. Determine the temperature, in degrees Celsius, at which the volume of the gas would be 1.0 L at the same pressure.

Combined Gas Law (Section 7.6)

7.15 Rearrange the standard form of the combined gas law equation so that each of the following variables is by itself on one side of the equation.

a. T_1 b. P_2 c. V_1

7.16 Rearrange the standard form of the combined gas law equation so that each of the following variables is by itself on one side of the equation.

a. V_2 b. T_2 c. P_1

7.17 A sample of carbon dioxide (CO_2) gas has a volume of 15.2 L at a pressure of 1.35 atm and a temperature of 33°C. Determine the following for this gas sample.

a. Volume, in liters, at $T = 35°C$ and $P = 3.50$ atm

b. Pressure, in atmospheres, at $T = 42°C$ and $V = 10.0$ L

c. Temperature, in degrees Celsius, at $P = 7.00$ atm and $V = 0.973$ L

d. Volume, in milliliters, at $T = 97°C$ and $P = 6.70$ atm

7.18 A sample of carbon monoxide (CO) gas has a volume of 7.31 L at a pressure of 735 mm Hg and a temperature of 45°C. Determine the following for this gas sample.

a. Pressure, in millimeters of mercury, at $T = 357°C$ and $V = 13.5$ L

b. Temperature, in degrees Celsius, at $P = 1275$ mm Hg and $V = 0.800$ L

c. Volume, in liters, at $T = 45°C$ and $P = 325$ mm Hg

d. Pressure, in atmospheres, at $T = 325°C$ and $V = 2.31$ L

Ideal Gas Law (Section 7.7)

7.19 What is the temperature, in degrees Celsius, of 5.23 moles of helium (He) gas confined to a volume of 5.23 L at a pressure of 5.23 atm?

7.20 What is the temperature, in degrees Celsius, of 1.50 moles of neon (Ne) gas confined to a volume of 2.50 L at a pressure of 1.00 atm?

7.21 Calculate the volume, in liters, of 0.100 mole of O_2 gas at 0°C and 2.00 atm pressure.

7.22 Calculate the pressure, in atmospheres, of 0.100 mole of O_2 gas in a 2.00-L container at a temperature of 75°C.

7.23 Determine the following for a 0.250-mole sample of CO_2 gas.

a. Volume, in liters, at 27°C and 1.50 atm

b. Pressure, in atmospheres, at 35°C in a 2.00-L container

c. Temperature, in degrees Celsius, at 1.20 atm pressure in a 3.00-L container.

d. Volume, in milliliters, at 125°C and 0.500 atm pressure

7.24 Determine the following for a 0.500-mole sample of CO gas.

a. Pressure, in atmospheres, at 35°C in a 1.00-L container

b. Temperature, in degrees Celsius, at 5.00 atm pressure in a 5.00-L container

c. Volume, in liters, at 127°C and 3.00 atm

d. Pressure, in millimeters of mercury, at 25°C in a 2.00-L container

Dalton's Law of Partial Pressures (Section 7.8)

7.25 The total pressure exerted by a mixture of O_2, N_2, and He gases is 1.50 atm. What is the partial pressure, in atmospheres, of the O_2, given that the partial pressures of the N_2 and He are 0.75 and 0.33 atm, respectively?

7.26 The total pressure exerted by a mixture of He, Ne, and Ar gases is 2.00 atm. What is the partial pressure, in atmospheres, of Ne, given that the partial pressures of the other gases are both 0.25 atm?

7.27 A gas mixture contains O_2, N_2, and Ar at partial pressures of 125, 175, and 225 mm Hg, respectively. If CO_2 gas is added to the mixture until the total pressure reaches 623 mm Hg, what is the partial pressure, in millimeters of mercury, of CO_2?

7.28 A gas mixture contains He, Ne, and H_2S at partial pressures of 125, 175, and 225 mm Hg, respectively. If all of the H_2S is removed from the mixture, what will be the partial pressure, in millimeters of mercury, of Ne?

Changes of State (Section 7.9)

7.29 Indicate whether each of the following is an exothermic or an endothermic change of state.

a. Sublimation b. Melting

c. Condensation

7.30 Indicate whether each of the following is an exothermic or an endothermic change of state.

a. Freezing b. Evaporation

c. Deposition

7.31 Indicate whether the liquid state is involved in each of the following changes of state.

a. Sublimation b. Melting

c. Condensation

7.32 Indicate whether the solid state is involved in each of the following changes of state.

a. Freezing b. Deposition

c. Evaporation

Properties of Liquids (Sections 7.10 through 7.12)

7.33 Match each of the following statements to the appropriate term: *vapor, vapor pressure, volatile, boiling,* or *boiling point.*

a. This is a temperature at which the liquid vapor pressure is equal to the external pressure on a liquid.

b. This property can be measured by allowing a liquid to evaporate in a closed container.

c. In this process, bubbles of vapor form within a liquid.

d. This temperature changes appreciably with changes in atmospheric pressure.

7.34 Match each of the following statements to the appropriate term: *vapor, vapor pressure, volatile, boiling,* or *boiling point.*

a. This state involves gaseous molecules of a substance at a temperature and pressure at which we would ordinarily think of the substance as a liquid.

b. This term describes a substance that readily evaporates at room temperature because of a high vapor pressure.

c. This process is a special form of evaporation.

d. This property always increases in magnitude with increasing temperature.

7.35 Offer a clear, concise explanation for each of the following observations.

a. All liquids do not have the same vapor pressure at a given temperature.

b. The boiling point of a liquid decreases as atmospheric pressure decreases.

c. A person emerging from an outdoor swimming pool on a breezy day gets the shivers.

d. Food will cook just as fast in boiling water with the stove set at low heat as in boiling water with the stove set at high heat.

7.36 Offer a clear, concise explanation for each of the following observations.

a. Increasing the temperature of a liquid increases its vapor pressure.

b. It takes more time to cook an egg in boiling water on a mountaintop than at sea level.

c. Food cooks faster in a pressure cooker than in an open pan.

d. Evaporation is a cooling process.

Intermolecular Forces in Liquids (Section 7.13)

7.37 Describe the molecular conditions necessary for the existence of a dipole–dipole interaction.

7.38 Describe the molecular conditions necessary for the existence of a London force.

7.39 In liquids, what is the relationship between boiling point and the strength of intermolecular forces?

7.40 In liquids, what is the relationship between vapor pressure magnitude and the strength of intermolecular forces?

7.41 For liquid-state samples of the following diatomic substances, classify the dominant intermolecular forces present as London forces, dipole–dipole interactions, or hydrogen bonds.

a. H_2 b. HF c. CO d. F_2

7.42 For liquid-state samples of the following diatomic substances, classify the dominant intermolecular forces present as London forces, dipole–dipole interactions, or hydrogen bonds.

a. O_2 b. HCl c. Cl_2 d. BrCl

7.43 In which of the following substances, in the pure liquid state, would hydrogen bonding occur?

a. 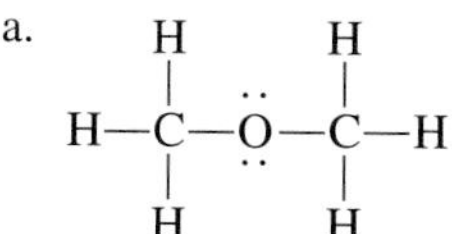

b. 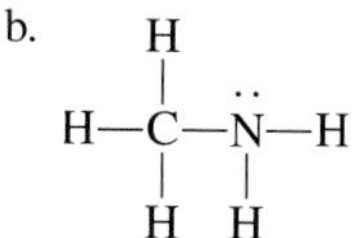

c. 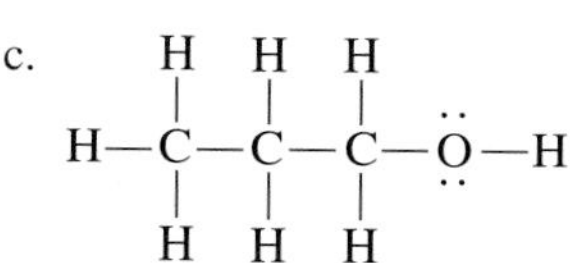

d. H—I:

7.44 In which of the following substances, in the pure liquid state, would hydrogen bonding occur?

a. $H_3C—CH_3$ (H—C—C—H with three H on each C)

b. Cl—N—H with H on N

c. H—N—N—H with H on each N

d. H—C—C—H with H above and below the first C and O double-bonded to the second C

7.45 How many hydrogen bonds can form between a single water molecule and other water molecules?

7.46 How many hydrogen bonds can form between a single ammonia molecule (NH_3) and other ammonia molecules?

Additional Problems

7.47 A sample of NO_2 gas in a 575-mL container at a pressure of 1.25 atm and a temperature of 125°C is transferred to a new container with a volume of 825 mL.

a. What is the new pressure, in atmospheres, if no change in temperature occurs?

b. What is the new temperature, in degrees Celsius, if no change in pressure occurs?

c. What is the new temperature, in degrees Celsius, if the pressure is increased to 2.50 atm?

7.48 A sample of NO_2 gas in a nonrigid container, at a temperature of 24°C, occupies a certain volume at a certain pressure. What will be its temperature, in degrees Celsius, in each of the following situations?

a. Both pressure and volume are doubled.

b. Both pressure and volume are cut in half.

c. The pressure is doubled and the volume is cut in half.

d. The pressure is cut in half and the volume is tripled.

7.49 What is the pressure, in atmospheres, inside a 4.00-L container that contains the following amounts of O_2 gas at 40.0°C?

a. 0.72 mole b. 4.5 moles

c. 0.72 g d. 4.5 g

7.50 How many molecules of hydrogen sulfide (H_2S) gas are contained in 2.00 L of H_2S at 0.0°C and 1.00 atm pressure?

7.51 A 1.00-mole sample of dry ice (solid CO_2) is placed in a flexible sealed container and allowed to sublime. After complete sublimation, what will be the container volume, in liters, at 23°C and 0.983 atm pressure?

7.52 A piece of Ca metal is placed in a 1.00-L container with pure N_2. The N_2 is at a pressure of 1.12 atm and a temperature of 26°C. One hour later, the pressure has dropped to 0.924 atm and the temperature has dropped to 24°C. Calculate the number of grams of N_2 that reacted with the Ca.

7.53 The vapor pressure of PBr_3 reaches 400 mm Hg at 150°C. The vapor pressure of PI_3 reaches 400 mm Hg at 57°C.

a. Which substance should evaporate at the slower rate at 100°C?

b. Which substance should have the lower boiling point?

c. Which substance should have the weaker intermolecular forces?

7.54 In each of the following pairs of molecules, predict which member of the pair would be expected to have the higher boiling point.

a. Cl_2 and Br_2 b. H_2O and H_2S

c. O_2 and CO d. C_3H_8 and CO_2

Grid Problems

7.55

1. definite shape	2. indefinite volume	3. high density
4. cohesive force dominance	5. kinetic energy dominance	6. small compressibility

Select from the grid *all* correct responses for each of the following situations.

a. Characterizations that apply to the solid state

b. Characterizations that apply to the gaseous state

c. Characterizations that apply to both the solid state and the liquid state

d. Characterizations that apply to only one of the three states of matter

7.56

1. $P_1V_1 = P_2V_2$	2. $\dfrac{V_1}{T_1} = \dfrac{V_2}{T_2}$	3. $V_1T_2 = V_2T_1$
4. $\dfrac{P_1V_1}{T_1} = \dfrac{P_2V_2}{T_2}$	5. $P_1V_1T_2 = P_2V_2T_1$	6. $PV = nRT$

Select from the grid *all* correct responses for each of the following situations.

a. Expression that requires a constant number of moles of gas
b. Expression that requires a constant pressure
c. Expression that requires that the temperature be expressed in terms of the Kelvin scale
d. Expression that carries the name of a scientist

7.57

1. Boyle's law	2. Charles's law	3. ideal gas constant
4. combined gas law	5. ideal gas law	6. Dalton's law of partial pressures

Select from the grid *all* correct responses for each of the following situations.

a. Gas law that involves two variables and two sets of conditions
b. Gas law that involves four variables and one set of conditions
c. Gas law that involves two variables and a direct proportion
d. Gas law that involves a fixed quantity of gas; that is, *n* is a constant

7.58

1. H—H	2. :C̤̈l—C̤̈l:	3. :C≡O:
4. H—F̤̈:	5. H—Ö: with H bonded below O	6. H—P̈—H with H bonded below P

Select from the grid *all* correct responses for each of the following situations.

a. London forces are present in liquid-state samples of the substance.
b. Dipole–dipole interactions are present in liquid-state samples of the substance.
c. Hydrogen bonding is present in liquid-state samples of the substance.
d. Pairs of substances that can hydrogen-bond to each other when both substances are in the liquid state

Solutions

CHAPTER OUTLINE

8.1 Characteristics of Solutions 179
8.2 Solubility and Saturated Solutions 181
8.3 Solution Formation 183
8.4 Solubility Rules 184
8.5 Solution Concentration Units 186
Chemistry at a Glance: **Solutions** 193
8.6 Colloidal Dispersions 194
8.7 Colligative Properties of Solutions 195
8.8 Osmosis and Osmotic Pressure 196
Chemistry at a Glance: **Colligative Properties of Solutions** 199
8.9 Dialysis 201
Chemical Connections
8.1 Factors Affecting Gas Solubility 182
8.2 Solubility of Vitamins 186
8.3 Controlled-Release Drugs: Regulating Concentration, Rate, and Location of Release 192
8.4 The Artificial Kidney: A Hemodialysis Machine 202

Ocean water is a solution in which many different substances are dissolved.

Solutions are common in nature, and they represent an abundant form of matter. Solutions carry nutrients to the cells of our bodies and carry away waste products. The ocean is a solution of water, sodium chloride, and many other substances (even gold). A large percentage of all chemical reactions take place in solution, including most of those discussed in later chapters in this text.

8.1 Characteristics of Solutions

All samples of matter are either *pure substances* or *mixtures* (Section 1.5). Pure substances are of two types: *elements* and *compounds.* Mixtures are of two types: *homogeneous* (uniform properties throughout) and *heterogeneous* (different properties in different regions).

Where do solutions fit in this classification scheme? The term *solution* is just an alternative way of saying *homogeneous mixture.* A **solution** *is a homogeneous combination of two or more substances in which each substance retains its own chemical identity.*

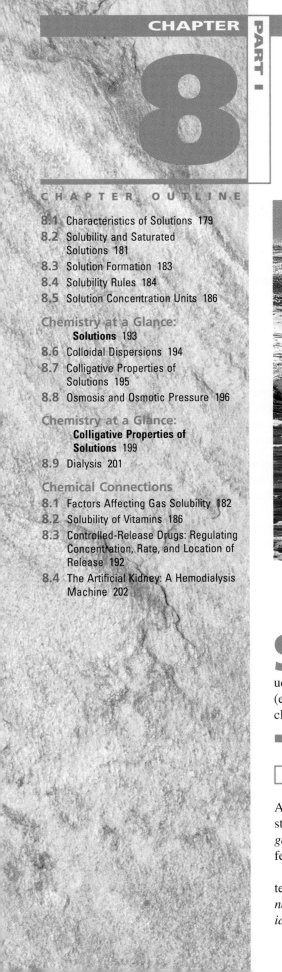

Figure 8.1
The colored crystals are the solute, and the clear liquid is the solvent. Stirring produces the solution.

• "All solutions are mixtures" is a valid statement. However, the reverse statement, "All mixtures are solutions," is not valid. Only those mixtures that are *homogeneous* are solutions.

• Generally, solutions are *transparent;* that is, you can see through them. A synonym for *transparent* is *clear.* Clear solutions may be colorless or colored. A solution of potassium dichromate is a clear yellow-orange solution.

Jewelry often involves solid solutions in which one metal has been dissolved in another metal.

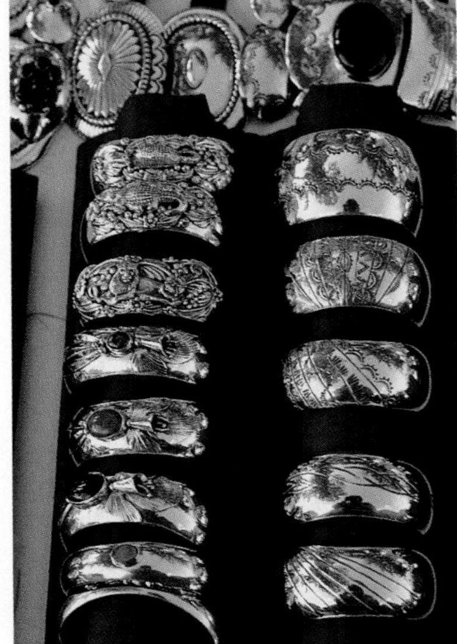

It is often convenient to call one component of a solution the solvent and other components that are present solutes (Figure 8.1). A **solvent** *is the component of a solution that is present in the greatest amount.* A solvent can be thought of as the medium in which the other substances present are dissolved. A **solute** *is a solution component that is present in a small amount relative to that of the solvent.* More than one solute can be present in the same solution. For example, both sugar and salt (two solutes) can be dissolved in a container of water (solvent) to give salty sugar water.

In most of the situations we will encounter, the solutes present in a solution will be of more interest to us than to the solvent. The solutes are the active ingredients in the solution. They are the substances that undergo reaction when solutions are mixed.

The general properties of a solution (homogeneous mixture) were outlined in Section 1.5. These properties, restated using the concepts of solvent and solute, are as follows:

1. A solution contains two or more components: a solvent (the substance present in the greatest amount) and one or more solutes.
2. A solution has a variable composition; that is, the ratio of solute to solvent may be varied.
3. The properties of a solution change as the ratio of solute to solvent is changed.
4. The dissolved solutes are present as individual particles (molecules, atoms, or ions). Intermingling of components at the particle level is a requirement for homogeneity.
5. The solutes remain uniformly distributed throughout the solution and will not settle out with time. Every part of a solution has exactly the same properties and composition as every other part.
6. The solute(s) generally can be separated from the solvent by physical means such as evaporation.

• Types of Two-Component Solutions

Solutions used in laboratories and clinical settings are most often liquids, and the solvent is almost always water. However, gaseous solutions, solid solutions, and liquid solutions in which water is not the solvent are also possible and are relatively common.

Two-component solutions can be classified into nine types according to the physical states of the solvent and solute before mixing. These types, along with an example of each, are listed in Table 8.1. In most solutions encountered in this text, the final state of the solution components is liquid.

The physical state of a solute becomes that of the solvent when a solution is formed. For example, solid naphthalene (moth repellent) must be sublimed (Section 7.13) in order for it to dissolve in air. Merely finely pulverizing a solid and dispersing it in air does not produce a solution. (Dust particles in air, for example, are not in solution.) The particles of the solid must be subdivided to the molecular level; the solid must sublime. Similarly, fog is a suspension of water droplets in air; the droplets are large enough to reflect

Table 8.1
Examples of Various Types of Solutions

Solution type (solute listed first)	Example
Gaseous Solutions	
gas dissolved in gas	dry air (oxygen and other gases dissolved in nitrogen)
liquid dissolved in gas[a]	wet air (water vapor in air)
solid dissolved in gas[a]	moth repellent (or moth balls) sublimed into air
Liquid Solutions	
gas dissolved in liquid	carbonated beverage (carbon dioxide in water)
liquid dissolved in liquid	vinegar (acetic acid dissolved in water)
solid dissolved in liquid	salt water
Solid Solutions	
gas dissolved in solid	hydrogen in platinum
liquid dissolved in solid	dental filling (mercury dissolved in silver)
solid dissolved in solid	sterling silver (copper dissolved in silver)

[a]An alternative viewpoint is that liquid-in-gas and solid-in-gas solutions do not actually exist as true solutions. From this perspective, water vapor or moth repellent in air is considered a gas-in-gas solution because the water or moth repellent must evaporate or sublime first in order to enter the air.

light, a fact that becomes evident when we drive an automobile on a foggy night. Thus fog is not a solution. Water vapor, however, is present in solution form in air (we experience it not as fog but as humidity). When hydrogen gas dissolves in platinum metal (a gas-in-solid solution), the gas molecules take up fixed positions in the metal structure. The gas is "solidified" as a result.

8.2 Solubility and Saturated Solutions

In addition to *solvent* and *solute,* several other terms are used to describe characteristics of solutions. The **solubility** *of a solute is the maximum amount of solute that will dissolve in a given amount of solvent.* Many factors affect the numerical value of a solute's solubility in a given solvent, including the nature of the solvent itself, the temperature, and, in some cases, the pressure and presence of other solutes.

• Most solutes are more soluble in hot solvent than in cold solvent.

Solubility is commonly expressed as grams of solute per 100 g of solvent. The temperature of the solution must also be specified. Table 8.2 gives the solubilities of selected solutes in the solvent water at three different temperatures.

A **saturated solution** *is a solution that contains the maximum amount of solute that can be dissolved under the conditions at which the solution exists.* A saturated solution containing excess undissolved solute is an equilibrium situation where an amount of undissolved solute is continuously dissolving while an equal amount of dissolved solute is

Table 8.2
Solubilities of Various Compounds in Water at 0°C, 50°C, and 100°C

Solute	Solubility (g solute/100 g H_2O)		
	0°C	50°C	100°C
lead(II) bromide ($PbBr_2$)	0.455	1.94	4.75
silver sulfate (Ag_2SO_4)	0.573	1.08	1.41
copper(II) sulfate ($CuSO_4$)	14.3	33.3	75.4
sodium chloride (NaCl)	35.7	37.0	39.8
silver nitrate ($AgNO_3$)	122	455	952
cesium chloride (CsCl)	161.4	218.5	270.5

Chemical CONNECTIONS

8.1 Factors Affecting Gas Solubility

Both temperature and pressure affect the solubility of a gas in water. The effects are opposite. Increased temperature decreases gas solubility, and increased pressure increases gas solubility. The following table quantifies such effects for carbon dioxide (CO_2), a gas we often encounter dissolved in water.

Solubility of CO_2 (g/100 mL water)

Temperature effect (at 1 atm pressure)		Pressure effect (at 0°C)	
0°C	0.348	1 atm	0.348
20°C	0.176	2 atm	0.696
40°C	0.097	3 atm	1.044
60°C	0.058		

The effect of temperature on gas solubility has important environmental consequences because of the use of water from rivers and lakes for industrial cooling. Water used for cooling and then returned to its source at higher than ambient temperatures contains less oxygen and is less dense than when it was diverted. This lower-density "oxygen-deficient" water tends to "float" on colder water below, which blocks normal oxygen adsorption processes. This makes it more difficult for fish and other aquatic forms to obtain the oxygen they need to sustain life. This overall situation is known as *thermal pollution.*

Thermal pollution is sometimes unrelated to human activities. On hot summer days, the temperature of shallow water sometimes reaches the point where dissolved oxygen levels are insufficient to support some life. Under these conditions, suffocated fish may be found on the surface.

A flat taste is often associated with boiled water. This is due in part to the removal of dissolved gases during the boiling process. The removal of dissolved carbon dioxide particularly affects the taste.

The effect of pressure on gas solubility is observed every time a can or bottle of carbonated beverage is opened. The fizzing that occurs results from the escape of gaseous CO_2. The atmospheric pressure associated with an open container is much lower than the pressure used in the bottling process.

Carbon dioxide escaping from an opened bottle of a carbonated beverage.

Deep-sea divers usually breathe a mixture of helium and oxygen instead of air in order to avoid solubility-pressure problems associated with air.

Pressure is a factor in the solubility of gases in the bloodstream. In hospitals, persons who are having difficulty obtaining oxygen are given supplementary oxygen. The result is an oxygen pressure greater than that in air. Hyperbaric medical procedures involve the use of pure oxygen. Oxygen pressure is sufficient to cause it to dissolve directly into the bloodstream, bypassing the body's normal mechanism for oxygen uptake (hemoglobin). Treatment of carbon monoxide poisoning is a situation where hyperbaric procedures are often needed.

Deep-sea divers can experience solubility-pressure problems. For every 30 feet that divers descend, the pressure increases by 1 atm. As a result, the air they breathe (particularly the N_2 component) dissolves to a greater extent in the blood. If a diver returns to the surface too quickly after a deep dive, the dissolved gases form bubbles in the blood (in the same way CO_2 does in a freshly opened can of carbonated beverage). This bubble formation may interfere with nerve impulse transmission and restrict blood flow. This painful condition, known as the *bends,* can cause paralysis or death.

Divers can avoid the bends by returning to the surface slowly and by using helium–oxygen gas mixtures, instead of air, in their breathing apparatus. Helium is less soluble in blood than N_2 and, because of its small atomic size, can escape from body tissues. Nitrogen must be removed via normal respiration.

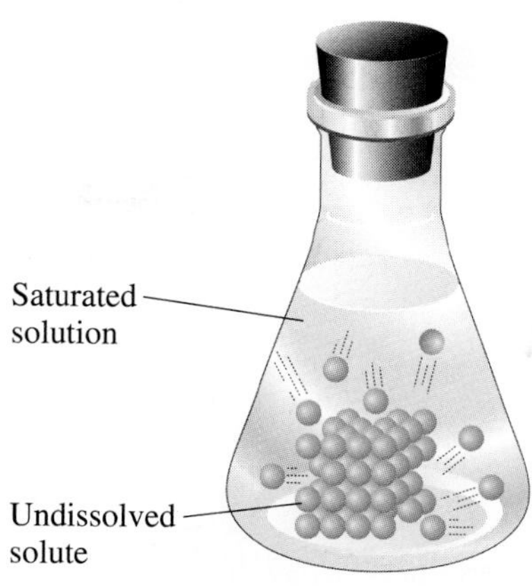

Figure 8.2
In a saturated solution, the dissolved solute is in a dynamic equilibrium with the undissolved solute. Solute enters and leaves the solution at the same rate.

continuously crystallizing. Consider the process of adding table sugar (sucrose) to a container of water. Initially, the added sugar dissolves as the solution is stirred. Finally, as we add more sugar, we reach a point where no amount of stirring will cause the added sugar to dissolve. The last-added sugar remains as a solid on the bottom of the container; the solution is saturated. Although it appears to the eye that nothing is happening once the saturation point is reached, this is not the case on the molecular level. Solid sugar from the bottom of the container is continuously dissolving in the water, and an equal amount of sugar is coming out of solution. Accordingly, the net number of sugar molecules in the liquid remains the same. The equilibrium situation in the saturated solution is somewhat similar to the evaporation of a liquid in a closed container (Section 7.15). Figure 8.2 illustrates the dynamic equilibrium process occurring in a saturated solution that contains undissolved excess solute.

• When the amount of dissolved solute in a solution corresponds to the solute's solubility in the solvent, the solution formed is a saturated solution.

An **unsaturated solution** *is a solution in which less solute than the maximum amount possible is dissolved in the solution.* Most solutions we encounter fall into this category.

The terms *dilute* and *concentrated* are also used to convey qualitative information about the degree of saturation of a solution. A **dilute solution** *is a solution that contains a small amount of solute relative to the amount that could dissolve.* A **concentrated solution** *is a solution that contains a large amount of solute relative to the amount that could dissolve.* A concentrated solution does not have to be a saturated solution.

Another term commonly encountered in solution discussions is *aqueous solution.* An **aqueous solution** *is a solution in which water is the solvent.*

• When the term *solution* is used, it is generally assumed that "aqueous solution" is meant, unless the context makes it clear that the solvent is not water.

8.3 Solution Formation

In a solution, solute particles are uniformly dispersed throughout the solvent. Considering what happens at the molecular level during the solution process will help us understand how this is accomplished.

In order for a solute to dissolve in a solvent, two types of interparticle attractions must be overcome: (1) attractions between solute particles (solute–solute attractions) and (2) attractions between solvent particles (solvent–solvent attractions). Only when these attractions are overcome can particles in both pure solute and pure solvent separate from one another and begin to intermingle. A new type of interaction, which does not exist prior to solution formation, arises as a result of the mixing of solute and solvent. This new interaction is the attraction between solute and solvent particles (solute–solvent attractions). These attractions are the primary driving force for solution formation.

An important type of solution process is one in which an ionic solid dissolves in water. Let us consider in detail the process of dissolving sodium chloride, a typical ionic solid, in water (Figure 8.3). The polar water molecules become oriented in such a way that the negative oxygen portion points toward positive sodium ions and the positive hydrogen portions point toward negative chloride ions. As the polar water molecules begin

• The fact that water molecules are polar is very important in the dissolving of an ionic solid in water.

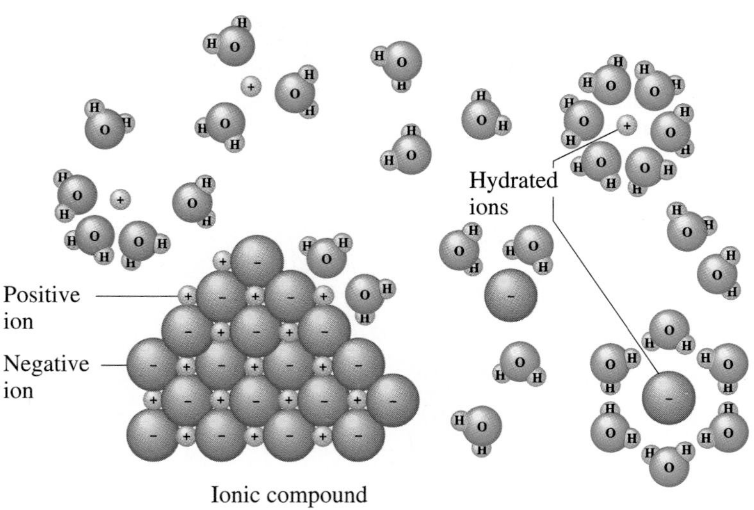

Figure 8.3
When an ionic solid, such as sodium chloride, dissolves in water, the water molecules *hydrate* the ions. The positive ions are bound to the water molecules by their attraction for the partial negative charge on the water's oxygen atom, and the negative ions are bound to the water molecules by their attraction for the partial positive charge on the water's hydrogen atoms.

to surround ions on the crystal surface, they exert sufficient attraction to cause these ions to break away from the crystal surface. After leaving the crystal, an ion retains its surrounding group of water molecules; it has become a *hydrated ion.* As each hydrated ion leaves the surface, other ions are exposed to the water, and the crystal is picked apart ion by ion. Once in solution, the hydrated ions are uniformly distributed either by stirring or by random collisions with other molecules or ions.

The random motion of solute ions in solutions causes them to collide with one another, with solvent molecules, and occasionally with the surface of any undissolved solute. Ions undergoing the latter type of collision occasionally stick to the solid surface and thus leave the solution. When the number of ions in solution is low, the chances for collision with the undissolved solute are low. However, as the number of ions in solution increases, so do the chances for collisions, and more ions are recaptured by the undissolved solute. Eventually, the number of ions in solution reaches a level where ions return to the undissolved solute at the same rate at which other ions leave. At this point, the solution is saturated, and the equilibrium process discussed in the previous section is in operation.

Oil spills can be contained to some extent using trawlers and a boom apparatus because oil and water, having different polarities, are relatively insoluble in each other. The oil, which is of lower density, floats on top of the water.

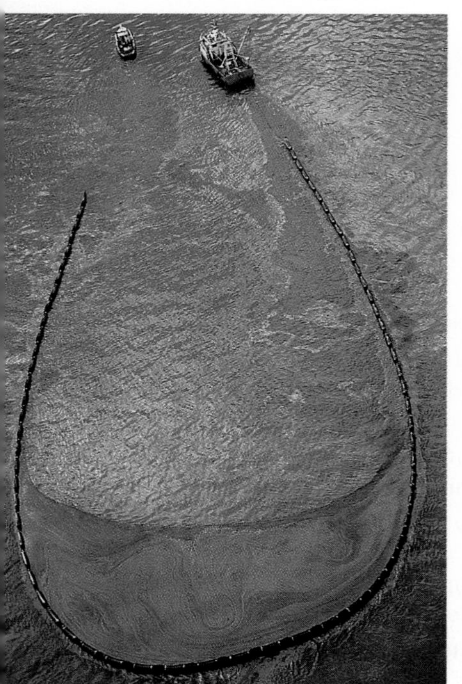

8.4 Solubility Rules

In this section, we will present some rules for qualitatively predicting solute solubilities. These rules summarize in a concise form the results of thousands of experimental determinations of solute–solvent solubility.

A very useful generalization that relates polarity to solubility is that *substances of like polarity tend to be more soluble in each other than substances that differ in polarity.* This conclusion is often expressed as the simple phrase "*like dissolves like.*" Polar substances, in general, are good solvents for other polar substances but not for nonpolar substances. Similarly, nonpolar substances exhibit greater solubility in nonpolar solvents than in polar solvents.

The generalization "like dissolves like" is a useful tool for predicting solubility behavior in many, but not all, solute–solvent situations. Results that agree with this generalization are nearly always obtained in the cases of gas-in-liquid and liquid-in-liquid solutions and for solid-in-liquid solutions in which the solute is not an ionic compound. For example, NH_3 gas (a polar gas) is much more soluble in H_2O (a polar liquid) than is O_2 gas (a nonpolar gas).

In the common case of solid-in-liquid solutions in which the solute is an ionic compound, the rule "like dissolves like" is not adequate. Their polar nature would suggest that all ionic compounds are soluble in a polar solvent such as water, but this is not the case.

Table 8.3
Solubility Guidelines for Ionic Compounds in Water

Ion contained in the compound	Solubility	Exceptions
Group IA (Li^+, Na^+, K^+, etc.)	soluble	
ammonium (NH_4^+)	soluble	
acetates ($C_2H_3O_2^-$)	soluble	
nitrates (NO_3^-)	soluble	
chlorides (Cl^-), bromides (Br^-), and iodides (I^-)	soluble	Ag^+, Pb^{2+}, Hg_2^{2+}
sulfates (SO_4^{2-})	soluble	Ca^{2+}, Sr^{2+}, Ba^{2+}, Pb^{2+}
carbonates (CO_3^{2-})	insoluble[a]	Group IA and NH_4^+
phosphates (PO_4^{3-})	insoluble	Group IA and NH_4^+
sulfides (S^{2-})	insoluble	Groups IA and IIA and NH_4^+
hydroxides (OH^-)	insoluble	Group IA, Ca^{2+}, Sr^{2+}, Ba^{2+}

[a]All ionic compounds, even the least soluble ones, dissolve to some slight extent in water. Thus the "insoluble" classification really means ionic compounds that have very limited solubility in water.

• The generalization "like dissolves like" is not adequate for predicting the solubilities of *ionic compounds* in water. More detailed solubility guidelines are needed (see Table 8.3).

The failure of the generalization for ionic compounds is related to the complexity of the factors involved in determining the magnitude of the solute–solute (ion–ion) and solvent–solute (solvent–ion) interactions. Among other things, both the charge and the size of the ions in the solute must be considered. Changes in these factors affect both types of interactions, but not to the same extent.

Some guidelines concerning the solubility of ionic compounds in water, which should be used in place of "like dissolves like," are given in Table 8.3.

Example 8.1

Predicting Solute Solubility Using Polarity Considerations

Predict the solubility of the following solutes in the solvent indicated.

a. CH_4 (a nonpolar gas) in water
b. Ethyl alcohol (a polar liquid) in chloroform (a polar liquid)
c. AgCl (an ionic solid) in water
d. Na_2SO_4 (an ionic solid) in water
e. $AgNO_3$ (an ionic solid) in water

Solution

a. Insoluble. They are of unlike polarity because water is polar.
b. Soluble. Both substances are polar, so they should be relatively soluble in one another—like dissolves like.
c. Insoluble. Table 8.3 indicates that all chlorides except those of silver, lead, and mercury(I) are soluble. Thus AgCl is one of the exceptions.
d. Soluble. Table 8.3 indicates that all ionic sodium-containing compounds are soluble.
e. Soluble. Table 8.3 indicates that all compounds containing the nitrate ion (NO_3^-) are soluble.

Practice Exercise 8.1

Predict the solubility of the following solutes in the solvent indicated.

a. NO_2 (a polar gas) in water
b. CCl_4 (a nonpolar liquid) in benzene (a nonpolar liquid)
c. NaBr (an ionic solid) in water
d. $MgCO_3$ (an ionic solid) in water
e. $(NH_4)_3PO_4$ (an ionic solid) in water

• *Answers.* **a.** soluble; **b.** soluble; **c.** soluble; **d.** insoluble; **e.** soluble

Chemical CONNECTIONS

8.2 Solubility of Vitamins

Polarity plays an important role in the solubility of many substances in the fluids and tissues of the human body. For example, consider vitamin solubilities. The 13 known vitamins fall naturally into two classes: fat-soluble and water-soluble. The fat-soluble vitamins are A, D, E, and K. Water-soluble vitamins are vitamin C and the eight B vitamins (thiamine, riboflavin, niacin, vitamin B_6, folic acid, vitamin B_{12}, pantothenic acid, and biotin). Water-soluble vitamins have polar molecular structures, as does water. By contrast, fat-soluble vitamins have nonpolar molecular structures that are compatible with the nonpolar nature of fats.

Vitamin C is water-soluble. Because of this, vitamin C is not stored in the body and must be ingested in our daily diet. Unused vitamin C is eliminated rapidly from the body via body fluids. Vitamin A, on the other hand, is fat-soluble. It can be, and is, stored by the body in fat tissue for later use. If vitamin A is consumed in excess quantities (from excessive vitamin supplements), illness can result. Because of its limited water solubility, vitamin A cannot be rapidly eliminated from the body by body fluids.

The water-soluble vitamins can be easily leached out of foods as they are prepared. As a rule of thumb, you should eat foods every day that are rich in the water-soluble vitamins. Taking megadose vitamin supplements of water-soluble vitamins is seldom effective. The extra amounts of these vitamins are usually picked up by the extracellular fluids, carried away by blood, and excreted in the urine. As one person aptly noted, "If you take supplements of water-soluble vitamins, you may have the most expensive urine in town."

8.5 Solution Concentration Units

Both solutions contain the same amount of solute. A concentrated solution (left) contains a relatively large amount of solute compared with the amount of solvent. A dilute solution (right) contains a relatively small amount of solute compared with the amount of solvent.

Because solutions are mixtures (Section 8.1), they have a variable composition. Specifying what the composition of a solution is involves specifying solute concentrations. In general, the **concentration** *of a solution is the amount of solute present in a specified amount of solution.* Many methods of expressing concentration exist, and certain methods are better suited for some purposes than others. In this section we consider two methods: *percent concentration* and *molarity.*

• Percent Concentration

There are three different ways of representing percent concentration:

1. Percent by mass (or mass–mass percent)
2. Percent by volume (or volume–volume percent)
3. Mass–volume percent

Percent by mass (or mass–mass percent) is the percentage unit most often used in chemical laboratories. **Percent by mass** *is equal to the mass of solute divided by the total mass of solution, multiplied by 100 (to put the value in terms of percentage).*

$$\text{Percent by mass} = \frac{\text{mass of solute}}{\text{mass of solution}} \times 100$$

The solute and solution masses must be measured in the same unit, which is usually grams. The mass of the solution is equal to the mass of the solute plus the mass of the solvent.

$$\text{Mass of solution} = \text{mass of solute} + \text{mass of solvent}$$

• The concentration of butterfat in milk is expressed in terms of percent by mass. When you buy 1% milk, you are buying milk that contains 1 g of butterfat per 100 g of milk.

A solution whose mass percent concentration is 5.0% would contain 5.0 g of solute per 100.0 g of solution (5.0 g of solute and 95.0 g of solvent). Thus percent by mass directly gives the number of grams of solute in 100 g of solution. The percent-by-mass concentration unit is often abbreviated as %(m/m).

Example 8.2

Calculating the Percent-by-Mass Concentration of a Solution

What is the percent-by-mass, %(m/m), concentration of sucrose (table sugar) in a solution made by dissolving 7.6 g of sucrose in 83.4 g of water?

Solution

Both the mass of solute and the mass of solvent are known. Substituting these numbers into the percent-by-mass equation

$$\%(\text{m/m}) = \frac{\text{mass of solute}}{\text{mass of solution}} \times 100$$

gives

$$\%(\text{m/m}) = \frac{7.6 \text{ g sucrose}}{7.6 \text{ g sucrose} + 83.4 \text{ g water}} \times 100$$

Remember that the denominator of the preceding equation (mass of solution) is the combined mass of the solute and the solvent.

Doing the mathematics gives

$$\%(\text{m/m}) = \frac{7.6 \cancel{\text{g}}}{91.0 \cancel{\text{g}}} \times 100 = 8.4\%$$

Practice Exercise 8.2

What is the percent-by-mass, %(m/m), concentration of Na_2SO_4 in a solution made by dissolving 7.6 g of Na_2SO_4 in enough water to give 87.3 g of solution?

- *Answer.* 8.7%(m/m)

Example 8.3

Calculating the Mass of Solute Needed to Produce a Solution of a Given Percent-by-Mass Concentration

How many grams of sucrose must be added to 375 g of water to prepare a 2.75%(m/m) solution of sucrose?

Solution

Often, when a solution concentration is given as part of a problem statement, the concentration information is used in the form of a conversion factor when you solve the problem. That will be the case in this problem.

The given quantity is 375 g of H_2O (grams of solvent), and the desired quantity is grams of sucrose (grams of solute).

$$375 \text{ g } H_2O = ? \text{ g sucrose}$$

The conversion factor relating these two quantities (solvent and solute) is obtained from the given concentration. In a 2.75%-by-mass sucrose solution, there are 2.75 g of sucrose for every 97.25 g of water.

$$100.00 \text{ g solution} - 2.75 \text{ g sucrose} = 97.25 \text{ g } H_2O$$

The relationship between grams of solute and grams of solvent (2.75 to 97.25) gives us the needed conversion factor.

$$\frac{2.75 \text{ g sucrose}}{97.25 \text{ g } H_2O}$$

The problem is set up and solved, using dimensional analysis, as follows:

$$375 \cancel{\text{g } H_2O} \times \left(\frac{2.75 \text{ g sucrose}}{97.25 \cancel{\text{g } H_2O}}\right) = 10.6 \text{ g sucrose}$$

Practice Exercise 8.3

How many grams of $LiNO_3$ must be added to 25.0 g of water to prepare a 5.00%(m/m) solution of $LiNO_3$?

• *Answer.* 1.32 g $LiNO_3$

The second type of percentage unit, percent by volume (or volume–volume percent), which is abbreviated %(v/v), is used as a concentration unit in situations where the solute and solvent are both liquids or both gases. In these cases, it is more convenient to measure volumes than masses. **Percent by volume** *is equal to the volume of solute divided by the total volume of solution, multiplied by 100.*

$$\text{Percent by volume} = \frac{\text{volume of solute}}{\text{volume of solution}} \times 100$$

Solute and solution volumes must always be expressed in the same units when you use percent by volume.

When the numerical value of a concentration is expressed as a percent by volume, it directly gives the number of milliliters of solute in 100 mL of solution. Thus a 100-mL sample of a 5.0%(v/v) alcohol-in-water solution contains 5.0 mL of alcohol dissolved in enough water to give 100 mL of solution. Note that such a 5.0%(v/v) solution could not be made by adding 5 mL of alcohol to 95 mL of water, because the volumes of two liquids are not usually additive. Differences in the way molecules are packed, as well as differences in distances between molecules, almost always result in the volume of the solution being different from the sum of the volumes of solute and solvent. For example, the final volume resulting from the addition of 50.0 mL of ethyl alcohol to 50.0 mL of water is 96.5 mL of solution (see Figure 8.4). Working problems involving percent by volume entails the same kinds of steps as those used for problems involving percent by mass.

• The proof system for specifying the alcoholic content of beverages is twice the percent by volume. Hence 40 proof is 20%(v/v) alcohol; 100 proof is 50%(v/v) alcohol.

The third type of percentage unit in common use is mass–volume percentage. This unit, which is often encountered in clinical and hospital settings, is particularly convenient to use when you work with a solid solute, which is easily weighed, and a liquid solvent. Solutions of drugs for internal and external use, intravenous and intramuscular injectables, and reagent solutions for testing are usually labeled in mass–volume percent.

Mass–volume percent, which is abbreviated %(m/v), *is equal to the mass of solute (in grams) divided by the total volume of solution (in milliliters), multiplied by* 100.

$$\text{Mass–volume percent} = \frac{\text{mass of solute (g)}}{\text{volume of solution (mL)}} \times 100$$

When volumes of two different liquids are combined, the volumes are not additive. This process is somewhat analogous to pouring marbles and golf balls together. The marbles can fill in the spaces between the golf balls. This results in the "mixed" volume being less than the sum of the "premixed" volumes.

Figure 8.4
Identical volumetric flasks are filled to the 50.0-mL mark with ethanol and with water. When the two liquids are poured into a 100-mL volumetric flask, the volume is seen to be less than the expected 100.0 mL; it is only about 96.5 mL.

• For dilute aqueous solutions, %(m/m) and %(m/v) are almost the same, because mass in grams of the solution equals the volume in milliliters when the density is close to 1.00 g/mL, as it is for pure water and for most dilute solutions.

Note that in the definition of mass–volume percent, specific mass and volume units are given. This is necessary because the units do not cancel, as was the case with mass percent and volume percent.

Mass–volume percent indicates the number of grams of solute dissolved in each 100 mL of solution. Thus a 2.3%(m/v) solution of any solute contains 2.3 g of solute in each 100 mL of solution, and a 5.4%(m/v) solution contains 5.4 g of solute in each 100 mL of solution.

Example 8.4

Calculating the Mass of Solute Needed to Produce a Solution of a Given Mass–Volume Percent Concentration

Normal saline solution that is used to dissolve drugs for intravenous use is 0.92%(m/v) NaCl in water. How many grams of NaCl are required to prepare 35.0 mL of normal saline solution?

Solution

The given quantity is 35.0 mL of solution, and the desired quantity is grams of NaCl.

$$35.0 \text{ mL solution} = ? \text{ g NaCl}$$

The given concentration, 0.92%(m/v), which means 0.92 g of NaCl per 100 mL of solution, is used as a conversion factor to go from milliliters of solution to grams of NaCl. The setup for the conversion is

$$35.0 \text{ }\cancel{\text{mL solution}} \times \left(\frac{0.92 \text{ g NaCl}}{100 \text{ }\cancel{\text{mL solution}}}\right)$$

Doing the arithmetic after canceling the units gives

$$\left(\frac{35.0 \times 0.92}{100}\right) \text{g NaCl} = 0.32 \text{ g NaCl}$$

Practice Exercise 8.4

How many grams of glucose ($C_6H_{12}O_6$) are needed to prepare 500.0 mL of a 4.50%(m/v) glucose–water solution?

• *Answer.* 22.5 g $C_6H_{12}O_6$

• Molarity

The **molarity** (M) *of a solution is a ratio giving the number of moles of solute per liter of solution.* The mathematical equation for molarity is

$$\text{Molarity (M)} = \frac{\text{moles of solute}}{\text{liters of solution}}$$

A solution containing 1 mole of KBr in 1 L of solution has a molarity of 1 and is said to be a 1 M (1 *molar*) solution.

This concentration unit is often used in laboratories where chemical reactions are being studied. Because chemical reactions occur between molecules and atoms, the mole—

Example 8.5

Calculating the Molarity of a Solution

Determine the molarities of the following solutions.

a. 4.35 moles of $KMnO_4$ are dissolved in enough water to give 750 mL of solution.
b. 20.0 g of NaOH is dissolved in enough water to give 1.50 L of solution.

Solution

a. The number of moles of solute is given in the problem statement.

$$\text{Moles of solute } (KMnO_4) = 4.35$$

The volume of the solution is also given in the problem statement, but not in the right units. Molarity requires liters for the volume units, and we are given milliliters of solution. Making the unit change yields

$$750\ \cancel{\text{mL}} \times \left(\frac{10^{-3}\ \text{L}}{1\ \cancel{\text{mL}}}\right) = 0.750\ \text{L}$$

The molarity of the solution is obtained by substituting the known quantities into the equation

$$\text{M} = \frac{\text{moles of solute}}{\text{liters of solution}}$$

which gives

$$\text{M} = \frac{4.35\ \text{moles } KMnO_4}{0.750\ \text{L solution}} = 5.80\ \frac{\text{moles } KMnO_4}{\text{L solution}}$$

Note that the units for molarity are always moles per liter.

b. This time, the volume of solution is given in liters.

$$\text{Volume of solution} = 1.50\ \text{L}$$

The moles of solute must be calculated from the grams of solute (given) and the solute's molar mass, which is 40.00 g/mole (calculated from atomic masses).

$$20.0\ \text{g}\ \cancel{\text{NaOH}} \times \left(\frac{1\ \text{mole NaOH}}{40.00\ \text{g}\ \cancel{\text{NaOH}}}\right) = 0.500\ \text{mole NaOH}$$

Substituting the known quantities into the defining equation for molarity gives

$$\text{M} = \frac{0.500\ \text{mole NaOH}}{1.50\ \text{L solution}} = 0.333\ \frac{\text{mole NaOH}}{\text{L solution}}$$

Practice Exercise 8.5

Determine the molarities of the following solutions.

a. 2.37 moles of KNO_3 are dissolved in enough water to give 650.0 mL of solution.
b. 40.0 g of KCl is dissolved in enough water to give 0.850 L of solution.

• *Answers.* **a.** 3.65 M KNO_3; **b.** 0.631 M KCl.

• In preparing 100 mL of a solution of a specific molarity, enough solvent is added to a weighed amount of solute to give a *final* volume of 100 mL. The weighed solute is not added to a *starting* volume of 100 mL; this would produce a final volume greater than 100 mL, because the solute volume increases the total volume.

• When you perform molarity concentration calculations, you need the *identity* of the solute. You cannot calculate moles of solute without knowing the chemical identity of the solute. When you perform percent concentration calculations, the *identity* of the solute is not used in the calculation; all you need is the *amount* of solute.

a unit that counts particles—is desirable. Equal volumes of two solutions of the same molarity contain the same number of solute molecules.

In order to find the molarity of a solution, we need to know the solution volume in liters and the number of moles of solute present. An alternative to knowing the number of moles of solute is knowing the number of grams of solute present and the solute's formula mass. The number of moles can be calculated by using these two quantities (Section 6.4).

The mass of solute present in a known volume of solution is an easily calculable quantity if the molarity of the solution is known. When we do such a calculation, molarity serves as a conversion factor that relates liters of solution to moles of solute. In a similar manner, the volume of solution needed to supply a given amount of solute can be calculated by using the solution's molarity as a conversion factor.

Example 8.6

Calculating the Amount of Solute Present in a Given Amount of Solution

How many grams of sucrose (table sugar, $C_{12}H_{22}O_{11}$) are present in 185 mL of a 2.50 M sucrose solution?

Solution

The given quantity is 185 mL of solution, and the desired quantity is grams of $C_{12}H_{22}O_{11}$.

$$185 \text{ mL of solution} = ?\text{ g } C_{12}H_{22}O_{11}$$

The pathway used to solve this problem is

$$\text{mL solution} \longrightarrow \text{L solution} \longrightarrow \text{moles } C_{12}H_{22}O_{11} \longrightarrow \text{g } C_{12}H_{22}O_{11}$$

The given molarity (2.50 M) serves as the conversion factor for the second unit change; the formula mass of sucrose (which is not given and must be calculated) is used to accomplish the third unit change.

The dimensional-analysis setup for this pathway is

$$185 \cancel{\text{mL solution}} \times \left(\frac{10^{-3}\ \cancel{\text{L solution}}}{1\ \cancel{\text{mL solution}}}\right) \times \left(\frac{2.50\ \cancel{\text{moles } C_{12}H_{22}O_{11}}}{1\ \cancel{\text{L solution}}}\right) \times \left(\frac{342.34\ \text{g } C_{12}H_{22}O_{11}}{1\ \cancel{\text{mole } C_{12}H_{22}O_{11}}}\right)$$

Canceling the units and doing the arithmetic, we find that

$$\left(\frac{185 \times 10^{-3} \times 2.50 \times 342.34}{1 \times 1 \times 1}\right) \text{g } C_{12}H_{22}O_{11} = 158 \text{ g } C_{12}H_{22}O_{11}$$

Practice Exercise 8.6

How many grams of silver nitrate ($AgNO_3$) are present in 375 mL of 1.50 M silver nitrate solution?

• *Answer:* 95.6 g $AgNO_3$

Example 8.7

Calculating the Amount of Solution Needed to Supply a Given Amount of Solute

A typical dose of iron(II) sulfate ($FeSO_4$) used in the treatment of iron-deficiency anemia is 0.35 g. How many milliliters of a 0.10 M iron(II) sulfate solution would be needed to supply this dose?

Solution

The given quantity is 0.35 g of $FeSO_4$, and the desired quantity is milliliters of $FeSO_4$ solution.

(continued)

$$0.35 \text{ g FeSO}_4 = ? \text{ mL FeSO}_4 \text{ solution}$$

The pathway used to solve this problem is

$$\text{g FeSO}_4 \longrightarrow \text{moles FeSO}_4 \longrightarrow \text{L FeSO}_4 \text{ solution} \longrightarrow \text{mL FeSO}_4 \text{ solution}$$

We accomplish the first unit conversion by using the formula mass of $FeSO_4$ (which must be calculated) as a conversion factor. The second unit conversion involves the use of the given molarity as a conversion factor.

$$0.35 \text{ g FeSO}_4 \times \left(\frac{1 \text{ mole FeSO}_4}{151.92 \text{ g FeSO}_4}\right) \times \left(\frac{1 \text{ L solution}}{0.10 \text{ mole FeSO}_4}\right) \times \left(\frac{1 \text{ mL solution}}{10^{-3} \text{ L solution}}\right)$$

Canceling units and doing the arithmetic, we find that

$$\left(\frac{0.35 \times 1 \times 1 \times 1}{151.92 \times 0.10 \times 10^{-3}}\right) \text{mL solution} = 23 \text{ mL solution}$$

Practice Exercise 8.7

How many milliliters of a 0.100 M NaOH solution would be needed to provide 15.0 g of NaOH for a chemical reaction?

• *Answer:* 3750 mL

See Chemistry at a Glance on the following page for a review of solutions and the ways in which we represent their concentration.

Chemical CONNECTIONS

8.3 Controlled-Release Drugs: Regulating Concentration, Rate, and Location of Release

In the use of both prescription and over-the-counter drugs, body concentration levels of the drug are obviously of vital importance. All drugs have an optimum concentration range where they are most effective. Below this optimum concentration range, a drug is ineffective, and above it the drug may have adverse side effects. Hence, the much-repeated warning "Take as directed."

Ordinarily, in the administration of a drug, the body's concentration level of the drug rapidly increases toward the higher end of the effective concentration range and then gradually declines and falls below the effective limit. The period of effectiveness of the drug can be extended by using the drug in a controlled-release form. This causes the drug to be released in a regulated, continuous manner over a longer period of time. The accompanying graph contrasts "ordinary-release" and "controlled-release" modes of drug action.

The use of controlled-release medication began in the early 1960s with the introduction of the decongestant Contac. Contac's controlled-release mechanism, which is now found in many drugs and used by all drug manufacturers, involves drug particles encapsulated within a slowly dissolving coating that *varies in thickness* from particle to particle. Particles of the drug with a thinner coating dissolve first. Those particles with a thicker coating dissolve more slowly, extending the period of drug release. The number of particles of various thicknesses, within a formulation, is predetermined by the manufacturer.

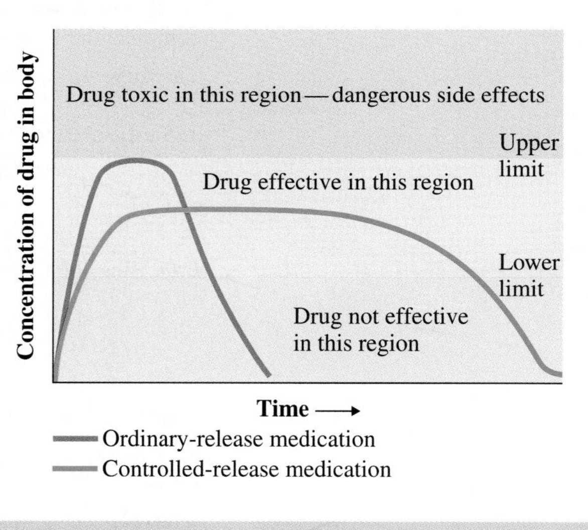

When drugs are taken orally, they first encounter the acidic environment of the stomach. Two problems can occur here: (1) The drug itself may damage the stomach lining. (2) The drug may be rendered inactive by the gastric acid present in the stomach. Controlled-release techniques are useful in overcoming these problems. Drug particle coatings are now available that are acid-resistant; that is, they do not dissolve in acidic solution. Drugs with such coatings pass from the stomach into the small intestine in undissolved form. Within the nonacidic (basic) environment of the small intestine, the dissolving process then begins.

CHEMISTRY AT A GLANCE

SOLVENT

The component of a solution present in the greatest quantity

SOLUTE

The component of a solution present in the lesser quantity

SOLUTION

A homogeneous mixture of two or more substances in which each substance retains its chemical identity

CONCENTRATION OF A SOLUTION

The amount of solute in a specified amount of solution

PERCENT BY MASS	PERCENT BY VOLUME	MASS–VOLUME PERCENT	MOLARITY
$\%(m/m) = \frac{\text{mass of solute}}{\text{mass of solution}} \times 100$	$\%(v/v) = \frac{\text{volume of solute}}{\text{volume of solution}} \times 100$	$\%(m/v) = \frac{\text{mass of solute (g)}}{\text{volume of solution (mL)}} \times 100$	$M = \frac{\text{moles of solute}}{\text{liters of solution}}$

1%(m/m)
milk

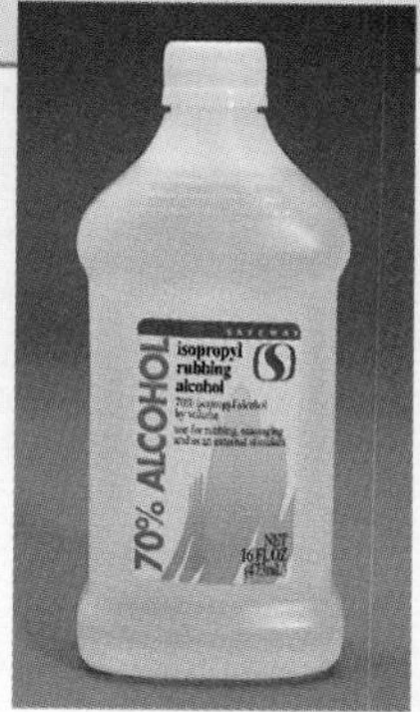

70%(v/v)
rubbing alcohol

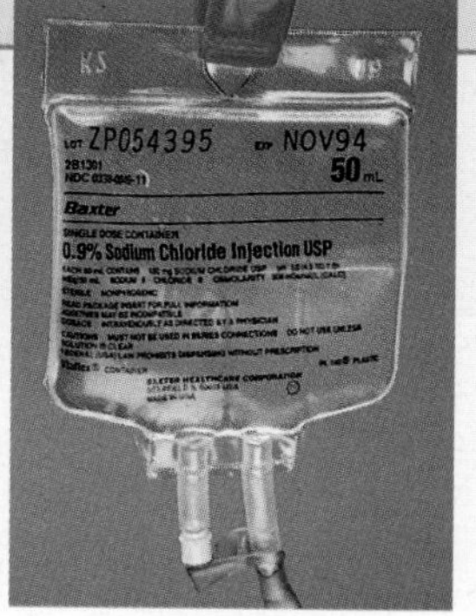

0.9%(m/v)
physiological saline solution

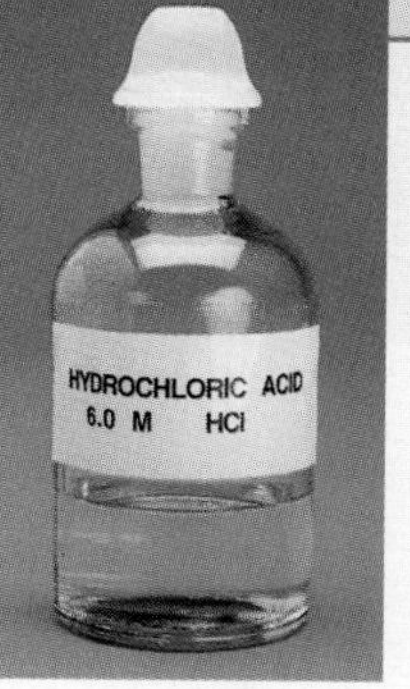

6.0 M
hydrochloric acid

• Dilution Calculations

A common problem encountered when we work with solutions is that of diluting a solution of known concentration (usually called a stock solution) to a lower concentration. **Dilution** *is the process in which more solvent is added to a solution in order to lower its concentration.* The same amount of solute is present, but it is now distributed in a larger amount of solvent (the original solvent plus the added solvent).

Often, we prepare a solution of a specific concentration by adding a predetermined volume of solvent to a specific volume of stock solution. A simple relationship exists

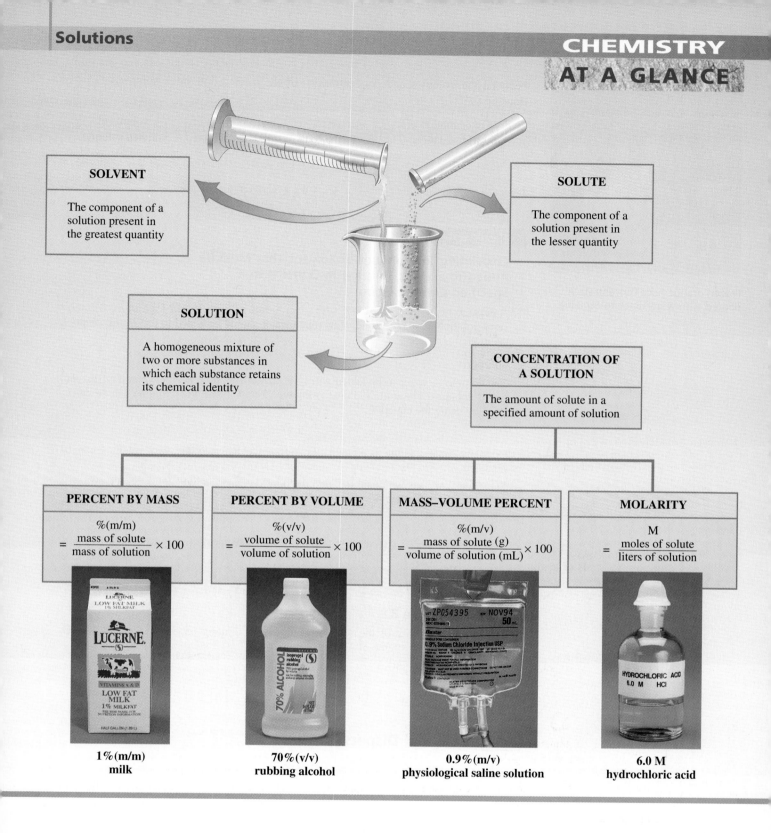

Frozen orange juice concentrate is diluted with water prior to drinking.

between the volumes and concentrations of the diluted and stock solutions. This relationship is

$$\begin{pmatrix}\text{Concentration of}\\ \text{stock solution}\end{pmatrix} \times \begin{pmatrix}\text{volume of}\\ \text{stock solution}\end{pmatrix} = \begin{pmatrix}\text{concentration of}\\ \text{diluted solution}\end{pmatrix} \times \begin{pmatrix}\text{volume of}\\ \text{diluted solution}\end{pmatrix}$$

or

$$C_s \times V_s = C_d \times V_d$$

Example 8.8

Calculating the Amount of Solvent That Must Be Added to a Stock Solution to Dilute It to a Specified Concentration

A nurse wants to prepare a 1.0%(m/v) silver nitrate solution from 24 mL of a 3.0%(m/v) stock solution of silver nitrate. How much water should be added to the 24 mL of stock solution?

Solution

The volume of water to be added will be equal to the difference between the final and initial volumes. The initial volume is known (24 mL). The final volume can be calculated by using the equation

$$C_s \times V_s = C_d \times V_d$$

Once the final volume is known, the difference between the two volumes can be obtained.

Substituting the known quantities into the dilution equation, which has been rearranged to isolate V_d on the left side, gives

$$V_d = \frac{C_s \times V_s}{C_d} = \frac{3.0\cancel{\%(\text{m/v})} \times 24 \text{ mL}}{1.0\cancel{\%(\text{m/v})}} = 72 \text{ mL}$$

The solvent added is

$$V_d - V_s = (72 - 24) \text{ mL} = 48 \text{ mL}$$

Practice Exercise 8.8

What is the molarity of the solution prepared by diluting 65 mL of 0.95 M Na_2SO_4 solution to a final volume of 135 mL?

- *Answer.* 0.46 M Na_2SO_4

8.6 Colloidal Dispersions

Colloidal dispersions are heterogeneous mixtures that have many properties similar to those of solutions, although they are not true solutions. In a broad sense, colloidal dispersions may be thought of as mixtures in which a material is *dispersed* rather than *dissolved.* A **colloidal dispersion** *is a dispersion of small particles of one substance in another substance.* The terms *solute* and *solvent* are not used to indicate the components of a colloidal dispersion. Instead, the particles dispersed in a colloidal dispersion are called the *dispersed phase,* and the material in which they are dispersed is called the *dispersing medium.*

• Some chemists use the term *colloid* instead of colloidal dispersion.

Particles of the dispersed phase in a colloidal dispersion are so small that (1) they are not usually discernible by the naked eye, (2) they do not settle out under the influence of gravity, and (3) they cannot be filtered out using filter paper that has relatively large pores. In these respects, the dispersed phase behaves similarly to a solute in a solution. However, the dispersed-phase particle size is sufficiently large to make the dispersion nonhomogeneous.

Figure 8.5
A beam of light travels through a true solution (the yellow liquid) without being scattered—that is, its path cannot be seen. This is not the case for a colloidal dispersion (the red liquid), where scattering of light by the dispersed phase makes the light pathway visible.

• Milk is a colloidal dispersion. If you shine a flashlight through a glass of salt water and a glass of milk, you can duplicate the experiment illustrated in Figure 8.5. (For the best effect, dilute the milk with some water until it just looks cloudy.)

• Particle size for the dispersed phase in a colloidal dispersion is larger than that for solutes in a true solution.

A beam of light reveals the nonhomogeneity of a colloidal dispersion. When we shine a beam of light through a true solution, we cannot see the track of the light. However, a beam of light passing through a colloidal dispersion can be observed, because the light is scattered by the dispersed phase (Figure 8.5). This scattered light is reflected into our eyes.

The diameters of the dispersed particles in a colloidal dispersion are in the range of 10^{-7} cm to 10^{-5} cm. This compares with diameters of less than 10^{-7} cm for particles such as ions, atoms, and molecules. Thus colloidal particles are up to 1000 times larger than those present in a true solution. The dispersed particles are usually aggregates of molecules, but this is not always the case. Some protein molecules are large enough to form colloidal dispersions that contain single molecules in suspension. Colloidal dispersions that contain particles with diameters larger than 10^{-5} cm are usually not encountered. Suspended particles of this size usually settle out under the influence of gravity.

Many different biochemical colloidal dispersions occur within the human body. Foremost among them is blood, which has numerous components that are colloidal in size. Fat is transported in the blood and lymph systems as colloidal-sized particles.

8.7 Colligative Properties of Solutions

Adding a solute to a pure solvent causes its physical properties to change. A special group of physical properties that change when a solute is added are called colligative properties. **Colligative properties** *are the physical properties of a solution that depend only on the number (concentration) of solute particles (molecules or ions) in a given quantity of solvent and not on their chemical identities.* Examples of colligative properties include vapor-pressure lowering, boiling-point elevation, freezing-point depression, and osmotic pressure. The first three of these colligative properties are discussed in this section. The fourth, osmotic pressure, will be considered in Section 8.8.

A water–antifreeze mixture has a higher boiling point and a lower freezing point than pure water.

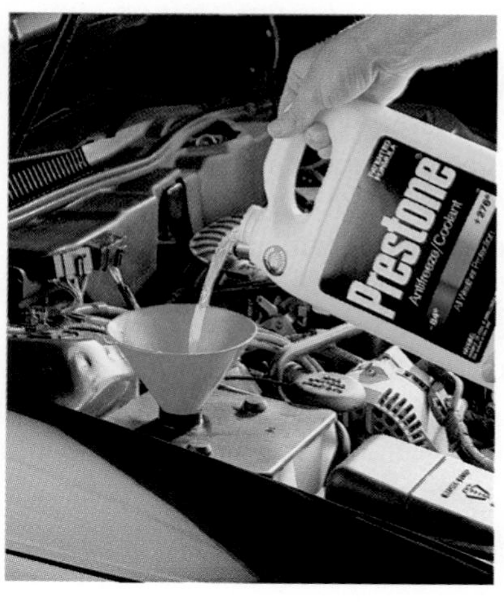

Adding a nonvolatile solute to a solvent *lowers* the vapor pressure of the resulting solution below that of the pure solvent at the same temperature. (A nonvolatile solute is one that has a low vapor pressure and therefore a low tendency to vaporize.) This lowering of vapor pressure is a direct consequence of some of the solute molecules or ions occupying positions on the surface of the liquid. Their presence decreases the probability of solvent molecules escaping; that is, the number of surface-occupying solvent molecules has been decreased. Figure 8.6 illustrates the decrease in surface concentration of solvent molecules when a solute is added. As the *number* of solute particles increases, the reduction in vapor pressure also increases; thus vapor pressure is a colligative property. What is important is not the identity of the solute molecules but the fact that they take up room on the surface of the liquid.

Adding a nonvolatile solute to a solvent *raises* the boiling point of the resulting solution above that of the pure solvent. This is logical when we remember that the vapor pressure of the solution is lower than that of pure solvent and that the boiling point is dependent on vapor pressure (Section 7.16). A higher temperature will be needed to raise

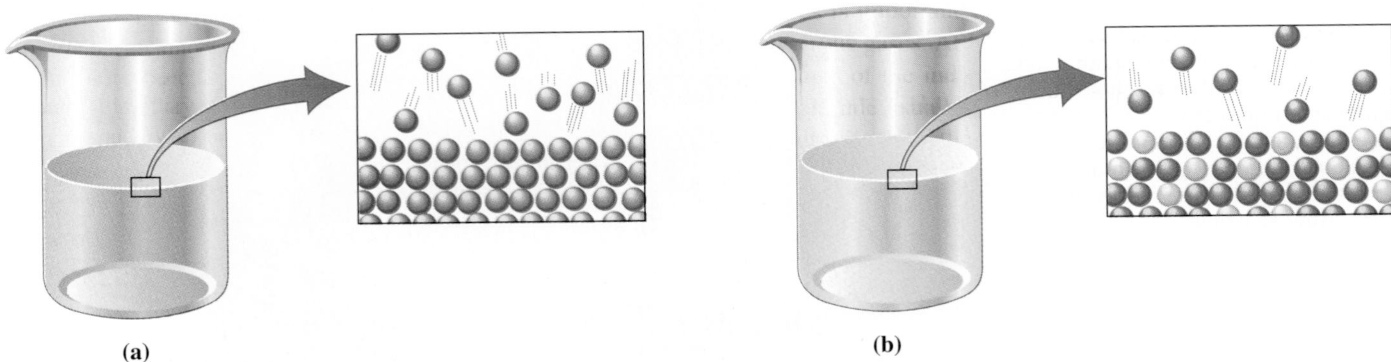

Figure 8.6
Close-ups of the surface of a liquid solvent (a) before and (b) after solute has been added. There are fewer solvent molecules on the surface of the liquid after solute has been added. This results in a decreased vapor pressure for the solution compared with pure solvent.

the depressed vapor pressure of the solution to atmospheric pressure; this is the condition required for boiling.

A common application of the phenomenon of boiling point elevation involves automobiles. The coolant ethylene glycol (a nonvolatile solute) is added to car radiators to prevent boilover in hot weather. The engine may not run any cooler, but the coolant–water mixture will not boil until it reaches a temperature well above the normal boiling point of water.

Adding a nonvolatile solute to a solvent *lowers* the freezing point of the resulting solution below that of the pure solvent. The presence of the solute particles within the solution interferes with the tendency of solvent molecules to line up in an organized manner, a condition necessary for the solid state. A lower temperature is necessary before the solvent molecules will form the solid.

• In the making of homemade ice cream, the function of the rock salt added to the ice is to depress the freezing point of the ice–water mixture surrounding the ice cream mix sufficiently to allow the mix (which contains sugar and other solutes and thus has a freezing point below 0°C) to freeze.

Applications of freezing-point depression are even more numerous than those for boiling-point elevation. In climates where the temperature drops below 0°C in the winter, it is necessary to protect water-cooled automobile engines from freezing. This is done by adding antifreeze (usually ethylene glycol) to the radiator. The addition of this nonvolatile material causes the vapor pressure and freezing point of the resulting solution to be much lower than those of pure water. Also in the winter, salt, usually NaCl or $CaCl_2$, is spread on roads and sidewalks to melt ice or prevent it from forming. The salt dissolves in the water to form a solution that will not freeze until the temperature drops much lower than 0°C, the normal freezing point of water.

8.8 Osmosis and Osmotic Pressure

The process of osmosis and the colligative property of osmotic pressure are extremely important phenomena when we consider biological solutions. These phenomena govern many of the processes important to a functioning human body.

• Osmosis

• The term *osmosis* comes from the Greek *osmos,* which means "push."

• An osmotic semipermeable membrane contains very small pores (holes)—too small to see—that are big enough to let small solvent molecules through but not big enough to let larger solute molecules pass through.

Osmosis *is the passage of a solvent from a dilute solution (or pure solvent) through a semipermeable membrane into a more concentrated solution.* The simple apparatus shown in Figure 8.7a is helpful in explaining, at the molecular level, what actually occurs during the osmotic process. The apparatus consists of a tube containing a concentrated salt–water solution that has been immersed in a dilute salt–water solution. The immersed end of the tube is covered with a semipermeable membrane. A **semipermeable membrane** *is a thin layer of material that allows certain types of molecules to pass through but prohibits the passage of others.* The selectivity of the membrane is based on size differences between molecules. The particles that are allowed to pass through (usually just solvent molecules like water) are relatively small. Thus, the membrane functions somewhat like a sieve. Using the experimental setup of Figure 8.7a, we can observe a net flow of solvent from the dilute to the concentrated solution over the course of time. This is indicated by a rise in the level of the solution in the tube and a drop in the level of the dilute solution, as shown in Figure 8.7b.

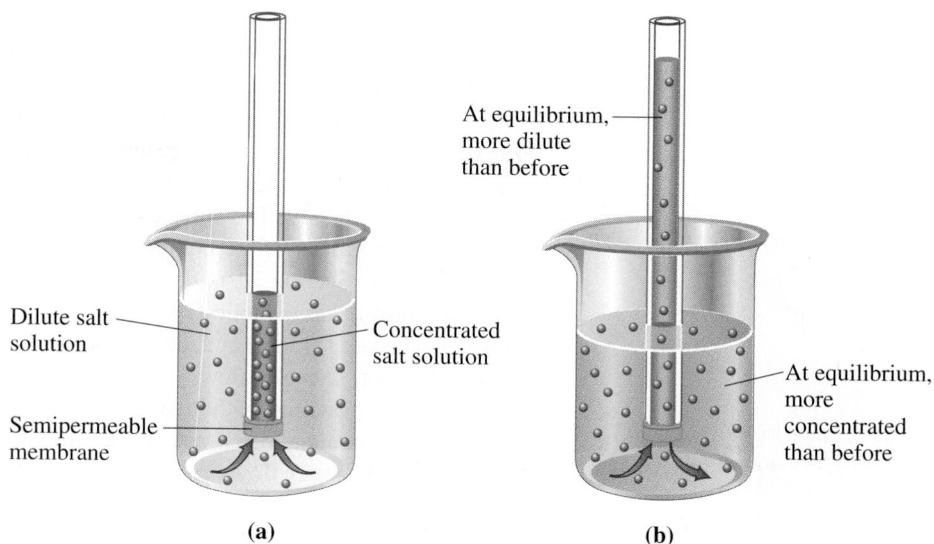

Figure 8.7
(a) Osmosis, the flow of solvent through a semipermeable membrane from a dilute to a more concentrated solution, can be observed with this apparatus. (b) At equilibrium, the molecules move back and forth at equal rates.

What is actually happening on a molecular level as the process of osmosis occurs? Water is flowing in both directions through the membrane. However, the rate of flow into the concentrated solution is greater than the rate of flow in the other direction (see Figure 8.8). Why? The presence of solute molecules diminishes the ability of water molecules to cross the membrane. The solute molecules literally get in the way; they occupy some of the surface positions next to the membrane. Because there is a greater concentration of solute molecules on one side of the membrane than on the other, the flow rates differ. The flow rate is diminished to a greater extent on the side of the membrane where the greater concentration of solute is present.

The net transfer of solvent across the membrane continues until (1) the concentrations of solute particles on both sides of the membrane become equal or (2) the hydrostatic pressure on the concentrated side of the membrane (from the difference in liquid levels) becomes sufficient to counterbalance the greater escaping tendency of molecules from the dilute side. From here on, there is an equal flow of solvent in both directions across the membrane, and the volume of liquid on each side of the membrane remains constant.

• Osmotic Pressure

Osmotic pressure *is the amount of pressure that must be applied to prevent the net flow of solvent through a semipermeable membrane from a solution of lower solute concentration to a solution of higher solute concentration.* In terms of Figure 8.7, osmotic pressure is the pressure required to prevent water from rising in the tube. Figure 8.9 shows how this pressure can be measured. The greater the concentration difference between the separated solutions, the greater the magnitude of the osmotic pressure.

• A process called *reverse osmosis* is used in the desalination of seawater to make drinking water. Pressure greater than the osmotic pressure is applied on the salt water side of the membrane to force solvent water across the membrane from the salt water side to the "pure" water side.

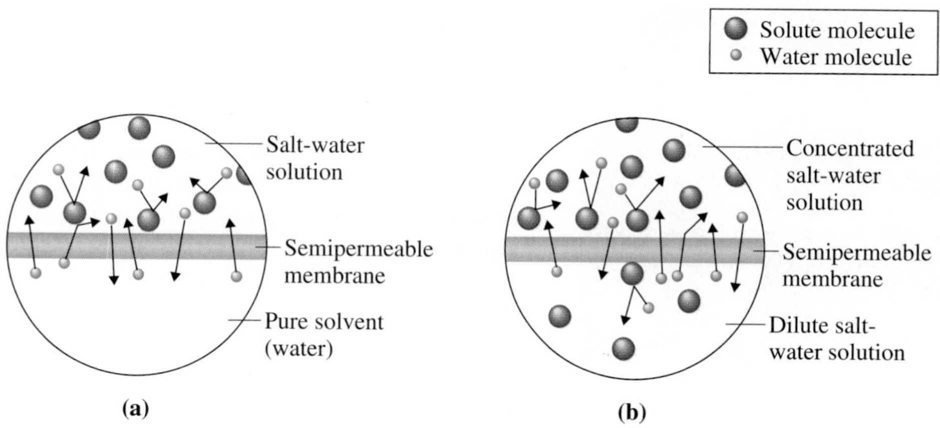

Figure 8.8
Enlarged views of a semipermeable membrane separating (a) pure water and a salt–water solution, and (b) a dilute salt–water solution and a concentrated salt–water solution. In both cases, water moves from the area of lower solute concentration to the area of higher solute concentration.

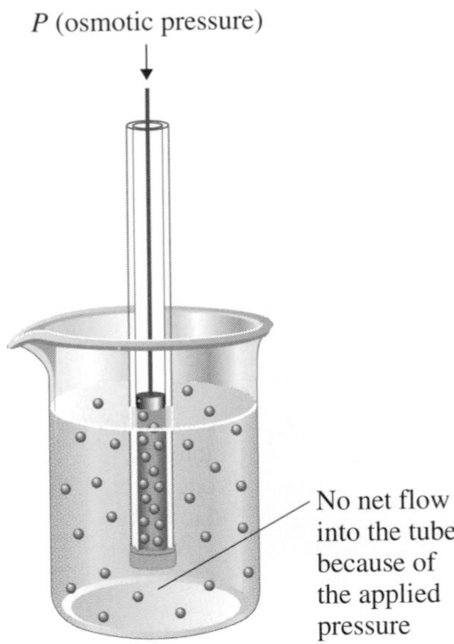

Figure 8.9
Osmotic pressure is the amount of pressure needed to prevent the solution in the tube from rising as a result of the process of osmosis.

• Osmolarity is greater for ionic solutes than for molecular solutes (solutes that do not separate into ions, such as glucose and sucrose) if the concentrations of the solutions are equal, because ionic solutes dissociate to form more than 1 mole of particles per mole of compound.

Figure 8.10
The dissolved substances in tree sap create a more concentrated solution than the surrounding ground water. Water enters membranes in the roots and rises in the tree, creating an osmotic pressure that can exceed 20 atm in extremely tall trees.

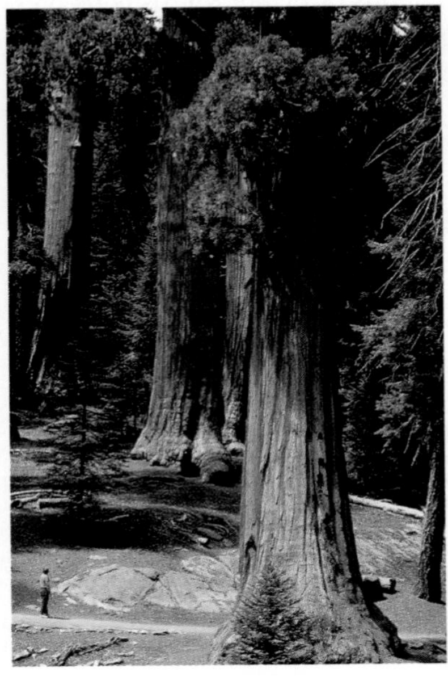

Cell membranes in both plants and animals are semipermeable in nature (see Figure 8.10). The selective passage of fluid materials through these membranes governs the balance of fluids in living systems. Thus osmotic-type phenomena are of prime importance for life. We say "osmotic-type phenomena" instead of "osmosis" because the semipermeable membranes found in living cells usually permit the passage of small solute molecules (nutrients and waste products) in addition to solvent. The term *osmosis* implies the passage of solvent only. The substances prohibited from passing through the membrane in osmotic-type processes are colloidal-sized molecules and insoluble suspended materials (see Dialysis, Section 8.9).

It is because of an osmotic-type process that plants will die if they are watered with salt water. The salt solution outside the root membranes is more concentrated than the solution in the root, so water flows out of the roots; then the plant becomes dehydrated and dies. This same principle is the reason for not drinking excessive amounts of salt water, even if you are stranded on a raft in the middle of the ocean. When salt water is taken into the stomach, water flows out of the stomach wall membranes and into the stomach; then the tissues become dehydrated. Drinking seawater will cause greater thirst because the body will lose water rather than absorb it.

• Osmolarity

The osmotic pressure of a solution depends on the number of solute particles present. This in turn depends on the solute concentration and on whether the solute forms ions once it is in solution. Note that two factors are involved in determining osmotic pressure.

The fact that some solutes dissociate into ions in solution is of utmost importance in osmotic pressure considerations. For example, the osmotic pressure of a 1 M NaCl solution is twice that of a 1 M glucose solution, despite the fact that both solutions have equal concentrations (1 M). Sodium chloride is an ionic solute, and it dissociates in solution to give two particles (a Na^+ and a Cl^- ion) per formula unit; however, glucose is a molecular solute and does not dissociate. It is the number of particles present that determines osmotic pressure.

The concentration unit osmolarity is used to compare the osmotic pressures of solutions. The **osmolarity** *of a solution is the product of its molarity and the number of particles produced per formula unit when the solute dissociates.* The equation for osmolarity is

$$\text{Osmolarity} = \text{molarity} \times i$$

where i is the number of particles produced from the dissociation of one formula unit of solute.

Solutions of equal osmolarity have equal osmotic pressures. If the osmolarity of one solution is three times that of another, then the osmotic pressure of the first solution is three times that of the second solution. A solution with high osmotic pressure will take up more water than a solution of lower osmotic pressure; thus more pressure must be applied to prevent osmosis.

Chemistry at a Glance on the following page summarizes this chapter's discussion of colligative properties of solutions.

• Isotonic, Hypertonic, and Hypotonic Solutions

The terms *isotonic solution, hypertonic solution,* and *hypotonic solution* are used repeatedly when we deal with osmotic-type phenomena in the human body. A consideration of what happens to red blood cells when they are placed in three different solutions will help us understand the differences in meaning of these three terms. The solution media are distilled water, concentrated sodium chloride solution, and physiological saline solution.

When red blood cells are placed in pure water, they swell up (enlarge in size) and finally rupture (burst); this process is called *hemolysis* (Figure 8.11a). Hemolysis is caused by an increase in the amount of water entering the cells compared with the amount of wa-

COLLIGATIVE PROPERTIES OF SOLUTIONS

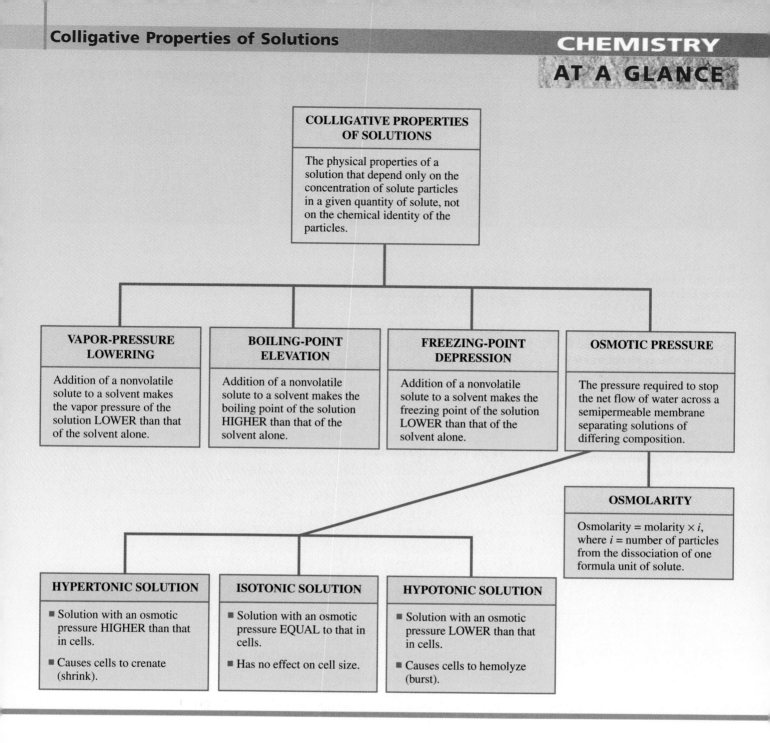

The physical properties of a solution that depend only on the concentration of solute particles in a given quantity of solute, not on the chemical identity of the particles.

VAPOR-PRESSURE LOWERING

Addition of a nonvolatile solute to a solvent makes the vapor pressure of the solution LOWER than that of the solvent alone.

BOILING-POINT ELEVATION

Addition of a nonvolatile solute to a solvent makes the boiling point of the solution HIGHER than that of the solvent alone.

FREEZING-POINT DEPRESSION

Addition of a nonvolatile solute to a solvent makes the freezing point of the solution LOWER than that of the solvent alone.

OSMOTIC PRESSURE

The pressure required to stop the net flow of water across a semipermeable membrane separating solutions of differing composition.

OSMOLARITY

Osmolarity = molarity $\times i$, where i = number of particles from the dissociation of one formula unit of solute.

HYPERTONIC SOLUTION

- Solution with an osmotic pressure HIGHER than that in cells.
- Causes cells to crenate (shrink).

ISOTONIC SOLUTION

- Solution with an osmotic pressure EQUAL to that in cells.
- Has no effect on cell size.

HYPOTONIC SOLUTION

- Solution with an osmotic pressure LOWER than that in cells.
- Causes cells to hemolyze (burst).

• The pickling of cucumbers and salt curing of meat are practical applications of the concept of crenation. A concentrated salt solution (brine) is used to draw water from the cells of the cucumber to produce a pickle. Salt on the surface of the meat preserves the meat by crenation of bacterial cells.

• The word *tonicity* refers to the tone, or firmness, of a biological cell.

ter leaving the cells. This is the result of cell fluid having a greater osmotic pressure than pure water.

When red blood cells are placed in a concentrated sodium chloride solution, a process opposite to hemolysis occurs. This time, water moves from the cells to the solution, causing the cells to shrivel (shrink in size); this process is called *crenation* (Figure 8.11b). Crenation occurs because the osmotic pressure of the concentrated salt solution surrounding the red cells is greater than that of the fluid within the cells. Water always moves in the direction of greater osmotic pressure.

Finally, when red blood cells are placed in physiological saline solution, a 0.9%(m/v) sodium chloride solution, water flow is balanced and neither hemolysis nor crenation occurs (Figure 8.11c). The osmotic pressure of physiological saline solution is the same as that of red blood cell fluid. Thus the rates of water flow into and out of the red blood cells are the same.

We will now define the terms *isotonic, hypotonic,* and *hypertonic.* An **isotonic solution** *is a solution whose osmotic pressure is equal to that within cells.* Red blood cell

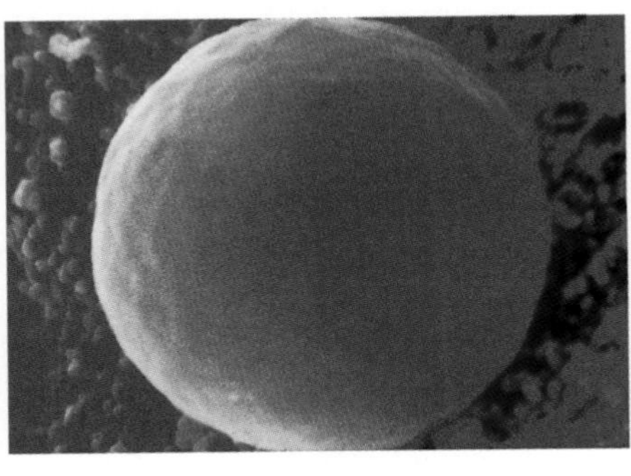

(a)

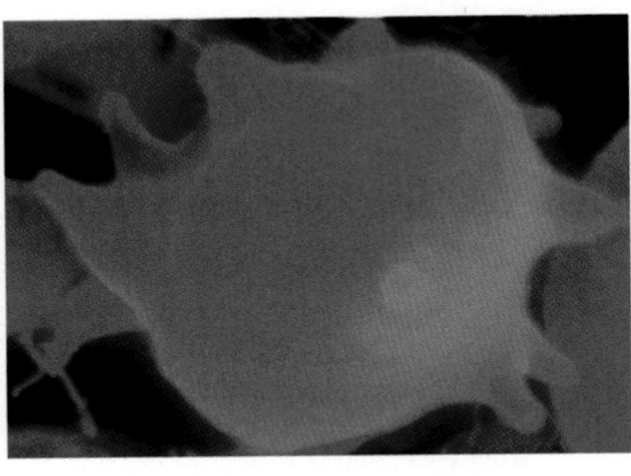

(b)

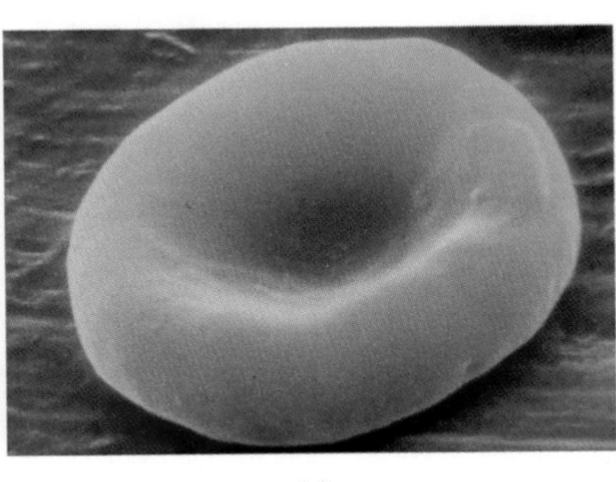

(c)

Figure 8.11
Effects of bathing red blood cells in various types of solutions.
(a) Hemolysis in pure water (a hypotonic solution). (b) Crenation in concentrated sodium chloride solution (a hypertonic solution).
(c) Cells neither swell nor shrink in physiological saline solution (an isotonic solution).

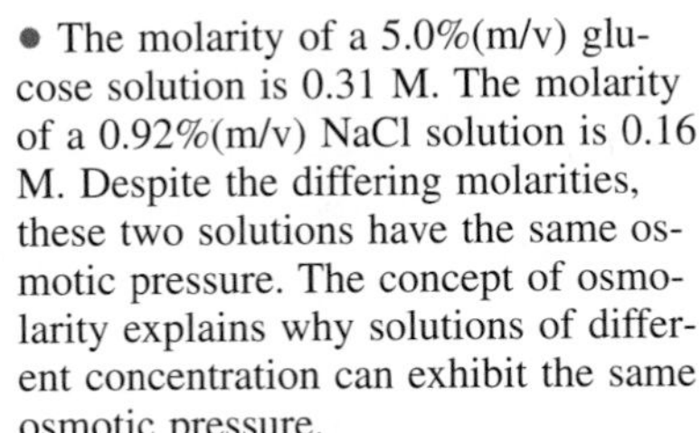

• The molarity of a 5.0%(m/v) glucose solution is 0.31 M. The molarity of a 0.92%(m/v) NaCl solution is 0.16 M. Despite the differing molarities, these two solutions have the same osmotic pressure. The concept of osmolarity explains why solutions of different concentration can exhibit the same osmotic pressure.

Example 8.9

Calculating the Osmolarity of Various Solutions

What is the osmolarity of each of the following solutions?

a. 2 M NaCl **b.** 2 M $CaCl_2$ **c.** 2 M glucose
d. 2 M in both NaCl and glucose
e. 2 M in NaCl and 1 M in glucose

Solution

The general equation for osmolarity that was previously given will be applicable in each of the parts of the problem.

$$\text{Osmolarity} = \text{molarity} \times i$$

a. Two particles per dissociation are produced when NaCl dissociates in solution.

$$NaCl \longrightarrow Na^+ + Cl^-$$

The value of *i* is 2, and the osmolarity is twice the molarity.

$$\text{Osmolarity} = 2\text{ M} \times 2 = 4$$

b. For $CaCl_2$, the value of *i* is 3, because three ions are produced from the dissociation of one $CaCl_2$ unit.

$$CaCl_2 \longrightarrow Ca^{2+} + 2\,Cl^-$$

The osmolarity will therefore be triple the molarity:

$$\text{Osmolarity} = 2\text{ M} \times 3 = 6$$

c. Glucose is a nondissociating solute. Thus the value of *i* is 1, and the molarity and osmolarity will be the same—two molar and two osmolar.

d. With two solutes present, we must consider the collective effects of both solutes. For NaCl, $i = 2$; and for glucose, $i = 1$. The osmolarity is calculated as follows:

$$\text{Osmolarity} = \underbrace{2\text{ M} \times 2}_{\text{NaCl}} + \underbrace{2\text{ M} \times 1}_{\text{glucose}} = 6$$

e. This problem differs from the previous one in that the two solutes are not present in equal concentrations. This does not change the way we work the problem. The *i* values are the same as before, and the osmolarity is

$$\text{Osmolarity} = \underbrace{2\text{ M} \times 2}_{\text{NaCl}} + \underbrace{1\text{ M} \times 1}_{\text{glucose}} = 5$$

Practice Exercise 8.9

What is the osmolarity of each of the following solutions?

a. 3 M $NaNO_3$ **b.** 3 M $Ca(NO_3)_2$ **c.** 3 M sucrose
d. 3 M in both $Ca(NO_3)_2$ and sucrose
e. 3 M in both $NaNO_3$ and $Ca(NO_3)_2$

• *Answers.* **a.** 6; **b.** 9; **c.** 3; **d.** 12; **e.** 15

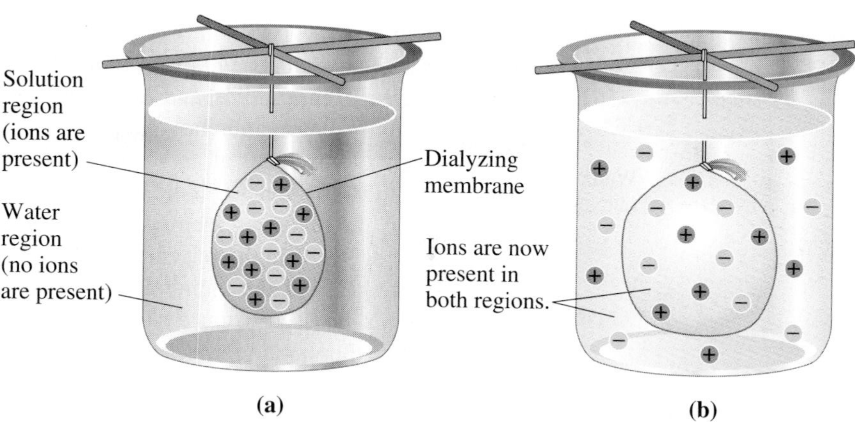

Figure 8.12
In dialysis, there is a net movement of ions from a region of higher concentration to a region of lower concentration. (a) Before dialysis. (b) After dialysis.

• The terminology "D5W," often heard in television shows involving doctors and paramedics, refers to a 5.5%(m/v) solution of glucose (also called dextrose, D) in water (W).

• The use of 5%(m/v) glucose solution for intravenous feeding has a shortcoming. A patient can accommodate only about 3 L of water in a day. Three liters of 5%(m/v) glucose water will supply only about 640 kcal of energy, an inadequate amount of energy. A resting patient requires about 1400 kcal/day.

This problem is solved by using solutions that are about 6 times as concentrated as isotonic solutions. They are administered, through a tube, directly into a large blood vessel leading to the heart (the superior vena cava) rather than through a small vein in the arm or leg. The large volume of blood flowing through this vein quickly dilutes the solution to levels that do not upset the osmotic balance in body fluids. Using this technique, patients can be given up to 5000 kcal/day of nourishment.

• In *osmosis,* only solvent passes through the membrane. In *dialysis,* both solvent and small solute particles (ions and small molecules) pass through the membrane.

fluid, physiological saline solution, and 5%(m/v) glucose water are all isotonic with respect to one another. The processes of replacing body fluids and supplying nutrients to the body intravenously require the use of isotonic solutions such as physiological saline and glucose water. If isotonic solutions were not used, the damaging effects of hemolysis or crenation would occur.

A **hypotonic solution** *is a solution with a lower osmotic pressure than that within cells.* The prefix *hypo-* means "under" or "less than normal." Distilled water is hypotonic with respect to red blood cell fluid, and these cells will hemolyze when placed in it (Figure 8.12a). A **hypertonic solution** *is a solution with a higher osmotic pressure than that within cells.* The prefix *hyper-* means "over" or "more than normal." Concentrated sodium chloride solution is hypertonic with respect to red blood cell fluid, and these cells undergo crenation when placed in it (Figure 8.12b).

It is sometimes necessary to introduce a hypertonic or hypotonic solution, under controlled conditions, into the body to correct an improper "water balance" in a patient. A hypertonic solution will cause the net transfer of water from tissues to blood; then the kidneys will remove the water. Some laxatives, such as Epsom salts, act by forming hypertonic solutions in the intestines. A hypotonic solution can be used to cause water to flow from the blood into surrounding tissue; blood pressure can be decreased in this manner. Table 8.4 summarizes the differences in meaning among the terms *isotonic, hypertonic,* and *hypotonic.*

8.9 Dialysis

Dialysis is closely related to osmosis. It is the osmotic-type process that occurs in living systems. Osmosis, you recall (Section 8.8), occurs when a solution and a solvent are separated by a semipermeable membrane that allows solvent but not solute to pass through it. There is a net transfer of solvent from the dilute solution (or pure solvent) into the more concentrated solution. **Dialysis** *is the process in which a semipermeable membrane*

Table 8.4
Characteristics of Isotonic, Hypertonic, and Hypotonic Solutions

	Type of Solution		
	Isotonic	**Hypertonic**	**Hypotonic**
osmolarity relative to body fluids	equal	greater than	less than
osmotic pressure relative to body fluids	equal	greater than	less than
osmotic effect on cells	equal water flow into and out of cells	net flow of water out of cells	net flow of water into cells

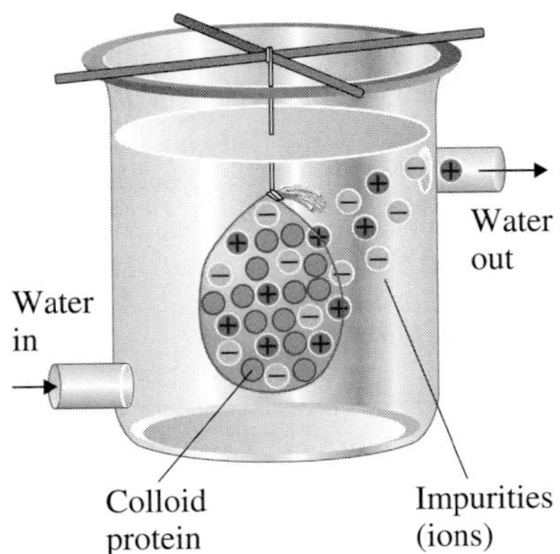

Figure 8.13
Impurities (ions) can be removed from a colloidal dispersion by using a dialysis procedure.

permits the passage of solvent, dissolved ions, and small molecules but blocks the passage of colloidal particles and large molecules. Thus dialysis allows for the separation of small particles from colloids and large molecules. Many plant and animal membranes function as dialyzing membranes.

Consider the placement of an aqueous solution of sodium chloride in a dialyzing bag that is surrounded by water (Figure 8.12a). What happens? Sodium ions and chloride ions move through the dialyzing membrane into the water; that is, there is a net movement of ions from a region of high concentration to a region of low concentration. This will occur until both sides of the membrane have equal concentrations of ions (Figure 8.12b).

Dialysis can be used to purify a colloidal solution containing protein molecules and solute. The smaller solute molecules pass through the dialyzing membrane and leave the solution. The larger protein molecules remain behind. The result is a purified protein colloidal dispersion (Figure 8.13).

The human kidneys are a complex dialyzing system that is responsible for removing waste products from the blood. The removed products are then eliminated in urine. When the kidneys fail, these waste products build up and eventually poison the body.

When a person goes into shock, there is a sudden increase in the permeability of the membranes of the blood capillaries. Large colloidally dispersed molecules, such as proteins, leave the bloodstream and leak into the space between cells. This damage disrupts the normal chemistry of the blood. If a patient in shock is left untreated, death can occur.

Chemical CONNECTIONS

8.4 The Artificial Kidney: A Hemodialysis Machine

Individuals who once would have died of kidney failure can now be helped through the use of artificial-kidney machines, which clean the blood of toxic waste products. In these machines, the blood is pumped through tubing made of dialyzing membrane. The tubing passes through a water bath that collects the impurities from the blood. Blood proteins and other important large molecules remain in the blood. This procedure to cleanse the blood is called *hemodialysis.*

In hemodialysis, a catheter is attached to a major artery of one arm, and the patient's blood is passed through a collection of tiny tubes with a carefully selected pore size. These tubes are immersed in a bath (dialyzing solution) that is isotonic in the normal components of blood. The isotonic solution consists of 0.6%(m/v) NaCl, 0.04%(m/v) KCl, 0.2%(m/v) $NaHCO_3$, and 1.5%(m/v) glucose. This solution does not contain urea or other wastes, which diffuse from the blood through the membrane and into the dialyzing solution.

The accompanying figure illustrates a typical hollow-fiber dialysis device.

Typically, an artificial-kidney patient must receive dialysis treatment two or three times a week, for 4 hours per treatment, in order to maintain proper health. A kidney transplant is preferable to many years on hemodialysis. However, kidney transplants are possible only when the donor kidneys are close tissue matches to the recipient.

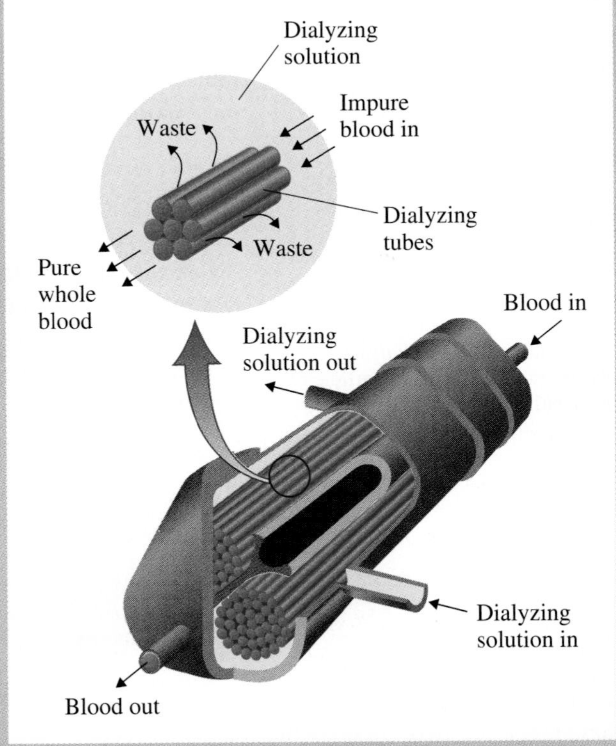

Concepts to Remember

Solution components. The component of a solution that is present in the greatest amount is the *solvent.* A *solute* is a solution component that is present in a small amount relative to the solvent.

Solution characteristics. A solution is a homogeneous (uniform) mixture. Its composition and properties are dependent on the ratio of solute(s) to solvent. Dissolved solutes are present as individual particles (molecules, atoms, or ions).

Solubility. The solubility of a solute is the maximum amount of solute that will dissolve in a given amount of solvent. The extent to which a solute dissolves in a solvent depends on the structure of solute and solvent, the temperature, and the pressure. Molecular polarity is a particularly important factor in determining solubility. A saturated solution contains the maximum amount of solute that can be dissolved under the conditions at which the solution exists.

Solution concentration. Solution concentration is the amount of solute present in a specified amount of solution. Percent solute and molarity are commonly encountered concentration units. Percent concentration units include percent by mass, percent by volume, and mass–volume percent. Molarity gives the moles of solute per liter of solution.

Colloidal dispersion. A colloidal dispersion is a dispersion (suspension) of small particles of one substance in another substance. Colloidal dispersions differ from true solutions in that the dispersed particles are large enough to scatter light even though they cannot be seen with the naked eye. Many different biochemical colloidal dispersions occur within the human body.

Colligative properties of solutions. Properties of a solution that depend on the number of solute particles in solution, not on their identity, are called colligative properties. Vapor pressure lowering, boiling-point elevation, freezing-point depression, and osmotic pressure are all colligative properties.

Osmosis and osmotic pressure. Osmosis involves the passage of a solvent from a dilute solution (or pure solvent) through a semipermeable membrane into a more concentrated solution. Osmotic pressure is the amount of pressure needed to prevent the net flow of solvent across the membrane in the direction of the more concentrated solution.

Dialysis. Dialysis is the process in which a semipermeable membrane permits the passage of solvent, dissolved ions, and small molecules but blocks the passage of large molecules. Many plant and animal membranes function as dialyzing membranes.

Key Reactions and Equations

1. Percent by mass (Section 8.5)

$$\%(\text{m/m}) = \frac{\text{mass of solute}}{\text{mass of solution}} \times 100$$

2. Percent by volume (Section 8.5)

$$\%(\text{v/v}) = \frac{\text{volume of solute}}{\text{volume of solution}} \times 100$$

3. Mass–volume percent (Section 8.5)

$$\%(\text{m/v}) = \frac{\text{mass of solute (g)}}{\text{volume of solution (mL)}} \times 100$$

4. Molarity (Section 8.5)

$$\text{M} = \frac{\text{moles of solute}}{\text{liters of solution}}$$

5. Dilution of stock solution to make less-concentrated solution (Section 8.5)

$$C_s \times V_s = C_d \times V_d$$

6. Osmolarity (Section 8.8)

$$\text{Osmolarity} = \text{molarity} \times i$$

Key Terms

Aqueous solution (8.2)
Colligative properties (8.7)
Colloidal dispersions (8.6)
Concentrated solution (8.2)
Concentration (8.5)
Dialysis (8.9)
Dilute solution (8.2)
Dilution (8.5)
Hypertonic solution (8.8)
Hypotonic solution (8.8)
Isotonic solution (8.8)
Mass–volume percent (8.2)
Molarity (8.5)
Osmolarity (8.8)
Osmosis (8.8)
Osmotic pressure (8.8)
Percent by mass (8.5)
Percent by volume (8.5)
Saturated solution (8.2)
Semipermeable membrane (8.8)
Solubility (8.2)
Solute (8.1)
Solution (8.1)
Solvent (8.1)
Unsaturated solution (8.2)

Exercises and Problems

The members of each pair of problems in this section test similar material.

Solution Characteristics and Terminology (Sections 8.1, 8.2)

8.1 Indicate which substance is the *solvent* in each of the following solutions.

a. A solution containing 20.0 g of NaBr and 200.0 mL of water

b. A solution containing 20.0 mL of ethyl alcohol and 15.0 mL of water

8.2 Indicate which substance is the *solvent* in each of the following solutions.

a. A solution containing 1.00 g of $AgNO_3$ and 50.0 mL of water

b. A solution containing 45.0 mL of acetone and 15.0 mL of water

8.3 Give an example of each of the following types of solutions.

a. Solute is a gas and solvent is a gas.

b. Solute is a solid and solvent is a liquid.

c. Solute is a liquid and solvent is a solid.

d. Solute is a gas and solvent is a liquid.

8.4 Give an example of each of the following types of solutions.

a. Solute is a liquid and solvent is a liquid.

b. Solute is a liquid and solvent is a gas.

c. Solute is a solid and solvent is a solid.

d. Solute is a solid and solvent is a gas.

8.5 Use Table 8.2 to determine whether each of the following solutions is saturated or unsaturated.

a. 1.94 g of $PbBr_2$ in 100 g of H_2O at 50°C

b. 34.0 g of NaCl in 100 g of H_2O at 0°C

c. 75.4 g of $CuSO_4$ in 200 g of H_2O at 100°C

d. 0.540 g of Ag_2SO_4 in 50 g of H_2O at 50°C

8.6 Use Table 8.2 to determine whether each of the following solutions is saturated or unsaturated.

a. 175 g of CsCl in 100 g of H_2O at 100°C

b. 455 g of $AgNO_3$ in 100 g of H_2O at 50°C

c. 2.16 g of Ag_2SO_4 in 200 g of H_2O at 50°C

d. 0.97 g of $PbBr_2$ in 50 g of H_2O at 50°C

8.7 Use Table 8.2 to determine whether each of the following solutions is dilute or concentrated.

a. 0.20 g of $CuSO_4$ in 100 g of H_2O at 100°C

b. 1.50 g of $PbBr_2$ in 100 g of H_2O at 50°C

c. 61 g of $AgNO_3$ in 100 g of H_2O at 50°C

d. 0.50 g of Ag_2SO_4 in 100 g of H_2O at 0°C

8.8 Use Table 8.2 to determine whether each of the following solutions is dilute or concentrated.

a. 255 g of $AgNO_3$ in 100 g of H_2O at 100°C

b. 35.0 g of NaCl in 100 g of H_2O at 0°C

c. 1.87 g of $PbBr_2$ in 100 g of H_2O at 50°C

d. 1.87 g of $CuSO_4$ in 100 g of H_2O at 50°C

Solubility Rules (Section 8.4)

8.9 Predict whether the following solutes are very soluble or slightly soluble in water.

a. O_2 (a nonpolar gas) b. CH_3OH (a polar liquid)
c. CBr_4 (a nonpolar liquid) d. AgCl (an ionic solid)

8.10 Predict whether the following solutes are very soluble or slightly soluble in water.

a. NH_3 (a polar gas) b. N_2 (a nonpolar gas)
c. C_6H_6 (a nonpolar liquid) d. Na_3PO_4 (an ionic solid)

8.11 In each of the following sets of ionic compounds, identify the members of the set that are soluble in water.

a. NaCl, Na_2SO_4, $NaNO_3$, Na_2CO_3

b. $AgNO_3$, KNO_3, $Ca(NO_3)_2$, $Cu(NO_3)_2$

c. CaS, $Ca(OH)_2$, $CaCl_2$, $CaSO_4$

d. $NiSO_4$, $Ni_3(PO_4)_2$, $Ni(C_2H_3O_2)_2$, $NiCO_3$

8.12 In each of the following sets of ionic compounds, identify the members of the set that are soluble in water.

a. K_2S, KOH, KI, K_3PO_4

b. NaCl, AgCl, $BeCl_2$, $CuCl_2$

c. $Ba(OH)_2$, $BaSO_4$, BaS, $Ba(NO_3)_2$

d. $CoBr_2$, $Co(C_2H_3O_2)_2$, $Co(OH)_2$, $CoSO_4$

Percent Concentration Units (Section 8.5)

8.13 Calculate the mass percent of solute in the following solutions.

a. 6.50 g of NaCl dissolved in 85.0 g of H_2O

b. 2.31 g of LiBr dissolved in 35.0 g of H_2O

c. 12.5 g of KNO_3 dissolved in 125 g of H_2O

d. 0.003 g of NaOH dissolved in 1.2 g of H_2O

8.14 Calculate the mass percent of solute in the following solutions.

a. 2.13 g of $AgNO_3$ dissolved in 30.0 g of H_2O

b. 135 g of CsCl dissolved in 455 g of H_2O

c. 10.3 g of K_2SO_4 dissolved in 93.7 g of H_2O

d. 10.3 g of KBr dissolved in 125 g of H_2O

8.15 How many grams of glucose must be added to 275 g of water in order to prepare each of the following percent-by-mass concentrations of aqueous glucose solution?

a. 1.30% b. 5.00% c. 20.0% d. 31.0%

8.16 How many grams of lactose must be added to 655 g of water in order to prepare each of the following percent-by-mass concentrations of aqueous lactose solution?

a. 0.50% b. 2.00% c. 10.0% d. 25.0%

8.17 Calculate the mass, in grams, of K_2SO_4 needed to prepare 32.00 g of 2.000%(m/m) K_2SO_4 solution.

8.18 Calculate the mass, in grams, of KCl needed to prepare 200.0 g of 5.000%(m/m) KCl solution.

8.19 How many grams of water must be added to 20.0 g of NaOH in order to prepare a 6.75%(m/m) solution?

8.20 How many grams of water must be added to 10.0 g of $Ca(NO_3)_2$ in order to prepare a 12.0%(m/m) solution?

8.21 Calculate the volume percent of solute in each of the following solutions.

a. 20.0 mL of methyl alcohol in enough water to give 475 mL of solution

b. 4.00 mL of bromine in enough carbon tetrachloride to give 87.0 mL of solution

8.22 Calculate the volume percent of solute in each of the following solutions.

a. 60.0 mL of water in enough ethylene glycol to give 970.0 mL of solution

b. 455 mL of ethyl alcohol in enough water to give 1375 mL of solution

8.23 What is the percent by volume of isopropyl alcohol in an aqueous solution made by diluting 22 mL of pure isopropyl alcohol with water to give a volume of 125 mL of solution?

8.24 What is the percent by volume of acetone in an aqueous solution made by diluting 75 mL of pure acetone with water to give a volume of 785 mL of solution?

8.25 Calculate the mass–volume percent of magnesium chloride in each of the following solutions.

a. 5.0 g of $MgCl_2$ in enough water to give 250 mL of solution

b. 85 g of $MgCl_2$ in enough water to give 580 mL of solution

8.26 Calculate the mass–volume percent of sodium nitrate in each of the following solutions.

a. 1.00 g of $NaNO_3$ in enough water to give 75.0 mL of solution

b. 100.0 g of $NaNO_3$ in enough water to give 1250 mL of solution

8.27 How many grams of Na_2CO_3 are needed to prepare 25.0 mL of a 2.00%(m/v) Na_2CO_3 solution?

8.28 How many grams of $Na_2S_2O_3$ are needed to prepare 50.0 mL of a 5.00%(m/v) $Na_2S_2O_3$ solution?

8.29 How many grams of NaCl are present in 50.0 mL of a 7.50%(m/v) NaCl solution?

8.30 How many grams of glucose are present in 250.0 mL of a 10.0%(m/v) glucose solution?

Molarity (Section 8.5)

8.31 Calculate the molarity of the following solutions.

a. 3.0 moles of potassium nitrate (KNO_3) in 0.50 L of solution

b. 12.5 g of sucrose ($C_{12}H_{22}O_{11}$) in 80.0 mL of solution

c. 25.0 g of sodium chloride (NaCl) in 1250 mL of solution

d. 0.00125 mole of baking soda ($NaHCO_3$) in 2.50 mL of solution

8.32 Calculate the molarity of the following solutions.

a. 2.0 moles of ammonium chloride (NH_4Cl) in 2.50 L of solution

b. 14.0 g of silver nitrate ($AgNO_3$) in 1.00 L of solution

c. 0.025 mole of potassium chloride (KCl) in 50.0 mL of solution

d. 25.0 g of glucose ($C_6H_{12}O_6$) in 1.25 L of solution

8.33 Calculate the number of grams of solute in each of the following solutions.

a. 2.50 L of a 3.00 M HCl solution

b. 10.0 mL of a 0.500 M KCl solution

c. 875 mL of a 1.83 M $NaNO_3$ solution

d. 75 mL of a 12.0 M H_2SO_4 solution

8.34 Calculate the number of grams of solute in each of the following solutions.

a. 3.00 L of a 2.50 M HCl solution

b. 50.0 mL of a 12.0 M HNO_3 solution

c. 50.0 mL of a 12.0 M $AgNO_3$ solution

d. 1.20 L of a 0.032 M Na_2SO_4 solution

8.35 Calculate the volume, in milliliters, of solution required to supply each of the following.

a. 1.00 g of sodium chloride (NaCl) from a 0.200 M sodium chloride solution

b. 2.00 g of glucose ($C_6H_{12}O_6$) from a 4.20 M glucose solution

c. 3.67 moles of silver nitrate ($AgNO_3$) from a 0.400 M silver nitrate solution

d. 0.0021 mole of sucrose ($C_{12}H_{22}O_{11}$) from an 8.7 M sucrose solution

8.36 Calculate the volume, in milliliters, of solution required to supply each of the following.

a. 4.30 g of lithium chloride (LiCl) from a 0.089 M lithium chloride solution

b. 429 g of lithium nitrate ($LiNO_3$) from an 11.2 M lithium nitrate solution

c. 2.25 moles of potassium sulfate (K_2SO_4) from a 0.300 M potassium sulfate solution

d. 0.103 mole of potassium hydroxide (KOH) from an 8.00 M potassium hydroxide solution

Dilution (Section 8.5)

8.37 What is the molarity of the solution prepared by diluting 25.0 mL of 0.220 M NaCl to each of the following final volumes?

a. 30.0 mL b. 75.0 mL c. 457 mL d. 2.00 L

8.38 What is the molarity of the solution prepared by diluting 35.0 mL of 1.25 M $AgNO_3$ to each of the following final volumes?

a. 50.0 mL b. 95.0 mL c. 975 mL d. 3.60 L

8.39 For each of the following solutions, how many milliliters of water should be added to yield a solution that has a concentration of 0.100 M?

a. 50.0 mL of 3.00 M NaCl b. 2.00 mL of 1.00 M NaCl

c. 1.45 L of 6.00 M NaCl d. 75.0 mL of 0.110 M NaCl

8.40 For each of the following solutions, how many milliliters of water should be added to yield a solution that has a concentration of 0.125 M?

a. 25.0 mL of 1.00 M $AgNO_3$
b. 5.00 mL of 10.0 M $AgNO_3$
c. 2.50 L of 2.50 M $AgNO_3$
d. 75.0 mL of 0.130 M $AgNO_3$

8.41 Determine the final concentration of each of the following solutions after 20.0 mL of water has been added.

a. 30.0 mL of 5.0 M NaCl solution
b. 30.0 mL of 5.0 M $AgNO_3$ solution
c. 30.0 mL of 7.5 M NaCl solution
d. 60.0 mL of 2.0 M NaCl solution

8.42 Determine the final concentration of each of the following solutions after 30.0 mL of water has been added.

a. 20.0 mL of 5.0 M NaCl solution
b. 20.0 mL of 5.0 M $AgNO_3$ solution
c. 20.0 mL of 0.50 M NaCl solution
d. 60.0 mL of 3.0 M NaCl solution

Colligative Properties of Solutions (Section 8.7)

8.43 Why is the vapor pressure of a solution that contains a nonvolatile solute always less than that of pure solvent?

8.44 How are the boiling point and freezing point of water affected by the addition of solute?

8.45 Why is the freezing point of a 0.10 M KCl solution lower than that of a 0.10 M glucose solution?

8.46 Why are the freezing point of a 0.30 M KCl solution and that of a 0.30 M NaCl solution the same?

Osmosis and Osmotic Pressure (Section 8.8)

8.47 Indicate whether the osmotic pressure of a 0.1 M NaCl solution will be less than, the same as, or greater than that of each of the following solutions.

a. 0.1 M NaBr
b. 0.050 M $MgCl_2$
c. 0.1 M $MgCl_2$
d. 0.1 M glucose

8.48 Indicate whether the osmotic pressure of a 0.1 M $NaNO_3$ solution will be less than, the same as, or greater than that of each of the following solutions.

a. 0.1 M NaCl
b. 0.1 M KNO_3
c. 0.1 M Na_2SO_4
d. 0.1 M glucose

8.49 What is the ratio of the osmotic pressures of 0.30 M NaCl and 0.10 M $CaCl_2$?

8.50 What is the ratio of the osmotic pressures of 0.20 M NaCl and 0.30 M $CaCl_2$?

8.51 Would red blood cells swell, remain the same size, or shrink when placed in each of the following solutions?

a. 0.9%(m/v) glucose solution
b. 0.9%(m/v) NaCl solution
c. 2.3%(m/v) glucose solution
d. 5.0%(m/v) NaCl solution

8.52 Would red blood cells swell, remain the same size, or shrink when placed in each of the following solutions?

a. distilled water
b. 0.5%(m/v) NaCl solution
c. 3.3%(m/v) glucose solution
d. 5.0%(m/v) glucose solution

8.53 Classify each of the solutions in Problem 8.51 as isotonic, hypertonic, or hypotonic.

8.54 Classify each of the solutions in Problem 8.52 as isotonic, hypertonic, or hypotonic.

Dialysis (Section 8.9)

8.55 What happens in each of the following situations?

a. A dialyzing bag containing a 1 M solution of potassium chloride is immersed in pure water.
b. A dialyzing bag containing colloidal-sized protein, 1 M potassium chloride, and 1 M glucose is immersed in pure water.

8.56 What happens in each of the following situations?

a. A dialyzing bag containing a 1 M solution of potassium chloride is immersed in a 1 M sodium chloride solution.
b. A dialyzing bag containing colloidal-sized protein is immersed in a 1 M glucose solution.

Additional Problems

8.57 With the help of Table 8.3, determine in which of the following pairs of compounds both members of the pair have like solubility in water (both soluble or both insoluble).

a. $Be(C_2H_3O_2)_2$ and $AgNO_3$
b. $ZnCl_2$ and $Mg(OH)_2$
c. BaS and $NiCO_3$
d. AgCl and $Al(OH)_3$

8.58 How many grams of solute are dissolved in the following amounts of solution?

a. 134 g of 3.00%(m/m) KNO_3 solution
b. 75.02 g of 9.735%(m/m) NaOH solution
c. 1576 g of 0.800%(m/m) HI solution
d. 1.23 g of 12.0%(m/m) NH_4Cl solution

8.59 What volume of water, in quarts, is contained in 3.50 qt of a 2.00%(v/v) solution of water in acetone?

8.60 How many liters of 0.10 M solution can be prepared from 60.0 g of each of the following solutes?

a. $NaNO_3$
b. HNO_3
c. KOH
d. LiCl

8.61 What is the molarity of the solution prepared by concentrating, by evaporation of solvent, 2212 mL of 0.400 M potassium sulfate (K_2SO_4) solution to each of the following final volumes?

a. 1875 mL
b. 1.25 L
c. 853 mL
d. 553 mL

8.62 After all the water is evaporated from 10.0 mL of a CsCl solution, 3.75 of CsCl remains. Express the original concentration of the CsCl solution in each of the following units.

a. mass–volume percent
b. molarity

8.63 Find the molarity of a solution obtained when 352 mL of 4.00 M sodium bromide (NaBr) solution is mixed with

a. 225 mL of 4.00 M NaBr solution
b. 225 mL of 2.00 M NaBr solution

8.64 Which of the following aqueous solutions would give rise to a greater osmotic pressure?

a. 8.00 g of NaCl in 375 mL of solution or 4.00 g of NaBr in 155 mL of solution

b. 6.00 g of NaCl in 375 mL of solution or 6.00 g of $MgCl_2$ in 225 mL of solution

Grid Problems

8.65

1. NO_3^-	2. Cl^-	3. CO_3^{2-}
4. OH^-	5. SO_4^{2-}	6. PO_4^{3-}

Select from the grid *all* correct responses for each of the following situations.

a. Combines with Ba^{2+} ion to produce a soluble ionic compound

b. Combines with Ca^{2+} ion to produce an insoluble ionic compound

c. Combines with all positive ions to produce soluble ionic compounds

d. Pairs of ions that have identical solubility guidelines

8.66

1. 1 M NaCl	2. 2 M $CaCl_2$	3. 3 M NH_4Cl
4. 1 M $NaNO_3$	5. 2 M K_3PO_4	6. 3 M glucose

Select from the grid *all* correct responses for each of the following situations.

a. Solutions that have a freezing point lower than that of pure water

b. Solutions that have a vapor pressure lower than that of pure water

c. Solutions that have an osmotic pressure greater than that of 1 M Na_2SO_4

d. Pairs of solutions that have the same osmolarity

8.67

1. 0.9%(m/v) NaCl	2. 3.5%(m/v) NaCl	3. 5.0%(m/v) NaCl
4. 0.9%(m/v) glucose	5. 3.5%(m/v) glucose	6. 5.0%(m/v) glucose

Select from the grid *all* correct responses for each of the following situations.

a. Solution in which crenation of red blood cells occurs

b. Solution in which hemolysis of red blood cells occurs

c. Hypotonic solution for red blood cells

d. Isotonic solution for red blood cells

8.68

1. molarity	2. mass–volume percent	3. mass–mass percent
4. percent by volume	5. percent by mass	6. volume–volume percent

Select from the grid *all* correct responses for each of the following situations.

a. The numerator of the defining equation is moles of solute.

b. The number 100 is present in the defining equation.

c. The numerator of the defining equation has volume units.

d. The denominator of the defining equation has mass units.

PART I

Chemical Reactions

CHAPTER OUTLINE

9.1 Types of Chemical Reactions 208
9.2 Redox and Nonredox Reactions 211

Chemistry at a Glance: **Types of Chemical Reactions** 212

9.3 Terminology Associated with Redox Processes 214
9.4 Collision Theory and Chemical Reactions 216
9.5 Exothermic and Endothermic Reactions 218
9.6 Factors That Influence Reaction Rates 218

Chemistry at a Glance: **Factors That Influence Reaction Rates** 221

9.7 Chemical Equilibrium 221
9.8 Equilibrium Constants 224
9.9 Altering Equilibrium Conditions: Le Châtelier's Principle 227

Chemical Connections

9.1 Smog Formation: A Set of Simple Reactions 210
9.2 Stratospheric Ozone: An Equilibrium Situation 223
9.3 Oxygen, Hemoglobin, Equilibrium, and Le Châtelier's Principle 230

Acid-rain-caused corrosion of churches and statues involves a double-replacement reaction involving sulfuric acid (H_2SO_4). The reaction is $CaCO_3 + H_2SO_4 \longrightarrow CaSO_4 + H_2CO_3$

In the previous two chapters, we considered the properties of matter in various pure and mixed states. Nearly all of the subject matter dealt with interactions and changes of a *physical* nature. We now concern ourselves with the *chemical* changes that occur when various types of matter interact.

We first consider important categories of chemical changes and then discuss important fundamentals common to all chemical changes. Of particular concern to us will be how fast chemical changes occur (chemical reaction rates) and how far chemical changes go (chemical equilibrium).

9.1 Types of Chemical Reactions

A **chemical reaction** *is a process in which at least one new substance is produced as a result of chemical change.* An almost inconceivable number of chemical reactions is possible. The majority of chemical reactions (but not all) fall into four major categories:

combination reactions, *decomposition* reactions, *single-replacement* reactions, and *double-replacement* reactions.

When a hot nail is stuck into a pile of zinc and sulfur, a fiery combination reaction occurs and zinc sulfide forms.

$$Zn + S \longrightarrow ZnS$$

• Combination Reactions

A **combination reaction** *is a reaction in which a single product is produced from two (or more) reactants.* The general equation for a combination reaction involving two reactants is

$$X + Y \longrightarrow XY$$

In such a combination reaction, two substances join together to form a more complicated product. The reactants X and Y can be elements or compounds or an element and a compound. The product of the reaction (XY) is always a compound. Some representative combination reactions that have elements as the reactants are

$$Ca + S \longrightarrow CaS$$
$$N_2 + 3H_2 \longrightarrow 2NH_3$$
$$2Na + O_2 \longrightarrow Na_2O_2$$

Some examples of combination reactions in which compounds are involved as reactants are

$$SO_3 + H_2O \longrightarrow H_2SO_4$$
$$2NO + O_2 \longrightarrow 2NO_2$$
$$2NO_2 + H_2O_2 \longrightarrow 2HNO_3$$

• In organic chemistry (Chapters 12–17), combination reactions are called *addition reactions.* One reactant, usually a small molecule, is considered to be added to a larger reactant molecule to produce a single product.

• Decomposition Reactions

A **decomposition reaction** *is a reaction in which a single reactant is converted into two (or more) simpler substances (elements or compounds).* Thus a decomposition reaction is the opposite of a combination reaction. The general equation for a decomposition reaction in which there are two products is

$$XY \longrightarrow X + Y$$

Although the products may be elements or compounds, the reactant is *always* a compound.

At sufficiently high temperatures, all compounds can be broken down (decomposed) into their constituent elements. Examples of such reactions include

$$2CuO \longrightarrow 2Cu + O_2$$
$$2H_2O \longrightarrow 2H_2 + O_2$$

At lower temperatures, compound decomposition often produces other compounds as products.

$$CaCO_3 \longrightarrow CaO + CO_2$$
$$2KClO_3 \longrightarrow 2KCl + 3O_2$$
$$4HNO_3 \longrightarrow 4NO_2 + 2H_2O + O_2$$

• In organic chemistry, decomposition reactions are often called *elimination reactions.* In many reactions, including some metabolic reactions that occur in the human body, either H_2O or CO_2 is eliminated from a molecule (a decomposition).

Decomposition reactions are easy to recognize in that they are the only type of reaction in which there is only one reactant.

• Single-Replacement Reactions

A **single-replacement reaction** *is a reaction in which an atom or molecule replaces an atom or group of atoms from a compound.* There are always two reactants and two products in a single-replacement reaction. The general equation for a single-replacement reaction is

$$X + YZ \longrightarrow Y + XZ$$

Chemical CONNECTIONS

9.1 Smog Formation: A Set of Simple Reactions

A simple set of reactions—three combination reactions and one decomposition reaction—generates the air pollution condition we call smog. The atmosphere is the site for all of these reactions except the first one, which occurs at locations where coal and petroleum products are burned (automobile engines, power plants, and so on).

At the high temperatures associated with the burning of fossil fuels, the two major components of air (N_2 and O_2) become slightly reactive toward each other. Thus, when fuel is burned, small amounts of N_2 and O_2 react, in a combination reaction, to produce nitric oxide (NO), a colorless, odorless gas.

$$N_2(g) + O_2(g) \longrightarrow 2NO(g)$$

The nitric oxide so produced reacts further, with oxygen in another combination reaction, to produce nitrogen dioxide (NO_2), a reddish brown gas with a strong odor.

$$2NO(g) + O_2(g) \longrightarrow 2NO_2(g)$$

This reaction does not occur at the fuel combustion site because NO_2 is unstable at the high temperatures associated with such sites. It readily occurs, however, once the NO enters the atmosphere, where temperatures are much lower.

Smog production occurs only during daylight hours. This is because the next step in the process, a decomposition reaction, requires sunlight as an activator.

$$NO_2(g) + \text{sunlight} \longrightarrow NO(g) + O(g)$$

The products are regenerated NO and atomic oxygen (O). The latter is a very reactive form of oxygen, much more reactive than diatomic oxygen (O_2).

The atomic oxygen (O) reacts with "normal" atmospheric oxygen (O_2), in another combination reaction, to give ozone (O_3), a third form of oxygen.

$$O(g) + O_2(g) \longrightarrow O_3(g)$$

Ozone is the most abundant ingredient in smog. It is an irritant to the mucous membranes of the nose and throat and can cause eye irritation. It also has an effect on lung function and is particularly dangerous for individuals with chronic lung disease.

Other "active ingredients" in smog are formed when ozone reacts with other air pollutants present in air. Particularly important are reactions between ozone and organic compounds such as those found in gasoline.

Atmospheric ozone levels go up and down with weather conditions. Concentration levels are higher during the summer months, when there is more sunlight, than during winter months. High ozone levels can occur, however, during wintertime as a result of air stagnation (temperature inversions).

A common type of single-replacement reaction is one in which an element and a compound are reactants and an element and a compound are products. Examples of this type of single-replacement reaction include

$$Fe + CuSO_4 \longrightarrow Cu + FeSO_4$$
$$Mg + Ni(NO_3)_2 \longrightarrow Ni + Mg(NO_3)_2$$
$$Cl_2 + NiI_2 \longrightarrow I_2 + NiCl_2$$
$$F_2 + 2NaCl \longrightarrow Cl_2 + 2NaF$$

The first two equations illustrate one metal replacing another metal from its compound. The latter two equations illustrate one nonmetal replacing another nonmetal from its compound. A more complicated example of a single-replacement reaction, in which all reactants and products are compounds, is

$$4PH_3 + Ni(CO)_4 \longrightarrow 4CO + Ni(PH_3)_4$$

• Double-Replacement Reactions

A **double-replacement reaction** *is a reaction in which two substances exchange parts with one another and form two different substances.* The general equation for a double-replacement reaction is

$$AX + BY \longrightarrow AY + BX$$

Such reactions can be thought of as involving "partner switching." The AX and BY partnerships are dissolved, and new AY and BX partnerships are formed in their place.

• In organic chemistry, replacement reactions (both single and double) are often called *substitution* reactions. Substitution reactions of the single-replacement type are seldom encountered. However, double-replacement reactions are common in organic chemistry.

A double-replacement reaction involving solutions of potassium iodide and lead(II) nitrate (both colorless solutions) produces yellow, insoluble lead(II) iodide as one of the products.

$$2KI(aq) + Pb(NO_3)_2(aq) \longrightarrow 2KNO_3(aq) + PbI_2(s)$$

When the reactants in a double-replacement reaction are ionic compounds in solution, the parts exchanged are the positive and negative ions of the compounds present.

$$AgNO_3(aq) + NaCl(aq) \longrightarrow AgCl(s) + NaNO_3(aq)$$
$$2KI(aq) + Pb(NO_3)_2(aq) \longrightarrow 2KNO_3(aq) + PbI_2(s)$$

In most reactions of this type, one of the product compounds is in a different physical state (solid or gas) from that of the reactants. Insoluble solids formed from such a reaction are called *precipitates;* AgCl and PbI_2 are precipitates in the foregoing reactions.

Chemistry at a Glance on the following page summarizes much of what we have just learned about reactions.

9.2 Redox and Nonredox Reactions

Chemical reactions can also be classified, in terms of whether transfer of electrons occurs, as either oxidation–reduction (redox) or nonoxidation–reduction (nonredox) reactions. An **oxidation–reduction (redox) reaction** *is a reaction in which there is a transfer of electrons from one reactant to another reactant.* A **nonoxidation–reduction (nonredox) reaction** *is a reaction in which there is no transfer of electrons from one reactant to another reactant.*

A "bookkeeping system" known as oxidation numbers is used to identify whether electron transfer occurs in a chemical reaction. The **oxidation number** *of an atom is a number that represents the charge that an atom appears to have when the electrons in each bond it is participating in are assigned to the more electronegative of the two atoms involved in the bond.*

There are several rules for determining oxidation numbers.

• *Oxidation numbers* are also sometimes called *oxidation states.*

1. *The oxidation number of an element in its elemental state is zero.* For example, the oxidation number of copper in Cu is zero, and the oxidation number of chlorine in Cl_2 is zero.
2. *The oxidation number of a monatomic ion is equal to the charge on the ion.* For example, the Na^+ ion has an oxidation number of $+1$, and the S^{2-} ion has an oxidation number of -2.
3. *The oxidation numbers of Groups IA and IIA metals are always $+1$, and $+2$, respectively.*
4. *The oxidation number of hydrogen is $+1$ in most hydrogen-containing compounds.*
5. *The oxidation number of oxygen is -2 in most oxygen-containing compounds.*
6. *In binary molecular compounds, the more electronegative element is assigned a negative oxidation number equal to its charge in binary ionic compounds.* For example, in CCl_4 the element Cl is the more electronegative, and its oxidation number is -1 (the same as in the simple Cl^- ion).

Types of Chemical Reactions

CHEMISTRY AT A GLANCE

COMBINATION REACTION

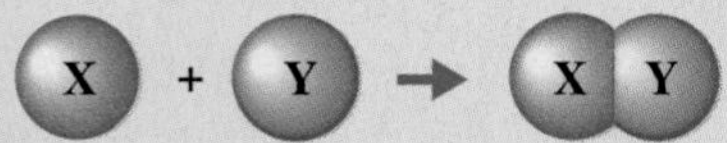

$2Al + 3I_2 \longrightarrow 2AlI_3$

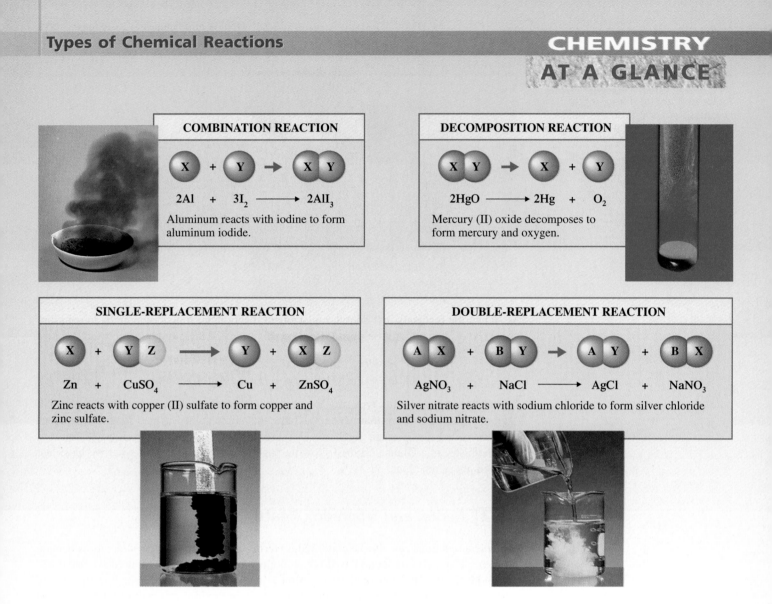

Aluminum reacts with iodine to form aluminum iodide.

DECOMPOSITION REACTION

$2HgO \longrightarrow 2Hg + O_2$

Mercury (II) oxide decomposes to form mercury and oxygen.

SINGLE-REPLACEMENT REACTION

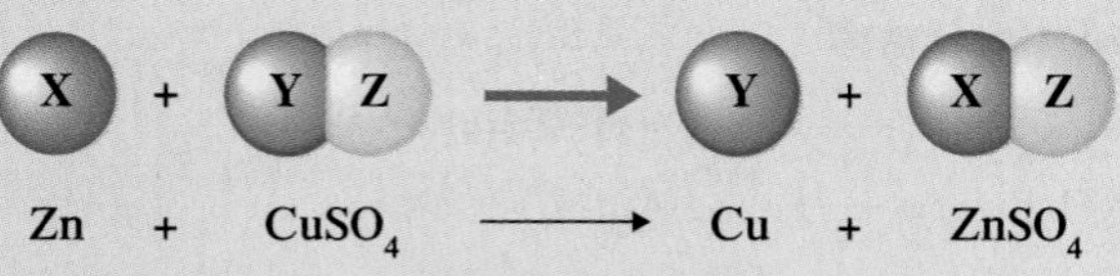

$Zn + CuSO_4 \longrightarrow Cu + ZnSO_4$

Zinc reacts with copper (II) sulfate to form copper and zinc sulfate.

DOUBLE-REPLACEMENT REACTION

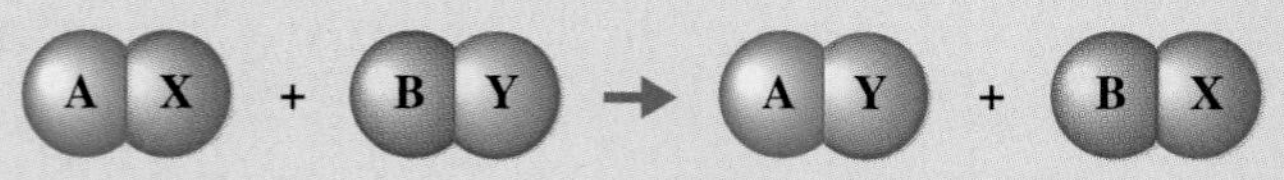

$AgNO_3 + NaCl \longrightarrow AgCl + NaNO_3$

Silver nitrate reacts with sodium chloride to form silver chloride and sodium nitrate.

7. *For a compound, the sum of the individual oxidation numbers is equal to zero; for a polyatomic ion, the sum is equal to the charge on the ion.*

Example 9.1 illustrates the use of these rules.

Example 9.1

Assigning Oxidation Numbers to Elements in a Compound or Polyatomic Ion

Assign an oxidation number to each element in the following compounds or polyatomic ions.

a. P_2O_5 **b.** $KMnO_4$ **c.** NO_3^-

Solution

a. The sum of the oxidation numbers of all the atoms present must add to zero (rule 7).

$$2(\text{oxid. no. P}) + 5(\text{oxid. no. O}) = 0$$

The oxidation number of oxygen is −2 (rule 5 or rule 6). Substituting this value into the previous equation enables us to calculate the oxidation number of phosphorus.

$$2(\text{oxid. no. P}) + 5(-2) = 0$$
$$2(\text{oxid. no. P}) = +10$$
$$(\text{oxid no. P}) = +5$$

Thus the oxidation numbers for the elements involved in this compound are

$$P = +5 \quad \text{and} \quad O = -2$$

Note that the oxidation number of phosphorus is not +10; that is the calculated charge associated with two phosphorus atoms. Oxidation number is always specified on a *per-atom* basis.

b. The sum of the oxidation numbers of all the atoms present must add to zero (rule 7).

$$(\text{oxid. no. K}) + (\text{oxid. no. Mn}) + 4(\text{oxid. no. O}) = 0$$

The oxidation number of potassium, a Group IA element, is +1 (rule 3), and the oxidation number of oxygen is −2 (rule 5). Substituting these two values into the rule 7 equation enables us to calculate the oxidation number of manganese.

$$(+1) + (\text{oxid. no. Mn}) + 4(-2) = 0$$

$$(\text{oxid. no. Mn}) = 8 - 1 = +7$$

Thus the oxidation numbers for the elements involved in this compound are

$$K = +1 \quad Mn = +7 \quad \text{and} \quad O = -2$$

Note that all the oxidation numbers add to zero when it is taken into account that there are four oxygen atoms.

$$(+1) + (+7) + 4(-2) = 0$$

c. The species NO_3^- is a polyatomic ion rather than a neutral compound. Thus the second part of rule 7 applies: The oxidation numbers must add to −1, the charge on the ion.

$$(\text{oxid. no. N}) + 3(\text{oxid. no. O}) = -1$$

The oxidation number of oxygen is −2 (rule 5). Substituting this value into the sum equation gives

$$(\text{oxid. no. N}) + 3(-2) = -1$$

$$(\text{oxid. no. N}) = -1 + 6 = +5$$

Thus the oxidation numbers for the elements involved in the polyatomic ion are

$$N = +5 \quad \text{and} \quad O = -2$$

Practice Exercise 9.1

Assign oxidation numbers to each element in the following compounds or polyatomic ions.

a. N_2O_4 **b.** $K_2Cr_2O_7$ **c.** NH_4^+

• *Answers:* **a.** N = +4, O = −2; **b.** K = +1, Cr = +6, O = −2; **c.** N = −3, H = +1

The burning of calcium metal in chlorine is a redox reaction. The burning calcium emits a red-orange flame.

Many elements display a range of oxidation numbers in their various compounds. For example, nitrogen exhibits oxidation numbers ranging from −3 to +5. Selected examples are

NH_3	N_2O	NO	N_2O_3	NO_2	HNO_3
−3	+1	+2	+3	+4	+5

As shown in this listing of nitrogen-containing compounds, the oxidation number of an atom is written *underneath* the atom in the formula. This convention is used to avoid confusion with the charge on an ion.

To determine whether a reaction is a redox reaction or a nonredox reaction, we look for changes in the oxidation number of elements involved in the reaction. Changes in oxidation number are a requirement for a redox reaction. The reaction

$$\underset{0}{Ca} + \underset{0}{Cl_2} \longrightarrow \underset{+2\ -1}{CaCl_2}$$

is a redox reaction; the oxidation number of Ca changes from zero to +2, and the oxidation number of Cl changes from zero to −1. The reaction

$$\underset{+2+4-2}{CaCO_3} \longrightarrow \underset{+2-2}{CaO} + \underset{+4-2}{CO_2}$$

is a nonredox reaction because there are no changes in oxidation number.

Example 9.2

Using Oxidation Numbers to Determine Whether a Chemical Reaction Is a Redox Reaction

By using oxidation numbers, determine whether the following reaction is a redox reaction or a nonredox reaction.

$$4NH_3 + 3O_2 \longrightarrow 2N_2 + 6H_2O$$

Solution

For the reactant NH_3, H has an oxidation number of +1 (rule 4) and N an oxidation number of −3 (rule 7). The other reactant, O_2, is an element and thus has an oxidation number of zero (rule 1). The product N_2 also has an oxidation number of zero, because it is an element. In H_2O, the other product, H has an oxidation number of +1 (rule 4) and oxygen an oxidation number of −2 (rule 5).

The overall oxidation number analysis is

$$\underset{-3+1}{4NH_3} + \underset{0}{3O_2} \longrightarrow \underset{0}{2N_2} + \underset{+1-2}{6H_2O}$$

This reaction is a redox reaction because the oxidation numbers of both N and O change.

Practice Exercise 9.2

By using oxidation numbers, determine whether the following reaction is a redox reaction or a nonredox reaction.

$$SO_3 + H_2O \longrightarrow H_2SO_4$$

• *Answer:* nonredox reaction

9.3 Terminology Associated with Redox Processes

Four key terms are used in describing redox processes. These terms are *oxidation, reduction, oxidizing agent,* and *reducing agent.* The definitions for these terms are closely tied to the concepts of "electron transfer" and "oxidation number change"—concepts considered in Section 9.2. It is electron transfer that links all redox processes together. Change in oxidation number is a direct consequence of electron transfer.

In a redox reaction, one reactant undergoes oxidation and another reactant undergoes reduction. **Oxidation** *is the process whereby a substance in a chemical reaction loses one*

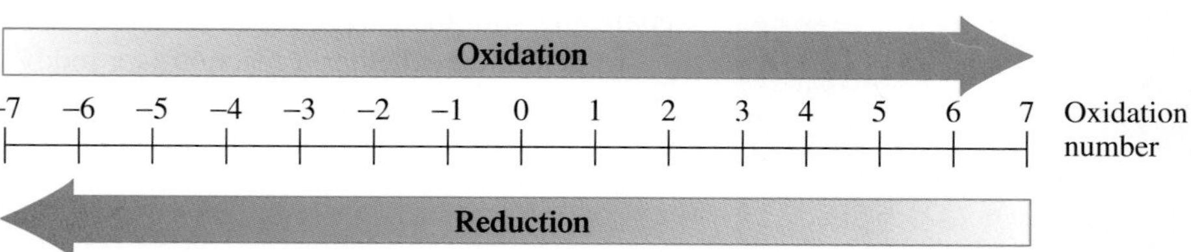

Figure 9.1 An increase in oxidation number is associated with the process of oxidation, a decrease with the process of reduction.

Table 9.1
Summary of Redox Terminology in Terms of Electron Transfer

Term	Electron transfer
oxidation	loss of electron(s)
reduction	gain of electron(s)
oxidizing agent (substance reduced)	electron(s) gained
reducing agent (substance oxidized)	electron(s) lost

or more electrons. **Reduction** *is the process whereby a substance in a chemical reaction gains one or more electrons.*

• Oxidation involves the *loss* of electrons, and reduction involves the *gain* of electrons. Students often have trouble remembering which is which. Two helpful mnemonic devices follow.

LEO the lion says *GER*

Loss of *E*lectrons: *O*xidation.
Gain of *E*lectrons: *R*eduction.

OIL RIG

*O*xidation *I*s *L*oss (of electrons).
*R*eduction *I*s *G*ain (of electrons).

Oxidation and reduction are complementary processes that always occur together. When electrons are lost by one species, they do not disappear; rather, they are always gained by another species. Thus electron transfer always involves both oxidation and reduction.

Electron loss (oxidation) always leads to an increase in oxidation number. Conversely, electron gain (reduction) always leads to a decrease in oxidation number. These generalizations are consistent with the rules for monatomic ion formation (Section 4.5); electron loss produces positive ions (increase in oxidation number), and electron gain produces negative ions (decrease in oxidation number). Figure 9.1 summarizes the relationship between change in oxidation number and the processes of oxidation and reduction.

• Oxidizing Agents and Reducing Agents

• The terms *oxidizing agent* and *reducing agent* sometimes cause confusion, because the oxidizing agent is not oxidized (it is reduced) and the reducing agent is not reduced (it is oxidized). A simple analogy is that a travel agent is not the one who takes a trip; he or she is the one who plans (causes) the trip that is taken.

There are two different ways of looking at the reactants in a redox reaction. First, the reactants can be viewed as being "acted on." From this viewpoint, one reactant is *oxidized* (the one that loses electrons) and one is *reduced* (the one that gains electrons). Second, the reactants can be looked on as "bringing about" the reaction. In this approach, the terms *oxidizing agent* and *reducing agent* are used. An **oxidizing agent** *causes oxidation by accepting electrons from the other reactant.* This acceptance of electrons means that the oxidizing agent itself is reduced. Similarly, the **reducing agent** *causes reduction by providing electrons for the other reactant to accept.* Thus the reducing agent and the substance oxidized are one and the same, as are the oxidizing agent and the substance reduced:

Substance oxidized = reducing agent
Substance reduced = oxidizing agent

Table 9.1 summarizes the redox terminology presented in this section in terms of electron transfer.

Example 9.3

Identifying the Oxidizing Agent and Reducing Agent in a Redox Reaction

For the redox reaction

$$FeO + CO \longrightarrow Fe + CO_2$$

identify the following.

a. The substance oxidized **b.** The substance reduced
c. The oxidizing agent **d.** The reducing agent

Solution

Oxidation numbers are calculated using the methods illustrated in Example 9.1.

$$\underset{+2-2}{FeO} + \underset{+2-2}{CO} \longrightarrow \underset{0}{Fe} + \underset{+4-2}{CO_2}$$

(continued)

a. Oxidation involves an increase in oxidation number. The oxidation number of C has increased from +2 to +4. Therefore, the reactant that contains C, which is CO, is the substance that has been oxidized.
b. Reduction involves a decrease in oxidation number. The oxidation number of Fe has decreased from +2 to zero. Therefore, the reactant that contains Fe, which is FeO, is the substance that has been reduced.
c. The oxidizing agent and the substance reduced are always one and the same. Therefore, FeO is the oxidizing agent.
d. The reducing agent and the substance oxidized are always one and the same. Therefore, CO is the reducing agent.

Practice Exercise 9.3

For the redox reaction

$$3MnO_2 + 4Al \longrightarrow 2Al_2O_3 + 3Mn$$

identify the following.

a. The substance oxidized **b.** The substance reduced
c. The oxidizing agent **d.** The reducing agent

• *Answers:* **a.** Al; **b.** MnO_2; **c.** MnO_2; **d.** Al

9.4 Collision Theory and Chemical Reactions

What causes a chemical reaction, either redox or nonredox, to take place? A set of three generalizations, developed after study of thousands of different reactions, helps answer this question. Collectively these generalizations are known as collision theory. **Collision theory** *is a set of statements that give the conditions that must be met before a chemical reaction will take place.* Central to collision theory are the concepts of molecular collisions, activation energy, and collision orientation. The statements of collision theory are

1. *Molecular collisions.* Reactant particles must interact (that is, collide) with one another before any reaction can occur.
2. *Activation energy.* Colliding particles must possess a certain minimum total amount of energy, called the activation energy, if the collision is to be effective (that is, result in reaction).
3. *Collision orientation.* Colliding particles must come together in the proper orientation unless the particles involved are single atoms or small, symmetrical molecules.

Let's look at these statements in the context of a reaction between two molecules or ions.

• Molecular Collisions

When reactions involve two or more reactants, collision theory assumes (statement 1) that the reactant molecules, ions, or atoms must come in contact (collide) with one another in order for any chemical change to occur. The validity of this statement is fairly obvious. Reactants cannot react if they are separated from each other.

Most reactions are carried out either in liquid solution or in the gaseous phase, wherein reacting particles are more free to move around, and thus it is easier for the reactants to come in contact with one another. Reactions in which reactants are solids can and do occur; however, the conditions for molecular collisions are not as favorable as they are for liquids and gases. Reactions of solids usually take place only on the solid surface and thus include only a small fraction of the total particles present in the solid. As the reaction proceeds and products dissolve, diffuse, or fall from the surface, fresh solid is exposed. In this way, the reaction can eventually consume all of the solid. The rusting of iron is an example of this type of process.

Rubbing a match head against a rough surface provides the activation energy needed for it to ignite.

• Activation Energy

The collisions between reactant particles do not always result in the formation of reaction products. Sometimes, reactant particles rebound unchanged from a collision. Statement 2 of collision theory indicates that in order for a reaction to occur, particles must collide with a certain minimum energy; that is, the kinetic energies of the colliding particles must add to a certain minimum value. **Activation energy** *is the minimum combined kinetic energy that reactant particles must possess in order for their collision to result in a reaction.* Every chemical reaction has a different activation energy. In a slow reaction, the activation energy is far above the average energy content of the reacting particles. Only those few particles with above-average energy undergo collisions that result in reaction; this is the reason for the overall slowness of the reaction.

It is sometimes possible to start a reaction by providing activation energy and then have the reaction continue on its own. Once the reaction is started, enough energy is released to activate other molecules and keep the reaction going. The striking of a kitchen match is an example of such a situation. Activation energy is initially provided by rubbing the match head against a rough surface; heat is generated by friction. Once the reaction is started, the match continues to burn.

• Collision Orientation

Reaction rates are sometimes very slow because reactant molecules must be oriented in a certain way in order for collisions to lead successfully to products. For nonspherical molecules and polyatomic ions, orientation relative to one another at the moment of collision is a factor that determines whether a collision produces a reaction.

As an illustration of the importance of proper collision orientation, consider the chemical reaction between NO_2 and CO to produce NO and CO_2.

$$NO_2(g) + CO(g) \longrightarrow NO(g) + CO_2(g)$$

• Many reactions in the human body do not occur unless specialized proteins called *enzymes* (Chapter 21) are present. One of the functions of these enzymes is to hold reactant molecules in the orientation required for a reaction to occur.

In this reaction, an O atom is transferred from an NO_2 molecule to a CO molecule. The collision orientation most favorable for this to occur is one that puts an O atom from NO_2 near a C atom from CO at the moment of collision. Such an orientation is shown in Figure 9.2a. In Figure 9.2b–d, three undesirable NO_2–CO orientations are shown, where the likelihood of successful reaction is very low.

Figure 9.2
In the reaction of NO_2 with CO to produce NO and CO_2, the most favorable collision orientation is one that puts an O atom from NO_2 in close proximity to the C atom of CO.

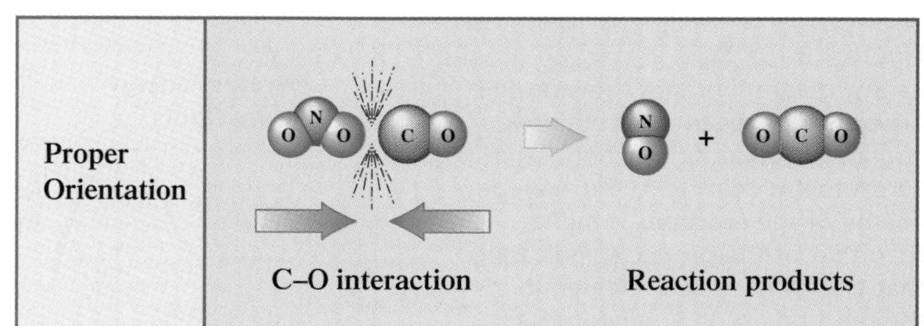

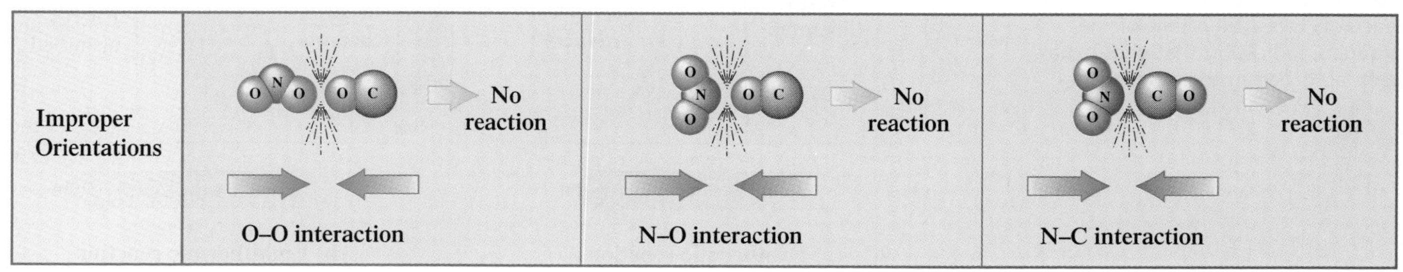

9.5 Exothermic and Endothermic Reactions

In Section 7.9, the terms *exothermic* and *endothermic* were used to classify changes of state. Melting, sublimation, and evaporation are endothermic changes of state, and freezing, condensation, and deposition are exothermic changes of state. The terms *exothermic* and *endothermic* are also used to classify chemical reactions. An **exothermic chemical reaction** *is one in which energy is released as the reaction occurs.* The burning of a fuel (reaction of the fuel with oxygen) is an exothermic process. An **endothermic chemical reaction** *is one that requires the continuous input of energy as the reaction occurs.* The photosynthesis process that occurs in plants is an example of an endothermic reaction. Light is the energy source for photosynthesis. Light energy must be continuously supplied in order for photosynthesis to occur; a green plant that is kept in the dark will die.

• *Exothermic* means energy is released; energy is a "product" of the chemical reaction. *Endothermic* means energy is absorbed; energy is a "reactant" in the reaction.

What determines whether a chemical reaction is exothermic or endothermic? The answer to this question is related to the strength of chemical bonds—that is, the energy stored in chemical bonds. Different types of bonds, such as oxygen–hydrogen bonds and fluorine–nitrogen bonds, have different energies associated with them. In a chemical reaction, bonds are broken within reactant molecules, and new bonds are formed within product molecules. The energy balance between this bond-breaking and bond-forming determines whether there is a net loss or a net gain of energy.

An exothermic reaction (release of energy) occurs when the energy required to break bonds in the reactants is less than the energy released by bond formation in the products. The opposite situation applies for an endothermic reaction. There is more energy stored in product molecule bonds than in reactant molecule bonds. The necessary additional energy must be supplied from external sources as the reaction proceeds. Figure 9.3 illustrates the energy relationships associated with exothermic and endothermic chemical reactions. Note that both of these diagrams contain a "hill" or "hump." The height of this "hill" corresponds to the activation energy needed for reaction between molecules to occur. This activation energy is independent of whether a given reaction is exothermic or endothermic.

9.6 Factors That Influence Reaction Rates

The **rate of a chemical reaction** *is the rate at which reactants are consumed or products produced in a given time period.* Natural processes have a wide range of reaction rates (see Figure 9.4). In this section we consider four different factors that affect reaction rate: (1) the physical nature of the reactants, (2) reactant concentrations, (3) reaction temperature, and (4) the presence of catalysts.

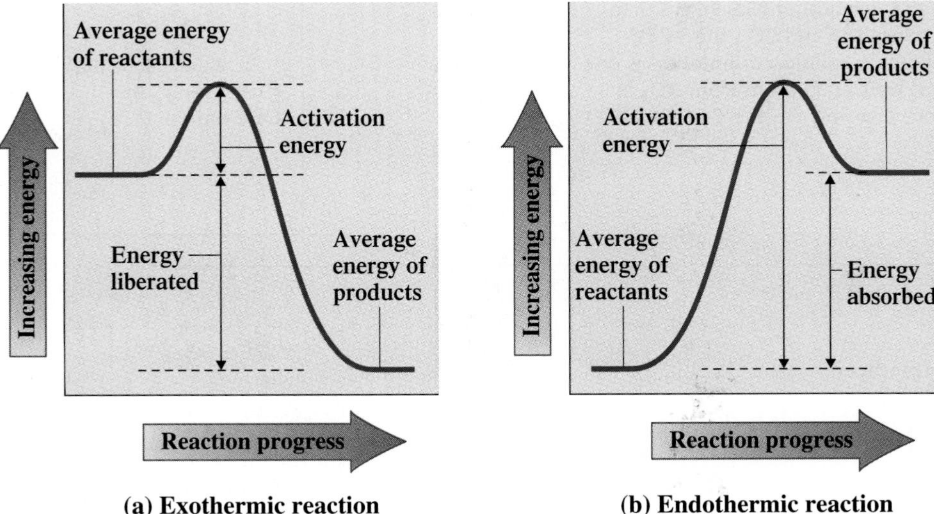

Figure 9.3
Energy diagram graphs showing the difference between an exothermic and an endothermic reaction. (a) In an exothermic reaction, the average energy of the reactants is higher than that of the products, indicating that energy has been released in the reaction. (b) In an endothermic reaction, the average energy of the reactants is less than that of the products, indicating that energy has been absorbed in the reaction.

Figure 9.4
Natural processes occur at a wide range of reaction rates. A fire (a) is a much faster reaction than the ripening of fruit (b), which is much faster than the process of rusting (c), which is much faster than the process of aging (d).

Physical Nature of Reactants

The physical nature of reactants includes not only the physical state of each reactant (solid, liquid, or gas) but also the particle size. In reactions where reactants are all in the same physical state, the reaction rate is generally faster between liquid-state reactants than between solid reactants and is fastest between gaseous reactants. Of the three states of matter, the gaseous state is the one where there is the most freedom of movement; hence, collisions between reactants are the most frequent in this state.

• For solid reactants, reaction rate increases as subdivision of the solid increases.

In the solid state, reactions occur at the boundary surface between reactants. The reaction rate increases as the amount of boundary surface area increases. Subdividing a solid into smaller particles increases surface area and thus increases reaction rate.

When the particle size of a solid is extremely small, reaction rates can be so fast that an explosion results. Although a lump of coal is difficult to ignite, the spontaneous ignition of coal dust is a real threat to underground coal-mining operations.

Reactant Concentrations

An increase in the concentration of a reactant causes an increase in the rate of the reaction. Combustible substances burn much more rapidly in pure oxygen than in air (21% oxygen). A person with a respiratory problem such as pneumonia or emphysema is often given air enriched with oxygen because an increased partial pressure of oxygen facilitates the absorption of oxygen in the alveoli of the lungs and thus expedites all subsequent steps in respiration (Section 7.8).

• Reaction rate increases as the concentration of reactants increases.

Increasing the concentration of a reactant means that there are more molecules of that reactant present in the reaction mixture; thus collisions between this reactant and other reactant particles are more likely. An analogy can be drawn to the game of billiards. The more billiard balls there are on the table, the greater the probability that a moving cue ball will strike one of them.

When the concentration of reactants is increased, the actual quantitative change in reaction rate is determined by the specific reaction. The rate usually increases, but not to the same extent in all cases. Sometimes the rate doubles with a doubling of concentration, but not always.

You may have noticed that pictures from an "instant camera" develop more rapidly on warm days than on cold days. The chemical reactions involved in the development process occur faster at higher temperatures.

• Reaction Temperature

The effect of temperature on reaction rates can also be explained by using the molecular-collision concept. An increase in the temperature of a system results in an increase in the average kinetic energy of the reacting molecules. The increased molecular speed causes more collisions to take place in a given time. Because the average energy of the colliding molecules is greater, a larger fraction of the collisions will result in reaction from the point of view of activation energy. As a rule of thumb, chemists have found that for the temperature ranges we normally encounter, the rate of a chemical reaction doubles for every 10°C increase in temperature.

• Reaction rate increases as the temperature of the reactants increases.

• Presence of Catalysts

A **catalyst** *is a substance that increases a reaction rate without being consumed in the reaction.* Catalysts enhance reaction rates by providing alternative reaction pathways that have lower activation energies than the original, uncatalyzed pathway. This lowering of activation energy is diagrammatically shown in Figure 9.5.

Catalysts exert their effects in varying ways. Some catalysts provide a lower-energy pathway by entering into a reaction and forming an "intermediate," which then reacts further to produce the desired products and regenerate the catalyst. The following equations, where C is the catalyst, illustrate this concept.

• Catalysts lower the activation energy for a reaction. Lowered activation energy increases the rate of a reaction.

Uncatalyzed reaction: $X + Y \longrightarrow XY$

Catalyzed reaction: *Step 1:* $X + C \longrightarrow XC$

Step 2: $XC + Y \longrightarrow XY + C$

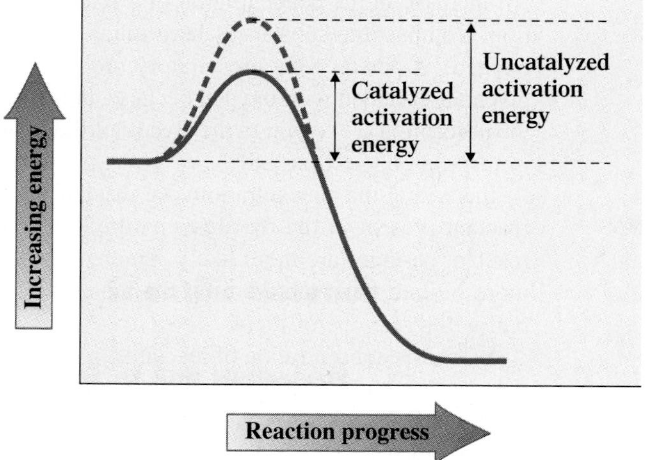

Figure 9.5
Catalysts lower the activation energy for chemical reactions. Reactions proceed more rapidly with the lowered activation energy.

CHEMISTRY AT A GLANCE

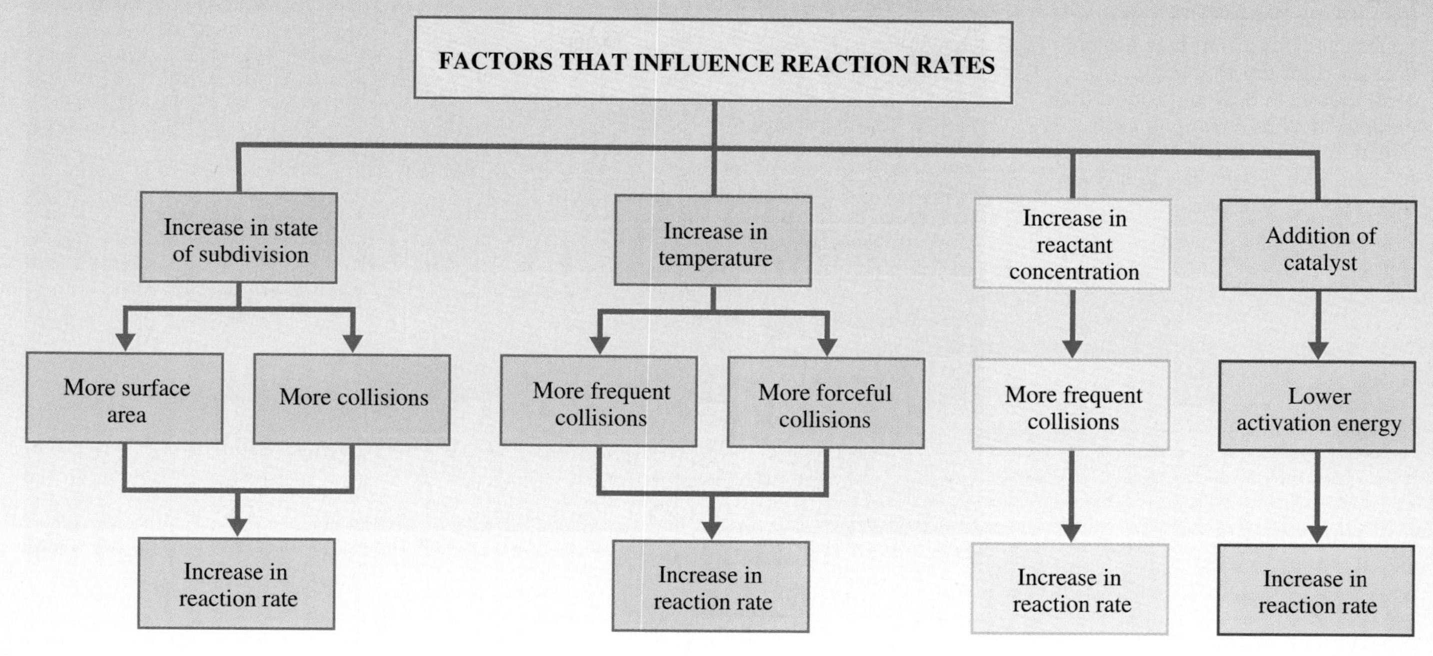

• Catalysts are extremely important for the proper functioning of the human body and other biological systems. Enzymes, which are proteins, are the catalysts within the human body (Chapter 21). They cause many reactions to take place rapidly under mild conditions and at body temperature. Without these enzymes, the reactions would proceed very slowly and then only under harsher conditions.

Solid-state catalysts often act by providing a surface to which reactant molecules are physically attracted and on which they are held with a particular orientation. These "held" reactants are sufficiently close to and favorably oriented toward one another that the reaction takes place. The products of the reaction then leave the surface and make it available to catalyze other reactants.

Chemistry at a Glance summarizes the factors that influence reaction rates.

9.7 Chemical Equilibrium

In our discussions of chemical reactions up to this point, we have assumed that chemical reactions go to completion; that is, reactions continue until one or more of the reactants are used up. This assumption is valid as long as product concentrations are not allowed to build up in the reaction mixture. If one or more products are gases that can escape from the reaction mixture or insoluble solids that can be removed from the reaction mixture, no product buildup occurs.

When product buildup does occur, reactions do not go to completion. This is because product molecules begin to react with one another to re-form reactants. With time, a steady-state situation results wherein the rate of formation of products and the rate of re-formation of reactants are equal. At this point, the concentrations of all reactants and all products remain constant, and a state of *chemical equilibrium* is reached. **Chemical equilibrium** *is the process wherein two opposing chemical reactions occur simultaneously at the same rate.* We discussed equilibrium situations in Sections 7.11 (vapor pressure) and 8.2 (saturated solutions), but the previous examples involved physical equilibrium rather than chemical equilibrium. Figure 9.6 illustrates an "everyday" equilibrium situation.

• A chemical reaction is in a state of chemical equilibrium when the rates of the forward and reverse reactions are equal. At this point, the concentrations of reactants and products no longer change.

The conditions that exist in a system in a state of chemical equilibrium can best be seen by considering an actual chemical reaction. Suppose equal molar amounts of gaseous H_2 and I_2 are mixed together in a closed container and allowed to react.

$$H_2 + I_2 \longrightarrow 2HI$$

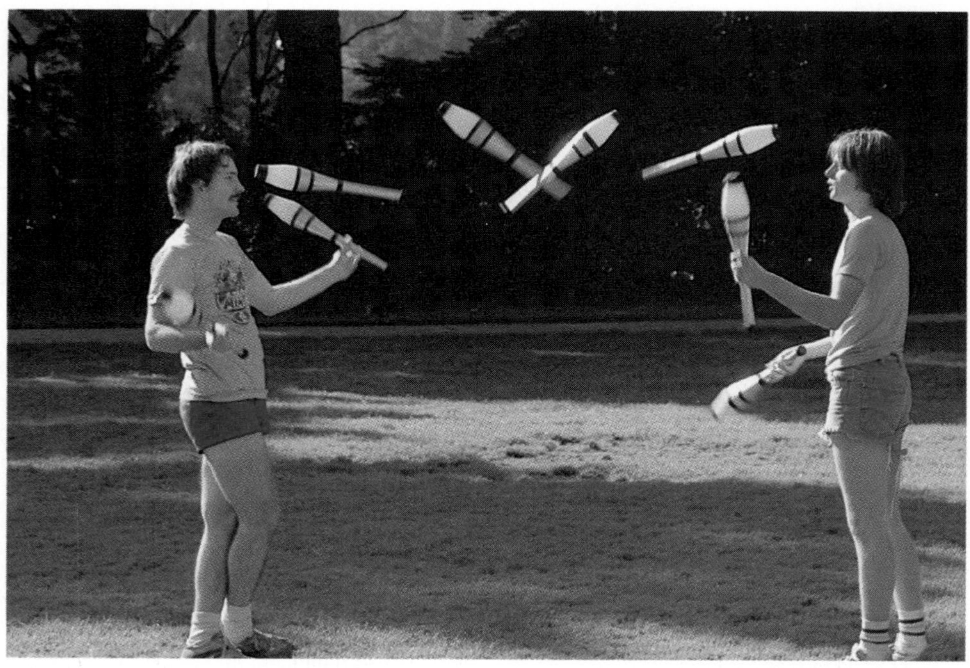

Figure 9.6
These jugglers provide an illustration of equilibrium. Each throws clubs to the other at the same rate at which he receives clubs from that person. Because clubs are thrown continuously in both directions, the number of clubs moving in each direction is constant, and the number of clubs each juggler has at a given time remains constant.

Initially, no HI is present, so the only reaction that can occur is that between H_2 and I_2. However, as the HI concentration increases, some HI molecules collide with one another in a way that causes a reverse reaction to occur:

$$2HI \longrightarrow H_2 + I_2$$

The initially low concentration of HI makes this reverse reaction slow at first, but as the concentration of HI increases, the reaction rate also increases. At the same time that the reverse-reaction rate is increasing, the forward-reaction rate (production of HI) is decreasing as the reactants are used up. Eventually, the concentrations of H_2, I_2, and HI in the reaction mixture reach a level at which the rates of the forward and reverse reactions become equal. At this point, a state of chemical equilibrium has been reached.

Figure 9.7a illustrates the behavior of reaction rates over time for both the forward and reverse reactions in the H_2–I_2–HI system. Figure 9.7b illustrates the important point that the reactant and product concentrations are usually not equal at the point at which equilibrium is reached.

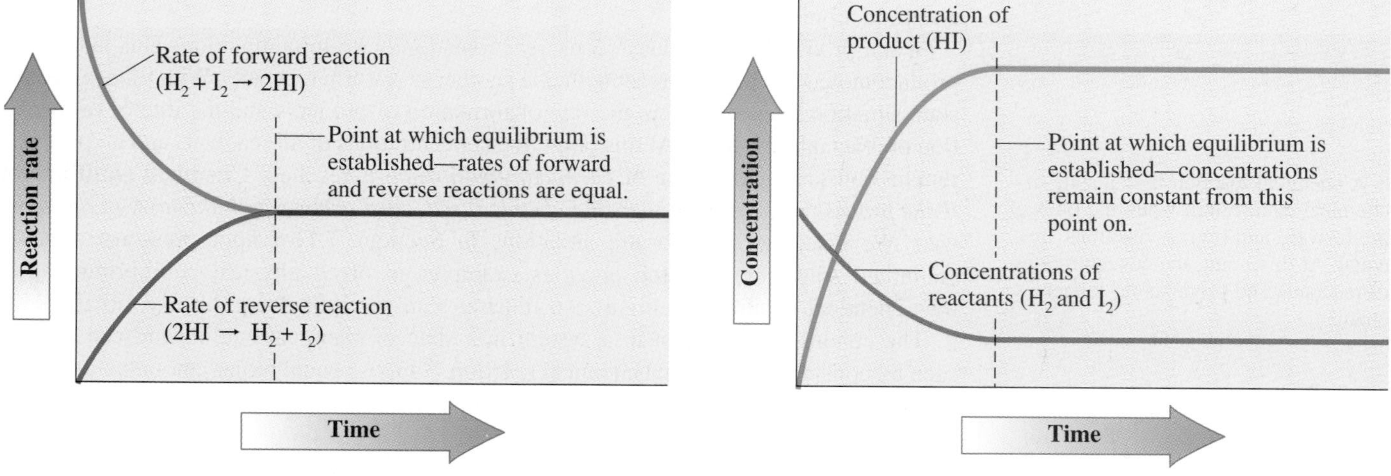

Figure 9.7
Graphs showing how reaction rates and reactant concentrations vary with time for the chemical system H_2–I_2–HI. (a) At equilibrium, rates of reaction are equal. (b) At equilibrium, concentrations of reactants remain constant but are not equal.

Chemical CONNECTIONS

9.2 Stratospheric Ozone: An Equilibrium Situation

Ozone is oxygen that has undergone conversion from its normal diatomic form (O_2) to a triatomic form (O_3). The presence of ozone in the *lower atmosphere* is considered undesirable, because its production contributes to air pollution; it is the major "active ingredient" in smog.

Los Angeles smog.

In the *upper atmosphere* (stratosphere), ozone is a naturally occurring species whose presence is not only desirable but absolutely essential to the well-being of humans on Earth. Stratospheric ozone screens out 95% to 99% of the ultraviolet radiation that comes from the sun. It is ultraviolet light that causes sunburn and that can be a causative factor in some types of skin cancer. The upper region of the stratosphere, where ozone concentrations are greatest, is often called the ozone layer. This ozone maximization occurs at altitudes of 25 to 30 miles (see the accompanying graph).

Within the ozone layer, ozone is continually being consumed and formed through the equilibrium process

$$3O_2(g) \rightleftharpoons 2O_3(g)$$

The source for ozone is thus diatomic oxygen. It is estimated that on any given day, 300 million tons of stratospheric ozone is formed and an equal amount destroyed in this equilibrium process.

In the early 1970s, concern began to be expressed about the possibility of very unreactive chemical species such as chlorofluorocarbons (CFCs), put into the air through human activities, reaching the stratosphere and negatively affecting the ozone layer. Of particular concern was the possibility that such species would catalyze the ozone decomposition reaction, causing it to proceed faster than the ozone formation reaction. The production of CFCs has been phased out because of their potential impact on the ozone layer.

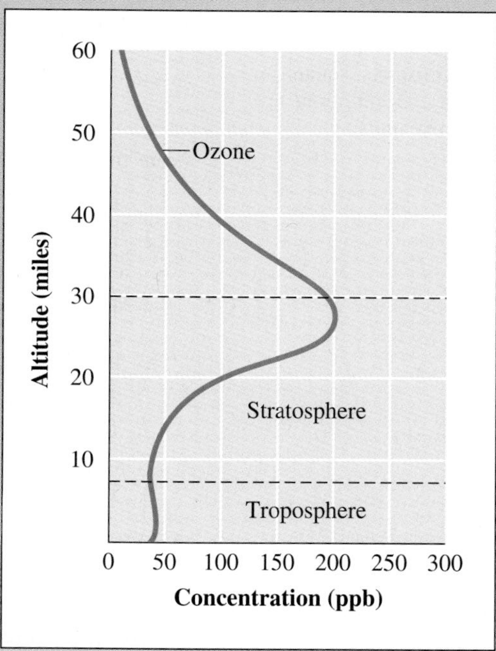

• At chemical equilibrium, forward and reverse reaction rates are equal. Reactant and product concentrations, although constant, do not have to be equal.

• Theoretically, all reactions are reversible (can go in either direction). Sometimes, the reverse reaction is so slight, however, that we say the reaction has "gone to completion" because no detectable reactants remain.

The equilibrium involving H_2, I_2, and HI could have been established just as easily by starting with pure HI and allowing it to change into H_2 and I_2 (the reverse reaction). The final position of equilibrium does not depend on the direction from which equilibrium is approached.

It is normal procedure to represent an equilibrium by using a single equation and two half-headed arrows pointing in opposite directions. Thus the reaction between H_2 and I_2 at equilibrium is written as

$$H_2 + I_2 \rightleftharpoons 2HI$$

The half-headed arrows denote a chemical system at equilibrium.

The term *reversible* is often used to describe a reaction like the one we have just discussed. A **reversible reaction** *is a chemical reaction in which the products formed can* *react to yield the original reactants.* When the half-headed arrow notation is used in an equation, it means that a reaction is reversible.

9.8 Equilibrium Constants

As noted in Section 9.7, the concentrations of reactants and products are constant (not changing) in a system at chemical equilibrium. This constancy allows us to describe the extent of reaction in a given equilibrium system by a single number called an equilibrium constant. An **equilibrium constant** *is a numerical value that characterizes the relationship between the concentrations of reactants and products in a system at chemical equilibrium.*

The equilibrium constant is obtained by writing an *equilibrium expression* and then evaluating it numerically. For a hypothetical chemical reaction, where A and B are reactants, C and D are products, and w, x, y, and z are equation coefficients,

$$wA + xB \rightleftharpoons yC + zD$$

the equilibrium expression is

• In equilibrium constants, square brackets mean that concentrations are expressed in molarity units.

$$K_{eq} = \frac{[C]^y[D]^z}{[A]^w[B]^x}$$

Note the following points about this general equilibrium expression:

1. The square brackets refer to molar (moles/liter) concentrations.
2. The concentrations of the products always appear in the numerator of the equilibrium expression.
3. The concentrations of the reactants always appear in the denominator of the equilibrium expression.
4. When more than one concentration appears in the numerator or denominator, the terms are multiplied.
5. The coefficients in the balanced equation give us the powers to which the concentrations are raised.
6. The abbreviation K_{eq} is used to denote an equilibrium constant.

An additional convention in writing equilibrium constants, which is not apparent from the equilibrium constant definition, is that only *concentrations of gases and substances in solution* are written in an equilibrium constant expression. The reason for this convention is that other substances (pure solids and pure liquids) have constant concentrations. These constant concentrations are incorporated into the equilibrium constant itself. For example, pure water in the liquid state has a concentration of 55.5 moles/L. It does not matter whether we have 1.00, 50.0, or 750 mL of liquid water. The concentration will be the same. In the liquid state, pure water is pure water, and it has only one concentration. Similar reasoning applies to other pure liquids and pure solids. All such substances have constant concentrations.

• The concentrations of *pure liquids* and *pure solids,* which are constants, are never included in an equilibrium constant expression.

The only information we need to write an equilibrium expression is a balanced chemical equation, which includes physical state information. Using the preceding generalizations about equilibrium expressions, for the reaction

$$4NH_3(g) + 7O_2(g) \rightleftharpoons 4NO_2(g) + 6H_2O(g)$$

we write the equilibrium expression as

Coefficient of NO_2 ↘ ↙ Coefficient of H_2O

$$K_{eq} = \frac{[NO_2]^4[H_2O]^6}{[NH_3]^4[O_2]^7}$$

Coefficient of NH_3 ↗ ↖ Coefficient of O_2

Example 9.4

Writing the Equilibrium Constant Expression for a Chemical Reaction from the Chemical Equation for the Reaction

Write the equilibrium constant expression for each of the following reactions.

a. $I_2(g) + Cl_2(g) \rightleftharpoons 2ICl(g)$ **b.** $C(s) + H_2O(g) \rightleftharpoons CO(g) + H_2(g)$

Solution

a. All of the substances involved in this reaction are gases. Therefore, each reactant and product will appear in the equilibrium constant expression.

The numerator of an equilibrium expression always contains product concentrations. There is only one product, ICl. Write its concentration in the numerator and square it, because the coefficient of ICl in the equation is 2.

$$[ICl]^2$$

Next, place the concentrations of the reactants in the denominator. Their powers will be an understood (not written) 1, because the coefficient of each reactant is 1.

$$K_{eq} = \frac{[ICl]^2}{[I_2][Cl_2]}$$

The equilibrium expression is now complete.

b. The reactant carbon (C) is a solid and thus will not appear in the equilibrium expression. Therefore,

$$K_{eq} = \frac{[CO][H_2]}{[H_2O]}$$

Note that all of the powers in this expression are 1 as a result of all the coefficients in the balanced equation being equal to unity.

Practice Exercise 9.4

Write the equilibrium constant expression for each of the following reactions.

a. $2Cl_2(g) + 2H_2O(g) \rightleftharpoons 4HCl(g) + O_2(g)$
b. $NH_4Cl(s) \rightleftharpoons HCl(g) + NH_3(g)$

• *Answers:* **a.** $K_{eq} = \frac{[O_2][HCl]^4}{[H_2O]^2[Cl_2]^2}$; **b.** $K_{eq} = [NH_3][HCl]$

If the concentrations of all reactants and products are known at equilibrium, the numerical value of the equilibrium constant can be calculated by using the equilibrium expression.

Example 9.5

Calculating the Value of an Equilibrium Constant from Equilibrium Concentrations

Calculate the value of the equilibrium constant for the equilibrium system

$$2NO(g) \rightleftharpoons N_2(g) + O_2(g)$$

at 1000°C, given that the equilibrium concentrations are 0.0026 M for NO, 0.024 M for N_2, and 0.024 M for O_2.

(continued)

Solution

First, write the equilibrium constant expression.

$$K_{eq} = \frac{[N_2][O_2]}{[NO]^2}$$

Next, substitute the equilibrium concentrations into the equilibrium expression and solve the equation.

$$K_{eq} = \frac{[0.024][0.024]}{[0.0026]^2}$$

$$K_{eq} = 85$$

In doing the mathematics, remember that the number 0.0026 must be squared.

Practice Exercise 9.5

Calculate the value of the equilibrium constant for the equilibrium system

$$N_2(g) + 3H_2(g) \rightleftharpoons 2NH_3(g)$$

at 532°C, given that the equilibrium concentrations are 0.079 M for N_2, 0.12 M for H_2, and 0.0051 M for NH_3.

- *Answer:* 0.19

• Temperature Dependence of Equilibrium Constants

The value of K_{eq} for a reaction depends on the reaction temperature. If the temperature changes, the value of K_{eq} also changes, because differing amounts of reactants and products will be present. Note that the equilibrium constant calculated in Example 9.5 is for a temperature of 1000°C. The equilibrium constant for this reaction would have a different value at a lower or a higher temperature.

Does the value of an equilibrium constant increase or decrease when reaction temperature is increased? For reactions where the forward reaction is *exothermic,* the equilibrium constant *decreases* with increasing temperature. For reactions where the forward reaction is *endothermic,* the equilibrium constant *increases* with increasing temperature (Section 9.5).

The space shuttle. Reactions in which the products escape from the reaction mixture, such in burning rocket fuel, never reach equilibrium. An equilibrium constant consideration is not applicable to such reactions.

• Equilibrium Constant Values and Reaction Completeness

The magnitude of an equilibrium constant value conveys information about how far a reaction has proceeded toward completion. If the equilibrium constant value is large (10^3 or greater), the equilibrium system contains more products than reactants. Conversely, if the equilibrium constant value is small (10^{-3} or less), the equilibrium system contains more reactants than products. Table 9.2 further compares equilibrium constant values and the extent to which a chemical reaction has occurred.

The **position of equilibrium** *specifies, in a qualitative way, the relative amounts of reactants and products present at equilibrium for a chemical reaction.* As shown in the last column of Table 9.2, the terms *far to the right, to the right, neither to right nor to left, to the left,* and *far to the left* are used in describing equilibrium position. In equilibrium situations where the concentrations of products are greater than those of reactants, the equilibrium position is said to lie to the *right* because products are always listed on the right side of an equation.

Position of equilibrium can also be indicated by varying the length of the arrows in the half-headed arrow notation for a reversible reaction. The longer arrow indicates the direction of the predominant reaction. For example, the arrow notation in the equation

$$CO_2 + H_2O \rightleftharpoons H_2CO_3$$

indicates that the equilibrium position lies to the right.

Table 9.2
Equilibrium Constant Values and the Extent to Which a Chemical Reaction Has Taken Place

Value of K_{eq}	Relative amounts of products and reactants	Description of equilibrium position
very large (10^{30})	essentially all products	far to the right
large (10^{10})	more products than reactants	to the right
near unity (between 10^3 and 10^{-3})	significant amounts of both reactants and products	neither to right nor to left
small (10^{-10})	more reactants than products	to the left
very small (10^{-30})	essentially all reactants	far to the left

9.9 Altering Equilibrium Conditions: Le Châtelier's Principle

A chemical system at equilibrium is very susceptible to disruption from outside forces. A change in temperature or a change in pressure can upset the balance within the equilibrium system. Changes in the concentrations of reactants or products upset an equilibrium also.

• Products are written on the right side of a chemical equation. A *shift to the right* means more products are produced. Conversely, because reactants are written on the left side of an equation, a *shift to the left* means more reactants are produced.

Disturbing an equilibrium has one of two results: Either the forward reaction speeds up (to produce more products), or the reverse reaction speeds up (to produce additional reactants). Over time, the forward and reverse reactions again become equal, and a new equilibrium, different from the previous one, is established. If more products have been produced as a result of the disruption, the equilibrium is said to have *shifted to the right.* Similarly, when disruption causes more reactants to form, the equilibrium has *shifted to the left.*

• The surname Le Châtelier is pronounced "le-SHOT-lee-ay."

An equilibrium system's response to disrupting influences can be predicted by using a principle introduced by the French chemist Henry Louis Le Châtelier (1850–1936). **Le Châtelier's principle** *states that if a stress (change of conditions) is applied to a system in equilibrium, the system will readjust (change the position of equilibrium) in the direction that best reduces the stress imposed on it.* We will use this principle to consider how four types of changes affect equilibrium position. The changes are (1) concentration changes, (2) temperature changes, (3) pressure changes, and (4) addition of catalysts.

Henri Louis Le Châtelier (1850–1936), although most famous for the principle that bears his name, was amazingly diverse in his interests. He worked on metallurgical processes, cements, glasses, fuels, and explosives and was also noted for his skills in industrial management.

• Concentration Changes

Adding a reactant or product to or removing it from a reaction mixture at equilibrium always upsets the equilibrium. If an additional amount of any reactant or product has been *added* to the system, the stress is relieved by shifting the equilibrium in the direction that *consumes* (uses up) some of the added reactant or product. Conversely, if a reactant or product is *removed* from an equilibrium system, the equilibrium shifts in a direction that will *produce* more of the substance that was removed.

Let us consider the effect that concentration changes will have on the gaseous equilibrium

$$N_2(g) + 3H_2(g) \rightleftharpoons 2NH_3(g)$$

Suppose some additional H_2 is added to the equilibrium mixture. The stress of "added H_2" causes the equilibrium to shift to the right; that is, the forward reaction rate increases in order to use up some of the additional H_2.

Stress: Too much H_2
Response: Use up "extra" H_2

$$N_2(g) + 3H_2(g) \rightleftharpoons 2NH_3(g)$$

Shift to the right →

$[N_2]$ decreases, $[H_2]$ decreases, $[NH_3]$ increases

Figure 9.8
Concentration changes that result when H_2 is added to an equilibrium mixture involving the system
$N_2(g) + 3H_2(g) \rightleftharpoons 2NH_3(g)$

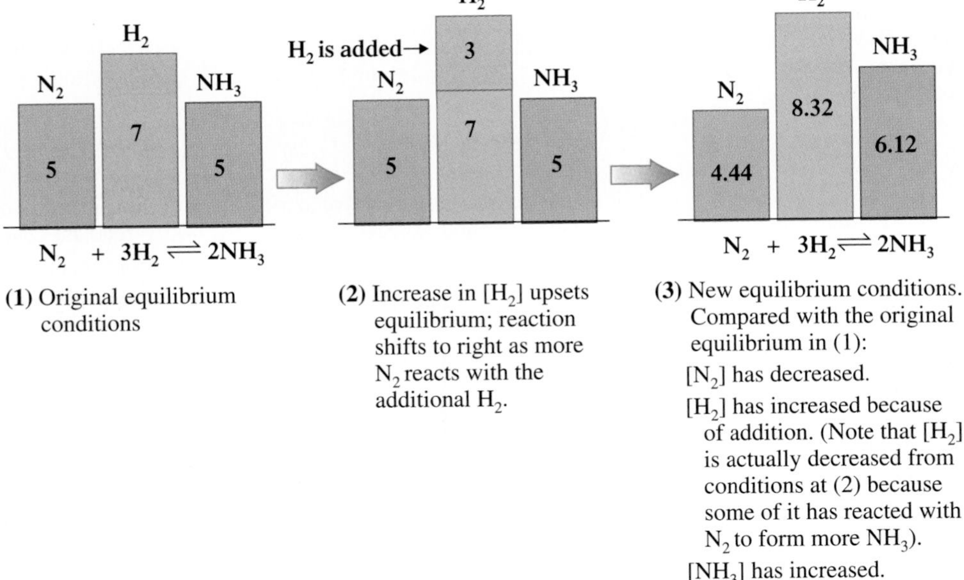
N2
H2
NH3
7
5
5
N2 + 3H2 ⇌ 2NH3
(1) Original equilibrium conditions
H2
H2 is added→
3
N2
NH3
7
5
5
(2) Increase in [H2] upsets equilibrium; reaction shifts to right as more N2 reacts with the additional H2.
H2
NH3
N2
8.32
6.12
4.44
N2 + 3H2 ⇌ 2NH3
(3) New equilibrium conditions. Compared with the original equilibrium in (1):
[N2] has decreased.
[H2] has increased because of addition. (Note that [H2] is actually decreased from conditions at (2) because some of it has reacted with N2 to form more NH3).
[NH3] has increased.

As the H_2 reacts, the amount of N_2 also decreases (it reacts with the H_2) and the amount of NH_3 increases (it is formed as H_2 and N_2 react).

With time, the equilibrium shift to the right caused by the addition of H_2 will cease because a new equilibrium condition (not identical to the original one) has been reached. At this new equilibrium condition, most (but not all) of the added H_2 will have been converted to NH_3. Necessary accompaniments to this change are a decreased N_2 concentration (some of it reacted with the H_2) and an increased NH_3 concentration (that produced from the N_2–H_2 reaction). Figure 9.8 quantifies the changes that occur in the N_2–H_2–NH_3 equilibrium system when it is upset by the addition of H_2 for a specific set of concentrations.

Consider again the reaction between N_2 and H_2 to form NH_3.

$$N_2(g) + 3H_2(g) \rightleftharpoons 2NH_3(g)$$

Le Châtelier's principle applies in the same way to removing a reactant or product from the equilibrium mixture as it does to adding a reactant or product at equilibrium. Suppose that at equilibrium we remove some NH_3. The equilibrium position shifts to the right to replenish the NH_3.

• Thousands of chemical equilibria simultaneously exist in biological systems. Many of them are interrelated. When the concentration of one substance changes, many equilibria are affected.

Within the human body, numerous equilibrium situations exist that shift in response to a concentration change. Consider, for example, the equilibrium between glucose in the blood and stored glucose (glycogen) in the liver:

$$\text{Glucose in blood} \rightleftharpoons \text{stored glucose} + H_2O$$

Strenuous exercise or hard work causes our blood glucose level to decrease. Our bodies respond to this stress (not enough glucose in the blood) by the liver converting glycogen into glucose. Conversely, when an excess of glucose is present in the blood (after a meal), the liver converts the excess glucose in the blood to its storage form (glycogen).

• Temperature Changes

Le Châtelier's principle can be used to predict the influence of temperature changes on an equilibrium, provided we know whether the reaction is exothermic or endothermic. For *exothermic reactions,* heat can be treated as one of the *products;* for *endothermic reactions,* heat can be treated as one of the *reactants.*

Consider the exothermic reaction

$$H_2(g) + F_2(g) \rightleftharpoons 2HF(g) + \text{heat}$$

Figure 9.9
Effect of temperature change on the equilibrium mixture

$$\underset{\text{Blue}}{CoCl_4^{2-}} + 6H_2O \rightleftharpoons \underset{\text{Pink}}{Co(H_2O)_6^{2+}} + 4Cl^- + \text{heat}$$

At room temperature, the equilibrium mixture is blue from $CoCl_4^{2-}$. When cooled by the ice bath, the equilibrium mixture turns pink from $Co(H_2O)_6^{2+}$. The temperature decrease causes the equilibrium position to shift to the right.

Heat is produced when the reaction proceeds to the right. Thus, if we add heat to an exothermic system at equilibrium (by raising the temperature), the system will shift to the left in an "attempt" to decrease the amount of heat present. When equilibrium is reestablished, the concentrations of H_2 and F_2 will be higher and the concentration of HF will have decreased. Lowering the temperature of an exothermic reaction mixture causes the reaction to shift to the right as the system acts to replace the lost heat (Figure 9.9).

The behavior, with temperature change, of an equilibrium system involving an endothermic reaction, such as

$$\text{Heat} + 2CO_2(g) \rightleftharpoons 2CO(g) + O_2(g)$$

is opposite to that of an exothermic reaction, because a shift to the left produces heat. Consequently, an increase in temperature will cause the equilibrium to shift to the right (to decrease the amount of heat present), and a decrease in temperature will produce a shift to the left (to generate more heat).

• Pressure Changes

Pressure changes affect systems at equilibrium only when gases are involved, and then only in cases where the chemical reaction is such that a change in the total number of moles in the gaseous state occurs. This latter point can be illustrated by considering the following two gas-phase reactions:

$$\underbrace{2H_2(g) + O_2(g)}_{\text{3 moles of gas}} \longrightarrow \underbrace{2H_2O(g)}_{\text{2 moles of gas}}$$

$$\underbrace{H_2(g) + Cl_2(g)}_{\text{2 moles of gas}} \longrightarrow \underbrace{2HCl(g)}_{\text{2 moles of gas}}$$

In the first reaction, the total number of moles of gaseous reactants and products decreases as the reaction proceeds to the right. This is because 3 moles of reactants combine to give only 2 moles of products. In the second reaction, there is no change in the total number of moles of gaseous substances present as the reaction proceeds. This is because 2 moles of reactants combine to give 2 moles of products. Thus a pressure change will shift the position of the equilibrium in the first reaction but not in the second reaction.

Pressure changes are usually brought about through volume changes. A pressure increase results from a volume decrease, and a pressure decrease results from a volume increase (Section 7.4). Le Châtelier's principle correctly predicts the direction of the equilibrium position shift resulting from a pressure change only when the pressure change is

Chemical CONNECTIONS

9.3 Oxygen, Hemoglobin, Equilibrium, and Le Châtelier's Principle

Hemoglobin, an iron-containing protein found in red blood cells, facilitates the transport of oxygen to body tissues. Without the hemoglobin of red blood cells, 1 L of arterial blood could dissolve and carry only about 3 mL of oxygen (the solubility of oxygen in water). With hemoglobin present, 1 L of blood can dissolve and carry about 250 mL of oxygen.

The transport of oxygen by hemoglobin from the lungs, through the bloodstream, and to the tissues involves a reversible reaction in which the O_2 becomes chemically bonded to hemoglobin:

$$\underset{\text{Hemoglobin}}{Hb(aq)} + O_2(g) \rightleftharpoons \underset{\text{Oxyhemoglobin}}{HbO_2(aq)}$$

In the hemoglobin–oxyhemoglobin reaction, the relative amounts of the two compounds are controlled by the partial pressure of oxygen. High oxygen pressure (in the lungs) causes hemoglobin (Hb) to be converted to oxyhemoglobin (HbO_2). (This is a shift to the right in terms of Le Châtelier's principle; the equilibrium system shifts to decrease the pressure of oxygen.) Decreasing oxygen pressure, on the other hand, accelerates oxygen dissociation from oxyhemoglobin (a shift to the left in terms of Le Châtelier's principle). Decreased oxygen pressure is encountered as arterial blood flows through tissue capillaries and cells.

The partial pressure of oxygen in the lungs is 105 mm Hg. This is sufficient to cause 97% *oxygen saturation;* that is, 97% of the blood's hemoglobin has united with oxygen by the time blood leaves the lung capillaries to return to the heart. The accompanying figure shows the variance in oxygen saturation as a function of oxygen partial pressure. You can see from the figure that oxygen saturation is still high (90%) at an oxygen partial pressure of 60 mm Hg. Thus, a safety zone exists relative to drops in oxygen pressure as a result of doing severe exercise, being at high altitudes, or having heart or lung disease.

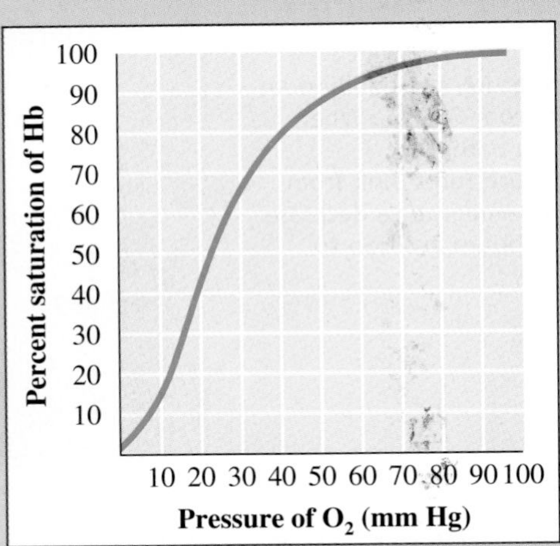

A graph of hemoglobin saturation as a function of oxygen partial pressure.

An individual accustomed to living at sea level will tire more easily at higher altitudes, because the partial pressure of oxygen is lower at the higher altitudes. This translates into fewer molecules of oxyhemoglobin (a shift to the left in terms of Le Châtelier's principle). However, if a person stays at the high altitude for a few days, the body adapts through increased red blood cell (hemoglobin) production. Conversely, individuals living at high altitudes who travel to locations at sea level tire less readily than usual.

Control of oxygen partial pressure is important in the practice of medicine. Increasing the partial pressure of oxygen above that normal for air is often used in the treatment of patients with diseased lungs.

caused by a change in volume. It does not apply to pressure increases caused by the addition of a nonreactive (inert) gas to the reaction mixture. This addition has no effect on the equilibrium position. The partial pressure (Section 7.8) of each of the gases involved in the reaction remains the same.

• Increasing the pressure associated with an equilibrium system by adding an inert gas (a gas that is not a reactant or a product in the reaction) does not affect the position of the equilibrium.

According to Le Châtelier's principle, the stress of increased pressure is relieved by decreasing the number of moles of gaseous substances in the system. This is accomplished by the reaction shifting in the direction of the fewer moles; that is, it shifts to the side of the equation that contains the fewer moles of gaseous substances. For the reaction

$$2NO_2(g) + 7H_2(g) \rightleftharpoons 2NH_3(g) + 4H_2O(g)$$

an increase in pressure would shift the equilibrium position to the right, because there are 9 moles of gaseous reactants and only 6 moles of gaseous products. On the other hand, the stress of decreased pressure causes an equilibrium system to produce more moles of gaseous substances.

• Addition of Catalysts

Catalysts cannot change the position of an equilibrium. A catalyst functions by lowering the activation energy for a reaction. It speeds up both the forward and the reverse reactions, so it has no net effect on the position of the equilibrium. However, the lowered activation energy allows equilibrium to be established more quickly than if the catalyst were absent.

Example 9.6

Using Le Châtelier's Principle to Predict How Various Changes Affect an Equilibrium System

How will the gas-phase equilibrium

$$CH_4(g) + 2H_2S(g) + \text{heat} \rightleftharpoons CS_2(g) + 4H_2(g)$$

be affected by the following?

a. The removal of $H_2(g)$
b. The addition of $CS_2(g)$
c. An increase in the temperature
d. An increase in the volume of the container (a decrease in pressure)

Solution

a. The equilibrium will *shift to the right,* according to Le Châtelier's principle, in an "attempt" to replenish the H_2 that was removed.
b. The equilibrium will *shift to the left* in an attempt to use up the extra CS_2 that has been placed in the system.
c. Raising the temperature means that heat energy has been added. In an attempt to minimize the effect of this extra heat, the position of the equilibrium will *shift to the right,* which is a direction that consumes heat; heat is one of the reactants in an endothermic reaction.
d. The system will *shift to the right,* the direction that produces more moles of gaseous substances (an increase of pressure). In this way, the reaction produces 5 moles of gaseous products for every 3 moles of gaseous reactants consumed.

Practice Exercise 9.6

How will the gas-phase equilibrium

$$CO(g) + 3H_2(g) \rightleftharpoons CH_4(g) + H_2O(g) + \text{heat}$$

be affected by the following?

a. The removal of $CH_4(g)$
b. The addition of $H_2O(g)$
c. A decrease in the temperature
d. A decrease in the volume of the container (an increase in pressure)

• *Answers:* **a.** shift to the right; **b.** shift to the left; **c.** shift to the right; **d.** shift to the right

Concepts to Remember

Chemical reaction. A process in which at least one new substance is produced as a result of chemical change.

Combination reaction. A chemical reaction in which a single product is produced from two or more reactants.

Decomposition reaction. A chemical reaction in which a single reactant is converted into two or more simpler substances (elements or compounds).

Single-replacement reaction. A chemical reaction in which an atom or a molecule replaces an atom or a group of atoms from a compound.

Double-replacement reaction. A chemical reaction in which two substances exchange parts with one another and form two different substances.

Redox reaction. A chemical reaction in which there is a transfer of electrons from one reactant to another reactant.

Nonredox reaction. A chemical reaction in which there is no transfer of electrons from one reactant to another reactant.

Oxidation number. An oxidation number for an atom is a number that represents the charge that an atom appears to have when the electrons in each bond it is participating in are assigned to the more electronegative of the two atoms involved in the bond. Oxidation numbers are used to identify the electron transfer that occurs in a redox reaction.

Oxidation–reduction terminology. Oxidation is the loss of electrons; reduction is the gain of electrons. An oxidizing agent causes oxidation by accepting electrons from the other reactant. A reducing agent causes reduction by providing electrons for the other reactant to accept.

Collision theory. Collision theory summarizes the conditions required for a chemical reaction to take place. The three basic tenets of collision theory are as follows: (1) Reactant molecules must collide with each other. (2) The collision must involve a certain minimum of energy. (3) In some cases, colliding molecules must be oriented in a specific way if reaction is to occur.

Exothermic and endothermic chemical reactions. An exothermic chemical reaction releases energy as the reaction occurs. An endothermic chemical reaction requires an input of energy as the reaction occurs.

Reaction rates. Reaction rate is the speed at which reactants are converted to products. Four factors affect the rates of all reactions: (1) the physical nature of the reactants, (2) reactant concentrations, (3) reaction temperature, and (4) the presence of catalysts.

Chemical equilibrium. Chemical equilibrium is the state wherein the rate of the forward reaction is equal to the rate of the reverse reaction. Equilibrium is indicated in chemical equations by writing half-headed arrows pointing in both directions between reactants and products.

Equilibrium constant. The equilibrium constant relates the concentrations of reactants and products at equilibrium. The value of an equilibrium constant is obtained by writing an equilibrium expression and then numerically evaluating it. Equilibrium expressions can be obtained from the balanced chemical equations for reactions.

Position of equilibrium. The relative amounts of reactants and products present in a system at equilibrium define the position of equilibrium. The equilibrium position is toward the right when a large amount of product is present and is toward the left when a large amount of reactant is present.

Le Châtelier's principle. Le Châtelier's principle states that when a stress (change of conditions) is applied to a system in equilibrium, the system will readjust (change the position of the equilibrium) in the direction that best reduces the stress imposed on it. Stresses known to change an equilibrium position include (1) changes in amount of reactants and/or products, (2) changes in temperature, and (3) changes in pressure.

Key Reactions and Equations

1. Combination reaction (Section 9.1)

$$X + Y \longrightarrow XY$$

2. Decomposition reaction (Section 9.1)

$$XY \longrightarrow X + Y$$

3. Single-replacement reaction (Section 9.1)

$$X + YZ \longrightarrow Y + XZ$$

4. Double-replacement reaction (Section 9.1)

$$AX + BY \longrightarrow AY + BX$$

5. Equilibrium expression equation for a general reaction (Section 9.8)

$$wA + xB \rightleftharpoons yC + zD$$

$$K_{eq} = \frac{[C]^y[D]^z}{[A]^w[B]^x}$$

Key Terms

Activation energy (9.4)
Catalyst (9.6)
Chemical equilibrium (9.7)
Chemical reaction (9.1)
Collision theory (9.4)
Combination reaction (9.1)
Decomposition reaction (9.1)
Double-replacement reaction (9.1)
Endothermic chemical reaction (9.5)
Equilibrium constant (9.8)
Exothermic chemical reaction (9.5)
Le Châtelier's principle (9.9)
Oxidation (9.3)
Oxidation number (9.2)
Oxidation–reduction reaction (9.2)
Oxidizing agent (9.3)
Nonoxidation–reduction reaction (9.2)
Position of equilibrium (9.8)
Rate of a chemical reaction (9.6)
Reducing agent (9.3)
Reduction (9.3)
Reversible reaction (9.7)
Single-replacement reaction (9.1)

Exercises and Problems

The members of each pair of problems in this section test similar material.

Types of Chemical Reactions (Section 9.1)

9.1 Classify each of the following reactions as a combination, decomposition, single-replacement, or double-replacement reaction.

a. $3CuSO_4 + 2Al \longrightarrow Al_2(SO_4)_3 + 3Cu$

b. $K_2CO_3 \longrightarrow K_2O + CO_2$

c. $2AgNO_3 + K_2SO_4 \longrightarrow Ag_2SO_4 + 2KNO_3$

d. $2SO_2 + O_2 \longrightarrow 2SO_3$

9.2 Classify each of the following reactions as a combination, decomposition, single-replacement, or double-replacement reaction.

a. $2NaHCO_3 \longrightarrow Na_2CO_3 + CO_2 + H_2O$

b. $2Ag_2CO_3 \longrightarrow 4Ag + 2CO_2 + O_2$

c. $2Fe + 3Cl_2 \longrightarrow 2FeCl_3$
d. $Mg + 2HCl \longrightarrow MgCl_2 + H_2$

Oxidation Numbers (Section 9.2)

9.3 Determine the oxidation number of
a. Ba in Ba^{2+} b. S in SO_3
c. F in F_2 d. P in PO_4^{3-}

9.4 Determine the oxidation number of
a. Al in Al^{3+} b. N in NO_2
c. O in O_3 d. S in SO_4^{2-}

9.5 Determine the oxidation number of Cr in each of the following chromium-containing species.
a. Cr_2O_3 b. CrO_2 c. CrO_3
d. Na_2CrO_4 e. $BaCrO_4$ f. $BaCr_2O_7$
g. $Na_2Cr_2O_7$ h. CrF_5

9.6 Determine the oxidation number of Cl in each of the following chlorine-containing species.
a. $BeCl_2$ b. $Ba(ClO)_2$ c. ClF_4^+ d. Cl_2O_7
e. NCl_3 f. $AlCl_4^-$ g. ClF h. ClO^-

9.7 What is the oxidation number of each element in each of the following substances?
a. PF_3 b. NaOH c. Na_2SO_4 d. CO_3^{2-}

9.8 What is the oxidation number of each element in each of the following substances?
a. H_2S b. H_2 c. N^{3-} d. MnO_4^-

Oxidation–Reduction Reactions (Sections 9.2 and 9.3)

9.9 Classify each of the following reactions as a redox reaction or a nonredox reaction.
a. $2Cu + O_2 \longrightarrow 2CuO$
b. $K_2O + H_2O \longrightarrow 2KOH$
c. $2KClO_3 \longrightarrow 2KCl + 3O_2$
d. $CH_4 + 2O_2 \longrightarrow CO_2 + 2H_2O$

9.10 Classify each of the following reactions as a redox reaction or a nonredox reaction.
a. $2NO + O_2 \longrightarrow 2NO_2$
b. $CO_2 + H_2O \longrightarrow H_2CO_3$
c. $Zn + 2AgNO_3 \longrightarrow Zn(NO_3)_2 + 2Ag$
d. $HNO_3 + NaOH \longrightarrow NaNO_3 + H_2O$

9.11 Identify which substance is oxidized and which substance is reduced in each of the following redox reactions.
a. $N_2 + 3H_2 \longrightarrow 2NH_3$
b. $Cl_2 + 2KI \longrightarrow 2KCl + I_2$
c. $Sb_2O_3 + 3Fe \longrightarrow 2Sb + 3FeO$
d. $3H_2SO_3 + 2HNO_3 \longrightarrow 2NO + H_2O + 3H_2SO_4$

9.12 Identify which substance is oxidized and which substance is reduced in each of the following redox reactions.
a. $2Al + 3Cl_2 \longrightarrow 2AlCl_3$
b. $Zn + CuCl_2 \longrightarrow ZnCl_2 + Cu$
c. $2NiS + 3O_2 \longrightarrow 2NiO + 2SO_2$
d. $3H_2S + 2HNO_3 \longrightarrow 3S + 2NO + 4H_2O$

9.13 Identify which substance is the oxidizing agent and which substance is the reducing agent in each of the redox reactions of Problem 9.11.

9.14 Identify which substance is the oxidizing agent and which substance is the reducing agent in each of the redox reactions of Problem 9.12.

Collision Theory (Section 9.4)

9.15 What is collision theory, and what are the three basic statements of collision theory?

9.16 Draw a diagram of the preferred orientation of molecules at the time of collision for the reaction of X_2 and Y_2 molecules to produce XY.

Exothermic and Endothermic Reactions (Section 9.5)

9.17 Which of the following reactions are endothermic and which are exothermic?
a. $C_2H_4 + 3O_2 \longrightarrow 2CO_2 + 2H_2O$ + heat
b. $N_2 + 2O_2$ + heat $\longrightarrow 2NO_2$
c. $2H_2O$ + heat $\longrightarrow 2H_2 + O_2$
d. $2KClO_3 \longrightarrow 2KCl + 3O_2$ + heat

9.18 Which of the following reactions are endothermic and which are exothermic?
a. $CaCO_3$ + heat $\longrightarrow CaO + CO_2$
b. $N_2 + 3H_2 \longrightarrow 2NH_3$ + heat
c. $CO + 3H_2$ + heat $\longrightarrow CH_4 + H_2O$
d. $2N_2 + 6H_2O$ + heat $\longrightarrow 4NH_3 + 3O_2$

9.19 Sketch an energy diagram graph representing an exothermic reaction, and label the following:
a. Average energy of reactants
b. Average energy of products
c. Activation energy
d. Amount of energy liberated during the reaction

9.20 Sketch an energy diagram graph representing an endothermic reaction, and label the following:
a. Average energy of reactants
b. Average energy of products
c. Activation energy
d. Amount of energy absorbed during the reaction

Factors That Influence Reaction Rates (Section 9.6)

9.21 Using collision theory, indicate why each of the following factors influences the rate of a reaction.
a. Temperature of reactants
b. Presence of a catalyst

9.22 Using collision theory, indicate why each of the following factors influences the rate of a reaction.
a. Physical nature of reactants
b. Reactant concentrations

9.23 Substances burn more rapidly in pure oxygen than in air. Explain why.

9.24 Milk will sour in a couple of days when left at room temperature, yet it can remain unspoiled for 2 weeks when refrigerated. Explain why.

9.25 Draw an energy diagram graph for an exothermic reaction where no catalyst is present. Then draw an energy diagram graph for the same reaction when a catalyst is present. Indicate the similarities and differences between the two diagrams.

9.26 Draw an energy diagram graph for an endothermic reaction where no catalyst is present. Then draw an energy diagram graph for the same reaction when a catalyst is present. Indicate the similarities and differences between the two diagrams.

9.27 The characteristics of four reactions, each of which involves only two reactants, are as follows.

Reaction	Activation energy	Temperature	Concentration of reactants
1	low	low	1 mole/L of each
2	high	low	1 mole/L of each
3	low	high	1 mole/L of each
4	low	low	1 mole/L of first reactant and 4 moles/L of second reactant

For each of the following pairs of the preceding reactions, compare the reaction rates when the two reactants are first mixed. Indicate which reaction is faster.

a. 1 and 2 b. 1 and 3 c. 1 and 4 d. 2 and 3

9.28 The characteristics of four reactions, each of which involves only two reactants, are as follows.

Reaction	Activation energy	Temperature	Concentration of reactants
1	high	low	1 mole/L of each
2	high	high	1 mole/L of each
3	low	low	1 mole/L of first reactant and 4 moles/L of second reactant
4	low	low	4 moles/L of each

For each of the following pairs of the preceding reactions, compare the reaction rates when the two reactants are first mixed. Indicate which reaction is faster.

a. 1 and 2 b. 1 and 3 c. 1 and 4 d. 3 and 4

Chemical Equilibrium (Section 9.7)

9.29 What condition must be met in order for a system to be in a state of chemical equilibrium?

9.30 What relationship exists between the rates of the forward and reverse reactions for a system in a state of chemical equilibrium?

9.31 Sketch a graph showing how the concentrations of the reactants and products of a typical reversible chemical reaction vary with time.

9.32 Sketch a graph showing how the rates of the forward and reverse reactions for a typical reversible chemical reaction vary with time.

Equilibrium Constants (Section 9.8)

9.33 Write equilibrium constant expressions for the following reactions.

a. $N_2O_4(g) \rightleftharpoons 2NO_2(g)$
b. $COCl_2(g) \rightleftharpoons CO(g) + Cl_2(g)$
c. $CS_2(g) + 4H_2(g) \rightleftharpoons CH_4(g) + 2H_2S(g)$
d. $2SO_2(g) + O_2(g) \rightleftharpoons 2SO_3(g)$

9.34 Write equilibrium constant expressions for the following reactions.

a. $3O_2(g) \rightleftharpoons 2O_3(g)$
b. $2NOCl(g) \rightleftharpoons 2NO(g) + Cl_2(g)$
c. $4NH_3(g) + 5O_2(g) \rightleftharpoons 4NO(g) + 6H_2O(g)$
d. $CO(g) + H_2O(g) \rightleftharpoons CO_2(g) + H_2(g)$

9.35 Write equilibrium constant expressions for the following reactions.

a. $H_2SO_4(l) \rightleftharpoons SO_3(g) + H_2O(l)$
b. $2Ag(s) + Cl_2(g) \rightleftharpoons 2AgCl(s)$
c. $BaCl_2(aq) + Na_2SO_4(aq) \rightleftharpoons 2NaCl(aq) + BaSO_4(s)$
d. $2Na_2O(s) \rightleftharpoons 4Na(l) + O_2(g)$

9.36 Write equilibrium constant expressions for the following reactions.

a. $2KClO_3(s) \rightleftharpoons 2KCl(s) + 3O_2(g)$
b. $PCl_5(s) \rightleftharpoons PCl_3(l) + Cl_2(g)$
c. $AgNO_3(aq) + NaCl(aq) \rightleftharpoons AgCl(s) + NaNO_3(aq)$
d. $2FeBr_3(s) \rightleftharpoons 2FeBr_2(s) + Br_2(g)$

9.37 Calculate the value of the equilibrium constant for the reaction

$$N_2O_4(g) \rightleftharpoons 2NO_2(g)$$

if the concentrations of the species at equilibrium are $[N_2O_4] = 0.213$ and $[NO_2] = 0.0032$.

9.38 Calculate the value of the equilibrium constant for the reaction

$$N_2(g) + 2O_2(g) \rightleftharpoons 2NO_2(g)$$

if the concentrations of the species at equilibrium are $[N_2] = 0.0013$, $[O_2] = 0.0024$, and $[NO_2] = 0.00065$.

9.39 Use the given K_{eq} value and the terminology in Table 9.2 to describe the relative amounts of reactants and products present in each of the following equilibrium situations.

a. $H_2(g) + Br_2(g) \rightleftharpoons 2HBr(g)$ K_{eq} (25°C) = 2.0×10^9
b. $2HCl(g) \rightleftharpoons H_2(g) + Cl_2(g)$ K_{eq} (25°C) = 3.2×10^{-34}
c. $SO_2(g) + NO_2(g) \rightleftharpoons NO(g) + SO_3(g)$ K_{eq} (460°C) = 85.0
d. $COCl_2(g) \rightleftharpoons CO(g) + Cl_2(g)$ K_{eq} (395°C) = 0.046

9.40 Use the given K_{eq} value and the terminology in Table 9.2 to describe the relative amounts of reactants and products present in each of the following equilibrium situations.

a. $2NO(g) \rightleftharpoons N_2(g) + O_2(g)$ K_{eq} (25°C) = 1×10^{30}
b. $N_2(g) + 3H_2(g) \rightleftharpoons 2NH_3(g)$ K_{eq} (25°C) = 1×10^9
c. $PCl_5(g) \rightleftharpoons PCl_3(g) + Cl_2(g)$ K_{eq} (127°C) = 1×10^{-2}
d. $2Na_2O(s) \rightleftharpoons 4Na(l) + O_2(g)$ K_{eq} (427°C) = 1×10^{-25}

Le Châtelier's Principle (Section 9.9)

9.41 For the reaction

$$2Cl_2(g) + 2H_2O(g) \rightleftharpoons 4HCl(g) + O_2(g)$$

determine in what direction the equilibrium will be shifted by each of the following changes.

a. Increase in Cl_2 concentration
b. Increase in O_2 concentration
c. Decrease in H_2O concentration
d. Decrease in HCl concentration

9.42 For the reaction

$$2Cl_2(g) + 2H_2O(g) \rightleftharpoons 4HCl(g) + O_2(g)$$

determine in what direction the equilibrium will be shifted by each of the following changes.

a. Increase in H_2O concentration

b. Increase in HCl concentration

c. Decrease in O_2 concentration

d. Decrease in Cl_2 concentration

9.43 For the reaction

$$C_6H_6(g) + 3H_2(g) \rightleftharpoons C_6H_{12}(g) + \text{heat}$$

determine in what direction the equilibrium will be shifted by each of the following changes.

a. Increasing the concentration of C_6H_{12}

b. Decreasing the concentration of C_6H_6

c. Increasing the temperature

d. Decreasing the pressure by increasing the volume of the container

9.44 For the reaction

$$C_6H_6(g) + 3H_2(g) \rightleftharpoons C_6H_{12}(g) + \text{heat}$$

determine in what direction the equilibrium will be shifted by each of the following changes.

a. Decreasing the concentration of H_2

b. Increasing the concentration of C_6H_6

c. Decreasing the temperature

d. Increasing the pressure by decreasing the volume of the container

9.45 Consider the following chemical system at equilibrium.

$$CO(g) + H_2O(g) + \text{heat} \rightleftharpoons CO_2(g) + H_2(g)$$

For each of the following adjustments of conditions, indicate the effect (shifts left, shifts right, or no effect) on the position of equilibrium.

a. Refrigerating the equilibrium mixture

b. Adding a catalyst to the equilibrium mixture

c. Adding CO to the equilibrium mixture

d. Increasing the size of the reaction container

9.46 Consider the following chemical system at equilibrium.

$$CO(g) + H_2O(g) + \text{heat} \rightleftharpoons CO_2(g) + H_2(g)$$

For each of the following adjustments of conditions, indicate the effect (shifts left, shifts right, or no effect) on the position of equilibrium.

a. Heating the equilibrium mixture

b. Increasing the pressure on the equilibrium mixture by adding a nonreactive gas

c. Adding H_2 to the equilibrium mixture

d. Decreasing the size of the container holding the mixture

Additional Problems

9.47 Characterize each of the following reactions using one selection from the choices *redox* and *nonredox* combined with one selection from the choices *combination, decomposition, single replacement,* and *double replacement.*

a. $Zn + Cu(NO_3)_2 \longrightarrow Zn(NO_3)_2 + Cu$

b. $2SO_2 + O_2 \longrightarrow 2SO_3$

c. $2CuO \longrightarrow 2Cu + O_2$

d. $NaCl + AgNO_3 \longrightarrow AgCl + NaNO_3$

9.48 Classify each of the following reactions as (1) a redox reaction, (2) a nonredox reaction, or (3) "can't classify" because of insufficient information.

a. A combination reaction in which one reactant is an element and the other is a compound

b. A decomposition reaction in which the products are all elements

c. A decomposition reaction in which one of the products is an element

d. A single-replacement reaction in which both of the reactants are compounds

9.49 In each of the following statements, choose the word in parentheses that best completes the statement.

a. The process of reduction is associated with the (loss, gain) of electrons.

b. The oxidizing agent in a redox reaction is the substance that undergoes (oxidation, reduction).

c. Reduction always results in an (increase, decrease) in the oxidation number of an element.

d. A reducing agent in a redox reaction is the substance that contains the element that undergoes an (increase, decrease) in oxidation number.

9.50 Indicate whether each of the following substances undergoes an increase in oxidation number or a decrease in oxidation number in a redox reaction.

a. The oxidizing agent

b. The reducing agent

c. The substance undergoing oxidation

d. The substance undergoing reduction

9.51 Which of the following changes would affect the *value* of a system's equilibrium constant?

a. Removal of a reactant or product from an equilibrium mixture

b. Decrease in the system's total pressure

c. Increase in the system's temperature

d. Addition of a catalyst to the equilibrium mixture

9.52 Write a balanced chemical equation for a totally gaseous equilibrium system that would lead to the following expression for the equilibrium constant.

$$K_{eq} = \frac{[CH_4][H_2S]^2}{[CS_2][H_2]^4}$$

9.53 For which of the following reactions is product formation favored by high temperature?

a. $N_2(g) + 2O_2(g) + \text{heat} \rightleftharpoons 2NO_2(g)$

b. $2N_2(g) + 6H_2O(g) + \text{heat} \rightleftharpoons 4NH_3(g) + 3O_2(g)$

c. $C_2H_4(g) + 3O_2(g) \rightleftharpoons 2CO_2(g) + 2H_2O(g) + \text{heat}$

d. $2KClO_3(s) + \text{heat} \rightleftharpoons 2KCl(s) + 3O_2(g)$

9.54 Predict the direction in which each of the following equilibria will shift if the pressure within the system is increased by reducing volume, using the choices *left, right,* and *no effect.*

a. $H_2(g) + C_2N_2(g) \rightleftharpoons 2HCN(g)$

b. $CO(g) + Br_2(g) \rightleftharpoons COBr_2(g)$

c. $CS_2(g) + 4H_2(g) \rightleftharpoons CH_4(g) + 2H_2S(g)$

d. $Ni(s) + 4CO(g) \rightleftharpoons Ni(CO)_4(g)$

Grid Problems

9.55

1. CH_4	2. SO_2	3. O_3
4. Na_2CO_3	5. NH_4^+	6. PO_4^{3-}

Select from the grid *all* correct responses for each of the following situations.

a. Substances in which one element has a +4 oxidation number
b. Substances in which the algebraic sum of the oxidation numbers is zero
c. Substances in which two elements have positive oxidation numbers
d. Substances in which all atoms have the same oxidation number

9.56

1. $S \longrightarrow SO_2$	2. $SO_2 \longrightarrow SO_4^{2-}$	3. $S^{2-} \longrightarrow S$
4. $S^{2-} \longrightarrow SO_4^{2-}$	5. $SO_2 \longrightarrow S^{2-}$	6. $SO_3 \longrightarrow SO_4^{2-}$

Select from the grid *all* correct responses for each of the following situations.

a. A change in which the reactant is oxidized
b. A change in which the reactant is reduced
c. A change in which the reactant is neither oxidized nor reduced
d. A change in which the reactant is an oxidizing agent

9.57

1. increase in temperature	2. increase in reactant surface area	3. addition of catalyst
4. decrease in collision frequency	5. increase in reactant concentration	6. decrease in activation energy

Select from the grid *all* correct responses for each of the following situations.

a. Factors that increase the reaction rate
b. Factors that increase the collision frequency
c. Factors that alter the reaction pathway
d. Factors that change the value of the equilibrium constant

9.58

1. increase in temperature	2. addition of a catalyst	3. increase in concentration of D
4. increase in reaction container size	5. addition of an inert gas	6. decrease in concentration of B

Select from the grid *all* correct responses for each of the following situations.

a. Changes that cause a shift to the right for the equilibrium
$$A(g) + B(g) \rightleftharpoons 2C(g) + D(g) + \text{heat}$$
b. Changes that cause a shift to the right for the equilibrium
$$2A(g) + C(g) \rightleftharpoons 2B(g) + D(g) + \text{heat}$$
c. Changes that cause a shift to the left for the equilibrium
$$A(g) + B(g) + C(g) \rightleftharpoons 3D(g) + \text{heat}$$
d. Changes that do not affect the position of equilibrium for the equilibrium
$$3A(g) + D(g) + \text{heat} \rightleftharpoons 2B(g) + C(g)$$

Acids, Bases, and Salts

CHAPTER OUTLINE

10.1 Arrhenius Acid–Base Theory 237
10.2 Brønsted–Lowry Acid–Base Theory 238
10.3 Mono-, Di-, and Triprotic Acids 241
10.4 Strengths of Acids and Bases 242
10.5 Ionization Constants for Acids and Bases 243
10.6 Salts 244
10.7 Acid–Base Neutralization Reactions 245
10.8 Self-Ionization of Water 246
10.9 The pH Concept 248
10.10 pK_a Method for Expressing Acid Strength 251

Chemistry at a Glance: **Acids and Acidic Solutions** 252

10.11 The pH of Aqueous Salt Solutions 252
10.12 Buffers 256

Chemistry at a Glance: **Buffer Systems** 262

10.13 The Henderson–Hasselbalch Equation 261
10.14 Electrolytes 262
10.15 Acid–Base Titrations 264

Chemical Connections

10.1 Acid Rain: Excess Acidity 254
10.2 Blood Plasma pH and Hydrolysis 256
10.3 Buffering Action in Human Blood 260
10.4 Electrolytes and Body Fluids 263

Fish are very sensitive to the acidity of the water present in an aquarium.

Acids, bases, and salts are among the most common and important compounds known. In the form of aqueous solutions, these compounds are key materials in both biological systems and the chemical industry. A major ingredient of gastric juice in the stomach is hydrochloric acid. Quantities of lactic acid are produced when the human body is subjected to strenuous exercise. The lye used in making soap contains the base sodium hydroxide. Bases are ingredients in many stomach antacid formulations. The white crystals you sprinkle on your food to make it taste better represent only one of many hundreds of salts that exist.

10.1 Arrhenius Acid–Base Theory

In 1884, the Swedish chemist Svante August Arrhenius (1859–1927) proposed that acids and bases be defined in terms of the chemical species they form when they dissolve in water. An **Arrhenius acid** *is a hydrogen-containing compound that, in water, produces hydrogen ions* (H^+). The acidic species in Arrhenius theory is thus the hydrogen ion. An **Arrhenius base** *is a hydroxide-containing compound that, in water,*

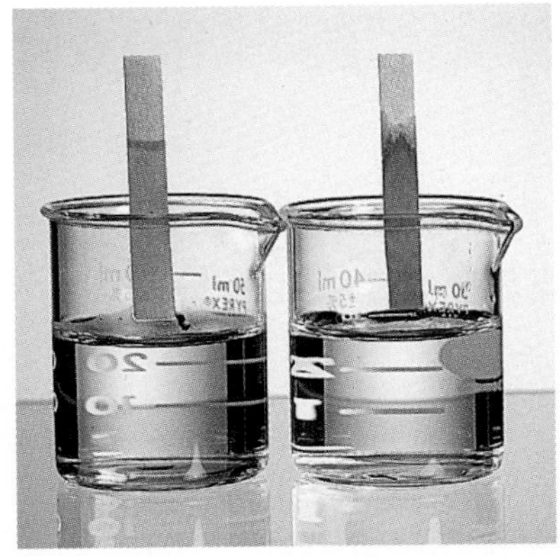

Litmus is a vegetable dye obtained from certain lichens found principally in the Netherlands. Paper treated with this dye turns from blue to red in acids (left) and from red to blue in bases (right).

• Arrhenius acids have a sour taste, change blue litmus paper to red, and are corrosive to many materials. Arrhenius bases have a bitter taste, change red litmus paper to blue, and are slippery (soapy) to the touch. (The bases themselves are not slippery but react with the fats in the skin to form new slippery or soapy compounds.)

produces hydroxide ions (OH^-). The basic species in Arrhenius theory is thus the hydroxide ion. For this reason, Arrhenius bases are also called *hydroxide bases.*

Two common examples of Arrhenius acids are HNO_3 (nitric acid) and HCl (hydrochloric acid).

$$HNO_3(l) \xrightarrow{H_2O} H^+(aq) + NO_3^-(aq)$$
$$HCl(g) \xrightarrow{H_2O} H^+(aq) + Cl^-(aq)$$

When Arrhenius acids are in the pure state (not in solution), they are covalent compounds; that is, they do not contain H^+ ions. This ion is formed through an interaction between water and the acid when they are mixed. **Ionization** *is the process in which individual positive and negative ions are produced from a molecular compound that is dissolved in solution.*

Two common examples of Arrhenius bases are NaOH (sodium hydroxide) and KOH (potassium hydroxide).

$$NaOH(s) \xrightarrow{H_2O} Na^+(aq) + OH^-(aq)$$
$$KOH(s) \xrightarrow{H_2O} K^+(aq) + OH^-(aq)$$

In direct contrast to acids, Arrhenius bases are ionic compounds in the pure state. When these compounds dissolve in water, the ions separate to yield the OH^- ions. **Dissociation** *is the process in which individual positive and negative ions are released from an ionic compound that is dissolved in solution.* Figure 10.1 contrasts the processes of dissociation (acids) and ionization (bases).

10.2 Brønsted–Lowry Acid–Base Theory

Although it is widely used, Arrhenius acid–base theory has some shortcomings. It is restricted to aqueous solution, and it cannot explain why compounds like ammonia (NH_3), which do not contain hydroxide ion, produce a basic water solution.

In 1923, Johannes Nicolaus Brønsted (1879–1947), a Danish chemist, and Thomas Martin Lowry (1874–1936), a British chemist, independently and almost simultaneously proposed broadened definitions for acids and bases that applied in both aqueous and nonaqueous solutions and that also explained how some nonhydroxide-containing substances, when added to water, produce basic solutions.

A **Brønsted–Lowry acid** *is any substance that can donate a proton* (H^+) *to some other substance.* A **Brønsted–Lowry base** *is any substance that can accept a proton* (H^+) *from some other substance.* In short, a Brønsted–Lowry acid is a *proton donor* (or hydrogen ion donor), and a Brønsted–Lowry base is a *proton acceptor* (or hydrogen ion acceptor). The terms *proton* and *hydrogen ion* are used interchangeably in acid–base discussions. Remember that an H^+ ion is a hydrogen atom (proton plus electron) that has lost its electron; hence it is a proton.

Figure 10.1
The difference between the aqueous solution processes of ionization (Arrhenius acids) and dissociation (Arrhenius bases). Ionization is the production of ions from a *molecular compound* that has been dissolved in solution. Dissociation is the production of ions from an *ionic compound* that has been dissolved in solution.

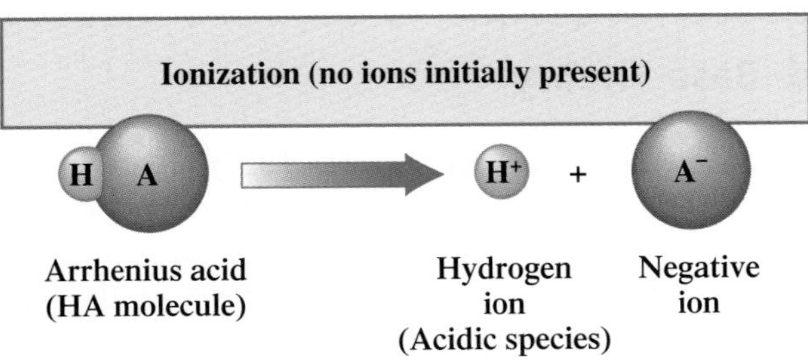

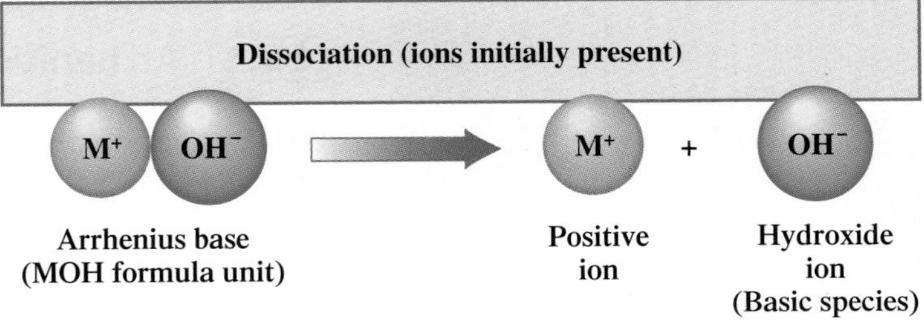

Table 10.1
Summary of Acid–Base Definitions

Arrhenius acid:	hydrogen-containing species that produces H^+ ion in aqueous solution
Arrhenius base:	hydroxide-containing species that produces OH^- ion in aqueous solution
Brønsted–Lowry acid: proton (H^+) donor	
Brønsted–Lowry base: proton (H^+) acceptor	

• The terms *hydrogen ion* and *proton* are used synonymously in acid–base discussions. Why? The predominant hydrogen isotope, ${}^1_1H$, is unique in that no neutrons are present; it consists of a proton and an electron. Thus the ion ${}^1_1H^+$, a hydrogen atom that has lost its only electron, is simply a proton.

Any chemical reaction involving a Brønsted–Lowry acid must also involve a Brønsted–Lowry base. You cannot have one without the other. Proton donation (from an acid) cannot occur unless an acceptor (a base) is present.

Brønsted–Lowry acid–base theory also includes the concept that hydrogen ions in an aqueous solution do not exist in the free state but, rather, react with water to form *hydronium ions.* The attraction between a hydrogen ion and polar water molecules is sufficiently strong to bond the hydrogen ion to a water molecule to form a hydronium ion (H_3O^+). The bond between them is a coordinate covalent bond (Section 5.5), because both electrons are furnished by the oxygen atom.

$$H^+ + :\ddot{O}\text{—}H \longrightarrow \left[H:\ddot{O}\text{—}H \right]^+$$

(with H bonded below O in each structure)

Hydronium ion

• A Brønsted–Lowry base—a proton acceptor—must contain an atom that possesses a pair of nonbonding electrons that can be used in forming a coordinate covalent bond to an incoming proton (from a Brønsted–Lowry acid).

When gaseous hydrogen chloride dissolves in water, it forms hydrochloric acid. This is a simple Brønsted–Lowry acid–base reaction. The chemical equation for this process is

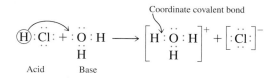

Coordinate covalent bond

$$\underset{\text{Acid}}{(H):\ddot{\underset{..}{Cl}}:} + \underset{\text{Base}}{:\ddot{O}:H} \longrightarrow \left[H:\ddot{O}:H \right]^+ + \left[:\ddot{\underset{..}{Cl}}: \right]^-$$

(with H below O in the water and hydronium structures)

The hydrogen chloride behaves as an acid by donating a proton to a water molecule. Because the water molecule accepts the proton, to become H_3O^+, it is the base.

It is not necessary that a water molecule be one of the reactants in a Brønsted–Lowry acid–base reaction; the reaction does not have to take place in the liquid state. Brønsted–Lowry acid–base theory can be used to describe gas-phase reactions. The white solid haze that often covers glassware in a chemistry laboratory results from the gas-phase reaction between HCl and NH_3:

$$\underset{\text{Acid}}{(H):\ddot{\underset{..}{Cl}}:} + \underset{\text{Base}}{:\underset{H}{\overset{H}{N}}:H} \longrightarrow \left[H:\underset{H}{\overset{H}{N}}:H \right]^+ + \left[:\ddot{\underset{..}{Cl}}: \right]^-$$

A white cloud of finely divided solid NH_4Cl is produced by the acid–base reaction that results when the colorless gases HCl and NH_3 mix. (The gases in this photograph escaped from the concentrated solutions of HCl and NH_3.)

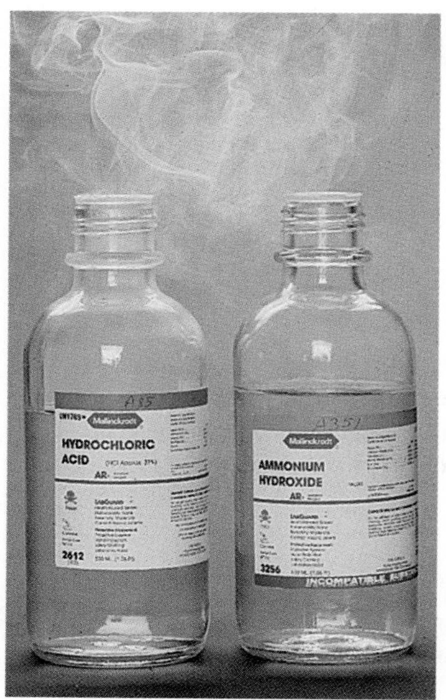

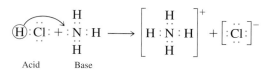

This is a Brønsted–Lowry acid–base reaction because the HCl molecules donate protons to the NH_3, forming NH_4^+ and Cl^- ions. These ions instantaneously combine to form the white solid NH_4Cl.

All the acids and bases included in the Arrhenius theory are also acids and bases according to the Brønsted–Lowry theory. However, the converse is not true; some substances that are not considered Arrhenius bases are Brønsted–Lowry bases. Table 10.1 summarizes these two types of acid–base definitions.

• Conjugate Acid–Base Pairs

For most Brønsted–Lowry acid–base reactions, 100% proton transfer does not occur. Instead, an equilibrium situation (Section 9.7) is reached in which a forward and a reverse reaction are occurring at the same rate.

The equilibrium mixture for a Brønsted–Lowry acid–base reaction always has two acids and two bases present. Consider the acid–base reaction involving hydrogen fluoride and water:

$$HF(aq) + H_2O(l) \rightleftharpoons H_3O^+(aq) + F^-(aq)$$

For the forward reaction, the HF molecules donate protons to water molecules. Thus the HF is functioning as an acid, and the H_2O is functioning as a base.

$$\underset{\text{Acid}}{HF(aq)} + \underset{\text{Base}}{H_2O(l)} \longrightarrow H_3O^+(aq) + F^-(aq)$$

For the reverse reaction, the one going from right to left, a different picture emerges. Here, H_3O^+ is functioning as an acid (by donating a proton), and F^- behaves as a base (by accepting the proton).

$$\underset{\text{Acid}}{H_3O^+(aq)} + \underset{\text{Base}}{F^-(aq)} \longrightarrow HF(aq) + H_2O(l)$$

• *Conjugate* means "coupled" or "joined together" (as in a pair).

The two acids and two bases involved in a Brønsted–Lowry acid–base equilibrium mixture can be grouped into two conjugate acid–base pairs. A **conjugate acid–base pair** *is two species that differ by one proton.* The two conjugate acid–base pairs in our example are

Conjugate pair (HF / F^-)

$$\underset{\text{Acid}}{HF(aq)} + \underset{\text{Base}}{H_2O(l)} \rightleftharpoons \underset{\text{Acid}}{H_3O^+(aq)} + \underset{\text{Base}}{F^-(aq)}$$

Conjugate pair (H_2O / H_3O^+)

• Every acid has a conjugate base, and every base has a conjugate acid. In general terms, these relationships can be diagrammed as follows:

$$\underset{\text{Acid}}{HA} + \underset{\text{Base}}{B} \rightleftharpoons \underset{\text{Conjugate acid}}{HB^+} + \underset{\text{Conjugate base}}{A^-}$$

The **conjugate base** *of an acid is the species that remains when an acid loses a proton.* The conjugate base of HF is F^-. The **conjugate acid** *of a base is the species formed when a base accepts a proton.* The H_3O^+ ion is the conjugate acid of H_2O. A slash mark separating an acid (on the left) from its conjugate base (on the right) is often used in representing conjugate acid–base pairs: for example, HF/F^- and H_3O^+/H_2O.

Example 10.1

Determining the Formula of One Member of a Conjugate Acid–Base Pair When Given the Other Member

Write the formula of each of the following.

a. The conjugate base of HSO_4^- **b.** The conjugate acid of NO_3^-
c. The conjugate base of H_3PO_4 **d.** The conjugate acid of $HC_2O_4^-$

Solution

a. A conjugate base can always be found by removing one H^+ from a given acid. Removing one H^+ (both the atom and the charge) from HSO_4^- leaves SO_4^{2-}. Thus SO_4^{2-} is the conjugate base of HSO_4^-.

b. A conjugate acid can always be found by adding one H^+ to a given base. Adding one H^+ (both the atom and the charge) to NO_3^- produces HNO_3. Thus HNO_3 is the conjugate acid of NO_3^-.

c. Proceeding as in part **a**, the removal of a H^+ ion from H_3PO_4 produces the $H_2PO_4^-$ ion. Thus $H_2PO_4^-$ is the conjugate base of H_3PO_4.

d. Proceeding as in part **b**, the addition of a H^+ ion to $HC_2O_4^-$ produces the $H_2C_2O_4$ molecule. Thus $H_2C_2O_4$ is the conjugate acid of $HC_2O_4^-$.

Practice Exercise 10.1

Write the formula of each of the following.

a. The conjugate acid of ClO_3^- **b.** The conjugate base of NH_3
c. The conjugate acid of PO_4^{3-} **d.** The conjugate base of HS^-

• *Answers:* **a.** $HClO_3$; **b.** NH_2^-; **c.** HPO_4^{2-}; **d.** S^{2-}

• Amphoteric Substances

• The term *amphoteric* comes from the Greek *amphoteres,* which means "partly one and partly the other." Just as an amphibian is an animal that lives partly on land and partly in the water, an amphoteric substance is sometimes an acid and sometimes a base.

Some molecules or ions are able to function as either an acid or a base, depending on the kind of substance with which they react. Such molecules are said to be amphoteric. An **amphoteric substance** *can either lose or accept a proton and thus can function as either an acid or a base.*

Water is the most common amphoteric substance. Water functions as a base in the first of the following two reactions and as an acid in the second.

$$\underset{\text{Acid}}{HNO_3(aq)} + \underset{\text{Base}}{H_2O(l)} \rightleftharpoons H_3O^+(aq) + NO_3^-(aq)$$

$$\underset{\text{Base}}{NH_3(aq)} + \underset{\text{Acid}}{H_2O(l)} \rightleftharpoons NH_4^+(aq) + OH^-(aq)$$

10.3 Mono-, Di-, and Triprotic Acids

Acids can be classified according to the number of hydrogen ions they can transfer per molecule during an acid–base reaction. A **monoprotic acid** *is an acid that transfers one H^+ ion (proton) per molecule during an acid–base reaction.* Hydrochloric acid (HCl) and nitric acid (HNO_3) are both monoprotic acids.

A **diprotic acid** *is an acid that can transfer two H^+ ions (two protons) per molecule during an acid–base reaction.* Carbonic acid (H_2CO_3) is a diprotic acid. The transfer of protons for a diprotic acid always occurs in steps. For H_2CO_3, the two steps are

• If the double arrows in the equation for a system at equilibrium are of unequal length, the longer arrow indicates the direction in which the equilibrium is displaced.
$\leftharpoondown\!\!\!\rightharpoonup$ Equilibrium displaced toward reactants
$\rightleftharpoons$ Equilibrium displaced toward products

$$H_2CO_3(aq) + H_2O(l) \rightleftharpoons H_3O^+(aq) + HCO_3^-(aq)$$
$$HCO_3^-(aq) + H_2O(l) \rightleftharpoons H_3O^+(aq) + CO_3^{2-}(aq)$$

A few triprotic acids exist. A **triprotic acid** *is an acid that can transfer three H^+ ions (three protons) per molecule during an acid–base reaction.* Phosphoric acid, H_3PO_4, is the most common triprotic acid. The three proton-transfer steps for this acid are

$$H_3PO_4(aq) + H_2O(l) \rightleftharpoons H_3O^+(aq) + H_2PO_4^-(aq)$$
$$H_2PO_4^-(aq) + H_2O(l) \rightleftharpoons H_3O^+(aq) + HPO_4^{2-}(aq)$$
$$HPO_4^{2-}(aq) + H_2O(l) \rightleftharpoons H_3O^+(aq) + PO_4^{3-}(aq)$$

The general term **polyprotic acid** *describes acids that can transfer two or more H^+ ions (protons) during an acid–base reaction.*

The number of hydrogen atoms present in one molecule of an acid *cannot* always be used to classify the acid as mono-, di-, or triprotic. For example, a molecule of acetic acid contains four hydrogen atoms, and yet it is a monoprotic acid. Only one of the hydrogen atoms in acetic acid is *acidic;* that is, only one of the hydrogen atoms leaves the molecule when it is in solution.

The sour taste of limes and other citrus fruit is due to the citric acid present in the fruit juice.

Whether a hydrogen atom is acidic is related to its location in a molecule—that is, to which other atom it is bonded. From a structural viewpoint, the acidic behavior of acetic acid can be represented by the equation

$$H{-}\underset{\underset{H}{|}}{\overset{\overset{H}{|}}{C}}{-}\overset{\overset{O}{\|}}{C}{-}O{-}H + H_2O \rightleftharpoons H_3O^+ + \left[H{-}\underset{\underset{H}{|}}{\overset{\overset{H}{|}}{C}}{-}\overset{\overset{O}{\|}}{C}{-}O\right]^-$$

Note that one hydrogen atom is bonded to an oxygen atom and the other three hydrogen atoms are bonded to a carbon atom. The hydrogen atom bonded to the oxygen atom is the only acidic hydrogen; the hydrogen atoms that are bonded to carbon atoms are too tightly held to be removed by reaction with water molecules. Water has very little effect on a carbon–hydrogen bond, because that bond is only slightly polar. On the other hand, the hydrogen bonded to oxygen is involved in a very polar bond because of oxygen's large electronegativity (Section 5.9). Water, which is a polar molecule, readily attacks this bond.

Table 10.2
Commonly Encountered Strong Acids

HCl	hydrochloric acid
HBr	hydrobromic acid
HI	hydroiodic acid
HNO_3	nitric acid
$HClO_4$	perchloric acid
H_2SO_4	sulfuric acid

Writing the formula for acetic acid as $HC_2H_3O_2$ instead of $C_2H_4O_2$ emphasizes that there are two different kinds of hydrogen. One of the hydrogen atoms is acidic and the other three are not. When some hydrogen atoms are acidic and others are not, we write the acidic hydrogens first, thus separating them from the other hydrogen atoms in the formula. Citric acid, the principal acid in citrus fruits, is another example of an acid that contains both acidic and nonacidic hydrogens. Its formula, $H_3C_6H_5O_7$, indicates that three of the eight hydrogen atoms present in a molecule are acidic.

10.4 Strengths of Acids and Bases

Brønsted–Lowry acids vary in their ability to transfer protons and produce hydronium ions in aqueous solution. Acids can be classified as strong or weak on the basis of the extent to which proton transfer occurs in aqueous solution. A **strong acid** *is a substance that transfers 100%, or very nearly 100%, of its protons to water.* Thus if an acid is strong, nearly all of the acid molecules present give up protons to water. This extensive transfer of protons produces many hydronium ions (the acidic species) within the solution. A **weak acid** *is a substance that transfers only a small percentage of its protons to water.* The extent of proton transfer for weak acids is usually less than 5%.

- Learn the names and formulas of the six commonly encountered strong acids, and then assume that all other acids you encounter are weak unless you are told otherwise.

The vast majority of acids are weak rather than strong. The six most commonly encountered strong acids are listed in Table 10.2.

The extent to which an acid transfers protons to water depends on the molecular structure of the acid; molecular polarity and the strength and polarity of individual bonds are particularly important factors in determining whether an acid is strong or weak.

- It is important not to confuse the terms *strong* and *weak* with the terms *concentrated* and *dilute. Strong* and *weak* apply to the *extent of proton transfer,* not to the concentration of acid or base. *Concentrated* and *dilute* are relative concentration terms. Stomach acid (gastric juice) is a dilute (not weak) solution of a strong acid (HCl); it is 5% by mass hydrochloric acid.

The difference between a strong acid and a weak acid can also be stated in terms of the position of equilibrium (Section 9.7). Consider the reaction wherein HA represents the acid and H_3O^+ and A^- are the products from the proton transfer to H_2O. For strong acids, the equilibrium lies far to the right (100% or almost 100%):

$$HA + H_2O \rightleftharpoons H_3O^+ + A^-$$

For weak acids, the equilibrium position lies far to the left:

$$HA + H_2O \rightleftharpoons H_3O^+ + A^-$$

Thus, in solutions of strong acids, the predominant species are H_3O^+ and A^-. In solutions of weak acids, the predominant species is HA; very little proton transfer has occurred. The differences between strong and weak acids, in terms of species present in solution, are illustrated in Figure 10.2.

Just as there are strong acids and weak acids, there are also strong bases and weak bases. As with acids, there are only a few strong bases. Strong bases are limited to the hydroxides of Groups IA and IIA listed in Table 10.3. Of the strong bases, only NaOH and KOH are commonly encountered in a chemical laboratory.

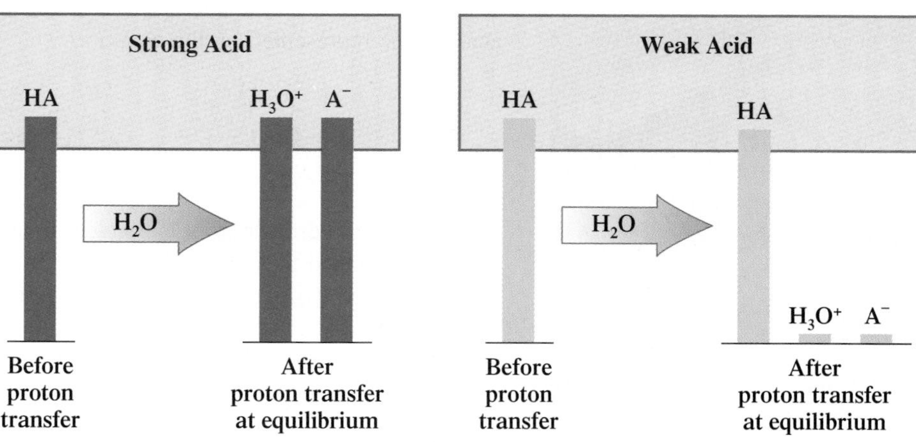

Figure 10.2
A comparison of the number of H_3O^+ ions (the acidic species) present in strong acid and weak acid solutions of the same concentration.

Table 10.3
Commonly Encountered Strong Hydroxide Bases

Group IA hydroxides	Group IIA hydroxides
LiOH	
NaOH	
KOH	$Ca(OH)_2$
RbOH	$Sr(OH)_2$
CsOH	$Ba(OH)_2$

Only one of the many weak bases that exist is fairly common—aqueous ammonia. In this solution of ammonia gas (NH_3) in water, small amounts of OH^- ions are produced through the reaction of NH_3 molecules with water.

$$NH_3(g) + H_2O(l) \rightleftharpoons NH_4^+(aq) + OH^-(aq)$$

A solution of aqueous ammonia is sometimes erroneously called ammonium hydroxide. Aqueous ammonia is the preferred designation because most of the NH_3 present has not reacted with water; the equilibrium position lies far to the left. Only a few ammonium ions (NH_4^+) and hydroxide ions (OH^-) are present.

10.5 Ionization Constants for Acids and Bases

The strengths of various acids and bases can be quantified by use of ionization constants, which are forms of equilibrium constants (Section 9.8).

• HA is a frequently used *general* notation for a monoprotic acid. Similarly, H_2A denotes a diprotic acid.

An **acid ionization constant** *is the equilibrium constant corresponding to the ionization for the acid.* For an acid with the general formula HA, the acid ionization constant is obtained by writing the equilibrium constant expression for the reaction

$$HA(aq) + H_2O(l) \rightleftharpoons H_3O^+(aq) + A^-(aq)$$

which is

$$K_a = \frac{[H_3O^+][A^-]}{[HA]}$$

• Note the following relationships among acid strength, percent ionization, and K_a magnitude.

- Acid strength increases as percent ionization increases.
- Acid strength increases as the magnitude of K_a increases.
- Percent ionization increases as the magnitude of K_a increases.

The concentration of water is not included in the equilibrium constant expression because water is a pure liquid (Section 9.8). The symbol K_a is used to denote an acid ionization constant.

Table 10.4 gives K_a values, as well as percent ionization values (which can be calculated from K_a values), for selected weak acids. Acid strength increases as the K_a value increases. The actual value for K_a for a given acid must be determined by experimentally measuring the concentrations of HA, H_3O^+, and A^- in the acid solution and then using these values to calculate K_a.

Example 10.2

Calculating the Acid Ionization Constant for an Acid When Given Its Concentration and Percent Ionization

A 0.0100 M solution of an acid, HA, is 15% ionized. Calculate the acid ionization constant for this acid.

Solution

To calculate K_a for the acid, we need the concentrations of H_3O^+, A^-, and HA in the aqueous solution.

The concentration of H_3O^+ will be 15% of the original HA concentration. Thus the concentration of hydronium ion is

$$H_3O^+ = (0.15) \times (0.0100\ M) = 0.0015\ M$$

The ionization of a monoprotic acid produces hydronium ions and the conjugate base of the acid (A^- ions) in a 1 : 1 ratio. Thus the concentration of A^- will be the same as that of hydronium ion—that is, 0.0015 M.

The concentration of HA is equal to the original concentration diminished by that which ionized (15%, or 0.0015 M):

$$HA = 0.0100\ M - 0.0015\ M = 0.0085\ M$$

Substituting these values in the equilibrium expression gives

$$K_a = \frac{[H_3O^+][A^-]}{[HA]} = \frac{[0.0015][0.0015]}{[0.0085]} = 2.6 \times 10^{-4}$$

Table 10.4
Ionization Constant Values (K_a) and Percent Ionization Values for 1.0 M Solutions, at 24°C, of Selected Weak Acids

Name	Formula	K_a	Percent ionization
phosphoric acid	H_3PO_4	7.5×10^{-3}	8.3
nitrous acid	HNO_2	7.2×10^{-4}	2.7
hydrofluoric acid	HF	6.6×10^{-4}	2.5
acetic acid	$HC_2H_3O_2$	1.8×10^{-5}	0.42
carbonic acid	H_2CO_3	4.3×10^{-7}	0.065
dihydrogen phosphate ion	$H_2PO_4^-$	6.2×10^{-8}	0.025
hydrocyanic acid	HCN	4.4×10^{-10}	0.0020
hydrogen carbonate ion	HCO_3^-	5.6×10^{-11}	0.00075
hydrogen phosphate ion	HPO_4^{2-}	2.2×10^{-13}	0.000047

Practice Exercise 10.2

A 0.100 M solution of an acid, HA, is 6.0% ionized. Calculate the acid ionization constant for this acid.

- *Answer:* 3.8×10^{-4}

In Section 10.3, we noted that dissociation of a polyprotic acid occurs in a stepwise manner. In general, each successive step of proton transfer for a polyprotic acid occurs to a lesser extent than the previous step. For the dissociation series

$$H_2CO_3(aq) + H_2O(l) \rightleftharpoons H_3O^+(aq) + HCO_3^-(aq)$$
$$HCO_3^-(aq) + H_2O(l) \rightleftharpoons H_3O^+(aq) + CO_3^{2-}(aq)$$

the second proton is not so easily transferred as the first, because it must be pulled away from a negatively charged particle, HCO_3^-. (Remember that particles with opposite charge attract one another.) Accordingly, HCO_3^- is a weaker acid than H_2CO_3. The K_a values for these two acids (Table 10.4) are 5.6×10^{-11} and 4.3×10^{-7}, respectively.

Base strength follows the same principle as acid strength. Here, however, we deal with a base ionization constant, K_b. A **base ionization constant** *is the equilibrium constant corresponding to the ionization for the base.* The general expression for K_b is

$$K_b = \frac{[BH^+][OH^-]}{[B]}$$

where the reaction is

$$B(aq) + H_2O(l) \rightleftharpoons BH^+(aq) + OH^-(aq)$$

For the reaction involving the weak base NH_3,

$$NH_3(aq) + H_2O(l) \rightleftharpoons NH_4^+(aq) + OH^-(aq)$$

the base ionization constant expression is

$$K_b = \frac{[NH_4^+][OH^-]}{[NH_3]}$$

The K_b value for NH_3, the only common weak base, is 1.8×10^{-5}.

10.6 Salts

To a nonscientist, the term *salt* denotes a white granular substance that is used as a seasoning for food. To the chemist, the term *salt* has a much broader meaning; sodium chloride (table salt) is only one of thousands of salts known to a chemist. From a chem-

The acid–base reaction between sulfuric acid and barium hydroxide produces the insoluble salt barium sulfate.

ical viewpoint, a **salt** *is an ionic compound containing a metal or polyatomic ion as the positive ion and a nonmetal or polyatomic ion (except hydroxide) as the negative ion.* (Ionic compounds that contain hydroxide ion are bases rather than salts.)

Much information about salts has been presented in previous chapters, although the term *salt* was not explicitly used in these discussions. Formula writing and nomenclature for binary ionic compounds (salts) were covered in Sections 4.7 and 4.9. Many salts contain polyatomic ions such as nitrate and sulfate; these ions were discussed in Section 4.10. The solubility of ionic compounds (salts) in water was the topic of Section 8.4.

All common soluble salts are *completely* dissociated into ions in solution (Section 8.3). Even if a salt is only slightly soluble, the small amount that does dissolve completely dissociates. Thus the terms *weak* and *strong,* which are used to denote qualitatively the percent dissociation of acids and bases, are not applicable to salts. We do not use the terms *weak salt* and *strong salt.*

Acids, bases, and salts are related in that a salt is one of the products that results from the reaction of an acid with a hydroxide base. This particular type of reaction will be discussed in Section 10.7.

10.7 Acid–Base Neutralization Reactions

When acids and hydroxide bases are mixed, they react with one another and their acidic and basic properties disappear; we say they have neutralized each other. **Neutralization** *is the reaction between an acid and a hydroxide base to form a salt and water.* The neutralization process can be viewed as either a double-replacement reaction or a proton transfer reaction.

From a *double-replacement viewpoint* (Section 9.1),

$$AX + BY \longrightarrow AY + BX$$

we have, for the HCl–KOH neutralization,

$$\underset{\text{Acid}}{HCl} + \underset{\text{Base}}{KOH} \longrightarrow \underset{\text{Water}}{HOH} + \underset{\text{Salt}}{KCl}$$

The salt that is formed contains the negative ion from the acid ionization and the positive ion from the base dissociation.

From a *proton transfer viewpoint,* the formation of water results from the transfer of protons from H_3O^+ ions (the acidic species in aqueous solution) to OH^- ions (the basic species) (Figure 10.3).

Any time an acid is completely reacted with a base, neutralization occurs. It does not matter whether the acid and base are strong or weak. Sodium hydroxide (a strong base) and nitric acid (a strong acid) react as follows:

$$HNO_3 + NaOH \longrightarrow NaNO_3 + H_2O$$

The equation for the reaction of potassium hydroxide (a strong base) with hydrocyanic acid (a weak acid) is

$$HCN + KOH \longrightarrow KCN + H_2O$$

• Hydrochloric acid (HCl), which is necessary for proper digestion of food, is present in the gastric juices of the human stomach. Overeating and emotional factors can cause the stomach to produce too much hydrochloric acid, a condition often called acid indigestion or heartburn. Substances known as *antacids* provide symptomatic relief from this condition. Over-the-counter antacids such as Maalox, Tums, and Alka-Seltzer contain one or more *basic* substances that are capable of neutralizing the hydrochloric acid in gastric juice.

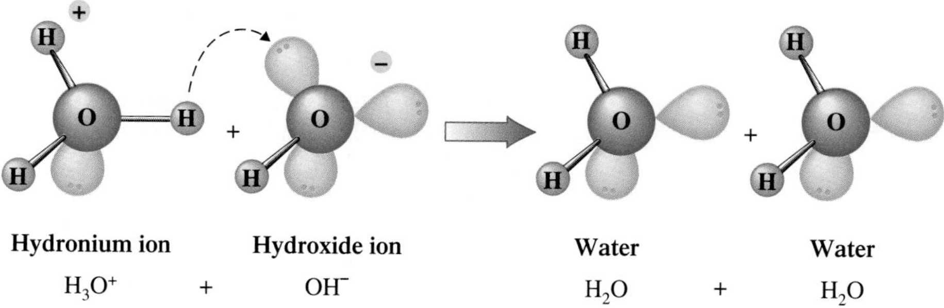

Figure 10.3
Formation of water by the transfer of protons from H_3O^+ ions to OH^- ions.

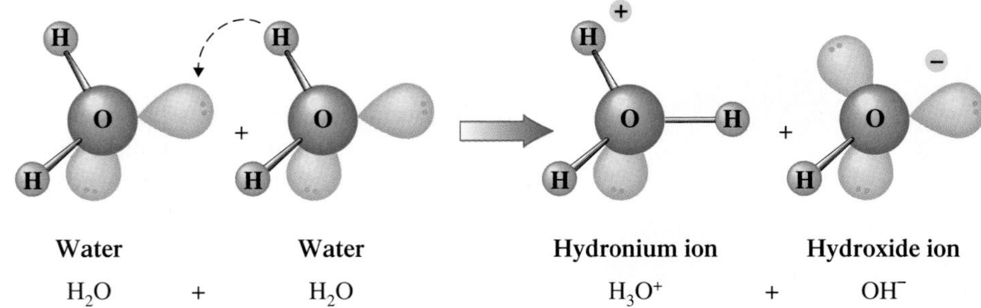

Figure 10.4
Self-ionization of water through proton transfer between water molecules.

Note that in both reactions, the products are a salt ($NaNO_3$ in the first reaction and KCN in the second) and water.

10.8 Self-Ionization of Water

Although we usually think of water as a covalent substance, experiments show that an *extremely small* percentage of water molecules in pure water interact with one another to form ions, a process that is called *self-ionization* (Figure 10.4). This interaction can be thought of as the transfer of protons between water molecules (Brønsted–Lowry theory, Section 10.2):

$$H_2O + H_2O \rightleftharpoons H_3O^+ + OH^-$$

The net effect of this transfer is the formation of *equal amounts* of hydronium and hydroxide ion. Such behavior for water should not seem surprising; we have already discussed the fact that water is an amphoteric substance (Section 10.2)—one that can either gain or lose protons. We have already seen several reactions in which H_2O acts as an acid and others where it acts as a base.

At any given time, the number of H_3O^+ and OH^- ions present in a sample of pure water is always extremely small. At equilibrium and 24°C, the H_3O^+ and OH^- concentrations are 1.00×10^{-7} M (0.000000100 M).

• Ion Product Constant for Water

The constant concentration of H_3O^+ and OH^- ions present in pure water at 24°C can be used to calculate a very useful relationship called the ion product constant for water. The **ion product constant for water** *is the numerical value* 1.00×10^{-14}, *obtained by multiplying together the molar concentrations of* H_3O^+ *ion and* OH^- *ion present in pure water.* We have the following equation for the ion product constant for water:

$$\begin{aligned} \text{Ion product constant for water} &= [H_3O^+] \times [OH^-] \\ &= (1.00 \times 10^{-7}) \times (1.00 \times 10^{-7}) \\ &= 1.00 \times 10^{-14} \end{aligned}$$

Remember that square brackets mean concentration in moles per liter (molarity).

The ion product constant expression for water is valid not only in pure water but also in water with solutes present. At all times, the product of the hydronium ion and hydroxide ion molarities in an aqueous solution at 24°C must equal 1.00×10^{-14}. Thus, if $[H_3O^+]$ is increased by the addition of an acidic solute, then $[OH^-]$ must decrease so that their product will still be 1.00×10^{-14}. Similarly, if additional OH^- ions are added to the water, then $[H_3O^+]$ must correspondingly decrease.

We can easily calculate the concentration of either H_3O^+ ion or OH^- ion present in an aqueous solution, if we know the concentration of the other ion, by simply rearranging the ion product expression $[H_3O^+] \times [OH^-] = 1.00 \times 10^{-14}$.

$$[H_3O^+] = \frac{1.00 \times 10^{-14}}{[OH^-]} \quad \text{or} \quad [OH^-] = \frac{1.00 \times 10^{-14}}{[H_3O^+]}$$

Example 10.3

Calculating the Hydroxide Ion Concentration of a Solution from a Given Hydronium Ion Concentration

Sufficient acidic solute is added to a quantity of water to produce $[H_3O^+] = 4.0 \times 10^{-3}$. What is $[OH^-]$ in this solution?

Solution

$[OH^-]$ can be calculated by using the ion product expression for water, rearranged in the form

$$[OH^-] = \frac{1.00 \times 10^{-14}}{[H_3O^+]}$$

Substituting into this expression the known $[H_3O^+]$ and doing the arithmetic give

$$[OH^-] = \frac{1.00 \times 10^{-14}}{4.0 \times 10^{-3}} = 2.5 \times 10^{-12}$$

Practice Exercise 10.3

Sufficient acidic solute is added to a quantity of water to produce $[H_3O^+] = 5.7 \times 10^{-6}$. What is $[OH^-]$ in this solution?

- *Answer:* 1.8×10^{-9} M

- If we know $[H_3O^+]$, we can always calculate $[OH^-]$, and vice versa, because of the ion product constant for water:

  $[H_3O^+] \times [OH^-] = 1.00 \times 10^{-14}$

- Neither $[H_3O^+]$ nor $[OH^-]$ is ever zero in an aqueous solution.

The relationship between $[H_3O^+]$ and $[OH^-]$ is that of an inverse proportion; when one increases, the other decreases. If $[H_3O^+]$ increases by a factor of 10^2, then $[OH^-]$ decreases by the same factor, 10^2. A graphical portrayal of this increase–decrease relationship for $[H_3O^+]$ and $[OH^-]$ is given in Figure 10.5.

• Acidic, Basic, and Neutral Solutions

Since a small amount of H_3O^+ ion and OH^- ion are present in all aqueous solutions, what determines whether a given solution is acidic or basic? It is the relative amounts of these two ions present. An **acidic solution** *is one in which the concentration of* H_3O^+ *ion is higher than that of* OH^- *ion.* A **basic solution** *is one in which the concentration of the* OH^- *ion is higher than that of the* H_3O^+ *ion.* A basic solution is also often referred to as

Figure 10.5
The relationship between $[H_3O^+]$ and $[OH^-]$ in aqueous solution is an inverse proportion; when $[H_3O^+]$ is increased, $[OH^-]$ decreases, and vice versa.

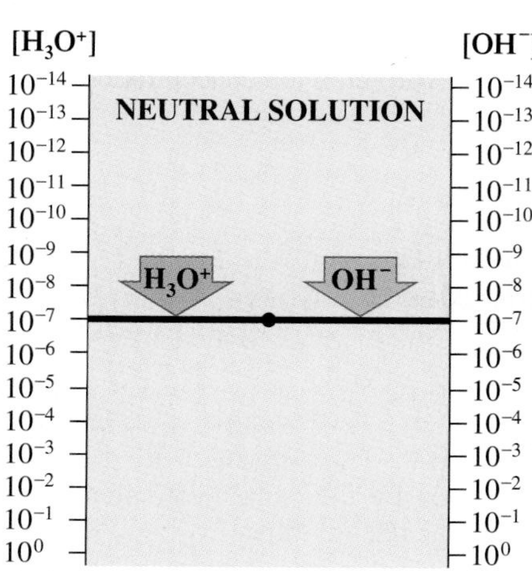

(a) In pure water the concentration of hydronium ions, $[H_3O^+]$, and that of hydroxide ions, $[OH^-]$, are equal. Both are 1.00×10^{-7} M at 24°C.

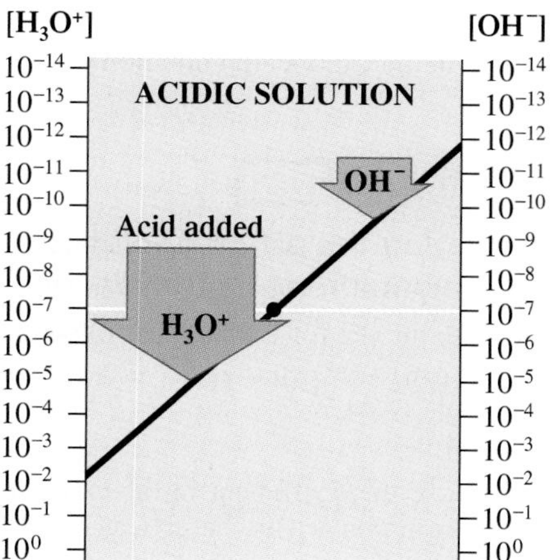

(b) If $[H_3O^+]$ is increased by a factor of 10^5 (from 10^{-7} M to 10^{-2} M), then $[OH^-]$ is decreased by a factor of 10^5 (from 10^{-7} M to 10^{-12} M).

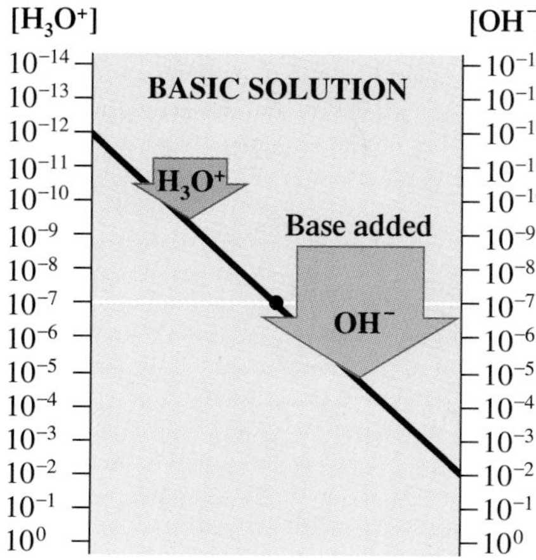

(c) If $[OH^-]$ is increased by a factor of 10^5 (from 10^{-7} M to 10^{-2} M), then $[H_3O^+]$ is decreased by a factor of 10^5 (from 10^{-7} M to 10^{-12} M).

Table 10.5
Relationship Between $[H_3O^+]$ and $[OH^-]$ in Neutral, Acidic, and Basic Solutions

neutral solution	$[H_3O^+] = [OH^-] = 1.00 \times 10^{-7}$
acidic solution $[H_3O^+] > [OH^-]$	$[H_3O^+]$ is greater than 1.00×10^{-7}; $[OH^-]$ is less than 1.00×10^{-7}
basic solution $[OH^-] > [H_3O^+]$	$[H_3O^+]$ is less than 1.00×10^{-7}; $[OH^-]$ is greater than 1.00×10^{-7}

an *alkaline solution.* It is possible to have an aqueous solution that is neither acidic nor basic but is, rather, a neutral solution. A **neutral solution** *is one in which the concentrations of* H_3O^+ *and* OH^- *ions are equal.* Table 10.5 summarizes the relationships between $[H_3O^+]$ and $[OH^-]$ that we have just considered.

10.9 The pH Concept

• The pH scale is a compact method for representing solution acidity.

Hydronium ion concentrations in aqueous solution range from relatively high values (10 M) to extremely small ones (10^{-14} M). It is inconvenient to work with numbers that extend over such a wide range; a hydronium ion concentration of 10 M is 1000 trillion times larger than a hydronium ion concentration of 10^{-14} M. The pH scale was developed as a more practical way to handle such a wide range of numbers. The **pH scale** *is a scale of small numbers that is used to specify molar hydronium ion concentration in an aqueous solution.*

Mathematically, pH is defined as the negative logarithm (log) of the hydronium ion concentration.

$$pH = -\log[H_3O^+]$$

• The *p* in pH comes from the German word *potenz,* which means "power," as in "power of 10."

(The letter *p*, as in pH, has been chosen to mean "negative logarithm of.")

• Integral pH Values

For any hydronium ion concentration expressed in exponential notation in which the coefficient is 1.0, the pH is given directly by the negative of the exponent value of the power of 10:

$$[H_3O^+] = 1.0 \times 10^{-x}$$
$$pH = x$$

• The rule for the number of significant figures in a logarithm is: The number of digits after the decimal place in a logarithm is equal to the number of significant figures in the original number.

$$[H_3O^+] = \underbrace{6.3}_{\text{Two significant figures}} \times 10^{-5}$$
$$pH = 4.\underbrace{20}_{\text{Two digits}}$$

Thus, if the hydronium ion concentration is 1.0×10^{-9}, then the pH will be 9.00. This simple relationship between pH and hydronium ion concentration is valid only when the coefficient in the exponential notation expression for the hydronium ion concentration is 1.0.

Example 10.4

Calculating the pH of a Solution When Given Its Hydronium Ion or Hydroxide Ion Concentration

Calculate the pH for each of the following solutions.

a. $[H_3O^+] = 1.0 \times 10^{-6}$ **b.** $[OH^-] = 1.0 \times 10^{-6}$

Solution

a. Because the coefficient in the exponential expression for the molar hydronium ion concentration is 1.0, the pH can be obtained from the relationships

$$[H_3O^+] = 1.0 \times 10^{-x}$$
$$pH = x$$

The power of 10 is -6 in this case, so the pH will be 6.00.

b. The given quantity involves hydroxide ion rather than hydronium ion. Thus we must calculate the hydronium ion concentration first and then calculate the pH.

$$[H_3O^+] = \frac{1.00 \times 10^{-14}}{1.0 \times 10^{-6}} = 1.0 \times 10^{-8}$$

A solution with a hydronium ion concentration of 1.0×10^{-8} M will have a pH of 8.00.

Practice Exercise 10.4

Calculate the pH for each of the following solutions.

a. $[H_3O^+] = 1.0 \times 10^{-3}$
b. $[OH^-] = 1.0 \times 10^{-8}$

• *Answers:* **a.** 3.00; **b.** 6.00

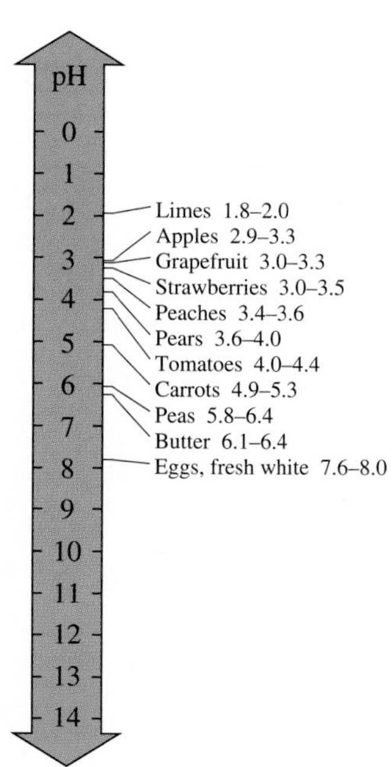

Most fruits and vegetables are acidic. Tart or sour taste is an indication that such is the case. Nonintegral pH values for selected foods are as shown here.

• Nonintegral pH Values

If the coefficient in the exponential expression for the molar hydronium ion concentration is *not* 1.0, then the pH will have a nonintegral value; that is, it will not be a whole number. For example, consider the following nonintegral pH values.

$$[H_3O^+] = 6.3 \times 10^{-5} \qquad pH = 4.20$$
$$[H_3O^+] = 5.3 \times 10^{-5} \qquad pH = 4.28$$
$$[H_3O^+] = 2.2 \times 10^{-4} \qquad pH = 3.66$$

The easiest way to obtain nonintegral pH values such as these involves using an electronic calculator that allows for the input of exponential numbers and that has a base-10 logarithm key (LOG).

In using such an electronic calculator, you can obtain logarithm values simply by pressing the LOG key after having entered the number whose log is desired. For pH, you must remember that after obtaining the log value, you must change signs because of the negative sign in the defining equation for pH.

Example 10.5

Calculating the pH of a Solution When Given Its Hydronium Ion Concentration

Calculate the pH for each of the following solutions.

a. $[H_3O^+] = 7.23 \times 10^{-8}$
b. $[H_3O^+] = 5.70 \times 10^{-3}$

Solution

a. Using an electronic calculator, first enter the number 7.23×10^{-8} into the calculator. Then use the LOG key to obtain the logarithm value, -7.1408617. Changing the sign of this number (because of the minus sign in the definition of pH) and adjusting for significant figures yield a pH value of 7.141.

b. Entering the number 5.70×10^{-3} into the calculator and then using the LOG key give a logarithm value of -2.2441251. This value translates into a pH value, after rounding, of 2.244.

Practice Exercise 10.5

Calculate the pH for each of the following solutions.

a. $[H_3O^+] = 4.44 \times 10^{-11}$
b. $[H_3O^+] = 8.92 \times 10^{-6}$

• *Answers:* **a.** 10.353; **b.** 5.050

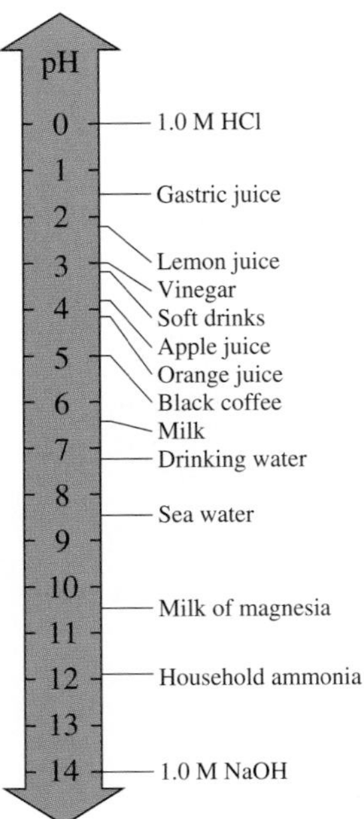

The pH values of selected common liquids. The lower the numerical value of the pH, the more acidic the substance is.

• pH Values and Hydronium Ion Concentration

It is often necessary to calculate the hydronium ion concentration for a solution from its pH value. This type of calculation, which is the reverse of that illustrated in Examples 10.4 and 10.5, is shown in Example 10.6.

Example 10.6

Calculating the Molar Hydronium Ion Concentration of a Solution from the Solution's pH

The pH of a solution is 6.80. What is the molar hydronium ion concentration for this solution?

Solution

From the defining equation for pH, we have

$$pH = -\log [H_3O^+] = 6.80$$
$$\log [H_3O^+] = -6.80$$

To find $[H_3O^+]$, we need to determine the *antilog* of -6.80.

How an antilog is obtained using a calculator depends on the type of calculator. Many calculators have an antilog function (sometimes labeled INV log) that performs this operation. If this key is present, then

1. Enter the number -6.80. Note that it is the *negative* of the pH that is entered into the calculator.
2. Press the INV log key (or an inverse key and then a log key). The result is the desired hydronium ion concentration.

$$\log [H_3O^+] = -6.80$$
$$\text{antilog } [H_3O^+] = 1.5848931 \times 10^{-7}$$

Rounded off, this value translates into a hydronium ion concentration of 1.6×10^{-7} M.

Some calculators use a 10^x key to perform the antilog operation. Use of this key is based on the mathematical identity

$$\text{antilog } x = 10^x$$

In our case, this means

$$\text{antilog } -6.80 = 10^{-6.80}$$

If the 10^x key is present, then

1. Enter the number -6.80 (the negative of the pH).
2. Press the function key 10^x. The result is the desired hydronium ion concentration.

$$[H_3O^+] = 10^{-6.80} = 1.6 \times 10^{-7}$$

Practice Exercise 10.6

The pH of a solution is 3.44. What is the molar hydronium ion concentration for this solution?

• *Answer:* 3.6×10^{-4} M

Figure 10.6
Relationships among pH values, $[H_3O^+]$, and $[OH^-]$ at 24°C.

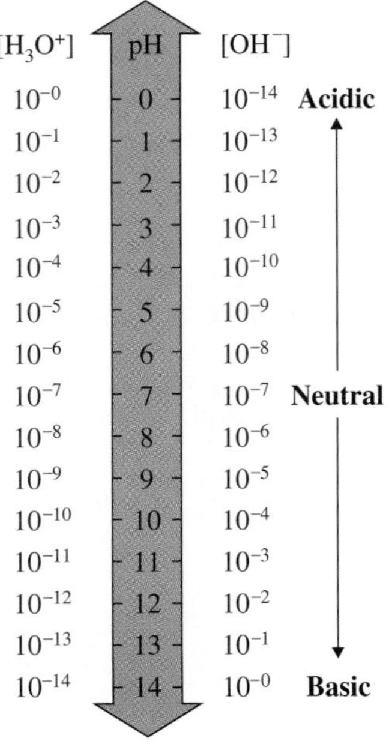

• Interpreting pH Values

The pH is simply a way of expressing hydronium ion concentration and thus identifies a solution as acidic, basic, or neutral. At 24°C, a neutral solution has a pH value of 7. Values of pH that are less than 7 correspond to acidic solutions, and values of pH that are greater than 7 are associated with basic solutions. The relationships among $[H_3O^+]$, $[OH^-]$, and pH are summarized in Figure 10.6. Note the following trends from the information presented in this figure.

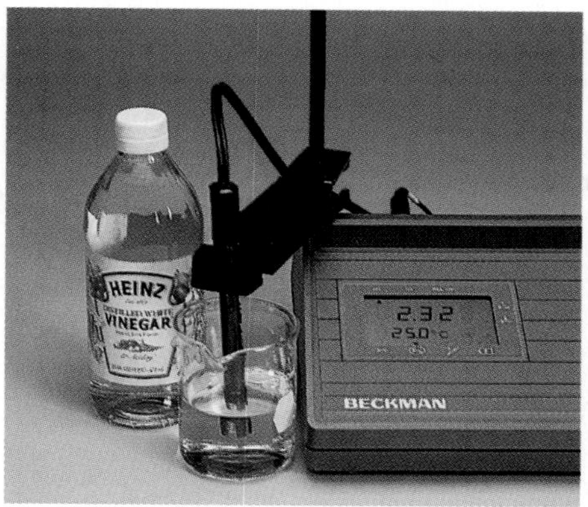

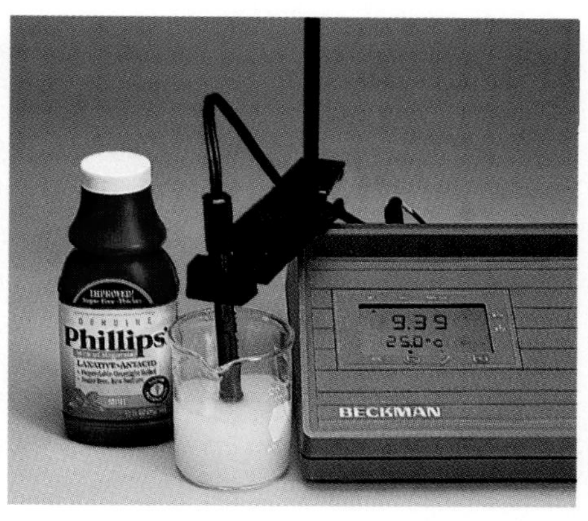

Figure 10.7
A pH meter gives an accurate measurement of pH values. The pH of vinegar is 2.32 (left). The pH of milk of magnesia in water is 9.39 (right).

• Solutions of low pH are more acidic than solutions of high pH; conversely, solutions of high pH are more basic than solutions of low pH.

1. The higher the concentration of hydronium ion, the lower the pH value. Another statement of this same trend is that lowering the pH always corresponds to increasing the hydronium ion concentration.
2. A change of 1 unit in pH always corresponds to a tenfold change in hydronium ion concentration. For example,

$$\text{Difference of 1}\begin{cases}\text{pH} = 1, \text{ then } [H_3O^+] = 0.1\,M \\ \text{pH} = 2, \text{ then } [H_3O^+] = 0.01\,M\end{cases}\text{tenfold difference}$$

In a laboratory, solutions of any pH can be created. The range of pH values that are displayed by natural solutions is more limited than that of prepared solutions, but solutions corresponding to most pH values can be found. A pH meter (Figure 10.7) helps chemists determine accurate pH values.

The pH values of several human body fluids are given in Table 10.6. Most human body fluids except gastric juices and urine have pH values within one unit of neutrality. Both blood plasma and spinal fluid are always slightly basic.

Chemistry at a Glance summarizes what we have said about acids and acidity.

10.10 pK_a Method for Expressing Acid Strength

In Section 10.5 ionization constants for acids and bases were introduced. These constants give an indication of the strengths of acids and bases. An additional method for expressing the strength of acids is in terms of pK_a units. The definition for pK_a is

$$pK_a = -\log K_a$$

• Like pH, pK_a is a positive number. The lower the pK_a value, the stronger the acid.

The pK_a for an acid is calculated from K_a in exactly the same way that pH is calculated from hydronium ion concentration.

Example 10.7

Calculating the pK_a of an Acid from the Acid's K_a Value

Determine the pK_a for acetic acid, $HC_2H_3O_2$, given that K_a for this acid is 1.8×10^{-5}.

Solution

Because the K_a value is 1.8×10^{-5} and pK_a = $-\log K_a$, we have

$$pK_a = -\log(1.8 \times 10^{-5}) = 4.74$$

The logarithm value 4.74 was obtained using an electronic calculator, as explained in Example 10.5.

Practice Exercise 10.7

Determine the pK_a for hydrocyanic acid, HCN, given that K_a for this acid is 4.4×10^{-10}.

• *Answer:* 9.36

Table 10.6
The Normal pH Range of Selected Body Fluids

Type of fluid	pH value
bile	6.8–7.0
blood plasma	7.3–7.5
gastric juices	1.0–3.0
milk	6.6–7.6
saliva	6.5–7.5
spinal fluid	7.3–7.5
urine	4.8–8.4

Chemical CONNECTIONS

10.1 Acid Rain: Excess Acidity

Rainfall, even in a pristine environment, has always been and will always be acidic. This acidity results from the presence of carbon dioxide in the atmosphere, which dissolves in water to produce carbonic acid (H_2CO_3), a weak acid.

$$CO_2(g) + H_2O(l) \longrightarrow H_2CO_3(aq)$$

This reaction produces rainwater with a pH between 5.6 and 6.2.

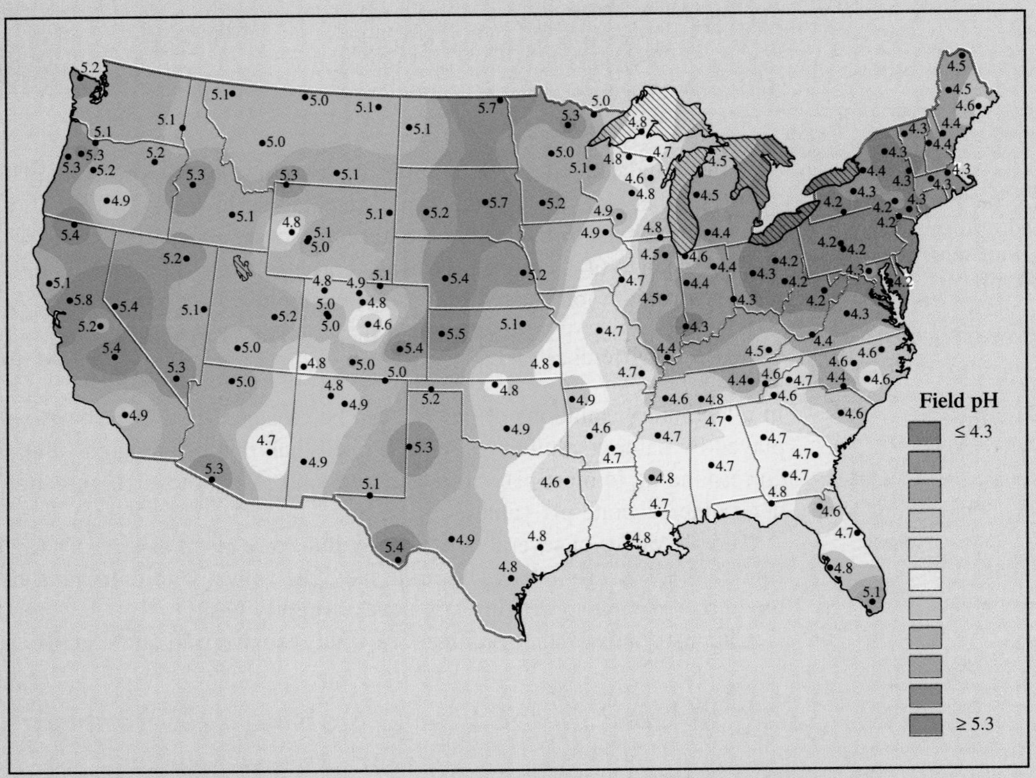

Average annual pH of precipitation in the United States

10.11 The pH of Aqueous Salt Solutions

The addition of an acid to water produces an acidic solution. The addition of a base to water produces a basic solution. What type of solution is produced when a salt is added to water? Because salts are the products of acid–base neutralizations, a logical supposition would be that salts dissolve in water to produce neutral (pH = 7) solutions. Such is the case for a *few* salts. Aqueous solutions of *most* salts, however, are either acidic or basic rather than neutral. Let us consider why this is so.

When a salt is dissolved in water, it completely ionizes; that is, it completely breaks up into the ions of which it is composed (Section 8.3). For many salts, one or more of the ions so produced are reactive toward water. The ensuing reaction, which is called hydrolysis, causes the solution to have a non-neutral pH. **Hydrolysis** *is the reaction of a substance with water to produce hydronium ion or hydroxide ion or both.*

• The term *hydrolysis* comes from the Greek *hydro,* which means "water," and *lysis,* which means "splitting."

• Types of Salt Hydrolysis

Not all salts hydrolyze. Which ones do and which ones do not? Of those salts that do hydrolyze, which produce acidic solutions and which produce basic solutions? The fol-

Acid rain is a generic term used to describe rainfall (or snowfall) whose pH is lower than the naturally produced value of 5.6. Acid rain has been observed with increasing frequency in many areas of the world. Rainfall with pH values between 4 and 5 is now common, and occasionally rainfall with a pH as low as 2 is encountered. Within the United States, the lowest acid rain pH values are encountered in the northeastern states (see the accompanying map). The maritime provinces of Canada have also been greatly affected.

Acid rain originates from the presence of sulfur oxides (SO_2 and SO_3), and to a lesser extent nitrogen oxides (NO and NO_2), in the atmosphere. After being discharged into the atmosphere, these pollutants can be chemically converted into sulfuric acid (H_2SO_4) and nitric acid (HNO_3) through oxidation processes. Several complicated pathways exist by which these two strong acids are produced. Which pathway is actually taken depends on numerous factors, including the intensity of sunlight and the amount of ammonia present in the atmosphere.

Small amounts of sulfur oxides and nitrogen oxides arise naturally from volcanic activity, lightning, and forest fires, but their major sources are human-related. The major source of sulfur oxide emissions is the combustion of coal associated with power plant operations. (The sulfur content of coal can be a high as 5% by mass.) Automobile exhaust is the major source of nitrogen oxides.

An important factor in determining the impact of acid rain on the environment is the ability of the natural ecosystem to neutralize incoming acidity. Generally speaking, the most sensitive areas overlie crystalline rock, whereas the least sensitive overlie limestone rock. Calcium carbonate and other basic substances associated with limestone rock are good neutralizing agents. Fortunately, low-pH rainfall directly entering lakes and streams does not automatically cause a severe decrease in pH. A large dilution factor accompanies rain falling directly into a large body of water.

The most observable effect of acid rain is the corrosion of building materials. Sulfuric acid (acid rain) readily attacks carbonate-based building materials (limestone, marble); the calcium carbonate is slowly converted into calcium sulfate.

$$CaCO_3(s) + H_2SO_4(aq) \longrightarrow CaSO_4(s) + CO_2(g) + H_2O(l)$$

The $CaSO_4$, which is more soluble than $CaCO_3$, is gradually eroded away. Many stone monuments show distinctly discernible erosion damage.

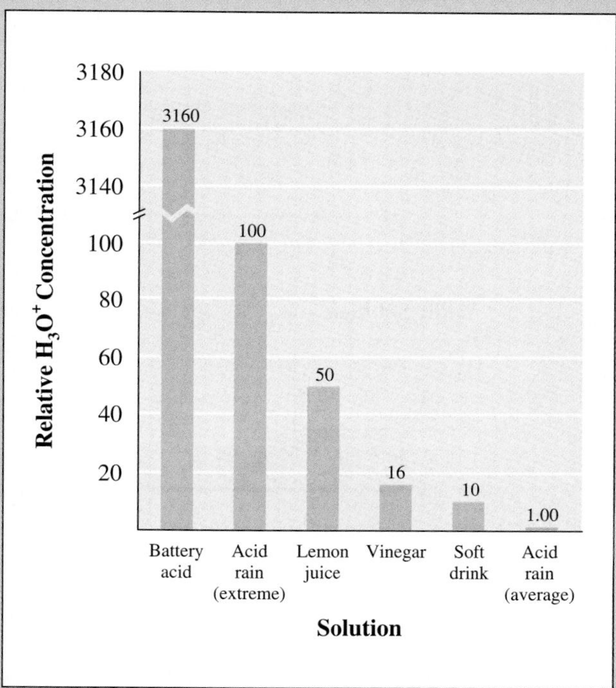

The relative hydronium concentrations in acid rain and some other acidic solutions compared with acid rain of pH 4.5, which is assigned a value of 1.00.

lowing guidelines, based on the neutralization "parentage" of a salt—that is, on the acid and base that produce the salt through neutralization—can be used to answer these questions.

1. The salt of a *strong acid* and a *strong base* does not hydrolyze, so the solution is neutral.
2. The salt of a *strong acid* and a *weak base* hydrolyzes to produce an acidic solution.
3. The salt of a *weak acid* and a *strong base* hydrolyzes to produce a basic solution.
4. The salt of a *weak acid* and a *weak base* hydrolyzes to produce a slightly acidic, neutral, or slightly basic solution, depending on the relative weaknesses of the acid and base.

These guidelines are summarized in Table 10.7.

The first prerequisite for using these guidelines is the ability to classify a salt into one of the four categories mentioned in the guidelines. This classification is accomplished by writing the neutralization equation (Section 10.7) that produces the salt and then

Acids and Acidic Solutions

CHEMISTRY AT A GLANCE

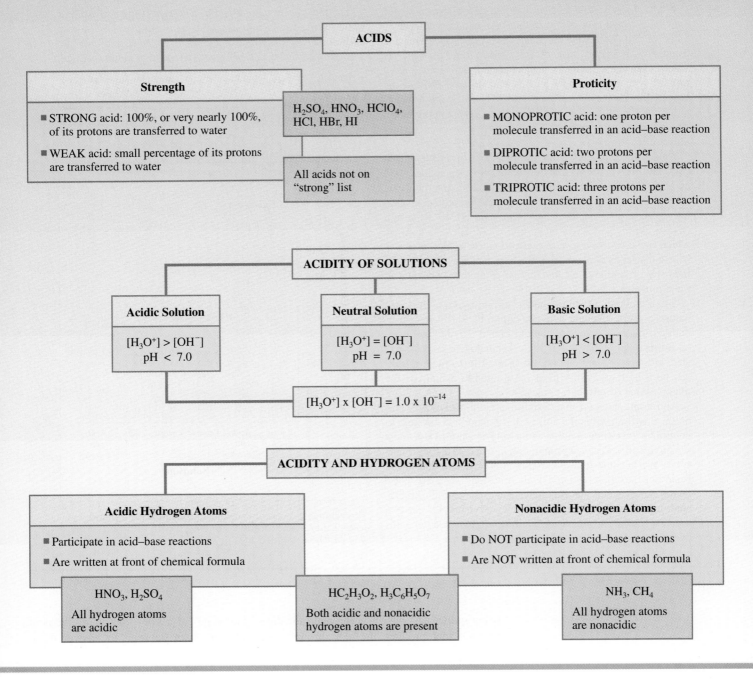

ACIDS

Strength

- STRONG acid: 100%, or very nearly 100%, of its protons are transferred to water
  - H_2SO_4, HNO_3, $HClO_4$, HCl, HBr, HI
- WEAK acid: small percentage of its protons are transferred to water
  - All acids not on "strong" list

Proticity

- MONOPROTIC acid: one proton per molecule transferred in an acid–base reaction
- DIPROTIC acid: two protons per molecule transferred in an acid–base reaction
- TRIPROTIC acid: three protons per molecule transferred in an acid–base reaction

ACIDITY OF SOLUTIONS

Acidic Solution

$[H_3O^+] > [OH^-]$
$pH < 7.0$

Neutral Solution

$[H_3O^+] = [OH^-]$
$pH = 7.0$

Basic Solution

$[H_3O^+] < [OH^-]$
$pH > 7.0$

$[H_3O^+] \times [OH^-] = 1.0 \times 10^{-14}$

ACIDITY AND HYDROGEN ATOMS

Acidic Hydrogen Atoms

- Participate in acid–base reactions
- Are written at front of chemical formula

HNO_3, H_2SO_4
All hydrogen atoms are acidic

$HC_2H_3O_2$, $H_3C_6H_5O_7$
Both acidic and nonacidic hydrogen atoms are present

Nonacidic Hydrogen Atoms

- Do NOT participate in acid–base reactions
- Are NOT written at front of chemical formula

NH_3, CH_4
All hydrogen atoms are nonacidic

specifying the strength (strong or weak) of the acid and base involved. The "parent" acid and base for the salt are identified by pairing the negative ion of the salt with H^+ (to form the acid) and pairing the positive ion of the salt with OH^- (to form the base). The following two equations illustrate the overall procedure.

$$\underset{\text{Strong base}}{NaOH} + \underset{\text{Strong acid}}{HCl} \longrightarrow H_2O + \underset{\text{Strong acid–strong base salt}}{NaCl}$$

$$\underset{\text{Strong base}}{KOH} + \underset{\text{Weak acid}}{HCN} \longrightarrow H_2O + \underset{\text{Weak acid–strong base salt}}{KCN}$$

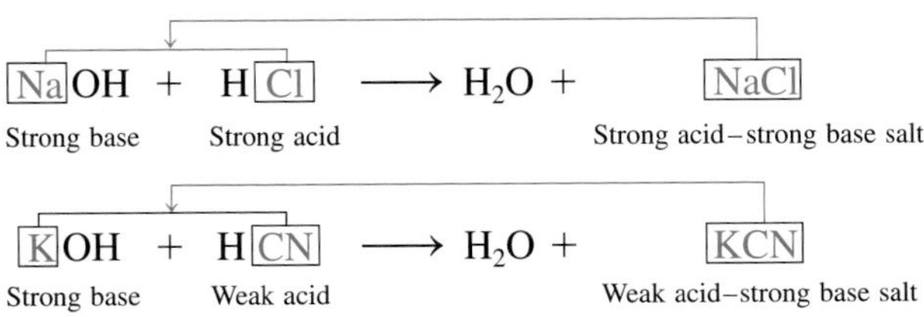

Table 10.7
Neutralization "Parentage" of Salts and the Nature of the Aqueous Solutions They Form

Type of salt	Nature of aqueous solution	Examples
strong acid–strong base	neutral	NaCl, KBr
strong acid–weak base	acidic	NH_4Cl, NH_4NO_3
weak acid–strong base	basic	$NaC_2H_3O_2$, K_2CO_3
weak acid–weak base	depends on the salt	$NH_4C_2H_3O_2$, NH_4NO_2

Knowing which acids and bases are strong and which are weak (Section 10.4) is a necessary part of the classification process. Once the salt classification has been determined, the guideline that is appropriate for the situation is easily selected.

Example 10.8

Predicting Whether a Salt's Aqueous Solution Will Be Acidic, Basic, or Neutral

Determine the acid–base "parentage" of each of the following salts, and then use this information to predict whether each salt's aqueous solution is acidic, basic, or neutral

a. Sodium acetate, $NaC_2H_3O_2$
b. Ammonium chloride, NH_4Cl
c. Potassium chloride, KCl
d. Ammonium fluoride, NH_4F

Solution

a. The ions present are Na^+ and $C_2H_3O_2^-$. The "parent" base of Na^+ is NaOH, a strong base. The "parent" acid of $C_2H_3O_2^-$ is $HC_2H_3O_2$, a weak acid. Thus the acid–base neutralization that produces this salt is

$$\underset{\text{Strong base}}{NaOH} + \underset{\text{Weak acid}}{HC_2H_3O_2} \longrightarrow H_2O + \underset{\text{Weak acid–strong base salt}}{NaC_2H_3O_2}$$

The solution of a weak acid–strong base salt (guideline 3) produces a basic solution.

b. The ions present are NH_4^+ and Cl^-. The "parent" base of NH_4^+ is NH_3, a weak base. The "parent" acid of Cl^- is HCl, a strong acid. This "parentage" will produce a strong acid–weak base salt through neutralization. Such a salt gives an acidic solution upon hydrolysis (guideline 2).

c. The ions present are K^+ and Cl^-. The "parent" base is KOH (a strong base), and the "parent" acid is HCl (a strong acid). The salt produced from neutralization involving this acid–base pair will be a strong acid–strong base salt. Such salts do not hydrolyze. The aqueous solution is neutral (guideline 1).

d. The ions present are NH_4^+ and F^-. Both ions are of weak "parentage"; NH_3 is a weak base and HF is a weak acid. Thus NH_4F is a weak acid–weak base salt. This is a guideline 4 situation. In this situation, you cannot predict the effect of hydrolysis unless you know the relative strengths of the weak acid and weak base (which is the weaker of the two). HF has a K_a of 6.6×10^{-4} (Table 10.4). NH_3 has a K_b of 1.8×10^{-5} (Section 10.5). Thus, NH_3 is the weaker of the two and will hydrolyze to the greater extent, causing the solution to be acidic.

Practice Exercise 10.8

Predict whether solutions of each of the following salts will be acidic, basic, or neutral.

a. Sodium bromide, NaBr
b. Potassium cyanide, KCN
c. Ammonium iodide, NH_4I
d. Barium chloride, $BaCl_2$

• *Answers:* **a.** neutral; **b.** basic; **c.** acidic; **d.** neutral

Chemical CONNECTIONS

10.2 Blood Plasma pH and Hydrolysis

Blood plasma has a slightly basic pH (7.35–7.45), as shown in Table 10.6. The reason for this is related to hydrolysis and becomes apparent when the identity of the ions present in blood plasma is specified.

The most abundant positive ion present in blood plasma is Na^+, an ion associated with a strong base (NaOH). Thus it does not hydrolyze. The predominant negative ion present is Cl^-, an ion that comes from a strong acid (HCl). Thus it also does not hydrolyze. Together, these two ions, Na^+ and Cl^-, produce a neutral solution because neither hydrolyzes.

The third most abundant ion in blood plasma is the hydrogen carbonate ion, HCO_3^-, which comes from the weak acid H_2CO_3. Hydrolysis of this ion ties up hydronium ions and leaves hydroxide ions in excess. Thus the plasma fluid has a slightly basic character. Other negative ions present in blood plasma, such as HPO_4^{2-}, also hydrolyze and add to the basic character. However, because of their lower concentrations, their effect on the pH is not so great as that of HCO_3^-.

The pH of numerous other body fluids besides blood plasma is also directly influenced by hydrolysis reactions.

• Chemical Equations for Salt Hydrolysis Reactions

Salt hydrolysis reactions are Brønsted–Lowry acid–base (proton transfer) reactions (Section 10.2). Such reactions are of the following two general types.

1. *Basic hydrolysis:* The reaction of the *negative ion* from a salt with water to produce the ion's conjugate acid and hydroxide ion.

• The only *negative ions* that undergo hydrolysis are those of "weak acid parentage." The driving force for the reaction is the formation of the weak acid "parent."

Conjugate acid–base pair (CN^- / HCN)

$$\underset{\text{Proton acceptor}}{CN^-} + \underset{\text{Proton donor}}{H_2O} \longrightarrow \underset{\text{Weak acid}}{HCN} + \underset{\text{Makes solution basic}}{OH^-}$$

Conjugate acid–base pair (F^- / HF)

$$\underset{\text{Proton acceptor}}{F^-} + \underset{\text{Proton donor}}{H_2O} \longrightarrow \underset{\text{Weak acid}}{HF} + \underset{\text{Makes solution basic}}{OH^-}$$

2. *Acidic hydrolysis:* The reaction of the *positive ion* from a salt with water to produce the ion's conjugate base and hydronium ion. The most common ion to undergo this type of reaction is the NH_4^+ ion.

• The only *positive ions* that undergo hydrolysis are those of "weak base parentage." The driving force for the reaction is the formation of the weak base "parent."

Conjugate acid–base pair (NH_4^+ / NH_3)

$$\underset{\text{Proton donor}}{NH_4^+} + \underset{\text{Proton acceptor}}{H_2O} \longrightarrow \underset{\text{Weak base}}{NH_3} + \underset{\text{Makes solution acidic}}{H_3O^+}$$

Hydrolysis reactions do not go 100% to completion. They occur only until equilibrium conditions are reached (Section 9.7). At the equilibrium point, solution pH can differ from neutrality by 2 to 4 pH units, as shown in Table 10.8.

10.12 Buffers

A **buffer** *is a solution that resists major changes in pH when small amounts of acid or base are added to it.* Buffers are used in a laboratory setting to maintain optimum pH conditions for chemical reactions. Many commercial products contain buffers, which are needed to maintain optimum pH conditions for product behavior. Examples include buffered aspirin (Bufferin) and pH-controlled hair shampoos. Most human body fluids are highly buffered. For example, a buffer system maintains blood's pH at a value close to 7.4, an optimum pH for oxygen transport.

Table 10.8
Approximate pH of Selected 0.1 M Aqueous Salt Solutions at 24°C

Name of salt	Formula of salt	pH	Category of salt
ammonium nitrate	NH_4NO_3	5.1	strong acid–weak base
ammonium nitrite	NH_4NO_2	6.3	weak acid–weak base
ammonium acetate	$NH_4C_2H_3O_2$	7.0	weak acid–weak base
sodium chloride	NaCl	7.0	strong acid–strong base
sodium fluoride	NaF	8.1	weak acid–strong base
sodium acetate	$NaC_2H_3O_2$	8.9	weak acid–strong base
ammonium cyanide	NH_4CN	9.3	weak acid–weak base
sodium cyanide	NaCN	11.1	weak acid–strong base

• A less common type of buffer involves a weak base and its conjugate acid. We will not consider this type of buffer here.

Buffers contain two chemical species: (1) a substance to react with and remove added base, and (2) a substance to react with and remove added acid. Typically, a buffer system is composed of a weak acid *and* its conjugate base—that is, a conjugate acid–base pair (Section 10.2). Such acid–base pairs that are commonly employed as buffers include $HC_2H_3O_2/C_2H_3O_2^-$, $H_2PO_4^-/HPO_4^{2-}$, and H_2CO_3/HCO_3^-.

Example 10.9

Recognizing Pairs of Chemical Substances That Can Function as a Buffer in Aqueous Solution

Predict whether each of the following pairs of substances could function as a buffer system in aqueous solution.

a. HCl and NaCl **b.** HCN and KCN
c. HCl and HCN **d.** NaCN and KCN

Solution

Buffer solutions contain either a weak acid and a salt of that weak acid or a weak base and a salt of that weak base.

a. No. We have an acid and the salt of that acid. However, the acid is a strong acid rather than a weak acid.
b. Yes. HCN is a weak acid, and KCN is a salt of that weak acid.
c. No. Both HCl and HCN are acids. No salt is present.
d. No. Both NaCN and KCN are salts. No weak acid is present.

Practice Exercise 10.9

Predict whether each of the following pairs of substances could function as a buffer system in aqueous solution.

a. HCl and NaOH **b.** $HC_2H_3O_2$ and $KC_2H_3O_2$
c. NaCl and NaCN **d.** HCN and $HC_2H_3O_2$

• *Answers:* **a.** No; **b.** Yes; **c.** No; **d.** No

As an illustration of buffer action, consider a buffer solution containing approximately equal concentrations of acetic acid (a weak acid) and sodium acetate (a salt of this weak acid). This solution resists pH change by the following mechanisms:

1. When a small amount of a strong acid such as HCl is added to the solution, the newly added H_3O^+ ions react with the acetate ions from the sodium acetate to give acetic acid.

$$H_3O^+ + C_2H_3O_2^- \longrightarrow HC_2H_3O_2 + H_2O$$

Most of the added H_3O^+ ions are tied up in acetic acid molecules, and the pH changes very little.

2. When a small amount of a strong base such as NaOH is added to the solution, the newly added OH^- ions react with the acetic acid (neutralization) to give acetate ions and water.

$$OH^- + HC_2H_3O_2 \longrightarrow C_2H_3O_2^- + H_2O$$

Most of the added OH^- ions are converted to water, and the pH changes only slightly.

• To resist both increases and decreases in pH effectively, a weak acid buffer must contain significant amounts of both the weak acid and its conjugate base. If a solution has a large amount of weak acid but very little conjugate base, it will be unable to consume much added acid. Consequently, the pH tends to drop significantly when acid is added. Conversely, a solution that contains a large amount of conjugate base but very little weak acid will provide very little protection against added base. Addition of just a little base will cause a big change in pH.

The reactions that are responsible for the buffering action in the acetic acid/acetate ion system can be summarized as follows:

$$C_2H_3O_2^- \underset{OH^-}{\overset{H_3O^+}{\rightleftharpoons}} HC_2H_3O_2$$

Note that one member of the buffer pair (acetate ion) removes excess H_3O^+ ion and that the other (acetic acid) removes excess OH^- ion. The buffering action always results in the active species being converted to its partner species.

Example 10.10

Writing Equations for Reactions That Occur in a Buffered Solution

Write an equation for each of the following buffering actions.

a. The response of $H_2PO_4^-/HPO_4^{2-}$ buffer to the addition of H_3O^+ ions
b. The response of HCN/CN^- buffer to the addition of OH^- ions

Solution

a. The base in a conjugate acid–base pair is the species that responds to the addition of acid. (Recall, from Section 10.2, that the base in a conjugate acid–base pair always has one less hydrogen than the acid.) The base for this reaction is HPO_4^{2-}. The equation for the buffering action is

$$H_3O^+ + HPO_4^{2-} \longrightarrow H_2PO_4^- + H_2O$$

In the buffering response, the base is always converted into its conjugate acid.

b. The acid in a conjugate acid–base pair is the species that responds to the addition of base. The acid for this reaction is HCN. The equation for the buffering action is

$$HCN + OH^- \longrightarrow CN^- + H_2O$$

Water will always be one of the products of buffering action.

Practice Exercise 10.10

Write an equation for each of the following buffering actions.

a. The response of H_2CO_3/HCO_3^- buffer to the addition of H_3O^+ ions
b. The response of $H_2PO_4^-/HPO_4^{2-}$ buffer to the addition of OH^- ions

• *Answers* **a.** $H_3O^+ + HCO_3^- \longrightarrow H_2CO_3 + H_2O$;
b. $H_2PO_4^- + OH^- \longrightarrow HPO_4^{2-} + H_2O$

• A common misconception about buffers is that a buffered solution is always a neutral (pH 7.0) solution. This is false. One can buffer a solution at any desired pH. A pH 7.4 buffer will hold the pH of the solution near pH 7.4, whereas a pH 9.3 buffer will tend to hold the pH of a solution near pH 9.3. The pH of a buffer is determined by the degree of weakness of the weak acid used and by the concentrations of the acid and its conjugate base.

A false notion about buffers is that they will hold the pH of a solution *absolutely* constant. The addition of even small amounts of a strong acid or a strong base to any solution, buffered or not, will lead to a change in pH. The important concept is that the shift in pH will be much less when an effective buffer is present (see Table 10.9).

Buffer systems have their limits. If large amounts of H_3O^+ or OH^- are added to a buffer, the buffer capacity can be exceeded; then the buffer system is overwhelmed and the pH changes (Figure 10.8). For example, if large amounts of H_3O^+ were added to the acetate/acetic acid buffer previously discussed, the H_3O^+ ion would react with acetate ion until the acetate was depleted. Then the pH would begin to drop as free H_3O^+ ions accumulated in the solution.

(a)

(b)

Figure 10.8
(a) The buffered solution on the left and the unbuffered solution on the right have the same pH (pH 8). They are basic solutions. (b) After the addition of 1 mL of a 0.01 M HCl solution, the pH of the buffered solution has not perceptibly changed, but the unbuffered solution has become acidic, as indicated by the change in the color of the acid–base indicator present.

Additional insights into the workings of buffer systems are obtained by considering buffer action within the framework of Le Châtelier's principle and an equilibrium system. Let us again consider an acetic acid/acetate ion buffer system. An equilibrium is established in solution between the acetic acid and the acetate ion.

$$HC_2H_3O_2(aq) + H_2O(l) \rightleftharpoons H_3O^+(aq) + C_2H_3O_2^-(aq)$$

This equilibrium system functions in accordance with *Le Chatelier's principle* (Section 9.9), which states that an equilibrium system, when stressed, will shift its position in such a way as to counteract the stress. Stresses for the buffer will be (1) addition of base (hydroxide ion) and (2) addition of acid (hydronium ion). Further details concerning these two stress situations are as follows.

Addition of base [OH^- ion] *to the buffer.* The addition of base causes the following changes to occur in the solution:

1. The added OH^- ion reacts with H_3O^+ ion, producing water (neutralization).
2. The neutralization reaction produces the stress of *not enough* H_3O^+ ion, because H_3O^+ ion was consumed in the neutralization.
3. The equilibrium shifts to the right, in accordance with Le Châtelier's principle, to produce more H_3O^+ ion, which maintains the pH close to its original level.

Addition of acid [H_3O^+ ion] *to the buffer.* The addition of acid causes the following changes to occur in the solution:

1. The added H_3O^+ ion increases the overall amount of H_3O^+ ion present.
2. The stress on the system is *too much* H_3O^+ ion.
3. The equilibrium shifts to the left, in accordance with Le Châtelier's principle, consuming most of the excess H_3O^+ ion and resulting in a pH close to the original level.

Table 10.9
A Comparison of pH Shifts in Buffered and Unbuffered Solutions

Unbuffered Solution	
1 liter water	pH = 7.0
1 liter water + 0.01 mole strong base (NaOH)	pH = 12.0
1 liter water + 0.01 mole strong acid (HCl)	pH = 2.0
Buffered Solution	
1 liter buffer[a]	pH = 7.2
1 liter buffer[a] + 0.01 mole strong base (NaOH)	pH = 7.3
1 liter buffer[a] + 0.01 mole strong acid (HCl)	pH = 7.1

[a]Buffer = 0.1 M HPO_4^{2-} + 0.1 M $H_2PO_4^-$

Chemical CONNECTIONS

10.3 Buffering Action in Human Blood

Carbonate, phosphate, and protein buffer systems play an important role in maintaining proper blood pH. Even small departures from blood's normal pH range of 7.35–7.45 can cause serious illness, and death can result from pH variations that exceed a few tenths of a unit. Many of the key reactions that occur in blood are enzyme-catalyzed and reach optimum conditions only within blood's normal pH range. Outside this normal range, enzyme action slows down or even stops.

The *carbonate buffer system* involves the species H_2CO_3 and HCO_3^-. Any excess acid formed in the blood reacts with the HCO_3^- ion, and any excess base reacts with H_2CO_3.

$$H_3O^+ + HCO_3^- \longrightarrow H_2CO_3 + H_2O$$

$$OH^- + H_2CO_3 \longrightarrow HCO_3^- + H_2O$$

The ratio of $[H_2CO_3]$ to $[HCO_3^-]$ in blood is approximately 1 to 10, which means this buffer has a greater ability to interact with acid than with base. Significant amounts of acids (up to 10 moles a day) are produced in the human body as a result of normal metabolic reactions. For example, lactic acid ($HC_3H_5O_2$) is produced in muscle tissue during exercise.

This 1-to-10 ratio of buffering species is also needed to maintain the blood at a pH of 7.4. A 1-to-1 ratio buffer would produce a pH of 6.4. The 1-to-10 ratio is easily maintained.

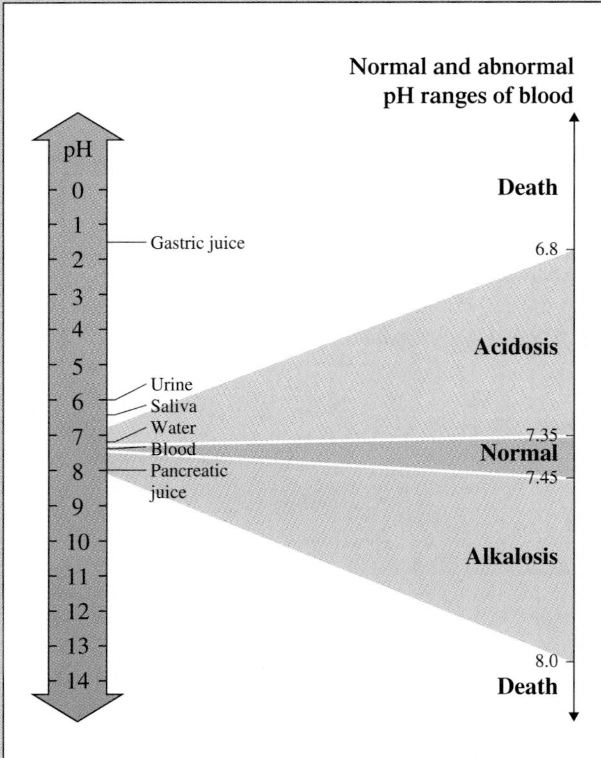

Excess H_2CO_3 decomposes to CO_2 and H_2O and is removed from the blood by the lungs.

$$H_2CO_3 \longrightarrow CO_2 + H_2O$$

Excess HCO_3^- can be eliminated from the body through the kidneys.

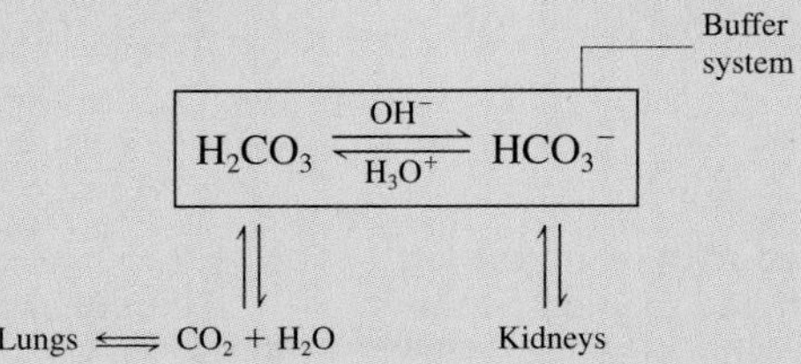

The *phosphate buffer system* involves the species dihydrogen phosphate ($H_2PO_4^-$) and hydrogen phosphate (HPO_4^{2-}). The control of pH using this system occurs by OH^- ion reacting with $H_2PO_4^-$ and by H_3O^+ reacting with HPO_4^{2-}.

$$H_2PO_4^- \underset{H_3O^+}{\overset{OH^-}{\rightleftharpoons}} HPO_4^{2-}$$

A 3-to-5 ratio of $[H_2PO_4^-]$ to $[HPO_4^{2-}]$ is required to maintain a pH of 7.4.

Under certain stress conditions, the blood's buffer systems can be overwhelmed. **Acidosis** *is a body condition in which the pH of blood drops from its normal value of* 7.4 *to* 7.1–7.2. **Alkalosis** *is a body condition in which the pH of blood increases from its normal value of* 7.4 *to a value of* 7.5. Both can be life-threatening if not properly taken care of; both can be caused by either metabolic processes or changes in breathing patterns (respiration).

Metabolic acidosis is seen in diabetics, who accumulate acidic substances from the metabolism of fats. Excessive loss of bicarbonate ion in cases of severe diarrhea is another cause. A temporary metabolic acidosis condition can result from prolonged intensive exercise. Exercise generates lactic acid (a weak acid) in the muscles. Some of the lactic acid ionizes, and this produces an influx of H_3O^+ ions into the bloodstream.

Metabolic alkalosis is less common than metabolic acidosis. It results from elevated HCO_3^- ion levels. Causes include prolonged vomiting and the side effects of certain drugs that change the concentrations of sodium, potassium, and chloride ions in the blood.

Respiratory acidosis results from higher than normal levels of CO_2 in the blood; inefficient CO_2 removal is usually the origin of this problem. Hypoventilation (a lowered breathing rate), caused by lung diseases such as emphysema and asthma or obstructed air passages, produces respiratory acidosis.

Respiratory alkalosis is caused by hyperventilation (an elevated breathing rate). Causes include hysteria and anxiety (occasioned by chemistry tests) and the rapid breathing associated with extremely high fevers.

10.13 The Henderson–Hasselbalch Equation

Buffers may be prepared from any ratio of concentrations of a weak acid and the salt of its conjugate base. However, a buffer is most effective in counteracting pH change when the acid-to-conjugate-base ratio is 1:1. If a buffer contains considerably more acid than the conjugate base, it is less efficient in handling an acid. Conversely, a buffer with considerably more of the conjugate base than the acid is less efficient in handling added base.

When the concentrations of an acid and its conjugate base are equal in a buffer solution, the solution's hydronium ion concentration is equal to the acid ionization constant of the weak acid—or, stated more concisely, the pH of the solution is equal to the pK_a of the weak acid. The mathematical basis for this equality is as follows.

For the weak acid,

$$K_a = \frac{[H_3O^+][A^-]}{[HA]}$$

If HA and A^- are equal, then they cancel from the equation and we have

$$K_a = [H_3O^+]$$

Taking the negative logarithm of both sides of this equation gives

$$pK_a = pH$$

The relationship between pK_a and pH for buffer solutions where the conjugate acid–base pair concentration ratio is something other than 1:1 is given by the equation

$$pH = pK_a + \log\frac{[A^-]}{[HA]}$$

This equation is called the *Henderson–Hasselbalch equation.* The Henderson–Hasselbalch equation indicates that if there is more A^- than HA in a solution, the pH is higher than pK_a; and if there is more HA than A^-, the pH is lower than pK_a.

Example 10.11

Calculating the pH of a Buffer Solution Using the Henderson–Hasselbalch Equation

What is the pH of a buffer solution that is 0.5 M in formic acid ($HCHO_2$) and 1.0 M in sodium formate ($NaCHO_2$). The pK_a for formic acid is 3.74.

Solution

The concentrations for the buffering species are

$$HCHO_2 = 0.5\,M \qquad CHO_2^- = 1.0\,M$$

Substituting these values into the Henderson-Hasselbalch equation gives

$$pH = pK_a + \log\frac{[CHO_2^-]}{[HCHO_2]} = 3.74 + \log\frac{1.0}{0.5}$$
$$= 3.74 + \log 2 = 3.74 + 0.30$$
$$= 4.04$$

Practice Exercise 10.11

What is the pH of a buffer solution that is 0.6 M in acetic acid ($HC_2H_3O_2$) and 1.5 M in sodium acetate ($NaC_2H_3O_2$). The pK_a for acetic acid is 4.74.

- *Answer:* 5.14

Buffer Systems

CHEMISTRY AT A GLANCE

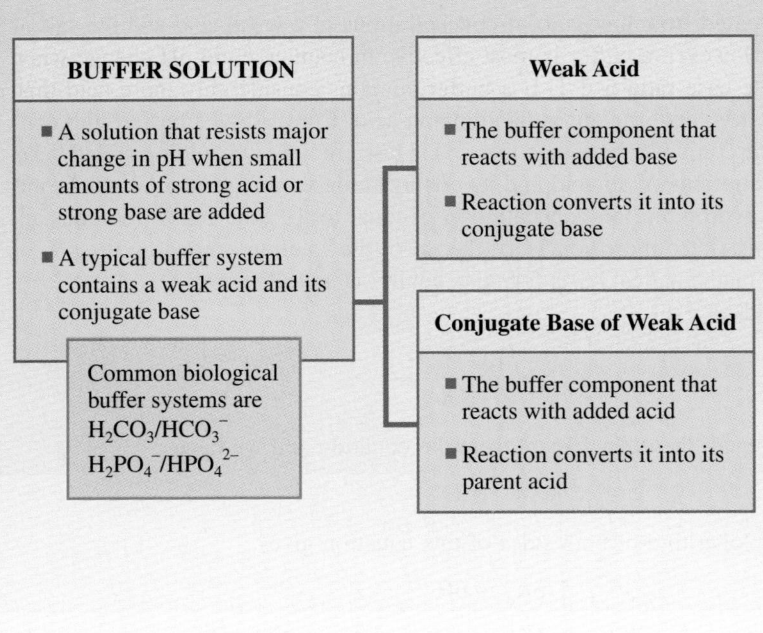

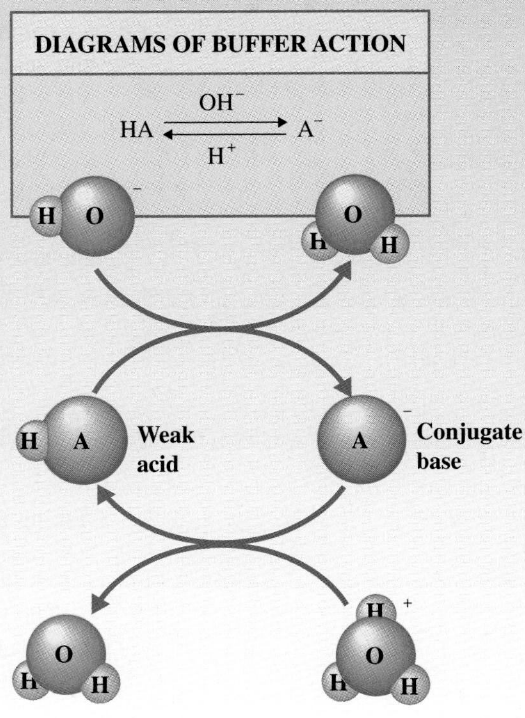

Chemistry at a Glance reviews some of what we have said about buffer systems.

10.14 Electrolytes

Aqueous solutions in which ions are present are good conductors of electricity, and the greater the number of ions present, the better the solution conducts electricity. Acids, bases, and soluble salts all produce ions in solution; thus they all produce solutions that conduct electricity. All three types of compounds are said to be electrolytes. An **electrolyte** *is a substance that forms a solution in water that conducts electricity.* The presence of ions (charged particles) explains the electrical conductivity.

Some substances, such as table sugar (sucrose), glucose, and isopropyl alcohol, do not produce ions in solution. These substances are called nonelectrolytes. A **nonelectrolyte** *is a substance that forms a solution in water that does not conduct electricity.*

Electrolytes can be divided into two groups—strong electrolytes and weak electrolytes. A **strong electrolyte** *is a substance that completely (or almost completely) dissociates into ions in solution.* Strong electrolytes produce strongly conducting solutions. All strong acids and strong bases and all soluble salts are strong electrolytes. A **weak electrolyte** *is a substance that only partially ionizes into ions in solution.* These electrolytes give solutions that are intermediate between those containing strong electrolytes and those containing nonelectrolytes in their ability to conduct an electric current. Weak acids, weak bases, and slightly soluble salts constitute the weak electrolytes.

You can determine whether a substance is an electrolyte in solution by testing the ability of the solution to conduct an electric current. A device such as that shown in Figure 10.9 can be used to distinguish among strong electrolytes, weak electrolytes, and nonelectrolytes. If the medium between the electrodes (the solution) is a conductor of electricity, the light bulb glows. A strong glow indicates a strong electrolyte. A faint glow occurs for a weak electrolyte, and there is no glow for a nonelectrolyte.

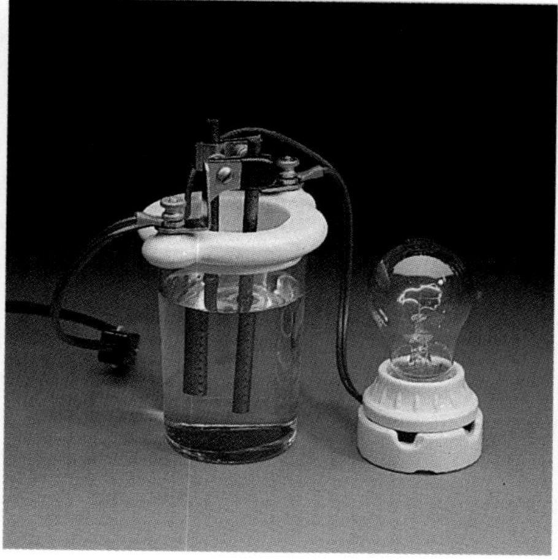

Figure 10.9
This simple device can be used to distinguish among strong electrolytes, weak electrolytes, and nonelectrolytes. The light bulb glows strongly for strong electrolytes (left), weakly for weak electrolytes (center), and not at all for nonelectrolytes (right).

Chemical CONNECTIONS

10.4 Electrolytes and Body Fluids

Structurally speaking, there are three types of body fluids: blood plasma, which is the liquid part of the blood; interstitial fluid, which is the fluid in tissues between and around cells; and intracellular fluids, which are the fluids within cells. For every kilogram of its mass, the body contains about 400 mL of intracellular fluid, 160 mL of interstitial fluid, and 40 mL of plasma.

Water is the main component of any type of body fluid. In addition, all body fluids contain electrolytes. It is the electrolytes present in body fluids that govern numerous body processes. The chemical makeup of the three types of body fluids, in terms of electrolytes (ions present), is shown in the accompanying figure.

Chemically, two of the body fluids (plasma and interstitial fluid) are almost identical. Intracellular fluid, on the other hand, shows striking differences. For example, K^+ is the dominant positive ion in intracellular fluid, and Na^+ dominates in the other two fluids. A similar situation occurs with negative ions. A different ion dominates in intracellular fluid (HPO_4^{2-}) than in the other two fluids (Cl^-).

The electrolytes present in body fluids (1) govern the movement of water between body fluid compartments and (2) maintain acid–base balance within the body fluids. Osmotic pressure, a major factor in controlling water movement, is directly related to electrolyte concentration gradients.

The fact that the presence of ions causes a solution to conduct electricity is of extreme biological significance. For example, messages to and from the brain are sent in the form of electrical signals. Ions in intracellular and interstitial fluids are often the carriers of these signals. The presence of electrolytes (ions) is essential to the proper functioning of the human body.

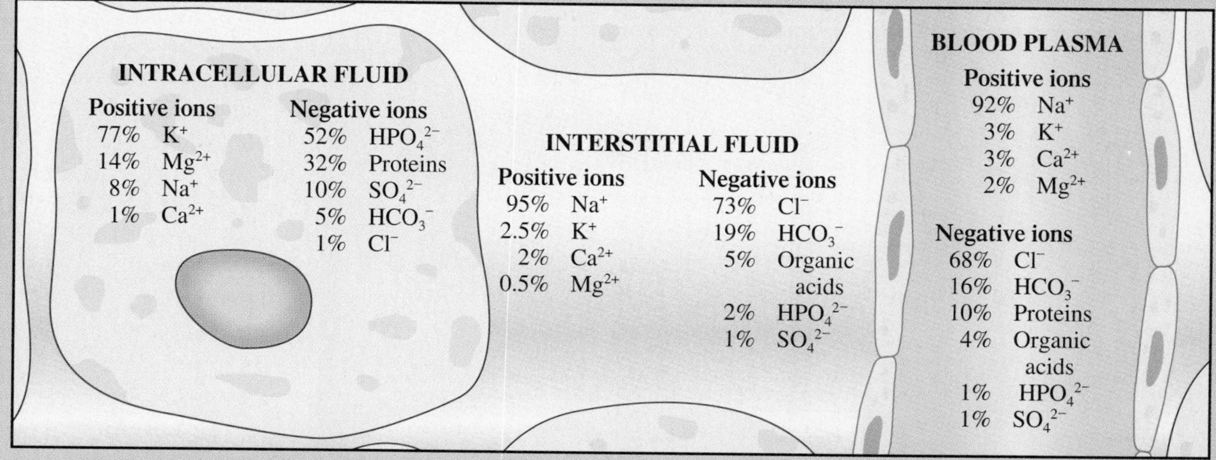

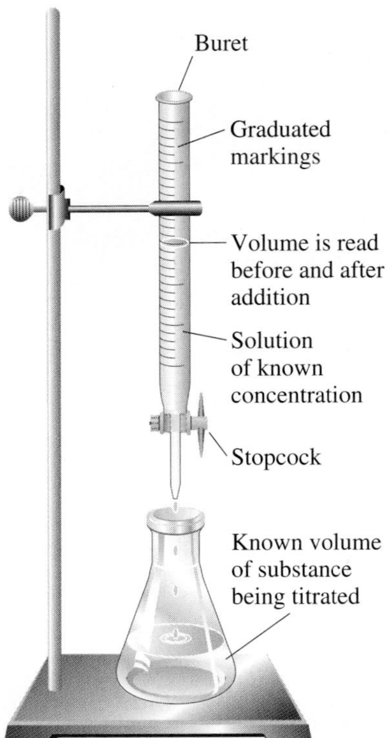

Figure 10.10
A schematic diagram showing the setup used for titration procedures.

10.15 Acid–Base Titrations

The analysis of solutions to determine the concentration of acid or base present is performed regularly in many laboratories. Such activity is different from determining a solution's pH value. The pH of a solution gives information about the *concentration* of hydronium ions in solution. Only ionized molecules influence the pH value. The concentration of an acid or a base gives information about the *total number* of acid or base molecules present; both dissociated and undissociated molecules are counted. Thus acid or base concentration is a measure of total acidity or total basicity.

The procedure most frequently used to determine the concentration of an acid or a base solution is an acid–base titration. In an **acid–base titration,** *a measured volume of an acid or a base of known concentration is exactly reacted with a measured volume of a base or an acid of unknown concentration.*

Suppose we want to determine the concentration of an acid solution by means of titration. We first measure out a known volume of the acid solution into a beaker or flask. Then we slowly add a solution of base of known concentration to the flask or beaker by means of a buret (Figure 10.10). We continue to add base until all the acid has completely reacted with the added base. The volume of base needed to reach this point is obtained from the buret readings. When we know the original volume of acid, the concentration of the base, and the volume of added base, we can calculate the concentration of the acid, as will be shown in Example 10.12.

In order to complete a titration successfully, we must be able to detect when the reaction between acid and base is complete. Neither the acid nor the base gives any outward sign that the reaction is complete. Thus an indicator is always added to the reaction mixture (Figure 10.11). An **indicator** *is a compound that exhibits different colors depending on the* pH *of its surroundings.* Typically, an indicator is one color in basic solutions and another color in acidic solutions. An indicator is selected that changes color at a pH that corresponds as nearly as possible to the pH of the solution when the titration is complete. If the acid and base are both strong, the pH at that point is 7. However, because of hydrolysis (Section 10.11), the pH is not 7 if a weak acid or weak base is part of the titration system.

Titration of a weak acid by a strong base requires an indicator that changes color above pH 7, because the salt formed in the titration will hydrolyze to form a basic solution. Conversely, titration of a weak base by a strong acid requires an indicator that changes color below pH 7.

Example 10.12 shows how titration data are used to calculate the molarity of an acid solution of unknown concentration.

Figure 10.11
An acid–base titration using an indicator that is yellow in acidic solution and red in basic solution.

Example 10.12

Calculating an Unknown Molarity from Acid–Base Titration Data

In a sulfuric acid (H_2SO_4)–sodium hydroxide (NaOH) acid–base titration, 17.3 mL of 0.126 M NaOH is needed to neutralize 25.0 mL of H_2SO_4 of unknown concentration. Find the molarity of the H_2SO_4 solution, given that the neutralization reaction that occurs is

$$H_2SO_4(aq) + 2NaOH(aq) \longrightarrow Na_2SO_4(aq) + 2H_2O(l)$$

Solution

First, we calculate the number of moles of H_2SO_4 that reacted with the NaOH. The pathway for this calculation, using dimensional analysis (Section 6.8), is

$$\text{mL of NaOH} \longrightarrow \text{L of NaOH} \longrightarrow \text{moles of NaOH} \longrightarrow \text{moles of } H_2SO_4$$

The sequence of conversion factors that effects this series of unit changes is

$$17.3\,\cancel{\text{mL NaOH}} \times \left(\frac{10^{-3}\,\cancel{\text{L NaOH}}}{1\,\cancel{\text{mL NaOH}}}\right) \times \left(\frac{0.126\,\cancel{\text{mole NaOH}}}{1\,\cancel{\text{L NaOH}}}\right) \times \left(\frac{1\text{ mole } H_2SO_4}{2\,\cancel{\text{moles NaOH}}}\right)$$

The first conversion factor derives from the definition of a milliliter, the second conversion factor derives from the definition of molarity (Section 8.5), and the third conversion factor uses the coefficients in the balanced chemical equation for the titration reaction (Section 6.7).

The number of moles of H_2SO_4 that react is obtained by combining all the numbers in the dimensional analysis setup in the manner indicated.

$$\left(\frac{17.3 \times 10^{-3} \times 0.126 \times 1}{1 \times 1 \times 2}\right) \text{mole } H_2SO_4 = 0.00109 \text{ mole } H_2SO_4$$

Now that we know how many moles of H_2SO_4 reacted, we calculate the molarity of the H_2SO_4 solution using the definition for molarity.

$$\text{Molarity } H_2SO_4 = \frac{\text{moles } H_2SO_4}{\text{L } H_2SO_4 \text{ solution}} = \frac{0.00109 \text{ mole}}{0.0250 \text{ L}}$$

$$= 0.0436 \, \frac{\text{mole}}{\text{L}}$$

Note that the units in the denominator of the molarity equation must be liters (0.0250) rather than milliliters (25.0).

Practice Exercise 10.12

In a nitric acid (HNO_3)–potassium hydroxide (KOH) acid–base titration, 32.4 mL of 0.352 M KOH is required to neutralize 50.0 mL of HNO_3 of unknown concentration. Find the molarity of the HNO_3 solution, given that the neutralization reaction that occurs is

$$HNO_3(aq) + KOH(aq) \longrightarrow KNO_3(aq) + H_2O(l)$$

- *Answer:* 0.228 M HNO_3

Concepts to Remember

Arrhenius acid–base theory. An Arrhenius acid is a hydrogen-containing compound that, in water, produces hydrogen ions. An Arrhenius base is a hydroxide-containing compound that, in water, produces hydroxide ions.

Brønsted–Lowry acid–base theory. A Brønsted–Lowry acid is any substance that can donate a proton (H^+) to some other substance. A Brønsted–Lowry base is any substance that can accept a proton from some other substance. Proton donation (from an acid) cannot occur unless an acceptor (a base) is present.

Conjugate acids and bases. A conjugate acid–base pair is two species that differ by one proton. The conjugate base of an acid is the species that remains when the acid loses a proton. The conjugate acid of a base is the species formed when the base accepts a proton.

Polyprotic acids. Polyprotic acids are acids that can transfer two or more hydrogen ions during an acid–base reaction.

Strengths of acids and bases. Acids can be classified as strong or weak in terms of the extent to which proton transfer occurs in aqueous solution. A strong acid completely transfers its protons to water. A weak acid transfers only a small percentage of its protons to water.

Acid ionization constant. The acid ionization constant quantitatively describes the degree of ionization of an acid. It is the equilibrium constant expression that corresponds to the ionization of the acid.

Salts. Salts are ionic compounds containing a metal or polyatomic ion as the positive ion and a nonmetal or polyatomic ion (except hydroxide ion) as the negative ion. Ionic compounds containing hydroxide ion are bases rather than salts.

Neutralization. Neutralization is the reaction between an acid and a hydroxide base to form a salt and water.

Self-ionization of water. In pure water, a small number of water molecules (1.0×10^{-7} mole/L) donate protons to other water molecules to produce small concentrations (1.0×10^{-7} mole/L) of hydronium and hydroxide ions.

The pH scale. The pH scale is a scale of small numbers that are used to specify molar hydronium ion concentration in an aqueous solution. Mathematically, the pH is the negative logarithm of the hydronium ion concentration. Solutions with a pH lower than 7 are acidic, those with a pH higher than 7 are basic, and those with a pH equal to 7 are neutral.

Hydrolysis of salts. Salt hydrolysis is a reaction in which a salt interacts with water to produce an acidic or a basic solution. Only salts that contain the conjugate base of a weak acid and/or the conjugate acid of a weak base hydrolyze.

Buffer solutions. A buffer solution is a solution that resists pH change when small amounts of acid or base are added to it. The resistance to pH change in most buffers is caused by the presence of a weak acid and a salt of its conjugate base.

Electrolytes. An electrolyte is a substance that forms a solution in water that conducts electricity. Strong acids, strong bases, and soluble salts are strong electrolytes. Weak acids, weak bases, and slightly soluble salts are weak electrolytes.

Acid–base titrations. Titration is a procedure in which an acid–base neutralization reaction is used in determining an unknown concentration. A measured volume of an acid or a base of known concentration is exactly reacted with a measured volume of a base or an acid of unknown concentration.

Key Reactions and Equations

1. Weak-acid equilibrium and acid ionization constant (K_a) expression (Section 10.5)

$$HA(aq) + H_2O(l) \rightleftharpoons H_3O^+(aq) + A^-(aq)$$

$$K_a = \frac{[H_3O^+][A^-]}{[HA]}$$

2. Weak-base equilibrium and base ionization constant (K_b) expression (Section 10.5)

$$B(aq) + H_2O(l) \rightleftharpoons BH^+(aq) + OH^-(aq)$$

$$K_b = \frac{[BH^+][OH^-]}{[B]}$$

3. Ion product constant for water (Section 10.8)

$$[H_3O^+][OH^-] = 1.0 \times 10^{-14}$$

4. Relationship between $[H_3O^+]$ and pH (Section 10.9)

$$pH = -\log[H_3O^+]$$

5. Henderson–Hasselbalch equation

$$pH = pK_a + \log\frac{[A^-]}{[HA]}$$

Key Terms

Acid–base titration (10.13)
Acid ionization constant (10.5)
Acidic solution (10.8)
Amphoteric substance (10.2)
Arrhenius acid (10.1)
Arrhenius base (10.1)
Base ionization constant (10.5)
Basic solution (10.8)
Brønsted–Lowry acid (10.2)
Brønsted–Lowry base (10.2)
Buffer (10.11)
Conjugate acid (10.2)
Conjugate acid–base pair (10.2)
Conjugate base (10.2)
Diprotic acid (10.3)
Dissociation (10.1)
Electrolyte (10.12)
Hydrolysis (10.10)
Ion product constant for water (10.8)
Ionization (10.1)
Monoprotic acid (10.3)
Neutral solution (10.8)
Neutralization (10.7)
Nonelectrolyte (10.12)
pH scale (10.9)
Polyprotic acid (10.3)
Salt (10.6)
Strong acid (10.4)
Strong electrolyte (10.12)
Triprotic acid (10.3)
Weak acid (10.4)
Weak electrolyte (10.12)

Exercises and Problems

The members of each pair of problems in this section test similar material.

Arrhenius Acid–Base Theory (Section 10.1)

10.1 Write equations depicting the behavior of the following Arrhenius acids and bases in water.

a. HI (hydroiodic acid)

b. HClO (hypochlorous acid)

c. LiOH (lithium hydroxide)

d. CsOH (cesium hydroxide)

10.2 Write equations depicting the behavior of the following Arrhenius acids and bases in water.

a. HBr (hydrobromic acid)

b. HCN (hydrocyanic acid)

c. RbOH (rubidium hydroxide)

d. KOH (potassium hydroxide)

Brønsted–Lowry Acid–Base Theory (Section 10.2)

10.3 In each of the following reactions, decide whether the underlined species is functioning as a Brønsted–Lowry acid or base.

a. $\underline{HF} + H_2O \rightarrow H_3O^+ + F^-$

b. $H_2O + \underline{S^{2-}} \rightarrow HS^- + OH^-$

c. $H_2O + \underline{H_2CO_3} \rightarrow H_3O^+ + HCO_3^-$

d. $\underline{HCO_3^-} + H_2O \rightarrow H_3O^+ + CO_3^{2-}$

10.4 In each of the following reactions, decide whether the underlined species is functioning as a Brønsted–Lowry acid or base.

a. $\underline{HClO_2} + H_2O \rightarrow H_3O^+ + ClO_2^-$

b. $\underline{OCl^-} + H_2O \rightarrow HOCl + OH^-$

c. $NH_3 + \underline{HNO_2} \rightarrow NH_4^+ + NO_2^-$

d. $HCl + \underline{H_2PO_4^-} \rightarrow H_3PO_4 + Cl^-$

10.5 Write equations to illustrate the acid–base reactions that can take place between the following Brønsted–Lowry acids and bases.

a. Acid: HClO; base: H_2O

b. Acid: $HClO_4$; base: NH_3

c. Acid: H_3O^+; base: OH^-

d. Acid: H_3O^+; base: NH_2^-

10.6 Write equations to illustrate the acid–base reactions that can take place between the following Brønsted–Lowry acids and bases.

a. Acid: $H_2PO_4^-$; base: NH_3

b. Acid: H_2O; base: ClO_4^-

c. Acid: HCl; base: OH^-

d. Acid: $HC_2H_3O_2$; base: H_2O

10.7 Write the formula of each of the following.

a. Conjugate base of H_2SO_3

b. Conjugate acid of CN^-

c. Conjugate base of $HC_2O_4^-$

d. Conjugate acid of HPO_4^{2-}

10.8 Write the formula of each of the following.

a. Conjugate base of NH_4^+

b. Conjugate acid of OH^-

c. Conjugate base of H_2S

d. Conjugate acid of NO_2^-

10.9 For each of the following amphoteric substances, write the two equations needed to describe its behavior in aqueous solution.

a. HS^- b. HPO_4^{2-} c. NH_3 d. OH^-

10.10 For each of the following amphoteric substances, write the two equations needed to describe its behavior in aqueous solution.

a. $H_2PO_4^-$ b. HSO_4^- c. $HC_2O_4^-$ d. PH_3

Polyprotic Acids (Section 10.3)

10.11 Classify the following acids as monoprotic, diprotic, or triprotic.

a. $HClO_4$ (perchloric acid) b. $H_2C_2O_4$ (oxalic acid)

c. $HC_2H_3O_2$ (acetic acid) d. H_2SO_4 (sulfuric acid)

10.12 Classify the following acids as monoprotic, diprotic, or triprotic.

a. $HC_4H_7O_2$ (butyric acid) b. H_3PO_4 (phosphoric acid)

c. HNO_3 (nitric acid) d. $H_2C_4H_4O_4$ (succinic acid)

10.13 Write equations showing all steps in the dissociation of citric acid ($H_3C_6H_5O_7$).

10.14 Write equations showing all steps in the dissociation of arsenic acid (H_3AsO_4).

10.15 The formula for lactic acid is preferably written as $HC_3H_5O_3$, rather than as $C_3H_6O_3$. Explain why.

10.16 The formula for tartaric acid is preferably written as $H_2C_4H_4O_6$ rather than as $C_4H_6O_6$. Explain why.

10.17 Pyruvic acid, which is produced in metabolic reactions, has the structure

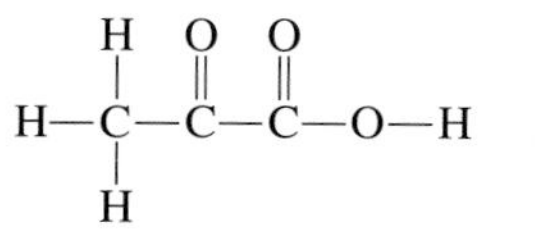

```
    H  O  O
    |  ‖  ‖
H—C—C—C—O—H
    |
    H
```

Would you predict that this acid is a mono-, di-, tri-, or tetraprotic acid? Give your reasoning.

10.18 Oxaloacetic acid, which is produced in metabolic reactions, has the structure

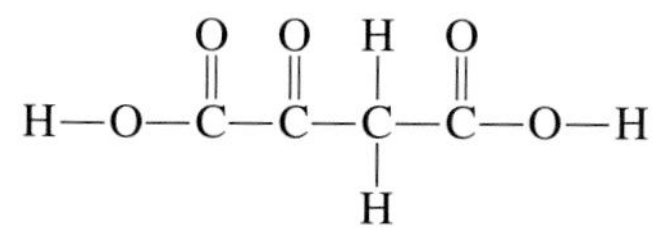

```
       O  O  H  O
       ‖  ‖  |  ‖
H—O—C—C—C—C—O—H
             |
             H
```

Would you predict that this acid is a mono-, di-, tri-, or tetraprotic acid? Give your reasoning.

Strengths of Acids and Bases (Section 10.4)

10.19 Classify each of the acids in Problem 10.11 as a strong acid or a weak acid.

10.20 Classify each of the acids in Problem 10.12 as a strong acid or a weak acid.

Ionization Constants for Acids and Bases (Section 10.5)

10.21 Write the acid ionization constant expression for the ionization of each of the following monoprotic acids.

a. HF (hydrofluoric acid)

b. $HC_2H_3O_2$ (acetic acid)

10.22 Write the acid ionization constant expression for the ionization of each of the following monoprotic acids.

a. HCN (hydrocyanic acid)

b. $HC_6H_7O_6$ (ascorbic acid)

10.23 Write the base ionization constant expression for the ionization of each of the following bases. In each case, the nitrogen atom accepts the proton.

a. NH_3 (ammonia)

b. $C_6H_5NH_2$ (aniline)

10.24 Write the base ionization constant expression for the ionization of each of the following bases. In each case, the nitrogen atom accepts the proton.

a. CH_3NH_2 (methylamine)

b. $C_2H_5NH_2$ (ethylamine)

10.25 Using the acid ionization constant information given in Table 10.4, indicate which acid is the stronger in each of the following acid pairs.

a. H_3PO_4 and HNO_2 b. HCN and HF

c. H_2CO_3 and HCO_3^- d. HNO_2 and HCN

10.26 Using the acid ionization constant information given in Table 10.4, indicate which acid is the stronger in each of the following acid pairs.

a. H_3PO_4 and $H_2PO_4^-$

b. H_3PO_4 and H_2CO_3

c. HPO_4^{2-} and $H_2PO_4^-$

d. $HC_2H_3O_2$ and HCN

10.27 A 0.00300 M solution of an acid is 12% ionized. Calculate the acid ionization constant K_a.

10.28 A 0.0500 M solution of a base is 7.5% ionized. Calculate the base ionization constant K_b.

Salts (Section 10.6)

10.29 Classify each of the following substances as an acid, a base, or a salt.

a. HBr b. NaI c. NH_4NO_3 d. $Ba(OH)_2$

e. $AlPO_4$ f. KOH g. HNO_3 h. $HC_2H_3O_2$

10.30 Classify each of the following substances as an acid, a base, or a salt.

a. KBr b. NH_4I c. H_2SO_4 d. $Ba_3(PO_4)_2$

e. $Ca(OH)_2$ f. HCN g. NaOH h. HCl

10.31 Write a balanced equation for the dissociation into ions of each of the following soluble salts in aqueous solution.

a. $Ba(NO_3)_2$ b. Na_2SO_4 c. $CaBr_2$ d. K_2CO_3

10.32 Write a balanced equation for the dissociation into ions of each of the following soluble salts in aqueous solution.

a. CaS b. $BeSO_4$ c. $MgCl_2$ d. $NaC_2H_3O_2$

Acid–Base Neutralization Reactions (Section 10.7)

10.33 Write a balanced molecular equation to represent each of the following acid–base neutralization reactions.

a. Acid: HCl; base: NaOH

b. Acid: HNO_3; base: KOH

c. Acid: H_2SO_4; base: LiOH

d. Acid: H_3PO_4; base: $Ba(OH)_2$

10.34 Write a balanced molecular equation to represent each of the following acid–base neutralization reactions.

a. Acid: HCl; base: LiOH

b. Acid: HNO_3; base: $Ba(OH)_2$

c. Acid: H_2SO_4; base: NaOH

d. Acid: H_3PO_4; base: KOH

10.35 Write a balanced molecular equation for the preparation of each of the following salts, using an acid–base neutralization reaction.

a. Li_2SO_4 (lithium sulfate)

b. NaCl (sodium chloride)

c. KNO_3 (potassium nitrate)

d. $Ba_3(PO_4)_2$ (barium phosphate)

10.36 Write a balanced molecular equation for the preparation of each of the following salts, using an acid–base neutralization reaction.

a. $LiNO_3$ (lithium nitrate)

b. $BaCl_2$ (barium chloride)

c. K_3PO_4 (potassium phosphate)

d. Na_2SO_4 (sodium sulfate)

Hydronium Ion and Hydroxide Ion Concentrations (Section 10.8)

10.37 Calculate the molar H_3O^+ ion concentration of a solution if the OH^- ion concentration is

a. 3.0×10^{-3} M b. 6.7×10^{-6} M

c. 9.1×10^{-8} M d. 1.2×10^{-11} M

10.38 Calculate the molar H_3O^+ ion concentration of a solution if the OH^- ion concentration is

a. 5.0×10^{-4} M b. 7.5×10^{-7} M

c. 2.3×10^{-12} M d. 1.1×10^{-10} M

10.39 Indicate whether the following solutions are acidic, basic, or neutral.

a. $[H_3O^+] = 1.0 \times 10^{-3}$ b. $[H_3O^+] = 3.0 \times 10^{-11}$

c. $[OH^-] = 4.0 \times 10^{-6}$ d. $[OH^-] = 2.3 \times 10^{-10}$

10.40 Indicate whether the following solutions are acidic, basic, or neutral.

a. $[H_3O^+] = 2.0 \times 10^{-4}$ b. $[H_3O^+] = 2.0 \times 10^{-8}$

c. $[OH^-] = 1.0 \times 10^{-7}$ d. $[OH^-] = 5.0 \times 10^{-9}$

pH Scale (Section 10.9)

10.41 Calculate the pH of the following solutions.

a. $[H_3O^+] = 1.0 \times 10^{-4}$ b. $[H_3O^+] = 1.0 \times 10^{-11}$

c. $[OH^-] = 1.0 \times 10^{-3}$ d. $[OH^-] = 1.0 \times 10^{-7}$

10.42 Calculate the pH of the following solutions.

a. $[H_3O^+] = 1.0 \times 10^{-6}$ b. $[H_3O^+] = 1.0 \times 10^{-2}$

c. $[OH^-] = 1.0 \times 10^{-9}$ d. $[OH^-] = 1.0 \times 10^{-5}$

10.43 Calculate the pH of the following solutions.

a. $[H_3O^+] = 2.1 \times 10^{-8}$ b. $[H_3O^+] = 4.0 \times 10^{-8}$
c. $[OH^-] = 7.2 \times 10^{-11}$ d. $[OH^-] = 7.2 \times 10^{-3}$

10.44 Calculate the pH of the following solutions.

a. $[H_3O^+] = 3.3 \times 10^{-5}$ b. $[H_3O^+] = 7.6 \times 10^{-5}$
c. $[OH^-] = 8.2 \times 10^{-10}$ d. $[OH^-] = 8.2 \times 10^{-4}$

10.45 What is the molar hydronium ion concentration in solutions with each of the following pH values?

a. 2.0 b. 6.0 c. 8.0 d. 10.0

10.46 What is the molar hydronium ion concentration in solutions with each of the following pH values?

a. 3.0 b. 5.0 c. 9.0 d. 12.0

10.47 What is the molar hydronium ion concentration in solutions with each of the following pH values?

a. 3.67 b. 5.09 c. 7.35 d. 12.45

10.48 What is the molar hydronium ion concentration in solutions with each of the following pH values?

a. 2.05 b. 4.88 c. 6.75 d. 11.33

pK_a Values (Section 10.10)

10.49 Calculate the pK_a value for each of the following acids.

a. Nitrous acid (HNO_2), $K_a = 7.2 \times 10^{-4}$
b. Carbonic acid (H_2CO_3), $K_a = 4.3 \times 10^{-7}$
c. Dihydrogen phosphate ion ($H_2PO_4^-$), $K_a = 6.2 \times 10^{-8}$
d. Sulfurous acid (H_2SO_3), $K_a = 1.5 \times 10^{-2}$

10.50 Calculate the pK_a value for each of the following acids.

a. Phosphoric acid (H_3PO_4), $K_a = 7.5 \times 10^{-3}$
b. Hydrofluoric acid (HF), $K_a = 6.6 \times 10^{-4}$
c. Hydrogen phosphate ion (HPO_4^{2-}), $K_a = 2.2 \times 10^{-13}$
d. Propanoic acid ($HC_3H_5O_2$), $K_a = 1.3 \times 10^{-5}$

10.51 Acid A has a pK_a value of 4.23, and acid B has a pK_a value of 3.97. Which of the two acids is the stronger?

10.52 Acid A has a pK_a value of 5.71, and acid B has a pK_a value of 5.30. Which of the two acids is the weaker?

Hydrolysis of Salts (Section 10.11)

10.53 Classify each of the following salts as a "strong acid–strong base salt," a "strong acid–weak base salt," a "weak acid–strong base salt," or a "weak acid–weak base salt."

a. NaCl b. $KC_2H_3O_2$ c. NH_4Br d. $Ba(NO_3)_2$

10.54 Classify each of the following salts as a "strong acid–strong base salt," a "strong acid–weak base salt," a "weak acid–strong base salt," or a "weak acid–weak base salt."

a. K_3PO_4 b. $NaNO_3$ c. KCl d. $Na_2C_2O_4$

10.55 Identify the ion (or ions) present in each of the salts in Problem 10.53 that will undergo hydrolysis in aqueous solution.

10.56 Identify the ion (or ions) present in each of the salts in Problem 10.54 that will undergo hydrolysis in aqueous solution.

10.57 Predict whether solutions of each of the salts in Problem 10.53 will be acidic, basic, or neutral.

10.58 Predict whether solutions of each of the salts in Problem 10.54 will be acidic, basic, or neutral.

Buffers (Section 10.12)

10.59 Predict whether each of the following pairs of substances could function as a buffer system in aqueous solution.

a. HNO_3 and $NaNO_3$
b. HF and NaF
c. KCl and KCN
d. H_2CO_3 and $NaHCO_3$

10.60 Predict whether each of the following pairs of substances could function as a buffer system in aqueous solution.

a. HNO_3 and HCl
b. HNO_2 and KNO_2
c. $NaC_2H_3O_2$ and $KC_2H_3O_2$
d. $HC_2H_3O_2$ and $NaNO_3$

10.61 Write an equation for each of the following buffering actions.

a. The response of a HF/F^- buffer to the addition of H_3O^+ ions
b. The response of a H_2CO_3/HCO_3^- buffer to the addition of OH^- ions
c. The response of a HCO_3^-/CO_3^{2-} buffer to the addition of H_3O^+ ions
d. The response of a $H_3PO_4/H_2PO_4^-$ buffer to the addition of OH^- ions

10.62 Write an equation for each of the following buffering actions.

a. The response of a HPO_4^{2-}/PO_4^{3-} buffer to the addition of OH^- ions
b. The response of a HF/F^- buffer to the addition of OH^- ions
c. The response of a HCN/CN^- buffer to the addition of H_3O^+ ions
d. The response of a $H_3PO_4/H_2PO_4^-$ buffer to the addition of H_3O^+ ions

The Henderson–Hasselbalch Equation (Section 10.13)

10.63 What is the pH of a buffer that is 0.230 M in a weak acid and 0.500 M in the acid's conjugate base? The pK_a for the acid is 6.72.

10.64 What is the pH of a buffer that is 0.250 M in a weak acid and 0.260 M in the acid's conjugate base? The pK_a for the acid is 5.53.

10.65 What is the pH of a buffer that is 0.150 M in a weak acid and 0.150 M in the acid's conjugate base? The acid's ionization constant in 6.8×10^{-6}.

10.66 What is the pH of a buffer that is 0.175 M in a weak acid and 0.200 M in the acid's conjugate base? The acid's ionization constant in 5.7×10^{-4}.

Electrolytes (Section 10.14)

10.67 Classify each of the following compounds as a strong electrolyte or a weak electrolyte.

a. H_2CO_3 b. KOH c. NaCl d. H_2SO_4

10.68 Classify each of the following compounds as a strong electrolyte or a weak electrolyte.

a. H_3PO_4 b. HNO_3 c. KNO_3 d. NaOH

Titration Calculations (Section 10.15)

10.69 Determine the molarity of a NaOH solution when each of the following amounts of acid neutralizes 25.0 mL of the NaOH solution.

a. 5.00 mL of 0.250 M HNO_3
b. 20.00 mL of 0.500 M H_2SO_4
c. 23.76 mL of 1.00 M HCl
d. 10.00 mL of 0.100 M H_3PO_4

10.70 Determine the molarity of a KOH solution when each of the following amounts of acid neutralizes 25.0 mL of the KOH solution.

a. 5.00 mL of 0.500 M H_2SO_4
b. 20.00 mL of 0.250 M HNO_3
c. 13.07 mL of 0.100 M H_3PO_4
d. 10.00 mL of 1.00 M HCl

Additional Problems

10.71 In which of the following pairs of substances do the two members of the pair constitute a conjugate acid–base pair?

a. HN_3 and N_3^-
b. H_2SO_4 and SO_4^{2-}
c. H_2CO_3 and $HClO_3$
d. NH_3 and NH_2^-

10.72 For which of the following pairs of acids are both members of the pair of "like strength"—that is, both strong or both weak?

a. HNO_3 and HNO_2
b. HCl and HBr
c. H_3PO_4 and $HClO_4$
d. H_2CO_3 and $H_2C_2O_4$

10.73 Solution A has $[OH^-] = 2.5 \times 10^{-3}$. Solution B has $[H_3O^+] = 3.5 \times 10^{-4}$.

a. Which solution is more acidic?
b. Which solution has the higher pH?

10.74 What would be the pH of a solution that contains 0.1 mole of each of the solutes NaCl, NaOH, and HCl in enough water to give 2.00 L of solution?

10.75 Arrange the following 0.1 M aqueous solutions in order of increasing pH: HCl, HCN, NaOH, and KCl.

10.76 Identify the buffer systems (conjugate acid–base pairs) present in a solution containing equal molar amounts of HCN, KCN, NaCN, and NaCl.

10.77 Both ions in each of the salts NH_4CN and $NH_4C_2H_3O_2$ undergo hydrolysis in aqueous solution. Upon hydrolysis, the first listed salt gives a basic solution and the second a neutral solution. Explain how this is possible.

10.78 How many grams of NaOH are needed to make 875 mL of a solution with a pH of 10.00?

Grid Problems

10.79

1.	2.	3.
H_2S	HS^-	S^{2-}
4. H_2CO_3	5. HCO_3^-	6. CO_3^{2-}

Select from the grid *all* correct responses for each of the following situations.

a. Species whose conjugate acid is also present in the grid
b. Species whose conjugate base is also present in the grid
c. Species that would be expected to exhibit amphoteric behavior
d. Pairs of species whose reaction would produce S^{2-}

10.80

1.	2.	3.
HNO_3 H_2SO_4	H_3PO_4 H_2CO_3	HCN HCl
4. $HC_2H_3O_2$ $HC_3H_5O_2$	5. $HClO_4$ $HClO_2$	6. $H_3C_6H_5O_7$ H_3BO_3

Select from the grid *all* correct responses for each of the following situations.

a. Both members of the pair are strong acids.
b. Both members of the pair are weak acids.
c. Both members of the pair are polyprotic acids.
d. All hydrogen atoms are acidic for both members of the pair.

10.81

1.	2.	3.
pH = 6.0	$[OH^-] = 10^{-8}$	$[H_3O^+] = [OH^-]$
4. $[OH^-] = 10^{-3}$	5. $[H_3O^+] = 10^{-11}$	6. pH = 7.0

Select from the grid *all* correct responses for each of the following situations.

a. Solutions that are neither acidic nor basic
b. Solutions that are basic
c. Pairs of solutions with the same pH
d. Pairs of solutions with the same hydroxide ion concentration

10.82

1.	2.	3.
NaCl	K_2CO_3	NH_4Br
4. Na_2SO_4	5. K_3PO_4	6. NH_4NO_3

Select from the grid *all* correct responses for each of the following situations.

a. Salts that will not hydrolyze in aqueous solution
b. Salts whose negative ion will hydrolyze in aqueous solution
c. Salts whose aqueous solutions have a pH greater than 7.0
d. Salts whose aqueous solutions are neutral

CHAPTER 11

PART I

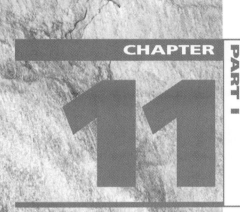

Nuclear Chemistry

CHAPTER OUTLINE

11.1 Stable and Unstable Nuclides 271
11.2 The Nature of Radioactivity 272
11.3 Radioactive Decay 273
11.4 Rate of Radioactive Decay 276
11.5 Transmutation and Bombardment Reactions 279
11.6 Radioactive Decay Series 280
11.7 Ionizing Effects of Radiation 282

Chemistry at a Glance:
Terminology Associated with Nuclear Reactions 283

11.8 Biological Effects of Radiation 283

Chemistry at a Glance
Ionizing Radiation 285

11.9 Detection of Radiation 285
11.10 Sources of Radiation Exposure 286
11.11 Nuclear Medicine 287
11.12 Nuclear Fission and Nuclear Fusion 290
11.13 Nuclear and Chemical Reactions Compared 293

Chemical Connections
11.1 Tobacco Radioactivity and the Uranium-238 Decay Series 281
11.2 The PET Scan—Seeing the Brain in Action 289

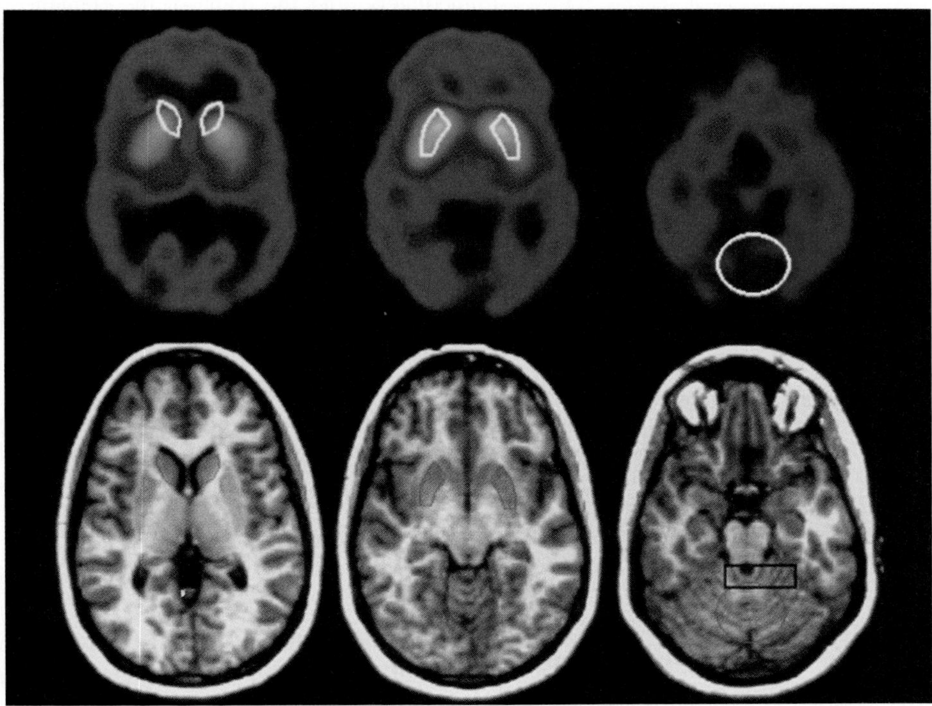

Associated with brain-scan technology is the use of small amounts of radioactive substances.

In this chapter we consider nuclear reactions. It is in the study of such reactions that we encounter the terms *radioactivity, nuclear power plants, nuclear weapons,* and *nuclear medicine.* The electricity produced by a nuclear power plant is generated through the use of heat energy obtained from nuclear reactions. In modern medicine, nuclear reactions are used in the diagnosis and treatment of numerous diseases. Despite some controversy concerning the use of nuclear reactions in weapons and power plants, it is important to remember that it is far more likely that your life will be extended by nuclear medicine than that your life will be taken by nuclear weapons.

11.1 Stable and Unstable Nuclides

A **nuclear reaction** *is a reaction in which changes occur in the nucleus of an atom.* Nuclear reactions are not considered to be ordinary chemical reactions. The governing principles for ordinary chemical reactions deal with the rearrangement of electrons; this rearrangement occurs as the result of electron transfer or electron sharing (Section 4.1). In nuclear reactions, it is nuclei rather than electron arrangements that undergo change.

The term *nuclide* is used extensively in discussions about nuclear reactions. A **nuclide** *is an atom of an element that has a specific number of protons and neutrons in its nucleus.* All atoms of a given nuclide must have the same number of protons and the same number of neutrons. The term *nuclide* is used to describe atomic forms of different elements. The species $^{12}_{6}C$ and $^{13}_{6}C$ are *isotopes* of the element carbon. The species $^{12}_{6}C$ and $^{16}_{8}O$ are *nuclides* of different elements.

In order for us to identify a nucleus or atom uniquely, both the atomic number and the mass number must be specified. Two different notation systems exist for doing this. Consider a nuclide of nitrogen that has seven protons and eight neutrons. This nuclide can be denoted as $^{15}_{7}N$ or nitrogen-15. In the first notation, the superscript is the mass number and the subscript is the atomic number. In the second notation, the mass number is placed immediately after the name of the element. Both types of notation will be used in this chapter. Note that both notations give the mass number.

Nuclides may be either stable or unstable. A **stable nuclide** *has a nucleus that does not easily undergo change.* Conversely, an **unstable nuclide** *has a nucleus that spontaneously undergoes change.*

• The term *radioactive isotope* is sometimes used in place of *radioactive nuclide.* We will use *radioactive nuclide* in this text.

A mechanism exists through which unstable nuclides can achieve stability. All unstable nuclides *spontaneously* emit energy (radiation). Such radiation is called radioactivity. **Radioactivity** *is the radiation spontaneously emitted from the nucleus of an unstable nuclide.* Of the 88 elements found in nature (Section 1.7), 29 have at least one naturally occurring isotope that has an unstable nucleus and is, therefore, a radioactive nuclide. A **radioactive nuclide** *(radionuclide) is an atom with an unstable nucleus that spontaneously emits energy (radiation).*

Radioactive nuclides are known for *all* 115 elements even though they occur naturally for only 29 elements. This is because laboratory procedures have been developed by which scientists convert nonradioactive nuclides (stable nuclei) into radioactive nuclides (unstable nuclei). Such procedures are considered in Section 11.5.

Marie Curie (1859–1934), one of the pioneers in the study of radioactivity, is the first person to have been awarded two Nobel Prizes for scientific work. In 1903, she, her husband Pierre, and Henri Becquerel were corecipients of the Nobel Prize in physics. In 1911, she received the Nobel Prize in chemistry. In 1934, Marie, now respectfully called Madame Curie, died of leukemia caused by overexposure to radiation.

11.2 The Nature of Radioactivity

The fact that unstable nuclei spontaneously emit radiation was accidentally discovered by the French physicist and engineer Antoine Henri Becquerel (1852–1908) in 1896. While working on an experiment involving rocks that phosphoresce, Becquerel discovered that a particular uranium-containing rock gave off radiation. Soon other scientists, such as the French chemists Marie (1867–1934) and Pierre (1859–1906) Curie and the British chemist Ernest Rutherford (1871–1937), began their own investigations into this strange phenomenon—a phenomenon that Marie Curie named radioactivity.

The first information concerning the nature of the radiation emanating from naturally radioactive materials was obtained by Rutherford in 1898–1899. Using an apparatus similar to that shown in Figure 11.1, he found that if radiation from uranium is passed between electrically charged plates, it is split into three components. This finding indicates the presence of three different types of emissions from naturally radioactive materials. A closer analysis of Rutherford's experiment reveals that one radiation component is positively charged (it is attracted to the negative plate); a second component is negatively charged (it is attracted to the positive plate); and the third component carries no charge (it is unaffected by either charged plate). Rutherford chose to call the three radiation components alpha rays (α rays)—the positive component; beta rays (β rays)—the negative component; and gamma rays (γ rays)—the uncharged component. (Alpha, beta, and gamma are the first three letters of the Greek alphabet.) We still use these names for these radiation types today, even though we know much more about their identity. Additional research has substantiated Rutherford's conclusion that three distinct types of radiation are present in the emissions from naturally radioactive substances. This research has also supplied the information necessary for the complete characterization of each type of radiation.

Alpha rays *consist of streams of positively charged particles (alpha particles), each of which is made up of two protons and two neutrons.* The notation used to represent an

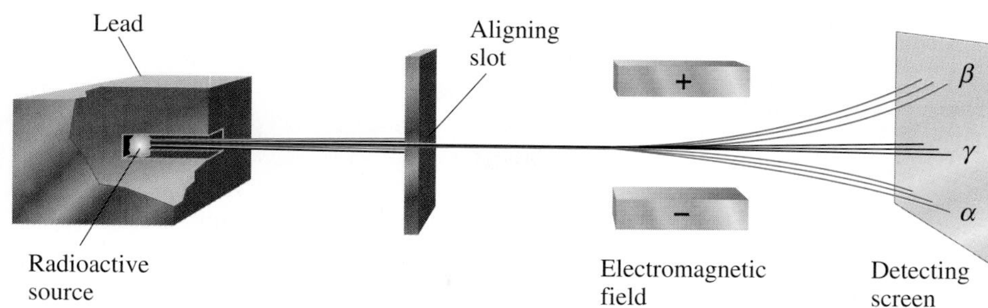

Figure 11.1
The effect of an electromagnetic field on alpha, beta, and gamma radiation. Alpha and beta particles are deflected in opposite directions, whereas gamma radiation is not affected.

alpha particle is ${}^{4}_{2}\alpha$. The numerical subscript indicates that the charge on the particle is +2 (from the two protons). The numerical superscript indicates a mass of 4 amu. Alpha particles are identical to the nuclei of helium-4 (${}^{4}_{2}He$) atoms; because of this, an alternative designation for an alpha particle is ${}^{4}_{2}He$.

• There are two key concepts for understanding the phenomenon of radioactivity: (1) certain nuclides possess unstable nuclei, and (2) nuclides with unstable nuclei spontaneously emit energy (radiation).

Beta rays *consist of streams of negatively charged particles (beta particles) whose charge and mass are identical to those of an electron.* However, beta particles are not extranuclear electrons; they are particles that have been produced inside the nucleus and then ejected. We will discuss this process in Section 11.3. The symbol used to represent a beta particle is ${}^{0}_{-1}\beta$. The numerical subscript indicates that the charge on the beta particle is −1; it is the same as that of an electron. The use of the superscript zero for the mass of a beta particle should be interpreted as meaning not that a beta particle has no mass but, rather, that the mass is very close to zero amu. The actual mass of a beta particle is 0.00055 amu.

• Early researchers in the field of radioactivity were hampered by the fact that many of the details of atomic structure were not yet known. For example, the neutron was not identified until 1932.

Gamma rays *are not considered to be particles; they are pure energy without charge or mass.* They are very high-energy radiation, somewhat like X rays. The symbol for gamma rays is ${}^{0}_{0}\gamma$.

• Alpha, beta, and gamma radiation are designated with the notations ${}^{4}_{2}\alpha$, ${}^{0}_{-1}\beta$, and ${}^{0}_{0}\gamma$, respectively.

11.3 Radioactive Decay

Alpha, beta, and gamma emissions come from the nucleus of an atom. These spontaneous emissions alter nuclei; obviously, if a nucleus loses an alpha particle (two protons and two neutrons), it will not be the same as it was before the departure of the particle. In the case of alpha and beta emissions, the nuclear alteration causes the identity of atoms to change, forming a new element. Thus nuclear reactions differ dramatically from ordinary chemical reactions, where the identities of the elements are always maintained. **Radioactive decay** *is the process whereby a radionuclide is transformed into a nuclide of another element as a result of the emission of radiation.* The terms *parent nuclide* and *daughter nuclide* are often used when describing radioactive decay processes. A **parent nuclide** *is the nuclide that undergoes decay in a radioactive decay process.* A **daughter nuclide** *is the nuclide that is produced as a result of a radioactive decay process.*

• In a *chemical* reaction, element identity is maintained. Atoms are rearranged to form new substances involving the same elements. In a *nuclear* reaction, element identity is not maintained. An element changes into a different element.

Alpha Particle Decay

Alpha particle decay, which is the emission of an alpha particle from a nucleus, always results in the formation of a nuclide of a different element. The product nucleus has an atomic number that is 2 less than that of the original nucleus and a mass number that is 4 less than that of the original nucleus. We can represent alpha particle decay in general terms by the equation

$$ {}^{A}_{Z}X \longrightarrow {}^{4}_{2}\alpha + {}^{A-4}_{Z-2}Y $$

where X is the symbol for the nucleus of the original element undergoing decay and Y is the symbol of the element formed as a result of the decay.

• Loss of an alpha particle from an unstable nucleus results in (1) a decrease of 4 units in the mass number (*A*) and (2) a decrease of 2 units in the atomic number (*Z*).

• Note that the symbols in nuclear equations stand for *nuclei* rather than atoms. We do not worry about electrons when writing nuclear equations.

Let us write equations for two alpha particle decay processes. Both $^{211}_{83}Bi$ and $^{238}_{92}U$ are radionuclides that undergo alpha particle decay. The nuclear equations for these two decay processes are

$$^{211}_{83}Bi \longrightarrow {}^{4}_{2}\alpha + {}^{207}_{81}Tl$$
$$^{238}_{92}U \longrightarrow {}^{4}_{2}\alpha + {}^{234}_{90}Th$$

where thallium-207 and thorium-234 are the daughter nuclides.

The procedures for balancing nuclear equations are different from those used for ordinary chemical equations. In a **balanced nuclear equation,** *the sums of the subscripts (atomic numbers or particle charges) on both sides of the equation are equal, and the sums of the superscripts (mass numbers) on both sides of the equation are equal.* Both of our example equations are balanced. In the alpha decay of $^{211}_{83}Bi$, the subscripts on both sides total 83, and the superscripts total 211.

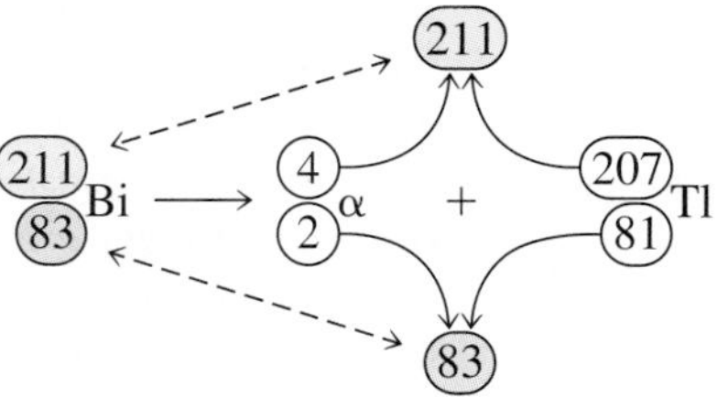

• The rules for balancing nuclear equations are

1. The sum of the subscripts must be the same on both sides of the equation.
2. The sum of the superscripts must be the same on both sides of the equation.

For the alpha decay of $^{238}_{92}U$, the subscripts total 92 on both sides, and the superscripts total 238.

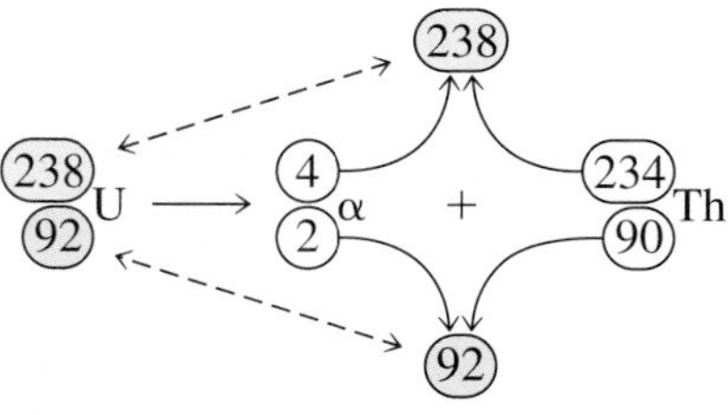

• Beta Particle Decay

Beta particle decay always results in the formation of a nuclide of a different element. The mass number of the new nuclide is the same as that of the original atom. However, the atomic number has increased by 1 unit. The general equation for beta decay is

$$^{A}_{Z}X \longrightarrow {}^{0}_{-1}\beta + {}_{Z+1}^{A}Y$$

• Loss of a beta particle from an unstable nucleus results in (1) no change in the mass number (*A*) and (2) an increase of 1 unit in the atomic number (*Z*).

Specific examples of beta particle decay are

$$^{10}_{4}Be \longrightarrow {}^{0}_{-1}\beta + {}^{10}_{5}B$$
$$^{234}_{90}Th \longrightarrow {}^{0}_{-1}\beta + {}^{234}_{91}Pa$$

Both of these nuclear equations are balanced; superscripts and subscripts add to the same sums on both sides of the equation.

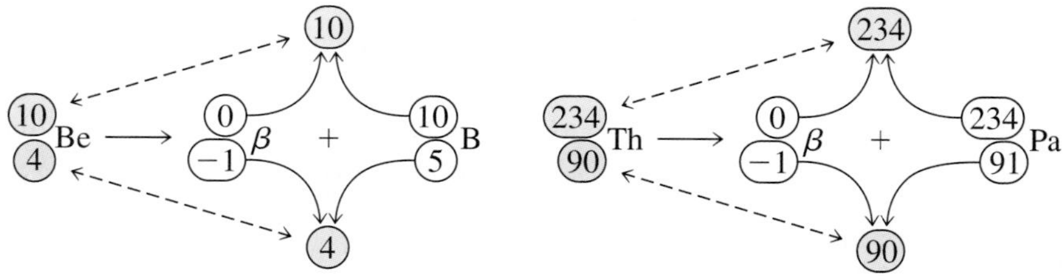

At this point in the discussion, you may be wondering how a nucleus, which is composed only of neutrons and protons, ejects a negative particle (beta particle) when no such particle is present in the nucleus. Explained simply, a neutron in the nucleus is transformed into a proton and a beta particle through a complex series of steps; that is,

$$\text{Neutron} \longrightarrow \text{proton} + \text{beta particle}$$

$$^{1}_{0}\text{n} \longrightarrow {}^{1}_{1}\text{p} + {}^{0}_{-1}\beta$$

Once it is formed within the nucleus, the beta particle is ejected with a high velocity. Note the symbols used to denote a neutron ($^{1}_{0}\text{n}$; no charge and a mass of 1 amu) and a proton ($^{1}_{1}\text{p}$; a +1 charge and a mass of 1 amu).

• Gamma Ray Emission

For naturally occurring radionuclides, gamma ray emission almost always takes place in conjunction with an alpha or a beta decay process; it never occurs independently. These gamma rays are often not included in the nuclear equation because they do not affect the balancing of the equation or the identity of the daughter nuclide. This can be seen from the following two nuclear equations.

• Gamma rays are to nuclear reactions what heat is to ordinary chemical reactions.

• Among *synthetically* produced radionuclides (Section 11.4), pure "gamma emitters," radionuclides that give off gamma rays but no alpha or beta particles, occur. These radionuclides are important in diagnostic nuclear medicine (Section 11.11). Pure "gamma emitters" are not found among naturally occurring radionuclides.

$$^{226}_{88}\text{Ra} \longrightarrow {}^{222}_{86}\text{Rn} + {}^{4}_{2}\alpha + {}^{0}_{0}\gamma$$

Balanced nuclear equation with gamma radiation included

$$^{226}_{88}\text{Ra} \longrightarrow {}^{222}_{86}\text{Rn} + {}^{4}_{2}\alpha$$

Balanced nuclear equation with gamma radiation omitted

The fact that gamma rays are often left out of balanced nuclear equations should not be interpreted to mean that such rays are not important in nuclear chemistry. On the contrary, gamma rays are more important than alpha and beta particles when the effects of external radiation exposure on living organisms are considered (Section 11.8).

Example 11.1

Writing Balanced Nuclear Equations, Given the Parent Nuclide and Its Mode of Decay

Write a balanced nuclear equation for the decay of each of the following radioactive nuclides. The mode of decay is indicated in parentheses.

a. $^{70}_{31}\text{Ga}$ (beta emission) **b.** $^{144}_{60}\text{Nd}$ (alpha emission)

c. $^{248}_{100}\text{Fm}$ (alpha emission) **d.** $^{113}_{47}\text{Ag}$ (beta emission)

Solution

In each case, the atomic and mass numbers of the daughter nucleus are obtained by writing the symbols of the parent nucleus and the particle emitted by the nucleus (alpha or beta). Then the equation is balanced.

a. Let X represent the product of the radioactive decay, the daughter nuclide. Then

$$^{70}_{31}\text{Ga} \longrightarrow {}^{0}_{-1}\beta + \text{X}$$

The sums of the superscripts on both sides of the equation must be equal, so the superscript for X must be 70. In order for the sums of the subscripts on both sides of the equation to be equal, the subscript for X must be 32. Then $31 = (-1) + (32)$. As soon as we determine the subscript of X, we can obtain the identity of X by looking at a periodic table. The element with an atomic number of 32 is Ge (germanium). Therefore,

$$^{70}_{31}\text{Ga} \longrightarrow {}^{0}_{-1}\beta + {}^{70}_{32}\text{Ge}$$

(*continued*)

b. Letting X represent the product of the radioactive decay, we have, for the alpha decay of $^{144}_{60}Nd$,

$$^{144}_{60}\text{Nd} \longrightarrow {}^{4}_{2}\alpha + \text{X}$$

We balance the equation by making the superscripts on each side of the equation total 144 and the subscripts total 60. We get

$$^{144}_{60}\text{Nd} \longrightarrow {}^{4}_{2}\alpha + {}^{140}_{58}\text{Ce}$$

c. Similarly, we write

$$^{248}_{100}\text{Fm} \longrightarrow {}^{4}_{2}\alpha + \text{X}$$

Balancing superscripts and subscripts, we get

$$^{248}_{100}\text{Fm} \longrightarrow {}^{4}_{2}\alpha + {}^{244}_{98}\text{Cf}$$

In alpha emission, the atomic number of the daughter nuclide always decreases by 2, and the mass number of the daughter nuclide always decreases by 4.

d. Finally, we write

$$^{113}_{47}\text{Ag} \longrightarrow {}^{0}_{-1}\beta + \text{X}$$

In beta emission, the atomic number of the daughter nuclide always increases by 1, and the mass number does not change from that of the parent. The balancing procedure gives us the result

$$^{113}_{47}\text{Ag} \longrightarrow {}^{0}_{-1}\beta + {}^{113}_{48}\text{Cd}$$

Practice Exercise 11.1

Write a balanced nuclear equation for the decay of the following radioactive nuclides. The mode of decay is indicated in parentheses.

a. $^{245}_{97}Bk$ (alpha emission) **b.** $^{89}_{38}Sr$ (beta emission)
c. $^{230}_{90}Th$ (alpha emission) **d.** $^{40}_{19}K$ (beta emission)

• *Answers:* **a.** $^{245}_{97}\text{Bk} \rightarrow {}^{4}_{2}\alpha + {}^{241}_{95}\text{Am}$; **b.** $^{89}_{38}\text{Sr} \rightarrow {}^{0}_{-1}\beta + {}^{89}_{39}\text{Y}$; **c.** $^{230}_{90}\text{Th} \rightarrow {}^{4}_{2}\alpha + {}^{226}_{88}\text{Ra}$; **d.** $^{40}_{19}\text{K} \rightarrow {}^{0}_{-1}\beta + {}^{40}_{20}\text{Ca}$

11.4 Rate of Radioactive Decay

All radioactive nuclides do not decay at the same rate. Some decay very rapidly; others undergo disintegration at extremely slow rates. This indicates that all radionuclides are not equally unstable. The faster the decay rate, the lower the stability of the nuclide.

The concept of *half-life* is used to express nuclear stability quantitatively. The **half-life** ($t_{1/2}$) *is the time required for one-half of any given quantity of a radioactive substance to undergo decay.* For example, if a radionuclide's half-life is 12 days and you have a 4.00-g sample of it, then after 12 days (one half-life), only 2.00 g of the sample (one-half of the original amount) will remain undecayed; the other half will have decayed into some other substance. Similarly, during the next half-life, one-half of the 2.00 g remaining will decay, leaving one-fourth of the original atoms (1.00 g) unchanged. After three half-lives, one-eighth ($\frac{1}{2} \times \frac{1}{2} \times \frac{1}{2}$) of the original sample will remain undecayed. Figure 11.2 illustrates the radioactive decay curve for a radionuclide.

• The faster the decay rate for a radionuclide, the shorter its half-life.

There is a wide range of half-lives for radionuclides. Half-lives as long as billions of years and as short as a fraction of a second have been determined (Table 11.1). Most naturally occurring radionuclides have long half-lives. However, some radionuclides with *short* half-lives are also found in nature. Naturally occurring mechanisms exist for the continual production of the short-lived species.

• Most radionuclides used in diagnostic medicine have short half-lives. This limits to a short time interval the exposure of the human body to radiation.

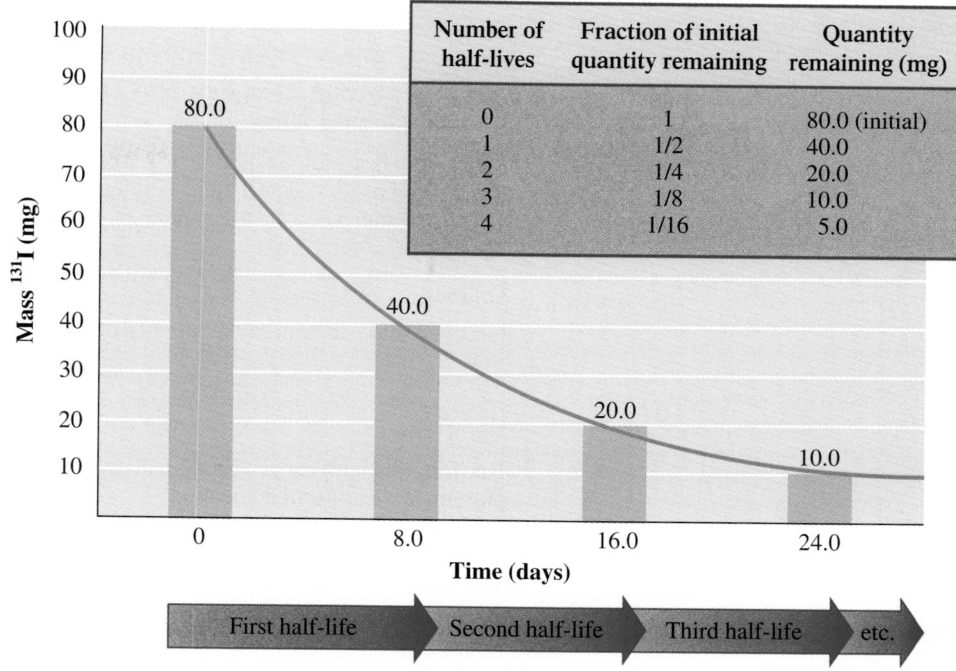

Figure 11.2
Decay of 80.0 mg of ^{131}I, which has a half life of 8.0 days. After each half-life period, the quantity of material present at the beginning of the period is reduced by half.

• The half-life for a radionuclide is independent of external conditions such as temperature, pressure, and state of chemical combination.

The decay rate (half-life) of a radionuclide is constant. It is independent of physical conditions such as temperature, pressure, and state of chemical combination. It depends only on the identity of the radionuclide. For example, radioactive sodium-24 decays at the same rate whether it is incorporated in NaCl, NaBr, Na_2SO_4, or $NaC_2H_3O_2$. If a nuclide is radioactive, nothing will stop it from decaying and nothing will increase or decrease its decay rate.

Calculations involving amounts of radioactive material decayed, amounts remaining undecayed, and time elapsed can be carried out by using the following equation:

$$\begin{pmatrix}\text{Amount of radionuclide}\\ \text{undecayed after } n \text{ half-lives}\end{pmatrix} = \begin{pmatrix}\text{original amount}\\ \text{of radionuclide}\end{pmatrix} \times \left(\frac{1}{2^n}\right)$$

Table 11.1
Range of Half-lives Found for Naturally Occurring Radionuclides

Element	Half-life ($t_{1/2}$)
vanadium-50	6×10^{15} yr
platinum-190	6.9×10^{11} yr
uranium-238	4.5×10^{9} yr
uranium-235	7.1×10^{8} yr
thorium-230	7.5×10^{4} yr
lead-210	22 yr
bismuth-214	19.7 min
polonium-212	3.0×10^{-7} sec

Example 11.2

Using Half-life to Calculate the Amount of Radioisotope That Remains Undecayed After a Certain Time

Iodine-131 is a radionuclide that is frequently used in nuclear medicine. Among other things, it is used to detect fluid buildup in the brain. The half-life of iodine-131 is 8.0 days. How much of a 0.16-g sample of iodine-131 will remain undecayed after a period of 32 days?

Solution

First, we must determine the number of half-lives that have elapsed.

$$32\ \text{days} \times \left(\frac{1\ \text{half-life}}{8.0\ \text{days}}\right) = 4\ \text{half-lives}$$

Knowing the number of elapsed half-lives and the original amount of radioactive iodine present, we can use the equation

$$\begin{pmatrix}\text{Amount of radionuclide}\\ \text{undecayed after } n \text{ half-lives}\end{pmatrix} = \begin{pmatrix}\text{original amount}\\ \text{of radionuclide}\end{pmatrix} \times \left(\frac{1}{2^n}\right)$$

$$= 0.16\,\text{g} \times \frac{1}{2^4} \longleftarrow \text{4 half-lives}$$

$$= 0.16\,\text{g} \times \frac{1}{16} = 0.010\,\text{g}$$

Constructing a tabular summary of the amount of sample remaining after each of the elapsed half-lives yields

Half-lives	0	1	2	3	4
Number of days	0	8	16	24	32
Amount remaining	0.16 g	0.080 g	0.040 g	0.020 g	0.010 g

Practice Exercise 11.2

The half-life of cobalt-60 is 5.3 years. If 2.0 g of cobalt-60 is allowed to decay for a period of 15.9 years, how many grams of cobalt-60 remain?

- *Answer:* 0.25 g

Example 11.3

Using Half-life to Calculate the Time Needed to Reduce Radioactivity to a Specific Level

Strontium-90 is a nuclide found in radioactive fallout from nuclear weapon explosions. Its half-life is 28.0 years. How long will it take for 94% (15/16) of the strontium-90 atoms present in a sample of material to undergo decay?

Solution

If 15/16 of the sample has decayed, then 1/16 of the sample remains undecayed. In terms of $1/2^n$, 1/16 is equal to $1/2^4$; that is,

$$\frac{1}{2} \times \frac{1}{2} \times \frac{1}{2} \times \frac{1}{2} = \frac{1}{2^4} = \frac{1}{16}$$

Thus 4 half-lives have elapsed in reducing the amount of strontium-90 to 1/16 of its original amount.

The half-life of strontium-90 is 28 years, so the total time elapsed will be

$$4\ \text{half-lives} \times \left(\frac{28.0\ \text{years}}{1\ \text{half-life}}\right) = 112\ \text{years}$$

Ernest Rutherford (1871–1937), the first person to carry out a bombardment reaction, was a "world-class" researcher. Earlier, he discovered that an atom has a nucleus (see Section 3.3), and he was the discoverer of the alpha and beta radiation associated with radioactivity.

Practice Exercise 11.3

Iodine-135 is a nuclide found in radioactive fallout from nuclear weapon explosions. Its half-life is 6.70 hr. How long will it take for 75.0% (three-quarters) of the iodine-135 atoms in a "fallout" sample to undergo decay?

- *Answer:* 13.4 hr

11.5 Transmutation and Bombardment Reactions

Radioactive decay, discussed in the previous two sections, is an example of a natural transmutation process. A **transmutation process** *is a nuclear reaction in which a nuclide of one element is changed into a nuclide of another element.* It is also possible to cause transmutation to occur in a laboratory setting by means of a bombardment reaction. A **bombardment reaction** *is a nuclear reaction in which small particles traveling at very high speeds are collided with stable nuclei, causing them to undergo nuclear change.*

The first successful bombardment reaction was carried out in 1919, 25 years after the discovery of radioactive decay. The reaction involved bombarding nitrogen gas with alpha particles from a natural source (radium). In this process, a new stable nuclide was formed: oxygen-17. The nuclear equation for this initial bombardment reaction is

$$^{14}_{7}\text{N} + ^{4}_{2}\alpha \longrightarrow ^{17}_{8}\text{O} + ^{1}_{1}\text{p}$$

Further research carried out by many investigators has shown that numerous nuclei experience change under the stress of bombardment by small, high-energy particles. In most cases, the new nuclide that is produced is radioactive (unstable). Two examples of bombardment reactions now carried out in laboratories in which the product nuclide is radioactive are

$$^{44}_{20}\text{Ca} + ^{1}_{1}\text{p} \longrightarrow ^{44}_{21}\text{Sc} + ^{1}_{0}\text{n}$$
$$^{23}_{11}\text{Na} + ^{2}_{1}\text{H} \longrightarrow ^{21}_{10}\text{Ne} + ^{4}_{2}\alpha$$

- In bombardment reactions, there are always two reactants (the target nuclide and the small, high-energy bombarding particle) and also two products (the daughter nuclide and another small particle such as a neutron or proton).

Radioactive nuclides produced by bombardment reactions obey the same laws as naturally occurring radionuclides. In many cases, the previously discussed alpha and beta modes of decay occur (Section 11.3).

- Production of the small, *high-energy* bombarding particles needed to effect a bombardment reaction requires use of a cyclotron or a linear accelerator (both very expensive pieces of equipment). Both use magnetic fields to accelerate charged particles to velocities at which the energy is sufficient to allow the particle to penetrate the nucleus and induce a nuclear reaction.

Synthetic Elements

Over 2000 bombardment-produced radionuclides that do not occur naturally are now known. This number is seven times greater than the number of naturally occurring nuclides

The synthetic element americium (element 95) is a component of nearly all standard smoke detectors. Small amounts of Am-241, which has a half-life of 458 years, produces radiation that ionizes air within the detector, causing it to conduct electricity. The presence of smoke in the detector causes a drop in electrical conductivity, and the alarm sounds.

Table 11.2
The Transuranium Elements

Name	Symbol	Atomic number	Mass number of most stable nuclide	Half-life of most stable nuclide	Discovery year for first isotope
neptunium	Np	93	237	2.14×10^6 yr	1940
plutonium	Pu	94	244	7.6×10^7 yr	1940
americium	Am	95	243	8.0×10^3 yr	1944
curium	Cm	96	247	1.6×10^7 yr	1944
berkelium	Bk	97	247	1400 yr	1950
californium	Cf	98	251	900 yr	1950
einsteinium	Es	99	252	472 days	1952
fermium	Fm	100	257	100 days	1953
mendelevium	Md	101	258	52 days	1955
nobelium	No	102	259	58 min	1958
lawrencium	Lr	103	262	3.6 hr	1961
rutherfordium	Rf	104	261	65 sec	1969
dubnium	Db	105	262	34 sec	1970
seaborgium	Sg	106	266	20 sec	1974
bohrium	Bh	107	264	1.4 sec	1980
hassium	Hs	108	277	16.5 min	1984
meitnerium	Mt	109	268	0.072 sec	1982
element 110	—	110	281	1.6 min	1994
element 111	—	111	272	0.002 sec	1994
element 112	—	112	285	15.4 min	1996
element 114	—	114	289	30.4 sec	1999
element 116	—	116	289	0.0006 sec	1999
element 118	—	118	293	0.0001 sec	1999

(Section 3.3). In this total is at least one radionuclide of every naturally occurring element. In addition, nuclides of 27 elements that do not occur in nature have been produced in small quantities as the result of bombardment reactions. Four of these "synthetic" elements, produced between 1937 and 1941, filled gaps in the periodic table for which no naturally occurring element had been found. These four elements are technetium (Tc, element 43), an element with numerous uses in nuclear medicine (Section 11.11); promethium (Pm, element 61); astatine (At, element 85); and francium (Fr, element 87). The remainder of the "synthetic" elements, elements 93 to 112, 114, 116, and 118 are called the *transuranium elements* because they occur immediately following uranium in the periodic table. (Uranium is the naturally occurring element with the highest atomic number.) All nuclides of all of the transuranium elements are radioactive. Table 11.2 gives information about the stability of the transuranium elements. Note the extremely short half-lives of the more recently produced elements.

• All nuclides of all elements beyond bismuth ($Z = 83$) in the periodic table are radioactive.

Most radioisotopes used in the field of medicine are "synthetic" radionuclides. For example, the synthetic radionuclides cobalt-60, yttrium-90, iodine-131, and gold-198 are used in radiotherapy treatments for cancer. Section 11.11 provides more information about the medical uses for radionuclides.

11.6 Radioactive Decay Series

Radioactive nuclides with high atomic numbers attain nuclear stability through a series of decay steps. When such nuclides decay, they produce nuclei that are also radioactive. These daughter nuclei in turn decay to a third radioactive product, and so on. Eventually, a stable nucleus is produced. Such a sequence of decay products is called a radioactive

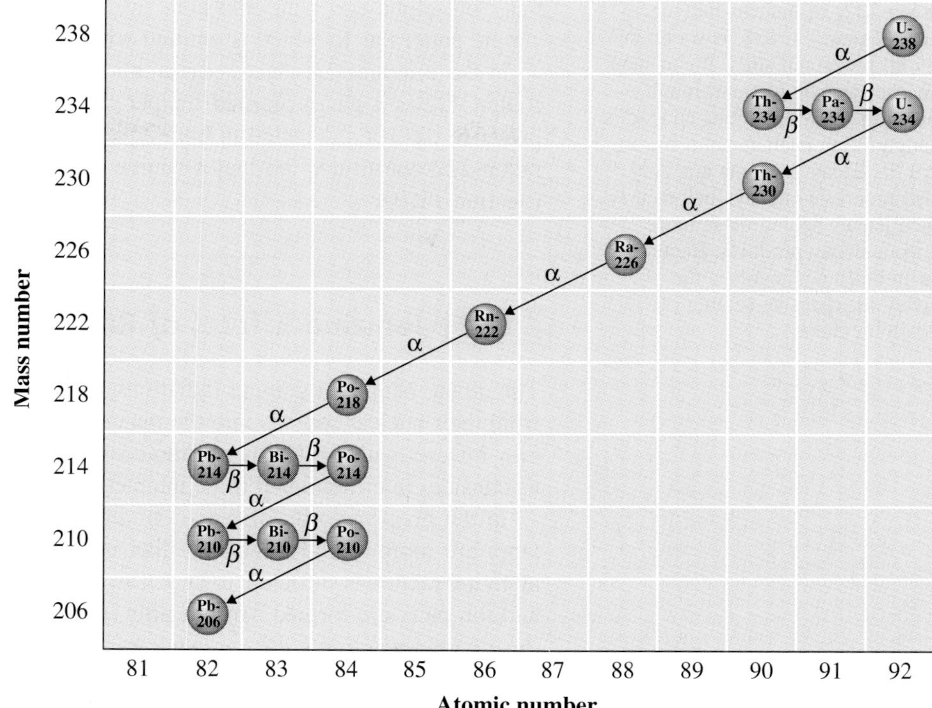

Figure 11.3
In the $^{238}_{92}U$ decay series, each nuclide except $^{206}_{82}Pb$ (the stable end product) is unstable; the successive transformations continue until this stable product is formed.

• In a decay series such as the one involving uranium-238, gamma rays are emitted at each step (even though they are not shown) in addition to the alpha or beta particle. Such gamma rays are very important in the effects of radiation exposure on health (Section 11.8).

decay series. A **radioactive decay series** *is a sequence of nuclear reactions beginning with a very long-lived radionuclide and ending with a stable nuclide of lower atomic number.*

Uranium-238, the most abundant isotope of uranium (99.2%), is the beginning nuclide for an important naturally occurring decay series. As shown in Figure 11.3, 14 steps are needed for uranium-238 to reach lead-206, its stable end product. Note from Figure 11.3

Chemical CONNECTIONS

11.1 Tobacco Radioactivity and the Uranium-238 Decay Series

About 362,000 persons die annually in the United States from tobacco use; 62,000 deaths are caused by lung disease, 130,000 are caused by cancer, and 170,000 result from heart and vascular diseases. One's life is shortened 14 minutes for every cigarette smoked. A 30- to 40-year cigarette smoker who smokes two packs of cigarettes per day loses an estimated 8 years of life.

The link between cigarette smoke and cancer is definitely established. The causative agents for the cancer involve many of the more than 2000 compounds identified in cigarette tar. And radioactivity has also been implicated.

The link between radioactivity and tobacco involves the following sequence of events. The soil in which tobacco is grown is heavily treated with phosphate fertilizers. The source for phosphate fertilizers is ultimately phosphate rock. Nearly all phosphate rock contains small amounts of uranium and its decay products as impurities. Hence small amounts of radioactive nuclides are present in fertilized tobacco-growing soil (as well as in many other crop soils).

Radon-222, one of the intermediate products from the uranium-238 decay series (Section 11.6), is present at relatively high concentrations in soil gas and in the surface air layer under the vegetation canopy provided by a field of growing tobacco plants. The decay products from radon-222 (see Figure 11.3), which include the solids polonium-218 and bismuth-214, often become firmly attached to the surface and interior of tobacco leaves. These short-lived isotopes decay further to lead-210, which has a half-life of 20.4 years. Gradually, lead-210 levels build in tobacco leaves.

During the burning of a cigarette, small insoluble particles (particulates) are produced in addition to gaseous products. Many of these particulates, some of which have a lead-210 content, are inhaled and deposited in the respiratory tract of the smoker and are eventually transported to storage sites in the liver, spleen, and bone marrow. With time (years of smoking), lead-210 concentrations (and decay products) continue to build within the body. The results are constant added exposure of organs and bone marrow to alpha and beta particles and an increased probability of cancer development in the smoker as compared with the nonsmoker.

• Radon-222, an intermediate decay product of uranium-238, is found in nearly all rocks and soils. It can seep into homes and buildings through openings in the foundations and walls. Although radon itself is not a health hazard, its decay products are. Any inhaled radon that decays before it is exhaled leaves behind these more dangerous decay products. Reducing radon in buildings involves the relatively inexpensive sealing of foundation cracks.

that both alpha and beta emissions are part of the decay sequence and that there is no simple pattern as to which is emitted when.

In the uranium-238 decay series, all the intermediate products are solids except one. Radon-222 is a gas at normal temperatures and is therefore a very mobile species. Its presence has been detected in both aqueous and atmospheric environments. Exposure to radon-222 constitutes the major source of radiation exposure for the average American (Section 11.10).

11.7 Ionizing Effects of Radiation

The alpha, beta, and gamma radiations produced from radioactive decay travel outward from their nuclear sources into the material surrounding the radioactive substance. There, they interact with the atoms and molecules of the material and thus lose their energy. Let us consider in closer detail these interactions between radiation and atoms and molecules.

In the great majority of radiation–atom interactions, the extranuclear electrons of an atom are more directly involved than is the nucleus. Energy transfer during radiation–atom interaction is sufficient to knock away electrons from atoms; that is, ionization occurs and ion pairs are formed. An **ion pair** *is the electron and positive ion that are produced during an ionization collision between an atom and radiation.* This ionization process is not the transfer of electrons that occurs during ionic compound formation but, rather, a nonchemical removal of electrons from atoms to form ions. Figure 11.4 diagrammatically shows ion pair formation. Many ion pairs are produced by a single "particle" of radiation, because such a particle must undergo many collisions before its energy is reduced to the level of the surrounding material. The electrons ejected from an atom frequently have enough energy to bombard neighboring molecules and cause additional ionization.

Free-radical formation is another effect caused by alpha, beta, and gamma radiation. A **free radical** *is a highly reactive* uncharged *molecular fragment—that is, an uncharged piece of a molecule.* Free radicals are very reactive entities because they contain an unpaired electron. (Recall from Section 5.2 that electrons occur in pairs in normal bonding situations.)

The ionizing effect of radiation (ion pair formation and free-radical formation) is what makes radiation harmful to both living matter and inert materials. In living matter, the formation of ions and free radicals disrupts cellular function.

The accompanying Chemistry at a Glance summarizes what we have learned about radioactivity thus far in the chapter.

Figure 11.4
Ion pair formation. When radiation interacts with an atom, electrons are often knocked away from the atom. The atoms that lose electrons become ions. An ion so produced and its "free electron" constitute an ion pair.

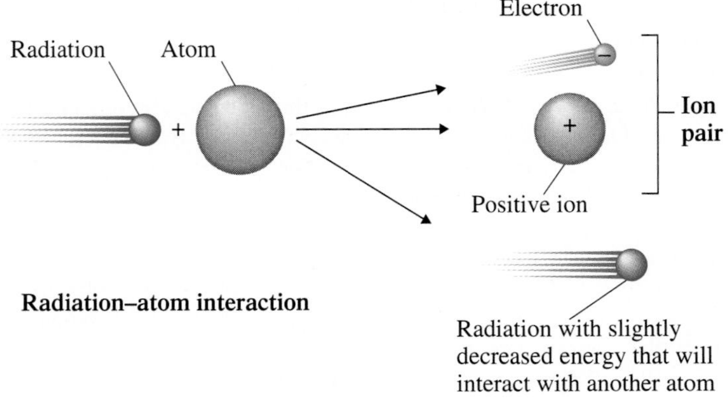

CHEMISTRY AT A GLANCE

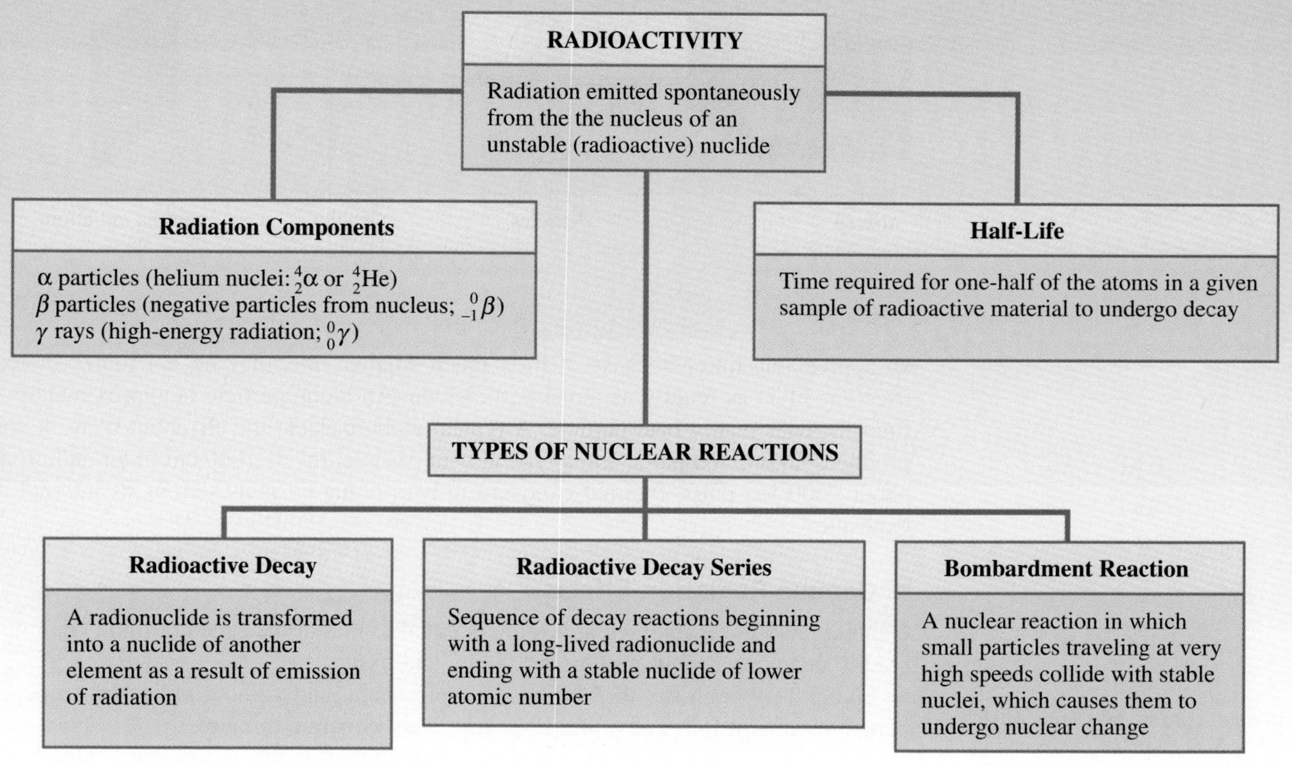

RADIOACTIVITY
Radiation emitted spontaneously from the the nucleus of an unstable (radioactive) nuclide
Radiation Components
α particles (helium nuclei: ${}^{4}_{2}\alpha$ or ${}^{4}_{2}He$)
β particles (negative particles from nucleus; ${}^{0}_{-1}\beta$)
γ rays (high-energy radiation; ${}^{0}_{0}\gamma$)
Half-Life
Time required for one-half of the atoms in a given sample of radioactive material to undergo decay
TYPES OF NUCLEAR REACTIONS
Radioactive Decay
A radionuclide is transformed into a nuclide of another element as a result of emission of radiation
Radioactive Decay Series
Sequence of decay reactions beginning with a long-lived radionuclide and ending with a stable nuclide of lower atomic number
Bombardment Reaction
A nuclear reaction in which small particles traveling at very high speeds collide with stable nuclei, which causes them to undergo nuclear change

11.8 Biological Effects of Radiation

The three types of naturally occurring radioactive emissions—alpha particles, beta particles, and gamma rays—differ in their ability to penetrate matter and cause ionization. Consequently, the extent of the biological effects of radiation depends on the type of radiation involved.

• Alpha Particle Effects

Alpha particles are the most massive and also the slowest particles involved in natural radioactive decay processes. Maximum alpha particle velocities are on the order of one-tenth of the speed of light. For a given alpha-emitting radionuclide, all alpha particles have the same energy; different alpha-emitting radionuclides, however, produce alpha particles of differing energies.

• The speed of light, 3.0×10^{8} m/sec (186,000 miles/sec), is the maximum limit of velocity. Objects cannot travel faster than the speed of light.

Because of their "slowness," alpha particles have low penetrating power and cannot penetrate the body's outer layers of skin. The major damage from alpha radiation occurs when alpha-emitting radionuclides are ingested—for example, in contaminated food. There are no protective layers of skin within the body.

• Beta Particle Effects

Unlike alpha particles, which are all emitted with the same discrete energy from a given radionuclide, beta particles emerge from a beta-emitting substance with a continuous range of energies up to a specific limit that is characteristic of the particular radionuclide. Maximum beta particle velocities are on the order of nine-tenths of the speed of light.

With their greater velocity, beta particles can penetrate much deeper than alpha particles and can cause severe skin burns if their source remains in contact with the skin for

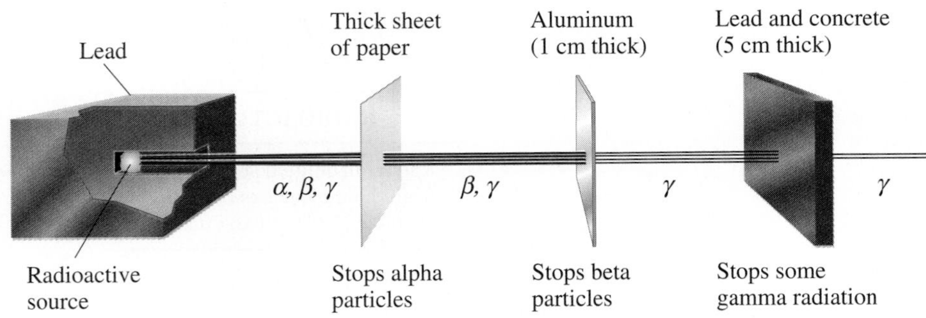

Figure 11.5
Alpha, beta, and gamma radiation differ in penetrating ability.

an appreciable time. Because of their much smaller size, they do not ionize molecules (Section 11.7) as readily as alpha particles do. An alpha particle is approximately 8000 times heavier than a beta particle. A typical alpha particle travels about 6 cm in air and produces 40,000 ion pairs, and a typical beta particle travels 1000 cm in air and produces about 2000 ion pairs. Internal exposure to beta radiation is as serious as internal alpha exposure.

• Gamma Radiation Effects

Gamma radiation is released at a velocity equal to that of the speed of light. Gamma rays readily penetrate deeply into organs, bone, and tissue.

Figure 11.5 contrasts the abilities of alpha, beta, and gamma radiations to penetrate paper, aluminum foil, and a thin layer of a lead–concrete mixture.

The minimum radiation dosage that causes human injury is unknown. However, the effects of larger doses have been studied (Table 11.3). As you can see, very serious damage or death can result from large doses of ionizing radiation.

• X rays and gamma rays are similar except that X rays are of lower energy. X rays used in diagnostic medicine have energies approximately 10% of that of gamma rays.

Table 11.3
The Effects of Short-Term Whole-Body Radiation Exposure on Humans

Dose (rems)[a]	Effects
0–25	No detectable clinical effects.
25–100	Slight short-term reduction in number of some blood cells; disabling sickness not common.
100–200	Nausea and fatigue, vomiting if dose is greater than 125 rems; longer-term reduction in number of some blood cells.
200–300	Nausea and vomiting first day of exposure; up to a 2-week latent period followed by appetite loss, general malaise, sore throat, pallor, diarrhea, and moderate emaciation. Recovery in about 3 months, unless complicated by infection or injury.
300–600	Nausea, vomiting, and diarrhea in first few hours. Up to a 1-week latent period followed by loss of appetite, fever, and general malaise in the second week, followed by hemorrhage, inflammation of mouth and throat, diarrhea, and emaciation. Some deaths in 2 to 6 weeks. Eventual death for 50% if exposure is above 450 rems; others recover in about 6 months.
600 or more	Nausea, vomiting, and diarrhea in first few hours. Rapid emaciation and death as early as second week. Eventual death of nearly 100%.

[a]A rem is the quantity of ionizing radiation that must be absorbed by a human to produce the same biological effect as 1 roentgen of high-penetration X rays. A roentgen is the quantity of high-penetration X rays that produces approximately 2×10^9 ion pairs per cubic centimeter of dry air at 0°C and 1 atm.

CHEMISTRY AT A GLANCE

Parent Nuclide

A nuclide that undergoes radioactive decay

α Particle Decay

$^{A}_{Z}X \rightarrow {}^{4}_{2}\alpha + {}^{A-4}_{Z-2}Y$

Decrease of 4 units in mass number and 2 units in atomic number

α particles

γ rays

β Particle Decay

$^{A}_{Z}X \rightarrow {}^{0}_{-1}\beta + {}_{Z+1}^{A}Y$

Mass number unchanged; atomic number increased by 1 unit

β particles

γ rays

α particles and β particles interacting with other matter

Ion Pair Formation

Atom → Ion pair (+, e^-)

Free-Radical Formation

Molecule → Free radicals

Daughter Nuclide

A nuclide produced as a result of radioactive decay

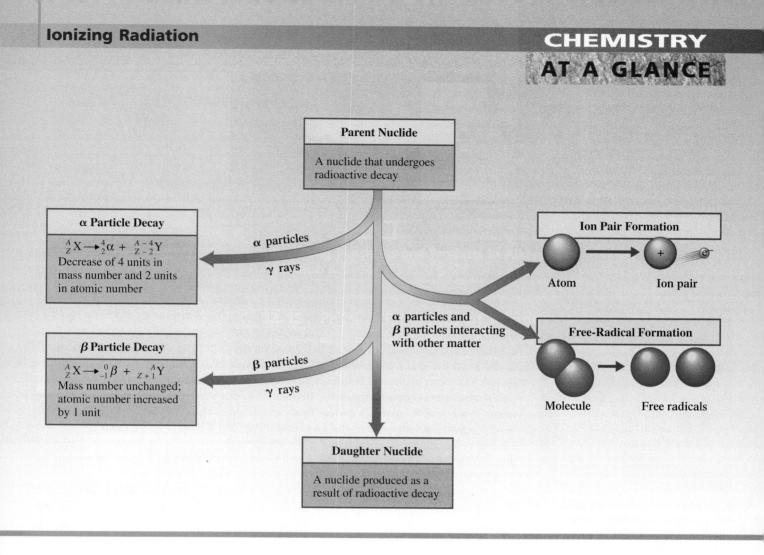

Figure 11.6
Film badges, such as the one shown here, are worn by technicians and others who work around radiation. The extent of exposure of all personnel is carefully monitored to help prevent overexposure.

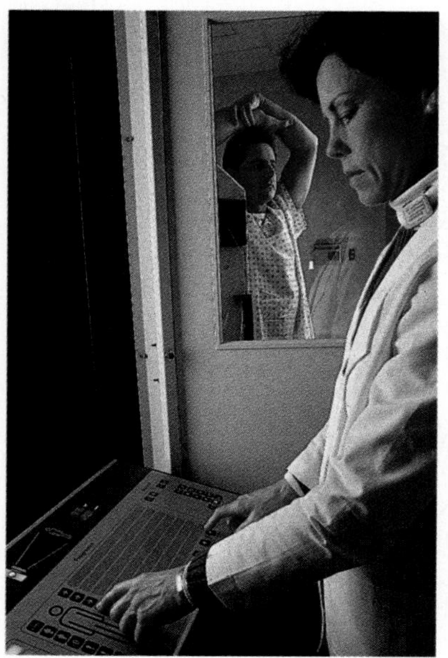

The accompanying Chemistry at a Glance reviews some of the things we have said about ionizing radiation.

11.9 Detection of Radiation

You cannot hear, feel, taste, see, or smell low levels of radiation. However, there are numerous methods for detecting its presence. Becquerel's initial discovery of radioactivity (Section 11.2) was a result of the effect of radiation on photographic plates. Radiation affects photographic film as ordinary light does; it exposes the film. Technicians and others who work around radiation usually wear film badges (see Figure 11.6) to record the extent of their exposure to radiation. When the film from the badge is developed, the degree of darkening of the film negative indicates the extent of radiation exposure. Different filters are used, so various parts of the film register exposures to the different types of radiations (alpha, beta, gamma, and X rays).

Radiation can also be detected by making use of the fact that it ionizes atoms and molecules (Section 11.7). The Geiger counter operates on this principle. The basic components of a Geiger counter are shown in Figure 11.7. The detection part of such a counter is a metal tube filled with a gas (usually argon). The tube has a thin-walled window made of a material that can be penetrated by alpha, beta, or gamma rays. In the center of the tube is a wire attached to the positive terminal of an electrical power source. The metal tube is attached to the negative terminal of the same source. Radiation entering the tube ionizes the gas, which allows a pulse of electricity to flow. This pulse of electricity is then amplified and displayed on a meter or some other type of readout display.

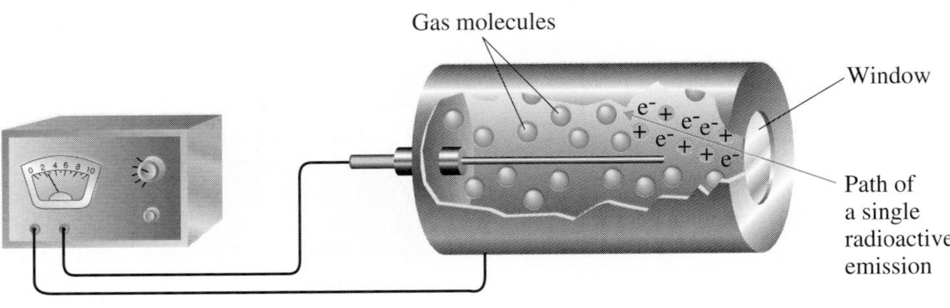

Figure 11.7
Radiation passing through the window of a Geiger counter ionizes one or more gas atoms, producing ion pairs. The electrons from the ion pairs are attracted to the central wire, and the positive ions are drawn to the metal tube. This constitutes a pulse of electric current, which is amplified and displayed on a meter or other readout.

11.10 Sources of Radiation Exposure

A commercially available kit to test for radon gas in the home. Radon gas is an intermediate in the decay of uranium. Because nearly all soils contain trace amounts of uranium, radon contamination of homes can occur. The radon gas seeps into homes through pores in concrete block walls, through cracks in basement floors or walls, or around pipes.

Most of us will never come in contact with the radiation dosage necessary to cause the effects listed in Table 11.3. Nevertheless, *low-level* exposure to ionizing radiation is something we constantly encounter. In fact, there is no way we can totally avoid this low-level exposure, because much of it results from naturally occurring environmental processes.

Both natural and human-made sources of low-level radiation exist. Natural sources include (1) radon seepage in buildings, (2) rocks and soils, (3) minerals in the body (mostly potassium), and (4) cosmic radiation. Radiation resulting from human activities may arise from (1) medical X rays, (2) nuclear medicine, (3) consumer products, and (4) miscellaneous sources, including occupational exposure, nuclear fallout from weapons testing, and nuclear power plants. The estimates of per capita radiation exposure given in Figure 11.8 are averages for all Americans; the actual exposure of individuals varies according to where they live and work, their medical history, and other factors.

A comparison of the values in Figure 11.8 with those of Table 11.3, taking into account that the units in the former are millirems and those in the latter are rems, shows that the current dosage levels received by the general population are very small, compared with those known to cause serious radiation sickness.

With low-level radiation exposure, chromosome damage rather than cell death can occur. If the damaged genetic material repairs itself improperly, then new, abnormal cells are produced when the cells replicate.

Cells that reproduce at a rapid rate, such as those in bone marrow, lymph nodes, and embryonic tissue, are the most sensitive to radiation damage. The sensitivity of embryonic tissue to radiation damage is the reason why pregnant women need to be protected from radiation exposure. One of the first signs of overexposure to radiation is a drop in

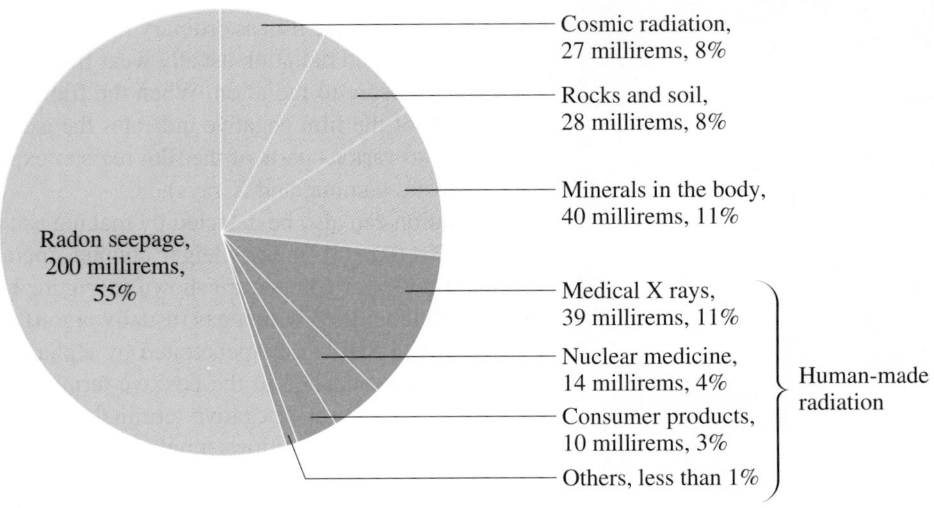

Figure 11.8
Components of the estimated annual radiation exposure (in millirems) of an average American. Individual exposures vary widely, but most such radiation comes from natural sources, the largest single contributor being radon gas.

red blood count. This is a direct consequence of the sensitivity of bone marrow, the site of red blood cell formation, to radiation.

11.11 Nuclear Medicine

• An additional use for radionuclides in medicine, besides diagnostic and therapeutic uses, is as a source of power (Section 11.12). Cardiac pacemakers powered by plutonium-238 can remain in a patient for longer periods than those powered by chemical batteries without the additional surgery required to replace batteries.

In medicine, radionuclides are used both diagnostically and therapeutically. In diagnostic applications, technicians use small amounts of radionuclides whose progress through the body or localization in specific organs can be followed. Larger quantities of radionuclides are used in therapeutic applications.

• Diagnostic Uses for Radioisotopes

The fundamental chemical principle behind the use of radionuclides in diagnostic medical work is the fact that a radioactive nuclide of an element has the same chemical properties as a nonradioactive nuclide of the element. Thus the body chemistry is not upset by the presence of a small amount of a radioactive substance whose nonradioactive form is already present in the body.

The criteria used in selecting radionuclides for diagnostic procedures include the following:

1. At low concentrations (to minimize radiation damage), the radionuclide must be detectable by instrumentation placed outside the body. Almost all diagnostic radionuclides are gamma emitters, because the penetrating power of alpha and beta particles is too low.
2. The radionuclide must have a short half-life so that the intensity of the radiation is sufficiently great to be detected. A short half-life also limits the time period of radiation exposure.
3. The radionuclide must have a known mechanism for elimination from the body so that the material does not remain in the body indefinitely.
4. The chemical properties of the radionuclide must be such that it is compatible with normal body chemistry. It must be able to be selectively transmitted to the part or system of the body that is under study.

The circulation of blood in the body can be followed by using radioactive sodium-24. A small amount of this isotope is injected into the bloodstream in the form of a sodium chloride solution. The movement of this radionuclide through the circulatory system can be followed easily with radiation detection equipment. If it takes longer than normal for the nuclide to show up at a particular part of the body, this is an indication that the circulation is impaired at that spot.

Radiologists evaluate the functioning of the thyroid gland by administering iodine-131, usually in the form of a sodium iodide (NaI) solution. The radioactive iodine behaves in the same manner as ordinary iodine and is absorbed by the thyroid at a rate related to the activity of the gland. If a hypothyroid condition exists, then the amount accumulated is less than normal; and if a hyperthyroid condition exists, then a greater-than-average amount accumulates.

The size and shape of organs, as well as the presence of tumors, can be determined in some situations by scanning the organ in which a radionuclide tends to concentrate. Iodine-131 and technetium-99 are used to generate thyroid and brain scans, respectively. In the brain, technetium-99, in the form of a polyatomic ion (TcO_4^-), concentrates in brain tumors more than in normal brain tissue; this helps radiologists determine the presence, size, and location of brain tumors. Figure 11.9 shows a brain scan obtained by using the radionuclide technetium-99. In this figure, the bright spot at the upper right indicates a tumor that has absorbed a greater amount of radioactive material than the normal brain tissue.

Table 11.4 lists a number of radionuclides that are used in diagnostic procedures. The half-life of the radionuclide, the body locations wherein it concentrates, and its diagnostic function are also given.

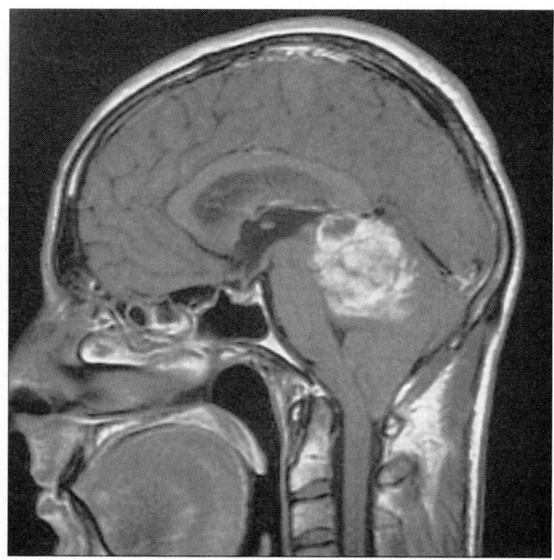

Figure 11.9
Brain scans, such as this one, are obtained using radioactive technetium-99, a laboratory-produced radionuclide.

• Therapeutic Uses for Radioisotopes

The objectives in therapeutic radionuclide use are entirely different from those for diagnostic procedures. The main objective in the therapeutic use of radionuclides is to *selectively destroy* abnormal (usually cancerous) cells. The radionuclide is often, but not always, placed within the body. Therapeutic radionuclides implanted in the body are usually alpha or beta emitters, because an intense dose of radiation in a small localized area is needed.

A commonly used implantation radionuclide that is effective in the localized treatment of tumors is yttrium-90, a beta emitter with a half-life of 64 hr. Yttrium-90 salts are implanted by inserting small, hollow needles into the tumor.

External, high-energy beams of gamma radiation are also extensively used in the treatment of certain cancers. Cobalt-60 is frequently used for this purpose; a beam of radiation is focused on the small area of the body where the tumor is located (see Figure 11.10).

Table 11.4
Selected Radionuclides Used in Diagnostic Procedures

Nuclide	Half-life	Part of body affected	Use in diagnosis
barium-131	11.6 days	bone	detection of bone tumors
chromium-51	27.8 days	blood	determination of blood volume and red blood cell lifetime
		kidney	assessment of kidney activity
iodine-131	8.05 days	brain	detection of fluid buildup in the brain
		kidney	location of cysts
		lung	location of blood clots
		thyroid	assessment of iodine uptake by thyroid
iron-59	45 days	blood	evaluation of iron metabolism in blood
phosphorus-32	14.3 days	blood	blood studies
		breast	assessment of breast carcinoma
potassium-42	12.4 hours	tissue	determination of intercellular spaces in fluids
sodium-24	15.0 hours	blood	detection of circulatory problems; assessment of peripheral vascular disease
technetium-99	6.0 hours	brain	detection of brain tumors, hemorrhages, or blood clots
		spleen	measurement of size and shape of spleen
		thyroid	measurement of size and shape of thyroid
		lung	location of blood clots

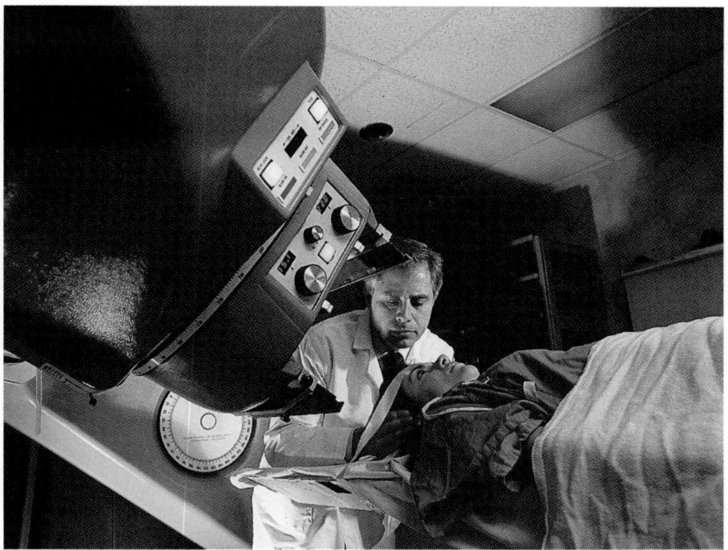

Figure 11.10
Cobalt-60 is used as a source of gamma radiation in radiation therapy.

Chemical CONNECTIONS

11.2 The PET Scan—Seeing the Brain in Action

An interesting combination of nuclear chemistry and advanced electronics now allows researchers to look into the brain safely and noninvasively. The technique, called positron emission tomography (PET), can provide minute-by-minute pictures of brain activities. Mapping such activities leads to a clearer understanding of normal brain functions. Comparing these pictures with those taken from patients with brain dysfunctions may point to dramatic new treatments for patients suffering from neurological disorders. One major goal, among many, is a cure for Parkinson's disease.

The nuclear chemistry involves the radioisotope fluorine-18 (half-life = 110 min), which is tagged to (incorporated into the molecular structure of) dopamine, a compound essential for brain function. The tagged compound then serves as a "beacon" within the functioning brain; the light is the positron emitted when the fluorine-18 undergoes decay. (Positrons are particles that have the same small mass as an electron but carry 1 unit of positive charge. A number of synthetic radioisotopes, but no naturally occurring ones, undergo positron decay.) Positrons, when formed, travel no more than a few millimeters in tissue before they collide with electrons. The result of such positron–electron collisions is the production of gamma rays. This gamma ray production is picked up by detectors that encircle the subject's head. A computer uses the detection data to reconstruct the spatial distribution of the gamma ray production. From this, the computer creates cross-sectional images of the distribution of fluorine-18 in the brain. Metabolically active regions of the brain can easily be distinguished from inactive ones.

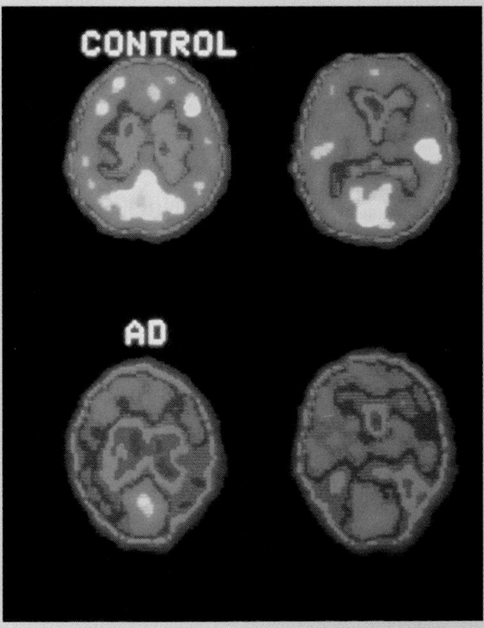

(Top) This PET scan shows normal brain activity. (Bottom) This PET scan shows inactive areas in the brain.

Research using ^{18}F-labeled compounds has shown Parkinson's disease patients to be deficient in the neurotransmitter dopamine, a substance that is used by nerve cells in the brain to communicate with one another. Such deficiency causes the symptoms of Parkinson's disease, which include difficulty in doing anything that requires a moderate amount of muscular control and (ultimately) uncontrollable shaking of the head and limbs.

Table 11.5
Some Radionuclides Used in Radiation Therapy

Nuclide	Half-life	Type of emitter	Use in therapy
cobalt-60	5.3 years	gamma	external source of radiation in treatment of cancer
iodine-131	8 days	beta, gamma	cancer of thyroid
phosphorus-32	14.3 days	beta, gamma	treatment of some types of leukemia and widespread carcinomas
radium-226	1620 years	alpha, gamma	used in implantation cancer therapy
radon-222	3.8 days	alpha, gamma	used in treatment of uterine, cervical, oral, and bladder cancers
yttrium-90	64 hours	beta, gamma	implantation therapy

• Abnormal cells are more susceptible to radiation damage than normal cells, because abnormal cells divide more frequently.

This therapy usually causes some radiation sickness because normal cells are also affected, but to a lesser extent. The operating principle here is that abnormal cells are more susceptible to radiation damage than normal cells. Radiation sickness is the price paid for abnormal-cell destruction. Table 11.5 lists some radionuclides that are used in therapy.

11.12 Nuclear Fission and Nuclear Fusion

Our glimpse into the world of nuclear chemistry would not be complete without a brief mention of two additional types of nuclear reactions that are used as sources of energy: nuclear fission and nuclear fusion.

• Fission Reactions

A nuclear power plant. The cooling tower at the Trojan nuclear power plant in Oregon dominates the landscape. The nuclear reactor is housed in the dome-shaped enclosure.

Nuclear fission *is the process in which a large nucleus (high atomic number) splits into two medium-sized nuclei with an accompanying release of several free neutrons and a large amount of energy.* The most important nucleus that undergoes fission is uranium-235. Bombardment of this nucleus with neutrons causes it to split into two fragments. Characteristics of the uranium-235 fission reaction include the following:

1. There is no unique way in which the uranium-235 nucleus splits. Thus, many different, lighter elements are produced during uranium-235 fission reactions. The following are examples of the ways in which this fission process proceeds.

$$^{235}_{92}U + ^{1}_{0}n \rightarrow \begin{cases} ^{135}_{53}I + ^{97}_{39}Y + 4\,^{1}_{0}n \\ ^{139}_{56}Ba + ^{94}_{36}Kr + 3\,^{1}_{0}n \\ ^{131}_{50}Sn + ^{103}_{42}Mo + 2\,^{1}_{0}n \\ ^{139}_{54}Xe + ^{95}_{38}Sr + 2\,^{1}_{0}n \end{cases}$$

2. Very large amounts of energy, which are many times greater than that released by ordinary radioactive decay, are emitted during the fission process. It is this large release of energy that makes nuclear fission of uranium-235 the important process that it is. In general, the term *nuclear energy* is used to refer to the energy released during a nuclear fission process. An older term for this energy is *atomic energy.*
3. Neutrons, which are reactants in the fission process, are also produced as products. The number of neutrons produced per fission depends on the way in which the nucleus splits; it ranges from 2 to 4 (as can be seen from the foregoing fission equations). On the average, 2.4 neutrons are produced per

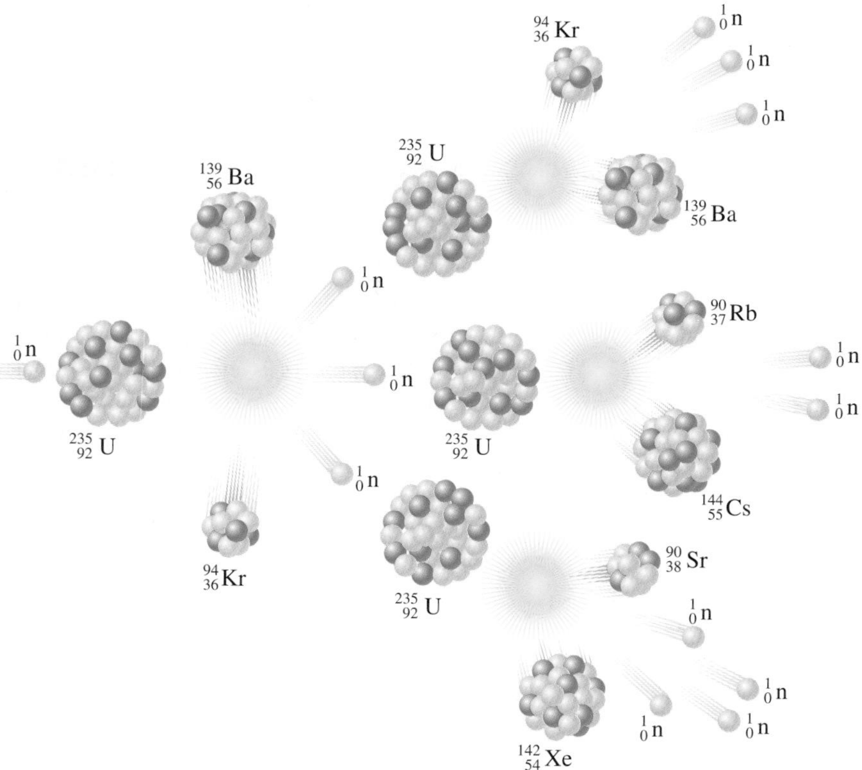

Figure 11.11
A fission chain reaction is caused by further reaction of the neutrons produced during fission.

fission. The significance of the neutrons that are produced is that they can cause the fission process to continue by colliding with further uranium-235 nuclei. Figure 11.11 shows the chain reaction that can occur once the fission process is started.

The process of nuclear fission—or "splitting the atom," as it is called in popularized science—can be carried out in both an uncontrolled and a controlled manner. The key to this control lies in what happens to the neutrons produced during fission. Do they react further, causing further fission, or do they escape into the surroundings? If the majority of produced neutrons react further (Figure 11.11), an uncontrolled nuclear reaction (an

Enormous amounts of energy are released in the explosion of a nuclear fission bomb.

atomic bomb) results. When only a few neutrons react further (on the average, one per fission), the fission reaction self-propagates in a controlled manner.

The process of *controlled* nuclear fission is the basis for the operation of nuclear power plants that are used to produce electricity. The reaction is controlled with rods that absorb excess neutrons (so that they cannot cause unwanted fissions) and with moderating substances that decrease the speed of the neutrons. The energy produced during the fission process, which appears as heat, is used to operate steam-powered electricity-generating equipment.

• Fusion Reactions

Another type of nuclear reaction, nuclear fusion, produces even more energy than nuclear fission. **Nuclear fusion** *is the process in which small nuclei are put together to make large ones.* This process is essentially the opposite of nuclear fission. In order for fusion to occur, a very high temperature, several hundred million degrees, is required.

• At the high temperature of fusion reactions, electrons completely separate from nuclei. Neutral atoms cannot exist. This high-temperature, gas-like mixture of nuclei and electrons is called a *plasma* and is considered by some scientists to represent a fourth state of matter.

Nuclear fusion is the process by which the sun generates its energy. Within the sun, in a three-step reaction, hydrogen-1 nuclei are converted to helium-4 nuclei with the release of extraordinarily large amounts of energy.

The use of nuclear fusion on Earth might seem impossible because of the high temperatures required. It has, however, been accomplished in a hydrogen bomb. In such a weapon, a *fission* bomb is used to achieve the high temperatures needed to start the following process:

$$^{3}_{1}H + ^{2}_{1}H \longrightarrow ^{4}_{2}He + ^{1}_{0}n$$

The use of nuclear fusion as a *controlled* (peaceful) energy source is a very active area of current scientific research. "Harnessing" this type of nuclear reaction would have numerous advantages:

1. Unlike the by-products of fission reactions, the by-products of fusion reactions are stable (nonradioactive) nuclides. Thus the problem of storing radioactive wastes does not arise.
2. The major fuel under study for controlled fusion is $^{2}_{1}H$ (called deuterium), a hydrogen isotope that can be readily extracted from ocean water (0.015% of all hydrogen atoms are $^{2}_{1}H$). Just $0.005\,km^3$ of ocean water contains enough $^{2}_{1}H$ to supply the United States with all the energy it needs for 1 year!

However, difficult scientific and engineering problems still remain to be solved before controlled fusion is a reality.

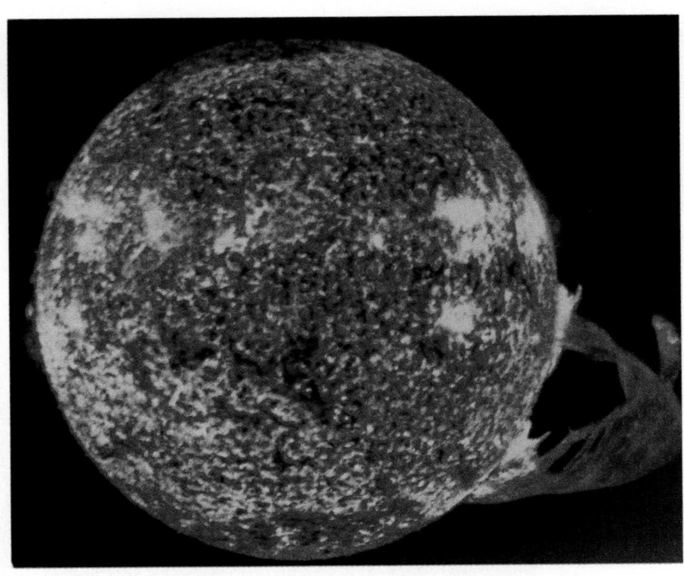

The process of nuclear fusion maintains the interior of the sun at a temperature of approximately 15 million degrees.

Table 11.6
Differences Between Nuclear and Chemical Reactions

Chemical reaction	Nuclear reaction
1. Different isotopes of an element have identical chemical processes.	1. Different isotopes of an element have different properties in nuclear properties.
2. The chemical reactivity of an element depends on the element's state of combination (free element, compound, etc.).	2. The nuclear reactivity of an element is independent of the state of chemical combination.
3. Elements retain their identity in chemical reactions.	3. Elements may be changed into other elements during nuclear reactions.
4. Energy changes that accompany chemical reactions are relatively small.	4. Nuclear reactions involve energy changes a number of orders of magnitude larger than those in chemical reactions.

11.13 Nuclear and Chemical Reactions Compared

As can be seen from the discussions in the previous sections of this chapter, nuclear chemistry is quite different from ordinary chemistry. Many of the laws of chemistry must be modified when we consider nuclear reactions. The major differences between nuclear reactions and ordinary chemical reactions are listed in Table 11.6. This table serves as a summary of many of the concepts presented in this chapter.

Concepts to Remember

Radioactivity. Some atoms possess nuclei that are unstable. To achieve stability, these unstable nuclei spontaneously emit energy (radiation). Such atoms are said to be radioactive.

Emissions from radioactive nuclei. The types of radiation emitted by naturally occurring radioactive nuclei are alpha, beta, and gamma. These radiations can be characterized by mass and charge values. Alpha particles carry a positive charge, beta particles a negative charge, and gamma radiation no charge.

Balanced nuclear equations. The procedures for balancing nuclear equations are different from those for balancing ordinary chemical equations. In nuclear equations, mass numbers and atomic numbers (rather than atoms) balance on both sides.

Half-life. Every radionuclide decays at a characteristic rate given by its half-life. One half-life is the time required for half of any given quantity of a radioactive substance to undergo decay.

Bombardment reactions. A bombardment reaction is a nuclear reaction in which small particles traveling at very high speeds are collided with stable nuclei; this causes these nuclei to undergo nuclear change (become unstable). Over 2000 synthetically produced radionuclides that do not occur naturally have been produced by using bombardment reactions.

Radioactive decay series. The product of the radioactive decay of an unstable nuclide is a nuclide of another element, which may or may not be stable. If it is not stable, it will decay and produce still another nuclide. Further decay will continue until a stable nuclide is formed. Such a sequence of reactions is called a radioactive decay series.

Biological effects of radiation. The biological effects of radiation depend on the energy, ionizing ability, and penetrating ability of the radiation. Alpha particles exhibit the greatest ionizing effect, and gamma rays have the greatest penetrating ability.

Detection of radiation. Radiation can be detected by making use of the fact that radiation ionizes atoms and molecules. The Geiger counter operates on this principle. Radiation also affects photographic film in the same way as ordinary light; the film is exposed. Hence film badges are used to record the extent of radiation exposure.

Sources of radiation exposure. Both natural and human-generated sources of low-level radiation exposure exist, with natural sources accounting for almost 80% of exposure (on the average). Current dosage levels received by the general population are very small, compared with those known to cause serious radiation sickness.

Nuclear medicine. Radionuclides are used in medicine for both diagnosis and therapy. The choice of radionuclide is dictated by the purpose as well as the target organ. Bombardment reactions are used to produce the nuclides used in medicine; they all have short half-lives.

Nuclear fission. Nuclear fission occurs when fissionable nuclides are bombarded with neutrons. The nuclides split into two fragments of about the same size. Also, more neutrons and large amounts of energy are produced. Nuclear fission is the process by which nuclear power plants generate energy.

Nuclear fusion. In nuclear fusion, small nuclei fuse to make heavier nuclei. Nuclear fusion is the process by which the sun generates its energy.

Key Reactions and Equations

1. General equation for alpha decay (Section 11.3)

$$^{A}_{Z}X \longrightarrow {}^{A-4}_{Z-2}Y + {}^{4}_{2}\alpha$$

2. General equation for beta decay (Section 11.3)

$$^{A}_{Z}X \longrightarrow {}_{Z+1}^{A}Y + {}^{0}_{-1}\beta$$

3. Half-life and amount of undecayed radionuclide (Section 11.4)

$$\begin{pmatrix}\text{Amount of radionuclide}\\ \text{undecayed after } n \text{ half-lives}\end{pmatrix} = \begin{pmatrix}\text{original amount}\\ \text{of radionuclide}\end{pmatrix} \times \left(\frac{1}{2^n}\right)$$

Key Terms

Alpha rays (11.2)
Balanced nuclear equation (11.3)
Beta rays (11.2)
Bombardment reaction (11.5)
Daughter nuclide (11.3)
Free radical (11.7)
Gamma rays (11.2)
Half-life (11.4)
Ion pair (11.7)
Nuclear fission (11.12)
Nuclear fusion (11.12)
Nuclear reaction (11.1)
Nuclide (11.1)
Parent nuclide (11.3)
Radioactive decay (11.3)
Radioactive decay series (11.6)
Radioactive nuclide (11.1)
Radioactivity (11.1)
Stable nuclide (11.1)
Transmutation process (11.5)
Unstable nuclide (11.1)

Exercises and Problems

The members of each pair of problems in this section test the same material.

Stable and Unstable Nuclides (Section 11.1)

11.1 Use two different notations to denote each of the following nuclides.
- a. Contains 4 protons, 4 electrons, and 6 neutrons
- b. Contains 11 protons, 11 electrons, and 14 neutrons
- c. Contains 41 protons, 41 electrons, and 55 neutrons
- d. Contains 103 protons, 103 electrons, and 154 neutrons

11.2 Use two different notations to denote each of the following nuclides.
- a. Contains 20 protons, 20 electrons, and 18 neutrons
- b. Contains 37 protons, 37 electrons, and 43 neutrons
- c. Contains 51 protons, 51 electrons, and 74 neutrons
- d. Contains 99 protons, 99 electrons, and 157 neutrons

11.3 Use a notation different from that given to designate each of the following nuclides.
- a. nitrogen-14 b. gold-197
- c. $^{121}_{50}Sn$ d. $^{10}_{5}B$

11.4 Use a notation different from that given to designate each of the following nuclides.
- a. oxygen-17 b. lead-212
- c. $^{92}_{37}Rb$ d. $^{201}_{83}Bi$

The Nature of Radioactivity (Section 11.2)

11.5 Supply a complete symbol, with superscript and subscript, for each of the following types of radiation.
- a. alpha particle b. beta particle c. gamma ray

11.6 Give the charge and mass (in amu) of each of the following types of radiation.
- a. alpha particle b. beta particle c. gamma ray

11.7 State the composition of an alpha particle in terms of protons and neutrons.

11.8 What is the relationship between a beta particle and an electron?

Radioactive Decay (Section 11.3)

11.9 Write balanced nuclear equations for the alpha decay of the following nuclides.
- a. $^{200}_{84}Po$ b. curium-240
- c. $^{244}_{96}Cm$ d. uranium-238

11.10 Write balanced nuclear equations for the alpha decay of the following nuclides.
- a. $^{229}_{90}Th$ b. bismuth-210
- c. $^{152}_{64}Gd$ d. americium-243

11.11 Write balanced nuclear equations for the beta decay of the following nuclides.
- a. $^{10}_{4}Be$ b. carbon-14
- c. $^{21}_{9}F$ d. sodium-25

11.12 Write balanced nuclear equations for the beta decay of the following nuclides.
- a. $^{77}_{32}Ge$ b. uranium-235
- c. $^{16}_{7}N$ d. iron-60

11.13 What is the effect on the mass number and atomic number of the parent nuclide when alpha particle decay occurs?

11.14 What is the effect on the mass number and atomic number of the parent nuclide when beta particle decay occurs?

11.15 Supply the missing symbol in each of the following radioactive decay equations.
- a. $^{34}_{14}Si \rightarrow {}^{34}_{15}P + ?$ b. $? \rightarrow {}^{28}_{13}Al + {}^{0}_{-1}\beta$
- c. $^{252}_{99}Es \rightarrow {}^{248}_{97}Bk + ?$ d. $^{204}_{82}Pb \rightarrow ? + {}^{4}_{2}\alpha$

11.16 Supply the missing symbol in each of the following radioactive decay equations.
- a. $? \rightarrow {}^{230}_{92}U + {}^{4}_{2}\alpha$ b. $^{192}_{78}Pt \rightarrow ? + {}^{4}_{2}\alpha$
- c. $^{84}_{35}Br \rightarrow ? + {}^{0}_{-1}\beta$ d. $^{10}_{4}Be \rightarrow {}^{10}_{5}B + ?$

11.17 Identify the mode of decay for each of the following parent radionuclides, given the identity of the daughter nuclide.
- a. parent = platinum-190; daughter = osmium-186
- b. parent = oxygen-19; daughter = fluorine-19

11.18 Identify the mode of decay for each of the following parent radionuclides, given the identity of the daughter nuclide.

a. parent = uranium-238; daughter = thorium-234

b. parent = rhodium-104; daughter = palladium-104

Rate of Radioactive Decay (Section 11.4)

11.19 Technetium-99 has a half-life of 6.0 hr. What fraction of the technetium-99 atoms in a sample will remain undecayed after the following times?

a. 12 hr b. 36 hr

c. 3 half-lives d. 6 half-lives

11.20 Copper-66 has a half-life of 5.0 min. What fraction of the copper-66 atoms in a sample will remain undecayed after the following times?

a. 20 min b. 30 min

c. 3 half-lives d. 8 half-lives

11.21 Determine the half-life of a radionuclide if after 5.4 days the fraction of undecayed nuclides present is

a. 1/16 b. 1/64 c. 1/256 d. 1/1024

11.22 Determine the half-life of a radionuclide if after 3.2 days the fraction of undecayed nuclides present is

a. 1/8 b. 1/128 c. 1/32 d. 1/512

11.23 The half-life of sodium-24 is 15.0 hr. How many grams of this nuclide in a 4.00-g sample will remain after 60.0 hr?

11.24 The half-life of strontium-90 is 28 years. How many grams of this nuclide in a 4.00-g sample will remain after 112 years?

Bombardment Reactions (Section 11.5)

11.25 Approximately how many laboratory-produced radionuclides are known?

11.26 How does the number of laboratory-produced radionuclides compare with the number of naturally occurring nuclides?

11.27 What is the highest-atomic-numbered naturally occurring element?

11.28 What is the highest-atomic-numbered element for which nonradioactive isotopes exist?

11.29 Identify the missing symbol in each of the following equations for bombardment reactions.

a. ${}^{24}_{12}\text{Mg} + ? \rightarrow {}^{27}_{14}\text{Si} + {}^{1}_{0}\text{n}$

b. ${}^{27}_{13}\text{Al} + {}^{2}_{1}\text{H} \rightarrow ? + {}^{4}_{2}\alpha$

c. ${}^{9}_{4}\text{Be} + ? \rightarrow {}^{12}_{6}\text{C} + {}^{1}_{0}\text{n}$

d. ${}^{6}_{3}\text{Li} + ? \rightarrow {}^{4}_{2}\text{He} + {}^{3}_{2}\text{He}$

11.30 Identify the missing symbol in each of the following equations for bombardment reactions.

a. $? + {}^{4}_{2}\alpha \rightarrow {}^{250}_{99}\text{Es} + {}^{3}_{1}\text{H}$

b. ${}^{14}_{7}\text{N} + {}^{4}_{2}\alpha \rightarrow ? + {}^{1}_{1}\text{H}$

c. ${}^{12}_{6}\text{C} + {}^{2}_{1}\text{H} \rightarrow {}^{13}_{7}\text{N} + ?$

d. ${}^{27}_{13}\text{Al} + ? \rightarrow {}^{30}_{15}\text{P} + {}^{1}_{0}\text{n}$

Radioactive Decay Series (Section 11.6)

11.31 The uranium-235 decay series terminates with lead-207. Would you expect lead-207 to be a stable or an unstable nuclide? Explain your answer.

11.32 A textbook erroneously indicates that the uranium-235 decay series terminates with radon-222. Explain why such a situation cannot be.

11.33 In the thorium-232 natural decay series, the thorium-232 initially undergoes alpha decay, the resulting daughter emits a beta particle, and the succeeding daughters emit a beta and an alpha particle in that order. Write four nuclear equations, one to represent each of the first four steps in the thorium-232 decay series.

11.34 In the uranium-235 natural decay series, the uranium-235 initially undergoes alpha decay, the resulting daughter emits a beta particle, and the succeeding daughters emit an alpha and a beta particle in that order. Write four nuclear equations, one to represent each of the first four steps in the uranium-235 decay series.

Effects of Radiation (Sections 11.7 and 11.8)

11.35 Contrast the abilities of alpha, beta, and gamma radiations to penetrate a thick sheet of paper.

11.36 Contrast the abilities of alpha, beta, and gamma radiation to penetrate human skin.

11.37 Contrast the velocities with which alpha, beta, and gamma radiations are emitted by nuclei.

11.38 Contrast the ionizing ability of alpha and beta radiations.

Radiation Exposure (Sections 11.8 and 11.10)

11.39 What would be the expected effect of each of the following short-term, whole-body radiation exposures?

a. 10 rems b. 150 rems

11.40 What would be the expected effect of each of the following short-term, whole-body radiation exposures?

a. 50 rems b. 250 rems

11.41 Contrast the radiation exposure that an average American receives from natural sources with that which an average American receives from human-made sources.

11.42 What are the five major sources of low-level radiation exposure, in terms of millirems per year, for the average American?

Detection of Radiation (Section 11.9)

11.43 Why do technicians who work around radiation usually wear film badges?

11.44 Explain the principle of operation of a Geiger counter.

Nuclear Medicine (Section 11.11)

11.45 Why are the radionuclides used for diagnostic procedures usually gamma emitters?

11.46 Why do the radionuclides used in diagnostic procedures almost always have short half-lives?

11.47 Explain how each of the following radionuclides is used in diagnostic medicine.

a. barium-131 b. sodium-24

c. iron-59 d. potassium-42

11.48 Explain how each of the following radionuclides is used in diagnostic medicine.

a. iodine-131 b. phosphorus-32

c. technetium-99 d. chromium-51

11.49 How do the radionuclides used for therapeutic purposes differ from the radionuclides used for diagnostic purposes?

11.50 Contrast the different ways in which cobalt-60 and yttrium-90 are used in radiation therapy.

Nuclear Fission and Nuclear Fusion (Section 11.12)

11.51 Identify which of the following characteristics apply to the fission process, which to the fusion process, and which to both processes.

a. An extremely high temperature is required to start the process.

b. An example of the process occurs on the sun.

c. Transmutation of elements occurs.

d. Neutrons are needed to start the process.

11.52 Identify which of the following characteristics apply to the fission process, which to the fusion process, and which to both processes.

a. Large amounts of energy are released in the process.

b. Energy released in the process is called *nuclear energy.*

c. The process is now used to generate some electrical power in the United States.

d. A fourth state of matter called *plasma* is encountered in studying this process.

11.53 Identify each of the following nuclear reactions as fission, fusion, or neither.

a. ${}^{3}_{2}He + {}^{3}_{2}He \rightarrow {}^{4}_{2}He + 2\,{}^{1}_{1}H$

b. ${}^{235}_{92}U + {}^{1}_{0}n \rightarrow {}^{144}_{55}Cs + {}^{90}_{37}Rb + 2\,{}^{1}_{0}n$

c. ${}^{209}_{83}Bi + {}^{4}_{2}\alpha \rightarrow {}^{210}_{85}At + 3\,{}^{1}_{0}n$

d. ${}^{238}_{92}U \rightarrow {}^{234}_{90}Th + {}^{4}_{2}\alpha$

11.54 Identify each of the following nuclear reactions as fission, fusion, or neither.

a. ${}^{239}_{92}U \rightarrow {}^{239}_{93}Np + {}^{0}_{-1}\beta$

b. ${}^{230}_{90}Th + {}^{1}_{1}p \rightarrow {}^{223}_{87}Fr + 2\,{}^{4}_{2}\alpha$

c. ${}^{3}_{1}H + {}^{2}_{1}H \rightarrow {}^{4}_{2}He + {}^{1}_{0}n$

d. ${}^{235}_{92}U + {}^{1}_{0}n \rightarrow {}^{139}_{54}Xe + {}^{95}_{38}Sr + 2\,{}^{1}_{0}n$

Additional Problems

11.55 Write nuclear equations for each of the following radioactive-decay processes.

a. Thallium-206 is formed by beta emission.

b. Palladium-109 undergoes beta emission.

c. Plutonium-241 is formed by alpha emission.

d. Fermium-249 undergoes alpha emission.

11.56 Cobalt-55 has a half-life of 18 hours. How long will it take, in hours, for the following fractions of nuclides in a cobalt-55 sample to decay?

a. 7/8 b. 31/32

c. 63/64 d. 127/128

11.57 Write equations for the following nuclear bombardment processes.

a. Bombardment of a radionuclide with an alpha particle produces curium-242 and one neutron.

b. Bombardment of curium-246 with a small particle produces nobelium-254 and four neutrons.

c. Aluminum-27 is bombarded with an alpha particle and produces a neutron.

d. Bombardment of sodium-23 with hydrogen-2 produces neon-21.

11.58 The second artificially produced element was promethium. Write the equation for the production of ${}^{143}Pm$ by the bombardment of ${}^{142}Nd$ with neutrons.

11.59 Using Table 11.2 as your source of information, determine for how many of the transuranium elements the most stable isotope has a half-life that is less than 1.0 day.

11.60 How many neutrons are produced as a result of the following fission reaction?

$$\text{Neutron} + \text{U-235} \longrightarrow \text{I-135} + \text{Y-97} + \text{? neutrons}$$

11.61 Consider the decay series

$$E \longrightarrow F \longrightarrow G \longrightarrow H$$

where E, F, and G are radioactive, with half-lives of 10.0 sec, 1.2 min and 12.5 days, respectively, and H is nonradioactive. Starting with 1000 atoms of E, and none of F, G, and H, estimate the numbers of atoms of E, F, G, and H that are present after 50 days.

11.62 Fill in the blanks in the following segment of a radioactive decay series.

$$\underline{\ ?\ } \xrightarrow{\beta} \underline{\ ?\ } \xrightarrow{\alpha} {}^{224}Ra \xrightarrow{\beta} \underline{\ ?\ }$$

Grid Problems

11.63

1. alpha particle	2. beta particle	3. gamma radiation
4. neutron	5. proton	6. electron

Select from the grid *all* correct responses for each of the following situations.

a. Its symbol contains a zero as a subscript.

b. Its symbol contains a zero as a superscript.

c. The subscript and superscript in its symbol are identical.

d. The subscript and superscript in its symbol are both greater than zero.

11.64

1. ${}^{218}_{86}Rn \longrightarrow {}^{214}_{84}Po$	2. ${}^{117}_{48}Cd \longrightarrow {}^{117}_{49}In$	3. ${}^{238}_{92}U \longrightarrow {}^{239}_{92}U$
4. ${}^{9}_{4}Be \longrightarrow {}^{12}_{6}C$	5. ${}^{136}_{53}I \longrightarrow {}^{136}_{54}Xe$	6. ${}^{244}_{94}Pu \longrightarrow {}^{240}_{92}U$

Select from the grid *all* correct responses for each of the following situations.

a. Parent–daughter relationship is consistent with the equation X → Y + alpha.

b. Parent–daughter relationship is consistent with the equation X → Y + beta.

c. Parent–daughter relationship is consistent with the equation X + alpha → Y + neutron.

d. Parent–daughter relationship is consistent with the equation X + neutron → Y.

11.65

1. bombardment α in, n out	2. bombardment n in, α out	3. bombardment p in, n out
4. decay series α, β, β	5. decay series β, α, α	6. decay series β, β, β

Select from the grid *all* correct responses for each of the following situations.

a. Change that increases both atomic number and mass number
b. Change that decreases both atomic number and mass number
c. Change that increases atomic number and holds mass number constant
d. Change that holds atomic number constant and decreases mass number

11.66

1. sample is 50% decayed	2. sample is $\frac{3}{4}$ decayed	3. sample is $\frac{15}{16}$ decayed
4. sample is $\frac{1}{4}$ undecayed	5. sample is 12.5% undecayed	6. sample is $\frac{1}{8}$ undecayed

Select from the grid *all* correct responses for each of the following situations.

a. Consistent with the elapse of 1 half-life
b. Consistent with the elapse of 2 half-lives
c. Consistent with the elapse of 3 half-lives
d. Consistent with the elapse of 4 half-lives

Saturated Hydrocarbons

CHAPTER OUTLINE

12.1 Organic and Inorganic Compounds 298
12.2 Bonding Characteristics of the Carbon Atom 299
12.3 Hydrocarbons and Hydrocarbon Derivatives 300
12.4 Structural Characteristics of Simple Alkanes 300
12.5 Structural Formulas 301
12.6 Structural Isomerism 303
12.7 Conformations of Alkanes 304
12.8 IUPAC Nomenclature for Alkanes 306
12.9 Classification of Carbon Atoms 311
12.10 Branched-Chain Alkyl Groups 311
12.11 Cycloalkanes 312
12.12 IUPAC Nomenclature for Cycloalkanes 313
12.13 Isomerism in Cycloalkanes 314
12.14 Sources of Alkanes and Cycloalkanes 316
12.15 Physical Properties of Alkanes and Cycloalkanes 317
12.16 Chemical Properties of Alkanes and Cycloalkanes 320
12.17 Nomenclature and Properties of Halogenated Alkanes 322

Chemistry at a Glance: **Properties of Alkanes and Cycloalkanes** 323

Chemical Connections
12.1 The Occurrence of Methane 302
12.2 The Physiological Effects of Alkanes 319
12.3 Chlorofluorocarbons and the Ozone Layer 324

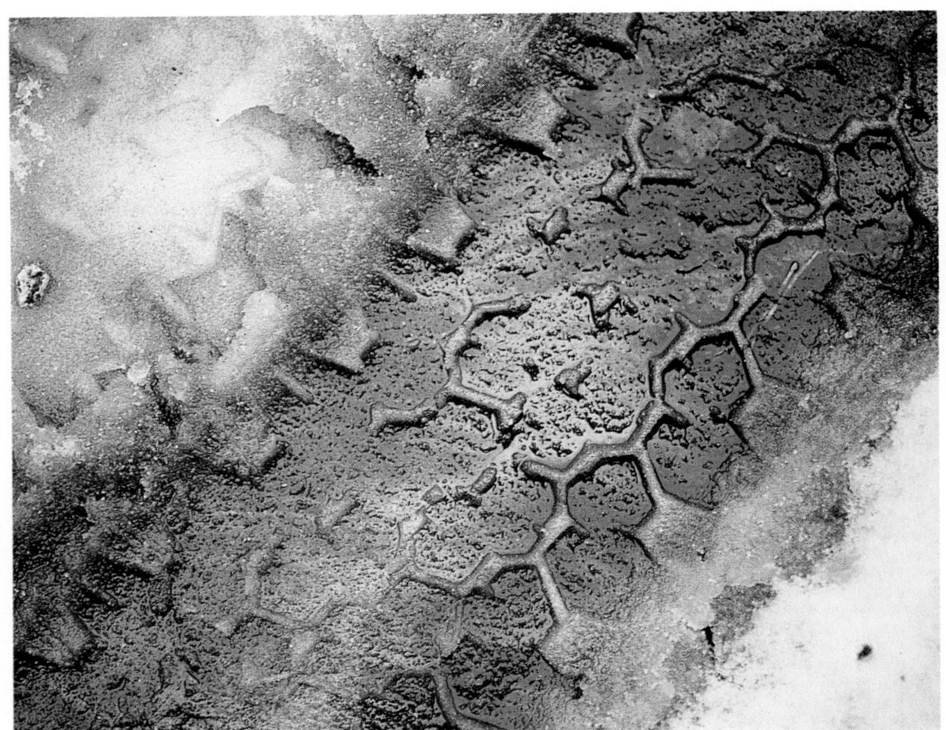

A tire tread in snow with oil (hydrocarbons) causing the iridescent pattern.

This chapter is the first of six that deal with the subject of organic chemistry and organic compounds. Organic compounds are the chemical basis for life itself, as well as an important component of the basis for our current high standard of living. Proteins, carbohydrates, enzymes, and hormones are organic molecules. Organic compounds also include natural gas, petroleum, coal, gasoline, and many synthetic materials such as dyes, plastics, and clothing fibers.

12.1 Organic and Inorganic Compounds

During the latter part of the eighteenth century and the early part of the nineteenth century, chemists began to categorize compounds into two types: organic and inorganic. Compounds obtained from living organisms were called *organic* compounds, and compounds obtained from mineral constituents of the earth were called *inorganic* compounds.

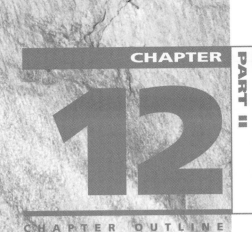

• The historical origins of the terms *organic* and *inorganic* involve the following conceptual pairings:

*org*anic—living *org*anism
*in*organic—*in*animate materials

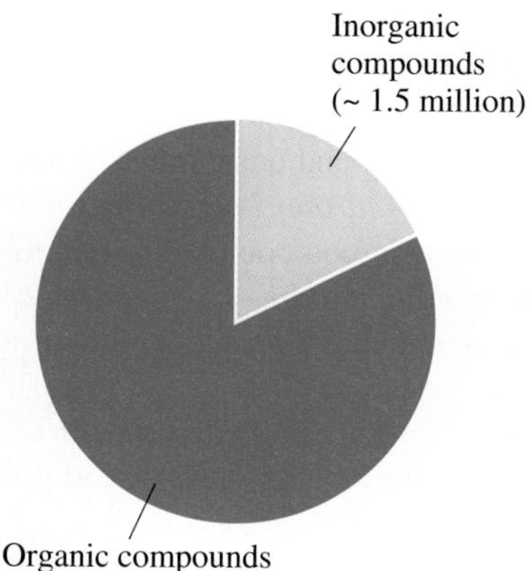

Sheer numbers is one reason why organic chemistry is a separate field of chemical study. Approximately 7 million organic compounds are known, compared to "just" 1.5 million inorganic compounds.

• Some textbooks define organic chemistry as the study of carbon-containing compounds. Almost all carbon-containing compounds qualify as organic compounds. However, the oxides of carbon, carbonates, cyanides, and metallic carbides are classified as inorganic rather than organic compounds. Inorganic carbon compounds involve carbon atoms that are not bonded to hydrogen atoms (CO, CO_2, Na_2CO_3, and so on).

• Carbon atoms in organic compounds, in accordance with the octet rule, always form four covalent bonds.

During this early period, chemists believed that a special "vital force," supplied by a living organism, was necessary for the formation of an organic compound. This concept was proved incorrect in 1828 by the German chemist Friedrick Wöhler. Wöhler heated an aqueous solution of two inorganic compounds, ammonium chloride and silver cyanate, and obtained urea (a component of urine).

$$NH_4Cl + AgNCO \longrightarrow \underset{\text{urea}}{(NH_2)_2CO} + AgCl$$

Soon other chemists had successfully synthesized organic compounds from inorganic starting materials. As a result, the vital force theory was completely abandoned. The terms *organic* and *inorganic* continue to be used, but the definitions of these terms have changed.

Today, **organic chemistry** *is defined as the study of hydrocarbons (compounds of hydrogen and carbon) and their derivatives.* Nearly all compounds found in living organisms are still classified as organic compounds, as are many compounds that have been synthesized in the laboratory that have never been found in a living organism.

In essence, organic chemistry is the study of the compounds of one element (carbon), and inorganic chemistry is the study of the compounds of the other 114 elements. This unequal partitioning occurs because there are approximately 7 million organic compounds and only an estimated 1.5 million inorganic compounds. This is an approximately 5:1 ratio between organic and inorganic compounds.

12.2 Bonding Characteristics of the Carbon Atom

What is unique about carbon that causes it to form 5 times as many compounds as all the other elements combined? The answer is that carbon atoms can be covalently bonded to other carbon atoms and to atoms of other elements in a wide variety of ways. Many of the bonding patterns result in molecules that contain chains or rings of carbon atoms. Sometimes, both chains and rings are present in the same molecule.

The variety of covalent bonding "behaviors" possible for carbon atoms is related to carbon's electron configuration. Carbon is a member of Group IVA of the periodic table, so carbon atoms possess four valence electrons (Section 4.2). In compound formation, four additional valence electrons are needed to give carbon atoms an octet of valence electrons (the octet rule, Section 4.3). These additional electrons are obtained by electron sharing (covalent bond formation). The sharing of *four* valence electrons requires the formation of *four* covalent bonds.

Carbon can meet this four-bond requirement in three different ways:

1. *By bonding to four other atoms.* This situation requires the presence of four single bonds.

$$-\overset{|}{\underset{|}{C}}-$$

Four single bonds

2. *By bonding to three other atoms.* This situation requires the presence of two single bonds and one double bond.

$$-\overset{|}{C}=$$

Two single bonds and one double bond

3. *By bonding to two other atoms.* This situation requires the presence of either two double bonds or a triple bond and a single bond.

$$=C= \qquad -C\equiv$$

Two double bonds — One triple bond and one single bond

12.3 Hydrocarbons and Hydrocarbon Derivatives

The field of organic chemistry encompasses the study of hydrocarbons and hydrocarbon derivatives (Section 12.1). A **hydrocarbon** *is a compound that contains only carbon and hydrogen atoms.* Thousands of hydrocarbons are known. A **hydrocarbon derivative** *is a compound that contains carbon and hydrogen and one or more additional elements.* Additional elements commonly found in hydrocarbon derivatives include O, N, S, P, F, Cl, and Br. Millions of hydrocarbon derivatives are known.

• The term *saturated* has the general meaning that there is no more room for something. Its use with hydrocarbons comes from early studies in which chemists tried to add hydrogen atoms to various hydrocarbon molecules. Compounds to which no more hydrogen atoms could be added (because they already contained the maximum number) were called saturated, and those to which hydrogen could be added were called unsaturated.

Hydrocarbons may be divided into two large classes: saturated and unsaturated. A **saturated hydrocarbon** *is a hydrocarbon in which all carbon–carbon bonds are single bonds.* Saturated hydrocarbons are the simplest type of organic compound. An **unsaturated hydrocarbon** *is a hydrocarbon that contains one or more carbon–carbon multiple bonds: double bonds, triple bonds, or both.* In general, saturated and unsaturated hydrocarbons undergo distinctly different chemical reactions.

Saturated hydrocarbons are the subject of this chapter. Unsaturated hydrocarbons are considered in the next chapter. Figure 12.1 summarizes the terminology presented in this section.

12.4 Structural Characteristics of Simple Alkanes

• Some saturated hydrocarbons have structures containing rings of carbon atoms. Such compounds, called *cycloalkanes,* are considered in Sections 12.11 and 12.12.

An **alkane** *is a saturated hydrocarbon in which the carbon atom arrangement is acyclic.* Thus an alkane is a hydrocarbon that contains only carbon–carbon single bonds (saturated) and has no rings of carbon atoms (acyclic). Note that the term *acyclic* means "not cyclic."

The molecular formulas of all alkanes fit the general formula C_nH_{2n+2}, where n is the number of carbon atoms present. The number of hydrogen atoms present in an alkane is always twice the number of carbon atoms plus two more, as in C_4H_{10}, C_5H_{12}, and C_8H_{18}.

The three simplest alkanes are methane (CH_4), ethane (C_2H_6), and propane (C_3H_8). Ball-and-stick and space-filling models showing the molecular structures of these three alkanes are given in Figure 12.2. Note, from these models, how each carbon atom in each of the models participates in four bonds (Section 12.2). Note also that the geometrical

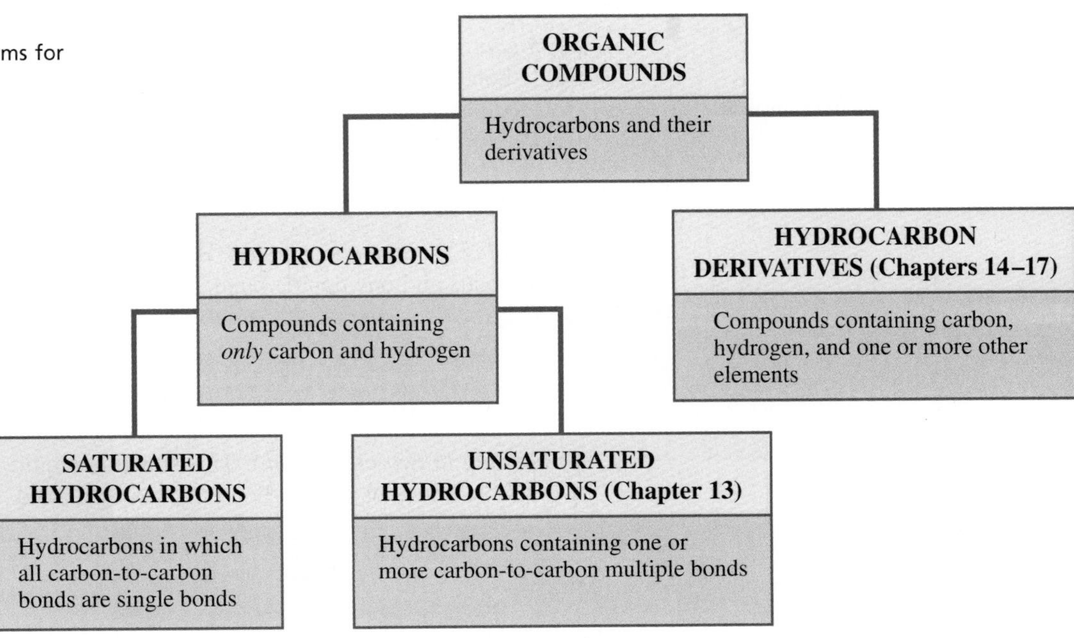

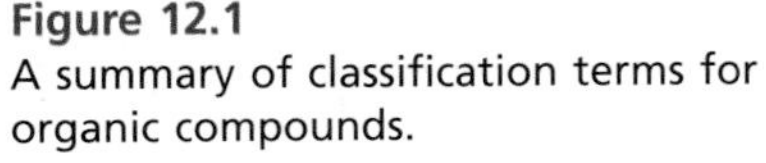
Figure 12.1
A summary of classification terms for organic compounds.

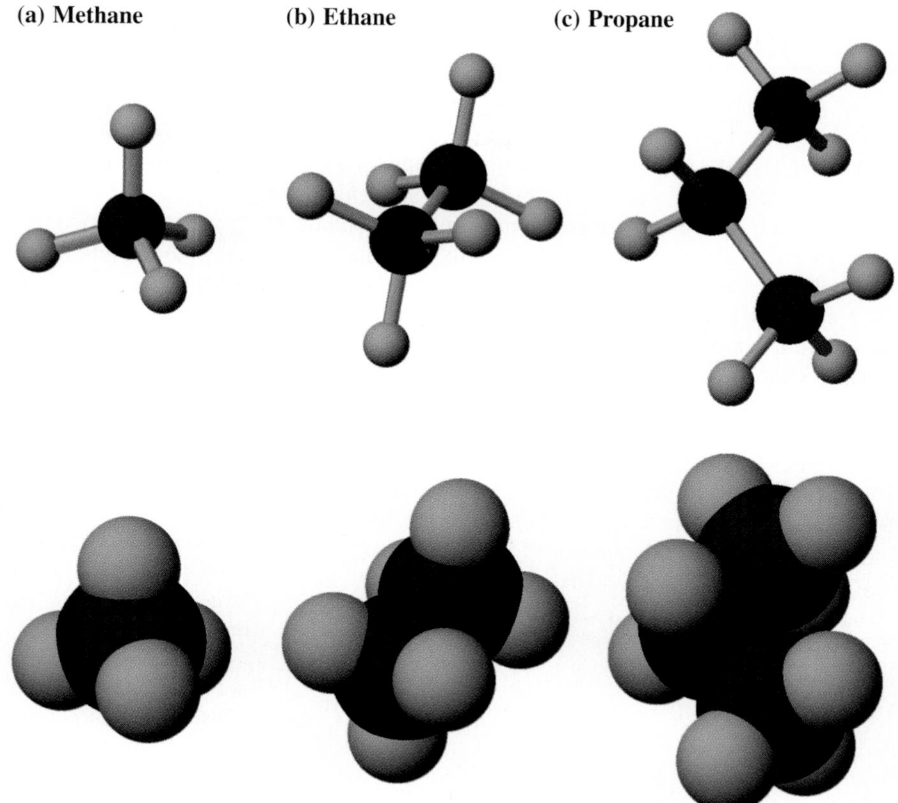

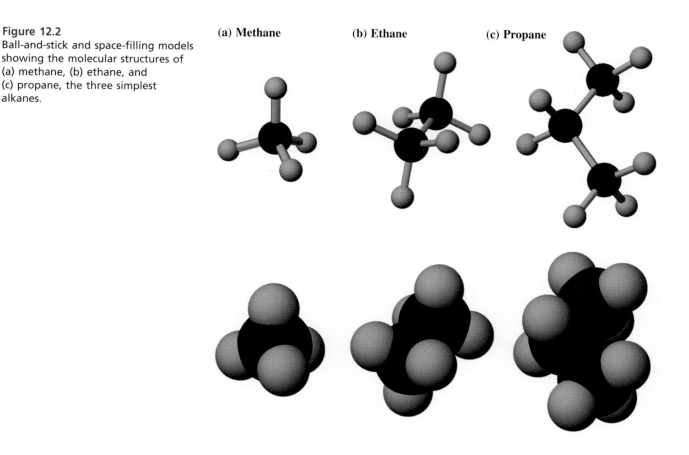

Figure 12.2
Ball-and-stick and space-filling models showing the molecular structures of (a) methane, (b) ethane, and (c) propane, the three simplest alkanes.

arrangement of atoms about each carbon atom is tetrahedral, an arrangement consistent with the principles of VSEPR theory (Section 5.8). The tetrahedral arrangement of the atoms bonded to alkane carbon atoms is fundamental to understanding the structural aspects of organic chemistry.

12.5 Structural Formulas

The structures of alkanes and other types of organic compounds are generally represented in two dimensions rather than three (Figure 12.2) because of the difficulty in drawing the latter. These two-dimensional structural representations make no attempt to portray accurately the bond angles or molecular geometry of molecules. Their purpose is to convey information about which atoms in a molecule are bonded to which other atoms.

Two-dimensional structural representations for organic molecules are of two types: expanded structural formulas and condensed structural formulas. An **expanded structural formula** *shows, in two dimensions, all atoms in a molecule and all the bonds connecting them.* When written out, expanded structural formulas generally occupy a lot of space, and condensed structural formulas represent a shorthand method for conveying the same information. A **condensed structural formula** *uses groupings of atoms, in which central atoms and the atoms connected to them are written as a group, to convey molecular structural information.* The expanded and condensed structural formulas for methane, ethane, and propane follow.

• Structural formulas, whether expanded or condensed, do not show the geometry (shape) of the molecule. That information can be conveyed only by 3-D drawings or models such as those in Figure 12.2.

$$\text{Expanded structural formula:}\quad \begin{array}{c} \text{H} \\ | \\ \text{H—C—H} \\ | \\ \text{H} \end{array} \qquad \begin{array}{c} \text{H}\ \ \text{H} \\ |\ \ \ \ | \\ \text{H—C—C—H} \\ |\ \ \ \ | \\ \text{H}\ \ \text{H} \end{array} \qquad \begin{array}{c} \text{H}\ \ \text{H}\ \ \text{H} \\ |\ \ \ \ |\ \ \ \ | \\ \text{H—C—C—C—H} \\ |\ \ \ \ |\ \ \ \ | \\ \text{H}\ \ \text{H}\ \ \text{H} \end{array}$$

	Methane	Ethane	Propane
Condensed structural formula	CH_4	$CH_3—CH_3$	$CH_3—CH_2—CH_3$

Chemical CONNECTIONS

12.1 The Occurrence of Methane

Methane (CH_4), the simplest of all hydrocarbons, is a major component of the atmospheres of Jupiter, Saturn, Uranus, and Neptune but only a minor component of Earth's atmosphere (see the accompanying table). Earth's gravitational field, being weaker than that of the large outer planets, cannot retain enough hydrogen (H_2) in its atmosphere to permit the formation of large amounts of methane; H_2 molecules (the smallest and fastest-moving of all molecules) escape from it into outer space.

The small amount of methane present in Earth's atmosphere comes from terrestrial sources. The decomposition of animal and plant matter in an oxygen-deficient environment—swamps, marshes, bogs, and the sediments of lakes—produces methane. A common name for methane, marsh gas, refers to the production of methane in this manner.

Decomposition of plant and animal matter in marshes is a source of methane gas.

Composition of Earth's Atmosphere (in parts per million by volume)

Major components		Minor components	
nitrogen	780,800	argon	9340
oxygen	209,500	carbon dioxide	314
		neon	18
		helium	5
		methane	2
		krypton	1

Bacteria that live in termites and in the digestive tracts of plant-eating animals have the ability to produce methane from plant materials (cellulose). The methane output of a large cow (belching and flatulence) can reach 20 liters per day.

Methane entering the atmosphere from terrestrial sources presents an environmental problem. Methane is a "greenhouse gas" that contributes to global warming. Methane is 15 to 30 times more efficient than carbon dioxide (the primary greenhouse gas) in trapping heat radiated from Earth. Fortunately, its atmospheric level of 2.0 ppm by volume is much lower than that of carbon dioxide (over 300 ppm).

Methane gas is also found associated with coal and petroleum deposits. Methane associated with coal mines is considered a hazard. If left to accumulate, it can form pockets where air is not present and asphyxiation of miners can occur. When mixed with air in certain ratios, it can also present an explosion hazard. Methane associated with petroleum deposits is most often recovered, processed, and marketed as *natural gas.* The processed natural gas used in the heating of homes is 85% to 95% methane by volume. Because methane is odorless, an odorant (smelly compound) must be added to the processed natural gas used in home heating. Otherwise, natural gas leaks could not be detected.

The condensed structural formulas of hydrocarbons in which a long chain of carbon atoms is present are often condensed even more. The formula

$$CH_3—CH_2—CH_2—CH_2—CH_2—CH_2—CH_2—CH_3$$

can be further abbreviated as

$$CH_3—(CH_2)_6—CH_3$$

in which parentheses and a subscript are used to denote the number of $—CH_2—$ groups in the chain.

In situations where the focus is on the arrangement of carbon atoms in a molecule, *skeletal formulas* that omit the hydrogen atoms are often used.

```
C—C—C—C—C      means the same as      CH3—CH2—CH—CH2—CH3
    |                                         |
    C                                         CH3
Skeletal formula                      Condensed structural formula
```

The skeletal formula still represents a unique compound, because we know that each carbon atom shown must have enough hydrogen atoms attached to it to give the carbon four bonds.

12.6 Structural Isomerism

The molecular formulas CH_4, C_2H_6, and C_3H_8 represent the alkanes methane, ethane, and propane, respectively. Next in the alkane molecular formula sequence is C_4H_{10}, which would be expected to be the molecular formula of the four-carbon alkane. A new phenomenon arises, however, when an alkane has four or more carbon atoms. There is more than one structural formula that is consistent with the molecular formula. Consequently, more than one compound is consistent with the molecular formula. This situation brings us to the topic of structural isomerism.

• The word *isomer* comes from the Greek *isos*, which means "the same," and *meros*, which means "parts." Isomers have the same parts put together in different ways.

Structural isomers *are compounds with the same molecular formula but different structural formulas—that is, different connections between atoms.* Structural isomers, with their differing structural formulas, always have different properties and are different compounds.

There are two four-carbon alkane structural isomers, the compounds *butane* and *isobutane.* Both have the molecular formula C_4H_{10}.

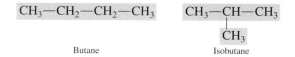

$$CH_3{-}CH_2{-}CH_2{-}CH_3 \qquad \begin{array}{c} CH_3{-}CH{-}CH_3 \\ | \\ CH_3 \end{array}$$

Butane Isobutane

• It is the existence of structural isomers that necessitates the use of structural formulas in organic chemistry. Structural isomers always have the same molecular formula and different structural formulas.

The one C_4H_{10} isomer, butane, has a chain of four carbon atoms. It is an example of a continuous-chain alkane. A **continuous-chain alkane** *is an alkane in which all carbon atoms are connected in a continuous nonbranching chain.* The other C_4H_{10} isomer, isobutane, has a chain of three carbon atoms with the fourth carbon attached as a branch on the middle carbon of the three-carbon chain. It is an example of a branched-chain alkane. A **branched-chain alkane** *is an alkane with one or more branches (of carbon atoms) attached to a continuous chain of carbon atoms.*

There are three structural isomers for alkanes with five carbon atoms (C_5H_{12}):

• Structural isomers of the type we are now considering are also called *skeletal isomers.* Such isomers differ in carbon atom arrangements; that is, the carbon skeletons differ. For alkanes, skeletal isomers are possible whenever more than three carbon atoms are present.

$$CH_3{-}CH_2{-}CH_2{-}CH_2{-}CH_3 \qquad \begin{array}{c} CH_3{-}CH{-}CH_2{-}CH_3 \\ | \qquad\qquad \\ CH_3 \qquad\qquad \end{array} \qquad \begin{array}{c} CH_3 \\ | \\ CH_3{-}C{-}CH_3 \\ | \\ CH_3 \end{array}$$

Pentane Isopentane Neopentane

Figure 12.3 shows space-filling models for the three isometric C_5 alkanes. Note how neopentane, the most branched isomer, has the most compact, most spherical three-dimensional shape.

Table 12.1 contrasts the differing physical properties of the two C_4 alkanes and the three C_5 alkanes.

The number of possible alkane structural isomers increases dramatically with increasing number of carbon atoms in the alkane, as shown in Table 12.2. Structural isomerism is one of the major reasons for the existence of so many organic compounds.

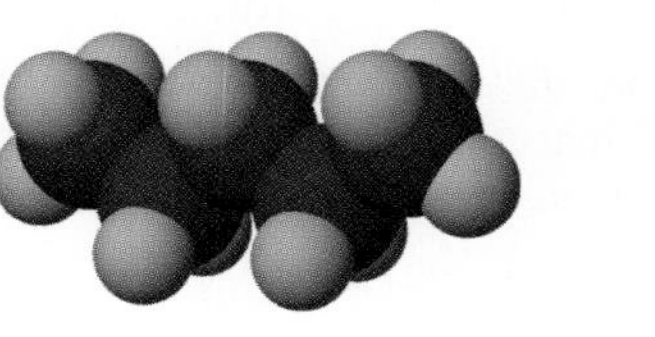

(a) pentane

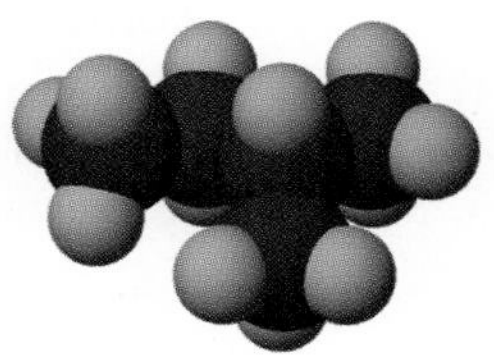
(b) isopentane

(c) neopentane

Figure 12.3 Space-filling models for the three isomeric C_5H_{12} alkanes: (a) pentane, (b) isopentane, and (c) neopentane.

Table 12.1
Selected Physical Properties of C_4 and C_5 Alkanes

Alkane	Boiling point	Density (at 20°C)
butane (C_4H_{10})	−0.5°C	0.579 g/mL
isobutane (C_4H_{10})	−11.6°C	0.549 g/mL
pentane (C_5H_{12})	36.1°C	0.626 g/mL
isopentane (C_5H_{12})	27.8°C	0.620 g/mL
neopentane (C_5H_{12})	9.5°C	0.614 g/mL

Table 12.2
Number of Structural Isomers Possible for Alkanes of Various Carbon-Chain Lengths

Molecular formula	Possible number of structural isomers
CH_4	1
C_2H_6	1
C_3H_8	1
C_4H_{10}	2
C_5H_{12}	3
C_6H_{14}	5
C_7H_{16}	9
C_8H_{18}	18
C_9H_{20}	35
$C_{10}H_{22}$	75
$C_{15}H_{32}$	4,347
$C_{20}H_{42}$	336,319
$C_{30}H_{62}$	4,111,846,763

12.7 Conformations of Alkanes

Conformations *are differing orientations of a molecule made possible by rotations about single bonds.* Rotation about carbon–carbon single bonds is an important property of alkane molecules. Two groups of atoms in an alkane connected by a carbon–carbon single bond can rotate with respect to one another around that bond, much as a wheel rotates around an axle.

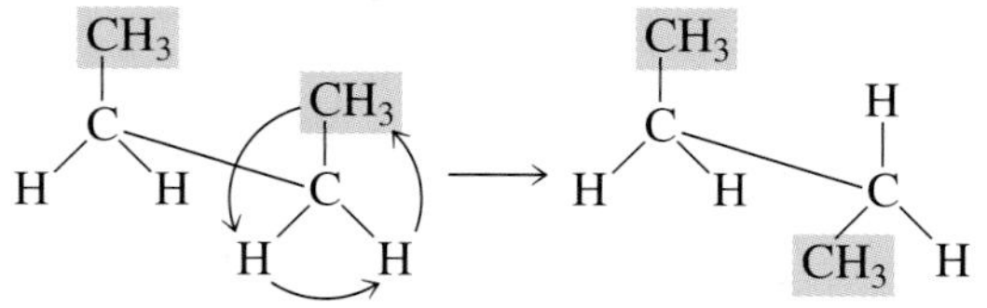

CH_3 / C / H H — C(CH_3) / H H → CH_3 / C / H H — C(H) / CH_3 H

As a result of rotation around single bonds, alkane molecules (except for methane) can exist in infinite numbers of orientations, or conformations. Three different conformations for a C_6H_{14} continuous-chain alkane molecule are shown in Figure 12.4.

The structures in Figure 12.4 illustrate several important concepts. The most *extended* arrangement possible for a chain of six carbon atoms is shown in Figure 12.4a. This molecular model shows that carbon atoms in a chain do not lie in a straight line but rather have a zigzag arrangement. A straight-line arrangement is not possible because of the tetrahedral orientation of bonds about each carbon atom in the molecule (Section 12.4). (Despite this zigzag pattern, carbon atoms are usually drawn in a straight line in structural formulas; see the right side of Figure 12.4a.) The transition from the extended conformation of Figure 12.4a to the conformation of Figure 12.4b is accomplished by rotation of groups about the single bond between carbons 2 and 3. A second rotation, involving carbons 4 and 5, produces the conformation of Figure 12.4c.

The right side of Figure 12.4 shows how different conformations of a molecule are denoted in two dimensions with structural formulas. All three structural formulas represent the same molecule; that is, they are different conformations of the same molecule. In all three cases, a continuous chain of six carbon atoms is present. In the second and third structures, the chain is "bent," but the bends do not disrupt the continuity of the chain.

• You should learn to recognize molecules drawn in several different ways (conformations). Like friends, they can be recognized whether they are sitting, reclining, or standing.

Example 12.1

Recognizing Different Conformations of a Molecule and Structural Isomers

Determine whether the members of each of the following pairs of structural formulas represent (1) different conformations of the same molecule, (2) different compounds that are structural isomers, or (3) different compounds that are not structural isomers.

a. $CH_3—CH_2—CH_2—CH_3$ and $CH_2—CH_2$ (each CH_2 bonded to a CH_3 below: CH_3 CH_3)

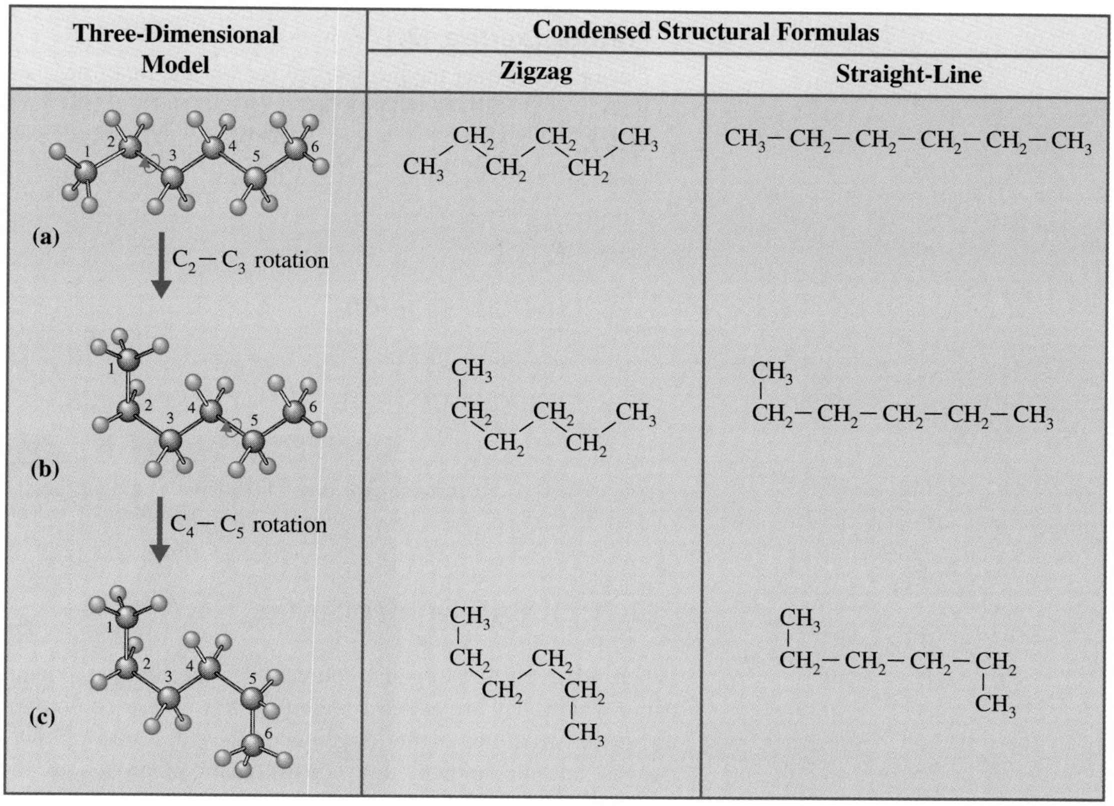

Figure 12.4 Various methods for denoting different conformations for a C_6H_{14} continuous-chain alkane.

Three-Dimensional Model	Condensed Structural Formulas: Zigzag	Condensed Structural Formulas: Straight-Line
(a)	CH_3 – CH_2 – CH_2 – CH_2 – CH_2 – CH_3 (zigzag)	$CH_3-CH_2-CH_2-CH_2-CH_2-CH_3$
C_2-C_3 rotation ↓		
(b)	CH_3 above CH_2, then CH_2 – CH_2 – CH_2 – CH_3 (zigzag)	CH_3 above; $CH_2-CH_2-CH_2-CH_2-CH_3$
C_4-C_5 rotation ↓		
(c)	CH_3 above CH_2, then CH_2 – CH_2 – CH_2, with CH_3 below	CH_3 above; $CH_2-CH_2-CH_2-CH_2$; CH_3 below last CH_2

b. $CH_2-CH_2-CH_3$ (with CH_3 below first CH_2) and $CH_2-CH_2-CH_2-CH_3$ (with CH_3 below first CH_2)

c. $CH_3-CH-CH_3$ (with CH_3 below CH) and $CH_3-CH_2-CH_2$ (with CH_3 below last CH_2)

Solution

a. Both molecules have the molecular formula C_4H_{10}. The connectivity of carbon atoms is the same for both molecules: a continuous chain of four carbon atoms. For the second structural formula, we need to go around two corners to get a four-carbon-atom chain, which is fine because of the free rotation associated with single bonds in alkanes.

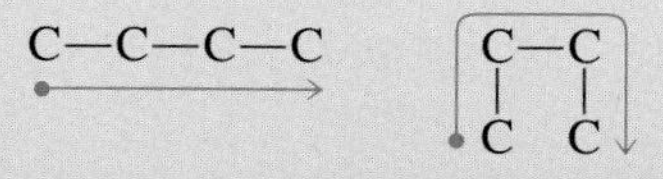

C—C—C—C C—C (with C below each)

With the same molecular formula and the same connectivity of atoms, these two structural formulas are conformations of the same molecule.

b. The molecular formula of the first compound is C_4H_{10}, and that of the second compound is C_5H_{12}. Thus the two structural formulas represent different compounds that are not structural isomers.

c. Both molecules have the same molecular formula, C_4H_{10}. The connectivity of atoms is different. In the first case, we have a chain of three carbon atoms with a branch off the chain. In the second case, a continuous chain of four carbon atoms is present.

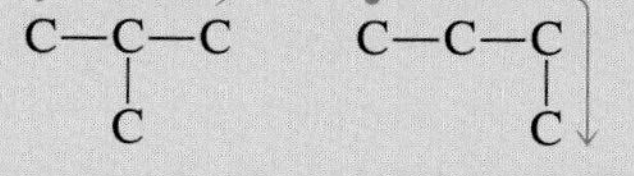

C—C—C (C below middle) C—C—C (C below last)

These two structural formulas are those of structural isomers.

Practice Exercise 12.1

Determine whether the members of each of the following pairs of structural formulas represent: (1) different conformations of the same molecule, (2) different compounds that are structural isomers, or (3) different compounds that are not structural isomers.

a. $CH_3-CH_2-CH_2-CH_2-CH_3$ and

$$\begin{array}{lll} CH_3-CH_2 & & \\ \quad\quad\;\; | & & \\ \quad\quad CH_2-CH_2 & & \\ \quad\quad\quad\quad\quad\; | & & \\ \quad\quad\quad\quad\quad CH_3 & & \end{array}$$

b.

$$\begin{array}{c} CH_3-\underset{\displaystyle CH_3}{\underset{|}{CH}}-CH_2-CH_3 \end{array} \quad \text{and} \quad \begin{array}{c} CH_3-\underset{\displaystyle CH_3}{\underset{|}{CH}}-\underset{\displaystyle CH_3}{\underset{|}{CH_2}} \end{array}$$

c.

$$\begin{array}{c} CH_3-\underset{\displaystyle CH_3}{\underset{|}{CH}}-CH_2-CH_3 \end{array} \quad \text{and} \quad \begin{array}{c} \underset{\displaystyle CH_3}{\underset{|}{CH_2}}-CH_2-\underset{\displaystyle CH_3}{\underset{|}{CH_2}} \end{array}$$

• *Answers:* **a.** different conformations; **b.** different conformations; **c.** structural isomers

12.8 IUPAC Nomenclature for Alkanes

When relatively few organic compounds were known, chemists arbitrarily named them using what today are called *common names.* These common names gave no information about the structures of the compounds they described. However, as more organic compounds became known, this nonsystematic approach to naming compounds became unwieldy.

Today, formal systematic rules exist for generating names for organic compounds. These rules, which were formulated, and are updated periodically, by the International Union of Pure and Applied Chemistry (IUPAC), are known as *IUPAC rules.* The advantage of the IUPAC naming system is that it assigns each compound a name that not only identifies it but also enables one to draw its structural formula.

• IUPAC is pronounced "eye-you-pack."

• Continuous-chain alkanes are also frequently called *straight-chain alkanes* and *normal-chain alkanes.*

• You need to memorize the prefixes in column 2 of Table 12.3. This is the way to count from 1 to 10 in "organic chemistry language."

IUPAC names for the first 10 *continuous-chain* alkanes are given in Table 12.3. Note that all of these names end in *-ane*, the characteristic ending for all alkane names. Note also that beginning with the five-carbon alkane, Greek numerical prefixes are used to denote the actual number of carbon atoms in the continuous chain.

To name *branched-chain* alkanes, we must be able to name the branch or branches that are attached to the main carbon chain. These branches are formally called substituents.

Table 12.3
IUPAC Names for the First Ten Continuous-Chain Alkanes[a]

Molecular formula	IUPAC prefix	IUPAC name	Structural formula
CH_4	meth-	methane	CH_4
C_2H_6	eth-	ethane	CH_3-CH_3
C_3H_8	prop-	propane	$CH_3-CH_2-CH_3$
C_4H_{10}	but-	butane	$CH_3-CH_2-CH_2-CH_3$
C_5H_{12}	pent-	pentane	$CH_3-CH_2-CH_2-CH_2-CH_3$
C_6H_{14}	hex-	hexane	$CH_3-CH_2-CH_2-CH_2-CH_2-CH_3$
C_7H_{16}	hept-	heptane	$CH_3-CH_2-CH_2-CH_2-CH_2-CH_2-CH_3$
C_8H_{18}	oct-	octane	$CH_3-CH_2-CH_2-CH_2-CH_2-CH_2-CH_2-CH_3$
C_9H_{20}	non-	nonane	$CH_3-CH_2-CH_2-CH_2-CH_2-CH_2-CH_2-CH_2-CH_3$
$C_{10}H_{22}$	dec-	decane	$CH_3-CH_2-CH_2-CH_2-CH_2-CH_2-CH_2-CH_2-CH_2-CH_3$

[a]The IUPAC naming system also includes prefixes for naming continuous-chain alkanes that have more than 10 carbon atoms, but we will not consider them in this text.

Table 12.4
Names for the First Six Continuous-Chain Alkyl Groups

Number of carbons	Structural formula	Stem of alkane name	Suffix	Alkyl group name
1	$—CH_3$	meth-	-yl	methyl
2	$—CH_2—CH_3$	eth-	-yl	ethyl
3	$—CH_2—CH_2—CH_3$	prop-	-yl	propyl
4	$—CH_2—CH_2—CH_2—CH_3$	but-	-yl	butyl
5	$—CH_2—CH_2—CH_2—CH_2—CH_3$	pent-	-yl	pentyl
6	$—CH_2—CH_2—CH_2—CH_2—CH_2—CH_3$	hex-	-yl	hexyl

A **substituent** *is an atom or group of atoms attached to a chain (or ring) of carbon atoms.*

For branched-chain alkanes, the substituents are specifically called *alkyl groups.* An **alkyl group** *is the group of atoms that would be obtained by removing a hydrogen atom from an alkane.*

The two most commonly encountered alkyl groups are the two simplest: the one-carbon and two-carbon alkyl groups. Their formulas and names are

$—CH_3$ Methyl group $—CH_2—CH_3$ Ethyl group

The extra long bond in these formulas (on the left) denotes the point of attachment to the carbon chain. Note that alkyl groups do not lead a stable, independent existence; that is, they are not molecules. They are always found attached to another entity (usually a carbon chain).

• The ending *-yl*, as in meth*yl*, eth*yl*, prop*yl*, and but*yl*, appears in the names of all alkyl groups.

Alkyl groups are named by taking the stem of the name of the alkane that contains the same number of carbon atoms and adding the ending *-yl*. Table 12.4 gives the names for small continuous-chain alkyl groups.

We are now ready for the IUPAC rules for naming branched-chain alkanes.

• An additional guideline for identifying the longest continuous carbon chain: If two different carbon chains in a molecule have the same largest number of carbon atoms, select as the parent chain the one with the largest number of substituents (alkyl groups) attached to the chain.

Rule 1: *Identify the longest continuous carbon chain (the parent chain), which may or may not be shown in a straight line, and name the chain.*

$CH_3—CH_2—CH_2—CH(CH_3)—CH_3$ — The parent chain name is *pentane,* because it has five carbon atoms.

$CH_3—CH(CH_2—CH_3)—CH_2—CH_2—CH_3$ — The parent chain name is *hexane,* because it has six carbon atoms.

• Additional guidelines for numbering carbon atom chains:

• 1. If each end of the chain has a substituent the same distance in, number from the end closest to the second-encountered substituent.

• 2. If there are substituents equidistant from each end of the chain and there is no third substituent to use as the "tie-breaker," begin numbering nearest the substituent that has alphabetical priority—that is, the substituent whose name occurs first in the alphabet.

Rule 2: *Number the carbon atoms in the parent chain from the end of the chain nearest a substituent (alkyl group).*

There always are two ways to number the chain (either from left to right or from right to left). This rule gives the first-encountered alkyl group the lowest possible number.

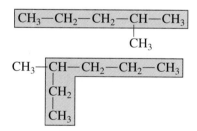

$\overset{5}{C}H_3—\overset{4}{C}H_2—\overset{3}{C}H_2—\overset{2}{C}H(CH_3)—\overset{1}{C}H_3$ — Right-to-left numbering system

$CH_3—\overset{3}{C}H(\overset{2}{C}H_2—\overset{1}{C}H_3)—\overset{4}{C}H_2—\overset{5}{C}H_2—\overset{6}{C}H_3$ — Left-to-right numbering system

Rule 3: *If only one alkyl group is present, name and locate it (by number), and attach the number and name to that of the parent carbon chain.*

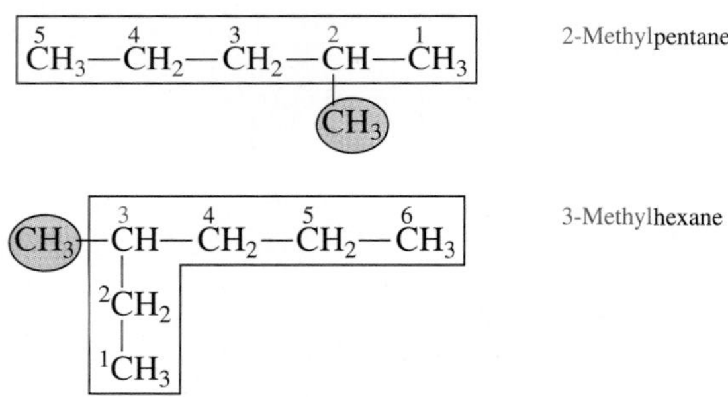

Note that the name is written as one word, with a hyphen between the number (location) and the name of the alkyl group.

Rule 4: *If two or more of the* same kind *of alkyl group are present in a molecule, indicate the number with a Greek numerical prefix (*di-, tri-, tetra-, penta-, *and so forth). In addition, a number specifying the location of each identical group must be included. These position numbers, separated by commas, precede the numerical prefix. Numbers are separated from words by hyphens.*

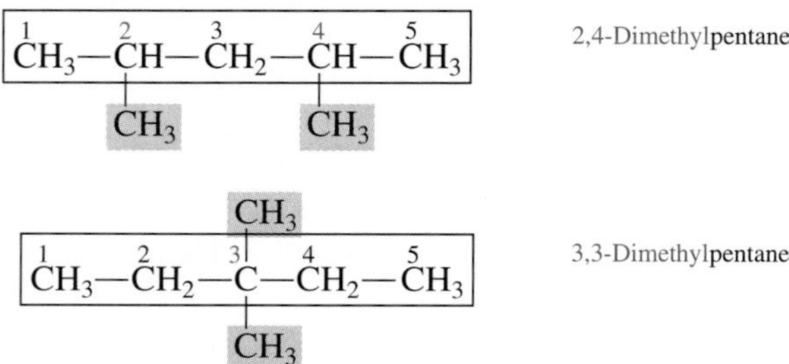

• There must be as many numbers as there are alkyl groups in the IUPAC name of a branched-chain alkane.

Note that the numerical prefix *di-* must always be accompanied by two numbers, *tri-* by three, and so on, even if the same number is written twice, as in 3,3-dimethylpentane.

Rule 5: *When two kinds of alkyl groups are present on the same carbon chain, number each group separately, and list the names of the alkyl groups in alphabetical order.*

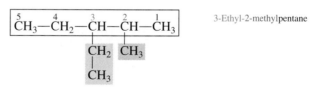

$\overset{5}{CH_3}-\overset{4}{CH_2}-\overset{3}{CH}(CH_2-CH_3)-\overset{2}{CH}(CH_3)-\overset{1}{CH_3}$

3-Ethyl-2-methylpentane

Note that ethyl is named first in accordance with the alphabetical rule.

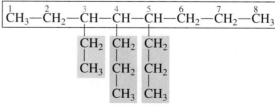

$\overset{1}{CH_3}-\overset{2}{CH_2}-\overset{3}{CH}(CH_2-CH_3)-\overset{4}{CH}(CH_2-CH_2-CH_3)-\overset{5}{CH}(CH_2-CH_2-CH_3)-\overset{6}{CH_2}-\overset{7}{CH_2}-\overset{8}{CH_3}$

3-Ethyl-4,5-dipropyloctane

• Numerical prefixes that designate numbers of alkyl groups, such as *di-*, *tri-*, and *tetra-*, are not considered when determining alphabetical priority for alkyl groups.

Note that the prefix *di-* does not affect the alphabetical order; ethyl precedes propyl.

Rule 6: *Follow IUPAC punctuation rules, which include the following: (1) Separate numbers from each other by commas. (2) Separate numbers from letters by hyphens. (3) Do not add a hyphen or space between the last-named substituent and the name of the parent alkane that follows.*

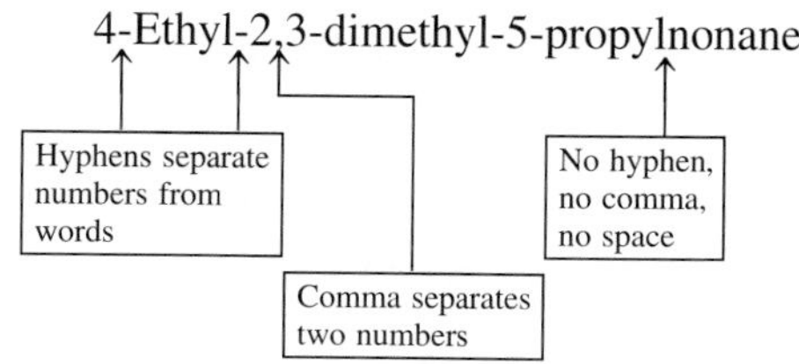

Example 12.2

Determining IUPAC Names for Branched-Chain Alkanes

Give the IUPAC name for each of the following branched-chain alkanes.

a. 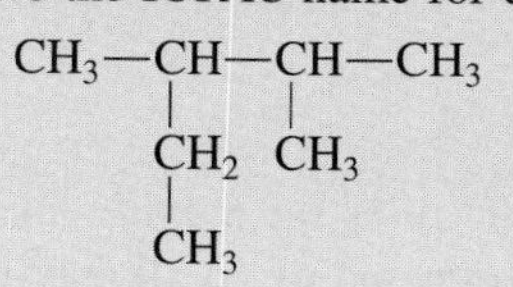
```
CH3—CH—CH—CH3
    |   |
    CH2 CH3
    |
    CH3
```

b. 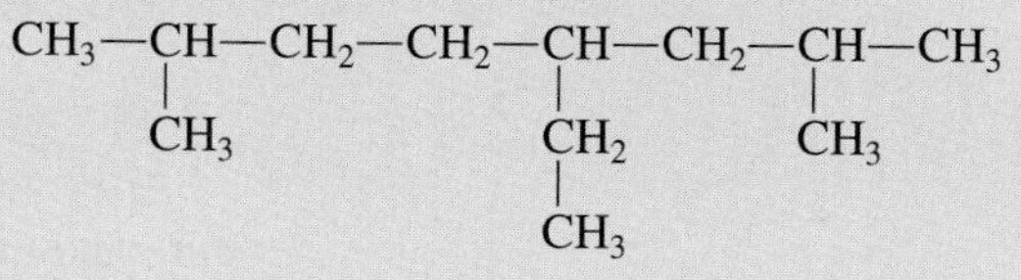
```
CH3—CH—CH2—CH2—CH—CH2—CH—CH3
    |              |       |
    CH3            CH2     CH3
                   |
                   CH3
```

Solution

a. The longest carbon chain possesses five carbon atoms. Thus the parent chain name is pentane.

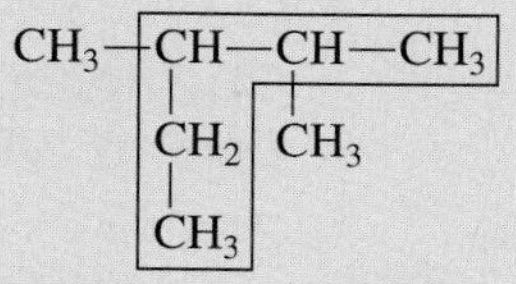

```
CH3—CH—CH—CH3
    |   |
    CH2 CH3
    |
    CH3
```

This parent chain is numbered from right to left, because an alkyl substituent is closer to the right end of the chain than to the left end.

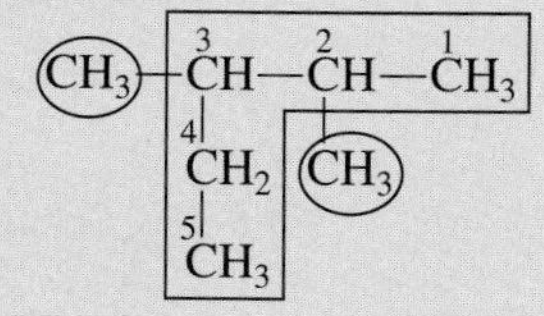

```
      3    2    1
(CH3)—CH—CH—CH3
     4|   |
      CH2 (CH3)
     5|
      CH3
```

There are two methyl group substituents (circled). One methyl group is located on carbon 2 and the other on carbon 3. The IUPAC name for the compound is 2,3-dimethylpentane.

b. There are eight carbon atoms in the longest carbon chain, so the parent name is octane. There are three alkyl groups present (circled).

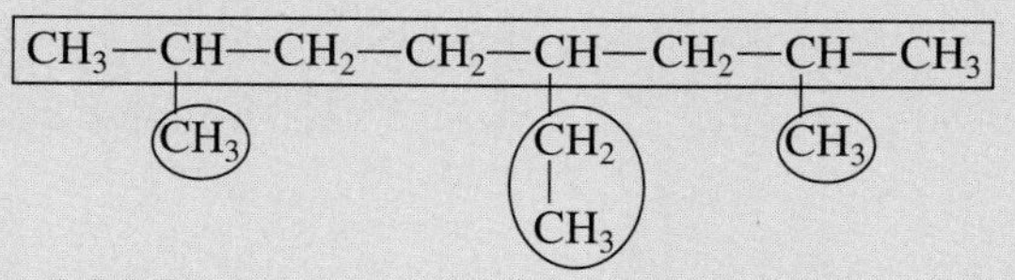

```
CH3—CH—CH2—CH2—CH—CH2—CH—CH3
    |              |       |
   (CH3)          (CH2    (CH3)
                   |
                   CH3)
```

Selection of the numbering system to be used cannot be made based on the "first-encountered-alkyl-group rule" because an alkyl group is equidistant from each end of the chain. Thus the second-encountered alkyl group is used as the "tie-breaker." It is closer to the right end of the parent chain (carbon 4) than to the left end (carbon 5). Thus we use the right-to-left numbering system.

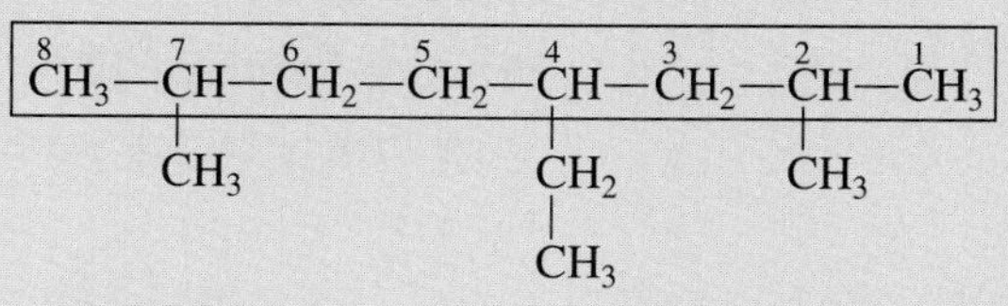

```
8   7   6    5    4   3    2   1
CH3—CH—CH2—CH2—CH—CH2—CH—CH3
    |              |       |
    CH3            CH2     CH3
                   |
                   CH3
```

Two different kinds of alkyl groups are present: ethyl and methyl. Ethyl has alphabetical priority over methyl and precedes methyl in the IUPAC name. The IUPAC name is 4-ethyl-2,7-dimethyloctane.

• Always compare the total number of carbon atoms in the name with the number in the structure to make sure they match. The name 4-ethyl-2,7-dimethyloctane indicates the presence of $2 + 2(1) + 8 = 12$ carbon atoms. The structure does have 12 carbon atoms.

Practice Exercise 12.2

Give the IUPAC name for each of the following alkanes.

a.

$$
\begin{array}{cccccccccccc}
CH_3 & - & CH & - & CH_2 & - & CH_2 & - & CH & - & CH_3 \\
 & & | & & & & & & | & & \\
 & & CH_2 & & & & & & CH_2 & & \\
 & & | & & & & & & | & & \\
 & & CH_3 & & & & & & CH_3 & &
\end{array}
$$

b.

$$
\begin{array}{ccccccccccccccc}
 & & & & & & CH_3 & & & & & & & & \\
 & & & & & & | & & & & & & & & \\
CH_3 & - & CH_2 & - & CH & - & C & - & CH & - & CH_2 & - & CH_2 & - & CH_3 \\
 & & & & | & & | & & | & & & & & & \\
 & & & & CH_3 & & CH_3 & & CH_3 & & & & & &
\end{array}
$$

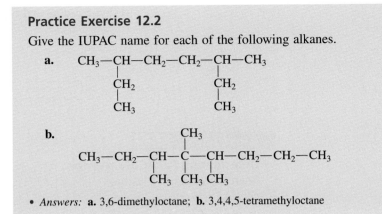

• *Answers:* **a.** 3,6-dimethyloctane; **b.** 3,4,4,5-tetramethyloctane

After you learn the rules for naming alkanes, it is relatively easy to reverse the procedure and translate the name of an alkane into a structural formula. Example 12.3 shows how this is done.

• A few smaller branched alkanes have common names—that is, non-IUPAC names—that still have widespread use. They make use of the prefixes *iso* and *neo,* as in isobutane, isopentane, and neohexane. These prefixes specifically denote particular carbon atom arrangements.

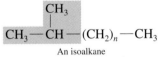

$$
\begin{array}{ccccccc}
 & & CH_3 & & & & \\
 & & | & & & & \\
CH_3 & - & CH & - & (CH_2)_n & - & CH_3
\end{array}
$$

An isoalkane
(e.g., $n = 1$, Isopentane)

$$
\begin{array}{ccccccc}
 & & CH_3 & & & & \\
 & & | & & & & \\
CH_3 & - & C & - & (CH_2)_n & - & CH_3 \\
 & & | & & & & \\
 & & CH_3 & & & &
\end{array}
$$

A neoalkane
(e.g., $n = 1$, Neohexane)

Example 12.3

Generating the Structural Formula of an Alkane from Its IUPAC Name

Draw the condensed structural formula for 3-ethyl-2,3-dimethylpentane.

Solution

Step 1: The name of this compound ends in *pentane,* so the longest continuous chain has five carbon atoms. Draw this chain of five carbon atoms and number it.

$$
\overset{1}{C}-\overset{2}{C}-\overset{3}{C}-\overset{4}{C}-\overset{5}{C}
$$

Step 2: Complete the carbon skeleton by attaching alkyl groups as they are specified in the name. An ethyl group goes on carbon 3, and methyl groups are attached to carbons 2 and 3.

$$
\begin{array}{ccccccccc}
 & & & & C & & & & \\
 & & & & | & & & & \\
\overset{1}{C} & - & \overset{2}{C} & - & \overset{3}{C} & - & \overset{4}{C} & - & \overset{5}{C} \\
 & & | & & | & & & & \\
 & & C & & C & & & & \\
 & & & & | & & & & \\
 & & & & C & & & &
\end{array}
$$

Step 3: Add hydrogen atoms to the carbon skeleton so that each carbon atom has four bonds.

$$
\begin{array}{ccccccccc}
 & & & & CH_3 & & & & \\
 & & & & | & & & & \\
\overset{1}{CH_3} & - & \overset{2}{CH} & - & \overset{3}{C} & - & \overset{4}{CH_2} & - & \overset{5}{CH_3} \\
 & & | & & | & & & & \\
 & & CH_3 & & CH_2 & & & & \\
 & & & & | & & & & \\
 & & & & CH_3 & & & &
\end{array}
$$

Practice Exercise 12.3

Draw the condensed structural formula for 4,5-diethyl-3,4,5-trimethyloctane.

• *Answer:*

$$
\begin{array}{ccccccccccccccc}
 & & & & & & CH_3 & & CH_3 & & & & & & \\
 & & & & & & | & & | & & & & & & \\
CH_3 & - & CH_2 & - & CH & - & C & - & C & - & CH_2 & - & CH_2 & - & CH_3 \\
 & & & & | & & | & & | & & & & & & \\
 & & & & CH_3 & & CH_2 & & CH_2 & & & & & & \\
 & & & & & & | & & | & & & & & & \\
 & & & & & & CH_3 & & CH_3 & & & & & &
\end{array}
$$

12.9 Classification of Carbon Atoms

Each of the carbon atoms within a hydrocarbon structure can be classified as a primary, secondary, tertiary, or quaternary carbon atom. A **primary carbon atom** *is bonded to only one other carbon atom.* Each of the "end" carbon atoms in the three-carbon propane structure is a primary carbon atom, whereas the middle carbon atom of propane is a secondary carbon atom. A **secondary carbon atom** *is bonded to two other carbon atoms.*

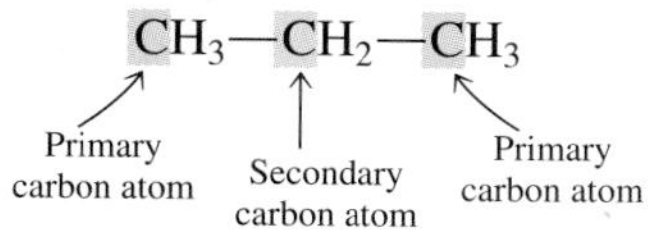

$CH_3—CH_2—CH_3$

Primary carbon atom, Secondary carbon atom, Primary carbon atom

• The notations 1°, 2°, 3°, and 4° are often used as designations for the terms *primary, secondary, tertiary,* and *quaternary.* Thus we can write
1° carbon atom
2° carbon atom
3° carbon atom
4° carbon atom

A **tertiary carbon atom** *is bonded to three other carbon atoms.* The molecule 2-methylpropane contains a tertiary carbon atom.

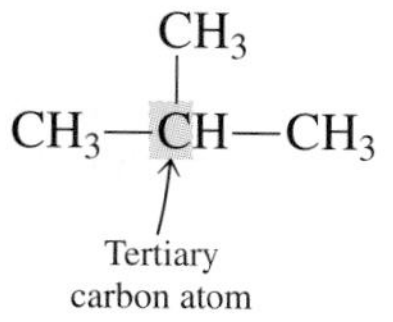

$$\begin{array}{c} CH_3 \\ | \\ CH_3—CH—CH_3 \end{array}$$

Tertiary carbon atom

A **quaternary carbon atom** *is bonded to four other carbon atoms.* The molecule 2,2-dimethylpropane contains a quaternary carbon atom.

$$\begin{array}{c} CH_3 \\ | \\ CH_3—C—CH_3 \\ | \\ CH_3 \end{array}$$

Quaternary carbon atom

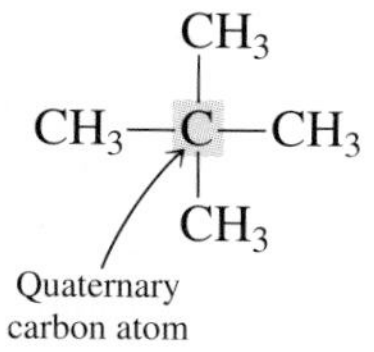

12.10 Branched-Chain Alkyl Groups

Just as there are continuous-chain and branched-chain alkanes, there are continuous-chain and branched-chain alkyl groups. Four branched-chain alkyl groups, shown in Figure 12.5, are so common that you should know their names and structures.

A secondary-butyl group (often abbreviated as *sec*-butyl or *s*-butyl) possesses, before it is attached to a parent chain, a secondary carbon atom (Section 12.9). This secondary carbon atom is the point of attachment of the group to the parent chain. Similarly, a tertiary-butyl group (often abbreviated as *tert*-butyl or *t*-butyl) possesses, before it is attached to a parent chain, a tertiary carbon atom. This tertiary carbon atom is the point of attachment of the group to the parent chain.

Figure 12.5
The four most common branched-chain alkyl groups and their IUPAC names.

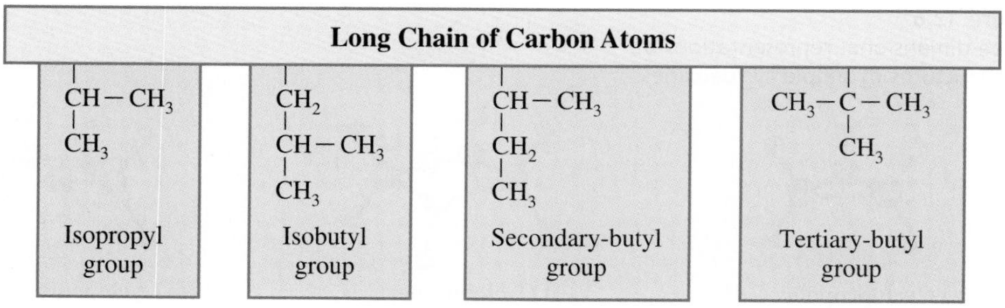

Long Chain of Carbon Atoms

$\begin{array}{l} CH—CH_3 \\ \vert \\ CH_3 \end{array}$	$\begin{array}{l} CH_2 \\ \vert \\ CH—CH_3 \\ \vert \\ CH_3 \end{array}$	$\begin{array}{l} CH—CH_3 \\ \vert \\ CH_2 \\ \vert \\ CH_3 \end{array}$	$\begin{array}{c} CH_3—C—CH_3 \\ \vert \\ CH_3 \end{array}$
Isopropyl group	Isobutyl group	Secondary-butyl group	Tertiary-butyl group

• You need to be able to recognize various conformations of branched-chain alkyl groups. For example, these structures all represent an isopropyl group:

```
        |
CH3—CH—CH3

     |            |
CH3—CH           CH
     |          /  \
    CH3       CH3  CH3
```

In each case, you have a chain of three carbon atoms with an attachment point (the long bond) involving the middle carbon atom of the chain.

Two examples of alkanes containing branched-chain alkyl groups follow.

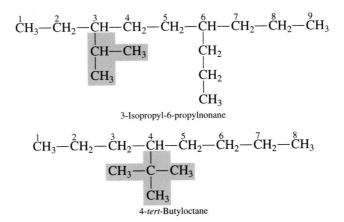

```
 1     2     3    4     5     6    7     8     9
CH3—CH2—CH—CH2—CH2—CH—CH2—CH2—CH3
             |               |
             CH—CH3          CH2
             |               |
             CH3             CH2
                             |
                             CH3
```

3-Isopropyl-6-propylnonane

```
 1     2     3     4    5     6     7     8
CH3—CH2—CH2—CH—CH2—CH2—CH2—CH3
                   |
              CH3—C—CH3
                   |
                   CH3
```

4-*tert*-Butyloctane

In IUPAC naming, hyphenated prefixes, such as *sec*- and *tert*-, are not considered when alphabetizing. The prefixes *iso* and *neo* are not hyphenated prefixes and are included when alphabetizing. The following IUPAC name is thus correct:

5-*sec*-Butyl-4-isopropyl-3-methyldecane

12.11 Cycloalkanes

• It takes a minimum of three carbon atoms to form a cyclic arrangement of carbon atoms.

A **cycloalkane** *is a saturated hydrocarbon in which the carbon atoms are connected to one another in a cyclic (ring) arrangement.* The simplest cycloalkane is cyclopropane, which contains a cyclic arrangement of three carbon atoms. Figure 12.6 shows a three-dimensional model of cyclopropane's structure and those of the four-, five-, and six-carbon cycloalkanes.

Cyclopropane's three carbon atoms lie in a flat ring. In all other cycloalkane molecules, some puckering of the ring occurs; that is, the ring systems are nonplanar, as shown in Figure 12.6.

The general formula for cycloalkanes is C_nH_{2n}. Thus a given cycloalkane contains two fewer hydrogen atoms than an alkane with the same number of hydrogen atoms (C_nH_{2n+2}). Butane (C_4H_{10}) and cyclobutane (C_4H_8) are not structural isomers; structural isomers must have the same molecular formula (Section 12.6).

Line-angle drawings are often used to represent cycloalkane structures. A **line-angle drawing** *is an abbreviated structural formula representation in which an angle represents a carbon atom and a line represents a bond.* The line angle drawing for cyclopropane is a triangle, that for cyclobutane a square, that for cyclopentane a pentagon, and that for cyclohexane a hexagon.

Figure 12.6
Three-dimensional representations of the structures of simple cycloalkanes.

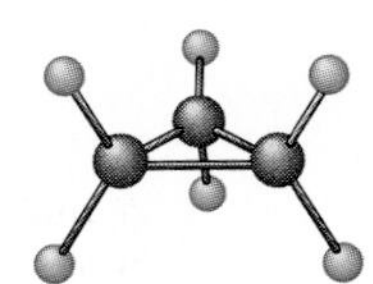

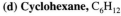

(a) Cyclopropane, C_3H_6

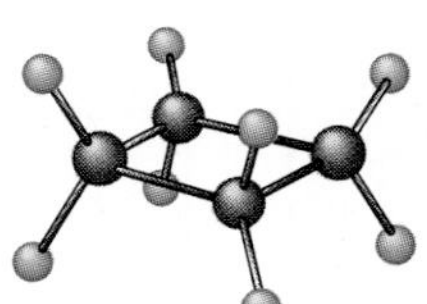

(b) Cyclobutane, C_4H_8

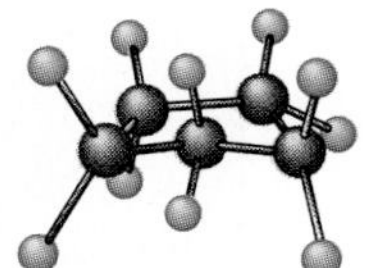

(c) Cyclopentane, C_5H_{10}

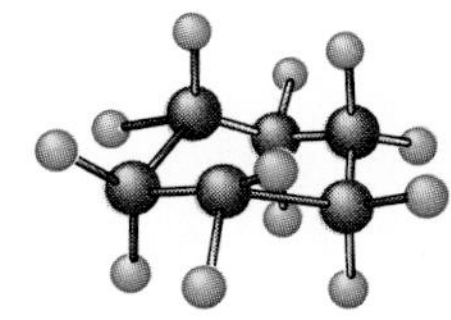

(d) Cyclohexane, C_6H_{12}

• Line-angle drawings, involving a sawtooth pattern of lines, can be used to denote chains of carbon atoms. In this notation, a carbon atom (with its associated hydrogen atoms) is understood to be present at every point where two lines meet and at the ends of lines. An unbranched chain of eight carbon atoms would be represented as

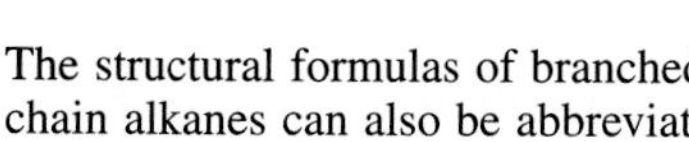

The structural formulas of branched-chain alkanes can also be abbreviated by using this notation. Such notation for 2,3,4-trimethylpentane is

Example 12.4

Generating Molecular Formulas from Line-Angle Drawings

Determine the molecular formula for each of the following cycloalkanes and alkanes.

a. b.

c. d.

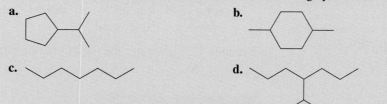

Solution

a. First replace each angle and line terminus with a carbon atom, and then add hydrogens as necessary to give each carbon four bonds. The molecular formula of this compound is C_8H_{16}.

$\Longrightarrow C_8H_{16}$

b. Similarly, we have

$C\text{—}C\text{(ring)}\text{—}C \Longrightarrow CH_3\text{—}CH(\text{ring})CH\text{—}CH_3 \Longrightarrow C_8H_{16}$

c. $C\text{—}C\text{—}C\text{—}C\text{—}C\text{—}C\text{—}C \Longrightarrow CH_3\text{—}CH_2\text{—}CH_2\text{—}CH_2\text{—}CH_2\text{—}CH_2\text{—}CH_3 \Longrightarrow C_7H_{16}$

d. $\Longrightarrow CH_3\text{—}CH_2\text{—}CH_2\text{—}CH(CH(CH_3)_2)\text{—}CH_2\text{—}CH_2\text{—}CH_3 \Longrightarrow C_{10}H_{22}$

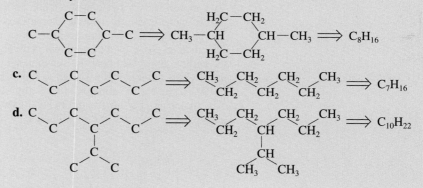

Practice Exercise 12.4

Determine the molecular formula for each of the following cycloalkanes and alkanes.

a. b.

c. d.

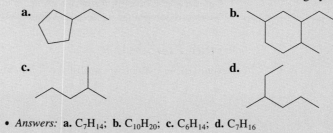

• *Answers:* **a.** C_7H_{14}; **b.** $C_{10}H_{20}$; **c.** C_6H_{14}; **d.** C_7H_{16}

12.12 IUPAC Nomenclature for Cycloalkanes

IUPAC naming procedures for cycloalkanes are similar to those for alkanes. The ring portion of a cycloalkane molecule serves as the name base, and the prefix *cyclo-* is used to indicate the presence of the ring. Alkyl group substituents are named in the same manner as in alkanes. Numbering conventions used in locating substituents include the following:

1. If there is just one ring substituent, it is not necessary to locate it by number.
2. When two ring substituents are present, the carbon atoms in the ring are numbered beginning with the substituent of higher alphabetical priority and

• Cycloalkanes of ring sizes ranging from 3 to over 30 are found in nature, and in principle, there is no limit to ring size. Five-membered rings (cyclopentanes) and six-membered rings (cyclohexanes) are especially abundant in nature and therefore have received special attention.

proceeding in the direction (clockwise or counterclockwise) that gives the other substituent the lower number.

3. When three or more ring substituents are present, ring numbering begins at the substituent that leads to the lowest set of location numbers. When two or more equivalent numbering sets exist, alphabetical priority among substituents determines the set used.

Example 12.5 illustrates the use of the ring-numbering guidelines.

Example 12.5

Determining IUPAC Names for Cycloalkanes

Assign IUPAC names to each of the following cycloalkanes.

a. CH_3

b.

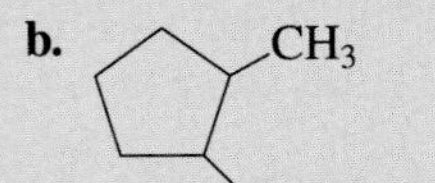

c.

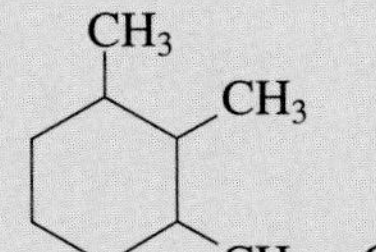

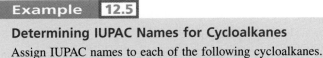

Solution

a. This molecule is a cyclobutane (four-carbon ring) with a methyl substituent. The IUPAC name is simply methylcyclobutane. No number is needed to locate the methyl group, because all four ring positions are equivalent.

b. This molecule is a cyclopentane with ethyl and methyl substituents. The numbers for the carbon atoms that bear the substituents are 1 and 2. On the basis of alphabetical priority, the number 1 is assigned to the carbon atom that bears the ethyl group. The IUPAC name for the compound is 1-ethyl-2-methylcyclopentane.

c. This molecule is a dimethylpropylcyclohexane. Two different 1,2,3 numbering systems exist for locating the substituents.

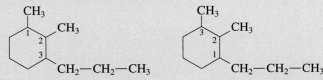

CH_3 (1), CH_3 (2), $CH_2—CH_2—CH_3$ (3) $\qquad$ CH_3 (3), CH_3 (2), $CH_2—CH_2—CH_3$ (1)

On the basis of alphabetical priority, we use the first numbering system. This system has carbon 1 bearing a methyl group; methyl has alphabetical priority over propyl. Thus the compound name is 1,2-dimethyl-3-propylcyclohexane.

Practice Exercise 12.5

Assign IUPAC names to each of the following cycloalkanes.

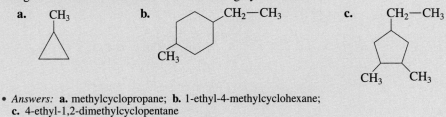

a. CH_3 $\qquad$ **b.** $CH_2—CH_3$, CH_3 $\qquad$ **c.** $CH_2—CH_3$, CH_3, CH_3

• *Answers:* **a.** methylcyclopropane; **b.** 1-ethyl-4-methylcyclohexane; **c.** 4-ethyl-1,2-dimethylcyclopentane

12.13 Isomerism in Cycloalkanes

Structural isomers are possible for cycloalkanes that contain four or more carbon atoms. For example, there are five cycloalkane structural isomers that have the formula C_5H_{10}: one based on a five-membered ring, one based on a four-membered ring, and three based on a three-membered ring. These isomers are

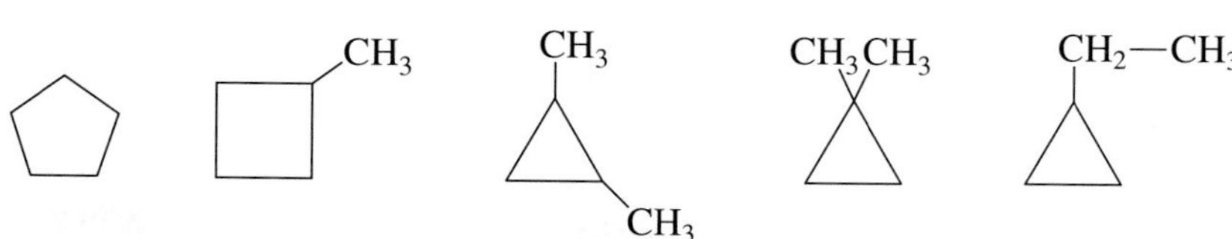

A second type of isomerism is possible for some *substituted* cycloalkanes. This new type of isomerism is *cis–trans* isomerism. ***Cis–trans* isomers** *are compounds that have the same molecular and structural formulas but different arrangements of atoms in space because of restricted rotation around bonds.*

• *Cis–trans isomers* have the same molecular formula and the same structural formula. The only difference between them is the orientation of atoms in space. *Structural isomers* have the same molecular formula but different structural formulas.

In alkanes, there is free rotation about all carbon–carbon bonds (Section 12.7). In cycloalkanes, the ring structure restricts rotation for the carbon atoms in the ring. The consequence of this lack of rotation in a cycloalkane is the creation of "top" and "bottom" positions for the two attachments on each of the ring carbon atoms. This "top–bottom" situation may lead to *cis–trans* isomerism in cycloalkanes with two or more substituents on the ring.

Consider the following two structures for the molecule 1,2-dimethylcyclopentane.

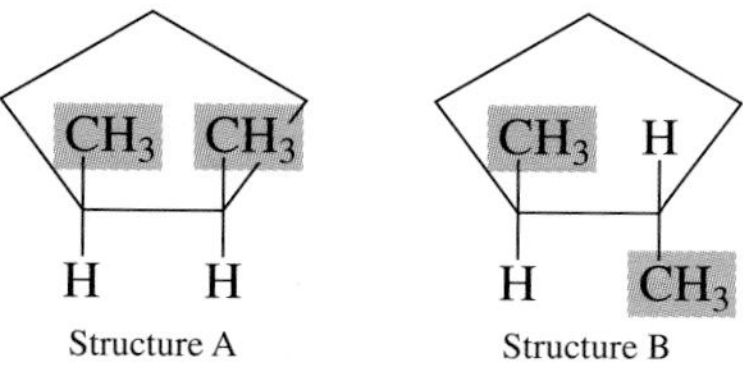

In structure A, both methyl groups are above the plane of the ring (the "top" side). In structure B, one methyl group is above the plane of the ring (the "top" side) and the other below it (the "bottom" side). Structure A cannot be converted into structure B without breaking bonds. Hence structures A and B are isomers; there are two 1,2-dimethylcyclopentanes. The first isomer is called *cis*-1,2-dimethylcyclopentane and the second *trans*-1,2-dimethylcyclopentane.

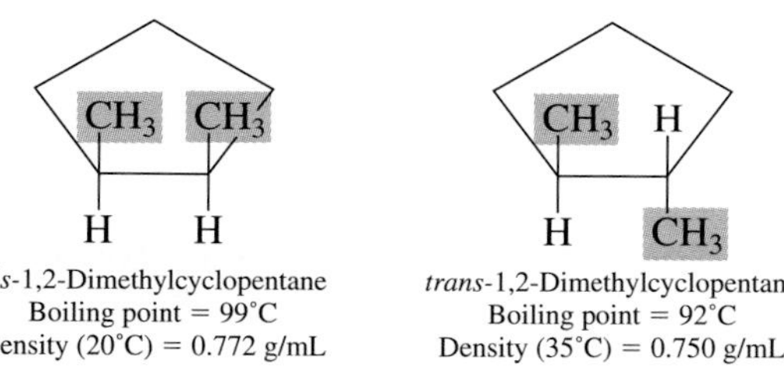

cis-1,2-Dimethylcyclopentane
Boiling point = 99°C
Density (20°C) = 0.772 g/mL

trans-1,2-Dimethylcyclopentane
Boiling point = 92°C
Density (35°C) = 0.750 g/mL

A ***cis* isomer** *is an isomer in which two atoms or groups are on the same side of a restricted rotation "barrier" in a molecule.* In our current discussion, the restricted rotation barrier is the ring of carbon atoms. A ***trans* isomer** *is an isomer in which two atoms or groups are on different sides of a restricted rotation "barrier" in a molecule.*

• The Latin *cis* means "on the same side," and the Latin *trans* means "across." Consider the use of the prefix *trans-* in the phrase "transatlantic voyage."

Cis–trans isomerism can occur in rings of all sizes. The presence of a substituent on each of two carbon atoms in the ring is the minimum requirement for its occurrence. In biochemistry, we will find that the human body often selectively distinguishes between the *cis* and *trans* isomers of a compound. One isomer will be active in the body and the other inactive.

Example 12.6

Identifying and Naming Cycloalkane *Cis–Trans* Isomers

Determine whether *cis–trans* isomerism is possible for each of the following cycloalkanes. If so, then draw structural formulas for the *cis* and *trans* isomers.

a. Methylcyclohexane
b. 1,1-Dimethylcyclohexane
c. 1,3-Dimethylcyclobutane
d. 1-Ethyl-2-methylcyclobutane

(*continued*)

• In cycloalkanes, *cis–trans* isomerism can also be denoted by using wedges and dotted lines. A heavy wedge-shaped bond to a ring structure indicates a bond *above* the plane of the ring, and a broken dotted line indicates a bond *below* the plane of the ring.

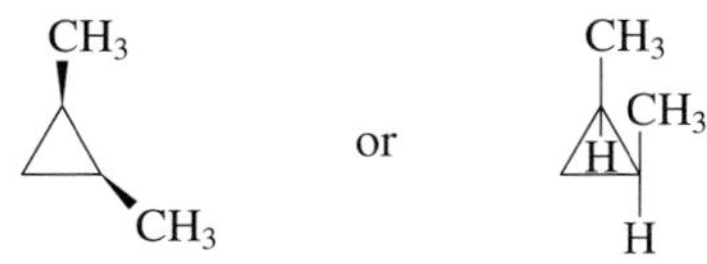

cis-1,2-Dimethylcyclopropane

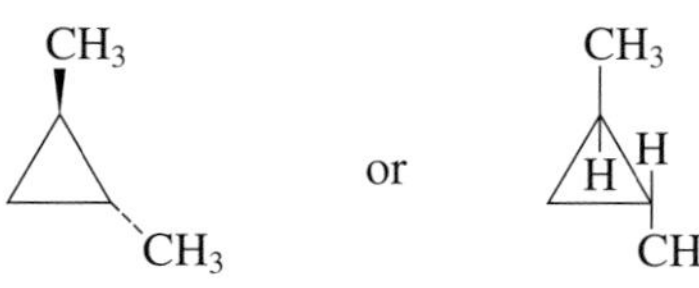

trans-1,2-Dimethylcyclopropane

Solution

a. *Cis–trans* isomerism is not possible because we do not have two substituents on the ring.
b. *Cis–trans* isomerism is not possible. We have two substituents on the ring, but they are on the same carbon atom. Each of two different carbons must bear substituents.
c. *Cis–trans* isomerism does exist.

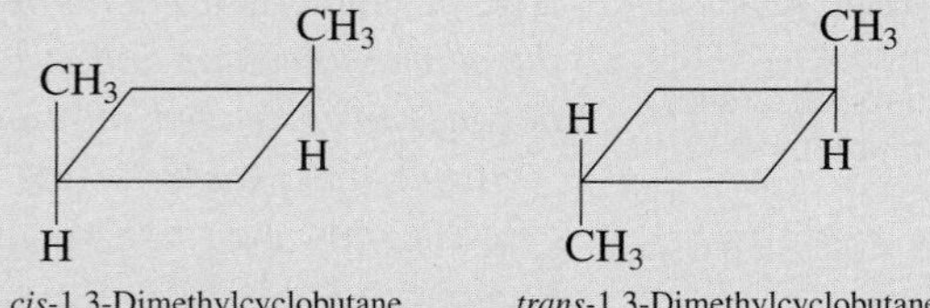

cis-1,3-Dimethylcyclobutane *trans*-1,3-Dimethylcyclobutane

d. *Cis–trans* isomerism does exist.

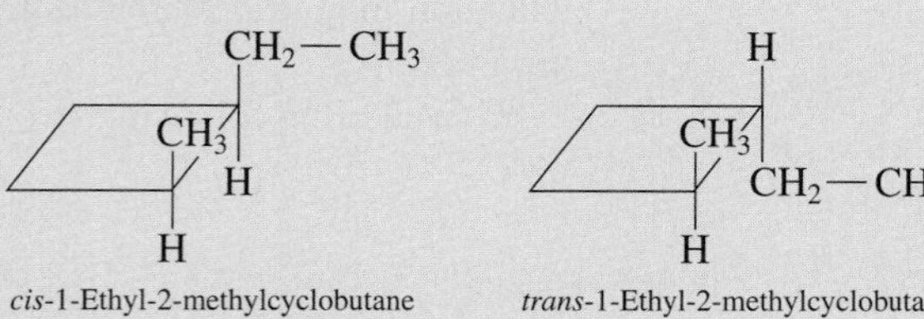

cis-1-Ethyl-2-methylcyclobutane *trans*-1-Ethyl-2-methylcyclobutane

Practice Exercise 12.6

Determine whether *cis–trans* isomerism is possible for each of the following cycloalkanes. If so, then draw structural formulas for the *cis* and *trans* isomers.

a. 1-Ethyl-1-methylcyclopentane
b. Ethylcyclohexane
c. 1,4-Dimethylcyclohexane
d. 1,1-Dimethylcyclooctane

• *Answers:* **a.** not possible; **b.** not possible;

c.

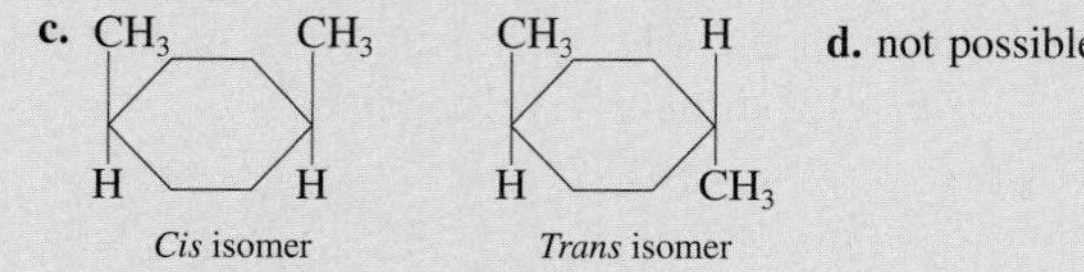

Cis isomer *Trans* isomer

d. not possible

12.14 Sources of Alkanes and Cycloalkanes

Natural gas and petroleum (oil) constitute the largest and most important sources of alkanes and cycloalkanes. Deposits of these substances are usually associated with underground dome-shaped rock formations (Figure 12.7). When a hole is drilled into such a rock formation, it is possible to recover some of the trapped hydrocarbons. Petroleum and natural gas do not occur in the earth in the form of liquid pools but rather are dispersed throughout a porous rock formation.

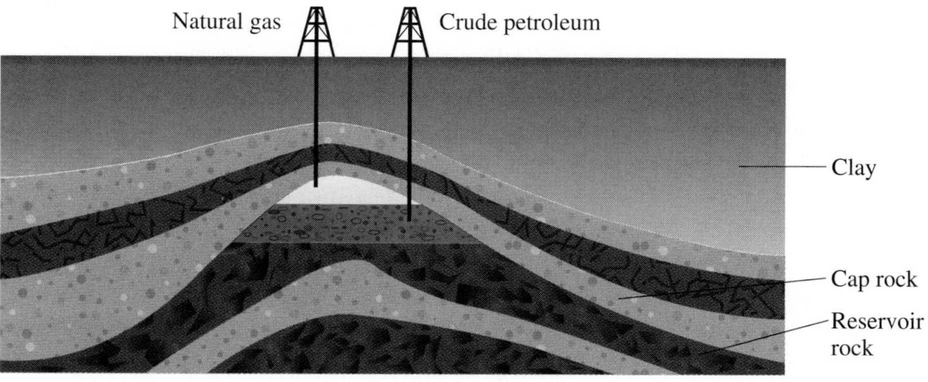

Figure 12.7
A rock formation such as this is necessary for the accumulation of petroleum and natural gas.

An oil rig pumping oil from an underground rock formation.

• The word *petroleum* comes from the Latin *petra,* which means "rock," and *oleum,* which means "oil."

Unprocessed natural gas contains 50%–90% methane, 1%–10% ethane, and up to 8% higher-molecular-mass alkanes (predominantly propane and butanes). The higher alkanes found in crude natural gas are removed prior to release of the gas into the pipeline distribution systems. Because the removed alkanes can be liquefied by the use of moderate pressure, they are stored as liquids under pressure in steel cylinders and are marketed as bottled gas.

Crude petroleum is a complex mixture of hydrocarbons (both cyclic and acyclic) that can be separated into useful fractions through refining. During refining, the physical separation of the crude into component fractions is accomplished by fractional distillation, a process that takes advantage of boiling-point differences between the components of the crude petroleum. Each fraction contains hydrocarbons within a specific boiling-point range. The fractions obtained from a typical fractionation process are shown in Figure 12.8.

12.15 Physical Properties of Alkanes and Cycloalkanes

In this section, we consider a number of generalizations about the physical properties of alkanes and cycloalkanes.

1. *Alkanes and cycloalkanes are insoluble in water.* Water molecules are polar, and alkane and cycloalkane molecules are nonpolar. Molecules of unlike

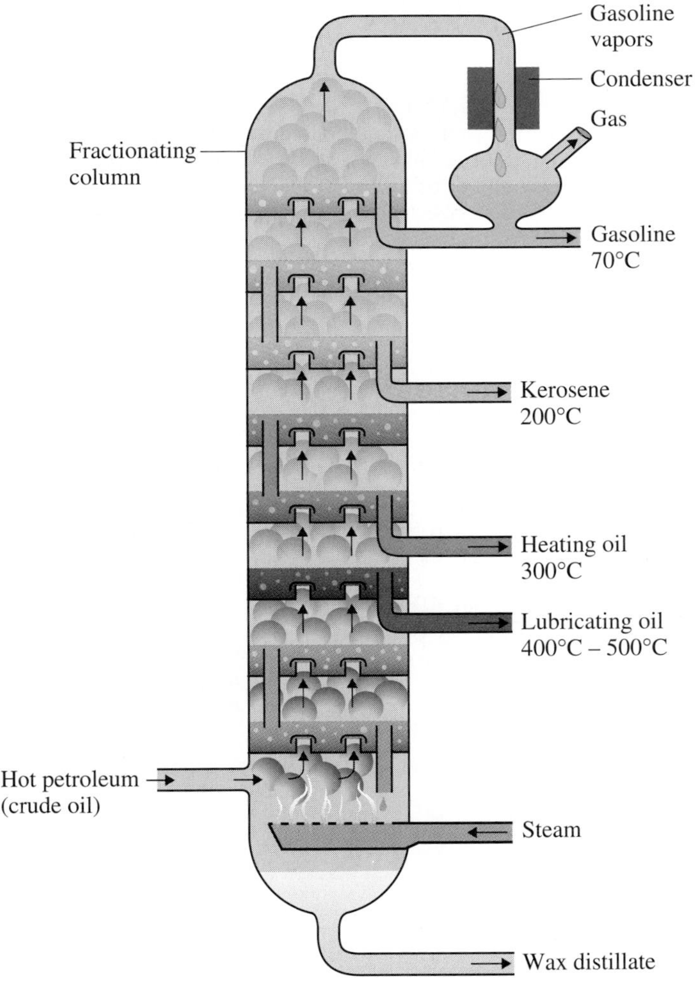

Figure 12.8
The complex hydrocarbon mixture present in petroleum is separated into simpler mixtures by means of a fractionating column.

The insolubility of alkanes in water is used to advantage by many plants, which produce unbranched long-chain alkanes that serve as protective coatings on leaves and fruits. Such protective coatings minimize water loss for plants. Apples can be "polished" because of the long-chain alkane coating on their skin, which involves the unbranched alkanes $C_{27}H_{56}$ and $C_{29}H_{60}$. The leaf wax of cabbage and broccoli is mainly unbranched $C_{29}H_{60}$.

polarity have limited solubility in one another (Section 8.4). The water insolubility of alkanes makes them good preservatives for metals. They prevent water from reaching the metal surface and causing corrosion.

2. *Alkanes and cycloalkanes have densities lower than that of water.* Alkane and cycloalkane densities fall in the range 0.6 g/mL to 0.8 g/mL, compared with water's density of 1.0 g/mL. When alkanes and cycloalkanes are mixed with water, two layers form (because of insolubility), with the hydrocarbon layer on top (because of its lower density). This density difference between alkanes/cycloalkanes and water explains why oil spills in aqueous environments spread so quickly. The *floating* oil follows the movement of the water.
3. *The boiling points of continuous-chain alkanes and cycloalkanes increase with an increase in carbon chain length or ring size.* For continuous-chain alkanes, the boiling point increases roughly 30°C for every carbon atom added to the chain. This trend, shown in Figure 12.9, is the result of increasing London force strength (Section 7.13). London forces become stronger as molecular surface area increases. Short, continuous-chain alkanes (1 to 4 carbon atoms) are gases at room temperature. Continuous-chain alkanes containing 5 to 17 carbon atoms are liquids, and alkanes that have carbon chains longer than this are solids at room temperature.

Branching on a carbon chain lowers the boiling point of an alkane. A comparison of the boiling points of unbranched alkanes and their 2-methyl-branched isomers is given in Figure 12.9. Branched alkanes are more compact, with smaller surface areas than their straight-chain isomers.

Cycloalkanes have higher boiling points than their noncyclic counterparts with the same number of carbon atoms (Figure 12.9). These differences are due in large part to cyclic systems having more rigid and more symmetrical

Figure 12.9
Trends in normal boiling points for continuous-chain alkanes, 2-methyl branched alkanes, and unsubstituted cycloalkanes as a function of the number of carbon atoms present. Although you are not expected to memorize specific boiling points and specific melting points of hydrocarbons, you should develop a feel for general trends in these values. For example, in a series of alkanes, melting point increases as carbon chain length increases.

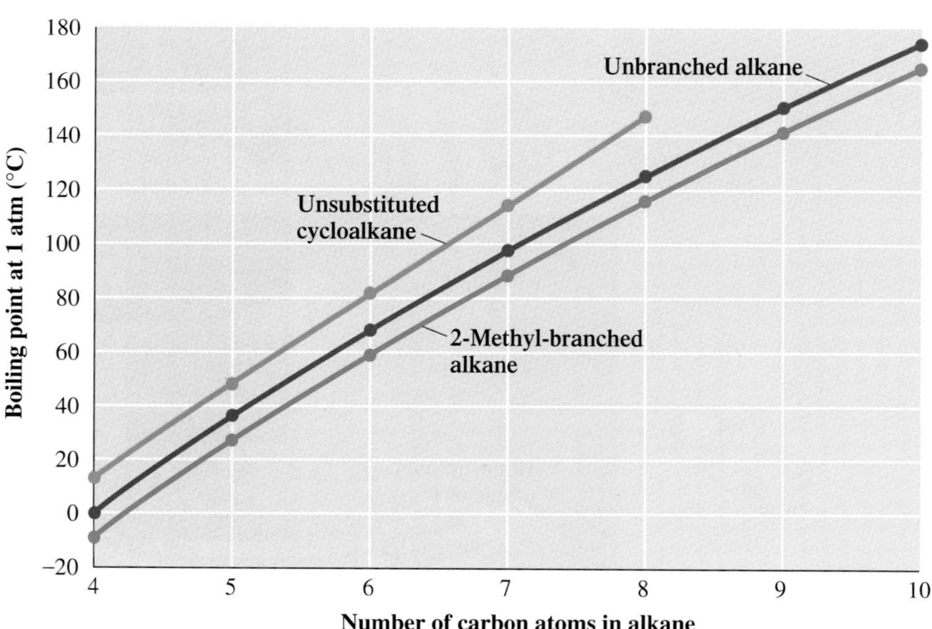

Chemical CONNECTIONS

12.2 The Physiological Effects of Alkanes

The simplest alkanes (methane, ethane, propane, and butane) are gases at room temperature and pressure. Methane and ethane are difficult to liquefy, so they are usually handled as compressed gases. Propane and butane are easily liquefied at room temperature under a moderate pressure. They are stored in low-pressure cylinders in a liquefied form. These four gases are colorless, odorless, and nontoxic, and they have limited physiological effects. The danger in inhaling them lies in potential suffocation due to lack of oxygen. The immediate major danger associated with a natural gas leak is the potential formation of an explosive air–alkane mixture rather than the formation of a toxic air–alkane mixture.

The C_5 to C_8 alkanes, of which there are many isomeric forms, are free-flowing, nonpolar, volatile liquids. They are the primary constituents of gasoline. These compounds are not particularly toxic, but gasoline should not be swallowed because (1) some of the additives present are harmful and (2) liquid alkanes can damage lung tissue because of physical rather than chemical effects. Physical effects include the dissolving of lipid molecules of cell membranes (see Chapter 19), causing pneumonia-like symptoms. Liquid alkanes can also affect the skin for related reasons. These alkanes dissolve natural body oils, causing the skin to dry out. (This "drying out" effect is easily noticed when paint thinner, a mixture of hydrocarbons, is used to remove paint from the hands.)

In direct contrast to liquid-state alkanes, solid-state alkanes are used to protect the skin. Pharmaceutical-grade *petrolatum* and *mineral oil* (also called liquid petrolatum), obtained as products from petroleum distillation, have such a function. Petrolatum is a mixture of C_{25} to C_{30} alkanes, and mineral oil involves alkanes in the C_{18}-to-C_{24} range.

Petrolatum (Vaseline is a well-known brand name) is a semi-solid hydrocarbon mixture that is useful both as a skin softener and as a skin protector. Many moisturizing hand lotions and some medicated salves contain petrolatum. Neither water nor water solutions (for example, urine) will penetrate protective petrolatum coatings. This explains why petrolatum products protect a baby's bottom from diaper rash.

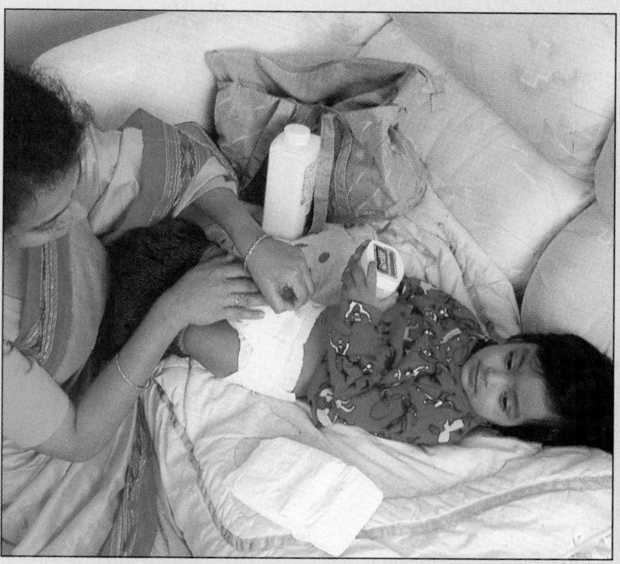

A semi-solid alkane mixture, such as Vaseline, is useful as a skin protector since neither water nor water solutions will penetrate a coating of it. Here, Vaseline is applied to a baby's bottom as a protection against diaper rash.

Mineral oil is often used to replace natural skin oils washed away by frequent bathing and swimming. Too much mineral oil, however, can be detrimental; it will dissolve nonpolar skin materials. Mineral oil has some use as a laxative; it effectively softens and lubricates hard stools. When taken by mouth, it passes through the gastrointestinal tract unchanged and is excreted chemically intact. Loss of fat-soluble vitamins (A, D, E, and K) can occur if mineral oil is consumed while these vitamins are in the digestive tract. Using a mineral oil enema instead avoids this drawback.

structures. Cyclopropane and cyclobutane are gases at room temperature, and cyclopentane through cyclooctane are liquids at room temperature.

The alkanes and cycloalkanes whose boiling points are compared in Figure 12.9 constitute *homologous series* of organic compounds. In a homologous series, the members differ structurally only in the number of —CH_2— groups present. Members exhibit gradually changing physical properties and usually have very similar chemical properties.

The existence of homologous series of organic compounds gives organization to organic chemistry in the same way that the periodic table gives organization to the chemistry of the elements. Knowing something about a few members of a homologous series usually enables us to deduce the properties of other members in the series.

Unbranched Alkanes			
C_1	C_3	C_5	C_7
C_2	C_4	C_6	C_8

Unsubstituted Cycloalkanes			
	C_3	C_5	C_7
	C_4	C_6	C_8

☐ Gas ☐ Liquid

◀ A physical-state summary for unbranched alkanes and unsubstituted cycloalkanes at room temperature and pressure.

12.16 Chemical Properties of Alkanes and Cycloalkanes

• The term *paraffins* is an older name for the alkane family of compounds. This name comes from the Latin *parum affinis,* which means "little activity." That is a good summary of the general chemical properties of alkanes.

Alkanes are the least reactive type of organic compound. They can be heated for long periods of time in strong acids and bases with no appreciable reaction. Strong oxidizing agents and reducing agents have little effect on alkanes.

Alkanes are not absolutely unreactive. Two important reactions that they undergo are combustion, which is reaction with oxygen, and halogenation, which is reaction with halogens.

• Combustion

A **combustion reaction** *is a reaction between a substance and oxygen (usually from air) that proceeds with the evolution of heat and light (usually as a flame).* Alkanes readily undergo combustion when ignited. When sufficient oxygen is present to support total combustion, carbon dioxide and water are the products.

$$CH_4 + 2O_2 \longrightarrow CO_2 + 2H_2O + \text{energy}$$

$$2C_6H_{14} + 19O_2 \longrightarrow 12CO_2 + 14H_2O + \text{energy}$$

The exothermic nature (Section 9.5) of alkane combustion reactions explains the extensive use of alkanes as fuels. Natural gas is predominantly methane. Propane is used for home heating. Butane fuels portable camping stoves. Gasoline is a complex mixture of many alkanes and other types of hydrocarbons.

Incomplete combustion can occur if insufficient oxygen is present during the combustion process. When this is the case, some carbon monoxide (CO) and/or elemental carbon are reaction products along with carbon dioxide (CO_2). In a chemical laboratory setting, incomplete combustion is often observed. The appearance of deposits of carbon black (soot) on the bottom of glassware is physical evidence that incomplete combustion is occurring. The problem is that the air-to-fuel ratio for the Bunsen burner is not correct. It is too rich; it contains too much fuel and not enough oxygen (air).

• Halogenation

Propane fuel tank on a home barbecue unit.

The **halogens** are the elements in Group VIIA of the periodic table: fluorine (F_2), chlorine (Cl_2), bromine (Br_2), and iodine (I_2) (Section 3.4). A **halogenation reaction** *is a reaction between a substance and a halogen in which one or more halogen atoms are incorporated into molecules of the substance.*

Halogenation of an alkane produces a hydrocarbon derivative in which one or more halogen atoms have been substituted for hydrogen atoms. An example of an alkane halogenation reaction is

$$CH_3\text{—}CH_3 + Br_2 \xrightarrow[\text{light}]{\text{Heat or}} CH_3\text{—}CH_2\text{—}Br + HBr$$

Alkane halogenation is an example of a substitution reaction, a type of reaction that occurs often in organic chemistry. A **substitution reaction** *is a reaction in which part of a small reacting molecule replaces an atom or a group of atoms on a hydrocarbon or hydrocarbon derivative.* A diagrammatic representation of a substitution reaction is shown in Figure 12.10.

A *general* equation for the substitution of a single halogen atom for one of the hydrogen atoms of an alkane is

$$\underset{\text{Alkane}}{R\text{—}H} + \underset{\text{Halogen}}{X_2} \xrightarrow[\text{light}]{\text{Heat or}} \underset{\text{Halogenated alkane}}{R\text{—}X} + \underset{\text{Hydrogen halide}}{H\text{—}X}$$

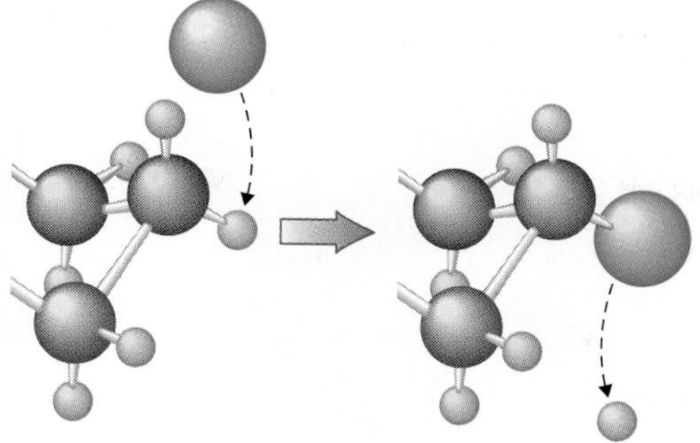

Figure 12.10
In an alkane substitution reaction, an incoming atom or group of atoms (represented by the orange sphere) replaces a hydrogen atom in the alkane molecule.

• Occasionally, it is useful to represent alkyl groups in a nonspecific way. The symbol R is used for this purpose. Just as *city* is a generic term for Chicago, New York, or San Francisco, the symbol R is a generic designation for any alkyl group. The symbol R comes from the German word *radikal*, which means, in a chemical context, "molecular fragment."

Note the following features of this general equation:

1. The notation R—H is a general formula for an alkane. R— is the general symbol for an alkyl group. Addition of a hydrogen atom to an alkyl group produces the parent hydrocarbon of the alkyl group.
2. The notation R—X on the product side is the general formula for a halogenated alkane. X is the general symbol for a halogen atom.
3. Reaction conditions are noted by placing these conditions on the equation arrow that separates reactants from products. Halogenation of an alkane requires the presence of heat or light.

In halogenation of an alkane, the alkane is said to undergo *fluorination, chlorination, bromination,* or *iodination,* depending on the identity of the halogen reactant. Chlorination and bromination are the two widely used alkane halogenation reactions. Fluorination reactions generally proceed too quickly to be useful, and iodination reactions go too slowly.

Halogenation usually results in the formation of a mixture of products rather than a single product. More than one product results because more than one hydrogen atom on an alkane can be replaced with halogen atoms. To illustrate this concept, let us consider the chlorination of methane, the simplest alkane.

Methane and chlorine, when heated to a high temperature or in the presence of light, react as follows:

$$CH_4 + Cl_2 \xrightarrow[\text{light}]{\text{Heat or}} CH_3Cl + HCl$$

The reaction does not stop at this stage, however, because the chlorinated methane product can react with additional chlorine to produce polychlorinated products.

$$CH_3Cl + Cl_2 \xrightarrow[\text{light}]{\text{Heat or}} CH_2Cl_2, + HCl$$

$$CH_2Cl_2 + Cl_2 \xrightarrow[\text{light}]{\text{Heat or}} CHCl_3 + HCl$$

$$CHCl_3 + Cl_2 \xrightarrow[\text{light}]{\text{Heat or}} CCl_4 + HCl$$

By controlling the reaction conditions and the ratio of chlorine to methane, it is possible to *favor* formation of one or another of the possible chlorinated methane products.

The chemical properties of cycloalkanes are similar to those of alkanes. Cycloalkanes readily undergo combustion as well as chlorination and bromination. With unsubstituted cycloalkanes, monohalogenation produces a single product, because all hydrogen atoms present in the cycloalkane are equivalent to one another.

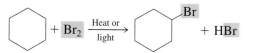

$$\text{cyclohexane} + Br_2 \xrightarrow[\text{light}]{\text{Heat or}} \text{bromocyclohexane (Br)} + HBr$$

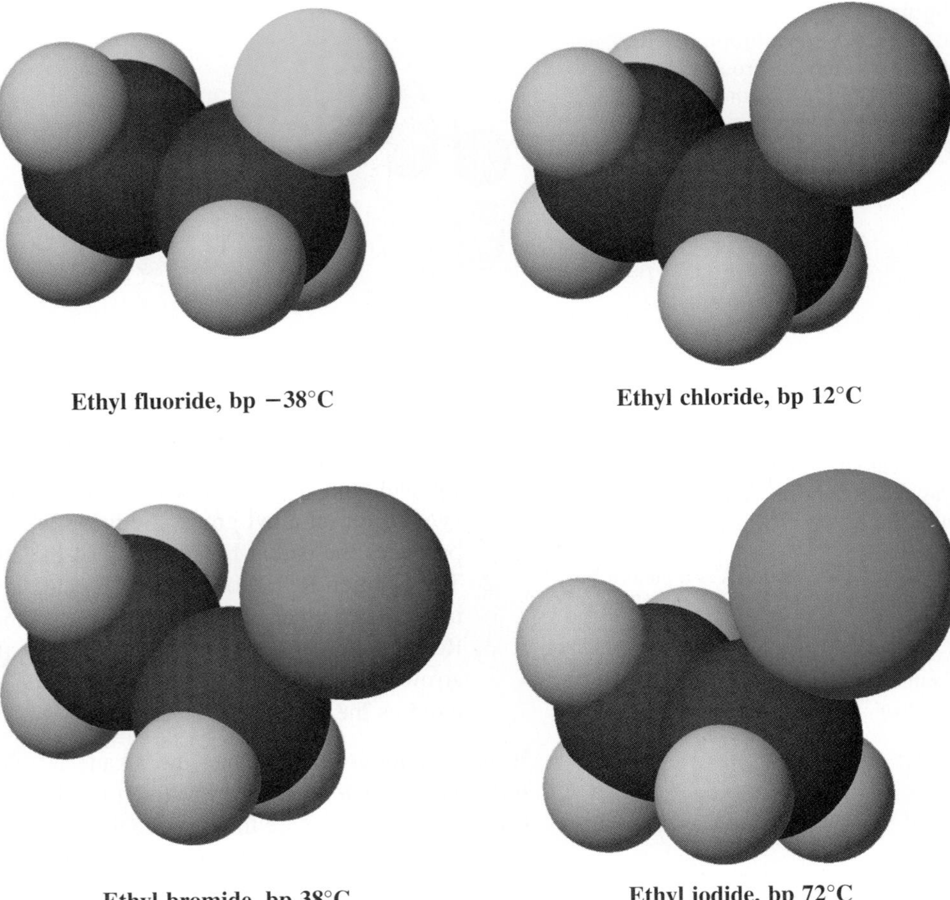

Figure 12.11
Space-filling models of the four ethyl halides. Melting point increases as the size of the halogen atom increases.

Figure 12.11 shows space-filling models of the four ethyl halides. Note the large differences in size associated with the various halogen atoms. Fluorine is the smallest and iodine the largest.

The accompanying Chemistry at a Glance summarizes the properties of alkanes and cycloalkanes.

12.17 Nomenclature and Properties of Halogenated Alkanes

• The contrast between IUPAC and common names for halogenated hydrocarbons is as follows:

IUPAC (one word)
haloalkane
chloromethane
Common (two words)
alkyl halide
methyl chloride

• An alternative designation for a halogenated alkane is *alkyl halide.*

• Halogenated alkane boiling points are generally higher than those of the corresponding alkane. An important factor contributing to this effect is the polarity of carbon–halogen bonds, which results in increased dipole–dipole interactions.

A **halogenated alkane** *is an alkane derivative in which one or more halogen atoms are present.* Similarly, a **halogenated cycloalkane** *is a cycloalkane derivative in which one or more halogen atoms are present.* Produced by halogenation reactions (Section 12.16), these two types of compounds represent the first class of hydrocarbon derivatives (Section 12.3) that we formally consider in this text.

Nomenclature of Halogenated Alkanes

The IUPAC rules for naming halogen derivatives of alkanes are similar to those for naming branched alkanes, with the following modifications:

1. Halogen atoms, treated as substituents on a carbon chain, are called *fluoro-, chloro-, bromo-,* and *iodo-*.
2. When a carbon chain bears both a halogen and an alkyl substituent, the two substituents are considered of equal rank in determining the numbering system for the chain. The chain is numbered from the end closer to a substituent, whether it be a *halo-* or an alkyl group.
3. Alphabetical priority determines the order in which all substituents present are listed.

Properties of Alkanes and Cycloalkanes

CHEMISTRY AT A GLANCE

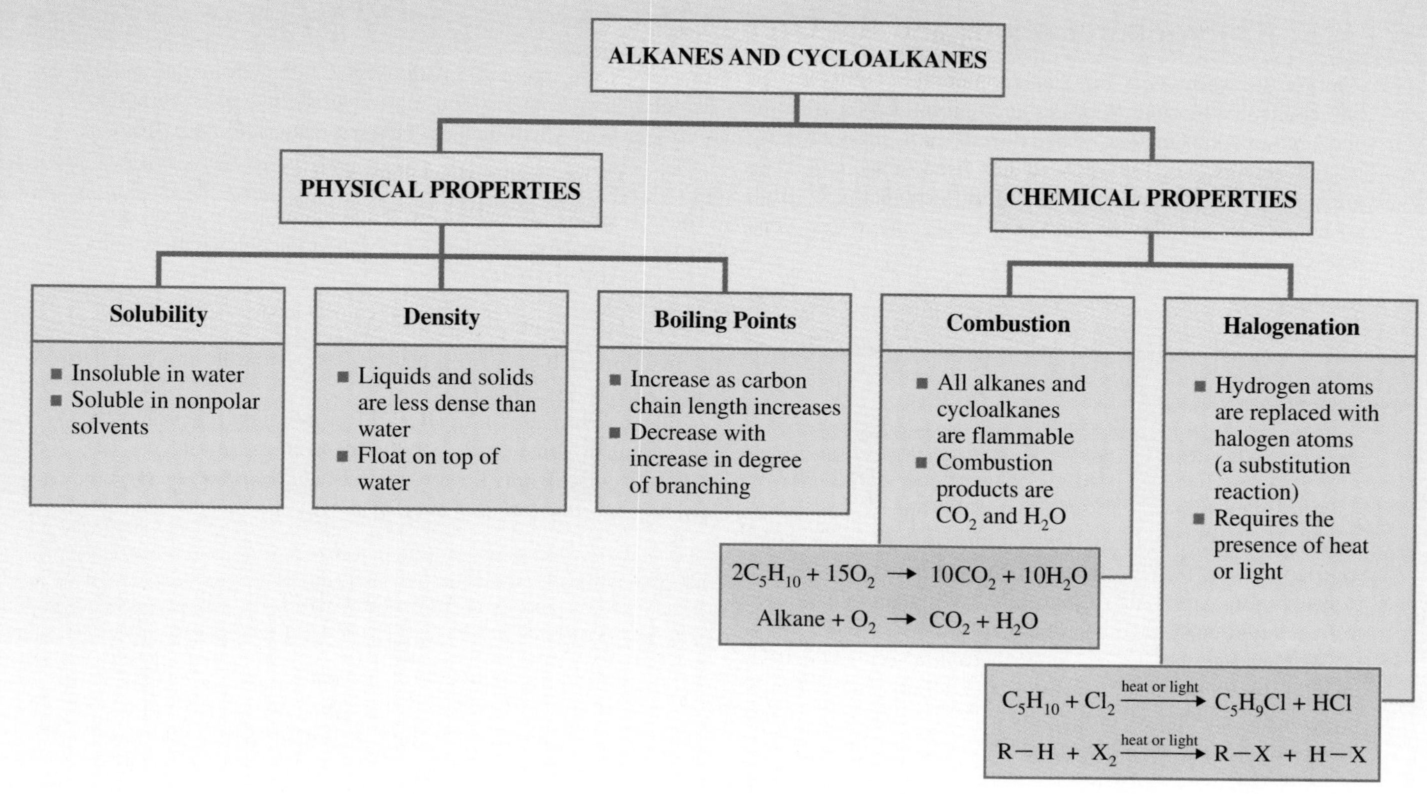

• Some halogenated alkanes have densities that are greater than that of water, a situation not common for organic compounds. Chloroalkanes containing two or more chlorine atoms, bromoalkanes, and iodoalkanes are all more dense than water.

The following names are derived using these rule adjustments.

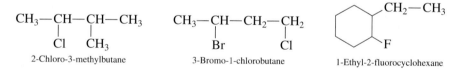

$CH_3{-}CH(Cl){-}CH(CH_3){-}CH_3$
2-Chloro-3-methylbutane

$CH_3{-}CH(Br){-}CH_2{-}CH_2(Cl)$
3-Bromo-1-chlorobutane

(cyclohexane ring with $CH_2{-}CH_3$ and F on adjacent carbons)
1-Ethyl-2-fluorocyclohexane

A common (non-IUPAC) system also exists for naming simple halogenated alkanes. These common names have two parts. The first part is the name of the hydrocarbon portion of the molecule (the alkyl group). The second part (as a separate word) identifies the halogen portion, which is named as though it were an ion (chloride, bromide, and so on), even though no ions are present (all bonds are covalent bonds). The following examples contrast the IUPAC names and the common names (in parentheses).

$CH_3{-}CH_2{-}Cl$
Chloroethane (ethyl chloride)

$CH_3{-}CH_2{-}CH_2{-}Br$
1-Bromopropane (propyl bromide)

$CH_3{-}CH(Cl){-}CH_3$
2-Chloropropane (isopropyl chloride)

Table 12.5
Additional Common Names for Selected Polyhalogenated Methanes

CH_2Cl_2	methylene chloride
$CHCl_3$	chloroform
CCl_4	carbon tetrachloride
CCl_3F	Freon-11
CCl_2F_2	Freon-12

A number of polyhalogenated methanes have acquired additional common names that are usually not clearly related to their structures. Table 12.5 gives five important examples of this additional nomenclature. The compounds Freon-11 and Freon-12, the last two entries in Table 12.5, are examples of chlorofluorocarbons (CFCs). CFCs are synthetic compounds that have been heavily used as refrigerants and as air conditioning chemicals. We now know that CFCs are factors in the destruction of stratospheric (high-altitude) ozone.

Chemical CONNECTIONS

12.3 Chloroflurocarbons and the Ozone Layer

Chlorofluorocarbons (CFCs) are compounds composed of the elements chlorine, fluorine, and carbon. CFCs are synthetic compounds that have been developed primarily for use as refrigerants. The two most widely used of the CFCs are trichlorofluoromethane and dichlorodifluoromethane. Both of these compounds are marketed under the trade name Freon.

```
     Cl                 F
     |                  |
 Cl—C—F            Cl—C—F
     |                  |
     Cl                 Cl
```

Trichlorofluoromethane (Freon-11) Dichlorodifluoromethane (Freon-12)

Freon-11 and Freon-12 possess ideal properties for use as a refrigerant gas. Both are inert, nontoxic, and easily compressible. Prior to their development, ammonia was used in refrigeration. Ammonia is toxic, and leaking ammonia-based refrigeration units have been fatal.

We now know that CFCs contribute to a serious environmental problem: destruction of the stratospheric (high-altitude) ozone that we commonly call the ozone layer. Once released into the atmosphere, CFCs persist for long periods without reaction. Consequently, they slowly drift upward in the atmosphere, finally reaching the stratosphere.

It is in the stratosphere, the location of the "ozone layer," that environmental problems occur. At these high altitudes, the CFCs are exposed to ultraviolet light (from the sun), which activates them. The ultraviolet light breaks carbon–chlorine bonds within the CFCs, releasing chlorine atoms.

$$CCl_2F_2 + \text{ultraviolet light} \longrightarrow CClF_2 + Cl$$

The Cl atoms so produced (called atomic chlorine) are extremely reactive species. One of the molecules with which they react is ozone (O_3).

$$Cl + O_3 \longrightarrow ClO + O_2$$

A reaction such as this upsets the O_3–O_2 equilibrium in the stratosphere (Section 9.6).

Recent international treaties (this is a worldwide problem) limit and in some cases ban future production and use of CFCs. Replacements for the phased-out CFCs will be HFCs (hydrogenfluorocarbons) such as

```
    F  F
    |  |
 F—C—C—H
    |  |
    F  H
```

1,1,1,2-Tetrafluoroethane

Haloalkanes with some carbon–hydrogen bonds are more reactive than CFCs and are generally destroyed at lower altitudes before they reach the stratosphere. Unfortunately, however, their refrigeration properties are not as good as those of the CFCs.

Concepts to Remember

Carbon atom bonding characteristics. Carbon atoms in organic compounds must have four bonds.

Types of hydrocarbons. Hydrocarbons, binary compounds of carbon and hydrogen, are of two types: saturated and unsaturated. In saturated hydrocarbons, all carbon–carbon bonds are single bonds. Unsaturated hydrocarbons have one or more carbon–carbon multiple bonds—double bonds, triple bonds, or both.

Alkanes. Alkanes are saturated hydrocarbons in which the carbon atom arrangement is that of an unbranched or branched chain. The formulas of all alkanes can be represented by the general formula C_nH_{2n+2}, where n is the number of carbon atoms present.

Structural formulas. Structural formulas are two-dimensional representations of the arrangement of the atoms in molecules. These formulas give complete information about the arrangement of the atoms in a molecule but not the spatial orientation of the atoms. Two types of structural formulas are commonly encountered: expanded and condensed.

Structural isomerism. Structural isomers are two or more compounds that have the same molecular formula but different structural formulas—that is, different arrangements of atoms within the molecule.

Conformations. Conformations are differing orientations of the same molecule made possible by free rotation about single bonds in the molecule.

Alkane nomenclature. The IUPAC name for an alkane is based on the longest continuous chain of carbon atoms in the molecule. A group of carbon atoms attached to the chain is an alkyl group. Both the position and the identity of the alkyl group are prefixed to the name of the longest carbon chain.

Cycloalkanes. Cycloalkanes are saturated hydrocarbons in which at least one cyclic arrangement of carbon atoms is present. The formulas of all cycloalkanes can be represented by the general formula C_nH_{2n}, where n is the number of carbon atoms present.

Cycloalkane nomenclature. The IUPAC name for a cycloalkane is obtained by placing the prefix *cyclo-* before the alkane name that corresponds to the number of carbon atoms in the ring. Alkyl groups attached to the ring are located by using a ring-numbering system.

***Cis–trans* isomerism.** For certain disubstituted cycloalkanes, *cis–trans* isomers exist. *Cis–trans* isomers are compounds that have the same molecular and structural formulas but different arrangements of atoms in space because of restricted rotation about bonds.

Natural sources of saturated hydrocarbons. Natural gas and petroleum are the largest and most important natural sources of both alkanes and cycloalkanes.

Physical properties of saturated hydrocarbons. Saturated hydrocarbons are not soluble in water and have lower densities than water. Melting and boiling points increase with increasing carbon chain length or ring size.

Chemical properties of saturated hydrocarbons. Two important reactions that saturated hydrocarbons undergo are combustion and halogenation. In combustion, saturated hydrocarbons burn in air to produce CO_2 and H_2O. Halogenation is a substitution reaction in which one or more hydrogen atoms of the hydrocarbon are replaced by halogen atoms.

Halogenated alkanes. Halogenated alkanes are hydrocarbon derivatives in which one or more halogen atoms have replaced hydrogen atoms of the alkane.

Halogenated alkane nomenclature. Halogenated alkanes are named by using the rules that apply to branched-chain alkanes, with halogen substituents being treated the same as alkyl groups.

Key Reactions and Equations

1. Combustion (rapid reaction with O_2) of alkanes (Section 12.16)

$$\text{Alkane} + O_2 \longrightarrow CO_2 + H_2O$$

2. Halogenation of alkanes (Section 12.16)

$$R{-}H + X_2 \xrightarrow[\text{light}]{\text{Heat or}} R{-}X + H{-}X$$

Key Terms

Alkane (12.4)
Alkyl group (12.8)
Branched-chain alkane (12.6)
***Cis* isomer** (12.13)
***Cis–trans* isomers** (12.13)
Combustion reaction (12.16)
Condensed structural formula (12.5)
Conformations (12.7)
Continuous-chain alkane (12.6)
Cycloalkane (12.11)
Expanded structural formula (12.5)
Halogen (12.16)
Halogenated alkane (12.17)
Halogenated cycloalkane (12.17)
Halogenation reaction (12.16)
Hydrocarbon (12.3)
Hydrocarbon derivative (12.3)
Line-angle drawing (12.10)
Organic chemistry (12.1)
Primary carbon atom (12.9)
Quaternary carbon atom (12.9)
Saturated hydrocarbon (12.3)
Secondary carbon atom (12.9)
Structural isomers (12.6)
Substituent (12.8)
Substitution reaction (12.16)
Tertiary carbon atom (12.9)
***Trans* isomer** (12.13)
Unsaturated hydrocarbon (12.3)

Exercises and Problems

The members of each pair of problems in this section test similar material.

Hydrocarbon Terminology and Notation (Sections 12.3–12.5)

12.1 Classify each of the following hydrocarbons as saturated or unsaturated.

a.
```
   H  H
   |  |
H—C—C—H
   |  |
   H  H
```

b.
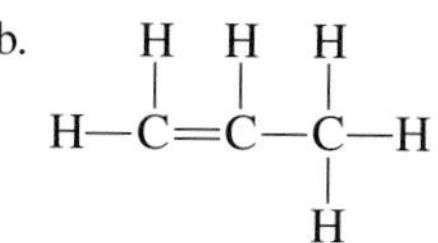

c.
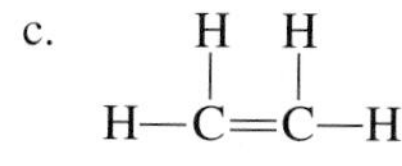

d.
```
            H  H
            |  |
H—C≡C—C—C—H
            |  |
            H  H
```

12.2 Classify each of the following hydrocarbons as saturated or unsaturated.

a. H—C≡C—H

b.
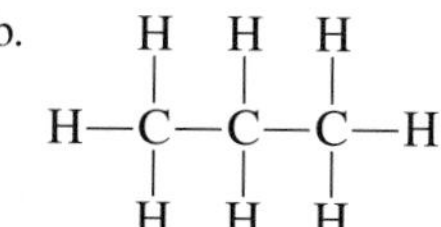

c.
```
   H  H  H  H
   |  |  |  |
H—C=C—C=C—H
```

d.
```
   H  H  H  H
   |  |  |  |
H—C—C—C—C—H
   |  |  |  |
   H  H  H  H
```

12.3 Using the general formula for an alkane, derive the following for specific alkanes.

a. Number of hydrogen atoms present when 8 carbon atoms are present

b. Number of carbon atoms present when 10 hydrogen atoms are present

c. Number of carbon atoms present when 41 total atoms are present

d. Total number of covalent bonds present in the molecule when 7 carbon atoms are present

12.4 Using the general formula for an alkane, derive the following for specific alkanes.

a. Number of carbon atoms present when 14 hydrogen atoms are present

b. Number of hydrogen atoms present when 6 carbon atoms are present

c. Number of hydrogen atoms present when 32 total atoms are present

d. Total number of covalent bonds present in the molecule when 16 hydrogen atoms are present

12.5 Convert each of the following expanded structural formulas into a condensed structural formula.

a.
```
   H  H  H  H
   |  |  |  |
H—C—C—C—C—H
   |  |  |  |
   H  H  H  H
```

b.
```
   H  H  H  H  H
   |  |  |  |  |
H—C—C—C—C—C—H
   |  |  |  |  |
   H  H  |  H  H
        H—C—H
           |
           H
```

c.
```
                 H
                 |
              H—C—H
   H  H  H  H   |  H
   |  |  |  |   |  |
H—C—C—C—C—C—C—H
   |  |  |  |   |  |
   H  H  |  H   H  H
        H—C—H
           |
           H
```

d.
```
   H  H  H  H  H
   |  |  |  |  |
H—C—C—C—C—C—H
   |  |  |  |  |
   H  H  |  H  H
        H—C—H
           |
        H—C—H
           |
           H
```

12.6 Convert each of the following expanded structural formulas into a condensed structural formula.

a.
```
   H  H  H  H  H  H
   |  |  |  |  |  |
H—C—C—C—C—C—C—H
   |  |  |  |  |  |
   H  H  H  H  H  H
```

b.
```
   H  H  H  H
   |  |  |  |
H—C—C—C—C—H
   |  |  |  |
   H  |  H  H
  H—C—H
     |
     H
```

c.
```
   H  H     H     H  H
   |  |     |     |  |
H—C—C————C————C—C—H
   |  |     |     |  |
   H  |     H     |  H
  H—C—H      H—C—H
     |           |
     H           H
```

d.
```
   H  H  H  H  H  H
   |  |  |  |  |  |
H—C—C—C—C—C—C—H
   |  |  |  |  |  |
   H  H  |  H  H  H
        H—C—H
           |
        H—C—H
           |
           H
```

12.7 The following structural formulas for alkanes are incomplete in that the hydrogen atoms attached to each carbon are not shown. Complete each of these formulas by writing in the correct number of hydrogen atoms attached to each carbon atom. That is, rewrite each of these formulas as a condensed structural formula such as $CH_3—CH_2—CH_3$.

a.
```
C—C—C—C
  |
  C
```

b.
```
C—C—C—C—C—C
  |  |  |
  C  C  C
```

c.
```
C—C—C—C—C—C
```

d.
```
  C
  |
C—C—C—C
  |
  C
```

12.8 The following structural formulas for alkanes are incomplete in that the hydrogen atoms attached to each carbon are not shown. Complete each of these formulas by writing in the correct number of hydrogen atoms attached to each carbon atom. That is, rewrite each of these formulas as a condensed structural formula such as $CH_3—CH_2—CH_3$.

a.
```
C—C—C—C—C
  |     |
  C     C
```

b.
```
  C
  |
C—C—C
  |
  C
```

c.
```
C—C—C—C—C
```

d.
```
C—C—C—C—C
  |  |
  C  C
     |
     C
```

12.9 Draw the indicated type of formula for the following alkanes.

a. The expanded structural formula for a continuous-chain alkane with the formula C_5H_{12}

b. The expanded structural formula for $CH_3—(CH_2)_6—CH_3$

c. The condensed structural formula, using parentheses for the $—CH_2—$ groups, for the continuous-chain alkane $C_{10}H_{22}$

d. The molecular formula for the alkane $CH_3—(CH_2)_4—CH_3$

12.10 Draw the indicated type of formula for the following alkanes.

a. The expanded structural formula for a continuous-chain alkane with the molecular formula C_6H_{14}

b. The condensed structural formula, using parentheses for the $—CH_2—$ groups, for the straight-chain alkane $C_{12}H_{26}$

c. The molecular formula for the alkane $CH_3—(CH_2)_6—CH_3$

d. The expanded structural formula for $CH_3—(CH_2)_3—CH_3$

Structural Isomers and Molecular Conformations (Sections 12.6 and 12.7)

12.11 For each of the following pairs of structures, determine whether they are

1. Different conformations of the same molecule
2. Different compounds that are structural isomers
3. Different compounds that are not structural isomers

a.
```
CH3—CH2—CH2—CH—CH3
             |
             CH3
```
and
```
CH3—CH—CH2—CH3
     |
     CH3
```

b. $CH_3—CH_2—CH_2—CH_2—CH_3$

and $CH_3—CH—CH_3$ (with $—CH_2—CH_3$ on the middle CH)

c. $CH_3—CH_2—CH_2—CH_3$ (terminal CH_2 bearing CH_3) and $CH_3—CH_2—CH_2—CH_3$ (terminal CH_2 bearing $CH_2—CH_3$)

$CH_3—CH_2—\underset{\displaystyle CH_3}{CH_2}$ and $CH_3—\underset{\displaystyle CH_2—CH_3}{CH_2}$

d. $CH_3—\overset{\displaystyle CH_3}{\underset{\displaystyle CH_3}{C}}—CH_3$ and $CH_3—\underset{\displaystyle CH_3}{CH}—CH_2—CH_3$

12.12 For each of the following pairs of structures, determine whether they are

1. Different conformations of the same molecule
2. Different compounds that are structural isomers
3. Different compounds that are not structural isomers

a. $CH_3—\underset{\displaystyle CH_2—CH_3}{CH}—CH_3$

and $CH_3—\underset{\displaystyle CH_3}{CH}—CH_2—CH_3$

b. $CH_3—\underset{\displaystyle CH_3}{CH}—CH_2—CH_3$

and $CH_3—CH_2—\underset{\displaystyle CH_3}{CH}—CH_3$

c. $CH_3—\underset{\displaystyle \underset{\displaystyle CH_3}{CH_2}}{CH}—CH_2—CH_3$

and $CH_3—\underset{\displaystyle CH_3}{CH}—\underset{\displaystyle CH_3}{CH}—CH_3$

d. $CH_3—CH_2—CH_2—CH_2—CH_2—CH_3$

and $CH_3—\overset{\displaystyle CH_3}{\underset{\displaystyle CH_3}{C}}—CH_3$

IUPAC Nomenclature for Alkanes (Section 12.8)

12.13 The first step in naming an alkane is to identify the longest continuous chain of carbon atoms. For each of the following skeletal carbon arrangements, how many carbon atoms are present in the longest continuous chain?

a. $C—C—\underset{\displaystyle \underset{\displaystyle C}{C}}{C}—C—\overset{\displaystyle C}{C}—C—C$

b. $C—\underset{\displaystyle \underset{\displaystyle C}{C}}{C}—C—C—C—\underset{\displaystyle C}{C}$

c. $C—C—C—C$ (the fourth C bonded above to $C—C—C$ and below to $C—C—C—C$)

$$\begin{array}{l} \qquad\qquad\quad C—C—C \\ \qquad\qquad\quad | \\ C—C—C—C \\ \qquad\qquad\quad | \\ \qquad\qquad\quad C—C—C—C \end{array}$$

d. $C—C—C$ (third C bonded below to $C—C—C—C$, whose first C continues down $C—C—C$)

$$\begin{array}{l} C—C—C \\ \qquad\quad | \\ \quad C—C—C—C \\ \qquad\quad | \\ \qquad\quad C \\ \qquad\quad | \\ \qquad\quad C—C \end{array}$$

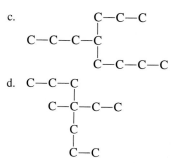

12.14 The first step in naming an alkane is to identify the longest continuous chain of carbon atoms. For each of the following skeletal carbon arrangements, how many carbon atoms are present in the longest continuous chain?

a. $C—\underset{\displaystyle \underset{\displaystyle C}{C}}{C}—C—\overset{\displaystyle C}{\underset{\displaystyle C}{C}}—C—C$

b. $\underset{\displaystyle C}{C}—C—\underset{\displaystyle \underset{\displaystyle C}{C}}{C}—C—\underset{\displaystyle C}{C}$

c.

$$\begin{array}{l} \qquad\qquad\quad C—C—C—C \\ \qquad\qquad\quad | \\ C—C—C—C \\ \qquad\qquad\quad | \\ C—C—C—C \end{array}$$

d. $C—C—\underset{\displaystyle \underset{\displaystyle \underset{\displaystyle C}{C}}{C}}{C}—C—\underset{\displaystyle \underset{\displaystyle \underset{\displaystyle C}{C}}{C}}{C}—C—C$

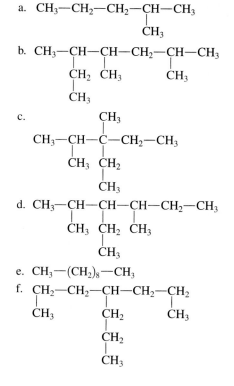

12.15 Give the IUPAC name for each of the following alkanes.

a. $CH_3—CH_2—CH_2—\underset{\displaystyle CH_3}{CH}—CH_3$

b. $CH_3—\underset{\displaystyle \underset{\displaystyle CH_3}{CH_2}}{CH}—\underset{\displaystyle CH_3}{CH}—CH_2—\underset{\displaystyle CH_3}{CH}—CH_3$

c. $CH_3—\underset{\displaystyle CH_3}{CH}—\overset{\displaystyle CH_3}{\underset{\displaystyle \underset{\displaystyle CH_3}{CH_2}}{C}}—CH_2—CH_3$

d. $CH_3—\underset{\displaystyle CH_3}{CH}—\underset{\displaystyle \underset{\displaystyle CH_3}{CH_2}}{CH}—\underset{\displaystyle CH_3}{CH}—CH_2—CH_3$

e. $CH_3—(CH_2)_8—CH_3$

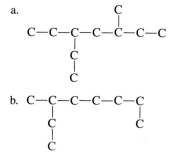

f. $\underset{\displaystyle CH_3}{CH_2}—CH_2—\underset{\displaystyle \underset{\displaystyle \underset{\displaystyle CH_3}{CH_2}}{CH_2}}{CH}—CH_2—\underset{\displaystyle CH_3}{CH_2}$

12.16 Give the IUPAC name for each of the following alkanes.

a.
```
CH3—CH—CH—CH2—CH3
    |   |
    CH3 CH3
```

b.
```
        CH3       CH3
        |         |
   CH3—C—CH2—C—CH3
        |         |
        CH3       CH3
```

c.
```
CH3—CH—CH2—CH—CH3
    |       |
    CH3     CH2
            |
            CH3
```

d.
```
CH3—CH2—CH2—CH—CH3
            |
    CH3—CH2—CH3
```

e. $CH_3—(CH_2)_7—CH_3$

f.
```
                        CH3
                        |
CH3—(CH2)3—CH—CH—(CH2)3—CH3
            |
            CH2
            |
            CH2
            |
            CH3
```

12.17 Two different carbon chains of eight atoms can be located in the following alkane.

```
                    CH2—CH2—CH3
                    |
CH3—CH2—CH2—CH2—CH—CH—CH2—CH3
                        |
                        CH3
```

Which of these chains should be used in determining the IUPAC name for the alkane? Explain your answer.

12.18 Two different carbon chains of seven atoms can be located in the following alkane.

```
CH3—CH—CH2—CH—CH2—CH—CH3
    |       |       |
    CH3     CH2     CH3
            |
            CH2—CH3
```

Which of these chains should be used in determining the IUPAC name for the alkane? Explain your answer.

12.19 Draw a condensed structural formula for each of the following alkanes.

a. 2-Methylbutane
b. 3,4-Dimethylhexane
c. 3-Ethyl-3-methylpentane
d. 2,3,4,5-Tetramethylheptane
e. 3,5-Diethyloctane
f. 4-Propylnonane

12.20 Draw a condensed structural formula for each of the following alkanes.

a. 3-Methylhexane
b. 2,4-Dimethylhexane
c. 5-Propyldecane
d. 2,3,4-Trimethyloctane
e. 3-Ethyl-3-methylheptane
f. 3,3,4,4-Tetramethylheptane

12.21 For each of the alkanes in Problem 12.19, determine (a) the number of alkyl groups present and (b) the number of substituents present.

12.22 For each of the alkanes in Problem 12.20 determine (a) the number of alkyl groups present and (b) the number of substituents present.

12.23 Explain why the name given for the following alkanes is not the correct IUPAC name. Then give the correct IUPAC name for the compound.

a. 4-Methylpentane
b. 2-Ethyl-2-methylpropane
c. 2,3,3-Trimethylbutane
d. 1,2,2-Trimethylpentane
e. 3-Methyl-4-ethylhexane
f. 2-Methyl-4-methylhexane

12.24 Explain why the name given for the following alkanes is not the correct IUPAC name. Then give the correct IUPAC name for the compound.

a. 1,6-Dimethylhexane
b. 2-Ethylpentane
c. 3,3,4-Trimethylpentane
d. 2-Ethyl-4-methylhexane
e. 3-Ethyl-4-ethylhexane
f. 3,4,5,5,6-Pentamethylhexane

Classification of Carbon Atoms (Section 12.9)

12.25 For each of the alkane structures in Problem 12.15, give the number of (a) primary, (b) secondary, (c) tertiary, and (d) quaternary carbon atoms present.

12.26 For each of the alkane structures in Problem 12.16, give the number of (a) primary, (b) secondary, (c) tertiary, and (d) quaternary carbon atoms present.

Branched-Chain Alkyl Groups (Section 12.10)

12.27 Give the name of the branched alkyl group attached to each of the following carbon chains, where the carbon chain is denoted by a horizontal line.

a.
```
———————
  |
  CH—CH3
  |
  CH3
```

b.
```
———————
   |
   CH2
   |
CH3—CH—CH3
```

c.
```
———————
   |
CH3—CH—CH3
```

d.
```
———————
  |
  CH—CH3
  |
  CH2
  |
  CH3
```

12.28 Give the name of the branched alkyl group attached to each of the following carbon chains, where the carbon chain is denoted by a horizontal line.

a.
```
———————
  |
  CH2
  |
  CH—CH3
  |
  CH3
```

b.
```
———————
   |
CH3—C—CH3
   |
   CH3
```

c. $CH_3—CH_2—CH—CH_3$ (attached through the CH carbon)

d. CH with two CH_3 groups (attached through the CH carbon)

12.29 Draw condensed structural formulas for the following branched alkanes.

a. 5-(*Sec*-butyl)decane
b. 4,4-Diisopropyloctane
c. 5-Isobutyl-2,3-dimethylnonane
d. 4-(*tert*-Butyl)-3-methylheptane

12.30 Draw condensed structural formulas for the following branched alkanes.

a. 5-Isobutylnonane
b. 4,4-Di(*sec*-butyl)decane
c. 4-(*tert*-Butyl)-3,3-diethylheptane
d. 4-Isopropyl-2,3-dimethyloctane

Cycloalkanes (Sections 12.11 through 12.13)

12.31 Using the general formula for a cycloalkane, derive the following for specific cycloalkanes.

a. Number of hydrogen atoms present when 8 carbon atoms are present
b. Number of carbon atoms present when 12 hydrogen atoms are present
c. Number of carbon atoms present when a total of 15 atoms are present
d. Number of covalent bonds present when 5 carbon atoms are present

12.32 Using the general formula for a cycloalkane, derive the following for specific cycloalkanes.

a. Number of hydrogen atoms present when 4 carbon atoms are present
b. Number of carbon atoms present when 6 hydrogen atoms are present
c. Number of hydrogen atoms present when a total of 18 atoms are present
d. Number of covalent bonds present when 8 hydrogen atoms are present

12.33 What is the molecular formula for each of the following cycloalkane molecules?

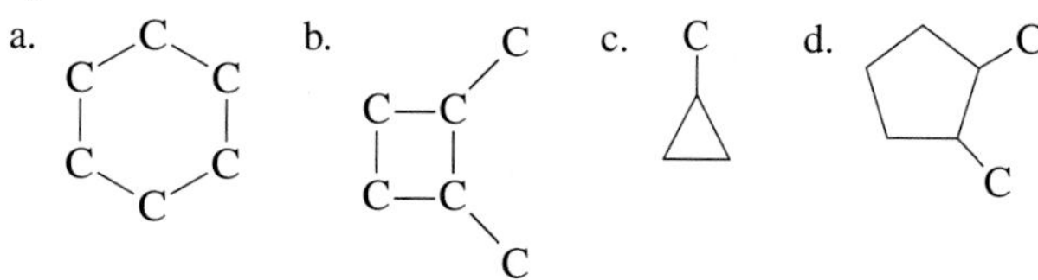

12.34 What is the molecular formula for each of the following cycloalkane molecules?

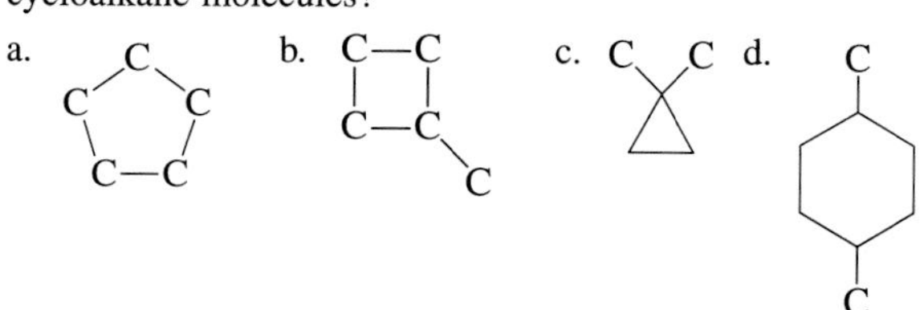

12.35 Assign an IUPAC name to each of the cycloalkanes in Problem 12.33.

12.36 Assign an IUPAC name to each of the cycloalkanes in Problem 12.34.

12.37 What is wrong with each of the following attempts to name a cycloalkane using IUPAC rules?

a. Dimethylcyclohexane
b. 3,4-Dimethylcyclohexane
c. 1-Ethylcyclobutane
d. 2-Ethyl-1-methylcyclopentane

12.38 What is wrong with each of the following attempts to name a cycloalkane using IUPAC rules?

a. Dimethylcyclopropane
b. 1-Methylcyclohexane
c. 2,5-Dimethylcyclobutane
d. 1-Propyl-2-ethylcyclohexane

12.39 Write structural formulas, with line-angle drawings denoting ring structure, for the following cycloalkanes.

a. Propylcyclobutane
b. Isopropylcyclobutane
c. *cis*-1,2-Diethylcyclohexane
d. *trans*-1-Ethyl-3-propylcyclopentane

12.40 Write structural formulas, with line-angle drawings denoting ring structure, for the following cycloalkanes.

a. Butylcyclopentane
b. Isobutylcyclopentane
c. *cis*-1,3-Diethylcyclopentane
d. *trans*-1-Ethyl-4-methylcyclohexane

12.41 Determine whether *cis–trans* isomerism is possible for each of the following cycloalkanes. If it is, then draw structural formulas for the *cis* and *trans* isomers.

a. Isopropylcyclobutane
b. 1,2-Diethylcyclopropane
c. 1-Ethyl-1-propylcyclopentane
d. 1,3-Dimethylcyclohexane

12.42 Determine whether *cis–trans* isomerism is possible for each of the following cycloalkanes. If it is, then draw structural formulas for the *cis* and *trans* isomers.

a. *Sec*-butylcyclohexane
b. 1-Ethyl-3-methylcyclobutane
c. 1,1-Dimethylcyclohexane
d. 1,3-Dipropylcyclopentane

Sources of Alkanes and Cycloalkanes (Section 12.14)

12.43 What physical property of hydrocarbons is the basis for the fractional distillation process for separating hydrocarbons?

12.44 Describe the process by which crude petroleum is separated into simpler mixtures (fractions).

Physical Properties of Alkanes and Cycloalkanes (Section 12.15)

12.45 Which member in each of the following pairs of compounds has the higher boiling point?

a. Hexane and octane
b. Cyclobutane and cyclopentane
c. Pentane and 1-methylbutane
d. Pentane and cyclopentane

12.46 Which member in each of the following pairs of compounds has the higher boiling point?

a. Methane and ethane

b. Cyclohexane and hexane

c. Butane and methylpropane

d. Pentane and 2,2-dimethylpropane

12.47 For which of the following pairs of compounds do both members of the pair have the same physical state (solid, liquid, or gas) at room temperature and pressure?

a. Ethane and hexane

b. Cyclopropane and butane

c. Octane and 3-methyloctane

d. Pentane and decane

12.48 For which of the following pairs of compounds do both members of the pair have the same physical state (solid, liquid, or gas) at room temperature and pressure?

a. Methane and butane

b. Cyclobutane and cyclopentane

c. Hexane and 2,3-dimethylbutane

d. Pentane and octane

Chemical Properties of Alkanes and Cycloalkanes (Section 12.16)

12.49 Write the formulas of the products from the complete combustion of each of the following alkanes or cycloalkanes.

a. C_3H_8 b. Butane

c. Cyclobutane d. $CH_3—(CH_2)_{15}—CH_3$

12.50 Write the formulas of the products from the complete combustion of each of the following alkanes or cycloalkanes.

a. C_4H_{10} b. 2-Methylpentane

c. Cyclopentane d. $CH_3—(CH_2)_7—CH_3$

12.51 Write molecular formulas for all the possible halogenated hydrocarbon products from the bromination of methane.

12.52 Write molecular formulas for all the possible halogenated hydrocarbon products from the fluorination of methane.

12.53 Write structural formulas for all the possible halogenated hydrocarbon products from the monochlorination of the following alkanes or cycloalkanes.

a. Ethane b. Butane

c. 2-Methylpropane d. Cyclopentane

12.54 Write structural formulas for all the possible halogenated hydrocarbon products from the monobromination of the following alkanes or cycloalkanes.

a. Propane b. Pentane

c. 2-Methylbutane d. Cyclohexane

Nomenclature of Halogenated Alkanes (Section 12.17)

12.55 Give both IUPAC and common names to each of the following halogenated hydrocarbons.

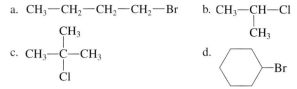

a. $CH_3—I$ b. $CH_3—CH_2—CH_2—Cl$

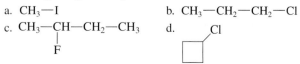

c. $CH_3—CH(F)—CH_2—CH_3$ d. (cyclobutane with Cl)

12.56 Give both IUPAC and common names to each of the following halogenated hydrocarbons.

a. $CH_3—CH_2—CH_2—CH_2—Br$ b. $CH_3—CH(CH_3)—Cl$

c. $CH_3—C(CH_3)(Cl)—CH_3$ d. (cyclohexane with Br)

12.57 Draw structural formulas for the following halogenated hydrocarbons.

a. Trichloromethane

b. 1,2-Dichloro-1,1,2,2-tetrafluoroethane

c. Isopropyl bromide

d. *trans*-1-Bromo-3-chlorocyclopentane

12.58 Draw structural formulas for the following halogenated hydrocarbons.

a. Trifluorochloromethane

b. Pentafluoroethane

c. Isobutyl chloride

d. *cis*-1,2-Dichlorocyclohexane

Additional Problems

12.59 Indicate whether the members of each of the following pairs of compounds are structural isomers.

a. Hexane and 2-methylhexane

b. Hexane and 2,2-dimethylbutane

c. Hexane and methylcyclopentane

d. Hexane and cyclohexane

12.60 Draw structural formulas for the following compounds.

a. *trans*-1,4-Difluorocyclohexane

b. *cis*-1-Chloro-2-methylcyclobutane

c. *Tert*-butyl bromide

d. Isobutyl iodide

12.61 What is the IUPAC name of the compound obtained by attaching a *tert*-butyl group to carbon 4 and a *sec*-butyl group to carbon 5 of the alkane 2,2-dimethylheptane?

12.62 Give the molecular formula for each of the following compounds.

a. an eighteen-carbon alkane

b. a seven-carbon cycloalkane

c. a seven-carbon difluorinated alkane

d. a six-carbon dibrominated cycloalkane

12.63 Draw the structural formula of each of the following compounds.

a. Neooctane b. Isobutane

c. Methylene chloride d. Freon-12

12.64 Classify each of the following molecular formulas as representing an alkane, a cycloalkane, a halogenated alkane, or a halogenated cycloalkane.

a. C_6H_{14} b. $C_6H_{10}Cl_2$

c. $C_4H_8Cl_2$ d. C_4H_8

12.65 Write skeletal formulas (without hydrogen atoms) and assign IUPAC names to all saturated hydrocarbon structural isomers (ignore *cis–trans* isomers) with the following molecular formulas.

a. C_6H_{12} (twelve isomers are possible)
b. C_6H_{14} (five isomers are possible)
c. $C_3H_6Br_2$ (four isomers are possible)

12.66 Assign an IUPAC name to each of the following hydrocarbons, whose structural formulas are given using line-angle notation.

a. 
b. 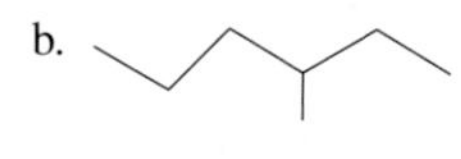
c. 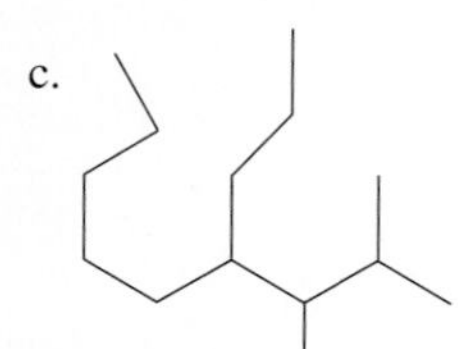
d. 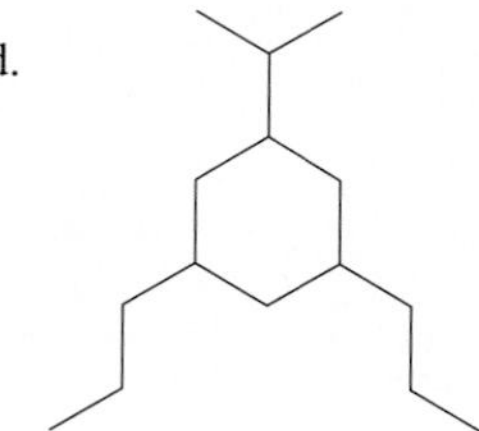

Grid Problems

12.67

1. 2-methylpentane	2. 3-methylpentane	3. 3-ethylpentane
4. cyclopentane	5. methylcyclopentane	6. ethylcyclopentane

Select from the grid *all* correct responses for each situation.

a. Compounds that contain six carbon atoms
b. Compounds whose molecular formula is C_7H_{14}
c. Compounds whose molecules contain five $-CH_2-$ units
d. Pairs of compounds that are structural isomers

12.68

1. $-CH_2-CH_3$	2. $-CH-CH_3$ $\quad|$ $\quad CH_3$	3. $-CH_2-CH-CH_3$ $\qquad\quad|$ $\qquad\quad CH_3$
4. $-CH_2-CH_2-CH_3$	5. $-CH-CH_2-CH_3$ $\quad|$ $\quad CH_3$	6. $\quad CH_3$ $\quad|$ $-C-CH_3$ $\quad|$ $\quad CH_3$

Select from the grid *all* correct responses for each situation.

a. Alkyl groups whose name contains the prefix *iso-*
b. Alkyl groups whose name contains the prefix *sec-*
c. Alkyl groups whose point of attachment to a carbon chain is a tertiary carbon atom
d. Alkyl groups whose point of attachment to a carbon chain is a primary carbon atom

12.69

1. butane	2. cyclopropane	3. bromomethane
4. pentane	5. cyclobutane	6. chloromethane

Select from the grid *all* correct responses for each situation.

a. Compounds that are insoluble in water
b. Compounds that have a density greater than that of water
c. Pairs of compounds that are homologs—that is, members of a homologous series
d. Compounds that are hydrocarbon derivatives

12.70

1. C_3H_8	2. $C_3H_4Cl_2$	3. C_3H_7Cl
4. C_3H_5Cl	5. C_3H_6	6. $C_3H_6Cl_2$

Select from the grid *all* correct responses for each situation.

a. Molecular formulas that represent alkanes or alkane derivatives
b. Molecular formulas that represent cycloalkanes or cycloalkane derivatives
c. Molecular formulas for which structural isomers exist
d. Molecular formulas for which *cis–trans* isomers exist

Unsaturated Hydrocarbons

CHAPTER OUTLINE

13.1 Unsaturated Hydrocarbons 332
13.2 Characteristics of Alkenes and Cycloalkenes 333
13.3 Names for Alkenes and Cycloalkenes 334
13.4 The Nature of Carbon–Carbon Multiple Bonds 337
13.5 Isomerism in Alkenes 339
13.6 Naturally Occurring Alkenes 343
13.7 Physical and Chemical Properties of Alkenes 344
13.8 Polymerization of Alkenes: Addition Polymers 350
13.9 Alkynes 352

Chemistry at a Glance: **Chemical Reactions of Alkenes** 353

13.10 Aromatic Hydrocarbons 354
13.11 Names for Aromatic Hydrocarbons 356
13.12 Aromatic Hydrocarbons: Physical Properties and Sources 359
13.13 Chemical Reactions of Aromatic Hydrocarbons 360

Chemistry at a Glance: **Benzene Substitution Reactions** 361

13.14 Fused-Ring Aromatic Compounds 361

Chemical Connections

13.1 Ethene: A Plant Hormone and High-Volume Industrial Chemical 337
13.2 *Cis–Trans* Isomerism and Vision 342
13.3 Carotenoids: A Source of Color 345
13.4 Fused-Ring Aromatic Hydrocarbons and Cancer 362

When acetylene, an unsaturated hydrocarbon, is burned with oxygen in an oxyacetylene welding torch, a temperature high enough to cut metals is produced.

Two general types of hydrocarbons exist, *saturated* and *unsaturated.* Saturated hydrocarbons, discussed in Chapter 12, include the alkanes and cycloalkanes. All bonds in saturated hydrocarbons are single bonds. Unsaturated hydrocarbons, the topic of this chapter, contain one or more carbon–carbon multiple bonds: double bonds, triple bonds, or both.

13.1 Unsaturated Hydrocarbons

Unsaturated hydrocarbons have *physical* properties similar to those of saturated hydrocarbons. However, their *chemical* properties are much different. Unsaturated hydrocarbons are chemically more reactive than their saturated counterparts. The increased reactivity of unsaturated hydrocarbons is related to the presence of the carbon–carbon multiple bond(s) in such compounds. These multiple bonds serve as locations where chemical reactions can occur.

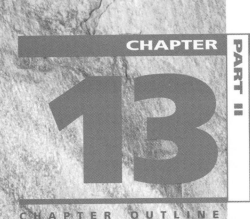

Whenever a specific portion of a molecule governs its chemical properties, that portion of the molecule is called a functional group. A **functional group** *is the part of a molecule where most of its chemical reactions occur.* Carbon–carbon multiple bonds are the functional group for an unsaturated hydrocarbon.

- The field of organic chemistry is organized in terms of functional groups.

The study of various functional groups and their respective reactions provides the organizational structure for organic chemistry. Each of the organic chemistry chapters that follow introduces new functional groups that characterize families of hydrocarbon derivatives.

- Alkanes and cycloalkanes (Chapter 12) lack functional groups; as a result, they are relatively unreactive.

Unsaturated hydrocarbons are subdivided into three groups on the basis of the type of multiple bond(s) present: (1) *alkenes,* which contain one or more carbon–carbon double bonds, (2) *alkynes,* which contain one or more carbon–carbon triple bonds, and (3) *aromatic hydrocarbons,* which exhibit a special type of multiple bonding that involves a six-membered carbon ring.

We begin our consideration of unsaturated hydrocarbons with an in-depth discussion of alkenes. Information about alkynes and aromatic hydrocarbons then follows.

13.2 Characteristics of Alkenes and Cycloalkenes

An **alkene** *is an acyclic unsaturated hydrocarbon that contains one or more carbon–carbon double bonds.* Note the close similarity between the family names *alkene* and *alkane* (Section 12.4); they differ only in their endings: *-ene* versus *-ane.* The *-ene* ending means a double bond is present.

- The general formula for an alkene with one double bond, C_nH_{2n}, is the same as that for a cycloalkane (Section 12.11). Thus such alkenes and cycloalkanes with the same number of carbon atoms are isomeric with one another.

The simplest type of alkene contains only one carbon–carbon double bond. Such compounds have the general formula C_nH_{2n}. Thus alkenes with one double bond have two fewer hydrogen atoms than do alkanes (C_nH_{2n+2}).

The two simplest alkenes are ethene (C_2H_4) and propene (C_3H_6).

$$\underset{\text{Ethene}}{CH_2{=}CH_2} \qquad \underset{\text{Propene}}{CH_3{-}CH{=}CH_2}$$

- An older but still widely used name for alkenes is *olefins,* pronounced "oh-la-fins." The term *olefin* means "oil-forming." Many alkenes react with Cl_2 to form "oily" compounds.

Comparing the geometrical shape of ethene with that of methane (the simplest alkane) reveals a major difference. The arrangement of bonds about the carbon atom in methane is tetrahedral (Section 12.4), whereas the carbon atoms in ethene have a trigonal planar arrangement of bonds; that is, they form a flat, triangle-shaped arrangement (see Figure 13.1). The two carbon atoms participating in a double bond and the four other atoms attached to these two carbon atoms always lie in a plane with a trigonal planar arrangement of

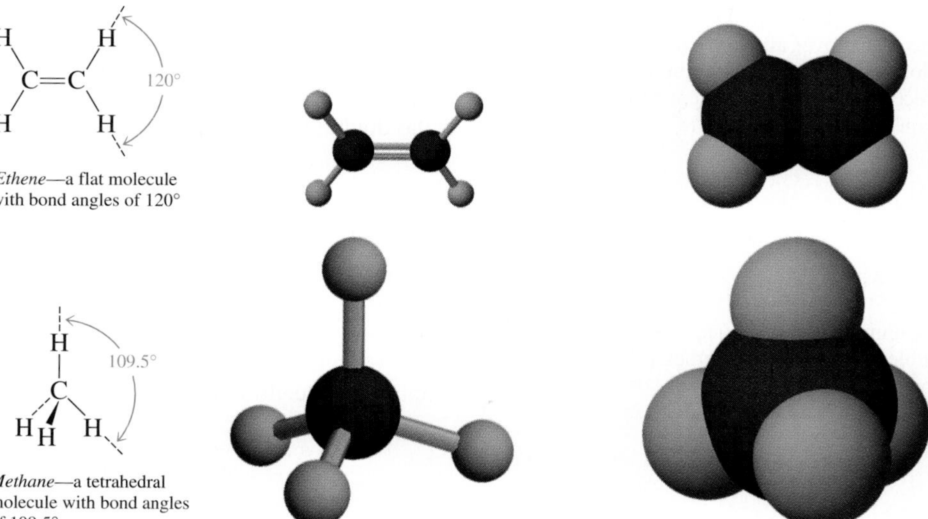

Ethene—a flat molecule with bond angles of 120°

Methane—a tetrahedral molecule with bond angles of 109.5°

Figure 13.1
Three-dimensional representations of the structures of ethene and methane. In ethene, the atoms are in a flat (planar) rather than a tetrahedral arrangement. Bond angles are 120°.

atoms about each carbon atom of the double bond. Such an arrangement of atoms is consistent with the principles of VSEPR theory (Section 5.7).

A **cycloalkene** *is a cyclic unsaturated hydrocarbon that contains one or more carbon–carbon double bonds within the ring system.* Cycloalkenes in which there is only one double bond have the general formula C_nH_{2n-2}. The general formula reflects the loss of four hydrogen atoms from that of an alkane (C_nH_{2n+2}). Note that two hydrogen atoms are lost because of the double bond and two because of the ring structure.

The simplest cycloalkene is the compound cyclopropene (C_3H_4), a three-membered carbon ring system containing one double bond.

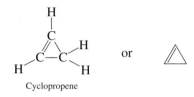

Cyclopropene

Alkenes with more than one carbon–carbon double bond are relatively common. When two double bonds are present, the compounds are often called *dienes.* Cycloalkenes that contain more than one double bond are possible but are not common.

13.3 Names for Alkenes and Cycloalkenes

The rules previously presented for naming alkanes and cycloalkanes (Sections 12.8 and 12.12) can be used, with some modification, to name alkenes and cycloalkenes.

1. Replace the alkane suffix *-ane* with the suffix *-ene,* which is used to indicate the presence of a carbon–carbon double bond.
2. Select as the parent carbon chain the longest continuous chain of carbon atoms that *contains both carbon atoms of the double bond.* For example, select

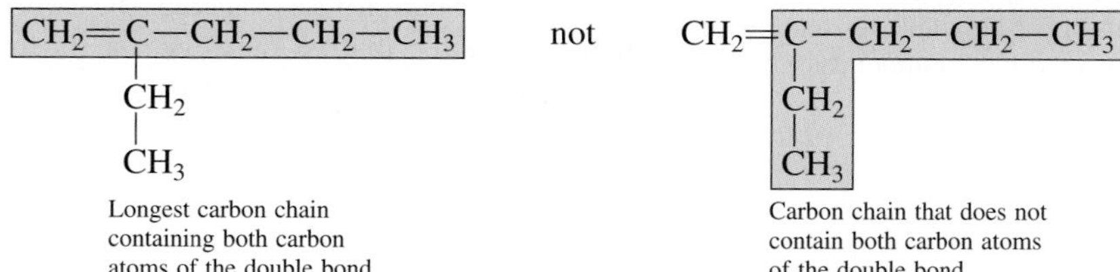

Longest carbon chain containing both carbon atoms of the double bond — not — Carbon chain that does not contain both carbon atoms of the double bond

• Carbon–carbon double bonds take precedence over alkyl groups and halogen atoms in determining the direction in which the parent carbon chain is numbered.

3. Number the parent carbon chain beginning at the end nearest the double bond.

$$\overset{1}{CH_3}-\overset{2}{CH}=\overset{3}{CH}-\overset{4}{CH_2}-\overset{5}{CH_3} \quad \text{not} \quad \overset{5}{CH_3}-\overset{4}{CH}=\overset{3}{CH}-\overset{2}{CH_2}-\overset{1}{CH_3}$$

If the double bond is equidistant from both ends of the parent chain, begin numbering from the end closer to a substituent.

$$\overset{4}{CH_3}-\overset{3}{CH}=\overset{2}{CH}-\underset{Cl}{\overset{1}{CH_2}} \quad \text{not} \quad \overset{1}{CH_3}-\overset{2}{CH}=\overset{3}{CH}-\underset{Cl}{\overset{4}{CH_2}}$$

• A number is not needed to specify double bond position in ethene and propene, because there is only one way of positioning the double bond in these molecules.

4. Give the position of the double bond in the chain as a *single* number, which is the lower-numbered carbon atom participating in the double bond. This number is placed immediately before the name of the parent carbon chain.

$$\overset{1}{CH_3}-\overset{2}{CH}=\overset{3}{CH}-\overset{4}{CH_3} \qquad \overset{1}{CH_2}=\overset{2}{CH}-\underset{CH_3}{\overset{3}{CH}}-\overset{4}{CH_3}$$

2-Butene — 3-Methyl-1-butene

• Line-angle drawings for the simpler noncyclic 1-alkenes are as follows.

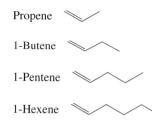

Propene

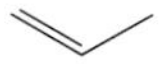

1-Butene

1-Pentene

1-Hexene

5. Use the suffixes *-diene, -triene, -tetrene,* and so on when more than one double bond is present in the molecule. A separate number must be used to locate each double bond.

$$\overset{1}{C}H_2{=}\overset{2}{C}H{-}\overset{3}{C}H{=}\overset{4}{C}H_2$$

1,3-Butadiene

$$\overset{1}{C}H_2{=}\overset{2}{C}H{-}\underset{CH_3}{\overset{3}{C}H}{-}\overset{4}{C}H{=}\overset{5}{C}H_2$$

3-Methyl-1,4-pentadiene

6. Do not use a number to locate the double bond in unsubstituted cycloalkenes with only one double bond, because that bond is assumed to be between carbons 1 and 2.

7. In substituted cycloalkenes with only one double bond, the double-bonded carbon atoms are numbered 1 and 2 in the direction (clockwise or counterclockwise) that gives the first-encountered substituent the lower number. Again, no number is used in the name to locate the double bond.

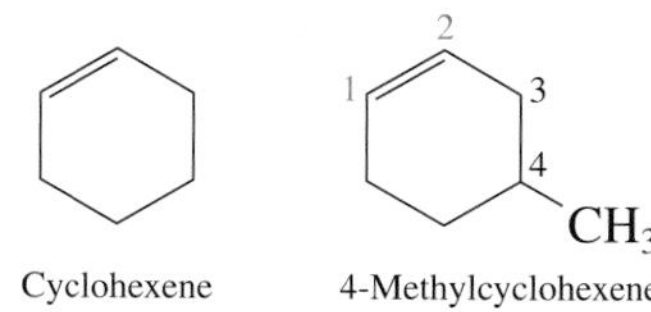

Cyclohexene 4-Methylcyclohexene

8. In cyclic hydrocarbon systems with more than one double bond within the ring, assign one double bond the numbers 1 and 2 and the other double bonds the lowest numbers possible.

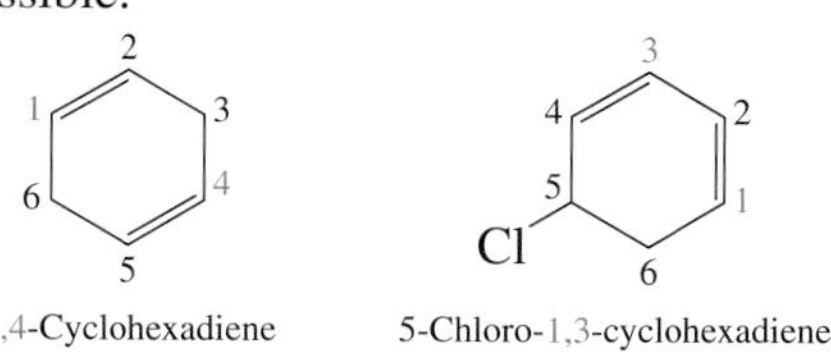

1,4-Cyclohexadiene 5-Chloro-1,3-cyclohexadiene

Example 13.1

Assigning IUPAC Names to Alkenes and Cycloalkenes

Assign IUPAC names to the following alkenes and cycloalkenes.

a. $CH_3{-}CH{=}CH{-}CH_2{-}CH_2{-}CH_3$

b. $CH_3{-}CH_2{-}\underset{\underset{CH_3}{|}}{\underset{CH_2}{|}}{C}{=}CH_2$

c. 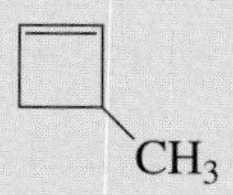 CH_3

d. 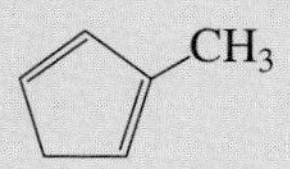 CH_3

Solution

a. The carbon chain in this hexene is numbered from the end closest to the double bond.

$$\overset{1}{C}H_3{-}\overset{2}{C}H{=}\overset{3}{C}H{-}\overset{4}{C}H_2{-}\overset{5}{C}H_2{-}\overset{6}{C}H_3$$

The complete IUPAC name is 2-hexene.

b. The longest carbon chain containing *both* carbons of the double bond has four carbon atoms. Thus we have a butene.

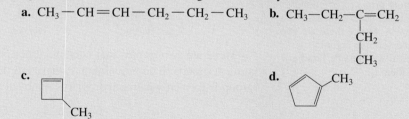

$$\boxed{CH_3{-}CH_2{-}C{=}CH_2}$$
$$\quad |$$
$$CH_2$$
$$\quad |$$
$$CH_3$$

(continued)

The chain is numbered from the end closest to the double bond. The complete IUPAC name is 2-ethyl-1-butene.

c. This compound is a methylcyclobutene. The numbers 1 and 2 are assigned to the carbon atoms of the double bond, and the ring is numbered clockwise, which results in a carbon 3 location for the methyl group. (Counterclockwise numbering would have placed the methyl group on carbon 4.) The complete IUPAC name of the cycloalkene is 3-methylcyclobutene. The double bond is understood to involve carbons 1 and 2.

d. A ring system containing five carbon atoms, two double bonds, and a methyl substituent on the ring is called a methylcyclopentadiene. Two different numbering systems produce the same locations (carbons 1 and 3) for the double bonds.

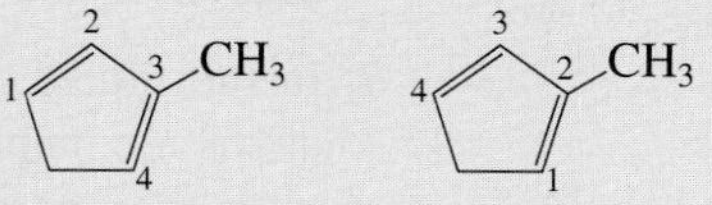

The counterclockwise numbering system assigns the lower number to the methyl group. The complete IUPAC name of the compound is 2-methyl-1,3-cyclopentadiene.

Practice Exercise 13.1

Assign IUPAC names to the following alkenes and cycloalkenes.

a. $CH_3—CH{=}CH—CH_2—CH(CH_3)—CH_3$

b. 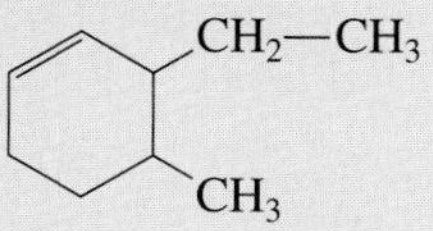 (cyclohexene ring with $CH_2—CH_3$ and CH_3 substituents)

c. $CH_2{=}CH—CH{=}CH_2$

d. 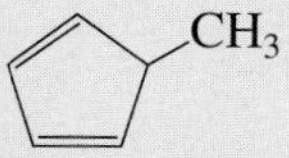 (cyclopentadiene ring with CH_3 substituent)

• *Answers:* **a.** 5-methyl-2-hexene; **b.** 3-ethyl-4-methylcyclohexene; **c.** 1,3-butadiene; **d.** 5-methyl-1,3-cyclopentadiene

• The simpler members of most families of organic compounds have common names. Often these common names predate the advent of systematic rules for naming organic compounds. It would be nice if such names did not exist, but they do. Hence there are two names for many simple organic compounds, and we must be familiar with both of them.

The simpler alkenes have common names. The most frequently encountered of these common names are

$CH_2{=}CH_2$ Ethylene

$CH_2{=}CH—CH_3$ Propylene

$CH_2{=}C(CH_3)—CH_3$ Isobutylene

There are two groups that contain carbon–carbon double bonds that are used as substituent groups in common names; these are the *vinyl group* and the *allyl group*. A vinyl group is formed when a hydrogen is removed from ethylene.

$CH_2{=}CH_2$ Ethylene

$CH_2{=}CH—$ Vinyl group

An allyl group is formed when a hydrogen is removed from the CH_3 carbon atom of propylene.

$CH_2{=}CH—CH_3$ Propylene

$CH_2{=}CH—CH_2—$ Allyl group

The terms *vinyl* and *allyl* find particular use in the common names of monohalogenated alkene derivatives.

$CH_2{=}CH—Cl$ Vinyl chloride (IUPAC name: chloroethene)

$CH_2{=}CH—CH_2—Br$ Allyl bromide (IUPAC name: 3-bromopropene)

Chemical CONNECTIONS

13.1 Ethene: A Plant Hormone and High-Volume Industrial Chemical

Ethene (ethylene), the simplest unsaturated hydrocarbon (C_2H_4), is a colorless, flammable gas with a slightly sweet odor. It occurs naturally in *small* amounts in plants where it functions as a plant hormone. A few parts per million ethene (less than 10 parts per million) stimulates the fruit-ripening process.

Ethene is the hormone that causes tomatoes to ripen.

The commercial fruit industry uses ethene's ripening property to advantage. Bananas, tomatoes, and some citrus fruits are picked green to prevent spoiling and bruising during transportation to markets. At its destination, the fruit is exposed to small amounts of ethene gas, which stimulates the ripening process.

Despite having no large natural source, ethene is an exceedingly important industrial chemical. Indeed, industrial production of ethene exceeds that of every other organic compound. Petrochemicals (substances found in natural gas and petroleum) are the starting materials for ethene production.

In one process, ethane (from natural gas) is *dehydrogenated* at a high temperature to produce ethene.

$$\underset{\text{Ethane}}{CH_3{-}CH_3} \xrightarrow{750°C} \underset{\text{Ethene}}{CH_2{=}CH_2} + H_2$$

In another process, called *thermal cracking,* hydrocarbons from petroleum are heated to a high temperature in the absence of air (to prevent combustion), which causes the cleavage of carbon–carbon bonds. Ethene is one of the smaller molecules produced by this process.

Industrially produced ethene serves as a starting material for the production of many plastics and fibers. Almost one-half of ethene production is used in the production of the well-known plastic polyethylene (Sec. 13.8). Polyvinyl chloride (PVC) and polystyrene (styrofoam) are two other important ethene-based materials. About one-sixth of ethene production is converted to ethylene glycol, the principal component of most brands of antifreeze for automobile radiators (Sec. 14.4).

13.4 The Nature of Carbon–Carbon Multiple Bonds

In a carbon–carbon multiple bond, either a double bond or a triple bond, there are two types of bonds present that are distinguished by the manner in which the atomic orbitals involved in the bonding overlap. Designations for these two bond types involve the Greek letters sigma (σ) and pi (π). A **sigma bond** *(σ-bond) is a covalent bond in which atomic orbital overlap occurs along the axis joining the two bonded atoms.* Both of the examples of atomic orbital overlap in Figure 13.2a give rise to σ-bonds.

• Rather than spelling out the words *sigma* and *pi* when referring to bonds, we use the Greek letters sigma (σ) and pi (π).

A **pi bond** *(π-bond) is a covalent bond in which atomic orbital overlap occurs above and below (but not on) the internuclear axis.* Figure 13.2b illustrates the *sideways* overlap of two *p* orbitals, which represents a π-bonding situation. The two overlap regions collectively constitute *one* π-bond. The sideways overlap of two *p* orbitals, as shown in the figure, is the type of π-bond present in all carbon–carbon multiple bonds.

Now let us relate the concepts of σ-bonds and π-bonds to carbon–carbon single, double, and triple bonds.

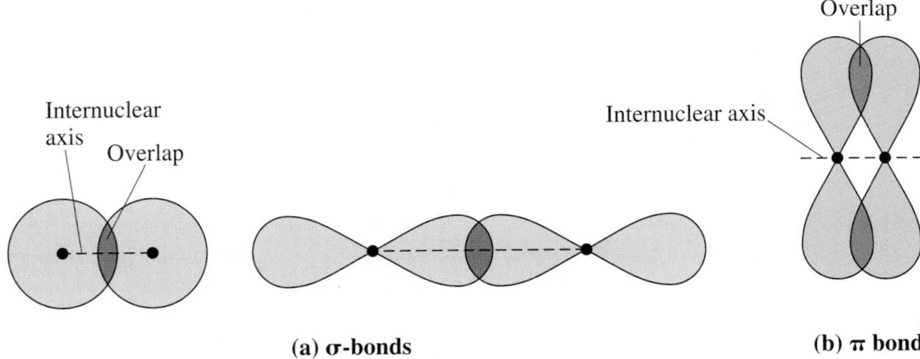
Internuclear
axis
Overlap
Overlap
Internuclear axis
(a) σ-bonds
(b) π bond

Figure 13.2
σ- and π-bonds. (a) In a σ-bond, the orbitals overlap along the axis joining the two atoms. (b) In a π-bond, the orbitals overlap above and below the axis joining the two atoms.

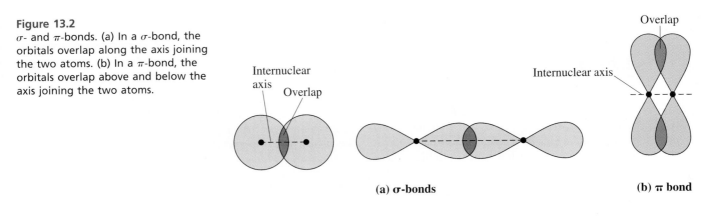

• A σ-bond involves an end-to-end overlap of atomic orbitals. A π-bond involves a side-to-side overlap of atomic orbitals.

• There can never be more than one σ-bond between a pair of atoms. All additional bonds between the pair of atoms must be π-bonds.

Single bond = σ-bond
Double bond = σ-bond + π-bond
Triple bond = σ-bond + 2 π-bonds

1. When a single bond is present between two atoms, that bond is always a σ-bond.
2. When a double bond is present between two atoms, that bond always consists of one σ-bond and one π-bond. Figure 13.3a shows the σ and π components of the double bond in the C_2H_4 molecule.
3. When a triple bond is present between two atoms, that bond always consists of one σ-bond and two π-bonds. Figure 13.3b shows the σ and π components of the triple bond in the C_2H_2 molecule.

Why is it important to know that π-bonding is present in carbon–carbon double bonds or carbon–carbon triple bonds? There are two important reasons.

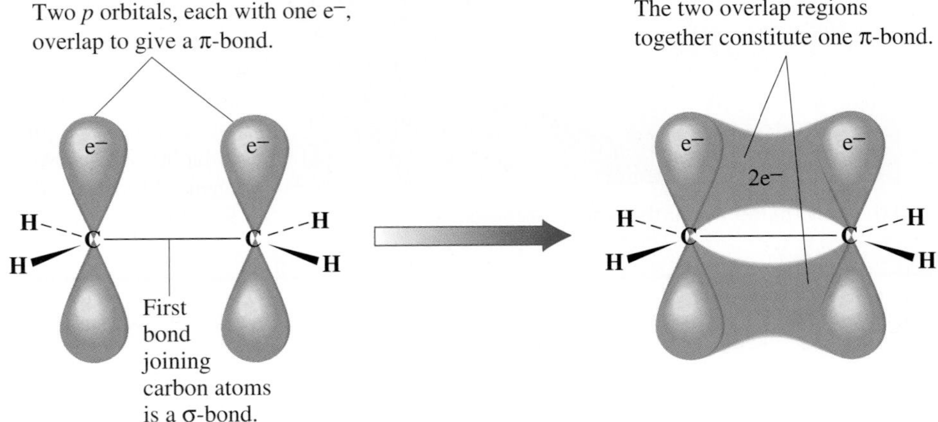
Two p orbitals, each with one e⁻, overlap to give a π-bond.
The two overlap regions together constitute one π-bond.
e⁻
e⁻
2e⁻
H
H
C
C
First bond joining carbon atoms is a σ-bond.
(a) C₂H₄, a hydrocarbon with a carbon–carbon double bond.

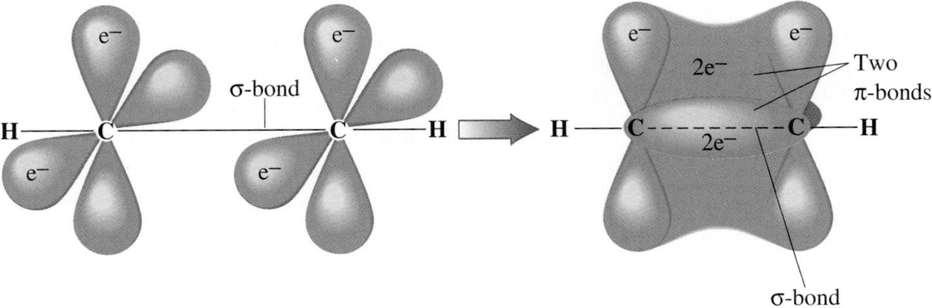
Two p orbitals on one carbon, each with one e⁻, overlap "sideways" with corresponding p orbitals on the other carbon to give two π-bonds.
e⁻
σ-bond
H
C
2e⁻
Two π-bonds
σ-bond
(b) C₂H₂, a hydrocarbon with a carbon–carbon triple bond.

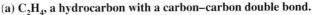

Figure 13.3
Carbon–carbon double and triple bonds. (a) A carbon–carbon double bond is made up of a σ-bond and a π-bond. (b) A carbon–carbon triple bond contains a σ-bond and two π-bonds.

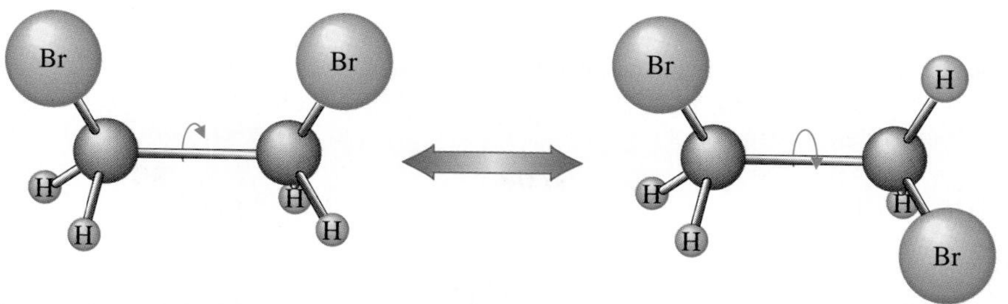

(a) The molecule 1,2-dibromoethane can rapidly interconvert between the two orientations shown, because a carbon–carbon single bond is not structurally rigid.

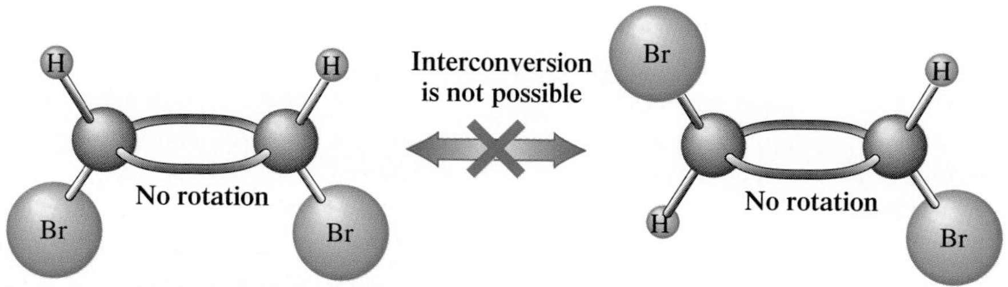

(b) The molecule 1,2-dibromoethene cannot interconvert between the two forms shown, because a carbon–carbon double bond is structurally rigid.

Figure 13.4 Rotation around carbon–carbon bonds. (a) A carbon–carbon single bond is not structurally rigid, allowing for free rotation of groups about the bond. (b) A carbon–carbon double bond is structurally rigid, preventing free rotation of groups about the bond.

• In structural formulas, no distinction is made between the σ-bond and the π-bond(s) present in a carbon–carbon multiple bond. Each bond is represented by a dash.

C—C	C=C	C≡C
Single bond	Double bond	Triple bond

1. *A carbon–carbon π-bond is weaker, and consequently more reactive (Section 13.7), than a carbon–carbon σ-bond.* The strength of a carbon–carbon π-bond is about 70% that of a carbon–carbon σ-bond.
2. *The presence of π-bonding causes a bond to be structurally rigid, which prevents free rotation of the carbon atoms involved in the bond.* Rotation is allowed in a carbon–carbon single bond (Figure 13.4a) but not in a carbon–carbon multiple bond (Figure 13.4b). If the two carbon atoms in a double bond did rotate, the π-bond would be broken.

13.5 Isomerism in Alkenes

• Structural isomers that differ only in the location of the functional group are often called *positional isomers.* The compounds 1-butene and 2-butene are positional isomers.

$$CH_2{=}CH{-}CH_2{-}CH_3$$
1-Butene

$$CH_3{-}CH{=}CH{-}CH_3$$
2-Butene

Structural isomerism is possible for alkenes, just as it was for alkanes (Section 12.6). Table 13.1 compares structural isomer possibilities for four-and five-carbon alkane and alkene systems. The potential for structural isomers is greater for alkenes than for alkanes, because there is more than one location where the double bond can be placed in systems containing four or more carbon atoms.

Cis–trans isomerism (Section 12.13) is possible for some alkenes. Such isomerism results from the structural rigidity associated with carbon–carbon double bonds (Section 13.4).

To determine whether an alkene has *cis* and *trans* isomers, draw the alkene structure in a manner that emphasizes the four attachments to the double-bonded carbon atoms.

$$\gt C{=}C \lt$$

• The double bond of alkenes, like the ring of cycloalkanes, imposes rotational restrictions.

If *each* of the two carbons of the double bond has two *different* groups attached to it, *cis* and *trans* isomers exist.

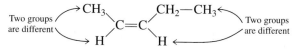

$$\begin{matrix} CH_3 & & & & CH_2{-}CH_3 \\ & \diagdown & & \diagup & \\ & & C{=}C & & \\ & \diagup & & \diagdown & \\ H & & & & H \end{matrix}$$

Two groups are different (left: CH_3 and H); Two groups are different (right: $CH_2{-}CH_3$ and H)

Table 13.1
A Comparison of Structural Isomerism Possibilities for Four- and Five-Carbon Alkane and Alkene Systems

Four-carbon alkanes (two isomers)	Four-carbon alkenes (three isomers)	
$CH_3-CH_2-CH_2-CH_3$ Butane	$CH_2=CH-CH_2-CH_3$ 1-Butene	$CH_2=\underset{\displaystyle CH_3}{\underset{\vert}{C}}-CH_3$ 2-Methyl-1-propene
$CH_3-\underset{\displaystyle CH_3}{\underset{\vert}{CH}}-CH_3$ 2-Methylpropane	$CH_3-CH=CH-CH_3$ 2-Butene	
Five-carbon alkanes (three isomers)	**Five-carbon alkenes (five isomers)**	
$CH_3-CH_2-CH_2-CH_2-CH_3$ Pentane	$CH_2=CH-CH_2-CH_2-CH_3$ 1-Pentene	$CH_2=CH-\underset{\displaystyle CH_3}{\underset{\vert}{CH}}-CH_3$ 3-Methyl-1-butene
$CH_3-\underset{\displaystyle CH_3}{\underset{\vert}{CH}}-CH_2-CH_3$ 2-Methylbutane	$CH_3-CH=CH-CH_2-CH_3$ 2-Pentene	$CH_3-\underset{\displaystyle CH_3}{\underset{\vert}{C}}=CH-CH_3$ 2-Methyl-2-butene
$CH_3-\overset{\displaystyle CH_3}{\overset{\vert}{\underset{\displaystyle CH_3}{\underset{\vert}{C}}}}-CH_3$ 2,2-Dimethylpropane	$CH_2=\underset{\displaystyle CH_3}{\underset{\vert}{C}}-CH_2-CH_3$ 2-Methyl-1-butene	

The simplest alkene for which *cis* and *trans* isomers exist is 2-butene.

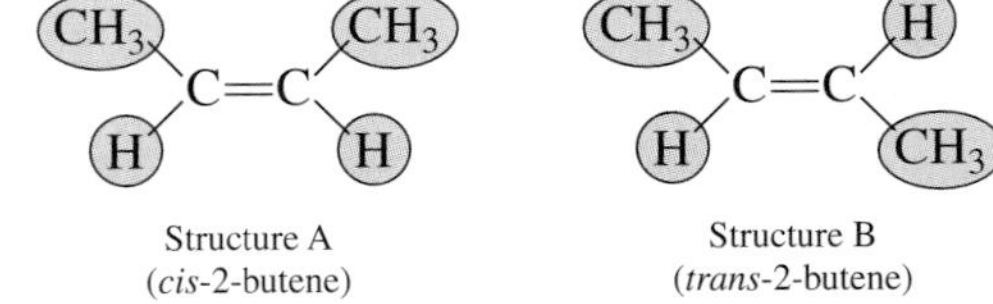

Structure A (*cis*-2-butene)

Structure B (*trans*-2-butene)

• Recall, from Section 12.13, that *cis* means "on the same side" and *trans* means "across."

In structure A, the *cis* isomer, both methyl groups are on the same side of the double bond. In structure B, the *trans* isomer, the methyl groups are on opposite sides of the double bond. The only way to convert structure A to structure B is to break the π-bond. At room temperature, such bond breaking does not occur. Hence these two structures represent two different compounds (*cis–trans* isomers) that differ in boiling point, density, and so on. Figure 13.5 shows three-dimensional representations of the *cis* and *trans* isomers of 2-butene.

Cis–trans isomerism is not possible when one of the double-bonded carbons bears two identical groups. Thus neither 1-butene nor 2-methylpropene is capable of existing in *cis* and *trans* forms.

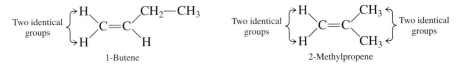

Two identical groups { H, H } C=C (CH_2-CH_3, H)
1-Butene

Two identical groups { H, H } C=C { CH_3, CH_3 } Two identical groups
2-Methylpropene

When alkenes contain more than one double bond, *cis–trans* considerations are more complicated. Orientation about each double bond must be considered independently of that at other sites. For example, for the molecule 2,4-heptadiene (two double bonds) there

Figure 13.5
Cis–trans isomers: Different representations of the *cis* and *trans* isomers of 2-butene.

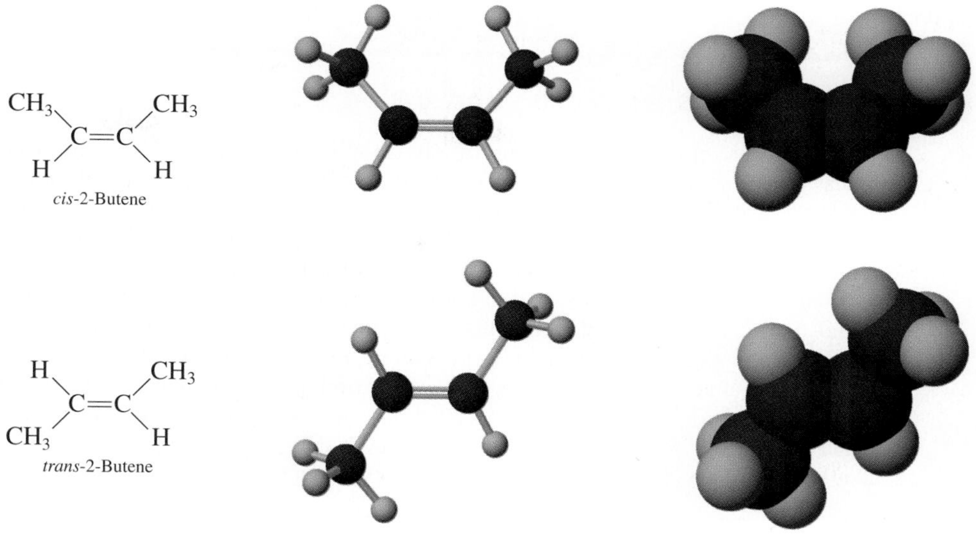

$$\begin{array}{ccc} CH_3 & & CH_3 \\ & C{=}C & \\ H & & H \end{array}$$
cis-2-Butene

$$\begin{array}{ccc} H & & CH_3 \\ & C{=}C & \\ CH_3 & & H \end{array}$$
trans-2-Butene

are four different *cis–trans* isomers (*trans–trans, trans–cis, cis–trans,* and *cis–cis*). The structures of two of these isomers are

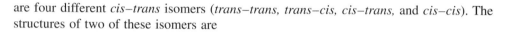

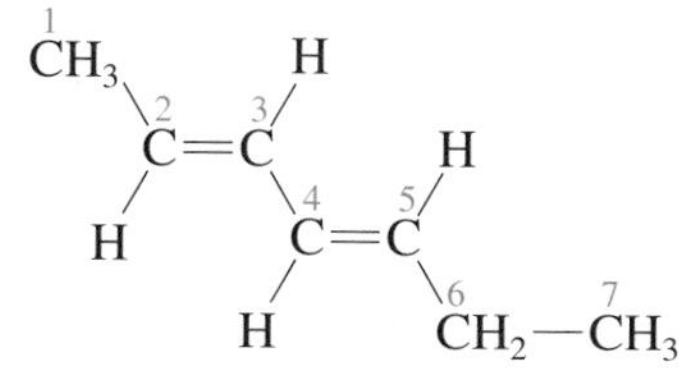

$$\begin{array}{ccccc} \overset{1}{C}H_3 & & H & & \\ & \overset{2}{C}{=}\overset{3}{C} & & H & \\ H & & \overset{4}{C}{=}\overset{5}{C} & & \\ & H & & \overset{6}{C}H_2{-}\overset{7}{C}H_3 & \end{array}$$
trans-trans-2,4-Heptadiene

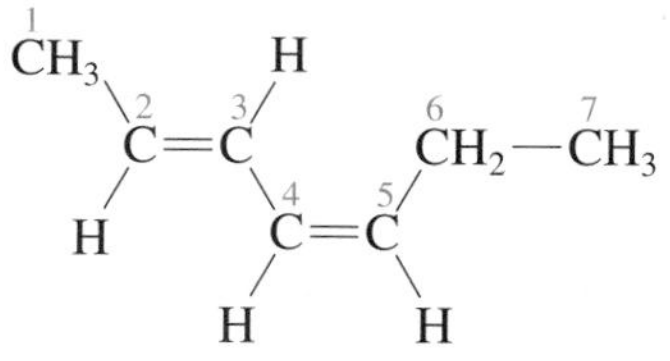

$$\begin{array}{ccccc} \overset{1}{C}H_3 & & H & & \\ & \overset{2}{C}{=}\overset{3}{C} & & \overset{6}{C}H_2{-}\overset{7}{C}H_3 & \\ H & & \overset{4}{C}{=}\overset{5}{C} & & \\ & H & & H & \end{array}$$
trans-cis-2,4-Heptadiene

Example 13.2

Determining Whether *Cis–Trans* Isomerism Is Possible in Substituted Alkenes

Determine whether each of the following substituted alkenes can exist in *cis–trans* isomeric forms.

a. 1-Bromo-1-chloroethene **b.** 2-Chloro-2-butene

Solution

a. The condensed structural formula for this compound is

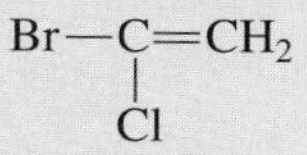

$$\begin{array}{l} Br{-}C{=}CH_2 \\ \quad\;\; | \\ \quad\;\, Cl \end{array}$$

Redrawing this formula to emphasize the four attachments to the double-bonded carbon atoms gives

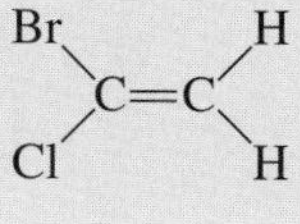

$$\begin{array}{ccc} Br & & H \\ & C{=}C & \\ Cl & & H \end{array}$$

(continued)

Chemical CONNECTIONS

13.2 *Cis–Trans* Isomerism and Vision

Cis–trans isomerism plays an important role in many biochemical processes, including the reception of light by the retina of the eye. Within the retina, microscopic structures called rods and cones contain a compound called *retinal,* which absorbs light. Retinal contains a carbon chain with five carbon–carbon double bonds, four in a *trans* configuration and one in a *cis* configuration. This arrangement of double bonds gives retinal a shape that fits the protein *opsin,* to which it is attached, as shown in the accompanying diagram.

When light strikes retinal, the π-bond in the *cis* double bond is broken for a brief instant, allowing free rotation about the σ-bond. The free rotation switches the *cis* bond to a *trans* bond before the π-bond re-forms. The resulting *trans*-retinal no longer fits the protein opsin and is subsequently released. Accompanying this release is an electrical impulse, which is sent to the brain. Receipt of such impulses by the brain is what enables us to see.

In order to trigger nerve impulses again, *trans*-retinal must be converted back to *cis*-retinal. This occurs in the membranes of the rods and cones, where enzymes change *trans*-retinal back to *cis*-retinal.

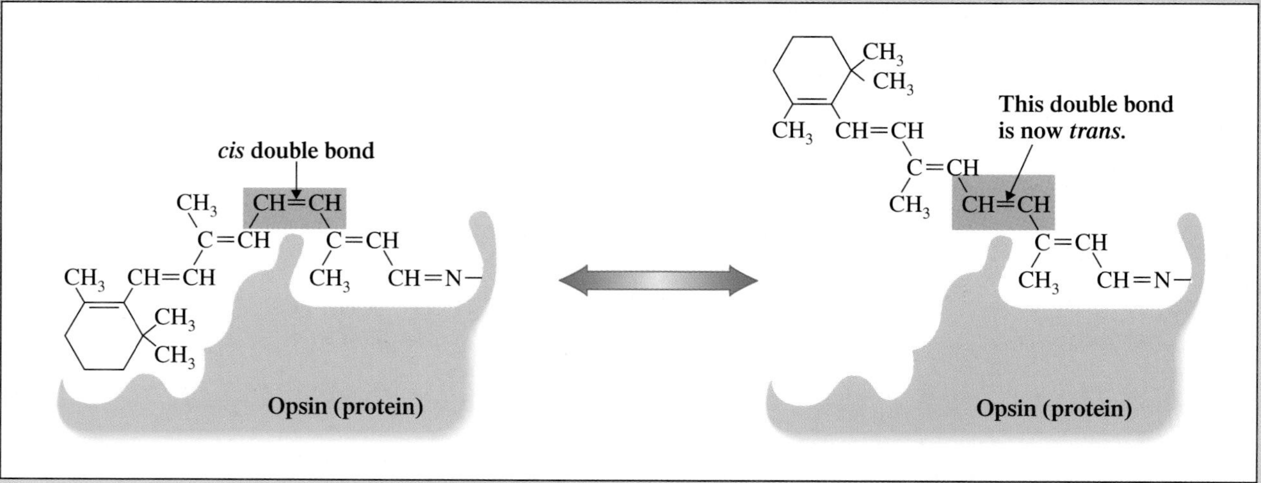

The carbon atom on the right has two identical attachments. Hence *cis–trans* isomerism is not possible.

b. The condensed structural formula for this compound is

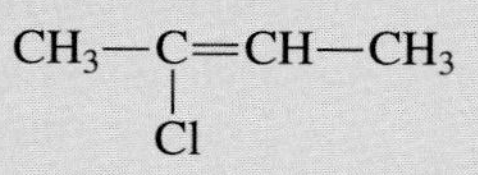

$$CH_3-\underset{\displaystyle Cl}{\underset{|}{C}}=CH-CH_3$$

Redrawing this formula to emphasize the four attachments to the double-bonded carbon atoms gives

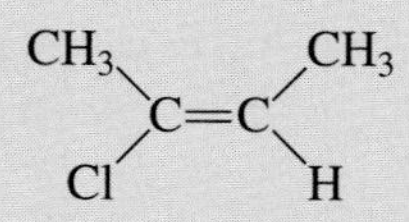

$$\begin{matrix} CH_3 & & CH_3 \\ & C=C & \\ Cl & & H \end{matrix}$$

Because both carbon atoms of the double bond bear two different attachments, *cis–trans* isomers are possible.

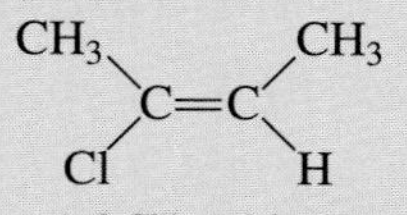

$$\begin{matrix} CH_3 & & CH_3 \\ & C=C & \\ Cl & & H \end{matrix} \qquad \begin{matrix} Cl & & CH_3 \\ & C=C & \\ CH_3 & & H \end{matrix}$$

cis-2-Chloro-2-butene 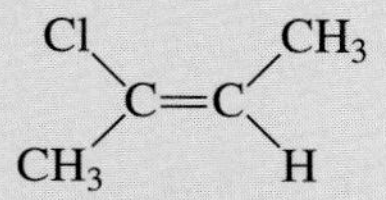 *trans*-2-Chloro-2-butene

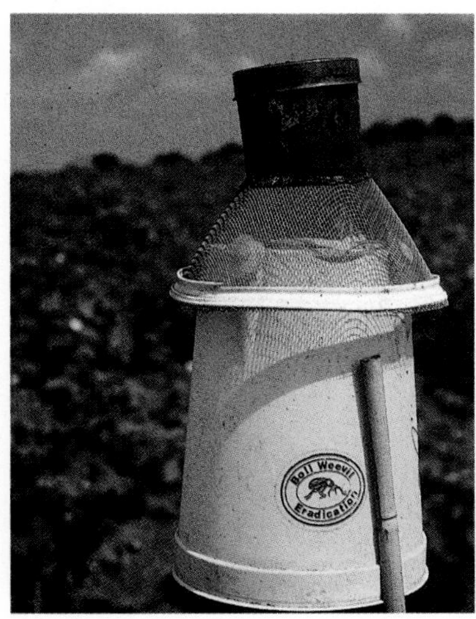

• Insect sex pheromones are useful in insect control. A small amount of synthetically produced sex pheromone is used to lure male insects of a single species into a trap. The trapped males are either killed or sterilized. Releasing sterilized males has proved effective in some situations. A sterile male can mate many times, preventing fertilization in many females, who usually mate only once. Sex attractant pheromones are now used to control the gypsy moth and the Mediterranean fruit fly.

Practice Exercise 13.2

Determine whether each of the following substituted alkenes can exist in *cis–trans* isomeric forms.

a. 1-Chloropropene **b.** 2-Chloropropene

• *Answers:* **a.** yes; **b.** no

13.6 Naturally Occurring Alkenes

Pheromones *are compounds used by insects (and some animals) to transmit messages to other members of the same species.* They are often alkenes or alkene derivatives. The biological activity of alkene-type pheromones is usually highly dependent on whether the double bonds present are in a *cis* or a *trans* arrangement (Section 13.5).

The sex attractant of the female silkworm is a 16-carbon alkene derivative containing an —OH group. Two double bonds are present, *trans* at carbon 10 and *cis* at carbon 12.

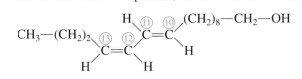

$$CH_3—(CH_2)_2—\overset{13}{C}H=\overset{12}{C}H—\overset{11}{C}H=\overset{10}{C}H—(CH_2)_8—CH_2—OH$$

This compound is 10 billion times more effective in eliciting a response from the male silkworm than is the 10-*cis*–12-*trans* isomer and 10 trillion times more effective than the isomer wherein both bonds are in a *trans* configuration.

Terpenes *are compounds whose carbon skeleton can be divided into two or more units that are identical to the carbon skeleton of isoprene.* Isoprene (2-methyl-1,3-butadiene) is a five-carbon diene.

$$\overset{1}{C}H_2=\overset{2}{C}(CH_3)—\overset{3}{C}H=\overset{4}{C}H_2$$

2-Methyl-1,3-butadiene
(Isoprene)

$$\text{tail} \rightarrow C—C(C)—C—C \leftarrow \text{head}$$

Isoprene unit

The molecule β-carotene is responsible for the yellow-orange color of carrots, apricots, and yams.

Terpenes are formed by joining the tail of one isoprene unit to the head of another unit. The isoprene unit maintains its isopentyl structure in a terpene, usually with modification of the isoprene double bonds.

Terpenes are among the most widely distributed compounds in the biological world, with over 22,000 structures known. Such compounds are responsible for the odors of many trees and many characteristic plant fragrances.

The number of carbon atoms present in a terpene is always a multiple of the number 5 (10, 15, and so on). Parts (a) and (b) of Figure 13.6 give the structures of selected 10- and 15-carbon terpenes found in plants. Beta-carotene is a terpene whose structure has 40 carbon atoms present in 8 isoprene units (Figure 13.6c).

In the human body, dietary beta-carotene (obtained by eating yellow-colored vegetables) serves as a precursor for vitamin A; splitting of a beta-carotene molecule produces two vitamin A molecules (Section 21.15). An additional role of beta-carotene in the body, independent of its vitamin A function, is that of antioxidant. An antioxidant is a substance that helps protect cells from damage from reactive oxygen-derived species called free radicals.

Figure 13.6
Selected terpenes containing two or three and eight isoprene units. Dashed lines in the structures separate the individual isoprene units.

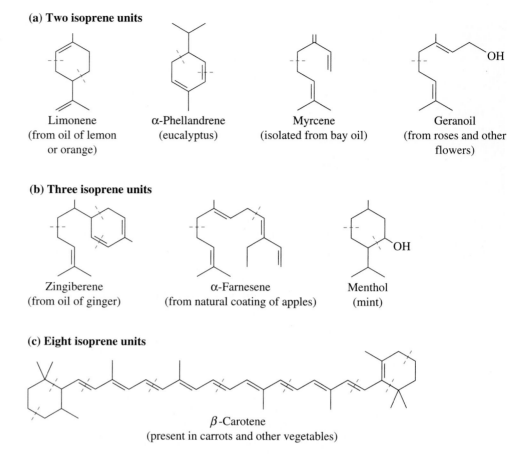

(a) Two isoprene units

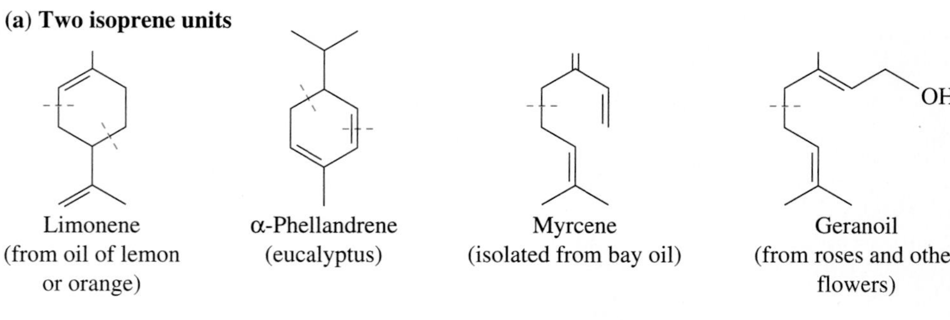

Limonene (from oil of lemon or orange)

α-Phellandrene (eucalyptus)

Myrcene (isolated from bay oil)

Geranoil (from roses and other flowers)

(b) Three isoprene units

OH

Zingiberene (from oil of ginger)

α-Farnesene (from natural coating of apples)

Menthol (mint)

(c) Eight isoprene units

β-Carotene (present in carrots and other vegetables)

• In later chapters, we will encounter additional isoprene-based molecules important in the functioning of the human body. They include vitamin K (Section 21.14), coenzyme Q (Section 23.7), and cholesterol (Section 19.10).

13.7 Physical and Chemical Properties of Alkenes

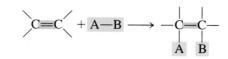

A physical-state summary for unbranched 1-alkenes and unsubstituted cycloalkenes with one double bond at room temperature and pressure.

Unbranched 1-Alkenes			
	C_3	C_5	C_7
C_2	C_4	C_6	C_8

Unsubstituted Cycloalkenes			
	C_3*	C_5	C_7
	C_4*	C_6	C_8

Gas Liquid

*Cyclopropene and cyclobutene are relatively unstable compounds, readily converting to other hydrocarbons, because of severe bond angle strain associated with a small ring containing a double bond.

The general physical properties of alkenes include insolubility in water, solubility in nonpolar solvents, and densities lower than that of water. Thus alkenes have physical properties similar to those of alkanes (Section 12.15). The melting point of an alkene is usually lower than that of the alkane with the same number of carbon atoms.

Alkenes with 2 to 4 carbon atoms are gases at room temperature. Unsubstituted alkenes with 5 to 17 carbon atoms and one double bond are liquids, and those with still more carbon atoms are solids.

Alkenes, like alkanes, are very flammable. The combustion products, as with any hydrocarbon, are carbon dioxide and water.

$$\underset{\text{Ethene}}{C_2H_4} + 3O_2 \longrightarrow 2CO_2 + 2H_2O$$

Aside from combustion, almost all other reactions of alkenes take place at the carbon–carbon double bond(s). Such reactions are called *addition reactions* because a substance is *added* to the double bond. This behavior contrasts with that of alkanes, where the most common reaction type, aside from combustion, is *substitution* (Section 12.16).

An **addition reaction** *is a reaction in which atoms or groups of atoms are added to each carbon atom of a carbon–carbon multiple bond.* A general equation for an alkene addition reaction is

$$>C=C< + A{-}B \longrightarrow -\underset{A}{\overset{|}{C}}-\underset{B}{\overset{|}{C}}-$$

Chemical CONNECTIONS

13.3 Carotenoids: A Source of Color

Carotenoids are the most widely distributed of the substances that give color to our world; they occur in flowers, fruits, plants, insects, and animals. These compounds are terpenes (Section 13.6) in which eight isoprene units are present. Structural formulas for two members of the carotenoid family, β-carotene and lycopene, follow.

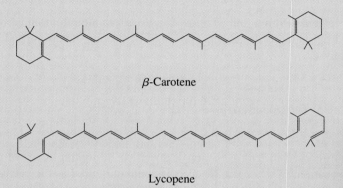

β-Carotene

Lycopene

Present in both of these carotenoid structures is a *conjugated* system of 11 double bonds. (Conjugated double bonds are double bonds separated from each other by one single bond.) Color is frequently caused by the presence of compounds that contain extended conjugated double bond systems. When *visible* light strikes these compounds, certain wavelengths of the visible light are absorbed by the electrons in the conjugated bond system. The unabsorbed wavelengths of visible light are reflected and are perceived as color.

The molecule β-carotene is responsible for the yellow-orange color in carrots, apricots, and yams. The yellowish tint of animal fat comes from β-carotene present in animal diets. The yellow-orange color of autumn leaves comes from β-carotene. Leaves contain chlorophyll (green pigment) and β-carotene (yellow pigment) in a ratio of approximately 3 to 1. The yellow-orange β-carotene color is masked by the chlorophyll until autumn, when the chlorophyll molecules decompose, as a result of lower temperatures and less sunlight, and are not replaced.

The molecule lycopene is the red pigment in tomatoes, paprika, and watermelon. Lycopene's structure differs from that of β-carotene in that the two rings in β-carotene have been opened. The ripening of a green tomato involves the gradual decomposition of chlorophyll with an associated unmasking of the red color of the lycopene present.

Carotenoids, molecules that contain eight isoprene units, are responsible for the yellow-orange color of autumn leaves.

In this reaction, the A part of the reactant A—B becomes attached to one carbon atom of the double bond, the B part to the other carbon atom (see Figure 13.7). As this occurs, the carbon–carbon double bond simultaneously becomes a carbon–carbon single bond.

A carbon–carbon double bond has a σ-bond component and a π-bond component (Section 13.4). The π-bond is weaker than the σ-bond and can be broken while the σ-bond

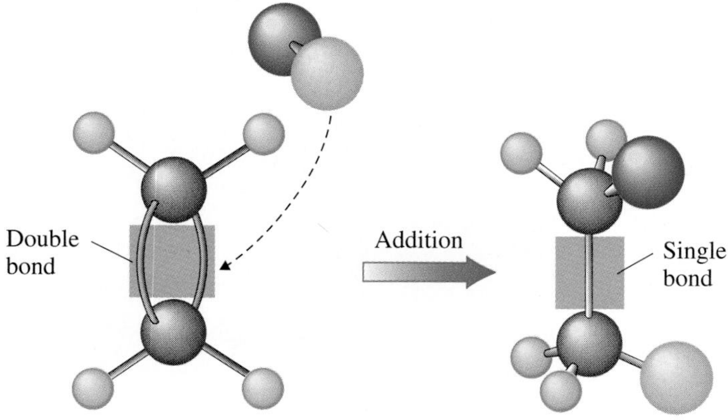

Figure 13.7
In an alkene addition reaction, the atoms provided by an incoming molecule are attached to the carbon atoms originally joined by a double bond. In the process, the double bond becomes a single bond.

• The following word associations are important to remember:

alkane—substitution reaction

alkene—addition reaction

An analogy can be drawn to a basketball team. When a *substitution* is made, one player leaves the game as another enters. The number of players on the court remains at five per team. If *addition* were allowed during a basketball game, two players could enter the game and no one would leave; there would be seven players per team on the court rather than five.

remains intact. This is what occurs in an alkene addition reaction. With the π-bond broken, each of the involved carbon atoms bonds to an atom (or group of atoms) supplied by the other reactant.

Addition reactions can be classified as symmetrical or unsymmetrical. In a **symmetrical addition reaction,** *identical atoms (or groups of atoms) are added to each carbon of the multiple bond.* In an **unsymmetrical addition reaction,** *different atoms (or groups of atoms) are added to the carbon atoms of the multiple bond.*

• Symmetrical Addition Reactions

The two most common examples of symmetrical addition reactions are hydrogenation and halogenation.

Hydrogenation of an alkene involves the addition of a hydrogen atom to each carbon of the double bond. Hydrogenation is usually accomplished by heating the alkene and H_2 in the presence of a catalyst.

$$\underset{\text{Propene}}{CH_2{=}CH{-}CH_3} + H_2 \xrightarrow[\substack{150°C \\ 12\text{–}15\text{ atm} \\ \text{pressure}}]{\text{Ni or Pt}} \underset{\text{Propane}}{\overset{H}{\overset{|}{C}}H_2{-}\overset{H}{\overset{|}{C}}H{-}CH_3}$$

• Hydrogenation of an alkene requires a catalyst. No reaction occurs if the catalyst is not present.

The identity of the catalyst used in hydrogenation is specified by writing it above the arrow in the chemical equation. In this case, Ni or Pt is commonly used as the catalyst. In general terms, hydrogenation of an alkene can be written as

$$\underset{\text{Alkene}}{\gt C{=}C\lt} + H_2 \xrightarrow[\substack{\text{Heat,} \\ \text{pressure}}]{\text{Ni or Pt}} \underset{\text{Alkane}}{-\overset{H}{\overset{|}{\underset{|}{C}}}-\overset{H}{\overset{|}{\underset{|}{C}}}-}$$

The hydrogenation of vegetable oils is a very important commercial process today. Vegetable oils from sources such as soybeans and cottonseeds are composed of long-chain organic molecules that contain several double bonds. When these oils are hydrogenated, they are converted to low-melting solids that are used in margarines and shortenings.

Halogenation of an alkene involves the addition of a halogen atom to each carbon of the double bond. Halogenation of alkenes most often involves Cl_2 or Br_2. No catalyst is needed. The product is always a dihalogenated alkene derivative.

$$\underset{\text{2-Butene}}{CH_3{-}CH{=}CH{-}CH_3} + Cl_2 \longrightarrow \underset{\text{2,3-Dichlorobutane}}{CH_3{-}\overset{Cl}{\overset{|}{C}}H{-}\overset{Cl}{\overset{|}{C}}H{-}CH_3}$$

A bromine in water solution is reddish brown (left). When a small amount of such a solution is added to an unsaturated hydrocarbon, the added solution is decolorized as the bromine adds to the hydrocarbon to form colorless dibromo compounds (right).

In general terms, halogenation of an alkene can be written as

$$\underset{\text{Alkene}}{>C{=}C<} + \underset{\text{Halogen}}{X_2} \longrightarrow \underset{\text{Dihalogenated alkane}}{-\overset{X}{\underset{|}{\overset{|}{C}}}-\overset{X}{\underset{|}{\overset{|}{C}}}-} \qquad (X = Cl, Br)$$

Bromination is often used to test for the presence of carbon–carbon double bonds in organic substances. Bromine in water or carbon tetrachloride is reddish brown. The dibromo compound(s) formed from the symmetrical addition of bromine to an organic compound is(are) colorless. Thus the decolorization of a Br_2 solution indicates the presence of carbon–carbon double bonds.

Unsymmetrical Addition Reactions

Two important types of unsymmetrical addition reactions are hydrohalogenation and hydration.

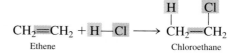

Hydrohalogenation of an alkene involves the addition of the reactant HCl, HBr, or HI to a carbon–carbon double bond. Hydrohalogenation reactions require no catalyst. For *symmetrical* alkenes, such as ethene, only one product results from hydrohalogenation.

$$\underset{\text{Ethene}}{CH_2{=}CH_2} + H{-}Cl \longrightarrow \underset{\text{Chloroethane}}{\overset{H}{\overset{|}{C}}H_2{-}\overset{Cl}{\overset{|}{C}}H_2}$$

Hydration of an alkene involves the addition of the reactant H_2O to a carbon–carbon double bond. Hydration requires a small amount of H_2SO_4 (sulfuric acid) as a catalyst. For *symmetrical* alkenes, only one product results from hydration.

• The addition of water to carbon–carbon double bonds occurs in many biochemical reactions that take place in the human body—for example, in the citric acid cycle (Section 23.6) and in the oxidation of fatty acids (Section 25.4).

$$\underset{\text{Ethene}}{CH_2{=}CH_2} + H{-}OH \xrightarrow{H_2SO_4} \underset{\text{Ethyl alcohol}}{\overset{H}{\overset{|}{C}}H_2{-}\overset{OH}{\overset{|}{C}}H_2}$$

In this equation, the water (H_2O) is written as H—OH to emphasize how this molecule adds to the double bond (the H goes to one carbon atom and the OH to the other carbon atom). Note also that the product of this hydration reaction contains an —OH group. Hydrocarbon derivatives of this type are called *alcohols.* Such compounds are the subject of Chapter 14.

When the alkene involved in a hydrohalogenation or hydration reaction is itself *unsymmetrical,* more than one product is possible. (An unsymmetrical alkene is one in which the two carbon atoms of the double bond are not equivalently substituted.) For example, the addition of HCl to propene (an unsymmetrical alkene) could produce either 1-chloropropane or 2-chloropropane, depending on whether the H from the HCl attaches itself to carbon 2 or carbon 1.

Vladimir Vasilevich Markovnikov (1837–1904). A professor of chemistry at several Russian universities, Markovnikov (pronounced Mar-cove-na-coff) synthesized rings containing four carbon atoms and seven carbon atoms, thereby disproving the notion of the day that carbon could form only five- and six-membered rings.

$$\underset{\text{Propene}}{CH_2{=}CH{-}CH_3} + HCl \longrightarrow \underset{\text{1-Chloropropane}}{\overset{Cl}{\overset{|}{C}}H_2{-}\overset{H}{\overset{|}{C}}H{-}CH_3}$$

or

$$\underset{\text{Propene}}{CH_2{=}CH{-}CH_3} + HCl \longrightarrow \underset{\text{2-Chloropropane}}{\overset{H}{\overset{|}{C}}H_2{-}\overset{Cl}{\overset{|}{C}}H{-}CH_3}$$

When two isomeric products are possible, one product often predominates. The dominant product can be predicted by using Markovnikov's rule, named after the Russian chemist Vladimir V. Markovnikov (1837–1904). **Markovnikov's rule** states that *when an unsymmetrical molecule of the form* HQ *adds to an unsymmetrical alkene, the hydrogen*

• Two catchy summaries of Markovnikov's rule are "Hydrogen goes where hydrogen is" and "The rich get richer" (in terms of hydrogen).

atom from the HQ *becomes attached to the unsaturated carbon atom that already has the most hydrogen atoms.* Thus the major product in our example involving propene is 2-chloropropane.

Example 13.3

Predicting Products in Alkene Addition Reactions Using Markovnikov's Rule

Using Markovnikov's rule, predict the predominant product in each of the following addition reactions.

a. $CH_3-CH_2-CH_2-CH=CH_2 + HBr \rightarrow$

b. 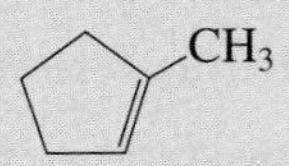 (1-methylcyclopentene) $CH_3 + HCl \rightarrow$

c. $CH_3-CH=CH-CH_2-CH_3 + HBr \rightarrow$

Solution

a. The hydrogen atom will add to carbon 1, because carbon 1 already contains more hydrogen atoms than carbon 2. The predominant product of the addition will be 2-bromopentane.

$$CH_3-CH_2-CH_2-\overset{②}{C}H=\overset{①}{C}H_2 + HBr \longrightarrow CH_3-CH_2-CH_2-\underset{}{\overset{Br}{\overset{|}{C}}}H-\overset{H}{\overset{|}{C}}H_2$$

b. Carbon 1 of the double bond does not have any H atoms directly attached to it. Carbon 2 of the double bond has one H atom (H atoms are not shown in the structure but are implied) attached to it. The H atom from the HCl will add to carbon 2, giving 1-chloro-1-methylcyclopentane as the product.

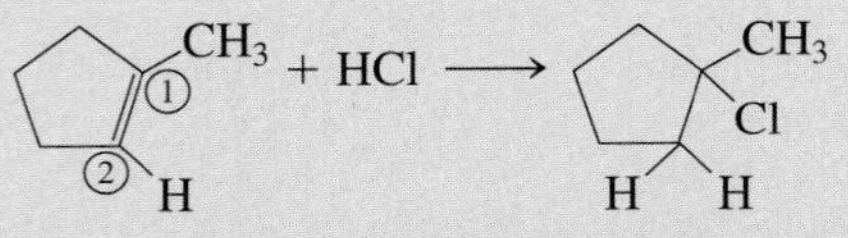

(cyclopentene ring with CH_3 on carbon ①, H on carbon ②) $+ HCl \longrightarrow$ (cyclopentane ring with CH_3 and Cl on one carbon, H H on adjacent carbon)

c. Each carbon atom of the double bond in this molecule has one hydrogen atom. Thus Markovnikov's rule does not favor either carbon atom. The result is two isomeric products that are formed in almost equal quantities.

$$CH_3-\underset{Br}{\underset{|}{C}}H-CH_2-CH_2-CH_3 \quad \text{and} \quad CH_3-CH_2-\underset{Br}{\underset{|}{C}}H-CH_2-CH_3$$

2-Bromopentane — 3-Bromopentane

Practice Exercise 13.3

Using Markovnikov's rule, predict the predominant product in each of the following addition reactions.

a. $CH_2=CH-CH_2-CH_3 + HCl \rightarrow$

b. 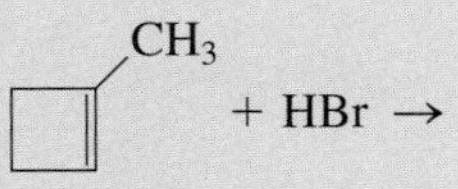 (1-methylcyclobutene) $CH_3 + HBr \rightarrow$

• *Answers:* **a.** $CH_3-\underset{Cl}{\underset{|}{C}}H-CH_2-CH_3$; **b.** 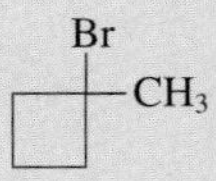 (cyclobutane ring with Br and CH_3 on the same carbon)

In compounds that contain more than one carbon–carbon double bond, such as dienes and trienes, addition can occur at each of the double bonds. In the complete hydrogenation of a diene and in that of a triene, the amount of hydrogen needed is twice as much and three times as much, respectively, as that needed for the hydrogenation of an alkene with one double bond.

$$CH_2{=}CH{-}CH_2{-}CH_2{-}CH_2{-}CH_3 + H_2 \xrightarrow{Ni} CH_3{-}(CH_2)_4{-}CH_3$$

$$CH_2{=}CH{-}CH{=}CH{-}CH_2{-}CH_3 + 2H_2 \xrightarrow{Ni} CH_3{-}(CH_2)_4{-}CH_3$$

$$CH_2{=}CH{-}CH{=}CH{-}CH{=}CH_2 + 3H_2 \xrightarrow{Ni} CH_3{-}(CH_2)_4{-}CH_3$$

Example 13.4

Predicting Reactants and Products in Alkene Addition Reactions

Supply the structural formula of the missing substance in each of the following addition reactions.

a. $CH_3{-}CH_2{-}CH{=}CH_2 + H_2O \xrightarrow{H_2SO_4} ?$

b. ? + Br_2 → (cyclohexane ring bearing Br, Br, and CH_3)

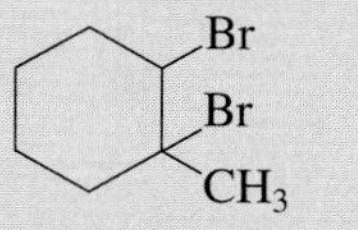

c. (1-methylcyclohexene, CH_3) + ? → (cyclohexane ring bearing Br and CH_3)

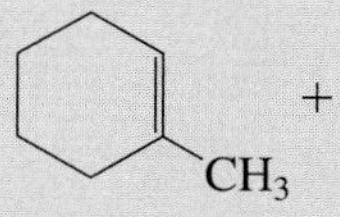

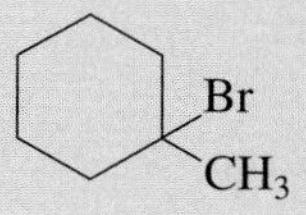

d. $CH_3{-}CH{=}CH{-}CH{=}CH_2 + 2H_2 \xrightarrow{Ni} ?$

Solution

a. This is a hydration reaction. Using Markovnikov's rule, we determine that the H will become attached to carbon 1, which has more hydrogen atoms than carbon 2, and that the —OH group will be attached to carbon 2.

$$CH_3{-}CH_2{-}\underset{}{\overset{OH}{\overset{|}{C}}}H{-}CH_3$$

b. The reactant alkene will have to have a double bond between the two carbon atoms that bromine atoms are attached to in the product.

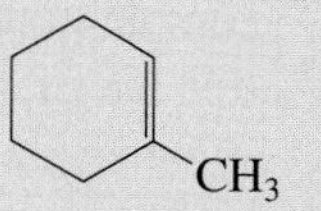

(1-methylcyclohexene, CH_3)

c. The small reactant molecule that adds to the double bond is HBr. The added Br atom from the HBr is explicitly shown in the product's structural formula, but the added H atom is not shown.

d. Hydrogen will add at each of the double bonds. The product hydrocarbon is pentane.

$$CH_3{-}CH_2{-}CH_2{-}CH_2{-}CH_3$$

Practice Exercise 13.4

Supply the structural formula of the missing substance in each of the following addition reactions.

a. $CH_3{-}CH_2{-}CH{=}CH_2 + HBr \rightarrow ?$

b. $? + H_2O \xrightarrow{H_2SO_4}$ (cyclopentane ring bearing OH)

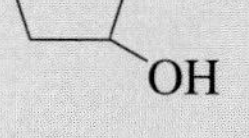

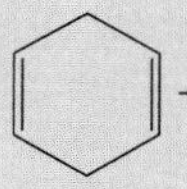

c. (cyclohexene) + 2? $\xrightarrow{Ni}$ (cyclohexane)

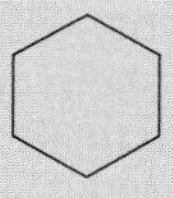

d. $CH_3{-}\underset{CH_3}{\underset{|}{C}}{=}CH{-}CH_3 + Cl_2 \longrightarrow ?$

• *Answers:* **a.** $CH_3{-}CH_2{-}\underset{Br}{\underset{|}{C}}H{-}CH_3$ **b.** (cyclopentene) **c.** H_2 **d.** $CH_3{-}\overset{Cl}{\overset{|}{\underset{CH_3}{\underset{|}{C}}}}{-}\overset{Cl}{\overset{|}{C}}H{-}CH_3$

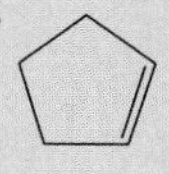

13.8 Polymerization of Alkenes: Addition Polymers

A **polymerization reaction** *is a reaction in which the repetitious combining of many small molecules produces a very large molecule.* The large molecule produced in this manner is called a polymer. A **polymer** *is a very large molecule composed of many identical repeating units.* The small repeating units are called *monomers.*

• The word *polymer* comes from the Greek *poly,* which means "many," and *meros,* which means "parts."

All the polymers discussed in this section are addition polymers. An **addition polymer** *is a polymer in which the monomers simply "add together," with no other products formed besides the polymer.* The monomers for addition polymers are most often alkenes or substituted alkenes. In an addition polymerization reaction, the π-bonds present in the alkene monomers are broken. The electrons that previously formed the π-bond now become involved in carbon–carbon bonds between the monomers.

• We will consider polymer types other than addition polymers in Sections 15.10 and 16.16.

The simplest alkene addition polymer has ethylene (ethene) as the monomer. With appropriate catalysts, ethylene readily adds to itself to produce polyethylene.

$$5\ \mathrm{CH_2{=}CH_2} \xrightarrow{\text{Catalyst}} \mathrm{-CH_2-CH_2-CH_2-CH_2-CH_2-CH_2-CH_2-CH_2-CH_2-CH_2-}$$

Polyethylene

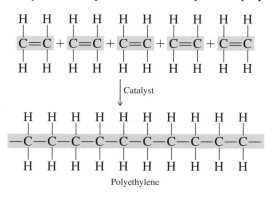

An *exact* formula for a polymer such as polyethylene cannot be written, because the length of the carbon chain varies from polymer molecule to polymer molecule. In recognition of this "inexactness" of formula, the notation used for denoting polymer formulas is independent of carbon chain length. We write the formula of the simplest repeating unit (the monomer with the double bond changed to a single bond) in parentheses and then add the subscript n after the parentheses, with n being understood to represent a very large number. Using this notation, we have, for the formula of polyethylene,

$$\left(\mathrm{-CH_2-CH_2-}\right)_n$$

This notation clearly identifies the basic repeating unit found in the polymer.

The properties of an ethene-based polymer depend not only on monomer identity but also on the average size (length) of polymer molecules and the extent of polymer branching. For example, for polyethylene, both high-density (HDPE) and low-density (LDPE) forms exist. HPDE, which has nonbranched polymer molecules, is a rigid material used in threaded bottle caps, toys, bottles, and milk jugs. LDPE, which has highly branched polymer molecules, is a flexible material used in plastic bags, plastic film, and squeeze bottles. Objects made of HDPE hold their shape in boiling water, while those made of LDPE become severely deformed at this temperature.

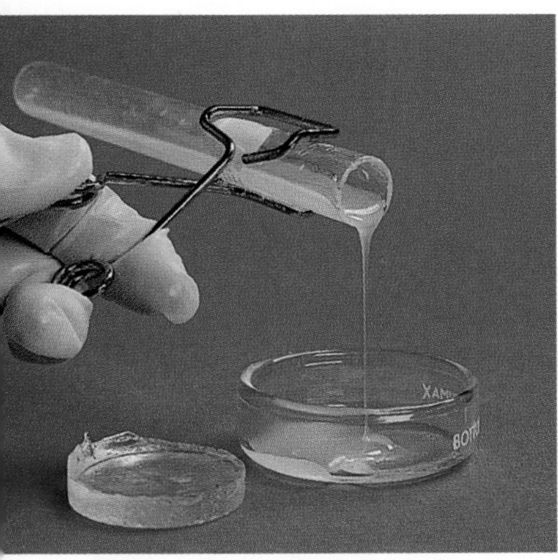

Preparation of polystyrene. When styrene, $C_6H_5CH{=}CH_2$, is heated with a catalyst (benzoyl peroxide), it yields a viscous liquid. After some time, this liquid sets to a hard plastic (sample shown at left).

Many substituted alkenes undergo polymerization in a manner similar to that for ethene when treated with the proper catalyst. For a monosubstituted-ethene monomer, the general polymerization equation is

$$H_2C{=}CH(Z) \xrightarrow{\text{Polymerization}} \left(CH_2{-}CH(Z)\right)_n$$

Variation in the substituent group Z can change polymer properties dramatically, as is shown by the entries in Table 13.2, a listing of formulas and uses for selected substituted-ethene-based polymers. Figure 13.8 contrasts the structures of polyethylene, polypropylene, and polyvinylchloride via space-filling models.

When dienes such as 1,3-butadiene are used as the monomers in addition polymerization reactions, the resulting polymers contain double bonds and are thus still unsaturated.

$$\underset{\text{1,3-Butadiene}}{CH_2{=}CH{-}CH{=}CH_2} \xrightarrow{\text{Polymerization}} \underset{\text{Polybutadiene}}{\left(CH_2{-}CH{=}CH{-}CH_2\right)_n}$$

Table 13.2
Some Common Polymers Obtained from Ethene-Based Monomers

Polymer formula and name	Monomer formula and name	Uses of polymer
$\left(CH_2{-}CH_2\right)_n$ polyethylene	$CH_2{=}CH_2$ ethylene	bottles, plastic bags, toys, electrical insulation
$\left(CH_2{-}CH(CH_3)\right)_n$ polypropylene	$CH_2{=}CH(CH_3)$ propylene	indoor-outdoor carpeting, bottles, molded parts (including heart valves)
$\left(CH_2{-}CH(Cl)\right)_n$ poly(vinyl chloride) (PVC)	$CH_2{=}CH(Cl)$ vinyl chloride	plastic wrap, bags for intravenous drugs, garden hose, plastic pipe, simulated leather (Naugahyde)
$\left(CF_2{-}CF_2\right)_n$ Teflon	$CF_2{=}CF_2$ tetrafluoroethylene	cooking utensil coverings, electrical insulation, component of artificial joints in body parts replacement
$\left(CH_2{-}CH(C_6H_5)\right)_n$ polystyrene	$CH_2{=}CH(C_6H_5)$ styrene	toys, styrofoam packaging, cups, simulated wood furniture

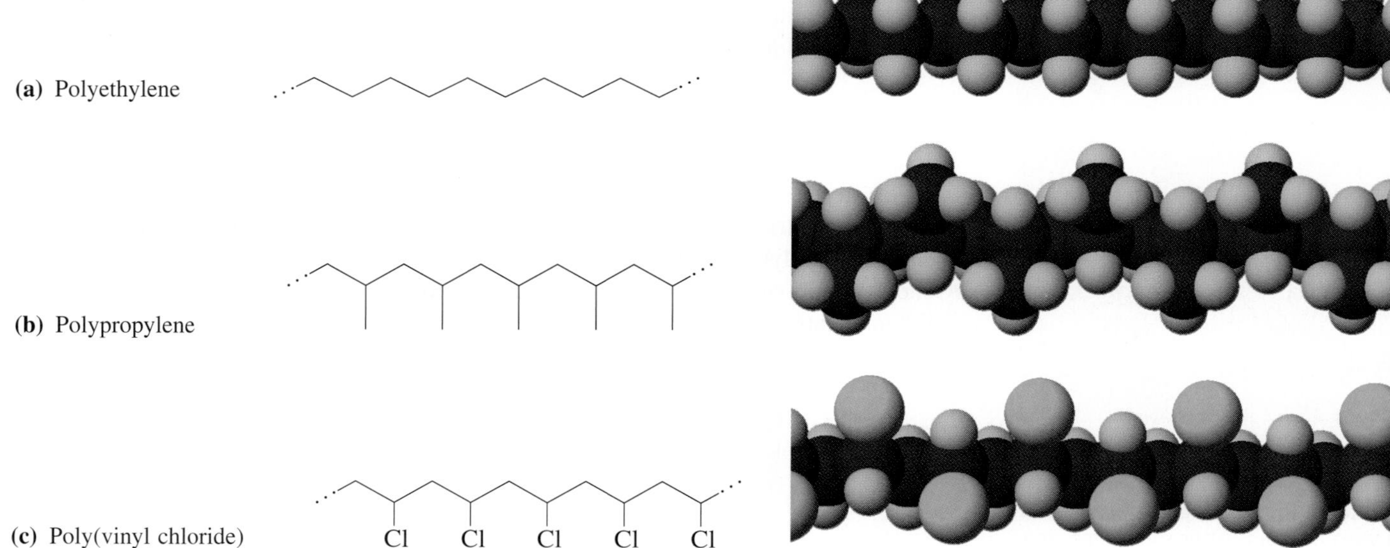

Figure 13.8
Line-angle drawings and space-filling models of segments of the ethene-based polymers (a) polyethylene, (b) polypropylene, and (c) polyvinylchloride.

In general, unsaturated polymers are much more flexible than the ethene-based saturated polymers listed in Table 13.2. Natural rubber is a flexible addition polymer whose repeating unit is isoprene (Section 13.6)—that is, 2-methyl-1,3-butadiene.

$$CH_2{=}\underset{\displaystyle CH_3}{\underset{|}{C}}{-}CH{=}CH_2 \xrightarrow{\text{Polymerization}} \left(CH_2{-}\underset{\displaystyle CH_3}{\underset{|}{C}}{=}CH{-}CH_2\right)_n$$

Isoprene (2-methyl-1,3-butadiene) → Polyisoprene (natural rubber)

When two or more different monomers are reacted to form a single polymer, the resulting polymer is called a *copolymer.* The commercial product Saran Wrap is a copolymer of chloroethene (vinyl chloride) and 1,1-dichloroethene.

$$\begin{matrix} H & & H \\ | & & | \\ C & = & C \\ | & & | \\ Cl & & H \end{matrix} + \begin{matrix} Cl & & H \\ | & & | \\ C & = & C \\ | & & | \\ Cl & & H \end{matrix} \xrightarrow{\text{Polymerization}} \left(\begin{matrix} H & & H & & Cl & & H \\ | & & | & & | & & | \\ C & - & C & - & C & - & C \\ | & & | & & | & & | \\ Cl & & H & & Cl & & H \end{matrix} \right)_n$$

Vinyl chloride + 1,1-Dichloroethene → Saran Wrap (1st monomer, 2nd monomer)

Natural rubber being harvested in Malaysia.

Another important copolymer is styrene–butadiene rubber, the leading synthetic rubber in use today. It contains the monomers 1,3-butadiene and styrene in a 3 : 1 ratio.

The accompanying Chemistry at a Glance summarizes much of what we have said about alkenes.

13.9 Alkynes

Alkynes represent a second type of unsaturated hydrocarbon. An **alkyne** *is an acyclic unsaturated hydrocarbon in which one or more carbon–carbon triple bonds are present.* As the family name *alkyne* indicates, the characteristic "ending" associated with a triple bond is *-yne.*

The general formula for an alkyne with one triple bond is C_nH_{2n-2}. Thus the simplest member of this type of alkyne has the formula C_2H_2, and the next member, with $n = 3$, has the formula C_3H_4.

CHEMISTRY AT A GLANCE

ALKENE

Combustion
- Products are CO_2 and H_2O
- Does not involve the process of addition

Addition

Addition of a small symmetrical molecule

Addition of a small unsymmetrical molecule

Polymerization
- Alkene molecules undergo an addition reaction with one another
- Specific catalysts are needed

H_2 Hydrogenation	Br_2 or Cl_2 Halogenation	HBr or HCl Hydrohalogenation	H_2O (H–OH) Hydration
$H_2 + \text{C=C} \xrightarrow{\text{Ni or Pt catalyst}}$ –C(H)–C(H)–	$X_2 + \text{C=C} \longrightarrow$ –C(X)–C(X)–	$HX + \text{C=C} \longrightarrow$ –C(H)–C(X)–	$H{-}OH + \text{C=C} \xrightarrow{H_2SO_4 \text{ catalyst}}$ –C(H)–C(OH)–
Alkane	**Dihaloalkane**	**Monohaloalkane***	**Alcohol***

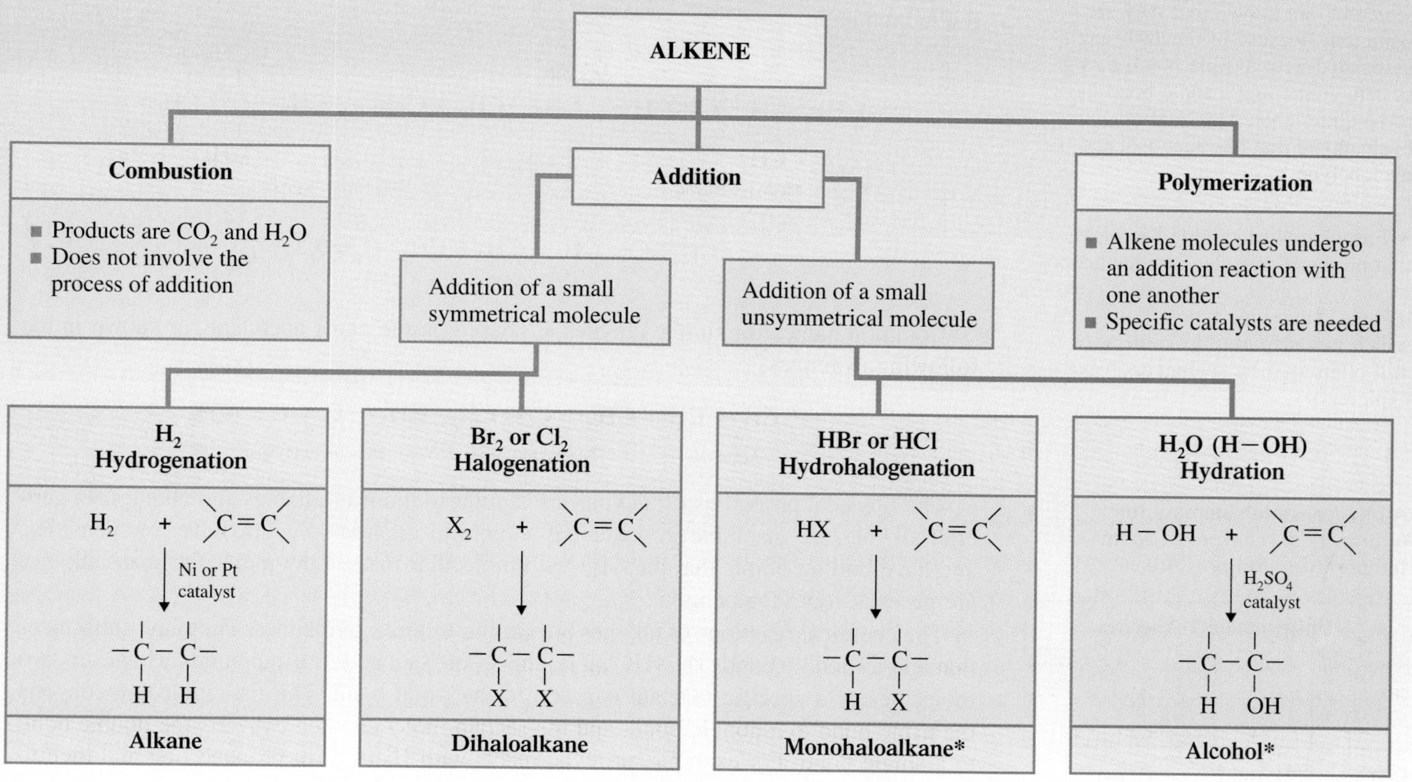

*Markovnikov's rule is needed to predict the product's exact structure if the alkene is unsymmetrical.

- Line-angle drawings for the simpler 1-alkynes:

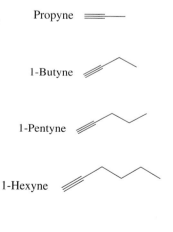

Propyne

1-Butyne

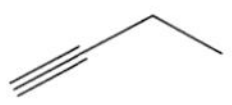

1-Pentyne

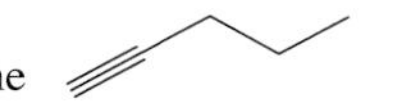

1-Hexyne

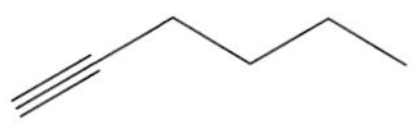

$CH{\equiv}CH$ — Ethyne $CH{\equiv}C{-}CH_3$ — Propyne

The presence of a carbon–carbon triple bond in a molecule always results in a linear arrangement for the two atoms attached to the carbons of the triple bond. Thus, ethyne is a linear molecule (see Figure 13.9).

Because of the linearity (180° angles) about the triple bond, *cis–trans* isomerism, like that found for 2-butene ($CH_3{-}CH{=}CH{-}CH_3$), is not possible for alkynes such as 2-butyne ($CH_3{-}C{\equiv}C{-}CH_3$).

The simplest alkyne, ethyne (C_2H_2), has the common name *acetylene*. In industry, acetylene is the most important of all alkynes. This colorless gas, which has a boiling point of −84°C, has traditionally been prepared by the reaction between calcium carbide and water.

$$CaC_2(s) + 2H_2O(l) \longrightarrow C_2H_2(g) + Ca(OH)_2(aq)$$

When acetylene is burned with oxygen in an oxyacetylene welding torch, a very high temperature is produced. These torches are used extensively for cutting and welding metals.

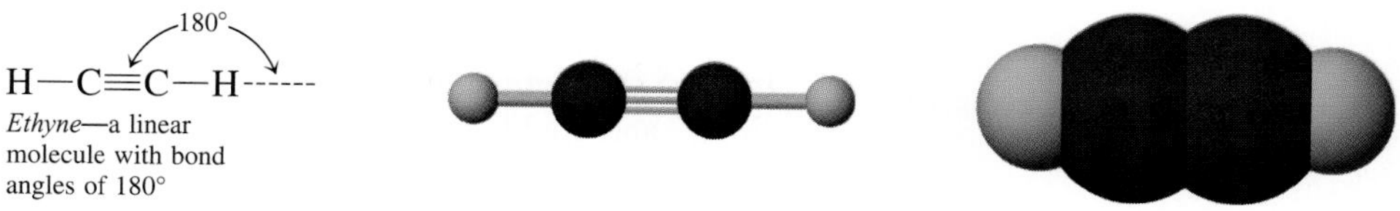

$H{-}C{\equiv}C{-}H$ (180°)

Ethyne—a linear molecule with bond angles of 180°

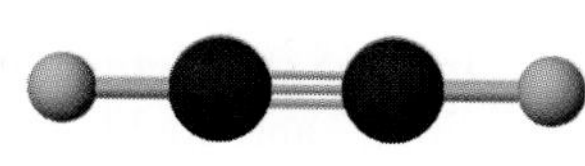

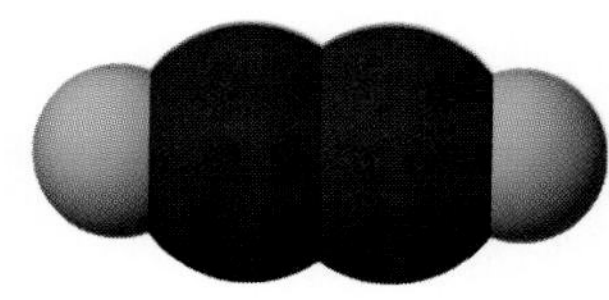

Figure 13.9 Structural representations of ethyne (acetylene), the simplest alkyne. The molecule is linear—that is, all four atoms lie in a straight line.

• *Cycloalkynes,* molecules that contain a triple bond as part of a ring structure, are known, but they are not common. Because of the 180° angle associated with a triple bond, a ring system containing a triple bond has to be quite large. The smallest cycloalkyne that has been isolated is cyclooctyne.

• Early cars had carbide headlights that produced acetylene by the action of slowly dripping water on calcium carbide. This same type of lamp, which was also used by miners, is still often used by spelunkers (cave explorers).

A physical-state summary for unbranched 1-alkynes at room temperature and pressure.

Unbranched 1-Alkynes			
	C_3	C_5	C_7
C_2	C_4	C_6	C_8

□ Gas ■ Liquid

• Students often ask whether it is possible to have hydrocarbons in which both double and triple bonds are present. The answer is yes. Immediately, another question is asked. How do we name such compounds? Such compounds are called *alkenynes*. An example is

$$CH\equiv C-CH=CH_2$$

1-Buten-3-yne

A double bond has priority over a triple bond in numbering the chain when numbering systems are equivalent. Otherwise, the chain is numbered from the end closest to a multiple bond.

• The name Kekulé is pronounced "Keck-u-la."

The rules for naming alkynes are identical to those used to name alkenes (Section 13.3), except the ending *-yne* is used instead of *-ene*. Consider the following structures and their IUPAC names.

$$\overset{4}{C}H_3-\overset{3}{C}H(CH_3)-\overset{2}{C}\equiv\overset{1}{C}H$$

3-Methyl-1-butyne

$$\overset{1}{C}H_3-\overset{2}{C}H_2-\overset{3}{C}\equiv\overset{4}{C}-\overset{5}{C}H_2-\overset{6}{C}(CH_3)_2-\overset{7}{C}H_3$$

6,6-Dimethyl-3-heptyne

$$\overset{1}{C}H\equiv\overset{2}{C}-\overset{3}{C}H_2-\overset{4}{C}H_2-\overset{5}{C}H_2-\overset{6}{C}\equiv\overset{7}{C}H$$

1,6-Heptadiyne

Common names for simple alkynes are based on the name acetylene, as shown in the following examples.

$$CH\equiv CH \qquad CH_3-C\equiv CH \qquad CH_3-C\equiv C-CH_3$$

Acetylene · Methylacetylene · Dimethylacetylene

The physical properties of alkynes are similar to those of alkenes and alkanes. In general, alkynes are insoluble in water but soluble in organic solvents, have low densities, and have boiling points that increase with molecular mass. Low-molecular-mass alkynes are gases at room temperature.

The chemical reactions of alkynes are similar to those of alkenes. The same substances that add to double bonds (H_2, HCl, Cl_2, and so on) can add to triple bonds. However, two molecules of a specific reactant can add to the triple bond. The first molecule converts the triple bond to a double bond, and the second molecule then converts the double bond to a single bond. For example, propyne reacts with H_2 to form propene first and then to form propane.

$$\underset{\text{An alkyne (propyne)}}{CH\equiv C-CH_3} \xrightarrow[Ni]{H_2} \underset{\text{An alkene (propene)}}{CH_2=CH-CH_3} \xrightarrow[Ni]{H_2} \underset{\text{An alkane (propane)}}{CH_3-CH_2-CH_3}$$

13.10 Aromatic Hydrocarbons

Aromatic hydrocarbons *are unsaturated cyclic compounds that do not readily undergo addition reactions.* This reaction behavior, which is almost opposite to that of alkenes and alkynes, explains the separate classification of aromatic hydrocarbons.

The distinctiveness of aromatic hydrocarbons is related to the chemical bonding present in such compounds. A discussion of the bonding features present in benzene, the simplest aromatic hydrocarbon, will be used to illustrate important features of the aromatic hydrocarbon bonding model.

Properties of benzene that must be accounted for through bonding considerations include

1. A molecular structure that involves a six-membered ring of carbon atoms in which all carbon–carbon bonds are identical.
2. A molecular structure in which each carbon atom has one hydrogen atom attached to it, with all hydrogen atoms equivalent to one another.
3. A molecular formula of C_6H_6, indicating a high degree of unsaturation because of the low hydrogen-to-carbon ratio.
4. A chemical reactivity different from that normally associated with an *unsaturated* compound. Substitution reactions, rather than addition reactions, occur.

In 1865 the German chemist August Kekulé (1829–1896) proposed a cyclic, alternating single- and double-bonded structure for benzene.

Friedrich August Kekulé (1829–1896) originally entered the German University of Giessen to study architecture but switched to chemistry after taking a chemistry course. In 1890, on the 25th anniversary of his proposal on the cyclic structure of benzene, he gave a speech in which he stated that the idea for benzene's structure came as a result of his dozing off in front of a fire while working on a textbook. He dreamed of chains of carbon atoms twisting and turning in a snake-like motion, when suddenly the head of one snake seized hold of its own tail and formed a spinning ring. This concept—that chains of carbon atoms could have cyclic structures—was a new idea at the time.

- The use of the resonance concept to describe bonding in a compound is not unique to organic chemistry. Many inorganic compounds also exhibit resonance.

- We know that neither of the Kekulé structures for benzene is, by itself, correct, because if it were, benzene would have three separate double bonds and would undergo addition reactions. Benzene does not undergo addition reactions.

or

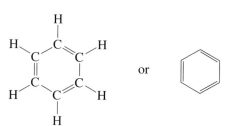

Kekulé further proposed that the double and single bonds in the carbon ring of benzene were oscillating rapidly around the ring. In order to indicate the oscillation of bonds around the ring, Kekulé drew two equivalent structures for benzene that differed only in the location of the double bonds (1,3,5 positions versus 2,4,6 positions) and connected the two structures by using a double-headed arrow.

Such bond oscillation would make all carbon–carbon bonds equivalent.

The only difference between Kekulé's two structures for benzene is the arrangement of bonding electrons. Structures that differ in this manner are called resonance structures. **Resonance structures** *are two or more Lewis structures for a molecule that differ only in the arrangement of bonding electrons.* When resonance structures are used to denote a molecule, it is understood that the "true structure" of the molecule is obtained by combining the properties of the resonance structures. In the case of benzene, the real structure is intermediate between the two Kekulé structures.

An alternative notation for denoting the bonding in benzene—a notation that involves a single structure—is

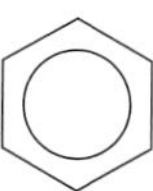

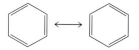

In this "circle-in-the-ring" structure for benzene, the circle denotes a system of π-bonds in which each of the carbon atoms participates. Each carbon atom in benzene participates in three σ-bonds and one π-bond. Two of each carbon's σ-bonds are to adjacent carbon atoms, and the third is to a hydrogen atom. This σ-bonding is shown in the following incomplete structure for benzene.

The fourth valence electron of each carbon atom is in a p orbital (Section 3.6) that is perpendicular to the plane of the benzene ring (Figure 13.10a). It is these electrons, six in number (one from each carbon atom), that interact to produce the π-bonding in a benzene molecule. The six mutually parallel p orbitals overlap (both above and below the plane of the ring) to produce a donut-shaped π-bond that involves all six carbon atoms (Figure 13.10b). The six electrons involved in the π-bond are shared by all six carbon atoms.

The π-bonding present in benzene is an example of delocalized bonding. A **delocalized bond** *is a covalent bond in which electrons are shared among three or more atoms.*

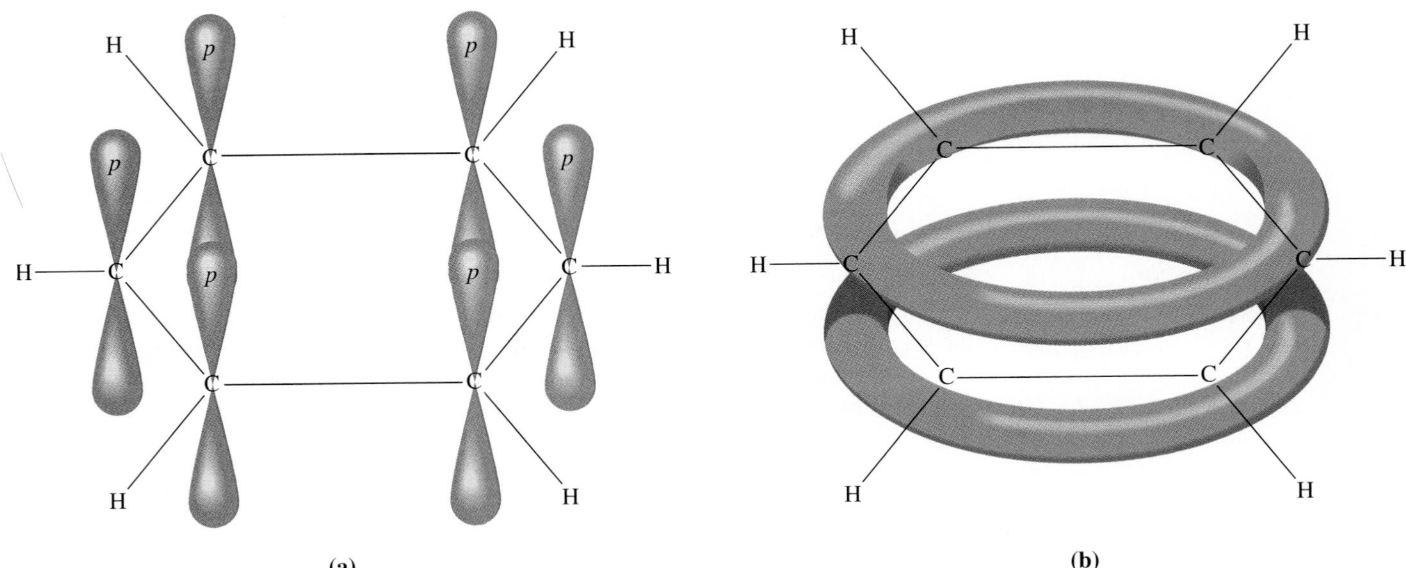

Figure 13.10
(a) The fourth valence electron of each carbon atom is in a *p* orbital that is perpendicular to the plane of the benzene ring. (b) The π-bonding present in a benzene molecule involves six mutually parallel *p* orbitals that interact to form a donut-shaped π-bond that involves all six carbon atoms of the ring.

• The novel bonding feature in benzene, compared to other types of hydrocarbons, is the side-to-side overlap (π-overlap) of *p* orbitals from all six carbon atoms to create a delocalized π-bond that encompasses the entire carbon ring structure.

• In 1825 Michael Faraday isolated a new hydrocarbon, which he called "bicarburet of hydrogen," from illuminating gas, a gaseous material obtained from coal. Nine years later, the same substance was prepared by heating benzoic acid with lime. Eventually, because of its relationship to benzoic acid, this compound became known as *benzin* and then later as *benzene,* the IUPAC name by which it is known today.

(All bonds we have considered up to this point have been localized bonds, covalent bonds in which electrons are shared between *two* atoms.)

The delocalized π-bonding present in the benzene molecule imparts additional stability to it. This phenomenon, which is known as *aromaticity,* is what causes benzene and other aromatic molecules to be resistant to addition reactions. Addition reactions would require breaking up the delocalized bonding within the ring so that additional bonds could form between carbon and the new, added species.

13.11 Names for Aromatic Hydrocarbons

Replacement of one or more of the hydrogen atoms on benzene with other groups produces benzene derivatives. Often the new group or groups attached to the benzene ring are alkyl groups (methyl, ethyl, and so on) or halogens. Other commonly encountered groups are hydroxyl ($-OH$), amino ($-NH_2$), nitro ($-NO_2$), and carboxyl ($-COOH$) groups.

The IUPAC system of naming monosubstituted benzene derivatives uses the name of the substituent as a prefix to the name of benzene. Examples of this type of nomenclature include

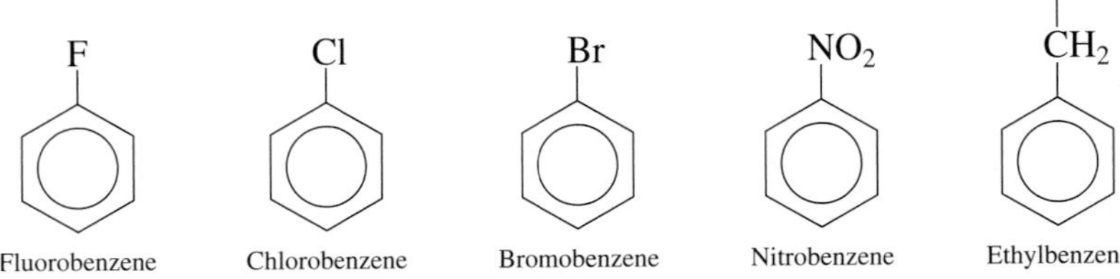

A few monosubstituted benzenes have names wherein the substituent and the benzene ring taken together constitute a new parent name. Four important examples of such nomenclature are

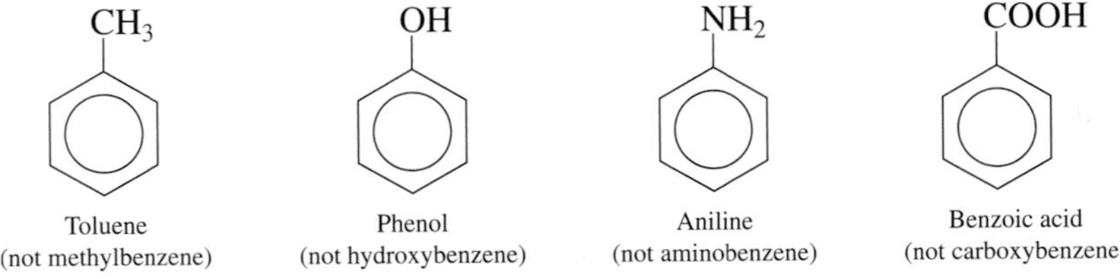

Space-filling and ball-and-stick models for the structure of benzene.

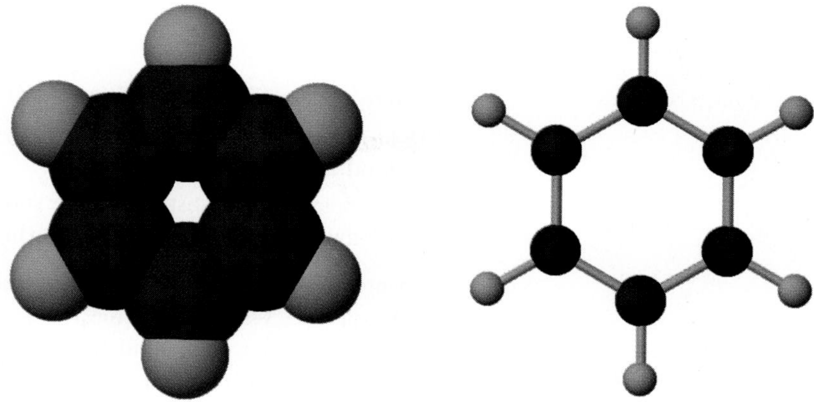

Because all the hydrogen atoms in benzene are equivalent, it does not matter at which carbon of the ring the substituted group is located. Each of the following formulas represents toluene.

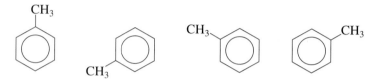

CH_3 CH_3 CH_3 CH_3

For monosubstituted benzene rings that have a group attached that is not easily named as a substituent, the benzene ring is often treated as a group attached to this substituent. In this reversed approach, the benzene ring attachment is called a *phenyl* group, and the compound is named according to the rules for naming alkanes, alkenes, and alkynes.

• The word *phenyl* comes from "phene," a European term used during the 1800s for benzene. The word is pronounced *fen*-nil.

$$CH_2{=}CH{-}CH{-}CH_3$$

3-Phenyl-1-butene

When two substituents, either the same or different, are attached to a benzene ring, three isomeric structures are possible.

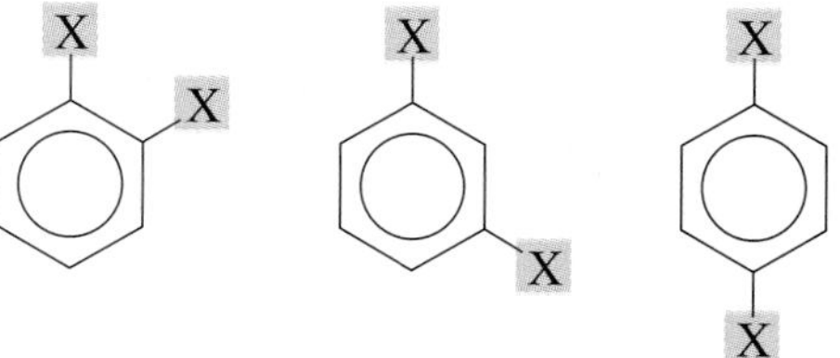

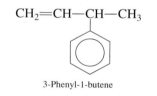

To distinguish among these three isomers, we must specify the positions of the substituents relative to one another. This can be done in either of two ways: by using numbers and by using nonnumerical prefixes.

When numbers are used, the three isomeric dichlorobenzenes have the first-listed set of names:

• *Cis–trans* isomerism is not possible for disubstituted benzenes. All 12 atoms of benzene are in the same plane—that is, benzene is a flat molecule. When a substituent group replaces an H atom, the atom that bonds the group to the ring is also in the plane of the ring.

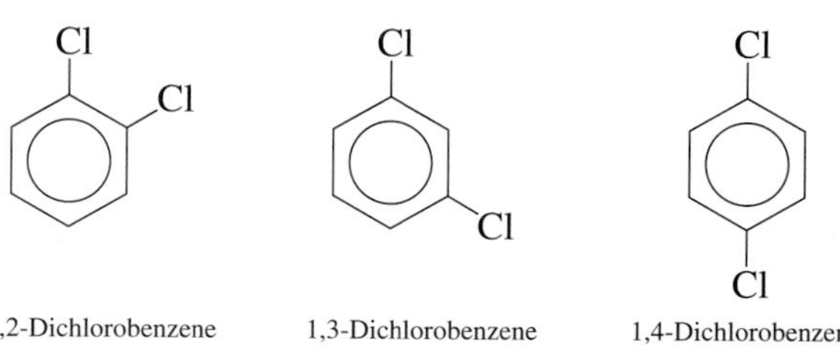

1,2-Dichlorobenzene (*ortho*-dichlorobenzene) 1,3-Dichlorobenzene (*meta*-dichlorobenzene) 1,4-Dichlorobenzene (*para*-dichlorobenzene)

• Learn the meaning of the prefixes *ortho-*, *meta-*, and *para-*. These prefixes are extensively used in naming disubstituted benzenes.

X

← *ortho* to X

← *meta* to X

para to X

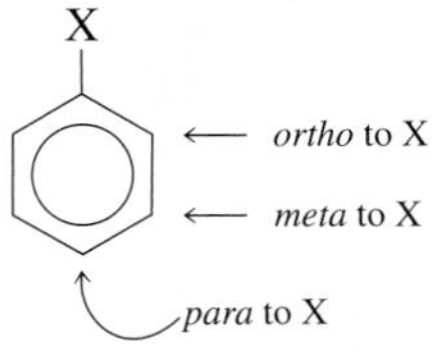

• The use of *ortho-*, *meta-*, and *para-* in place of position numbers is reserved exclusively for disubstituted benzenes. The system is never used with cyclohexanes or other ring systems.

• The dimethylbenzenes have the common name *xylene.*

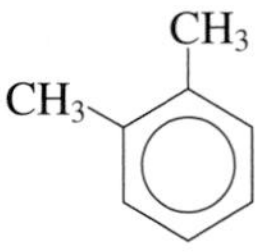

CH_3 CH_3

o-Xylene

The xylenes are good solvents for grease and oil and are used for cleaning microscope slides and optical lenses and for removing wax from skis.

• When parent names such as *toluene, xylene,* and *phenol* are used, additional substituents present cannot be the same as those included in the parent name. If such is the case, name the compound as a substituted benzene. The compound

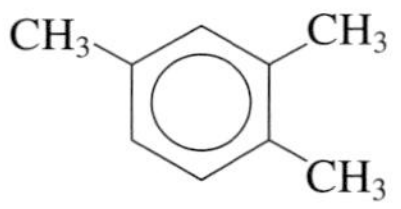

CH_3 CH_3 CH_3

is named as a trimethylbenzene and not as a methylxylene or a dimethyltoluene.

The prefix system uses the prefixes *ortho-*, *meta-*, and *para-* (abbreviated *o-*, *m-*, and *p-*).

Ortho- means 1,2 disubstitution; the substituents are on adjacent carbon atoms.

Meta- means 1,3 disubstitution; the substituents are one carbon removed from each other.

Para- means 1,4 disubstitution; the substituents are two carbons removed from each other (on opposite sides of the ring).

When prefixes are used, the three isomeric dichlorobenzenes have the second-listed set of names above. When one of the two substituents in a disubstituted benzene imparts a special name to the compound (as, for example, toluene or phenol), the compound is named as a derivative of that parent molecule. The special substituent is assumed to be at ring position one.

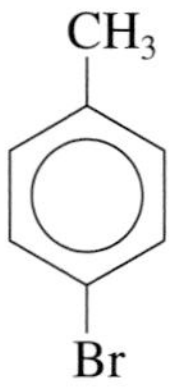

CH_3 Br — 4-Bromotoluene (not 1-bromo-4-methylbenzene)

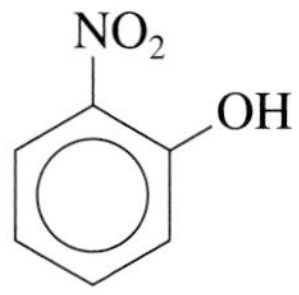

NO_2 OH — 2-Nitrophenol (not 1-hydroxy-2-nitrobenzene)

When neither substituent group imparts a special name, the substituents are cited in alphabetical order before the ending *-benzene.* The carbon of the benzene ring bearing the substituent with alphabetical priority becomes carbon 1.

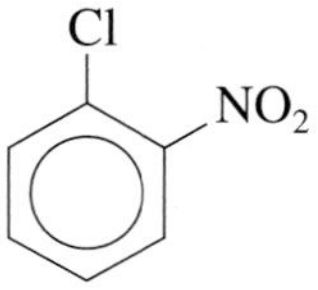

Cl NO_2 — 1-Chloro-2-nitrobenzene (not 2-chloro-1-nitrobenzene)

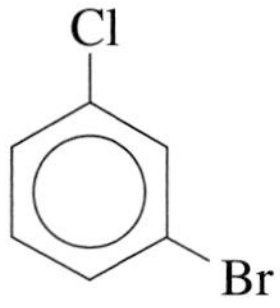

Cl Br — 1-Bromo-3-chlorobenzene (not 3-bromo-1-chlorobenzene)

When more than two groups are present on the benzene ring, their positions are indicated with *numbers.* The ring is numbered so as to obtain the lowest possible numbers for the carbon atoms that have substituents. If there is a choice of numbering systems (two systems give the same lowest set), then the group that comes first alphabetically is given the lower number.

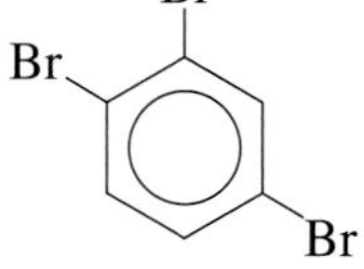

Br Br Br — 1,2,4-Tribromobenzene

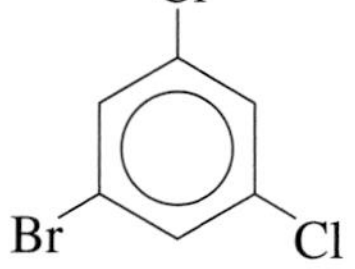

Cl Br Cl — 1-Bromo-3,5-dichlorobenzene

Example 13.5

Assigning IUPAC Names to Benzene Derivatives

Assign IUPAC names to the following benzene derivatives.

a. Cl, $CH_2—CH_3$

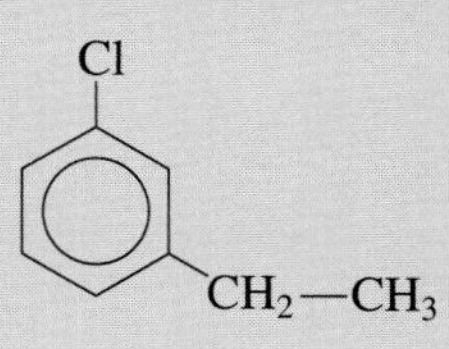

b. Br, Cl, NO_2

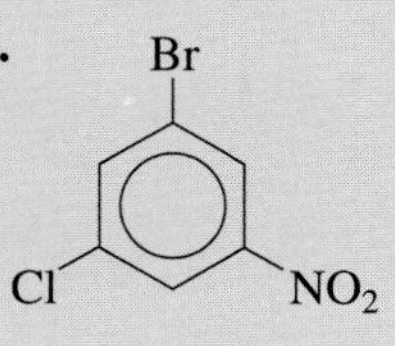

c. 2-bromo-3-phenylbutane structure: $CH_3-CH(Br)-CH(C_6H_5)-CH_3$

d. 2-chlorotoluene structure: benzene ring with CH_3 and Cl on adjacent carbons

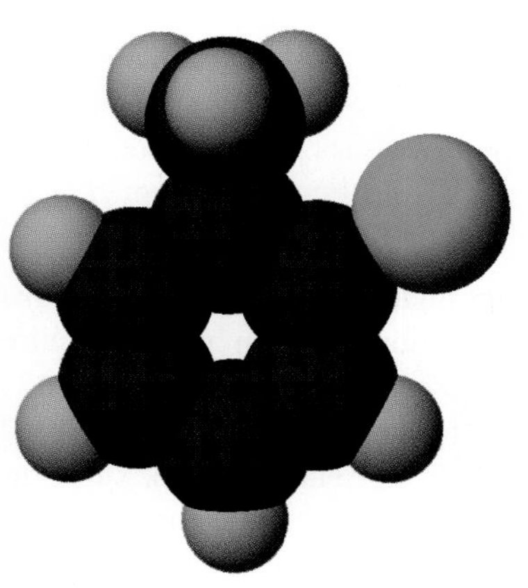

Space-filling model for the compound 2-chlorotoluene.

Solution

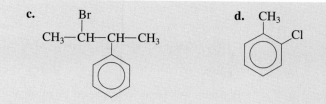

a. No substituents that will change the parent name from benzene are present on the ring. Alphabetical priority dictates that the chloro group is on carbon 1 and the ethyl group on carbon 3. The compound is named 1-chloro-3-ethylbenzene (or *m*-chloroethylbenzene).

b. Again, no substituents that will change the parent name from benzene are present on the ring. Alphabetical priority among substituents dictates that the bromo group is on carbon 1, the chloro group on carbon 3, and the nitro group on carbon 5. The compound is named 1-bromo-3-chloro-5-nitrobenzene.

c. This compound is named with the benzene ring treated as a substituent—that is, as a phenyl group. The compound is named 2-bromo-3-phenylbutane.

d. The methyl group present on the benzene ring changes the parent name from benzene to toluene. Carbon 1 bears the methyl group. Numbering clockwise, we obtain the name 2-chlorotoluene.

Practice Exercise 13.5

Assign IUPAC names to the following benzene derivatives.

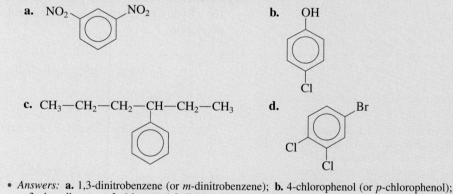

a. benzene ring with two NO_2 groups (meta)

b. benzene ring with OH and Cl (para)

c. $CH_3-CH_2-CH_2-CH(C_6H_5)-CH_2-CH_3$

d. benzene ring with Br, Cl, Cl

- *Answers:* **a.** 1,3-dinitrobenzene (or *m*-dinitrobenzene); **b.** 4-chlorophenol (or *p*-chlorophenol); **c.** 3-phenylhexane; **d.** 4-bromo-1,2-dichlorobenzene

13.12 Aromatic Hydrocarbons: Physical Properties and Sources

In general, aromatic hydrocarbons resemble other hydrocarbons in physical properties. They are insoluble in water, are good solvents for other nonpolar materials, and are less dense than water.

Benzene, monosubstituted benzenes, and many disubstituted benzenes are liquids at room temperature. Benzene itself is a colorless, flammable liquid that burns with a sooty flame because of incomplete combustion.

At one time, coal tar was the main source of aromatic hydrocarbons. Petroleum is now the primary source of such compounds. At high temperatures, with special catalysts, saturated hydrocarbons obtained from petroleum can be converted to aromatic hydrocarbons. The production of toluene from heptane is representative of such a conversion.

$$CH_3-CH_2-CH_2-CH_2-CH_2-CH_2-CH_3 \xrightarrow[\text{High temperature}]{\text{Catalyst}} C_6H_5-CH_3 + 4H_2$$

• Two common situations in which a person can be exposed to low-level benzene vapors are:
1. Inhaling of gasoline vapors while refueling an automobile. Gasoline contains about 2% (v/v) benzene.
2. Being around a cigarette smoker. Benzene is a combustion product present in cigarette smoke. For smokers themselves, inhaled cigarette smoke is a serious benzene-exposure source.

Benzene was once widely used as an organic solvent. Such use has been discontinued because of now known short- and long-term toxic effects. Benzene inhalation can cause nausea and respiratory problems.

13.13 Chemical Reactions of Aromatic Hydrocarbons

We have noted that aromatic hydrocarbons do not readily undergo the addition reactions characteristic of other unsaturated hydrocarbons. An addition reaction would require breaking up the delocalized π-bonding (Section 13.10) present in the ring system.

If benzene is so unresponsive to addition reactions, what reactions does it undergo? Benzene undergoes *substitution* reactions. As you recall from Section 12.16, substitution reactions are characterized by different atoms or groups of atoms replacing hydrogen atoms in a hydrocarbon molecule. There are four important types of substitution reactions for benzene and other aromatic hydrocarbons: alkylation, halogenation, nitration, and sulfonation.

1. *Alkylation:* An alkyl group (R—) from an alkyl chloride (R—Cl) substitutes for a hydrogen atom on the benzene ring. A catalyst, $AlCl_3$, is needed for alkylation.

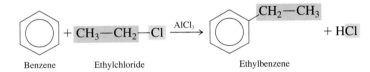

$$C_6H_6 + CH_3-CH_2-Cl \xrightarrow{AlCl_3} C_6H_5-CH_2-CH_3 + HCl$$

Benzene, Ethylchloride, Ethylbenzene

In general terms, the alkylation of benzene can be written as

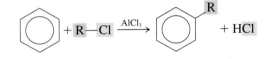

$$C_6H_6 + R-Cl \xrightarrow{AlCl_3} C_6H_5-R + HCl$$

• Alkylation, the reaction that attaches an alkyl group to an aromatic ring, is also known as a *Friedel–Crafts reaction,* named after Charles Friedel and James Mason Crafts, the French and American chemists responsible for its discovery in 1877.

Alkylation is the most important industrial reaction of benzene.

2. *Halogenation* (bromination or chlorination): A hydrogen atom on a benzene ring can be replaced by bromine or chlorine if benzene is treated with Br_2 or Cl_2 in the presence of a catalyst. The catalyst is usually $FeBr_3$ for bromination and $FeCl_3$ for chlorination.

$$C_6H_6 + Br_2 \xrightarrow{FeBr_3} C_6H_5-Br + HBr$$

$$C_6H_6 + Cl_2 \xrightarrow{FeCl_3} C_6H_5-Cl + HCl$$

Aromatic halogenation differs from alkane halogenation (Section 12.16) in that light is not required to initiate aromatic halogenation.

3. *Nitration:* Nitric acid (HNO_3) reacts with benzene to form nitrobenzene. Nitric acid is often written as HO—NO_2 to emphasize the —NO_2 group that becomes a substituent on the ring. Concentrated sulfuric acid (H_2SO_4) is required as a catalyst in this reaction, and the reaction mixture must be heated.

CHEMISTRY AT A GLANCE

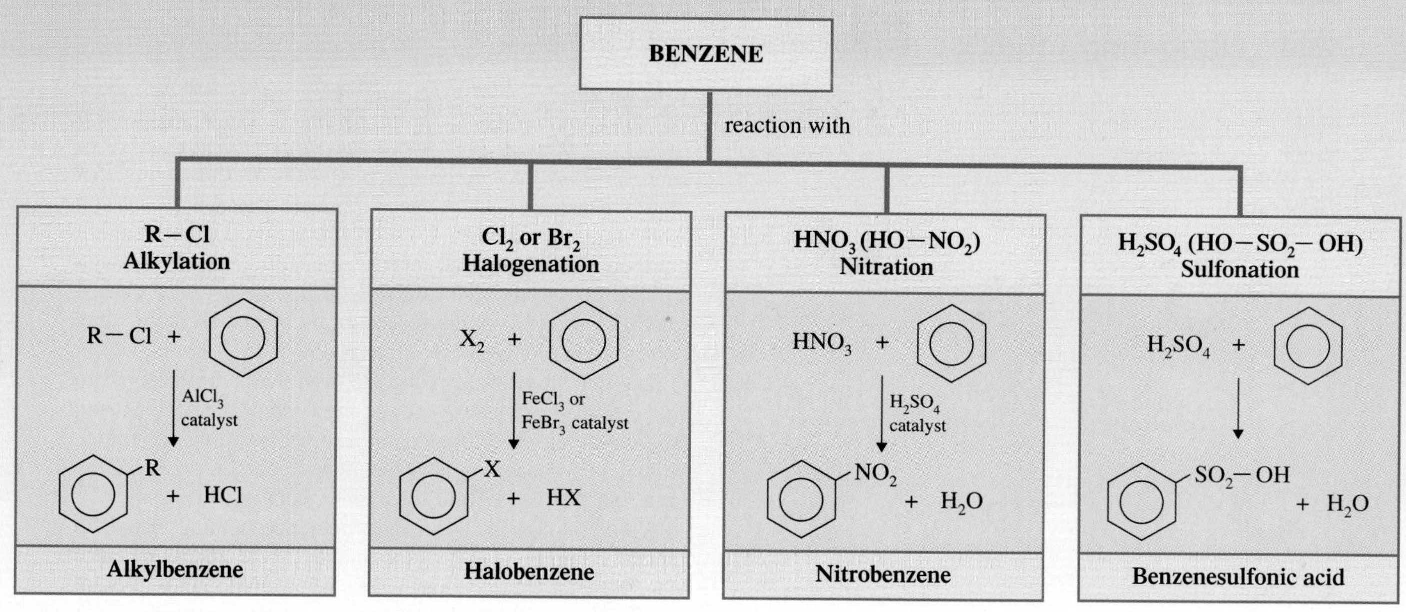

(The heating requirement is indicated in the equation by placing a small triangle under the reaction equation arrow.)

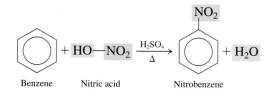

$$\text{Benzene} + HO\text{—}NO_2 \xrightarrow[\Delta]{H_2SO_4} \text{Nitrobenzene} + H_2O$$

Benzene Nitric acid Nitrobenzene

4. *Sulfonation:* Concentrated sulfuric acid (H_2SO_4) reacts with benzene to form benzenesulfonic acid.

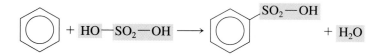

$$\text{Benzene} + HO\text{—}SO_2\text{—}OH \longrightarrow C_6H_5\text{—}SO_2\text{—}OH + H_2O$$

A sulfonic acid is a strong acid, about as strong as sulfuric acid. A better understanding of the structure of benzenesulfonic acid comes from writing the structure of the reacting sulfuric acid as HO—SO_2—OH. One of the —OH groups of sulfuric acid has been replaced with the aromatic ring of benzene. Aromatic-ring sulfonation is a key step in the synthesis of the family of antibiotics known as the sulfa drugs (Section 21.10).

The accompanying Chemistry at a Glance summarizes the benzene substitution reactions.

13.14 Fused-Ring Aromatic Compounds

Benzene and its substituted derivatives are not the only type of aromatic hydrocarbon that exists. Another large class of aromatic hydrocarbons is the fused-ring aromatic hydrocarbons. **Fused-ring aromatic hydrocarbons** *are aromatic compounds whose structures*

Chemical CONNECTIONS

13.4 Fused-Ring Aromatic Hydrocarbons and Cancer

A number of fused-ring aromatic hydrocarbons are known to be carcinogens—that is, to cause cancer. Three of the most potent carcinogens are

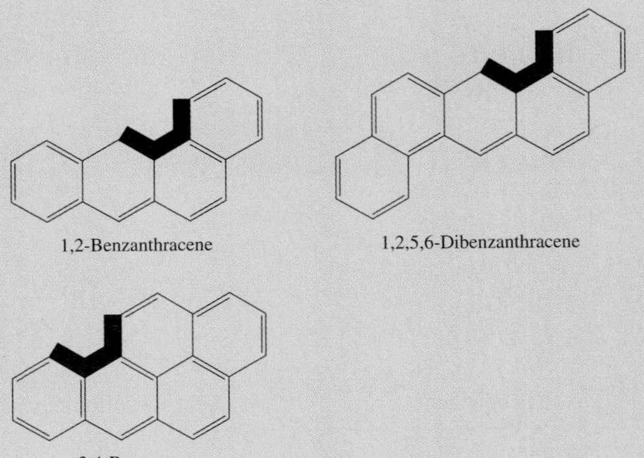

Very small amounts of these substances, when applied to the skin of mice, cause cancer.

Carcinogenic fused-ring aromatic hydrocarbons share some structural features. They all contain four or more fused rings, and they all have the same "angle" in the series of rings (the dark area in the structures shown). Fused-ring aromatic hydrocarbons are often formed when hydrocarbon materials are heated to high temperatures. These resultant compounds are present in low concentrations in tobacco smoke, in automobile exhaust, and sometimes in burned (charred) food. The charred portions of a well-done steak cooked over charcoal are a likely source.

Angular, fused-ring hydrocarbon systems are believed to be partially responsible for the high incidence of lung and lip cancer among cigarette smokers, because tobacco smoke contains 3,4-benzpyrene. The more a person smokes, the greater his or her risk of developing cancer.

We now know that the high incidence of lung cancer in British chimney sweeps (documented over 200 years ago) was caused by fused-ring hydrocarbon compounds present in the chimney soot that the sweeps inhaled regularly.

contain two or more rings fused together. Two carbon rings that share a pair of carbon atoms are said to be *fused.*

The three simplest fused-ring aromatic compounds are naphthalene, anthracene, and phenanthrene. All three are solids at room temperature.

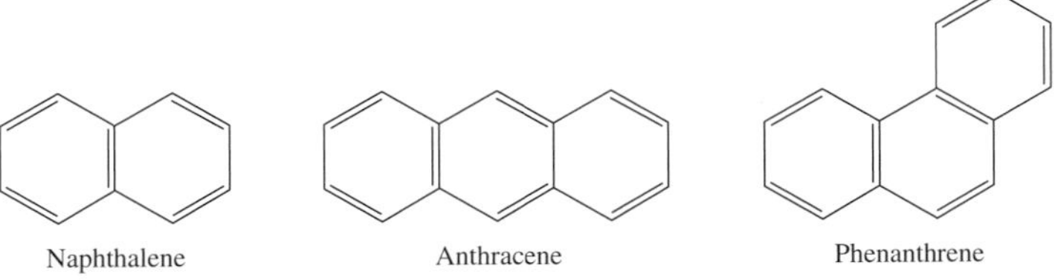

Concepts to Remember

Unsaturated hydrocarbons. An unsaturated hydrocarbon is a hydrocarbon that contains one or more carbon–carbon multiple bonds. Three main classes of unsaturated hydrocarbons exist: alkenes, alkynes, and aromatic hydrocarbons.

Alkenes and cycloalkenes. An alkene is an acyclic unsaturated hydrocarbon in which one or more carbon–carbon double bonds are present. A cycloalkene is a cyclic unsaturated hydrocarbon that contains one or more carbon–carbon double bonds within the ring system.

Alkene nomenclature. Alkenes and cycloalkenes are given IUPAC names using rules similar to those for alkanes and cycloalkanes, except that the ending *-ene* is used. Also, the double bond takes precedence both in selecting and in numbering the main chain or ring.

Nature of carbon–carbon multiple bonds. A carbon–carbon double bond consists of a σ-bond and a π-bond. A carbon–carbon triple bond consists of a σ-bond and two π-bonds. A σ-bond is a covalent bond in which the overlap between atomic orbitals lies along the axis joining the two bonded atoms. A π-bond is a covalent bond in which the overlap between atomic orbitals is above and below (but not on) the internuclear axis.

***Cis–trans* isomers.** Because rotation about a carbon–carbon double bond is restricted, some alkenes exist in two isomeric (*cis–trans*) forms. *Cis–trans* isomerism is possible when each carbon of the double bond is attached to two different groups.

Physical properties of alkenes. Alkenes and alkanes have similar physical properties. They are nonpolar, insoluble in water, less dense than water, and soluble in nonpolar solvents.

Addition reactions of alkenes. Numerous substances, including H_2, Cl_2, Br_2, HCl, HBr, and H_2O, add to the carbon–carbon double bond.

When both the alkene and the reactant are unsymmetrical, the addition proceeds according to Markovnikov's rule: The carbon atom of the double bond that already has the greater number of H atoms gets one more.

Addition polymers. Addition polymers are formed from alkene monomers that undergo repeated addition reactions with each other. Many familiar and widely used materials, such as fibers and plastics, are addition polymers.

Alkynes and cycloalkynes. Alkynes and cycloalkynes are unsaturated hydrocarbons that contain one or more carbon–carbon triple bonds. They are named in the same way as alkenes and cycloalkenes, except that their parent names end in *-yne*. Like alkenes, alkynes undergo addition reactions. These occur in two steps, an alkene forming first and then an alkane.

Aromatic hydrocarbons. Benzene, the simplest aromatic hydrocarbon, and other members of this family of compounds contain a six-membered ring with a cyclic, delocalized bond containing six π electrons. This aromatic ring is often drawn as a hexagon containing a circle, which represents the six electrons of the π-bond that move freely around the ring.

Nomenclature of aromatic hydrocarbons. Monosubstituted benzene compounds are named by adding the substituent name to the word *benzene.* Positions of substituents in disubstituted benzenes are indicated by using the numbering system or the *ortho-* (1,2), *meta-* (1,3), and *para-* (1,4) prefix system.

Reactions of aromatic hydrocarbons. Aromatic hydrocarbons undergo substitution reactions rather than addition reactions. Important substitution reactions are alkylation, halogenation, nitration, and sulfonation.

Key Reactions and Equations

1. Halogenation of an alkene (Section 13.7)

$$>C=C< + Br-Br \longrightarrow -\overset{|}{\underset{Br}{C}}-\overset{|}{\underset{Br}{C}}-$$

2. Hydrogenation of an alkene (Section 13.7)

$$>C=C< + H-H \xrightarrow{Ni} -\overset{|}{\underset{H}{C}}-\overset{|}{\underset{H}{C}}-$$

3. Hydrohalogenation of an alkene (Section 13.7)

$$>C=C< + H-Cl \longrightarrow -\overset{|}{\underset{H}{C}}-\overset{|}{\underset{Cl}{C}}-$$

4. Hydration of an alkene (Section 13.7)

$$>C=C< + H-OH \xrightarrow{H_2SO_4} -\overset{|}{\underset{H}{C}}-\overset{|}{\underset{OH}{C}}-$$

5. Hydrogenation of an alkyne (Section 13.9)

$$-C\equiv C- + H_2 \xrightarrow{Ni} -\overset{H}{C}=\overset{H}{C}- \xrightarrow[Ni]{H_2} -\overset{H}{\underset{H}{C}}-\overset{H}{\underset{H}{C}}-$$

6. Halogenation of an alkyne (Section 13.9)

$$-C\equiv C- + Br_2 \longrightarrow -\overset{Br}{C}=\overset{Br}{C}- \xrightarrow{Br_2} -\overset{Br}{\underset{Br}{C}}-\overset{Br}{\underset{Br}{C}}-$$

7. Hydrohalogenation of an alkyne (Section 13.9)

$$-C\equiv C- + HBr \longrightarrow -\overset{H}{C}=\overset{Br}{C}- \xrightarrow{HBr} -\overset{H}{\underset{H}{C}}-\overset{Br}{\underset{Br}{C}}-$$

8. Alkylation of benzene (Section 13.13)

$$C_6H_6 + R-Cl \xrightarrow{AlCl_3} C_6H_5R + HCl$$

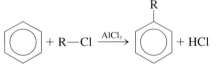

9. Halogenation of benzene (Section 13.13)

$$C_6H_6 + Br_2 \xrightarrow{FeBr_3} C_6H_5Br + HBr$$

10. Nitration of benzene (Section 13.13)

$$C_6H_6 + HO-NO_2 \xrightarrow[Heat]{H_2SO_4} C_6H_5NO_2 + HOH$$

11. Sulfonation of benzene (Section 13.13)

$$C_6H_6 + HO-SO_2-OH \longrightarrow C_6H_5SO_2-OH + HOH$$

Key Terms

Addition polymer (13.8)
Addition reaction (13.7)
Alkene (13.2)
Alkyne (13.9)
Aromatic hydrocarbons (13.10)
Cycloalkene (13.2)
Delocalized bonding (13.10)
Functional group (13.1)
Fused-ring aromatic hydrocarbons (13.14)
Markovnikov's rule (13.7)
Pheromones (13.6)
Pi bond (13.4)
Polymer (13.8)
Polymerization reaction (13.8)
Resonance structures (13.10)
Sigma bond (13.4)
Symmetrical addition reaction (13.7)
Terpenes (13.6)
Unsymmetrical addition reaction (13.7)

Exercises and Problems

The members of each pair of problems in this section test similar material.

Unsaturated Hydrocarbons with Double Bonds (Section 13.2)

13.1 Classify each of the following hydrocarbons as saturated or unsaturated. Further classify any unsaturated hydrocarbons as alkenes with one double bond, dienes, or trienes.

a. $CH_3-CH_2-CH=CH-CH_3$

b. $CH_3-CH_2-CH_2-CH_2-CH_3$

c.
$$\begin{array}{c} CH_3 \\ | \\ CH_2=C-C-CH_3 \\ || \\ CH_3CH_3 \end{array}$$

d. $CH_2=CH-CH_2-\underset{\underset{CH_2}{\|}}{C}-CH_3$

e. $CH_2=CH-CH=CH-CH=CH_2$

f. $CH_3-CH=C=CH-CH_3$

13.2 Classify each of the following hydrocarbons as saturated or unsaturated. Further classify any unsaturated hydrocarbons as alkenes with one double bond, dienes, or trienes.

a. $CH_3-CH=CH-CH=CH_2$

b. $CH_3-CH=CH-CH_3$

c. $CH_2=\underset{\underset{CH_3}{|}}{C}-CH_2-CH_3$

d. $CH_2=\underset{\underset{CH_3}{|}}{C}-CH=CH-CH=CH_2$

e. $CH_3-CH_2-CH_2-CH=CH_2$

f. $CH_3-\underset{\underset{CH_2}{\|}}{C}-CH_2-CH_2-\underset{\underset{CH_2}{\|}}{C}-CH_3$

13.3 Write the *molecular formula* for hydrocarbons with each of the following structural features.

a. Acyclic, four carbon atoms, no multiple bonds

b. Acyclic, five carbon atoms, one double bond

c. Cyclic, five carbon atoms, one double bond

d. Cyclic, seven carbon atoms, two double bonds

13.4 Write the *molecular formula* for hydrocarbons with each of the following structural features.

a. Acyclic, six carbon atoms, two double bonds

b. Acyclic, six carbon atoms, three double bonds

c. Cyclic, five carbon atoms, no multiple bonds

d. Cyclic, eight carbon atoms, four double bonds

13.5 Write the *general* molecular formula (C_nH_{2n}, and so on) for each of the following families of compounds.

a. Cycloalkene with one double bond

b. Alkadiene

c. Diene

d. Cycloalkatriene

13.6 Write the *general* molecular formula (C_nH_{2n}, and so on) for each of the following families of compounds.

a. Cycloalkadiene

b. Alkene with one double bond

c. Triene

d. Alkatriene

Names for Hydrocarbons Containing Double Bonds (Section 13.3)

13.7 Assign an IUPAC name to each of the following unsaturated hydrocarbons.

a. $CH_3-CH=CH-CH_3$

b. $CH_3-\underset{\underset{CH_3}{|}}{C}=CH-\underset{\underset{CH_3}{|}}{CH}-CH_3$

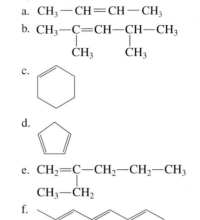

c.

d.

e. $CH_2=\underset{\underset{CH_3-CH_2}{|}}{C}-CH_2-CH_2-CH_3$

f.

13.8 Assign an IUPAC name to each of the following unsaturated hydrocarbons.

a. $CH_3-CH_2-CH=CH-CH_3$

b. $CH_3-CH_2-\underset{\underset{CH_3}{|}}{C}=CH-CH_3$

c.

d. CH_3

e. $CH_3-CH_2-\underset{\underset{CH_2}{\|}}{C}-CH_2-CH_3$

f.

13.9 Assign an IUPAC name to each of the hydrocarbons in Problem 13.1.

13.10 Assign an IUPAC name to each of the hydrocarbons in Problem 13.2.

13.11 Draw a condensed structural formula for each of the following unsaturated hydrocarbons.

a. 3-Methyl-1-pentene

b. 3-Methylcyclopentene

c. 1,3-Butadiene

d. 3-Ethyl-1,4-pentadiene

e. 4-Propyl-2-heptene

f. 3,6-Diethyl-1,4-cyclohexadiene

13.12 Draw a condensed structural formula for each of the following unsaturated hydrocarbons.

a. 4-Methyl-1-hexene

b. 4-Methylcyclohexene

c. 1,3-Pentadiene

d. 2-Ethyl-1,4-pentadiene
e. 1,3-Cyclohexadiene
f. 4,4,5-Trimethyl-2-heptene

13.13 The following names are *incorrect* by IUPAC rules. Determine the correct IUPAC name for each compound.
a. 2-Ethyl-2-pentene
b. 4,5-Dimethyl-4-hexene
c. 3,5-Cyclopentadiene
d. 1,2-Dimethyl-4-cyclohexene

13.14 The following names are *incorrect* by IUPAC rules. Determine the correct IUPAC name for each compound.
a. 2-Methyl-4-pentene
b. 3-Methyl-2,4-pentadiene
c. 3-Methyl-3-cyclopentene
d. 1,2-Dimethyl-3-cyclohexene

σ-Bonds and π-Bonds (Section 13.4)

13.15 Explain the differences between a σ-bond and a π-bond.

13.16 Characterize each of the following chemical bonds in terms of σ- and π-components.
a. Carbon–carbon double bond
b. Carbon–carbon single bond
c. Carbon–hydrogen single bond
d. Carbon–carbon triple bond

13.17 Determine the number of carbon–carbon σ-bonds and the number of carbon–carbon π-bonds in each of the following molecules.
a. $CH_3—CH_2—CH_2—CH_2—CH_3$
b. $CH_3—CH{=}CH—CH_3$
c. $CH_2{=}CH—CH{=}CH_2$
d. $CH{\equiv}CH$
e. CH_3
f.

13.18 Determine the number of carbon–carbon σ-bonds and the number of carbon–carbon π-bonds in each of the following molecules.
a. $CH_3—CH_2—CH{=}CH_2$
b. $CH_3—CH(CH_3)—CH(CH_3)—CH(CH_3)—CH_3$
c. $CH{\equiv}C—CH_2—CH_3$
d. $CH{\equiv}C—CH_2—CH{=}CH_2$
e. CH_3
f.

13.19 Determine the total number of σ-bonds and the total number of π-bonds in each of the molecules in Problem 13.17.

13.20 Determine the total number of σ-bonds and the total number of π-bonds in each of the molecules in Problem 13.18.

Isomerism in Alkenes (Section 13.5)

13.21 Draw skeletal formulas showing only carbon atoms, and give the IUPAC names, for the 13 possible alkene structural isomers with the formula C_6H_{12}. (Three of the structural isomers are hexenes, six are methylpentenes, three are dimethylbutenes, and one is an ethylbutene.)

13.22 Draw skeletal formulas showing only carbon atoms, and give the IUPAC names, for the 16 possible alkadiene structural isomers with the formula C_6H_{10}. (Six of the structural isomers are hexadienes, eight are methylpentadienes, one is a dimethylbutadiene, and one is an ethylbutadiene.)

13.23 For each molecule, tell whether *cis–trans* isomers exist. If they do, draw the two isomers and label them as *cis* and *trans*.
a. $CH_2{=}CH—CH_3$
b. $CH_2{=}CH—CH_2Cl$ (Cl on the terminal CH_2)
c. $CH_3—C(CH_3){=}CH—CH_3$
d. 3-Hexene
e. 4-Methyl-2-pentene
f. 1,2-Dimethylcyclopentane

13.24 For each molecule, tell whether *cis–trans* isomers exist. If they do, draw the two isomers and label them as *cis* and *trans*.
a. $CH_3—CH_2—CH{=}CH_2$
b. $CH_3—CH_2—C(Cl){=}CH_2$
c. $CH_3—CH_2—CH{=}CH—Cl$ (Cl on the terminal CH)
d. 2-Pentene
e. 1,2-Dichloroethene
f. 1,3-Dichlorocyclobutane

13.25 Assign an IUPAC name to each of the following molecules. Include the prefix *cis-* or *trans-* when appropriate.
a. CH_3, H on one carbon; $CH_2—CH_3$, H on the other: $(CH_3)(H)C{=}C(CH_2—CH_3)(H)$
b. $(I)(H)C{=}C(H)(Br)$
c. 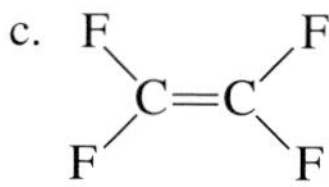
d. $(H)(CH_3)C{=}C(CH_3)(CH_3)$

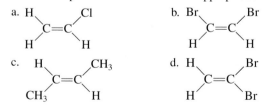

13.26 Assign an IUPAC name to each of the following molecules. Include the prefix *cis-* or *trans-* when appropriate.
a. $(H)(H)C{=}C(Cl)(H)$
b. $(Br)(H)C{=}C(Br)(H)$
c. 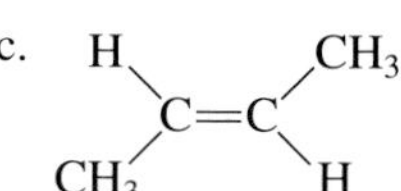
d. $(H)(H)C{=}C(Br)(Br)$

13.27 Draw a structural formula for each of the following compounds.
a. *trans*-3-Methyl-3-hexene
b. *cis*-2-Pentene
c. *trans*-5-Methyl-2-heptene
d. *trans*-1,3-Pentadiene

13.28 Draw a structural formula for each of the following compounds.
a. *trans*-2-Hexene
b. *cis*-4-Methyl-2-pentene
c. *cis*-1-Chloro-1-pentene
d. *cis*-1,3-Pentadiene

Alkene Addition Reactions (Section 13.7)

13.29 Which of the following reactions are addition reactions?

a. $C_4H_8 + Cl_2 \rightarrow C_4H_8Cl_2$
b. $C_6H_6 + Cl_2 \rightarrow C_6H_5Cl + HCl$
c. $C_3H_6 + HCl \rightarrow C_3H_7Cl$
d. $C_7H_{16} \rightarrow C_7H_8 + 4H_2$

13.30 Which of the following reactions are addition reactions?

a. $C_3H_6 + Cl_2 \rightarrow C_3H_6Cl_2$
b. $C_8H_{10} \rightarrow C_8H_8 + H_2$
c. $C_6H_6 + C_2H_5Cl \rightarrow C_8H_{10} + HCl$
d. $C_4H_8 + HCl \rightarrow C_4H_9Cl$

13.31 Write a chemical equation, showing reactants, products, and catalysts needed (if any), for the reaction of ethene with each of the following substances.

a. Cl_2 b. HCl
c. H_2 d. HBr

13.32 Write a chemical equation, showing reactants, products, and catalysts needed (if any), for the reaction of ethene with each of the following substances.

a. H_2O b. Br_2
c. HI d. I_2

13.33 Write a chemical equation, showing reactants, products, and catalysts needed (if any), for the reaction of propene with each of the reactants in Problem 13.31. Use Markovnikov's rule as needed.

13.34 Write a chemical equation, showing reactants, products, and catalysts needed (if any), for the reaction of propene with each of the reactants in Problem 13.32. Use Markovnikov's rule as needed.

13.35 Supply the structural formula of the product in each of the following alkene addition reactions.

a. $CH_3-CH=CH-CH_3 + Cl_2 \rightarrow ?$
b. $CH_3-C(CH_3)=CH_2 + HBr \rightarrow ?$
c. $CH_3-CH_2-CH=CH_2 + HCl \rightarrow ?$
d. 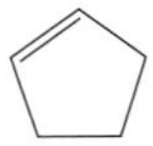 (cyclopentene) $+ H_2 \xrightarrow[\text{catalyst}]{\text{Ni}} ?$
e. 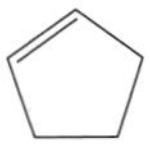 (cyclopentene) $+ H_2 \xrightarrow[\text{catalyst}]{\text{no}} ?$
f. 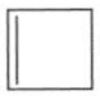 (cyclobutene) $+ H_2O \xrightarrow{H_2SO_4} ?$

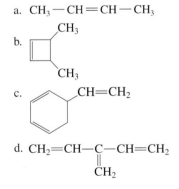

13.36 Supply the structural formula of the product in each of the following alkene addition reactions.

a. $CH_3-CH_2-CH=CH_2 + Cl_2 \rightarrow ?$
b. $CH_3-CH(CH_3)-CH=CH_2 + HBr \rightarrow ?$
c. $CH_3-C(CH_3)=C(CH_3)-CH_3 + HCl \rightarrow ?$
d. 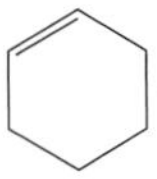 (cyclohexene) $+ H_2 \xrightarrow[\text{catalyst}]{\text{Ni}} ?$
e. 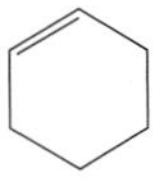 (cyclohexene) $+ H_2 \xrightarrow[\text{catalyst}]{\text{no}} ?$
f. $CH_3-CH=CH_2 + H_2O \xrightarrow{H_2SO_4} ?$

13.37 What reactant would you use to prepare each of the following from cyclohexene?

a. 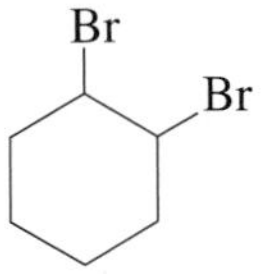 (1,2-dibromocyclohexane: Br, Br)
b. 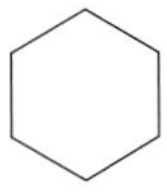 (cyclohexane)
c. 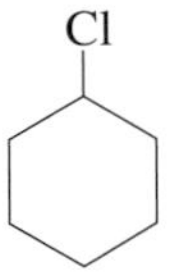 (chlorocyclohexane: Cl)
d. 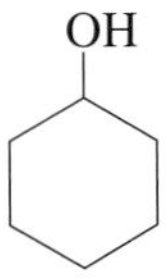 (cyclohexanol: OH)

13.38 What reactant would you use to prepare each of the following from cyclopentene?

a. 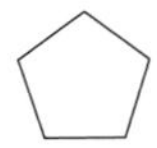 (cyclopentane)
b. 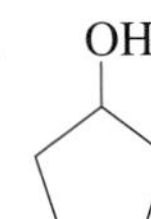

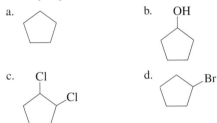

c. 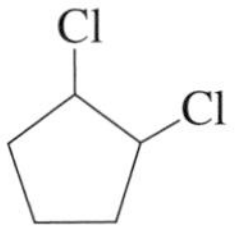 (1,2-dichlorocyclopentane: Cl, Cl)
d. 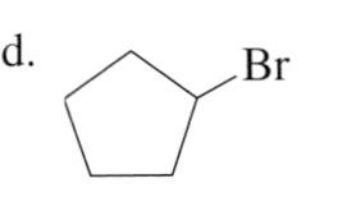

13.39 How many molecules of H_2 gas will react with 1 molecule of each of the following unsaturated hydrocarbons?

a. $CH_3-CH=CH-CH=CH-CH_3$
b. (methylcyclobutadiene: CH_3)

c. 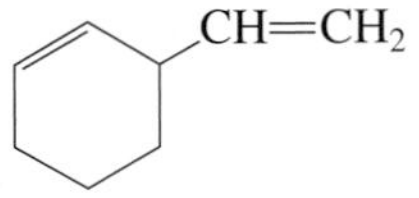 (vinylcyclohexadiene: $CH=CH_2$)
d. $CH_3-CH=C=C(CH_3)-CH=CH_2$

13.40 How many molecules of H_2 gas will react with 1 molecule of each of the following unsaturated hydrocarbons?

a. $CH_3-CH=CH-CH_3$
b. 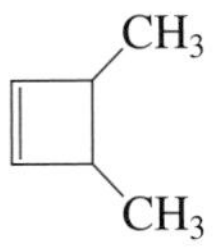 (dimethylcyclobutene: CH_3, CH_3)
c. 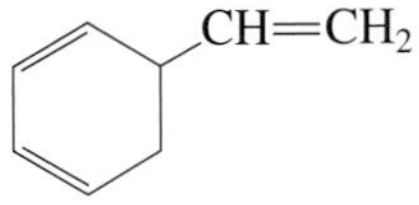 (vinylcyclohexadiene: $CH=CH_2$)
d. $CH_2=CH-C(=CH_2)-CH=CH_2$

Polymerization of Alkenes (Section 13.8)

13.41 Draw the structural formula of the monomer(s) from which each of the following polymers was made.

a. 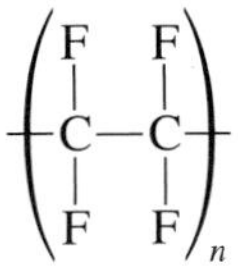 $\left(\begin{array}{cc} F & F \\ | & | \\ -C- & -C- \\ | & | \\ F & F \end{array}\right)_n$

b. 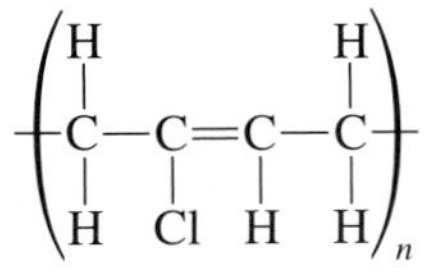 $\left(\begin{array}{cccc} H & & & H \\ | & & & | \\ -C- & C= & C- & C- \\ | & | & | & | \\ H & Cl & H & H \end{array}\right)_n$

c. $\left(\begin{array}{cc} H & H \\ | & | \\ -C- & C- \\ | & | \\ H & Cl \end{array}\right)_n$

d. 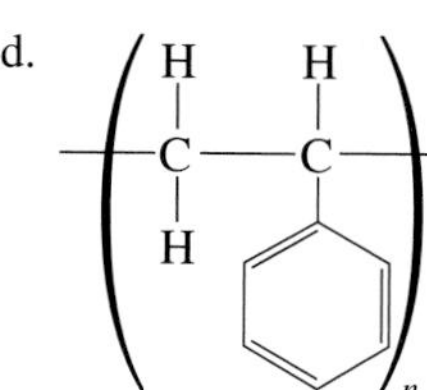

13.42 Draw the structural formula of the monomer(s) from which each of the following polymers was made.

a. $\left(\begin{array}{cc} H & F \\ | & | \\ -C- & C- \\ | & | \\ H & F \end{array}\right)_n$

b. $\left(\begin{array}{cccc} H & & & H \\ | & & & | \\ -C- & C= & C- & C- \\ | & | & | & | \\ H & Cl & Cl & H \end{array}\right)_n$

c. $\left(\begin{array}{cc} H & H \\ | & | \\ -C- & C- \\ | & | \\ Cl & CH_3 \end{array}\right)_n$

d. $\left(\begin{array}{cc} H & H \\ | & | \\ -C- & C- \\ | & | \\ Cl & C_6H_5 \end{array}\right)_n$

13.43 Draw the "start" (the first three repeating units) of the structural formula of the addition polymers made from the following monomers.

a. Ethylene
b. Vinyl chloride
c. 1,2-Dichloroethene
d. 1-Chloroethene

13.44 Draw the "start" (the first three repeating units) of the structural formula of the addition polymers made from the following monomers.

a. Propylene
b. 1,1,2,2-Tetrafluoroethene
c. 2-Methyl-1-propene
d. 1,2-Dichloroethylene

Alkynes (Section 13.9)

13.45 Assign an IUPAC name to each of the following unsaturated hydrocarbons.

a. $CH_3-CH_2-CH_2-CH_2-C\equiv CH$

b. $CH_3-C\equiv C-\underset{\displaystyle CH_3}{\underset{|}{CH}}-CH_3$

c. $CH_3-\underset{\displaystyle CH_3}{\underset{|}{\overset{\displaystyle CH_3}{\overset{|}{C}}}}-C\equiv C-CH_2-CH_2-CH_3$

d. $CH_3-C\equiv C-CH_3$

e. $CH\equiv C-\underset{\displaystyle CH_3}{\underset{|}{CH}}-C\equiv C-CH_3$

f. $CH_3-C\equiv C-\underset{\displaystyle CH_2-CH_3}{\underset{|}{CH}}-CH_3$

13.46 Assign an IUPAC name to each of the following unsaturated hydrocarbons.

a. $CH_3-\underset{\displaystyle CH_3}{\underset{|}{CH}}-C\equiv CH$

b. $CH_3-C\equiv C-CH_3$

c. $CH_3-\underset{\displaystyle CH_3}{\underset{|}{CH}}-C\equiv C-\underset{\displaystyle CH_3}{\underset{|}{CH}}-CH_3$

d. $CH_3-\underset{\displaystyle CH_3}{\underset{|}{\underset{\displaystyle CH_2}{\underset{|}{CH}}}}-CH_2-\underset{\displaystyle CH}{\underset{|||}{C}}$

e. $CH\equiv C-C\equiv CH$

f. $CH_3-CH_2-\underset{\displaystyle CH_2-CH_2-CH_3}{\underset{|}{CH}}-C\equiv CH$

13.47 Supply the structural formula of the product in each of the following alkyne addition reactions.

a. $CH\equiv CH + 2H_2 \xrightarrow{Ni} ?$

b. $CH_3-C\equiv CH + 2Br_2 \rightarrow ?$

c. $CH_3-C\equiv CH + 2HBr \rightarrow ?$

d. $CH\equiv CH + 1HCl \rightarrow ?$

e. C≡CH $+ 3H_2 \xrightarrow{Ni} ?$

f. $CH_3-CH_2-C\equiv CH + 1HBr \rightarrow ?$

13.48 Supply the structural formula of the product in each of the following alkyne addition reactions.

a. $CH_3-C\equiv C-CH_3 + 2Br_2 \rightarrow ?$

b. $CH_3-C\equiv C-CH_3 + 2HBr \rightarrow ?$

c. $CH\equiv C-CH_2-CH_3 + 1H_2 \xrightarrow{Ni} ?$

d. $CH\equiv C-CH_3 + 1HCl \rightarrow ?$

e. C≡CH $+ 4H_2 \xrightarrow{Ni} ?$

f. $CH_3-CH_2-C\equiv CH + 1HBr \rightarrow ?$

Nomenclature for Aromatic Compounds (Section 13.10)

13.49 Assign an IUPAC name to each of the following disubstituted benzenes. Use numbers rather than prefixes to locate the substituents on the benzene ring.

a. 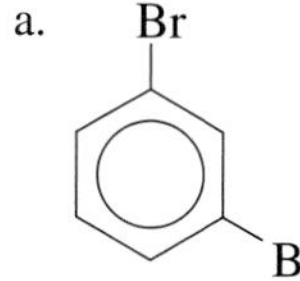

b. 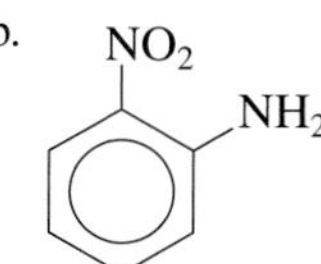

c. 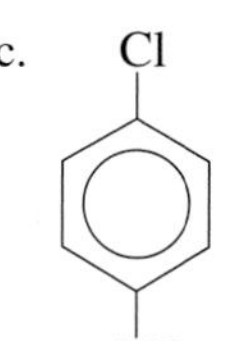

d. 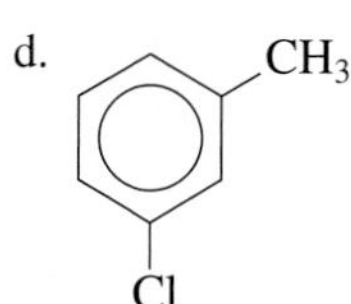

e. 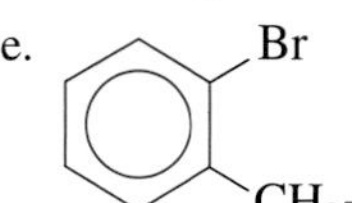

f. 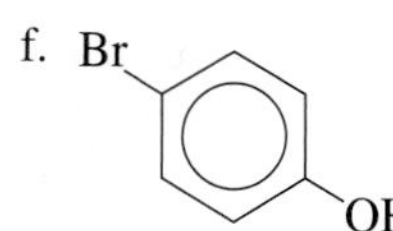

13.50 Assign an IUPAC name to each of the following disubstituted benzenes. Use numbers rather than prefixes to locate the substituents on the benzene ring.

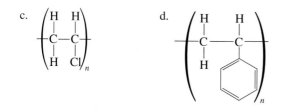

a. 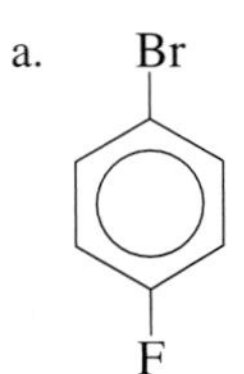

b. OH
CH_2-CH_3

c. 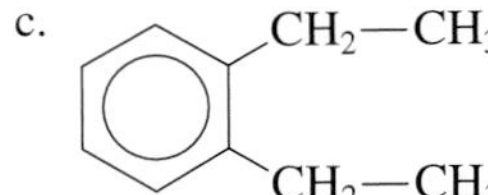

d. CH_3
NO_2

e. Br NO_2

f. Cl
NH_2

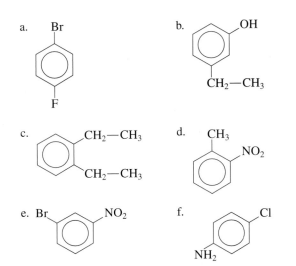

13.51 Assign, to each of the compounds in Problem 13.49, an IUPAC name in which the substituents on the benzene ring are located using the *ortho*-, *meta*-, *para*- prefix system.

13.52 Assign, to each of the compounds in Problem 13.50, an IUPAC name in which the substituents on the benzene ring are located using the *ortho*-, *meta*-, *para*- prefix system.

13.53 Assign an IUPAC name to each of the following substituted benzenes.

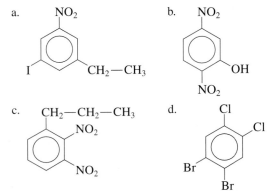

a. 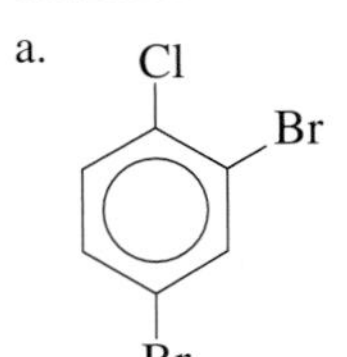

b. 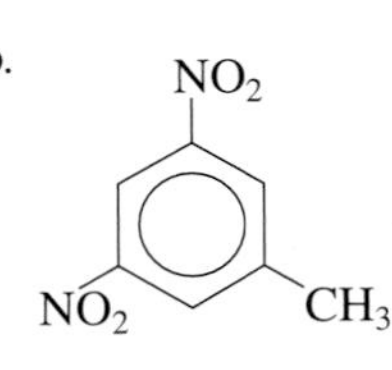

c. 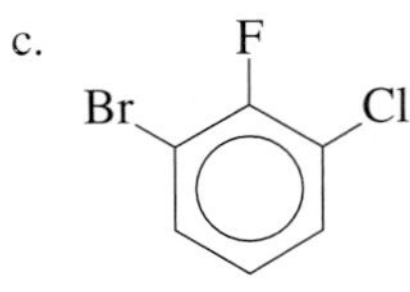

d. 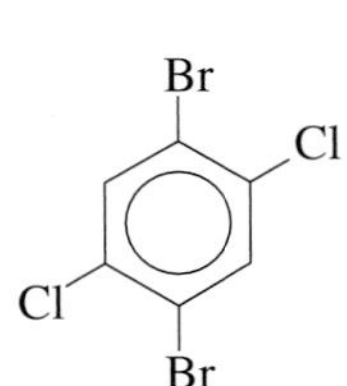

13.54 Assign an IUPAC name to each of the following substituted benzenes.

a. NO_2
I CH_2-CH_3

b. NO_2
OH
NO_2

c. $CH_2-CH_2-CH_3$
NO_2
NO_2

d. Cl
Cl
Br
Br

13.55 Assign an IUPAC name to each of the following compounds, in which the benzene ring is treated as a substituent.

a. $CH_3-CH-CH=CH_2$

b. $CH_3-CH-CH=CH_2$

c. $CH_3-CH-CH_2-CH_2$
CH_3

d. $CH_3-CH-CH_2-CH-CH_3$

13.56 Assign an IUPAC name to each of the following compounds, in which the benzene ring is treated as a substituent.

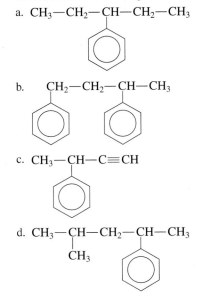

a. $CH_3-CH_2-CH-CH_2-CH_3$

b. $CH_2-CH_2-CH-CH_3$

c. $CH_3-CH-C\equiv CH$

d. $CH_3-CH-CH_2-CH-CH_3$
CH_3

13.57 Write a structural formula for each of the following compounds.

a. 1,3-Diethylbenzene
b. *o*-Xylene
c. *p*-Ethyltoluene
d. 4-Chlorobenzoic acid
e. Aniline
f. 3-Methyl-3-phenylpentane

13.58 Write a structural formula for each of the following compounds.

a. *o*-Ethylpropylbenzene
b. *m*-Xylene
c. 2-Bromotoluene
d. 3-Nitrobenzoic acid
e. Phenol
f. Triphenylmethane

Chemical Reactions of Aromatic Hydrocarbons (Section 13.13)

13.59 For each of the following classes of compounds, indicate whether addition or substitution is the most characteristic reaction for the class.

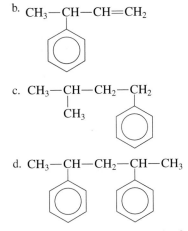

a. Alkanes b. Dienes
c. Alkylbenzenes d. Cycloalkenes

13.60 For each of the following classes of compounds, indicate whether addition or substitution is the most characteristic reaction for the class.

a. Alkynes b. Cycloalkanes
c. Aromatic hydrocarbons d. Saturated hydrocarbons

13.61 Complete the following reaction equations by supplying the formula of the missing reactant or product.

a. $C_6H_6 + ? \xrightarrow{FeBr_3} C_6H_5{-}Br + HBr$

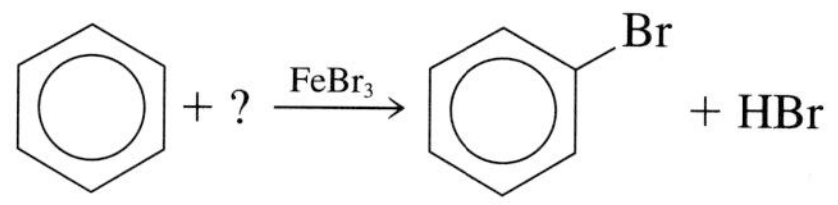

b. $C_6H_6 + ? \xrightarrow[\Delta]{H_2SO_4} C_6H_5{-}NO_2 + H_2O$

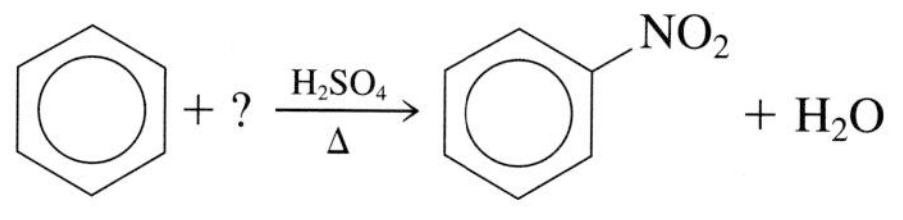

c. $C_6H_6 + CH_3{-}\underset{\displaystyle CH_3}{\underset{|}{CH}}{-}Cl \xrightarrow{AlCl_3} ? + HCl$

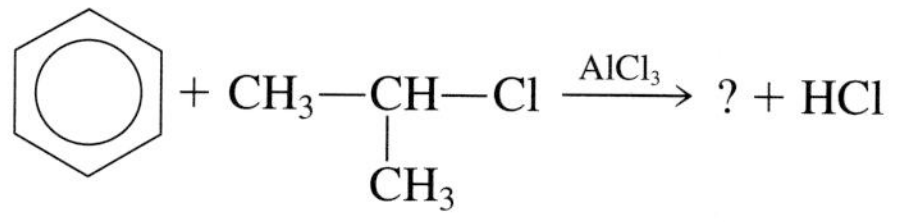

d. $C_6H_6 + ? \xrightarrow{AlBr_3} C_6H_5{-}CH_2{-}CH_3 + HBr$

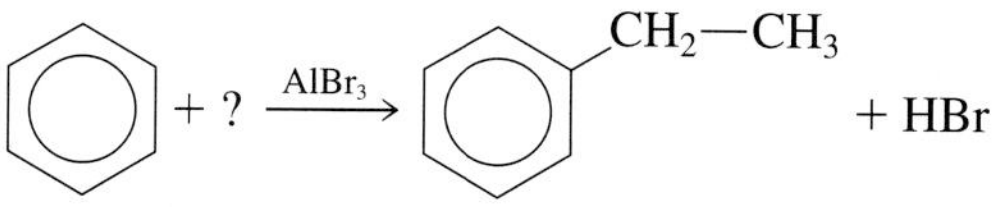

13.62 Complete the following reaction equations by supplying the formula of the missing reactant, product, or catalyst.

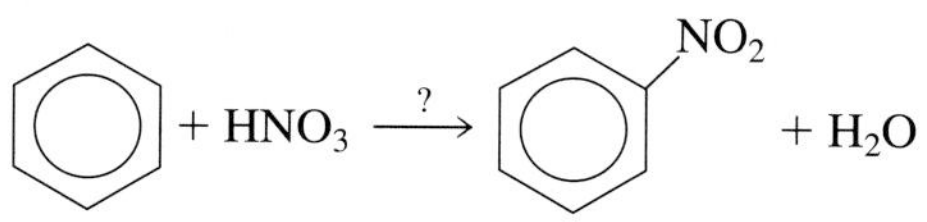

a. $C_6H_6 + Cl_2 \xrightarrow{FeCl_3} C_6H_5{-}Cl + ?$

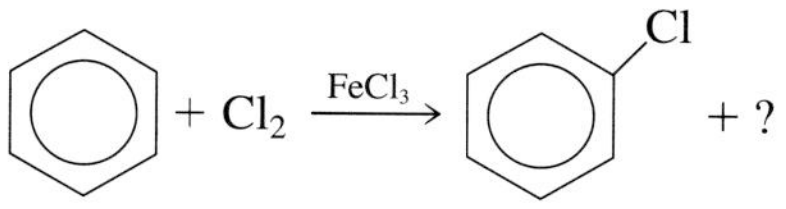

b. $C_6H_6 + HNO_3 \xrightarrow{?} C_6H_5{-}NO_2 + H_2O$

c. $C_6H_6 + ? \xrightarrow{AlBr_3} C_6H_5{-}CH_3 + HBr$

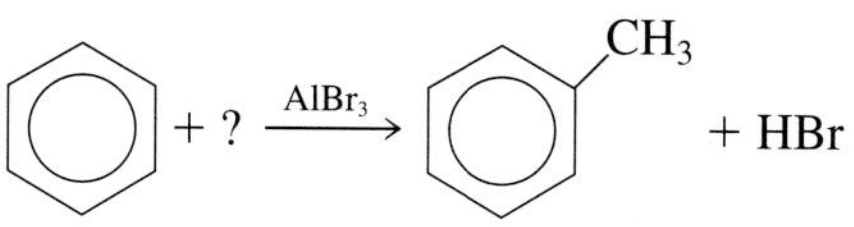

d. $C_6H_6 + CH_3{-}\overset{\displaystyle CH_3}{\overset{|}{\underset{\displaystyle CH_3}{\underset{|}{C}}}}{-}Br \xrightarrow{AlBr_3} ? + HBr$

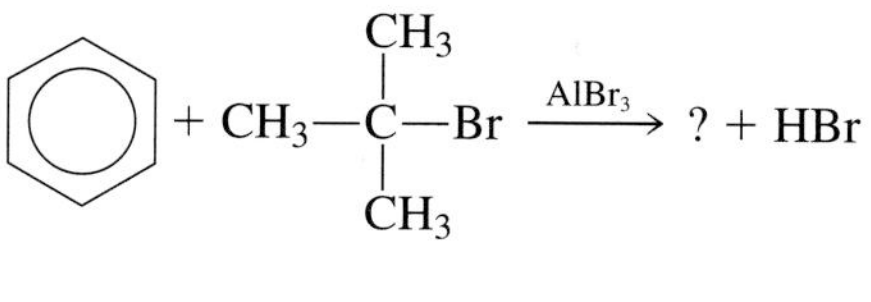

Additional Problems

13.63 Draw a condensed structural formula for each of the following unsaturated hydrocarbons.

a. 5-Methyl-2-hexyne
b. 1-Chloro-2-butene
c. 5,6-Dimethyl-2-heptyne
d. 3-Isopropyl-1-hexene
e. 1,6-Heptadiene
f. 3-Methyl-1,4-pentadiyne

13.64 How many molecules of H_2 will react with 1 molecule of each of the compounds in Problem 13.63 when the appropriate catalyst is present?

13.65 Draw a condensed structural formula for each of the following compounds.

a. Vinylbenzene b. Allyl chloride
c. Propylacetylene d. Dipropylacetylene
e. 3-Ethylbenzoic acid f. *m*-Phenyltoluene

13.66 The compound 2-methyl-1-propene is a well-known substance. The compound 2,2-dimethyl-1-propene does not exist. Explain why this is so.

13.67 The compound 1,2-dichlorocyclohexane exists in *cis–trans* forms. However, *cis–trans* isomerism is not possible for the compound 1,2-dichlorobenzene. Explain why this is so.

13.68 Hydrocarbons with the formula C_5H_{10} can be either alkenes or cycloalkanes. Draw the nine possible structural isomers that fit this formula; five are alkenes and four are cycloalkanes. Then indicate which of these nine isomers exist in *cis–trans* forms.

13.69 There are eight isomeric substituted benzenes that have the formula C_9H_{12}. What are the IUPAC names for these eight structural isomers?

13.70 How many different compounds are there that fit each of the following descriptions?

a. Bromochlorobenzenes
b. Trichlorobenzenes
c. Dibromodichlorobenzenes
d. Monobromoanthracenes

Grid Problems

13.71

1.	2.	3.
$\underset{\displaystyle CH_3}{\underset{\mid}{CH}}{=}CH{-}\underset{\displaystyle CH_3}{\underset{\mid}{CH_2}}$	$CH_3{-}\underset{\displaystyle CH_3}{\underset{\mid}{C}}{=}CH_2$	$CH_2{=}\underset{\displaystyle CH_3}{\underset{\mid}{C}}{-}CH_2{-}CH_3$
4.	**5.**	**6.**
$CH_3{-}\underset{\displaystyle CH_3}{\underset{\mid}{C}}{=}CH{-}CH_3$	$\underset{\displaystyle CH_3}{\underset{\mid}{CH_2}}{-}CH{=}CH{-}CH_3$	$\underset{\displaystyle CH_3}{\underset{\mid}{CH}}{=}CH{-}CH_3$

Select from the grid *all* correct responses for each of the following situations.

a. Compounds that are named as unsubstituted alkenes
b. Compounds that are named as 2-methyl-1-alkenes
c. Compounds that can exist in *cis–trans* forms
d. Compounds that contain 11 σ-bonds

13.72

1.	2.	3.
alkene (one double bond)	diene	alkyne (one triple bond)
4.	**5.**	**6.**
alkane	cycloalkene (one double bond)	cycloalkane

Select from the grid *all* correct responses for each of the following situations.

a. Compounds that undergo substitution reactions
b. Compounds that undergo addition reactions
c. Compounds that have the general molecular formula C_nH_{2n}
d. Compounds for which *cis–trans* isomerism is possible in selected situations

13.73

1. Toluene	2. Phenol	3. Aniline
4. Chlorophenol	5. Xylene	6. Nitroaniline

Select from the grid *all* correct responses for each of the following situations.

a. Compounds that are monosubstituted benzenes
b. Compounds that exist in three isomeric forms
c. Compounds that contain six carbon atoms
d. Compounds that are hydrocarbons

13.74

1. 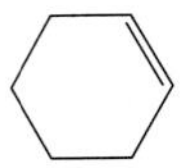	2. $CH_2{=}CH_2$	3. $CH_2{=}CH{-}CH_3$
4. 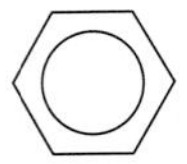	5. $CH{\equiv}CH$	6. $CH_3{-}CH{=}CH{-}CH_3$

Select from the grid *all* correct responses for each of the following situations.

a. Reaction with one Br_2 molecule gives a monobrominated organic product.
b. Reaction with one Br_2 molecule gives a dibrominated organic product.
c. Reaction with two Br_2 molecules gives a tetrabrominated organic product.
d. Reaction with one HBr molecule gives only one brominated organic product.

CHAPTER 14

PART II

Alcohols, Phenols, and Ethers

CHAPTER OUTLINE

14.1 Bonding Characteristics of Oxygen Atoms in Organic Compounds 371
14.2 Structural Features of Alcohols, Phenols, and Ethers 372
14.3 Nomenclature of Alcohols and Phenols 373
14.4 Important Commonly Encountered Alcohols 376
14.5 Physical Properties of Alcohols 380
14.6 Preparation of Alcohols 383
14.7 Reactions of Alcohols 383
14.8 Polymeric Alcohols 388

Chemistry at a Glance:
Summary of Reactions Involving Alcohols 389

14.9 Properties and Uses of Phenols 389
14.10 Nomenclature for Ethers 392
14.11 Physical and Chemical Properties of Ethers 394
14.12 Cyclic Ethers 395
14.13 Sulfur Analogs of Alcohols and Ethers 396

Chemical Connections
14.1 Menthol: A Useful Naturally Occurring Terpene Alcohol 380
14.2 Marijuana: The Most Commonly Used Illicit Drug 391
14.3 Ethers as General Anesthetics 393

The physiological effects of poison ivy are caused by certain phenol compounds present in the leaves.

This chapter is the first of three that consider hydrocarbon derivatives with *oxygen-containing functional groups.* Many biochemically important molecules contain carbon atoms bonded to oxygen atoms.

In this chapter we consider hydrocarbon derivatives whose functional group contains one oxygen atom participating in two single bonds (alcohols, phenols, and ethers). Chapter 15 focuses on derivatives whose functional groups have one oxygen atom participating in a double bond (aldehydes and ketones), and in Chapter 16 we examine functional groups that contain two oxygen atoms, one participating in single bonds and the other in a double bond (carboxylic acids and esters).

14.1 Bonding Characteristics of Oxygen Atoms in Organic Compounds

An understanding of the bonding characteristics of the oxygen atom is a prerequisite to our study of compounds with oxygen-containing functional groups. Normal bonding behavior for oxygen atoms in such functional groups is the formation of two covalent bonds. Oxygen is a member of Group VIA of the periodic table and thus possesses six valence electrons. To complete its octet by electron sharing, an oxygen atom can form either two single bonds or a double bond.

Two single bonds — One double bond

Thus, in organic chemistry, carbon forms four bonds, hydrogen forms one bond, and oxygen forms two bonds.

—C—	H—	:O—
4 valence electrons, 4 covalent bonds, no nonbonding electron pairs	1 valence electron, 1 covalent bond, no nonbonding electron pairs	6 valence electrons, 2 covalent bonds, 2 nonbonding electron pairs

14.2 Structural Features of Alcohols, Phenols, and Ethers

An **alcohol** *is a hydrocarbon derivative in which a hydroxyl group (—OH) is attached to a saturated carbon atom.* Even though alcohols contain an —OH group, they are not hydroxides. A *hydroxide* is a compound in which the polyatomic OH^- ion (Section 4.10) is present. Alcohols are not ionic compounds. In an alcohol, the —OH group is *covalently* bonded to a carbon atom.

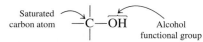

Saturated carbon atom → —C—OH ← Alcohol functional group

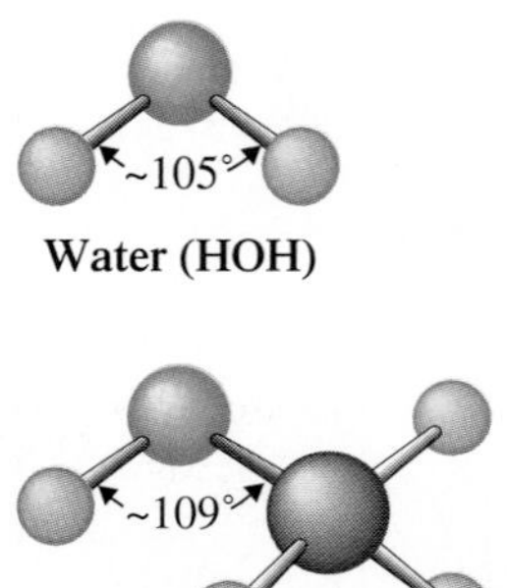

Figure 14.1
The similar shapes of water and methanol. Methyl alcohol may be viewed structurally as an alkyl derivative of water.

The generalized formula for an alcohol is R—OH, where R represents the hydrocarbon portion of the molecule.

Alcohols may be viewed structurally in two ways: (1) as hydroxyl derivatives of hydrocarbons in which a hydrogen atom has been replaced by a hydroxyl group, and (2) as alkyl derivatives of water in which a hydrogen atom has been replaced by an alkyl group (see Figure 14.1).

$$\underset{\text{Alkane}}{\text{R—H}} \xrightarrow[+\text{OH}]{-\text{H}} \underset{\text{Alcohol}}{\text{R—OH}} \xleftarrow[+\text{R}]{-\text{H}} \underset{\text{Water}}{\text{H—O—H}}$$

Examples of formulas for simple alcohols include

$$CH_3\text{—OH} \qquad CH_3\text{—}CH_2\text{—OH} \qquad CH_3\text{—}CH_2\text{—}CH_2\text{—OH}$$

A **phenol** *is a compound in which a hydroxyl group (—OH) is attached to a carbon atom that is part of an aromatic carbon ring system.* The general formula for phenols is

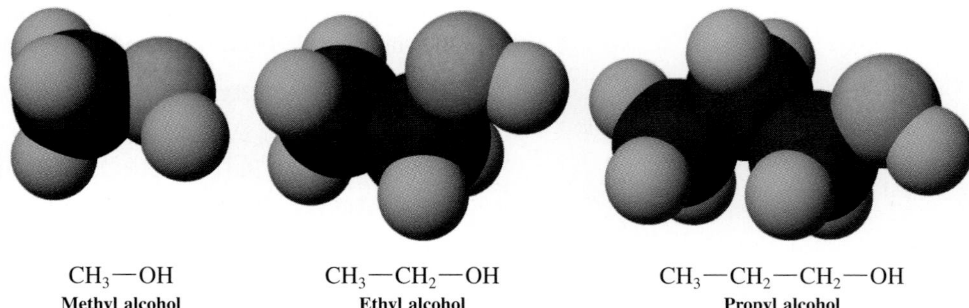

Space-filling models for the three simplest unbranched-chain alcohols: methyl alcohol, ethyl alcohol, and propyl alcohol.

• The generic term *aryl group* (Ar) is the aromatic counterpart of the nonaromatic generic term *alkyl group* (R).

Ar—OH, where Ar represents an aryl group. An **aryl group** *is an aromatic carbon ring system from which one hydrogen atom has been removed.*

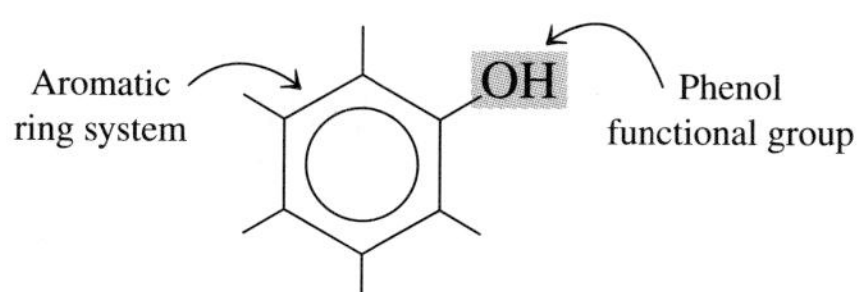

A hydroxyl group is the functional group for both phenols and alcohols.

The following compounds are examples of simple phenols.

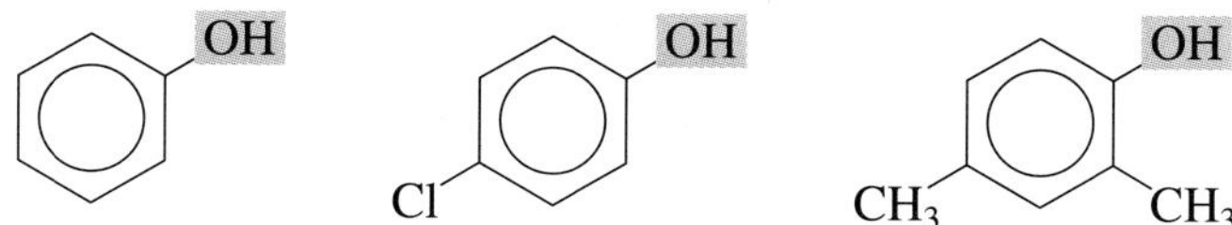

An **ether** *is an organic compound in which an oxygen atom is bonded to two carbon atoms by single bonds.* In an ether, the carbon atoms that are attached to the oxygen atom can be part of alkyl, cycloalkyl, or aryl groups. Examples of ethers include

$CH_3—O—CH_3$ $\quad$ $CH_3—CH_2—O—$(cyclopentyl) $\quad$ $CH_3—O—$(phenyl)

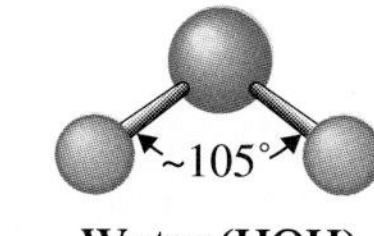

Water (HOH)

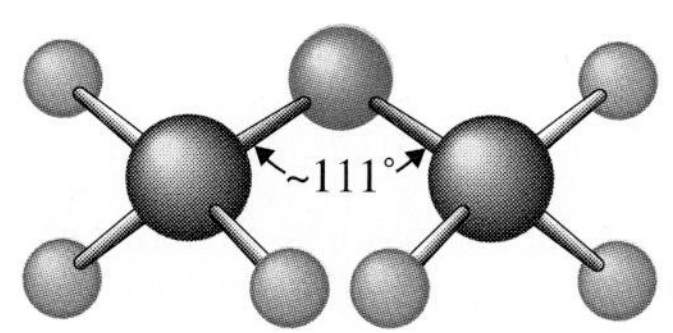

Dimethyl ether (CH_3OCH_3)

The similar shapes of water and dimethyl ether. Diethyl ether may be viewed structurally as a dialkyl derivative of water.

The two groups attached to the oxygen atom of an ether can be the same (first structure), but they need not be so (second and third structures).

All ethers contain a C—O—C unit, which is the ether functional group.

Ether functional group
---C—O—C---

Generalized formulas for ethers, which depend on the types of groups attached to the oxygen atom (alkyl or aryl), include R—O—R, R—O—R′ (where R′ is an alkyl group different from R), R—O—Ar, and Ar—O—Ar.

Structurally, an ether can be visualized as a derivative of water in which both hydrogen atoms have been replaced by hydrocarbon groups. Note that unlike alcohols and phenols, ethers do not possess a hydroxyl (—OH) group.

14.3 Nomenclature of Alcohols and Phenols

• Line-angle drawings for selected simple alcohols:

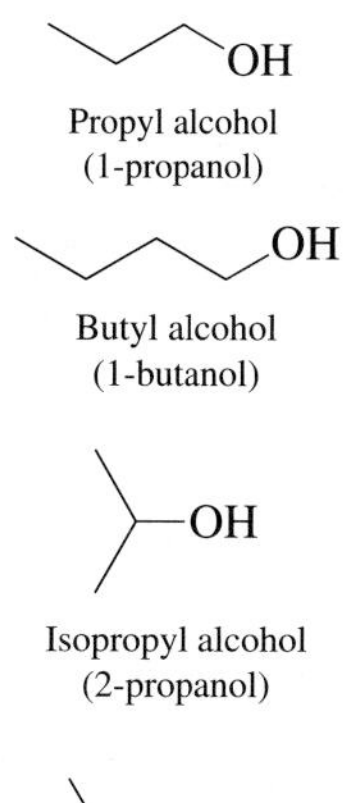

Propyl alcohol (1-propanol)

Butyl alcohol (1-butanol)

Isopropyl alcohol (2-propanol)

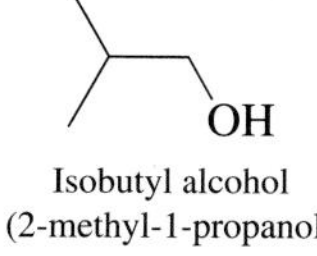

Isobutyl alcohol (2-methyl-1-propanol)

• Alcohols

Common names exist for alcohols with simple (generally C_1 through C_4) alkyl groups. The word *alcohol,* as a separate word, is placed after the name of the alkyl or cycloalkyl group present.

$CH_3—OH$	$CH_3—CH_2—OH$	$CH_3—CH_2—CH_2—OH$
Methyl alcohol	Ethyl alcohol	Propyl alcohol

$$CH_3—\underset{\displaystyle CH_3}{\underset{|}{CH}}—OH$$

Isopropyl alcohol

(cyclobutane ring)—OH

Cyclobutyl alcohol

IUPAC rules for naming alcohols that contain a single hydroxyl group follow.

Rule 1: *Name the longest carbon chain to which the hydroxyl group is attached.* The chain name is obtained by dropping the final *-e* from the alkane name and adding the suffix *-ol.*

Rule 2: *Number the chain starting at the end nearest the hydroxyl group, and use the appropriate number to indicate the position of the —OH group.* (In numbering of the longest carbon chain, the hydroxyl group has priority over double and triple bonds, as well as over alkyl, cycloalkyl, and halogen substituents.)

Rule 3: *Name and locate any other substituents present.*

Rule 4: *In alcohols where the —OH group is attached to a carbon atom in a ring, the hydroxyl group is assumed to be on carbon 1.*

Example 14.1

Determining IUPAC Names for Alcohols

Name the following alcohols, utilizing IUPAC nomenclature rules.

a. $CH_3—CH_2—C(CH_3)(OH)—CH_2—CH_2—CH_3$

b. $CH_3—CH_2—CH(CH_2—OH)—CH_2—CH_3$

c. (cyclohexane ring bearing —OH, with CH_3 groups on the two adjacent ring carbons)

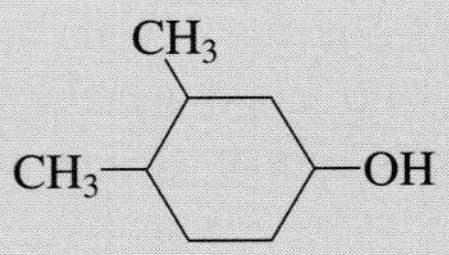

Solution

a. The longest carbon chain that contains the alcohol functional group has six carbons. When we change the *-e* to *-ol,* hexane becomes *hexanol.* Numbering the chain from the end nearest the —OH group identifies carbon number 3 as the location of both the —OH group and a methyl group. The complete name is 3-methyl-3-hexanol.

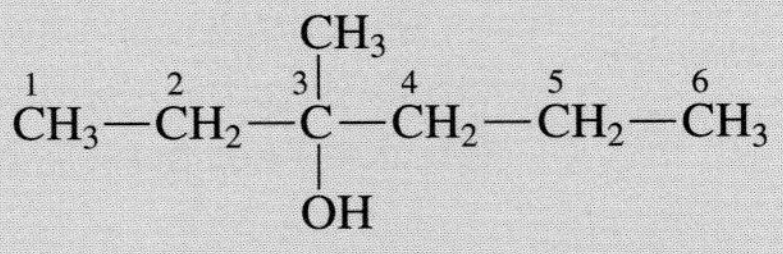

$$\overset{1}{CH_3}—\overset{2}{CH_2}—\overset{3}{C}(CH_3)(OH)—\overset{4}{CH_2}—\overset{5}{CH_2}—\overset{6}{CH_3}$$

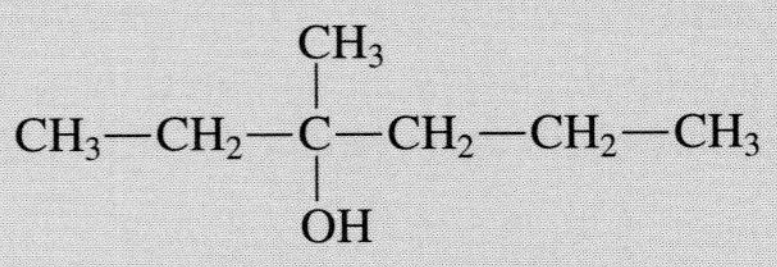

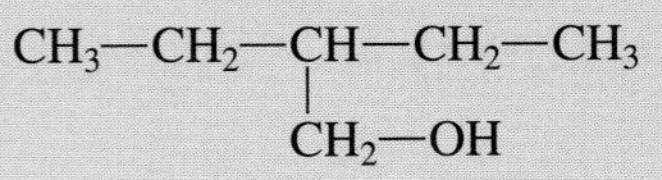

b. The longest carbon chain containing the —OH group has four carbon atoms. It is numbered from the end closest to the —OH group as follows:

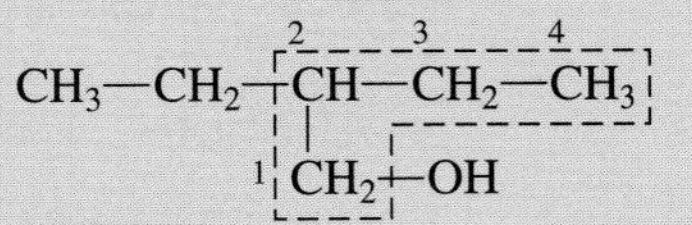

$$CH_3—CH_2—\overset{2}{CH}(\overset{1}{CH_2}—OH)—\overset{3}{CH_2}—\overset{4}{CH_3}$$

The base name is 1-butanol. The complete name is 2-ethyl-1-butanol.

c. This alcohol is a cyclohexanol. The carbon to which the —OH group is attached is assigned the number 1. The complete name for this alcohol is 3,4-dimethylcyclohexanol.

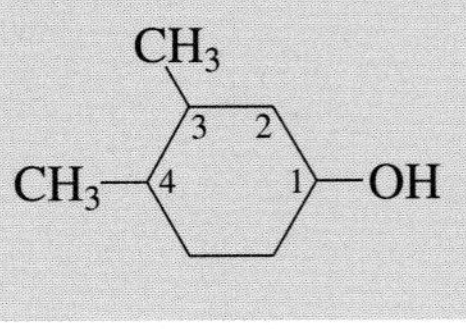

(ring numbered: C1 bears —OH, C2, C3 bears CH_3, C4 bears CH_3)

Practice Exercise 14.1

Name the following alcohols, utilizing IUPAC nomenclature rules.

a. $CH_3—CH(CH_3)—CH(OH)—CH_2—CH(CH_3)—CH_3$

• In the naming of alcohols with *unsaturated* carbon chains, two endings are needed: one for the double or triple bond and one for the hydroxyl group. The *-ol* suffix always comes last in the name; that is, unsaturated alcohols are named as *alkenols* or *alkynols.*

$$\overset{3}{CH_2}=\overset{2}{CH}—\overset{1}{CH_2}—OH$$

2-Propen-1-ol
(common name: allyl alcohol)

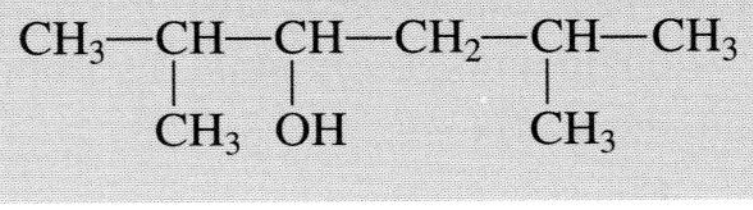

b. $CH_3-CH_2-CH(CH_2-CH_2-OH)-CH_3$

c. (cyclopentane ring bearing CH_3 and OH on one carbon and CH_3 on the adjacent carbon)

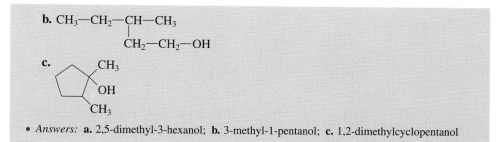

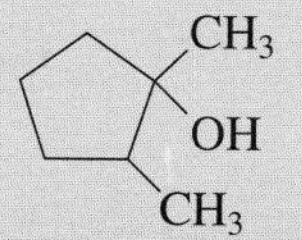

• *Answers:* **a.** 2,5-dimethyl-3-hexanol; **b.** 3-methyl-1-pentanol; **c.** 1,2-dimethylcyclopentanol

• The contrast between IUPAC and common names for alcohols is as follows:

IUPAC (one word)
alkanol
ethanol
Common (two words)
alkyl alcohol
ethyl alcohol

• Structural isomers that differ only in the location of the functional group are often called *positional isomers.* The compounds 1-butanol and 2-butanol are positional isomers (see Table 14.1).

• A hydroxyl group as a substituent in a molecule is called a hydroxy group; an *-oxy* rather than an *-oxyl* ending is used.

Table 14.1 gives both IUPAC and common names for monohydroxy alcohols that contain four or fewer carbon atoms. This table also shows that structural isomerism is possible for alcohols containing three or more carbon atoms. Indeed, the structural isomerism possibilities for alcohols far exceed those for alkanes. There are 75 possible $C_{10}H_{22}$ structural isomers and 507 possible $C_{10}H_{21}OH$ structural isomers.

Polyhydroxy alcohols—alcohols that possess more than one hydroxyl group—can be named with only a slight modification of the preceding IUPAC rules. An alcohol in which two hydroxyl groups are present is named as a *diol,* one containing three hydroxyl groups is named as a *triol,* and so on. In these names for diols, triols, and so forth, the final *-e* of the parent alkane name is retained for pronunciation reasons.

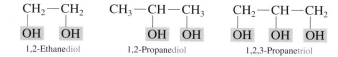

$CH_2(OH)-CH_2(OH)$ — 1,2-Ethanediol

$CH_3-CH(OH)-CH_2(OH)$ — 1,2-Propanediol

$CH_2(OH)-CH(OH)-CH_2(OH)$ — 1,2,3-Propanetriol

• Phenols

Besides being the name for a family of compounds, *phenol* is also the IUPAC-approved name of the simplest member of the phenol family of compounds (Section 13.11).

Table 14.1
IUPAC and Common Names of Monohydroxy Alcohols That Contain Up to Four Carbon Atoms

Formula	IUPAC name	Common name
One carbon atom (CH_3OH)		
CH_3-OH	methanol	methyl alcohol
Two carbon atoms (C_2H_5OH)		
CH_3-CH_2-OH	ethanol	ethyl alcohol
Three carbon atoms (C_3H_7OH); two structural isomers exist		
$CH_3-CH_2-CH_2-OH$	1-propanol	propyl alcohol
$CH_3-CH(OH)-CH_3$	2-propanol	isopropyl alcohol
Four carbon atoms (C_4H_9OH); four structural isomers exist		
$CH_3-CH_2-CH_2-CH_2-OH$	1-butanol	butyl alcohol
$CH_3-CH(CH_3)-CH_2-OH$	2-methyl-1-propanol	isobutyl alcohol
$CH_3-CH_2-CH(CH_3)-OH$	2-butanol	*sec*-butyl alcohol
$CH_3-C(CH_3)_2-OH$	2-methyl-2-propanol	*tert*-butyl alcohol

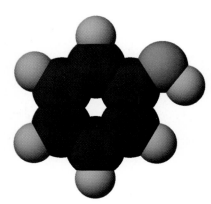

A space-filling model for *phenol,* a compound that has an —OH group bonded directly to a benzene (aromatic) ring.

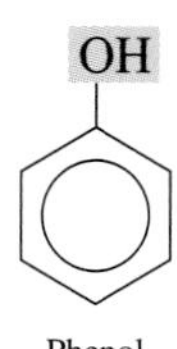

Phenol

Substituted phenols are named as derivatives of phenol. Ring numbering begins with the hydroxyl group and proceeds in the direction that gives the lower number to the next carbon atom bearing a substituent.

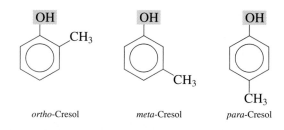

3-Chlorophenol (or *meta*-chlorophenol) 4-Ethyl-2-methylphenol 2,5-Dibromophenol

Methyl and hydroxy derivatives of phenol have IUPAC-accepted common names. Methylphenols are called cresols. The name *cresol* applies to all three isomeric methylphenols.

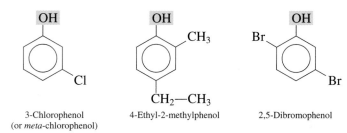

ortho-Cresol *meta*-Cresol *para*-Cresol

• Several neurotransmitters in the human body (Section 17.9), including norepinephrine, epinephrine (adrenaline), and dopamine, are catechol derivatives.

For hydroxyphenols, each of the three isomers has a different common name.

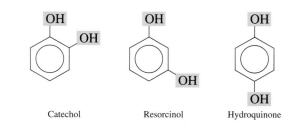

Catechol Resorcinol Hydroquinone

14.4 Important Commonly Encountered Alcohols

In this section we consider the properties and uses of six commonly encountered alcohols: methyl, ethyl, and isopropyl alcohols (all monohydroxy alcohols), ethylene glycol and propylene glycol (both diols), and glycerin (a triol).

• Methyl Alcohol (Methanol)

Racing cars at the Indianapolis Speedway are fueled with methyl alcohol.

Methyl alcohol, with one carbon atom and one —OH group, is the simplest alcohol. This colorless liquid is a good fuel for internal combustion engines. Since 1965 all racing cars at the Indianapolis Speedway have been fueled with methyl alcohol. (Methyl alcohol fires are easier to put out than gasoline fires, because water mixes with and dilutes methyl alcohol.) Methyl alcohol also has excellent solvent properties, and it is the solvent of choice for paints, shellacs, and varnishes.

Methyl alcohol is sometimes called *wood alcohol,* terminology that draws attention to an early method for its preparation—the heating of wood to a high temperature in the absence of air. Today, almost all methyl alcohol is produced via the reaction between H_2 and CO.

$$CO + 2H_2 \xrightarrow[300°C - 400°C,\ 200\ atm]{ZnO-Cr_2O_3} CH_3-OH$$

• Methyl alcohol poisoning is treated with ethyl alcohol, which ties up the enzyme that oxidizes methyl alcohol to its toxic metabolites. Ethyl alcohol has 10 times the affinity for the alcohol dehydrogenase enzyme that methyl alcohol does.

Drinking methyl alcohol is very dangerous. Within the human body, methyl alcohol is oxidized by the liver enzyme *alcohol dehydrogenase* to the toxic metabolites formaldehyde and formic acid.

$$CH_3-OH \xrightarrow[\text{dehydrogenase}]{\text{Alcohol}} \underset{\text{Formaldehyde}}{H-\overset{\overset{\large O}{\|}}{C}-H} \xrightarrow[\text{oxidation}]{\text{Further}} \underset{\text{Formic acid}}{H-\overset{\overset{\large O}{\|}}{C}-OH}$$

Formaldehyde is toxic to the eye and can cause blindness (temporary or permanent). Formic acid causes acidosis (Chemical Connections 10.3). Ingesting as little as 1 oz (30 mL) of methyl alcohol can cause optic nerve damage.

• Many people imagine ethanol to be relatively nontoxic and methanol to be extremely toxic. Actually, their toxicities differ by a factor of only 2. Typical fatal doses for adults are about 100 mL for methanol and about 200 mL for ethanol, although smaller doses of methanol may damage the optic nerve.

• Ethyl Alcohol (Ethanol)

Ethyl alcohol, the two-carbon monohydroxy alcohol, is the alcohol present in alcoholic beverages and is commonly referred to as simply alcohol or *drinking alcohol*. Like methyl alcohol, ethyl alcohol is oxidized in the human body by the liver enzyme *alcohol dehydrogenase*.

$$CH_3-CH_2-OH \xrightarrow[\text{dehydrogenase}]{\text{Alcohol}} \underset{\text{Acetaldehyde}}{CH_3-\overset{\overset{\large O}{\|}}{C}-H} \xrightarrow[\text{oxidation}]{\text{Further}} \underset{\text{Acetic acid}}{CH_3-\overset{\overset{\large O}{\|}}{C}-OH}$$

Acetaldehyde, the first oxidation product, is largely responsible for the symptoms of hangover. The odors of both acetaldehyde and acetic acid are detected on the breath of someone who has consumed a large amount of alcohol. Ethyl alcohol oxidation products are less toxic than those of methyl alcohol.

Long-term excessive use of ethyl alcohol may cause undesirable effects such as cirrhosis of the liver, loss of memory, and strong physiological addiction. Links have also been established between certain birth defects and the ingestion of ethyl alcohol by women during pregnancy (fetal alcohol syndrome).

Ethyl alcohol can be produced by yeast fermentation of sugars found in plant extracts (see Figure 14.2). The synthesis of ethyl alcohol in this manner, from grains such as corn, rice, and barley, is the reason why ethyl alcohol is often called *grain alcohol.*

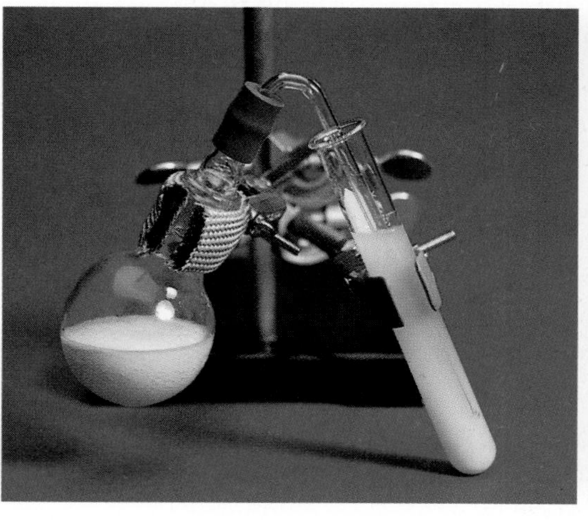

(a)

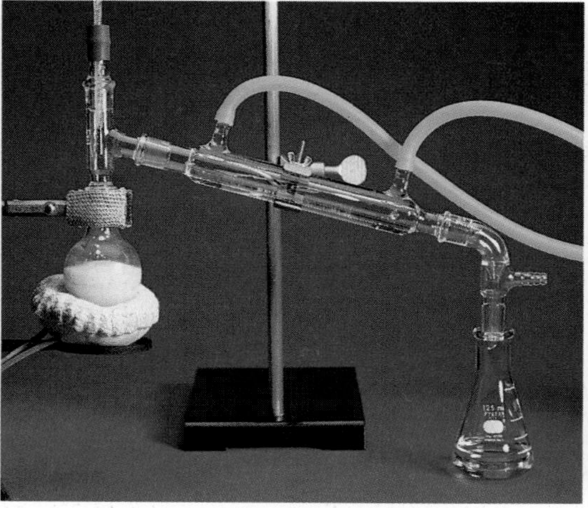

(b)

(c)

Figure 14.2
An experimental setup for preparing ethyl alcohol by fermentation. (a) A small amount of yeast has been added to the aqueous sugar solution in the flask. Yeast enzymes catalyze the decomposition of sugar to ethanol and carbon dioxide, CO_2. The CO_2 is bubbling through lime water, $Ca(OH)_2$, producing calcium carbonate, $CaCO_3$. (b) More concentrated ethanol is produced from the solution in the flask by collecting the fraction that boils at about 78°C. (c) Concentrated ethanol (50% v/v) burns when it is ignited.

• The alcohol content of strong alcoholic beverages is often stated in terms of proof. *Proof* is twice the percentage of alcohol. This system dates back to the seventeenth century and is based on the fact that a 50% (v/v) alcohol–water mixture will burn. Its flammability was *proof* that a liquor had not been watered down.

Fermentation is the process by which ethyl alcohol for alcoholic beverages is produced. The maximum concentration of ethyl alcohol obtainable by fermentation is about 18% (v/v), because yeast enzymes cannot function in stronger alcohol solutions. Alcoholic beverages with a higher concentration of alcohol than this are prepared by either distillation or fortification with alcohol obtained by the distillation of another fermentation product. Table 14.2 lists the alcohol content of common alcoholic beverages and of selected common household products and over-the-counter drug products.

Denatured alcohol is ethyl alcohol that has been rendered unfit to drink by the addition of small amounts of toxic substances (denaturing agents). Almost all of the ethyl alcohol used for industrial purposes is denatured alcohol.

Most ethyl alcohol used in industry is prepared from ethene via a hydration reaction (Section 13.7).

$$CH_2{=}CH_2 + H_2O \xrightarrow{\text{Catalyst}} CH_3{-}CH_2{-}OH$$

The reaction produces a product that is 95% alcohol and 5% water. In applications where water does interfere with use, the mixture is treated with a dehydrating agent to produce 100% ethyl alcohol. Such alcohol, with all traces of water removed, is called *absolute alcohol.*

• Isopropyl Alcohol (2-Propanol)

• The "medicinal" odor associated with doctors' offices is usually that of isopropyl alcohol.

Isopropyl alcohol is one of two three-carbon monohydroxy alcohols; the other is propyl alcohol. A 70% isopropyl alcohol–30% water solution is marketed as *rubbing alcohol.* Isopropyl alcohol's rapid evaporation rate creates a dramatic cooling effect when it is applied to the skin, hence its use for alcohol rubs to combat high body temperature.

Isopropyl alcohol has a bitter taste. Its toxicity is twice that of ethyl alcohol but causes few fatalities because it often induces vomiting and thus doesn't stay down long enough to kill you. In the body it is oxidized to acetone.

$$\underset{\text{Isopropyl alcohol}}{CH_3{-}\overset{\overset{\displaystyle OH}{|}}{C}H{-}CH_3} \xrightarrow[\text{dehydrogenase}]{\text{Alcohol}} \underset{\text{Acetone}}{CH_3{-}\overset{\overset{\displaystyle O}{\|}}{C}{-}CH_3}$$

Table 14.2
Ethyl Alcohol Content (volume percent) of Common Alcoholic Beverages, Household Products, and Over-the-Counter Drugs

Product type	Product	% Ethyl alcohol
Alcoholic Beverages	Beer	3.2–9
	Wine (unfortified)	12
	Brandy	40–45
	Whiskey	45–55
	Rum	45
Flavorings	Vanilla extract	35
	Almond extract	50
Cough and Cold Remedies	Pertussin Plus	25
	Nyquil	25
	Dristan	12
	Vicks 44	10
	Robitussin, DM	1.4
Mouthwashes	Listerine	25
	Scope	18
	Colgate 100	17
	Cepacol	14
	Lavoris	5

Large amounts, about 150 mL, of ingested isopropyl alcohol can be fatal; death occurs from paralysis of the central nervous system.

Ethylene Glycol (1,2-Ethanediol) and Propylene Glycol (1,2-Propanediol)

• The ethylene glycol and propylene glycol used in antifreeze formulations are colorless and odorless; the color and odor of antifreezes come from additives for rust protection and the like.

• Ethylene glycol and propylene glycol are synthesized from ethylene and propylene, respectively, hence their common names.

Ethylene glycol and propylene glycol are the two simplest alcohols possessing two —OH groups. Besides being diols, they are also classified as glycols. A **glycol** *is a diol in which the two —OH groups are on adjacent carbon atoms.*

$$\underset{\text{Ethylene glycol}}{\underset{|\quad\;\;|}{CH_2{-}CH_2}\atop{OH\;\;OH}} \qquad \underset{\text{Propylene glycol}}{CH_3{-}\underset{OH}{\underset{|}{CH}}{-}\underset{OH}{\underset{|}{CH_2}}}$$

Both of these glycols are colorless, odorless, high-boiling liquids that are completely miscible with water. Their major uses are as the main ingredient in automobile "year-round" antifreeze and airplane "de-icers" and as a starting material for the manufacture of polyester fibers (Section 17.16).

Ethylene glycol is extremely toxic when ingested. In the body, liver enzymes oxidize it to oxalic acid.

$$\underset{\text{Ethylene glycol}}{HO{-}CH_2{-}CH_2{-}OH} \xrightarrow[\text{enzymes}]{\text{Liver}} \underset{\text{Oxalic acid}}{HO{-}\overset{O}{\overset{\|}{C}}{-}\overset{O}{\overset{\|}{C}}{-}OH}$$

Oxalic acid, as a calcium salt, crystallizes in the kidneys, which leads to renal problems.

On the other hand, propylene glycol is essentially nontoxic and has been used as a solvent for drugs. Like ethylene glycol, it is oxidized by liver enzymes; however, pyruvic acid, its oxidation product, is a compound normally found in the human body, being an intermediate in carbohydrate metabolism (Chapter 24).

$$\underset{\text{Propylene glycol}}{CH_3{-}\underset{OH}{\underset{|}{CH}}{-}\underset{OH}{\underset{|}{CH_2}}} \xrightarrow[\text{enzymes}]{\text{Liver}} \underset{\text{Pyruvic acid}}{CH_3{-}\overset{O}{\overset{\|}{C}}{-}\overset{O}{\overset{\|}{C}}{-}OH}$$

Glycerol (1,2,3-Propanetriol)

Glycerol is a clear, thick liquid that has the consistency of honey. Its molecular structure involves three —OH groups on three different carbon atoms.

$$\underset{OH}{\underset{|}{CH_2}}{-}\underset{OH}{\underset{|}{CH}}{-}\underset{OH}{\underset{|}{CH_2}}$$

Space-filling molecular model for the dihydroxy alcohol *ethylene glycol,* which is the major ingredient in airplane "de-icers."

$$\underset{OH}{\underset{|}{CH_2}}{-}\underset{OH}{\underset{|}{CH_2}}$$

Chemical CONNECTIONS

14.1 Menthol: A Useful Naturally Occurring Terpene Alcohol

Menthol is a naturally occurring terpene (Section 13.6) alcohol with a pleasant, minty odor. Its IUPAC name is 2-isopropyl-5-methylcyclohexanol.

OH

$CH—CH_3$

CH_3

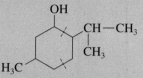

H_3C

In the pure state, menthol is a white crystalline solid with a melting point of 41°C to 43°C. It can be obtained from peppermint oil and can also be prepared synthetically.

Topical application of menthol to the skin causes a refreshing, cooling sensation followed by a slight burning-and-prickling sensation. Its mode of action is that of a *differential* anesthetic. It stimulates the receptor cells in the skin that normally respond to cold to give a sensation of coolness that is unrelated to body temperature. (This cooling sensation is particularly noticeable in the respiratory tract when low concentrations of menthol are inhaled.) At the same time as cooling is perceived, menthol can depress the nerves for pain reception.

Numerous products contain menthol.

- Throat sprays and lozenges containing menthol temporarily soothe inflamed mucous surfaces of the nose and throat. Lozenges contain 2–20 milligrams of menthol per wafer.
- Cough drops and cigarettes of the "mentholated" type use menthol for its counterirritant effect.
- Pre-electric shave preparations and aftershave lotions often contain menthol. A concentration of only 0.1% (m/v) gives ample cooling to allay the irritation of a "close" shave.
- Many dermatologic preparations contain menthol as an antipruritic (anti-itching agent).
- Chest-rub preparations containing menthol include Ben Gay [7% (m/v)] and Mentholatum [6% (m/v)].
- Artificial mint flavors have menthol as an ingredient. Several toothpastes and mouthwashes use menthol as a flavoring agent.

Glycerol is normally present in the human body because it is a product of fat metabolism. It is present, in combined form, in all animal fats and vegetable oils (Section 19.3).

Because glycerol has a great affinity for water vapor (moisture), it is often added to pharmaceutical preparations such as skin lotions and soap. Florists sometimes use glycerol on cut flowers to help retain water and maintain freshness. Its lubricative properties also make it useful in shaving creams and applications such as glycerin suppositories for rectal administration of medicines. It is used in candies and icings as a retardant for preventing sugar crystallization.

14.5 Physical Properties of Alcohols

Alcohol molecules have both polar and nonpolar character. The hydroxyl groups present are polar, and the alkyl (R) group present is nonpolar.

Nonpolar portion ↘ ↙ Polar portion

$(CH_3—CH_2—CH_2)—(OH)$

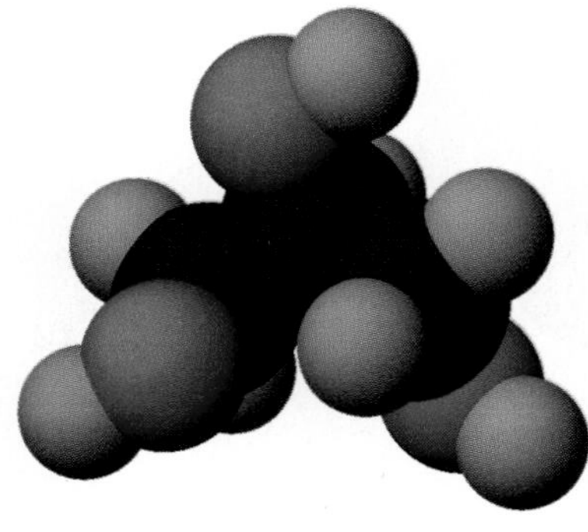

Space-filling molecular model for the trihydroxy alcohol *glycerol*, which is often called biological antifreeze. For survival in Arctic and northern winters, many fish and insects, including the common housefly, produce large amounts of glycerol that dissolve in their blood, thereby lowering the freezing point of the blood.

$$\underset{OH}{CH_2}—\underset{OH}{CH}—\underset{OH}{CH_2}$$

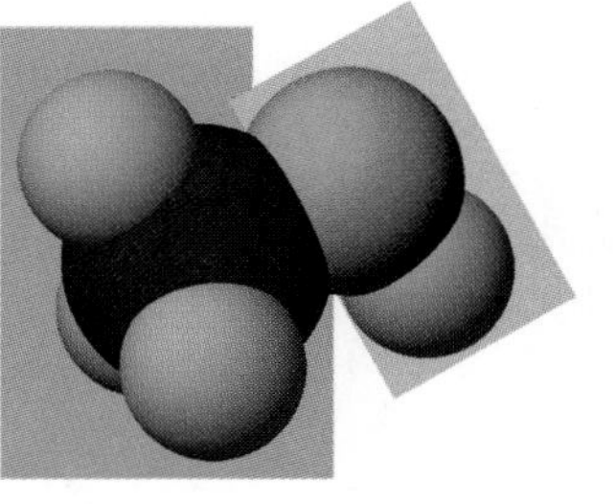
Methanol

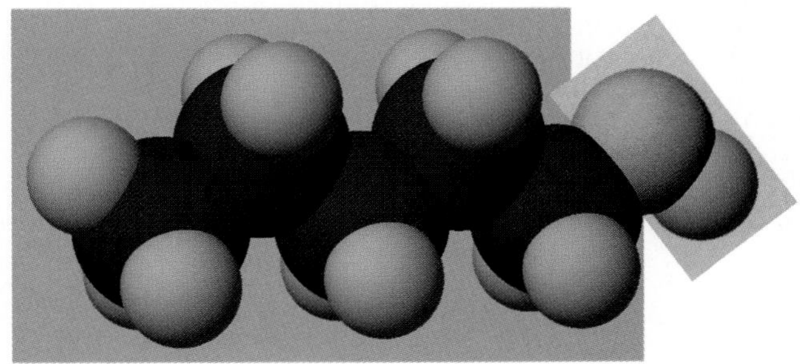
1-Pentanol

Figure 14.3
Space-filling molecular models showing the nonpolar (green) and polar (pink) parts of methanol and 1-pentanol. The polar hydroxyl functional group dominates the physical properties of methanol. The molecule is completely soluble in water (polar) but only partially so in hexane (nonpolar). Conversely, the nonpolar portion of 1-pentanol dominates its physical properties; it is infinitely soluble in hexane and has limited solubility in water.

Unbranched 1-Alcohols			
C_1	C_3	C_5	C_7
C_2	C_4	C_6	C_8

Unsubstituted Cycloalcohols			
	C_3	C_5	C_7
	C_4	C_6	C_8

 Liquid

A physical-state summary for unbranched 1-alcohols and unsubstituted cycloalcohols at room temperature and pressure.

The physical properties of an alcohol depend on whether the polar or the nonpolar portion of its structure "dominates." Factors that determine this include the *length* of the nonpolar carbon chain present and the *number* of polar hydroxyl groups present (see Figure 14.3).

Figure 14.4a shows that the boiling point for unbranched-chain alcohols with an —OH group on an end carbon, 1-alcohols, increases as the length of the carbon chain increases. This trend results from increasing London forces (Sections 7.13 and 12.15) with increasing carbon chain length. Alcohols with more than one hydroxyl group present have significantly higher boiling points (bp) than their monohydroxy counterparts.

$CH_3—CH_2—CH_2—OH$ (OH on C3), bp = 97°C

$CH_3—CH(OH)—CH_2(OH)$, bp = 188°C

$CH_2(OH)—CH(OH)—CH_2(OH)$, bp = 290°C

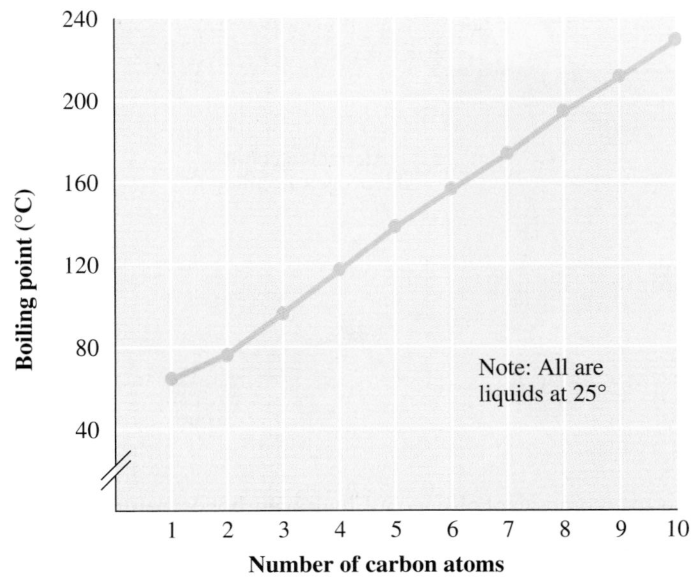

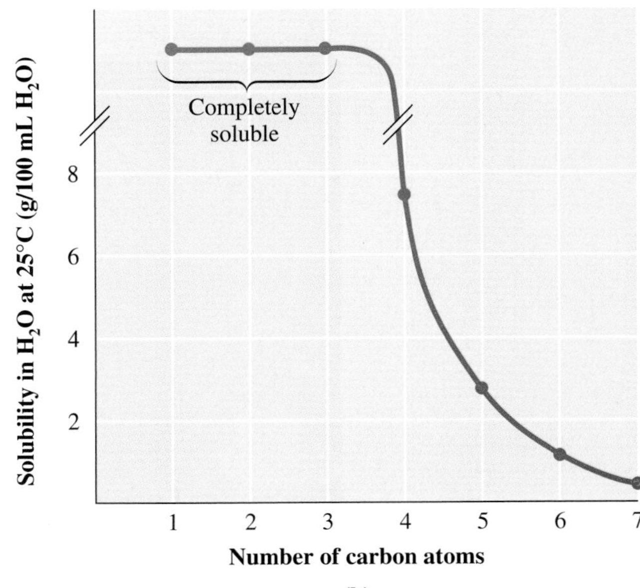

Figure 14.4
(a) Boiling points and (b) solubilities in water of selected 1-alcohols.

Table 14.3
A Comparison of Selected Physical Properties of Alcohols with Alkane Counterparts of Similar Molecular Mass

Type of compound	Compound	Structure	Molecular mass	Boiling point (°C)	Solubility in water
alkane	ethane	CH_3-CH_3	30	−89	slight solubility
alcohol	methanol	CH_3-OH	32	65	unlimited solubility
alkane	propane	$CH_3-CH_2-CH_3$	44	−42	slight solubility
alcohol	ethanol	CH_3-CH_2-OH	46	78	unlimited solubility
alkane	butane	$CH_3-CH_2-CH_2-CH_3$	58	−1	slight solubility
alcohol	1-propanol	$CH_3-CH_2-CH_2-OH$	60	97	unlimited solubility
alcohol	2-propanol	$CH_3-CH(CH_3)-OH$	60	83	unlimited solubility

This boiling-point trend is related to increased hydrogen bonding between alcohol molecules (to be discussed shortly).

Small monohydroxy alcohols are soluble in water in all proportions. As carbon chain length increases beyond three carbons, solubility in water rapidly decreases (Figure 14.4b) because of the increasingly nonpolar character of the alcohol. Alcohols with two —OH groups present are more soluble in water than their counterparts with only one —OH group. Increased hydrogen bonding is responsible for this. Diols containing as many as seven carbon atoms show appreciable solubility in water.

A comparison of the properties of alcohols with their alkane counterparts (Table 14.3) shows that

1. Alcohols have *higher* boiling points than alkanes of similar molecular mass.
2. Alcohols have much *higher* solubility in water than alkanes of similar molecular mass.

The differences in physical properties between alcohols and alkanes are related to hydrogen bonding. Because of their hydroxyl group(s), alcohols can participate in hydrogen bonding, whereas alkanes cannot. Hydrogen bonding between alcohol molecules is similar to that which occurs between water molecules (Section 7.13).

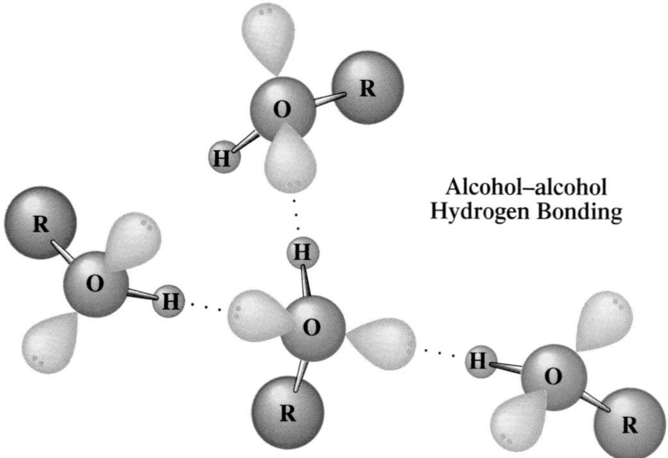

Extra energy is needed to overcome alcohol–alcohol hydrogen bonds before alcohol molecules can enter the vapor phase. Hence alcohol boiling points are higher than those for the corresponding alkanes (where no hydrogen bonds are present).

Alcohol molecules can also hydrogen-bond to water molecules. The formation of such hydrogen bonds explains the solubility of small alcohol molecules in water. As the alcohol chain length increases, alcohols become more alkane-like (nonpolar), and solubility decreases.

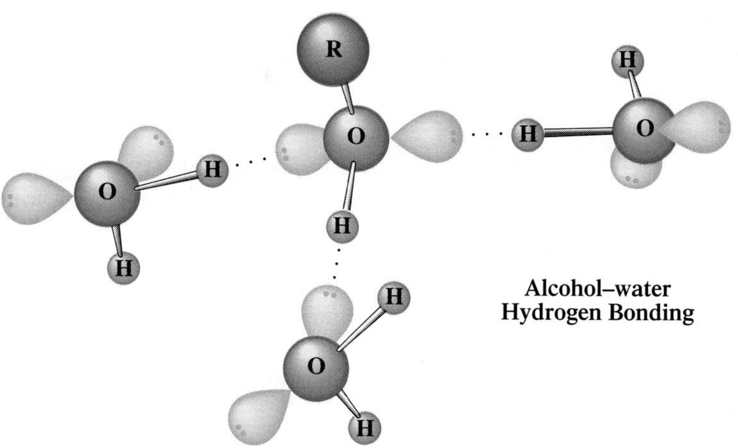

14.6 Preparation of Alcohols

One general method for preparing alcohols—the hydration of alkenes—was discussed in the previous chapter (Section 13.7). Alkenes react with water (an unsymmetrical addition agent) in the presence of sulfuric acid (the catalyst) to form an alcohol. Markovnikov's rule is used to determine the predominant alcohol product.

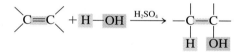

$$\gt C{=}C\lt + H{-}OH \xrightarrow{H_2SO_4} -\underset{H}{\overset{|}{\underset{|}{C}}}-\underset{OH}{\overset{|}{\underset{|}{C}}}-$$

• Alcohols are intermediate products in the metabolism of both carbohydrates (Chapter 24) and fats (Chapter 25). In these metabolic processes, both addition of water to a carbon–carbon double bond and addition of hydrogen to a carbon–oxygen double bond lead to the introduction of the alcohol functional group into a biomolecule.

Another method of synthesizing alcohols involves the addition of H_2 to a carbon–oxygen double bond (a carbonyl group, $\gt C{=}O$). (The carbonyl group is a functional group that will be discussed in detail in Chapter 15.) A carbonyl group behaves very much like a carbon–carbon double bond when it reacts with H_2 under the proper conditions. As a result of H_2 addition, the oxygen of the carbonyl group is converted to an —OH group.

$$R-\overset{\overset{O}{\|}}{C}-H + H_2 \xrightarrow{\text{Catalyst}} R-\underset{H}{\overset{OH}{\underset{|}{\overset{|}{C}}}}-H$$

Aldehyde (Section 15.2) Alcohol

$$R-\overset{\overset{O}{\|}}{C}-R' + H_2 \xrightarrow{\text{Catalyst}} R-\underset{H}{\overset{OH}{\underset{|}{\overset{|}{C}}}}-R'$$

Ketone (Section 15.2) Alcohol

14.7 Reactions of Alcohols

• Pronounce 1° as "primary," 2° as "secondary," and 3° as "tertiary."

Alcohols are classified as primary (1°), secondary (2°), or tertiary (3°), depending on the number of carbon atoms bonded to the carbon atom that bears the hydroxyl group. A **primary alcohol** *is an alcohol in which the hydroxyl-bearing carbon atom is attached to*

• Methyl alcohol, CH_3—OH, an alcohol in which the hydroxyl-bearing carbon atom is attached to three hydrogen atoms, does not fit any of the alcohol classification definitions. It is usually grouped with the primary alcohols because its reactions are similar to those of these alcohols.

only one other carbon atom. A **secondary alcohol** *is an alcohol in which the hydroxyl-bearing carbon atom is attached to two other carbon atoms.* A **tertiary alcohol** *is an alcohol in which the hydroxyl-bearing carbon atom is attached to three other carbon atoms.* Chemical reactions of alcohols often depend on alcohol class (1°, 2°, or 3°).

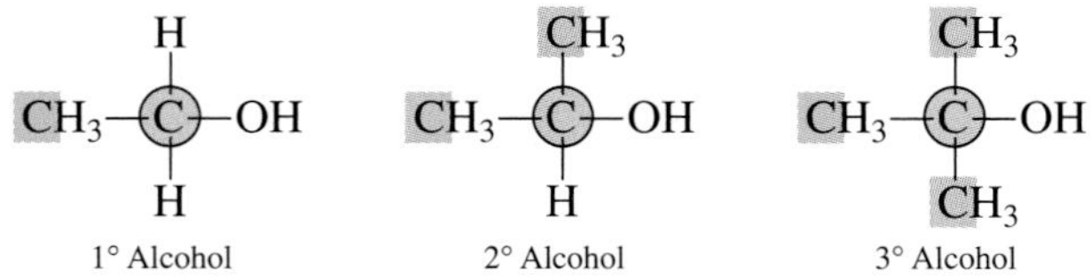

$$\underset{\text{1° Alcohol}}{CH_3-\overset{H}{\underset{H}{C}}-OH} \qquad \underset{\text{2° Alcohol}}{CH_3-\overset{CH_3}{\underset{H}{C}}-OH} \qquad \underset{\text{3° Alcohol}}{CH_3-\overset{CH_3}{\underset{CH_3}{C}}-OH}$$

In general, alcohols (1°, 2°, and 3°) are very flammable substances that, when burned, produce carbon dioxide and water. Additional important reactions of alcohols besides combustion include

1. Intramolecular dehydration to produce an alkene
2. Intermolecular dehydration to produce an ether
3. Oxidation to produce aldehydes, ketones, and carboxylic acids
4. Substitution reactions to produce alkyl halides

• Intramolecular Alcohol Dehydration

A **dehydration reaction** *is a reaction in which the components of water (H and OH) are removed from a single reactant or from two reactants (H from one and OH from the other).* In *intramolecular* dehydration, both water components are removed from the same molecule.

Reaction conditions for the intramolecular dehydration of an alcohol are a temperature of 180°C and the presence of sulfuric acid (H_2SO_4) as a catalyst. The dehydration product is an alkene.

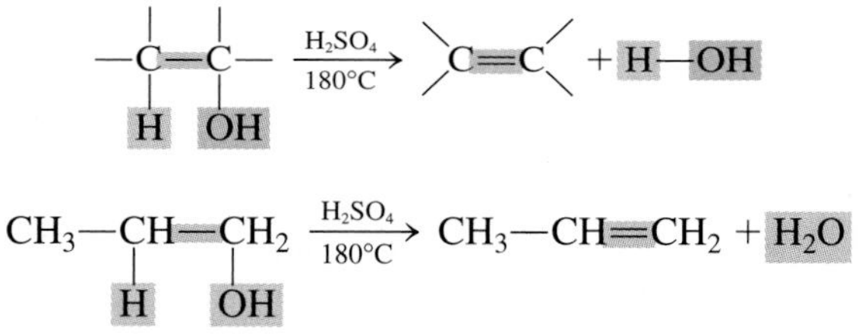

$$-\underset{H}{\overset{|}{C}}-\underset{OH}{\overset{|}{C}}- \xrightarrow[180°C]{H_2SO_4} \;\rangle C=C\langle \; + H-OH$$

$$CH_3-\underset{H}{CH}-\underset{OH}{CH_2} \xrightarrow[180°C]{H_2SO_4} CH_3-CH=CH_2 + H_2O$$

Intramolecular alcohol dehydration is an example of an *elimination reaction* (see Figure 14.5), as contrasted to a substitution reaction (Section 12.16) and an addition reaction (Section 13.7). An **elimination reaction** *is a reaction in which two groups or two atoms on neighboring carbon atoms are removed, or eliminated, from a molecule, leaving a multiple bond between the carbon atoms.*

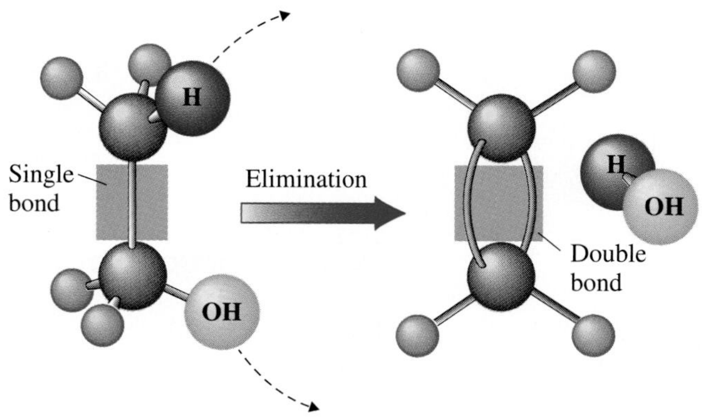

Figure 14.5
In an intramolecular alcohol dehydration (elimination reaction), the components of water (H and OH) are removed from neighboring carbon atoms with the resultant introduction of a double bond into the molecule.

$$-\overset{|}{\underset{|}{C}}-\overset{|}{\underset{\boxed{A}}{C}}\!\!\!\!\underset{\boxed{B}}{}- \longrightarrow \; \rangle C{=}C\langle \; + \; A{-}B$$

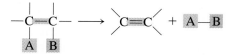

What occurs in an elimination reaction is the reverse of what occurs in an addition reaction.

Dehydration of an alcohol can result in the production of more than one alkene product. This happens when there is more than one neighboring carbon atom from which hydrogen loss can occur. Dehydration of 2-butanol produces two alkenes.

> • Dehydration of alcohols to form carbon–carbon double bonds occurs in several metabolic pathways in living systems, such as the citric acid cycle (Section 23.6) and the fatty acid spiral (Section 25.4). In these biochemical dehydrations, enzymes serve as catalysts instead of acids, and the reaction temperature is 37°C instead of the elevated temperatures required in the laboratory.

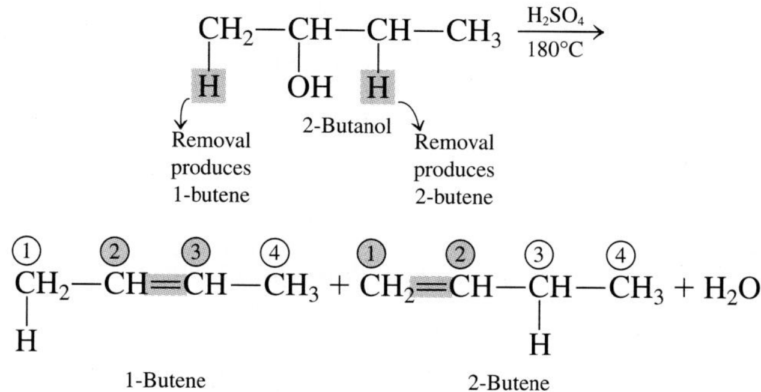

The dominant product can be predicted using Zaitsev's rule, named after the Russian chemist Alexander Zaitsev. **Zaitsev's rule** states that *the major product in an intramolecular alcohol dehydration reaction is the alkene that has the greater number of alkyl groups attached to the carbon atoms of the double bond.* In the preceding reaction, 2-butene (with two alkyl groups) is favored over 1-butene (with one alkyl group).

> • Alexander Zaitsev (1841–1910), a nineteenth-century Russian chemist, studied at the University of Paris and then returned to his native Russia to become a professor of chemistry at the University of Kazan. His surname is pronounced "zait-zeff."

Two alkyl groups on double-bonded carbons: $(CH_3)-CH{=}CH-(CH_3)$ 2-Butene

$CH_2{=}CH-(CH_2-CH_3)$ 1-Butene — One alkyl group on double-bonded carbons

> • An alternative way of expressing Zaitsev's rule is "Hydrogen atom loss, during intramolecular alcohol dehydration to form an alkene, will occur preferentially from the carbon atom (adjacent to the hydroxyl-bearing carbon) that already has the fewest hydrogen atoms."

• Intermolecular Alcohol Dehydration

At a lower temperature (140°C) than that required for alkene formation (180°C), an *inter*molecular rather than an *intra*molecular alcohol dehydration process can occur to produce an ether. In ether formation, two alcohol molecules interact, an H atom being lost from one and an —OH group from the other. The resulting "leftover" portions of the two alcohol molecules join to form the ether. This reaction, which gives useful yields only for primary alcohol reactants (2° and 3° alcohols yield predominantly alkenes), can be written as

> • Alkene formation from alcohol dehydration is the "reverse" of the hydration reaction for alkenes discussed in Section 13.7.
>
> $$\text{Alkene} \underset{\text{Dehydration}}{\overset{\text{Hydration}}{\rightleftharpoons}} \text{alcohol}$$
>
> Many organic reactions go both forward and backward.

$$-\overset{|}{\underset{|}{C}}-O-H + H-O-\overset{|}{\underset{|}{C}}- \xrightarrow[140°C]{H_2SO_4} -\overset{|}{\underset{|}{C}}-O-\overset{|}{\underset{|}{C}}- + H-O-H$$

$$\underset{\text{Ethanol}}{CH_3-CH_2-O-H} + \underset{\text{Ethanol}}{H-O-CH_2-CH_3} \xrightarrow[140°C]{H_2SO_4}$$

$$CH_3-CH_2-O-CH_2-CH_3 + H_2O$$

The preceding reaction is an example of *condensation.* A **condensation reaction** *is a reaction in which two molecules combine to form a larger one while liberating a small molecule, usually water.* In this case, two alcohol molecules combine to give an ether and water.

• The following is a summary of products obtained from alcohol dehydration reactions using H_2SO_4 as a catalyst.

Primary alcohol $\xrightarrow{180°C}$ alkene; $\xrightarrow{140°C}$ ether

Secondary alcohol $\xrightarrow{180°C}$ alkene; $\xrightarrow{140°C}$ alkene

Tertiary alcohol $\xrightarrow{180°C}$ alkene; $\xrightarrow{140°C}$ alkene

Example 14.2

Predicting the Reactant in an Alcohol Dehydration Reaction When Given the Product

Identify the alcohol reactant needed to produce each of the following compounds as the *major product* of an alcohol dehydration reaction.

a. Alcohol $\xrightarrow[180°C]{H_2SO_4}$ $CH_3—CH=CH—CH_3$

b. Alcohol $\xrightarrow[180°C]{H_2SO_4}$ $CH_2=CH—CH(CH_3)—CH_3$

c. Alcohol $\xrightarrow[140°C]{H_2SO_4}$ $CH_3—CH(CH_3)—CH_2—O—CH_2—CH(CH_3)—CH_3$

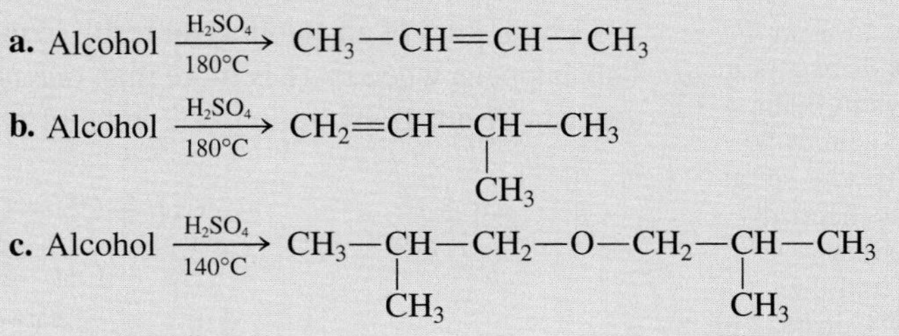

Solution

a. Both carbon atoms of the double bond are equivalent to each other. Add an H atom to one carbon atom of the double bond and an OH group to the other carbon atom of the double bond. It does not matter which goes where; you get the same molecule either way.

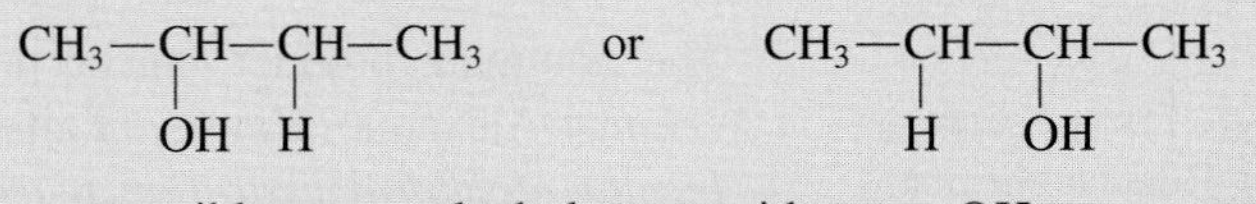

$CH_3—CH(OH)—CH(H)—CH_3$ or $CH_3—CH(H)—CH(OH)—CH_3$

b. There are two possible parent alcohols: one with an —OH group on carbon 1 and the other with an —OH group on carbon 2.

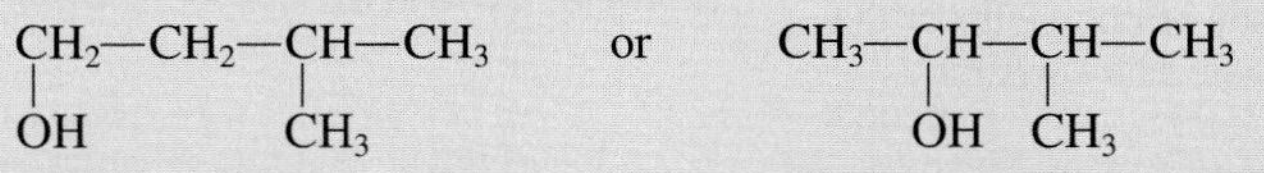

$CH_2(OH)—CH_2—CH(CH_3)—CH_3$ or $CH_3—CH(OH)—CH(CH_3)—CH_3$

Using the reverse of Zaitsev's rule, we find that the hydrogen atom will go back on the double-bonded carbon that bears the most alkyl groups.

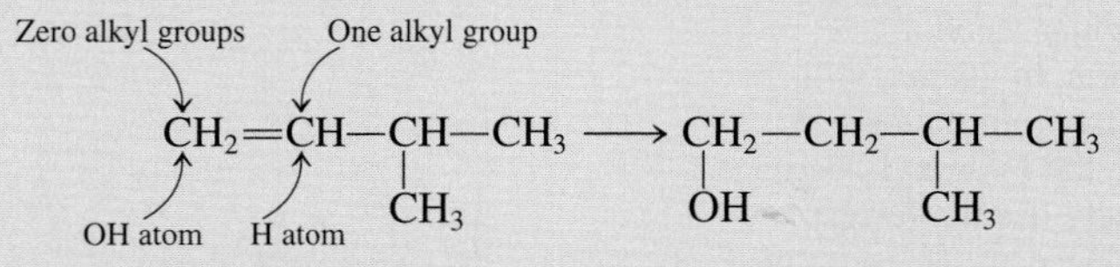

Zero alkyl groups; One alkyl group

$CH_2=CH—CH(CH_3)—CH_3 \longrightarrow CH_2(OH)—CH_2—CH(CH_3)—CH_3$

OH atom; H atom

c. This is an ether. The primary alcohol from which the ether was formed will have the same alkyl group present as is in the ether. Thus the alcohol is

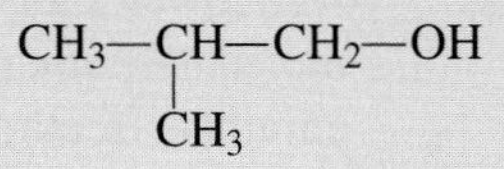

$CH_3—CH(CH_3)—CH_2—OH$

Practice Exercise 14.2

Identify the starting alcohol from which each of the following products was obtained by an alcohol dehydration reaction.

a. Alcohol $\xrightarrow[140°C]{H_2SO_4}$ $CH_2=CH—CH_2—CH_3$

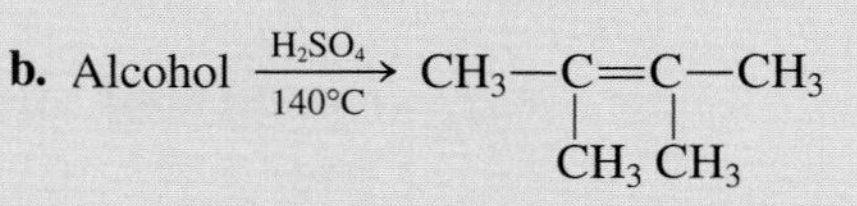

b. Alcohol $\xrightarrow[140°C]{H_2SO_4}$ $CH_3—C(CH_3)=C(CH_3)—CH_3$

c. Alcohol $\xrightarrow[140°C]{H_2SO_4}$ $CH_3—CH_2—CH_2—O—CH_2—CH_2—CH_3$

• *Answers:* **a.** $CH_2(OH)—CH_2—CH_2—CH_3$ **b.** $CH_3—C(OH)(CH_3)—CH(CH_3)—CH_3$

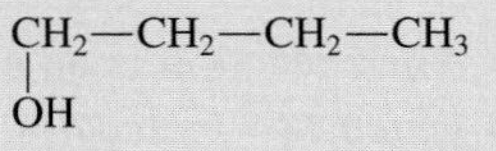

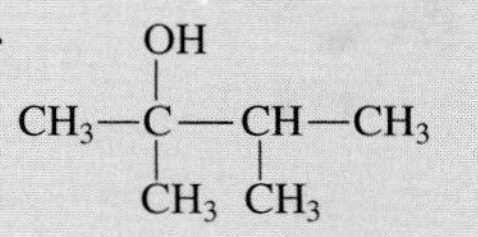

c. $CH_3—CH_2—CH_2—OH$

Oxidation

Primary and secondary alcohols readily undergo oxidation in the presence of mild oxidizing agents to produce compounds that contain a carbon–oxygen double bond (aldehydes, ketones, and carboxylic acids). A number of different oxidizing agents can be used for the oxidation, including potassium permanganate ($KMnO_4$), potassium dichromate ($K_2Cr_2O_7$), and chromic acid (H_2CrO_4).

The net effect of the action of a mild oxidizing agent on a primary or secondary alcohol is the removal of two hydrogen atoms from the alcohol. One hydrogen comes from the —OH group, the other from the carbon atom to which the —OH group is attached. This H removal generates a carbon–oxygen double bond.

$$\underset{\text{An alcohol}}{-\overset{\overset{\text{O—H}}{|}}{\underset{|}{\text{C}}}\text{—H}} \xrightarrow[\text{agent}]{\text{Mild oxidizing}} \underset{\text{Compound containing a carbon–oxygen double bond}}{-\overset{\overset{\text{O}}{\|}}{\underset{|}{\text{C}}}} + 2\text{H}$$

The two "removed" hydrogen atoms combine with oxygen supplied by the oxidizing agent to give H_2O.

The three classes of alcohols behave differently toward mild oxidizing agents.

$$\text{Primary alcohol} \xrightarrow[\text{ox. agent}]{\text{Mild}} \text{aldehyde} \xrightarrow[\text{ox. agent}]{\text{Mild}} \text{carboxylic acid}$$

$$\text{Secondary alcohol} \xrightarrow[\text{ox. agent}]{\text{Mild}} \text{ketone}$$

$$\text{Tertiary alcohol} \xrightarrow[\text{ox. agent}]{\text{Mild}} \text{no reaction}$$

The general reaction for the oxidation of a primary alcohol is

$$\underset{\text{1° Alcohol}}{\text{R—}\overset{\overset{\text{O—(H)}}{|}}{\underset{\underset{\text{(H)}}{|}}{\text{C}}}\text{—H}} \xrightarrow{[O]} \underset{\text{Aldehyde}}{\text{R—}\overset{\overset{\text{O}}{\|}}{\text{C}}\text{—H}} \xrightarrow{[O]} \underset{\text{Carboxylic acid}}{\text{R—}\overset{\overset{\text{O}}{\|}}{\text{C}}\text{—OH}}$$

In this equation, the symbol [O] represents the mild oxidizing agent. The immediate product of the oxidation of a primary alcohol is an aldehyde. Because aldehydes themselves are readily oxidized by the same oxidizing agents that oxidize alcohols, aldehydes are further converted to carboxylic acids. A specific example of a primary alcohol oxidation reaction is

$$CH_3\text{—}CH_2\text{—OH} \xrightarrow{[O]} CH_3\text{—}\overset{\overset{\text{O}}{\|}}{\text{C}}\text{—H} \xrightarrow{[O]} CH_3\text{—}\overset{\overset{\text{O}}{\|}}{\text{C}}\text{—OH}$$

The general reaction for the oxidation of a secondary alcohol is

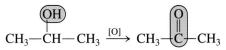

$$\underset{\text{2° Alcohol}}{\text{R—}\overset{\overset{\text{O—(H)}}{|}}{\underset{\underset{\text{(H)}}{|}}{\text{C}}}\text{—R}} \xrightarrow{[O]} \underset{\text{Ketone}}{\text{R—}\overset{\overset{\text{O}}{\|}}{\text{C}}\text{—R}}$$

As with primary alcohols, oxidation involves the removal of two hydrogen atoms. Unlike aldehydes, ketones are resistant to further oxidation. A specific example of the oxidation of a secondary alcohol is

$$CH_3\text{—}\overset{\overset{\text{OH}}{|}}{\text{CH}}\text{—}CH_3 \xrightarrow{[O]} CH_3\text{—}\overset{\overset{\text{O}}{\|}}{\text{C}}\text{—}CH_3$$

• The oxidation of ethanol is the basis for the "breathalyzer test" that law enforcement officers use to determine whether an individual suspected of driving under the influence (DUI) has a blood alcohol level exceeding legal limits.

The DUI suspect is required to breathe into an apparatus containing a solution of potassium dichromate ($K_2Cr_2O_7$). The unmetabolized alcohol in the person's breath is oxidized by the dichromate ion ($Cr_2O_7^{2-}$), and the extent of the reaction gives a measure of the amount of alcohol present.

The dichromate ion is a yellow-orange color in solution. As oxidation of the alcohol proceeds, the dichromate ions are converted to Cr^{3+} ions, which have a green color in solution. The intensity of the green color that develops is measured and is proportional to the amount of ethanol in the suspect's breath, which in turn has been shown to be proportional to the person's blood alcohol level.

Tertiary alcohols do not undergo oxidation with mild oxidizing agents. This is because they do not have hydrogen on the —OH-bearing carbon atom.

$$\begin{array}{c} \text{OH} \\ | \\ \text{R—C—R} \\ | \\ \text{R} \end{array} \xrightarrow{[O]} \text{no reaction}$$

3° Alcohol

In inorganic chemistry (Section 9.2), oxidation numbers are used to characterize oxidation–reduction processes. This same technique could be used in characterizing oxidation–reduction processes involving organic compounds, but it is not. Formal use of the oxidation number rules with organic compounds is usually cumbersome because of the many carbon and hydrogen atoms present; often, fractional oxidation numbers for carbon result.

A better approach is to use the following set of operational rules, instead of oxidation numbers, in characterizing oxidation–reduction processes in organic chemistry.

1. A carbon atom in an organic compound is *oxidized* if it *loses hydrogen atoms* or *gains oxygen atoms* in a chemical reaction.
2. A carbon atom in an organic compound is *reduced* if it *gains hydrogen atoms* or *loses oxygen atoms* in a chemical reaction.

Substitution Reactions

Alcohols also undergo substitution reactions. An important reaction of this type is replacement of the hydroxyl group by a halogen atom, giving an alkyl halide (Section 12.17).

Several different halogen-containing reactants, including phosphorus trihalides (PX_3; X is Cl or Br), are useful in producing alkyl halides from alcohols.

$$3\text{R—OH} + \text{PX}_3 \xrightarrow{\Delta} 3\text{R—X} + \text{H}_3\text{PO}_3$$

Notice that heating of the reactants (Δ) is required.

Using such a substitution reaction for making alkyl halides is more convenient than halogenation of an alkane (Section 12.17), because mixtures of products are *not* obtained. In the product molecules, the halogen is found only where the —OH groups were originally located.

The accompanying Chemistry at a Glance summarizes the reactions that involve alcohols.

14.8 Polymeric Alcohols

It is possible to make polymeric alcohols with structures similar to those of substituted polyethylenes (Section 13.8). Two of the simplest such compounds are poly(vinyl alcohol) (PVA) and poly(ethylene glycol) (PEG).

$$\left(\begin{array}{cc} \text{H} & \text{H} \\ | & | \\ \text{—C—} & \text{C—} \\ | & | \\ \text{H} & \text{OH} \end{array} \right)_n \qquad \left(\begin{array}{cc} \text{H} & \text{H} \\ | & | \\ \text{—C—} & \text{C—} \\ | & | \\ \text{OH} & \text{OH} \end{array} \right)_n$$

PVA PEG

Poly(vinyl alcohol) is a tough, whitish polymer that can be formed into strong films, tubes, and fibers that are highly resistant to hydrocarbon solvents. Unlike most organic polymers, PVA is water-soluble. PVA has oxygen-barrier properties under dry conditions that are superior to those of any other polymer. PVA can be rendered insoluble in water, if needed, by use of chemical agents that cross-link individual polymer strands.

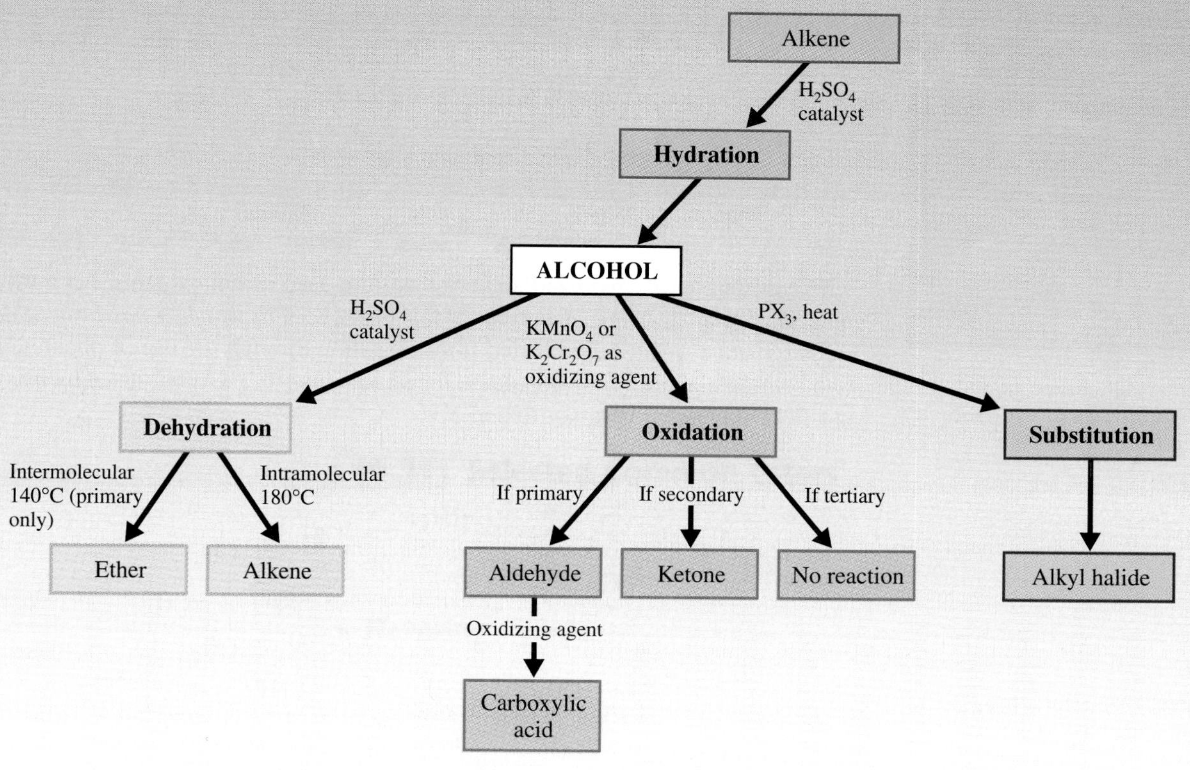

Aqueous solutions of PEG are very viscous (thick) because of the great solubility of PEG in water. PEG is used as an additive in many shampoos. It contributes little to the cleansing action of the shampoo, but it gives the shampoo texture or richness.

14.9 Properties and Uses of Phenols

• An *antiseptic* is a substance that kills microorganisms on living tissue. A *disinfectant* is a substance that kills microorganisms on inanimate objects.

Phenols are generally low-melting solids or oily liquids at room temperature. Most of them are only slightly soluble in water. Many phenols have antiseptic and disinfectant properties.

The simplest phenol, phenol itself, is a colorless solid with a medicinal odor. Its melting point is 41°C, and it is more soluble in water than are most other phenols. Dilute (2%) solutions of phenol have long been used as antiseptics. Concentrated phenol solutions, however, can cause severe skin burns.

Today, phenol has been largely replaced by more effective phenol derivatives such as 4-hexylresorcinol. The compound 4-hexylresorcinol is an ingredient in many mouthwashes and throat lozenges.

• The "parent" name for a benzene ring bearing two hydroxyl groups "meta" to each other is resorcinol.

OH

OH

$CH_2—(CH_2)_4—CH_3$

4-Hexylresorcinol

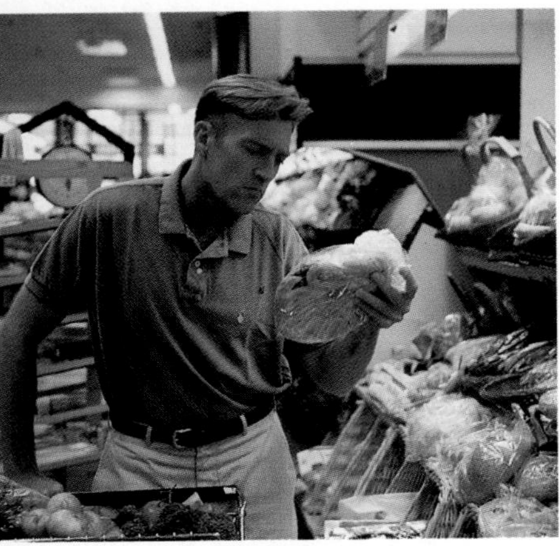

Many commercially baked goods contain the antioxidants BHA and BHT to help prevent spoilage.

The phenol derivatives *o*-phenylphenol and 2-benzyl-4-chlorophenol are the active ingredients in Lysol, a disinfectant for walls, floors, and furniture in homes and hospitals.

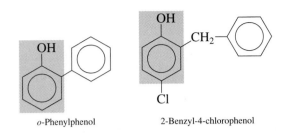

o-Phenylphenol 2-Benzyl-4-chlorophenol

Phenols are easily oxidized and are therefore used as antioxidants. An **antioxidant** *is an additive that is oxidized instead of the substances to which it has been added.* Many foods sensitive to air are protected from oxidation through the use of phenolic additives. Two commercial phenolic antioxidant food additives are BHA (butylated hydroxy anisole) and BHT (butylated hydroxy toluene).

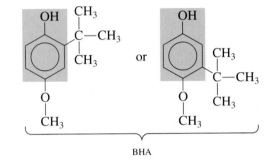

• The "parent" name for a benzene ring bearing a methoxy group is anisole.

BHA

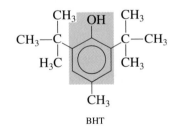

BHT

A naturally occurring phenolic antioxidant that is important in the functioning of the human body is vitamin E (Section 21.14).

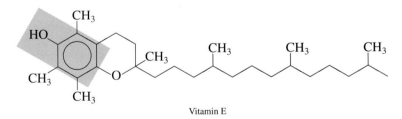

Vitamin E

Unlike alcohols, phenols are weak acids in solution. As acids, phenols have K_a values (Section 10.5) of about 10^{-10}. Such K_a values are higher than that of water (10^{-14}) but lower than those of most weak inorganic acids (10^{-5} to 10^{-10}). The acid dissociation reaction for phenol itself is

$$C_6H_5OH + H_2O \rightleftharpoons C_6H_5O^- + H_3O^+$$

Chemical CONNECTIONS

14.2 Marijuana: The Most Commonly Used Illicit Drug

Prepared from the leaves, flowers, seeds, and small stems of a hemp plant called *Cannabis sativa,* marijuana, which is also called pot or grass, is the most commonly used illicit drug in the United States. The most active ingredient of the many in marijuana is the molecule *tetrahydrocannabinol,* called THC for short. Three different functional groups are present in a THC molecule; it is a phenol, a cyclic ether, and a cycloalkene.

cycloalkene; phenol; cyclic ether

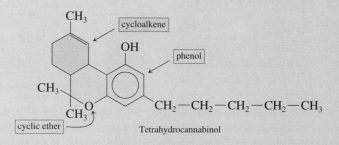

$CH_2—CH_2—CH_2—CH_2—CH_3$

Tetrahydrocannabinol

The THC content of marijuana varies considerably. Most marijuana sold in the North American illegal drug market has a THC content of 1% to 2%.

Marijuana has a pharmacology unlike that of any other drug. A marijuana "high" is a combination of sedation, tranquilization, and mild hallucination. THC readily penetrates the brain. The portions of the brain that involve memory and motor control contain the receptor sites where THC molecules interact. Even moderate doses of marijuana cause short-term memory loss. Marijuana unquestionably impairs driving ability, even after ordinary social use. THC readily crosses the placental barrier and reaches the fetus. Heavy marijuana users experience inflammation of the bronchi, sore throat, and inflamed sinuses. Increased heart rate, to as high as 160 beats per second, can occur with marijuana use.

The onset of action of THC is usually within minutes after smoking begins, and peak concentration in plasma occurs in 10 to 30 minutes. Unless more is smoked, the effects seldom last longer than 2 to 3 hours. Because THC is only slightly soluble in water, it tends to be deposited in fatty tissues. Unlike alcohol, THC persists in the bloodstream for several days, and the products of its breakdown remain in the blood for as long as 8 days.

New research indicates that physical dependence on THC can develop. Drug withdrawal symptoms are seen in some individuals who have been exposed repeatedly to high doses.

A number of phenols found in plants are used as flavoring agents and/or antibacterials. Included among these phenols are

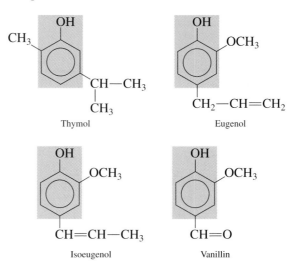

Thymol

Eugenol

Isoeugenol

Vanillin

Nutmeg tree fruit. A phenolic compound, isoeugenol, is responsible for the odor associated with nutmeg.

Thymol, obtained from the herb thyme, possesses both flavorant and antibacterial properties. It is used as an ingredient in several mouthwash formulations.

Eugenol is responsible for the flavor of cloves. Dentists traditionally used clove oil as an antiseptic because of eugenol's presence; they use it to a limited extent even today.

Isoeugenol, which differs in structure from eugenol only in the location of the double bond in the hydrocarbon side chain, is responsible for the odor associated with nutmeg.

Vanillin, which gives vanilla its flavor, is extracted from the dried seed pods of the vanilla orchid. Natural supplies of vanillin are inadequate to meet demand for this flavoring agent. Synthetic vanillin is produced by oxidation of eugenol. Vanillin is an unusual substance in that even though its odor can be perceived at extremely low concentrations, the strength of its odor does not increase greatly as its concentration is increased.

Certain phenols exert profound physiological effects. For example, the irritating constituents of poison ivy and poison oak are derivatives of catechol (Section 14.3). These skin irritants have 15-carbon alkyl side chains with varying degrees of unsaturation (zero to three double bonds).

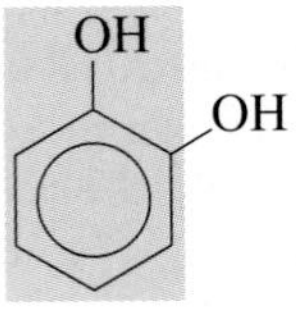

OH OH

Catechol

OH OH $C_{15}H_{25-31}$

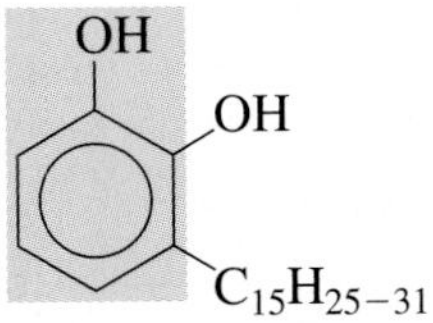

Poison ivy irritants

14.10 Nomenclature for Ethers

Ethers are compounds with the general formula R—O—R′ (Section 14.2). Common names for ethers are formed by naming the two hydrocarbon groups attached to the oxygen atom and adding the word *ether.* The hydrocarbon groups are listed in alphabetical order. When both hydrocarbon groups are the same, the prefix *di-* is placed before the name of the hydrocarbon group. In this system, ether names consist of two or three separate words.

- Line-angle drawings for selected simple ethers:

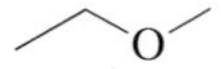

O — Methyl ethyl ether (methoxyethane)

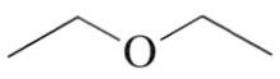

O — Diethyl ether (ethoxyethane)

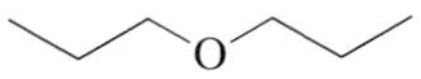

O — Dipropyl ether (1-propoxypropane)

$CH_3—O—CH_2—CH_3$ — Ethyl methyl ether

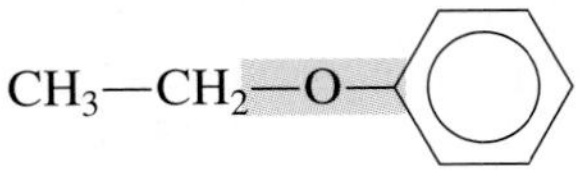

$CH_3—CH_2—O—C_6H_5$ — Ethyl phenyl ether

$CH_3—O—CH_3$ — Dimethyl ether

In the IUPAC nomenclature system, ethers are named as substituted hydrocarbons. The smaller hydrocarbon attachment and the oxygen atom are called an *alkoxy group,* and this group is considered a substituent on the larger hydrocarbon group. An **alkoxy group** *is an alkyl group to which an oxygen atom has been added.* Simple alkoxy groups include the following:

$CH_3—O—$ — Methoxy group

$CH_3—CH_2—O—$ — Ethoxy group

$CH_3—CH_2—CH_2—O—$ — Propoxy group

The general symbol for an alkoxy group is —O—R (or —OR).

The following examples illustrate IUPAC ether nomenclature. In each structure, the alkoxy groups present have been highlighted.

- It is possible to have compounds that contain both ether and alcohol functional groups, such as

$CH_3—CH(OH)—CH_2—CH_2—O—CH_3$

4-Methoxy-2-butanol

(The alcohol functional group has higher priority in IUPAC nomenclature, so the compound is named as an alcohol rather than as an ether.)

$CH_3—O—CH_2—CH_2—CH_2—CH_3$ — 1-Methoxybutane

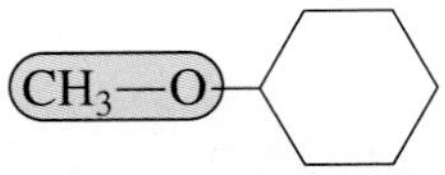

$CH_3—O—C_6H_{11}$ — Methoxycyclohexane

$CH_3—CH_2—CH_2—O—CH_2—CH_2—CH_3$ — 1-Propoxypropane

$CH_3—CH(O—CH_3)—CH_2—CH_3$ — 2-Methoxybutane

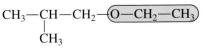

$CH_3—CH(CH_3)—CH_2—O—CH_2—CH_3$ — 1-Ethoxy-2-methylpropane

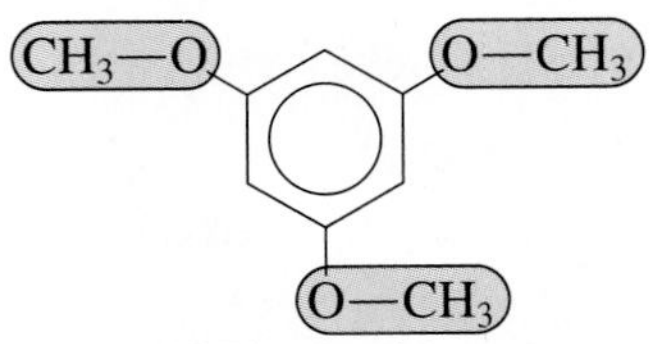

$C_6H_3(O—CH_3)_3$ — 1,3,5-Trimethoxybenzene

Chemical CONNECTIONS

14.3 Ethers as General Anesthetics

For many people, the word *ether* evokes thoughts of hospital operating rooms and anesthesia. This response derives from the *former* large-scale use of diethyl ether as a general anesthetic. In 1846, the Boston dentist William Morton was the first to demonstrate publicly the use of diethyl ether as a surgical anesthetic.

In many ways, diethyl ether is an ideal general anesthetic. It is relatively easy to administer, it is readily made in pure form, and it causes excellent muscle relaxation. There is less danger of an overdose with diethyl ether than with almost any other anesthetic, because there is a large gap between the effective level for anesthesia and the lethal dose.

Despite these ideal properties, diethyl ether is rarely used today because of two drawbacks: (1) It causes nausea and irritation of the respiratory passage, and (2) it is a highly flammable substance, forming explosive mixtures with air, which can be set off by a spark.

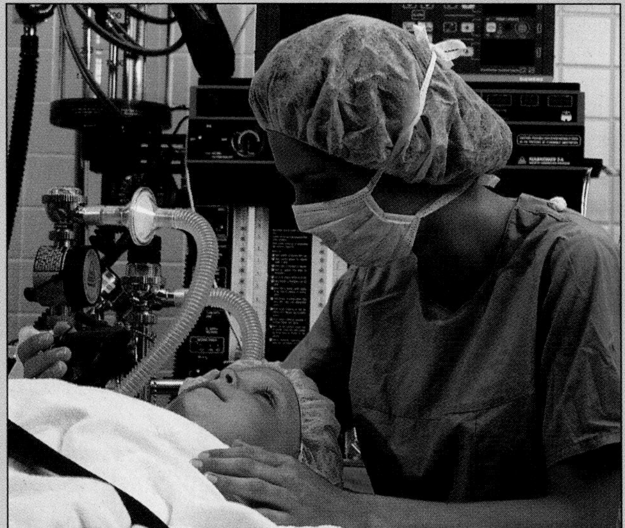

By the 1930s, nonether anesthetics had been developed that solved the problems of nausea and irritation. They also, however, were extremely flammable compounds. The simple hydrocarbon cyclopropane was the most widely used of these newer compounds.

It was not until the late 1950s and early 1960s that nonflammable general anesthetics became available. Anesthetic nonflammability was achieved by incorporating halogen atoms into anesthetic molecules. Three of the most used of these "halogenated" anesthetics are enflurane, isoflurane, and halothane.

Enflurane and isoflurane, which are structural isomers, are hexahalogenated ethers.

$$\text{H}-\underset{\text{F}}{\overset{\text{Cl}}{\text{C}}}-\underset{\text{F}}{\overset{\text{F}}{\text{C}}}-\text{O}-\underset{\text{F}}{\overset{\text{F}}{\text{C}}}-\text{H}$$

Enflurane

$$\text{F}-\underset{\text{F}}{\overset{\text{F}}{\text{C}}}-\underset{\text{Cl}}{\overset{\text{H}}{\text{C}}}-\text{O}-\underset{\text{F}}{\overset{\text{F}}{\text{C}}}-\text{H}$$

Isoflurane

With these compounds, induction of anesthesia can be achieved in less than 10 minutes with an inhaled concentration of 3% in oxygen.

Halothane, which is potent at relatively low doses, and whose effects wear off quickly, is a hexahalogenated ethane derivative rather than an ether.

$$\text{F}-\underset{\text{F}}{\overset{\text{F}}{\text{C}}}-\underset{\text{Cl}}{\overset{\text{Br}}{\text{C}}}-\text{H}$$

Halothane

• The contrast between IUPAC and common names for ethers is as follows:

IUPAC (one word)
alkoxyalkane
2-methoxybutane
Common (three or two words)
alkyl alkyl ether
ethyl methyl ether
or
dialkyl ether
dipropyl ether

In the United States, the hydrocarbon ethene, $CH_2{=}CH_2$, is the organic chemical produced in the largest quantities. An ether, methyl *tert*-butyl ether (MTBE), is the number-two organic chemical.

$$CH_3-O-\underset{CH_3}{\overset{CH_3}{C}}-CH_3$$

Methyl *tert*-butyl ether (MTBE)

Since 1979 small amounts of MTBE have been used in gasoline to raise octane levels, but in recent years it has been added in much larger amounts as a clean-burning oxygenate to produce EPA-mandated reformulated gasoline used to improve air quality in polluted areas. The amount of MTBE in gasoline is likely to be reduced in the future in response to a growing problem: contamination of water supplies by small amounts of MTBE from leaking

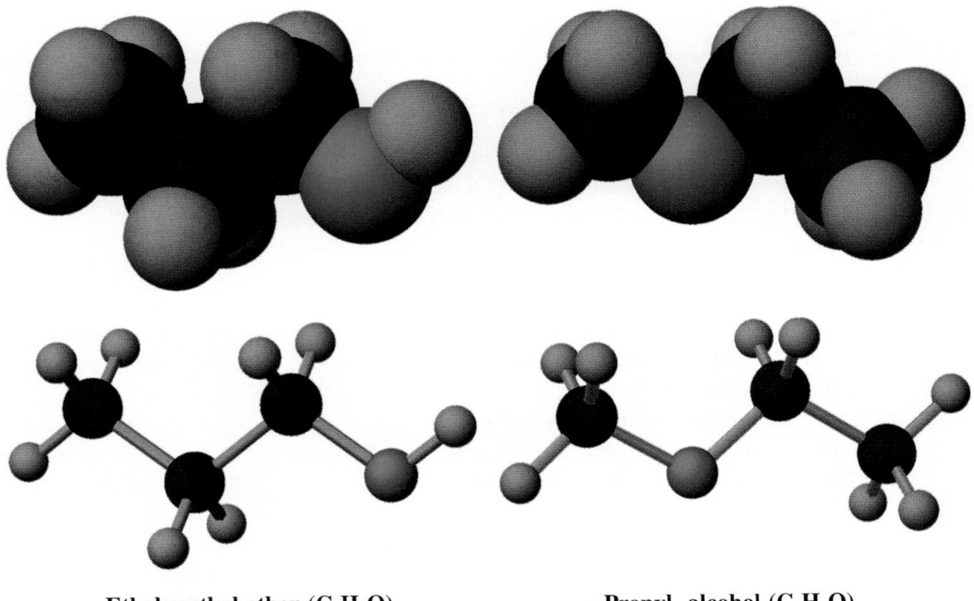

Figure 14.6
Alcohols and ethers with the same number of carbon atoms and the same degree of saturation are structural isomers, as is illustrated here for propyl alcohol and ethyl methyl ether.

gasoline tanks and from spills. MTBE in the water supplies is not a health-and-safety issue at this time, but its presence does affect taste and odor in contaminated supplies.

Compounds with ether functional groups occur in a variety of plants. The phenolic flavoring agents eugenol, isoeugenol, and vanillin (Section 14.9) are also ethers; each has a methoxy attachment on the ring.

• Structural isomers that differ in the type of functional group present, such as ethyl alcohol and dimethyl ether, are often called *functional group isomers.*

Alcohols and ethers with the same number of carbon atoms and the same degree of saturation are structural isomers. For example, both propyl alcohol and ethyl methyl ether have the molecular formula C_3H_8O and are thus structural isomers (see Figure 14.6.)

14.11 Physical and Chemical Properties of Ethers

The boiling points of ethers are similar to those of alkanes of comparable molecular mass and are much lower than those of alcohols of comparable molecular mass.

Alkane

$$CH_3{-}CH_2{-}CH_2{-}CH_2{-}CH_3$$

Mol. mass = 72 amu
bp = 36°C

Ether

$$CH_3{-}CH_2{-}O{-}CH_2{-}CH_3$$

Mol. mass = 74 amu
bp = 35°C

Alcohol

$$CH_3{-}CH_2{-}CH_2{-}CH_2{-}OH$$

Mol. mass = 74 amu
bp = 117°C

A physical-state summary for unbranched alkyl alkyl ethers at room temperature and pressure.

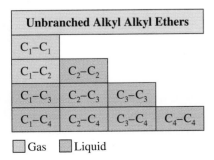

Unbranched Alkyl Alkyl Ethers			
C_1–C_1			
C_1–C_2	C_2–C_2		
C_1–C_3	C_2–C_3	C_3–C_3	
C_1–C_4	C_2–C_4	C_3–C_4	C_4–C_4

☐ Gas ☐ Liquid

The much higher boiling point of the alcohol results from hydrogen bonding between alcohol molecules. Ether molecules, like alkanes, cannot hydrogen-bond to one another. Ether oxygen atoms have no hydrogen atom attached directly to them.

Ethers, in general, are more soluble in water than are alkanes of similar molecular mass, because ether molecules are able to form hydrogen bonds with water.

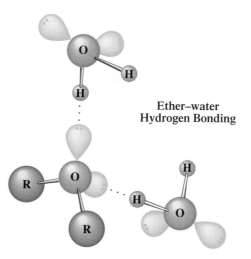

Ether–water
Hydrogen Bonding

• The term *ether* comes from the Latin *aether,* which means "to ignite." This name is given to these compounds because of their high vapor pressure at room temperature, which makes them very flammable.

Ethers have water solubilities similar to those of alcohols of the same molecular mass. For example, diethyl ether and butyl alcohol have the same solubility in water. Because ethers can also hydrogen-bond to alcohols, alcohols and ethers tend to be mutually soluble. Nonpolar substances tend to be more soluble in ethers than in alcohols because ethers have no hydrogen-bonding network that has to be broken up for solubility to occur.

Two chemical properties of ethers are especially important.

1. *Ethers are flammable.* Special care must be exercised in laboratories where ethers are used. Diethyl ether, whose boiling point of 35°C is only a few degrees above room temperature, is a particular flash-fire hazard.
2. *Ethers react slowly with oxygen from the air to form unstable hydroperoxides and peroxides.*

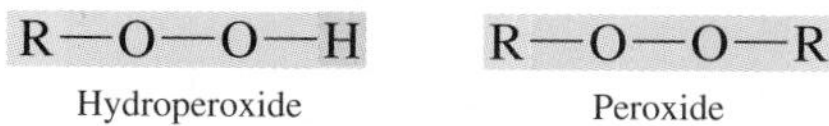

Such compounds, when concentrated, represent an explosion hazard and must be removed before *stored* ethers are used.

Ethers are unreactive towards acids, bases, and oxidizing agents. Like alkanes, they do undergo halogenation reactions.

The general chemical unreactivity of ethers, coupled with the fact that most organic compounds are ether-soluble, makes ethers excellent solvents in which to carry out organic reactions. Their relatively low boiling points simplify their separation from the reaction products.

A chemical reaction for the preparation of ethers has been previously considered. In Section 14.7 we noted that the intermolecular dehydration of a primary alcohol will produce an ether. Although additional methods exist for ether preparation, we will not consider them in this text.

14.12 Cyclic Ethers

Cyclic ethers contain ether functional groups as part of a ring system. Some examples of such cyclic ethers, along with their common names, follow.

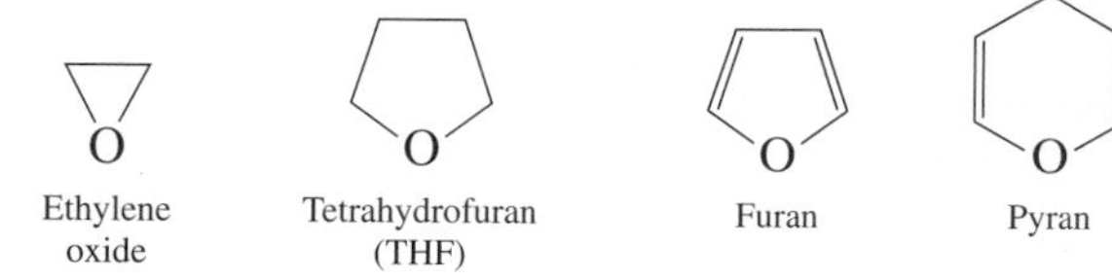

Ethylene oxide is used extensively in industrial chemical processes. THF is a particularly useful solvent in that it dissolves many organic compounds and yet is miscible with water. We will encounter the furan and pyran ring systems when we discuss carbohydrates (Chapter 18). The vitamin E molecule and the THC molecule in marijuana both have structures in which there is a cyclic-ether component.

Cyclic ethers are our first encounter with heterocyclic organic compounds. A **heterocyclic organic compound** *is a cyclic compound in which one or more of the carbon atoms in the ring have been replaced with other kinds of atoms.* The hetero atom is usually oxygen or nitrogen.

We have just seen that *cyclic ethers*—compounds in which the ether functional group is part of a ring system—exist. In contrast, cyclic alcohols—compounds in which the alcohol functional group is part of a ring system—do not exist. To incorporate an alcohol functioning group into a ring system would require an oxygen atom with three bonds, and oxygen atoms form only two bonds.

H

O

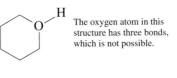

The oxygen atom in this structure has three bonds, which is not possible.

Compounds such as

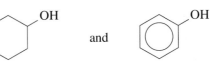

OH and OH

which do exist, are not cyclic alcohols in the sense in which we are using the term, because the alcohol functional group is attached to a ring system rather than part of it.

14.13 Sulfur Analogs of Alcohols and Ethers

Many organic compounds containing oxygen have sulfur analogs, in which a sulfur atom has replaced an oxygen atom. Sulfur is in the same group of the periodic table as oxygen, so the two elements have similar electron configurations (Section 3.8).

• The root *thio-* indicates that a sulfur atom has replaced an oxygen atom in a compound. It originates from the Greek *theion,* meaning "brimstone," which is an older name for the element sulfur.

Sulfur analogs of alcohols are called *thiols.* These compounds contain —SH groups instead of —OH groups. The —SH group is known as the *sulfhydryl group.* An older term used for thiols was *mercaptans.* The general structures for alcohols and thiols are

Hydroxyl group ↙ R—OH An alcohol

Sulfhydryl group ↙ R—SH A thiol

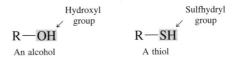

Thiols are named in the same way as alcohols in the IUPAC system, except that the *-ol* becomes *-thiol.* The prefix *thio-* indicates the substitution of a sulfur atom for an oxygen atom in a compound.

$CH_3—CH(OH)—CH_2—CH_3$ 2-Butanol

$CH_3—CH(SH)—CH_2—CH_3$ 2-Butanethiol

• Even though thiols have a higher molecular mass than alcohols with the same number of carbon atoms, they have a much lower boiling point, because they do not exhibit hydrogen bonding as alcohols do.

As in the case of diols and triols, the *-e* at the end of the alkane name is also retained for thiols.

Two important properties of thiols are low boiling points (because of reduced hydrogen bonding) and a strong, disagreeable odor. The familiar odor of natural gas results from the addition of thiols to the gas. The exceptionally low threshold of detection for thiols enables

Thiols are responsible for the strong odor of "essence of skunk." Their odor is an effective defense mechanism.

consumers to smell a gas leak long before the gas reaches dangerous levels. Thiols are responsible for the odor of skunks. Many of the strong odors associated with foods are caused by small thiol molecules (Table 14.4).

Thiols are easily oxidized but yield different products than their alcohol analogs. Thiols form *disulfides.* Each of two thiol groups loses a hydrogen atom, thus linking the two sulfur atoms together via a disulfide group, —S—S—.

$$R{-}SH + HS{-}R \underset{\text{Reduction}}{\overset{\text{Oxidation}}{\rightleftharpoons}} \underset{\text{A disulfide}}{R{-}S{-}S{-}R} + 2H$$

Table 14.4
Selected Thiols That Are Present in Foods

Flavor or odor of	Structures of compounds involved
oysters	$CH_3{-}SH$
cheddar cheese	$CH_3{-}SH$
freshly chopped onions	$CH_3{-}CH_2{-}CH_2{-}SH$
garlic	$CH_2{=}CH{-}CH_2{-}SH$

Sulfur analogs of ethers exist. They are known as sulfides (or thioethers). Like thiols, sulfides have strong characteristic odors.

Sulfides are named in the same way as ethers, with *sulfide* used in place of *ether* in common names and *alkylthio* used in place of *alkoxy* in IUPAC names.

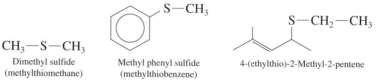

$CH_3{-}S{-}CH_3$
Dimethyl sulfide (methylthiomethane)

$C_6H_5{-}S{-}CH_3$
Methyl phenyl sulfide (methylthiobenzene)

$S{-}CH_2{-}CH_3$ (skeletal structure)
4-(ethylthio)-2-Methyl-2-pentene

• Bacteria in the mouth interact with saliva and leftover food to produce such compounds as hydrogen sulfide, methanethiol (a thioalcohol), and dimethyl sulfide (a thioether). These compounds, which have odors detectable in air at concentrations of parts per billion, are responsible for "morning breath."

In general, thiols are more reactive than their alcohol counterparts, and sulfides are more reactive than their corresponding ethers. The larger size of a sulfur atom compared to an oxygen atom (see Figure 14.7) results in a carbon–sulfur covalent bond that is weaker than a carbon–oxygen covalent bond. An added factor is that sulfur's electronegativity (2.5) is significantly lower than that of oxygen (3.5).

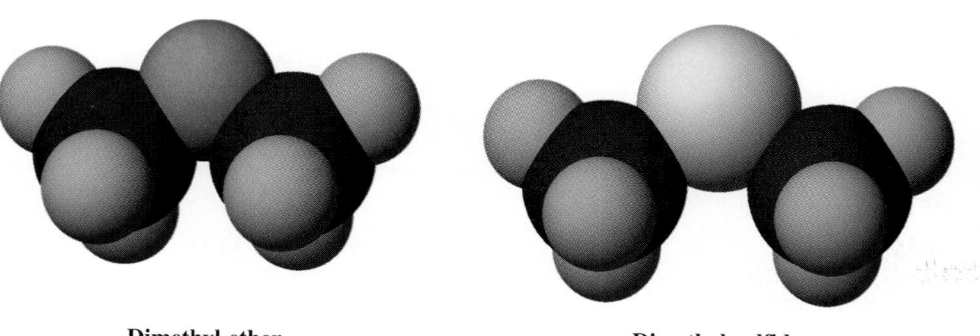

Figure 14.7
A comparison involving space-filling models for dimethyl ether and dimethyl sulfide (dimethyl thioether). A sulfur atom is much larger than an oxygen atom. This results in a carbon–sulfur bond being weaker than a carbon–oxygen bond.

Concepts to Remember

Alcohols and phenols. Alcohols are organic compounds that contain an —OH group attached to a saturated carbon atom. The general formula for an alcohol is R—OH, where R is an alkyl group. Phenols have the general formula Ar—OH, where Ar represents an aryl group derived from an aromatic compound.

Nomenclature of alcohols and phenols. The IUPAC names of simple alcohols end in *-ol,* and their chains are numbered to give precedence to the location of the —OH group. The common names have the word *alcohol* preceded by the name of the alkyl group. Phenols are named as derivatives of the parent compound phenol.

Physical properties of alcohols. Alcohol molecules hydrogen-bond to each other and to water molecules. They thus have higher-than-normal boiling points, and the low-molecular-mass alcohols are soluble in water.

Classifications of alcohols. Alcohols are classified on the basis of the number of carbon atoms bonded to the carbon attached to the —OH group. In primary alcohols, the —OH group is bonded to a carbon atom bonded to only one other C atom. In secondary alcohols, the —OH-containing C atom is attached to two other C atoms. In tertiary alcohols, it is attached to three other C atoms.

Alcohol dehydration. Alcohols can be dehydrated in the presence of sulfuric acid to form alkenes or ethers. At 180°C, an alkene is produced; at 140°C, primary alcohols produce an ether.

Alcohol oxidation. Oxidation of primary alcohols first produces an aldehyde, which is then further oxidized to a carboxylic acid. Secondary alcohols are oxidized to ketones, and tertiary alcohols are resistant to oxidation.

Ethers. The general formula for an ether is R—O—R′, where R and R′ are alkyl, cycloalkyl, or aryl groups. In the IUPAC system, ethers are named as alkoxy derivatives of alkanes. Common names are obtained by giving the R group names in alphabetical order and adding the word *ether.*

Properties of ethers. Ethers have lower boiling points than alcohols because ether molecules do not hydrogen-bond to each other. Ethers are slightly soluble in water because water forms hydrogen bonds with ethers.

Thiols and disulfides. Thiols are the sulfur analogs of alcohols. They have the general formula R—SH. The —SH group is called the sulfhydryl group. Oxidation of thiols forms disulfides, which have the general formula R—S—S—R. The most distinctive physical property of thiols is their foul odor.

Key Reactions and Equations

1. Intramolecular dehydration of alcohols to give alkenes (Section 14.7)

$$\underset{\text{H}\quad\ \text{OH}}{-\overset{|\quad |}{\text{C}-\text{C}}-} \xrightarrow[180°\text{C}]{H_2SO_4} \text{>C=C<} + H_2O$$

2. Intermolecular dehydration of primary alcohols to give ethers (Section 14.7)

$$\text{R—O—H} + \text{H—O—R} \xrightarrow[140°\text{C}]{H_2SO_4} \text{R—O—R} + H_2O$$

3. Oxidation of a primary alcohol to give an aldehyde and then a carboxylic acid (Section 14.7)

$$\text{R}-\underset{\text{H}}{\overset{\text{OH}}{\text{C}}}-\text{H} \xrightarrow{[O]} \underset{\text{Aldehyde}}{\text{R}-\overset{\text{O}}{\overset{\|}{\text{C}}}-\text{H}} + H_2O \xrightarrow[\text{oxidation}]{[O]\ \text{Further}} \underset{\text{Carboxylic acid}}{\text{R}-\overset{\text{O}}{\overset{\|}{\text{C}}}-\text{OH}}$$

4. Oxidation of a secondary alcohol to give a ketone (Section 14.7)

$$\text{R}-\underset{\text{H}}{\overset{\text{OH}}{\text{C}}}-\text{R}' \xrightarrow{[O]} \text{R}-\overset{\text{O}}{\overset{\|}{\text{C}}}-\text{R}' + 2\text{H}$$

5. Attempted oxidation of a tertiary alcohol, which gives no reaction (Section 14.7)

$$\text{R}-\underset{\text{R}''}{\overset{\text{OH}}{\text{C}}}-\text{R}' \xrightarrow{[O]} \text{no reaction}$$

6. Production of an alkyl halide from an alcohol by substitution using PX_3 (X is Cl or Br) (Section 14.7)

$$\text{R—OH} \xrightarrow[\Delta]{PX_3} \text{R—X}$$

7. Oxidation of a thiol to give a disulfide (Section 14.13)

$$\text{R—SH} + \text{HS—R} \xrightarrow{[O]} \text{R—S—S—R} + 2\text{H}$$

8. Reduction of a disulfide to give a thiol (Section 14.13)

$$\text{R—S—S—R} + 2\text{H} \longrightarrow \text{R—SH} + \text{HS—R}$$

Key Terms

Alcohol (14.2)
Alkoxy group (14.10)
Antioxidant (14.9)
Aryl group (14.2)
Condensation reaction (14.7)
Dehydration reaction (14.7)
Elimination reaction (14.7)
Ether (14.2)
Glycol (14.4)
Heterocyclic organic compound (14.11)
Phenol (14.2)
Primary alcohol (14.7)
Secondary alcohol (14.7)
Tertiary alcohol (14.7)
Zaitsev's rule (14.7)

Exercises and Problems

The members of each pair of problems in this section test similar material.

Structural Characteristics of Alcohols, Phenols, and Ethers (Section 14.2)

14.1 Classify each of the following compounds as an alcohol, a phenol, or an ether.

a. 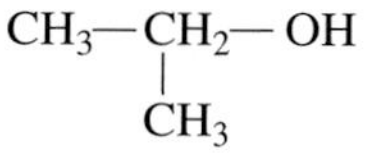 $CH_3{-}CH(CH_3){-}OH$

b. $CH_3{-}O{-}CH_2{-}CH_2{-}CH_3$

c. 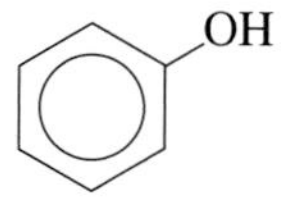 $C_6H_5{-}OH$

d. 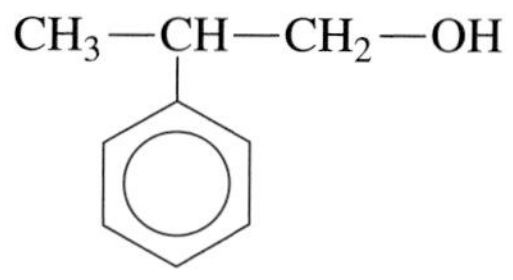 $CH_3{-}CH(C_6H_5){-}CH_2{-}OH$

e. 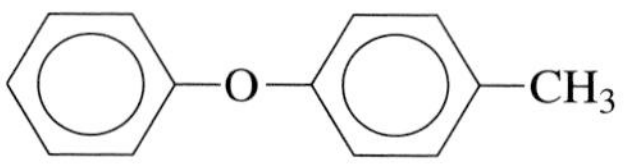 $C_6H_5{-}O{-}C_6H_4{-}CH_3$

f. 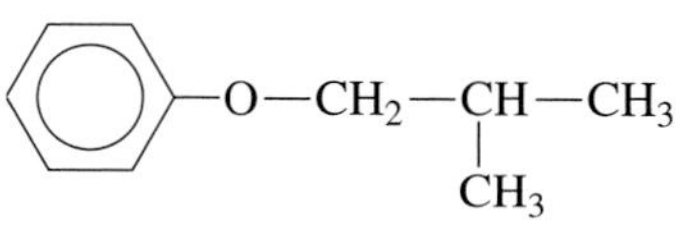 $C_6H_5{-}O{-}CH_2{-}CH(CH_3){-}CH_3$

14.2 Classify each of the following compounds as an alcohol, a phenol, or an ether.

a. $CH_3{-}CH_2{-}O{-}CH_2{-}CH_3$

b. 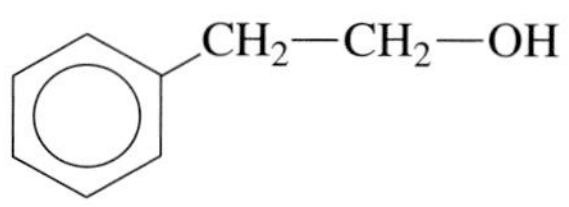 $C_6H_5{-}CH_2{-}CH_2{-}OH$

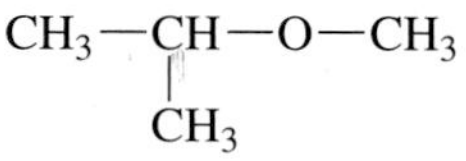

c. $CH_3{-}CH(CH_3){-}O{-}CH_3$

d. 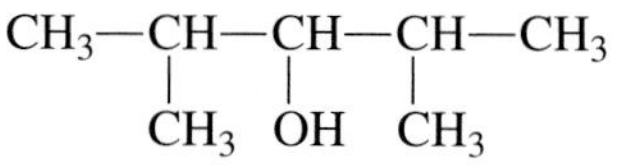 $CH_3{-}CH(CH_3){-}CH(OH){-}CH(CH_3){-}CH_3$

e. 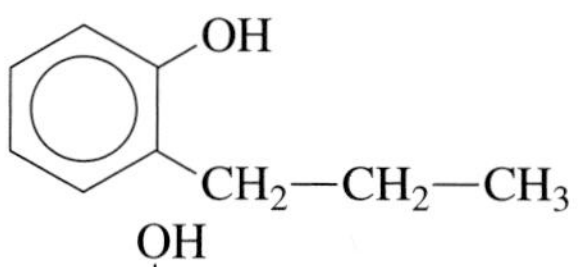 $C_6H_4(OH){-}CH_2{-}CH_2{-}CH_3$

f. 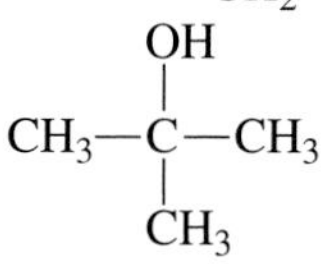 $CH_3{-}C(OH)(CH_3){-}CH_3$

Nomenclature of Alcohols (Section 14.3)

14.3 Assign an IUPAC name to each of the following alcohols.

a. 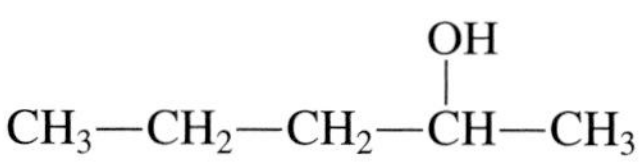 $CH_3{-}CH_2{-}CH_2{-}CH(OH){-}CH_3$

b. $CH_3{-}CH_2{-}OH$

c. 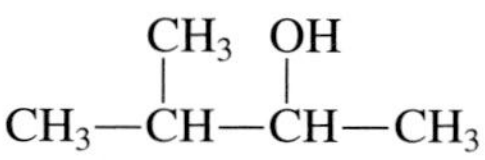 $CH_3{-}CH(CH_3){-}CH(OH){-}CH_3$

d. 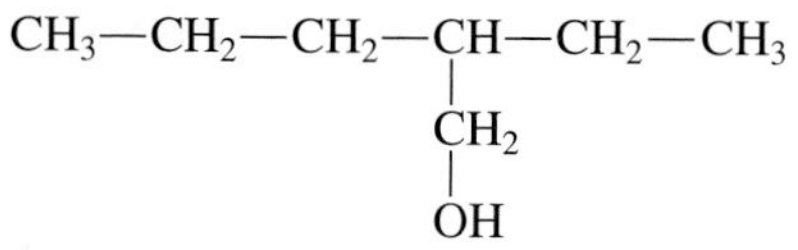 $CH_3{-}CH_2{-}CH_2{-}CH(CH_2OH){-}CH_2{-}CH_3$

e. 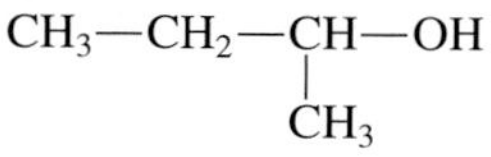 $CH_3{-}CH_2{-}CH(CH_3){-}OH$

f. 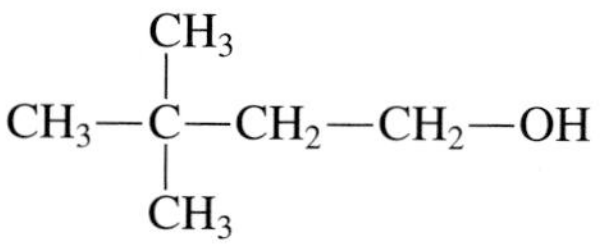 $CH_3{-}C(CH_3)_2{-}CH_2{-}CH_2{-}OH$

14.4 Assign an IUPAC name to each of the following alcohols.

a. 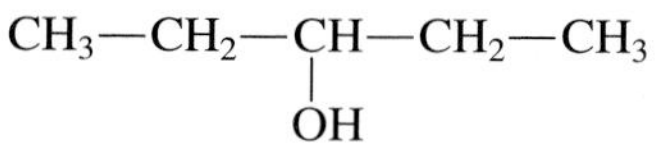 $CH_3{-}CH_2{-}CH(OH){-}CH_2{-}CH_3$

b. 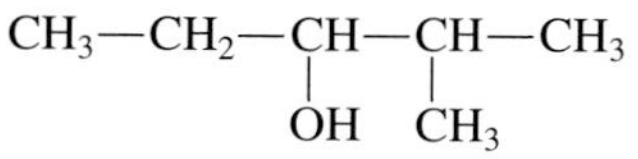 $CH_3{-}CH_2{-}CH(OH){-}CH(CH_3){-}CH_3$

c. 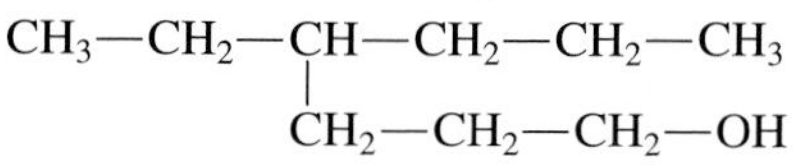 $CH_3{-}CH_2{-}CH(CH_2{-}CH_2{-}CH_2{-}OH){-}CH_2{-}CH_2{-}CH_3$

d. 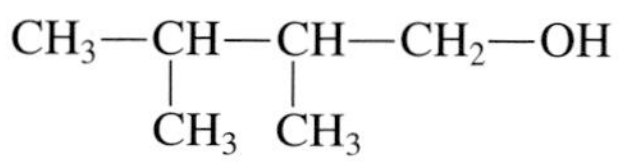 $CH_3{-}CH(CH_3){-}CH(CH_3){-}CH_2{-}OH$

e. 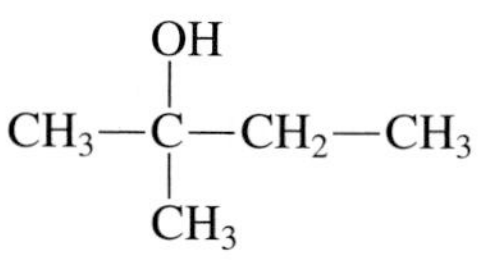 $CH_3{-}C(OH)(CH_3){-}CH_2{-}CH_3$

f. 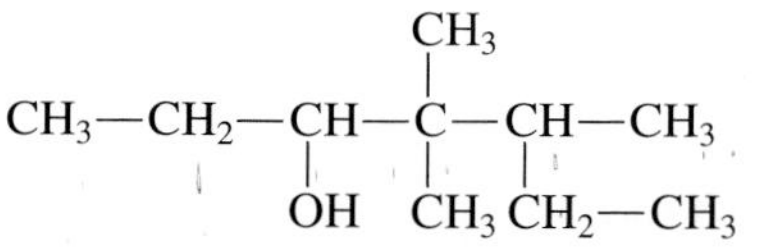 $CH_3{-}CH_2{-}CH(OH){-}C(CH_3)_2{-}CH(CH_2{-}CH_3){-}CH_3$

14.5 Write a structural formula for each of the following alcohols.

a. 3-Pentanol
b. 3-Ethyl-3-hexanol
c. 2-Methyl-1-propanol
d. 4-Methyl-2-pentanol
e. 2-Phenyl-2-propanol
f. 2-Methylcyclobutanol

14.6 Write a structural formula for each of the following alcohols.

a. 2-Heptanol
b. 2,2-Dimethyl-1-hexanol
c. 2-Methyl-2-heptanol
d. 3-Ethyl-2-pentanol
e. 3-Phenyl-1-butanol
f. 3,5-Dimethylcyclohexanol

14.7 Write a structural formula for, and assign an IUPAC name to, each of the following alcohols.

a. Pentyl alcohol
b. Propyl alcohol
c. Isobutyl alcohol
d. *sec*-Butyl alcohol

14.8 Write a structural formula for, and assign an IUPAC name to, each of the following alcohols.

a. Butyl alcohol
b. Hexyl alcohol
c. Isopropyl alcohol
d. *tert*-Butyl alcohol

14.9 Assign an IUPAC name to each of the following polyhydroxy alcohols.

a. 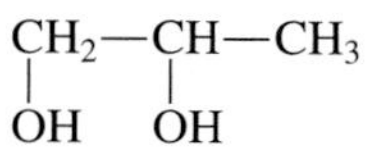 $CH_2(OH){-}CH(OH){-}CH_3$

b. 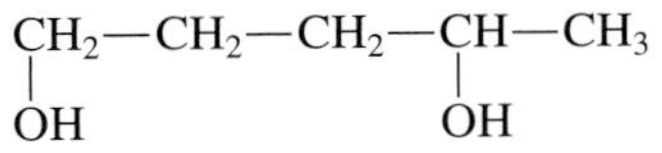 $CH_2-CH_2-CH_2-CH-CH_3$
OH OH

c. 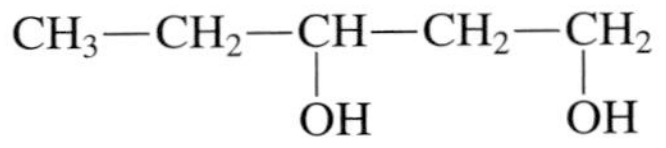 $CH_3-CH_2-CH-CH_2-CH_2$
OH OH

d. 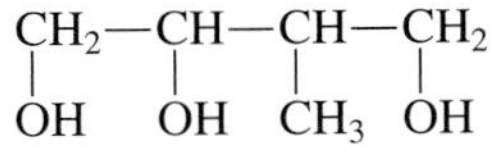 $CH_2-CH-CH-CH_2$
OH OH CH_3 OH

14.10 Assign an IUPAC name to each of the following polyhydroxy alcohols.

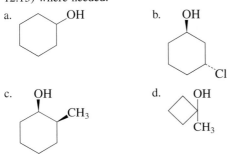

a. 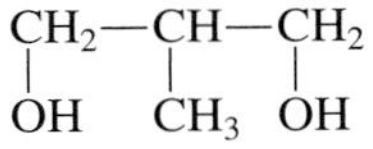 $CH_2-CH-CH_2$
OH CH_3 OH

b. 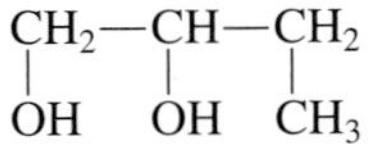 $CH_2-CH-CH_2$
OH OH CH_3

c. 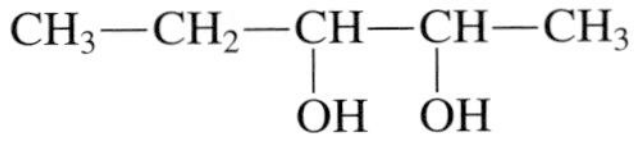 $CH_3-CH_2-CH-CH-CH_3$
OH OH

d. 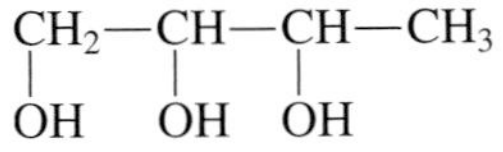 $CH_2-CH-CH-CH_3$
OH OH OH

14.11 Utilizing IUPAC rules, name each of the following compounds. Don't forget to use *cis* and *trans* prefixes (Section 12.13) where needed.

a. OH

b. OH Cl

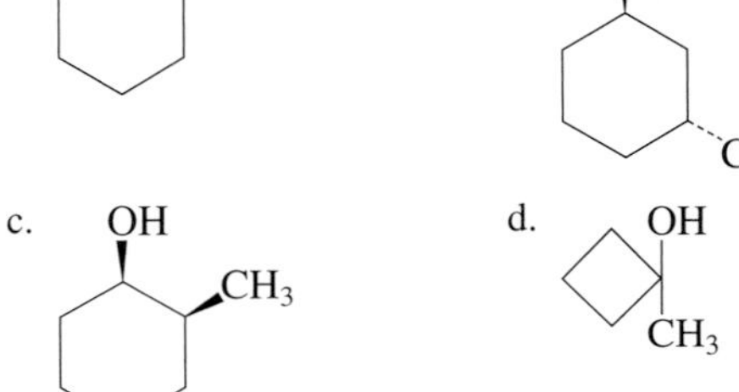

14.12 Utilizing IUPAC rules, name each of the following compounds. Don't forget to use *cis* and *trans* prefixes (Section 12.13) where needed.

a. 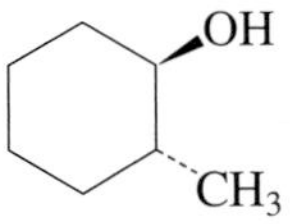 OH CH_3

b. 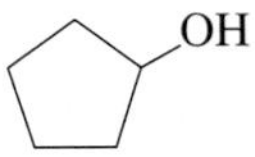 OH

c. 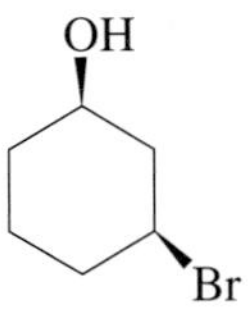 OH Br

d.  HO CH_3

14.13 Write a structural formula for each of the following *unsaturated* alcohols.

a. 4-Penten-2-ol
b. 1-Pentyn-3-ol
c. 3-Methyl-3-buten-2-ol
d. *cis*-2-Buten-1-ol

14.14 Write a structural formula for each of the following *unsaturated* alcohols.

a. 1-Penten-3-ol
b. 3-Butyn-1-ol
c. 2-Methyl-3-buten-1-ol
d. *trans*-3-Penten-1-ol

14.15 Each of the following alcohols is named incorrectly. However, the names give correct structural formulas. Draw structural formulas for the compounds, and then write the correct IUPAC name for each alcohol.

a. 2-Ethyl-1-propanol
b. 2,4-Butanediol
c. 2-Methyl-3-butanol
d. 1,4-Cyclopentanediol

14.16 Each of the following alcohols is named incorrectly. However, the names give correct structural formulas. Draw structural formulas for the compounds, and then write the correct IUPAC name for each alcohol.

a. 3-Ethyl-2-butanol
b. 3,4-Pentanediol
c. 3-Methyl-3-butanol
d. 1,1-Dimethyl-1-butanol

Nomenclature of Phenols (Section 14.3)

14.17 Name the following phenols.

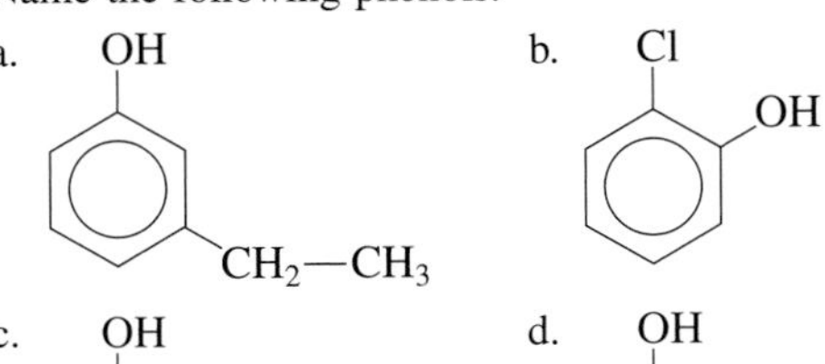

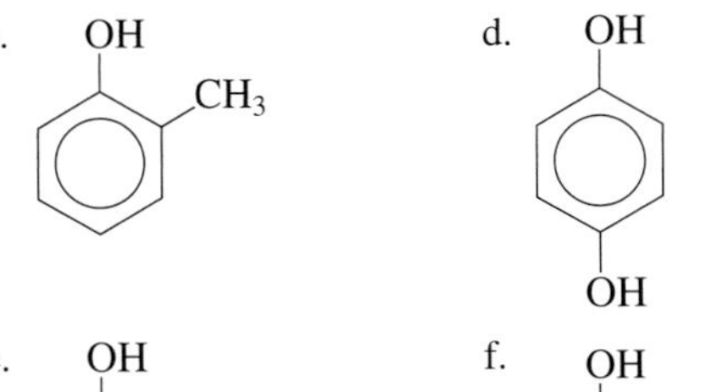

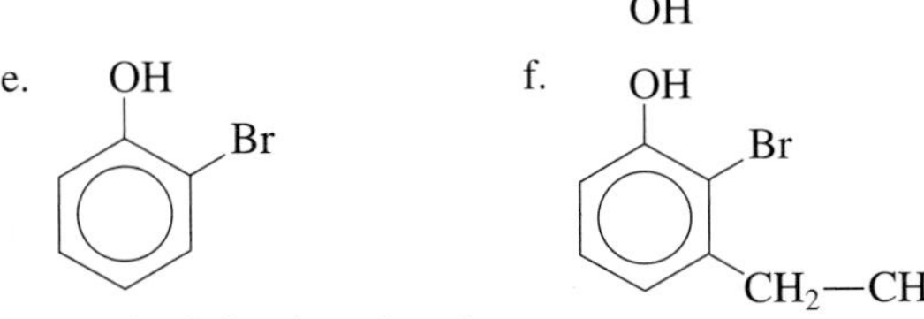

14.18 Name the following phenols.

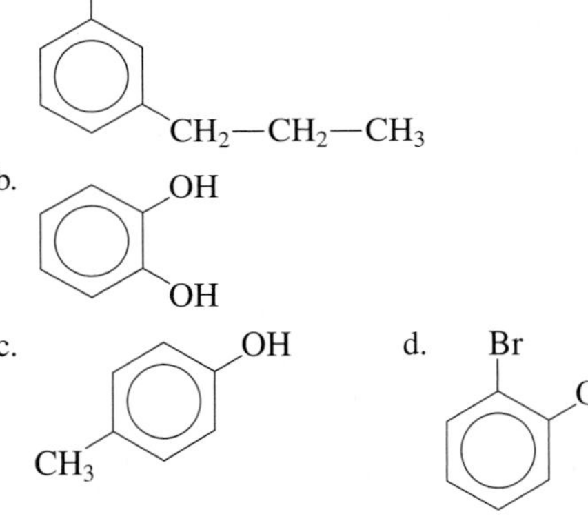

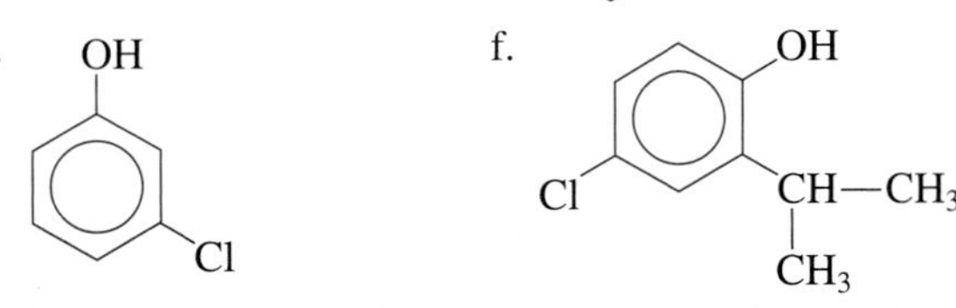

14.19 Draw a structural formula for each of the following phenols.

a. 4-Chlorophenol
b. 2-Ethylphenol
c. 2,4-Dinitrophenol
d. *m*-Cresol
e. Resorcinol
f. 2,6-Diethyl-4-methylphenol

14.20 Draw a structural formula for each of the following phenols.

a. 3-Bromophenol
b. *m*-Ethylphenol
c. Hydroquinone
d. *o*-Cresol
e. Catechol
f. 2,6-Dichlorophenol

14.21 Explain why the first of the following two compounds is a phenol and the second is not.

$$CH_3-\langle\bigcirc\rangle-OH$$

$$CH_3-\langle\bigcirc\rangle-CH_2-OH$$

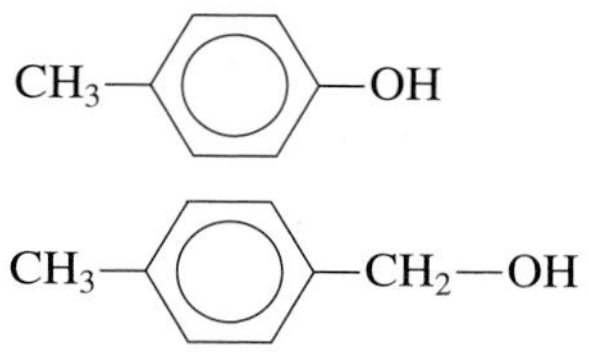

14.22 Explain why the first of the following two compounds is a phenol and the second is not.

$$CH_3-\langle\bigcirc\rangle-OH \qquad CH_3-\langle\;\;\rangle-OH$$

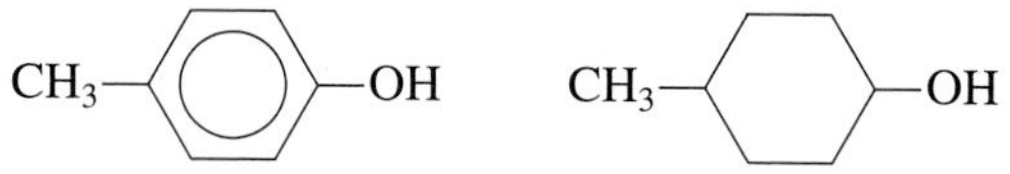

Important Common Alcohols (Section 14.4)

14.23 What does each of the following terms mean?

a. Absolute alcohol b. Grain alcohol
c. Rubbing alcohol d. Drinking alcohol

14.24 What does each of the following terms mean?

a. Wood alcohol b. Denaturated alcohol
c. 70-Proof alcohol d. "Alcohol"

14.25 Give the name of the alcohol that fits each of the following descriptions.

a. Thick liquid that has the consistency of honey
b. Often produced via a fermentation process
c. Used as a race car fuel
d. Industrially produced from CO and H_2

14.26 Give the name of the alcohol that fits each of the following descriptions.

a. Sometimes used as a skin coolant for the human body
b. Antifreeze ingredient
c. Active ingredient in alcoholic beverages
d. Moistening agent in many cosmetics

Physical Properties of Alcohols (Section 14.5)

14.27 Explain why the boiling points of alcohols are much higher than those of alkanes with similar molecular masses.

14.28 Explain why the water solubilities of alcohols are much higher than those of alkanes with similar molecular masses.

14.29 Which member of each of the following pairs of compounds would you expect to have the higher boiling point?

a. 1-Butanol and 1-heptanol
b. Butane and 1-propanol
c. Ethanol and 1,2-ethanediol

14.30 Which member of each of the following pairs of compounds would you expect to have the higher boiling point?

a. 1-Octanol and 1-pentanol
b. Pentane and 1-butanol
c. 1,3-Propanediol and 1-propanol

14.31 Which member of each of the following pairs of compounds would you expect to be more soluble in water?

a. Butane and 1-butanol
b. 1-Octanol and 1-pentanol
c. 1,2-Butanediol and 1-butanol

14.32 Which member of each of the following pairs of compounds would you expect to be more soluble in water?

a. 1-Pentanol and 1-butanol
b. 1-Propanol and 1-hexanol
c. 1,2,3-Propanetriol and 1-hexanol

14.33 Determine the maximum number of hydrogen bonds that can form between an ethanol molecule and

a. other ethanol molecules b. water molecules
c. methanol molecules d. 1-propanol molecules

14.34 Determine the maximum number of hydrogen bonds that can form between a methanol molecule and

a. other methanol molecules b. water molecules
c. 1-propanol molecules d. 2-propanol molecules

Preparation of Alcohols (Section 14.6)

14.35 Write the structure of the expected predominant organic product formed in each of the following reactions.

a. $CH_2{=}CH_2 + H_2O \xrightarrow{H_2SO_4}$

b. $CH_3-CH_2-\overset{\overset{\displaystyle O}{\|}}{C}-H + H_2 \xrightarrow{\text{Catalyst}}$

c. $CH_3-CH_2-\underset{\underset{\displaystyle CH_3}{|}}{C}{=}CH_2 + H_2O \xrightarrow{H_2SO_4}$

d. $CH_3-CH_2-\overset{\overset{\displaystyle O}{\|}}{C}-CH_2-CH_3 + H_2 \xrightarrow{\text{Catalyst}}$

14.36 Write the structure of the expected predominant organic product formed in each of the following reactions.

a. $CH_3-CH{=}CH-CH_3 + H_2O \xrightarrow{H_2SO_4}$

b. $CH_3-CH_2-\overset{\overset{\displaystyle O}{\|}}{C}-CH_3 + H_2 \xrightarrow{\text{Catalyst}}$

c. $CH_3-\overset{\overset{\displaystyle O}{\|}}{C}-H + H_2 \xrightarrow{\text{Catalyst}}$

d. $CH_3-\underset{\underset{\displaystyle CH_3}{|}}{CH}-CH{=}CH-CH_3 + H_2O \xrightarrow{H_2SO_4}$

Reactions of Alcohols (Section 14.7)

14.37 Draw the structure of the expected predominant organic product when each of the following alcohols is dehydrated using sulfuric acid at the temperature indicated.

a. $CH_3-\underset{\underset{\displaystyle OH}{|}}{CH}-CH_3 \xrightarrow[180°C]{H_2SO_4}$

b. $CH_3-\underset{\underset{\displaystyle CH_3}{|}}{CH}-CH_2-OH \xrightarrow[180°C]{H_2SO_4}$

c. $CH_3-\underset{\underset{\displaystyle CH_3}{|}}{CH}-OH \xrightarrow[140°C]{H_2SO_4}$

d. $CH_3-CH_2-CH_2-OH \xrightarrow[140°C]{H_2SO_4}$

14.38 Draw the structure of the expected predominant organic product when each of the following alcohols is dehydrated using sulfuric acid at the temperature indicated.

a. $CH_3-\underset{CH_3}{\underset{|}{CH}}-CH_2-OH \xrightarrow[140°C]{H_2SO_4}$

b. $CH_3-\underset{CH_3}{\underset{|}{CH}}-CH_2-OH \xrightarrow[180°C]{H_2SO_4}$

c. $CH_3-\underset{OH}{\underset{|}{CH}}-CH_2-CH_3 \xrightarrow[180°C]{H_2SO_4}$

d. $CH_3-\underset{OH}{\underset{|}{CH}}-CH_2-CH_3 \xrightarrow[140°C]{H_2SO_4}$

14.39 Identify the alcohol reactant from which each of the following products was obtained by an alcohol dehydration reaction.

a. Alcohol $\xrightarrow[180°C]{H_2SO_4} CH_3-CH=\underset{CH_3}{\underset{|}{C}}-CH_3$

b. Alcohol $\xrightarrow[180°C]{H_2SO_4} CH_3-CH=CH_2$

c. Alcohol $\xrightarrow[140°C]{H_2SO_4} CH_3-CH_2-O-CH_2-CH_3$

d. Alcohol $\xrightarrow[140°C]{H_2SO_4}$
$CH_3-\underset{CH_3}{\underset{|}{CH}}-CH_2-O-CH_2-\underset{CH_3}{\underset{|}{CH}}-CH_3$

14.40 Identify the alcohol reactant from which each of the following products was obtained by an alcohol dehydration reaction.

a. Alcohol $\xrightarrow[180°C]{H_2SO_4} CH_2=\underset{CH_3}{\underset{|}{C}}-CH_2-CH_3$

b. Alcohol $\xrightarrow[180°C]{H_2SO_4} CH_3-CH_2-CH=CH_2$

c. Alcohol $\xrightarrow[140°C]{H_2SO_4} CH_3-O-CH_3$

d. Alcohol $\xrightarrow[140°C]{H_2SO_4} CH_3-CH_2-O-CH_2-CH_3$

14.41 Draw the structure of the alcohol that could be used to prepare each of the following compounds in an oxidation reaction.

a. $CH_3-CH_2-\overset{O}{\overset{\|}{C}}-CH_3$

b. $CH_3-CH_2-\overset{O}{\overset{\|}{C}}-OH$

c. $CH_3-CH_2-\overset{O}{\overset{\|}{C}}-H$

d. [cyclopentane ring]$-\overset{O}{\overset{\|}{C}}-H$

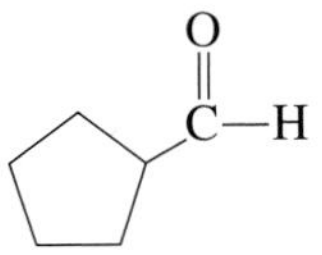

14.42 Draw the structure of the alcohol that could be used to prepare each of the following compounds in an oxidation reaction.

a. $CH_3-\underset{CH_3}{\underset{|}{CH}}-\overset{O}{\overset{\|}{C}}-H$

b. $CH_3-\underset{CH_3}{\underset{|}{CH}}-\overset{O}{\overset{\|}{C}}-CH_3$

c. $CH_3-\underset{CH_3}{\underset{|}{CH}}-CH_2-\overset{O}{\overset{\|}{C}}-OH$

d. [cyclohexane ring]$-\overset{O}{\overset{\|}{C}}-H$

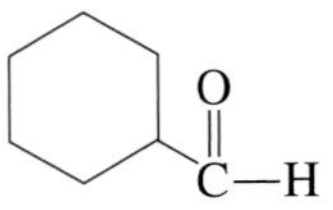

14.43 Draw the structure of the expected predominant organic product formed in each of the following reactions.

a. $CH_3-CH_2-CH_2-OH \xrightarrow[\Delta]{PCl_3}$

b. [cyclopentane ring with CH_3 and OH on adjacent carbons] $\xrightarrow[180°C]{H_2SO_4}$

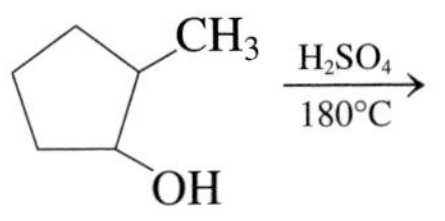

c. $CH_3-\underset{OH}{\underset{|}{CH}}-CH_2-CH_3 \xrightarrow{K_2Cr_2O_7}$

d. $CH_3-CH_2-\underset{OH}{\underset{|}{CH}}-CH_2-CH_3 \xrightarrow[\Delta]{PCl_3}$

e. $CH_3-CH_2-OH \xrightarrow[140°C]{H_2SO_4}$

f. $\underset{OH}{\underset{|}{CH_2}}-\underset{OH}{\underset{|}{CH_2}} \xrightarrow[\Delta]{PBr_3}$

14.44 Draw the structure of the expected predominant organic product formed in each of the following reactions.

a. $CH_3-CH_2-CH_2-OH \xrightarrow[\Delta]{PBr_3}$

b. [cyclopentane ring with CH_3 and OH on adjacent carbons] $\xrightarrow[\Delta]{PCl_3}$

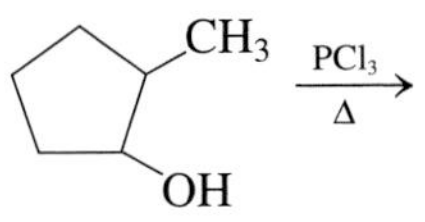

c. $\underset{OH}{\underset{|}{CH_2}}-CH_2-CH_2-CH_3 \xrightarrow[140°C]{H_2SO_4}$

d. $CH_3-CH_2-\underset{OH}{\underset{|}{CH}}-CH_2-CH_3 \xrightarrow{K_2Cr_2O_7}$

e. $CH_3-\underset{CH_3}{\underset{|}{CH}}-OH \xrightarrow[180°C]{H_2SO_4}$

f. $\underset{OH}{\underset{|}{CH_2}}-\underset{OH}{\underset{|}{CH_2}} \xrightarrow[\Delta]{PCl_3}$

Polymeric Alcohols (Section 14.8)

14.45 Draw a structural representation for the polymeric alcohol PEG [poly(ethylene glycol)].

14.46 Draw a structural representation for the polymeric alcohol PVA [poly(vinyl alcohol)].

Properties and Uses of Phenols (Section 14.9)

14.47 Phenolic compounds are frequently used as antiseptics and disinfectants. What is the difference between an antiseptic and a disinfectant?

14.48 Phenolic compounds are frequently used as antioxidants. What is an antioxidant?

14.49 Phenols are weak acids. Write an equation for the acid dissociation of the compound phenol.

14.50 How does the acidity of phenols compare with that of inorganic weak acids?

Nomenclature for Ethers (Section 14.10)

14.51 Assign an IUPAC name to each of the following ethers.

a. $CH_3—O—CH_2—CH_2—CH_3$

b. $CH_3—CH_2—CH_2—O—CH_2—CH_3$

c. 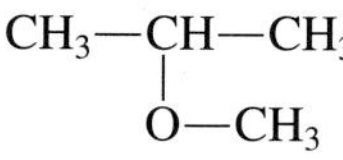 $CH_3—CH—CH_3$ (with $O—CH_3$ on the CH)

d. 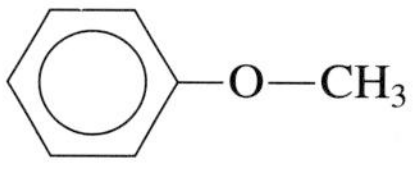 (benzene ring)$—O—CH_3$

e. 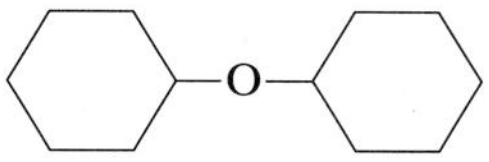 (cyclohexane ring)$—O—$(cyclohexane ring)

f. 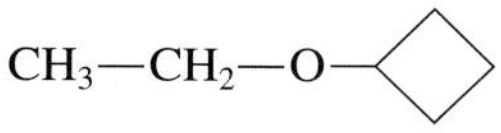 $CH_3—CH_2—O—$(cyclobutane ring)

14.52 Assign an IUPAC name to each of the following ethers.

a. $CH_3—CH_2—O—CH_2—CH_3$

b. 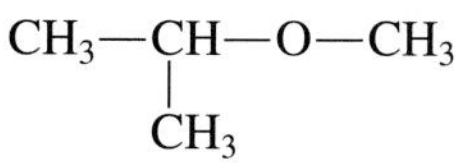 $CH_3—CH—O—CH_3$ (with CH_3 on the CH)

c. 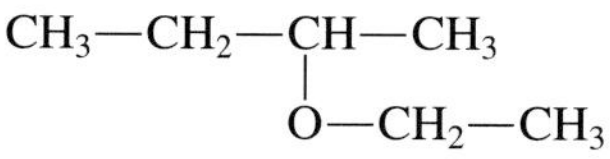 $CH_3—CH_2—CH—CH_3$ (with $O—CH_2—CH_3$ on the CH)

d. 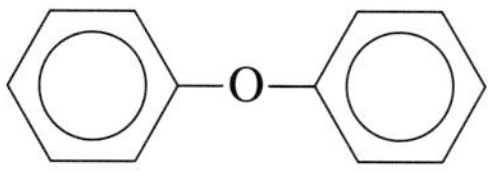 (benzene ring)$—O—$(benzene ring)

e. 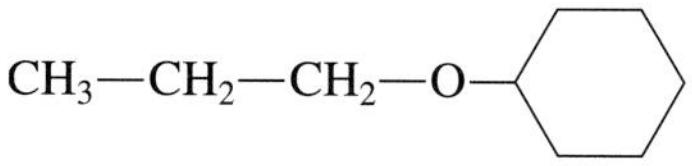 $CH_3—CH_2—CH_2—O—$(cyclohexane ring)

f. 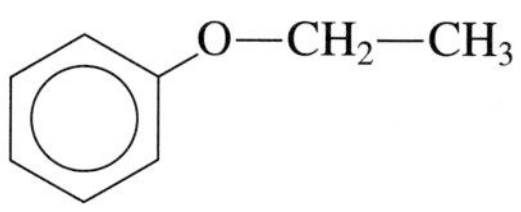 (benzene ring)$—O—CH_2—CH_3$

14.53 Assign a common name to each of the ethers in Problem 14.51.

14.54 Assign a common name to each of the ethers in Problem 14.52.

14.55 Draw the structure of each of the following ethers.

a. Isopropyl propyl ether

b. Ethyl phenyl ether

c. Isobutyl methyl ether

d. 2-Ethoxypentane

e. Ethoxycyclobutane

f. 1-Methoxy-2,2-dimethylpropane

14.56 Draw the structure of each of the following ethers.

a. Butyl methyl ether

b. Diethyl ether

c. Phenyl propyl ether

d. 3-Propoxyheptane

e. 1,3-Dimethoxybenzene

f. 3-Methoxy-2-methylhexane

Properties of Ethers (Section 14.11)

14.57 Dimethyl ether and ethanol have the same molecular mass. Dimethyl ether is a gas at room temperature, and ethanol is a liquid at room temperature. Explain these observations.

14.58 Compare the solubility in water of ethers and alcohols that have similar molecular masses.

14.59 What are the two chemical hazards associated with ether use?

14.60 How do the chemical reactivities of ethers compare with those of (a) alkanes and (b) alcohols?

14.61 Explain why ether molecules cannot hydrogen-bond to each other.

14.62 How many hydrogen bonds can form between a single ether molecule and water molecules?

Cyclic Ethers (Section 14.12)

14.63 Classify each of the following molecular structures as that of a cyclic ether, a noncyclic ether, or a nonether.

a. 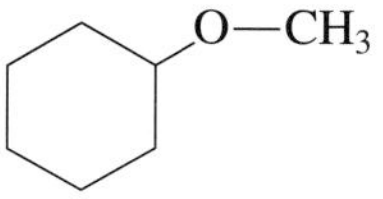 (cyclohexane ring)$—O—CH_3$

b. 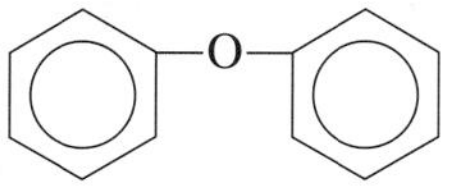 (benzene ring)$—O—$(benzene ring)

c. 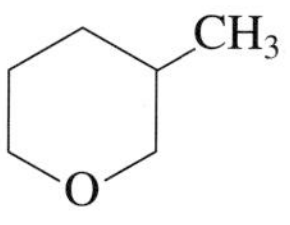 (tetrahydropyran ring with O)$—CH_3$

d. 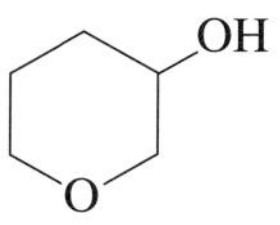 (tetrahydropyran ring with O)$—OH$

e. 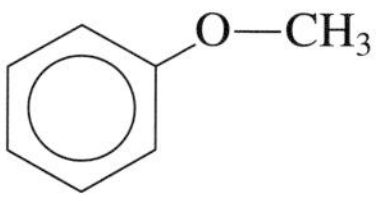 (benzene ring)$—O—CH_3$

f. 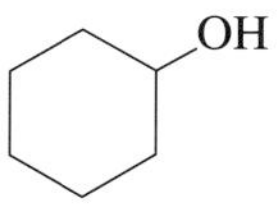 (cyclohexane ring)$—OH$

14.64 Classify each of the following molecular structures as that of a cyclic ether, a noncyclic ether, or a nonether.

a. 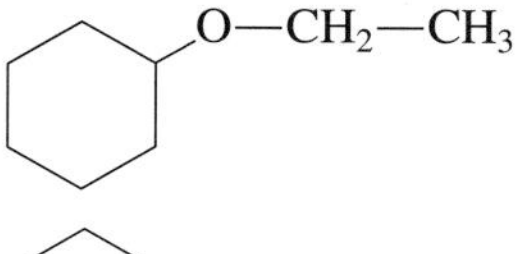 (cyclohexane ring)$—O—CH_2—CH_3$

b. 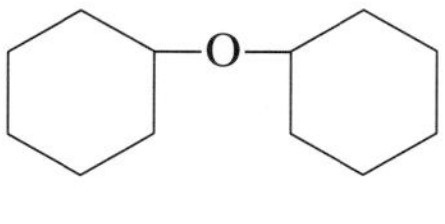 (cyclohexane ring)$—O—$(cyclohexane ring)

c. 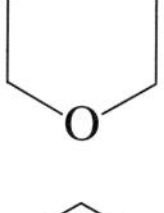 (tetrahydropyran ring with O)

d. 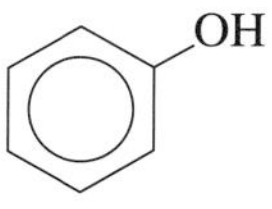 (benzene ring)$—OH$

e. 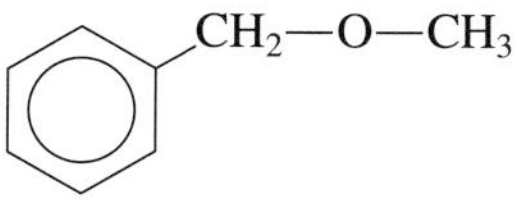 (benzene ring)$—CH_2—O—CH_3$

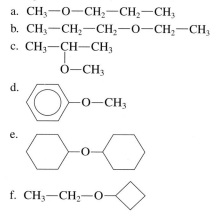

f. 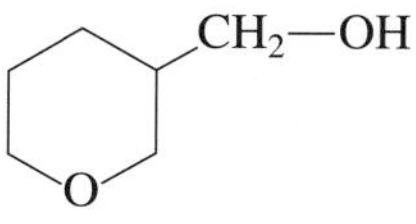 (tetrahydropyran ring with O)$—CH_2—OH$

Sulfur Analogs of Alcohols and Ethers (Section 14.13)

14.65 What are the structural similarities and differences between a thiol and an alcohol?

14.66 What are the structural similarities and differences between a disulfide and an ether?

14.67 Draw a structural formula for each of the following thiols.

a. Methanethiol
b. 2-Propanethiol
c. 1-Butanethiol
d. 3-Methyl-1-pentanethiol
e. Cyclopentanethiol
f. 1,2-Ethanedithiol

14.68 Draw a structural formula for each of the following thiols.

a. 1-Propanethiol
b. Ethanethiol
c. 1,3-Pentanedithiol
d. 3-Methyl-3-pentanethiol
e. 2-Methylcyclopentanethiol
f. 2,2-Dimethyl-1-hexanethiol

14.69 Contrast the products that result from the oxidation of an alcohol and the oxidation of a thiol.

14.70 Write the formulas for the sulfur-containing organic products of the following reactions.

a. $2CH_3-CH_2-SH \xrightarrow[\text{agent}]{\text{Oxidizing}}$

b. $CH_3-CH_2-S-S-CH_2-CH_3 \xrightarrow[\text{agent}]{\text{Reducing}}$

14.71 Assign an IUPAC name to each of the following thioethers.

a. $CH_3-CH_2-S-CH_3$

b. $CH_3-\underset{\underset{CH_3}{|}}{CH}-S-CH_3$

c. 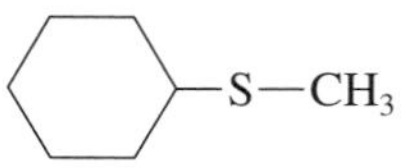

d. 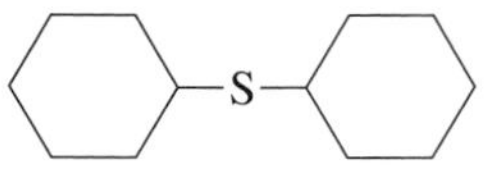

e. $CH_3-CH{=}CH-S-CH_3$

f. $CH_3-CH_2-\underset{\underset{\underset{CH_3}{|}}{\underset{S}{|}}}{CH}-CH_3$

14.72 Assign an IUPAC name to each of the following thioethers.

a. $CH_3-CH_2-S-CH_2-CH_3$

b. $CH_3-\underset{\underset{CH_3}{|}}{CH}-S-CH_2-CH_3$

c. 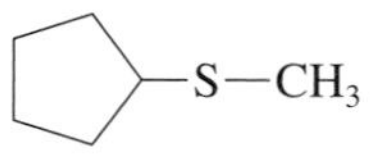

d. 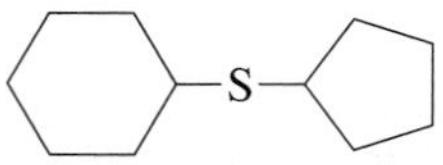

e. $CH_2{=}CH-CH_2-S-CH_3$

f. $CH_3-\underset{\underset{\underset{CH_3}{|}}{\underset{S}{|}}}{CH}-CH_3$

Additional Problems

14.73 Classify each of the alcohols in Problem 14.4 as a primary, secondary, or tertiary alcohol.

14.74 Classify each of the alcohol functional groups in the compounds in Problem 14.10 as a primary-, secondary-, or tertiary-alcohol functional group.

14.75 Assign an IUPAC name to each of the following compounds.

a. 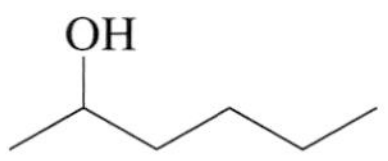

b. 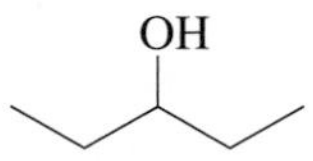

c. 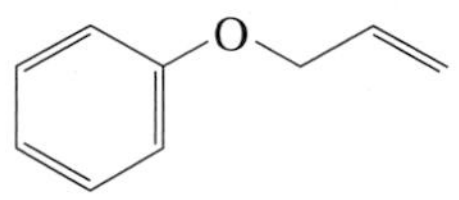

d. 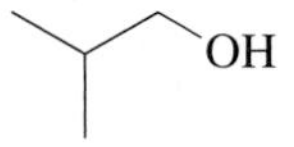

e. 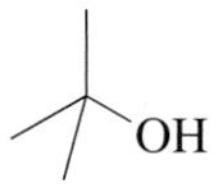

f. 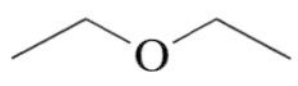

14.76 Draw structural formulas for the eight isomeric alcohols and six isomeric ethers that have the molecular formula $C_5H_{10}O$.

14.77 Three isomeric pentanols with unbranched carbon chains exist. Which of these, upon dehydration at 180°C, yields only 1-pentene as a product?

14.78 A mixture of methanol, 1-propanol, and H_2SO_4 (catalyst) is heated to 140°C. After reaction, the solution contains three different ethers. Draw a structural formula for each of the ethers.

14.79 Which of the terms *ether, alcohol, diol, thiol, thioether, thioalcohol, disulfide, sulfide,* and *peroxide* characterizes each of the following compounds? Note that more than one term may apply to a given compound.

a. $CH_3-S-S-CH_3$

b. CH_3-CH_2-SH

c. $CH_3-\underset{\underset{OH}{|}}{CH}-CH_2-CH_3$

d. $CH_3-CH_2-O-O-CH_3$

e. $HO-CH_2-CH_2-SH$

f. $CH_3-O-CH_2-S-CH_3$

14.80 Assign IUPAC names to the following compounds.

a. $HS-CH_2-CH_2-SH$

b. $CH_3-O-CH_2-CH_2-CH_2-OH$

c. $HO-CH_2-CH_2-CH_3$

d. $CH_3-CH_2-O-CH_2-CH_2-O-CH_2-CH_3$

e. $CH_3-S-CH_2-CH_3$

f. $CH_3-O-CH_2-CH_2-S-CH_2-CH_3$

Grid Problems

14.81

1.	2.	3.
$\mathrm{CH_3{-}CH_2{-}OH}$	$\mathrm{CH_3{-}CH_2{-}CH_2{-}OH}$	$\mathrm{CH_3{-}CH{-}OH}$ $\mathrm{CH_3}$ (on CH)
4.	**5.**	**6.**
$\mathrm{CH_2{-}CH{-}OH}$ $\mathrm{CH_3}$ (on $\mathrm{CH_2}$), $\mathrm{CH_3}$ (on CH)	$\mathrm{CH_3{-}CH{-}CH_2{-}OH}$ $\mathrm{CH_3}$ (on CH)	$\mathrm{CH_3}$ $\mathrm{CH_3{-}C{-}OH}$ $\mathrm{CH_3}$

Select from the grid *all* correct responses for each of the following situations.

a. Compounds that are secondary alcohols
b. Compounds that do not undergo oxidation with mild oxidizing agents
c. Compounds that produce ketones when oxidized by mild oxidizing agents
d. Compounds that undergo intermolecular dehydration at 140°C with H_2SO_4 as a catalyst to produce an ether

14.82

1.	2.	3.
o-cresol	catechol	resorcinol
4.	**5.**	**6.**
2-bromophenol	*m*-cresol	phenol

Select from the grid *all* correct responses for each of the following situations.

a. Compounds whose structures involve a benzene ring with two substituents
b. Compounds that are phenols
c. Compounds with structures that involve at least one —OH group attached to a benzene ring
d. Pairs of compounds that are structural isomers

14.83

1.	2.	3.
$\mathrm{CH_2{-}OH}$ $\mathrm{CH_3}$	$\mathrm{CH_2{-}O{-}CH_3}$ $\mathrm{CH_2{-}O{-}CH_3}$	$\mathrm{CH_3}$ $\mathrm{CH{-}OH}$ $\mathrm{CH_3}$
4.	**5.**	**6.**
$\mathrm{CH_2{-}O{-}CH_3}$ $\mathrm{CH_3}$	$\mathrm{CH_2{-}OH}$ $\mathrm{CH_2{-}OH}$	$\mathrm{CH_2{-}O{-}CH_3}$ $\mathrm{CH{-}OH}$ $\mathrm{CH_3}$

Select from the grid *all* correct responses for each of the following situations.

a. Compounds that contain both ether and alcohol functional groups
b. Compounds that are named, in the IUPAC system, as alcohols
c. Compounds that are named, in the IUPAC system, as substituted alkanes
d. Pairs of compounds that are structural isomers

14.84

1.	2.	3.
$\mathrm{CH_3{-}SH}$	$\mathrm{CH_3{-}S{-}S{-}CH_3}$	$\mathrm{CH_3{-}O{-}CH_3}$
4.	**5.**	**6.**
$\mathrm{CH_3{-}OH}$	$\mathrm{CH_3{-}O{-}O{-}CH_3}$	$\mathrm{CH_3{-}S{-}CH_3}$

Select from the grid *all* correct responses for each of the following situations.

a. Compounds that contain a sulfhydryl group
b. Compounds that are thioethers
c. Compounds whose oxidation produces a disulfide
d. Compounds whose slow reaction with O_2 can produce a peroxide

PART II

Aldehydes and Ketones

CHAPTER OUTLINE

15.1 The Carbonyl Group 406
15.2 Structure of Aldehydes and Ketones 407
15.3 Nomenclature for Aldehydes 408
15.4 Nomenclature for Ketones 410
15.5 Selected Common Aldehydes and Ketones 413
15.6 Physical Properties of Aldehydes and Ketones 414
15.7 Preparation of Aldehydes and Ketones 417
15.8 Oxidation and Reduction of Aldehydes and Ketones 418
15.9 Reaction of Aldehydes and Ketones with Alcohols 420

Chemistry at a Glance:
Reactions Involving Aldehydes and Ketones 425

15.10 Formaldehyde-Based Polymers 425
15.11 Sulfur-Containing Carbonyl Groups 426

Chemical Connections
15.1 Lachrymatory Aldehydes and Ketones 412
15.2 Suntan, Sunburn, and Ketones 415
15.3 Diabetes, Aldehyde Oxidation, and Glucose Testing 418

Benzaldehyde is the main flavor component in almonds. Aldehydes and ketones are responsible for the odor and taste of numerous nuts and spices.

In this chapter, we continue our discussion of hydrocarbon derivatives that contain the element oxygen. In the previous chapter, the functional groups considered (alcohols, phenols, and ethers) have the common feature of carbon–oxygen *single* bonds. Carbon–oxygen *double* bonds are also possible in hydrocarbon derivatives. We will now consider the simplest types of compounds that contain this structural feature: aldehydes and ketones.

15.1 The Carbonyl Group

Both aldehydes and ketones contain a carbonyl group. A **carbonyl group** *consists of a carbon atom and an oxygen atom joined by a double bond.*

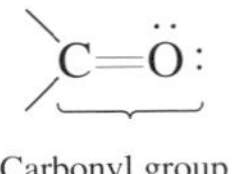

Carbonyl group

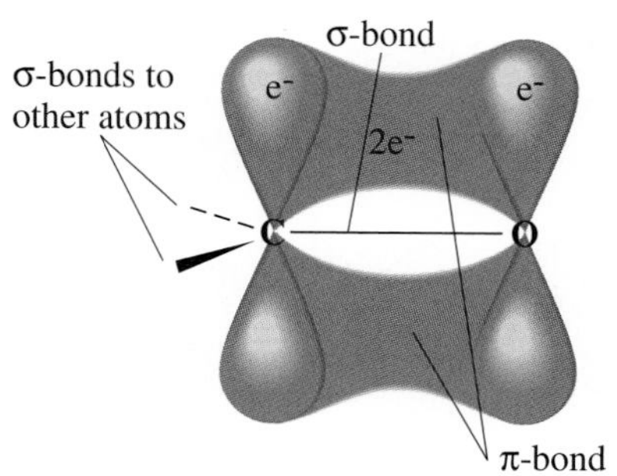

Figure 15.1
The double bond of a carbonyl group has both a σ-bond and a π-bond component.

The carbon–oxygen double bond in a carbonyl group, like any double bond, consists of a σ-bond and a π-bond (Section 13.4), the π-bond being weaker than the σ-bond (Figure 15.1). In this respect, the carbon–carbon double bonds found in alkenes and carbonyl double bonds are similar.

Carbon–carbon and carbon–oxygen double bonds differ in a major way. A carbon–carbon double bond is *nonpolar,* and a carbon–oxygen double bond is *polar.* The electronegativity (Section 5.9) of oxygen (3.5) is much greater than that of carbon (2.5). Hence the carbon–oxygen double bond is polarized, the oxygen atom acquiring a partial negative charge (δ^-) and the carbon atom acquiring a partial positive charge (δ^+).

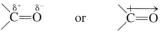

$$\overset{\delta^+}{C}{=}\overset{\delta^-}{O} \quad \text{or} \quad \overset{+\!\!\longrightarrow}{C{=}O}$$

Polar nature of carbon–oxygen double bond

• The word *carbonyl* is pronounced "carbon-EEL."

• The difference in electronegativity between oxygen and carbon causes a carbon–oxygen double bond to be polar.

• The word *aldehyde* is pronounced "AL-da-hide."

15.2 Structure of Aldehydes and Ketones

An **aldehyde** *is a compound that has at least one hydrogen atom attached to the carbon atom of a carbonyl group.* The remaining group attached to the carbonyl carbon atom can be hydrogen, an alkyl group (R), a cycloalkyl group, or an aryl group (Ar).

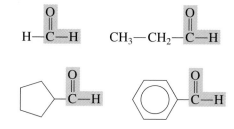

$$H{-}\overset{O}{\overset{\|}{C}}{-}H \qquad CH_3{-}CH_2{-}\overset{O}{\overset{\|}{C}}{-}H$$

$$\text{cyclopentyl}{-}\overset{O}{\overset{\|}{C}}{-}H \qquad \text{phenyl}{-}\overset{O}{\overset{\|}{C}}{-}H$$

The aldehyde functional group, the structural feature common to all the preceding compounds, is

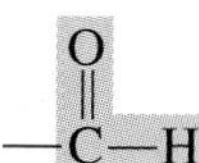

• Remember that for general *condensed* functional group structures such as RCHO, carbon always has four bonds and hydrogen always has only one. In RCHO, you know one of carbon's bonds goes to the R group and one to H; therefore, two bonds must go to O.

Linear notations for an aldehyde functional group and for an aldehyde itself are —CHO and RCHO, respectively. Note that the ordering of the symbols H and O in these notations is HO, not OH (which denotes a hydroxyl group).

Cyclic aldehyde structures are not possible. For a carbonyl carbon atom to be part of a ring system, it would have to form two bonds to ring atoms. This is not possible, because three of the four bonds from an aldehyde carbonyl carbon atom *must* go to oxygen and hydrogen. This means that an aldehyde group must always be at the end of a carbon chain.

• The word *ketone* is pronounced "KEY-tone."

A **ketone** *is a compound that has two carbon atoms attached to the carbon atom of a carbonyl group.* The groups containing these bonded carbon atoms may be alkyl, cycloalkyl, or aryl.

$$CH_3{-}\overset{O}{\overset{\|}{C}}{-}CH_2{-}CH_3 \qquad \text{cyclopentyl}{-}\overset{O}{\overset{\|}{C}}{-}CH_3 \qquad \text{phenyl}{-}\overset{O}{\overset{\|}{C}}{-}\text{phenyl}$$

• In an aldehyde, the carbonyl group is always located at the end of a hydrocarbon chain.

$$CH_3{-}CH_2{-}CH_2{-}CH_2{-}\boxed{\overset{O}{\overset{\|}{C}}}{-}H$$

In a ketone, the carbonyl group is always at a nonterminal (interior) position on the hydrocarbon chain.

$$CH_3{-}CH_2{-}CH_2{-}\boxed{\overset{O}{\overset{\|}{C}}}{-}CH_2{-}CH_3$$

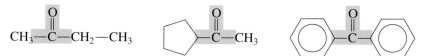

The ketone functional group, the structural feature common to all the preceding compounds, is

The general condensed formula for a ketone is RCOR, in which the oxygen atom is understood to be double-bonded to the carbonyl carbon at the left of it in the formula.

Unlike aldehydes, ketones can form cyclic structures, such as

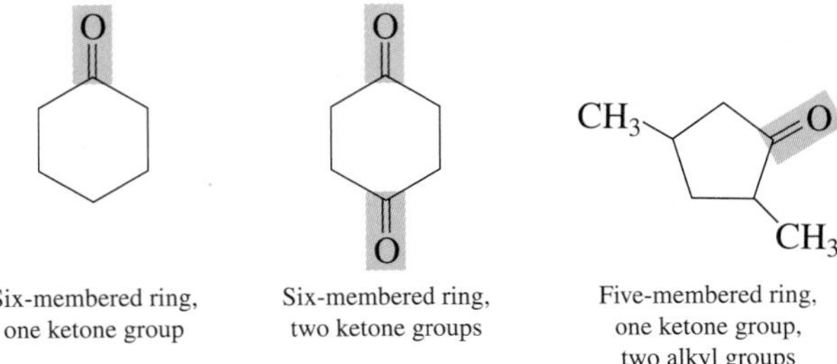

Cyclic ketones are *not* heterocyclic ring systems as were cyclic ethers (Section 14.13).

• This is the second time we have encountered functional group isomerism. Alcohols and ethers (Section 14.10) also can exhibit such isomerism.

Aldehydes and ketones with the same number of carbon atoms and the same degree of saturation are isomeric (functional group isomerism). For example, there are both an aldehyde and a ketone that have the molecular formula C_3H_6O. (See Figure 15.2.)

$$CH_3-CH_2-\overset{\overset{\displaystyle O}{\|}}{C}-H \quad \text{and} \quad CH_3-\overset{\overset{\displaystyle O}{\|}}{C}-CH_3$$

An aldehyde — A ketone

15.3 Nomenclature for Aldehydes

The IUPAC rules for naming aldehydes are as follows:

• The carbonyl carbon atom in an aldehyde cannot have any number but 1, so we do not have to include this number in the aldehyde's IUPAC name.

1. Select as the parent carbon chain the longest chain that *includes* the carbon atom of the carbonyl group.
2. Name the parent chain by changing the *-e* ending of the corresponding alkane name to *-al*.
3. Number the parent chain by assigning the number 1 to the carbonyl carbon atom of the aldehyde group.
4. Determine the identity and location of any substituents, and append this information to the front of the parent chain name.

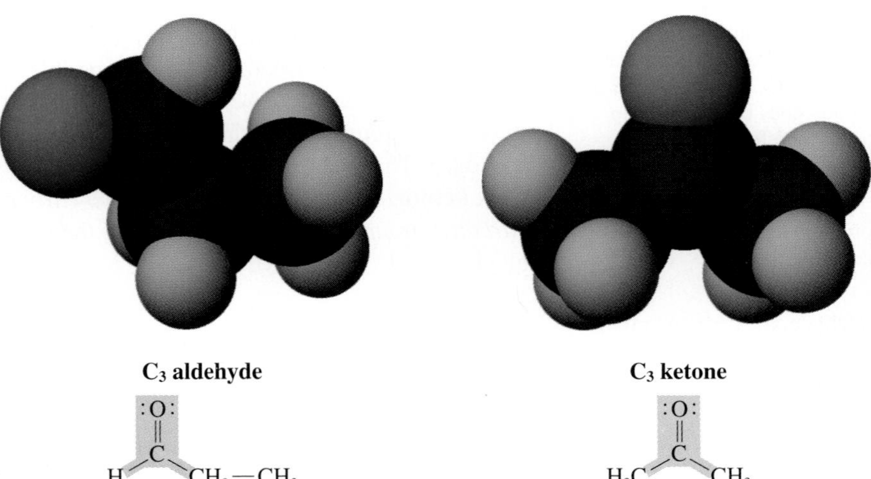

Figure 15.2
Aldehydes and ketones with the same number of carbon atoms and the same degree of saturation are structural isomers, as is illustrated here for the three-carbon aldehyde (propanal) and the three-carbon ketone (2-propanone). Both have the molecular formula C_3H_6O.

- Line-angle drawings for the simpler unbranched-chain aldehydes:

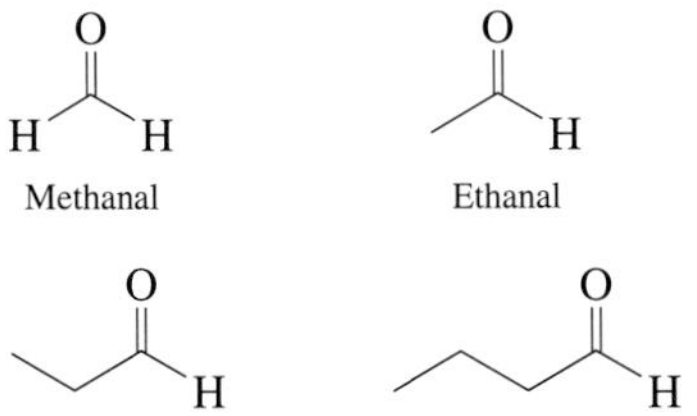

Methanal

Ethanal

Propanal

Butanal

- Be careful about the endings *-al* and *-ol*. They are easily confused. The suffix *-al* (pronounced like the man's name Al) denotes an aldehyde; the suffix *-ol* (pronounced like the *ol* in old) denotes an alcohol.

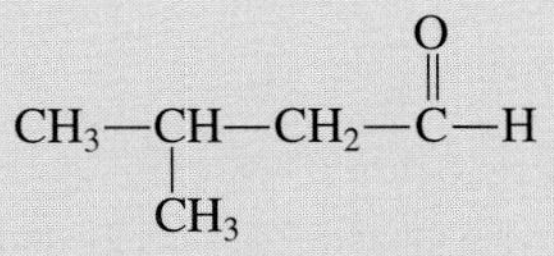

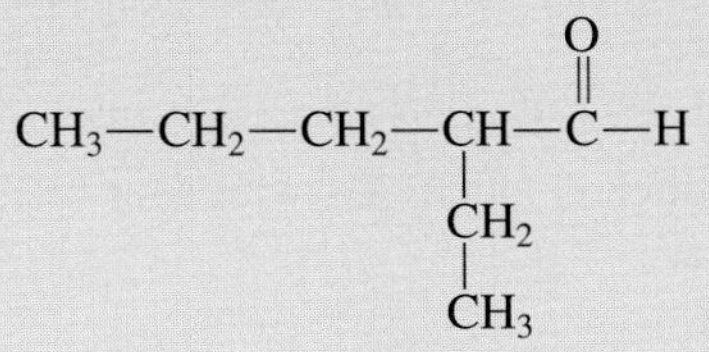

- When a compound contains more than one type of functional group, the suffix for only one of them can be used as the ending of the name. The IUPAC rules establish priorities that specify which suffix is used. For the functional groups we have discussed up to this point in the text, the IUPAC priority system is

Increasing priority (top = highest):
aldehyde
ketone
alcohol
alkene
alkyne
alkoxy
halogen
alkyl
(alkoxy, halogen, alkyl: Equal-priority substituents (listed in alphabetical order))

aldehyde–ether

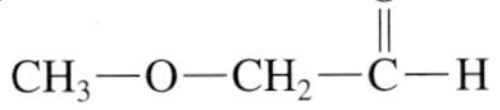

$CH_3-O-CH_2-\overset{O}{\overset{\|}{C}}-H$

2-methoxyethanal

aldehyde–alkene

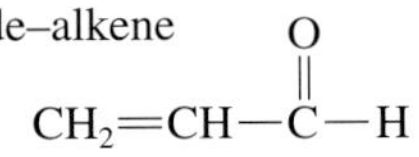

$CH_2{=}CH-\overset{O}{\overset{\|}{C}}-H$

2-propenal

Example 15.1

Determining IUPAC Names for Aldehydes

Assign IUPAC names to the following aldehydes.

a. $CH_3-CH_2-CH_2-CH_2-\overset{O}{\overset{\|}{C}}-H$

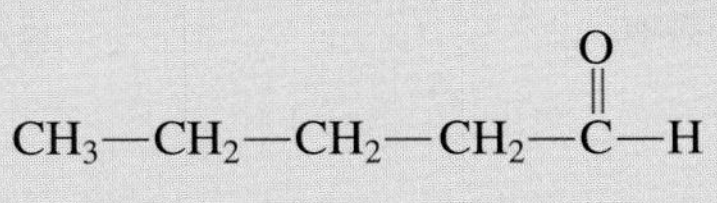

b. $CH_3-\underset{CH_3}{\underset{|}{CH}}-CH_2-\overset{O}{\overset{\|}{C}}-H$

c. $CH_3-CH_2-CH_2-\underset{CH_3}{\underset{|}{\underset{CH_2}{\underset{|}{CH}}}}-\overset{O}{\overset{\|}{C}}-H$

d. $CH_3-CH_2-\underset{OH}{\underset{|}{CH}}-CH_2-\overset{O}{\overset{\|}{C}}-H$

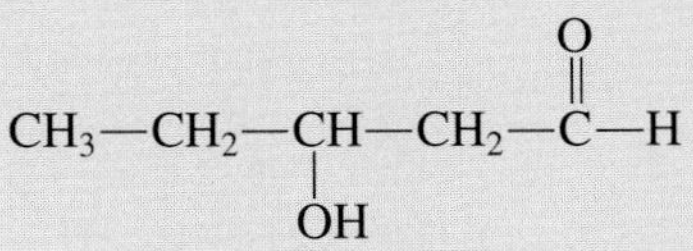

Solution

a. The parent chain name comes from pentane. Remove the *-e* ending and add the aldehyde suffix *-al*. The name becomes *pentanal.* The location of the carbonyl carbon atom need not be specified, because this carbon atom is always number 1. The complete name is simply *pentanal.*

b. The parent chain name is *butanal.* To locate the methyl group, we number the carbon chain beginning with the carbonyl carbon atom. The complete name of the aldehyde is *3-methylbutanal.*

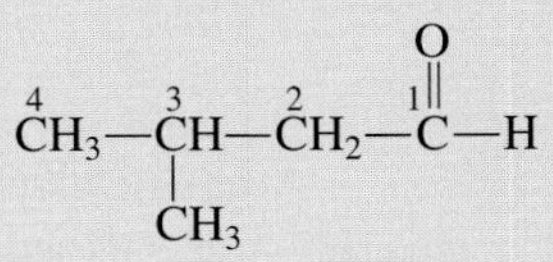

$$\overset{4}{C}H_3-\underset{CH_3}{\underset{|}{\overset{3}{C}H}}-\overset{2}{C}H_2-\overset{O}{\overset{\|}{\overset{1}{C}}}-H$$

c. The longest chain containing the carbonyl carbon atom is five carbons long, giving a parent chain name of *pentanal.* An ethyl group is present on carbon 2. Thus the complete name is *2-ethylpentanal.*

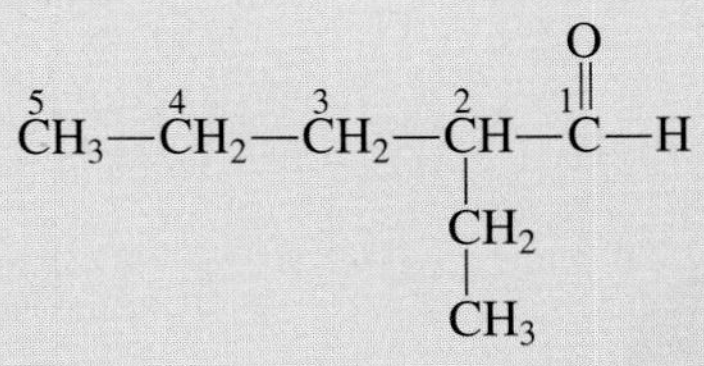

$$\overset{5}{C}H_3-\overset{4}{C}H_2-\overset{3}{C}H_2-\underset{CH_3}{\underset{|}{\underset{CH_2}{\underset{|}{\overset{2}{C}H}}}}-\overset{O}{\overset{\|}{\overset{1}{C}}}-H$$

d. This is a hydroxyaldehyde, with the hydroxyl group located on carbon 3.

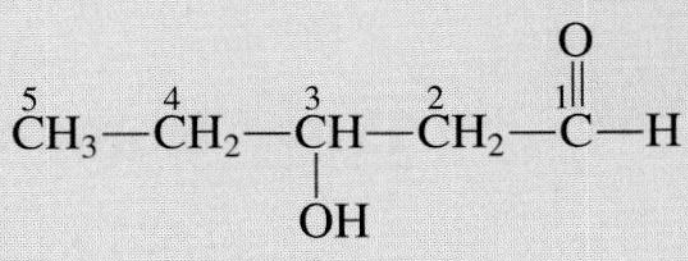

$$\overset{5}{C}H_3-\overset{4}{C}H_2-\underset{OH}{\underset{|}{\overset{3}{C}H}}-\overset{2}{C}H_2-\overset{O}{\overset{\|}{\overset{1}{C}}}-H$$

The complete name of the compound is *3-hydroxypentanal.* An aldehyde functional group has priority over an alcohol functional group in IUPAC nomenclature. An alcohol group named as a substituent is a *hydroxy* group.

Practice Exercise 15.1

Assign IUPAC names to the following aldehydes.

a. $CH_3—CH(CH_3)—C(=O)—H$

b. $CH_3—CH_2—CH(CH_2—CH_2—CH_3)—C(=O)—H$

c. $CH_3—CH(Cl)—CH(Cl)—C(=O)—H$

d. $CH_3—(CH_2)_4—C(=O)—H$

• *Answers:* **a.** 2-methylpropanal; **b.** 2-ethylpentanal; **c.** 2,3-dichlorobutanal; **d.** hexanal

• The common names for simple aldehydes illustrate a second method for counting from one to four: *form-*, *acet-*, *propion-*, and *butyr-*. We will use this method again in the next chapter when we consider the common names for carboxylic acids and esters. (The first method for counting from one to four, with which you are now thoroughly familiar, is *meth-*, *eth-*, *prop-*, and *but-*, as in methane, ethane, propane, and butane.)

Unbranched aldehydes with four or fewer carbon atoms have common names:

$H—C(=O)—H$
Formaldehyde

$CH_3—C(=O)—H$
Acetaldehyde

$CH_3—CH_2—C(=O)—H$
Propionaldehyde

$CH_3—CH_2—CH_2—C(=O)—H$
Butyraldehyde

• The contrast between IUPAC names and common names for aldehydes is as follows:

IUPAC (one word)
alkanal
pentanal
Common (one word)
(prefix) aldehyde*
butyraldehyde

*The common-name prefixes are related to natural sources for carboxylic acids with the same number of carbon atoms (see 16.3)

Unlike the common names for alcohols and ethers, the common names for aldehydes are *one* word rather than two or three.

In assigning common names to branched aldehyde systems, we name and locate substituents (branches) in the same way as in the IUPAC system.

$\overset{4}{C}H_3—\overset{3}{C}H(CH_3)—\overset{2}{C}H_2—\overset{1}{C}(=O)—H$
3-Methylbutyraldehyde

$Cl—\overset{3}{C}H_2—\overset{2}{C}H_2—\overset{1}{C}(=O)—H$
3-Chloropropionaldehyde

Aromatic aldehydes—compounds in which an aldehyde group is attached to a benzene ring—are named as derivatives of benzaldehyde, the parent compound.

• Aromatic aldehydes are not cyclic aldehydes (which do not exist). The carbonyl carbon atom in an aromatic aldehyde is not part of the ring system.

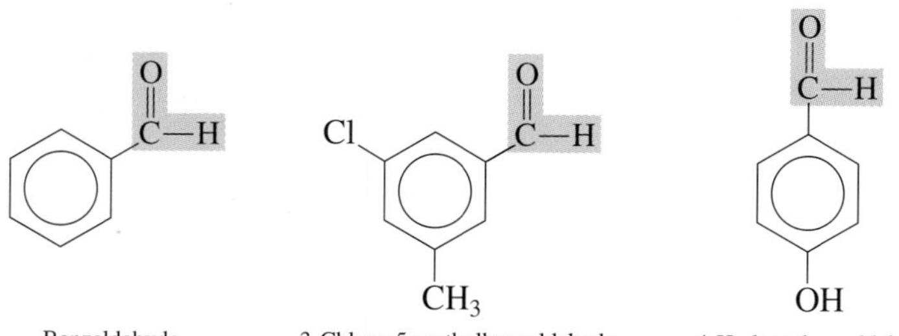

Benzaldehyde 3-Chloro-5-methylbenzaldehyde 4-Hydroxybenzaldehyde

The last of these compounds is named as a benzaldehyde rather than as a phenol because the aldehyde group has priority over the hydroxyl group in the IUPAC naming system.

Space-filling model for benzaldehyde, the simplest aromatic aldehyde.

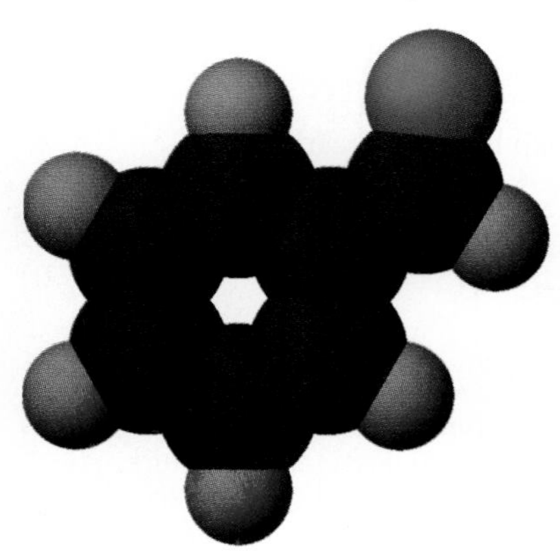

15.4 Nomenclature for Ketones

Assigning IUPAC names to ketones is similar to naming aldehydes except that the ending *-one* is used instead of *-al*. The rules for IUPAC ketone nomenclature follow.

1. Select as the parent carbon chain the longest carbon chain that *includes* the carbon atom of the carbonyl group.

• 2-Propanone is the simplest possible ketone. One- and two-carbon ketones cannot exist. A minimum of three carbon atoms is required for a ketone: one C atom for the carbonyl group and one C atom for each of the groups attached to the carbonyl carbon atom.

• Line-angle drawings for the simpler unbranched-chain 2-ketones:

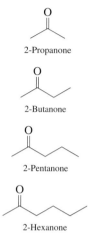

2-Propanone

2-Butanone

2-Pentanone

2-Hexanone

• In IUPAC nomenclature, the ketone functional group has precedence over all groups we have discussed so far except the aldehyde group. When both aldehyde and ketone groups are present in the same molecule, the ketone group is named as a substituent (the *oxo-* group).

$$CH_3-\overset{O}{\overset{\|}{C}}-CH_2-CH_2-\overset{O}{\overset{\|}{C}}-H$$

4-Oxopentanal

2. Name the parent chain by changing the *-e* ending of the corresponding alkane name to *-one*.
3. Number the carbon chain such that the carbonyl carbon atom receives the lowest possible number. The position of the carbonyl carbon atom is noted by placing a number immediately before the parent chain name.
4. Determine the identity and location of any substituents, and append this information to the front of the parent chain name.
5. Cyclic ketones are named by assigning the number 1 to the carbon atom of the carbonyl group. The ring is then numbered to give the lowest number(s) to the atom(s) bearing substituents.

Example 15.2

Determining IUPAC Names for Ketones

Assign IUPAC names to the following ketones.

a. $CH_3-\overset{O}{\overset{\|}{C}}-CH_2-CH_2-CH_3$

b. $CH_3-\underset{CH_3}{\underset{|}{CH}}-\underset{CH_3}{\underset{|}{CH}}-\overset{O}{\overset{\|}{C}}-CH_3$

c. 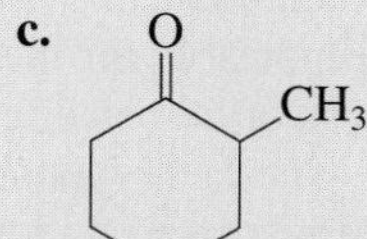

d. 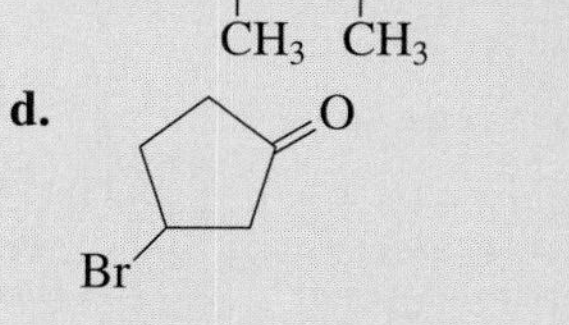

Solution

a. The parent chain name is *pentanone.* We number the chain from the end closest to the carbonyl carbon atom. Locating the carbonyl carbon at carbon 2 completes the name: *2-pentanone.*

b. The longest carbon chain of which the carbonyl carbon is a member is five carbons long. The parent chain name is again *pentanone.* There are two methyl groups attached, and the numbering system is from right to left.

$$\overset{5}{C}H_3-\underset{CH_3}{\underset{|}{\overset{4}{C}H}}-\underset{CH_3}{\underset{|}{\overset{3}{C}H}}-\overset{O}{\overset{\|}{\overset{2}{C}}}-\overset{1}{C}H_3$$

The complete name for the compound is *3,4-dimethyl-2-pentanone.*

c. The base name is *cyclohexanone.* The methyl group is bonded to carbon 2 because we begin numbering at the carbonyl carbon; the name is *2-methylcyclohexanone.*

d. This ketone has a base name of *cyclopentanone.* Numbering clockwise from the carbonyl carbon atom locates the bromo group on carbon 3. The complete name is *3-bromocyclopentanone.*

Practice Exercise 15.2

Assign IUPAC names to the following ketones.

a. $CH_3-CH_2-\overset{O}{\overset{\|}{C}}-CH_2-CH_3$

b. $CH_3-\underset{CH_3}{\underset{|}{CH}}-\overset{O}{\overset{\|}{C}}-\underset{CH_3}{\underset{|}{CH}}-CH_3$

(continued)

c. (cyclobutane ring with $=O$)

d. (cyclohexane ring with $=O$, CH_3, OH)

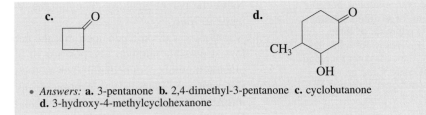

- *Answers:* **a.** 3-pentanone **b.** 2,4-dimethyl-3-pentanone **c.** cyclobutanone **d.** 3-hydroxy-4-methylcyclohexanone

Chemical CONNECTIONS

15.1 Lachrymatory Aldehydes and Ketones

A lachrymator, pronounced "lack-ra-mater," is a compound that causes the production of tears. A number of aldehydes and ketones have lachrymatic properties.

Two lachrymal ketones are 2-chloroacetophenone and bromoacetone.

$$C_6H_5-\overset{\overset{\displaystyle O}{\|}}{C}-CH_2-Cl$$

2-Chloroacetophenone
(2-chloro-1-phenylethanone)

$$CH_3-\overset{\overset{\displaystyle O}{\|}}{C}-CH_2-Br$$

Bromoacetone
(1-bromo-2-propanone)

2-Chloroacetophenone is a component of the tear gas used by police and the military. It is also the active ingredient in MACE canisters now marketed for use by individuals to protect themselves from attackers. The compound bromoacetone has been used as a chemical war gas.

Smoke contains compounds that cause the eyes to tear. A predominant lachrymator in wood smoke is formaldehyde, the one carbon aldehyde. The smoke associated with an outdoor barbecue contains the unsaturated aldehyde *acrolein.*

$$CH_2{=}CH-\overset{\overset{\displaystyle O}{\|}}{C}-H$$

Acrolein
(2-propenal)

Acrolein is formed as fats that are present in meat break down when heated. (Besides being a lachrymator, acrolein is responsible for the "pleasant" odor associated with the process of barbecuing meat.)

The lachrymatory compound associated with onions is a derivative of thiopropionaldehyde.

$$CH_3-CH_2-\overset{\overset{\displaystyle S}{\|}}{C}-H$$

Thiopropionaldehyde
(propanethial)

$$CH_3-CH_2-\overset{\overset{\displaystyle S{=}O}{\|}}{C}-H$$

Thiopropionaldehyde-S-oxide
(propanethial-S-oxide)
[lachrymator in chopped onions]

Onions do not cause tear production until they are chopped or sliced. The onion cells damaged by these actions release the enzyme *allinase,* which converts an odorless compound naturally present in onions to the lachrymatic compound.

Scientists are not sure why thiopropionaldehyde-S-oxide causes tear production, but it is known that it is an unstable molecule that is readily broken up by water into propanal, hydrogen sulfide, and sulfuric acid.

$$CH_3-CH_2-\overset{\overset{\displaystyle S{=}O}{\|}}{C}-H \xrightarrow{H_2O} CH_3-CH_2-\overset{\overset{\displaystyle O}{\|}}{C}-H + H_2S + H_2SO_4$$

Sulfuric acid may be responsible for the eye irritation.

Many people can peel onions under water, or peel cold ones from the refrigerator, without crying. Water washes away the soluble lachrymator and also breaks it down chemically. If the onion is cold, the enzymatic reaction making the lachrymator is slower, so less is formed. Also the vapor pressure of the lachrymator is greatly reduced at the lower temperature, so its concentration in air is reduced.

• The procedure for coining common names for ketones is the same as that used for ether common names (Section 14.11).

• The contrast between IUPAC names and common names for ketones is as follows:

IUPAC (one word)
alkanone
2-butanone
Common (three or two words)
alkyl alkyl ketone
ethyl methyl ketone
or
dialkyl ketone
dipropyl ketone

Common names for ketones are constructed by giving, in alphabetical order, the names of the alkyl or aryl groups attached to the carbonyl functional group and then adding the word *ketone.* Unlike aldehyde common names, which are one word, those for ketones are two or three words.

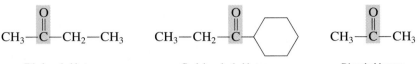

$CH_3-\overset{O}{\overset{\|}{C}}-CH_2-CH_3$ Ethyl methyl ketone

$CH_3-CH_2-\overset{O}{\overset{\|}{C}}-C_6H_{11}$ Cyclohexyl ethyl ketone

$CH_3-\overset{O}{\overset{\|}{C}}-CH_3$ Dimethyl ketone

Three ketones have additional common names besides those obtained with the preceding procedures. These three ketones are

$CH_3-\overset{O}{\overset{\|}{C}}-CH_3$ Acetone (dimethyl ketone)

$C_6H_5-\overset{O}{\overset{\|}{C}}-CH_3$ Acetophenone (methyl phenyl ketone)

$C_6H_5-\overset{O}{\overset{\|}{C}}-C_6H_5$ Benzophenone (diphenyl ketone)

15.5 Selected Common Aldehydes and Ketones

• The oxidation of methanol to formaldehyde has been previously mentioned (Section 14.7). Ingested methanol is oxidized in the human body to formaldehyde, and it is the formaldehyde that causes blindness.

Formaldehyde, the simplest aldehyde, with only one carbon atom, is manufactured on a large scale by the oxidation of methanol.

$$CH_3-OH \xrightarrow[600°C-700°C]{\text{Ag catalyst}} H-\overset{O}{\overset{\|}{C}}-H + H_2$$

Its major use is in the manufacture of polymers (Section 15.10). At room temperature and pressure, formaldehyde is an irritating gas. When this colorless gas is bubbled through water, a highly concentrated aqueous formaldehyde solution called *formalin* is produced. Formalin is used as a germicide for disinfecting surgical instruments and as a preservative that hardens tissues. Anyone who has experience in a biology laboratory is familiar with the pungent odor of formalin.

Formalin, a 35%–40% by mass solution of formaldehyde in water, is used to preserve biological specimens. The formaldehyde reacts with proteins to change their structure, causing them to resist decay. This coelacanth, a "prehistoric" fish, is preserved in formaldehyde.

The ketone used in the largest volume is acetone, the simplest ketone (Section 15.4). Acetone is an excellent solvent because it is miscible with both water and nonpolar solvents. Acetone is the main ingredient in gasoline treatments that are designed to solubilize water in the gas tank and allow it to pass through the engine in miscible form. Acetone can also be used to remove water from glassware in the laboratory. And it is a major component of some nail polish removers.

Acetone is normally present in the human body in small amounts; it is a product of the synthesis of ketone bodies in the liver (Section 25.6). Abnormal metabolic conditions, such as those associated with diabetes, can result in elevated blood acetone levels. Such excess acetone is excreted in the urine, and its presence in urine is used as a diagnostic test for diabetes. The sweetish odor of acetone can also often be detected on the breath of an untreated diabetic.

Aldehydes and ketones occur widely in nature. Naturally occurring compounds of these types, with higher molecular masses, usually have pleasant odors and flavors and are often used for these properties in consumer products (perfumes, air fresheners, and the like). Table 15.1 gives the structures and uses for selected naturally occurring aldehydes and ketones.

Many important steroid hormones (Chapter 19) are ketones, including testosterone, the hormone that controls the development of male sex characteristics; progesterone, the hormone secreted at the time of ovulation in females; and cortisone, a hormone from the adrenal glands that is used medicinally to relieve inflammation.

Table 15.1
Selected Aldehydes and Ketones Whose Uses Are Based on Their Odor or Flavor

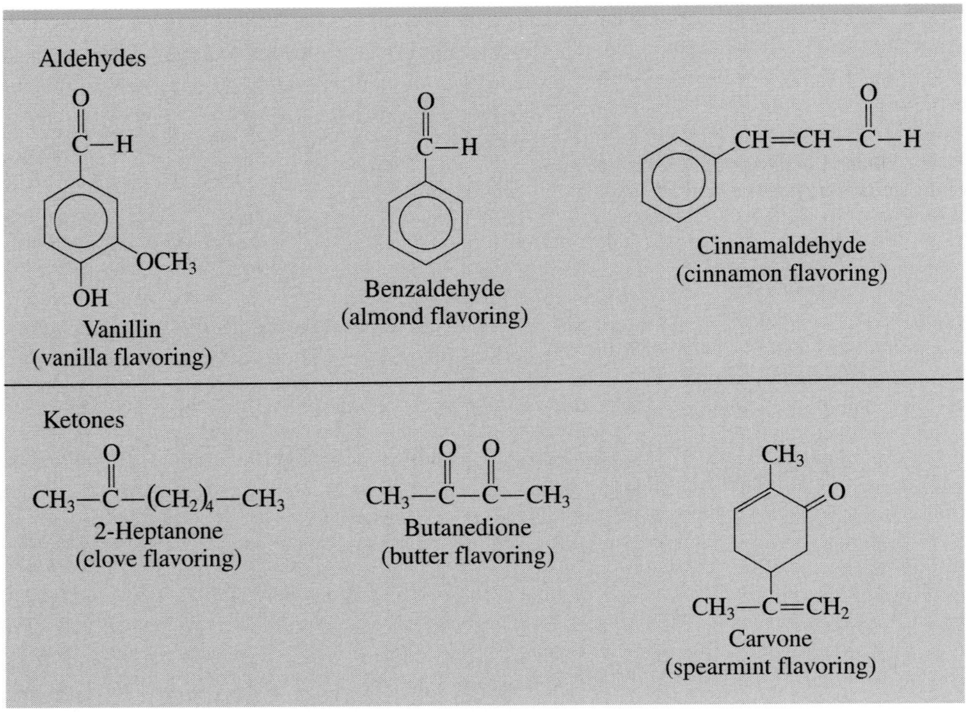

Aldehydes

Vanillin (vanilla flavoring)

Benzaldehyde (almond flavoring)

Cinnamaldehyde (cinnamon flavoring)

Ketones

$CH_3-\overset{O}{\overset{\|}{C}}-(CH_2)_4-CH_3$
2-Heptanone (clove flavoring)

$CH_3-\overset{O}{\overset{\|}{C}}-\overset{O}{\overset{\|}{C}}-CH_3$
Butanedione (butter flavoring)

Carvone (spearmint flavoring)

The delightful odor of melted butter is largely due to butanedione.

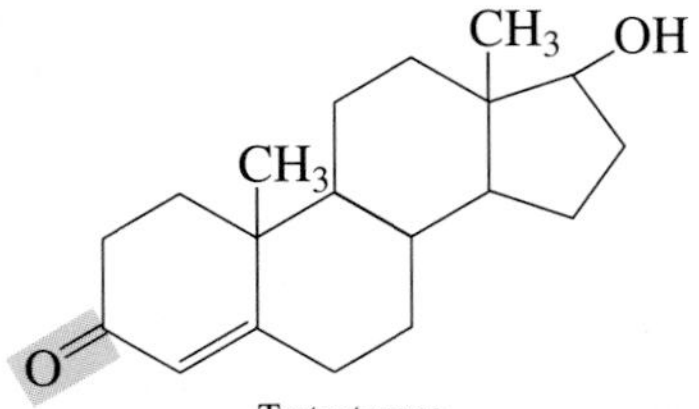

Testosterone

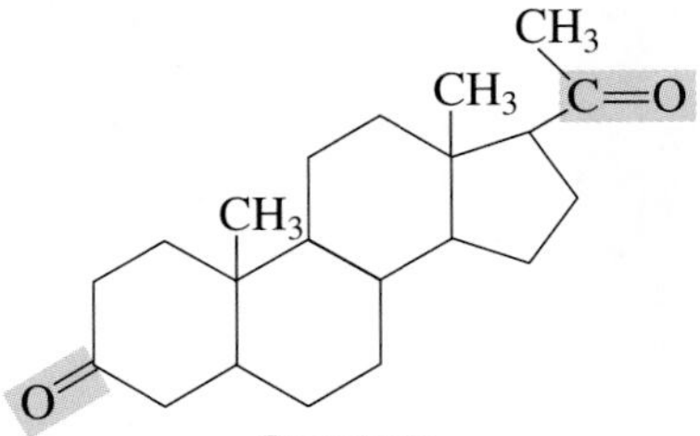

Progesterone

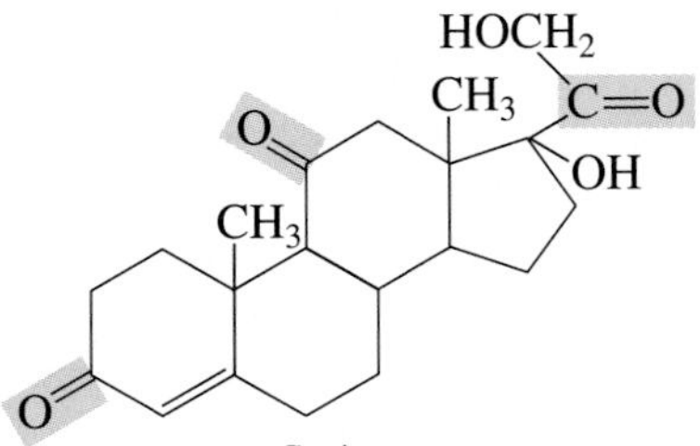

Cortisone

15.6 Physical Properties of Aldehydes and Ketones

The C_1 and C_2 aldehydes are gases at room temperature. The C_3 through C_{11} straight-chain saturated aldehydes are liquids, and the higher aldehydes are solids. The presence of alkyl groups tends to lower both boiling points and melting points, as does the presence of

Chemical CONNECTIONS

15.2 Suntan, Sunburn, and Ketones

Both sunburn and suntan are caused by the interaction of ultraviolet (UV) radiation from the sun with skin. Sudden high-level UV radiation exposure can cause the skin to burn. Steady low-level exposure to UV radiation can have a different effect—a suntan.

Human skin has a built-in defense system to protect itself against ultraviolet radiation. The brown-colored skin pigment called melanin acts as a protective barrier by absorbing and scattering UV light. Structurally, melanin is a polymeric substance involving many interconnected *cyclic ketone* units. The following is a representation of a portion of its structure. A dark-skinned person has more melanin molecules in the upper layers of the skin (and more protection against sunburn) than a light-skinned person.

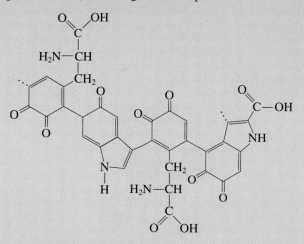

When melanin-producing cells deep in the skin are exposed to UV radiation, melanin production increases. The presence of this extra melanin in the skin gives the skin an appearance that we call a "tan." The larger the melanin molecules so produced, the deeper the tan. People who tan readily have skin that can produce a large amount of melanin.

When a person experiences a sunburn, the skin peels. When peeling occurs, any tan that has been built up (excess melanin) sloughs off with the dead skin. Thus, the tanning process must begin anew.

Sunscreen products contain substances that absorb UV radiation in a manner similar to that of melanin. Sunscreen use enables sunbathers to control UV radiation dosage and thus avoid sunburn (too much UV radiation) and yet develop a suntan (a small amount of UV radiation). Light-skinned people must use stronger sunscreen protection, because they have less natural protection (melanin). The UV-absorbing agent in many sunscreen formulations is benzophenone or one of its derivatives.

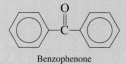

Benzophenone

Many people believe that a suntan is desirable. Studies show, however, that sunbathing, especially when sunburn results, ages the skin prematurely and increases the risk of skin cancer.

unsaturation in the carbon chain. Lower-molecular-mass ketones are colorless liquids at room temperature.

• The ordering of boiling points for carbonyl compounds (aldehydes and ketones), alcohols, and alkanes of similar molecular mass is

$$\text{Alcohol} > \begin{matrix}\text{carbonyl}\\ \text{compounds}\end{matrix} > \text{alkane}$$

The physical properties of aldehydes and ketones are intermediate between those of alcohols and alkanes of similar molecular mass. Aldehydes and ketones have higher boiling points than alkanes because of dipole–dipole attractions between molecules. Carbonyl group polarity (Section 15.1) makes these dipole–dipole interactions possible. Aldehydes and ketones have lower boiling points than the corresponding alcohols, because no hydrogen bonding occurs as it does with alcohols. Dipole–dipole attractions are weaker forces than hydrogen bonds (Section 7.14). Table 15.2 provides boiling-point information for selected aldehydes, alcohols, and alkanes.

Table 15.2
Boiling Points of Some Alkanes, Aldehydes, and Alcohols of Similar Molecular Mass

Type of compound	Compound	Structure	Molecular mass	Boiling point (°C)
alkane	ethane	$CH_3—CH_3$	30	−89
aldehyde	methanal	H—CHO	30	−21
alcohol	methanol	$CH_3—OH$	32	65
alkane	propane	$CH_3—CH_2—CH_3$	44	−42
aldehyde	ethanal	$CH_3—CHO$	44	20
alcohol	ethanol	$CH_3—CH_2—OH$	46	78
alkane	butane	$CH_3—CH_2—CH_2—CH_3$	58	−1
aldehyde	propanal	$CH_3—CH_2—CHO$	58	49
alcohol	1-propanol	$CH_3—CH_2—CH_2—OH$	60	97

Water molecules can hydrogen-bond with aldehyde and ketone molecules.

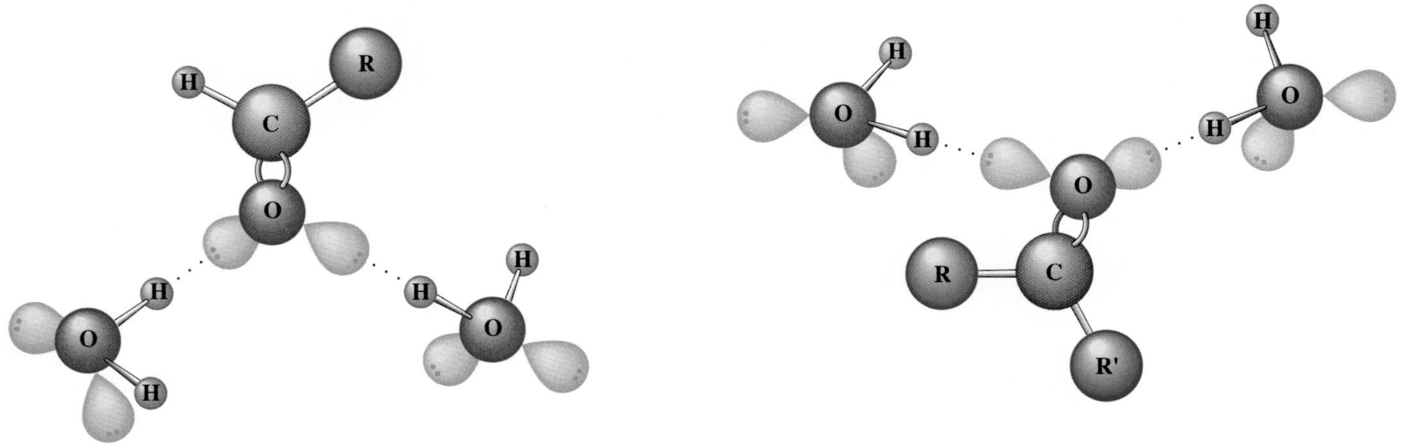

Aldehyde–water Hydrogen Bonding

Ketone–water Hydrogen Bonding

This hydrogen bonding causes low-molecular-mass aldehydes and ketones to be water-soluble. As the hydrocarbon portions get larger, the water solubility of aldehydes and ketones decreases. Table 15.3 gives data on solubility in water for selected aldehydes and ketones.

Although low-molecular-mass aldehydes have pungent, penetrating, unpleasant odors, higher-molecular-mass aldehydes (above C_8) and ketones are more fragrant, especially benzaldehyde derivatives.

Table 15.3
Water-Solubility Data (g/100 g H_2O) for Various Aldehydes and Ketones

Number of carbon atoms	Aldehyde	Water-solubility of aldehyde	Ketone	Water-solubility of ketone
1	methanal	very soluble		
2	ethanal	infinite		
3	propanal	16	propanone	infinite
4	butanal	7	2-butanone	26
5	pentanal	4	2-pentanone	5
6	hexanal	1	2-hexanone	1.6
7	heptanal	0.1	2-heptanone	0.4
8	octanal	insoluble	2-octanone	insoluble

Unbranched Aldehydes			
C_1	C_3	C_5	C_7
C_2	C_4	C_6	C_8

Unbranched 2-Ketones			
	C_3	C_5	C_7
	C_4	C_6	C_8

Gas Liquid

A physical-state summary for unbranched aldehydes and unbranched 2-ketones at room temperature and pressure.

15.7 Preparation of Aldehydes and Ketones

Aldehydes and ketones can be produced by the oxidation of primary and secondary alcohols, respectively (Section 14.7).

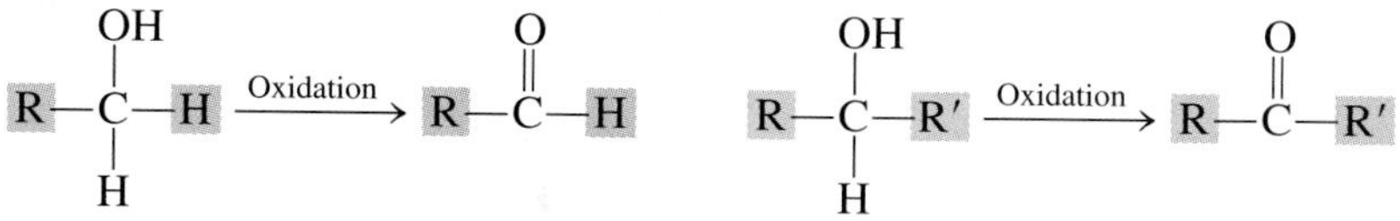

$$\underset{\text{Primary alcohol}}{R-\overset{OH}{\underset{H}{|}}{C}-H} \xrightarrow{\text{Oxidation}} \underset{\text{Aldehyde}}{R-\overset{O}{\overset{\|}{C}}-H} \qquad \underset{\text{Secondary alcohol}}{R-\overset{OH}{\underset{H}{|}}{C}-R'} \xrightarrow{\text{Oxidation}} \underset{\text{Ketone}}{R-\overset{O}{\overset{\|}{C}}-R'}$$

When this type of reaction is used for aldehyde preparation, reaction conditions must be sufficiently mild to avoid further oxidation of the aldehyde to a carboxylic acid (Section 14.7). Ketones do not undergo the further oxidation that aldehydes do.

Example 15.3

Predicting Products in Alcohol Oxidation Reactions

Draw the structure of the aldehyde or ketone formed from the oxidation of each of the following alcohols. Assume that reaction conditions are sufficiently mild that any aldehydes produced are not oxidized further.

a. $CH_3-CH_2-CH_2-OH$

b. 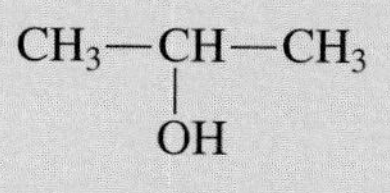 $CH_3-\underset{OH}{\underset{|}{CH}}-CH_3$

c. 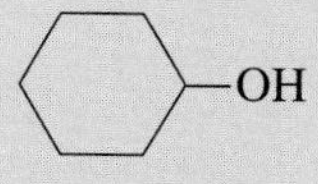 (cyclohexane ring)—OH

d. 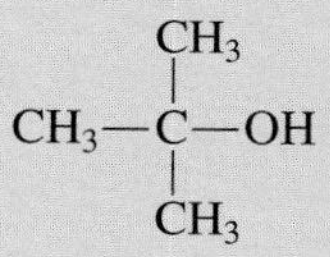 $CH_3-\overset{CH_3}{\overset{|}{\underset{CH_3}{\underset{|}{C}}}}-OH$

Solution

a. This is a primary alcohol that will give the aldehyde *propanal* as the oxidation product.

$$CH_3-CH_2-\overset{O}{\overset{\|}{C}}-H$$

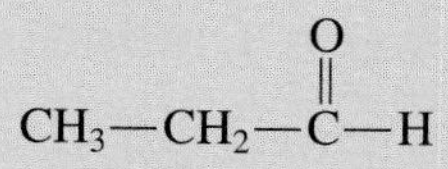

Propanal

b. This is a secondary alcohol. Upon oxidation, secondary alcohols are converted to ketones.

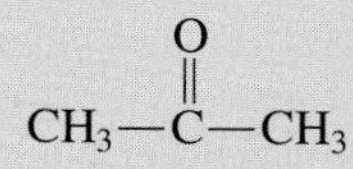

$$CH_3-\overset{O}{\overset{\|}{C}}-CH_3$$

2-Propanone

c. This cyclic alcohol is a secondary alcohol; hence a ketone is the oxidation product.

(cyclohexane ring with =O)

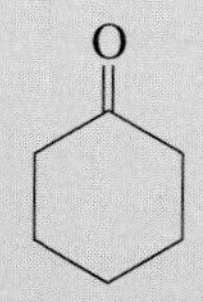

Cyclohexanone

d. This is a tertiary alcohol. Tertiary alcohols do not undergo oxidation (Section 14.7).

Practice Exercise 15.3

Draw the structure of the aldehyde or ketone formed from the oxidation of each of the following alcohols. Assume that reaction conditions are sufficiently mild that any aldehydes produced are not oxidized further to carboxylic acids.

(continued)

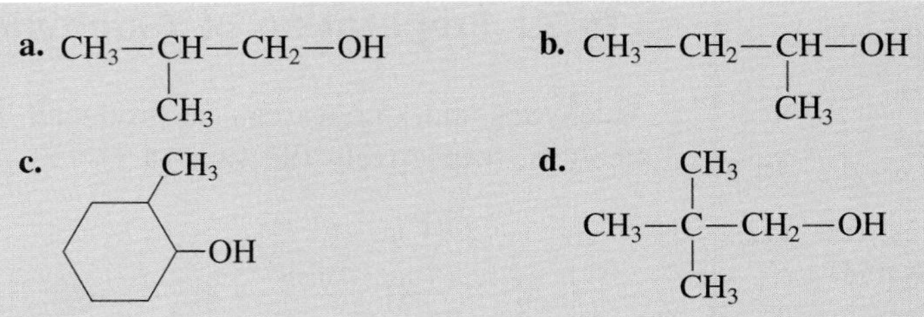

a. $CH_3—CH(CH_3)—CH_2—OH$

b. $CH_3—CH_2—CH(CH_3)—OH$

c. 2-methylcyclohexanol (cyclohexane ring with CH_3 and —OH on adjacent carbons)

d. $CH_3—C(CH_3)_2—CH_2—OH$

• *Answers:*

a. $CH_3—CH(CH_3)—\overset{\displaystyle O}{\overset{\|}{C}}—H$

b. $CH_3—CH_2—\overset{\displaystyle O}{\overset{\|}{C}}—CH_3$

c. 2-methylcyclohexanone (cyclohexane ring with CH_3 and =O on adjacent carbons)

d. $CH_3—C(CH_3)_2—\overset{\displaystyle O}{\overset{\|}{C}}—H$

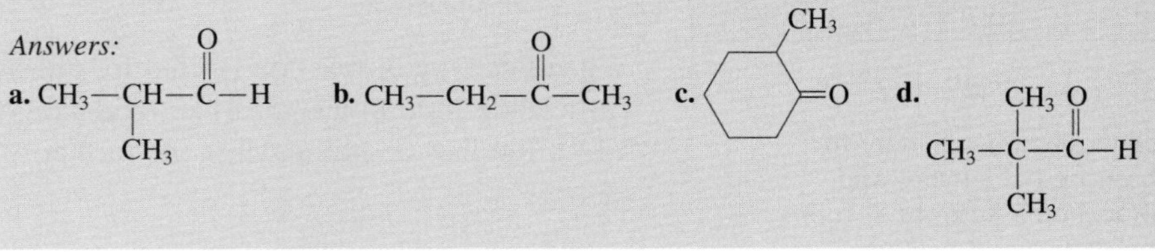

15.8 Oxidation and Reduction of Aldehydes and Ketones

• Oxidation of Aldehydes and Ketones

Aldehydes readily undergo oxidation to carboxylic acids (Section 15.7), and ketones are resistant to oxidation.

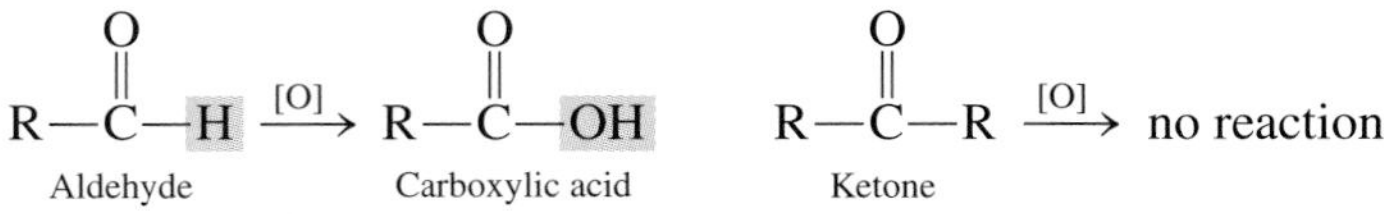

$$\underset{\text{Aldehyde}}{R—\overset{\displaystyle O}{\overset{\|}{C}}—H} \xrightarrow{[O]} \underset{\text{Carboxylic acid}}{R—\overset{\displaystyle O}{\overset{\|}{C}}—OH} \qquad \underset{\text{Ketone}}{R—\overset{\displaystyle O}{\overset{\|}{C}}—R} \xrightarrow{[O]} \text{no reaction}$$

Chemical CONNECTIONS

15.3 Diabetes, Aldehyde Oxidation, and Glucose Testing

Diabetes mellitus is a disease that involves the hormone insulin, a substance necessary to control blood-sugar (glucose) levels. There are two forms of diabetes. In one form, the pancreas does not produce insulin at all. Patients with this condition require injections of insulin to control glucose levels. In the second form, the body cannot make proper use of insulin. Patients with this form of diabetes can often control glucose levels through their diet but may require medication. If the blood-sugar level of a diabetic becomes too high, serious kidney damage can result.

Normal urine does not contain glucose. When the kidneys become overloaded with glucose (the blood-glucose level is too high), glucose is excreted in the urine. Benedict's test (Section 15.8) can be used to detect glucose in urine, because glucose has an aldehyde group present in its structure.

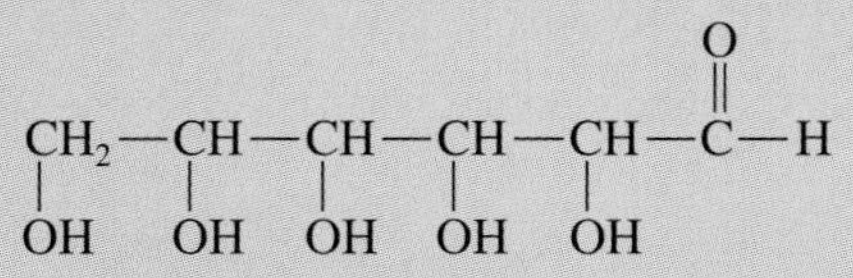

$$\underset{\text{Glucose}}{\underset{OH}{CH_2}—\underset{OH}{CH}—\underset{OH}{CH}—\underset{OH}{CH}—\underset{OH}{CH}—\overset{\displaystyle O}{\overset{\|}{C}}—H}$$

A urine glucose test is carried out using either plastic test strips coated with Benedict's solution or Clinitest tablets (a convenient solid form of Benedict's reagent). A few drops of urine are added to the plastic strip or tablet, and the degree of coloration is used to estimate the blood-glucose level. The solution turns greenish at a low glucose level, then yellow-orange, and finally a dark orange-red.

Tests are also available for directly measuring glucose concentration in blood. These tests involve placing a drop of blood (from a finger prick) on a plastic strip containing a dye and an enzyme that will oxidize glucose's aldehyde group. A two-step reaction sequence occurs. First, the enzyme causes glucose oxidation to a carboxylic acid, with hydrogen peroxide (H_2O_2) being another product of the reaction.

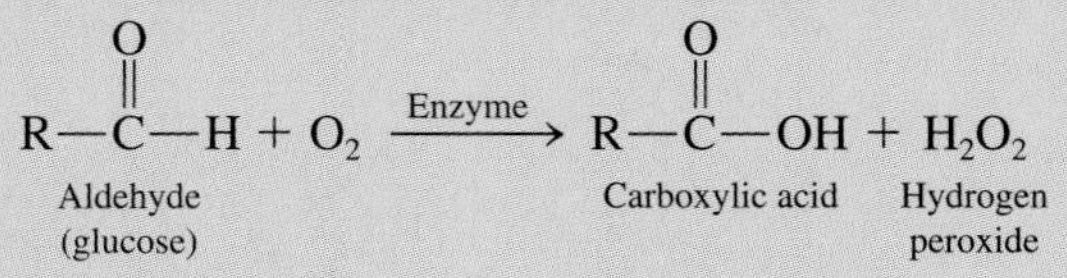

$$\underset{\substack{\text{Aldehyde}\\\text{(glucose)}}}{R—\overset{\displaystyle O}{\overset{\|}{C}}—H} + O_2 \xrightarrow{\text{Enzyme}} \underset{\text{Carboxylic acid}}{R—\overset{\displaystyle O}{\overset{\|}{C}}—OH} + \underset{\substack{\text{Hydrogen}\\\text{peroxide}}}{H_2O_2}$$

Then the H_2O_2 reacts with the dye to produce a colored product. The amount of color produced, measured by comparison with a color chart or by an electronic monitor, is proportional to the blood-glucose concentration.

(a)

(b)

(c)

Figure 15.3
A positive Tollens test for aldehydes involves the formation of a silver mirror. (a) An aqueous solution of ethanal is added to a solution of silver nitrate in aqueous ammonia and stirred. (b) The solution darkens as ethanal is oxidized to ethanoic acid, and Ag^+ ion is reduced to silver. (c) The inside of the beaker becomes coated with metallic silver.

Because both aldehydes and ketones contain carbonyl groups, we might expect similar reactions for the two types of compounds. Oxidation of an aldehyde involves breaking a carbon–hydrogen bond, and oxidation of a ketone involves breaking a carbon–carbon bond. The former is much easier to accomplish than the latter. For ketones to be oxidized, strenuous reaction conditions must be employed.

The difference in the tendency of aldehydes and ketones to undergo oxidation enables us to use simple tests to distinguish between these two types of compounds. To determine whether a compound is an aldehyde or a ketone, we treat it with a mild oxidizing agent: If oxidation occurs, it is an aldehyde; if no oxidation occurs, it is a ketone. Two tests commonly used for this purpose are the Tollens test and Benedict's test. These tests are particularly important in carbohydrate chemistry (Section 18.12).

The Tollens test, also called the silver mirror test, involves a solution that contains silver nitrate ($AgNO_3$) and ammonia (NH_3) in water. When Tollens solution is added to an aldehyde, Ag^+ ion (the oxidizing agent) is reduced to silver metal, which deposits on the inside of the test tube, forming a silver mirror. The appearance of this silver mirror (see Figure 15.3) is a positive test for the presence of the aldehyde group.

$$\underset{\text{Aldehyde}}{R-\overset{\overset{\displaystyle O}{\|}}{C}-H} + Ag^+ \xrightarrow[\Delta]{NH_3,\ H_2O} \underset{\text{Carboxylic acid}}{R-\overset{\overset{\displaystyle O}{\|}}{C}-OH} + \underset{\text{Silver metal}}{Ag}$$

The Ag^+ ion will not oxidize ketones.

Benedict's test is similar to the Tollens test in that a metal ion is the oxidizing agent. With this test, Cu^{2+} ion is reduced to Cu^+ ion, which precipitates from solution as Cu_2O (a brick red solid; Figure 15.4).

$$\underset{\text{Aldehyde}}{R-\overset{\overset{\displaystyle O}{\|}}{C}-H} + Cu^{2+} \longrightarrow \underset{\text{Carboxylic acid}}{R-\overset{\overset{\displaystyle O}{\|}}{C}-OH} + \underset{\text{Brick red solid}}{Cu_2O}$$

Benedict's solution is made by dissolving copper sulfate, sodium citrate, and sodium carbonate in water.

Figure 15.4
Benedict's solution, which is blue in color, turns brick red when an aldehyde reacts with it.

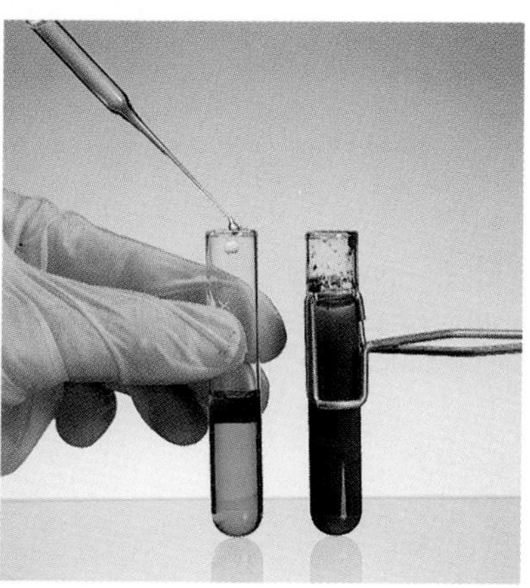

• Reduction of Aldehydes and Ketones

Aldehydes and ketones are easily reduced by hydrogen gas (H_2), in the presence of a catalyst (Ni, Pt, or Cu), to form alcohols. The reduction of aldehydes produces primary alcohols, and the reduction of ketones yields secondary alcohols.

General Equation		*Specific Example*	
$R-\overset{O}{\overset{\parallel}{C}}-H + H_2 \xrightarrow{Ni}$	$R-\underset{H}{\overset{OH}{C}}-H$	$CH_3-\overset{O}{\overset{\parallel}{C}}-H + H_2 \xrightarrow{Ni}$	$CH_3-\underset{H}{\overset{OH}{C}}-H$
Aldehyde	Primary alcohol	Ethanal	Ethanol
$R-\overset{O}{\overset{\parallel}{C}}-R' + H_2 \xrightarrow{Ni}$	$R-\underset{H}{\overset{OH}{C}}-R'$	$CH_3-\overset{O}{\overset{\parallel}{C}}-CH_3 + H_2 \xrightarrow{Ni}$	$CH_3-\underset{H}{\overset{OH}{C}}-CH_3$
Ketone	Secondary alcohol	2-Propanone	2-Propanol

• The term *aldehyde* stems from <u>*al*</u>*cohol* <u>*dehyd*</u>*rogenation,* indicating that aldehydes are related to alcohols by the loss of hydrogen.

Aldehyde and ketone reductions by H_2 gas involve the addition of the H_2 to the carbon–oxygen double bond, a process that is similar to the addition of H_2 to a carbon–carbon double bond (Section 13.7).

$$>C=C< + H_2 \xrightarrow{\text{Catalyst}} >\underset{H}{C}-\underset{H}{C}<$$

$$>C=O + H_2 \xrightarrow{\text{Catalyst}} >\underset{H}{C}-\underset{H}{O}$$

When alcohols are *oxidized* to produce aldehydes and ketones (Section 14.8), they *lose H atoms*. When aldehydes and ketones are *reduced* to produce alcohols, they *gain H atoms*. The two processes are "opposites."

$$1^\circ \text{ Alcohol} \underset{\text{Reduction}}{\overset{\text{Oxidation}}{\rightleftharpoons}} \text{aldehyde} \qquad 2^\circ \text{ Alcohol} \underset{\text{Reduction}}{\overset{\text{Oxidation}}{\rightleftharpoons}} \text{ketone}$$

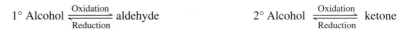

15.9 Reaction of Aldehydes and Ketones with Alcohols

Many different substances can add to the carbon–oxygen double bond of a carbonyl group. In fact, because of its polarity, a carbon–oxygen double bond is even more susceptible to addition reactions than a carbon–carbon double bond.

Polarity is used to predict locations for the entering groups from addition. In a carbon–oxygen double bond, due to electronegativity differences, the carbon atom possesses a partial positive charge (δ^+), and the oxygen atom possesses a partial negative charge (δ^-).

$$>\overset{\delta^+}{C}=\overset{\delta^-}{O}$$

An unsymmetrical addition agent also has partial charges.

$$\overset{\delta^+}{X}-\overset{\delta^-}{Y}$$

The atom (or group of atoms) in the addition agent with the positive partial charge (δ^+) bonds to the carbonyl oxygen atom (δ^-), and the atom (or group of atoms) with the partial negative charge (δ^-) bonds to the carbonyl carbon atom (δ^+).

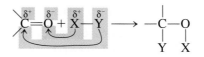

$$>\overset{\delta^+}{C}=\overset{\delta^-}{O} + \overset{\delta^+}{X}-\overset{\delta^-}{Y} \longrightarrow -\underset{Y}{\overset{|}{C}}-\underset{X}{O}$$

• Hemiacetals and Hemiketals

When an alcohol molecule adds across the carbon–oxygen double bond of an aldehyde or ketone, the H atom from the alcohol adds to the carbonyl oxygen atom, and the R—O portion of the alcohol (an alkoxy group; Section 14.10) adds to the carbonyl carbon atom.

The product of the addition of *one* molecule of an alcohol to an aldehyde is called a *hemiacetal.* Similarly, the addition of *one* molecule of alcohol to a ketone produces a *hemiketal.*

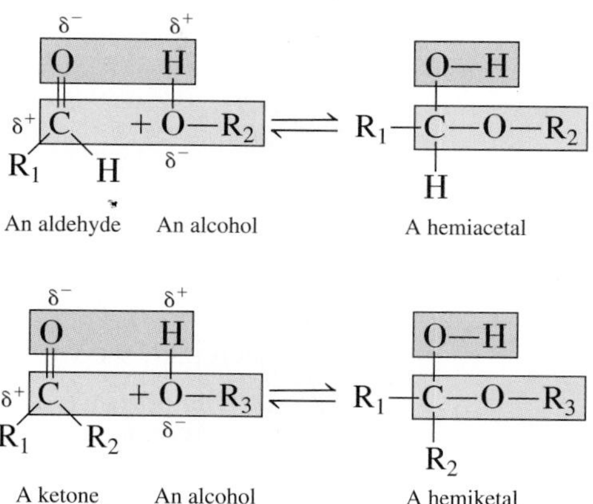

• Hemiacetals and hemiketals both contain an alcohol group (hydroxyl group) and an ether group (alkoxy group) on the same carbon atom. What differentiates them is the presence or lack of a hydrogen atom on this same carbon atom. In a hemiacetal a hydrogen atom is present. In a hemiketal no hydrogen atom is present.

Hemiacetal   Hemiketal

Formally defined, a **hemiacetal** *is a compound that has a carbon atom to which a hydroxyl group (—OH), an alkoxy group (—OR), and a hydrogen atom (—H) are attached.* A hemiketal differs from a hemiacetal in that an R group has replaced the H atom. A **hemiketal** *is a compound that has a carbon atom to which a hydroxyl group (—OH) and an alkoxy group (—OR), but no hydrogen atom (—H), are attached.*

Reaction mixtures containing hemiacetals and hemiketals are always in equilibrium with the alcohol and carbonyl compounds from which they are made, and the equilibrium lies to the carbonyl compound side of the reaction (Section 9.7).

$$\text{Alcohol} + \text{aldehyde} \rightleftharpoons \text{hemiacetal}$$
$$\text{Alcohol} + \text{ketone} \rightleftharpoons \text{hemiketal}$$

This situation makes isolation of the hemiacetal or hemiketal difficult; in practice, it usually cannot be done.

An important exception to this isolation difficulty is the case where the —OH and —C=O functional groups that react to form the hemiacetal or hemiketal *come from the same molecule.* This produces a *cyclic* hemiacetal or *cyclic* hemiketal rather than a noncyclic one, and the cyclic species are more stable than the noncyclic ones and can be isolated.

Illustrative of intramolecular hemiacetal formation is the reaction

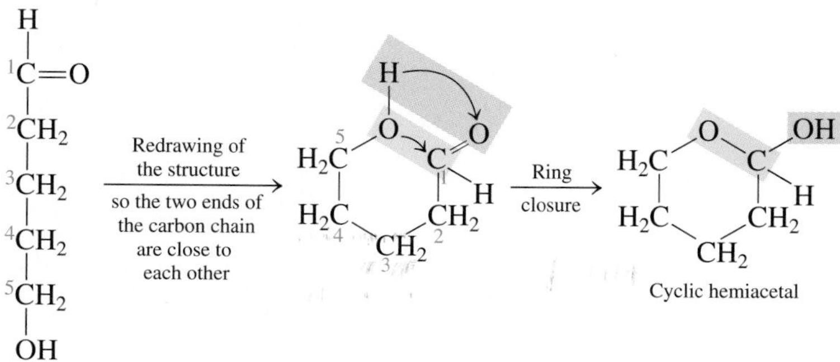

Cyclic hemiacetals are very important compounds in carbohydrate chemistry, the topic of Chapter 18.

Example 15.4

Recognizing Hemiacetal and Hemiketal Structures

Identify each of the following compounds as a hemiacetal, a hemiketal, or neither of these two types of compounds.

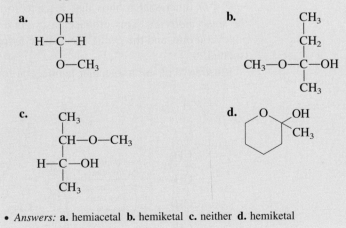

a. 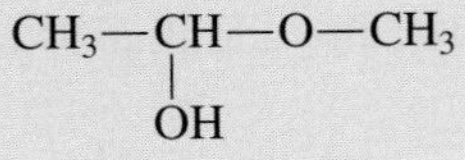 $CH_3—CH(OH)—O—CH_3$

b. 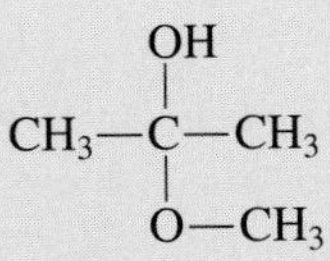 $CH_3—C(OH)(O—CH_3)—CH_3$

c. $CH_3—CH(OH)—CH(CH_3)—O—CH_3$

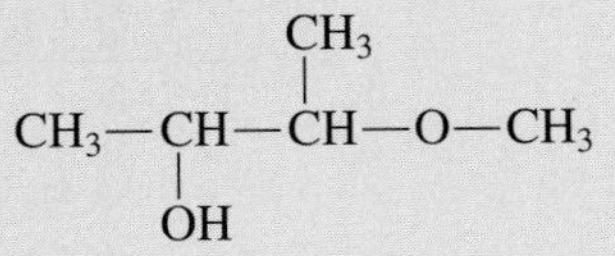

d. 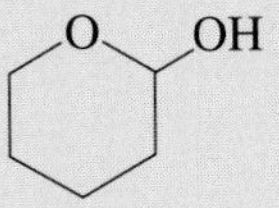 Six-membered ring with ring O; the carbon next to O bears OH

Solution

In each part, we will be looking for the following structural features:

1. The presence of an —OH group and an —OR group attached to the same carbon atom.
2. The presence of an H atom on the carbon to which the —OH and —OR are bonded (a hemiacetal) or the lack of such an H atom (a hemiketal).

a. We have an —OH group and an —OR group attached to the same carbon atom. Because there is also an H atom on this oxygen-bearing carbon, the compound is a *hemiacetal.*

b. We have an —OH group and an —OR group attached to the same carbon atom. Because this oxygen-bearing carbon atom bears no H atoms, the compound is a *hemiketal.*

c. The —OH and —OR groups present in this molecule are attached to *different* carbon atoms. Therefore, the molecule is *neither a hemiacetal nor a hemiketal.*

d. We have a ring carbon atom bonded to two oxygen atoms: one oxygen atom in an —OH substituent and the other oxygen atom bonded to the rest of the ring (the same as an R group). This is a *hemiacetal* because the oxygen-bearing carbon atom also has an H atom attached to it.

• The carbon atom in a hemiacetal or a hemiketal that bears the hydroxyl and alkoxy groups is the original carbonyl carbon atom.

Practice Exercise 15.4

Identify each of the following compounds as a hemiacetal, a hemiketal, or neither of these two types of compounds.

a. 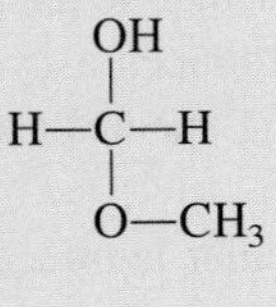 $H—C(H)(OH)—O—CH_3$

b. $CH_3—O—C(OH)(CH_3)—CH_2—CH_3$

c. $CH_3—CH(O—CH_3)—CH(OH)—CH_3$

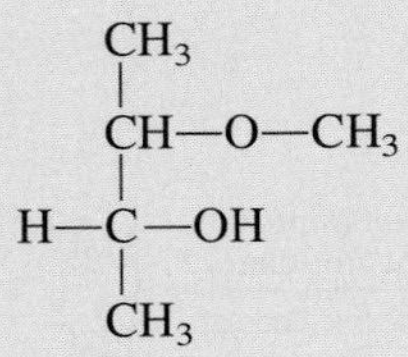

d. 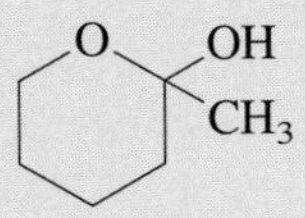 Six-membered ring with ring O; the carbon next to O bears OH and CH_3

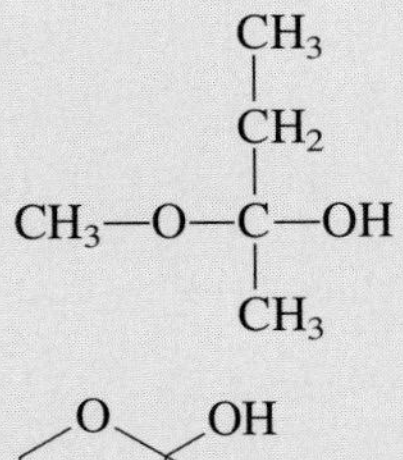

• *Answers:* **a.** hemiacetal **b.** hemiketal **c.** neither **d.** hemiketal

• Acetals and Ketals

If a small amount of acid catalyst is added to a hemiacetal reaction mixture, then the hemiacetal reacts with a second alcohol molecule to form an acetal.

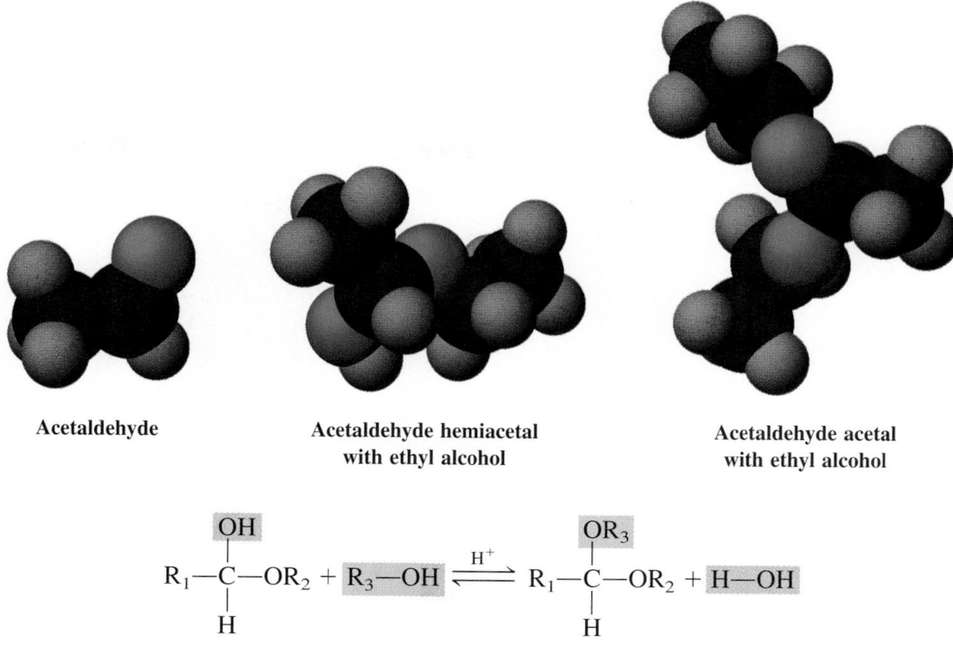

Acetaldehyde | Acetaldehyde hemiacetal with ethyl alcohol | Acetaldehyde acetal with ethyl alcohol

Molecular models for acetaldehyde and its hemiacetal and acetal formed by reaction with ethyl alcohol.

$$\underset{\text{A hemiacetal}}{R_1-\overset{\overset{\displaystyle OH}{|}}{\underset{\underset{\displaystyle H}{|}}{C}}-OR_2} + \underset{\text{An alcohol}}{R_3-OH} \overset{H^+}{\rightleftharpoons} \underset{\text{An acetal}}{R_1-\overset{\overset{\displaystyle OR_3}{|}}{\underset{\underset{\displaystyle H}{|}}{C}}-OR_2} + H-OH$$

An **acetal** *is a compound that has a carbon atom to which two alkoxy groups* (—*OR*) and a hydrogen atom (—*H*) *are attached.* An example of acetal formation is

$$CH_3-\overset{\overset{\displaystyle OH}{|}}{\underset{\underset{\displaystyle O-CH_3}{|}}{CH}} + CH_3-CH_2-OH \overset{H^+}{\rightleftharpoons} CH_3-\overset{\overset{\displaystyle O-CH_2-CH_3}{|}}{\underset{\underset{\displaystyle O-CH_3}{|}}{CH}} + H-OH$$

• Both acetals and ketals have two alkoxy groups (—OR) attached to the same carbon atom. What differentiates them is the presence or lack of a hydrogen atom on this same carbon atom. In an acetal, a hydrogen atom is present. In a ketal, no hydrogen atom is present.

$$\underset{\text{Acetal}}{R_1-\overset{\overset{\displaystyle OR_3}{|}}{\underset{\underset{\displaystyle H}{|}}{C}}-OR_2} \qquad \underset{\text{Ketal}}{R_1-\overset{\overset{\displaystyle OR_4}{|}}{\underset{\underset{\displaystyle R_2}{|}}{C}}-OR_3}$$

Similarly, in the presence of an acid catalyst, the reaction of a second alcohol molecule with a hemiketal produces a ketal.

$$\underset{\text{A hemiketal}}{R_1-\overset{\overset{\displaystyle OH}{|}}{\underset{\underset{\displaystyle R_2}{|}}{C}}-OR_3} + \underset{\text{An alcohol}}{R_4-OH} \rightleftharpoons \underset{\text{A ketal}}{R_1-\overset{\overset{\displaystyle OR_4}{|}}{\underset{\underset{\displaystyle R_2}{|}}{C}}-OR_3} + H-OH$$

A **ketal** *is a compound that has a carbon atom to which two alkoxy groups* (—*OR*) *and no hydrogen atom* (—*H*) *are attached.*

Acetals and ketals, unlike hemiacetals and hemiketals, are easily isolated from reaction mixtures. They are stable in basic solution but undergo *hydrolysis* in acidic solution. A **hydrolysis reaction** *is the reaction of a compound with water, in which the compound splits into two or more fragments as the elements of water (H— and —OH) are added.* The products of such hydrolysis are the aldehyde or ketone and alcohols that originally reacted to form the acetal or ketal.

• In Section 24.1, we will find that the enzyme-catalyzed hydrolysis of acetals is an important step in the digestion of carbohydrates.

$$\underset{\text{Acetal or ketal}}{-\overset{\overset{\displaystyle O-R_1}{|}}{\underset{|}{C}}-O-R_2} + H-OH \overset{\text{Acid catalyst}}{\rightleftharpoons} \underset{\text{Aldehyde or ketone}}{\overset{\overset{\displaystyle O}{\|}}{C}} + R_1-OH + R_2-O-H$$

For example,

$$CH_3-\overset{\overset{\displaystyle O-CH_2-CH_3}{|}}{\underset{\underset{\displaystyle CH_3}{|}}{C}}-O-CH_3 + H-OH \overset{\text{Acid catalyst}}{\rightleftharpoons} CH_3-\overset{\overset{\displaystyle O}{\|}}{C}-CH_3 + CH_3-OH + CH_3-CH_2-OH$$

Example 15.5

Predicting Products in Acetal and Ketal Hydrolysis Reactions

Draw structures for the aldehyde (or ketone) and alcohols produced when the following acetal and ketal undergo acid hydrolysis.

a. $CH_3-CH_2-CH(O-CH_3)(O-CH_2-CH_3)$

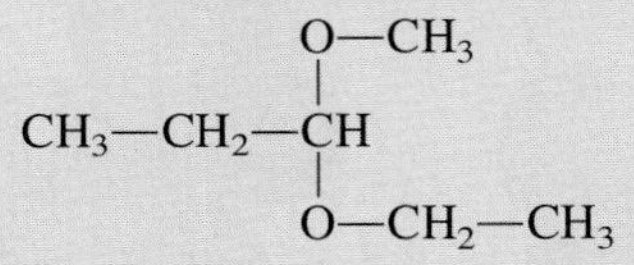

b. $CH_3-C(CH_3)(O-CH(CH_3)-CH_3)-O-C(CH_3)_2-CH_3$

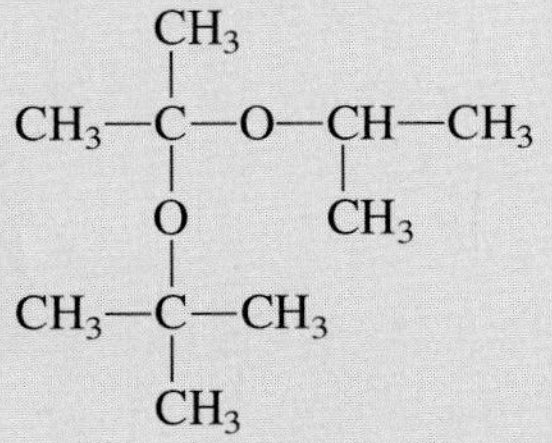

Solution

a. This compound is an acetal because there are two —OR groups and the oxygen-bearing carbon also bears a hydrogen atom. Each of the alkoxy (—OR) groups will be converted to an alcohol during the hydrolysis. The remainder of the molecule becomes an aldehyde, with the carbon atom to which the alkoxy groups were attached becoming the carbonyl carbon atom.

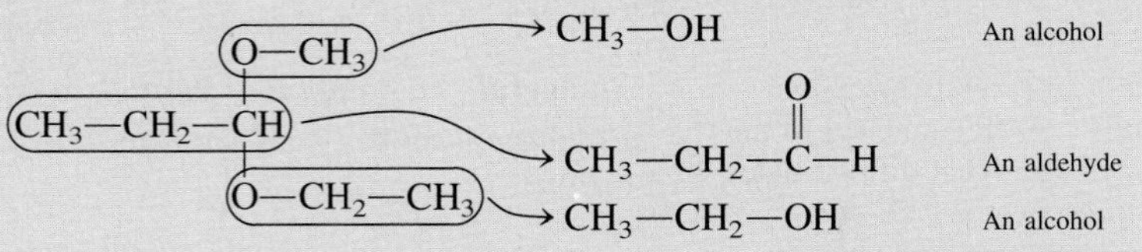

$O-CH_3 \rightarrow CH_3-OH$ An alcohol

$CH_3-CH_2-CH \rightarrow CH_3-CH_2-C(=O)-H$ An aldehyde

$O-CH_2-CH_3 \rightarrow CH_3-CH_2-OH$ An alcohol

b. This compound is a ketal because there is no H atom on the carbon atom that bears the two —OR groups. The hydrolysis products will be a ketone and two alcohols.

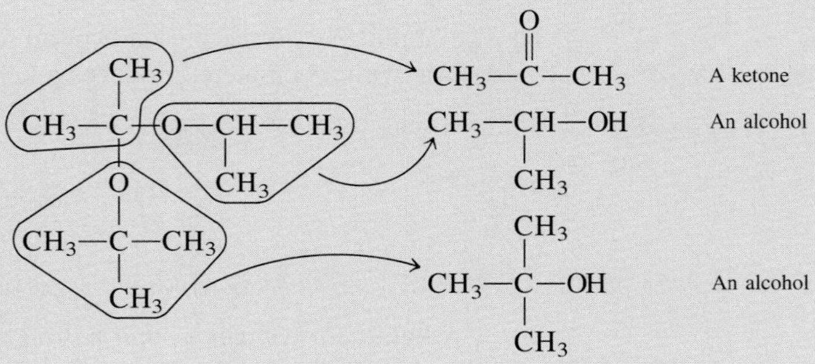

$CH_3-C(=O)-CH_3$ A ketone

$CH_3-CH(OH)-CH_3$ An alcohol

$CH_3-C(CH_3)_2-OH$ An alcohol

Practice Exercise 15.5

Draw the structures of the aldehyde (or ketone) and the two alcohols produced when the following compound undergoes hydrolysis in acidic solution.

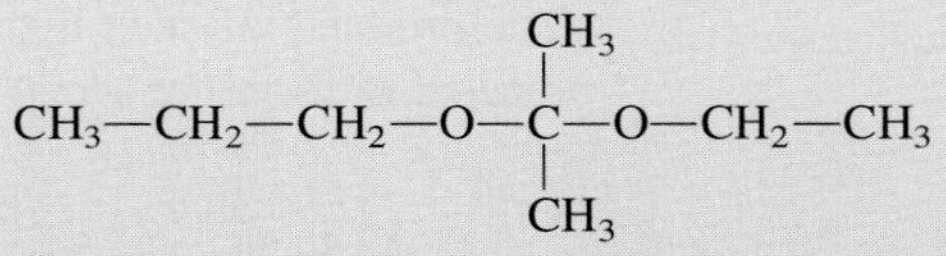

$$CH_3-CH_2-CH_2-O-C(CH_3)_2-O-CH_2-CH_3$$

- *Answers:* $CH_3-CH_2-CH_2-OH$ (alcohol), CH_3-CH_2-OH (alcohol), $CH_3-C(=O)-CH_3$ (ketone)

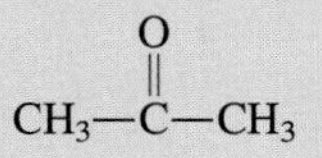

The accompanying Chemistry at a Glance summarizes reactions that involve aldehydes and ketones.

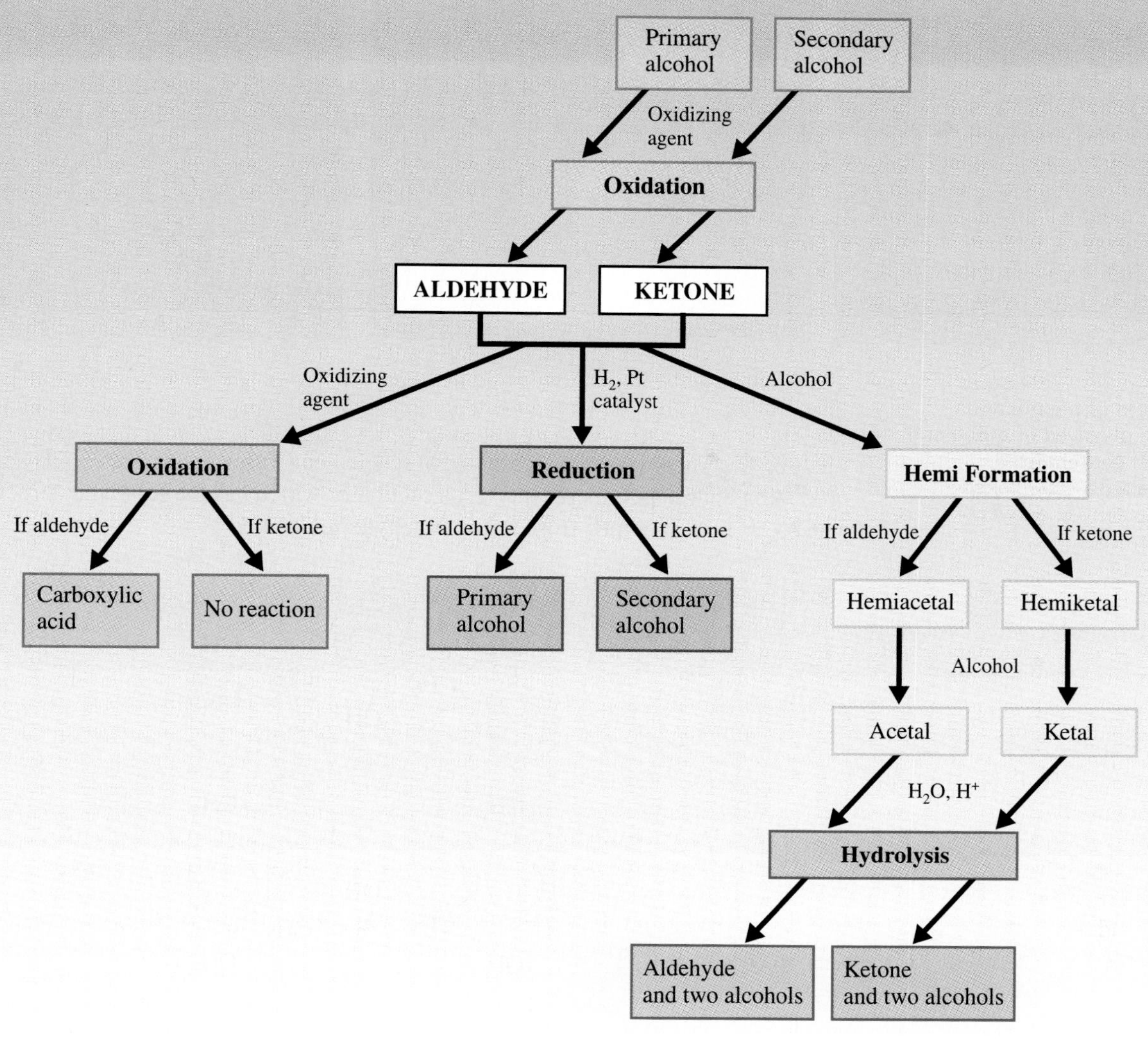

15.10 Formaldehyde-Based Polymers

Many types of organic compounds can serve as reactants (monomers) for polymerization reactions, including ethylenes (Section 13.8), alcohols (Section 14.9), and carbonyl compounds.

Formaldehyde, the simplest aldehyde, is a prolific "polymer former." As representative of its polymer reactions, let us consider the reaction between formaldehyde and phenol to form a phenol–formaldehyde network polymer (see Figure 15.5). A **network polymer** *is a polymer in which monomers are connected in a three-dimensional cross-linked network.*

As a first step in the polymerization process, a formaldehyde molecule reacts with two phenol molecules with the production of one water molecule.

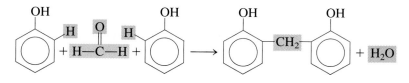

$$\text{phenol (H)} + H{-}\overset{O}{\overset{\|}{C}}{-}H + \text{(H) phenol} \longrightarrow \text{phenol}{-}CH_2{-}\text{phenol} + H_2O$$

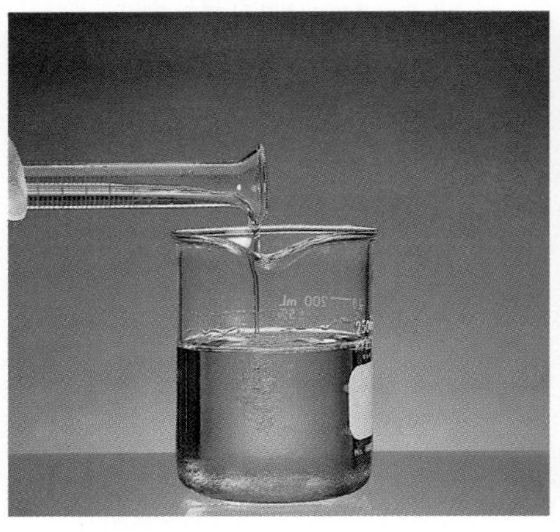

Figure 15.5
When a mixture of phenol and formaldehyde dissolved in acetic acid is treated with concentrated hydrochloric acid, a cross-linked phenol–formaldehyde network polymer is formed.

Each of the phenol groups in this new species can form two more $—CH_2—$ bridges, producing a large polymeric species, a section of which has the following structure. Each $—CH_2—$ bridge comes from a formaldehyde molecule.

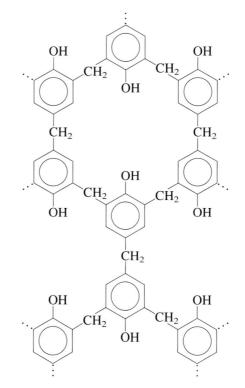

The first synthetic plastic, Bakelite, produced in 1907, was a phenol–formaldehyde polymer. One of its first uses was in the manufacture of billiard balls. Modern phenol–formaldehyde polymers, called phenolics, are adhesives used in the production of plywood and particle board.

15.11 Sulfur-Containing Carbonyl Groups

The introduction of sulfur into a carbonyl group produces two different classes of compounds, depending on whether the sulfur atom replaces the carbonyl oxygen atom or the carbonyl carbon atom.

Bakelite jewelry in use during the 1930–1950 time period.

Replacement of the carbonyl oxygen atom with sulfur produces *thiocarbonyl compounds*—thioaldehydes (thials) and thioketones (thiones)—the simplest of which are

$$\underset{\text{(methanethial)}}{\underset{\text{Thioformaldehyde}}{\mathrm{H{-}\overset{\overset{\displaystyle S}{\|}}{C}{-}H}}} \qquad \underset{\text{(propanethione)}}{\underset{\text{Thioacetone}}{\mathrm{CH_3{-}\overset{\overset{\displaystyle S}{\|}}{C}{-}CH_3}}}$$

Thiocarbonyl compounds, such as these, are unstable and readily decompose.

Replacement of the carbonyl carbon atom with sulfur produces *sulfoxides,* compounds that are much more stable than thiocarbonyl compounds. The oxidation of a sulfide constitutes the most common route to a sulfoxide.

$$\underset{\text{Sulfide}}{\mathrm{R{-}S{-}R}} \xrightarrow{[O]} \underset{\text{Sulfoxide}}{\mathrm{R{-}\overset{\overset{\displaystyle O}{\|}}{S}{-}R}}$$

A highly interesting sulfoxide is DMSO (dimethyl sulfoxide), a sulfur analog of acetone, the simplest ketone.

$$\underset{\text{DMSO}}{\mathrm{CH_3{-}\overset{\overset{\displaystyle O}{\|}}{S}{-}CH_3}} \qquad \underset{\text{Acetone}}{\mathrm{CH_3{-}\overset{\overset{\displaystyle O}{\|}}{C}{-}CH_3}}$$

DMSO is an odorless liquid with unusual properties. Because of the presence of the highly polar sulfur–oxide bond, DMSO is miscible with water and also quite soluble in less polar organic solvents. When rubbed on the skin, DMSO has remarkable penetrating power and is quickly absorbed into the body, where it relieves pain and inflammation. For many years it has been heralded as a "miracle drug" for arthritis, sprains, burns, herpes, infections, and high blood pressure. However, the FDA has steadfastly refused to approve it for general medical use. For example, the FDA says that DMSO's powerful penetrating action could cause an insecticide on a gardener's skin to be carried accidentally into his or her bloodstream. Another complication is that DMSO is reduced in the body to dimethyl sulfide, a compound with a strong garlic-like odor that soon appears on the breath.

$$\mathrm{CH_3{-}\overset{\overset{\displaystyle O}{\|}}{S}{-}CH_3} \xrightarrow{\text{Reduction}} \mathrm{CH_3{-}S{-}CH_3}$$

The FDA has approved DMSO for use in certain bladder conditions and as a veterinary drug for topical use in nonbreeding dogs and horses. For example, DMSO is used as an anti-inflammatory rub for race horses.

Concepts to Remember

The carbonyl group. A carbonyl group consists of a carbon atom bonded to an oxygen atom through a double bond. Aldehydes and ketones are compounds that contain a carbonyl functional group. The carbonyl carbon in an aldehyde has at least one hydrogen attached to it, and the carbonyl carbon in a ketone has no hydrogens.

Nomenclature of aldehydes and ketones. The IUPAC names of aldehydes and ketones are based on the longest carbon chain that contains the carbonyl group. The chain numbering is done from the end that results in the lowest number for the carbonyl group. The names of aldehydes end in *-al,* those of ketones in *-one.*

Physical properties of aldehydes and ketones. The boiling points of aldehydes and ketones are intermediate between those of alcohols and alkanes. The polarity of the carbonyl groups enables aldehyde and ketone molecules to interact with each other through dipole–dipole interactions. They cannot, however, hydrogen-bond to each other. Lower-molecular-mass aldehydes and ketones are soluble in water.

Preparation of aldehydes and ketones. Oxidation of primary and secondary alcohols, using mild oxidizing agents, produces aldehydes and ketones, respectively.

Oxidation and reduction of aldehydes and ketones. Aldehydes are easily oxidized to carboxylic acids; ketones do not readily undergo oxidation. Reduction of aldehydes and ketones produces primary and secondary alcohols, respectively.

Hemiacetals, hemiketals, acetals, and ketals. A characteristic reaction of aldehydes and ketones is the addition of an alcohol across the carbonyl double bond to produce hemiacetals and hemiketals. The reaction of a second alcohol molecule with a hemiacetal or hemiketal produces an acetal or a ketal.

Key Reactions and Equations

1. Oxidation of an aldehyde to give a carboxylic acid (Section 15.8)

$$R{-}\overset{\overset{\Large O}{\|}}{C}{-}H \xrightarrow{[O]} R{-}\overset{\overset{\Large O}{\|}}{C}{-}OH$$

2. Attempted oxidation of a ketone (Section 15.8)

$$R{-}\overset{\overset{\Large O}{\|}}{C}{-}R' \xrightarrow{[O]} \text{no reaction}$$

3. Reduction of an aldehyde to give a primary alcohol (Section 15.8)

$$R{-}\overset{\overset{\Large O}{\|}}{C}{-}H + H_2 \xrightarrow{\text{Catalyst}} R{-}\underset{\underset{\Large H}{|}}{\overset{\overset{\Large OH}{|}}{C}}{-}H$$

4. Reduction of a ketone to give a secondary alcohol (Section 15.8)

$$R{-}\overset{\overset{\Large O}{\|}}{C}{-}R' + H_2 \xrightarrow{\text{Catalyst}} R{-}\underset{\underset{\Large H}{|}}{\overset{\overset{\Large OH}{|}}{C}}{-}R'$$

5. Addition of an alcohol to an aldehyde to form a hemiacetal and then an acetal (Section 15.9)

$$\underset{\text{Aldehyde}}{R_1{-}\overset{\overset{\Large O}{\|}}{C}{-}H + R_2{-}O{-}H} \overset{H^+}{\rightleftharpoons} \underset{\text{Hemiacetal}}{R_1{-}\underset{\underset{\Large H}{|}}{\overset{\overset{\Large OH}{|}}{C}}{-}OR_2}$$

$$\underset{\text{Hemiacetal}}{R_1{-}\underset{\underset{\Large H}{|}}{\overset{\overset{\Large OH}{|}}{C}}{-}OR_2} + R_3{-}OH \overset{H^+}{\rightleftharpoons} \underset{\text{Acetal}}{R_1{-}\underset{\underset{\Large H}{|}}{\overset{\overset{\Large OR_3}{|}}{C}}{-}OR_2} + H_2O$$

6. Addition of an alcohol to a ketone to form a hemiketal and a ketal (Section 15.9)

$$\underset{\text{Ketone}}{R_1{-}\overset{\overset{\Large O}{\|}}{C}{-}R_2} + R_3{-}O{-}H \overset{H^+}{\rightleftharpoons} \underset{\text{Hemiketal}}{R_1{-}\underset{\underset{\Large R_2}{|}}{\overset{\overset{\Large OH}{|}}{C}}{-}OR_3}$$

$$\underset{\text{Hemiketal}}{R_1{-}\underset{\underset{\Large R_2}{|}}{\overset{\overset{\Large OH}{|}}{C}}{-}OR_3} + R_4{-}OH \overset{H^+}{\rightleftharpoons} \underset{\text{Ketal}}{R_1{-}\underset{\underset{\Large R_2}{|}}{\overset{\overset{\Large OR_4}{|}}{C}}{-}OR_3} + H_2O$$

7. Hydrolysis of an acetal to yield an aldehyde and two alcohols (Section 15.9)

$$R_1{-}\underset{\underset{\Large H}{|}}{\overset{\overset{\Large OR_3}{|}}{C}}{-}OR_2 + H_2O \overset{H^+}{\rightleftharpoons} R_1{-}\overset{\overset{\Large O}{\|}}{C}{-}H + R_2{-}OH + R_3{-}OH$$

8. Hydrolysis of a ketal to yield a ketone and two alcohols (Section 15.9)

$$R_1{-}\underset{\underset{\Large R_2}{|}}{\overset{\overset{\Large OR_4}{|}}{C}}{-}OR_3 + H_2O \overset{H^+}{\rightleftharpoons} R_1{-}\overset{\overset{\Large O}{\|}}{C}{-}R_2 + R_3{-}OH + R_4{-}OH$$

Key Terms

Acetal (15.9)
Aldehyde (15.2)
Carbonyl group (15.1)
Hemiacetal (15.9)
Hemiketal (15.9)
Hydrolysis reaction (15.9)
Ketal (15.9)
Ketone (15.2)

Exercises and Problems

The members of each pair of problems in this section test similar material.

The Carbonyl Group (Section 15.1)

15.1 Indicate which of the following compounds contain a carbonyl group.

a. $CH_3-CH_2-CH_2-\overset{O}{\overset{\|}{C}}-H$

b. CH_3-O-CH_3

c. $CH_3-\overset{O}{\overset{\|}{C}}-CH_2-CH_3$

d. $O{=}\overset{CH_3}{\overset{|}{C}}-H$

e. $CH_3-O-CH_2-O-CH_3$

f. $CH_3-\underset{O-CH_3}{\underset{|}{CH}}-CH_3$

15.2 Indicate which of the following compounds contain a carbonyl group.

a. $CH_3-CH_2-\overset{O}{\overset{\|}{C}}-CH_2-CH_3$

b. $CH_3-CH_2-O-CH_2-CH_3$

c. $CH_3-CH_2-CH_2-O-H$

d. $CH_3-CH_2-\overset{O}{\overset{\|}{C}}-H$

e. $CH_3-O-CH_2-CH_2-O-H$

f. $CH_3-CH_2-\overset{CH_3}{\overset{|}{C}}{=}O$

15.3 What are the similarities and differences between the bonding in a carbon–oxygen double bond and that in a carbon–carbon double bond?

15.4 Use δ^+ and δ^- notation to show the polarity in a carbon–oxygen double bond.

Structure of Aldehydes and Ketones (Section 15.2)

15.5 Classify each of the following structures as an aldehyde, a ketone, or neither.

a. $CH_3-CH_2-CH_2-\overset{O}{\overset{\|}{C}}-OH$

b. $CH_3-CH_2-CH_2-CH_2-\overset{O}{\overset{\|}{C}}-H$

c. $CH_3-\overset{O}{\overset{\|}{C}}-CH_3$

d. $CH_3-O-CH_2-CH_3$

e. $H-\underset{O}{\underset{\|}{C}}-CH_2-CH_2-CH_3$

f. CH_3-CHO

15.6 Classify each of the following structures as an aldehyde, a ketone, or neither.

a. $CH_3-\overset{O}{\overset{\|}{C}}-CH_2-CH_3$

b. $CH_3-CH_2-CH_2-\overset{O}{\overset{\|}{C}}-O-CH_3$

c. $CH_3-CH_2-\underset{CH_3}{\underset{|}{C}}{=}O$

d. $CH_3-CH_2-\overset{O}{\overset{\|}{C}}-H$

e. $CH_3-CH_2-\underset{CH_3}{\underset{|}{CH}}-\overset{O}{\overset{\|}{C}}-H$

f. $CH_3-\overset{CH_3}{\underset{CH_3}{\overset{|}{\underset{|}{C}}}}-CH_2-CHO$

15.7 Draw the structures of the two simplest aldehydes and the two simplest ketones.

15.8 One- and two-carbon ketones do not exist. Explain why.

15.9 Classify each of the following structures as an aldehyde, a ketone, or neither.

a. (cyclohexane ring bearing OH and CH_3 on adjacent carbons)

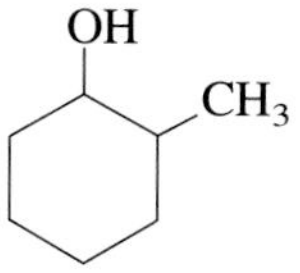

b. (benzene ring)$-\overset{O}{\overset{\|}{C}}-H$

c. (benzene ring)$-\overset{O}{\overset{\|}{C}}-O-CH_3$

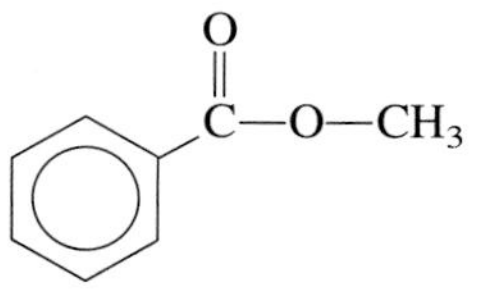

d. (cyclohexane ring)$=O$

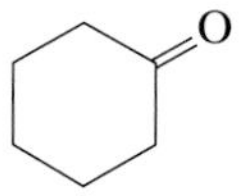

e. (benzene ring)$-\overset{O}{\overset{\|}{C}}-CH_3$

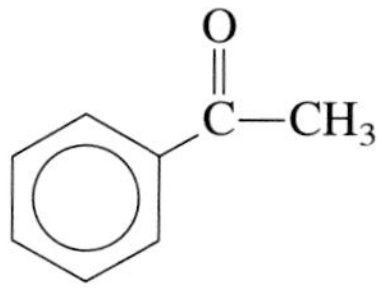

f. (cyclohexane ring)$-\overset{O}{\overset{\|}{C}}-H$, with adjacent ring carbon bearing CH_3

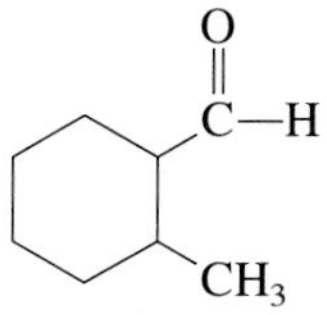

15.10 Classify each of the following structures as an aldehyde, a ketone, or neither.

a. (cyclohexane ring)$-\overset{O}{\overset{\|}{C}}-CH_2-CH_3$

b. (benzene ring)$-\overset{O}{\overset{\|}{C}}-OH$

c. (cyclohexane ring)$-\overset{O}{\overset{\|}{C}}-H$

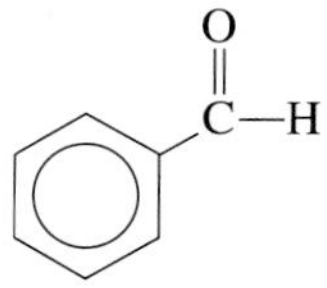

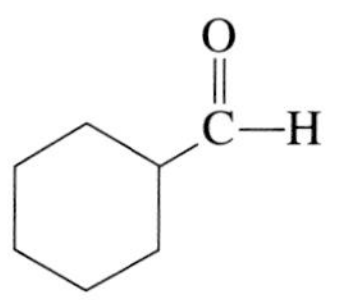

d. (cyclohexane ring with $=O$ at two opposite carbons)

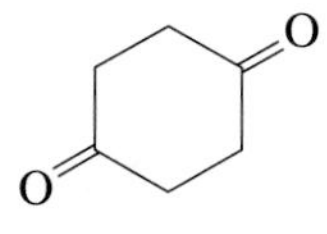

e. (benzene ring)$-\overset{O}{\overset{\|}{C}}-O-CH_3$

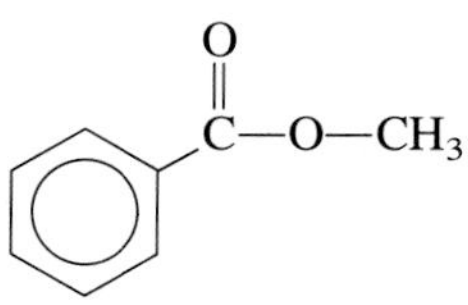

f. (benzene ring bearing two CH_3 groups)$-\overset{O}{\overset{\|}{C}}-CH_3$

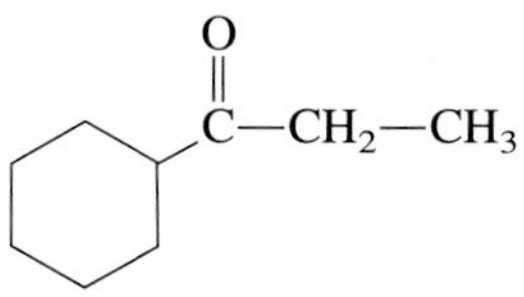

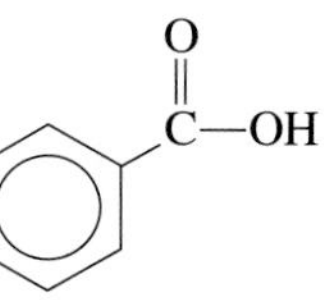

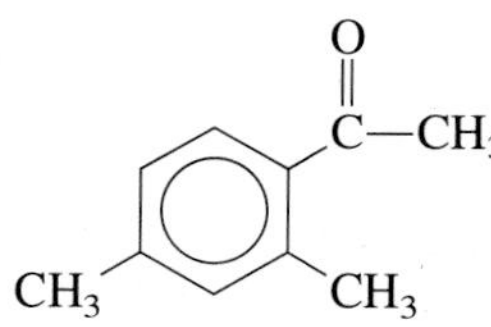

Nomenclature for Aldehydes (Section 15.3)

15.11 Assign an IUPAC name to each of the following aldehydes.

a. $CH_3-CH_2-CH_2-\overset{\overset{O}{\|}}{C}-H$

b. $CH_3-CH_2-\underset{\underset{CH_3}{|}}{CH}-\overset{\overset{O}{\|}}{C}-H$

c. $CH_3-\underset{\underset{CH_2-CH_2-CH_3}{|}}{CH}-CH_2-CH_2-\overset{\overset{O}{\|}}{C}-H$

d. $C_6H_5-CH_2-CH_2-\overset{\overset{O}{\|}}{C}-H$

e. CH_3-CH_2-CHO

f. $CH_3-\underset{\underset{CH_3}{|}}{\overset{\overset{CH_3}{|}}{C}}-CH_2-\overset{\overset{O}{\|}}{C}-H$

15.12 Assign an IUPAC name to each of the following aldehydes.

a. $CH_3-\underset{\underset{CH_3}{|}}{CH}-CH_2-CH_2-\overset{\overset{O}{\|}}{C}-H$

b. $CH_3-CH_2-\underset{\underset{CH_2-CH_3}{|}}{CH}-\overset{\overset{O}{\|}}{C}-H$

c. $\underset{\underset{CH_3}{|}}{CH_2}-CH_2-\underset{\underset{CH_2-CH_2-CH_3}{|}}{CH}-CH_2-\overset{\overset{O}{\|}}{C}-H$

d. $C_6H_5-CH_2-\overset{\overset{O}{\|}}{C}-H$

e. $CH_3-CH_2-\underset{\underset{CH_3}{|}}{\overset{\overset{CH_3}{|}}{C}}-CH_2-\overset{\overset{O}{\|}}{C}-H$

f. $CH_3-CH_2-CH_2-CHO$

15.13 Draw a structural formula for each of the following aldehydes.

a. 3-Methylpentanal
b. 2-Ethylhexanal
c. 3,4-Dimethylheptanal
d. 2,2-Dichloropropanal
e. 2,4,5-Trimethylheptanal
f. 4-Hydroxy-2-methyloctanal

15.14 Draw a structural formula for each of the following aldehydes.

a. 2-Methylpentanal
b. 4-Ethylhexanal
c. 3,3-Dimethylhexanal
d. 2,3-Dibromopropanal
e. 2-Bromo-4-methylhexanal
f. 2,4-Dichloroheptanal

15.15 Draw a structural formula for each of the following aldehydes.

a. Formaldehyde
b. Propionaldehyde
c. Chloroacetaldehyde
d. 2-Chlorobutyraldehyde
e. *o*-Methylbenzaldehyde
f. 2,4-Dimethylbenzaldehyde

15.16 Draw a structural formula for each of the following aldehydes.

a. Acetaldehyde
b. Butyraldehyde
c. 2-Chloropropionaldehyde
d. 2-Methylbutyraldehyde
e. Benzaldehyde
f. *p*-Bromobenzaldehyde

15.17 Assign a common name to each of the following aldehydes.

a. $CH_3-CH_2-\overset{\overset{O}{\|}}{C}-H$

b. $CH_3-\underset{\underset{CH_3}{|}}{CH}-\overset{\overset{O}{\|}}{C}-H$

c. $\underset{\underset{CH_3}{|}}{CH_2}-CH_2-\overset{\overset{O}{\|}}{C}-H$

d. $Cl-\underset{\underset{Cl}{|}}{CH}-\overset{\overset{O}{\|}}{C}-H$

e. $C_6H_4(Cl)-\overset{\overset{O}{\|}}{C}-H$

f. $C_6H_3(Cl)(HO)-\overset{\overset{O}{\|}}{C}-H$

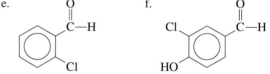

15.18 Assign a common name to each of the following aldehydes.

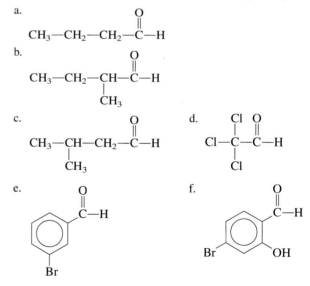

a. $CH_3-CH_2-CH_2-\overset{\overset{O}{\|}}{C}-H$

b. $CH_3-CH_2-\underset{\underset{CH_3}{|}}{CH}-\overset{\overset{O}{\|}}{C}-H$

c. $CH_3-\underset{\underset{CH_3}{|}}{CH}-CH_2-\overset{\overset{O}{\|}}{C}-H$

d. $Cl-\underset{\underset{Cl}{|}}{\overset{\overset{Cl}{|}}{C}}-\overset{\overset{O}{\|}}{C}-H$

e. $C_6H_4(Br)-\overset{\overset{O}{\|}}{C}-H$

f. $C_6H_3(Br)(OH)-\overset{\overset{O}{\|}}{C}-H$

Nomenclature for Ketones (Section 15.4)

15.19 Using IUPAC nomenclature, name each of the following ketones.

a. $CH_3-CH_2-\overset{\overset{O}{\|}}{C}-CH_3$

b. $CH_3-\underset{\underset{CH_3}{|}}{CH}-\underset{\underset{CH_3}{|}}{CH}-\overset{\overset{O}{\|}}{C}-\underset{\underset{CH_3}{|}}{CH}-CH_3$

c. 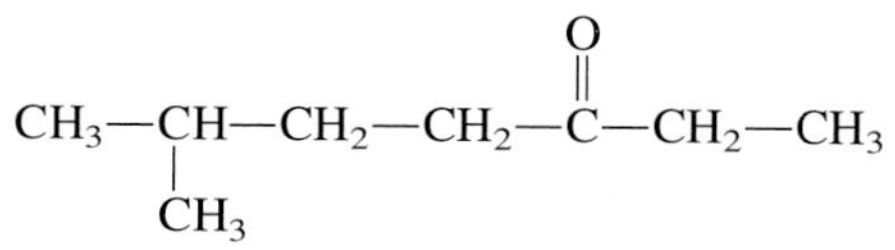 $CH_3-CH(CH_3)-CH_2-CH_2-C(=O)-CH_2-CH_3$

d. 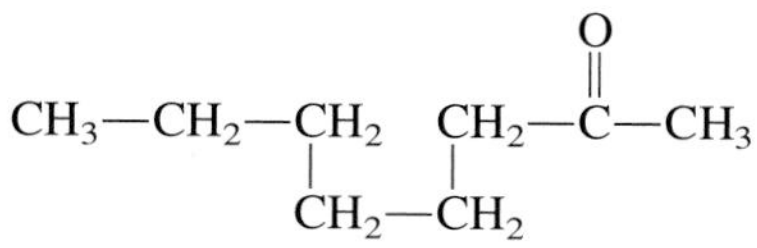 $CH_3-CH_2-CH_2-CH_2-CH_2-CH_2-C(=O)-CH_3$

e. 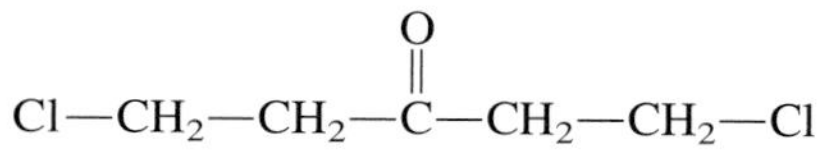 $Cl-CH_2-CH_2-C(=O)-CH_2-CH_2-Cl$

f. 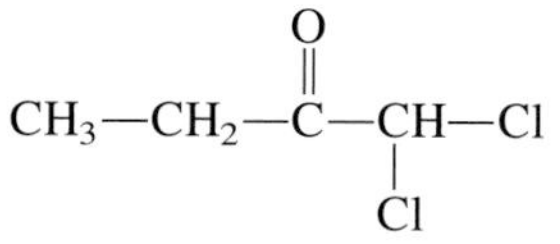 $CH_3-CH_2-C(=O)-CH(Cl)-Cl$

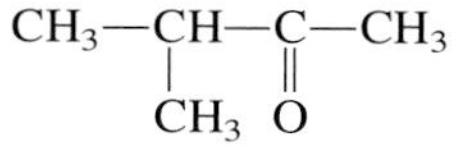

15.20 Using IUPAC nomenclature, name each of the following ketones.

a. 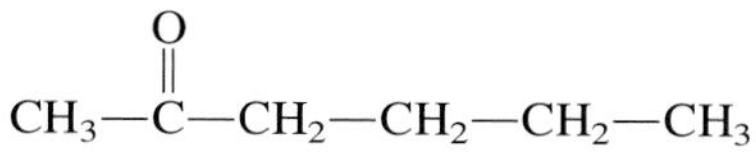 $CH_3-C(=O)-CH_2-CH_2-CH_2-CH_3$

b. $CH_3-CH(CH_3)-C(=O)-CH_3$

c. $CH_3-CH_2-C(=O)-CH_2-CH(CH_2-CH_3)-CH_3$

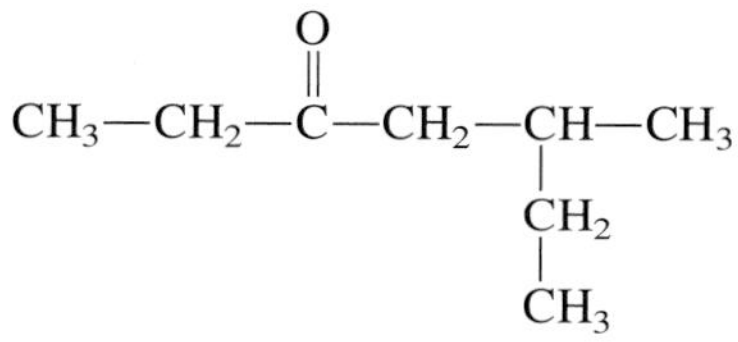

d. 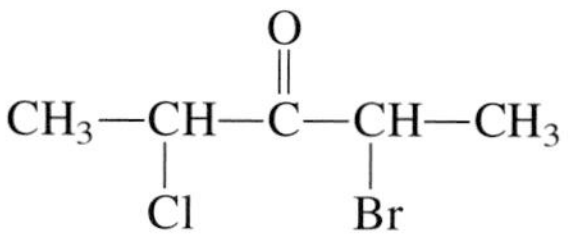 $CH_3-CH(Cl)-C(=O)-CH(Br)-CH_3$

e. 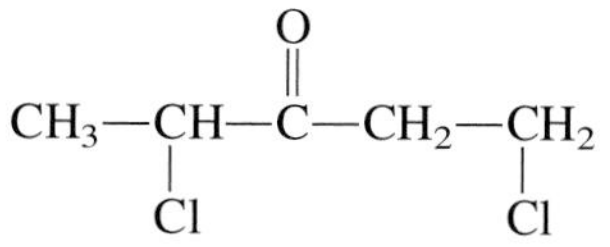 $CH_3-CH(Cl)-C(=O)-CH_2-CH_2-Cl$

f. 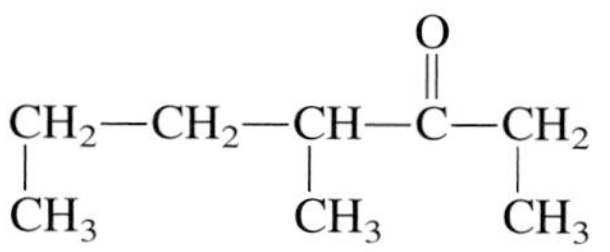 $CH_3-CH_2-CH_2-CH(CH_3)-C(=O)-CH_2-CH_3$

15.21 Using IUPAC nomenclature, name each of the following ketones.

a. 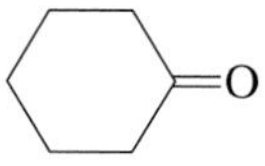 O

b. 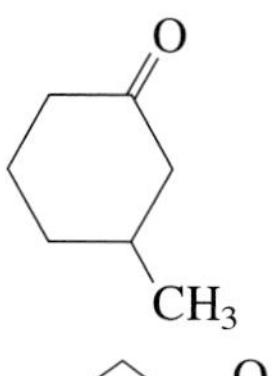 O, CH_3

c. 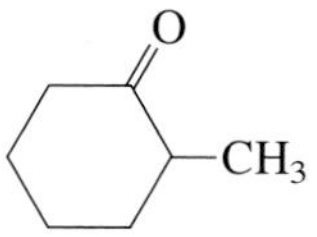 O, CH_3

d. 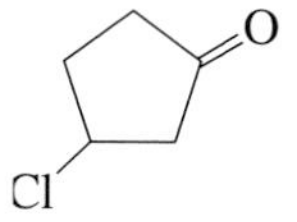 O, Cl

15.22 Using IUPAC nomenclature, name each of the following ketones.

a. 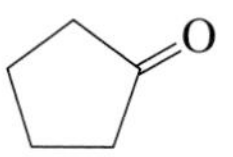 O

b. 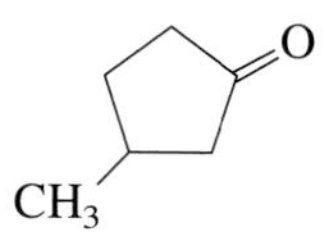 O, CH_3

c. 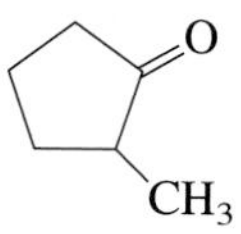 O, CH_3

d. 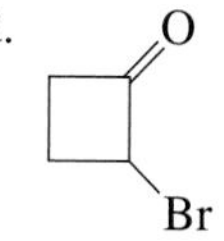 O, Br

15.23 Draw a structural formula for each of the following ketones.

a. 3-Methyl-2-pentanone
b. 3-Hexanone
c. Cyclobutanone
d. 2,4-Dimethyl-3-pentanone
e. Chloropropanone
f. 1,3-Dichloropropanone

15.24 Draw a structural formula for each of the following ketones.

a. 2-Methyl-3-pentanone
b. 2-Pentanone
c. 2,2-Dimethyl-4-octanone
d. Bromopropanone
e. 1,1-Dibromopropanone
f. Cyclopentanone

15.25 Draw a structural formula for each of the following ketones.

a. Diethyl ketone
b. Acetone
c. Isopropyl propyl ketone
d. Chloromethyl methyl ketone
e. Acetophenone
f. Methyl phenyl ketone

15.26 Draw a structural formula for each of the following ketones.

a. Dimethyl ketone
b. Phenyl propyl ketone
c. Methyl *tert*-butyl ketone
d. Dichloromethyl ethyl ketone
e. Benzophenone
f. Diphenyl ketone

Physical Properties of Aldehydes and Ketones (Section 15.6)

15.27 Aldehydes and ketones have higher boiling points than alkanes of similar molecular mass. Explain why.

15.28 Aldehydes and ketones have lower boiling points than alcohols of similar molecular mass. Explain why.

15.29 How many hydrogen bonds can form between an acetone molecule and water molecules?

15.30 How many hydrogen bonds can form between an acetaldehyde molecule and water molecules?

15.31 Would you expect ethanal or octanal to be more soluble in water? Explain your answer.

15.32 Would you expect ethanal or octanal to have the more fragrant odor? Explain your answer.

Preparation of Aldehydes and Ketones (Section 15.7)

15.33 Draw the structure of the aldehyde or ketone formed from oxidation of each of the following alcohols. Assume that reaction conditions are sufficiently mild that any aldehydes produced are not oxidized further to carboxylic acids.

a. $CH_3-CH_2-CH_2-CH_2-CH_2-OH$

b. 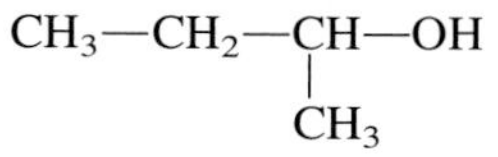 $CH_3-CH_2-CH(CH_3)-OH$

c. 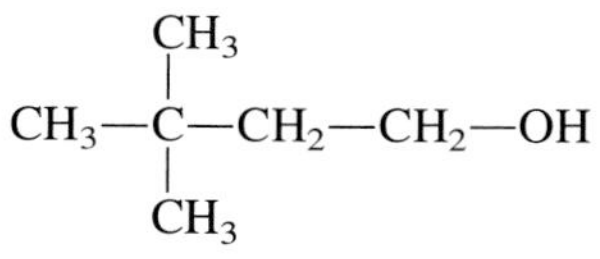 $CH_3-C(CH_3)_2-CH_2-CH_2-OH$

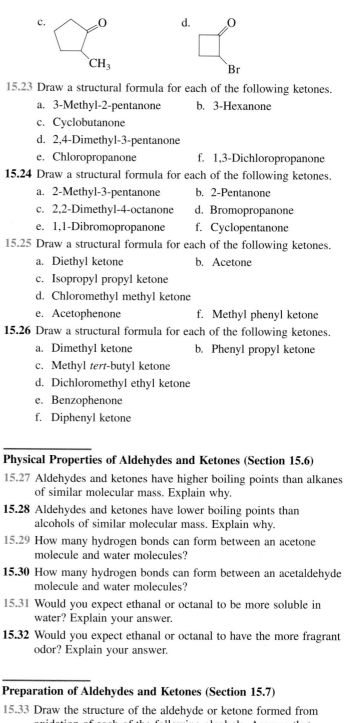

d. 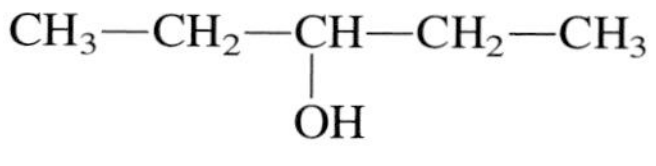 $CH_3-CH_2-CH(OH)-CH_2-CH_3$

e. (cyclopentane ring)—OH

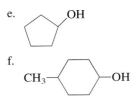

f. CH_3—(cyclohexane ring)—OH

15.34 Draw the structure of the aldehyde or ketone formed from oxidation of each of the following alcohols. Assume that reaction conditions are sufficiently mild that any aldehydes formed are not oxidized further to carboxylic acids.

a. $CH_3-CH_2-CH-CH_2-OH$
  (with CH_3 on the CH)

b. $CH_3-CH-CH-OH$
  (with CH_3 CH_3 on the two CH)

c. CH_3 above, CH_3-C-OH, CH_3 below

d. $CH_3-CH_2-CH_2-CH-CH_3$
  (with OH on the CH)

e. (cyclohexane ring)—OH

f. (cyclohexane ring) with CH_2-CH_3 and —OH

15.35 Draw the structure of the alcohol needed to prepare each of the following aldehydes or ketones by alcohol oxidation.

a. Ethanal b. Diethyl ketone
c. Phenylpropanone d. Acetaldehyde
e. Acetone f. 2-Ethylhexanal

15.36 Draw the structure of the alcohol needed to prepare each of the following aldehydes or ketones by alcohol oxidation.

a. Propanal b. Dipropyl ketone
c. 3-Phenyl-2-butanone d. Chloroacetone
e. Formaldehyde f. Cyclohexanone

Oxidation and Reduction of Aldehydes and Ketones (Section 15.8)

15.37 Draw the structural formula of the organic product when each of the following aldehydes is oxidized to a carboxylic acid.

a. Ethanal b. Pentanal
c. Formaldehyde d. 3,4-Dichlorohexanal

15.38 Draw the structural formula of the organic product when each of the following aldehydes is oxidized to a carboxylic acid.

a. Butanal b. 2-Methylpentanal
c. Acetaldehyde d. Benzaldehyde

15.39 What are the characteristics of a positive Tollens test for aldehydes?

15.40 What are the characteristics of a positive Benedict's test for aldehydes?

15.41 What is the oxidizing agent in Benedict's solution?

15.42 What is the oxidizing agent in Tollens solution?

15.43 Which of the following compounds would react with Tollens solution?

a. $CH_3-CH_2-CH_2-C(=O)-CH_3$

b. $CH_3-CH_2-CH_2-C(=O)-H$

c. $CH_3-CH-CH_2-C(=O)-H$
  (with OH on the CH)

d. $CH_3-CH-C(=O)-CH_3$
  (with OH on the CH)

15.44 Which of the following compounds would react with Benedict's solution?

a. $CH_3-CH_2-C(=O)-H$

b. $CH_3-C(=O)-CH_3$

c. $CH_3-CH_2-CH-C(=O)-CH_2-CH_3$
  (with OH on the CH)

d. $CH_3-CH-CH-CH_2-C(=O)-H$
  (with CH_3 CH_3 on the two CH)

15.45 Draw the structure of the major organic compound produced when each of the following compounds is reduced using molecular H2 and a Ni catalyst.

a. $CH_3-CH_2-CH_2-C(=O)-H$

b. $CH_3-CH_2-C(=O)-CH_2-CH_3$

c. $CH_3-CH-CH_2-C(=O)-H$
  (with CH_3 on the CH)

d. $CH_3-CH-C(=O)-CH_2-CH_2-CH_3$
  (with CH_3 on the CH)

15.46 Draw the structure of the major organic compound produced when each of the following compounds is reduced using molecular H_2 and a Ni catalyst.

a. $CH_3-CH_2-CH_2-C(=O)-CH_3$

b. $CH_3-CH_2-CH_2-CH_2-C(=O)-H$

c. CH_3 above, $CH_3-C-CH_2-CH_2-C(=O)-H$, CH_3 below

d. $CH_3-CH_2-C(=O)-CH_3$

Hemiacetal and Hemiketal Formation (Section 15.9)

15.47 When an alcohol molecule (R—O—H) adds across a carbon–oxygen double bond, into what "fragments" is the alcohol split?

15.48 When an alcohol molecule (R—O—H) adds across a carbon–oxygen double bond, which part of the alcohol molecule adds to the carbonyl oxygen atom?

15.49 Identify each of the following compounds as a hemiacetal, a hemiketal, or neither a hemiacetal nor a hemiketal.

a. $CH_3-CH_2-O-CH_3$

b. CH_3-O
   |
   $CH_3-CH-OH$

c. $CH_3-O-CH_2-CH_2-OH$

d.
   $O-H$
   |
   CH_3-C-CH_3
   |
   $O-CH_3$

e. (five-membered ring containing O, with OH on the carbon adjacent to the ring O)

f. (five-membered ring containing O, with $O-CH_3$ on the carbon adjacent to the ring O)

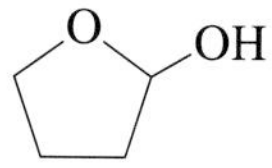

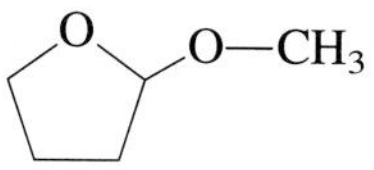

15.50 Identify each of the following compounds as a hemiacetal, a hemiketal, or neither a hemiacetal nor a hemiketal.

a.
   OH
   |
   $CH_3-CH_2-C-CH_3$
   |
   OH

b.
   OH
   |
   $CH_3-CH_2-C-CH_3$
   |
   $O-CH_3$

c.
   $O-CH_3$
   |
   $CH_3-CH_2-C-CH_3$
   |
   $O-CH_3$

d. $CH_3-CH_2-O-CH_3$

e. 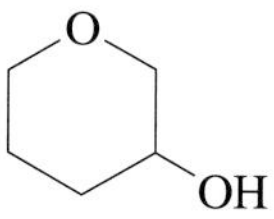 (six-membered ring containing O, with OH on a ring carbon not adjacent to the ring O)

f. 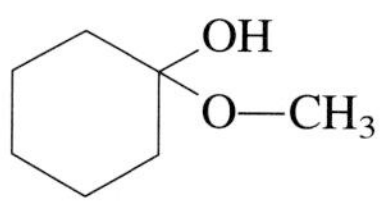 (cyclohexane ring with OH and $O-CH_3$ on the same carbon)

15.51 Draw the structural formula of the hemiacetal or hemiketal formed from each of the following pairs of reactants.

a. Acetaldehyde and ethyl alcohol
b. 2-Pentanone and methanol
c. Butanal and ethanol
d. Acetone and isopropyl alcohol

15.52 Draw the structural formula of the hemiacetal or hemiketal formed from each of the following pairs of reactants.

a. Acetaldehyde and methanol
b. 2-Pentanone and ethyl alcohol
c. Butanal and isopropyl alcohol
d. Acetone and ethanol

15.53 Draw the structural formula of the missing compound in each of the following reactions.

a.
   O
   ‖
   $CH_3-(CH_2)_2-C-H + CH_3-CH_2-OH \overset{H^+}{\rightleftharpoons} ?$

b.
   OH
   |
   $? + CH_3-OH \overset{H^+}{\rightleftharpoons} CH_3-CH_2-CH$
   |
   $O-CH_3$

c.
   O
   ‖
   $CH_3-CH_2-C-CH_3 + CH_3-OH \overset{H^+}{\rightleftharpoons} ?$

d. 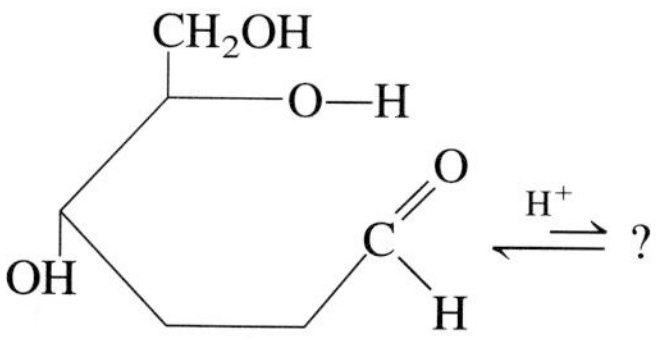 (open chain: CH_2OH carbon bearing O—H, chain ending with OH and $C(=O)H$) $\overset{H^+}{\rightleftharpoons} ?$

15.54 Draw the structural formula of the missing compound in each of the following reactions.

a.
   O
   ‖
   $CH_3-CH_2-C-H + CH_3-OH \overset{H^+}{\rightleftharpoons} ?$

b.
   OH
   |
   $? + CH_3-CH_2-OH \overset{H^+}{\rightleftharpoons} CH_3-CH_2-CH$
   |
   $O-CH_2-CH_3$

c.
   O
   ‖
   $CH_3-CH_2-CH_2-C-CH_3 + CH_3-CH_2-OH \overset{H^+}{\rightleftharpoons} ?$

d. 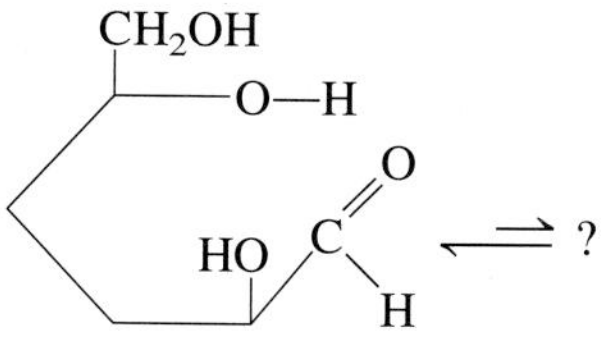 (open chain: CH_2OH carbon bearing O—H, chain ending with HO and $C(=O)H$) $\rightleftharpoons ?$

Acetal and Ketal Formation (Section 15.9)

15.55 Identify each of the following compounds as an acetal, a ketal, or neither an acetal nor a ketal.

a.
   $O-CH_3$
   |
   $CH_3-CH_2-CH_2-CH$
   |
   $O-CH_3$

b.
   $O-CH_3$
   |
   $CH_3-CH_2-CH_2-O-C-CH_3$
   |
   CH_3

c.
   CH_3
   |
   $CH_3-CH_2-CH_2-O-C-OH$
   |
   CH_3

d.
   $O-CH_2-CH_2-CH_3$
   |
   $CH_3-C-CH_2-CH_2-CH_3$
   |
   $O-CH_2-CH_2-CH_3$

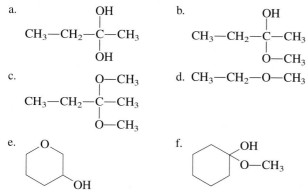

15.56 Identify each of the following compounds as an acetal, a ketal, or neither an acetal nor a ketal.

a.
   $O-CH_2-CH_3$
   |
   CH_3-CH_2-CH
   |
   $O-CH_3$

b.
   $O-CH_3$
   |
   $CH_3-CH_2-O-C-CH_3$
   |
   CH_3

c. 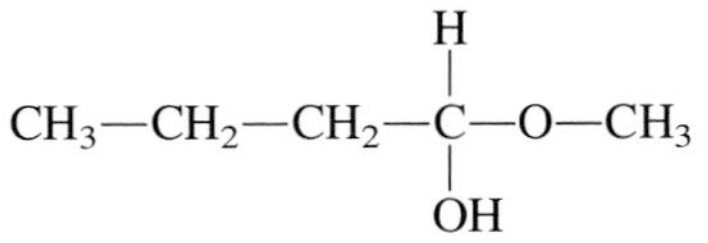
```
                      H
                      |
CH3—CH2—CH2—C—O—CH3
                      |
                      OH
```

d. 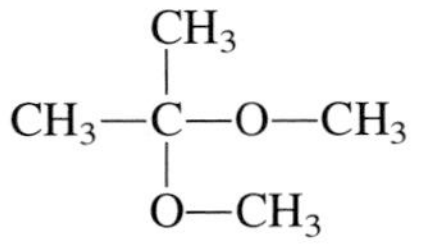
```
       CH3
       |
CH3—C—O—CH3
       |
       O—CH3
```

15.57 Draw the structural formula of the missing compound(s) in each of the following reactions.

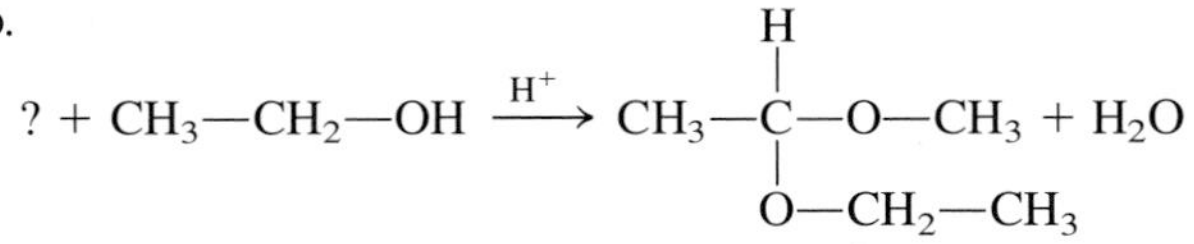

a. 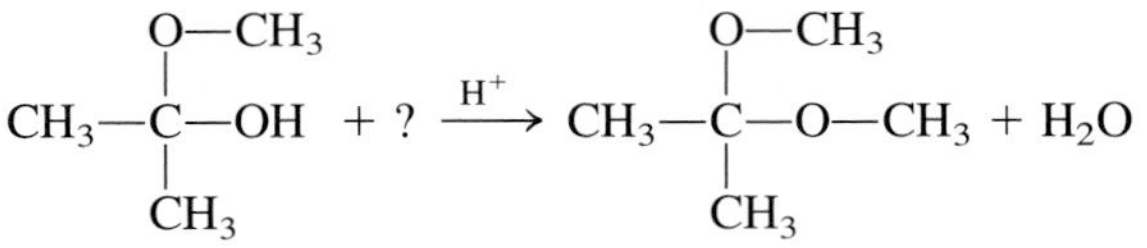
```
       O—CH3                       O—CH3
       |                 H+         |
CH3—C—OH  + ?  ——→  CH3—C—O—CH3 + H2O
       |                            |
       CH3                          CH3
```

b.
```
                                 H
                      H+         |
? + CH3—CH2—OH  ——→  CH3—C—O—CH3 + H2O
                                 |
                                 O—CH2—CH3
```

c.
```
             OH
             |                               H+
CH3—CH2—C—O—CH3 + CH3—CH—OH  ——→
             |                    |
             H                    CH3
```

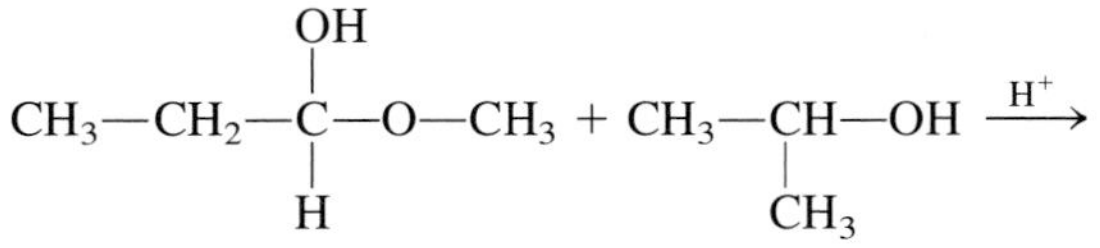

d. 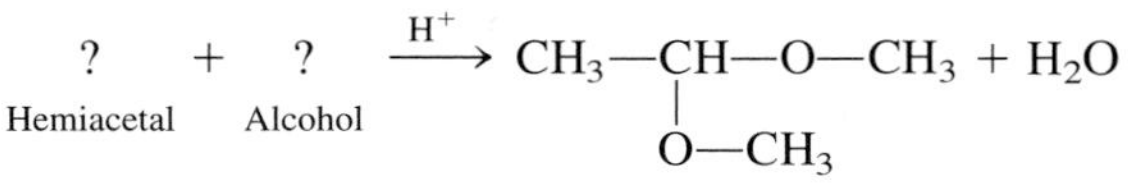
```
                          H+
   ?        +    ?       ——→  CH3—CH—O—CH3 + H2O
Hemiacetal   Alcohol                 |
                                     O—CH3
```

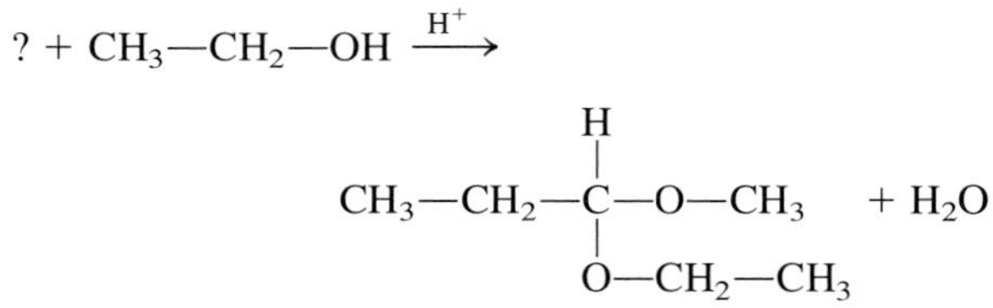

15.58 Draw the structural formula of the missing compound(s) in each of the following reactions.

a. 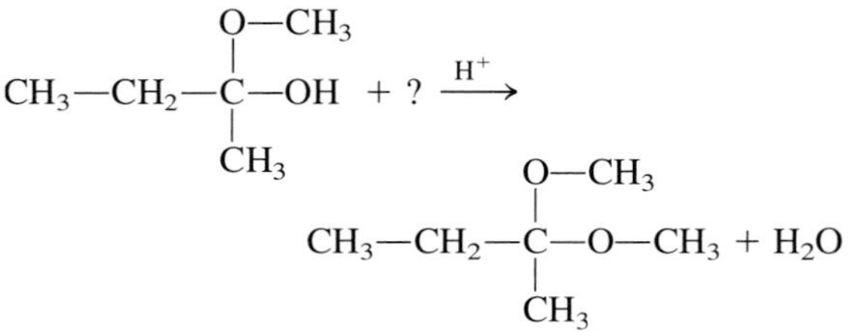
```
             O—CH3
             |              H+
CH3—CH2—C—OH  + ?  ——→
             |
             CH3                 O—CH3
                                 |
                       CH3—CH2—C—O—CH3 + H2O
                                 |
                                 CH3
```

b.
```
                      H+
? + CH3—CH2—OH  ——→

                                 H
                                 |
                       CH3—CH2—C—O—CH3   + H2O
                                 |
                                 O—CH2—CH3
```

c.
```
             OH
             |                               H+
CH3—CH2—C—O—CH3 + CH3—CH—OH  ——→
             |                    |
             H                    CH3
                                              ? + H2O
```

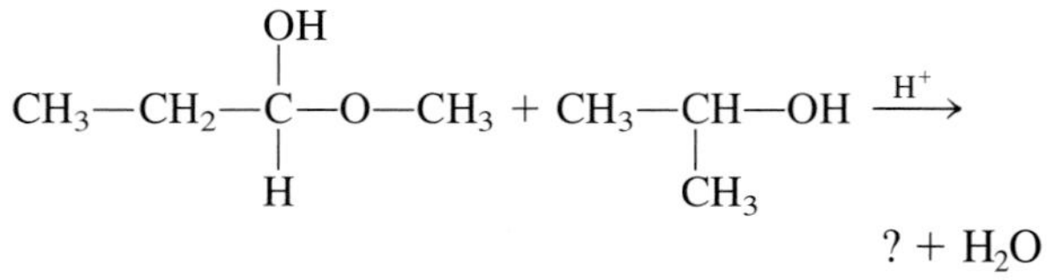

d. 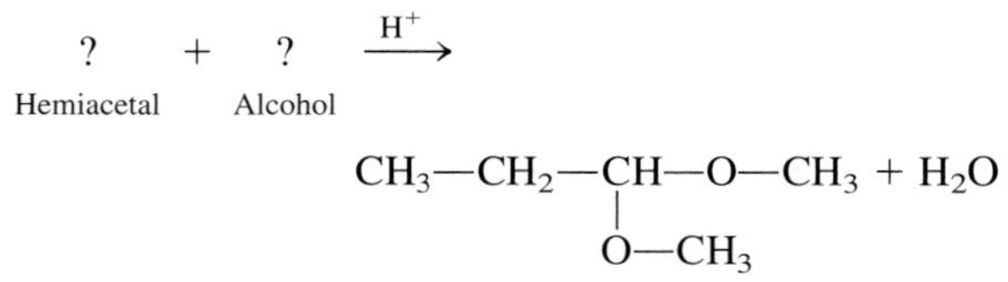
```
                          H+
   ?        +    ?       ——→
Hemiacetal   Alcohol

                       CH3—CH2—CH—O—CH3 + H2O
                                |
                                O—CH3
```

15.59 Draw the structural formulas of the aldehyde (or ketone) and the two alcohols produced when the following compounds undergo hydrolysis in acidic solution.

a. 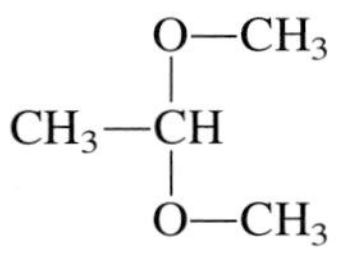
```
        O—CH3
        |
CH3—CH
        |
        O—CH3
```

b. 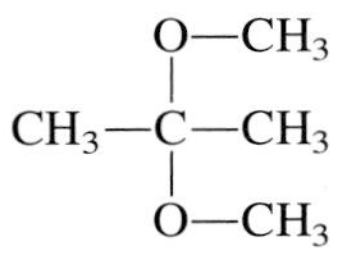
```
       O—CH3
       |
CH3—C—CH3
       |
       O—CH3
```

c.
```
             CH2—CH3
             |
CH3—O—C—O—CH2—CH3
             |
             CH2—CH3
```

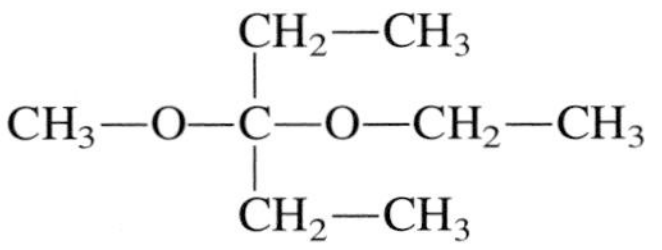

d. 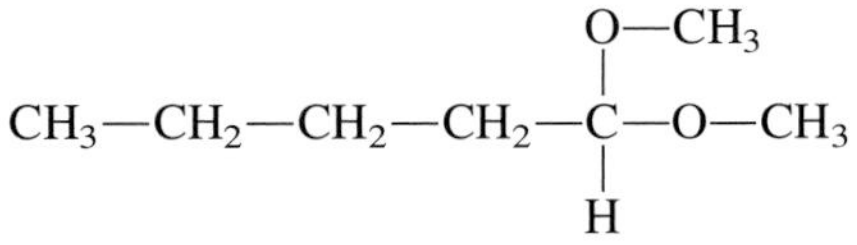
```
                                O—CH3
                                |
CH3—CH2—CH2—CH2—C—O—CH3
                                |
                                H
```

15.60 Draw the structural formulas of the aldehyde (or ketone) and the two alcohols produced when the following compounds undergo hydrolysis in acidic solution.

a. 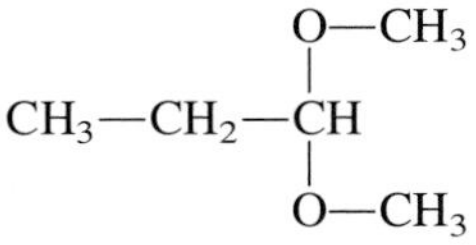
```
                 O—CH3
                 |
CH3—CH2—CH
                 |
                 O—CH3
```

b. 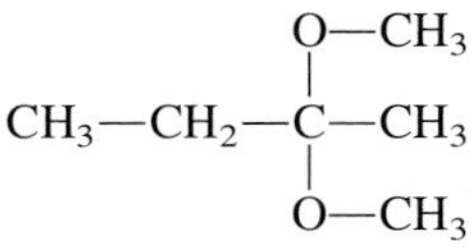
```
                 O—CH3
                 |
CH3—CH2—C—CH3
                 |
                 O—CH3
```

c.
```
                      H
                      |
CH3—CH2—O—C—O—CH2—CH3
                      |
                      CH3
```

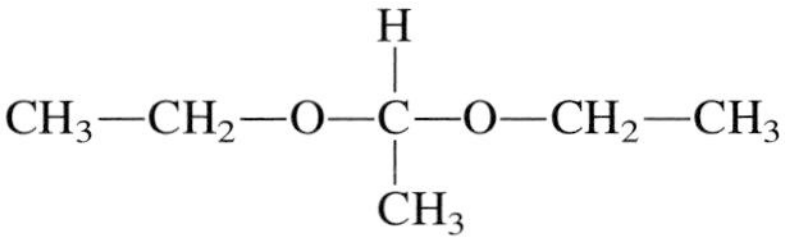

d. 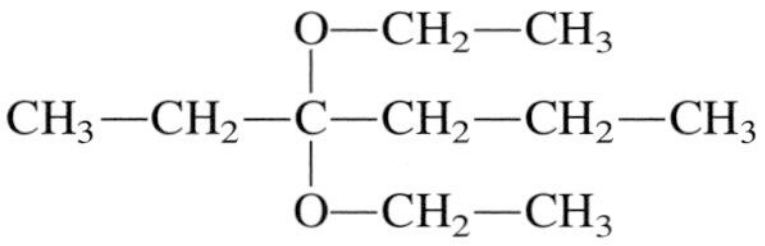
```
                 O—CH2—CH3
                 |
CH3—CH2—C—CH2—CH2—CH3
                 |
                 O—CH2—CH3
```

Additional Problems

15.61 The IUPAC name for the three-carbon aldehyde is propanal rather than 1-propanal. Explain why.

15.62 Each of the following compound names represents an impossible structure. In each case, explain why.

a. Methanone
b. 1-Chlorobutanal
c. 3-Methyl-3-pentanone
d. Cyclohexanal

15.63 Give IUPAC names for all possible saturated unbranched-chain heptanals.

15.64 How many ketone structures can be drawn that are isomeric with the aldehyde pentanal?

15.65 Draw structural formulas and assign IUPAC names to all aldehydes and ketones with the molecular formula $C_6H_{12}O$. There are eight aldehyde isomers and six ketone isomers.

15.66 The compound 4-hydroxybutanal can form an intramolecular cyclic hemiacetal. Draw the structural formula of this cyclic hemiacetal.

15.67 Name the functional groups present in each of the following polyfunctional compounds.

a. 4-Octen-2-one
b. 2-Methoxy-4-hydroxypentanal
c. 3-Hexyn-2-one
d. 4-Oxohexanal

15.68 Assign an IUPAC name to each of the following compounds.

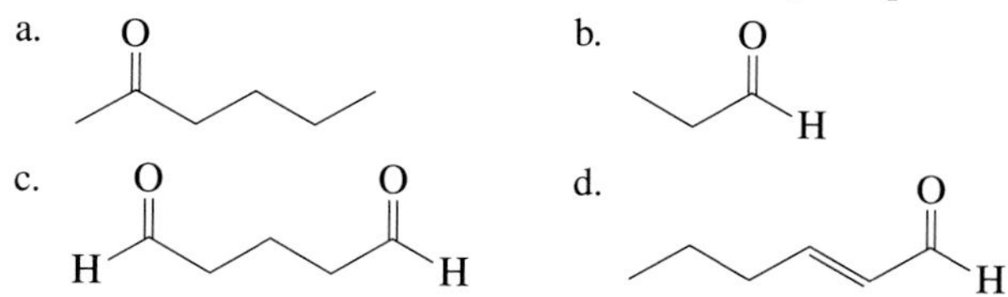

Grid Problems

15.69

1. $\overset{CH_3}{\overset{\vert}{CH_2}}-CH_2-\overset{O}{\overset{\Vert}{C}}-H$	2. $CH_3-\overset{O}{\overset{\Vert}{C}}-CH_3$	3. $CH_3-\overset{CH_3}{\overset{\vert}{CH}}-\overset{O}{\overset{\Vert}{C}}-H$
4. $CH_3-\overset{O}{\overset{\Vert}{C}}-\overset{CH_3}{\overset{\vert}{CH_2}}$	5. $CH_3-\overset{O}{\overset{\Vert}{C}}-CH_2-CHO$	6. $\overset{OH}{\overset{\vert}{CH_2}}-CH_2-\overset{O}{\overset{\Vert}{C}}-CH_3$

Select from the grid *all* correct responses for each of the following situations.

a. IUPAC name ends in *-one.*

b. IUPAC name ends in *-al.*

c. IUPAC name does not contain any numbers.

d. IUPAC name requires the use of two or more numbers.

15.70

1. acetone	2. acetaldehyde	3. formaldehyde
4. acetophenone	5. benzaldehyde	6. benzophenone

Select from the grid *all* correct responses for each of the following situations.

a. Compounds that contain three or fewer carbon atoms

b. Compounds that contain a carbonyl group

c. Compounds with two identical groups attached to a carbonyl carbon atom

d. Compounds in which one or more aromatic carbon ring systems are present.

15.71

1. aldehyde	2. hemiacetal	3. acetal
4. ketone	5. hemiketal	6. ketal

Select from the grid *all* correct responses for each of the following situations.

a. Compounds that contain a carbonyl group

b. Compounds that contain two oxygen atoms

c. Compounds that contain two alkoxy groups

d. Compounds that contain three carbon–oxygen single bonds

15.72

1. aldehyde	2. hemiacetal	3. acetal
4. ketone	5. hemiketal	6. ketal

Select from the grid *all* correct responses for each of the following situations.

a. Compound undergoes acidic hydrolysis to produce two alcohol molecules and an additional type of organic compound

b. Compound undergoes reduction with H_2 gas, with a Ni catalyst, to produce an alcohol

c. Compound gives a positive Tollens test

d. Compound reacts with an alcohol to give a new type of organic compound and a water molecule

PART II

Carboxylic Acids and Esters

CHAPTER OUTLINE

16.1 Structure of Carboxylic Acids 436
16.2 IUPAC Nomenclature for Carboxylic Acids 437
16.3 Common Names for Carboxylic Acids 440
16.4 Polyfunctional Carboxylic Acids 443
16.5 Physical Properties of Carboxylic Acids 445
16.6 Preparation of Carboxylic Acids 446
16.7 Acidity of Carboxylic Acids 447
16.8 Carboxylic Acid Salts 448
16.9 Structure of Esters 450
16.10 Nomenclature for Esters 451
16.11 Selected Common Esters 453
16.12 Physical Properties of Esters 455
16.13 Chemical Reactions of Esters 456
16.14 Sulfur Analogs of Esters 458

Chemistry at a Glance:
Summary of Reactions Involving Carboxylic Acids and Esters 459

16.15 Polyesters 459
16.16 Esters of Inorganic Acids 461

Chemical Connections
16.1 Nonprescription Pain Relievers Derived from Propanoic Acid 442
16.2 Aspirin 455
16.3 Nitroglycerin: An Inorganic Triester 462

Red ants in a defensive position. The irritant in the "bite" of a red ant is methanoic acid (formic acid), the simplest carboxylic acid.

In Chapter 15, we discussed the carbonyl group and two families of compounds—aldehydes and ketones—that contain this group. In this chapter, we discuss two more families of compounds in which the carbonyl group is present: carboxylic acids and esters. In these compounds, however, another oxygen-containing entity is present in addition to the carbonyl group.

16.1 Structure of Carboxylic Acids

A **carboxylic acid** *is a compound whose characteristic functional group is the carboxyl group.* Because of their wide distribution and abundance in natural products, these compounds were some of the first organic compounds studied in detail. A **carboxyl group** *is a carbonyl group (C=O) with a hydroxyl group (—OH) bonded to the carbonyl carbon atom.* A general structural representation for a carboxyl group is

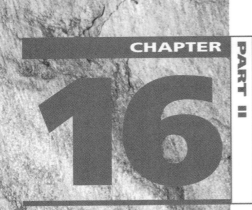

$$—\overset{\overset{\large O}{\|}}{C}—OH$$

• The term *carboxyl* is a contraction of the words *carb*onyl and hydr*oxyl*.

Abbreviated linear designations for the carboxyl group are

—COOH and —CO_2H

Although we see within a carboxyl group both a carbonyl group (C═O) and a hydroxyl group (—OH), the carboxyl group does not show characteristic behavior of either an alcohol or a carbonyl compound (aldehyde or ketone). Rather, it is a unique functional group with a set of characteristics different from those of its component parts.

The double bond present in the C═O portion of a carboxyl group has polarity associated with it and consists of a σ-bond and a π-bond, as was the case with aldehyde and ketone carbonyl groups (Section 15.1).

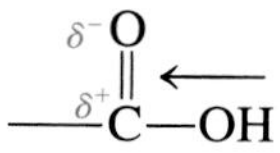

δ^-O, δ^+C—OH ← Polar carbon–oxygen bond

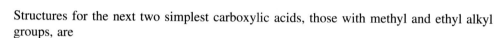

The simplest carboxylic acid has a hydrogen atom attached to the carboxyl group carbon atom.

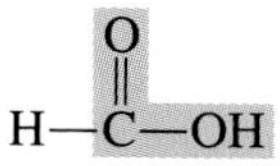

H—C(═O)—OH

Structures for the next two simplest carboxylic acids, those with methyl and ethyl alkyl groups, are

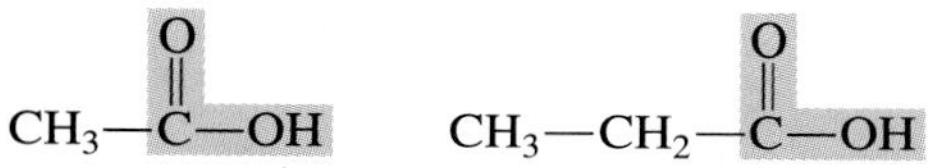

CH_3—C(═O)—OH CH_3—CH_2—C(═O)—OH

• General formulas for carboxylic acids containing alkyl and aryl groups, respectively, are
R—COOH and Ar—COOH

The structure of the simplest aromatic carboxylic acid involves a benzene ring to which a carboxyl group is attached.

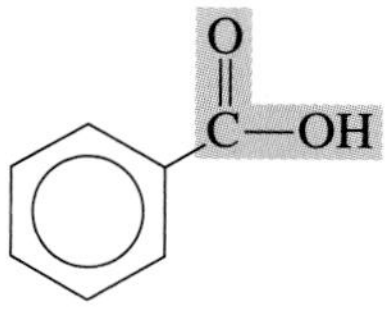

C_6H_5—C(═O)—OH

Cyclic carboxylic acids do not exist; having the carboxyl carbon atom as part of a ring system creates a situation where the carboxyl carbon atom would have five bonds. The nonexistence of cyclic carboxylic acids parallels the nonexistence of cyclic aldehydes (Section 15.2).

16.2 IUPAC Nomenclature for Carboxylic Acids

IUPAC rules for naming carboxylic acids resemble those for naming aldehydes (Section 15.3).

1. Select as the parent carbon chain the longest carbon chain that *includes* the carbon atom of the carboxyl group.
2. Name the parent chain by changing the *-e* ending of the corresponding alkane to *-oic acid.*
3. Number the parent chain by assigning the number 1 to the carboxyl carbon atom.
4. Determine the identity and location of any substituents in the usual manner, and append this information to the front of the parent chain name.

• A carboxyl group must occupy a terminal (end) position in a carbon chain because there can be only one other bond to it.

• Line-angle drawings for the simpler unbranched-chain carboxylic acids:

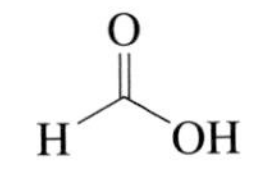

Methanoic acid

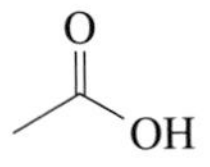

Ethanoic acid

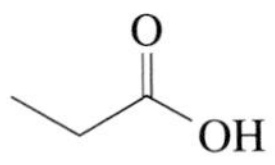

Propanoic acid

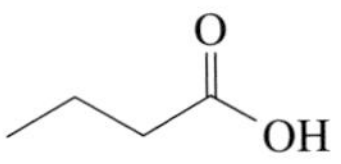

Butanoic acid

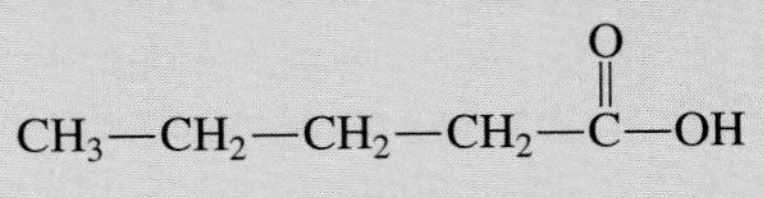

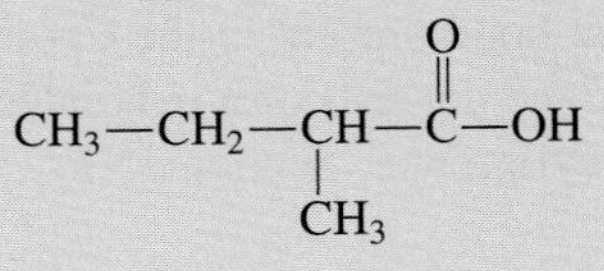

Example 16.1

Determining IUPAC Names for Carboxylic Acids

Assign IUPAC names to the following carboxylic acids.

a. $CH_3—CH_2—CH_2—CH_2—\overset{O}{\overset{\|}{C}}—OH$

b. $CH_3—CH_2—\underset{CH_3}{\underset{|}{CH}}—\overset{O}{\overset{\|}{C}}—OH$

c. $CH_3—\underset{Br}{\underset{|}{CH}}—\underset{CH_2—CH_3}{\underset{|}{CH}}—\overset{O}{\overset{\|}{C}}—OH$

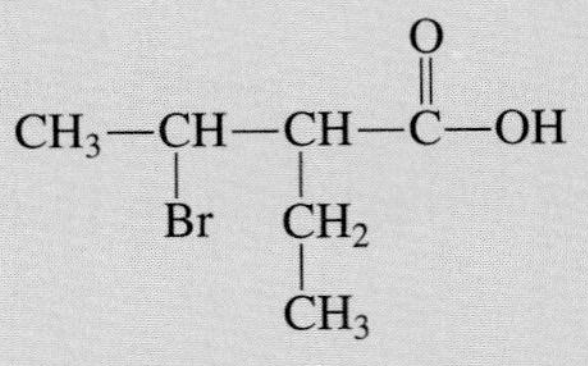

Solution

a. The parent chain name is based on pentane. Removing the *-e* ending from pentane and replacing it with the ending *-oic acid* gives *pentanoic acid.* The location of the carboxyl group need not be specified, because by definition the carboxyl carbon atom is always carbon 1.

b. The parent chain name is *butanoic acid.* To locate the methyl group substituent, we number the carbon chain beginning with the carboxyl carbon atom. The complete name of the acid is *2-methylbutanoic acid.*

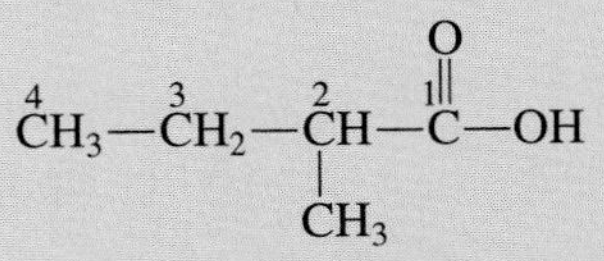

$$\overset{4}{C}H_3—\overset{3}{C}H_2—\underset{CH_3}{\underset{|}{\overset{2}{C}H}}—\overset{O}{\overset{\|}{\overset{1}{C}}}—OH$$

c. The longest carboxyl-carbon-containing chain has four carbon atoms. The parent chain name is thus *butanoic acid.* There are two substituents present, an ethyl group on carbon 2 and a bromo group on carbon 3. The complete name is *3-bromo-2-ethylbutanoic acid.*

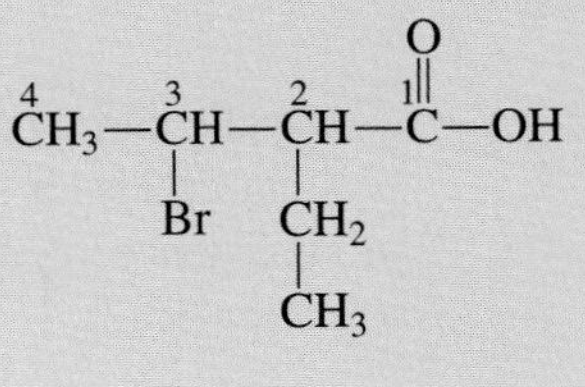

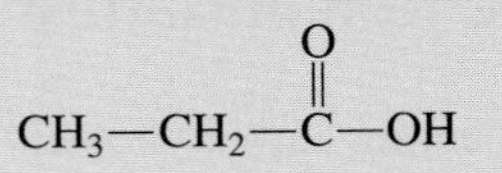

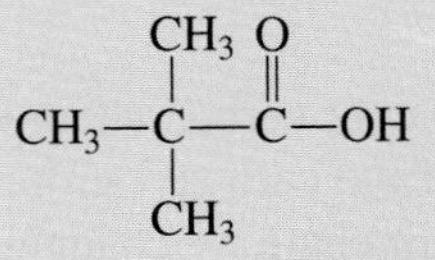

$$\overset{4}{C}H_3—\underset{Br}{\underset{|}{\overset{3}{C}H}}—\underset{CH_2—CH_3}{\underset{|}{\overset{2}{C}H}}—\overset{O}{\overset{\|}{\overset{1}{C}}}—OH$$

Practice Exercise 16.1

Assign IUPAC names to the following carboxylic acids.

a. $CH_3—CH_2—\overset{O}{\overset{\|}{C}}—OH$

b. $CH_3—\underset{CH_3}{\underset{|}{\overset{CH_3}{\overset{|}{C}}}}—\overset{O}{\overset{\|}{C}}—OH$

c. $CH_3—CH_2—\underset{CH_3—CH_2—CH_2}{\underset{|}{CH}}—\overset{O}{\overset{\|}{C}}—OH$

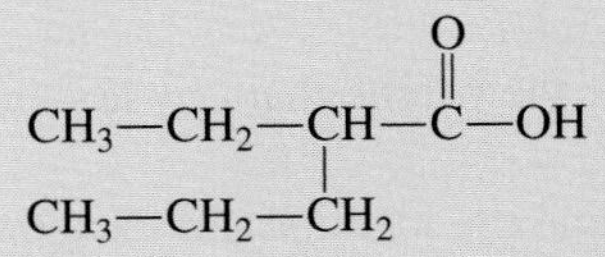

• *Answers:* **a.** propanoic acid **b.** 2,2-dimethylpropanoic acid **c.** 2-ethylpentanoic acid

• The carboxyl functional group has the highest priority in the IUPAC naming system of all functional groups considered so far. When both a carboxyl group and a carbonyl group (aldehyde, ketone) are present in the same molecule, the prefix *oxo-* is used to denote the carbonyl group.

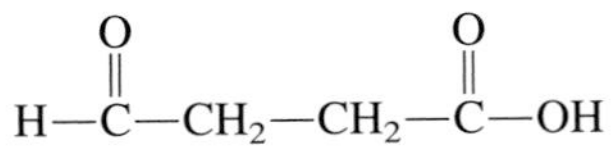

$$H—\overset{O}{\overset{\|}{C}}—CH_2—CH_2—\overset{O}{\overset{\|}{C}}—OH$$

4-Oxobutanoic acid

Space-filling models for the three simplest carboxylic acids: methanoic acid, ethanoic acid, and propanoic acid.

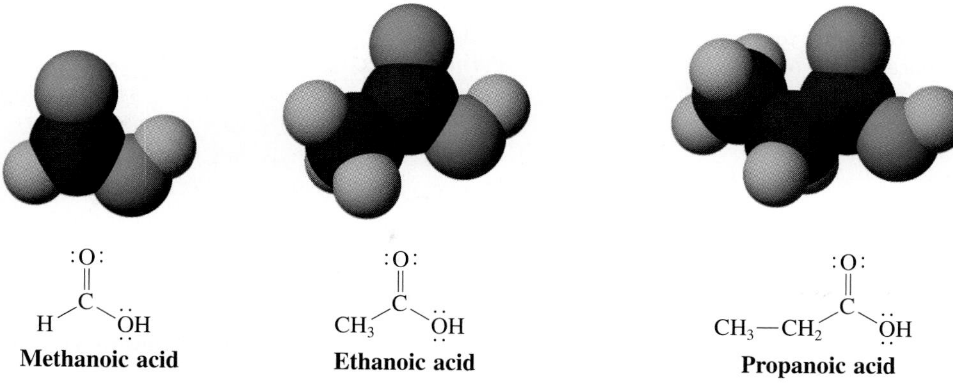

• Dicarboxylic Acids

A **Dicarboxylic acid** *is a compound that contains two carboxyl groups, one at each end of a carbon chain.* Saturated acids of this type are named by appending the suffix *-dioic acid* to the corresponding alkane name (the *-e* is retained to facilitate pronunciation). Both carboxyl carbon atoms must be part of the parent carbon chain, and the carboxyl locations need not be specified with numbers because they will always be at the two ends of the chain.

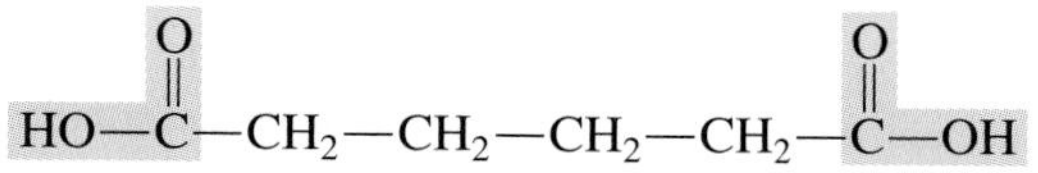

$$HO-\overset{\overset{\displaystyle O}{\|}}{C}-CH_2-CH_2-CH_2-CH_2-\overset{\overset{\displaystyle O}{\|}}{C}-OH$$

Hexanedioic acid

$$HO-\overset{\overset{\displaystyle O}{\|}}{C}-\underset{\underset{\displaystyle CH_3}{|}}{CH}-CH_2-\overset{\overset{\displaystyle O}{\|}}{C}-OH$$

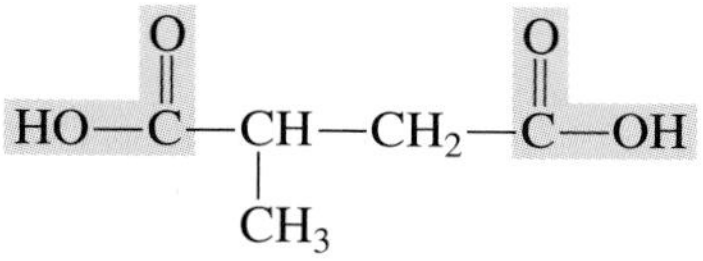

2-Methylbutanedioic acid

• Aromatic Carboxylic Acids

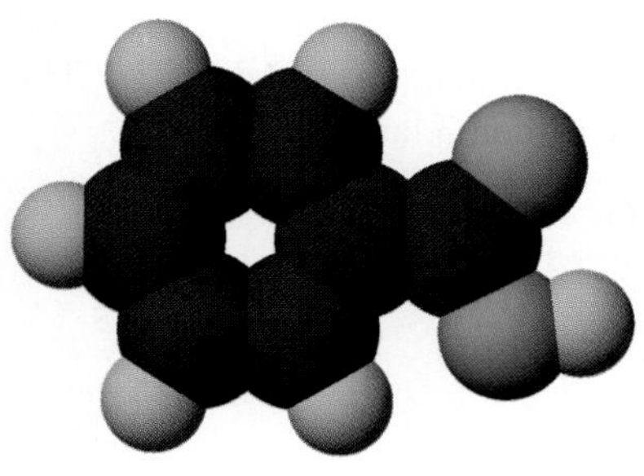

Space-filling model for benzoic acid, the simplest aromatic carboxylic acid.

The simplest aromatic carboxylic acid is called benzoic acid (Section 13.13).

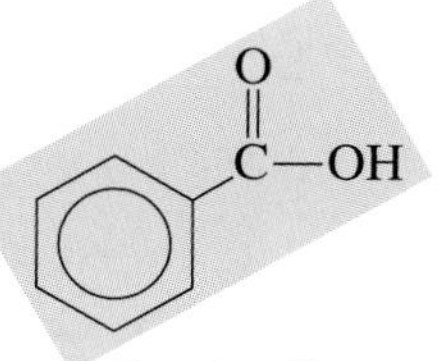

Benzoic acid

Other simple aromatic acids are named as derivatives of benzoic acid.

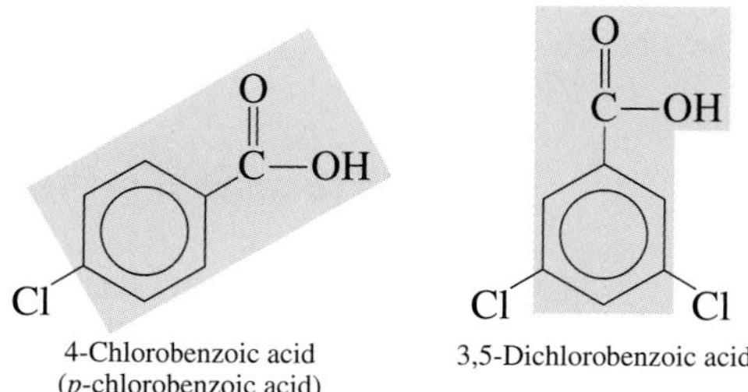

4-Chlorobenzoic acid (*p*-chlorobenzoic acid)

3,5-Dichlorobenzoic acid

In substituted benzoic acids, the ring carbon atom bearing the carboxyl group is always carbon 1.

• Methyl benzoic acids go by the name *toluic acid.* (This situation parallels methyl benzene being called toluene.)

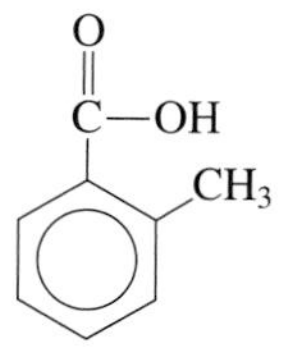

o-Toluic acid

Table 16.1
Common Names for the First Six Monocarboxylic Acids

Length of carbon chain	Structural formula	Common name[a]	IUPAC name
C_1 monoacid	H—COOH	formic acid	methanoic acid
C_2 monoacid	CH_3—COOH	acetic acid	ethanoic acid
C_3 monoacid	CH_3—CH_2—COOH	propionic acid	propanoic acid
C_4 monoacid	CH_3—$(CH_2)_2$—COOH	butyric acid	butanoic acid
C_5 monoacid	CH_3—$(CH_2)_3$—COOH	valeric acid	pentanoic acid
C_6 monoacid	CH_3—$(CH_2)_4$—COOH	caproic acid	hexanoic acid

[a]The mnemonic "*Frogs are polite, being very courteous*" is helpful in remembering, in order, the first letters of the common names of these six simple saturated monocarboxylic acids.

16.3 Common Names for Carboxylic Acids

The use of common names is more prevalent for carboxylic acids than for any other family of organic compounds.

The common names of monocarboxylic acids (Table 16.1) indicate the natural sources from which these acids were isolated. The stinging sensation associated with red ant bites is due in part to formic acid (Latin, *formica,* "ant"). Acetic acid gives vinegar its tartness (sour taste); vinegar is a 4%–8% (v/v) acetic acid solution (Latin, *acetum,* "sour"). Propionic acid is the smallest acid that can be obtained from fats (Greek, *protos,* "first," and *pion,* "fat"). Rancid butter contains butyric acid (Latin, *butyrum,* "butter"). Valeric acid, found in valerian root (an herb), has a strong odor (Latin, *valere,* "to be strong"). The skin secretions of goats contain caproic acid, which contributes to the odor associated with these animals (Latin, *caper,* "goats").

- The common names of monocarboxylic acids are the basis for aldehyde common names (Section 15.3).
  - C_1: formic acid and formaldehyde
  - C_2: acetic acid and acetaldehyde
  - C_3: propionic acid and propionaldehyde
  - C_4: butyric acid and butyraldehyde

Acetic acid is the most common carboxylic acid. One of its primary uses is as an *acidulant,* a substance that gives the proper acidic conditions for a chemical reaction. Pure acetic acid is known as *glacial* acetic acid, because it freezes on a moderately cold day (fp = 17°C), producing icy-looking crystals.

When using common names for carboxylic acids, we designate the positions (locations) of substituents by using letters of the Greek alphabet rather than numbers. (The first four letters of the Greek alphabet are alpha (α), beta (β), gamma (γ), and delta (δ). The alpha-carbon atom is carbon 2, the beta-carbon atom is carbon 3, and so on.

- The alpha-carbon atom in a carboxylic acid is the carbon atom to which the carboxyl group is attached. It is never the carboxyl carbon atom itself.

$$\cdots\cdots \text{C—C—C—C—}\overset{\text{O}}{\overset{\|}{\text{C}}}\text{—OH}$$

IUPAC: 5 4 3 2 1
Greek letter: δ γ β α

- Greek letters are *never* used for specifying substituent location in the IUPAC system.

With the Greek letter system, the compound

$$CH_3\text{—}CH_2\text{—}\underset{CH_3}{\underset{|}{CH}}\text{—}CH_2\text{—}\overset{O}{\overset{\|}{C}}\text{—}OH$$

β carbon (arrow to CH), α carbon (arrow to CH_2)

would be called β-methylvaleric acid.

• Dicarboxylic Acids

Common names for the first six dicarboxylic acids are given in Table 16.2. Oxalic acid, the simplest dicarboxylic acid, is found in plants of the genus *Oxalis,* which includes rhubarb and spinach, and in cabbage. This acid and its salts are poisonous in *high*

Table 16.2
Common Names for the First Six Dicarboxylic Acids

Length of carbon chain	Structural formula	Common name[a]	IUPAC name
C_2 diacid	$HOOC-COOH$	oxalic acid	ethanedioic acid
C_3 diacid	$HOOC-CH_2-COOH$	malonic acid	propanedioic acid
C_4 diacid	$HOOC-(CH_2)_2-COOH$	succinic acid	butanedioic acid
C_5 diacid	$HOOC-(CH_2)_3-COOH$	glutaric acid	pentanedioic acid
C_6 diacid	$HOOC-(CH_2)_4-COOH$	adipic acid	hexanedioic acid
C_7 diacid	$HOOC-(CH_2)_5-COOH$	pimelic acid	heptanedioic acid

[a]The mnemonic "*Oh my, such good apple pie*" is helpful in remembering, in order, the first letters of the common names of these six simple dicarboxylic acids.

The C_2 dicarboxylic acid, oxalic acid, contributes to the tart taste of rhubarb stalks. Although the stalks are edible, rhubarb leaves are not. They have too high an oxalic acid content to be safe for human consumption. (They are also toxic to sheep and horses.)

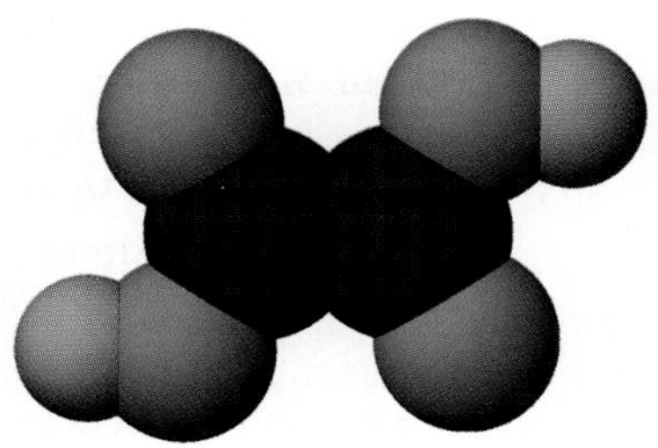

$$HO-\overset{\overset{\displaystyle O}{\|}}{C}-\overset{\overset{\displaystyle O}{\|}}{C}-OH$$

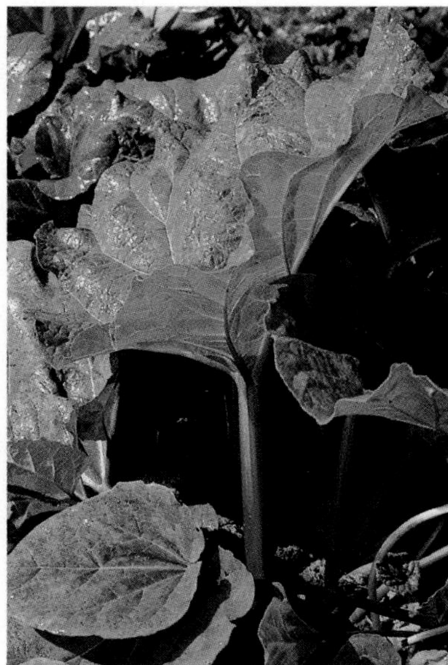

concentrations. The amount of oxalic acid present in spinach and cabbage is not harmful. However, rhubarb *leaves* should not be eaten, because they contain much higher levels of oxalic acid. Oxalic acid is used to remove rust, bleach straw and leather, and remove ink stains. Succinic and glutaric acid and their derivatives play important roles in biological reactions that occur in the human body (Section 23.6).

Example 16.2

Generating the Structural Formulas of Carboxylic Acids from Their Names

Draw a structural formula for each of the following carboxylic acids.

a. Caproic acid　　**b.** Glutaric acid
c. α-Phenylsuccinic acid　　**d.** β-Chlorobutyric acid

Solution

a. Caproic acid is the six-carbon unsubstituted monocarboxylic acid. Its structural formula is

$$CH_3-CH_2-CH_2-CH_2-CH_2-\overset{\overset{\displaystyle O}{\|}}{C}-OH$$

b. Glutaric acid is the five-carbon unsubstituted dicarboxylic acid, with a carboxyl group at each end of the carbon chain.

$$HO-\overset{\overset{\displaystyle O}{\|}}{C}-CH_2-CH_2-CH_2-\overset{\overset{\displaystyle O}{\|}}{C}-OH$$

c. Succinic acid is the four-carbon unsubstituted dicarboxylic acid. A phenyl group (Section 13.11) is present on the alpha-carbon atom.

$$HO-\overset{\overset{\displaystyle O}{\|}}{C}-\underset{\underset{\displaystyle C_6H_5}{|}}{\overset{\alpha}{C}H}-\overset{\beta}{C}H_2-\overset{\overset{\displaystyle O}{\|}}{C}-OH$$

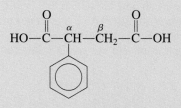

d. Butyric acid is the four-carbon unsubstituted monocarboxylic acid. A chloro group is attached to the beta-carbon atom (carbon 3).

$$\overset{\gamma}{C}H_3-\underset{\underset{\displaystyle Cl}{|}}{\overset{\beta}{C}H}-\overset{\alpha}{C}H_2-\overset{\overset{\displaystyle O}{\|}}{C}-OH$$

• The contrast between IUPAC names and common names for mono- and dicarboxylic acids is as follows:

Monocarboxylic Acids

IUPAC (two words)

alkanoic acid

Common (two words)

(prefix)ic acid*

Dicarboxylic Acids

IUPAC (two words)

alkanedioic acid

Common (two words)

(prefix)ic acid*

*The common-name prefixes are related to natural sources for the acids.

Practice Exercise 16.2

Draw a structural formula for each of the following carboxylic acids.

a. Adipic acid
b. β-Chlorovaleric acid
c. Malonic acid
d. Phenylacetic acid

• *Answers:* **a.** $HO{-}\overset{\overset{O}{\|}}{C}{-}CH_2{-}CH_2{-}CH_2{-}CH_2{-}\overset{\overset{O}{\|}}{C}{-}OH$

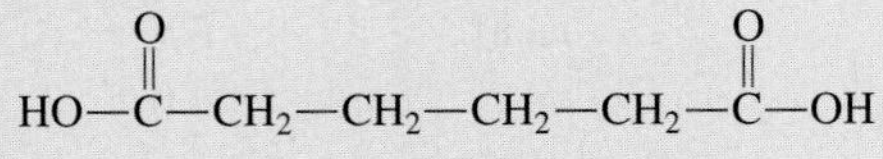

b. $CH_3{-}CH_2{-}\underset{\underset{Cl}{|}}{CH}{-}CH_2{-}\overset{\overset{O}{\|}}{C}{-}OH$

c. $HO{-}\overset{\overset{O}{\|}}{C}{-}CH_2{-}\overset{\overset{O}{\|}}{C}{-}OH$

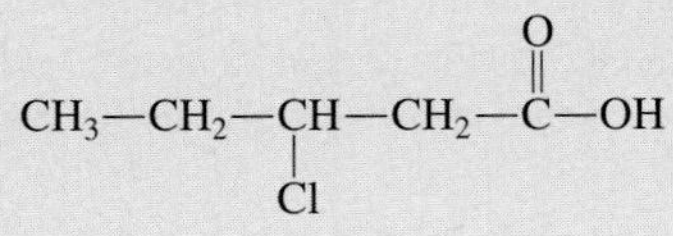

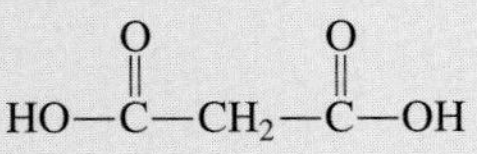

d. $C_6H_5{-}CH_2{-}\overset{\overset{O}{\|}}{C}{-}OH$ (benzene ring attached to CH_2)

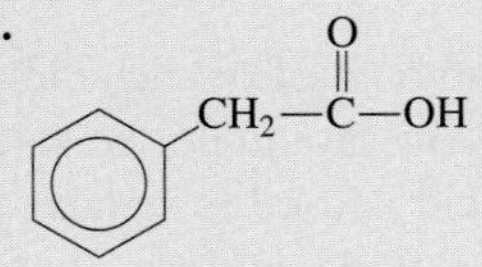

Chemical CONNECTIONS

16.1 Nonprescription Pain Relievers Derived from Propanoic Acid

Consumers are faced with a shelf-full of choices when looking for an over-the-counter medicine to treat aches, pains, and fever. The multitude of brands available, however, represent only five chemical formulations. Besides the long-available aspirin and acetaminophen, consumers can now purchase products that contain ibuprofen, naproxen, or ketoprofen.

These three newer entrants to the over-the-counter pain-reliever market are all derivatives of propanoic acid, the three-carbon monocarboxylic acid.

Ibuprofen, marketed under the brand names Advil, Motrin-IB, and Nuprin, was cleared by the FDA in 1984 for nonprescription sales. Numerous studies have shown that nonprescription-strength ibuprofen relieves minor pain and fever as well as aspirin or acetaminophen. Like aspirin, ibuprofen reduces inflammation. (Prescription-strength ibuprofen has extensive use as an anti-inflammatory agent for the treatment of rheumatoid arthritis.) There is evidence that ibuprofen is more effective than either aspirin or acetaminophen in reducing dental pain and menstrual pain. Both aspirin and ibuprofen can cause stomach bleeding in some people, although ibuprofen seems to cause fewer problems. Ibuprofen is more expensive than either aspirin or acetaminophen.

Naproxen, marketed under the brand names Aleve and Anaprox, was cleared by the FDA in 1994 for nonprescription use. The effects of naproxen last longer in the body (8–12 hr per dose) than the effects of ibuprofen (4–6 hr per dose) and of aspirin and acetaminophen (4 hr per dose). Naproxen is more likely to cause slight intestinal bleeding and stomach upset than is ibuprofen. It is also not recommended for use by children under 12.

Ketoprofen, marketed under the brand names Orudis KT and Actron, became available for nonprescription sales in 1996. Consumers take ketoprofen in 12.5-mg doses, compared with the dosages of 200 to 500 mg for other over-the-counter pain relievers. The easy-to-swallow size of this tiny product is touted by its manufacturers. Ketoprofen should not be taken by people allergic to aspirin or ibuprofen, because its side effects are similar.

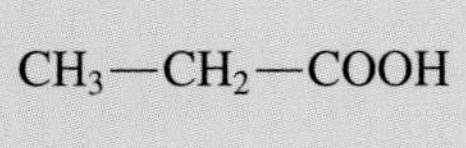

$CH_3{-}CH_2{-}COOH$

Propanoic acid

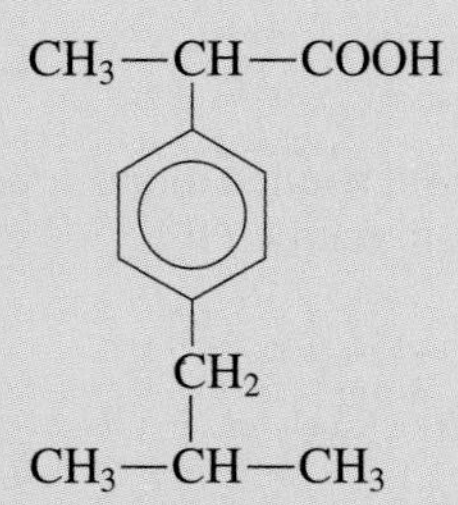

$CH_3{-}CH{-}COOH$ (on benzene ring, para to $CH_2{-}CH(CH_3){-}CH_3$)

Ibuprofen

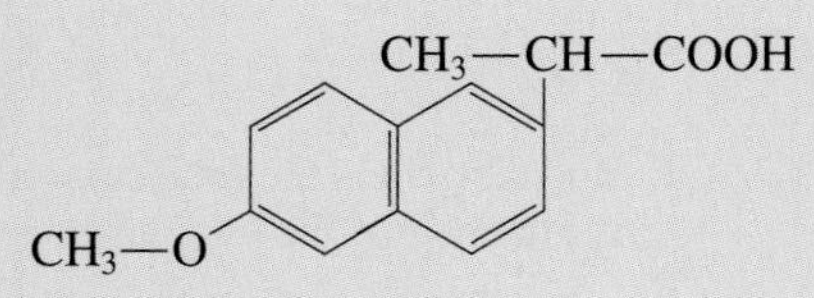

$CH_3{-}CH{-}COOH$ on naphthalene ring bearing $CH_3{-}O$

Naproxen

$CH_3{-}CH{-}COOH$ on benzene ring bearing $\overset{\overset{O}{\|}}{C}$ linked to a second benzene ring

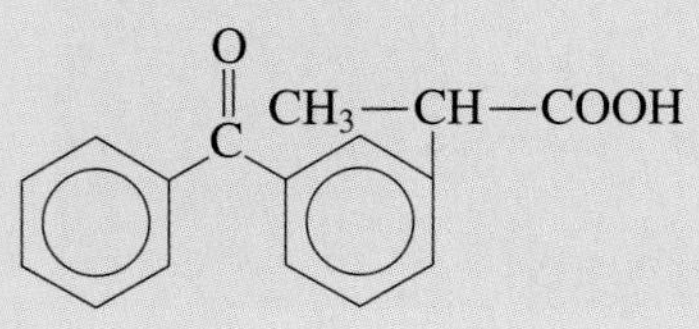

Ketoprofen

16.4 Polyfunctional Carboxylic Acids

Polyfunctional carboxylic acids are carboxylic acids that contain one or more additional functional groups besides the carboxyl group. Such acids occur naturally, are important in the normal functioning of the human body, and find use in cosmetics and prescription drugs. Three commonly encountered types of polyfunctional carboxylic acids are unsaturated acids, keto acids, and hydroxy acids.

Eight key intermediate compounds in the metabolic reactions (Section 23.1) that occur in the human body are polyfunctional carboxylic acids. These compounds, which we will meet repeatedly in the biochemistry chapters of this text, are

• It is important to be familiar with the structures of the eight "metabolic" carboxylic acids.

1. Fumaric acid (an unsaturated acid)
2. Pyruvic acid (a keto acid)
3. Oxaloacetic acid (a keto acid)
4. α-Ketoglutaric acid (a keto acid)
5. Lactic acid (a hydroxy acid)
6. Malic acid (a hydroxy acid)
7. Glyceric acid (a dihydroxy acid)
8. Citric acid (a carboxyhydroxy acid)

Structures for these acids, along with commentary on their occurrence in human body processes, are given in Figure 16.1. Interestingly, these carboxylic acids that are important in the functioning of the human body are derived from just three of the simple carboxylic acids: propionic acid, succinic acid, and glutaric acid.

• An unsaturated monocarboxylic acid with the structure

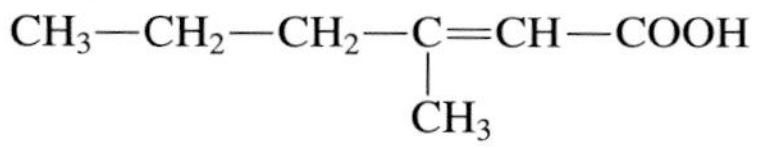

$$CH_3—CH_2—CH_2—\underset{\displaystyle CH_3}{\underset{|}{C}}=CH—COOH$$

3-Methyl-2-hexenoic acid

has been found to be largely responsible for "body odor." It is produced by skin bacteria, particularly those found in armpits.

The simplest *unsaturated mono*carboxylic acid is propenoic acid (acrylic acid), a substance used in the manufacture of several polymeric materials. Two forms exist for the simplest *unsaturated di*carboxylic acid, butenedioic acid. The two isomers have separate common names, fumaric acid (*trans*) and maleic acid (*cis*), a naming procedure seldom encountered.

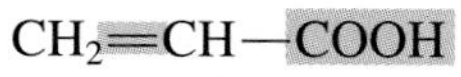

$CH_2=CH—COOH$

Acrylic acid

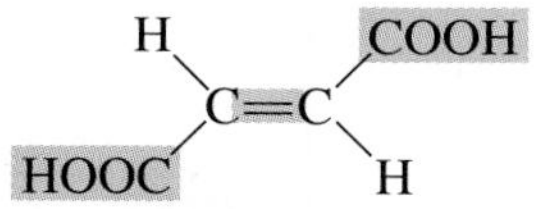

H, COOH / C=C / HOOC, H

Fumaric acid (*trans* isomer)

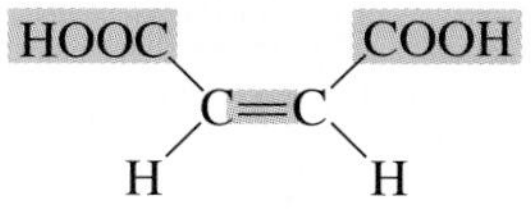

HOOC, COOH / C=C / H, H

Maleic acid (*cis* isomer)

Fumaric acid is a "metabolic" acid. Some antihistamines (Section 17.9) are salts of maleic acid. The addition of small amounts of maleic acid to fats and oils prevents them from becoming rancid (Section 19.5).

*Poly*unsaturated carboxylic acids with structures related to that of vitamin A (Section 21.14) are used extensively in the treatment of severe acne. The prescription drugs Tretinoin and Accutane are such compounds.

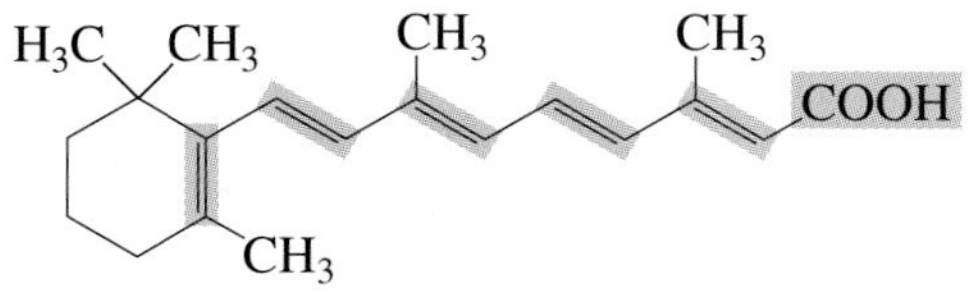

Tretinoin

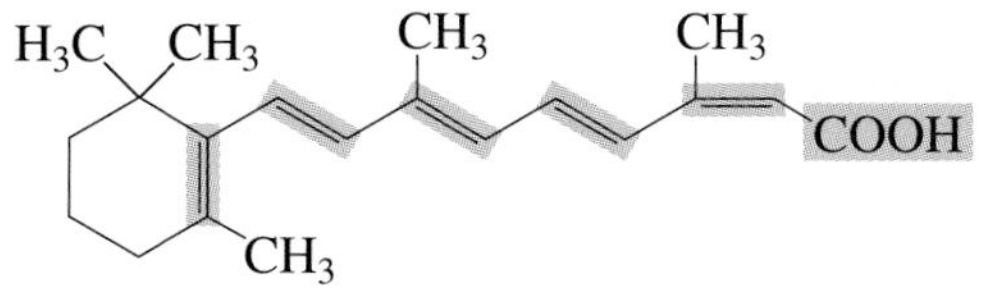

Accutane

Tretinoin has an all-*trans* double-bond configuration. Accutane has a *cis* double bond in the position closest to the carboxyl group, with the rest of the double bonds in *trans* configurations.

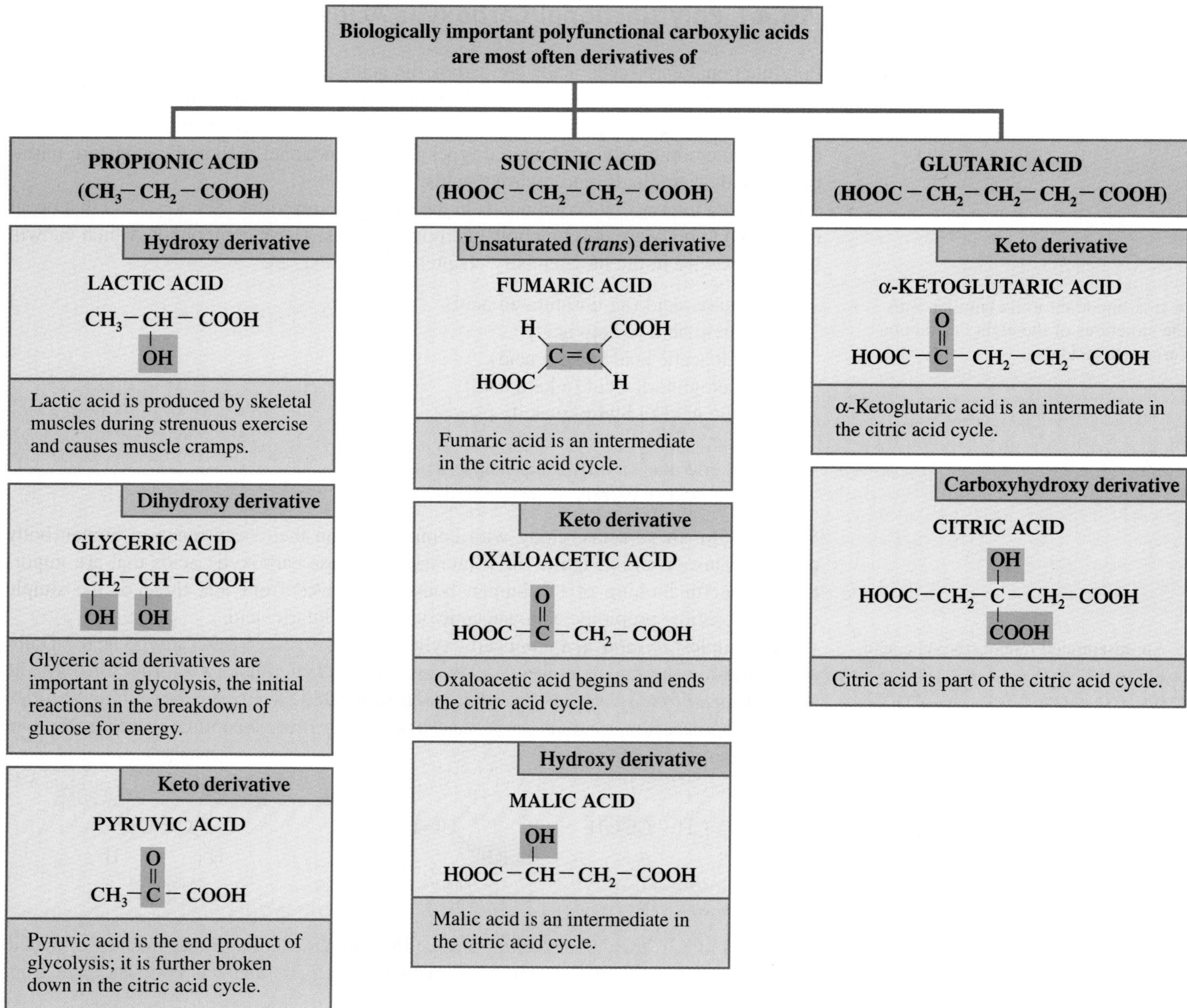

Figure 16.1
Structures of polyfunctional carboxylic acids that are important in metabolic reactions in the human body.

● An alpha-hydroxy acid is a carboxylic acid that has a hydroxyl group attached to the alpha-carbon atom, which is the carbon atom adjacent to the carboxyl carbon atom. (See Figure 16.2.)

Three of the eight "metabolic" acids (Figure 16.1) are *keto* acids: pyruvic, oxaloacetic, and α-ketoglutaric. Pyruvic acid, with three carbon atoms, is the simplest keto acid that can exist. In the pure state, it is a liquid with an odor resembling that of vinegar (acetic acid; Section 16.3).

The simplest *hydroxy* acid is the two-carbon glycolic acid, a substance that occurs naturally in sugar cane juice. It finds use in cosmetic products that address problems such as dryness, flaking, and itchiness of skin. A number of α-hydroxy acids, of which glycolic acid is the simplest, are used as ingredients in these products. Other "skin care acids" include malic, tartaric, lactic, and citric acids.

$$\underset{\text{Glycolic acid}}{\underset{|\;\;\;\;\;\;\;\;\;\;\;\;\;}{CH_2{-}COOH} \atop OH\;\;\;\;\;\;\;\;\;\;\;\;\;\;} \qquad \underset{\text{Tartaric acid}}{HOOC{-}\underset{OH}{\underset{|}{CH}}{-}\underset{OH}{\underset{|}{CH}}{-}COOH}$$

The structures of lactic, malic, glyceric, and citric acids, all hydroxy "metabolic" acids, are found in Figure 16.1.

Both malic and tartaric acids occur naturally in fruits. The sharp taste of apples (fruit of trees of the genus *Malus*) is due to malic acid. Tartaric acid is particularly abundant in

Figure 16.2
Two skin care products containing alpha-hydroxy acids.

grapes. It is also a component of tartar sauce and an acidic ingredient in many baking powders. Lactic acid is present in sour milk, sauerkraut, and dill pickles. Citric acid, which is perhaps the best known of all carboxylic acids, gives citrus fruits their "sharp" taste; lemon juice contains 4%–8% citric acid, and orange juice is about 1% citric acid. Citric acid is used widely in beverages and in foods. In jams, jellies, and preserves, it produces tartness and pH adjustment to optimize conditions for gelation. In fresh salads, citric acid prevents enzymatic browning reactions, and in frozen fruits, it prevents deterioration of color and flavor. Addition of citric acid to seafood retards microbial growth by lowering pH.

16.5 Physical Properties of Carboxylic Acids

Unsubstituted saturated monocarboxylic acids containing up to nine carbon atoms are liquids that have strong, sharp odors. Acids with 10 or more carbon atoms in an unbranched chain are waxy solids that are odorless (because of low volatility). Aromatic carboxylic acids, as well as dicarboxylic acids, are also odorless solids.

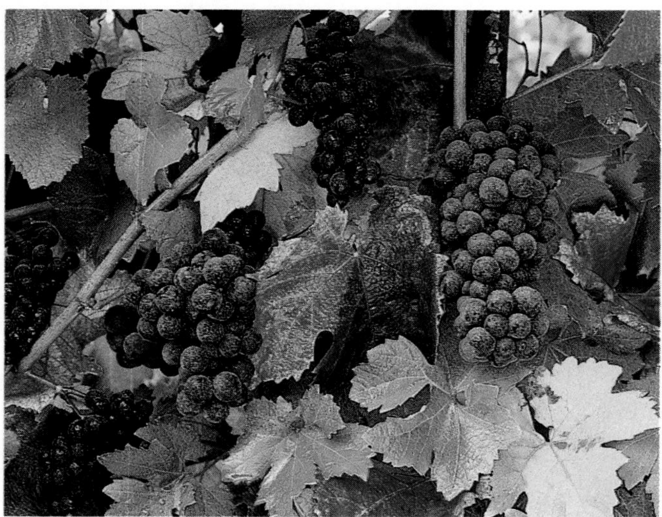

Tartaric acid, the dihydroxy derivative of succinic acid, is particularly abundant in ripe grapes.

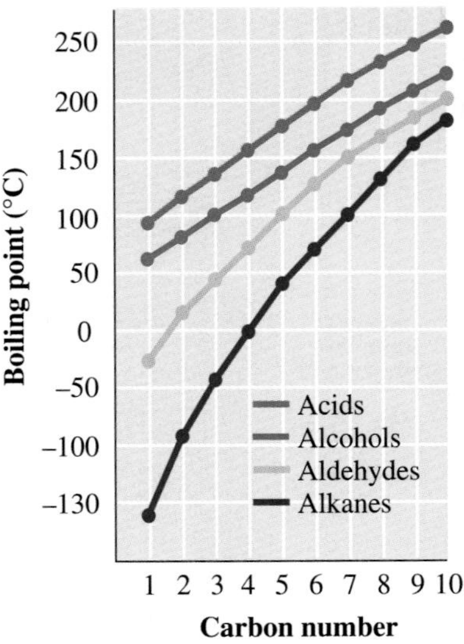

Figure 16.3
The boiling points of monocarboxylic acids compared to those of other types of compounds. All compounds in the comparison have unbranched carbon chains.

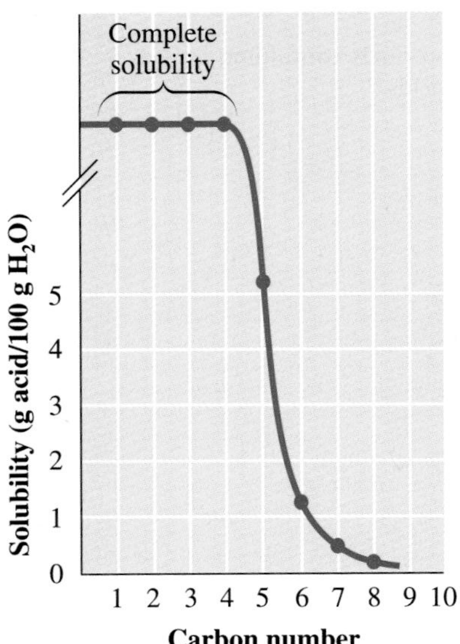

Figure 16.4
The solubility of saturated unbranched-chain carboxylic acids.

Carboxylic acids have the highest boiling points of all organic compounds we have considered so far (Figure 16.3), because hydrogen-bonding opportunities are more extensive for carboxylic acids. A given carboxylic acid molecule can form two hydrogen bonds with another carboxylic acid molecule.

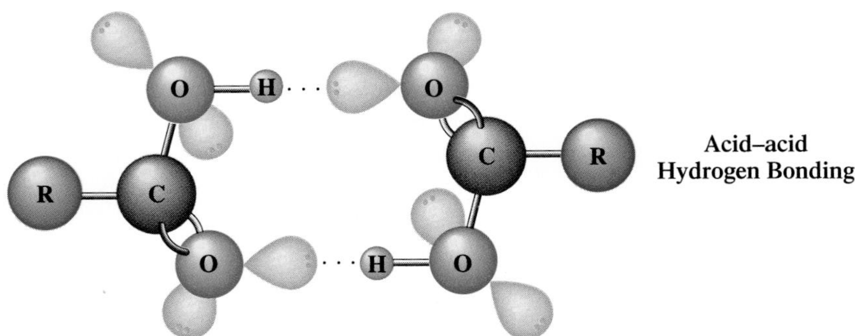

Dogs can differentiate among the odors of individual humans because of differing proportions of various carboxylic acids in human sweat.

"Drug-sniffing" dogs used by narcotics agents can find hidden heroin by detecting the odor of acetic acid (vinegar odor). Acetic acid is a by-product of the final step in illicit heroin production, and trace amounts remain in the heroin.

Carboxylic acids readily hydrogen-bond to water molecules. Such hydrogen bonding contributes to water solubility for short-chain carboxylic acids. The unsubstituted C_1 to C_4 monocarboxylic acids are completely miscible with water. Solubility then rapidly decreases with carbon number, as shown in Figure 16.4. Short-chain dicarboxylic acids are also water-soluble. In general, aromatic acids are not water-soluble.

16.6 Preparation of Carboxylic Acids

Oxidation of primary alcohols or aldehydes, using an oxidizing agent such as CrO_3 or $K_2Cr_2O_7$, produces carboxylic acids.

$$\text{Primary alcohol} \xrightarrow{[O]} \text{aldehyde} \xrightarrow{[O]} \text{carboxylic acid}$$

Unbranched Monocarboxylic Acids			
C_1	C_3	C_5	C_7
C_2	C_4	C_6	C_8

Unbranched Dicarboxylic Acids			
	C_3	C_5	C_7
C_2	C_4	C_6	C_8

Liquid Solid

A physical-state summary for unbranched mono- and dicarboxylic acids at room temperature and pressure.

Aromatic acids can be prepared by oxidizing a carbon side chain (alkyl group) on a benzene derivative. In this process, all the carbon atoms of the alkyl group except the one attached to the ring are lost. The remaining carbon becomes part of a carboxyl group.

$$C_6H_5-CH_2-CH_2-CH_3 \xrightarrow[H_2SO_4]{K_2Cr_2O_7} C_6H_5-\overset{O}{\overset{\|}{C}}-OH + 2CO_2 + 3H_2O$$

16.7 Acidity of Carboxylic Acids

Carboxylic acids, as the name implies, are *acidic*. When a carboxylic acid is placed in water, hydrogen ion transfer (proton transfer; Section 10.2) occurs to produce hydronium ion (the acidic species in water; Section 10.2) and carboxylate ion.

$$R-COOH + H_2O \longrightarrow \underset{\text{Hydronium ion}}{H_3O^+} + \underset{\text{Carboxylate ion}}{R-COO^-}$$

A **carboxylate ion** *is the negative ion produced when a carboxylic acid loses one or more acidic hydrogen atoms.*

Carboxylate ions formed from monocarboxylic acids always carry a -1 charge; only one acidic hydrogen atom is present in such molecules. Dicarboxylic acids, which possess two acidic hydrogen atoms (one in each carboxyl group), can produce carboxylate ions bearing a -2 charge.

Carboxylate ions are named by dropping the *-ic acid* ending from the name of the parent acid and replacing it with *-ate.*

$$\underset{\substack{\text{Acetic acid}\\\text{(ethanoic acid)}}}{CH_3-\overset{O}{\overset{\|}{C}}-OH} + H_2O \longrightarrow H_3O^+ + \underset{\substack{\text{Acetate ion}\\\text{(ethanoate ion)}}}{CH_3-\overset{O}{\overset{\|}{C}}-O^-}$$

$$\underset{\substack{\text{Oxalic acid}\\\text{(ethanedioic acid)}}}{HO-\overset{O}{\overset{\|}{C}}-\overset{O}{\overset{\|}{C}}-OH} + 2H_2O \longrightarrow 2H_3O^+ + \underset{\substack{\text{Oxalate ion}\\\text{(ethanedioate ion)}}}{{}^-O-\overset{O}{\overset{\|}{C}}-\overset{O}{\overset{\|}{C}}-O^-}$$

Carboxylic acids are weak acids (Section 10.4). The extent of proton transfer is usually less than 5%; that is, an equilibrium situation exists in which the equilibrium lies far to the left.

$$\underset{\substack{\text{More than 95\%}\\\text{of molecules in this form}}}{R-COOH + H_2O} \rightleftharpoons \underset{\substack{\text{Less than 5\%}\\\text{of molecules in this form}}}{H_3O^+ + R-COO^-}$$

Table 16.3 gives K_a values (Section 10.5) and percent ionization in 0.100M solution for selected monocarboxylic acids.

Table 16.3
Acid Strength for Selected Monocarboxylic Acids

Acid	K_a	Percent ionization (0.100 M solution)
Formic	1.8×10^{-4}	4.2%
Acetic	1.8×10^{-5}	1.3%
Propionic	1.3×10^{-5}	1.2%
Butyric	1.5×10^{-5}	1.2%
Valeric	1.5×10^{-5}	1.2%
Caproic	1.4×10^{-5}	1.2%

16.8 Carboxylic Acid Salts

• Carboxylic acid salt formation involves an acid–base neutralization reaction (Section 10.7).

In a manner similar to that of inorganic acids (Section 10.6), carboxylic acids react completely with strong bases to produce water and a carboxylic acid salt.

$$\underset{\text{Carboxylic acid}}{CH_3-\overset{\overset{\large O}{\|}}{C}-OH} + \underset{\text{Strong base}}{NaOH} \longrightarrow \underset{\text{Carboxylic acid salt}}{CH_3-\overset{\overset{\large O}{\|}}{C}-O^-\,Na^+} + \underset{\text{Water}}{H_2O}$$

A **carboxylic acid salt** *is an ionic compound in which the negative ion is a carboxylate ion.*

Carboxylic acid salts are named similarly to other ionic compounds (Section 4.9): *The positive ion is named first, followed by a separate word giving the name of the negative ion.* The salt formed in the preceding reaction contains sodium ions and acetate ions (from acetic acid); hence the salt's name is sodium acetate.

Example 16.3

Writing Equations for the Formation of Carboxylic Acid Salts

Using an acid–base neutralization reaction, write a chemical equation for the formation of each of the following carboxylic acid salts.

a. Sodium propionate

b. Potassium oxalate

Solution

a. This salt contains sodium ion (Na^+) and propionate ion, the three-carbon monocarboxylate ion.

$$CH_3-CH_2-\overset{\overset{\large O}{\|}}{C}-O^-Na^+$$

From a neutralization standpoint, the sodium ion's source is the base sodium hydroxide, NaOH, and the negative ion's source is the acid propanoic acid. The acid–base neutralization equation is

$$\underset{\text{Propionic acid}}{CH_3-CH_2-\overset{\overset{\large O}{\|}}{C}-OH} + \underset{\text{Sodium hydroxide}}{NaOH} \longrightarrow \underset{\text{Sodium propionate}}{CH_3-CH_2-\overset{\overset{\large O}{\|}}{C}-O^-Na^+} + \underset{\text{Water}}{H_2O}$$

b. This salt contains potassium ions (K^+) whose source would be the base potassium hydroxide, KOH. The salt also contains oxalate ions, whose source would be the acid oxalic acid.

$$\underset{\text{Oxalic acid}}{HO-\overset{\overset{\large O}{\|}}{C}-\overset{\overset{\large O}{\|}}{C}-OH} + \underset{\text{Potassium hydroxide}}{2KOH} \longrightarrow \underset{\text{Potassium oxalate}}{K^{+-}O-\overset{\overset{\large O}{\|}}{C}-\overset{\overset{\large O}{\|}}{C}-O^-K^+} + \underset{\text{Water}}{2H_2O}$$

Note that two molecules of base are needed to react completely with one molecule of acid because the acid is a dicarboxylic acid.

Practice Exercise 16.3

Using an acid–base neutralization reaction, write a chemical equation for the formation of each of the following carboxylic acid salts.

a. Sodium formate

b. Potassium malonate

• *Answers:* **a.**

$$H{-}\overset{\overset{\displaystyle O}{\|}}{C}{-}OH + NaOH \rightarrow H{-}\overset{\overset{\displaystyle O}{\|}}{C}{-}O^-Na^+ + H_2O$$

b.

$$HO{-}\overset{\overset{\displaystyle O}{\|}}{C}{-}CH_2{-}\overset{\overset{\displaystyle O}{\|}}{C}{-}OH + 2KOH \rightarrow K^{+-}O{-}\overset{\overset{\displaystyle O}{\|}}{C}{-}CH_2{-}\overset{\overset{\displaystyle O}{\|}}{C}{-}O^-K^+ + 2H_2O$$

• The interconversion reactions between carboxylic acid salts and the carboxylic acids themselves are so easy to carry out that organic chemists consider these two types of compounds interchangeable.

Converting the carboxylic acid salt back to the carboxylic acid itself is very simple. React the salt with a solution of a strong acid such as hydrochloric acid (HCl) or sulfuric acid (H_2SO_4).

$$CH_3{-}\overset{\overset{\displaystyle O}{\|}}{C}{-}O^-\,Na^+ + HCl \longrightarrow CH_3{-}\overset{\overset{\displaystyle O}{\|}}{C}{-}OH + NaCl$$

Sodium acetate — Hydrochloric acid — Acetic acid — Sodium chloride

• The solubility of benzoic acid in water at 25°C is 3.4 g/L. The solubility of sodium benzoate, the sodium salt of benzoic acid in water at 25°C is 550 g/L.

The solubility of carboxylic acid salts in water is much greater than that of the carboxylic acids from which they are derived. Drugs and medicines that contain acid groups are usually marketed as the sodium or potassium salt of the acid. This greatly enhances the solubility of the medication, increasing the ease of its absorption by the body.

Many *antimicrobials,* compounds used as food preservatives, are carboxylic acid salts. Particularly important are the salts of benzoic, sorbic, and propionic acids.

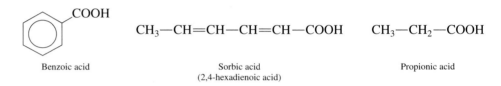

$$C_6H_5{-}COOH \qquad CH_3{-}CH{=}CH{-}CH{=}CH{-}COOH \qquad CH_3{-}CH_2{-}COOH$$

Benzoic acid — Sorbic acid (2,4-hexadienoic acid) — Propionic acid

The benzoate salts of sodium and potassium are effective against yeast and mold in beverages, jams and jellies, pie fillings, ketchup, and syrups. Concentrations of up to 0.1% (m/m) benzoate are found in such products.

$$C_6H_5{-}\overset{\overset{\displaystyle O}{\|}}{C}{-}O^-\,Na^+ \qquad C_6H_5{-}\overset{\overset{\displaystyle O}{\|}}{C}{-}O^-\,K^+$$

Sodium benzoate — Potassium benzoate

Sodium and potassium sorbates inhibit mold and yeast growth in dairy products, dried fruits, sauerkraut, and some meat and fish products. Sorbate preservative concentrations range from 0.02% to 0.2% (m/m).

$$CH_3{-}CH{=}CH{-}CH{=}CH{-}\overset{\overset{\displaystyle O}{\|}}{C}{-}O^-\,Na^+$$

Sodium sorbate

$$CH_3{-}CH{=}CH{-}CH{=}CH{-}\overset{\overset{\displaystyle O}{\|}}{C}{-}O^-\,K^+$$

Potassium sorbate

Propionates, salts of propionic acid, extend the shelf life of bread by preventing the formation of mold.

Calcium and sodium propionates are used in baked products and also in cheese foods and spreads. Benzoates and sorbates cannot be used in yeast-leavened baked goods, because they affect the activity of the yeast.

$$(CH_3{-}CH_2{-}\overset{\overset{\displaystyle O}{\|}}{C}{-}O^-)_2\,Ca^{2+} \qquad CH_3{-}CH_2{-}\overset{\overset{\displaystyle O}{\|}}{C}{-}O^-\,Na^+$$

Calcium propionate — Sodium propionate

Carboxylate salts do not directly kill microorganisms present in food. Rather, they prevent further growth and proliferation of these organisms by increasing the pH of the foods in which they are used.

16.9 Structure of Esters

Esters are carboxylic acid derivatives in which the —OH group of the carboxylic acid has been replaced with an —OR group.

$$\underset{\text{Carboxylic acid}}{R-\overset{\overset{\displaystyle O}{\|}}{C}-O-H} \qquad \underset{\text{Ester}}{R-\overset{\overset{\displaystyle O}{\|}}{C}-O-R}$$

An **ester** *is an organic compound whose characteristic functional group is*

$$-\overset{\overset{\displaystyle O}{\|}}{C}-O-R$$

In linear form, the ester functional group can be represented as —COOR or $-CO_2R$.

The reaction of a carboxylic acid with an alcohol, using a strong-acid catalyst (generally H_2SO_4), produces an ester.

$$\underset{\text{Carboxylic acid}}{R-\overset{\overset{\displaystyle O}{\|}}{C}-O-H} + \underset{\text{Alcohol}}{H-O-R'} \overset{H^+}{\rightleftharpoons} \underset{\text{Ester}}{R-\overset{\overset{\displaystyle O}{\|}}{C}-O-R'} + \underset{\text{Water}}{H_2O}$$

• Esterification is a *condensation reaction.* This is the second time we have encountered this type of reaction. The first encounter involved the preparation of hemiacetals and acetals and their ketone counterparts (Section 15.9).

In this reaction, called *esterification,* a —OH group is lost from the carboxylic acid, a —H atom is lost from the alcohol, and water is formed as a by-product. The net effect of this reaction is substitution of the —OR group of the alcohol for the —OH group of the acid.

$$R-\overset{\overset{\displaystyle O}{\|}}{C}-O-H + H-O-R' \overset{H^+}{\rightleftharpoons} R-\overset{\overset{\displaystyle O}{\|}}{C}-O-R' + H_2O$$

A specific example of esterification is the reaction of acetic acid with methyl alcohol.

• Studies show that in ester formation, the hydroxyl group of the acid (not of the alcohol) becomes part of the water molecule.

$$CH_3-\overset{\overset{\displaystyle O}{\|}}{C}-O-H + H-O-CH_3 \overset{H^+}{\rightleftharpoons} CH_3-\overset{\overset{\displaystyle O}{\|}}{C}-O-CH_3 + H_2O$$

Esterification reactions are equilibrium processes, with the position of equilibrium (Section 9.8) usually only slightly favoring products. That is, at equilibrium, substantial amounts of both reactants and products are present. The amount of ester formed can be

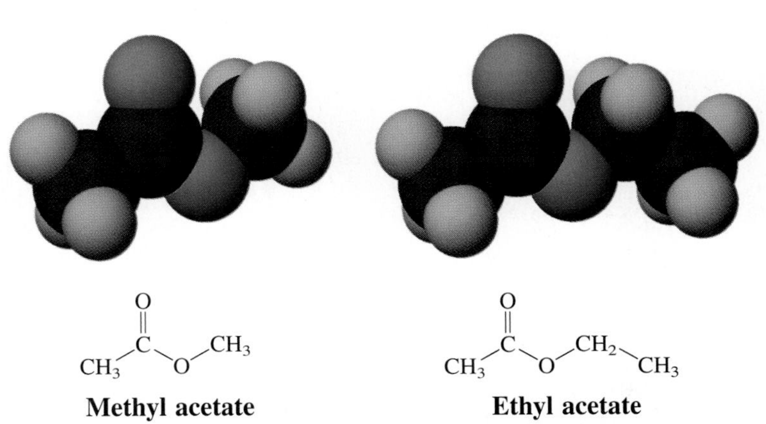

Space-filling models for the methyl and ethyl esters of acetic acid.

increased by using an excess of alcohol or constantly removing one of the products. According to Le Châtelier's principle (Section 9.9), both of these techniques will shift the position of equilibrium to the right (the product side of the equation). This equilibrium problem explains the use of the "double-arrow notation" in all the esterification equations in this section.

It is often useful to think of the structure of an ester in terms of its "parent" alcohol and acid molecules; the ester has an acid part and an alcohol part.

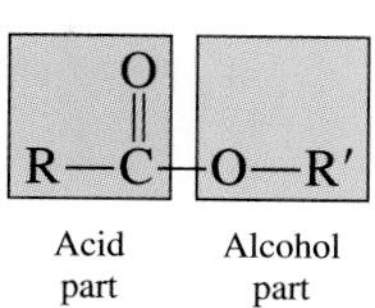

• Esters and carboxylic acids with the same number of carbon atoms and the same degree of saturation are structural isomers (functional group isomers). This is the third time we have encountered functional group isomerism. The other examples are alcohol–ether and aldehyde–ketone isomers.

In this context, it is easy to identify the acid and alcohol from which a given ester can be produced; just add a —OH group to the acid part of the ester and a —H atom to the alcohol part to generate the parent molecules.

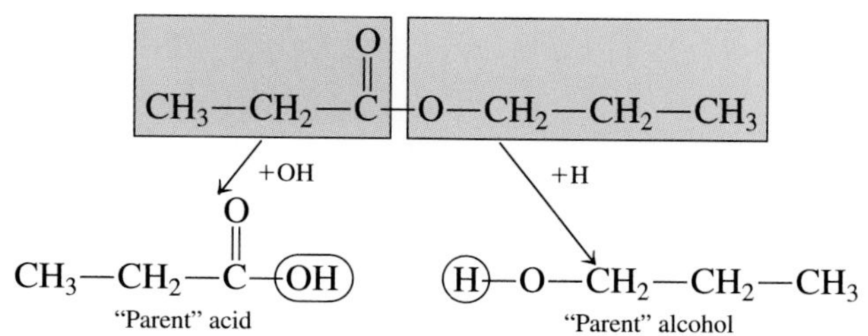

Hydroxyacids (Section 16.4) contain both functional groups needed for an esterification reaction (acid + alcohol). If through "twisting" of the carbon chain, the acid and alcohol group can come into contact with each other, an internal reaction can occur to produce a cyclic ester.

• Cyclic esters formed from hydroxyacids are called *lactones.*

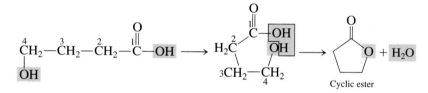

$$\underset{\text{OH}}{\overset{4}{\text{CH}_2}}-\overset{3}{\text{CH}_2}-\overset{2}{\text{CH}_2}-\overset{\text{O}}{\overset{\|}{\overset{1}{\text{C}}}}-\text{OH} \longrightarrow \text{(ring-shaped intermediate: } {}^2\text{H}_2\text{C},\ {}^3\text{CH}_2-{}^4\text{CH}_2,\ \overset{\text{O}}{\overset{\|}{\text{C}}}-\text{OH},\ \text{OH)} \longrightarrow \text{cyclic ester} + \text{H}_2\text{O}$$

Cyclic ester

Such a reaction easily takes place in situations where a five-or six-membered ring can be formed.

16.10 Nomenclature for Esters

• Salts and esters of carboxylic acids are named in the same way. The name of the positive ion (in the case of a salt) or the name of the organic group attached to the oxygen of the carbonyl group (in the case of an ester) precedes the name of the acid. The *-ic acid* part of the name of the acid is converted to *-ate.*

$$CH_3-CH_2-CH_2-\overset{O}{\overset{\|}{C}}-O^-\,Na^+$$

IUPAC: Sodium butanoate
Common: Sodium butyrate

$$CH_3-CH_2-CH_2-\overset{O}{\overset{\|}{C}}-O-CH_3$$

IUPAC: Methyl butanoate
Common: Methyl butyrate

Visualizing esters as having an "alcohol part" and an "acid part" (Section 16.9) is the key to naming them in both the common and the IUPAC systems of nomenclature. The name of the alcohol part of the ester appears first and is followed by a separate word giving the name for the acid part of the ester. The name for the alcohol part of the ester is simply the name of the R group (alkyl, cycloalkyl, or aryl) present in the —OR portion of the ester. The name for the acid part of the ester is obtained by dropping the *-ic acid* ending from the acid's name and adding the suffix *-ate.*

Consider the ester derived from ethanoic acid (acetic acid) and methanol (methyl alcohol). Its name will be *methyl ethanoate* (IUPAC) or *methyl acetate* (common).

$$CH_3-\overset{O}{\overset{\|}{C}}-OH + HO-CH_3 \longrightarrow CH_3-\overset{O}{\overset{\|}{C}}-O-CH_3 + H_2O$$

IUPAC:	Ethanoic acid	Methanol	Methyl ethanoate
Common:	Acetic acid	Methyl alcohol	Methyl acetate

Dicarboxylic acids can form diesters, with each of the carboxyl groups undergoing esterification. An example of such a molecule and how it is named is

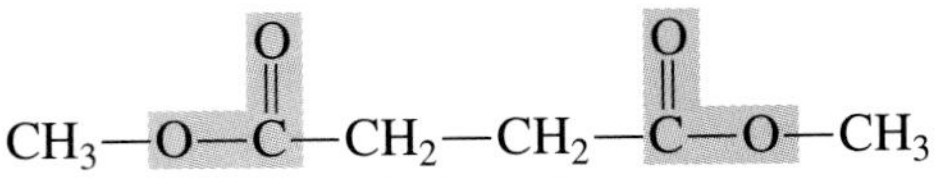

$$CH_3{-}O{-}\overset{\overset{\displaystyle O}{\|}}{C}{-}CH_2{-}CH_2{-}\overset{\overset{\displaystyle O}{\|}}{C}{-}O{-}CH_3$$

IUPAC: Dimethyl butanedioate
Common: Dimethyl succinate

Further examples of ester nomenclature, for compounds in which substituents are present, are

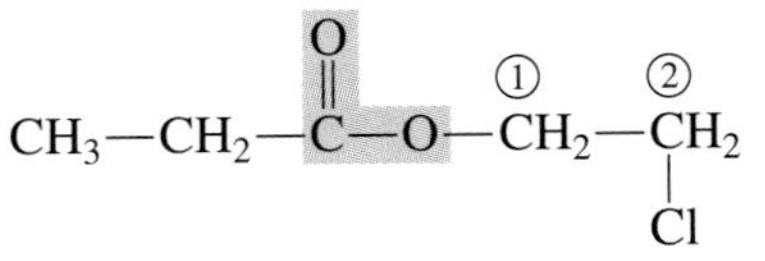

$$CH_3{-}CH_2{-}\overset{\overset{\displaystyle O}{\|}}{C}{-}O{-}\overset{①}{CH_2}{-}\underset{\underset{\displaystyle Cl}{|}}{\overset{②}{CH_2}} \qquad \overset{③}{CH_3}{-}\underset{\underset{\displaystyle CH_3}{|}}{\overset{②}{CH}}{-}\overset{\overset{\displaystyle O}{\|}}{C}{-}O{-}CH_2{-}CH_3$$

IUPAC: 2-Chloroethyl propanoate — Ethyl 2-methylpropanoate
Common: 2-Chloropropionate — Ethyl 2-methylpropionate

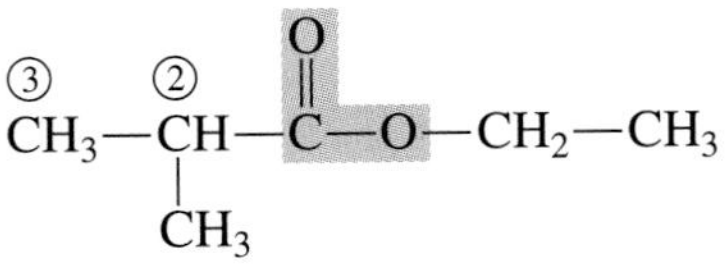

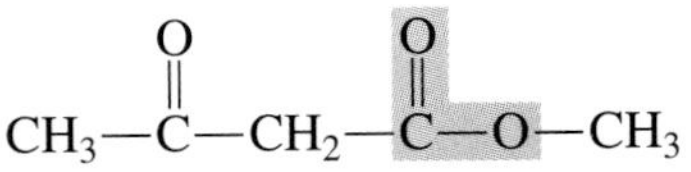

$$CH_3{-}\overset{\overset{\displaystyle O}{\|}}{C}{-}CH_2{-}\overset{\overset{\displaystyle O}{\|}}{C}{-}O{-}CH_3$$

IUPAC: Methyl 3-oxobutanoate
Common: Methyl 3-oxobutyrate

• Line-angle drawings for the simpler unbranched-chain methyl esters:

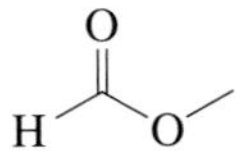

Methyl methanoate

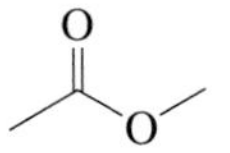

Methyl ethanoate

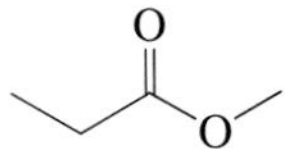

Methyl propanoate

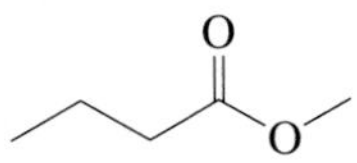

Methyl butanoate

Example 16.4

Determining IUPAC and Common Names for Esters

Assign both IUPAC and common names to the following esters.

a. $CH_3{-}CH_2{-}\overset{\overset{\displaystyle O}{\|}}{C}{-}O{-}CH_2{-}CH_3$

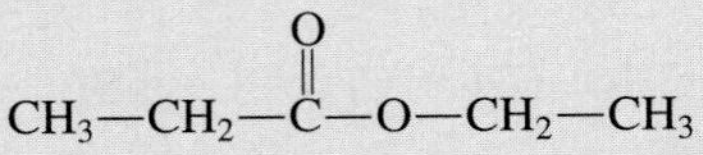

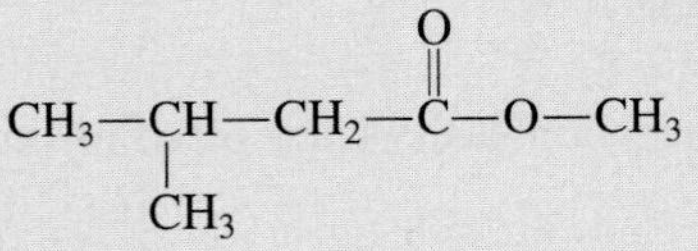

b. $CH_3{-}\underset{\underset{\displaystyle CH_3}{|}}{CH}{-}CH_2{-}\overset{\overset{\displaystyle O}{\|}}{C}{-}O{-}CH_3$

c. $C_6H_5{-}\overset{\overset{\displaystyle O}{\|}}{C}{-}O{-}CH_2{-}CH_2{-}CH_3$ (benzene ring attached to the carbonyl carbon)

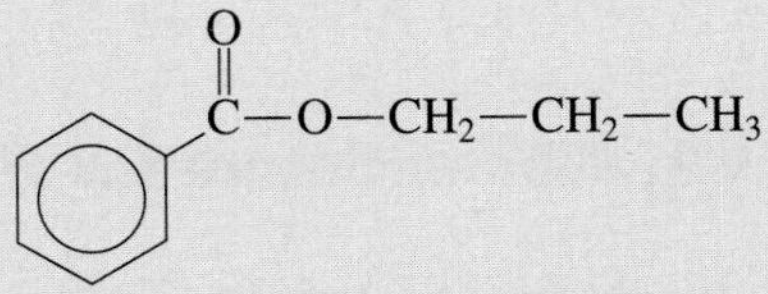

Solution

a. The name *ethyl* characterizes the alcohol part of the molecule. The name of the acid is propanoic acid (IUPAC) or propionic acid (common). Deleting the *-ic acid* ending and adding *-ate* give the name *ethyl propanoate* (IUPAC) or *ethyl propionate* (common).

b. The name of the alcohol part of the molecule is methyl (from methanol or methyl alcohol). The name of the five-carbon acid is 3-methylbutanoic acid or β-methylbutyric acid. Hence the ester name is *methyl 3-methylbutanoate* (IUPAC) or *methyl β-methylbutyrate* (common).

c. The name *propyl* characterizes the alcohol part of the molecule. The acid part of the molecule is derived from benzoic acid (both IUPAC and common name). Hence the ester name in both systems is *propyl benzoate.*

• The contrast between IUPAC names and common names for unbranched esters of carboxylic acids is as follows:

IUPAC (two words)

alkyl alkanoate

methyl propanoate

Common (two words)

alkyl (prefix)ate*

methyl acetate

*The common-name prefixes are related to natural sources for the "parent" carboxylic acids.

Practice Exercise 16.4

Assign both IUPAC and common names to the following esters.

a. $CH_3-\overset{\overset{\displaystyle O}{\|}}{C}-O-CH_2-CH_3$

b. $CH_3-CH_2-CH_2-CH_2-\overset{\overset{\displaystyle O}{\|}}{C}-O-CH_3$

c. $H-\overset{\overset{\displaystyle O}{\|}}{C}-O-CH_2-CH_2-CH_3$

• *Answers:* **a.** ethyl ethanoate (IUPAC), ethyl acetate (common); **b.** methyl pentanoate (IUPAC), methyl valerate (common); **c.** propyl methanoate (IUPAC), propyl formate (common)

16.11 Selected Common Esters

In this section we consider selected esters that function as flavoring agents, pheromones, and medications.

• Flavoring Agents

Esters are largely responsible for the flavor and fragrance of fruits and flowers. Generally, a natural flavor or odor is caused by a mixture of esters, with one particular compound being dominant. The synthetic production of these "dominant" compounds is the basis for the flavoring agents used in ice cream, gelatins, soft drinks, and so on. Table 16.4 gives the structures of selected esters used as flavoring agents. What is surprising about the structures in Table 16.4 is how closely some of them resemble each other. For example, the apple and pineapple flavoring agents differ by one carbon atom (methyl

Table 16.4
Selected Esters That Are Used as Flavoring Agents

IUPAC name	Structural formula	Characteristic flavor and odor
isobutyl methanoate	$H-\overset{\overset{\displaystyle O}{\|}}{C}-O-CH_2-\overset{\overset{\displaystyle CH_3}{\vert}}{CH}-CH_3$	raspberry
propyl ethanoate	$CH_3-\overset{\overset{\displaystyle O}{\|}}{C}-O-(CH_2)_2-CH_3$	pear
pentyl ethanoate	$CH_3-\overset{\overset{\displaystyle O}{\|}}{C}-O-(CH_2)_4-CH_3$	banana
octyl ethanoate	$CH_3-\overset{\overset{\displaystyle O}{\|}}{C}-O-(CH_2)_7-CH_3$	orange
pentyl propanoate	$CH_3-CH_2-\overset{\overset{\displaystyle O}{\|}}{C}-O-(CH_2)_4-CH_3$	apricot
methyl butanoate	$CH_3-(CH_2)_2-\overset{\overset{\displaystyle O}{\|}}{C}-O-CH_3$	apple
ethyl butanoate	$CH_3-(CH_2)_2-\overset{\overset{\displaystyle O}{\|}}{C}-O-CH_2-CH_3$	pineapple

versus ethyl); a five-carbon chain versus an eight-carbon chain makes the difference between banana and orange flavor.

Cats of all types (from lions to house cats) are strongly attracted to the catnip plant. The attractant in the catnip plant, which is not a pheromone because different species are involved, is the compound nepetalactone, a cyclic ester.

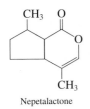

CH_3 O O CH_3

Nepetalactone

• Pheromones

A number of pheromones (Section 13.6) contain ester functional groups. The compound isoamyl acetate,

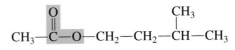

$$CH_3-\overset{\overset{\displaystyle O}{\|}}{C}-O-CH_2-CH_2-\overset{\overset{\displaystyle CH_3}{|}}{CH}-CH_3$$

is an alarm pheromone for the honey bee. The compound methyl *p*-hydroxybenzoate,

$$HO-C_6H_4-\overset{\overset{\displaystyle O}{\|}}{C}-O-CH_3$$

is a sexual attractant for canine species. It is secreted by female dogs in heat and evokes attraction and sexual arousal in male dogs.

• Medications

Numerous esters have medicinal value, including benzocaine (a local anesthetic), aspirin, and oil of wintergreen (a counterirritant). The structure of benzocaine, an amino ester, is

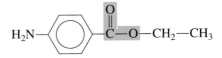

$$H_2N-C_6H_4-\overset{\overset{\displaystyle O}{\|}}{C}-O-CH_2-CH_3$$

Both aspirin and oil of wintergreen are esters of salicylic acid, an aromatic hydroxy-acid.

$$C_6H_4(OH)-\overset{\overset{\displaystyle O}{\|}}{C}-OH$$

Salicylic acid

Because this acid has both an acid group and a hydroxyl group, it can form two different types of esters: one by reaction of its acid group with an alcohol, the other by reaction of its alcohol group with a carboxylic acid.

Reaction of acetic acid with the alcohol group of salicylic acid produces aspirin.

$$C_6H_4(O-H)-\overset{\overset{\displaystyle O}{\|}}{C}-OH + HO-\overset{\overset{\displaystyle O}{\|}}{C}-CH_3 \xrightarrow[\text{Heat}]{H^+} C_6H_4\left(O-\overset{\overset{\displaystyle O}{\|}}{C}-CH_3\right)-\overset{\overset{\displaystyle O}{\|}}{C}-OH + H_2O$$

Salicylic acid — Acetic acid — Aspirin

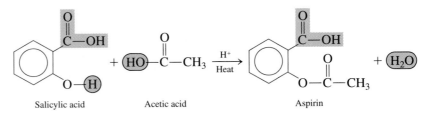

Aspirin's mode of action in the human body is considered in Chemical Connections 16.2.

Reaction of methanol with the acid group of salicylic acid produces oil of wintergreen.

$$C_6H_4(OH)-\overset{\overset{\displaystyle O}{\|}}{C}-OH + H-O-CH_3 \xrightarrow[\text{Heat}]{H^+} C_6H_4(OH)-\overset{\overset{\displaystyle O}{\|}}{C}-O-CH_3 + H_2O$$

Salicylic acid — Methanol — Oil of wintergreen

Chemical CONNECTIONS

16.2 Aspirin

Aspirin, an ester of salicylic acid (Section 16.11), is a drug that has the ability to decrease pain (analgesic properties), to lower body temperature (antipyretic properties), and to reduce inflammation (anti-inflammatory properties). It is most frequently taken in tablet form, and the tablet usually contains 325 mg of aspirin held together with an inert starch binder.

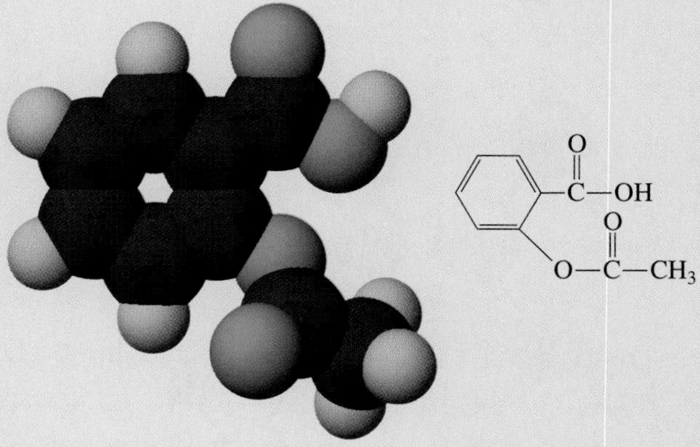

After ingestion, aspirin undergoes hydrolysis to produce salicylic acid and acetic acid. Salicylic acid is the active ingredient of aspirin—the substance that has analgesic, antipyretic, and anti-inflammatory effects.

Salicylic acid is capable of irritating the lining of the stomach, inducing a small amount of bleeding. Breaking (or chewing) an aspirin tablet, rather than taking it whole, reduces the chance of bleeding by eliminating drug concentration on one part of the stomach lining. Buffered aspirin products contain alkaline chemicals (such as aluminum glycinate or aluminum hydroxide) to neutralize the acidity of the aspirin when it contacts the stomach lining.

Aspirin—that is, salicylic acid—inhibits the synthesis of a class of hormones called prostaglandins (Section 19.11), molecules that cause pain, fever, and inflammation when present in the bloodstream in higher-than-normal levels. Salicylic acid's mode of action is irreversible inhibition (Section 21.7) of *cyclooxygenase,* an enzyme necessary for the production of prostaglandins.

Recent studies show that aspirin also increases the time it takes blood to coagulate (clot). For blood to coagulate, platelets must first be able to aggregate, and prostaglandins (which aspirin inhibits) appear to be necessary for platelet aggregation to occur. One study suggests that healthy men can cut their risk of heart attacks nearly in half by taking one baby aspirin per day (85 mg compared to the 325 mg in a regular tablet). Aspirin acts by making the blood less likely to clot. Heart attacks usually occur when clots form in the coronary arteries, cutting off blood supply to the heart.

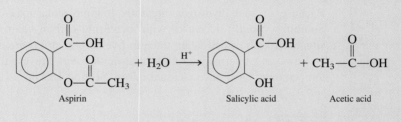

$$\text{Aspirin} + H_2O \xrightarrow{H^+} \text{Salicylic acid} + CH_3-\overset{O}{\overset{\|}{C}}-OH \ (\text{Acetic acid})$$

Aspirin Salicylic acid Acetic acid

Oil of wintergreen, also called methyl salicylate, is used in skin rubs and liniments to help decrease the pain of sore muscles. It is absorbed through the skin, where it is hydrolyzed to produce salicylic acid. Salicylic acid is the actual pain reliever.

16.12 Physical Properties of Esters

Ester molecules cannot form hydrogen bonds to each other because they do not have a hydrogen atom bonded to an oxygen atom. Consequently, the boiling points of esters are much lower than those of alcohols and acids of comparable molecular mass. Esters are more like ethers in their physical properties. Table 16.5 gives boiling-point data for compounds of similar molecular mass that contain different functional groups.

Water molecules can hydrogen-bond to esters through the oxygen atoms present in the ester functional group.

Table 16.5
Boiling Points of Compounds of Similar Molecular Mass That Contain Different Functional Groups

Name	Functional-group class	Molecular mass	Boiling point (°C)
diethyl ether	ether	74	34
ethyl formate	ester	74	54
methyl acetate	ester	74	57
butanal	aldehyde	72	76
1-butanol	alcohol	74	118
propionic acid	acid	74	141

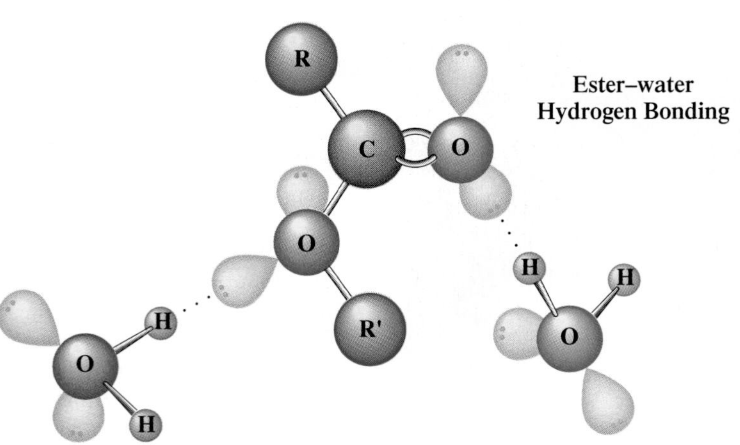

A physical-state summary for methyl and ethyl esters of unbranched-chain carboxylic acids at room temperature and pressure.

Methyl Esters			
☒	C_3	C_5	C_7
C_2	C_4	C_6	C_8

Ethyl Esters			
☒	C_3	C_5	C_7
☒	C_4	C_6	C_8

□ Liquid

Because of such hydrogen bonding, low-molecular-mass esters are soluble in water. Solubility rapidly decreases with increasing carbon chain length; borderline solubility situations are reached when three to five carbon atoms are in a chain.

Low- and intermediate-molecular-mass esters are usually colorless liquids at room temperature. Most have pleasant odors (Section 16.11).

16.13 Chemical Reactions of Esters

The most important reaction of esters involves breaking the carbon–oxygen single bond that holds the "alcohol part" and the "acid part" of the ester together. This reaction process is called either ester hydrolysis or ester saponification, depending on reaction conditions.

• Ester Hydrolysis

In ester hydrolysis, an ester reacts with water, producing the carboxylic acid and alcohol from which the ester was formed. Ester hydrolysis is thus the reverse of esterification (Section 16.9).

• The breaking of a bond within a molecule and the attachment of the components of water to the fragments are characteristics of all hydrolysis reactions.

• This is our second encounter with hydrolysis reactions. The first encounter involved the hydrolysis of acetals and ketals (Section 15.9).

$$\mathrm{R{-}\overset{\displaystyle O}{\overset{\|}{C}}{-}O{-}R' + \textcircled{H}{-}\boxed{OH} \xrightarrow{H^+} R{-}\overset{\displaystyle O}{\overset{\|}{C}}{-}\boxed{OH} + R'{-}O{-}\textcircled{H}}$$

$$\underset{\text{Methyl acetate}}{\mathrm{CH_3{-}\overset{\displaystyle O}{\overset{\|}{C}}{-}O{-}CH_3}} + \underset{\text{Water}}{\mathrm{\textcircled{H}{-}\boxed{OH}}} \xrightarrow{H^+} \underset{\text{Acetic acid}}{\mathrm{CH_3{-}\overset{\displaystyle O}{\overset{\|}{C}}{-}\boxed{OH}}} + \underset{\text{Methyl alcohol}}{\mathrm{CH_3{-}O{-}\textcircled{H}}}$$

Ester hydrolysis requires the presence of a strong-acid catalyst or enzymes.

• In both ester hydrolysis and ester saponification, an alcohol is produced. Under acidic conditions (ester hydrolysis), the other product is a carboxylic acid; under basic conditions (ester saponification), the other product is a carboxylic acid salt.

• Ester Saponification

Ester saponification *is the base-catalyzed hydrolysis of an ester.* Either of the strong bases NaOH and KOH is used. The products of ester saponification are an alcohol and a carboxylic acid salt. (Any carboxylic acid product formed is converted to its salt because of the basic reaction conditions.)

$$\underset{\text{An ester}}{R{-}\overset{\overset{\displaystyle O}{\|}}{C}{-}O{-}R'} + \underset{\text{A strong base}}{NaOH} \xrightarrow{H_2O} \underset{\text{A carboxylate salt}}{R{-}\overset{\overset{\displaystyle O}{\|}}{C}{-}O^-\ Na^+} + \underset{\text{An alcohol}}{R'{-}OH}$$

A specific example of ester saponification is

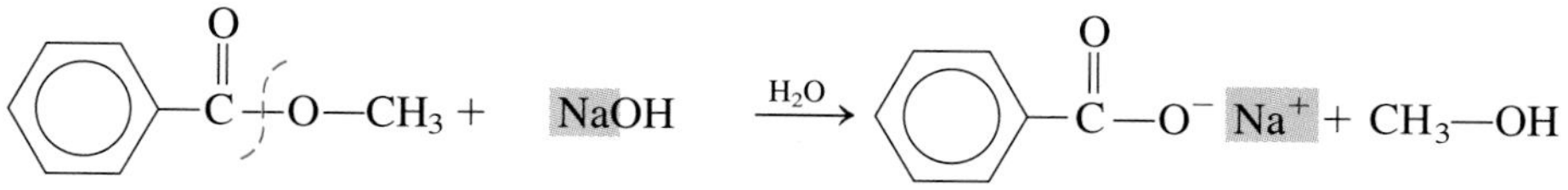

$$\underset{\text{Methyl benzoate}}{C_6H_5{-}\overset{\overset{\displaystyle O}{\|}}{C}{-}O{-}CH_3} + \underset{\text{Sodium hydroxide}}{NaOH} \xrightarrow{H_2O} \underset{\text{Sodium benzoate}}{C_6H_5{-}\overset{\overset{\displaystyle O}{\|}}{C}{-}O^-\ Na^+} + \underset{\text{Methyl alcohol}}{CH_3{-}OH}$$

Example 16.5

Structural Equations for Reactions That Involve Esters

Write structural equations for each of the following reactions.

a. Hydrolysis, with an acidic catalyst, of ethyl acetate
b. Saponification, with NaOH, of methyl formate
c. Esterification of propionic acid using isopropyl alcohol

Solution

a. Hydrolysis, under acidic conditions, cleaves an ester to produce its "parent" carboxylic acid and alcohol.

$$\underset{\text{Ethyl acetate}}{CH_3{-}\overset{\overset{\displaystyle O}{\|}}{C}{-}O{-}CH_2{-}CH_3} + H_2O \xrightarrow{H^+} \underset{\text{Acetic acid}}{CH_3{-}\overset{\overset{\displaystyle O}{\|}}{C}{-}OH} + \underset{\text{Ethyl alcohol}}{CH_3{-}CH_2{-}OH}$$

b. Saponification cleaves an ester to produce its "parent" alcohol and the *salt* of its "parent" carboxylic acid.

$$\underset{\text{Methyl formate}}{H{-}\overset{\overset{\displaystyle O}{\|}}{C}{-}O{-}CH_3} + \underset{\text{Sodium hydroxide}}{NaOH} \xrightarrow{H_2O} \underset{\text{Sodium formate}}{H{-}\overset{\overset{\displaystyle O}{\|}}{C}{-}O^-\ Na^+} + \underset{\text{Methyl alcohol}}{CH_3{-}OH}$$

c. Esterification is the reaction in which a carboxylic acid and an alcohol react to produce an ester.

$$\underset{\text{Propionic acid}}{CH_3{-}CH_2{-}\overset{\overset{\displaystyle O}{\|}}{C}{-}OH} + \underset{\text{Isopropyl alcohol}}{CH_3{-}\underset{\underset{\displaystyle CH_3}{|}}{CH}{-}OH} \overset{H^+}{\rightleftharpoons}$$

$$\underset{\text{Isopropyl propionate}}{CH_3{-}CH_2{-}\overset{\overset{\displaystyle O}{\|}}{C}{-}O{-}\underset{\underset{\displaystyle CH_3}{|}}{CH}{-}CH_3} + H_2O$$

Practice Exercise 16.5

Write structural equations for each of the following reactions.

a. Hydrolysis, with an acidic catalyst, of propyl propanoate
b. Saponification, with KOH, of ethyl propanoate
c. Esterification of acetic acid with propyl alcohol

• *Answers:* **a.**

$$CH_3{-}CH_2{-}\overset{\overset{\displaystyle O}{\|}}{C}{-}O{-}CH_2{-}CH_2{-}CH_3 + H_2O \xrightarrow{H^+} CH_3{-}CH_2{-}\overset{\overset{\displaystyle O}{\|}}{C}{-}OH + CH_3{-}CH_2{-}CH_2{-}OH$$

b.

$$CH_3{-}CH_2{-}\overset{\overset{\displaystyle O}{\|}}{C}{-}O{-}CH_2{-}CH_3 + KOH \xrightarrow{H_2O} CH_3{-}CH_2{-}\overset{\overset{\displaystyle O}{\|}}{C}{-}O^-K^+ + CH_3{-}CH_2{-}OH$$

c.

$$CH_3{-}\overset{\overset{\displaystyle O}{\|}}{C}{-}OH + CH_3{-}CH_2{-}CH_2{-}OH \overset{H^+}{\rightleftharpoons} CH_3{-}\overset{\overset{\displaystyle O}{\|}}{C}{-}O{-}CH_2{-}CH_2{-}CH_3 + H_2O$$

The accompanying Chemistry at a Glance summarizes reactions that involve carboxylic acids and esters.

16.14 Sulfur Analogs of Esters

Just as alcohols react with carboxylic acids to produce esters, thiols (Section 14.13) react with carboxylic acids to produce thioesters. A **thioester** *is a sulfur-containing analog of an ester in which an —SR group has replaced the —OR group.*

$$\underset{\text{A carboxylic acid}}{CH_3{-}\overset{\overset{\displaystyle O}{\|}}{C}{-}OH} + \underset{\text{A thiol}}{CH_3{-}CH_2{-}S{-}H} \longrightarrow \underset{\text{A thioester}}{CH_3{-}\overset{\overset{\displaystyle O}{\|}}{C}{-}S{-}CH_2{-}CH_3} + H_2O$$

The thioester methyl thiobutanoate is used as an artificial flavoring agent. It generates the taste we call "strawberry."

$$CH_3{-}CH_2{-}CH_2{-}\overset{\overset{\displaystyle O}{\|}}{C}{-}S{-}CH_3$$

Methyl thiobutanoate

The most important naturally occurring thioester is acetyl coenzyme A, whose abbreviated structure is

$$CH_3{-}\overset{\overset{\displaystyle O}{\|}}{C}{-}S{-}CoA$$

Acetyl coenzyme A

Acetyl coenzyme A plays a central role in the metabolic cycles through which the body obtains energy to "run itself" (Section 23.6).

Coenzyme A, the parent molecule for acetyl coenzyme A, is a large complex *thiol* whose structure, for simplicity, is usually abbreviated as CoA—S—H. The formation of acetyl coenzyme A (acetyl CoA) from coenzyme A can be envisioned as a thioesterification reaction between acetic acid and coenzyme A.

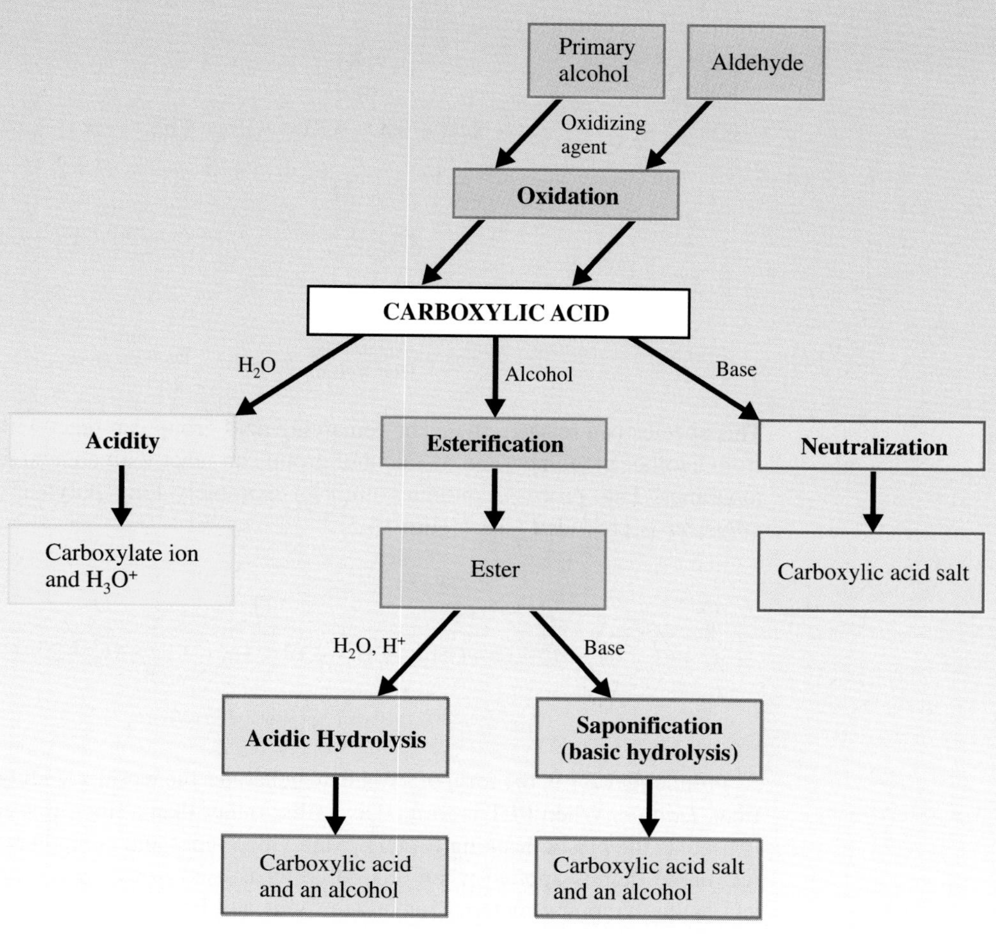

$$\underset{\text{Acetic acid}}{CH_3-\overset{\overset{\displaystyle O}{\|}}{C}-OH} + \underset{\substack{\text{Coenzyme A}\\ \text{(a thiol)}}}{CoA-S-H} \longrightarrow \underset{\substack{\text{Acetyl coenzyme A}\\ \text{(acetyl CoA)}}}{CH_3-\overset{\overset{\displaystyle O}{\|}}{C}-S-CoA} + H_2O$$

The complete structure of acetyl CoA is given in Section 23.3.

16.15 Polyesters

A **polyester** *is a polymer in which the monomers are joined through ester linkages.* Dicarboxylic acids and dialcohols are the monomers used to form polyesters.

The best known of the many polyesters now marketed is *poly(ethylene terephthalate),* which is also known by the acronym *PET.* The monomers used to produce PET are terephthalic acid (a diacid) and ethylene glycol (a dialcohol).

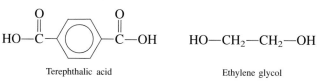

$$\underset{\text{Terephthalic acid}}{HO-\overset{\overset{\displaystyle O}{\|}}{C}-C_6H_4-\overset{\overset{\displaystyle O}{\|}}{C}-OH} \qquad \underset{\text{Ethylene glycol}}{HO-CH_2-CH_2-OH}$$

The reaction of one acid group of the diacid with one alcohol group of the dialcohol initially produces an ester molecule, with an acid group left over on one end and an alcohol group left over on the other end.

$$HO-\overset{O}{\overset{\|}{C}}-C_6H_4-\overset{O}{\overset{\|}{C}}-OH + HO-CH_2-CH_2-OH \longrightarrow$$

$$HO-\overset{O}{\overset{\|}{C}}-C_6H_4-\overset{O}{\overset{\|}{C}}-O-CH_2-CH_2-OH + H_2O$$

Leftover acid group that can react further

Ester functional group

Leftover alcohol group that can react further

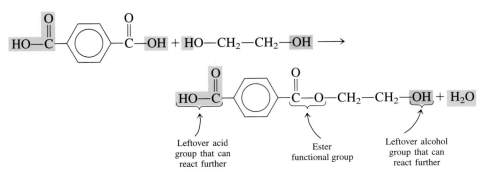

This species can react further. The remaining acid group can react with an alcohol group from another monomer, and the alcohol group can react with an acid group from another monomer. This process continues until an extremely long polymer molecule called a *polyester* is produced (see Figure 16.5).

$$\cdots\overset{O}{\overset{\|}{C}}-C_6H_4-\overset{O}{\overset{\|}{C}}-O-CH_2-CH_2-O-\overset{O}{\overset{\|}{C}}-C_6H_4-\overset{O}{\overset{\|}{C}}-O-CH_2-CH_2-O\cdots$$

Ester linkage

Ester linkage

Poly(ethylene terephthalate), a polyester

Producing PET in the form of a filament generates the world's leading synthetic clothing fiber, *Dacron.* When PET is formed as a film, rather than a fiber, it is called *Mylar,* which is used as the plastic backing for audio and video tapes and computer diskettes. Its chemical name PET is applied when this polyester is used in clear, flexible soft-drink bottles and as the wrapping material for frozen foods and boil-in bags for foods.

PET is also used in medicine. Because it is physiologically inert, PET is used in the form of a mesh to replace diseased sections of arteries. It has also been used in synthetic heart valves.

PET is an example of a condensation polymer. A **condensation polymer** *is a polymer formed by reacting bifunctional monomers to give a polymer and some small molecule, such as water.* Condensation polymers differ from addition polymers (Section 13.8), where the polymer is the *only* product.

A variation of the diacid–dialcohol monomer formulation for polyesters involves using hydroxyacids as monomers. In this situation, both of the functional groups required are present in the same molecule.

A copolymer (Section 13.8) in which lactic acid and glycolic acid (both hydroxyacids, Section 16.4) are monomers produces a biodegradable material (trade name *Lactomer*) that is used as surgical staples in several types of surgery. Traditional suture materials

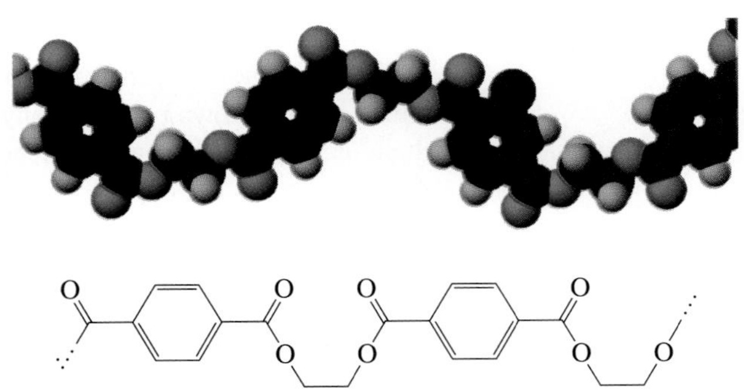

Figure 16.5 Space-filling model of a segment of the polyester condensation polymer known as poly(ethylene terephthalate), or PET.

must be removed later on, after they have served their purpose. Lactomer staples start to dissolve (hydrolyze) after a period of several weeks. The hydrolysis products are the starting monomers, lactic acid and glycolic acid, both of which are normally present in the human body. By the time tissue has fully healed, the staples have fully degraded.

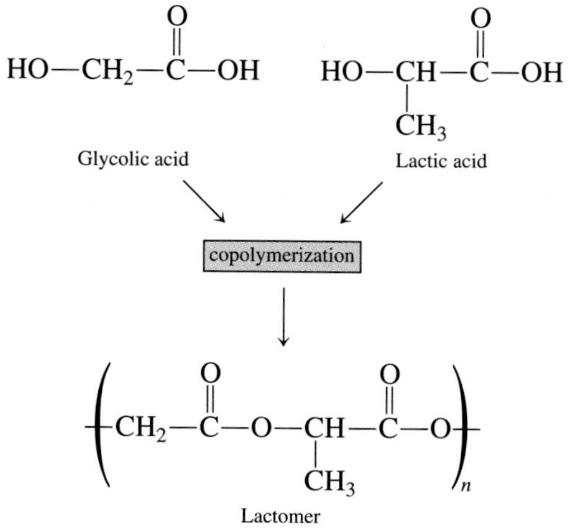

$$HO-CH_2-\overset{\overset{O}{\|}}{C}-OH \qquad HO-\underset{\underset{CH_3}{|}}{CH}-\overset{\overset{O}{\|}}{C}-OH$$

Glycolic acid Lactic acid

copolymerization

$$\left(CH_2-\overset{\overset{O}{\|}}{C}-O-\underset{\underset{CH_3}{|}}{CH}-\overset{\overset{O}{\|}}{C}-O\right)_n$$

Lactomer

16.16 Esters of Inorganic Acids

Inorganic acids such as sulfuric, phosphoric, and nitric acids react with alcohols to form esters in a manner similar to that for carboxylic acids.

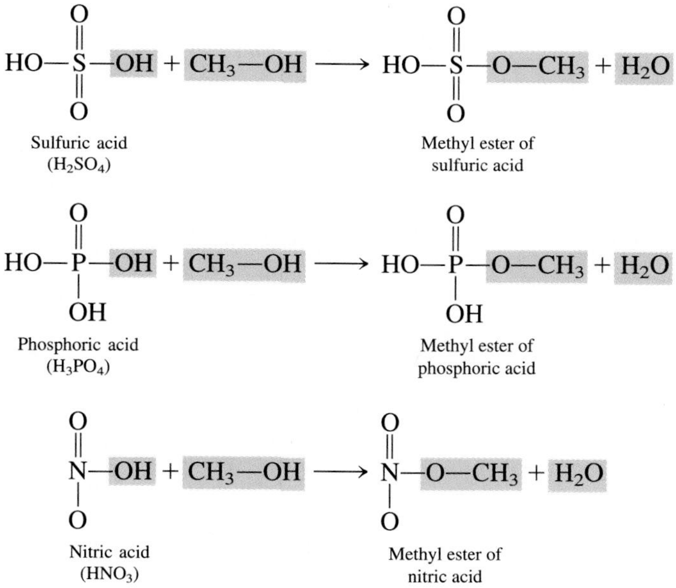

$$HO-\underset{\underset{O}{\|}}{\overset{\overset{O}{\|}}{S}}-OH + CH_3-OH \longrightarrow HO-\underset{\underset{O}{\|}}{\overset{\overset{O}{\|}}{S}}-O-CH_3 + H_2O$$

Sulfuric acid (H_2SO_4) Methyl ester of sulfuric acid

$$HO-\underset{\underset{OH}{|}}{\overset{\overset{O}{\|}}{P}}-OH + CH_3-OH \longrightarrow HO-\underset{\underset{OH}{|}}{\overset{\overset{O}{\|}}{P}}-O-CH_3 + H_2O$$

Phosphoric acid (H_3PO_4) Methyl ester of phosphoric acid

$$\underset{\underset{O}{|}}{\overset{\overset{O}{\|}}{N}}-OH + CH_3-OH \longrightarrow \underset{\underset{O}{|}}{\overset{\overset{O}{\|}}{N}}-O-CH_3 + H_2O$$

Nitric acid (HNO_3) Methyl ester of nitric acid

• Esters of inorganic acids undergo hydrolysis reactions in a manner similar to that for esters of carboxylic acids (Section 16.13).

The most important inorganic esters, from a biochemical standpoint, are those of phosphoric acid—that is, phosphate esters. A **phosphate ester** *is a compound formed by reaction of an alcohol with phosphoric acid.* Because phosphoric acid has three hydroxyl groups, it can form mono-, di-, and triesters by reaction with one, two, and three molecules of alcohol, respectively.

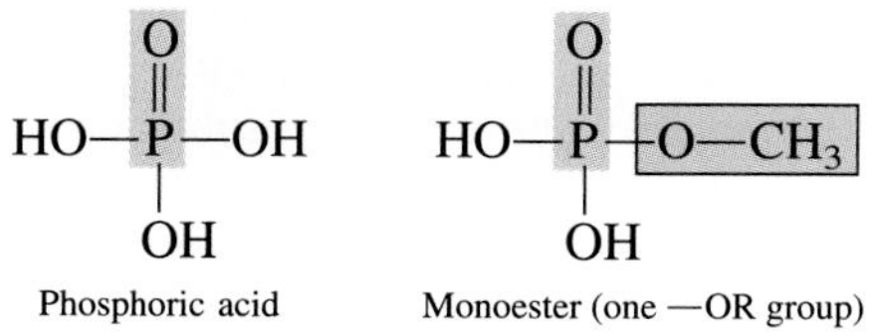

$$HO-\underset{\underset{OH}{|}}{\overset{\overset{O}{\|}}{P}}-OH \qquad HO-\underset{\underset{OH}{|}}{\overset{\overset{O}{\|}}{P}}-O-CH_3$$

Phosphoric acid Monoester (one —OR group)

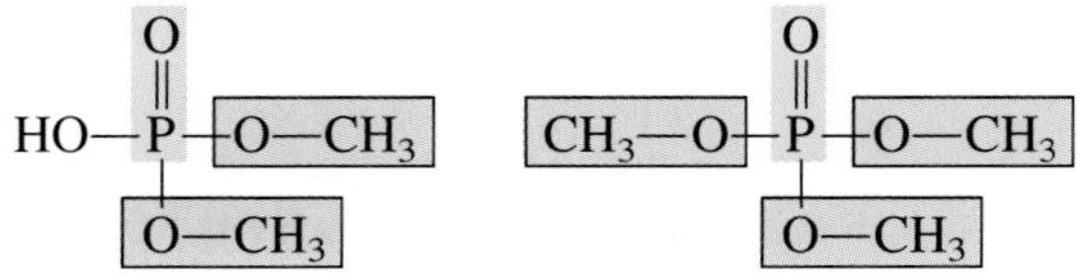

$$HO{-}\overset{\overset{\Large O}{\|}}{\underset{\underset{\Large O-CH_3}{|}}{P}}{-}O{-}CH_3 \qquad CH_3{-}O{-}\overset{\overset{\Large O}{\|}}{\underset{\underset{\Large O-CH_3}{|}}{P}}{-}O{-}CH_3$$

Diester (two —OR groups) Triester (three —OR groups)

Phosphoric acid molecules have the ability to react with each other to produce diphosphoric acid and triphosphoric acid.

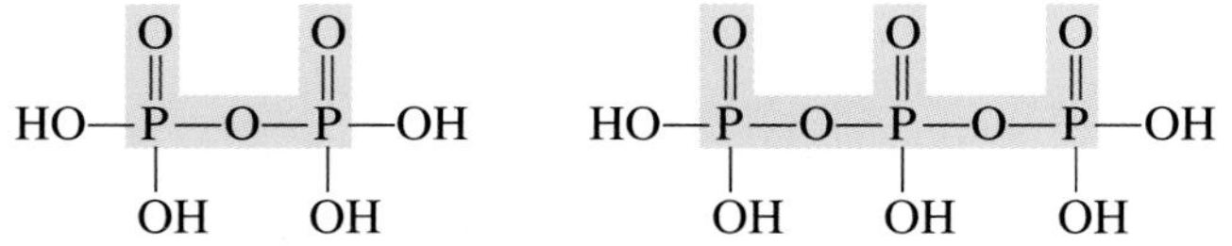

$$HO{-}\overset{\overset{\Large O}{\|}}{\underset{\underset{\Large OH}{|}}{P}}{-}O{-}\overset{\overset{\Large O}{\|}}{\underset{\underset{\Large OH}{|}}{P}}{-}OH \qquad HO{-}\overset{\overset{\Large O}{\|}}{\underset{\underset{\Large OH}{|}}{P}}{-}O{-}\overset{\overset{\Large O}{\|}}{\underset{\underset{\Large OH}{|}}{P}}{-}O{-}\overset{\overset{\Large O}{\|}}{\underset{\underset{\Large OH}{|}}{P}}{-}OH$$

Diphosphoric acid Triphosphoric acid

These acids also undergo esterification reactions with alcohols, producing species such as those shown below.

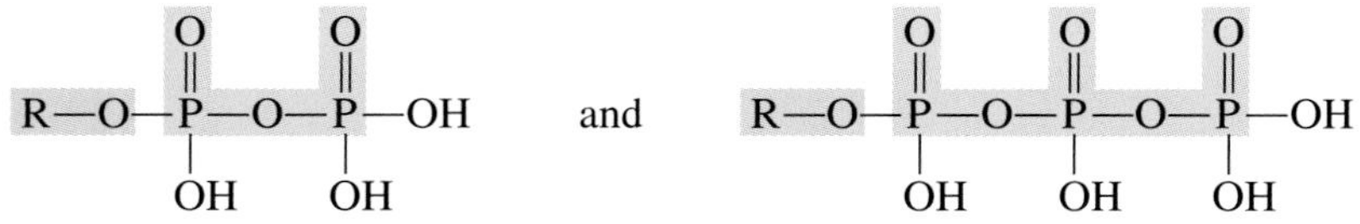

$$R{-}O{-}\overset{\overset{\Large O}{\|}}{\underset{\underset{\Large OH}{|}}{P}}{-}O{-}\overset{\overset{\Large O}{\|}}{\underset{\underset{\Large OH}{|}}{P}}{-}OH \quad \text{and} \quad R{-}O{-}\overset{\overset{\Large O}{\|}}{\underset{\underset{\Large OH}{|}}{P}}{-}O{-}\overset{\overset{\Large O}{\|}}{\underset{\underset{\Large OH}{|}}{P}}{-}O{-}\overset{\overset{\Large O}{\|}}{\underset{\underset{\Large OH}{|}}{P}}{-}OH$$

Diphosphate monoester Triphosphate monoester

We will encounter esters of these types in Chapter 23 when we consider the biochemical production of energy in the human body. Adenosine diphosphate (ADP) and adenosine triphosphate (ATP) are important examples of such compounds.

Chemical CONNECTIONS

16.3 Nitroglycerin: An Inorganic Triester

The reaction of one molecule of glycerol (a trihydroxyalcohol) with three molecules of nitric acid produces the trinitrate ester called nitroglycerin.

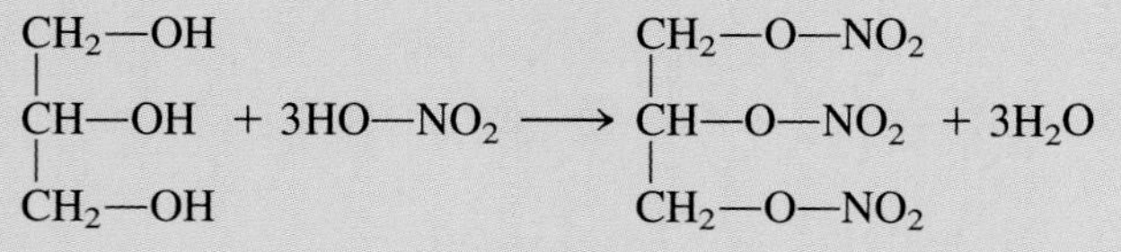

$$\begin{array}{l} CH_2{-}OH \\ | \\ CH{-}OH \\ | \\ CH_2{-}OH \end{array} + 3HO{-}NO_2 \longrightarrow \begin{array}{l} CH_2{-}O{-}NO_2 \\ | \\ CH{-}O{-}NO_2 \\ | \\ CH_2{-}O{-}NO_2 \end{array} + 3H_2O$$

Besides being a component of dynamite explosives, nitroglycerin has medicinal value. It is used in treating patients with angina pectoris—sharp chest pains caused by an insufficient supply of oxygen reaching heart muscle. Its effect on the human body is that of a vasodilator, a substance that increases blood flow by relaxing constricted muscles around blood vessels.

Nitroglycerin medication is available in several forms: (1) as a liquid diluted with alcohol to render it nonexplosive, (2) as a liquid adsorbed to a tablet for convenience of sublingual (under the tongue) administration, (3) in ointments for topical use, and (4) as "skin patches" that release the drug continuously through the skin over a 24-hr period. Nitroglycerin is rapidly absorbed through the skin, enters the bloodstream, and finds its way to heart muscle within seconds.

In the pure state, nitroglycerin is a shock-sensitive liquid that can decompose to produce large volumes of gases (N_2, CO_2, H_2O, and O_2). When used in dynamite, it is adsorbed on clay-like materials, giving products that will not explode without a formal ignition system.

Another compound used for the same medicinal purposes as nitroglycerin is isopentyl nitrite. It is a monoester involving nitrous acid and isopentyl alcohol.

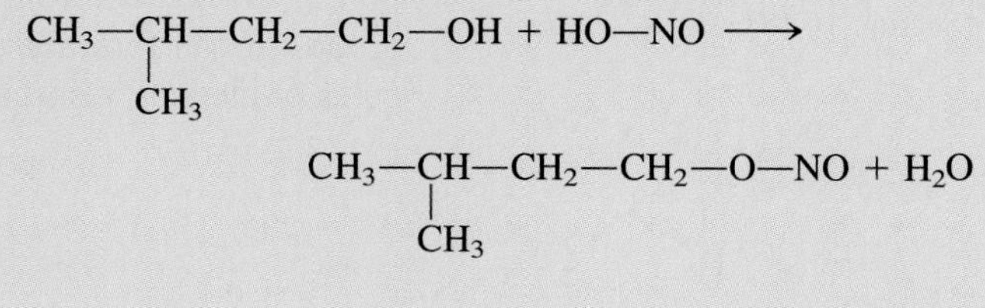

$$CH_3{-}\underset{\underset{\Large CH_3}{|}}{CH}{-}CH_2{-}CH_2{-}OH + HO{-}NO \longrightarrow$$

$$CH_3{-}\underset{\underset{\Large CH_3}{|}}{CH}{-}CH_2{-}CH_2{-}O{-}NO + H_2O$$

Concepts to Remember

The carboxyl group. The functional group present in carboxylic acids is the carboxyl group. A carboxyl group is composed of a hydroxyl group bonded to a carbonyl carbon atom. It thus contains two oxygen atoms directly bonded to the same carbon atom.

Nomenclature of carboxylic acids. The IUPAC name for a monocarboxylic acid is formed by replacing the final *-e* of the hydrocarbon parent name with *-oic acid.* As with previous IUPAC nomenclature, the longest carbon chain containing the functional group is identified, and it is numbered starting with the carboxyl carbon atom. Common-name usage is more prevalent for carboxylic acids than for any other type of organic compound.

Types of carboxylic acids. Carboxylic acids are classified by the number of carboxyl groups present (monocarboxylic, dicarboxylic, etc.), by the degree of saturation (saturated, unsaturated, aromatic), and by additional functional groups present (hydroxy, keto, etc.).

Physical properties of carboxylic acids. Low-molecular-mass carboxylic acids are liquids at room temperature and have sharp or unpleasant odors. Long-chain acids are wax-like solids. The carboxyl group is polar and forms hydrogen bonds to other carboxyl groups or other molecules. Thus carboxylic acids have relatively high boiling points, and those with lower molecular masses are soluble in water.

Preparation of carboxylic acids. Carboxylic acids are synthesized through oxidation of primary alcohols or aldehydes using strong oxidizing agents. Aromatic carboxylic acids can be prepared by oxidizing a carbon side chain on a benzene derivative using a strong oxidizing agent.

Acidity of carboxylic acids. Soluble carboxylic acids behave as weak acids, donating protons to water molecules. The portion of the acid molecule left after proton loss is called a carboxylate ion.

Carboxylic acid salts. Carboxylic acids are neutralized by bases to produce carboxylic acid salts. Such salts are usually more soluble in water than the acids from which they were derived. Carboxylic acid salts are named by changing the *-ic* ending of the acid to *-ate.*

Esters. Esters are formed by the reaction of an acid with an alcohol. In such reactions, the —OR group from the alcohol replaces the —OH group in the carboxylic acid. Esters are polar compounds, but they cannot form hydrogen bonds to each other. Therefore, their boiling points are lower than those of alcohols and acids of similar molecular mass.

Nomenclature of esters. An ester is named as an alkyl (from the name of the alcohol reactant) carboxylate (from the name of the acid reactant).

Reactions of esters. Esters can be converted back to carboxylic acids and alcohols under either acidic or basic conditions. Under acidic conditions, the process is called hydrolysis, and the products are the acid and alcohol. Under basic conditions, the process is called saponification, and the products are the acid salt and alcohol.

Thioesters. Thioesters are sulfur-containing analogs of esters in which a —SR group has replaced the —OR group.

Polyesters. Polyesters are polymers in which the monomers (diacids and dialcohols) are joined through ester linkages.

Esters of inorganic acids. Alcohols can react with inorganic acids, such as nitric, sulfuric, and phosphoric acids, to form esters. Phosphate esters are an important class of biological compounds.

Key Reactions and Equations

1. Oxidation of a primary alcohol to an acid (Section 16.6)

$$R{-}CH_2{-}OH \xrightarrow{[O]} R{-}\overset{\overset{\large O}{\|}}{C}{-}H \xrightarrow{[O]} R{-}\overset{\overset{\large O}{\|}}{C}{-}OH$$

2. Oxidation of an alkylbenzene to an acid (Section 16.6)

$$C_6H_5{-}CH_2{-}R \xrightarrow{[O]} C_6H_5{-}COOH$$

3. Ionization of a carboxylic acid to give a carboxylate ion and a hydronium ion (Section 16.7)

$$R{-}\overset{\overset{\large O}{\|}}{C}{-}OH + H_2O \rightleftharpoons R{-}\overset{\overset{\large O}{\|}}{C}{-}O^- + H_3O^+$$

4. Reaction of a carboxylic acid with a base to produce a carboxylic acid salt plus water (Section 16.8)

$$R{-}\overset{\overset{\large O}{\|}}{C}{-}OH + NaOH \longrightarrow R{-}\overset{\overset{\large O}{\|}}{C}{-}O^-Na^+ + H_2O$$

5. Preparation of an ester from an acid and an alcohol (Section 16.9)

$$R{-}\overset{\overset{\large O}{\|}}{C}{-}OH + R'{-}OH \overset{H^+}{\rightleftharpoons} R{-}\overset{\overset{\large O}{\|}}{C}{-}O{-}R' + H_2O$$

6. Ester hydrolysis to produce a carboxylic acid and an alcohol (Section 16.13)

$$R{-}\overset{\overset{\large O}{\|}}{C}{-}O{-}R' + H{-}OH \overset{H^+}{\rightleftharpoons} R{-}\overset{\overset{\large O}{\|}}{C}{-}OH + R'{-}OH$$

7. Ester saponification to give a carboxylic acid salt and alcohol (Section 16.13)

$$R{-}\overset{\overset{\large O}{\|}}{C}{-}O{-}R' + NaOH \xrightarrow{H_2O} R{-}\overset{\overset{\large O}{\|}}{C}{-}O^-Na^+ + R'{-}OH$$

8. Preparation of a thioester (Section 16.14)

$$R{-}\overset{\overset{\large O}{\|}}{C}{-}OH + R'{-}SH \xrightarrow{H^+} R{-}\overset{\overset{\large O}{\|}}{C}{-}S{-}R' + H_2O$$

9. Phosphate ester formation (Section 16.16)

$$R{-}OH + HO{-}\underset{\underset{\large OH}{|}}{\overset{\overset{\large O}{\|}}{P}}{-}OH \longrightarrow R{-}O{-}\underset{\underset{\large OH}{|}}{\overset{\overset{\large O}{\|}}{P}}{-}OH + H_2O$$

Key Terms

Carboxylate ion (16.7)
Carboxylic acid (16.1)
Carboxylic acid salt (16.8)
Condensation polymer (16.15)
Dicarboxylic acid (16.2)
Ester (16.9)
Ester saponification (16.13)
Phosphate ester (16.16)
Polyester (16.15)
Thioester (16.14)

Exercises and Problems

The members of each pair of problems in this section test similar material.

The Carboxyl Functional Group (Section 16.1)

16.1 In which of the following compounds is a carboxyl group present?

a. 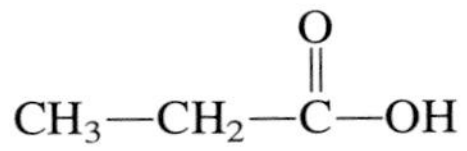 $CH_3-CH_2-\overset{O}{\overset{\|}{C}}-OH$

b. $CH_3-CH_2-CH_2-\overset{O}{\overset{\|}{C}}-CH_3$

c. $C_6H_5-\overset{O}{\overset{\|}{C}}-OH$

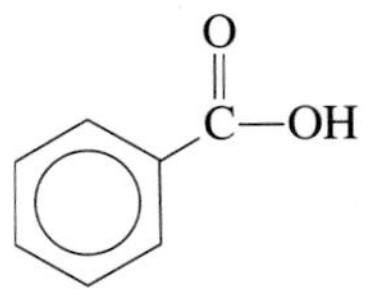

d. 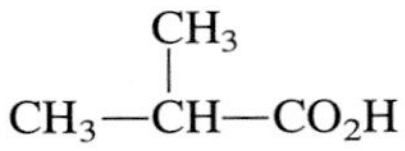 $CH_3-\overset{CH_3}{\overset{|}{CH}}-CO_2H$

e. 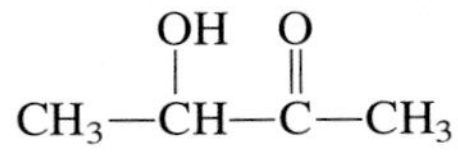 $CH_3-\overset{OH}{\overset{|}{CH}}-\overset{O}{\overset{\|}{C}}-CH_3$

f. CH_3-CH_2-COOH

16.2 In which of the following compounds is a carboxyl group present?

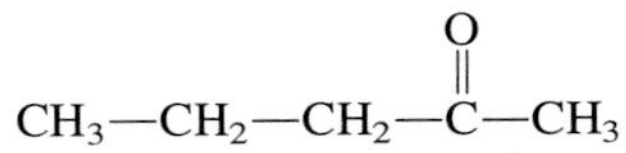

a. 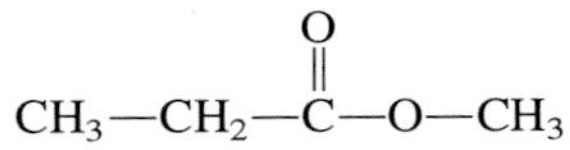 $CH_3-CH_2-\overset{O}{\overset{\|}{C}}-O-CH_3$

b. 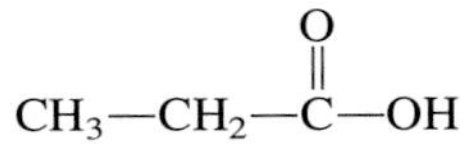 $CH_3-CH_2-\overset{O}{\overset{\|}{C}}-OH$

c. $HOOC-CH_2-CH_3$

d. 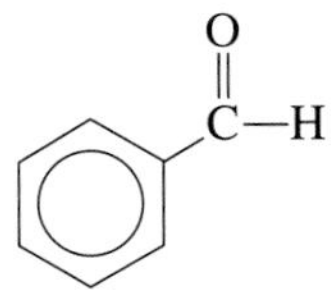 $C_6H_5-\overset{O}{\overset{\|}{C}}-H$

e. 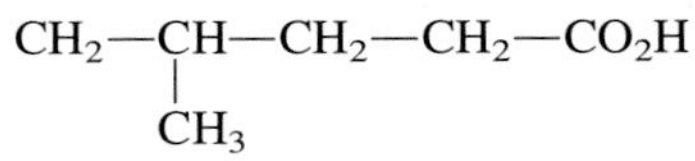 $CH_2-\underset{CH_3}{\underset{|}{CH}}-CH_2-CH_2-CO_2H$

f. 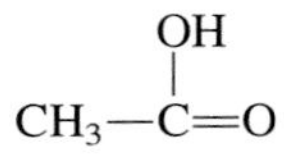 $CH_3-\overset{OH}{\overset{|}{C}}=O$

IUPAC Nomenclature for Carboxylic Acids (Section 16.2)

16.3 Give the IUPAC name for each of the following carboxylic acids.

a. 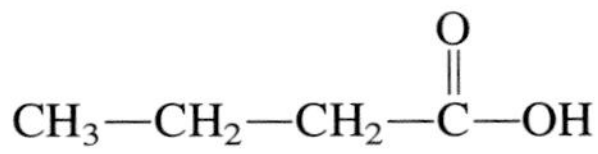 $CH_3-CH_2-CH_2-\overset{O}{\overset{\|}{C}}-OH$

b. $CH_3-CH_2-CH_2-CH_2-CH_2-CH_2-\overset{O}{\overset{\|}{C}}-OH$

c. $CH_3-CH_2-\overset{CH_3}{\overset{|}{CH}}-\overset{CH_3}{\overset{|}{CH}}-\overset{O}{\overset{\|}{C}}-OH$

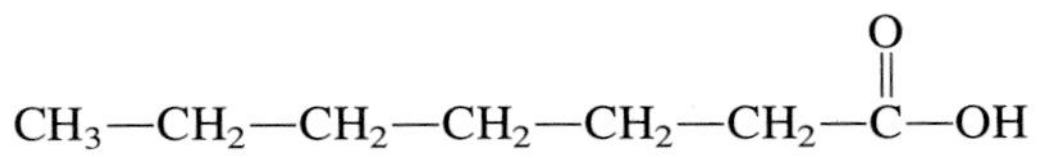

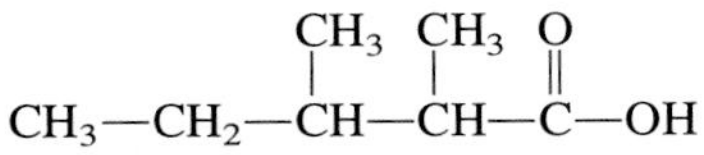

d. 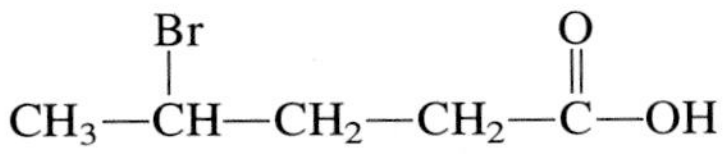 $CH_3-\overset{Br}{\overset{|}{CH}}-CH_2-CH_2-\overset{O}{\overset{\|}{C}}-OH$

e. 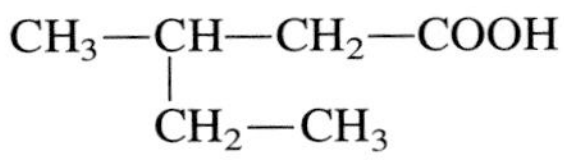 $CH_3-\underset{CH_2-CH_3}{\underset{|}{CH}}-CH_2-COOH$

f. 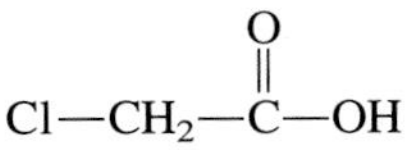 $Cl-CH_2-\overset{O}{\overset{\|}{C}}-OH$

16.4 Give the IUPAC name for each of the following carboxylic acids.

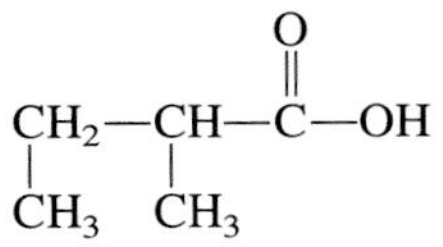

a. 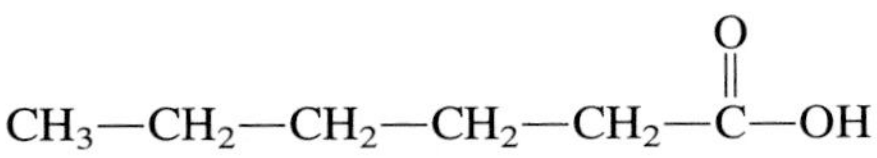 $CH_3-CH_2-CH_2-CH_2-CH_2-\overset{O}{\overset{\|}{C}}-OH$

b. $\underset{CH_3}{\underset{|}{CH_2}}-\underset{CH_3}{\underset{|}{CH}}-\overset{O}{\overset{\|}{C}}-OH$

c. $CH_3-\underset{Cl}{\underset{|}{\overset{CH_3}{\overset{|}{C}}}}-\overset{O}{\overset{\|}{C}}-OH$

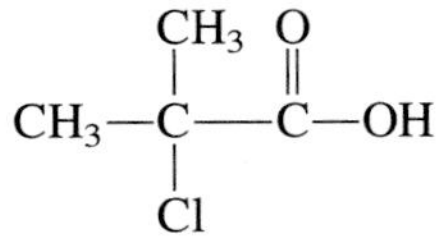

d. 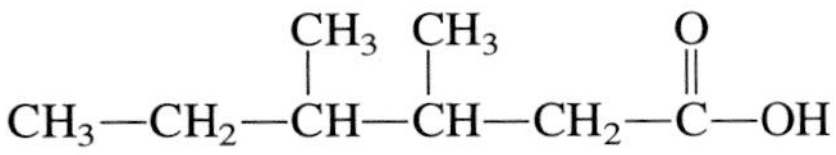 $CH_3-CH_2-\overset{CH_3}{\overset{|}{CH}}-\overset{CH_3}{\overset{|}{CH}}-CH_2-\overset{O}{\overset{\|}{C}}-OH$

e. $HOOC-CH_3$

f. 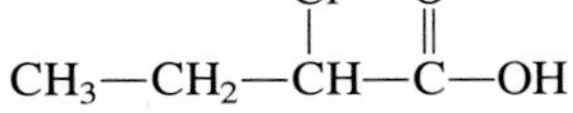 $CH_3-CH_2-\overset{Cl}{\overset{|}{CH}}-\overset{O}{\overset{\|}{C}}-OH$

16.5 Draw a condensed structural formula that corresponds to each of the following carboxylic acids.

a. 2-Ethylbutanoic acid
b. 2,5-Dimethylhexanoic acid
c. Methylpropanoic acid
d. Dichloroethanoic acid
e. 3-Bromo-5-chlorooctanoic acid
f. 2,3-Dimethylbutanoic acid

16.6 Draw a condensed structural formula that corresponds to each of the following carboxylic acids.

a. 3,3-Dimethylheptanoic acid
b. 4-Methylpentanoic acid
c. 3-Chloropropanoic acid

d. Trichloroethanoic acid
e. 3-Isopropylhexanoic acid
f. 4-Ethyl-3,5-dimethylhexanoic acid

16.7 Give the IUPAC name for each of the following carboxylic acids.

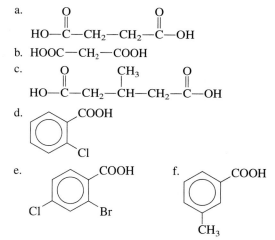

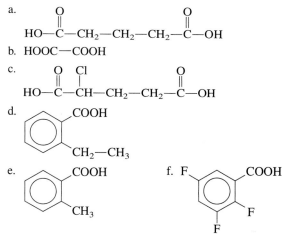

a. $$HO-\overset{\overset{\displaystyle O}{\|}}{C}-CH_2-CH_2-\overset{\overset{\displaystyle O}{\|}}{C}-OH$$

b. $HOOC-CH_2-COOH$

c. $$HO-\overset{\overset{\displaystyle O}{\|}}{C}-CH_2-\overset{\overset{\displaystyle CH_3}{|}}{CH}-CH_2-\overset{\overset{\displaystyle O}{\|}}{C}-OH$$

d. COOH, Cl (benzene ring)

e. 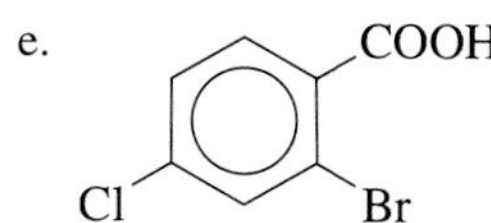

f. COOH, CH_3 (benzene ring)

16.8 Give the IUPAC name for each of the following carboxylic acids.

a. $$HO-\overset{\overset{\displaystyle O}{\|}}{C}-CH_2-CH_2-CH_2-\overset{\overset{\displaystyle O}{\|}}{C}-OH$$

b. $HOOC-COOH$

c. $$HO-\overset{\overset{\displaystyle O}{\|}}{C}-\overset{\overset{\displaystyle Cl}{|}}{CH}-CH_2-CH_2-\overset{\overset{\displaystyle O}{\|}}{C}-OH$$

d. COOH, CH_2-CH_3 (benzene ring)

e. 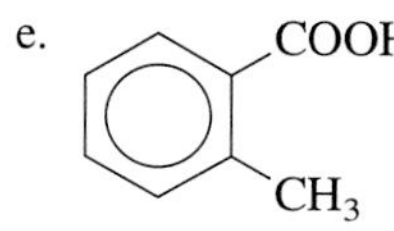

f. F, COOH, F, F (benzene ring)

16.9 Draw a condensed structural formula that corresponds to each of the following carboxylic acids.

a. 2,2-Dimethylbutanoic acid
b. 2,2-Dimethylbutanedioic acid
c. 2,2-Dimethylpentanedioic acid
d. *o*-Bromobenzoic acid
e. 2,4-Dichlorobenzoic acid
f. *p*-Toluic acid

16.10 Draw a condensed structural formula that corresponds to each of the following carboxylic acids.

a. 2,3-Dichlorohexanoic acid
b. 2,3-Dichlorohexanedioic acid
c. 2,3-Dichloroheptanedioic acid
d. *m*-Bromobenzoic acid
e. 3,5-Dichlorobenzoic acid
f. *o*-Toluic acid

Common Names for Carboxylic Acids (Section 16.3)

16.11 Draw a condensed structural formula that corresponds to each of the following carboxylic acids.

a. Valeric acid b. Propionic acid
c. Acetic acid d. α-Chlorobutyric acid
e. β-Bromocaproic acid
f. γ-Chloro-α-methylvaleric acid

16.12 Draw a condensed structural formula that corresponds to each of the following carboxylic acids.

a. Butyric acid b. Caproic acid
c. Formic acid d. Chloroacetic acid
e. α-Methylpropionic acid
f. β-Chloro-β-iodocaproic acid

16.13 Draw a condensed structural formula that corresponds to each of the following carboxylic acids.

a. Malonic acid b. Succinic acid
c. Adipic acid d. γ-Bromopimelic acid
e. α-Methylglutaric acid
f. α-Bromo-α-chlorosuccinic acid

16.14 Draw a condensed structural formula that corresponds to each of the following carboxylic acids.

a. Oxalic acid b. Glutaric acid
c. Pimelic acid d. Chloromalonic acid
e. α-Methyladipic acid
f. α,β-Dichlorosuccinic acid

16.15 Classify the two carboxylic acids in each of the following pairs as (1) both dicarboxylic acids, (2) both monocarboxylic acids, or (3) one dicarboxylic and one monocarboxylic acid.

a. Glutaric acid and valeric acid
b. Adipic acid and oxalic acid
c. Caproic acid and formic acid
d. Succinic acid and malonic acid

16.16 Classify the two carboxylic acids in each of the following pairs as (1) both dicarboxylic acids, (2) both monocarboxylic acids, or (3) one dicarboxylic and one monocarboxylic acid.

a. Formic acid and acetic acid
b. Butyric acid and succinic acid
c. Pimelic acid and caproic acid
d. Malonic acid and adipic acid

Polyfunctional Carboxylic Acids (Section 16.4)

16.17 Each of the following acids contains an additional type of functional group besides the carboxyl group. For each acid, specify the noncarboxyl functional group present.

a. Acrylic acid b. Lactic acid
c. Oxaloacetic acid d. Glyceric acid

16.18 Each of the following acids contains an additional type of functional group besides the carboxyl group. For each acid, specify the noncarboxyl functional group present.

a. Fumaric acid b. Pyruvic acid
c. Malic acid d. Tartaric acid

16.19 Give the IUPAC name for each of the acids in Problem 16.17.

16.20 Give the IUPAC name for each of the acids in Problem 16.18.

16.21 Draw a structural formula for each of the following acids.

a. 3-Oxopentanoic acid
b. 2-Hydroxybutanoic acid
c. *trans*-4-Hexenoic acid
d. α,β-Dihydroxyglutaric acid

16.22 Draw a structural formula for each of the following acids.

a. 3-Hydroxypentanoic acid
b. α,γ-Dihydroxyvaleric acid
c. 2-Oxobutanoic acid
d. *cis*-3-Heptenoic acid

Physical Properties of Carboxylic Acids (Section 16.5)

16.23 Determine the maximum number of hydrogen bonds that can form between an acetic acid molecule and

a. another acetic acid molecule
b. water molecules

16.24 Determine the maximum number of hydrogen bonds that can form between a butanoic acid molecule and

a. another butanoic acid molecule
b. water molecules

16.25 What is the physical state (solid, liquid, or gas) of the following carboxylic acids at room temperature?

a. Oxalic acid b. Decanoic acid
c. Hexanoic acid d. Benzoic acid

16.26 What is the physical state (solid, liquid, or gas) of the following carboxylic acids at room temperature?

a. Succinic acid b. Octanoic acid
c. Pentanoic acid d. *p*-Chlorobenzoic acid

Preparation of Carboxylic Acids (Section 16.6)

16.27 Draw a structural formula for the carboxylic acid expected to be formed when each of the following substances is oxidized using a strong oxidizing agent.

a. $CH_3—CH_2—OH$

b. $CH_3—\overset{\overset{O}{\|}}{C}—H$

c. $CH_3—CH_2—\underset{}{\overset{\overset{CH_3}{|}}{C}H}—CH_2—\overset{\overset{O}{\|}}{C}—H$

d. 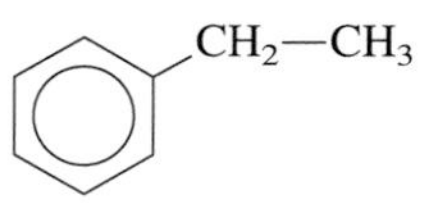 (benzene ring)$—CH_2—CH_3$

16.28 Draw a structural formula for the carboxylic acid expected to be formed when each of the following substances is oxidized using a strong oxidizing agent.

a. $CH_3—CH_2—\overset{\overset{O}{\|}}{C}—H$

b. $CH_3—CH_2—CH_2—OH$

c. $CH_3—\underset{\underset{CH_3}{|}}{CH}—\underset{\underset{CH_3}{|}}{CH}—\overset{\overset{O}{\|}}{C}—H$

d. 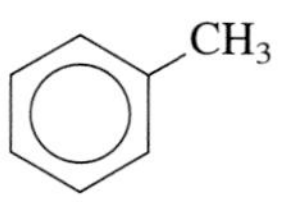 (benzene ring)$—CH_3$

Acidity of Carboxylic Acids (Section 16.7)

16.29 How many acidic hydrogen atoms are present in each of the following carboxylic acids?

a. Pentanoic acid b. Citric acid
c. Succinic acid d. Oxalic acid

16.30 How many acidic hydrogen atoms are present in each of the following carboxylic acids?

a. Acetic acid b. Benzoic acid
c. Propanoic acid d. Glutaric acid

16.31 What is the charge on the carboxylate ion formed when each of the acids in Problem 16.29 ionizes in water?

16.32 What is the charge on the carboxylate ion formed when each of the acids in Problem 16.30 ionizes in water?

16.33 What is the name of the carboxylate ion that forms when each of the acids in Problem 16.29 ionizes in water? (Use an IUPAC carboxylate name if the acid name is IUPAC; use a common name if the acid name is common.)

16.34 What is the name of the carboxylate ion that forms when each of the acids in Problem 16.30 ionizes in water? (Use an IUPAC carboxylate name if the acid name is IUPAC; use a common name if the acid name is common.)

16.35 Write a chemical equation for the formation of each of the following carboxylate ions, in aqueous solution, from its parent acid.

a. Acetate b. Citrate
c. Ethanoate d. 2-Methylbutanoate

16.36 Write a chemical equation for the formation of each of the following carboxylate ions, in aqueous solution, from its parent acid.

a. Butanoate b. Succinate
c. Benzoate d. α-Methylbutyrate

Carboxylic Acid Salts (Section 16.8)

16.37 Give the IUPAC name for each of the following carboxylic acid salts.

a. $CH_3—\overset{\overset{O}{\|}}{C}—O^-\ K^+$

b. $\left(CH_3—CH_2—\overset{\overset{O}{\|}}{C}—O^-\right)_2\ Ca^{2+}$

c. $K^+\ {}^-O—\overset{\overset{O}{\|}}{C}—CH_2—CH_2—\overset{\overset{O}{\|}}{C}—O^-\ K^+$

d. $CH_3—CH_2—CH_2—CH_2—\overset{\overset{O}{\|}}{C}—O^-\ Na^+$

16.38 Give the IUPAC name for each of the following carboxylic acid salts.

a. $\left(H—\overset{\overset{O}{\|}}{C}—O^-\right)_2\ Ca^{2+}$

b. $CH_3—CH_2—\overset{\overset{O}{\|}}{C}—O^-\ Na^+$

c. $CH_3—CH_2—CH_2—\overset{\overset{O}{\|}}{C}—O^-\ K^+$

d. $C_6H_5-\overset{O}{\overset{\|}{C}}-O^- Na^+$

16.39 Write a chemical equation for the preparation of each of the salts in Problem 16.37 using an acid–base neutralization reaction.

16.40 Write a chemical equation for the preparation of each of the salts in Problem 16.38 using an acid–base neutralization reaction.

16.41 Write a chemical equation for the conversion of each of the following carboxylic acid salts to its parent carboxylic acid. Let hydrochloric acid (HCl) be the source of the needed hydronium ions.

a. Sodium butanoate b. Potassium oxalate
c. Calcium malonate d. Sodium benzoate

16.42 Write a chemical equation for the conversion of each of the following carboxylic acid salts to its parent carboxylic acid. Let hydrochloric acid (HCl) be the source of the needed hydronium ions.

a. Calcium propanoate b. Sodium lactate
c. Magnesium succinate d. Potassium benzoate

Structure of Esters (Section 16.9)

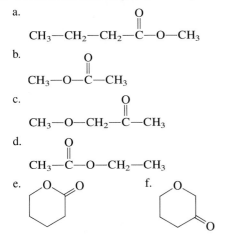

16.43 Which of the following structures represent esters?

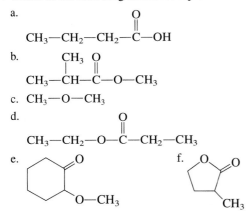

a. $CH_3-CH_2-CH_2-\overset{O}{\overset{\|}{C}}-O-CH_3$

b. $CH_3-O-\overset{O}{\overset{\|}{C}}-CH_3$

c. $CH_3-O-CH_2-\overset{O}{\overset{\|}{C}}-CH_3$

d. $CH_3-\overset{O}{\overset{\|}{C}}-O-CH_2-CH_3$

e. (six-membered ring containing O, with =O on the carbon adjacent to the ring O) f. (six-membered ring containing O, with =O on a carbon not adjacent to the ring O)

16.44 Which of the following structures represent esters?

a. $CH_3-CH_2-CH_2-\overset{O}{\overset{\|}{C}}-OH$

b. $CH_3-\overset{CH_3}{\overset{|}{CH}}-\overset{O}{\overset{\|}{C}}-O-CH_3$

c. CH_3-O-CH_3

d. $CH_3-CH_2-O-\overset{O}{\overset{\|}{C}}-CH_2-CH_3$

e. 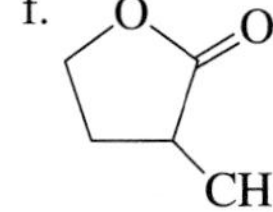

f. (five-membered ring containing O, with =O on the carbon adjacent to the ring O, and a CH_3 substituent)

16.45 Draw the structure of the ester produced when each of the following pairs of carboxylic acid and alcohol react.

a. Propanoic acid and methanol
b. Acetic acid and 1-propanol
c. 2-Methylbutanoic acid and 2-propanol
d. Valeric acid and *sec*-Butyl alcohol

16.46 Draw the structure of the ester produced when each of the following pairs of carboxylic acid and alcohol react.

a. Methanoic acid and 1-propanol
b. Propanoic acid and ethanol
c. 2-Methylpropanoic acid and 2-butanol
d. Valeric acid and isobutyl alcohol

16.47 For each of the following esters, draw the structural formula of the "parent" acid and the "parent" alcohol.

a. $CH_3-CH_2-\overset{O}{\overset{\|}{C}}-O-CH_2-CH_3$

b. $CH_3-CH_2-CH_2-\overset{O}{\overset{\|}{C}}-O-CH_3$

c. $CH_3-O-\overset{O}{\overset{\|}{C}}-CH_2-CH_2-CH_3$

d. $CH_3-\overset{O}{\overset{\|}{C}}-O-C_6H_5$

e. $C_6H_5-\overset{O}{\overset{\|}{C}}-O-CH_3$

f. $CH_3-\overset{Cl}{\overset{|}{CH}}-\overset{O}{\overset{\|}{C}}-O-CH_2-CH_3$

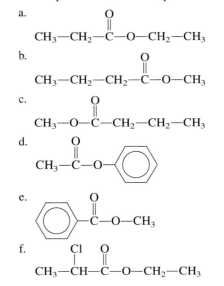

16.48 For each of the following esters, draw the structural formula of the "parent" acid and the "parent" alcohol.

a. $CH_3-\overset{O}{\overset{\|}{C}}-O-CH_2-CH_3$

b. $CH_3-CH_2-\overset{O}{\overset{\|}{C}}-O-CH_3$

c. $CH_3-O-\overset{O}{\overset{\|}{C}}-CH_2-CH_3$

d. $C_6H_5-\overset{O}{\overset{\|}{C}}-O-CH_3$

e. $CH_3-\overset{CH_3}{\overset{|}{CH}}-CH_2-\overset{O}{\overset{\|}{C}}-O-C_6H_5$

f. $CH_3-\underset{C_6H_5}{\underset{|}{CH}}-CH_2-\overset{O}{\overset{\|}{C}}-O-CH_3$

Nomenclature for Esters (Section 16.10)

16.49 Assign an IUPAC name to each of the following esters.

a. $CH_3-CH_2-C(=O)-O-CH_3$

b. $H-C(=O)-O-CH_3$

c. $CH_3-C(=O)-O-CH_3$

d. $CH_3-CH_2-CH_2-O-C(=O)-CH_3$

e. $CH_3-CH_2-C(=O)-O-CH(CH_3)-CH_3$

f. $C_6H_5-C(=O)-O-CH_2-CH_3$

16.50 Assign an IUPAC name to each of the following esters.

a. $CH_3-C(=O)-O-CH_2-CH_2-CH_2-CH_3$

b. $CH_3-CH_2-CH_2-C(=O)-O-CH_3$

c. $CH_3-CH_2-C(=O)-O-CH_2-CH_2-CH_3$

d. $CH_3-CH_2-O-C(=O)-H$

e. $CH_3-CH(CH_3)-CH(CH_3)-C(=O)-O-CH_2-CH_3$

f. $CH_3-CH_2-C(=O)-O-C_6H_5$

16.51 Assign a common name to each of the esters in Problem 16.49.

16.52 Assign a common name to each of the esters in Problem 16.50.

16.53 Draw a structural formula for each of the following esters.

a. Methyl formate
b. Propyl acetate
c. Octyl decanoate
d. Ethyl phenylacetate
e. Isopropyl acetate
f. 2-Bromopropyl ethanoate

16.54 Draw a structural formula for each of the following esters.

a. Ethyl butyrate
b. Butyl ethanoate
c. 2-Methylpropyl formate
d. Ethyl 2-methylpropanoate
e. Methyl valerate
f. Phenyl benzoate

16.55 Assign IUPAC names to the esters that are produced from the reaction of the following carboxylic acids and alcohols.

a. Acetic acid and ethanol
b. Ethanoic acid and methanol
c. Butyric acid and ethyl alcohol
d. Lactic acid and propyl alcohol
e. 1-Pentanol and pentanoic acid
f. 2-Butanol and caproic acid

16.56 Assign IUPAC names to the esters that are produced from the reaction of the following carboxylic acids and alcohols.

a. Ethanoic acid and propyl alcohol
b. Acetic acid and 1-pentanol
c. Acetic acid and 2-pentanol
d. Methyl alcohol and butyric acid
e. Ethanol and benzoic acid
f. Pyruvic acid and methyl alcohol

Physical Properties of Esters (Section 16.12)

16.57 Explain why ester molecules cannot form hydrogen bonds to each other.

16.58 How many hydrogen bonds can form between a methyl acetate molecule and two water molecules?

16.59 Explain why esters have lower boiling points than carboxylic acids of comparable molecular mass.

16.60 Explain why esters are less soluble in water than carboxylic acids of comparable molecular mass.

Chemical Reactions of Esters (Section 16.13)

16.61 Write the structural formulas of the reaction products when each of the following esters is hydrolyzed.

a. $CH_3-CH_2-C(=O)-O-CH_2-CH_3$

b. $CH_3-C(=O)-O-CH_2-CH_3$

c. $CH_3-CH(CH_3)-C(=O)-O-C_6H_5$

d. Methyl butanoate
e. Ethyl formate
f. Isopropyl benzoate

16.62 Write the structural formulas of the reaction products when each of the following esters is hydrolyzed.

a. $H-C(=O)-O-CH_2-CH_2-CH_3$

b. $CH_3-CH(CH_3)-C(=O)-O-CH(CH_3)-CH_3$

c. $C_6H_5-C(=O)-O-C_6H_5$

d. Ethyl valerate
e. Butyl butyrate
f. Pentyl benzoate

16.63 Write the structural formulas of the reaction products when each of the esters in Problem 16.61 is saponified using sodium hydroxide.

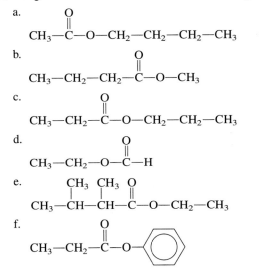

16.64 Write the structural formulas of the reaction products when each of the esters in Problem 16.62 is saponified using sodium hydroxide.

16.65 Draw structures of the reaction products in the following chemical reactions.

a. 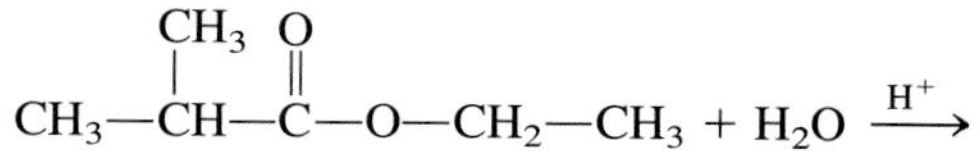 $$CH_3{-}CH(CH_3){-}C(=O){-}O{-}CH_2{-}CH_3 + H_2O \xrightarrow{H^+}$$

b. $$CH_3{-}CH(CH_3){-}C(=O){-}O{-}CH_2{-}CH_3 + NaOH \xrightarrow{H_2O}$$

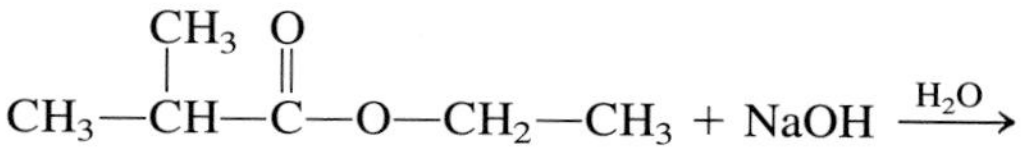

c. $$H{-}C(=O){-}O{-}CH_2{-}CH_2{-}CH_2{-}CH_3 + H_2O \xrightarrow{H^+}$$

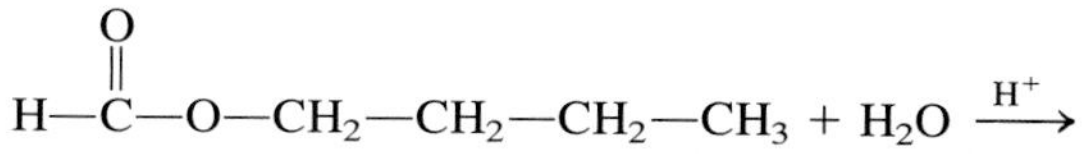

d. 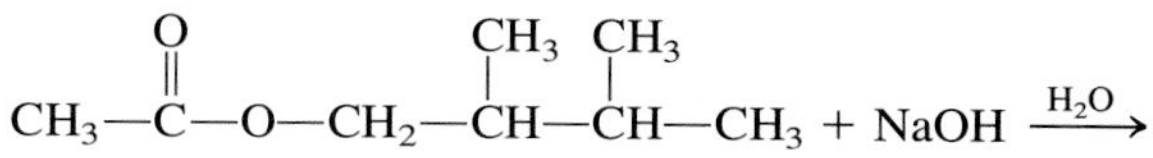 $$CH_3{-}C(=O){-}O{-}CH_2{-}CH(CH_3){-}CH(CH_3){-}CH_3 + NaOH \xrightarrow{H_2O}$$

16.66 Draw structures of the reaction products in the following chemical reactions.

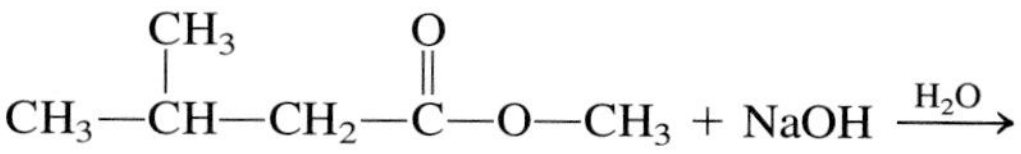

a. 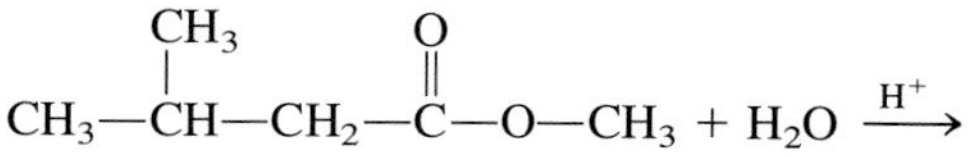 $$CH_3{-}CH(CH_3){-}CH_2{-}C(=O){-}O{-}CH_3 + H_2O \xrightarrow{H^+}$$

b. $$CH_3{-}CH(CH_3){-}CH_2{-}C(=O){-}O{-}CH_3 + NaOH \xrightarrow{H_2O}$$

c. $$CH_3{-}CH_2{-}C(=O){-}O{-}(CH_2)_5{-}CH_3 + H_2O \xrightarrow{H^+}$$

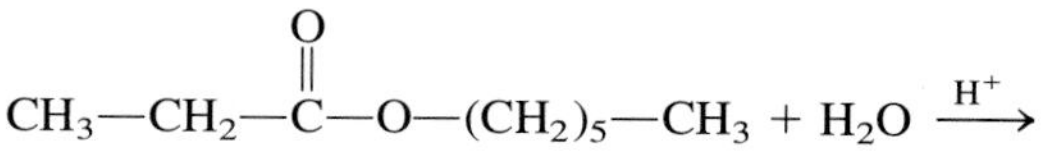

d. 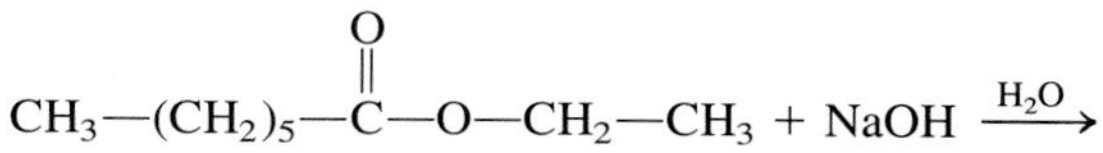 $$CH_3{-}(CH_2)_5{-}C(=O){-}O{-}CH_2{-}CH_3 + NaOH \xrightarrow{H_2O}$$

Sulfur Analogs of Esters (Section 16.14)

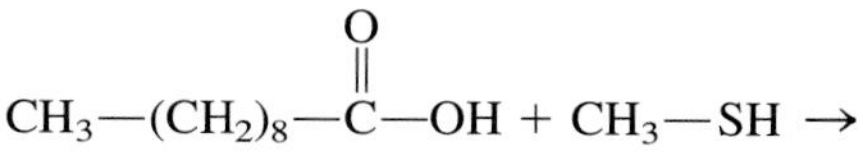

16.67 Draw the structures of the thioesters formed as a result of each of the following reactions between carboxylic acids and thiols.

a. 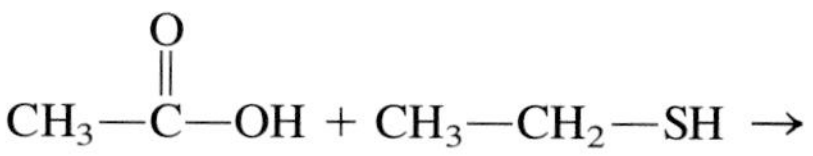 $$CH_3{-}C(=O){-}OH + CH_3{-}CH_2{-}SH \rightarrow$$

b. $$CH_3{-}(CH_2)_8{-}C(=O){-}OH + CH_3{-}SH \rightarrow$$

c. $$C_6H_5{-}COOH + CH_3{-}CH(CH_3){-}SH \rightarrow$$

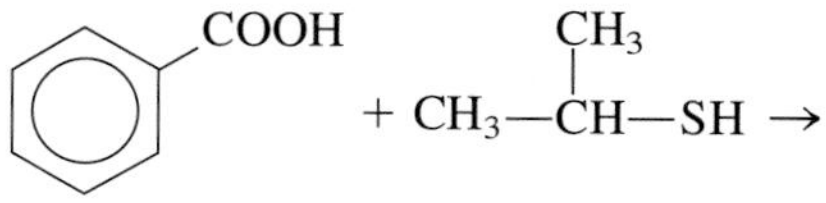

d. 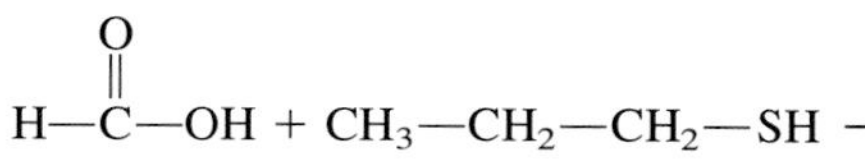 $$H{-}C(=O){-}OH + CH_3{-}CH_2{-}CH_2{-}SH \rightarrow$$

16.68 Draw the structures of the thioesters formed as a result of each of the following reactions between carboxylic acids and thiols.

a.  $$CH_3{-}CH_2{-}C(=O){-}OH + CH_3{-}CH_2{-}SH \rightarrow$$

b. $$CH_3{-}CH_2{-}CH_2{-}COOH + CH_3{-}SH \rightarrow$$

c. 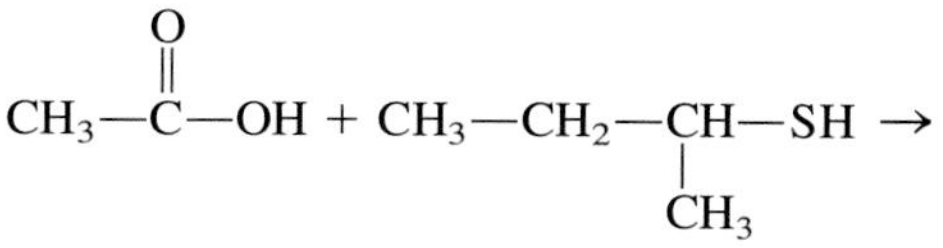 $$CH_3{-}C(=O){-}OH + CH_3{-}CH_2{-}CH(CH_3){-}SH \rightarrow$$

d. 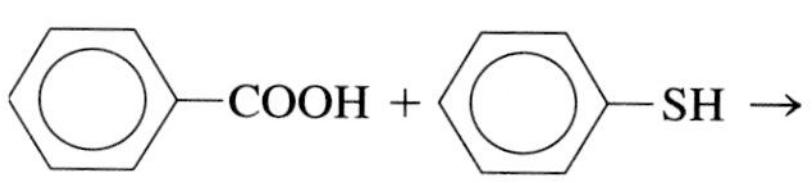 $$C_6H_5{-}COOH + C_6H_5{-}SH \rightarrow$$

Polyesters (Section 16.15)

16.69 Write the structure (two repeating units) of the polyester polymer formed from oxalic acid and 1,3-propanediol.

16.70 Write the structure (two repeating units) of the polyester polymer formed from malonic acid and ethylene glycol.

16.71 Draw the structural formulas of the monomers needed to form the following polyester.

$$\left[O{-}(CH_2)_3{-}O{-}C(=O){-}(CH_2)_2{-}C(=O){-}O\right]_n$$

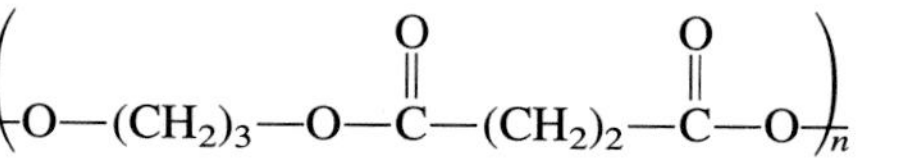

16.72 Draw the structural formulas of the monomers needed to form the following polyester.

$$\left[O{-}C(=O){-}(CH_2)_3{-}C(=O){-}O{-}(CH_2)_2{-}O\right]_n$$

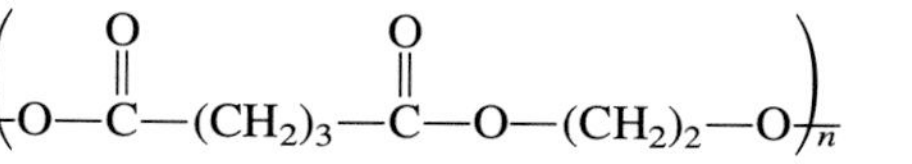

Esters of Inorganic Acids (Section 16.16)

16.73 Draw the structures of the esters formed by reacting the following substances.

a. 1 molecule methanol and 1 molecule phosphoric acid

b. 2 molecules methanol and 1 molecule phosphoric acid

c. 1 molecule methanol and 1 molecule nitric acid

d. 1 molecule ethylene glycol and 2 molecules nitric acid

16.74 Draw the structures of the esters formed by reacting the following substances.

a. 1 molecule ethanol and 1 molecule phosphoric acid

b. 2 molecules methanol and 1 molecule sulfuric acid

c. 1 molecule ethylene glycol and 1 molecule nitric acid

d. 1 molecule glycerol and 3 molecules nitric acid

16.75 Phosphoric acid can form triesters but sulfuric acid cannot. Explain why.

16.76 Sulfuric acid can form diesters but nitric acid cannot. Explain why.

Additional Problems

16.77 With the help of Figure 16.2 and IUPAC naming rules, specify the number of carbon atoms present and the number of carboxyl groups present in each of the following carboxylic acids.

a. Oxalic acid
b. Heptanoic acid
c. *cis*-3-Heptenoic acid
d. Citric acid
e. Pyruvic acid
f. Dichloroethanoic acid

16.78 Malonic, maleic, and malic acids are dicarboxylic acids with similar-sounding names. How do the structures of these acids differ from each other?

16.79 The general molecular formula for an alkane is C_nH_{2n+2}. What is the general molecular formula for an unsaturated unsubstituted monocarboxylic acid containing one carbon–carbon double bond?

16.80 Draw structural formulas and give IUPAC names for all possible saturated unsubstituted monocarboxylic acids that contain six carbon atoms. There are eight isomers.

16.81 Monocarboxylic acids and esters with the same number of carbon atoms and the same degree of unsaturation are isomeric. Give IUPAC names for all possible esters that are isomeric with butanoic acid. There are four isomers.

16.82 Assign IUPAC names to the following compounds.

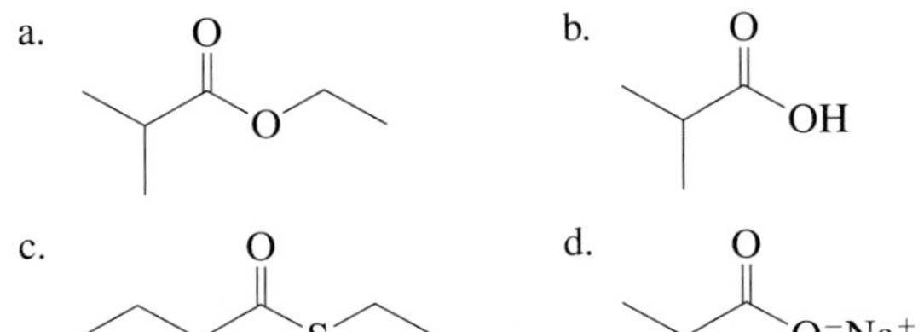

16.83 A sample of ethyl alcohol is divided into two portions. Portion A is added to an aqueous solution of a strong oxidizing agent and allowed to react. The organic product of this reaction is mixed with portion B of the ethyl alcohol. A trace of acid is added and the solution is heated. What is the structure of the final product of this reaction scheme?

16.84 For each of the following reactions, draw the structure(s) of the organic product(s).

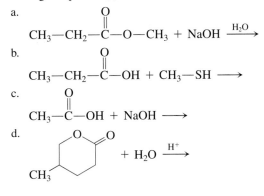

a. $CH_3-CH_2-\overset{\overset{\displaystyle O}{\|}}{C}-O-CH_3 + NaOH \xrightarrow{H_2O}$

b. $CH_3-CH_2-\overset{\overset{\displaystyle O}{\|}}{C}-OH + CH_3-SH \longrightarrow$

c. $CH_3-\overset{\overset{\displaystyle O}{\|}}{C}-OH + NaOH \longrightarrow$

d. (cyclic ester ring with O and =O, substituent CH_3) $+ H_2O \xrightarrow{H^+}$

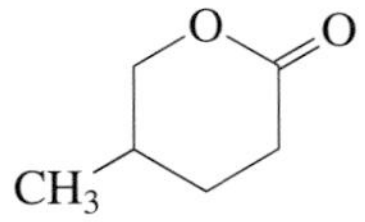

Grid Problems

16.85

1. succinic acid	2. acetic acid	3. butyric acid
4. oxalic acid	5. benzoic acid	6. formic acid

Select from the grid *all* correct responses for each of the following situations.

a. Compounds that are monocarboxylic acids
b. Compounds that are dicarboxylic acids
c. Compounds in which the number of carboxyl groups present and the number of carbon atoms present are the same
d. Pairs of compounds that have the same number of carbon atoms

16.86

1. $CH_3-\underset{\displaystyle CH_3}{\underset{\vert}{CH}}-COOH$	2. $CH_3-\underset{\displaystyle OH}{\underset{\vert}{CH}}-COOH$	3. $HOOC-CH_2-COOH$
4. $CH_2{=}CH-CH_2-COOH$	5. $CH_3-\underset{\displaystyle NH_2}{\underset{\vert}{CH}}-CH_2-COOH$	6. $CH_3-\overset{\overset{\displaystyle O}{\|}}{C}-CH_2-COOH$

Select from the grid *all* correct responses for each of the following situations.

a. Compounds that would be named as acids
b. Compounds that are keto acids
c. Compounds that contain two functional groups
d. Compounds that have a substituent on the alpha-carbon atom

16.87

1. methyl acetate	2. sodium ethanoate	3. ethanoic acid
4. ethyl formate	5. potassium acetate	6. methyl formate

Select from the grid *all* correct responses for each of the following situations.

a. Compounds that are esters
b. Compounds that are carboxylate salts
c. Compounds whose names are IUPAC names
d. Pairs of compounds that are structural isomers

16.88

1. $CH_3-\overset{\overset{\displaystyle O}{\|}}{C}-O-CH_3$	2. $CH_3-\overset{\overset{\displaystyle O}{\|}}{C}-O^-\ Na^+$	3. CH_3-CH_2-COOH
4. $CH_3-\overset{\overset{\displaystyle O}{\|}}{C}-S-CH_3$	5. $CH_3-\overset{\overset{\displaystyle O}{\|}}{C}-OH$	6. $CH_3-\overset{\overset{\displaystyle O}{\|}}{C}-O-\overset{\overset{\displaystyle CH_3}{\vert}}{CH_2}$

Select from the grid *all* correct responses for each of the following situations.

a. Compounds for which acetic acid is a hydrolysis product
b. Compounds that can be synthesized using acetic acid as a reactant
c. Compounds that will react with aqueous sodium hydroxide solution
d. Compounds whose aqueous solutions contain carboxylate ions

Amines and Amides

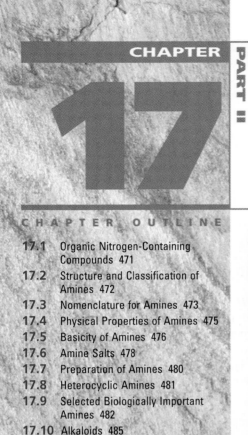

CHAPTER OUTLINE

17.1 Organic Nitrogen-Containing Compounds 471
17.2 Structure and Classification of Amines 472
17.3 Nomenclature for Amines 473
17.4 Physical Properties of Amines 475
17.5 Basicity of Amines 476
17.6 Amine Salts 478
17.7 Preparation of Amines 480
17.8 Heterocyclic Amines 481
17.9 Selected Biologically Important Amines 482
17.10 Alkaloids 485
17.11 Structure of and Nomenclature for Amides 486
17.12 Selected Amides and Their Uses 488
17.13 Properties of Amides 488
17.14 Preparation of Amides 490
17.15 Hydrolysis of Amides 492

Chemistry at a Glance:
Summary of Reactions Involving Amines and Amides 494
17.16 Polyamides and Polyurethanes 494

Chemical Connections
17.1 Caffeine: The Most Widely Used Central Nervous System Stimulant 482
17.2 Nicotine Addiction: A Widespread Example of Drug Dependence 483
17.3 Acetaminophen: A Substituted Amide 489

Parachutist with a parachute made of the polyamide nylon.

The four most abundant elements in living organisms are carbon, hydrogen, oxygen, and nitrogen. In previous chapters, we have discussed compounds containing the first three of these elements. Alkanes, alkenes, alkynes, and aromatic hydrocarbons are all carbon–hydrogen compounds. The carbon–hydrogen–oxygen compounds we have discussed include alcohols, phenols, ethers, aldehydes, ketones, carboxylic acids, and esters. We now extend our discussion to organic compounds that contain the element nitrogen.

17.1 Organic Nitrogen-Containing Compounds

Two types of organic nitrogen-containing compounds are the focus of this chapter: amines and amides. Amines are carbon–hydrogen–nitrogen compounds, and amides contain oxygen in addition to these elements. Amines and amides occur widely in nature in both plants and animals. Many of these naturally occurring compounds are very active physiologically. In addition, numerous drugs used for the treatment of mental illness, hay fever, heart problems, and other physical disorders are amines or amides.

An understanding of the bonding characteristics of the nitrogen atom is a prerequisite to our study of amines and amides. Nitrogen is a member of Group VA of the periodic table; it has five valence electrons (Section 4.2) and will form three covalent bonds to complete its octet of electrons (Section 4.3). Thus, in organic chemistry, carbon forms four bonds (Section 12.2), nitrogen forms three bonds, and oxygen forms two bonds (Section 14.1).

$-\overset{|}{\underset{|}{C}}-$	$-\ddot{N}-$ (one bond down)	$:\ddot{O}-$ (one bond down)
4 valence electrons 4 covalent bonds no nonbonding electron pairs	5 valence electrons 3 covalent bonds 1 nonbonding electron pair	6 valence electrons 2 covalent bonds 2 nonbonding electron pairs

17.2 Structure and Classification of Amines

Amines *are organic derivatives of ammonia (NH_3) in which one or more hydrogen atoms on the nitrogen have been replaced by an alkyl, a cycloalkyl, or an aryl group.* Amines bear the same relationship to ammonia that alcohols and ethers bear to water.

Amines can be classified as primary, secondary, or tertiary (Figure 17.1). Classification is by the number of carbon atoms directly attached to the nitrogen atom.

• Line-angle drawings for selected primary, secondary, and tertiary amines.

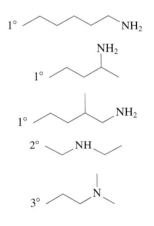

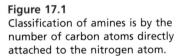

1° (line-angle chain) NH_2

1° (branched chain) NH_2

1° (branched chain) NH_2

2° (chain) NH (chain)

3° (chain) N (with two methyl branches)

The basis for this primary–secondary–tertiary classification system differs from that for alcohols.

1. For alcohols we look at how many R groups are on a *carbon* atom, the hydroxyl-bearing carbon atom.
2. For amines we look at how many R groups are on the *nitrogen* atom.

Tert-butyl alcohol is a *tertiary* alcohol, whereas *tert*-butylamine is a *primary* amine.

$$CH_3-\overset{CH_3}{\underset{CH_3}{C}}-OH \qquad CH_3-\overset{CH_3}{\underset{CH_3}{C}}-NH_2$$

Tertiary carbon atom → C; *tert*-Butyl alcohol (a tertiary alcohol)

Tertiary carbon atom → C; Primary nitrogen atom → N; *tert*-Butylamine (a primary amine)

The functional group present in a primary amine, the $-NH_2$ group, is called an *amino* group. Secondary and tertiary amines possess substituted amino groups.

$-NH_2$	$-\underset{R}{\underset{|}{N}}H$	$-\underset{R}{\underset{|}{N}}-R'$
Amino group	Monosubstituted amino group	Disubstituted amino group

Figure 17.1
Classification of amines is by the number of carbon atoms directly attached to the nitrogen atom.

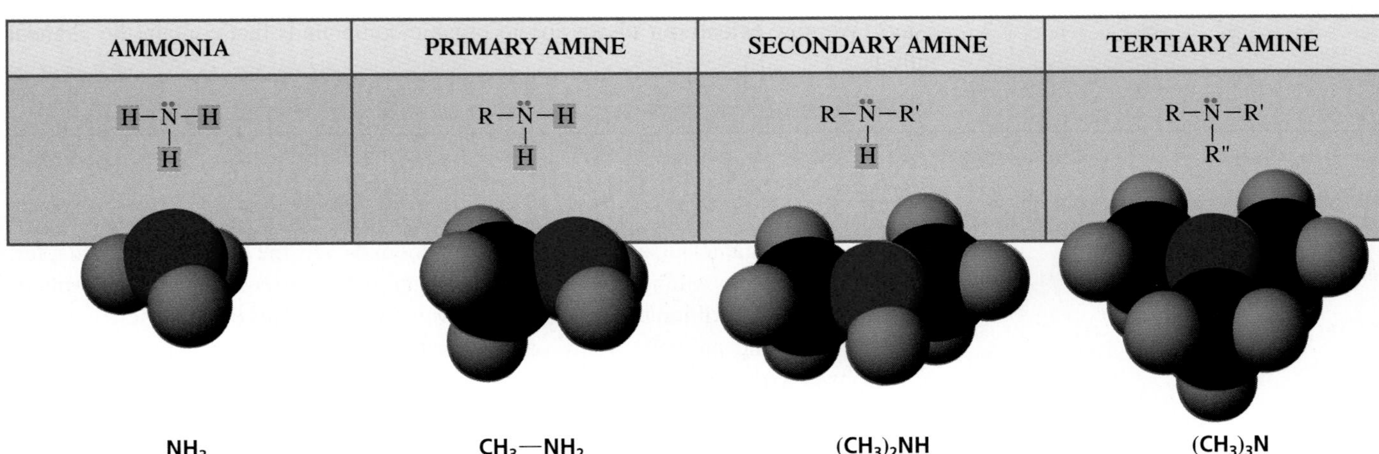

Example 17.1

Classifying Amines as Primary, Secondary, or Tertiary

Classify each of the following amines as a primary, secondary, or tertiary amine.

a. $CH_3—NH—C_6H_5$ (phenyl ring)

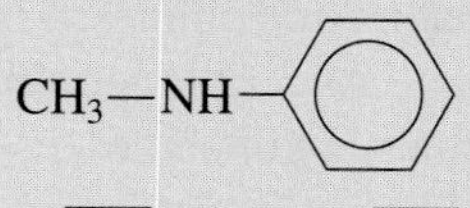

b. $CH_3—N(CH_3)—CH_3$

c. $C_6H_5—N(CH_3)—C_6H_5$

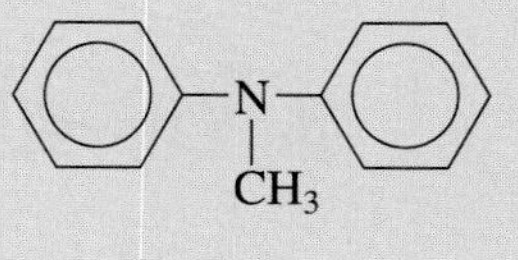

d. Cyclohexane ring with CH_3 and NH_2 substituents

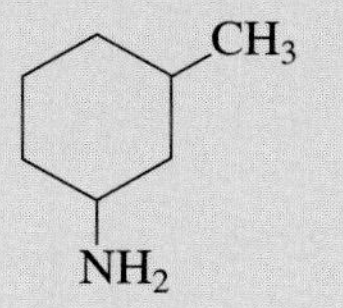

Solution

The number of carbon atoms directly bonded to the nitrogen atom determines the amine classification.

a. This is a secondary amine, because the nitrogen is bonded to both a methyl group and a phenyl group.
b. Here we have a tertiary amine, because the nitrogen atom is bonded to three methyl groups.
c. This is also a tertiary amine; the nitrogen atom is bonded to two phenyl groups and a methyl group.
d. This is a primary amine. The nitrogen atom is bonded to only one carbon atom.

Practice Exercise 17.1

Classify each of the following amines as a primary, secondary, or tertiary amine.

a. $CH_3—CH_2—CH_2—NH_2$

b. $CH_3—NH—CH_2—CH_3$

c. $C_6H_5—NH_2$ (phenyl ring)

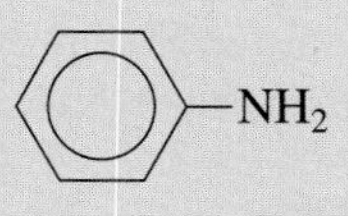

d. $CH_3—N(CH_3)—C_6H_5$ (phenyl ring)

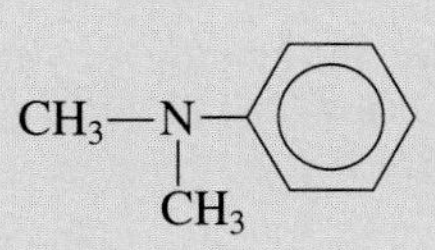

• *Answers:* **a.** primary **b.** secondary **c.** primary **d.** tertiary

Cyclic amines also occur. Such compounds are always either secondary or tertiary amines.

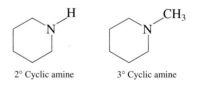

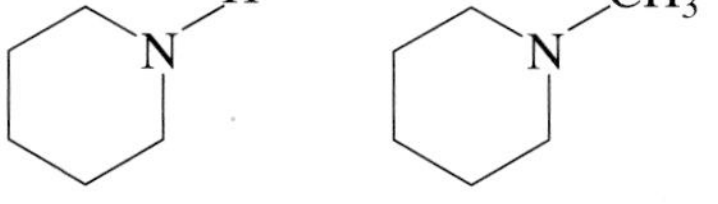

Piperidine ring with N—H: 2° Cyclic amine

Piperidine ring with N—CH_3: 3° Cyclic amine

Cyclic amines are heterocyclic compounds (Section 14.12). Numerous cyclic amine compounds are found in biological systems (Section 17.8).

17.3 Nomenclature for Amines

Both common and IUPAC names are extensively used for amines. In the common system of nomenclature, simple amines are named by listing the alkyl group or groups attached to the nitrogen atom in alphabetical order and adding the suffix *-amine;* all of this appears as one word. Prefixes such as *di-* and *tri-* are added when identical groups are bonded to the nitrogen atom.

$CH_3{-}CH_2{-}NH_2$ Ethylamine

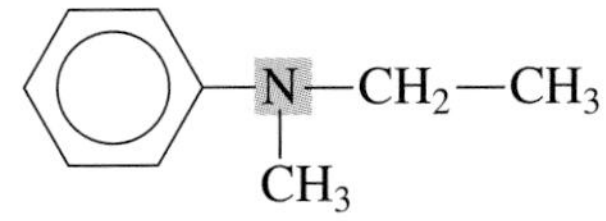

$CH_3{-}NH{-}CH_3$ Dimethylamine

$C_6H_5{-}N(CH_3){-}CH_2{-}CH_3$ Ethylmethylphenylamine

• The common names of amines, like those of aldehydes, are written as a single word, which is different from the common names of alcohols (two words), ethers (two or three words), ketones (two or three words), acids (two words), and esters (two words).

IUPAC nomenclature for amines is similar to that for alcohols, except the suffix is *-amine* rather than *-ol*. An $-NH_2$ group, like an $-OH$ group, has priority in numbering the parent carbon chain. In diamines, as with diols, the final *-e* of the hydrocarbon chain is retained.

$CH_3{-}CH_2{-}CH_2{-}CH_2{-}NH_2$ 1-Butanamine

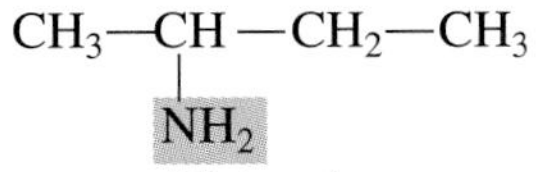

$CH_3{-}CH(NH_2){-}CH_2{-}CH_3$ 2-Butanamine

$CH_3{-}CH(CH_3){-}CH_2{-}CH_2{-}NH_2$ 3-Methyl-1-butanamine

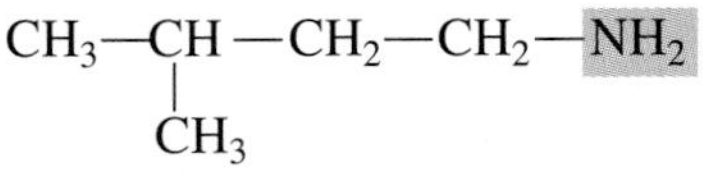

$H_2N{-}CH_2{-}CH_2{-}CH_2{-}CH_2{-}NH_2$ 1,4-Butanediamine

In secondary and tertiary amines, the prefix *N-* is used for each substituent on the nitrogen atom.

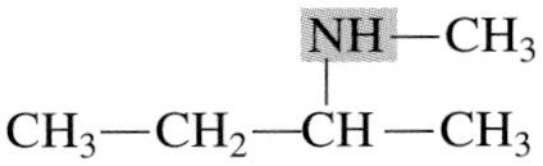

$CH_3{-}CH_2{-}CH(NH{-}CH_3){-}CH_3$ *N*-methyl-2-butanamine

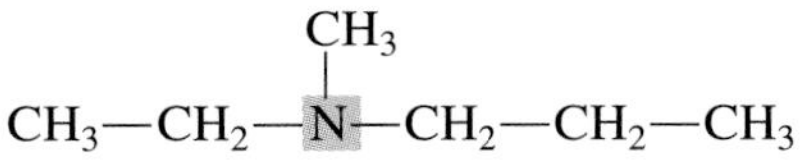

$CH_3{-}CH_2{-}N(CH_3){-}CH_2{-}CH_2{-}CH_3$ *N*-ethyl-*N*-methyl-1-propanamine

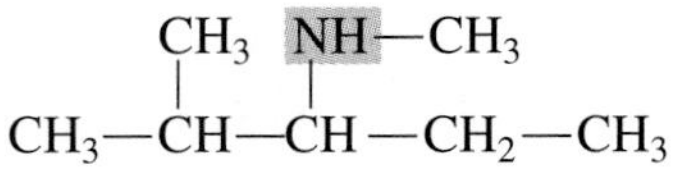

$CH_3{-}CH(CH_3){-}CH(NH{-}CH_3){-}CH_2{-}CH_3$ 2,*N*-dimethyl-3-pentanamine

• In IUPAC nomenclature, the amino group has a priority just below that of an alcohol. The priority list for functional groups is

carboxylic acid
aldehyde
ketone
alcohol
amine
(Increasing priority ↑)

In amines where additional functional groups are present, the amine group is treated as a substituent. As a substituent, an $-NH_2$ group is called an *amino* group.

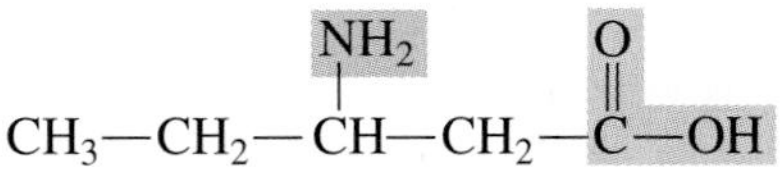

$CH_3{-}CH_2{-}CH(NH_2){-}CH_2{-}C({=}O){-}OH$ 3-Aminopentanoic acid

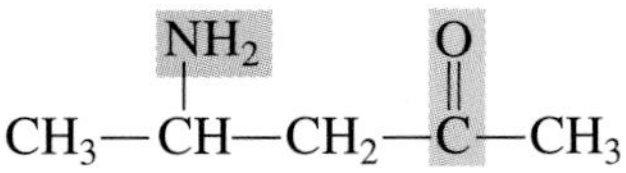

$CH_3{-}CH(NH_2){-}CH_2{-}C({=}O){-}CH_3$ 4-Amino-2-pentanone

$CH_3{-}NH{-}CH_2{-}CH_2{-}CH_2{-}OH$ 3-(*N*-methylamino)-1-propanol

The simplest aromatic amine, a benzene ring bearing an amino group, is called *aniline* (Section 13.11). Other simple aromatic amines are named as derivatives of aniline.

• A benzene ring with both an amino group and a methyl group as substituents is called *toluidine*. This name is a combination of the names *toluene* and *aniline*.

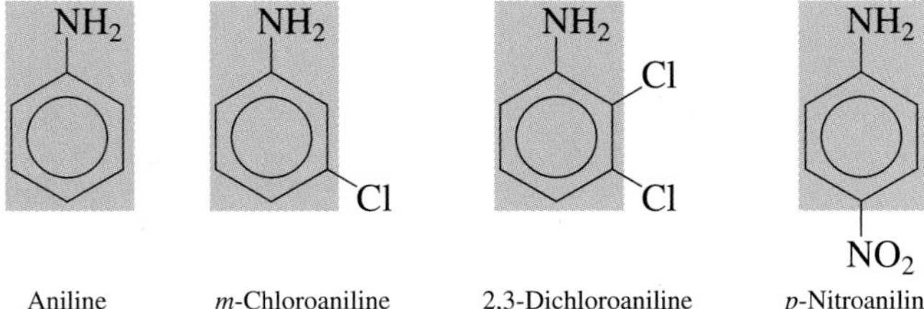

Aniline *m*-Chloroaniline 2,3-Dichloroaniline *p*-Nitroaniline

In secondary and tertiary aromatic amines, the additional group or groups attached to the nitrogen atom are located using a capital *N-*.

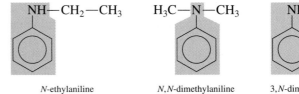

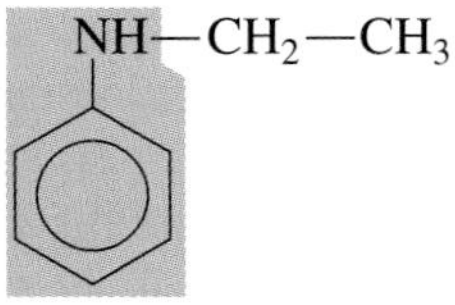

$C_6H_5{-}NH{-}CH_2{-}CH_3$ *N*-ethylaniline

$C_6H_5{-}N(CH_3){-}CH_3$ *N,N*-dimethylaniline

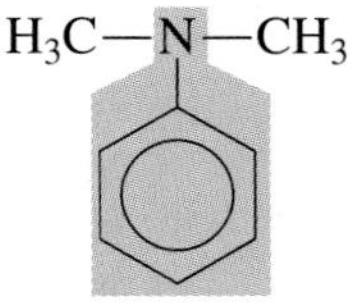

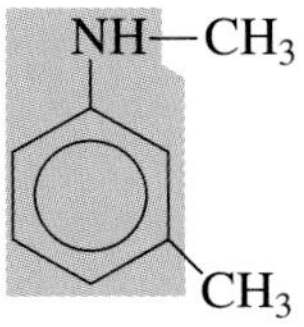

$3{-}CH_3C_6H_4{-}NH{-}CH_3$ 3,*N*-dimethylaniline

Example 17.2

Determining IUPAC Names for Amines

Assign IUPAC names to each of the following amines.

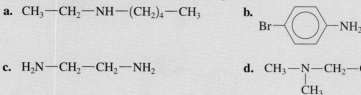

a. $CH_3—CH_2—NH—(CH_2)_4—CH_3$

b. Br—⟨benzene ring⟩—NH_2

c. $H_2N—CH_2—CH_2—NH_2$

d. $CH_3—N—CH_2—CH_3$ with CH_3 on N

Solution

a. The longest carbon chain has five carbons. The name of the compound is *N-ethyl-1-pentanamine.*

b. This compound is named as a derivative of aniline: *4-bromoaniline* (or *p*-bromoaniline). The carbon in the ring to which the $—NH_2$ is attached is carbon 1.

c. Two $—NH_2$ groups are present in this molecule. The name is *1,2-ethanediamine.*

d. This is a tertiary amine in which the longest carbon chain has two carbons (ethane). The base name is thus *ethanamine.* We also have two methyl groups attached to the nitrogen atom. The name of the compound is *N,N-dimethylethanamine.*

• The contrast between IUPAC names and common names for primary, secondary, and tertiary amines is as follows:

Primary Amines

IUPAC (one word)

alkanamine

Common (one word)

alkylamine

Secondary Amines

IUPAC (one word)

N-alkylalkanamine

Common (one word)

alkylalkylamine

Tertiary Amines

IUPAC (one word)

N-alkyl-*N*-alkylalkanamine

Common (one word)

alkylalkylalkylamine

Practice Exercise 17.2

Assign IUPAC names to each of the following amines.

a. $CH_3—CH_2—CH—CH_2—CH_2—CH_3$ with NH_2 on the third carbon

b. $CH_3—CH_2—CH_2—NH—CH_2—CH_2—CH_3$

c. $CH_3—N—CH_3$ with CH_3 on N

d. ⟨benzene ring⟩—$NH—CH_3$

• *Answers:* **a.** 3-hexanamine; **b.** *N*-propyl-1-propanamine; **c.** *N,N*-dimethylmethanamine; **d.** *N*-methylaniline

17.4 Physical Properties of Amines

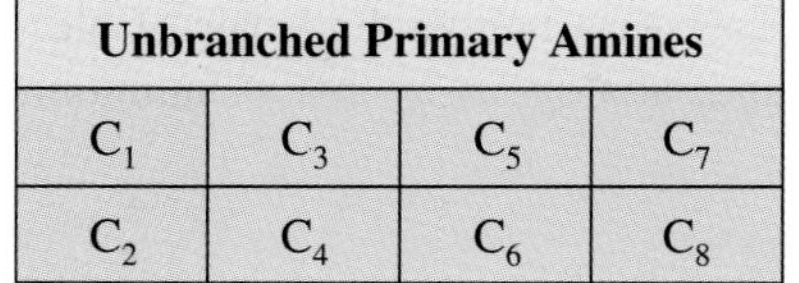

Unbranched Primary Amines			
C_1	C_3	C_5	C_7
C_2	C_4	C_6	C_8

☐ Gas ☐ Liquid

A physical-state summary for unbranched primary amides at room temperature and room pressure.

The methylamines (mono-, di-, and tri-) and ethylamine are gases at room temperature and have ammonia-like odors. Most other amines are liquids, and many have odors resembling that of raw fish. A few amines, particularly diamines, have strong, disagreeable odors. The foul odor arising from dead fish and decaying flesh is due to amines released by the bacterial decomposition of protein. Two of these "odoriferous" compounds are the diamines putrescine and cadaverine.

$H_2N—(CH_2)_4—NH_2$
Putrescine
(1,4-butanediamine)

$HN_2—(CH_2)_5—NH_2$
Cadaverine
(1,5-pentanediamine)

The simpler amines are irritating to the skin, eyes, and mucous membranes and are toxic by ingestion. Aromatic amines are generally toxic. Many are readily absorbed through the skin and affect both the blood and the nervous system.

The boiling points of amines are intermediate between those of alkanes and alcohols of similar molecular mass. They are higher than alkane boiling points, because hydrogen bonding is possible between amine molecules but not between alkane molecules. Intermolecular hydrogen bonding of amines involves the hydrogen atoms and nitrogen atoms of the amino groups (Figure 17.2).

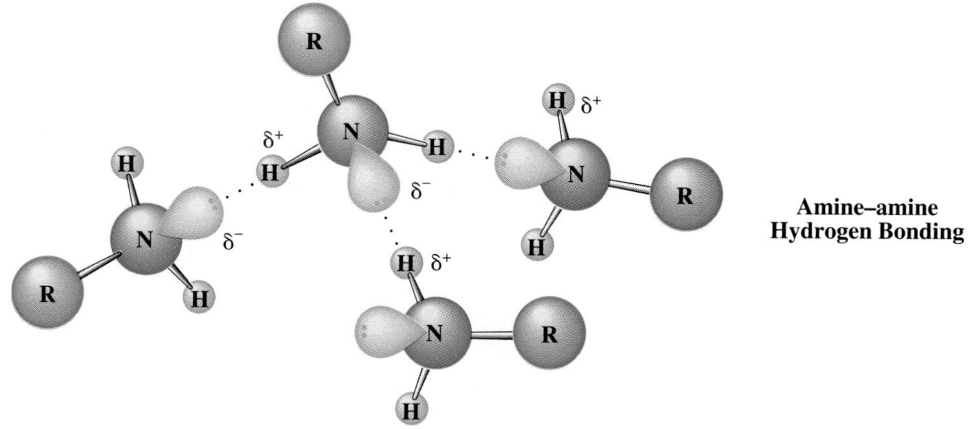

Figure 17.2
Hydrogen bonds link the hydrogen and nitrogen atoms of the amino groups of amines.

• Tertiary amines have lower boiling points than primary and secondary amines, because intermolecular hydrogen bonding is not possible in tertiary amines. Such amines have no hydrogen atoms directly bonded to the nitrogen atom.

The boiling points of amines are lower than those of corresponding alcohols (Figure 17.3), because N···H hydrogen bonds are weaker than O···H hydrogen bonds. [The difference in hydrogen-bond strength results from electronegativity differences; nitrogen is less electronegative than oxygen (Section 5.9).]

Amines of low molecular mass, with fewer than six carbon atoms, are infinitely soluble in water. This solubility results from hydrogen bonding between the amines and water. Even tertiary amines are water-soluble, because the amine nitrogen atom has a nonbonding electron pair that can form a hydrogen bond with a hydrogen atom of water.

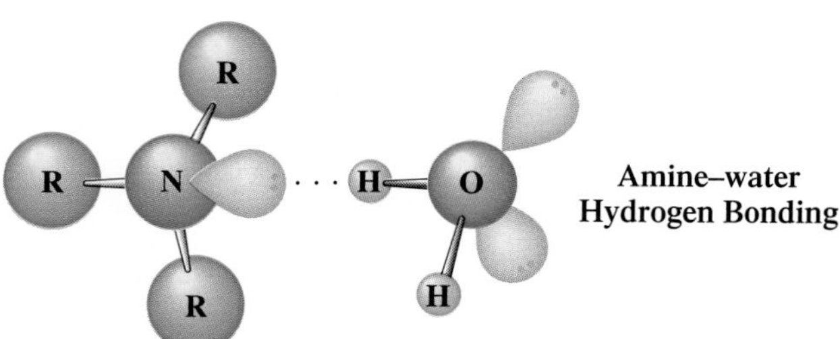

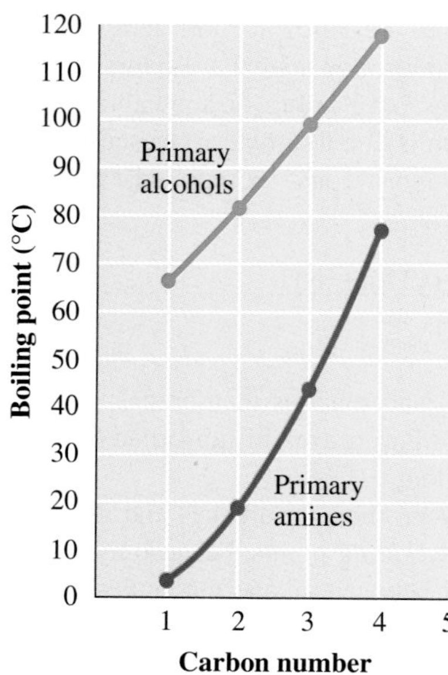

Figure 17.3
A comparison of boiling points of unbranched primary amines and unbranched primary alcohols.

17.5 Basicity of Amines

Like ammonia, amines have a nitrogen atom that possesses a nonbonding pair of electrons. Thus amines can interact with water in a manner analogous to that of ammonia (Section 10.4).

$$\underset{\text{Ammonia}}{\ddot{N}H_3} + HOH \rightleftharpoons \underset{\text{Ammonium ion}}{NH_4^+} + \underset{\text{Hydroxide ion}}{OH^-}$$

$$\underset{\text{Methylamine}}{CH_3{-}\ddot{N}H_2} + HOH \rightleftharpoons \underset{\text{Methylammonium ion}}{[CH_3{-}\overset{+}{N}H_3]} + \underset{\text{Hydroxide ion}}{OH^-}$$

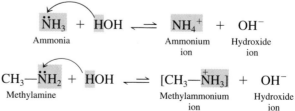

The result of the interaction of an amine with water is a basic solution containing substituted ammonium ions and hydroxide ions. A **substituted ammonium ion** *is an ammonium ion in which one or more alkyl, cycloalkyl, or aryl groups have been substituted for hydrogen atoms.*

Three important generalizations apply to substituted ammonium ions.

1. *Substituted ammonium ions are always charged species rather than neutral molecules.*

• Amines, like ammonia, have a pair of unshared electrons on the nitrogen atom present. These unshared electrons can accept a hydrogen ion from water. Thus both amines and ammonia produce basic aqueous solutions.

2. *The nitrogen atom in an ammonium ion or a substituted ammonium ion participates in four bonds.* In a neutral compound, nitrogen atoms form only three bonds. Four bonds about a nitrogen atom are possible, however, when the species is a positive ion.
3. *Substituted ammonium ions have common names derived from the names of the "parent" amines.* Replacement of the word *amine* in the name of the "parent" amine with the words *ammonium ion* generates the name of the substituted ammonium ion. The following two examples illustrate this nomenclature pattern.

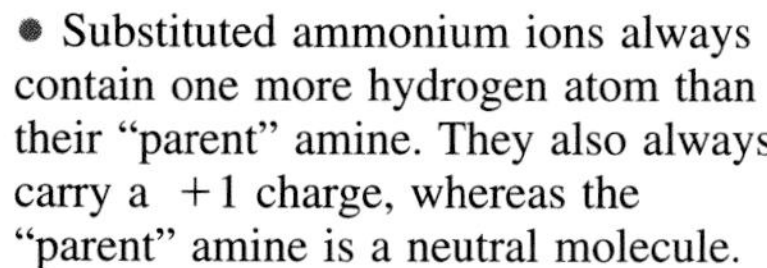

• Substituted ammonium ions always contain one more hydrogen atom than their "parent" amine. They also always carry a +1 charge, whereas the "parent" amine is a neutral molecule.

$$\underset{\text{Ethylamine}}{CH_3{-}CH_2{-}NH_2} \xrightarrow{H_2O} \underset{\text{Ethylammonium ion}}{CH_3{-}CH_2{-}\overset{+}{N}H_3} + OH^-$$

$$\underset{\text{Ethyldimethylamine}}{CH_3{-}\underset{\displaystyle CH_3}{\underset{|}{N}}{-}CH_2{-}CH_3} \xrightarrow{H_2O} \underset{\text{Ethyldimethylammonium ion}}{CH_3{-}\underset{\displaystyle CH_3}{\underset{|}{\overset{+}{N}H}}{-}CH_2{-}CH_3} + OH^-$$

Aromatic amines also exhibit basic behavior in water. With such compounds, the positive ion formed is called a substituted *anilinium ion.*

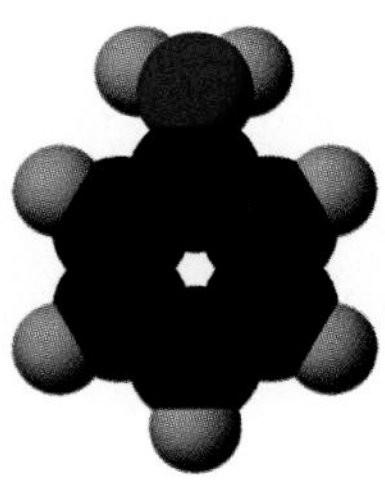

Space-filling model of aniline, the simplest aromatic amine. Aromatic amines, including aniline, are generally toxic; they are readily absorbed through the skin.

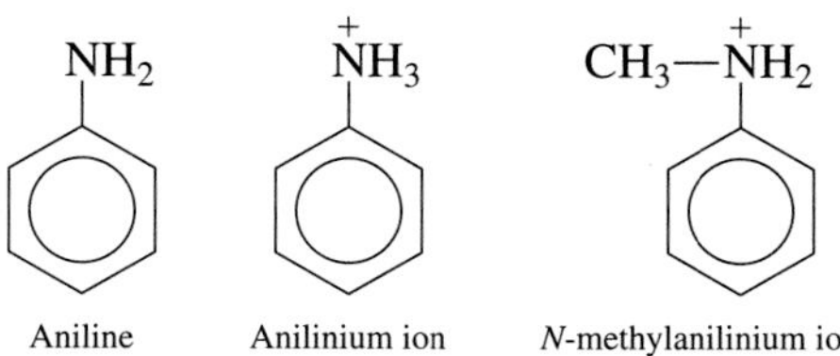

Example 17.3

Determining Names for Substituted Ammonium and Substituted Anilinium Ions

Name the following substituted ammonium or substituted anilinium ions.

a. $CH_3{-}CH_2{-}\overset{+}{N}H_2{-}CH_2{-}CH_3$

b. $CH_3{-}\underset{\displaystyle CH_3}{\underset{|}{CH}}{-}CH_2{-}\overset{+}{N}H_3$

c. $CH_3{-}\underset{\displaystyle CH_3}{\underset{|}{\overset{+}{N}H}}{-}CH_3$

d. $CH_3{-}\overset{+}{N}H{-}CH_3$ (with a benzene ring attached to N)

Solution

a. The parent amine is diethylamine. Replacing the word *amine* in the parent name with *ammonium ion* generates the name of the ion, *diethylammonium ion.*
b. The parent amine is isobutylamine. The name of the ion is *isobutylammonium ion.*
c. The parent amine is trimethylamine. The name of the ion is *trimethylammonium ion.*
d. The parent name is *N,N*-dimethylaniline. Replacing the word *aniline* in the parent name with *anilinium* ion generates the name of the ion, *N,N-dimethylanilinium ion.*

Practice Exercise 17.3

Name the following substituted ammonium or substituted anilinium ions.

a. $CH_3{-}CH_2{-}\overset{+}{N}H_2{-}CH_3$

b. $CH_3{-}\underset{\displaystyle CH_3}{\underset{|}{CH}}{-}\overset{+}{N}H_3$

(continued)

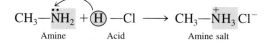

c. $CH_3—CH_2—\overset{+}{N}H(CH_3)—CH_2—CH_3$ d. $\overset{+}{N}H_2—CH_2—CH_2—CH_3$ (N attached to benzene ring)

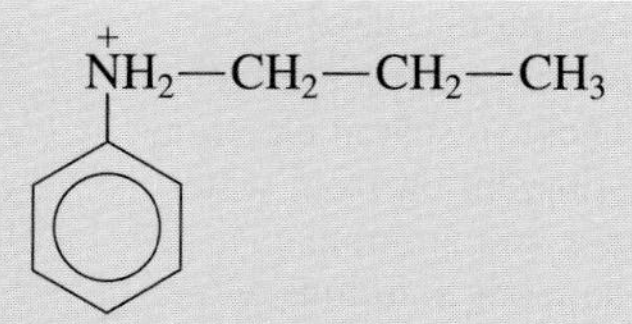

• *Answers:* **a.** ethylmethylammonium ion; **b.** isopropylammonium ion; **c.** diethylmethylammonium ion; **d.** *N*-propylanilinium ion

17.6 Amine Salts

The reaction of an acid with a base (neutralization) produces a salt (Section 10.7). Because amines are bases (Section 17.5), their reaction with an acid produces a salt, an amine salt.

$$CH_3—\ddot{N}H_2 + H—Cl \longrightarrow CH_3—\overset{+}{N}H_3\,Cl^-$$

Amine Acid Amine salt

Aromatic amines react with acids in a similar manner.

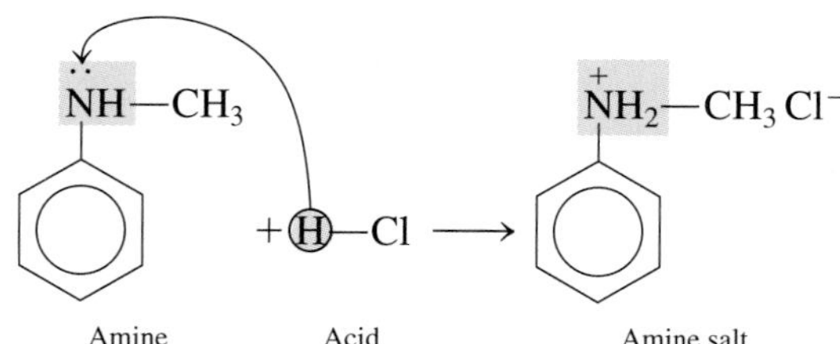

An **amine salt** *is an ionic compound in which the positive ion is a mono-, di-, or trisubstituted ammonium ion (RNH_3^+, $R_2NH_2^+$, or R_3NH^+) and the negative ion comes from the acid.* Amine salts can be obtained in crystalline form (odorless, white crystals) by evaporating the water from the acidic solutions in which amine salts are prepared.

Amine salts are named using standard nomenclature procedures for ionic compounds (Section 4.9). The name of the positive ion, the substituted ammonium or anilinium ion, is given first and is followed by a separate word for the name of the negative ion.

$$CH_3—CH_2—\overset{+}{N}H_3\,Cl^- \qquad CH_3—\overset{+}{N}H_2—CH_3\,Br^-$$

Ethylammonium chloride Dimethylammonium bromide

An older naming system for amine salts, still used in the pharmaceutical industry, treats amine salts as amine–acid complexes rather than as ionic compounds. In this system, the amine salt made from dimethylamine and hydrochloric acid is named and represented as

$$CH_3—NH(CH_3) \cdot HCl$$

Dimethylamine hydrochloride

rather than as

$$CH_3—\overset{+}{N}H_2(CH_3)\,Cl^-$$

Dimethylammonium chloride

Many medication labels refer to hydrochlorides or hydrogen sulfates (from sulfuric acid), indicating that they are in a water-soluble ionic (salt) form.

Many higher-molecular-mass amines are water-insoluble; however, virtually all amine salts are water-soluble. Thus amine salt formation, like carboxylic acid salt formation (Section 16.8), provides a means for converting water-insoluble compounds into water-soluble

• The cocaine molecule is both an amine and an ester; one amine and two ester functional groups are present. As an illegal street drug, cocaine is consumed as a water-soluble amine salt and in a water-insoluble, free-base form (nonsalt form—freed of the base required to make the salt). Cocaine hydrochloride, the amine salt, is a white powder that is snorted or injected intravenously. Free-base cocaine is heated and its vapors are inhaled. Cocaine users and dealers call the water-insoluble form of the drug "crack." "Snow" and "coke" are street names for the water-soluble form of the drug.

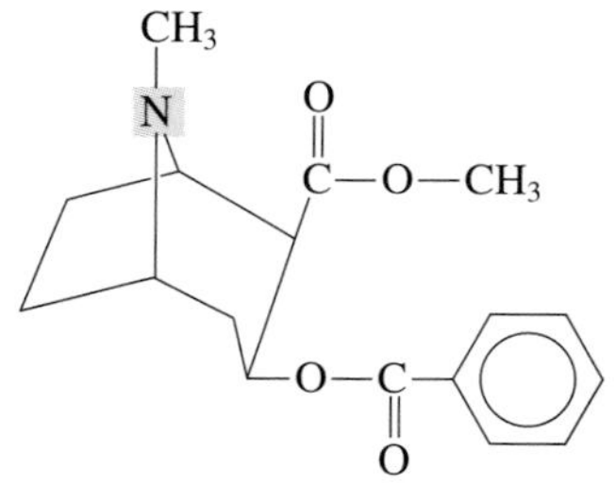

Cocaine

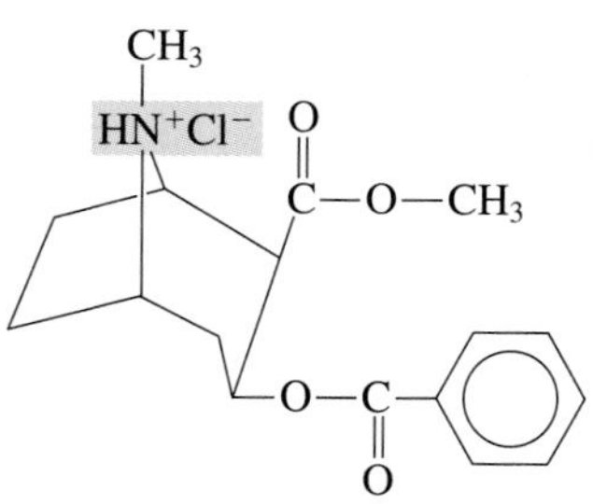

Cocaine hydrochloride

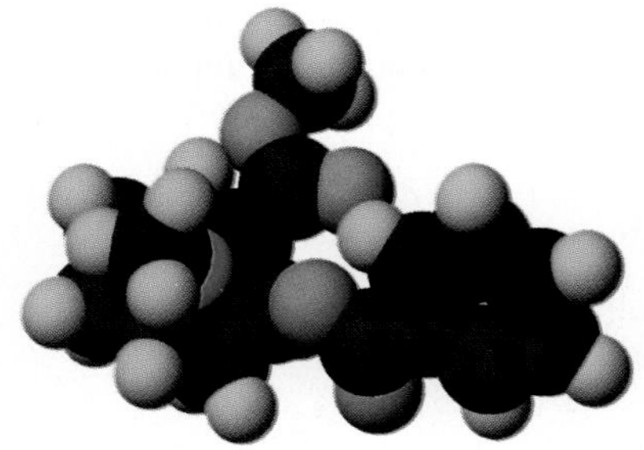

compounds. Many drugs that contain amine functional groups are administered to patients in the form of amine salts because of their increased solubility in water in this form.

Many people unknowingly use acids to form amine salts when they put vinegar or lemon juice on fish. Such action converts amines in fish (often smelly compounds) to salts, which are odorless.

The process of forming amine salts with acids is an easily reversed process. Treating an amine salt with a strong base such as NaOH regenerates the "parent" amine.

$$\underset{\text{Amine salt}}{CH_3{-}\overset{+}{N}H_3\ Cl^-} + \underset{\text{Base}}{NaOH} \longrightarrow \underset{\text{Amine}}{CH_3{-}NH_2} + NaCl + H_2O$$

Example 17.4

Writing Chemical Equations for Reactions That Involve Amine Salts

Write the structures of the products that form when each of the following reactions involving amines or amine salts takes place.

a. $CH_3{-}NH{-}CH_3 + HCl \longrightarrow$

b. $C_6H_5{-}NH{-}CH_3 + H_2SO_4 \longrightarrow$ (benzene ring bearing NH—CH₃)

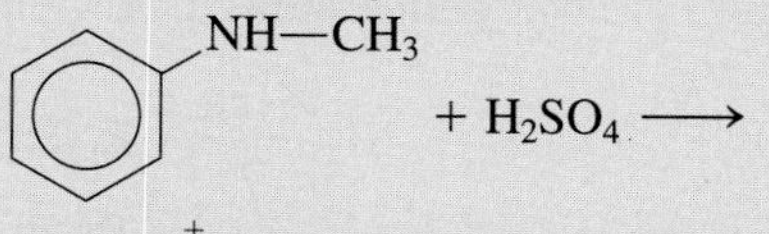

c. $CH_3{-}\overset{+}{N}H_2{-}CH_3\ Cl^- + NaOH \longrightarrow$

Solution

a. The reactants are an amine and a strong acid. Their interaction produces an amine salt.

$$CH_3{-}\ddot{N}H{-}CH_3 + \text{Ⓗ}Cl \longrightarrow CH_3{-}\overset{+}{N}H_2{-}CH_3\ Cl^-$$

b. Again, we have the reaction of an amine with a strong acid. A hydrogen ion is transferred from the acid to the amine.

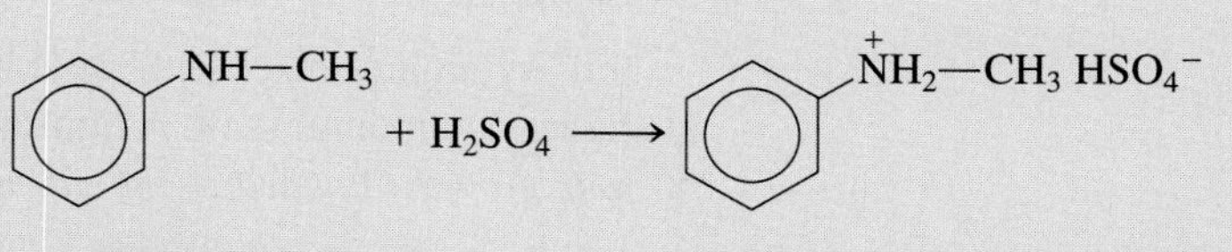

$$C_6H_5{-}NH{-}CH_3 + H_2SO_4 \longrightarrow C_6H_5{-}\overset{+}{N}H_2{-}CH_3\ HSO_4^-$$

c. The reactants are an amine salt and a strong base. Their interaction regenerates the "parent" amine.

$$CH_3{-}\overset{+}{N}H_2{-}CH_3\ Cl^- + NaOH \longrightarrow CH_3{-}NH{-}CH_3 + NaCl + H_2O$$

Practice Exercise 17.4

Write the structures of the products formed in the following reactions.

a. $CH_3{-}CH_2{-}\underset{\underset{CH_3}{|}}{N}{-}CH_3 + HCl \longrightarrow$

b. $CH_3{-}CH_2{-}NH_2 + H_2SO_4 \longrightarrow$

c. $CH_3{-}CH_2{-}\underset{\underset{CH_3}{|}}{\overset{+}{N}H}{-}CH_3\ Br^- + NaOH \longrightarrow$

• *Answers:* **a.** $CH_3{-}CH_2{-}\underset{\underset{CH_3}{|}}{\overset{+}{N}H}{-}CH_3\ Cl^-$ **b.** $CH_3{-}CH_2{-}\overset{+}{N}H_3\ HSO_4^-$

c. $CH_3{-}CH_2{-}\underset{\underset{CH_3}{|}}{N}{-}CH_3 + NaBr + H_2O$

17.7 Preparation of Amines

In a two-step process, ammonia reacts with an alkyl halide in the presence of a strong base such as NaOH to produce a primary amine. In the first step of the reaction, an amine salt is produced.

$$NH_3 + R{-}X \longrightarrow R{-}\overset{+}{N}H_3\ X^-$$

The second step, which involves the NaOH present, converts the amine salt to free amine.

$$R{-}\overset{+}{N}H_3\ X^- + NaOH \longrightarrow RNH_2 + NaX + H_2O$$

A specific example of the production of a primary amine from ammonia is the reaction of ethyl bromide with ammonia to produce ethylamine. The chemical equation (with both steps combined) is

$$NH_3 + CH_3{-}CH_2{-}Br + NaOH \longrightarrow CH_3{-}CH_2{-}NH_2 + NaBr + H_2O$$

If the newly formed primary amine produced in an ammonia alkylation reaction is not quickly removed from the reaction mixture, then the nitrogen atom of the amine may react with further alkyl halide molecules, giving, in succession, secondary and tertiary amines.

$$NH_3 \xrightarrow[OH^-]{RX} \underset{\text{Primary amine}}{RNH_2} \xrightarrow[OH^-]{RX} \underset{\text{Secondary amine}}{R_2NH} \xrightarrow[OH^-]{RX} \underset{\text{Tertiary amine}}{R_3N}$$

This reaction sequence indicates that primary and secondary amines are also appropriate starting materials for amine production through alkylation. A secondary amine can be produced from a primary amine, and a tertiary amine from a secondary amine.

$$\underset{\text{Primary amine}}{CH_3{-}CH_2{-}NH_2} + \underset{\text{Alkyl halide}}{CH_3{-}Br} + \underset{\text{Base}}{NaOH} \longrightarrow \underset{\text{Secondary amine}}{CH_3{-}CH_2{-}NH{-}CH_3} + NaBr + H_2O$$

$$\underset{\text{Secondary amine}}{CH_3{-}CH_2{-}NH{-}CH_3} + \underset{\text{Alkyl halide}}{CH_3{-}Br} + \underset{\text{Base}}{NaOH} \longrightarrow \underset{\text{Tertiary amine}}{CH_3{-}CH_2{-}\underset{\substack{|\\CH_3}}{N}{-}CH_3} + NaBr + H_2O$$

Tertiary amines react with alkyl halides in the presence of a strong base to produce a quaternary ammonium salt. A **quaternary ammonium salt** *is an ammonium salt in which all four groups attached to the nitrogen atom are organic groups.*

$$R{-}\underset{\substack{|\\R}}{N}{-}R + R{-}X \xrightarrow{OH^-} R{-}\overset{\substack{R\\|}}{\underset{\substack{|\\R}}{N^+}}{-}R\ X^-$$

Quaternary ammonium salts differ from amine salts in that addition of strong base does not convert quaternary ammonium salts back to their "parent" amines; there is no hydrogen atom on the nitrogen with which the OH^- can react. Quaternary ammonium salts are colorless, odorless, crystalline solids that have high melting points and are usually water soluble.

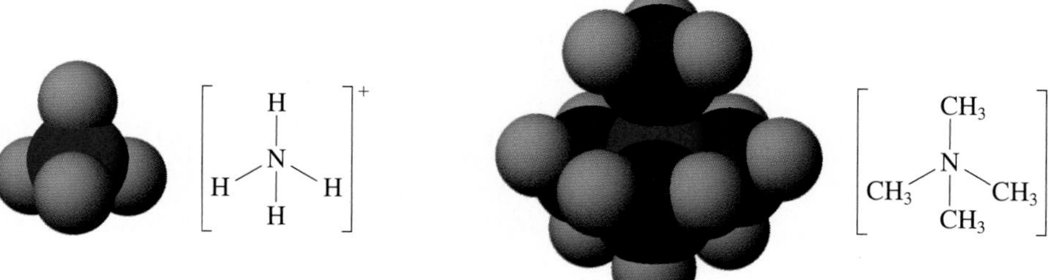

Space-filling models showing that the ammonium ion (NH_4^+) has a tetrahedral structure, as does the quaternary ammonium ion, in which four methyl groups are present $[(CH_3)_4N^+]$.

Compounds that contain quaternary ammonium ions are important in biological systems. Choline and acetylcholine are two important quaternary ammonium *ions* present in the human body. Choline has important roles in both fat transport and growth regulation. Acetylcholine is involved in the transmission of nerve impulses.

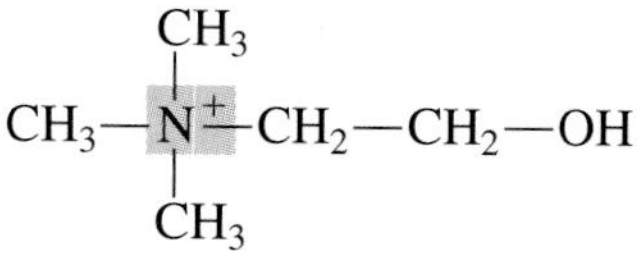

$$CH_3-\overset{CH_3}{\underset{CH_3}{N^+}}-CH_2-CH_2-OH$$

Choline

$$CH_3-\overset{CH_3}{\underset{CH_3}{N^+}}-CH_2-CH_2-O-\overset{O}{\overset{\|}{C}}-CH_3$$

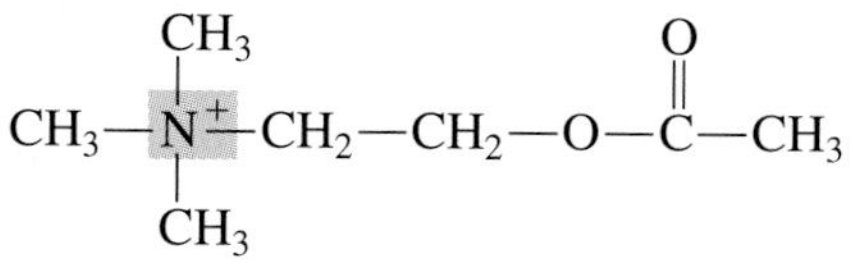

Acetylcholine

17.8 Heterocyclic Amines

A **heterocyclic amine** *is an organic compound in which nitrogen atoms of amine groups are part of either an aromatic or a nonaromatic ring system.* Heterocyclic amines are the most common type of heterocyclic organic compound (Section 14.12). Figure 17.4 gives structures for a number of "key" heterocyclic amines. These compounds are the "parent" compounds for numerous derivatives that are important in medicinal, agricultural, food, and industrial chemistry as well as in the functioning of the human body.

Study of the heterocyclic amine structures in Figure 17.4 shows that (1) ring systems may be saturated, unsaturated, or aromatic, (2) more than one nitrogen atom may be present in a given ring, and (3) fused ring systems often occur.

The two most widely used central nervous system stimulants in our society, caffeine and nicotine, are heterocyclic amine derivatives. Caffeine's structure is based on a purine ring system. Nicotine's structure contains one pyridine ring and one pyrrolidine ring.

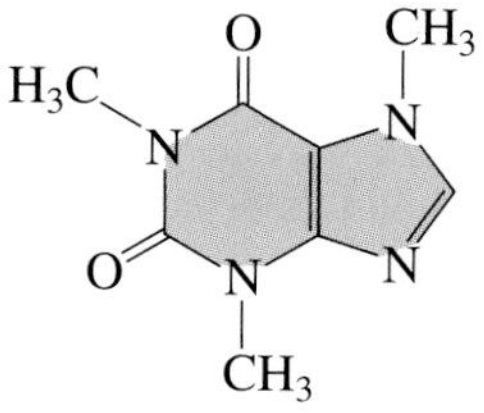

Caffeine

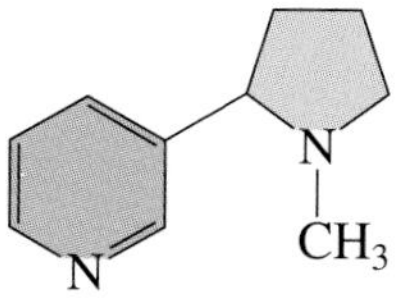

Nicotine

A large cyclic structure built on four pyrrole rings (Figure 17.4), called a porphyrin, is important in the chemistry of living organisms. Porphyrins form metal ion complexes in which the metal ion is located in the middle of the large ring structure. *Heme,* an iron–porphyrin complex present in the red blood pigment hemoglobin, is responsible for oxygen transport in the human body.

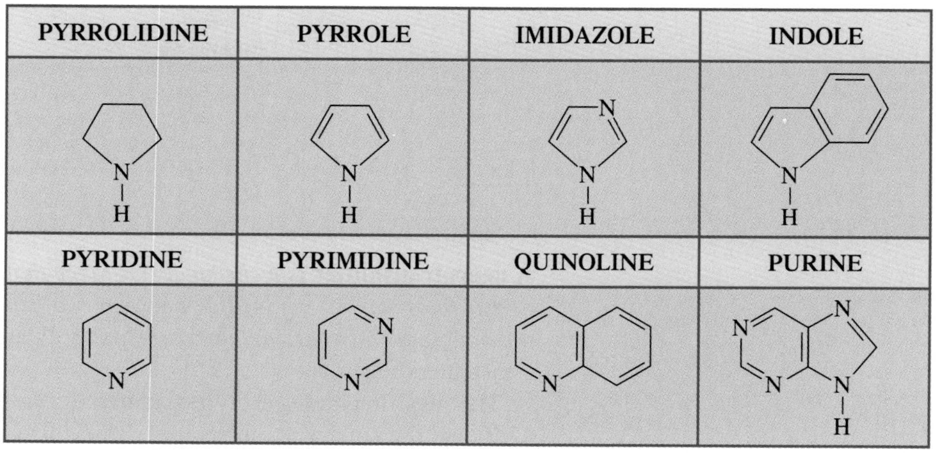

• Heterocyclic amines are our first encounter with heterocyclic compounds that have nitrogen heteroatoms. In previous chapters, we have encountered heterocyclic compounds with oxygen as the heteroatom: cyclic ethers (Section 14.12); the cyclic forms of hemiacetals, hemiketals, acetals, and ketals (Section 15.9); and cyclic esters (Section 16.9).

• Heterocyclic amines often have strong odors, some agreeable and others disagreeable. The "pleasant" aroma of many heat-treated foods is caused by heterocyclic amines formed during the heat treatment. The compounds responsible for the pervasive odors of popped popcorn and hot roasted peanuts are heterocyclic amines.

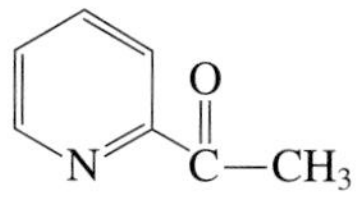

Methyl-2-pyridyl ketone (odor of popcorn)

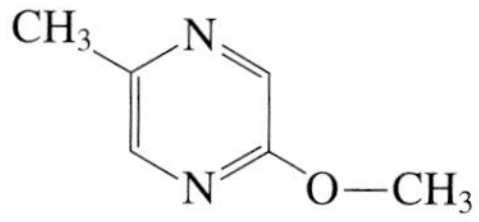

2-Methoxy-5-methylpyrazine (odor of peanuts)

PYRROLIDINE	PYRROLE	IMIDAZOLE	INDOLE
PYRIDINE	PYRIMIDINE	QUINOLINE	PURINE

Figure 17.4
Structural formulas for selected heterocyclic amines that serve as "parent" molecules for more complex amine derivatives.

Chemical CONNECTIONS

17.1 Caffeine: The Most Widely Used Central Nervous System Stimulant

Caffeine, found in coffee beans and tea leaves, is the most widely used nonprescription central nervous system stimulant today. Structurally, it is a derivative of the heterocyclic amine purine (see Figure 17.4).

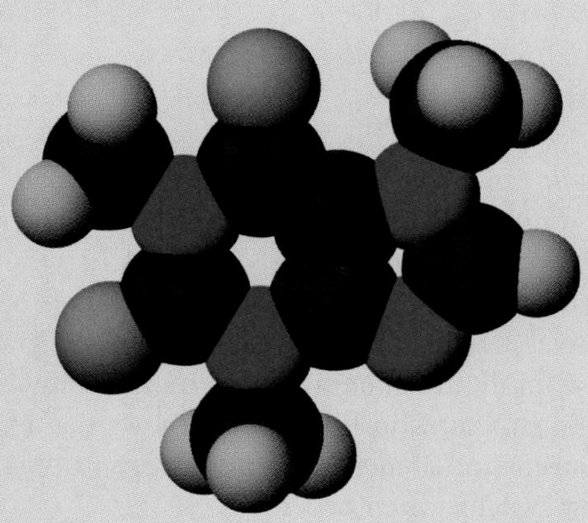

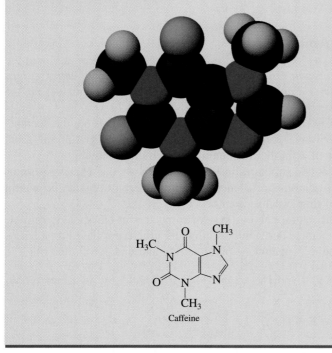

Caffeine

In addition to stimulating the central nervous system, caffeine can produce a variety of other effects. It increases heartbeat and basal metabolic rate, promotes secretion of stomach acid, and steps up production of urine. The overall effect an individual experiences is usually interpreted as a "lift."

Caffeine is mildly addicting. People who ordinarily consume substantial amounts of caffeine-containing beverages or drugs experience withdrawal symptoms if caffeine is eliminated. Such symptoms include headache and depression for a period of several days. As a result of caffeine dependence, many people need a cup of coffee before they feel good each morning.

Current scientific thought holds that caffeine's mode of action in the body is exerted through a chemical substance called cyclic adenosine monophosphate (cyclic AMP). Caffeine inhibits an enzyme that ordinarily breaks down cyclic AMP to its inactive end-product. The resulting increase in cyclic AMP leads to increased glucose production within cells and thus makes available more energy to allow higher rates of cellular activity.

Coffee is the major source of caffeine for most Americans. However, substantial amounts of caffeine may be consumed in soft drinks, tea, and numerous nonprescription medications, including combination pain relievers (Anacin, Midol, Empirin), cold remedies (Dristan, Triaminicin), and antisleep agents (No Doz, Vivarin).

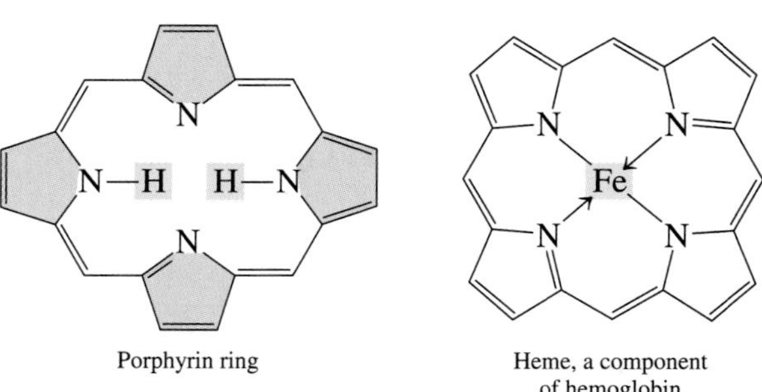

Porphyrin ring

Heme, a component of hemoglobin

17.9 Selected Biologically Important Amines

• Neurotransmitters

A **neurotransmitter** *is a chemical substance that is released at the end of a nerve, travels across the synaptic gap located between two nerves, and then bonds to a receptor site in the other nerve, triggering a nerve impulse.* Figure 17.5 shows schematically how neurotransmitters function.

The most important neurotransmitters in the human body are acetylcholine (Section 17.7) and the amines norepinephrine, dopamine, and serotonin.

Chemical CONNECTIONS

17.2 Nicotine Addiction: A Widespread Example of Drug Dependence

Next to caffeine, nicotine is the most widely used central nervous system stimulant in our society. It is found in smoking tobacco and chewing tobacco. Structurally, nicotine contains both a pyrrolidine and a pyridine ring system (see Figure 17.4); these rings are connected through a carbon–carbon bond rather than fused.

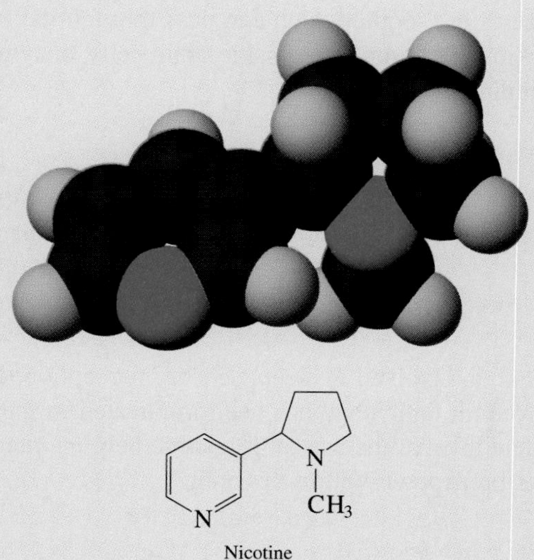

Nicotine

Nicotine is readily and completely absorbed from the stomach after oral administration and from the lungs upon inhalation. Its mild effect on the central nervous system is rather transient. After an initial response, depression follows. Most cigarettes contain between 0.5 mg and 2.0 mg of nicotine, of which approximately 20% (between 0.1 mg and 0.4 mg) will actually be inhaled and absorbed into the bloodstream.

Nicotine is quickly distributed throughout the body, rapidly penetrating the brain, all body organs, and, in general, all body fluids. There is now strong evidence that cigarette smoking affects a developing fetus.

Nicotine induces both physiological and psychological dependence. Withdrawal from nicotine is accompanied by headache, stomach pain, irritability, and insomnia. Full withdrawal can take six months or longer.

In large doses nicotine is a potent poison. Nicotine poisoning causes vomiting, diarrhea, nausea, and abdominal pain. Death occurs from respiratory paralysis. The lethal dose of the drug is considered to be approximately 60 mg.

Nicotine can influence the action of prescription drugs that the smoker is taking concurrently. For example, Darvon can lose its pain-killing effect in a smoker, and tranquilizers such as Valium may work with diminished effect. The carcinogenic effects associated with cigarette smoking are caused not by nicotine but by other substances present in tobacco, including fused-ring aromatic hydrocarbons (see Chemical Connections 13.2) and perhaps radon-222 decay products (see Chemical Connections 11.1).

Figure 17.5
Neurotransmitters are chemical messengers between nerve cells. Neurotransmitters released from one nerve cell stimulate (activate) an adjacent nerve cell. (a) Before the conduction of a nerve impulse. (b) An incoming nerve impulse triggers the release of neurotransmitter molecules. (c) Neurotransmitters bind to receptor sites, activating the receptor nerve cell.

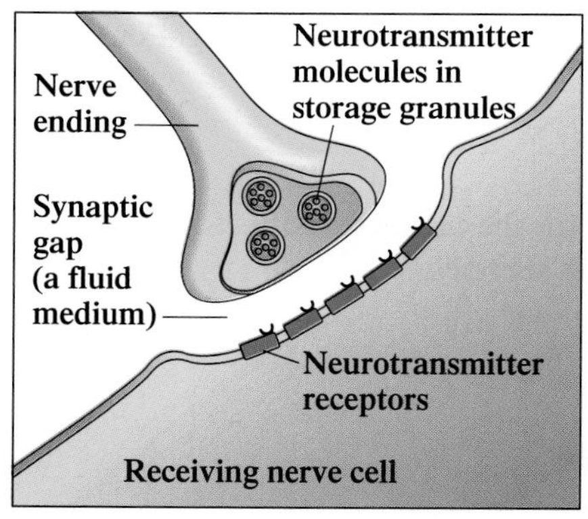

(a)

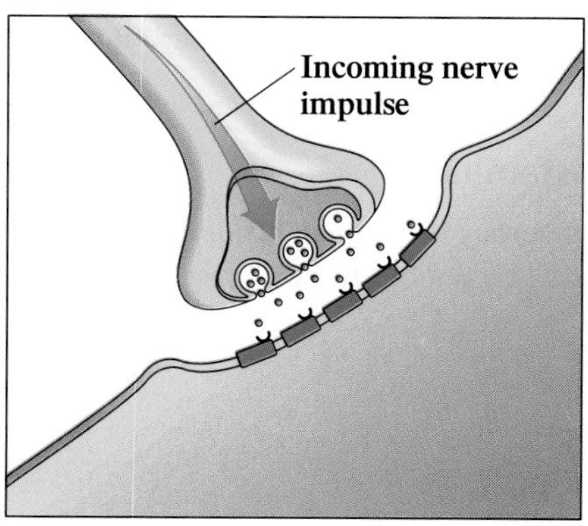

(b)

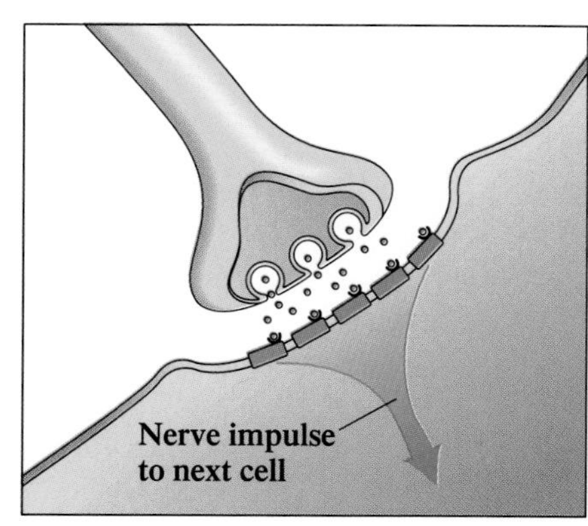

(c)

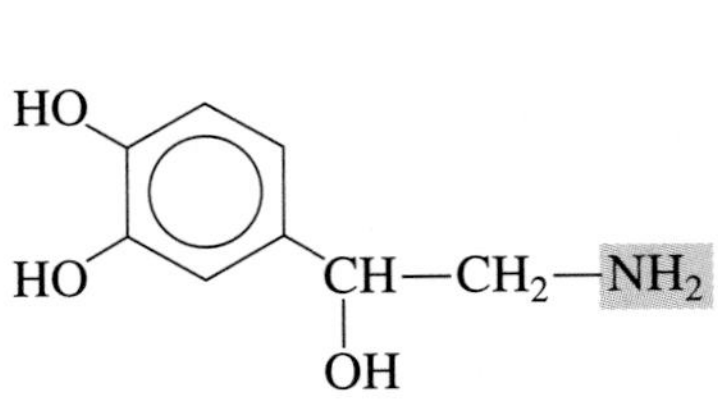

Norepinephrine

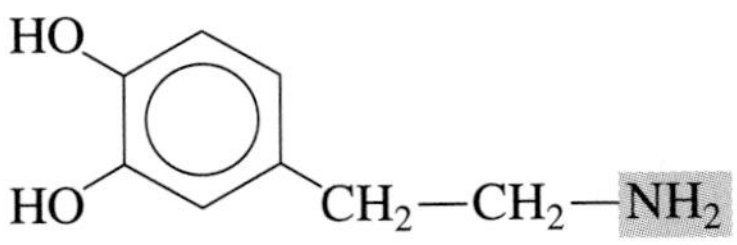

Dopamine

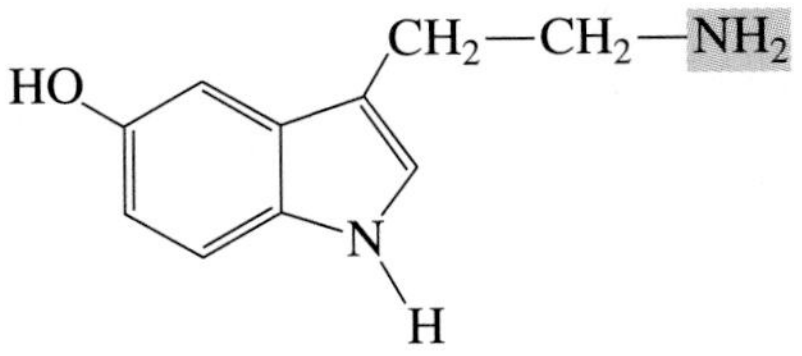

Serotonin

• The names of the amine neurotransmitters are pronounced nor-ep-in-NEFF-rin, SER-oh-tone-in, and DOPE-a-mean.

Norepinephrine, a compound secreted by the adrenal glands into the blood, helps maintain muscle tone in the blood vessels.

Dopamine is found in the brain. A deficiency of this neurotransmitter results in Parkinson's disease, a degenerative neurological disease. Administration of dopamine to a patient does not stop the symptoms of this disease, because dopamine in the blood cannot cross the blood–brain barrier. The drug L-dopa, which can pass through the blood–brain barrier, does give relief from Parkinson's symptoms. Inside brain cells, enzymes catalyze the conversion of L-dopa to dopamine.

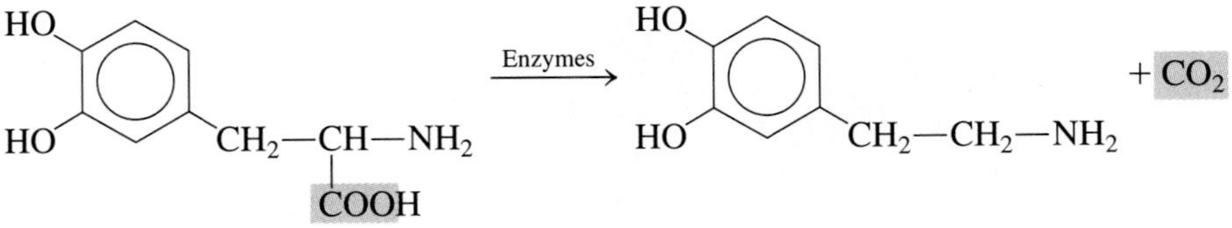

Enzymes

$+ CO_2$

L-Dopa

Dopamine

• Prozac, the most widely prescribed drug for mental depression, inhibits the reuptake of serotonin, thus maintaining serotonin levels. Chemically, Prozac is a methyl propyl secondary amine derivative with three fluorine atoms present in the structure. The element fluorine is very seldom encountered in biochemical molecules.

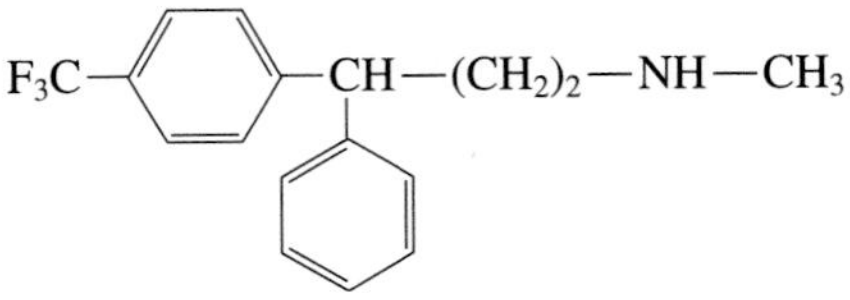

Serotonin, also a brain chemical, is involved in sleep, sensory perception, and the regulation of body temperature. Serotonin deficiency has been implicated in mental illness. Treatment of mental depression can involve the use of drugs that help maintain serotonin at normal levels by preventing its breakdown within the brain.

• Epinephrine

Epinephrine, also known as adrenaline, has some neurotransmitter functions but is more important as a central nervous system stimulant. Produced by the adrenal glands, epinephrine differs in structure from norepinephrine in that a methyl group substituent is present on the amine nitrogen atom.

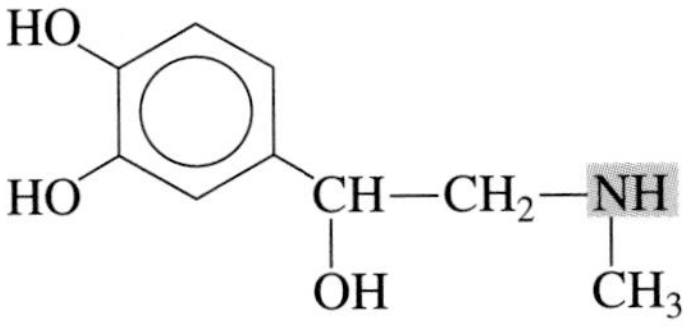

Epinephrine

Pain, excitement, and fear trigger the release of large amounts of epinephrine into the bloodstream. The effect is increased blood glucose levels, which in turn increase blood pressure, rate and force of heart contraction, and muscular strength. These changes cause the body to function at a "higher" level. Epinephrine is often called the "fight or flight" hormone.

• Histamine

The heterocyclic amine *histamine* is responsible for the unpleasant effects felt by individuals susceptible to hay fever and various pollen allergies.

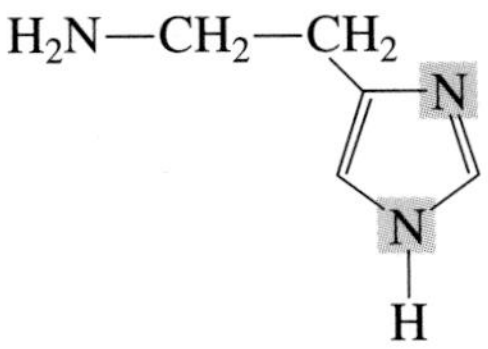

Histamine

• Antihistamines are drugs that counteract, to some extent, the effects of histamine release in the body. Antihistamines share a common structural feature with histamine—an ethanamine chain.

$$-CH_2-CH_2-N\langle$$

This structure allows antihistamines to occupy receptor sites in nerves normally occupied by histamine, thus blocking histamine from occupying the nerve sites.

Histamine is naturally present in the human body in a "stored" form; it is part of more complex molecules. A number of situations can trigger the release of histamine. Activators include (1) contact with pollen, dust, and other allergens, (2) substances released from damaged cells, and (3) contact with chemicals to which an individual has become sensitized.

The presence of "free" histamine causes the symptoms associated with hay fever, such as watery eyes and stuffy nose, and many of the symptoms associated with the common cold. A group of substances called *antihistamines* can be taken as medication to counteract the effects of the histamine.

17.10 Alkaloids

• The name *alkaloid,* which means "like a base," reflects the fact that alkaloids react with acids. Such is expected behavior for substances with amine functional groups, because amines are weak bases.

People in various parts of the world have known for centuries that physiological effects can be obtained by eating or chewing the leaves, roots, or bark of certain plants. Over 5000 different compounds that are physiologically active have been isolated from such plants. Almost all of these compounds, which are collectively called alkaloids, contain amine functional groups. An **alkaloid** *is a nitrogen-containing compound extracted from plant material.*

Three well-known compounds that we have considered previously are alkaloids. They are nicotine (tobacco plant), caffeine (coffee beans and tea leaves), and cocaine (coca plant).

A number of alkaloids are currently used in medicine. Quinine, which occurs in cinchona bark, is used to treat malaria. Atropine, which is isolated from the belladonna plant, is used to dilate the pupil of the eye in patients undergoing eye examinations. Atropine is also used as a preoperative drug to relax muscles and reduce the secretion of saliva in surgical patients.

OH, CH, CH_3-O, N, N, $CH=CH_2$

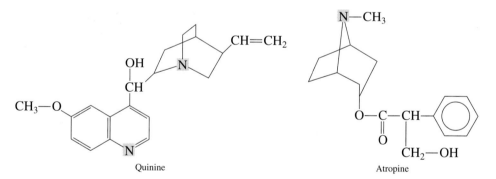

Quinine

$N-CH_3$, O—C—CH—, O, CH_2-OH

Atropine

Fruit of the belladonna plant; the alkaloid atropine is obtained from this plant.

Oriental poppy plants.

An extremely important family of alkaloids is the narcotic painkillers, a class of drugs derived from the resin (opium) of the oriental poppy plant. The most important drugs obtained from opium are morphine and codeine. Synthetic modification of morphine produces heroin.

These three compounds have similar chemical structures.

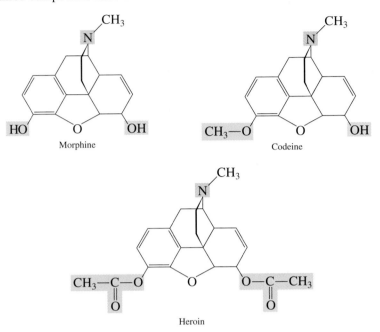

[Structures: Morphine (HO, O, OH, N–CH_3); Codeine (CH_3—O, O, OH, N–CH_3); Heroin (CH_3—C(=O)—O, O, O—C(=O)—CH_3, N–CH_3)]

Morphine

Codeine

Heroin

Morphine is one of the most effective painkillers known; its painkilling properties are about a hundred times greater than those of aspirin. Morphine acts by blocking the process in the brain that interprets pain signals coming from the peripheral nervous system. The major drawback to the use of morphine is that it is addictive.

Codeine is a methylmorphine. Almost all codeine used in modern medicine is produced by methylating the more abundant morphine. Codeine is less potent than morphine, having a painkilling effect about one sixth that of morphine.

Heroin is a synthetic compound, the diacetyl ester of morphine. This chemical modification increases painkilling potency; heroin has more than three times the painkilling effect of morphine. However, heroin is so addictive that it has no accepted medical use in the United States.

17.11 Structure of and Nomenclature for Amides

An **amide** *is a carboxylic acid derivative in which the carboxyl —OH group is replaced by an amino or a substituted amino group.* The amide functional group is thus

$$-\overset{\overset{\displaystyle O}{\|}}{C}-\underset{|}{N}-$$

Amides, like amines, can be classified as primary, secondary, or tertiary, depending on how many carbon atoms are attached to the nitrogen atom.

• Primary, secondary, and tertiary amides are also called unsubstituted, monosubstituted, and disubstituted amides, respectively.

$$R-\overset{\overset{\displaystyle O}{\|}}{C}-NH_2 \qquad R-\overset{\overset{\displaystyle O}{\|}}{C}-NH-R' \qquad R-\overset{\overset{\displaystyle O}{\|}}{C}-\underset{\underset{\displaystyle R''}{|}}{N}-R'$$

Primary amide Secondary amide Tertiary amide

• Line-angle drawings for selected primary, secondary, and tertiary amides:

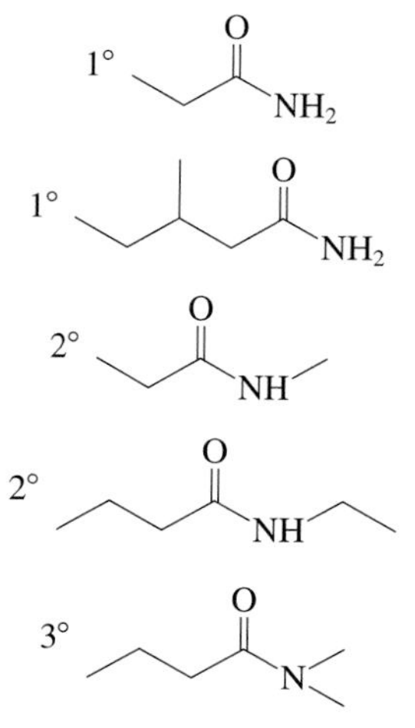

1° (line-angle structure, NH_2)

1° (line-angle structure, NH_2)

2° (line-angle structure, NH)

2° (line-angle structure, NH)

3° (line-angle structure, N)

• The contrast between IUPAC names and common names for unbranched unsubstituted amides of carboxylic acids is as follows:

IUPAC (one word)

alkanamide

ethanamide

Common (one word)

(prefix)amide*

acetamide

*The common-name prefixes are related to natural sources for the acids.

Cyclic amide structures are possible. Examples of such structures include

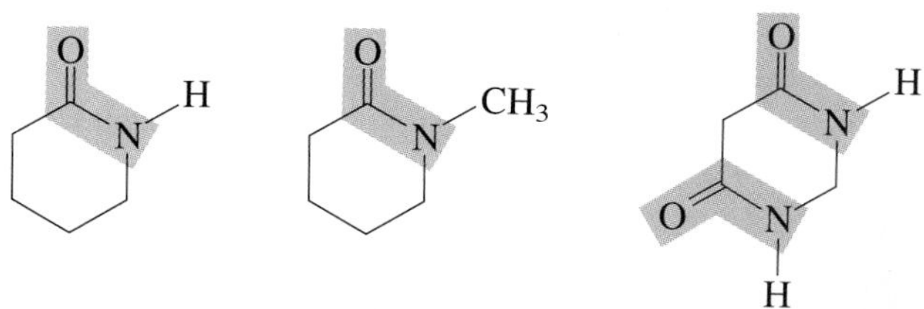

In the naming of amides, the name of the "parent" acid from which the amide can be considered to be derived (either common name or IUPAC name) is the base for the name. This base name is modified by replacing the *-ic acid* ending (common system) or the *-oic acid* ending (IUPAC system) with *-amide.* Alkyl groups attached to the nitrogen are included as prefixes, using a capital *N-* with each group to indicate location.

Example 17.5

Determining IUPAC and Common Names for Amides

Assign both common and IUPAC names to each of the following amides.

a. $CH_3{-}\overset{O}{\overset{\|}{C}}{-}NH_2$

b. $CH_3{-}\underset{}{\overset{Br}{\overset{|}{CH}}}{-}\overset{O}{\overset{\|}{C}}{-}NH{-}CH_3$

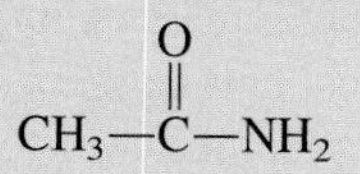

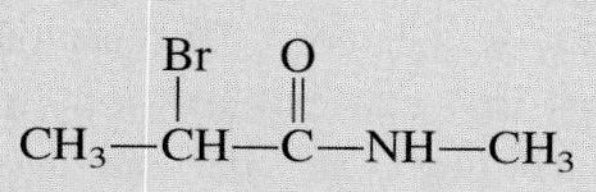

c. (benzene ring)$-\overset{O}{\overset{\|}{C}}-N-$(benzene ring), with a third benzene ring on N

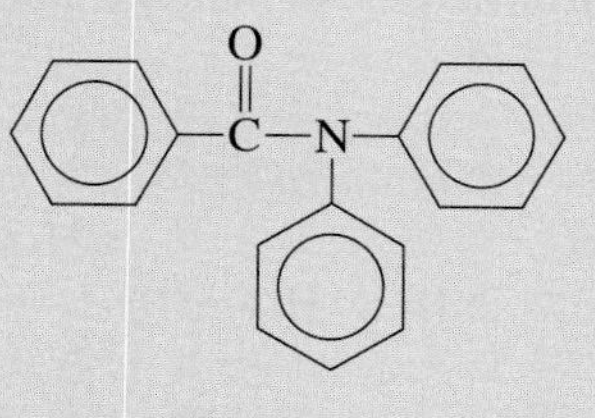

Solution

a. The parent acid for this amide is acetic acid (common) or ethanoic acid (IUPAC). The common name for this amide is *acetamide,* and the IUPAC name is *ethanamide.*

b. The common and IUPAC names of the acid are very similar; they are propionic acid and propanoic acid, respectively. The common name is *2-bromo-N-methylpropionamide,* and the IUPAC name is *2-bromo-N-methylpropanamide.* The prefix *N-* must be used with the methyl group to indicate that it is attached to the nitrogen atom.

c. In both the common and IUPAC systems of nomenclature, the name of the parent acid is the same: benzoic acid. The name of the amide is *N,N-diphenylbenzamide.*

Practice Exercise 17.5

Assign both common and IUPAC names to each of the following amides.

a. $CH_3{-}CH_2{-}\underset{Br}{\underset{|}{CH}}{-}\overset{O}{\overset{\|}{C}}{-}NH_2$

b. $CH_3{-}\overset{O}{\overset{\|}{C}}{-}NH{-}CH_3$

c. (benzene ring)$-\overset{O}{\overset{\|}{C}}-\underset{CH_3}{\underset{|}{N}}-CH_3$

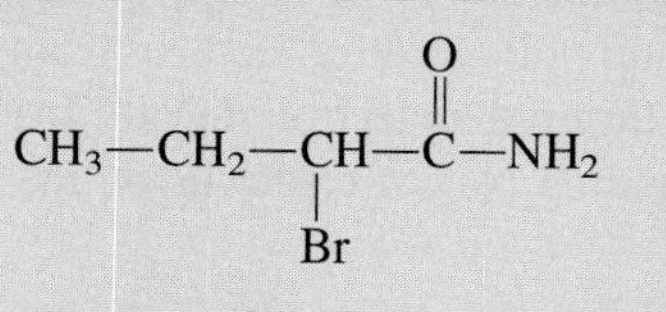

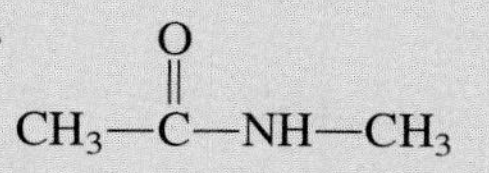

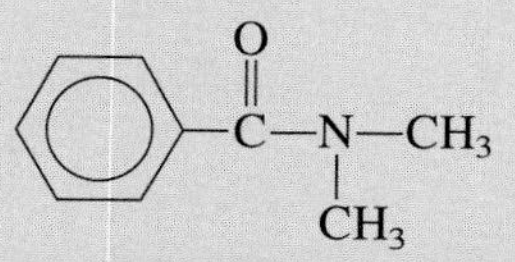

• *Answers:* **a.** 2-bromobutyramide, 2-bromobutanamide; **b.** *N*-methylacetamide, *N*-methylethanamide; **c.** *N,N*-dimethylbenzamide (both common and IUPAC)

Space-filling models for the simplest primary, secondary, and tertiary amides.

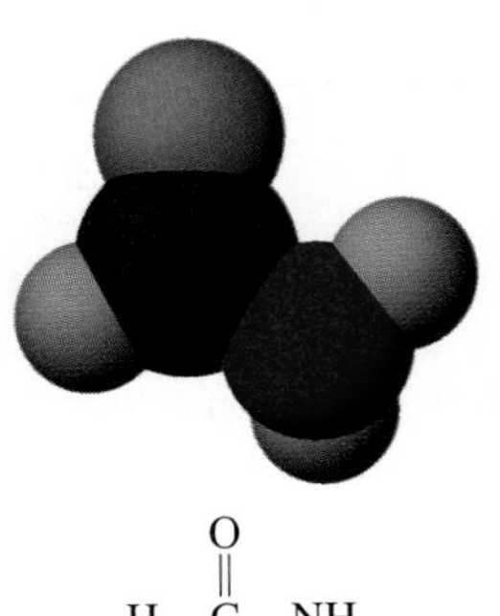

$$H-\overset{\overset{\Large O}{\|}}{C}-NH_2$$

Methanamide (a primary amide)

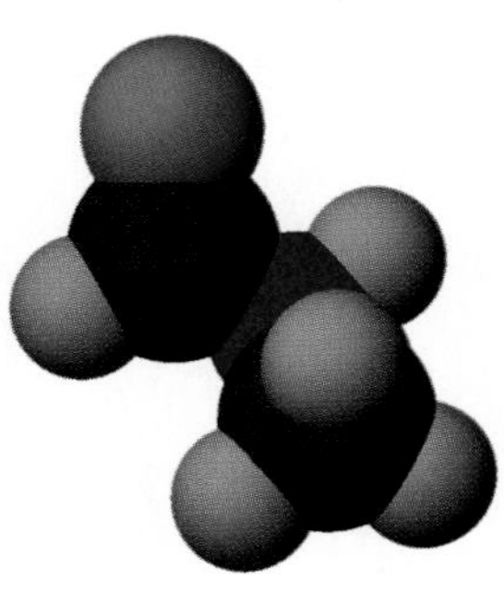

$$H-\overset{\overset{\Large O}{\|}}{C}-NH-CH_3$$

***N*-methyl methamide (a secondary amide)**

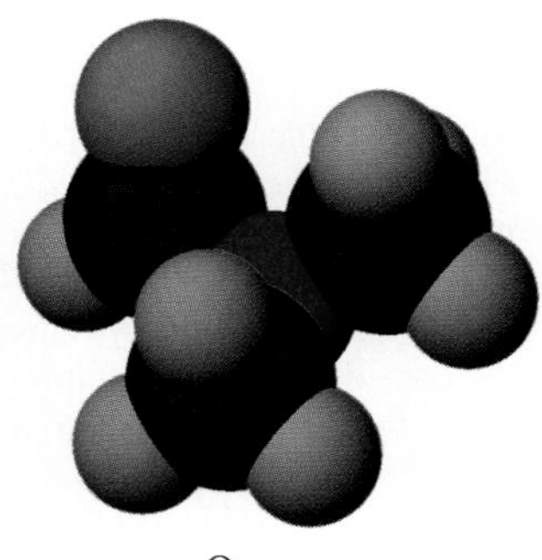

$$H-\overset{\overset{\Large O}{\|}}{C}-\underset{\underset{\Large CH_3}{|}}{N}-CH_3$$

***N,N*-dimethyl methamide (a tertiary amide)**

17.12 Selected Amides and Their Uses

The simplest naturally occurring amide is urea, a water-soluble white solid produced in the human body from carbon dioxide and ammonia through a complex series of metabolic reactions (Section 26.4).

$$CO_2 + 2NH_3 \longrightarrow (H_2N)_2CO + H_2O$$

Urea is a one-carbon diamide. Its molecular structure is

$$H_2N-\overset{\overset{\Large O}{\|}}{C}-NH_2$$

Urea formation is the human body's primary method for eliminating "waste" nitrogen. The kidneys remove urea from the blood and provide for its excretion in urine. With malfunctioning kidneys, urea concentrations in the body can build to toxic levels—a condition called *uremia.*

A number of synthetic amides exhibit physiological activity and are used as drugs in the human body. Foremost among them, in terms of use, is acetaminophen, which in 1992 replaced aspirin as the top-selling over-the-counter pain reliever. Acetaminophen is a derivative of acetamide (see Chemical Connections 17.3).

Barbiturates, which are cyclic amide compounds, are a heavily used group of prescription drugs that cause relaxation (tranquilizers), sleep (sedatives), and death (overdoses). All barbiturates are derivatives of barbituric acid, a cyclic amide that was first synthesized from urea and malonic acid.

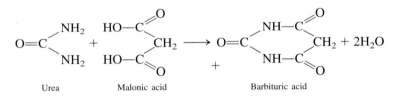

$$\underset{\text{Urea}}{O=C\begin{matrix}NH_2\\NH_2\end{matrix}} + \underset{\text{Malonic acid}}{\begin{matrix}HO-C{=}O\\ \quad CH_2\\ HO-C{=}O\end{matrix}} \longrightarrow \underset{\text{Barbituric acid}}{O=C\begin{matrix}NH-C{=}O\\ \quad CH_2\\ NH-C{=}O\end{matrix}} + 2H_2O$$

(The researcher who first synthesized this compound named it after his girlfriend Barbara.)

17.13 Properties of Amides

Amides do not exhibit basic properties in solution as amines do (Section 17.5). Although the nitrogen atom present in amides has a nonbonding pair of electrons, as in amines, these electrons are not available for bonding to a H^+ ion. The oxygen atom of the carbonyl group pulls electron density from the carbonyl carbon (an electronegativity effect), which in turn pulls electron density from the nitrogen atom. The net effect of this is that the nonbinding pair of electrons on the N atom does not exhibit basic behavior.

$$-\overset{\overset{\Large O^{\delta-}}{\|}}{C}-\ddot{N}^{\delta+}-$$

With the exception of formamide, which is a liquid, all unsubstituted amides are solids at room temperature. Numerous intermolecular hydrogen-bonding possibilities exist, between amide H atoms and carbonyl O atoms, in unsubstituted amides.

Chemical CONNECTIONS

17.3 Acetaminophen: A Substituted Amide

Often called the aspirin substitute, acetaminophen is the most widely used of all nonprescription pain relievers, accounting for over half of this market. Acetaminophen is a derivative of acetamide in which a hydroxyphenyl group has replaced one of the amide hydrogens.

$$CH_3-\overset{O}{\overset{\|}{C}}-NH_2$$

Acetamide

$$CH_3-\overset{O}{\overset{\|}{C}}-NH-C_6H_4-OH$$

Acetaminophen

Acetaminophen is the active ingredient in Tylenol, Datril, Tempra, and Anacin-3. Excedrin, which contains both acetaminophen and aspirin, is a combination pain reliever.

Acetaminophen is often used as an aspirin substitute because it has no irritating effect on the intestinal tract and yet has comparable analgesic and antipyretic effects. Unlike aspirin, however, it is not effective against inflammation and is of limited use for the aches and pains of arthritis. Also, acetaminophen does not inhibit platelet aggregation and therefore is not useful for preventing vascular clotting.

Acetaminophen is available in a liquid form that is used extensively for small children and other patients who have difficulty taking solid tablets. The extensive use of acetaminophen for children has a drawback; it is the drug most often involved in childhood poisonings. An overdose of acetaminophen is toxic to the liver.

Acetaminophen's mode of action in the body is similar to that of aspirin—inhibition of prostaglandin synthesis.

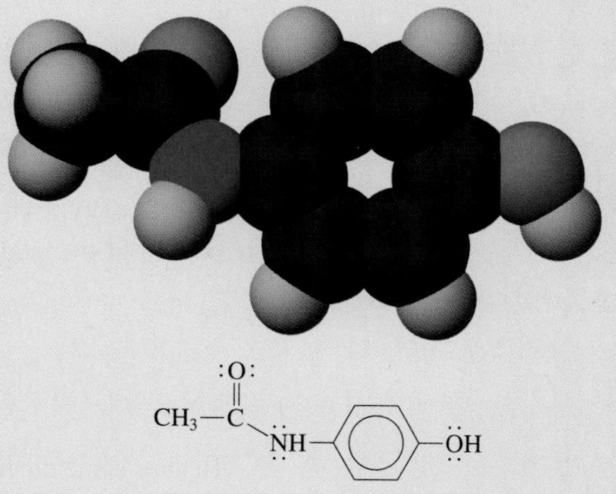

$$CH_3-\overset{:O:}{\overset{\|}{C}}-\ddot{N}H-C_6H_4-\ddot{O}H$$

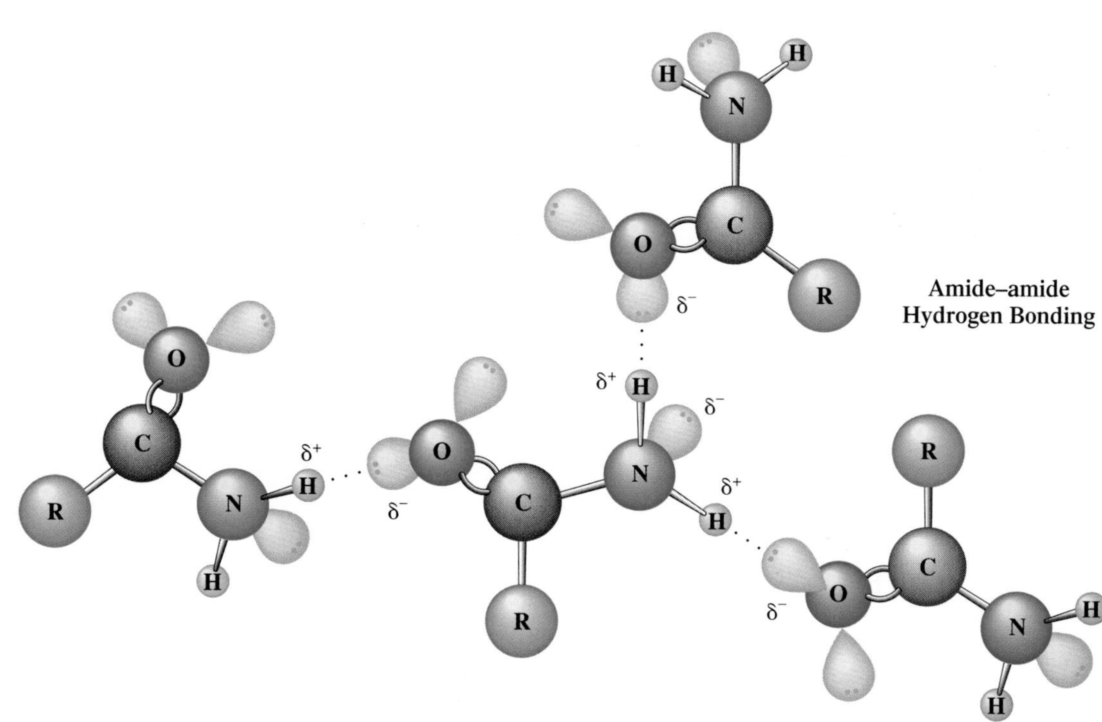

Amide–amide Hydrogen Bonding

Unbranched Primary Amides			
C_1	C_3	C_5	C_7
C_2	C_4	C_6	C_8

□ Liquid □ Solid

A physical-state summary for unbranched primary amines at room temperature and pressure.

Fewer hydrogen-bonding possibilities exist for monosubstituted amides, because the nitrogen atom now has only one hydrogen atom; hence lower melting points are the rule for such amides. Still lower melting points are observed for disubstituted amides, because no hydrogen bonding is possible. The disubstituted *N,N*-dimethylacetamide has a melting point of −20°C, which is about 100°C lower than that of the unsubstituted acetamide.

Amides of low molecular mass, up to five or six carbon atoms, are soluble in water. Again, numerous hydrogen-bonding possibilities exist between water and the amide. Even disubstituted amides can participate in such hydrogen bonding.

$$\mathrm{R{-}\overset{\overset{\displaystyle \ddot{O}:}{\|}}{C}{-}\ddot{N}{-}H} \quad (\text{N bonded to a second H})$$

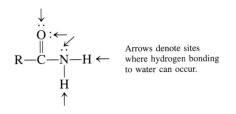

Arrows denote sites where hydrogen bonding to water can occur.

17.14 Preparation of Amides

The reaction of a carboxylic acid with ammonia produces a primary amide. The —OH portion of the acid's carboxyl group is replaced, and water is formed as a by-product.

$$\underset{\text{Acid}}{\mathrm{R{-}\overset{\overset{\displaystyle O}{\|}}{C}{-}OH}} + \underset{\text{Ammonia}}{NH_3} \xrightarrow{\Delta} \underset{\text{Amide}}{\mathrm{R{-}\overset{\overset{\displaystyle O}{\|}}{C}{-}NH_2}} + H_2O$$

Heat or some other means of activation is needed for the reaction to proceed.

Mono- and disubstituted amides can also be prepared from carboxylic acids. To obtain these products, primary amines and secondary amines, respectively, are used as the second reactant

$$\underset{\text{Butanoic acid}}{\mathrm{CH_3{-}CH_2{-}CH_2{-}\overset{\overset{\displaystyle O}{\|}}{C}{-}OH}} + \underset{\substack{\text{Methylamine}\\\text{(primary amine)}}}{\mathrm{H{-}\overset{\overset{\displaystyle H}{|}}{N}{-}CH_3}} \xrightarrow{\Delta} \underset{\substack{N\text{-methylbutanamide}\\\text{(a monosubstituted amine)}}}{\mathrm{CH_3{-}CH_2{-}CH_2{-}\overset{\overset{\displaystyle O}{\|}}{C}{-}\overset{\overset{\displaystyle H}{|}}{N}{-}CH_3}} + H_2O$$

$$\underset{\text{Benzoic acid}}{\mathrm{C_6H_5{-}\overset{\overset{\displaystyle O}{\|}}{C}{-}OH}} + \underset{\substack{\text{Diethylamine}\\\text{(secondary amine)}}}{\mathrm{H{-}\underset{\underset{\displaystyle CH_2{-}CH_3}{|}}{N}{-}CH_2{-}CH_3}} \xrightarrow{\Delta} \underset{\substack{N,N\text{-diethylbenzamide}\\\text{(a disubstituted amine)}}}{\mathrm{C_6H_5{-}\overset{\overset{\displaystyle O}{\|}}{C}{-}\underset{\underset{\displaystyle CH_2{-}CH_3}{|}}{N}{-}CH_2{-}CH_3}} + H_2O$$

Just as it is useful to think of the structure of an ester (Section 16.9) in terms of an "acid part" and an "alcohol part," it is useful to think of an amide in terms of an "acid part" and an "amine (or ammonia) part."

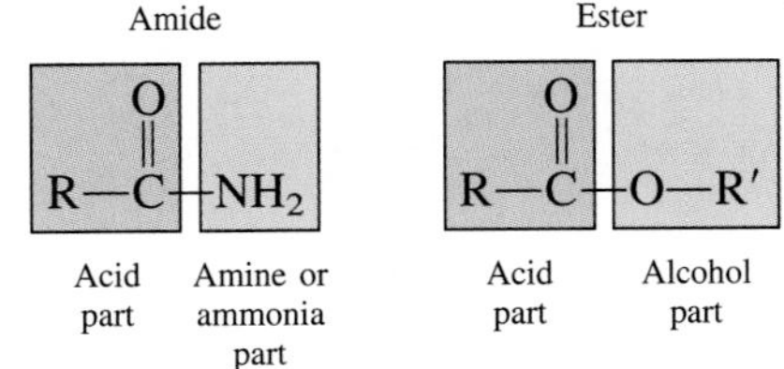

In this context, it is easy to identify the parent acid and amine from which a given amide can be produced; to generate the parent molecules, just add an —OH group to the acid part of the amide and an H atom to the amine part.

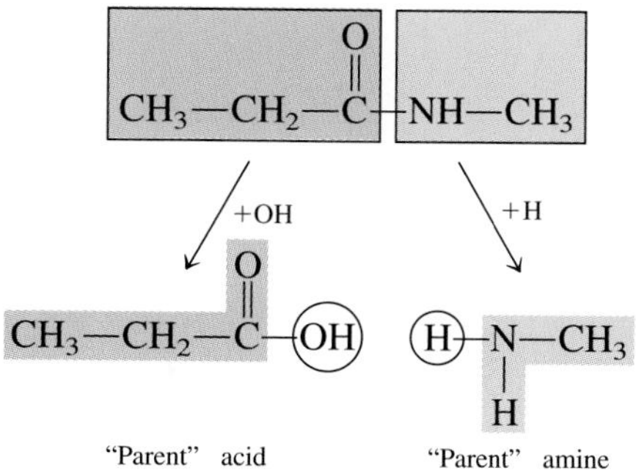

$$CH_3-CH_2-\overset{\overset{\large O}{\|}}{C}-NH-CH_3$$

+OH / +H

$$CH_3-CH_2-\overset{\overset{\large O}{\|}}{C}-OH \qquad H-\underset{\underset{\large H}{|}}{N}-CH_3$$

"Parent" acid "Parent" amine

Example 17.6

Predicting Reactants Needed to Prepare Specific Amides

What carboxylic acid and amine (or ammonia) are needed to prepare each of the following amides?

a. $CH_3-\overset{\overset{\large O}{\|}}{C}-NH-CH_2-CH_3$

b. $CH_3-CH_2-\overset{\overset{\large O}{\|}}{C}-NH_2$

c. $CH_3-CH_2-\overset{\overset{\large O}{\|}}{C}-\underset{\underset{\large CH_3}{|}}{N}-CH_3$

Solution

a. Viewing the molecule as having an acid part and an amine part, we obtain

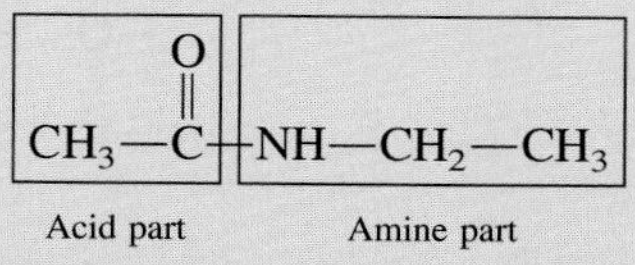

$$CH_3-\overset{\overset{\large O}{\|}}{C}-NH-CH_2-CH_3$$

Acid part Amine part

Adding an —OH group to the acid part and a H atom to the amine part, we obtain the "parent" molecules, which are

$$CH_3-\overset{\overset{\large O}{\|}}{C}-OH \quad \text{and} \quad CH_3-CH_2-NH_2$$

b. Proceeding as in part **a,** we find that the "parent" acid and amine molecules are, respectively,

$$CH_3-CH_2-\overset{\overset{\large O}{\|}}{C}-OH \quad \text{and} \quad NH_3$$

c. Proceeding again as in part **a,** we find that the "parent" acid and amine molecules are, respectively,

$$CH_3-CH_2-\overset{\overset{\large O}{\|}}{C}-OH \quad \text{and} \quad CH_3-NH-CH_3$$

Practice Exercise 17.6

What carboxylic acid and amine (or ammonia) are needed to prepare each of the following amides?

a. $CH_3-\overset{\overset{\large O}{\|}}{C}-\underset{\underset{\large CH_3}{|}}{N}-CH_2-CH_3$

b. $CH_3-CH_2-\overset{\overset{\large O}{\|}}{C}-NH-CH_3$

(continued)

c. $C_6H_5-\overset{O}{\overset{\|}{C}}-NH_2$

• *Answers:* **a.** $CH_3-\overset{O}{\overset{\|}{C}}-OH,\ CH_3-CH_2-NH-CH_3$

b. $CH_3-CH_2-\overset{O}{\overset{\|}{C}}-OH,\ CH_3-NH_2$ **c.** $C_6H_5-\overset{O}{\overset{\|}{C}}-OH,\ NH_3$

17.15 Hydrolysis of Amides

As was the case with esters (Section 16.13), the most important reaction of amides is hydrolysis. In amide hydrolysis, the bond between the carbonyl carbon atom and the nitrogen is broken, and free acid and free amine are produced. Amide hydrolysis is catalyzed by acids, bases, or certain enzymes; sustained heating is also often required.

$$\underset{\text{Amide}}{R-\overset{O}{\overset{\|}{C}}-NH-R'} + H_2O \xrightarrow{\Delta} \underset{\text{Carboxylic acid}}{R-\overset{O}{\overset{\|}{C}}-OH} + \underset{\text{Amine}}{R'-NH_2}$$

• Amide hydrolysis under basic conditions is also called amide saponification, just as ester hydrolysis under basic conditions is called ester saponification (Section 16.14).

Acidic or basic hydrolysis conditions have an effect on the products. *Acidic* conditions convert the product amine to an amine salt (Section 17.6). *Basic* conditions convert the product acid to a carboxylic acid salt (Section 16.8).

$$\underset{\text{Acidic hydrolysis of an amide}}{R-\overset{O}{\overset{\|}{C}}-NH-R' + H_2O + HCl} \xrightarrow{\Delta} \underset{\text{Carboxylic acid}}{R-\overset{O}{\overset{\|}{C}}-OH} + \underset{\text{Amine salt}}{R'-\overset{+}{N}H_3\ Cl^-}$$

$$\underset{\text{Basic hydrolysis of an amide}}{R-\overset{O}{\overset{\|}{C}}-NH-R' + NaOH} \xrightarrow{\Delta} \underset{\text{Carboxylic acid salt}}{R-\overset{O}{\overset{\|}{C}}-O^-\ Na^+} + \underset{\text{Amine}}{R'-NH_2}$$

Example 17.7

Predicting the Products of Amide Hydrolysis Reactions

Draw structural formulas for the organic products of each of the following amide hydrolysis reactions. Be sure to take into account whether the hydrolysis occurs under neutral, acidic, or basic conditions.

a. $CH_3-CH_2-\overset{O}{\overset{\|}{C}}-NH-CH_3 + H_2O \xrightarrow{\Delta}$

b. $CH_3-CH_2-\overset{O}{\overset{\|}{C}}-NH-CH_2-CH_3 + H_2O \xrightarrow[HCl]{\Delta}$

c. $CH_3-\overset{O}{\overset{\|}{C}}-\underset{CH_3}{\underset{|}{N}}-CH_3 + H_2O \xrightarrow[NaOH]{\Delta}$

d.

$$CH_3-\underset{\underset{CH_3}{|}}{CH}-\overset{\overset{O}{\|}}{C}-NH_2 + H_2O \xrightarrow{\Delta}$$

Solution

a. This reaction is hydrolysis under neutral conditions. The products will be the "parent" acid and amine for the amide. These "parents" are

$$CH_3-CH_2-\overset{\overset{O}{\|}}{C}-OH \quad \text{and} \quad CH_3-NH_2$$

b. This reaction is hydrolysis under acidic conditions. The acid is hydrochloric acid (HCl). The products will be the "parent" carboxylic acid and the chloride salt of the amine. The HCl converts the amine to its chloride salt.

$$CH_3-CH_2-\overset{\overset{O}{\|}}{C}-OH \quad \text{and} \quad CH_3-CH_2-\overset{+}{N}H_3\ Cl^-$$

c. This reaction is hydrolysis under basic conditions. The base is sodium hydroxide (NaOH). The products will be the "parent" amine and the salt of the carboxylic acid. The NaOH converts the carboxylic acid to its sodium salt.

$$CH_3-\overset{\overset{O}{\|}}{C}-O^-\ Na^+ \quad \text{and} \quad CH_3-NH-CH_3$$

d. This reaction is hydrolysis under neutral conditions. The products will be the "parent" acid and amine of the amide. Because the amide is unsubstituted, the parent amine is actually ammonia.

$$CH_3-\underset{\underset{CH_3}{|}}{CH}-\overset{\overset{O}{\|}}{C}-OH \quad \text{and} \quad NH_3$$

Practice Exercise 17.7

Draw structural formulas for the organic products of each of the following amide hydrolysis reactions. Be sure to take into account whether the hydrolysis occurs under neutral, acidic, or basic conditions.

a.

$$CH_3-\overset{\overset{O}{\|}}{C}-NH-CH_3 + H_2O \xrightarrow[NaOH]{\Delta}$$

b.

$$CH_3-\overset{\overset{O}{\|}}{C}-NH-CH_3 + H_2O \xrightarrow[HCl]{\Delta}$$

c.

$$CH_3-\overset{\overset{O}{\|}}{C}-NH-CH_3 + H_2O \xrightarrow{\Delta}$$

d.

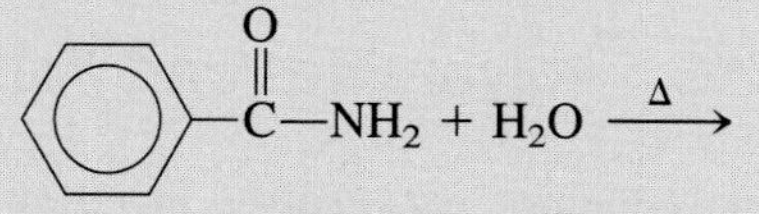

$$C_6H_5-\overset{\overset{O}{\|}}{C}-NH_2 + H_2O \xrightarrow{\Delta}$$

• *Answers:* **a.** $CH_3-\overset{\overset{O}{\|}}{C}-O^-\ Na^+,\ CH_3-NH_2$ **b.** $CH_3-\overset{\overset{O}{\|}}{C}-OH,\ CH_3-\overset{+}{N}H_3\ Cl^-$

c. $CH_3-\overset{\overset{O}{\|}}{C}-OH,\ CH_3-NH_2$ **d.** $C_6H_5-\overset{\overset{O}{\|}}{C}-OH,\ NH_3$

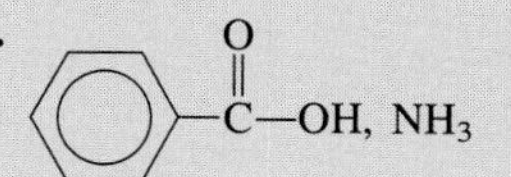

Summary of Reactions Involving Amines and Amides

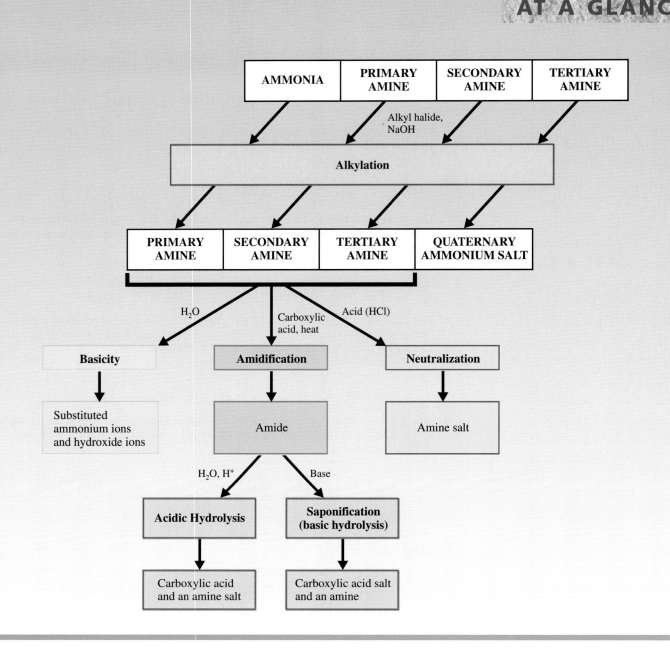

CHEMISTRY AT A GLANCE

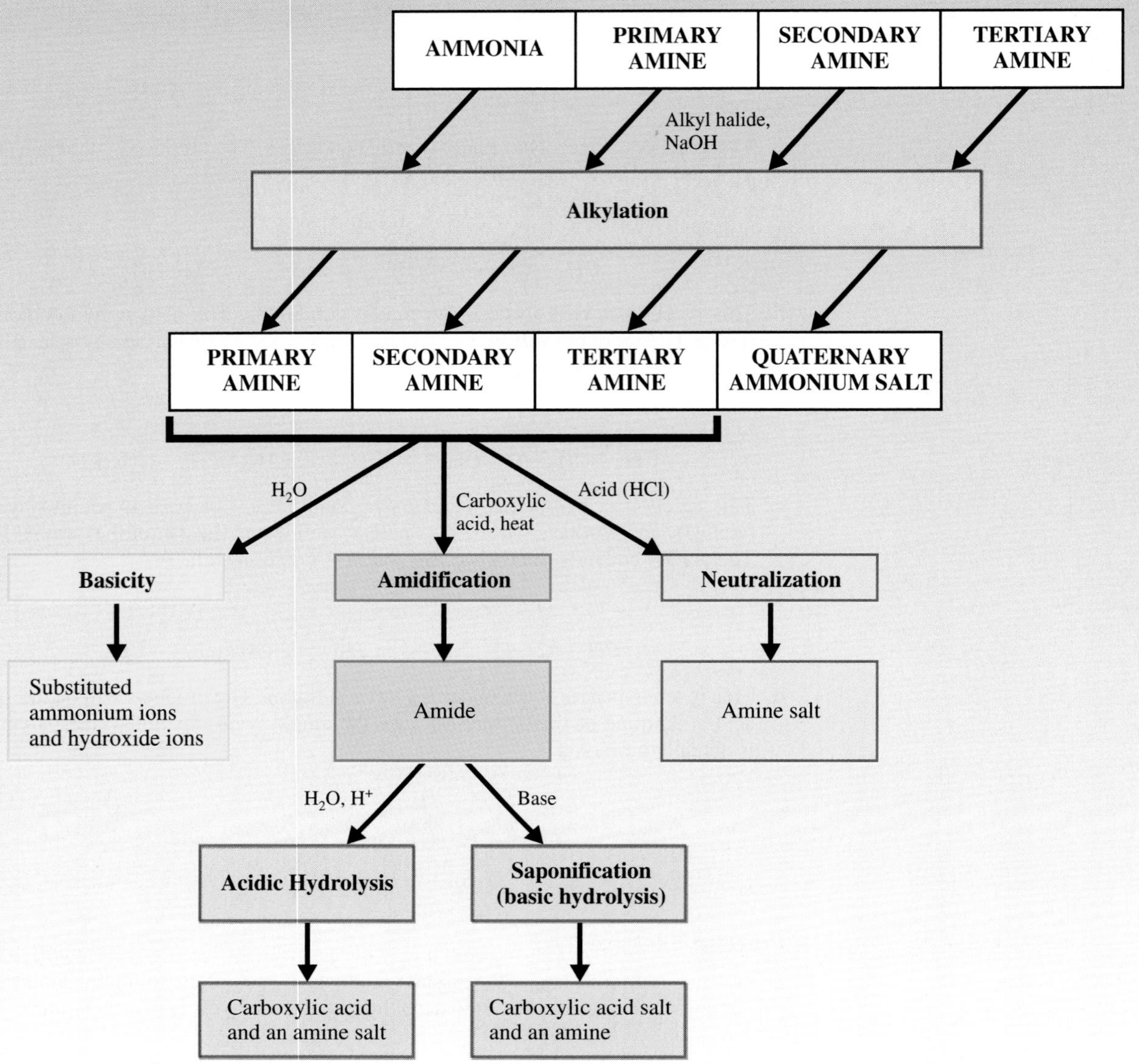

The accompanying Chemistry at a Glance summarizes the reactions that involve amines and amides.

17.16 Polyamides and Polyurethanes

Amide polymers, polyamides, are synthesized by combining diamines and dicarboxylic acids in a condensation polymerization reaction (Section 16.15). A **polyamide** *is a condensation polymer in which the monomers are linked together by amide functional groups.*

The most important synthetic polyamide is *nylon.* Nylon is used in clothing and hosiery, as well as in carpets, tire cord, rope, and parachutes. It also has nonfiber uses; for example, it is used in paint brushes, electrical parts, valves, and fasteners. It is a tough, strong, nontoxic, nonflammable material that is resistant to chemicals. Surgical suture is made of nylon because it is such a strong fiber.

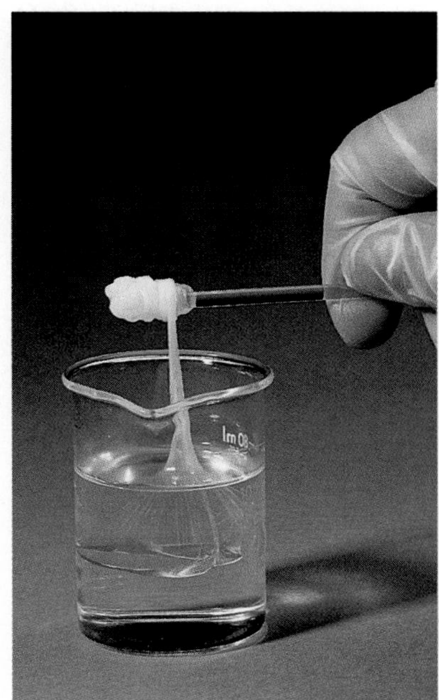

Figure 17.6
A white strand of a nylon polymer forms between two layers of a solution containing a diacid (bottom layer) and a diamine (top layer).

• The name *nylon 66* comes from the fact that each of the monomers has six carbon atoms.

There are actually many different types of nylon, all of which are based on diamine and diacid monomers (Figure 17.6). The most important nylon is nylon 66, which is made by using 1,6-hexanediamine and hexanedioic acid as monomers.

$$H{-}N(H){-}(CH_2)_6{-}N(H){-}H$$

1,6-Hexanediamine

$$HO{-}C(=O){-}(CH_2)_4{-}C(=O){-}OH$$

Hexanedioic acid

The reaction of one acid group of the diacid with one amine group of the diamine initially produces an amide molecule; an acid group is left over on one end, and an amine group is left over on the other end.

$$HO{-}C(=O){-}(CH_2)_4{-}C(=O){-}OH + H{-}N(H){-}(CH_2)_6{-}N(H){-}H \longrightarrow$$

$$HO{-}C(=O){-}(CH_2)_4{-}C(=O){-}N(H){-}(CH_2)_6{-}N(H){-}H + H_2O$$

Leftover acid group that can react further; Amide linkage; Leftover amine group that can react further

This species then reacts further, and the process continues until a long polymeric molecule, nylon, has been produced.

$$\cdots C(=O){-}(CH_2)_4{-}C(=O){-}N(H){-}(CH_2)_6{-}N(H){-}C(=O){-}(CH_2)_4{-}C(=O){-}N(H){-}(CH_2)_6{-}N(H){-}C(=O){-}(CH_2)_4{-}C(=O)\cdots$$

Amide linkages; Amide linkages

A portion of the polyamide nylon 66

Additional stiffness and toughness are imparted to polyamides if aromatic rings are present in the polymer "backbone." The polyamide Kevlar is now used in place of steel in bullet-resistant vests. The polymeric repeating unit in Kevlar is

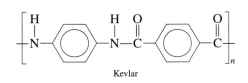

$$\left[N(H){-}C_6H_4{-}N(H){-}C(=O){-}C_6H_4{-}C(=O) \right]_n$$

Kevlar

Fire fighters with flame-resistant clothing containing Nomex.

Nomex is a polymer whose structure is a variation of that of Kevlar. With Nomex, the monomers are *meta* isomers rather than *para* isomers. Nomex is used in flame-resistant clothing for fire fighters and race car drivers.

Silk and wool are examples of *naturally occurring* polyamide polymers. Silk and wool are proteins, and proteins are polyamide polymers. Because much of the human body is protein material, much of the human body is polyamide polymer. Proteins are the subject of Chapter 20 of this text.

Polyurethanes are polymers related to polyesters and polyamides. The backbone of a polyurethane polymer contains aspects of both ester and amide functional groups. The following is a portion of the structure of a typical polyurethane polymer.

$$---C(=O){-}N(H){-}(CH_2)_6{-}N(H){-}C(=O){-}O{-}(CH_2)_4{-}O{-}C(=O){-}N(H){-}(CH_2)_6{-}N(H){-}C(=O){-}O{-}(CH_2)_4{-}O---$$

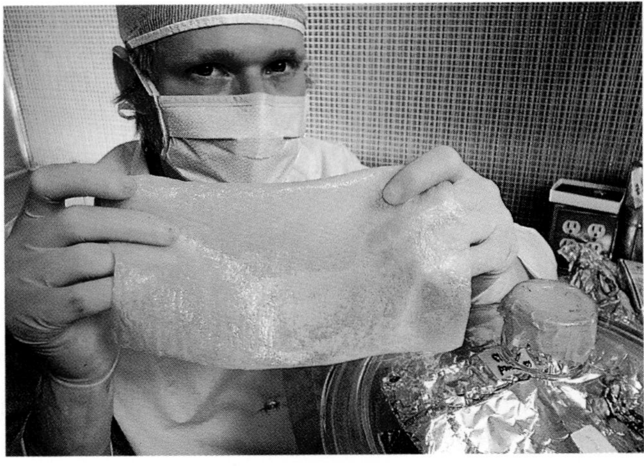

Figure 17.7
Polyurethanes have medical applications. For example, polyurethane membranes are used as skin substitutes for severe burn victims. Because they pass only oxygen and water, these membranes help patients recover more rapidly.

Foam rubber in furniture upholstery, packaging materials, life preservers, elastic fibers, and many other products contain polyurethane polymers (Figure 17.7).

Concepts to Remember

Structural characteristics of amines. Amines are derivatives of ammonia (NH_3) in which one or more hydrogen atoms have been replaced by an alkyl, a cycloalkyl, or an aryl group.

Classification of amines. Amines are classified as primary, secondary or tertiary, depending on the number of carbon atoms (one, two, or three) directly attached to the nitrogen atom. The functional group present in a primary amine, the $—NH_2$ group, is called an *amino* group. Secondary and tertiary amines contain substituted amino groups.

Nomenclature for amines. Common names for amines are formed by listing the hydrocarbon groups attached to the nitrogen atom in alphabetical order, followed by the suffix *-amine.* In the IUPAC system, the *-e* ending of the name of the longest carbon chain present is changed to *-amine,* and a number is used to locate the position of the amino group. Carbon-chain substituents are given numbers to designate their locations.

Properties of amines. The methylamines and ethylamine are gases at room temperature; amines of higher molecular mass are usually liquids and smell like raw fish. Primary and secondary, but not tertiary, amines can participate in hydrogen bonding to other amine molecules.

Basicity of amines. Amines are weak bases because of the ability of the unshared electron pair on the amine nitrogen atom to accept a proton in acidic solution.

Amine salts. The reaction of a strong acid with an amine forms an amine salt. Such salts are more soluble in water than is the parent amine.

Alkylation of ammonia and amines. Alkylation of ammonia, primary amines, secondary amines, and tertiary amines produces primary amines, secondary amines, tertiary amines, and quaternary ammonium salts, respectively.

Heterocyclic amines. In a heterocyclic amine, the nitrogen atoms of amino groups present are part of either an aromatic or a nonaromatic ring system. Numerous heterocyclic amines are important biological compounds.

Classification of amides. An amide is derived from a carboxylic acid by replacing the hydroxyl group with an amino or a substituted amino group. Amides, like amines, can be classified as primary, secondary, or tertiary, depending on how many nonhydrogen atoms are attached to the nitrogen atom.

Nomenclature for amides. The nomenclature for amides is derived from that for carboxylic acids by changing the *-oic acid* ending to *-amide.* Groups attached to the nitrogen atom of the amide are included as prefixes, using a capital *N-* with each group to indicate location.

Properties of amides. Amides do not exhibit basic properties in solution. The electronegative oxygen atom in the carbonyl group draws electron density away from the nitrogen, leaving very little electron density on the nitrogen to bond to an incoming proton. Most unbranched amides are solids at room temperature and have correspondingly high boiling points because of strong hydrogen bonds between molecules.

Preparation of amides. Reaction of carboxylic acids with ammonia, primary amines, and secondary amines produces unsubstituted, monosubstituted, and disubstituted amides, respectively.

Hydrolysis of amides. In amide hydrolysis, the bond between the carbonyl carbon atom and the nitrogen is broken, and free acid and free amine are produced. Acidic hydrolysis conditions convert the product amine to an amine salt. Basic hydrolysis conditions convert the product acid to an acid salt.

Polyamides. Polyamides are condensation polymers with monomers joined together by amide linkages. The monomers for polyamides are diacids and diamines.

Key Terms

Alkaloid (17.10)
Amide (17.11)
Amine salt (17.6)
Amines (17.2)
Heterocyclic amine (17.8)
Neurotransmitter (17.9)
Polyamide (17.16)
Quaternary ammonium salt (17.7)
Substituted ammonium ion (17.5)

Key Reactions and Equations

1. Reaction of amines with water to give a basic solution (Section 17.5)

$$R—NH_2 + H_2O \rightleftharpoons R—\overset{+}{N}H_3 + OH^-$$

2. Reaction of amines with acids to produce amine salts (Section 17.6)

$$R—NH_2 + HCl \longrightarrow R—\overset{+}{N}H_3\ Cl^-$$

3. Conversion of an amine salt to an amine (Section 17.6)

$$R—\overset{+}{N}H_3Cl^- + NaOH \longrightarrow R—NH_2 + NaCl + H_2O$$

4. Alkylation of ammonia to produce a primary amine (Section 17.7)

$$NH_3 + R—X + NaOH \longrightarrow R—NH_2 + NaX + H_2O$$

5. Alkylation of primary and secondary amines to produce, respectively, secondary and tertiary amines (Section 17.7)

$$R—NH_2 + R—X + NaOH \longrightarrow R_2NH + NaX + H_2O$$

$$R_2NH + R—X + NaOH \longrightarrow R_3N + NaX + H_2O$$

6. Alkylation of a tertiary amine to produce a quaternary ammonium salt (Section 17.7)

$$R_3N + R—X \longrightarrow R_4\overset{+}{N}\ X^-$$

7. Reaction of amines with carboxylic acids to form amides (Section 17.14)

$$R—\overset{\displaystyle O}{\overset{\|}{C}}—OH + R'—NH_2 \xrightarrow{\Delta} R—\overset{\displaystyle O}{\overset{\|}{C}}—NH—R' + H_2O$$

8. Acid hydrolysis of amides to produce a carboxylic acid and an amine salt (Section 17.15)

$$R—\overset{\displaystyle O}{\overset{\|}{C}}—NH—R' + H_2O + HCl \longrightarrow R—\overset{\displaystyle O}{\overset{\|}{C}}—OH + R'—\overset{+}{N}H_3\ Cl^-$$

9. Basic hydrolysis of amides to produce a carboxylic acid salt and an amine (Section 17.15)

$$R—\overset{\displaystyle O}{\overset{\|}{C}}—NH—R' + NaOH \longrightarrow R—\overset{\displaystyle O}{\overset{\|}{C}}—O^-\ Na^+ + R'—NH_2$$

Exercises and Problems

The members of each pair of problems in this section test similar material.

The Amine Functional Group (Section 17.1)

17.1 In which of the following compounds is an amine functional group present?

a. $CH_3—\underset{\displaystyle NH_2}{\underset{|}{CH}}—CH_3$

b. $CH_3—NH—CH_3$

c. $CH_3—CH_2—\overset{\displaystyle O}{\overset{\|}{C}}—NH_2$

d. $CH_3—CH_2—\underset{\displaystyle CH_3}{\underset{|}{N}}—CH_2—CH_3$

e. $CH_3—CH_2—\overset{\displaystyle O}{\overset{\|}{C}}—\underset{\displaystyle CH_3}{\underset{|}{N}}—CH_3$

f. (benzene ring)—NH—(benzene ring)

17.2 In which of the following compounds is an amine functional group present?

a. $CH_3—CH_2—CH_2—NH_2$

b. $CH_3—CH_2—\underset{\displaystyle CH_3}{\underset{|}{N}}—CH_3$

c. $CH_3—NH—$(benzene ring)

d. $CH_3—\overset{\displaystyle O}{\overset{\|}{C}}—NH—CH_3$

e. 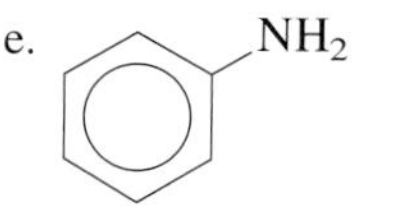

f. $CH_3—CH_2—CH_2—\overset{\displaystyle O}{\overset{\|}{C}}—NH_2$

Classification of Amines (Section 17.2)

17.3 Classify each of the following amines as a primary, secondary, or tertiary amine.

a. $CH_3—NH_2$

b. $CH_3—\underset{\displaystyle NH_2}{\underset{|}{CH}}—CH_3$

c. $CH_3—NH—CH_3$

d. $CH_3—CH_2—CH_2—\underset{\displaystyle H}{\underset{|}{N}}—CH_3$

e. $CH_3—CH_2—\underset{\displaystyle CH_3}{\underset{|}{CH}}—NH_2$

f. $CH_3—CH_2—\underset{\displaystyle CH_3}{\underset{|}{N}}—CH_2—\underset{\displaystyle CH_3}{\underset{|}{CH}}—CH_3$

17.4 Classify each of the following amines as a primary, secondary, or tertiary amine.

a. $CH_3—CH_2—\underset{\displaystyle NH_2}{\underset{|}{CH}}—CH_2—CH_3$

b. $CH_3—\overset{\displaystyle CH_3}{\overset{|}{\underset{\displaystyle NH_2}{\underset{|}{C}}}}—CH_3$

c. $CH_3—\underset{\displaystyle CH_3}{\underset{|}{N}}—CH_3$

d. $CH_3—\underset{\displaystyle CH_3}{\underset{|}{CH}}—NH—\underset{\displaystyle CH_3}{\underset{|}{CH}}—CH_2$

e. $CH_3-CH_2-CH_2-CH_2-NH_2$

f. 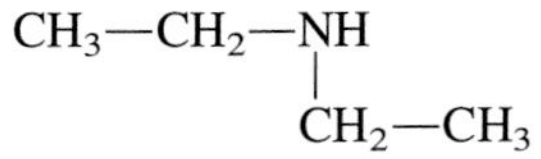 CH_3-CH_2-NH
$\qquad\qquad\quad |$
$\qquad\quad CH_2-CH_3$

17.5 Classify each of the following amines as a primary, secondary, or tertiary amine.

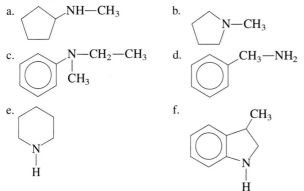

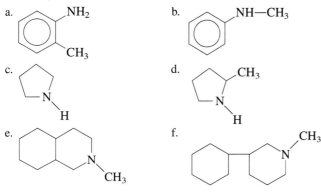

a. 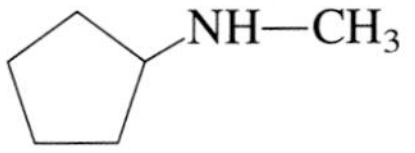 (cyclopentane ring)$-NH-CH_3$ b. 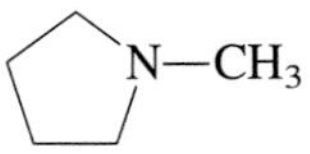 (pyrrolidine ring)$N-CH_3$

c. 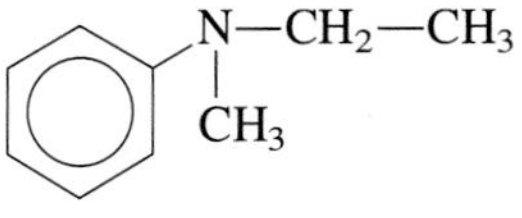 (benzene ring)$-N-CH_2-CH_3$, with CH_3 on N d. 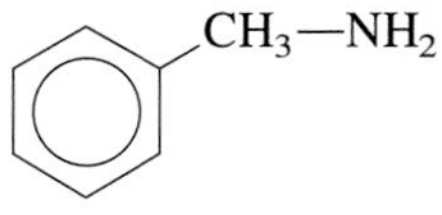 (benzene ring)$-CH_3-NH_2$

e.

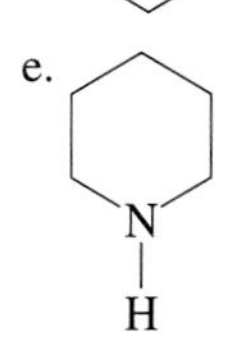

f. 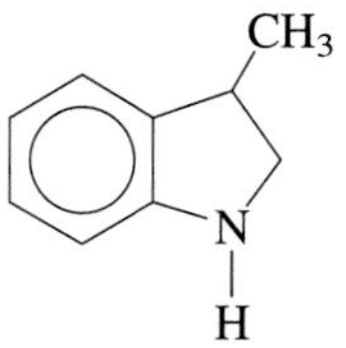 (ring with CH_3, N, H)

17.6 Classify each of the following amines as a primary, secondary, or tertiary amine.

a. 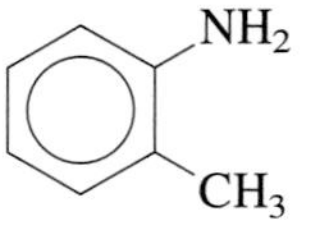 (benzene ring with NH_2 and CH_3) b. 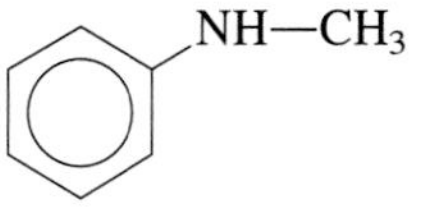 (benzene ring)$-NH-CH_3$

c.

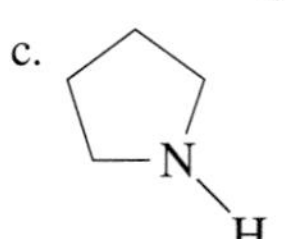

d. 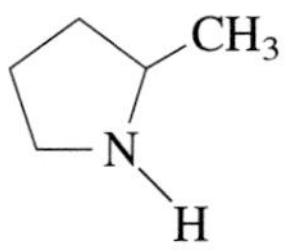 (ring with CH_3, N, H)

e.

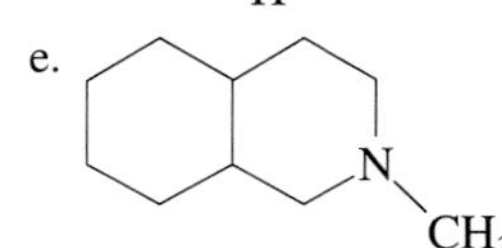

f. 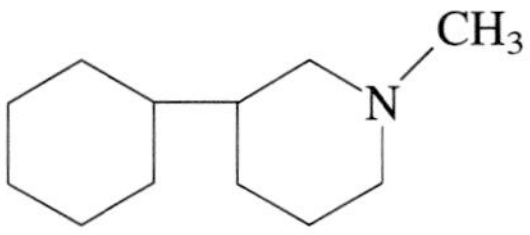 (cyclohexyl ring attached to ring with N$-CH_3$)

Nomenclature for Amines (Section 17.3)

17.7 Assign a common name to each of the following amines.

a. $CH_3-NH-CH_2-CH_3$

b. $CH_3-CH_2-CH_2-NH_2$

c. 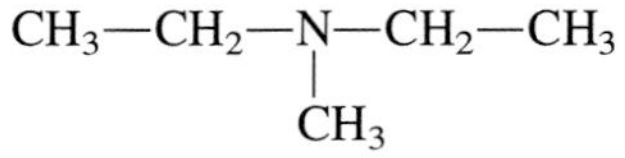 $CH_3-CH_2-N-CH_2-CH_3$, with CH_3 on N

d.

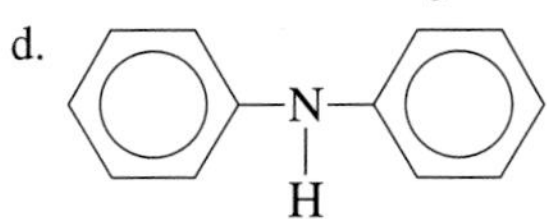

e. 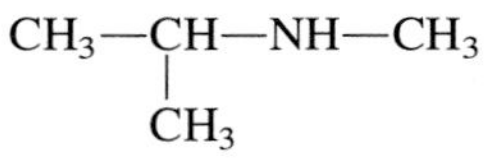 $CH_3-CH-NH-CH_3$, with CH_3 on CH

f. 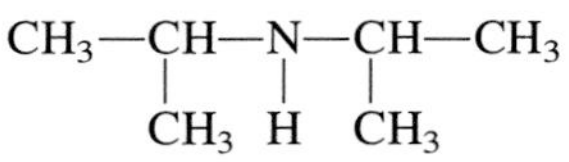 $CH_3-CH-N-CH-CH_3$, with CH_3, H, CH_3 below

17.8 Assign a common name to each of the following amines.

a. 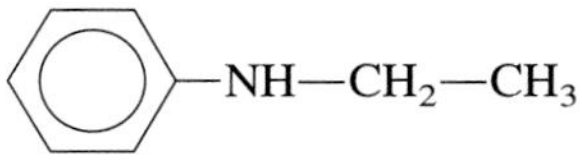 (benzene ring)$-NH-CH_2-CH_3$

b. 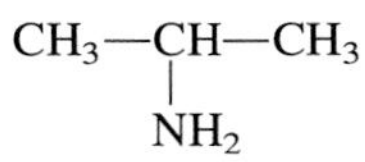 $CH_3-CH-CH_3$, with NH_2 on CH

c. $H_2N-CH_2-CH_2-CH_2-CH_3$

d. 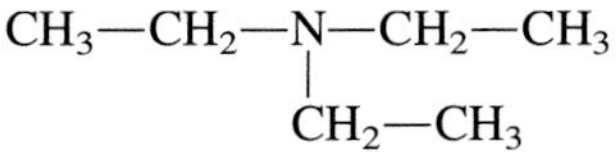 $CH_3-CH_2-N-CH_2-CH_3$, with CH_2-CH_3 on N

e. 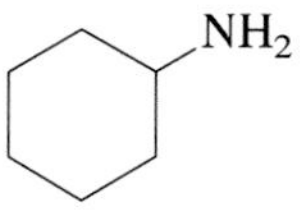 (cyclohexane ring)$-NH_2$

f. 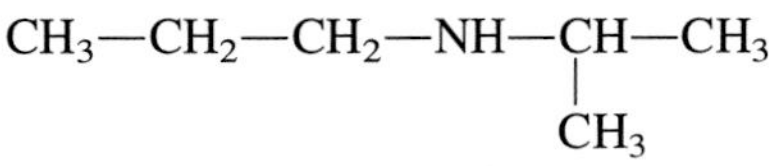 $CH_3-CH_2-CH_2-NH-CH-CH_3$, with CH_3 on CH

17.9 Assign an IUPAC name to each of the following amines.

a. 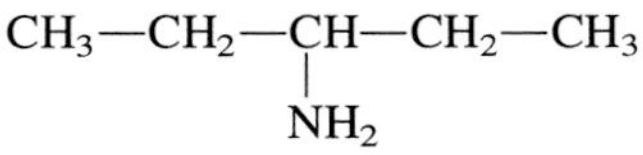 $CH_3-CH_2-CH-CH_2-CH_3$, with NH_2 on CH

b. 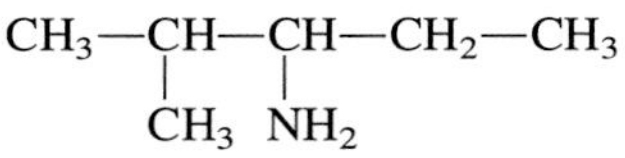 $CH_3-CH-CH-CH_2-CH_3$, with CH_3 and NH_2 below

c. 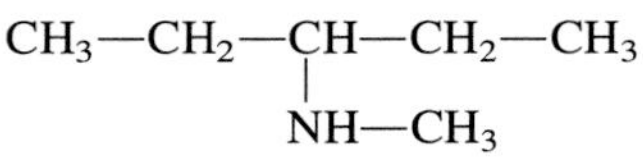 $CH_3-CH_2-CH-CH_2-CH_3$, with $NH-CH_3$ on CH

d. $H_2N-CH_2-CH_2-CH_2-CH_2-CH_2-NH_2$

e. 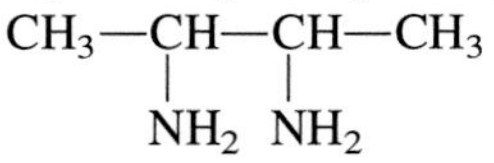 $CH_3-CH-CH-CH_3$, with NH_2 and NH_2 below

f. 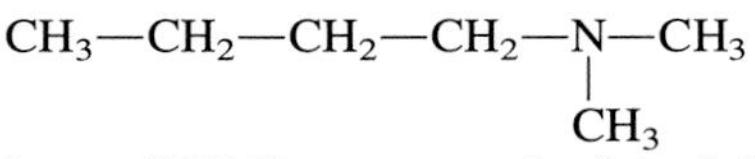 $CH_3-CH_2-CH_2-CH_2-N-CH_3$, with CH_3 on N

17.10 Assign an IUPAC name to each of the following amines.

a. $CH_3-CH_2-CH_2-NH_2$

b. 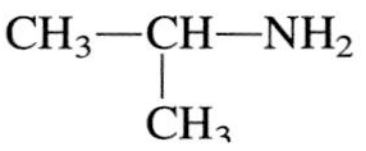 $CH_3-CH-NH_2$, with CH_3 on CH

c. 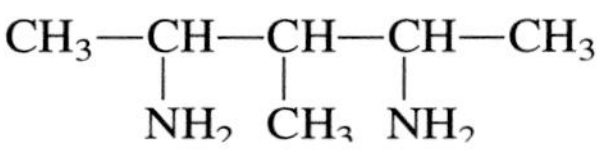 $CH_3-CH-CH-CH-CH_3$, with NH_2, CH_3, NH_2 below

d. $CH_3-CH_2-CH_3-NH-CH_3$

e. 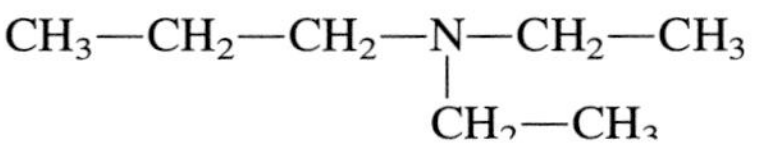 $CH_3-CH_2-CH_2-N-CH_2-CH_3$, with CH_2-CH_3 on N

f. $CH_3-CH_2-CH_2-NH-CH_2-CH_3$

17.11 Name each of the following aromatic amines as a derivative of aniline.

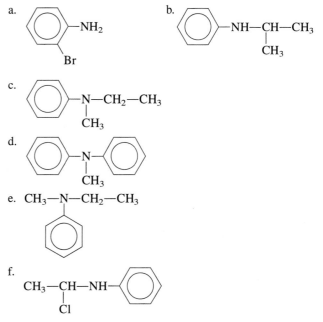

a. 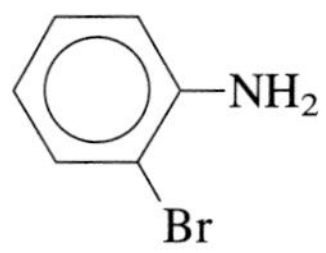 (benzene ring)$-NH_2$, with Br b. 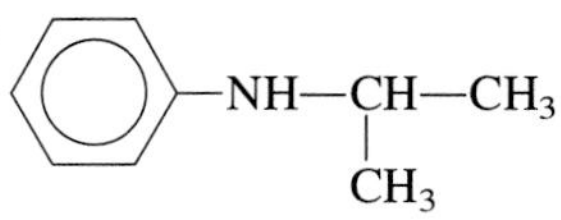 (benzene ring)$-NH-CH-CH_3$, with CH_3 on CH

c. 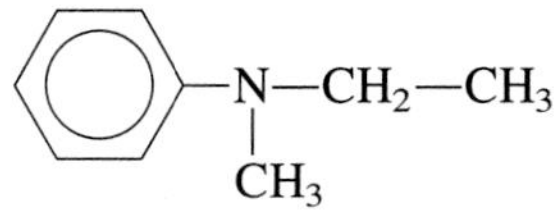 (benzene ring)$-N-CH_2-CH_3$, with CH_3 on N

d. 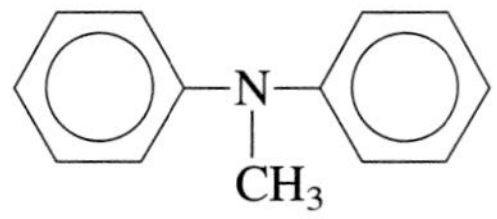 (benzene ring)$-N-$(benzene ring), with CH_3 on N

e. $CH_3-N-CH_2-CH_3$, with benzene ring on N

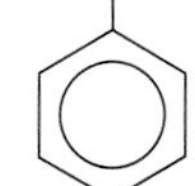

f. 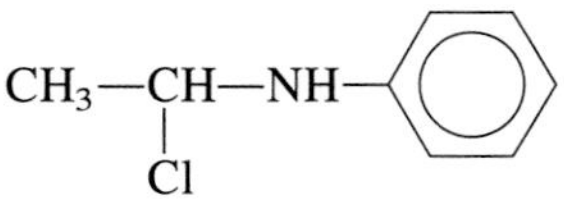 $CH_3-CH-NH-$(benzene ring), with Cl on CH

17.12 Name each of the following aromatic amines as a derivative of aniline.

a. 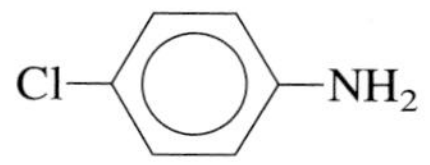

b. 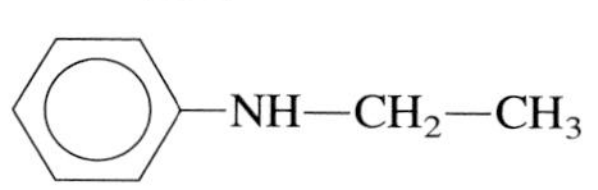

c. 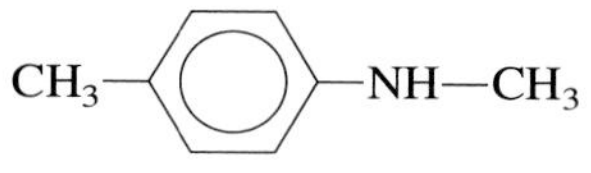

d. 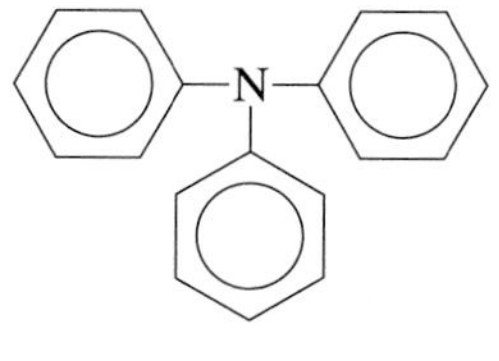

e. 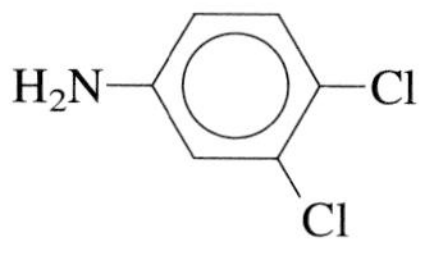

f. 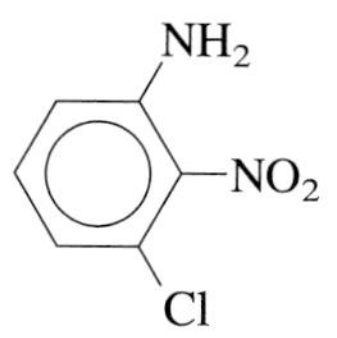

17.13 Draw a structural formula for each of the following amines.

a. Ethylamine
b. Triisopropylamine
c. *o*-Methylaniline
d. *N*-methylaniline
e. 2-Methyl-2-butanamine
f. 1,6-Hexanediamine
g. 2-Amino-3-pentanone
h. 2-Aminopropanoic acid

17.14 Draw a structural formula for each of the following amines.

a. Ethylmethylamine
b. Diethylpropylamine
c. *p*-Nitroaniline
d. *N,N*-dimethylaniline
e. 2-Methyl-3-ethyl-1-hexanamine
f. 1,3-Pentanediamine
g. 3-Amino-2-pentanol
h. *N,N*-dimethyl-1-butanamine

Physical Properties of Amines (Section 17.4)

17.15 Indicate whether each of the following amines is a liquid or a gas at room temperature.

a. Butylamine b. Dimethylamine
c. Ethylamine d. Dibutylamine

17.16 Indicate whether each of the following amines is a liquid or a gas at room temperature.

a. Methylamine b. Propylamine
c. Trimethylamine d. Pentylamine

17.17 Determine the maximum number of hydrogen bonds that can form between a methylamine molecule and

a. other methylamine molecules
b. water molecules

17.18 Determine the maximum number of hydrogen bonds that can form between a dimethylamine molecule and

a. other dimethylamine molecules b. water molecules

17.19 Although they have similar molecular masses (73 and 72 amu, respectively), the boiling point of butylamine is much higher (78°C) than that of pentane (36°C). Explain why.

17.20 Although they have similar molecular masses (73 and 74 amu, respectively), the boiling point of 1-butanamine is much lower (78°C) than that of 1-butanol (118°C). Explain why.

17.21 Which compound in each of the following pairs of amines would you expect to be more soluble in water. Justify each answer.

a. $CH_3—CH_2—NH_2$ and $CH_3—CH_2—CH_2—CH_2—CH_2—NH_2$
b. $CH_3—CH_2—CH_2—NH_2$ and $H_2N—CH_2—CH_2—CH_2—NH_2$

17.22 Which compound in each of the following pairs of amines would you expect to be more soluble in water. Justify each answer.

a. $CH_3—CH_2—CH_2—NH_2$ and $CH_3—CH_2—CH_2—CH_2—NH_2$
b. $CH_3—CH_2—NH—CH_3$ and $CH_3—\underset{\displaystyle CH_3}{\underset{|}{N}}—CH_3$

Basicity of Amines (Section 17.5)

17.23 Show the structures of the missing substances in the following acid–base equilibria.

a. $CH_3—CH_2—NH_2 + H_2O \rightleftharpoons ? + OH^-$

b. 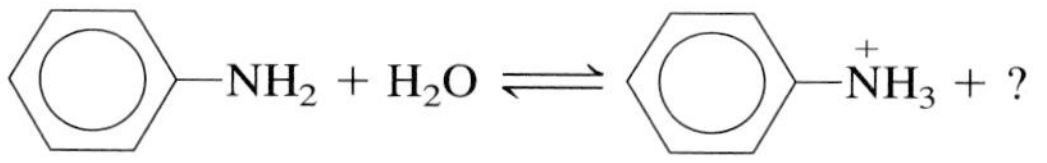

c. $? + H_2O \rightleftharpoons CH_3—\underset{\displaystyle CH_3}{\underset{|}{CH}}—\overset{+}{N}H_2—CH_3 + OH^-$

d. Diethylamine $+ H_2O \longrightarrow ? + ?$

17.24 Show the structures of the missing substances in the following acid–base equilibria.

a. $CH_3—CH_2—CH_2—NH_2 + H_2O \rightleftharpoons CH_3—CH_2—CH_2—\overset{+}{N}H_3 + ?$

b. 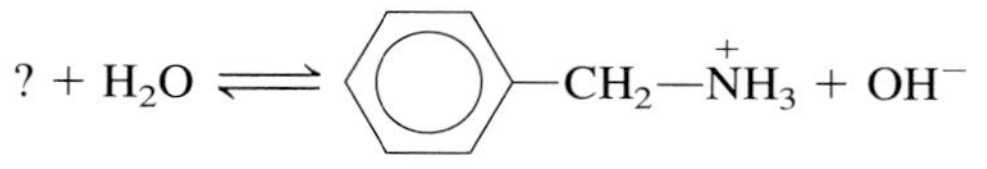

c. $CH_3—\underset{\displaystyle CH_3}{\underset{|}{CH}}—CH_2—NH—CH_3 + H_2O \rightleftharpoons ? + OH^-$

d. Trimethylamine $+ H_2O \rightleftharpoons ? + ?$

17.25 Name each of the following substituted ammonium and substituted anilinium ions.

a. $CH_3—\overset{+}{N}H_2—CH_3$

b. $CH_3—CH_2—\underset{\displaystyle CH_2—CH_3}{\underset{|}{\overset{+}{N}H}}—CH_2—CH_3$

c. 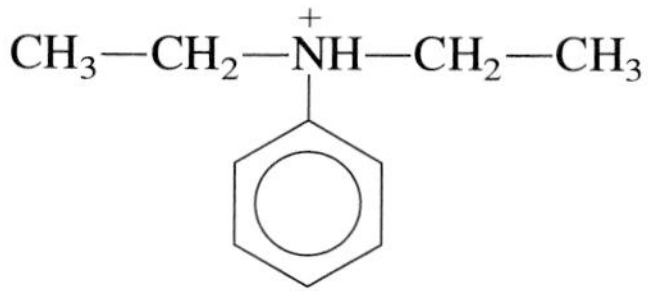

d. $CH_3-CH_2-CH_2-\overset{+}{N}H(CH_3)-CH_3$

e. $CH_3-CH_2-CH_2-\overset{+}{N}H_3$

f. $C_6H_5-\overset{+}{N}H_2-CH(CH_3)-CH_3$

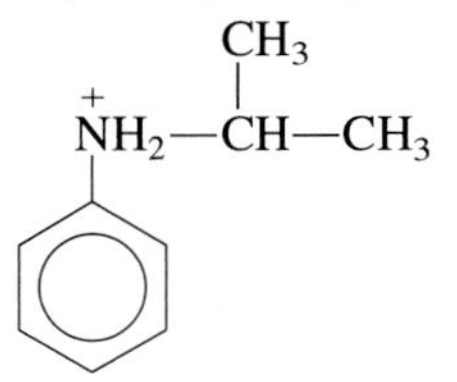

17.26 Name each of the following substituted ammonium and substituted anilinium ions.

a. $CH_3-\overset{+}{N}H_3$

b. $CH_3-CH_2-CH_2-\overset{+}{N}H_2-CH_3$

c. $C_6H_5-\overset{+}{N}H_2-CH_2-CH_3$

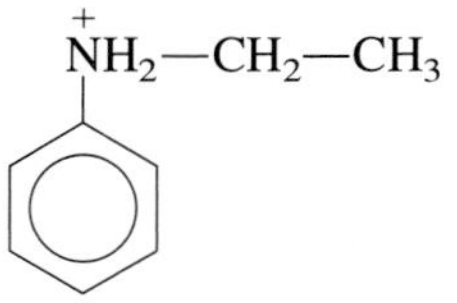

d. $CH_3-CH_2-CH_2-\overset{+}{N}H(CH_3)-CH_2-CH_3$

e. $CH_3-CH_2-CH_2-\overset{+}{N}H_2-CH_2-CH_2-CH_3$

f. $CH_3-CH_2-\overset{+}{N}H(C_6H_5)-CH_2-CH_3$

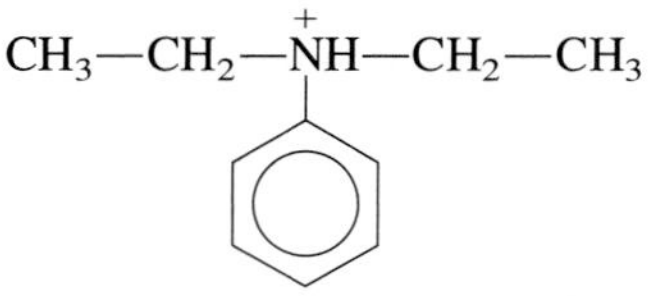

17.27 Draw a structural formula for the "parent" amine of each of the substituted ammonium and substituted anilinium ions in Problem 17.25.

17.28 Draw a structural formula for the "parent" amine of each of the substituted ammonium and substituted anilinium ions in Problem 17.26.

Amine Salts (Section 17.6)

17.29 Draw the structure of the missing substance in each of the following reactions involving amine salts.

a. $CH_3-CH_2-NH_2 + HCl \longrightarrow ?$

b. $C_6H_5-NH_2 + HBr \longrightarrow ?$

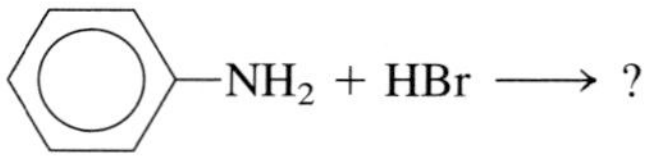

c. $? + HBr \longrightarrow CH_3-C(CH_3)_2-\overset{+}{N}H_3\ Br^-$

d. $CH_3-CH_2-NH-CH_3 + ? \longrightarrow CH_3-CH_2-\overset{+}{N}H_2-CH_3\ Cl^-$

17.30 Draw the structure of the missing substance in each of the following reactions involving amine salts.

a. $CH_3-CH_2-NH-CH_2-CH_3 + HBr \longrightarrow ?$

b. $CH_3-NH_2 + ? \longrightarrow CH_3-\overset{+}{N}H_3\ Cl^-$

c. $? + HBr \longrightarrow CH_3-CH(CH_3)-\overset{+}{N}H(CH_3)-CH_3\ Br^-$

d. $C_6H_5-NH-CH_3 + HCl \longrightarrow ?$

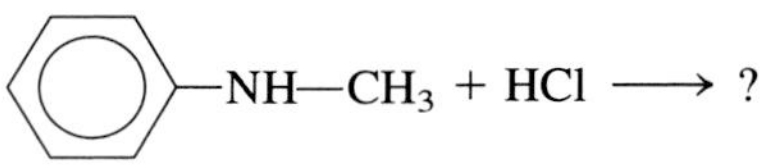

17.31 Draw the structures of the missing substances in each of the following reactions involving amine salts.

a. $CH_3-CH(CH_3)-\overset{+}{N}H_3\ Cl^- + NaOH \longrightarrow ? + NaCl + H_2O$

b. $? + NaOH \longrightarrow CH_3-NH-CH_3 + NaCl + H_2O$

c. $C_6H_5-\overset{+}{N}H(CH_3)-CH_3\ Br^- + NaOH \longrightarrow ? + ? + H_2O$

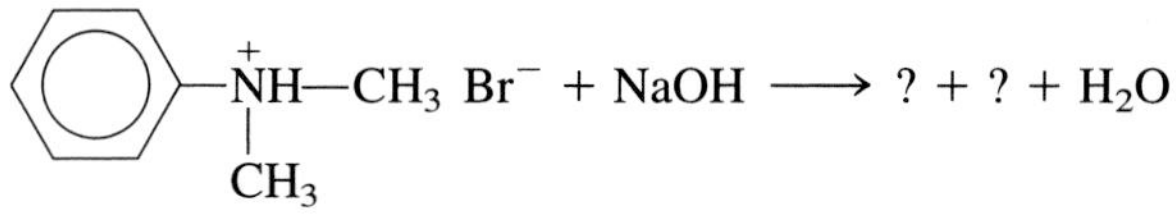

d. $CH_3-\overset{+}{N}H_2-CH_3\ Cl^- + NaOH \longrightarrow ? + NaCl + H_2O$

17.32 Draw the structures of the missing substances in each of the following reactions involving amine salts.

a. $CH_3-CH_2-CH_2-\overset{+}{N}H_3\ Br^- + NaOH \longrightarrow ? + NaBr + H_2O$

b. $? + NaOH \longrightarrow CH_3-N(CH_3)-CH_3 + NaBr + H_2O$

c. $CH_3-CH(CH_3)-\overset{+}{N}H_2-CH_3\ Cl^- + NaOH \longrightarrow ? + NaCl + H_2O$

d. $C_6H_5-\overset{+}{N}H_3\ Cl^- + NaOH \longrightarrow ? + NaCl + ?$

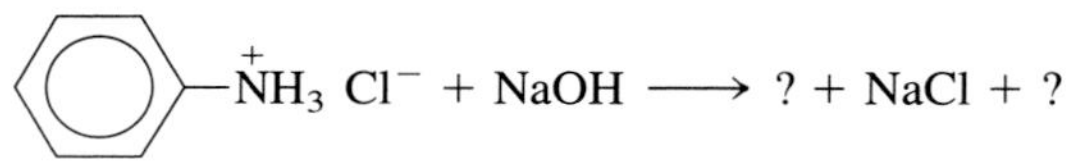

17.33 Name each of the following amine salts.

a. $CH_3-CH_2-CH_2-\overset{+}{N}H_3\ Cl^-$

b. $CH_3-CH_2-CH_2-\overset{+}{N}H_2(CH_3)\ Cl^-$

c. $CH_3-CH_2-\overset{+}{N}H(CH_3)-CH_3\ Br^-$

d. $C_6H_5-\overset{+}{N}H(CH_3)-CH_3\ Br^-$

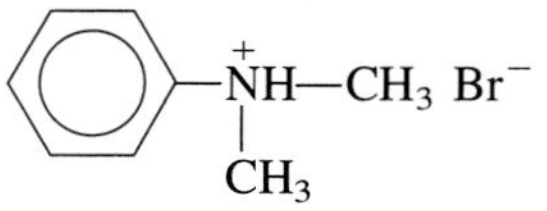

17.34 Name each of the following amine salts.

a. $CH_3-CH_2-\overset{+}{N}H_2(CH_3)\ Cl^-$

b. $CH_3-CH_2-CH_2-CH_2-\overset{+}{N}H_3\ Cl^-$

c. $CH_3-CH(CH_3)-\overset{+}{N}H(CH_3)-CH_3\ Br^-$

d. $C_6H_5-\overset{+}{N}H_2(CH_3)\ Cl^-$

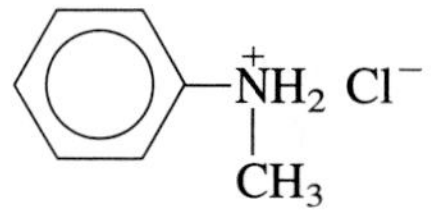

17.35 Why are drugs that contain the amine functional group most often administered to patients in the form of amine chloride or hydrogen sulfate salts?

17.36 Both heptylamine and heptyl alcohol are insoluble in water. If you were given a mixture of these two liquids, how could you separate them without heating (distilling) them?

17.37 How would the structure and name of the amine salt ethylmethylammonium chloride probably be written by someone in the pharmaceutical industry?

17.38 A student looking in an old chemistry book found the following name and structure for a compound.

$$CH_3-CH_2-NH_2 \cdot HBr$$

Ethylamine hydrobromide

What are the modern name and structure for this compound?

Alkylation of Ammonia and Amines (Section 17.7)

17.39 Identify the three products in each of the following reactions.

a. $NH_3 + CH_3-CH_2-CH_2-Cl + NaOH \longrightarrow$

b. 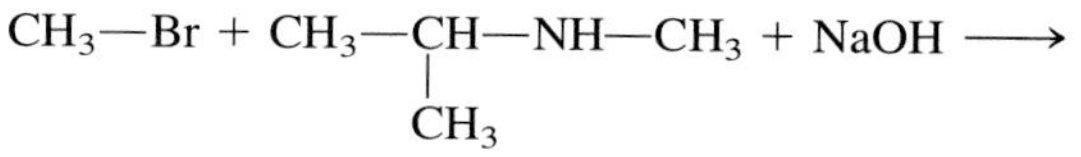 $CH_3-Br + CH_3-\underset{CH_3}{CH}-NH-CH_3 + NaOH \longrightarrow$

c. $CH_3-CH_2-NH_2 + CH_3-CH_2-Cl + NaOH \longrightarrow$

d. 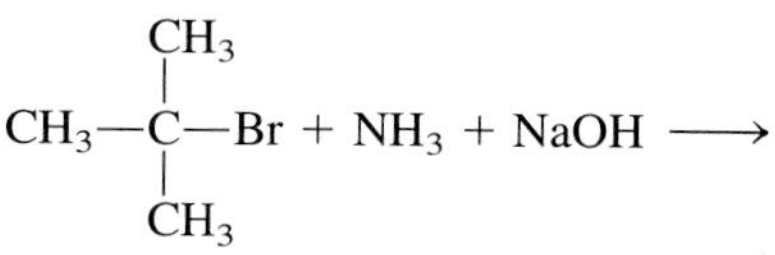 $CH_3-\overset{CH_3}{\underset{CH_3}{C}}-Br + NH_3 + NaOH \longrightarrow$

17.40 Identify the three products in each of the following reactions.

a. 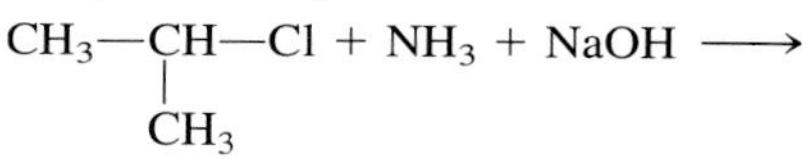 $CH_3-\underset{CH_3}{CH}-Cl + NH_3 + NaOH \longrightarrow$

b. $CH_3-NH-CH_3 + CH_3-Br + NaOH \longrightarrow$

c. $CH_3-CH_2-CH_2-NH_2 + CH_3-CH_2-Br + NaOH \longrightarrow$

d. 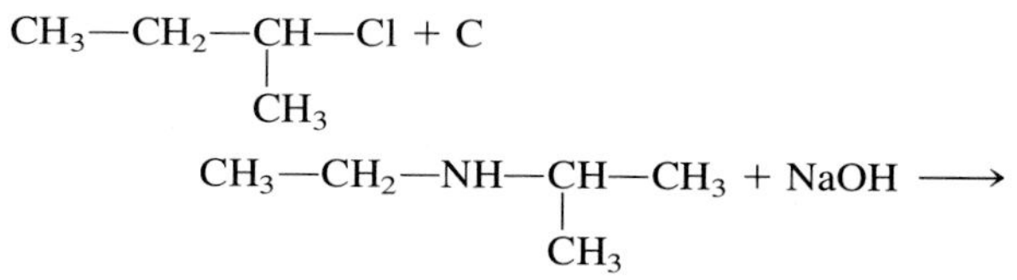 $CH_3-CH_2-\underset{CH_3}{CH}-Cl + C$
$CH_3-CH_2-NH-\underset{CH_3}{CH}-CH_3 + NaOH \longrightarrow$

17.41 List three different sets of alkyl chloride–secondary amine reactants that could be used to prepare the tertiary amine ethylmethylpropylamine.

17.42 List three different sets of alkyl chloride–secondary amine reactants that could be used to prepare the tertiary amine butylethylpropylamine.

17.43 Draw the structure of the amine or quaternary ammonium salt produced when each of the following pairs of compounds react in the presence of strong base.

a. Trimethylamine and ethyl bromide

b. Diisopropylamine and methyl bromide

c. Ethylmethylpropylamine and methyl chloride

d. Ethylamine and ethyl chloride

17.44 Draw the structure of the amine or quaternary ammonium salt produced when each of the following pairs of compounds reacts in the presence of strong base.

a. Dimethylamine and propyl bromide

b. Diethylmethylamine and isopropyl chloride

c. Methylpropylamine and ethyl chloride

d. Tripropylamine and propyl chloride

17.45 Classify each of the following salts as an amine salt or a quaternary ammonium salt.

a. 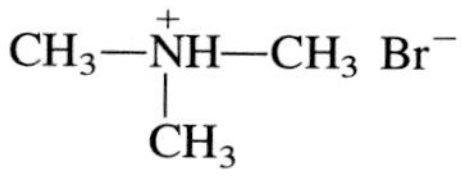 $CH_3-\underset{CH_3}{\overset{+}{N}H}-CH_3\ Br^-$

b. 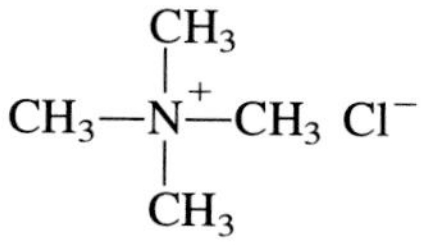 $CH_3-\overset{CH_3}{\underset{CH_3}{\overset{+}{N}}}-CH_3\ Cl^-$

c. $CH_3-CH_2-\overset{+}{N}H_2-CH_3\ Br^-$

d. 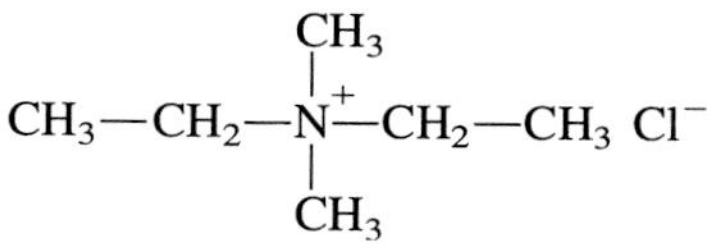 $CH_3-CH_2-\overset{CH_3}{\underset{CH_3}{\overset{+}{N}}}-CH_2-CH_3\ Cl^-$

17.46 Classify each of the following salts as an amine salt or a quaternary ammonium salt.

a. 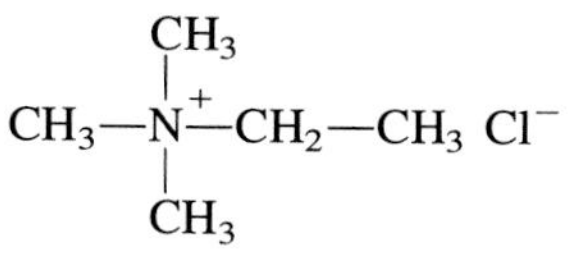 $CH_3-\overset{CH_3}{\underset{CH_3}{\overset{+}{N}}}-CH_2-CH_3\ Cl^-$

b. 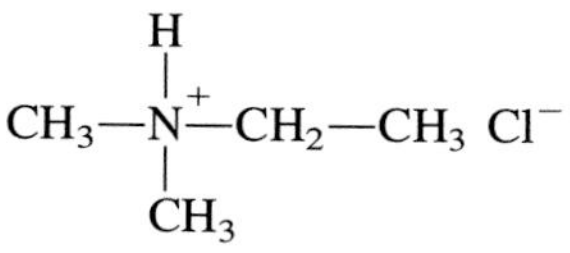 $CH_3-\overset{H}{\underset{CH_3}{\overset{+}{N}}}-CH_2-CH_3\ Cl^-$

c. 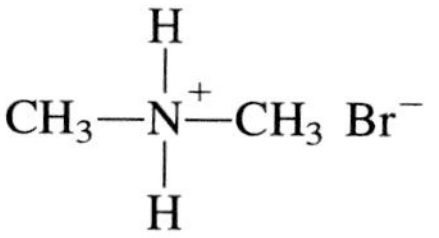 $CH_3-\overset{H}{\underset{H}{\overset{+}{N}}}-CH_3\ Br^-$

d. 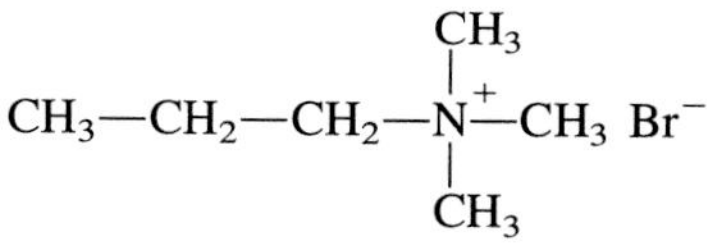 $CH_3-CH_2-CH_2-\overset{CH_3}{\underset{CH_3}{\overset{+}{N}}}-CH_3\ Br^-$

17.47 Name each of the salts in Problem 17.45.

17.48 Name each of the salts in Problem 17.46.

Selected Important Amines (Sections 17.8 through 17.10)

17.49 With the help of Figure 17.4, identify the heterocyclic amine ring system or systems present in each of the following substances.

a. Caffeine
b. Heme
c. Histamine
d. Serotonin

17.50 With the help of Figure 17.4, identify the heterocyclic amine ring system or systems present in each of the following.

a. Nicotine
b. Quinine
c. Odor of popcorn
d. Porphyrin ring

17.51 Indicate whether each of the following statements about biologically important amines is true or false.

a. Both caffeine and nicotine are alkaloids.

b. The alkaloid quinine is used medically to dilate the pupil of the eye.

c. Structurally, morphine and codeine differ by a methyl group

d. Heroin is a naturally occurring substance.

e. Serotonin deficiency is associated with Parkinson's disease.

f. Adrenaline is another name for norepinephrine.

17.52 Indicate whether each of the following statements about biologically important amines is true or false.

a. Epinephrine and norepinephrine are hormones produced by the adrenal glands.
b. Both serotonin and dopamine are found in the brain and are heterocyclic amines.
c. The alkaloid atropine is used in the treatment of malaria.
d. "Free" histamine in tissues and the blood causes the symptoms associated with hay fever.
e. Nicotine is the chemical directly responsible for the high incidence of lung cancer in heavy smokers.
f. Structurally, epinephrine and norepinephrine differ by a methyl group.

Structure of and Nomenclature for Amides (Section 17.11)

17.53 Which of the following compounds contain an amide functional group?

a. $CH_3-CH_2-\overset{O}{\overset{\|}{C}}-NH_2$

b. $C_6H_5-\overset{O}{\overset{\|}{C}}-\underset{CH_3}{\underset{|}{N}}-CH_2-CH_3$ (benzene ring)

c. $CH_3-\overset{O}{\overset{\|}{C}}-CH_2-CH_2-NH_2$

d. $CH_3-\overset{O}{\overset{\|}{C}}-NH-C_6H_5$ (benzene ring)

e. $CH_3-\overset{NH_2}{\overset{|}{CH}}-\overset{O}{\overset{\|}{C}}-OH$

f. six-membered ring containing N—H, with C=O adjacent to N

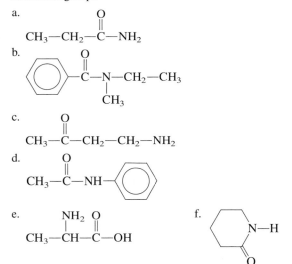

17.54 Which of the following compounds contain an amide functional group?

a. $CH_3-\overset{O}{\overset{\|}{C}}-NH-CH_3$

b. $CH_3-\overset{O}{\overset{\|}{C}}-\overset{CH_2-CH_3}{\overset{|}{N}}-CH_2-CH_2-CH_3$

c. $\overset{O}{\overset{\|}{C}}(-NH_2)-CH_2-CH_3$

d. $H_2N-\overset{O}{\overset{\|}{C}}-CH_2-CH_3$

e. $CH_3-CH_2-\overset{CH_3}{\overset{|}{CH}}-\overset{O}{\overset{\|}{C}}-NH-CH_3$

f. six-membered ring containing N—H, with C=O adjacent to N and CH_3 on the other carbon adjacent to N

17.55 Classify each of the following amides as unsubstituted, monosubstituted, or disubstituted.

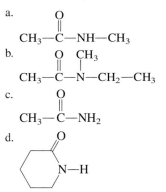

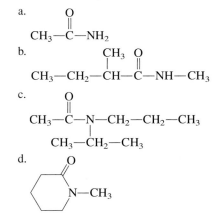

a. $CH_3-\overset{O}{\overset{\|}{C}}-NH-CH_3$

b. $CH_3-\overset{O}{\overset{\|}{C}}-\overset{CH_3}{\overset{|}{N}}-CH_2-CH_3$

c. $CH_3-\overset{O}{\overset{\|}{C}}-NH_2$

d. six-membered ring containing N—H, with C=O adjacent to N

17.56 Classify each of the following amides as unsubstituted, monosubstituted, or disubstituted.

a. $CH_3-\overset{O}{\overset{\|}{C}}-NH_2$

b. $CH_3-CH_2-\overset{CH_3}{\overset{|}{CH}}-\overset{O}{\overset{\|}{C}}-NH-CH_3$

c. $CH_3-\overset{O}{\overset{\|}{C}}-\underset{CH_3-CH_2-CH_3}{\underset{|}{N}}-CH_2-CH_2-CH_3$

d. six-membered ring containing N—CH_3, with C=O adjacent to N

17.57 Classify each of the amides in Problem 17.55 as a primary, secondary, or tertiary amide.

17.58 Classify each of the amides in Problem 17.56 as a primary, secondary, or tertiary amide.

17.59 Assign an IUPAC name to each of the following amides.

a. $CH_3-\overset{O}{\overset{\|}{C}}-NH-CH_2-CH_3$

b. $CH_3-CH_2-\overset{O}{\overset{\|}{C}}-\overset{CH_3}{\overset{|}{N}}-CH_3$

c. $H_2N-\overset{O}{\overset{\|}{C}}-CH_2-CH_2-CH_3$

d. $H-\overset{O}{\overset{\|}{C}}-\overset{CH_3}{\overset{|}{N}}-H$

e. $Cl-\overset{CH_3}{\overset{|}{CH}}-\overset{O}{\overset{\|}{C}}-NH_2$

f. $CH_3-\overset{CH_3}{\overset{|}{CH}}-\overset{O}{\overset{\|}{C}}-NH-CH_3$

17.60 Assign an IUPAC name to each of the following amides.

a. $CH_3-CH_2-\overset{O}{\overset{\|}{C}}-NH-CH_2-CH_3$

b. $CH_3-CH_2-CH_2-CH_2-\overset{O}{\overset{\|}{C}}-NH_2$

c. $\mathrm{CH_3{-}CH_2{-}CH_2{-}\overset{\overset{O}{\|}}{C}{-}\overset{\overset{CH_3}{|}}{N}{-}CH_3}$

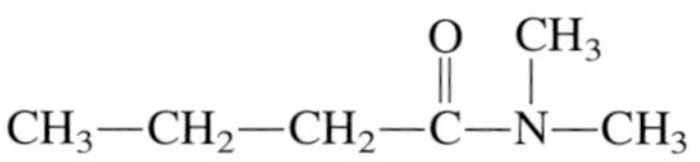

d. $\mathrm{H_2N{-}\overset{\overset{CH_3}{|}}{C}{=}O}$

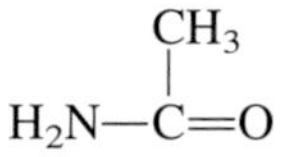

e. $\mathrm{CH_3{-}\overset{\overset{CH_3}{|}}{CH}{-}\overset{\overset{CH_3}{|}}{CH}{-}\overset{\overset{O}{\|}}{C}{-}NH_2}$

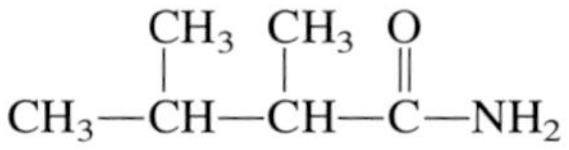

f. $\mathrm{CH_3{-}\overset{\overset{CH_3}{|}}{CH}{-}\overset{\overset{CH_3}{|}}{CH}{-}\overset{\overset{O}{\|}}{C}{-}NH{-}CH_3}$

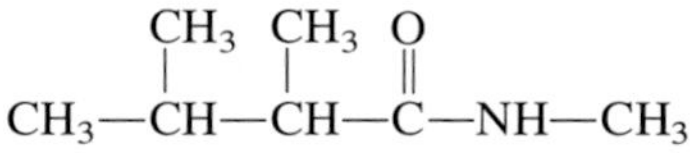

17.61 Assign a common name to each of the amides in Problem 17.59.

17.62 Assign a common name to each of the amides in Problem 17.60.

17.63 Write a structural formula for each of the following amides.

a. *N,N*-dimethylacetamide
b. 2-Methylbutyramide
c. 3,*N*-dimethylbutanamide
d. Methanamide
e. *N*-phenylbenzamide
f. Formamide

17.64 Write a structural formula for each of the following amides.

a. *N,N*-diethylpropanamide
b. Propionamide
c. 3-Methylbutyramide
d. *N*-methylbenzamide
e. 3,3,*N*-trimethylbutyramide
f. *N*-methyl-*N*-phenylpentanamide

Properties of Amides (Section 17.13)

17.65 Although amides contain a nitrogen atom, they are not bases like amines. Explain why.

17.66 Would you expect *N*-ethylacetamide or *N,N*-diethylacetamide to have the higher boiling point? Explain.

17.67 Determine the maximum number of hydrogen bonds that can form between an acetamide molecule and

a. other acetamide molecules
b. water molecules

17.68 Determine the maximum number of hydrogen bonds that can form between a propanamide molecule and

a. other propanamide molecules
b. water molecules

Preparation of Amides (Section 17.14)

17.69 Draw the structures of the missing substances in each of the following reactions involving amides.

a. $\mathrm{CH_3{-}CH_2{-}\overset{\overset{O}{\|}}{C}{-}OH} + ? \xrightarrow{\Delta} \mathrm{CH_3{-}CH_2{-}\overset{\overset{O}{\|}}{C}{-}NH{-}CH_3} + H_2O$

b. $\mathrm{CH_3{-}\underset{\underset{CH_3}{|}}{\overset{\overset{CH_3}{|}}{C}}{-}\overset{\overset{O}{\|}}{C}{-}OH} + \mathrm{CH_3{-}NH{-}CH_3} \xrightarrow{\Delta} ? + H_2O$

c. $\mathrm{CH_3{-}\overset{\overset{O}{\|}}{C}{-}OH} + ? \xrightarrow{\Delta} \mathrm{CH_3{-}\overset{\overset{O}{\|}}{C}{-}NH_2} + H_2O$

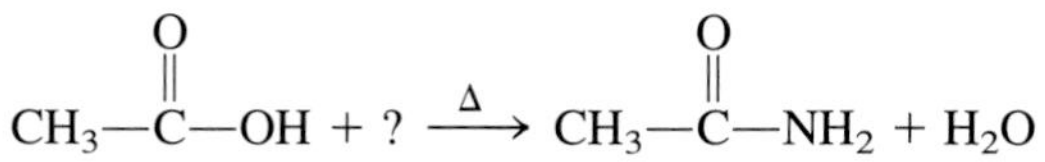

d. $? + \mathrm{C_6H_5{-}NH_2} \xrightarrow{\Delta} \mathrm{C_6H_5{-}\overset{\overset{O}{\|}}{C}{-}NH{-}C_6H_5} + H_2O$

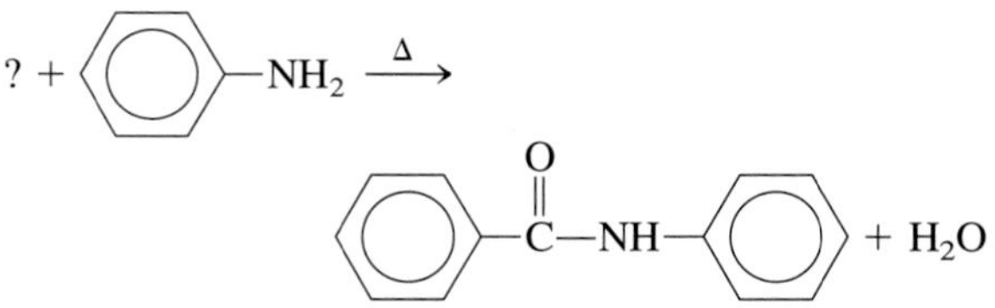

17.70 Draw the structures of the missing substances in each of the following reactions involving amides.

a. $\mathrm{CH_3{-}CH_2{-}CH_2{-}\overset{\overset{O}{\|}}{C}{-}OH} + \mathrm{CH_3{-}CH_2{-}NH_2} \xrightarrow{\Delta} ? + H_2O$

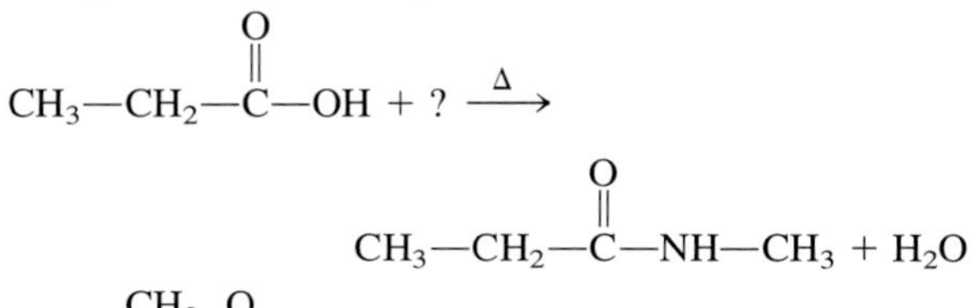

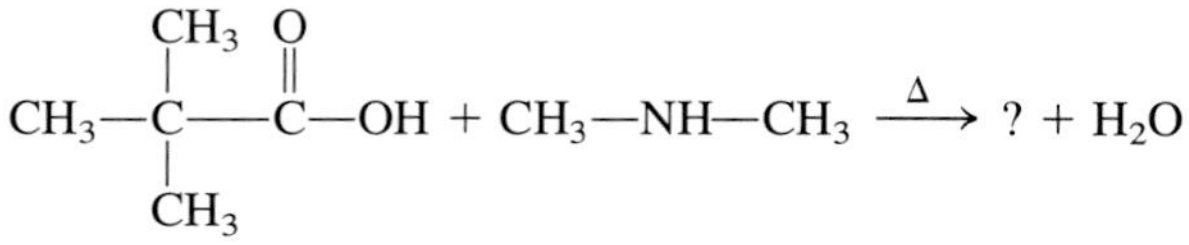

b. $? + NH_3 \xrightarrow{\Delta} \mathrm{H{-}\overset{\overset{O}{\|}}{C}{-}NH_2} + H_2O$

c. $\mathrm{CH_3{-}\overset{\overset{CH_3}{|}}{CH}{-}\overset{\overset{O}{\|}}{C}{-}OH} + ? \xrightarrow{\Delta} \mathrm{CH_3{-}\overset{\overset{CH_3}{|}}{CH}{-}\overset{\overset{O}{\|}}{C}{-}\overset{\overset{CH_3}{|}}{N}{-}CH_3} + H_2O$

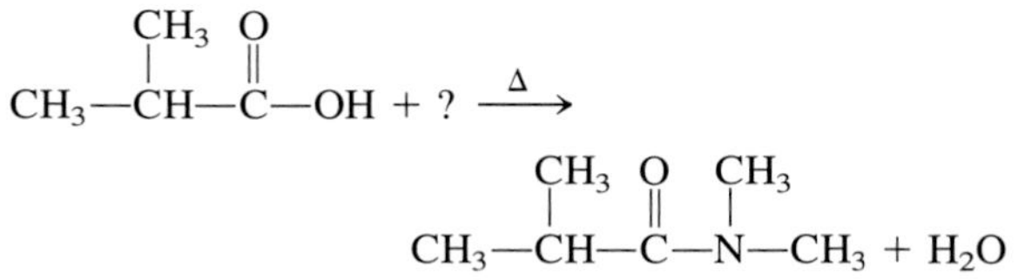

d. $? + \mathrm{CH_3{-}NH_2} \xrightarrow{\Delta} \mathrm{(CH_3)C_6H_4{-}\overset{\overset{O}{\|}}{C}{-}NH{-}CH_3}$ (benzene ring with CH_3 meta to the amide group)

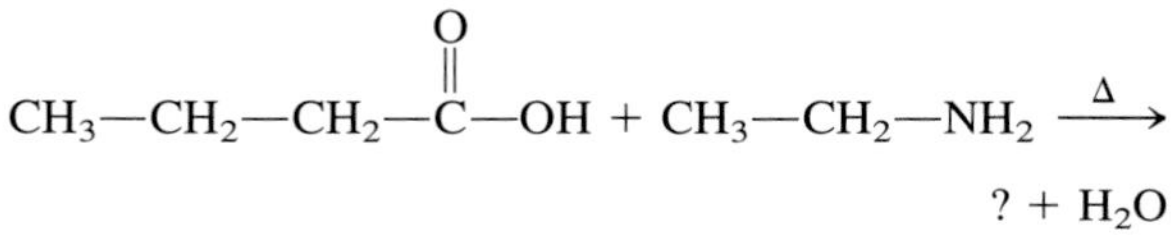

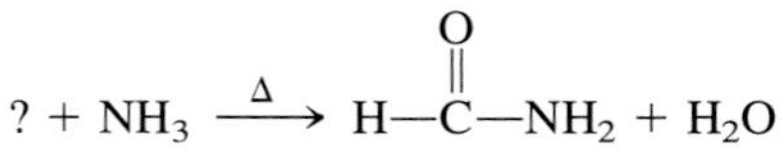

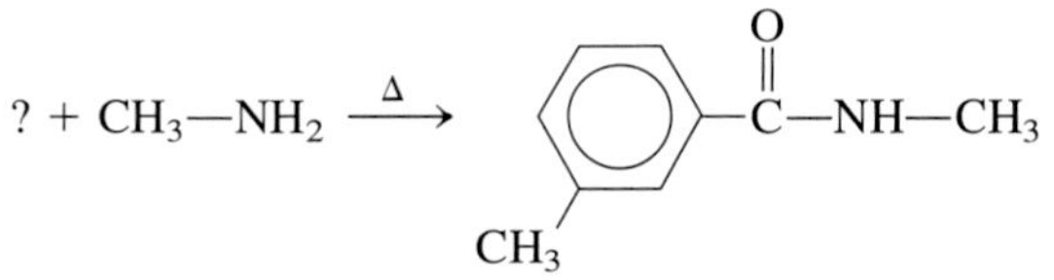

17.71 Draw the structures of the carboxylic acid and the amine from which each of the following amides could be formed.

a. $\mathrm{CH_3{-}\overset{\overset{O}{\|}}{C}{-}\overset{\overset{CH_3}{|}}{N}{-}\overset{\overset{CH_3}{|}}{CH}{-}CH_3}$

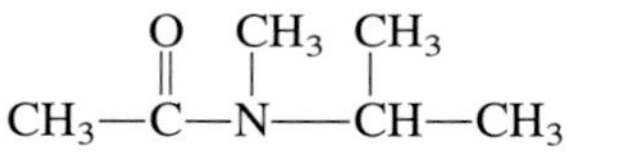

b. *N*-methylpentanamide

c. $\mathrm{CH_3{-}\overset{\overset{CH_3}{|}}{CH}{-}\overset{\overset{O}{\|}}{C}{-}NH{-}CH_3}$

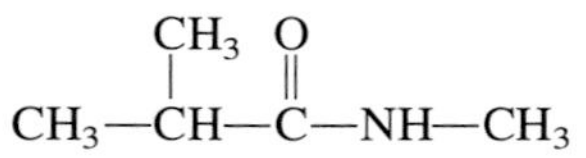

d. 2,3,*N*-trimethylbutanamide

17.72 Draw the structures of the carboxylic acid and the amine from which each of the following amides could be formed.

a. $\mathrm{CH_3{-}CH_2{-}\overset{\overset{O}{\|}}{C}{-}\underset{\underset{CH_3}{|}}{N}{-}CH_3}$

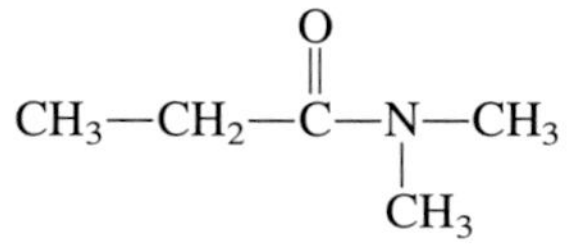

b. 2-Methylpentanamide

c. $\mathrm{CH_3{-}\underset{\underset{CH_3}{|}}{\overset{\overset{CH_3}{|}}{C}}{-}\overset{\overset{O}{\|}}{C}{-}NH{-}CH_2{-}CH_3}$

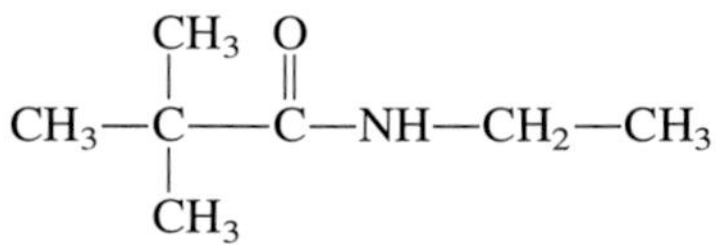

d. *N,N*-diethylacetamide

Hydrolysis of Amides (Section 17.15)

17.73 Draw the structures of the organic products in each of the following hydrolysis reactions.

a. $\mathrm{CH_3{-}CH_2{-}CH_2{-}\overset{\overset{O}{\|}}{C}{-}NH{-}CH_3} + H_2O \xrightarrow{\Delta}$

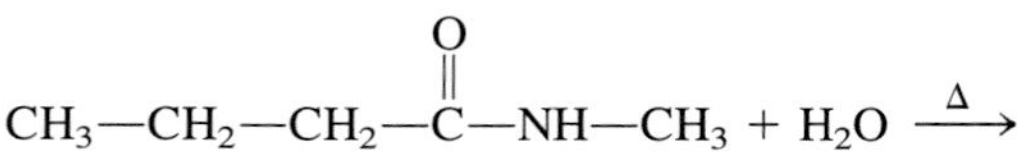

b. $CH_3—CH_2—CH_2—\overset{O}{\overset{\|}{C}}—NH—CH_3 + H_2O \xrightarrow[HCl]{\Delta}$

c. $CH_3—CH_2—CH_2—\overset{O}{\overset{\|}{C}}—NH—CH_3 + H_2O \xrightarrow[NaOH]{\Delta}$

d. $C_6H_5—\overset{O}{\overset{\|}{C}}—\underset{CH_3}{\underset{|}{N}}—C_6H_5 + H_2O \xrightarrow{\Delta}$

17.74 Draw the structures of the organic products in each of the following hydrolysis reactions.

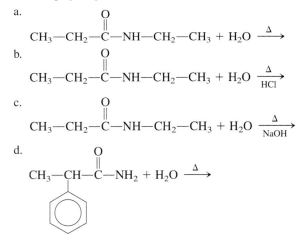

a. $CH_3—CH_2—\overset{O}{\overset{\|}{C}}—NH—CH_2—CH_3 + H_2O \xrightarrow{\Delta}$

b. $CH_3—CH_2—\overset{O}{\overset{\|}{C}}—NH—CH_2—CH_3 + H_2O \xrightarrow[HCl]{\Delta}$

c. $CH_3—CH_2—\overset{O}{\overset{\|}{C}}—NH—CH_2—CH_3 + H_2O \xrightarrow[NaOH]{\Delta}$

d. $CH_3—\underset{C_6H_5}{\underset{|}{CH}}—\overset{O}{\overset{\|}{C}}—NH_2 + H_2O \xrightarrow{\Delta}$

Polyamides and Polyurethanes (Section 17.16)

17.75 List the general characteristics of the monomers needed to produce a polyamide.

17.76 Contrast the monomers needed to produce a polyamide with those needed to produce a polyester.

17.77 Draw a structural representation of the polyamide formed from the reaction of succinic acid and 1,4-butanediamine.

17.78 Draw a structural representation of the polyamide formed from the reaction of adipic acid and 1,2-ethanediamine.

Additional Problems

17.79 Draw structural formulas for the following compounds.

a. Formamide
b. 2-Pentamine
c. 2-Methylpentanamide
d. *N*-Isopropylethanamide
e. Diethylammonium chloride
f. Trimethylanilinium chloride

17.80 What is the structure of the organic product (or products) in each of the following reactions?

a. $CH_3—CH_2—\overset{+}{N}H_2—CH_3\ Br^- + NaOH \longrightarrow$

b. $CH_3—CH_2—\overset{CH_3}{\overset{|}{CH}}—\overset{O}{\overset{\|}{C}}—OH + CH_3—NH—CH_3 \xrightarrow{\Delta}$

c. $CH_3—\overset{CH_3}{\overset{|}{N}}—CH_3 + CH_3—Cl \xrightarrow{NaOH}$

d. $CH_3—CH_2—\overset{CH_3}{\overset{|}{CH}}—\overset{O}{\overset{\|}{C}}—NH_2 + H_2O \xrightarrow[\Delta]{NaOH}$

e. $CH_3—Br + CH_3—NH_2 \xrightarrow{NaOH}$

f. $CH_3—CH_2—NH_2 + H_2O \longrightarrow$

17.81 Draw structural formulas and assign IUPAC names to the eight amine structural isomers with the formula $C_4H_{11}N$.

17.82 Draw structural formulas and assign IUPAC names to the four amide structural isomers with the formula C_3H_7ON.

17.83 Draw the structural formula of the quaternary ammonium salt with the formula $C_5H_{14}NCl$.

17.84 Classify each of the following amines or amides as unsubstituted, monosubstituted, or disubstituted.

a. *o*-Methylbenzamide
b. *N*-Methylbenzamide
c. Cyclopentylmethylamine
d. *N,N*,-dimethylhexanamide
e. Isopropylamine
f. 4-Methylheptanamine

17.85 Indicate whether each of the following compounds is an amine, an amide, both, or neither.

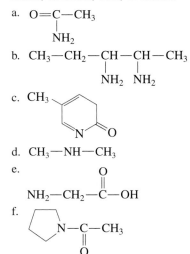

a. $O{=}\underset{NH_2}{\underset{|}{C}}—CH_3$

b. $CH_3—CH_2—\underset{NH_2}{\underset{|}{CH}}—\underset{NH_2}{\underset{|}{CH}}—CH_3$

c. CH_3 (ring with N and =O)

d. $CH_3—NH—CH_3$

e. $NH_2—CH_2—\overset{O}{\overset{\|}{C}}—OH$

f. (pyrrolidine ring) $N—\underset{O}{\underset{\|}{C}}—CH_3$

17.86 Assign IUPAC names to each of the following compounds.

a. NH_2

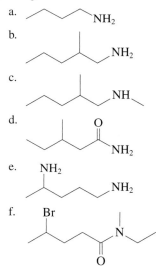

b. NH_2

c. NH

d. O, NH_2

e.

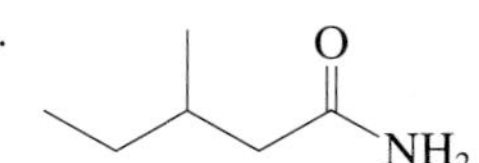

f.

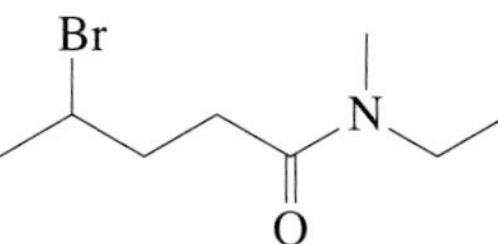

Grid Problems

17.87

1. $R—NH_2$	2. $R—NH—R$	3. $R—\overset{O}{\overset{\|}{C}}—NH—R$
4. $R\overset{+}{N}H_3\ Cl^-$	5. $R_3\overset{+}{N}H\ Cl^-$	6. $R_4\overset{+}{N}\ Cl^-$

Select from the grid *all* correct responses for each of the following situations.

a. Compounds classified as amines
b. Compounds classified as amides
c. Compounds classified as amine salts
d. Compounds classified as quaternary ammonium salts

17.88

1. $CH_3—CH_2—NH_2$	2. $CH_3—CH_2—\overset{CH_3}{\overset{\|}{N}H}$	3. $CH_3—\overset{CH_3}{\overset{\|}{N}}—CH_3$
4. $CH_3—\overset{O}{\overset{\|}{C}}—NH_2$	5. $CH_3—\overset{O}{\overset{\|}{C}}—\overset{CH_3}{\overset{\|}{N}H}$	6. $H—\overset{O}{\overset{\|}{C}}—\overset{CH_3}{\overset{\|}{N}}—CH_3$

Select from the grid *all* correct responses for each of the following situations.

a. Compounds that are disubstituted amides
b. Compounds that are disubstituted amines
c. Compounds that are primary amines
d. Compounds that are secondary amides

17.89

1. methanamine	2. benzamide	3. *o*-methylaniline
4. formamide	5. 1,2-ethanediamine	6. N-methylaniline

Select from the grid *all* correct responses for each of the following situations.

a. The structure of the molecule includes a benzene ring.
b. A —NH— structural unit is present in the molecule.
c. The number of carbon atoms present and the number of nitrogen atoms present are the same.
d. A $—\overset{O}{\overset{\|}{C}}—NH_2$ structural unit is present in the molecule.

17.90

1. $CH_3—\overset{O}{\overset{\|}{C}}—NH—CH_3$	2. $CH_3—\overset{O}{\overset{\|}{C}}—NH_2$	3. $CH_3—\overset{+}{N}H_3\ Cl^-$
4. $CH_3—CH_2—NH_2$	5. $CH_3—\overset{CH_3}{\overset{\|}{N}}—CH_3$	6. $CH_3—\underset{CH_3}{\underset{\|}{\overset{CH_3}{\overset{\|}{\overset{+}{N}}}}}—CH_3\ Cl^-$

Select from the grid *all* correct responses for each of the following situations.

a. Compounds that can be synthesized by reacting a carboxylic acid with ammonia under appropriate conditions
b. Compounds that can be synthesized by reacting an amine with an alkyl halide in the presence of a strong base
c. Compounds that can be synthesized by reacting an amine with a strong acid
d. Compounds for which a carboxylic acid salt is a hydrolysis product when the hydrolysis is carried out under basic conditions

Carbohydrates

CHAPTER OUTLINE

18.1 Biochemistry—An Overview 506
18.2 Occurrence and Functions of Carbohydrates 507
18.3 Carbohydrate Classifications 508
18.4 Chirality: Handedness in Molecules 509
18.5 Stereoisomerism: Enantiomers and Diastereomers 511
18.6 Fischer Projections 513
18.7 Properties of Enantiomers 517
Chemistry at a Glance: **Isomers** 519
18.8 Classification of Monosaccharides 520
18.9 Biologically Important Monosaccharides 523
18.10 Cyclic Forms of Monosaccharides 525
18.11 Haworth Projection Formulas 527
18.12 Reactions of Monosaccharides 528
18.13 Disaccharides 532
Chemistry at a Glance: **Reactions of Monosaccharides** 539
18.14 Polysaccharides 539
18.15 Mucopolysaccharides 543
18.16 Glycolipids and Glycoproteins 544
Chemical Connections
18.1 Blood Types and Monosaccharides 530
18.2 Lactose Intolerance and Galactosemia 535
18.3 Artificial Sweeteners 537
18.4 Sucrose Derivatives and Flatulence 538

Carbohydrates in the form of cotton and linen may be woven into clothing materials.

In previous chapters, we have considered numerous examples of the importance of chemistry in biological systems. Beginning with this chapter on carbohydrates, we will focus almost exclusively on biochemistry, the chemistry of living systems. Like organic chemistry, biochemistry is a vast subject, and we can discuss only a few of its facets. Our approach to biochemistry will be similar to our approach to organic chemistry. We will devote individual chapters to each of the major classes of biochemical compounds, which are carbohydrates, lipids, proteins, and nucleic acids. Then we will examine the major types of chemical reactions in living organisms. In this first "biochapter," carbohydrates are considered.

18.1 Biochemistry—An Overview

Biochemistry *is the study of the chemical substances found in living systems and the chemical interactions of these substances with each other.* Biochemistry is a field in which new discoveries are made almost daily about how cells manufacture the molecules needed for life and how the chemical reactions by which life is maintained occur.

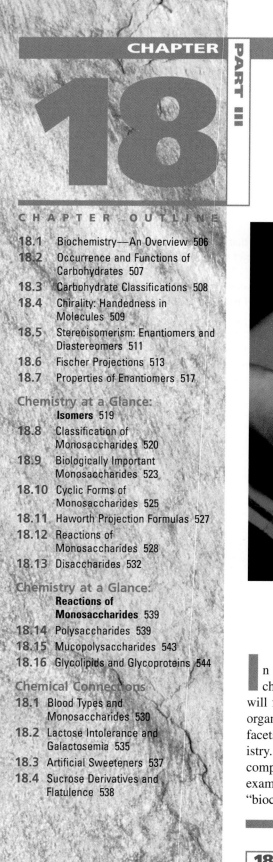

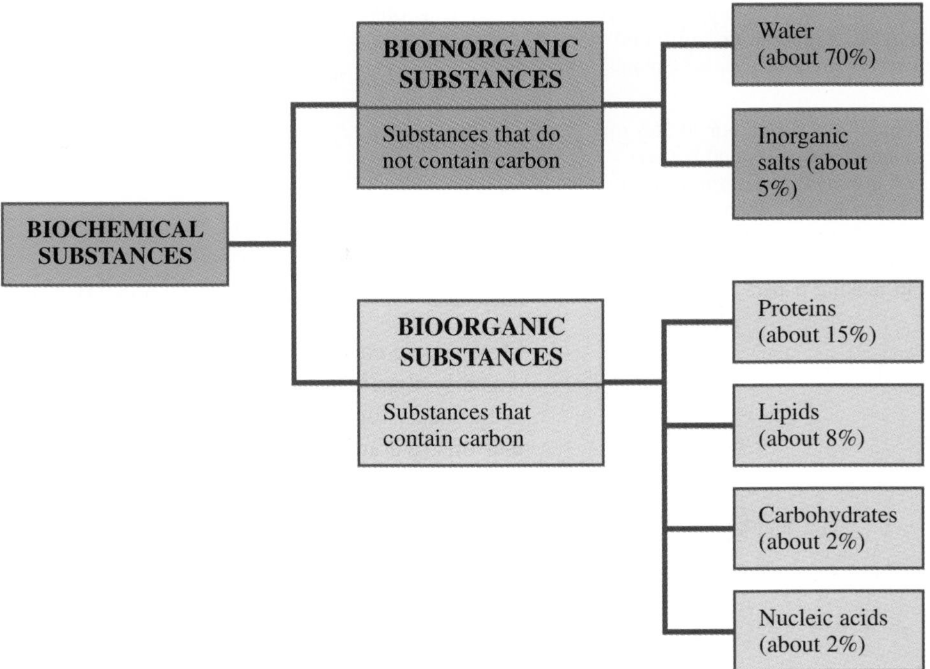

Figure 18.1
Mass composition data for the human body in terms of major types of biochemical substances.

• The functional groups present in bioorganic substances are the same ones found in the organic compounds studied in previous chapters. The major difference between simple organic and bioorganic compounds is structural complexity; bioorganic substances generally have several different functional groups present.

• As isolated compounds, bioinorganic and bioorganic substances have no life in and of themselves. Yet when these substances are gathered together in a cell, their chemical interactions are able to sustain life.

The knowledge explosion that has occurred in the field of biochemistry during the last decades of the twentieth century is truly phenomenal.

Biochemical substances *are the chemical substances found within a living organism.* They are divided into two groups: bioinorganic substances and bioorganic substances. *Bioinorganic substances* include water and inorganic salts. *Bioorganic substances* include carbohydrates, lipids, proteins, and nucleic acids. Figure 18.1 gives an approximate mass composition for the human body in terms of types of biochemical substances present.

Although we tend to think of the human body as made up of organic substances, bioorganic molecules make up only about one-fourth of body mass. The bioinorganic substance water constitutes over two-thirds of the mass of the human body, and another 4%–5% of body mass comes from inorganic salts (Section 10.6).

18.2 Occurrence and Functions of Carbohydrates

Most of the matter in plants, except water, is carbohydrate material. Photosynthesis, the process by which carbohydrates are made, requires sunlight.

Carbohydrates are the most abundant class of bioorganic molecules on planet Earth. Although their abundance in the human body is relatively low (Section 18.1), carbohydrates constitute about 75% by mass of dry plant materials.

Green (chlorophyll-containing) plants produce carbohydrates via *photosynthesis.* In this process, carbon dioxide from the air and water from the soil are the reactants, and sunlight absorbed by chlorophyll is the energy source.

$$CO_2 + H_2O + \text{solar energy} \xrightarrow[\text{Plant enzymes}]{\text{Chlorophyll}} \text{carbohydrates} + O_2$$

Plants have two main uses for the carbohydrates they produce. In the form of *cellulose,* carbohydrates serve as structural elements, and in the form of *starch,* they provide energy reserves for the plants.

Dietary intake of plant materials is the major carbohydrate source for humans and animals. The average human diet should ideally be about two-thirds carbohydrate by mass. Carbohydrates have the following functions in humans:

1. Carbohydrate oxidation provides energy.
2. Carbohydrate storage, in the form of glycogen, provides a short-term energy reserve.

• It is estimated that more than half of all organic carbon atoms are found in the carbohydrate materials of plants.

• Human uses for carbohydrates of the plant kingdom extend beyond food. Carbohydrates in the form of cotton, rayon, and linen are used as clothing. Carbohydrates in the form of wood are used for shelter and heating and in making paper.

3. Carbohydrates supply carbon atoms for the synthesis of other biochemical substances (proteins, lipids, and nucleic acids).
4. Carbohydrates form part of the structural framework of DNA and RNA molecules.
5. Carbohydrate "markers" on cell surfaces play key roles in cell–cell recognition processes.

18.3 Carbohydrate Classifications

Most simple carbohydrates have empirical formulas that fit the general formula $C_nH_{2n}O_n$. An early observation by scientists that this general formula can also be written as $C_n(H_2O)_n$ is the basis for the term *carbohydrate*—that is, "hydrate of carbon." It is now known that this hydrate viewpoint is not correct, but the term *carbohydrate* still persists. Today the term is used to refer to an entire family of compounds, only some of which have the formula $C_nH_{2n}O_n$.

Carbohydrates *are polyhydroxy aldehydes, polyhydroxy ketones, or compounds that yield such substances upon hydrolysis.* The carbohydrate glucose is a polyhydroxy aldehyde, and the carbohydrate fructose is a polyhydroxy ketone.

```
Aldehyde ——→ CHO                        CH2OH
group         |                           |
          H—C—OH                        C=O ←—— Ketone
              |                           |          group
         HO—C—H                     HO—C—H
              |                           |
          H—C—OH                     H—C—OH
              |                           |
          H—C—OH                     H—C—OH
              |                           |
            CH2OH                       CH2OH
```

Glucose (a polyhydroxy aldehyde) — Fructose (a polyhydroxy ketone)

A striking structural feature of carbohydrates is the large number of functional groups present. In glucose and fructose there is a functional group attached to each carbon atom.

Carbohydrates are classified on the basis of molecular size as monosaccharides, oligosaccharides, and polysaccharides.

• The term *monosaccharide* is pronounced "mon-oh-SACK-uh-ride."

Monosaccharides *are carbohydrates that contain a single polyhydroxy aldehyde or polyhydroxy ketone unit.* Monosaccharides cannot be broken down into simpler units by hydrolysis reactions. Both glucose and fructose are monosaccharides. Naturally occurring monosaccharides have from three to seven carbon atoms; five- and six-carbon species are especially common. Pure monosaccharides are water-soluble, white, crystalline solids.

• The *oligo* in the term *oligosaccharides* comes from the Greek *oligos,* which means "small" or "few."

Oligosaccharides *are carbohydrates that contain from two to ten monosaccharide units.* Disaccharides are the most common type of oligosaccharide. **Disaccharides** *are carbohydrates composed of two monosaccharide units covalently bonded to each other.* Like monosaccharides, disaccharides are crystalline, water-soluble substances. Sucrose (table sugar) and lactose (milk sugar) are disaccharides.

Within the human body, oligosaccharides are often found associated with proteins and lipids in complexes that have both structural and regulatory functions. Free oligosaccharides, other than disaccharides, are seldom encountered in biological systems.

Complete hydrolysis of an oligosaccharide produces monosaccharides. Upon hydrolysis, a disaccharide produces two monosaccharides, a trisaccharide three monosaccharides, a hexasaccharide six monosaccharides, and so on.

• Types of carbohydrates are related to each other through hydrolysis.

Polysaccharides
↓ Hydrolysis
Oligosaccharides
↓ Hydrolysis
Monosaccharides

Polysaccharides *are carbohydrates made up of many monosaccharide units.* Polysaccharides, which are polymers, often consist of tens of thousands of monosaccharide units. Both cellulose and starch are polysaccharides. We encounter these two substances everywhere. The paper on which this book is printed is mainly cellulose, as are the cotton in our clothes and the wood in our houses. Starch is a component of many types of foods, including bread, pasta, potatoes, rice, corn, beans, and peas.

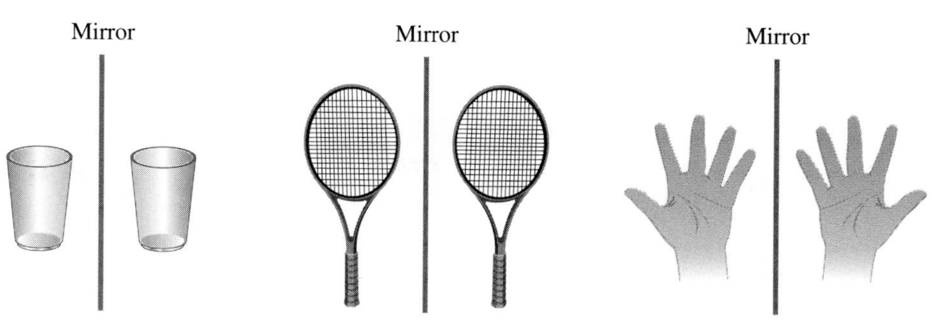

Figure 18.2
Some common objects and their mirror images.

18.4 Chirality: Handedness in Molecules

• The property of handedness is not restricted to carbohydrates. It is a general phenomenon found in all classes of organic compounds.

Before considering structures for and reactions of specific carbohydrates, we will consider *handedness,* a biologically important structural property exhibited by most carbohydrates. Most carbohydrate molecules can exist in two forms—a left-handed form and a right-handed form. Significantly, these different forms often elicit different responses within the human body.

• Mirror Images

The concept of *mirror images* is the key to understanding molecular handedness. All objects, including all molecules, have mirror images. The **mirror image** *of an object is the object's reflection in a mirror.* Figure 18.2 shows the mirror images of a water glass, a tennis racket, and a human hand.

Objects can be divided into two classes on the basis of their mirror images: (1) objects that *can* be superimposed on their mirror image, and (2) objects that *cannot* be superimposed on their mirror image. Superimposable objects have parts that coincide exactly at all points when the objects are laid upon each other. The glass and tennis racket of Figure 18.2 are superimposable upon their mirror images, whereas the human hand is not. Figure 18.3 is a closer look at the *nonsuperimposable* human hand. It shows that when human hands are superimposed on each other, the two thumbs point in opposite directions and the fingers do not align correctly; the index fingers are on opposite sides.

• The term *chiral* (rhymes with *spiral*) comes from the Greek word *cheir,* which means "hand." Chiral objects are said to possess "handedness."

• Chirality

Objects that *cannot* be superimposed upon their mirror image are said to be chiral objects. A **chiral object** *is an object that is not identical to its mirror image.* Your hands and feet are chiral objects, as are gloves and shoes. Objects that *can* be superimposed upon their mirror images are achiral. An **achiral object** *is identical to its mirror image.* Achiral objects include tube socks, solid-colored ties and T-shirts.

Figure 18.3
A person's left and right hands are not superimposable upon each other.

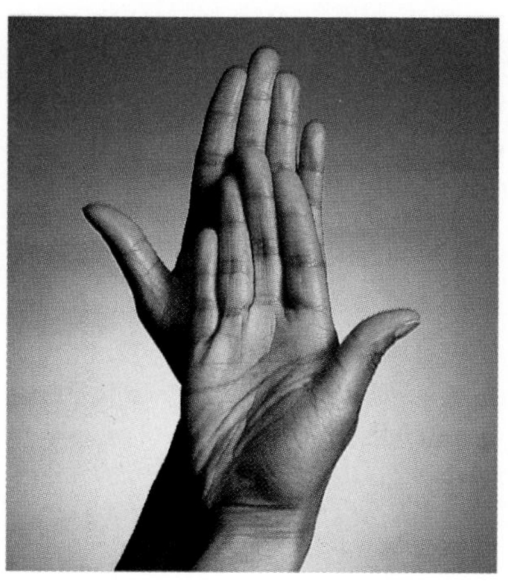

Molecules, like larger objects, can be chiral or achiral. A simple example of a chiral molecule is the trisubstituted methane bromochloroiodomethane.

```
      H
      |
  Br—C—Cl
      |
      I
```

Figure 18.4a shows the nonsuperimposability of the mirror-image forms of this molecule. The simplest example of a chiral carbohydrate is the three-carbon molecule glyceraldehyde.

```
     CHO
      |
  H—C—OH
      |
     CH2OH
```

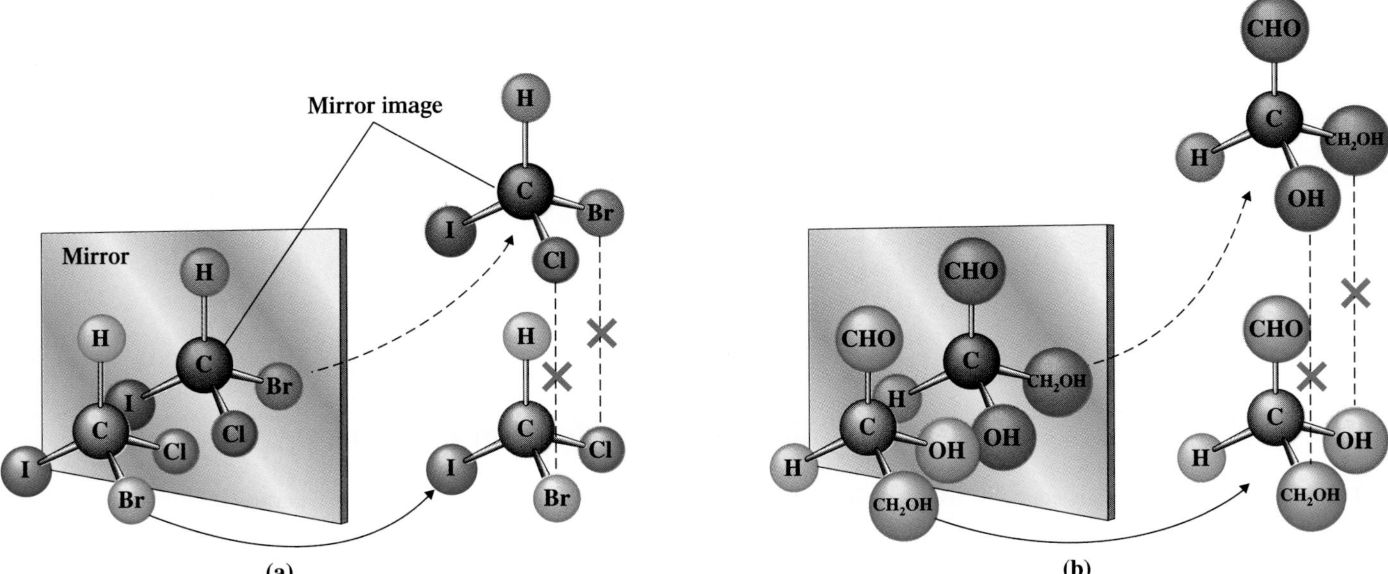

Figure 18.4
Examples of simple molecules that are chiral. (a) The mirror-image forms of the molecule bromochloroiodomethane are nonsuperimposable. (b) The mirror-image forms of the molecule glyceraldehyde are nonsuperimposable.

• There are a few chiral molecules known that do not have a chiral center. Such exceptions are not important for the applications that we will make of the chirality concept in this text.

The nonsuperimposability of the mirror-image forms of glyceraldehyde is shown in Figure 18.4b.

Trying to superimpose the mirror image of a molecule on that molecule visually, as is done in Figure 18.4, is one way to determine molecular chirality. Another method, which is much easier to apply, makes use of the observation that *generally, whenever a carbon atom in a molecule is bonded to four* different *groups, the molecule as a whole is chiral.*

Any organic molecule containing a *single* carbon atom with four *different* groups attached to it exhibits chirality. Such a carbon atom is called a chiral center. A **chiral center** *is an atom in a molecule that has four different groups tetrahedrally bonded to it.* Note that each of the molecules in Figure 18.4 has a chiral center.

Chiral centers within molecules are often denoted by a small asterisk. Note the chiral centers in the following molecules.

$$CH_3-CH_2-{}^*\overset{\displaystyle H}{\underset{\displaystyle OH}{C}}-CH_3 \qquad H-{}^*\overset{\displaystyle Cl}{\underset{\displaystyle I}{C}}-CH_3 \qquad CH_3-CH_2-CH_2-{}^*\overset{\displaystyle CH_3}{\underset{\displaystyle H}{C}}-CH_2-CH_3$$

2-Butanol — 1-Chloro-1-iodoethane — 3-Methylhexane

Example 18.1

Identifying Chiral Centers in Molecules

Indicate whether the circled carbon atom in each of the following molecules is a chiral center.

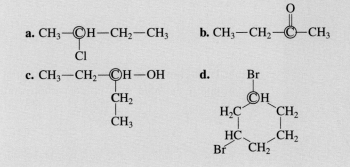

a. $CH_3-\underset{\displaystyle Cl}{\text{Ⓒ}H}-CH_2-CH_3$

b. $CH_3-CH_2-\overset{\displaystyle O}{\overset{\|}{\text{Ⓒ}}}-CH_3$

c. $CH_3-CH_2-\text{Ⓒ}H(-OH)-CH_2-CH_3$ (circled carbon bonded to H, OH, CH_3-CH_2- and $-CH_2-CH_3$ below)

d. Ring: Br–ⒸH bonded to H_2C and CH_2; H_2C–HC(Br); CH_2–CH_2; HC–CH_2–CH_2

Solution

a. This is a chiral center. The four different groups attached to the carbon atom are $-CH_3$, $-Cl$, $-CH_2-CH_3$, and $-H$.
b. No chiral center is present. The carbon atom is attached to only *three* groups.
c. No chiral center is present. Two of the groups attached to the carbon atom are identical.
d. The chirality rules for ring carbon atoms are the same as for acyclic carbon atoms. A chiral center is present. Two of the groups are $-H$ and $-Br$. The third group, obtained by proceeding clockwise around the ring, is $-CH_2-CH_2-CH_2$. The fourth group, obtained by proceeding counterclockwise around the ring, is $-CH_2-CHBr-CH_2$.

Practice Exercise 18.1

Indicate whether the circled carbon atom in each of the following molecules is a chiral center.

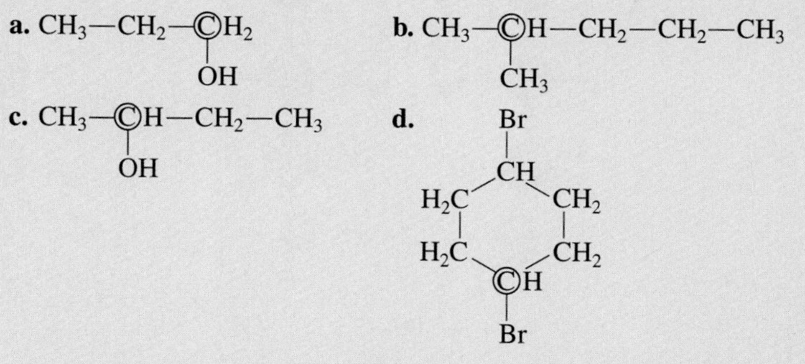

a. CH_3-CH_2-ⒸH_2 (with OH on the circled carbon)

b. CH_3-Ⓒ$H-CH_2-CH_2-CH_3$ (with CH_3 on the circled carbon)

c. CH_3-Ⓒ$H-CH_2-CH_3$ (with OH on the circled carbon)

d. Ring: Br–CH, H_2C, CH_2, H_2C, CH_2, ⒸH–Br

• *Answers:* **a.** no chiral center; **b.** no chiral center; **c.** chiral center; **d.** no chiral center

• Remember the meaning of the structural notations —CHO and $-CH_2OH$.

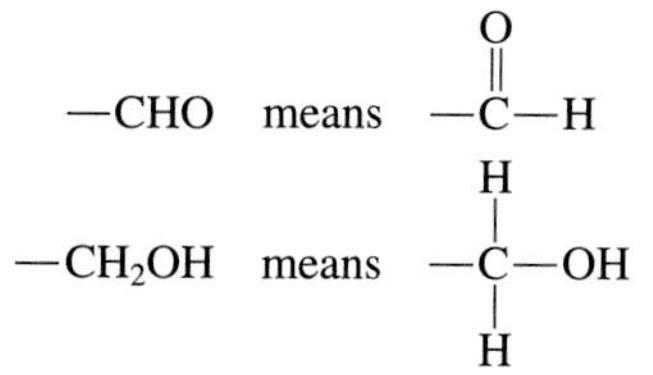

—CHO means —C(=O)—H

$-CH_2OH$ means —C(H)(H)—OH

Organic molecules, especially carbohydrates, may contain more than one chiral center. For example, the following carbohydrate has two chiral centers.

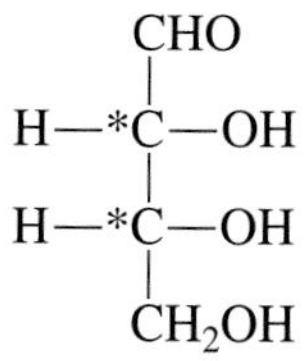

$$\begin{array}{c} CHO \\ | \\ H-{}^*C-OH \\ | \\ H-{}^*C-OH \\ | \\ CH_2OH \end{array}$$

18.5 Stereoisomerism: Enantiomers and Diastereomers

Stereoisomers *are isomers whose atoms are connected in the same way but differ in their arrangement in space.* The two nonsuperimposable mirror-image forms of a chiral molecule are stereoisomers.

There are two major causes of stereoisomerism: (1) the presence of a chiral center in a molecule, and (2) the presence of "structural rigidity" in a molecule. Structural rigidity is caused by restricted rotation about chemical bonds. It is the basis for *cis–trans* stereoisomerism, a phenomenon found in some substituted cycloalkanes (Section 12.13) and some alkenes (Section 13.5).

• The term *enantiomer* comes from the Greek *enantios,* which means "opposite." It is pronounced "en-AN-tee-o-mer."

• Some textbooks use the term *diastereoisomers* instead of *diastereomers.* The pronunciation for *diastereomer* is "dye-a-STEER-ee-o-mer."

Stereoisomers can be subdivided into two types: *enantiomers* and *diastereomers* (Figure 18.5). **Enantiomers** *are stereoisomers whose molecules are nonsuperimposable mirror images of each other.* Left- and right-handed forms of a molecule with a single chiral center are enantiomers.

Diastereomers *are stereoisomers whose molecules are not mirror images of each other. Cis–trans* isomers (of both the alkene and the cycloalkane types) are diastereomers. We

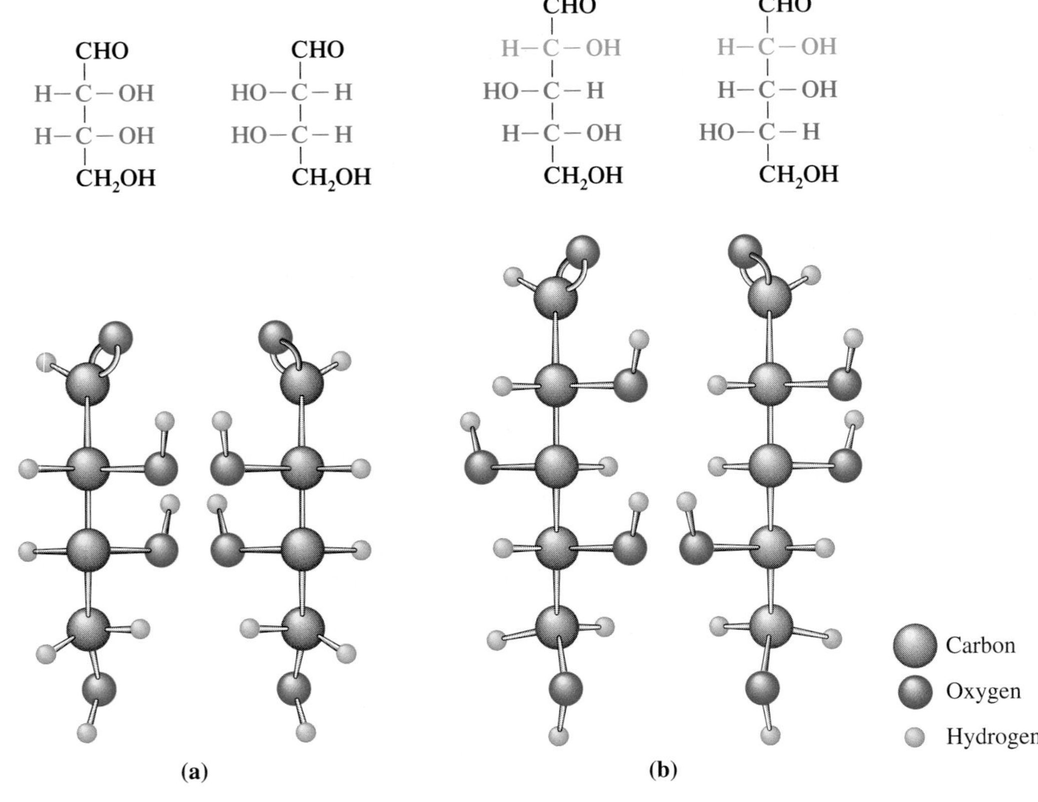

Figure 18.5
(a) Enantiomers are stereoisomers whose molecules are nonsuperimposable mirror images of each other, as in left-handed and right-handed forms of a molecule. (b) Diastereomers are stereoisomers whose molecules are not mirror images of each other.

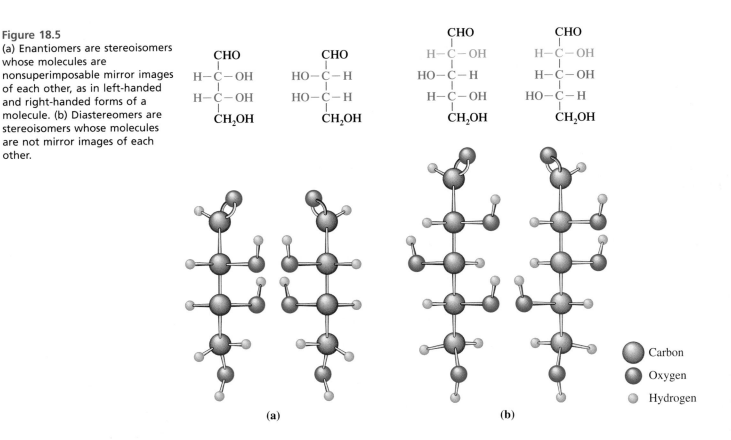

will see additional examples of carbohydrate diastereomers in the next section. Stereoisomers that are not enantiomers are diastereomers; they must be one or the other (see Figure 18.6).

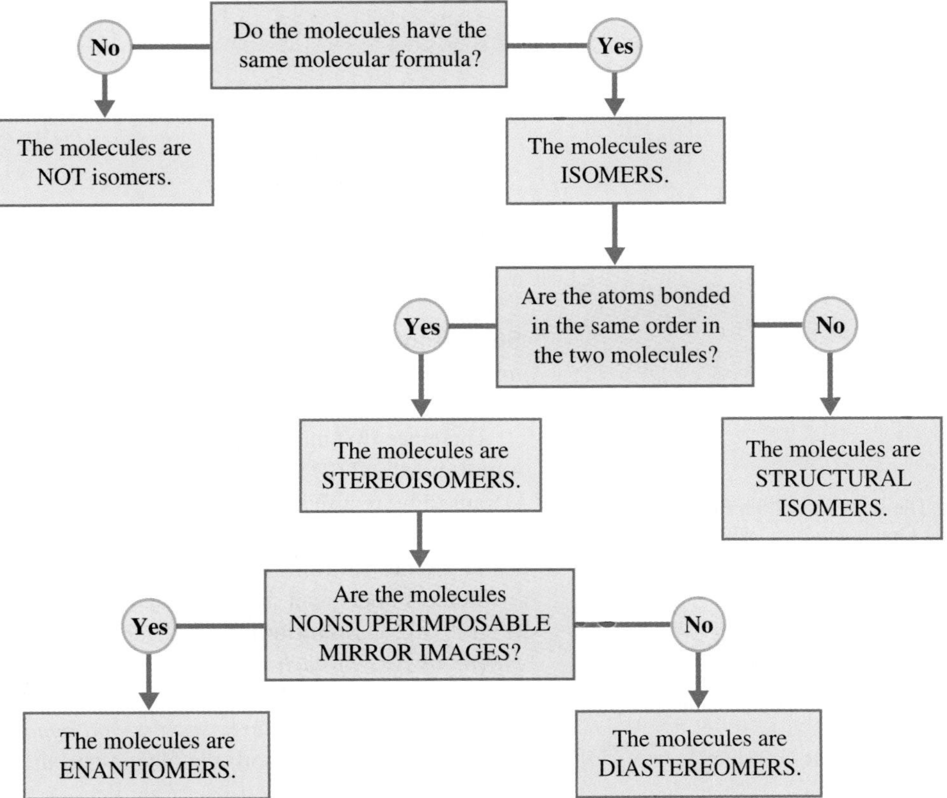

Figure 18.6
A summary of the "thought process" used in classifying molecules as enantiomers or diastereomers.

The German chemist Emil Fischer (1852–1919), the developer of the two-dimensional system for specifying chirality, was one of the early greats in organic chemistry. He made many fundamental discoveries about carbohydrates, proteins, and other natural products and in 1902 was awarded the second Nobel Prize in chemistry.

18.6 Fischer Projections

Drawing *three*-dimensional representations of chiral molecules, as in Figure 18.5, can be both time-consuming and awkward. Fischer projections represent a method for giving molecular chirality specifications in *two* dimensions. A **Fischer projection** *is a two-dimensional notation for showing the spatial arrangement of groups about chiral centers in molecules.*

In a Fischer projection, a chiral center is represented as the intersection of vertical and horizontal lines. The atom at the chiral center, which is almost always carbon, is not explicitly shown.

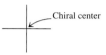

Chiral center

The tetrahedral arrangement of the four groups attached to the atom at the chiral center is governed by the following conventions: (1) Vertical lines from the chiral center represent bonds to groups directed into the printed page. (2) Horizontal lines from the chiral center represent bonds to groups directed out of the printed page.

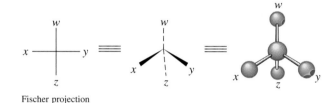

w, x, y, z ≡ w, x, y, z ≡ w, x, y, z

Fischer projection

Our immediate concern is Fischer projections for monosaccharides (the simplest type of carbohydrate; Section 18.2). Such projections have the monosaccharide *carbon chain* positioned vertically with the carbonyl group (aldehyde or ketone) at or near the top. The smallest monosaccharide that has a chiral center is the compound glyceraldehyde (2,3-dihydroxypropanal). The structural formula and Fischer projections for the two enantiomers of glyceraldehyde are

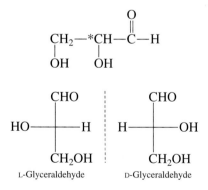

$$\underset{\displaystyle OH}{CH_2}—\underset{\displaystyle OH}{{}^*CH}—\overset{\displaystyle O}{\overset{\|}{C}}—H$$

	L-Glyceraldehyde		D-Glyceraldehyde	
	CHO		CHO	
HO	—	H	H	—	OH
	CH_2OH		CH_2OH	

• The D and L designations for the handedness of the two members of an enantiomeric pair come from the Latin words *dextro,* which means "right," and *levo,* which means "left."

The handedness (right and left) of these two enantiomers is specified by using the designations D and L. The enantiomer with the chiral center —OH group on the right in the Fischer projection is by definition the right-handed isomer (D-glyceraldehyde), and the enantiomer with the chiral center —OH group on the left in the Fischer projection is by definition the left-handed isomer (L-glyceraldehyde).

We now consider Fischer projections for the compound 2,3,4-trihydroxybutanal, a monosaccharide with four carbons and *two* chiral centers.

$$\underset{\displaystyle OH}{CH_2}—\underset{\displaystyle OH}{{}^*CH}—\underset{\displaystyle OH}{{}^*CH}—\overset{\displaystyle O}{\overset{\|}{C}}—H$$

There are four stereoisomers for this compound—two pairs of enantiomers.

A	B	C	D
CHO	CHO	CHO	CHO
H—OH	HO—H	HO—H	H—OH
H—OH	HO—H	H—OH	HO—H
CH_2OH	CH_2OH	CH_2OH	CH_2OH

First enantiomeric pair (A, B) Second enantiomeric pair (C, D)

• Any given molecular structure can have only one mirror image. Hence enantiomers always come in pairs; there can never be more than two.

In the first enantiomeric pair, both chiral center —OH groups are on the same side of the Fischer projection, and in the second enantiomeric pair, the chiral center —OH groups are on opposite sides of the Fischer projection. These are the only —OH group arrangements possible.

The D,L system used to designate the handedness of glyceraldehyde enantiomers is extended to monosaccharides with more than one chiral center in the following manner. The carbon chain is numbered, starting at the carbonyl group end of the molecule, and the highest-numbered chiral center is used to determine D or L configuration.

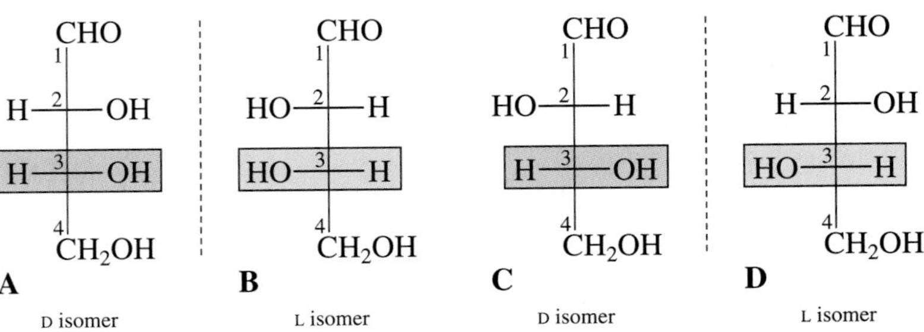

• Diastereomers that have two chiral centers must have the same handedness (both left or both right) at one chiral center and opposite handedness (one left and one right) at the other chiral center.

The D,L nomenclature gives the configuration (handedness) only at the highest-numbered chiral center. The configuration at other chiral centers in a molecule is accounted for by assigning a different common name to each pair of D,L enantiomers. In our present example, compounds A and B (the first enantiomeric pair) are D-erythrose and L-erythrose; compounds C and D (the second enantiomeric pair) are D-threose and L-threose.

What is the relationship between compounds A and C in our present example? They are diastereomers (Section 18.5), stereoisomers that are not mirror images of each other. Other diastereomeric pairs in our example are A and D, B and C, and B and D. These four pairs are epimers. **Epimers** *are diastereomers that differ only in the configuration at one chiral center.*

Example 18.2

Drawing Fischer Projections for Monosaccharides

Draw a Fischer projection for the enantiomer of each of the following monosaccharides.

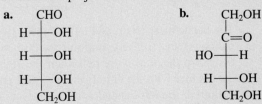

a.
CHO
H—OH
H—OH
H—OH
CH_2OH

b.
CH_2OH
C=O
HO—H
H—OH
CH_2OH

Solution

Given the Fischer projection of one member of an enantiomeric pair, we draw the other enantiomer's Fischer projection by reversing the substituents that are in horizontal positions *at each chiral center.*

a. Three chiral centers are present in this polyhydroxy aldehyde. Reversing the positions of the —H and —OH groups at each chiral center produces the Fischer projection of the other enantiomer.

The given enantiomer		The other enantiomer
CHO		CHO
H—|—OH	→	HO—|—H
H—|—OH		HO—|—H
H—|—OH		HO—|—H
CH_2OH		CH_2OH

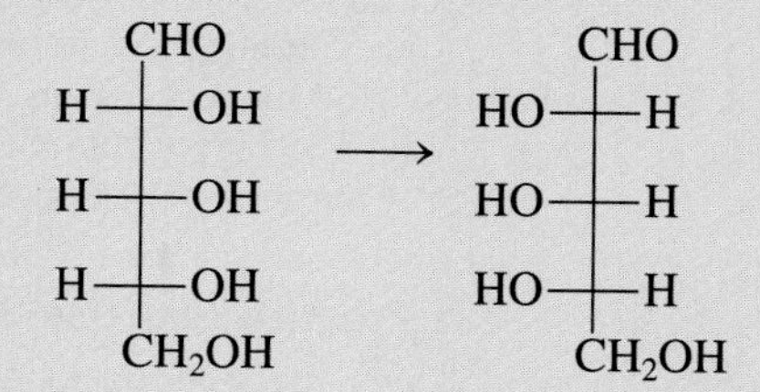

b. This monosaccharide is a polyhydroxy ketone with two chiral centers. Reversing the positions of the —H and —OH groups at both chiral centers generates the Fischer projection of the other enantiomer.

The given enantiomer		The other enantiomer
CH_2OH		CH_2OH
C=O		C=O
HO—|—H	→	H—|—OH
H—|—OH		HO—|—H
CH_2OH		CH_2OH

Practice Exercise 18.2

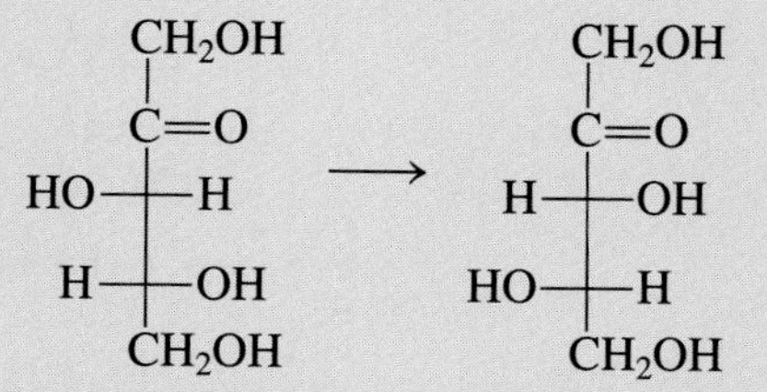

Draw a Fischer projection for the enantiomer of each of the following monosaccharides.

a.	b.
CHO	CH_2OH
H—|—OH	C=O
HO—|—H	H—|—OH
HO—|—H	H—|—OH
CH_2OH	CH_2OH

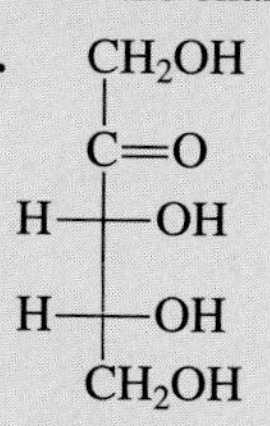

• *Answers:*

a.	b.
CHO	CH_2OH
HO—|—H	C=O
H—|—OH	HO—|—H
H—|—OH	HO—|—H
CH_2OH	CH_2OH

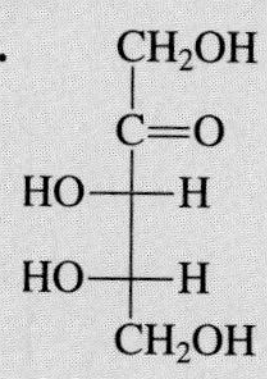

Example 18.3

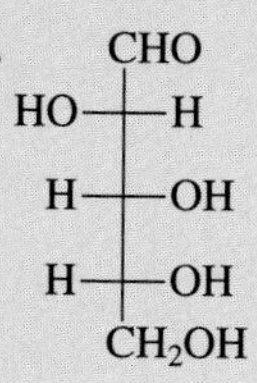

Classifying Monosaccharides as D or L Enantiomers

Classify each of the following monosaccharides as a D enantiomer or an L enantiomer.

a.	b.
^{1}CHO	$^1CH_2OH$
HO—^{2}—H	^{2}C=O
H—^{3}—OH	H—^{3}—OH
H—^{4}—OH	HO—^{4}—H
$^5CH_2OH$	HO—^{5}—H
	$^6CH_2OH$

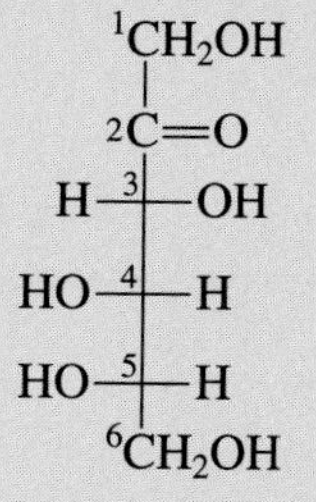

(continued)

Solution

D or L configuration for a monosaccharide is determined by the highest-numbered chiral center, the one farthest from the carbonyl carbon atom.

a. The highest-numbered chiral center, which involves carbon-4, has the —OH group on the right. Thus, this monosaccharide is a D enantiomer.
b. The highest-numbered chiral center, which involves carbon-5, has the —OH group on the left. Thus, this monosaccharide is an L enantiomer.

Practice Exercise 18.3

Classify each of the following monosaccharides as a D enantiomer or an L enantiomer.

a.

$$\begin{array}{c} CH_2OH \\ | \\ C{=}O \\ HO-\!\!-H \\ H-\!\!-OH \\ H-\!\!-OH \\ CH_2OH \end{array}$$

b.

$$\begin{array}{c} CHO \\ H-\!\!-OH \\ HO-\!\!-H \\ H-\!\!-OH \\ HO-\!\!-H \\ CH_2OH \end{array}$$

• *Answers:* **a.** D enantiomer; **b.** L enantiomer

Example 18.4

Recognizing Enantiomers and Diastereomers

Characterize each of the following pairs of structures as enantiomers, diastereomers, or neither enantiomers nor diastereomers.

a.

$$\begin{array}{c} CHO \\ H-\!\!-OH \\ HO-\!\!-H \\ H-\!\!-OH \\ CH_2OH \end{array} \quad \text{and} \quad \begin{array}{c} CHO \\ H-\!\!-OH \\ H-\!\!-OH \\ HO-\!\!-H \\ CH_2OH \end{array}$$

b.

$$\begin{array}{c} CHO \\ H-\!\!-OH \\ HO-\!\!-H \\ H-\!\!-OH \\ CH_2OH \end{array} \quad \text{and} \quad \begin{array}{c} CHO \\ HO-\!\!-H \\ H-\!\!-OH \\ HO-\!\!-H \\ CH_2OH \end{array}$$

c.

$$\begin{array}{c} CHO \\ H-\!\!-OH \\ HO-\!\!-H \\ H-\!\!-OH \\ CH_2OH \end{array} \quad \text{and} \quad \begin{array}{c} CHO \\ | \\ H-C-H \\ H-\!\!-OH \\ HO-\!\!-H \\ CH_2OH \end{array}$$

Solution

a. These two structures represent *diastereomers*—the arrangement of —H and —OH substituents is identical for at least one chiral center, whereas the arrangement of —H and —OH substituents at remaining chiral centers is that of mirror images. The —H and —OH substituent arrangement is the same at the first chiral center and is that of mirror images at the second and third chiral centers.
b. These two structures represent *enantiomers*—a mirror-image substituent relationship exists between the two isomers at *every* chiral center.
c. These two structures are *neither enantiomers nor diastereomers.* The connectivity of atoms differs in the two structures at carbon-2. Stereoisomers (enantiomers and

diastereomers) must have the same connectivity throughout both structures. (The two structures are not even structural isomers because the first structure contains one more oxygen atom than the second.)

Practice Exercise 18.4

Characterize the following pairs of structures as enantiomers, diastereomers, or neither enantiomers nor diastereomers.

a.

```
      CHO                  CHO
   H──┼──OH            HO──┼──H
   H──┼──OH    and     HO──┼──H
   H──┼──OH             H──┼──OH
     CH2OH                CH2OH
```

b.

```
      CHO                  CHO
   H──┼──OH            HO──┼──H
   H──┼──OH    and     HO──┼──H
   H──┼──OH            HO──┼──H
     CH2OH                CH2OH
```

c.

```
      CHO                  CHO
   H──┼──OH            HO──┼──H
  HO──┼──H     and      H──┼──OH
  HO──┼──H             HO──┼──H
     CH2OH                CH2OH
```

• *Answers:* **a.** diastereomers; **b.** enantiomers; **c.** diastereomers

• We calculate 2^n to predict the maximum possible number of stereoisomers for a molecule with n chiral atoms. In a few cases, the actual number of stereoisomers is less than the maximum because of symmetry considerations that make some mirror images superimposable.

In general, a compound that has n chiral centers may exist in a *maximum* of 2^n stereoisomeric forms. For example, when three chiral centers are present, at most eight stereoisomers ($2^3 = 8$) are possible (four pairs of enantiomers).

Figure 18.7
Vibrational characteristics of ordinary (unpolarized) light (a), and polarized light (b). The direction of travel of the light is toward the reader.

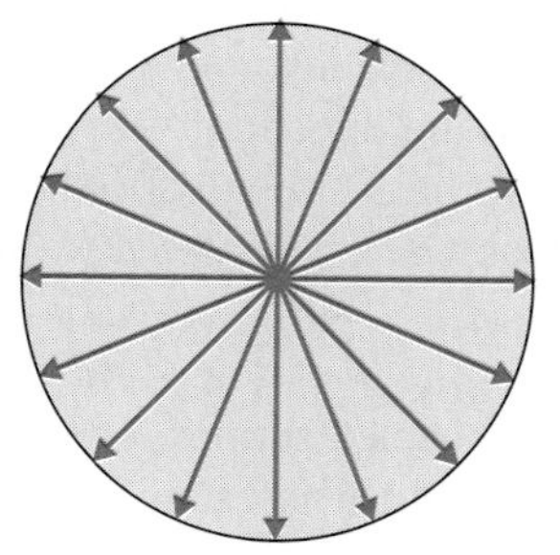

(a) Ordinary (unpolarized) light

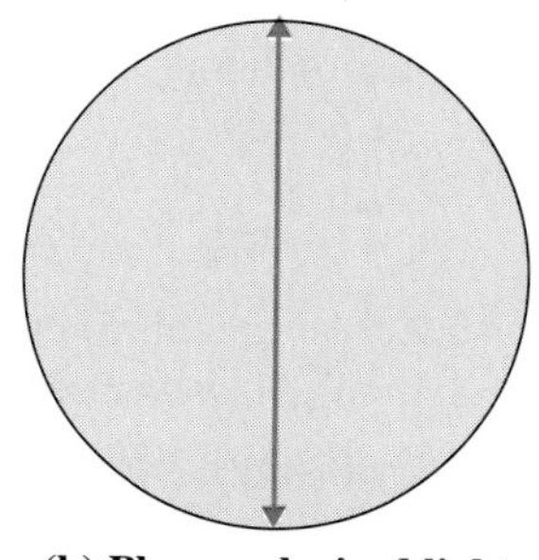

(b) Plane-polarized light

18.7 Properties of Enantiomers

Structural isomers differ in most chemical and physical properties. For example, structural isomers have different boiling points and melting points. Diastereomers also differ in most chemical and physical properties. They also have different boiling points and freezing points. In contrast, nearly all the properties of a pair of enantiomers are the same; for example, they have identical boiling points and freezing points. Enantiomers exhibit property differences in only two areas: their interaction with plane-polarized light and their interaction with other chiral substances.

• Interaction of Enantiomers with Plane-Polarized Light

All light moves through space with a wave motion. Ordinary light waves—that is, unpolarized light waves—vibrate in *all* planes at right angles to their direction of travel. Plane-polarized light waves, by contrast, vibrate in *only one* plane at right angles to their direction of travel. Figure 18.7 contrasts the vibrational behavior of ordinary light with that of plane-polarized light.

Ordinary light can be converted to plane-polarized light by passing it through a *polarizer,* an instrument with lenses or filters containing special types of crystals. When plane-polarized light is passed through a solution containing a *single* enantiomer, the plane of the polarized light is rotated counterclockwise (to the left) or clockwise (to the right),

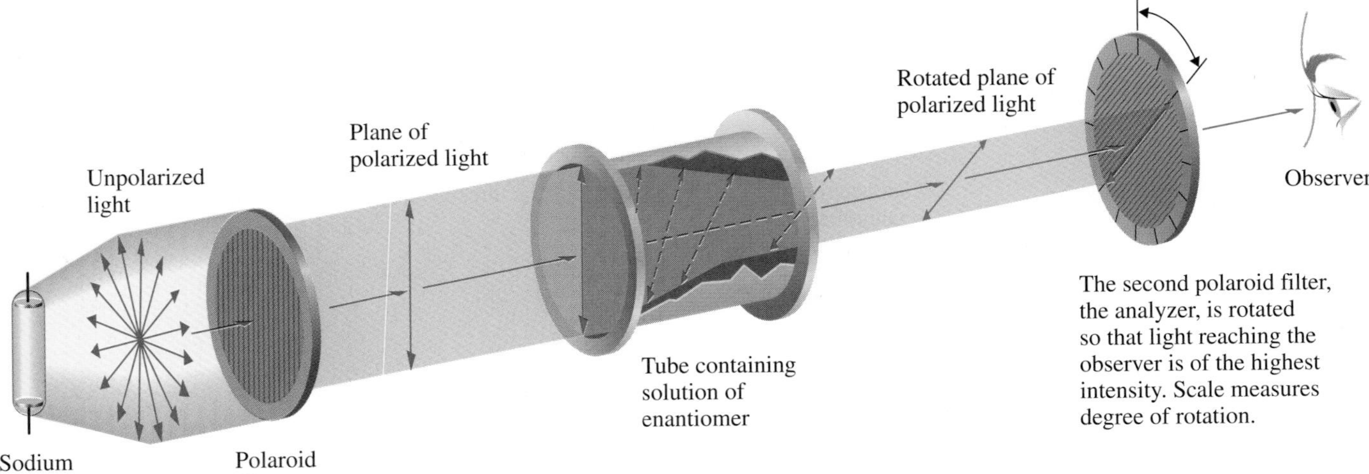

Figure 18.8
Schematic depiction of how a polarimeter works.

depending on the enantiomer. The extent of rotation depends on the concentration of the enantiomer as well as on its identity. Furthermore, the two enantiomers of a pair rotate the plane-polarized light the same number of degrees, but in opposite directions. If a 0.50 M solution of one enantiomer rotates the light 30° to the right, then a 0.50 M solution of the other enantiomer rotates the light 30° to the left.

Instruments used to measure the degree of rotation of plane-polarized light by enantiomeric compounds are called *polarimeters.* The schematic diagram in Figure 18.8 shows the basis for these instruments.

• Dextrorotatory and Levorotatory Compounds

• Achiral molecules are optically *inactive.* Chiral molecules are optically *active.*

• Because of their ability to rotate the plane of polarized light, enantiomers are sometimes referred to as *optical isomers.*

• In any pair of enantiomers, one, the (+)-enantiomer, always rotates the plane of polarized light to the right, and the other, the (−)-enantiomer, to the left.

• Both handedness and direction of rotation of plane-polarized light can be incorporated into the name of an enantiomer. For example, the notation D-(+)-mannose specifies that the right-handed isomer of the monosaccharide mannose rotates plane-polarized light in a clockwise direction (to the right).

Enantiomers are said to be optically active because of the way they interact with plane-polarized light. An **optically active compound** *is a compound that rotates the plane of polarized light.*

An enantiomer that rotates plane-polarized light to the *right* is said to be dextrorotatory (the Latin *dexter* means "right"). A **dextrorotatory compound** *is a chiral compound that rotates the plane of polarized light in a clockwise (to the right) direction.* An enantiomer that rotates plane-polarized light to the *left* is said to be levorotatory (the Latin *laevus* means "left"). A **levorotatory compound** *is a chiral compound that rotates the plane of polarized light in a counterclockwise (to the left) direction.* If one member of an enantiomeric pair is dextrorotatory, then the other member must be levorotatory.

A plus or minus sign inside parentheses is used to denote the direction of rotation of plane-polarized light by a chiral compound. The notation (+) means rotation to the right (clockwise), and (−) means rotation to the left (counterclockwise). Thus the dextrorotatory enantiomer of glucose is (+)-glucose.

The handedness of enantiomers (D or L, Section 18.6) and the direction of rotation of plane-polarized light by enantiomers [(+) or (−)] are not connected entities. There is no way of knowing which way an enantiomer will rotate light until it is examined with a polarimeter. Not all D enantiomers rotate plane-polarized light in one direction, nor do all L enantiomers rotate plane-polarized light in the opposite direction. Some D enantiomers are dextrorotatory; others are levorotatory.

The accompanying Chemistry at a Glance summarizes information about the various kinds of isomers we have encountered so far in the text.

• Interactions Between Chiral Compounds

A left-handed baseball player (chiral) and a right-handed baseball player (chiral) can use the same baseball bat (achiral) or wear the same baseball hat (achiral). However, left- and right-handed baseball players (chiral) cannot use the same baseball glove (chiral). This nonchemical example illustrates that the chirality of an object becomes important when the object interacts with another chiral object.

CHEMISTRY AT A GLANCE

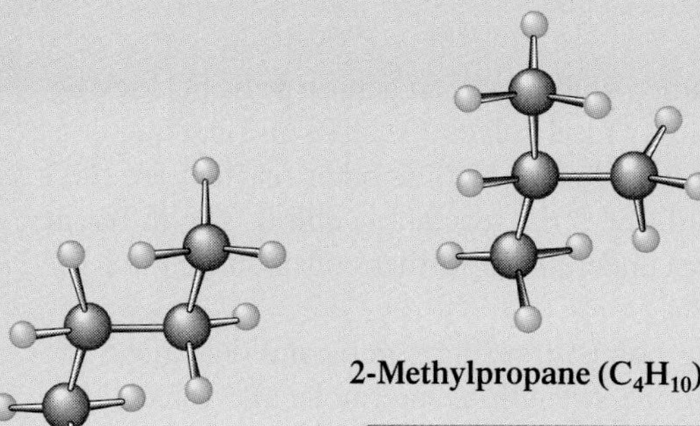
2-Methylpropane (C_4H_{10})

Butane (C_4H_{10})

ISOMERS

Different compounds with the same molecular formula.

STRUCTURAL ISOMERS (Sections 12.6, 13.5)

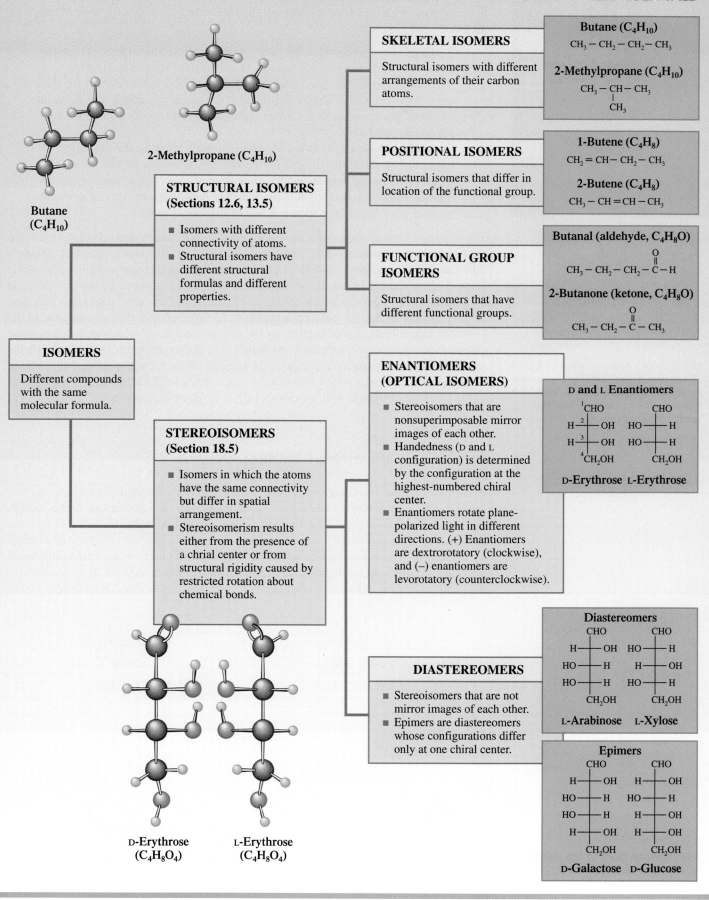

- Isomers with different connectivity of atoms.
- Structural isomers have different structural formulas and different properties.

SKELETAL ISOMERS

Structural isomers with different arrangements of their carbon atoms.

Butane (C_4H_{10})

$CH_3-CH_2-CH_2-CH_3$

2-Methylpropane (C_4H_{10})

$CH_3-CH(CH_3)-CH_3$

POSITIONAL ISOMERS

Structural isomers that differ in location of the functional group.

1-Butene (C_4H_8)

$CH_2=CH-CH_2-CH_3$

2-Butene (C_4H_8)

$CH_3-CH=CH-CH_3$

FUNCTIONAL GROUP ISOMERS

Structural isomers that have different functional groups.

Butanal (aldehyde, C_4H_8O)

$CH_3-CH_2-CH_2-C(=O)-H$

2-Butanone (ketone, C_4H_8O)

$CH_3-CH_2-C(=O)-CH_3$

STEREOISOMERS (Section 18.5)

- Isomers in which the atoms have the same connectivity but differ in spatial arrangement.
- Stereoisomerism results either from the presence of a chrial center or from structural rigidity caused by restricted rotation about chemical bonds.

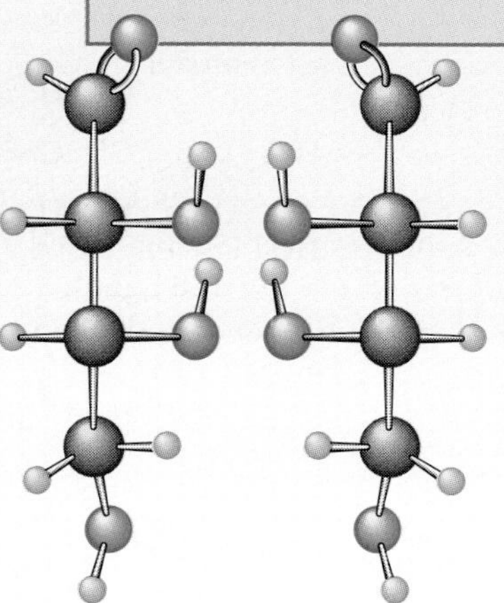
D-Erythrose ($C_4H_8O_4$) L-Erythrose ($C_4H_8O_4$)

ENANTIOMERS (OPTICAL ISOMERS)

- Stereoisomers that are nonsuperimposable mirror images of each other.
- Handedness (D and L configuration) is determined by the configuration at the highest-numbered chiral center.
- Enantiomers rotate plane-polarized light in different directions. (+) Enantiomers are dextrorotatory (clockwise), and (–) enantiomers are levorotatory (counterclockwise).

D and L Enantiomers

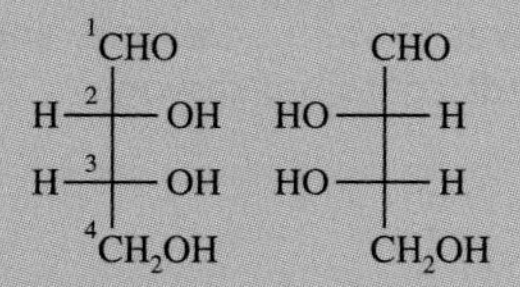

D-Erythrose	L-Erythrose
1 CHO	CHO
H–2–OH	HO–H
H–3–OH	HO–H
4 CH_2OH	CH_2OH

DIASTEREOMERS

- Stereoisomers that are not mirror images of each other.
- Epimers are diastereomers whose configurations differ only at one chiral center.

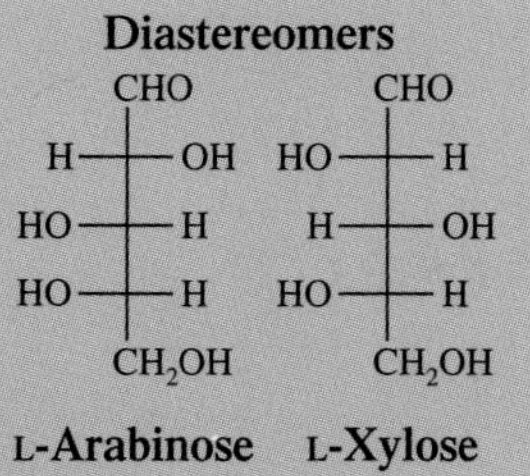

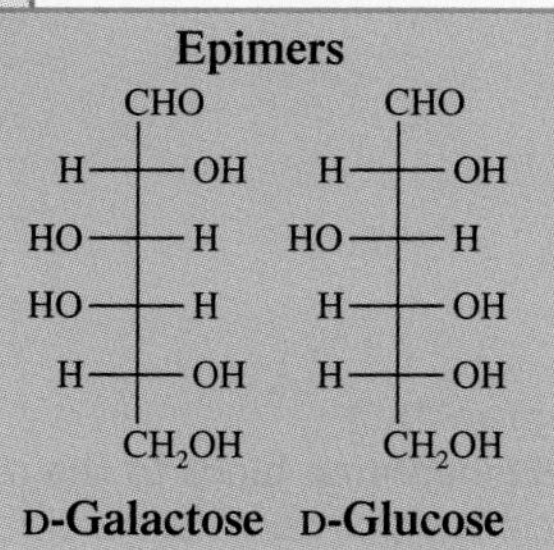

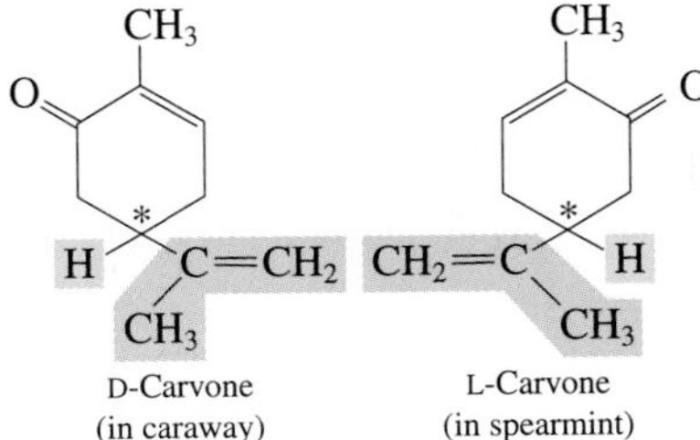

The distinctly different natural flavors of spearmint and caraway are caused by enantiomeric molecules. Spearmint leaves contain L-carvone, and caraway seeds contain D-carvone.

Applying this generalization to molecules, we find that the two members of an enantiomeric pair, because of their differing chirality, interact differently with other chiral molecules. We find that

1. A pair of enantiomers have the same solubility in an achiral solvent, such as ethanol, but differing solubilities in a chiral solvent, such as D-2-butanol.
2. The rate and extent of reaction of enantiomers with another reactant are the same if the reactant is achiral but differ if the reactant is chiral. The different reactions that different enantiomers undergo are further considered in the paragraph that follows.
3. Enantiomers have identical boiling points, freezing points, and densities, because such properties depend on the strength of intermolecular forces (Section 7.13), and intermolecular force strength does not depend on chirality. Intermolecular force strength is the same for both forms of a chiral molecule, because both forms have identical sets of functional groups.

The two enantiomeric forms of a chiral molecule often generate different responses from the human body. Sometimes both enantiomers are biologically active, each form giving a different response; sometimes both give the same response, but one isomer's response is many times greater than that of the other; and sometimes only one of the two forms is biologically active, the other form giving no response. For example, the body's response to the D isomer of the hormone epinephrine is 20 times greater than its response to the L isomer. Epinephrine exerts its effect by binding to specialized *receptors.* It binds to the receptor site by means of a three-point contact as is shown in Figure 18.9, D-epinephrine makes a perfect three-point contact with the receptor surface, but the biologically weaker L-epinephrine can make only a two-point contact. Because of the poorer fit, the binding of the L isomer is weaker, and less physiological response is observed.

18.8 Classification of Monosaccharides

Now that we have considered molecular chirality and its consequences (Sections 18.4 through 18.7), we return to the subject of carbohydrates by considering further details about monosaccharides, the simplest carbohydrates (Section 18.3).

Although there is no limit to the number of carbon atoms that can be present in a monosaccharide, only monosaccharides with three to seven carbon atoms are commonly found in nature. A three-carbon monosaccharide is called a *triose,* and those that contain four, five, and six carbon atoms are called *tetroses, pentoses,* and *hexoses,* respectively.

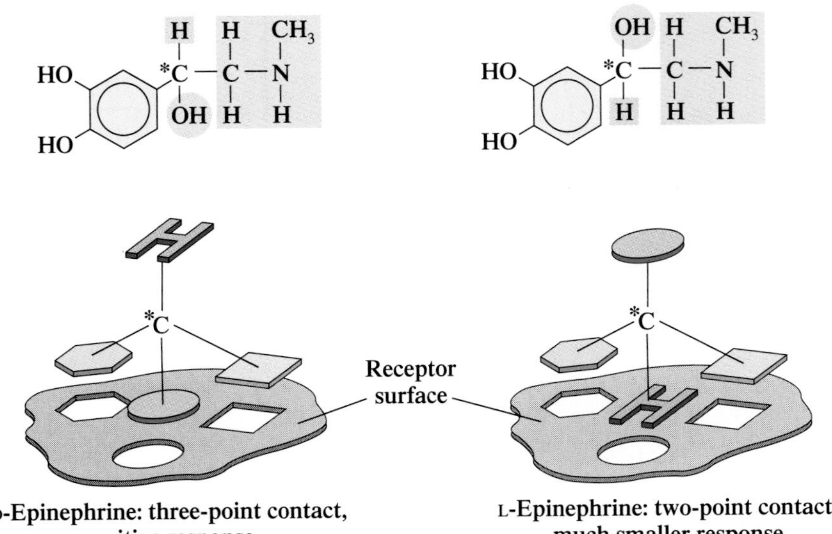

Figure 18.9
D-Epinephrine binds to the receptor at three points, whereas the biologically weaker L-epinephrine binds at only two sites.

Monosaccharides are classified as *aldoses* or *ketoses* on the basis of type of carbonyl group (Section 15.1) present. **Aldoses** *are monosaccharides that contain an aldehyde group.* **Ketoses** *are monosaccharides that contain a ketone group.*

Monosaccharides are often classified by both their number of carbon atoms and their functional group. A six-carbon monosaccharide with an aldehyde functional group is an *aldohexose;* a five-carbon monosaccharide with a ketone functional group is a *ketopentose.*

• The Latin word for "sugar" is *saccharum,* from whence comes the term *saccharide.*

Monosaccharides are also often called sugars. Hexoses are six-carbon sugars, pentoses five-carbon sugars, and so on. The word *sugar* is associated with "sweetness," and most (but not all) monosaccharides have a sweet taste. The designation *sugar* is also applied to disaccharides, many of which also have a sweet taste. Thus **sugar** *is a general designation for either a monosaccharide or a disaccharide.*

Example 18.5

Classifying Monosaccharides by Using Their Structural Characteristics

Classify each of the following monosaccharides according to both the number of carbon atoms and the type of carbonyl group present.

a. 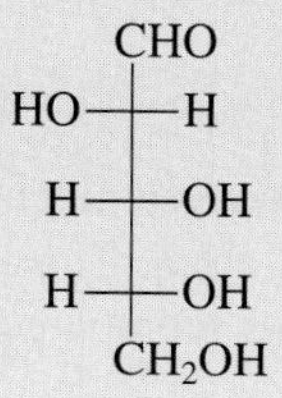
CHO
HO—H
H—OH
H—OH
CH_2OH

b. 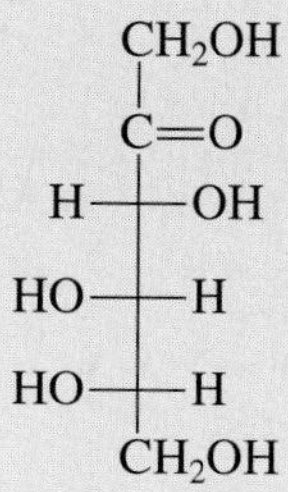
CH_2OH
C=O
H—OH
HO—H
HO—H
CH_2OH

c.
CHO
HO—H
H—OH
HO—H
HO—H
CH_2OH

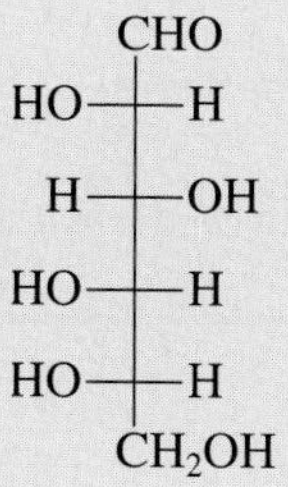

d. 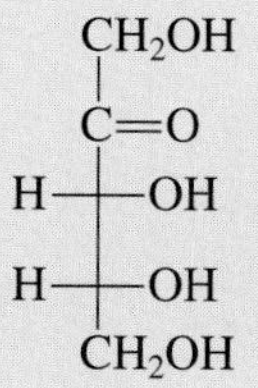
CH_2OH
C=O
H—OH
H—OH
CH_2OH

Solution

a. An aldehyde functional group is present as well as five carbon atoms. This monosaccharide is thus an *aldopentose.*

b. This monosaccharide contains a ketone group and six carbon atoms, so it is a *ketohexose.*

c. Six carbon atoms and an aldehyde group in a monosaccharide are characteristic of an *aldohexose.*

d. This monosaccharide is a *ketopentose.*

Practice Exercise 18.5

Classify each of the following monosaccharides according to both the number of carbon atoms and the type of carbonyl group present.

a. 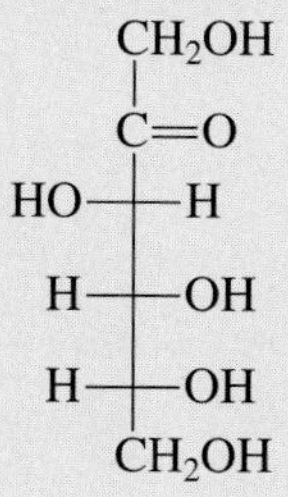
CH_2OH
C=O
HO—H
H—OH
H—OH
CH_2OH

b. 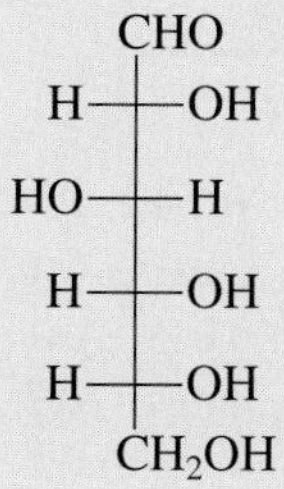
CHO
H—OH
HO—H
H—OH
H—OH
CH_2OH

(continued)

c.
CHO
H—OH
H—OH
CH_2OH

d.
CH_2OH
C=O
H—OH
HO—H
CH_2OH

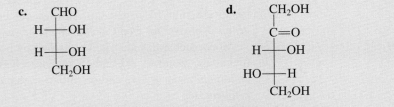

• *Answers:* **a.** ketohexose; **b.** aldohexose; **c.** aldotetrose; **d.** ketopentose

• Nearly all naturally occurring monosaccharides are D isomers. These D monosaccharides are important energy sources for the human body. L Monosaccharides, which can be produced in the laboratory, cannot be used by the body as energy sources. Body enzymes are specific for D isomers.

The D and L designations specify the configuration at the highest-numbered chiral center in a monosaccharide (Section 18.6). The configurations about other chiral centers are accounted for by assigning a different common name to each pair of D and L enantiomers. This naming system, for simple aldoses, is given in Figure 18.10. Only the D form of each aldose is shown in Figure 18.10. The L forms are mirror images of the molecules shown.

The simplest aldose and ketose are the trioses glyceraldehyde and dihydroxyacetone.

CHO
H—OH
CH_2OH
D-Glyceraldehyde

CH_2OH
C=O
CH_2OH
Dihydroxyacetone

A major difference between glyceraldehyde and dihydroxyacetone is that the latter does not possess a chiral carbon atom. Thus, D and L forms are not possible for dihydroxyacetone. This reduces by half (compared with aldoses) the number of stereoisomers pos-

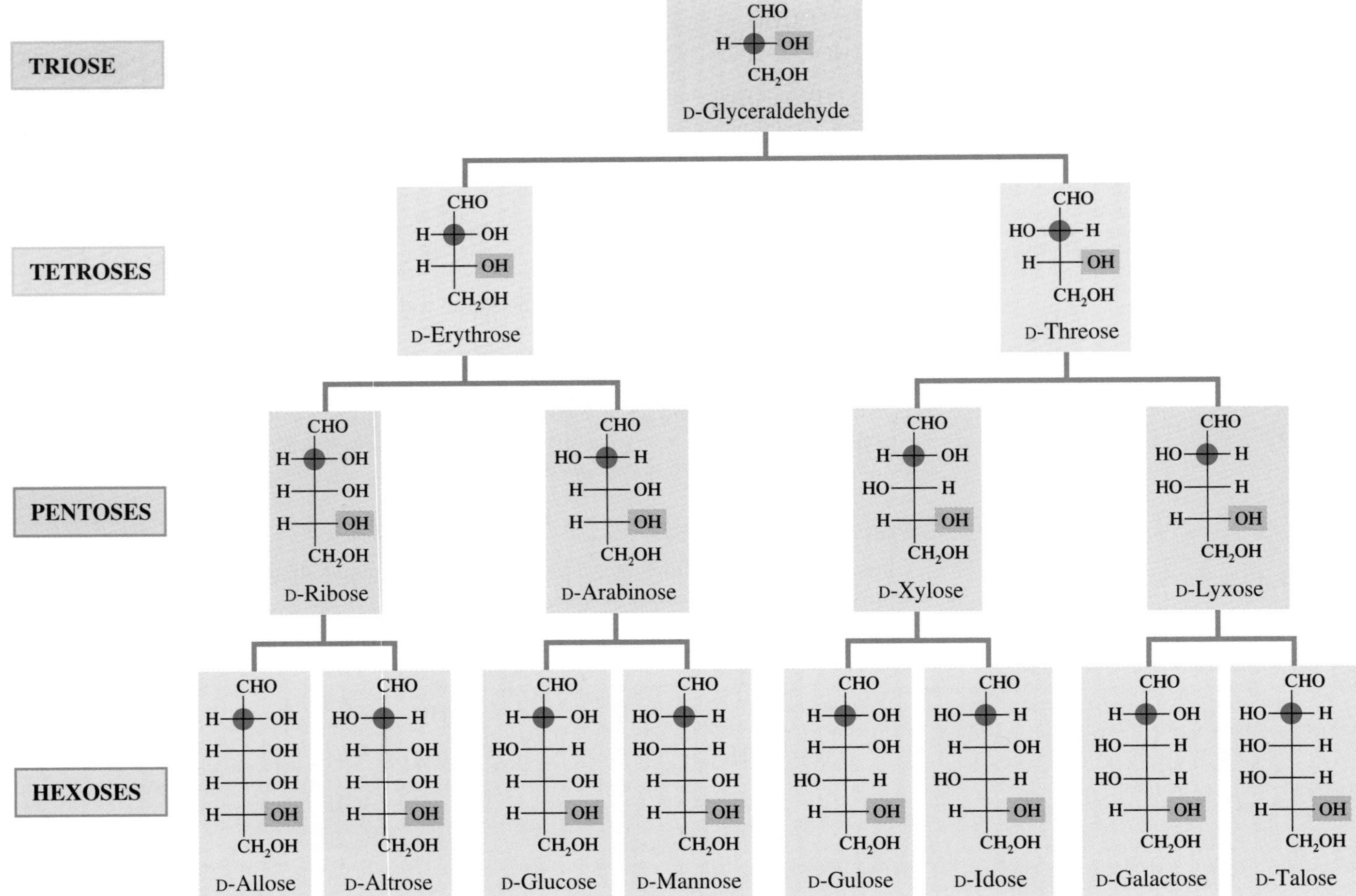

Figure 18.10
Fischer projections and common names for D aldoses containing three, four, five, and six carbon atoms. The new chiral-center carbon atom added in going from triose to tetrose to pentose to hexose is marked in color. This new chiral center can have the hydroxyl group at the right or left in the Fischer projection, which doubles the number of stereoisomers. The hydroxyl group that specifies the D configuration is highlighted in red.

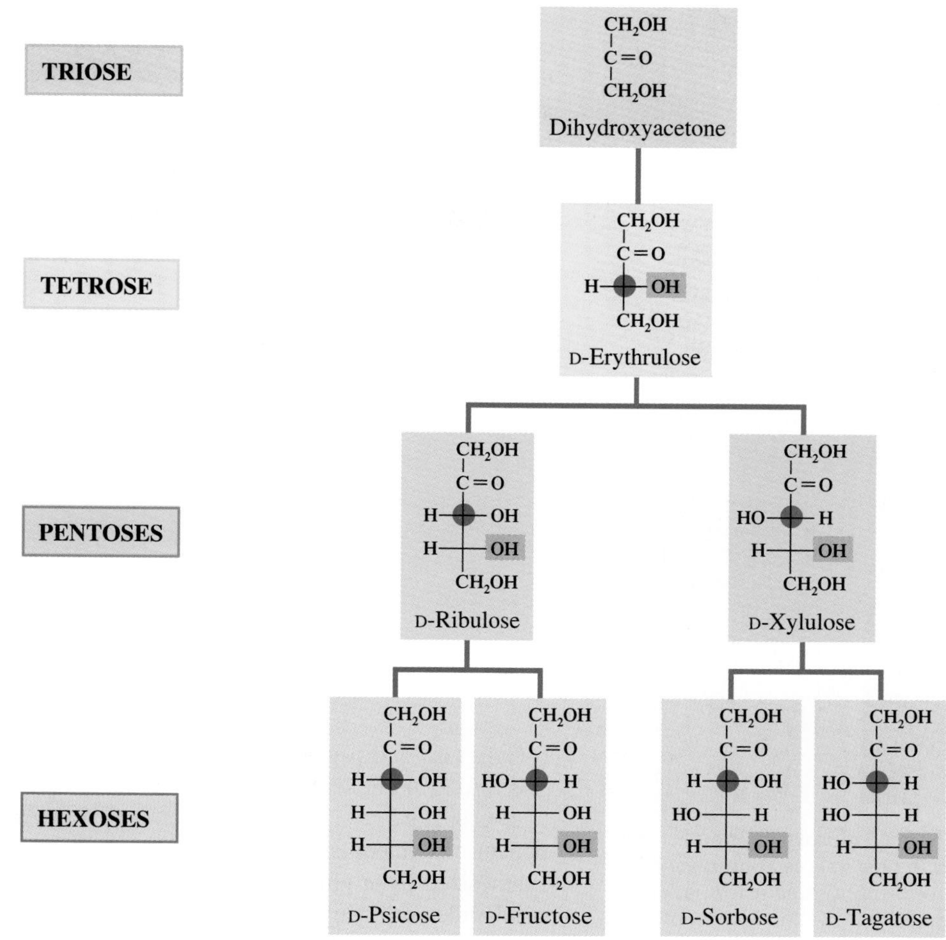

Figure 18.11 Fischer projections and common names for D ketoses containing three, four, five, and six carbon atoms. The new chiral-center carbon atom added in going from triose to tetrose to pentose to hexose is marked in color. This new chiral center can have the hydroxyl group at the right or left in the Fischer projection, which doubles the number of stereoisomers. The hydroxyl group that specifies the D configuration is highlighted in red.

• All monosaccharides have names that end in *-ose* except the trioses glyceraldehyde and dihydroxyacetone.

sible for ketotetroses, ketopentoses, and ketohexoses. An aldohexose has four chiral carbon atoms, but a ketohexose has only three. Figure 18.11 gives the projection formulas and common names for the D forms of ketoses containing three, four, five, and six carbon atoms.

18.9 Biologically Important Monosaccharides

• You should memorize the structures of the four monosaccharides considered in this section.

Of the many monosaccharides, the most important in the human body are the D-forms of glucose, galactose, fructose, and ribose. Glucose and galactose are aldohexoses, fructose is a ketohexose, and ribose is an aldopentose. All four of these monosaccharides are water-soluble, white, crystalline solids.

• D-Glucose

• D-Glucose tastes sweet, is nutritious, and is an important component of the human diet. L-Glucose, on the other hand, is tasteless, and the body cannot use it.

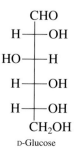

Of all monosaccharides, D-glucose is the most abundant in nature and the most important from a nutritional standpoint. Its Fischer projection is

CHO
H—OH
HO—H
H—OH
H—OH
CH_2OH

D-Glucose

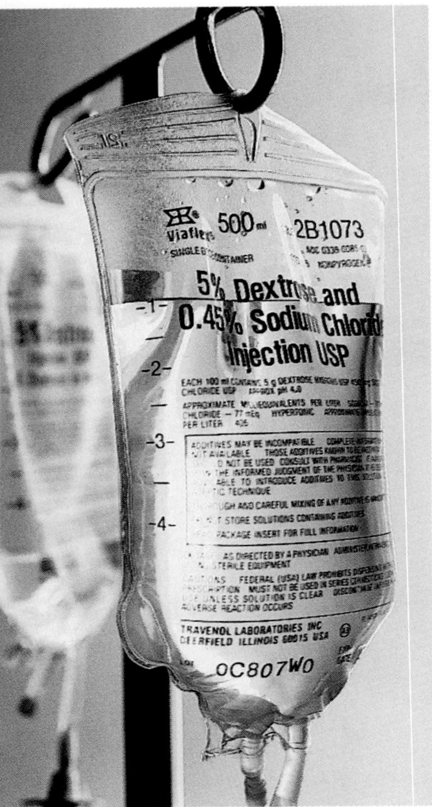

A 5% (m/v) glucose solution is often used in hospitals as an intravenous source of nourishment for patients who cannot take food by mouth. The body can use it as an energy source without digesting it.

Ripe fruits, particularly ripe grapes (20%–30% glucose by mass), are a good source of glucose, which is often referred to as *grape sugar.* Two other names for D-glucose are dextrose and blood sugar. The name *dextrose* draws attention to the fact that the optically active D-glucose, in aqueous solution, rotates plane-polarized light to the right. The term *blood sugar* draws attention to the fact that blood contains dissolved glucose. The concentration of glucose in human blood is fairly constant; it is in the range of 70–100 mg per 100 mL of blood. Cells use this glucose as a primary energy source.

• D-Galactose

A comparison of the Fischer projections for D-galactose and D-glucose shows that these two compounds differ only in the configuration of the —OH group and —H group on carbon-4.

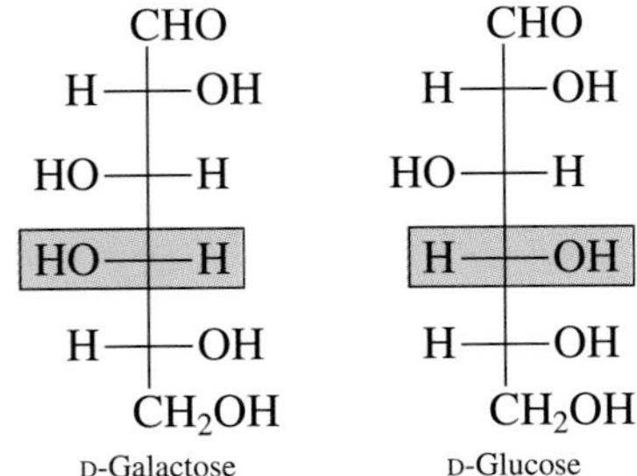

D-Galactose and D-glucose are epimers (diastereomers that differ only in the configuration at one chiral center; Section 18.6).

D-Galactose is seldom encountered as a free monosaccharide. It is, however, a component of numerous important biochemical substances. In the human body, galactose is synthesized from glucose in the mammary glands for use in lactose (milk sugar), a disaccharide consisting of a glucose unit and a galactose unit (Section 18.13). D-Galactose is sometimes called *brain sugar* because it is a component of glycoproteins (protein–carbohydrate compounds; Section 18.15) found in brain and nerve tissue. D-Galactose is also present in the chemical markers that distinguish various types of blood—A, B, AB, and O (see Chemical Connections 18.1).

• D-Fructose

D-Fructose is the most important ketohexose. It is also known as *levulose* and *fruit sugar.* Aqueous solutions of naturally occurring D-fructose rotate plane-polarized light to the left; hence the name *levulose.* The sweetest-tasting of all sugars, D-fructose is found in many fruits and is present in honey in equal amounts with glucose. It is sometimes used as a dietary sugar, not because it has fewer calories per gram than other sugars but because less is needed for the same amount of sweetness.

From the third to the sixth carbon, the structure of D-fructose is identical to that of D-glucose. Differences at carbons 1 and 2 are related to the presence of a ketone group in fructose and an aldehyde group in glucose.

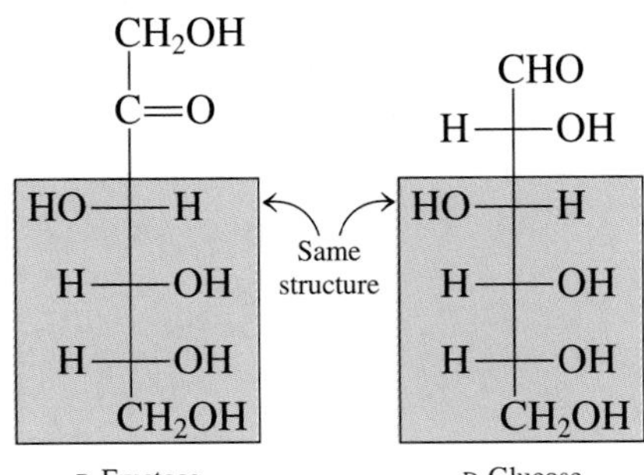

• D-Ribose

The three monosaccharides previously discussed in this section have all been hexoses. D-Ribose is a pentose. If carbon-3 and its accompanying —H and —OH groups were eliminated from the structure of D-glucose, the remaining structure would be that of D-ribose.

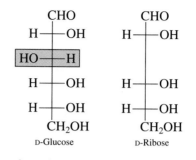

D-Glucose	D-Ribose
CHO	CHO
H—OH	H—OH
HO—H	
H—OH	H—OH
H—OH	H—OH
CH_2OH	CH_2OH

D-Ribose is a component of a variety of complex molecules, including ribonucleic acids (RNAs) and energy-rich compounds such as adenosine triphosphate (ATP). The compound 2-deoxy-D-ribose is also important in nucleic acid chemistry. This monosaccharide is a component of DNA molecules. The prefix *deoxy-* means "minus an oxygen"; the structures of ribose and 2-deoxyribose differ in that the latter compound lacks an oxygen atom at carbon-2.

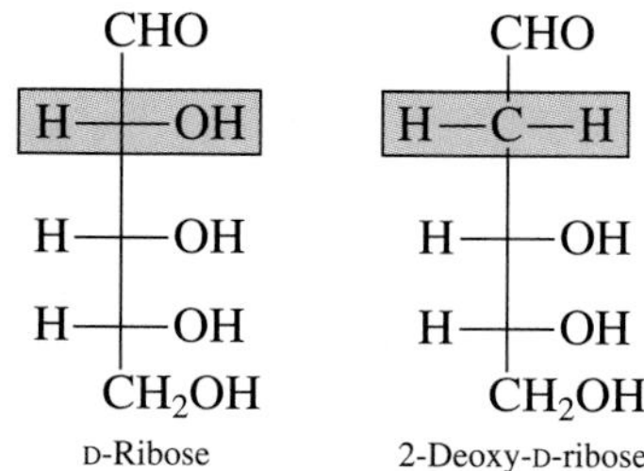

D-Ribose 2-Deoxy-D-ribose

18.10 Cyclic Forms of Monosaccharides

So far in this chapter, the structures of monosaccharides have been depicted as open-chain polyhydroxy aldehydes or ketones. However, experimental evidence indicates that for monosaccharides containing five or more carbon atoms, such open-chain structures are actually in equilibrium with two cyclic structures, and the cyclic structures are the dominant forms at equilibrium.

• Recall from Section 15.9 that hemiacetals and hemiketals have both an —OH group and an —OR group attached to the same carbon atom. In the cyclic hemiacetals and hemiketals that monosaccharides form, it is the carbonyl carbon atom that bears the —OH and —OR groups.

The cyclic forms of monosaccharides result from the ability of their carbonyl group to react intramolecularly with a hydroxyl group. The result is a cyclic hemiacetal or cyclic hemiketal (Section 15.9). Such an intramolecular cyclization reaction for D-glucose is shown in Figure 18.12.

In Figure 18.12, structure 2 is a rearrangement of the projection formula for D-glucose in which the carbon atoms have locations similar to those found for carbon atoms in a six-membered ring. All hydroxyl groups drawn to the right in the original projection formula appear below the ring. Those to the left in the projection formula appear above the ring.

Structure 3 in Figure 18.12 is obtained by rotating the groups attached to carbon-5 in a counterclockwise direction so that they are in the positions where it is easiest to visualize intramolecular hemiacetal formation. The intramolecular reaction occurs between the hydroxyl group on carbon-5 and the carbonyl group (carbon-1). The —OH group adds across the carbon–oxygen double bond, producing a heterocyclic ring that contains five carbon atoms and one oxygen atom.

Addition across the carbon–oxygen double bond with its accompanying ring formation produces a chiral center at carbon-1, so two stereoisomers are possible. See Figure

Figure 18.12
The cyclic hemiacetal forms of D-glucose result from the intramolecular reaction between the carbonyl group and the hydroxyl group on carbon-5.

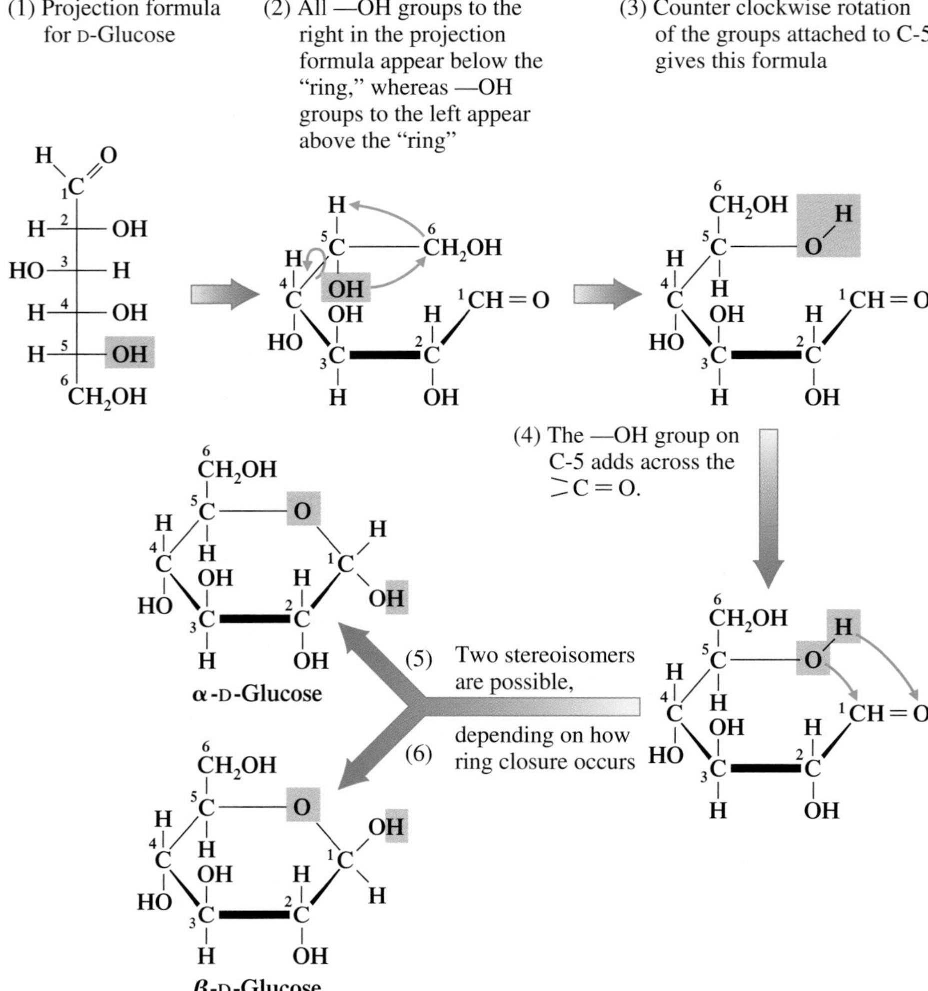

(1) Projection formula for D-Glucose
(2) All —OH groups to the right in the projection formula appear below the "ring," whereas —OH groups to the left appear above the "ring"
(3) Counter clockwise rotation of the groups attached to C-5 gives this formula
(4) The —OH group on C-5 adds across the C=O.
(5) Two stereoisomers are possible,
(6) depending on how ring closure occurs
α-D-Glucose
β-D-Glucose

• Cyclization of glucose (hemiacetal formation) creates a new chiral center at carbon-1, and the presence of this new chiral center produces two stereoisomers, called α and β isomers.

18.12, structures 4–6. These two forms differ in the orientation of the —OH group on the hemiacetal carbon atom (carbon-1). In α-D-glucose, the —OH group is on the opposite side of the ring from the CH_2OH group attached to carbon-5. In β-D-glucose, the CH_2OH group on carbon-5 and the —OH group on carbon-1 are on the same side of the ring.

In an aqueous solution of D-glucose, a dynamic equilibrium exists among the α, β, and open-chain forms, and there is continual interconversion among them. For example, a freshly mixed solution of pure α-D-glucose slowly converts to a mixture of both α- and β-D-glucose by an opening and a closing of the cyclic structure. When equilibrium is established, 63% of the molecules are β-D-glucose, 37% are α-D-glucose, and less than 0.01% are in the open-chain form (see Figure 18.13).

Intramolecular cyclic hemiacetal formation and the equilibrium between forms associated with it is not restricted to glucose. All aldoses with five or more carbon atoms establish similar equilibria, but with different percentages of the alpha, beta, and open-chain forms. Fructose and other ketoses with a sufficient number of carbon atoms also cyclize; here, cyclic hemi*ketal* formation occurs.

Galactose, like glucose, forms a six-membered ring, but both D-fructose and D-ribose form a five-membered ring.

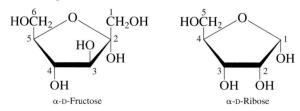

α-D-Fructose   α-D-Ribose

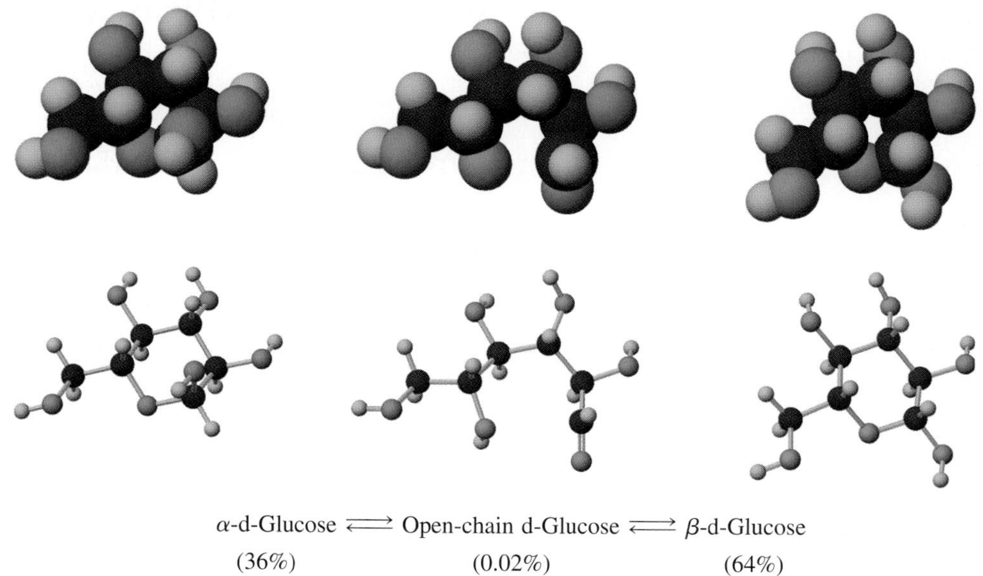

Figure 18.13
In an aqueous solution of D-glucose, a dynamic equilibrium exists among the alpha, beta, and open-chain forms, the beta form being favored.

D-Fructose cyclization involves carbon-2 (the keto group) and carbon-5, which results in two CH_2OH groups being outside the ring (carbons 1 and 6). D-Ribose cyclization involves carbon-1 (the aldehyde group) and carbon-4.

18.11 Haworth Projection Formulas

The structural representations of the cyclic forms of monosaccharides found in the previous section are examples of Haworth projection formulas. A **Haworth projection** *is a two-dimensional notation that specifies the three-dimensional structure of a cyclic form of a carbohydrate.*

In a Haworth projection, the hemiacetal ring system is viewed "edge on" with the oxygen ring atom at the upper right (six-membered ring) or at the top (five-membered ring).

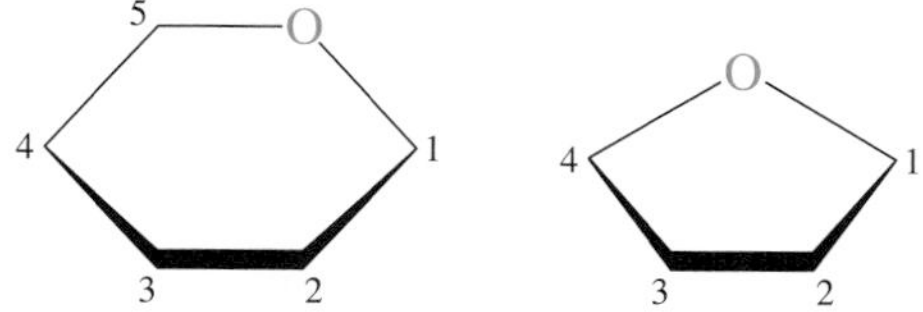

Walter Norman Haworth (1883–1950), the developer of Haworth projection formulas, was a British carbohydrate chemist. He helped determine the structures of the cyclic forms of glucose, was the first to synthesize vitamin C, and was a corecipient of the 1937 Nobel Prize in chemistry.

The D or L form of a monosaccharide is determined by the position of the terminal CH_2OH group on the highest-numbered ring carbon atom. In the D form, this group is positioned above the ring. In the L form, which is not usually encountered in biological systems, the terminal CH_2OH group is positioned below the ring.

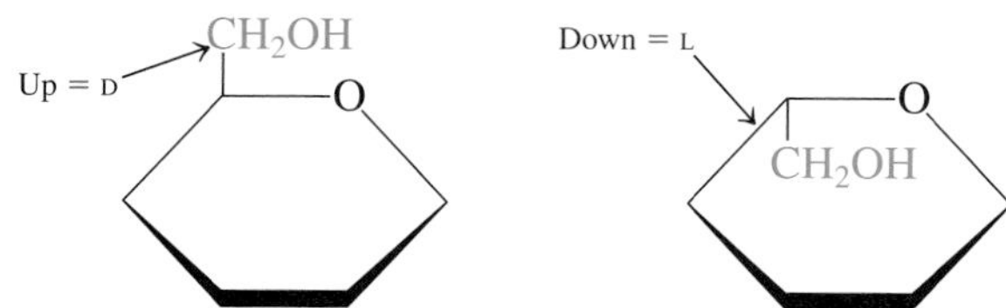

α or β configuration is determined by the position of the —OH group on carbon-1 relative to the CH_2OH group that determines D or L series. In a β configuration, both of these groups point in the same direction; in an α configuration, the two groups point in opposite directions.

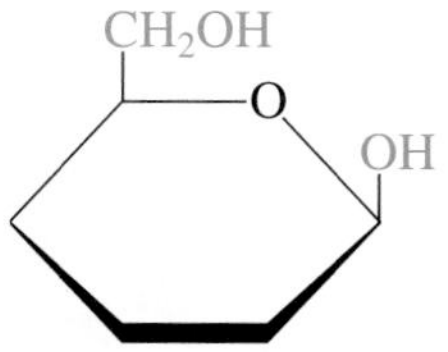

β-D-Monosaccharide 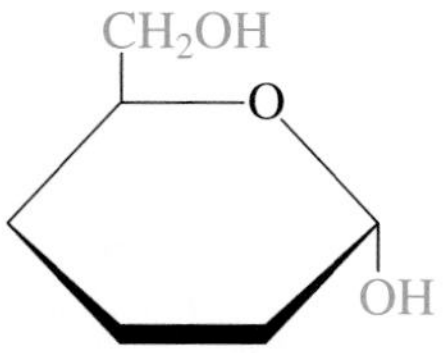 α-D-Monosaccharide 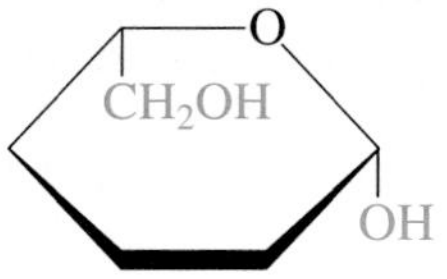 β-L-Monosaccharide

Where α or β configuration does not matter, the —OH group on carbon-1 is placed in a horizontal position, and a wavy line is used as the bond that connects it to the ring.

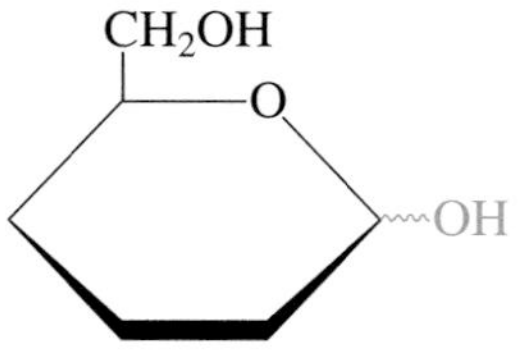

The specific identity of a monosaccharide is determined by the positioning of the other —OH groups in the Haworth projection. Any —OH group at a chiral center that is to the right in a Fischer projection formula points down in the Haworth projection. Any group to the left in a Fischer projection points up in the Haworth projection. The following is a matchup between Haworth projection and Fischer projection.

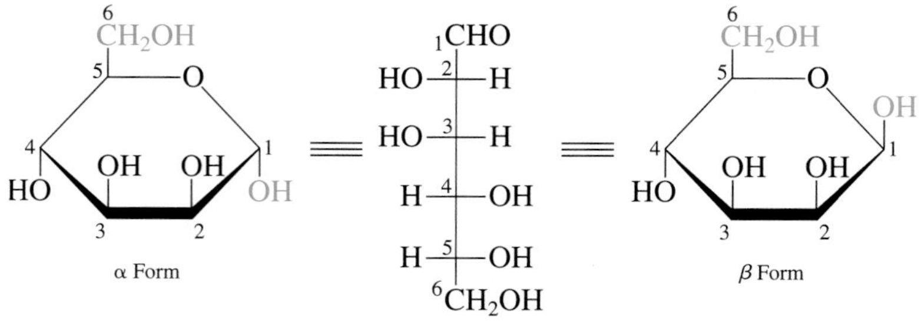

α Form β Form

Comparison of this Fischer projection with those given in Figure 18.10 reveals that the monosaccharide is D-mannose.

18.12 Reactions of Monosaccharides

Five important reactions of monosaccharides are oxidation, reduction, glycoside formation, phosphate ester formation, and amino sugar formation. In considering these reactions, we will use glucose as the monosaccharide reactant. Remember, however, that other aldoses as well as ketoses undergo similar reactions.

• Oxidation

Monosaccharide oxidation can yield three different types of oxidation products. The oxidizing agent used determines the product.

Weak oxidizing agents, such as Tollens, Fehling's, and Benedict's solutions (Section 15.8), oxidize the carbonyl group end of a monosaccharide to give an *-onic acid.* Oxidation of the aldehyde end of glucose produces gluconic acid, and oxidation of the aldehyde end of galactose produces galactonic acid. The structures involved in the glucose reaction are

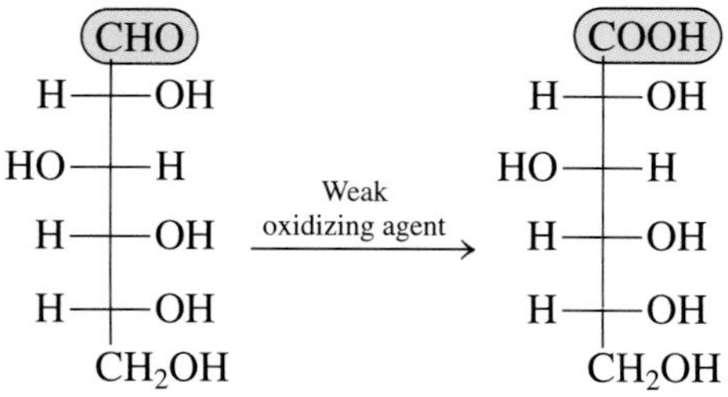

Weak oxidizing agent →

D-Glucose D-Gluconic acid

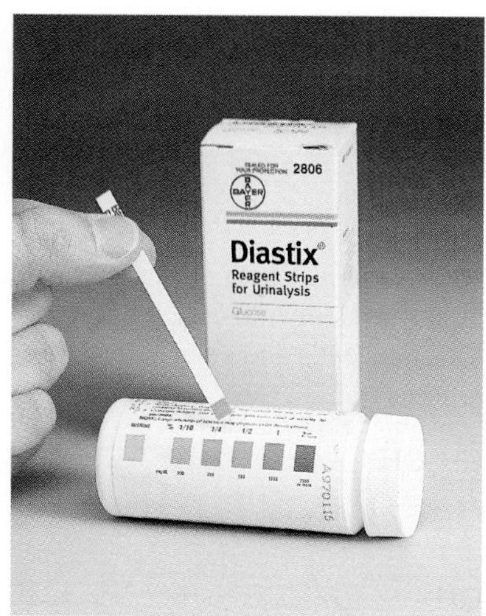

Figure 18.14
The glucose content of urine can be determined by dipping a plastic strip treated with oxidizing agents into the urine sample and comparing the color change of the strip to a color chart that indicates glucose concentration.

Because monosaccharides act as reducing agents in such reactions, they are called *reducing sugars.* With Tollens solution, glucose reduces Ag^+ ion to Ag, and with Benedict's and Fehling's solutions, glucose reduces Cu^{2+} ion to Cu^+ ion (see Section 15.8). A **reducing sugar** *is a carbohydrate that gives a positive test with Tollens, Fehling's, and Benedict's solutions.* All monosaccharides are reducing sugars.

Tollens, Fehling's, and Benedict's solutions can be used to test for glucose in urine, a symptom of diabetes. For example, using Benedict's solution, we observe that if no glucose is present in the urine (a normal condition), the Benedict's solution remains blue. The presence of glucose is indicated by the formation of a red precipitate. Testing for the presence of glucose in urine is such a common laboratory procedure that much effort has been put into the development of easy-to-use test methods (Figure 18.14).

Strong oxidizing agents can oxidize both ends of a monosaccharide at the same time (the carbonyl group and the terminal primary alcohol group) to produce a dicarboxylic acid. Such polyhydroxy dicarboxylic acids are known as *-aric acids.* For glucose, such an oxidation produces glucaric acid.

CHO
H—OH
HO—H
H—OH
H—OH
CH_2OH
D-Glucose

Strong oxidizing agent →

COOH
H—OH
HO—H
H—OH
H—OH
COOH
D-Glucaric acid

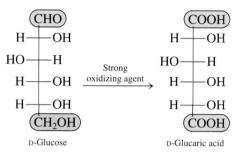

Although it is difficult to do in the laboratory, in biological systems enzymes can oxidize the primary alcohol end of an aldose such as glucose, without oxidation of the aldehyde group, to produce a *-uronic acid.* For glucose, such an oxidation produces D-glucuronic acid.

CHO
H—OH
HO—H
H—OH
H—OH
CH_2OH
D-Glucose

Enzymes →

CHO
H—OH
HO—H
H—OH
H—OH
COOH
D-Glucuronic acid

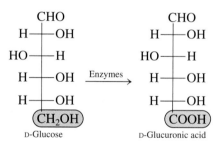

• Reduction

The carbonyl group present in a monosaccharide (either an aldose or a ketose) can be reduced to a hydroxyl group, using hydrogen as the reducing agent. For aldoses and ketoses, the product of the reduction is the corresponding polyhydroxy alcohol, which is sometimes called a *sugar alcohol.* For example, the reduction of D-glucose gives D-glucitol.

CHO
H—C—OH
HO—C—H
H—C—OH
H—C—OH
CH_2OH
D-Glucose

H_2, catalyst →

CH_2OH
H—C—OH
HO—C—H
H—C—OH
H—C—OH
CH_2OH
D-Glucitol

D-Glucitol is also known by the common name D-sorbitol. Hexahydroxy alcohols such as D-sorbitol have properties similar to those of the trihydroxy alcohol *glycerol* (Section

• D-Sorbitol accumulation in the eye is a major factor in the formation of cataracts due to diabetes.

14.4). These alcohols are used as moisturizing agents in foods and cosmetics because of their affinity for water. D-Sorbitol is also used as a sweetening agent in chewing gum; bacteria that cause tooth decay cannot use polyalcohols as food sources, as they can glucose and many other monosaccharides.

• Glycoside Formation

• Remember, from Section 15.9, that acetals and ketals have two —OR groups attached to the same carbon atom.

In Section 15.9 we learned that hemiacetals and hemiketals can react with alcohols in acid solution to produce acetals and ketals. Because the cyclic forms of monosaccharides are hemiacetals and hemiketals, they react with alcohols to form acetals and ketals, as is illustrated for the reaction of β-D-glucose with methyl alcohol.

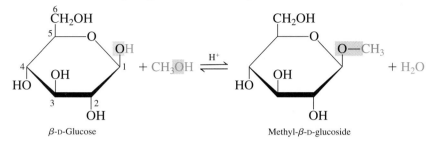

$$\beta\text{-D-Glucose} + CH_3OH \overset{H^+}{\rightleftharpoons} \text{Methyl-}\beta\text{-D-glucoside} + H_2O$$

β-D-Glucose

Methyl-β-D-glucoside

Chemical CONNECTIONS

18.1 Blood Types and Monosaccharides

Human blood is classified into four types: A, B, AB, and O. If a blood transfusion is necessary and the patient's own blood is not available, the donor's blood must be matched to that of the patient. Blood of one type cannot be given to a recipient with blood of another type unless the two types are compatible. A transfusion of the wrong blood type can cause the blood cells to form clumps, a potentially fatal reaction. The table opposite shows compatibility relationships. People with type O blood are universal donors, and those with type AB blood are universal recipients.

Human Blood Group Compatibilities

	Recipient Blood Type			
Donor blood type	**A**	**B**	**AB**	**O**
A	+	−	+	−
B	−	+	+	−
AB	−	−	+	−
O	+	+	+	+

+ = compatible; − = incompatible

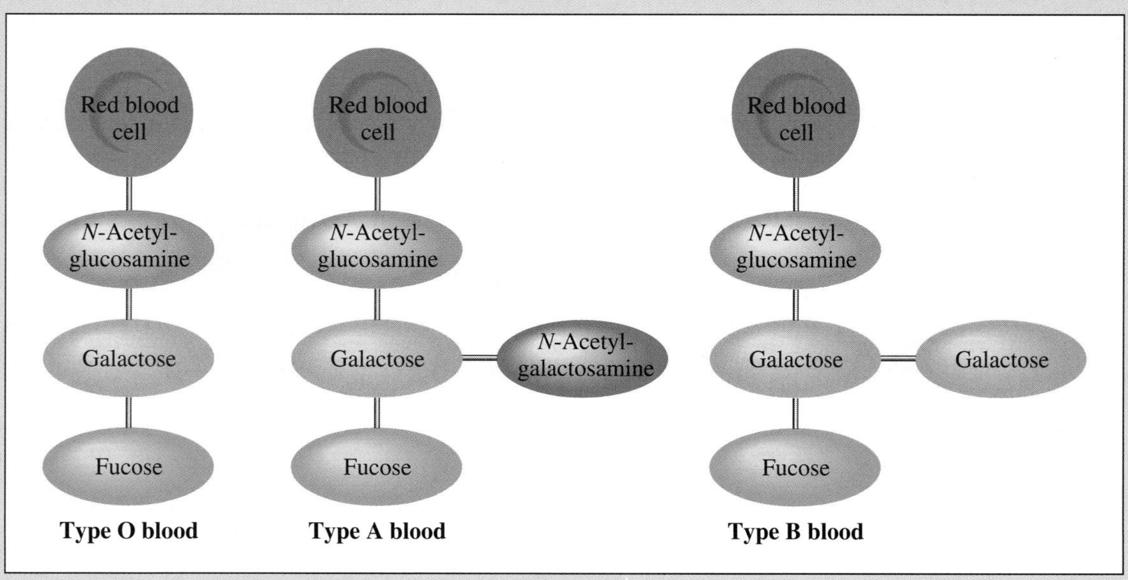

The general name for monosaccharide acetals and ketals is glycoside. A **glycoside** *is an acetal or a ketal formed from a cyclic monosaccharide.* More specifically, a glycoside produced from glucose is a glucoside, from galactose a galactoside, and so on. Glycosides, like the hemiacetals and hemiketals from which they are formed, can exist in both α and β forms. Glycosides are named by listing the alkyl or aryl group attached to the oxygen, followed by the name of the monosaccharide involved, with the suffix *-ide* appended to it.

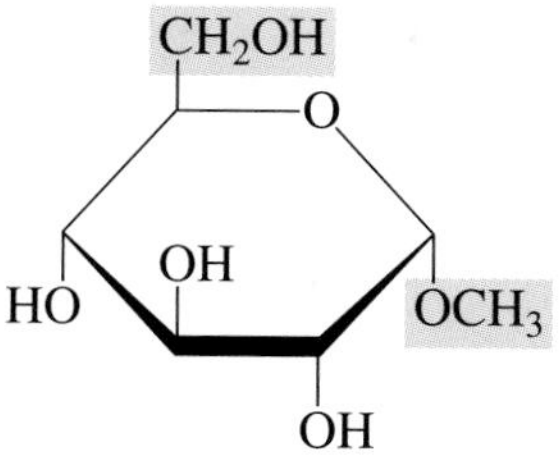

Methyl-α-D-glucoside

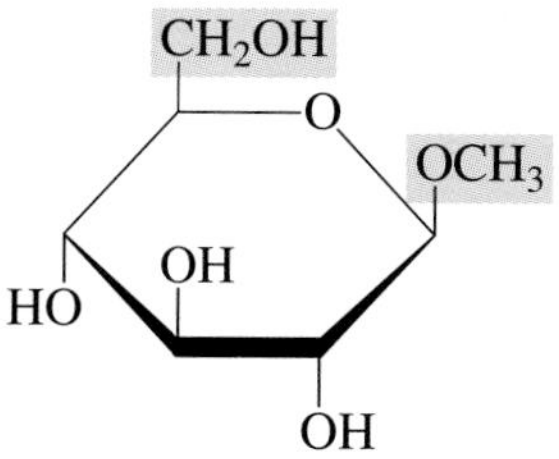

Methyl-β-D-glucoside

• Phosphate Ester Formation

The hydroxyl groups of a monosaccharide can react with inorganic oxyacids to form inorganic esters (Section 16.16). Phosphate esters, formed from phosphoric acid and various

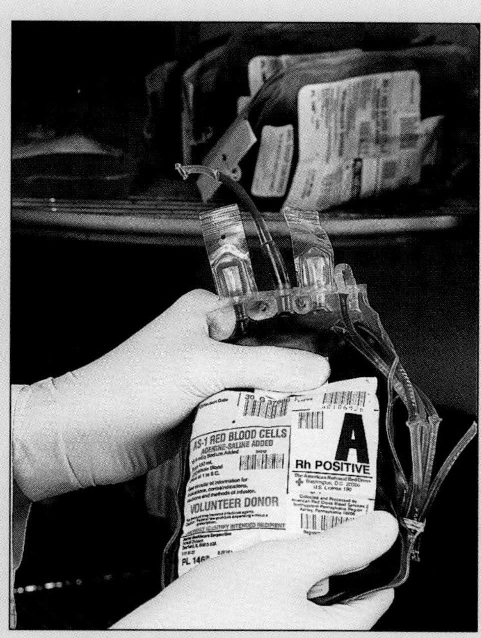

A unit of blood obtained from a blood bank.

In the United States, sampling studies show that 41% of the population has type A blood, 10% type B, 4% type AB, and 45% type O.

The biochemical basis for the various blood types involves monosaccharides. The plasma membranes of red blood cells carry biochemical markers made up of monosaccharides. Four monosaccharides are involved in the "marking system." One is the simple monosaccharide D-galactose, and the other three are monosaccharide derivatives. Two of these are *N*-acetyl amino derivatives (Section 18.12), those of D-glucose and D-galactose. The third is L-fucose (6-deoxy-L-galactose), an L-galactose derivative in which the oxygen atom at carbon- 6 has been removed (converting the $—CH_2OH$ group to a $—CH_3$ group). The L configuration of this derivative is unusual in that L-monosaccharides are seldom found in the human body. The arrangement of these monosaccharides in the biochemical marker determines blood type.

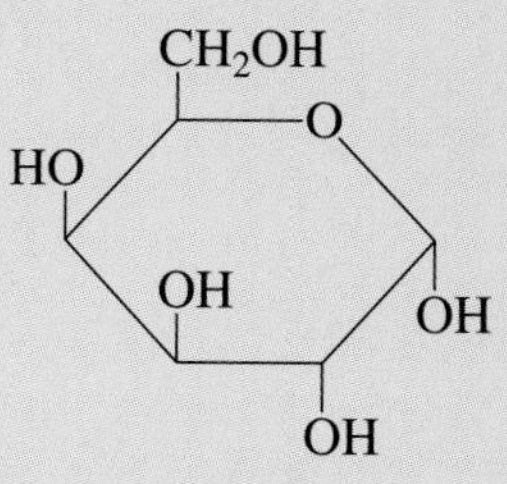

α-D-galactose

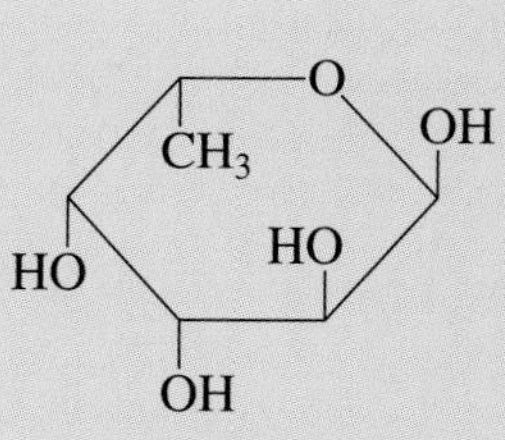

α-L-fucose (α-6-deoxy-L-galactose)

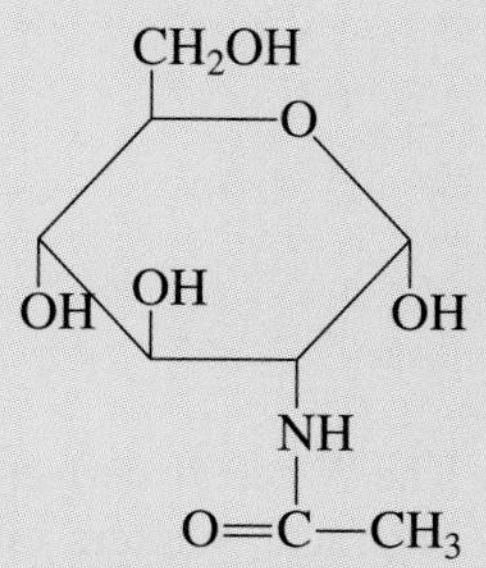

α-*N*-acetyl-D-glucosamine

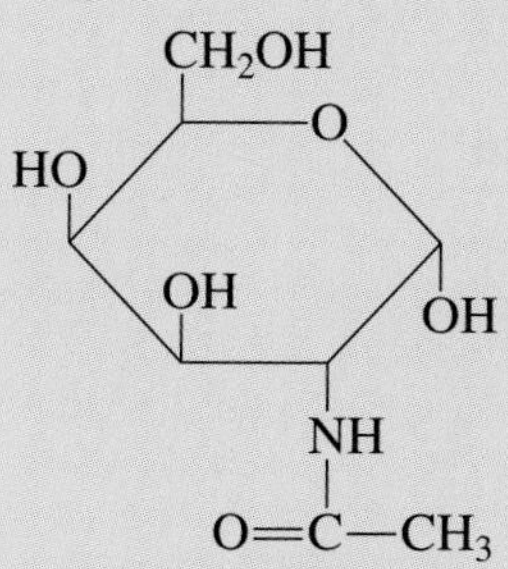

α-*N*-acetyl-D-galactosamine

Note that all of the biochemical markers have a common structural portion that involves three monosaccharide units. Type A markers differ from type O markers in that an *N*-acetyl galactosamine unit is also present. In type B markers, a second galactose unit is present. Type AB blood contains both type A and type B markers.

monosaccharides, are commonly encountered in biological systems. For example, specific enzymes in the human body catalyze the esterification of the carbonyl group (carbon-1) and the primary alcohol group (carbon-6) in glucose to produce the compounds glucose 1-phosphate and glucose 6-phosphate, respectively.

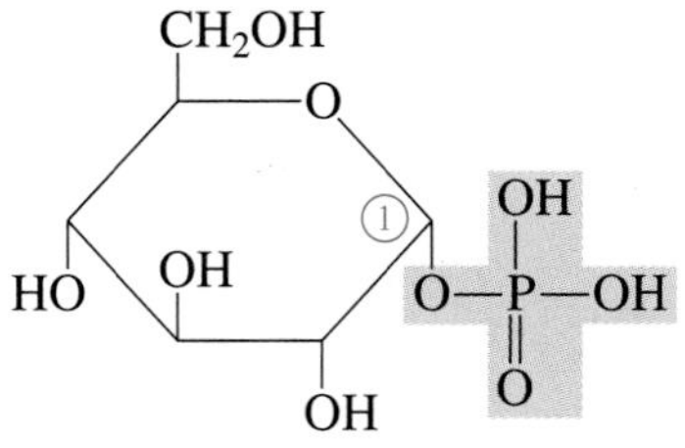

α-D-Glucose 1-phosphate

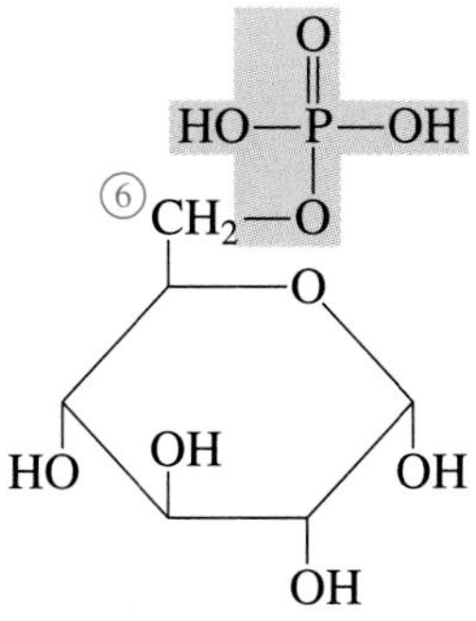

α-D-Glucose 6-phosphate

These phosphate esters of glucose are stable in aqueous solution and play important roles in the metabolism of carbohydrates.

• Amino Sugar Formation

Amino sugars of glucose, mannose, and galactose are common in nature. Such sugars are produced by replacing the hydroxyl group on carbon-2 on the monosaccharide with an amino group. Amino sugars and their *N*-acetyl derivatives are important building blocks of polysaccharides found in cartilage.

• An *acetyl group* has the structure

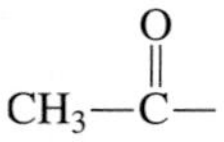

$CH_3-\overset{\overset{\displaystyle O}{\|}}{C}-$

It can be considered to be derived from acetic acid by removal of the —OH portion of that structure.

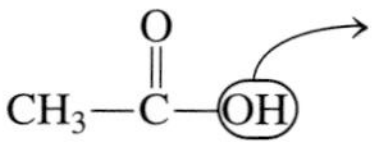

$CH_3-\overset{\overset{\displaystyle O}{\|}}{C}-OH$

Acetic acid

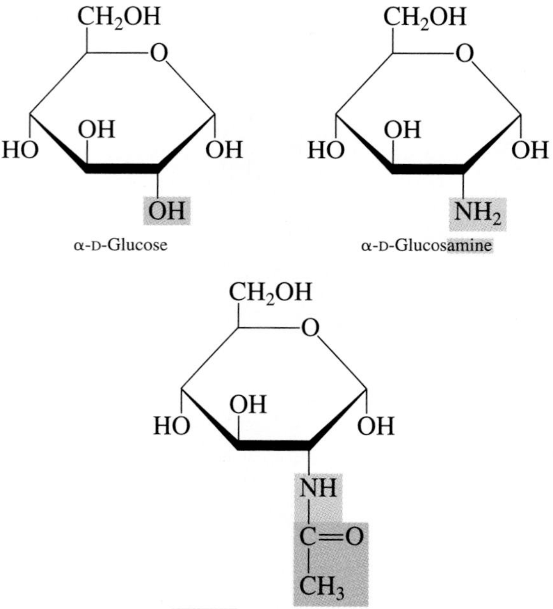

α-D-Glucose

α-D-Glucosamine

N-acetyl-α-D-glucosamine

The *N*-acetyl derivatives of D-glucosamine and D-galactosamine are present in the biochemical markers on red blood cells, which distinguish the various blood types. (See Chemical Connections 18.1 on pages 530–531.)

18.13 Disaccharides

A monosaccharide that has cyclic forms (hemiacetal or hemiketal) can react with an alcohol to form a glycoside (acetal or ketal), as we noted in Section 18.12. This same type of reaction can be used to produce a *disaccharide,* a carbohydrate in which two monosaccharides are bonded together (Section 18.3). In disaccharide formation, one of the monosaccharide reactants functions as a hemiacetal or hemiketal, and the other functions as an alcohol.

Monosaccharide + monosaccharide ⟶ disaccharide + H_2O
(Functioning as a hemiacetal or a hemiketal) (Functioning as an alcohol) (Glycoside)

Glycosidic linkage

CH₂OH … + H_2O

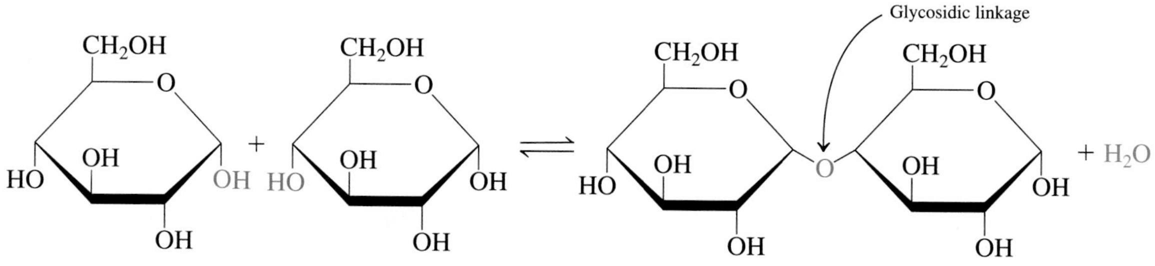

The bond that links the two monosaccharides of a disaccharide together is called a glycosidic linkage. A **glycosidic linkage** *is the carbon–oxygen–carbon bond that joins the two components of a glycoside together.*

We now examine the structures and properties of four important disaccharides: maltose, cellobiose, lactose, and sucrose. As we consider details of the structures of these compounds, we will find that the configuration (α or β) at carbon-1 of the reacting monosaccharides is often of prime importance.

• Maltose

Maltose, often called *malt sugar,* is produced whenever the polysaccharide starch (Section 18.14) breaks down, as happens in plants when seeds germinate and in human beings during starch digestion. It is a common ingredient in baby foods and is found in malted milk. Malt (germinated barley that has been baked and ground) contains maltose; hence the name *malt sugar.*

Structurally, maltose is made up of two D-glucose units, one of which must be α-D-glucose. The formation of maltose from two glucose molecules is as follows:

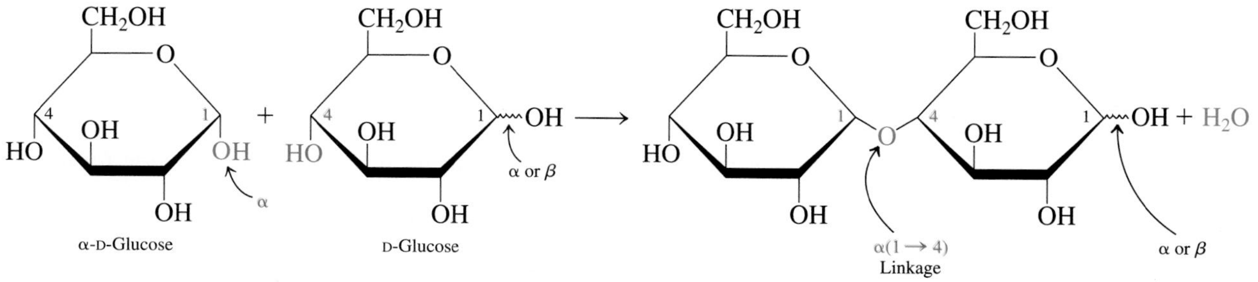

α-D-Glucose + D-Glucose (α or β) ⟶ α(1 → 4) Linkage, α or β + H_2O

The glycosidic linkage between the two glucose units is called an $\alpha(1 \rightarrow 4)$ linkage. The two —OH groups that form the linkage are attached, respectively, to carbon-1 of the first glucose unit (in an α configuration) and to carbon-4 of the second.

Maltose is a reducing sugar (Section 18.12), because the glucose unit on the right has a hemiacetal carbon atom (C-1). Thus this glucose unit can open and close; it is in equilibrium with its open-chain aldehyde form (Section 18.10). This means there are actually three forms of the maltose molecule: α-maltose, β-maltose, and the open-chain form. Structures for these three maltose forms are shown in Figure 18.15. In the solid state, the β form is dominant.

The most important chemical reaction of maltose is that of hydrolysis. Hydrolysis of D-maltose, whether in a laboratory flask or in a living organism, produces two molecules of D-glucose.

$$\text{D-Maltose} \xrightarrow[H_2O]{H^+ \text{ or maltase}} 2\ \text{D-Glucose}$$

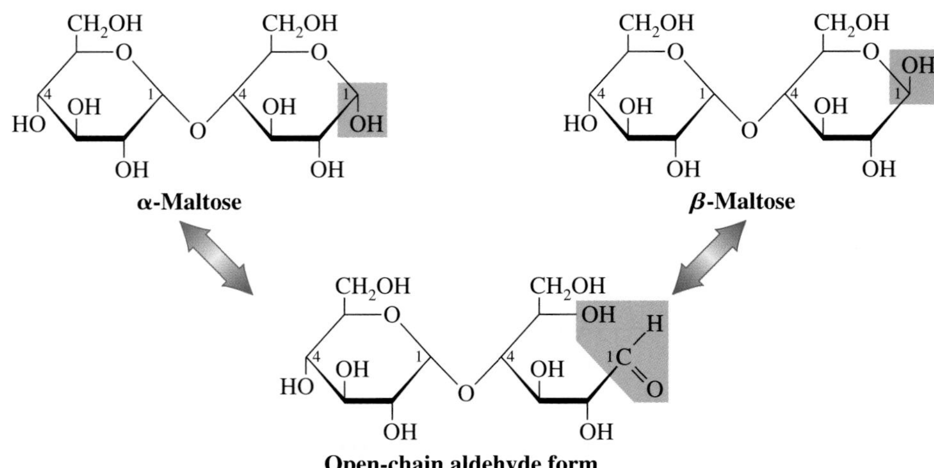

Figure 18.15
The three forms of maltose present in aqueous solution.

• Cellobiose

Cellobiose is produced as an intermediate in the hydrolysis of the polysaccharide cellulose (Section 18.14). Like maltose, cellobiose contains two D-glucose monosaccharide units. It differs from maltose in that one of the D-glucose units—the one functioning as a hemiacetal—must have a β configuration instead of the α configuration for maltose. This change in configuration results in a $\beta(1 \rightarrow 4)$ glycosidic linkage.

β-D-Glucose + D-Glucose (α or β) → $\beta(1 \rightarrow 4)$ Linkage (α or β) + H_2O

Like maltose, cellobiose is a reducing sugar, has three isomeric forms in aqueous solution, and upon hydrolysis produces two D-glucose molecules.

$$\text{D-Cellobiose} \xrightarrow[H_2O]{H^+ \text{ or cellobiase}} 2\ \text{D-glucose}$$

Despite these similarities, maltose and cellobiose have different biological behaviors. These differences are related to the stereochemistry of their glycosidic linkages. Maltase, the enzyme that breaks the glucose–glucose $\alpha(1 \rightarrow 4)$ linkage present in maltose, is found both in the human body and in yeast. Consequently, maltose is digested easily by humans and is readily fermented by yeast. Both the human body and yeast lack the enzyme cellobiase needed to break the glucose–glucose $\beta(1 \rightarrow 4)$ linkage of cellobiose. Thus cellobiose cannot be digested by humans or fermented by yeast.

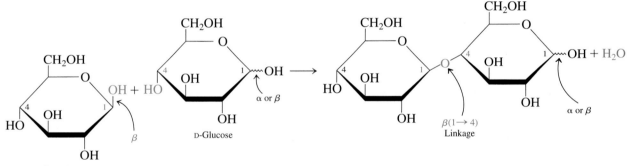

• It is important to distinguish between the notation used for an $\alpha(1 \rightarrow 4)$ glycosidic linkage and that used for a $\beta(1 \rightarrow 4)$ glycosidic linkage.

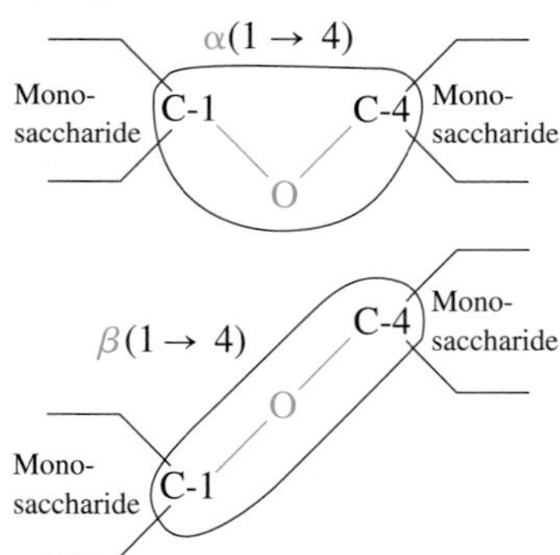

• Lactose

In maltose and cellobiose, the two units of the disaccharide are identical—two glucose units in each case. However, the two monosaccharide units in a disaccharide need not be identical. *Lactose* is made up of a β-D-galactose unit and a D-glucose unit joined by a $\beta(1 \rightarrow 4)$ glycosidic linkage.

β(1 → 4) Glycosidic linkage

β-D-Galactose — D-Glucose

Lactose

Chemical CONNECTIONS

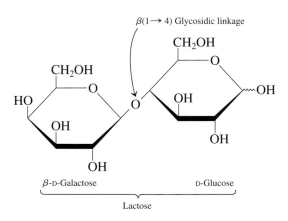

18.2 Lactose Intolerance and Galactosemia

Lactose is the principal carbohydrate in milk. Human mother's milk obtained by nursing infants contains 7%–8% lactose, almost double the 4%–5% lactose found in cow's milk.

For many people, the digestion and absorption of lactose is a problem. This problem, called *lactose intolerance,* is a condition in which people lack the enzyme *lactase,* which is needed to hydrolyze lactose to galactose and glucose.

$$\text{Lactose} \xrightarrow{\text{Lactase}} \text{Glucose} + \text{galactose}$$

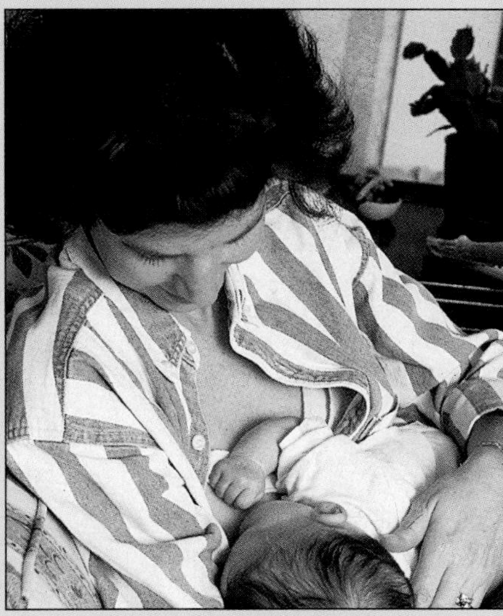

Deficiency of lactase can be caused by a genetic defect, by physiological decline with age, or by injuries to the mucosa lining the intestines. When lactose molecules remain in the intestine undigested, they attract water to themselves, causing fullness, discomfort, cramping, nausea, and diarrhea. Bacterial fermentation of the lactose further along the intestinal tract produces acid (lactic acid) and gas, adding to the discomfort.

The level of the enzyme lactase in humans varies with age. Most children have sufficient lactase during the early years of their life when milk is a much-needed source of calcium in their diet. In adulthood, the enzyme level decreases, and lactose intolerance develops. This explains the change in milk-drinking habits of many adults. Some researchers estimate that as many as one of three adult Americans exhibits a degree of lactose intolerance.

The level of the enzyme lactase in humans varies widely among ethnic groups, indicating that the trait is genetically determined (inherited). The occurrence of lactose intolerance is lowest among Scandinavians and other northern Europeans and highest among native North Americans, Southeast Asians, Africans, and Greeks. The estimated prevalence of lactose intolerance is as follows:

80% Asian Americans	60% Inuits
80% Native Americans	50% Hispanics
75% African Americans	20% Caucasians
70% Mediterranean peoples	10% Northern Europeans

After lactose has been degraded into glucose and galactose, the galactose has to be converted into glucose before it can be used by cells. In humans, the genetic condition called *galactosemia* is caused by the absence of one or more of the enzymes needed for this conversion. In people with this condition, galactose and its toxic metabolic derivative dulcitol accumulate in the blood.

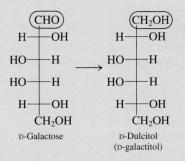

D-Galactose → D-Dulcitol (D-galactitol)

If not treated, galactosemia can cause mental retardation in infants and even death. Treatment involves exclusion of milk and milk products from the diet.

The glucose hemiacetal center is unaffected when galactose bonds to glucose in the formation of lactose, so lactose is a reducing sugar (the glucose ring can open to give an aldehyde).

Lactose is the major sugar found in milk. This accounts for its common name, *milk sugar.* Enzymes in mammalian mammary glands take glucose from the bloodstream and synthesize lactose in a four-step process. Epimerization (Section 18.9) of glucose yields galactose, and then the $\beta(1 \rightarrow 4)$ linkage forms between a galactose and a glucose unit. Lactose is an important ingredient in commercially produced infant formulas that are designed to simulate mother's milk. Souring of milk is caused by the conversion of lactose to lactic acid by bacteria in the milk. Pasteurization of milk is a quick-heating process that kills most of the bacteria and retards the souring process.

Lactose can be hydrolyzed by acid or by the enzyme lactase, forming an equimolar mixture of galactose and glucose.

$$\text{D-Lactose} \xrightarrow[H_2O]{H^+ \text{ or lactase}} \text{D-galactose} + \text{D-glucose}$$

In the human body, the galactose so produced is then converted to glucose by other enzymes. The genetic condition *lactose intolerance,* an inability of the human digestive system to hydrolyze lactose, is considered in Chemical Connections 18.2 on the preceding page.

• The α form of lactose is sweeter to the taste and more soluble in water than the β form. The β form can be found in ice cream that has been stored for a long time; it crystallizes and gives the ice cream a gritty texture.

• The enzyme needed to break a $\beta(1 \rightarrow 4)$ linkage between a galactose and a glucose molecule (lactose) has a different structure from the enzyme needed to break a $\beta(1 \rightarrow 4)$ linkage between two glucose molecules (cellobiose). These enzyme structures are dictated by the differing molecular structures of lactose and cellobiose.

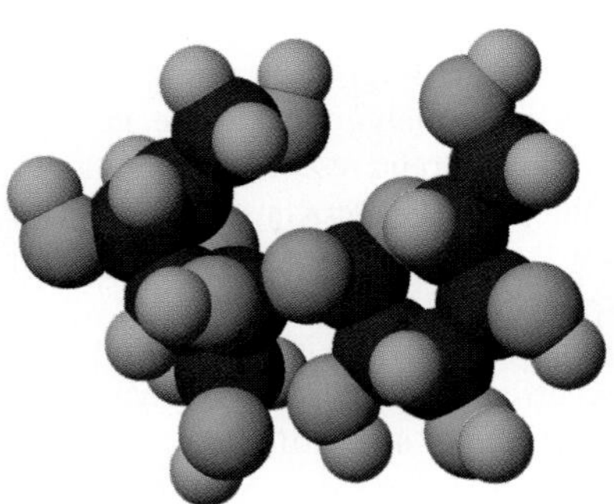

Space-filling model of the disaccharide sucrose. Average per capita consumption of sucrose in the United States is approximately 100 pounds per year. Two-thirds of this is sucrose that is added to food for extra sweetening.

• Sucrose

Sucrose, common *table sugar,* is the most abundant of all disaccharides and occurs throughout the plant kingdom. It is produced commercially from the juice of sugar cane and sugar beets. Sugar cane contains up to 20% by mass sucrose, and sugar beets contain up to 17% by mass sucrose.

The two monosaccharide units present in a D-sucrose molecule are α-D-glucose and β-D-fructose. The glycosidic linkage is not a $(1 \rightarrow 4)$ linkage, as was the case for maltose, cellobiose, and lactose. It is instead an $\alpha,\beta(1 \rightarrow 2)$ glycosidic linkage. The —OH group on carbon-2 of D-fructose (the hemiketal carbon) reacts with the —OH group on carbon-1 of D-glucose (the hemiacetal carbon).

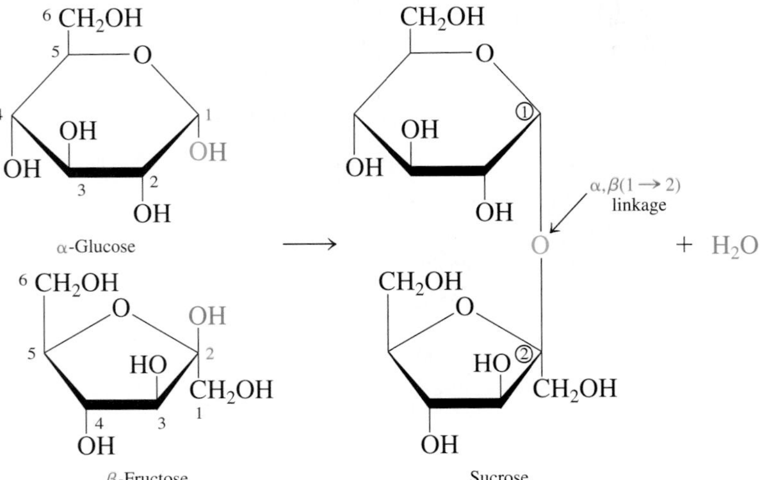

• The glycosidic linkage in sucrose is very different from that in maltose, cellobiose, and lactose. The linkages in the latter three compounds can be characterized as "head-to-tail" linkages—that is, the front end (carbon-1) of one monosaccharide is linked to the back end (carbon-4) of the other monosaccharide. Sucrose has a "head-to-head" glycosidic linkage; the front ends of the two monosaccharides (carbon-1 for glucose and carbon-2 for fructose) are linked.

Sucrose, unlike maltose, cellobiose, and lactose, is a *nonreducing sugar.* No hemiacetal or hemiketal center is present in the molecule, because the glycosidic linkage involves the reducing ends of both monosaccharides. Sucrose, in the solid state and in solution, exists in only one form—there are no α and β isomers, and an open-chain form is not possible.

Sucrase, the enzyme needed to break the $\alpha,\beta(1 \rightarrow 2)$ linkage in sucrose, is present in the human body. Hence sucrose is an easily digested substance. Sucrose hydrolysis (digestion) produces an equimolar mixture of glucose and fructose called *invert sugar.*

• The term *invert sugar* comes from the observation that the direction of rotation of plane-polarized light (Section 18.7) changes from positive (clockwise) to negative (counterclockwise) when sucrose is hydrolyzed to invert sugar. The rotation is +66° for sucrose. The *net* rotation for the invert sugar mixture of fructose (−92°) and glucose (+52°) is −40°.

Chemical CONNECTIONS

18.3 Artificial Sweeteners

Because of the high caloric value of sucrose, it is often difficult to satisfy a demanding "sweet tooth" with sucrose without adding pounds to the body frame or inches to the waistline. Artificial sweeteners, which provide virtually no calories, are now used extensively as a solution to the "sucrose problem."

Three artificial sweeteners that have been widely used are saccharin, sodium cyclamate, and aspartame.

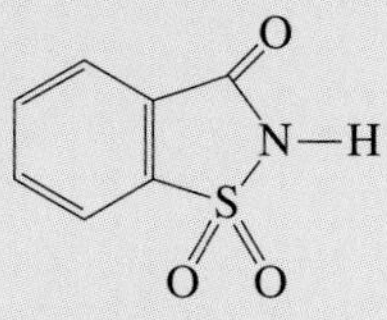

Saccharin 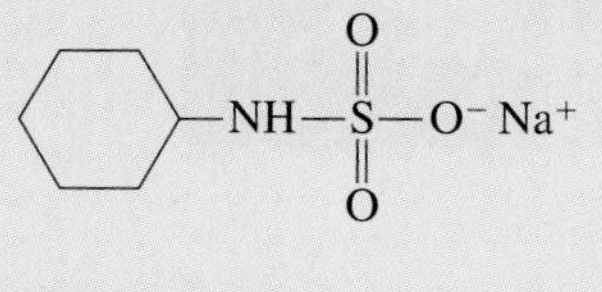 Sodium cyclamate

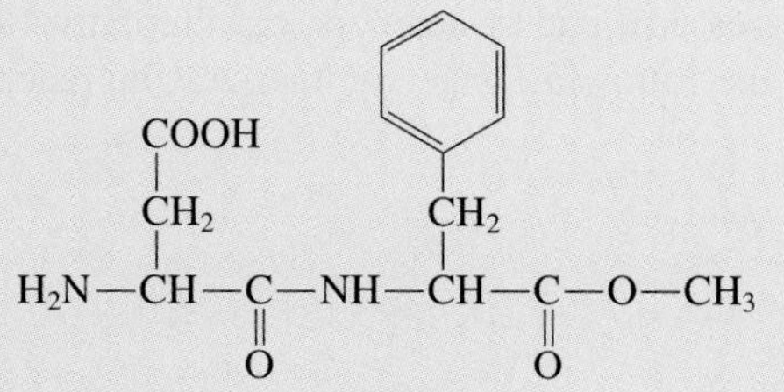

Aspartame

All three of these artificial sweeteners have received wide publicity because of concern about their safety.

Saccharin is the oldest of the artificial sweeteners, having been in use since before 1900. Questions about its safety arose in 1977 from a study that suggested that large doses of saccharin caused bladder tumors in rats. The FDA proposed banning saccharin as a result, but public support for its use caused Congress to impose a moratorium on the ban. In 1991, on the basis of many further studies, the FDA withdrew its proposal to ban saccharin.

Sodium cyclamate, approved by the FDA in 1949, dominated the artificial sweetener market for 20 years. In 1969, principally on the basis of one study suggesting that it caused cancer in laboratory animals, the FDA banned its use. Further studies have shown that neither sodium cyclamate nor its metabolites cause cancer in animals. Reapproval of sodium cyclamate has been suggested, but there has been little action by the FDA. Interestingly, Canada has approved sodium cyclamate use but banned the use of saccharin.

Aspartame (Nutra-Sweet), approved by the FDA in 1981, is used in both the United States and Canada and accounts for three-fourths of current artificial sweetener use. It tastes like sucrose but is 180 times sweeter. It provides 4 kcal/g, as does sucrose, but because so little is used, its calorie contribution is negligible. Aspartame has quickly found its way into almost every diet food on the market today.

The safety of aspartame lies with its hydrolysis products: the amino acids aspartic acid and phenylalanine. These amino acids are identical to those obtained from digestion of proteins. The only danger aspartame poses is that it contains phenylalanine, an amino acid that can lead to mental retardation among young children suffering from PKU (phenylketonuria). Labels on all products containing aspartame warn phenylketonurics of this potential danger.

Sucralose, a derivative of sucrose, is a new low-calorie entry into the artificial sweetener market. It is synthesized from sucrose by substitution of three chlorine atoms for hydroxyl groups.

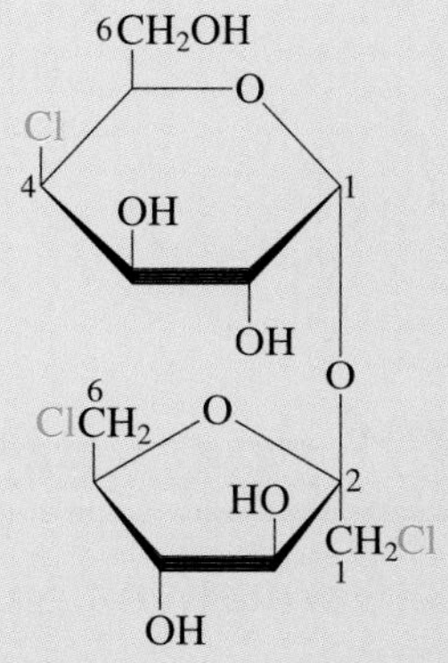

An advantage of sucralose over aspartame is that it is heat-stable and can therefore be used in cooked food. Aspartame loses its sweetness when heated. Sucralose is 600 times sweeter than sucrose and has a similar taste. It is calorie-free because it cannot be hydrolyzed as it passes through the digestive tract.

Name	Type	Sweetness
lactose	disaccharide	16
glucose	monosaccharide	74
sucrose	disaccharide	100
fructose	monosaccharide	173
sodium cyclamate	noncarbohydrate	3,000
aspartame	noncarbohydrate	15,000
saccharin	noncarbohydrate	35,000
sucralose	disaccharide derivative	60,000

$$\text{D-Sucrose} + H_2O \xrightarrow{H^+ \text{ or sucrase}} \underbrace{\text{D-glucose} + \text{D-fructose}}_{\text{Invert sugar}}$$

When sucrose is cooked with acid-containing foods such as fruits or berries, partial hydrolysis takes place, forming some invert sugar. Jams and jellies prepared in this manner

• Honeybees and many other insects possess an enzyme called *invertase* that hydrolyzes sucrose to invert sugar. Thus honey is predominantly a mixture of D-glucose and D-fructose, with some unhydrolyzed sucrose. Honey also contains flavoring agents obtained from the particular flowers whose nectars are collected. Whether a person eats monosaccharides individually, as in honey, or linked together, as in sucrose, they end up the same way in the human body: as glucose and fructose.

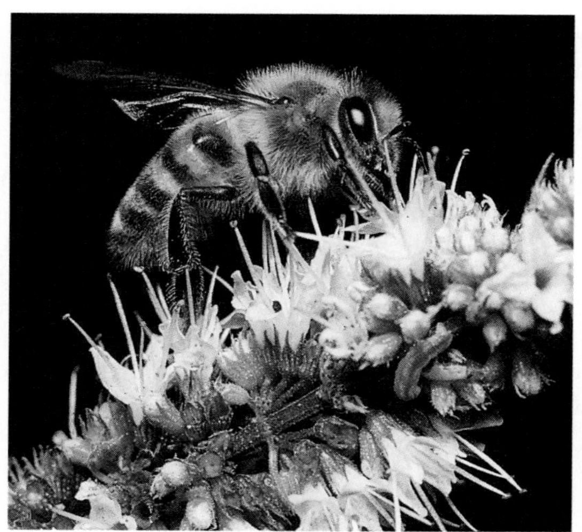

are actually sweeter than the pure sucrose added to the original mixture, because one-to-one mixtures of glucose and fructose taste sweeter than sucrose. Chemical Connections 18.3 on the preceding page discusses artificial sweeteners, an alternative to sucrose use.

The Chemistry at a Glance on the following page summarizes the reactions of monosaccharides.

Chemical CONNECTIONS

18.4 Sucrose Derivatives and Flatulence

Flatulence (or flatus) is the gases produced in the stomach and intestines by the degradation of food constituents. Ingestion of some foods, particularly certain members of the bean and pea families, leads to increased flatus production in most individuals. The compounds in beans and peas that are responsible for increased production of intestinal gas are the sucrose derivatives *raffinose, stachyose,* and *verbascose.* In these compounds, one or more galactose units are attached to a sucrose molecule.

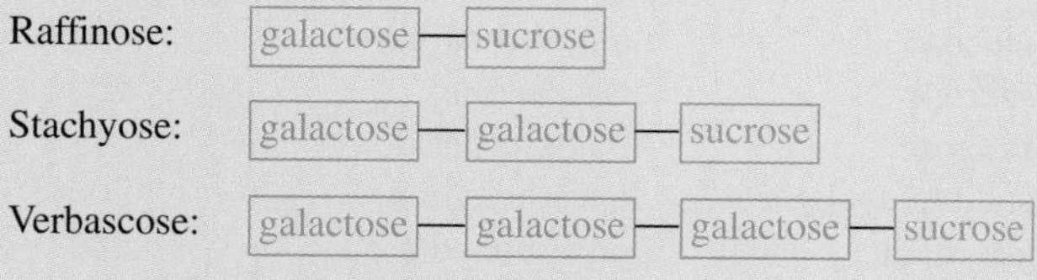

Thus raffinose is a trisaccharide, stachyose a tetrasaccharide, and verbascose a pentasaccharide.

The human body lacks enzymes that can hydrolyze the glycosidic linkages in these oligosaccharides, so they pass into the intestine undigested. Intestinal bacteria and their enzymes cause fermentation of these compounds, which generates large amounts of gases, primarily H_2 and CO_2. After the ingestion of a meal of beans, the volume of flatus generated can increase by a factor of 10.

The volume and composition of the flatus of an individual vary, depending mainly on the food ingested. The following table gives flatus composition, in volume percent, for 11 subjects.

	Volume Percent	
Gas	Mean	Range
nitrogen	64	26–88
carbon dioxide	14	5.5–27
hydrogen	19	0.17–49
methane	3.2	0–20
oxygen	0.7	0.1–1.8

Nitrogen gas, from air swallowed with food, is the most abundant flatus component. Most of the rest is hydrogen and carbon dioxide—products of carbohydrate metabolism by intestinal bacteria. Methane is not formed in all individuals. The possession of methane-producing bacteria appears to be an environmentally acquired familial trait. Very little oxygen is found in flatus, because any that is produced is rapidly consumed by the bacteria. All of the gases so far mentioned are *odorless.* The odor-generating gases of flatus, which account for less than 1% of total volume, include ammonia and hydrogen sulfide. Their source is protein digestion.

CHEMISTRY AT A GLANCE

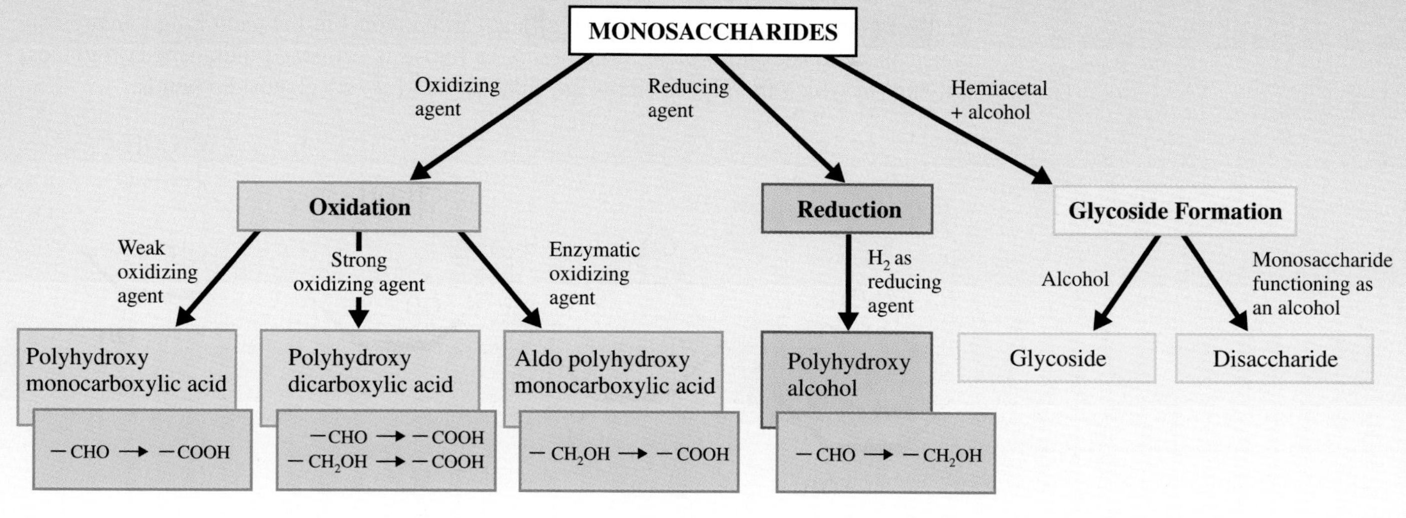

18.14 Polysaccharides

A **polysaccharide** *contains many monosaccharide units bonded to each other by glycosidic linkages.* The number of monosaccharide units varies with the polysaccharide from a few hundred to hundreds of thousands. Polysaccharides are polymers (Section 13.8). In some, the monosaccharides are bonded together in a linear (unbranched) chain. In others, there is extensive branching of the chains (Figure 18.16).

• In nutrition discussions, monosaccharides and disaccharides are called *simple carbohydrates,* and polysaccharides are called *complex carbohydrates.*

Unlike monosaccharides and most disaccharides, polysaccharides are not sweet and do not test positive in Tollens, Benedict's, and Fehling's solutions. They have limited water solubility because of their size. However, the —OH groups present can individually become hydrated by water molecules. The result is usually a thick colloidal suspension of the polysaccharide in water. Polysaccharides, such as flour and cornstarch, are often used as thickening agents in sauces, desserts, and gravy.

Although there are many naturally occurring polysaccharides, in this section we will focus on only four of them: cellulose, starch, glycogen, and chitin. All play vital roles in living systems—cellulose and starch in plants, glycogen in humans and other animals, and chitin in arthropods.

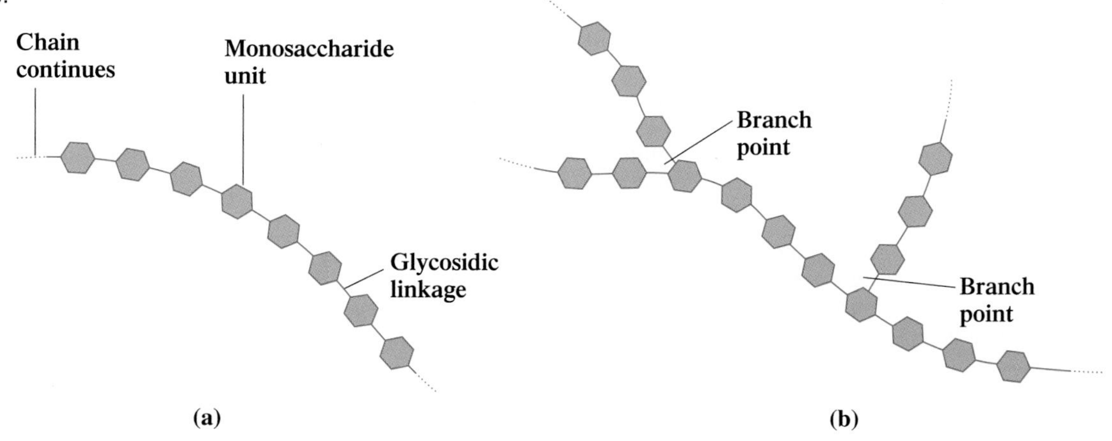

Figure 18.16
Some polysaccharides have a linear chain structure (a); others have a branched chain structure (b).

• Cellulose

Cellulose is the most abundant polysaccharide. It is the structural component of the cell walls of plants. Approximately half of all the carbon atoms in the plant kingdom are contained in cellulose molecules. Structurally, cellulose is a linear (unbranched) D-glucose polymer in which the glucose units are linked by $\beta(1 \rightarrow 4)$ glycosidic bonds.

A sandwhich such as this is high in dietary fiber; that is, it is a cellulose-rich "meal."

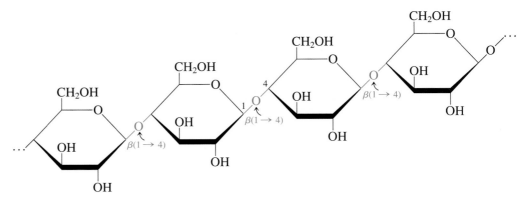

Typically, cellulose chains contain about 5000 glucose units, which gives macromolecules with molecular masses of about 900,000 amu. Cotton is almost pure cellulose (95%), and wood is about 50% cellulose.

Even though it is a glucose polymer, cellulose is not a source of nutrition for human beings. Humans lack the enzymes capable of catalyzing the hydrolysis of $\beta(1 \rightarrow 4)$ linkages in cellulose. Even grazing animals lack the enzymes necessary for cellulose digestion. However, the intestinal tracts of animals such as horses, cows, and sheep contain bacteria that produce *cellulase,* an enzyme that can hydrolyze $\beta(1 \rightarrow 4)$ linkages and produce free glucose from cellulose. Thus grasses and other plant materials are a source of nutrition for grazing animals. The intestinal tracts of termites contain the same microorganisms, which enable termites to use wood as their source of food. Microorganisms in the soil can also metabolize cellulose, which makes possible the biodegradation of dead plants.

Despite its nondigestibility, cellulose is still an important component of a balanced diet. It serves as dietary fiber. Dietary fiber provides the digestive tract with "bulk" that helps move food through the intestinal tract and facilitates the excretion of solid wastes. Cellulose readily absorbs water, leading to softer stools and frequent bowel action. Links have been found between the length of time stools spend in the colon and possible colon cancer.

High-fiber food may also play a role in weight control. Obesity is not seen in parts of the world where people eat large amounts of fiber-rich foods. Many of the weight-loss products on the market are composed of bulk-inducing fibers such as methylcellulose.

Some fibers bind lipids such as cholesterol (Section 19.10) and carry them out of the body with the feces. This lowers blood lipid concentrations and possibly the risk of heart and artery disease.

About 20–35 grams of dietary fiber daily is a desirable intake. This is two to three times higher than the average intake in the United States.

• Starch

Starch, like cellulose, is a polysaccharide containing only glucose units. It is the storage polysaccharide in plants. If excess glucose enters a plant cell, it is converted to starch and stored for later use. When the cell cannot get enough glucose from outside the cell, it hydrolyzes starch to release glucose.

• Amylose and cellulose are both linear chains of D-glucose molecules. They are stereoisomers that differ in the configuration at carbon-1 of each D-glucose unit. In amylose, α-D-glucose is present; in cellulose, β-D-glucose.

Iodine is often used to test for the presence of starch in solution. Starch-containing solutions turn a dark blue-black when iodine is added. As starch is broken down through acid or enzymatic hydrolysis to glucose monomers, the blue-black color disappears.

Two different polyglucose polysaccharides can be isolated from most starches: amylose and amylopectin. *Amylose,* a straight-chain glucose polymer, usually accounts for

Use of iodine to test for starch. Starch-containing solutions turn dark blue-black when iodine is added.

15%–20% of the starch; *amylopectin,* a highly branched glucose polymer, accounts for the remaining 80%–85% of the starch.

In amylose's structure, the glucose units are connected by $\alpha(1 \rightarrow 4)$ glycosidic linkages.

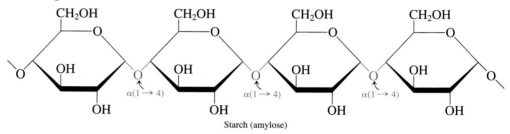

Starch (amylose)

The number of glucose units present in an amylose chain depends on the source of the starch; 300–500 monomer units are usually present.

Amylopectin, the other polysaccharide in starch, is similar to amylose in that all linkages are α linkages. It is different in that there is a high degree of branching in the polymer. A branch occurs about once every 25–30 glucose units. The branch points involve $\alpha(1 \rightarrow 6)$ linkages (Figure 18.17). Because of the branching, amylopectin has a larger

Figure 18.17
Two perspectives on the structure of the polysaccharide amylopectin. (left) Molecular structure of amylopectin. (right) An overview of the branching that occurs in the amylopectin structure. Each circle is a glucose unit.

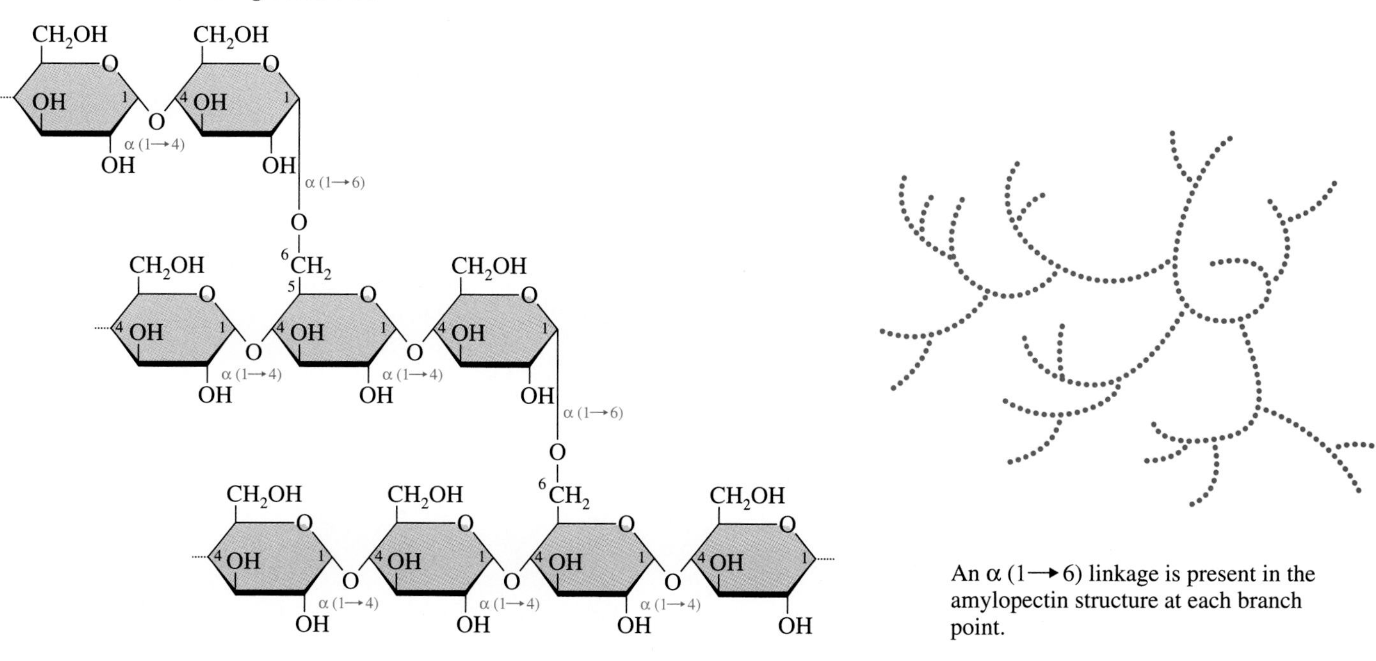
An α (1→6) linkage is present in the amylopectin structure at each branch point.
(a)
(b)

average molecular mass than the linear amylose. The average molecular mass of amylose is 50,000 amu or more; it is 300,000 or more for amylopectin.

Note that all of the glycosidic linkages in starch (both amylose and amylopectin) are of the α type. In amylose, they are all $(1 \rightarrow 4)$; in amylopectin, both $(1 \rightarrow 4)$ and $(1 \rightarrow 6)$ linkages are present. Because α linkages can be broken through hydrolysis within the human digestive tract (with the help of the enzyme *amylase*), starch has nutritional value for humans. The starches present in potatoes and cereal grains (wheat, rice, corn, etc.) account for approximately two-thirds of the world's food consumption.

• Glycogen

• The glucose polymers amylose, amylopectin, and glycogen compare as follows in molecular size and degree of branching.

Amylose:	Up to 1000 glucose units; no branching
Amylopectin:	Up to 100,000 glucose units; branch points every 24–30 glucose units
Glycogen:	Up to 1,000,000 glucose units; branch points every 8–12 glucose units

Glycogen, like cellulose and starch, is a polysaccharide containing only glucose units. It is the glucose storage polysaccharide in humans and animals. Its function is thus similar to that of starch in plants, and it is sometimes referred to as *animal starch.* Liver cells and muscle cells are the storage sites for glycogen in humans.

Glycogen has a structure similar to that of amylopectin; all glycosidic linkages are of the α type, and both $(1 \rightarrow 4)$ and $(1 \rightarrow 6)$ linkages are present. Glycogen and amylopectin differ in the number of glucose units between branches and the total number of glucose units present in a molecule. Glycogen is about three times more highly branched than amylopectin, and it is much larger, with a molar mass of up to 3,000,000 amu.

• The amount of stored glycogen in the human body is relatively small. Muscle tissue is approximately 1% glycogen, and liver tissue 2%–3%. However, this amount is sufficient to take care of normal-activity glucose demands for about 15 hours. During strenuous exercise, glycogen supplies can be exhausted rapidly. At this point, the body begins to oxidize fat as a source of energy.

Many marathon runners eat large quantities of starch foods the day before a race. This practice, called *carbohydrate loading,* maximizes body glycogen reserves.

When excess glucose is present in the blood (normally from eating too much starch), the liver and muscle tissue convert the excess glucose to glycogen, which is then stored in these tissues. Whenever the glucose blood level drops (from exercise, fasting, or normal activities), some stored glycogen is hydrolyzed back to glucose. These two opposing processes are called *glycogenesis* and *glycogenolysis,* the formation and decomposition of glycogen, respectively.

$$\text{Glucose} \underset{\text{Glycogenolysis}}{\overset{\text{Glycogenesis}}{\rightleftharpoons}} \text{glycogen}$$

Glycogen is an ideal storage form for glucose. The large size of these macromolecules prevents them from diffusing out of cells. Also, conversion of glucose to glycogen reduces osmotic pressure (Section 9.9). Cells would burst because of increased osmotic pressure if all of the glucose in glycogen were present in cells in free form. High concentrations of glycogen in a cell sometimes precipitate or crystallize into *glycogen granules.* These granules are discernible in photographs of cells under electron microscope magnification (Figure 18.18).

Figure 18.18
The small, dense particles within this electron micrograph of a liver cell are glycogen granules.

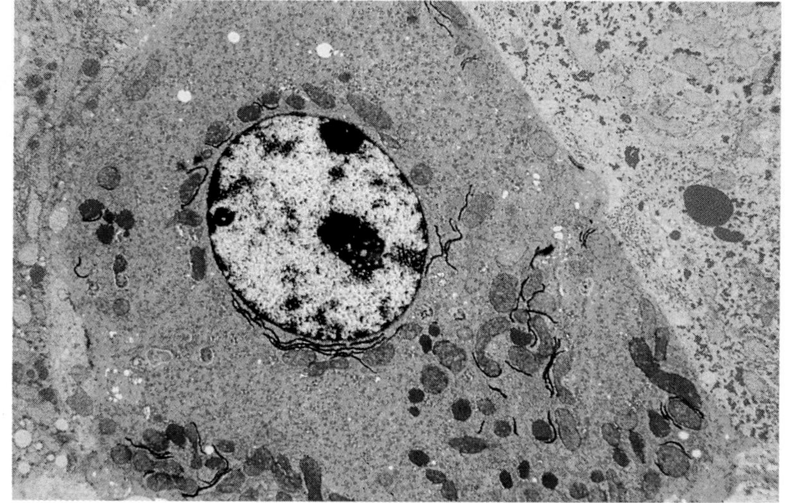

Chitin, a linear $\beta(1 \rightarrow 4)$ polysaccharide produces the rigidity in the exoskeletons of crabs and other arthropods.

• The word *chitin* is pronounced "kye-ten"; it rhymes with *Titan.*

• Chitin

Chitin is a polysaccharide that is similar to cellulose in both function and structure. Its function is to give rigidity to the exoskeletons of crabs, lobsters, shrimp, insects, and other arthropods. It also occurs in the cell walls of fungi.

Structurally, chitin is a linear polymer (no branching) with all $\beta(1 \rightarrow 4)$ glycosidic linkages, as is cellulose. Chitin differs from cellulose in that the monosaccharide present is an *N*-acetyl amino derivative of D-glucose (Section 18.12). Figure 18.19 contrasts the structures of chitin and cellulose.

18.15 Mucopolysaccharides

Mucopolysaccharides are compounds that occur in connective tissue associated with joints in animals and humans. Their function is primarily that of lubrication, a necessary requirement if movement is to occur. The name *muco*polysaccharide comes from the highly viscous, gelatinous (mucus-like) consistency of these substances in aqueous solution.

Unlike all the polysaccharides we have discussed up to this point, mucopolysaccharides are *hetero*polysaccharides rather than *homo*polysaccharides. A **homopolysaccharide** *is a polysaccharide in which only one type of monosaccharide unit is present.* Cellulose, starch, glycogen, and chitin are all homopolysaccharides. A **heteropolysaccharide** *is a polysaccharide in which more than one (usually two) type of monosaccharide unit is present.*

One of the most common mucopolysaccharides is hyaluronic acid, a heteropolysaccharide in which the following two glucose derivatives alternate in the structure.

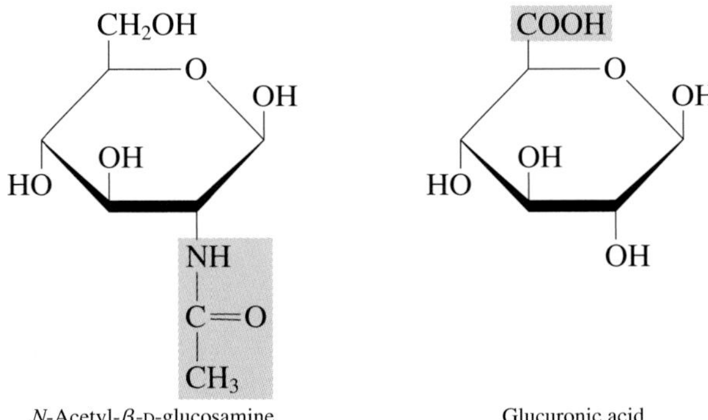

Both of these monosaccharide derivatives have been encountered previously. *N*-acetyl-β-D-glucosamine is the repeating unit in chitin (see Section 18.14). Glucuronic acid is derived from glucose by oxidation of the —OH group at carbon-6 to an acid group

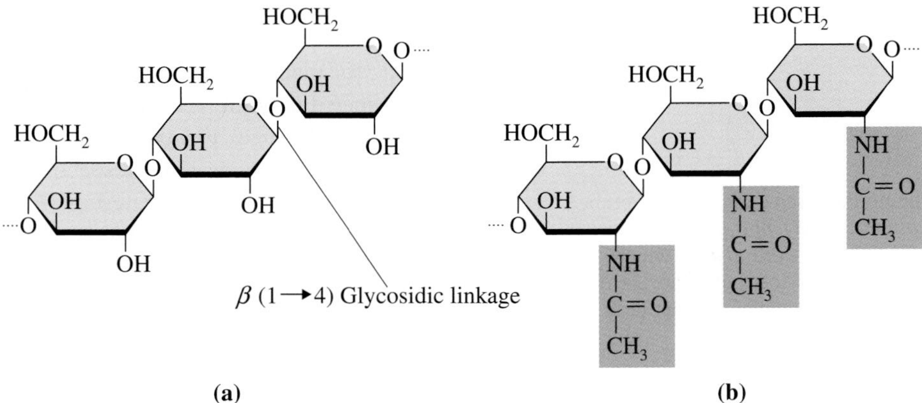

Figure 18.19
The structures of cellulose (a) and chitin (b). In both substances, all glycosidic linkages are of the $\beta(1 \rightarrow 4)$ type.

Mucopolysaccharides associated with the connective tissue of joints give hurdlers such as these the flexibility needed to accomplish their task.

(see Section 18.12). A section of the polymeric structure for hyaluronic acid is

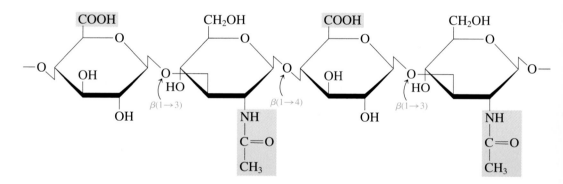

COOH CH$_2$OH COOH CH$_2$OH
O O O O
—O OH O HO O OH O HO O—
$\beta(1 \rightarrow 3)$ $\beta(1 \rightarrow 4)$ $\beta(1 \rightarrow 3)$
OH NH OH NH
C=O C=O
CH$_3$ CH$_3$

Note also, in this structure, the alternating pattern of glycosidic bond types, $\beta(1 \rightarrow 3)$ and $\beta(1 \rightarrow 4)$.

18.16 Glycolipids and Glycoproteins

Before 1960, the biochemistry of carbohydrates was thought to be rather simple. These compounds served (1) as energy sources for plants, humans, and animals and (2) as structural materials for plants and arthropods.

Research since that time has shown that carbohydrates (oligosaccharides) attached through glycosidic linkages to lipids (Chapter 19) and proteins (Chapter 20), which are called *glycolipids* and *glycoproteins,* respectively, have a wide variety of cellular functions.

• The prefix *glyco-,* used in the terms *glycolipid* and *glycoprotein,* is derived from the Greek word *glykys,* which means "sweet." Most monosaccharides and disaccharides have a sweet taste.

It is now known that carbohydrate units on cell surfaces (glycolipids and glycoproteins) govern how individual cells interact with other cells, with invading bacteria, and with viruses. In the human reproductive process, fertilization begins with the binding of a sperm to a specific oligosaccharide on the surface of an egg. The oligosaccharide markers on cell surfaces that are the basis for blood types were considered in Chemical Connections 18.1.

Concepts to Remember

Biochemistry. Biochemistry is the study of the chemical substances found in living systems and the chemical interactions of these substances with each other.

Carbohydrates. Carbohydrates are polyhydroxy aldehydes, polyhydroxy ketones, or compounds that yield such substances upon hydrolysis. Plants contain large quantities of carbohydrates produced via photosynthesis.

Carbohydrate classification. Carbohydrates are classified into three groups: monosaccharides, oligosaccharides, and polysaccharides.

Chirality and achirality. A chiral object is not identical to its mirror image. An achiral object is identical to its mirror image.

Chiral center. A chiral center is an atom in a molecule that has four different groups tetrahedrally bonded to it. Molecules that contain a single chiral center exist in a left-handed and a right-handed form.

Stereoisomerism. The atoms of stereoisomers are connected in the same way but are arranged differently in space. The major causes of stereoisomerism in molecules are structural rigidity and the presence of a chiral center.

Enantiomers and diastereomers. Two types of stereoisomers exist: enantiomers and diastereomers. Enantiomers have structures that are nonsuperimposable mirror images of each other. Enantiomers have identical achiral properties but different chiral properties. Diastereomers have structures that are not mirror images of each other.

Fischer projections. Fischer projections are two-dimensional structural formulas used to depict the three-dimensional shapes of molecules with chiral centers.

Chirality of monosaccharides. Monosaccharides are classified as D or L stereoisomers on the basis of the configuration of the chiral center farthest from the carbonyl group.

Optical activity. Chiral compounds are optically active—that is, they rotate the plane of polarized light. Enantiomers rotate the plane of polarized light in opposite directions. The prefix (+) indicates that the compound rotates the plane of polarized light in a clockwise direction, whereas compounds that rotate the plane of polarized light in a counterclockwise direction have the prefix (−).

Classification of monosaccharides. Monosaccharides are classified as aldoses or ketoses on the basis of the type of carbonyl group present. They are further classified as trioses, tetroses, pentoses, etc. on the basis of the number of carbon atoms present.

Important monosaccharides. Important monosaccharides include glucose, galactose, fructose, and ribose. Glucose and galactose are aldohexoses, fructose is a ketohexose, and ribose is an aldopentose.

Cyclic monosaccharides. Cyclic monosaccharides form through an intramolecular reaction between the carbonyl group and an alcohol group of an open-chain monosaccharide. These cyclic forms predominate in solution.

Reactions of monosaccharides. Five important reactions of monosaccharides are (1) oxidation to a polyhydroxy acid, (2) reduction to a polyhydroxy alcohol, (3) glycoside formation, (4) phosphate ester formation, and (5) amino sugar formation.

Disaccharides. Disaccharides are glycosides formed from the linkage of two monosaccharides. The most important disaccharides are maltose, cellobiose, lactose, and sucrose. Each of these has at least one glucose unit in its structure.

Polysaccharides. Polysaccharides are polymers of monosaccharides. Cellulose and starch are polymers of glucose, but the linkages between glucose units are different. The α linkages in starch are easily broken by enzymatic hydrolysis in humans, thus providing great sources of dietary energy. However, the β linkages in cellulose prevent enzymatic hydrolysis in humans, so cellulose provides no usable energy in our diets. Glycogen is a highly branched glucose polymer utilized for energy storage within the body.

Key Reactions and Equations

1. Monosaccharide oxidation (Section 18.12)

$$\text{Aldose} + \text{weak oxidizing agent} \longrightarrow \text{carboxylic acid}$$

2. Monosaccharide reduction (Section 18.12)

$$\text{Aldose or ketose} + H_2 \xrightarrow{\text{Catalyst}} \text{polyhydroxy alcohol}$$

3. Glycoside (acetal or ketal) formation (Section 18.12)

$$\text{Cyclic monosaccharide} + \text{alcohol} \longrightarrow \text{glycoside (acetal or ketal)} + H_2O$$

4. Monosaccharide ester formation (Section 18.12)

$$\text{Monosaccharide} + \text{oxyacid} \longrightarrow \text{ester} + H_2O$$

5. Hydrolysis of disaccharide (Section 18.13)

$$\text{Disaccharide} + H_2O \xrightarrow{\text{Catalyst}} \text{two monosaccharides}$$

6. Hydrolysis of maltose (Section 18.13)

$$\text{D-Maltose} + H_2O \xrightarrow{H^+ \text{ or maltase}} 2\ \text{D-glucose}$$

7. Hydrolysis of cellobiose (Section 18.13)

$$\text{D-Cellobiose} + H_2O \xrightarrow{H^+ \text{ or cellobiase}} 2\ \text{D-glucose}$$

8. Hydrolysis of lactose (Section 18.13)

$$\text{D-Lactose} + H_2O \xrightarrow{H^+ \text{ or lactose}} \text{D-galactose} + \text{D-glucose}$$

9. Hydrolysis of sucrose (Section 18.13)

$$\text{D-Sucrose} + H_2O \xrightarrow{H^+ \text{ or sucrase}} \text{D-fructose} + \text{D-glucose}$$

10. Complete hydrolysis of polysaccharide (Section 18.14)

$$\text{Polysaccharide} + H_2O \xrightarrow{H^+ \text{ or enzymes}} \text{many monosaccharides}$$

11. Complete hydrolysis of starch (Section 18.14)

$$\text{Starch} + H_2O \xrightarrow{H^+ \text{ or enzymes}} \text{many D-glucose}$$

12. Complete hydrolysis of glycogen (Section 18.14)

$$\text{Glycogen} + H_2O \xrightarrow{H^+ \text{ or enzymes}} \text{many D-glucose}$$

Key Terms

Achiral object (18.4)
Aldoses (18.8)
Biochemical substance (18.1)
Biochemistry (18.1)
Carbohydrates (18.3)
Chiral center (18.4)
Chiral object (18.4)
Dextrorotatory compound (18.7)
Diastereomers (18.5)
Disaccharides (18.3)
Enantiomers (18.5)
Epimers (18.6)
Fischer projection (18.6)
Glycoside (18.12)
Glycosidic linkage (18.13)
Haworth projection (18.11)
Heteropolysaccharide (18.15)
Homopolysaccharide (18.15)
Ketoses (18.8)
Levorotatory compound (18.7)
Mirror image (18.4)
Monosaccharides (18.3)
Oligosaccharides (18.3)
Optically active compound (18.7)
Polysaccharides (18.3, 18.14)
Reducing sugar (18.12)
Stereoisomers (18.5)
Sugar (18.8)

Exercises and Problems

The members of each pair of problems in this section test similar material.

Biochemical Substances (Section 18.1)

18.1 Define the term *biochemistry.*

18.2 What are the two general groups of biochemical substances?

18.3 What are the four major types of bioorganic substances?

18.4 What are the two most abundant types of bioorganic substances present in the human body?

Occurrence of Carbohydrates (Section 18.2)

18.5 Write a general chemical equation for photosynthesis.

18.6 What role does chlorophyll play in photosynthesis?

18.7 What are the two major functions of carbohydrates in the plant kingdom?

18.8 What are the major functions of carbohydrates in the human body?

Structural Characteristics of Carbohydrates (Section 18.3)

18.9 Define the term *carbohydrate.*

18.10 What functional group is present in all carbohydrates?

18.11 Explain the difference between

a. a monosaccharide and an oligosaccharide
b. a disaccharide and a tetrasaccharide

18.12 Explain the difference between

a. an oligosaccharide and a polysaccharide
b. a trisaccharide and an oligosaccharide

Chirality (Section 18.4)

18.13 Explain what the term *superimposable* means.

18.14 Explain what the term *nonsuperimposable* means.

18.15 In each of the following lists of objects, identify those that are chiral.

a. Nail, hammer, screwdriver, drill bit
b. Your hand, your foot, your ear, your nose
c. The words TOT, TOOT, POP, PEEP

18.16 In each of the following lists of objects, identify those that are chiral.

a. Baseball cap, glove, shoe, scarf
b. Pliers, scissors, spoon, fork
c. The words MOM, DAD, AHA, WAX

18.17 Indicate whether the circled carbon atom in each of the following molecules is a chiral center.

a. $CH_3-\textcircled{C}H_2-OH$
b. $CH_3-\textcircled{C}H(CH_3)-OH$
c. $CH_3-\textcircled{C}H(Cl)-OH$
d. $CH_3-CH_2-\textcircled{C}H(CH_3)-OH$

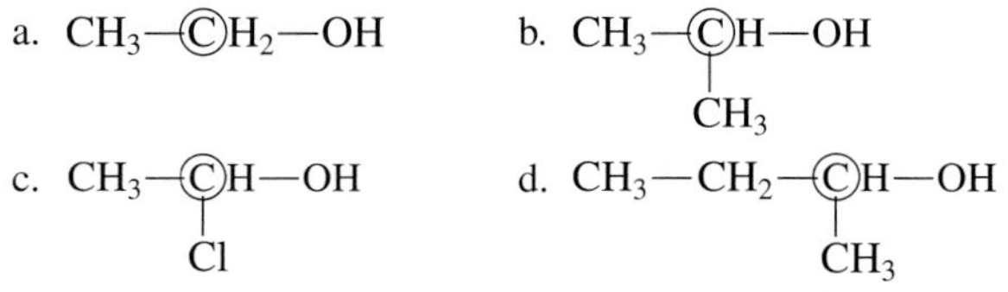

18.18 Indicate whether the circled carbon atom in each of the following molecules is a chiral center.

a. $CH_3-\textcircled{C}H_2-NH_2$
b. $CH_3-\textcircled{C}H(NH_2)-CH_3$
c. $CH_3-\textcircled{C}H(CH_3)-NH_2$
d. $CH_3-\textcircled{C}H(Cl)-NH_2$

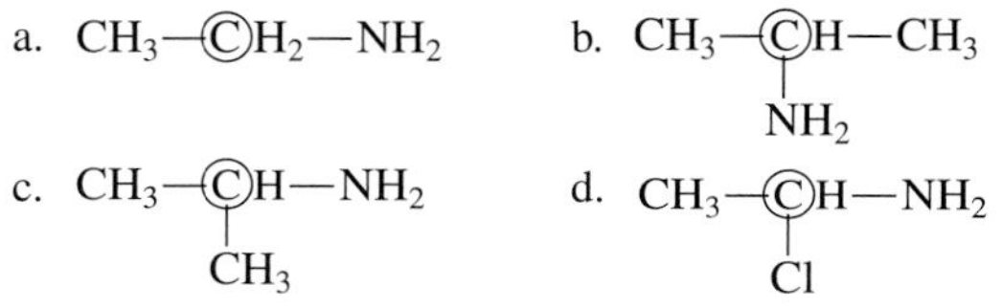

18.19 Use asterisks to show the chiral center(s) in the following structures.

a. $Cl-C(H)(H)-C(H)(Cl)-Br$
b. $Br-C(H)(H)-C(Cl)(Br)-C(H)(Br)-Cl$
c. $CH_2(OH)-CH(OH)-CH(OH)-CH(OH)-C(=O)-H$
d. $CH_2(OH)-CH(OH)-CH(OH)-CH(OH)-CH(OH)-CH_2(OH)$

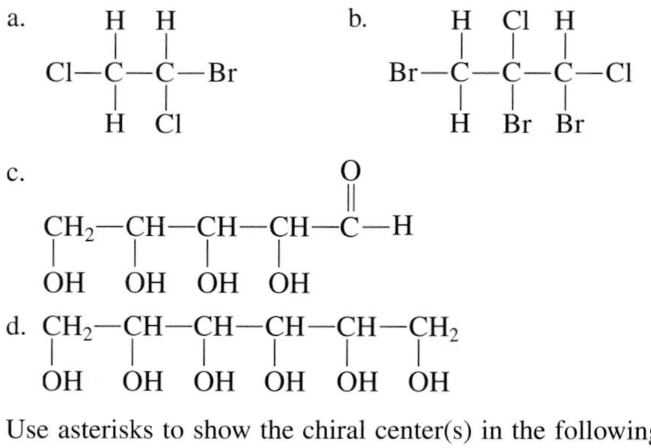

18.20 Use asterisks to show the chiral center(s) in the following structures.

a. $Cl-C(H)(Br)-C(Cl)(Br)-Cl$
b. $Cl-C(H)(Br)-C(H)(OH)-C(H)(Br)-Cl$
c. $CH_3-CH(OH)-CH(OH)-CH(OH)-C(=O)-H$
d. $CH_2(OH)-CH(OH)-CH(OH)-CH(OH)-CH_2(OH)$

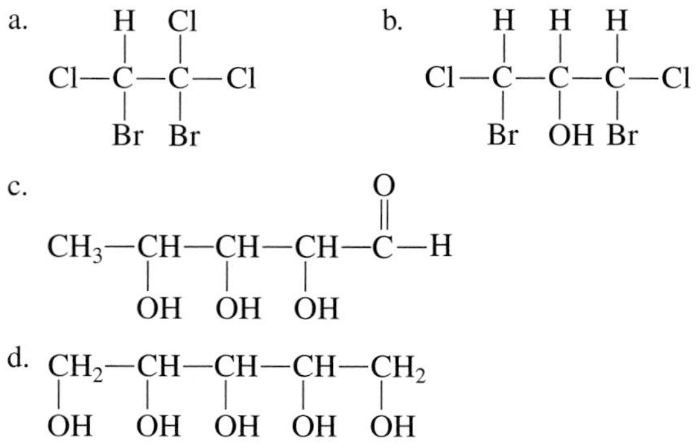

18.21 How many chiral centers are present in each of the following molecular structures?

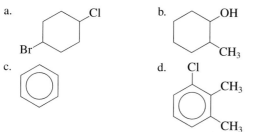

a. 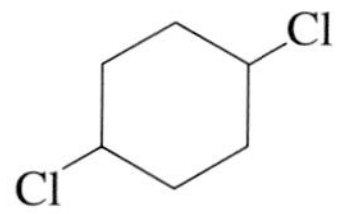 (cyclohexane ring with Cl at positions 1 and 4) b. 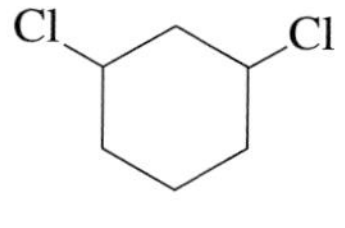 (cyclohexane ring with Cl at positions 1 and 3)

c. 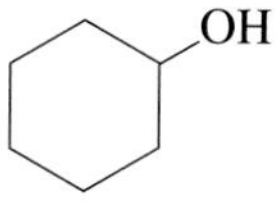 (cyclohexane ring with OH) d. 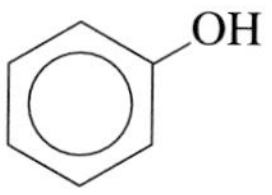 (benzene ring with OH)

18.22 How many chiral centers are present in each of the following molecular structures?

a. 

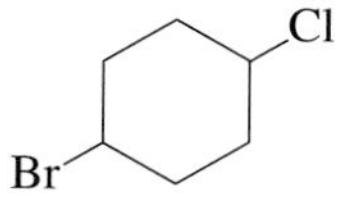

b. 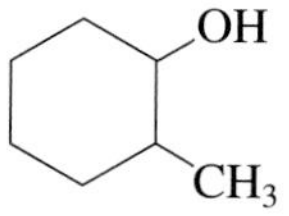 (cyclohexane ring with OH and CH_3)

c. (benzene ring)

d. 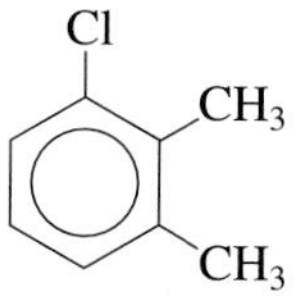 (benzene ring with Cl, CH_3, CH_3)

Stereoisomerism: Enantiomers and Diastereomers (Section 18.5)

18.23 What is the difference between structural isomers and stereoisomers?

18.24 Both enantiomers and diastereomers are stereoisomers. How do they differ?

Fischer Projections (Section 18.6)

18.25 Draw the Fischer projection for each of the following molecules.

a. 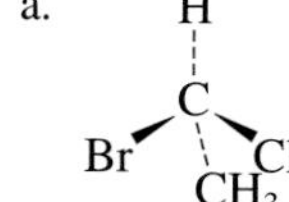

b. 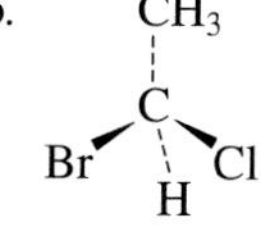

c. 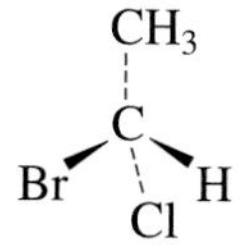 CH_3, C, Br, H, Cl

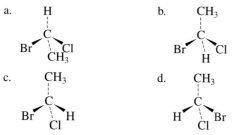

d. 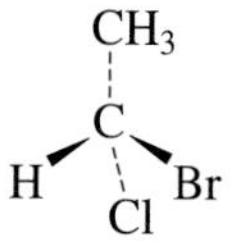 CH_3, C, H, Br, Cl

18.26 Draw the Fischer projection for each of the following molecules.

a. 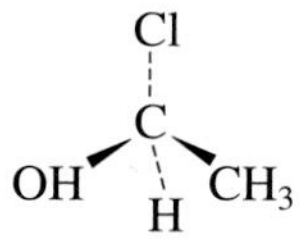 Cl, C, OH, CH_3, H

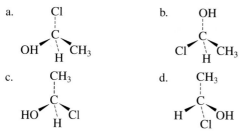

b. 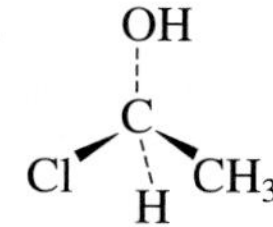

c. 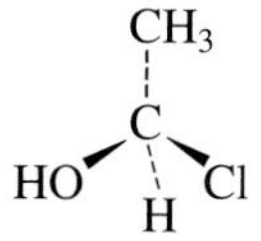 CH_3, C, HO, Cl, H

d. 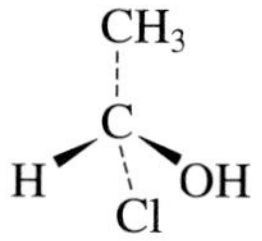 CH_3, C, H, OH, Cl

18.27 Draw a Fischer projection for the enantiomer of each of the following monosaccharides.

a. 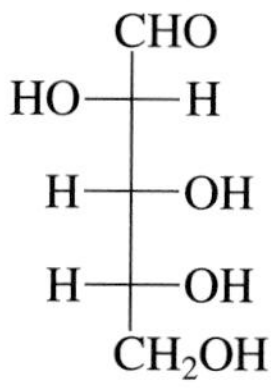
CHO
HO—H
H—OH
H—OH
CH_2OH

b. 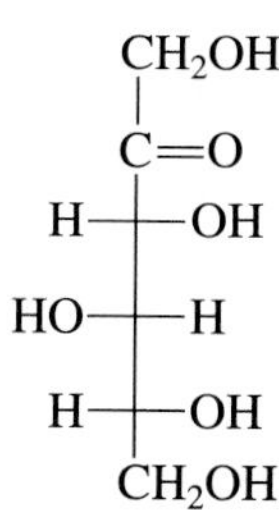

c. 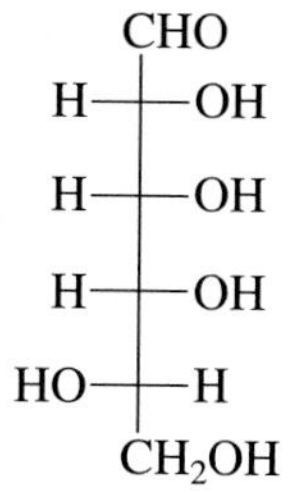
CHO
H—OH
H—OH
H—OH
HO—H
CH_2OH

d. 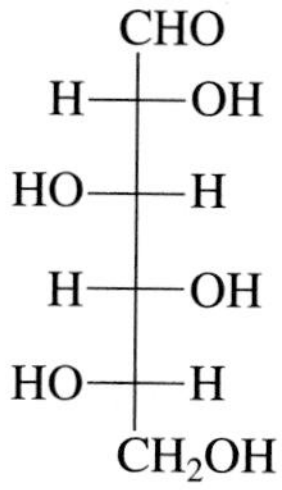
CHO
H—OH
HO—H
H—OH
HO—H
CH_2OH

18.28 Draw a Fischer projection for the enantiomer of each of the following monosaccharides.

a. 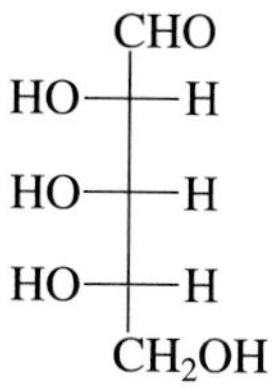
CHO
HO—H
HO—H
HO—H
CH_2OH

b. 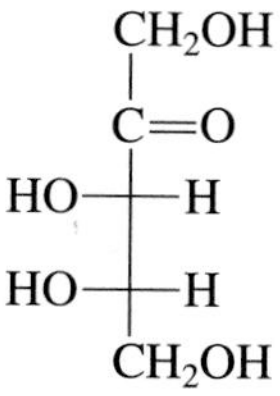
CH_2OH
C=O
HO—H
HO—H
CH_2OH

c.
CHO
HO—H
HO—H
H—OH
H—OH
CH_2OH

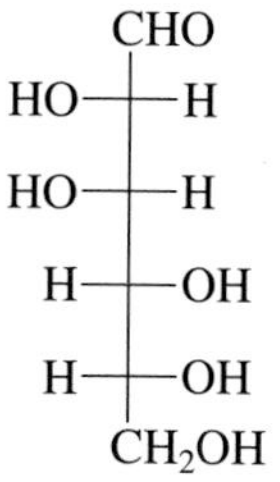

d. 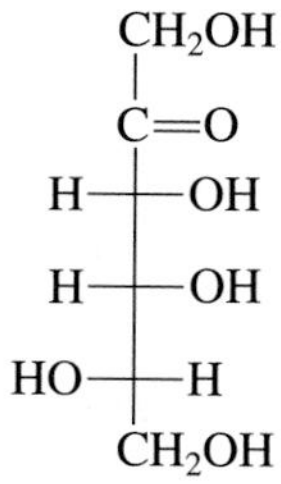
CH_2OH
C=O
H—OH
H—OH
HO—H
CH_2OH

18.29 Classify each of the molecules in Problem 18.27 as a D enantiomer or an L enantiomer.

18.30 Classify each of the molecules in Problem 18.28 as a D enantiomer or an L enantiomer.

18.31 Characterize the members of each of the following pairs of structures as (1) enantiomers, (2) diastereomers, or (3) neither enantiomers nor disastereomers.

a. 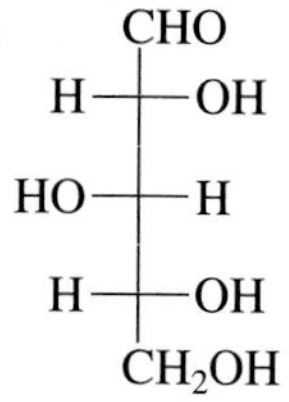
CHO / H—OH / HO—H / H—OH / CH_2OH
and
CHO / H—OH / H—OH / H—OH / CH_2OH

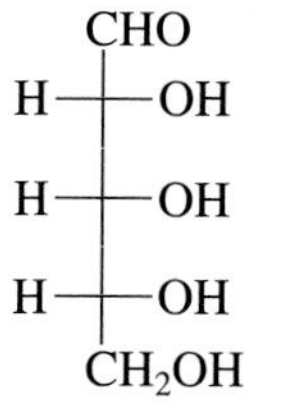

b. 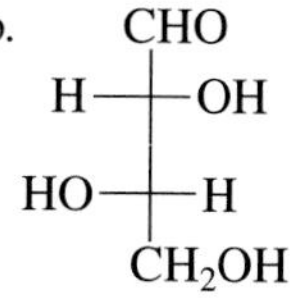
CHO / H—OH / HO—H / CH_2OH
and
CHO / H—H / HO—H / CH_2OH

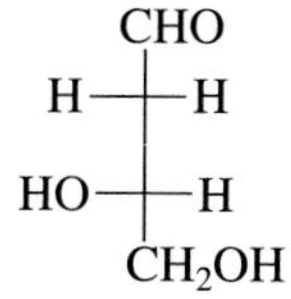

c. 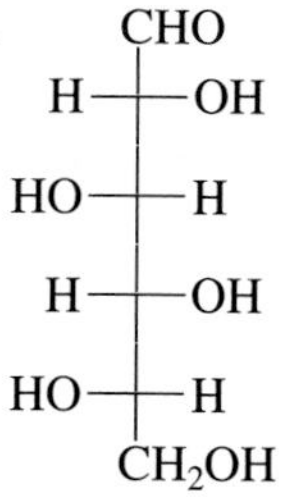
CHO / H—OH / HO—H / H—OH / HO—H / CH_2OH
and
CHO / HO—H / H—OH / HO—H / H—OH / CH_2OH

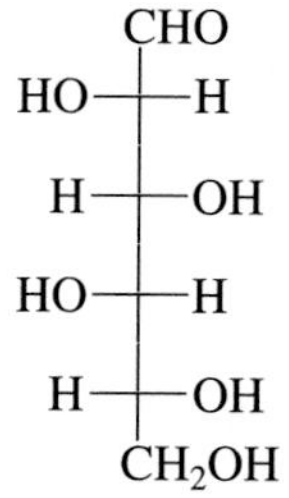

d. 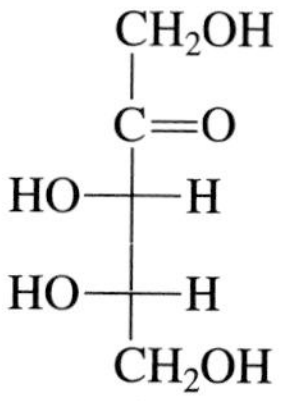
CH_2OH / C=O / HO—H / HO—H / CH_2OH
and
CH_2OH / C=O / HO—H / H—OH / CH_2OH

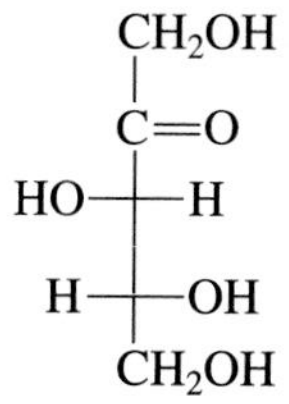

18.32 Characterize the members of each of the following pairs of structures as (1) enantiomers, (2) diastereomers, or (3) neither enantiomers nor disastereomers.

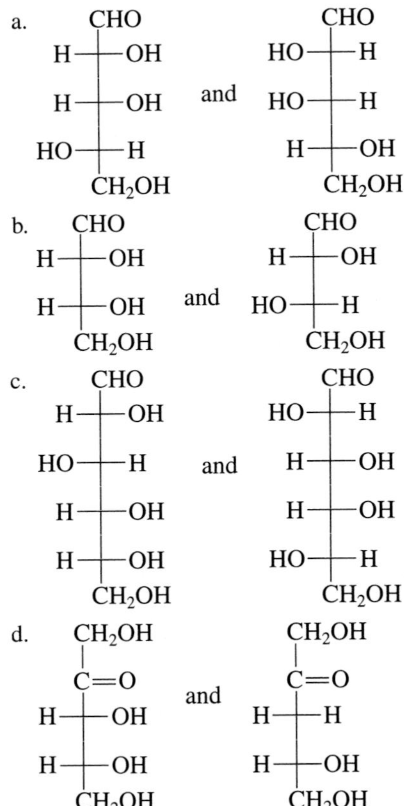

Properties of Enantiomers (Section 18.7)

18.33 The members of an enantiomeric pair would be expected to show differences in which of the following properties?

a. Solubility in an achiral solvent
b. Density
c. Melting point
d. Effect on plane-polarized light

18.34 The members of an enantiomeric pair would be expected to show differences in which of the following properties?

a. Solubility in a chiral solvent
b. Freezing point
c. Reaction with ethanol
d. Reaction with (+)-lactic acid

18.35 Compare (+)-lactic acid and (−)-lactic acid with respect to each of the following properties.

a. Boiling point
b. Optical activity
c. Solubility in water
d. Reaction with (+)-2,3-butanediol

18.36 Compare (+)-glyceraldehyde and (−)-glyceraldehyde with respect to each of the following properties.

a. Freezing point
b. Rotation of plane-polarized light
c. Reaction with ethanol
d. Reaction with (−)-2,3-butanediol

Classification of Monosaccharides (Section 18.8)

18.37 Classify each of the following monosaccharides as an aldose or a ketose.

a. 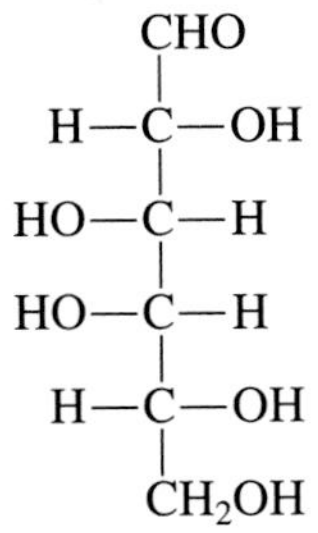 CHO
H—C—OH
HO—C—H
HO—C—H
H—C—OH
CH_2OH

b. CH_2OH
C=O
H—C—OH
H—C—OH
H—C—OH
CH_2OH

c. CH_2OH
C=O
CH_2OH

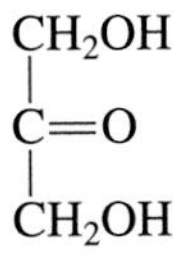

d. 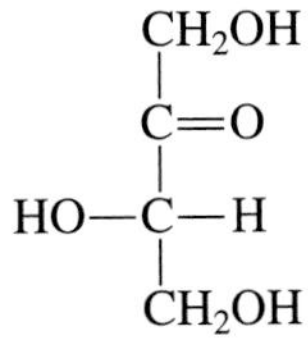 CH_2OH
C=O
HO—C—H
CH_2OH

18.38 Classify each of the following monosaccharides as an aldose or a ketose.

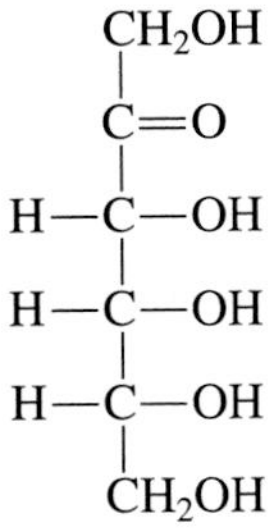

a. 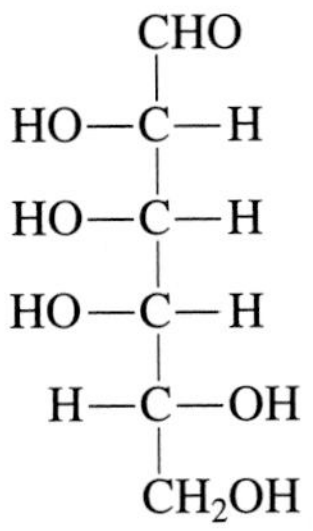 CHO
HO—C—H
HO—C—H
HO—C—H
H—C—OH
CH_2OH

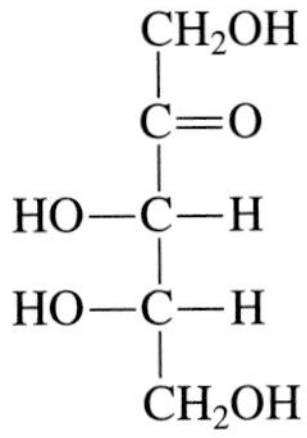

b. CH_2OH
C=O
HO—C—H
HO—C—H
CH_2OH

c. CHO
HO—C—H
H—C—OH
CH_2OH

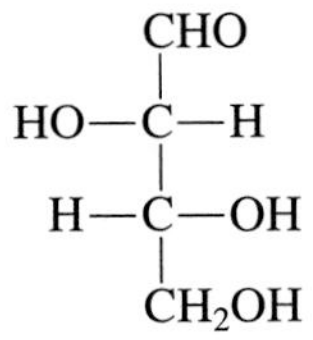

d. 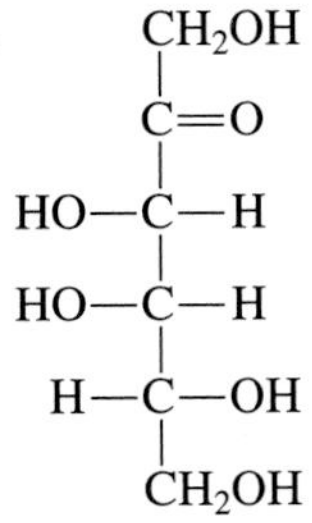 CH_2OH
C=O
HO—C—H
HO—C—H
H—C—OH
CH_2OH

18.39 Classify each monosaccharide in Problem 18.37 by its number of carbon atoms and its type of carbonyl group.

18.40 Classify each monosaccharide in Problem 18.38 by its number of carbon atoms and its type of carbonyl group.

18.41 Using the information in Figures 18.10 and 18.11, assign a name to each of the monosaccharides in Problem 18.37.

18.42 Using the information in Figures 18.10 and 18.11, assign a name to each of the monosaccharides in Problem 18.38.

Biologically Important Monosaccharides (Section 18.9)

18.43 What differences exist between the structures of D-glucose and D-fructose?

18.44 What differences exist between the structures of D-glucose and D-galactose?

Cyclic Forms of Monosaccharides (Section 18.10)

18.45 The intramolecular reaction that produces the cyclic forms of D-glucose involves a reaction between the carbonyl group and the —OH group on which carbon atom?

18.46 The intramolecular reaction that produces the cyclic forms of D-galactose involves a reaction between the carbonyl group and the —OH group on which carbon atom?

18.47 What is the structural difference between the α hemiacetal form of an aldose and its β form?

18.48 What is the structural difference between the α hemiketal form of a ketose and its β form?

18.49 Fructose contains six carbon atoms, and ribose has only five carbon atoms. Why do both of these monosaccharides have cyclic forms that contain the same number of carbon atoms (four)?

18.50 Fructose and glucose both contain six carbon atoms. Why do the cyclic forms of fructose have a five-membered ring instead of the six-membered ring found in the cyclic forms of glucose?

18.51 The structure of glucose is sometimes written in an open-chain form and sometimes as a cyclic hemiacetal structure. Explain why either form is acceptable.

18.52 When pure α-D-glucose is dissolved in water, β-D-glucose and α-D-glucose are both soon present. Explain how this is possible.

Haworth Projection Formulas (Section 18.11)

18.53 Identify each of the following structures as an α-D-monosaccharide or a β-D-monosaccharide.

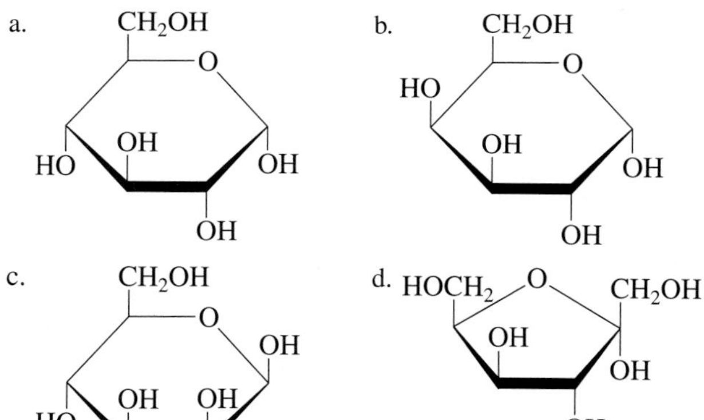

18.54 Identify each of the following structures as an α-D-monosaccharide or a β-D-monosaccharide.

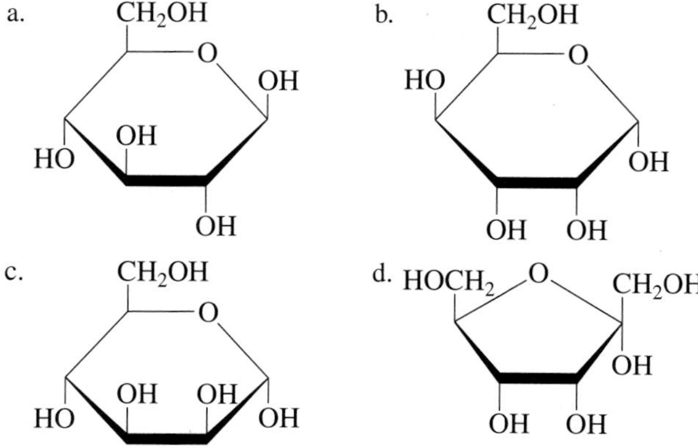

18.55 Identify each of the structures in Problem 18.53 as a hemiacetal or a hemiketal.

18.56 Identify each of the structures in Problem 18.54 as a hemiacetal or a hemiketal.

18.57 Draw the open-chain form for each of the monosaccharides in Problem 18.53.

18.58 Draw the open-chain form for each of the monosaccharides in Problem 18.54.

18.59 Using the information in Figures 18.10 and 18.11, assign a name to each of the monosaccharides in Problem 18.53.

18.60 Using the information in Figures 18.10 and 18.11, assign a name to each of the monosaccharides in Problem 18.54.

18.61 Draw the Haworth projection for each of the following monosaccharides.

a. α-D-Galactose b. β-D-Galactose
c. α-L-Galactose d. β-L-Galactose

18.62 Draw the Haworth projection for each of the following monosaccharides.

a. α-D-Mannose b. β-D-Mannose
c. α-L-Mannose d. β-L-Mannose

Reactions of Monosaccharides (Section 18.12)

18.63 Which of the following monosaccharides is a *reducing sugar?*

a. D-Glucose b. D-Galactose
c. D-Fructose d. D-Ribose

18.64 Which of the following monosaccharides will give a positive test with Benedict's solution?

a. D-Glucose b. D-Galactose
c. D-Fructose d. D-Ribose

18.65 In terms of oxidation and reduction, explain what occurs to both D-glucose and Tollens solution when they react with each other.

18.66 Describe the chemical reaction that is used to detect glucose in urine that involves Benedict's solution.

18.67 Draw structures for the following compounds.

a. Galactonic acid b. Galactaric acid
c. Galacturonic acid d. Galactitol

18.68 Draw structures for the following compounds.

a. Mannonic acid b. Mannaric acid
c. Mannuronic acid d. Mannitol

18.69 Identify each of the following compounds as an acetal or a ketal.

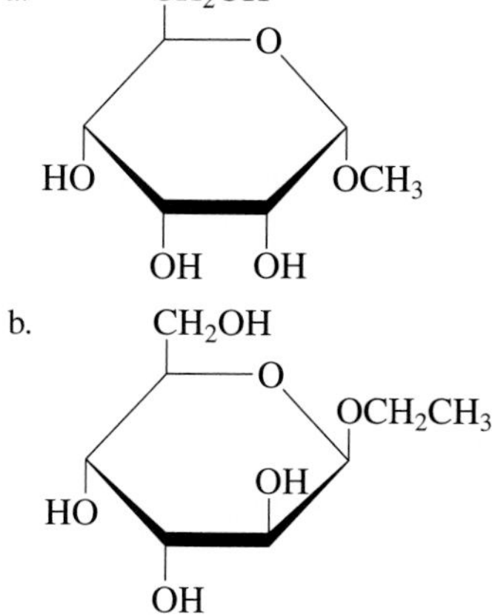

c. 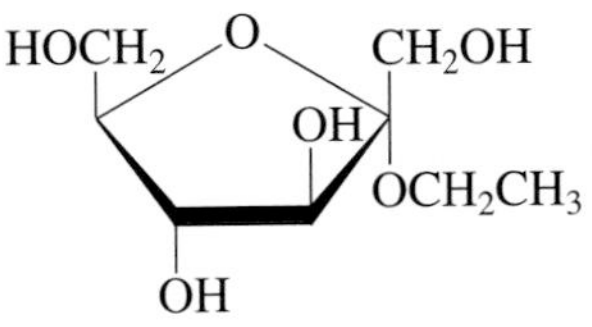 $HOCH_2$ O CH_2OH OH OCH_2CH_3 OH

d. 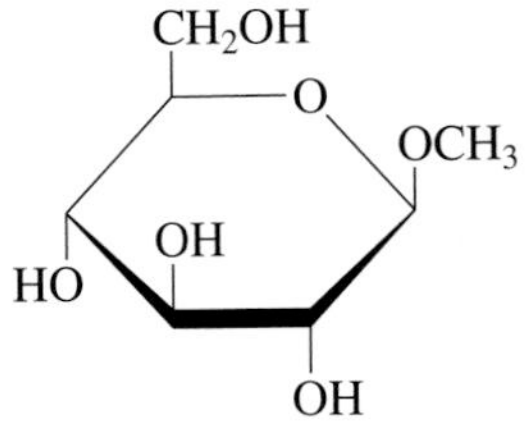 CH_2OH O OCH_3 OH HO OH

18.70 Identify each of the following compounds as an acetal or a ketal.

a. 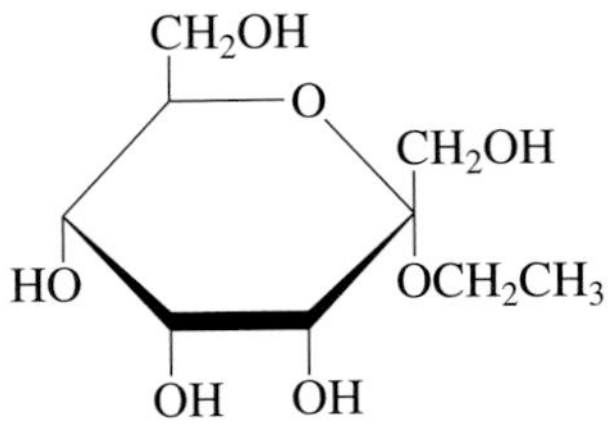 CH_2OH O CH_2OH HO OCH_2CH_3 OH OH

b. 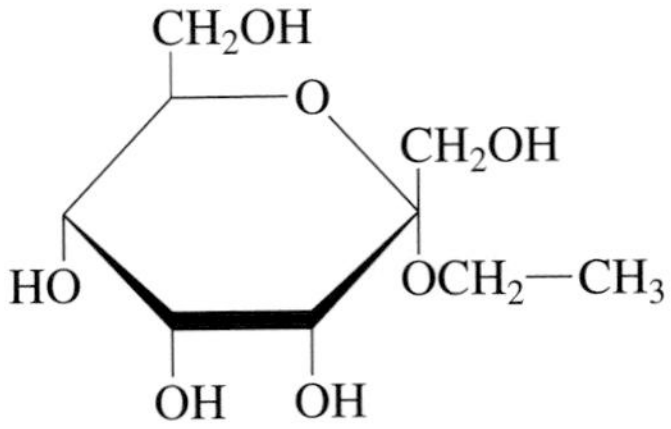 CH_2OH O CH_2OH HO $OCH_2—CH_3$ OH OH

c. 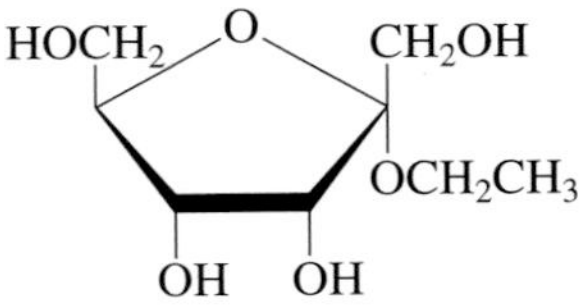 $HOCH_2$ O CH_2OH OCH_2CH_3 OH OH

d. 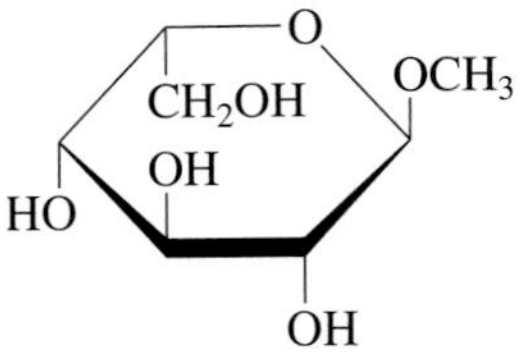 O OCH_3 CH_2OH OH HO OH

18.71 For each structure in Problem 18.69, identify the configuration at the acetal or ketal carbon atom as α or β.

18.72 For each structure in Problem 18.70, identify the configuration at the acetal or ketal carbon atom as α or β.

18.73 Identify the alcohol needed to produce each of the compounds in Problem 18.69 by reaction of the alcohol with the appropriate monosaccharide.

18.74 Identify the alcohol needed to produce each of the compounds in Problem 18.70 by reaction of the alcohol with the appropriate monosaccharide.

18.75 What is the difference in meaning between the terms *glycoside* and *glucoside?*

18.76 What is the difference in meaning between the terms *glycoside* and *galactoside?*

18.77 Draw structures for the following compounds.

a. Ethyl-β-D-glucoside

b. Methyl-α-D-galactoside

18.78 Draw structures for the following compounds.

a. Ethyl-α-D-galactoside

b. Methyl-β-D-glucoside

18.79 Draw structures for the following compounds.

a. α-D-Galactose 6-phosphate

b. *N*-acetyl-α-D-galactosamine

18.80 Draw structures for the following compounds.

a. α-D-Mannose 6-phosphate

b. *N*-acetyl-α-D-mannosamine

Disaccharides (Section 18.13)

18.81 What monosaccharides are produced from the hydrolysis of the following disaccharides?

a. Sucrose b. Maltose

c. Lactose d. Cellobiose

18.82 What type of glycosidic linkage [α(1 → 4), etc.] is present in the following disaccharides?

a. Sucrose b. Maltose

c. Lactose d. Cellobiose

18.83 Explain why lactose is a reducing sugar.

18.84 Explain why sucrose is not a reducing sugar.

18.85 Indicate whether each of the following disaccharides gives a positive or a negative Benedict's test.

a. Sucrose b. Maltose

c. Lactose d. Cellobiose

18.86 Indicate whether each of the following disaccharides gives a positive or a negative Tollens test.

a. Maltose b. Lactose

c. Cellobiose d. Sucrose

18.87 What type of glycosidic linkage [α(1 → 4), etc.] is present in each of the following disaccharides?

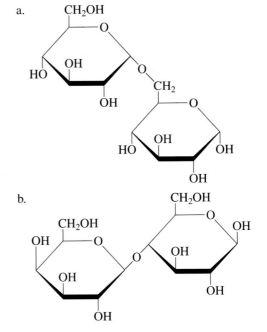

a. 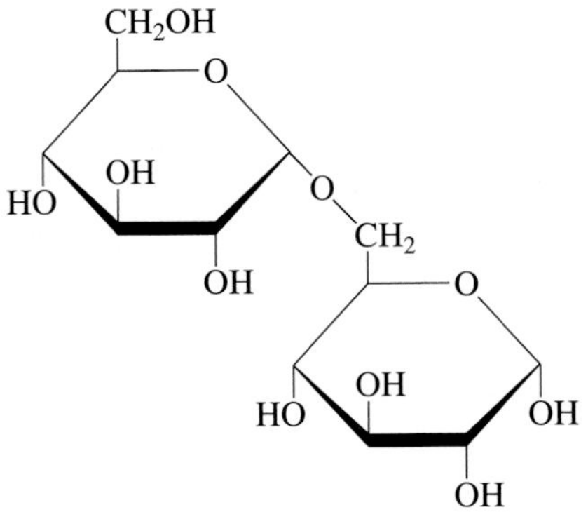 CH_2OH O OH HO OH O CH_2 O OH HO OH OH

b.

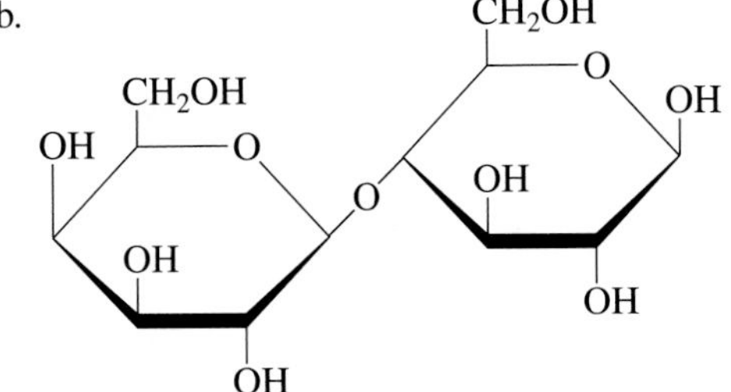

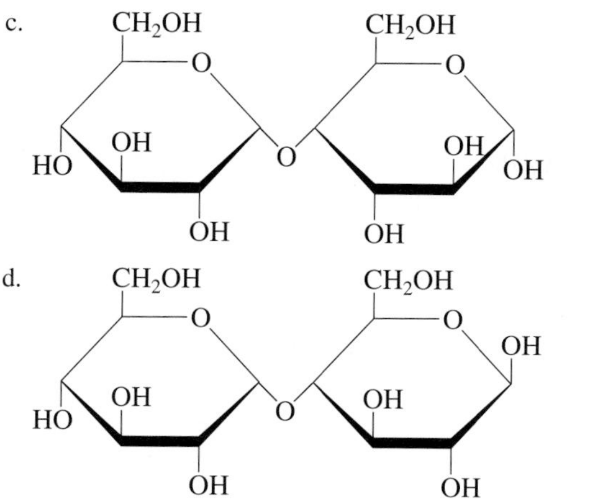

18.88 What type of glycosidic linkage [$\alpha(1 \rightarrow 4)$, etc.] is present in each of the following disaccharides?

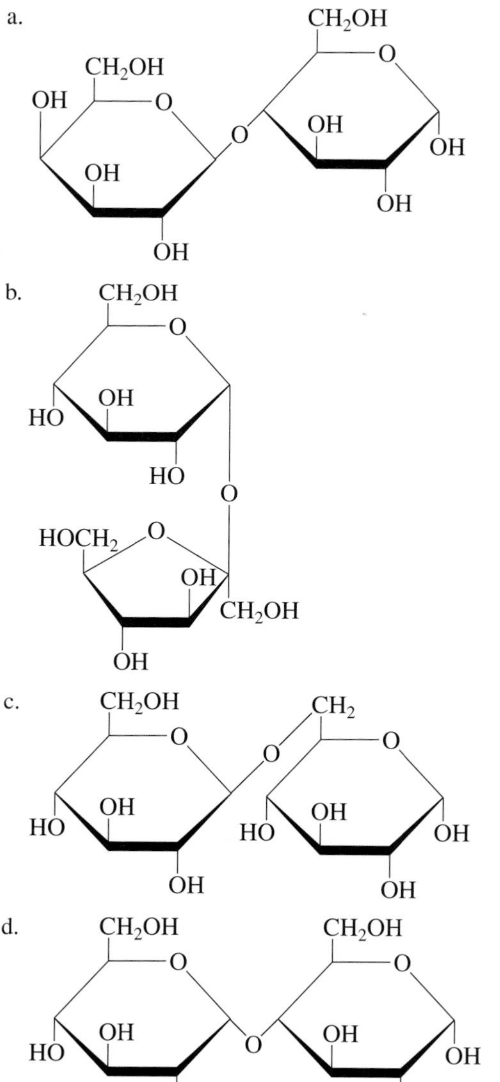

18.89 For each of the structures in Problem 18.87, specify whether the disaccharide is in an α or a β configuration, or neither.

18.90 For each of the structures in Problem 18.88, specify whether the disaccharide is in an α or a β configuration, or neither.

18.91 Identify each of the structures in Problem 18.87 as a reducing sugar or a nonreducing sugar.

18.92 Identify each of the structures in Problem 18.88 as a reducing sugar or a nonreducing sugar.

18.93 Give the identity of the monosaccharides present in each of the structures in Problem 18.87 (see Figures 18.10 and 18.11).

18.94 Give the identity of the monosaccharides present in each of the structures in Problem 18.88 (see Figures 18.10 and 18.11).

Polysaccharides (Section 18.14)

18.95 Describe the structural differences and similarities between the following pairs of polysaccharides.

a. Glycogen and amylopectin
b. Amylose and cellulose

18.96 Describe the structural differences and similarities between the following pairs of polysaccharides.

a. Amylose and glycogen
b. Amylose and amylopectin

18.97 Match each of the following structural characteristics to the polysaccharides amylopectin, amylose, glycogen, cellulose, and chitin. A specific characteristic may apply to more than one polysaccharide.

a. Contains both $\alpha(1 \rightarrow 4)$ and $\alpha(1 \rightarrow 6)$ glycosidic linkages
b. Composed of glucose monosaccharide units
c. Composed of unbranched molecular chains
d. Contains only $\beta(1 \rightarrow 4)$ glycosidic linkages

18.98 Match each of the following structural characteristics to the polysaccharides amylopectin, amylose, glycogen, cellulose, and chitin. A specific characteristic may apply to more than one polysaccharide.

a. Contains acetal linkages between monosaccharide units
b. Contains only $\alpha(1 \rightarrow 4)$ glycosidic linkages
c. Monosaccharide units are derivatives of glucose
d. Composed of highly branched molecular chains

18.99 Why, when both contain D-glucose, can humans digest starch but not cellulose?

18.100 What is the difference between plant starch and animal starch?

Additional Problems

18.101 Indicate whether each of the following compounds is chiral or achiral.

a. 1-Chloro-2-methylpentane b. 2-Chloro-2-methylpentane
c. 2-Chloro-3-methylpentane d. 3-Chloro-2-methylpentane

18.102 Which of the following compounds is (are) optically active?

a. H / Cl—C—Cl / Cl

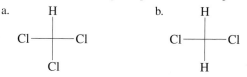

b. H / Cl—C—Cl / H

c. 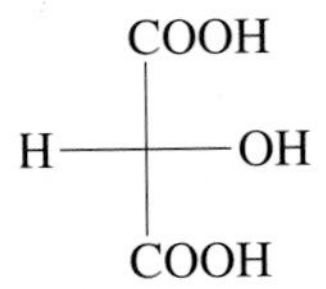 COOH
H—C—OH
COOH

d. 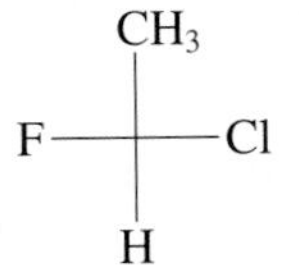 CH_3
F—C—Cl
H

18.103 In which of the following pairs of monosaccharides do both members of the pair contain the same number of carbon atoms?

a. Glyceraldehyde and glucose
b. Dihydroxyketone and ribose
c. Ribose and deoxyribose
d. Glyceraldehyde and dihydroxyketone

18.104 Draw Fischer projections for the four stereoisomers of the molecule

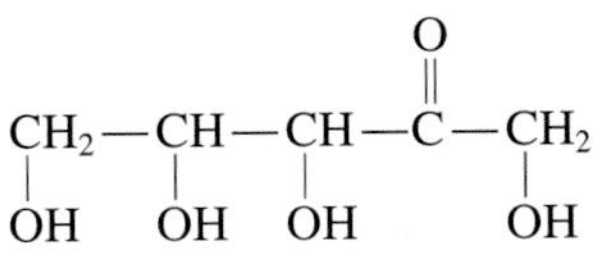

$$\underset{\text{OH}}{CH_2}-\underset{\text{OH}}{CH}-\underset{\text{OH}}{CH}-\overset{\text{O}}{\overset{\|}{C}}-\underset{\text{OH}}{CH_2}$$

18.105 What is the alkane of lowest molecular mass that is a chiral compound?

18.106 What monosaccharide(s) is (are) obtained from the hydrolysis of each of the following?

a. Sucrose
b. Glycogen
c. Starch
d. Amylose

18.107 Classify each of the following carbohydrates as a homopolysaccharide or a heteropolysaccharide.

a. Chitin
b. Amylopectin
c. Hyaluronic acid
d. Glycogen

18.108 List the reactant(s) necessary to effect the following chemical changes.

a. 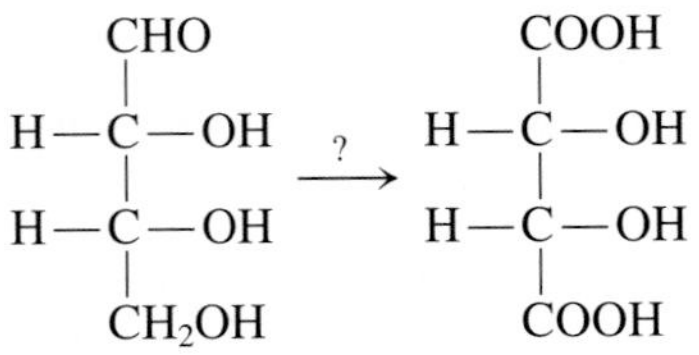 CHO / H—C—OH / H—C—OH / CH_2OH $\xrightarrow{?}$ COOH / H—C—OH / H—C—OH / COOH

b. ring with CH_2OH, HO, OH, OH, OH $\xrightarrow{?}$ ring with CH_2OH, HO, OH, OH, O—CH_2—CH_3

c. disaccharide (CH_2OH, HO, OH, OH; O linkage; CH_2OH, OH, OH, OH) $\xrightarrow{?}$ ring (CH_2OH, HO, OH, OH, OH) + ring (CH_2OH, HO, OH, OH, OH)

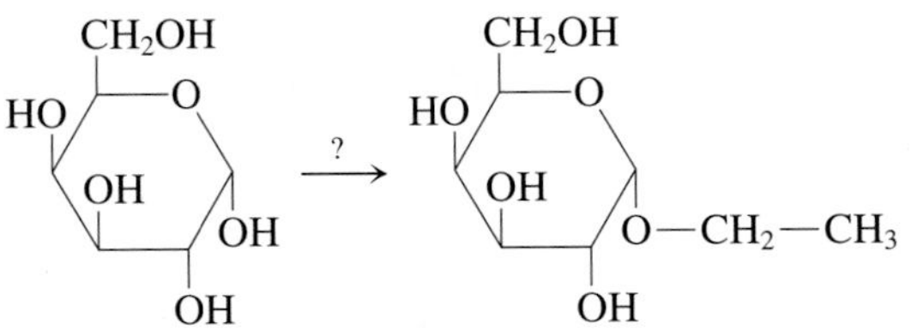

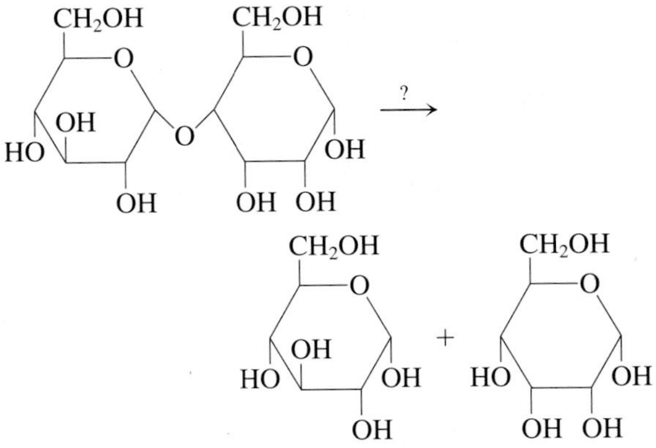

d. 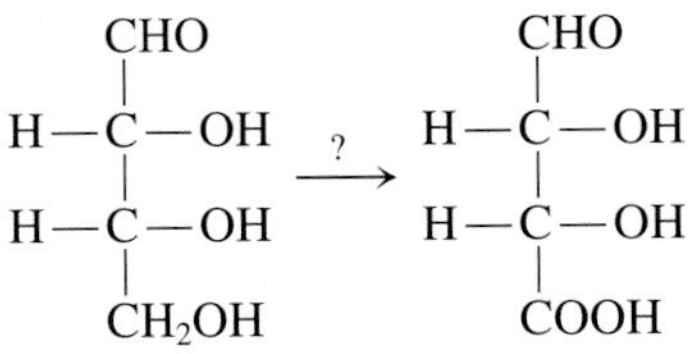 CHO / H—C—OH / H—C—OH / CH_2OH $\xrightarrow{?}$ CHO / H—C—OH / H—C—OH / COOH

Grid Problems

18.109

1. glucose	2. galactose	3. fructose
4. ribose	5. mannose	6. xylose

Select from the grid *all* correct responses for each of the following situations.

a. Compounds that are monosaccharides, aldoses, and pentoses
b. Compounds that can exist in D or L forms
c. Compounds that contain four —OH groups
d. Compounds that are chiral

18.110

1. α-maltose	2. α-lactose	3. α-cellobiose
4. β-maltose	5. β-lactose	6. sucrose

Select from the grid *all* correct responses for each of the following situations.

a. Compounds that contain a $\beta(1 \rightarrow 4)$ glycosidic linkage
b. Compounds that yield one product on hydrolysis
c. Disaccharides in which one of the structural units is glucose
d. Reducing sugars

18.111

1. amylose	2. amylopectin	3. glycogen
4. cellulose	5. chitin	6. starch

Select from the grid *all* correct responses for each of the following situations.

a. Homopolysaccharides
b. Polysaccharides containing some $\alpha(1 \rightarrow 6)$ linkages
c. Unbranched polysaccharides
d. Polysaccharides that produce only glucose upon complete hydrolysis

18.112

1. D-glucose	2. D-fructose	3. D-ribose
4. L-glucose	5. L-fructose	6. L-galactose

Select from the grid *all* correct responses for each of the following situations.

a. Compounds with three chiral centers
b. Pairs of compounds that are enantiomers
c. Pairs of compounds that are diastereomers
d. Compounds that are optically active

CHAPTER

19

PART III

Lipids

CHAPTER OUTLINE

19.1 Characteristics of Lipids 554
19.2 Fatty Acids: Saponifiable Lipid Building Blocks 555
19.3 Physical Properties of Fatty Acids 558

Chemistry at a Glance: **Classification Schemes for Fatty Acids** 559

19.4 Fats and Oils 559
19.5 Chemical Reactions of Triacylglycerols 563
19.6 Phosphoacylglycerols 567
19.7 Waxes 569
19.8 Sphingolipids 570

Chemistry at a Glance: **Saponifiable Lipids** 572

19.9 Nonsaponifiable Lipids 572
19.10 Steroids 573
19.11 Eicosanoids 577
19.12 Plasma Membranes 578
19.13 Transport Across Cell Membranes 580

Chemical Connections

19.1 Artificial Fat Substitutes 563
19.2 The Cleansing Action of Soap 565
19.3 *Trans* Fatty Acids and Blood Cholesterol Levels 566
19.4 Steroid Drugs in Sports 576

Fats and oils are the most widely occurring types of lipids. Thick layers of fat help insulate polar bears against the effects of low temperatures.

In the previous chapter, we noted that there are four major classes of bioorganic substances: carbohydrates, lipids, proteins, and nucleic acids. We now turn our attention to the second of the bioorganic classes, the compounds we call lipids. Lipids are a structurally heterogeneous class of bioorganic compounds and include such diverse types of molecules as fats and oils, waxes, cholesterol and its derivatives, some vitamins, and prostaglandins. What these substances have in common is limited solubility in water.

19.1 Characteristics of Lipids

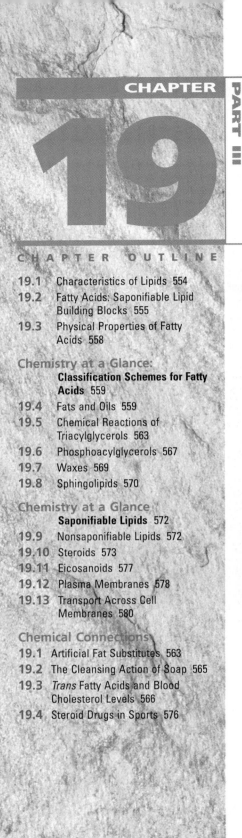

Lipids form a large class of relatively water-insoluble bioorganic compounds. In humans and many animals, excess carbohydrates and other energy-yielding foods are converted to, and stored in the body in the form of, lipids called fats. These fat reservoirs constitute a major way of storing chemical energy and carbon atoms in the body. Fats and other lipids also surround and insulate vital body organs, providing protection from mechanical shock and helping to maintain correct body temperature. Lipids function as coverings for nerve fibers and as the basic structural components of all cell membranes. Many chemical messengers in the human body, substances called hormones, are lipids.

Lipids, unlike carbohydrates and most other classes of compounds, cannot be defined from a structural viewpoint. A variety of functional groups and structural features are found in molecules classified as lipids. What lipids share are their solubility properties. **Lipids** *are a structurally heterogeneous group of substances of biological origin that are only sparingly soluble, if at all, in water but are soluble in nonpolar organic solvents.* When biological material (animal or plant tissue) is homogenized in a blender and mixed with a nonpolar organic solvent, the substances that dissolve in the solvent are the lipids.

• The term *lipid* comes from the Greek *lipos,* which means "fat" or "lard."

Lipids can be divided into two major classes on the basis of whether they undergo hydrolysis reactions in alkaline (basic) solution. **Saponifiable lipids** *can be hydrolyzed under alkaline conditions to yield salts of fatty acids.* **Nonsaponifiable lipids** *do not undergo hydrolysis reactions in alkaline solution.*

• The term *saponification* has been encountered previously. In Section 16.13 we learned that saponification of an ester involves breaking of the ester linkage to produce an alcohol and a carboxylic acid salt. In Section 17.16 we learned that amide linkages are also subject to saponification, producing an amine and a carboxylic acid salt. Ester and/or amide linkages are present in all saponifiable lipids.

We begin our consideration of lipids with an in-depth discussion of saponifiable lipids. Information about nonsaponifiable lipids follows.

19.2 Fatty Acids: Saponifiable Lipid Building Blocks

We begin our discussion of saponifiable lipids by considering fatty acids, compounds that are a building block in all saponifiable lipid structures.

• Fatty acids were first isolated from naturally occurring fats; hence the designation *fatty acids.*

Fatty acids *are naturally occurring carboxylic acids with an unbranched carbon chain and an even number of carbon atoms.* They are rarely found free in nature but rather occur mostly in esterified form in the structures of saponifiable lipids. Because of the pathway by which fatty acids are biosynthesized (Section 25.7) they almost always contain an even number of carbon atoms. *Long-chain* fatty acids (12 to 26 carbon atoms) are found in meats and fish; *medium-chain* fatty acids (6 to 10 carbon atoms) and *short-chain* fatty acids (fewer than 6 carbon atoms) occur primarily in dairy products.

• Saturated and Unsaturated Fatty Acids

Fatty acids are further classified as saturated, monounsaturated, or polyunsaturated. **Saturated fatty acids** *have a carbon chain in which all carbon–carbon bonds are single bonds.* An example is hexadecanoic acid, a 16-carbon acid whose common name is palmitic acid.

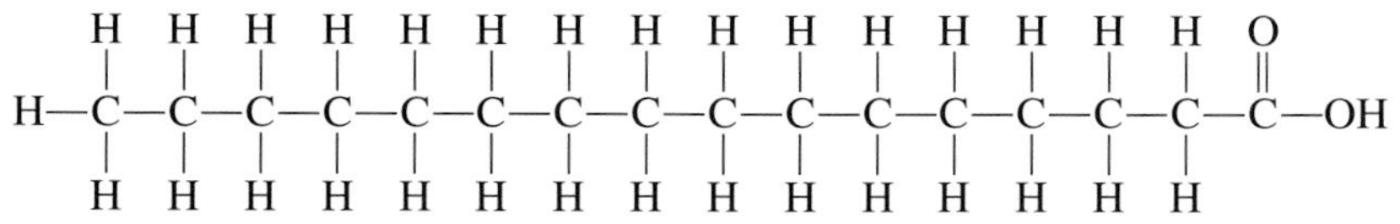

$$\mathrm{H{-}\overset{H}{\underset{H}{C}}{-}\overset{H}{\underset{H}{C}}{-}\overset{H}{\underset{H}{C}}{-}\overset{H}{\underset{H}{C}}{-}\overset{H}{\underset{H}{C}}{-}\overset{H}{\underset{H}{C}}{-}\overset{H}{\underset{H}{C}}{-}\overset{H}{\underset{H}{C}}{-}\overset{H}{\underset{H}{C}}{-}\overset{H}{\underset{H}{C}}{-}\overset{H}{\underset{H}{C}}{-}\overset{H}{\underset{H}{C}}{-}\overset{H}{\underset{H}{C}}{-}\overset{H}{\underset{H}{C}}{-}\overset{H}{\underset{H}{C}}{-}\overset{O}{\overset{\|}{C}}{-}OH}$$

Hexadecanoic acid (palmitic acid)

The structural formulas of fatty acids are usually written in a more condensed form than the preceding structural formula. Two alternative notations for palmitic acid's structure are

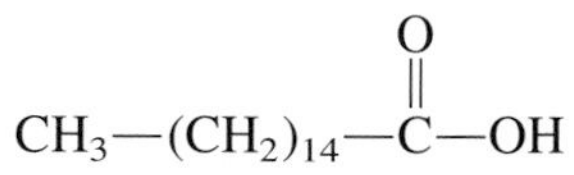

$$CH_3{-}(CH_2)_{14}{-}\overset{O}{\overset{\|}{C}}{-}OH$$

and

[line-angle zigzag chain]–COOH

(We first encountered "line-angle notation" in Section 12.11.)

Monounsaturated fatty acids *have a carbon chain in which one carbon–carbon double bond is present.* In naturally occurring and biologically important unsaturated fatty acids, the configuration about the double bond is almost always *cis* (Section 14.5). Different ways of depicting the structure of a monounsaturated fatty acid are as follows; the molecule shown is the 18-carbon acid with one double bond (*cis*-9-octadecenoic acid, or oleic acid).

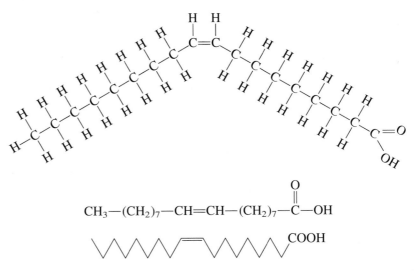

$$CH_3-(CH_2)_7-CH{=}CH-(CH_2)_7-\overset{\displaystyle O}{\overset{\|}{C}}-OH$$

COOH

The first of these structures correctly emphasizes that the presence of a *cis* double bond in the carbon chain puts a rigid 30° bend in the chain. Such a bend affects the physical properties of a fatty acid, as we will see in Section 19.3.

• More than 500 different fatty acids have been isolated from the lipids of microorganisms, plants, animals, and humans. These fatty acids differ from one another in the length of their carbon chains, their degree of unsaturation (number of double bonds), and the positions of the double bonds in the chains.

Polyunsaturated fatty acids *have a carbon chain in which two or more carbon–carbon double bonds are present.* Up to six double bonds are found in biologically important unsaturated fatty acids.

Table 19.1 gives structures and names for the naturally occurring fatty acids most often found in saponifiable lipid structures. Fatty acids are almost always referred to by their common names. IUPAC names, although easily understandable, are usually quite long. Selected IUPAC names are given as a footnote to Table 19.1.

• Omega-3 and Omega-6 Fatty Acids

Not only do the fatty acids come in three categories (saturated, monounsaturated, and polyunsaturated), but the polyunsaturates also belong to families, two of which are particularly important in human body chemistry. These two important families are the omega-6 and the omega-3 fatty acids.

The basis for the omega classification system involves the following considerations. A fatty acid has two ends, designated as the methyl (CH_3) end and the carboxyl (COOH) end.

$$\underbrace{CH_3-(CH_2)_4}_{\text{Methyl end}}-(CH{=}CH-CH_2)_4-\underbrace{(CH_2)_2-C\overset{\diagup O}{\underset{\diagdown OH}{}}}_{\text{Carboxyl end}}$$

Fish that live in deep, cold water—mackerel, herring, tuna, salmon, and bluefish—are better sources of omega-3 fatty acids than other fish.

In the omega classification system, the carbon chain is numbered beginning at the methyl end, which is the reverse of the usual way carbon chains are numbered in the naming of simple acids. The "reverse" numbering system is used because of the mechanism by which fatty acid carbon chains are lengthened during biotransformations within the body. Lengthening involves adding carbon atoms, two at a time, at the carboxyl end of the chain. Thus, by numbering polyunsaturated fatty acid chains from the methyl end, which does not change, chemists ease the task of keeping track of fatty acid identities. When an omega-3 acid is lengthened, the new acid is still an omega-3 acid.

An **omega-3 fatty acid** *is a polyunsaturated fatty acid with its endmost double bond three carbons away from its methyl end.* An **omega-6 fatty acid** *is a polyunsaturated fatty*

Table 19.1
Selected Fatty Acids of Biological Importance

Structure notation[a]	Common name[b]	Structure
Saturated Fatty Acids		
12:0	lauric acid	COOH
14:0	myristic acid	COOH
16:0	palmitic acid	COOH
18:0	stearic acid	COOH
20:0	arachidic acid	COOH
Monounsaturated Fatty Acids		
16:1	palmitoleic acid	COOH
18:1	oleic acid	COOH
Polyunsaturated Fatty Acids		
18:2	linoleic acid	COOH
18:3	linolenic acid	COOH
20:4	arachidonic acid	COOH
20:5	EPA (eicosapentaenoic acid)	COOH
22:6	DHA (docosahexaenoic acid)	COOH

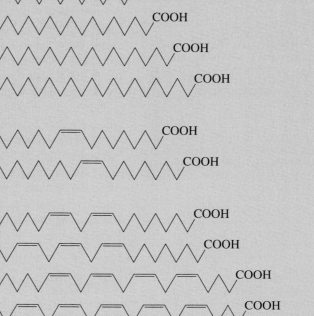

[a]In this shorthand notation for identifying fatty acids, the number to the left of the colon gives the number of carbon atoms present, and the number to the right of the colon gives the number of carbon–carbon double bonds present. The notation can be further extended to give double-bond locations by adding a Δ symbol and numbers as superscripts. With this extended notation, linoleic acid would be denoted as $18{:}2^{\Delta 9,12}$ and linolenic acid as $18{:}3^{\Delta 9,12,15}$. Carbon atoms are numbered beginning with the carboxyl carbon atom.

[b]Common names are used instead of IUPAC names in most situations because of the length of the IUPAC names. Selected examples of IUPAC names follow.

16:1 acid *cis*-9-hexadecenoic acid
18:2 acid *cis,cis*-9,12-octadecadienoic acid
20:5 acid *cis,cis,cis,cis,cis*-5,8,11,14,17-eicosapentaenoic acid

acid with its endmost double bond six carbons away from its methyl end. Table 19.2 lists the omega-3 and omega-6 fatty acids that are most important from a nutritional viewpoint.

• Essential Fatty Acids

• Humans lack the enzymes needed to introduce double bonds at carbon atoms beyond carbon-9 in a fatty acid carbon chain. Hence linoleic and linolenic acids cannot be synthesized within the body.

Essential fatty acids *are fatty acids that are needed by the human body and must be obtained from dietary sources because they cannot be synthesized within the body from other substances.* There are two essential fatty acids: *linoleic acid* and *linolenic acid.* Both are 18-carbon polyunsaturated acids; the former is an omega-6 acid, and the latter is an omega-3 acid. Their structures are given in Table 19.1.

Both of the essential fatty acids contribute to proper membrane structure. When linoleic acid is missing from the diet, the skin reddens and becomes irritated, infections and dehydration are likely to occur, and the liver may develop abnormalities. If the fatty acid is restored, then the conditions reverse themselves. Infants are especially in need

Table 19.2
Nutritionally Important Omega-3 and Omega-6 Fatty Acids

Omega-3 acids
linolenic acid (lin-oh-LEN-ic) (18:3)
eicosapentaenoic acid (EYE-cossa-PENTA-ee-NO-ic) (20:5)
docosahexaenoic acid (DOE-cossa-HEXA-ee-NO-ic) (22:6)
Omega-6 acids
linoleic acid (lin-oh-LAY-ic) (18:2)
arachidonic acid (a-RACK-ih-DON-ic) (20:4)

of linoleic acid for their growth. Human breast milk has a much higher percentage of it than cow's milk.

Linoleic acid and linolenic acid furnished by the diet are also the starting points for the synthesis of a variety of other longer-chain polyunsaturated acids, including arachidonic acid, EPA, and DHA (see Table 19.1). Arachidonic acid is the major precursor of eicosanoids (Section 19.11), substances that help regulate blood pressure, clotting, and several other important body functions. EPA and DHA are important constituents of the communication membranes of the brain and are necessary for normal brain development. EPA and DHA are also active in the retina of the eye.

Normally, vegetable oils and meats supply enough linoleic acid and other omega-6 acids to meet the body's needs. Optimal amounts of linolenic acid and other omega-3 acids are often lacking in the typical American diet. Because fish is a good source of omega-3 acids, nutritionists recommend adding more fish to the diet.

In 1985, researchers reported that the Inuit people of Greenland have a low incidence of heart disease despite a diet very high in fat. Studies on the U.S. population show correlation between a high-fat diet and a high incidence of heart disease. What accounts for the difference between the two peoples?

A U.S. resident consumes about double the amount of omega-6 acids and half the amount of omega-3 acids that an Inuit consumes. Researchers speculate that this difference may explain why the heart attack rate is so much higher in the United States. It may also explain why the people in Greenland have prolonged bleeding times and a high incidence of stroke. Studies such as this show the importance of a well-balanced diet. Ideally, the ratio between omega-6 and omega-3 fatty acids in the diet should be 4–10 grams of omega-6 acids to 1 gram of omega-3 acids.

19.3 Physical Properties of Fatty Acids

The melting points of fatty acids depend on both the length of their hydrocarbon chains and their degree of unsaturation (number of double bonds per molecule). The graph in Figure 19.1 shows melting-point variation as a function of both of these variables. A trend of particular significance is that saturated acids have higher melting points than unsaturated acids with the same number of carbon atoms. The greater the degree of unsaturation, the greater the reduction in melting point. Figure 19.1 shows this effect for the 18-carbon acids with zero, one, two, and three double bonds.

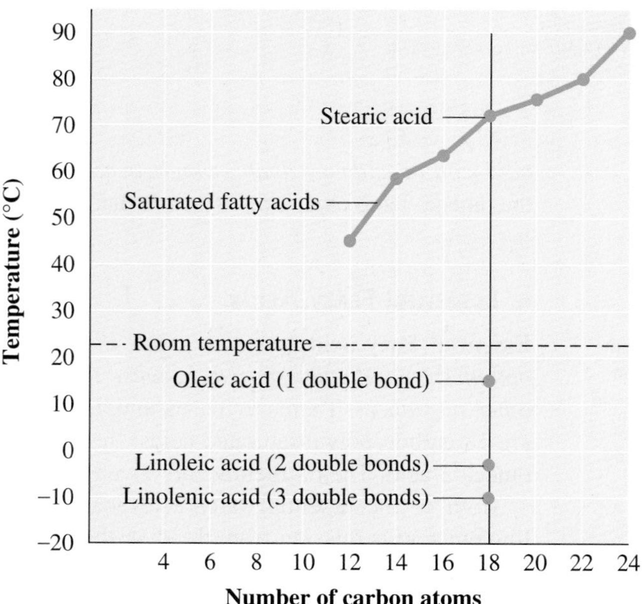

Figure 19.1
The melting point of a fatty acid depends on the length of the carbon chain and on the number of double bonds present in the carbon chain.

CHEMISTRY AT A GLANCE

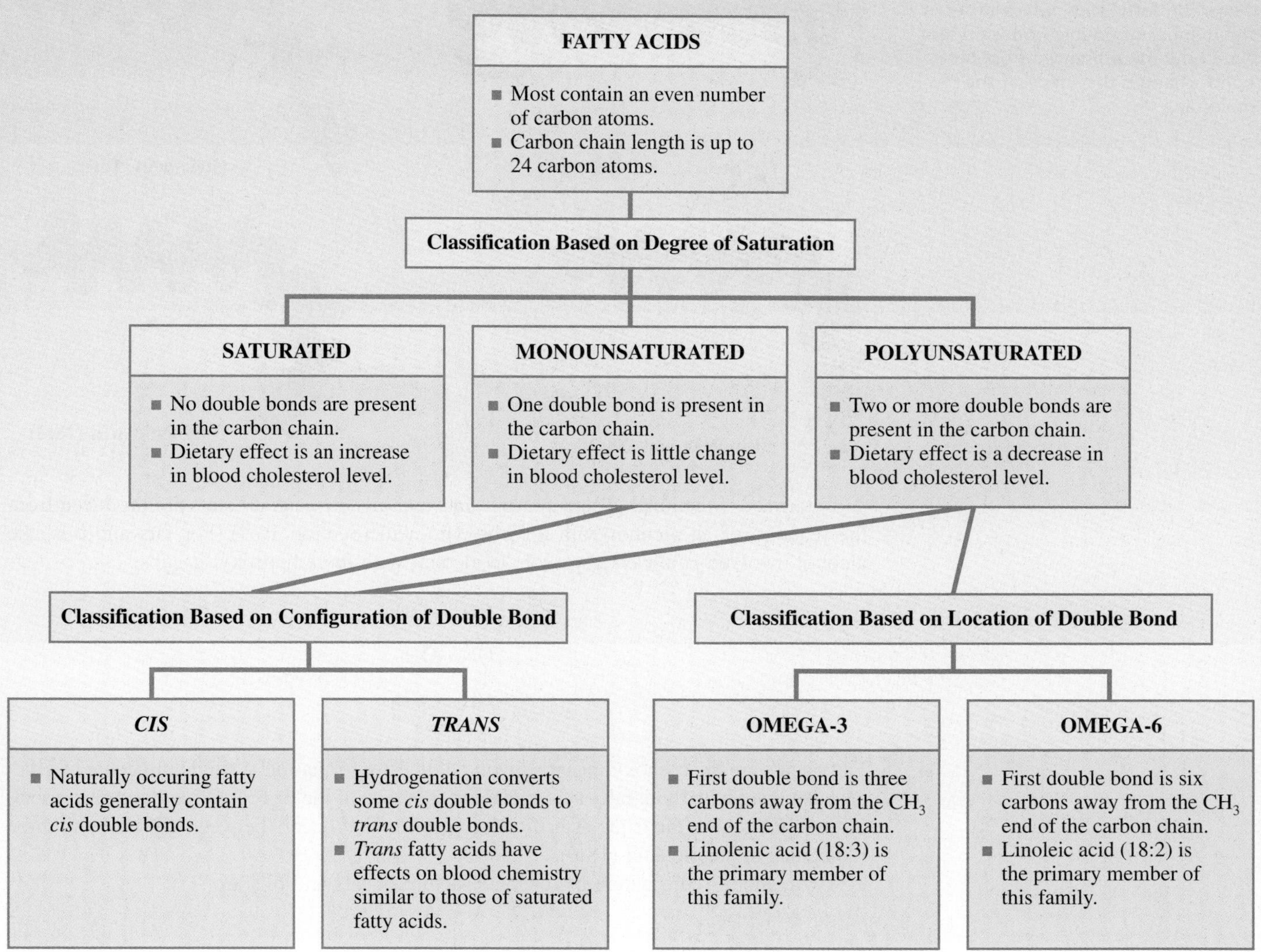

FATTY ACIDS
■ Most contain an even number of carbon atoms.
■ Carbon chain length is up to 24 carbon atoms.
Classification Based on Degree of Saturation
SATURATED
■ No double bonds are present in the carbon chain.
■ Dietary effect is an increase in blood cholesterol level.
MONOUNSATURATED
■ One double bond is present in the carbon chain.
■ Dietary effect is little change in blood cholesterol level.
POLYUNSATURATED
■ Two or more double bonds are present in the carbon chain.
■ Dietary effect is a decrease in blood cholesterol level.
Classification Based on Configuration of Double Bond
Classification Based on Location of Double Bond
CIS
■ Naturally occuring fatty acids generally contain cis double bonds.
TRANS
■ Hydrogenation converts some cis double bonds to trans double bonds.
■ Trans fatty acids have effects on blood chemistry similar to those of saturated fatty acids.
OMEGA-3
■ First double bond is three carbons away from the CH_3 end of the carbon chain.
■ Linolenic acid (18:3) is the primary member of this family.
OMEGA-6
■ First double bond is six carbons away from the CH_3 end of the carbon chain.
■ Linoleic acid (18:2) is the primary member of this family.

• Fatty acids have low water solubilities, which decrease with increasing carbon chain length; at 30°C, lauric acid (12:0) has a water solubility of 0.063 g/L and stearic acid (18:0) a solubility of 0.0034 g/L. Contrast this with glucose's solubility in water at the same temperature, 1100 g/L.

The decreasing melting point associated with increasing degree of unsaturation in fatty acids is explained by decreased molecular attractions between carbon chains. The double bonds in unsaturated fatty acids, which generally have the *cis* configuration, produce "bends" in the carbon chains of these molecules (see Figure 19.2). These "bends" prevent unsaturated fatty acids from packing together as tightly as fully saturated fatty acids. The greater the number of double bonds, the less efficient the packing. As a result, unsaturated fatty acids always have fewer intermolecular attractions and therefore lower melting points than their saturated counterparts.

The accompanying Chemistry at a Glance is a review of the classification of fatty acids.

19.4 Fats and Oils

The most abundant fatty-acid-containing (saponifiable) lipids are the *fats* and *oils*. Because these two types of compounds have the same general chemical structure, we will consider them at the same time.

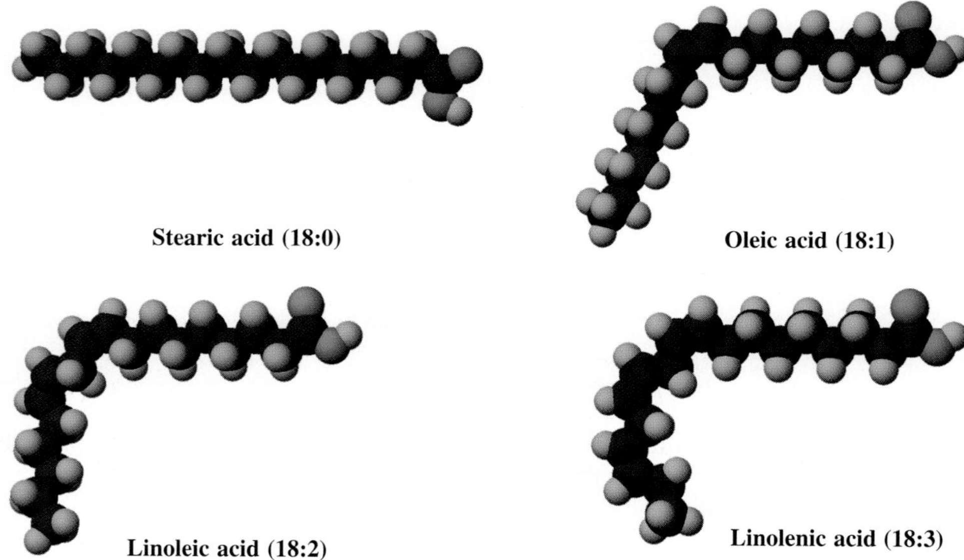

Figure 19.2 Space-filling models of four 18-carbon fatty acids, which differ in the number of double bonds present. Note how the presence of double bonds changes the shape of the molecule.

In terms of functional groups present, fats and oils are esters. Esters are produced from the reaction of an alcohol with a carboxylic acid (Section 16.9). For fats and oils, the alcohol involved is always glycerol, an alcohol with three hydroxyl groups.

$$\begin{array}{l} CH_2\text{—}OH \\ | \\ CH\text{—}OH \\ | \\ CH_2\text{—}OH \end{array}$$

Glycerol

Fatty acids are the carboxylic acids involved. In ester formation, a single molecule of glycerol can react with three fatty acid molecules because of the three hydroxyl groups present. Figure 19.3 shows the triple esterification reaction that occurs between glycerol and three molecules of stearic acid (18:0).

Two general representations for the structure of fats and oils are

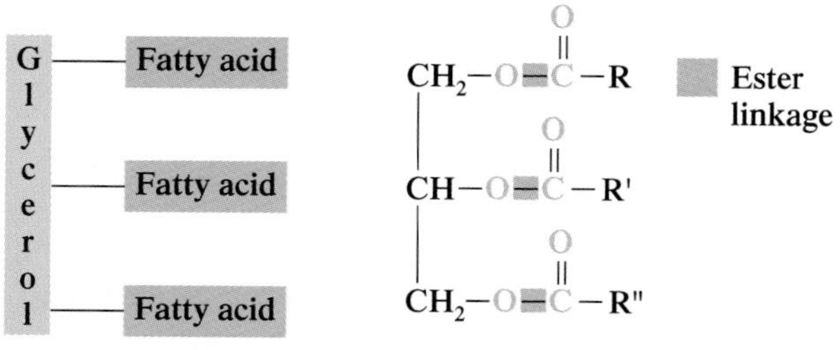

The first representation, a block diagram, shows that four structural components are present in the structure: glycerol and three fatty acids. The second representation, a general structural formula, shows the triester nature of fats and oils. Each of the fatty acids is attached to glycerol through an ester linkage.

• Triacylglycerols do not actually contain glycerol and three fatty acids, as the block diagram for a triacylglycerol implies. They actually contain a glycerol *residue* and three fatty acid *residues.* In the formation of the triacylglycerol, three molecules of water have been removed from the structural components of the fat or oil, leaving residues of the reacting molecules.

Fats and oils are formally called triacylglycerols. A **triacylglycerol** *is a compound formed by esterification of three fatty acids to glycerol.* An *acyl group* is the group that remains after the —OH group is removed from a fatty acid. Thus, as the name implies, triacylglycerols contain three fatty acid residues (acyl groups) esterified to glycerol. An older name that is still frequently used for a triacylglycerol is *triglyceride.*

The triacylglycerol produced from glycerol and three molecules of stearic acid (see Figure 19.3) is an example of a simple triacylglycerol. A **simple triacylglycerol** *is a triester formed from the reaction of glycerol with three identical fatty acid molecules.* If the

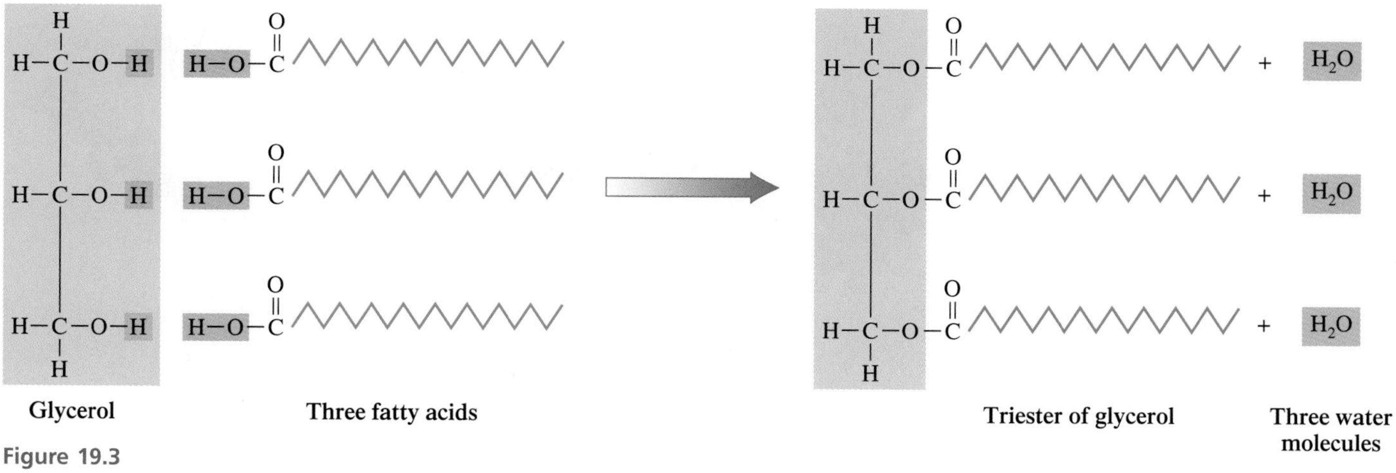

Glycerol

Three fatty acids

Triester of glycerol

Three water molecules

Figure 19.3
Triple esterification reaction between glycerol and three molecules of stearic acid (18:0 acid). Three molecules of water are a by-product of this reaction.

reacting fatty acid molecules are not all identical, then the result is a mixed triacylglycerol. A **mixed triacylglycerol** *is a triester formed from the reaction of glycerol with more than one kind of fatty acid molecule.* Figure 19.4 shows the structure of a mixed triacylglycerol in which one fatty acid is saturated, another monounsaturated, and the third polyunsaturated.

In nature, simple triacylglycerols are rare. Most naturally occurring triacylglycerols are mixed triacylglycerols. Biologically important fats and oils are complex mixtures of mixed triacylglycerols. They cannot be represented by a single specific chemical formula because many different fatty acids are present in the mixed triacylglycerol molecules. The variation is wide because the composition of any given fat or oil depends not only on the species involved but also on dietary and climatic factors. For example, fat from corn-fed hogs has a different overall composition than fat from peanut-fed hogs. Flax seed grown in warm climates gives oil with a different composition than that from flax seed grown in colder climates.

Fats are mixtures of triacylglycerols. Oils are mixtures of triacylglycerols. What, then, distinguishes a fat from an oil? The answer is physical state at room temperature. **Fats** *are triacylglycerol mixtures that are solids or semi-solids at room temperature* (25°C). **Oils** *are triacylglycerol mixtures that are liquids at room temperature.* Here are some broad generalizations relative to this distinction, in terms of physical state, between fats and oils:

1. The melting points of triacylglycerols are directly related to the identity of the fatty acids present within them. Fats are composed largely of triacylglycerols in which saturated fatty acids predominate, although some unsaturated fatty acids are present. Such triacylglycerols can pack closely together because of

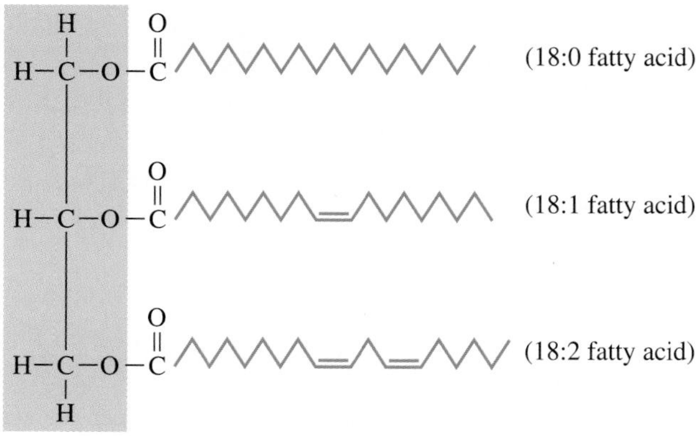

(18:0 fatty acid)

(18:1 fatty acid)

(18:2 fatty acid)

Figure 19.4
Structure of a mixed triacylglycerol in which three different fatty acid residues are present.

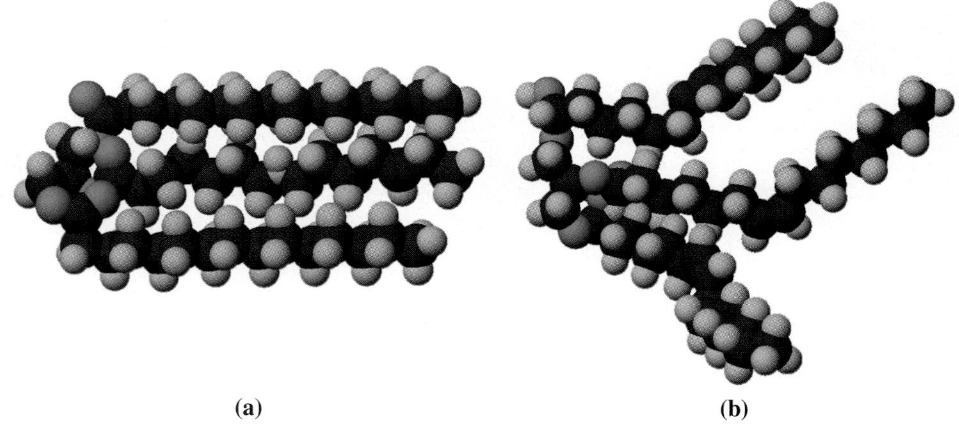

Figure 19.5
Representative triacylglycerols from (a) a fat and (b) an oil.

• *Petroleum oils* (Section 12.14) are structurally different from *lipid oils.* The former are mixtures of alkanes and cycloalkanes. The latter are triesters of glycerol.

the "linearity" of their fatty acid chains (Figure 19.5a), causing the higher melting points associated with fats (Section 19.3). Oils contain triacylglycerols with larger amounts of mono- and polyunsaturated fatty acids than those in fats. Such triacylglycerols cannot pack as tightly together because of the "bends" in their fatty acid chains (Figure 19.5b). The result is lower melting points.

Figure 19.6 gives the percentages of saturated, monounsaturated, and polyunsaturated fatty acids found in common dietary oils and fats. In general, a higher degree of fatty acid unsaturation is associated with oils than with fats. A notable exception to this generalization is coconut oil, which is highly saturated. This oil is a liquid not because it contains many double bonds within the fatty acids but because it is rich in *shorter-chain* fatty acids, particularly lauric acid (12:0).

2. Fats are generally obtained from animals; hence the term *animal fat.* Although fats are solids at room temperature, the warmer body temperature of the living animal keeps the fat somewhat liquid (semi-solid) and thus allows for movement. Oils typically come from plants, although there are also fish oils. A fish would have some serious problems if its triacylglycerols "solidified" when it encountered cold waters.

• All oils, even polyunsaturated oils, contain some *saturated* fatty acids. All fats, even highly saturated fats, contain some *unsaturated* fatty acids.

3. Pure fats and pure oils are colorless, odorless, and tasteless. The tastes, odors, and colors associated with dietary plant oils are caused by small amounts of

Dietary Oil/Fat	Saturated	Monounsaturated	Polyunsaturated
Canola oil	6%	58%	36%
Safflower oil	9%	13%	78%
Sunflower oil	11%	20%	69%
Corn oil	13%	25%	62%
Olive oil	14%	77%	9%
Soybean oil	15%	24%	61%
Peanut oil	18%	48%	34%
Cottonseed oil	27%	19%	54%
Lard	41%	47%	12%
Palm oil	51%	39%	10%
Beef tallow	52%	44%	4%
Butterfat	66%	30%	4%
Coconut oil	92%	6%	2%

Figure 19.6
Percentages of saturated, monounsaturated, and polyunsaturated fatty acids in the triacylglycerols of various dietary fats and oils.

other naturally occurring substances present in the plant that have been carried along during processing. The presence of these "other" compounds is usually considered desirable.

When a person consumes too much food, much of the excess energy (calories) is stored as fat. This fat is concentrated in special cells (adipocytes) that are nearly filled with large fat droplets. Adipose tissue containing these cells is found in various parts of the body—under the skin, in the abdominal cavity, in the mammary glands, and around various organs (Figure 19.7).

In the past decade, considerable attention has been paid to the role of dietary factors as a cause of disease (obesity, diabetes, cancer, hypertension, and atherosclerosis). Several organizations have proposed new dietary guidelines on the basis of these studies. The guidelines strongly suggest that total fat intake should be reduced, with particular emphasis on reduction of saturated fatty acid intake.

A new development involving fats is the creation, by food scientists, of fat substitutes called *artificial fats.* Chemical Connections 19.1 considers this rapidly evolving area of food science.

Figure 19.7
An electron micrograph of fat cells. Note their bulging spherical shape.

19.5 Chemical Reactions of Triacylglycerols

The chemical properties of triacylglycerols (fats and oils) are typical of esters and alkenes because these are the two functional groups present in triacylglycerols. Four important triacylglycerol reactions are hydrolysis, saponification, hydrogenation, and rancidity.

Chemical CONNECTIONS

19.1 Artificial Fat Substitutes

Artificial sweeteners (sugar substitutes) have been an accepted part of the diet of most people for many years. New since the 1990s are artificial fats, substances that create the sensations of "richness" of taste and "creaminess" of texture in food without the negative effects associated with dietary fats (heart disease and obesity).

Food scientists have been trying to develop fat substitutes since the 1960s. Now available for consumer use are two types of fat substitutes: *calorie-reduced* fat substitutes and *calorie-free* substitutes. They differ in their chemical structures and therefore in how the body handles them.

Simplesse, the best-known calorie-reduced fat substitute, received FDA marketing approval in 1990. It is made from the protein of fresh egg whites and milk by a procedure called microparticulation. This procedure produces tiny, round protein particles so fine that the tongue perceives them as a fluid rather than as the solid they are. Their fineness creates a sensation of smoothness, richness, and creaminess on the tongue.

In the body, Simplesse is digested and absorbed, contributing to energy intake. But 1 gram of Simplesse provides $1\frac{1}{3}$ cal, compared with the 9 cal provided by 1 g of fat. Simplesse is used only to replace fats in *formulated* foods such as salad dressings, cheeses, sour creams, and other dairy products. Simplesse is unsuitable for frying or baking, because it turns rubbery or rigid (gels) when heated. Consequently, it is not available for home use.

Olestra, a sucrose polyester and the best-known calorie-free fat substitute, received FDA marketing approval in 1996. It is produced by heating soybean oil with sucrose in the presence of methyl alcohol. The resulting product has a structure somewhat similar to that of a triacylglycerol; sucrose takes the place of the glycerol molecule, and six to eight fatty acids are attached to it rather than the three in a triacylglycerol. Unlike triacylglycerols, however, olestra cannot be hydrolyzed by the body's enzymes and therefore passes through the digestive tract undigested.

Olestra looks, feels, and tastes like dietary fat and can substitute for fats and oils in foods such as shortenings, oils, margarines, snacks, ice creams, and other desserts. It has the same cooking properties as fats and oils.

In the digestive tract, Olestra interferes with the absorption of both dietary and body-produced cholesterol; thus it may lower total cholesterol levels. A concern about its use revolves around its interference with vitamin E absorption. All products containing Olestra must carry the following label: "Olestra may cause abdominal cramping and loose stools. Olestra inhibits the absorption of some vitamins and other nutrients. Vitamins A, D, E, and K have been added." Despite these drawbacks, Olestra is considered safe for use.

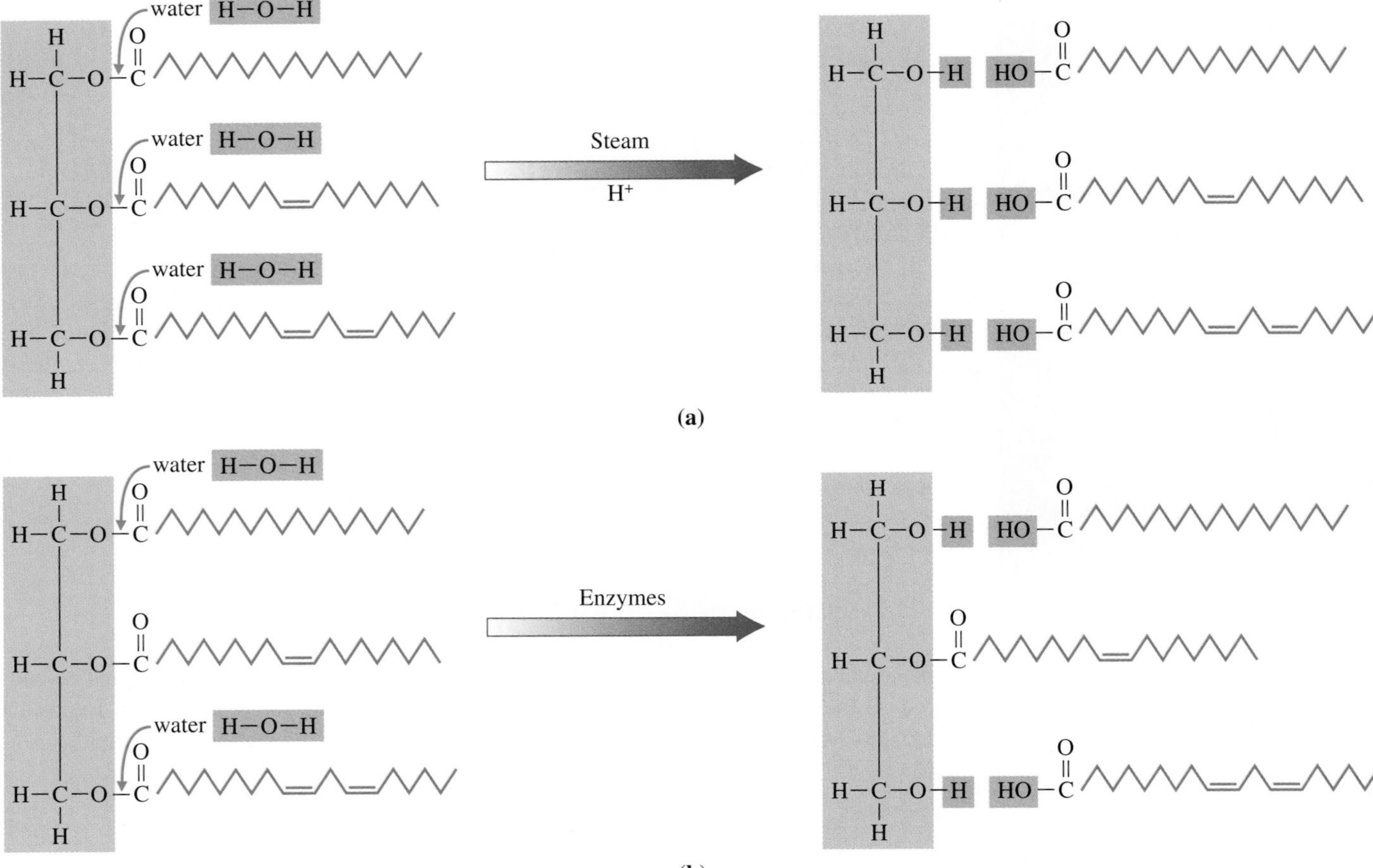

Figure 19.8
Complete and partial hydrolysis of a triacylglycerol. (a) Complete hydrolysis of a triacylglycerol produces glycerol and three fatty acid molecules. (b) Partial hydrolysis (during digestion) of a triacylglycerol produces a monoacylglycerol and two fatty acid molecules.

• Naturally occurring mono- and diacylglycerols are seldom encountered. *Synthetic* mono- and diacylglycerols are used as emulsifiers in many food products. Emulsifiers prevent suspended particles in colloidal solutions (Section 8.6) from coalescing and settling. Emulsifiers are usually present in so-called fat-free cakes and other fat-free products.

• Recall, from Section 16.8, the structural difference between a carboxylic acid and a carboxylic acid salt.

$$R-\overset{\overset{\displaystyle O}{\|}}{C}-OH \qquad R-\overset{\overset{\displaystyle O}{\|}}{C}-O^{-}\ Na^{+}$$

Carboxylic acid   Carboxylic acid salt

• Hydrolysis

Hydrolysis of a triacylglycerol is the reverse of the esterification reaction by which it was formed (see Figure 19.3a). *Complete* hydrolysis of a triacylglycerol molecule always gives one glycerol molecule and three fatty acid molecules as products (Figure 19.8a).

Triacylglycerol hydrolysis within the human body requires the help of enzymes (protein catalysts; Section 21.1) produced by the pancreas. These enzymes cause the triacylglycerol to be hydrolyzed in a *stepwise* fashion. First one of the outer fatty acids is removed, then the other outer one, leaving a monoacylglycerol. In most cases this is the end product of the initial hydrolysis (digestion) of the triacylglycerol (Figure 19.8b). Sometimes, enzymes remove all three fatty acids, leaving a free molecule of glycerol.

• Saponification

Saponification (Section 16.13) is a hydrolysis reaction carried out in an alkaline (basic) solution. For fats and oils, the products of saponification are glycerol and fatty acid *salts.*

The overall reaction of triacylglycerol saponification can be thought of as occurring in two steps. The first step is the hydrolysis of the ester linkages to produce glycerol and three fatty acid molecules:

$$\text{Fat or oil} + 3H_2O \longrightarrow 3 \text{ fatty acids} + \text{glycerol}$$

The second step involves a reaction between the acid molecules and the base (usually NaOH) in the alkaline solution. This is an acid–base reaction that produces water plus salts:

$$3 \text{ Fatty acids} + 3NaOH \longrightarrow 3 \text{ fatty acid salts} + 3H_2O$$

Chemical CONNECTIONS

19.2 The Cleansing Action of Soap

The cleansing action of soap is directly related to the structure of the carboxylate ions present in soap as fatty acid salts. Their structure is such that they exhibit a "dual polarity." The hydrocarbon portion of the carboxylate ion is nonpolar, and the carboxyl portion is polar. This dual polarity for the fatty acid salt *sodium stearate,* which is representative of all fatty acid salts present in soap, is as follows:

$COO^- \; Na^+$

Nonpolar portion — Polar portion

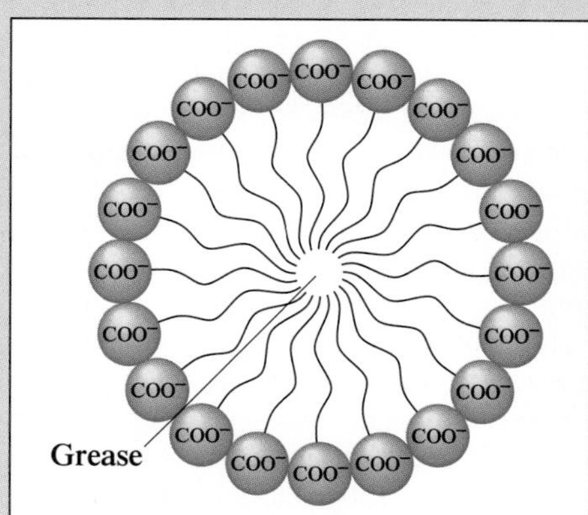

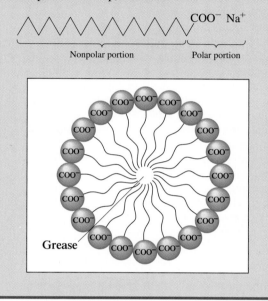

Soap solubilizes oily and greasy materials in the following manner: The nonpolar portion of the carboxylate ion dissolves in the nonpolar oil or grease, but the polar carboxyl portion maintains its solubility in the polar water.

The penetration of the oil or grease by the nonpolar end of the carboxylate ion is followed by the formation of micelles. A **micelle** *is a small, spherical grease–carboxylate ion droplet that is soluble in water as a result of the polar groups (carboxyl groups) on its surface* (see the accompanying diagram). The carboxyl groups and water molecules are attracted to each other, causing the solubilizing of the micelle.

The micelles do not combine into larger drops, because their surfaces are all negatively charged, and like charges repel each other. The water-soluble micelles are subsequently rinsed away, leaving a material devoid of oil and grease.

For many cleansing purposes, synthetic detergents have largely replaced soaps. The basis of the cleansing action of synthetic detergents is very similar to that of soaps, because their structures are very similar. The structure of the sodium salt of a benzene sulfonic acid is typical of the types of molecules used in detergents.

$$-SO_2-O^- \; Na^+$$

Saponification of animal fat is the process by which soap was made in pioneer times. Soap making involved heating lard (fat) with lye (ashes of wood, an impure form of KOH). Today most soap is prepared by hydrolyzing fats and oils (animal fat and coconut oil) under high pressure and high temperature. Sodium carbonate is used as the base.

The cleansing action of soap is related to the structure of the fatty acid salts that are present, as is explained in Chemical Connections 19.2.

• Hydrogenation

Hydrogenation is a reaction we encountered in Section 13.7. It involves hydrogen addition across carbon–carbon multiple bonds, which increases the degree of saturation as some double bonds are converted to single bonds. With this change, there is a corresponding increase in the melting point of the substance.

The carbon–carbon double bonds in vegetable oils are partially hydrogenated (some, but not all, of the double bonds are converted to single bonds) to produce semi-solid rather than liquid products. Many food products are produced in this way. The peanut oil in many popular brands of peanut butter has been partially hydrogenated to convert the oil into a solid that does not separate out of the mixture. Hydrogenation is used to produce solid cooking shortenings or margarines from liquid vegetable oils.

Soft-spread margarines are partially hydrogenated oils. The extent of hydrogenation is carefully controlled to make the margarine soft at refrigerated temperatures (4°C). Recently, a health concern has developed about the hydrogenation of food products (see Chemical Connections 19.3).

Chemical CONNECTIONS

19.3 *Trans* Fatty Acids and Blood Cholesterol Levels

All current dietary recommendations stress reducing saturated fat intake. In accordance with such recommendations, many people have switched from butter to margarine and now use partially hydrogenated vegetable oils rather than animal fat for cooking. However, recent studies suggest that partially hydrogenated products also have roles in raising blood cholesterol levels. Why would this be so?

During hydrogenation (Section 19.5) some of the *unreacted* double bonds that are naturally present in the *cis* configuration are isomerized to the *trans* configuration.

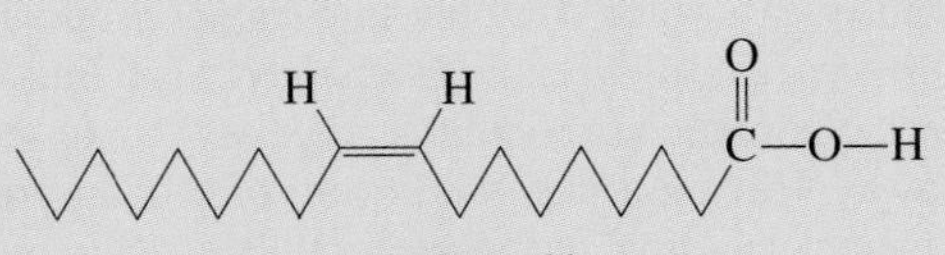

H H O ‖ C—O—H

Cis fatty acid

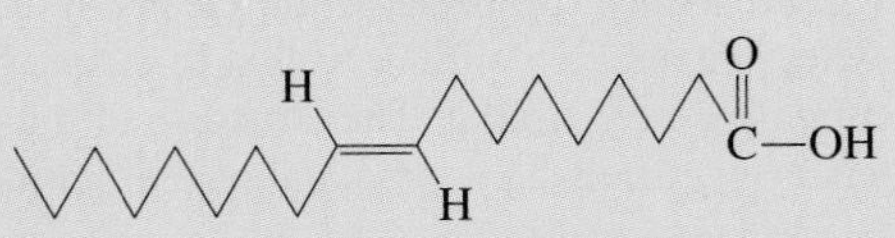

H O ‖ C—OH H

Trans fatty acid

Studies show that fatty acids with *trans* double bonds (an unnatural situation) affect blood cholesterol levels in a manner similar to saturated fatty acids.

Trans fatty acids (*trans* fat) make up approximately 5% of the fat intake in the typical diet in the United States, and the amount of *trans* fat a person consumes depends on the amount of fat eaten and on the types of foods selected. The best example of a *trans*-fat food may be stick margarine, but it is also found in crackers, cookies, pastries, and deep-fried fast foods. Spreadable margarine in tubs, though, contains little if any *trans* fat.

Until recently, the only way consumers could determine whether a food included *trans* fat was to look for "hydrogenated" on the list of ingredients. A food that lists partially hydrogenated oils among its first three ingredients usually contains substantial amounts of *trans* fatty acids as well as some saturated fat. In the year 2000, the U.S. Food and Drug Administration (FDA) adopted rules requiring that the *trans*-fat content of a food be included in the Nutrition Facts panel found on all food products. When *trans* fatty acids are present, an asterisk (or other symbol) placed after the heading "Saturated Fat" refers the consumer to a footnote that lists the amount of grams of *trans* fat in the product.

The health implications of *trans* fatty acids remain an area of active research; many answers are yet to be found. At present, eating saturated fatty acids still appears to pose greater health risks than eating *trans* fatty acids. There should be concern, however, about the consumption of excessive amounts of either type of fatty acid.

• Rancidity

Fats and oils often develop unpleasant odors and/or flavors upon exposure to moist air at room temperature. When this occurs, fats and oils are said to have become *rancid.* Rancidity results from two kinds of unwanted reactions: hydrolysis of triacylglycerol ester linkages and oxidation of carbon–carbon double bonds in the fatty acid chains of triacylglycerols.

Hydrolytic rancidity results from the exposure of fats and oils to *moist* air. (All air contains some moisture.) Microorganisms in the air supply necessary enzymes to catalyze the hydrolysis of ester bonds within triacyglycerols. Volatile short-chain fatty acid molecules are freed as the result of such triacylglycerol hydrolysis, and they contribute to the disagreeable odors of rancid foods. For example, butyric acid is the source of the characteristic odor of rancid butter.

Oxidative rancidity, in most cases, is a more important reaction than hydrolytic rancidity. In oxidative rancidity, double bonds in the unsaturated fatty acid components of triacylglycerols are cleaved, producing low-molecular-mass aldehydes with objectionable odors. In addition, these aldehydes can be further oxidized to give equally offensive-smelling, low-molecular-mass carboxylic acids. Warmth and exposure to atmospheric oxygen induce oxidative rancidity. To avoid this unwanted oxidation, the food industry adds antioxidants to foods. Two naturally occurring antioxidants are vitamin C (ascorbic acid) and vitamin E (α-tocopherol). Two synthetic oxidation inhibitors are BHA and BHT (Section 14.9). In the presence of air, antioxidants, rather than food, are oxidized.

• Antioxidants are compounds that are easily oxidized. When added to foods, they are more easily oxidized than the food. Thus they prevent the food from being oxidized.

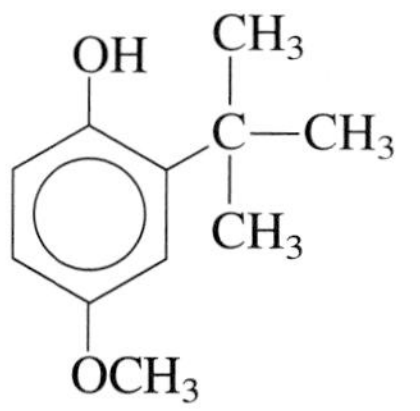

Butylated hydroxyanisole (BHA)

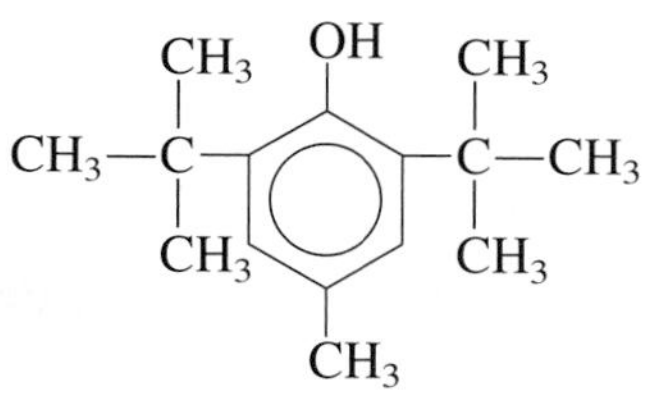

Butylated hydroxytoluene (BHT)

19.6 Phosphoacylglycerols

• Phosphoacylglycerols differ from fats and oils in that one of the fatty acids has been replaced with phosphoric acid (a triprotic acid), which is further esterified with an alcohol.

A second group of glycerol esters besides fats and oils exists—the phosphoacylglycerols. **Phosphoacylglycerols** *are triesters of glycerol in which two —OH groups are esterified with fatty acids and the third is esterified with phosphoric acid, which in turn is esterified to an alcohol.* The block diagram for a phosphoacylglycerol has the following general structure:

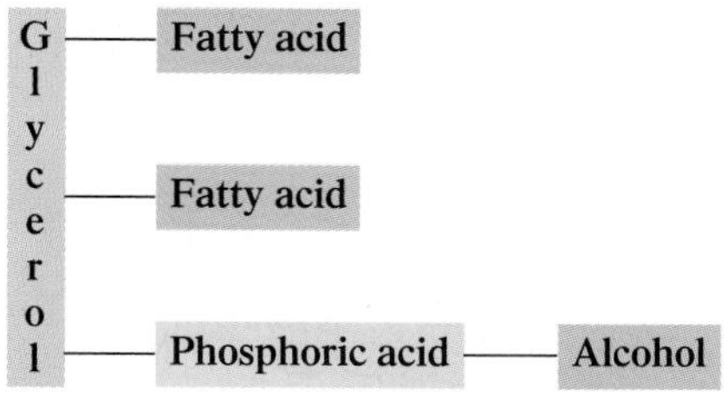

• An amino alcohol molecule contains both a hydroxyl group, —OH, and an amino group, $—NH_2$.

The most abundant phosphoacylglycerols have an amino alcohol (choline, ethanolamine, or serine) attached to the phosphate group. The structures of these three amino alcohols, given in terms of the charged forms (Sections 17.8 and 18.6) that they adopt in neutral solution, are

$$HO—CH_2—CH_2—\overset{+}{N}(CH_3)_3$$

Choline (a quaternary ammonium ion)

$$HO—CH_2—CH_2—\overset{+}{N}H_3$$

Ethanolamine (positive-ion form)

$$HO—CH_2—\underset{\displaystyle COO^-}{\underset{|}{CH}}—\overset{+}{N}H_3$$

Serine (two ionic groups present)

Phosphoacylglycerols containing these three amino alcohols are respectively known as phosphatidylcholines, phosphatidylethanolamines, and phosphatidylserines. The fatty acid, glycerol, and phosphoric acid portions of a phosphoacylglycerol structure constitute a *phosphatidyl* group.

Although the general structural features of phosphoacylglycerols are similar in many respects to those of triacylglycerols, these two types of lipids have quite different biological functions. Triacylglycerols serve as storage molecules for metabolic fuel. Phosphoacylglycerols function almost exclusively as components of cell membranes (Section 19.12) and are not stored. A major structural difference between the two types of lipids, that of polarity, is related to their differing biological functions. Triacylglycerols are a nonpolar class of lipids, whereas phosphoacylglycerols are polar. In general, membrane lipids have polarity associated with their structures.

Further consideration of general phosphoacylglycerol structure reveals an additional structural characteristic of most membrane lipids. Let us consider a phosphatidylcholine containing stearic and oleic acids as our example. The chemical structure of this molecule is shown in Figure 19.9a.

A molecular model for this compound, which gives the orientation of groups in space, is illustrated in Figure 19.9b. There are two important things to notice about this model: (1) There is a "head" part, the choline and phosphate, and (2) there are two "tails," the

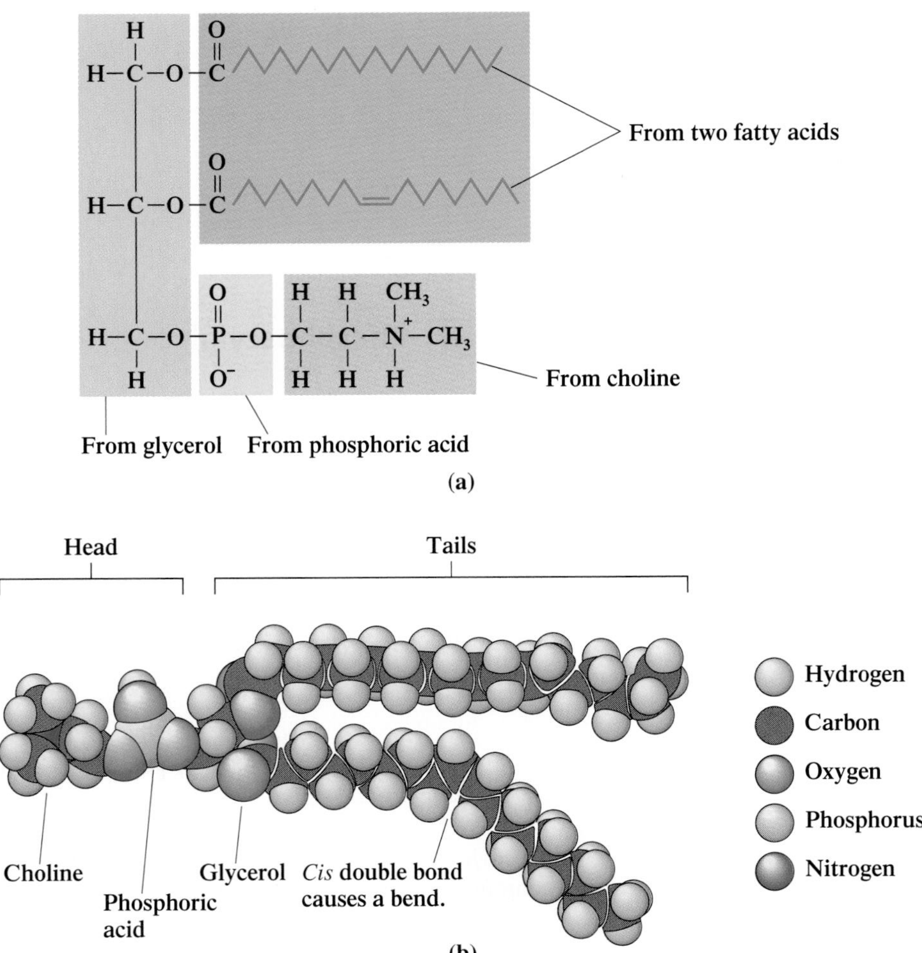

Figure 19.9 (a) Structural formula and (b) molecular model showing the "head and two tails" structure of a phosphatidylcholine molecule containing stearic acid (18:0) and oleic acid (18:1).

two fatty acid carbon chains. The head part is polar. The two tails, the carbon chains, are nonpolar.

All phosphoacylglycerols have structures similar to that shown in Figure 19.9. All have a "head" and two "tails." A simplified representation for this structure uses a circle to represent the polar head and two wavy lines to represent the nonpolar tails.

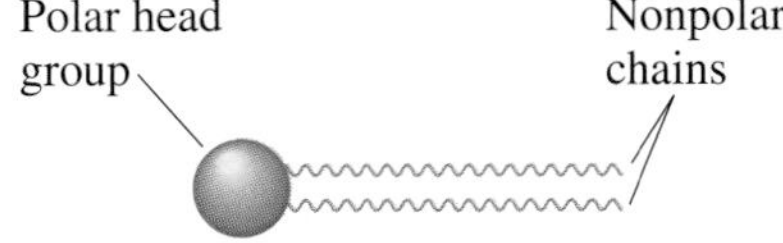

The polar head group of a phosphoacylglycerol is soluble in water. The nonpolar tail chains are insoluble in water but soluble in nonpolar substances. This dual-polarity feature, previously encountered when soaps were discussed (Chemical Connections 19.2), is a structural characteristic of most membrane lipids.

- Phosphoacylglycerols have a *hydrophobic* ("water-hating") portion, the nonpolar fatty acid groups, and a *hydrophilic* ("water-loving") portion, the polar phosphate-amino alcohol–ester group.

• Phosphatidylcholines

Phosphatidylcholines are also known as lecithins. There are a number of different phosphatidylcholines because different fatty acids may be bonded to the glycerol portion of the phosphatidylcholine structure. In general, phosphatidylcholines are waxy solids that form colloidal suspensions in water. Egg yolks and soybeans are good dietary sources of these lipids. Within the body, phosphatidylcholines are prevalent in cell membranes.

- The amino alcohol in phosphatidylcholines (pronounced fahs-fuh-TIDE-ul-KOH-leen) is choline.

Plant leaves often have a wax coating to prevent excessive loss of water.

Periodically, claims arise that phosphatidylcholine should be taken as a nutritive supplement; some claims indicate it will improve memory. There is no evidence that these supplements are useful. The enzyme lecithinase in the intestine hydrolyzes most of the phosphatidylcholine taken orally before it passes into body fluids, so it does not reach body tissues. The phosphatidylcholine present in cell membranes is made by the liver; thus phosphatidylcholines are not essential nutrients.

The food industry uses phosphatidylcholines as emulsifiers to promote the mixing of otherwise immiscible materials. Mayonnaise, ice cream, and custards are some of the products they are found in. It is the polar–nonpolar (head–tail) structure of phosphatidylcholines that enables them to function as emulsifiers.

Phosphatidylethanolamines and Phosphatidylserines

Phosphatidylethanolamines and phosphatidylserines are also known as cephalins. These compounds are found in heart and liver tissue and in high concentrations in the brain. They are important in blood clotting. Much is yet to be learned about how these compounds function within the human body.

19.7 Waxes

Two categories of saponifiable lipids exist—glycerol esters and nonglycerol esters. Fats and oils (Section 19.4) and phosphoacylglycerols (Section 19.6) are glycerol esters. With this section, we begin consideration of nonglycerol esters, of which waxes (this section) and sphingolipids (Section 19.8) are important representatives.

• The general structural formula for a wax is the same as that for a simple ester:

$$R-\overset{\overset{\displaystyle O}{\|}}{C}-O-R'$$

However, for waxes both R and R' must be long carbon chains (usually 20–30 carbon atoms).

Everyday use of the term *wax* differs from the chemical definition for a wax. To the layperson, a wax is a substance that is insoluble in water, hard when cold, and pliable when warm. Household paraffin "wax," a mixture of long-chain hydrocarbons obtained from the processing of petroleum (Section 12.14), is the most familiar substance with such properties. However, the chemical definition of a wax is based on structure rather than properties. A **wax** *is a monoester formed from the reaction of a long-chain monohydroxy alcohol with a fatty acid molecule.* The block diagram for a wax is

Fatty acid ——— **Long-chain alcohol**

Molecules with this structure have the properties commonly associated with a wax. There are also substances, such as paraffin "wax," that have wax-like properties but do not fit the structural definition of a wax.

Many naturally occurring waxes are known. In most cases, they function as protective coatings. Many plants that grow in dry regions have leaves coated with wax to minimize water loss. Bees secrete wax and use it as a structural material.

$$CH_3-(CH_2)_{14}-\overset{\overset{\displaystyle O}{\|}}{C}-O-(CH_2)_{29}-CH_3$$

Beeswax

• Human ear wax, which acts as a protective barrier against infection by capturing airborne particles, is not a true wax—that is, it is not a mixture of simple esters. Human ear wax is a gland secretion that is a mixture of triacylglycerols, phospholipids, and esters of cholesterol.

Waxes also coat the hair of many animals and the feathers of birds. Commercially, waxes are marketed as protective coatings (automobile and floor waxes; lotions) and are used in manufacturing candles.

The structure of waxes, like that of phosphoacylglycerols, involves a head and two tails (Figure 19.10). The polar head of waxes is very small compared with the polar head of phosphoacylglycerols. Consequently, the hydrocarbon tail portions of waxes totally overwhelm the head (ester linkage), and this gives waxes hydrocarbon-like (nonpolar) properties. Waxes are completely insoluble in water; as noted previously, they serve as biosynthesized water repellents. They have no function as membrane lipids.

• Waxes are not so easily hydrolyzed as phosphoacylglycerols, and therefore are useful as protective coatings.

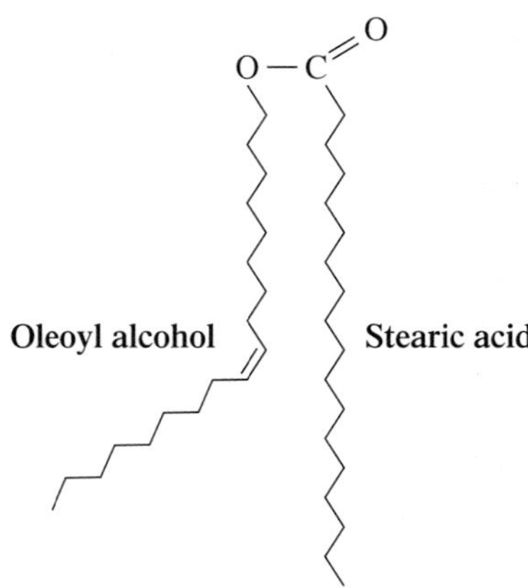

Figure 19.10
A wax has a structure with a very small "head" and two "tails."

Many synthetic materials are now available with properties that closely match those of waxes. Such synthetic materials, which are generally polymers, are used in cosmetics, ointments, and the like. The synthetic *carbowax,* for example, is a polyether.

19.8 Sphingolipids

• A sphingolipid is similar to a phosphoacylglycerol in that it has two nonpolar tails and a *relatively large* polar head. Waxes, on the other hand, have a very small polar head associated with their two nonpolar tails.

Sphingolipids are the second major class of nonglycerol-based saponifiable lipids. Like phosphoacylglycerols (Section 19.6), they are polar lipids and are major constituents of cell membranes.

Sphingolipids have structures based on the long-chain amino dialcohol *sphingosine.* A **sphingolipid** *is a saponifiable lipid derived from the amino dialcohol sphingosine.* The structure of sphingosine is

$$CH_3—(CH_2)_{12}—CH{=}CH—\underset{\displaystyle OH}{\underset{|}{CH}}—\underset{\displaystyle NH_2}{\underset{|}{CH}}—\underset{\displaystyle OH}{\underset{|}{CH_2}}$$

Sphingosine

All lipids derived from sphingosine have (1) a fatty acid connected to the $—NH_2$ group via an *amide* linkage, and (2) a group attached to the —OH group on the terminal carbon atom via an *ester* linkage. The general block diagram for a sphingolipid is shown in Figure 19.11a.

• When sphingolipids were discovered over a century ago by the physician-chemist Johann Thudichum (1829–1901), their biological role seemed as enigmatic as the Sphinx, for which he therefore named them.

A molecular model showing orientation of groups in space for a sphingolipid is given in Figure 19.11b. Note again, as in phosphoacylglycerols and waxes, the structural features of a head and two tails. For sphingolipids, the fatty acid is one of the tails, and the long carbon chain of sphingosine itself is the other tail. The "additional component" is the head, and it is a phosphoric acid–choline group for the particular model shown.

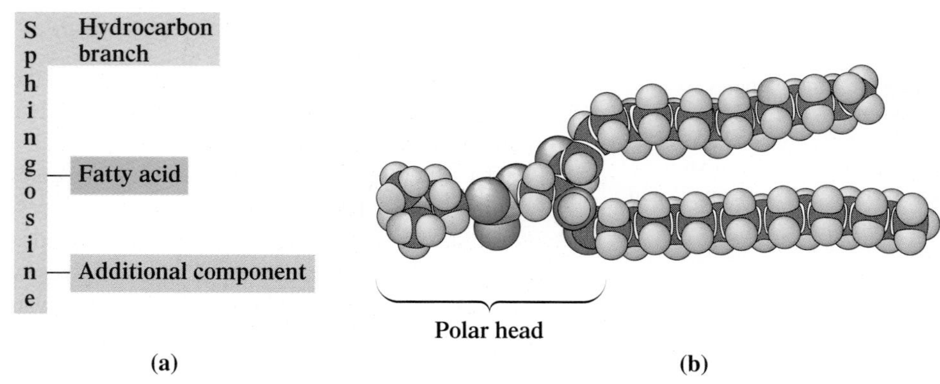

Figure 19.11
(a) A block diagram for a sphingolipid. (b) A molecular model of a sphingolipid shows a structure with a polar "head" and two nonpolar "tails." This particular sphingolipid contains a phosphoric acid–choline entity as its additional component.

• The term *phospholipid* applies to all lipids that contain a phosphate group as part of their structure. Sphingomyelins are a type of phospholipid, as are the phosphoacylglycerols discussed in Section 19.6.

• Sphingomyelins

Sphingolipids in which the esterifying group is phosphoric acid to which choline is attached are called *sphingomyelins.* Sphingomyelins are found in all cell membranes and are important structural components of the myelin sheath, the protective and insulating coating that surrounds nerves. The block diagram for the structure of a sphingomyelin is

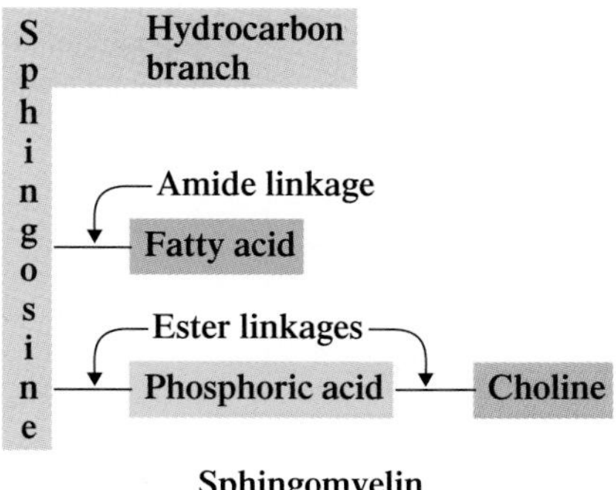

Sphingomyelin

• Cerebrosides and Gangliosides

Some sphingosine-based membrane lipids have a small carbohydrate as the head group. *Cerebrosides,* the simplest of such carbohydrate-containing lipids, usually have a glucose or galactose as the carbohydrate unit. The cerebrosides, as the name suggests, occur primarily in the brain (7% of dry mass) and in the myelin sheath of nerves. *Gangliosides* contain more complex carbohydrate heads; up to seven monosaccharide units are present. These substances occur in the gray matter of the brain as well as in the myelin sheath. The block diagrams for the structures of cerebrosides and gangliosides are

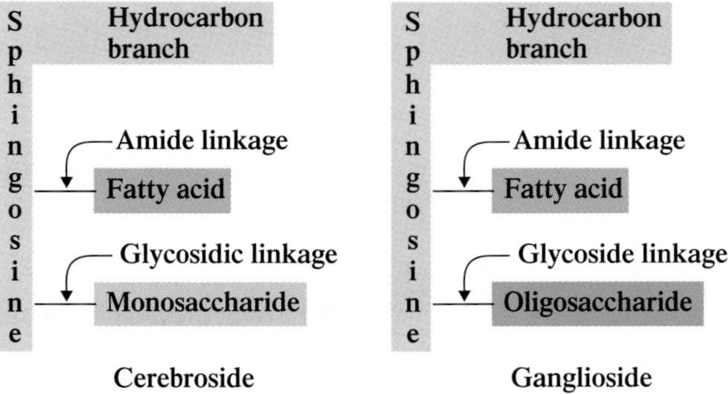

• The term *glycolipid* (Section 18.16) applies to all lipids that contain a carbohydrate group as part of their structure. Both cerebrosides and gangliosides are glycolipids.

The specific structure of the cerebroside in which stearic acid (18:0) is the fatty acid and galactose is the monosaccharide is

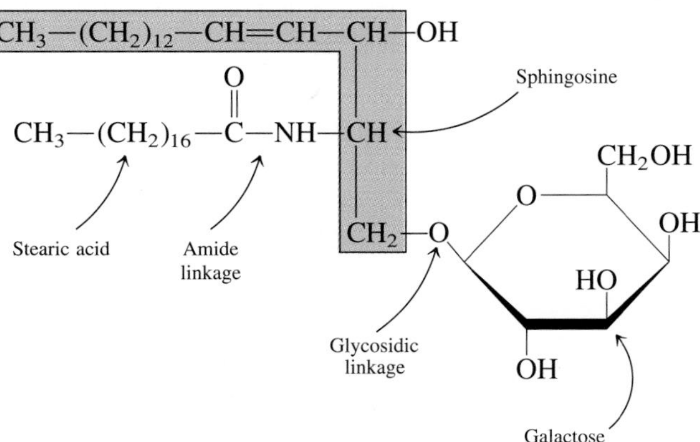

Saponifiable Lipids

CHEMISTRY AT A GLANCE

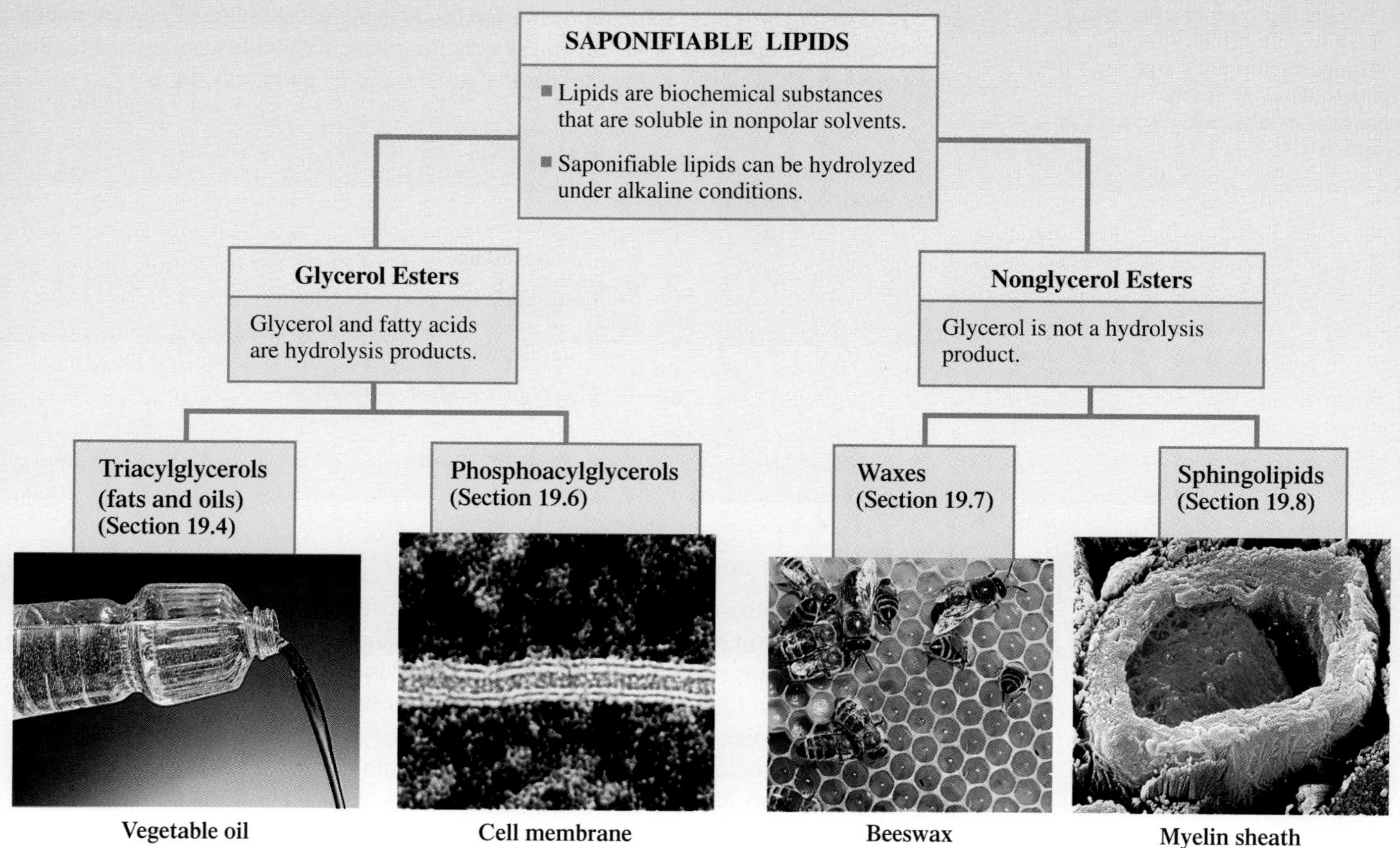

SAPONIFIABLE LIPIDS
Lipids are biochemical substances that are soluble in nonpolar solvents.
Saponifiable lipids can be hydrolyzed under alkaline conditions.
Glycerol Esters
Glycerol and fatty acids are hydrolysis products.
Nonglycerol Esters
Glycerol is not a hydrolysis product.
Triacylglycerols (fats and oils) (Section 19.4)
Phosphoacylglycerols (Section 19.6)
Waxes (Section 19.7)
Sphingolipids (Section 19.8)
Vegetable oil
Cell membrane
Beeswax
Myelin sheath

The accompanying Chemistry at a Glance briefly outlines the classification of saponifiable lipids.

19.9 Nonsaponifiable Lipids

Nonsaponifiable lipids do not undergo hydrolysis in alkaline solution (Section 19.1). Their structures are much different from those of the saponifiable lipids; neither ester nor amide linkages are present. Of the five types of nonsaponifiable lipids listed in Table 19.3, we will discuss only two in this chapter—steroids and eicosanoids. The other three types have been or will be discussed elsewhere: terpenes in Section 13.6, pheromones in Section 13.6, and fat-soluble vitamins in Section 21.14.

Table 19.3
Nonsaponifiable Lipids

Steroids	(Section 19.10)
Eicosanoids	(Section 19.11)
Terpenes	(Section 13.6)
Pheromones	(Section 13.6)
Fat-soluble vitamins	(Section 21.14)

19.10 Steroids

Steroids *are lipids with structures that are based on a fused-ring system involving three 6-membered rings and one 5-membered ring.* The fused-ring system of steroids, called a steroid nucleus, has the following structure:

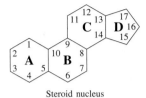

Steroid nucleus

Note that the rings are customarily labeled with letters and each carbon atom is labeled with a number.

Numerous steroids have been isolated from plants, animals, and human beings. Location of double bonds within the fused-ring system and the nature and location of substituents distinguish one steroid from another. Most steroids have an oxygen functional group (=O or —OH) at carbon 3 and some kind of side chain at carbon 17. Many also have a double bond from carbon 5 to either carbon 4 or carbon 6.

The steroids discussed in this section are cholesterol, bile salts, and steroid hormones.

Cholesterol contributes to many gallbladder attacks caused by gallstones. A large percentage of gallstones are almost pure crystallized cholesterol that has precipitated from bile solution.

• Cholesterol

Cholesterol is the most abundant steroid in the human body. The name *cholesterol* has the *-ol* ending because it is an alcohol, with an —OH group on carbon 3 of the steroid nucleus. In addition, cholesterol has methyl groups bonded to carbon atoms 10 and 13 and a small branched hydrocarbon chain on carbon 17.

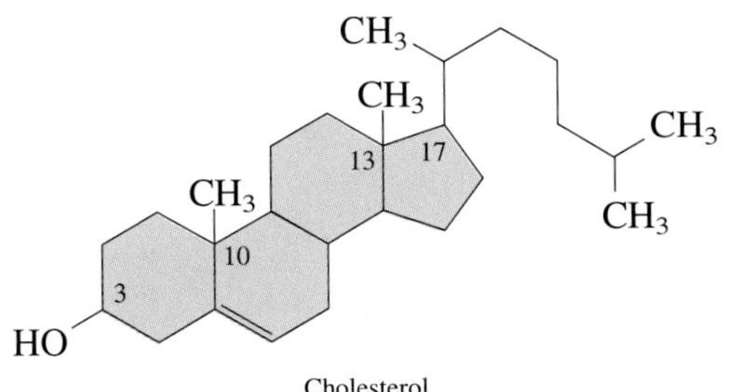

Cholesterol

Within the human body, cholesterol is found in cell membranes (up to 25% by mass), nerve tissue, and brain tissue (about 10% by dry mass), and it is the main component of gallstones. Human blood plasma contains about 50 mg of free cholesterol per 100 mL and about 170 mg of cholesterol esterified with various fatty acids.

A space-filling model of the cholesterol molecule (Figure 19.12) shows the rather compact nature of this molecule. The "head and two tails" arrangement found in many lipids is not present. The lack of a large polar head causes cholesterol to have limited water solubility. The —OH group on carbon 3 is considered the head of the molecule.

Food	Cholesterol (mg)
liver (3 oz)	410
egg (1 large)	213
shrimp (3 oz)	166
pork chop (3 oz)	83
chicken (3 oz)	75
beef steak (3 oz)	70
fish fillet (3 oz)	54
whole milk (1 cup)	33
cheddar cheese (1 oz)	30
Swiss cheese (1 oz)	26
low-fat milk (1 cup)	22

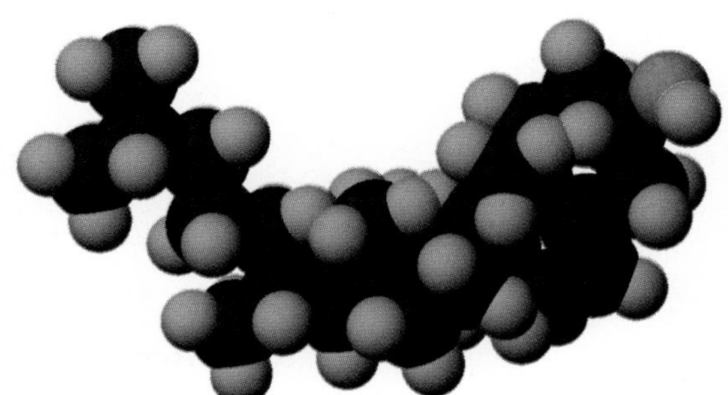

Figure 19.12
A molecular model showing the compact nature of the cholesterol molecule.

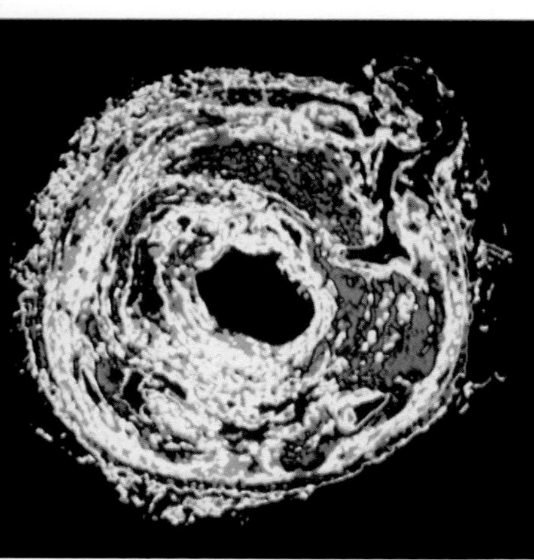

Occluded artery, cholesterol.

Cholesterol plays a vital biological role in chemical synthesis within the human body. It is the starting material for the synthesis of numerous steroid hormones, vitamin D, and bile salts. Its presence in the body is essential to life.

The human body, mainly within the liver, synthesizes about 1 gram of cholesterol each day, an amount sufficient to meet the body's biosynthetic needs. Therefore, cholesterol is not necessary in the diet. When we ingest cholesterol, the amount synthesized by the body is reduced. However, the reduction is less than the amount ingested. Therefore, total body cholesterol level increases with dietary cholesterol level.

Medical science now considers high blood cholesterol, along with high blood pressure and smoking, as the major risk factors for cardiovascular disease (CVD). High blood cholesterol contributes to atherosclerosis, the main form of CVD, which is characterized by the buildup of plaque along the inner walls of the arteries. Plaque is a mound of lipid material mixed with smooth muscle cells and calcium. Much of the lipid material in plaque is cholesterol. Extensive plaque formation leads to hardening of the arteries. Plaque deposits in the arteries that serve the heart reduce blood flow to the heart muscle and can lead to a heart attack.

People who want to reduce their level of dietary cholesterol should reduce the amount of animal products they eat (meat, dairy products, etc.) and eat more fruit and vegetables. Plant foods contain no cholesterol; it is found only in foods of animal origin.

• Bile Salts

• The source for bile salts is *bile,* a fluid secreted by the liver, stored in the gallbladder, and released into the small intestine during digestion. Besides bile salts, bile also contains bile pigments (breakdown products of hemoglobin; Section 26.7) and electrolytes such as bicarbonate ion.

Bile salts *are emulsifying agents that make dietary lipids soluble in the aqueous environment of the digestive tract.* During digestion, bile salts are released into the intestine from the gallbladder, where they help digestion by emulsifying (solubilizing) fats and oils. Their mode of action is much like that of soap during washing.

Bile salts are cholesterol oxidation products. They are trihydroxy cholesterol derivatives in which the carbon-17 side chain has been oxidized to a carboxylic acid. This acid side chain is then bonded to an amino acid through an amide linkage. The two principal bile salts are sodium glycocholate (glycine is the amino acid) and sodium taurocholate (taurine is the amino acid).

Glycine portion

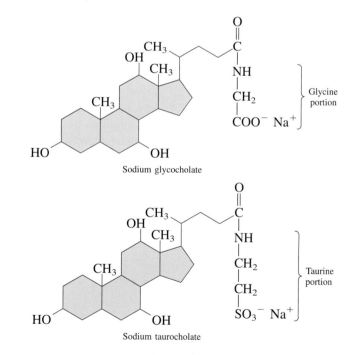

Sodium glycocholate

Taurine portion

Sodium taurocholate

• Steroid Hormones

Hormones *are chemical messengers produced by ductless glands.* They serve as a means of communication between various tissues. Many, but not all, hormones in the human body are steroids. Cholesterol is the ultimate starting material for the production of all steroid hormones, so they contain its characteristic system of four fused rings. Steroid hormone synthesis is always a multistep process.

There are two major classes of steroid hormones: the *sex hormones,* which control reproduction and secondary sex characteristics, and the *adrenocortical hormones,* which regulate numerous biochemical processes in the body.

The sex hormones can be classified into three major groups:

1. Estrogens—the female sex hormones
2. Androgens—the male sex hormones
3. Progestins—the pregnancy hormones

• Estrogens are a class of molecules rather than a single molecule. Statements like "the estrogen level is high" should be rephrased as "a high level of estrogens."

Estrogens are synthesized in the ovaries and adrenal cortex and are responsible for the development of female secondary sex characteristics at the onset of puberty and for regulation of the menstrual cycle. They also stimulate the development of the mammary glands during pregnancy and induce estrus (heat) in animals.

Androgens are synthesized in the testes and adrenal cortex and promote the development of secondary male characteristics. They also promote muscle growth.

Progestins are synthesized in the ovaries and the placenta and prepare the lining of the uterus for implantation of the fertilized ovum. They also suppress ovulation.

The upper part of Figure 19.13 gives the structure of the primary hormone in each of the three subclasses of sex hormones. Other members of these hormone families are metabolized forms of the primary hormone.

Note, in Figure 19.13a, how similar the structures are for these principal hormones, and yet how different their functions. The fact that seemingly minor changes in structure effect great changes in biofunction points out, again, the extreme specificity (Section 21.5) of the enzymes that control biochemical reactions.

Increased knowledge of the structures and functions of sex hormones has led to the development of a number of *synthetic* steroids whose actions often mimic those of the natural hormones. Among the best known of the synthetic steroids are oral contraceptives and anabolic agents.

Figure 19.13
Structures of selected sex hormones and synthetic compounds that have similar actions.

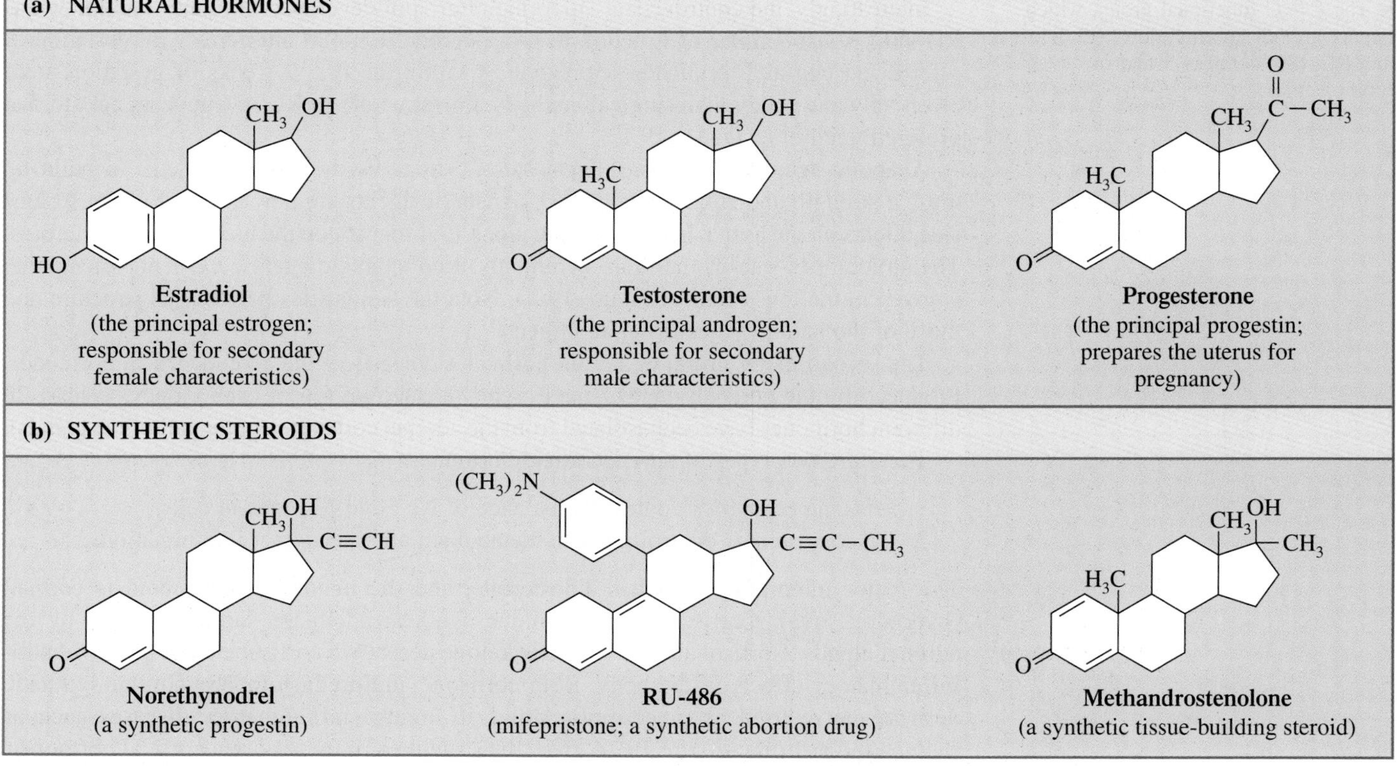

Chemical CONNECTIONS

19.4 Steroid Drugs in Sports

The steroid hormone testosterone is the principal male sex hormone. It has masculinizing (androgenic) effects and muscle-building (anabolic) effects. Masculinizing effects of testosterone include the growth of facial and body hair, deepening of the voice, and maturation of the male sex organs. Testosterone's anabolic effects are responsible for the muscle development that boys experience at puberty.

Some of the many synthetic testosterone derivatives exert primarily androgenic effects, whereas others exert primarily anabolic effects. Androgenic compounds can be used to correct hormonal imbalances in the body. Anabolic steroids can be used to prevent the withering of muscle in persons recovering from major surgery or serious injuries.

Anabolic steroids have also been "discovered" by athletes, who have found that these compounds can be used to help build muscle mass and reduce the healing time for muscle injuries. Often the net result of anabolic hormone use by an athlete is enhanced athletic performance.

However, anabolic steroid use by athletes has been banned by most sports organizations, including the International Olympic Committee. Why? There are two reasons: (1) Their use is considered a form of cheating, because it confers an unfair advantage. And (2) their "beneficial effects" are far outweighed by serious negative side effects.

Medical evidence now indicates that the use of anabolic steroids is a dangerous practice. The list of adverse effects is long and continues to grow.

The blood lipid profiles of anabolic steroid users are indicative of a very high risk of heart disease. The high-risk profile still remains 6 months after cessation of steroid use. Oral-dose steroids are toxic to the liver and impair its function. The male reproductive system is altered, causing testicular shrinkage and decreased sperm production.

In women, anabolic steroids produce the additional effects of acne, oily skin, facial hair, male-pattern baldness, deepening of the voice, and changes in the menstrual cycle. Many of these changes are not reversible.

Oral contraceptives are used to suppress ovulation as a method of birth control. Generally, a mixture of a synthetic estrogen and a synthetic progestin is used. The synthetic estrogen regulates the menstrual cycle, and the synthetic progestin prevents ovulation, thus creating a false state of pregnancy. The structure of norethynodrel (Enovid), a synthetic progestin, is given in the lower half of Figure 19.13. Compare its structure to that of progesterone (the real hormone); the structures are very similar.

• The C≡C functional group, which occurs in both norethynodrel (Enovid) and RU-486, is rarely found in biomolecules.

Interestingly, the controversial "morning after" pill developed in France and known as RU-486, is also similar in structure to progesterone. RU-486 interferes with gestation of a fertilized egg and terminates a pregnancy within the first 9 weeks of gestation more effectively and safely than surgical methods. The structure of RU-486 appears next to that of norethynodrel in Figure 19.13.

Anabolic agents include the illegal steroid drugs used by some athletes to build up muscle strength and enhance endurance. Anabolic agents are now known to have serious side effects on the user. Chemical Connections 19.4 focuses on the use of anabolic steroids. The structure of one of the more commonly used anabolic agents, methandrostenolone, is given in the lower half of Figure 19.13. Note the similarities between its structure and those of the naturally occurring androgens.

The second major group of steroid hormones consists of the adrenocortical hormones. Produced by the adrenal glands, small organs located on top of each kidney, at least 28 different hormones have been isolated from the adrenal cortex (the outer part of the glands).

There are two types of adrenocortical hormones.

1. *Mineralocorticoids* control the balance of Na^+ and K^+ ions in cells.
2. *Glucocorticoids* control glucose metabolism and counteract inflammation.

The major mineralocorticoid is aldosterone, and the major glucocorticoid is cortisol (hydrocortisone). Cortisol is the hormone synthesized in the largest amount by the adrenal glands. Cortisol and its synthetic ketone derivative cortisone exert powerful anti-inflammatory effects in the body. Both cortisone and prednisolone, a similar synthetic derivative, are used as prescription drugs to control inflammatory diseases such as rheumatoid arthritis. Figure 19.14 gives the structures of these adrenocortical hormones.

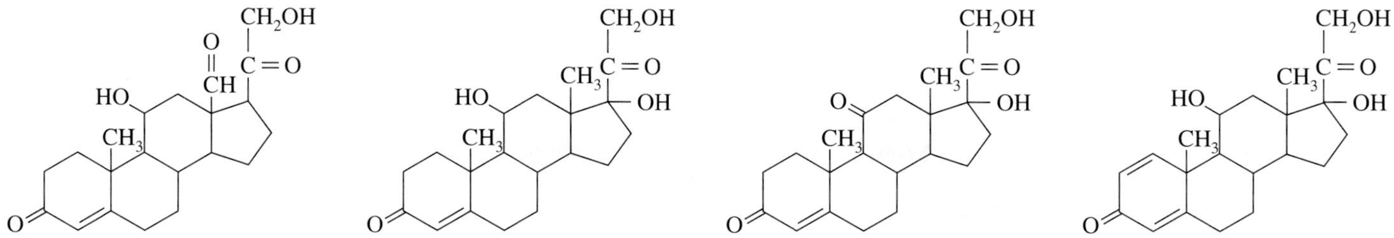

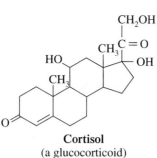

Aldosterone (a mineralocorticoid)

Cortisol (a glucocorticoid)

Cortisone (an anti-inflammatory drug)

Prednisolone (an anti-inflammatory drug)

Figure 19.14
Structures of selected adrenocortical hormones and related synthetic compounds.

19.11 Eicosanoids

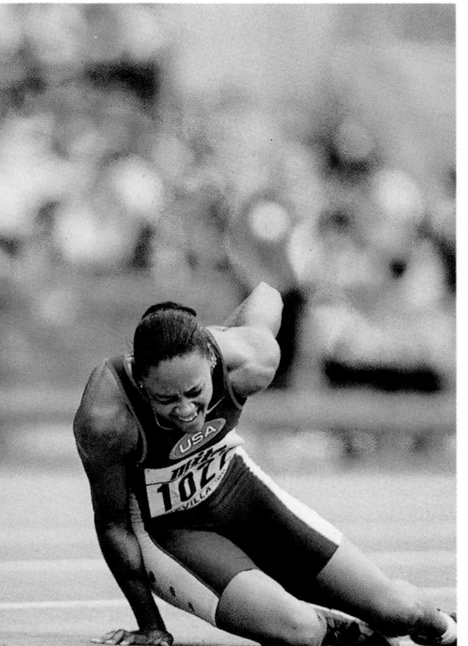

Molecules called eicosanoids are involved in the pain and inflammation response associated with "muscle injuries," which athletes occasionally experience.

Eicosanoids *are oxygenated derivatives of polyunsaturated 20-carbon fatty acids.* The metabolic precursor of most eicosanoids is arachidonic acid, the 20:4 fatty acid. The name *eicosanoid* is derived from the Greek word *eikos,* which means "twenty."

Almost all cells, except red blood cells, produce eicosanoids. These substances, like hormones, have profound physiological effects at extremely low concentrations. Eicosanoids are hormone-like molecules rather than true hormones, because they are not transported in the bloodstream to their site of action, as are true hormones; instead, they exert their effects in the tissues where they are synthesized. The physiological effects of eicosanoids include mediation of

1. The inflammatory response, a normal response to tissue damage
2. The production of pain and fever
3. The regulation of blood pressure
4. The induction of blood clotting
5. The control of reproduction functions, such as induction of labor
6. The regulation of the sleep/wake cycle

There are three principal types of eicosanoids: prostaglandins, thromboxanes, and leukotrienes.

• Prostaglandins

Prostaglandins *are 20-carbon fatty acid derivatives that contain a cyclopentane ring and oxygen-containing functional groups.* Twenty-carbon fatty acids are converted into a prostaglandin structure when the eighth and twelfth carbon atoms of the fatty acid become connected to form a five-membered ring (Figure 19.15a, b).

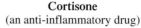

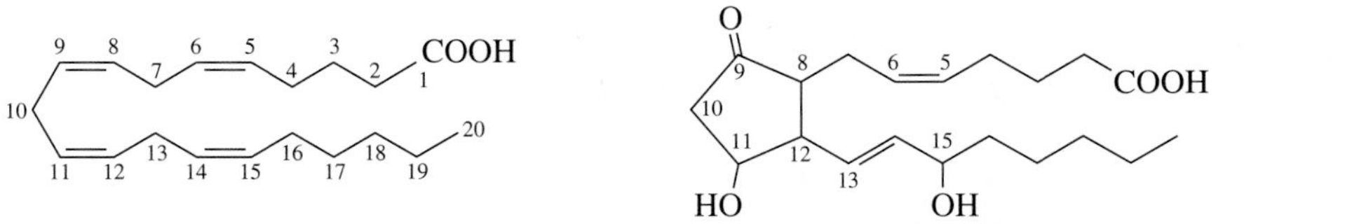

(a) Arachidonic acid

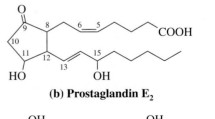

(b) Prostaglandin E_2

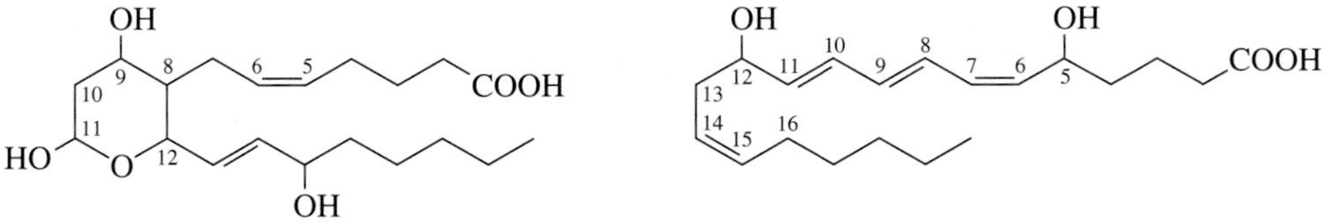

(c) Thromboxane B_2

(d) Leukotriene B_4

Figure 19.15
Relationship of the structures of various eicosanoids to their precursor, arachidonic acid.

Prostaglandins are named after the prostate gland, which was first thought to be their only source. Today, more than 20 prostaglandins have been discovered in a variety of tissues in both males and females.

Within the human body, prostaglandins are involved in many regulatory functions, including raising body temperature, inhibiting the secretion of gastric juices, relaxing and contracting smooth muscle, directing water and electrolyte balance, intensifying pain, and enhancing inflammation responses. Aspirin reduces inflammation and fever because it inactivates the enzyme needed for prostaglandin synthesis.

• Thromboxanes

Thromboxanes *are 20-carbon fatty acid derivatives that contain a cyclic ether ring and oxygen-containing functional groups.* As with prostaglandins, the cyclic structure involves a bond between carbons 8 and 12 (Figure 19.15c). An important function of thromboxanes is to promote the formation of blood clots. Thromboxanes are produced by blood platelets and promote platelet aggregation.

• Leukotrienes

Leukotrienes *are 20-carbon fatty acid derivatives that contain three conjugated double bonds and hydroxy groups.* Fatty acids and their derivatives do not normally contain *conjugated* double bonds (see Chemical Connections 13.3), as is the case in leukotrienes (Figure 19.15d). Leukotrienes are found in leukocytes (white blood cells). Their source and the presence of the three conjugated double bonds account for their name. Various inflammatory and hypersensitivity (allergy) responses are associated with elevated levels of leukotrienes. The development of drugs that inhibit leukotriene synthesis has been an active area of research.

19.12 Plasma Membranes

• Although triacylglycerols are the most abundant type of lipid in the human body, they are not structural components of membranes because they lack the needed "head and two tails" structure and are nonpolar. The most abundant lipids in membranes are phosphoacylglycerols.

All cells are surrounded by a plasma membrane that confines their contents. A **plasma membrane** *separates the aqueous interior of a cell from the aqueous environment surrounding the cell.* We discuss plasma membranes in this chapter because up to 80% of the mass of a plasma membrane is lipid material.

The fundamental structure of a plasma membrane is a lipid bilayer predominantly made up of various phospholipids (phosphoacylglycerols and sphingolipids). A **lipid bilayer** *is a two-layer-thick structure of lipid molecules in which the nonpolar tails of the lipids are in the middle and the polar heads are on the outside surface.* Figure 19.16 shows an electron micrograph of a plasma membrane and a diagram of a section of a lipid bilayer.

Such a bilayer is 6 to 9 billionths of a meter thick—that is, 6 to 9 nanometers thick. There are three distinct parts to such a layer: the exterior polar "heads," the interior polar "heads," and the central nonpolar "tails." These three distinct regions can be seen in the electron micrograph of a cell membrane.

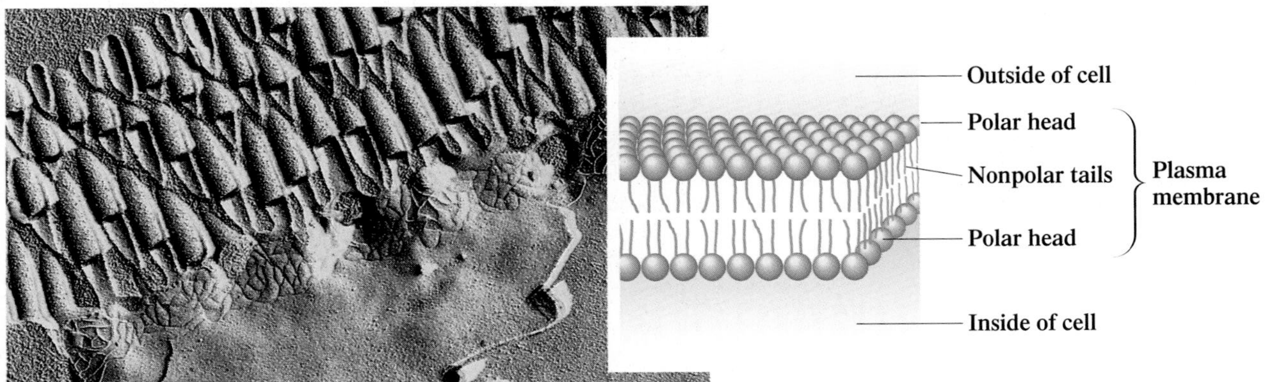

Figure 19.16
(a) Electron micrograph of a plasma membrane. (b) Lipid bilayer model of plasma membrane structure. Lipids with "head and two tails" structures form a double-layered sheet.

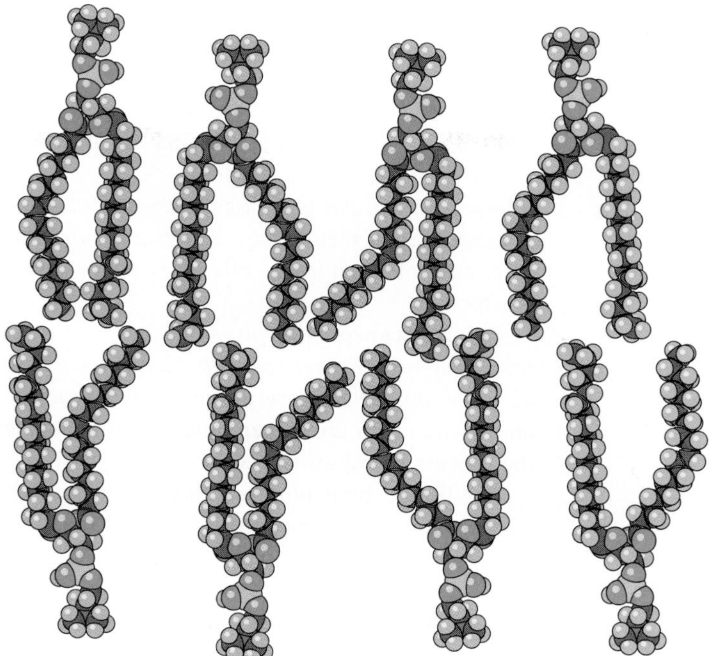

Figure 19.17
The kinks associated with *cis* double bonds in fatty acid chains prevent tight packing of the lipid molecules in a lipid bilayer.

• The percentage of lipid and protein components in a plasma membrane is related to the function of the cell. The lipid/protein ratio ranges from about 80% lipid/20% protein by mass in the myelin sheath of nerve cells to the unique 20% lipid/80% protein ratio for the inner mitochondrial membrane (Section 23.2). Red blood cell membranes contain approximately equal amounts of lipid and protein. A typical membrane also has a carbohydrate content that varies between 2% and 10% by mass.

The lipid bilayer is held together by dipole–dipole interactions, not by covalent bonds. This means each phospholipid or sphingolipid is free to diffuse laterally within the lipid bilayer. Most lipid molecules in the bilayer contain at least one unsaturated fatty acid. The presence of such acids, with the kinks in their carbon chains (Section 19.3), prevent tight packing of fatty acid chains (Figure 19.17). The open packing imparts a liquid-like character to the membrane—a necessity because numerous types of biochemicals must pass into and out of a cell.

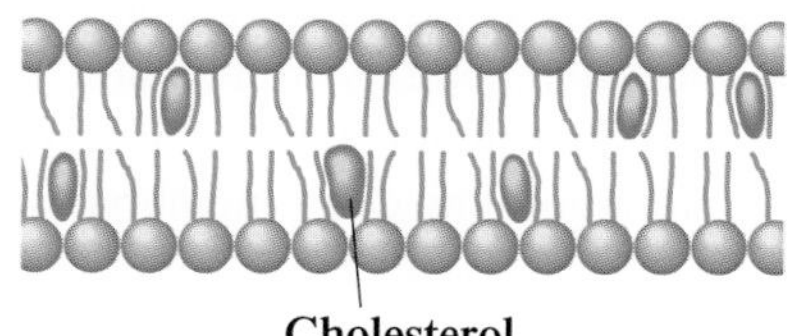

Figure 19.18
Cholesterol molecules fit between fatty acid chains in a lipid bilayer.

Cholesterol molecules are also components of plasma membranes. They regulate membrane fluidity. Because of their compact shape (Section 19.10; Figure 19.13), cholesterol molecules fit between the fatty acid chains of the lipid bilayer (Figure 19.18), restricting movement of the fatty acid chains and making the bilayer more rigid.

Proteins are also components of lipid bilayers. The proteins are responsible for moving substances such as nutrients and electrolytes across the membrane, and they also act as receptors that bind hormones and neurotransmitters (Figure 19.19).

Small carbohydrate molecules are also components of plasma membranes. They are found on the *outer* membrane surface covalently bonded to protein molecules (a glycoprotein) or lipid molecules (a glycolipid). The carbohydrate portions of glycoproteins and

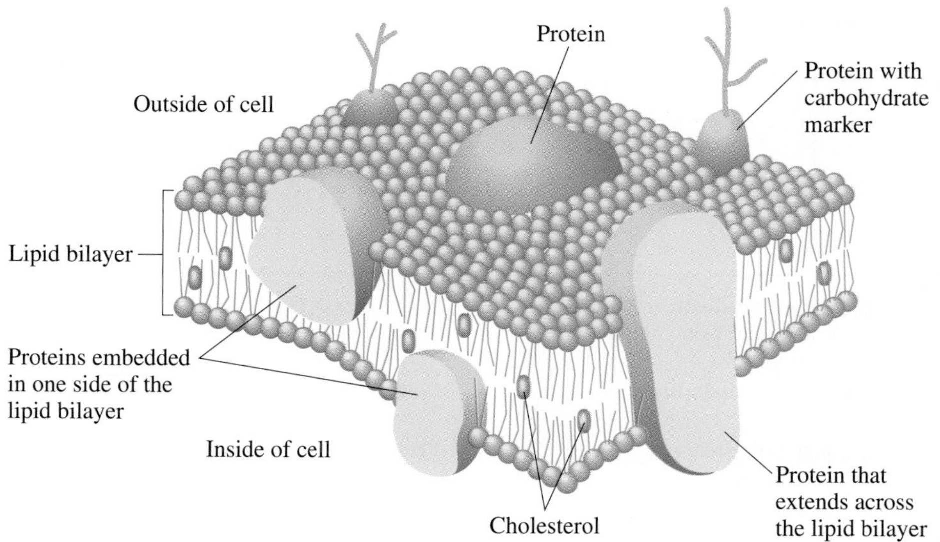

Figure 19.19
Proteins are important structural components of plasma membranes.

glycolipids function as *markers,* substances that play key roles in the process by which different cells recognize each other.

19.13 Transport Across Cell Membranes

In order for cellular processes to be maintained, molecules of various types must be able to cross plasma membranes. Three common transport mechanisms exist by which molecules can enter and leave cells. They are passive transport, facilitated transport, and active transport.

Passive transport *involves movement of a substance across a membrane by diffusion from a region of higher concentration to a region of lower concentration without the expenditure of any cellular energy.* Only a few types of molecules, including O_2, N_2, H_2O, urea, and ethanol, can cross membranes in this manner. Passive transport is closely related to the process of osmosis (Section 8.8).

Facilitated transport *involves movement of a substance across a membrane by diffusion, with the aid of membrane proteins, from a region of higher concentration to a region of lower concentration without the expenditure of cellular energy.* The specific protein molecules involved in the process are called *carriers* or *transporters.* A carrier protein forms a complex with a specific molecule at one surface of the membrane. Formation of the complex induces a conformational change in the protein that allows the molecule to move through a "gate" to the other side of the membrane. Once the molecule is released, the protein returns to its original conformation. Glucose, chloride ion, and bicarbonate ion cross membranes in this manner.

Active transport *involves movement of a substance across a membrane, with the aid of membrane proteins, against a concentration gradient with the expenditure of cellular energy.* Proteins involved in active transport are called "pumps," because they require energy much as a water pump requires energy in order to function. The needed energy is supplied by molecules such as ATP (Section 23.3). The need for energy expenditure is related to the molecules moving against a concentration gradient—from lower to higher concentration. It is essential to life processes to have some solutes "permenantly" at different concentrations on the two sides of a membrane, a situation contrary to the natural tendency (osmosis) to establish equal concentrations on both sides of a membrane. Hence the need for active transport. Sodium, potassium, and hydronium ions cross membranes through active transport.

Figure 19.20 contrasts the processes of passive transport, facilitated transport, and active transport.

Figure 19.20
Three processes by which substances can cross plasma membranes: (a) passive transport, (b) facilitated transport, and (c) active transport.

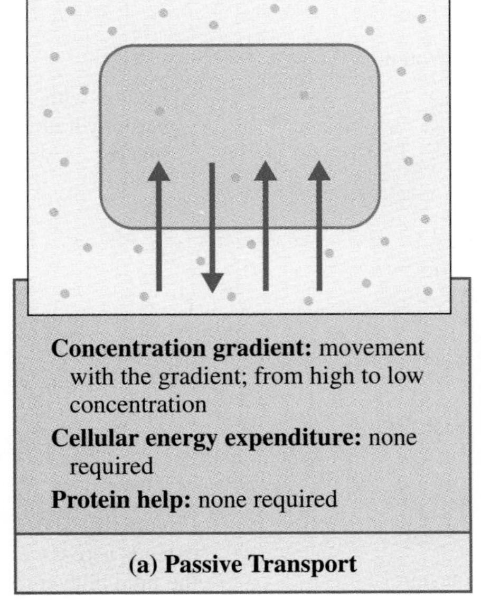

Concentration gradient: movement with the gradient; from high to low concentration
Cellular energy expenditure: none required
Protein help: none required
(a) Passive Transport

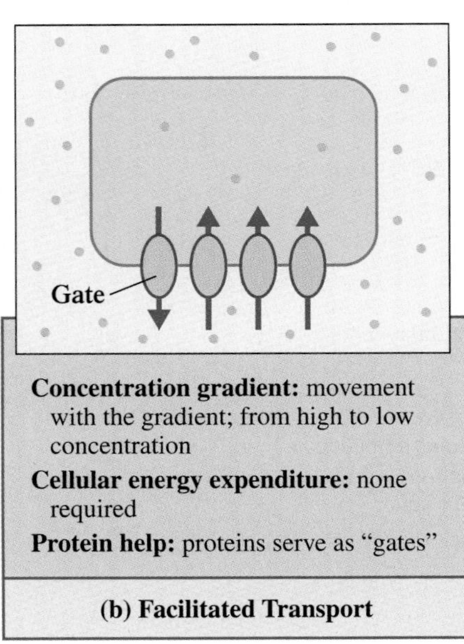

Gate
Concentration gradient: movement with the gradient; from high to low concentration
Cellular energy expenditure: none required
Protein help: proteins serve as "gates"
(b) Facilitated Transport

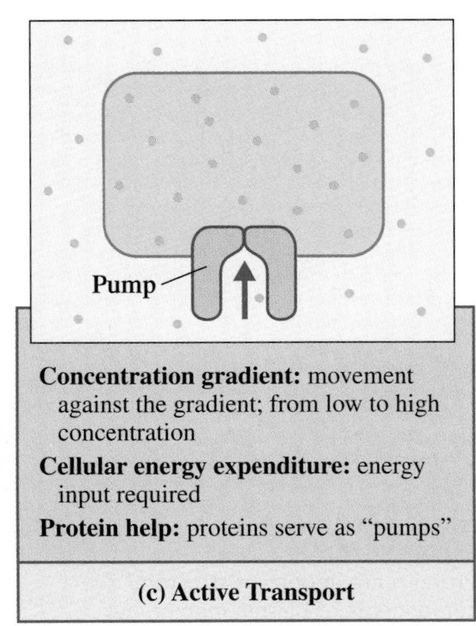

Pump
Concentration gradient: movement against the gradient; from low to high concentration
Cellular energy expenditure: energy input required
Protein help: proteins serve as "pumps"
(c) Active Transport

Concepts to Remember

Lipids. Lipids are a structurally heterogeneous group of compounds that are soluble in nonpolar organic solvents and insoluble in water. Lipids are divided into two major classes, saponifiable and nonsaponifiable, on the basis of whether they undergo hydrolysis reactions in alkaline (basic) solution.

Fatty acids. Fatty acids are monocarboxylic acids that contain long unbranched hydrocarbon chains. The carbon chain may be saturated, monounsaturated, or polyunsaturated. Length of carbon chain, degree of unsaturation, and location of the unsaturation influence the properties of fatty acids. Fatty acid residues are present in all saponifiable lipids.

Fats and oils. Fats and oils, also called triacylglycerols, are triesters formed from the reaction between three fatty acid molecules and a glycerol molecule. Fats, which are solids at room temperature, have a relatively high percentage of saturated fatty acid residues. Oils, which are liquids at room temperature, have a relatively high percentage of unsaturated fatty acid residues.

Phosphoacylglycerols. Phosphoacylglycerols are triesters of glycerol in which two —OH groups are esterified with fatty acids and one —OH group is esterified with phosphoric acid, which in turn is linked to an alcohol. Phosphoacylglycerols, which have a "head and two tails" structure, are major components of cell membranes. Phosphatidylcholines, phosphatidylethanolamines, and phosphatidylserines are phosphoacylglycerols.

Waxes. Waxes are monoesters formed from the reaction of a long-chain monohydroxy alcohol with a fatty acid molecule. Their structure involves a head and two tails; the two tails are the alcohol chain and the fatty acid chain.

Sphingolipids. Sphingolipids have structures based on the long-chain amino dialcohol sphingosine rather than on glycerol. A fatty acid is bonded to the sphingosine via the amino group (an amide linkage), and an additional group is bonded to the terminal hydroxyl group of sphingosine (an ester linkage). Sphingolipids have a "head and two tails" structure; the fatty acid and the carbon chain of sphingosine itself are the tails. Sphingomyelins, cerebrosides, and gangliosides are sphingolipids.

Steroids. Steroids are nonsaponifiable lipids that have structures involving a fused-ring system containing three 6-membered rings and one 5-membered ring. Cholesterol is the most abundant steroid in the human body.

Eicosanoids. Eicosanoids are hormone-like molecules derived from 20-carbon fatty acids. Prostaglandins are derivatives in which a cyclopentane ring is present; thromboxanes are derivatives in which a cyclic ether ring is present; and leukotrienes are derivatives in which three conjugated double bonds are present.

Lipid bilayers. A lipid bilayer is the fundamental structure associated with a plasma membrane. It is a two-layer-thick structure of lipid molecules (mostly phosphoacylglycerols and sphingolipids) in which nonpolar tails of the lipids are in the middle and the polar heads are on the outside surfaces.

Membrane transport mechanisms. Transport mechanisms by which molecules enter and leave cells are *passive* transport, *facilitated* transport, and *active* transport. Passive and facilitated transport follow a concentration gradient and do not involve cellular energy expenditure. Active transport involves movement against a concentration gradient and requires cellular energy expenditure.

Key Reactions and Equations

1. Formation of a triacylglycerol (Section 19.5)

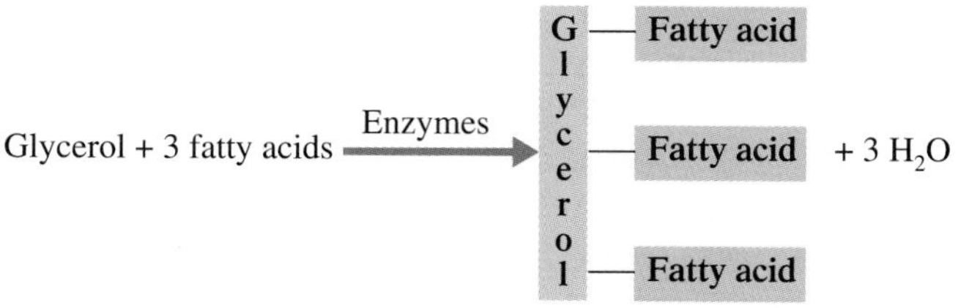

Glycerol + 3 fatty acids —Enzymes→ Glycerol (—Fatty acid, —Fatty acid, —Fatty acid) + 3 H_2O

2. Hydrolysis of a triacylglycerol to produce glycerol and fatty acids (Section 19.5)

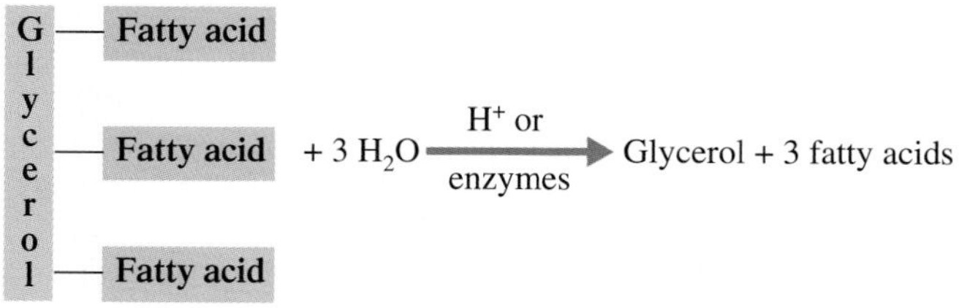

Glycerol (—Fatty acid, —Fatty acid, —Fatty acid) + 3 H_2O —H^+ or enzymes→ Glycerol + 3 fatty acids

3. Saponification of a triacylglycerol to produce glycerol and fatty acid salts (Section 19.5)

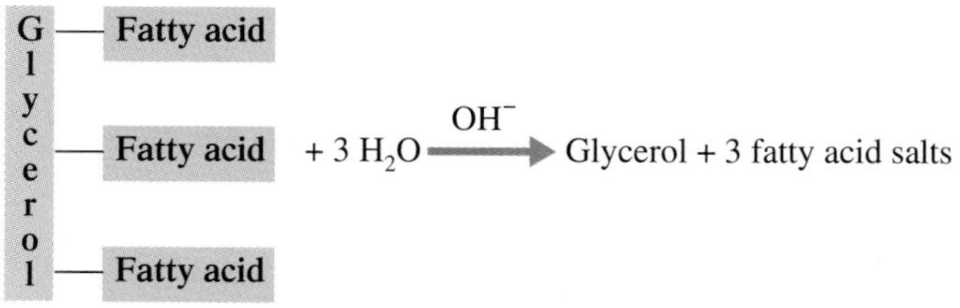

Glycerol (—Fatty acid, —Fatty acid, —Fatty acid) + 3 H_2O —OH^-→ Glycerol + 3 fatty acid salts

4. Hydrogenation of a triacylglycerol to reduce the unsaturation of its fatty acid components (Section 19.5)

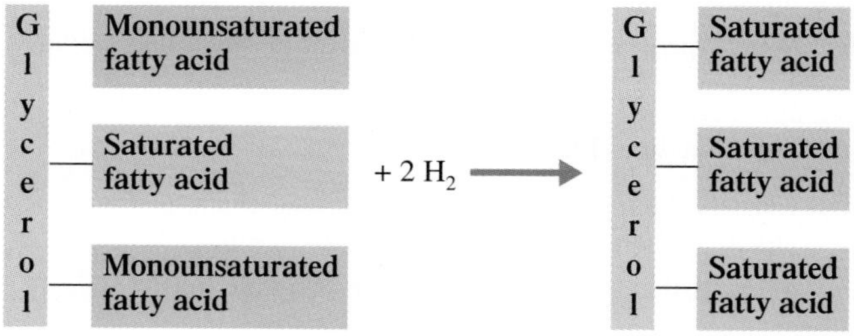

Glycerol (—Monounsaturated fatty acid, —Saturated fatty acid, —Monounsaturated fatty acid) + 2 H_2 → Glycerol (—Saturated fatty acid, —Saturated fatty acid, —Saturated fatty acid)

5. Formation of a phosphoacylglycerol (Section 19.6)

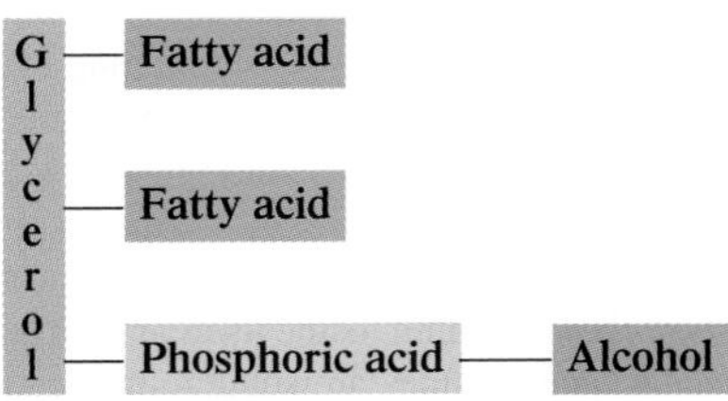

6. Formation of a wax (Section 19.7)

7. Formation of a sphingolipid (Section 19.8)

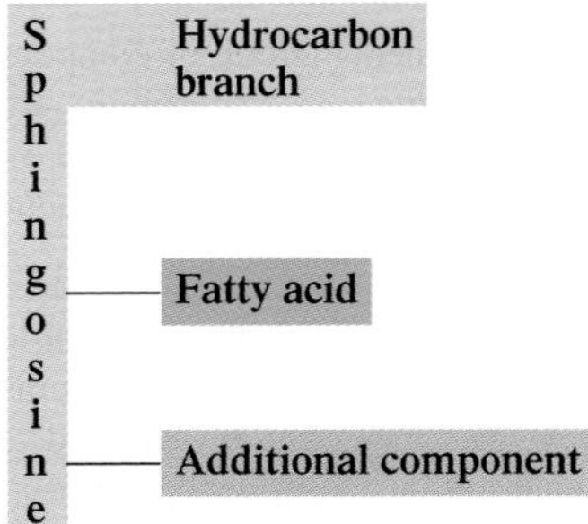

Key Terms

Active transport (19.13)
Bile salts (19.10)
Eicosanoids (19.11)
Essential fatty acid (19.2)
Facilitated transport (19.13)
Fats (19.4)
Fatty acids (19.2)
Hormones (19.10)
Leukotrienes (19.11)
Lipid bilayer (19.12)
Lipids (19.1)
Mixed triacylglycerol (19.4)
Monounsaturated fatty acids (19.2)
Nonsaponifiable lipids (19.1)
Oils (19.4)
Omega-3 fatty acid (19.2)
Omega-6 fatty acid (19.2)
Passive transport (19.13)
Phosphoacylglycerols (19.6)
Plasma membrane (19.12)
Polyunsaturated fatty acids (19.2)
Prostaglandins (19.11)
Saponifiable lipids (19.1)
Saturated fatty acids (19.2)
Simple triacylglycerol (19.4)
Sphingolipid (19.8)
Steroids (19.10)
Thromboxanes (19.11)
Triacylglycerol (19.4)
Wax (19.7)

Exercises and Problems

The members of each pair of problems in this section test similar material.

Characteristics of Lipids (Section 19.1)

19.1 Give a formal definition for the class of biochemicals called lipids.

19.2 What structural feature, if any, do all lipid molecules have in common? Explain your answer.

19.3 Would you expect lipids to be soluble or insoluble in each of the following solvents?

a. H_2O (polar)
b. $CH_3—CH_2—O—CH_2—CH_3$ (nonpolar)
c. $CH_3—OH$ (polar)
d. $CH_3—CH_2—CH_2—CH_2—CH_3$ (nonpolar)

19.4 Would you expect lipids to be soluble or insoluble in each of the following solvents?

a. $CH_3—(CH_2)_7—CH_3$ (nonpolar)
b. $CH_3—Cl$ (polar)
c. CCl_4 (nonpolar)
d. $CH_3—CH_2—OH$ (polar)

19.5 What is the difference between a saponifiable lipid and a nonsaponifiable lipid?

19.6 What type of chemical bond is present in saponifiable lipids?

Fatty Acids (Sections 19.2 and 19.3)

19.7 Classify the following fatty acids as long-chain, medium-chain, or short-chain.

a. Myristic (14:0) b. Caproic (6:0)
c. Arachidic (20:0) d. Capric (10:0)

19.8 Classify the following fatty acids as long-chain, medium-chain, or short-chain.

a. Lauric (12:0) b. Oleic (18:1)
c. Butyric (4:0) d. Stearic (18:0)

19.9 Classify the following fatty acids as saturated, monounsaturated, or polyunsaturated.

a. Stearic (18:0) b. Linolenic (18:3)
c. Docosahexaenoic (22:6) d. Oleic (18:1)

19.10 Classify the following fatty acids as saturated, monounsaturated, or polyunsaturated.

a. Palmitic (16:0) b. Linoleic (18:2)
c. Arachidonic (20:4) d. Palmitoleic (16:1)

19.11 With the help of Table 19.1, classify each of the acids in Problem 19.9 as an omega-3 acid, an omega-6 acid, or neither an omega-3 nor an omega-6 acid.

19.12 With the help of Table 19.1, classify each of the acids in Problem 19.10 as an omega-3 acid, an omega-6 acid, or neither an omega-3 nor an omega-6 acid.

19.13 Classify each of the acids in Problem 19.9 as an essential fatty acid or a nonessential fatty acid.

19.14 Classify each of the acids in Problem 19.10 as an essential fatty acid or a nonessential fatty acid.

19.15 Why does the introduction of double bonds into a fatty acid molecule lower its melting point?

19.16 What effect does a *cis* double bond have on the shape of a fatty acid molecule?

19.17 In each of the following pairs, select the fatty acid that has the lower melting point.

a. 18:0 acid and 18:1 acid

b. 18:2 acid and 18:3 acid

c. 14:0 acid and 16:0 acid

d. 18:1 acid and 20:0 acid

19.18 In each of the following pairs, select the fatty acid that has the higher melting point.

a. 14:0 acid and 18:0 acid

b. 20:4 acid and 20:5 acid

c. 18:3 acid and 20:3 acid

d. 16:0 acid and 16:1 acid

19.19 On the basis of the information given at the bottom of Table 19.1, what would be the IUPAC name for linolenic acid?

19.20 On the basis of the information given at the bottom of Table 19.1, what would be the IUPAC name for oleic acid?

Fats and Oils (Section 19.4)

19.21 What is the difference between a fat and an oil in terms of chemical structure?

19.22 What is the difference between a fat and an oil in terms of melting point?

19.23 In terms of structural features, define the class of lipids called triacylglycerols.

19.24 What is the difference between a simple triacylglycerol molecule and a mixed triacylglycerol molecule?

19.25 Draw the condensed structural formula of a triacylglycerol formed from glycerol and three molecules of palmitic acid.

19.26 Draw the condensed structural formula of a triacylglycerol formed from glycerol and three molecules of stearic acid.

19.27 Draw block diagram structures for the four different triacylglycerols that can be produced from glycerol, stearic acid, and linolenic acid.

19.28 Draw block diagram structures for the three different triacylglycerols that can be produced from glycerol, palmitic acid, stearic acid, and linolenic acid.

19.29 Identify the fatty acids present in the following triacylglycerols.

a.

$$CH_2—O—\overset{O}{\overset{\|}{C}}—(CH_2)_{14}—CH_3$$
$$CH—O—\overset{O}{\overset{\|}{C}}—(CH_2)_{12}—CH_3$$
$$CH_2—O—\overset{O}{\overset{\|}{C}}—(CH_2)_7—CH{=}CH—(CH_2)_7—CH_3$$

b.

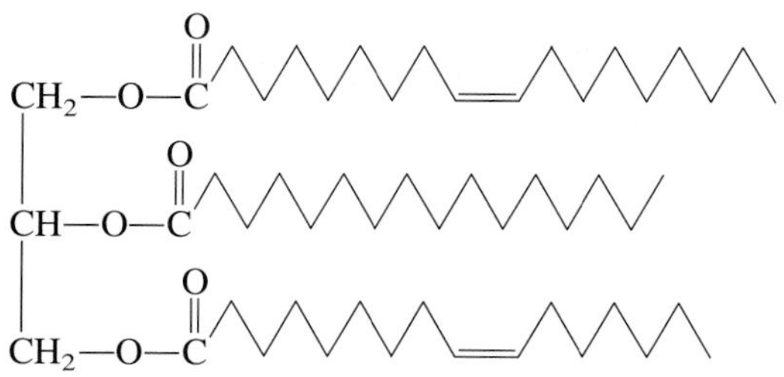

$$CH_2—O—\overset{O}{\overset{\|}{C}}$$
$$CH—O—\overset{O}{\overset{\|}{C}}$$
$$CH_2—O—\overset{O}{\overset{\|}{C}}$$

19.30 Identify the fatty acids present in the following triacylglycerols.

a.

$$CH_2—O—\overset{O}{\overset{\|}{C}}—(CH_2)_{16}—CH_3$$
$$CH—O—\overset{O}{\overset{\|}{C}}—(CH_2)_7—CH{=}CH—(CH_2)_7—CH_3$$
$$CH_2—O—\overset{O}{\overset{\|}{C}}—(CH_2)_{12}—CH_3$$

b.

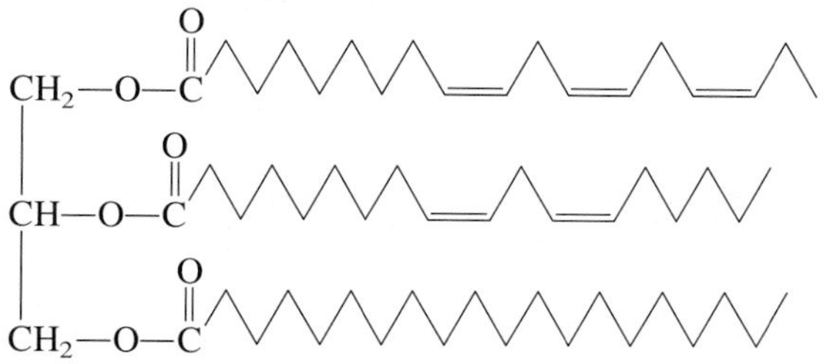

$$CH_2—O—\overset{O}{\overset{\|}{C}}$$
$$CH—O—\overset{O}{\overset{\|}{C}}$$
$$CH_2—O—\overset{O}{\overset{\|}{C}}$$

Chemical Reactions of Triacylglycerols (Section 19.5)

19.31 Name, in general terms, the products of the complete

a. hydrolysis of a fat b. saponification of an oil

19.32 Name, in general terms, the products of the complete

a. saponification of a fat b. hydrolysis of an oil

19.33 Draw condensed structural formulas for all products you would obtain from the complete hydrolysis of the following triacylglycerol.

$$CH_2—O—\overset{O}{\overset{\|}{C}}—(CH_2)_{14}—CH_3$$
$$CH—O—\overset{O}{\overset{\|}{C}}—(CH_2)_{12}—CH_3$$
$$CH_2—O—\overset{O}{\overset{\|}{C}}—(CH_2)_7—CH{=}CH—(CH_2)_7—CH_3$$

19.34 Draw condensed structural formulas for all products you would obtain from the complete hydrolysis of the following triacylglycerol.

$$CH_2—O—\overset{O}{\overset{\|}{C}}—(CH_2)_{16}—CH_3$$
$$CH—O—\overset{O}{\overset{\|}{C}}—(CH_2)_{12}—CH_3$$
$$CH_2—O—\overset{O}{\overset{\|}{C}}—(CH_2)_6—(CH_2—CH{=}CH)_3—CH_2—CH_3$$

19.35 With the help of Table 19.1, determine the names of each of the products obtained in Problem 19.33.

19.36 With the help of Table 19.1, determine the names of each of the products obtained in Problem 19.34.

19.37 Draw condensed structural formulas for all products you would obtain from the saponification with NaOH of the triacylglycerol in Problem 19.33.

19.38 Draw condensed structural formulas for all products you would obtain from the saponification with KOH of the triacylglycerol in Problem 19.34.

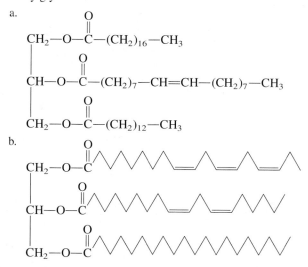

19.39 With the help of Table 19.1, name each of the products obtained in Problem 19.37.

19.40 With the help of Table 19.1, name each of the products obtained in Problem 19.38.

19.41 Why can only unsaturated triacylglycerols undergo hydrogenation?

19.42 A food package label lists an oil as "partially hydrogenated." What does this mean?

19.43 How many molecules of H_2 will react with one molecule of the following triacylglycerol?

$$CH_2{-}O{-}\overset{\overset{\displaystyle O}{\|}}{C}{-}(CH_2)_6{-}(CH_2{-}CH{=}CH)_2{-}(CH_2)_4{-}CH_3$$
$$CH{-}O{-}\overset{\overset{\displaystyle O}{\|}}{C}{-}(CH_2)_7{-}CH{=}CH{-}(CH_2)_7{-}CH_3$$
$$CH_2{-}O{-}\overset{\overset{\displaystyle O}{\|}}{C}{-}(CH_2)_6{-}(CH_2{-}CH{=}CH)_3{-}CH_2{-}CH_3$$

19.44 How many molecules of H_2 will react with one molecule of the following triacylglycerol?

$$CH_2{-}O{-}\overset{\overset{\displaystyle O}{\|}}{C}{-}(CH_2)_7{-}CH{=}CH{-}(CH_2)_7{-}CH_3$$
$$CH{-}O{-}\overset{\overset{\displaystyle O}{\|}}{C}{-}(CH_2)_{16}{-}CH_3$$
$$CH_2{-}O{-}\overset{\overset{\displaystyle O}{\|}}{C}{-}(CH_2)_6{-}(CH_2{-}CH{=}CH)_3{-}CH_2{-}CH_3$$

19.45 Draw block diagram structures for all possible products of the partial hydrogenation with two molecules of H_2 of the following molecules.

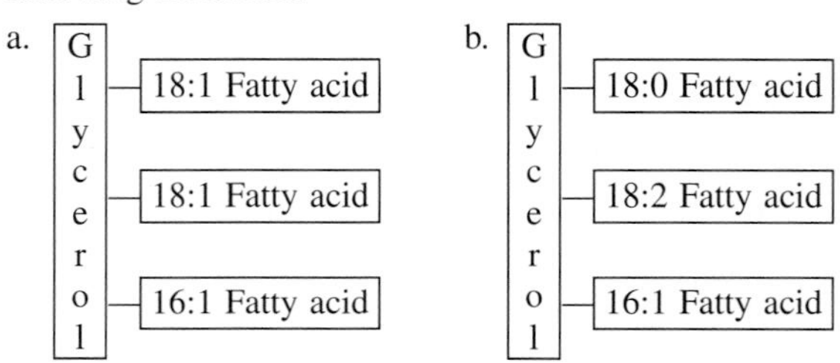

19.46 Draw block diagram structures for all possible products of the partial hydrogenation with two molecules of H_2 of the following molecules.

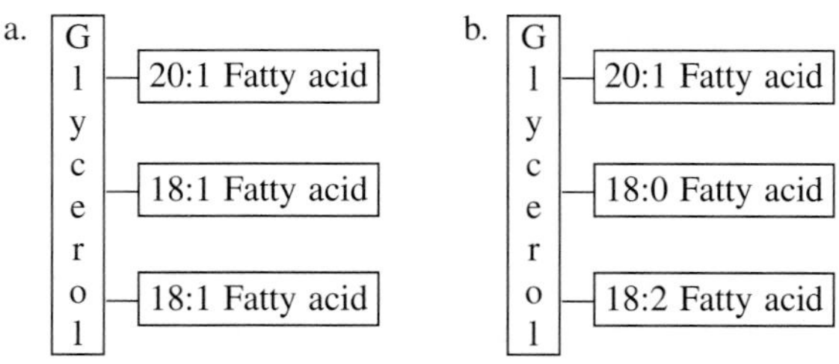

19.47 Why do animal fats and vegetable oils become rancid when exposed to moist warm air?

19.48 Why are the compounds BHA and BHT often added to foods that contain fats and oils?

Phosphoacylglycerols (Section 19.6)

19.49 Draw the general block diagram structure of a phosphoacylglycerol.

19.50 How do phosphoacylglycerols differ structurally from triacylglycerols?

19.51 Draw the structures of the three amino alcohols commonly esterified to the phosphate group in phosphoacylglycerols.

19.52 What is the structure of a phosphatidyl group?

19.53 Phosphoacylglycerols have a "head and two tails" structure. What is the chemical identity of the head and of each of the two tails?

19.54 Phosphoacylglycerols have a hydrophobic portion and a hydrophilic portion in their structure. What does this statement mean?

Waxes (Section 19.7)

19.55 Draw the general block diagram structure of a wax.

19.56 Why is paraffin wax not considered a "true" wax?

19.57 Draw the condensed structural formula of a wax formed from palmitic acid (Table 19.1) and cetyl alcohol, $CH_3{-}(CH_2)_{14}{-}CH_2{-}OH$.

19.58 Name the two classes of organic compounds produced when a wax is hydrolyzed.

19.59 How do waxes differ structurally from triacylglycerols?

19.60 Waxes have a "head and two tails" structure. What is the chemical identity of the head and of each of the two tails?

Sphingolipids (Section 19.8)

19.61 Draw the general block diagram structure of a sphingolipid.

19.62 Draw the structure of the sphingolipid that could form from sphingosine, stearic acid, phosphoric acid, and ethanolamine.

19.63 Describe the location of the *amide* linkage in a sphingolipid.

19.64 Describe the location(s) of the *ester* linkage(s) in a sphingolipid.

19.65 Sphingolipids have a "head and two tails" structure. What is the chemical identity of the head and of each of the two tails?

19.66 Are phosphoric acid and fatty acid components a necessity in all sphingolipid structures? Explain.

19.67 What is the major structural difference between a sphingomyelin and a cerebroside?

19.68 What is the major structural difference between a cerebroside and a ganglioside?

Steroids (Section 19.10)

19.69 Draw and number the fused hydrocarbon ring system characteristic of all steroids.

19.70 What positions in the steroid nucleus are particularly likely to bear substituents?

19.71 Describe the structure of cholesterol in terms of substituents attached to the steroid nucleus.

19.72 Explain the fallacy in the statement that a cholesterol-free diet would eventually reduce body cholesterol levels to near zero.

19.73 What is the role of bile salts in the human digestive process?

19.74 Contrast the general structural differences between cholesterol and bile salts.

19.75 What are the two major classes of steroid hormones?

19.76 Describe the general function of each of the following types of steroid hormones.

a. Estrogens b. Androgens
c. Progestins d. Mineralocorticoids

19.77 How do the sex hormones estradiol and testosterone differ in structure?

19.78 What functional groups are present in each of the following steroid hormones?

a. Estradiol b. Testosterone
c. Progesterone d. Cortisone

Eicosanoids (Section 19.11)

19.79 What is the major structural difference between a prostaglandin and its parent fatty acid?

19.80 What is the major structural difference between a leukotriene and its parent fatty acid?

19.81 List six physiological processes that are regulated by eicosanoids.

19.82 What is the biochemical basis for the effectiveness of aspirin in decreasing inflammation?

Plasma Membranes (Section 19.12)

19.83 What are the two major types of lipids present in a plasma membrane?

19.84 What are the structural characteristics common to lipids found in plasma membranes?

19.85 What is a lipid bilayer?

19.86 What is the basic structure of a plasma membrane?

19.87 What is the function of unsaturation in the hydrocarbon tails of membrane lipids?

19.88 How would a plasma membrane with a large percentage of linoleic acid differ from one with a large percentage of palmitic acid?

Transport Across Cell Membranes (Section 19.13)

19.89 What is the difference between *passive transport* and *facilitated transport*?

19.90 What is the difference between *facilitated transport* and *active transport*?

19.91 Match each of the following statements relating to membrane transport processes to the appropriate term: *passive transport, facilitated transport, active transport.* More than one term may apply in a given situation.

a. Movement across the membrane is against the concentration gradient.
b. Proteins serve as "gates."
c. Cellular energy expenditure is required.
d. Movement across the membrane is from a high to a low concentration.

19.92 Match each of the following statements relating to membrane transport processes to the appropriate term: *passive transport, facilitated transport, active transport.* More than one term may apply in a given situation.

a. Movement across the membrane is with the concentration gradient.
b. Proteins serve as "pumps."
c. Cellular energy expenditure is not required.
d. Movement across the membrane is from a low to a high concentration.

Additional Problems

19.93 Classify each of the following types of lipids as (1) glycerol-based, (2) sphingosine-based, or (3) neither glycerol-based nor sphingosine-based.

a. Bile salts b. Fats
c. Thromboxanes d. Gangliosides
e. Waxes f. Leukotrienes

19.94 Indicate whether or not each of the lipid types in Problem 19.93 has a "head and two tails" structure.

19.95 Identify the type of lipid that fits each of the following "structural component" characterizations.

a. Sphingosine + fatty acid + phosphoric acid + choline
b. Glycerol + three fatty acids
c. Fused-ring system with three 6-membered rings and one 5-membered ring
d. 20-carbon fatty acid + three conjugated double bonds
e. 20-carbon fatty acid + cyclopentane ring
f. Sphingosine + fatty acid + monosaccharide

19.96 Classify each of the following types of lipids as (1) a storage lipid, (2) a membrane lipid, (3) a polar lipid, or (4) a nonpolar lipid. More than one classification may apply to a given lipid type.

a. Fats b. Triacylglycerols
c. Phosphoacylglycerols d. Waxes
e. Sphingomyelins f. Cerebrosides

19.97 Indicate whether or not each of the lipid types in Problem 19.96 is a saponifiable lipid.

19.98 Indicate which of the lipid types in Problem 19.96 can be categorized as

a. phospholipids
b. sphingolipids
c. glycolipids
d. both phospholipids and sphingolipids

19.99 Specify the number of ester linkages, amide linkages, and glycosidic linkages present in each of the following types of lipids.

a. Oils b. Lecithins
c. Sphinogomyelins d. Waxes
e. Cerebrosides f. Phosphatidylcholines

19.100 Indicate whether or not the following types of lipids contain a "steroid nucleus" as part of structure.

a. Prostaglandins b. Cortisone
c. Cholesterol d. Bile salts
e. Estrogens f. Leukotrienes

Grid Problems

19.101

1. glycerol	2. fatty acid	3. monosaccharide
4. sphingosine	5. amino alcohol	6. phosphoric acid

Select from the grid *all* correct responses for each of the following situations.

a. Structural component of a wax
b. Structural component of a fat
c. Structural component of a phosphatidylserine
d. Structural component of a cerebroside

19.102

1. fats	2. oils	3. phosphatidyl-cholines
4. cephalins	5. sphingomyelins	6. cerebrosides

Select from the grid *all* correct responses for each of the following situations.

a. Types of compounds that are saponifiable
b. Types of compounds whose complete hydrolysis produces glycerol and other products
c. Types of compounds whose complete hydrolysis produces at least one fatty acid as well as other products
d. Types of compounds that contain both amide and ester linkages.

19.103

1. a lecithin	2. cholesterol	3. a cerebroside
4. a bile salt	5. a sphingomyelin	6. a wax

Select from the grid *all* correct responses for each of the following situations.

a. Substance present in plasma membranes
b. Substance classified as a phospholipid
c. Substance classified as a glycolipid
d. Substance classified as a sphingolipid

19.104

1. prostaglandins	2. estrogens	3. gangliosides
4. bile salts	5. androgens	6. leukotrienes

Select from the grid *all* correct responses for each of the following situations.

a. Types of compounds that are cholesterol derivatives
b. Types of compounds that are eicosanoids
c. Types of compounds that are steroids
d. Types of compounds that are nonsaponifiable lipids

CHAPTER 20

PART III

Proteins

CHAPTER OUTLINE

20.1 Characteristics of Proteins 587
20.2 Amino Acids: The Building Blocks for Proteins 588
20.3 Chirality and Amino Acids 590
20.4 Acid–Base Properties of Amino Acids 591
20.5 Peptide Formation 595
20.6 Levels of Protein Structure 598
20.7 Primary Structure of Proteins 598
20.8 Secondary Structure of Proteins 600
20.9 Tertiary Structure of Proteins 602
20.10 Quaternary Structure of Proteins 604
20.11 Globular and Fibrous Proteins 605

Chemistry at a Glance: **Protein Structure** 606

20.12 Simple and Conjugated Proteins 607
20.13 Protein Hydrolysis 608
20.14 Protein Denaturation 609
20.15 Glycoproteins 610
20.16 Lipoproteins 615

Chemical Connections

20.1 The Essential Amino Acids 590
20.2 Substitutes for Human Insulin 599
20.3 Denaturation and Human Hair 610
20.4 Cyclosporine: An Antirejection Drug 613
20.5 Lipoproteins and Heart Attack Risk 614

The protein made by spiders to produce a web is a form of silk that can be exceptionally strong.

In this chapter we consider the third of the bioorganic classes of molecules (Section 18.1), the compounds called proteins. An extraordinary number of different proteins, each with a different function, exist in the human body. A typical human cell contains about 9000 different kinds of proteins, and the human body contains about 100,000 different proteins. Proteins are needed for the synthesis of enzymes, certain hormones, and some blood components; for the maintenance and repair of existing tissues; for the synthesis of new tissue; and sometimes for energy.

20.1 Characteristics of Proteins

Next to water, proteins are the most abundant substances in most cells—from 10% to 20% of the cell's mass (Section 18.1). All proteins contain the elements carbon, hydrogen, oxygen, and nitrogen; most also contain sulfur. The presence of nitrogen in proteins sets them apart from carbohydrates and lipids, which generally do not contain nitrogen. The average nitrogen content of proteins is 15.4% by mass. Other elements, such as phosphorus and iron, are essential constituents of certain specialized proteins. Casein, the main protein of milk, contains phosphorus, an element very important in the diet of infants and children. Hemoglobin, the oxygen-transporting protein of blood, contains iron.

20.2 Amino Acids: The Building Blocks for Proteins

• The word *protein* comes from the Greek *proteios,* which means "of first importance." This derivation alludes to the key role that proteins play in life processes.

A **protein** *is a polymer in which the monomer units are amino acids.* Thus the starting point for a discussion of proteins is an understanding of the structures and chemical properties of amino acids.

An **amino acid** *is an organic compound that contains both an amino* ($-NH_2$) *group and a carboxyl* ($-COOH$) *group.* The amino acids found in proteins are always α-amino acids—that is, amino acids in which the amino group is attached to the α-carbon atom of the carboxylic acid carbon chain. The general structural formula for an α-amino acid is

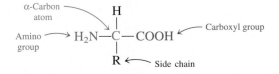

• In an α-amino acid, the carboxyl group and the amino group are attached to the same carbon atom.

$$H_2N-\underset{\substack{|\\R}}{\overset{\substack{H\\|}}{C}}-COOH$$

α-Carbon atom; Amino group; Carboxyl group; Side chain

• The nature of the side chain (R group) distinguishes α-amino acids from each other, both physically and chemically.

The R group present in an α-amino acid is called the amino acid *side chain.* The nature of this side chain distinguishes α-amino acids from each other. Side chains vary in size, shape, charge, acidity, functional groups present, hydrogen-bonding ability, and chemical reactivity.

Over 700 different naturally occurring amino acids are known, but only 20 of them, called standard amino acids, are normally present in proteins. A **standard amino acid** *is one of the 20 α-amino acids normally found in proteins.* The structures of the 20 standard amino acids are given in Table 20.1. Within Table 20.1, amino acids are grouped according to side-chain polarity. In this system there are four categories: (1) nonpolar amino acids, (2) polar neutral amino acids, (3) polar acidic amino acids, and (4) polar basic amino acids. This classification system gives insights into how various types of amino acid side chains help determine the properties of proteins (Section 20.9).

• The nonpolar amino acid *proline* has a structural feature not found in any other standard amino acid. Its side chain, a propyl group, is bonded to both the α-carbon atom and the amino nitrogen atom, giving a cyclic side chain.

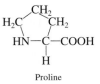

CH_2, H_2C, CH_2, $HN-C-COOH$, H

Proline

Nonpolar amino acids *contain one amino group, one carboxyl group, and a nonpolar side chain.* When incorporated into a protein, such amino acids are *hydrophobic* ("water-fearing"); that is, they are not attracted to water molecules. They are generally found in the interior of proteins, where there is limited contact with water. There are eight nonpolar amino acids.

The three types of polar amino acids have varying degrees of affinity for water. Within a protein, such amino acids are said to be *hydrophilic* ("water-loving"). Hydrophilic amino acids are often found on the surfaces of proteins.

Polar neutral amino acids *contain one amino group, one carboxyl group, and a side chain that is polar but neutral.* The side chain is neutral in that it is neither acidic nor basic in solution at physiological pH. There are seven polar neutral amino acids.

Polar acidic amino acids *contain one amino group and two carboxyl groups, the second carboxyl group being part of the side chain.* In solution at physiological pH, the side chain of a polar acidic amino acid bears a negative charge; the side-chain carboxyl group has lost its acidic hydrogen atom. There are two polar acidic amino acids: aspartic acid and glutamic acid.

Polar basic amino acids *contain two amino groups and one carboxyl group, the second amino group being part of the side chain.* In solution at physiological pH, the side chain of a polar basic amino acid bears a positive charge; the nitrogen atom of the amino group has accepted a proton (basic behavior; Section 17.5). There are three polar basic amino acids: lysine, arginine, and histidine.

The names of the standard amino acids are often abbreviated using three-letter codes. Except in four cases, these abbreviations are the first three letters of the amino acid's

Table 20.1
The 20 Standard Amino Acids, Grouped According to Side-Chain Polarity. Below each amino acid's structure are its name (with pronunciation), its three-letter abbreviation, and its one-letter abbreviation.

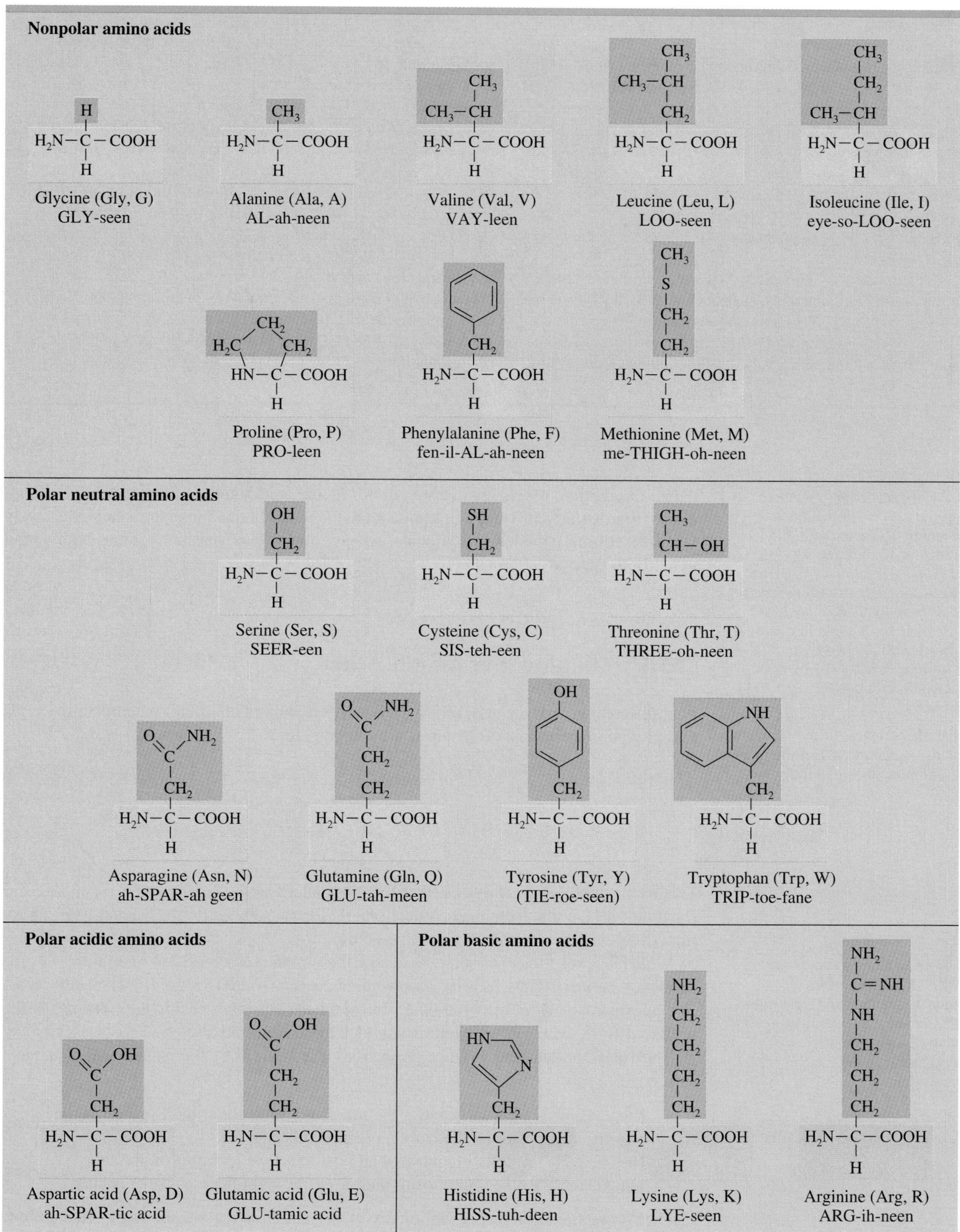

Chemical CONNECTIONS

20.1 The Essential Amino Acids

All of the amino acids in Table 20.1 are necessary constituents of human protein. Adequate amounts of 11 of the 20 amino acids can be synthesized from carbohydrates and lipids in the body if a source of nitrogen is also available. Because the human body is incapable of producing 9 of these 20 acids fast enough or in sufficient quantities to sustain normal growth, these 9 amino acids, called essential amino acids, must be obtained from food. **Essential amino acids** *are amino acids that must be obtained from food.* An adequate human diet must include foods that contain these essential amino acids.

The human body can synthesize small amounts of some of the essential amino acids, but not enough to meet its needs, especially in the case of growing children.

The 9 essential amino acids for adults are histidine, isoleucine, leucine, lysine, methionine, phenylalanine, threonine, tryptophan, and valine. (In addition, arginine is essential for children.)

A **complete dietary protein** *contains all the essential amino acids in the same relative amounts in which human beings require them.* A complete dietary protein may or may not contain all the nonessential amino acids. Most animal proteins, including casein from milk and proteins found in meat, fish, and eggs, are complete proteins, although gelatin is an exception (it lacks tryptophan). Proteins from plants (vegetables, grains, and legumes) have quite diverse amino acid patterns, and some tend to be limiting in one or more essential amino acids. Some plant proteins (for example, corn protein) are far from complete. Others (for example, soy protein) are complete. Thus vegetarians must eat a variety of plant foods to obtain all of the essential amino acids in appropriate quantities.

• A variety of functional groups are present in the side chains of the 20 standard amino acids: six have alkyl groups (Section 12.8), three have aromatic groups (Section 13.10), two have sulfur-containing groups (Section 14.13), two have hydroxyl (alcohol) groups (Section 14.2), three have amino groups (Section 17.2), two have carboxyl groups (Section 16.1), and two have amide groups (Section 17.11).

name. In addition, a new one-letter code for amino acid names is currently gaining popularity (particularly in computer applications). These abbreviations are used extensively when describing peptides and proteins, which contain tens and hundreds of amino acid units. Both types of abbreviations are given in Table 20.1.

20.3 Chirality and Amino Acids

• Glycine, the simplest of the standard amino acids, is achiral. All of the other standard amino acids are chiral.

Four different groups are attached to the α-carbon atom in all of the standard amino acids except glycine, where the R group is a hydrogen atom.

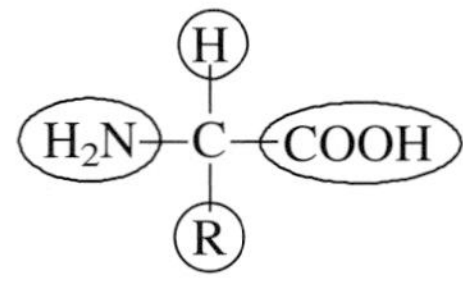

This means that the structures of 19 of the 20 standard amino acids possess a chiral center (Section 18.4) at this location, so enantiomeric forms (left- and right-handed forms; Section 18.5) exist for each of these amino acids.

• Because only L amino acids are constituents of proteins, the enantiomer designation of L or D will be omitted in subsequent amino acid and protein discussions. It is understood that it is the L isomer that is always present.

With few exceptions (in some bacteria), the amino acids found in nature and in proteins are L isomers. Thus, as is the case with monosaccharides (Section 18.8), nature favors one mirror-image form over the other. Interestingly, for amino acids the L isomer is the preferred form, whereas for monosaccharides the D isomer is preferred.

The rules for drawing Fischer projections (Section 18.6) for amino acid structures follow.

1. The —COOH group is put at the top of the projection, the R group at the bottom. This positions the carbon chain vertically.
2. The $-NH_2$ group is in a horizontal position. Positioning it on the left denotes the L isomer, and positioning it on the right denotes the D isomer.

Fischer projections for both enantiomers of the amino acids alanine and serine follow.

• Two of the 19 chiral standard amino acids, isoleucine and threonine, possess two chiral centers (see Table 20.1). With two chiral centers present, four stereoisomers are possible for these amino acids. However, only one of the L isomers is found in proteins.

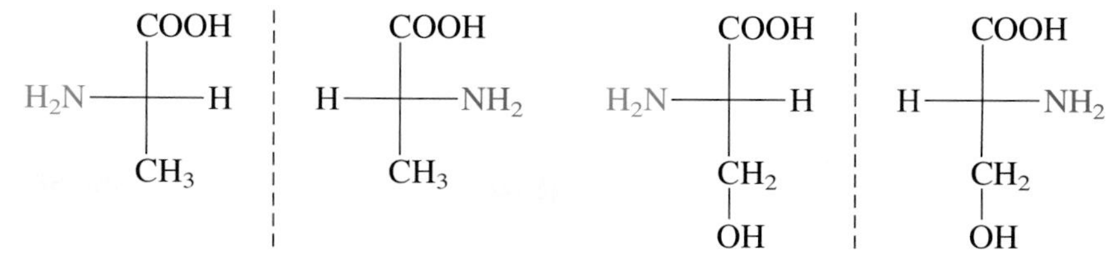

L-Alanine: COOH at top, H_2N on left, H on right, CH_3 at bottom.
D-Alanine: COOH at top, H on left, NH_2 on right, CH_3 at bottom.

L-Serine: COOH at top, H_2N on left, H on right, CH_2 at bottom, with OH bonded below the CH_2.
D-Serine: COOH at top, H on left, NH_2 on right, CH_2 at bottom, with OH bonded below the CH_2.

20.4 Acid–Base Properties of Amino Acids

In pure form, amino acids are white crystalline solids with relatively high decomposition points. (Most amino acids decompose before they melt.) Also most amino acids are *not* very soluble in water because of strong intermolecular forces within their crystal structures. Such properties are those often exhibited by compounds in which charged species are present. Studies of amino acids confirm that they are charged species both in the solid state and in solution. Why is this so?

Both an acidic group ($-COOH$) and a basic group ($-NH_2$) are present on the same carbon in an α-amino acid.

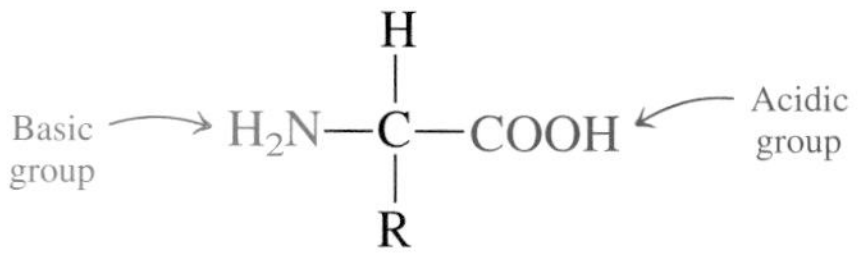

$$\text{Basic group} \rightarrow H_2N-\overset{\displaystyle H}{\underset{\displaystyle R}{C}}-COOH \leftarrow \text{Acidic group}$$

In Section 16.7, we learned that in neutral solution, carboxyl groups have a tendency to lose protons (H^+), producing a negatively charged species:

$$-COOH \longrightarrow -COO^- + H^+$$

In Section 17.5, we learned that in neutral solution, amino groups have a tendency to accept protons (H^+), producing a positively charged species:

$$-NH_2 + H^+ \longrightarrow -\overset{+}{N}H_3$$

As is consistent with the behavior of these groups, in neutral solution, the $-COOH$ group of an amino acid donates a proton to the $-NH_2$ of the same amino acid. We can characterize this behavior as an *internal* acid–base reaction. The net result is that in neutral solution, amino acid molecules have the structure

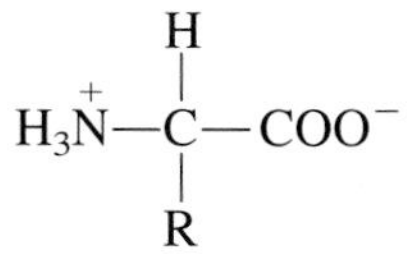

$$H_3\overset{+}{N}-\overset{\displaystyle H}{\underset{\displaystyle R}{C}}-COO^-$$

• Strong intermolecular forces between the positive and negative centers of zwitterions are the cause of the high melting points of amino acids.

Such a molecule is known as a zwitterion, from the German term meaning "double ion." A **zwitterion** *is a molecule that has a positive charge on one atom and a negative charge on another atom.* Note that the net charge on a zwitterion is zero even though parts of the molecule carry charges. In solution and also in the solid state, α-amino acids are zwitterions.

• From this point on in the text, the structures of amino acids will be drawn in their zwitterion form unless information given about the pH of the solution indicates otherwise.

Zwitterion structure changes when the pH of a solution containing an amino acid is changed from neutral either to acidic (low pH) by adding an acid such as HCl or to basic (high pH) by adding a base such as NaOH. In an acidic solution, the zwitterion accepts a proton (H^+) to form a positively charged ion.

$$H_3\overset{+}{N}-\overset{\displaystyle H}{\underset{\displaystyle R}{C}}-\boxed{COO^-} + H_3O^+ \longrightarrow H_3\overset{+}{N}-\overset{\displaystyle H}{\underset{\displaystyle R}{C}}-\boxed{COOH} + H_2O$$

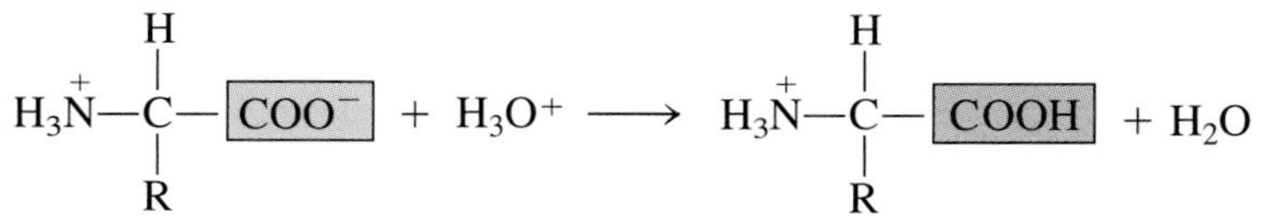

Zwitterion (no net charge) — Positively charged ion

In basic solution, the $-\overset{+}{N}H_3$ of the zwitterion loses a proton, and a negatively charged species is formed.

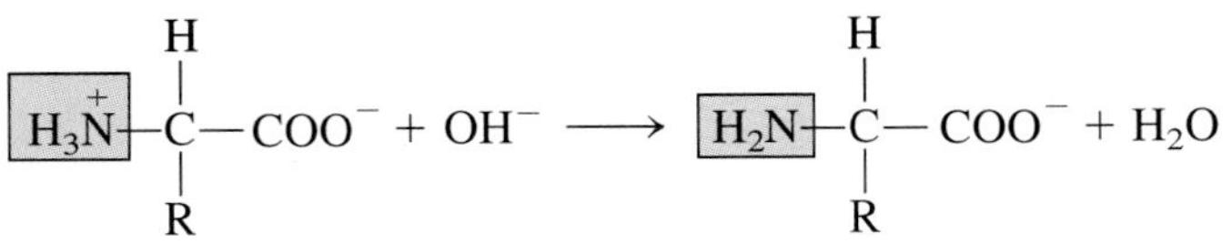

$$H_3\overset{+}{N}-\underset{R}{\overset{H}{C}}-COO^- + OH^- \longrightarrow H_2N-\underset{R}{\overset{H}{C}}-COO^- + H_2O$$

Zwitterion (no net charge) — Negatively charged ion

Thus, in solution, three different amino acid forms can exist (zwitterion, negative ion, and positive ion). The three species are actually in equilibrium with each other, and the equilibrium shifts with pH change. The overall equilibrium process can be represented as follows:

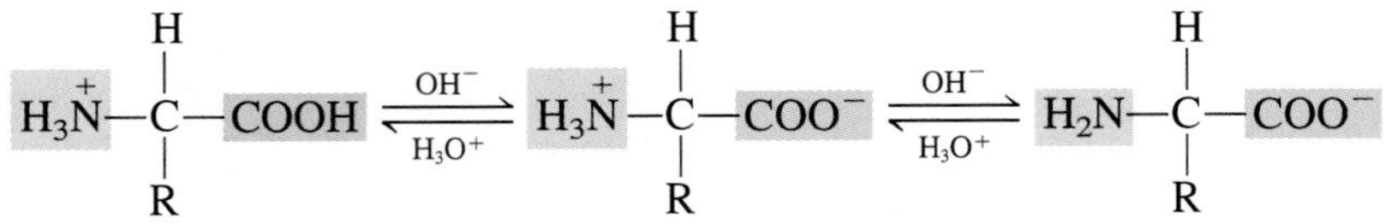

$$H_3\overset{+}{N}-\underset{R}{\overset{H}{C}}-COOH \underset{H_3O^+}{\overset{OH^-}{\rightleftharpoons}} H_3\overset{+}{N}-\underset{R}{\overset{H}{C}}-COO^- \underset{H_3O^+}{\overset{OH^-}{\rightleftharpoons}} H_2N-\underset{R}{\overset{H}{C}}-COO^-$$

Acidic solution (low pH) — Neutral solution (pH = 7) — Basic solution (high pH)

In acidic solution, the positively charged species on the left predominates; nearly neutral solutions have the middle species (the zwitterion) as the dominant species; in basic solution, the negatively charged species on the right predominates.

Example 20.1

Determining Amino Acid Form in Solutions of Various pH

Draw an appropriate structural form for the amino acid alanine that predominates in solution at each of the following pH values.

a. pH = 1
b. pH = 7
c. pH = 11

Solution

At low pH, both amino and carboxyl groups are protonated. At high pH, both groups have lost their protons. At neutral pH, the zwitterion is present.

a. $H_3\overset{+}{N}-\underset{CH_3}{\overset{H}{C}}-COOH$ pH = 1 (net charge of +1)

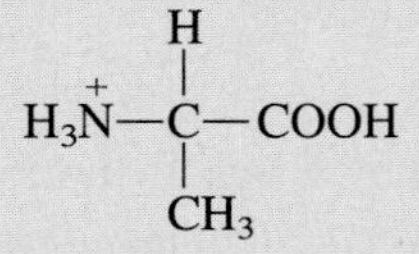

b. $H_3\overset{+}{N}-\underset{CH_3}{\overset{H}{C}}-COO^-$ pH = 7 (no net charge)

c. $H_2N-\underset{CH_3}{\overset{H}{C}}-COO^-$ pH = 11 (net charge of −1)

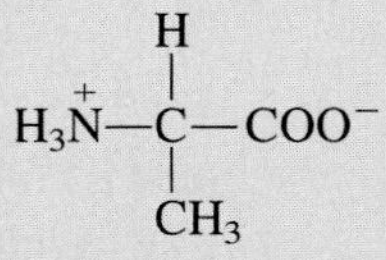

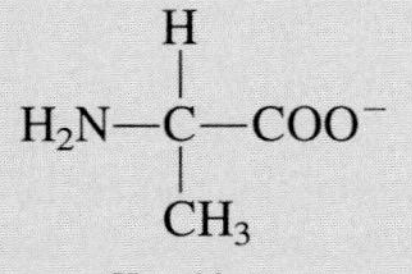

Practice Exercise 20.1

Draw an appropriate structural form for the amino acid valine that predominates in solution at each of the following pH values.

a. pH = 7 **b.** pH = 12 **c.** pH = 2

• *Answers:* **a.** $H_3\overset{+}{N}-\overset{H}{C}(-CH(CH_3)-CH_3)-COO^-$ **b.** $H_2N-\overset{H}{C}(-CH(CH_3)-CH_3)-COO^-$ **c.** $H_3\overset{+}{N}-\overset{H}{C}(-CH(CH_3)-CH_3)-COOH$

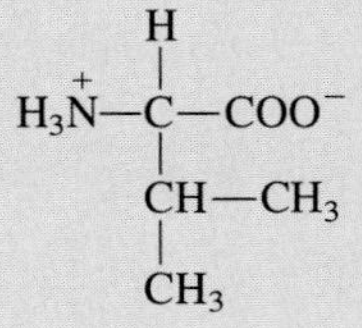

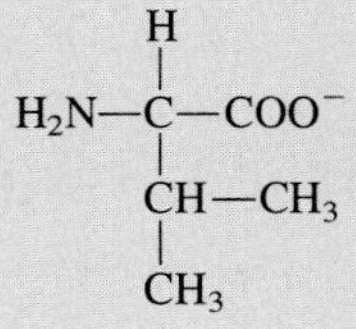

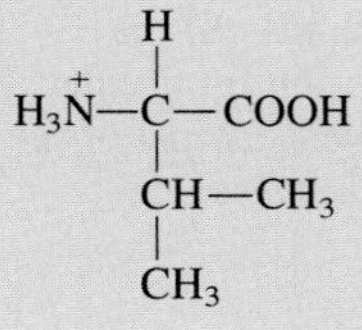

The previous discussion assumed that the side chain (R group) of an amino acid remains unchanged in solution as the pH is varied. This is the case for neutral amino acids but not for acidic or basic ones. For these latter compounds, the side chain can also acquire a charge, becuase it contains an amino or a carboxyl group that can, respectively, gain or lose a proton.

• The term *protonated* denotes gain of a H^+ ion, and the term *deprotonated* denotes loss of a H^+ ion.

Because of the extra site that can be protonated or deprotonated, acidic and basic amino acids have four charged forms in solution. These four forms for aspartic acid, one of the acidic amino acids are

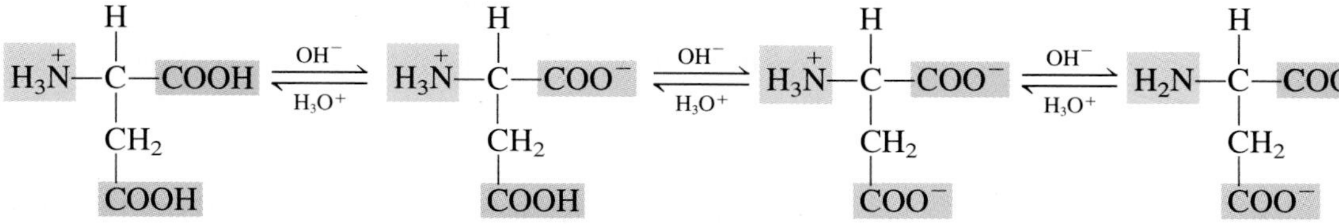

$$H_3\overset{+}{N}-C(H)(CH_2COOH)-COOH \underset{H_3O^+}{\overset{OH^-}{\rightleftharpoons}} H_3\overset{+}{N}-C(H)(CH_2COOH)-COO^- \underset{H_3O^+}{\overset{OH^-}{\rightleftharpoons}} H_3\overset{+}{N}-C(H)(CH_2COO^-)-COO^- \underset{H_3O^+}{\overset{OH^-}{\rightleftharpoons}} H_2N-C(H)(CH_2COO^-)-COO^-$$

Low-pH form (+1 charge) | Moderately-low-pH form (no net charge) (zwitterion) | Neutral-pH form (−1 net charge) | High-pH form (−2 net charge)

• Guidelines for amino acid form as a function of solution pH are

Low pH:	All acid groups are protonated (—COOH). All amino groups are protonated (—$\overset{+}{N}H_3$).
High pH:	All acid groups are deprotonated (—COO^-). All amino groups are deprotonated (—NH_2).
Neutral pH:	All acid groups are deprotonated (—COO^-). All amino groups are protonated (—$\overset{+}{N}H_3$).

• The isoelectric point is a pH value. For neutral amino acids, it is the pH at which the amount of the zwitterion form of the amino acid is maximized.

The existence of two low-pH forms for aspartic acid results from the two carboxyl groups being deprotonated at different pH values. For basic amino acids, two high-pH forms exist because deprotonation of the amino groups does not occur simultaneously. The side-chain amino group deprotonates before the α-amino group.

The **isoelectric point** *for an amino acid is the pH at which the total charge on the amino acid is zero.* Every amino acid has a different isoelectric point. Fifteen of the 20 amino acids, those with nonpolar or polar neutral side chains (Table 20.1), have isoelectric points in the range of 4.8–6.3. The three basic amino acids have higher isoelectric points (His = 7.59, Lys = 9.74, Arg = 10.76), and the two acidic amino acids have lower ones (Asp = 2.77, Glu = 3.22).

A pH below the isoelectric point favors the positively charged form of the amino acid. Conversely, a pH above the isoelectric point favors the negatively charged form of the amino acid.

When two electrodes (one positively charged and one negatively charged) are immersed in a solution containing an amino acid, molecules with a net positive charge are attracted to the negatively charged electrode, and negatively charged amino acid molecules migrate toward the positively charged electrode. The zwitterion form exhibits no net migration toward either electrode. This behavior is the basis for the measurement of isoelectric points. The pH of the solution is adjusted until no net migration occurs.

Mixtures of amino acids in solution can be separated by using their different migration patterns at various pH values. This type of analytical separation is called electrophoresis. **Electrophoresis** *is the process of separating charged molecules on the basis of their migration toward charged electrodes.*

Figure 20.1 schematically shows the separation of the amino acids Lys, Phe, and Glu by electrophoresis. At a pH of 5.5, these amino acids exist in the following forms:

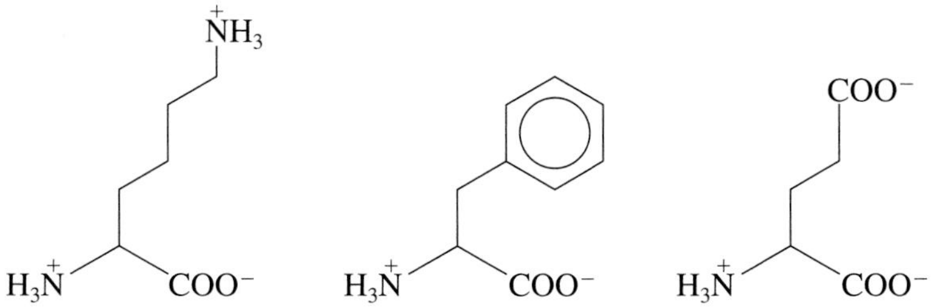

$\overset{+}{N}H_3$ … $H_3\overset{+}{N}$ COO^- — Lys (+1 charge)

$H_3\overset{+}{N}$ COO^- — Phe (no net charge)

COO^- … $H_3\overset{+}{N}$ COO^- — Glu (−1 charge)

When a current is applied, Phe will not move (it has no charge); Lys, because of its positive charge, will migrate toward the negative electrode; and Glu, with a negative charge, will move toward the positive electrode.

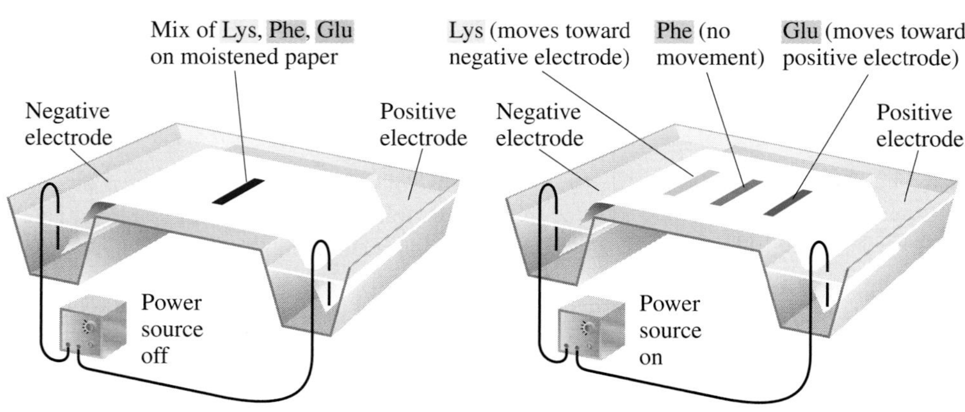

Figure 20.1
Separation, at a pH of 5.5 of the three amino acids Lys, Phe, and Glu using electrophoresis.

Proteins, which are amino acid polymers (Section 20.6), also have isoelectric points and also can be separated via electrophoresis techniques.

Example 20.2

Migration Patterns of Amino Acids at Various pH Values

Predict the direction (if any) of migration toward the positively or negatively charged electrode for the following amino acids in solutions of the specified pH. Write "isoelectric" if no migration occurs.

a. Lysine at pH = 7
b. Glutamic acid at pH = 7
c. Serine at pH = 1

Solution

a. Lysine at pH = 7

$$H_3\overset{+}{N}-CH(COO^-)-CH_2-CH_2-CH_2-CH_2-\overset{+}{N}H_3$$

Net positive charge (2 "+" and 1 "−"); migrates toward negatively charged electrode

b. Glutamic acid at pH = 7

$$H_3\overset{+}{N}-CH(COO^-)-CH_2-CH_2-COO^-$$

Net negative charge (1 "+" and 2 "−"); migrates toward positively charged electrode

c. Serine at pH = 1

$$H_3\overset{+}{N}-CH(COOH)-CH_2-OH$$

One positive charge; migrates toward negatively charged electrode

Practice Exercise 20.2

Predict the direction (if any) of migration toward the positively or negatively charged electrode for the following amino acids in solutions of the specified pH. Write "isoelectric" if no migration occurs.

a. Lysine at pH = 12
b. Glutamic acid at pH = 2
c. Serine at pH = 7

• *Answers:* **a.** toward positively charged electrode; **b.** toward negatively charged electrode; **c.** isoelectric

20.5 Peptide Formation

In Section 17.14, we learned that a carboxylic acid and an amine can react to produce an amide. The general equation for this reaction is

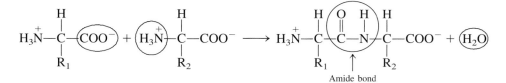

$$\underset{\text{Acid}}{R-\overset{\overset{\displaystyle O}{\|}}{C}-OH} + \underset{\text{Amine}}{H-\overset{\overset{\displaystyle H}{|}}{N}-R} \longrightarrow \underset{\text{Amide}}{R-\overset{\overset{\displaystyle O}{\|}}{C}-\overset{\overset{\displaystyle H}{|}}{N}-R} + H_2O$$

Two amino acids can react in a similar way—the carboxyl group of one amino acid reacts with the amino group of the other amino acid. The products are a molecule of water and a molecule containing the two amino acids linked by an amide bond.

• Peptide bond formation is an example of a condensation reaction.

$$H_3\overset{+}{N}-\overset{\overset{\displaystyle H}{|}}{\underset{\underset{\displaystyle R_1}{|}}{C}}-COO^- + H_3\overset{+}{N}-\overset{\overset{\displaystyle H}{|}}{\underset{\underset{\displaystyle R_2}{|}}{C}}-COO^- \longrightarrow H_3\overset{+}{N}-\overset{\overset{\displaystyle H}{|}}{\underset{\underset{\displaystyle R_1}{|}}{C}}-\overset{\overset{\displaystyle O}{\|}}{C}-\overset{\overset{\displaystyle H}{|}}{N}-\overset{\overset{\displaystyle H}{|}}{\underset{\underset{\displaystyle R_2}{|}}{C}}-COO^- + H_2O$$

Amide bond

Removal of the elements of water from the reacting carboxyl and amino groups and the ensuing formation of the amide bond are better visualized when expanded structural formulas for the reacting groups are used.

$$\underset{\substack{\text{Carboxyl}\\\text{group}\\(-COO^-)}}{-\overset{\overset{\displaystyle O}{\|}}{C}-O} + \underset{\substack{\text{Amino}\\\text{group}\\(H_3\overset{+}{N}-)}}{H-\overset{\overset{\displaystyle H}{|}}{\underset{\underset{\displaystyle H}{|}}{N}}-} \longrightarrow \underset{\substack{\text{Amide}\\\text{bond}}}{-\overset{\overset{\displaystyle O}{\|}}{C}-\overset{\overset{\displaystyle H}{|}}{N}-} + H_2O$$

In amino acid chemistry, amide bonds that link amino acids together are given the specific name of peptide bond. A **peptide bond** *is a bond between the carboxyl group of one amino acid and the amino group of another amino acid.*

Under proper conditions, many amino acids can bond together to give chains of amino acids containing numerous peptide bonds. For example, four peptide bonds are present in a chain of five amino acids.

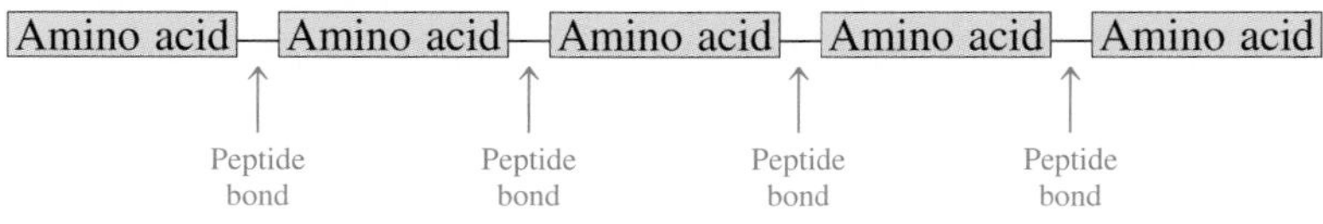

Short to medium-sized chains of amino acids are known as peptides. A **peptide** *is a sequence of amino acids, of up to 50 units, in which the amino acids are joined together*

through amide (peptide) bonds. A compound containing two amino acids joined by a peptide bond is specifically called a *dipeptide;* three amino acids in a chain constitute a *tripeptide;* and so on. The name *oligopeptide* is loosely used to refer to peptides with 10 to 20 amino acid residues and *polypeptide* to larger peptides.

• A peptide chain has *directionality* because its two ends are different. There is an N-terminal end and a C-terminal end. By convention, the direction of the peptide chain is always

N-terminal end ⟶ C-terminal end

The N-terminal end is always on the left, and the C-terminal end is always on the right.

In all peptides, the amino acid at one end of the amino acid sequence has a free $H_3\overset{+}{N}$ group, and the amino acid at the other end of the sequence has a free COO^- group. The end with the free $H_3\overset{+}{N}$ group is called the *N-terminal end,* and the end with the free COO^- group is called the *C-terminal end.* By convention, the sequence of amino acids in a peptide is written with the N-terminal end amino acid at the left. The individual amino acids within a peptide chain are called *amino acid residues.*

The structural formula for a polypeptide may be written out in full, or the sequence of amino acids present may be indicated by using the standard three-letter amino acid abbreviations. The abbreviated formula for the tripeptide

N-terminal end ⟶ $H_3\overset{+}{N}-CH(H)-C(=O)-NH-CH(CH_3)-C(=O)-NH-CH(CH_2OH)-COO^-$ ⟵ C-terminal end

Glycine — Alanine — Serine

which contains the amino acids glycine, alanine, and serine, is Gly–Ala–Ser. When we use this abbreviated notation, by convention, the amino acid at the N-terminal end of the peptide is always written on the left.

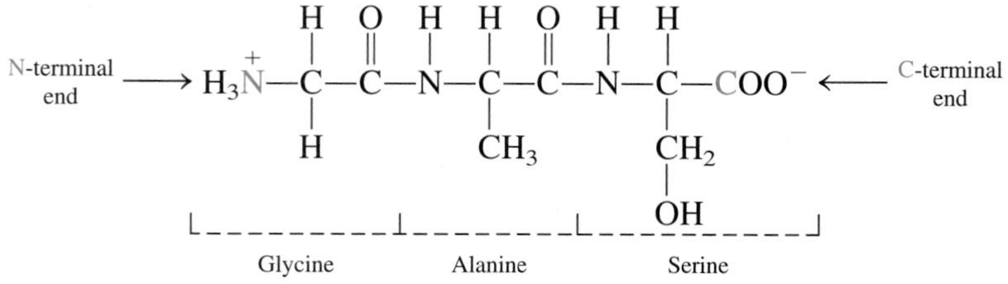

• The *backbone* of a protein is the alternating sequence of peptide bonds and α-carbon atoms.

$-C-C(=O)-N(H)-C-C(=O)-N(H)-C-$

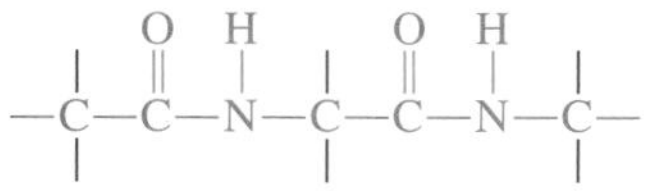

R group side chains are not part of the backbone.

The repeating chain of peptide bonds and α-carbon atoms in a peptide is referred to as the *backbone* of the peptide. The R group side chains are substituents on the backbone.

Example 20.3

Converting an Abbreviated Peptide Formula to a Structural Peptide Formula

Draw the structural formula for the tripeptide Ala–Gly–Val.

Solution

Step 1: The N-terminal end of the peptide involves alanine. Its structure is written first.

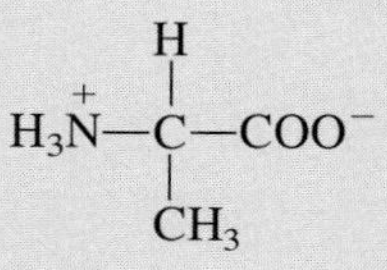

$$H_3\overset{+}{N}-CH(CH_3)-COO^-$$

Step 2: The structure of glycine is written to the right of the alanine structure, and a peptide bond is formed between the two amino acids by removing the elements of H_2O and bonding the N of glycine to the carboxyl C of alanine.

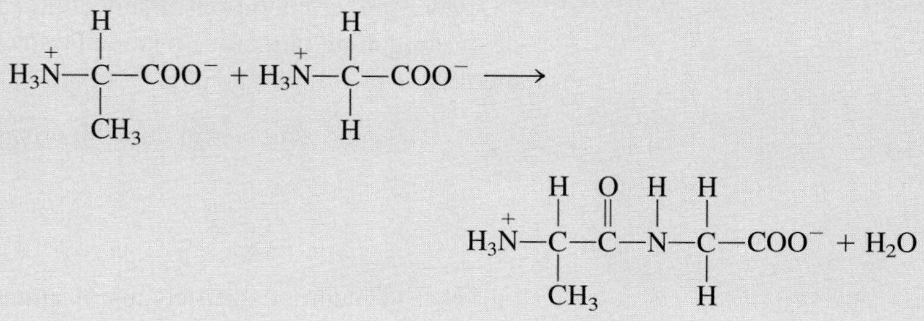

$$H_3\overset{+}{N}-CH(CH_3)-COO^- + H_3\overset{+}{N}-CH(H)-COO^- \longrightarrow$$

$$H_3\overset{+}{N}-CH(CH_3)-C(=O)-NH-CH(H)-COO^- + H_2O$$

Step 3: To the right of the just-formed dipeptide, draw the structure of valine. Then repeat Step 2 to form the desired tripeptide.

$$\begin{array}{ccccccccccccccc}
 & & \text{H} & & \text{O} & & \text{H} & & \text{H} & & & & \text{H} & & \\
 & & | & & \| & & | & & | & & & & | & & \\
\overset{+}{\text{H}_3\text{N}} & — & \text{C} & — & \text{C} & — & \text{N} & — & \text{C} & —\text{COO}^- & + & \overset{+}{\text{H}_3\text{N}} & —\text{C} & —\text{COO}^- & \longrightarrow \\
 & & | & & & & & & | & & & & | & & \\
 & & \text{CH}_3 & & & & & & \text{H} & & & & \text{CH—CH}_3 & & \\
 & & & & & & & & & & & & | & & \\
 & & & & & & & & & & & & \text{CH}_3 & &
\end{array}$$

$$\begin{array}{cccccccccccccccc}
 & & \text{H} & & \text{O} & & \text{H} & & \text{H} & & \text{O} & & \text{H} & & \text{H} & \\
 & & | & & \| & & | & & | & & \| & & | & & | & \\
\overset{+}{\text{H}_3\text{N}} & — & \text{C} & — & \text{C} & — & \text{N} & — & \text{C} & — & \text{C} & — & \text{N} & — & \text{C} & —\text{COO}^- + \text{H}_2\text{O} \\
 & & | & & & & & & | & & & & & & | & \\
 & & \text{CH}_3 & & & & & & \text{H} & & & & & & \text{CH—CH}_3 & \\
 & & & & & & & & & & & & & & | & \\
 & & & & & & & & & & & & & & \text{CH}_3 &
\end{array}$$

Practice Exercise 20.3

Draw the structural formula for the tripeptide Val–Ala–Gly.

- *Answer:*

$$\begin{array}{cccccccccccccccc}
 & & \text{H} & & \text{O} & & \text{H} & & \text{H} & & \text{O} & & \text{H} & & \text{H} & \\
 & & | & & \| & & | & & | & & \| & & | & & | & \\
\overset{+}{\text{H}_3\text{N}} & — & \text{C} & — & \text{C} & — & \text{N} & — & \text{C} & — & \text{C} & — & \text{N} & — & \text{C} & —\text{COO}^- \\
 & & | & & & & & & | & & & & & & | & \\
 & & \text{CH—CH}_3 & & & & & & \text{CH}_3 & & & & & & \text{H} & \\
 & & | & & & & & & & & & & & & & \\
 & & \text{CH}_3 & & & & & & & & & & & & &
\end{array}$$

Peptides that contain the same amino acids but in different order are different molecules (structural isomers) with different properties. For example, two different dipeptides can be formed from one molecule of alanine and one molecule of glycine.

Ala–Gly

$$\begin{array}{cccccccccc}
 & & \text{H} & & \text{O} & & \text{H} & & \text{H} & \\
 & & | & & \| & & | & & | & \\
\overset{+}{\text{H}_3\text{N}} & — & \text{C} & — & \text{C} & — & \text{N} & — & \text{C} & —\text{COO}^- \\
 & & | & & & & & & | & \\
 & & \text{CH}_3 & & & & & & \text{H} &
\end{array}$$

Gly–Ala

$$\begin{array}{cccccccccc}
 & & \text{H} & & \text{O} & & \text{H} & & \text{H} & \\
 & & | & & \| & & | & & | & \\
\overset{+}{\text{H}_3\text{N}} & — & \text{C} & — & \text{C} & — & \text{N} & — & \text{C} & —\text{COO}^- \\
 & & | & & & & & & | & \\
 & & \text{H} & & & & & & \text{CH}_3 &
\end{array}$$

In the first dipeptide, the alanine is the N-terminal residue, and in the second molecule, it is the C-terminal residue. These two compounds are isomers with different chemical and physical properties.

• Amino acid sequence in a peptide has biological importance. Isomeric peptides give different biological responses; that is, they have different biological specificities.

The number of isomeric peptides possible increases rapidly as the length of the peptide chain increases. Let us consider the tripeptide Ala–Ser–Cys as another example. In addition to this sequence, five other arrangements of these three components are possible, each representing another isomeric tripeptide: Ala–Cys–Ser, Ser–Ala–Cys, Ser–Cys–Ala, Cys–Ala–Ser, and Cys–Ser–Ala. For a pentapeptide containing 5 different amino acids, 120 isomers are possible.

More than two hundred peptides have been isolated and identified as essential to the proper functioning of the human body. In general, these substances serve as hormones or neurotransmitters. Their functions range from controlling pain to controlling muscle contraction or kidney fluid excretion.

Two important hormones produced by the pituitary gland are oxytocin and vasopressin. Each hormone is a nonapeptide (nine amino acid residues) with six of the residues held in the form of a loop by a disulfide bond formed from the interaction of two cysteine residues.

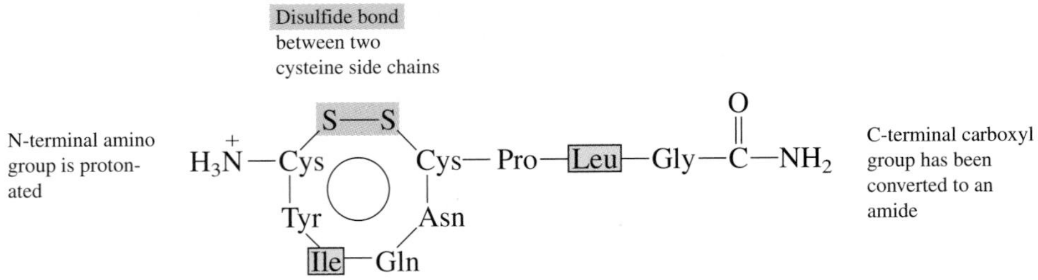

Disulfide bond between two cysteine side chains

N-terminal amino group is protonated

$\overset{+}{H_3N}$—Cys(—S—S—)Cys—Pro—Leu—Gly—C(=O)—NH_2
Tyr, Ile—Gln, Asn (loop)

C-terminal carboxyl group has been converted to an amide

Oxytocin

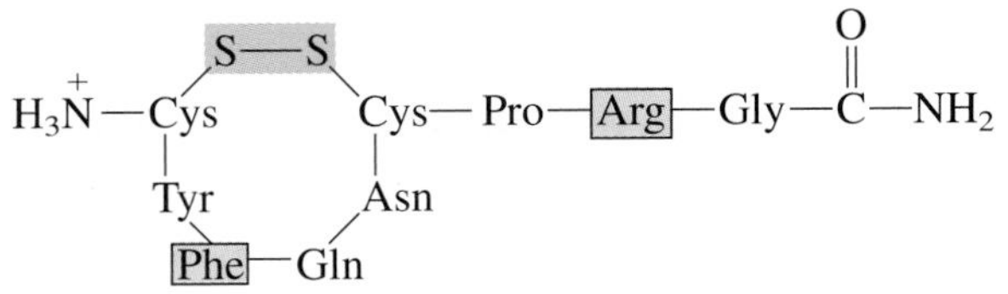

$\overset{+}{H_3N}$—Cys(—S—S—)Cys—Pro—Arg—Gly—C(=O)—NH_2
Tyr, Phe—Gln, Asn (loop)

Vasopressin

Oxytocin regulates uterine contractions and lactation. Vasopressin regulates the excretion of water by the kidneys; it also affects blood pressure. The structure of vasopressin differs from that of oxytocin at only two amino acid positions: the third and eighth amino acid residues. The result of these variations is a significant difference in physiological action.

Endorphins are peptides that bind at receptor sites in the brain to reduce pain. These compounds are synthesized by the brain itself. A subclass of such molecules, the *enkephalins,* are simple pentapeptides. Two important enkephalins are methionine enkephalin (Tyr–Gly–Gly–Phe–Met) and leucine enkephalin (Tyr–Gly–Gly–Phe–Leu). The action of the prescription painkillers morphine and codeine is based on their binding at the same receptor sites in the brain as naturally occurring enkephalins.

20.6 Levels of Protein Structure

• Proteins are the second type of biological polymer we have encountered; the other was polysaccharides (Section 18.14). Protein monomers are amino acids, whereas polysaccharide monomers are monosaccharides.

Proteins *are polypeptides that contain more than 50 amino acid units.* The dividing line between a polypeptide and a protein is arbitrary. The important point is that proteins are polymers containing a large number of amino acid units linked by peptide bonds. Polypeptides are shorter chains of amino acids. Some proteins have molecular masses in the millions. Some proteins also contain more than one polypeptide chain.

To aid us in describing protein structure, we will consider four levels of substructure: primary, secondary, tertiary, and quaternary. Even though we consider these structure levels one by one, remember that it is the combination of all four levels of structure that controls protein function.

20.7 Primary Structure of Proteins

The **primary structure of a protein** *is the sequence of amino acids present in its peptide chain or chains.* Knowledge of primary structure tells us which amino acids are present, the number of each, their sequence, and the length and number of polypeptide chains.

The first protein whose primary structure was determined was insulin, the hormone that regulates blood-glucose level; a deficiency of insulin leads to diabetes. The sequencing of insulin, which took over 8 years, was completed in 1953. Today, thousands of proteins

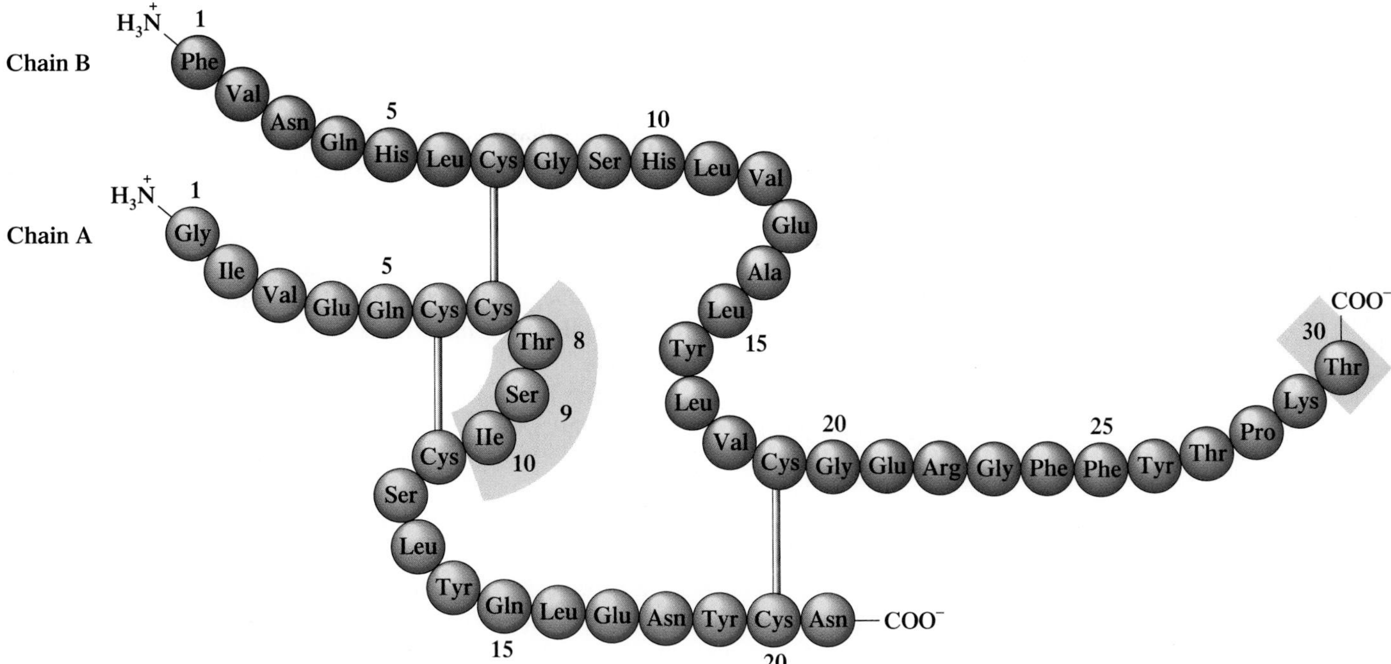

Figure 20.2
The amino acid sequence (primary structure) in the protein insulin in humans. The locations where variance occurs between human insulin and insulin of certain animals are highlighted.

• When a protein consists of more than one polypeptide chain, as in insulin, each chain is called a *subunit* of the protein.

have been sequenced; that is, researchers have determined the order of amino acids within the polypeptide chain or chains.

Figure 20.2 shows the primary structure of human insulin. Insulin's molecular structure involves two polypeptide chains designated A and B. Chain A contains 21 amino acids, and chain B has 30 amino acids. The A and B chains are bonded together (cross-linked) in two places, and there is an additional cross link within chain A.

The primary structure of a specific protein is always the same, regardless of where the protein is found within an organism. The structures of certain proteins are even similar among different species of animals. For example, the primary structures of insulin in cows, pigs, sheep, and horses are very similar both to each other and to human insulin. Until recently, this similarity was particularly important for diabetics who required supplemental injections of insulin. (See Chemical Connections 20.2.)

Chemical CONNECTIONS

20.2 Substitutes for Human Insulin

In humans, an insufficient production of insulin results in the disease *diabetes mellitus*. Treatment of this disease involves giving the patient extra insulin via subcutaneous injection. For many years, because of the limited availability of human insulin, most insulin used by diabetics was obtained from the pancreases of slaughter-house animals. Such animal insulin, primarily from cows and pigs, was used by most diabetics without serious side effects because it is structurally very similar to human insulin. Immunological reactions gradually do increase over time, however, because the animal insulin is foreign to the human body.

A comparison of the primary structure of human insulin with pig and cow insulins shows differences at only 4 of the 51 amino acid positions: positions 8, 9, and 10 on chain A and position 30 on chain B (see Figure 20.2 and the accompanying table).

	Chain A			Chain B
Species	**#8**	**#9**	**#10**	**#30**
human	Thr	Ser	Ile	Thr
pig (porcine)	Thr	Ser	Ile	Ala
cow (bovine)	Ala	Ser	Val	Ala

The dependence of diabetics on animal insulin has declined because of the availability of human insulin produced by genetically engineered bacteria (Section 22.14). These bacteria carry a gene that directs the synthesis of human insulin. Such bacteria-produced insulin is fully functional. All diabetics will soon be able to use human insulin for injection instead of animal insulin.

An analogy is often drawn between the primary structure of proteins and words. Words, which convey information, are formed when the 26 letters of the English alphabet are properly sequenced. Proteins, which function biologically, are formed from the proper sequence of 20 amino acids. The proper sequence of letters in a word is necessary for it to make sense, just as the proper sequence of amino acids is necessary to make biologically active protein. Furthermore, the letters that form a word are written from left to right, as are amino acids in protein formulas. As any dictionary of the English language will document, a tremendous variety of words can be formed by different letter sequences. Imagine the number of amino acid sequences possible for a large protein. There are 1.55×10^{66} sequences possible for the 51 amino acids found in insulin! From these possibilities, the body reliably produces only *one,* illustrating the remarkable precision of life processes. From the simplest bacterium to the human brain cell, only those amino acid sequences needed by the cell are produced. The fascinating process of protein biosynthesis and the way in which genes in DNA direct this process will be discussed in Chapter 22.

• The primary structure of a protein is the *sequence* of amino acids in a protein chain—that is, the order in which the amino acids are connected to each other.

20.8 Secondary Structure of Proteins

The **secondary structure of a protein** *is the arrangement in space of the atoms in the backbone of the protein.* Three major types of protein secondary structure are known: the *alpha helix,* the *beta pleated sheet,* and the *triple helix.* The major force responsible for all three types of secondary structure is hydrogen bonding (Section 7.13) between a carbonyl oxygen atom of a peptide linkage and the hydrogen atom of an amino group (—NH) of another peptide linkage farther along the backbone. This hydrogen-bonding interaction may be diagrammed as follows:

• The Alpha Helix

The *alpha helix* (α helix) structure resembles a coiled helical spring, with the coil configuration maintained by hydrogen bonds between >N—H and >C=O groups of every fourth amino acid, as is shown diagrammatically in Figure 20.3.

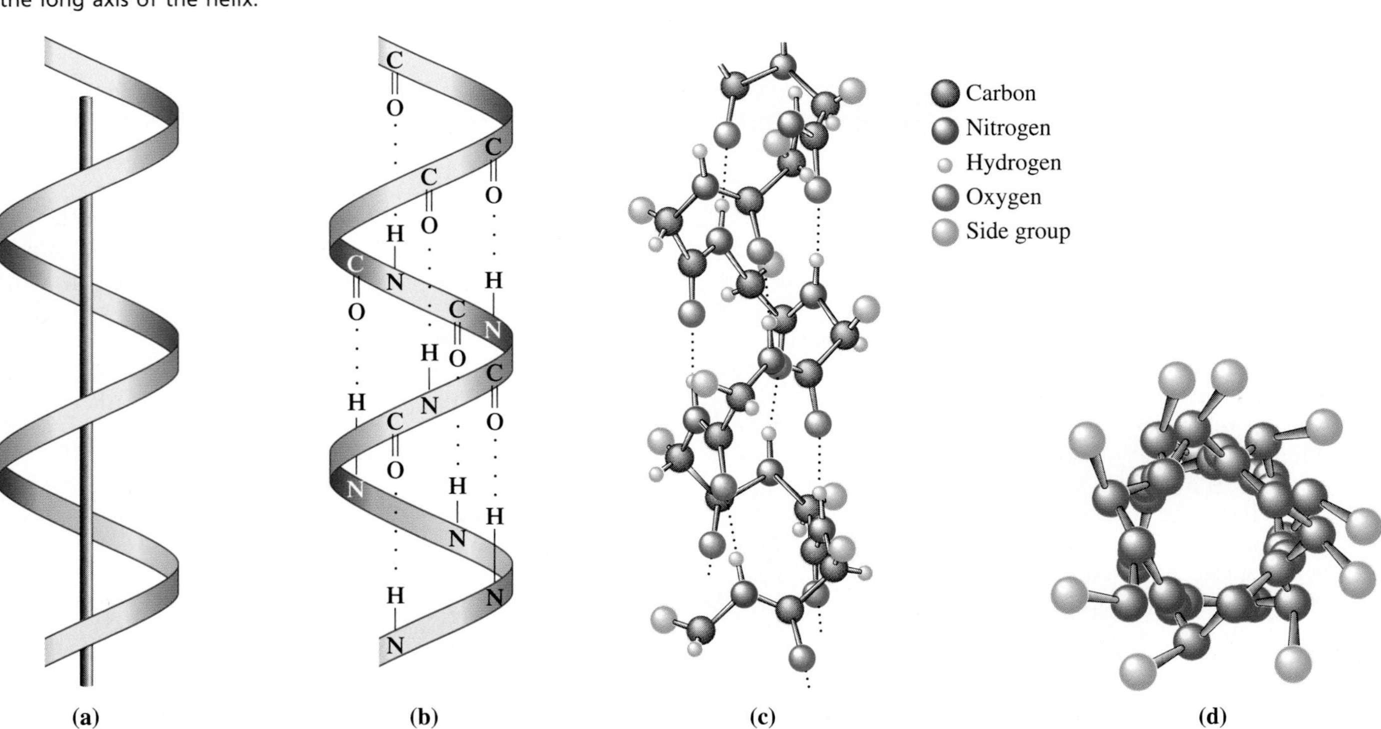

Figure 20.3
Three representations of (a) the α helix protein structure. Hydrogen bonds between amide groups (peptide linkages) are shown in (b) and (c). (d) The top view of an α helix shows that amino acid side chains (R groups) point away from the long axis of the helix.

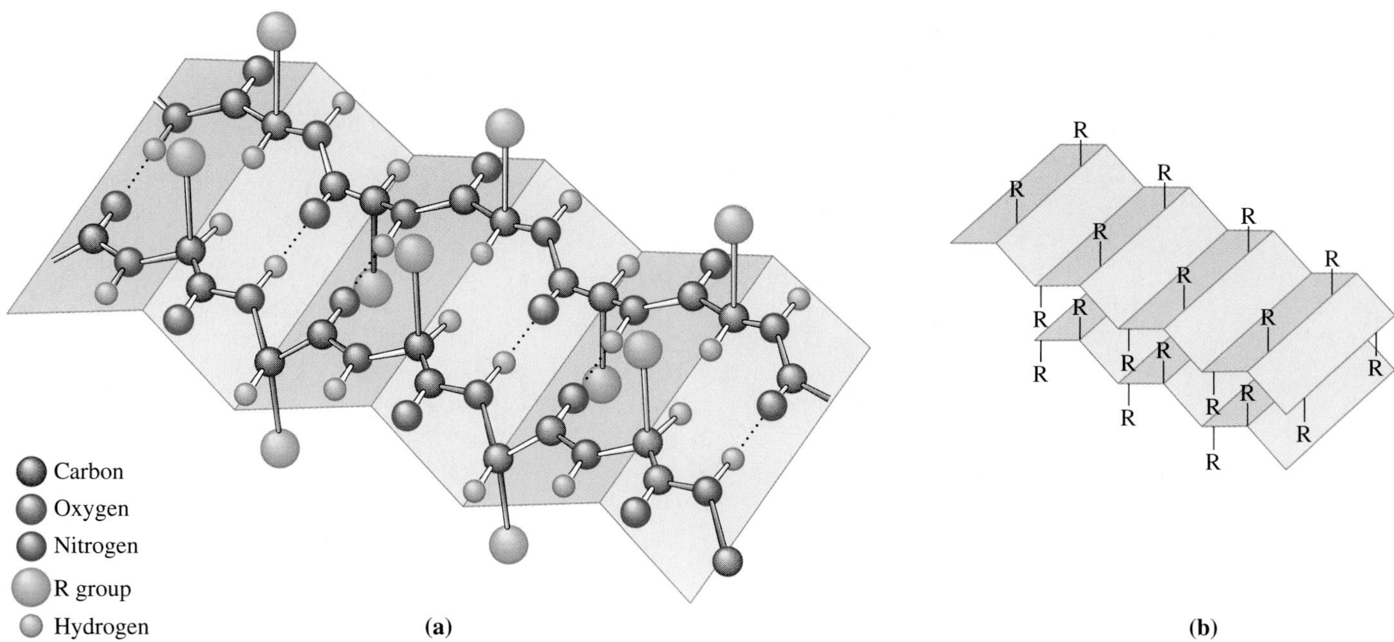

Figure 20.4
Two representations of the β pleated sheet protein structure. (a) A representation emphasizing the hydrogen bonds between protein chains. (b) A representation emphasizing the pleats and the location of the R groups.

Proteins have varying amounts of α-helical secondary structure, ranging from a few percent to nearly 100%. In an α helix, all of the amino acid side chains (R groups) lie outside the helix; there is not enough room for them in the interior. Figure 20.3d illustrates this situation. This structural feature of the α helix is the basis for protein tertiary structure (Section 20.9).

The Beta Pleated Sheet

• The hydrogen bonding present in an α helix is *intra*molecular. In a β pleated sheet, the hydrogen bonding can be *inter*molecular (between two different chains) or *intra*molecular (a single chain folding back on itself).

The *beta pleated sheet* (β pleated sheet) secondary structure involves amino acid chains that are almost completely extended. Hydrogen bonds form between two different side-by-side protein chains (interchain bonds) as shown in Figure 20.4, or between different parts of a single chain that folds back on itself (intrachain bonds), as shown in Figure 20.5. The term *pleated sheet* arises from the repeated zigzag pattern in the structure (Figure 20.4b). Amino acid side chains are located above and below the plane of the sheet.

• The term *random* in "random structure" is really a misnomer because identical irregular structure is found in all molecules of a given protein.

Very few proteins have entirely α helix or β pleated sheet structures. Instead, most proteins have only certain portions of their molecules in these conformations. The rest of the molecule assumes a "random structure." It is possible to have both α helix and β pleated sheet structures within the same protein. Figure 20.5 is a diagram of a protein that has more than one secondary structural feature. Only sections of the protein where the side chains (R groups) are relatively small can have the helical or pleated sheet structures.

The Triple Helix

The *triple helix* is another secondary structure for proteins. This structure involves three coiled polypeptide chains wound around each other about a common axis to give a rope-like arrangement (Figure 20.6). The intertwining of the three polypeptide chains and some cross-linking between chains involving covalent bonds hold the triple helix together.

• Collagen, pronounced "kahl-uh-jen," is the most abundant protein in the human body.

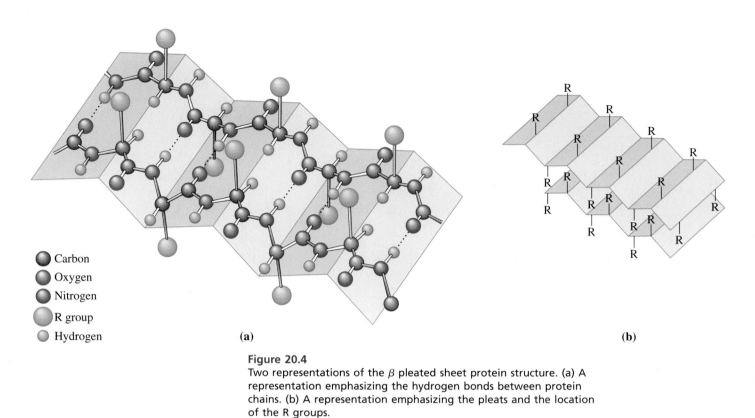

Collagen, the structural protein of connective tissue (cartilage, tendon, and skin), has a triple-helix structure. Collagen molecules are very long, thin, and rigid. Many such molecules, lined up alongside each other, combine to make collagen fibers. Cross-linking gives

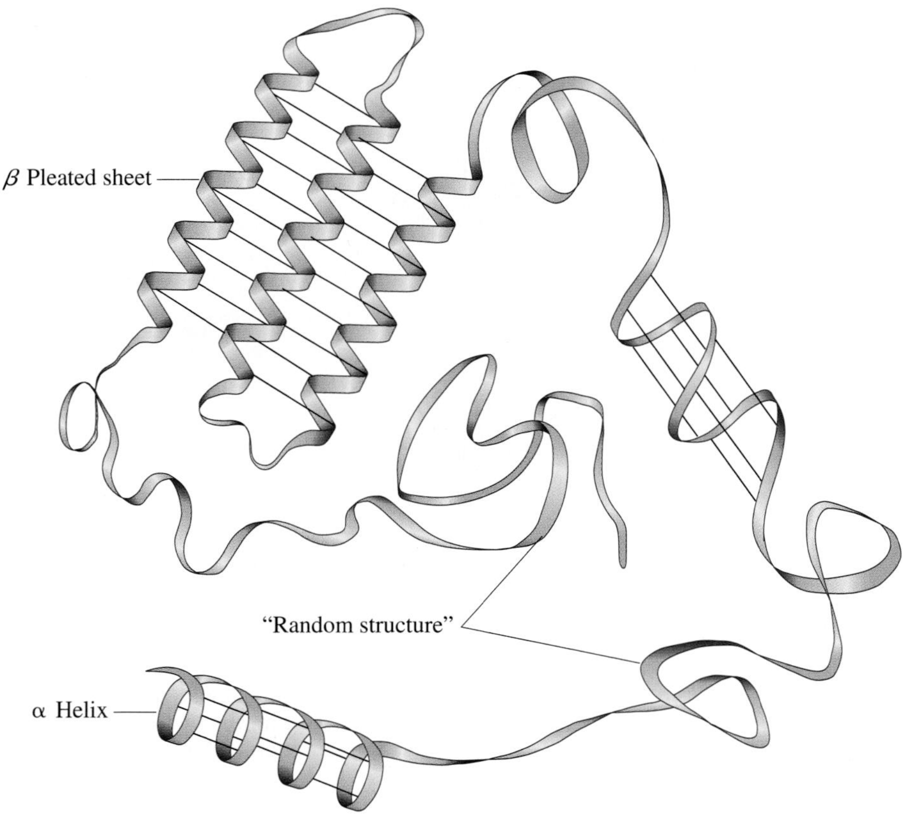

Figure 20.5
The secondary structure of a single protein often shows areas of α helix and β pleated sheet configurations, as well as areas of random coiling.

the fibers extra strength. Figure 20.7 shows collagen fibers viewed through an electron microscope. (Additional details about the structure of collagen molecules will be given in Section 20.14.)

20.9 Tertiary Structure of Proteins

The **tertiary structure of a protein** *is the overall three-dimensional shape that results from the attractive forces between amino acid side chains (R groups) that are widely separated from each other within the chain.*

Figure 20.6
A schematic diagram emphasizing how three helical polypeptide chains intertwine to form a triple helix. The chains are partially unwound and cut away to show their structure.

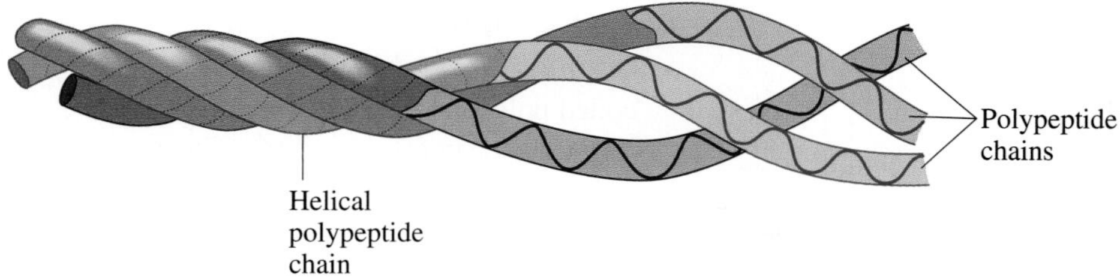

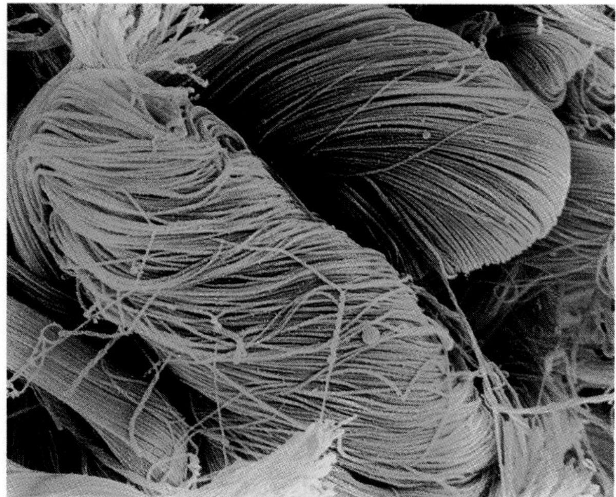

Figure 20.7
Electron micrograph of collagen fibers.

A good analogy for the relationships among the primary, secondary, and tertiary structures of a protein is that of a telephone cord (Figure 20.8). The primary structure is the long, straight cord. The coiling of the cord into a helical arrangement gives the secondary structure. The supercoiling arrangement the cord adopts after you hang up the receiver is the tertiary structure.

• Interactions Responsible for Tertiary Structure

Four types of attractive interactions contribute to the tertiary structure of a protein: (1) covalent disulfide bonds, (2) electrostatic attractions (salt bridges), (3) hydrogen bonds, and (4) hydrophobic attractions. All four of these interactions are interactions between amino acid R groups. This is a major distinction between tertiary-structure interactions and secondary-structure interactions. Tertiary-structure interactions involve the R groups of amino acids; secondary-structure interactions involve the peptide linkages between amino acid units.

• Cysteine is the only α-amino acid that contains a sulfhydryl group (—SH).

Disulfide bonds, the strongest of the tertiary-structure interactions, result from the —SH groups of two cysteine molecules reacting with each other to form a covalent disulfide. This type of interaction is the only one of the four tertiary-structure interactions that involves a covalent bond. Recall, from Section 14.14, that —SH groups are readily oxidized to give a disulfide bond, —S—S—. Figure 20.9a shows a disulfide bond. Disulfide bonds may involve two cysteine units in the same chain or in different chains. Referring to Figure 20.2, we note the presence of both types of disulfide bonds in the structure of insulin. A disulfide linkage is also present in the structures of oxytocin and vasopressin (Section 20.5).

Electrostatic interactions, also called *salt bridges,* always involve amino acids with charged side chains. These amino acids are the acidic and basic amino acids. The two R

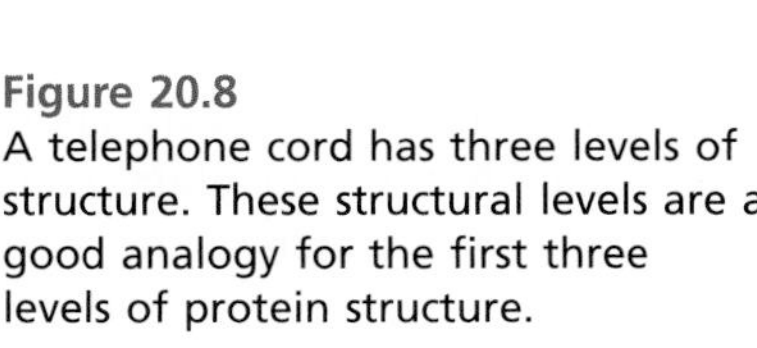

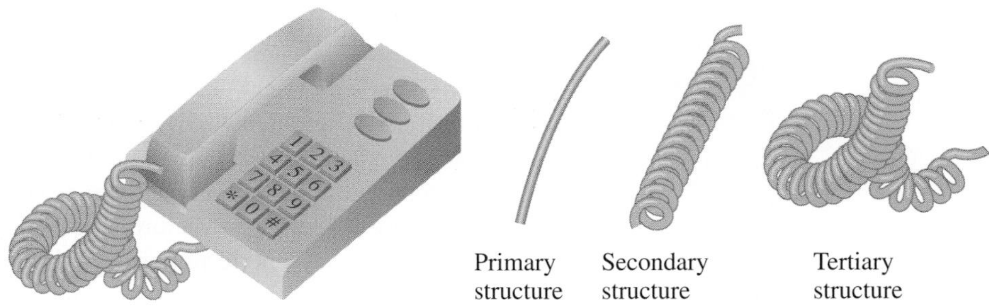

Figure 20.8
A telephone cord has three levels of structure. These structural levels are a good analogy for the first three levels of protein structure.

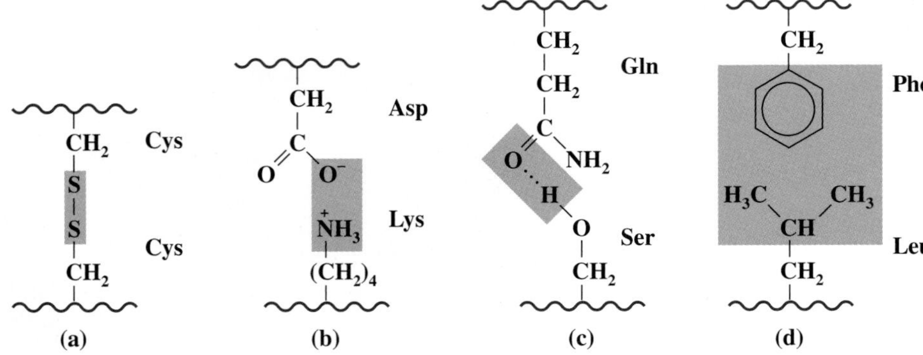

Figure 20.9
Four types of interactions between amino acid R groups produce the tertiary structure of a protein. (a) Disulfide bonds. (b) Electrostatic interactions (salt bridges). (c) Hydrogen bonds. (d) Hydrophobic interactions.

groups, one acidic and one basic, interact through ion–ion attractions. Figure 20.9b shows an electrostatic interaction.

Hydrogen bonds can occur between amino acids with polar R groups. A variety of polar side chains can be involved, especially those that possess the following functional groups:

$$-\mathrm{OH} \qquad -\mathrm{NH_2} \qquad -\overset{\overset{\displaystyle \mathrm{O}}{\|}}{\mathrm{C}}-\mathrm{OH} \qquad -\overset{\overset{\displaystyle \mathrm{O}}{\|}}{\mathrm{C}}-\mathrm{NH_2}$$

Hydrogen bonds are relatively weak and are easily disrupted by changes in pH and temperature. Figure 20.9c shows the hydrogen-bonding interactions between the R groups of glutamine and serine.

Hydrophobic interactions result when two nonpolar side chains are close to each other. In aqueous solution, many proteins have their polar R groups outward, toward the aqueous solvent (which is also polar), and their nonpolar R groups inward (away from the polar water molecules). The nonpolar R groups then interact with each other. Hydrophobic interactions are common between phenyl rings and alkyl side chains. Although hydrophobic interactions are weaker than hydrogen bonds or electrostatic interactions, they are a significant force in some proteins because there are so many of them; their cumulative effect can be greater in magnitude than the effects of hydrogen bonding. Figure 20.9d shows the hydrophobic interactions between the R groups of phenylalanine and leucine.

In 1959, a protein tertiary structure was determined for the first time. The determination involved myoglobin, a protein whose function is oxygen storage in muscle tissue. Figure 20.10 shows myoglobin's tertiary structure. It involves a single chain of 153 amino acids with numerous α helix segments within the chain. The structure also contains a heme group, an iron-containing group with the ability to bind molecular oxygen.

Figure 20.10
A schematic diagram showing the tertiary structure of the single-chain protein myoglobin.

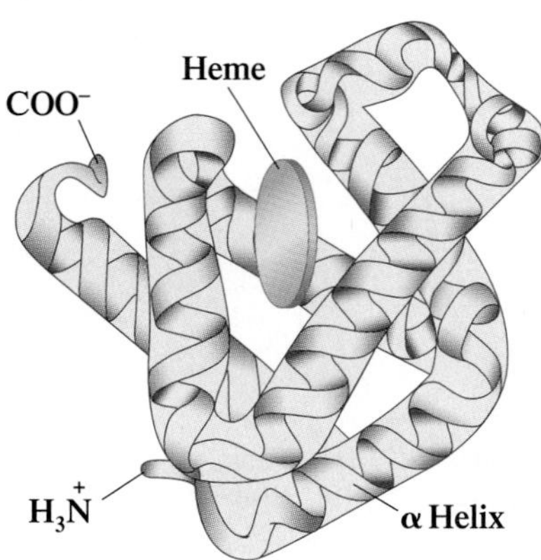

20.10 Quaternary Structure of Proteins

Quaternary structure is the highest level of protein organization. It is found only in proteins that have structures involving two or more polypeptide chains that are independent of each other—that is, are not covalently bonded to each other. These multichain proteins are often called *oligomeric proteins.* The **quaternary structure of a protein** *involves the associations among the separate chains in an oligomeric protein.*

Most oligomeric proteins contain an even number of subunits (two subunits = a dimer, four subunits = a tetramer, and so on). The subunits are held together mainly by hydrophobic interactions between amino acid R groups.

The hydrophobic interactions maintaining quaternary structure are more easily interrupted than those for tertiary structure. For example, only small changes in cellular conditions can cause a tetrameric protein to fall apart, dissociating into dimers or perhaps four separate subunits, with a resulting temporary loss of protein activity. As conditions change back, the oligomer automatically re-forms, and normal protein function is restored.

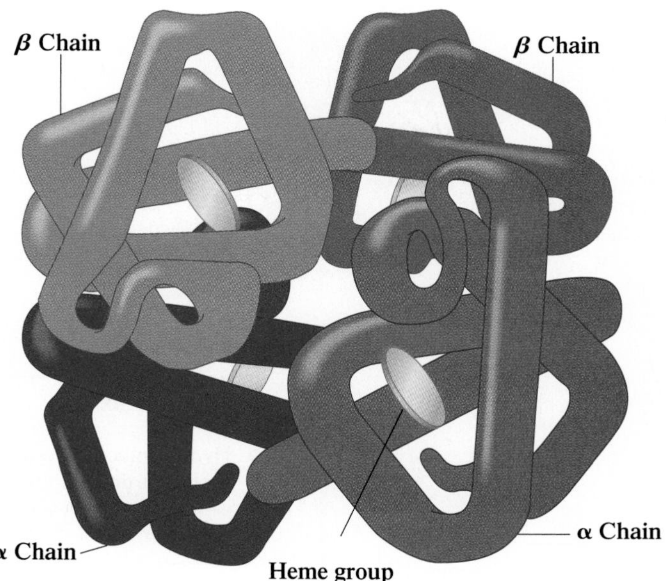

Figure 20.11
A schematic diagram showing the quaternary structure of the oxygen-carrying protein hemoglobin.

An example of a protein with quaternary structure is hemoglobin, the oxygen-carrying protein in blood (Figure 20.11). It is a tetramer in which there are two identical α chains and two identical β chains. Each chain enfolds a heme group, the site where oxygen binds to the protein.

Chemistry at a Glance reviews what we have said about protein structure.

20.11 Globular and Fibrous Proteins

On the basis of structural shape, proteins can be classified into two major types: fibrous proteins and globular proteins. A **fibrous protein** *is a protein that has a long, thin, fibrous shape.* Such proteins are made up of long rod-shaped or string-like molecules that can intertwine with one another and form strong fibers. They are water-insoluble and generally have structural functions within the human body. A **globular protein** *is a protein whose overall shape is roughly spherical or globular.* Globular proteins either dissolve in water or form stable suspensions in water, which allows them to travel through the blood and other body fluids to sites where their activity is needed. Table 20.2 gives examples of selected common fibrous and globular proteins.

• Secondary structure is primarily responsible for the nature of fibrous proteins. By contrast, tertiary structure is of prime importance in determining the overall shapes of globular proteins.

Table 20.2
Some Common Fibrous and Globular Proteins

Name	Occurrence and function
Fibrous proteins (insoluble)	
keratins	found in wool, feathers, hooves, silk, and fingernails
collagens	found in tendons, bone, and other connective tissue
elastins	found in blood vessels and ligaments
myosins	found in muscle tissue
fibrin	found in blood clots
Globular proteins (soluble)	
insulin	regulatory hormone for controlling glucose metabolism
myoglobin	protein involved in oxygen transport in muscles
hemoglobin	protein involved in oxygen transport in blood
transferrin	protein involved in iron transport in blood
immunoglobulins	proteins involved in immune response

CHEMISTRY AT A GLANCE

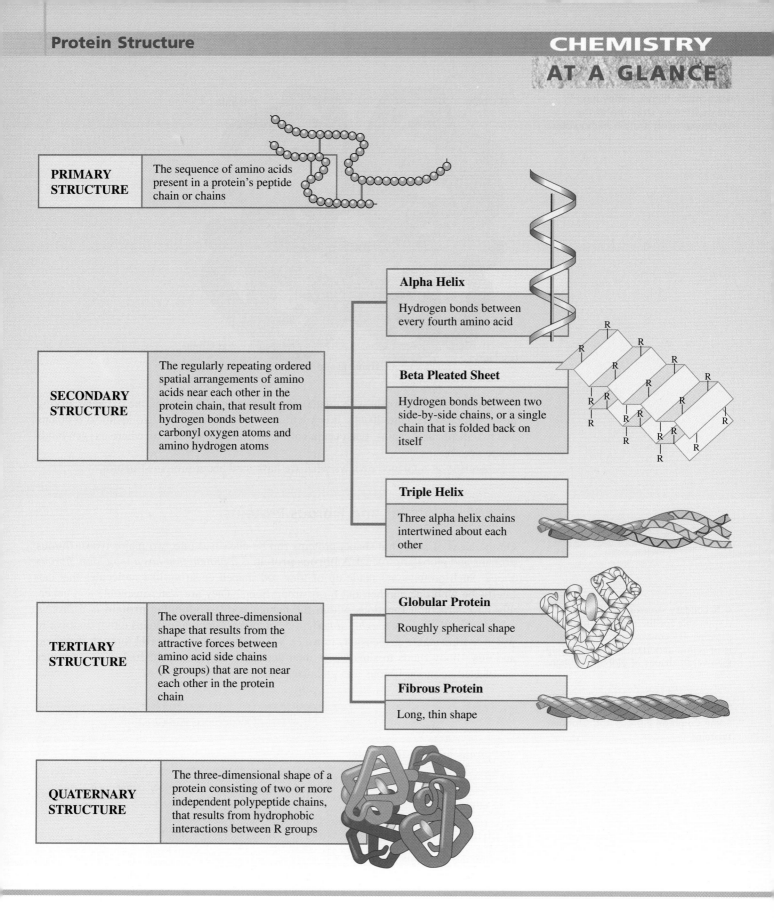

PRIMARY STRUCTURE	The sequence of amino acids present in a protein's peptide chain or chains

SECONDARY STRUCTURE	The regularly repeating ordered spatial arrangements of amino acids near each other in the protein chain, that result from hydrogen bonds between carbonyl oxygen atoms and amino hydrogen atoms

Alpha Helix
Hydrogen bonds between every fourth amino acid

Beta Pleated Sheet
Hydrogen bonds between two side-by-side chains, or a single chain that is folded back on itself

Triple Helix
Three alpha helix chains intertwined about each other

TERTIARY STRUCTURE	The overall three-dimensional shape that results from the attractive forces between amino acid side chains (R groups) that are not near each other in the protein chain

Globular Protein
Roughly spherical shape

Fibrous Protein
Long, thin shape

QUATERNARY STRUCTURE	The three-dimensional shape of a protein consisting of two or more independent polypeptide chains, that results from hydrophobic interactions between R groups

The tail feathers of a peacock contain the fibrous protein *alpha keratin.*

The fibrous protein *alpha keratin* is particularly abundant in nature, where it is found in protective coatings for organisms. It is the major protein constituent of hair, feathers, wool, fingernails and toenails, claws, scales, horns, turtle shells, quills, and hooves.

The structure of a typical alpha keratin, that of hair, is depicted in Figure 20.12. The individual molecules are almost wholly alpha-helical (Figure 20.12a). Pairs of these helices twine about one another to produce a coiled coil (Figure 20.12b). In hair, two of the coiled coils then further twist together to form a four-molecule protofilament (Figure 20.12c). Finally, eight protofilaments combine in either a circular or a square arrangement to make a microfilament, which is the basic structural unit. Introduction of disulfide links (between cysteine residues) within the several levels of structure determine the "hardness" of the alpha keratin. "Hard" keratins, such as those found in horns and nails, have considerably more disulfide bridges than their softer counterparts found in hair, wool, and feathers.

20.12 Simple and Conjugated Proteins

Proteins are classified as either simple proteins or conjugated proteins. A **simple protein** *is made up entirely of amino acid residues.* More than one polypeptide chain may be

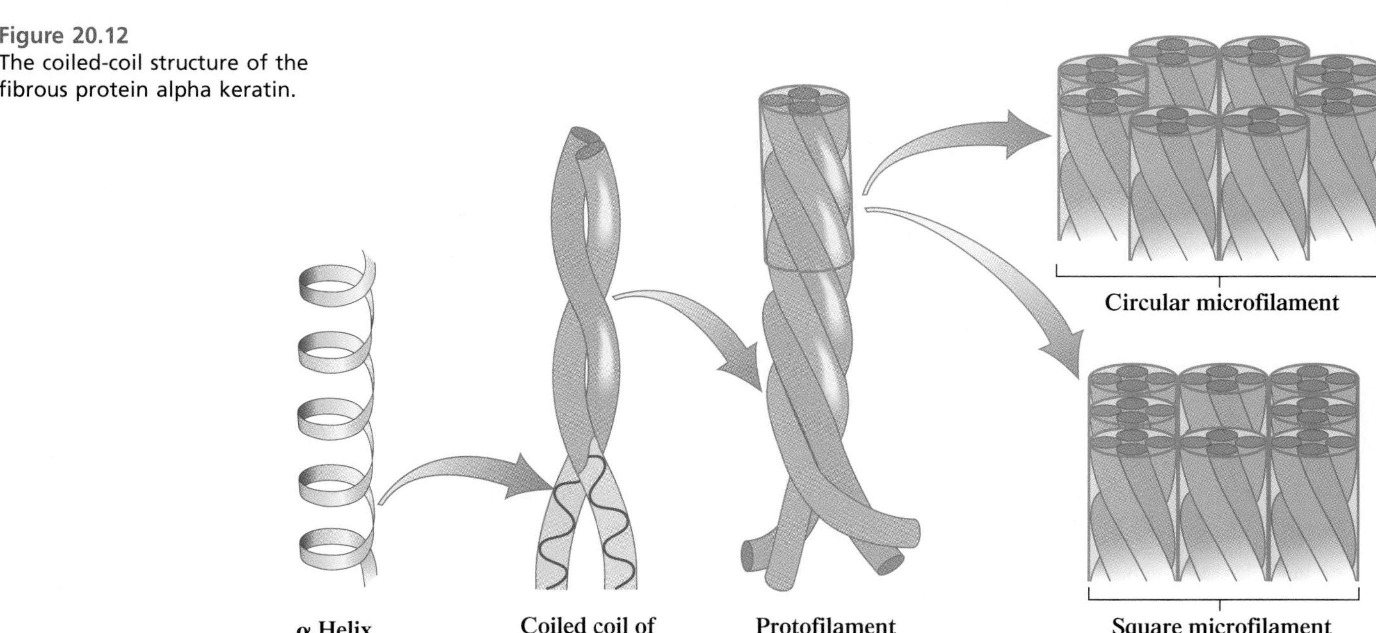

Figure 20.12
The coiled-coil structure of the fibrous protein alpha keratin.

Table 20.3
Types of Conjugated Proteins

Class	Prosthetic group	Specific example	Function of example
hemoproteins	heme unit	hemoglobin	carrier of O_2 in blood
		myoglobin	oxygen binder in muscles
lipoproteins	lipid	low-density lipoprotein (LDL)	lipid carrier
		high-density lipoprotein (HDL)	lipid carrier
glycoproteins	carbohydrate	gamma globulin	antibody
		mucin	lubricant in mucous secretions
		interferon	antiviral protection
phosphoproteins	phosphate group	glycogen phosphorylase	enzyme in glycogen phosphorylation
nucleoproteins	nucleic acid	ribosomes	site for protein synthesis in cells
		viruses	self-replicating, infectious complex
metalloproteins	metal ion	iron–ferritin	storage complex for iron
		zinc–alcohol dehydrogenase	enzyme in alcohol oxidation

present, but all chains contain only amino acids. A **conjugated protein** *has other chemical components in addition to amino acids.* These additional components, which may be organic or inorganic, are called prosthetic groups. A **prosthetic group** *is a non-amino acid unit permanently associated with a protein.*

Conjugated proteins may be further classified according to the nature of the prosthetic group. For example, proteins containing lipids, those containing carbohydrates, and those containing metal ions are called lipoproteins, glycoproteins, and metalloproteins, respectively. Table 20.3 gives further examples of the types of conjugated proteins. Several examples of glycoproteins and lipoproteins are discussed in Sections 20.15 and 20.16, respectively.

20.13 Protein Hydrolysis

When a protein or polypeptide in a solution of strong acid or strong base is heated, the peptide bonds of the amino acid chain are hydrolyzed and free amino acids are produced. The hydrolysis reaction is the reverse of the formation reaction for a peptide bond. Amine and carboxylic acid functional groups are regenerated.

Let us consider the hydrolysis of the tripeptide Ala–Gly–Cys under acidic conditions. Complete hydrolysis produces one unit each of the amino acids alanine, glycine, and cysteine. The equation for the hydrolysis is

$$\overset{+}{H_3N}-\underset{CH_3}{\overset{H}{C}}-\overset{O}{\overset{\|}{C}}-\overset{H}{N}-\underset{H}{\overset{H}{C}}-\overset{O}{\overset{\|}{C}}-\overset{H}{N}-\underset{\underset{SH}{CH_2}}{\overset{H}{C}}-COOH \xrightarrow[H^+,\ \Delta]{H_2O} \overset{+}{H_3N}-\underset{CH_3}{\overset{H}{C}}-COOH + \overset{+}{H_3N}-\underset{H}{\overset{H}{C}}-COOH + \overset{+}{H_3N}-\underset{\underset{SH}{CH_2}}{\overset{H}{C}}-COOH$$

Ala–Gly–Cys → Ala + Gly + Cys

Note that the product amino acids in this reaction are written in positive ion form because of the acidic reaction conditions.

• Protein hydrolysis produces free amino acids. This process is the reverse of protein synthesis, where free amino acids are combined.

Protein digestion is simply enzyme-catalyzed hydrolysis of ingested protein. The free amino acids produced from this process are absorbed through the intestinal wall into the bloodstream and transported to the liver. Here they become the raw materials for the synthesis of new protein tissue. Also, the hydrolysis of cellular proteins to amino acids is an ongoing process, as the body resynthesizes needed molecules and tissue.

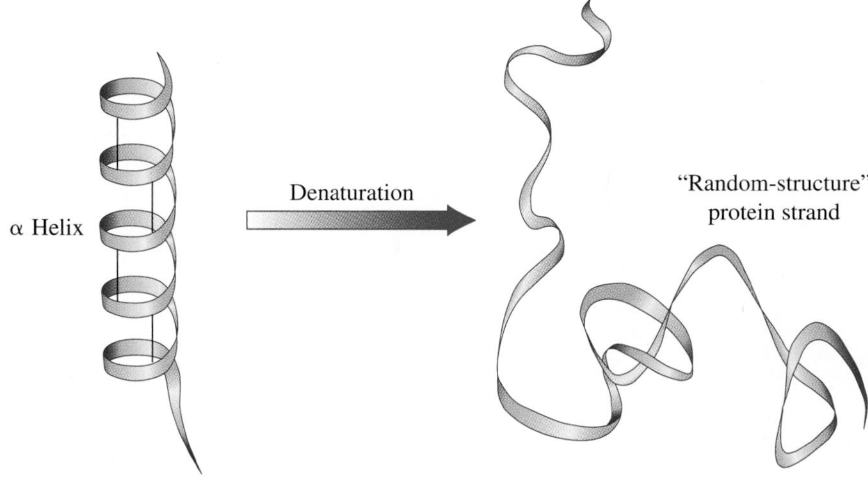

Figure 20.13
Protein denaturation involves loss of the protein's three-dimensional structure. Complete loss of such structure produces a random-coil protein strand.

20.14 Protein Denaturation

• A consequence of protein denaturation, the partial or complete loss of a protein's three-dimensional structure, is loss of biological activity for the protein.

Protein denaturation *is the partial or complete disorganization of a protein's characteristic three-dimensional shape as a result of disruption of its secondary, tertiary, and quaternary structural interactions.* Because the biological function of a protein depends on its three-dimensional shape, the result of denaturation is loss of biological activity. Protein denaturation does not affect the primary structure of a protein.

Although some proteins lose all of their three-dimensional structural characteristics upon denaturation (Figure 20.13), most proteins maintain some three-dimensional structure. For a few small proteins, it is possible to find conditions under which the effects of denaturation can be reversed; this restoration process in which the protein is "refolded" is called *renaturation.* Denaturation is irreversible, however, for most proteins.

Loss of water solubility is a frequent physical consequence of protein denaturation. The precipitation out of biological solution of denatured protein is called *coagulation.*

A most dramatic example of protein denaturation occurs when egg white (a concentrated solution of the protein albumin) is poured onto a hot surface. The clear albumin solution immediately changes into a white solid with a jelly-like consistency. A similar process occurs when hamburger juices encounter a hot surface. A brown, jelly-like solid forms.

Cooking food denatures the protein present but does not alter the protein's nutritional value. The primary structure of the protein remains intact. Above, the denaturation of protein in egg whites.

One of the reasons why many foods are cooked is to denature the protein present so that it is more easily digested. It is easier for digestive enzymes to "work on" denatured (unraveled) protein. Cooking foods also kills microorganisms through protein denaturation. For example, ham and bacon can harbor parasites that cause trichinosis. Cooking the ham or bacon denatures parasite protein.

In surgery, heat is often used to seal small blood vessels. This process is called *cauterization.* Small wounds can also be sealed by cauterization. Heat-induced denaturation is used in sterilizing surgical instruments and in canning foods; bacteria are destroyed when the heat denatures their protein.

The body temperature of a patient with fever may rise to 102°F, 103°F, or even 104°F without serious consequences. A temperature above 106°F (41°C) is extremely dangerous, for at this level, the enzymes of the body begin to be inactivated. Enzymes, which function as catalysts for almost all body reactions, are protein. Inactivation of enzymes, through denaturation, can have lethal effects on body chemistry.

The effect of ultraviolet radiation from the sun, a nonionizing radiation (Section 11.7), is similar to that of heat. Denatured skin proteins cause most of the problems associated with sunburn.

Storage room for cheese; during storage cheese "matures" as bacteria and enzymes ferment the cheese, giving it a stronger flavor.

A curdy precipitate of casein, the principal protein in milk, is formed in the stomach when the hydrochloric acid of gastric juice denatures milk. The curdling of milk that takes place when milk sours or cheese is made results from the presence of lactic acid, a by-product of bacterial growth. Yogurt is prepared by growing lactic acid–producing bacteria in skim milk. The coagulated denatured protein gives yogurt its semi-solid consistency.

Serious eye damage can result from eye tissue contact with acids or bases, when irreversibly denatured and coagulated protein causes a clouded cornea. This reaction is part of the basis for the rule that students wear protective eyewear in the chemistry laboratory.

Alcohols are an important type of denaturing agent. Denaturation of bacterial protein takes place when isopropyl or ethyl alcohol is used as a disinfectant. This accounts for the common practice of swabbing the skin with alcohol before giving an injection. Interestingly, pure isopropyl or ethyl alcohol is less effective than the commonly used 70% alcohol solution. Pure alcohol quickly denatures and coagulates the bacterial surface, thereby forming an effective barrier to further penetration by the alcohol. The 70% solution denatures more slowly and allows complete penetration to be achieved before coagulation of the surface proteins takes place.

20.15 Glycoproteins

Glycoproteins *are conjugated proteins that contain carbohydrates or carbohydrate derivatives in addition to amino acids.* The carbohydrate content of glycoproteins is variable (from a few percent up to 85%), but it is fixed for any specific glycoprotein.

Glycoproteins include a number of very important substances; two of these, collagen and immunoglobulins, are described in this section. Many of the proteins in plasma (cell)

Chemical CONNECTIONS

20.3 Denaturation and Human Hair

The process used in waving hair—that is, in a hair permanent—involves reversible denaturation. Hair is protein, in which many disulfide ($-S-S-$) linkages occur as part of its tertiary structure; 16%–18% of hair is the amino acid cysteine. It is these disulfide linkages that give hair protein its overall shape. When a permanent is administered, hair is first treated with a reducing agent (ammonium thioglycolate) that breaks the disulfide linkages in the hair, producing two sulfhydryl ($-SH$) groups:

$$\text{Disulfide bridges} \xrightarrow[\text{agents}]{\text{Reducing}} \text{sulfhydryl groups}$$

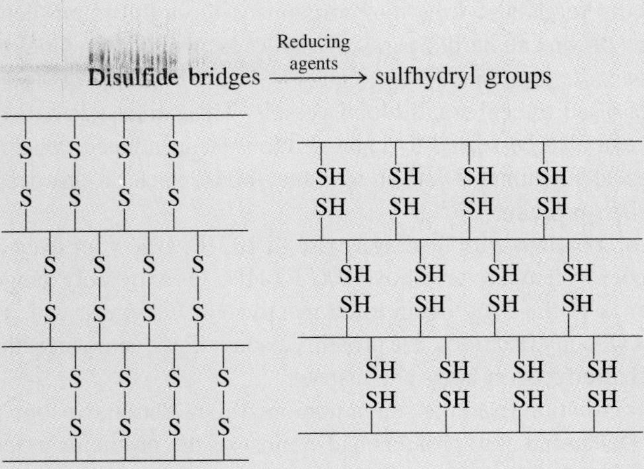

The "reduced" hair, whose tertiary structure has been disrupted, is then wound on curlers to give it a new configuration. Finally, the reduced and rearranged hair is treated with an oxidizing agent (potassium bromate) to form disulfide linkages at new locations within the hair:

$$\text{Sulfhydryl groups} \xrightarrow[\text{agents}]{\text{Oxidizing}} \text{re-formed disulfide bridges}$$

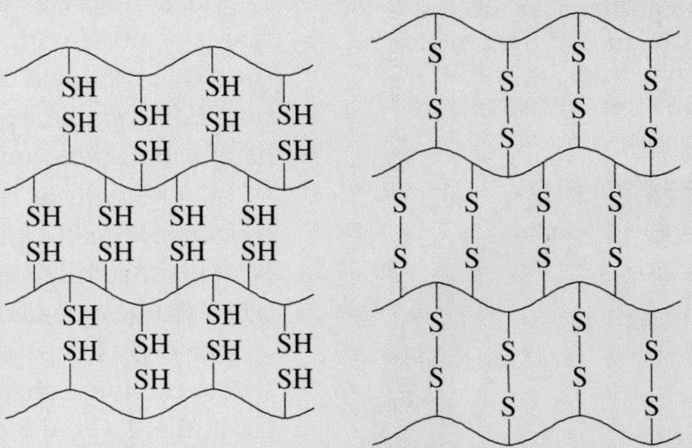

The new shape and curl of the hair are maintained by the newly formed disulfide bonds and the resulting new tertiary structure accompanying their formation. Of course, as new hair grows in, the "permanent" process has to be repeated.

Table 20.4
The Collagen Content of Selected Body Tissues

Tissue	Collagen (% dry mass)
Achilles tendon	86
aorta	12–24
bone (mineral-free)	88
cartilage	46–63
cornea	68
ligament	17
skin	72

membranes (lipid bilayers; see Section 19.12) are actually glycoproteins. The blood group markers of the ABO system (see Chemical Connections 18.1) are also glycoproteins in which the carbohydrate content can reach 85%.

• Collagen

Collagen, the most abundant of all proteins in humans (30% of total body protein), is a major structural material in tendons, ligaments, blood vessels, and skin; it is also the organic component of bones and teeth. Table 20.4 lists the collagen content of selected body tissues. The predominant structural feature within collagen molecules, three chains of amino acids wrapped (wound) into a triple helix, has already been considered (Section 20.8).

The rich content of the amino acid proline (up to 20%) in collagen is one reason why it has a triple-helix conformation rather than the simpler α helix structure (Section 20.8). Proline amino acid residues do not fit into regular α helices because of the cyclic nature of the side chain present and its accompanying different "geometry."

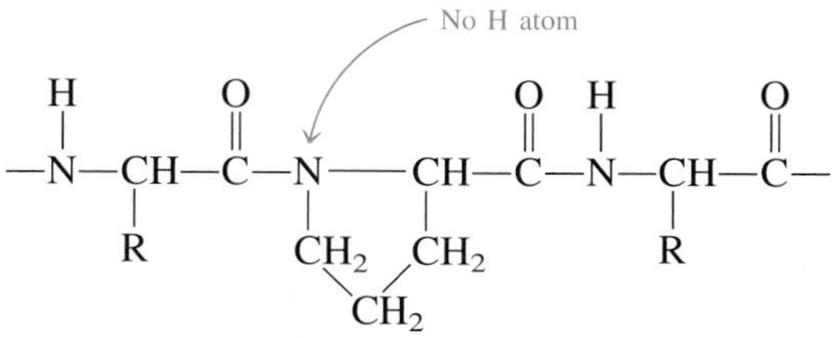

No H atom

—N(H)—CH(R)—C(=O)—N—CH—C(=O)—N(H)—CH(R)—C(=O)—
(ring: N—CH_2—CH_2—CH_2—CH)

Portion of a collagen chain

• Nonstandard amino acids consist of amino acid residues that have been chemically modified after their incorporation into a protein (as is the case with 4-hydroxyproline and 5-hydroxylysine) and amino acids that occur in living organisms but are not found in proteins.

An additional structural feature of collagen is the presence of the *nonstandard* amino acids 4-hydroxyproline (5%) and 5-hydroxylysine (1%)—derivatives of the standard amino acids proline and lysine (Table 20.1).

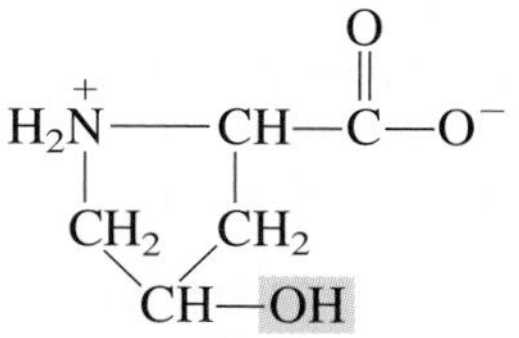

$H_2\overset{+}{N}$—CH—C(=O)—O^- (ring: CH_2—CH—OH—CH_2)

4-Hydroxyproline

$H_3\overset{+}{N}$—CH(—CH_2—CH_2—CH(OH)—CH_2—$^+NH_3$)—C(=O)—O^-

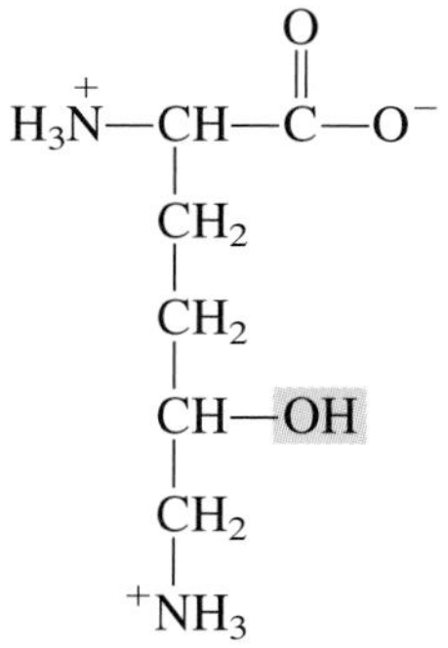

5-Hydroxylysine

• The primary biological function of vitamin C involves the hydroxylation of proline and lysine during collagen formation. These hydroxylation processes require the enzymes proline hydroxylase and lysine hydroxylase. These enzymes can function only in the presence of vitamin C.

The presence of carbohydrate units (mostly glucose, galactose, and their disaccharides) attached by glycosidic linkages (Section 18.13) to collagen at its 5-hydroxylysine residues causes collagen to be classified as a *glycoprotein.* The function of the carbohydrate groups in collagen is related to cross-linking; they direct the assembly of collagen triple helices into more complex aggregations called *collagen fibrils.*

Collagen molecules (triple helices) are very long, thin, and rigid. Many such molecules, lined up alongside each other, combine to make collagen fibrils. Cross-linking between helices gives the fibrils extra strength. The greater the number of cross links, the more rigid the fibril is. The stiffening of skin and other tissues associated with aging is thought to result, at least in part, from an increasing amount of cross-linking between collagen molecules. The process of tanning, which converts animal hides to leather, involves increasing the degree of cross-linking.

When collagen is boiled in water, under basic conditions, it is converted to the water-soluble protein gelatin. This process involves both denaturation (Section 20.14) and hydrolysis (Section 20.13). Heat acts as a denaturant, causing rupture of the hydrogen bonds supporting collagen's triple-helix structure. Regions in the amino acid chains where proline and hydroxyproline concentrations are high are particularly susceptible to hydrolysis, which breaks up the polypeptide chains. Meats become more tender when cooked because of the conversion of some collagen to gelatin. Tougher cuts of meat (more cross-linking), such as stew meat, need longer cooking times.

• Immunoglobulins

Immunoglobulins are among the most important and interesting of the soluble proteins in the human body. **Immunoglobulins** *are glycoprotein molecules produced by an organism as a protective response to the invasion of microorganisms or foreign molecules.* Different classes of immunoglobulins, identified by differing carbohydrate content and molecular mass, exist.

Immunoglobulins serve as *antibodies* to combat invasion of the body by *antigens.* **Antigens** *are foreign substances, such as bacteria and viruses, that invade the body.* **Antibodies** *are molecules that counteract specific antigens.* The immune system of the human body has the capability to produce immunoglobulins that respond to several thousand different antigens.

All types of immunoglobulin molecules have a similar basic structure, which includes the following features:

1. Four polypeptide chains are present: two identical heavy (H) chains and two identical light (L) chains.
2. The H chains, which usually contain 400–500 amino acid residues, are approximately twice as long as the L chains.
3. Both the H and L chains have constant and variable regions. The constant regions have the same amino acid sequence from immunoglobulin to immunoglobulin, and the variable regions have a different amino acid sequence in each immunoglobulin.
4. The carbohydrate content of various immunoglobulins varies from 1% to 12% by mass.
5. The secondary and tertiary structures are similar for all immunoglobulins. They involve a Y-shaped conformation (Figure 20.14), with disulfide linkages between H and L chains stabilizing the structure.

The interaction of an immunoglobulin molecule with an antigen occurs at the "tips" (uppermost part) of the Y structure. These tips are the variable-composition region of the immunoglobulin structure. It is here that the antigen binds specifically, and it is here that the amino acid sequence differs from one immunoglobulin to another.

Each immunoglobulin has two identical active sites and can thus bind to two molecules of the antigen it is "designed for." The action of many such immunoglobulins of a given type in concert with each other creates an antigen–antibody complex that precipi-

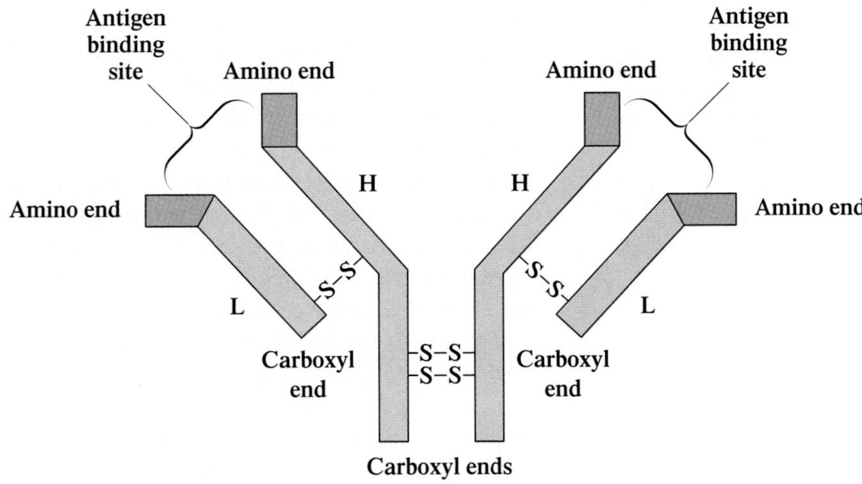

Figure 20.14 This schematic diagram shows the structure of an immunoglobulin. Two heavy (H) polypeptide chains and two light (L) polypeptide chains are cross-linked by disulfide bridges. The purple areas are the constant amino acid regions, and the areas shown in red are the variable amino acid regions of each chain. Carbohydrate molecules attached to the heavy chains aid in determining the destinations of immunoglobulins in the tissues.

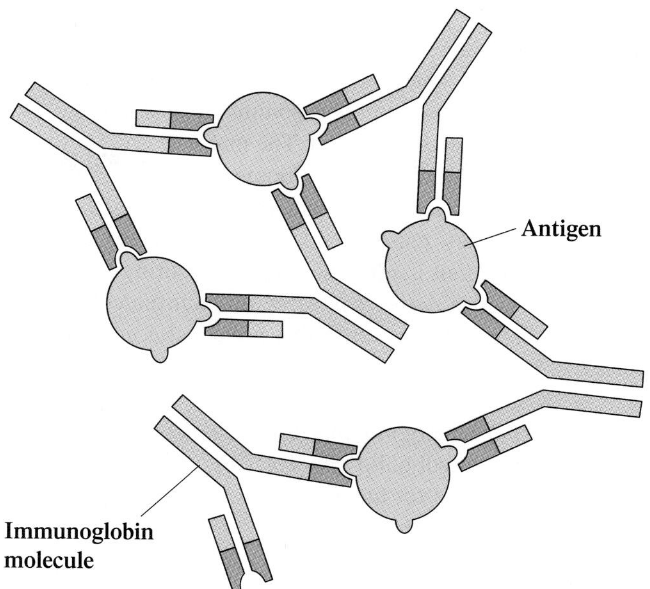

Figure 20.15
In this immunoglobulin–antigen complex, note that more than one immunoglobulin molecule can attach itself to a given antigen. Also, any given immunoglobulin has only two sites where antigen can bind.

tates from solution (Figure 20.15). Eventually, an invading antigen can be eliminated from the body through such precipitation. The bonding of an antigen to the variable region of an immunoglobulin occurs through dipole–dipole interactions and hydrogen bonds rather than covalent bonds.

The importance of immunoglobulins is amply and tragically demonstrated by the effects of AIDS (acquired immune deficiency syndrome). The AIDS virus upsets the body's

Chemical CONNECTIONS

20.4 Cyclosporine: An Antirejection Drug

The survival rate for patients undergoing human organ transplant operations, such as heart, liver, or kidney replacement, has risen dramatically since 1985. This increased success coincides with the introduction of a new drug for controlling transplant rejection by a patient's own immune system. This new immunosuppressive agent (antirejection drug) is cyclosporine, a substance obtained from a particular type of soil fungus.

The primary structure of cyclosporine is that of a cyclic peptide containing 11 amino acid units. Ten of these are amino acids with simple side chains (four or fewer carbon atoms). The eleventh amino acid, which is the key to cyclosporine's pharmacological activity, had not been previously reported. This novel amino acid has a 7-carbon branched, unsaturated, hydroxylated side chain with the structure

$$-\underset{\displaystyle OH}{\underset{|}{CH}}-\underset{\displaystyle CH_3}{\underset{|}{CH}}-CH_2-CH{=}CH-CH_3$$

The diagram shows the amino acid sequence within the cyclosporine ring. Seven of the amino acid units, denoted by

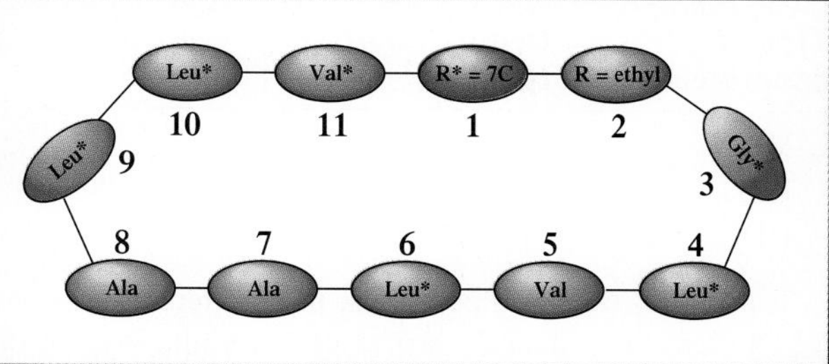

asterisks in the preceding structure, have their nitrogen atom methylated; that is, a methyl group has replaced the hydrogen atom. This unique structural feature makes cyclosporine water-insoluble but fat-soluble.

The fat solubility of cyclosporine allows it to cross cell membranes readily and to be widely distributed in the body. It is administered either intravenously or orally. Because of its low water solubility, the drug is supplied in olive oil for oral administration. Cyclosporine has a narrow therapeutic index. When the blood concentration is too low, inadequate immunosuppression occurs. On the other hand, a high cyclosporine concentration can lead to kidney problems.

normal production of immunoglobulins and leaves the body susceptible to what would otherwise not be debilitating and deadly infections.

Individuals who receive organ transplants must be given drugs to suppress the production of immunoglobulins against foreign proteins in the new organ, thus preventing rejection of the organ. The major reason for the increasing importance of organ transplants is the successful development of drugs that can properly manipulate the body's immune system (see Chemical Connections 20.4).

Many reasons exist for a mother to breast-feed a newborn infant. One of the most important is immunoglobulins. During the first two or three days of lactation, the breasts produce *colostrum,* a premilk substance containing immunoglobulins from the mother's blood. Colostrum helps protect the newborn infant from those infections to which the mother has developed immunity. These diseases are the ones in her environment—precisely those the infant needs protection from. Breast milk, once it is produced, is a source of immunoglobulins for the infant for a short time. (After the first week of nursing, immunoglobulin concentrations in the milk decrease rapidly.) Infant formula used as a substitute for breast milk is almost always nutritionally equivalent, but it does not contain immunoglobulins.

Chemical CONNECTIONS

20.5 Lipoproteins and Heart Attack Risk

The lipoproteins present in blood serum are classified according to their density, which is related to the fractions of protein and lipid present. The more protein in the lipoprotein, the higher its density. On a density basis, there are three general categories of blood serum lipoproteins: (1) very-low-density lipoprotein (VLDL), (2) low-density lipoprotein (LDL), and (3) high-density lipoprotein (HDL). Characteristics of these three types of lipoproteins are given in the accompanying table.

Type of lipoprotein	Density range (g/mL)	Approximate percent-by-mass protein
VLDL	1.006–1.019	5
LDL	1.019–1.063	25
HDL	1.063–1.21	50

The various types of lipoproteins have different functions. VLDLs are the principal carriers of triacylglycerols in the blood. (As cells remove triacylglycerols as needed from VLDLs, the VLDLs become LDLs.) Both LDLs and HDLs are involved in cholesterol transport. LDLs carry approximately 80% of this substance, and HDLs the remainder.

Of significance, LDLs and HDLs carry cholesterol for different purposes. LDLs carry cholesterol to cells for their use, whereas HDLs carry excess cholesterol away from cells to the liver for processing and excretion from the body.

Studies show that LDL levels correlate *directly* with heart disease, whereas HDL levels correlate *inversely* with heart disease risk. HDL is sometimes referred to as "good" cholesterol (*H*DL = *H*ealthy) and LDL as "bad" cholesterol (*L*DL = *L*ess healthy).

The goal of dietary measures to slow the advance of atherosclerosis is to reduce LDL cholesterol levels. Reduction in the dietary intake of saturated fat appears to be a key action (see Chemical Connections 19.3).

High HDL levels are desirable, because they give the body an efficient means of removing excess cholesterol. Low HDL levels can result in excess cholesterol depositing within the circulatory system.

In general, women have higher HDL levels than men—an average of 55 mg per 100 mL of blood serum versus 45 mg per 100 mL. This may explain in part why proportionately fewer women have heart attacks than men. Nonsmokers have uniformly higher HDL levels than smokers. Exercise on a regular basis tends to increase HDL levels. This discovery has increased the popularity of walking and running exercise. It appears that genetics also plays a role in establishing HDL as well as other lipoprotein concentrations in the blood.

Guidelines from the American Heart Association relative to HDL levels are given in the accompanying table.

	HDL Level	
Guideline	Male (mg/100 mL)	Female (mg/100 mL)
desirable	greater than 45	greater than 55
acceptable	38–44	47–54
moderate risk	33–37	41–46
high risk	less than 33	less than 41

20.16 Lipoproteins

Lipoproteins *are conjugated proteins composed of both lipids and amino acids.* The major function of such proteins is to help suspend lipids and transport them through the bloodstream. Lipids, in general, are insoluble in blood (an aqueous medium) because of their nonpolar nature (Section 19.1).

The presence or absence of various types of lipoproteins in the blood appears to have implications for the health of the heart and blood vessels. Lipoprotein levels in the blood are now used as an indicator of heart attack risk (see Chemical Connections 20.5).

Concepts to Remember

Protein. A protein is a polymer in which the monomer units are amino acids.

α-Amino acid. An α-amino acid is a compound in which the amino group is attached to the α-carbon atom of the carboxylic acid carbon chain.

Standard amino acid. A standard amino acid is one of the 20 α-amino acids that are normally present in protein.

Amino acid classifications. Amino acids are classified as nonpolar, polar neutral, polar basic, or polar acidic, depending on the nature of the side chain (R group) present.

Chirality of amino acids. Amino acids found in proteins are always left-handed (L isomer).

Zwitterion. A zwitterion is a molecule that has a positive charge on one atom and a negative charge on another atom. In neutral solution and in the solid state, amino acids exist as zwitterions. For neutral amino acids in solution, the isoelectric point is the pH at which the concentration of zwitterions is a maximum.

Peptide bond. A peptide bond is an amide bond involving the carboxyl group of one amino acid and the amino group of another amino acid. In a protein, the amino acids are linked to each other through peptide bonds.

Protein primary structure. The primary structure of a protein is the sequence of amino acids present in the peptide chain or chains of the protein.

Protein secondary structure. The secondary structure of a protein is the arrangement in space of the atoms in the backbone of the protein. The three major types of protein secondary structure are the alpha helix, the beta pleated sheet, and the triple helix.

Protein tertiary structure. The tertiary structure of a protein is the overall three-dimensional shape that results from the attractive forces among amino acid side chains (R groups).

Protein quaternary structure. The quaternary structure of a protein involves the associations among various polypeptide chains present in the protein.

Fibrous protein. Fibrous proteins have a long, thin, fibrous shape. They are generally insoluble in water.

Globular protein. Globular proteins have an overall shape that is roughly spherical or globular. They are generally soluble in water.

Conjugated Proteins. A conjugated protein contains one or more other chemical components, called prosthetic groups, in addition to amino acids.

Protein hydrolysis. Protein hydrolysis is a chemical reaction in which peptide bonds within a protein are broken through reaction with water. Complete hydrolysis produces free amino acids.

Protein denaturation. Protein denaturation is the partial or complete disorganization of a protein's characteristic three-dimensional shape as a result of disruption of its secondary, tertiary, and quaternary structural interactions.

Glycoproteins. Glycoproteins are conjugated proteins that contain carbohydrates or carbohydrate derivatives in addition to amino acids. Collagen and immunoglobulins are important glycoproteins.

Lipoproteins. Lipoproteins are conjugated proteins that are composed of both lipids and amino acids. Lipoproteins are classified on the basis of their density.

Key Reactions and Equations

1. Formation of a zwitterion at pH 7 (Section 20.4)

$$H_2N-\underset{\underset{R}{|}}{CH}-\overset{\overset{O}{\|}}{C}-OH \longrightarrow H_3\overset{+}{N}-\underset{\underset{R}{|}}{CH}-\overset{\overset{O}{\|}}{C}-O^-$$

2. Conversion of a zwitterion to a positive ion in acidic solution (Section 20.4)

$$H_3\overset{+}{N}-\underset{\underset{R}{|}}{CH}-\overset{\overset{O}{\|}}{C}-O^- + H_3O^+ \longrightarrow H_3\overset{+}{N}-\underset{\underset{R}{|}}{CH}-\overset{\overset{O}{\|}}{C}-OH + H_2O$$

3. Conversion of a zwitterion to a negative ion in basic solution (Section 20.4)

$$H_3\overset{+}{N}-\underset{\underset{R}{|}}{CH}-\overset{\overset{O}{\|}}{C}-O^- + OH^- \longrightarrow H_2N-\underset{\underset{R}{|}}{CH}-\overset{\overset{O}{\|}}{C}-O^- + H_2O$$

4. Formation of a peptide bond (Section 20.5)

$$H_3\overset{+}{N}-\underset{\underset{R}{|}}{CH}-COO^- + H_3\overset{+}{N}-\underset{\underset{R'}{|}}{CH}-COO^- \longrightarrow$$

$$H_3\overset{+}{N}-\underset{\underset{R}{|}}{CH}-\overset{\overset{O}{\|}}{C}-\overset{\overset{H}{|}}{N}-\underset{\underset{R'}{|}}{CH}-COO^- + H_2O$$

5. Hydrolysis of a protein in acidic solution (Section 20.13)

$$\text{Protein} + H_2O \xrightarrow{H^+} \text{smaller peptides} \xrightarrow{H^+} \text{amino acids}$$

6. Denaturation of a protein (Section 20.14)

$$\text{Protein with } 1°, 2°, \text{ and } 3° \text{ structure} \xrightarrow[\text{agent}]{\text{Denaturing}} \text{Protein with } 1° \text{ structure only}$$

Key Terms

Amino acid (20.2)
Antibodies (20.15)
Antigens (20.15)
Complete dietary protein (20.2)
Conjugated protein (20.12)
Electrophoresis (20.4)
Essential amino acid (20.2)
Fibrous protein (20.11)
Globular protein (20.11)
Glycoproteins (20.15)
Immunoglobulins (20.15)
Isoelectric point (20.4)
Lipoproteins (20.16)
Nonpolar amino acid (20.2)
Peptide (20.5)
Peptide bond (20.5)
Polar acidic amino acid (20.2)
Polar basic amino acid (20.2)
Polar neutral amino acid (20.2)
Primary structure of a protein (20.7)
Prosthetic group (20.12)
Protein (20.2 and 20.6)
Protein denaturation (20.14)
Quaternary structure of a protein (20.10)
Secondary structure of a protein (20.8)
Simple protein (20.12)
Standard amino acid (20.2)
Tertiary structure of a protein (20.9)
Zwitterion (20.4)

Exercises and Problems

The members of each pair of problems in this section test similar material.

Amino Acid Structural Characteristics (Section 20.2)

20.1 Which of the following structures represent α-amino acids?

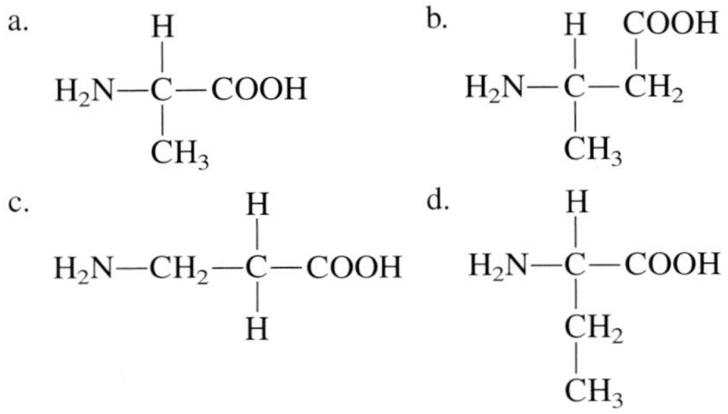

a. $H_2N—C(H)(CH_3)—COOH$

b. $H_2N—C(H)(CH_3)—CH_2—COOH$ (H and COOH above, CH_3 below)

c. $H_2N—CH_2—C(H)(H)—COOH$

d. $H_2N—C(H)(CH_2CH_3)—COOH$

20.2 What is the significance of the prefix α in the designation α-amino acid?

20.3 What is the major structural difference among the various standard amino acids?

20.4 On the basis of polarity, what are the four types of side chains found in the standard amino acids?

20.5 With the help of Table 20.1, determine which of the standard amino acids have a side chain with the following characteristics.

a. Contains an aromatic group
b. Contains the element sulfur
c. Contains a carboxyl group
d. Contains a hydroxyl group

20.6 With the help of Table 20.1, determine which of the standard amino acids have a side chain with the following characteristics.

a. Contains only carbon and hydrogen
b. Contains an amino group
c. Contains an amide group
d. Contains more than four carbon atoms

20.7 What is the distinguishing characteristic of a polar basic amino acid?

20.8 What is the distinguishing characteristic of a polar acidic amino acid?

20.9 In what way is the structure of the amino acid proline different from that of the other 19 standard amino acids?

20.10 Which two of the standard amino acids are structural isomers?

Amino Acid Nomenclature (Section 20.2)

20.11 What amino acids do these abbreviations stand for?

a. Ala b. Leu c. Met d. Trp

20.12 What amino acids do these abbreviations stand for?

a. Asp b. Cys c. Phe d. Val

20.13 Which four standard amino acids have three-letter abbreviations that are not the first three letters of their common names?

20.14 What are the three-letter abbreviations for the three polar basic amino acids?

20.15 Classify each of the following amino acids as nonpolar, polar neutral, polar acidic, or polar basic.

a. Asn b. Glu c. Pro d. Ser

20.16 Classify each of the following amino acids as nonpolar, polar neutral, polar acidic, or polar basic.

a. Gly b. Thr c. Tyr d. His

Chirality and Amino Acids (Section 20.3)

20.17 To which family of mirror-image isomers do nearly all naturally occurring amino acids belong?

20.18 In what way is the structure of glycine different from that of the other 19 common amino acids?

20.19 Draw Fischer projection formulas for the following amino acids.

a. L-Serine b. D-Serine
c. D-Alanine d. L-Leucine

20.20 Draw Fischer projection formulas for the following amino acids.

a. L-Cysteine b. D-Cysteine
c. L-Alanine d. L-Valine

Acid–Base Properties of Amino Acids (Section 20.4)

20.21 Amino acids are solids at room temperature with relatively high decomposition points. Explain why.

20.22 Amino acids exist as zwitterions in the solid state. Explain why.

20.23 Draw the zwitterion structure for each of the following amino acids.

a. Leucine   b. Isoleucine

c. Cysteine   d. Glycine

20.24 Draw the zwitterion structure for each of the following amino acids.

a. Serine   b. Methionine

c. Threonine   d. Phenylalanine

20.25 Draw the structure of serine at each of the following pH values.

a. 7   b. 1   c. 12   d. 3

20.26 Draw the structure of glycine at each of the following pH values.

a. 7   b. 13   c. 2   d. 11

20.27 Explain what is meant by the term *isoelectric point.*

20.28 Most amino acids have isoelectric points between 5 and 6, but the isoelectric point of lysine is 9.7. Explain why lysine has such a high value for its isoelectric point.

20.29 Glutamic acid exists in two low-pH forms instead of the usual one. Explain why.

20.30 Arginine exists in two high-pH forms instead of the usual one. Explain why.

20.31 Predict the direction of movement of each of the following amino acids in a solution at the pH value specified. Indicate the direction as toward the positive electrode or toward the negative electrode. Write "isoelectric" if no net movement occurs.

a. Alanine at pH = 12   b. Valine at pH = 7

c. Aspartic acid at pH = 1   d. Arginine at pH = 13

20.32 Predict the direction of movement of each of the following amino acids in a solution at the pH value specified. Indicate the direction as toward the positive electrode or toward the negative electrode. Write "isoelectric" if no net movement occurs.

a. Alanine at pH = 2   b. Valine at pH = 12

c. Aspartic acid at pH = 13   d. Arginine at pH = 1

20.33 A direct current was passed through a solution containing valine, histidine, and aspartic acid at a pH of 6.0. One amino acid migrated to the positive electrode, one migrated to the negative electrode, and one did not migrate to either electrode. Which amino acids went where?

20.34 A direct current was passed through a solution containing alanine, arginine, and glutamic acid at a pH of 6.0. One amino acid migrated to the positive electrode, one migrated to the negative electrode, and one did not migrate to either electrode. Which amino acids went where?

Peptide Formation (Section 20.5)

20.35 What is a peptide?

20.36 What is a peptide bond?

20.37 What two functional groups are involved in the formation of a peptide bond?

20.38 What is meant by the N-terminal end and the C-terminal end of a peptide?

20.39 Write out the full structure of the tripeptide Val–Phe–Cys.

20.40 Write out the full structure of the tripeptide Glu–Ala–Leu.

20.41 Explain why the notations Ser–Cys and Cys–Ser represent two different molecules rather than the same molecule.

20.42 Explain why the notations Ala–Gly–Val–Ala and Ala–Val–Gly–Ala represent two different molecules rather than the same molecule.

20.43 There are a total of six different sequences for a tripeptide containing one molecule each of serine, valine, and glycine. Using three-letter abbreviations for the amino acids, draw the six possible sequences of amino acids.

20.44 There are a total of six different sequences for a tetrapeptide containing two molecules each of serine and valine. Using three-letter abbreviations for the amino acids, draw the six possible sequences of amino acids.

20.45 Identify the amino acids contained in each of the following tripeptides.

a.
```
            O  H        O  H
     +      ‖  |        ‖  |
  H3N—CH—C—N—CH—C—N—CH—COO⁻
      |         |         |
      CH2       CH3       CH2
      |                   |
      OH                  SH
```

b.
```
            O  H        O  H
     +      ‖  |        ‖  |
  H3N—CH—C—N—CH—C—N—CH—COO⁻
      |         |         |
      CH2       CH—OH     CH2
      |         |         |
      COO⁻      CH3       C—NH2
                          ‖
                          O
```

20.46 Identify the amino acids contained in each of the following tripeptides.

a.
```
              O  H        O  H
     +        ‖  |        ‖  |
  H3N—CH2—C—N—CH—C—N—CH2—COO⁻
                  |
                  CH—CH3
                  |
                  CH3
```

b.
```
            O  H        O  H
     +      ‖  |        ‖  |
  H3N—CH—C—N—CH—C—N—CH—COO⁻
      |         |         |
      CH2       CH2       CH2
      |         |         |
      OH        CH—CH3    CH2
                |         |
                CH3       COO⁻
```

20.47 How many peptide bonds are present in each of the molecules in Problem 20.45?

20.48 How many peptide bonds are present in each of the molecules in Problem 20.46?

20.49 What is the difference between an amide bond and a peptide bond?

20.50 What is the difference between a polypeptide and a protein?

Levels of Protein Structure (Sections 20.6–20.10)

20.51 What is the primary structure of a protein?

20.52 Two proteins with the same amino acid composition do not have to have the same primary structure. Explain why.

20.53 What are the three major types of secondary protein structure?

20.54 Hydrogen bonding between which functional groups stabilizes protein secondary structure arrangements?

20.55 The β pleated sheet secondary structure can be formed through either intramolecular hydrogen bonding or intermolecular hydrogen bonding. Explain why.

20.56 The α helix secondary structure always involves intramolecular hydrogen bonding and never involves intermolecular hydrogen bonding. Explain why.

20.57 Can more than one type of secondary structure be present in the same protein molecule? Explain your answer.

20.58 What is meant by the statement that a section of a protein has a "random structure" arrangement?

20.59 What is the difference between the types of hydrogen bonding that occur in secondary and tertiary protein structures?

20.60 State the four types of attractive interactions that give rise to tertiary structure for proteins.

20.61 Specify the nature of each of the following tertiary-structure interactions, using the choices (1) hydrophobic, (2) electrostatic, (3) hydrogen bonding, and (4) disulfide bond.

a. Phenylalanine and leucine
b. Arginine and glutamic acid
c. Two cysteines
d. Serine and tyrosine

20.62 Specify the nature of each of the following tertiary-structure interactions, using the choices (1) hydrophobic, (2) electrostatic, (3) hydrogen bonding, and (4) disulfide bond.

a. Lysine and aspartic acid
b. Threonine and tyrosine
c. Alanine and valine
d. Leucine and isoleucine

20.63 What is an oligomeric protein?

20.64 What is the quaternary structure of a protein?

Protein Classifications (Sections 20.11 and 20.12)

20.65 Contrast the properties of fibrous and globular proteins in terms of

a. basic structural shape
b. solubility characteristics in water

20.66 What is meant by each of the following "structural terms" associated with the structure of the fibrous protein alpha keratin?

a. Coiled coil
b. Four-molecule protofilament
c. Microfilament

20.67 What is the major difference between a simple protein and a conjugated protein?

20.68 What is the prosthetic group in each of the following types of conjugated proteins?

a. Lipoproteins
b. Glycoproteins
c. Phosphoproteins
d. Metalloproteins

Protein Hydrolysis (Section 20.13)

20.69 Will hydrolysis of the dipeptides Ala–Val and Val–Ala yield the same products? Explain your answer.

20.70 A shampoo bottle lists "partially hydrolyzed protein" as one of its ingredients. What is the difference between partially hydrolyzed protein and completely hydrolyzed protein?

20.71 Drugs that are proteins, such as insulin, must always be injected rather than taken by mouth. Explain why.

20.72 Which structural levels of a protein are affected by hydrolysis?

20.73 Identify the primary structure of a hexapeptide containing six different amino acids if the following smaller peptides are among the partial-hydrolysis products:
Ala–Gly, His–Val–Arg, Ala–Gly–Met, and Gly–Met–His.

20.74 Identify the primary structure of a hexapeptide containing five different amino acids if the following smaller peptides are among the partial-hydrolysis products:
Gly–Cys, Ala–Ser, Ala–Gly, and Cys–Val–Ala.

20.75 How many different di- and tripeptides could be present in a solution of partially hydrolyzed Ala–Gly–Ser–Tyr?

20.76 How many different di- and tripeptides could be present in a solution of partially hydrolyzed Ala–Gly–Ala–Gly?

Protein Denaturation (Section 20.14)

20.77 Which structural levels of a protein are affected by denaturation?

20.78 Suppose a sample of protein is completely hydrolyzed and another sample of the same protein is denatured. Compare the final products of these processes.

20.79 In what way is the protein in a cooked egg the same as that in a raw egg?

20.80 Why is 70% ethanol rather than pure ethanol preferred for use as an antiseptic agent?

Glycoproteins (Section 20.15)

20.81 What is the dominant structural feature within collagen molecules?

20.82 What two nonstandard amino acids are present in collagen?

20.83 Where are the carbohydrate units located in collagen?

20.84 What is the role of vitamin C in the biosynthesis of collagen?

20.85 What is the difference between an antigen and an antibody?

20.86 What is an immunoglobulin?

20.87 Describe the structural features of a typical immunoglobulin molecule.

20.88 Describe the process by which blood immunoglobulins help protect the body from invading bacteria and viruses.

Lipoproteins (Section 20.16)

20.89 What is the major biological function of lipoproteins?

20.90 What is the basis for the classification of blood serum lipoproteins into groups?

Additional Problems

20.91 State whether each of the following statements apply to primary, secondary, tertiary, or quaternary protein structure.

a. A disulfide bond forms between two cysteine residues in different protein chains.

b. A salt bridge forms between amino acids with acidic and basic side chains.

c. Hydrogen bonding between carbonyl oxygen atoms and nitrogen atoms of amino groups causes a peptide to coil into a helix.

d. Peptide linkages hold amino acids together in a polypeptide chain.

20.92 What is the common name for each of the following IUPAC-named standard amino acids?

a. 2-Aminopropanoic acid

b. 2-Amino-4-methylbutanoic acid

c. 2-Amino-3-hydroxybutanoic acid

d. 2-Aminobutanedioic acid

20.93 What is the net charge at a pH of 1 for each of the following peptides?

a. Val–Ala–Leu b. Tyr–Trp–Thr

c. Asp–Asp–Glu–Gly d. His–Arg–Ser–Ser

20.94 What is the net charge at a pH of 13 for each of the peptides in Problem 20.93?

20.95 The amino acid isoleucine possesses two chiral centers. Draw Fischer projection formulas for the four stereoisomers that are possible for this amino acid.

20.96 Indicate how many structurally isomeric tetrapeptides are possible for a tetrapeptide in which

a. four different amino acids are present

b. three different amino acids are present

c. two different amino acids are present

20.97 Contrast the lower-level structural characteristics of the fibrous proteins collagen and alpha keratin.

20.98 Draw the structures of the three hydrolysis products obtained when the tripeptide in part a of Problem 20.45 undergoes hydrolysis under

a. low-pH (acidic) conditions

b. high-pH (basic) conditions

Grid Problems

20.99

1. alanine	2. arginine	3. cysteine
4. lysine	5. serine	6. tyrosine

Select from the grid *all* correct responses for each of the following situations.

a. A standard amino acid b. An essential amino acid

c. A nonpolar amino acid d. A polar basic amino acid

20.100

1. Ala (nonpolar)	2. Asp (polar acidic)	3. His (polar basic)
4. Lys (polar basic)	5. Tyr (polar neutral)	6. Val (nonpolar)

Select from the grid *all* correct responses for each of the following situations.

a. Possesses a charge of zero at a pH of 7

b. Possesses a charge of +1 at a pH of 7

c. Possesses a charge of +2 at a pH of 1

d. Possesses a charge of −1 at a pH of 13

20.101

1. α helix	2. β pleated sheet	3. triple helix
4. globular protein	5. fibrous protein	6. amino acid sequence

Select from the grid *all* correct responses for each of the following situations.

a. Interactions between amino acid side chains are responsible for this protein structural characteristic.

b. Interactions between carbonyl oxygen atoms and amino hydrogen atoms are responsible for this protein structural characteristic.

c. Chemical reactions between carboxyl groups and amino groups are responsible for this protein structural characteristic.

d. Hydrogen bonding is solely or partially responsible for this protein structural characteristic.

20.102

1. Gly–Val–Gly	2. Gly–Val–Val	3. Gly–Gly–Val–Val
4. Val–Gly–Gly	5. Val–Gly–Val	6. Gly–Val–Gly–Val

Select from the grid *all* correct responses for each of the following situations.

a. Val–Gly is a possible dipeptide product when partial hydrolysis occurs.

b. Three different dipeptide products are possible when partial hydrolysis occurs.

c. Two different amino acids are produced in equal numbers when complete hydrolysis occurs.

d. These are pairs of peptides for which possible dipeptide products are the same when partial hydrolysis occurs.

Enzymes, Vitamins, and Minerals

CHAPTER OUTLINE

21.1 General Characteristics of Enzymes 621
21.2 Enzyme Nomenclature 621
21.3 Enzyme Structure 622
21.4 Models of Enzyme Action 622
21.5 Enzyme Specificity 625
21.6 Factors That Affect Enzyme Activity 626

Chemistry at a Glance: **Enzyme Activity** 629

21.7 Enzyme Inhibition 630

Chemistry at a Glance: **Enzyme Inhibition** 632

21.8 Regulation of Enzyme Activity: Allosteric Enzymes 632
21.9 Regulation of Enzyme Activity: Zymogens 633
21.10 Antibiotics That Inhibit Enzyme Activity 634
21.11 Medical Uses of Enzymes 636
21.12 Vitamins 636
21.13 Water-Soluble Vitamins 638
21.14 Fat-Soluble Vitamins 639
21.15 Minerals 642

Chemical Connections

21.1 Enzymatic Browning: Discoloration of Fruits and Vegetables 627
21.2 Heart Attacks and Enzyme Analysis 637
21.3 Iron: The Most Abundant Trace Mineral in the Human Body 645

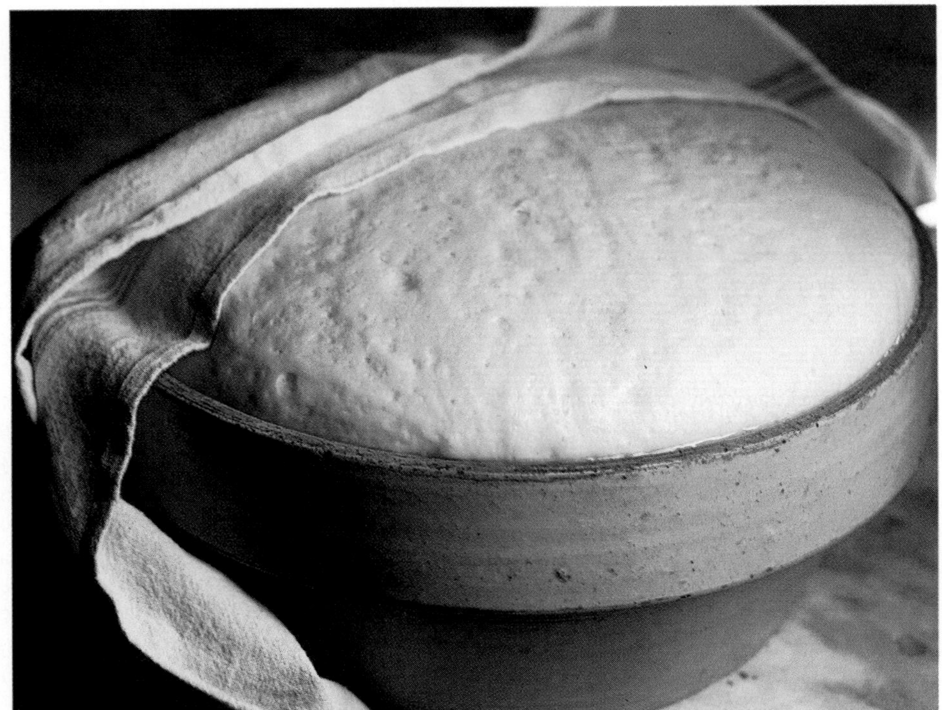

Bread dough rises as the result of the action of enzymes.

In this chapter we consider two topics: (1) enzymes, and (2) vitamins and minerals. Enzymes govern all chemical reactions in living orgnisms. They are specialized proteins that, with fascinating precision and selectivity, catalyze biochemical reactions that store and release energy, make pigments in our hair and eyes, digest the food we eat, synthesize cellular building materials, and protect us by repairing cellular damage or clotting our blood. Enzymes are sensitive to their environment, responding automatically to changes in the cell. The deficiency or excess of particular enzymes can cause certain diseases or signal problems such as heart attacks and other organ damage. Our knowledge of protein structure (Chapter 20) can help us appreciate and better understand how enzymes function in living cells.

Vitamins and minerals, necessary components of a healthful diet, play important roles in cellular metabolism. In most cases, they function as enzyme cofactors or carriers of functional groups during biosynthesis. Vitamins are organic substances, and minerals are inorganic substances.

Long before their chemical nature was understood, yeast enzymes were used in the production of bread and alcoholic beverages. The action of yeast on sugars produces the carbon dioxide gas that causes bread to rise. Fermentation of sugars in fruit juices with the same yeast enzymes produces alcoholic beverages. Above, a vat of fermenting grapes in a winery.

21.1 General Characteristics of Enzymes

Enzymes *are catalysts for biological reactions.* Each cell in the human body contains thousands of different enzymes, because almost every reaction in a cell requires its own specific enzyme. Enzymes cause cellular reactions to occur millions of times faster than corresponding uncatalyzed reactions. As catalysts (Section 9.6), enzymes are not consumed during the reaction but merely help the reaction occur more rapidly.

Most enzymes are globular proteins (Section 20.11). Some are simple proteins, consisting entirely of amino acid chains. Others are more complex, containing additional chemical components (Section 21.3). Until the 1980s, it was thought that *all* enzymes were proteins. A few enzymes are now known that are made of ribonucleic acids (RNA; Section 22.7) and catalyze cellular reactions involving nucleic acids. In this chapter, we will consider only enzymes that are proteins.

Enzymes undergo all the reactions of proteins, including *denaturation* (Section 20.14). Slight alterations in pH, temperature, or other protein denaturants affect enzyme activity dramatically. Good cooks realize that overheating yeast kills the action of the yeast. A person suffering from a high fever (greater than 106°F) runs the risk of denaturing certain enzymes. The biochemist must exercise extreme caution in handling enzymes to avoid the loss of their activity. Even vigorous shaking of an enzyme solution can destroy enzyme activity.

• The word *enzyme* comes from the Greek words *en,* which means "in," and *zyme,* which means "yeast."

Many enzymes function only when needed, automatically turning on or off. Most laboratory catalysts need to be removed from the reaction to stop their catalytic action; not so with enzymes. In some cases, if a certain chemical is needed in the cell, then the enzyme responsible for its production automatically "switches on." When a sufficient quantity has been produced, the enzyme "switches off." In other situations, the cell may produce more or less enzyme as required. Because individual enzymes catalyze each reaction in the cell, certain necessary reactions can be accelerated without affecting the rest of the cellular metabolism.

21.2 Enzyme Nomenclature

• Enzymes, the most efficient catalysts known, increase the rates of biological reactions by factors of up to 10^{20} over uncatalyzed reactions. Nonenzymatic catalysts, on the other hand, typically enhance the rate of a reaction by factors of 10^2 to 10^4.

Enzymes are most commonly named by using a system that attempts to provide information about the *function* (rather than the structure) of the enzyme. Type of reaction catalyzed and *substrate* identity are focal points for the nomenclature. A **substrate** *is the reactant in an enzyme-catalyzed reaction.* The substrate is the substance upon which the enzyme "acts."

Three important aspects of the naming process are the following:

1. The suffix *-ase* identifies a substance as an enzyme. Thus ure*ase,* sucr*ase,* and lip*ase* are all *enzyme designations.* The suffix *-in* is still found in the names of some of the first enzymes studied, many of which are digestive enzymes. Such names include *trypsin, chymotrypsin,* and *pepsin.*
2. The type of reaction catalyzed by an enzyme is often noted with a prefix. An *oxidase* enzyme catalyzes an oxidation reaction, and a *hydrolase* enzyme catalyzes a hydrolysis reaction.
3. The identity of the substrate is often noted in addition to the type of reaction. Enzyme names of this type include *glucose oxidase, pyruvate carboxylase,* and *succinate dehydrogenase.* Infrequently, the substrate but not the reaction type is given, as in the names *urease* and *lactase.* In such names, the reaction involved is hydrolysis; *urease* catalyzes the hydrolysis of urea, *lactase* the hydrolysis of lactose.

Example 21.1

Predicting Enzyme Function from an Enzyme's Name

Predict the function of the following enzymes.

a. Cellulase **b.** Sucrase
c. L-Amino acid oxidase **d.** Aspartate aminotransferase

Solution

a. Cellulase catalyzes the hydrolysis of cellulose.
b. Sucrase catalyzes the hydrolysis of the disaccharide sucrose.
c. L-Amino acid oxidase catalyzes the oxidation of L-amino acids.
d. Aspartate aminotransferase catalyzes the transfer of an amino group from aspartate to a different molecule.

Practice Exercise 21.1

Predict the function of the following enzymes.

a. Maltase **b.** Lactate dehydrogenase
c. Fructose oxidase **d.** Maleate isomerase

• *Answers:* **a.** hydrolysis of maltose; **b.** removal of hydrogen from lactate ion; **c.** oxidation of fructose; **d.** rearrangement (isomerization) of maleate ion.

Enzymes are grouped into six major categories based on the types of reactions they catalyze: oxidoreductases, transferases, hydrolases, lyases, isomerases, and ligases. Table 21.1 gives further information about these six types of enzymes.

21.3 Enzyme Structure

• *Holoenzyme* is an alternative name for a conjugated enzyme. A holoenzyme, a biologically active entity, results from the combination of an apoenzyme and a cofactor.
Apoenzyme + cofactor = holoenzyme

Enzymes can be divided into two general structural classes: simple enzymes and conjugated enzymes. **Simple enzymes** *are composed only of protein (amino acid chains).* **Conjugated enzymes** *have a nonprotein portion in addition to a protein portion.* By itself, neither the protein part nor the nonprotein portion of a conjugated enzyme has catalytic properties. An **apoenzyme** *is the protein portion of a conjugated enzyme.* A **cofactor** *is the nonprotein portion of a conjugated enzyme.* Only the combination of apoenzyme and cofactor—the conjugated enzyme—shows biological activity.

Why do many enzymes need cofactors? Cofactors provide additional chemically reactive functional groups besides those present in the amino acid side chains of apoenzymes.

A cofactor is generally either a small organic molecule or an inorganic ion (usually a metal ion). A **coenzyme** *is a small organic molecule that serves as a cofactor in a conjugated protein.* Many vitamins (Section 21.12) have coenzyme functions in the human body.

Typical inorganic ion cofactors include Zn^{2+}, Mg^{2+}, Mn^{2+}, and Fe^{2+}. The nonmetallic Cl^- ion occasionally acts as a cofactor. Dietary minerals are an important source of inorganic ion cofactors.

21.4 Models of Enzyme Action

Explanations of *how* enzymes function as catalysts in biological systems are based on the concepts of an enzyme active site and enzyme–substrate complex formation.

Table 21.1
Main Classes and Subclasses of Enzymes

Main classes	Selected subclasses	Type of reaction catalyzed
Oxidoreductases catalyze oxidation–reduction reactions involving substrate molecules	oxidases	oxidation of a substrate
	reductases	reduction of a substrate
	dehydrogenases	introduction of double bond (oxidation) by formal removal of two H atoms from substrate, the H being accepted by a coenzyme
Transferases catalyze transfer of functional groups between two substrates	transaminases	transfer of an amino group between substrates
	kinases	transfer of a phosphate group between substrates
Hydrolases catalyze substrate hydrolysis reactions, the addition of a water molecule to a bond, causing the bond to break	lipases	hydrolysis of ester linkages in lipids
	proteases	hydrolysis of amide linkages in proteins
	nucleases	hydrolysis of sugar–phosphate ester bonds in nucleic acids
	carbohydrases	hydrolysis of glycosidic bonds in carbohydrates
	phosphatases	hydrolysis of phosphate–ester bonds
Lyases catalyze the addition of a group to a double bond or the removal of a group to create a double bond in a manner that does not involve hydrolysis or oxidation	dehydratases	removal of H_2O from substrate
	decarboxylases	removal of CO_2 from substrate
	deaminases	removal of NH_3 from substrate
	hydratases	addition of H_2O to a substrate
Isomerases catalyze the conversion of a substrate into another compound that is isomeric with it, such as *cis* $\rightleftharpoons$ *trans*, L $\rightleftharpoons$ D, or aldehyde $\rightleftharpoons$ ketone	racemases	conversion of D to L isomer, or vice versa
	mutases	conversion of one structural isomer into another
Ligases catalyze the bonding together of two substrate molecules, with ATP hydrolysis to provide energy	synthetases	formation of new bond between two substrates, with participation of ATP
	carboxylases	formation of new bond between substrate and CO_2, with participation of ATP

• Enzyme Active Site

Studies show that only a small portion of an enzyme molecule, called the active site, participates in the interaction with a substrate or substrates during a reaction. The **active site** *is the relatively small part of an enzyme that is actually involved in catalysis.*

• It is possible for an enzyme to have more than one active site.

The active site in an enzyme is a three-dimensional entity formed by groups that come from different parts of the protein chain(s); these groups are brought together by the folding and bending (secondary and tertiary structure; Sections 20.8 and 20.9) of the protein. The active site is usually a "crevice-like" location in the enzyme (see Figure 21.1).

• Enzyme–Substrate Complex

Catalysts offer an alternative pathway with lower activation energy through which a reaction can occur (Section 9.6). In enzyme-controlled reactions, this alternative pathway involves the formation of an enzyme–substrate complex as an intermediate species in the reaction. An **enzyme–substrate complex** *is the intermediate reaction species that is formed when a substrate binds to the active site of an enzyme.* Within the enzyme–substrate complex, proximity effects and orientation effects create more favorable reaction conditions than if substrates were free. The result is faster formation of products.

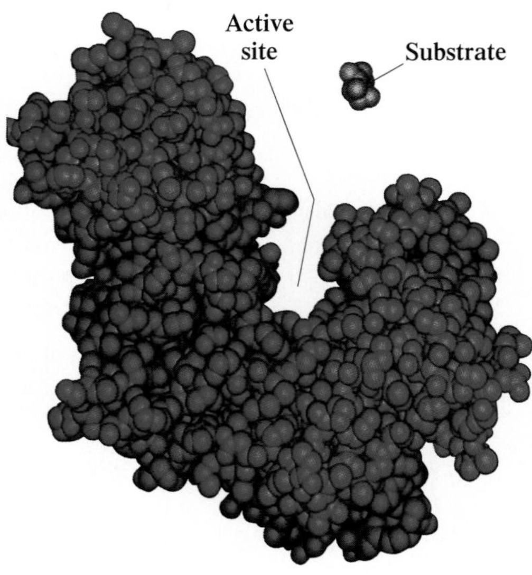

Figure 21.1
The active site of an enzyme is usually a crevice-like region formed as the result of the protein's secondary and tertiary structural characteristics.

• Lock-and-Key Model

• The lock-and-key model is more than just a "shape fit." In addition, there are weak binding forces (R group interactions) between parts.

To account for the highly specific way an enzyme selects a substrate and binds it to the active site, researchers have proposed several models. The simplest of these models is the lock-and-key model.

In the lock-and-key model, the active site in the enzyme has a fixed, rigid geometrical conformation. Only substrates with a complementary geometry can be accommodated at such a site, much as a lock accepts only certain keys. Figure 21.2 illustrates the lock-and-key concept of substrate–enzyme interaction.

• Induced-Fit Model

The lock-and-key model explains the action of many enzymes. It is, however, too restrictive for the action of other enzymes. Experimental evidence indicates that many enzymes have flexibility in their shapes. They are not rigid and static; there is constant change in their shape. The induced-fit model is used for this type of situation.

The induced-fit model allows for small changes in the shape or geometry of the active site of an enzyme in order to accommodate a substrate. An analogy would be the changes that occur in the shape of a glove when a hand is inserted into it. The induced fit is a result of the enzyme's flexibility; it adapts to accept the incoming substrate. This model, illustrated in Figure 21.3, is a more thorough explanation for the active-site properties of

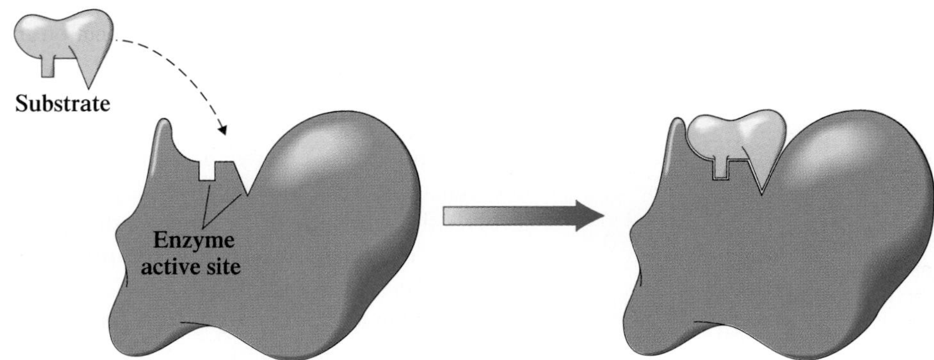

Figure 21.2
The lock-and-key model for enzyme activity. Only a substrate whose shape and chemical nature are complementary to those of the active site can interact with the enzyme.

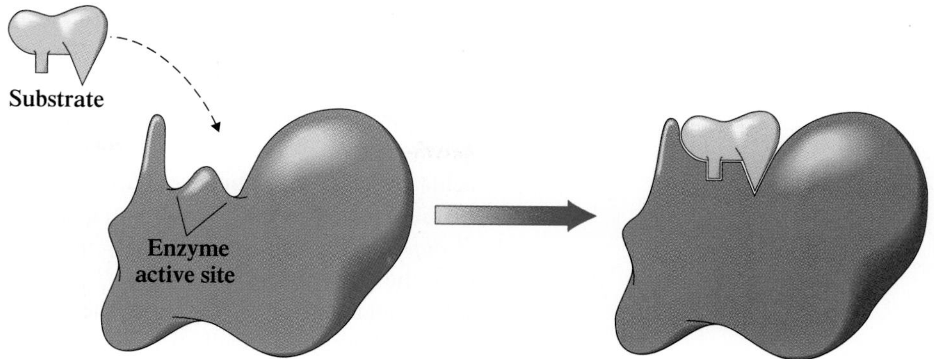

Figure 21.3
The induced-fit model for enzyme activity. The enzyme active site, although not exactly complementary in shape to that of the substrate, is flexible enough that it can adapt to the shape of the substrate.

an enzyme because it includes the specificity of the lock-and-key model coupled with the flexibility of the enzyme protein.

The forces that draw the substrate into the active site are many of the same forces that maintain tertiary structure in the folding of polypeptide chains. Ionic attractions, hydrogen bonds, and hydrophobic interactions all help attract and bind substrate molecules. For example, a protonated (positively charged) amino group in a substrate could be attracted and held at the active site by a negatively charged aspartate or glutamate residue. Alternatively, cofactors such as positively charged metal ions often help bind substrate molecules. Figure 21.4 is a schematic representation of the amino acid R group interactions that bind a substrate to an enzyme active site.

21.5 Enzyme Specificity

Enzymes exhibit different levels of selectivity, or specificity, for substrates. The degree of enzyme specificity is determined by the active site. Some active sites accommodate only one particular compound, whereas others can accommodate a "family" of closely related compounds. Types of enzyme specificity include

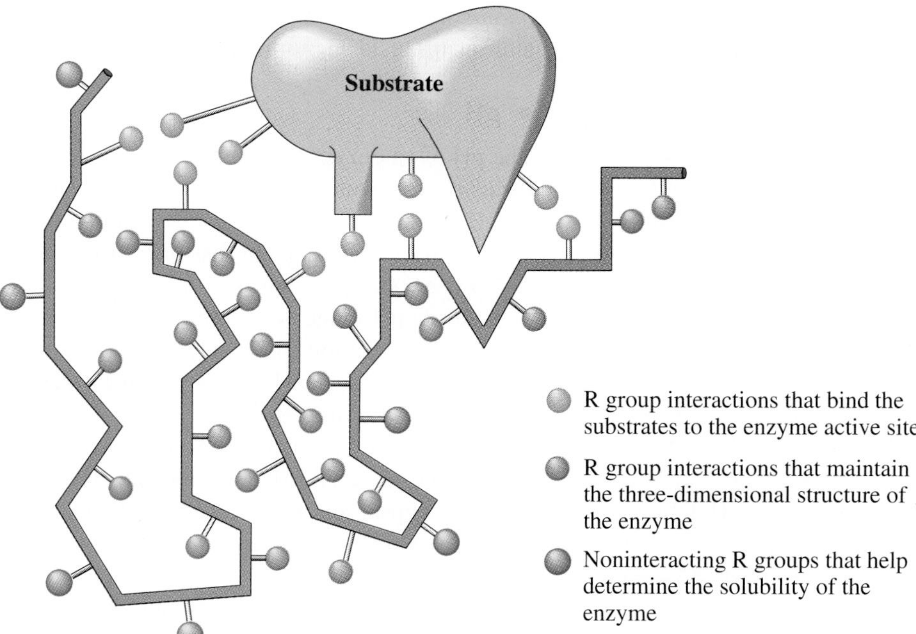

Figure 21.4
A schematic diagram representing amino acid R group interactions that bind a substrate to an enzyme active site. The R group interactions that maintain the three-dimensional structure of the enzyme (secondary and tertiary structure) are also shown.

1. *Absolute Specificity.* Such specificity means an enzyme will catalyze a particular reaction for *only one* substrate. This most restrictive of all specificities is not common. Urease is an enzyme with absolute specificity.
2. *Stereochemical Specificity.* Such specificity means an enzyme can distinguish between stereoisomers. Chirality is inherent in an active site, because amino acids are chiral compounds. L-Amino acid oxidase will catalyze reactions of L-amino acids but not of D-amino acids.
3. *Group Specificity.* Such specificity involves structurally similar compounds that have the same functional groups. Carboxypeptidase is group-specific; it cleaves amino acids, one at a time, from the carboxyl end of the peptide chain.
4. *Linkage Specificity.* Such specificity involves a particular type of bond, irrespective of the structural features in the vicinity of the bond. Phosphatases hydrolyze phosphate–ester bonds in all types of phosphate esters. Linkage specificity is the most general of the specificities considered.

21.6 Factors That Affect Enzyme Activity

Enzyme activity *is a measure of the rate at which an enzyme converts substrate to products.* Four factors affect enzyme activity: temperature, pH, substrate concentration, and enzyme concentration.

• Temperature

Temperature is a measure of the kinetic energy (energy of motion) of molecules. Higher temperatures mean molecules are moving faster and colliding more frequently. This concept applies to collisions between substrate molecules and enzymes. As the temperature of an enzymatically catalyzed reaction increases, so does the rate (velocity) of the reaction.

However, when the temperature increases beyond a certain point, the increased energy begins to cause disruptions in the tertiary structure of the enzyme; denaturation is occurring. Tertiary-structure change at the active site impedes catalytic action, and the enzyme activity quickly decreases as the temperature climbs past this point (Figure 21.5). The temperature that produces maximum activity for an enzyme is known as the optimum temperature for that enzyme. **Optimum temperature** *is the temperature at which the rate of an enzyme-catalyzed reaction is the greatest.* For human enzymes, the optimum temperature is often 37°C, normal body temperature.

Figure 21.5
Effect of temperature on the rate of an enzymatic reaction.

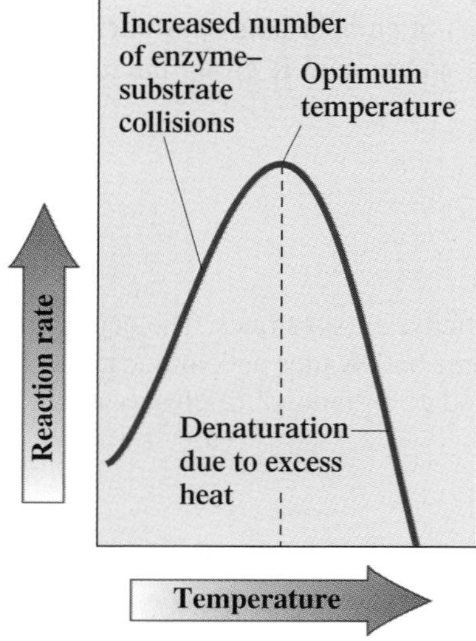

• pH

The pH of an enzyme's environment can affect its activity. This is not surprising, because the *charge* on acidic and basic amino acids (Section 20.2) located at the active site depends on pH. Small changes in pH (less than one unit) can result in enzyme denaturation (Section 20.14) and subsequent loss of catalytic activity.

Most enzymes exhibit maximum activity over a very narrow pH range. Only within this narrow pH range do the enzyme's amino acids exist in properly charged forms (Section 20.4). **Optimum pH** *is the pH at which an enzyme has maximum activity.* Figure 21.6 shows the effect of pH on an enzyme's activity. Biological buffers help maintain the optimum pH for an enzyme.

Each enzyme has a characteristic optimum pH, which usually falls within the physiological pH range of 7.0–7.5. Notable exceptions to this generalization are the digestive enzymes pepsin and trypsin. Pepsin, which is active in the stomach, functions best at a pH of 2.0. On the other hand, trypsin, which operates in the small intestine, functions best at a pH of 8.0.

A variation from normal pH can also affect substrates, causing either protonation or deprotonation of groups on the substrate. The interaction between the altered substrate and the enzyme active site may be less efficient than normal—or even impossible.

Figure 21.6
Effect of pH on an enzyme's activity.

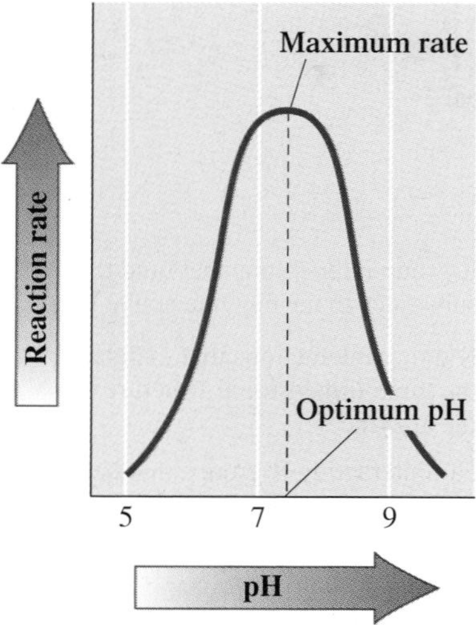

Chemical CONNECTIONS

21.1 Enzymatic Browning: Discoloration of Fruits and Vegetables

Everyone is familiar with the way fruits such as apples, pears, peaches, apricots, and bananas and vegetables such as potatoes quickly turn brown when their tissue is exposed to oxygen. Such oxygen exposure occurs when the food is sliced or bitten into or when it has sustained bruises, cuts, or other injury to the peel. This "browning reaction" is related to the work of an enzyme called phenolase (or polyphenoloxidase), a conjugated enzyme in which copper is present.

Phenolase is classified as an oxidoreductase. The substrates for phenolase are phenolic compounds present in the tissues of the fruits and vegetables. Phenolase hydroxylates monophenols to *o*-diphenols and oxidizes *o*-diphenols to *o*-quinones. The *o*-quinones then enter into a number of other reactions, which produce the "undesirable" brown discolorations. Quinone formation is enzyme- and oxygen-dependent. Once the quinones have formed, the subsequent reactions occur spontaneously and no longer depend on the presence of phenolase or oxygen.

Enzymatic browning can be prevented or slowed in several ways. Immersing the "injured" food (for example, apple slices) in cold water slows the browning process. The lower temperature decreases enzyme activity, and the water limits the enzyme's access to oxygen. Refrigeration slows enzyme activity even more, and boiling temperatures destroy (denature) the enzyme. A long-used method for preventing browning involves lemon juice. Phenolase works very slowly in the acidic environment created by the lemon juice's presence. In addition, the vitamin C (ascorbic acid) present in lemon juice functions as an antioxidant. It is more easily oxidized than the phenolic-derived compounds, and its oxidation products are colorless.

Monophenols $\xrightarrow[\text{Phenolase}]{O_2}$ *o*-Diphenols $\xrightarrow[\text{Phenolase}]{O_2}$ *o*-Quinones $\longrightarrow$ Brownish oxidation products

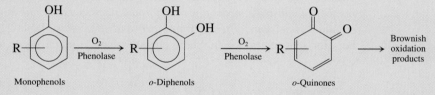

At left, a freshly cut apple. Brownish oxidation products form in a few minutes (at right).

• Substrate Concentration

When the concentration of an enzyme is kept constant and the concentration of substrate is increased, the enzyme activity pattern shown in Figure 21.7 is obtained. This activity pattern is called a *saturation curve*. Enzyme activity increases up to a certain substrate concentration and thereafter remains constant.

What limits enzymatic activity to a certain maximum value? As substrate concentration increases, the point is eventually reached where enzyme capabilities are used to their

Table 21.2
Turnover Numbers for Selected Enzymes

Enzyme	Turnover number (molecules of substrate per second per enzyme molecule)	Turnover time (seconds per molecule of product)
carbonic anhydrase	600,000	2×10^{-6}
α-glucosidase	17,000	6×10^{-5}
glutamate dehydrogenase	500	2×10^{-3}
phosphoglucomutase	21	5×10^{-2}
chymotrypsin	2	5×10^{-1}

maximum extent. The rate remains constant from this point on (Figure 21.7). Each substrate must occupy an enzyme active site for a finite amount of time, and the products must leave the site before the cycle can be repeated. When each enzyme molecule is working at full capacity, the incoming substrate molecules must "wait their turn" for an empty active site. At this point, the enzyme is said to be under saturating conditions.

The rate at which an enzyme accepts and releases substrate molecules at substrate saturation is given by its turnover number. An enzyme's **turnover number** *is equal to the number of substrate molecules transformed per second by one molecule of enzyme under optimum conditions of temperature, pH, and saturation.* Table 21.2 gives turnover numbers for selected enzymes. Some enzymes have a much faster mode of operation than others.

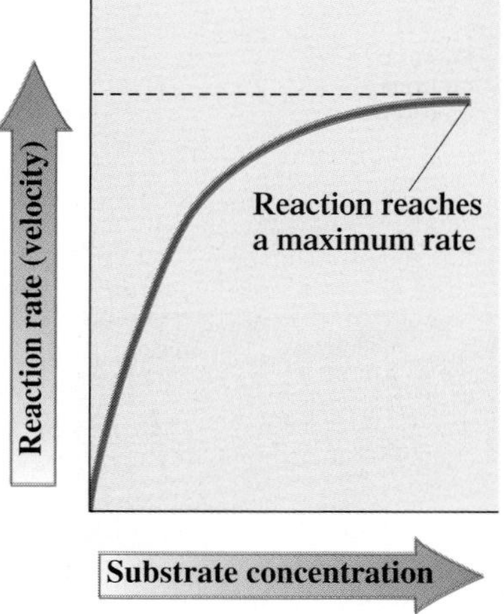

Figure 21.7
A graph showing the change in enzyme activity with a change in substrate concentration at constant temperature, pH, and enzyme concentration. Enzyme activity remains constant after a certain substrate concentration is reached.

• Enzyme Concentration

Because enzymes are not consumed in the reactions they catalyze, the cell usually keeps the number of enzymes low compared with the number of substrate molecules. This is efficient; the cell avoids paying the energy costs of synthesizing and maintaining a large work force of enzyme molecules. Thus, in general, the concentration of substrate in a reaction is much higher than that of the enzyme.

If the amount of substrate present is kept constant and the enzyme concentration is increased, the reaction rate increases because more substrate molecules can be accommodated in a given amount of time. A plot of enzyme activity versus enzyme concentration, at a constant substrate concentration that is high relative to enzyme concentration, is shown in Figure 21.8. The greater the enzyme concentration, the greater the reaction rate.

The accompanying Chemistry at a Glance reviews what we have said about enzyme activity.

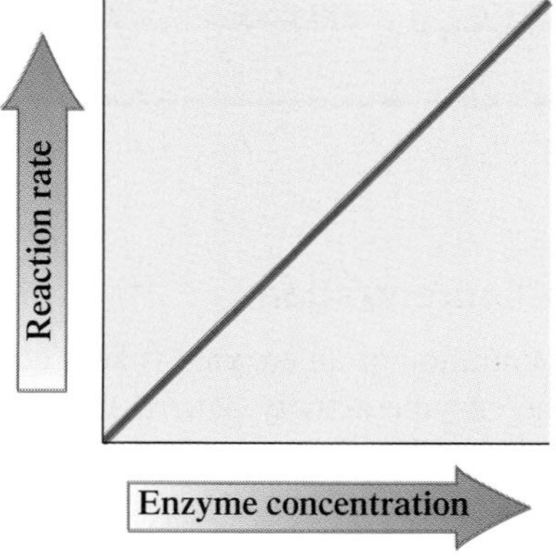

Figure 21.8
A graph showing the change in reaction rate with a change in enzyme concentration for an enzymatic reaction. Temperature, pH, and substrate concentration are constant. The substrate concentration is high relative to enzyme concentration.

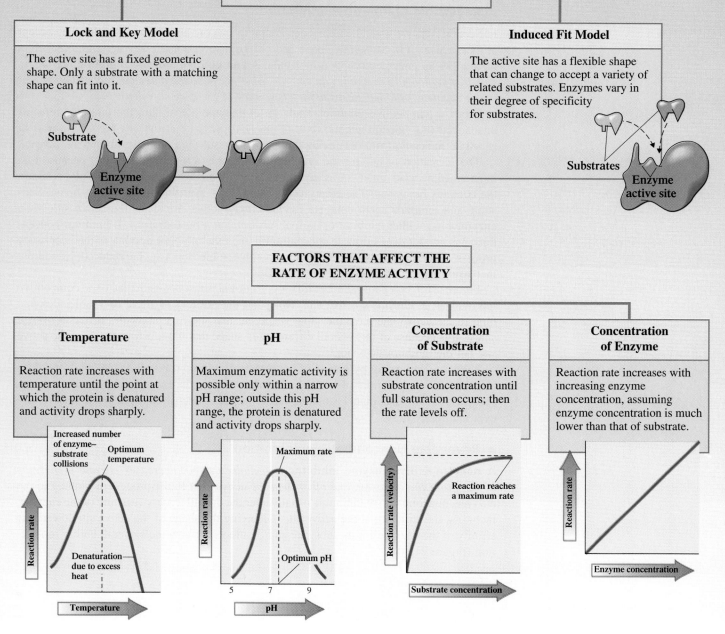

THE MECHANISM OF ENZYME ACTION

Formation of an enzyme–substrate complex as an intermediate species provides an alternative pathway, with lower activation energy, through which a reaction can occur.

Lock and Key Model

The active site has a fixed geometric shape. Only a substrate with a matching shape can fit into it.

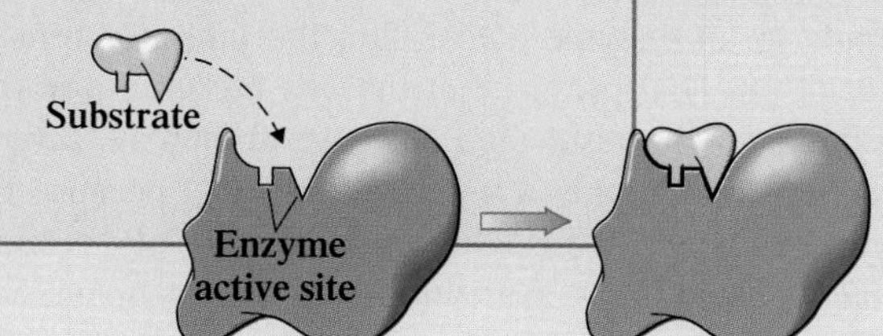

Induced Fit Model

The active site has a flexible shape that can change to accept a variety of related substrates. Enzymes vary in their degree of specificity for substrates.

FACTORS THAT AFFECT THE RATE OF ENZYME ACTIVITY

Temperature

Reaction rate increases with temperature until the point at which the protein is denatured and activity drops sharply.

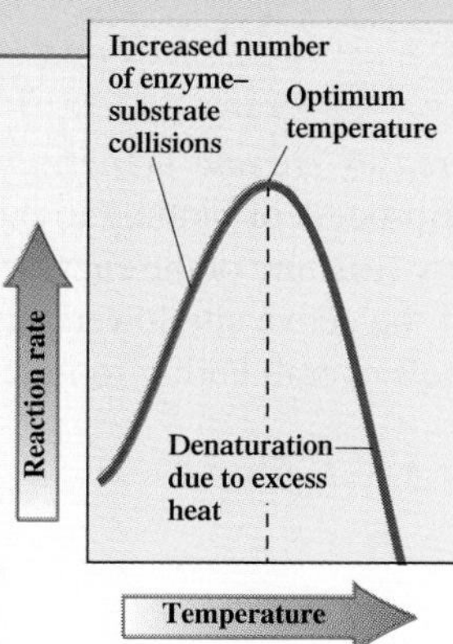

pH

Maximum enzymatic activity is possible only within a narrow pH range; outside this pH range, the protein is denatured and activity drops sharply.

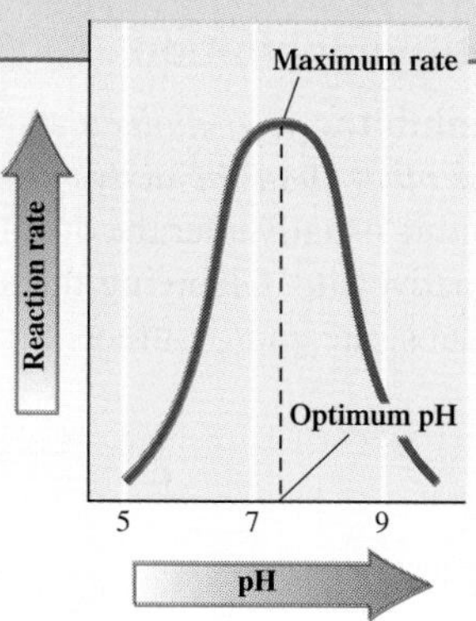

Concentration of Substrate

Reaction rate increases with substrate concentration until full saturation occurs; then the rate levels off.

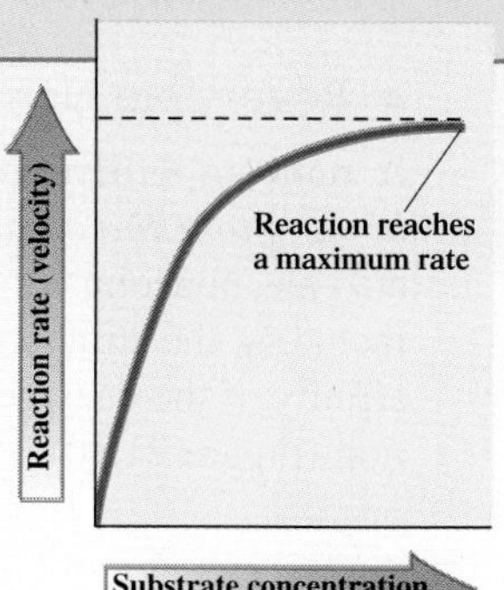

Concentration of Enzyme

Reaction rate increases with increasing enzyme concentration, assuming enzyme concentration is much lower than that of substrate.

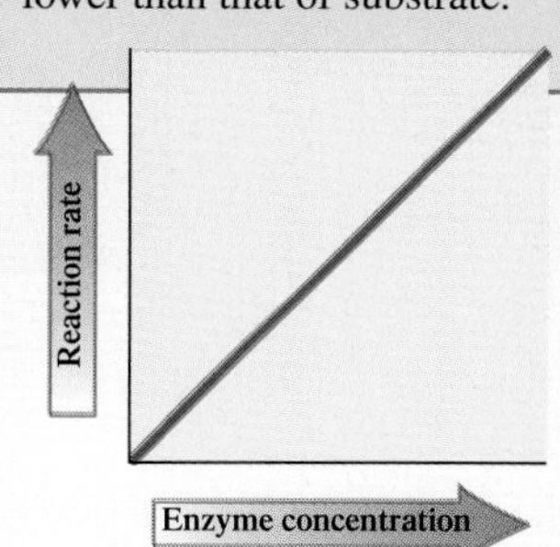

21.7 Enzyme Inhibition

The rates of enzyme-catalyzed reactions can be *decreased* by a group of substances called inhibitors. An **enzyme inhibitor** *is a substance that slows or stops the normal catalytic function of an enzyme by binding to it.* In this section, we consider three modes by which inhibition takes place: reversible competitive inhibition, reversible noncompetitive inhibition, and irreversible inhibition.

• Reversible Competitive Inhibition

In Section 21.5 we noted that enzymes are quite specific about the molecules they accept at their active sites. Molecular shape and charge distribution are key determining factors in whether an enzyme accepts a molecule. A **competitive enzyme inhibitor** *is a molecule that sufficiently resembles an enzyme substrate in shape and charge distribution that it can compete with the substrate for occupancy of the enzyme's active site.*

When a competitive inhibitor binds to an enzyme active site, the inhibitor remains unchanged (no reaction occurs), but its physical presence at the site prevents a normal substrate molecule from occupying the site. The result is a decrease in enzyme activity.

The formation of an enzyme–inhibitor complex is a reversible process, because it is maintained by weak interactions (hydrogen bonds, etc.). With time (fractions of a second), the complex breaks up. The empty active site is now available for a new occupant. Substrate and inhibitor again compete for the empty active site. Thus the active site of an enzyme binds either inhibitor or normal substrate on a random basis. If inhibitor concentration is greater than substrate concentration, then the inhibitor dominates the occupancy process. The reverse is also true. Competitive inhibition can be reduced by simply increasing the concentration of the substrate.

Figure 21.9 compares the binding of a normal substrate and that of a competitive inhibitor at an enzyme's active site. Note that the portions of these two molecules that bind to the active site have the same shape but that the two molecules differ in *overall* shape. It is because of this overall difference in shape that the substrate reacts at the active site but the inhibitor does not.

Numerous drugs act by means of competitive inhibition. For example, antihistamines are competitive inhibitors of histidine decarboxylation, the enzymatic reaction that converts histidine to histamine. Histamine causes the usual allergy and cold symptoms: watery eyes and runny nose.

• Reversible Noncompetitive Inhibition

A **noncompetitive enzyme inhibitor** *is a molecule that decreases enzyme activity by binding to a site on an enzyme other than the active site.* In the process of binding to the enzyme, noncompetitive inhibitors usually alter the overall tertiary structure of the enzyme, including the structure at the active site. Changing the shape of the active site lowers the affinity of the enzyme for its substrate and/or affects the enzyme's overall ability to function (Figure 21.10).

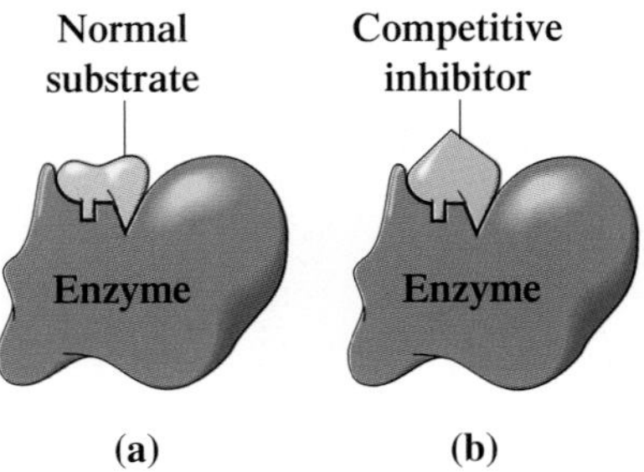

Figure 21.9
A comparison of an enzyme with a substrate at its active site (a) and an enzyme with a competitive inhibitor at its active site (b).

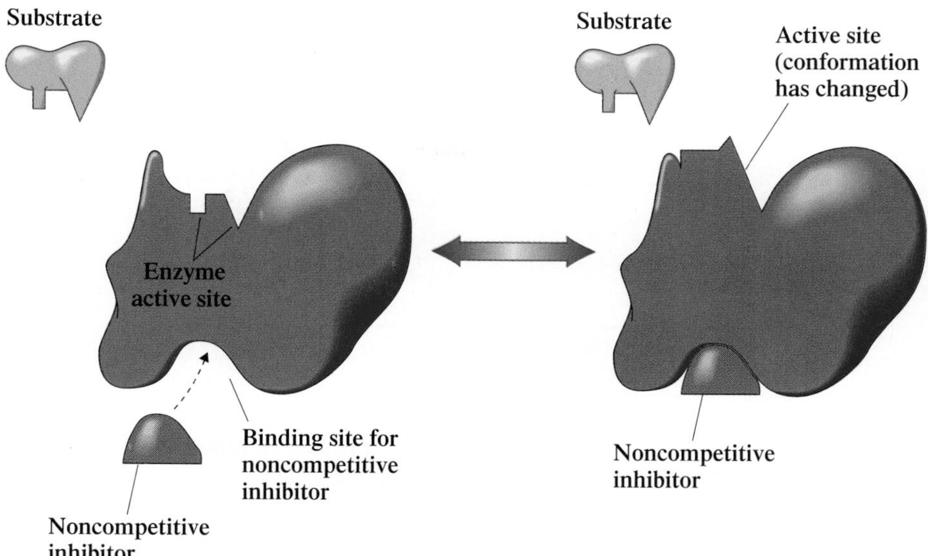

Figure 21.10
The interaction of an enzyme and a noncompetitive inhibitor induces conformation changes in the enzyme's active site.

Unlike the situation in competitive inhibition, increasing the concentration of substrate does not completely overcome the inhibitory effect in this case. However, lowering the concentration of a noncompetitive inhibitor sufficiently does free many enzymes, which return to normal activity.

Examples of noncompetitive inhibitors include the heavy metal ions Pb^{2+}, Ag^{+}, and Hg^{2+}. The binding sites for these ions are sulfhydryl ($-SH$) groups located away from the active site. Metal disulfide linkages are formed (Figure 21.11), an effect that disrupts secondary and tertiary structure.

• Irreversible Inhibition

An **irreversible enzyme inhibitor** *is a molecule that inactivates enzymes by forming a strong covalent bond to an amino acid side-chain group at the enzyme active site.* In general, such inhibitors do *not* have structures similar to that of the enzyme's normal substrate. The inhibitor–active site bond is sufficiently strong that addition of excess substrate does not reverse the inhibition process. Thus the enzyme is permanently deactivated. The actions of chemical warfare agents (nerve gases) and organophosphate insecticides are based on irreversible inhibition.

The accompanying Chemistry at a Glance summarizes what we have said about enzyme inhibition.

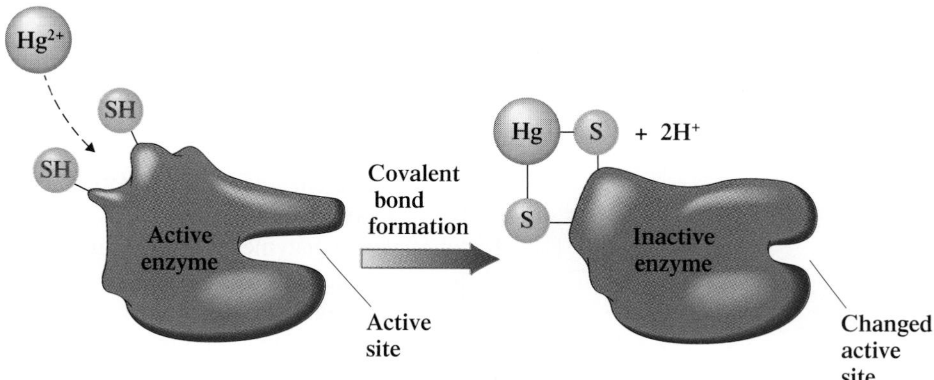

Figure 21.11
Heavy metal poisoning is an example of noncompetitive inhibition of an enzyme.

CHEMISTRY AT A GLANCE

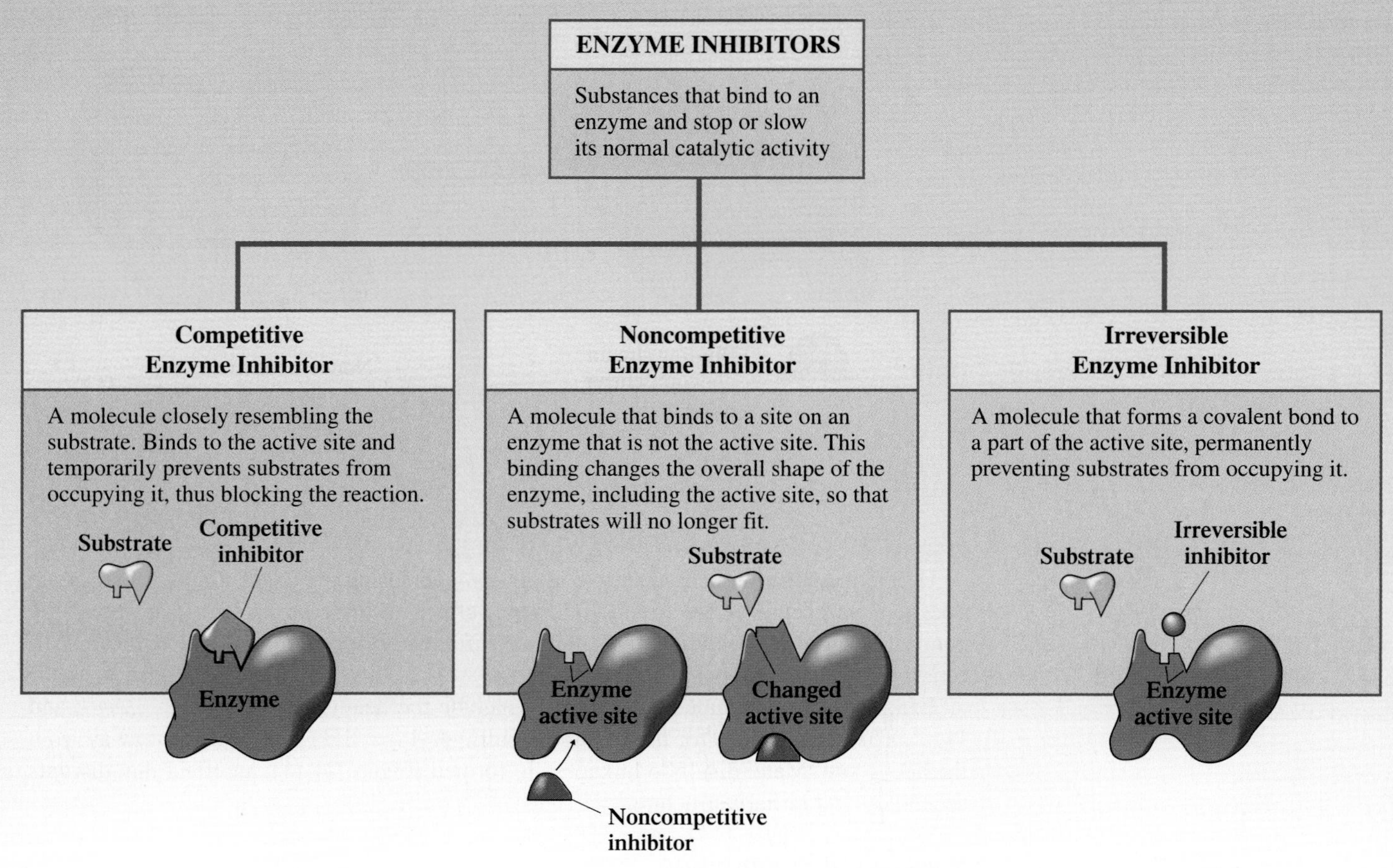

ENZYME INHIBITORS
Substances that bind to an enzyme and stop or slow its normal catalytic activity
Competitive Enzyme Inhibitor
A molecule closely resembling the substrate. Binds to the active site and temporarily prevents substrates from occupying it, thus blocking the reaction.
Substrate
Competitive inhibitor
Enzyme
Noncompetitive Enzyme Inhibitor
A molecule that binds to a site on an enzyme that is not the active site. This binding changes the overall shape of the enzyme, including the active site, so that substrates will no longer fit.
Substrate
Enzyme active site
Changed active site
Noncompetitive inhibitor
Irreversible Enzyme Inhibitor
A molecule that forms a covalent bond to a part of the active site, permanently preventing substrates from occupying it.
Substrate
Irreversible inhibitor
Enzyme active site

21.8 Regulation of Enzyme Activity: Allosteric Enzymes

In the previous section, we looked at the decrease in enzyme activity caused by inhibiting agents that were "foreign" to normal cells. In this section, we consider the regulation of enzyme activity by substances produced within a cell—that is, regulation by "normal" cell components. The concept of noncompetitive inhibition developed in the previous section will be part of our discussion.

• Allosteric Enzymes

Many, but not all, of the molecules responsible for regulating cellular processes are a special group of enzymes called *allosteric enzymes.* Characteristics of such enzymes are as follows:

1. All allosteric enzymes have quaternary structure; that is, they are composed of two or more protein chains.
2. All allosteric enzymes have two kinds of binding sites: those for substrate and those for regulators.
3. Active and regulatory binding sites are distinct from each other in both location and shape. Often the regulatory site is on one protein chain and the active site on another.

4. Binding of a molecule at the regulatory site causes changes in the overall three-dimensional structure of the enzyme, including structural changes at the active site.

• The term *allosteric* comes from the Greek *allo,* which means "other," and *stereos,* which means "site or space."

Thus an **allosteric enzyme** *is a molecule with two or more protein chains (quaternary structure) and two kinds of binding sites (substrate and regulator).*

• Some regulators of allosteric enzyme function are inhibitors (negative regulators), and some increase enzyme activity (positive regulators).

Substances that bind at regulatory sites of allosteric enzymes are called *regulators.* The binding of a *positive regulator* increases enzyme activity; the shape of the active site is changed such that it can more readily accept substrate. The binding of a *negative regulator* (a noncompetitive inhibitor) decreases enzyme activity; changes to the active site are such that substrate is less readily accepted.

• Feedback Control

One of the mechanisms by which allosteric enzyme activity is regulated is feedback control. **Feedback control** *is a process in which activation or inhibition of the first reaction in a reaction sequence is controlled by a product of the reaction sequence.*

• Most biochemical processes within cells take place in several steps rather than in a single step. A different enzyme is required for each step of the process.

To illustrate the feedback control mechanism, let us consider a biochemical process within a cell that occurs in several steps, each step catalyzed by a different enzyme.

$$\text{A} \xrightarrow{\text{Enzyme 1}} \text{B} \xrightarrow{\text{Enzyme 2}} \text{C} \xrightarrow{\text{Enzyme 3}} \text{D}$$

The product of each step is the substrate for the next enzyme.

What will happen in this reaction series if the final product (D) is a negative regulator of the first enzyme (enzyme 1)? At low concentrations of D, the reaction sequence proceeds rapidly. At higher concentrations of D, the activity of enzyme 1 becomes inhibited (by feedback), and eventually the activity stops. At the stopping point, there is sufficient D present in the cell to meet its needs. Later, when the concentration of D decreases through use in other cell reactions, the activity of enzyme 1 increases and more D is produced.

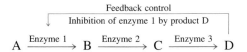

Feedback control

Inhibition of enzyme 1 by product D

$$\text{A} \xrightarrow{\text{Enzyme 1}} \text{B} \xrightarrow{\text{Enzyme 2}} \text{C} \xrightarrow{\text{Enzyme 3}} \text{D}$$

• The general term *allosteric control* is often used to describe a process in which a regulatory molecule that binds at one site in an enzyme influences substrate binding at the active site in the enzyme.

Feedback control is not the only mechanism by which an allosteric enzyme can be regulated; it is just one of the more common ways. Regulators of a particular allosteric enzyme may be products of entirely different pathways of reaction within the cell, or they may even be compounds produced outside the cell (hormones).

21.9 Regulation of Enzyme Activity: Zymogens

Because they would otherwise destroy the tissues that produce them, some enzymes are generated in an inactive form and then converted to an active form when it is needed. Most digestive and blood-clotting enzymes are of this type. These inactive enzyme forms are called zymogens or proenzymes. A **zymogen,** or **proenzyme,** *is an inactive precursor of an enzyme.*

• The names of zymogens can be recognized by the suffix *-ogen* or the prefix *pre-* or *pro-*.

Activation of a zymogen requires an enzyme-controlled reaction that either adds to the zymogen structure or removes some part of it. Such modification changes the three-dimensional structure (secondary and tertiary structure) of the zymogen protein, which affects active site conformation. For example, the zymogen pepsinogen is converted to the active enzyme pepsin in the stomach, where it then functions as a digestive enzyme. Pepsin would digest the tissues of the stomach wall if it were prematurely generated in active form. Pepsinogen activation involves removal of a polypeptide fragment from its structure (Figure 21.12).

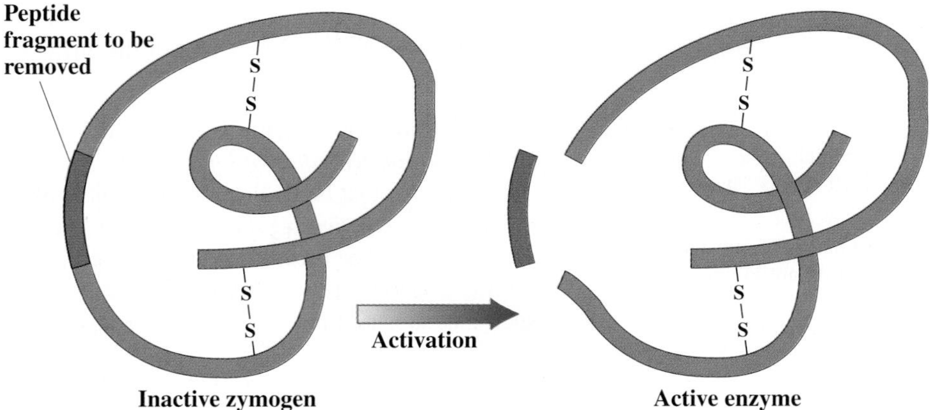

Figure 21.12
Conversion of a zymogen to an active enzyme usually involves removal of a polypeptide chain segment from the zymogen structure.

21.10 Antibiotics That Inhibit Enzyme Activity

Antibiotics *are substances that kill bacteria or inhibit their growth.* A good antibiotic exerts its action selectively on bacteria and does not affect the normal metabolism of the host organism. The two best-known families of antibiotics, the sulfa drugs and the penicillins, are inhibitors of specific enzymes essential to the life processes of bacteria.

• Sulfa Drugs

The antibiotic activity of a compound called sulfanilamide was first discovered in 1932 by a German chemist who was synthesizing sulfur-containing dyes, some of which were sulfanilamide derivatives. Since that time, scientists have synthesized many biologically active derivatives of sulfanilamide, collectively called sulfa drugs.

Sulfanilamide inhibits bacterial growth because it is structurally similar to PABA (*p*-aminobenzoic acid).

$$H_2N-C_6H_4-\overset{\overset{\displaystyle O}{\|}}{\underset{\underset{\displaystyle O}{\|}}{S}}-NH_2 \qquad\qquad H_2N-C_6H_4-\overset{\overset{\displaystyle O}{\|}}{C}-OH$$

Sulfanilamide

p-Aminobenzoic acid
(PABA)

Many bacteria need PABA in order to produce an important coenzyme, folic acid. Sulfanilamide acts as a competitive inhibitor to enzymes in the biosynthetic pathway for converting PABA into folic acid in these bacteria. Folic acid deficiency retards growth of the bacteria and can eventually kill them.

Sulfa drugs selectively inhibit only bacteria metabolism and growth, because humans absorb folic acid from their diet and thus do not use PABA for its synthesis. A few of the most common sulfa drugs and their structures are shown in Figure 21.13.

• Penicillins

Penicillin, one of the most widely used antibiotics, was accidentally discovered by Alexander Fleming in 1928 while he was working with cultures of an infectious staphylococcus bacterium. A decade later, the scientists Howard Flory and Ernst Chain isolated penicillin in pure form and proved its effectiveness as an antibiotic.

Several naturally occurring penicillins have now been isolated, and numerous derivatives of these substances have been synthetically produced. All have structures containing a four-membered β-lactam ring fused with a five-membered thiazolidine ring (Figure 21.14). As with sulfa drugs, derivatives of the basic structure differ from each other in the identity of a particular R group.

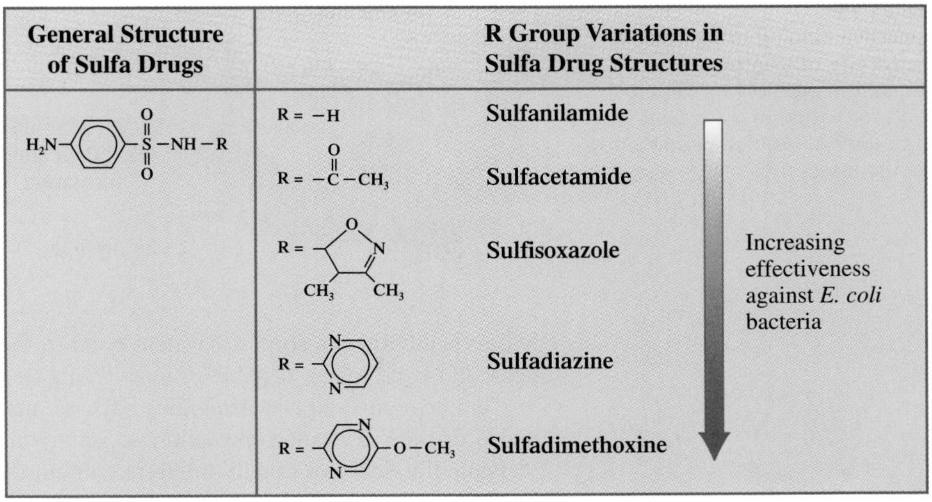

Figure 21.13
Structures of selected sulfa drugs in use today as antibiotics.

General Structure of Sulfa Drugs	R Group Variations in Sulfa Drug Structures	
$H_2N-C_6H_4-SO_2-NH-R$	R = –H	Sulfanilamide
	R = $-\overset{O}{\overset{\parallel}{C}}-CH_3$	Sulfacetamide
	R = dimethylisoxazole ring (CH_3, CH_3; O, N)	Sulfisoxazole
	R = pyrimidine ring (N, N)	Sulfadiazine
	R = pyrimidine ring (N, N)–O–CH_3	Sulfadimethoxine

Increasing effectiveness against *E. coli* bacteria (arrow from Sulfanilamide down to Sulfadimethoxine)

Penicillins inhibit *transpeptidase,* an enzyme that catalyzes the formation of peptide cross links between polysaccharide strands in bacteria cell walls. These cross links strengthen cell walls. A strong cell wall is necessary to protect the bacterium from lysis (breaking open). By inhibiting transpeptidase, penicillin prevents the formation of a strong cell wall. Any osmotic or mechanical shock then causes lysis, killing the bacterium.

Penicillin's unique action depends on two aspects of enzyme deactivation that we have discussed before: structural similarity to the enzyme's natural substrate and irreversible inhibition. Penicillin is *highly specific* in binding to the active site of transpeptidase. In this sense, it acts as a very selective competitive inhibitor. However, unlike a normal competitive inhibitor, once bound to the active site, the β-lactam ring opens as the highly reactive amide bond forms a covalent linkage bond to a critical serine residue required for normal catalytic action. The result is an irreversibly inhibited transpeptidase enzyme (Figure 21.15).

Some bacteria produce the enzyme *penicillinase,* which protects them from penicillin. Penicillinase selectively binds penicillin and catalyzes the opening of the β-lactam ring

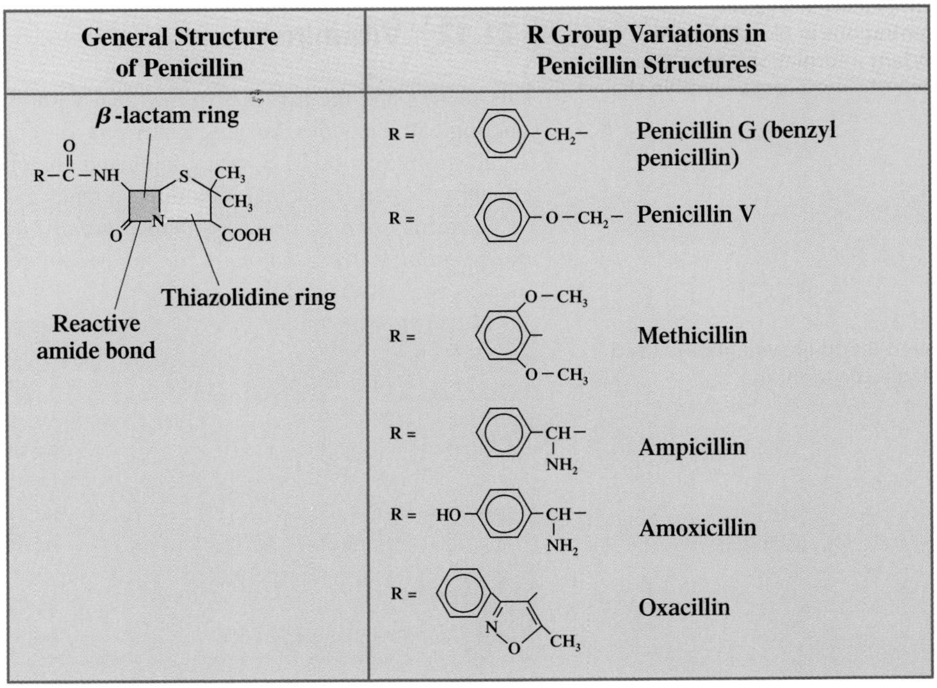

General Structure of Penicillin	R Group Variations in Penicillin Structures	
$R-\overset{O}{\overset{\parallel}{C}}-NH-$ (β-lactam ring fused to thiazolidine ring: S, CH_3, CH_3, N, O, COOH). Labels: β-lactam ring; Thiazolidine ring; Reactive amide bond	R = $C_6H_5-CH_2-$	Penicillin G (benzyl penicillin)
	R = $C_6H_5-O-CH_2-$	Penicillin V
	R = 2,6-dimethoxyphenyl (O–CH_3, O–CH_3)	Methicillin
	R = $C_6H_5-\underset{NH_2}{CH}-$	Ampicillin
	R = HO–$C_6H_4-\underset{NH_2}{CH}-$	Amoxicillin
	R = phenyl-isoxazole ring (N, O, CH_3)	Oxacillin

Figure 21.14
Structures of selected penicillins in use today as antibiotics.

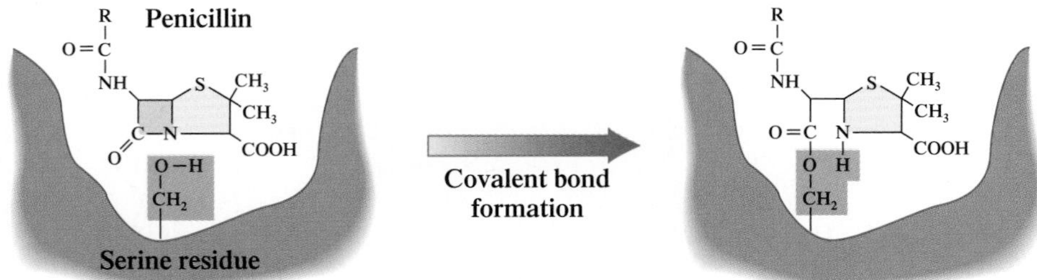

Figure 21.15
The selective binding of penicillin to the active site of transpeptidase. Subsequent irreversible inhibition through formation of a covalent bond to a serine residue permanently blocks the active site.

before penicillin can form a covalent bond to the enzyme. Once the ring is opened, the penicillin is no longer capable of inactivating transpeptidase.

Certain semi-synthetic penicillins such as methicillin and amoxicillin have been produced that are resistant to penicillinase activity and are thus clinically important.

Penicillin does not usually interfere with normal metabolism in humans because of its highly selective binding to bacterial transpeptidase. This selectivity makes penicillin an extremely useful antibiotic.

21.11 Medical Uses of Enzymes

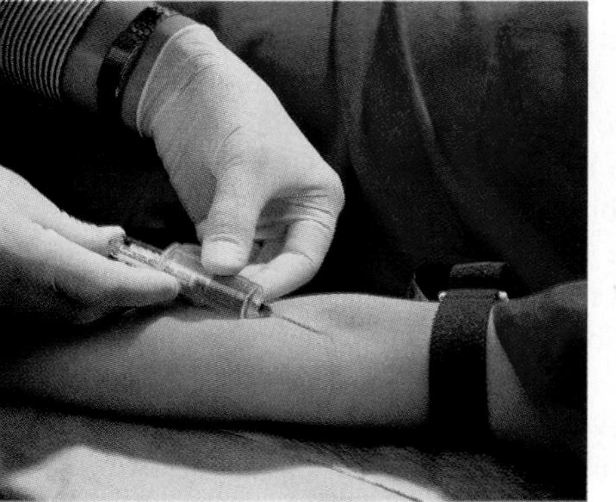

Drawing of a blood sample. Determination of enzyme concentrations in blood provides important information about the "state" of various organs within the body.

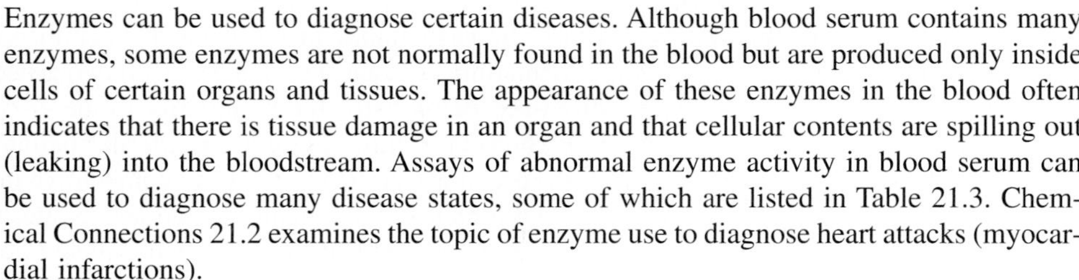

Enzymes can be used to diagnose certain diseases. Although blood serum contains many enzymes, some enzymes are not normally found in the blood but are produced only inside cells of certain organs and tissues. The appearance of these enzymes in the blood often indicates that there is tissue damage in an organ and that cellular contents are spilling out (leaking) into the bloodstream. Assays of abnormal enzyme activity in blood serum can be used to diagnose many disease states, some of which are listed in Table 21.3. Chemical Connections 21.2 examines the topic of enzyme use to diagnose heart attacks (myocardial infarctions).

Scientists have recently attempted to use enzymes in the treatment of diseases. A recent advance in treating heart attacks is the use of tissue plasminogen activator (TPA), which activates the enzyme plasminogen. When so activated, this enzyme dissolves blood clots in the heart and often provides immediate relief.

21.12 Vitamins

This section and the two that follow deal with vitamins. Vitamins are considered in conjunction with enzymes, because many enzymes contain vitamins as part of their structure. Recall, from Section 21.3, that conjugated enzymes have a protein part (apoenzyme) and a nonprotein part (cofactor). Vitamins, in many cases, are cofactors in conjugated enzymes.

Vitamins *are organic compounds that must be obtained from dietary sources and that are essential in trace amounts for the proper functioning of the human body.* Vitamins

Table 21.3
Selected Blood Enzyme Assays Used in Diagnostic Medicine

Enzyme	Condition indicated by abnormal level
lactate dehydrogenase (LDH)	heart disease, liver disease
creatine phosphokinase (CPK)	heart disease
aspartate transaminase (AST)	heart disease, liver disease, muscle damage
alanine transaminase (ALT)	heart disease, liver disease, muscle damage
gamma-glutamyl transpeptidase (GGTP)	heart disease, liver disease
alkaline phosphatase (ALP)	bone disease, liver disease

Chemical CONNECTIONS

21.2 Heart Attacks and Enzyme Analysis

The symptoms of a heart attack—that is, a *myocardial infarction (MI)*—include irregular breathing and pain in the left chest that may radiate to the neck, left shoulder, and arm. An initial diagnosis of an MI is based on these and other physical symptoms, and treatment is initiated on this basis.

Physicians then use enzyme analysis to confirm the diagnosis and to monitor the course of treatment. The blood levels of three enzymes are commonly assayed in MI situations: creatine phosphokinase (CPK), aspartate transaminase (AST), and lactate dehydrogenase (LDH). The CKP level rises and falls relatively rapidly after a heart attack, maximizing after about 30 hours at a level approximately six times normal. The AST level triples after about 40 hours. LDH, whose concentrations rise slowly, is used to monitor the later stages of the MI and to assess the extent of heart damage. The accompanying graph shows blood levels of these three enzymes as a function of time in an MI situation.

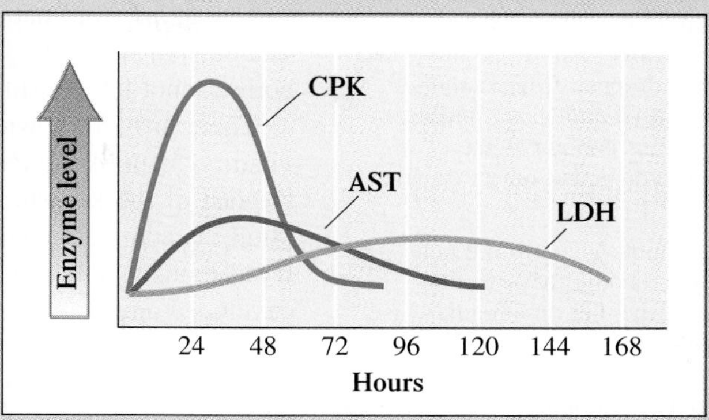

Further information concerning the seriousness of an MI is obtained by studying isoenzymes. **Isoenzymes** *are forms of the same enzyme with slightly different amino acid sequences.* Lactate dehydrogenase is a mixture of five isoenzymes, denoted LDH_1 to LDH_5. Creatine phosphokinase is a mixture of three isoenzymes, denoted CK-MM, CK-MB, and CK-BB.

Consideration of further details about the LDH isoenzymes shows how isoenzymes give information about whether a heart attack has occurred or not. Note from the following table that heart tissue is particularly high in LDH_1 and that liver and skeletal muscle are particularly high in LDH_5.

Tissue	LDH_1	LDH_2	LDH_3	LDH_4	LDH_5
brain	23	34	30	10	3
heart	50	36	9	3	2
kidney	28	34	21	11	6
liver	4	6	17	16	57
lung	10	20	30	25	15
serum	28	41	19	7	5
skeletal muscle	5	5	10	22	58

When a heart attack occurs, some heart muscle cells are damaged (destroyed), and their enzymes "leak" into the bloodstream. This changes the ratios among the various LDHs present in blood, because heart muscle is particularly high in LDH_1. The accompanying graph of an LDH isoenzyme assay for a heart attack victim shows that the LDH_1/LDH_2 ratio, which is normally less than 1 in blood serum, is now greater than 1.

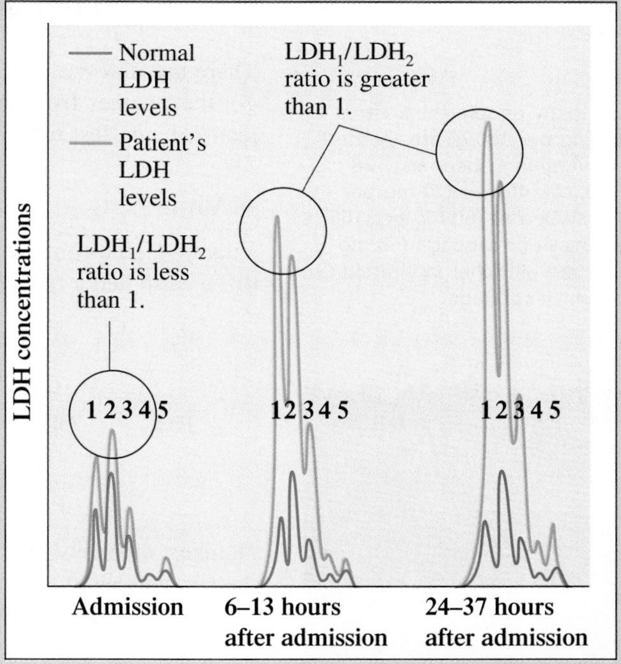

must be obtained from the diet, because the human body cannot synthesize them in the required amounts.

Vitamins differ from the major classes of foods (carbohydrates, lipids, and proteins) in the amount required; for vitamins it is *micro*gram or *milli*gram quantities per day compared with 50–200 grams per day for the major food categories. To illustrate the small amount of vitamins needed by the human body, consider the recommended daily allowance (RDA) of vitamin B_{12}, which is 2.0 micrograms per day for an adult. Just 1.0 gram of this vitamin could theoretically supply the daily needs of 500,000 people.

• The spelling of the term *vitamin* was originally *vitamine,* a word derived from the Latin *vita,* meaning "life," and from the fact that these substances were all thought to contain the *amine* functional group. When this supposition was found to be false, the final *e* was dropped from *vitamine,* and the term *vitamin* came into use. Some vitamins contain amine functional groups, but others do not.

• Water-soluble vitamins are not stored by the body; they must be provided in the diet on a regular basis. Fat-soluble vitamins are stored by the body, in fat tissues, and thus are needed only in periodic doses.

• Why is vitamin C called ascorbic *acid* when there is no carboxyl group (acid group) present in its structure? Vitamin C is a cyclic ester in which a carbon-1 carboxyl group has reacted with a carbon-4 hydroxyl group, forming the ring structure.

A well-balanced diet usually meets all the body's vitamin requirements. Supplemental vitamins are often required for women during pregnancy and for people recovering from certain illnesses.

One of the most common myths associated with the nutritional aspects of vitamins is that vitamins from natural sources are superior to synthetic vitamins. In truth, synthetic vitamins, manufactured in the laboratory, are identical to the vitamins found in foods. The body cannot tell the difference and gets the same benefits from either source.

There are 13 known vitamins, and scientists believe that the discovery of additional vitamins is unlikely. Despite searches for new vitamins, it has been over 50 years since the last of the known vitamins (B_{12}) was discovered. Strong evidence that the vitamin family is complete comes from the fact that many people have lived for years being fed, intravenously, solutions containing the known vitamins and nutrients, and they have not developed any known vitamin deficiency disease.

Solubility characteristics divide the vitamins into two major classes: the water-soluble vitamins and the fat (lipid)-soluble vitamins. Water-soluble vitamins must be constantly replenished in the body, because they are rapidly eliminated from the body in the urine. They are carried in the bloodstream, are needed in frequent, small doses, and are unlikely to be toxic except when taken in unusually large doses. The fat-soluble vitamins are found dissolved in lipid materials. They are, in general, carried in the blood by protein carriers, are stored in fat tissues, are needed in periodic doses, and are more likely to be toxic when consumed in excess of need.

21.13 Water-Soluble Vitamins

There are nine water-soluble vitamins: vitamin C and eight B vitamins. These vitamins got their names from the labels B and C on the test tubes in which they were first collected. Later, test tube B was found to contain more than one vitamin.

Although many people think citrus fruits (50 mg per 100 g) are the best source of vitamin C, peppers (128 mg per 100 g), cauliflower (70 mg per 100 g), strawberries (60 mg per 100 g), and spinach or cabbage (60 mg per 100 g) are all richer in vitamin C. Below, rows of cabbage.

• Vitamin C

Vitamin C, the compound ascorbic acid, has two active forms in the human body: an oxidized form and a reduced form.

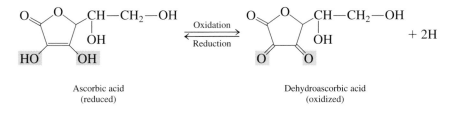

O, O, CH—CH_2—OH, OH, HO, OH ⇌ (Oxidation / Reduction) O, O, CH—CH_2—OH, OH, O, O + 2H

Ascorbic acid (reduced) — Dehydroascorbic acid (oxidized)

Humans, monkeys, apes, and guinea pigs are among the relatively few species that require dietary sources of vitamin C. Other species synthesize vitamin C from glucose.

The most completely characterized role of vitamin C is its function as a cosubstrate in the formation of the structural protein collagen (Section 20.15), which makes up much of the skin, ligaments, and tendons and also serves as the matrix on which bone and teeth are formed. Specifically, biosynthesis of the amino acids hydroxyproline and hydroxylysine (important in binding collagen fibers together) from proline and lysine requires the presence of both vitamin C and iron. Iron serves as a cofactor in the reaction, and vitamin C maintains iron in the oxidation state that allows it to function.

Vitamin C is also involved in the metabolism of several amino acids that end up being converted to the hormones norepinephrine and thyroxine. The adrenal glands contain a higher concentration of vitamin C than any other organ in the body.

An intake of 100 mg/day of vitamin C saturates all body tissues with the compound. After the tissues are saturated, all additional vitamin C is excreted. The RDA for vitamin C varies from country to country. It is 30 mg/day in Great Britain, 60 mg/day in the United States and Canada, and 75 mg/day in Germany.

• Pronunciation guidelines for the standard names of the B vitamins:
THIGH-ah-min
RYE-boh-flay-vin
NIGH-a-sin
FOLL-ate
PAN-toe-THEN-ick acid
BY-oh-tin

• Vitamin B

There are eight B vitamins. Much confusion exists concerning B vitamin nomenclature; many have both word names and number designations. A single set of standard names for these vitamins has recently been agreed (see Table 21.4). All of the B vitamins function as coenzymes. The conjugated enzymes in which they are present function in various steps of the metabolism of carbohydrates, lipids, and proteins. Table 21.4 gives the structures and enzyme involvement for the eight B vitamins.

21.14 Fat-Soluble Vitamins

There are four fat-soluble vitamins, denoted by the letters A, D, E, and K. Many of the functions of the fat-soluble vitamins involve processes that occur in plasma membranes. The structures of the fat-soluble vitamins are more hydrocarbon-like, with fewer functional groups, than the water-soluble vitamins. Their structures as a whole are nonpolar, which enhances their solubility in plasma membranes.

• Vitamin A

• Beta-carotene is a deep yellow (almost orange) compound. If a plant food is white or colorless, it possesses little or no vitamin A activity. Potatoes, pasta, and rice are foods in this category.

Three forms of vitamin A are active in the body: retinol (an alcohol), retinal (an aldehyde), and retinoic acid (an acid). In addition, the plant pigment beta-carotene can be converted to vitamin A by the liver. Figure 21.16 shows the structural similarities and differences among the various vitamin A species.

The best-known vitamin A function in the body involves retinal and its role in vision (see Chemical Connections 13.2). However, only one-thousandth of the vitamin A in the body is in the retina. Most vitamin A, as retinol and retinoic acid, is in the body's skin and linings. Vitamin A helps to maintain the integrity (smoothness) of mucous membranes.

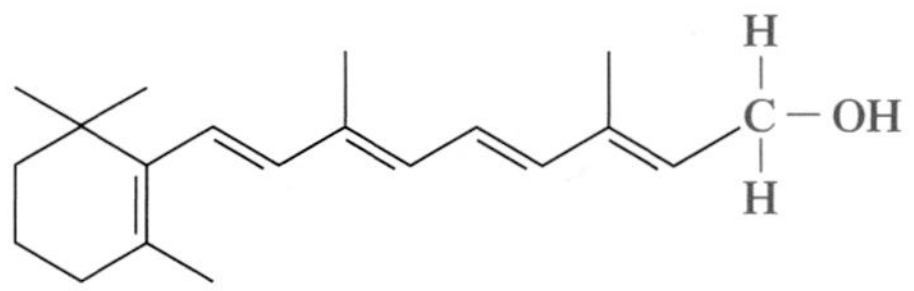

H
C—OH
H
Retinol

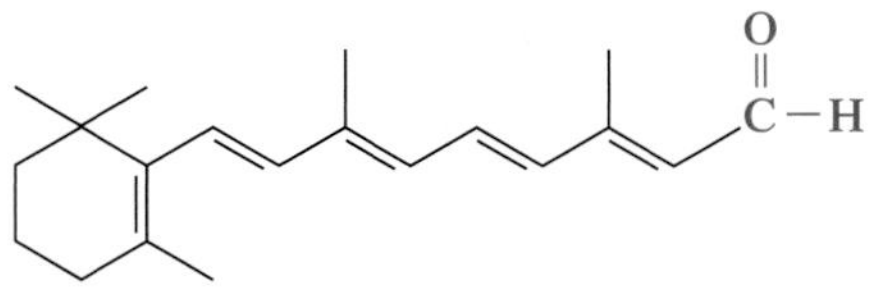

O
C—H
Retinal

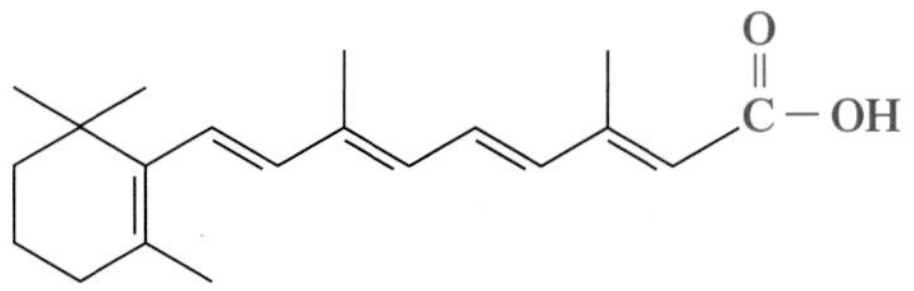

O
C—OH
Retinoic acid

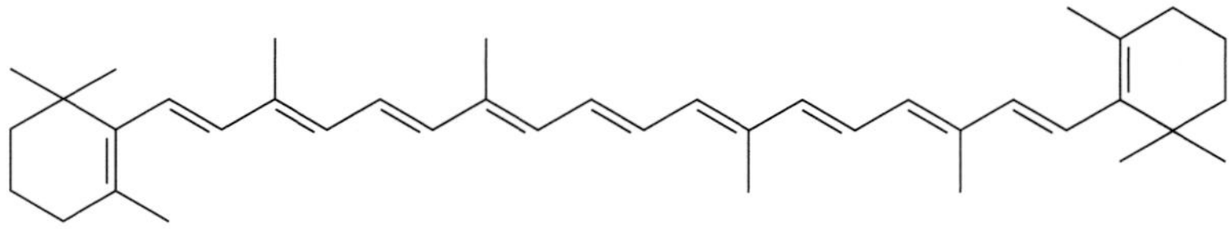

Beta-carotene

Figure 21.16
Structures of the three forms of vitamin A and the vitamin A precursor beta-carotene. Side-chain double bonds are *cis* double bonds.

Table 21.4
Structures and Main Functions of B Vitamins

Name	Main function in the body	Structure (active forms)
thiamin	part of TPP (thiamin pyrophosphate), a coenzyme used in energy metabolism; supports normal appetite and nervous system function	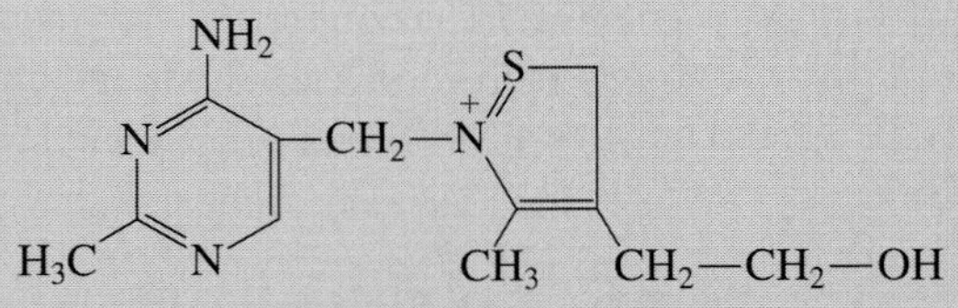
riboflavin	part of FMN (flavin mononucleotide) and FAD (flavin adenine dinucleotide), coenzymes used in energy metabolism; supports normal vision and skin health	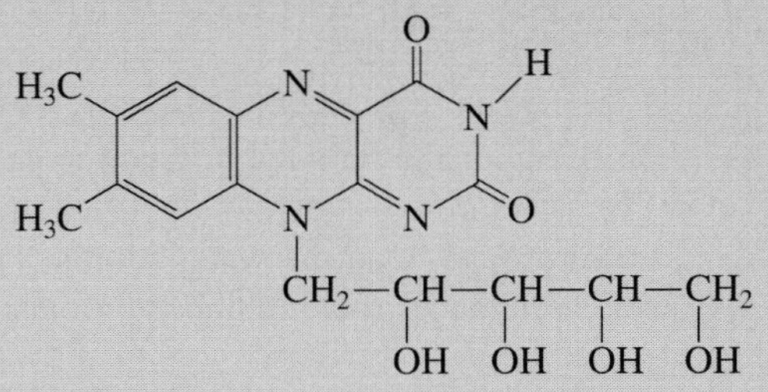
niacin[a] (2 compounds)	part of NAD (nicotinamide adenine dinucleotide) and NADP (its phosphorylated form), coenzymes used in energy production; supports health of skin, nervous system, and digestive system	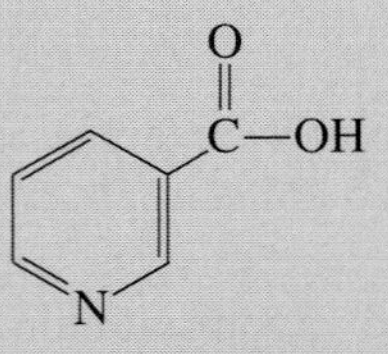 Nicotinic acid; 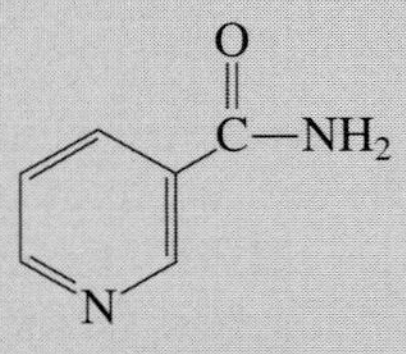 Nicotinamide
vitamin B_6[a] (3 compounds)	part of PLP (pyridoxal phosphate) and PMP (pyridoxamine phosphate), coenzymes used in amino acid, glycogen, and fatty acid metabolism; helps convert tryptophan to niacin; helps make red blood cells	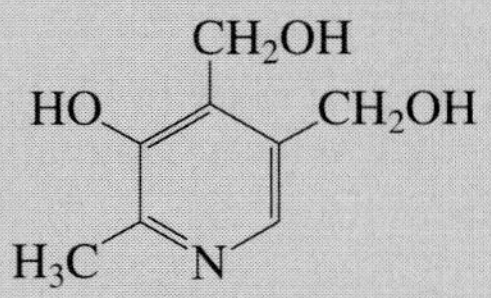 Pyridoxine; 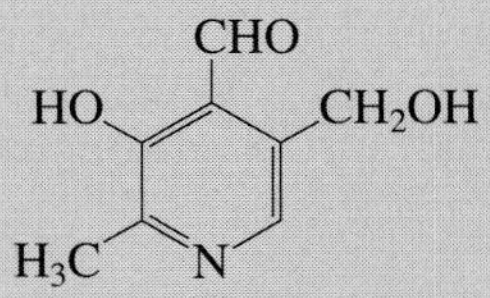 Pyridoxal; 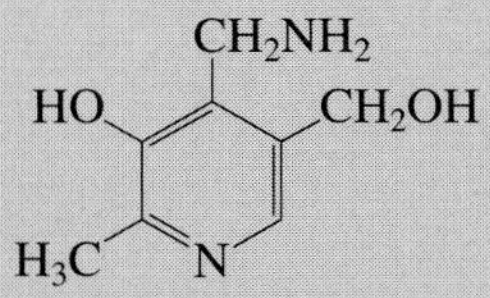 Pyridoxamine
folate	part of THF (tetrahydrofolate) and DHF (dihydrofolate), coenzymes used in new cell synthesis	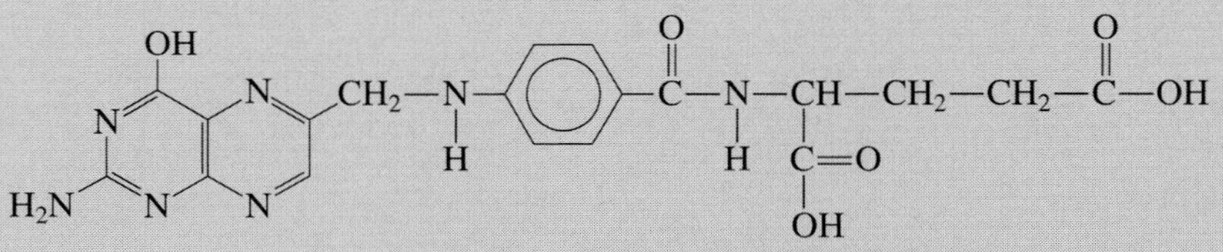
vitamin B_{12}	part of methylcobalamin and deoxyadenocobalamin, coenzymes used in new cell synthesis; helps to maintain nerve cells	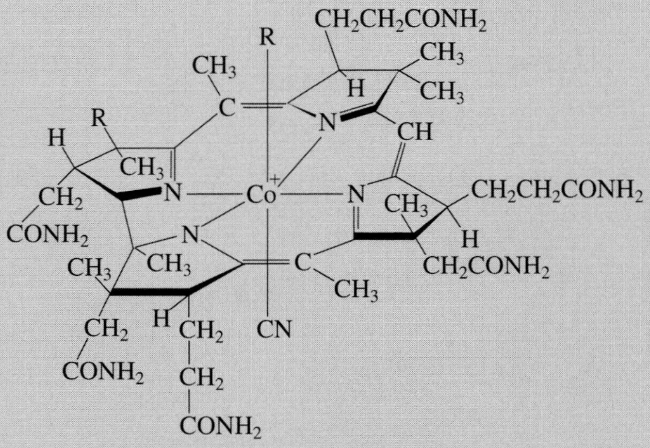
pantothenic acid	part of coenzyme A, which is used in energy production.	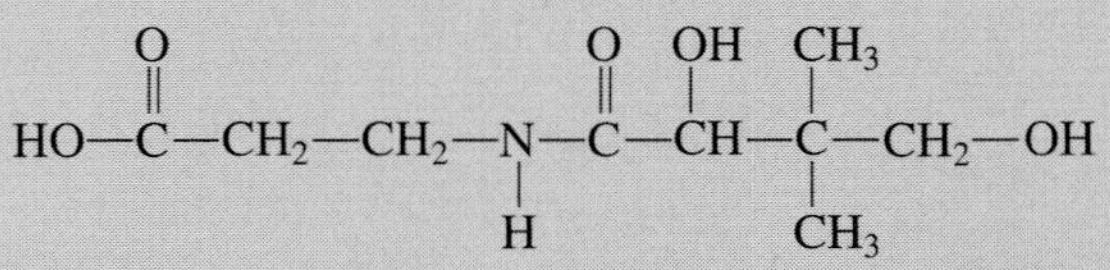 $HO-C(=O)-CH_2-CH_2-NH-C(=O)-CH(OH)-C(CH_3)_2-CH_2-OH$

[a] The different forms of niacin and vitamin B_6 are interconverted in the human body.

Table 21.4 (continued)
Structures and Main Functions of B Vitamins

Name	Main function in the body	Structure (active forms)
biotin	part of coenzymes used in energy production, fat synthesis, amino acid metabolism, and glycogen synthesis	bicyclic ring (S; two N—H; C=O) —CH_2—CH_2—CH_2—CH_2—C(=O)—OH

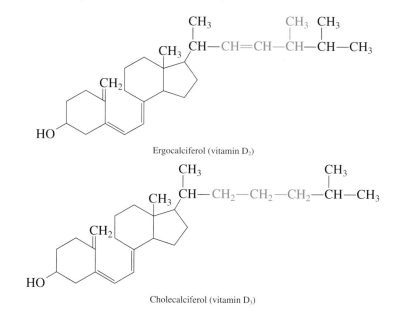

• Vitamin D

Two forms of vitamin D are active in the body: ergocalciferol (vitamin D_2) and cholecalciferol (vitamin D_3).

Ergocalciferol (vitamin D_2)

Cholecalciferol (vitamin D_3)

Cholecalciferol is sometimes called the "sunshine vitamin," because it can be synthesized in the skin by sunlight irradiation of 7-dehydrocholesterol (a normal metabolite of cholesterol present in the skin).

The quantity of vitamin D synthesized by exposure of the skin to sunlight (ultraviolet radiation) varies with latitude, the length of exposure time, and skin pigmentation. (Darker-skinned people synthesize less vitamin D because the pigmentation filters out ultraviolet light.)

• Milk is enriched in vitamin D by exposure to ultraviolet light. Cholesterol in milk is converted to cholecalciferol (vitamin D) by ultraviolet light.

The special function of vitamin D is to promote normal bone mineralization. Its presence facilitates correct calcium and phosphorus concentrations in the blood that bathes bones, to be deposited as the bones harden (mineralize). Acting as a hormone, it promotes phosphorus and calcium absorption in the intestine.

• Vitamin E

The compound alpha-tocopherol is the most biologically active of several different forms of vitamin E.

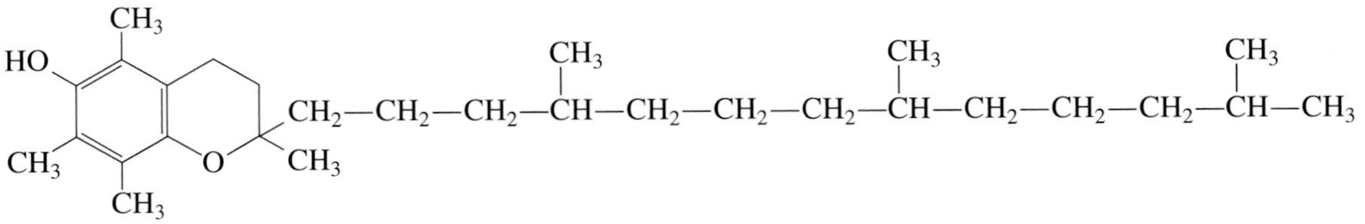

HO, CH_3, CH_3, CH_3 (ring substituents); O; CH_3

$CH_2—CH_2—CH_2—CH(CH_3)—CH_2—CH_2—CH_2—CH(CH_3)—CH_2—CH_2—CH_2—CH(CH_3)—CH_3$

Other forms of vitamin E are beta-, gamma-, and delta-tocopherol, which differ from alpha-tocopherol only in the number and position of methyl groups on the ring.

Plant oils (margarine, salad dressings, shortenings), green and leafy vegetables, and whole-grain products are sources of vitamin E.

• A most important location in the human body where vitamin E exerts its antioxidant effect is in the lungs, where the exposure of cells to oxygen (and air pollutants) is greatest.

The primary function of vitamin E in the body is as an antioxidant, a compound that protects other compounds from oxidation by being oxidized itself. Vitamin E is particularly important in preventing the oxidation of polyunsaturated fatty acids (Section 19.2) in phospholipids present in plasma membranes. It also protects vitamin A from oxidation.

• Vitamin K

Vitamin K occurs in several forms that differ only in the length and unsaturation of their side chains. The K_1 form is found naturally in dark green, leafy vegetables, and the K_2 form is synthesized by bacteria that normally grow in the colon. Typically, about half of the body's vitamin K comes from the diet and half from synthesis by intestinal bacteria.

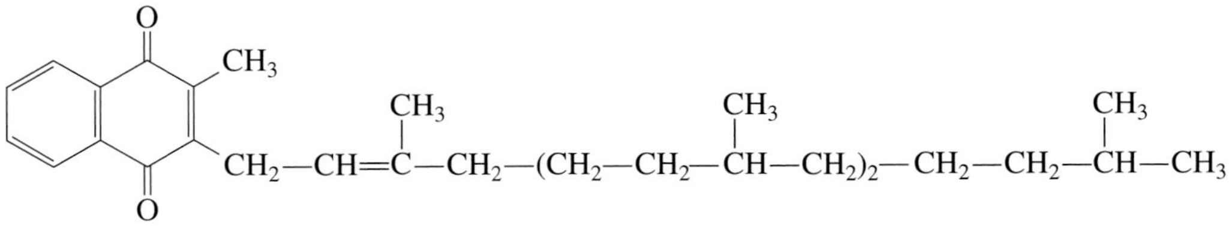

O, O, CH_3

$CH_2—CH{=}C(CH_3)—CH_2—(CH_2—CH_2—CH(CH_3)—CH_2)_2—CH_2—CH_2—CH(CH_3)—CH_3$

Vitamin K_1

O, O, CH_3

$CH_2—CH{=}C(CH_3)—CH_2—(CH_2—CH{=}C(CH_3)—CH_2)_5—CH_2—CH{=}C(CH_3)—CH_3$

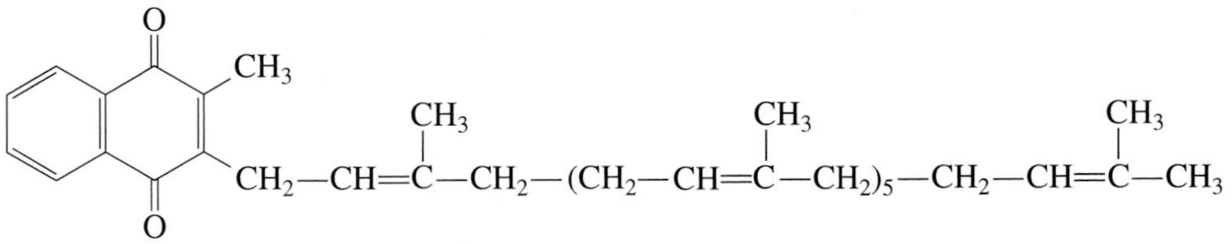

Vitamin K_2

Compounds with vitamin K activity are essential for the formation of prothrombin and at least five other proteins involved in the regulation of blood clotting. Vitamin K is sometimes given to presurgical patients to ensure adequate prothrombin levels and prevent hemorrhaging. Vitamin K is also required for the biosynthesis of some other proteins found in the plasma, bone, and kidney.

21.15 Minerals

Proteins, carbohydrates, and lipids in the diet supply the body with adequate amounts of carbon, hydrogen, oxygen, nitrogen, and sulfur. The amino acids of protein are the primary source of nitrogen and sulfur. All three types of dietary compounds are good sources of carbon, hydrogen, and oxygen. For proper functioning, the human body also needs a number of other elements. Our need for these additional elements, which is *small but absolute,* is met by dietary intake of minerals. In a dietary context, **minerals** *are inorganic compounds containing elements other than C, H, O, N, and S that are needed for the proper functioning of the human body.*

• A more general definition of a mineral is any compound produced naturally as a result of inorganic (rather than organic) processes.

Three important functions of dietary minerals are as (1) cofactors needed to complete active enzymes (Section 21.3), (2) principal components of bone structure, and (3) regulators of fluid movement in the body.

• Vitamins are dietary *organic* substances needed by the body in small amounts. Minerals are dietary *inorganic* substances needed by the body in small amounts. One of the functions of both vitamins and minerals is to act as enzyme cofactors.

Mineral nutrients differ from organic nutrients in that the body uses them, in general, in the form of ions (monatomic or polyatomic—Section 4.10), the form in which they are usually ingested.

Dietary minerals may be divided into two classes: major minerals and trace minerals. The distinction between these two types of minerals is based on quantity needed rather than on importance—all are absolutely essential. **Major minerals** *are dietary minerals needed by the body in amounts greater than 100 mg/day.* There are six major minerals: calcium, phosphorus, magnesium, sodium, potassium, and chlorine. **Trace minerals** *are dietary minerals needed by the body in amounts less than 20 mg/day.* Figure 21.17 shows actual gram amounts of various minerals present in the human body.

• Major Minerals

• Minerals are not present in the body in "elemental" form. The statement "red meats are a good source of iron" does not mean that red meats contain metallic (elemental) iron. Rather, the meats contain compounds in which iron is present.

Four of the six major minerals are metals (Ca, K, Na, and Mg) and two are nonmetals (P and Cl). All six are present in body fluids as either monatomic ions (Ca^{2+}, K^+, Na^+, Mg^{2+}, Cl^-) or polyatomic ions (HPO_4^{2-} or $H_2PO_4^-$). One of the functions of these ions in body fluids is to act as electrolytes (Section 10.14). Chemical Connections 10.4 contains information on cellular electrolyte concentrations and electrolytic function. Table 21.5 summarizes additional functions of major minerals and also gives their dietary sources.

• Trace Minerals

Some information about trace minerals was presented previously (see Chemical Connections 3.3). If all the trace minerals were extracted from a human body, the material obtained would not even fill a teaspoon (5 g). Yet each of the trace elements performs some vital role for which no substitute is known. The trace minerals may be divided into the following three categories.

1. Trace minerals for which recommended dietary allowances (RDA) have been established. These minerals, which have been extensively studied, involve the elements iron, zinc, iodine, and selenium.

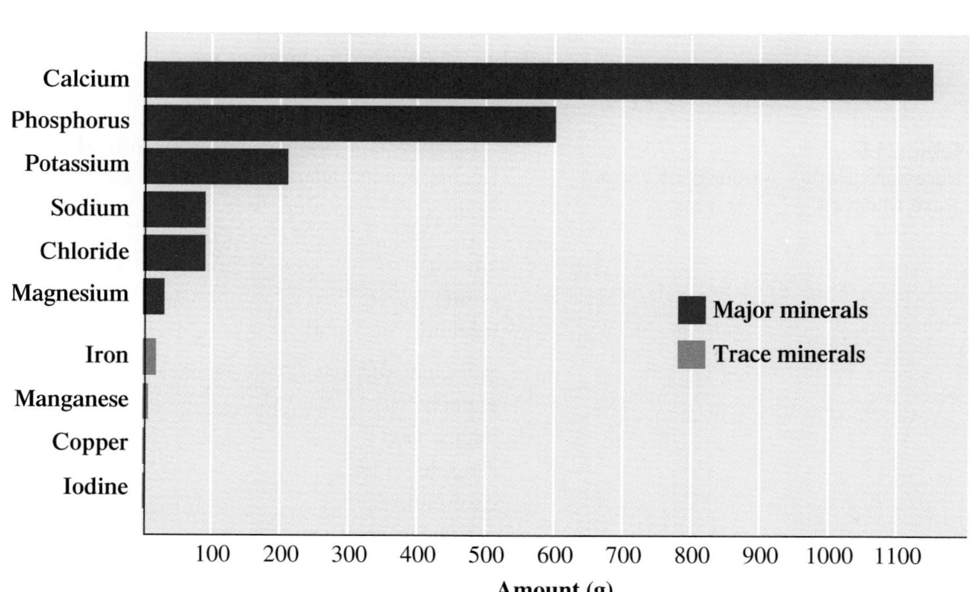

Figure 21.17
Typical amounts, in grams, of minerals present in a 60-kg (132-lb) person. Major minerals are those present in amounts greater than 5 grams (a teaspoon); only four trace minerals are shown. Calcium and phosphorus are the only two minerals present in amounts greater than 1 lb (454 g).

Table 21.5
Nonelectrolyte Functions of Major Minerals and Their Dietary Sources

Micronutrient	Nonelectrolyte functions in the body	Significant sources
calcium	an important mineral of bones and teeth; also involved in normal muscle (including heart muscle) contraction and relaxation, proper nerve functioning, blood clotting, blood pressure, and immune defenses	milk and milk products, small fish (with bones), tofu (bean curd), greens (broccoli, chard), legumes
phosphorus	an important mineral of bones and teeth; part of every plasma membrane; important in genetic material (DNA and RNA), as part of phospholipids, in energy transfer, and in buffer systems that maintain acid–base balance	all animal tissues
potassium	facilitates many reactions, including the making of protein; supports cell integrity; assists in the transmission of nerve impulses and the contraction of muscles, including the heart	all whole foods: meats, milk, fruits, vegetables, grains, legumes
sodium	assists in nerve impulse transmission	table salt, soy sauce; moderate quantities in whole, unprocessed foods; large amounts in processed foods
chlorine	part of the hydrochloric acid found in the stomach	table salt, soy sauce; moderate quantities in whole, unprocessed foods; large amounts in processed foods
magnesium	involved in bone mineralization, the building of protein, enzyme action, normal muscular contraction, transmission of nerve impulses, and maintenance of teeth	nuts, legumes, whole grains, dark green vegetables, seafood, chocolate, cocoa

2. Trace minerals for which tentative ranges for safe and adequate daily dietary intakes have been published. These minerals, which still need more study before RDAs are established, involve the elements copper, manganese, fluorine, chromium, and molybdenum.
3. Trace minerals that are recognized as essential for some animals but have not been definitely proved to be required for human beings. These minerals involve the elements arsenic, nickel, silicon, and boron. Many other trace elements are presently under study to determine whether they, too, play indispensable roles in the body.

Table 21.6 gives the amount needed in the diet of the various trace minerals.

Table 21.6
Necessary Dietary Amounts of Various Trace Minerals

Recommended Dietary Allowances (RDA)	
iron	10 mg (men), 15 mg (women before menopause) 10 mg (women after menopause)
zinc	15 mg (men), 12 mg (women)
iodine	0.15 mg
selenium	0.070 mg (men), 0.055 mg (women)
Safe and Adequate Daily Dietary Intakes	
copper	1.5 to 3.0 mg
manganese	2.5 to 5.0 mg
fluorine	1.5 to 4.0 mg
chromium	0.050 to 0.200 mg
molybdenum	0.075 to 0.250 mg

Chemical CONNECTIONS

21.3 Iron: The Most Abundant Trace Mineral in the Human Body

By far the most abundant of the trace minerals is iron. Most of the body's iron is found as a component of the proteins hemoglobin and myoglobin, where it functions in the transport of oxygen. Hemoglobin is the oxygen carrier in red blood cells, and myoglobin is the oxygen reservoir in muscle cells. Iron-deficient blood has less oxygen-carrying capacity and often cannot completely meet the body's energy needs. Energy deficiency—tiredness and apathy—is one of the symptoms of iron deficiency.

Iron deficiency is a worldwide problem. Millions of people are unknowingly deficient. Even in the United States and Canada, about 20% of women and 3% of men have this problem; some 8% of women and 1% of men are anemic, experiencing fatigue, weakness, apathy, and headaches.

Inadequate intake of iron, either from malnutrition or from high consumption of the wrong foods, is the usual cause of iron deficiency. In the Western world, the cause is often displacement of iron-rich foods by foods high in sugar and fat.

About 80% of the iron in the body is in the blood, so iron losses are greatest whenever blood is lost. Blood loss from menstruation makes a woman's need for iron nearly twice as great as a man's. Also, women usually consume less food than men do. These two factors—lower intake and higher loss—cause iron deficiency to be likelier in women than in men. The iron RDA (recommended dietary allowance) is 10 mg per day for adult males and older women. For women of childbearing age, the RDA is 15 mg. This amount is necessary to replace menstrual loss and to provide the extra iron needed during pregnancy.

Iron deficiency may also be caused by poor absorption of ingested iron. A normal, healthy person absorbs about 2%–10% of the iron in vegetables and about 10%–30% in meats. About 40% of the iron in meat, fish, and poultry is bound into molecules of heme, the iron-containing part of hemoglobin and myoglobin. Heme iron is much more readily absorbed (23%) than nonheme iron (2%–10%). (See the accompanying graphs.)

Cooking utensils can enhance the amount of iron delivered by the diet. The iron content of 100 g of spaghetti sauce simmered in a glass dish is 3 mg, but it is 87 mg when the sauce is cooked in an unenameled iron skillet. Even in the short time it takes to scramble eggs, their iron content can be tripled by cooking them in an iron pan.

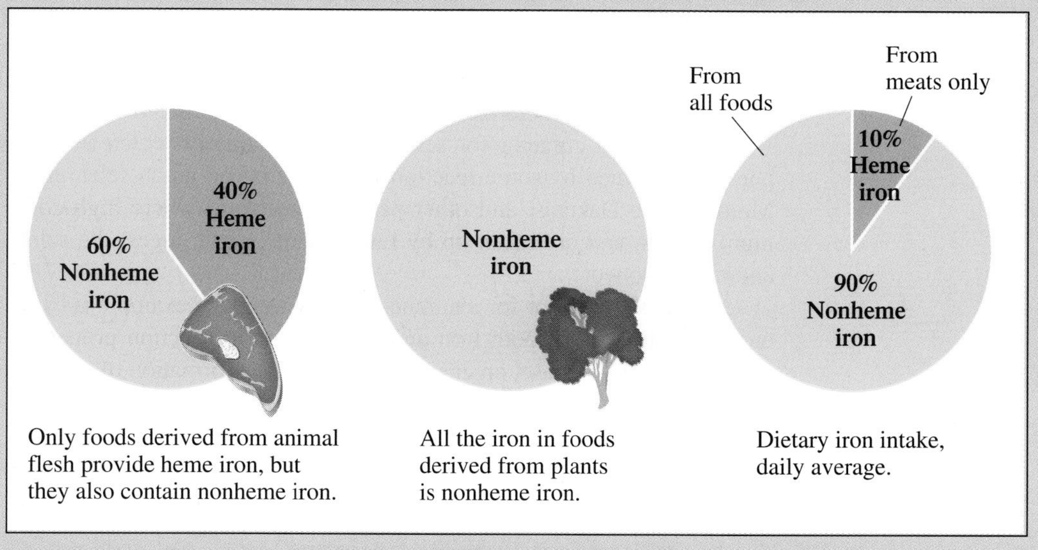

Only foods derived from animal flesh provide heme iron, but they also contain nonheme iron.

All the iron in foods derived from plants is nonheme iron.

Dietary iron intake, daily average.

● Iron is absorbed by the body as Fe^{2+}. Most iron in foods is present as Fe^{3+}. To increase absorption of iron, the body must reduce the Fe^{3+} to Fe^{2+}. Vitamin C aids in this reduction. Dietary iron supplements frequently contain vitamin C as an added ingredient.

Iron is the best-known and most abundant trace mineral. In addition to its involvement in oxygen transport, iron acts as a cofactor for enzymes that contain heme groups, such as cytochrome oxidase, catalase, and peroxidase. These enzymes participate in oxidation–reduction reactions, the iron atoms acting as electron acceptors or donors as they alternate between ferrous (Fe^{2+}) and ferric (Fe^{3+}) states. The heme structure, in which iron is present as Fe^{2+}, is

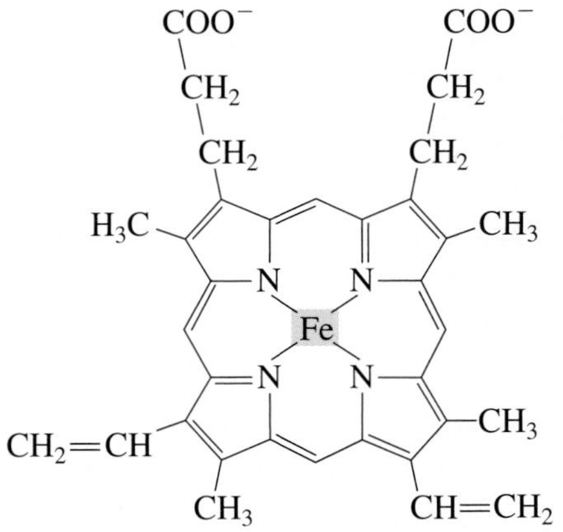

Heme

Chemical Connections 21.3 gives further information about iron in the human body.

Zinc is an essential cofactor for over 100 enzymes, including carboxypeptidase, alcohol dehydrogenase, RNA and DNA polymerases, and carbonic anhydrase. Zinc also is required for the proper formation of taste buds and smell-sensing receptors of the nasal passages.

Iodine is found in the hormone thyroxine. Each thyroxine molecule contains four iodine atoms.

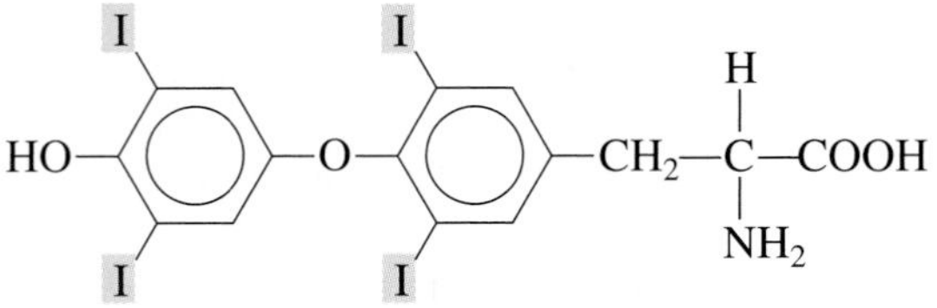

• Seafood is a good natural source of iodine. Seaweed is particularly rich in iodine. Although iodine is found in cereals, vegetables, and milk, the amount varies widely, depending on the amount of iodine in the soil.

Thyroxine is a key regulator of basal metabolic rate; the more thyroxine produced, the higher the basal metabolic rate. The best source of iodine (iodide ions) is iodized salt (which includes small amounts of KI or NaI). The clinical manifestation of iodine deficiency is *goiter,* characterized by an enlarged thyroid gland.

Selenium is a cofactor for several enzymes, one of which functions to destroy toxic peroxides formed by some reactions in the cell. Too much selenium is toxic. In parts of Montana, the Dakotas, and other parts of the world where high concentrations of selenium salts in soil are taken up by range plants, cattle ingest the selenium while grazing and are poisoned.

Copper is a cofactor for a number of enzymes. One example is cytochrome oxidase, an enzyme that requires both iron and copper ions to function properly. Another important enzyme is lysyl oxidase, an enzyme that catalyzes formation of cross links that strengthen collagen protein. Animals lacking copper in their diets suffer from weak arterial walls that tend to rupture easily. Very few cases of copper deficiency are seen in the population at large. However, an *excess* of copper in the tissues is associated with *Wilson's disease,* which can be treated with penicillamine, an agent that promotes the excretion of copper.

Concepts to Remember

Enzyme. Enzymes are highly specialized protein molecules that act as biological catalysts. Enzymes have common names that provide information about their function rather than their structure. The suffix *-ase* is characteristic of an enzyme name.

Enzyme structure. Simple enzymes are composed only of protein (amino acids). Conjugated enzymes have a nonprotein portion (cofactor) in addition to a protein portion (apoenzyme). Cofactors may be small organic molecules (coenzymes) or inorganic ions.

Enzyme classification. There are six classes of enzymes based on function: oxidoreductases, transferases, hydrolases, lyases, isomerases, and ligases.

Enzyme active site. An enzyme active site is the relatively small part of the enzyme that is actually involved in catalysis. It is where substrate binds to the enzyme.

Lock-and-key model of enzyme activity. The active site in an enzyme has a fixed, rigid geometrical conformation. Only substrates with a complementary geometry can be accommodated at the active site.

Induced-fit model of enzyme activity. The active site in an enzyme can undergo small changes in shape or geometry in order to accommodate a substrate.

Enzyme activity. Enzyme activity is a measure of the rate at which an enzyme converts substrate to products. Four factors that affect enzyme activity are temperature, pH, substrate concentration, and enzyme concentration.

Enzyme inhibition. An enzyme inhibitor slows or stops the normal catalytic function of an enzyme by binding to it. Three modes of inhibition are reversible competitive inhibition, reversible noncompetitive inhibition, and irreversible inhibition.

Allosteric enzyme. An allosteric enzyme is an enzyme with two or more protein chains and two kinds of binding sites (for substrate and regulator).

Zymogen. A zymogen is an inactive precursor of an enzyme; the zymogen is activated by a chemical reaction that removes or adds to part of its structure.

Vitamin. A vitamin is an organic compound necessary in small amounts for the normal growth of humans and some animals. Vitamins must be obtained from dietary sources, because they cannot be synthesized in the body.

Water-soluble vitamins. Vitamin C and the eight B vitamins are the water-soluble vitamins. Vitamin C is essential for the proper formation of bones and teeth. All eight B vitamins function as coenzymes.

Fat-soluble vitamins. The four fat-soluble vitamins are vitamins A, D, E, and K. The best-known function of vitamin A is its role in vision. Vitamin D is essential for the proper use of calcium and phosphorus to form bones and teeth. The primary function of vitamin E is as an antioxidant. Vitamin K is essential in the regulation of blood clotting.

Minerals. Minerals are inorganic substances necessary in small quantities for the proper functioning of the human body; they must be obtained from dietary sources. Minerals needed in relatively large amounts are called major minerals, and those needed in very small amounts are called trace minerals.

Key Reactions and Equations

1. Conversion of an apoenzyme to an active enzyme (Section 21.3)

Apoenzyme + cofactor ⟶ active enzyme (holoenzyme)

2. Mechanism of enzyme action (Section 21.4)

Enzyme + substrate ⟶ enzyme–substrate complex

Enzyme–substrate complex ⟶ enzyme + product

3. Enzyme inhibition (Section 21.7)

Enzyme + inhibitor ⟶ inactive enzyme

4. Conversion of a zymogen to an active enzyme (Section 21.9)

$$\text{Zymogen} \xrightarrow{\text{Enzyme}} \text{active enzyme} + \text{peptide fragment}$$

Key Terms

Active site (21.4)
Allosteric enzyme (21.8)
Antibiotics (21.10)
Apoenzyme (21.3)
Coenzyme (21.3)
Cofactor (21.3)
Competitive enzyme inhibitor (21.7)
Conjugated enzymes (21.3)
Enzyme activity (21.6)
Enzyme inhibitor (21.7)
Enzyme–substrate complex (21.4)
Enzymes (21.1)
Feedback control (21.8)
Irreversible enzyme inhibitor (21.7)
Optimum pH (21.6)
Optimum temperature (21.6)
Major minerals (21.15)
Minerals (21.15)
Noncompetitive enzyme inhibitor (21.7)
Proenzyme (21.9)
Simple enzymes (21.3)
Substrate (21.2)
Trace minerals (21.15)
Turnover number (21.6)
Vitamins (21.12)
Zymogen (21.9)

Exercises and Problems

The members of each pair of problems in this section test similar material.

Importance of Enzymes (Section 21.1)

21.1 What is the general role of enzymes in the human body?

21.2 Why does the body need so many different enzymes?

21.3 List two ways in which enzymes differ from inorganic laboratory catalysts.

21.4 Occasionally, we refer to the "delicate" nature of enzymes. Explain why this adjective is appropriate.

Enzyme Nomenclature (Section 21.2)

21.5 Which of the following substances are enzymes?

a. Sucrase b. Galactose
c. Trypsin d. Xylulose reductase

21.6 Which of the following substances are enzymes?

a. Sucrose b. Pepsin
c. Glutamine synthetase d. Cellulase

21.7 Predict the function of the following enzymes.

a. Pyruvate carboxylase

b. Alcohol dehydrogenase

c. L-Amino acid reductase

d. Maltase

21.8 Predict the function of the following enzymes.

a. Cytochrome oxidase

b. *Cis–trans* isomerase

c. Succinate dehydrogenase

d. Lactase

21.9 Suggest a name for an enzyme that catalyzes each of the following reactions.

a. Hydrolysis of sucrose

b. Decarboxylation of pyruvate

c. Isomerization of glucose

d. Removal of hydrogen from lactate

21.10 Suggest a name for an enzyme that catalyzes each of the following reactions.

a. Hydrolysis of lactose

b. Oxidation of nitrite

c. Decarboxylation of citrate

d. Reduction of oxalate

21.11 Give the name of the substrate on which each of the following enzymes acts.

a. Pyruvate carboxylase b. Galactase

c. Alcohol dehydrogenase d. L-Amino acid reductase

21.12 Give the name of the substrate on which each of the following enzymes acts.

a. Cytochrome oxidase b. Lactase

c. Succinate dehydrogenase d. Tyrosine kinase

21.13 To which of the six major classes of enzymes does each of the following belong?

a. Mutase b. Dehydratase

c. Carboxylase d. Kinase

21.14 To which of the six major classes of enzymes does each of the following belong?

a. Protease b. Racemase

c. Dehydrogenase d. Synthase

21.15 To which of the six major classes of enzymes does the enzyme that catalyzes each of the following reactions belong?

a. A *cis* double bond is converted to a *trans* double bond.

b. An alcohol is dehydrated to form a compound with a double bond.

c. An amino group is transferred from one substrate to another.

d. An ester linkage is hydrolyzed.

21.16 To which of the six major classes of enzymes does the enzyme that catalyzes each of the following reactions belong?

a. An L isomer is converted to a D isomer.

b. A phosphate group is transferred from one substrate to another.

c. An amide linkage is hydrolyzed.

d. Hydrolysis of a carbohydrate to monosaccharides occurs.

21.17 Identify the enzyme needed in each of the following reactions as an isomerase, a decarboxylase, a dehydrogenase, a lipase, or a phosphatase.

a.
$$CH_3-\overset{\overset{O}{\|}}{C}-COOH \rightarrow CH_3-\overset{\overset{O}{\|}}{C}-H + CO_2$$

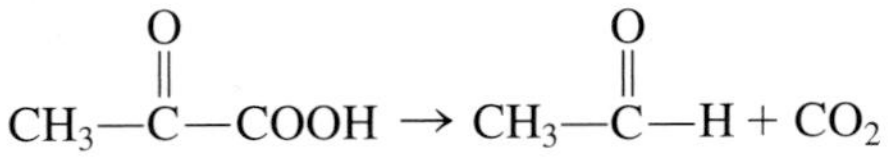

b.
$$\begin{array}{l} CH_2-O-\overset{\overset{O}{\|}}{C}-R \\ | \\ CH-O-\overset{\overset{O}{\|}}{C}-R + 3H_2O \rightarrow \\ | \\ CH_2-O-\overset{\overset{O}{\|}}{C}-R \end{array} \quad \begin{array}{l} CH_2-OH \\ | \\ CH-OH \\ | \\ CH_2-OH \end{array} + 3R-COOH$$

c.
$$\begin{array}{l} \overset{+}{H_3N}-CH-COO^- + H_2O \rightarrow \\ \qquad\quad | \\ \qquad\;\; CH_2 \\ \qquad\quad | \\ \qquad OPO_3^{2-} \end{array} \quad \begin{array}{l} \overset{+}{H_3N}-CH-COO^- + HPO_4^{2-} \\ \qquad\quad | \\ \qquad\;\; CH_2 \\ \qquad\quad | \\ \qquad\;\; OH \end{array}$$

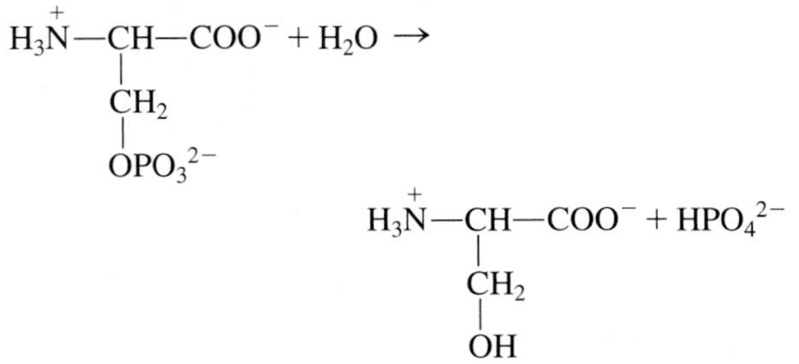

d.
$$\begin{array}{l} \qquad\;\; OH \\ \qquad\;\; | \\ CH_3-CH-COOH + NAD^+ \rightarrow \end{array} \quad CH_3-\overset{\overset{O}{\|}}{C}-COOH + NADH + H^+$$

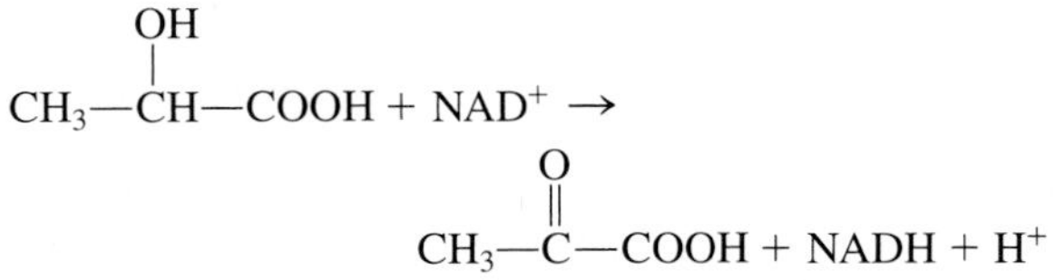

21.18 Identify the enzyme needed in each of the following reactions as an isomerase, a decarboxylase, a dehydrogenase, a protease, or a phosphatase.

a.
$$CH_3-\overset{\overset{O}{\|}}{C}-\overset{\overset{O}{\|}}{C}-OH \rightarrow CH_3-\overset{\overset{O}{\|}}{C}-H + CO_2$$

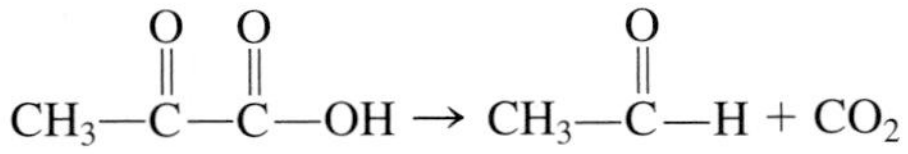

b.
$$\begin{array}{r} CHO \\ | \\ HC-OH \\ | \\ HO-CH \\ | \\ HC-OH \\ | \\ HO-CH \\ | \\ CH_2OPO_3^{2-} \end{array} \rightarrow \begin{array}{r} CH_2OH \\ | \\ C{=}O \\ | \\ HO-CH \\ | \\ HC-OH \\ | \\ HO-CH \\ | \\ CH_2OPO_3^{2-} \end{array} + 2H$$

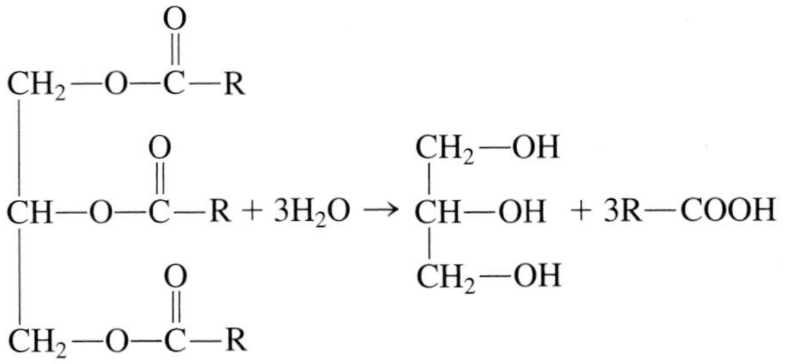

c.
$$HO-\overset{\overset{O}{\|}}{C}-CH_2-CH_2-\overset{\overset{O}{\|}}{C}-OH \rightarrow HO-\overset{\overset{O}{\|}}{C}-CH{=}CH-\overset{\overset{O}{\|}}{C}-OH + 2H$$

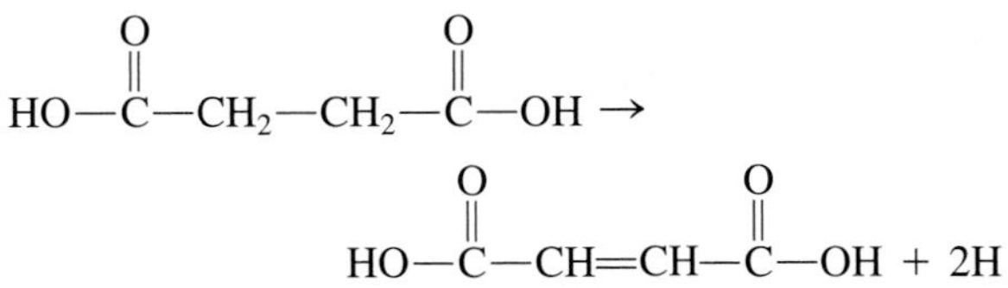

d.
$$\begin{array}{l} \overset{+}{H_3N}-CH-\overset{\overset{O}{\|}}{C}-NH-CH-COO^- + H_2O \rightarrow \\ \qquad\quad | \qquad\qquad\quad\; | \\ \qquad\; CH_3 \qquad\qquad CH_3 \end{array} \quad \begin{array}{l} 2\ \overset{+}{H_3N}-CH-COO^- \\ \qquad\qquad | \\ \qquad\quad\; CH_3 \end{array}$$

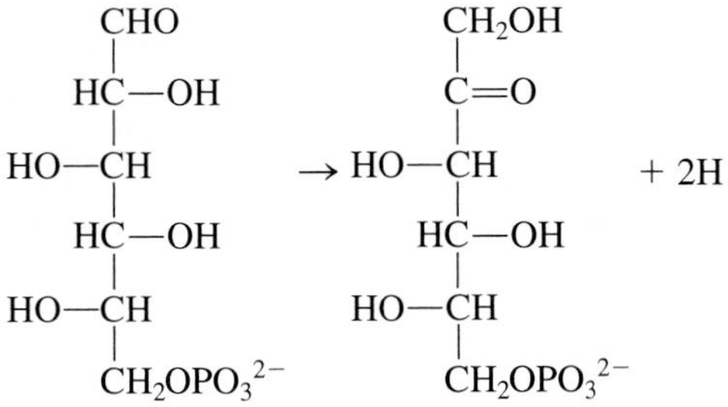

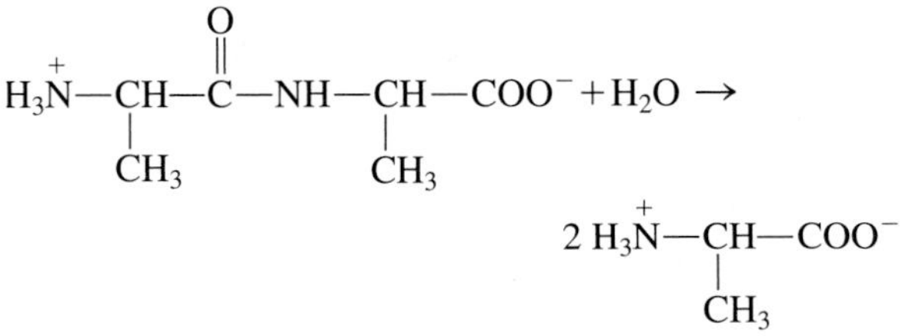

Enzyme Structure (Section 21.3)

21.19 Indicate whether each of the following phrases describes a simple or a conjugated enzyme.

a. An enzyme that has both a protein and a nonprotein portion

b. An enzyme that requires Mg^{2+} ion for activity

c. An enzyme in which only amino acids are present

d. An enzyme in which a cofactor is present

21.20 Indicate whether each of the following phrases describes a simple or a conjugated enzyme.

a. An enzyme that contains a carbohydrate portion

b. An enzyme that contains only protein

c. A holoenzyme

d. An enzyme that has a vitamin as part of its structure

21.21 What is the difference between a cofactor and a coenzyme?

21.22 All coenzymes are cofactors, but not all cofactors are coenzymes. Explain this statement.

21.23 Why are cofactors present in most enzymes?

21.24 What is the difference between an apoenzyme and a holoenzyme?

Models of Enzyme Action (Section 21.4)

21.25 What is an enzyme active site?

21.26 What is an enzyme–substrate complex?

21.27 How does the lock-and-key model of enzyme action explain the highly specific way some enzymes select a substrate?

21.28 How does the induced-fit model of enzyme action explain the broad specificities of some enzymes?

21.29 What types of forces hold a substrate at an enzyme active site?

21.30 The forces that hold a substrate at an enzyme active site are not covalent bonds. Explain why not.

Enzyme Specificity (Section 21.5)

21.31 Define the following terms dealing with enzyme specificity.

a. Absolute specificity

b. Linkage specificity

21.32 Define the following terms dealing with enzyme specificity.

a. Group specificity

b. Stereochemical specificity

21.33 Which type(s) of enzyme specificity are best accounted for by the lock-and-key model of enzyme action?

21.34 Which type(s) of enzyme specificity are best accounted for by the induced-fit model of enzyme action?

21.35 Which enzyme in each of the following pairs would be most limited in its catalytic scope?

a. Absolute-specific enzyme or group-specific enzyme

b. Stereochemical-specific enzyme or linkage-specific enzyme

21.36 Which enzyme in each of the following pairs would be most limited in its catalytic scope?

a. Linkage-specific enzyme or absolute-specific enzyme

b. Group-specific enzyme or stereochemical-specific enzyme

Factors That Affect Enzyme Activity (Section 21.6)

21.37 Temperature affects enzymatic reaction rates in two ways. An increase in temperature can accelerate the rate of a reaction or it can stop the reaction. Explain each of these effects.

21.38 Define the optimum temperature for an enzyme.

21.39 Explain why all enzymes do not possess the same optimum pH.

21.40 Why does an enzyme lose activity when the pH is drastically changed from the optimum pH?

21.41 Draw a graph that shows the effect of increasing substrate concentration on the rate of an enzyme-catalyzed reaction (at constant temperature, pH, and enzyme concentration).

21.42 Draw a graph that shows the effect of increasing enzyme concentration on the rate of an enzyme-catalyzed reaction (at constant temperature, pH, and substrate concentration).

21.43 In an enzyme-catalyzed reaction, all of the enzyme active sites are saturated by substrate molecules at a certain substrate concentration. What happens to the rate of the reaction when the substrate concentration is doubled?

21.44 What is an enzyme turnover number?

Enzyme Inhibition (Section 21.7)

21.45 In competitive inhibition, can both the inhibitor and the substrate bind to an enzyme at the same time? Explain your answer.

21.46 Compare the sites where competitive and noncompetitive inhibitors bind to enzymes.

21.47 Indicate whether each of the following statements describes a reversible competitive inhibitor, a reversible noncompetitive inhibitor, or an irreversible inhibitor. More than one answer may apply.

a. Both inhibitor and substrate bind at the active site on a random basis.

b. The inhibitor effect cannot be reversed by the addition of more substrate.

c. Inhibitor structure does not have to resemble substrate structure.

d. The inhibitor can bind to the enzyme at the same time as substrate.

21.48 Indicate whether each of the following statements describes a reversible competitive inhibitor, a reversible noncompetitive inhibitor, or an irreversible inhibitor. More than one answer may apply.

a. It bonds covalently to the enzyme active site.

b. The inhibitor effect can be reversed by the addition of more substrate.

c. Inhibitor structure must be somewhat similar to that of substrate.

d. The inhibitor cannot bind to the enzyme at the same time as substrate.

Regulation of Enzyme Activity (Sections 21.8 and 21.9)

21.49 What is an allosteric enzyme?

21.50 What is a regulator molecule?

21.51 What is feedback control?

21.52 What is the difference between positive and negative feedback to an allosteric enzyme?

21.53 What is a zymogen?

21.54 What is a proenzyme?

21.55 Why are some digestive enzymes produced in a zymogen form?

21.56 How is a zymogen activated?

Antibiotics That Inhibit Enzyme Activity (Section 21.10)

21.57 By what mechanism do sulfa drugs kill bacteria?

21.58 By what mechanism do penicillins kill bacteria?

21.59 Why is penicillin toxic to bacteria but not to higher organisms?

21.60 What amino acid in transpeptidase forms a covalent bond to penicillin?

Vitamins (Sections 21.12–21.14)

21.61 What is a vitamin?

21.62 List a way in which vitamins differ from carbohydrates, fats, and proteins (the major classes of food).

21.63 How many known vitamins are there?

21.64 Scientists do not expect to discover any additional vitamins. Explain why.

21.65 Indicate whether each of the following is a fat-soluble or a water-soluble vitamin.

a. Vitamin K
b. Vitamin B_{12}
c. Vitamin C
d. Thiamin

21.66 Indicate whether each of the following is a fat-soluble or a water-soluble vitamin.

a. Vitamin A
b. Vitamin B_6
c. Vitamin E
d. Riboflavin

21.67 Indicate whether each of the vitamins in Problem 21.65 would be likely or unlikely to be toxic when consumed in excess.

21.68 Indicate whether each of the vitamins in Problem 21.66 would be likely or unlikely to be toxic when consumed in excess.

21.69 Describe the most completely characterized role in the body for vitamin C.

21.70 Structurally, how do the oxidized and reduced forms of vitamin C differ?

21.71 What is the dominant function within the body of the B vitamins as a group?

21.72 With the help of Table 21.4, indicate whether each of the following B vitamins exists in more than one structural form.

a. Folate
b. Niacin
c. Vitamin B_6
d. Biotin

21.73 With the help of Table 21.4, identify the B vitamin or vitamins to which each of the following characterizations applies.

a. Is water-soluble
b. Has the most complex structure
c. Contains a metal atom as part of its structure
d. Contains sulfur as part of its structure

21.74 With the help of Table 21.4, identify the B vitamin or vitamins to which each of the following characterizations applies.

a. Is part of the coenzymes NAD and NADP
b. Is part of coenzyme A
c. Contains a fused ring system in its structure
d. Contains both a five-membered and a six-membered ring in its structure

21.75 Describe the structural differences among the three active forms of vitamin A.

21.76 What is the relationship between the plant pigment beta-carotene and vitamin A?

21.77 What is the major function of vitamins K and E in the human body?

21.78 What is the major function of vitamins D and A in the human body?

Minerals (Section 21.15)

21.79 What are three important functions of dietary minerals in the human body?

21.80 What is the difference between a dietary mineral and a vitamin?

21.81 List the major minerals that have each of the following characteristics.

a. A nonmetal
b. Present in body fluids primarily as monatomic ions
c. Principal component of bones and teeth
d. Important in muscle contraction

21.82 List the major minerals that have each of the following characteristics.

a. A metal
b. Present in body fluids primarily as polyatomic ions
c. Assists in nerve impulse transmission
d. Important in the clotting of blood

21.83 List a major biological function for each of the following trace minerals.

a. Iron
b. Iodine

21.84 List a major biological function for each of the following trace minerals.

a. Copper
b. Zinc

21.85 Using Table 21.6, find the RDA for each of the following trace minerals.

a. Iodine
b. Selenium

21.86 Using Table 21.6, find the RDA for each of the following trace minerals.

a. Iron
b. Zinc

Additional Problems

21.87 Explain the difference between the following types of enzymes.

a. Apoenzyme and proenzyme
b. Simple enzyme and allosteric enzyme
c. Coenzyme and isoenzyme
d. Conjugated enzyme and holoenzyme

21.88 Identify the functional groups present in a molecule of each of the following vitamins.

a. Vitamin C b. Vitamin A (retinol)
c. Vitamin D d. Vitamin K

21.89 Indicate whether each of the following vitamins functions as a coenzyme.

a. Vitamin C b. Vitamin A
c. Vitamin D d. Niacin
e. Riboflavin f. Biotin

21.90 Classify each of the following as a major mineral or a trace mineral.

a. Potassium b. Chlorine
c. Zinc d. Iodine
e. Copper f. Phosphorus

21.91 What general kinds of reactions do the following types of enzymes catalyze?

a. Oxidoreductases b. Lyases
c. Isomerases d. Ligases
e. Hydrolases f. Transferases

21.92 Explain what is meant by the equation

$$E + S \rightleftharpoons ES \rightarrow E + P$$

given that ES stands for enzyme–substrate complex.

21.93 Alcohol dehydrogenase catalyzes the conversion of ethanol to acetaldehyde. This enzyme, in its active state, consists of a protein molecule and a zinc ion. On the basis of this information, identify the following for this chemical system.

a. Substrate b. Cofactor
c. Apoenzyme d. Holoenzyme

21.94 Each of the following is an abbreviation for an enzyme used in the diagnosis and/or treatment of heart attacks. What does each abbreviation stand for?

a. TPA b. LDH
c. CPK d. AST

Grid Problems

21.95

1. sucrase	2. hydrolase	3. phosphatase
4. kinase	5. lyase	6. decarboxylase

Select from the grid *all* correct responses for each of the following situations.

a. Enzyme action involves a hydrolysis reaction.
b. Enzyme action involves transfer of a group between substrates.
c. Enzyme action is restricted to one substrate.
d. Enzyme action involves removal of a group to create a double bond in the substrate.

21.96

1. conjugated enzyme	2. apoenzyme	3. coenzyme
4. holoenzyme	5. proenzyme	6. isoenzyme

Select from the grid *all* correct responses for each of the following situations.

a. Entity that is a part of a conjugated enzyme
b. Entity that by itself is biologically inactive
c. Substance that contains protein
d. Pairs of terms with the same meaning

21.97

1. vitamin A	2. vitamin B_{12}	3. vitamin D
4. vitamin B_6	5. vitamin C	6. vitamin E

Select from the grid *all* correct responses for each of the following situations.

a. Is fat-soluble
b. Has a structure that contains one or more metal atoms
c. Has two or more structural forms
d. Has a carbon chain of six or more atoms

21.98

1. phosphorus	2. calcium	3. potassium
4. iodine	5. copper	6. zinc

Select from the grid *all* correct responses for each of the following situations.

a. Trace mineral
b. Major mineral
c. Trace mineral for which an RDA has been established
d. Major mineral that exists in solution as a monatomic ion

CHAPTER 22

PART III

Nucleic Acids

CHAPTER OUTLINE

22.1 Types of Nucleic Acids 652
22.2 Nucleotides: Building Blocks of Nucleic Acids 653
22.3 Primary Structure of Nucleic Acids 656
22.4 The DNA Double Helix 658
22.5 Replication of DNA Molecules 660

Chemistry at a Glance: **DNA Replication** 665

22.6 Overview of Protein Synthesis 665
22.7 Ribonucleic Acids 666
22.8 Transcription: RNA Synthesis 667
22.9 The Genetic Code 669
22.10 Anticodons and tRNA Molecules 671
22.11 Translation: Protein Synthesis 672
22.12 Mutations 676

Chemistry at a Glance: **Protein Synthesis** 677

22.13 Nucleic Acids and Viruses 678
22.14 Recombinant DNA and Genetic Engineering 678

Chemical Connections
22.1 Use of Synthetic Nucleic Acid Bases in Medicine 662
22.2 Chromosomes in Cell Division and Reproduction 664
22.3 Antibiotics That Inhibit Bacterial Protein Synthesis 676
22.4 Polymerase Chain Reaction and DNA Sequencing 680

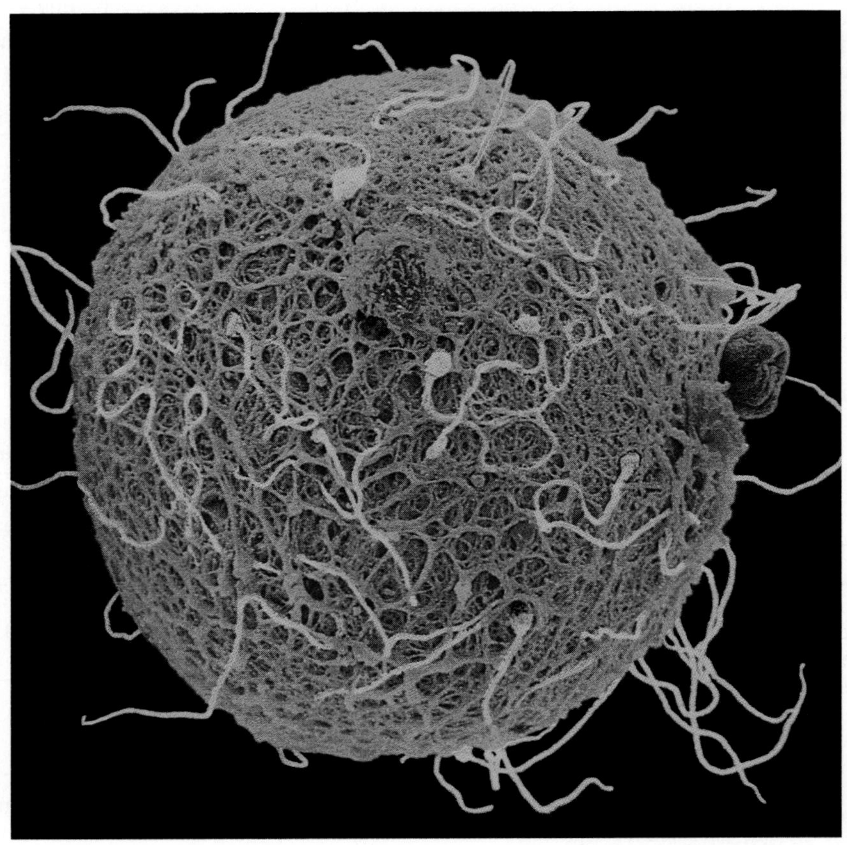

Human egg and sperms.

A most remarkable property of living cells is their ability to produce exact replicas of themselves. Furthermore, cells contain all the instructions needed for making the complete organism of which they are a part. The molecules within a cell that are responsible for these amazing capabilities are nucleic acids.

The Swiss physiologist Friedrich Miescher (1844–1895) discovered nucleic acids in 1869 while studying the nuclei of white blood cells. The fact that they were initially found in cell nuclei and are acidic accounts for the name *nucleic acid.* Although we now know that nucleic acids are found throughout a cell, not just in the nucleus, the name is still used for such materials.

22.1 Types of Nucleic Acids

Two types of nucleic acids are found within cells of higher organisms: *deoxyribonucleic acid* (DNA) and *ribonucleic acid* (RNA). Nearly all the DNA is found within the

• It was not until 1944, 75 years after the discovery of nucleic acids, that scientists obtained the first evidence that these molecules are responsible for the storage and transfer of genetic information.

cell nucleus. Its primary function is the storage and transfer of genetic information. This information is used (indirectly) to control many functions of a living cell. In addition, DNA is passed from existing cells to new cells during cell division. RNA occurs in all parts of a cell. It functions primarily in synthesis of proteins, the molecules that carry out essential cellular functions. The structural distinctions between DNA and RNA molecules are considered in Section 22.2.

22.2 Nucleotides: Building Blocks of Nucleic Acids

• Proteins are polypeptides, many carbohydrates are polysaccharides, and nucleic acids are polynucleotides.

Like polysaccharides and proteins, nucleic acids are polymers. The repeating unit for the nucleic acid structure is a nucleotide. **Nucleic acids** *are polymeric molecules in which the repeating unit is a nucleotide.*

The monomers for nucleic acid polymers, nucleotides, have a more complex structure than polysaccharide monomers (monosaccharides) or protein monomers (amino acids). Within each nucleotide monomer are three subunits. A **nucleotide** *is a molecule composed of a pentose sugar bonded to both a phosphate group and a nitrogen-containing heterocyclic base.*

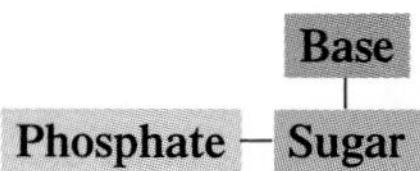

Pentose Sugars

The sugar unit of a nucleotide is either the pentose *ribose* or the pentose *2-deoxyribose.*

• The systems for numbering the atoms in the pentose and nitrogen-containing base subunits of a nucleotide are important and will be used extensively in later sections of this chapter. The convention is that
1. Pentose ring atoms are designated with *primed* numbers.
2. Nitrogen-containing base ring atoms are designated with *unprimed* numbers.

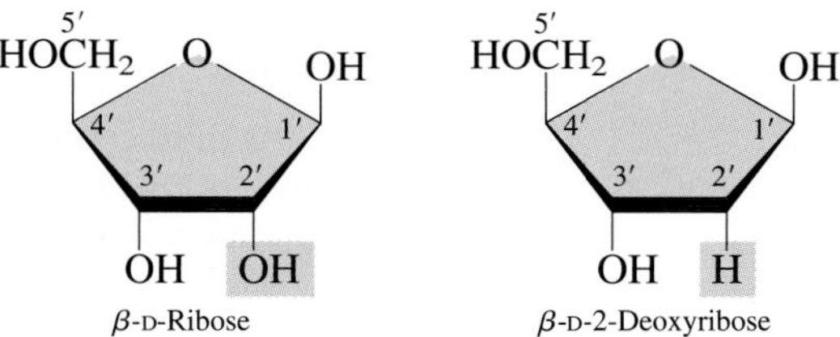

β-D-Ribose β-D-2-Deoxyribose

Structurally, the only difference between these two sugars occurs at carbon 2′. The —OH group present on this carbon in ribose becomes a —H atom in 2-deoxyribose. (The prefix *deoxy-* means "without oxygen.")

RNA and DNA differ in the identity of the sugar unit in their nucleotides. In RNA the sugar unit is *r*ibose—hence the *R* in RNA. In DNA the sugar unit is 2-*d*eoxyribose—hence the *D* in DNA.

Nitrogen-Containing Bases

• A pyrimidine derivative that we have encountered previously is thiamin (see Section 21.13).

• Caffeine, the most widely used nonprescription central nervous system stimulant, is the 3,7-dimethyl-2,6-dioxo derivative of purine (Section 17.8).

Five nitrogen-containing bases are nucleotide components. Three of them are derivatives of pyrimidine, a monocyclic base with a six-membered ring, and two are derivatives of purine, a bicyclic base with fused five- and six-membered rings.

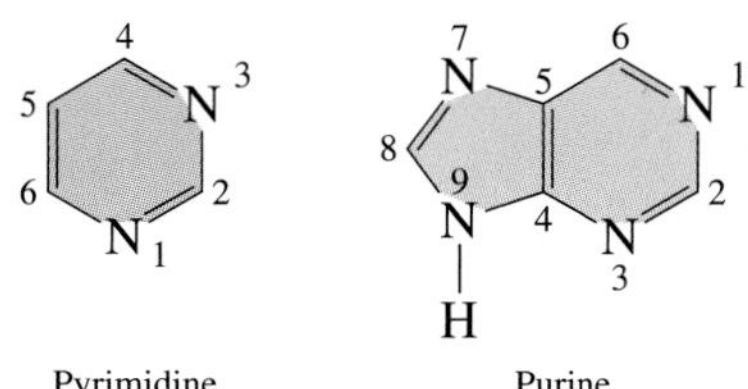

Pyrimidine Purine

These two parent bases (Section 17.8) and their derivatives are bases because their nitrogen atoms, which possess a nonbonding pair of electrons, can accept protons (Brønsted-Lowry base behavior; Section 17.5).

The three pyrimidine derivatives found in nucleotides are thymine (T), cytosine (C), and uracil (U).

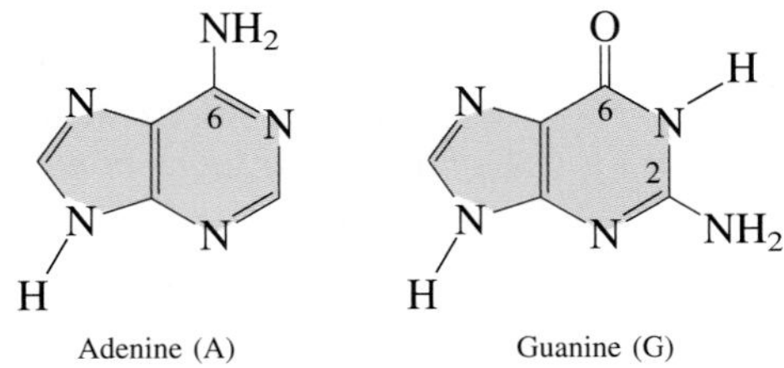

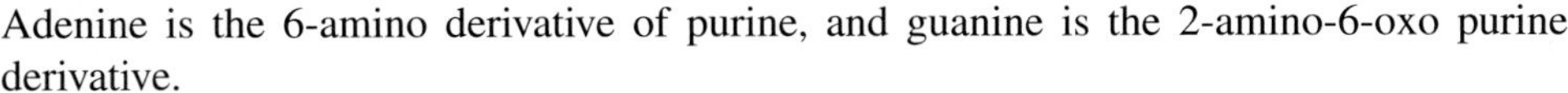

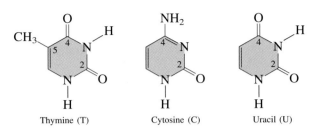

Thymine is the 5-methyl-2,4-dioxo derivative, cytosine the 4-amino-2-oxo derivative, and uracil the 2,4-dioxo derivative of pyrimidine.

The two purine derivatives found in nucleotides are adenine (A) and guanine (G).

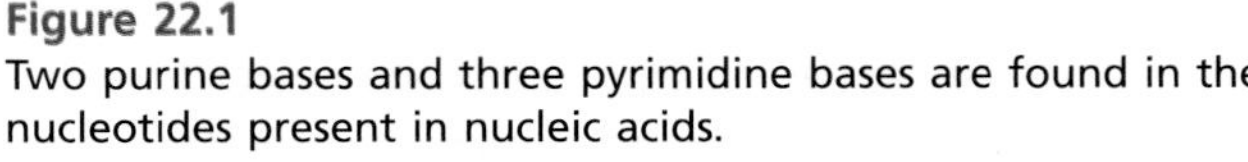

Adenine is the 6-amino derivative of purine, and guanine is the 2-amino-6-oxo purine derivative.

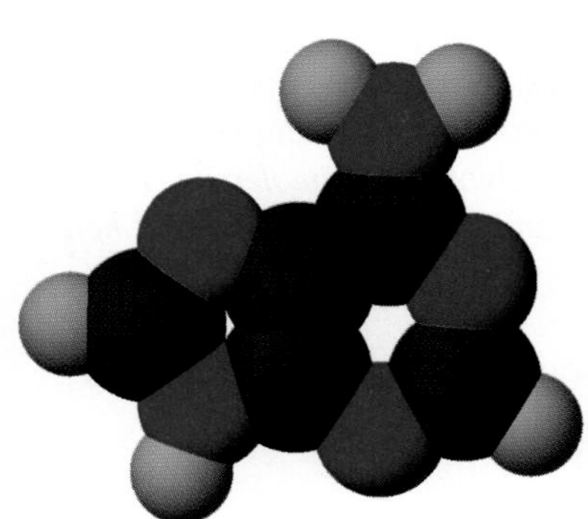

Space-filling model of the molecule adenine, a nucleotide present in both DNA and RNA.

Adenine, guanine, and cytosine are found in both DNA and RNA. Uracil is found only in RNA, and thymine usually occurs only in DNA. Figure 22.1 summarizes the occurrences of nitrogen-containing bases in nucleic acids.

Phosphate

Phosphate, the third component of a nucleotide, is derived from phosphoric acid (H_3PO_4). Under cellular pH conditions, the phosphoric acid loses two of its hydrogen atoms to give a -2-charged hydrogen phosphate ion.

• To remember which two of the five nucleotide bases are the purine derivatives (fused rings), use the phrase "pure silver" and the chemical symbol for silver, which is Ag.

pure Ag
purine A and G

Figure 22.1
Two purine bases and three pyrimidine bases are found in the nucleotides present in nucleic acids.

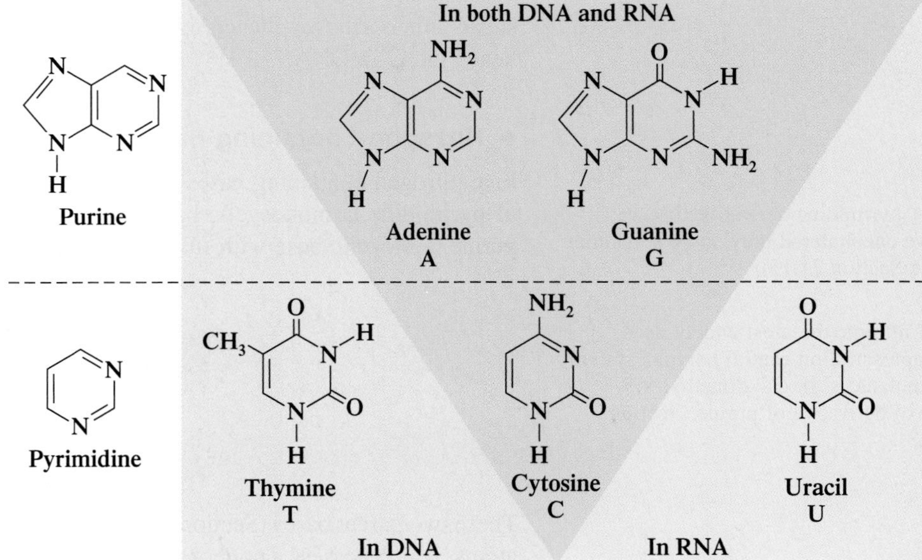

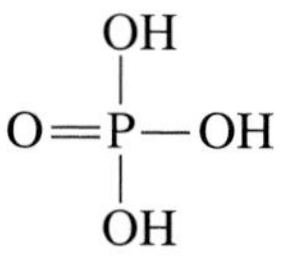

Phosphoric acid

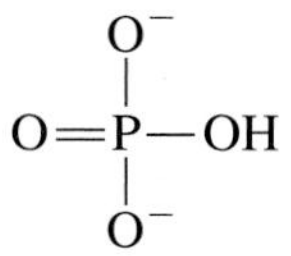

Hydrogen phosphate ion under cellular pH conditions

• Nucleotide Formation

The formation of a nucleotide from sugar, base, and phosphate can be visualized as occurring in the following manner:

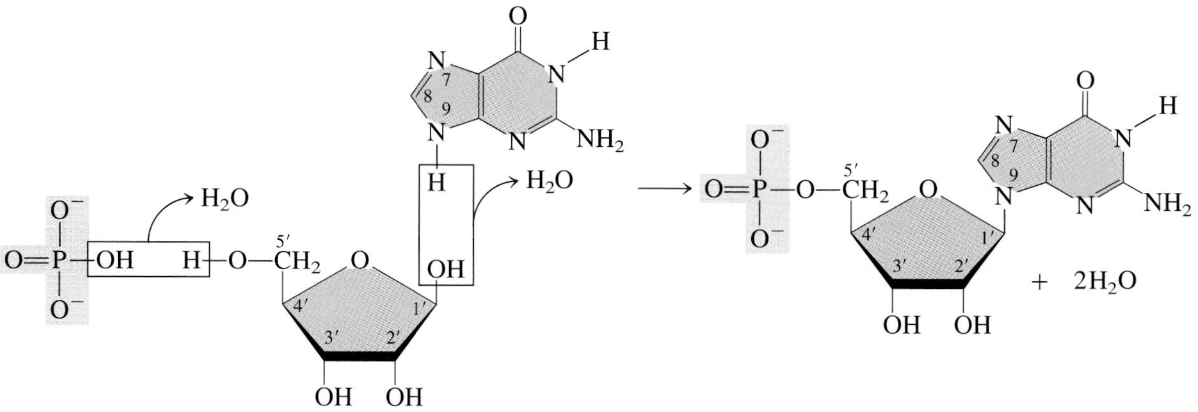

Phosphate

Sugar

Nucleotide

Important characteristics of this combining of three molecules into one molecule (the nucleotide) are that

1. Condensation, with formation of a water molecule, occurs at two locations: between sugar and base, and between sugar and phosphate.
2. The base is always attached at the C-1′ position of the sugar. For purine bases, attachment is through N-9; for pyrimidine bases, N-1 is involved. The C-1′ carbon atom of the ribose unit is always in a β configuration (Section 18.10), and the bond connecting the sugar and base is a β-*N*-glycosidic linkage.
3. The phosphate group is attached to the sugar at the C-5′ position through a phosphate–ester linkage.

There are four possible RNA nucleotides, differing in the base present (A, C, G, or U) and four possible DNA nucleotides, differing in the base present (A, C, G, or T).

• Nucleotide Nomenclature

The eight nucleotides of DNA and RNA molecules have common names and abbreviations (see Table 22.1). It is important to be familiar with them, because they are frequently encountered in biochemistry.

We can make several generalizations about the nomenclature given in Table 22.1.

1. All of the names end in 5′-monophosphate, which signifies the presence of a phosphate group attached to the 5′ carbon atom of ribose or deoxyribose. (In Chapter 23 we will encounter nucleotides that contain two or three phosphate groups—diphosphates and triphosphates.)
2. Preceding the monophosphate ending is the name of the base present in a modified form. The suffix *-osine* is used with purine bases, the suffix *-idine* with pyrimidine bases.
3. The prefix *deoxy-* at the start of the name signifies that the sugar present is deoxyribose. When no prefix is present, the sugar is ribose.

Table 22.1
The Names of the Eight Nucleotides Found in DNA and RNA

Base	Sugar	Nucleotide name	Nucleotide abbreviation
DNA Nucleotides			
adenine	deoxyribose	deoxyadenosine 5′-monophosphate	dAMP
guanine	deoxyribose	deoxyguanosine 5′-monophosphate	dGMP
cytosine	deoxyribose	deoxycytidine 5′-monophosphate	dCMP
thymine	deoxyribose	deoxythymidine 5′-monophosphate	dTMP
RNA Nucleotides			
adenine	ribose	adenosine 5′-monophosphate	AMP
guanine	ribose	guanosine 5′-monophosphate	GMP
cytosine	ribose	cytidine 5′-monophosphate	CMP
uracil	ribose	uridine 5′-monophosphate	UMP

4. The abbreviations in Table 22.1 for the nucleotides come from the one-letter symbols for the bases (A, C, G, T, and U), the use of MP for monophosphate, and a lower-case *d* at the start of the abbreviation whenever deoxyribose is the sugar.

22.3 Primary Structure of Nucleic Acids

• Nucleotides are related to nucleic acids in the same way that amino acids are related to proteins.

• The backbone of a nucleic acid structure is always an alternating sequence of phosphate and sugar groups. The sugar is ribose in RNA and deoxyribose in DNA.

Nucleic acids are polymers in which the repeating units, the monomers, are nucleotides (Section 22.2). The nucleotide units within a nucleic acid molecule are linked to each other through sugar–phosphate bonds. The resulting molecular structure (Figure 22.2) involves a chain of alternating sugar and phosphate groups with a base group protruding from the chain at regular intervals.

The alternating sugar–phosphate chain in a nucleic acid structure is often called the *nucleic acid backbone*. This backbone is constant throughout the entire nucleic acid structure. For DNA molecules, the backbone consists of alternating phosphate and *deoxyribose* sugar units; for RNA molecules, the backbone consists of alternating phosphate and *ribose* sugar units.

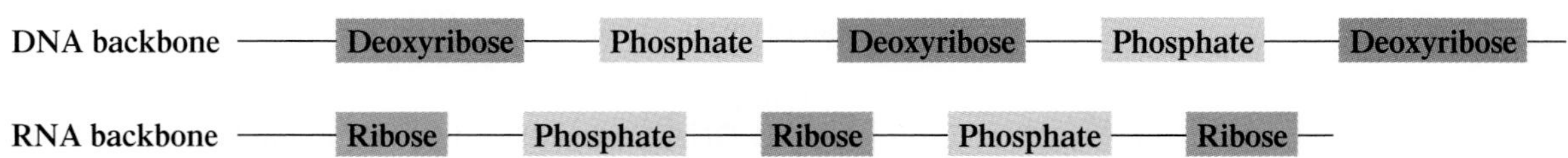

The variable portion of nucleic acid structure is the sequence of bases attached to the sugar units of the backbone. The sequence of these base side chains distinguishes various DNAs from each other and various RNAs from each other. Only four types of bases are found in any given nucleic acid structure. This situation is much simpler than that for proteins, where 20 side-chain entities (amino acids) are available (Section 20.2). In both RNA and DNA, adenine, guanine, and cytosine are encountered as side-chain components; thymine is found mainly in DNA, and uracil only in RNA (Figure 22.1).

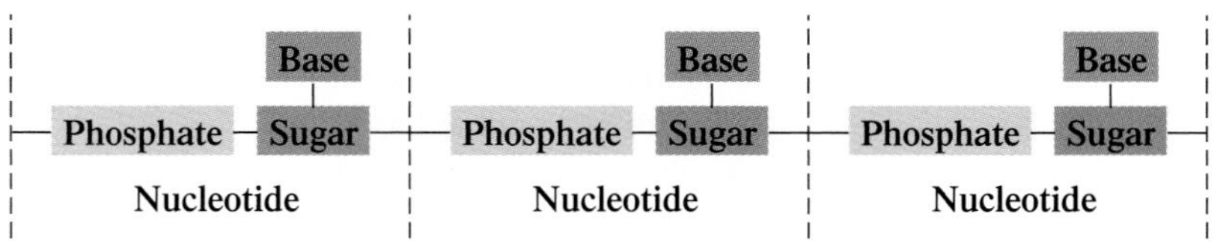

Figure 22.2
The general structure of a nucleic acid in terms of nucleotide subunits.

Primary nucleic acid structure *is the sequence of nucleotides in the molecule.* Because the sugar–phosphate backbone of a given nucleic acid does not vary, the primary structure of the nucleic acid depends only on the sequence of bases present. Further information about nucleic acid structure can be obtained by considering the detailed four-nucleotide segment of a DNA molecule shown in Figure 22.3.

The following list describes some important points of DNA structure that are illustrated in Figure 22.3.

• For both nucleic acids and proteins, a distinction is made between the two ends of the polymer chain. For nucleic acids there is a 5′ end and a 3′ end; for proteins there is an N-terminal end and a C-terminal end (Section 20.5).

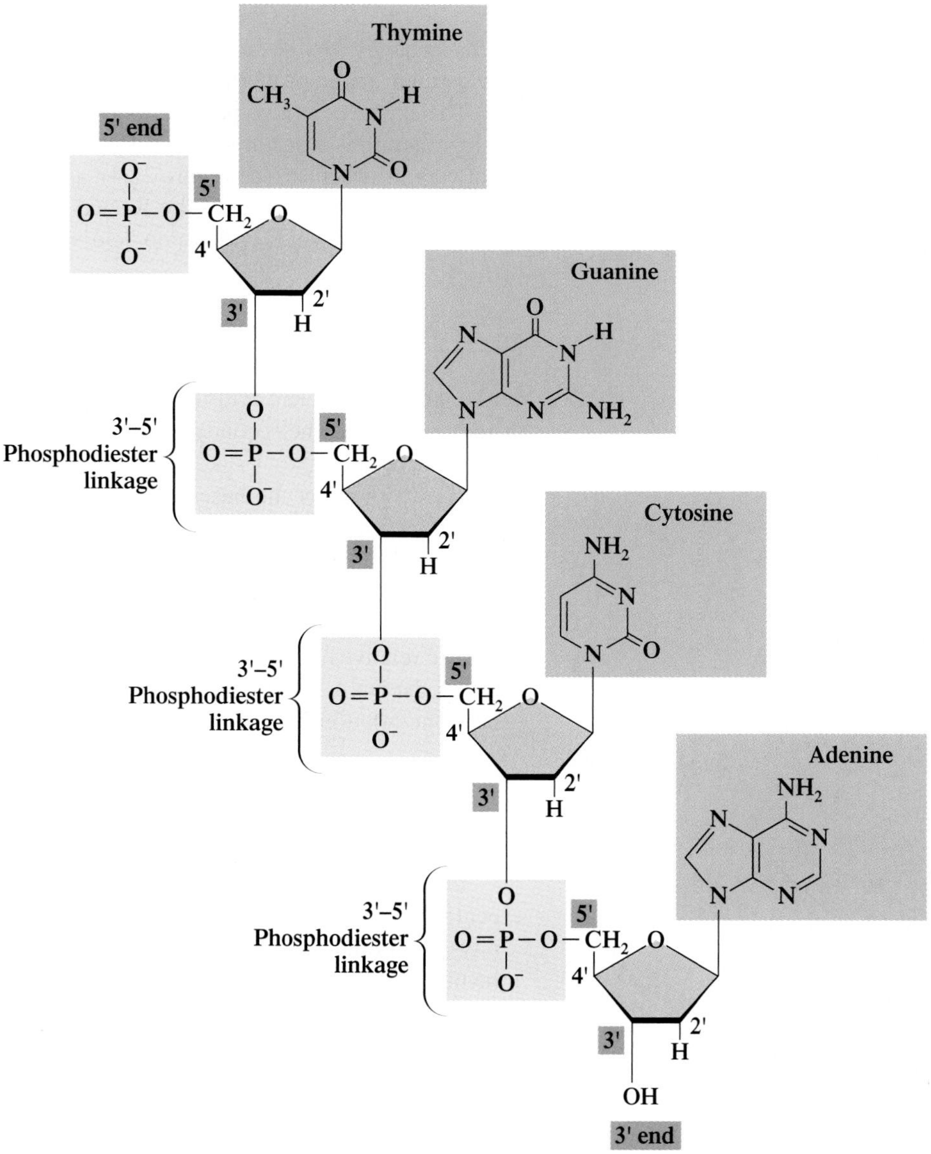

1. Each nonterminal phosphate group of the sugar–phosphate backbone is bonded to two sugar molecules through a *3′,5′-phosphodiester linkage.* There is a phosphoester bond to the 5′ carbon of one sugar unit and a phosphoester bond to the 3′ carbon of the other sugar.
2. A nucleotide chain has *directionality.* One end of the nucleotide chain, the *5′ end,* normally carries a free phosphate group attached to the 5′ carbon atom. The other end of the nucleotide chain, the *3′ end,* normally has a free hydroxyl group attached to the 3′ carbon atom. By convention, the sequence of bases of a nucleic acid strand is read from the 5′ end to the 3′ end.
3. Each nonterminal phosphate group in the backbone of a nucleic acid carries a −1 charge. The parent phosphoric acid molecule from which the phosphate

Figure 22.3
A four-nucleotide-long segment of DNA. (The choice of bases was arbitrary.)

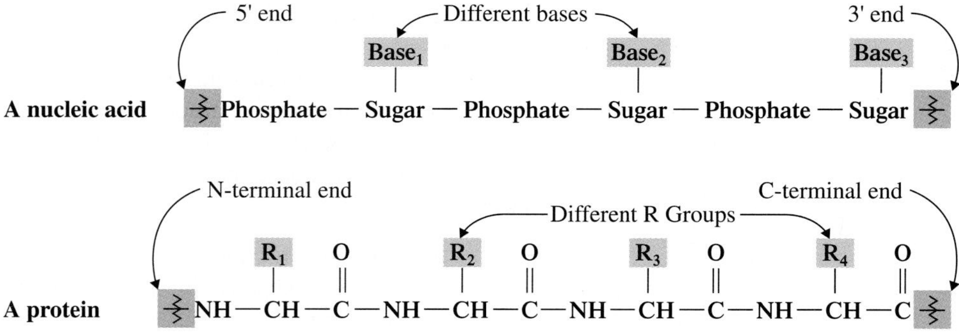

Figure 22.4
A comparison of the general primary structures of nucleic acids and proteins.

was derived originally had three —OH groups (Section 22.2). Two of these become involved in the 3′,5′-phosphodiester linkage. The remaining —OH group is free to exhibit acidic behavior—that is, to produce an H^+ ion.

$$\text{---O—}\underset{\text{OH}}{\overset{\text{O}}{\overset{\|}{\text{P}}}}\text{—O---} \rightleftharpoons \text{---O—}\underset{\text{O}^-}{\overset{\text{O}}{\overset{\|}{\text{P}}}}\text{—O---} + H^+$$

This behavior by the many phosphate groups in a nucleic acid backbone gives nucleic acids their acidic properties.

The existence of two parallels between primary nucleic acid structure and primary protein structure is worth noting: (1) Both nucleic acids and proteins have backbones that do not vary in structure (see Figure 22.4), and (2) the differences among various nucleic acids and among various proteins are related to the order of groups attached to the backbone: bases in nucleic acids and amino acid side chains (R groups) in proteins.

22.4 The DNA Double Helix

Like proteins, nucleic acids have secondary, or three-dimensional, structure as well as primary structure. The secondary structures of DNAs and RNAs differ, and we will discuss them separately.

The amounts of the bases A, T, G, and C present in DNA molecules were the key to determination of the general three-dimensional structure of DNA molecules. Base composition data for DNA molecules from many different organisms revealed a definite pattern of base occurrence. The amounts of A and T were always equal, and the amounts of C and G were always equal, as were the amounts of total purines and total pyrimidines.

The relative amounts of these base pairs in DNA vary depending on the life form from which the DNA is obtained. (Each animal or plant has a unique base composition.) However, the relationships

$$\%A = \%T \qquad \text{and} \qquad \%C = \%G$$

always hold true. For example, human DNA contains 30% adenine, 30% thymine, 20% guanine, and 20% cytosine.

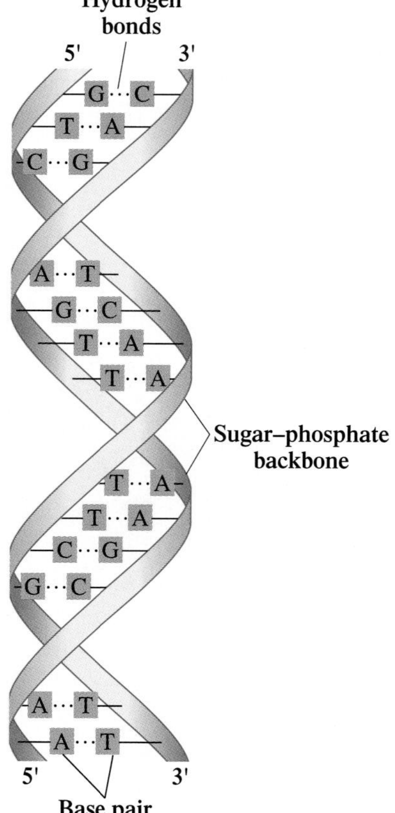

Figure 22.5
A schematic drawing of the DNA double helix that emphasizes the hydrogen bonding between bases on the two chains.

In 1953, an explanation for the base composition patterns associated with DNA molecules was proposed by the American microbiologist James Watson and the English biophysicist Francis Crick. Their model, which has now been validated in numerous ways, involves a double-helix structure that accounts for the equality of bases present, as well as for all other known DNA structural data.

The DNA double helix involves two polynucleotide strands coiled around each other in a helical fashion, much like two entwined rails from a spiral staircase. The sugar–phosphate backbones of the two DNA strands form the outside of the helix (Figure 22.5). The bases (side chains) of each backbone extend inward toward the bases of the other

• The *antiparallel* nature of the two polynucleotide chains in the DNA double helix means that there is a 5′ end and a 3′ end at both ends of the double helix.

strand. The two strands are connected by *hydrogen bonds* (Section 7.16) between their bases. Additionally, the two strands of the double helix are *antiparallel*—that is, they run in opposite directions. One strand runs in the 5′-to-3′ direction, and the other is oriented in the 3′-to-5′ direction.

• Base Pairing

A physical restriction, the size of the interior of the DNA double helix, limits the base pairs that can hydrogen-bond to one another. Only pairs involving one small base (a pyrimidine) and one large base (a purine) correctly "fit" within the helix interior. There is not enough room for two large purine bases to fit opposite each other (they overlap), and two small pyrimidine bases are too far apart to hydrogen-bond to one another effectively. Of the four possible purine–pyrimidine combinations (A–T, A–C, G–T, and G–C), hydrogen-bonding possibilities are *most favorable* for the A–T and G–C pairings, and these two combinations are the *only two* that normally occur in DNA. Figure 22.6 shows the specific hydrogen-bonding interactions for the four possible purine–pyrimidine base-pairing combinations.

• A mnemonic device for recalling base-pairing combinations in DNA and RNA involves listing the base abbreviations in alphabetical order. Then the first and last bases pair, and so do the middle two bases.

DNA: A C G T

RNA: A C G U

Figure 22.6
Hydrogen-bonding possibilities are more favorable when A–T and G–C base pairing occurs than when A–C and G–T base pairing occurs. (a) Two and three hydrogen bonds can form, respectively, between A–T and G–C base pairs. These combinations are present in DNA molecules. (b) Only one hydrogen bond can form between G–T and A–C base pairs. These combinations are not present in DNA molecules.

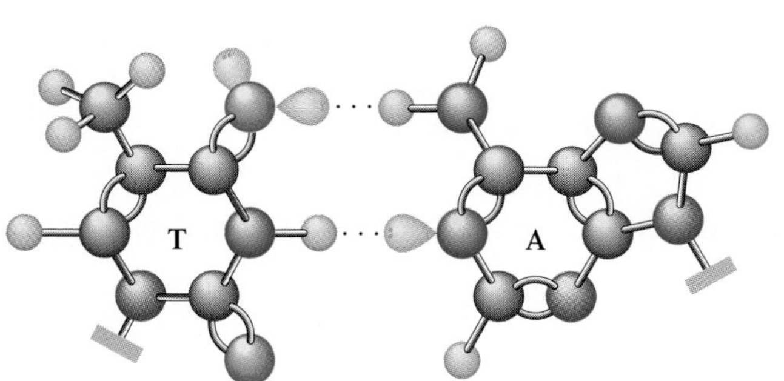

Thymine–Adenine Base Pairing (two hydrogen bonds form)

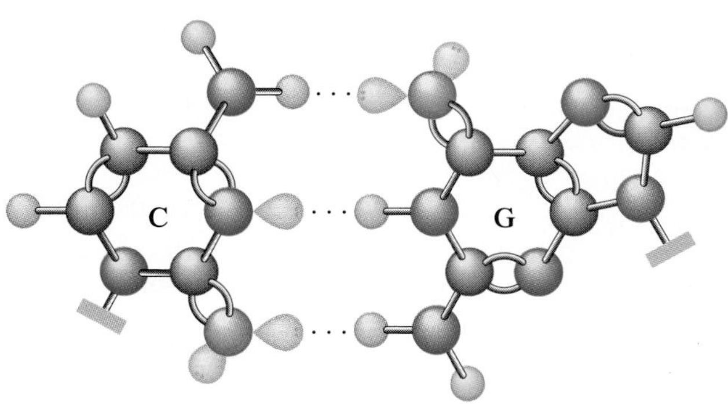

Cytosine–Guanine Base Pairing (three hydrogen bonds form)

(a)

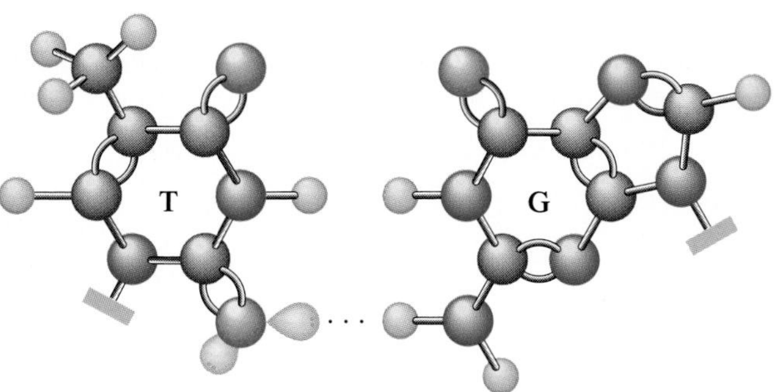

Thymine–Guanine Base Pairing (only one hydrogen bond forms)

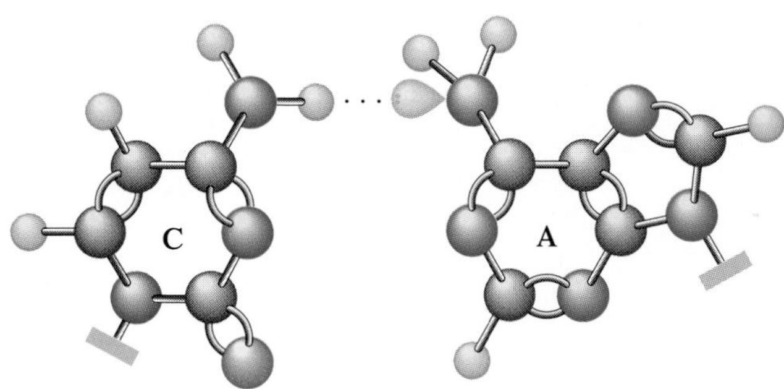

Cytosine–Adenine Base Pairing (only one hydrogen bond forms)

(b)

Carbon | Oxygen | Lone pair
Nitrogen | Hydrogen | Attachment to backbone

• Hydrogen bonding is responsible for the secondary structure (double helix) of DNA. Hydrogen bonding is also responsible for secondary structure in proteins (Section 20.8).

• The two strands of DNA in a double helix are complementary. This means that if you know the order of bases in one strand, you can predict the order of bases in the other strand.

The pairing of A with T and of G with C is said to be *complementary.* A and T are complementary bases, as are G and C. **Complementary bases** *are specific pairs of bases in nucleic acid structures that hydrogen-bond to each other.* The fact that complementary base pairing occurs in DNA molecules explains, very simply, why the amounts of the bases A and T present are always equal, as are the amounts of G and C.

Hydrogen bonding holds the two strands of the DNA double helix together. Although hydrogen bonds are relatively weak forces, each DNA molecule contains so many base pairs that the hydrogen-bonding attractions are sufficient in magnitude, collectively, to prevent the two entwined DNA strands from separating spontaneously under normal physiological conditions.

The two strands of DNA in a double helix are *not identical*—they are complementary. Wherever G occurs in one strand, there is a C in the other strand; wherever T occurs in one strand, there is an A in the other strand. An important ramification of this complementary relationship is that knowing the base sequence of one strand of DNA enables us to predict the base sequence of the complementary strand.

In specifying the base sequence of a segment of a strand of DNA (or RNA), we list the bases in sequential order (using their one-letter abbreviations) in the direction from the 5′ end to the 3′ end of the segment.

5′–A–A–G–C–T–A–G–C–T–T–A–C–T–3′

Example 22.1

Predicting Base Sequence in a Complementary DNA Strand

Predict the sequence of bases in the DNA strand that is complementary to the single DNA strand shown.

5′–C–G–A–A–T–C–C–T–A–3′

Solution

Because only A forms a complementary base pair with T, and only G with C, the complementary strand is as follows:

Given: 5′–C–G–A–A–T–C–C–T–T–A–3′

Complementary strand: 3′–G–C–T–T–A–G–G–A–T–5′

Note the reversal of the numbering of the ends of the complementary strand compared to the given strand. This is due to the antiparallel nature of the two strands in a DNA double helix.

Practice Exercise 22.1

Predict the sequence of bases in the DNA strand complementary to the single DNA strand shown.

5′–A–A–T–G–C–A–G–C–T–3′

• *Answer:* 3′–T–T–A–C–G–T–C–G–A–5′

22.5 Replication of DNA Molecules

DNA molecules are the carriers of the genetic information within a cell; that is, they are the molecules of heredity. Each time a cell divides, an exact copy of the DNA of the parent cell is needed for the new daughter cell. The process by which new DNA molecules are generated is DNA replication. **DNA replication** *is the process by which DNA molecules*

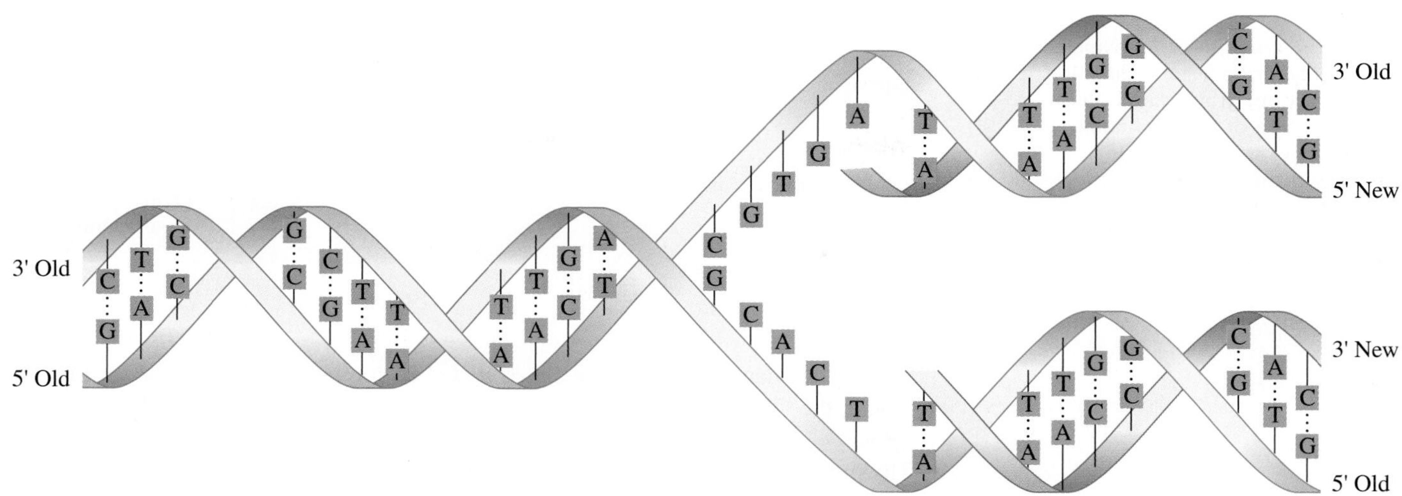

Figure 22.7
In DNA replication, the two strands of the DNA double helix unwind, the separated strands serving as templates for the formation of new DNA strands. Free nucleotides pair with the complementary bases on the separated strands of DNA. This process ultimately results in the complete replication of the DNA molecule.

produce exact duplicates of themselves. The key concept in understanding DNA replication is the base pairing associated with the DNA double helix.

• DNA Replication Overview

To understand DNA replication, we must regard the two strands of the DNA double helix as a pair of *templates,* or patterns. During replication, the strands separate. Each can then act as a template for the synthesis of a new, complementary strand. The result is two daughter DNA molecules with base sequences identical to those of the parent double helix. Let us consider details of this replication.

Under the influence of the enzyme *DNA helicase,* the DNA double helix unwinds, and the hydrogen bonds between complementary bases are broken. This unwinding process, as shown in Figure 22.7, is somewhat like opening a zipper.

The bases of the separated strands are no longer connected by hydrogen bonds. They can pair with *free* individual nucleotides present in the cell's nucleus. As shown in Figure 22.7, the base pairing always involves C pairing with G and A pairing with T. After the free nucleotides have formed hydrogen bonds with the old strand (the template), the enzyme *DNA polymerase* verifies that the base pairing is correct and catalyzes the formation of new phosphodiester linkages between nucleotides (represented by colored ribbons in Figure 22.7).

Each of the two daughter molecules of double-stranded DNA formed in the DNA replication process contains one strand from the original parent molecule and one newly formed strand.

• The Replication Process in Finer Detail

Though simple in principle, the DNA replication process has many intricacies.

1. The enzyme *DNA polymerase* can operate on a forming DNA daughter strand only in the 5′-to-3′ direction. Because the two strands of parent DNA run in opposite directions (one is 5′ to 3′ and the other 3′ to 5′; Section 22.4), only one strand can grow continuously in the 5′-to-3′ direction. The other strand must be formed in short segments, called *Okazaki fragments* (after their discoverer, Reiji Okazaki), as the DNA unwinds (see Figure 22.8). The breaks or

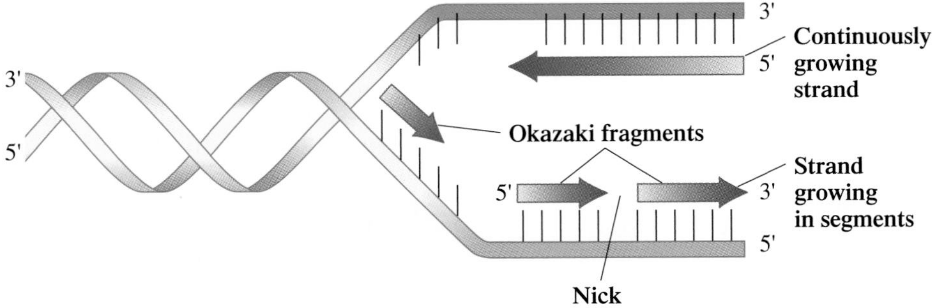

Figure 22.8
Because the enzyme DNA polymerase can act only in the 5′-to-3′ direction, one strand (top) grows continuously in the direction of the unwinding, and the other strand grows in segments in the opposite direction. The segments in this latter chain are then connected by a different enzyme, DNA ligase.

gaps in this daughter strand are called *nicks.* To complete the formation of this strand, the Okazaki fragments are connected by action of the enzyme *DNA ligase.*

2. The process of DNA unwinding does not have to begin at an end of the DNA molecule. It may occur at any location within the molecule. Indeed, studies show that unwinding usually occurs at several interior locations simultaneously and that DNA replication is bidirectional for these locations; that is, it proceeds in both directions from the unwinding sites. As shown in Figure 22.9, the result of this multiple-site replication process is formation of "bubbles" of newly synthesized DNA. The bubbles grow larger and eventually coalesce, giving rise to two complete daughter DNAs. Multiple-site replication enables large DNA molecules to be replicated rapidly.

The use of synthetic nucleic acid bases in medicine is considered in Chemical Connections 22.1.

Chemical CONNECTIONS

22.1 Use of Synthetic Nucleic Acid Bases in Medicine

Many hundreds of modified nucleic acid bases have been prepared in laboratories, and their effects on nucleic acid synthesis investigated. Several of them are now in clinical use as drugs for controlling, at the cellular level, cancers and other related disorders.

The theory behind the use of these modified bases involves their masquerading as legitimate nucleic acid building blocks. The enzymes associated with the DNA replication process incorporate the modified bases into growing nucleic acid chains. The presence of these "pseudonucleotides" in the chain stops further growth of the chain, thus interfering with nucleic acid synthesis.

Examples of drugs now in use include 5-fluorouracil, which is employed against a variety of cancers, especially those of the breast and digestive tract, and 6-mercaptopurine, which is used in the treatment of leukemia.

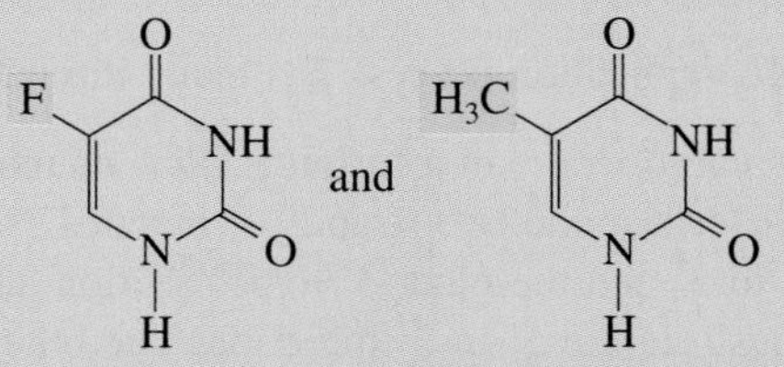

5-Fluorouracil (a modified thymine) and Thymine

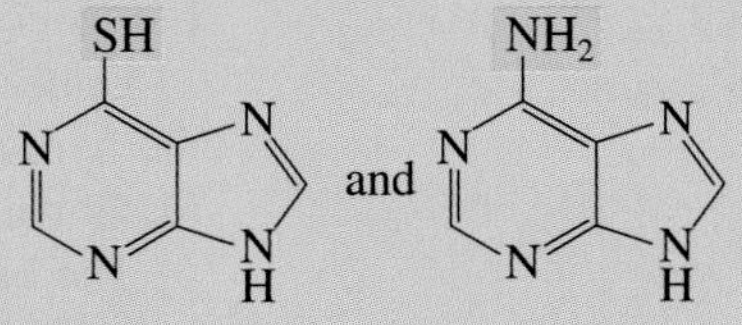

6-Mercaptopurine (a modified adenine) and Adenine

The rapidly dividing cells that are characteristic of cancer require large quantities of DNA. Anticancer drugs based on modified nucleic acid bases block DNA synthesis and therefore block the increase in the number of cancer cells. Cancer cells are generally affected to a greater extent than normal cells because of the rapid growth factor. Eventually, the normal cells are affected to such a degree that use of the drugs must be discontinued. 5-Fluorouracil inhibits the formation of thymine-containing nucleotides required for DNA synthesis. 6-Mercaptopurine, which substitutes for adenine, inhibits the synthesis of nucleotides that incorporate adenine and guanine.

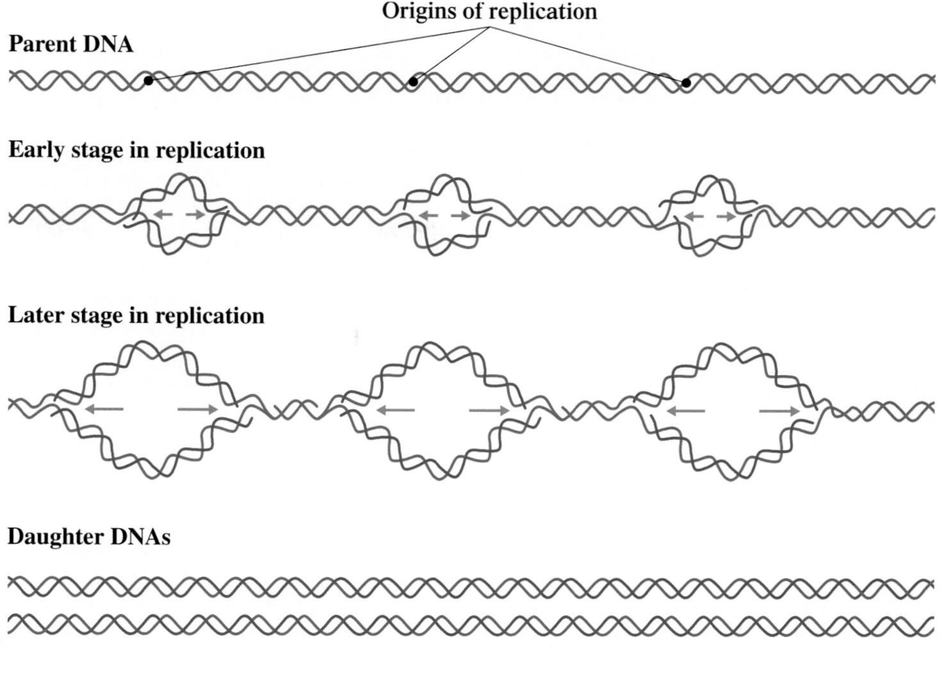

Figure 22.9
DNA replication usually occurs at multiple sites within a molecule, and the replication is bidirectional from these sites.

• Chromosomes

Once the DNA within a cell has been replicated, it interacts with specific proteins in the cell to form structural units called chromosomes. A **chromosome** *is an individual DNA molecule bound to a group of proteins.* Typically, a chromosome is about 15% by mass DNA and 85% by mass protein.

Cells from different kinds of organisms have different numbers of chromosomes. A normal human has 46 chromosomes per cell, a mosquito 6, a frog 26, a dog 78, and a turkey 82.

• Chromosomes are *nucleoproteins.* They are a combination of nucleic acid (DNA) and various proteins.

Chromosomes occur in matched (*homologous*) pairs. The 46 chromosomes of a human cell constitute 23 homologous pairs. One member of each homologous pair is derived from a chromosome inherited from the father, and the other is a copy of one of the chromosomes inherited from the mother. Homologous chromosomes have similar, but not identical, DNA base sequences; both code for the same traits but for different forms of the trait (for example, blue eyes versus brown eyes). Further details concerning the production of homologous chromosome pairs are considered in Chemical Connections 22.2.

The accompanying Chemistry at a Glance summarizes the steps in DNA replication.

Twins sharing identical physical characteristics.

Chemical CONNECTIONS

22.2 Chromosomes in Cell Division and Reproduction

Two human cell division processes are known: mitosis and meiosis. In mitosis, the two daughter cells resulting from cell division receive DNA identical to that in the parent cell. Most human cells divide by mitosis. Meiosis is a special type of cell division that occurs in certain cells of the ovaries and testes of sexually mature people. In meiosis, the newly produced cells contain 50% less DNA than parent cells.

During mitosis (see the accompanying illustration), the 46 chromosomes (23 homologous pairs) in the nucleus of the original cell duplicate through the DNA replication process (Section 22.5). This produces a parent cell containing 92 chromosomes (2 sets of 23 homologous pairs), which then divides into two daughter cells, each daughter cell receiving an identical set of 46 chromosomes (23 homologous pairs).

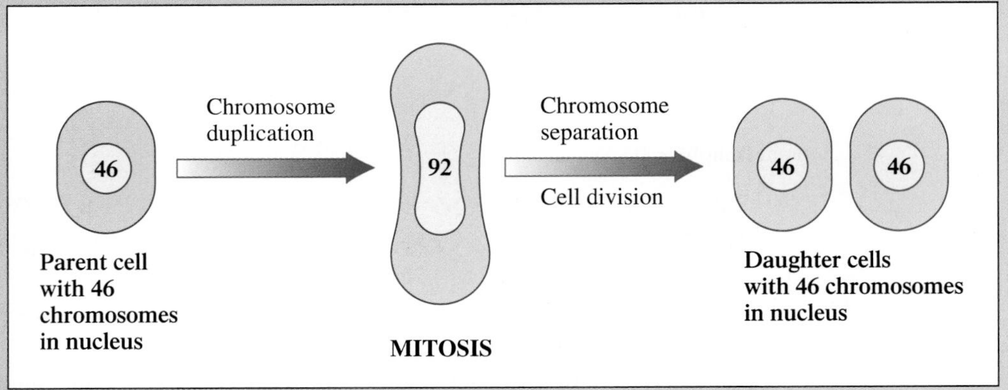

Meiosis, like mitosis, begins with DNA replication, which increases the number of chromosomes to 92—that is, 23 *pairs* of *doubled* chromosomes. Then, stepwise, two cell divisions occur, compared with the one cell division in mitosis. The first meiotic division (see the accompanying illustration) separates the two members of each pair of homologous chromosomes. Each of the two daughter cells of this first division contains *two copies* of one member of each homologous pair of chromosomes. There is no DNA replication before the second meiotic division, and in the course of this second division, the two copies of each of the 23 chromosomes are separated. Because both daughter cells of the first division divide, the result of the second division is four cells, each containing 23 chromosomes. These cells containing 23 chromosomes (no homologous pairs) are found in the sperm and ovum of the human reproductive process. When the sperm and ovum unite, a new cell containing 46 chromosomes (23 from the father and 23 from the mother) is formed, an event that marks the beginning of a new individual. The 23 chromosomes from each parent join together to form 23 homologous pairs. This first cell and its daughter cells reproduce by mitosis, eventually forming the complete new individual.

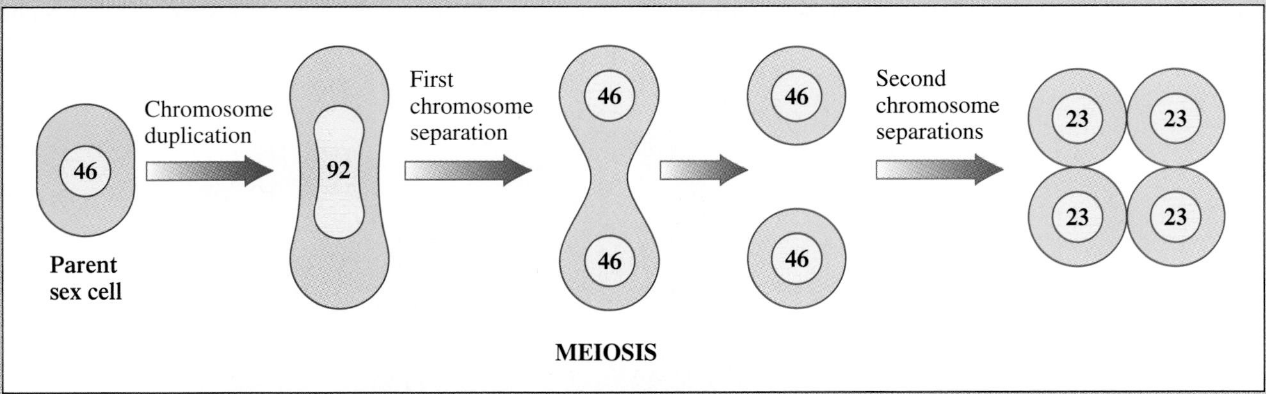

CHEMISTRY AT A GLANCE

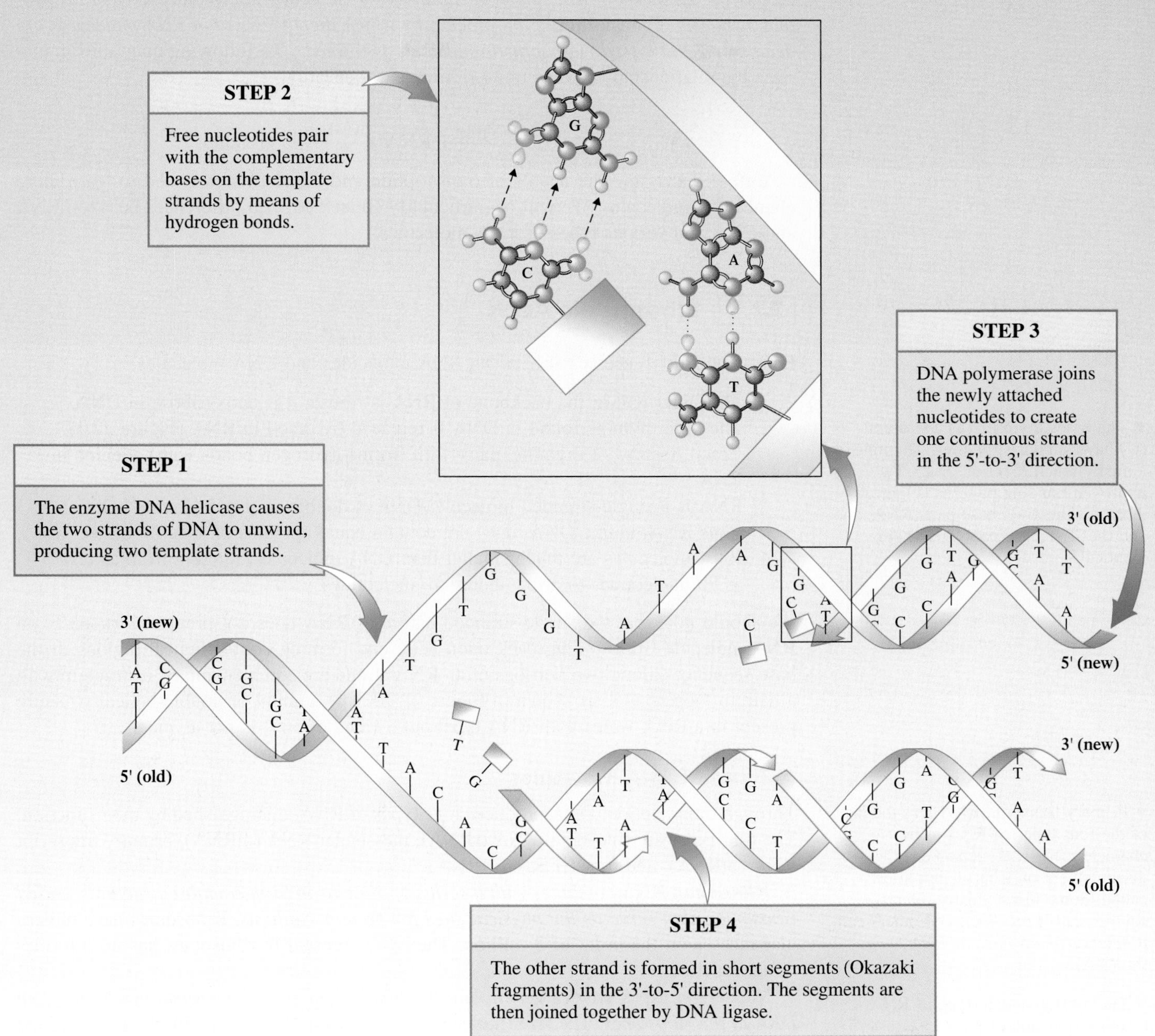

STEP 2
Free nucleotides pair with the complementary bases on the template strands by means of hydrogen bonds.
STEP 3
DNA polymerase joins the newly attached nucleotides to create one continuous strand in the 5'-to-3' direction.
STEP 1
The enzyme DNA helicase causes the two strands of DNA to unwind, producing two template strands.
3' (old)
3' (new)
5' (new)
5' (old)
3' (new)
5' (old)
STEP 4
The other strand is formed in short segments (Okazaki fragments) in the 3'-to-5' direction. The segments are then joined together by DNA ligase.

22.6 Overview of Protein Synthesis

We saw in the previous section how the replication of DNA makes it possible for a new cell to contain the same genetic information as its parent cell. We will now consider how the genetic information contained in a cell is expressed in cell operation. This brings us to the topic of protein synthesis. The synthesis of proteins (skin, hair, enzymes, hormones, and so on) is under the direction of DNA molecules. It is this role of DNA that establishes the similarities between parent and offspring that we regard as hereditary characteristics.

We can divide the overall process of protein synthesis into two steps. The first step is called transcription and the second translation. **Transcription** *is the process by which DNA directs the synthesis of RNA molecules that carry the coded information needed for protein synthesis.* **Translation** *is the process by which the codes within RNA molecules are deciphered and a particular protein molecule is formed.* The following diagram summarizes the relationship between transcription and translation.

$$\text{DNA} \xrightarrow{\text{Transcription}} \text{RNA} \xrightarrow{\text{Translation}} \text{protein}$$

Before discussing the details of transcription and translation, we need to learn more about RNA molecules. We will be particularly concerned with differences between RNA and DNA and various types of RNA molecules.

22.7 Ribonucleic Acids

Four major differences exist between RNA molecules and DNA molecules.

1. The sugar unit in the backbone of RNA is ribose; it is deoxyribose in DNA.
2. The base thymine found in DNA is replaced by uracil in RNA (Figure 22.1). Uracil, instead of thymine, pairs with (forms hydrogen bonds with) adenine in RNA.
3. RNA is a single-stranded molecule; DNA is double-stranded (double helix). Thus RNA, unlike DNA, does not contain equal amounts of specific bases.
4. RNA molecules are much smaller than DNA molecules, ranging from as few as 75 nucleotides to a few thousand nucleotides.

• The bases thymine (T) and uracil (U) have similar structures. Thymine is a methyluracil (Section 22.2). The hydrogen-bonding patterns (Figure 22.5) for the A–U base pair (RNA) and the A–T base pair (DNA) are identical.

We should note that the single-stranded nature of RNA does not prevent *portions* of an RNA molecule from folding back upon itself and forming double-helical regions. If the base sequences along two portions of an RNA strand are complementary, a structure with a hairpin loop results, as shown in Figure 22.10. The amount of double-helical structure present in a RNA varies with RNA type, but a value of 50% is not atypical.

• Types of RNA Molecules

Through transcription, DNA produces four types of RNA, distinguished by their function. The four types are ribosomal RNA (rRNA), messenger RNA (mRNA), primary transcript RNA (ptRNA), and transfer RNA (tRNA).

Ribosomal RNA *combines with a series of proteins to form complex structures, called ribosomes, that serve as the physical sites for protein synthesis.* Ribosomes have molecular masses on the order of 3 million. The rRNA present in ribosomes has no informational function.

Messenger RNA *carries genetic information (instructions for protein synthesis) from DNA to the ribosomes.* The size (molecular mass) of mRNA varies with the length of the protein whose synthesis it will direct. Each kind of protein in the body has its own mRNA.

Primary transcript RNA *is the material from which messenger RNA is made.*

Transfer RNA *delivers specific individual amino acids to the ribosomes, the sites of protein synthesis.* These RNAs are the smallest of the RNAs, possessing only 75–90 nucleotide units.

• Primary transcript RNA was the last of the four types of RNA to be characterized. There is some disagreement on what it should be called. Some biochemistry reference sources call it *pre-RNA,* and others call it *heterogeneous nuclear RNA* (hnRNA).

• The most abundant type of RNA in a cell is ribosomal RNA (75% to 80% by mass). Transfer RNA constitutes 10%–15% of cellular RNA; messenger RNA and its precursor, primary transcript RNA, make up the remaining 5%–10% of RNA material in the cell.

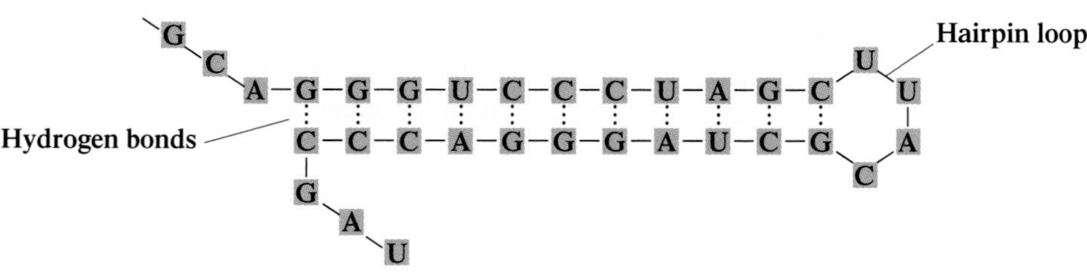

Figure 22.10
A hairpin loop is produced when single-stranded RNA doubles back on itself and complementary base pairing occurs.

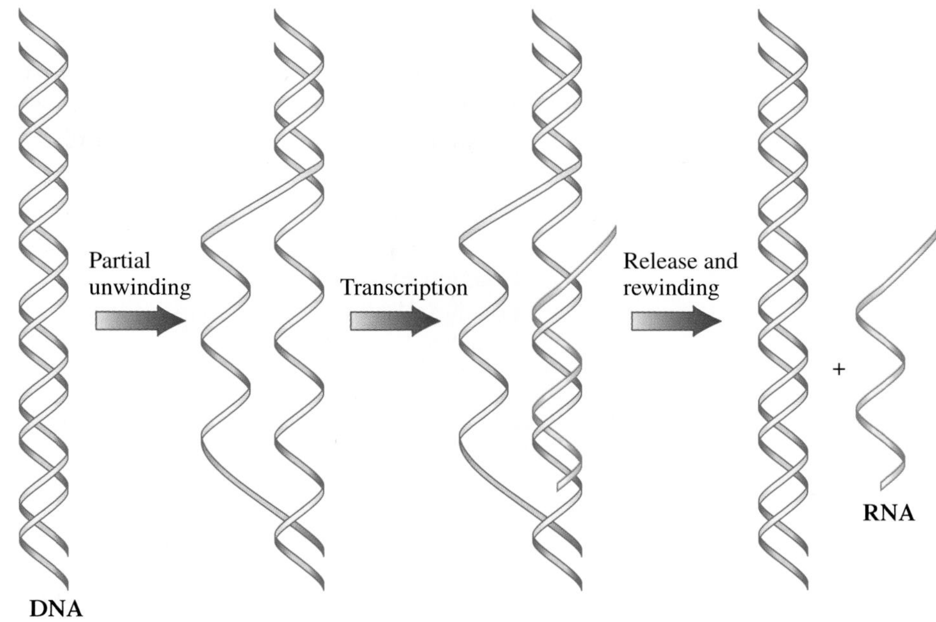

Figure 22.11
The transcription of DNA to form RNA involves an unwinding of a portion of the DNA double helix. Only one strand of the DNA is copied during transcription.

22.8 Transcription: RNA Synthesis

The mechanics of transcription are in many ways similar to those of DNA replication. Four steps are involved.

1. A *portion* of the DNA double helix unwinds, exposing some bases. The unwinding process is governed by the enzyme *RNA polymerase* rather than by *DNA helicase* (replication enzyme).
2. Free *ribo*nucleotides align along *one* of the exposed strands of DNA bases, forming new base pairs. In this process, U rather than T aligns with A in the base-pairing process. Because ribonucleotides rather than deoxyribonucleotides are involved in the base pairing, ribose, rather than deoxyribose, becomes incorporated into the new nucleic acid backbone.
3. RNA polymerase links the aligned ribonucleotides.
4. Transcription ends when the RNA polymerase enzyme encounters a sequence of bases that is "read" as a stop signal. The newly formed RNA molecule and the RNA polymerase enzyme are released, and the DNA then rewinds to re-form the original double helix.

• In DNA–RNA base pairing, the complementary base pairs are

DNA	RNA
A ——	U
G ——	C
C ——	G
T ——	A

RNA molecules contain the base U instead of the base T.

Figure 22.11 shows the overall process of transcription of DNA to form RNA. All types of RNA molecules (ptRNA/mRNA, rRNA, and tRNA) are synthesized in the nucleus of a cell via this general transcription process.

Example 22.2

Base Pairing Associated with the Transcription Process

Determine the RNA base sequence that is complementary to the following DNA "template."

$$5'\text{–A–T–G–C–C–C–G–A–G–T–T–}3'$$

(continued)

Solution

An RNA molecule cannot contain the base T. The base U is present instead. Therefore, U pairs with DNA A units instead of T. The transcription process will therefore produce the RNA sequence

3′–U–A–C–G–G–G–C–U–C–A–A–5′

Note that the DNA and RNA strands are antiparallel.

Practice Exercise 22.2

Determine the RNA base sequence that is complementary to the following DNA "template."

5′–G–A–T–T–C–A–G–C–T–A–3′

• *Answer:* 3′–C–U–A–A–G–U–C–G–A–U–5′

• Genes

The primary function of mRNA molecules is to direct the synthesis of the many different proteins needed for cellular function. Within a DNA strand are instructions for the synthesis of numerous mRNA molecules, which in turn direct the synthesis of numerous specific proteins. During transcription, the DNA molecule unwinding is controlled by RNA polymerase and occurs only at the particular spot where the appropriate base sequence is found for the mRNA (and protein) of concern. Such short segments of DNA, containing instructions for the formation of particular mRNAs, are called genes. A **gene** *is a segment of a DNA molecule that contains the base sequence for the production of a single, specific RNA molecule, which in turn produces a single, specific protein molecule.*

In humans, most genes are composed of 1000–3500 nucleotide units. Hundreds of genes can exist along a DNA molecule strand. The DNA found in one human cell contains an estimated 5.5 billion nucleotide base pairs, making up approximately 80,000 genes.

The gene content of a human cell is an extraordinary amount of information. To put the informational content of a human cell in perspective, consider the following example: Let us assume you have "special abilities" such that with your naked eye you can see DNA molecules with their base pairs. You write down the characteristics (base sequence) of all genes in a human cell. You use one page of a lab notebook for each gene, and it takes 5 minutes (a very rapid speed) to record each gene's characteristics. When you finish the project, your notebook, 80,000 pages long, will be equivalent to one hundred 800-page textbooks. With a recording time of 5 minutes per gene, it would take you—working 8 hours a day, 5 days a week—over 3 years to complete your work. All this information is found in each cell.

• Exons and Introns

Though simple in principle, transcription is a complex process. A consideration of one of the finer details of transcription illustrates this and leads us to the concepts of exons and introns.

It is now known that not all bases in a gene convey genetic information. Instead, a gene is *segmented* in that it has portions called *exons* that contain genetic information and portions called *introns* that do not convey genetic information.

An **exon** *is a DNA segment that conveys (codes for) genetic information. Ex*ons are DNA segments that help *ex*press a genetic message. An **intron** *is a DNA segment that does not convey (code for) genetic information. Int*rons are DNA segments that *int*errupt a genetic message. A gene consists of alternating exon and intron segments (Figure 22.12). At present, it is not known why introns occur in genes, but determining their function is an active area of biochemical research.

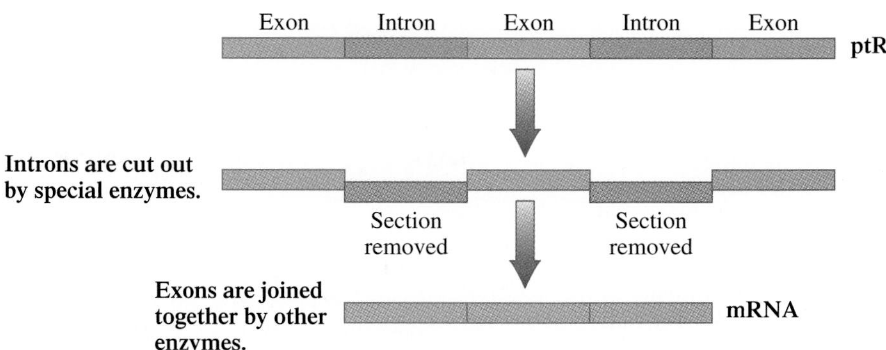

Figure 22.12 Primary transcript RNA contains both exons and introns. Messenger RNA is primary transcript RNA from which the introns have been excised.

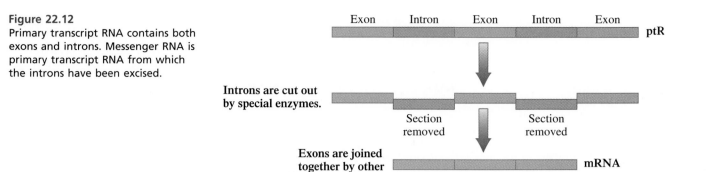

Both the exons and the introns of a gene are transcribed during production of *primary transcript RNA* (ptRNA). The ptRNA is then "edited," under the direction of enzymes, to remove the introns. The remaining exons are joined together to form a shortened RNA strand that carries the genetic information of the transcribed gene. This "edited" RNA is the messenger RNA (mRNA) that serves as a blueprint for protein assembly.

22.9 The Genetic Code

The nucleotide (base) sequence of an mRNA molecule is the informational part of such a molecule. This base sequence in a given mRNA determines the amino acid sequence for the protein synthesized under that mRNA's direction.

How can the base sequence of an mRNA molecule (which involves only *4* different bases—A, C, G, and U) encode enough information to direct proper sequencing of *20* amino acids in proteins? If each base encoded for a particular standard amino acid, then only 4 amino acids would be specified out of the 20 needed for protein synthesis, a clearly inadequate number. If two-base sequences were used to code amino acids, then there would be $4^2 = 16$ possible combinations, so 16 amino acids could be represented uniquely. This is still an inadequate number. If three-base sequences were used to code for amino acids, there would be $4^3 = 64$ possible combinations, which is more than enough combinations for uniquely specifying each of the 20 standard amino acids found in proteins.

Research has verified that sequences of three nucleotides in mRNA molecules specify the amino acids that go into synthesis of a protein. Such three-nucleotide sequences are called codons. A **codon** *is a sequence of three nucleotides in an mRNA molecule that codes for a specific amino acid.*

Which amino acid is specified by which codon? (We have 64 codons to choose from.) Researchers deciphered codon–amino acid relationships by adding different *synthetic* mRNA molecules (whose base sequences were known) to cell extracts and then determining the structure of any newly formed protein. After many such experiments, researchers finally matched all 64 possible codons with their functions in protein synthesis. It was found that 61 of the 64 codons formed by various combinations of the bases A, C, G, and U were related to specific amino acids; the other 3 combinations were termination codons ("stop" signals) for protein synthesis. Collectively, these relationships between three-nucleotide sequences in mRNA and amino acid identities are known as the genetic code. The **genetic code** *gives the assignment of the 64 mRNA codons to specific amino acids (or stop signals).* The determination of this code is one of the most remarkable of twentieth-century scientific achievements. The 1968 Noble Prize in chemistry was awarded to Marshall Nirenberg and Gobind Khovana for their work in illuminating this essential code.

The complete genetic code is given in Table 22.2. Examination of this table indicates that the genetic code has several remarkable features.

Table 22.2
The Universal Genetic Code. The code is composed of 64 three-nucleotide sequences (codons), which can be read from the table. The left-hand column indicates the nucleotide base found in the first (5′) position of the codon. The nucleotides in the second (middle) position of the codon are in the middle columns. The right-hand column indicates the nucleotide found in the third (3′) position. Thus the codon ACG encodes for the amino acid Thr, and the codon GGG encodes for the amino acid Gly.

First position (5′ end)	Second Position				Third position (3′ end)
	U	C	A	G	
U	Phe	Ser	Tyr	Cys	U
	Phe	Ser	Tyr	Cys	C
	Leu	Ser	Stop	Stop	A
	Leu	Ser	Stop	Trp	G
C	Leu	Pro	His	Arg	U
	Leu	Pro	His	Arg	C
	Leu	Pro	Gln	Arg	A
	Leu	Pro	Gln	Arg	G
A	Ile	Thr	Asn	Ser	U
	Ile	Thr	Asn	Ser	C
	Ile	Thr	Lys	Arg	A
	Met	Thr	Lys	Arg	G
G	Val	Ala	Asp	Gly	U
	Val	Ala	Asp	Gly	C
	Val	Ala	Glu	Gly	A
	Val	Ala	Glu	Gly	G

• There is a rough correlation between the number of codons for a particular amino acid and that amino acid's frequency of occurrence in proteins. For example, the two amino acids that have a single codon, Met and Trp, are two of the least common amino acids in proteins.

1. *The genetic code is highly degenerate; that is, many amino acids are designated by more than one codon.* Three amino acids (Arg, Leu, and Ser) are represented by six codons. Two or more codons exist for all other amino acids except Met and Trp, which have only a single codon. Codons that specify the same amino acid are called *synonyms.*
2. *There is a pattern to the arrangement of synonyms in the genetic code table.* All synonyms for an amino acid fall within a single box in Table 22.2, unless there are more than four synonyms, where two boxes are needed. The significance of the "single box" pattern is that with synonyms, the first two bases of the codon are the same—they differ only in the third base. For example, the four synonyms for the amino acid Pro are CCU, CCC, CCA, and CCG.
3. *The genetic code is almost universal.* Although Table 22.2 does not show this feature, studies of many organisms indicate that with minor exceptions, the code is the same in all of them. The same codon specifies the same amino acid whether the cell is a bacterial cell, a corn plant cell, or a human cell.
4. *An initiation codon exists.* The existence of "stop" codons (UAG, UAA, and UGA) suggests the existence of "start" codons. There is one initiation codon. Besides coding for the amino acid methionine, the codon AUG functions as an initiator of protein synthesis when it occurs as the first codon in an amino acid sequence.

Example 22.3

Using the Genetic Code and mRNA Codons to Predict Amino Acid Sequences

Using the genetic code in Table 22.2, determine the sequence of amino acids encoded by the mRNA codon sequence

5′–GCC–AUG–GUA–AAA–UGC–GAC–CCA–3′

Solution

Matching the codons with the right amino acids yields

mRNA: 5′–GCC–AUG–GUA–AAA–UGC–GAC–CCA–3′

Peptide: Ala- Met- Val- Lys- Cys- Asp- Pro-

Practice Exercise 22.3

Using the genetic code in Table 22.2, determine the sequence of amino acids encoded by the mRNA codon sequence

5′–CAU–CCU–CAC–ACU–GUU–UGU–UGG–3′

• *Answer:* His- Pro- His- Thr- Val- Cys- Trp-

22.10 Anticodons and tRNA Molecules

The amino acids used in protein synthesis do not directly interact with the codons of a mRNA molecule. Instead, tRNA molecules function as intermediaries that deliver amino acids to the mRNA. At least one type of tRNA molecule exists for each of the 20 amino acids found in proteins.

All tRNA molecules have the same general shape, and this shape is crucial to how they function. Figure 22.13 shows the general *two-dimensional* "cloverleaf" shape of a tRNA molecule, a shape produced by the molecule's folding and twisting into regions of parallel strands and regions of hairpin loops. (The actual three-dimensional shape of a tRNA molecule involves considerable additional twisting of the "cloverleaf" shape.)

Two features of the tRNA structure are of particular importance.

1. The 3′ end of the open part of the cloverleaf structure is where an amino acid becomes *covalently* bonded to the tRNA molecule through an ester bond. Each

Figure 22.13
A tRNA molecule. The amino acid attachment site is at the open end of the cloverleaf (the 3′ end), and the anticodon is located in the loop opposite the open end.

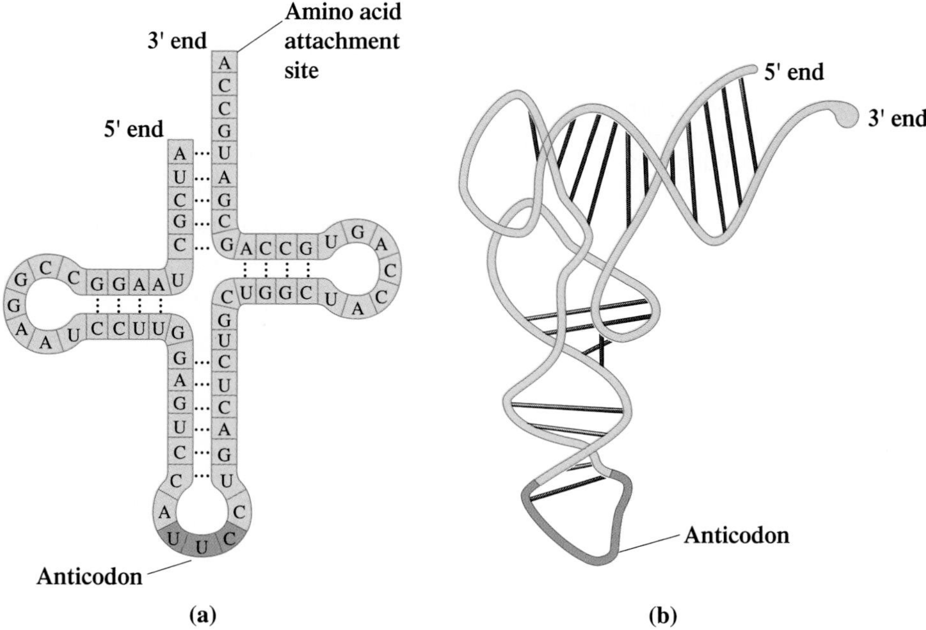

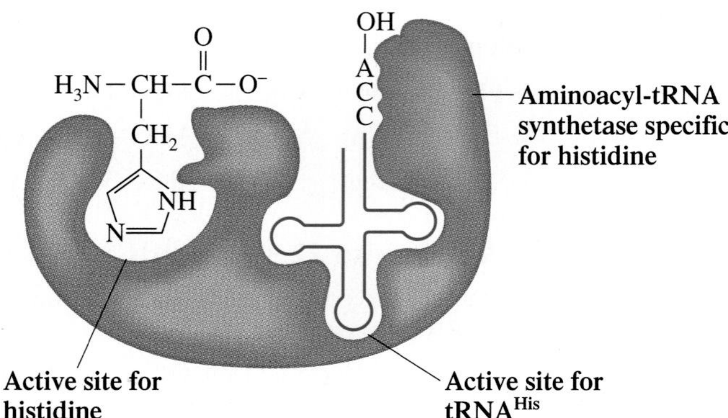

Figure 22.14
An aminoacyl–tRNA synthetase has an active site for tRNA and a binding site for the particular amino acid that is to be attached to that tRNA.

of the different tRNA molecules is specifically recognized by an *aminoacyl synthetase* enzyme. These enzymes also recognize the one kind of amino acid that "belongs" with the particular tRNA and facilitates its bonding to the tRNA (see Figure 22.14).

2. The loop *opposite* the open end of the cloverleaf is the site for a sequence of three bases called an anticodon. An **anticodon** *is a three-nucleotide sequence in tRNA that is complementary to the mRNA codon for the amino acid that bonds to the tRNA.*

The interaction between the anticodon of the tRNA and the codon of the mRNA leads to the proper placement of an amino acid into a growing peptide chain during protein synthesis. This interaction, which involves complementary base pairing, is shown in Figure 22.15.

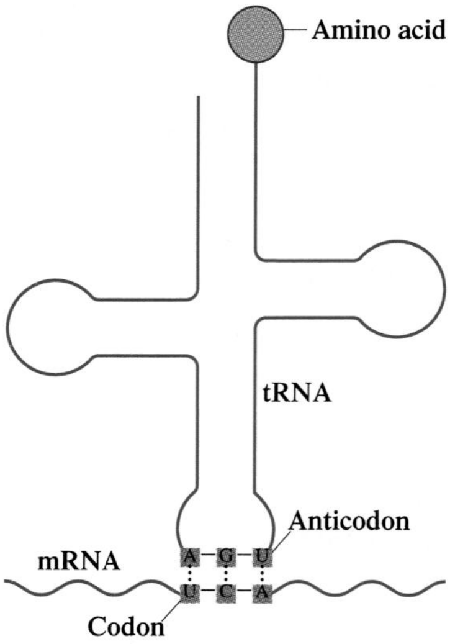

Amino acid
tRNA
Anticodon
mRNA
A G U
U C A
Codon

Figure 22.15
The interaction between anticodon (tRNA) and codon (mRNA), which involves complementary base pairing, governs the proper placement of amino acids in a protein.

22.11 Translation: Protein Synthesis

The substances needed for the translation phase of protein synthesis are mRNA molecules, tRNA molecules, amino acids, ribosomes, and a number of different enzymes. Ribosomes, which serve as sites for protein synthesis, have structures involving two subunits—a large subunit and a small subunit (see Figure 22.16). Each subunit has a composition of approximately 65% rRNA and 35% protein. The rRNA helps maintain the structure of the ribosome and also provides sites where mRNA can attach itself to the ribosome.

There are five general steps to the translation process: (1) activation of tRNA, (2) initiation, (3) elongation, (4) termination, and (5) post-translational processing.

• Activation of tRNA

There are two steps involved in tRNA activation. First, an amino acid interacts with an activator molecule (ATP; Section 23.3) to form a highly energetic complex. This complex

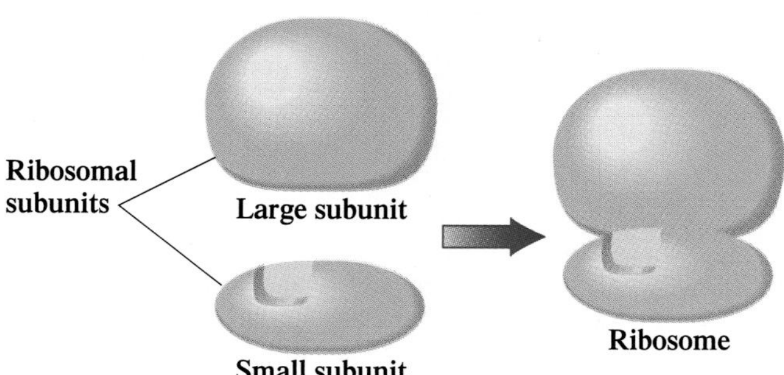

Figure 22.16
Ribosomes, which contain both rRNA and protein, have structures that contain two subunits. One subunit is much larger than the other.

then reacts with the appropriate tRNA molecule to produce an *activated tRNA molecule,* a tRNA molecule that has an amino acid covalently bonded to it at its 3′ end through an ester linkage.

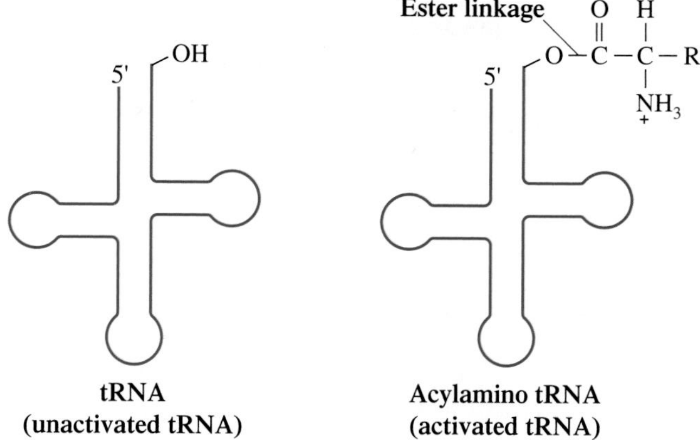

• Initiation

The initiation of protein synthesis begins when mRNA attaches itself to the surface of a small ribosomal subunit such that its first codon, which is always the initiating codon AUG, occupies a site called the P site (peptidyl site). (See Figure 22.17a.) An activated tRNA molecule with anticodon complementary to the codon AUG attaches itself, through complementary base pairing, to the AUG codon (Figure 22.17b). The resulting complex then interacts with a large ribosomal subunit to complete the formation of an initiation complex (Figure 22.17c).

When functioning as the initiating codon, AUG codes for a derivative of methionine, *N*-formyl methionine, rather than for methionine itself. This derivative, designated as f-MET, has the structure

$$\underbrace{\mathrm{H{-}\overset{O}{\overset{\|}{C}}}}_{\text{Formyl group}}\mathrm{{-}NH{-}\underset{\underset{\displaystyle CH_2{-}S{-}CH_3}{|}}{\underset{CH_2}{\underset{|}{CH}}}{-}\overset{O}{\overset{\|}{C}}{-}O^-}$$

Figure 22.17
Initiation of protein synthesis begins with the formation of an initiation complex.

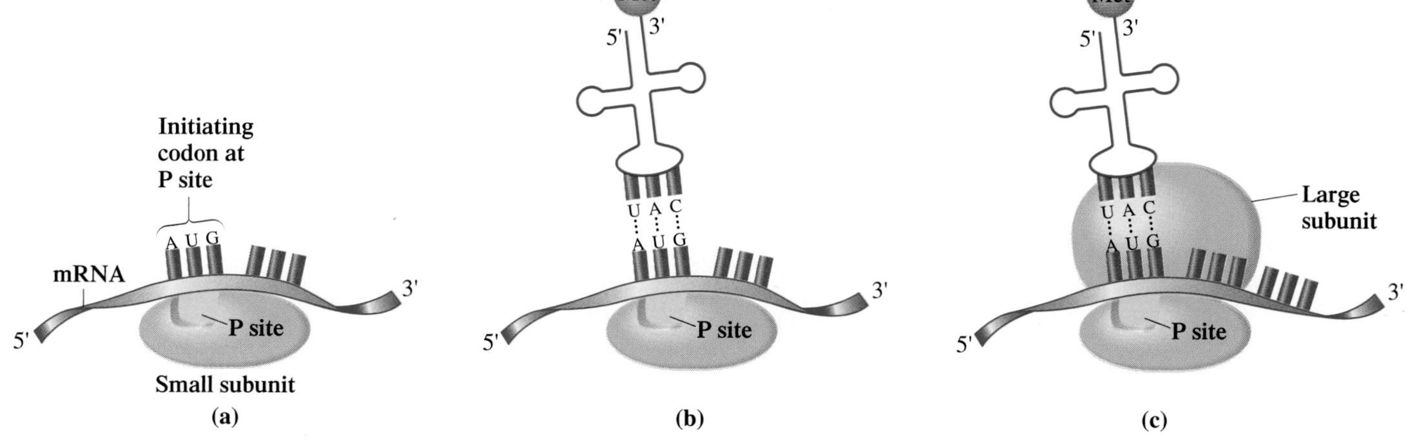

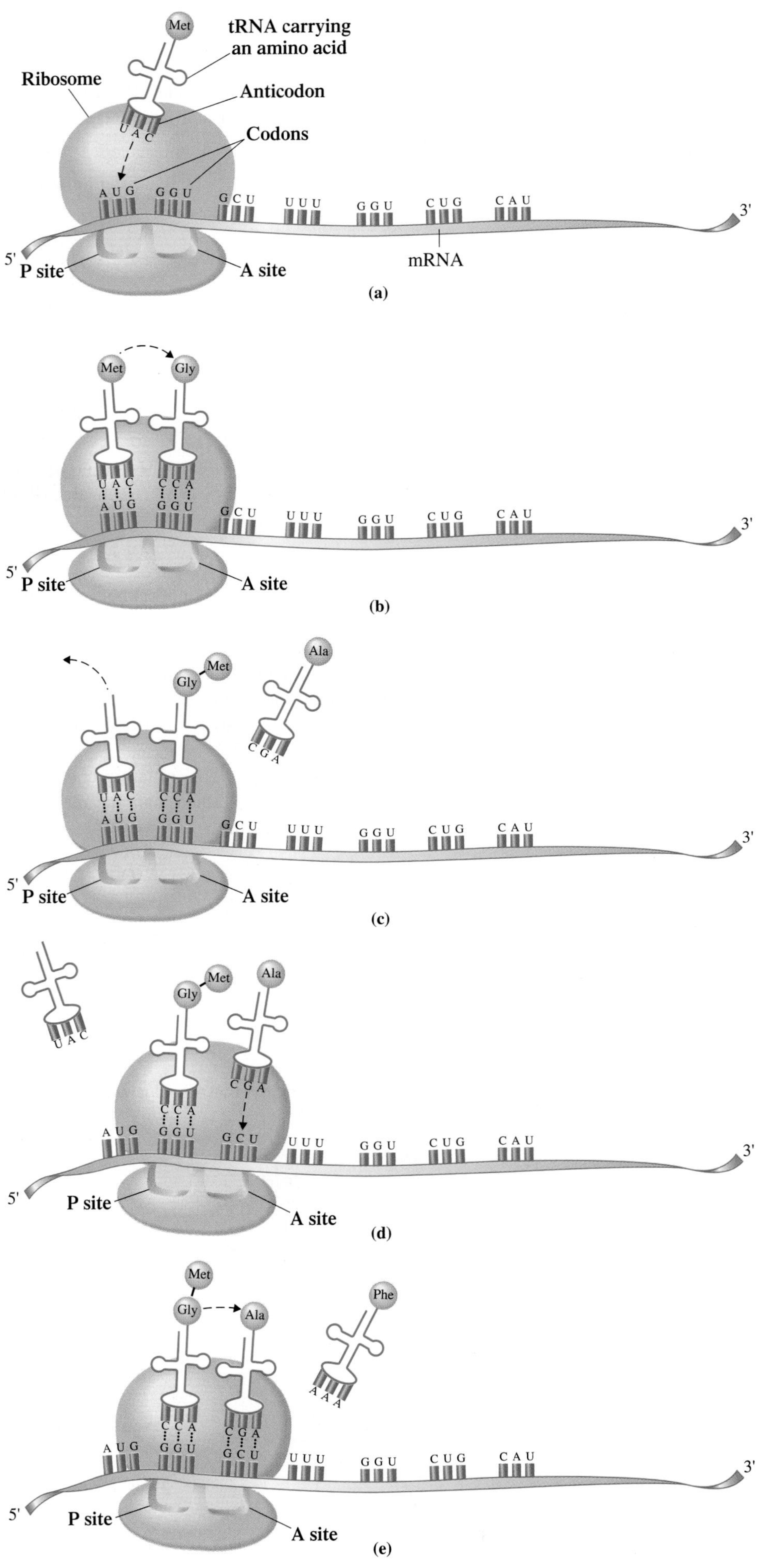

Figure 22.18
The process of translation that occurs during protein synthesis. The anticodons of tRNA molecules are paired with the codons of an mRNA molecule to bring the appropriate amino acids into sequence for protein formation.

• Elongation

Next to the P site in an mRNA–ribosome complex is a second binding site called the A site (aminoacyl site). (See Figure 22.18a). At this second site the next mRNA codon is exposed, and a tRNA with the appropriate anticodon binds to it (Figure 22.18b). With amino acids in place at both P and the A sites, the enzyme *peptidyl transferase* effects the linking of the P site amino acid to the A site amino acid to form a dipeptide. Such peptide bond formation leaves the tRNA at the P site empty and the tRNA at the A site bearing the dipeptide (Figure 22.18c).

The empty tRNA at the P site now leaves that site and is free to pick up another molecule of its specific amino acid. Simultaneously with the release of tRNA from the P site, the ribosome shifts along the mRNA. This shift puts the newly formed dipeptide at the P site, and the third codon of mRNA is now available, at site A, to accept a tRNA molecule whose anticodon complements this codon (see Figure 22.18d).

• In elongation, the polypeptide chain grows one amino acid at a time.

Now a repetitious process begins. The third codon, now at the A site, accepts an incoming tRNA with its accompanying amino acid; and then the entire dipeptide at the P site is transferred and bonded to the A site amino acid to give a tripeptide (see Figure 22.18e). The empty tRNA at the P site is released, the ribosome shifts along the mRNA, and the process continues.

• Termination

The polypeptide continues to grow by way of translation until all necessary amino acids are in place and bonded to each other. Appearance in the mRNA codon sequence of one of the three stop codons (UAA, UAG, or UGA) terminates the process. No tRNA has an anticodon that can base-pair with these stop codons. The polypeptide is then cleaved from the tRNA through hydrolysis.

• Post-Translation Processing

Some modification of proteins usually occurs after translation. For example, most proteins do not have f-Met (the initiation codon) as their first amino acid. Cleavage of N-terminal f-Met is part of post-translation processing. Formation of S—S bonds between cysteine units is another example of post-translation processing.

• Efficiency of mRNA Utilization

Many ribosomes can move simultaneously along a single mRNA molecule (Figure 22.19). In this highly efficient arrangement, many identical protein chains can be synthesized almost at the same time from a single strand of mRNA. This multiple use of mRNA molecules reduces the amount of resources and energy expended by the cell to synthesize

Figure 22.19
Several ribosomes can simultaneously proceed along a single strand of mRNA one after another. Such a complex of mRNA and ribosomes is called a polysome.

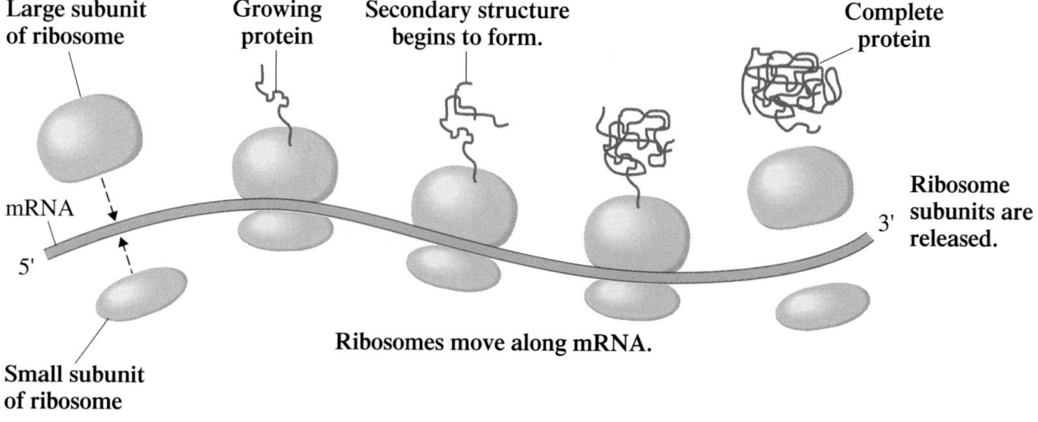

Chemical CONNECTIONS

22.3 Antibiotics That Inhibit Bacterial Protein Synthesis

Some antibiotics work because they inhibit protein synthesis in bacteria but not in humans. They inhibit one specific enzyme or another in the bacterial ribosomes. These antibiotics are useful in treating disease and in studying protein synthesis mechanisms in bacteria. The accompanying table lists a few of the most commonly encountered antibiotics and their modes of action.

Antibiotic	Biological action
chloramphenicol	inhibits an important enzyme (peptidyl transferase) in the large ribosomal subunit
erythromycin	binds to the large subunit and stops the ribosome from moving along the mRNA from one codon to the next
puromycin	induces premature polypeptide chain termination
streptomycin	inhibits initiation of protein synthesis and also causes the mRNA codons to be read incorrectly
tetracycline	binds to the small ribosomal subunit and inhibits the binding of incoming tRNA molecules

needed protein. Such complexes of several ribosomes and mRNA are called polyribosomes or polysomes. A **polysome** *is a complex of mRNA and several ribosomes.*

The accompanying Chemistry at a Glance summarizes the steps in protein synthesis.

22.12 Mutations

Mutations *are changes in the base sequence in DNA molecules.* These changes alter the genetic information that is passed on during transcription. The altered information can cause changes in amino acid sequence during protein synthesis. Sometimes, such changes have a profound effect on an organism.

Mutagens *are the substances or agents that cause a change in the structure of a DNA molecule.* Radiation and chemical agents are two important types of mutagens. Radiation, in the form of ultraviolet light, X rays, radioactivity (Chapter 11), and cosmic rays, has the potential to be mutagenic. Ultraviolet light from the sun is the radiation that causes sunburn and can cause changes in the DNA of the skin cells. Sustained exposure to ultraviolet light can lead to serious problems such as skin cancer.

Chemical agents can also have mutagenic effects. Nitrous acid (HNO_2) is a mutagen that causes deamination of heterocyclic nitrogen bases. For example, HNO_2 can convert cytosine to uracil.

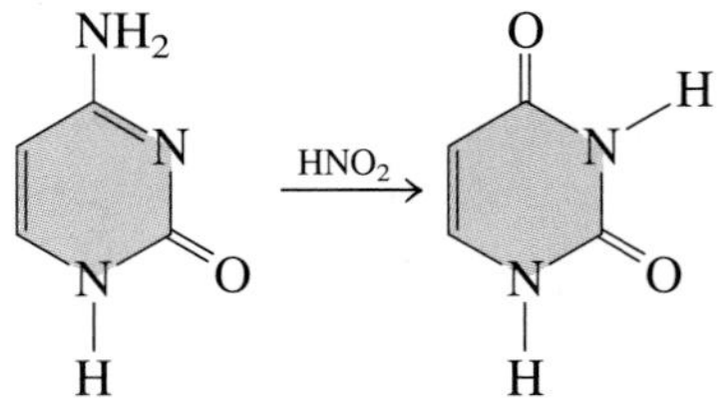

Cytosine $\xrightarrow{HNO_2}$ Uracil

Deamination of a cytosine that was part of an mRNA codon would change the codon; for example, CGG would become UGG.

A variety of chemicals—including nitrites, nitrates, and nitrosamines—can form nitrous acid in the body. The use of nitrates and nitrites as preservatives in foods such as bologna and hot dogs is a cause of concern because of their conversion to nitrous acid in the body and possible damage to DNA.

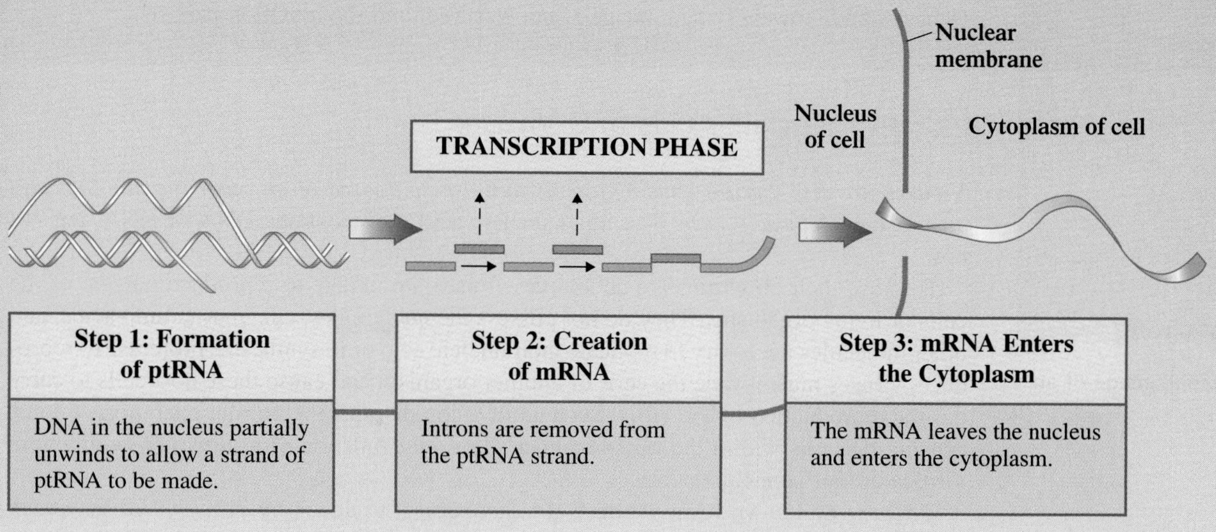

Step 1: Formation of ptRNA

DNA in the nucleus partially unwinds to allow a strand of ptRNA to be made.

Step 2: Creation of mRNA

Introns are removed from the ptRNA strand.

Step 3: mRNA Enters the Cytoplasm

The mRNA leaves the nucleus and enters the cytoplasm.

TRANSLATION PHASE

Met
3' end
5' end
Anticodon

Step 4: Activation of tRNA

An amino acid interacts with ATP to become highly energized. It then forms a covalent bond with the 3' end of a tRNA molecule. Amino acid–tRNA pairing is governed by enzymes.

Ribosome
Codons
mRNA
P site
A site

Step 5: Initiaton

The mRNA attaches to a ribosome so that the first codon (AUG) is at the P site. A tRNA carrying N-formyl methionine, an amino acid derivitave, attaches to the first codon.

Step 6: Elongation

Another tRNA with the second amino acid binds at the A site. The methionine derivative transfers from the P site to the A site. The ribosome shifts to the next codon, making its A site available for the tRNA carrying the third amino acid.

Steps 7 and 8: Termination and Post-Translation Processing

The polypeptide chain continues to lengthen until a stop codon appears on the mRNA. The new protein is cleaved from the last tRNA.

During post-translation processing, cleavage of f-met (the initiation codon) usually occurs. S—S bonds between cys units also can form.

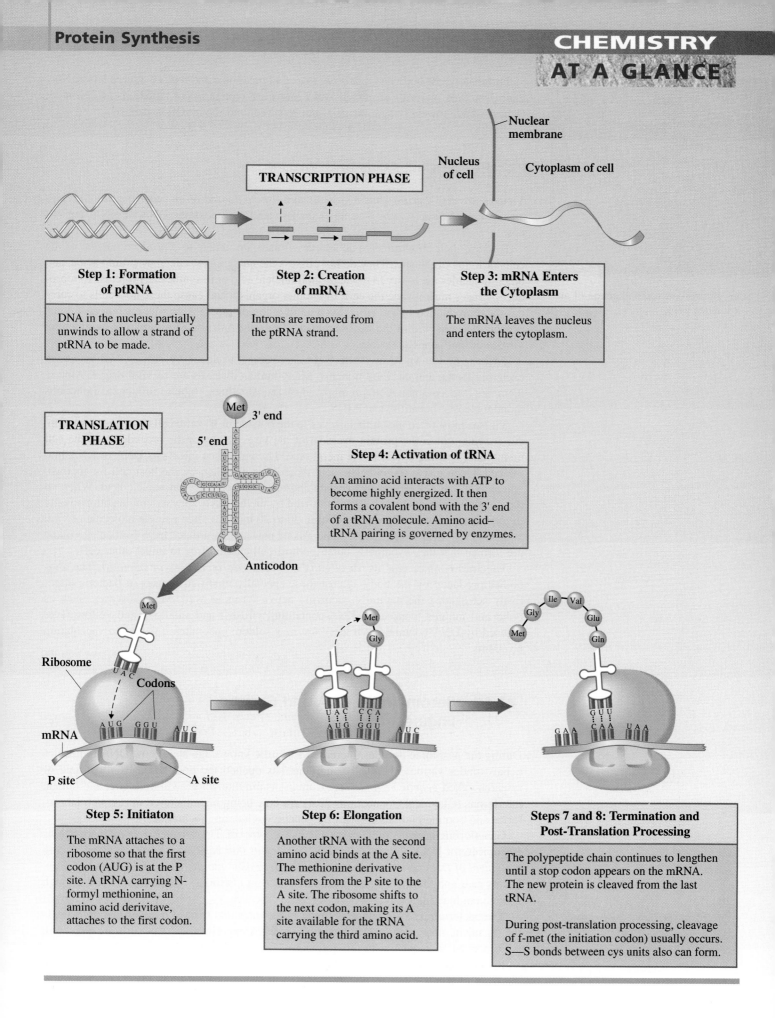

An electron microscope image of an influenza virus.

Fortunately, the body has *repair enzymes* that recognize and replace altered bases. Normally, the vast majority of altered DNA bases are repaired, and mutations are avoided. Occasionally, however, the damage is not repaired, and the mutation persists.

22.13 Nucleic Acids and Viruses

Viruses *are tiny disease-causing agents that are composed of an outer protein coat and an inner nucleic acid core.* The inner nucleic acid core contains DNA or RNA, but not both.

Because their structures are so simple, viruses are unable to reproduce outside of the cells of living organisms. They do not possess the nucleotides, enzymes, amino acids, and other molecules necessary to replicate their nucleic acid or to synthesize proteins. To reproduce, viruses must invade the cells of another organism and cause these host cells to carry out the reproduction of the virus. Such an invasion disrupts the normal operation of cells, causing diseases within the host organism. The only function of a virus is reproduction; viruses do not generate energy.

There is no known form of life that is not subject to attack by viruses. Viruses attack bacteria, plants, animals, and humans. Many human diseases are of viral origin. Among them are the common cold, mumps, measles, smallpox, rabies, influenza, infectious mononucleosis, hepatitis, and AIDS.

Viruses most often attach themselves to the outside of specific cells in a host organism. An enzyme within the protein overcoat of the virus catalyzes the breakdown of the cell membrane, opening a hole in the membrane. The virus then injects its DNA or RNA into the cell. Once inside, this nucleic acid material is mistaken by the host cell for its own, whereupon that cell begins to translate and/or transcribe the viral nucleic acid. When all the virus components have been synthesized by the host cell, they assemble automatically to form many new virus particles. Within 20 to 30 minutes after a single molecule of viral nucleic acid enters the host cell, hundreds of new virus particles have formed. So many are formed that they eventually burst the host cell and are free to infect other cells.

Vaccines *contain inactive or slightly altered forms of viruses or bacteria.* The antibodies produced by the body against these specially modified viruses or bacteria effectively act against the naturally occurring active forms as well. Many diseases, such as polio and mumps (caused by RNA-containing viruses) and smallpox and yellow fever (caused by DNA-containing viruses), are now seldom encountered thanks to vaccination programs.

22.14 Recombinant DNA and Genetic Engineering

During the past three decades, increased scientific knowledge about how DNA molecules behave under various chemical conditions has opened the door to an exciting field of research called *genetic engineering.* Human insulin, human growth hormone, and human interferons (Chemical Connections 22.4) are now being manufactured in research laboratories and commercial laboratories by genetic engineering techniques.

Genetic engineering procedures involve a type of DNA called recombinant DNA. **Recombinant DNA** *molecules are DNA molecules that have been synthesized by splicing a segment of DNA (usually a gene) from one organism into the DNA of another organism.* Let us examine the theory and procedures used in obtaining recombinant DNA through genetic engineering.

The bacterium *E. coli,* which is found in the intestinal tract of humans and animals, is the organism most often used in recombinant DNA experiments. Yeast cells are also used, with increasing frequency, in this research.

In addition to their chromosomal DNA, *E. coli* (and other bacteria) contain DNA in the form of small, circular, double-stranded molecules called *plasmids.* These plasmids, which carry only a few genes, replicate independently of the chromosome. Also, they are transferred relatively easily from one cell to another. Plasmids from *E. coli* are used in recombinant DNA work.

Let us consider the procedure used to obtain *E. coli* cells that contain recombinant DNA (see Figure 22.20).

Step 1: *E. coli* cells of a specific strain are placed in a solution that dissolves plasma membranes, thus releasing the contents of the cells.

Step 2: The released cell components are separated into fractions, one fraction being the plasmids. The isolated plasmid fraction is the material used in further steps.

Step 3: A special enzyme, called a *restriction enzyme,* is used to cleave the double-stranded DNA of a circular plasmid. The result is a linear (noncircular) DNA molecule.

Step 4: The same restriction enzyme is then used to remove a desired gene from a chromosome of another organism.

Step 5: The gene (from Step 4) and the opened plasmid (from Step 3) are mixed in the presence of the enzyme *DNA ligase,* which splices the two together. This splicing, which attaches one end of the gene to one end of the opened plasmid and attaches the other end of the gene to the other end of the plasmid, results in an altered circular plasmid (the recombinant DNA).

Step 6: The altered plasmids (recombinant DNA) are placed in a live *E. coli* culture, where they are taken up by the *E. coli* bacteria. The *E. coli* culture into which the plasmids are placed need not be identical to that from which the plasmids were originally obtained.

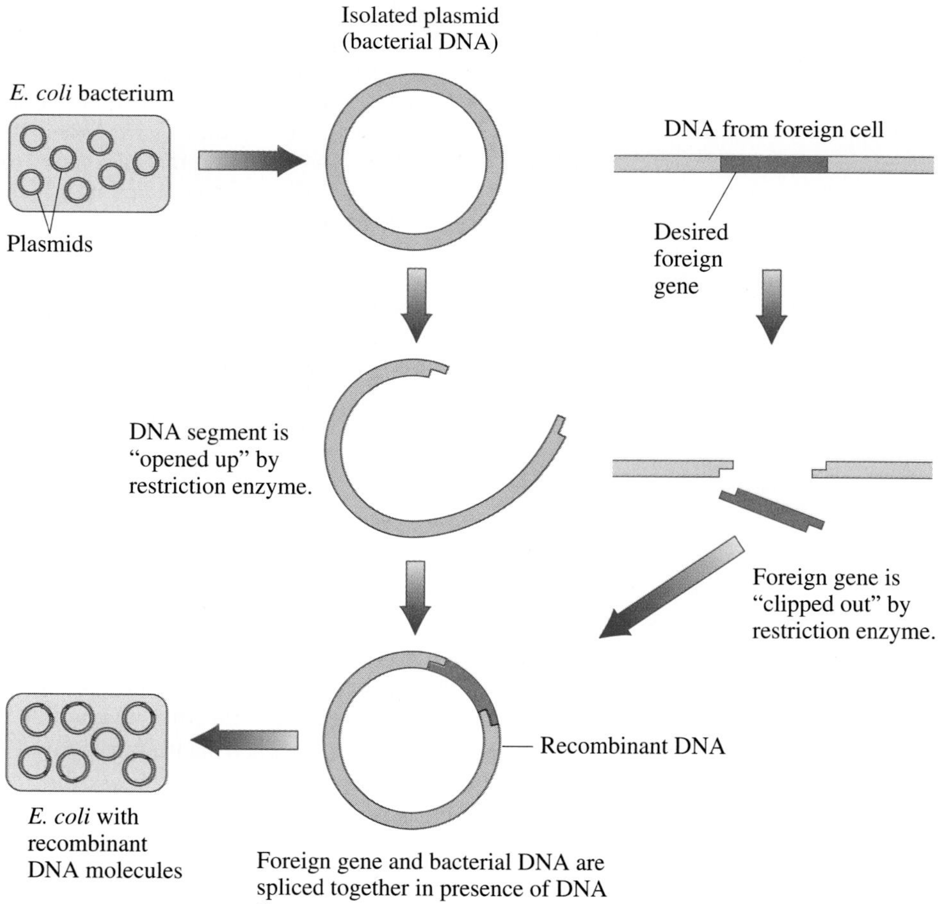

Figure 22.20
Recombinant DNA is made by inserting a gene obtained from a cell of one kind of organism into the DNA of another kind of organism.

Chemical CONNECTIONS

22.4 Polymerase Chain Reaction and DNA Sequencing

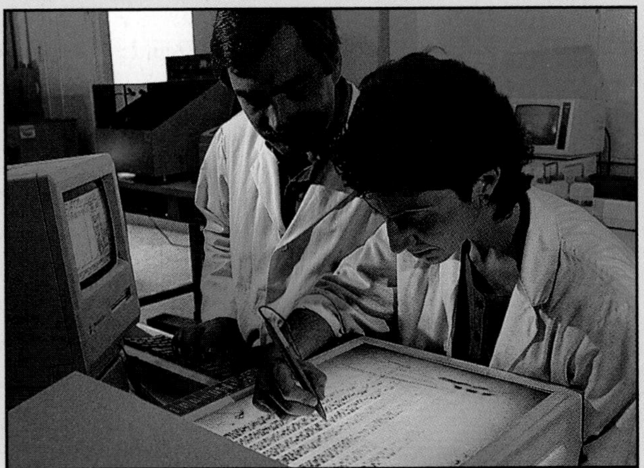

An understanding of DNA replication has helped scientists develop new DNA-based procedures with applications from diagnosing disease to convicting criminals. Two of the most exciting techniques are polymerase chain reaction and DNA sequencing.

The polymerase chain reaction (PCR) *is a method for rapidly producing multiple copies of a DNA nucleotide sequence.* Billions of copies of a specific DNA sequence (gene) can be produced in a few hours. The PCR is easy to carry out, requiring only a few chemicals, a test tube, and a source of heat.

With PCR, DNA that is available only in very small quantities can be amplified to quantities large enough to analyze. The process, devised in 1983, has become a valuable tool for diagnosing diseases and detecting pathogens in the body. It is now used in prenatal diagnosis of a number of genetic diseases, including muscular dystrophy and cystic fibrosis, and in the identification of bacterial pathogens. It is also the definitive way to detect the AIDS virus.

The PCR process has also proved useful in certain types of forensic investigations. A DNA sample may be obtained from a single drop of blood or semen or a single strand of hair at a crime scene and amplified by the PCR process. A forensic chemist can then compare the amplified samples with DNA samples taken from suspects. Work with DNA in the forensic area is often referred to as *DNA fingerprinting.*

The basis of the PCR technique is DNA polymerase, an enzyme present in all living organisms. DNA polymerase can attach additional nucleotides to a short starter nucleotide chain, called the primer, when the primer is bound to a complementary strand of DNA that functions as a template. The original DNA is heated to separate its strands, and then primers, DNA polymerase, and deoxyribonucleotides are added so that the polymerase can replicate the original strand. The process is repeated until, in a short time, millions of DNA molecules have been produced.

Another technique, discovered in 1977, is also based on our understanding of DNA replication. **DNA sequencing** *is a process in which the exact base sequence in a DNA molecule (or a portion of it) is determined.* Because DNA molecules are too large to sequence as a whole, restriction enzymes are used to cut the molecules into smaller fragments, which are sequenced. By identifying the points of overlap of the fragments, it is possible to determine the sequence of the entire original molecule.

DNA sequencing is the basis for a large-scale research project now under way called the Human Genome Project (HGP). A **genome** *is all of the genetic material in the chromosomes of an organism,* Its size is given as the number of base pairs. The HGP is a massive attempt to sequence the entire human genome, some 3.3 billion base pairs spread over 23 pairs of chromosomes and 50,000 to 100,000 genes. This project, started formally in the early 1990s, is a worldwide effort that is expected to be completed by the year 2005 and is proceeding on schedule.

The goal of the HGP is to determine the location and base sequence of each of the genes in the human genome. It is hoped that the results of this project will lead to increased understanding of the nearly 4000 genetic disorders known to afflict humans, as well as an increased understanding of the many "normal" human genetic traits.

In Step 3 we noted that the conversion of a circular plasmid into a linear DNA molecule requires a restriction enzyme. **Restriction enzymes** *are enzymes that recognize specific base sequences in DNA and cleave the DNA in a predictable manner at these sequences.* The discovery of restriction enzymes in the late 1960s and early 1970s made genetic engineering possible.

Restriction enzymes occur naturally in numerous types of bacterial cells. Their function is to protect the bacteria from invasion by foreign DNA by catalyzing the cleavage of the invading DNA. Their names are derived from their placing a "restriction" on the type of DNA allowed into the bacterial cells.

To understand how a restriction enzyme works, let us consider one that cleaves DNA between G and A bases in the 5′-to-3′ direction in the sequence G–A–A–T–T–C. This enzyme will cleave the double helix structure of a DNA molecule in the manner shown in Figure 22.21.

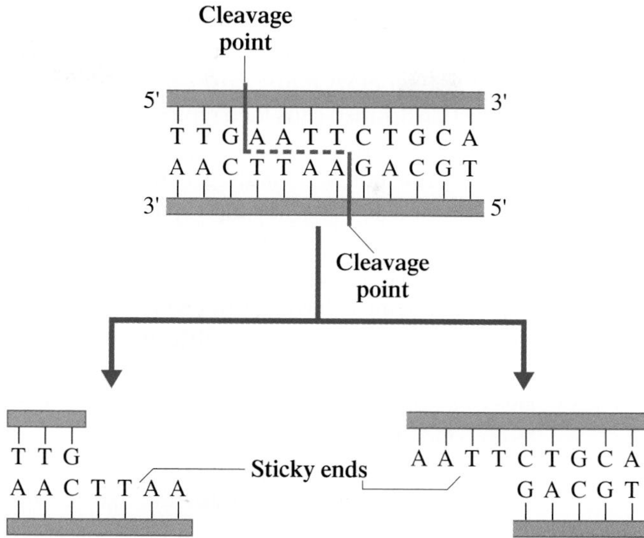

Figure 22.21
Cleavage pattern resulting from the use of a restriction enzyme that cleaves DNA between G and A bases in the 5′-to-3′ direction in the sequence G–A–A–T–T–C. The double helix structure is not cut straight across.

Note that the double helix is not cut straight across; the individual strands are cut at different points, giving a staircase cut. (Both cuts must be between G and A in the 5′-to-3′ direction.) This staircase cut leaves unpaired bases on each cut strand. These ends with unpaired bases are called "sticky ends" because they are ready to "stick to" (pair up with) a complementary section of DNA if they can find one.

If the same restriction enzyme used to cut a plasmid is also used to cut a gene from another DNA molecule, the sticky ends of the gene will be complementary to those of the plasmid. This enables the plasmid and gene to combine readily, forming a new, modified plasmid molecule. This modified plasmid molecule is called recombinant DNA. In addition to the newly spliced gene, the recombinant DNA plasmid contains all of the genes and characteristics of the original plasmid. Figure 22.22 shows diagrammatically the match between sticky ends that occurs when plasmid and gene combine.

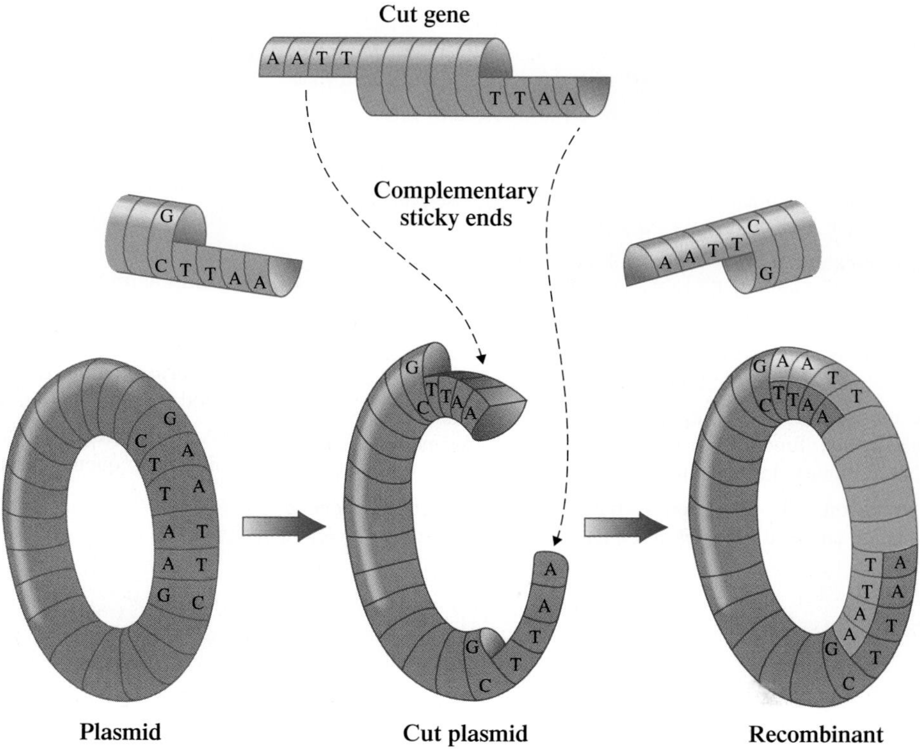

Figure 22.22
The "sticky ends" of the cut plasmid and the cut gene are complementary and combine to form recombinant DNA.

Step 6 involves inserting the recombinant DNA (modified plasmids) back into *E. coli* cells. The process is called transformation. **Transformation** *is the process of incorporating foreign DNA into a host cell.*

The transformed cells then reproduce, resulting in large numbers of identical cells called clones. **Clones** *are cells that have descended from a single cell and have identical DNA.* Within a few hours, a single genetically altered bacterial cell can give rise to thousands of clones. Each clone has the capacity to synthesize the protein directed by the foreign gene it carries.

Researchers are not limited to selection of naturally occurring genes for transforming bacteria. Chemists have developed nonenzymatic methods of linking nucleotides together such that they can construct artificial genes of any sequence they desire. In fact, benchtop instruments are now available that can be programmed by a microprocessor to synthesize any DNA base sequence *automatically.* The operator merely enters a sequence of desired bases, starts the instrument, and returns later to obtain the product. This synthesis requires only about 15 to 20 minutes per nucleotide. Such flexibility in manufacturing DNA has opened many doors, accelerated the pace of recombinant DNA research, and redefined the term *designer genes!*

Concepts to Remember

Nucleic acids. Nucleic acids are polymeric molecules in which the repeating units are nucleotides. Cells contain two kinds of nucleic acids—deoxyribonucleic acids (DNA) and ribonucleic acids (RNA). The major biological functions of DNA and RNA are, respectively, transfer of genetic information and synthesis of proteins.

Nucleotides. Nucleotides, the monomers of nucleic acid polymers, are molecules composed of a pentose sugar bonded to both a phosphate group and a nitrogen-containing heterocyclic base. The pentose sugar must be either ribose or deoxyribose. Five nitrogen-containing bases are found in nucleotides: adenine (A), guanine (G), cytosine (C), thymine (T), and uracil (U).

Primary structure of nucleic acids. The “backbone” of a nucleic acid molecule is a constant alternating sequence of sugar and phosphate groups. Each sugar unit has a nitrogen-containing base attached to it.

Secondary structure of DNA. A DNA molecule exists as two polynucleotide chains coiled around each other in a double helix arrangement. The double helix is held together by hydrogen bonding between complementary pairs of bases. Only two base-pairing combinations occur: A with T, and C with G.

DNA replication. DNA replication occurs when the two strands of a parent DNA double helix separate and act as templates for the synthesis of new chains using the principle of complementary base pairing.

Chromosome. A chromosome is a cell structure that consists of an individual DNA molecule bound to a group of proteins.

RNA molecules. Four important types of RNA molecules, distinguished by their function, are ribosomal RNA (rRNA), messenger RNA (mRNA), primary transcript RNA (ptRNA), and transfer RNA (tRNA).

Transcription. Transcription is the process in which the genetic information encoded in the base sequence of DNA is copied into RNA molecules.

Complementary bases. Complementary bases are specific pairs of bases in nucleic acid structures that hydrogen-bond to each other.

Gene. A gene is a portion of a DNA molecule that contains the base sequences needed for the production of a specific protein. Genes are segmented, with portions called exons that contain genetic information and portions called introns that do not convey genetic information.

Codon. A codon is a three-nucleotide sequence in mRNA that codes for a specific amino acid needed during the process of protein synthesis.

Genetic code. The genetic code consists of all the mRNA codons that specify either a particular amino acid or the termination of protein synthesis.

Anticodon. An anticodon is a three-nucleotide sequence in tRNA that binds to a complementary sequence (a codon) in mRNA.

Translation. Translation is the stage of protein synthesis in which the codons in mRNA are translated into amino acid sequences of new proteins. Translation occurs at the ribosome and involves interactions between the codons of mRNA and the anticodons of tRNA.

Mutations. Mutations are changes in the base sequence in DNA molecules.

Recombinant DNA. Recombinant DNA molecules are synthesized by splicing a segment of DNA, usually a gene, from one organism into the DNA of another organism.

Polymerase chain reaction. The polymerase chain reaction is a method for rapidly producing many copies of a DNA sequence.

DNA sequencing. DNA sequencing is a multistep process for determining the sequence of bases in a DNA segment.

Key Reactions and Equations

1. Formation of a nucleotide (Section 22.2)

   Pentose sugar (ribose or deoxyribose) + phosphate group + nitrogen-containing heterocyclic base $\longrightarrow$

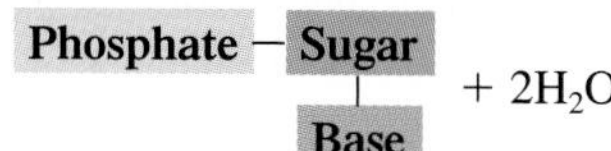

$+ 2H_2O$

2. Formation of a nucleic acid (Section 22.3)

   Many deoxyribose-containing nucleotides $\longrightarrow$ DNA

   Many ribose-containing nucleotides $\longrightarrow$ RNA

3. Protein synthesis (Section 22.6)

   $\text{DNA} \xrightarrow{\text{Transcription}} \text{RNA} \xrightarrow{\text{Translation}} \text{Protein}$

Key Terms

Anticodon (22.10)
Chromosome (22.5)
Clones (22.14)
Codon (22.9)
Complementary bases (22.4)
DNA replication (22.5)
DNA sequencing (22.14)
Exon (22.8)
Gene (22.8)
Genetic code (22.9)
Genome (22.14)
Interferons (22.14)
Intron (22.8)
Messenger RNA (22.7)
Mutagen (22.12)
Mutations (22.12)
Nucleic acids (22.2)
Nucleotide (22.2)
Polymerase chain reaction (22.14)
Primary nucleic acid structure (22.3)
Primary transcript RNA (22.7)
Polysome (22.12)
Recombinant DNA (22.14)
Restriction enzymes (22.14)
Ribosomal RNA (22.7)
Transcription (22.6)
Transfer RNA (22.7)
Transformation (22.14)
Translation (22.6)
Vaccines (22.13)
Viruses (22.13)

Exercises and Problems

The members of each pair of problems in this section test similar material.

Nucleotides (Section 22.2)

22.1 What is the structural difference between the pentose sugars ribose and 2-deoxyribose?

22.2 What are the names of the pentose sugars present, respectively, in DNA and RNA molecules?

22.3 Characterize each of the following nitrogen-containing bases as a purine derivative or a pyrimidine derivative.

a. Thymine b. Cytosine
c. Adenine d. Guanine

22.4 Characterize each of the following nitrogen-containing bases as a component of (1) both DNA and RNA, (2) DNA but not RNA, or (3) RNA but not DNA.

a. Adenine b. Thymine
c. Uracil d. Cytosine

22.5 How many different choices are there for each of the following subunits in the specified type of nucleotide?

a. Pentose sugar subunit in DNA nucleotides
b. Nitrogen-containing base subunit in RNA nucleotides
c. Phosphate subunit in DNA nucleotides

22.6 How many different choices are there for each of the following subunits in the specified type of nucleotide?

a. Pentose sugar subunit in RNA nucleotides
b. Nitrogen-containing base subunit in DNA nucleotides
c. Phosphate subunit in RNA nucleotides

22.7 Which nitrogen-containing base is present in each of the following nucleotides?

a. AMP b. dGMP
c. dTMP d. UMP

22.8 Which nitrogen-containing base is present in each of the following nucleotides?

a. GMP b. dAMP
c. CMP d. dCMP

22.9 Which pentose sugar is present in each of the nucleotides in Problem 22.7?

22.10 Which pentose sugar is present in each of the nucleotides in Problem 22.8?

22.11 Characterize as true or false each of the following statements about the given nucleotide.

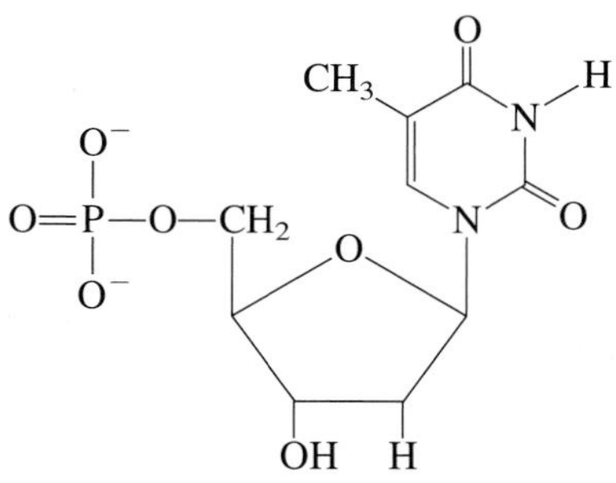

a. The nitrogen-containing base is a purine derivative.
b. The phosphate group is attached to the sugar unit at carbon 3′.
c. The sugar unit is ribose.
d. The nucleotide could be a component of both DNA and RNA.

22.12 Characterize as true or false each of the following statements about the given nucleotide.

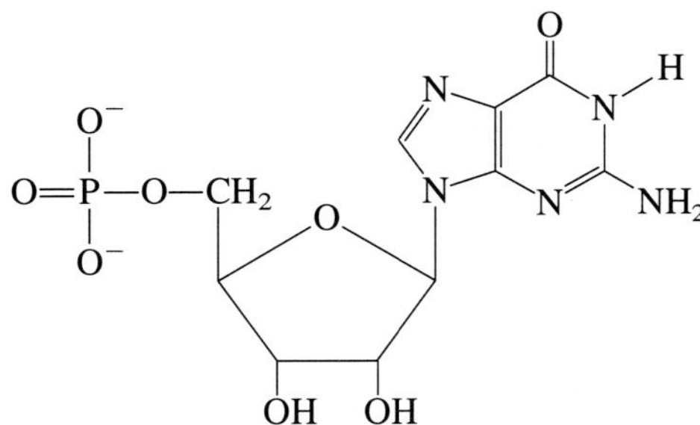

a. The sugar unit is 2-deoxyribose.
b. The sugar unit is attached to the nitrogen-containing base at nitrogen 3.
c. The nitrogen-containing base is a pyrimidine derivative.
d. The nucleotide could be a component of both DNA and RNA.

22.13 Draw the structures of the three products produced when the nucleotide in Problem 22.11 undergoes hydrolysis.

22.14 Draw the structures of the three products produced when the nucleotide in Problem 22.12 undergoes hydrolysis.

Primary Structure of Nucleic Acids (Section 22.3)

22.15 What is meant by the phrase "backbone of a nucleic acid"?

22.16 How does the "backbone" for a DNA molecule differ from that for an RNA molecule?

22.17 What distinguishes various DNA molecules from each other?

22.18 What distinguishes various RNA molecules from each other?

22.19 What is the difference between a nucleic acid's 3′ end and its 5′ end?

22.20 In the lengthening of a polynucleotide chain, which type of nucleotide subunit would bond to the 3′ end of the polynucleotide chain?

22.21 Draw the structure of the dinucleotide product obtained by combining the nucleotides of Problems 22.11 and 22.12 such that the Problem 22.11 nucleotide is the 5′ end of the dinucleotide.

22.22 Draw the structure of the dinucleotide product obtained by combining the nucleotides of Problems 22.11 and 22.12 such that the Problem 22.11 nucleotide is the 3′ end of the dinucleotide.

The DNA Double Helix (Section 22.4)

22.23 Describe the DNA double helix in terms of general shape.

22.24 Describe the DNA double helix in terms of what is on the outside of the helix and what is within the interior of the helix.

22.25 The base content of a particular DNA molecule is 36% thymine. What is the percentage of each of the following bases in the molecule?

a. Adenine b. Guanine c. Cytosine

22.26 The base content of a particular DNA molecule is 24% guanine. What is the percentage of each of the following bases in the molecule?

a. Adenine b. Cytosine c. Thymine

22.27 In terms of hydrogen bonding, a G–C base pair is more stable than an A–T base pair. Explain why this is so.

22.28 What structural consideration prevents the following bases from forming complementary base pairs?

a. A and G b. T and C

22.29 Identify the 3′ and 5′ ends of the DNA base sequence TAGCC.

22.30 The two-base DNA sequences TA and AT represent different dinucleotides. Explain why this is so.

22.31 Using the concept of complementary base pairing, write the complementary DNA strands, with their 5′ and 3′ ends labeled, for each of the following DNA base sequences.

a. 5′–ACGTAT–3′ b. 5′–TTACCG–3′
c. 3′–GCATAA–5′ d. AACTGG

22.32 Using the concept of complementary base pairing, write the complementary DNA strands, with their 5′ and 3′ ends labeled, for each of the following DNA base sequences.

a. 5′–CCGGTA–3′ b. 5′–CACAGA–3′
c. 3′–TTTAGA–5′ d. CATTAC

22.33 How many total hydrogen bonds would exist between the DNA strand 5′–AGTCCTCA–3′ and its complementary strand?

22.34 How many total hydrogen bonds would exist between the DNA strand 5′–CCTAGGAT–3′ and its complementary strand?

Replication of DNA Molecules (Section 22.5)

22.35 What is the function of the enzyme DNA helicase in the DNA replication process?

22.36 What are two functions of the enzyme DNA polymerase in the DNA replication process?

22.37 In the replication of a DNA molecule, two daughter molecules, Q and R, are formed. The following base sequence is part of the newly formed strand in daughter molecule Q.

5′–ACTTAG–3′

a. What is the corresponding base sequence in the newly formed strand in daughter molecule R?
b. What is the corresponding base sequence in the "parent" strand in daughter molecule Q?
c. What is the corresponding base sequence in the "parent" strand in daughter molecule R?

22.38 In the replication of a DNA molecule, two daughter molecules, S and T, are formed. The following base sequence is part of the "parent" strand in daughter molecule S.

5′–TTCAGAG–3′

a. What is the corresponding base sequence for the newly formed strand in daughter molecule T?
b. What is the corresponding base sequence for the newly formed strand in daughter molecule S?
c. What is the corresponding base sequence in the "parent" strand in daughter molecule T?

22.39 During DNA replication, one of the newly formed strands grows continuously, whereas the other grows in segments that are later connected together. Explain why this is so.

22.40 DNA replication is most often a bidirectional process. Explain why this is so.

22.41 What is a chromosome?

22.42 Chromosomes are nucleoproteins. Explain.

RNA Molecules (Section 22.7)

22.43 What are the four major differences between RNA molecules and DNA molecules?

22.44 What are the names and abbreviations for the four major types of RNA molecules?

22.45 State whether each of the following phrases applies to ptRNA, mRNA, tRNA, or rRNA.

a. Material from which messenger RNA is made
b. Delivers amino acids to protein synthesis sites
c. Smallest of the RNAs in terms of nucleotide units present
d. Type of RNA most recently discovered

22.46 State whether each of the following phrases applies to ptRNA, mRNA, tRNA, or rRNA.

a. Associated with a series of enzymes (proteins) in a complex structure
b. Contains genetic information needed for protein synthesis
c. Most abundant type of RNA in a cell
d. Some disagreement exists on the name of this RNA

Transcription: RNA Synthesis (Section 22.8)

22.47 Describe the transcription process by listing its four steps.

22.48 Which types of RNA are involved in transcription?

22.49 What are the complementary base pairs in DNA–RNA interactions?

22.50 In DNA–DNA interactions there are two complementary base pairs, and in DNA–RNA interactions there are three complementary base pairs. Explain.

22.51 Write the base sequence of the ptRNA formed by transcription of the following DNA base sequence.

5′–ATGCTTA–3′

22.52 Write the base sequence of the ptRNA formed by transcription of the following DNA base sequence.

5′–TAGTGAT–3′

22.53 From what DNA base sequence was the following ptRNA sequence transcribed?

5′–UUCGCAG–3′

22.54 From what DNA base sequence was the following ptRNA sequence transcribed?

5′–GCUUAUC–3′

22.55 What is the relationship between an exon and a gene?

22.56 What is the relationship between an intron and a gene?

22.57 What mRNA base sequence would be obtained from the following portion of a gene?

exon intron exon
5′–TCAG–TAGC–TTCA–3′

22.58 What mRNA base sequence would be obtained from the following portion of a gene?

intron exon intron
5′–TTAC–AACG–GCAT–3′

The Genetic Code (Section 22.9)

22.59 What is a codon?

22.60 On what type of RNA molecule are codons found?

22.61 Using the information in Table 22.2, determine the amino acid that is coded for by each of the following codons.

a. CUU b. AAU c. AGU d. GGG

22.62 Using the information in Table 22.2, determine the amino acid that is coded for by each of the following codons.

a. GUA b. CCC c. CAC d. CCA

22.63 Using the information in Table 22.2, determine the synonyms, if any, of each of the codons in Problem 22.61.

22.64 Using the information in Table 22.2, determine the synonyms, if any, of each of the codons in Problem 22.62.

22.65 Explain why the base sequence ATC could not be a codon.

22.66 Explain why the base sequence AGAC could not be a codon.

22.67 Predict the sequence of amino acids coded by the mRNA sequence

5′–AUG–AAA–GAA–GAC–CUA–3′

22.68 Predict the sequence of amino acids coded by the mRNA sequence

5′–GGA–GGC–ACA–UGG–GAA–3′

Anticodons and tRNA Molecules (Section 22.10)

22.69 Describe the general structure of a tRNA molecule.

22.70 Where is the anticodon site on a tRNA molecule?

22.71 By what type of bond is an amino acid attached to a tRNA molecule?

22.72 What principle governs the codon–anticodon interaction that leads to proper placement of amino acids in proteins?

22.73 What is the anticodon that would interact with each of the following codons?

a. AGA b. CGU c. UUU d. CAA

22.74 What is the anticodon that would interact with each of the following codons?

a. CCU b. GUA c. AUC d. GCA

22.75 Identify the amino acid carried by tRNA molecules with the following anticodons.

a. UGG b. GAC c. GGA d. AGA

22.76 Identify the amino acid carried by tRNA molecules with the following anticodons.

a. UGU b. ACG c. AGU d. CAC

Translation: Protein Synthesis (Section 22.11)

22.77 What are the five major steps in translation (protein synthesis)?

22.78 What is a ribosome, and what role do ribosomes play in protein synthesis?

22.79 In the growth step of protein synthesis, at which site in the ribosome does new peptide bond formation actually take place?

22.80 What two changes occur at a ribosome during protein synthesis immediately after peptide bond formation?

22.81 Write a possible mRNA base sequence that would lead to the production of this pentapeptide. (There is more than one correct answer.)

Gly–Ala–Cys–Val–Tyr

22.82 Write a possible mRNA base sequence that would lead to the production of this pentapeptide. (There is more than one correct answer.)

Lys–Met–Thr–His–Phe

Mutations (Section 22.12)

22.83 For the codon sequence

5′–GGC–UAU–AGU–AGC–CCC–3′

write the amino acid sequence produced in the following ways.

a. Translation proceeds in a normal manner.
b. A mutation changes CCC to CCU.
c. A mutation changes CCC to ACC.

22.84 For the codon sequence

5′–GGA–AUA–UGG–UUC–CUA–3′

write the amino acid sequence produced in the following ways.

a. Translation proceeds in a normal manner.
b. A mutation changes GGA to GGG.
c. A mutation changes GGA to CGA.

Viruses and Vaccines (Section 22.13)

22.85 Describe the general structure of a virus.

22.86 What is the only function of a virus?

22.87 What is the most common method by which viruses invade cells?

22.88 Why must a virus infect another organism in order to reproduce?

Recombinant DNA and Genetic Engineering (Section 22.14)

22.89 How does recombinant DNA differ from normal DNA?

22.90 Give two reasons why bacterial cells are used for recombinant DNA procedures.

22.91 What role do plasmids play in recombinant DNA procedures?

22.92 Describe what occurs when a particular restriction enzyme operates on a segment of double-stranded DNA.

22.93 Describe what happens during transformation.

22.94 How are plasmids obtained from *E. coli* bacteria?

22.95 A particular restriction enzyme will cleave DNA between A and A in the sequence AAGCTT in the 5′-to-3′ direction. Draw a diagram showing the structural details of the "sticky ends" that result from cleavage of the following DNA segment.

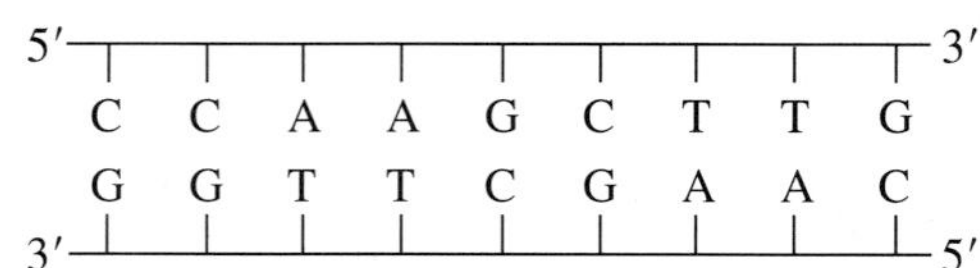

22.96 A particular restriction enzyme will cleave DNA between A and A in the sequence AAGCTT in the 5′-to-3′ direction. Draw a diagram showing the structural details of the "sticky ends" that result from cleavage of the following DNA segment.

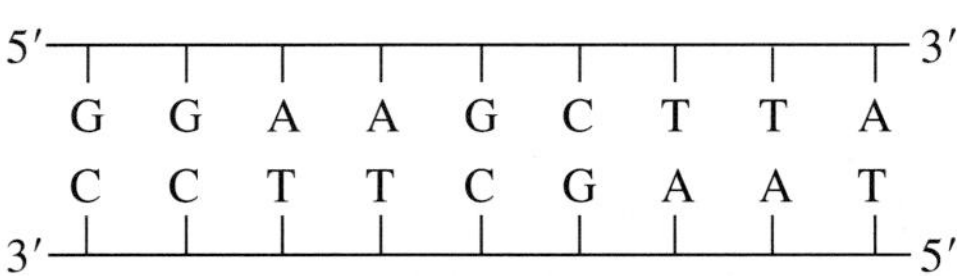

Additional Problems

22.97 With the help of the structures given in Section 22.2, describe the structural differences between the following pairs of nucleotide bases.

a. Thymine and uracil b. Adenine and guanine

22.98 The following is a sequence of bases for an exon portion of a strand of a gene.

5′–CATACAGCCTGGAAGCTA–3′

a. What is the sequence of bases on the strand of DNA complementary to this segment?
b. What is the sequence of bases on the mRNA molecule synthesized from this strand?
c. What codons are present on the mRNA molecule from part b?
d. What anticodons will be found on the tRNA molecules that interact with the codons from part c?
e. What is the sequence of amino acids in the peptide formed using these protein synthesis instructions?

22.99 Which of these RNA types, (1) mRNA, (2) ptRNA, (3) rRNA, or (4) tRNA, is most closely associated with each of the following terms?

a. Codon b. Anticodon
c. Intron d. Amino acid carrier

22.100 Which of these processes, (1) translation phase of protein synthesis, (2) transcription phase of protein synthesis, (3) replication of DNA, or (4) formation of recombinant DNA, is associated with each of the following events?

a. Complete unwinding of a DNA molecule occurs
b. Partial unwinding of a DNA molecule occurs
c. An mRNA–ribosome complex is formed
d. Okazaki fragments are formed

22.101 Which of these base-pairing situations, (1) between two DNA segments, (2) between two RNA segments, (3) between a DNA segment and an RNA segment, or (4) between a codon and an anticodon, fits each of the following base-pairing sequences? More than one response may apply to a given base-pairing situation.

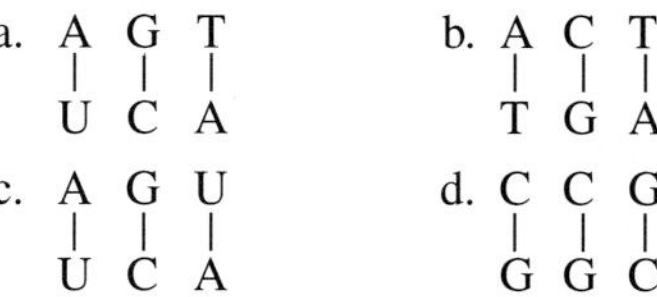

22.102 Which of these characterizations, (1) found in DNA but not RNA, (b) found in RNA but not DNA, (3) found in both DNA and RNA, or (4) not found in DNA or RNA, fits each of the following di- or trinucleotides?

a. 5′–dAMP–dAMP–3′

b. 5′–AMP–AMP–CMP–3′

c. 5′–dAMP–CMP–3′

d. 5′–GGA–3′

22.103 Suppose that 28% of the nucleotides of a DNA molecule are deoxythymidine 5′-monophosphate, and during replication the relative amounts of available nucleotide bases are 22% A, 28% T, 22% C, and 28% G. What base would be depleted first in the replication process?

22.104 Describe the extent (size) of the human genome.

Grid Problems

22.105

1. dAMP	2. dCMP	3. UMP
4. dTMP	5. CMP	6. GMP

Select from the grid *all* correct responses for each of the following situations.

a. A nucleotide in which ribose is present

b. A nucleotide that is never found in RNA molecules

c. A pair of nucleotides that differ only in the identity of the nitrogen-containing base present

d. A pair of nucleotides that differ only in the identity of the sugar present

22.106

1. adenine	2. guanine	3. ribose
4. thymine	5. deoxyribose	6. uracil

Select from the grid *all* correct responses for each of the following situations.

a. A base that could be present in both DNA and RNA molecules

b. A base that is a pyrimidine derivative

c. A base that bonds to the carbon-1′ position of a sugar during nucleotide formation

d. A base that bonds through its nitrogen-9 position to a sugar during nucleotide formation

22.107

1. TCA AGT	2. ATG UAC	3. AGU UCA
4. UCA AGT	5. UUU AAA	6. CCC GGG

Select from the grid *all* correct responses for each of the following situations.

a. A possible base-pairing combination between two DNA strands

b. A possible base-pairing combination between two RNA strands

c. A possible base-pairing combination between a DNA strand and an RNA strand

d. A possible base-pairing combination between a codon and an anticodon

22.108

1. AUG	2. CUC	3. GGA
4. GGG	5. UAA	6. UUA

Select from the grid *all* correct responses for each of the following situations.

a. Codons that are part of the genetic code

b. Codons that initiate protein synthesis

c. Codons that do not specify a particular amino acid

d. Pairs of codons that are synonyms

Biochemical Energy Production

CHAPTER OUTLINE

23.1 Metabolism 688
23.2 Metabolism and Cell Structure 689
23.3 Important Intermediate Compounds in Metabolic Pathways 691
23.4 High-Energy Phosphate Compounds 695
23.5 An Overview of Biochemical Energy Production 696

Chemistry at a Glance:
Catabolic Pathways 697
23.6 The Citric Acid Cycle 698
23.7 The Electron Transport Chain 702
23.8 Oxidative Phosphorylation 706
23.9 ATP Production for the Common Metabolic Pathway 707
23.10 The Importance of ATP 708

Chemical Connections
23.1 Cyanide Poisoning 706
23.2 Brown Fat, Newborn Babies, and Hibernating Animals 709

The energy consumed by these scarlet ibises in flight is generated by numerous sequences of biochemical reactions that occur within their bodies.

This chapter is the first of four dealing with the chemical reactions that occur in a living organism. In this first chapter, we consider those molecules that are repeatedly encountered in biological reactions, as well as those reactions that are common to the processing of carbohydrates, lipids, and proteins. The three following chapters consider the unique reactions associated with carbohydrate, lipid, and protein processing, respectively.

23.1 Metabolism

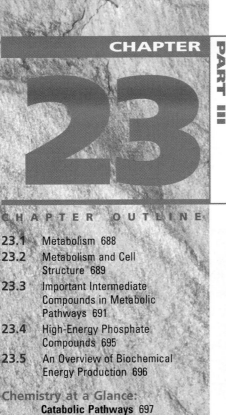

Metabolism *is the sum total of all the chemical reactions that take place in a living organism.* Human metabolism is quite remarkable. An average human adult whose weight remains the same for 40 years processes about 6 *tons* of solid food and 10,000 gallons of water, during which time the composition of the body is essentially constant. Just as we must put gasoline in a car to make it go or plug in a kitchen appliance to make it run, we also need a source of energy to think, breathe, exercise, or

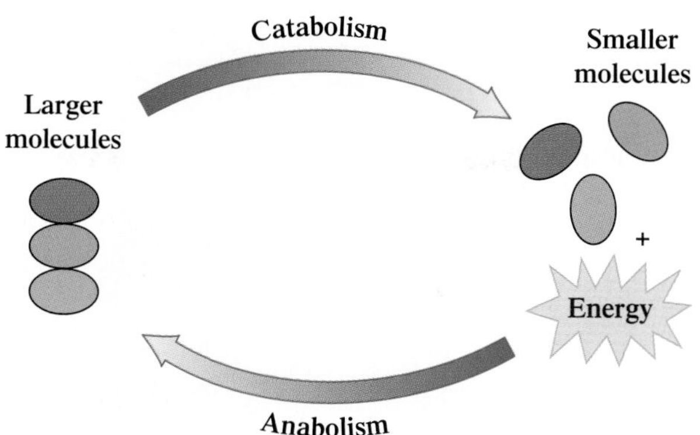

Figure 23.1
The processes of catabolism and anabolism are opposite in nature. The first usually produces energy, and the second usually consumes energy.

work. As we have seen in previous chapters, even the simplest living cell is continually carrying on energy-demanding processes such as protein synthesis, DNA replication, RNA transcription, and membrane transport.

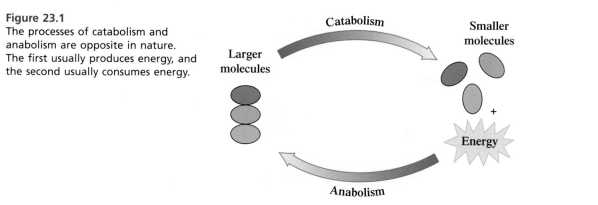

• *Catabolism* is pronounced ca-TAB-o-lism, and *anabolism* is pronounced an-ABB-o-lism. *Catabolic* is pronounced CAT-a-bol-ic, and *anabolic* is pronounced AN-a-bol-ic.

Metabolic reactions fall into one of two subtypes: catabolism and anabolism. **Catabolism** *includes all metabolic reactions in which large molecules are broken down to smaller ones.* Catabolic reactions usually release energy. The reactions involved in the oxidation of glucose are catabolic. **Anabolism** *includes all metabolic reactions in which small molecules are put together to form larger ones.* Anabolic reactions usually need energy in order to proceed. The synthesis of proteins from amino acids is an anabolic process. Figure 23.1 contrasts catabolic and anabolic processes.

The metabolic reactions that occur in a cell are usually organized into sequences called *metabolic pathways.* A **metabolic pathway** *is a series of consecutive biochemical reactions used to convert a starting material into an end product.* Such pathways may be *linear,* in which a series of reactions generates a final product, or *cyclic,* in which a series of reactions regenerates the first reactant.

Linear metabolic pathway: $A \longrightarrow B \longrightarrow C \longrightarrow D$

Cyclic metabolic pathway:

$$\begin{array}{ccc} A & \longrightarrow & B \\ \uparrow & & \downarrow \\ D & \longleftarrow & C \end{array}$$

The major metabolic pathways for all life forms are similar. This enables scientists to study metabolic reactions in simpler life forms and use the results to help understand the corresponding metabolic reactions in more complex organisms, including humans.

23.2 Metabolism and Cell Structure

Knowledge of the major structural features of a cell is a prerequisite to understanding *where* metabolic reactions take place.

• The term *eukaryotic,* pronounced you-KAHR-ee-ah-tic, is from the Greek *eu,* meaning "true," and *karyon,* meaning "nucleus." The term *prokaryotic,* which contains the Greek *pro,* meaning "before," literally means "before the nucleus."

• Eukaryotic and prokaryotic cells differ in that the former contain a well-defined nucleus, set off from the rest of the cell by a membrane.

Cells are of two types: prokaryotic and eukaryotic. *Prokaryotic cells* have no nucleus and are found only in bacteria. The DNA that governs the reproduction of prokaryotic cells is usually a single circular molecule found near the center of the cell in a region called the *nucleoid.* In *eukaryotic cells,* the DNA is found in the membrane-enclosed nucleus. Cells of this type, which are found in all higher organisms, are about 1000 times larger than bacterial cells. Our focus in the remainder of this section will be on eukaryotic cells, the type present in humans.

The *cytoplasm* of a eukaryotic cell is the material that lies between the nucleus and the outer plasma membrane of the cell. Within the water-based cytoplasm are small bodies called *organelles.* An **organelle** *is a minute structure within the cell cytoplasm that carries out a specific cellular function.* Among the organelles are *ribosomes, lysosomes,* and *mitochondria.* Ribosomes are the sites of protein synthesis (Section 22.11). Lysosomes

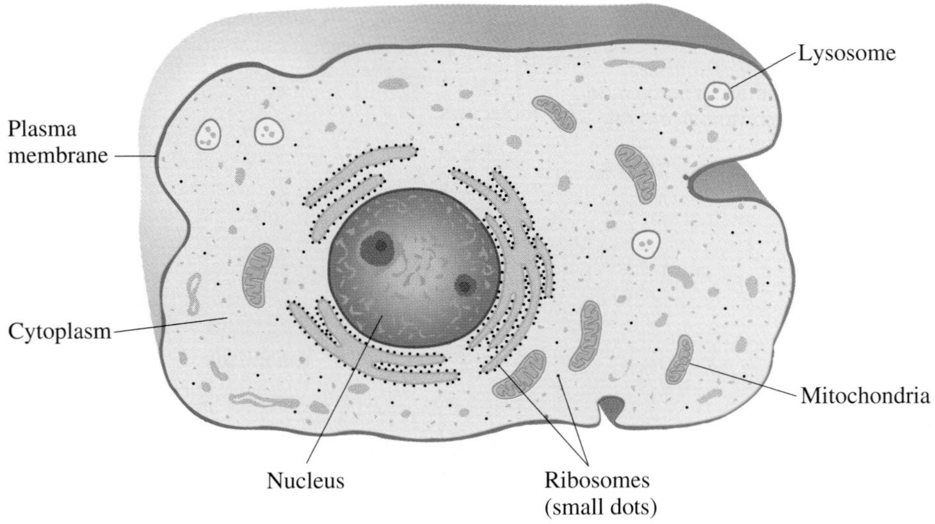

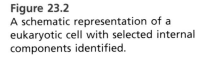

Figure 23.2
A schematic representation of a eukaryotic cell with selected internal components identified.

contain hydrolytic enzymes, which are needed for cellular rebuilding, repair, and degradation. Mitochondria house some of the enzymes required for energy generation in the cell. **Mitochondria** *are organelles that have a central role in the production of energy.* Figure 23.2 shows the general internal structure of a eukaryotic cell.

• *Mitochondria,* pronounced my-toe-KON-dree-ah, is plural. The singular form of the term is *mitochondrion.* The thread-like shape of mitochondria is responsible for this organelle's name; *mitos* is Greek for "thread," and *chondrion* is Greek for "granule."

Mitochondria are sausage-shaped organelles containing both an *outer membrane* and a *multifolded inner membrane* (see Figure 23.3). The outer membrane, which is about 50% lipid and 50% protein, is freely permeable to small molecules. The inner membrane, which is about 20% lipid and 80% protein, is highly impermeable to most substances. The nonpermeable nature of the inner membrane divides a mitochondrion into two separate compartments—an interior region called the *matrix* and the region between the inner and outer membranes, called the *intermembrane space.* The folds of the inner membrane that protrude into the matrix are called *cristae.*

The invention of high-resolution electron microscopes allowed researchers to see the interior structure of the mitochondrion more clearly and led to the discovery, in 1962, of small spherical knobs attached to the cristae called *ATP synthase complexes.* As their name implies, these relatively small knobs, which are located on the matrix side of the inner membrane, are responsible for ATP synthesis, and their association with the inner membrane is critically important for this task. More will be said about ATP in the next section.

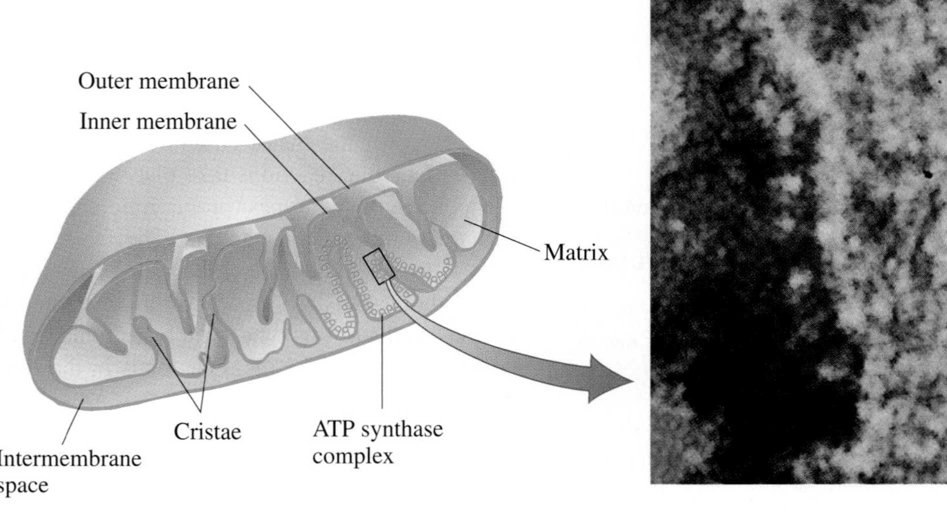

Figure 23.3
(a) A schematic representation of a mitochondrion, showing key features of its internal structure. (b) An electron micrograph of a single mitochondrial crista, showing the ATP synthase knobs extending into the matrix.

23.3 Important Intermediate Compounds in Metabolic Pathways

• Nucleotides are the monomer units from which nucleic acids are made, and they are also present in several *nonpolymeric* molecules that are important in energy production in living things.

As a prelude to an overview presentation (Section 23.5) of the metabolic processes by which our food is converted to energy, we now consider several compounds that repeatedly function as key intermediates in these metabolic pathways. Knowing about these compounds will make it easier to understand the details of metabolic pathways. The compounds to be discussed all have nucleotides (Section 22.2) as part of their structures.

• Adenosine Phosphates

Adenosine (Section 22.2) exists in several phosphate forms: *mono*phosphate, *di*phosphate, *tri*phosphate, and *cyclic mono*phosphate. Figure 23.4 shows the structural relationships among these phosphate forms.

Adenosine monophosphate (AMP) was discussed in detail in Section 22.2 as a structural component of RNA. Adenosine diphosphate (ADP) and adenosine triphosphate (ATP) are key compounds in numerous metabolic pathways. Structurally, ADP and ATP differ from AMP in the number of phosphate groups present. The additional phosphates are connected to AMP through *strained* bonds (Section 23.4), which, when hydrolyzed, require less-than-normal amounts of energy to break. The hydrolysis equations are

$$\text{ATP} + H_2O \longrightarrow \text{ADP} + HPO_4^{2-} + H^+$$
$$\text{ADP} + H_2O \longrightarrow \text{AMP} + HPO_4^{2-} + H^+$$

The net energy produced from these reactions is used to drive cellular processes that require energy input.

• A certain amount of stimulation is associated with drinking a cup of coffee, especially late at night. Why? The caffeine present in the coffee inhibits the action of the enzyme phosphodiesterase, whose job it is to inactivate molecules of cAMP. The molecule cAMP is involved in the release of glucose into the bloodstream. Deactivation of phosphodiesterase by caffeine allows cAMP to do its job. More glucose is produced and the coffee drinker feels more energetic.

Cyclic AMP (cAMP) is a molecule with several functions. One function is associated with the activity of the hormone epinephrine (Section 17.9). This hormone stimulates cAMP synthesis within a cell, and the cAMP then helps regulate the release of glucose from glycogen (Section 24.9).

ATP is not the only nucleotide triphosphate present in cells, although it is the most prevalent. The other nitrogen-containing bases associated with nucleotides (Section 22.2) are also present in triphosphate form. Uridine triphosphate (UTP) is involved in carbohydrate metabolism, guanosine triphosphate (GTP) participates in protein and carbohydrate metabolism, and cytidine triphosphate (CTP) is involved in lipid metabolism.

Figure 23.4
Structures of the various phosphate forms of adenosine.

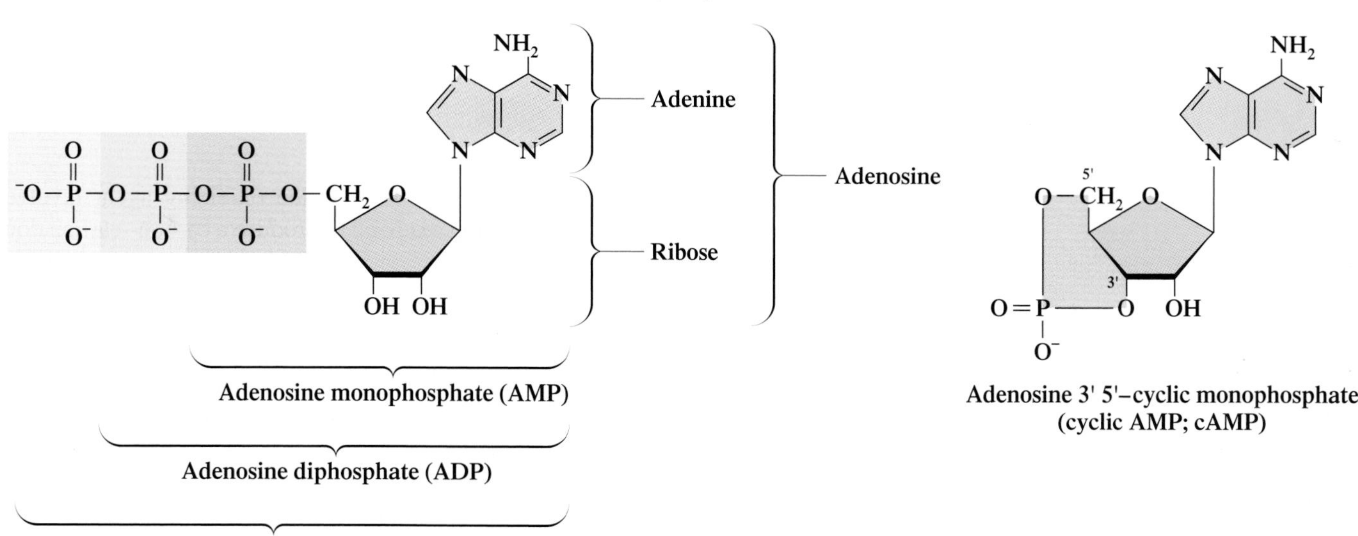

• Flavin Adenine Dinucleotide

Flavin adenine dinucleotide (FAD) is a coenzyme (Section 21.3) required in numerous metabolic redox reactions. Structurally, FAD can be visualized as containing either three subunits or six subunits. A block diagram of FAD from the three-subunit viewpoint is

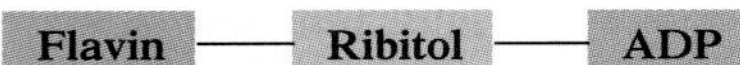

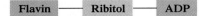

Flavin and ribitol, the two components attached to the ADP unit, together constitute the B vitamin riboflavin (Section 21.13). The block diagram for FAD from the six-subunit viewpoint is

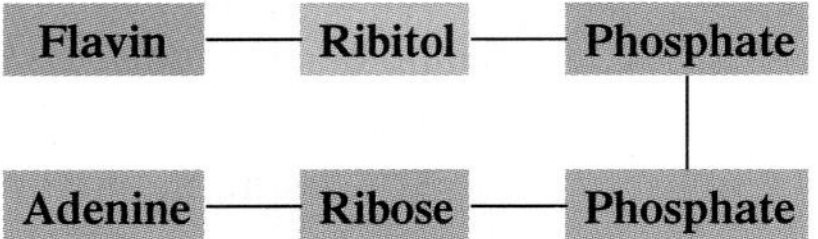

This block diagram shows the basis for the name *f*lavin *a*denine *d*inucleotide. Ribitol is a reduced form of ribose; a —CH_2OH group is present in place of the —CHO group (Section 18.12).

D-Ribose	D-Ribitol
CHO	CH_2OH
H—C—OH	H—C—OH
H—C—OH	H—C—OH
H—C—OH	H—C—OH
CH_2OH	CH_2OH

The complete structural formula of FAD is given in Figure 23.5a.

The active portion of FAD in redox reactions is the flavin subunit of the molecule. The flavin is reduced, converting the FAD to $FADH_2$, a molecule with two additional hydrogen atoms. Thus FAD is the *oxidized* form of the molecule, $FADH_2$ the *reduced* form.

$$\text{FAD (oxidized form)} + 2H^+ + 2e^- \rightleftharpoons \text{FADH}_2\ \text{(reduced form)}$$

FAD
(oxidized form)

$FADH_2$
(reduced form)

R = —Ribitol—ADP

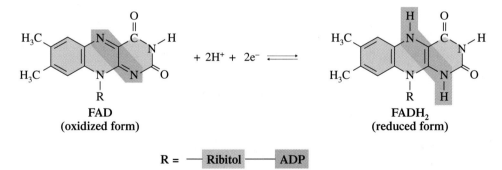

A typical cellular reaction in which FAD serves as the oxidizing agent involves a —CH_2—CH_2— portion of a substrate being oxidized to produce a carbon–carbon double bond.

$$\text{R—CH}_2\text{—CH}_2\text{—R} + \text{FAD} \longrightarrow \text{R—CH=CH—R} + \text{FADH}_2$$

Saturated (reduced) → Unsaturated (oxidized)

For an enzyme-catalyzed redox reaction involving removal of two hydrogen atoms, such as this, each removed hydrogen atom is equivalent to a hydrogen *ion,* H^+, plus an electron, e^-.

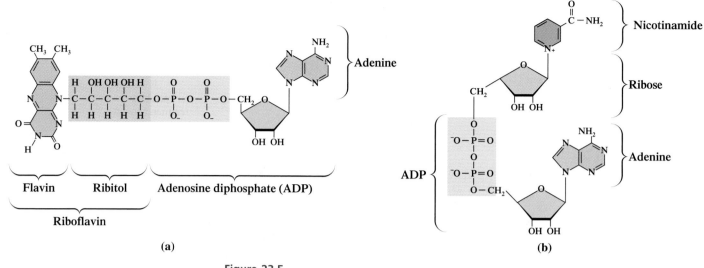

Figure 23.5
Structural formulas of the molecules flavin adenine dinucleotide, FAD (a) and nicotinamide adenine dinucleotide, NAD^+ (b).

• In metabolic pathways in which it is involved, flavin adenine dinucleotide continually changes back and forth between its oxidized and reduced forms.

$$2H^+ + 2e^- + FAD \rightleftharpoons FADH_2$$

$$2\text{ H atoms (removed)} \longrightarrow 2H^+ + 2e^-$$

On the basis of this equivalency, the summary equation relating the oxidized and reduced forms of flavin adenine dinucleotide is usually written as

$$\underbrace{2H^+ + 2e^-}_{\text{2 H atoms}} + FAD \rightleftharpoons FADH_2$$

• Nicotinamide Adenine Dinucleotide

Several parallels exist between the characteristics of nicotinamide adenine dinucleotide (NAD^+) and FAD. Both have coenzyme functions in metabolic redox pathways, both have a B vitamin as a structural component, and both can be represented structurally by using a three-subunit or a six-subunit formulation. In the case of NAD^+, the B vitamin present is nicotinamide (Section 21.13).

A block diagram of the structure of NAD^+, in which we have used the three-subunit formulation, is

The six-subunit block diagram, which emphasizes the dinucleotide nature of the coenzyme as well as the origin of its name, is

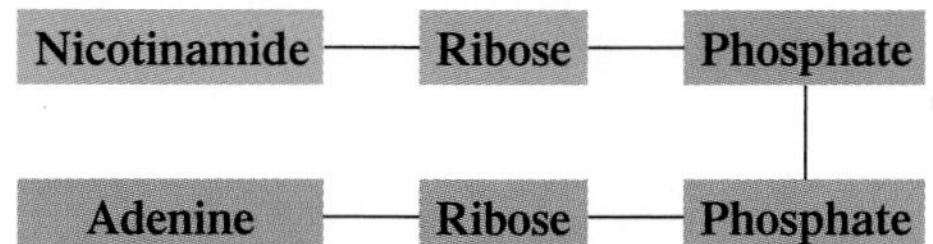

Examination of the detailed structure of NAD^+ (Figure 23.5b) reveals the basis for the positive electrical charge. The + sign refers to the positive charge on the nitrogen atom in the nicotinamide component of the structure; this nitrogen atom has four bonds instead of the usual three (Section 17.7).

The active portion of NAD^+ in redox reactions is the nicotinamide subunit of the molecule. The nicotinamide is reduced, converting the NAD^+ to NADH, a molecule with one additional hydrogen atom and two additional electrons. Thus NAD^+ is the *oxidized* form of the molecule, NADH the *reduced* form.

$$\text{NAD}^+ + \text{H}^+ + 2e^- \rightleftharpoons \text{NADH}$$

NAD⁺ (oxidized form) NADH (reduced form)

R = —Ribose—ADP

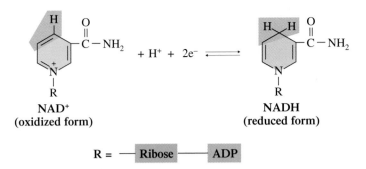

A typical cellular reaction in which NAD^+ serves as the oxidizing agent is the oxidation of a secondary alcohol to give a ketone.

$$\underset{\text{2° alcohol}}{R-\overset{OH}{\underset{H}{C}}-R} + NAD^+ \longrightarrow \underset{\text{Ketone}}{R-\overset{O}{\overset{\|}{C}}-R} + NADH + H^+$$

• In metabolic pathways, nicotinamide adenine dinucleotide continually changes back and forth between its oxidized and reduced forms.
$2H^+ + 2e^- + NAD^+ \rightleftharpoons NADH + H^+$

In this reaction, one hydrogen atom of the alcohol substrate is directly transferred to NAD^+, whereas the other appears in solution as H^+ ion. Both electrons lost by the alcohol go to the nicotinamide ring in $NADH^+$. (Two electrons are required, rather than one, because of the original positive charge on NAD^+.) Thus the summary equation relating the oxidized and reduced forms of flavin adenine dinucleotide is written as

$$\underbrace{2H^+ + 2e^-}_{\text{2 H atoms}} + NAD^+ \rightleftharpoons NADH + H^+$$

• Coenzyme A

Another important coenzyme in metabolic pathways is coenzyme A, a derivative of the B vitamin pantothenic acid (Section 21.13). The three-subunit and six-subunit block diagrams for coenzyme A are

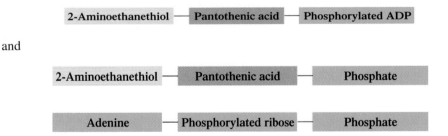

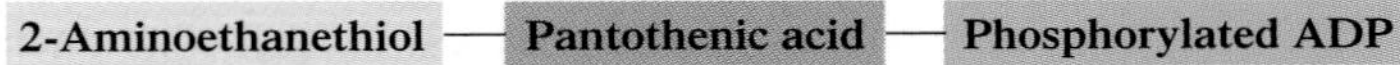

and

2-Aminoethanethiol — Pantothenic acid — Phosphate

Adenine — Phosphorylated ribose — Phosphate

Note, in the three-subunit block diagram, that the ADP subunit present is phosphorylated. As shown in the following complete structural formula for coenzyme A, the phosphorylated version of ADP carries an extra phosphate group attached to carbon 3′ of its ribose.

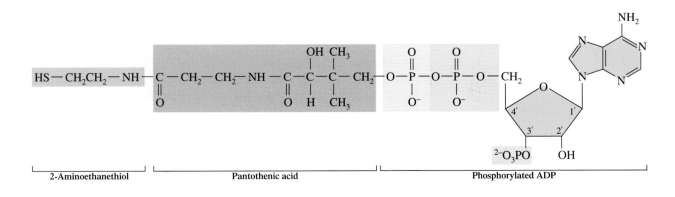

$$HS-CH_2CH_2-NH-\overset{\|}{\underset{O}{C}}-CH_2-CH_2-NH-\underset{O}{C}-\overset{OH}{\underset{H}{C}}-\overset{CH_3}{\underset{CH_3}{C}}-CH_2-O-\overset{O}{\underset{O^-}{P}}-O-\overset{O}{\underset{O^-}{P}}-O-CH_2-(\text{ribose, } 3'\text{-}{}^{2-}O_3PO,\ 2'\text{-}OH)-\text{adenine}(NH_2)$$

2-Aminoethanethiol | Pantothenic acid | Phosphorylated ADP

The active portion of coenzyme A is the sulfhydryl group (—SH group; Section 14.13) in the ethanethiol subunit of the coenzyme. For this reason, the abbreviation CoA—SH is used for coenzyme A.

Think of the letter A in the name *coenzyme A* as reflecting a general metabolic function of this substance; it is the transfer of acetyl groups in metabolic pathways. Such groups bond to CoA—SH through a thioester bond (Section 16.14) to give acetyl CoA.

• An acetyl group, which can be considered to be derived from acetic acid, has the structure

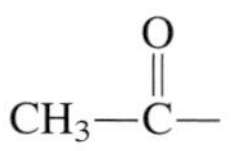

$$\underset{\text{Acetyl group}}{CH_3-\overset{\overset{\large O}{\|}}{C}-} \qquad \underset{\text{Acetic acid}}{CH_3-\overset{\overset{\large O}{\|}}{C}-OH}$$

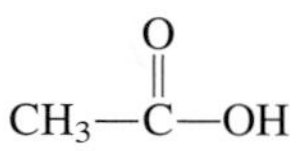

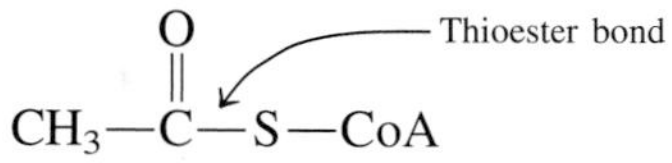

$$\underset{\text{Acetyl CoA}}{CH_3-\overset{\overset{\large O}{\|}}{C}-S-CoA}$$

Thioester bond

23.4 High-Energy Phosphate Compounds

In the previous section, we noted that knowing about several key intermediate compounds in metabolic reactions makes it easier to understand the details of metabolic processes, which are yet to come. In like manner, knowing about a particular type of bond present in certain phosphate-containing metabolic intermediates makes the details of metabolic processes easier to understand.

Several phosphate-containing compounds found in metabolic pathways are known as high-energy compounds. A **high-energy compound** *has a greater free energy of hydrolysis than a typical compound.* High-energy compounds differ from other compounds in that they contain one or more *very reactive* bonds, often called *strained bonds.* The energy required to break these strained bonds, during hydrolysis, is less than that generally required to break a chemical bond. Consequently, the hydrolysis of high-energy compounds produces greater-than-normal energy (high energy) because the bond-formation energy released from product formation will exceed the lower-than-normal bond-breaking energy.

• In the definition of a high-energy compound, the term *free energy* rather than simply *energy* was used. Free energy is the amount of energy released by a chemical reaction that is actually available for further use at a given temperature and pressure. In reality, the energy released in a chemical reaction is divisible into two parts. One part, lost as heat, is not available for further use. The other part, the free energy, is available for further use; in cells, it can be used to "drive" reactions that require energy.

Greater-than-normal electron–electron repulsive forces at specific locations within a molecule are a cause of bond strain. Highly electronegative atoms and/or highly charged atoms occurring together in a molecule cause increased repulsive forces and thus increase bond strain.

Let us specifically consider bond strain as it is related to phosphate-containing organic molecules involved in metabolic pathways. The parent molecule for phosphate groups is phosphoric acid, H_3PO_4, a weak triprotic inorganic acid (Section 10.3). This acid exists in aqueous solution in several forms, the dominant form at cellular pH being

• In a chemical reaction, the energy balance between bond breaking among reactants (energy-input) and new bond formation among products (energy-release) determines whether there is a net loss or net gain of energy (Section 9.5).

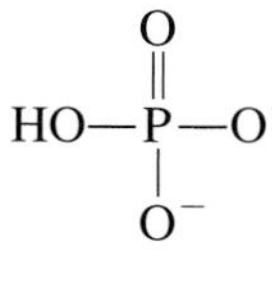

$$HO-\overset{\overset{\large O}{\|}}{\underset{\underset{\large O^-}{|}}{P}}-O^-$$

Diphosphate and triphosphate ions can also exist in cellular fluids.

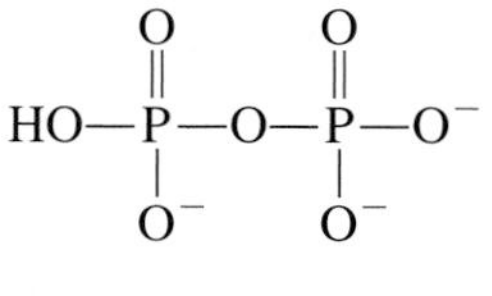

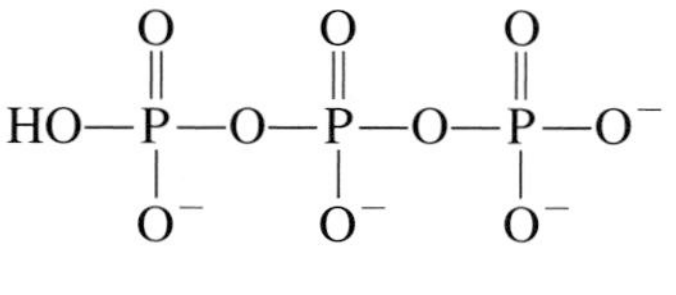

$$HO-\overset{\overset{\large O}{\|}}{\underset{\underset{\large O^-}{|}}{P}}-O-\overset{\overset{\large O}{\|}}{\underset{\underset{\large O^-}{|}}{P}}-O^- \qquad HO-\overset{\overset{\large O}{\|}}{\underset{\underset{\large O^-}{|}}{P}}-O-\overset{\overset{\large O}{\|}}{\underset{\underset{\large O^-}{|}}{P}}-O-\overset{\overset{\large O}{\|}}{\underset{\underset{\large O^-}{|}}{P}}-O^-$$

Note the presence in these three phosphate structures of highly electronegative oxygen atoms, many of which bear negative charges. The factors that can produce bond strain are present when phosphates (mono-, di-, and tri-) are bonded to certain organic molecules.

• The designation *high-energy compound* does not mean that a compound is different from other compounds in terms of bonding. High-energy compounds obey the normal rules for chemical bonding. The only difference between such compounds and other compounds is the presence of one or more *strained bonds.* The breaking of such bonds requires lower-than-normal amounts of energy.

Table 23.1 gives the structures of commonly encountered phosphate-containing compounds, as well as a numerical parameter—the free energy of hydrolysis—that can be considered a measure of the extent of bond strain in the molecules. The more negative the free energy of hydrolysis, the greater the bond strain. A free-energy release greater than 6.0 kcal/mole is generally considered indicative of bond strain. In Table 23.1, strained bonds within the molecules are noted with a squiggle (~), a notation often employed to denote strained bonds.

Table 23.1
Free Energies of Hydrolysis of Common Phosphate-Containing Metabolic Compounds

Type	Example	Free energy of hydrolysis (kcal/mole)
enol phosphates $R(O{\sim}PO_3^{2-})C{=}CH_2$	phosphoenolpyruvate	−14.8
acyl phosphates $R{-}C(=O){-}O{\sim}PO_3^{2-}$	1,3-bisphosphoglycerate	−11.8
	acetyl phosphate	−11.3
guanidine phosphates $R{-}NH{-}C(=NH){-}NH{\sim}PO_3^{2-}$	creatine phosphate	−10.3
	arginine phosphate	−9.1
triphosphates $R{-}O{-}PO_2^-{-}O{\sim}PO_2^-{-}O{\sim}PO_3^{2-}$	$ATP \longrightarrow AMP + PP_i$[a]	−7.7
	$ATP \longrightarrow ADP + P_i$[a]	−7.5
diphosphates $R{-}O{-}PO_2^-{-}O{\sim}PO_3^{2-}$	$PP_i \longrightarrow 2P_i$	−7.8
	$ADP \longrightarrow AMP + P_i$	−7.5
sugar phosphates $R{-}O{-}PO_3^{2-}$	glucose 1-phosphate	−5.0
	fructose 6-phosphate	−3.8
	$AMP \longrightarrow adenosine + P_i$	−3.4
	glucose 6-phosphate	−3.3
	glycerol 3-phosphate	−2.2

[a] The notation P_i is used as a general designation for any free monophosphate species present in cellular fluid. Free diphosphate ions are designated as PP_i ("i" stands for *inorganic*).

• The $—PO_3^{2-}$ group as part of a larger organic phosphate molecule is referred to as a *phosphoryl group.*

23.5 An Overview of Biochemical Energy Production

The energy needed to run the human body is obtained from ingested food through a multistep process that involves several different catabolic pathways. There are four general stages in the biochemical energy production process, and numerous reactions are associated with each stage. See the accompanying Chemistry at a Glance.

• The first stage of biochemical energy production, digestion, is not considered part of metabolism because it is extracellular. Metabolic processes are intracellular.

Stage 1: The first stage, *digestion,* begins in the mouth (saliva contains starch-digesting enzymes), continues in the stomach (gastric juices), and is completed in the small intestine (the majority of digestive enzymes and bile

Catabolic Pathways

CHEMISTRY AT A GLANCE

STAGE 1

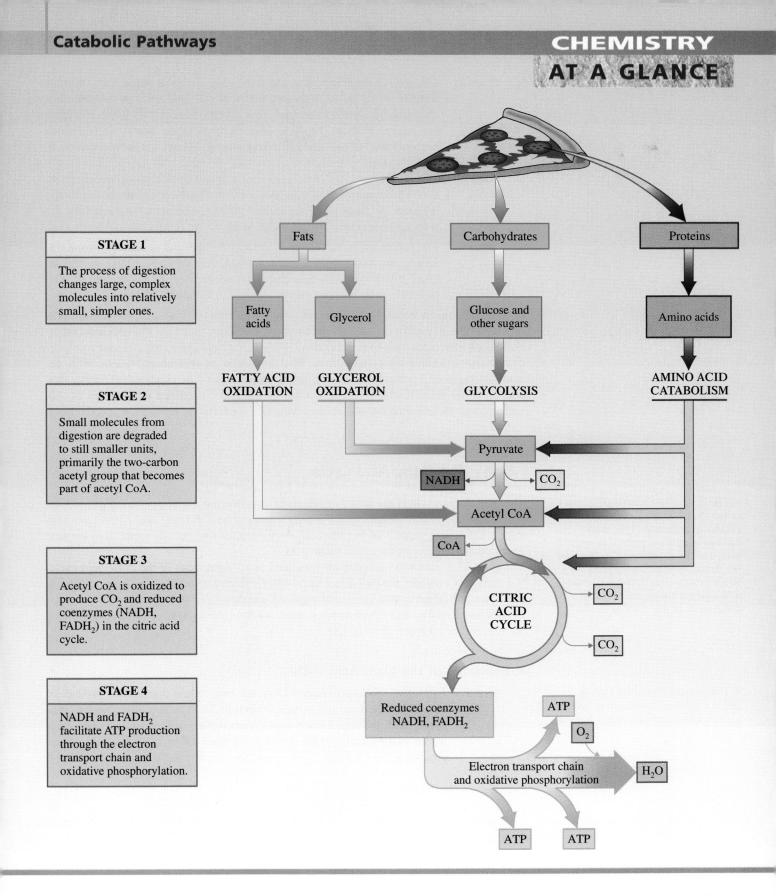

The process of digestion changes large, complex molecules into relatively small, simpler ones.

STAGE 2

Small molecules from digestion are degraded to still smaller units, primarily the two-carbon acetyl group that becomes part of acetyl CoA.

STAGE 3

Acetyl CoA is oxidized to produce CO_2 and reduced coenzymes (NADH, $FADH_2$) in the citric acid cycle.

STAGE 4

NADH and $FADH_2$ facilitate ATP production through the electron transport chain and oxidative phosphorylation.

salts). The end products of digestion—glucose and other monosaccharides from carbohydrates, amino acids from proteins, and fatty acids and glycerol from fats and oils—are small enough to pass across intestinal membranes and into the blood, where they are transported to the body's cells (see the upper portion of the Chemistry at a Glance).

Stage 2: The second stage, *acetyl group formation,* involves numerous reactions, some of which occur in the cytoplasm of cells and some in cellular mitochondria. The small molecules from digestion are further oxidized during this stage. Primary products include two-carbon acetyl units (which become attached to coenzyme A to give acetyl CoA) and the reduced coenzyme NADH.

Stage 3: The third stage, the *citric acid cycle,* occurs inside mitochondria. Here acetyl groups are oxidized to produce CO_2 and energy. Some of the energy released by these reactions is lost as heat, and some is carried by the reduced coenzymes NADH and $FADH_2$ to the fourth stage.

Stage 4: The fourth stage, the *electron transport chain and oxidative phosphorylation,* also occurs inside mitochondria. NADH and $FADH_2$ supply the "fuel" (hydrogen ions and electrons) needed for the production of ATP molecules, the primary energy carriers in metabolic pathways.

The reactions in stages 3 and 4 are the same for all types of foods (carbohydrates, fats, proteins). These reactions constitute the common metabolic pathway. The **common metabolic pathway** *is the sum of the reactions of the citric acid cycle, the electron transport chain, and oxidative phosphorylation.* The remainder of this chapter deals with the common metabolic pathway. The reactions of stages 1 and 2 of biochemical energy production differ for different types of foodstuffs. They are discussed in Chapters 24–26, which cover the metabolism of carbohydrates, fats (lipids), and proteins, respectively.

23.6 The Citric Acid Cycle

• The citric acid cycle (CAC) is also called the *tricarboxylic acid cycle* (TCA), in reference to the three carboxylate groups present in citric acid, and the *Krebs cycle,* in honor of the British biochemist Sir Hans A. Krebs (1900–1981), who established relationships among the different compounds in the cycle in 1937.

The **citric acid cycle** *is the series of reactions in which the acetyl portion of acetyl CoA is oxidized to carbon dioxide and the reduced coenzymes $FADH_2$ and NADH are produced.* This cycle, stage 3 of biochemical energy production, gets its name from the first intermediate product in the cycle, citric acid.

Figure 23.6 shows the compounds produced in all eight steps of the citric acid cycle. We shall now consider the individual steps of the cycle in detail. As we go through these steps, we will observe two important types of reactions: (1) oxidation, which produces NADH and $FADH_2$, and (2) decarboxylation, wherein the carbon chain is shortened by the removal of a carbon atom as CO_2.

• Reactions of the Citric Acid Cycle

• The formation of citryl CoA is called a condensation reaction because a new carbon–carbon bond is formed.

Step 1: *Formation of Citrate.* Acetyl CoA, the two-carbon degradation product of carbohydrates, fats, and proteins (Section 23.5), enters the cycle by combining with the four-carbon keto dicarboxylate species oxaloacetate. This results in the transfer of the acetyl group from coenzyme A to oxaloacetate, producing the C_6 citrate species and free coenzyme A.

$$\underset{\text{Oxaloacetate}}{\begin{array}{c} COO^- \\ | \\ O{=}C \\ | \\ CH_2 \\ | \\ COO^- \end{array}} + \underset{\text{Acetyl CoA}}{CH_3{-}\overset{\overset{O}{\|}}{C}{-}S{-}CoA} \xrightarrow[\text{synthase}]{\text{Citrate}} \underset{\text{Citryl CoA}}{\begin{array}{c} O \\ \| \\ C{-}S{-}CoA \\ | \\ CH_2 \\ | \\ HO{-}C{-}COO^- \\ | \\ CH_2 \\ | \\ COO^- \end{array}} \xrightarrow{H_2O} \underset{\text{Citrate}}{\begin{array}{c} COO^- \\ | \\ CH_2 \\ | \\ HO{-}C{-}COO^- \\ | \\ CH_2 \\ | \\ COO^- \end{array}} + CoA{-}SH + H^+$$

There are two parts to the reaction: (1) the condensation of acetyl CoA and oxaloacetate to form citryl CoA, a process catalyzed by the enzyme

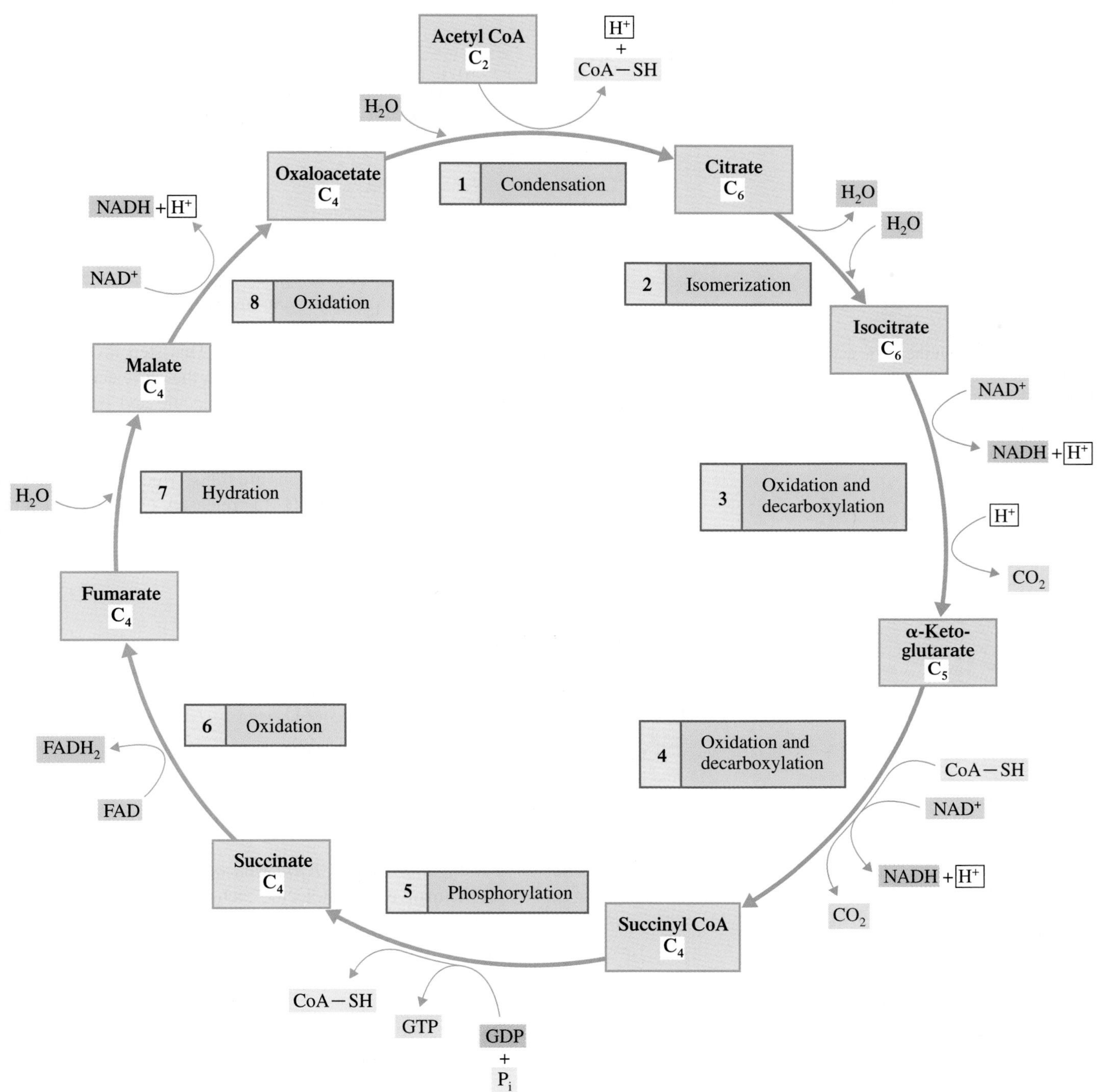

Figure 23.6
The citric acid cycle. Details of the numbered steps are given in the text.

• The formation of citryl CoA involves addition across the carbon–oxygen double bond. A —CH_3 hydrogen atom of acetyl CoA adds to the O atom, and the remainder of the acetyl CoA adds to the carbon atom.

citrate synthase, and (2) hydrolysis of the thioester bond in citryl CoA to produce CoA—SH and citrate, also catalyzed by the enzyme citrate synthase.

Step 2: *Formation of Isocitrate.* Citrate is converted to its less symmetrical isomer isocitrate in an isomerization process that involves a dehydration followed by a hydration, both catalyzed by the enzyme *aconitase*. The net result of these reactions is that the —OH group from citrate ends up on a different carbon atom.

$$\underset{\text{Citrate}}{\begin{array}{c} COO^- \\ | \\ CH_2 \\ | \\ HO-C-COO^- \\ | \\ H-C-H \\ | \\ COO^- \end{array}} \xrightarrow[\text{Aconitase}]{H_2O} \underset{\textit{cis}\text{-Aconitate}}{\begin{array}{c} COO^- \\ | \\ CH_2 \\ | \\ C-COO^- \\ \| \\ C-H \\ | \\ COO^- \end{array}} \xrightarrow[\text{Aconitase}]{H_2O} \underset{\text{Isocitrate}}{\begin{array}{c} COO^- \\ | \\ CH_2 \\ | \\ H-C-COO^- \\ | \\ HO-C-H \\ | \\ COO^- \end{array}}$$

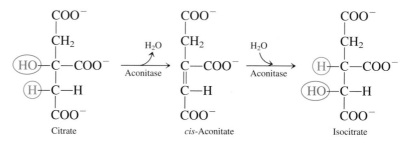

Citrate is a tertiary alcohol and isocitrate a secondary alcohol. Tertiary alcohols are not readily oxidized; secondary alcohols are easier to oxidize (Section 14.7). The next step in the cycle involves oxidation.

• All acids found in the citric acid cycle exist as negative ions (carboxylate ions; Section 16.7) at cellular pH.

Step 3: *Oxidation of Isocitrate and Formation of CO_2.* This step involves oxidation-reduction (the first of four redox reactions in the citric acid cycle) and decarboxylation. The reactants are a NAD^+ molecule and isocitrate. The reaction, catalyzed by isocitrate dehydrogenase, is complex: (1) Isocitrate is oxidized to a ketone (oxalosuccinate) by NAD^+, releasing two hydrogens. (2) One hydrogen and two electrons are transferred to NAD^+ to form NADH; the remaining hydrogen ion (H^+) is released. (3) The oxalosuccinate remains bound to the enzyme and undergoes decarboxylation (loses CO_2), which produces the C_5 species α-ketoglutarate.

$$\underset{\text{Isocitrate}}{\begin{array}{c} COO^- \\ | \\ CH_2 \\ | \\ H-C-COO^- \\ | \\ HO-C-H \\ | \\ COO^- \end{array}} \xrightarrow[\text{Isocitrate dehydrogenase}]{NAD^+ \quad NADH + H^+} \underset{\text{Oxalosuccinate}}{\begin{array}{c} COO^- \\ | \\ CH_2 \\ | \\ H-C-COO^- \\ | \\ C=O \\ | \\ COO^- \end{array}} \xrightarrow{H^+ \quad CO_2} \underset{\alpha\text{-Ketoglutarate}}{\begin{array}{c} COO^- \\ | \\ CH_2 \\ | \\ H-C-H \\ | \\ C=O \\ | \\ COO^- \end{array}}$$

This step yields the first molecules of CO_2 and NADH in the cycle.

• The CO_2 molecules produced in Steps 3 and 4 of the citric acid cycle are the CO_2 molecules we exhale in the process of respiration.

Step 4: *Oxidation of α-Ketoglutarate and Formation of CO_2.* This second redox reaction of the cycle involves one molecule each of NAD^+, CoA—SH, and α-ketoglutarate. The catalyst is an aggregate of three enzymes called the α-ketoglutarate dehydrogenase complex. As in Step 3, both oxidation and decarboxylation occur. There are three products: CO_2, NADH, and the C_4 species succinyl CoA.

$$\underset{\alpha\text{-Ketoglutarate}}{\begin{array}{c} COO^- \\ | \\ CH_2 \\ | \\ CH_2 \\ | \\ C=O \\ | \\ COO^- \end{array}} + NAD^+ + CoA-SH \xrightarrow{\substack{\alpha\text{-Ketoglutarate} \\ \text{dehydrogenase} \\ \text{complex}}} \underset{\text{Succinyl CoA}}{\begin{array}{c} COO^- \\ | \\ CH_2 \\ | \\ CH_2 \\ | \\ C=O \\ | \\ S-CoA \end{array}} + NADH + CO_2 + H^+$$

• The thioester bond is a strained bond. Its hydrolysis releases energy, which is trapped by GTP formation. The function of the GTP produced is similar to that of ATP: to store energy in the form of a high-energy phosphate bond (Section 23.4).

Step 5: *Thioester bond cleavage in Succinyl CoA and Phosphorylation of GDP.* Two molecules react with succinyl CoA—a molecule of GDP (similar to ADP; Section 23.3) and a free phosphate group (P_i). The enzyme succinyl CoA synthase removes coenzyme A by thioester bond cleavage. The energy released is used to combine GDP and P_i to form GTP. Succinyl CoA has been converted to succinate.

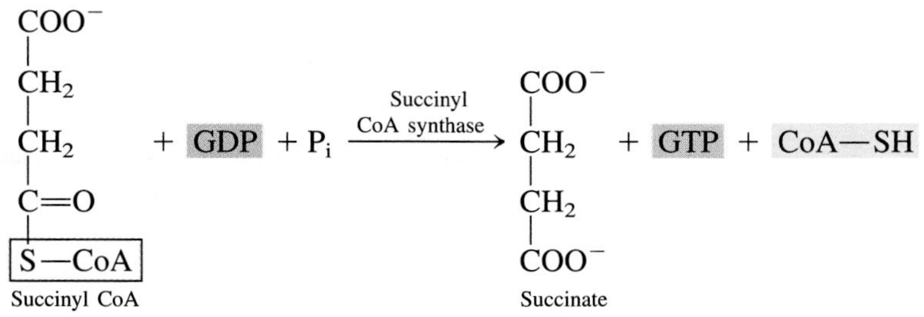

$$\begin{array}{c} COO^- \\ | \\ CH_2 \\ | \\ CH_2 \\ | \\ C{=}O \\ | \\ \boxed{S{-}CoA} \end{array} + \text{GDP} + P_i \xrightarrow[\text{CoA synthase}]{\text{Succinyl}} \begin{array}{c} COO^- \\ | \\ CH_2 \\ | \\ CH_2 \\ | \\ COO^- \end{array} + \text{GTP} + \text{CoA—SH}$$

Succinyl CoA — Succinate

Steps 6 through 8 involve a sequence of functional-group changes that we have encountered many times in the organic sections of the text. The reaction sequence is

$$\text{Alkane} \xrightarrow[\text{(dehydrogenation)}]{\text{① Oxidation}} \text{alkene} \xrightarrow{\text{② Hydration}} \begin{array}{c}\text{secondary}\\\text{alcohol}\end{array} \xrightarrow[\text{(dehydrogenation)}]{\text{③ Oxidation}} \text{ketone}$$

Step 6: *Oxidation of Succinate.* This is the third redox reaction of the cycle. The enzyme involved is succinate dehydrogenase, and the oxidizing agent is FAD rather than NAD^+. Two hydrogen atoms are removed from the succinate to produce fumarate, a C_4 species with a *trans* double bond. FAD is reduced to $FADH_2$ in the process.

• Fumarate, with its *trans* double bond, is an essential metabolic intermediate in both plants and animals. Its isomer, with a *cis* double bond, is called maleate, and it is toxic and irritating to tissues. Succinate dehydrogenase produces only the *trans* isomer of this unsaturated diacid.

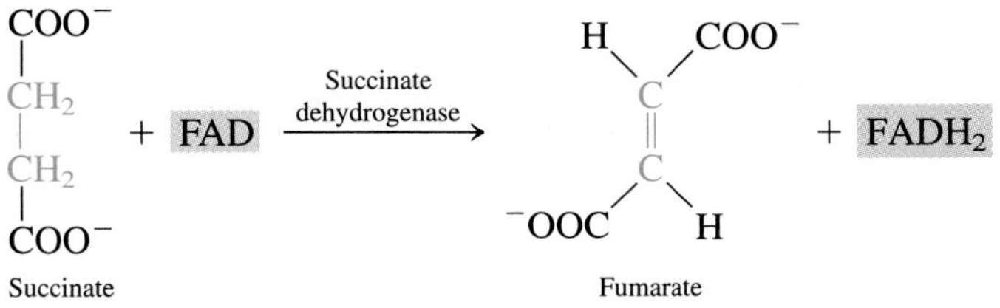

$$\begin{array}{c} COO^- \\ | \\ CH_2 \\ | \\ CH_2 \\ | \\ COO^- \end{array} + \text{FAD} \xrightarrow{\text{Succinate dehydrogenase}} \begin{array}{c} H \quad COO^- \\ \backslash \;\; / \\ C \\ \| \\ C \\ / \;\; \backslash \\ {}^-OOC \quad H \end{array} + FADH_2$$

Succinate — Fumarate

Step 7: *Hydration of Fumarate.* The enzyme fumarase catalyzes the addition of water to the double bond of fumarate. The enzyme is very stereospecific, so only the L isomer of the product malate is produced.

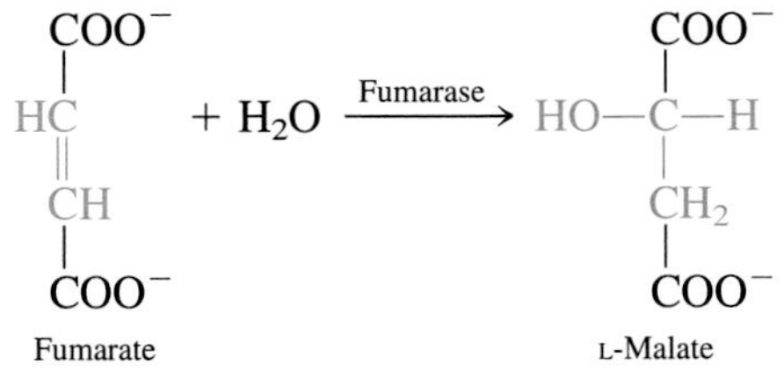

$$\begin{array}{c} COO^- \\ | \\ HC \\ \| \\ CH \\ | \\ COO^- \end{array} + H_2O \xrightarrow{\text{Fumarase}} \begin{array}{c} COO^- \\ | \\ HO{-}C{-}H \\ | \\ CH_2 \\ | \\ COO^- \end{array}$$

Fumarate — L-Malate

Step 8: *Oxidation of L-Malate to Regenerate Oxaloacetate.* In the fourth oxidation–reduction reaction of the cycle, a molecule of NAD^+ reacts with malate, picking up two hydrogen atoms with their associated energy to form $NADH + H^+$. The product of this reaction is oxaloacetate, so we are back where we started. The oxaloacetate formed in this step can combine with another molecule of acetyl CoA (Step 1), and the cycle can begin again.

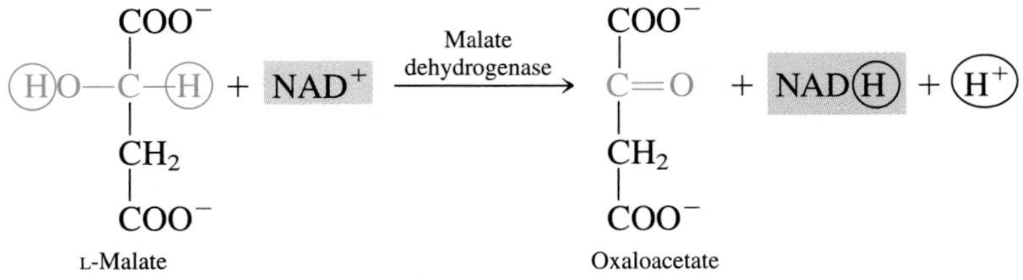

$$\begin{array}{c} COO^- \\ | \\ HO{-}C{-}H \\ | \\ CH_2 \\ | \\ COO^- \end{array} + NAD^+ \xrightarrow{\text{Malate dehydrogenase}} \begin{array}{c} COO^- \\ | \\ C{=}O \\ | \\ CH_2 \\ | \\ COO^- \end{array} + NADH + H^+$$

L-Malate — Oxaloacetate

• Summary of the Citric Acid Cycle

An overall summary equation for the citric acid cycle is obtained by adding together the individual reactions of the cycle:

$$\text{Acetyl CoA} + 3\text{NAD}^+ + \text{FAD} + \text{GDP} + \text{P}_i + 2\text{H}_2\text{O} \longrightarrow 2\text{CO}_2 + \text{CoA—SH} + 3\text{NADH} + 2\text{H}^+ + \text{FADH}_2 + \text{GTP}$$

Important features of the cycle include the following:

1. The reactions of the cycle take place in the mitochondrial matrix, except the succinate dehydrogenase reaction that involves FAD. The enzyme that catalyzes this reaction is an integral part of the inner mitochondrial membrane.
2. The "fuel" for the cycle is acetyl CoA, obtained from the breakdown of carbohydrates, fats, and proteins.
3. Four of the cycle reactions involve oxidation and reduction. The oxidizing agent is either NAD^+ (three times) or FAD (once). The operation of the cycle depends on the availability of these oxidizing agents.
4. In redox reactions, NAD^+ is the oxidizing agent when a carbon–oxygen double bond is formed; FAD is the oxidizing agent when a carbon–carbon double bond is formed.
5. The three NADH and one $FADH_2$ that are formed during the cycle carry electrons and H^+ to the electron transport chain (Section 25.6) through which ATP is synthesized.
6. Two carbon atoms enter the cycle as the acetyl unit of acetyl CoA, and two carbon atoms leave the cycle as two molecules of CO_2. The carbon atoms that enter and leave are not the same ones. The carbon atoms that leave during one turn of the cycle are carbon atoms that entered during the previous turn of the cycle.
7. Four B vitamins are necessary for the proper functioning of the cycle: riboflavin (in both FAD and the α-ketoglutarate dehydrogenase complex), nicotinamide (in NAD^+), pantothenic acid (in CoA—SH), and thiamin (in the α-ketoglutarate dehydrogenase complex).
8. One high-energy GTP molecule is produced by phosphorylation.

• The eight B vitamins and their structures were discussed in Section 21.13.

• Regulation of the Citric Acid Cycle

The rate at which the citric acid cycle operates is controlled by the body's need for energy (ATP). When the body's ATP supply is high, the ATP present inhibits the activity of citrate synthase, the enzyme in Step 1 of the cycle. When energy is being used at a high rate, a state of low ATP and high ADP concentrations, the ADP activates citrate synthase and the cycle speeds up. A similar control mechanism exists at Step 3, which involves isocitrate dehydrogenase; here NADH acts as an inhibitor and ADP as an activator.

23.7 The Electron Transport Chain

The NADH and $FADH_2$ produced in the citric acid cycle pass to the electron transport chain. The **electron transport chain** *is a series of reactions in which electrons and hydrogen ions from NADH and $FADH_2$ are passed to intermediate carriers and ultimately react with molecular oxygen to produce water.* The oxidation reactions for NADH and $FADH_2$ are

$$\text{NADH} + \text{H}^+ \longrightarrow \text{NAD}^+ + 2\text{H}^+ + 2e^-$$

$$\text{FADH}_2 \longrightarrow \text{FAD} + 2\text{H}^+ + 2e^-$$

Water is formed when the electrons and hydrogen ions that originate from these reactions react with molecular oxygen.

$$\text{O}_2 + 4e^- + 4\text{H}^+ \longrightarrow 2\text{H}_2\text{O}$$

• The oxygen involved in the water formation associated with the electron transport chain is the oxygen we breathe.

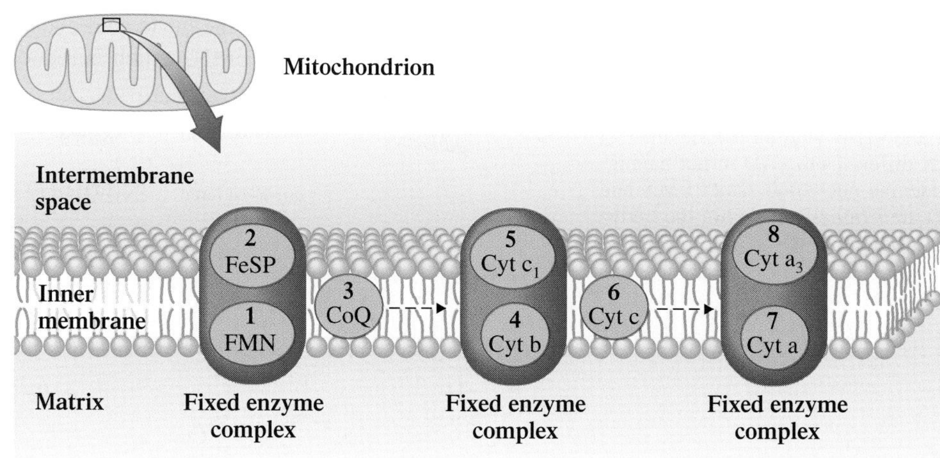

Figure 23.7
Six of the eight electron carriers (enzymes) of the electron transport chain are found at fixed sites within the inner mitochondrial membrane. The other two electron carriers are mobile, moving between fixed sites.

The electrons that pass through the various steps of the electron transport chain (ETC) lose some energy with each transfer along the chain. Some of this "lost" energy is used to make ATP from ADP (oxidative phosphorylation), as we will see in Section 23.8.

The enzyme complexes of the ETC are found in the inner mitochondrial membrane in the sequence in which they are needed. The product from one enzyme-mediated reaction is passed directly to the next enzyme. Each of the enzyme complexes in the pathway is alternately reduced and oxidized as it first accepts electrons and then passes them on to the next enzyme complex in the ETC.

• The *electron transport chain* is also frequently called the *respiratory chain.*

The electron carriers in the enzyme complexes of the ETC that are actually involved in the electron-transfer process are eight in number. Listed in the order in which they are encountered, they are

1. Flavin mononucleotide (FMN)
2. Iron–sulfur protein (FeSP)
3. Coenzyme Q (CoQ)
4. Cytochrome b (cyt b)
5. Cytochrome c_1 (cyt c_1)
6. Cytochrome c (cyt c)
7. Cytochrome a (cyt a)
8. Cytochrome a_3 (cyt a_3)

With the exception of coenzyme Q and cytochrome c, these electron carriers occur as part of enzyme complexes that have *fixed* locations within the inner mitochondrial membrane (Figure 23.7). Coenzyme Q and cytochrome c are *mobile*—free to move between specific fixed enzyme complexes.

Figure 23.8 is an overview of the redox reactions of the ETC. The individual reactions of the chain will now be considered in detail.

Step 1: *Flavin Mononucleotide (FMN).* The initial step of the ETC involves NADH/H^+ (from the citric acid cycle) and the enzyme complex containing the coenzyme FMN. The NADH/H^+ is oxidized to NAD^+ (which can again

Figure 23.8
An overview of the reactions in the electron transport chain. The oxidized product of each reaction is shown in red; the reduced product in blue.

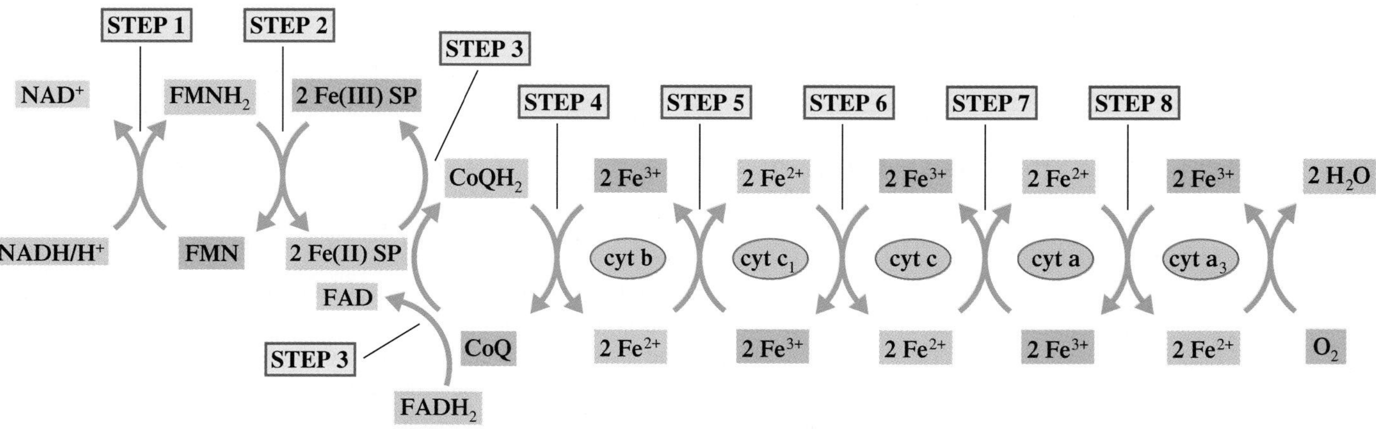

• The FMN/$FMNH_2$ pair is the third situation we have encountered in which a flavin molecule is present. The other two are the FAD/$FADH_2$ pair and the B vitamin riboflavin. FMN differs from FAD in not having an adenine nucleotide. Both FMN and FAD are synthesized within the body from riboflavin.

participate in the citric acid cycle) as it passes two hydrogen ions and two electrons to FMN, which is reduced to $FMNH_2$.

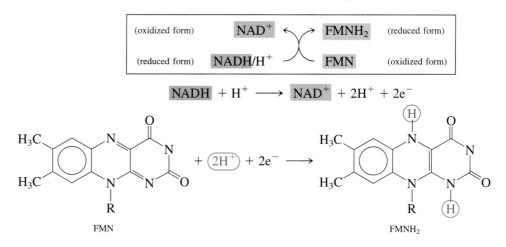

(oxidized form)	NAD^+		$FMNH_2$	(reduced form)
(reduced form)	NADH/H^+		FMN	(oxidized form)

$$NADH + H^+ \longrightarrow NAD^+ + 2H^+ + 2e^-$$

$$FMN + (2H^+) + 2e^- \longrightarrow FMNH_2$$

FMN FMNH$_2$

Step 2: *Iron/Sulfur Protein (FeSP).* $FMNH_2$ transfers the electrons it receives to an iron/sulfur protein (FeSP) located in the same enzyme complex where Step 1 occurred. The iron present in FeSP is Fe^{3+}, which is reduced in the reaction to Fe^{2+}.

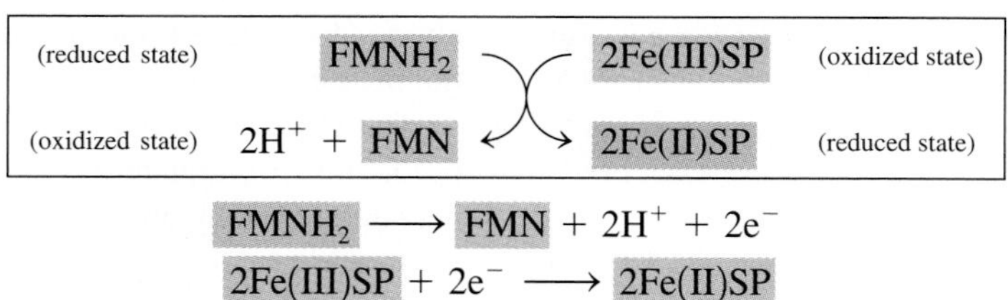

$$FMNH_2 \longrightarrow FMN + 2H^+ + 2e^-$$
$$2Fe(III)SP + 2e^- \longrightarrow 2Fe(II)SP$$

The two H atoms of $FMNH_2$ are released as two H^+ ions. Two FeSP units are needed to accommodate the two electrons released by $FMNH_2$, because an Fe^{3+}/Fe^{2+} change involves only one electron.

• The molecule quinone, a cyclic ketone (Section 15.2), has the structure

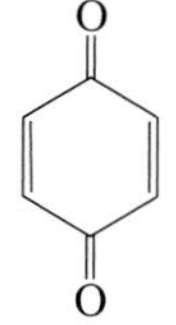

Step 3: *Coenzyme Q.* Two FeSP units next pass an electron pair to the mobile coenzyme Q. This coenzyme is a quinone derivative, hence the designation Q for the coenzyme. In its most common form, coenzyme Q has a long carbon chain containing 10 isoprene (Section 13.6) units.

(oxidized state)	2Fe(III)SP		$CoQH_2$	(reduced state)
(reduced state)	2Fe(II)SP		CoQ	(oxidized state)

$$2Fe(II)SP \longrightarrow 2Fe(III)SP + 2e^-$$

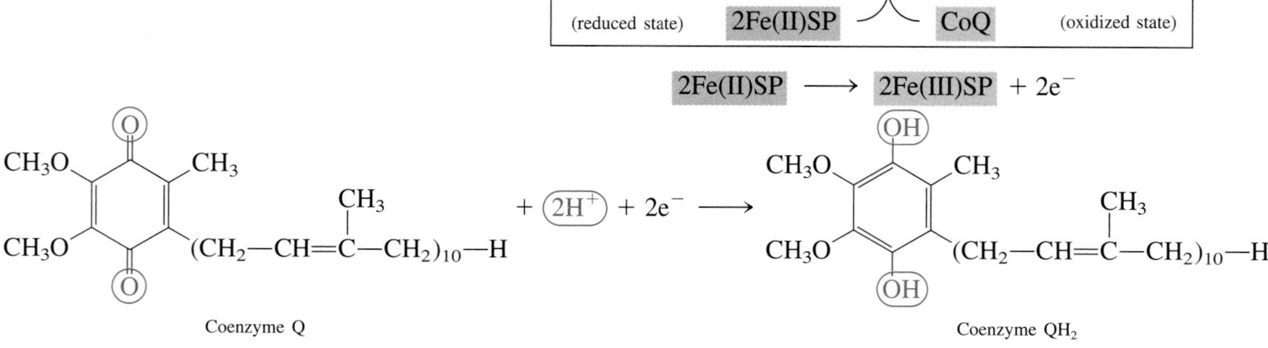

Coenzyme Q Coenzyme QH_2

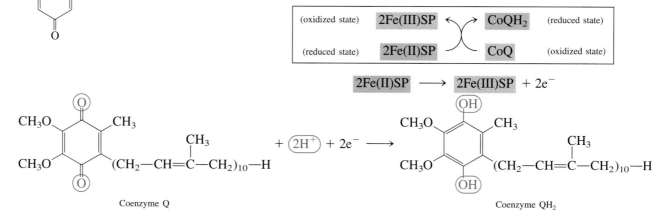

Coenzyme Q picks up two H^+ ions in solution, in addition to the two electrons, in forming coenzyme QH_2.

Coenzyme Q is also the entry point into the ETC for electrons and H^+ ions available from $FADH_2$, the other citric acid cycle (CAC) electron carrier besides NADH. The interaction between $FADH_2$ and coenzyme Q is

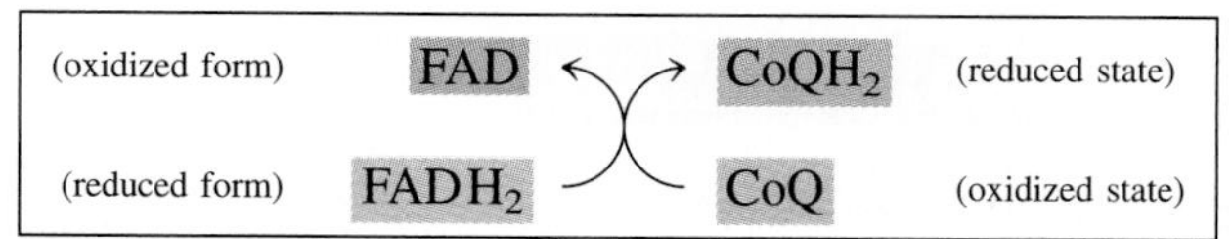

The $FADH_2$ is oxidized to FAD (which can again participate in the CAC) as it passes two hydrogen ions and two electrons to CoQ.

The formation of $CoQH_2$ is the last step of the ETC in which H^+ ions directly participate in the redox process.

• All H^+ ions required for the reactions of NADH, CoQ, and O_2 in the ETC come from the matrix side of the inner mitochondrial membrane.

Steps 4–8: *Cytochromes.* Next follows a series of steps involving only electrons (no H^+ ions) and protein molecules called cytochromes. **Cytochromes** *are heme-containing proteins that undergo reversible oxidation and reduction of their iron atoms.* Heme, a compound also present in hemoglobin and myoglobin (Section 20.10), has the structure

• In cytochromes the iron of the heme group does not bind to oxygen; instead, the iron is involved in redox reactions in which the iron changes back and forth between the +2 and +3 oxidation states.

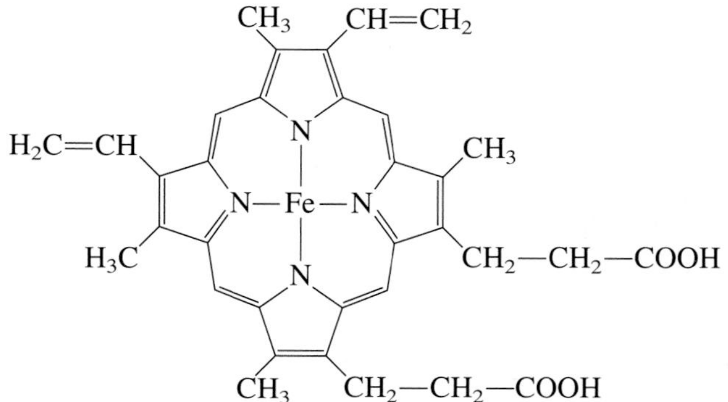

• Iron/sulfur protein (FeSP) is a *nonheme iron protein.* Most proteins of this type contain sulfur, as is the case with FeSP. Often the iron is bound to the sulfur atom in the amino acid cysteine.

Heme-containing proteins function similarly to FeSP; iron changes back and forth between the +3 and +2 oxidation states.

Various cytochromes, abbreviated cyt a, cyt b, cyt c, and so on, differ from each other in (1) their protein constituents, (2) the manner in which the heme is bound to the protein, and (3) attachments to the heme ring. Again, because the Fe^{3+}/Fe^{2+} system involves only a one-electron change, two cytochrome molecules are needed in these steps of the ETC to move two electrons along the chain.

	Step 4		Step 5		Step 6		Step 7		Step 8			
$CoQH_2$		$2Fe^{3+}$		$2Fe^{2+}$		$2Fe^{3+}$		$2Fe^{2+}$		$2Fe^{3+}$		H_2O
		cyt b		cyt c_1		cyt c		cyt a		cyt a_3		
CoQ		$2Fe^{2+}$		$2Fe^{3+}$		$2Fe^{2+}$		$2Fe^{3+}$		$2Fe^{2+}$		$\frac{1}{2}O_2 + 2H^+$

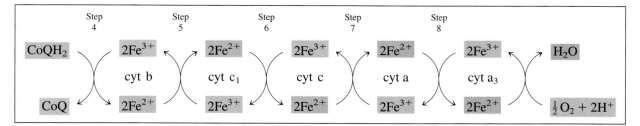

• A feature all steps in the ETC share is that as each electron carrier passes electrons along the chain, it becomes reoxidized and thus able to accept more electrons.

The last cytochrome in the ETC (Step 8), cyt a_3, has the ability to bind molecular oxygen to itself and effect its reduction. The reduced oxygen atoms combine with H^+ ions to produce water.

$$\frac{1}{2}O_2 + 2H^+ + 2e^- \longrightarrow H_2O$$

It is estimated that 95% of the oxygen used by cells serves as the final electron acceptor for the ETC.

The fixed enzyme complex that contains cytochromes a and a_3 is called cytochrome oxidase. Its structure is known to include two different metals, iron and copper. The role that copper plays in the activity of cytochrome oxidase is not definitely known, although it appears that electrons are transferred from copper to iron to oxygen.

Chemical CONNECTIONS

23.1 Cyanide Poisoning

Inhalation of hydrogen cyanide gas (HCN) or ingestion of solid potassium cyanide (KCN) rapidly inhibits the electron transport chain in all tissues, making cyanide one of the most potent and rapidly acting poisons known. The attack point for the cyanide ion (CN^-) is cytochrome oxidase, the last fixed-site enzyme in the electron transport chain. Cyanide inactivates this enzyme by bonding itself to the Fe^{3+} in the enzyme's heme portions. As a result, Fe^{3+} is unable to transfer electrons to oxygen, blocking the cell's use of oxygen. Death results from tissue asphyxiation, particularly of the central nervous system. Cyanide also binds to the heme group in hemoglobin, blocking oxygen transport in the bloodstream.

One treatment for cyanide poisoning is to administer various nitrites (NO_2^-), which oxidize the iron atoms of hemoglobin to Fe^{3+}. This form of hemoglobin helps draw CN^- back into the bloodstream, where it can be converted to thiocyanate (SCN^-) by thiosulfate ($S_2O_3^{2-}$), which is administered along with the nitrite (see the accompanying figure).

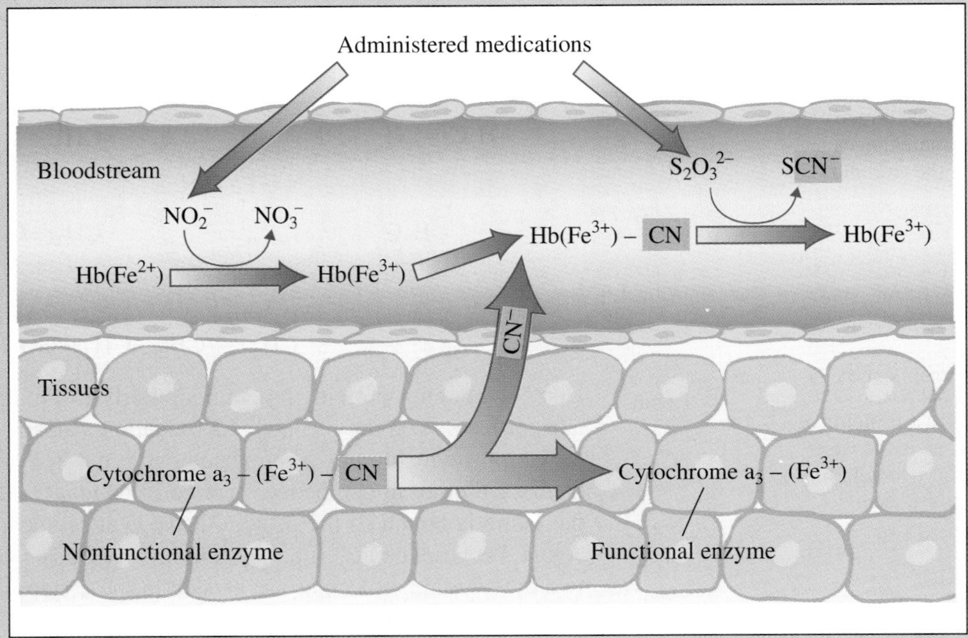

23.8 Oxidative Phosphorylation

Oxidative phosphorylation *is the process by which ATP is synthesized from ADP using energy released in the electron transport chain.* A key aspect of this process is coupled reactions. **Coupled reactions** *are pairs of chemical reactions in which energy released from one reaction changes the equilibrium position of a second reaction.*

The interdependence (coupling) of ATP synthesis with the reactions of the ETC is related to the movement of protons (H^+ ions) across the inner mitochondrial membrane. The three fixed enzyme complexes involved in the ETC chain have a second function besides that of electron transfer down the chain. They also serve as "proton pumps," transferring protons from the matrix side of the inner mitochondrial membrane to the intermembrane space (Figure 23.9).

• *Oxidative phosphorylation* is not the only process by which ATP is produced in cells. A second process, *substrate phosphorylation* (Section 24.2), can also be an ATP source. However, the amount of ATP produced by this second process is much less than that produced by oxidative phosphorylation.

Some of the H^+ ions crossing the inner mitochondrial membrane come from the reduced electron carriers, and some come from the matrix; the details of how the H^+ ions cross the inner mitochondrial membrane are not fully understood.

For every two electrons passed through the ETC, four protons cross the inner mitochondrial membrane at the first fixed enzyme site, two at the second fixed enzyme site, and four more at the third fixed enzyme site. This proton flow causes a buildup of H^+

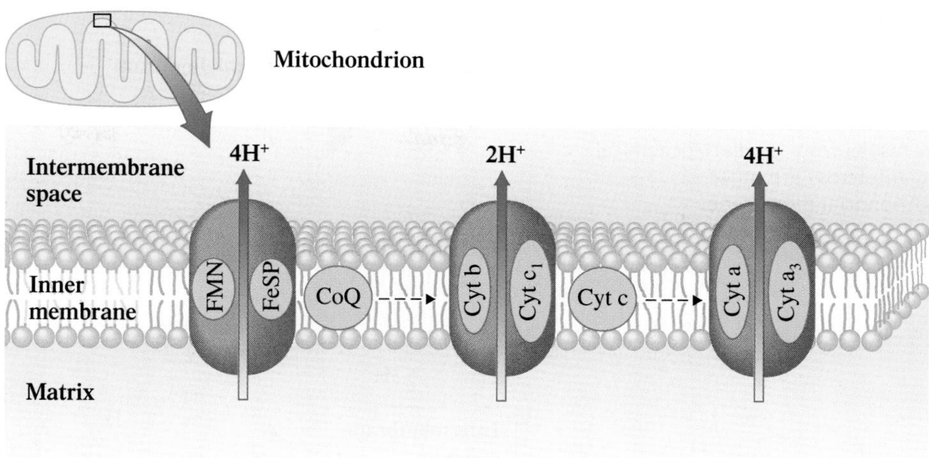

Figure 23.9
A second function for the fixed enzyme sites involved in the electron transport chain is that of proton pumps. For every two electrons passed through the ETC, 10 protons are transferred from the mitochondrial matrix to the intermembrane space.

ions (protons) in the intermembrane space; this high concentration of protons becomes the basis for ATP synthesis.

The "proton flow" explanation for ATP–ETC coupling is formally called chemiosmotic coupling. **Chemiosmotic coupling** *is an explanation for the coupling of ATP synthesis with electron transport chain reactions that requires a proton gradient across the inner mitochondrial membrane.* The main concepts in this explanation for coupling follow.

• Some of the energy released at each of the three fixed enzyme complex sites is consumed in the movement of H^+ ions across the inner membrane from the matrix into the intermembrane space. Movement of ions from a region of lower concentration (the matrix) to one of higher concentration (the intermembrane space) requires the expenditure of energy because it opposes the natural tendency, as exhibited in the process of osmosis (Section 8.8), to equalize concentrations.

1. The result of the pumping of protons from the mitochondrial matrix across the inner mitochondrial membrane is a higher concentration of protons in the intermembrane space than in the matrix. This concentration difference constitutes an *electrochemical (proton) gradient.* A chemical gradient exists whenever a substance has a higher concentration in one region than in another. Because the proton has an electrical charge (is an ion), an electrical gradient also exists. Potential energy (Section 7.2) is always associated with an electrochemical gradient.
2. A spontaneous flow of protons from the region of high concentration to the region of low concentration occurs because of the electrochemical gradient. This proton flow is not through the membrane itself (it is not permeable to H^+ ions) but rather through enzyme complexes called *ATP synthases* located on the inner mitochondrial membrane (Section 23.2). These enzymes catalyze the conversion of ADP to ATP.

$$\text{ADP} + \text{P}_i \xrightarrow{\text{ATP synthase}} \text{ATP} + \text{H}_2\text{O}$$

3. ATP synthase has two subunits, the F_0 and F_1 subunits (Figure 23.10). The F_0 part of the synthase is the channel for proton flow, whereas the formation of ATP takes place in the F_1 subunit. As protons return to the mitochondrial matrix through the F_0 subunit, the potential energy associated with the electrochemical gradient is released and used in the F_1 subunit for the synthesis of ATP.

23.9 ATP Production for the Common Metabolic Pathway

• The difference in H^+ ion concentration between the two sides of the inner mitochondrial membrane causes a pH difference of about 1.4 units. A pH difference of 1.4 units means that the intermembrane space, the more acidic region, has 25 times more protons than the matrix.

Three molecules of ATP are formed for every NADH molecule that enters the ETC. $FADH_2$, which does not enter the ETC at its start, produces only two ATP molecules. $FADH_2$'s entrance point into the chain, coenzyme Q, is beyond the site of the first fixed enzyme complex site—that is, beyond the first "proton-pumping" site. Hence one fewer ATP molecule is produced than for NADH.

The energy yield, in terms of ATP production, can now be totaled for the common metabolic pathway (Section 23.5). Every acetyl CoA entering the CAC produces three

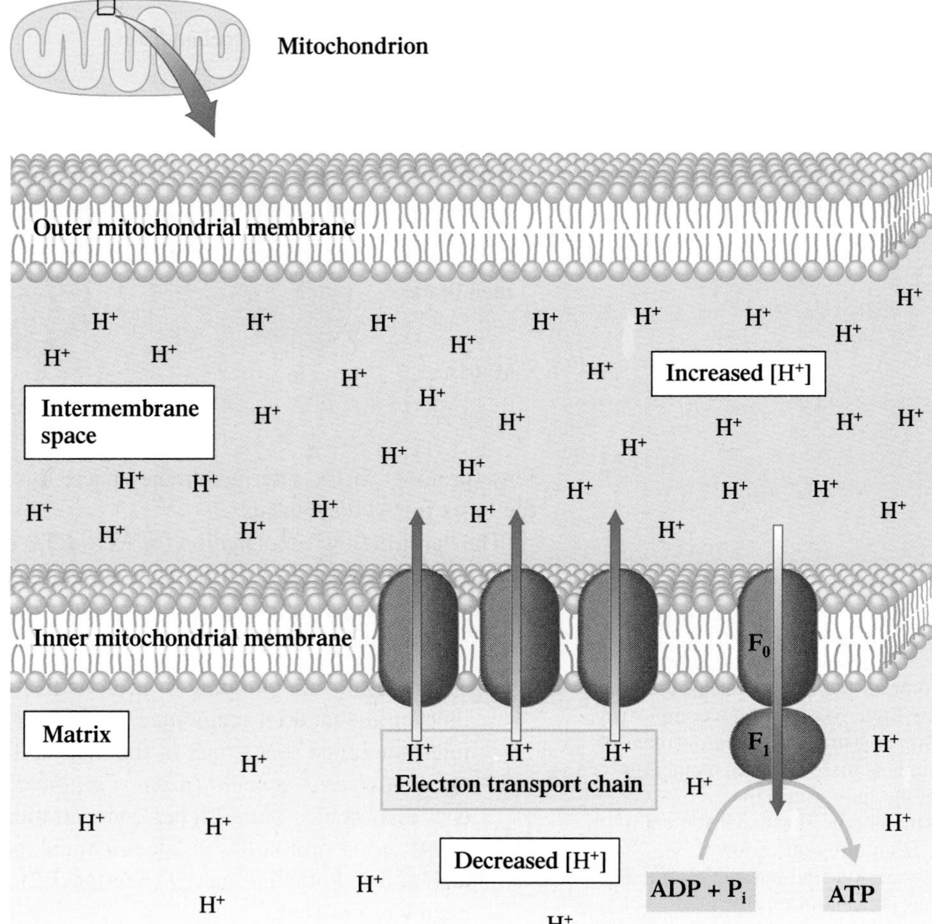

Figure 23.10
Formation of ATP accompanies the flow of protons from the intermembrane space back into the mitochondrial matrix. The proton flow results from an electrochemical gradient across the inner mitochondrial membrane.

NADH, one $FADH_2$, and one GTP (which is equivalent in energy to ATP; Section 23.6). Thus 12 molecules of ATP are produced for each acetyl CoA catabolized.

$$\begin{aligned} 3\ \text{NADH} &\longrightarrow 9\ \text{ATP} \\ 1\ \text{FADH}_2 &\longrightarrow 2\ \text{ATP} \\ 1\ \text{GTP} &\longrightarrow \underline{1\ \text{ATP}} \\ & \quad\ 12\ \text{ATP} \end{aligned}$$

23.10 The Importance of ATP

The cycling of ATP and ADP in metabolic processes is the principal medium for energy exchange in biological processes. The conversion

$$\text{ATP} \longrightarrow \text{ADP} + \text{P}_i$$

powers life processes (the biosynthesis of essential compounds, muscle contraction, nutrient transport, and so on). The conversion

$$\text{P}_i + \text{ADP} \longrightarrow \text{ATP}$$

which occurs in food catabolism cycles, regenerates the ATP expended in cell operation. Figure 23.11 summarizes the ATP–ADP cycling process.

ATP is a high-energy phosphate compound (Section 23.4). Its hydrolysis to ADP produces an *intermediate* amount of free energy (−7.5 kcal/mole; Table 23.1) when

Chemical CONNECTIONS

23.2 Brown Fat, Newborn Babies, and Hibernating Animals

Ordinarily, metabolic processes generate enough heat to maintain normal body temperature. In certain cases, however, including newborn infants and hibernating animals, normal metabolism is not sufficient to meet the body's heat requirements. In these cases, a supplemental method of heat generation, which involves *brown fat tissue,* occurs.

Brown fat tissue, as the name implies, is darker in color than ordinary fat tissue, which is white. Brown fat is specialized for heat production. It contains many more blood vessels and mitochondria than white fat. (The increased number of mitochondria gives brown fat its color.)

Another difference between the two types of fat is that the mitochondria in brown fat cells contain a protein called *thermogenin,* which functions as an *uncoupling agent.* This protein "uncouples" the ATP production associated with the electron transport chain. The ETC reactions still take place, but the energy that would ordinarily be used for ATP synthesis is simply released as heat.

Brown fat tissue is of major importance for newborn infants. Newborns are immediately faced with a temperature-regulation problem. They leave an environment of constant 37°C temperature and enter a much colder environment (25°C). A supply of *active* brown fat, present at birth, helps the baby adapt to the cooler environment.

Very limited amounts of brown fat are present in most adults. However, stores of brown fat increase in adults who are regularly exposed to cold environs. Thus the production of brown fat is one of the body's mechanisms for cold adaptation.

Hibernating bears rely on brown fat tissue to help meet their body's heat requirements.

Thermogenin, the uncoupling agent in brown fat, is a protein bound to the inner mitochondrial membrane. When activated, it functions as a proton channel through the inner membrane. The proton gradient produced by the electron transport chain is dissipated through this "new" proton channel, and less ATP synthesis occurs because the normal proton channel, ATP synthase, has been bypassed. The energy of the proton gradient, no longer useful for ATP synthesis, is released as heat.

• ATP molecules in cells have a high turnover rate. Normally, a given ATP molecule in a cell does not last more than a minute before it is converted to ADP. The concentration of ATP in a cell varies from 0.5 to 2.5 mg/mL of cell fluid.

compared with hydrolysis energies for other organophosphate compounds (Table 23.1). Of major importance, the energy derived from ATP hydrolysis is a *biologically useful* amount of energy. It is larger than the amount of energy needed by compounds to which ATP donates energy, and yet it is smaller than that available in compounds used to form ATP. If the ATP hydrolysis energy were *unusually high,* the body would not be able to

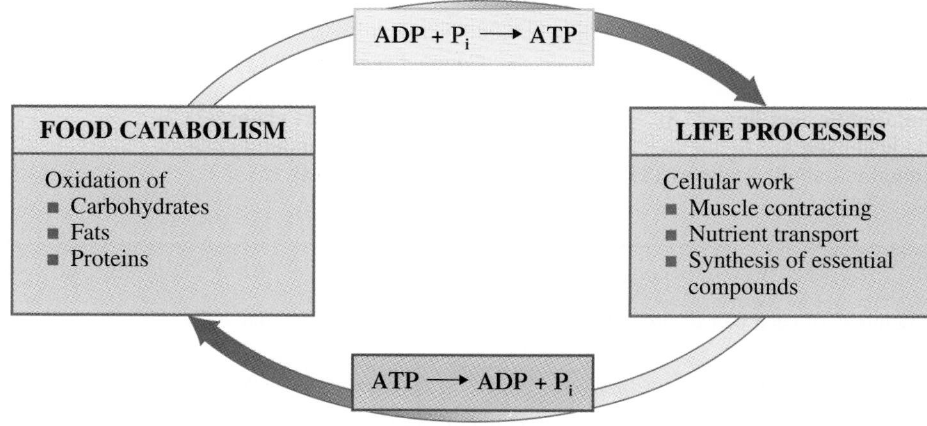

Figure 23.11
The interconversion of ATP and ADP is the principal medium for energy exchange in biological processes.

convert ADP back to ATP, because ATP synthesis requires an energy input equal to or greater than the hydrolysis energy, and such an unusually high amount of energy would not be available.

Concepts to Remember

Metabolism. Metabolism is the sum total of all the chemical reactions that take place in a living organism. Metabolism consists of catabolism and anabolism. Catabolic reactions involve the breakdown of large molecules into smaller fragments. Anabolic reactions synthesize large molecules from smaller ones.

Mitochondria. Mitochondria are membrane-enclosed subcellular structures that are the site of energy production in the form of ATP molecules. Enzymes for both the citric acid cycle and the electron transport chain are housed in the mitochondria.

Important coenzymes. Three very important coenzymes involved in catabolism are NAD^+, FAD, and CoA. NAD^+ and FAD are oxidizing agents that participate in the oxidation reactions of the citric acid cycle. They transport hydrogen atoms and electrons from the citric acid cycle to the electron transport chain. CoA interacts with acetyl groups produced from food degradation to form acetyl CoA. Acetyl CoA is the "fuel" for the citric acid cycle.

High-energy compounds. A high-energy compound liberates a larger-than-normal amount of free energy upon hydrolysis, because structural features in the molecule contribute to repulsive strain in one or more bonds. Most high-energy biochemical molecules contain phosphate groups.

Common catabolic pathway. The common catabolic pathway includes the reactions of the citric acid cycle and those of the electron transport chain and oxidative phosphorylation. The degradation products from all types of foods (carbohydrates, fats, and proteins) participate in these reactions.

Citric acid cycle. The citric acid cycle is a cyclic series of eight reactions that oxidize the acetyl portion of acetyl CoA, resulting in the production of two molecules of CO_2. The complete oxidation of one acetyl group produces three molecules of NADH, one of $FADH_2$, and one of GTP.

Electron transport chain. The electron transport chain is a series of reactions that passes electrons from NADH and $FADH_2$ to molecular oxygen. Each electron carrier that participates in the chain has an increasing affinity for electrons. Upon accepting the electrons and hydrogen ions, the O_2 is reduced to H_2O.

Oxidative phosphorylation. In the electron transport chain, released energy is used to convert ADP to ATP. One molecule of NADH produces three molecules of ATP. $FADH_2$, which enters the chain later than NADH, produces only two molecules of ATP.

Chemiosmotic coupling. Chemiosmotic coupling explains how the energy needed for ATP synthesis is obtained. Synthesis takes place because of a flow of protons across the inner mitochondrial membrane.

Importance of ATP. ATP is the link between energy production and energy use in cells. The conversion of ATP to ADP powers life processes, and the conversion of ADP back to ATP regenerates the energy expended in cell operation.

Key Reactions and Equations

1. Oxidation by NAD^+ (Section 23.3)

$$NAD^+ + 2H^+ + 2e^- \longrightarrow NADH + H^+$$

2. Oxidation by FAD (Section 23.3)

$$FAD + 2H^+ + 2e^- \longrightarrow FADH_2$$

3. The citric acid cycle (Section 23.6)

$$\text{Acetyl CoA} + 3NAD^+ + FAD + GDP + P_i + 2H_2O \longrightarrow 2CO_2 + CoA + 3NADH + 2H^+ + FADH_2 + GTP$$

4. The electron transport chain (Section 23.7)

$$NADH + H^+ \longrightarrow NAD^+ + 2H^+ + 2e^-$$

$$FADH_2 \longrightarrow FAD + 2H^+ + 2e^-$$

$$O_2 + 4H^+ + 4e^- \longrightarrow 2H_2O$$

5. Oxidative phosphorylation (Section 23.8)

$$ADP + P_i \xrightarrow[\text{from ETC}]{\text{Energy}} ATP$$

Key Terms

Anabolism (23.1)
Catabolism (23.1)
Chemiosmotic coupling (23.8)
Citric acid cycle (23.6)
Common metabolic pathway (23.5)
Coupled reactions (23.8)
Cytochromes (23.7)
Electron transport chain (23.7)
High-energy compound (23.4)
Metabolic pathway (23.1)
Metabolism (23.1)
Mitochondria (23.2)
Organelle (23.2)
Oxidative phosphorylation (23.8)

Exercises and Problems

The members of each pair of problems in this section test similar material.

Metabolism (Section 23.1)

23.1 Classify anabolism and catabolism as synthetic or degradative processes.

23.2 Classify anabolism and catabolism as energy-producing or energy-consuming processes.

23.3 What is a metabolic pathway?

23.4 What is the difference between a linear and a cyclic metabolic pathway?

23.5 What general characteristics are associated with a catabolic pathway?

23.6 What general characteristics are associated with an anabolic pathway?

Cell Structure (Section 23.2)

23.7 List several differences between prokaryotic cells and eukaryotic cells.

23.8 What kinds of organisms have prokaryotic cells and what kinds have eukaryotic cells?

23.9 What is an organelle?

23.10 What is a mitochondrion?

23.11 In a mitochondrion, what separates the matrix from the intermembrane space?

23.12 In what major way do the inner and outer mitochondrial membranes differ?

23.13 What is the intermembrane space of a mitochondrion?

23.14 Where are ATP synthase complexes located in a mitochondrion?

Intermediate Compounds in Metabolic Pathways (Section 23.3)

23.15 What does each letter stand for in ATP?

23.16 What does each letter stand for in ADP?

23.17 Draw a block diagram structure for ATP.

23.18 Draw a block diagram structure for ADP.

23.19 What is the structural difference between ATP and AMP?

23.20 What is the structural difference between AMP and cAMP?

23.21 What is the structural difference between ATP and GTP?

23.22 What is the structural difference between ATP and CTP?

23.23 In terms of hydrolysis, what is the relationship between ATP and ADP?

23.24 In terms of hydrolysis, what is the relationship between ADP and AMP?

23.25 What does each letter stand for in FAD?

23.26 What does each letter stand for in NAD^+?

23.27 Draw a block diagram structure for FAD based on the presence of an ADP core (3-block diagram).

23.28 Draw a block diagram structure for FAD based on the presence of two nucleotides (6-block diagram).

23.29 Draw a block diagram structure for NAD^+ based on the presence of two nucleotides (6-block diagram).

23.30 Draw a block diagram structure for NAD^+ based on the presence of an ADP core (3-block diagram).

23.31 Which part of an NAD^+ molecule is the active participant in redox reactions?

23.32 Which part of an FAD molecule is the active participant in redox reactions?

23.33 Give the letter designation for

a. the reduced form of FAD

b. the oxidized form of NADH

23.34 Give the letter designation for

a. the oxidized form of $FADH_2$

b. the reduced form of NAD^+

23.35 Name the vitamin B molecule that is part of the structure of

a. NAD^+ b. FAD

23.36 In which of the following molecules is the vitamin B portion the "active" portion in redox processes?

a. NAD^+ b. FAD

23.37 Draw the 3-block diagram structure for coenzyme A.

23.38 Which part of a coenzyme A molecule is the active participant in a redox reaction?

High-Energy Phosphate Compounds (Section 23.4)

23.39 What is a high-energy compound?

23.40 What factors contribute to a strained bond in high-energy phosphate compounds.

23.41 What does the designation P_i denote?

23.42 What does the designation PP_i denote?

23.43 With the help of Table 23.1, determine which compound in each of the following pairs of phosphate-containing compounds releases more free energy upon hydrolysis?

a. ATP and phosphoenolpyruvate

b. Creatine phosphate and ADP

c. Glucose 1-phosphate and 1,3-diphosphoglycerate

d. AMP and glycerol 3-phosphate

23.44 With the help of Table 23.1, determine which compound in each of the following pairs of phosphate-containing compounds releases more free energy upon hydrolysis?

a. ATP and creatine phosphate

b. Glucose 1-phosphate and glucose 6-phosphate

c. ADP and AMP

d. Phosphoenolpyruvate and PP_i

Biochemical Energy Production (Section 23.5)

23.45 Describe the four general stages of the process by which biochemical energy is obtained from food.

23.46 Of the four general stages of biochemical energy production from food, which are part of the common metabolic pathway?

The Citric Acid Cycle (Section 23.6)

23.47 What are two other names for the citric acid cycle?

23.48 What is the basis for the name *citric acid cycle?*

23.49 What is the "fuel" for the citric acid cycle?

23.50 What are the products of the citric acid cycle?

23.51 Consider the reactions that occur during *one turn* of the citric acid cycle in answering each of the following questions.

a. How many CO_2 molecules are formed?

b. How many molecules of $FADH_2$ are formed?

c. How many times is a secondary alcohol oxidized?

d. How many times does water add to a carbon–carbon double bond?

23.52 Consider the reactions that occur during *one turn* of the citric acid cycle in answering each of the following questions.

a. How many molecules of NADH are formed?

b. How many GTP molecules are formed?

c. How many decarboxylation reactions occur?

d. How many oxidation–reduction reactions occur?

23.53 There are eight steps in the citric acid cycle. List those steps that involve

a. oxidation

b. isomerization

c. hydration

23.54 There are eight steps in the citric acid cycle. List those steps that involve

a. oxidation and decarboxylation

b. phosphorylation

c. condensation

23.55 There are four C_4 dicarboxylic acid species in the citric acid cycle. What are their names and structures?

23.56 There are two simple keto carboxylic acid species in the citric acid cycle. What are their names and structures?

23.57 What type of reaction occurs in the citric acid cycle whereby a C_6 compound is converted to a C_5 compound?

23.58 What type of reaction occurs in the citric acid cycle whereby a C_5 compound is converted to a C_4 compound?

23.59 Identify the oxidized coenzyme (NAD^+ or FAD) that participates in each of the following citric acid cycle reactions.

a. Isocitrate $\longrightarrow$ α-ketoglutarate

b. Succinate $\longrightarrow$ fumarate

23.60 Identify the oxidized coenzyme (NAD^+ or FAD) that participates in each of the following citric acid cycle reactions.

a. Malate $\longrightarrow$ oxaloacetate

b. α-Ketoglutarate $\longrightarrow$ succinyl CoA

23.61 List the two citric acid cycle intermediates involved in the reaction governed by each of the following enzymes. List the reactant first.

a. Isocitrate dehydrogenase

b. Fumarase

c. Malate dehydrogenase

d. Aconitase

23.62 List the two citric acid cycle intermediates involved in the reaction governed by each of the following enzymes. List the reactant first.

a. α-Ketoglutarate dehydrogenase

b. Succinate dehydrogenase

c. Citrate synthase

d. Succinyl CoA synthase

The Electron Transport Chain (Section 23.7)

23.63 By what other name is the electron transport chain known?

23.64 Give a one-sentence summary of what occurs during the reactions known as the electron transport chain.

23.65 What is the final electron acceptor of the electron transport chain?

23.66 Which substances generated in the citric acid cycle participate in the electron transport chain?

23.67 What do the following abbreviations stand for?

a. FMN b. Cyt c. FeSP d. NADH

23.68 What do the following abbreviations stand for?

a. CoQ b. $FMNH_2$ c. $FADH_2$ d. $CoQH_2$

23.69 Classify each of the following electron transport chain electron carriers as mobile or fixed-site. For fixed-site electron carriers, further specify the location as first, second, or third fixed site.

a. Cyt c b. Cyt b

c. FMN d. CoQ

23.70 Classify each of the following electron transport chain electron carriers as mobile or fixed-site. For fixed-site electron carriers, further specify the location as first, second, or third fixed site.

a. Cyt c_1 b. FeSP

c. $CoQH_2$ d. Cyt a_3

23.71 Put the following substances in the correct order of their participation in the electron transport chain: cyt c_1, cyt a_3, FeSP, and NADH.

23.72 Put the following substances in the correct order of their participation in the electron transport chain: CoQ, $FADH_2$, FMN, and cyt b.

23.73 Indicate whether each of the following changes represents oxidation or reduction.

a. $CoQH_2 \longrightarrow CoQ$

b. $NAD^+ \longrightarrow NADH$

c. Cyt c (Fe^{2+}) $\longrightarrow$ cyt c (Fe^{3+})

d. Cyt b (Fe^{3+}) $\longrightarrow$ cyt b (Fe^{2+})

23.74 Indicate whether each of the following changes represents oxidation or reduction.

a. $FADH_2 \longrightarrow FAD$

b. $FMN \longrightarrow FMNH_2$

c. Fe(III)SP $\longrightarrow$ Fe(II)SP

d. Cyt c_1 (Fe^{3+}) $\longrightarrow$ cyt c_1 (Fe^{2+})

23.75 Fill in the missing substances in the following electron transport chain reaction sequences.

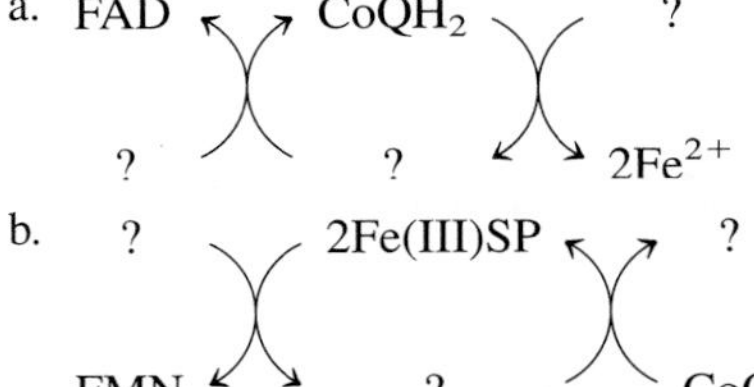

23.76 Fill in the missing substances in the following electron transport chain reaction sequences.

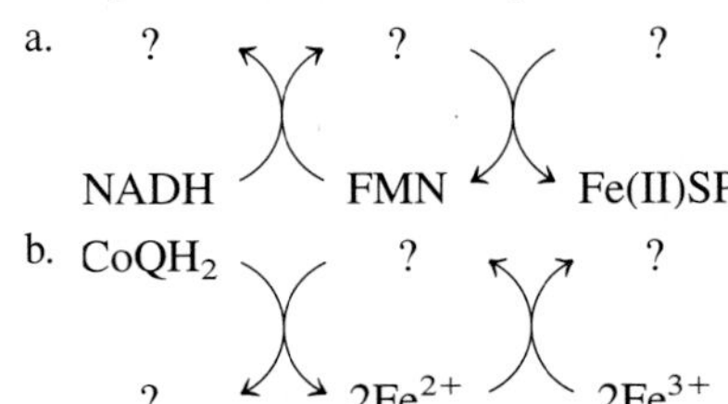

23.77 Complete the blanks in the following electron transport chain reactions.

a. $FMNH_2$ + 2Fe(III)SP $\longrightarrow$ ________ + ________

b. $FADH_2$ + ________ $\longrightarrow$ FAD + ________

23.78 Complete the blanks in the following electron transport chain reactions.

a. Cyt b (Fe^{2+}) + cyt c_1 (Fe^{3+}) $\longrightarrow$ ________ + ________

b. NADH + ________ $\longrightarrow$ NAD^+ + ________

Oxidative Phosphorylation (Section 23.8)

23.79 What is oxidative phosphorylation?

23.80 What are coupled reactions?

23.81 The coupling of ATP synthesis with the reactions of the ETC is related to the movement of what chemical species across the inner mitochrondial membrane?

23.82 At what enzyme location(s) in the electron transport chain does proton pumping occur?

23.83 At what mitochondrial location does H^+ ion buildup occur as the result of proton pumping?

23.84 How many protons cross the inner mitochondrial membrane for every two electrons that are passed through the electron transport chain?

23.85 What is the name of the enzyme that catalyzes ATP production during oxidative phosphorylation?

23.86 What is the location of the enzyme that uses stored energy in a proton gradient to drive the reaction that produces ATP.

23.87 How is the proton gradient associated with chemiosmotic coupling dissipated during ATP synthesis?

23.88 What are the "starting materials" from which ATP is synthesized as the proton gradient associated with chemiosmotic coupling is dissipated?

ATP Production (Section 23.9)

23.89 How many ATP molecules are formed for each NADH molecule that enters the electron transport chain?

23.90 How many ATP molecules are formed for each $FADH_2$ molecule that enters the electron transport chain?

23.91 NADH and $FADH_2$ molecules do not yield the same number of ATP molecules. Explain why.

23.92 What is the energy yield, in terms of ATP molecules, from one turn of the citric acid cycle, assuming that the products of the cycle enter the electron transport chain?

Additional Problems

23.93 Classify each of the following substances as (1) a reactant in the citric acid cycle, (2) a reactant in the electron transport chain, or (3) a reactant in both the CAC and the ETC.

a. NAD^+ b. NADH c. O_2
d. H_2O e. fumarate f. cytochrome a

23.94 Classify each of the following substances as (1) a product in the citric acid cycle, (2) a product in the electron transport chain, or (3) a product in both the CAC and the ETC.

a. $FADH_2$ b. FAD c. CO_2
d. H_2O e. malate f. flavin mononucleotide

23.95 Which of these substances, (1) ATP, (2) CoA, (3) FAD, and (4) NAD^+, contains the following subunits of structure? More than one choice may apply in a given situation.

a. Contains two ribose subunits
b. Contains two phosphate subunits
c. Contains one adenine subunit
d. Contains one ribitol subunit

23.96 Characterize, in terms of number of carbon atoms present, each of the following citric acid cycle changes as (a) a C_6 to C_6 change, (b) a C_6 to C_5 change, (c) a C_5 to C_4 change, or (d) a C_4 to C_4 change.

a. Citrate to isocitrate b. Succinate to fumarate
c. Malate to oxaloacetate
d. Isocitrate to alpha-ketoglutarate

23.97 At which of these locations, (1) first fixed enzyme site, (2) second fixed enzyme site, (3) third fixed enzyme site, or (4) mobile enzyme site, does each of the following ETC reactions occur?

a. $FADH_2$ + CoQ b. NADH + FMN
c. cyt a_3 + O_2 d. FeSP + CoQ

23.98 In what way are the processes of the citric acid cycle and the electron-transport chain interrelated?

23.99 Where within a cell does each of the following take place?

a. Citric acid cycle
b. Electron transport chain and oxidative phosphorylation

23.100 One of the oxidation steps that occurs when lipids are metabolized is

$$R{-}CH_2{-}CH_2{-}C{-}S{-}CoA \longrightarrow R{-}CH{=}CH{-}C{-}S{-}CoA$$

Would you expect this reaction to require FAD or NAD^+ as the oxidizing agent?

Grid Problems

23.101

1. B vitamin	2. ribose	3. ribitol
4. ADP	5. phosphorylated ADP	6. flavin

Select from the grid *all* correct responses for each of the following situations.

a. A structural subunit of FAD
b. A structural subunit of NAD^+
c. A structural subunit of CoA—SH
d. A structural subunit in FAD, NAD^+, and CoA—SH

23.102

1. citrate	2. malate	3. oxaloacetate
4. fumarate	5. α-ketoglutarate	6. isocitrate

Select from the grid *all* correct responses for each of the following situations.

a. A four-carbon intermediate in the citric acid cycle
b. A citric acid cycle intermediate that undergoes both oxidation and decarboxylation
c. A citric acid cycle intermediate that undergoes oxidation but not decarboxylation
d. A citric acid cycle intermediate that is a reactant in forming a six-carbon species

23.103

1. FMN	2. FeSP	3. CoQ
4. Cyt b	5. Cyt c	6. Cyt a_3

Select from the grid *all* correct responses for each of the following situations.

a. Part of an ETC fixed-site enzyme complex
b. A coenzyme that contains the element iron
c. A coenzyme in an ETC fixed-site enzyme complex with which a reduced coenzyme from the CAC interacts
d. A coenzyme in an ETC fixed-site enzyme complex with which molecular oxygen interacts

23.104

1. oxidative phosphorylation	2. electron transport chain	3. citric acid cycle
4. proton gradient dissipation	5. proton gradient formation	6. common metabolic pathway

Select from the grid *all* correct responses for each of the following situations.

a. The oxidized coenzymes NAD^+ and FAD are reactants in this process.
b. The reduced coenzymes NADH and $FADH_2$ are reactants in this process.
c. ATP is involved in this process.
d. Coenzyme A is involved in this process.

Carbohydrate Metabolism

CHAPTER OUTLINE

24.1 Digestion and Absorption of Carbohydrates 715
24.2 Glycolysis 717
24.3 Fates of Pyruvate 724
24.4 ATP Production from the Complete Oxidation of Glucose 727
24.5 Glycogen Synthesis and Degradation 728
24.6 Gluconeogenesis 730
24.7 Terminology for Glucose Metabolic Pathways 733
24.8 The Pentose Phosphate Pathway 733
Chemistry at a Glance: **Glucose Metabolism** 735
24.9 Hormonal Control of Carbohydrate Metabolism 736
Chemical Connections
24.1 Lactate Accumulation 726
24.2 Diabetes Mellitus 737

Carbohydrates are the major energy source for animals as well as human beings.

In this chapter we explore the relationship between carbohydrate metabolism and energy production in cells. The molecule glucose is the focal point of carbohydrate metabolism. Commonly called blood sugar, glucose is supplied to the body via the circulatory system and, after being absorbed by the cell, can be either oxidized to yield energy or stored as glycogen for future use. When sufficient oxygen is present, glucose is totally oxidized to CO_2 and H_2O. However, in the absence of oxygen, glucose is only partially oxidized to lactic acid. Besides supplying energy needs, glucose and other six-carbon sugars can be converted into a variety of different sugars (C_3, C_4, C_5, and C_7) needed for biosynthesis. Some of the oxidative steps in carbohydrate metabolism also produce NADH and NADPH, sources of reductive power in cells.

24.1 Digestion and Absorption of Carbohydrates

Digestion *is the breakdown of food molecules, through hydrolysis, into simpler chemical units that can be used by cells for their metabolic needs.* Digestion is the first stage in the processing of food products.

• Salivary α-amylase is a constituent of saliva, the fluid secreted by the salivary glands. Saliva is 99% water plus small amounts of several inorganic ions and organic molecules. Saliva secretion can be triggered by the taste, smell, sight, and even thought of food. Average saliva output is about 1.5 L per day.

The digestion of carbohydrates begins in the mouth, where the enzyme *salivary α-amylase* catalyzes the hydrolysis of α-glycosidic linkages (Section 18.13) in starch from plants and glycogen from meats to produce smaller polysaccharides and the disaccharide maltose.

Only a small amount of carbohydrate digestion occurs in the mouth, because food is swallowed so quickly. Although the food mass remains longer in the stomach, very little further carbohydrate digestion occurs there either, because salivary α-amylase is inactivated by the acidic environment of the stomach, and the stomach's own secretions do not contain any carbohydrate-digesting enzymes.

The primary site for carbohydrate digestion is within the lumen of the small intestine, where α-amylase, this time secreted by the pancreas, again begins to function. The *pancreatic α-amylase* breaks down polysaccharide chains into shorter and shorter segments until the disaccharide maltose (two glucose units; Section 18.13) and glucose itself are the dominant species.

The final step in carbohydrate digestion occurs on the outer membranes of intestinal mucosal cells, where the enzymes that convert disaccharides to monosaccharides are located. The important disaccharidase enzymes are *maltase, sucrase,* and *lactase.* These enzymes convert, respectively, maltose to two glucose units, sucrose to one glucose and one fructose unit, and lactose to one glucose and one galactose unit (Section 18.13). (The disaccharides sucrose and lactose present in food are not digested until they reach this point.)

The three major breakdown products from carbohydrate digestion are thus glucose, galactose, and fructose. These monosaccharides are absorbed into the bloodstream through the intestinal wall. The folds of the intestinal wall are lined with finger-like projections called *villi,* which are rich in blood capillaries (Figure 24.1). Absorption is by *active transport* (Section 19.13), which, unlike passive transport, is an energy-requiring process. In this case, ATP is needed. Protein carriers mediate the passage of the monosaccharides through plasma membranes. Figure 24.2 summarizes the different phases in the digestive process for carbohydrates.

After their absorption into the bloodstream, monosaccharides are transported to the liver, where fructose and galactose are rapidly converted into compounds that are metabolized by the same pathway as glucose. Thus the central focus of carbohydrate metabolism is the pathway by which glucose is further processed, a pathway called *glycolysis* (Section 24.2), a series of ten reactions, each of which involves a different enzyme.

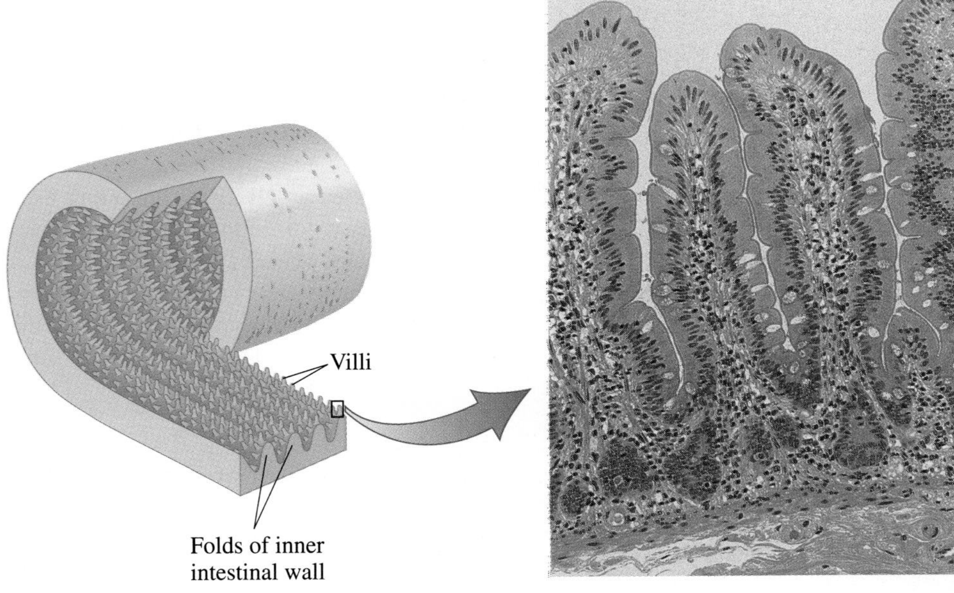

Figure 24.1
A section of the small intestine, showing its folds and the villi that cover the inner surface of the folds. Villi greatly increase the inner intestinal surface area.

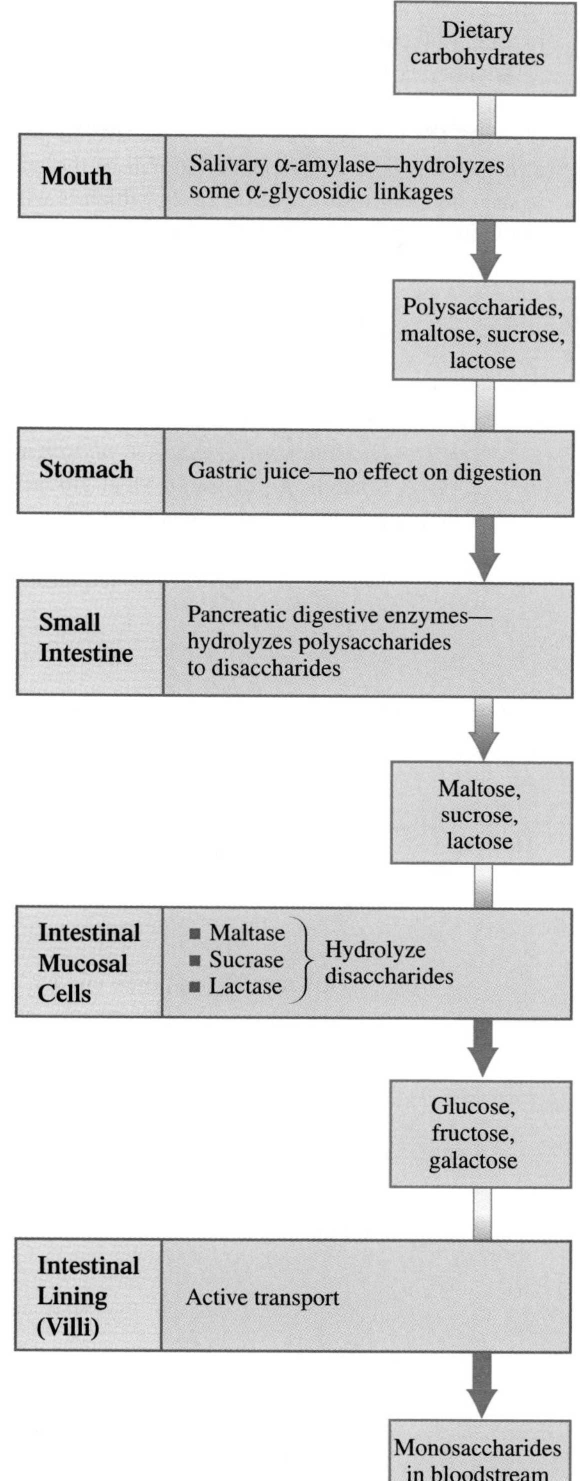

Figure 24.2
Summary of carbohydrate digestion in the human body.

• The term *glycolysis,* pronounced "gligh-KOLL-ih-sis," comes from the Greek *glyco,* meaning "sweet," and *lysis,* meaning "breakdown."

• *Pyruvate,* pronounced "PIE-roo-vate," is the carboxylate ion (Section 16.8) produced when pyruvic acid (a three-carbon keto acid) loses its acidic hydrogen atom.

$$CH_3-C(=O)-COOH \longrightarrow CH_3-C(=O)-COO^- + H^+$$

Pyruvic acid — Pyruvate ion

24.2 Glycolysis

Glycolysis *is the metabolic pathway by which glucose (a C_6 molecule) is converted into two molecules of pyruvate (a C_3 molecule).* This metabolic pathway functions in almost all cells.

The conversion of glucose to pyruvate is an oxidation process in which no molecular oxygen is utilized. The oxidizing agent is the coenzyme NAD^+. Metabolic pathways in

• *Anaerobic* is pronounced "AN-air-ROE-bic." *Aerobic* is pronounced "air-ROE-bic."

• Glycolysis is also called the *Embden–Meyerhof pathway* after the German chemists Gustav Embden and Otto Meyerhof, who discovered many of the details of the pathway.

which molecular oxygen is not a participant are called *anaerobic* pathways. Pathways that require molecular oxygen are called *aerobic* pathways. Glycolysis is an anaerobic pathway.

Glycolysis is a ten-step process (compared to the eight steps of the citric acid cycle; Section 23.6) in which every step is enzyme catalyzed. Figure 24.3 gives an overview of glycolysis. There are two stages in the overall process, a *six-carbon stage* (Steps 1–3) and a *three-carbon stage* (Steps 4–10). All of the enzymes needed for glycolysis are present in the cell cytoplasm (Section 23.2), which is where glycolysis takes place. Details of the individual steps within the glycolysis pathway are now considered.

• Six-Carbon Stage of Glycolysis (Steps 1–3)

The intermediates of the six-carbon stage of glycolysis are all either *glucose* or *fructose* derivatives in which phosphate groups are present.

• A *kinase* is an enzyme that catalyzes the transfer of a phosphoryl group (PO_3^{2-}) from ATP (or some other high-energy phosphate compound) to a substrate.

Step 1: *Formation of Glucose 6-phosphate.* Glycolysis begins with the phosphorylation of glucose to yield glucose 6-phosphate, a glucose molecule with a phosphate group attached to the hydroxyl oxygen on carbon 6 (the carbon atom outside the ring). The phosphate group is from an ATP molecule. *Hexokinase,* an enzyme that requires Mg^{2+} ion for its activity, catalyzes the reaction.

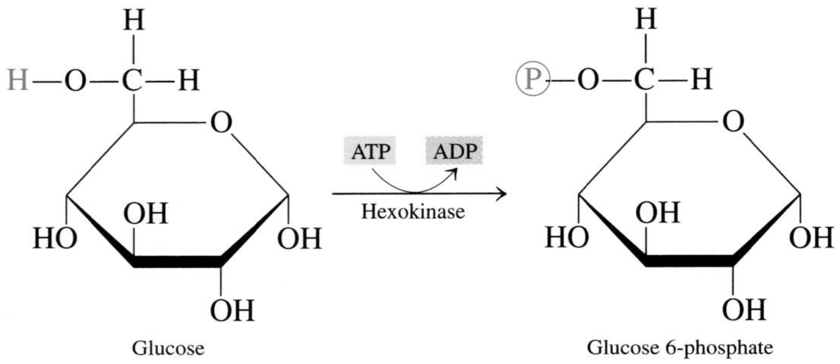

H—O—C(H)(H) ... Glucose —(ATP → ADP; Hexokinase)→ Ⓟ—O—C(H)(H) ... Glucose 6-phosphate

Glucose

Glucose 6-phosphate

The symbol Ⓟ is a shorthand notation for a PO_3^{2-} unit.

This reaction requires energy, which is provided by the breakdown of an ATP molecule. This energy expenditure will be recouped later in the cycle. Phosphorylation of glucose provides a way of "trapping" glucose within a cell. Glucose can cross cell membranes, but glucose 6-phosphate cannot.

Step 2: *Formation of Fructose 6-phosphate.* Glucose 6-phosphate is isomerized to fructose 6-phosphate by *phosphoglucoisomerase.*

Ⓟ—OH_2C ... Glucose 6-phosphate ⇌ (Phosphoglucoisomerase) Ⓟ—OH_2C ... CH_2OH Fructose 6-phosphate

Glucose 6-phosphate

Fructose 6-phosphate

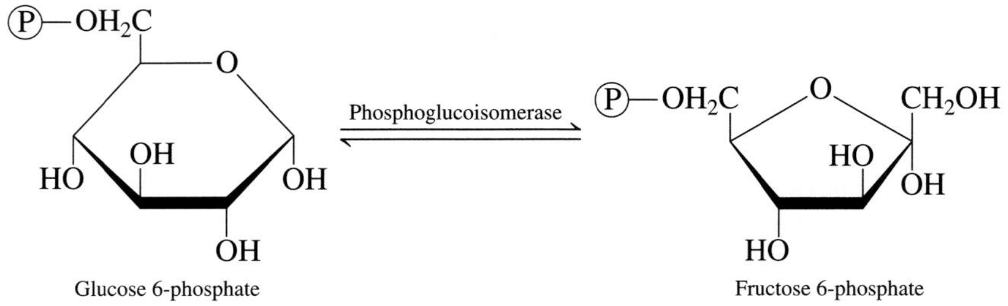

• Step 3 of glycolysis commits the original glucose molecule to the glycolysis pathway. Glucose 6-phosphate (Step 1) and fructose 6-phosphate (Step 2) can enter other metabolic pathways, but fructose 1,6-bisphosphate can only enter glycolysis.

The net result of this change is that carbon 1 of glucose is no longer part of the ring structure. (Glucose, an aldose, forms a six-membered ring, and fructose, a ketose, forms a five-membered ring [Section 18.10]; both sugars, however, contain six carbon atoms.)

Step 3: *Formation of Fructose 1,6-bisphosphate.* This step, like Step 1, is a phosphorylation reaction and therefore requires the expenditure of energy. ATP is the source of the phosphate and the energy. The enzyme involved, *phosphofructokinase,* is another enzyme that requires Mg^{2+} ion for its activity.

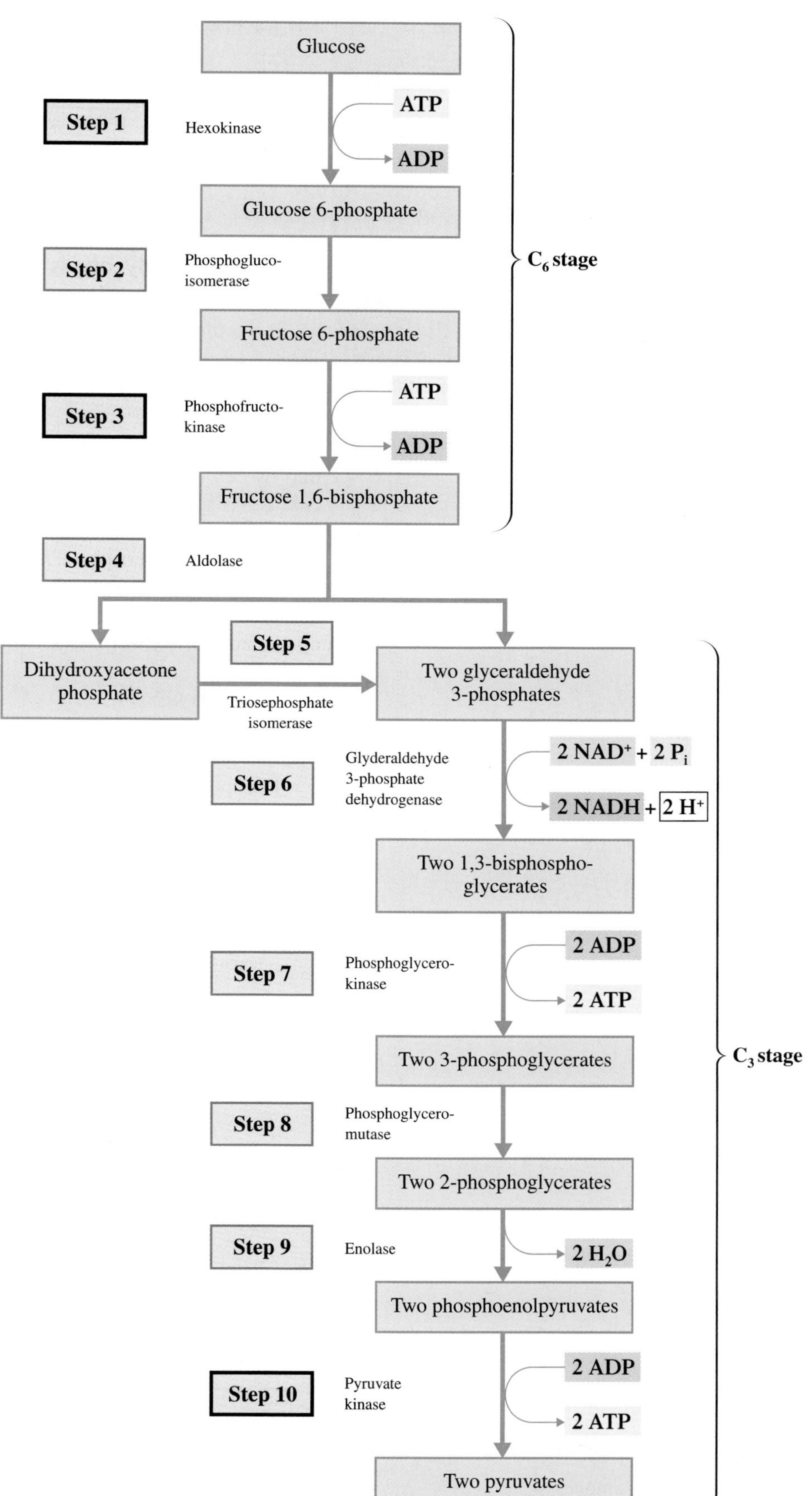

Figure 24.3
An overview of glycolysis.

The fructose molecule now contains two phosphate groups.

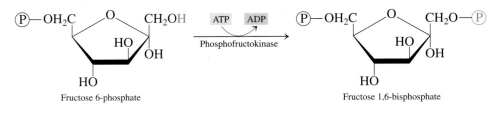

Ⓟ—OH_2C O CH_2OH HO OH HO → (ATP → ADP, Phosphofructokinase) Ⓟ—OH_2C O CH_2O—Ⓟ HO OH HO

Fructose 6-phosphate — Fructose 1,6-bisphosphate

• Three-Carbon Stage of Glycolysis (Steps 4–10)

All intermediates in the three-carbon stage of glycolysis are phosphorylated derivatives of *dihydroxyacetone, glyceraldehyde, glycerate,* or *pyruvate,* which in turn are derivatives of either glycerol or acetone. Figure 24.4 shows the structural relationships among these molecules.

Step 4: *Formation of Triose Phosphates.* In this step, the reacting C_6 species is split into two C_3 (triose) species. Because fructose 1,6-bisphosphate, the molecule being split, is unsymmetrical, the two trioses produced are not identical. One product is dihydroxyacetone phosphate, and the other is glyceraldehyde 3-phosphate. *Aldolase* is the enzyme that catalyzes this reaction. A better understanding of the structural relationships between reactant and products is obtained if the fructose 1,6-bisphosphate is written in its open-chain form (Section 18.10) rather than in its cyclic form.

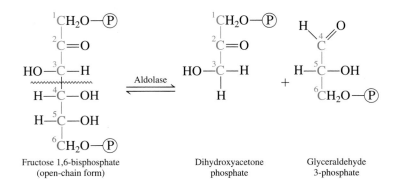

Fructose 1,6-bisphosphate (open-chain form): ${}^1CH_2O$—Ⓟ, ${}^2C{=}O$, HO—${}^3C$—H, H—${}^4C$—OH, H—${}^5C$—OH, ${}^6CH_2O$—Ⓟ

⇌ (Aldolase)

Dihydroxyacetone phosphate: ${}^1CH_2O$—Ⓟ, ${}^2C{=}O$, HO—${}^3C$—H, H

\+

Glyceraldehyde 3-phosphate: H, O, ${}^4C$, H—${}^5C$—OH, ${}^6CH_2O$—Ⓟ

Step 5: *Isomerization of Triose Phosphates.* Only one of the two trioses produced in Step 4, glyceraldehyde 3-phosphate, is a glycolysis intermediate. Dihydroxyacetone phosphate, the other triose, can, however, be readily converted into glyceraldehyde 3-phosphate. Dihydroxyacetone phosphate (a ketose)

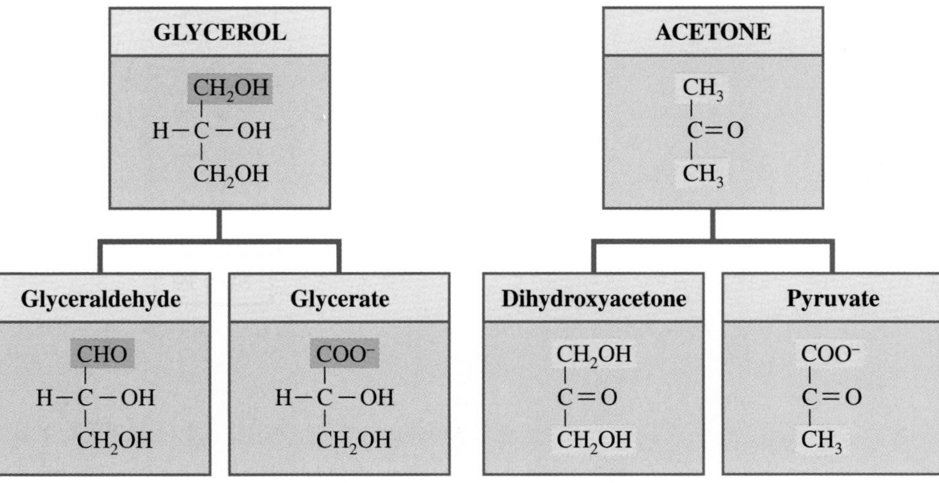

Figure 24.4 Structural relationships among glycerol and acetone and the C_3 intermediates in the process of glycolysis.

and glyceraldehyde 3-phosphate (an aldose) are isomers, and the isomerization process from ketose to aldose is catalyzed by the enzyme *triosephosphate isomerase.*

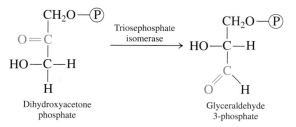

Dihydroxyacetone phosphate —(Triosephosphate isomerase)→ Glyceraldehyde 3-phosphate

Step 6: *Formation of 1,3-Bisphosphoglycerate.* In a reaction catalyzed by *glyceraldehyde 3-phosphate dehydrogenase,* a phosphate group is added to glyceraldehyde 3-phosphate to produce 1,3-bisphosphoglycerate. The hydrogen of the aldehyde group becomes part of NADH.

Glyceraldehyde 3-phosphate + NAD^+ + P_i —(Glyceraldehyde 3-phosphate dehydrogenase)→ 1,3-Bisphosphoglycerate + NADH + H^+

• Keep in mind that from Step 6 onward, two molecules of each of the C_3 compounds take part in every reaction for each original C_6 glucose molecule.

The newly added phosphate group in 1,3-bisphosphoglycerate is a high-energy phosphate group (Section 23.4). A high-energy phosphate group is produced when a phosphate group is attached to a carbon atom that is also participating in a carbon–carbon or carbon–oxygen double bond.

Note that a molecule of the reduced coenzyme NADH is a product of this reaction and also that the source of the added phosphate is inorganic phosphate (P_i).

Step 7: *Formation of 3-Phosphoglycerate.* In this step, the diphosphate species just formed is converted back to a monophosphate species. This is an ATP-producing step in which the C-1 phosphate group of 1,3-bisphosphoglycerate (the high-energy phosphate) is transferred to an ADP molecule to form the ATP. The enzyme involved is *phosphoglycerokinase.*

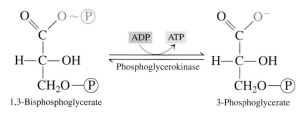

1,3-Bisphosphoglycerate ⇌ (ADP → ATP; Phosphoglycerokinase) 3-Phosphoglycerate

Remember that two ATP molecules are produced for each original glucose molecule, because both C_3 molecules produced from the glucose react.

ATP production in this step involves substrate-level phosphorylation. **Substrate-level phosphorylation** *is the direct transfer of a high-energy phosphate group from an intermediate compound (substrate) to an ADP molecule to produce ATP.* Substrate-level phosphorylation differs from oxidative phosphorylation (Section 23.8) in that the latter process involves the transfer of free phosphate ions in solution (P_i) to ADP molecules to form ATP.

Step 8: *Formation of 2-Phosphoglycerate.* In this isomerization step, the phosphate group of 3-phosphoglycerate is moved from carbon 3 to carbon 2. The

• A *mutase* is an enzyme that effects the shift of a phosphoryl group (PO_3^{2-}) from one oxygen atom to another within a molecule.

enzyme *phosphoglyceromutase* catalyzes the exchange of the phosphate group between the two carbons.

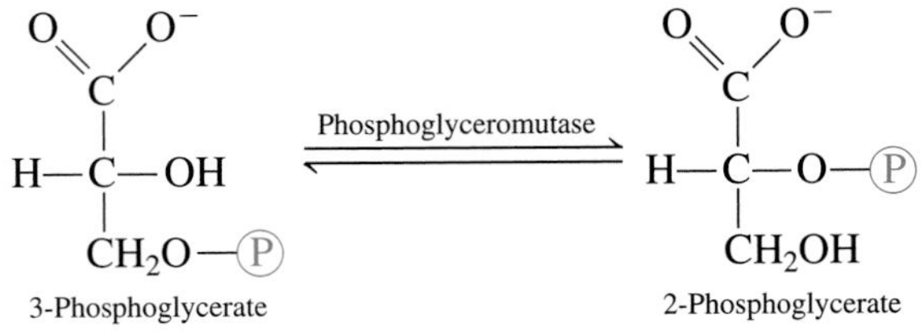

3-Phosphoglycerate ⇌ (Phosphoglyceromutase) 2-Phosphoglycerate

• An *enol* (from *ene* + *ol*), as in phospho*enol*pyruvate, is a compound in which an —OH group is attached to a carbon atom involved in a carbon–carbon double bond. Note that in phosphoenolpyruvate the —OH group has been phosphorylated.

Step 9: *Formation of Phosphoenolpyruvate.* This is an alcohol dehydration reaction that proceeds with the enzyme *enolase,* another Mg^{2+}-requiring enzyme. The result is another compound containing a high-energy phosphate group; the phosphate group is attached to a carbon atom that is involved in a carbon–carbon double bond.

2-Phosphoglycerate ⇌ (Enolase) Phosphoenolpyruvate + HOH

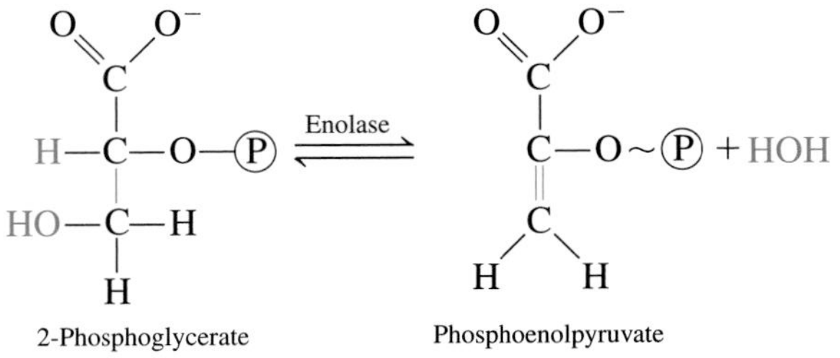

Step 10: *Formation of Pyruvate.* In this step, substrate-level phosphorylation again occurs. Phosphoenolpyruvate transfers its high-energy phosphate group to an ADP molecule to produce ATP and pyruvate.

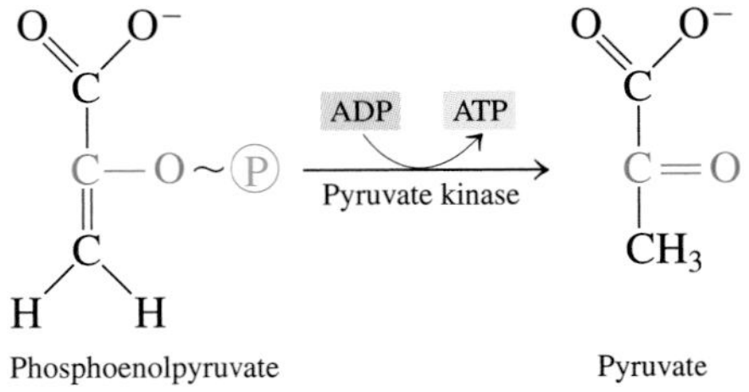

Phosphoenolpyruvate → (ADP → ATP; Pyruvate kinase) Pyruvate

The enzyme involved, *pyruvate kinase,* requires both Mg^{2+} and K^+ ions for its activity. Again, because two C_3 molecules are reacting, two ATP molecules are produced.

ATP molecules are involved in Steps 1, 3, 7, and 10 of glycolysis. Considering these steps collectively shows that there is a net gain of two ATP molecules for every glucose molecule converted into two pyruvates (Table 24.1). Though useful, this is a small amount of ATP compared to that generated in oxidative phosphorylation (Section 23.8).

Table 24.1
ATP Production and Consumption During Glycolysis

Step	Reaction	ATP change per glucose
1	Glucose → glucose 6-phosphate	−1
3	Fructose 6-phosphate → fructose 1,6-bisphosphate	−1
7	2(1,3-Bisphosphoglycerate → 3-phosphoglycerate)	+2
10	2(Phosphoenolpyruvate → pyruvate)	+2
		Net +2

Entry of Galactose and Fructose into Glycolysis

The breakdown products from carbohydrate digestion are glucose, fructose, and galactose (Section 24.1). Both fructose and galactose are converted, in the liver, to intermediates that enter into the glycolysis pathway.

The entry of fructose into the glycolytic pathway involves phosphorylation by ATP to produce fructose 1-phosphate, which is then split into two trioses—glyceraldehyde and dihydroxyacetone phosphate. Dihydroxyacetone phosphate enters glycolysis directly; glyceraldehyde must be phosphorylated by ATP to glyceraldehyde 3-phosphate before it enters the pathway (see Figure 24.5).

The entry of galactose into the glycolytic pathway begins with its conversion to glucose 1-phosphate (a four-step sequence), which is then converted to glucose 6-phosphate, a glycolysis intermediate (see Figure 24.5).

Figure 24.5
Entry points for fructose and galactose into the glycolysis pathway.

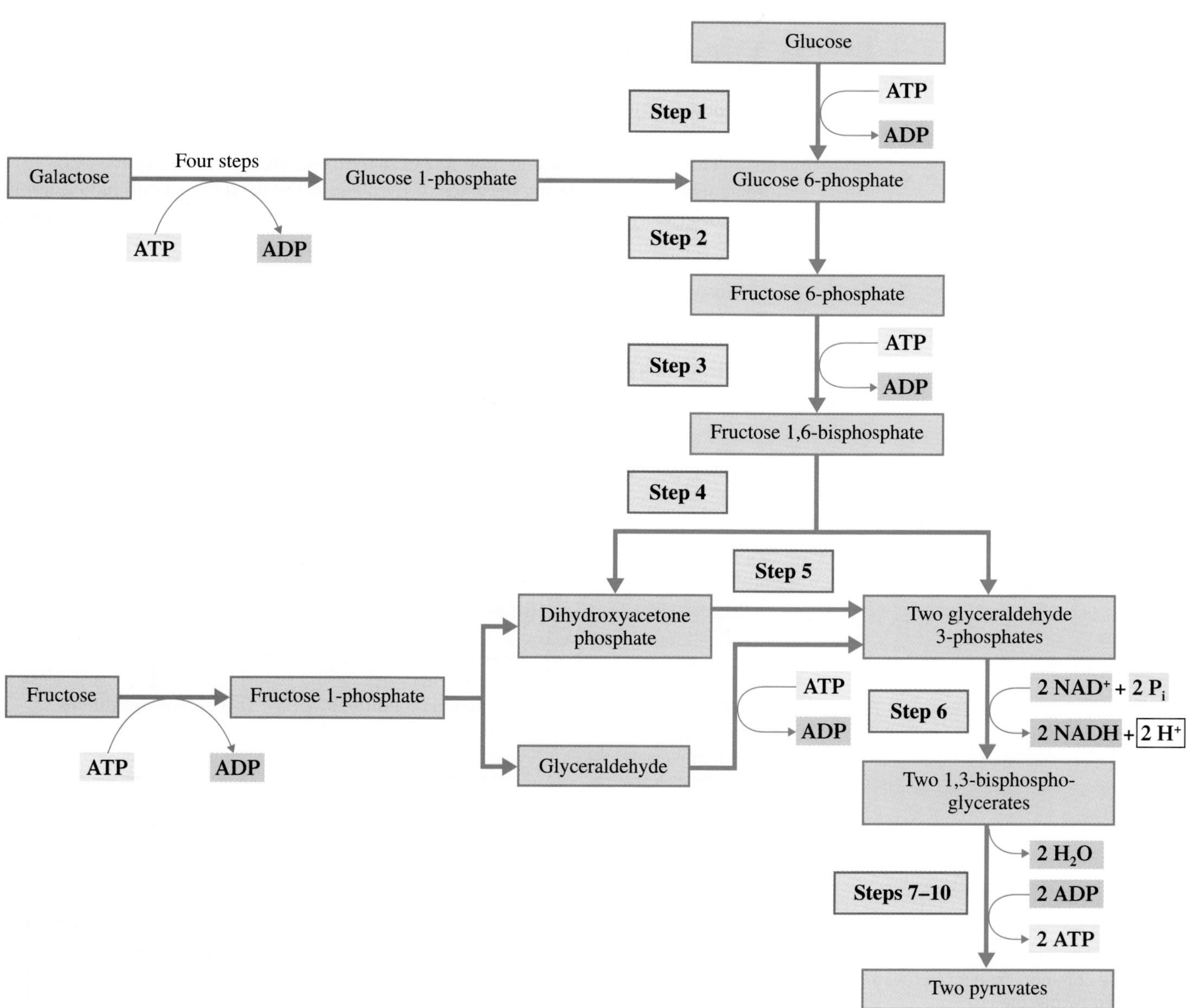

• Regulation of Glycolysis

Glycolysis, like all metabolic pathways, must have control mechanisms associated with it. In glycolysis, the control points are Steps 1, 3, and 10 (see Figure 24.3).

Step 1, the conversion of glucose to glucose 6-phosphate, involves the enzyme *hexokinase.* This particular enzyme is inhibited by glucose 6-phosphate, the substance produced by its action (feedback inhibition; Section 21.9).

At Step 3, where fructose 6-phosphate is converted to fructose 1,6-bisphosphate by the enzyme *phosphofructokinase,* high concentrations of ATP and citrate inhibit enzyme activity. A high ATP concentration, which is characteristic of a state of low energy consumption, thus stops glycolysis at the fructose 6-phosphate stage. This stoppage also causes increases in glucose 6-phosphate stores, because glucose 6-phosphate is in equilibrium with fructose 6-phosphate.

The third control point involves the last step of glycolysis, the conversion of phosphoenolpyruvate to pyruvate. *Pyruvate kinase,* the enzyme needed at this point, is inhibited by high ATP concentrations. Both pyruvate kinase (Step 10) and phosphofructokinase (Step 3) are allosteric enzymes (Section 21.9).

24.3 Fates of Pyruvate

The breakdown of glucose to pyruvate occurs in most living systems. What happens to the pyruvate thus produced, however, varies with cellular conditions and the nature of the organism. The three different common fates for pyruvate are shown in Figure 24.6.

A key concept in considering the fates of pyruvate is the need for a continuous supply of NAD^+ for glycolysis. As glucose is oxidized to pyruvate in glycolysis, NAD^+ is reduced to NADH.

$$\text{Glucose} + 2NAD^+ \xrightarrow[2ADP + 2P_i \;\to\; 2ATP]{} 2\text{ pyruvate} + 2NADH + 2H^+$$

It is significant that each pathway of pyruvate metabolism includes provisions for regeneration of NAD^+ from NADH so that glycolysis can continue.

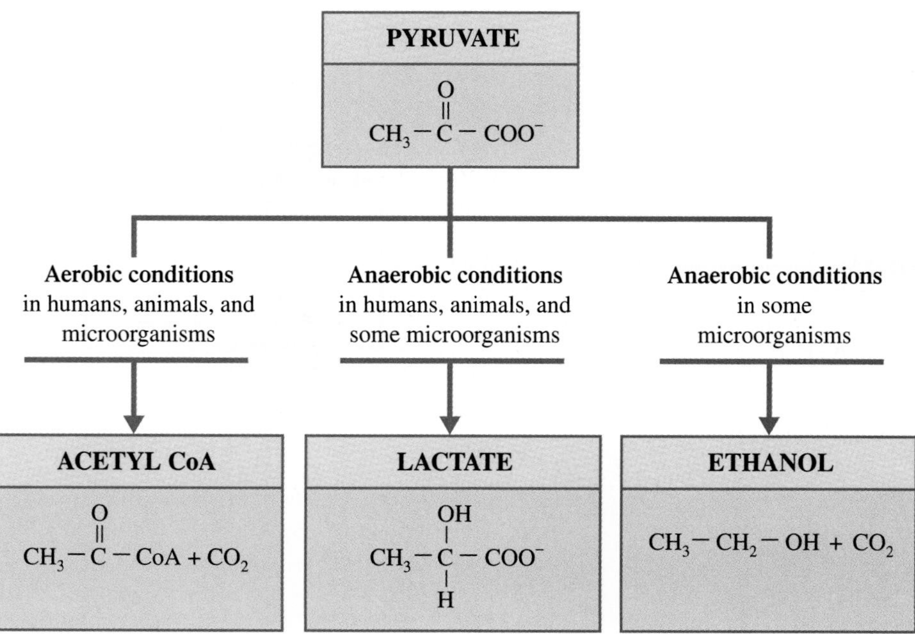

Figure 24.6
The three common fates of pyruvate generated by glycolysis.

Oxidation to Acetyl CoA

Under *aerobic* (oxygen-rich) conditions, pyruvate is oxidized to acetyl CoA. Pyruvate formed in the cytoplasm through glycolysis crosses the two mitochondrial membranes and enters the matrix, where the oxidation takes place. The overall reaction, in simplified terms, is

$$\underset{\text{Pyruvate}}{CH_3-\overset{O}{\overset{\|}{C}}-COO^-} + CoA-SH + NAD^+ \xrightarrow[\text{complex}]{\text{Pyruvate dehydrogenase}} \underset{\text{Acetyl CoA}}{CH_3-\overset{O}{\overset{\|}{C}}-S-CoA} + NADH + CO_2$$

This reaction, which involves both oxidation and decarboxylation (CO_2 is produced), is far more complex than the simple stoichiometry of the equation suggests. The enzyme complex involved contains three different enzymes, each with numerous subunits. The overall reaction process involves four separate steps and requires NAD^+, CoA—SH, FAD, and two other coenzymes (lipoic acid and thiamin pyrophosphate, the latter derived from the B vitamin thiamin).

Most acetyl CoA molecules produced from pyruvate enter the citric acid cycle. Citric acid cycle operations change more NAD^+ to its reduced form, NADH. The NADH from glycolysis, from the conversion of pyruvate to acetyl CoA, and from the citric acid cycle enters the electron transport chain directly or indirectly (Section 23.7). In the ETC, electrons from NADH are transferred to O_2, and the NADH is changed back to NAD^+. The NAD^+ needed for glycolysis, pyruvate–acetyl CoA conversion, and the citric acid cycle has been regenerated.

• Not all acetyl CoA produced from pyruvate enters the citric acid cycle. Particularly when high levels of acetyl CoA are produced (from excess ingestion of dietary carbohydrates), some acetyl CoA is used as the starting material for the production of the fatty acids needed for fat (triacylglycerol) formation (Section 25.7).

Reduction to Lactate

Acetyl CoA production is associated with oxygen-rich conditions. In an oxygen-poor environment (anaerobic conditions), pyruvate is reduced to lactate in humans and many other organisms. In humans, oxygen-poor conditions often accompany strenuous muscle activity.

In an anaerobic situation, NADH cannot be oxidized to NAD^+ by the electron transport chain, which is oxygen-dependent. Reduction of pyruvate to lactate is an alternative method for producing the NAD^+ needed for glycolysis.

$$\underset{\text{Pyruvate}}{CH_3-\overset{O}{\overset{\|}{C}}-COO^-} + NADH + H^+ \xrightarrow[\text{dehydrogenase}]{\text{Lactate}} \underset{\text{Lactate}}{CH_3-\overset{OH}{\overset{|}{C}}H-COO^-} + NAD^+$$

The accumulation of lactate in muscles and in the blood is what causes the fatigue associated with strenuous muscle activity (see Chemical Connections 24.1). The fate of the lactate produced from pyruvate will be considered in Section 24.6.

When the reaction for conversion of pyruvate to lactate is added to the net glycolysis reaction (Section 24.2), an overall reaction for the conversion of glucose to lactate is obtained.

$$\underset{\text{Glucose}}{C_6H_{12}O_6} + 2ADP + 2P_i \longrightarrow \underset{\text{Lactate}}{2CH_3-\overset{OH}{\overset{|}{C}}H-COO^-} + 2ATP$$

• Red blood cells have no mitochondria and therefore always form lactate as the end product of glycolysis.

Reduction to Ethanol

Under anaerobic conditions, several simple organisms, including yeast, possess the ability to regenerate NAD^+ through ethanol, rather than lactate, production. Such a process is called alcohol fermentation. **Alcohol fermentation** *is the enzymatic conversion of pyruvate to ethanol and carbon dioxide.* Fermentation causes bread and related products to rise as a result of CO_2 bubbles being released during baking. Beer, wine, and other alcoholic drinks are produced by fermentation of the sugars in grain and fruit products.

Chemical CONNECTIONS

24.1 Lactate Accumulation

During strenuous exercise, conditions in muscle cells can change from aerobic to anaerobic as the oxygen supply becomes inadequate to meet demand. Such conditions cause pyruvate to be converted to lactate rather than acetyl CoA. (Lactate production can also be high at the start of strenuous exercise before the delivery of oxygen is stepped up via an increased respiration rate.)

The resulting lactate begins to accumulate in the cytoplasm of cells where it is produced. Some lactate diffuses out of the cells into the blood, where it contributes to a slight decrease in blood pH. This lower pH triggers fast breathing, which helps supply more oxygen to the cells.

Lactate accumulation is the cause of muscle pain and cramping during prolonged, strenuous exercise. As a result of such cramping, muscles may be stiff and sore the next day. Regular, hard exercise increases the efficiency with which oxygen is delivered to the body. Thus athletes can function longer than nonathletes under aerobic conditions without lactate production.

Lactate accumulation can also occur in heart muscle if it experiences decreased oxygen supply (from artery blockage). The heart muscle experiences cramps and stops beating (cardiac arrest). Massage of heart muscle often reduces such cramps, just as it does for skeletal muscle, and it is sometimes possible to start the heart beating again by using such a technique.

Premature infants born with underdeveloped lungs are often given increased amounts of oxygen to minimize lactate accumulation. They are also given bicarbonate (HCO_3^-) solution to counteract the acidity change in blood that accompanies lactate buildup.

Strenuous muscular activity can result in lactate accumulation.

The first step in conversion of pyruvate to ethanol is a decarboxylation reaction to produce acetaldehyde.

$$\underset{\text{Pyruvate}}{CH_3-\overset{\overset{\large O}{\|}}{C}-COO^-} + H^+ \xrightarrow[\text{decarboxylase}]{\text{Pyruvate}} \underset{\text{Acetaldehyde}}{CH_3-\overset{\overset{\large O}{\|}}{C}-H} + CO_2$$

The second step involves acetaldehyde reduction to produce ethanol.

$$\underset{\text{Acetaldehyde}}{CH_3-\overset{\overset{\large O}{\|}}{C}-H} + NADH + H^+ \xrightarrow[\text{dehydrogenase}]{\text{Alcohol}} \underset{\text{Ethanol}}{CH_3-\overset{OH}{\underset{H}{C}}-H} + NAD^+$$

The overall equation for the conversion of pyruvate to ethanol (the sum of the two steps) is

$$\underset{\text{Pyruvate}}{CH_3-\overset{\overset{\large O}{\|}}{C}-COO^-} + 2H^+ + NADH \xrightarrow{\text{Two steps}} \underset{\text{Ethanol}}{CH_3-CH_2-OH} + NAD^+ + CO_2$$

An overall reaction for the production of ethanol from glucose is obtained by combining the reaction for the conversion of pyruvate with the net reaction for glycolysis (Section 24.2).

$$\underset{\text{Glucose}}{C_6H_{12}O_6} + 2ADP + 2P_i \longrightarrow \underset{\text{Ethanol}}{2CH_3-CH_2-OH} + 2CO_2 + 2ATP$$

24.4 ATP Production for the Complete Oxidation of Glucose

We now assemble energy production figures for glycolysis, oxidation of pyruvate to acetyl CoA, the citric acid cycle, and the electron transport chain. The result, with one added piece of information, gives the ATP yield for the *complete* oxidation of one molecule of glucose.

The new piece of information involves the NADH produced during Step 6 of glycolysis. This NADH, produced in the cytoplasm, cannot *directly* participate in the electron transport chain, because mitochondria are impermeable to NADH (and NAD^+). A transport system shuttles the electrons from NADH, but not NADH itself, across the membrane. This shuttle involves dihydroxyacetone phosphate (a glycolysis intermediate) and glycerol 3-phosphate.

The first step in the shuttle is the cytoplasmic reduction of dihydroxyacetone phosphate by NADH to produce glycerol 3-phosphate and NAD^+ (see Figure 24.7). Glycerol 3-phosphate then crosses the outer mitochondrial membrane, where it is reoxidized to dihydroxyacetone phosphate. The oxidizing agent is FAD rather than NAD^+. The regenerated dihydroxyacetone phosphate diffuses out of the mitochondrion and returns to the cytoplasm for participation in another "turn" of the shuttle. The $FADH_2$ coproduced in the mitochondrial reaction can participate in the electron transport chain reactions. The net reaction of this shuttle process is

$$\underset{\text{(cytoplasmic)}}{\text{NADH}} + H^+ + \underset{\text{(mitochondrial)}}{\text{FAD}} \longrightarrow \underset{\text{(cytoplasmic)}}{\text{NAD}^+} + \underset{\text{(mitochondrial)}}{\text{FADH}_2}$$

The consequence of this reaction is that only two, rather than three, molecules of ATP are formed for each cytoplasmic NADH, because $FADH_2$ yields one less ATP than does NADH in the electron transport chain.

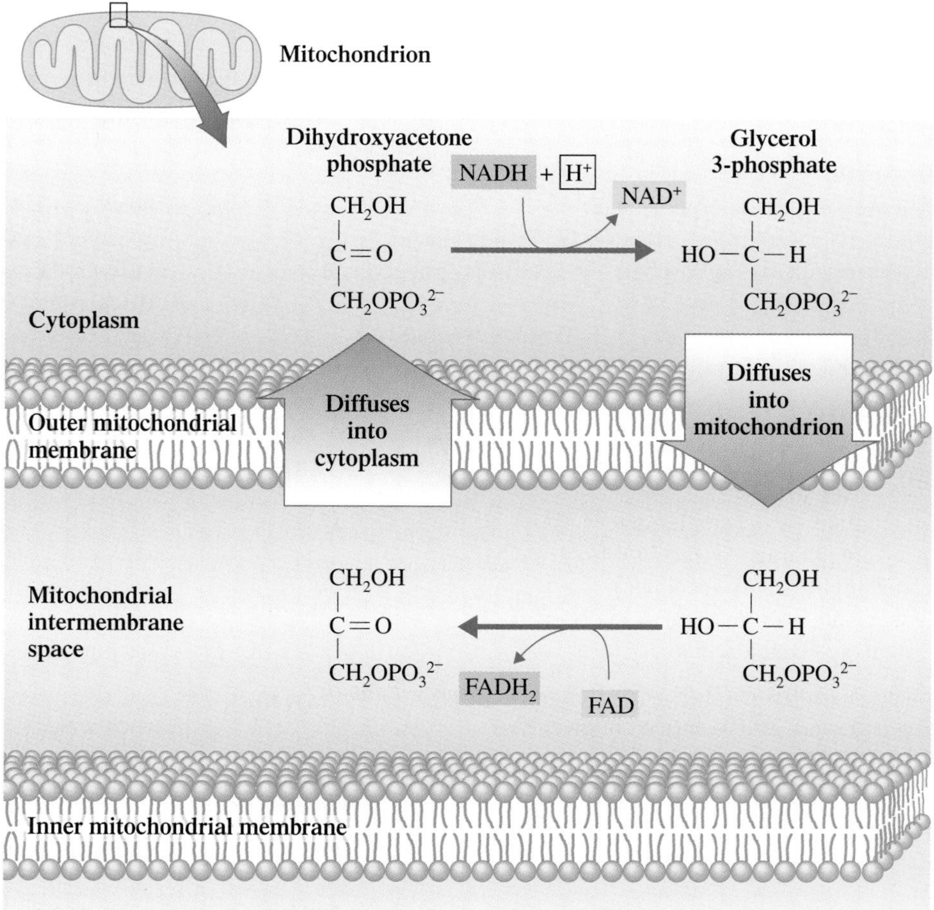

Figure 24.7
The dihydroxyacetone phosphate–glycerol 3-phosphate shuttle.

Table 24.2 shows ATP production for the complete oxidation of a molecule of glucose. The final number is 36 ATP, 32 of which come from the oxidative phosphorylation associated with the electron transport chain. This total of 36 ATP for complete oxidation contrasts markedly with a total of 2 ATP for oxidation of glucose to lactate and 2 ATP for oxidation of glucose to ethanol. Neither of these latter processes involves the citric acid cycle or the electron transport chain. Thus the aerobic oxidation of glucose is 18 times more efficient in the production of ATP than the anaerobic lactate and ethanol processes.

The yield of 36 ATP molecules per glucose (Table 24.2) is for those cells where the dihydroxyacetone phosphate–glycerol 3-phosphate shuttle operates (skeletal muscle and nerve cells). In certain other cells, particularly heart and liver cells, a more complex shuttle system called the malate–aspartate shuttle functions. In this shuttle, 3 ATP molecules result from 1 cytoplasmic NADH, which changes the total ATP yield to 38 molecules per glucose.

24.5 Glycogen Synthesis and Degradation

Glycogen, a branched polymeric form of glucose (Section 18.14), is the storage form of carbohydrates in humans and animals. It is found primarily in muscle and liver tissue. In muscles it is the source of glucose needed for glycolysis. In the liver, it is the source of glucose needed to maintain normal glucose levels in the blood.

• Glycogenesis

Glycogenesis *is the synthesis of glycogen from glucose.* Glycogenesis involves three reactions (steps).

Table 24.2
Yield of ATP from the Complete Oxidation of One Glucose Molecule in a Skeletal Muscle Cell

Reaction	Comments	Yield of ATP
Glycolysis		
glucose → glucose 6-phosphate	consumes 1 ATP	−1
glucose 6-phosphate → fructose 1,6-bisphosphate	consumes 1 ATP	−1
2(glyceraldehyde 3-phosphate → 1,3-bisphosphoglycerate)	each produces 1 cytoplasmic NADH	—
2(1,3-bisphosphoglycerate → 3-phosphoglycerate)	each produces 1 ATP	+2
2(phosphoenolpyruvate → pyruvate)	each produces 1 ATP	+2
Oxidation of Pyruvate		
2(pyruvate → acetyl CoA + CO_2)	each produces 1 NADH	—
Citric Acid Cycle		
2(isocitrate → α-ketoglutarate + CO_2)	each produces 1 NADH	—
2(α-ketoglutarate → succinyl CoA + CO_2)	each produces 1 NADH	—
2(succinyl CoA → succinate)	each produces 1 GTP	+2
2(succinate → fumarate)	each produces 1 $FADH_2$	—
2(malate → oxaloacetate)	each produces 1 NADH	—
Electron Transport Chain and Oxidative Phosphorylation		
2 cytoplasmic NADH formed in glycolysis	each yields 2 ATP	+4
2 NADH formed in the oxidation of pyruvate	each yields 3 ATP	+6
2 $FADH_2$ formed in the citric acid cycle	each yields 2 ATP	+4
6 NADH formed in the citric acid cycle	each yields 3 ATP	+18
	Net yield of ATP	+36

Step 1: *Formation of Glucose 1-phosphate.* The starting material for this step is not glucose itself but rather glucose 6-phosphate (available from the first step of glycolysis). The enzyme *phosphoglucomutase* effects the change from a 6-phosphate to a 1-phosphate.

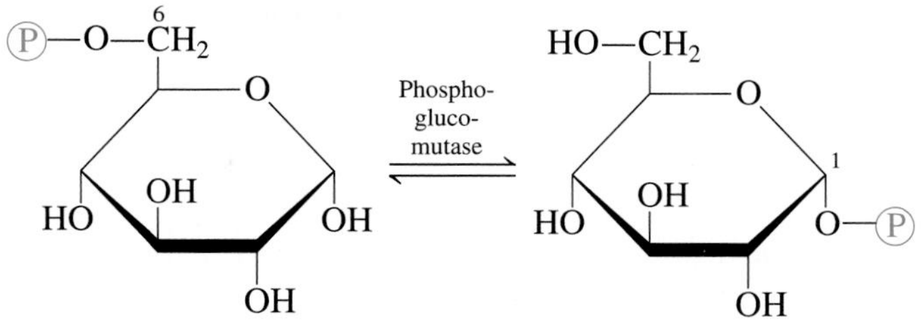

Phospho-gluco-mutase

Glucose 6-phosphate ⇌ Glucose 1-phosphate

Step 2: *Formation of UDP-glucose.* Glucose 1-phosphate from Step 1 must be activated before it can be added to a growing glycogen chain. The activator is the high-energy compound UTP (uridine triphosphate). The UTP is hydrolyzed to UMP and PP_i, and then the PP_i is further hydrolyzed to $2P_i$. The UMP that is formed bonds to the glucose 1-phosphate to form UDP-glucose.

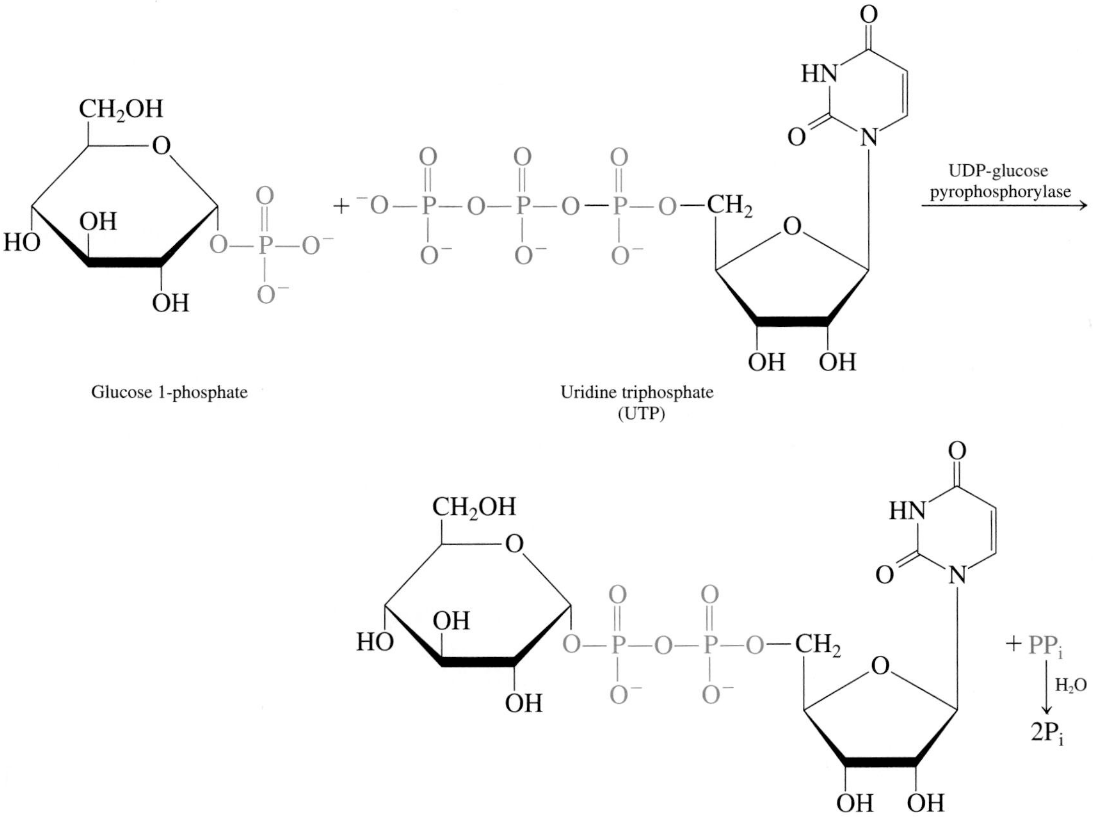

Glucose 1-phosphate + Uridine triphosphate (UTP) $\xrightarrow{\text{UDP-glucose pyrophosphorylase}}$ Uridine diphosphate glucose (UDP-glucose) + PP_i $\xrightarrow{H_2O}$ $2P_i$

Step 3: *Glucose Transfer to a Glycogen Chain.* The glucose unit of UDP-glucose is then attached to the end of a glycogen chain.

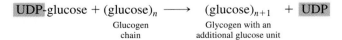

$$\text{UDP-glucose} + \underset{\text{Glucogen chain}}{(\text{glucose})_n} \longrightarrow \underset{\text{Glycogen with an additional glucose unit}}{(\text{glucose})_{n+1}} + \text{UDP}$$

In a subsequent reaction, the UDP produced in Step 3 is converted back to UTP, which can then react with another glucose 1-phosphate (Step 2). The conversion reaction requires ATP.

$$\text{UDP} + \text{ATP} \longrightarrow \text{UTP} + \text{ADP}$$

Adding a single glucose unit to a growing glycogen chain requires the investment of two ATP molecules: one in the formation of glucose 6-phosphate and one in the regeneration of UTP.

• Glycogenolysis

Glycogenolysis *is the breakdown of glycogen into free glucose units.* This process is not simply the reverse of glycogen synthesis (glycogenesis), because it does not require UTP or UDP molecules. Glycogenolysis, like glycogenesis, is a three-step process.

• A **phosphorylase** is an enzyme that catalyzes the cleavage of a bond by P_i (in contrast to hydrolysis, which refers to bond cleavage by water), such as removal of a glucose unit from glycogen to give glucose 1-phosphate.

Step 1: *Phosphorylation of a Glucose Residue.* The enzyme *glycogen phosphorylase* effects the removal of an end glucose unit from a glycogen molecule and the subsequent phosphorylation of this glucose to glucose 1-phosphate.

$$\underset{\text{Glycogen}}{(\text{Glucose})_n} + \text{P}_\text{i} \longrightarrow \underset{\substack{\text{Glycogen with one}\\ \text{fewer glucose unit}}}{(\text{glucose})_{n-1}} + \text{glucose 1-phosphate}$$

Step 2: *Glucose 1-phosphate Isomerization.* The enzyme *phosphoglucomutase* catalyzes the isomerization process whereby the phosphate group of glucose 1-phosphate is moved to the carbon-6 position.

$$\text{Glucose 1-phosphate} \rightleftharpoons \text{glucose 6-phosphate}$$

This process is the reverse of the first step of glycogenesis.

Step 3: *Hydrolysis of Glucose 6-phosphate to Glucose.* The reaction for this step is

$$\text{Glucose 6-phosphate} + \text{H}_2\text{O} \longrightarrow \text{glucose} + \text{P}_\text{i}$$

• A **phosphatase** is an enzyme that effects the removal of a phosphate group (P_i) from a molecule, such as converting glucose 6-phosphate to glucose.

The enzyme needed for this reaction, *glucose 6-phosphatase,* is found only in the liver, kidneys, and intestine. Thus *complete* glycogenolysis occurs only in these tissues.

Muscle and brain cells, which lack glucose 6-phosphatase, cannot form free glucose from glucose 6-phosphate. The first two steps of glycogenolysis do, however, occur in these tissues. The glucose 6-phosphate so produced can contribute to energy production through glycolysis and the common metabolic pathway, because glucose 6-phosphate is the first intermediate in the glycolytic pathway (see Figure 24.3).

Thus brain and muscle cells can use glycogen for energy production only. The liver, however, has the capacity to supply additional glucose to the blood.

• The fact that glycogen synthesis (glycogenesis) and glycogen degradation (glycogenolysis) are not totally reverse processes has significance. In fact, it is almost always the case in biochemistry that "opposite" biosynthetic and degradative pathways differ in some steps. This allows for separate control of the pathways.

24.6 Gluconeogenesis

Gluconeogenesis *is the synthesis of glucose from noncarbohydrate materials.* Glycogen stores in muscle and liver tissue are depleted within 12–18 hours from fasting or in even less time from heavy work or strenuous exercise. Without gluconeogenesis, the brain, which is dependent on glucose as a fuel, would have problems functioning if food intake were restricted for even one day.

The noncarbohydrate starting materials for gluconeogenesis are lactate (from hard-working muscles and from red blood cells), glycerol (from triacylglycerol hydrolysis), and certain amino acids (from dietary protein hydrolysis or from muscle protein during starvation). About 90% of gluconeogenesis takes place in the liver. Hence gluconeogenesis helps to maintain normal blood glucose levels in times of inadequate dietary carbohydrate intake (such as between meals).

Figure 24.8 gives the steps in gluconeogenesis. Although at first glance the steps appear to be the reverse of those for glycolysis, this is only partially true. The most obvious difference between these two "opposite" pathways is that 11 compounds are involved in gluconeogenesis and only 10 in glycolysis. Why the difference? The last step of glycolysis

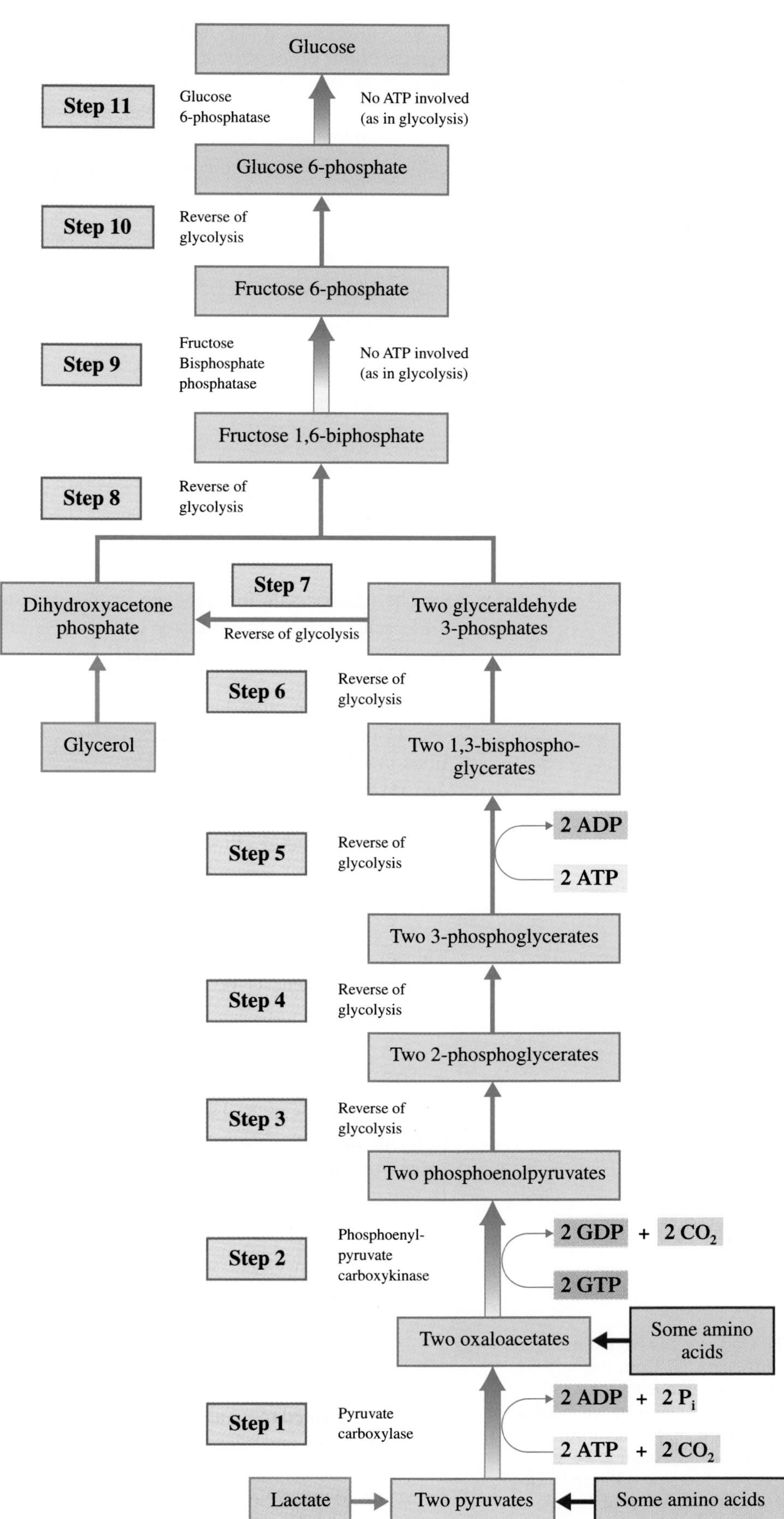

Figure 24.8
The pathway for gluconeogenesis is similar, but not identical, to the pathway for glycolysis. The distinctive reactions of gluconeogenesis are indicated by large arrows. Entry points of noncarbohydrate materials to the pathway are also shown.

is the conversion of the high-energy compound phosphoenolpyruvate to pyruvate. The reverse of this process, which is the beginning of gluconeogenesis, cannot be accomplished in a single step because of the large energy difference between the two compounds and the slow rate of the reaction. Instead, a two-step process by way of oxaloacetate is required to effect the change, and this adds an extra compound to the pathway. Both an ATP molecule and a GTP molecule are required to drive these two reactions.

$$\underset{\text{Pyruvate}}{\mathrm{COO^-{-}C(=O){-}CH_3}} + CO_2 + \text{ATP} + H_2O \xrightleftharpoons[\text{Biotin}]{\text{Pyruvate carboxylase}} \underset{\text{Oxaloacetate}}{\mathrm{COO^-{-}C(=O){-}CH_2{-}COO^-}} + \text{ADP} + P_i$$

$$\underset{\text{Oxaloacetate}}{\mathrm{COO^-{-}C(=O){-}CH_2{-}COO^-}} + \text{GTP} \xrightleftharpoons{\text{Phosphoenolpyruvate carboxykinase}} \underset{\text{Phosphoenolpyruvate}}{\mathrm{{}^-O(O{=})C{-}C(=CH_2){-}O{-}P(=O)(O^-){-}O^-}} + CO_2 + \text{GDP}$$

The oxaloacetate intermediate in this two-step process provides a connection to the citric acid cycle. In the first step of this cycle, oxaloacetate combines with acetyl CoA. If energy rather than glucose is needed, then oxaloacetate can go directly into the citric acid cycle.

There are two other locations where gluconeogenesis and glycolysis differ. In Steps 9 and 11 of gluconeogenesis (Steps 1 and 3 of glycolysis), the reactant–product combinations match between pathways. However, different enzymes are required for the forward and reverse processes. The new enzymes for gluconeogenesis are *fructose 1,6-bisphosphatase* and *glucose 6-phosphatase.*

The overall net reaction for gluconeogenesis is

$$2\text{ Pyruvate} + 4\text{ATP} + 2\text{GTP} + 2\text{NADH} + 2H_2O \longrightarrow \text{glucose} + 4\text{ADP} + 2\text{GDP} + 6P_i + 2\text{NAD}^+$$

• Glycolysis has a net production of 2 ATP (Section 24.2). Gluconeogenesis has a net expenditure of 4 ATP and 2 GTP, which is equivalent to the expenditure of 6 ATP.

Thus to reconvert pyruvate to glucose requires the expenditure of 4 ATP and 2 GTP. Whenever gluconeogenesis occurs, it is at the expense of other ATP-producing metabolic processes.

• The Cori Cycle

Gluconeogenesis using lactate as a source of pyruvate is particularly important because of lactate formation during strenuous exercise. The lactate so produced (Section 24.3) diffuses from muscle cells into the blood, where it is transported to the liver. Here the enzyme lactate dehydrogenase (the same enzyme that catalyzes lactate formation in muscle) converts lactate back to pyruvate.

$$\underset{\text{Lactate}}{\mathrm{COO^-{-}C(H)(OH){-}CH_3}} + \text{NAD}^+ \xrightarrow{\text{Lactate dehydrogenase}} \underset{\text{Pyruvate}}{\mathrm{COO^-{-}C(=O){-}CH_3}} + \text{NADH} + H^+$$

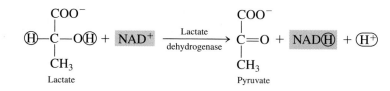

• The Cori cycle is named in honor of Gerty and Carl Cori, the husband-and-wife team who discovered it. They were awarded a Nobel Prize in 1947, the third husband-and-wife team to be so recognized. Marie and Pierre Curie were the first, Irene and Frederic Joliot-Curie the second.

The newly formed pyruvate is then converted via gluconeogenesis to glucose, which enters the bloodstream and goes to the muscles. This cyclic process, called the Cori cycle, is diagrammed in Figure 24.9. The **Cori cycle** *is a cyclic process in which glucose is converted to lactate in muscle tissue, the lactate is reconverted to glucose in the liver, and the glucose is returned to the muscle.*

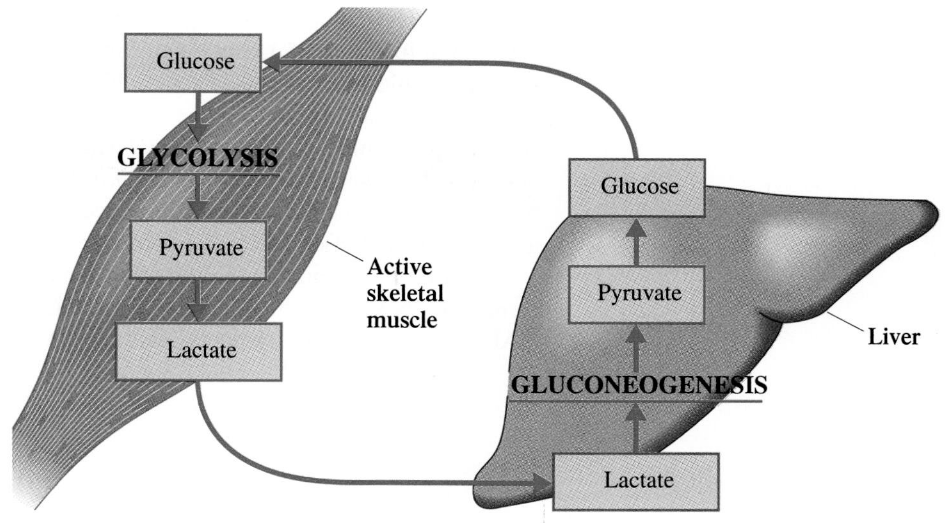

Figure 24.9
The Cori cycle. Lactate, formed from glucose under anaerobic conditions in muscle cells, is transferred to the liver, where it is reconverted to glucose, which is then transferred back to the muscle cells.

24.7 Terminology for Glucose Metabolic Pathways

In the preceding three sections we considered the processes of glycolysis, glycogenesis, glycogenolysis, and gluconeogenesis. Because of their like-sounding names, keeping the terminology for these four processes "straight" is often a problem. Figure 24.10 shows the relationships among these processes. Note that the glycogen degradation pathways (left side of Figure 24.10) have names ending in *-lysis,* which means "breakdown." The pathways associated with glycogen synthesis (right side of Figure 24.10) have names ending in *-genesis,* which means "making."

24.8 The Pentose Phosphate Pathway

Glycolysis is not the only pathway by which glucose may be degraded. Depending on the type of cell, various amounts of glucose are degraded by the pentose phosphate pathway,

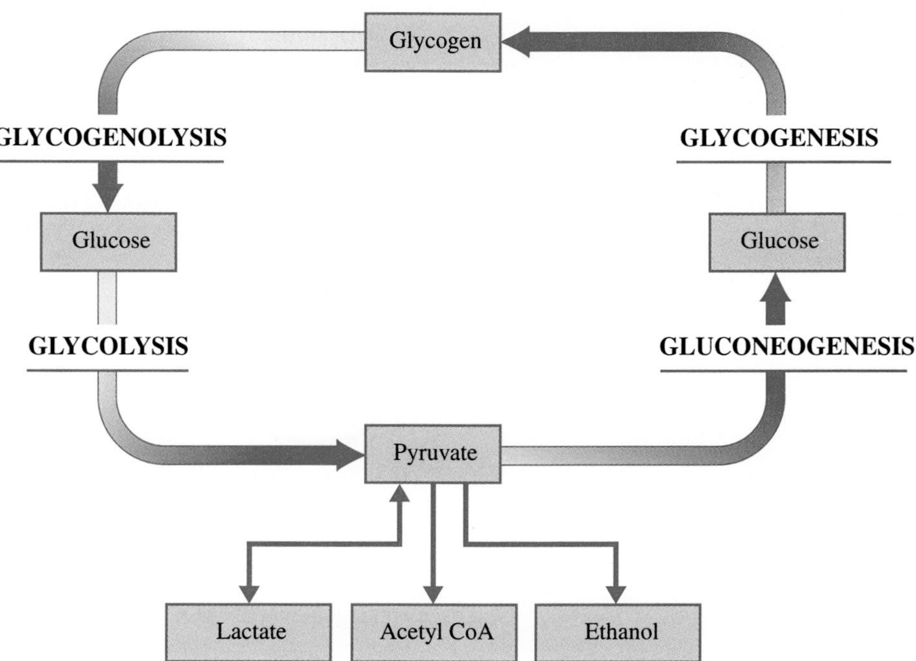

Figure 24.10
The relationships among four common metabolic pathways that involve glucose.

a pathway whose main focus is *not* subsequent ATP production as is the case for glycolysis. Major functions of this alternative pathway are (1) synthesis of the coenzyme NADPH needed in lipid biosynthesis (Section 25.7), and (2) production of ribose 5-phosphate, a pentose derivative needed for the synthesis of nucleic acids and many coenzymes. The **pentose phosphate pathway** *uses glucose to produce NADPH, ribose 5-phosphate (a pentose), and numerous other sugar phosphates.* The operation of the pentose phosphate pathway is significant in cells that produce lipids: fatty tissue, the liver, mammary glands, and the adrenal cortex (an active producer of steroid lipids).

NADPH, the coenzyme produced in the pentose phosphate pathway, is the reduced form of $NADP^+$ (nicotinamide adenine dinucleotide phosphate). Structurally, $NADP^+$/NADPH is a phosphorylated version of NAD^+/NADH.

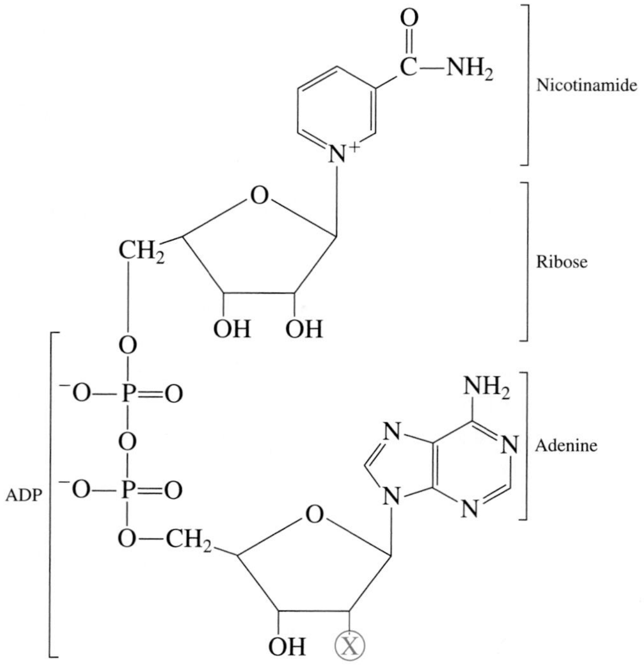

Nicotinamide

Ribose

Adenine

ADP

Nicotinamide adenine dinucleotide (NAD^+), X = —OH
Nicotinamide adenine dinucleotide phosphate ($NADP^+$), X = $—OPO_3^{2-}$

The nonphosphorylated and phosphorylated versions of this coenzyme have significantly different functions. The nonphosphorylated version is involved, mainly in its oxidized form (NAD^+/NADH), in the reactions of the common metabolic pathway (Section 23.5). The phosphorylated version is involved, mainly in its reduced form (NADPH/$NADP^+$), in biosynthetic reactions of lipids and nucleic acids.

There are two stages within the pentose phosphate pathway—an oxidative stage and a nonoxidative stage. The oxidative stage, which occurs first, involves three steps through which glucose 6-phosphate is converted to ribulose 5-phosphate and CO_2.

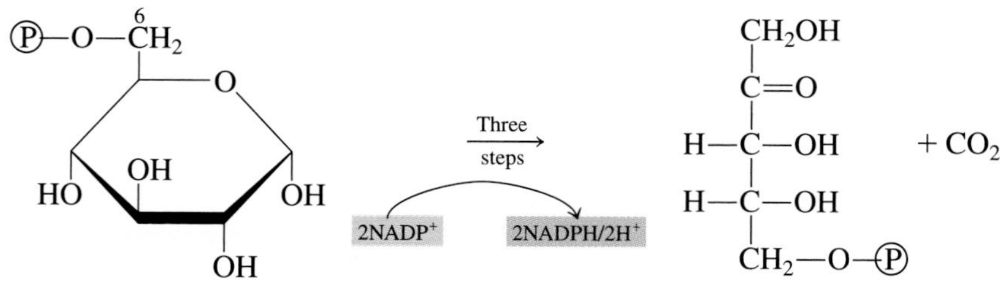

Glucose 6-phosphate —Three steps ($2NADP^+$ → $2NADPH/2H^+$)→ Ribulose 5-phosphate + CO_2

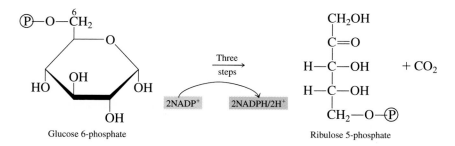

The net equation for the oxidative stage of the pentose phosphate pathway is

$$\text{Glucose 6-phosphate} + 2NADP^+ + H_2O \longrightarrow \text{ribulose 5-phosphate} + CO_2 + 2NADPH + 2H^+$$

CHEMISTRY AT A GLANCE

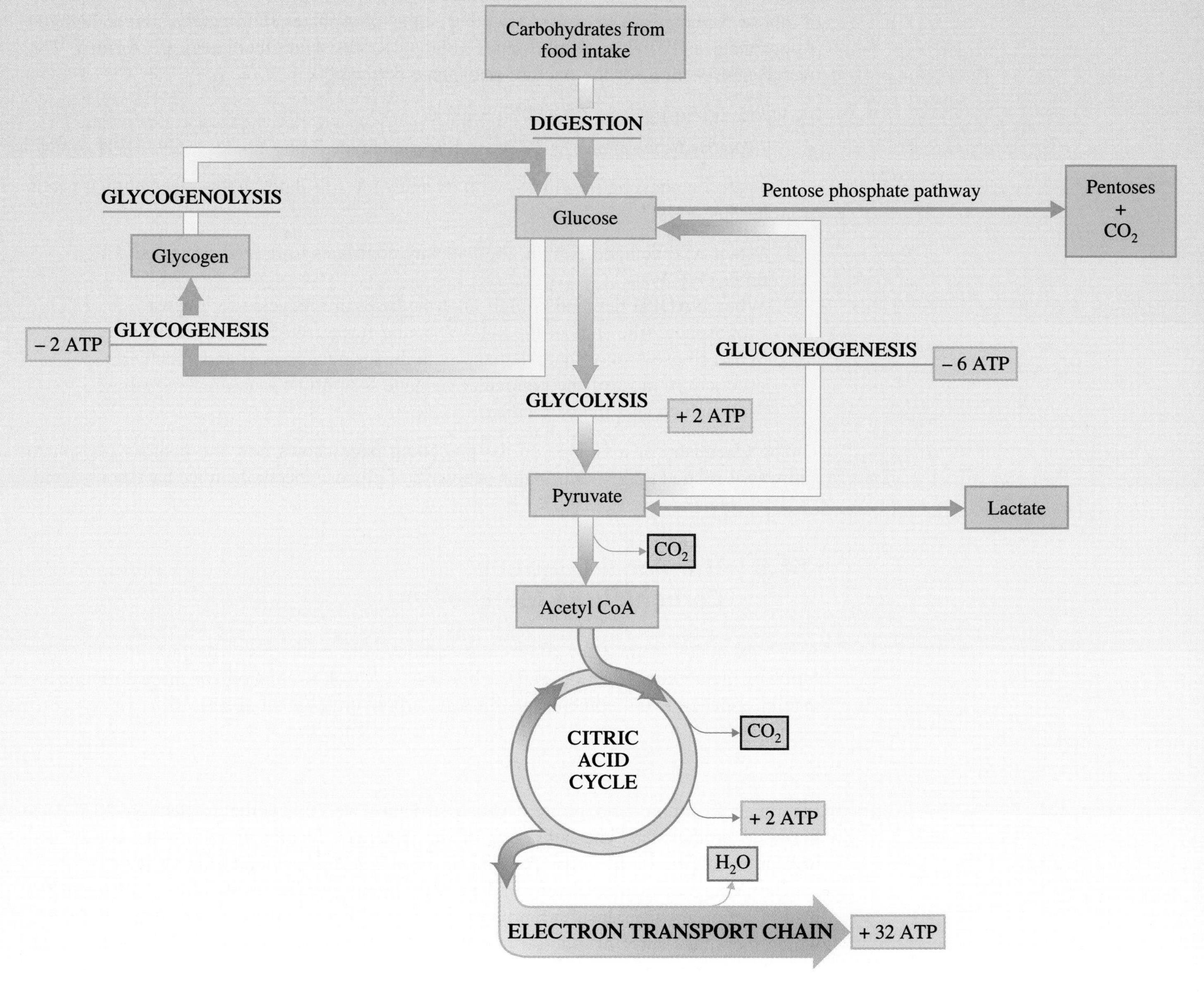

Note the production of two NADPH molecules per glucose 6-phosphate processed during this stage.

In the first step of the nonoxidative stage of the pentose phosphate pathway, ribulose 5-phosphate (a ketose) is isomerized to ribose 5-phosphate (an aldose).

$$\begin{array}{c} CH_2OH \\ | \\ C{=}O \\ | \\ H{-}C{-}OH \\ | \\ H{-}C{-}OH \\ | \\ CH_2{-}O{-}\text{Ⓟ} \end{array} \xrightarrow{\text{Phosphopentose isomerase}} \begin{array}{c} H \quad O \\ \backslash \;\; /\!/ \\ C \\ | \\ H{-}C{-}OH \\ | \\ H{-}C{-}OH \\ | \\ H{-}C{-}OH \\ | \\ CH_2{-}O{-}\text{Ⓟ} \end{array}$$

Ribulose 5-phosphate → Ribose 5-phosphate

The pentose ribose is a component of ATP, GTP, UTP, CoA, NAD^+/NADH, FAD/$FADH_2$, and RNA. Further steps in the nonoxidative stage contain provision for the conversion of ribose 5-phosphate to numerous other sugar phosphates. Ultimately, glyceraldehyde 3-phosphate and fructose 6-phosphate (both glycolysis intermediates) are formed. The overall net reaction for the pentose phosphate pathway is

$$\text{3 Glucose 6-phosphate} + 6NADP^+ + 3H_2O \longrightarrow$$
$$\text{2 fructose 6-phosphate} + 3CO_2 + \text{glyceraldehyde 3-phosphate} + 6NADPH + 6H^+$$

The pentose phosphate pathway, with its many intermediates, helps meet cellular needs in numerous ways.

1. When ATP demand is high, the pathway continues to its end products, which enter glycolysis.
2. When NADPH demand is high, intermediates are recycled to glucose 6-phosphate (the start of the pathway), and further NADPH is produced.
3. When ribose 5-phosphate demand is high, for nucleic acid and coenzyme production, most of the nonoxidative stage is nonfunctional, leaving ribose 5-phosphate as a major product.

The Chemistry at a Glance on the preceding page shows how the pentose phosphate pathway is related to the other major pathways of glucose metabolism we have considered.

24.9 Hormonal Control of Carbohydrate Metabolism

A second major method for controlling carbohydrate metabolism, besides enzyme inhibition by metabolites (Section 24.2), is hormonal control. Among others, three hormones—insulin, glucagon, and epinephrine—affect carbohydrate metabolism.

• Insulin

Insulin, a 51-amino-acid protein whose structure we considered in Section 20.7, is a hormone produced by the beta cells of the pancreas. Insulin promotes the uptake and utilization of glucose by cells. Thus its function is to lower blood glucose levels.

The release of insulin is triggered by *high* blood glucose levels. The mechanism for insulin action involves insulin binding to protein receptors on the outer surfaces of cells, which facilitates entry of glucose into the cells. Insulin also produces an increase in the rate of glycogen synthesis.

• Glucagon

Glucagon is a polypeptide hormone (29 amino acids) produced in the pancreas by alpha cells. It is released when blood glucose levels are *low.* Its principal function is to increase blood glucose concentrations by speeding up the conversion of glycogen to glucose (glycogenolysis) in the liver. Thus glucagon's effects are opposite to those of insulin.

• Epinephrine

Epinephrine (Section 17.9), also called adrenaline, is released by the adrenal glands in response to anger, fear, or excitement. Its function is similar to that of glucagon—stimulation of glycogenolysis, the release of glucose from glycogen. Its primary target is muscle cells, where energy is needed for quick action.

Epinephrine acts by binding to a receptor site on the outside of the cell membrane, stimulating the enzyme adenyl cyclase to begin production of a *secondary messenger,*

Chemical CONNECTIONS

24.2 Diabetes Mellitus

Diabetes mellitus is the best-known and most prevalent metabolic disease in humans, affecting approximately 4% of the population. There are two major forms of this disease: insulin-dependent (type I) and non–insulin-dependent (type II) diabetes.

Type I diabetes, which often appears in children, is the result of inadequate insulin production by the beta cells of the pancreas. Control of this condition involves insulin injections and special dietary programs. A risk associated with the insulin injections is that too much insulin can produce severe hypoglycemia (insulin shock); blackout or a coma can result. Treatment involves a quick infusion of glucose. Diabetics often carry candy bars (quick glucose sources) for use if they feel any of the symptoms that signal the onset of insulin shock.

In Type II diabetes, which usually occurs in overweight individuals more than 40 years old, body insulin production is normal, but the cells do not respond to it normally. Some of the insulin receptors on the plasma membranes are not functioning properly and fail to recognize the insulin. Treatment involves drugs that increase body insulin levels and a carefully regulated diet (to reduce obesity). More efficient use of undamaged receptors occurs at increased insulin levels.

The effects of both types of diabetes are the same—inadequate glucose uptake by cells. The result is blood glucose levels much higher than normal (hyperglycemia). With an inadequate glucose intake, cells must resort to other procedures for energy production, procedures that involve the breakdown of fats and protein.

A frequently used diagnostic test for diabetes is the glucose tolerance test. A patient who has fasted for 10–16 hours is given a single dose of glucose, typically in a fruit-flavored drink. Blood glucose levels are then monitored at regular intervals over several hours. As the accompanying diagram shows, glucose levels drop to a fasting level in a nondiabetic in about 2 hours but remain high in a diabetic.

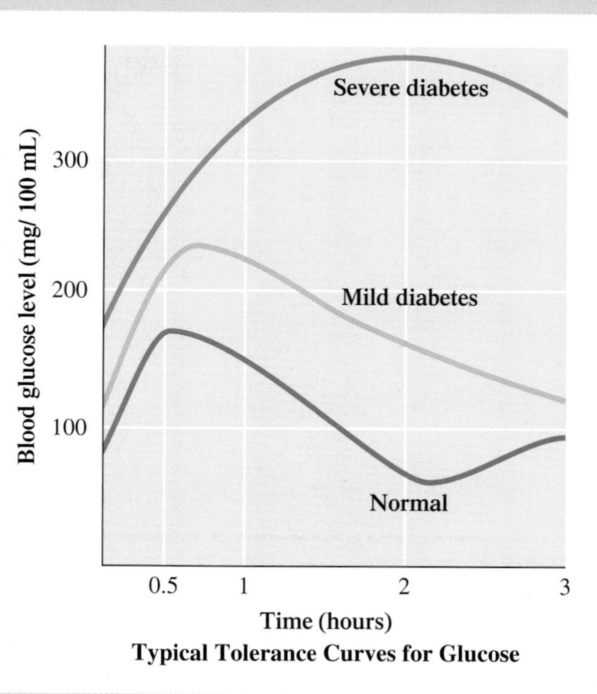

Typical Tolerance Curves for Glucose

A diabetic giving himself a blood glucose test.

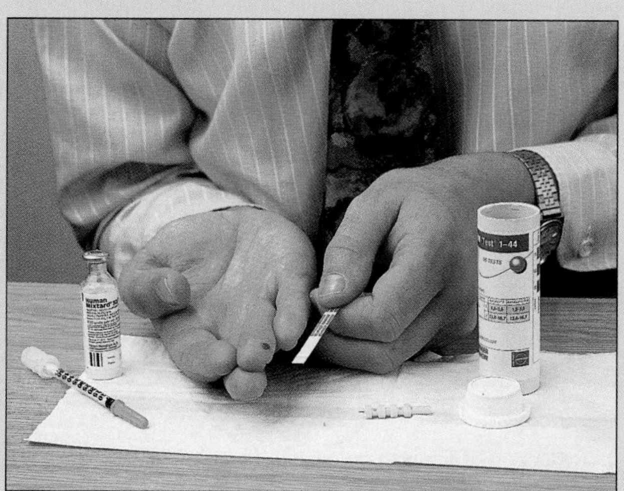

cyclic AMP (cAMP) from ATP. (The structure of cAMP was considered in Section 23.3.) The cAMP is released in the cell interior, where, in a series of reactions, it activates glycogen phosphorylase, the enzyme that initiates glycogenolysis. The glucose 6-phosphate that is produced from the glycogen breakdown provides a source of quick energy. Figure 24.11 shows the series of events initiated by the release of the hormone epinephrine. Cyclic AMP also inhibits glycogenesis, thus preventing glycogen production at the same time.

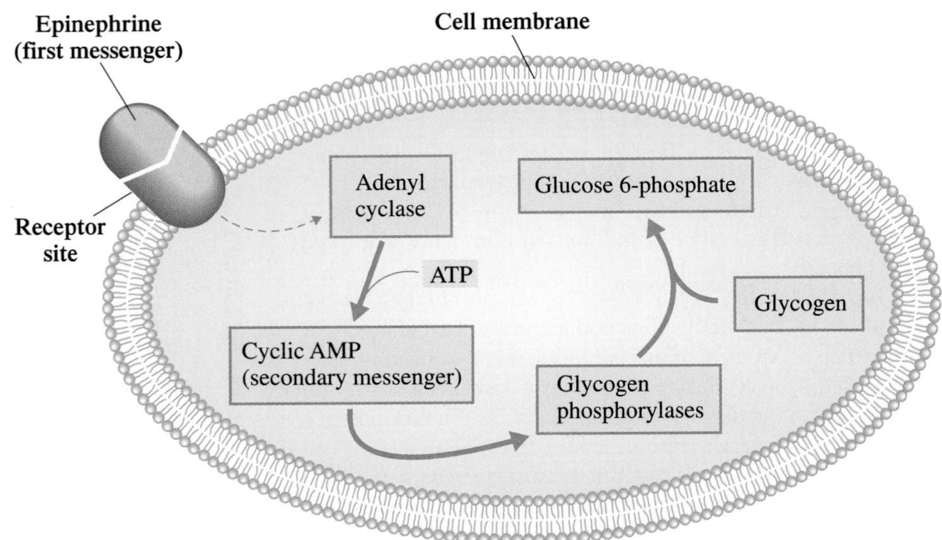

Figure 24.11 The series of events by which the hormone epinephrine stimulates glucose production.

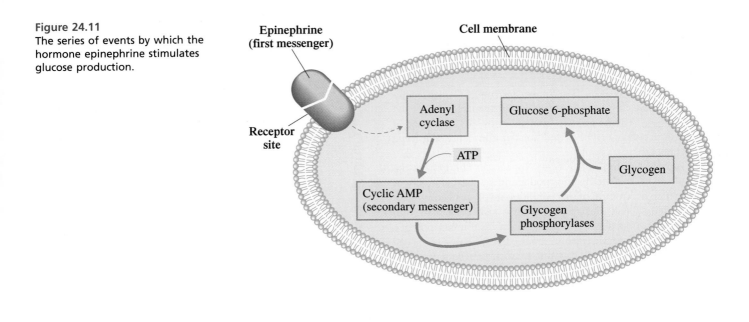

Concepts to Remember

Glycolysis. Glycolysis, a series of ten reactions that occur in the cytoplasm, is a process in which one glucose molecule is converted into two molecules of pyruvate. A net gain of two molecules of ATP and two molecules of NADH results from the metabolizing of glucose to pyruvate.

Fates of pyruvate. With respect to energy-yielding metabolism, the pyruvate produced by glycolysis can be converted to acetyl CoA under aerobic conditions or to lactate under anaerobic conditions. Some microorganisms convert pyruvate to ethanol, an anaerobic process.

Glycogenesis. Glycogenesis is the process whereby excess glucose is converted into glycogen. The glycogen is stored in the liver and in muscle tissue.

Glycogenolysis. Glycogenolysis is the breakdown of glycogen into glucose. This process occurs when muscles need energy and when the liver is restoring a low blood-sugar level to normal.

Gluconeogenesis. Gluconeogenesis is the formation of glucose from lactate and certain other substances. This process takes place in the liver when glycogen supplies are being depleted and when carbohydrate intake is low.

Cori cycle. The Cori cycle is the cyclic process involving the transport of lactate from muscle tissue to the liver, the resynthesis of glucose by gluconeogenesis, and the return of glucose to muscle tissue.

Pentose phosphate pathway. The pentose phosphate pathway metabolizes glucose to produce ribose (a pentose), NADPH, and other sugars needed for biosynthesis.

Carbohydrate metabolism and hormones. Insulin decreases blood glucose levels by promoting the uptake of glucose by cells. Glucagon increases blood glucose levels by promoting the conversion of glycogen to glucose. Epinephrine stimulates the release of glucose from glycogen in muscle cells.

Key Reactions and Equations

1. Glycolysis (Section 24.2)

$$\text{Glucose} + 2P_i + 2ADP + 2NAD^+ \longrightarrow 2\text{ pyruvate} + 2ATP + 2NADH + 2H^+ + 2H_2O$$

2. Oxidation of pyruvate to acetyl CoA (Section 24.3)

$$CH_3{-}\overset{\overset{\large O}{\|}}{C}{-}COO^- + CoA{-}SH + NAD^+ \xrightarrow[\text{steps}]{\text{Four}} CH_3{-}\overset{\overset{\large O}{\|}}{C}{-}S{-}CoA + NADH + CO_2$$

3. Reduction of pyruvate to lactate (Section 24.3)

$$CH_3{-}\overset{\overset{\large O}{\|}}{C}{-}COO^- + NADH + H^+ \longrightarrow CH_3{-}\overset{\overset{\large OH}{|}}{CH}{-}COO^- + NAD^+$$

4. Reduction of pyruvate to ethanol (Section 24.3)

$$CH_3{-}\overset{\overset{\large O}{\|}}{C}{-}COO^- + 2H^+ + NADH \xrightarrow[\text{steps}]{\text{Two}} CH_3{-}CH_2{-}OH + NAD^+ + CO_2$$

5. Glycogenesis (Section 24.5)

$$\text{Glucose} \longrightarrow \text{glycogen}$$

6. Glycogenolysis (Section 24.5)

$$\text{Glycogen} \longrightarrow \text{glucose}$$

7. Gluconeogenesis (Section 24.6)

$$\left.\begin{matrix}\text{Lactate, certain}\\ \text{amino acids,}\\ \text{citric acid cycle}\\ \text{intermediates}\end{matrix}\right\} \longrightarrow \text{pyruvate} \longrightarrow \text{glucose}$$

Key Terms

Alcohol fermentation (24.3)
Cori cycle (24.6)
Digestion (24.1)
Gluconeogenesis (24.6)
Glycogenesis (24.4)
Glycogenolysis (24.4)
Glycolysis (24.2)
Pentose phosphate pathway (24.8)
Substrate-level phosphorylation (24.2)

Exercises and Problems

The members of each pair of problems in this section test similar material.

Carbohydrate Digestion (Section 24.1)

24.1 Where does starch digestion begin in the body, and what is the name of the enzyme involved in this initial digestive process?

24.2 Very little digestion of starch occurs in the stomach. Why?

24.3 What is the primary site for carbohydrate digestion, and what organ produces the enzymes that are active at this location?

24.4 Where does the final step in carbohydrate digestion take place, and in what form are carbohydrates as they enter this final step?

24.5 Where does the digestion of sucrose begin, and what is the reaction that occurs?

24.6 Where does the digestion of lactose begin, and what is the reaction that occurs?

24.7 Identify the three major monosaccharides produced by digestion of carbohydrates.

24.8 The various stages of carbohydrate digestion all involve the same general type of reaction. What is this reaction type?

Glycolysis (Section 24.2)

24.9 What is the starting material for glycolysis?

24.10 What is the end product from glycolysis?

24.11 What coenzyme functions as the oxidizing agent in glycolysis?

24.12 What is meant by the statement that glycolysis is an anaerobic pathway?

24.13 What is the first step of glycolysis, and why is it important in retaining glucose inside the cell?

24.14 Step 3 of glycolysis is the commitment step. Explain.

24.15 What two C_3 fragments are formed by the splitting of fructose 1,6-bisphosphate?

24.16 In one step of the glycolysis pathway, a C_6 chain is broken into two C_3 fragments, only one of which can be further degraded. What happens to the other C_3 fragment?

24.17 How many pyruvate molecules are produced per glucose molecule during glycolysis?

24.18 How many molecules of ATP and NADH are produced per glucose molecule during glycolysis?

24.19 How many steps in the glycolysis pathway produce ATP?

24.20 How many steps in the glycolysis pathway consume ATP?

24.21 Of the 10 steps of glycolysis, which ones involve phosphorylation?

24.22 Of the 10 steps of glycolysis, which ones involve oxidation?

24.23 Where in a cell does glycolysis occur?

24.24 Do the reactions of glycolysis and the citric acid cycle occur at the same location in a cell? Explain.

24.25 Replace the question mark in each of the following word equations with the name of a substance.

a. Glucose + ATP $\xrightarrow{\text{Hexokinase}}$? + ADP + H^+

b. ? $\xrightarrow{\text{Enolase}}$ phosphoenolpyruvate + water

c. 3-Phosphoglycerate $\xrightarrow{?}$ 2-phosphoglycerate

d. 1,3-Bisphosphoglycerate + ? $\xrightarrow[\text{kinase}]{\text{Phosphoglycero-}}$ 3-phosphoglycerate + ATP

24.26 Replace the question mark in each of the following word equations with the name of a substance.

a. Glucose 6-phosphate $\xrightarrow[\text{isomerase}]{\text{Phosphogluco-}}$?

b. ? $\xrightarrow{\text{Aldolase}}$ dihydroxyacetone phosphate + glyceraldehyde 3-phosphate

c. Phosphoenolpyruvate + ? + H^+ $\xrightarrow[\text{kinase}]{\text{Pyruvate}}$ pyruvate + ATP

d. Dihydroxyacetone phosphate $\xrightarrow{?}$ glyceraldehyde 3-phosphate

24.27 In which step of glycolysis does each of the following occur?

a. Second substrate-level phosphorylation reaction
b. First ATP-consuming reaction
c. Third isomerization reaction
d. Use of NAD^+ as an oxidizing agent

24.28 In which step of glycolysis does each of the following occur?

a. First energy-producing reaction
b. First ATP-producing reaction
c. A dehydration reaction
d. First isomerization reaction

24.29 What is the net ATP production when each of the following molecules is processed through the glycolysis pathway?

a. One glucose molecule
b. One sucrose molecule

24.30 What is the net ATP production when each of the following molecules is processed through the glycolysis pathway?

a. One lactose molecule
b. One maltose molecule

24.31 Draw structural formulas for each of the following pairs of molecules.

a. Pyruvic acid and pyruvate
b. Dihydroxyacetone and dihydroxyacetone phosphate
c. Fructose 6-phosphate and fructose 1,6-bisphosphate
d. Glyceric acid and glyceraldehyde

24.32 Draw structural formulas for each of the following pairs of molecules.

a. Glyceric acid and glycerate

b. Glycerate and pyruvate

c. Glucose 6-phosphate and fructose 6-phosphate

d. Dihydroxyacetone and glyceric acid

24.33 Number the carbon atoms of fructose 1,6-bisphosphate 1 through 6, and show the location of each carbon in the two trioses produced during Step 4 of glycolysis.

24.34 Number the carbon atoms of glucose 1 through 6, and show the location of each carbon in the two molecules of pyruvate produced by glycolysis.

Fates of Pyruvate (Section 24.3)

24.35 What are the three common possible fates for pyruvate produced from glycolysis?

24.36 Compare the fates of pyruvate in the body under aerobic and under anaerobic conditions.

24.37 What is the overall reaction equation for the conversion of pyruvate to acetyl CoA?

24.38 What is the overall reaction equation for the conversion of pyruvate to acetyl CoA?

24.39 Explain how lactate production allows glycolysis to continue under anaerobic conditions.

24.40 How is the alcohol fermentation in yeast similar to lactate production in skeletal muscle?

24.41 In alcohol fermentation, a C_3 pyruvate molecule is changed to a C_2 ethanol molecule. What is the fate of the third pyruvate carbon?

24.42 What are the structural differences between pyruvate and lactate ions?

24.43 What is the net reaction for the conversion of one glucose molecule to two lactate molecules?

24.44 What is the net reaction for the conversion of one glucose molecule to two ethanol molecules?

Complete Oxidation of Glucose (Section 24.4)

24.45 How does the fact that cytoplasmic $NADH/H^+$ cannot cross the mitochondrial membranes affect ATP production from cytoplasmic $NADH/H^+$?

24.46 What is the net reaction for the shuttle mechanism involving glycerol 3-phosphate by which NADH electrons are shuttled across the mitochondrial membrane.

24.47 Contrast, in terms of ATP production, the oxidation of glucose to CO_2 and H_2O with the oxidation of glucose to pyruvate.

24.48 Contrast, in terms of ATP production, the oxidation of glucose to CO_2 and H_2O with the oxidation of glucose to ethanol.

24.49 How many of the 36 ATP molecules produced from the complete oxidation of 1 glucose molecule are produced during glycolysis?

24.50 How many of the 36 ATP molecules produced from the complete oxidation of 1 glucose molecule are the result of the oxidation of pyruvate to acetyl CoA?

Glycogen Metabolism (Section 24.5)

24.51 Compare the meanings of the terms *glycogenesis* and *glycogenolysis.*

24.52 Where is most of the body's glycogen stored?

24.53 Glucose 1-phosphate is the product of the first step of glycogenesis. What is the reactant?

24.54 Glucose 1-phosphate is the product of the first step of glycogenolysis. What are the reactants?

24.55 What is the source of the PP_i produced during the second step of glycogenesis?

24.56 What is the function of the PP_i produced during the second step of glycogenesis?

24.57 How is ATP involved in glycogenesis?

24.58 How many ATP molecules are needed to attach a single glucose molecule to a growing glycogen chain?

24.59 Which step of glycogenolysis is the reverse of Step 1 of glycogenesis?

24.60 What reaction determines whether glucose formed by glycogenolysis can leave a cell?

24.61 What is the difference between glycogenolysis in liver cells and in muscle cells?

24.62 The liver, but not the brain or muscle cells, has the capacity to supply free glucose to the blood. Explain.

24.63 In what form does glycogen enter the glycolysis pathway?

24.64 Explain why one more ATP is produced when glucose is obtained from glycogen rather than directly from the blood.

Gluconeogenesis (Section 24.6)

24.65 What organ is primarily responsible for gluconeogenesis?

24.66 What is the physiological function of gluconeogenesis?

24.67 How does gluconeogenesis get around the three irreversible steps of glycolysis?

24.68 Although gluconeogenesis and glycolysis are "reverse" processes, there are 11 steps in gluconeogenesis and only 10 steps in glycolysis. Explain.

24.69 What intermediate in gluconeogenesis is also an intermediate in the citric acid cycle?

24.70 What are the sources of high-energy bonds in gluconeogenesis?

24.71 What is the fate of lactate formed by muscular activity?

24.72 What is the physiological function of the Cori cycle?

The Pentose Phosphate Pathway (Section 24.8)

24.73 What is the starting material for the pentose phosphate pathway?

24.74 What are two major functions of the pentose phosphate pathway?

24.75 How do the biochemical functions of NADH and NADPH differ?

24.76 How do the structures of NADH and NADPH differ?

24.77 Write a general equation for the oxidative stage of the pentose phosphate pathway.

24.78 Write a general equation for the entire pentose phosphate pathway.

24.79 What compound contains the carbon atom lost from glucose (a hexose) in its conversion to ribose (a pentose)?

24.80 How many molecules of NADPH are produced per glucose 6-phosphate in the pentose phosphate pathway?

Control of Carbohydrate Metabolism (Section 24.9)

24.81 What effect does insulin have on glycogen metabolism?

24.82 What effect does insulin have on blood glucose levels?

24.83 What effect does glucagon have on blood glucose levels?

24.84 What effect does glucagon have on glycogen metabolism?

24.85 What organ is the source of insulin?

24.86 What organ is the source of glucagon?

24.87 The hormone epinephrine generates a "second messenger." Explain.

24.88 What is the relationship between cAMP and the hormone epinephrine?

24.89 Compare the target tissues for glucagon and epinephrine.

24.90 Compare the biological functions of glucagon and epinephrine.

Additional Problems

24.91 Indicate in which of the four processes *glycolysis, glycogenesis, glycogenolysis,* and *gluconeogenesis* each of the following compounds is encountered. There may be more than one correct answer for a given compound.

a. Glucose 6-phosphate
b. Glucose 1-phosphate
c. Dihydroxyacetone phosphate
d. Oxaloacetate

24.92 Indicate in which of the four processes *glycolysis, glycogenesis, glycogenolysis,* and *gluconeogenesis* each of the following situations is encountered. There may be more than one correct answer for a given situation.

a. NAD^+ is consumed
b. ATP is produced
c. ATP is consumed
d. UDP is involved

24.93 Indicate in which of the four processes *glycolysis, glycogenesis, glycogenolysis,* and *glyconeogenesis* each of the following characterizations applies.

a. Glucose is converted to two pyruvates.
b. Glycogen is synthesized from glucose.
c. Glycogen is broken down into free glucose units.
d. Glucose is synthesized from pyruvate.

24.94 What is the ATP yield *per glucose molecule* in each of the following processes?

a. Glycolysis
b. Glycolysis, acetyl CoA formation, and the common metabolic pathway
c. Glycolysis plus oxidation of pyruvate to acetyl CoA
d. Glycolysis plus reduction of pyruvate to lactate

24.95 Which one of these characterizations, (1) Cori cycle, (2) an anaerobic process, (3) oxidative stage of pentose phosphate pathway, or (4) nonoxidative stage of pentose phosphate pathway, applies to each of the following chemical changes?

a. Pyruvate to lactate
b. Pyruvate to ethanol
c. Glucose 6-phosphate to ribulose 5-phosphate
d. Ribulose 5-phosphate to ribose 5-phosphate

24.96 What condition or conditions determine that pyruvate is involved in each of the following?

a. Gluconeogenesis
b. Converted to lactate
c. Citric acid cycle
d. Converted to ethanol

24.97 In the complete metabolism of 1 mole of sucrose, how many moles of each of the following are produced?

a. CO_2
b. Pyruvate
c. Acetyl CoA
d. ATP

24.98 Under what conditions does glucose 6-phosphate enter each of the following pathways?

a. Glycogenesis
b. Glycolysis
c. Pentose phosphate pathway
d. Hydrolysis to free glucose

Grid Problems

24.99

1. pyruvate	2. lactate	3. acetyl part of acetyl CoA
4. dihydroxyacetone	5. glycerate	6. glyceraldehyde

Select from the grid *all* correct responses for each of the following situations.

a. Substances that are C_3 species
b. Substances for which one or more phosphoderivatives are intermediates in glycolysis
c. Substances for which one or more bisphosphoderivatives are intermediates in glycolysis
d. Substances that are not part of the glycolysis process

24.100

1. glucose 6-phosphate	2. dihydroxyacetone phosphate	3. 3-phospho-glycerate
4. glyceraldehyde 3-phosphate	5. fructose 6-phosphate	6. phosphoenol-pyruvate

Select from the grid *all* correct responses for each of the following situations.

a. Production of this substance in glycolysis requires expenditure of ATP.
b. Production of this substance in glycolysis occurs in two different ways.
c. Reaction of this substance in glycolysis generates ATP.
d. This is a molecule that contains a high-energy phosphate group.

24.101

1. 0 ATP	2. 2 ATP	3. 4 ATP
4. 12 ATP	5. 18 ATP	6. 36 ATP

Select from the grid *all* correct responses for each of the following situations.

a. ATP yield per glucose for glycolysis
b. ATP yield per glucose for glycolysis plus the common metabolic pathway
c. ATP yield per glucose for glycolysis plus oxidation of pyruvate to acetyl CoA
d. ATP yield per glucose for glycolysis plus reduction of pyruvate to lactate

24.102

1. glycogenolysis	2. glycogenesis	3. pentose phosphate pathway
4. glycolysis	5. gluconeogenesis	6. alcohol fermentation

Select from the grid *all* correct responses for each of the following situations.

a. Pyruvate is the starting material or end product.
b. Glucose is the starting material or end product.
c. Glycogen is the starting material or end product.
d. Carbon dioxide is a product.

Lipid Metabolism

CHAPTER OUTLINE

25.1 Digestion and Absorption of Lipids 743
25.2 Triacylglycerol Storage and Mobilization 744
25.3 Glycerol Metabolism 746
25.4 Oxidation of Fatty Acids 747
25.5 ATP Production from Fatty Acid Oxidation 751
25.6 Ketone Bodies 753
25.7 Biosynthesis of Fatty Acids: Lipogenesis 756
25.8 Biosynthesis of Cholesterol 759
25.9 Relationships Between Lipid and Carbohydrate Metabolism 762

Chemistry at a Glance:
Interrelationships Between Carbohydrate and Lipid Metabolism 763

Chemical Connections
25.1 High-Intensity Versus Low-Intensity Workouts 752
25.2 Statins: Drugs That Lower Plasma Levels of Cholesterol 761

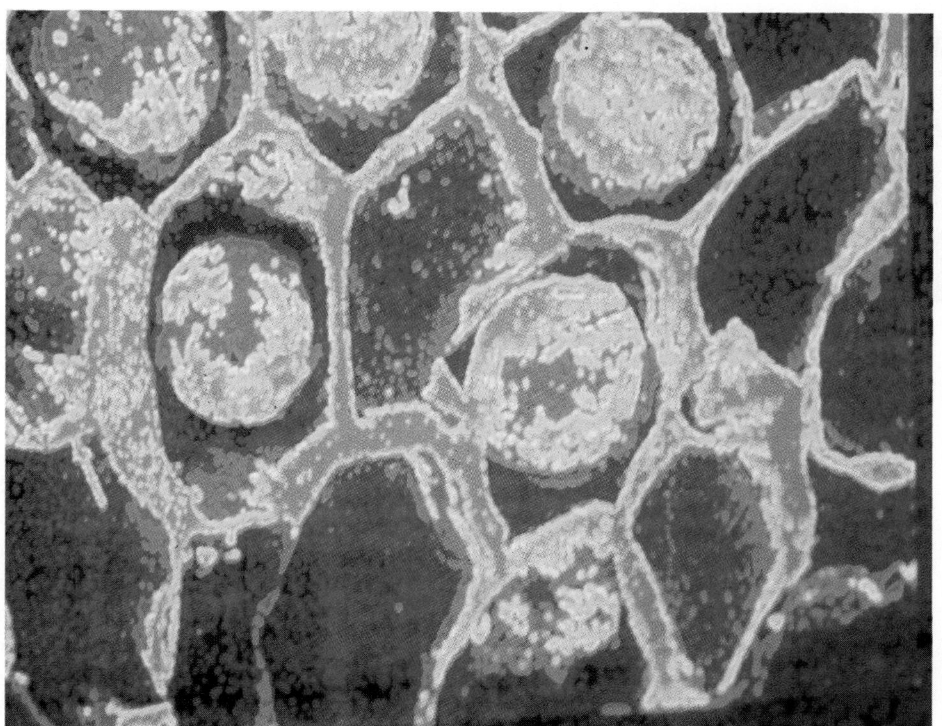

■ Fat cells.

Certain classes of lipids play an extremely important role in cellular metabolism, because they represent an energy-rich "fuel" that can be stored in large amounts in adipose (fat) tissue. Between one-third and one-half of the calories present in the diet of the average U.S. resident are supplied by lipids. Furthermore, excess energy derived from carbohydrates and proteins beyond normal daily needs is stored in lipid molecules (in adipose tissue), later to be mobilized and used when needed.

25.1 Digestion and Absorption of Lipids

Because 98% of total *dietary* lipids are triacylglycerols (fats and oils; Section 19.4), this chapter focuses on triacylglycerol metabolism. Like all lipids, triacylglycerols (TAGs) are insoluble in water. Hence, water-based salivary enzymes in the mouth have little effect on them. In the stomach, TAGs are changed physically but not chemically. The churning action of the stomach breaks up triacylglycerol materials into small globules, or droplets, which float as a layer above the other components of swallowed food. The resulting material, called *chyme,* then enters the small intestine.

The arrival of triacylglycerol "globules" into the small intestine triggers, through the action of the hormone *cholecystokinin,* the release of bile stored in the gallbladder. The bile (Section 19.10), which contains no enzymes, acts as an emulsifier. An **emulsifier** *is a substance that can disperse and stabilize water-insoluble substances as colloidal particles in an aqueous solution.* Colloidal particle formation (Section 8.6) through bile emulsification "solubilizes" the triacylglycerol globules, and digestion of the TAGs can begin. The major enzymes involved are the *pancreatic lipases,* which hydrolyze the ester linkages between the glycerol and fatty acid units of the TAGs. *Complete* hydrolysis does not usually occur; only two of the three fatty acid units are liberated, producing a monoacylglycerol and two free fatty acids. Occasionally, enzymes remove all three fatty acid units, leaving a free glycerol molecule.

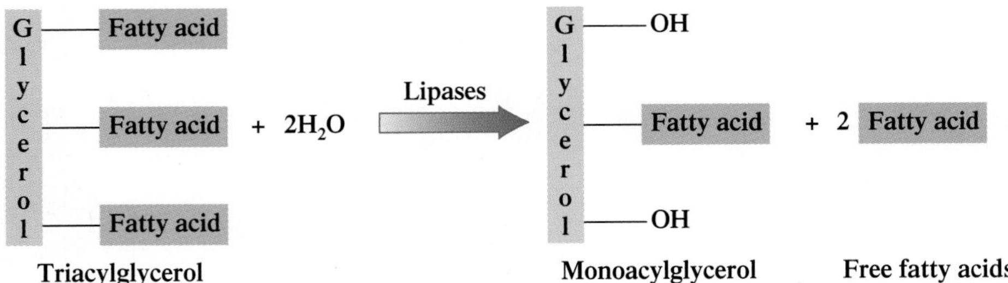

• The physical breakup of high-fat materials into small globules takes longer than the breakup of low-fat materials, so high-fat foods remain in the stomach longer. This is why a high-fat meal causes you to feel "full" for a longer period of time.

• Bile, a yellowish-green liquid produced in the liver and stored in the gallbladder is a water-based mixture of bile salts (Section 19.10), bile pigments from the breakdown of hemoglobin (Section 26.7), cholesterol, and inorganic ions. The bile salts are the emulsifiers.

• When freed of the triacylglycerol molecules they "transport" during digestion, bile salts are mostly recycled. Small amounts are excreted.

• Chylomicron is pronounced "kye-lo-MY-cron."

• Dietary triacylglycerol molecules undergo hydrolysis three times (to form free fatty acids and/or monoacylglycerols) and are repackaged twice (to re-form TAGs) before they become available for use by cells.

• Triacylglycerol reserves would enable the average person to survive starvation for about 30 days, given sufficient water. Glycogen reserves (stored glucose) would be depleted within 1 day.

With the help of bile, the free fatty acids and monoacylglycerols produced from hydrolysis are combined into tiny spherical droplets called fatty acid micelles. A **fatty acid micelle** *is a tiny spherical droplet containing about 20 fatty acids and/or monoacylglycerols and some bile.* Fatty acid micelles are very small compared to the original triacylglycerol globules, which contain thousands of triacylglycerol molecules.

Micelles, with their free fatty acid and monoacylglycerol components, are small enough to be readily absorbed through the membranes of intestinal cells. Within the intestinal cells, a "repackaging" occurs in which the free fatty acids and monoacylglycerols are reassembled into triacylglycerols. The newly formed triacylglycerols are then combined with water-soluble proteins to produce a type of lipoprotein (Section 20.16) called a chylomicron. A **chylomicron** *is a lipoprotein that transports triacylglycerols from intestinal cells to the bloodstream.* Chylomicrons are too large to pass through capillary walls directly into the bloodstream. Consequently, delivery of the chylomicrons to the bloodstream is accomplished through the body's lymphatic system. Chylomicrons enter the lymphatic system through tiny lymphatic vessels in the intestinal lining. They enter the bloodstream through the thoracic duct (a large lymphatic vessel just below the collarbone), where the fluid of the lymphatic system flows into a vein, joining the bloodstream. Figure 25.1 summarizes the events that must occur before triacylglycerols can reach the bloodstream through the digestive process.

Once the chylomicrons reach the bloodstream, the TAGs they carry are again hydrolyzed to produce glycerol and free fatty acids. The products from this hydrolysis are absorbed by the cells of the body and are either broken down to acetyl CoA for energy or stored as lipids (some are again repackaged as TAGs).

Soon after a meal heavily laden with TAGs is ingested, the chylomicron content of both blood and lymph increases dramatically. Chylomicron concentrations usually begin to rise within 2 hours after a meal (see Figure 25.2), reach a peak in 4–6 hours, and then drop rather rapidly to a normal level as they move into adipose cells (Section 25.2) or into the liver.

25.2 Triacylglycerol Storage and Mobilization

Most cells in the body have limited capability for storage of TAGs. However, this activity is the major function of specialized cells called adipocytes, found in adipose tissue.

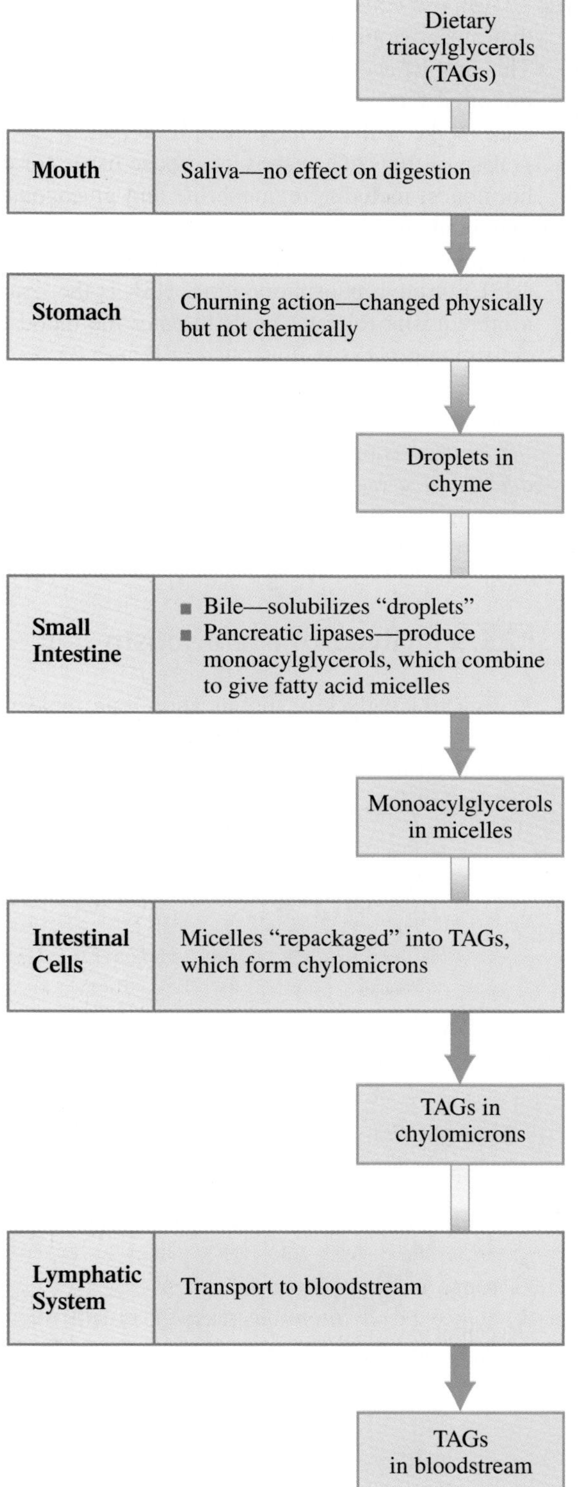

Figure 25.1
A summary of the events that must occur before triacylglycerols (TAGs) can reach the bloodstream through the digestive process.

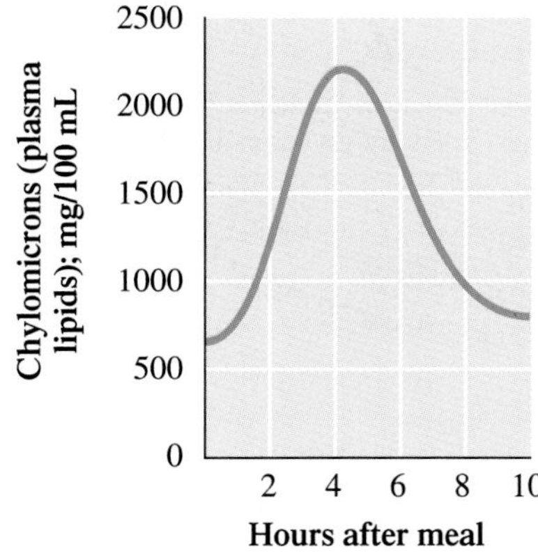

Figure 25.2
The rise in chylomicron concentration in blood after a meal with a high triacylglycerol content.

Adipocytes *are triacylglycerol-storing cells.* **Adipose tissue** *consists of large numbers of adipocyte cells.*

Adipose tissue is located primarily directly beneath the skin (subcutaneous), particularly in the abdominal region, and in areas around vital organs. Besides its function as a storage location for the chemical energy inherent in TAGs, subcutaneous adipose tissue also serves as an insulator against excessive heat loss to the environment and provides organs with protection against physical shock.

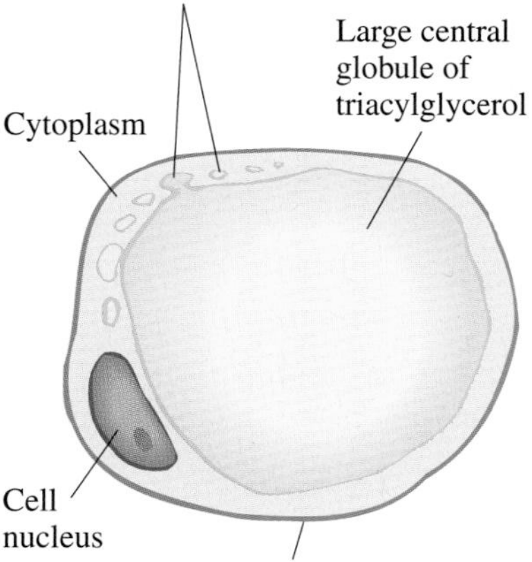

Figure 25.3 Structural characteristics of an adipose cell.

• Adult humans have approximately 30–40 billion adipose cells.

• Adipose tissue is the only tissue in which *free* TAGs occur in appreciable amounts. In other types of cells and in the bloodstream, TAGs are part of lipoprotein particles.

Adipose cells are among the largest cells in the body. They differ from other cells in that most cytoplasm has been replaced with a large triacylglycerol droplet (Figure 25.3). This droplet accounts for nearly the entire volume of the cell. As newly formed TAGs are imported into an adipose cell, they form small droplets at the periphery of the cell that later merge with the large central droplet.

Use of the TAGs stored in adipose tissue for energy production is triggered by several hormones, including epinephrine and glucagon. Hormonal interaction with adipose cell membrane receptors stimulates production of cAMP from ATP inside the adipose cell. In a series of enzymatic reactions, the cAMP (Section 23.3) activates hormone-sensitive lipase (HSL) through phosphorylation. HSL is the lipase needed for triacylglycerol hydrolysis, a prerequisite for fatty acids to enter the bloodstream from an adipose cell. This cAMP activation process is illustrated in Figure 25.4.

The overall process of tapping the body's triacylglycerol energy reserves (adipose tissue) for energy is called triacylglycerol mobilization. **Triacylglycerol mobilization** *is the hydrolysis of triacylglycerols stored in adipose tissue, followed by release of the fatty acids and glycerol produced into the bloodstream.* Triacylglycerol mobilization is an ongoing process. On the average, about 10% of the TAGs in adipose tissue are replaced daily by new triacylglycerol molecules.

25.3 Glycerol Metabolism

During triacylglycerol mobilization, one molecule of glycerol is produced for each triacylglycerol completely hydrolyzed. Glycerol metabolism primarily involves processes considered in the previous chapter. After entering the bloodstream, glycerol travels to the liver or kidneys, where it is converted, in a two-step process, to dihydroxyacetone phosphate.

$$\begin{array}{c} H_2C-OH \\ | \\ HC-OH \\ | \\ H_2C-OH \\ \text{Glycerol} \end{array} \xrightarrow[\text{ATP} \rightarrow \text{ADP}]{\text{Glycerol kinase}} \begin{array}{c} H_2C-OH \\ | \\ HC-OH \\ | \\ H_2C-O-Ⓟ \\ \text{Glycerol 3-phosphate} \end{array} \xrightarrow[\text{NAD}^+ \rightarrow \text{NADH/H}^+]{\text{Glycerol 3-phosphate dehydrogenase}} \begin{array}{c} H_2C-OH \\ | \\ C=O \\ | \\ H_2C-O-Ⓟ \\ \text{Dihydroxyacetone phosphate} \end{array}$$

Figure 25.4 Hydrolysis of stored triacylglycerols in adipose tissue is triggered by hormones that stimulate cAMP production within adipose cells.

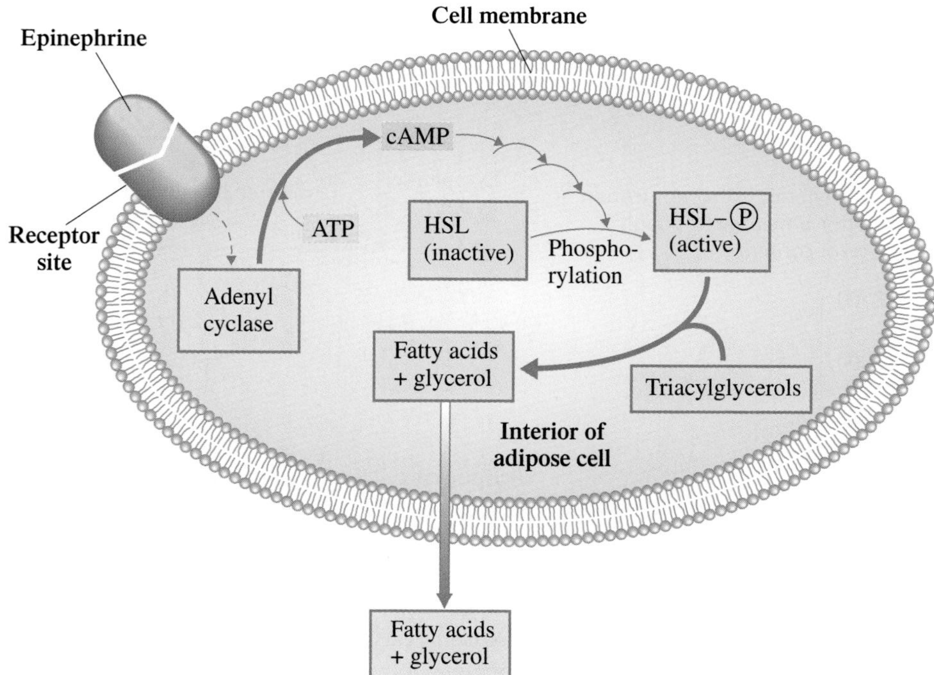

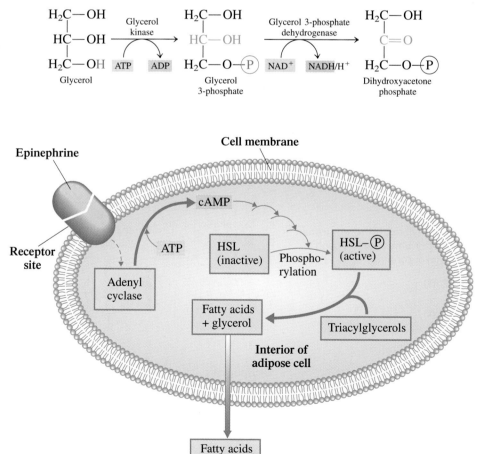

• The use of cAMP in the activation of hormone-sensitive lipase in adipose cells is similar to cAMP's role in the activation of the glycogenolysis process (Section 24.9).

• The stored TAGs in adipose tissue supply approximately 60% of the body's energy needs when the body is in a resting state.

The first step involves phosphorylation of a primary hydroxyl group of the glycerol. In the second step, glycerol's secondary alcohol group (C-2) is oxidized to a ketone.

Dihydroxyacetone phosphate is an intermediate in both glycolysis (Section 24.2) and gluconeogenesis (Section 24.6). It can be broken down to form pyruvate, then acetyl CoA, and finally carbon dioxide, or it can be used to form glucose. Dihydroxyacetone phosphate formation represents the first of several situations we will consider where carbohydrate and lipid metabolism are connected.

25.4 Oxidation of Fatty Acids

There are three parts to the process by which fatty acids are broken down to obtain energy.

1. The fatty acid must be *activated* by bonding to coenzyme A.
2. The fatty acid must be *transported* into the mitochondrial matrix by a shuttle mechanism.
3. The fatty acid must be repeatedly *oxidized,* cycling through a series of four reactions, to produce acetyl CoA, $FADH_2$, and NADH.

• Fatty Acid Activation

The outer mitochondrial membrane is the site of fatty acid activation, the first stage of fatty acid oxidation. Here the fatty acid is converted to a high-energy derivative of coenzyme A. Reactants are the fatty acid, coenzyme A, and a molecule of ATP.

$$\underset{\text{Free fatty acid}}{R-\overset{\overset{\large O}{\|}}{C}-O^-} + HS-CoA \xrightarrow[\text{ATP} \;\to\; \text{AMP} + 2P_i]{\text{Acyl CoA synthetase}} \underset{\text{Acyl CoA}}{R-\overset{\overset{\large O}{\|}}{C}-S-CoA}$$

This reaction requires the expenditure of two high-energy phosphate bonds from a single ATP molecule; the ATP is converted to AMP rather than ADP, and the resulting pyrophosphate (PP_i) is hydrolyzed to $2P_i$.

The activated fatty acid–CoA molecule is called *acyl* CoA. The difference between the designations *acyl* CoA and *acetyl* CoA is that *acyl* refers to a random-length fatty acid carbon chain that is covalently bonded to coenzyme A, whereas *acetyl* refers to a two-carbon chain covalently bonded to coenzyme A.

• *Acyl* is a generic term for

$$R-\overset{\overset{\large O}{\|}}{C}-$$

which is the species formed when the carboxyl —OH is removed from an organic acid. The R group can involve a carbon chain of any length.

$$\underset{\substack{\text{Acyl CoA}\\ \text{R = carbon chain of any length}}}{R-\overset{\overset{\large O}{\|}}{C}-S-CoA} \qquad \underset{\substack{\text{Acetyl CoA}\\ \text{R = CH}_3\text{ group}}}{CH_3-\overset{\overset{\large O}{\|}}{C}-S-CoA}$$

• Fatty Acid Transport

Acyl CoA is too large to pass through the inner mitochondrial membrane to the mitochondrial matrix, where the enzymes needed for fatty acid oxidation are located. A shuttle mechanism involving the molecule carnitine effects the entry of acyl CoA into the matrix (see Figure 25.5).

• The Fatty Acid Spiral

In the mitochondrial matrix, a sequence of four reactions *repeatedly* cleaves two-carbon units from the carboxyl end of the fatty acid. This process is called the *fatty acid spiral,* because of its repetitive nature, or *β oxidation spiral,* because the second, or beta, carbon from the carboxyl end of the chain is oxidized. The **fatty acid spiral** *is a repetitive series of four reactions in which each sequence produces acetyl CoA, $FADH_2$, NADH, and a fatty acid that is shorter than the previous fatty acid by two carbon atoms.*

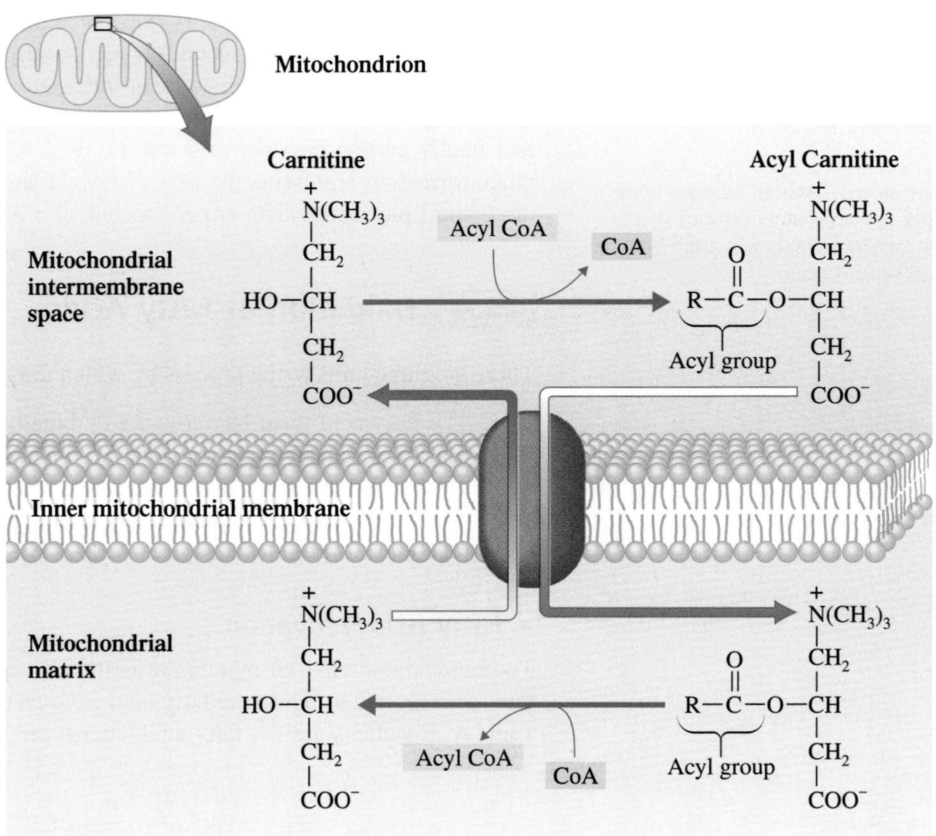

Figure 25.5
Fatty acids are transported across the inner mitochondrial membrane in the form of acyl carnitine.

For a *saturated* fatty acid, the fatty acid spiral involves the following functional group changes at the β carbon and the following reaction types.

• We have encountered this identical set of functional group changes before, in the back side of the citric acid cycle (Section 23.6).

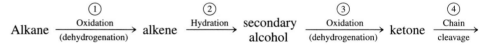

$$\text{Alkane} \xrightarrow[\text{(dehydrogenation)}]{\text{① Oxidation}} \text{alkene} \xrightarrow{\text{② Hydration}} \text{secondary alcohol} \xrightarrow[\text{(dehydrogenation)}]{\text{③ Oxidation}} \text{ketone} \xrightarrow[\text{cleavage}]{\text{④ Chain}}$$

Details about Steps 1–4 of the fatty acid spiral follow.

Step 1: *Oxidation (dehydrogenation).* Hydrogen atoms are removed from the α and β carbons, creating a double bond between these two carbon atoms. FAD is the oxidizing agent, and a $FADH_2$ molecule is a product.

$$\underset{\text{Acyl CoA}}{R{-}\overset{\beta}{C}H_2{-}\overset{\alpha}{C}H_2{-}\overset{O}{\overset{\|}{C}}{-}S{-}CoA} \xrightarrow[\text{FAD} \rightarrow \text{FADH}_2]{\text{Acyl CoA dehydrogenase}} \underset{\textit{trans}\text{-Enoyl CoA}}{R{-}\overset{\beta}{C}H{=}\overset{\alpha}{C}H{-}\overset{O}{\overset{\|}{C}}{-}S{-}CoA}$$

The enzyme involved is stereospecific in that only *trans* double bonds are produced.

Step 2: *Hydration.* A molecule of water is added across the *trans* double bond, producing a secondary alcohol at the β-carbon position. Again, the enzyme involved is stereospecific in that only the L-hydroxy isomer is produced from the *trans* double bond.

$$\underset{\textit{trans}\text{-Enoyl CoA}}{R{-}CH{=}CH{-}\overset{O}{\overset{\|}{C}}{-}S{-}CoA} \xrightarrow[H_2O]{\text{Enoyl CoA hydratase}} \underset{\text{L-}\beta\text{-Hydroxyacyl CoA}}{R{-}\overset{OH}{\overset{|}{C}}H{-}CH_2{-}\overset{O}{\overset{\|}{C}}{-}S{-}CoA}$$

The enzyme involved in this hydration will also hydrate a *cis* double bond, but the product then is the D isomer. We shall return to this point later in considering how unsaturated fatty acids are oxidized.

• The reaction sequence dehydrogenation–hydration–dehydrogenation in the fatty acid spiral has a parallel in Steps 6–8 of the citric acid cycle (Section 23.6), where succinate is dehydrogenated to fumarate, which is hydrated to malate, which is dehydrogenated to oxaloacetate.

Step 3: *Oxidation (dehydrogenation).* The β-hydroxy group is oxidized to a ketone functional group with NAD^+ serving as the oxidizing agent. The required enzyme exhibits absolute stereospecificity for the L isomer.

$$R{-}\underset{H}{\overset{OH}{C}}{-}\underset{H}{\overset{H}{C}}{-}\overset{O}{\overset{\|}{C}}{-}S{-}CoA \xrightarrow[NAD^+ \;\; NADH + H^+]{\beta\text{-Hydroxyacyl CoA dehydrogenase}} R{-}\overset{O}{\overset{\|}{C}}{-}CH_2{-}\overset{O}{\overset{\|}{C}}{-}S{-}CoA$$

L-β-Hydroxyacyl CoA $\qquad$ β-Ketoacyl CoA

It is now apparent why one of the names for this series of reactions is β oxidation spiral. The β-carbon atom has been oxidized from a $-CH_2-$ group to a ketone group.

Step 4: *Chain Cleavage.* The fatty acid chain is broken between the α and β carbons by reaction with a coenzyme A molecule. The result is an acetyl CoA molecule and a new acyl CoA molecule that is shorter by two carbon atoms than its predecessor.

$$R{-}\overset{O}{\overset{\|}{C}}{-}CH_2{-}\overset{O}{\overset{\|}{C}}{-}S{-}CoA \xrightarrow[CoA{-}SH]{\text{Thiolase}} R{-}\overset{O}{\overset{\|}{C}}{-}S{-}CoA + CH_3{-}\overset{O}{\overset{\|}{C}}{-}S{-}CoA$$

β-Ketoacyl CoA $\qquad$ Acyl CoA with two fewer carbon atoms $\qquad$ Acetyl CoA

The new acyl CoA molecule (now shorter by two carbons) is *recycled* through the same set of four reactions again. This yields another acetyl CoA, a two-carbon-shorter new acyl CoA, $FADH_2$, and NADH. Recycling occurs again and again, until the entire fatty acid is converted to acetyl CoA. Thus the fatty acid carbon chain is sequentially degraded, two carbons at a time.

• This sequence of reactions is called the fatty acid *spiral* rather than the fatty acid *cycle* because a different product results from each turn.

Figure 25.6 summarizes the reactions of the fatty acid spiral for stearic acid (18:0) as the starting fatty acid.

The fatty acids normally found in dietary triacylglycerols contain an *even* number of carbon atoms. Thus the number of acetyl CoA molecules produced in the fatty acid spiral is equal to half the number of carbon atoms in the fatty acid. The number of *turns* of the fatty acid spiral required to produce the acetyl CoA is always one less than the number of acetyl CoA molecules produced, because the last turn produces two acetyl CoA molecules as a C_4 unit splits into two C_2 units.

$$C_{18}\text{ fatty acid} \longrightarrow 9\text{ acetyl CoA (8 cycles)}$$
$$C_{14}\text{ fatty acid} \longrightarrow 7\text{ acetyl CoA (6 cycles)}$$

• Unsaturated Fatty Acids

Unsaturated fatty acids are common components of dietary triacylglycerols. Their oxidation through the fatty acid spiral requires two additional enzymes besides those needed for oxidation of saturated fatty acids. These two—an epimerase that can change a D configuration to an L configuration and a *cis–trans* isomerase—are needed for two reasons. First, the double bonds in naturally occurring unsaturated fatty acids are almost always *cis* double bonds, which yield on hydration a D-hydroxy product rather than the L-hydroxy product needed for Step 3 of the spiral. The epimerase enzyme effects a configuration change from the D form to the L form.

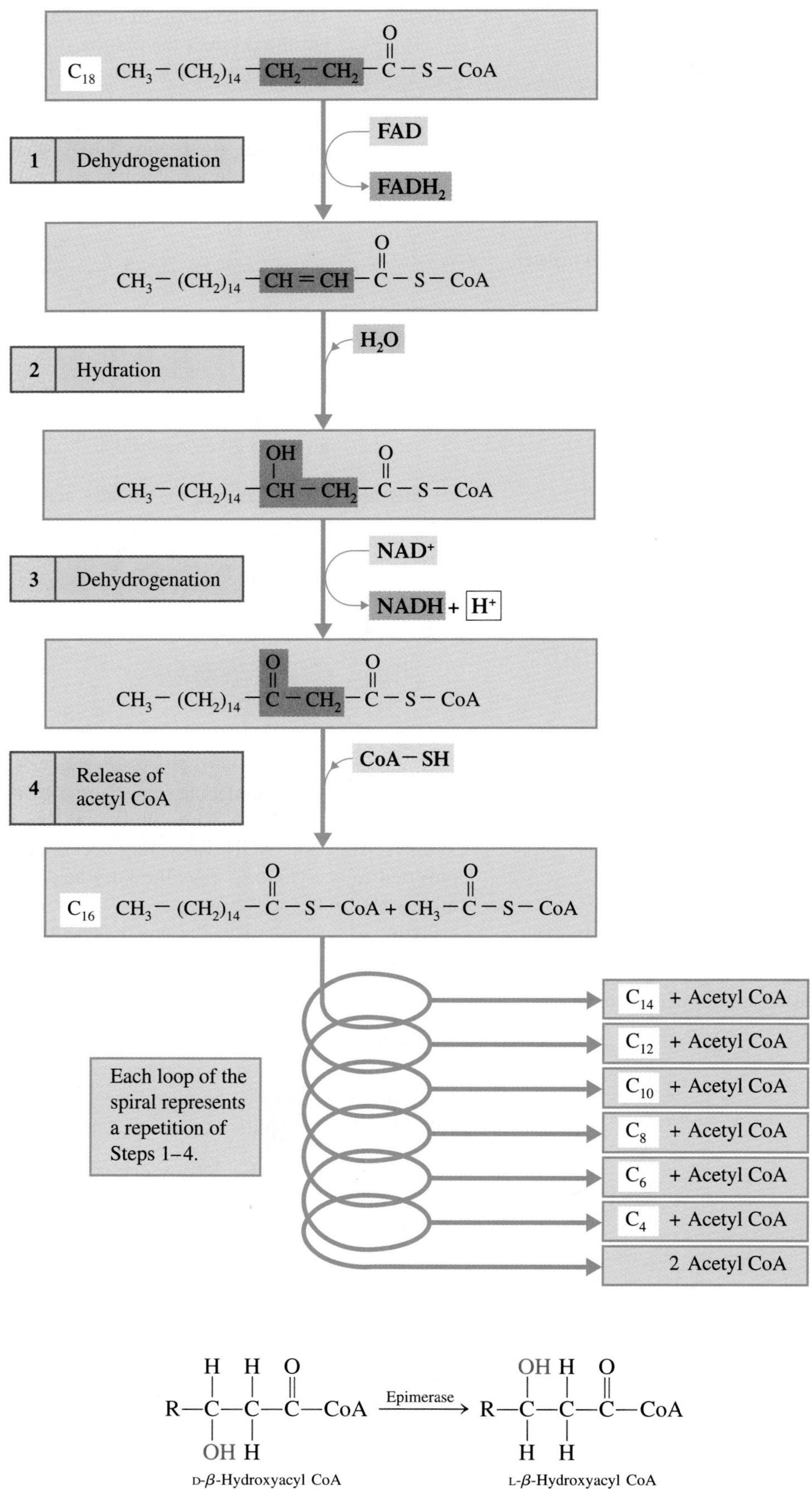

Figure 25.6
Reactions of the fatty acid spiral for an 18:0 fatty acid (stearic acid).

$$\underset{\text{D-}\beta\text{-Hydroxyacyl CoA}}{R-\underset{OH}{\overset{H}{C}}-\underset{H}{\overset{H}{C}}-\overset{O}{\overset{\|}{C}}-CoA} \xrightarrow{\text{Epimerase}} \underset{\text{L-}\beta\text{-Hydroxyacyl CoA}}{R-\underset{H}{\overset{OH}{C}}-\underset{H}{\overset{H}{C}}-\overset{O}{\overset{\|}{C}}-CoA}$$

Second, the double bonds in naturally occurring unsaturated fatty acids often occupy odd-numbered positions (Section 19.2). The hydratase in Step 2 of the fatty acid spiral can effect only an even-numbered double bond. The *cis–trans* isomerase produces a *trans*-(2,3) double bond from a *cis*-(3,4) double bond.

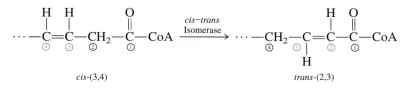

$$\cdots-\overset{\text{H}}{\overset{|}{\text{C}}}=\overset{\text{H}}{\overset{|}{\text{C}}}-\text{CH}_2-\overset{\text{O}}{\overset{\|}{\text{C}}}-\text{CoA} \xrightarrow[\text{Isomerase}]{\textit{cis–trans}} \cdots-\text{CH}_2-\underset{\text{H}}{\underset{|}{\text{C}}}=\overset{\text{H}}{\overset{|}{\text{C}}}-\overset{\text{O}}{\overset{\|}{\text{C}}}-\text{CoA}$$

cis-(3,4) *trans*-(2,3)

The Step 2 hydratase can then work on the *trans*-(2,3) double bond in the normal fashion.

25.5 ATP Production from Fatty Acid Oxidation

How does the total energy output from fatty acid oxidation compare to that of glucose oxidation? Let us calculate ATP production for the oxidation of a specific fatty acid molecule, stearic acid (18:0), and compare it with that from glucose.

Figure 25.6 shows that for all turns of the fatty acid spiral except the last turn, one $FADH_2$ molecule, one NADH molecule, and one acetyl CoA molecule are produced. In the final turn, two acetyl CoA molecules are produced in addition to the $FADH_2$ and NADH molecules.

Eight turns of the fatty acid spiral are required for the oxidation of stearic acid, an 18-carbon acid. These eight turns of the spiral produce 9 acetyl CoA molecules, 8 $FADH_2$ molecules, and 8 NADH molecules. Further processing of these products through the common metabolic pathway (citric acid cycle, electron transport chain, and oxidative phosphorylation) leads to ATP production as follows:

$$9 \text{ acetyl CoA} \times \frac{12 \text{ ATP}}{1 \text{ acetyl CoA}} = 108 \text{ ATP}$$

$$8 \text{ FADH}_2 \times \frac{2 \text{ ATP}}{1 \text{ FADH}_2} = 16 \text{ ATP}$$

$$8 \text{ NADH} \times \frac{3 \text{ ATP}}{1 \text{ NADH}} = 24 \text{ ATP}$$

$$148 \text{ ATP}$$

The conversion factors used in this calculation were first presented in Section 23.8.

This *gross* production of 148 ATP must be decreased by the ATP needed to activate the fatty acid before it enters the fatty acid spiral. The activation consumes two high-energy phosphate bonds of an ATP molecule. For accounting purposes, this is equivalent to hydrolyzing 2 ATP molecules to ADP. Thus the *net* ATP production from oxidation of stearic acid is 146 ATP (148 minus 2).

The comparison between complete fatty acid oxidation and complete glucose oxidation shows that a stearic acid molecule produces more than four times as much ATP as a glucose molecule.

$$1 \text{ glucose} \longrightarrow 36 \text{ ATP (Section 24.4)}$$

$$1 \text{ stearic acid} \longrightarrow 146 \text{ ATP}$$

Taking into account the fact that glucose has only 6 carbon atoms and stearic acid has 18 carbon atoms still shows more ATP production from the fatty acid.

$$3 \text{ glucose (18 C)} \longrightarrow 108 \text{ ATP}$$

$$1 \text{ stearic acid (18 C)} \longrightarrow 146 \text{ ATP}$$

On an equal-mass basis, fatty acids produce more than twice as much energy per gram as carbohydrates (glucose); this is shown by the following calculation involving 1.00 gram of stearic acid and 1.00 gram of glucose.

Chemical CONNECTIONS

25.1 High-Intensity Versus Low-Intensity Workouts

In a resting state, the human body burns more fat than carbohydrate. The fuel consumed is about one-third carbohydrate and two-thirds fat.

Information about fuel consumption ratios is obtainable from respiratory gas measurements, specifically from the respiratory exchange ratio (RER). The RER is the ratio of carbon dioxide to oxygen inhaled divided by the ratio of carbon dioxide to oxygen exhaled. For 100% fat burning, the RER would be 0.7; for 100% carbohydrate burning, the RER would be 1.0.

When a person at rest begins exercising, his or her body suddenly needs energy at a faster rate—more fuel and more oxygen are needed. It takes 0.7 L of oxygen to burn 1 gram of carbohydrate and 1.0 L of oxygen to burn 1 gram of fat. At the onset of exercise, the body is immediately short of oxygen. Also, there is a time delay in triacylglycerol mobilization. Triacylglycerols have to be broken down to fatty acids, which have to be attached to protein carriers before they can be carried in the bloodstream to working muscles. At their destination, they must be released from the carriers and then undergo energy-producing reactions. By contrast, glycogen is already present in muscle cells, and it can release glucose 6-phosphate as an instant fuel.

The initial stages of exercise are fueled primarily by glucose; in later stages, triacylglycerols become the primary fuel.

Consequently, the initial stages of exercise are fueled primarily by glucose—it requires less oxygen and can even be burned anaerobically (to lactate). During the first few minutes of exercise, up to 80% of the fuel used comes from glycogen.

With time, increased breathing rates increase oxygen supplies to muscles, and triacylglycerol use increases. Continued activity for three-quarters of an hour achieves a 50–50 balance of triacylglycerol and glucose use. Beyond an hour, triacylglycerol use may be as high as 80%.

Suppose a person is exercising at a moderate rate and decides to speed up. Immediately, body fuel and oxygen needs are increased. The response is increased use of glycogen supplies.

The accompanying table compares exercise on a stationary cycle at 45% and 70% of maximal oxygen uptake sufficient to burn 300 calories.

	Low-intensity exercise	High-intensity exercise
percent of maximum oxygen uptake	45%	70%
time required to burn 300 calories	48 min	30 min
calories obtained from fat	133 cal	65 cal
percent of calories from fat	44%	22%
rate of fat burning per minute	2.8 cal/min	2.1 cal/min

$$1.00\ \cancel{\text{g stearic acid}} \times \left(\frac{1\ \cancel{\text{mole stearic acid}}}{284\ \cancel{\text{g stearic acid}}}\right) \times \left(\frac{146\ \text{moles ATP}}{1\ \cancel{\text{mole stearic acid}}}\right) = 0.514\ \text{mole ATP}$$

$$1.00\ \cancel{\text{g glucose}} \times \left(\frac{1\ \cancel{\text{mole glucose}}}{180\ \cancel{\text{g glucose}}}\right) \times \left(\frac{36\ \text{moles ATP}}{1\ \cancel{\text{mole glucose}}}\right) = 0.200\ \text{mole ATP}$$

The fact that fatty acids (stearic acid) yield more than twice as much energy per gram as carbohydrates (glucose) means that the former "do twice as much damage" to a person on a diet.

In dietary calculations, we say that 1 gram of carbohydrate equals 4 kcal and that 1 gram of fat equals 9 kcal. We now know the basis for these numbers. The figure of 9 kcal for fat takes into account the fact that not all fatty acids present in fat contain 18 carbon atoms (the basis for our preceding calculations) and also the fact that fats contain glycerol, which produces ATP when degraded.

Is the preferred fuel for "running" the human body fatty acids, which yield twice as much energy per gram as glucose, or is it glucose? In a normally functioning human body, certain organs use both fuels, others prefer glucose, and still others prefer fatty acids. Here are some generalizations about "fuel" use:

1. Skeletal muscle uses glucose (from glycogen) when in an active state. In a resting state, it uses fatty acids.
2. Cardiac muscle depends first on fatty acids and secondarily on ketone bodies (Section 25.6), glucose, and lactate.
3. The liver uses fatty acids as the preferred fuel.
4. Brain function is maintained by glucose and ketone bodies (Section 25.6). Fatty acids cannot cross the blood–brain barrier and thus are unavailable.
5. Red blood cells use only glucose. Such cells lack mitochondria and thus cannot obtain energy via oxidative phosphorylation.

25.6 Ketone Bodies

Ordinarily, when there is adequate balance between lipid and carbohydrate metabolism, most of the acetyl CoA produced from the fatty acid spiral is further processed through the citric acid cycle. The first step of the citric acid cycle (Section 23.6) involves the reaction between oxaloacetate and acetyl CoA. Sufficient oxaloacetate must be present for the acetyl CoA to react with. Oxaloacetate concentration depends on pyruvate produced from glycolysis (Section 24.2); pyruvate can be converted to oxaloacetate by *pyruvate carboxylase* (Section 24.6).

Certain body conditions upset the lipid–carbohydrate balance required for acetyl CoA generated by fatty acids to be processed by the citric acid cycle. These conditions include (1) dietary intakes high in fat and low in carbohydrates, (2) diabetic conditions where the body cannot adequately process glucose even though it is present, and (3) *prolonged* fasting conditions, including starvation, where glycogen supplies are exhausted. Under these conditions, the problem of inadequate oxaloacetate arises, which is compounded by the body's using oxaloacetate that is present to produce glucose through gluconeogenesis (Section 24.6).

• Ketone bodies are produced when the amount of acetyl CoA is excessive compared with the amount of oxaloacetate available to react with it (Step 1 of the citric acid cycle).

What happens when oxaloacetate supplies are too low for all acetyl CoA present to be processed through the citric acid cycle? The excess acetyl CoA is diverted to the formation of ketone bodies. **Ketone bodies** *are the substances acetoacetate, β-hydroxybutyrate, and acetone.* The structures of these substances are

$$CH_3-\overset{\overset{O}{\|}}{C}-CH_2-\overset{\overset{O}{\|}}{C}-O^-$$

Acetoacetate

$$CH_3-\overset{\overset{OH}{|}}{CH}-CH_2-\overset{\overset{O}{\|}}{C}-O^-$$

β-Hydroxybutyrate

$$CH_3-\overset{\overset{O}{\|}}{C}-CH_3$$

Acetone

The structure of β-hydroxybutyrate does not actually include a ketone group, but it is still classified as a ketone body.

For a number of years, ketone bodies were thought of as degradation products that had little physiological significance. It is now known that ketone bodies can serve as sources of energy for various tissues and are very important energy sources in heart muscle and the renal cortex. During starvation, the brain adapts to the use of ketone bodies for energy production.

• Ketogenesis

• Even when ketogenic conditions are not present in the human body, the liver produces a *small amount* of ketone bodies.

Ketogenesis *is the synthesis of ketone bodies from acetyl CoA.* The primary site for ketogenesis is liver mitochondria. After they are produced, the ketone bodies diffuse from these structures into the bloodstream, where they are transported to peripheral tissues.

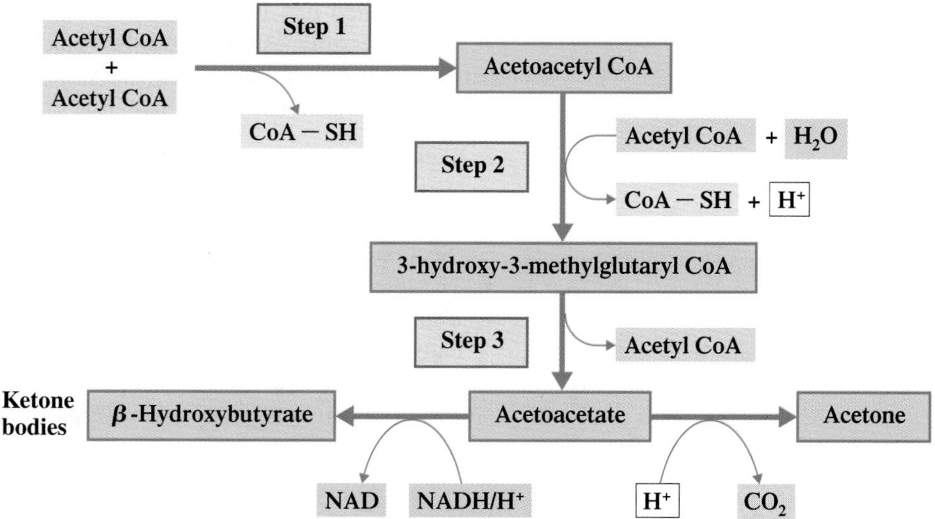

Figure 25.7 Ketogenesis involves the production of ketone bodies from acetyl CoA.

The reactions that constitute ketogenesis are shown in Figure 25.7. Ketogenesis begins as two acetyl CoA molecules combine to produce acetoacetyl CoA, a reversal of the last step of the fatty acid spiral (Section 25.4).

$$\underset{\text{Acetyl CoA}}{CH_3-\overset{O}{\overset{\|}{C}}-S-CoA} + \underset{\text{Acetyl CoA}}{CH_3-\overset{O}{\overset{\|}{C}}-S-CoA} \xrightarrow{\text{Thiolase}} \underset{\text{Acetoacetyl CoA}}{CH_3-\overset{O}{\overset{\|}{C}}-CH_2-\overset{O}{\overset{\|}{C}}-S-CoA} + CoA-SH$$

Acetoacetyl CoA then reacts with a third acetyl CoA and water to produce 3-hydroxy-3-methylglutaryl CoA (HMG-CoA) and CoA-SH.

$$\underset{\text{Acetoacetyl CoA}}{CH_3-\overset{O}{\overset{\|}{C}}-CH_2-\overset{O}{\overset{\|}{C}}-S-CoA} + \underset{\text{Acetyl CoA}}{CH_3-\overset{O}{\overset{\|}{C}}-S-CoA} + H_2O \xrightarrow[\text{synthase}]{\text{HMG-CoA}} \underset{\text{HMG-CoA}}{{}^-OOC-CH_2-\overset{OH}{\overset{|}{\underset{\underset{CH_3}{|}}{C}}}-CH_2-\overset{O}{\overset{\|}{C}}-S-CoA} + CoA-SH + H^+$$

HMG-CoA is then cleaved to acetyl CoA and acetoacetate.

$$\underset{\text{HMG-CoA}}{{}^-OOC-CH_2-\overset{OH}{\overset{|}{\underset{\underset{CH_3}{|}}{C}}}-CH_2-\overset{O}{\overset{\|}{C}}-S-CoA} \xrightarrow[\text{lyase}]{\text{HMG-CoA}} \underset{\text{Acetoacetate}}{{}^-OOC-CH_2-\overset{O}{\overset{\|}{C}}-CH_3} + \underset{\text{Acetyl CoA}}{CH_3-\overset{O}{\overset{\|}{C}}-S-CoA}$$

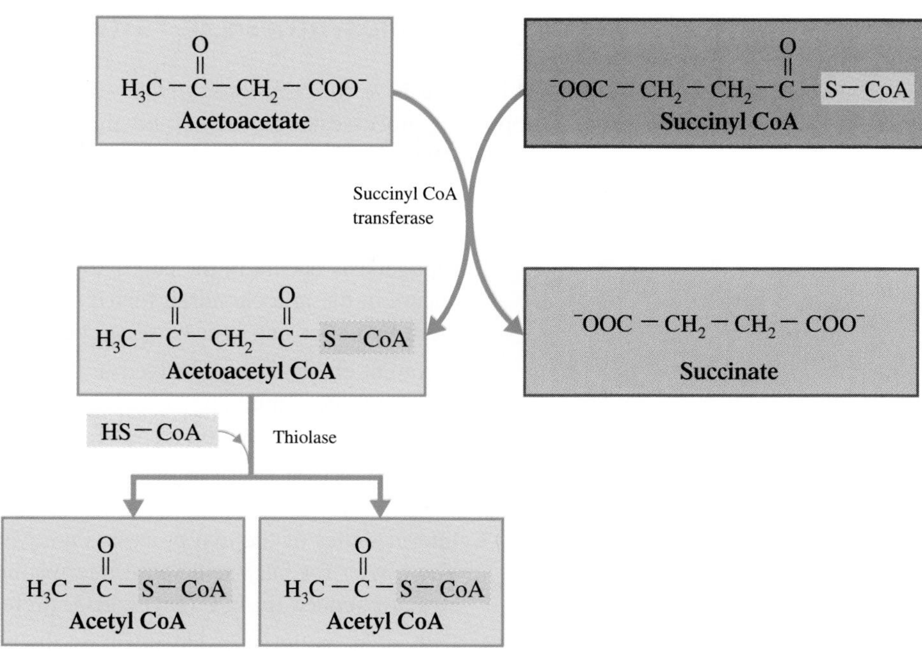

Figure 25.8
The pathway for utilization of acetoacetate as a fuel. The required succinyl CoA comes from the citric acid cycle.

Summing these three reactions to obtain the net reaction for ketogenesis yields

$$2 \text{ Acetyl CoA} + H_2O \longrightarrow \text{acetoacetate} + 2 \text{ CoA} + H^+$$

The ketone body acetoacetate is the "parent" compound for the other two ketone bodies. Acetone arises from acetoacetate by the loss of the carboxyl group (as CO_2). Reduction of the keto group of acetoacetate to a hydroxyl group by NADH produces β-hydroxybutyrate. The amount of acetone present is usually small compared with the other two species.

For acetoacetate to be used as a fuel—in heart muscle, for example—it must first be activated. Acetoacetate is activated by transfer of a CoA group from succinyl CoA (a citric acid cycle intermediate). The resulting acetoacetyl CoA is then cleaved to give two acetyl CoA molecules that can enter the citric acid cycle (see Figure 25.8). In effect, acetoacetate is a water-soluble transportable form of acetyl units.

Ketosis

Under normal metabolic conditions (an appropriate glucose–fatty acid balance), the concentration of ketone bodies in the blood is very low—about 1 mg/100 mL. Abnormal metabolic conditions, such as those mentioned at the start of this section, produce elevated blood ketone levels, levels 50–100 times greater than normal. Excess accumulation of ketone bodies in blood (20 mg/100 mL) is called *ketonemia.* At a level of 70 mg/100 mL, the renal threshold is exceeded, and ketone bodies are excreted in the urine, a condition called *ketonuria.* The overall accumulation of ketone bodies in the blood and urine is called *ketosis.* Ketosis is often detectable by the smell of acetone on a person's breath; acetone is very volatile and is excreted through the lungs.

Two of the three ketone bodies—acetoacetate and β-hydroxybutyrate — are acids. Their presence in blood causes a slight but significant decrease in blood pH. This can result in acidosis (Section 10.12) in severe ketosis situations. Symptoms include heavy breathing (because acidic blood can carry less oxygen) and increased urine output that can lead to dehydration. Ultimately, the condition can cause coma and death.

Acidosis from elevated ketone body levels is often called keto acidosis or metabolic acidosis to distinguish it from respiratory acidosis (Section 10.12), which is not linked to ketone bodies.

25.7 Biosynthesis of Fatty Acids: Lipogenesis

Lipogenesis *is the synthesis of fatty acids from acetyl CoA.* As was the case for the opposing processes of glycolysis and gluconeogenesis, lipogenesis is not simply a reversal of the steps for degradation of fatty acids (the fatty acid spiral). Before we look at the details of fatty acid synthesis, we will consider some differences between the synthesis and degradation of fatty acids.

1. Lipogenesis occurs in the cell cytoplasm, whereas degradation of fatty acids occurs in the mitochondrial matrix. Because they have different reaction sites, these two opposing processes can occur at the same time when necessary.
2. Different enzymes are involved in the two processes. Lipogenesis enzymes are collected into a multienzyme complex called *fatty acid synthase.* This enzyme complex ties the reaction steps of lipogenesis closely together. The enzymes involved in fatty acid degradation are not physically associated, so the reaction steps are independent.
3. Intermediates of the two processes are covalently bonded to different carriers. The carrier for fatty acid spiral intermediates is CoA. Lipogenesis intermediates are bonded to ACP (acyl carrier protein).
4. Fatty acid synthesis is dependent on the reducing agent NADPH. Fatty acid degradation is dependent on the oxidizing agents FAD and NAD^+.
5. Fatty acids are built up two carbons at a time during synthesis and are broken down two carbons at a time during degradation. The source of the two carbon units differs between the two processes. In lipogenesis, acetyl CoA is used to form malonyl ACP, which becomes the carrier of the two carbon units. CoA derivatives are involved in all steps of the fatty acid spiral.

In general, fatty acid biosynthesis (lipogenesis) occurs any time dietary intake provides more nutrients than are needed for energy requirements. The primary lipogenesis sites are the liver, adipose tissue, and mammary glands. The mammary glands show increased synthetic activity during periods of lactation.

• Formation of Malonyl CoA

Acetyl CoA is the starting material for fatty acid synthesis. Because acetyl CoA is generated in mitochondria and fatty acid synthesis occurs in cytoplasm, the acetyl CoA must first be transported to the cytoplasm. It exits the mitochondria through a transport system that involves citrate ion.

Mitochondrial acetyl CoA reacts with oxaloacetate (the first step of the citric acid cycle) to produce citrate, which is then transported through the mitochondrial membrane.

$$\text{Acetyl CoA} + \text{oxaloacetate} + H_2O \longrightarrow \text{citrate} + \text{coenzyme A}$$

Once in the cytoplasm, the citrate undergoes the reverse reaction to regenerate acetyl CoA; an ATP molecule is required to drive the shuttle process.

Cytoplasmic acetyl CoA is then converted to malonyl CoA in a carboxylation reaction that involves carbon dioxide (CO_2) and ATP.

• The parent compound for the malonyl group is malonic acid, the C_3 dicarboxylic acid

$$HO-\overset{O}{\overset{\|}{C}}-CH_2-\overset{O}{\overset{\|}{C}}-OH$$

$$\underset{\text{Acetyl CoA}}{CH_3-\overset{O}{\overset{\|}{C}}-S-CoA} + CO_2 \xrightarrow[\text{ATP} \quad \text{ADP} + P_i]{} \underset{\text{Malonyl CoA}}{{}^-O-\overset{O}{\overset{\|}{C}}-CH_2-\overset{O}{\overset{\|}{C}}-S-CoA}$$

This reaction occurs only when cellular ATP levels are high. It is catalyzed by *acetyl CoA carboxylase complex,* which requires both Mn^{2+} ion and the B vitamin biotin for its activity.

• ACP Complex Formation

Studies show that all intermediates in fatty acid synthesis are linked to acyl carrier proteins (ACP—SH) rather than to CoA—SH. Even the small C_2 acetyl and C_3 malonyl groups are bound to such carriers.

$$\text{Acetyl CoA} + \text{ACP—SH} \longrightarrow \text{acetyl ACP} + \text{CoA—SH}$$
$$\text{Malonyl CoA} + \text{ACP—SH} \longrightarrow \text{malonyl ACP} + \text{CoA—SH}$$

ACP—SH can be regarded as a "giant CoA—SH molecule." Involved in its structure are the 2-ethanethiol and pantothenic acid components of CoA—SH (Section 23.3) attached to a polypeptide chain containing 77 amino acid residues.

• Chain Elongation

Four reactions that occur in a cyclic pattern within the multienzyme *fatty acid synthase complex* constitute the chain elongation process used for fatty acid synthesis. The reactions of the *first* turn of the cycle, in general terms, are shown in Figure 25.9. Specific details about this series of reactions follow.

Step 1: *Condensation.* Acetyl ACP and malonyl ACP condense together to form acetoacetyl ACP.

$$\underset{\text{Acetyl ACP}}{CH_3\text{—}\overset{\overset{\displaystyle O}{\|}}{C}\text{—S—ACP}} + \underset{\text{Malonyl ACP}}{{}^{-}O\text{—}\overset{\overset{\displaystyle O}{\|}}{C}\text{—}CH_2\text{—}\overset{\overset{\displaystyle O}{\|}}{C}\text{—S—ACP}} \longrightarrow \underset{\text{Acetoacetyl ACP}}{CH_3\text{—}\overset{\overset{\displaystyle O}{\|}}{C}\text{—}CH_2\text{—}\overset{\overset{\displaystyle O}{\|}}{C}\text{—S—ACP}} + CO_2 + \text{ACP—SH}$$

Note that a C_2 species (acetyl) and a C_3 species (malonyl) react to produce a C_4 species (acetoacetyl) rather than a C_5 species. One carbon atom leaves the reaction in the form of a CO_2 molecule.

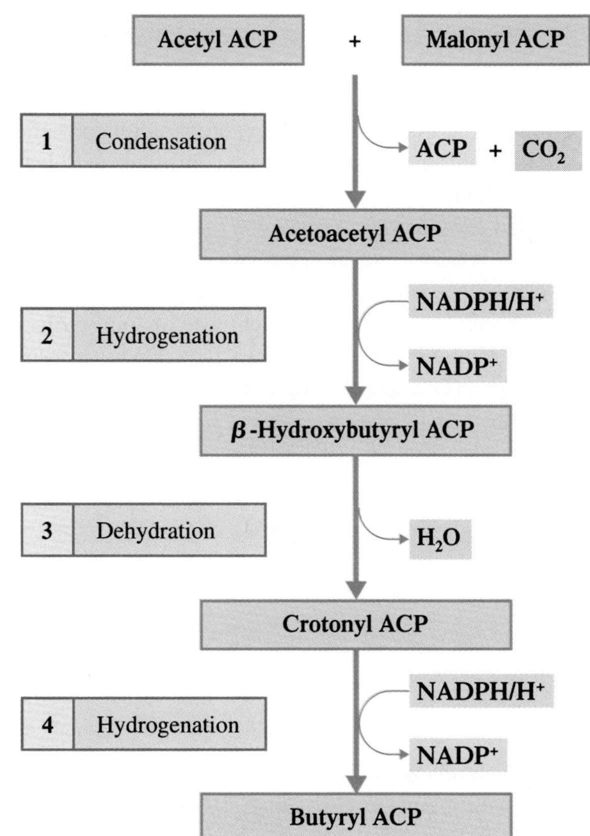

Figure 25.9
In the first cycle of the fatty acid biosynthetic pathway, acetyl ACP is converted to butyryl ACP. In the next cycle (not shown), the butyryl ACP reacts with another malonyl ACP to produce a 6-carbon acid. Continued cycles produce acids with 8, 10, 12, 14, and 16 carbon atoms.

Steps 2 through 4 involve a sequence of functional-group changes that we have encountered twice before—in the fatty acid spiral (Section 25.4) and in the citric acid cycle (Section 23.6). This time, however, the changes occur in the reverse sequence to that previously encountered. The functional-group changes are

$$\text{Ketone} \xrightarrow[\text{(hydrogenation)}]{\text{② Reduction}} \begin{array}{c}\text{secondary}\\\text{alcohol}\end{array} \xrightarrow{\text{③ Dehydration}} \text{alkene} \xrightarrow[\text{(hydrogenation)}]{\text{④ Reduction}} \text{alkane}$$

Step 2: *Hydrogenation.* The keto group of the acetoacetyl complex, which involves the β-carbon atom, is reduced to the corresponding alcohol by NADPH.

$$\underset{\text{Acetoacetyl ACP}}{CH_3-\overset{\overset{O}{\|}}{C}-CH_2-\overset{\overset{O}{\|}}{C}-S-ACP} \xrightarrow[\text{NADPH/H}^+ \;\; \text{NADP}^+]{} \underset{\beta\text{-Hydroxybutyryl ACP}}{CH_3-\overset{\overset{OH}{|}}{CH}-CH_2-\overset{\overset{O}{\|}}{C}-S-ACP}$$

Step 3: *Dehydration.* The alcohol produced in Step 2 is dehydrated to introduce a double bond into the molecule (between the α and β carbons).

$$\underset{\beta\text{-Hydroxybutyryl ACP}}{CH_3-\overset{\overset{OH}{|}}{CH}-CH_2-\overset{\overset{O}{\|}}{C}-S-ACP} \xrightarrow[H_2O]{} \underset{\text{Crotonyl ACP}}{CH_3-CH\overset{trans}{=}CH-\overset{\overset{O}{\|}}{C}-S-ACP}$$

Step 4: *Hydrogenation.* The double bond introduced in Step 3 is converted to a single bond through hydrogenation. As in Step 2, NADPH is the reducing agent.

$$\underset{\text{Crotonyl ACP}}{CH_3-CH\overset{trans}{=}CH-\overset{\overset{O}{\|}}{C}-S-ACP} \xrightarrow[\text{NADPH/H}^+ \;\; \text{NADP}^+]{} \underset{\text{Butyryl ACP}}{CH_3-CH_2-CH_2-\overset{\overset{O}{\|}}{C}-S-ACP}$$

• Steps 2, 3, and 4 of fatty acid biosynthesis accomplish the reverse of Steps 3, 2, and 1 of the fatty acid spiral.

Further cycles of the preceding four-step process convert the four-carbon acyl group to a six-carbon acyl group, then to an eight-carbon acyl group, and so on (see Figure 25.10). Elongation of the acyl group chain through this procedure, which is tied to the fatty acid synthase complex, stops upon formation of the C_{16} acyl group (palmitic acid). Different enzyme systems and different cellular locations are required for elongation of the chain beyond C_{16} and for introduction of double bonds into the acyl group (unsaturated fatty acids).

A relatively large input of energy is needed to biosynthesize a fatty acid molecule, as can be seen from the data in Table 25.1, which gives a net summary of the reactants and products involved in the synthesis of one molecule of palmitic acid, the 16:0 fatty acid.

Production of unsaturated fatty acids (insertion of double bonds) requires molecular oxygen (O_2). In an oxidation step, hydrogen is removed and combined with the O_2 to form water.

Table 25.1
Reactants and Products in the Biosynthesis of One Molecule of Palmitic Acid, the 16:0 Fatty Acid

Reactants	Products
8 acetyl CoA	1 palmitate
7 ATP	8 coenzyme A
14 NADPH	7 ADP
6 H^+	7 P_i
	14 $NADP^+$
	6 H_2O

$$R-\overset{\overset{H}{|}}{\underset{\underset{H}{|}}{C}}-\overset{\overset{H}{|}}{\underset{\underset{H}{|}}{C}}-(CH_2)_n-\overset{\overset{O}{\|}}{C}-O^- + O_2 \xrightarrow[\text{NADPH/H}^+ \;\; \text{NADP}^+]{} R-\underset{\underset{H}{|}}{C}=\underset{\underset{H}{|}}{C}-(CH_2)_n-\overset{\overset{O}{\|}}{C}-O^- + 2H_2O$$

In humans and animals, enzymes can introduce double bonds only between C-4 and C-5 and between C-9 and C-10. Thus the important unsaturated fatty acids linoleic (C_{18} with C-9 and C-12 double bonds) and linolenic (C_{18} with C-9, C-12, and C-15 double

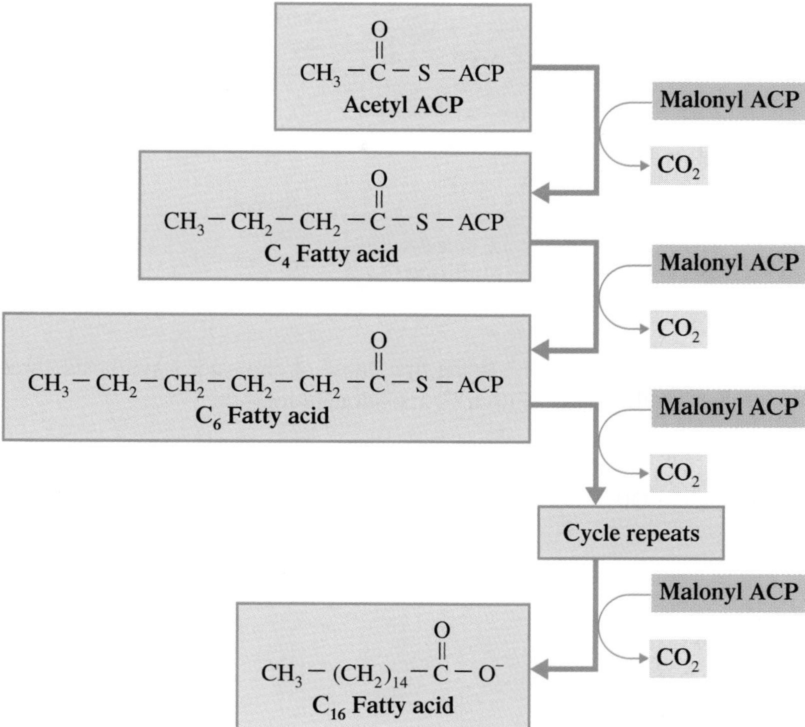

Figure 25.10
The sequence of cycles needed to produce a C_{16} fatty acid from acetyl ACP. Each loop represents one cycle.

bonds) cannot be biosynthesized. They must be obtained from the diet. (Plants have the enzymes necessary to synthesize these acids.)

Acids such as linoleic and linolenic, which cannot be synthesized by the body but are necessary for its proper functioning, are called *essential fatty acids* (Section 19.2). Linolenic acid is the starting material from which the body makes arachidonic acid, the precursor for prostaglandins, the hormone-like substances that regulate a wide range of body functions, including blood pressure (Section 19.11).

Lipogenesis can be used to convert glucose to fatty acids via acetyl CoA. The reverse process, conversion of fatty acids to glucose, is not possible within the human body. Fatty acids can be broken down to acetyl CoA, but there is no enzyme present for the conversion of acetyl CoA to pyruvate or oxaloacetate, starting materials for gluconeogenesis (Section 24.6). Plants and some bacteria do possess the needed enzymes and can thus convert fatty acids to carbohydrates.

25.8 Biosynthesis of Cholesterol

So far in this chapter, our discussion of lipid metabolism has focused on fats and oils (triacylglycerols) and their hydrolysis products, fatty acids and glycerol. We now consider another very important lipid—cholesterol.

Every membrane of every cell in the body has cholesterol as a necessary component. This substance is also the precursor for bile salts, sex hormones, and adrenal hormones (Section 19.10).

In today's health-conscious world, dietary intake of cholesterol is of great interest because of correlations between high serum cholesterol levels and coronary heart disease. Average daily dietary intake of cholesterol is approximately 0.3 gram. This amount, though important, is small compared to the 1.5–2.0 grams of cholesterol that the body synthesizes every day from acetyl CoA units.

The biosynthesis of cholesterol, a C_{27} molecule, occurs primarily in the liver. Its production consumes 15 molecules of acetyl CoA and involves at least 27 separate enzymatic steps. An overview of cholesterol synthesis is given in Figure 25.11.

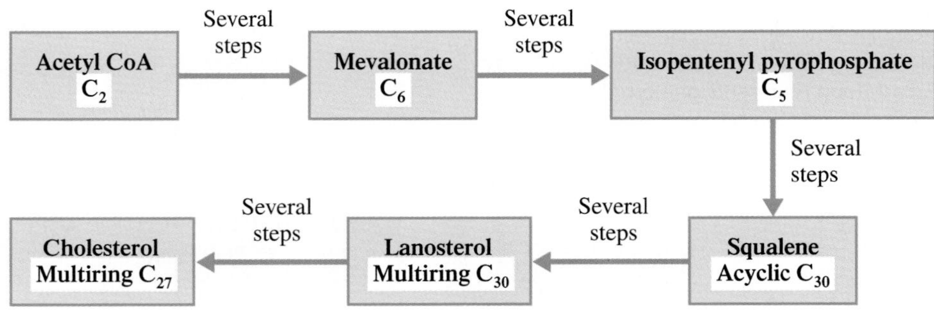

Figure 25.11 An overview of the biosynthetic pathway for cholesterol synthesis.

- The "parent" compound for mevalonate ion is mevalonic acid (3,5-dihydroxy-3-methylpentanoic acid).

$$\underset{\text{OH}}{\text{CH}_2}-\text{CH}_2-\underset{\text{OH}}{\overset{\text{CH}_3}{\text{C}}}-\text{CH}_2-\overset{\text{O}}{\overset{\|}{\text{C}}}-\text{OH}$$

In the first phase of cholesterol synthesis, three molecules of acetyl CoA are condensed into a C_6 mevalonate ion.

$$\underset{\text{Acetyl CoA}}{3\text{CH}_3-\overset{\text{O}}{\overset{\|}{\text{C}}}-\text{S}-\text{CoA}} \xrightarrow[\text{steps}]{\text{Several}} \underset{\text{Mevalonate}}{\text{COO}^- - \text{CH}_2-\text{C(OH)(CH}_3)-\text{CH}_2-\text{CH}_2\text{OH}}$$

The C_6 mevalonate undergoes a decarboxylation to yield a C_5 isoprene derivative called isopentenyl pyrophosphate and CO_2. Three ATP molecules are needed in accomplishing this process.

$$\underset{\text{Mevalonate}}{\text{COO}^- - \text{CH}_2-\text{C(OH)(CH}_3)-\text{CH}_2-\text{CH}_2\text{OH}} \xrightarrow[\text{steps}]{\text{Several}} \underset{\text{Isopentenyl pyrophosphate}}{\text{CH}_2{=}\text{C(CH}_3)-\text{CH}_2-\text{CH}_2\text{O}-(\text{P})-(\text{P})}$$

- Compounds whose structures are based on the five-carbon isoprene unit are called *terpenes* (Section 13.6).

The isoprene structural unit (Section 13.6), present in isoprene derivatives in a modified form, is a commonly used five-carbon building block in biosynthetic processes.

The next stage of cholesterol biosynthesis involves the condensation of six isoprene units to give the C_{30} squalene molecule.

$$\text{H}_3\text{C}-\overset{\text{CH}_3}{\text{C}}{=}\text{CH}-\text{CH}_2-(\text{CH}_2-\overset{\text{CH}_3}{\text{C}}{=}\text{CH}-\text{CH}_2)_2-(\text{CH}_2-\text{CH}{=}\overset{\text{CH}_3}{\text{C}}-\text{CH}_2)_2-\text{CH}_2-\text{CH}{=}\overset{\text{CH}_3}{\text{C}}-\text{CH}_3$$

Squalene

A "redrawing" of the squalene structure, with numerous twists and bends in it, is helpful in visualizing the next stage of cholesterol biosynthesis, the formation of the four-ring steroid nucleus (Section 19.10) associated with lanosterol (and cholesterol).

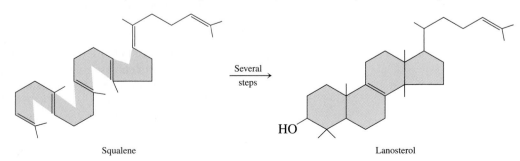

Squalene $\xrightarrow[\text{steps}]{\text{Several}}$ Lanosterol (HO)

Chemical CONNECTIONS

25.2 Statins: Drugs That Lower Plasma Levels of Cholesterol

Over half of all deaths in the United States are directly or indirectly related to heart disease, in particular to atherosclerosis. Atherosclerosis results from the buildup of plaque (fatty acid deposits) on the inner walls of arteries. Cholesterol, obtained from low-density-lipoproteins (LDL) that circulate in blood plasma, is also a major component of plaque.

Because most of the cholesterol in the human body is synthesized in the liver, from acetyl CoA, much research has focused on finding ways to inhibit its biosynthesis. The rate-determining step in cholesterol biosynthesis involves the conversion of 3-hydroxy-3-methylglutaryl CoA (HMG-CoA) to mevalonate, a process catalyzed by the enzyme HMG-CoA reductase.

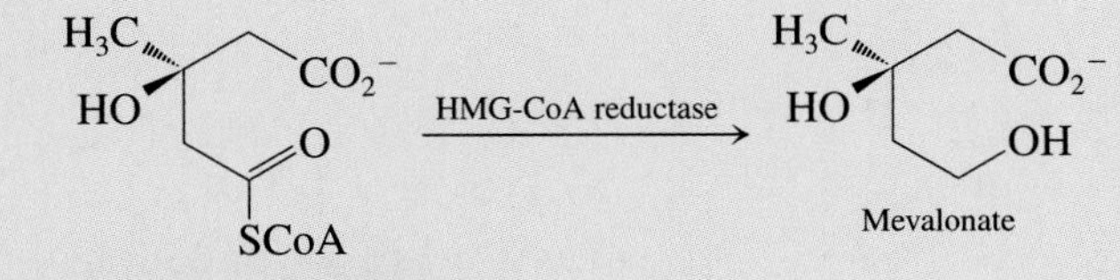

3-Hydroxy-3-methylglutaryl-CoA (HMG-CoA) —HMG-CoA reductase→ Mevalonate

In 1976, as the result of screening more than 8000 strains of microorganisms, a compound now called *mevastatin*—a potent inhibitor of HMG-CoA reductase—was isolated from culture broths of a fungus. Soon thereafter, a second, more active compound called *lovastatin* was isolated.

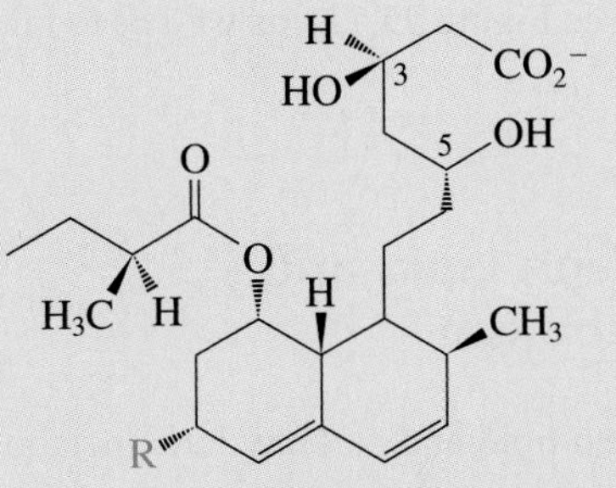

R = H, mevastatin
R = CH_3, lovastatin

These "statins" are very effective in lowering plasma concentrations of LDL by functioning as competitive inhibitors of HMG-CoA reductase.

After years of testing, the statins are now available as prescription drugs for lowering blood cholesterol levels. Clinical studies indicate that use of these drugs lowers the incidence of heart disease in individuals with mildly elevated blood cholesterol levels. Note the structural resemblance between part of the structure of lovastatin and mevalonate.

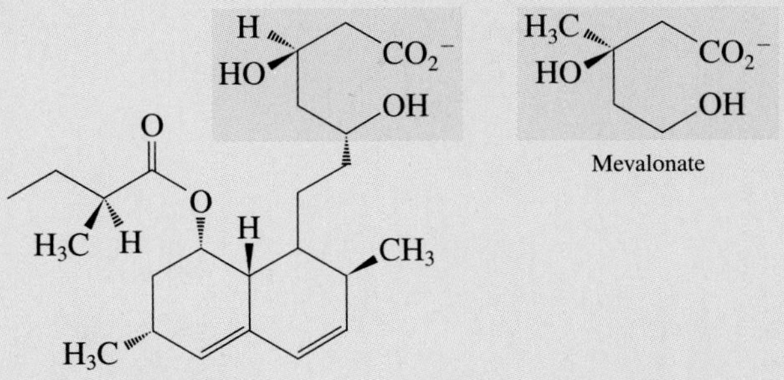

Lovastatin

Mevalonate

Recent research studies have unexpectedly shown that the cholesterol-lowering statins have two added benefits.

Laboratory studies with animals indicate that statins prompt growth cells to build new bone, replacing bone that has been leached away by osteoporosis ("brittle-bone disease"). A retrospective study of osteoporosis patients who also took statins shows evidence that their bones became more dense than did bones of osteoporosis patients who did not take the drugs.

Statins have also been shown to function as anti-inflammatory agents that counteract the effects of a common virus, cytomegalovirus, which is now believed to contribute to the development of coronary heart disease. Researchers believe that by age 65, more than 70% of all people have been exposed to this virus. The virus, along with other infecting agents in blood, may actually trigger the inflammation mechanism for heart disease.

The multistep squalene-to-lanosterol transition involves the formation of four ring systems, a decrease in double bonds from six to two, the migration of two methyl groups to new locations, and the addition of an —OH group to the C_{30} system. Addition of the —OH group requires the use of molecular oxygen; the O of the —OH group comes from the molecular O_2.

The transition from lanosterol to cholesterol involves removal of three methyl groups (C_{30} to C_{27}), reduction of the double bond in the side chain, and migration of the other double bond to a new location.

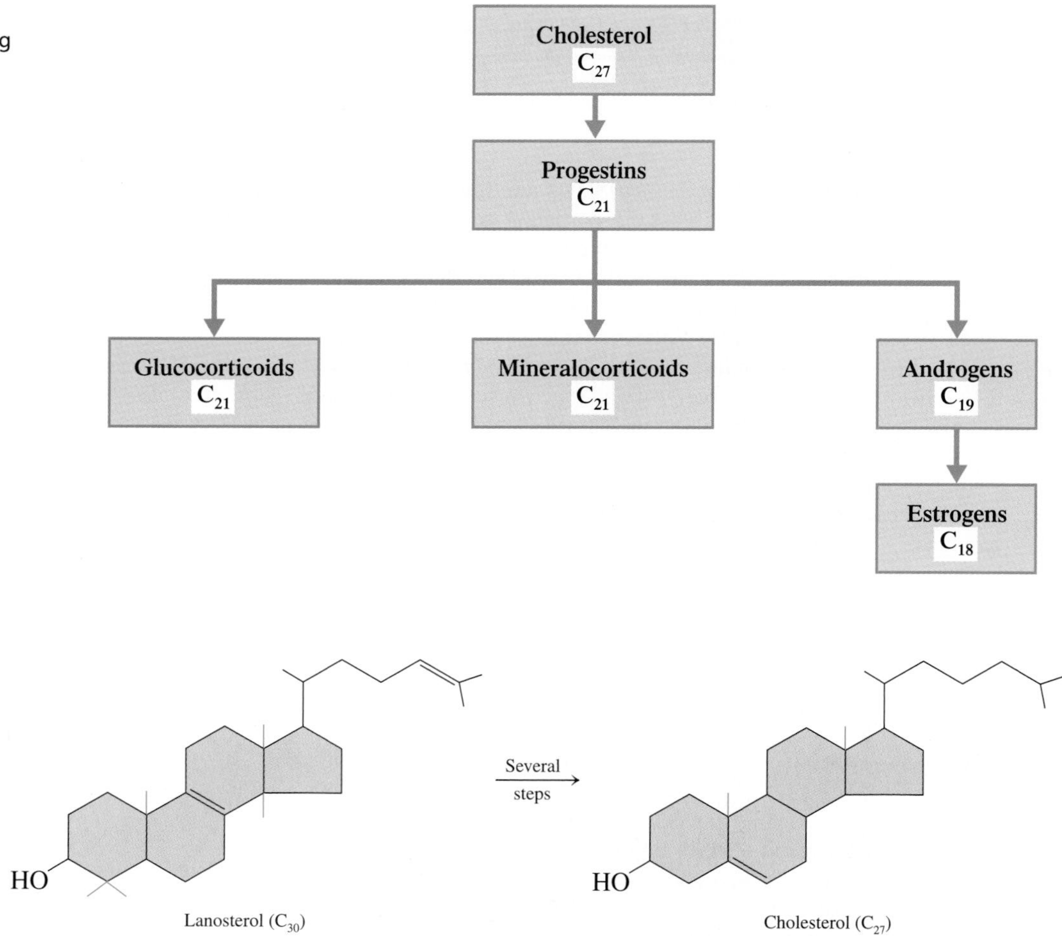

Figure 25.12 Biosynthetic relationships among steroid hormones.

Once cholesterol has been formed, biosynthetic pathways are available to convert it to each of the five major classes of steroid hormones: progestins, androgens, estrogens, glucocorticoids, and mineralocorticoids (see Figure 25.12), as well as to bile salts and vitamin D (Section 19.10).

25.9 Relationships Between Lipid and Carbohydrate Metabolism

Acetyl CoA is the primary link between lipid and carbohydrate metabolism. As shown in the accompanying Chemistry at a Glance, acetyl CoA is the degradation product for glucose, glycerol, and fatty acids, and it is also the starting material for the biosynthesis of fatty acids, cholesterol, and ketone bodies.

Note the four possible fates of acetyl CoA produced from fatty acid, glycerol, and glucose degradation processes.

1. *Oxidation in the citric acid cycle.* Both lipids (fatty acids and glycerol) and carbohydrates (glucose) supply acetyl CoA for the operation of this cycle.
2. *Ketone body formation.* This process is of major importance when there is imbalance between lipid and carbohydrate metabolic processes. The imbalance is caused by inadequate glucose metabolism during times of adequate lipid metabolism.
3. *Fatty acid biosynthesis.* The buildup of excess acetyl CoA when dietary intake exceeds energy needs leads to accelerated fatty acid biosynthesis.
4. *Cholesterol biosynthesis.* As with fatty acid biosynthesis, cholesterol biosynthesis occurs primarily when the body is in an acetyl CoA–rich state.

Interrelationships Between Carbohydrate and Lipid Metabolism

CHEMISTRY AT A GLANCE

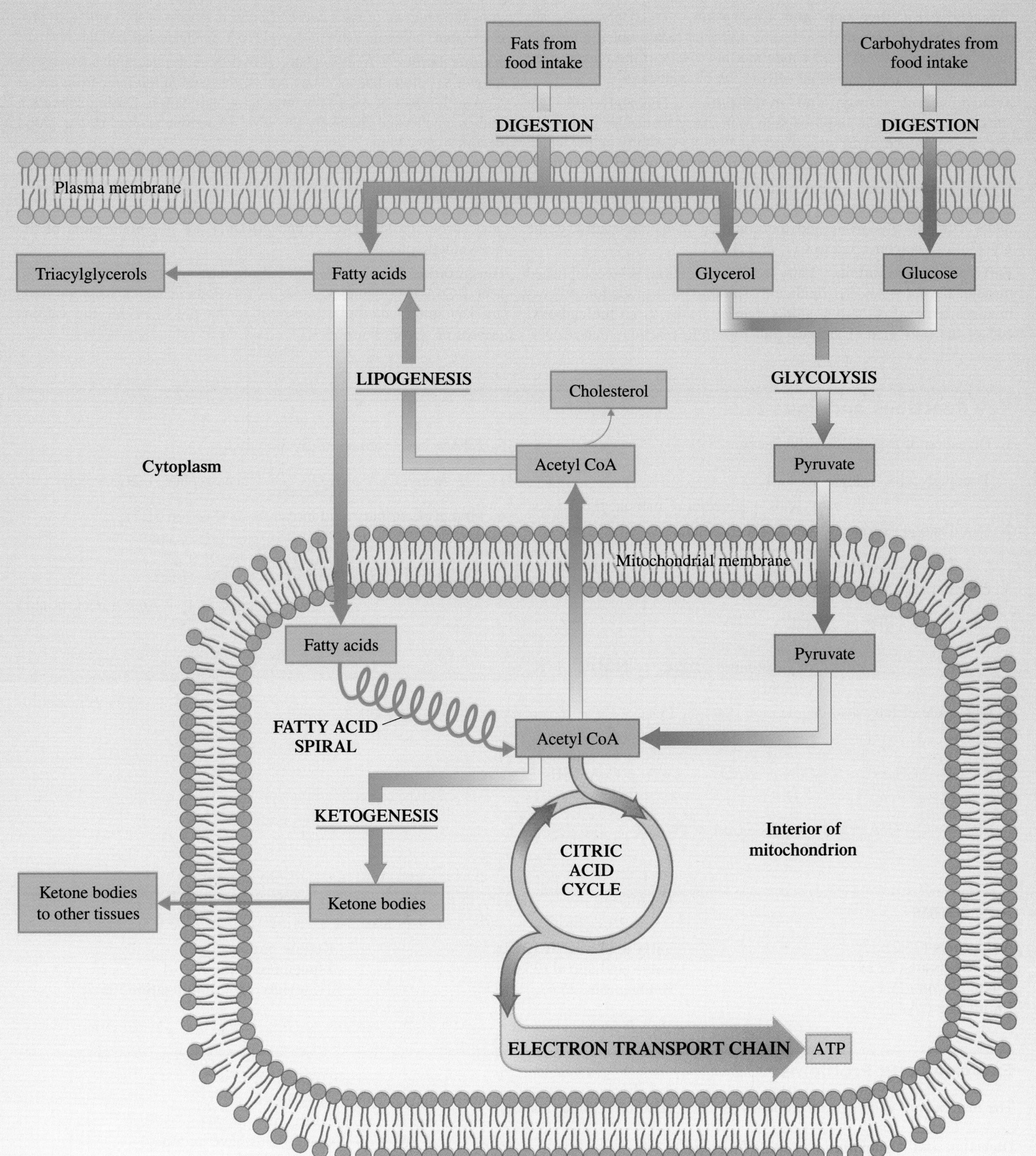

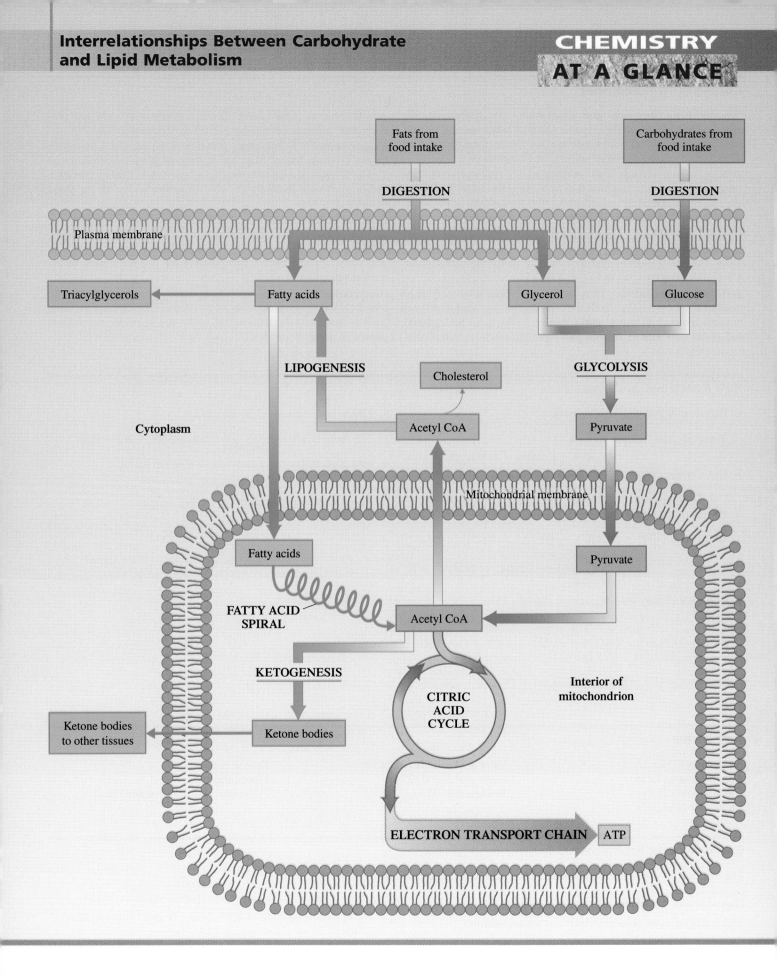

Concepts to Remember

Triacylglycerol digestion and absorption. Triacylglycerols are digested (hydrolyzed) in the intestine and then reassembled after passage into the intestinal wall. Chylomicrons transport the reassembled triacylglycerols from intestinal cells to the bloodstream.

Triacylglycerol storage and mobilization. Triacylglycerols are stored as fat droplets in adipose tissue. When they are needed for energy, enzyme-controlled hydrolysis reactions liberate the fatty acids, which then enter the bloodstream and travel to tissues where they are utilized.

Glycerol metabolism. Glycerol is first phosphorylated and then oxidized to dihydroxyacetone phosphate, a glycolysis pathway intermediate. Through glycolysis and the common metabolic pathway, the glycerol can be converted to CO_2 and H_2O.

Fatty acid degradation. Fatty acid degradation is accomplished through the fatty acid (β oxidation) spiral. The degradation process involves removal of carbon atoms, two at a time, from the carboxyl end of the fatty acid. There are four repeating reactions that accompany the removal of each two-carbon unit. A turn of the cycle also produces one molecule each of acetyl CoA, NADH, and $FADH_2$.

Ketone bodies. Acetoacetate, β-hydroxybutyrate, and acetone are known as ketone bodies. They are synthesized in the liver from acetyl CoA as a result of excessive fatty acid degradation. During starvation and in unchecked diabetes, the level of ketone bodies in the blood becomes very high.

Fatty acid biosynthesis. Fatty acids are biosynthesized through the addition of two-carbon units to a growing acyl chain. The added two-carbon units come from malonyl CoA. A multienzyme complex, an acyl carrier protein (ACP), and NADPH are important parts of the biosynthetic process.

Biosynthesis of cholesterol. Cholesterol is biosynthesized from acetyl CoA in a complex series of reactions in which isoprene units are key intermediates. Cholesterol is the precursor for the various classes of steroid hormones.

Key Reactions and Equations

1. Digestion of triacylglycerols (Section 25.1)

$$\text{Triacylglycerol} + H_2O \xrightarrow{\text{Lipase}} \text{fatty acids} + \text{glycerol} + \text{monoacylglycerols}$$

2. Mobilization of triacylglycerols (Section 25.2)

$$\text{Triacylglycerol} + 3H_2O \xrightarrow{\text{Lipase}} 3\text{ fatty acids} + \text{glycerol}$$

3. Glycerol metabolism (Section 25.3)

$$\text{Glycerol} + \text{ATP} + \text{NAD}^+ \xrightarrow[\text{steps}]{\text{Two}} \text{dihydroxyacetone phosphate} + \text{ADP} + \text{NADH} + H^+$$

4. One cycle of fatty acid degradation (Section 25.4)

$$R{-}CH_2{-}CH_2{-}\overset{\overset{\large O}{\|}}{C}{-}S{-}CoA + \text{NAD}^+ + \text{FAD} + CoA{-}SH \longrightarrow R{-}\overset{\overset{\large O}{\|}}{C}{-}S{-}CoA + CH_3{-}\overset{\overset{\large O}{\|}}{C}{-}S{-}CoA + \text{NADH} + FADH_2$$

5. Ketone body formation (Section 25.6)

$$2\text{ Acetyl CoA} + H_2O \xrightarrow[\text{steps}]{\text{Three}} \text{acetoacetate} + 2CoA{-}SH$$

6. First cycle of fatty acid biosynthesis (Section 25.7)

$$\text{Acetyl ACP} + \text{malonyl ACP} \xrightarrow{2NADPH/H^+ \;\; 2NADP^+} \text{butyryl ACP} + \text{ACP} + CO_2 + H_2O$$

Key Terms

Adipocytes (25.2)
Adipose tissue (25.2)
Chylomicron (25.1)
Emulsifier (25.1)
Fatty acid micelle (25.1)
Fatty acid spiral (25.4)
Ketogenesis (25.6)
Ketone bodies (25.6)
Lipogenesis (25.7)
Triacylglycerol mobilization (25.2)

Exercises and Problems

The members of each pair of problems in this section test similar material.

Digestion and Absorption of Lipids (Section 25.1)

25.1 What percent of dietary lipids are triacylglycerols?

25.2 What are the solubility characteristics of triacylglycerols?

25.3 What effect do salivary enzymes have on triacylglycerols?

25.4 What effect do stomach fluids have on triacylglycerols?

25.5 Why does ingestion of lipids make one feel "full" for a long time?

25.6 Where does the digestion of triacylglycerols begin?

25.7 What function does bile serve in lipid digestion?

25.8 What are the major products of triacylglycerol digestion?

25.9 *Complete* hydrolysis of triacylglycerols during digestion is unusual. Explain.

25.10 What is a fatty acid micelle?

25.11 What happens to the products of triacylglycerol digestion after they pass through the intestinal wall?

25.12 What is a chylomicron, and what is its function?

Triacylglycerol Storage and Mobilization (Section 25.2)

25.13 What is the distinctive structural feature of *adipocytes?*

25.14 What is the major metabolic function of adipose tissue?

25.15 What is triacylglycerol mobilization?

25.16 What situation signals the need for mobilization of triacylglycerols from adipose tissue?

25.17 What role does cAMP play in triacylglycerol mobilization?

25.18 Triacylglycerols in adipose tissue do not enter the bloodstream as triacylglycerols. Explain.

Glycerol Metabolism (Section 25.3)

25.19 In what order are the compounds glycerol 3-phosphate and dihydroxyacetone phosphate encountered in the degradation of glycerol?

25.20 How many reactions are needed to convert glycerol into a glycolysis intermediate?

25.21 How many ATP molecules are expended in the conversion of glycerol to a glycolysis intermediate?

25.22 What are the two fates of glycerol after it has been converted to a glycolysis intermediate?

Oxidation of Fatty Acids (Section 25.4)

25.23 Where in a cell does fatty acid activation take place?

25.24 What is the chemical form for an activated fatty acid?

25.25 Only one molecule of ATP is used to activate fatty acids before oxidation occurs, yet we count this expenditure as *two* ATP molecules for accounting purposes. Explain.

25.26 What is the difference between an acetyl CoA molecule and an acyl CoA molecule?

25.27 What is the function of carnitine in the fatty acid degradation process?

25.28 The locations in a cell for fatty acid activation and fatty acid oxidation differ. Explain.

25.29 Explain what functional-group change occurs, during one turn of the fatty acid spiral, in

a. Step 1 b. Step 2 c. Step 3

25.30 For one turn of the fatty acid spiral, arrange the following β-carbon functional groups in the order in which they are encountered: secondary alcohol, ketone, alkane, and alkene.

25.31 What is the configuration of the unsaturated enoyl CoA formed by dehydrogenation during a turn of the fatty acid spiral?

25.32 What is the configuration of the β-hydroxyacyl CoA formed by hydration during a turn of the fatty acid spiral?

25.33 In which step (of Steps 1 through 4) and which turn (first or second) of the fatty acid spiral is each of the following compounds encountered as a reactant if the fatty acid to be degraded is hexanoic acid?

a. $CH_3-CH_2-CH_2-\overset{\displaystyle OH}{\overset{|}{CH}}-CH_2-\overset{\displaystyle O}{\overset{\|}{C}}-S-CoA$

b. $CH_3-CH=CH-\overset{\displaystyle O}{\overset{\|}{C}}-S-CoA$

c. $CH_3-\overset{\displaystyle O}{\overset{\|}{C}}-CH_2-\overset{\displaystyle O}{\overset{\|}{C}}-S-CoA$

d. $CH_3-CH_2-CH_2-CH_2-CH_2-\overset{\displaystyle O}{\overset{\|}{C}}-S-CoA$

25.34 In which step (of Steps 1 through 4) and which turn (first or second) of the fatty acid spiral is each of the following compounds encountered as a reactant if the fatty acid to be degraded is hexanoic acid?

a. $CH_3-CH_2-CH_2-\overset{\displaystyle O}{\overset{\|}{C}}-CH_2-\overset{\displaystyle O}{\overset{\|}{C}}-S-CoA$

b. $CH_3-CH_2-CH_2-\overset{\displaystyle O}{\overset{\|}{C}}-S-CoA$

c. $CH_3-CH_2-CH_2-CH=CH-\overset{\displaystyle O}{\overset{\|}{C}}-S-CoA$

d. $CH_3-\overset{\displaystyle OH}{\overset{|}{CH}}-CH_2-\overset{\displaystyle O}{\overset{\|}{C}}-S-CoA$

25.35 Which compound(s) in Problem 25.33 undergo(es) a dehydrogenation reaction during a turn of the fatty acid spiral?

25.36 Which compound(s) in Problem 25.34 undergo(es) a chain-cleavage reaction during a turn of the fatty acid spiral?

25.37 How many turns of the fatty acid spiral would be needed to degrade each of the following fatty acids to acetyl CoA?

a. 16:0 fatty acid

b. 12:0 fatty acid

25.38 How many turns of the fatty acid spiral would be needed to degrade each of the following fatty acids to acetyl CoA?

a. 20:0 fatty acid

b. 10:0 fatty acid

25.39 The degradation of *cis*-3-hexenoic acid, a 6:1 acid, requires one more step than the degradation of hexanoic acid, a 6:0 acid. Describe the nature of this extra step.

25.40 The degradation of *cis*-4-hexenoic acid, a 6:1 acid, requires one more step than the degradation of hexanoic acid, a 6:0 acid. Describe the nature of this extra step.

ATP Production from Fatty Acid Oxidation (Section 25.5)

25.41 Identify the major fuel for skeletal muscle in

a. an active state b. a resting state

25.42 Explain why fatty acids cannot serve as fuel for the brain.

25.43 Consider the conversion of a C_{10} saturated acid entirely to acetyl CoA.

a. How many turns of the fatty acid spiral are required?

b. What is the yield of acetyl CoA?

c. What is the yield of NADH?

d. What is the yield of $FADH_2$?

e. How many high-energy ATP bonds are consumed?

25.44 Consider the conversion of a C_{14} saturated acid entirely to acetyl CoA.

a. How many turns of the fatty acid spiral are required?

b. What is the yield of acetyl CoA?

c. What is the yield of NADH?

d. What is the yield of $FADH_2$?

e. How many high-energy ATP bonds are consumed?

25.45 What is the net ATP production for the complete oxidation to CO_2 and H_2O of the fatty acid in Problem 25.43?

25.46 What is the net ATP production for the complete oxidation to CO_2 and H_2O of the fatty acid in Problem 25.44?

25.47 Which yield more $FADH_2$, saturated or unsaturated fatty acids? Explain.

25.48 Which yield more NADH, saturated or unsaturated fatty acids? Explain.

25.49 Compare the energy released when 1 g of carbohydrate and 1 g of lipid are completely degraded in the body.

25.50 Compare the net ATP produced from 1 molecule of glucose and 1 molecule of hexanoic acid when they are completely degraded in the body.

Ketone Bodies (Section 25.6)

25.51 What three body conditions are conducive to ketone body formation?

25.52 Why does a deficiency of carbohydrates in the diet lead to ketone body formation?

25.53 What is the relationship between oxaloacetate concentration and ketone body formation?

25.54 What is the relationship between pyruvate concentration and ketone body formation?

25.55 Draw the structures of the three compounds classified as ketone bodies.

25.56 Two of the three ketone bodies can be synthesized from the third one. Write equations for the formation of these two compounds.

25.57 What is the primary site for ketone body formation?

25.58 What is the first reaction step in the process of ketogenesis?

25.59 What reaction step is required to activate the ketone body acetoacetate before it can be used as a fuel?

25.60 In what order are the compounds acetoacetyl CoA and 3-hydroxy-3-methylglutaryl CoA encountered in the process of using ketone bodies as fuel. Explain.

25.61 What is ketosis?

25.62 Severe ketosis situations produce acidosis. Explain.

Biosynthesis of Fatty Acids (Section 25.7)

25.63 Compare the locations of the enzymes for fatty acid biosynthesis and fatty acid degradation.

25.64 How does the structure of fatty acid synthase differ from that of the enzymes that degrade fatty acids?

25.65 Coenzyme A plays an important role in fatty acid degradation. What is its counterpart in fatty acid biosynthesis, and how does its structure differ from that of coenzyme A?

25.66 What does the designation ACP stand for?

25.67 What are the primary locations within the human body where fatty acid biosynthesis occurs?

25.68 What is the starting material for fatty acid biosynthesis?

25.69 What is the role of oxaloacetate in fatty acid biosynthesis?

25.70 What is the role of citrate in fatty acid biosynthesis?

25.71 What is the role of malonyl ACP in fatty acid biosynthesis?

25.72 Write an equation for the reaction by which malonyl ACP is formed from acetyl ACP.

25.73 What type of reaction occurs in each of the four steps in the elongation of a fatty acid chain?

25.74 Why do almost all fatty acids in the human body contain an even number of carbon atoms?

25.75 In which step (of Steps 1 through 4) and which cycle (first or second turn) of fatty acid biosynthesis is each of the following compounds encountered as a product?

a. $CH_3—C(=O)—CH_2—C(=O)—S—ACP$

b. $CH_3—CH_2—CH_2—CH(OH)—CH_2—C(=O)—S—ACP$

c. $CH_3—CH_2—CH_2—CH=CH—C(=O)—S—ACP$

d. $CH_3—CH_2—CH_2—C(=O)—S—ACP$

25.76 In which step (of Steps 1 through 4) and which cycle (first or second turn) of fatty acid biosynthesis is each of the following compounds encountered as a product?

a. $CH_3—CH_2—CH_2—CH_2—CH_2—C(=O)—S—ACP$

b. $CH_3—CH=CH—C(=O)—S—ACP$

c. $CH_3—CH(OH)—CH_2—C(=O)—S—ACP$

d. $CH_3—CH_2—CH_2—C(=O)—CH_2—C(=O)—S—ACP$

25.77 Which of the compounds in Problem 25.75 is produced by a hydrogenation reaction?

25.78 Which of the compounds in Problem 25.76 is produced by a dehydration reaction?

25.79 What is the longest fatty acid that can be produced by the fatty acid synthase complex?

25.80 What central role does palmitic acid play in fatty acid biosynthesis?

25.81 What role does molecular oxygen, O_2, play in fatty acid biosynthesis?

25.82 What is the characteristic structural feature of an essential fatty acid?

25.83 Consider the biosynthesis of a C_{14} saturated fatty acid from acetyl CoA molecules.

a. How many turns of the fatty acid biosynthetic pathway are needed?

b. How many molecules of malonyl ACP must be formed?

c. How many high-energy ATP bonds are consumed?

d. How many NADPH molecules are needed?

25.84 Consider the biosynthesis of a C_{16} saturated fatty acid from acetyl CoA molecules.

a. How many turns of the fatty acid biosynthetic pathway are needed?

b. How many molecules of malonyl ACP must be formed?

c. How many high-energy ATP bonds are consumed?

d. How many NADPH molecules are needed?

Biosynthesis of Cholesterol (Section 25.8)

25.85 Approximately what percent of the total amount of cholesterol in your body is derived from the following?

a. Your diet   b. Biosynthesis

25.86 What is the starting material for the biosynthesis of cholesterol?

25.87 In each of the following pairs of intermediates in the biosynthetic pathway for cholesterol, specify which one is encountered first in the pathway.

a. Mevalonate and squalene

b. Isopentenyl pyrophosphate and lanosterol

c. Lanosterol and squalene

25.88 In each of the following pairs of intermediates in the biosynthetic pathway for cholesterol, specify which one is encountered first in the pathway.

a. Mevalonate and lanosterol

b. Isopentenyl pyrophosphate and squalene

c. Mevalonate and isopentenyl pyrophosphate

25.89 For each pair of compounds in Problem 25.87, tell whether the number of carbon atoms in the first compound is fewer than, the same as, or more than the number of carbon atoms in the second compound.

25.90 For each pair of compounds in Problem 25.88, tell whether the number of carbon atoms in the first compound is fewer than, the same as, or more than the number of carbon atoms in the second compound.

Additional Problems

25.91 With which of these processes, (1) glycerol catabolism, (2) fatty acid spiral, (3) lipogenesis, or (4) ketogenesis, is each of the following molecules associated?

a. Acyl CoA

b. Enoyl CoA

c. Malonyl ACP

d. Dihydroxyacetone phosphate

e. β-Hydroxybutyrate

f. Acetoacetyl CoA

25.92 With which of these processes, (1) fatty acid catabolism, (2) lipogenesis, (3) ketogenesis, or (4) consumption of molecular O_2, is each of the following situations associated?

a. Carnitine shuttle system

b. Citrate shuttle system

c. Fatty acid synthase complex

d. Conversion of acetoacetyl CoA to HMG-CoA

e. Conversion of squalene to cholesterol

f. Conversion of a saturated fatty acid to an unsaturated fatty acid

25.93 Identify the step (among Steps 1 through 4) of the fatty acid chain elongation process in lipogenesis to which each of the following characterizations applies.

a. Malonyl ACP is a reactant.

b. CO_2 is a product.

c. A dehydration reaction occurs.

d. A carbon–carbon double bond is converted to a carbon–carbon single bond.

25.94 Indicate in what order the following events occur in the digestion of triacylglycerols (TAGs).

(1) Bile emulsifies TAG "droplets."

(2) TAGs incorporated into chylomicrons enter the lymph system.

(3) TAGs are hydrolyzed to monoacylglycerols.

(4) Free fatty acids are "repackaged" into TAGs.

25.95 Indicate whether each of the following statements is *true* or *false*.

a. Chylomicrons are lipoproteins.

b. Acetoacetate is an intermediate in the conversion of glycerol to dihydroxyacetone phosphate.

c. The molecule carnitine is involved in fatty acid activation.

d. One turn of the fatty acid spiral produces two molecules of ATP.

25.96 Indicate whether each of the following pairings of terms is *correct* or *incorrect* for reactions in the fatty acid spiral.

a. Alkene functional group; dehydrogenation

b. Ketone functional group; chain cleavage

c. Alkane functional group; hydration

d. Secondary alcohol functional group; oxidation

25.97 Indicate whether each of the following pairings of terms is *correct* or *incorrect* for reactions in the chain elongation phase of lipogenesis.

a. Alkene functional group; hydrogenation

b. Secondary alcohol group; dehydration

c. Ketone group; reduction

d. Ketone group; hydrogenation

25.98 Arrange the four molecules (1) glucose, (2) sucrose, (3) C_8 unsaturated fatty acid, and (4) C_{14} unsaturated fatty acid in order of increasing biological energy content (ATP production) per mole.

Grid Problems

25.99

1. acyl CoA	2. acetyl CoA	3. CoA—SH
4. H_2O	5. FAD	6. NAD^+

Select from the grid *all* correct responses for each of the following situations.

a. Substances involved as reactants or products in Step 1 of the fatty acid spiral

b. Substances involved as reactants or products in Step 2 of the fatty acid spiral
c. Substances involved as reactants or products in Step 3 of the fatty acid spiral
d. Substances involved as reactants or products in Step 4 of the fatty acid spiral

25.100

1. acetoacetyl CoA	2. acetoacetate	3. acetone
4. β-hydroxy-butyrate	5. acetyl CoA	6. HMG-CoA

Select from the grid *all* correct responses for each of the following situations.

a. Substances classified as ketone bodies
b. Substances involved as reactants or products in the first step of ketogenesis
c. Substances involved as reactants or products in the second step of ketogenesis
d. Substances involved as reactants or products in the third step of ketogenesis

25.101

1. acetyl ACP	2. malonyl ACP	3. butyryl ACP
4. ACP—SH	5. CO_2	6. NADPH

Select from the grid *all* correct responses for each of the following situations.

a. Substances involved as reactants or products in Step 1 of a cycle of fatty acid biosynthesis
b. Substances involved as reactants or products in two or more steps of a cycle of fatty acid biosynthesis
c. Substances involved as reactants or products in all cycles of fatty acid biosynthesis
d. Substances involved as reactants or products in only one cycle of fatty acid biosynthesis

25.102

1. fatty acid spiral	2. fatty acid activation	3. malonyl CoA formation
4. ketogenesis	5. lipogenesis	6. cholesterol biosynthesis

Select from the grid *all* correct responses for each of the following situations.

a. Processes that require the expenditure of ATP molecules
b. Processes in which CO_2 is produced
c. Processes in which acetyl CoA is a starting reactant or an end product
d. Processes in which molecular oxygen (O_2) is often a

PART III

CHAPTER 26

Protein Metabolism

CHAPTER OUTLINE

26.1 Protein Digestion and Absorption 769
26.2 Amino Acid Utilization 770
26.3 Transamination and Oxidative Deamination 772
26.4 The Urea Cycle 776
26.5 Amino Acid Carbon Skeletons 781
26.6 Amino Acid Biosynthesis 782
26.7 Hemoglobin Catabolism 783

Chemistry at a Glance:
Interrelationships Among Lipid, Carbohydrate, and Protein Metabolism 787
26.8 Interrelationships Among Metabolic Pathways 789

Chemical Connections
26.1 The Chemical Composition of Urine 780
26.2 Arginine, Citrulline, and the Chemical Messenger Nitric Oxide 781

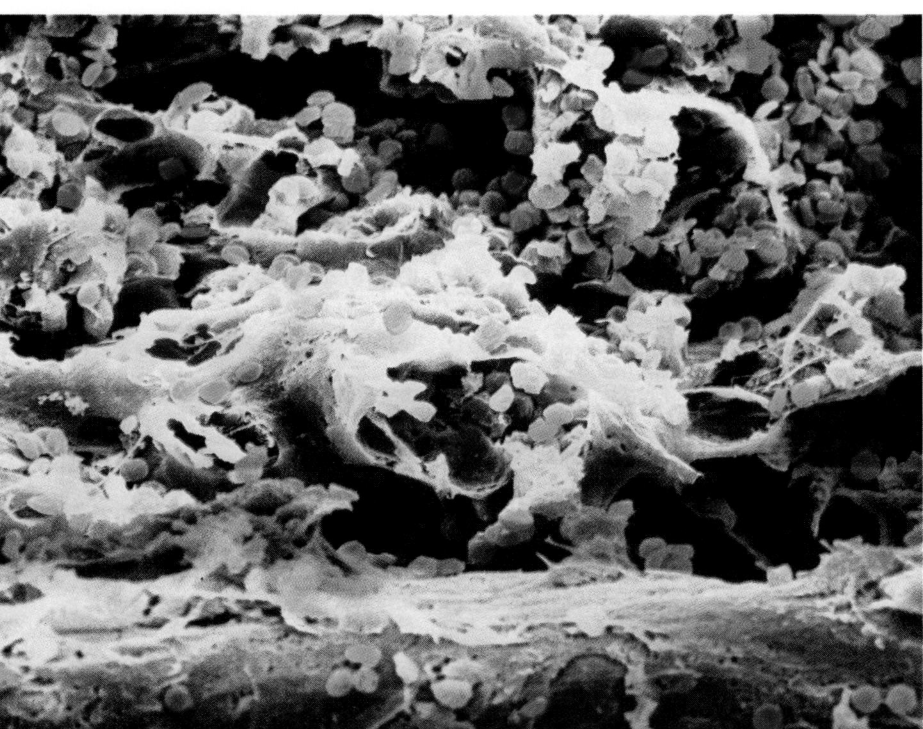

False color SEM of bone marrow with developing red blood cells.

From an energy production standpoint, proteins supply only a small portion of the body's needs. With a normal diet, carbohydrates and fats supply 90% of the body's energy, with only 10% of our daily calories coming from proteins. However, protein metabolism plays an important role in maintaining good health despite its minor role in energy production. The amino acids obtained from proteins are needed for both protein synthesis and synthesis of other nitrogen-containing compounds in the cell. In this chapter, we examine protein digestion, the oxidative degradation of amino acids, and amino acid biosynthesis.

26.1 Protein Digestion and Absorption

Protein digestion begins in the stomach rather than in the mouth, because saliva contains no enzymes that affect proteins. Both protein denaturation (Section 20.14) and protein hydrolysis (Section 20.13) occur in the stomach. The partially digested protein (large polypeptides) passes from the stomach into the small intestine, where digestion is completed (Figure 26.1).

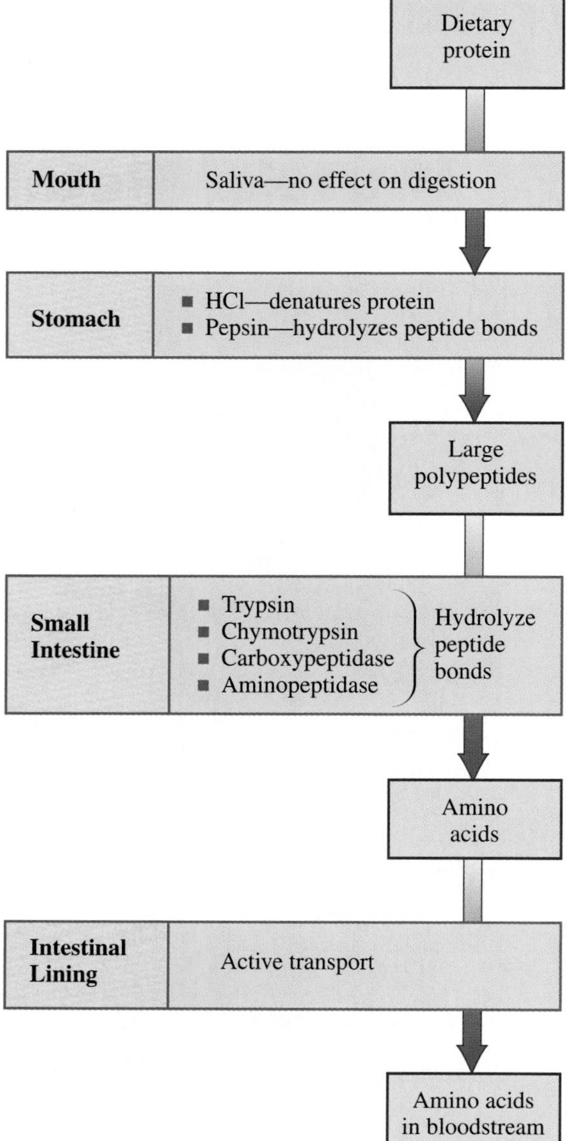

Figure 26.1 Summary of protein digestion in the human body.

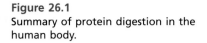

• Some digestive enzymes are secreted in inactive forms called *zymogens* (Section 21.10). Pepsin is formed from the zymogen pepsinogen, which is activated by hydrogen ions from hydrochloric acid. Trypsin and chymotrypsin are formed from trypsinogen and chymotrypsinogen, respectively.

• The passage of polypeptide chains and small proteins across the intestinal wall is uncommon in adults. However, in infants, such transport allows the passage of antibodies (proteins) in colostral milk from a mother to a nursing infant to build up immunologic protection in the infant.

Proteins are denatured in the stomach by the hydrochloric acid present in gastric juice. The acid gives gastric juice a pH of between 1.5 and 2.0. The enzyme pepsin effects the hydrolysis of about 10% of peptide bonds in proteins, producing a variety of polypeptides. In the small intestine, trypsin, chymotrypsin, and carboxypeptidase in pancreatic juice attack peptide bonds. The pH of pancreatic juice is between 7 and 8, and it neutralizes the acidity of the material from the stomach. Aminopeptidase, secreted by intestinal mucosal cells, also attacks peptide bonds.

The net result of protein digestion is the release of the protein's constituent amino acids. Absorption of these "free" amino acids through the intestinal wall requires active transport with the expenditure of energy (Section 19.13). Different transport systems exist for the various kinds of amino acids. After passing through the intestinal wall, the free amino acids enter the bloodstream, which distributes them throughout the body.

26.2 Amino Acid Utilization

Amino acids produced from the digestion of proteins enter the amino acid pool of the body. The **amino acid pool** *is the body's total supply of free amino acids available for* use. Dietary protein is one of three sources that contributes amino acids to the amino acid pool. The other two sources are *protein turnover* and *biosynthesis* of amino acids in the liver.

• The rate of protein turnover varies from a few minutes to several hours. Proteins with short turnover rates include many enzymes and regulatory hormones. In a healthy adult, about 2% of the body's protein is broken down and resynthesized every day.

Within the human body, proteins are continually being degraded (hydrolyzed) to amino acids and resynthesized. Disease, injury, and "wear and tear" are all causes of degradation. The degradation–resynthesis process is called protein turnover. **Protein turnover** *is the repetitive process in which body proteins are degraded and resynthesized.*

Biosynthesis of amino acids by the liver also supplies the amino acid pool with amino acids. However, only the *nonessential* amino acids (Sections 20.2 and 26.6) can be produced in this manner.

In a healthy adult, the amount of nitrogen taken into the body each day (dietary proteins) equals the amount of nitrogen excreted from the body. Such a person is said to be in a state of nitrogen balance. **Nitrogen balance** *is the state that results when the amount of nitrogen taken into the body (as protein) equals the amount of nitrogen excreted from the body.*

Two types of nitrogen imbalance can occur. When protein degradation exceeds protein synthesis, the amount of nitrogen in the urine exceeds the amount of nitrogen ingested (dietary protein). This condition of *negative nitrogen balance* accompanies a state of "tissue wasting," because more tissue proteins are being catabolized than are being replaced by protein synthesis. Protein-poor diets, starvation, and wasting illnesses, for example, produce a negative nitrogen balance.

A *positive nitrogen balance* (nitrogen intake exceeds nitrogen output) indicates that the rate of protein anabolism (synthesis) exceeds that of protein catabolism. This state indicates that large amounts of tissue are being synthesized, such as during growth, pregnancy, and convalescence from an emaciating illness.

Although the overall nitrogen balance in the body often varies, the relative concentrations of amino acids within the amino acid pool remain essentially constant. No specialized storage forms for amino acids exist in the body, as is the case for glucose (glycogen) and fatty acids (triacylglycerols). Therefore, the body needs a relatively constant source of amino acids to maintain normal metabolism. During negative nitrogen balance, the body must resort to degradation of proteins that were synthesized for other functions. As we have seen throughout our discussion of metabolism, a well-balanced diet consisting of carbohydrates, lipids, and proteins is essential for good health and normal metabolism.

The amino acids from the body's amino acid pool are used in four different ways.

1. *Protein synthesis.* It is estimated that about 75% of the free amino acids in a healthy, well-nourished adult go into protein synthesis. Proteins are continually needed to replace old tissue (protein turnover) and also to build new tissue (growth). The subject of protein synthesis was considered in Section 22.11.
2. *Synthesis of nonprotein nitrogen-containing compounds.* Amino acids are regularly withdrawn from the amino acid pool for the synthesis of nonprotein nitrogen-containing compounds. Such molecules include the purines and pyrimidines of nucleic acids, the heme of hemoglobin, neurotransmitters such as acetylcholine and serotonin, the choline and ethanolamine of phosphoglycerides, and hormones such as thyroxine and epinephrine.
3. *Synthesis of nonessential amino acids.* When required, the body draws on the amino acid pool for raw materials for the production of nonessential amino acids that are in short supply. The "roadblock" preventing the synthesis of the essential amino acids is not lack of nitrogen but lack of a correct carbon skeleton upon which enzymes can work. In general, the essential amino acids contain carbon chains or aromatic rings not present in other amino acids or the intermediates of carbohydrate or lipid metabolism. Table 26.1 lists the essential amino acids and the nonessential amino acids with the precursors needed to form the latter.
4. *Production of energy.* Because excess amino acids cannot be stored for later use, the body's response is to degrade them. The degradation process is

• Higher plants and certain microorganisms are capable of synthesizing all the protein amino acids from carbon dioxide, water, and inorganic salts.

• There are approximately 100 grams of free amino acids present in the amino acid pool. Two amino acids, glutamic acid and glutamine, account for half of the amino acids present in the pool. The essential amino acids constitute approximately 10 grams of the pool.

Table 26.1
Essential and Nonessential Amino Acids

Nutritionally Essential Amino Acids	Nutritionally Nonessential	
	Amino acid	**Precursor**
histidine	alanine	pyruvate
isoleucine	arginine	glutamate
leucine	asparagine	aspartate
lysine	aspartic acid	oxaloacetate
methionine	cysteine	serine
phenylalanine	glutamic acid	α-ketoglutarate
threonine	glutamine	glutamate
tryptophan	glycine	serine
valine	proline	glutamate
	serine	3-phosphoglycerate
	tyrosine	phenylalanine

complex, because each of the 20 standard amino acids has a different degradation pathway.

In all the degradation pathways, the amino nitrogen atom is removed and converted to ammonium ion, which ultimately is excreted from the body as urea. The remaining carbon skeleton is then converted to pyruvate, acetyl CoA, or a citric acid cycle intermediate, depending on its makeup, with the resulting energy production or energy storage. Figure 26.2 shows the various pathways available for the products of amino acid catabolism. Subsequent sections of this chapter give further details about these processes.

26.3 Transamination and Oxidative Deamination

Degradation of an amino acid has two stages: (1) the removal of the α-amino group and (2) the degradation of the remaining carbon skeleton. In this section and the next, we consider what happens to the amino group; in Section 26.5, the fate of the carbon skeleton is considered.

Figure 26.2
Possible fates for amino acid degradation products.

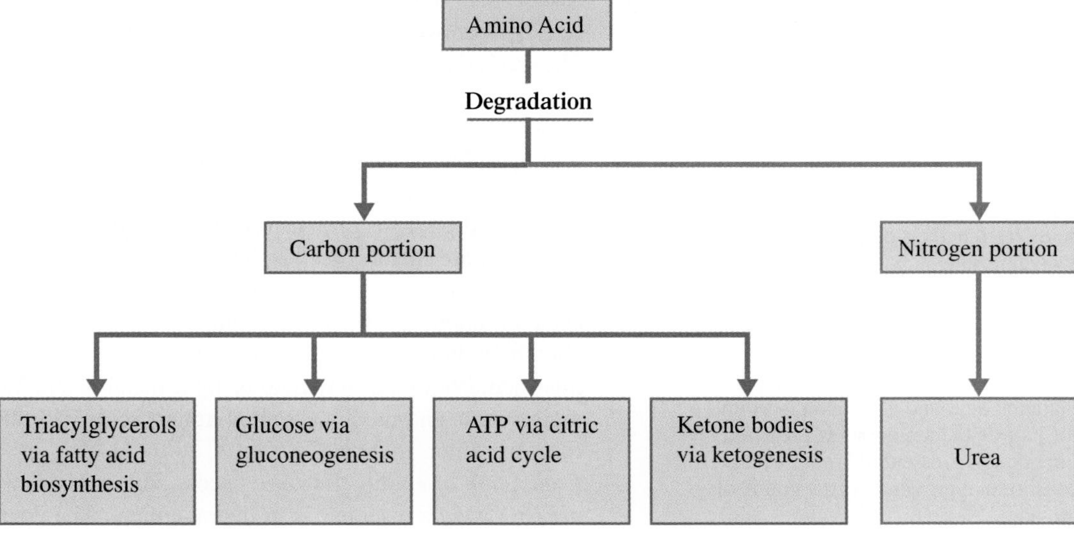

The release of an amino group from most amino acids requires a two-step process involving *transamination* followed by *oxidative deamination.* The following two procedures will make these processes easier to visualize.

1. Draw amino acid structures in the general format

$$\mathrm{R{-}\overset{\overset{+}{N}H_3}{\overset{|}{C}H}{-}COO^-}$$

   Remember that the ordering of the four groups attached to the carbon in an amino acid is not critical except in stereochemical considerations (Fischer projections; Section 20.3).
2. Review the structural relationships among six molecules—three pairs of keto/amino acids.

In Section 16.4, we noted that the derivatives of three carboxylic acids — propionic (a three-carbon monoacid), succinic (a four-carbon diacid) and glutaric (a five-carbon diacid) — are particularly important in metabolic reactions. It is the α-keto and α-amino derivatives of these three acids that are the "key players" in the transamination/oxidative deamination process. Figure 26.3 gives the structural relationships among these compounds.

Figure 26.3
Key compounds in the transamination/oxidative deamination process include three keto acid/amino acid pairs.

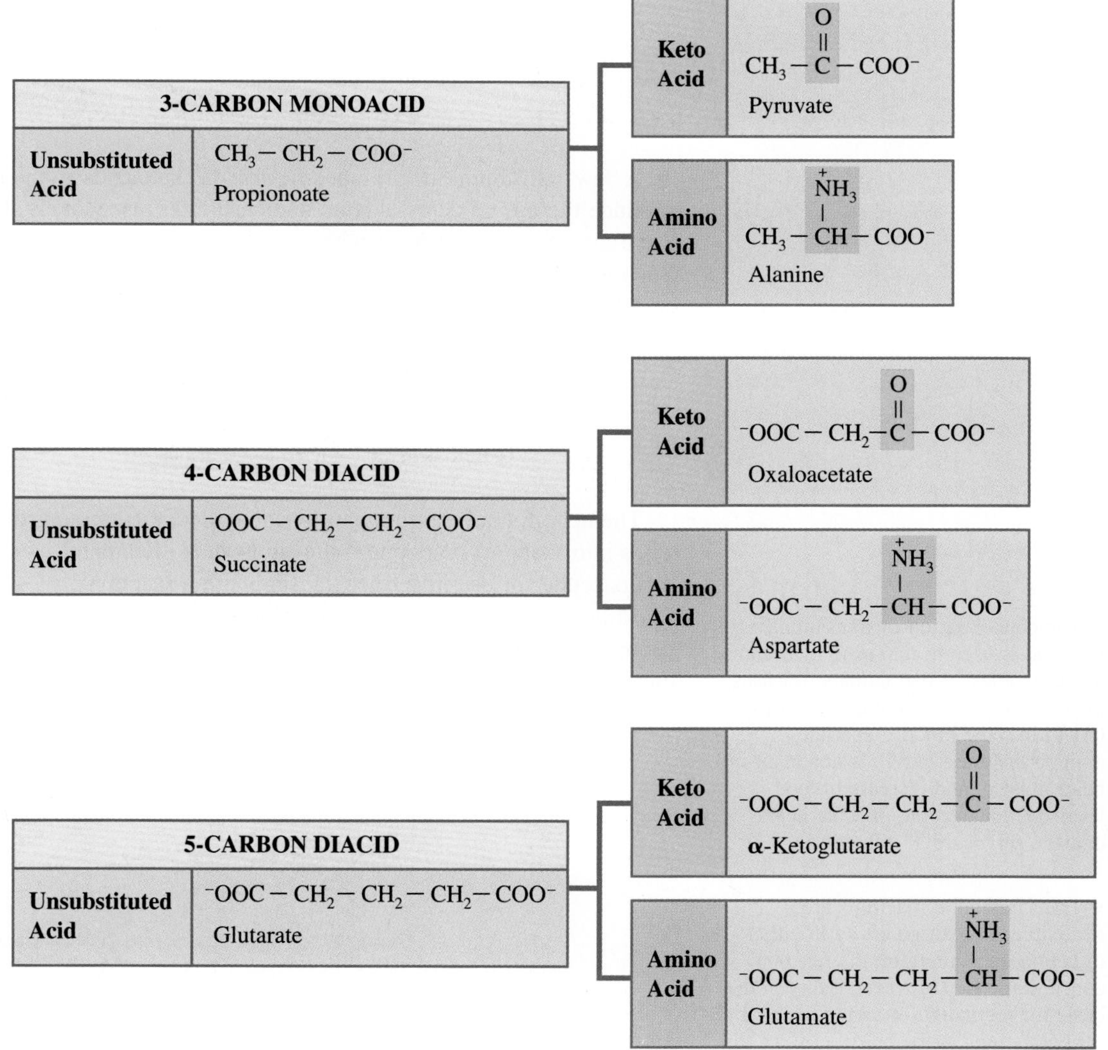

• Transamination

Transamination *is the interchange of the amino group of an amino acid with the keto group of an α-keto acid.* A general equation for the transamination process is

$$\underset{\text{Amino acid}}{R-\overset{\overset{+}{N}H_3}{\overset{|}{C}H}-COO^-} + \underset{\alpha\text{-Keto acid}}{R'-\overset{O}{\overset{\|}{C}}-COO^-} \longrightarrow \underset{\text{New }\alpha\text{-keto acid}}{R-\overset{O}{\overset{\|}{C}}-COO^-} + \underset{\text{New amino acid}}{R'-\overset{\overset{+}{N}H_3}{\overset{|}{C}H}-COO^-}$$

• The purpose of transamination is to remove amino groups from the various α-amino acids and collect them in a single amino acid, glutamate. Glutamate then acts as the source of amino groups for continued nitrogen metabolism (excretion or biosynthesis).

There are at least 50 transaminase enzymes associated with the transamination process. Most have a specificity for α-ketoglutarate as the amino group acceptor. Glutamate is the product from the action of these enzymes.

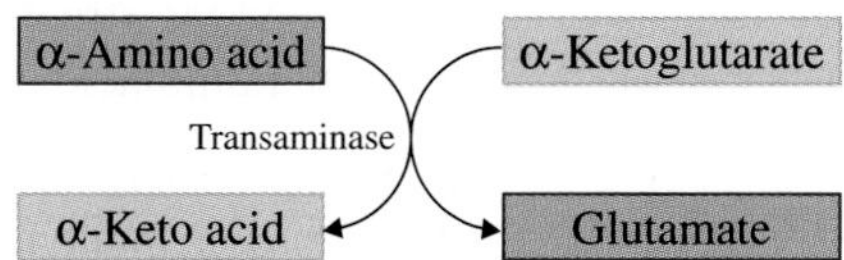

A specific example of this type of reaction is

$$\underset{\text{Alanine}}{CH_3-\overset{\overset{+}{N}H_3}{\overset{|}{C}H}-COO^-} + \underset{\alpha\text{-Ketoglutarate}}{^-OOC-CH_2-CH_2-\overset{O}{\overset{\|}{C}}-COO^-} \xrightarrow{\text{Trans-aminase}}$$

$$\underset{\text{Pyruvate}}{CH_3-\overset{O}{\overset{\|}{C}}-COO^-} + \underset{\text{Glutamate}}{^-OOC-CH_2-CH_2-\overset{\overset{+}{N}H_3}{\overset{|}{C}H}-COO^-}$$

A few transaminases are specific for the ketoacids pyruvate and oxaloacetate. They produce the amino acids alanine and aspartate, respectively. Ultimately, the alanine and aspartate so produced react with α-ketoglutarate, via transamination, to give glutamate. Such a "double" transamination sequence involving oxaloacetate would be diagrammed as follows:

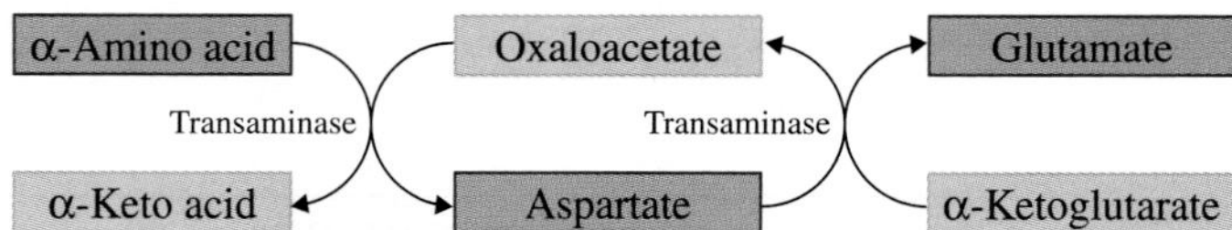

The net effect of transamination is to collect the amino groups from a variety of amino acids into a single compound—the amino acid glutamate—and to regenerate pyruvate and oxaloacetate for use in further transamination reactions.

• The concentration of transaminases in blood is used to diagnose liver and heart disorders. Liver damage releases the enzyme alanine aminotransferase (ALT) into the blood. Aspartate aminotransferase (AST) is abundant in heart muscle, and increased blood levels of this enzyme indicate heart damage (myocardial infarction).

Although the transamination reaction appears to involve the simple transfer of a $-\overset{+}{N}H_3$ group between two molecules, the reaction involves several steps and requires the presence of pyridoxal phosphate, a coenzyme produced from pyridoxine (vitamin B_6).

[Structures: Pyridoxine (vitamin B_6): pyridinium ring (N–H, +) bearing $HOCH_2$, CH_2OH, OH, CH_3. Pyridoxal phosphate (coenzyme): pyridinium ring (N–H, +) bearing Ⓟ–OCH_2, CHO, OH, CH_3.]

Pyridoxine
(vitamin B_6)

Pyridoxal phosphate
(coenzyme)

• Transamination reactions are reversible and can go easily in either direction, depending on the reactant concentrations. This reversibility is the basis for regulation of amino acid concentrations in the body.

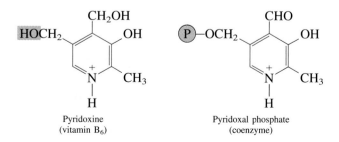

This coenzyme is an integral part of the transamination process. The amino group of the amino acid is transferred first to the pyridoxal phosphate and then from the pyridoxal

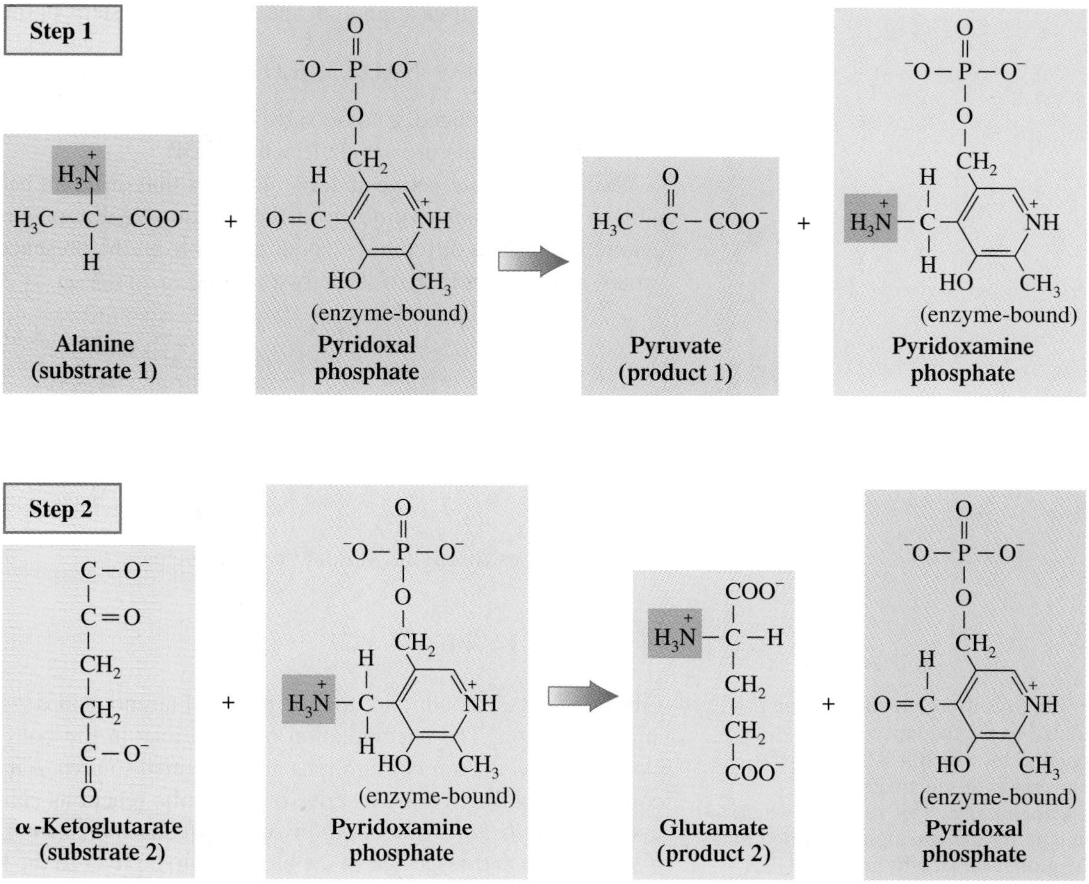

Step 1

Alanine (substrate 1) + Pyridoxal phosphate (enzyme-bound) → Pyruvate (product 1) + Pyridoxamine phosphate (enzyme-bound)

Step 2

α-Ketoglutarate (substrate 2) + Pyridoxamine phosphate (enzyme-bound) → Glutamate (product 2) + Pyridoxal phosphate (enzyme-bound)

Figure 26.4
The role of pyridoxal phosphate in the process of transamination.

phosphate to the α-keto acid. Figure 26.4 shows the role of this coenzyme in the transamination process, where alanine is the amino acid and α-ketoglutarate is the α-keto acid.

• Oxidative Deamination

In the second step of amino acid degradation, ammonium ion (NH_4^+) is liberated from the glutamate formed by transamination. This step involves oxidative deamination. **Oxidative deamination** *is the conversion of an amino acid into a keto acid with the release of ammonium ion.* Oxidative deamination occurs primarily in the liver and kidney mitochondria.

Oxidative deamination of glutamate requires the enzyme *glutamate dehydrogenase.* This enzyme is unusual in that it can function with either $NADP^+$ or NAD^+ as a coenzyme. With NAD^+ as the coenzyme, the reaction is

$$\underset{\text{Glutamate}}{^{-}OOC—CH_2—CH_2—\overset{\overset{+}{N}H_3}{|}{CH}—COO^-} + NAD^+ + H_2O \xrightarrow[\text{dehydrogenase}]{\text{Glutamate}}$$

$$NH_4^+ + \underset{\alpha\text{-Ketoglutarate}}{^{-}OOC—CH_2—CH_2—\overset{\overset{O}{\|}}{C}—COO^-} + NADH + H^+$$

Note that α-ketoglutarate is a product of this process. It can be reused in the transamination process (first step). The $NADH^+/H^+$ formed can undergo oxidative phosphorylation to produce three ATP molecules.

The sum of the transamination and deamination steps of the degradation of amino acids is

$$\alpha\text{-Amino acid} + NAD^+ + H_2O \longrightarrow \alpha\text{-keto acid} + NH_4^+ + NADH^+ + H^+$$

The NH_4^+ so produced, a toxic substance if left to accumulate in the body, is then converted to urea in the urea cycle (Section 26.4).

Two amino acids, serine and threonine, exhibit different behavior from the others. They undergo *direct deamination* by a dehydration–hydration process rather than *oxidative deamination.* This different behavior results from the presence of a side chain β-hydroxyl group, a feature unique to these two acids.

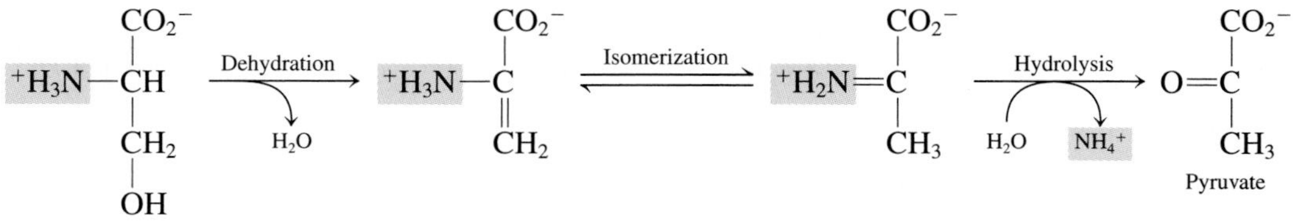

$$\underset{\text{Serine}}{{}^+H_3N-\underset{\underset{OH}{|}}{\underset{CH_2}{|}}\overset{CO_2^-}{\overset{|}{CH}}} \xrightarrow[H_2O]{\text{Dehydration}} {}^+H_3N-\underset{\underset{CH_2}{\|}}{\overset{CO_2^-}{\overset{|}{C}}} \xrightleftharpoons{\text{Isomerization}} {}^+H_2N{=}\underset{\underset{CH_3}{|}}{\overset{CO_2^-}{\overset{|}{C}}} \xrightarrow[H_2O \;\; NH_4^+]{\text{Hydrolysis}} \underset{\text{Pyruvate}}{O{=}\underset{\underset{CH_3}{|}}{\overset{CO_2^-}{\overset{|}{C}}}}$$

Threonine goes through a similar series of steps.

26.4 The Urea Cycle

• The toxicity of ammonium ion is related to the oxidative deamination reaction by which it is formed, the conversion of glutamate to α-ketoglutarate. This reaction, which is an equilibrium situation, is shifted to the glutamate side by increased ammonium ion levels. This shift decreases α-ketoglutarate levels significantly, which affects the citric acid cycle of which α-ketoglutarate is an intermediate. Cellular ATP production drops, and the lack of ATP causes central nervous system problems.

From a nitrogen standpoint, the net effect of amino acid degradation is the production of ammonium ion. The accumulation of these ions in the body has potential toxic effects. Consequently, the ammonium ions are converted to urea, a less toxic nitrogen-containing compound, in the liver by a series of metabolic reactions called the urea cycle. The **urea cycle** *is a cyclic biochemical pathway that produces urea from ammonium ions and carbon dioxide.* Urea is transported in the blood from the liver to the kidneys and eliminated from the body in urine.

In the pure state, urea is a white solid with a melting point of 133°C. Its structure is

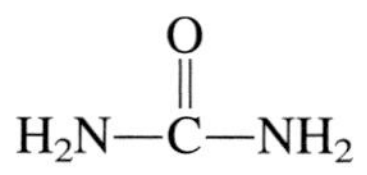

$$H_2N-\overset{\overset{O}{\|}}{C}-NH_2$$

Urea is very soluble in water (1 g per 1 mL), is odorless and colorless, and has a salty taste. (Urea does not contribute to the odor or color of urine.) With normal metabolism, an adult excretes about 30 g of urea daily in urine, although the exact amount varies with the protein content of the diet.

Three amino acids are involved as intermediates in the conversion of ammonium ions to urea through the urea cycle. These acids are arginine, ornithine, and citrulline, the latter two of which are nonstandard amino acids—that is, amino acids not found in protein. Structurally, all three of these amino acids have the same carbon chain.

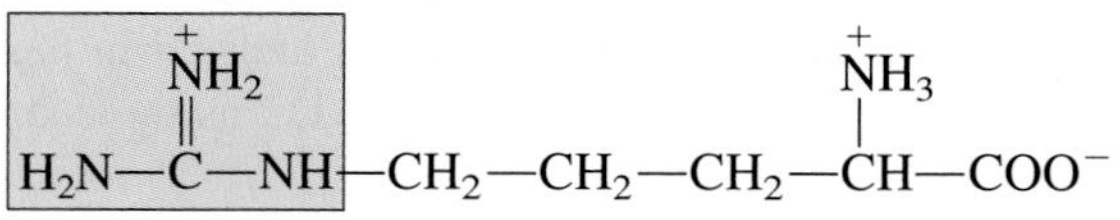

$$\boxed{H_2N-\overset{\overset{\overset{+}{NH_2}}{\|}}{C}-NH}-CH_2-CH_2-CH_2-\overset{\overset{\overset{+}{NH_3}}{|}}{CH}-COO^-$$

Arginine
(standard amino acid)

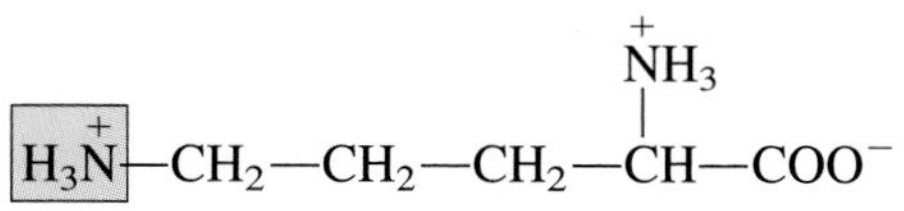

$$\boxed{H_3\overset{+}{N}}-CH_2-CH_2-CH_2-\overset{\overset{\overset{+}{NH_3}}{|}}{CH}-COO^-$$

Ornithine
(nonstandard amino acid)

$$\boxed{H_2N-\overset{\overset{O}{\|}}{C}-NH}-CH_2-CH_2-CH_2-\overset{\overset{\overset{+}{NH_3}}{|}}{CH}-COO^-$$

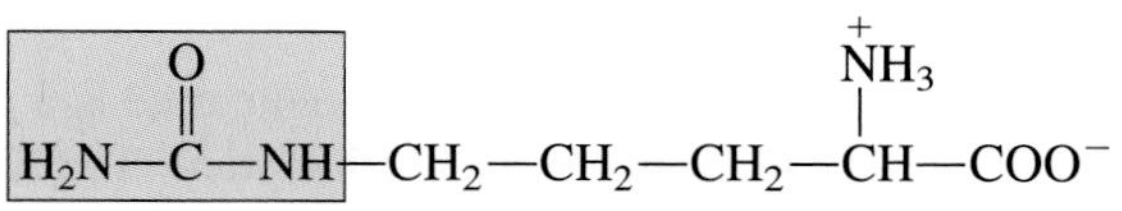

Citrulline
(nonstandard amino acid)

• The functional group attached to the phosphate in carbamoyl phosphate is the simple amide functional group

$$-\overset{\overset{\displaystyle O}{\|}}{C}-NH_2$$

The term *carbamoyl* is the *prefix* that denotes an amide group. Most often, amide groups are named by using the *suffix* system, in which case the terminology is *amide.*

• Carbamoyl Phosphate

Ammonium ions enter the four-step urea cycle in the form of carbamoyl phosphate rather than as simple ammonium ions. The reactants for the formation of this substance are ammonium ions (from oxidative deamination), carbon dioxide (from the citric acid cycle), water, and two ATP molecules. The reaction equation is

$$NH_4^+ + CO_2 + H_2O + 2ATP \longrightarrow H_2N-\overset{\overset{\displaystyle O}{\|}}{C}-O-\overset{\overset{\displaystyle O}{\|}}{\underset{\underset{\displaystyle O^-}{|}}{P}}-O^- + 2ADP + P_i + 3H^+$$

Carbamoyl phosphate

Note that two ATP molecules are expended in the formation of carbamoyl phosphate.

• Steps of the Urea Cycle

Figure 26.5 shows in outline form the four-step urea cycle. Note that the urea cycle occurs partially in the mitochondria and partially in the cytoplasm and that ornithine and citrulline are transported across the mitochondrial membrane by specific transport systems. We will now consider in detail the individual steps of the urea cycle.

Figure 26.5
The four-step urea cycle in which carbamoyl phosphate is converted to urea.

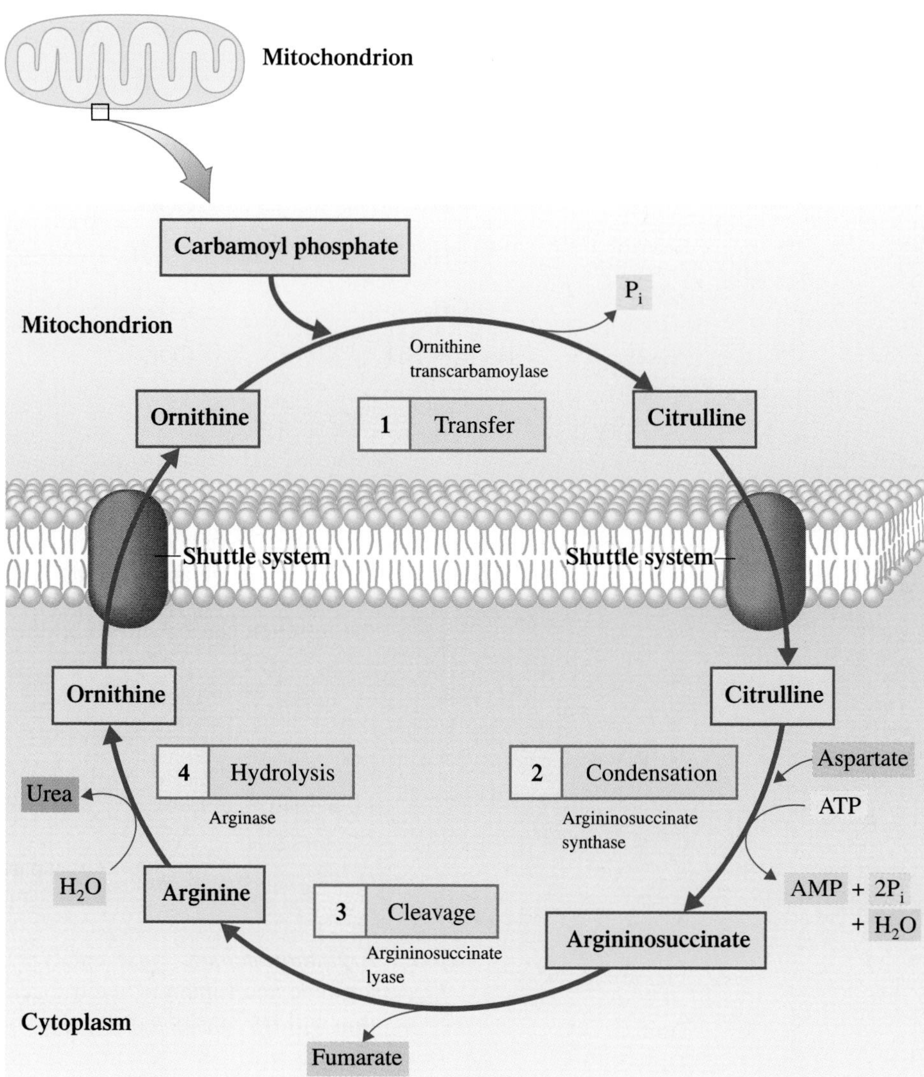

• The standard amino acid lysine and the nonstandard amino acid ornithine, both basic amino acids, have closely related structures.

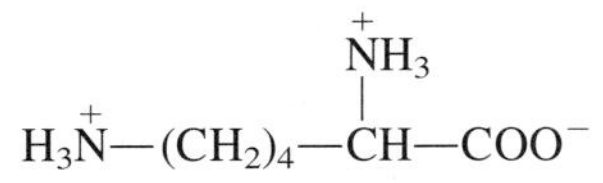

$$H_3\overset{+}{N}-(CH_2)_4-CH(\overset{+}{N}H_3)-COO^-$$

Lysine

$$H_3\overset{+}{N}-(CH_2)_3-CH(\overset{+}{N}H_3)-COO^-$$

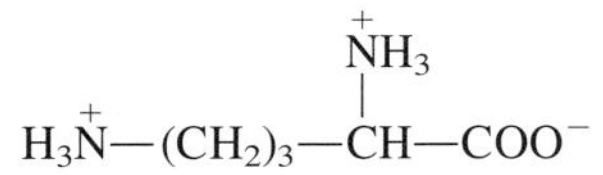

Ornithine

Step 1: The carbamoyl group of carbamoyl phosphate is transferred to ornithine to form citrulline in a reaction catalyzed by *ornithine transcarbamoylase.*

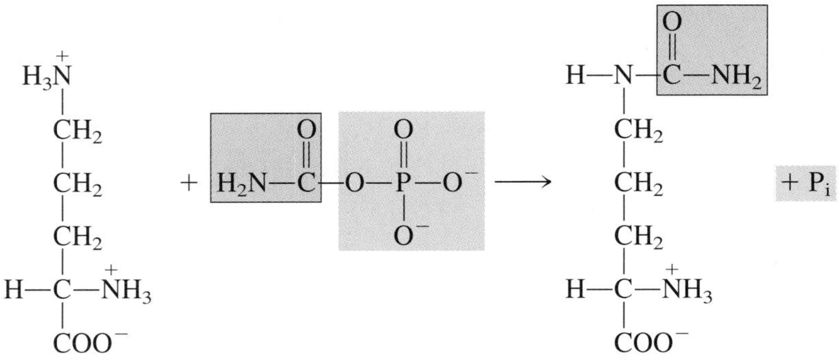

$$\text{Ornithine} + H_2N-C(=O)-O-P(=O)(O^-)-O^- \longrightarrow \text{Citrulline} + P_i$$

Ornithine: $H_3\overset{+}{N}-CH_2-CH_2-CH_2-CH(\overset{+}{N}H_3)-COO^-$; Carbamoyl phosphate; Citrulline: $H_2N-C(=O)-NH-CH_2-CH_2-CH_2-CH(\overset{+}{N}H_3)-COO^-$

The breaking of the high-energy phosphate bond in carbamoyl phosphate drives the transfer process. With the carbamoyl transfer, the first of the two nitrogen atoms and the carbon atom needed for the formation of urea have been introduced into the cycle.

Step 2: A condensation reaction between citrulline and aspartate (a standard amino acid) produces argininosuccinate. This condensation, catalyzed by *argininosuccinate synthase,* is driven by the expenditure of ATP.

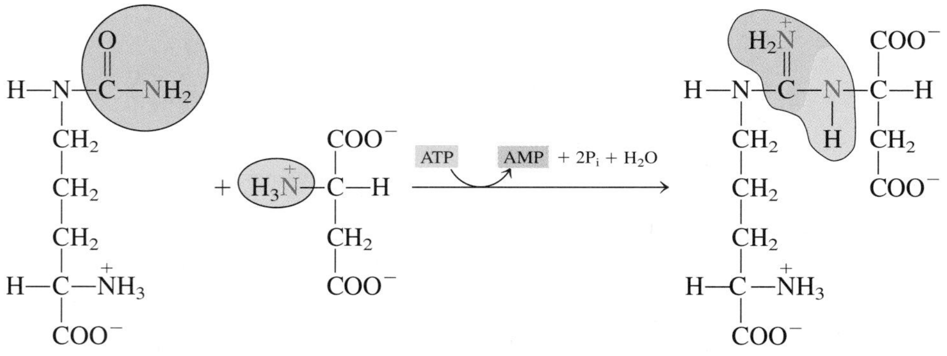

$$\text{Citrulline} + \overset{+}{H_3N}-CH(COO^-)-CH_2-COO^- \xrightarrow{ATP \;\to\; AMP + 2P_i + H_2O} \text{Argininosuccinate}$$

Citrulline; Aspartate; Argininosuccinate: $H-N(-C(=\overset{+}{N}H_2)-N(H)-CH(COO^-)-CH_2-COO^-)-CH_2-CH_2-CH_2-CH(\overset{+}{N}H_3)-COO^-$

With this reaction, the second of the two nitrogen atoms that will be part of the end-product urea has been introduced into the cycle. One nitrogen atom comes from carbamoyl phosphate, the other from aspartate.

However, the original source of both nitrogens is glutamate. The flow of the nitrogen can be shown by these reactions:

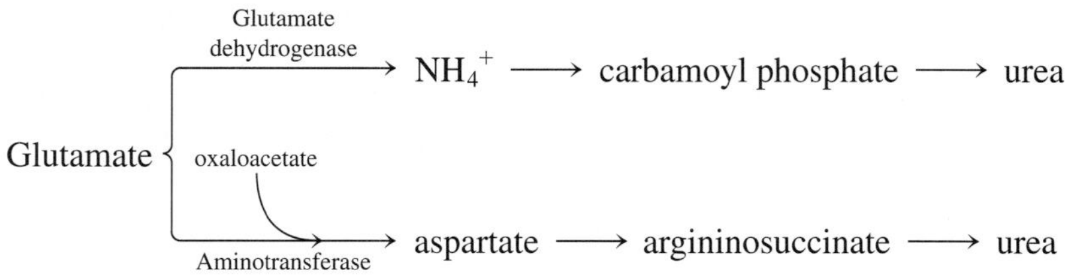

$$\text{Glutamate} \xrightarrow{\text{Glutamate dehydrogenase}} NH_4^+ \longrightarrow \text{carbamoyl phosphate} \longrightarrow \text{urea}$$

$$\text{Glutamate} \xrightarrow[\text{Aminotransferase}]{\text{oxaloacetate}} \text{aspartate} \longrightarrow \text{argininosuccinate} \longrightarrow \text{urea}$$

Step 3: *Argininosuccinate lyase* catalyzes the cleavage of argininosuccinate into arginine and fumarate, a citric acid cycle intermediate. The significance of this will be considered shortly.

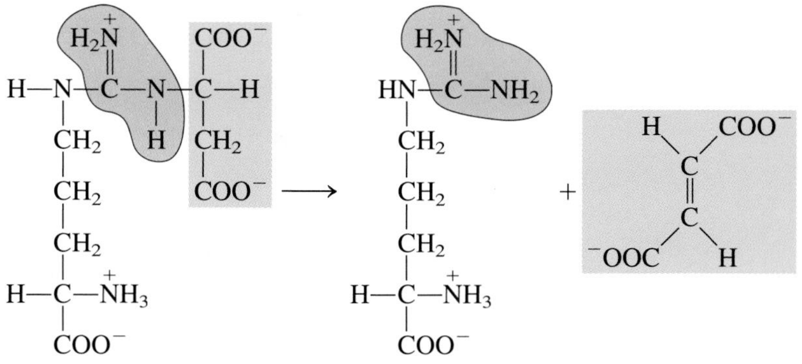

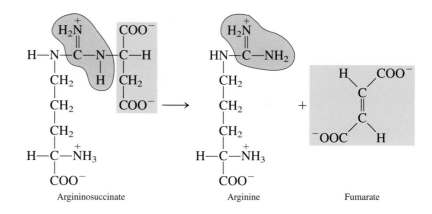

Argininosuccinate → Arginine + Fumarate

Step 4: Hydrolysis of arginine produces urea and regenerates ornithine, one of the cycle's starting materials. The enzyme involved is *arginase*.

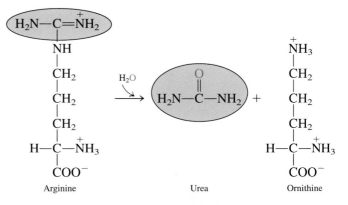

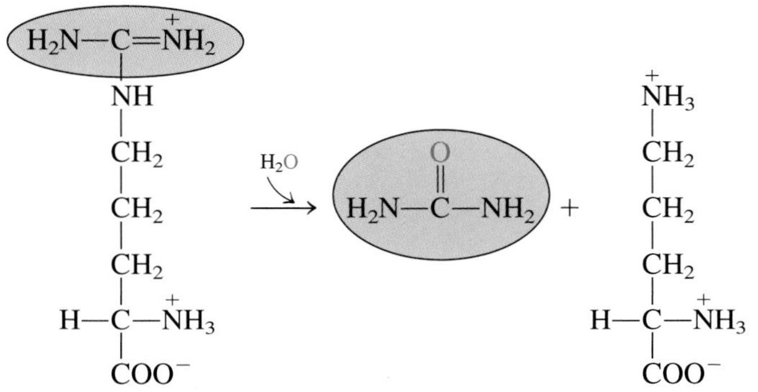

Arginine $\xrightarrow{H_2O}$ Urea + Ornithine

The oxygen atom present in the urea comes from the water involved in the hydrolysis.

• Urea Cycle Net Reaction

The net reaction for urea formation, in which all of the urea cycle intermediates cancel out of the equation, is

$$NH_4^+ + CO_2 + 3ATP + 2H_2O + \text{aspartate} \longrightarrow \text{urea} + 2ADP + AMP + 4P_i + \text{fumarate}$$

The equivalent of a total of four ATP molecules is expended in the production of one urea molecule. Two ATP molecules are consumed in the production of carbamoyl phosphate, and the equivalent of two ATP molecules is consumed in Step 2 of the urea cycle, where an ATP is hydrolyzed to AMP and PP_i and the PP_i is then further hydrolyzed to two P_i.

• Sulfur-containing amino acids (cysteine and methionine) contain both sulfur and nitrogen. The nitrogen-containing group is lost through transamination and processed to urea. The sulfur-containing group is processed to sulfur dioxide (SO_2), which is then oxidized to sulfate (SO_4^{2-}). The SO_4^{2-} ion, the negative ion from sulfuric acid, is eliminated in urine.

• Linkage Between the Urea and Citric Acid Cycles

The net equation for urea formation shows fumarate, a citric acid cycle intermediate, as a product. This fumarate enters the citric acid cycle, where it is converted to malate and then to oxaloacetate, which can then be converted to aspartate through transamination. The aspartate then re-enters the urea cycle at Step 2 (see Figure 26.6).

Besides undergoing transamination, the oxaloacetate produced from fumarate of the urea cycle can be (1) converted to glucose via gluconeogenesis, (2) condensed with acetyl CoA to form citrate, or (3) converted to pyruvate.

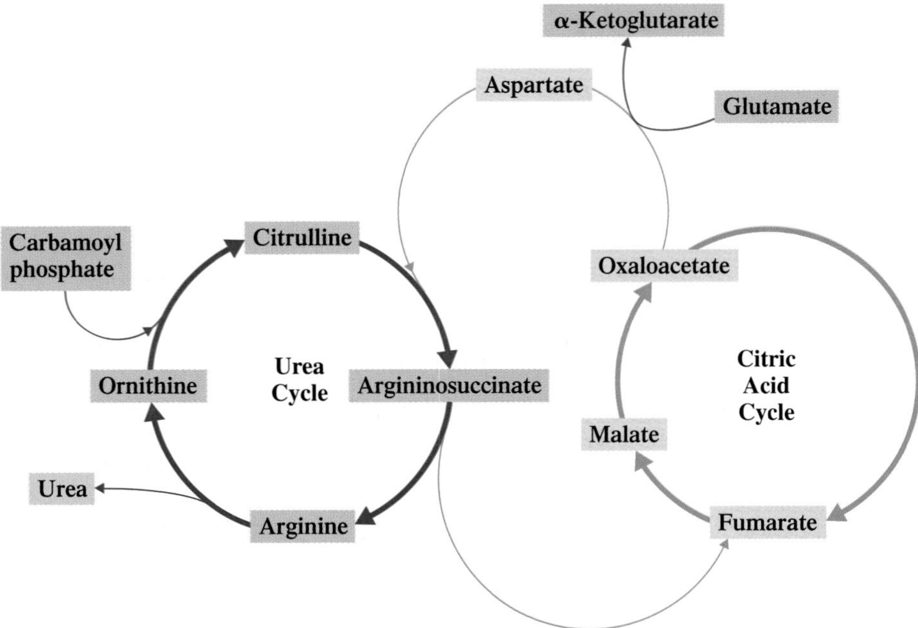

Figure 26.6
Fumarate from the urea cycle enters the citric acid cycle, and aspartate produced from oxaloacetate of the citric acid cycle enters the urea cycle.

Chemical CONNECTIONS

26.1 The Chemical Composition of Urine

Urine is a dilute aqueous solution containing many solutes whose concentrations are dependent on the diet and state of health of the individual. On average, about 4 g of solutes is present in a 100-g urine sample; thus urine is an approximately 4%-by-mass aqueous solution of materials eliminated from the body.

The solutes present in urine are of two general types: organic compounds and inorganic ions. Generally, the organic compounds are more abundant because of the dominance of urea, as shown in the following composition data.

Major Constituents of Urine (for a 1400-mL specimen obtained over a 24-hour period)

Organic constituents		Inorganic constituents	
urea	25.0 g	chloride (Cl^-)	6.3 g
creatinine	1.5 g	sodium (Na^+)	3.0 g
amino acids	0.8 g	potassium (K^+)	1.7 g
uric acid	0.7 g	sulfate (SO_4^{2-})	1.4 g
		dihydrogen phosphate ($H_2PO_4^-$)	1.2 g
		ammonium (NH_4^+)	0.8 g
		calcium (Ca^{2+})	0.2 g
		magnesium (Mg^{2+})	0.2 g

Urea, the solute present in the greatest quantity in urine, is odorless and colorless in solution (Section 26.4). (The pale yellow color of urine is due to small amounts of urobilin and related compounds, as discussed in Section 26.7.) Urea is the principal nitrogen-containing end product of protein metabolism.

Creatinine, the second most abundant organic product in urine, is produced from the amino acids arginine, methionine, and glycine. Uric acid is a product of the metabolism of purines from nucleic acids.

The most abundant inorganic constituent of urine is chloride ion. Its primary source is dietary table salt (NaCl). Correspondingly, the second most abundant ion present is sodium ion, the positive ion in table salt. The sulfate ion present in urine comes primarily from the metabolism of sulfur-containing amino acids. Ammonium ions come primarily from the hydrolysis of urea.

Urine is normally slightly acidic, having an average pH value of 6.6. However, the pH range is wide—from 4.5 to 8.0. Fruits and vegetables in the diet tend to raise urine pH, and high-protein foods tend to lower urine pH.

A normal adult excretes 1000–1500 mL of urine daily. Actual urine volume depends on liquid intake and weather. During hot weather, urine volume decreases as a result of increased water loss through perspiration.

Chemical CONNECTIONS

26.2 Arginine, Citrulline, and the Chemical Messenger Nitric Oxide

A somewhat startling biochemical discovery, made during the early 1990s, was the existence within the human body of a *gaseous* chemical messenger, the simple diatomic molecule nitric oxide (NO). Its production involves two of the amino acid intermediates of the urea cycle—arginine and citrulline. Arginine reacts with oxygen to produce citrulline and NO. The reaction requires NADPH and the enzyme nitric oxide synthase (NOS).

$$O_2 + \underset{\text{Arginine}}{H_2N{-}\overset{+}{C}({=}NH_2){-}NH{-}(CH_2)_3{-}CH(\overset{+}{N}H_3){-}COO^-} \xrightarrow[\text{NADPH}\ \rightarrow\ \text{NADP}^+]{\text{Nitric oxide synthase}} \underset{\text{Citrulline}}{H_2N{-}C({=}O){-}NH{-}(CH_2)_3{-}CH(\overset{+}{N}H_3){-}COO^-} + NO$$

Even though this reaction involves urea cycle intermediates, it is completely independent of the urea cycle.

Nitric oxide affects many kinds of cells and has particularly striking effects in the following areas:

1. NO helps maintain blood pressure by dilating blood vessels.
2. NO is a chemical messenger in the central nervous system.
3. NO is involved in the immune system's response to invasion by foreign organisms or materials.
4. NO is found in the brain and may be a major biochemical component of long-term memory.

In humans, nitric oxide is the first known biological messenger compound that is a gas. It can easily pass through biological membranes by diffusion. No specific receptor or transport system is needed. Because of its extreme reactivity, NO exists for less than 10 seconds before undergoing reaction. This high reactivity prevents it from getting more than 1 millimeter from its site of synthesis.

The action of nitroglycerin, when it is used as a heart medication (for angina pectoris), is now known to be related to NO. Nitric oxide is the active metabolite from nitroglycerin.

Before the discovery of nitric oxide's role as a biochemical messenger, this gas was thought of mainly as a noxious atmospheric gas found in cigarette smoke and smog, as a destroyer of ozone, and as a precursor of acid rain. The contrast between nitric oxide's roles in environmental pollution and in the human body as a chemical messenger is indeed startling.

26.5 Amino Acid Carbon Skeletons

The removal of the amino group of an amino acid by transamination or oxidative deamination (Section 26.3) produces an α-keto acid that contains the carbon skeleton from the amino acid. Each of the 20 amino acid carbon skeletons undergoes a different degradation process. For alanine and serine, the degradation requires a single step. For most carbon arrangements, however, multistep reaction sequences are required. We will not consider the details of these various degradation procedures in this text. It is important, however, to consider the products of these degradation sequences. There are only seven, and each is a compound that we have previously encountered in our discussions of metabolism. The seven degradation products are pyruvate, acetyl CoA, acetoacetyl CoA, α-ketoglutarate, succinyl CoA, fumarate, and oxaloacetate. The last four products are intermediates in the citric acid cycle. Figure 26.7 relates these seven degradation products to the amino acids from which they are obtained. Some amino acids appear in more than one box in Figure 26.7. This means either that there is more than one pathway for degradation or that some of the carbon atoms of the skeleton emerge as one product and others as another product.

Amino acids that are degraded to citric acid cycle intermediates can serve as glucose precursors and are called glucogenic. A **glucogenic amino acid** *is an amino acid whose carbon-containing degradation product(s) can be used to produce glucose via gluconeogenesis.*

Amino acids that are degraded to acetyl CoA or acetoacetyl CoA can contribute to the formation of fatty acids or ketone bodies and are called ketogenic. A **ketogenic amino acid** *is an amino acid whose carbon-containing degradation product(s) can be used to* *produce ketone bodies.* Even though acetyl CoA can enter the citric acid cycle, there can be no *net* production of glucose from it. Acetyl groups are C_2 species, and such species only maintain the carbon count in the cycle, because two CO_2 molecules exit the cycle (Section 23.6). Thus amino acids that are degraded to acetyl CoA (or acetoacetyl CoA) are not glucogenic.

Amino acids that are degraded to pyruvate can be either glucogenic or ketogenic. Pyruvate can be metabolized to either oxaloacetate (glucogenic) or acetyl CoA (ketogenic).

Only two amino acids are purely ketogenic: leucine and lysine. Nine amino acids are both glucogenic and ketogenic: those degraded to pyruvate (see Figure 26.7), as well as tyrosine, phenylalanine, and isoleucine (which have two degradation products). The remaining nine amino acids are purely glucogenic.

Our discussion of glucogenicity and ketogenicity for amino acids points out that ATP production (common metabolic pathway) is not the only fate for amino acid degradation products. They can also be converted to glucose, ketone bodies, or fatty acids (via acetyl CoA).

26.6 Amino Acid Biosynthesis

The classification of amino acids as essential or nonessential for humans (Section 20.2) roughly parallels the number of steps in their biosynthetic pathways and the energy

Figure 26.7
Fates of the carbon skeletons of amino acids. Glucogenic amino acids are shaded blue and ketogenic amino acids are shaded green. Some amino acids (marked with an asterisk) have more than one degradation pathway.

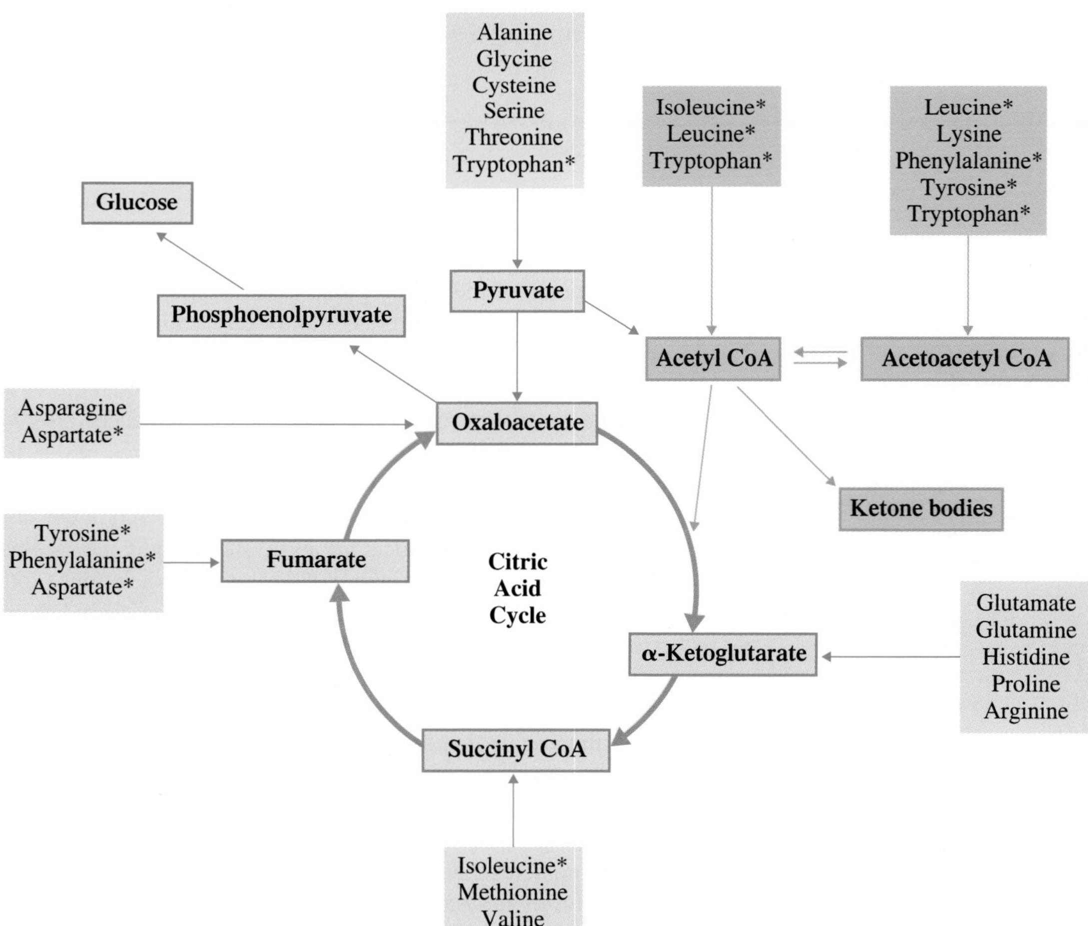

Figure 26.8
A summary of the starting materials for the biosynthesis of the 11 nonessential amino acids.

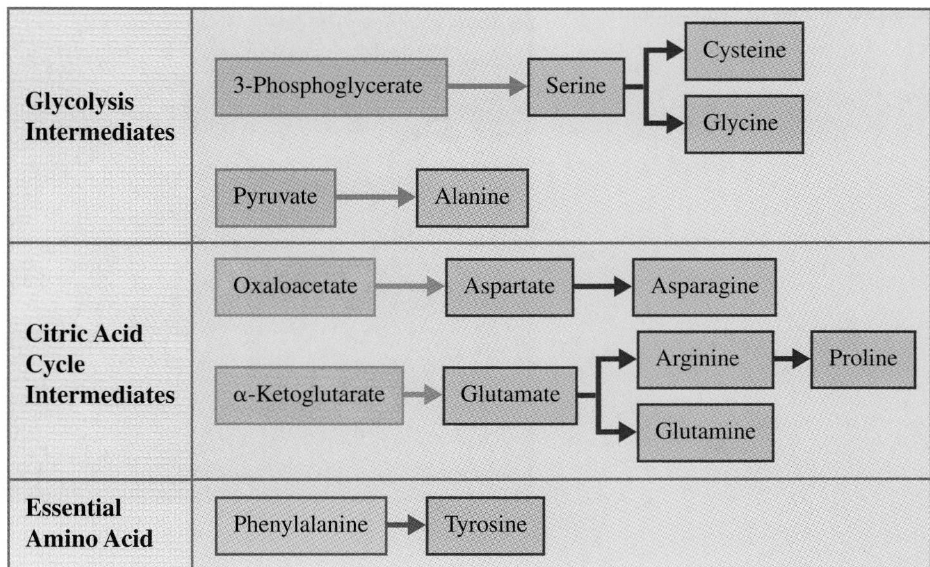

• There is considerable variation in biosynthetic pathways for amino acids among different species. By contrast, the basic pathways of carbohydrate and lipid metabolism are almost universal.

required for their synthesis. The nonessential amino acids can be made in 1–3 steps. The essential ones have biosynthetic pathways that require 7–10 steps, based on observations of their synthesis in microorganisms. Most bacteria and plants can synthesize all the amino acids by pathways not present in humans. Plants, consumed as food, are the major source of the essential amino acids in humans and animals.

The starting materials for the biosynthesis of the 11 nonessential amino acids are the glycolysis intermediates 3-phosphoglycerate and pyruvate and the citric acid cycle intermediates oxaloacetate and α-ketoglutarate (see Figure 26.8).

Three of the nonessential amino acids—alanine, aspartate, and glutamate—are biosynthesized by transamination (Section 26.3) of the appropriate α-keto acid starting material.

$$\underset{\text{Pyruvate}}{CH_3-\overset{\overset{O}{\|}}{C}-COO^-} \xrightarrow{\text{Transamination}} \underset{\text{Alanine}}{CH_3-\overset{\overset{+}{N}H_3}{\overset{|}{C}H}-COO^-}$$

$$\underset{\text{Oxaloacetate}}{{}^-OOC-CH_2-\overset{\overset{O}{\|}}{C}-COO^-} \xrightarrow{\text{Transamination}} \underset{\text{Aspartate}}{{}^-OOC-CH_2-\overset{\overset{+}{N}H_3}{\overset{|}{C}H}-COO^-}$$

$$\underset{\alpha\text{-Ketoglutaric acid}}{{}^-OOC-CH_2-CH_2-\overset{\overset{O}{\|}}{C}-COO^-} \xrightarrow{\text{Transamination}} \underset{\text{Glutamate}}{{}^-OOC-CH_2-CH_2-\overset{\overset{+}{N}H_3}{\overset{|}{C}H}-COO^-}$$

• PKU is characterized by elevated blood levels of phenylalanine and phenylpyruvate. The physical consequence of PKU (elevated phenylpyruvate levels) is damage to *developing* brain cells. In children up to six years old, PKU leads to retarded mental development. The major defense against PKU is mandatory screening of newborns to identify the one in every 20,000 who is afflicted and then restricting those children's dietary phenylalanine intake to that needed for protein synthesis until they are six years old. After that age, brain cells are not so susceptible to the toxic effect of phenylpyruvate.

The nonessential amino acid tyrosine is obtained from the essential amino acid phenylalanine in a one-step oxidation that involves molecular O_2, NADPH, and the enzyme *phenylalanine hydroxylase.* Lack of this enzyme causes the metabolic disease phenylketonuria (PKU).

26.7 Hemoglobin Catabolism

Red blood cells are highly specialized cells whose primary function is to deliver oxygen to, and remove carbon dioxide from, body tissues. Mature red cells have no nucleus or DNA. Instead, they are filled with the red pigment hemoglobin. Red blood cell formation

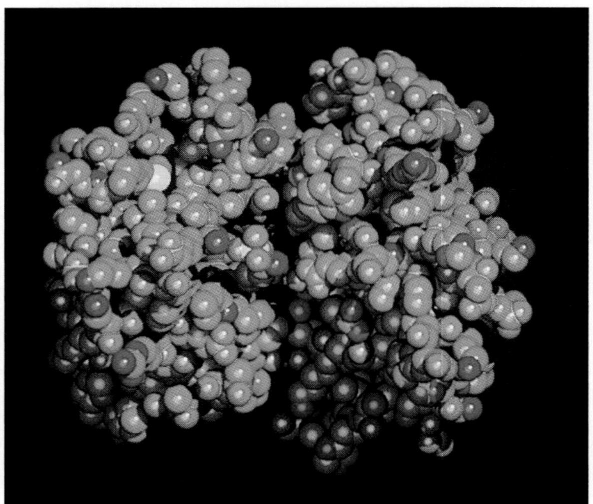
A molecular model of hemoglobin.

occurs in the bone marrow, and approximately 200 billion new red blood cells are formed daily. The life span of a red blood cell is about 4 months.

The oxygen-carrying ability of red blood cells is due to the molecule hemoglobin present in such cells. Hemoglobin is a conjugated protein (Section 20.11); the protein portion is called *globin,* and the prosthetic group (nonprotein portion) is *heme.* Heme contains four pyrrole groups (Section 17.8) joined together with an iron atom in the center.

• An excess of red blood cells is called *polycythemia,* and a shortage is called *anemia.*

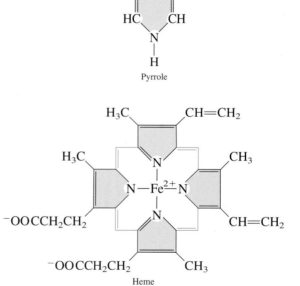

Pyrrole

Heme

It is the iron atom in heme that interacts with O_2, forming a reversible complex with it (Section 9.9). This complexation increases the amount of O_2 that the blood can carrry by a factor of 80 over that which simply "dissolves" in the blood.

• The tetrapyrrole heme ring is the only component of hemoglobin that is not reused by the body.

Old red blood cells are broken down in the spleen (primary site) and liver (secondary site). Part of this process is degradation of hemoglobin. The globin protein is hydrolyzed to amino acids, which become part of the amino acid pool (Section 26.2). The iron atom of heme becomes part of *ferritin,* an iron-storage protein, which saves the iron for use in the biosynthesis of new hemoglobin molecules. The tetrapyrrole carbon arrangement of heme is degraded to *bile pigments* that are eliminated in feces and to a lesser extent in urine.

Degradation of heme begins with a ring-opening reaction in which a single carbon atom is removed. The product is called *biliverdin.*

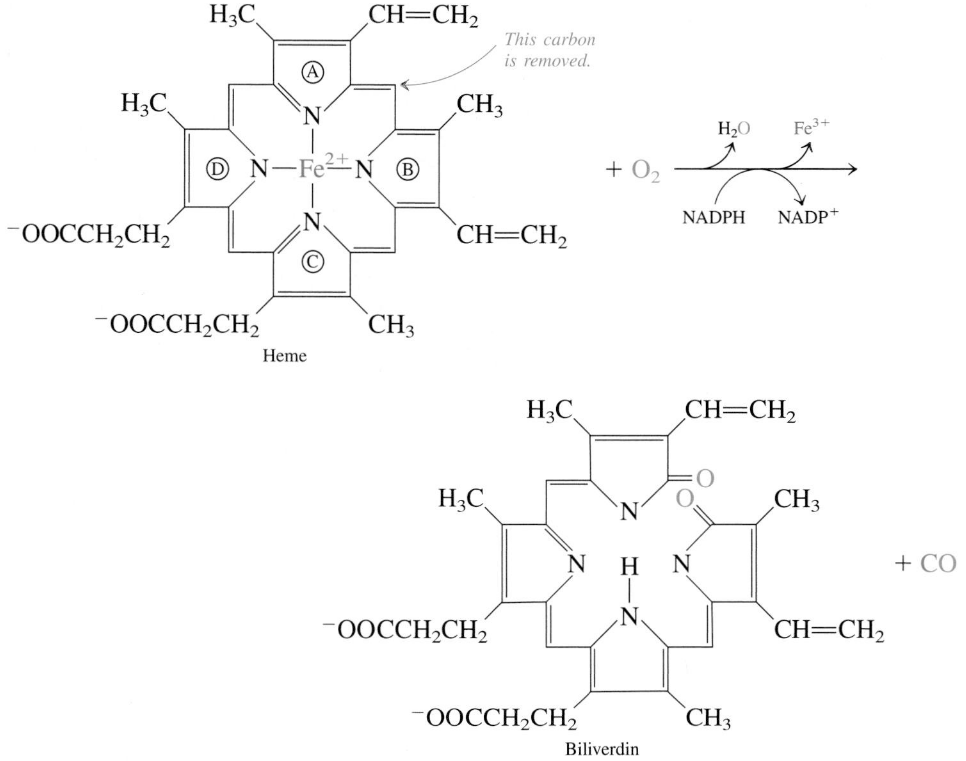

Heme

Biliverdin

• The level of carbon monoxide produced in the first step of hemoglobin degradation is sufficient to complex 1% of the oxygen-binding sites of the blood's hemoglobin.

This reaction has several important characteristics. (1) Molecular oxygen, O_2, is required as a reactant. (2) Ring opening releases the iron atom to be incorporated into ferritin. (3) The product containing the excised carbon atom is *carbon monoxide* (a substance toxic to the human body). The carbon monoxide so produced reacts with functioning hemoglobin, forming a CO–hemoglobin complex; this decreases the oxygen-carrying ability of the blood. CO–hemoglobin complexes are very stable; CO release to the lungs is a slow process.

An alternative rendering of the structure of biliverdin is

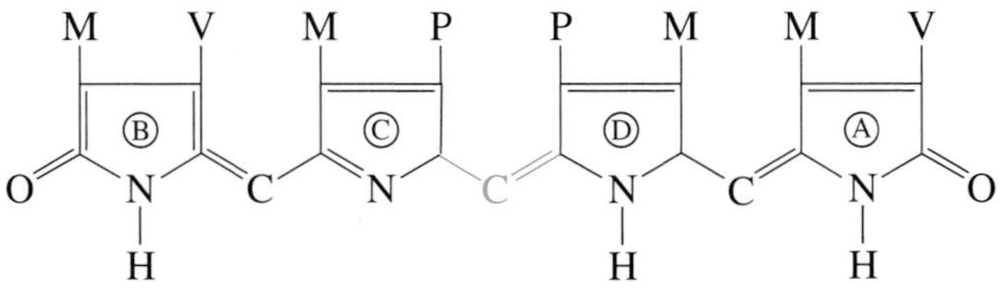

Biliverdin

M = —CH_3 (methyl)
V = —$CH{=}CH_2$ (vinyl)
P = —CH_2—CH_2—COO^- (propionate)

This structure correctly emphasizes the *linear* (straight-line) arrangement of the pyrrole rings that results from the ring opening. It also employs a notation, common in heme chemistry, in which letters are used to denote attachments to the pyrrole rings.

In the second step of heme degradation, biliverdin is converted to bilirubin. This change involves reduction of the central methylene bridge of biliverdin.

Biliverdin

$NADPH + H^+$

$NADP^+$

Bilirubin

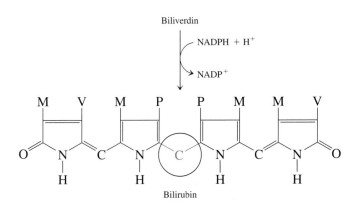

The change from heme to biliverdin to bilirubin usually occurs in the spleen. The bilirubin is then transported by serum albumin to the liver, where it is rendered more water-soluble by the attachment of sugar residues to its propionate side chains (P side chains). The solubilizing sugar is *glucuronide* (glucose with a $—COO^-$ group on C-6 instead of a $—CH_2OH$ group).

Bilirubin diglucuronide

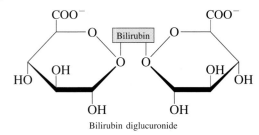

The solubilized bilirubin is excreted from the liver in bile, which flows into the small intestine. Here the bilirubin diglucuronide is changed, in a multistep process, to either stercobilin for excretion in feces or urobilin for excretion in urine. Both stercobilin and urobilin still have tetrapyrrole structures (Figure 26.9). Intestinal bacteria are primarily responsible for the changes that produce stercobilin and urobilin.

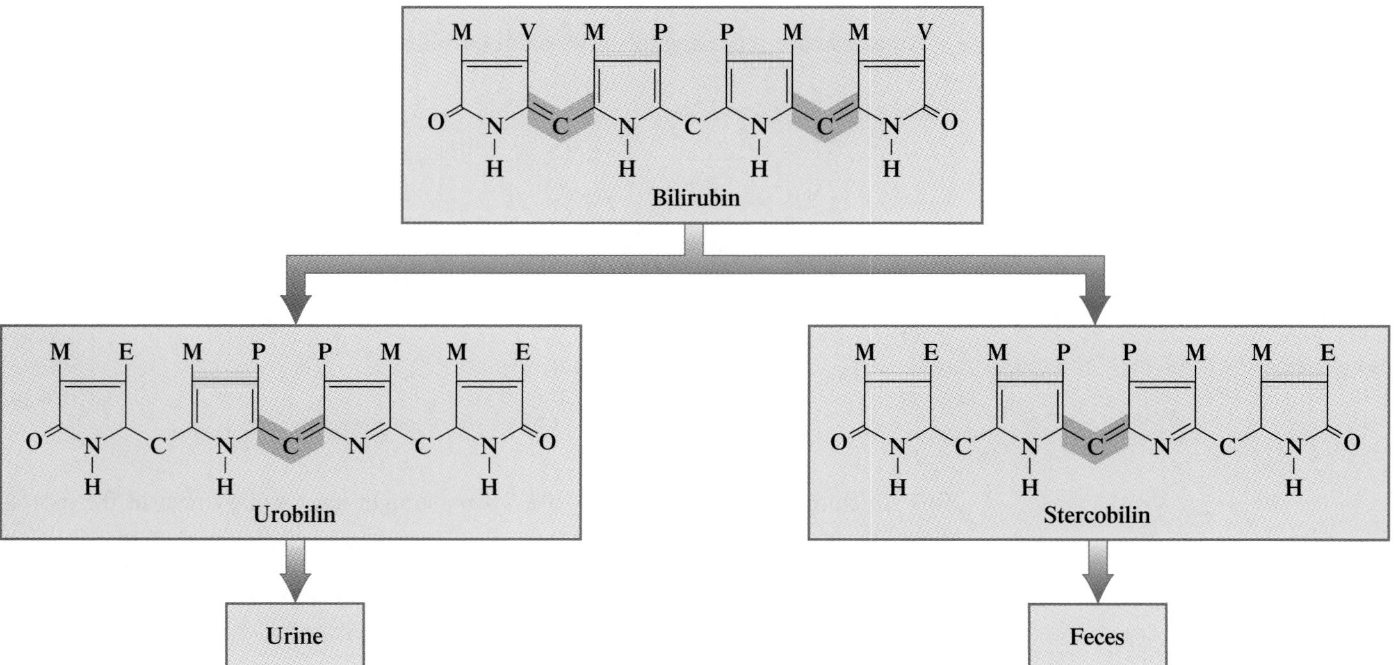

Figure 26.9
Stercobilin and urobilin have structures closely resembling that of bilirubin. Changes include reduction of vinyl (V) groups to ethyl (E) groups and reduction of the $—CH_2—$ bridge.

Interrelationships Among Lipid, Carbohydrate, and Protein Metabolism

CHEMISTRY AT A GLANCE

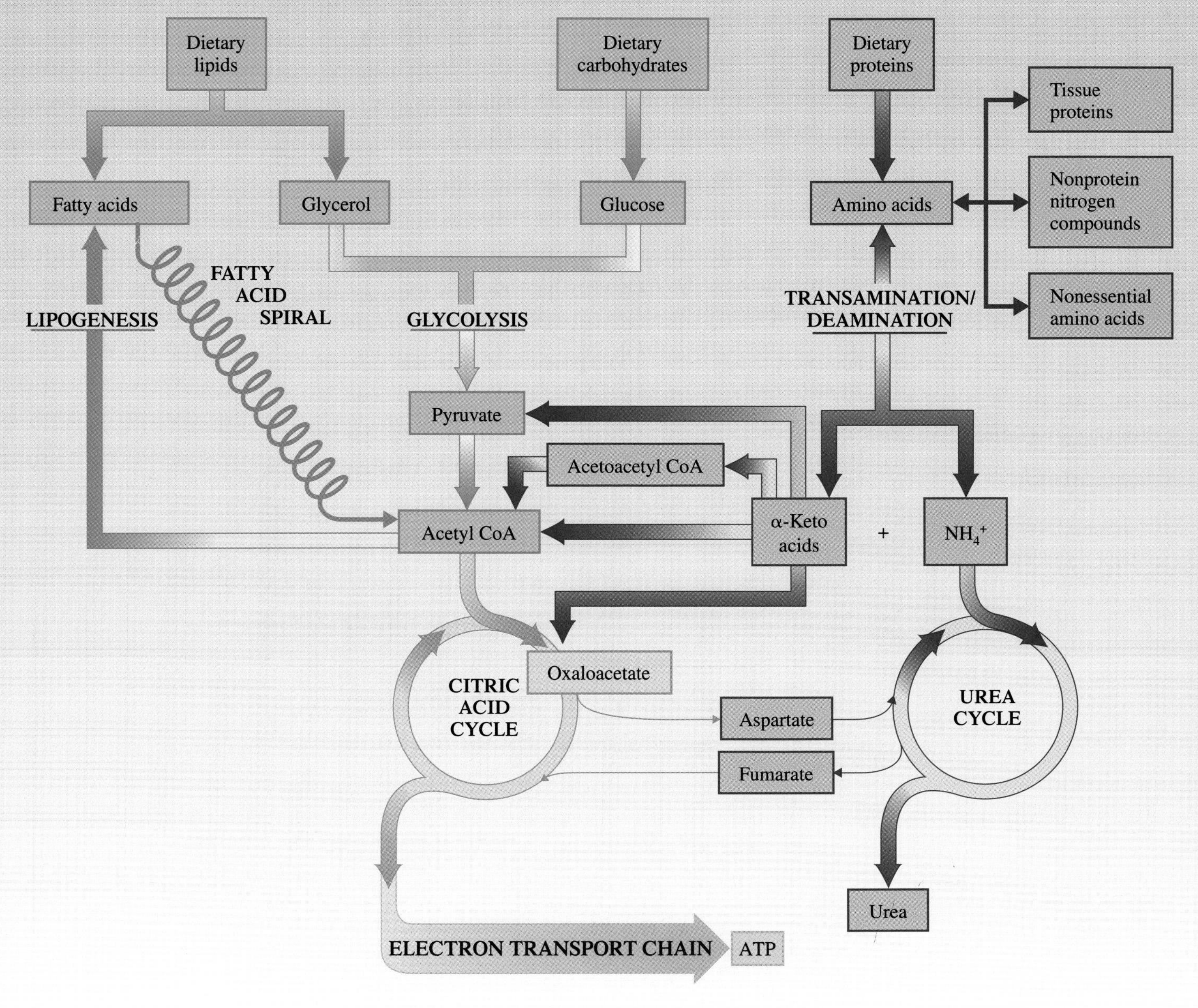

• The first part of the names *biliverdin* and *bilirubin* and the last part of the names *stercobilin* and *urobilin* all come from the Latin *bilis,* which means "bile." As for the other parts of the names:

1. Latin *virdis* means "green"; biliverdin = "green bile."
2. Latin *rubin* means "red"; bilirubin = "red bile."
3. Latin *urina* means "urine"; urobilin = "urine bile."
4. Latin *sterco* means "dung"; stercobilin = "dung bile."

• Bile Pigments

The tetrapyrrole degradation products obtained from heme are known as bile *pigments* because they are secreted with the bile (Section 25.1), and most of them are highly colored. Biliverdin and bilirubin are, respectively, green and reddish orange in color. Stercobilin has a brownish hue and is the compound that gives feces their characteristic color. Urobilin is the pigment that gives urine its characteristic yellow color. Normally, the body excretes 1–2 mg of bile pigments in urine daily and 250–350 mg of bile pigments in feces daily.

When the body is functioning properly, the degradation of heme in the spleen to bilirubin and the removal of bilirubin from the blood by the liver balance each other.

Jaundice is the condition that occurs when this balance is upset such that bilirubin concentrations in the blood become higher than normal. The skin and the white of the eyes

• The word *jaundice* comes from the French *jaune,* which means "yellow."

• A mild form of jaundice is common among premature infants because of underdeveloped liver function. Treatment involves the use of white or ultraviolet light, which breaks the bilirubin down to simpler compounds that are more easily excreted.

acquire a yellowish tint because of the excess bilirubin in the blood. Jaundice can occur as a result of liver diseases, such as infectious hepatitis and cirrhosis, that decrease the liver's ability to process bilirubin; spleen malfunction, in which heme is degraded faster than it can be adsorbed by the liver; and gallbladder malfunction, usually from an obstruction of the bile duct.

The local coloration associated with a deep bruise is also related to the pigmentation associated with heme, biliverdin, and bilirubin. The changing color of the bruise as it heals reflects the dominant degradation product present at the time as the tissue repairs itself.

Figure 26.10
The human body's response to feasting, to fasting, and to starvation.

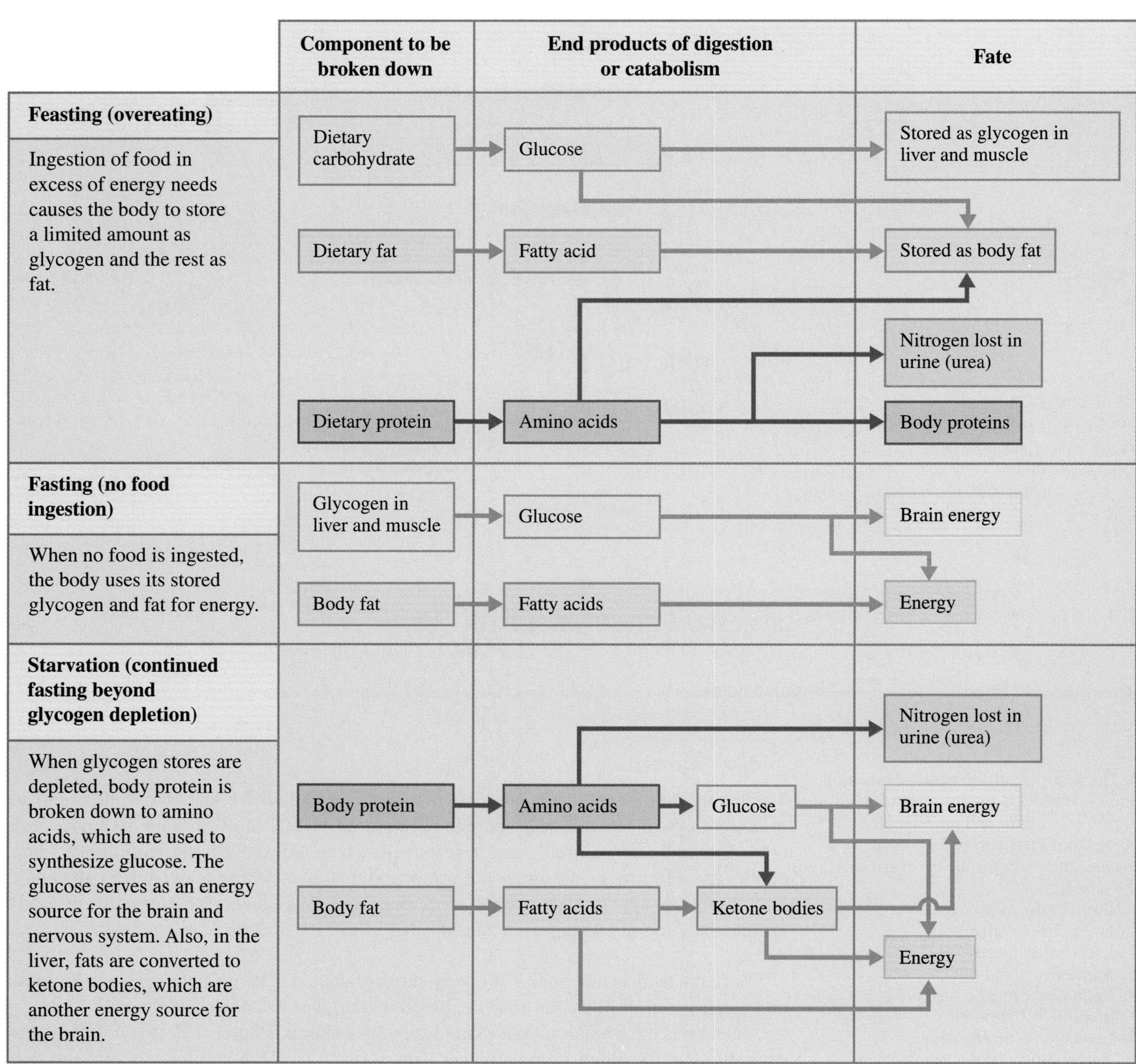

26.8 Interrelationships Among Metabolic Pathways

In this chapter and the previous two chapters, we have considered metabolic pathways of carbohydrates, lipids, and proteins. These pathways are not independent of each other but rather are integrally linked, as shown in the Chemistry at a Glance on page 787. The numerous connections between pathways mean that a change in one pathway can affect many other pathways.

A good illustration of the interrelationships among pathways emerges from comparing the processes of eating (feasting), not eating for a short period (fasting), and not eating for a prolonged period (starvation). Figure 26.10 shows how the body responds to each of these situations.

Concepts to Remember

Protein digestion and absorption. Digestion of proteins involves the hydrolysis of the peptide bonds that link amino acids to each other. This process begins in the stomach and is completed in the small intestine. The amino acids released by digestion are absorbed through the intestinal wall into the bloodstream.

Amino acid pool. The amino acid pool within cells consists of moderate amounts of each of the 20 standard amino acids found in proteins.

Amino acid utilization. Amino acids from the amino acid pool are used for protein synthesis, synthesis of nonprotein nitrogen compounds, synthesis of nonessential amino acids, and energy production.

Transamination. Transamination is an enzyme-catalyzed transfer of an amino group from an amino acid to an α-keto acid. Transamination is a step in obtaining energy from amino acids.

Oxidative deamination. Oxidative deamination is a reaction in which an amino acid is converted into a keto acid, accompanied by the release of a free ammonium ion. Deamination is a step in obtaining energy from amino acids.

Urea cycle. The urea cycle is the metabolic pathway that converts ammonium ions into urea. This cycle processes the ammonium ions in the form of carbamoyl phosphate, a compound formed from CO_2, NH_4^+, ATP, and H_2O.

Amino acid carbon skeletons. Amino acids are classified as glucogenic or ketogenic on the basis of their catabolic pathways. Glucogenic amino acids are degraded to intermediates of the citric acid cycle and can be used for glucose synthesis. Ketogenic amino acids are degraded into acetoacetyl CoA or acetyl CoA and can be used to make ketone bodies.

Amino acid biosynthesis. Amino acid biosynthesis is the process in which the body synthesizes amino acids from intermediates of the glycolysis pathway and the citric acid cycle. Eleven amino acids can be synthesized by the body. The other nine amino acids, called essential amino acids, must be obtained from the diet.

Hemoglobin catabolism. Hemoglobin from red blood cells undergoes a stepwise degradation to biliverdin, to bilirubin, and then to bile pigments that are excreted from the body.

Key Reactions and Equations

1. Digestion of protein (Section 26.1)

$$\text{Protein} \xrightarrow[\text{and HCl}]{\text{Pancreatic enzymes}} \text{amino acids}$$

2. Transamination (Section 26.3)

$$R_1\text{—}\underset{\overset{|}{\overset{+}{N}H_3}}{CH}\text{—}COO^- + R_2\text{—}\overset{\overset{O}{\|}}{C}\text{—}COO^- \longrightarrow R_1\text{—}\overset{\overset{O}{\|}}{C}\text{—}COO^- + R_2\text{—}\overset{\overset{\overset{+}{N}H_3}{|}}{CH}\text{—}COO^-$$

3. Oxidative deamination (Section 26.3)

$$R\text{—}\overset{\overset{NH_3^+}{|}}{CH}\text{—}COO^- + NAD^+ + H_2O \longrightarrow R\text{—}\overset{\overset{O}{\|}}{C}\text{—}COO^- + NH_4^+ + NADH + H^+$$

4. Formation of urea (Section 26.4)

$$NH_4^+ + CO_2 + 3ATP + 2H_2O + \text{aspartate} \longrightarrow \text{urea} + \text{fumarate} + 2ADP + AMP + 4P_i$$

Key Terms

Amino acid pool (26.2)
Glucogenic amino acid (26.5)
Ketogenic amino acid (26.5)
Nitrogen balance (26.2)
Oxidative deamination (26.3)
Protein turnover (26.2)
Transamination (26.3)
Urea cycle (26.4)

Exercises and Problems

The members of each pair of problems in this section test similar material.

Protein Digestion and Absorption (Section 26.1)

26.1 The first step in protein digestion is denaturation. Where does denaturation occur in the body, and what is the denaturant?

26.2 What is the first digestive enzyme that protein encounters, and where does this encounter take place?

26.3 What is the relationship between pepsinogen and pepsin?

26.4 What is the relationship between trypsinogen and trypsin?

26.5 Contrast gastric juice and pancreatic juice in terms of pH.

26.6 Contrast gastric juice and pancreatic juice in terms of enzymes present.

26.7 Absorption of amino acids through the intestinal wall requires a transport system. Explain.

26.8 The passage of small polypeptides through the intestinal wall is particularly important in infants. Explain.

26.9 What is the amino acid pool?

26.10 What are the three major sources of amino acids for the amino acid pool?

26.11 What is protein turnover?

26.12 The protein turnover rate is not the same for all proteins. Explain.

26.13 What is the difference between a positive nitrogen balance and a negative nitrogen balance?

26.14 What happens to the nitrogen balance during a period of fasting?

26.15 What happens to the nitrogen balance when the diet is lacking in one of the essential amino acids?

26.16 What happens to the nitrogen balance of a pregnant woman?

Amino Acid Utilization (Section 26.2)

26.17 What four types of processes draw amino acids out of the amino acid pool?

26.18 What percent of amino acid utilization from the amino acid pool is for protein synthesis?

26.19 Classify each of the following amino acids as essential or nonessential.

a. Lysine b. Arginine
c. Serine d. Tryptophan

26.20 Classify each of the following amino acids as essential or nonessential.

a. Proline b. Asparagine
c. Glutamic acid d. Tyrosine

26.21 Which of the amino acids in Problem 26.19 can be biosynthesized in the human body?

26.22 Which of the amino acids in Problem 26.20 can be biosynthesized in the human body?

Transamination and Oxidative Deamination (Section 26.3)

26.23 In general terms, what are the two reactants in a transamination reaction?

26.24 In general terms, what are the two products in a transamination reaction?

26.25 Write structural equations for the transamination reactions that involve the following pairs of reactants.

a. Threonine and pyruvate
b. Alanine and oxaloacetate
c. Glycine and α-ketoglutarate
d. Threonine and α-ketoglutarate

26.26 Write structural equations for the transamination reactions that involve the following pairs of reactants.

a. Threonine and oxaloacetate
b. Glycine and pyruvate
c. Alanine and oxaloacetate
d. Isoleucine and α-ketoglutarate

26.27 What are the three keto acids that are usually reactants in transamination reactions?

26.28 The net effect of transamination is to collect the amino groups from a variety of amino acids into the compound glutamate. Explain.

26.29 What is the function of pyridoxal phosphate in transamination processes?

26.30 Which one of the B vitamins is important in the process of transamination?

26.31 Describe the process of oxidative deamination.

26.32 What coenzyme is required for an oxidative deamination reaction?

26.33 How does oxidative deamination differ from transamination?

26.34 What do the processes of oxidative deamination and transamination have in common?

26.35 Draw the structure of the keto acid produced from the oxidative deamination of each of the following amino acids.

a. Glutamate b. Cysteine
c. Alanine d. Phenylalanine

26.36 Draw the structure of the keto acid produced from the oxidative deamination of each of the following amino acids.

a. Glycine b. Leucine
c. Aspartate d. Tyrosine

26.37 The following α-keto acid can be used as a substitute for a particular *essential* amino acid in the diet. Explain how this is possible, and draw the structure of the essential amino acid.

$$CH_3-\underset{\underset{CH_3}{|}}{CH}-CH_2-\overset{\overset{O}{\|}}{C}-COO^-$$

26.38 The following α-keto acid can be used as a substitute for a particular *essential* amino acid in the diet. Explain how this is possible, and draw the structure of the essential amino acid.

$$CH_3-CH_2-\underset{\underset{CH_3}{|}}{CH}-\overset{\overset{O}{\|}}{C}-COO^-$$

26.39 Give the *name* of the compound produced from each reactant or the reactant needed to produce each product using transamination.

a. Oxaloacetate ⟶ ? b. ? ⟶ α-ketoglutarate
c. Alanine ⟶ ? d. ? ⟶ glutamate

26.40 Give the *name* of the compound produced from each reactant or the reactant needed to produce each product using transamination.

a. Pyruvate ⟶ ? b. ? ⟶ oxaloacetate
c. Aspartate ⟶ ? d. ? ⟶ alanine

The Urea Cycle (Section 26.4)

26.41 Draw the chemical structure of urea.

26.42 What are some of the physical characteristics of urea?

26.43 In what chemical form do ammonium ions enter the urea cycle?

26.44 What are the chemical reactants for the formation of carbamoyl phosphate?

26.45 What is a carbamoyl group?

26.46 Draw the structure of the molecule carbamoyl phosphate.

26.47 How do the structures of the three amino acids involved as intermediates in the urea cycle differ from each other?

26.48 Three amino acids are involved as intermediates in the urea cycle. Name them and classify them as standard or nonstandard amino acids.

26.49 Name the compound that enters the urea cycle by combining with ornithine.

26.50 Name the compound that enters the urea cycle by combining with citrulline.

26.51 What substance is consumed and then regenerated in the urea cycle?

26.52 If the urea cycle were named in the same way as the citric acid cycle, what would the cycle's name be?

26.53 What is the first reaction of the urea cycle that occurs in the cytoplasm?

26.54 What is the first reaction of the urea cycle that occurs in the mitochondrial matrix?

26.55 In each of the following pairs of compounds associated with the urea cycle, specify which one is encountered first in the cycle.

a. Citrulline and arginine
b. Ornithine and aspartate
c. Argininosuccinate and fumarate
d. Carbamoyl phosphate and citrulline

26.56 In each of the following pairs of compounds associated with the urea cycle, specify which one is encountered first in the cycle.

a. Carbamoyl phosphate and fumarate
b. Argininosuccinate and arginine
c. Ornithine and aspartate
d. Citrulline and ATP

26.57 How much energy is expended in the synthesis of a molecule of urea?

26.58 What are the sources of the carbon atom and the two nitrogen atoms in urea?

26.59 What is the fate of the fumarate formed in the urea cycle?

26.60 Explain how the urea cycle is linked to the citric acid cycle.

Amino Acid Carbon Skeletons (Section 26.5)

26.61 What are the four possible degradation products of the carbon skeletons of amino acids that are citric acid cycle intermediates?

26.62 What are the three possible degradation products of the carbon skeletons of amino acids that are not citric acid cycle intermediates.

26.63 With the help of Figure 26.7, write the name of the compound (or compounds) to which each of the following amino acid carbon skeletons is metabolized.

a. Leucine b. Isoleucine
c. Aspartate d. Arginine

26.64 With the help of Figure 26.7, write the name of the compound (or compounds) to which each of the following amino acid carbon skeletons is metabolized.

a. Serine b. Tyrosine
c. Tryptophan d. Histidine

26.65 What degradation characteristics do all purely glucogenic amino acids share?

26.66 What degradation characteristics do all purely ketogenic amino acids share?

Amino Acid Biosynthesis (Section 26.6)

26.67 What compound is a major source of amino groups in amino acid biosynthesis?

26.68 How does transamination play a role in both catabolism and anabolism of amino acids?

26.69 What are the five starting materials for the biosynthesis of the 11 nonessential amino acids?

26.70 What is a major difference between the biosynthesis pathways for the essential and the nonessential amino acids?

Hemoglobin Catabolism (Section 26.7)

26.71 What happens to the globin produced from the breakdown of hemoglobin?

26.72 What happens to the iron (Fe^{2+}) produced from the breakdown of hemoglobin?

26.73 What are the structural differences between heme and biliverdin?

26.74 What are the structural differences between biliverdin and bilirubin?

26.75 Arrange the following substances in the order in which they appear during the catabolism of heme: bilirubin, urobilin, biliverdin, and bilirubin diglucuronide.

26.76 Carbon monoxide is a by-product of the degradation of heme. At what point in the degradation process is it formed, and what happens to it once it is formed?

26.77 Which bile pigment is responsible for the yellow color of urine?

26.78 Which bile pigment is responsible for the brownish-red color of feces?

26.79 What chemical condition is responsible for jaundice?

26.80 What physical conditions cause jaundice?

Interrelationships of Metabolic Pathways (Section 26.8)

26.81 Arrange the following compounds in the order in which they would be produced if carbons from the carbon skeleton of serine were to become part of one of the ketone bodies: acetoacetate, pyruvate, and acetyl CoA.

26.82 Arrange the following compounds in the order in which they would be produced if carbons from the carbon skeleton of serine were to become part of carbon dioxide: citrate, isocitrate, pyruvate.

26.83 Briefly explain how the carbon atoms from amino acids can end up in adipose tissue.

26.84 Briefly explain how the carbon atoms from amino acids can end up in glucose.

26.85 How are the amino acids from protein "processed" when they are present in amounts that exceed the body's needs?

26.86 How are the amino acids from protein "processed" when an individual is in a state of starvation?

Additional Problems

26.87 In which of the processes (1) urea cycle, (2) hemoglobin catabolism, and (3) transamination reactions would each of the following molecules be encountered.

a. Citrulline b. Carbon monoxide
c. Pyruvate d. Urobilin
e. Arginine f. Pyridoxal phosphate

26.88 Characterize each of the following molecules as a possible reactant, product, or enzyme of (1) transamination, (2) oxidative deamination, or (3) both transamination and oxidative deamination.

a. Arginine b. Glutamate
c. α-Ketoglutarate d. Ammonium ion
e. Oxaloacetate f. Glutamate dehydrogenase

26.89 Arrange the following events in the order in which they occur in the digestive process for proteins.

(1) Peptide bonds are hydrolyzed with the help of pepsin.
(2) Peptide bonds are hydrolyzed with the help of trypsin.
(3) Large polypeptides pass from the stomach into the small intestine.
(4) Amino acids pass through the intestinal wall into the bloodstream.

26.90 With the help of Figure 26.7 and the given conversion information, classify each of the following amino acids as (1) ketogenic but not glucogenic, (2) glucogenic but not ketogenic, (3) both ketogenic and glucogenic, or (4) neither ketogenic nor glucogenic.

a. Alanine is converted to pyruvate.
b. Aspartate is converted to either fumarate or oxaloacetate.
c. Lysine is converted to acetoacetyl CoA.
d. Isoleucine is converted to either succinyl CoA or acetyl CoA.

26.91 Indicate whether each of the following statements refers to *transamination* or *deamination.*

a. Both an amino acid and a keto acid are reactants.
b. An amino acid and water are reactants.
c. The ammonium ion is a product.
d. An amino acid is produced from a keto acid.

26.92 Which of the compounds (1) ornithine, (2) citrulline, (3) argininosuccinate, and (4) arginine is associated with each of the following urea cycle "occurrences"?

a. Reacts with carbamoyl phosphate.
b. Reacts with water to produce urea.
c. Reacts with aspartate.
d. Fumarate is a product of its "breakup."

26.93 Indicate whether each of the following statements is *true* or *false.*

a. Glutamate is the most abundant amino acid in the amino acid pool.
b. Pyruvate is a compound that participates in both the urea cycle and the citric acid cycle.
c. Citrulline, a participant in the urea cycle, is a nonstandard amino acid.
d. Glutamate is a reactant in oxidative deamination.

26.94 Which of the heme degradation products (1) bilirubin, (2) biliverdin, (3) stercobilin, and (4) urobilin is associated with each of the following heme degradation characterizations?

a. CO is produced at the same time as this substance.
b. The buildup of this substance in the blood produces jaundice.
c. Molecular O_2 is involved in the reaction that produces this substance.
d. This degradation product gives feces its characteristic color.

Grid Problems

26.95

1. carbamoyl phosphate	2. citrulline	3. ornithine
4. arginine	5. arginino-succinate	6. aspartate

Select from the grid *all* correct responses for each of the following situations.

a. Urea cycle intermediates
b. Amino acids
c. Substances that furnish nitrogen atoms for urea molecules
d. Substances involved in the first step of the urea cycle as reactants or products

26.96

1. acetyl CoA	2. succinyl CoA	3. α-ketoglutarate
4. fumarate	5. pyruvate	6. oxaloacetate

Select from the grid *all* correct responses for each of the following situations.

a. Substances to which amino acid carbon skeletons may be degraded
b. Amino acid degradation product that identifies the amino acid as glucogenic
c. Amino acid degradation product that identifies the amino acid as ketogenic
d. Degradation product that can be used to produce either glucose or ketone bodies

26.97

1. heme	2. urobilin	3. bilirubin
4. biliverdin	5. stercobilin	6. bilirubin diglucuronide

Select from the grid *all* correct responses for each of the following situations.

a. Substances whose structure is based on four pyrrole groups
b. Bile pigments
c. Substances that react with molecular oxygen (O_2) during heme catabolism
d. Substances from which carbon monoxide (CO) is extracted during heme catabolism

26.98

1. $CH_3-\overset{O}{\overset{\|}{C}}-COO^-$	2. $CH_3-\overset{\overset{+}{N}H_3}{\overset{\|}{C}H}-COO^-$	3. $H_2N-\overset{O}{\overset{\|}{C}}-NH_2$
4. $\overset{^-OOC}{\overset{\|}{C}H_2}-\overset{\overset{+}{N}H_3}{\overset{\|}{C}H}-COO^-$	5. $\overset{^-OOC}{\overset{\|}{(CH_2)_2}}-\overset{\overset{+}{N}H_3}{\overset{\|}{C}H}-COO^-$	6. $H_2N-\overset{O}{\overset{\|}{C}}-O-Ⓟ$

Select from the grid *all* correct responses for each of the following situations.

a. Starting materials for the biosynthesis of nonessential amino acids
b. Reactants in a transamination reaction
c. Products of a transamination reaction
d. Reactants in an oxidative deamination reaction

Chapter 1 **1.1** (a) matter (b) matter (c) energy (d) energy (e) matter (f) matter **1.3** (a) shape (b) volume **1.5** (a) no (b) no (c) yes (d) yes **1.7** (a) physical (b) chemical (c) chemical (d) physical **1.9** (a) chemical (b) physical (c) chemical (d) physical **1.11** (a) chemical (b) physical (c) physical (d) chemical **1.13** (a) physical (b) physical (c) chemical (d) physical **1.15** (a) false (b) true (c) false (d) true **1.17** (a) heterogeneous mixture (b) homogeneous mixture (c) pure substance (d) heterogeneous mixture **1.19** (a) homogeneous mixture, one phase (b) heterogeneous mixture, two phases (c) heterogeneous mixture, three phases (d) heterogeneous mixture, three phases **1.21** (a) compound (b) compound (c) classification not possible (d) classification not possible **1.23** (a) A, classification not possible; B, classification not possible; C, compound (b) D, compound; E, classification not possible; F, classification not possible; G, classification not possible **1.25** (a) true (b) false (c) false (d) false **1.27** (a) false (b) true (c) false (d) true **1.29** (a) silver (b) gold (c) calcium (d) sodium (e) phosphorus (f) sulfur **1.31** (a) Sn (b) Cu (c) Al (d) B (e) Ba (f) Ar **1.33** (a) no (b) yes (c) yes (d) no **1.35** (a) true (b) false; Triatomic molecules must contain at least one kind of atom. (c) true (d) false; Both homoatomic and heteroatomic molecules may contain three or more atoms.

1.37 (a)  (b) 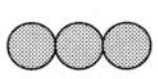 (c)  (d) 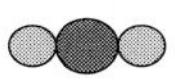 **1.39** (a) compound (b) compound (c) element (d) compound (e) compound (f) compound (g) element (h) element **1.41** (a) $C_8H_{10}N_4O_2$ (b) $C_{12}H_{22}O_{11}$ (c) HCN (d) H_2SO_4 **1.43** (a) 3 elements, 2 H atoms, 1 C atom, 3 O atoms (b) 4 elements, 1 N atom, 4 H atoms, 1 Cl atom, 4 O atoms (c) 3 elements, 1 Ca atom, 1 S atom, 4 O atoms (d) 2 elements, 4 C atoms, 10 H atoms **1.45** (a) solid is pulverized, granules are heated (b) discoloration, bursts in flame and burns **1.46** (a) element (b) compound (c) mixture (d) compound **1.47** (a) homogeneous mixture (b) heterogeneous mixture (c) compound (d) compound **1.48** (a) element (b) mixture (c) mixture (d) compound **1.49** (a) B-Ar-Ba-Ra (b) Eu-Ge-Ne (c) He-At-H-Er (d) Al-La-N **1.50** (a) same, both 4 (b) more, 6 and 5 (c) same, both 5 (d) fewer, 13 and 15 **1.51** (a) $1 + 2 + x = 6$; $x = 3$ (b) $2 + 3 + 3x = 17$; $x = 4$ (c) $1 + x + x = 5$; $x = 2$ (d) $x + 2x + x = 8$; $x = 2$ **1.52** (a) 2 (N_2, NH_3) (b) 4 (N, H, C, Cl) (c) 110; 5(2 + 6 + 4 + 5 + 5) (d) 56; 4(4 + 3 + 4 + 3) **1.53** (a) 2 (b) 1, 4, 5, 6 (c) 2, 3 (d) 1, 3, 4, 5, 6 **1.54** (a) 2, 3, 4 (b) 6 (c) 3, 5 (d) 1, 4 **1.55** (a) 6 (b) 4, 5 (c) 2, 5 (d) all choices **1.56** (a) 2 (b) 4, 5 (c) 1, 4 (d) 3 and 6

Chapter 2 **2.1** (1) kilo (b) milli (c) micro (d) deci **2.3** (a) centimeter (b) kiloliter (c) microliter (d) nanogram **2.5** (a) nanogram, milligram, centigram (b) kilometer, megameter, gigameter (c) picoliter, microliter, deciliter, (d) microgram, milligram, kilogram **2.7** (a) 0.1°C (b) 0.01 mL (c) 1 mL (d) 0.1 mm **2.9** (a) 4 (b) 2 (c) 4 (d) 3 (e) 5 (f) 4 **2.11** (a) same (b) different (c) same (d) same **2.13** (a) 0.351 (b) 653,900 (c) 22.556 (d) 0.2777 **2.15** (a) 2 (b) 2 (c) 2 (d) 2 **2.17** (a) 0.0080 (b) 0.0143 (c) 14 (d) 0.182 (e) 1.1 (f) 5.72 **2.19** (a) 162 (b) 9.3 (c) 1261 (d) 20.0 **2.21** (a) 1.207×10^2 (b) 3.4×10^{-3} (c) 2.3100×10^2 (d) 2.3×10^4 (e) 2.00×10^{-1} (f) 1.011×10^{-1} **2.23** (a) 10^8 (b) 10^2 (c) 10^{-8} (d) 10^8 (e) 10^{-8} (f) 10^{-2} **2.25** (a) 5.50×10^{12} (b) 4.14×10^{-2} (c) 1.5×10^4 (d) 2.0×10^{-7} (e) 1.5×10^{11} (f) 1.2×10^6

2.27 (a) $\frac{10^3\ \text{g}}{1\ \text{kg}}, \frac{1\ \text{kg}}{10^3\ \text{g}}$ (b) $\frac{10^{-9}\ \text{m}}{1\ \text{nm}}, \frac{1\ \text{nm}}{10^{-9}\ \text{m}}$ (c) $\frac{10^{-3}\ \text{L}}{1\ \text{mL}}, \frac{1\ \text{mL}}{10^{-3}\ \text{L}}$ (d) $\frac{454\ \text{g}}{1.00\ \text{lb}}, \frac{1.00\ \text{lb}}{454\ \text{g}}$ (e) $\frac{1.00\ \text{km}}{0.621\ \text{m}}, \frac{0.621\ \text{mi}}{1.00\ \text{km}}$ (f) $\frac{0.265\ \text{gal}}{1.00\ \text{L}}, \frac{1.00\ \text{L}}{0.265\ \text{gal}}$

2.29 (a) 1.6×10^2 m (b) 2.4×10^{-8} m (c) 3 m (d) 3.0×10^5 m **2.31** 2.5 L **2.33** 3.41 lb **2.35** 0.0066 gal **2.37** 183 lb, 6.30 ft (6 ft 4 in.) **2.39** 13.55 g/cm³ **2.41** 25.3 mL **2.43** 243 g **2.45** 274°C **2.47** −38.0°F **2.49** −10°C **2.51** 2.6 J/g · °C **2.53** 0.155 cal/g · °C **2.55** (a) 48 cal (b) 8.40×10^2 cal (c) 180 cal **2.57** (a) 4.720506 (b) 4.7205 (c) 4.721 (d) 4.7 **2.58** (a) 3.00×10^{-3} (b) 9.4×10^5 (c) 2.35×10^1 (d) 4.50000×10^8 **2.59** (a) smaller, 10^3 (b) larger, 10^9 (c) smaller, 10^8 (d) smaller, 10^8 **2.60** (a) four significant figures (b) four significant figures (c) four significant figures (d) exact **2.61** (a) 5.0×10^{-1} g/cm³ (b) 5.00×10^{-1} g/cm³ (b) 5.000×10^{-1} g/cm³ (d) 5.000×10^{-1} g/cm³ **2.62** (a) 1.3×10^2 mL (b) 81 mL (c) 9.88×10^4 mL (d) 5.51 mL **2.63** (a) 4.5×10^3 mg/L (b) 4.5×10^9 pg/mL (c) 4.5 g/L (d) 4.5 kg/m³ **2.64** 11.6 dollars **2.65** (a) 3, 5 (b) 1, 4, 6 (c) 1 and 5, 3 and 4 (d) 3 and 5, 1 and 4, 4 and 6 **2.66** (a) 1, 4, 6 (b) 2, 3, 5 (c) 1, 4 (d) 1, 2, 3, 4, 5, 6 **2.67** (a) 2, 5 (b) 1, 2, 3 (c) 2, 5 (d) 1, 2, 3, 4, 5, 6 **2.68** (a) 1 (b) 2 (c) 2, 3, 4, 5 (d) 4, 5

Chapter 3 **3.1** (a) electron (b) neutron (c) proton (d) proton **3.3** (a) false (b) false (c) false (d) true **3.5** (a) true (b) false (c) false (d) false **3.7** (a) 2 and 4 (b) 4 and 9 (c) 5 and 9 (d) 28 and 58 **3.9** 8, 8, and 8 (b) 8, 10, and 8 (c) 20, 24, and 20 (d) 100, 157, and 100 **3.11** (a) S, Cl, Ar, and K (b) Ar, K, Cl, and S (c) S, Cl, Ar, and K (d) S, Cl, K, and Ar **3.13** (a) 24, 29, 24, 53, and 77 (b) 101, 155, 101, 256, and 357 (c) 30, 37, 30, 67, and 97 (d) 20, 20, 20, 40, and 60 **3.15** $^{96}_{40}Zr$, $^{94}_{40}Zr$, $^{92}_{40}Zr$, $^{91}_{40}Zr$, and $^{90}_{40}Zr$ **3.17** (a) false (b) false (c) true (d) true **3.19** (a) not the same (b) same (c) not the same (d) same **3.21** 107 amu **3.23** (a) 6.95 amu (b) 24.31 amu **3.25** The exact number 12 applies only to ^{12}C, and the number 12.011 is an average obtained by considering *all* isotopes of C. **3.27** (a) Ca (b) Mo (c) Li (d) Sn **3.29** (a) K and Rb (b) P and As (c) F and I (d) Na and Cs **3.31** (a) group (b) periodic law (c) periodic law (d) group **3.33** (a) alkali metal (b) alkali metal (c) noble gas (d) alkaline earth metal (e) halogen (f) noble gas (g) alkaline earth metal (h) halogen **3.35** (a) no (b) no (c) yes (d) yes **3.37** (a) S (b) P (c) I (d) Cl **3.39** (a) orbital (b) orbital (c) shell (d) shell **3.41** (a) true (b) true (c) false (d) true **3.43** (a) 2 (b) 2 (c) 6 (d) 18 **3.45** (a) $1s^22s^22p^2$ (b) $1s^22s^22p^63s^1$ (c) $1s^22s^22p^63s^23p^4$ (d) $1s^22s^22p^63s^23p^6$ **3.47** (a) $1s^22s^22p^63s^23p^5$ (b) $1s^22s^22p^63s^23p^64s^23d^{10}4p^65s^24d^7$ (c) $1s^22s^22p^63s^23p^64s^2$ (d) $1s^22s^22p^63s^23p^64s^23d^1$ **3.49** (a) 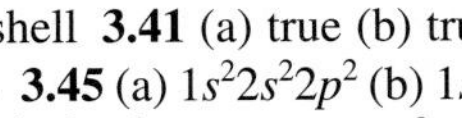 (↑↓) (↑↓) (↑)(↑)( )

(b) 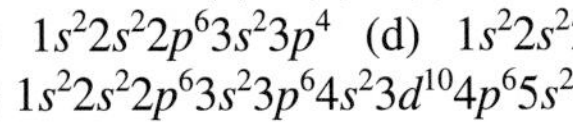 (↑↓) (↑↓) (↑↓)(↑↓)(↑↓) (↑↓)

(c) 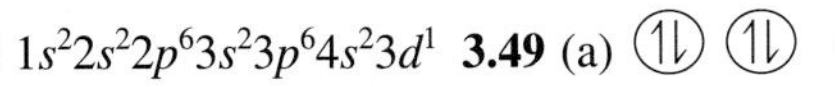 (↑↓) (↑↓) (↑↓)(↑↓)(↑↓) (↑↓) (↑)(↑)(↑)

(d) (↑↓) (↑↓) (↑↓)(↑↓)(↑↓) (↑↓) (↑↓)(↑↓)(↑↓) (↑↓) (↑↓)(↑↓)(↑)(↑)(↑)

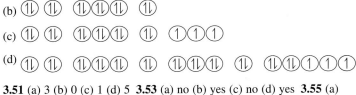

3.51 (a) 3 (b) 0 (c) 1 (d) 5 **3.53** (a) no (b) yes (c) no (d) yes **3.55** (a) p^1 (b) d^3 (c) s^2 (d) p^6 **3.57** (a) representative element (b) noble gas (c) transition element (d) inner transition element **3.59** (a) noble gas (b) representative element (c) transition element (d) representative element **3.61** (a) $^{44}_{20}Ca$ (b) $^{211}_{86}Rn$ (c) $^{110}_{47}Ag$ (d) $^{9}_{4}Be$ **3.62** (a) same number of neutrons, 7 (b) same number of neutrons, 10 (c) same total number of subatomic particles, 54 (d) same number of electrons, 17 **3.63** (a) $^{57}_{24}Cr$ (b) $^{50}_{24}Cr$ (c) $^{55}_{24}Cr$ (d) $^{65}_{24}Cr$ **3.64** 1638 electrons **3.65** (a) Be, Al (b) Be, Al, Ag, Au (the metals) (c) N, Be, Ar, Al, Ag, Au (d) Ag, Au **3.66** the same **3.67** (a) $1s^22s^22p^1$ (B) (b) $1s^22s^22p^63s^23p^1$ (Al) (c) $1s^22s^1$ (Li) (d) $1s^22s^22p^63s^23p^3$ (P) **3.68** (a) $_8O$ (b) $_{10}Ne$ (c) $_{30}Zn$ (d) $_{12}Mg$ **3.69** (a) all choices (b) 2, 3, 5, 6 (c) 1, 4 (d) (3, 4) and (5, 6) **3.70** (a) 1, 2 (b) 3, 5 (c) 1, 2 (d) (3, 4) and (5, 6) **3.71** (a) 2 (b) 1, 4 (c) 3, 4, 5, 6 (d) 1, 2, 3, 4, 5 **3.72** (a) 1, 4, 6 (b) 5, 6 (c) 5, 6 (d) 2

Chapter 4 **4.1** (a) 2 (b) 2 (c) 3 (d) 4 **4.3** (a) 1 (b) 8 (c) 2 (d) 7 **4.5** (a) $1s^22s^22p^2$ (b) $1s^22s^22p^5$ (c) $1s^22s^22p^63s^2$ (d) $1s^22s^22p^63s^23p^3$ **4.7** (a) Mg· (b) K· (c) :P̈· (d) :K̈r: **4.9** (a) Li (b) F (c) Be (d) N **4.11** (a) O^{2-} (b) Mg^{2+} (c) F^- (d) Al^{3+} **4.13** (a) Ca^{2+} (b) O^{2-} (c) Na^+ (d) Al^{3+} **4.15** (a) 15p, 18e (b) 7p, 10e (c) 12p, 10e (d) 3p, 2e **4.17** (a) 2+ (b) 3− (c) 1+ (d) 1− **4.19** (a) 2 lost (b) 1 gained (c) 2 lost (d) 2 gained **4.21**

(a) $1s^22s^22p^63s^23p^1$ (b) $1s^22s^22p^6$ **4.23** (a) Be· → Ö: (b) Mg: → S̈:

(c) 3 K· → N: (d) Ca: → 2 F: **4.25** (a) $BaCl_2$ (b) $BaBr_2$ (c) Ba_3N_2 (d) BaO **4.27** (a) MgF_2 (b) BeF_2 (c) LiF (d) AlF_3 **4.29** (a) Na_2S (b) CaI_2 (c) Li_3N (d) $AlBr_3$ **4.31** (a) potassium iodide (b) beryllium oxide (c) aluminum fluoride (d) sodium phosphide **4.33** (a) +1 (b) +2 (c) +4 (d) +2 **4.35** (a) iron(II) oxide (b) gold(III) oxide (c) copper(II) sulfide (d) cobalt(II) bromide **4.37** (a) gold(I) chloride (b) potassium chloride (c) silver chloride (d) copper(II) chloride **4.39** (a) KBr (b) Ag_2O (c) BeF_2 (d) Ba_3P_2 **4.41** (a) CoS (b) Co_2S_3 (c) SnI_4 (d) Pb_3N_2 **4.43** (a) SO_4^{2-} (b) ClO_3^- (c) OH^- (d) CN^- **4.45** (a) PO_4^{3-} and HPO_4^{2-} (b) NO_3^- and NO_2^- (c) H_3O^+ and OH^- (d) CrO_4^{2-} and $Cr_2O_7^{2-}$ **4.47** (a) $NaClO_4$ (b) $Fe(OH)_3$ (c) $Ba(NO_3)_2$ (d) $Al_2(CO_3)_3$ **4.49** (a) magnesium carbonate (b) zinc sulfate (c) beryllium nitrate (d) silver phosphate **4.51** (a) iron(II) hydroxide (b) copper(II) carbonate (c) gold(I) cyanide (d) manganese(II) phosphate **4.53** (a) $KHCO_3$ (b) $Au_2(SO_4)_3$ (c) $AgNO_3$ (d) $Cu_3(PO_4)_2$ **4.55** (a) Na^+ (b) F^- (c) S^{2-} (d) Ca^{2+} **4.56** (a) XZ_2 (b) X_2Z (c) XZ (d) XZ **4.57** (a) S (b) Mg (c) P (d) Al **4.58** (a) K^+, Cl^- (b) Ca^{2+}, S^{2-} (c) Be^{2+}, two F^- (d) two Al^{3+}, three S^{2-} **4.59** (a) tin(IV) chloride, tin(II) chloride (b) iron(II) sulfide, iron(III) sulfide (c) copper(I) nitride, copper(II) nitride (d) nickel(II) iodide, nickel(III) iodide **4.60** (a) same (3+) (b) different (1+ and 2+) (c) different (1+ and 3+) (d) different (2+ and 1+) **4.61** (a) copper(I) nitrate, copper(II) nitrate (b) lead(II) phosphate, lead(IV) phosphate (c) manganese(III) cyanide, manganese(II) cyanide (d) cobalt(II) chlorate, cobalt(III) chlorate **4.62** (a) Na_2S (b) Na_2SO_4 (c) Na_2SO_3 (d) $Na_2S_2O_3$ **4.63** (a) 1 and 2 (b) 4 (c) 1 and 3, 2 and 3, 6 and 5 (d) 1 and 5, 2 and 5, 6 and 3 **4.64** (a) 1 and 2 (b) 1 and 3 (c) 1, 2, and 4 (d) 2, 5, and 6 **4.65** (a) 4 and 6 (b) 1, 2, and 6 (c) 3 (d) 3 **4.66** (a) 1, 2, 4, and 5 (b) 3 and 4 (c) 1, 3, 5, and 6 (d) 1 and 5

Chapter 5 **5.1** (a) :B̈r:B̈r: (b) H:Ï: (c) :Ï:B̈r: (d) :B̈r:F̈: **5.3** (a) 2 (b) 2 (c) 0 (d) 6 **5.5** (a) :N≡N: (b) H—C=C—H (with H on each C) (c) H—C—H with =Ö: on C (d) H—Ö—Ö—H **5.7** (a) NF_3 (b) Cl_2O (c) H_2S (d) CH_4 **5.9** (a) N (b) C (c) N (d) C **5.11** Oxygen forms three bonds instead of the normal two. **5.13** (a) H:P̈:H with H on P (b) :C̈l:P̈:C̈l: with :C̈l: on P (c) :B̈r:Si:B̈r: with :B̈r: above and below Si (d) :F̈:Ö:F̈:

5.15 (a) :F̈:S̈:F̈: (b) :Ï:C:Ï: with :Ï: above and below C (c) :B̈r:N̈:B̈r: with :B̈r: on N (d) H:S̈e:H **5.17** (a) H:C::C::C:H with H on the first and last C (b) :F̈:N::N:F̈: (c) H:C:C:::N: with H above and below the first C (d) H:C:C:::C:H with H above and below the first C

5.19 (a) $[:\ddot{O}:H]^-$ (b) $[H:Be:H]^{2-}$ with H above and below Be (c) $[:\ddot{Cl}:Al:\ddot{Cl}:]^-$ with :C̈l: above and below Al (d) $[:\ddot{O}:N:\ddot{O}:]^-$ with :Ö on N **5.21** (a) $Na^+[:C:::N:]^-$ (b) $3[K]^+[:\ddot{O}:P:\ddot{O}:]^{3-}$ with :Ö: above and below P

5.23 (a) angular (b) angular (c) angular (d) linear **5.25** (a) trigonal pyramidal (b) trigonal planar (c) tetrahedral (d) tetrahedral **5.27** (a) trigonal pyramidal (b) tetrahedral (c) angular (d) angular **5.29** (a) trigonal planar about each carbon atom (b) tetrahedral about carbon atom and angular about oxygen atom **5.31** (a) Na, Mg, Al, P (b) I, Br, Cl, F (c) Al, P, S, O (d) Ca, Mg, C, O **5.33** (a) N, O, Cl, F, Br (b) Li, Na, K, Ca (c) F, O, Cl, N (d) 0.5 units **5.35** (a) $\overset{\delta+}{B}—\overset{\delta-}{N}$ (b) $\overset{\delta+}{Cl}—\overset{\delta-}{F}$ (c) $\overset{\delta-}{N}—\overset{\delta+}{C}$ (d) $\overset{\delta-}{F}—\overset{\delta+}{O}$ **5.37** (a) H—Br, H—Cl, H—O (b) O—F, P—O, Al—O (c) Br—Br, H—Cl, B—N (d) P—N, S—O, Br—F **5.39** (a) polar covalent (b) ionic (c) nonpolar covalent (d) polar covalent **5.41** (a) nonpolar (b) polar (c) polar (d) polar **5.43** (a) polar (b) polar (c) nonpolar (d) polar **5.45** (a) sulfur tetrafluoride (b) tetraphosphorus hexoxide (c) chlorine dioxide (d) hydrogen sulfide **5.47** (a) ICl (b) N_2O (c) NCl_3 (d) HBr **5.49** (a) H_2O_2 (b) CH_4 (c) NH_3 (d) PH_3 **5.51** (a) 26 (b) 24 (c) 14 (d) 8 **5.52** (a) same, both single (b) same, both triple (c) different, double and single (d) same, both single **5.53** (a) not enough electron dots (b) not enough electron dots (c) improper placement of a correct number of electron dots (d) too many electron dots **5.54** (a) tetrahedral; tetrahedral (b) tetrahedral; tetrahedral (c) trigonal planar; angular (d) trigonal planar; trigonal planar **5.55** (a) BrI (b) SO_2 (c) NF_3 (d) H_3CF **5.56** BA, CA, DB, DA **5.57** (a)

CH_4: H—C—H with H above and below C; CH_3F: H—C—F̈: with H above and below C; CH_2F_2: H—C—F̈: with :F̈: above and H below C; CHF_3: H—C—F̈: with :F̈: above and below C; CF_4: :F̈—C—F̈: with :F̈: above and below C

(b) all are tetrahedral (c) nonpolar, polar, polar, polar, nonpolar **5.58** (a) sodium chloride (b) bromine monochloride (c) potassium sulfide (d) dichlorine monoxide **5.59** (a) 4 (b) 5 (c) 4 and 1, 4 and 2, 4 and 3 (d) 1 and 2, 1 and 3, 2 and 3 **5.60** (a) 1, 3, and 6 (b) 1, 4, 5, and 6 (c) 2, 4, and 5 (d) 1 and 6, 2 and 4, 2 and 5, 4 and 5 **5.61** (a) 2 and 5 (b) 1, 2, and 6 (c) 5 (d) 1, 2, 4, and 5 **5.62** (a) 2 and 4 (b) 5 (c) 4 (d) 1, 2, 4, and 5

Chapter 6 **6.1** (a) 342.34 amu (b) 100.23 amu (c) 183.20 amu (d) 132.17 amu **6.3** (a) 1.20×10^{24} molecules (b) 1.96×10^{24} molecules (c) 2.14×10^{23} molecules (d) 6.02×10^{23} molecules **6.5** (a) 3.26×10^{23} atoms (b) 3.26×10^{23} atoms (c) 3.26×10^{23} molecules (d) 3.26×10^{23} molecules **6.7** (a) 28.0 g (b) 44.0 g (c) 58.4 g (d) 342 g **6.9** (a) 6.7 g (b) 3.7 g (c) 48.0 g (d) 96.0 g **6.11** (a) 0.179 mole (b) 0.114 mole (c) 0.0937 mole (d) 0.0210 mole **6.13** (a) $\frac{2 \text{ moles H}}{1 \text{ mole } H_2SO_4}$ $\frac{1 \text{ mole } H_2SO_4}{2 \text{ moles H}}$ $\frac{1 \text{ mole S}}{1 \text{ mole } H_2SO_4}$ $\frac{1 \text{ mole } H_2SO_4}{1 \text{ mole S}}$ $\frac{4 \text{ moles O}}{1 \text{ mole } H_2SO_4}$ $\frac{1 \text{ mole } H_2SO_4}{4 \text{ moles O}}$ (b) $\frac{1 \text{ mole P}}{1 \text{ mole } POCl_3}$ $\frac{1 \text{ mole } POCl_3}{1 \text{ mole P}}$ $\frac{1 \text{ mole O}}{1 \text{ mole } POCl_3}$ $\frac{1 \text{ mole } POCl_3}{1 \text{ mole O}}$ $\frac{3 \text{ moles Cl}}{1 \text{ mole } POCl_3}$ $\frac{1 \text{ mole } POCl_3}{3 \text{ moles Cl}}$ **6.15** (a) 2.00 moles S, 4.00 moles O (b) 2.00 moles S, 6.00 moles O (c) 3.00 moles N, 9.00 moles H (d) 6.00 moles N, 12.0 moles H **6.17** (a) 5.57×10^{23} atoms (b) 4.81×10^{23} atoms (c) 6.0×10^{22} atoms (d) 3.0×10^{23} atoms **6.19** (a) 63.6 g (b) 31.8 g (c) 5.88×10^{-20} g (d) 1.06×10^{-22} g **6.21** (a) 2.50 moles (b) 0.227 mole (c) 6.6×10^{-14} mole (d) 6.6×10^{-14} mole **6.23** (a) 6.14×10^{22} atoms S (b) 1.50×10^{23} atoms S (c) 3.61×10^{23} atoms S (d) 2.41×10^{24} atoms S **6.25** (a) 32.1 g S (b) 6.39×10^{-22} g S (c) 64.1 g S (d) 1150 g S **6.27** (a) $2H_2 + O_2 \rightarrow 2H_2O$ (b) $2NO + O_2 \rightarrow 2NO_2$ (c) $2Fe_2O_3 \rightarrow 4Fe + 3O_2$ (d) $NH_4NO_2 \rightarrow N_2 + 2H_2O$ **6.29** (a) $2Na + 2H_2O \rightarrow 2NaOH + H_2$ (b) $2Na + ZnSO_4 \rightarrow Na_2SO_4 + Zn$ (c) $2NaBr + Cl_2 \rightarrow 2NaCl + Br_2$ (d) $2ZnS + 3O_2 \rightarrow 2ZnO + 2SO_2$ **6.31** (a) $CH_4 + 2O_2 \rightarrow CO_2 + 2H_2O$ (b) $2C_6H_6 + 15O_2 \rightarrow 12CO_2 + 6H_2O$ (c) $C_4H_8O_2 + 5O_2 \rightarrow 4CO_2 + 4H_2O$ (d) $C_5H_{10}O + 7O_2 \rightarrow 5CO_2 + 5H_2O$ **6.33** (a) $3PbO + 2NH_3 \rightarrow 3Pb + N_2 + 3H_2O$ (b) $2Fe(OH)_3 + 3H_2SO_4 \rightarrow Fe_2(SO_4)_3 + 6H_2O$ **6.35** $\frac{2 \text{ moles } Ag_2CO_3}{4 \text{ moles Ag}}$ $\frac{2 \text{ moles } Ag_2CO_3}{2 \text{ moles } CO_2}$ $\frac{2 \text{ moles } Ag_2CO_3}{1 \text{ mole } O_2}$ $\frac{4 \text{ moles Ag}}{2 \text{ moles } CO_2}$ $\frac{4 \text{ moles Ag}}{1 \text{ mole } O_2}$ $\frac{2 \text{ moles } CO_2}{1 \text{ mole } O_2}$ The other six are the reciprocals of these six factors. **6.37** (a) 14.0 moles CO_2 (b) 1.00 mole CO_2 (c) 4.00 moles CO_2 (d) 2.00 moles CO_2 **6.39** (a) 24.3 g NH_3 (b) 1.80×10^2 g $(NH_4)_2Cr_2O_7$ (c) 22.9 g N_2H_4 (d) 24.3 g NH_3 **6.41** 5.09 g O_2 **6.43** 14.3 g O_2 **6.45** 5.63 g H_2O **6.47** $y = 8$ **6.48** (a) 1.00 mole S_8 (b) 28.0 g Al (c) 30.0 g Mg (d) 6.02×10^{23} atoms He **6.49** (a) 0.03560 mole SiH_4 (b) 2.139 g SiO_2 (c) 2.144×10^{22} molecules $(CH_3)_3SiCl$ (d) 2.144×10^{22} atoms Si **6.50** 59.0 g Si **6.51** C_4H_6 **6.52** 8.33 g N_2, 21.4 g H_2O, and 45.2 g Cr_2O_3 **6.53** 109 g Ag and 16.2 g S **6.54** 43.4 g Be **6.55** (a) 2 and 3, 5 and 6 (b) 2 and 3, 5 and 6 (c) 5 and 6 (d) 5 and 6 **6.56** (a) 1, 3, 5, 6 (b) 2, 6 (c) 1 (d) 6 **6.57** (a) 1, 3, 4 (b) 2, 6 (c) 2, 3, 4, 5, 6 (d) 1, 4, 6 **6.58** (a) 2, 3 (b) 2, 3, 6 (c) 2, 3, 4, 6 (d) 2, 3

Chapter 7 **7.1** (a) Velocity increases with increasing temperature. (b) potential energy (c) Increasing temperature increases disruptive force magnitude. (d) gaseous state **7.3** (a) Increase in vibrational movement is limited because particles are already close together; hence, there is little change in volume. (b) Particles of a gas are widely separated because disruptive forces are greater than cohesive forces. **7.5** (a) 0.967 atm (b) 403 mm Hg (c) 403 torr (d) 0.816 atm **7.7** 7.2 atm **7.9** 2.71 L **7.11** 3.64 L **7.13** 144°C **7.15** (a) $T_1 = \frac{P_1V_1T_2}{P_2V_2}$ (b) $P_2 = \frac{P_1V_1T_2}{V_2T_1}$ (c) $V_1 = \frac{P_2V_2T_1}{P_1T_2}$ **7.17** (a) 5.90 L (b) 2.11 atm (c) −171°C (d) 3.70×10^3 mL **7.19** −209°C **7.21** 1.12 L **7.23** (a) 4.11 L (b) 3.16 atm (c) −98°C (d) 16,300 mL **7.25** 0.42 atm **7.27** 98 mm Hg **7.29** (a) endothermic (b) endothermic (c) exothermic **7.31** (a) no (b) yes (c) yes **7.33** (a) boiling point (b) vapor pressure (c) boiling (d) boiling point **7.35** (a) They differ in the strength of inter-molecular forces. (b) Vapor pressure becomes equal to atmospheric pressure at a lower temperature. (c) Evaporation is a cooling process. (d) Low-heat and high-heat boiling water have the same temperature and the same heat content. **7.37** Molecules must be polar. **7.39** Boiling point increases as intermolecular force strength increases. **7.41** (a) London (b) hydrogen bonding (c) dipole–dipole (d) London **7.43** (a) no (b) yes (c) yes (d) no **7.45** four (see Figure 7.15) **7.47** (a) 0.871 atm (b) 298°C (c) 869°C **7.48** (a) 915°C (b) −199°C (c) 24°C (d) 172°C **7.49** (a) 4.6 atm (b) 29 atm (c) 0.14 atm (d) 0.90 atm **7.50** 5.37×10^{22} molecules H_2S **7.51** 24.7 L **7.52** 0.22 g N_2 **7.53** (a) PBr_3 (b) PI_3 (c) PI_3 **7.54** (a) Br_2, larger mass (b) H_2O, hydrogen bonding (c) CO, dipole–dipole (d) C_3H_8, larger size **7.55** (a) 1, 3, 4, and 6 (b) 2 and 5 (c) 3 and 6 (d) 1, 2, 4, and 5 **7.56** (a) 1, 2, 3, 4, 5 (b) 2 and 3 (c) 2, 3, 4, 5, 6 (d) 1, 2, and 3 **7.57** (a) 1 and 2 (b) 5 (c) 2 (d) 1, 2, 4, and 6 **7.58** (a) all (b) 3, 4, 5, and 6 (c) 4 and 5 (d) 4 and 5

Chapter 8 **8.1** (a) water (b) ethyl alcohol **8.3** See Table 8.1 **8.5** (a) saturated (b) unsaturated (c) unsaturated (d) saturated **8.7** (a) dilute (b) concentrated (c) dilute (d) concentrated **8.9** (a) slightly soluble (b) very soluble (c) slightly soluble (d) slightly soluble **8.11** (a) all (b) all (c) CaS, $Ca(OH)_2$, $CaCl_2$ (d) $NiSO_4$, $Ni(C_2H_3O_2)_2$ **8.13** (a) 7.10%(m/m) (b) 6.19%(m/m) (c) 9.06%(m/m) (d) 0.27%(m/m) **8.15** (a) 3.62 g (b) 14.5 g (c) 68.8 g (d) 124 g **8.17** 0.6400 g **8.19** 276 g **8.21** (a) 4.21%(v/v) (b) 4.60%(v/v) **8.23** 18%(v/v) **8.25** (a) 2.0%(m/v) (b) 15%(m/v) **8.27** 0.500 g **8.29** 3.75 g **8.31** (a) 6.0 M (b) 0.456 M (c) 0.342 M (d) 0.500 M **8.33** (a) 273 g (b) 0.373 g (c) 136 g (d) 88 g **8.35** (a) 85.6 mL (b) 2.64 mL (c) 9180 mL (d) 0.24 mL **8.37** (a) 0.183 M (b) 0.0733 M (c) 0.0120 M (d) 0.00275 M **8.39** (a) 1450 mL (b) 18.0 mL (c) 85,600 mL (d) 7.5 mL **8.41** (a) 3.0 M (b) 3.0 M (c) 4.5 M (d) 1.5 M **8.43** The presence of solute molecules decreases the ability of solvent molecules to escape. **8.45** KCl produces ions, so more solute particles are present. **8.47** (a) same as (b) greater than (c) less than (d) greater than **8.49** 2 to 1 **8.51** (a) swell (b) remain the same (c) swell (d) shrink **8.53** (a) hypotonic (b) isotonic (c) hypotonic (d) hypertonic **8.55** (a) K^+ and Cl^- leave the bag. (b) K^+, Cl^-, and glucose leave the bag. **8.57** (a) like (both soluble) (b) unlike (c) unlike (d) like (both insoluble) **8.58** (a) 4.02 g (b) 7.303 g (c) 12.6 g (d) 0.148 g **8.59** 0.0700 qt **8.60** (a) 7.1 L (b) 9.5 L (c) 11 L (d) 14 L **8.61** (a) 0.472 M (b) 0.708 M (c) 1.04 M (d) 1.60 M **8.62** (a) 37.5%(m/v) (b) 2.23 M **8.63** (a) 4.00 M (b) 3.22 M **8.64** (a) NaCl (b) $MgCl_2$ **8.65** (a) 1, 2, and 4 (b) 3, 5, and 6 (c) 1 (d) 3 and 6 **8.66** (a) all (b) all (c) 2, 3, and 5 (d) 1 and 4, 2 and 3 **8.67** (a) 2 and 3 (b) 4 and 5 (c) 4 and 5 (d) 1 and 6 **8.68** (a) 1 (b) 2, 3, 4, 5, and 6 (c) 4 and 6 (d) 3 and 5

Chapter 9 **9.1** (a) single replacement (b) decomposition (c) double replacement (d) combination **9.3** (a) +2 (b) +6 (c) 0 (d) +5 **9.5** (a) +3 (b) +4 (c) +6 (d) +6 (e) +6 (f) +6 (g) +6 (h) +5 **9.7** (a) +3P, −1F (b) +1Na, −2O, +1H (c) +1Na, +6S, −2O (d) +4C, −2O **9.9** (a) redox (b) nonredox (c) redox (d) redox **9.11** (a) H_2 oxidized, N_2 reduced (b) KI oxidized, Cl_2 reduced (c) Fe oxidized, Sb_2O_3 reduced (d) H_2SO_3 oxidized, HNO_3 reduced **9.13** (a) N_2 oxidizing agent, H_2 reducing agent (b) Cl_2 oxidizing agent, KI reducing agent (c) Sb_2O_3 oxidizing agent, Fe reducing agent (d) HNO_3 oxidizing agent, H_2SO_3 reducing agent **9.15** See listing in Section 9.4. **9.17** (a) exothermic (b) endothermic (c) endothermic (d) exothermic **9.19**

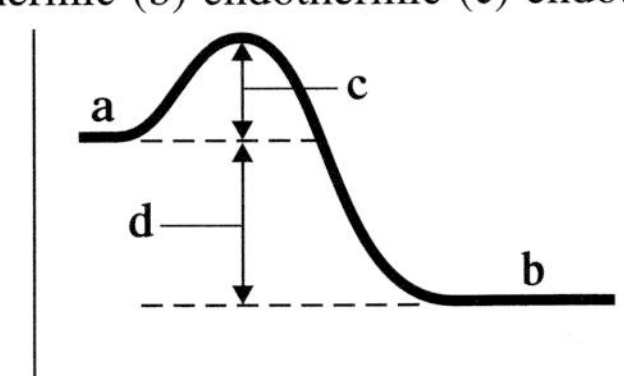

9.21 (a) As temperature increases, so does the number of collisions per second. (b) A catalyst lowers the activation energy. **9.23** The concentration of O_2 has increased from 21% to 100%.

9.25

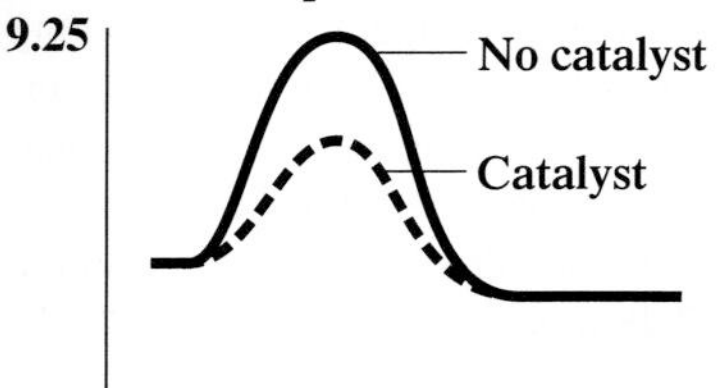

9.27 (a) 1 (b) 3 (c) 4 (d) 3 **9.29** rate of forward reaction = rate of reverse reaction

9.31

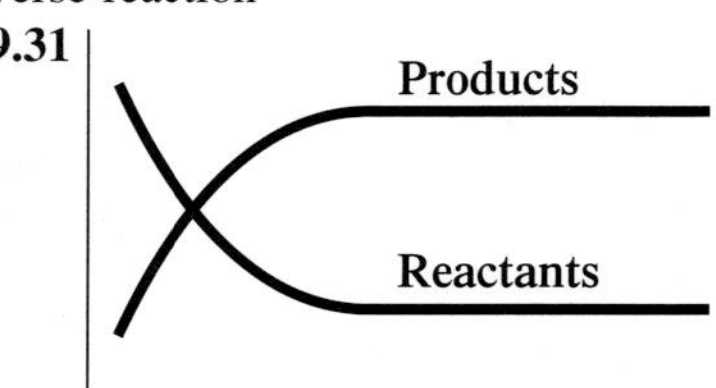

9.33 (a) $K_{eq} = \dfrac{[NO_2]^2}{[N_2O_4]}$ (b) $K_{eq} = \dfrac{[Cl_2][CO]}{[COCl_2]}$ (c) $K_{eq} = \dfrac{[CH_4][H_2S]^2}{[CS_2][H_2]^4}$ (d) $K_{eq} = \dfrac{[SO_3]^2}{[O_2][SO_2]^2}$ **9.35** (a) $K_{eq} = [SO_3]$ (b) $K_{eq} = \dfrac{1}{[Cl_2]}$ (c) $K_{eq} = \dfrac{[NaCl]^2}{[Na_2SO_4][BaCl_2]}$ (d) $K_{eq} = [O_2]$ **9.37** 4.8×10^{-5} **9.39** (a) more products than reactants (b) essentially all reactants (c) significant amounts of both reactants and products (d) significant amounts of both reactants and products **9.41** (a) right (b) left (c) left (d) right **9.43** (a) left (b) left (c) left (d) left **9.45** (a) left (b) no effect (c) right (d) no effect **9.47** (a) redox, single replacement (b) redox, combination (c) redox, decomposition (d) nonredox, double replacement **9.48** (a) redox (b) redox (c) redox (d) can't classify **9.49** (a) gain (b) reduction (c) decrease (d) increase **9.50** (a) decrease (b) increase (c) increase (d) decrease **9.51** (a) no (b) no (c) yes (d) no **9.52** $CH_4(g) + 2H_2S(g) \longrightarrow CS_2(g) + 4H_2(g)$ **9.53** (a) yes (b) yes (c) no (d) yes **9.54** (a) no effect (b) right (c) right (d) right **9.55** (a) 2 and 4 (b) 1, 2, 3, and 4 (c) 4 (d) 3 **9.56** (a) 1, 2, 3, and 4 (b) 5 (c) 6 (d) 5 **9.57** (a) 1, 2, 3, 5, and 6 (b) 1, 2, and 5 (c) 3 (d) 1 **9.58** (a) 4 (b) 6 (c) 1, 3, and 6 (d) 2 and 5

Chapter 10 **10.1** (a) $HI \xrightarrow{H_2O} H^+ + I^-$ (b) $HClO \xrightarrow{H_2O} H^+ + ClO^-$ (c) $LiOH \xrightarrow{H_2O} Li^+ + OH^-$ (d) $CsOH \xrightarrow{H_2O} Cs^+ + OH^-$ **10.3** (a) acid (b) base (c) acid (d) acid **10.5** (a) $HClO + H_2O \rightarrow H_3O^+ + ClO^-$ (b) $HClO_4 + NH_3 \rightarrow NH_4^+ + ClO_4^-$ (c) $H_3O^+ + OH^- \rightarrow H_2O + H_2O$ (d) $H_3O^+ + NH_2^- \rightarrow H_2O + NH_3$ **10.7** (a) HSO_3^- (b) HCN (c) $C_2O_4^{2-}$ (d) $H_2PO_4^-$ **10.9** (a) $HS^- + H_2O \rightarrow H_3O^+ + S^{2-}$, $HS^- + H_2O \rightarrow H_2S + OH^-$ (b) $HPO_4^{2-} + H_2O \rightarrow H_3O^+ + PO_4^{3-}$, $HPO_4^{2-} + H_2O \rightarrow H_2PO_4^- + OH^-$ (c) $NH_3 + H_2O \rightarrow H_3O^+ + NH_2^-$, $NH_3 + H_2O \rightarrow NH_4^+ + OH^-$ (d) $OH^- + H_2O \rightarrow H_3O^+ + O^{2-}$, $OH^- + H_2O \rightarrow OH^- + H_2O$ **10.11** (a) monoprotic (b) diprotic (c) monoprotic (d) diprotic **10.13** $H_3C_6H_5O_7 + H_2O \rightarrow H_3O^+ + H_2C_6H_5O_7^-$, $H_2C_6H_5O_7^- + H_2O \rightarrow H_3O^+ + HC_6H_5O_7^{2-}$, $HC_6H_5O_7^{2-} + H_2O \rightarrow H_3O^+ + C_6H_5O_7^{3-}$ **10.15** To show that it is a monoprotic acid **10.17** Monoprotic; only one H atom is involved in a polar bond. **10.19** (a) strong (b) weak (c) weak (d) strong **10.21** (a) $K_a = \dfrac{[H^+][F^-]}{[HF]}$ (b) $K_a = \dfrac{[H^+][C_2H_3O_2^-]}{[HC_2H_3O_2]}$ **10.23** (a) $K_b = \dfrac{[NH_4^+][OH^-]}{[NH_3]}$ (b) $K_b = \dfrac{[C_6H_5NH_3^+][OH^-]}{[C_6H_5NH_2]}$ **10.25** (a) H_3PO_4 (b) HF (c) H_2CO_3 (d) HNO_2 **10.27** 4.9×10^{-5} **10.29** (a) acid (b) salt (c) salt (d) base (e) salt (f) base (g) acid (h) acid **10.31** (a) $Ba(NO_3)_2 \xrightarrow{H_2O} Ba^{2+} + 2NO_3^-$ (b) $Na_2SO_4 \xrightarrow{H_2O} 2Na^+ + SO_4^{2-}$ (c) $CaBr_2 \xrightarrow{H_2O} Ca^{2+} + 2Br^-$ (d) $K_2CO_3 \xrightarrow{H_2O} 2K^+ + CO_3^{2-}$ **10.33** (a) $HCl + NaOH \rightarrow NaCl + H_2O$ (b) $HNO_3 + KOH \rightarrow KNO_3 + H_2O$ (c) $H_2SO_4 + 2LiOH \rightarrow Li_2SO_4 + 2H_2O$ (d) $2H_3PO_4 + 3Ba(OH)_2 \rightarrow Ba_3(PO_4)_2 + 6H_2O$ **10.35** (a) $H_2SO_4 + 2LiOH \rightarrow Li_2SO_4 + 2H_2O$ (b) $HCl + NaOH \rightarrow NaCl + H_2O$ (c) $HNO_3 + KOH \rightarrow KNO_3 + H_2O$ (d) $2H_3PO_4 + 3Ba(OH)_2 \rightarrow Ba_3(PO_4)_2 + 6H_2O$ **10.37** (a) 3.3×10^{-12}M (b) 1.5×10^{-9}M (c) 1.1×10^{-7}M (d) 8.3×10^{-4}M **10.39** (a) acidic (b) basic (c) basic (d) acidic **10.41** (a) 4.00 (b) 11.00 (c) 11.00 (d) 7.00 **10.43** (a) 7.68 (b) 7.40 (c) 3.85 (d) 11.85 **10.45** (a) 1×10^{-2} (b) 1×10^{-6} (c) 1×10^{-8} (d) 1×10^{-10} **10.47** (a) 2.1×10^{-4}M (b) 8.1×10^{-6}M (c) 4.5×10^{-8}M (d) 3.6×10^{-13}M **10.49** (a) 3.14 (b) 6.37 (c) 7.21 (d) 1.82 **10.50** (a) 2.12 (b) 3.18 (c) 12.66 (d) 4.89 **10.51** acid B **10.52** acid A **10.53** (a) strong acid–strong base salt (b) weak acid–strong base salt (c) strong acid–weak base salt (d) strong acid–strong base salt **10.55** (a) none (b) $C_2H_3O_2^-$ (c) NH_4^+ (d) none **10.57** (a) neutral (b) basic (c) acidic (d) neutral **10.59** (a) no (b) yes (c) no (d) yes **10.61** (a) $F^- + H_3O^+ \rightarrow HF + H_2O$ (b) $H_2CO_3 + OH^- \rightarrow HCO_3^- + H_2O$ (c) $CO_3^{2-} + H_3O^+ \rightarrow HCO_3^- + H_2O$ (d) $H_3PO_4 + OH^- \rightarrow H_2PO_4^- + H_2O$ **10.63** 7.06 **10.64** 5.55 **10.65** 5.17 **10.66** 3.30 **10.67** (a) weak (b) strong (c) strong (d) strong **10.69** (a) 0.0500 M (b) 0.800 M (c) 0.952 M (d) 0.120 M **10.71** (a) yes (b) no (c) no (d) yes **10.72** (a) no (b) yes, both strong (c) no (d) yes, both weak **10.73** (a) B (b) A **10.74** 7.0 **10.75** HCl, HCN, KCl, NaOH **10.76** HCN/CN^- **10.77** CN^- ion undergoes hydrolysis to a greater extent than NH_4^+ ion, resulting in a basic solution; NH_4^+ ion and $C_2H_3O_2^-$ ion hydrolyze to an equal extent, resulting in a neutral solution. **10.78** 0.0035 g NaOH **10.79** (a) 2, 3, 5, and 6 (b) 1, 2, 4, and 5 (c) 2 and 5 (d) 2 and 5, 2 and 6 **10.80** (a) 1 (b) 2, 4, and 6 (c) 2 and 6 (d) 1, 2, 3, and 5 **10.81** (a) 3 and 6 (b) 4 and 5 (c) 1 and 2, 3 and 6, 4 and 5 (d) 1 and 2, 3 and 6, 4 and 5 **10.82** (a) 1 and 4 (b) 2 and 5 (c) 2 and 5 (d) 1 and 4

Chapter 11 **11.1** (a) $^{10}_{4}Be$, Be-10 (b) $^{25}_{11}Na$, Na-25 (c) $^{96}_{41}Nb$, Nb-96 (d) $^{257}_{103}Lr$, Lr-257 **11.3** (a) $^{14}_{7}N$ (b) $^{197}_{79}Au$ (c) Sn-121 (d) B-10 **11.5** (a) $^{4}_{2}\alpha$ (b) $^{0}_{-1}\beta$ (c) $^{0}_{0}\gamma$ **11.7** 2 protons and 2 neutrons **11.9** (a) $^{200}_{84}Po \rightarrow {}^{4}_{2}\alpha + {}^{196}_{82}Pb$ (b) $^{240}_{96}Cm \rightarrow {}^{4}_{2}\alpha + {}^{236}_{94}Pu$ (c) $^{244}_{96}Cm \rightarrow {}^{4}_{2}\alpha + {}^{240}_{94}Pu$ (d) $^{238}_{92}U \rightarrow {}^{4}_{2}\alpha + {}^{234}_{90}Th$ **11.11** (a) $^{10}_{4}Be \rightarrow {}^{0}_{-1}\beta + {}^{10}_{5}B$ (b) $^{14}_{6}C \rightarrow {}^{0}_{-1}\beta + {}^{14}_{7}N$ (c) $^{21}_{9}F \rightarrow {}^{0}_{-1}\beta + {}^{21}_{10}Ne$ (d) $^{25}_{11}Na \rightarrow {}^{0}_{-1}\beta + {}^{25}_{12}Mg$ **11.13** $A \rightarrow A - 4, Z \rightarrow Z - 2$ **11.15** (a) $^{0}_{-1}\beta$ (b) $^{28}_{12}Mg$ (c) $^{4}_{2}\alpha$ (d) $^{200}_{80}Hg$ **11.17** (a) $^{4}_{2}\alpha$ (b) $^{0}_{-1}\beta$ **11.19** (a) $\frac{1}{4}$ (b) $\frac{1}{64}$ (c) $\frac{1}{8}$ (d) $\frac{1}{64}$ **11.21** (a) 1.4 days (b) 0.90 day (c) 0.68 day (d) 0.54 day **11.23** 0.250 g **11.25** 2000 **11.27** 92 **11.29** (a) $^{4}_{2}\alpha$ (b) $^{25}_{12}Mg$ (c) $^{4}_{2}\alpha$ (d) $^{1}_{1}p$ **11.31** Termination of a decay series requires a stable nuclide. **11.33** $^{232}_{90}Th \rightarrow {}^{4}_{2}\alpha + {}^{228}_{88}Ra$, $^{228}_{88}Ra \rightarrow {}^{0}_{-1}\beta + {}^{228}_{89}Ac$, $^{228}_{89}Ac \rightarrow {}^{0}_{-1}\beta + {}^{228}_{90}Th$, $^{228}_{90}Th \rightarrow {}^{4}_{2}\alpha + {}^{224}_{88}Ra$ **11.35** Alpha is stopped; beta and gamma go through. **11.37** alpha, 0.1 the speed of light; beta, up to 0.9 the speed of light; gamma, the speed of light **11.39** (a) no detectable effects (b) nausea, fatigue, lowered blood cell count **11.41** 19% human-made, 81% natural sources **11.43** to monitor the extent of radiation exposure **11.45** so radiation can be detected externally **11.47** (a) bone tumors (b) circulatory problems (c) iron metabolism in blood (d) intercellular space problems **11.49** They are usually α or β emitters. **11.51** (a) fusion (b) fusion (c) both

(d) fission **11.53** (a) fusion (b) fission (c) neither (d) neither **11.55** (a) $^{206}_{80}Hg \rightarrow ^{0}_{-1}\beta + ^{206}_{81}Tl$ (b) $^{109}_{46}Pd \rightarrow ^{0}_{-1}\beta + ^{109}_{47}Ag$ (c) $^{245}_{96}Cm \rightarrow ^{4}_{2}\alpha + ^{241}_{94}Pu$ (d) $^{249}_{100}Fm \rightarrow ^{4}_{2}\alpha + ^{245}_{98}Cf$ **11.56** (a) 54 hr (b) 90 hr (c) 108 hr (d) 126 hr **11.57** (a) $^{243}_{94}Pu + ^{4}_{2}\alpha \rightarrow ^{246}_{96}Cm + ^{1}_{0}n$ (b) $^{246}_{96}Cm + ^{12}_{6}C \rightarrow ^{254}_{102}No + 4^{1}_{0}n$ (c) $^{27}_{13}Al + ^{4}_{2}\alpha \rightarrow ^{1}_{0}n + ^{30}_{15}P$ (d) $^{23}_{11}Na + ^{2}_{1}H \rightarrow ^{21}_{10}Ne + ^{4}_{2}\alpha$ **11.58** $^{142}_{60}Nd + ^{1}_{0}n \rightarrow ^{143}_{61}Pm + ^{0}_{-1}\beta$ **11.59** 14 elements **11.60** 4 neutrons **11.61** A = 0 (negligible amount), B = 0 (negligible amount), C = 63 atoms, D = 937 atoms **11.62** ^{228}Ac, ^{228}Th, ^{224}Ac **11.63** (a) 3 and 4 (b) 2, 3, and 6 (c) 3 and 5 (d) 1 and 5 **11.64** (a) 1 and 6 (b) 2 and 5 (c) 4 (d) 3 **11.65** (a) 1 (b) 2 and 5 (c) 3 and 6 (d) 4 **11.66** (a) 1 (b) 2 and 4 (c) 5 and 6 (d) 3

Chapter 12 **12.1** (a) saturated (b) unsaturated (c) unsaturated (d) unsaturated **12.3** (a) 18 (b) 4 (c) 13 (d) 22 **12.5** (a) $CH_3—CH_2—CH_2—CH_3$

(b) 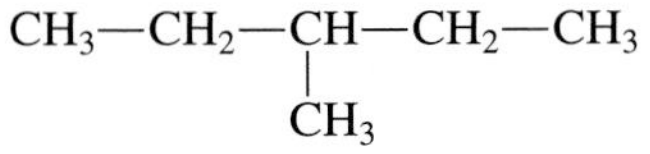
```
CH3—CH2—CH—CH2—CH3
        |
        CH3
```
(c) 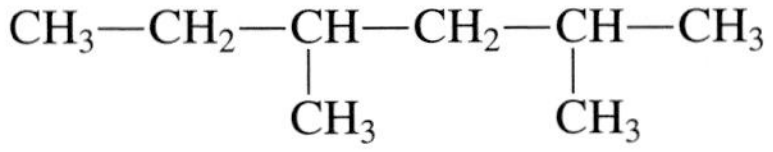
```
CH3—CH2—CH—CH2—CH—CH3
        |       |
        CH3     CH3
```
(d) 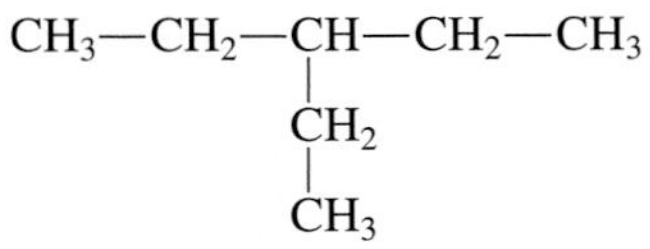
```
CH3—CH2—CH—CH2—CH3
        |
        CH2
        |
        CH3
```

12.7 (a) 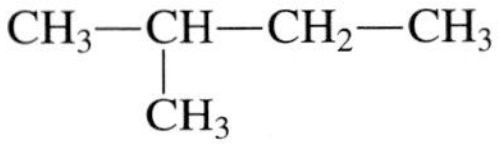
```
CH3—CH—CH2—CH3
    |
    CH3
```
(b) 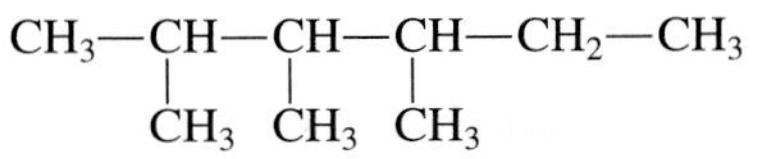
```
CH3—CH—CH—CH—CH2—CH3
    |   |   |
    CH3 CH3 CH3
```
(c) $CH_3—CH_2—CH_2—CH_2—CH_2—CH_3$

(d) 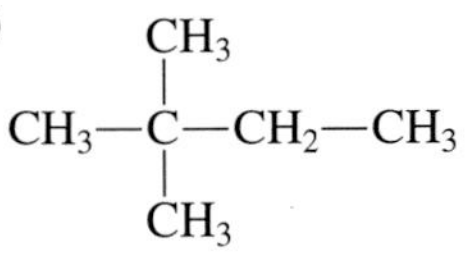
```
    CH3
    |
CH3—C—CH2—CH3
    |
    CH3
```

12.9 (a) 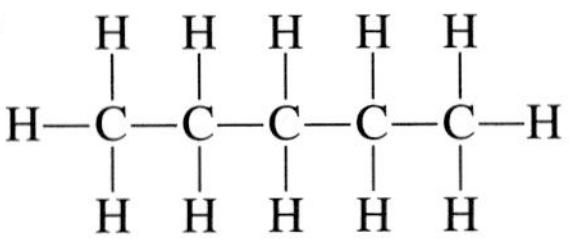
```
  H H H H H
  | | | | |
H—C—C—C—C—C—H
  | | | | |
  H H H H H
```
(b) 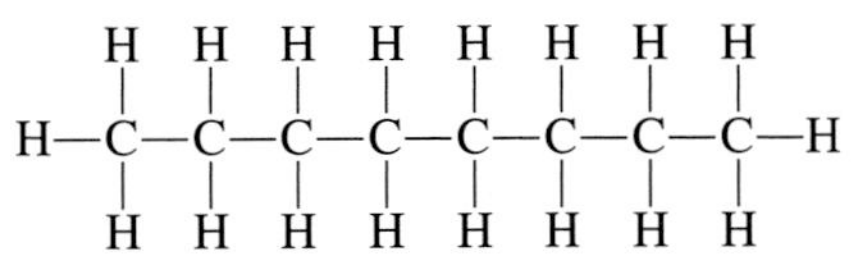
```
  H H H H H H H H
  | | | | | | | |
H—C—C—C—C—C—C—C—C—H
  | | | | | | | |
  H H H H H H H H
```
(c) $CH_3—(CH_2)_8—CH_3$ (d) C_6H_{14} **12.11** (a) different compounds that are not structural isomers (b) different compounds that are structural isomers (c) different conformations of the same molecule (d) different compounds that are structural isomers **12.13** (a) seven-carbon chain (b) eight-carbon chain (c) eight-carbon chain (d) seven-carbon chain **12.15** (a) 2-methylpentane (b) 2,4,5-trimethylheptane (c) 3-ethyl-2,3-dimethylpentane (d) 3-ethyl-2,4-dimethylhexane (e) decane (f) 4-propylheptane **12.17** horizontal chain, because it has more substituents (two)

12.19 (a) 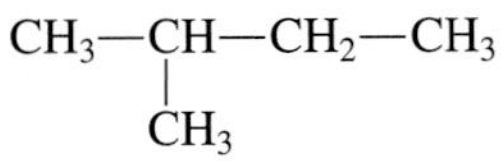
```
CH3—CH—CH2—CH3
    |
    CH3
```
(b) 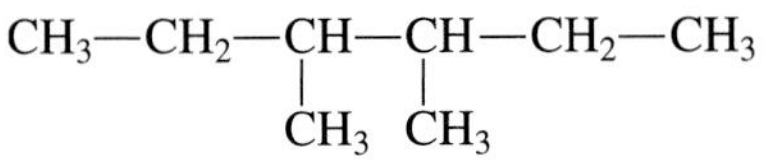
```
CH3—CH2—CH—CH—CH2—CH3
        |   |
        CH3 CH3
```
(c) 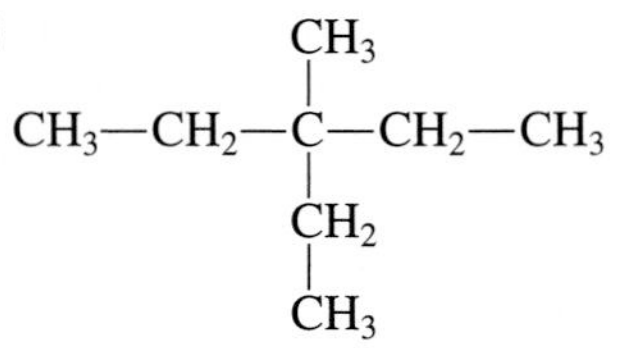
```
        CH3
        |
CH3—CH2—C—CH2—CH3
        |
        CH2
        |
        CH3
```
(d) 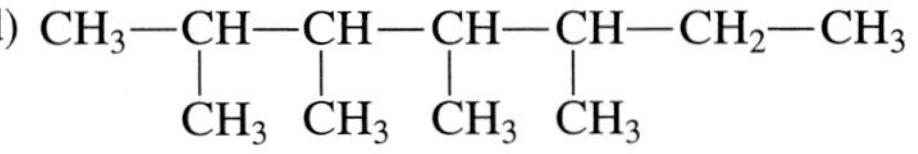
```
CH3—CH—CH—CH—CH—CH2—CH3
    |   |   |   |
    CH3 CH3 CH3 CH3
```
(e) 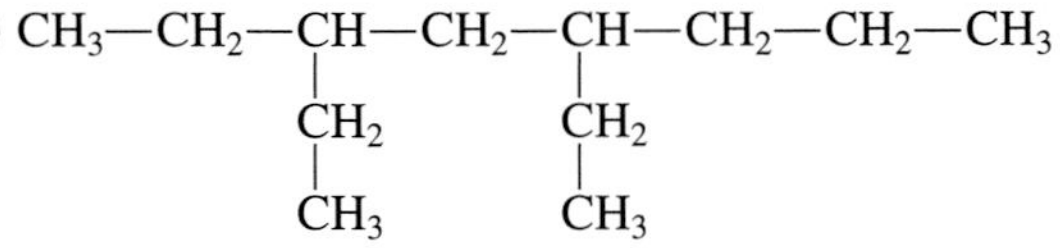
```
CH3—CH2—CH—CH2—CH—CH2—CH2—CH3
        |       |
        CH2     CH2
        |       |
        CH3     CH3
```
(f) 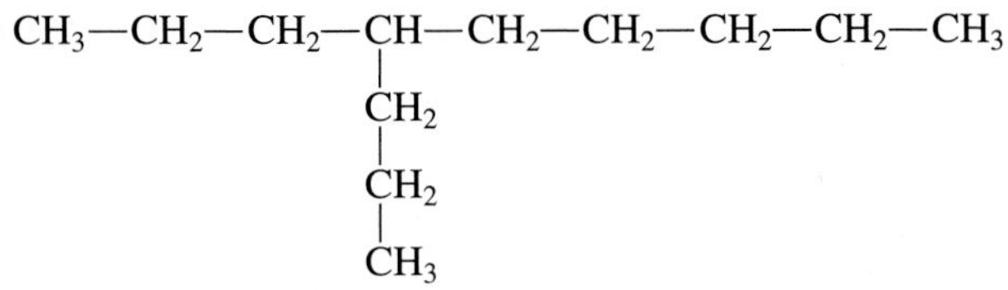
```
CH3—CH2—CH2—CH—CH2—CH2—CH2—CH2—CH3
            |
            CH2
            |
            CH2
            |
            CH3
```

12.21 (a) 1, 1 (b) 2, 2 (c) 2, 2, (d) 4, 4 (e) 2, 2 (f) 1, 1 **12.23** (a) carbon chain numbered from wrong end; 2-methylpentane (b) not based on longest carbon chain; 2,2-dimethylbutane (c) carbon chain numbered from wrong end; 2,2,3-trimethylbutane (d) not based on longest carbon chain; 3,3-dimethylhexane (e) carbon chain numbered from wrong end and alkyl groups not listed alphabetically; 3-ethyl-4-methylhexane (f) like alkyl groups listed separately; 2,4-dimethylhexane **12.25** (a) 3, 2, 1, 0 (b) 5, 2, 3, 0 (c) 5, 2, 1, 1 (d) 5, 2, 3, 0 (e) 2, 8, 0, 0 (f) 3, 6, 1, 0 **12.27** (a) isopropyl (b) isobutyl (c) isopropyl (d) *sec*-butyl **12.29**

(a) 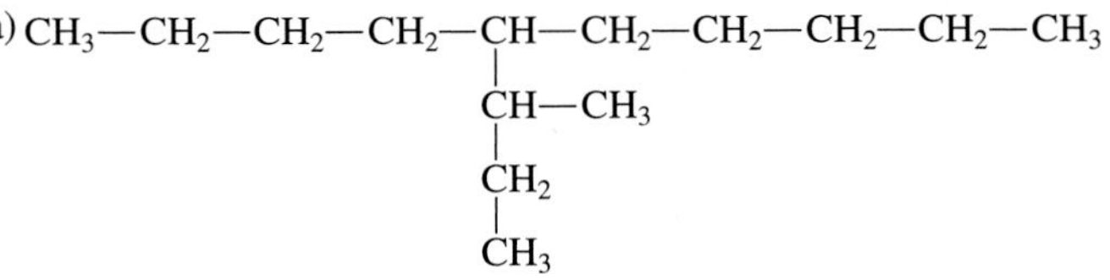
```
CH3—CH2—CH2—CH2—CH—CH2—CH2—CH2—CH2—CH3
                |
                CH—CH3
                |
                CH2
                |
                CH3
```
(b) 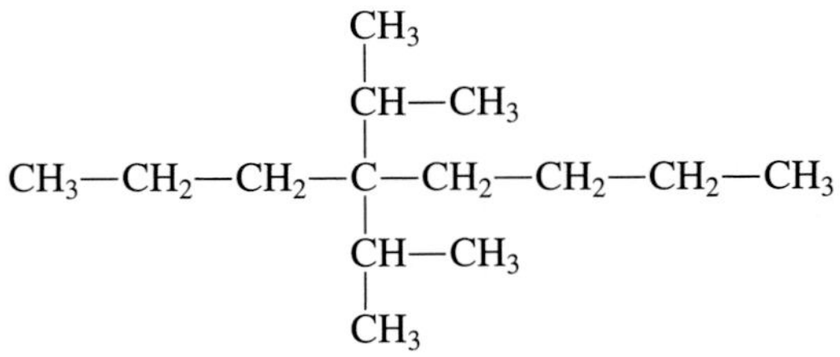
```
            CH3
            |
            CH—CH3
            |
CH3—CH2—CH2—C—CH2—CH2—CH2—CH3
            |
            CH—CH3
            |
            CH3
```
(c)
```
CH3—CH—CH—CH2—CH—CH2—CH2—CH2—CH3
    |   |       |
    CH3 CH3     CH2
                |
                CH—CH3
                |
                CH3
```
(d) 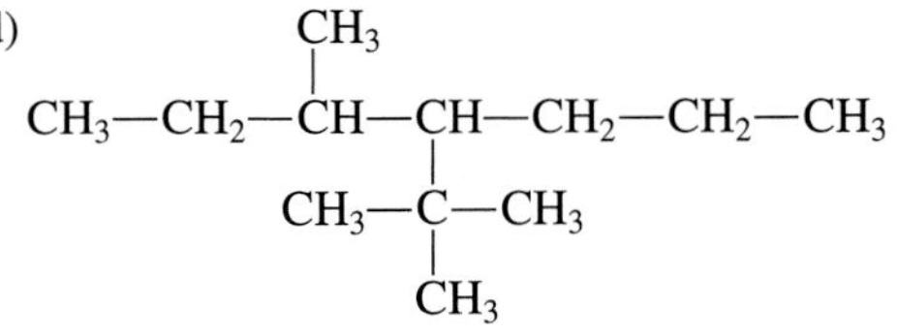
```
            CH3
            |
CH3—CH2—CH—CH—CH2—CH2—CH3
            |
        CH3—C—CH3
            |
            CH3
```

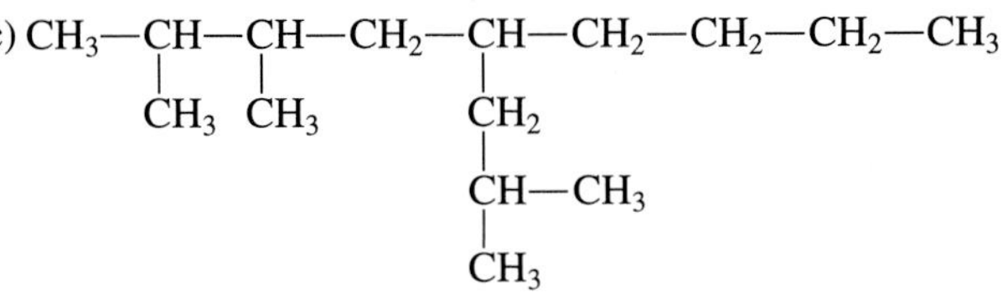

12.31 (a) 16 (b) 6 (c) 5 (d) 15 **12.33** (a) C_6H_{12} (b) C_6H_{12} (c) C_4H_8 (d) C_7H_{14} **12.35** (a) cyclohexane (b) 1,2-dimethylcyclobutane (c) methylcyclopropane (d) 1,2-dimethylcyclopentane **12.37** (a) must locate methyl groups with numbers (b) wrong numbering system for ring (c) no number needed (d) wrong numbering system for ring **12.39**

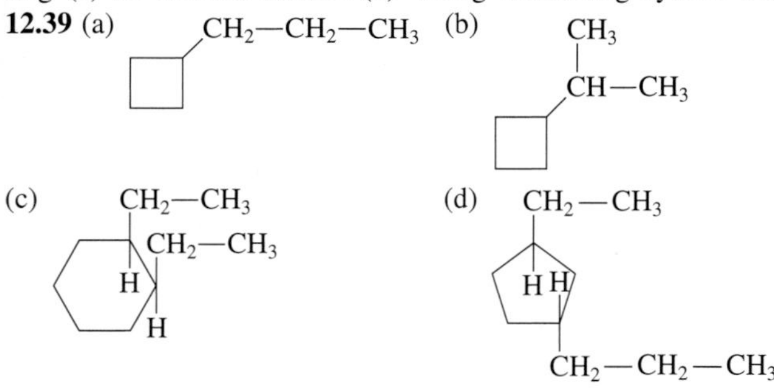

12.41 (a) not possible
(b) $CH_3—CH_2$ / $CH_2—CH_3$ on cyclopropane, H H (*cis*); $CH_3—CH_2$ / H and H / $CH_2—CH_3$ on cyclopropane (*trans*)

(c) not possible (d) cyclohexane with CH_3 and CH_3, H and H (*cis*); cyclohexane with CH_3 / H and H / CH_3 (*trans*) **12.43** boiling point

12.45 (a) octane (b) cyclopentane (c) pentane (d) cyclopentane **12.47** (a) different states (b) same states (c) same states (d) same states **12.49** (a) CO_2 and H_2O (b) CO_2 and H_2O (c) CO_2 and H_2O (d) CO_2 and H_2O **12.51** CH_3Br, CH_2Br_2, $CHBr_3$, CBr_4 **12.53** (a) $CH_3—CH_2—Cl$

(b) $ClCH_2—CH_2—CH_2—CH_3$ $\quad CH_3—CH(Cl)—CH_2—CH_3$

(c) $ClCH_2—CH(CH_3)—CH_3$ $\quad CH_3—C(Cl)(CH_3)—CH_3$ (d) cyclopentane—Cl

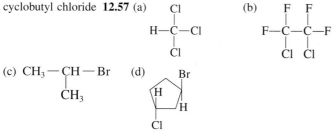

12.55 (a) iodomethane, methyl iodide (b) 1-chloropropane, propyl chloride (c) 2-fluorobutane, sec-butyl fluoride (d) chlorocyclobutane, cyclobutyl chloride **12.57** (a) $H—C(Cl)_2—Cl$ (b) $F—C(F)(Cl)—C(F)(Cl)—F$

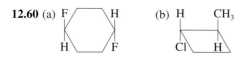

(c) $CH_3—CH(CH_3)—Br$ (d) cyclopentane with Br, H, H, Cl

12.59 (a) no (b) yes (c) no (d) no

12.60 (a) cyclohexane with F, H and H, F (b) cyclobutane with H, CH_3 and Cl, H

(c) $CH_3—C(CH_3)(Br)—CH_3$ (d) $CH_3—CH(CH_3)—CH_2—I$

12.61 4-tert-butyl-5-ethyl-2,2,6-trimethyloctane **12.62** (a) $C_{18}H_{38}$ (b) C_7H_{14} (c) $C_7H_{14}F_2$ (d) $C_6H_{10}Br_2$
12.63 (a) $CH_3—C(CH_3)_2—CH_2—CH_2—CH_2—CH_3$ (b) $CH_3—CH(CH_3)—CH_3$

(c) $Cl—C(Cl)(H)—H$ (d) $Cl—C(Cl)(F)—F$

12.64 (a) alkane (b) halogenated cycloalkane (c) halogenated alkane (d) cycloalkane

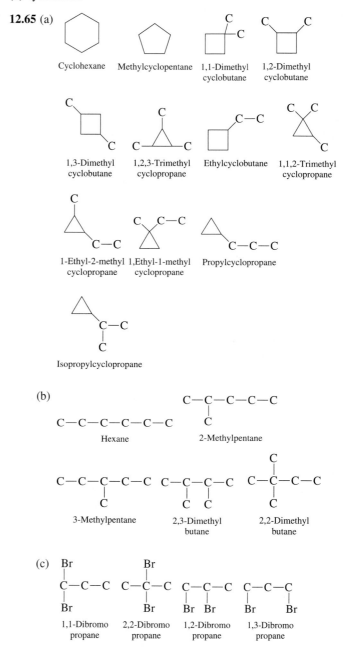

12.65 (a)

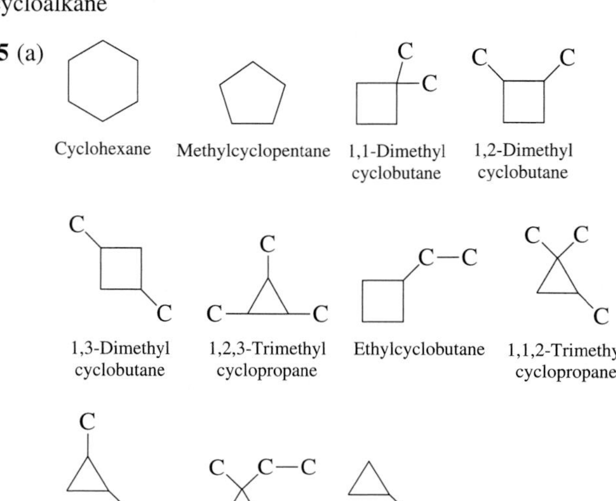

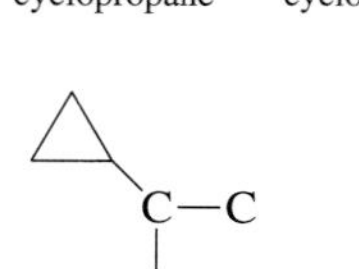

Isopropylcyclopropane

(b)

C—C—C—C—C—C
Hexane

C—C(C)—C—C—C
2-Methylpentane

C—C—C(C)—C—C
3-Methylpentane

C—C(C)—C(C)—C
2,3-Dimethyl butane

C—C(C)(C)—C—C
2,2-Dimethyl butane

(c)

C(Br)(Br)—C—C
1,1-Dibromo propane

C—C(Br)(Br)—C
2,2-Dibromo propane

C(Br)—C(Br)—C
1,2-Dibromo propane

C(Br)—C—C(Br)
1,3-Dibromo propane

12.66 (a) 1,2-diethylcyclohexane (b) 3-methylhexane (c) 2,3-dimethyl-3-propylnonane (d) 1-isopropyl-3,5-dipropylcyclohexane **12.67** (a) 1, 2, and 5 (b) 6 (c) 4 and 6 (d) 1 and 2 **12.68** (a) 2 and 3 (b) 5 (c) 6 (d) 1, 3, and 4 **12.69** (a) all choices (b) 3 (c) 1 and 4; 2 and 5 (d) 3 and 6 **12.70** (a) 1, 3, and 6 (b) 2, 4, and 5 (c) 2, 3, and 6 (d) 2

Chapter 13 **13.1** (a) unsaturated, alkene with one double bond (b) saturated (c) unsaturated, alkene with one double bond (d) unsaturated, diene (e) unsaturated, triene (f) unsaturated, diene **13.3** (a) C_4H_{10} (b) C_5H_{10} (c) C_5H_8 (d) C_7H_{10} **13.5** (a) C_nH_{2n-2} (b) C_nH_{2n-2} (c) C_nH_{2n-2} (d) C_nH_{2n-6} **13.7** (a) 2-butene (b) 2,4-dimethyl-2-pentene (c) cyclohexene (d) 1,3-cyclopentadiene (e) 2-ethyl-1-pentene (f) 2,4,6-octatriene **13.9** (a) 2-pentene (b) pentane (c) 2,3,3-trimethyl-1-butene (d) 2-methyl-1,4-pentadiene (e) 1,3,5-hexatriene (f) 2,3-pentadiene **13.11** (a) $CH_2{=}CH{-}CH(CH_3){-}CH_2{-}CH_3$

(b) cyclopentene ring with CH_3 substituent (c) $CH_2{=}CH{-}CH{=}CH_2$

(d) $CH_2{=}CH{-}CH(CH_2{-}CH_3){-}CH{=}CH_2$

(e) $CH_3{-}CH{=}CH{-}CH(CH_2{-}CH_2{-}CH_3){-}CH_2{-}CH_2{-}CH_3$

(f) cyclohexadiene ring with $CH_2{-}CH_3$ and $CH_3{-}CH_2$ substituents

13.13 (a) 3-methyl-3-hexene (b) 2,3-dimethyl-2-hexene (c) 1,3-cyclopentadiene (d) 4,5-dimethylcyclohexene **13.15** σ-bond: orbital overlap along internuclear axis; π-bond: orbital overlap above and below (but not on) internuclear axis **13.17** (a) 4 σ, 0 π (b) 3 σ, 1 π (c) 3 σ, 2 π (d) 1 σ, 2 π (e) 5 σ, 0 π (f) 10 σ, 2 π **13.19** (a) 16 σ, 0 π (b) 11 σ, 1 π (c) 9 σ, 2 π (d) 3 σ, 2 π (e) 15 σ, 0 π (f) 26 σ, 2 π

13.21
C=C—C—C—C—C (1-Hexene)
C—C=C—C—C—C (2-Hexene)
C—C—C=C—C—C (3-Hexene)
C=C(C)—C—C—C (2-Methyl-1-pentene)
C=C—C(C)—C (3-Methyl-1-pentene)
C=C—C—C(C)—C (4-Methyl-1-pentene)
C—C(C)=C—C—C (2-Methyl-2-pentene)
C—C=C(C)—C—C (3-Methyl-2-pentene)
C—C=C—C(C)—C (4-Methyl-2-pentene)
C=C(C)—C(C)—C (2,3-Dimethyl-1-butene)
C=C—C(C)(C)—C (3,3-Dimethyl-1-butene)
C—C(C)=C(C)—C (2,3-Dimethyl-2-butene)
C=C(C—C)—C—C (2-Ethyl-1-butene)

13.23 (a) no (b) no (c) no (d) $CH_3{-}CH_2(H)C{=}C(H)CH_2{-}CH_3$ *cis*; $CH_3{-}CH_2(H)C{=}C(H)CH_2{-}CH_3$ *trans*

(e) $CH_3(H)C{=}C(H)CH(CH_3){-}CH_3$ *cis*; $CH_3(H)C{=}C(H)CH(CH_3){-}CH_3$ *trans*

(f) 1,2-dimethylcyclopentane: *cis* (both CH_3 on same side); *trans* (CH_3 on opposite sides)

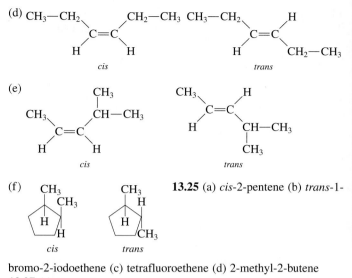

13.25 (a) *cis*-2-pentene (b) *trans*-1-bromo-2-iodoethene (c) tetrafluoroethene (d) 2-methyl-2-butene

13.27 (a) $CH_3{-}CH_2(CH_3)C{=}C(H)CH_2{-}CH_3$
(b) $CH_3(H)C{=}C(H)CH_2{-}CH_3$
(c) $CH_3(H)C{=}C(H)CH_2{-}CH(CH_3){-}CH_2{-}CH_3$
(d) $CH_2{=}CH(H)C{=}C(H)CH_3$

13.29 (a) yes (b) no (c) yes (d) no

13.31 (a) $CH_2{=}CH_2 + Cl_2 \longrightarrow CH_2Cl{-}CH_2Cl$
(b) $CH_2{=}CH_2 + HCl \longrightarrow CH_3{-}CH_2Cl$
(c) $CH_2{=}CH_2 + H_2 \xrightarrow{Ni} CH_3{-}CH_3$
(d) $CH_2{=}CH_2 + HBr \longrightarrow CH_3{-}CH_2Br$

13.33 (a) $CH_2{=}CH{-}CH_3 + Cl_2 \longrightarrow CH_2Cl{-}CHCl{-}CH_3$
(b) $CH_2{=}CH{-}CH_3 + HCl \longrightarrow CH_3{-}CHCl{-}CH_3$
(c) $CH_2{=}CH{-}CH_3 + H_2 \xrightarrow{Ni} CH_3{-}CH_2{-}CH_3$
(d) $CH_2{=}CH{-}CH_3 + HBr \longrightarrow CH_3{-}CHBr{-}CH_3$

13.35 (a) $CH_3{-}CHCl{-}CHCl{-}CH_3$ (b) $CH_3{-}CBr(CH_3){-}CH_3$ (c) $CH_3{-}CH_2{-}CHCl{-}CH_3$ (d) cyclopentane (e) cyclopentene (f) HO-cyclobutane

13.37 (a) Br_2 (b) H_2 + Ni catalyst (c) HCl (d) $H_2O + H_2SO_4$ catalyst

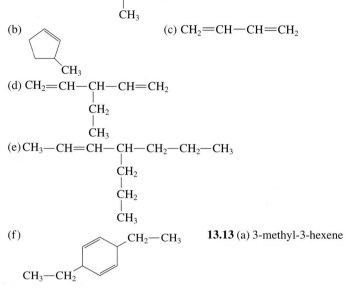

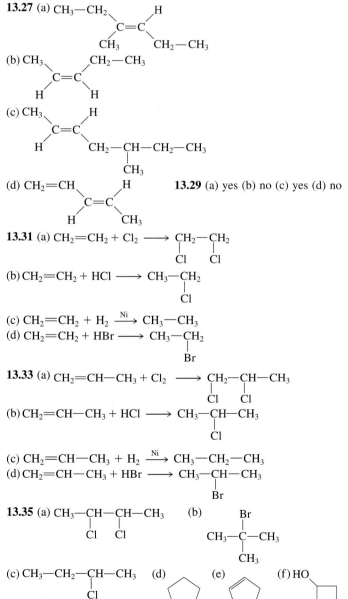

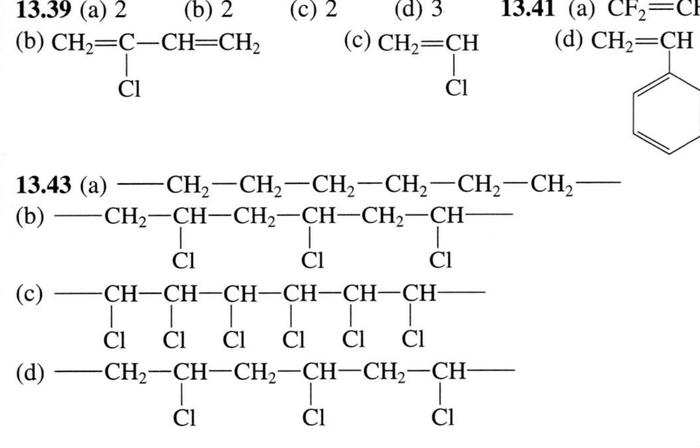

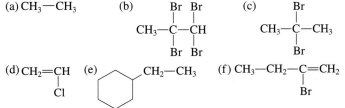

13.45 (a) 1-hexyne (b) 4-methyl-2-pentyne (c) 2,2-dimethyl-3-heptyne (d) 2-butyne (e) 3-methyl-1,4-hexadiyne (f) 4-methyl-2-hexyne **13.47** (a) $CH_3—CH_3$ (b) $CH_3—CBr_2—CHBr_2$ (c) $CH_3—CBr_2—CH_3$ (d) $CH_2=CH(Cl)$ (e) cyclohexyl—$CH_2—CH_3$ (f) $CH_3—CH_2—C(Br)=CH_2$

13.49 (a) 1,3-dibromobenzene (b) 2-nitroaniline (c) 1-chloro-4-nitrobenzene (d) 3-chlorotoluene (e) 1-bromo-2-ethylbenzene (f) 4-bromophenol **13.51** (a) *m*-dibromobenzene (b) *o*-nitroaniline (c) *p*-chloronitrobenzene (d) *m*-chlorotoluene (e) *o*-bromoethylbenzene (f) *p*-bromophenol **13.53** (a) 2,4-dibromo-1-chlorobenzene (b) 3,5-dinitrotoluene (c) 1-bromo-3-chloro-2-fluorobenzene (d) 1,4-dibromo-2,5-dichorobenzene **13.55** (a) 2-phenylbutane (b) 3-phenyl-1-butene (c) 3-methyl-1-phenylbutane (d) 2,4-diphenylpentane

13.57 (a) benzene ring with $CH_2—CH_3$ and $CH_2—CH_3$ (meta) (b) benzene ring with CH_3 and CH_3 (ortho) (c) benzene ring with CH_3 and $CH_2—CH_3$ (para) (d) benzene ring with COOH and Cl (para) (e) benzene ring with NH_2 (f) $CH_3—CH_2—C(CH_3)(C_6H_5)—CH_2—CH_3$

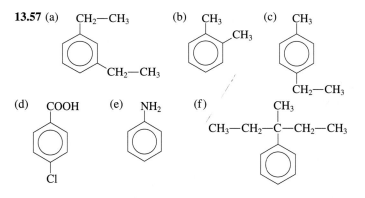

13.59 (a) substitution (b) addition (c) substitution (d) addition **13.61** (a) Br_2 (b) HNO_3 (c) benzene ring with $CH(CH_3)—CH_3$ (d) $CH_3—CH_2—Br$

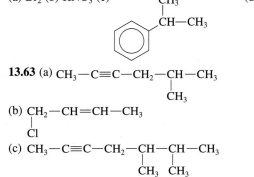

13.63 (a) $CH_3—C≡C—CH_2—CH(CH_3)—CH_3$
(b) $CH_2(Cl)—CH=CH—CH_3$
(c) $CH_3—C≡C—CH_2—CH(CH_3)—CH(CH_3)—CH_3$
(d) $CH_2=CH—CH(CH(CH_3)—CH_3)—CH_2—CH_2—CH_3$
(e) $CH_2=CH—CH_2—CH_2—CH_2—CH=CH_2$
(f) $CH≡C—CH(CH_3)—C≡CH$ **13.64** (a) two (b) one (c) two (d) one (e) two (f) four **13.65** (a) $CH_2=CH$—benzene ring
(b) $CH_2=CH—CH_2—Cl$ (c) $CH_3—CH_2—CH_2—C≡CH$
(d) $CH_3—CH_2—CH_2—C≡C—CH_2—CH_2—CH_3$
(e) benzene ring with COOH and $CH_2—CH_3$ (meta) (f) benzene ring with CH_3 and a second benzene ring (meta)

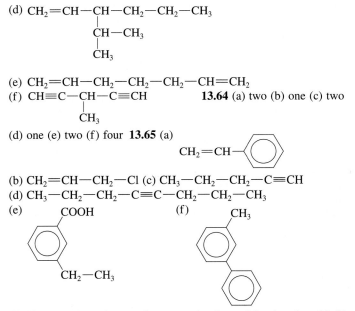

13.66 It would require a carbon atom that formed five bonds. **13.67** Each carbon atom in 1,2-dichlorobenzene has only one substituent. **13.68** $CH_2=CH—CH_2—CH_2—CH_3$ $CH_3—CH=CH—CH_2—CH_3$ (*cis–trans* forms)
$CH_2=C(CH_3)—CH_2—CH_3$ $CH_2=CH—CH(CH_3)—CH_3$
$CH_3—C(CH_3)=CH—CH_3$ cyclopentane methylcyclobutane
1,1-dimethylcyclopropane 1,2-dimethylcyclopropane (*cis–trans* forms)

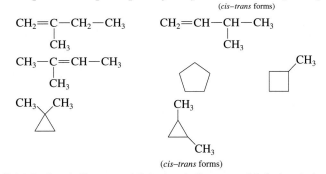

13.69 1,2,3-trimethylbenzene; 1,2,4-trimethylbenzene; 1,3,5-trimethylbenzene; 2-ethyltoluene; 3-ethyltoluene; 4-ethyltoluene; propylbenzene; isopropylbenzene **13.70** (a) 3 (b) 3 (c) 11 (d) 3 **13.71** (a) 1, 5, and 6 (b) 2 and 3 (c) 1, 5, and 6 (d) 2 and 6 **13.72** (a) 4 and 6 (b) 1, 2, 3, and 5 (c) 1 and 6 (d) 1, 2, 5, and 6 **13.73** (a) 1, 2, and 3 (b) 4, 5, and 6 (c) 2, 3, 4, and 6 (d) 1 and 5 **13.74** (a) 4 (b) 1, 2, 3, 5, and 6 (c) 5 (d) 1, 2, 4, 5, and 6

Chapter 14 **14.1** (a) alcohol (b) ether (c) phenol (d) alcohol (e) ether (f) ether **14.3** (a) 2-pentanol (b) ethanol (c) 3-methyl-2-butanol (d) 2-ethyl-1-pentanol (e) 2-butanol (f) 3,3-dimethyl-1-butanol
14.5 (a) $CH_3—CH_2—CH(OH)—CH_2—CH_3$
(b) $CH_3—CH_2—C(OH)(CH_2—CH_3)—CH_2—CH_2—CH_3$
(c) $CH_2(OH)—CH(CH_3)—CH_3$
(d) $CH_3—CH(OH)—CH_2—CH(CH_3)—CH_3$
(e) $CH_3—C(OH)(C_6H_5)—CH_3$

(f) [cyclobutane ring with OH and CH_3 on adjacent carbons]

OH

CH_3

14.7 (a) $CH_2—CH_2—CH_2—CH_2—CH_3$

OH

1-Pentanol

(b) $CH_2—CH_2—CH_3$

OH

1-Propanol

(c) $CH_3—CH—CH_2—OH$

CH_3

2-Methyl-1-Propanol

(d) $CH_3—CH_2—CH—OH$

CH_3

2-Butanol

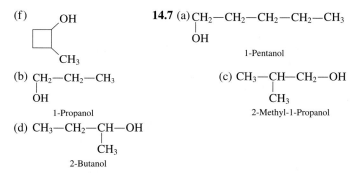

14.9 (a) 1,2-propanediol (b) 1,4-pentanediol (c) 1,3-pentanediol (d) 3-methyl-1,2,4-butanetriol **14.11** (a) cyclohexanol (b) *trans*-3-chlorocyclohexanol (c) *cis*-2-methylcyclohexanol (d) 1-methylcyclobutanol

14.13 (a) $CH_3—CH—CH_2—CH{=}CH_2$

OH

(b) $CH{\equiv}C—CH—CH_2—CH_3$

OH

(c) $CH_3—CH—C{=}CH_2$

OH CH_3

(d) $HO—CH_2$ CH_3

C=C

H H

14.15 (a) $CH_2—CH—CH_3$

OH CH_2

CH_3

2-Methyl-1-Butanol

(b) $CH_3—CH—CH_2—CH_2$

OH OH

1,3-Butanediol

(c) $CH_3—CH—CH—CH_3$

CH_3 OH

3-Methyl-2-Butanol

(d) [cyclopentane ring with OH at C-1 and HO at C-3]

OH

HO

1,3-Cyclopentanediol

14.17 (a) 3-ethylphenol (b) 2-chlorophenol (c) *o*-cresol (d) hydroquinone (e) 2-bromophenol (f) 2-bromo-3-ethylphenol **14.19** (a) [benzene ring with OH and, para to it, Cl]

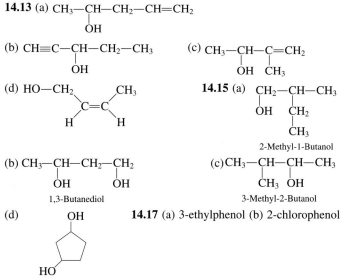

OH

Cl

(b) [benzene ring with OH and, ortho to it, $CH_2—CH_3$]

OH

$CH_2—CH_3$

(c) [benzene ring with OH, NO_2 ortho to the OH, and NO_2 para to the OH]

OH

NO_2

NO_2

(d) [benzene ring with OH and, meta to it, CH_3]

OH

CH_3

(e) [benzene ring with OH and, meta to it, a second OH]

OH

OH

(f) [benzene ring with OH, $CH_3—CH_2$ and $CH_2—CH_3$ on the two carbons ortho to the OH, and CH_3 para to the OH]

OH

$CH_3—CH_2$ $CH_2—CH_3$

CH_3

14.21 Phenols require the —OH groups to be attached directly to the benzene ring. **14.23** (a) ethanol with all traces of H_2O removed (b) ethanol (c) 70% solution of isopropyl alcohol (d) ethanol **14.25** (a) glycerol (b) ethanol (c) methanol (d) methanol **14.27** Alcohol molecules can hydrogen-bond to each other; alkane molecules cannot. **14.29** (a) 1-heptanol (b) 1-propanol (c) 1,2-ethanediol **14.31** (a) 1-butanol (b) 1-pentanol (c) 1,2-butanediol **14.33** (a) 3 (b) 3 (c) 3 (d) 3 **14.35** (a) $CH_2—CH_3$

OH

(b) $CH_3—CH_2—CH_2$

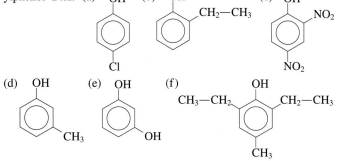

OH

(c) OH

$CH_3—CH_2—C—CH_3$

CH_3

(d) $CH_3—CH_2—CH—CH_2—CH_3$

OH

14.37 (a) $CH_2{=}CH—CH_3$ (b) $CH_3—C{=}CH_2$

CH_3

(c) $CH_3—CH{=}CH_2$ (d) $CH_3—CH_2—CH_2—O—CH_2—CH_2—CH_3$

14.39 (a) $CH_3—CH—CH—CH_3$

OH CH_3

(b) $CH_3—CH_2—CH_2$

OH

(c) $CH_3—CH_2—OH$ (d) $CH_3—CH—CH_2—OH$

CH_3

14.41 (a) $CH_3—CH_2—CH—CH_3$

OH

(b) $CH_3—CH_2—CH_2$

OH

(c) $CH_3—CH_2—CH_2$

OH

(d) [cyclopentane ring with $CH_2—OH$ substituent]

$CH_2—OH$

14.43 (a) $CH_3—CH_2—CH_2—Cl$ (b) [cyclopentene ring with CH_3 on a double-bonded carbon]

CH_3

(c) O

$CH_3—C—CH_2—CH_3$

(d) $CH_3—CH_2—CH—CH_2—CH_3$

Cl

(e) $CH_3—CH_2—O—CH_2—CH_3$ (f) $CH_2—CH_2$

Br Br

14.45 $\left(\begin{matrix}CH—CH\\ OH\quad OH\end{matrix}\right)_n$ **14.47** An antiseptic kills microorganisms on living tissue; disinfectant kills microorganisms on inanimate objects.

14.49 [phenol] + $H_2O \rightleftharpoons$ [phenoxide ion] + H_3O^+

OH O^-

14.51 (a) 1-methoxypropane (b) 1-ethoxypropane (c) 2-methoxypropane (d) methoxybenzene (e) cyclohexoxycyclohexane (f) ethoxycyclobutane **14.53** (a) methyl propyl ether (b) ethyl propyl ether (c) isopropyl methyl ether (d) methyl phenyl ether (e) dicyclohexyl ether (f) cyclobutyl ethyl ether **14.55** (a) $CH_3—CH—O—CH_2—CH_2—CH_3$

CH_3

(b) $CH_3—CH_2—O—$[benzene ring]

(c) $CH_3—CH—CH_2—O—CH_3$

CH_3

(d) $CH_3—CH—CH_2—CH_2—CH_3$

$O—CH_2—CH_3$

(e) [cyclobutane ring with $O—CH_2—CH_3$ substituent]

$O—CH_2—CH_3$

(f) CH_3

$CH_3—O—CH_2—C—CH_3$

CH_3

14.57 Dimethyl ether molecules cannot hydrogen-bond to each other; ethanol molecules can. **14.59** flammability and peroxide formation **14.61** No oxygen–hydrogen bonds are present. **14.63** (a) noncyclic ether (b) noncyclic ether (c) cyclic ether (d) cyclic ether (e) noncyclic ether (f) nonether **14.65** R—S—H versus R—O—H **14.67** (a) $CH_3—SH$ (b) $CH_3—CH—CH_3$

SH

(c) $CH_3—CH_2—CH_2—CH_2$

SH

(d) $CH_3—CH_2—CH—CH_2—CH_2$

CH_3 SH

(e) [cyclopentane ring with SH substituent]

SH

(f) $CH_2—CH_2$

SH SH

14.69 Alcohol oxidation produces aldehydes and ketones; thioalcohol oxidation produces disulfides. **14.71** (a) methylthioethane (b) 2-methylthiopropane (c) methylthiocyclohexane (d) cyclohexylthiocyclohexane (e) 1-methylthio-1-propene (f) 2-methylthiobutane **14.72**

(a) ethylthioethane (b) 2-ethylthiopropane (c) methylthiocyclopentane (d) cyclopentylthiocyclohexane (e) 1-methylthio-2-propene (f) 2-methylthiopropane **14.73** (a) secondary (b) secondary (c) primary (d) primary (e) tertiary (f) secondary **14.74** (a) primary, primary (b) primary, secondary (c) secondary, secondary (d) primary, secondary, secondary **14.75** (a) 2-hexanol (b) 3-pentanol (c) 3-phenyoxy-1-propene (d) 2-methyl-1-propanol (e) 2-methyl-2-propanol (f) ethoxyethane **14.76** $CH_2—CH_2—CH_2—CH_2—CH_3$ (with OH on C-1)

$CH_2—CH—CH_2—CH_3$ (OH on C-1, CH_3 on C-2)

$CH_3—C—CH_2—CH_3$ (CH_3 and OH on C-2)

$CH_3—CH—CH_2—CH_2—CH_3$ (OH on C-2)

$CH_2—CH_2—CH—CH_3$ (OH on C-1, CH_3 on C-3)

$CH_3—CH_2—CH—CH_2—CH_3$ (OH on C-3)

$CH_3—C—CH_2$ (CH_3 on C-2, OH on CH_2; CH_3 above C-2)

$CH_3—CH—CH—CH_3$ (OH on C-2, CH_3 on C-3)

14.77 1-pentanol

14.78 $CH_3—O—CH_3$, $CH_3—CH_2—CH_2—O—CH_2—CH_2—CH_3$, and $CH_3—O—CH_2—CH_2—CH_3$ **14.79** (a) disulfide (b) thiol, thioalcohol (c) alcohol (d) peroxide (e) alcohol, thiol, thioalcohol (f) ether, sulfide, thioether **14.80** (a) 1,2-ethanedithiol (b) 3-methoxy-1-propanol (c) 1-propanol (d) 1,2-dimethoxyethane (e) methylthioethane (f) 1-ethylthio-2-methoxyethane **14.81** (a) 3 and 4 (b) 6 (c) 3 and 4 (d) 1, 2, and 5 **14.82** (a) 1, 2, 3, 4, and 5 (b) all choices (c) all choices (d) 1 and 5, 2 and 3 **14.83** (a) 6 (b) 1, 3, 5, and 6 (c) 2 and 4 (d) 3 and 4, 2 and 6 **14.84** (a) 1 (b) 6 (c) 1 (d) 3

Chapter 15 **15.1** (a) yes (b) no (c) yes (d) yes (e) no (f) no **15.3** similarity: both have σ and π components; difference: C=O is polar, C=C is not polar **15.5** (a) neither (b) aldehyde (c) ketone (d) neither (e) aldehyde (f) aldehyde

15.7 $H—C(=O)—H$, $CH_3—C(=O)—H$, $CH_3—C(=O)—CH_3$,

$CH_3—CH_2—C(=O)—CH_3$ **15.9** (a) neither (b) aldehyde (c) neither (d) ketone (e) ketone (f) aldehyde **15.11** (a) butanal (b) 2-methylbutanal (c) 4-methylheptanal (d) 3-phenylpropanal (e) propanal (f) 3,3-dimethylbutanal **15.13** (a) $CH_3—CH_2—CH(CH_3)—CH_2—C(=O)—H$

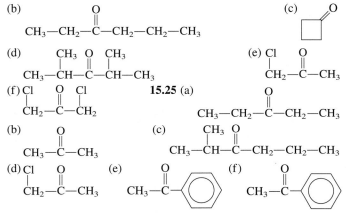

(b) $CH_3—CH_2—CH_2—CH_2—CH(CH_2—CH_3)—C(=O)—H$

(c) $CH_3—CH_2—CH_2—CH(CH_3)—CH(CH_3)—CH_2—C(=O)—H$

(d) $CH_3—C(Cl)(Cl)—C(=O)—H$

(e) $CH_3—CH_2—CH(CH_3)—CH(CH_3)—CH_2—CH(CH_3)—C(=O)—H$

(f) $CH_3—CH_2—CH_2—CH_2—CH(OH)—CH_2—CH(CH_3)—C(=O)—H$

15.15 (a) $H—C(=O)—H$ (b) $CH_3—CH_2—C(=O)—H$ (c) $CH_2(Cl)—C(=O)—H$

(d) $CH_3—CH_2—CH(Cl)—C(=O)—H$ (e) benzene ring bearing $C(=O)—H$ and, ortho, CH_3

(f) benzene ring bearing $C(=O)—H$, with CH_3 and CH_3 substituents

15.17 (a) propionaldehyde (b) 2-methylpropionaldehyde (c) butyraldehyde (d) dichloroacetaldehyde (e) *o*-chlorobenzaldehyde (f) 3-chloro-4-hydroxybenzaldehyde **15.19** (a) 2-butanone (b) 2,4,5-trimethyl-3-hexanone (c) 6-methyl-3-heptanone (d) 2-octanone (e) 1,5-dichloro-3-pentanone (f) 1,1-dichloro-2-butanone **15.21** (a) cyclohexanone (b) 3-methylcyclohexanone (c) 2-methylcyclohexanone (d) 3-chlorocyclopentanone **15.23** (a) $CH_3—C(=O)—CH(CH_3)—CH_2—CH_3$

(b) $CH_3—CH_2—C(=O)—CH_2—CH_2—CH_3$ (c) cyclobutanone (four-membered ring with =O)

(d) $CH_3—CH(CH_3)—C(=O)—CH(CH_3)—CH_3$ (e) $CH_2(Cl)—C(=O)—CH_3$

(f) $CH_2(Cl)—C(=O)—CH_2(Cl)$ **15.25** (a) $CH_3—CH_2—C(=O)—CH_2—CH_3$

(b) $CH_3—C(=O)—CH_3$ (c) $CH_3—CH(CH_3)—C(=O)—CH_2—CH_2—CH_3$

(d) $CH_2(Cl)—C(=O)—CH_3$ (e) $CH_3—C(=O)—C_6H_5$ (f) $CH_3—C(=O)—C_6H_5$

15.27 Dipole–dipole attractions between molecules raise the boiling point. **15.29** 2 **15.31** ethanal, because it has a shorter carbon chain **15.33** (a) $CH_3—CH_2—CH_2—CH_2—C(=O)—H$

(b) $CH_3—CH_2—C(=O)—CH_3$ (c) $CH_3—C(CH_3)_2—CH_2—C(=O)—H$

(d) $CH_3—CH_2—C(=O)—CH_2—CH_3$ (e) cyclopentanone (five-membered ring =O)

(f) CH_3—(cyclohexane ring)=O (4-methylcyclohexanone) **15.35** (a) $CH_3—CH_2—OH$

(b) $CH_3—CH_2—CH(OH)—CH_2—CH_3$ (c) $C_6H_5—CH_2—CH(OH)—CH_3$

(d) CH_3-CH_2-OH

(e) $CH_3-CH-CH_3$
OH

(f) $CH_3-CH_2-CH_2-CH_2-CH-CH_2$
CH_3-CH_2 OH

15.37 (a)
O
CH_3-C-OH

(b)
O
$CH_3-CH_2-CH_2-CH_2-C-OH$

(c)
O
$H-C-OH$

(d)
Cl Cl O
$CH_3-CH_2-CH-CH-CH_2-C-OH$

15.39 appearance of a silver mirror **15.41** Cu^{2+} ion **15.43** (a) no (b) yes (c) yes (d) no **15.45**

(a) $CH_3-CH_2-CH_2-CH_2$
OH

(b) $CH_3-CH_2-CH-CH_2-CH_3$
OH

(c) $CH_3-CH-CH_2-CH_2$
CH_3 OH

(d) $CH_3-CH-CH-CH_2-CH_2-CH_3$
CH_3 OH

15.47 R—O— and H— **15.49** (a) neither (b) hemiacetal (c) neither (d) hemiketal (e) hemiacetal (f) neither

15.51 (a) $CH_3-CH-O-CH_2-CH_3$
OH

(b)
OH
$CH_3-C-CH_2-CH_2-CH_3$
$O-CH_3$

(c)
OH
$CH_3-CH_2-CH_2-CH$
$O-CH_2-CH_3$

(d)
OH
CH_3-C-CH_3
$O-CH-CH_3$
CH_3

15.53 (a)
OH
$CH_3-CH_2-CH_2-CH$
$O-CH_2-CH_3$

(b)
O
CH_3-CH_2-C-H

(c)
OH
$CH_3-CH_2-C-CH_3$
$O-CH_3$

(d)
CH_2OH
O
OH
OH

15.55 (a) acetal (b) ketal (c) neither (d) ketal

15.57 (a) CH_3-OH

(b) $CH_3-CH-O-CH_3$
OH

(c)
CH_3
$O-CH-CH_3$
$CH_3-CH-O-CH_3$

(d) $CH_3-CH-O-CH_3$, CH_3-OH
OH

15.59 (a)
O
CH_3-C-H, $2CH_3-OH$

(b)
O
CH_3-C-CH_3, $2CH_3-OH$

(c)
O
$CH_3-CH_2-C-CH_2-CH_3$, CH_3-OH, CH_3-CH_2-OH

(d)
O
$CH_3-CH_2-CH_2-CH_2-C-H$, $2CH_3-OH$

15.61 A number is not needed because there is only one possible location for the double bond. **15.62** (a) A ketone carbonyl group cannot be on a terminal carbon atom. (b) It requires a carbon atom with five bonds. (c) It requires a carbon atom with five bonds. (d) It requires a carbon atom with five bonds. **15.63** heptanal **15.64** three **15.65**

O
$CH_3-CH_2-CH_2-CH_2-CH_2-C-H$
Hexanal

CH_3 O
$CH_3-CH_2-CH_2-CH-C-H$
2-Methylpentanal

CH_3 O
$CH_3-CH_2-CH-CH_2-C-H$
3-Methylpentanal

CH_3 O
$CH_3-CH-CH_2-CH_2-C-H$
4-Methylpentanal

CH_3 O
$CH_3-CH_2-C-C-H$
CH_3
2,2-Dimethylbutanal

CH_3 O
CH_3-C-CH_2-C-H
CH_3
3,3-Dimethylbutanal

CH_3 CH_3 O
$CH_3-CH-CH-C-H$
2,3-Dimethylbutanal

CH_3
CH_2 O
$CH_3-CH_2-CH-C-H$
2-Ethylbutanal

O
$CH_3-C-CH_2-CH_2-CH_2-CH_3$
2-Hexanone

O
$CH_3-CH_2-C-CH_2-CH_2-CH_3$
3-Hexanone

O CH_3
$CH_3-C-CH-CH_2-CH_3$
3-Methyl-2-pentanone

O CH_3
$CH_3-C-CH_2-CH-CH_3$
4-Methyl-2-pentanone

CH_3 O
$CH_3-CH-C-CH_2-CH_3$
2-Methyl-3-pentanone

O CH_3
$CH_3-C-C-CH_3$
CH_3
3,3-Dimethyl-2-butanone

15.66
O OH
CH_2 CH
CH_2-CH_2

15.67 (a) ketone, alkene (b) aldehyde, alcohol, ether (c) ketone, alkyne (d) aldehyde, ketone **15.68** (a) 2-hexanone (b) propanal (c) pentanedial (d) 2-hexenal **15.69** (a) 2, 4, and 6 (b) 1, 3, and 5 (c) 1, 2, 3, and 4 (d) 6 **15.70** (a) 1, 2, and 3 (b) all choices (c) 1, 3, and 6 (d) 4, 5, and 6 **15.71** (a) 1 and 4 (b) 2, 3, 5, and 6 (c) 3 and 6 (d) 2 and 5 **15.72** (a) 3 and 6 (b) 1 and 4 (c) 1 (d) 2 and 5

Chapter 16 **16.1** (a) yes (b) no (c) yes (d) yes (e) no (f) yes **16.3** (a) butanoic acid (b) heptanoic acid (c) 2,3-dimethylpentanoic acid (d)

4-bromopentanoic acid (e) 3-methylpentanoic acid (f) chloroethanoic acid **16.5** (a)

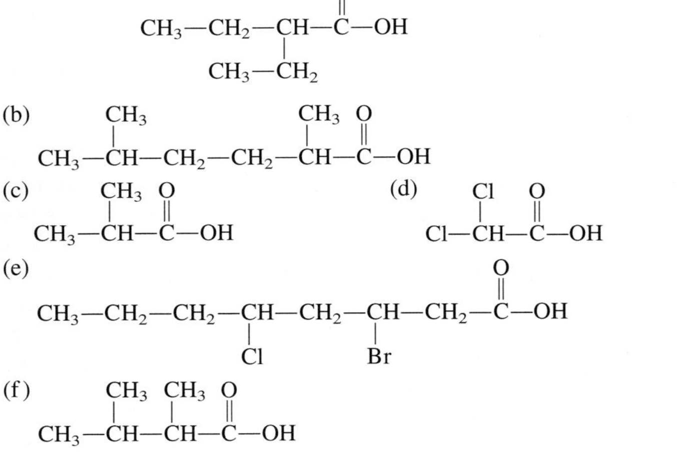

$$CH_3-CH_2-CH(CH_3-CH_2)-C(=O)-OH$$

(b) $$CH_3-CH(CH_3)-CH_2-CH_2-CH(CH_3)-C(=O)-OH$$

(c) $$CH_3-CH(CH_3)-C(=O)-OH$$

(d) $$Cl-CH(Cl)-C(=O)-OH$$

(e) $$CH_3-CH_2-CH_2-CH(Cl)-CH_2-CH(Br)-CH_2-C(=O)-OH$$

(f) $$CH_3-CH(CH_3)-CH(CH_3)-C(=O)-OH$$

16.7 (a) butanedioic acid (b) propanedioic acid (c) 3-methylpentanedioic acid (d) 2-chlorobenzoic acid (e) *m*-toluic acid (f) 2-bromo-4-chlorobenzoic acid

16.9 (a) $$CH_3-CH_2-C(CH_3)_2-C(=O)-OH$$

(b) $$HO-C(=O)-C(CH_3)_2-CH_2-C(=O)-OH$$

(c) $$HO-C(=O)-C(CH_3)_2-CH_2-CH_2-C(=O)-OH$$

(d) COOH
Br

(e) COOH
Cl
Cl

(f) COOH
CH_3

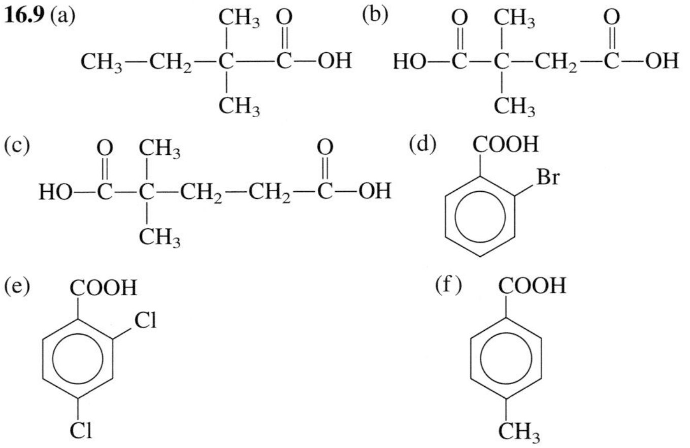

16.11 (a) $$CH_3-CH_2-CH_2-CH_2-C(=O)-OH$$

(b) $$CH_3-CH_2-C(=O)-OH$$

(c) $$CH_3-C(=O)-OH$$

(d) $$CH_3-CH_2-CH(Cl)-C(=O)-OH$$

(e) $$CH_3-CH_2-CH_2-CH(Br)-CH_2-C(=O)-OH$$

(f) $$CH_3-CH(Cl)-CH_2-CH(CH_3)-C(=O)-OH$$

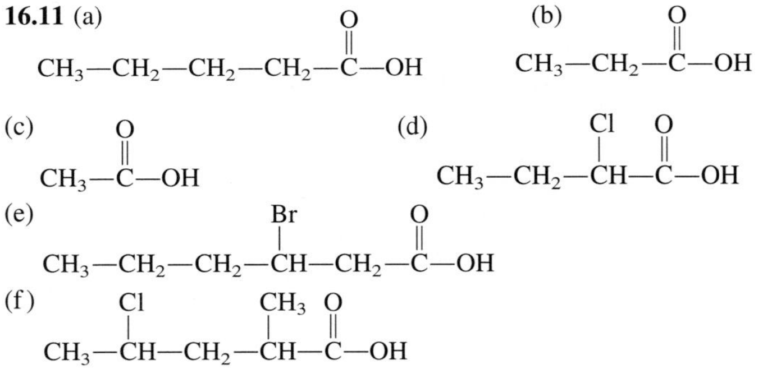

16.13 (a) $$HO-C(=O)-CH_2-C(=O)-OH$$

(b) $$HO-C(=O)-CH_2-CH_2-C(=O)-OH$$

(c) $$HO-C(=O)-CH_2-CH_2-CH_2-CH_2-C(=O)-OH$$

(d) $$HO-C(=O)-CH_2-CH_2-CH(Br)-CH_2-CH_2-C(=O)-OH$$

(e) $$HO-C(=O)-CH(CH_3)-CH_2-CH_2-C(=O)-OH$$

(f) $$HO-C(=O)-C(Cl)(Br)-CH_2-C(=O)-OH$$

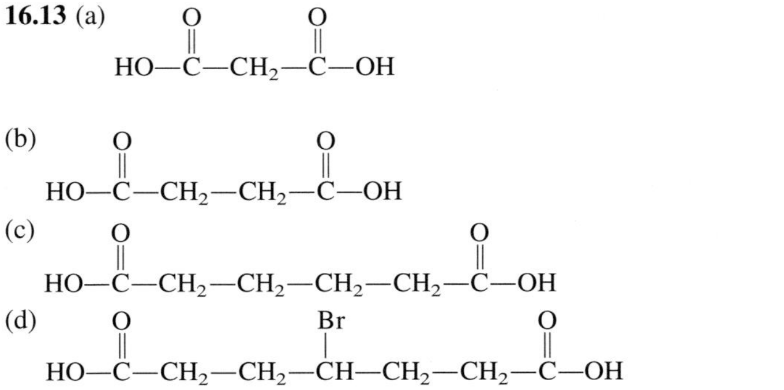

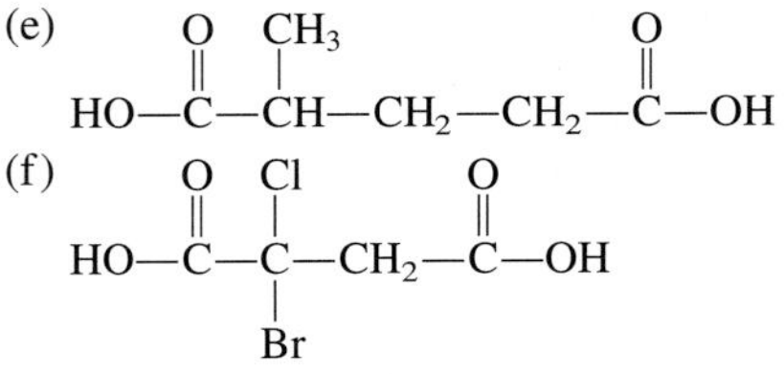

16.15 (a) 3 (b) 1 (c) 2 (d) 1 **16.17** (a) carbon–carbon double bond (b) hydroxyl group (c) keto group (d) two hydroxyl groups **16.19** (a) propenoic acid (b) 2-hydroxy-propanoic acid (c) 2-oxobutanedioic acid (d) 2,3-dihydroxypropanoic acid

16.21 (a) $$CH_3-CH_2-C(=O)-CH_2-C(=O)-OH$$

(b) $$CH_3-CH_2-CH(OH)-C(=O)-OH$$

(c) $$CH_3-CH=CH-CH_2-CH_2-C(=O)-OH$$

(d) $$HO-C(=O)-CH(OH)-CH(OH)-CH_2-C(=O)-OH$$

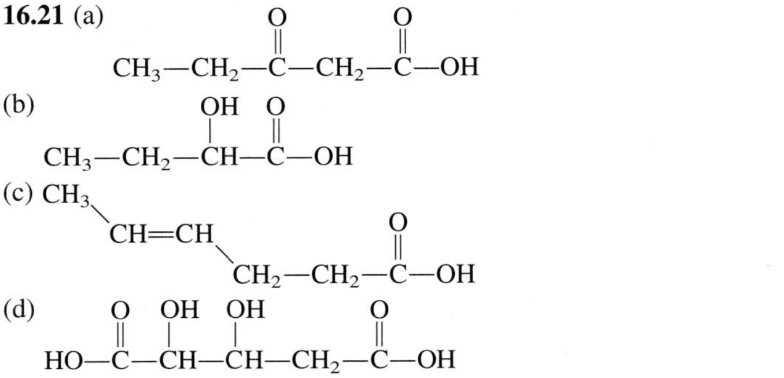

16.23 (a) 2 (b) 5 **16.25** (a) solid (b) solid (c) liquid (d) solid

16.27 (a) $$CH_3-C(=O)-OH$$

(b) $$CH_3-C(=O)-OH$$

(c) $$CH_3-CH_2-CH(CH_3)-CH_2-C(=O)-OH$$

(d) $$C_6H_5-C(=O)-OH$$

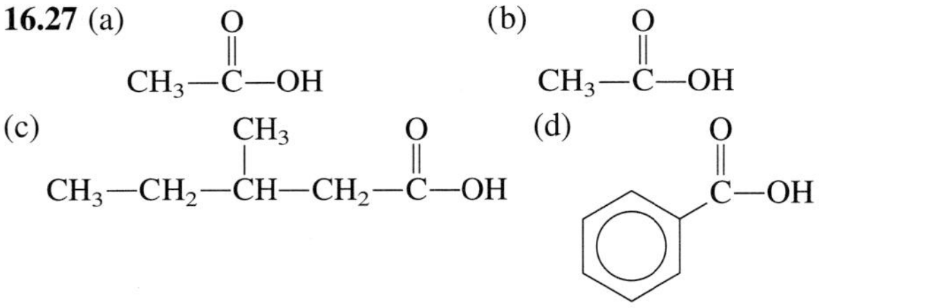

16.29 (a) 1 (b) 3 (c) 2 (d) 2 **16.31** (a) −1 (b) −3 (c) −2 (d) −2 **16.33** (a) pentanoate ion (b) citrate ion (c) succinate ion (d) oxalate ion **16.35**

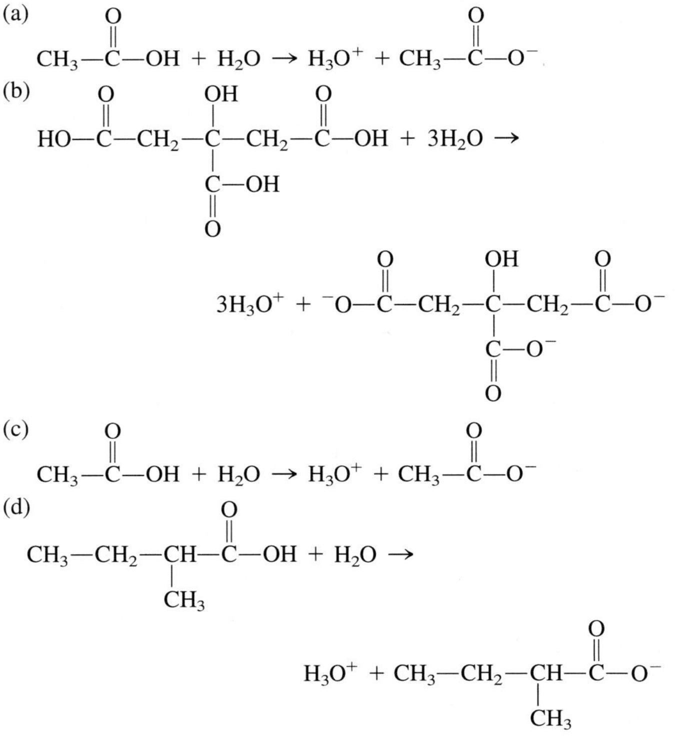

(a) $$CH_3-C(=O)-OH + H_2O \rightarrow H_3O^+ + CH_3-C(=O)-O^-$$

(b) $$HO-C(=O)-CH_2-C(OH)(C(=O)-OH)-CH_2-C(=O)-OH + 3H_2O \rightarrow$$
$$3H_3O^+ + {}^-O-C(=O)-CH_2-C(OH)(C(=O)-O^-)-CH_2-C(=O)-O^-$$

(c) $$CH_3-C(=O)-OH + H_2O \rightarrow H_3O^+ + CH_3-C(=O)-O^-$$

(d) $$CH_3-CH_2-CH(CH_3)-C(=O)-OH + H_2O \rightarrow$$
$$H_3O^+ + CH_3-CH_2-CH(CH_3)-C(=O)-O^-$$

16.37 (a) potassium ethanoate (b) calcium propanoate (c) potassium butanedioate (d) sodium pentanoate **16.39**

(a) $CH_3-\overset{\overset{O}{\|}}{C}-OH + KOH \rightarrow CH_3-\overset{\overset{O}{\|}}{C}-O^-\ K^+ + H_2O$

(b) $2CH_3-CH_2-\overset{\overset{O}{\|}}{C}-OH + Ca(OH)_2 \rightarrow$

$\left(CH_3-CH_2-\overset{\overset{O}{\|}}{C}-O^-\right)_2 Ca^{2+} + H_2O$

(c) $HO-\overset{\overset{O}{\|}}{C}-CH_2-CH_2-\overset{\overset{O}{\|}}{C}-OH + 2KOH \rightarrow$

$K^+\ {}^-O-\overset{\overset{O}{\|}}{C}-CH_2-CH_2-\overset{\overset{O}{\|}}{C}-O^-\ K^+ + 2H_2O$

(d) $CH_3-CH_2-CH_2-CH_2-\overset{\overset{O}{\|}}{C}-OH + NaOH \rightarrow$

$CH_3-CH_2-CH_2-CH_2-\overset{\overset{O}{\|}}{C}-O^-\ Na^+ + H_2O$

16.41 (a) $CH_3-CH_2-CH_2-\overset{\overset{O}{\|}}{C}-O^-\ Na^+ + HCl \rightarrow$

$CH_3-CH_2-CH_2-\overset{\overset{O}{\|}}{C}-OH + NaCl$

(b) $K^+\ {}^-O-\overset{\overset{O}{\|}}{C}-\overset{\overset{O}{\|}}{C}-O^-\ K^+ + 2HCl \rightarrow$

$HO-\overset{\overset{O}{\|}}{C}-\overset{\overset{O}{\|}}{C}-OH + 2KCl$

(c) $\left({}^-O-\overset{\overset{O}{\|}}{C}-CH_2-\overset{\overset{O}{\|}}{C}-O^-\right)_2 Ca^{2+} + 2HCl \rightarrow$

$HO-\overset{\overset{O}{\|}}{C}-CH_2-\overset{\overset{O}{\|}}{C}-OH + CaCl_2$

(d) $C_6H_5-\overset{\overset{O}{\|}}{C}-O^-\ Na^+ + HCl \rightarrow C_6H_5-\overset{\overset{O}{\|}}{C}-OH + NaCl$

16.43 (a) yes (b) yes (c) no (d) yes (e) yes (f) no **16.45**

(a) $CH_3-CH_2-\overset{\overset{O}{\|}}{C}-O-CH_3$

(b) $CH_3-\overset{\overset{O}{\|}}{C}-O-CH_2-CH_2-CH_3$

(c) $CH_3-CH_2-\overset{\overset{CH_3}{|}}{CH}-\overset{\overset{O}{\|}}{C}-O-\overset{\overset{CH_3}{|}}{CH}-CH_3$

(d) $CH_3-CH_2-CH_2-CH_2-\overset{\overset{O}{\|}}{C}-O-\overset{\overset{CH_3}{|}}{CH}-CH_2-CH_3$

16.47 (a) $CH_3-CH_2-\overset{\overset{O}{\|}}{C}-OH;\ CH_3-CH_2-OH$

(b) $CH_3-CH_2-CH_2-\overset{\overset{O}{\|}}{C}-OH;\ CH_3-OH$

(c) $CH_3-CH_2-CH_2-\overset{\overset{O}{\|}}{C}-OH;\ CH_3-OH$

(d) $CH_3-\overset{\overset{O}{\|}}{C}-OH;\ C_6H_5-OH$ (e) $C_6H_5-\overset{\overset{O}{\|}}{C}-OH;\ CH_3-OH$

(f) $CH_3-\overset{\overset{Cl}{|}}{CH}-\overset{\overset{O}{\|}}{C}-OH;\ CH_3-CH_2-OH$

16.49 (a) methyl propanoate (b) methyl methanoate (c) methyl ethanoate (d) propyl ethanoate (e) isopropyl propanoate (f) ethyl benzoate **16.51** (a) methyl propionate (b) methyl formate (c) methyl acetate (d) propyl acetate (e) isopropyl propionate (f) ethyl benzoate **16.53** (a) $H-\overset{\overset{O}{\|}}{C}-O-CH_3$ (b) $CH_3-\overset{\overset{O}{\|}}{C}-O-CH_2-CH_2-CH_3$

(c) $CH_3-(CH_2)_8-\overset{\overset{O}{\|}}{C}-O-(CH_2)_7-CH_3$

(d) $C_6H_5-CH_2-\overset{\overset{O}{\|}}{C}-O-CH_2-CH_3$

(e) $CH_3-\overset{\overset{O}{\|}}{C}-O-\overset{\overset{CH_3}{|}}{CH}-CH_3$ (f) $CH_3-\overset{\overset{O}{\|}}{C}-O-CH_2-\overset{\overset{Br}{|}}{CH}-CH_3$

16.55 (a) ethyl ethanoate (b) methyl ethanoate (c) ethyl butanoate (d) propyl 2-hydroxypropanoate (e) pentyl pentanoate (f) 1-methylpropyl hexanoate **16.57** No oxygen–hydrogen bonds are present. **16.59** There is no hydrogen bonding between ester molecules **16.61**

(a) $CH_3-CH_2-\overset{\overset{O}{\|}}{C}-OH;\ CH_3-CH_2-OH$ (b) $CH_3-\overset{\overset{O}{\|}}{C}-OH;$ CH_3-CH_2-OH (c) $CH_3-\overset{\overset{CH_3}{|}}{CH}-\overset{\overset{O}{\|}}{C}-OH;\ C_6H_5-OH$

(d) $CH_3-CH_2-CH_2-\overset{\overset{O}{\|}}{C}-OH;\ CH_3-OH$

(e) $H-\overset{\overset{O}{\|}}{C}-OH;\ CH_3-CH_2-OH$

(f) $C_6H_5-\overset{\overset{O}{\|}}{C}-OH;\ CH_3-\overset{\overset{CH_3}{|}}{CH}-OH$

16.63 (a) $CH_3-CH_2-\overset{\overset{O}{\|}}{C}-O^-\ Na^+;\ CH_3-CH_2-OH$

(b) $CH_3-\overset{\overset{O}{\|}}{C}-O^-\ Na^+;\ CH_3-CH_2-OH$

(c) $CH_3-\overset{\overset{CH_3}{|}}{CH}-\overset{\overset{O}{\|}}{C}-O^-\ Na^+;\ C_6H_5-OH$

(d) $CH_3-CH_2-CH_2-\overset{\overset{O}{\|}}{C}-O^-\ Na^+;\ CH_3-OH$

(e) $H-\overset{O}{\overset{\|}{C}}-O^-\ Na^+;\ CH_3-CH_2-OH$

(f) $C_6H_5-\overset{O}{\overset{\|}{C}}-O^-\ Na^+;\ CH_3-\overset{CH_3}{\overset{|}{CH}}-OH$

16.65 (a) $CH_3-\overset{CH_3}{\overset{|}{CH}}-\overset{O}{\overset{\|}{C}}-OH;\ CH_3-CH_2-OH$

(b) $CH_3-\overset{CH_3}{\overset{|}{CH}}-\overset{O}{\overset{\|}{C}}-O^-\ Na^+;\ CH_3-CH_2-OH$

(c) $H-\overset{O}{\overset{\|}{C}}-OH;\ CH_3-CH_2-CH_2-CH_2-OH$

(d) $CH_3-\overset{O}{\overset{\|}{C}}-O^-\ Na^+;\ CH_3-\overset{CH_3}{\overset{|}{CH}}-\overset{CH_3}{\overset{|}{CH}}-CH_2-OH$

16.67 (a) $CH_3-\overset{O}{\overset{\|}{C}}-S-CH_2-CH_3$

(b) $CH_3-(CH_2)_8-\overset{O}{\overset{\|}{C}}-S-CH_3$ (c) $C_6H_5-\overset{O}{\overset{\|}{C}}-S-\underset{CH_3}{\underset{|}{CH}}-CH_3$

(d) $H-\overset{O}{\overset{\|}{C}}-S-CH_2-CH_2-CH_3$

16.69 $-\overset{O}{\overset{\|}{C}}-\overset{O}{\overset{\|}{C}}-O-(CH_2)_3-O-\overset{O}{\overset{\|}{C}}-\overset{O}{\overset{\|}{C}}-O-(CH_2)_3-O-$

16.71 $HO-\overset{O}{\overset{\|}{C}}-CH_2-CH_2-\overset{O}{\overset{\|}{C}}-OH;\ HO-CH_2-CH_2-CH_2-OH$

16.73 (a) $HO-\underset{OH}{\underset{|}{\overset{O}{\overset{\|}{P}}}}-O-CH_3$ (b) $HO-\underset{O-CH_3}{\underset{|}{\overset{O}{\overset{\|}{P}}}}-O-CH_3$

(c) $O-\overset{O}{\overset{\|}{N}}-O-CH_3$ (d) $O-\overset{O}{\overset{\|}{N}}-O-CH_2-CH_2-O-\overset{O}{\overset{\|}{N}}-O$

16.75 H_3PO_4 is a triprotic acid, and H_2SO_4 is a diprotic acid. **16.77** (a) 2,2 (b) 7,1 (c) 7,1 (d) 6,3 (e) 3,1 (f) 2,1

16.78 $HO-\overset{O}{\overset{\|}{C}}-CH_2-\overset{O}{\overset{\|}{C}}-OH$

Malonic acid

$HO-\overset{O}{\overset{\|}{C}}-CH{=}CH-\overset{O}{\overset{\|}{C}}-OH$

Maleic acid (*cis* isomer)

$HO-\overset{O}{\overset{\|}{C}}-\overset{OH}{\overset{|}{CH}}-CH_2-\overset{O}{\overset{\|}{C}}-OH$

Malic acid

16.79 $C_nH_{2n-2}O_2$ **16.80**

$CH_3-CH_2-CH_2-CH_2-CH_2-\overset{O}{\overset{\|}{C}}-OH$

Hexanoic acid

$CH_3-\underset{CH_3}{\underset{|}{CH}}-CH_2-CH_2-\overset{O}{\overset{\|}{C}}-OH$

4-Methylpentanoic acid

$CH_3-CH_2-\underset{CH_3}{\underset{|}{CH}}-CH_2-\overset{O}{\overset{\|}{C}}-OH$

3-Methylpentanoic acid

$CH_3-CH_2-CH_2-\underset{CH_3}{\underset{|}{CH}}-\overset{O}{\overset{\|}{C}}-OH$

2-Methylpentanoic acid

$CH_3-\underset{CH_3}{\underset{|}{\overset{CH_3}{\overset{|}{C}}}}-CH_2-\overset{O}{\overset{\|}{C}}-OH$

3,3-Dimethylbutanoic acid

$CH_3-CH_2-\underset{CH_3}{\underset{|}{\overset{CH_3}{\overset{|}{C}}}}-\overset{O}{\overset{\|}{C}}-OH$

2,2-Dimethylbutanoic acid

$CH_3-\underset{CH_3}{\underset{|}{CH}}-\underset{CH_3}{\underset{|}{CH}}-\overset{O}{\overset{\|}{C}}-OH$

2,3-Dimethylbutanoic acid

$CH_3-CH_2-\underset{CH_3}{\underset{|}{\underset{CH_2}{\underset{|}{CH}}}}-\overset{O}{\overset{\|}{C}}-OH$

2-Ethylbutanoic acid

16.81 methyl propanoate; ethyl ethanoate; propyl methanoate; 1-methylethyl methanoate **16.82** (a) ethyl 2-methylpropanoate (b) 2-methylbutanoic acid (c) ethyl thiobutanoate (d) sodium propanoate **16.83**

$CH_3-\overset{O}{\overset{\|}{C}}-O-CH_2-CH_3$

16.84 (a) $CH_3-CH_2-\overset{O}{\overset{\|}{C}}-O^-\ Na^+;\ CH_3-OH$

(b) $CH_3-CH_2-\overset{O}{\overset{\|}{C}}-S-CH_3$ (c) $CH_3-\overset{O}{\overset{\|}{C}}-O^-\ Na^+$

(d) $HO-\overset{O}{\overset{\|}{C}}-CH_2-CH_2-\underset{CH_3}{\underset{|}{CH}}-CH_2-\overset{OH}{\overset{|}{CH_2}}$

16.85 (a) 2, 3, 5, and 6 (b) 1 and 4 (c) 4 and 6 (d) 1 and 3, 2 and 4 **16.86** (a) all choices (b) 6 (c) 2, 3, 4, 5, and 6 (d) 1 and 2 **16.87** (a) 1, 4, and 6 (b) 2 and 5 (c) 2 and 3 (d) 1 and 4, 3 and 6 **16.88** (a) 1, 4, and 6 (b) 1, 2, 4, and 6 (c) 1, 3, 4, 5, and 6 (d) 2, 3, and 5

Chapter 17 **17.1** (a) yes (b) yes (c) no (d) yes (e) no (f) yes **17.3** (a) 1° (b) 1° (c) 2° (d) 2° (e) 1° (f) 3° **17.5** (a) 2° (b) 3° (c) 3° (d) 1° (e) 2° (f) 2° **17.7** (a) ethylmethylamine (b) propylamine (c) diethylmethylamine (d) diphenylamine (e) isopropylmethylamine (f) diisopropylamine **17.9** (a) 3-pentanamine (b) 2-methyl-3-pentanamine (c) *N*-methyl-3-pentanamine (d) 1,5-pentanediamine (e) 2,3-butanediamine (f) *N,N*-dimethyl-1-butanamine **17.11** (a) 2-bromoaniline (b) *N*-isopropylaniline (c) *N*-ethyl-*N*-methylaniline (d) *N*-methyl-*N*-phenylaniline (e) *N*-ethyl-*N*-methylaniline (f) *N*-(1-chloroethyl)aniline

17.13 (a) $CH_3-CH_2-NH_2$ (b) $CH_3-CH(CH_3)-N(CH(CH_3)-CH_3)-CH(CH_3)-CH_3$

(c) $C_6H_4(NH_2)(CH_3)$ (ortho) (d) $C_6H_5-NH-CH_3$

(e) $CH_3-C(NH_2)(CH_3)-CH_2-CH_3$

(f) $H_2N-CH_2-CH_2-CH_2-CH_2-CH_2-CH_2-NH_2$

(g) $CH_3-CH(NH_2)-C(=O)-CH_2-CH_3$ (h) $CH_3-CH(NH_2)-C(=O)-OH$

17.15 (a) liquid (b) gas (c) gas (d) liquid **17.17** (a) 3 (b) 3

$R-N(R)-H\cdots N(H)(R)-R$

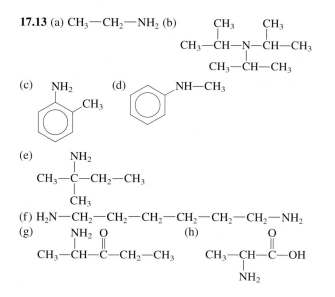

17.19 Hydrogen bonding is possible for the amine. **17.21** (a) $CH_3-CH_2-NH_2$; it has fewer carbon atoms. (b) $H_2N-CH_2-CH_2-CH_2-NH_2$; it has two amino groups rather than one. **17.23** (a) $CH_3-CH_2-NH_3$ (b) OH^-

(c) $CH_3-CH(CH_3)-NH-CH_3$

(d) $CH_3-CH_2-\overset{+}{N}H_2-CH_2-CH_3$; OH^- **17.25** (a) dimethyl-ammonium ion (b) triethylammonium ion (c) *N,N*-diethylanilinium ion (d) dimethylpropylammonium ion (e) propylammonium ion (f) *N*-isopropylanilinium ion **17.27** (a) $CH_3-NH-CH_3$

(b) $CH_3-CH_2-N(CH_2-CH_3)-CH_2-CH_3$

(c) $CH_3-CH_2-N(C_6H_5)-CH_2-CH_3$ (d) $CH_3-CH_2-CH_2-N(CH_3)-CH_3$

(e) $CH_3-CH_2-CH_2-NH_2$ (f) $C_6H_5-NH-CH(CH_3)-CH_3$

17.29 (a) $CH_3-CH_2-\overset{+}{N}H_3$ Cl^-

(b) $C_6H_5-\overset{+}{N}H_3$ Br^-

(c) $CH_3-C(CH_3)(CH_3)-NH_2$ (d) HCl **17.31** (a) $CH_3-CH(CH_3)-NH_2$

(b) $CH_3-\overset{+}{N}H_2(CH_3)$ Cl^- (c) $C_6H_5-N(CH_3)-CH_3$ (d) $CH_3-NH-CH_3$

17.33 (a) propylammonium chloride (b) methylpropylammonium chloride (c) ethyldimethylammonium bromide (d) *N,N*-dimethylanilinium bromide **17.35** to increase water solubility **17.37** ethylmethylamine hydrochloride **17.39** (a) $CH_3-CH_2-CH_2-NH_2$, NaCl, H_2O (b) $CH_3-CH(CH_3)-N(CH_3)-CH_3$, NaBr, H_2O

(c) $CH_3-CH_2-NH-CH_2-CH_3$, NaCl, H_2O

(d) $CH_3-C(CH_3)(CH_3)-NH_2$, NaBr, H_2O **17.41** ethylmethylamine and propyl chloride, ethylpropylamine and methyl chloride, methylpropylamine and ethyl chloride **17.43** (a) $CH_3-\overset{+}{N}(CH_3)(CH_3)-CH_2-CH_3$ Br^-

(b) $CH_3-CH(CH_3)-N(CH_3)-CH(CH_3)-CH_3$

(c) $CH_3-CH_2-\overset{+}{N}(CH_3)(CH_3)-CH_2-CH_2-CH_3$ Cl^-

(d) $CH_3-CH_2-NH-CH_2-CH_3$ **17.45** (a) amine salt (b) quarternary ammonium salt (c) amine salt (d) quarternary ammonium salt **17.47** (a) trimethylammonium bromide (b) tetramethylammonium chloride (c) ethylmethylammonium bromide (d) diethyldimethylammonium chloride **17.49** (a) purine (b) pyrrole (c) imidazole (d) indole **17.51** (a) true (b) false (c) true (d) false (e) false (f) false **17.53** (a) yes (b) yes (c) no (d) yes (e) no (f) yes **17.55** (a) monosubstituted (b) disubstituted (c) unsubstituted (d) monosubstituted **17.57** (a) secondary amide (b) tertiary amide (c) primary amide (d) secondary amide **17.59** (a) *N*-ethylethanamide (b) *N,N*-dimethylpropanamide (c) butanamide (d) *N*-methylmethanamide (e) 2-chloropropanamide (f) 2,*N*-dimethylpropanamide **17.61** (a) *N*-ethylacetamide (b) *N,N*-dimethylpropionamide (c) butyramide (d) *N*-methylformamide (e) 2-chloropropionamide (f) 2,*N*-dimethylpropionamide

17.63 (a) $CH_3-C(=O)-N(CH_3)-CH_3$ (b) $CH_3-CH_2-CH(CH_3)-C(=O)-NH_2$

(c) $CH_3-CH(CH_3)-CH_2-C(=O)-NH-CH_3$ (d) $H-C(=O)-NH_2$

(e) $C_6H_5-C(=O)-NH-C_6H_5$ (f) $H-C(=O)-NH_2$ **17.65** An electronegativity effect induced by the carbonyl oxygen atom makes the lone pair of electrons on the nitrogen atom unavailable. **17.67** (a) 5 (b) 5

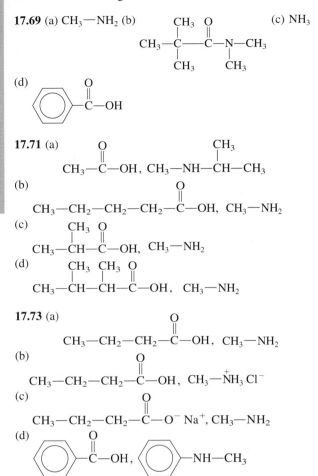

17.69 (a) $CH_3—NH_2$ (b)

$$\begin{array}{c} \quad CH_3\;\; O \\ \quad |\quad\;\; \| \\ CH_3—C——C—N—CH_3 \\ \quad |\qquad\;\; | \\ \quad CH_3\quad\; CH_3 \end{array}$$

(c) NH_3

(d)

$$\begin{array}{c} O \\ \| \\ \text{(benzene ring)}—C—OH \end{array}$$

17.71 (a)

$$\begin{array}{c} O \qquad\qquad\qquad\qquad CH_3 \\ \| \qquad\qquad\qquad\qquad\;\; | \\ CH_3—C—OH,\; CH_3—NH—CH—CH_3 \end{array}$$

(b)

$$\begin{array}{c} O \\ \| \\ CH_3—CH_2—CH_2—CH_2—C—OH,\; CH_3—NH_2 \end{array}$$

(c)

$$\begin{array}{c} CH_3\;\; O \\ |\quad\;\; \| \\ CH_3—CH—C—OH,\; CH_3—NH_2 \end{array}$$

(d)

$$\begin{array}{c} CH_3\;\; CH_3\;\; O \\ |\qquad\; |\qquad \| \\ CH_3—CH—CH—C—OH,\; CH_3—NH_2 \end{array}$$

17.73 (a)

$$\begin{array}{c} O \\ \| \\ CH_3—CH_2—CH_2—C—OH,\; CH_3—NH_2 \end{array}$$

(b)

$$\begin{array}{c} O \\ \| \\ CH_3—CH_2—CH_2—C—OH,\; CH_3—\overset{+}{N}H_3\; Cl^- \end{array}$$

(c)

$$\begin{array}{c} O \\ \| \\ CH_3—CH_2—CH_2—C—O^-\; Na^+,\; CH_3—NH_2 \end{array}$$

(d)

$$\begin{array}{c} O \\ \| \\ \text{(benzene ring)}—C—OH,\; \text{(benzene ring)}—NH—CH_3 \end{array}$$

17.75 diacid and diamine

17.77

$$\left(\begin{array}{c} O \qquad\qquad\qquad\quad O\;\; H \qquad\qquad\qquad\qquad\qquad\qquad H \\ \| \qquad\qquad\qquad\qquad \|\;\;\; | \qquad\qquad\qquad\qquad\qquad\qquad\; | \\ C—CH_2—CH_2—C—N—CH_2—CH_2—CH_2—CH_2—N \end{array} \right)_n$$

17.79 (a)

$$\begin{array}{c} O \\ \| \\ H—C—NH_2 \end{array}$$

(b)

$$\begin{array}{c} CH_3—CH—CH_2—CH_2—CH_3 \\ | \\ NH_2 \end{array}$$

(c)

$$\begin{array}{c} CH_3\;\; O \\ |\quad\;\; \| \\ CH_3—CH_2—CH_2—CH—C—NH_2 \end{array}$$

(d)

$$\begin{array}{c} O \qquad\quad CH_3 \\ \| \qquad\qquad | \\ CH_3—C—NH—CH—CH_3 \end{array}$$

(e) $CH_3—CH_2—\overset{+}{N}H_2—CH_2—CH_3\; Cl^-$

(f)

$$\begin{array}{c} \qquad\; CH_3 \\ CH_3—\overset{+}{N}H \qquad Cl^- \\ \qquad\; CH_3 \\ | \\ \text{(benzene ring)} \end{array}$$

17.80 (a) $CH_3—CH_2—NH—CH_3$

(b)

$$\begin{array}{c} CH_3\;\; O \\ |\quad\;\; \| \\ CH_3—CH_2—CH—C—N—CH_3 \\ | \\ CH_3 \end{array}$$

(c)

$$\begin{array}{c} CH_3 \\ | \\ CH_3—\overset{+}{N}—CH_3\; Cl^- \\ | \\ CH_3 \end{array}$$

(d)

$$\begin{array}{c} CH_3\;\; O \\ |\quad\;\; \| \\ CH_3—CH_2—CH—C—O^-\; Na^+ \end{array}$$

(e) $CH_3—NH—CH_3$ (f) $CH_3—CH_2—NH_3^+$

17.81 $CH_3—CH_2—CH_2—CH_2—NH_2$
1-Butanamine

$$\begin{array}{c} CH_3—CH—CH_2—CH_3 \\ | \\ NH_2 \end{array}$$

2-Butanamine

$$\begin{array}{c} CH_3—CH—CH_2—NH_2 \\ | \\ CH_3 \end{array}$$

2-Methyl-1-propanamine

$$\begin{array}{c} CH_3 \\ | \\ CH_3—C—NH_2 \\ | \\ CH_3 \end{array}$$

2-Methyl-2-propanamine

$CH_3—CH_2—CH_2—NH—CH_3$
N-methyl-1-propanamine

$$\begin{array}{c} CH_3—CH—NH—CH_3 \\ | \\ CH_3 \end{array}$$

N-methyl-2-propanamine

$CH_3—CH_2—NH—CH_2—CH_3$
N-ethylethanamine

$$\begin{array}{c} CH_3—N—CH_2—CH_3 \\ | \\ CH_3 \end{array}$$

N,N-dimethylethanamine

17.82

$$\begin{array}{c} O \\ \| \\ CH_3—CH_2—C—NH_2 \end{array}$$

Propanamide

$$\begin{array}{c} O \\ \| \\ CH_3—C—NH—CH_3 \end{array}$$

N-methylethanamide

$$\begin{array}{c} O \\ \| \\ H—C—NH—CH_2—CH_3 \end{array}$$

N-ethylmethanamide

$$\begin{array}{c} O\;\; CH_3 \\ \|\quad\; | \\ H—C—N—CH_3 \end{array}$$

N,N-dimethylmethanamide

17.83

$$\begin{array}{c} CH_3 \\ | \\ CH_3—\overset{+}{N}—CH_2—CH_3\; Cl^- \\ | \\ CH_3 \end{array}$$

17.84 (a) unsubstituted (b) monosubstituted (c) disubstituted (d) disubstituted (e) unsubstituted (f) unsubstituted **17.85** (a) amide (b) amine (c) amide (d) amine (e) amine (f) amide **17.86** (a) 1-butanamine (b) 2-methyl-1-pentanamine (c) *N*,2-dimethylpentanamine (d) 3-methylpentanamide (e) 1,4-pentandiamine (f) 4-bromo-*N*-ethyl-*N*-methylpentanamide **17.87** (a) 1 and 2 (b) 3 (c) 4 and 5 (d) 6 **17.88** (a) 6 (b) 3 (c) 1 (d) 5 **17.89** (a) 2, 3, and 6 (b) 6 (c) 1, 4, and 5 (d) 2 and 4 **17.90** (a) 2 (b) 5 (c) 3 (d) 1 and 2

Chapter 18 **18.1** Biochemistry is the study of the chemical substances found in living systems and the chemical interactions of these substances with each other. **18.3** proteins, lipids, carbohydrates, and nucleic acids

18.5 $CO_2 + H_2O + \text{solar energy} \xrightarrow[\text{Plant enzymes}]{\text{Chlorophyll}} \text{carbohydrates} + O_2$

18.7 serve as structural elements, provide energy reserves **18.9** Carbohydrates are polyhydroxy aldehydes, polyhydroxy ketones, or compounds that yield such substances upon hydrolysis. **18.11** (a) one unit versus a few units (b) two units versus four units **18.13** Superimposable objects have parts that coincide exactly at all points when the objects are laid upon each other. **18.15** (a) drill bit (b) hand, foot, ear (c) PEEP, POP **18.17** (a) no (b) no (c) yes (d) yes **18.19**

(a)

$$\begin{array}{c} CH_2—{}^*CH—Br \\ |\qquad\;\; | \\ Cl\qquad Cl \end{array}$$

(b)

$$\begin{array}{c} \qquad\; Cl\;\; Cl \\ \qquad\;\; |\quad\; | \\ CH_2—\overset{*}{C}—\overset{*}{C}H \\ |\qquad\; |\quad\; | \\ Br\qquad Br\;\; Br \end{array}$$

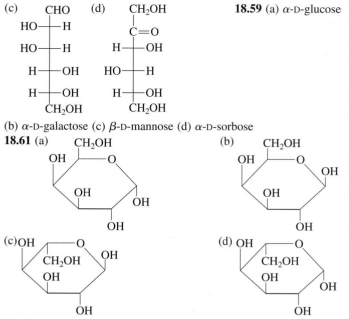

(c)

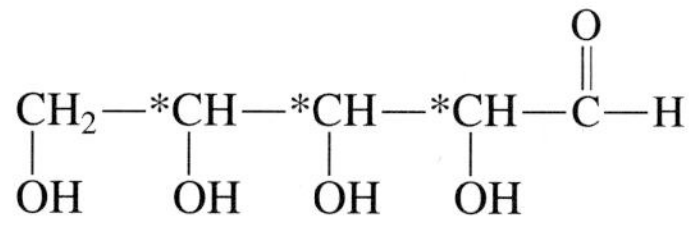

(d)

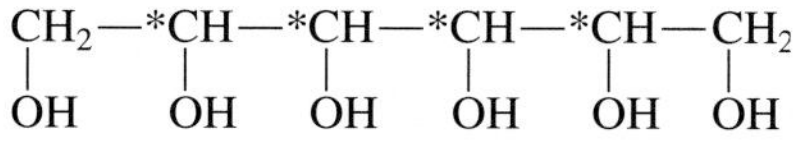

18.21 (a) zero (b) two (c) zero (d) zero **18.23** Structural isomers have different connectivity of atoms. Stereoisomers have the same connectivity of atoms with different arrangements of the atoms in space. **18.25**

(a)

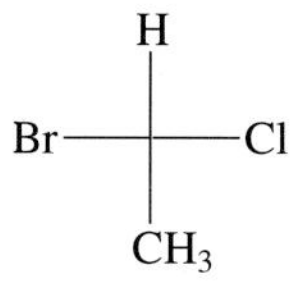

(b)

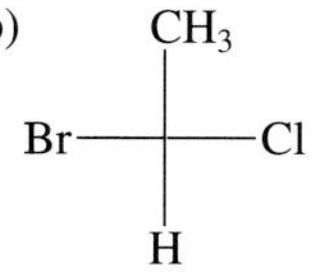

(c)

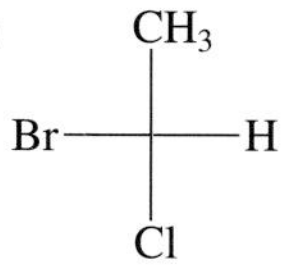

(d)

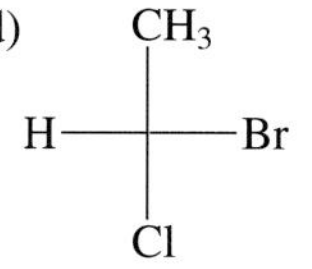

18.27 (a)

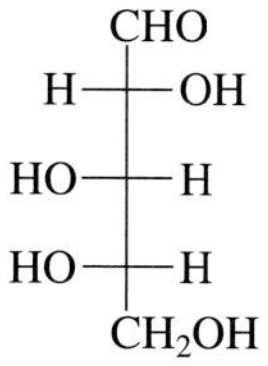

(b)

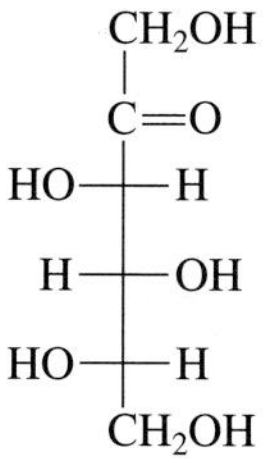

(c)

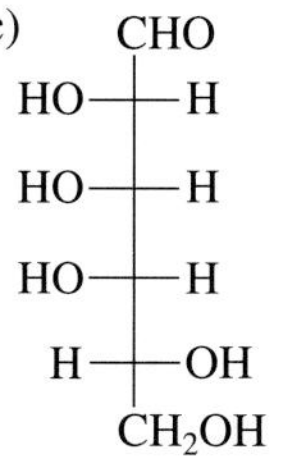

(d)

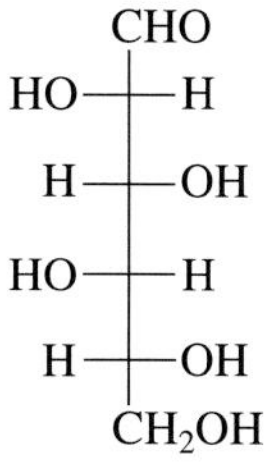

18.29 (a) D enantiomer (b) D enantiomer (c) L enantiomer (d) L enantiomer **18.31** (a) diastereomers (b) neither enantiomers nor diastereomers (c) enantiomers (d) diastereomers **18.33** (d) effect on plane-polarized light **18.35** (a) same (b) different (c) same (d) different **18.37** (a) aldose (b) ketose (c) ketose (d) ketose **18.39** (a) aldohexose (b) ketohexose (c) ketotriose (d) ketotetrose **18.41** (a) D-galactose (b) D-psicose (c) dihydroxyacetone (d) L-erythrulose **18.43** They differ at carbons 1 and 2; glucose has an aldehyde group and fructose has a ketone group. **18.45** carbon 5 **18.47** The difference involves the —OH group on the hemiacetal carbon atom. In the β form it is on the same side of the ring as the $—CH_2OH$ group, and in the α form it is on the opposite side of the ring from the $—CH_2OH$ group. **18.49** In fructose the cyclization involves carbons 2 and 5, and in ribose the cyclization involves carbons 1 and 4; both processes give five-membered rings. **18.51** The cyclic and noncyclic forms interconvert; an equilibrium exists between the forms. **18.53** (a) α-D-monosaccharide (b) α-D-monosaccharide (c) β-D-monosaccharide (d) α-D-monosaccharide **18.55** (a) hemiacetal (b) hemiacetal (c) hemiacetal (d) hemiketal **18.57** (a)

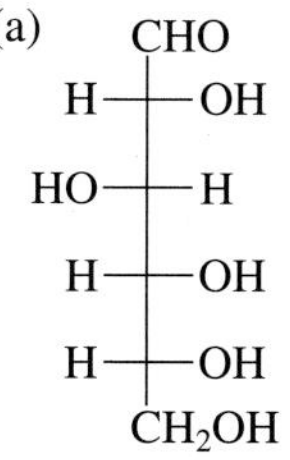

(b)

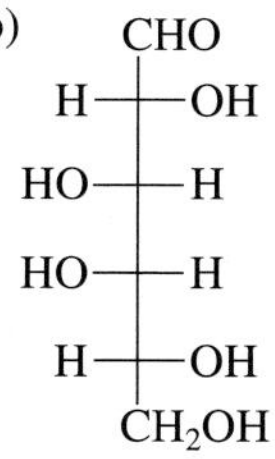

(c)

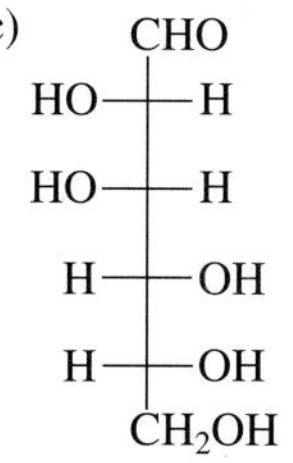

(d)

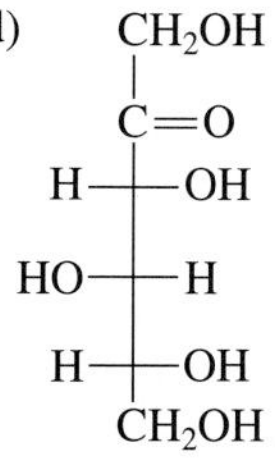

18.59 (a) α-D-glucose (b) α-D-galactose (c) β-D-mannose (d) α-D-sorbose

18.61 (a)

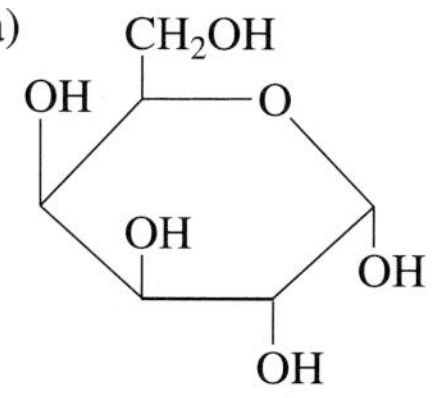

(b)

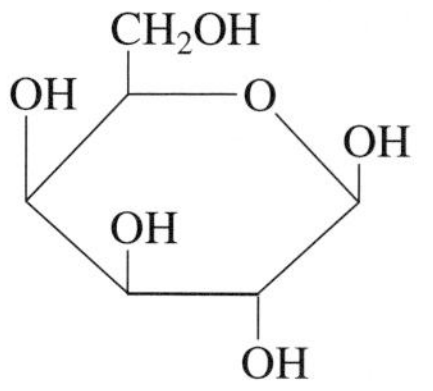

(c)

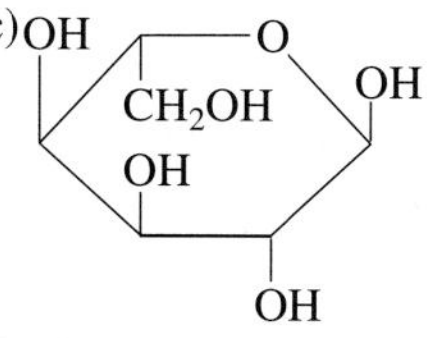

(d)

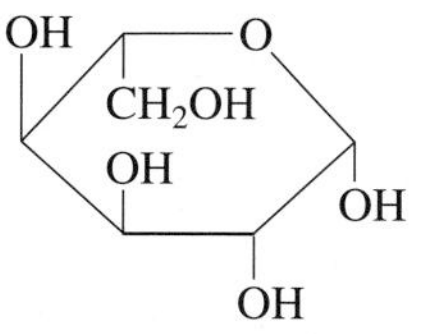

18.63 (a) reducing sugar (b) reducing sugar (c) reducing sugar (d) reducing sugar **18.65** The aldehyde group in glucose is oxidized to an acid group. The Ag^+ in Tollens solution is reduced to Ag. **18.67**

(a)

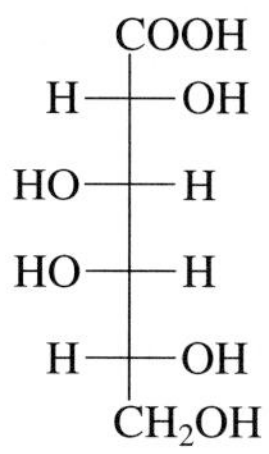

(b)

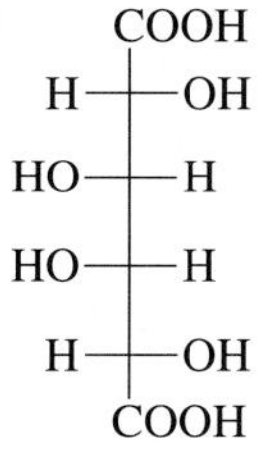

(c)

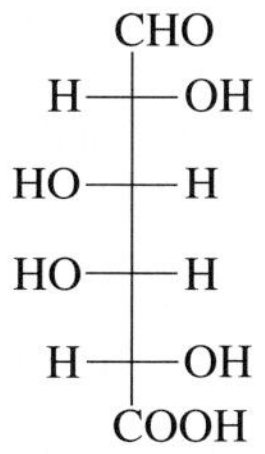

(d)

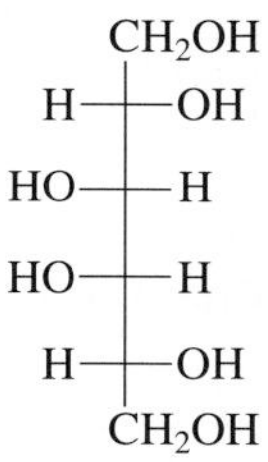

18.69 (a) acetal (b) acetal (c) ketal (d) acetal **18.71** (a) alpha (b) beta (c) alpha (d) beta **18.73** (a) methyl alcohol (b) ethyl alcohol (c) ethyl alcohol (d) methyl alcohol **18.75** A glycoside is an acetal or a ketal formed from a cyclic monosaccharide. A glucoside is a glycoside in which the monosaccharide is glucose. **18.77**

(a)

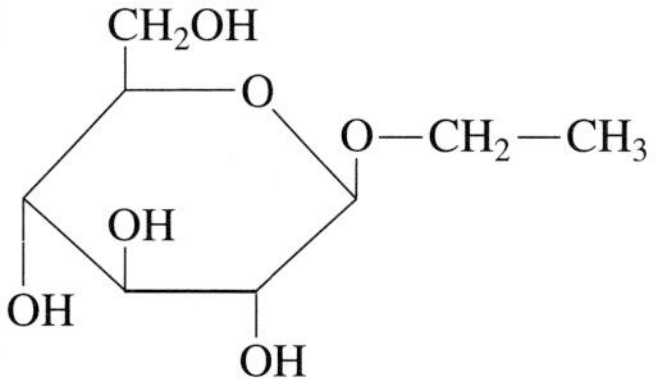

(b)

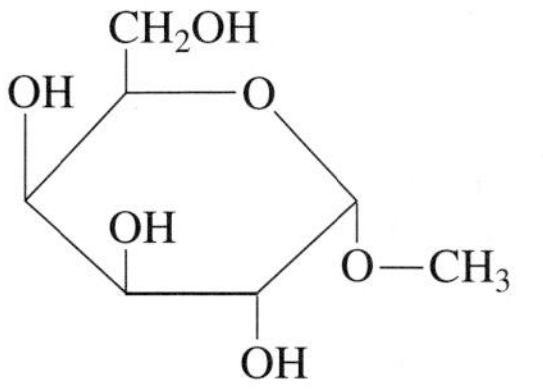

18.79 (a)

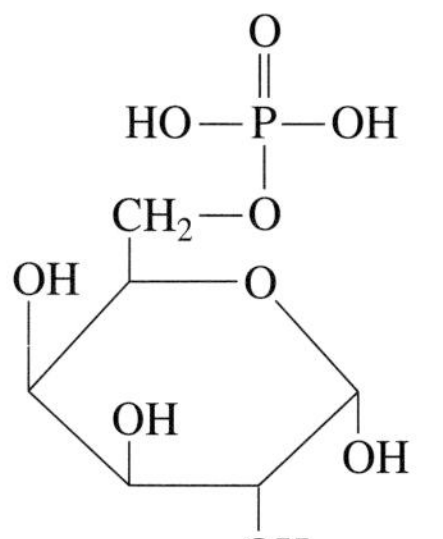

(b)

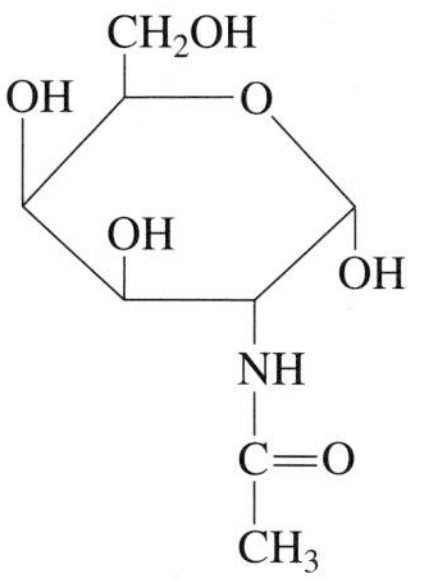

18.81 (a) glucose and fructose (b) glucose (c) glucose and galactose (d) glucose **18.83** The glucose part of the lactose structure has a hemiacetal carbon atom. **18.85** (a) negative (b) positive (c) positive (d) positive **18.87** (a) $\alpha(1 \rightarrow 6)$ (b) $\beta(1 \rightarrow 4)$ (c) $\alpha(1 \rightarrow 4)$ (d) $\alpha(1 \rightarrow 4)$ **18.89** (a) alpha (b) beta (c) alpha (d) beta **18.91** (a) reducing sugar (b) reducing sugar (c) reducing sugar (d) reducing sugar **18.93** (a)

glucose (b) galactose and glucose (c) glucose (d) glucose **18.95** (a) Both are glucose polymers with $\alpha(1 \rightarrow 4)$ and $\alpha(1 \rightarrow 6)$ linkages. Glycogen is more highly branched than amylopectin. (b) Both are unbranched glucose polymers. Amylose has $\alpha(1 \rightarrow 4)$ linkages, and cellulose has $\beta(1 \rightarrow 4)$ linkages. **18.97** (a) glycogen and amylopectin (b) amylopectin, amylose, glycogen, and cellulose (c) amylose, cellulose, and chitin (d) cellulose and chitin **18.99** The human body possesses enzymes for $\alpha(1 \rightarrow 4)$ linkages (starch) but not for $\beta(1 \rightarrow 4)$ linkages (cellulose). **18.101** (a) chiral (b) achiral (c) chiral (d) chiral **18.102** (a) no (b) no (c) no (d) yes **18.103** (a) no (b) no (c) yes (d) yes **18.104**

CH_2OH	CH_2OH	CH_2OH	CH_2OH
C=O	C=O	C=O	C=O
H—C—OH	HO—C—H	H—C—OH	HO—C—H
H—C—OH	HO—C—H	HO—C—H	H—C—OH
CH_2OH	CH_2OH	CH_2OH	CH_2OH

18.105 3-methyl hexane (hydrogen, methyl, ethyl, and propyl groups attached to a carbon atom) **18.106** (a) glucose, fructose (b) glucose (c) glucose (d) glucose **18.107** (a) homopolysaccharide (b) homopolysaccharide (c) heteropolysaccharide (d) homopolysaccharide **18.108** (a) strong oxidizing agent (b) ethyl alcohol, H^+ ion (c) water (H^+ ion or enzymes) (d) enzymes **18.109** (a) 4 and 6 (b) all choices (c) 3, 4, and 6 (d) all choices **18.110** (a) 2, 3, and 5 (b) 1, 3, and 4 (c) 2, 5, and 6 (d) 1, 2, 3, 4, and 5 **18.111** (a) all choices (b) 2, 3, and 6 (c) 1, 4, and 5 (d) 1, 2, 3, 4, and 6 **18.112** (a) 2, 3, and 5 (b) 1 and 4, 2 and 3 (c) 1 and 6, 4 and 6 (d) all choices

Chapter 19 **19.1** Lipids are organic compounds of biological origin that are water-insoluble but soluble in nonpolar solvents. **19.3** (a) insoluble (b) soluble (c) insoluble (d) soluble **19.5** Saponifiable lipids can be hydrolyzed under basic conditions, whereas nonsaponifiable lipids cannot. **19.7** (a) long-chain (b) medium-chain (c) long-chain (d) medium-chain **19.9** (a) saturated (b) polyunsaturated (c) polyunsaturated (d) monounsaturated **19.11** (a) neither (b) omega-3 (c) omega-3 (d) neither **19.13** (a) nonessential (b) essential (c) nonessential (d) nonessential **19.15** There are fewer attractions between fatty acid carbon chains because of bends in the chains caused by the presence of the double bonds. **19.17** (a) 18:1 acid (b) 18:3 acid (c) 14:0 acid (d) 18:1 acid **19.19** *cis,cis,cis*-9,12,15-octadecatrienoic acid **19.21** There are more saturated fatty acid residues present in a fat than in an oil. **19.23** Triacylglycerols are triesters formed from the reaction between three fatty acid molecules and a glycerol molecule. **19.25**

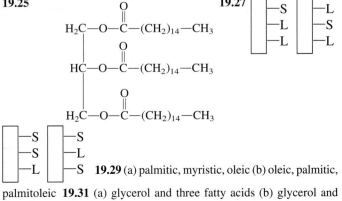

$H_2C-O-\overset{O}{\overset{\|}{C}}-(CH_2)_{14}-CH_3$
$HC-O-\overset{O}{\overset{\|}{C}}-(CH_2)_{14}-CH_3$
$H_2C-O-\overset{O}{\overset{\|}{C}}-(CH_2)_{14}-CH_3$

19.27 [—S, —L, —L] [—L, —S, —L] [—S, —S, —L] [—S, —L, —S] **19.29** (a) palmitic, myristic, oleic (b) oleic, palmitic, palmitoleic **19.31** (a) glycerol and three fatty acids (b) glycerol and three fatty acid salts **19.33** $CH_2(OH)-CH(OH)-CH_2(OH)$

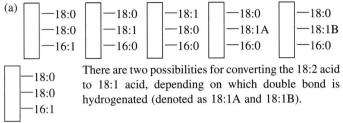

$CH_3-(CH_2)_{14}-COOH$ $CH_3-(CH_2)_{12}-COOH$ $CH_3-(CH_2)_7-CH{=}CH-(CH_2)_7-COOH$ **19.35** glycerol, palmitic acid, myristic acid, oleic acid. **19.37** $CH_2(OH)-CH(OH)-CH_2(OH)$

$CH_3-(CH_2)_{14}-COO^-\ Na^+$ $CH_3-(CH_2)_{12}-COO^-\ Na^+$ $CH_3-(CH_2)_7-CH{=}CH-(CH_2)_7-COO^-\ Na^+$ **19.39** glycerol, sodium palmitate, sodium myristate, sodium oleate **19.41** Carbon–carbon double bond(s) must be present. **19.43** six **19.45** (a) [—18:0, —18:0, —16:1] [—18:0, —18:1, —16:0] [—18:1, —18:0, —16:0] [—18:0, —18:1A, —16:0] [—18:0, —18:1B, —16:0]

[—18:0, —18:0, —16:1] There are two possibilities for converting the 18:2 acid to 18:1 acid, depending on which double bond is hydrogenated (denoted as 18:1A and 18:1B).

19.47 Rancidity results from hydrolysis of ester linkages and oxidation of carbon–carbon double bonds.

19.49

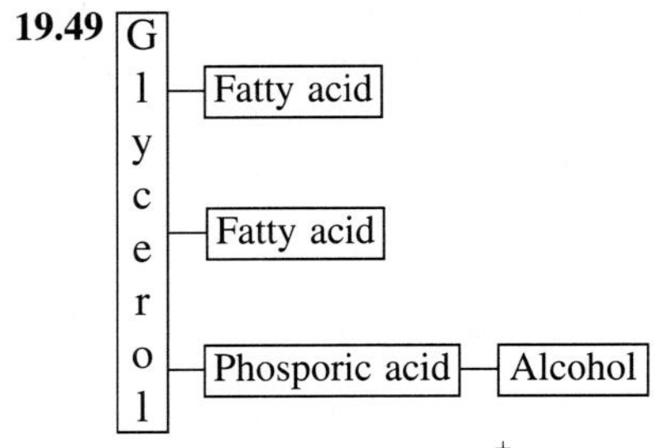

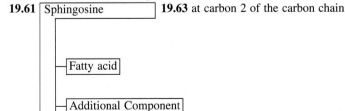

19.51 $HO-CH_2-CH_2-\overset{+}{N}-(CH_3)_3$ $HO-CH_2-CH_2-\overset{+}{N}H_3$ $HO-CH_2-CH(COO^-)-\overset{+}{N}H_3$ **19.53** The head is the combined glycerol–phosphoric acid–amino alcohol part, and the tails are the two fatty acids. **19.55** [Fatty acid]—[Long-chain alcohol]

19.57 $CH_3-(CH_2)_{14}-\overset{O}{\overset{\|}{C}}-O-(CH_2)_{15}-CH_3$ **19.59** A wax is a monoester (monoalcohol and fatty acid) and a triacylglycerol is a triester (trialcohol and 3 fatty acids).

19.61 [Sphingosine] with —[Fatty acid] and —[Additional Component] **19.63** at carbon 2 of the carbon chain

19.65 The head is the additional component, the fatty acid is one tail, and the sphingosine chain is the other tail. **19.67** The additional component is phosphoric acid–choline in sphingomyelins and it is a monosaccharide in cerebrosides. **19.69** [steroid ring system with carbons numbered 1–17]

19.71 —OH on carbon 3, $-CH_3$ on carbons 10 and 13, and a hydrocarbon chain on carbon 17 **19.73** solubilize lipids in an aqueous environment **19.75** sex hormones, adrenocortical hormones **19.77** estradiol has an —OH on carbon 3, while testosterone has a ketone group at this location; testosterone has an extra $-CH_3$ group at carbon

10 **19.79** Prostaglandins have a bond between carbons 8 and 12, which creates a cyclopentane ring structural feature. **19.81** inflammatory response, production of pain and fever, blood pressure regulation, induction of blood clotting, control of some reproductive functions, regulation of sleep/wake cycle **19.83** phosphoacylglycerols, sphingolipids **19.85** a two-layer structure of lipid molecules with nonpolar "tails" in the interior and polar "heads" on the exterior **19.87** creates "open" areas in the bilayer **19.89** Protein help is required in facilitated transport but not in passive transport. **19.90** Cellular energy expenditure is required in active transport but not in facilitated transport; movement is against the concentration gradient in active transport but not in facilitated transport. **19.91** (a) active transport (b) facilitated transport (c) active transport (d) passive transport and facilitated transport **19.92** (a) passive transport and facilitated transport (b) active transport (c) passive transport and facilitated transport (d) active transport **19.93** (a) neither (b) glycerol-based (c) neither (d) sphingosine-based (e) neither (f) neither **19.94** (a) no (b) no (c) no (d) yes (e) yes (f) no **19.95** (a) sphingomyelins (b) triacylglycerols (c) steroids (d) leukotrienes (e) prostaglandins (f) cerebrosides **19.96** (a) storage lipid, nonpolar lipid (b) storage lipid, nonpolar lipid (c) membrane lipid, polar lipid (d) nonpolar lipid (e) membrane lipid, polar lipid (f) membrane lipid, polar lipid **19.97** (a) yes (b) yes (c) yes (d) yes (e) yes (f) yes **19.98** (a) phosphoacylglycerols and sphingomyelins (b) sphingomyelins and cerebrosides (c) cerebrosides (d) sphingomyelins **19.99** (a) 3, 0, 0 (b) 4, 0, 0 (c) 2, 1, 0 (d) 1, 0, 0 (e) 0, 1, 1 (f) 4, 0, 0 **19.100** (a) no (b) yes (c) yes (d) yes (e) yes (f) no **19.101** (a) 2 (b) 1 and 2 (c) 1, 2, 5, and 6 (d) 2, 3, and 4 **19.102** (a) all choices (b) 1, 2, 3, and 4 (c) all choices (d) 5 **19.103** (a) 1, 2, 3, and 5 (b) 1 and 5 (c) 3 (d) 3 and 5 **19.104** (a) 2, 4, and 5 (b) 1 and 6 (c) 2, 4, and 5 (d) 1, 2, 4, 5, and 6

Chapter 20 **20.1** (a) yes (b) no (c) no (d) yes **20.3** the identity of the R group (side chain) **20.5** (a) phenylalanine, tyrosine, tryptophan (b) methionine, cysteine (c) aspartic acid, glutamic acid (d) serine, threonine, tyrosine **20.7** An amino group is part of the side chain **20.9** The side chain is part of a cyclic structure. **20.11** (a) alanine (b) leucine (c) methionine (d) tryptophan **20.13** asparagine, glutamine, isoleucine, tryptophan **20.15** (a) polar neutral (b) polar acidic (c) nonpolar (d) polar neutral **20.17** L family **20.19** (a)

```
(a)      COOH
          |
   H2N ---+--- H
          |
          CH2
          |
          OH

(b)      COOH          (c)      COOH          (d)      COOH
          |                      |                      |
   H  ----+--- NH2       H  -----+--- NH2      H2N  ----+--- H
          |                      |                      |
          CH2                    CH3                    CH2
          |                                             |
          OH                                            CH—CH3
                                                        |
                                                        CH3
```

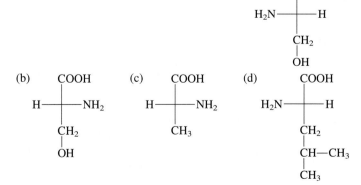

20.21 They exist as zwitterions. **20.23** (a)

```
(a)         H
            |
   +H3N ----+---- COO−
            |
            CH2
            |
            CH—CH3
            |
            CH3

(b)         H                  (c)         H
            |                              |
   +H3N ----+---- COO−            +H3N ----+---- COO−
            |                              |
            CH—CH3                         CH2
            |                              |
            CH2                            SH
            |
            CH3

(d)         H
            |
   +H3N ----+---- COO−
            |
            H
```

20.25

```
(a)         H                  (b)         H
            |                              |
   +H3N ----+---- COO−            +H3N ----+---- COOH
            |                              |
            CH2                            CH2
            |                              |
            OH                             OH

(c)         H                  (d)         H
            |                              |
    H2N ----+---- COO−            +H3N ----+---- COOH
            |                              |
            CH2                            CH2
            |                              |
            OH                             OH
```

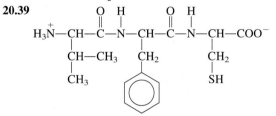

20.27 the pH at which zwitterion concentration in a solution is maximized **20.29** Two —COOH groups are present, which deprotonate at different times. **20.31** (a) toward positive electrode (b) isoelectric (c) toward negative electrode (d) toward positive electrode **20.33** Aspartic acid migrates toward the positive electrode, histidine migrates toward the negative electrode, and valine does not migrate. **20.35** a short sequence of amino acids linked through peptide bonds **20.37** —COOH and —NH_2

20.39

```
               O  H         O  H
               ‖  |         ‖  |
+H3N—CH—C—N—CH—C—N—CH—COO−
      |           |           |
      CH—CH3      CH2         CH2
      |           |           |
      CH3       (C6H5)        SH
```

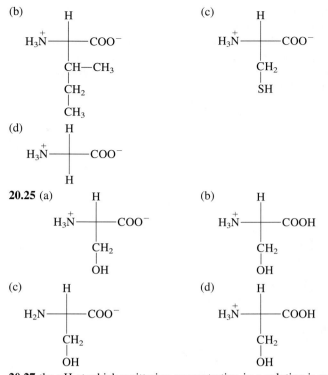

20.41 Ser is the N-terminal end of Ser–Cys and Cys is the N-terminal end of Cys–Ser. **20.43** Ser–Val–Gly, Val–Ser–Gly, Gly–Ser–Val, Ser–Gly–Val, Val–Gly–Ser, Gly–Val–Ser **20.45** (a) Ser–Ala–Cys (b) Asp–Thr–Asn **20.47** two in each **20.49** nothing; they are the same **20.51** the sequence of amino acids in the protein chain **20.53** α helix, β pleated sheet, triple helix **20.55** Intermolecular involves two separate chains and intramolecular involves a single chain bending back on itself. **20.57** Yes, both α helix and β pleated sheet can occur at different regions in the same chain. **20.59** Secondary-structure hydrogen bonding involves C=O···H—N interactions; tertiary-structure hydrogen bonding involves R group interactions. **20.61** (a) hydrophobic (b) electrostatic (c) disulfide bond (d) hydrogen bonding **20.63** a multi-chain protein **20.65** (a) fibrous protein: long, thin, fibrous shape; globular protein: a roughly spherical or globular shape (b) fibrous protein: water-insoluble; globular protein: dissolves in water or forms a stable suspension in water **20.66** (a) Pairs of alpha helices twine about one another. (b) Two coiled coils twist together. (c) Eight profilaments combine in either a circular or a square arrangement. **20.67** A simple protein contains only amino acids; a conjugated protein has one or more other chemical components besides amino acids. **20.69** Yes, both Ala and Val are products in each case. **20.71** Drug hydrolysis would occur in the stomach. **20.73** Ala–Gly–Met–His–Val–Arg **20.75** five: Ala–Gly–Ser, Gly–Ser–Tyr, Ala–Gly, Gly–Ser, Ser–Tyr **20.77** secondary, tertiary, and quaternary **20.79** same primary structure **20.81** triple helix **20.83** bonded to 5-hydroxylysine residues **20.85** An antigen is a substance foreign to the human body, and an antibody is a substance that defends against an invading antigen. **20.87** four polypeptide chains that have constant and variable amino acid regions; two chains are longer than the other two; 1%–12% carbohydrates present; long and short chains are connected through disulfide linkages **20.89** suspend and transport lipids in the bloodstream **20.91** (a) tertiary (b) tertiary (c) secondary (d) primary **20.92** (a) alanine (b) leucine (c) threonine (d) aspartic acid **20.93** (a) +1 (b) +1 (c) +1 (d) +3 **20.94** (a) −1 (b) −1 (c) −4 (d) −1 **20.95**

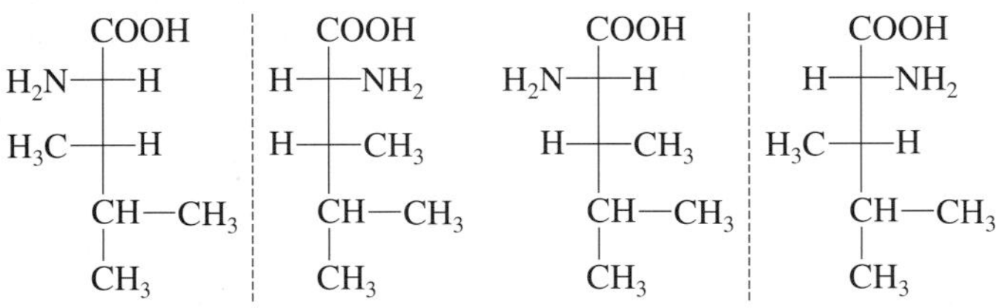

COOH	COOH	COOH	COOH
H_2N—+—H	H—+—NH_2	H_2N—+—H	H—+—NH_2
H_3C—+—H	H—+—CH_3	H—+—CH_3	H_3C—+—H
CH—CH_3	CH—CH_3	CH—CH_3	CH—CH_3
CH_3	CH_3	CH_3	CH_3

20.96 (a) 24 (b) 36 (c) 20 **20.97** Collagen has a triple-helix structure; alpha keratin has a double-double-helix structure. **20.98** (a)

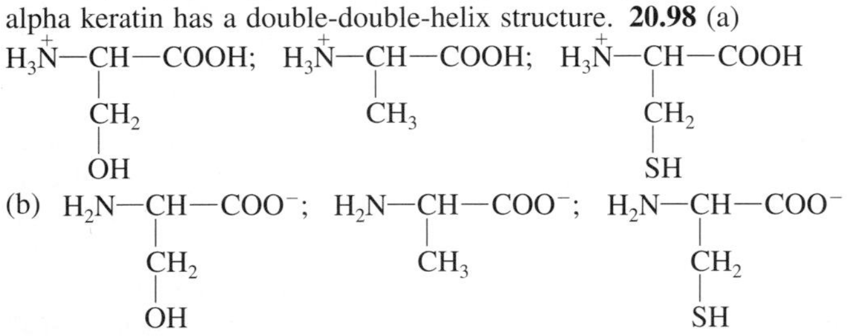

$H_3\overset{+}{N}$—CH(CH_2OH)—COOH; $H_3\overset{+}{N}$—CH(CH_3)—COOH; $H_3\overset{+}{N}$—CH(CH_2SH)—COOH

(b) H_2N—CH(CH_2OH)—COO^-; H_2N—CH(CH_3)—COO^-; H_2N—CH(CH_2SH)—COO^-

20.99 (a) all choices (b) 4 (c) 1 (d) 2 and 4 **20.100** (a) 1, 5, and 6 (b) 3 and 4 (c) 3 and 4 (d) 1, 3, 4, 5, and 6 **20.101** (a) 4 and 5 (b) 1, 2, and 3 (c) 6 (d) 1, 2, 3, 4, and 5 **20.102** (a) 1, 4, 5, and 6 (b) 3 (c) 3 and 6 (d) 1 and 5, 1 and 6, 5 and 6

Chapter 21 **21.1** catalyst **21.3** more efficient, more specific **21.5** (a) yes (b) no (c) yes (d) yes **21.7** (a) add a carboxylate group to pyruvate (b) remove H_2 from an alcohol (c) reduce an L-amino acid (d) hydrolyze maltose **21.9** (a) sucrase (or sucrose hydrolase) (b) pyruvate decarboxylase (c) glucose isomerase (d) lactate dehydrogenase **21.11** (a) pyruvate (b) galactose (c) an alcohol (d) an L-amino acid **21.13** (a) isomerase (b) lyase (c) ligase (d) transferase **21.15** (a) isomerase (b) lyase (c) transferase (d) hydrolase **21.17** (a) decarboxylase (b) lipase (c) phosphatase (d) dehydrogenase **21.19** (a) conjugated (b) conjugated (c) simple (d) conjugated **21.21** A coenzyme is a cofactor that is an organic substance. A cofactor can be an inorganic or an organic substance. **21.23** to provide additional functional groups **21.25** the portion of an enzyme actually involved in the catalysis process **21.27** the substrate must have the same shape as the active site. **21.29** interactions with amino acid R groups **21.31** (a) accepts only one substrate (b) accepts substrate with a particular type of bond **21.33** absolute specificity and stereochemical specificity **21.35** (a) absolute (b) stereochemical **21.37** Rate increases until enzyme denaturation occurs. **21.39** Enzymes vary in the number of acidic and base amino acids present. **21.41**

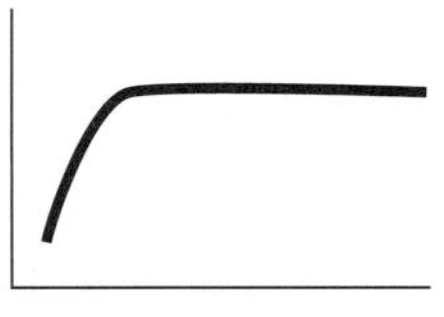

21.43 nothing; the rate remains constant **21.45** no; only one molecule may occupy the active site at a given time **21.47** (a) reversible competitive (b) reversible noncompetitive, irreversible (c) reversible noncompetitive, irreversible (d) reversible noncompetitive **21.49** enzyme that has quaternary structure and more than one binding site **21.51** The product of a subsequent reaction in a series of reactions inhibits a prior reaction. **21.53** inactive precursor of an enzyme **21.55** so that the enzymes do not "digest" the glands that produce them **21.57** competitive inhibition of the conversion of PABA to folic acid **21.59** has absolute specificity for bacterial transpeptidase **21.61** dietary organic compound needed by the body in trace amounts **21.63** 13 **21.65** (a) fat-soluble (b) water-soluble (c) water-soluble (d) water-soluble **21.67** (a) likely (b) unlikely (c) unlikely (d) unlikely **21.69** serves as a cosubstrate in the formation of collagen **21.71** coenzymes **21.73** (a) all B vitamins (b) B_{12} (c) B_{12} (d) thiamin and biotin **21.75** alcohol, aldehyde, acid **21.77** Vitamin K is necessary for the formation of blood-clotting agents; vitamin E functions as an antioxidant. **21.79** cofactors, bone component formation, regulation of fluid movement **21.81** (a) P, Cl (b) Na, K, Ca, Mg, Cl (c) Ca, P (d) Ca, K, Mg **21.83** (a) transport of oxygen (b) formation of the hormone thyroxin **21.85** (a) 0.15 mg (b) 0.070 mg (male), 0.055 mg (female) **21.87** (a) An apoenzyme is the protein portion of a conjugated enzyme; a proenzyme is an inactive precursor of an enzyme. (b) A simple enzyme is pure protein; an allosteric enzyme has two or more protein chains and two binding sites. (c) A coenzyme is an organic cofactor, and an isoenzyme is one of several similar forms of an enzyme. (d) A conjugated enzyme has both a protein and a nonprotein portion; holoenzyme is just another name for a conjugated enzyme. **21.88** (a) alcohol, ketone (b) double bond, alcohol (c) double bond, alcohol (d) double bond, ketone **21.89** (a) no (b) no (c) no (d) yes (e) yes (f) yes **21.90** (a) major mineral (b) major mineral (c) trace mineral (d) trace mineral (e) trace mineral (f) major mineral **21.91** (a) oxidation–reduction reactions (b) addition of a group to, or removal of a group from, a double bond in a manner that does not involve hydrolysis or oxidation–reduction (c) conversion of a compound into another isomeric with it (d) bonding together of two molecules with the involvement of ATP (e) hydrolysis reactions (f) transfer of functional groups between two molecules **21.92** (a) enzyme plus substrate produces an enzyme–substrate complex, which breaks apart to regenerate the enzyme and a product molecule **21.93** (a) ethanol (b) zinc ion (c) protein molecule (d) alcohol dehydrogenase **21.94** (a) tissue plasminogen activator (b) lactate dehydrogenase (c) creatine phosphokinase (d) aspartate transaminase **21.95** (a) 1, 2, and 3 (b) 4 (c) 1 (d) 6 **21.96** (a) 2 and 3 (b) 2, 3, and 5 (c) 1, 2, 4, 5, and 6 (d) 1 and 4 **21.97** (a) 1, 3, and 6 (b) 2 (c) 1, 3, 4, 5, and 6 (d) 1, 3, and 6 **21.98** (a) 4, 5, and 6 (b) 1, 2, and 3 (c) 4 and 6 (d) 2 and 3

Chapter 22 **22.1** Ribose has both an —H group and an —OH group on carbon 2; deoxyribose has 2 —H atoms on carbon 2. **22.3** (a) pyrimidine (b) pyrimidine (c) purine (d) purine **22.5** (a) one (b) four (c) one **22.7** (a) adenine (b) guanine (c) thymine (d) uracil **22.9** (a) ribose (b) deoxyribose (c) deoxyribose (d) ribose **22.11** (a) false (b) false (c) false (d) false **22.13**

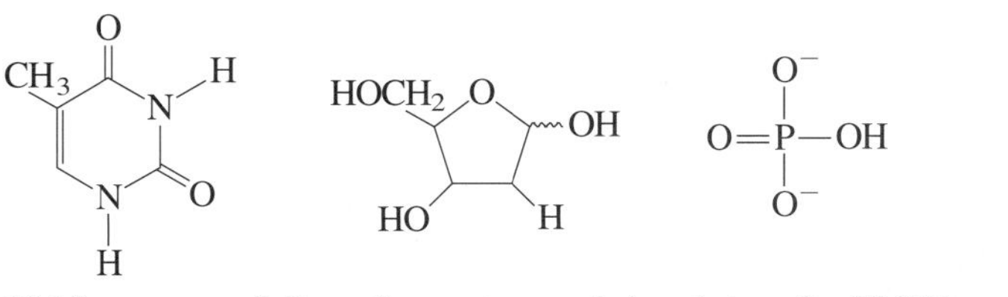

CH_3, O, N, H, O, N, H (thymine ring) $HOCH_2$, O, OH, HO, H (deoxyribose ring) O^-, O=P—OH, O^-

22.15 sequence of alternating pentose and phosphate units **22.17** base sequence **22.19** 5′ end has a phosphate group attached to the 5′ carbon; 3′ end has a hydroxyl group attached to the 3′ carbon

22.21

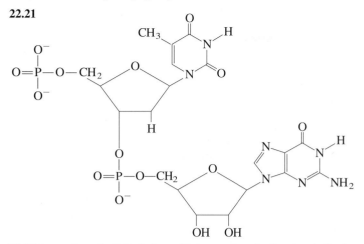

22.23 two polynucleotide chains coiled around each other in a helical fashion **22.25** (a) 36% (b) 14% (c) 14% **22.27** A G–C pairing involves 3 hydrogen bonds, and an A–T pairing involves 2 hydrogens bonds. **22.29** 5′–TAGCC–3′ **22.31** (a) 3′–TGCATA–5′ (b) 3′–AATGGC–5′ (c) 5′–CGTATT–3′ (d) 3′–TTGACC–5′ **22.33** 20 hydrogen bonds **22.35** catalyzes the unwinding of the double helix structure **22.37** (a) 3′–TGAATC–5′ (b) 3′–TGAATC–5′ (c) 5′–ACTTAG–3′ **22.39** The unwound strands are antiparallel (5′ → 3′ and 3′ → 5′). Only the 5′ → 3′ strand can grow continuously. **22.41** a DNA molecule bound to a group of small proteins **22.43** (1) RNA contains ribose instead of deoxyribose, (2) RNA contains the base U instead of T, (3) RNA is single-stranded rather than double-stranded, and (4) RNA has a lower molecular mass. **22.45** (a) ptRNA (b) tRNA (c) tRNA (d) ptRNA **22.47** DNA unwinding, base pairing, ribonucleotide linkage, RNA releases **22.49** T–A, A–U, G–C, C–G **22.51** 3′–UACGAAU–5′ **22.53** 3′–AAGCGTC–5′ **22.55** Exons convey genetic information whereas introns do not. **22.57** 3′–AGUCAAGU–5′ **22.59** A three-nucleotide sequence in mRNA that codes for a specific amino acid **22.61** (a) Leu (b) Asn (c) Ser (d) Gly **22.63** (a) CUC, CUA, CUG, UUA, UUG (b) AAC (c) AGC, UCU, UCC, UCA, UCG (d) GGU, GGC, GGA **22.65** The base T cannot be present in a codon. **22.67** Met–Lys–Glu–Asp–Leu **22.69** A cloverleaf shape with three hairpin loops and one open side. **22.71** covalent bond **22.73** (a) UCU (b) GCA (c) AAA (d) GUU **22.75** (a) Thr (b) Leu (c) Pro (d) Ser **22.77** (1) activation of tRNA, (2) initiation, (3) elongation, (4) termination, and (5) post-translational processing. **22.79** A site **22.81** Gly: GGU, GGC, GGA or GGG; Ala: GCU, GCC, GCA or GCG; Cys: UGU or UGC; Val: GUU, GUC, GUA or GUG; Tyr: UAU or UAC **22.83** (a) Gly–Tyr–Ser–Ser–Pro (b) Gly–Tyr–Ser–Ser–Pro (c) Gly–Tyr–Ser–Ser–Thr **22.85** a DNA or an RNA molecule with a protein coating **22.87** (1) attaches itself to cell membrane, (2) opens a hole in the membrane, and (3) injects itself into the cell **22.89** contains a "foreign" gene **22.91** host for a "foreign" gene **22.93** recombinant DNA is incorporated into a host cell

22.95

22.97 (a) Thymine has a methyl group on carbon-5 that uracil lacks. (b) Adenine is 6-aminopurine, and guanine is 2-amino-6-oxo purine. **22.98** (a) 3′–GTATGTCGGACCTTCGAT–5′ (b) 3′–GUAUGUCGGACCUUCGAU–5′ (c) 3′–GUA–UGU–CGG–ACC–UUC–GAU–5′ (d) 5′–CAU–ACA–GCC–UGG–AAG–CUA–3′ (e) Val–Cys–Arg–Thr–Phe–Asp **22.99** (a) 1 (b) 4 (c) 2 (d) 4 **22.100** (a) 3 (b) 2 (c) 1 (d) 3 **22.101** (a) 3 (b) 1 (c) 4 (d) 1, 2, 3, and 4 **22.102** (a) 1 (b) 2 (c) 4 (d) 3 **22.103** A **22.104** 3.3 million base pairs spread over 23 pairs of chromosomes and 50,000 to 100,000 genes **22.105** (a) 3, 5, 6 (b) 1, 2, 4 (c) 1 and 2, 1 and 4, 2 and 4, 3 and 5, 3 and 6, 5 and 6, (d) 2 and 5 **22.106** (a) 1 and 2 (b) 4 and 6 (c) 1, 2, 4, and 6 (d) 1 and 2 **22.107** (a) 1 and 6 (b) 3, 5, and 6 (c) 2, 4, 5, and 6 (d) 3, 5, and 6 **22.108** (a) all choices (b) 1 (c) 5 (d) 2 and 6; 3 and 4

Chapter 23 **23.1** anabolism—synthetic; catabolism—degradative **23.3** a series of consecutive biochemical reactions **23.5** Large molecules are broken down to smaller ones; energy is released. **23.7** Prokaryotic cells have no nucleus, and the DNA is usually a single circular molecule. Eukaryotic cells have their DNA in a membrane-enclosed nucleus. **23.9** An organelle is a small structure within the cell cytoplasm that carries out a specific cellular function. **23.11** inner membrane **23.13** region between inner and outer membranes **23.15** adenosine triphosphate **23.17** phosphate—phosphate—phosphate—ribose—adenine **23.19** three phosphates versus one phosphate **23.21** adenine versus guanine **23.23** ADP is produced from the hydrolysis of ATP. **23.25** flavin adenine dinucleotide **23.27** flavin—ribitol—ADP **23.29** nicotinamide—ribose—phosphate / adenine—ribose—phosphate (the two phosphates joined) **23.31** nicotinamide subunit **23.33** (a) $FADH_2$ (b) NAD^+ **23.35** (a)nicotinamide (b) riboflavin **23.37** 2-aminoethanethiol—pantothenic acid—phosphorylated ADP **23.39** a compound with a greater free energy of hydrolysis than is typical for a compound **23.41** free monophosphate species **23.43** (a) phosphoenolpyruvate (b) creatine phosphate (c) 1,3-diphosphoglycerate (d) AMP **23.45** (1) digestion, (2) acetyl group formation, (3) citric acid cycle, (4) electron transport chain and oxidative phosphorylation **23.47** tricarboxylic acid cycle, Krebs cycle **23.49** acetyl CoA **23.51** (a) 2 (b) 1 (c) 2 (Steps 3, 8) (d) 2 (Steps 2, 7) **23.53** (a) Steps 3, 4, 6, 8 (b) Step 2 (c) Step 7 **23.55** succinate ($^-OOC—CH_2—CH_2—COO^-$); fumarate ($^-OOC—CH=CH—COO^-$); malate $^-OOC—CH_2—CH(OH)—COO^-$; oxaloacetate $^-OOC—CH_2—C(=O)—COO^-$ **23.57** oxidation and decarboxylation **23.59** (a) NAD^+ (b) FAD **23.61** (a) isocitrate, α-ketoglutarate (b) fumarate, malate (c) malate, oxaloacetate (d) citrate, isocitrate **23.63** respiratory chain **23.65** O_2 **23.67** (a) flavin mononucleotide (b) cytochrome (c) iron-sulfur protein (d) reduced form of nicotinamide adenine dinucleotide **23.69** (a) mobile (b) fixed, 2nd site (c) fixed, 1st site (d) mobile **23.71** (a) NADH, FeSP, cyt c_1, cyt a_3 **23.73** (a) oxidation (b) reduction (c) oxidation (d) reduction **23.75** (a) $FADH_2$, CoQ, $2Fe^{3+}$ (b) $FMNH_2$, 2Fe(II)SP, $CoQH_2$ **23.77** (a) FMN + 2Fe(II)SP (b) CoQ, $CoQH_2$ **23.79** ATP synthesis from ADP using energy from the electron transport chain **23.81** protons (H^+ ions) **23.83** intermembrane space **23.85** ATP synthase **23.87** Protons flow through ATP synthase complex. **23.89** three **23.91** They enter the ETC at different stages. **23.93** (a) 1 (b) 2 (c) 2 (d) 1 (e) 1 (f) 2 **23.94** (a) 1 (b) 2 (c) 1 (d) 3 (e) 1 (f) 2 **23.95** (a) 4 (b) 2, 3, and 4 (c) 1, 2, 3, and 4 (d) 3 **23.96** (a) 1 (b) 4

(c) 4 (d) 2 **23.97** (a) 4 (b) 1 (c) 3 (d) 1 **23.98** The products from the CAS, which are FAD and NAD^+, are the starting reactants for the ETC. **23.99** (a) inside mitochondria (b) inside mitochondria **23.100** FAD is the oxidizing agent for carbon–carbon double bond formation. **23.101** (a) 1, 2, 3, 4, and 6 (b) 1, 2, and 4 (c) 1 and 5 (d) 1 **23.102** (a) 2, 3, and 4 (b) 5 and 6 (c) 2 (d) 3 **23.103** (a) 1, 2, 4, and 6 (b) 2, 4, 5, and 6 (c) 1 (d) 6 **23.104** (a) 3 and 6 (b) 2 and 6 (c) 1, 4, and 6 (d) 3 and 6

Chapter 24 **24.1** mouth, salivary α-amylase **24.3** small intestine, pancreas **24.5** outer membranes of intestinal mucosal cells, sucrose hydrolysis **24.7** glucose, galactose, fructose **24.9** glucose **24.11** NAD^+ **24.13** formation of glucose 6-phosphate, a species that cannot cross cell membranes **24.15** dihydroxyacetone phosphate, glyceraldehyde 3-phosphate **24.17** two **24.19** two **24.21** Steps 1, 3, and 6 **24.23** cytoplasm **24.25** (a) glucose 6-phosphate (b) 2-phosphoglycerate (c) phosphoglyceromutase (d) ADP **24.27** (a) Step 10 (b) Step 1 (c) Step 8 (d) Step 6 **24.29** (a) +2 (b) +4 **24.31**

(a) $CH_3-\overset{\overset{\large O}{\|}}{C}-OH$ and $CH_3-\overset{\overset{\large O}{\|}}{C}-O^-$

(b) CH_2-OH / $C=O$ / CH_2-OH and CH_2-O-Ⓟ / $C=O$ / CH_2-OH

(c) Ⓟ$-O-CH_2$ (furanose ring: O, CH_2OH, OH, OH, OH) and Ⓟ$-O-CH_2$ (furanose ring: O, CH_2-O-Ⓟ, OH, OH, OH)

(d) COOH / CH—OH / CH_2—OH and CHO / CH—OH / CH_2—OH

24.33 $^1CH_2-O-$Ⓟ / $^2C=O$ / $^3CH_2-OH$ and ^4CHO / ^5CH-OH / $^6CH_2-O-$Ⓟ

24.35 acetyl CoA, lactate, ethanol **24.37** pyruvate + CoA + $NAD^+ \rightarrow$ acetyl CoA + NADH + CO_2 **24.39** NADH is oxidized to NAD^+, a substance needed for glycolysis. **24.41** CO_2 **24.43** glucose + 2ADP + $2P_i \rightarrow$ 2 lactate + 2ATP **24.45** decreases ATP production by 2 **24.47** 36 ATP versus 2 ATP **24.49** two **24.51** Glycogenesis converts glucose to glycogen and glycogenolysis is the reverse process. **24.53** glucose 6-phosphate **24.55** UTP **24.57** UDP + ATP → UTP + ADP **24.59** Step 2 **24.61** In liver cells the product is glucose, and in muscle cells it is glucose 6-phosphate. **24.63** as glucose 6-phosphate **24.65** the liver **24.67** two-step pathway for Step 10; different enzymes for Steps 1 and 3 **24.69** oxaloacetate **24.71** goes to the liver, where it is converted to glucose **24.73** glucose 6-phosphate **24.75** NADPH is consumed in its reduced form; NADH is consumed in its oxidized form (NAD^+). **24.77** Glucose 6-phosphate + $2NADP^+$ + $H_2O \rightarrow$ ribulose 5-phosphate + CO_2 + 2NADPH + $2H^+$ **24.79** CO_2 **24.81** increases rate of glycogen synthesis **24.83** increases blood glucose levels **24.85** pancreas **24.87** Epinephrine attaches to cell membrane and stimulates the production of cAMP, which activates glycogen phosphorylase. **24.89** glucagon (liver cells) and epinephrine (muscle cells) **24.91** (a) all four (b) glycogenesis and glycogenolysis (c) glycolysis and gluconeogenesis (d) gluconeogenesis **24.92** (a) glycolysis (b) glycolysis (c) glycolysis, gluconeogenesis, and glycogenesis (d) glycogenesis **24.93** (a) glycolysis (b) glycogenesis (c) glycogenolysis (d) glyconeogenesis **24.94** (a) 2 ATP (b) 36 ATP (c) 2 ATP (d) 2 ATP **24.95** (a) 2 (b) 2 (c) 3 (d) 4 **24.96** (a) when the body requires free glucose (b) anaerobic conditions in muscle; red blood cells (c) when the body requires energy (d) anaerobic conditions in yeast **24.97** (a) 12 moles (b) 4 moles (c) 4 moles (d) 72 moles **24.98** (a) The glucose supply is adequate, and the body doesn't need energy. (b) The glucose supply is adequate, and the body needs energy. (c) Ribose 5-phosphate or NADPH is needed. (d) The free glucose supply is not adequate. **24.99** (a) 1, 2, 4, 5, and 6 (b) 1, 4, 5, and 6 (c) 5 (d) 2 and 3 **24.100** (a) 1 (b) 4 (c) 6 (d) 6 **24.101** (a) 2 (b) 6 (c) 2 (d) 2 **24.102** (a) 4, 5, and 6 (b) 1, 2, 3, 4, and 5 (c) 1 and 2 (d) 3 and 6

Chapter 25 **25.1** 98% **25.3** no effect **25.5** because lipids have a long residence time in the stomach **25.7** acts as an emulsifier **25.9** monoacylglycerols are the major product **25.11** reassembled into triacylglycerols; converted to chylomicrons **25.13** They have a large storage capacity for triacylglycerols. **25.15** hydrolysis of triacylglycerols in adipose tissue; entry of hydrolysis products into bloodstream **25.17** activates hormone-sensitive lipase **25.19** glycerol 3-phosphate, dihydroxyacetone phosphate **25.21** one **25.23** outer mitochondrial membrane **25.25** ATP is converted to AMP and $2P_i$ **25.27** shuttles acyl groups across the inner mitochondrial membrane **25.29** (a) alkane to alkene (b) alkene to 2° alcohol (c) 2° alcohol to ketone **25.31** *trans* isomer **25.33** (a) Step 3, turn 1 (b) Step 2, turn 2 (c) Step 4, turn 2 (d) Step 1, turn 1 **25.35** compounds a and d **25.37** (a) 7 turns (b) 5 turns **25.39** *Cis–trans* isomerase converts a *cis*-(3,4) double bond to a *trans*-(2,3) double bond. **25.41** (a) glucose (b) fatty acids **25.43** (a) 4 turns (b) 5 acetyl CoA (c) 4 NADH (d) 4 $FADH_2$ (e) 2 high-energy bonds **25.45** 78 ATP **25.47** yield the same amount **25.49** 4 kcal versus 9 kcal **25.51** (1) dietary intakes high in fat and low in carbohydrates, (2) inadequate processing of glucose present, and (3) prolonged fasting **25.53** Ketone body formation occurs when oxaloacetate concentrations are low.

25.55 $CH_3-\overset{\overset{\large O}{\|}}{C}-CH_2-\overset{\overset{\large O}{\|}}{C}-O^-$ $\quad$ $CH_3-\overset{\overset{\large OH}{|}}{CH}-CH_2-\overset{\overset{\large O}{\|}}{C}-O^-$ $\quad$ $CH_3-\overset{\overset{\large O}{\|}}{C}-CH_3$

25.57 liver mitochondria **25.59** formation of acetoacetyl CoA **25.61** accumulation of ketone bodies in blood and urine **25.63** cytoplasm versus mitochondrial matrix **25.65** acyl carrier protein; polypeptide chain replaces phosphorylated ADP **25.67** liver, adipose tissue, mammary glands **25.69** It is involved in the citrate shuttle system. **25.71** source of C_2 units for the growing fatty acid chain **25.73** (1) condensation, (2) hydrogenation, (3) dehydration, and (4) hydrogenation **25.75** (a) Step 1, cycle 1 (b) Step 2, cycle 2 (c) Step 3, cycle 2 (d) Step 4, cycle 1 **25.77** compounds b and d **25.79** C_{16} fatty acid **25.81** needed to convert saturated fatty acids to unsaturated fatty acids **25.83** (a) 6 rounds (b) 6 malonyl ACP (c) 6 ATP bonds (d) 12 NADPH **25.85** (a) 13–17% (b) 83–87% **25.87** (a) mevalonate (b) isopentenyl pyrophosphate (c) squalene **25.89** (a) fewer than (b) fewer than (c) same as **25.91** (a) fatty acid spiral (b) fatty acid spiral (c) lipogenesis (d) glycerol catabolism (e) ketogenesis (f) ketogenesis **25.92** (a) fatty acid catabolism (b) lipogenesis (c) lipogenesis (d) ketogenesis (e) consumption of molecular O_2 (f) consumption of molecular O_2 **25.93** (a) Step 1 (b) Step 1 (c) Step 3 (d) Step 4 **25.94** (1), (3), (4), and (2) **25.95** (a) true (b) false (c) false (d) false **25.96** (a) incorrect (b) correct (c) incorrect (d) correct **25.97** (a) incorrect (b) correct (c) correct (d) correct **25.98** glucose, C_8 fatty acid, sucrose, C_{14} fatty acid **25.99** (a) 1 and 5 (b) 4 (c) 6 (d) 3 **25.100** (a) 2, 3, and 4 (b) 1 and 5 (c) 1, 5, and 6 (d) 2, 5, and 6 **25.101** (a) 1, 2, 4, and 5 (b) 6 (c) 2, 4, 5, and 6 (d) 1 and 3 **25.102** (a) 2, 3, 5, and 6 (b) 4, 5, and 6 (c) 1, 3, 4, 5, and 6 (d) 5 and 6

Chapter 26 **26.1** Denaturation occurs in the stomach with gastric juice as the denaturant. **26.3** Pepsinogen is the inactive precursor of pepsin. **26.5** Gastric juice is acidic (1.5–2.0 pH) and pancreatic juice is basic (7–8 pH). **26.7** Shuttle molecules facilitate the passage of amino acids through the intestinal wall. **26.9** total supply of free amino acids available for use **26.11** cyclic process of protein degradation and resynthesis **26.13** A positive nitrogen balance has nitrogen intake exceeding nitrogen output; a negative nitrogen balance has nitrogen output exceeding nitrogen intake. **26.15** negative balance; proteins are degraded to get the needed amino acid **26.17** protein synthesis; synthesis of nonprotein nitrogen-containing compounds; nonessential amino acid synthesis; energy production **26.19** (a) essential (b) nonessential (c) nonessential (d) essential **26.21** b and c **26.23** an amino acid and an α-keto acid **26.25**

(a)
$$HO{-}CH(CH_3){-}CH(\overset{+}{N}H_3){-}COO^- + CH_3{-}C(=O){-}COO^- \longrightarrow$$
$$HO{-}CH(CH_3){-}C(=O){-}COO^- + CH_3{-}CH(\overset{+}{N}H_3){-}COO^-$$

(b)
$$CH_3{-}CH(\overset{+}{N}H_3){-}COO^- + {}^-OOC{-}CH_2{-}C(=O){-}COO^- \longrightarrow$$
$$CH_3{-}C(=O){-}COO^- + {}^-OOC{-}CH_2{-}CH(\overset{+}{N}H_3){-}COO^-$$

(c)
$$H{-}CH(\overset{+}{N}H_3){-}COO^- + {}^-OOC{-}CH_2{-}CH_2{-}C(=O){-}COO^- \longrightarrow$$
$$H{-}C(=O){-}COO^- + {}^-OOC{-}CH_2{-}CH_2{-}CH(\overset{+}{N}H_3){-}COO^-$$

(d)
$$HO{-}CH(CH_3){-}CH(\overset{+}{N}H_3){-}COO^- + {}^-OOC{-}CH_2{-}CH_2{-}C(=O){-}COO^- \longrightarrow$$
$$HO{-}CH(CH_3){-}C(=O){-}COO^- + {}^-OOC{-}CH_2{-}CH_2{-}CH(\overset{+}{N}H_3){-}COO^-$$

26.27 pyruvate, α-ketoglutarate, oxaloacetate **26.29** coenzyme that participates in the amino group transfer **26.31** conversion of an amino acid into a keto acid with the release of ammonium ion **26.33** Oxidative deamination produces ammonium ion, and transamination produces an amino acid. **26.35** (a)
$${}^-OOC{-}CH_2{-}CH_2{-}C(=O){-}COO^-$$

(b)
$$HS{-}CH_2{-}C(=O){-}COO^-$$

(c)
$$CH_3{-}C(=O){-}COO^-$$

(d)
$$C_6H_5{-}CH_2{-}C(=O){-}COO^-$$

26.37 transamination of the α-keto acid produces the amino acids
$$CH_3{-}CH(CH_3){-}CH_2{-}CH(\overset{+}{N}H_3){-}COO^-$$

26.39 (a) aspartate (b) glutamate (c) pyruvate (d) α-ketoglutarate **26.41**
$$H_2N{-}C(=O){-}NH_2$$

26.43 carbamoyl phosphate **26.45** an amide group, $-C(=O){-}NH_2$

26.47
$$H_3\overset{+}{N}{-},\quad H_2N{-}C(=O){-}NH{-},\quad H_2N{-}C(=\overset{+}{N}H_2){-}NH{-}$$

26.49 carbamoyl phosphate **26.51** ornithine **26.53** conversion of citrulline to argininosuccinate **26.55** (a) citrulline (b) ornithine (c) argininosuccinate (d) carbamoyl phosphate **26.57** equivalent of four ATP molecules **26.59** goes to the citric acid cycle where it is converted to oxaloacetate, which is then converted to aspartate **26.61** α-ketoglutarate, succinyl CoA, fumarate, oxaloacetate **26.63** (a) acetoacetyl CoA and acetyl CoA (b) succinyl CoA and acetyl CoA (c) fumarate and oxaloacetate (d) α-ketoglutarate **26.65** Degradation products can be used to make glucose. **26.67** glutamate **26.69** pyruvate, α-ketoglutarate, 3-phosphoglycerate, oxaloacetate, and phenylalanine **26.71** hydrolyzed to amino acids **26.73** In biliverdin the heme ring has been opened and one carbon atom has been lost (as CO). **26.75** biliverdin, bilirubin, bilirubin diglucuronide, urobilin **26.77** urobilin **26.79** excess bilirubin **26.81** pyruvate acetyl CoA, acetoacetate **26.83** Amino acids are converted to acetyl CoA, lipogenesis converts acetyl CoA to fatty acids, and then fatty acids are converted to triacylglycerols. **26.85** converted to body fat stores **26.87** (a) 1 (b) 2 (c) 3 (d) 2 (e) 1 and 3 (f) 3 **26.88** (a) 1 (b) 3 (c) 3 (d) 2 (e) 1 (f) 2 **26.89** (1), (3), (2), and (4) **26.90** (a) 3 (b) 2 (c) 1 (d) 3 **26.91** (a) transamination (b) deamination (c) deamination (d) transamination **26.92** (a) 1 (b) 4 (c) 2 (d) 3 **26.93** (a) true (b) false (c) true (d) true **26.94** (a) 2 (b) 1 (c) 2 (d) 3 **26.95** (a) 2, 3, 4, and 5 (b) 2, 3, 4, and 6 (c) 1 and 6 (d) 1, 2, and 3 **26.96** (a) all choices (b) 2, 3, 4, and 6 (c) 1 (d) 5 **26.97** (a) all choices (b) 2, 3, 4, 5, and 6 (c) 1 (d) 1 **26.98** (a) 1 (b) 1, 2, 4, and 5 (c) 1, 2, 4, and 5 (d) 2, 4, and 5

Glossary/Index

Key terms, which appear in **boldface,** *are followed by their definitions.*

Absolute alcohol, 378
Acetal(s) A compound that has a carbon atom to which two alkoxy groups and a hydrogen atom are attached, 423
  hydrolysis of, 423–424
  synthesis of, from cyclic monosaccharides, 530–531
  synthesis of, from hemiacetals, 423
Acetaminophen
  pharmacology of, 489
  structure of, 489
Acetoacetate
  ketone body, 753
  ketogenesis and, 754
  utilization of, as fuel, 755
Acetoacetyl CoA
  amino acid degradation and, 782
  ketogenesis and, 754
  lipogenesis and, 757
Acetone
  derivatives of, glycolysis and, 720
  ketone body function for, 753
  properties of, 413
  uses of, 413
Acetyl CoA
  amino acid degradation and, 782
  cholesterol biosynthesis and, 760
  fatty acid spiral and, 749
  ketogenesis and, 753–755
  lipogenesis and, 756
  possible fates of, 762
  pyruvate oxidation to, 725
Achiral object(s) An object that is identical to its mirror image, 509
  examples of, 509
Acid(s)
  amino; *see* Amino acids
  Arrhenius, 237–238
  Brønsted-Lowry, 238–241
  carboxylic; *see* Carboxylic acids
  diprotic, 241–242
  monoprotic, 241–242
  neutralization of, 245
  polyprotic, 241–242
  strength of, and extent of dissociation, 242
  strength of, and pK_a, 251
  strong, 242
  summary diagram concerning, 254
  triprotic, 241–242
  weak, 242
Acid-base pairs
  conjugate, Brønsted-Lowry theory and, 239–240
Acid-base titration(s) A measured volume of an acid or a base of known concentration is exactly reacted with a measured volume of a base or an acid of unknown concentration, 264
  indicator use in, 264
  use of data from, in calculations, 265
Acid-base theory
  Arrhenius, 237–238
  Brønsted-Lowry, 238–241
Acid ionization constant(s) The equilibrium constant corresponding to the ionization for the acid, 243
  calculating value of, 243–244
  listing of, selected acids, 244
  mathematical expression for, 243
Acid rain
  effects of, 253
  formation of, 253
  pH values for, 252
Acidic solution(s) A solution in which the concentration of H_3O^+ ion is higher than that of OH^- ion, 247
  hydronium ion concentration and, 247–248
  pH value and, 250–251
Acidosis A body condition in which the pH of blood drops from its normal value of 7.4 to 7.1–7.2, 260
  metabolic, 260
  respiratory, 260
ACP complex (acyl carrier protein complex)
  lipogenesis and, 757
Activation energy The minimum combined kinetic energy that reactant particles must possess in order for their collision to result in a reaction, 217
  potential energy diagrams and, 218
Active site The relatively small part of an enzyme that is actually involved in catalysis, 623
Active transport Movement of a substance across a membrane, with the aid of membrane proteins, against a concentration gradient with the expenditure of cellular energy, 580
  process of, characteristics for, 580
Acyl
  meaning of term, 747
Acyl CoA
  fatty acid spiral and, 748–749
Addition polymer(s) A polymer in which the monomers simply "add together," with no other products formed besides the polymer, 350
  butadiene-based, 352
  ethene-based, 350–352
  formulas of, notation for, 350–352
  line-angle drawings for, 352
Addition reaction(s) A reaction in which atoms or groups of atoms are added to each carbon atom of a carbon-carbon multiple bond, 344
  symmetrical, 346–347
  unsymmetrical, 347–349
Adenine
  as nucleotide subunit, 654
Adenosine diphosphate; *see* ADP
Adenosine monophosphate; *see* AMP
Adenosine phosphate(s)
  general structures of, 691
Adenosine triphosphate; *see* ATP
Adipocytes Triacylglycerol-storing cells, 745
Adipose tissue Tissue containing a large number of adipocyte cells, 745
  triacylglycerol storage and, 745–746
ADP (adenosine diphosphate)
  structure of, 691
Adrenaline; *see* Epinephrine
AIDS (acquired immune deficiency syndrome)
  immunoglobins and, 613–614
Alcohol(s) A hydrocarbon derivative in which a hydroxyl group (—OH) is attached to a saturated carbon atom, 372
  chemical reaction summary for, 389
  classifications of, 383–384
  commonly encountered, 376–380
  dehydration, intermolecular, 385–386
  dehydration, intramolecular, 384–386
  halogen substitution reactions, 388
  hydrogen-bonding and, 382–383
  IUPAC-common name contrast for, 375
  line-angle drawings for, 373
  nomenclature of, 373–375
  oxidation of, 387–388
  physical properties of, 380–383
  physical-state summary for, 381
  polyhydroxy, 375
  polymeric, 388–389
  primary, 383–384
  secondary, 383–384
  structural isomerism for, 375
  sugar, 529
  sulfur analogs of, 396–397
  synthesis of, from aldehydes, 383
  synthesis of, from alkenes, 383
  synthesis of, from ketones, 383
  tertiary, 383–384
Alcohol fermentation The enzymatic conversion of pyruvate to ethanol and carbon dioxide, 725
Aldehyde(s) A compound that has at least one hydrogen atom attached to the carbon atom of a carbonyl group, 407
  chemical reaction summary for, 425
  commonly encountered, 413–414
  diabetes testing and, 418
  hemiacetal formation and, 421–422
  IUPAC-common name contrast for, 410
  lachrymatory, 412
  line-angle drawings for, 409
  nomenclature of, 408–410
  notations for, 407
  oxidation of, 418–419
  physical properties of, 414–416
  physical-state summary for, 417
  reaction with alcohols, 420–422
  reaction with Benedict's solution, 419
  reaction with H_2, 419–420
  reaction with Tollens solution, 419

reduction of, 419–420
synthesis of, from alcohols, 387–388, 417–418
thio, 427
Aldose(s) Monosaccharides that contain an aldehyde group, 521
common, listing of, 522
Alkaloid(s) A nitrogen-containing compound extracted from plant material, 485
examples of, 485–486
Alkalosis A body condition in which the pH of blood increases from its normal value of 7.4 to a value of 7.5, 260
metabolic, 260
respiratory, 260
Alkane(s) A saturated hydrocarbon in which the carbon atom arrangement is acyclic, 300
branched-chain, 303
chemical reactions of, 320–321
combustion of, 320
continuous-chain, 303
general molecular formula for, 300
halogenation of, 320–321
natural sources of, 316–317
nomenclature of, 306–310
physical properties of, 317–319
physical-state summary for, 319
physiological effects of, 319
structural formulas for, 301–303
structural isomerism for, 303–304
Alkene(s) An acyclic unsaturated hydrocarbon that contains one or more carbon-carbon double bonds, 333
addition reactions and, 344–349
cis-trans isomerism for, 340–343
general molecular formula for, 333
halogenation of, 346–347
hydration of, 347
hydrogenation of, 346
hydrohalogenation of, 347
line-angle drawings for, 334
naturally occurring, 343–344
nomenclature for, 334–336
physical properties of, 344
physical-state summary for, 344
reaction-summary for, 353
structural isomerism for, 339–340
synthesis of, from alcohols, 384–386
Alkoxy group(s) An alkyl group to which an oxygen atom has been added, 392
Alkyl group(s) The group of atoms that would be obtained by removing a hydrogen atom from an alkane, 307
branched chain, 311–312
R symbol for, 321
Alkyl halide(s); *see* Halogenated alkane(s)
Alkylation
of ammonia and amines, 480–481
of aromatic hydrocarbons and, 360
Alkyne(s) An acyclic unsaturated hydrocarbon in which one or more carbon-carbon triple bonds are present, 352
chemical reactions of, 354
general molecular formula for, 352
line-angle drawings for, 353
nomenclature for, 354
physical properties of, 354
physical-state summary for, 354
Allosteric enzyme(s) An enzyme with two or more protein chains (quaternary structure) and two kinds of binding sites (substrate and regulator), 633
characteristics of, 632–633
Alpha particle decay
equations for, 273–274
Alpha particles
biological effects of, 283–284
Alpha rays Streams of positively charged particles (alpha particles), each of which is made up of two protons and two neutrons, 272
characterization of, 273
notation for, 273
Amide(s) A carboxylic acid derivative in which the carboxyl —OH group is replaced by an amino or substituted amino group, 486
chemical reaction summary for, 494
classification of, 486
commonly encountered, 488
cyclic, 487
hydrogen bonding and, 488–490
hydrolysis of, 492–493
IUPAC-common name contrast for, 487
line-angle drawings for, 487
nomenclature for, 487
physical properties of, 488–490
physical-state summary for, 490
primary, 486
saponification of, 492–493
secondary, 486
synthesis of, from carboxylic acids, 490–492
tertiary, 486
Amine(s) Organic derivatives of ammonia (NH_3) in which one or more hydrogen atoms on the nitrogen atom have been replaced by an alkyl, a cycloalkyl, or an aryl group, 472
basicity of, 476–477
biologically important, selected, 482–485
classification of, 472–473
heterocyclic, 481–482
hydrogen bonding and, 475–476
IUPAC-common names contrast for, 475
line-angle drawings for, 472
neurotransmitter function for, 482–484
nomenclature for, 473–475
physical properties of, 475–476
physical-state summary for, 475
primary, 472–473
secondary, 472–473
synthesis of, with alkyl halides, 480
tertiary, 472–473
Amine salt(s) An ionic compound in which the positive ion is a mono-, di-, or trisubstituted ammonium ion and the negative ion comes from an acid, 478
nomenclature of, 478
synthesis of, with acids, 478–479
uses for, 478–479
Amino acid(s) An organic compound that contains both an amino (—NH_2) group and a carboxyl (—COOH) group, 588
acid-base properties of, 591–595
alpha, 588
biosynthesis of, 782–783
carbon skeletons, degradation of, 781–782
chirality of, 590–591
codons and, 669–671
degradation of, stages in, 772
essential, 590, 772
Fischer projections and, 590–591
glucogenic, 781–782
ketogenic, 781–782
nonessential, synthesis of, 772
nonpolar, 588–589
peptide formation from, 595–596
polar acidic, 588–589
polar basic, 588–589
polar neutral, 588–589
separation of, using electrophoresis, 593–595
standard, name abbreviations for, 589
standard, structures of, 589
transamination and, 773–775
urea cycle and, 776–779
utilization of, 770–772
Amino acid pool The body's total supply of free amino acids available for use, 770–771
fate of acids in, 771–772
Amino sugar(s)
formation of, from cyclic monosaccharides, 532
Ammonium ion
carbamoyl phosphate and, 777
oxidative deamination and, 775–776
AMP (adenosine monophosphate)
structure of, 691
Amphoteric substance(s) A substance that can either lose or accept a proton and thus can function as either an acid or a base, 241
Brønsted-Lowry theory and, 241
examples of, 241
Amylopectin
structure of, 541–542
Amylose
structure of, 540–542
Anabolism All metabolic reactions in which small molecules are put together to form larger ones, 689
Androgens
biological functions of, 575–576
synthesis of, from cholesterol, 762
Anesthetics
ethers as, 393
Antibiotic(s) Substances that kill bacteria or inhibit their growth, 634
penicillins, 635–636
protein synthesis inhibition and, 676
sulfa drugs, 634–635
Antibodies Molecules that counteract specific antigens, 612
immunoglobulin response to, 612–613
Anticodon(s) A three-nucleotide sequence in tRNA that is complementary to the mRNA codon for the amino acid which bonds to the tRNA, 672
codon interaction with, 672
transfer RNA and, 671–672
Antigen Foreign substances, such as bacteria and viruses, that invade the body, 612
Antihistamine(s)
counteracting effects of histamines, 484–485
Antimicrobials
carboxylic acid salt use as, 449–450
Antioxidant(s) An additive that is oxidized instead of the substances to which it has been added, 390
phenols as, 390
rancidity and, 566–567

Antiseptics
phenols as, 389
Apoenzyme(s) The protein portion of a conjugated enzyme, 622
Aqueous solution(s) A solution in which water is the solvent, 183
Arginine
nitric oxide production and, 781
urea cycle and, 776–779
Argininosuccinate
urea cycle and, 777–779
Aromatic hydrocarbon(s) Unsaturated cyclic compounds that do not readily undergo addition reactions, 354
alkylation of, 360
bonding analysis for, 354–356
chemical reaction summary for, 361
fused-ring, 361–362
halogenation of, 360
nitration of, 360–361
nomenclature for, 356–359
physical properties of, 359
sources for, 359–360
sulfonation of, 361
Aromaticity
stability associated with, 356
Arrhenius acid(s) A hydrogen-containing compound that, in water, produces hydrogen ions (H^+), 237
characteristics of, 237–238
ionization of, 238
Arrhenius base(s) A hydroxide-containing compound that, in water, produces hydroxide ions (OH^-), 237–238
characteristics of, 237–238
dissociation of, 238
Arrhenius, Svante August, 237
Artificial fat substitutes, 563
Artificial sweeteners, 537
Aryl group An aromatic ring system from which one hydrogen atom has been removed, 373
Aspartame, 537
Aspartate
urea cycle and, 777–779
Aspirin
pharmacology of, 455
structure of, 454
Atmosphere
pressure unit of, 155
Atom(s) The smallest particle of an element that can exist and still have the properties of the element, 13
as building blocks of matter, 13–14
atomic number for, 52
charge neutrality and, 49
chemical properties of, and electrons, 53
electron configurations for, 65–69
excited electronic states for, 67
ground electronic states for, 67
ion formation, generalizations for, 84–86
limit of chemical subdivision and, 14
mass number for, 52
nucleus of, 49–52
orbital diagrams for, 65–69
relative mass scale for, 55–56
size of, 13
subatomic particle arrangement within, 49–50
Atomic energy, 291
Atomic mass(es) The calculated average mass for the isotopes of an element, expressed on a scale using atoms of ^{12}C as the reference, 55
amu unit for, 55
calculation of, procedure for, 55–56
relative mass scale for, 55
values for, listing of, inside front cover
weighted-average concept and, 55–56
Atomic mass unit
relationship to grams unit, 131
relative scale for, 55
Atomic number(s) The number of protons in an atom's nucleus, 52
electrons and, 52–53
element identification and, 53
for elements, listing of, inside front cover
informational value of, 52–53
protons and, 52–53
use of, with chemical symbols, 54–55
Atomic structure
summary of, 56
ATP (adenosine triphosphate)
consumption of, gluconeogenesis and, 732
consumption of, lipogenesis and, 758
importance of, biochemical energy production and, 708–710
production of, common metabolic pathway and, 707–708
production of, complete oxidation of glucose, 727–728
production of, fatty acid spiral and, 751–753
production of, glycolysis and, 722
structure of, 691
ATP synthase complexes (adenosine triphosphate synthase complexes)
mitochondria and, 690
oxidative phosphorylation and, 707
Atropine, 485
Avogadro, Amedeo, 129–130
Avogadro's number
magnitude of, examples illustrating, 129
use of, in calculations, 129–130

Balanced chemical equation(s) A chemical equation that has the same number of atoms of each element involved in the reaction on each side of the equation, 136
Balanced nuclear equation(s) The sums of the subscripts (atomic numbers or particle charges) on each side of the equation are equal, and the sums of the superscripts (mass numbers) on each side of the equation are equal, 274
examples of, 274–276
Barbiturates
properties of, 488
Barometer A device used to measure atmospheric pressure, 155
Base(s)
Arrhenius, 237–238
Brønsted-Lowry, 238–241
neutralization of, 245
strengths of, 242–243
strong, 242–243
weak, 242–243
Base ionization constant(s) The equilibrium constant corresponding to the ionization for the base, 244
mathematical expression for, 244
Base pairing
complementary nature of, 660
DNA double helix and, 659–660
DNA-RNA, 667
Basic solution(s) A solution in which the concentration of the OH^- ion is higher than that of the H_3O^+ ion, 247
hydronium ion concentration and, 247–248
pH value and, 250–251
Becquerel, Antoine Henri, 272
Benedict's test
aldehyde oxidation and, 419
polysaccharides and, 539
reducing sugars and, 529
Benzene
chemical reaction summary for, 361
Beta particle(s)
biological effects of, 283–284
Beta particle decay
equations for, 274–276
Beta rays Streams of negatively charged particles (beta particles) whose charge and mass are identical to those of an electron, 273
characterization of, 273
notation for, 273
Bile
chemical composition of, 574
Bile pigments
coloration of, 787–788
Bile salt(s) Emulsifying agents that make dietary lipids soluble in the aqueous environment of the digestive tract, 574
biological functions of, 574
structures of, 574
Bilirubin
diglucuronide of, 786
hemoglobin catabolism and, 786
jaundice and, 787–788
Biliverdin
hemoglobin catabolism and, 785
Binary ionic compound(s) An ionic compound in which only two elements are present, 90
naming of, 91–93
Binary molecular compound(s)
common names for, 120–121
naming of, 120–121
Biochemical energy production
acetyl group formation and, 697–698
citric acid cycle and, 697–698
digestion and, 696–697
electron transport chain and, 697–698
overview diagram for, 697
Biochemical substances The chemical substances found within a living organism, 507
bioinorganic, 507
bioorganic, 507
types of, 507
Biochemistry The study of the chemical substances found in living systems and the chemical interactions of these substances with each other, 506
Biotin
biological functions of, 641
structure of, 641
1,3-Bisphosphoglycerate
gluconeogenesis and, 731
glycolysis and, 721
Blood
buffer systems within, 260

lipoproteins within, 614–615
pH change and acidosis, 260
pH change and alkalosis, 260
types of, 530
Blood plasma
effects of hydrolysis reactions on pH, 256
electrolyte composition of, 263
Blood types
monosaccharide markers and, 530
Boiling A special form of evaporation where conversion from the liquid state to the vapor state occurs within the body of a liquid through bubble formation, 166–167
bubble formation and, 167
Boiling point(s) The temperature at which the vapor pressure of a liquid becomes equal to the external (atmospheric) pressure exerted on the liquid, 167
elevation of, 195–196
factors affecting magnitude of, 167–168
normal, 167
table of, at various pressures for water, 167
Bombardment reaction(s) A nuclear reaction in which small particles traveling at very high speeds are collided with stable nuclei, causing them to undergo nuclear change, 279
equations for, 279–280
synthetic elements and, 279–280
Bond(s); *see* Chemical bonds, Covalent bonds, Ionic bonds
Bond polarity
electronegativity differences and, 117
notations for, 116–117
types of chemical bonds and, 116–117
Bonding electrons Pairs of valence electrons that are shared between atoms in a covalent bond, 104
Boyle, Robert, 155–156
Boyle's law The volume of a sample of a gas is inversely proportional to the pressure applied to the gas if the temperature is kept constant, 155
kinetic molecular theory and, 157
mathematical form of, 155–156
use of, in calculations, 156
Branched-chain alkane An alkane with one or more branches (of carbon atoms) attached to a continuous chain of carbon atoms, 303
Branched-chain alkyl groups, 311–312
Brønsted, Johannes Nicolaus, 238
Brønsted-Lowry acid(s) Any substance that can donate a proton (H^+) to some other substance, 238
characteristics of, 238–239
conjugate bases of, 239–240
Brønsted-Lowry base(s) Any substance that can accept a proton (H^+) from some other substance, 238
characteristics of, 238–239
conjugate acids of, 239–240
Brown fat
heat generation and, 709
thermogenin and, 709
Buffer(s) A solution that resists major changes in pH when small amounts of acid or base are added to it, 256
chemical species required for, 257
extent of pH change for, 259
Henderson-Hasselbalch equation calculations for, 261
mode of action of, 257–259
Butyryl ACP
lipogenesis and, 757–758

Caffeine
pharmacology of, 482
structure of, 482
Calcium
functions of, in human body, 60
similar behavior of other group IIA elements to, 60
Calorie The amount of heat energy needed to raise the temperature of 1 gram of water by 1 degree Celsius, 41
dietetic, 41
relationship to joule, 41
cAMP (cyclic adenosine monophosphate)
structure of, 691
Canal rays
discharge tube experiments and, 51
Cancer
fused-ring aromatic hydrocarbons and, 362
Carbamoyl phosphate
urea cycle and, 777–779
Carbohydrate(s) Polyhydroxy aldehydes, polyhydroxy ketones, or compounds that yield such substances upon hydrolysis, 508
classifications of, 508
complex, 539
functions of, in humans, 507–508
occurrence of, 507
plasma membranes and, 579
photosynthesis and, 507
simple, 539
Carbohydrate metabolism
glycolysis and, 718–724
hormonal control of, 736–738
relationships between lipid metabolism and, 763
relationships between protein metabolism and, 787–789
Carbon atom(s)
bonding characteristics of, 299
classifications of, 311
primary, 311
quaternary, 311
secondary, 311
tertiary, 311
Carbon-carbon multiple bond(s)
nature of, 337–338
structural rigidity of, 338
Carbon dioxide
acetyl CoA formation and, 725
alcohol fermentation and, 725–726
carbamoyl phosphate formation and, 777
citric acid cycle and, 699–700
ketone body formation and, 754–755
lipogenesis and, 756–757
pentose phosphate pathway and, 734
respiration and, 162
Carbon monoxide
hemoglobin catabolism and, 785
properties of, 4
steel-making and, 4
toxicity of, 4
Carbonyl group(s) Consists of a carbon atom and an oxygen atom joined by a double bond, 406
bonding within, 407
polarity of, 407
sulfur-containing, 426–427
Carboxyl group(s) A carbonyl group with a hydroxyl group bonded to the carbonyl carbon atom, 436
notations for, 437
polarity of, 437
Carboxylate ion(s) The negative ion produced when a carboxylic acid loses one or more of its acidic hydrogen atoms, 447
nomenclature for, 447
Carboxylic acid(s) Compounds whose characteristic functional group is the carboxyl group, 436
acidity of, 447
aromatic, 438
biochemically important, listing of, 444
chemical reaction summary for, 459
di-, 438
hydrogen bonding and, 446
IUPAC-common name contrast for, 442
line-angle drawings for, 438
nomenclature for, 437–442
physical properties of, 445–446
physical-state summary for, 447
polyfunctional, 443–445
reactions of, with alcohols, 450–451
salts of, 448–450
synthesis of, from alcohols, 387–388
synthesis of, from alcohols, 446–447
synthesis of, from aldehydes, 446–447
Carboxylic acid salt(s) An ionic compound in which the negative ion is a carboxylate ion, 448
nomenclature of, 448
reaction of, with acids, 449
solubilities of, 449
synthesis of, 448–449
uses for, 449–450
β-Carotene,
Vitamin A and, 639
Carotenoids
color and, 345
Catabolism All metabolic reactions in which large molecules are broken down to smaller ones, 689
Catalyst(s) A substance that increases a reaction rate without being consumed in the reaction, 220
activation energy and, 220–221
position of equilibrium and, 231
reaction rates and. 220–221
Cathode rays
discharge tube experiments and, 50
Cell(s)
cytoplasm of, 689
eukaryotic, characteristics of, 689–690
organelles and, 689–690
prokaryotic, characteristics of, 689
Cellobiose, hydrolysis of, 534
occurrence of, 534
structure of, 534
Cellulose
dietary fiber and, 540
properties of, 540
structure of, 540
Cephalins, 569
Cerebrosides, 571
Chain, Ernest, 634

Change(s)
chemical, 3–4
in matter, examples of, 4–5
physical, 3–4
Change(s) of state A process in which a substance is transformed from one physical state to another, 164
endothermic, 164
exothermic, 164
summary diagram of, 164
Charles, Jacques, 157–158
Charles's law The volume of a sample of gas is directly proportional to its Kelvin temperature if the pressure is kept constant, 157
kinetic molecular theory and, 158–159
mathematical form of, 158
use of, in calculations, 158
Chemical
use of the term, 5
Chemical bond(s) The attractive forces that hold atoms together in more complex units, 80
conjugated, 345
covalent bond model for, 102–103
delocalized, 355–356
ionic bond model for, 83–84
Lewis structures and, 86–87
octet rule and, 82–83
pi, 337–338
polarity of, 116–118
resonance structures and, 355
sigma, 337–338
strained, 695–696
types of, 80
Chemical change A process that involves a change in the basic nature (chemical composition) of the substance, 5
characteristics of, 5
control of, 5
Chemical equation(s) A written statement that uses symbols and formulas instead of words to describe the changes that occur in a chemical reaction, 136
conventions used in writing, 136
general calculations involving, 140–145
macroscopic level interpretation of, 140
microscopic level interpretation of, 140
molar interpretation of, 140
procedures for balancing, 137–139
special symbols used in, 139–140
Chemical equilibrium The process wherein two opposing chemical reactions occur simultaneously at the same rate, 221
conditions necessary for, 221–223
contrasted with physical equilibrium, 221
equilibrium constants and, 224–227
Le Chatelier's principle and, 227–231
position of, 226–227
Chemical formula(s) A notation made up of the symbols of the elements present in a compound and the numerical subscripts (located to the right of each symbol) that indicate the number of atoms of each element present in a molecule of the compound, 14
atomic interpretation of, 14–15
general calculations involving, 133–135
macroscopic level interpretation of, 132
microscopic level interpretation of, 132
molar interpretation of, 132–133
parenthesis use and, 14–15
subscripts in, 14–15
Chemical nomenclature; *see* Nomenclature
Chemical property(ies) A characteristic of a substance that describes the way the substance undergoes or resists change to form a new substance, 3
conditions that affect, 4
electrons and, 53
examples of, 3–4
Chemical reaction(s) A process in which at least one new substance is produced as a result of a chemical change, 208
addition, 209, 344–349
combination, 209
combustion, 320
completeness of, 226–227
contrasted with nuclear reactions, 293
coupled, 706
decomposition, 209
dehydration, 384
double-replacement, 210–211
elimination, 209, 384
endothermic, 218
equilibrium constants for, 224–227
exothermic, 218
halogenation, 320–321
hydrolysis of salts, 252–256
nonredox, 211
oxidation-reduction, 211–216
rate of, factors affecting, 218–221
redox, 211–216
reversible, 223–224
single-replacement, 209–210
substitution, 210, 320
Chemical stoichiometry, 143
Chemical subdivision
limit of, 14
Chemical symbol(s) A one- or two-letter notation used to represent the name of an element, 12
capitalization rules for, 12
generalizations concerning, 12
listing of, 12
Chemiosmotic coupling An explanation for the coupling of ATP synthesis with reactions of the electron transport chain that requires a proton gradient across the inner mitochondrial membrane, 707
concepts involved in, 707
electrochemical gradient and, 707–708
Chemistry The field of study concerned with the characteristics, composition, and transformations of matter, 1
as study of matter, 1–2
scope of, 2
Chiral center(s) An atom in a molecule that has four different groups tetrahedrally bonded to it, 510
identification of, in molecules, 510–511
interactions between, 518–520
Chiral object(s) An object that is not identical to its mirror image, 509
examples of, 509
Chitin
properties of, 543
structure of, 543
Chlorofluorocarbons
ozone layer and, 324
Cholesterol
biological functions of, 574
biosynthesis of, 759–762
blood levels and "trans" fatty acids, 566
component of plasma membranes, 579
human body levels of, 573
plasma levels of, statins and, 761
structure of, 573
Chromosome(s) An individual DNA molecule bound to a group of proteins, 663
cell division and, 664
general properties of, 663
Chylomicron(s) A lipoprotein that transports triacylglycerols from intestinal cells to the bloodstream, 744
lipid digestion and, 744
Chyme
lipid digestion and, 743–745
***Cis* isomer** An isomer in which two atoms or groups are on the same side of a restricted rotation "barrier" in a molecule, 315
***Cis-trans* isomer(s)** Compounds that have the same molecular and structural formulas but different arrangements of atoms in space because of restricted rotation around bonds, 315
for alkenes, 340–343
for cycloalkanes, 315–316
vision and, 342
Citrate
citric acid cycle and, 698–700
Citric acid cycle The series of reactions in which the acetyl portion of acetyl CoA is oxidized to carbon dioxide and the reduced coenzymes $FADH_2$ and NADH are produced, 698
important features of, 702
linkage to urea cycle, 779–780
reactions of, 698–702
regulation of, 702
Citrulline
nitric oxide production and, 781
urea cycle and, 776–779
Clone(s) Cells that have descended from a single cell and have identical DNA, 682
recombinant DNA production and, 682
Codeine, 486
Codon(s) A sequence of three nucleotides in an mRNA that codes for a specific amino acid, 669
amino acids and, 669–671
anticodon interaction with, 672
listing of, 670
messenger RNA and, 669–671
Coefficient(s) A number that is placed to the left of the formula of a substance and that changes the amount, but not the identity of the substance, 137
equation, determination of, 137–139
in scientific notation, 29–30
Coenzyme A (CoA)
general structural characteristics of, 694–695
Coenzyme Q (CoQ)
electron transport chain and, 704–705
Coenzyme(s) A small organic molecule that serves as a cofactor in a conjugated enzyme, 622
Cofactor(s) The nonprotein portion of a conjugated enzyme, 622

Cohesive forces
potential energy and, 152–155
Collagen
gelatin from, 611–612
nonstandard amino acids and, 611
occurrence of, 611
secondary structure of, 601–602
Colligative properties Physical properties of a solution that depend only on the number (concentration) of solute particles (molecules or ions) in a given quantity of solvent and not of their chemical identities, 195
examples of use of, 195–196
selected, listing of, 195
summary diagram for, 199
Collision theory A set of statements that give the conditions that must be met before a chemical reaction will take place, 216
concepts involved in, 216–217
Colloidal dispersion(s) A dispersion of small particles of one substance in another substance, 194
characteristics of, 194–195
terminology associated with, 194–195
Colostrum
immunoglobin content of, 614
Combination reaction(s) A reaction in which a single product is produced from two or more reactants, 209
examples of, 209
Combined gas law An expression obtained by mathematically combining Boyle's and Charles's laws, 159
mathematical form of, 159
use of, in calculations, 159–160
Combustion
of alkanes, 320
of cycloalkanes, 321
Combustion reaction A reaction between a substance and oxygen (usually from air) that proceeds with the evolution of heat and light (usually as a flame), 320
Common metabolic pathway The sum of the reactions of the citric acid cycle, the electron transport chain, and oxidative phosphorylation, 698
ATP production and, 707–708
citric acid cycle and, 698–702
electron transport chain and, 702–705
oxidative phosphorylation and, 706–707
Competitive enzyme inhibitor A molecule that sufficiently resembles an enzyme substrate in shape and charge distribution that it can compete with the substrate for occupancy of the enzyme's active site, 630
mode of action, 630
Complementary base(s) Specific pairs of bases in nucleic acid structures that hydrogen-bond to each other, 660
structural characteristics of, 659–660
Compound(s) A pure substance that can be broken down into two or more simpler substances by chemical means, 8
binary ionic, 91–93
characteristics of, 7–9
classification of, for naming purposes, 121
comparison with mixtures, 8
dextrorotatory, 518
formula mass of, 127–128
heteroatomic molecules and, 13–14
high-energy, 695–696
inorganic, contrasted with organic, 298–299
levorotatory, 518
molar mass of, 130–132
number of known, 7
optically active, 518
organic contrasted with inorganic, 298–299
Compressibility
and states of matter, 150–151
Concentrated solution(s) A solution that contains a large amount of solute relative to the amount that could dissolve, 183
Concentration(s) The amount of solute present in a specified amount of solution, 186
units for, 186–193
Concentration units
mass-volume percent, 188–189
molarity, 190–193
percent by mass, 186–188
percent by volume, 188
Condensation
process of, 164
Condensation polymer(s) A polymer formed by reacting bifunctional monomers to give a polymer and some small molecule, such as water, 460
diacid-dialcohol, 460–461
diacid-diamine, 494–495
polyamides, 494–495
polyesters, 460–461
polyurethanes, 495–496
Condensation reaction(s)
esterification, 450
Condensed structural formula Uses groupings of atoms, in which central atoms and the atoms connected to them are written as a group, to convey molecular structural information, 301
Conformation(s) Differing orientations of a molecule made possible by rotation about single bonds, 304
of alkanes, 304–306
Conjugate acid(s) The species formed when a base accepts a proton, 240
Brønsted-Lowry theory and, 239–240
relationship to conjugate bases, 240
Conjugate acid-base pair(s) Two species that differ by one proton, 240
Brønsted-Lowry theory and, 239–240
determining members of, 240
Conjugate base(s) The species that remains when an acid loses a proton, 240
Brønsted-Lowry theory and, 239–240
relationship to conjugate acids, 240
Conjugated double bonds
color and, 345
Conjugated enzyme(s) Has a nonprotein portion in addition to a protein portion, 622
Conjugated protein(s) Has other chemical components in addition to amino acids, 608
Continuous-chain alkane An alkane in which all carbon atoms are connected in a continuous nonbranching chain, 303
Conversion factor(s) A ratio that specifies how one unit of measurement is related to another, 32
Avogadro's number, use of in, 129–130
characteristics of, 36
density, use of as a, 38–39
dimensional analysis, use of in, 34–35
English-English, 32
equation coefficients, use of in, 140–145
formula subscripts, use of in, 132–133
mass-volume percent, use of as a, 189
metric-English, listing of, 33
metric-metric, 32–33
molar mass, use of in, 130–131
molarity, use of as a, 191–192
percent by mass, use of as a, 187
significant figures and, 32–33
Coordinate covalent bond(s) A bond in which both electrons of a shared pair come from one of the two atoms involved in the bond, 106
general characteristics of, 106
molecules containing, examples of, 106
Copper
biological function of, 646
Cori, Carl, 732
Cori, Gerty, 732
Cori cycle A cyclic process in which glucose is converted to lactate in muscle tissue, the lactate is reconverted to glucose in the liver, and the glucose is returned to the muscle, 732
steps in, 732–733
Coupled reaction(s) Pairs of chemical reactions in which energy released from one reaction changes the equilibrium position of a second reaction, 706
Covalent bond(s) A bond that results from the sharing of one or more pairs of electrons between atoms, 80
contrasted with ionic bonds, 102–103
coordinate, 106
double, notation for, 104–105
Lewis structures, 103–105
multiple, 104
nonpolar, 116–118
polar, 116–118
single, notation for, 104
triple, notation for, 104–105
types of, summary diagram, 118
Crenation
hypotonic solutions and, 199
Crick, Francis, 658
Crotonyl ACP (Crotonyl acyl carrier protein)
lipogenesis and, 757–758
Curie, Marie, 272
Curie, Pierre, 272
Cyanide poisoning
electron transport chain and, 706
Cycloalkane(s) A saturated hydrocarbon in which the carbon atoms are connected to one another in a cyclic arrangement, 312
combustion of, 321
general formula for, 312
halogenation of, 321
natural sources of, 316–317
nomenclature of, 313–314
physical properties of, 317–319
physical-state summary for, 319
structural formulas for, 312–313
Cycloalkene(s) A cyclic unsaturated hydrocarbon that contains one or more carbon-carbon double bonds within the ring system, 334
general molecular formula for, 334
nomenclature for, 335
physical-state summary for, 344

Cyclosporine, 613
Cystosine
  as nucleotide subunit, 654
Cytochrome(s) Heme-containing proteins that undergo reversible oxidation and reduction of their iron atoms, 705
  electron transport chain and, 703–705
  structural characteristics of, 705

Dalton, John, 161
Dalton's law of partial pressures The total pressure exerted by a mixture of gases is the sum of the partial pressures of the individual gases, 161
  mathematical form of, 161
  use of, in calculations, 162–163
Daughter nuclide The nuclide that is produced as a result of a radioactive decay process, 273
Decomposition reaction(s) A reaction in which a single reactant is converted into two or more simpler substances (elements or compounds), 209
  examples of, 209
Dehydration reaction(s) A reaction in which the components of water (H and OH) are removed from a single reactant or from two reactants (H from one and OH from the other), 384
  for alcohols, 384–386
Delocalized bond A covalent bond in which electrons are shared among three or more atoms, 355
Denatured alcohol, 378
Density The ratio of the mass of an object to the volume occupied by that object, 36
  of human body, and percent body fat, 38
  terminology associated with, 37
  units for, 37
  use of, as a conversion factor, 38–39
  values, table of, 37
Deoxyribonucleic acid; *see* DNA
Deoxyribose
  as nucleotide subunit, 653
  occurrence of, 525
  structure of, 525
Deposition, 164
Detergents
  cleansing action of, 565
Deuterium, 54
Dextrorotatory compound(s) A chiral compound that rotates the plane of polarized light in a clockwise (to the right) direction, 518
  notation for, 518
Dextrose, 524
Diabetes
  glucose tolerance test for, 737
  testing, aldehyde oxidation and, 418
  testing, reducing sugars and, 529
  type I, 737
  type II, 737
Diacylglycerol(s)
  use of, as emulsifiers, 564
Dialysis The process in which a semipermeable membrane permits the passage of solvent, dissolved ions, and small molecules but blocks the passage of colloidal particles and large molecules, 201
  artificial-kidney machines and, 202
  relationship to osmosis, 201–202
Diastereomer(s) Stereoisomers whose molecules are not mirror images of each other, 511
  epimers and, 514
  examples of, 512
  Fischer projections and, 513–514
  recognizing, 516–517
Dicarboxylic acid(s) A carboxylic acid that contains two carboxyl groups, one at each end of a carbon chain, 439
  nomenclature for, 439–442
  physical-state summary for, 447
Dietary fiber
  cellulose as, 540
Digestion The breakdown of food molecules, through hydrolysis, into simpler chemical units that can be used by cells for their metabolic needs, 715
  carbohydrate, 715–717
  lipid, 743–745
  protein, 769–770
Dihydroxyacetone phosphate
  gluconeogenesis and, 731
  glycerol metabolism and, 746–747
  glycolysis and, 720–721
Dilute solution(s) A solution that contains a small amount of solute relative to the amount that could dissolve, 183
Dilution The process in which more solvent is added to a solution in order to lower its concentration, 193
  mathematical equations associated with, 194
  process of, 193–194
Dimensional analysis A general problem-solving method in which the units associated with numbers are used as a guide in setting up calculations, 34
  metric-English conversions and, 35
  metric-metric conversions and, 34–35
  procedural steps in, 34
Dimethylsulfoxide (DMSO)
  uses of, 427
Dipole-dipole interactions Electrostatic attractions between polar molecules, 168
  characteristics of, 168–169
Diprotic acid(s) An acid that can transfer two H^+ ions (two protons) per molecule during an acid-base reaction, 241
  examples of, 241
Disaccharide(s) Carbohydrates composed of two monosaccharide units covalently bonded to each other, 508
  biologically important, 532–538
  glycosidic linkage within, 533
Discharge tube experiments
  canal rays and, 51
  cathode rays and, 50
  electrons/protons and, 50–51
  experimental set-up for, 50–51
Disinfectants
  phenols as, 389
Dispersed phase
  colloidal dispersions and, 194
Dispersing medium
  colloidal dispersions and, 194
Disruptive forces
  kinetic energy and, 152–155
Dissociation The process in which individual positive and negative ions are produced from an ionic compound that is dissolved in solution, 238
  Arrhenius bases and, 238
Distinguishing electron(s) The last electron added to an electron configuration when subshells are filled in order of increasing energy, 70
  identity of, and periodic table, 71
Disulfide bonds
  protein tertiary structure and, 603
Disulfide(s)
  nomenclature for, 397
  reduction of, 397
  synthesis of, from thiols, 397
DNA (deoxyribonucleic acid)
  backbone of, structure for, 656
  complementary strands of, 660
  differences between RNA and, 666
  double helix, base-pairing and, 659–660
  double helix, general structure of, 658–659
  mutations and base sequence for, 676
  polymerase chain reaction and, 680
  predicting base sequence, complementary strands and, 660
  recombinant, 678–682
  replication of, 660–662
  sequencing of, 680
DNA replication The process by which DNA molecules produce exact duplicates of themselves, 660–661
  bidirectional nature of, 662–663
  enzymes for, 661–662
  key concepts in, 661–662
  Okazaki fragments and, 661–662
  summary-diagram for, 665
  synthetic bases and, 662
DNA sequencing A process in which the exact base sequence in a DNA molecule (or a portion of it) is determined, 680
  human genome project and, 680
Double covalent bond(s) A bond in which two atoms share two pairs of electrons, 104
  molecules containing, examples of, 105
  strength of, 105
Double-replacement reaction(s) A reaction in which two substances exchange parts with one another and form two different substances, 210
  examples of, 210–211
Drinking alcohol, 377

Earth's crust
  elemental composition of, 10–11
Eicosanoid(s) Oxygenated derivatives of polyunsaturated 20-carbon fatty acids, 577
  biological functions of, 577
  important classes of, 577
  leukotrienes as, 578
  prostaglandins as, 577–578
  thromboxanes as, 578
Electrochemical gradient
  chemiosmotic coupling and, 707–708
Electrolyte(s) A substance that forms a solution in water that conducts electricity, 262
  body fluids and, 263
  strength of, 262
  strong, 262
  weak, 262

Electron(s) Subatomic particles that possess a negative electrical charge, 49
bonding, 104
canal rays and, 50
discharge tube experiments and, 50
discovery of, 50
distinguishing, 70
excited states for, 67
ground state for, 67
location of, within atom, 49
nonbonding, 104
orbitals for, 63–64
properties of, 49
sharing of, 103–104
shells for, 61, 63–64
spin of, 64–65
subshells for, 61, 63–64
transfer of, 83–87
valence, 80–82
Electron cloud
concept of, 49
Electron configuration(s) A statement of how many electrons an atom has in each of its subshells, 65
interpretation of, 65–69
periodic law and, 69–70
periodic table location of elements and, 70–71
procedures for writing, 65–69
valence electrons and, 81
Electron excitation
practical uses of, 62
Electron orbital(s) A region of space within an electron subshell where an electron with a specific energy is most likely to be found, 63
maximum electron occupancy of, 63–64
occupancy of, rules for, 65
overlap of, in bond formation, 103
size/shape of, 63–64
Electron shell(s) A region of space about a nucleus that contains electrons that have approximately the same energy and that spend most of their time approximately the same distance from the nucleus, 61
energy order for, 61, 63–64
notation for, 61, 63–64
number of subshells within, 61, 63–64
outermost, 80
Electron spin, 64–65
Electron subshell(s) A region of space within an electron shell that contains electrons that have the same energy, 61
energy order for, 61, 63–65
maximum electron occupancy of, 63–64
notation for, 61, 63–64
number of orbitals within, 63–64
types of, 61, 63–64
Electron transport chain (ETC) A series of reactions in which electrons and hydrogen ions from NADH and $FADH_2$ are passed to intermediate carriers and ultimately react with molecular oxygen to produce water, 702
cyanide poisoning and, 706
electron-carriers for, 703
fixed enzyme complexes and, 703
reactions of, 702–705
Electronegativity A measure of the relative attraction that an atom has for the shared electrons in a bond, 115
use of, in determining bond type, 117
use of, in determining molecular polarity, 119–120
values, listing of, 116
values, periodic trends in, 116
Electrophoresis The process of separating charged molecules on the basis of their migration toward charged electrodes, 593
apparatus diagram, 594
Electrostatic interactions Attractions and repulsions that occur between charged particles, 151
as a form of potential energy, 151
protein tertiary structure and, 603–604
Element(s) A pure substance that cannot be broken down into simpler substances by ordinary chemical means such as a reaction, an electric current, heat, or a beam of light; a pure substance in which all atoms present have the same atom number, 8, 53
abundances of, in different realms, 9–11
characteristics of, 7–9
chemical symbols for, 12–13
classification systems for, 71–73
discovery of, 9
groups of, within periodic table, 58
homoatomic molecules and, 13
inner-transition, 71–73
isotopic forms for, 53–57
listing of, 12, inside front cover
metallic, 59–61, 72
naming of, 11
necessary for human life, listing of, 11
noble gases, 71–72
nonmetallic, 59–61, 72
number of known, 7–8
periods of, within periodic table, 58
representative, 71–72
synthetic (laboratory-produced), 9
synthetic, listing of, 280
synthetic, uses for, 279–280
trace, 62
transition, 71–73
transuranium, 280
Elimination reaction(s) A reaction in which two groups of two atoms on neighboring carbon atoms are removed, or eliminated, from a molecule, leaving a multiple bond between the carbon atoms, 384
for alcohols, 384–386
Emulsifier(s) A substance that can disperse and stabilize water-insoluble substances as colloidal particles in an aqueous solution, 744
bile salts as, 574
lipid digestion, need for, 744
mono- and diacylglycerols as, 564
phosphoacylglycerol(s) as, 569
Enantiomer(s) Stereoisomers whose molecules are nonsuperimposable mirror images of each other, 511
examples of, 512
Fischer projections and, 513–514
interaction with plane-polarized light, 518
properties of, 517–518
recognizing, 516–517
Endorphins, 598
Endothermic change(s) of state A change of state that requires the input (absorption) of heat energy, 164
examples of, 164
Endothermic chemical reaction(s) A chemical reaction that requires the continuous input of energy as the reaction occurs, 218
Energy
activation, 216–217
atomic, 290
contrasted with matter, 1–2
free, 695
heat, units for, 41
kinetic, 151–155
nuclear, 290–291
potential, 151–155
Enkephalins, 598
Enol
structural characteristics of, 722
Enzymatic browning
factors that control, 627
Enzyme(s) Catalysts for biological reactions, 621
absolute specificity of, 626
action of, induced-fit model for, 624–625
action of, lock-and-key model for, 624
active site of, 623
allosteric, 633–634
carbohydrate digestion and, 716
citric acid cycle and, 698–702
classes of, by function, 623
classes of, by structure, 622
conjugated, 622
DNA replication and, 661–662
enzyme-substrate complex for, 623
general characteristics of, 621
gluconeogenesis and, 730–732
glycogenesis and, 728–730
glycogenolysis and, 730
glycolysis and, 718–724
group specificity of, 626
linkage specificity of, 626
medical uses for, 636–637
models for action of, 623–625
nomenclature of, 621–622
oxidative deamination and, 775–776
protein digestion and, 769–770
restriction, 680–681
RNA synthesis and, 667
simple, 622
specificity of, types of, 625–626
stereochemical specificity of, 626
terminology associated with, 622
transamination and, 774
turnover number and, 628
Enzyme activity A measure of the rate at which an enzyme converts substrate to products, 626
enzyme concentration and, 628
factors that affect, 626–629
pH and, 626
regulation of, 632–633
saturation curve and, 627
substrate concentration and, 627–628
summary-diagram for, 629
temperature and, 626
Enzyme inhibition
antibiotics and, 634–636
irreversible, 631
reversible competitive, 630
reversible noncompetitive, 630–631
summary-diagram for, 633

Enzyme inhibitor(s) A substance that slows or stops the normal catalytic function of an enzyme by binding to it, 630
types of, 630–631
Enzyme regulation
allosteric enzymes and, 632–633
feedback control and, 633
zymogens and, 633–634
Enzyme specificity
types of, 626
Enzyme-substrate complex The intermediate reaction species that is formed when a substrate binds to the active site of an enzyme, 623
Epimer(s) Diastereomers that differ only in the configuration at one chiral center, 514
examples of, 514
Epinephrine
biological functions of, 736–738
CNS stimulant function for, 484
Equation coefficient(s)
determination of, 137–139
molar interpretation of, 140
use of, in chemical calculations, 140–145
Equilibrium; *see* Chemical equilibrium
Equilibrium constant(s) A numerical value that characterizes the relationship between the concentrations of reactants and products in a system at chemical equilibrium, 224
magnitude, calculation of, 226–227
rules for writing, 224–226
temperature dependence of, 226
value of, and reaction completeness, 226–227
Essential amino acids Amino acids that must be obtained from food, 590
listing of, 590
Essential fatty acid(s) Fatty acids that are needed by the human body and must be obtained from dietary sources because they cannot be synthesized within the body from other substances, 557
biological functions of, 759
importance of, 557–558
Ester(s) An organic compound whose characteristic functional group is RCOR, 450
chemical reaction summary for, 459
commonly encountered, 453–455
cyclic, 451
flavoring agent use, 453–454
hydrogen bonding and, 455–456
hydrolysis of, 456–458
inorganic acid, 461–462
IUPAC-common name contrast for, 453
line-angle drawings for, 452
nitrate, 461
nomenclature of, 451–453
notations for, 450
pheromone use, 454
phosphate, 461–462
physical properties of, 455–456
physical-state summary for, 456
saponification of, 457–458
sulfate, 461
sulfur analogs of, 458–459
synthesis of, from carboxylic acids, 450–451
Ester saponification The base catalyzed hydrolysis of an ester, 457
Esterification
condensation reaction of, 450
Estrogens
biological functions of, 575–576
synthesis of, from cholesterol, 762
Ethanol
amount in beverages, 378
fermentation of, 725
from pyruvate, 725
properties of, 377–378
uses of, 377–378
Ethene
industrial uses of, 337
plant hormone function, 337
Ether(s) An organic compound in which an oxygen atom is bonded to two carbon atoms by single bonds, 373
chemical reactions of, 395
cyclic, 395–396
hydrogen bonding and, 394–395
IUPAC-common name contrast, 393
line-angle drawings for, 392
nomenclature of, 392
physical properties of, 394–395
physical-state summary for, 394
sulfur analogs of, 397
synthesis of, from alcohols, 385–386
use of, as anesthetics, 393
use of, as gasoline additive (MTBE), 393–394
Ethylene glycol
properties of, 379
uses of, 379
Evaporation The process by which molecules escape from the liquid phase to the gas phase, 164
factors affecting rate of, 165
in closed container, equilibrium and, 165–166
kinetic molecular theory and, 165
process of, 164
Exercise
fuel consumption and, 752
high-intensity versus low-intensity, 752
Exon(s) A DNA segment that conveys (codes for) genetic information and helps express a genetic message, 668
primary transcript RNA and, 668–669
Exothermic change(s) of state A change that requires heat to be given up (released), 164
examples of, 164
Exothermic chemical reaction(s) One in which energy is released as the reaction occurs, 218
Expanded structural formula Shows, in two dimensions, all atoms in a molecule and all the bonds connecting them, 301
Exponent(s)
use of, in scientific notation, 29–30
Exponential notation; *see* Scientific notation

Facilitated transport Movement of a substance across a membrane by diffusion, with the aid of membrane proteins, from a region of higher concentration to a region of lower concentration without the expenditure of cellular energy, 580
process of, characteristics for, 580
FAD; *see* Flavin adenine dinucleotide
$FADH_2$; *see* Flavin adenine dinucleotide
Faraday, Michael, 356
Fat(s) Triacylglycerol mixtures that are solids or semi-solids at room temperature (25°C), 561
animal, 562
artificial, 563
brown, 709
chemical reactions of, 563–567
dietary considerations and, 563
general structure of, 560
hydrogenation of, 565–566
properties contrasted with oils, 561–563
trans, 566
Fatty acid(s) Monocarboxylic acids that contain long, unbranched hydrocarbon chains generally 12 to 26 carbon atoms in length, 555
biochemical oxidation of, 747–751
biosynthesis of, 756–759
characteristics of, 555
essential, 557–558
lipogenesis and, 756–759
monounsaturated, 555–557
omega-3, 556–558
omega-6, 556–558
physical properties of, 558–559
polyunsaturated, 556–557
saturated, 555–557
trans, and blood cholesterol levels, 566
types of, summary, 559
Fatty acid micelle(s) A tiny spherical droplet containing about 20 fatty acids and/or monoacylglycerols and some bile, 744
lipid digestion and, 744
Fatty acid oxidation
activation and, 747
fatty acid spiral and, 747–751
transport and, 747
Fatty acid spiral A repetitive series of four reactions in which each sequence produces acetyl CoA, $FADH_2$, NADH, and a fatty acid that is shorter than the reactant fatty acid by two carbon atoms, 747
steps in, 748–749
summary-diagram for, 750
unsaturated acids and, 750–751
Feedback control A process in which activation or inhibition of the first reaction in a reaction sequence is controlled by a product of the reaction sequence, 633
Fehling's test
polysaccharides and, 539
reducing sugars and, 529
Ferritin
hemoglobin catabolism and, 784
Fiber
dietary, 540
Fibrous protein(s) A protein that has a long, thin, fibrous shape, 605
occurrence and function, 605–607
Fischer, Emil, 513
Fischer projection(s) A two dimensional notation for showing the spatial arrangement of groups about chiral centers in molecules, 513
amino acids and, 590–591
conventions for drawing, 513–515
D and L designations for, 513–516
monosaccharides and, 513–515
Flatulence
sucrose derivatives and, 538
Flavin adenine dinucleotide ($FAD/FADH_2$)
citric acid cycle and, 698–702
electron transport chain and, 703–704

general structural characteristics of, 692–693
oxidized form of, 692
reduced form of, 692
Flavin mononucleotide (FMN)
electron transport chain and, 703–704
Fleming, Alexander, 634
Flory, Howard, 634
Folate
biological functions of, 640
structure of, 640
Formaldehyde
polymers involving, 425–426
properties of, 413
uses of, 413
Formula mass The sum of the atomic masses of the atoms in a formula, 127–128
calculation of, 128
relationship to molar mass, 130–131
relationship to molecular mass, 128
Formula unit(s)
ionic compounds and, 89
Formula(s)
chemical, 14–15
skeletal, 302–303
structural, 301–303
Free energy
contrasted with simple energy, 695
Free radical A highly reactive uncharged molecular fragment, that is, an uncharged piece of a molecule, 282
Freezing, 164
Freezing point
depression of, 195–196
Freon(s), 323–324
Fructose
from carbohydrate digestion, 716
glycolysis and, 723
occurrence of, 524
structure of, 524
Fructose 1,6-bisphosphate
gluconeogenesis and, 731
glycolysis and 718–720
Fructose 1-phosphate
glycolysis and, 723
Fructose 6-phosphate
gluconeogenesis and, 731
glycolysis and, 718–720
Fumarate
amino acid degradation and, 782
citric acid cycle and, 701
urea cycle and, 777–779
Functional group(s) The part of a molecule where most of its chemical reactions occur, 333
alcohol, 372
alkoxy, 392
amine, 472
amino, 472
carbon-carbon multiple bond, 333
carbonyl, 406
carboxyl, 436–437
carboxylic acid, 436–437
ester, 450–451
ether, 373
hydroxyl, 372
phenol, 372–373
sulfhydryl, 396
sulfoxide, 427
thiocarbonyl, 427
thioester, 458
thioether, 397
thiol, 396–397
Fused-ring aromatic hydrocarbon(s) Aromatic hydrocarbons whose structures contain two or more rings fused together, 361–362
cancer and, 362
examples of, 362

Galactose
carbohydrate digestion and, 716
glycolysis and, 723
occurrence of, 524
structure of, 524
Galactosemia, 535
Gamma rays Not considered to be particles; they are pure energy without charge of mass, 273
biological effects of, 284
characterization of, 273
notation for, 273
Gangliosides, 571
Gas(es) Matter that has an indefinite shape and an indefinite volume, 3
condensation of, 164
density of, common phenomena explained using, 154
deposition of, 164
distinguishing characteristics of, 150–151
distinguishing characteristics of, 3
factors affecting solubility of, 182
kinetic molecular theory of matter applied to, 153–155
respiratory, exchange of, 162
Gas law(s) Generalizations that describe in mathematical terms the relationships among the pressure, temperature, and volume of a specific quantity of gas, 155
Boyle's, 155–157
Charles's, 157–159
combined, 159–160
Dalton's, of partial pressures, 161–164
ideal, 160–161
summary diagram for, 163
variables in, 155
Gaseous state A state characterized by a complete dominance of kinetic energy (disruptive forces) over potential energy (cohesive forces), 153
Geiger counters
radiation detection and, 285–286
Gelatin
from collagen, 611–612
Gene(s) A segment of a DNA molecule that contains the base sequence for the production of a single, specific RNA molecule, which in turn produces a single, specific protein molecule, 668
DNA and, 668
messenger RNA and, 668
Genetic code Gives the assignment of the 64 mRNA codons to specific amino acids (or stop signals), 669
codons and, 669–670
degeneracy of, 670
synonyms and, 670
Genome(s) All of the genetic material in the chromosomes of an organism, 680
human genome project and, 680
Globular protein(s) A protein whose overall shape is roughly spherical or globular, 605
occurrence and function, 605
Glucagon
biological functions of, 736
Glucocorticoids
biological functions of, 576–577
synthesis of, from cholesterol, 762
Glucogenic amino acid(s) An amino acid whose carbon-containing degradation product(s) can be used to produce glucose via gluconeogenesis, 781
listing of, 782
Gluconeogenesis The synthesis of glucose from noncarbohydrate materials, 730
steps in, 730–732
Glucose
Cori cycle and, 732–733
carbohydrate digestion and, 716
gluconeogenesis and, 730–732
glycogenolysis and, 730
metabolism, summary-diagram for, 735
metabolism, terminology associated with, 733
occurrence of, 523–524
oxidation of, to pyruvate, 718–724
structure of, 523
synthesis of, noncarbohydrate materials and, 730–732
Glucose 1-phosphate
glycogenesis and, 729
glycogenolysis and, 730
glycolysis and, 723
Glucose 6-phosphate
gluconeogenesis and, 731
glycogenesis and, 729
glycogenolysis and, 730
glycolysis and, 718
pentose phosphate pathway and, 734–735
Glutamate
derivatives of, transamination and, 773
oxidative deamination and, 775–776
transamination and, 774
Glyceraldehyde
glycolysis and, 723
Glyceraldehyde 3-phosphate
gluconeogenesis and, 731
glycolysis and, 720–721
Glycerol
derivatives of, glycolysis and, 720
properties of, 379–380
uses of, 379–380
Glycerol 3-phosphate
glycerol metabolism and, 746–747
Glycogen
breakdown of, 730
properties of, 542
structure of, 542
synthesis of, 728–730
Glycogenesis The synthesis of glycogen from glucose, 728
steps in, 728–730
Glycogenolysis The breakdown of glycogen into free glucose molecules, 730
steps in, 730
Glycol A diol in which the two —OH groups are on adjacent carbon atoms, 379
Glycolipid(s)
cell membrane interactions and, 544
cerebrosides and gangliosides as, 571

Glycolysis The metabolic pathway by which glucose (a C_6 molecule) is converted into two molecules of pyruvate (a C_3 molecule), 717
reactions of, 718–724
regulation of, 724
six-carbon stage of, 718–720
summary-diagram for, 719
three-carbon state of, 720–724
Glycoprotein(s) Conjugated proteins that contain carbohydrates or carbohydrate derivatives in addition to amino acids, 610
cell membrane interactions and, 544
collagen, 611–612
immunoglobulins, 612–614
Glycoside(s) An acetal or ketal formed from a cyclic monosaccharide, 531
nomenclature of, 531
Glycosidic linkage(s) The carbon-oxygen-carbon bond that joins the two components of a glycoside together, 533
disaccharides and, 533
polysaccharides and, 539–544
types of, 534
Gold-foil experiment
discovery of nucleus and, 51–52
experimental set-up for, 51
Goldstein, Eugene, 51
Grain alcohol, 377
Gram The base unit of mass in the metric system, 23
compared to English system units, 23
Group(s) A vertical column of elements in the periodic table, 58
in periodic table, notation for, 58
Guanine
as nucleotide subunit, 654

Half life(ves) The time required for one-half of any given quantity of a radioactive substance to undergo decay, 276
selected values for, table of, 277
use of, in calculations, 278–279
Halogen(s) The elements in Group VIIA of the periodic table: fluorine, chlorine, bromine, and iodine, 320
periodic table location of, 58, 72
Halogenated alkane(s) An alkane derivative in which one or more halogen atoms are present, 322
IUPAC-common name contrast for, 322
nomenclature for, 322–323
physical properties of, 322
synthesis of, from alcohols, 388
Halogenated cycloalkane(s) A cycloalkane derivative in which one or more halogen atoms are present, 322
synthesis of, from alcohols, 388
Halogenation
of alkanes, 320–321
of alkenes, 346–347
of aromatic hydrocarbons, 360
of cycloalkanes, 321
Halogenation reaction A reaction between a substance and a halogen in which one or more halogen atoms are incorporated into molecules of the substance, 320
Handedness
molecular, notation for, 513–516
molecular, recognition of, 509–511
Haworth, Walter Norman, 527
Haworth projection(s) A two-dimensional notation that specifies the three-dimensional structure of a cyclic form of a carbohydrate, 527
conventions for drawing, 527–528
monosaccharides and, 527–528
HDL (high density lipoprotein)
biological functions of, 614
Heart attacks
enzyme analysis and, 637
Heat energy
units for, 41
Heme
catabolism of, 783–786
structure of, 784
Hemiacetal(s) A compound that has a carbon atom to which a hydroxy group, an alkoxy group, and a hydrogen atom are attached, 421
cyclic, 421
formation of, from monosaccharides, 526–528
reaction of, with alcohols, 423–424
synthesis of, from aldehydes, 421–422
Hemiketal(s) A compound that has a carbon atom to which a hydroxy group, an alkoxy group, but no hydrogen atom are attached, 421
formation of, from monosaccharides, 526–528
reaction of, with alcohols, 423–424
synthesis of, from ketones, 421–422
Hemoglobin
catabolism of, 783–786
Le Chatelier's principle and oxygen transport by, 230
Hemolysis
hypertonic solutions and, 198–199
Henderson-Hasselbalch equation
buffer systems and, 261
Heroin, 486
Heteroatomic molecule(s) A molecule in which two or more kinds of atoms are present, 13
examples of, 14
Heterocyclic amine(s) An organic compound in which nitrogen atoms of amine groups are part of either an aromatic or a nonaromatic ring system, 481
ring systems in, 481
Heterocylic organic compound(s) A cyclic organic compound in which one or more of the carbon atoms in the ring have been replaced with other kinds of atoms, 396
amides as, 487
amines as, 481–482
ethers as, 396
Heterogeneous mixture(s) Matter which contains visibly different parts, or phases, each of which has different physical and chemical properties, 6
characteristics of, 6–9
Heteropolysaccharide A polysaccharide in which more than one (usually two) type of monosaccharide unit is present, 543
High-energy compound(s) A compound with a greater free energy of hydrolysis than a typical compound, 695
phosphate containing, 695–696
strained bonds and, 695–696
Histamine
allergy response from, 484
HMG-CoA (3-hydroxy-3-methylglutaryl CoA)
ketogenesis and, 754
Holoenzyme(s), 622
Homoatomic molecule(s) Molecule in which all atoms present are of the same kind, 13
examples of, 13
Homogeneous mixture(s) Mixture that contains only one visibly distinct phase, which has uniform properties throughout, 6
characteristics of, 6–9
Homopolysaccharide Polysaccharide in which only one type of monosaccharide unit is present, 543
Hormone(s) Chemical messengers produced by ductless glands, 574
adrenocortical, 575–577
sex, 575–576
Human body
calcium-like elements and, 60
elemental composition of, 11
normal temperature for, 42
percent body fat, determination of, 38
response of, to fasting, feasting, and starvation, 788
trace elements, need for, 62
Human genome project, 680
Human hair
denaturation of, 610
Hyaluronic acid
structure of, 543–544
Hydration
of alkenes, 347
Hydrocarbon(s) A compound that contains only carbon and hydrogen atoms, 300
alkanes, 300–312
cycloalkanes, 312–315
derivatives of, 300
saturated, 300
unsaturated, 300
Hydrocarbon derivative(s) A compound that contains carbon, hydrogen, and one or more additional elements, 300
Hydrogen bond(s) An extra-strong dipole-dipole interaction between a hydrogen atom covalently bonded to a small, very electronegative element (F, O, or N) and a lone pair of electrons on another small, very electronegative element (F, O, or N), 169
alcohols and, 382–383
aldehydes and, 415–416
amides and, 488–489
amines and, 475–476
carboxylic acids and, 446
characteristics of, 168–171
DNA base-pairing and, 659–660
effects of, on properties of water, 171
esters and, 455–456
ethers and, 394–395
ketones and, 415–416
predicting occurrence of, 170
protein secondary structure and, 600–602
protein tertiary structure and, 604
RNA and, 666
Hydrogenation reaction(s)
alkenes and, 346
trans fats and, 566
triacylglycerols, 565–566

Hydrohalogenation
of alkenes, 347
Hydrolases
examples of, 623
Hydrolysis The reaction of a substance with water to produce hydronium ion or hydroxide ion or both, 252
Hydrolysis reaction(s) A reaction in which a water molecule is added to a reactant, thereby breaking the reactant into two product molecules, 423
acetals, 423–424
amides, 492–493
disaccharides, 533–538
effects of, on blood plasma pH, 256
esters, 456–458
ketals, 423–424
protein, 608
salts, chemical equations for, 256
salts, guidelines for predicting, 252–256
triacylglycerols, 564
Hydroperoxides
from ethers, 395
Hydrophobic attractions
protein tertiary structure and, 604
β-Hydroxyacyl CoA
fatty acid spiral and, 748–749
β-Hydroxybutyrate
as ketone body, 753
β-Hydroxybutyryl ACP
lipogenesis and, 757–758
Hydroxyl group
in alcohols, 372
Hypertonic solution(s) A solution with a higher osmotic pressure than that within cells, 201
hemolysis and, 198–199
Hypotonic solution(s) A solution with a lower osmotic pressure than that within cells, 201
crenation and, 199

Ideal gas constant
value of, 160
Ideal gas law An equation that includes the quantity of gas in a sample as well as the temperature, pressure and volume, 160
mathematical form of, 160
use of, in calculations, 160–161
Immunoglobulins Glycoproteins produced by an organism as a protective response to the invasion of microorganisms or foreign molecules, 612
general structure of, 612
mode of action, 612–613
Indicator(s) A compound that exhibits different colors depending on the pH of its surroundings, 264
need for, in acid-base titrations, 264
Induced-fit model
enzyme action and, 624
Inner transition elements All the elements of the f area of the periodic table, 73
electronic characteristics of, 73
periodic table positions of, 71–72
Insulin
biological functions of, 736
primary structure of, 599
substitutes for human, 599
Intermolecular force(s) Forces that act between a molecule and another molecule, 168
contrasted with intramolecular forces, 168
dipole-dipole interactions, 168–169
effects of, on physical properties, 173
hydrogen bonds, 168–171
London forces, 171–173
summary diagram of, 172
types of, 168–173
Interstitial fluid
electrolyte composition of, 263
Intracellular fluid
electrolyte composition of, 263
Intron(s) A DNA segment that does not convey (code for) genetic information, but rather interrupts a genetic message, 668
primary transcript RNA and, 668–669
Iodine
biological function of, 646
Iodine test
starch and, 540–541
Ion(s) An atom (or group of atoms) that is electrically charged as a result of the loss or gain of electrons, 83
Lewis structures for, 86–87
magnitude of charge on, 83–84
monoatomic, 94
notation for, 83–84
number of protons and electrons in, 84
polyatomic, 94
types of, 83
Ion pair The electron and positive ion that are produced during an ionization collision between an atom and radiation, 282
Ion product for water The numerical value 1.00×10^{-14} obtained by multiplying together the molar concentrations of H_3O^+ ion and OH^- ion present in pure water, 246
calculations involving, 246–247
numerical value of, 246
Ionic bond(s) A bond that results from the transfer of one or more electrons from one atom or group of atoms to another, 80
contrasted with covalent bonds, 102–103
electronegativity differences and, 117
formation of, 83–87
Lewis structures and, 86–87
Ionic compound(s), binary, naming of, 91–93
formation of, electron transfer and, 86–87
formula units and, 89
formulas for, 87–88, 95–97
general properties of, 79–80
Lewis structures for, 86–87
polyatomic-ion-containing, naming of, 96–97
structure of, 88–90
Ionization The process by which individual positive and negative ions are produced from a molecular compound that is dissolved in a solution, 238
Arrhenius acids and, 238
Iron
biological function of, 645–646
deficiency of, 645
heme and nonheme, 645
use of, in steel, 4
Iron-sulfur protein (FeSP)
electron transport chain and, 704
Irreversible enzyme inhibitor A molecule that inactivates enzymes by forming a strong covalent bond to an amino acid side-chain group at the enzyme active site, 631
mode of action, 631
Isocitrate
citric acid cycle and, 699–700
Isoelectric point The pH at which the concentration of zwitterion ion form of an amino acid is at a maximum, 593
Isoenzyme(s) Forms of the same enzyme with slightly different amino acid sequences, 637
heart attack diagnosis and, 637
Isomer(s)
alcohol-ether, 394
alcohols, 375
aldehyde-ketone, 408
alkane, 303–304
alkene, 339–343
cis-trans, for alkenes, 340–343
cis-trans, for cycloalkanes, 315–316
cycloalkane, 314–316
diastereoisomers, 511–517
enantiomers, 511–517
epimers, 514
ester-carboxylic acid, 451
functional group, 394, 408
positional, 339, 375
skeletal, 302–303
structural, 303–304
types of, summary diagram, 519
Isomerases
examples of, 623
Isopentenyl pyrophosphate
cholesterol biosynthesis and, 760
Isoprene
terpenes and, 343–344
Isopropyl alcohol
properties of, 378–379
uses of, 378–379
Isotonic solution(s) A solution whose osmotic pressure is equal to that within cells, 199
examples of, 200–201
Isotope(s) Atoms of an element that have the same number of protons and electrons but different numbers of neutrons, 53–54
chemical properties of, 54
distinguishing among, for an element, 54–55
hydrogen, properties of, 54
known number of, 55
notation for, 54–55
percentage abundances of, for selected elements, 57
physical properties of, 54
relative masses of, for selected elements, 57

Jaundice
bilirubin concentrations and, 787–788
Joule
metric unit, for heat energy, 41
relationship to calorie, 41

Kekulé, August, 354–355
Kelvar, 495
α-Keratin
structure of, 607

Ketal(s) A compound that has a carbon atom to which two alkoxy groups and no hydrogen atom are attached, 423
formation of, from cyclic monosaccharides, 530–531
hydrolysis of, 423–424
synthesis of, from hemiketals, 423
β-Ketoacyl CoA
fatty acid spiral and, 749
Ketogenesis The synthesis of ketone bodies from acetyl CoA, 753
Ketogenic amino acid(s) An amino acid whose carbon-containing degradation product(s) can be used to produce ketone bodies, 781
listing of, 782
α-Ketoglutarate
amino acid degradation and, 782
citric acid cycle and, 700
Ketone(s) A compound that has two carbon atoms attached to the carbon atom of a carbonyl group, 407
chemical reaction summary for, 425
commonly encountered, 413–414
cyclic, 408
hemiketal formation and, 421–422
IUPAC-common name contrast for, 413
lachrymatory, 412
line-angle drawings for, 411
melanin (sunburn) and, 415
nomenclature of, 410–413
notations for, 407
oxidation of, 418–419
physical properties of, 415–416
physical-state summary for, 417
reaction with alcohols, 420–422
reaction with H_2, 419–420
reduction of, 419–420
synthesis, from alcohols, 417–418
synthesis of, from alcohols, 387–388
thio, 427
Ketone bodies The substances acetoacetate, β-hydroxybutyrate, and acetone, 753
formation of, ketogenesis and, 753–755
ketosis and, 755
Ketose(s) Monosaccharides that contain a ketone group, 521
common, listing of, 523
Ketosis
ketone body formation and, 755
Kinetic energy Energy that matter possesses because of its motion, 151
disruptive forces and, 152–155
kinetic molecular theory of matter and, 151–152
Kinetic molecular theory of matter A set of five statements that are used to explain the physical behavior of the three states of matter (solids, liquids, and gases), 151
applied to gaseous state, 153–155
applied to liquid state, 153
applied to solid state, 152–153
Boyle's law and, 157
Charles's law and, 158–159
concepts associated with, 151
evaporation and, 165
kinetic energy and, 151–152
osmosis and, 197
potential energy and, 151–152
Krebs, Hans A., 698
Lachrymator(s)
aldehydes and ketones as, 412
Lactate
anaerobic accumulation of, 726
Cori cycle and, 732–733
from pyruvate, 725
Lactones
cyclic ester structure for, 451
Lactose
hydrolysis of, 536
occurrence of, 536
structure of, 535
Lactose intolerance, 535
Lanosterol
cholesterol biosynthesis and, 760–762
Law(s)
Boyle's, 155–157
Charles's, 157–159
combined gas, 159–160
conservation of mass, 136
Dalton's, of partial pressures, 161–164
ideal gas, 160–161
periodic, 57
Law of conservation of mass, 136
LDL (low density lipoprotein)
biological functions of, 614
Le Chatelier, Henri Louis, 227
Le Chatelier's principle Statement that if a stress (change of condition) is applied to a system in equilibrium, the system will readjust (change the position of equilibrium) in the direction that best reduces the stress imposed on it, 227
catalyst addition and, 231
concentration changes and, 227–228
pressure changes and, 229–230
temperature changes and, 228–229
Length, conversion factors involving, 33
metric units of, 23
Leukotriene(s) 20-carbon fatty acid derivatives that contain three conjugated double bonds and hydroxy groups, 578
biological functions of, 578
structure of, 577–578
Levorotatory compound(s) A chiral compound that rotates the plane of polarized light in a counterclockwise (to the left) direction, 518
notation for, 518
Levulose, 524
Lewis, Gilbert N., 81–82
Lewis structure(s) An element's symbol with one dot for each valence electron placed around the elemental symbol, 80–81
generalizations concerning, 81–82
ionic compounds and, 86–87
molecular compounds and, 103–105
notation used in, 80–82
systematic procedures for drawing, 106–109
valence electrons and, 80–82
Ligases
examples of, 623
Light
plane-polarized, 518–519
Line-angle drawing An abbreviated structural formula representation in which an angle represents a carbon atom and a line represents a bond, 312
alcohols, 373
aldehydes, 409
alkanes, 313
alkenes, 335
amides, 487
amines, 472
carboxylic acids, 438
cycloalkanes, 312–313
esters, 452
ethers, 392
ketones, 411
Lipid(s) A structurally heterogeneous group of substances of biological origin that are only sparingly soluble, if at all, in water but are soluble in nonpolar organic solvents, 555
characteristics of, 554–555
classes of, 555
eicosanoids, 577–578
fatty acids, building blocks for, 555–560
fats, 559–563
functions of, in humans, 554–555
membrane structure and, 578–580
metabolism of, 743–762
nonsaponifiable, 572
oils, 559–563
phosphoacylglycerols, 567–569
saponifiable, 555
sphingolipids, 570–571
steroids, 573–577
triacylglycerols, 559–566
waxes, 569–570
Lipid bilayer A two-layer thick structure of lipid molecules in which the nonpolar tails of the lipids are in the middle and the polar heads are on the outside surface, 578
plasma membrane structure and, 578–579
Lipid digestion
steps in, 745
Lipid metabolism
relationships between carbohydrate metabolism and, 763
relationships between protein metabolism and, 787–789
Lipogenesis The synthesis of fatty acids from acetyl CoA, 756
contrasted with fatty acid spiral, 756
steps in, 756–759
unsaturated fatty acids and, 758–759
Lipoprotein(s) Conjugated proteins composed of both lipids and proteins, 615
HDL and LDL as, 614
Liquid(s) Matter that has an indefinite shape and a definite volume, 2
boiling of, 166–167
distinguishing characteristics of, 2–3, 150–151
evaporation of, 164
freezing of, 164
intermolecular forces in, 168–173
kinetic molecular theory of matter applied to, 153
vapor pressures of, 166
Liquid state A state characterized by potential energy (cohesive forces) and kinetic energy (disruptive forces) of about the same magnitude, 153
Liter The base unit of volume in the metric system, 24
compared to English system units, 23
Lock-and-key-model
enzyme action and, 624

London, Fritz, 171
London force(s) Instantaneous dipole-dipole interactions that exist between all atoms and molecules, polar as well as nonpolar, 171
characteristics of, 171–173
Lone electron pairs (lone pairs); *see* Nonbonding electrons
Lovastatin
cholesterol plasma levels and, 761
Lowry, Thomas Martin, 238
Lyases
examples of, 623

Major mineral(s) Dietary minerals needed by the body in amounts greater than 100 mg/day, 643
amounts of, human body, 644
biological functions of, 644
listing of, 644
Malate
citric acid cycle and, 701
Malonyl CoA
formation of, 756
lipogenesis and, 756–759
Maltose
hydrolysis of, 533–534
occurrence of, 533
structure of, 533
Marijuana
pharmacology of, 391
structure of, 391
Markovnikov, Vladmir Vasilevich, 347
Markovnikov's rule When an unsymmetrical molecule of the form H-Q adds to an unsymmetrical alkene, the hydrogen atom from the H-Q becomes attached to the unsaturated carbon atom that already has the most hydrogen atoms, 347–348
use of, 348–349
Mass A measure of the total quantity of matter in an object, 23–24
calculation of, using density, 38–39
conversion factors involving, 33
distinction between weight and, 24
metric units of, 23–24
Mass number(s) The sum of the numbers of protons and neutrons in an atom's nucleus, 52
informational value of, 52–53
neutrons and, 52–53
non-uniqueness of, 54
use of, with chemical symbols, 54–55
Mass-volume percent The mass of solute (in grams) divided by the total volume of solution (in milliliters), multiplied by 100, 188
calculations involving, 189
mathematical equation for, 188
Matter Anything that has mass and occupies space, 1
changes in, 4–5
chemistry as study of, 1–2
classification procedure for, 8
classifications of, 6–8
contrasted with nonmatter, 1–2
physical states of, 2–3
properties of, 3–4
Measurement(s) The determination of the dimensions, capacity, quantity, or extent of something, 21
precision of, 25
rules for recording of, 24–25
significant figure guidelines for, 25–26
systems for, 22
uncertainty associated with, 24–25
Meiosis
cell division and, 664
Melanin
suntan/sunburn and, 415
Melting
process of, 164
Membranes; *see* Plasma membranes
Mendeleev, Dmitri Ivanovich, 57
Menthol
properties of, 380
uses of, 380
Messenger RNA (mRNA) Carries genetic information (instructions for protein synthesis) from DNA to the ribosomes, 666
codons and, 669–671
efficiency of utilization of, 675–676
Meta
prefix, meaning of, 358
Metabolic pathway(s) A series of consecutive biochemical reactions used to convert a starting material into an end product, 689
cyclic, 689
linear, 689
Metabolism The sum total of all the chemical reactions that take place in a living organism, 688
carbohydrate, 715–738
glucose, terminology associated with, 733
glycerol, 746–747
lipid, 743–763
protein, 769–789
reaction subtypes in, 688–689
triacylglycerol, 747–751
Metal(s) An element that has the characteristic properties of luster, thermal conductivity, electrical conductivity, and malleability, 59
alkali, periodic table location of, 58, 72
alkaline earth, periodic table location of, 58, 72
fixed-charge, naming compounds and, 91–93
general properties of, 59–60, 72
periodic table locations for, 59–61
types of, naming compounds and, 91–93
variable-charge, naming compounds and, 91–92
Metal-foil experiments
discovery of nucleus and, 51–52
Meter The base unit of length in the metric system, 23
compared to English system units, 23
Methane
"greenhouse effect" of, 302
natural gas and, 316–317
occurrence of, 302
Methanol
properties of, 376–377
uses of, 376–377
Metric system, compared with English system, 23
prefixes, table of, 23
units of length, 23
units of mass, 23–24
units of volume, 24
Mevalonate
cholesterol biosynthesis and, 760
Mevastatin
cholesterol plasma levels and, 761
Meyer, Julius Lothar, 57
Micelle(s) A small, spherical grease-carboxylate ion droplet that is soluble in water as a result of the polar groups (carboxylate groups) on its surface, 565
fatty acid, lipid digestion and, 744
formation of, 565
Mineral oil, 319
Mineral(s) Inorganic compounds containing elements other than C, H, O, N, and S that are needed for the proper functioning of the human body, 643
major, 643–644
trace, 644–646
Mineralocorticoids
biological functions of, 576–577
synthesis of, from cholesterol, 762
Mirror image(s) An object's reflection in a mirror, 509
nonsuperimposability of, 509–511
superimposability of, 509–511
Mitochondria Organelles that have a central role in the production of energy, 689
innermembrane of, 689
outermembrane of, 689
substructure within, 689
Mitosis
cell division and, 664
Mixed triacylglycerol(s) A triester formed from the reaction of glycerol with more than one kind of fatty acid molecule, 561
structural formula for, 562
Mixture(s) A physical combination of two or more pure substances in which each substance retains its own chemical identity, 6
characteristics of, 6–9
comparison with compounds, 8
heterogeneous, 6–9
homogeneous, 6–9
separation of, by physical means, 6
types of, 6–9
Molar mass The mass in grams that is numerically equal to the substance's formula mass, 130
relationship to formula mass, 130–131
use of, in calculations, 131
Molarity A concentration unit giving the number of moles of solute per liter of solution, 190
calculations involving, 190–192
mathematical equation for, 190
Mole(s) 6.02×10^{23} objects; the amount of substance in a system that contains as many elementary particles (atoms, molecules, or formula units) as there are C atoms in exactly 12 grams of C, 129, 132
as a counting unit, 129–130
Avogadro's number and, 129
chemical equations and, 140
chemical formulas and, 132–133
general calculations involving, 133–135, 140–145
important relationships involving, summary of, 141
mass of, 130–132
number of objects in, calculation of, 129–130

Molecular collisions
elastic, 151
inelastic, 151
Molecular compound(s)
binary, naming of, 120–121
general properties of, 79–80
Lewis structures for, 103–105
polarity of, 119–120
Molecular mass, calculation of, 128
relationship to formula mass, 128
Molecular polarity, diatomic molecules and, 119
factors which determine, 119–120
tetraatomic molecules and, 119–120
triatomic molecules and, 119
Molecular shape, effects of nonbonding electron pairs on, 112–115
electron pair repulsions and, 111–112
odor theory and, 113
prediction of, using VSEPR theory, 111–115
role of central atom in determining, 111–115
Molecule(s) A group of two or more atoms that function as a unit because the atoms are tightly bound together, 13
characteristics of, 13–14
chiral, interactions between, 518–520
chirality of, 509–511
classification of, by number of atoms, 13–14
classification of, by types of atoms, 13–14
conformations for, 304
handedness in, notation for, 513–516
handedness in, recognition of, 509–511
heteroatomic, 13–14
homoatomic, 13
ionic solids, and, 14
limit of physical subdivision and, 14
nonpolar, 119–121
polar, 119–121
shape of, 111–116
Monoacylglycerol(s)
use of, as emulsifiers, 564
Monoatomic ion(s) Ion formed from a single atom that has lost or gained electrons and thus acquired a charge, 94
charges on, 85–86
formulas for compounds containing, 87–88
Monoprotic acid(s) An acid that transfers one H^+ ion (proton) per molecule during an acid-base reaction, 241
examples of, 241
Monosaccharide(s) Carbohydrates that contain a single polyhydroxy aldehyde or polyhydroxy ketone unit, 508
amino sugar formation and, 532
-aric acid formation from, 529
biologically important, 523–525
chemical reaction summary for, 539
classification of, by functional group, 520–523
classification of, by number of carbon atoms, 520–523
cyclic forms, formation of, 525–528
cyclic forms, Haworth projections and, 527–528
disaccharide formation from, 532–538
glycoside formation and, 530–531
hemiacetal forms of, 526–528
markers determining blood types, 530
-onic acid formation from, 528
oxidation of, 528–529
phosphate ester formation and, 531–532
reactions of, 528–532
reduction of, 529–530
-uronic acid formation from, 529
Monounsaturated fatty acid(s) Fatty acid with a carbon chain in which one carbon-carbon double bond is present, 556
common, listing of, 557
structural formula notation for, 556–557
Morphine
use of, as narcotic pain killer, 486
MTBE (methyl *tert*-butyl ether)
use of, as gasoline additive, 393–394
Mucopolysaccharides, 543–544
Mutagen(s) Substances or agents that cause a change in the structure of a DNA molecule, 676
selected types of, 676
Mutase
function of, 722
Mutation(s) Changes in the base sequence in DNA molecules, 676
DNA base sequence and, 676

NAD^+, *see* nicotinamide adenine dinucleotide
NADH; *see* nicotinamide adenine dinucleotide
Naming; *see* Nomenclature
Natural gas
chemical composition of, 316–317
Network polymer(s) A polymer in which monomers are connected in a three-dimensional cross-linked network, 425
formaldehyde-based, 425–426
Neurotransmitter(s) A chemical substance that is released at the end of a nerve, travels across the synaptic gap located between two nerves, and then bonds to a receptor site in the other nerve, triggering a nerve impulse, 482
amines as, 482
Neutral solution Solution in which the concentrations of H_3O^+ and OH^- ions are equal, 248
Neutralization The reaction between an acid and a hydroxide base to form a salt and water, 245
acid-base, equations for, 245–246
Neutron(s) Subatomic particles that have no charge associated with them; that is, they are neutral, 49
location of, within atom, 49
properties of, 49
role of in nuclear fission, 290–291
Niacin
biological functions of, 640
presence of, in NAD^+, 693
structural forms of, 640
Nicotinamide adenine dinucleotide (NAD^+/NADH)
citric acid cycle and, 698–702
electron transport chain and, 703–704
general structural characteristics of, 693–694
oxidized form of, 693–694
reduced form of, 694
Nicotine
pharmacology of, 483
structure of, 483
Nitration
of aromatic hydrocarbons, 360–361
Nitric oxide
biochemical messenger functions of, 109, 781
Lewis structure for, 109
properties of, 109
Nitrogen atom(s)
bonding characteristics of, 471–472
Nitrogen balance The state that results when nitrogen taken into the body (as protein) equals the amount of nitrogen excreted from the body, 771
negative, causes of, 771
positive, causes of, 771
Nitrogen-containing bases
as nucleotide subunits, 653–654
Nitroglycerin
structure of, 462
uses for, 462
Noble gas(es) Elements found in the far right column of the periodic table, 72
electronic characteristics of, 72
periodic table location of, 58, 71–72
stability of electron configurations of, 82–83
Nomenclature
alcohols, 373–375
aldehydes, 408–410
alkanes, 306–310
amides, 487
amine salts, 478
binary ionic compounds, 91–93
binary molecular compounds, 120–121
carboxylate ions, 447
carboxylic acid salts, 448
carboxylic acids, 437–442
cycloalkanes, 313–314
enzymes, 621–622
esters, 451–453
ethers, 392
glycosides, 531
halogenated alkanes, 322–323
inorganic, common name use in, 120–121
IUPAC rules and, 306
ketones, 410–413
nucleotides, 655–656
ortho-meta-para, use of, 358
phenols, 375–376
polyatomic-ion containing compounds, 96–97
rules summary, for ionic compounds, 96
substituted ammonium ions, 476–478
Nonbonding electrons Pairs of valence electrons that are not involved in electron sharing, 104
Noncompetitive enzyme inhibitor A molecule that decreases enzyme activity by binding to a site on an enzyme other than the active site, 630
mode of action, 630–631
Nonelectrolyte(s) A substance that forms a solution in water that does not conduct electricity, 262
characteristics of, 262
Nonmetal(s) An element characterized by the absence of the properties of luster, thermal conductivity, electrical conductivity, and malleability, 60
general properties of, 60–61, 72
periodic table locations for, 60–61
Nonoxidation-reduction (nonredox) reaction(s) A reaction in which there is no transfer of electrons from one reactant to another reactant, 211
Nonpolar amino acid(s) Amino acids that

contain one amino group, one carboxyl group, and a nonpolar side chain, 588
structures of, 589
Nonpolar covalent bond(s) A covalent bond in which there is equal sharing of electrons, 116
electronegativity differences and, 117
Nonpolar molecule
characteristics of, 119–120
Nonsaponifiable lipid(s) Lipids that do not undergo hydrolysis reactions in alkaline solution, 555
common types of, 573
eicosanoids as, 577
fat-soluble vitamins as, 572
pheromones as, 572
steroids as, 573
terpenes as, 572
Normal boiling point The temperature at which a liquid boils under a pressure of 760 mm Hg, 167
Nuclear chemistry
equations for reactions, 273–276
natural radioactive emissions and, 272–273
terminology associated with, 272
Nuclear energy
generation of, 290–291
Nuclear equation(s)
alpha-particle decay, 273–274
balancing of, 274–276
beta-particle decay, 274–276
fission, 290
fusion, 292
gamma-ray emission, 275
Nuclear fission The process in which a large nucleus (high atom number) splits into two smaller nuclei accompanied by the release of several free neutrons and a large amount of energy, 290
characteristics of, 290–291
nuclear power plants and, 290–292
uranium-235 and, 290–292
Nuclear fusion The process in which small nuclei are put together to make large ones, 292
occurrence on the sun, 292
Nuclear medicine
diagnostic uses, 287
PET scan, 289
therapeutic uses, 288–290
Nuclear power plants, 290–292
Nuclear reaction(s) A reaction in which changes occur in the nucleus of an atom, 271
bombardment, 279–280
contrasted with chemical reactions, 293
equations for, 273–276
fission and fusion, 290–292
Nuclear weapons, 291–292
Nucleic acid(s) Polymeric molecules in which the repeating unit is a nucleotide, 653
backbone of, 656
directionality of structure of, 657
nucleotides within, 653–656
primary structure of, 656–658
types of, 652–653
Nucleic acid bases
synthetic, use of in medicine, 662
Nucleons, 49
Nucleotide(s) A molecule composed of a pentose sugar bonded to both a phosphate group and a nitrogen-containing heterocyclic base, 653
formation of, 655
nitrogen-containing base subunit of, 653–654
nomenclature for, 655–656
pentose sugar subunit of, 653
phosphate subunit of, 654–655
structural subunits of, 653–656
Nucleus The very small, dense, positively charged center of an atom, 49
discovery of, 51–52
metal-foil experiments and, 51–52
protons and neutrons with, 49
size of, relative to whole atom, 49
stability of, 271–272
Nuclide(s) An atom of an element that has a specific number of protons and neutrons in its nucleus, 272
daughter, 273–276
parent, 273–276
radioactive, 272
stable, 272
unstable, 272
Nylon
structure of, 494–495

Octet rule In compound formation, atoms of elements lose, gain, or share electrons in such a way that their electron configurations become identical to that of the noble gas nearest them in the periodic table, 83
valence electron configurations and, 82–83
Odor response
relationship to molecular shape, 113
Oil(s) Triacylglycerol mixtures that are liquids at room temperature (25°), 561
chemical reactions of, 563–567
dietary considerations and, 563
general structure of, 560
hydrogenation of, 565–566
property-contrast with fats, 561–563
Okazaki, Rejii, 661
Okazaki fragments
DNA replication and, 661–662
Olefins, 333
Olestra, 563
Oligomeric protein(s), 604
Oligosaccharide(s) Carbohydrates that contain from two to ten monosaccharide units, 508
Omega-3 fatty acid(s) A polyunsaturated fatty acid with its endmost double bond three carbons away from its methyl end, 556
biological significance of, 556–558
Omega-6 fatty acid(s) A polyunsaturated fatty acid with its endmost double bond three carbons away from its methyl end, 556–557
biological significance of, 556–558
Optically active compound(s) A compound that rotates the plane of polarized light, 518
types of, 518
Optimum pH The pH at which an enzyme has maximum activity, 626
Optimum temperature The temperature at which an enzyme has maximum activity, 626
Orbital(s), *see* Electron orbital(s)
Orbital diagram(s) A statement of how many electrons an atom has in each of its orbitals, 65
interpretation of, 65–69
procedures for writing, 65–69
Organelle(s) A minute structure within the cell cytoplasm that carries out a specific cellular function, 689
types of, 689–690
Organic chemistry The study of hydrocarbons (compounds of hydrogen and carbon) and their derivatives, 299
contrasted with inorganic chemistry, 299
subdivisions within, 300
Ornithine
urea cycle and, 776–779
Ortho
prefix, meaning of, 358
Osmolarity The product of molarity and the number of particles produced per formula unit when a solute dissociates, 198
calculations involving, 200
mathematical equation for, 198
Osmosis The passage of a solvent from a dilute solution (or pure solvent) through a semipermeable membrane into a more concentration solution, 196
kinetic molecular theory and, 197
process of, description of, 196–197
relationship to dialysis, 201–202
Osmotic pressure The amount of pressure that must be applied to prevent the flow of solvent through a semipermeable membrane from a solution of lower solute concentration to a solution of higher solute concentration, 197
biological importance of, 198
factors affecting, 198
measurement of, 197–198
terminology associated with, 198–201
Oxaloacetate
amino acid degradation and, 782
citric acid cycle and, 698–699, 701
gluconeogenesis and, 731–732
Oxidation The process whereby a substance in a chemical reaction loses one or more electrons, 214–215
loss of electrons and, 215
oxidation number increase and, 214–215
Oxidation number(s) A number that represents the charge that an atom appears to have when the electrons in each bond it is participating in are assigned to the more electronegative of the two atoms involved in the bond, 211
assigning, example of, 212–214
rules for determining, 211–212
Oxidation-reduction (redox) reaction(s) A reaction in which there is a transfer of electrons from one reactant to another reactant, 211
terminology associated with, 214–216
Oxidative deamination The conversion of an amino acid into a keto acid with the release of ammonium ion, 775
ammonium ion production and, 775–776

Oxidative phosphorylation The process by which ATP is synthesized from ADP using energy released in the electron transport chain, 706
chemiosmotic coupling and, 707
electron transport chain and, 706
Oxidizing agent(s) Reactant that causes oxidation by accepting electrons from the other reactant, 215
identification of, 215–216
Oxidoreductases
examples of, 623
Oxygen atom(s)
bonding characteristics of, 371–372
Oxygen (O_2)
electron transport chain and, 705
hemoglobin complex with, 230
hemoglobin degradation and, 785
respiration and, 162
unsaturated fatty acid synthesis and, 758
Oxytocin, 598
Ozone
stratospheric concentrations of, 223
Ozone layer
chlorofluorocarbons and, 324

Pain relievers
acetaminophen, 489
aspirin, 455
narcotic, 486
propanoic acid-based, 442
Pantothenic acid
biological functions of, 640
presence of, in coenzyme A, 694
structure of, 640
Para
prefix, meaning of, 358
Parent nuclide The nuclide that undergoes decay in a radioactive decay process, 273
Partial pressure(s) The pressure that a gas in a mixture would exert if it were present alone under the same conditions, 161
importance of, respiratory gases and, 162
use of, in calculations, 162–163
Passive transport Movement of a substance across a membrane by diffusion from a region of higher concentration to a region of lower concentration without the expenditure of any cellular energy, 580
process of, characteristics for, 580
Pauling, Linus Carl, 115
Penicillins
mode of action, 635–636
Pentose phosphate pathway Uses glucose to produce NADPH, ribose 5-phosphate (a pentose), and numerous other sugar phosphates, 734
needs met by, 736
stages in, 734–736
Peptide(s) A sequence of up to 50 amino acids, in which the amino acids are joined together through amide (peptide) bonds, 595
backbone of, 596
biological functions of, 597–598
directionality of, 596
endorphins, 598
enkephalins, 598
isomeric forms of, 597
oxytocin, 598
peptide bonds with, 596
structural formulas of, 596–597
vasopressin, 598
Peptide bond(s) A bond between the carboxyl group of one amino acid and the amino group of another amino acid, 595
Percent by mass The mass of solute divided by the total mass of solution, multiplied by 100, 186
calculations involving, 187
mathematical equation for, 186
Percent by volume The volume of solute divided by the total volume of solution, multiplied by 100, 188
mathematical equation for, 188
Period(s) A horizontal row of elements in the periodic table, 58
in periodic table, notation for, 58
Periodic law When elements are arranged in order of increasing atomic number, elements with similar properties occur at periodic (regularly recurring) intervals, 57
discovery of, 57
electron configurations and, 69–70
graphical representation of, 57
Periodic table A graphical display of the elements in order of increasing atomic number in which elements with similar properties fall in the same column of the display, 57
atomic number sequence within, 59
classification systems for elements and, 71–73
distinguishing electrons and, 70–71
electron configurations and element location, 70–71
groups within, 58
information shown on, 57–58
long form of, 59
most common form of, 58
periods within, 58
shape of, rationale for, 70–71
specifying element position within, 58–59
Peroxides
from ethers, 395
PET [poly(ethylene terephthalate)]
condensation polymer, properties of, 460
PET scan (positron emission tomography scan)
nuclear chemistry of, 289
Petrolatum, 319
Petroleum
chemical composition of, 316–317
refining of, 317
pH
calculations involving, 250
effect on zwitterion structure, 591–593
enzyme activity and, 626
integral values, calculation of, 248–249
mathematical expression for, 248
measurement of, 251
nonintegral values, calculation of, 249–250
optimum, for enzymes, 626
relationship to hydronium and hydroxide concentration, 248–250
values of, for aqueous salt solutions, 252–256
values of, for selected common substances, 249–250
values of, significant figures and, 248
pH scale A scale of small numbers that is used to specify molar hydronium ion concentration in aqueous solution, 248
interpreting values on, 250–251
Phenol(s) A compound in which a hydroxyl group is attached to a carbon atom that is part of an aromatic carbon ring system, 372–373
acidity of, 390
antioxidant properties of, 390
naturally occurring, 391
nomenclature for, 375–376
properties of, 389
uses of, 389–390
Phenyl group
structure of, 357
Pheromone(s) Compounds used by insects (and some animals) to transmit messages to other members of the same species, 343
alkenes as, 343
esters as, 454
Phosphatase
function of, 730
Phosphate
as nucleotide subunit, 654–655
Phosphate ester(s) A compound formed by reaction of an alcohol with phosphoric acid, 461
examples of, 461–462
formation of, from cyclic monosaccharides, 531–532
Phosphatidylcholines, 567–569
Phosphatidylethanolamines, 569
Phosphatidylserines, 569
Phosphoacylglycerol(s) Triesters of glycerol in which two —OH groups are esterified with fatty acids and the third is esterified with phosphoric acid, which in turn is esterified to an alcohol, 567
biological functions of, 567
general structure of, 567
phosphatidylcholines, 567–569
phosphatidylethanolamines, 569
phosphatidylserines, 569
types of, 567
Phosphoenolpyruvate
gluconeogenesis and, 731
glycolysis and, 722
2-Phosphoglycerate
gluconeogenesis and, 731
glycolysis and, 721–722
3-Phosphoglycerate
gluconeogenesis and, 731
glycolysis and, 721–722
Phosphorylase
function of, 730
Phosphorylation
oxidative, 706–707
substrate-level, 721
Photosynthesis
carbohydrates and, 507
Physical
use of the term, 5
Physical change A process that does not alter the basic nature (chemical composition) of the substance undergoing change, 4
changes of state and, 4–5
characteristics of, 4–5
Physical property(ies) A characteristic of a substance that can be observed without changing the basic identity of the substance, 3
examples of, 5
Physical states
of matter, 2–3

Physical subdivision
limit of, 14
Pi bond(s) A covalent bond in which atomic orbital overlap occurs above and below (but not on) the internuclear axis, 337
orbital overlap and, 338
pK_a
calculation of, 251
definition of, 251
Plane-polarized light, 518–519
Plasma, 292
Plasma membrane(s) Separates the aqueous interior of a cell from the aqueous environment surrounding the cell, 578
bilayer structure of, 578–579
bonding interactions within, 579
carbohydrate components of, 579
protein components of, 579
transport across, mechanisms for, 580
Plasmid(s)
recombinant DNA production and, 678–681
Polar acidic amino acids Amino acid that contains one amino group and two carboxyl groups, the second carboxyl group being part of the side chain, 588
structures of, 589
Polar basic amino acids Amino acid that contains two amino groups and one carboxyl group, the second amino group being part of the side chain, 588
structures for, 589
Polar covalent bond(s) A covalent bond in which there is unequal sharing of electrons, 116
electronegativity differences and, 117
Polar molecules
characteristics of, 119–120
Polar neutral amino acids Amino acid that contains one amino group, one carboxyl group, and a side chain that is polar but neutral, 588
structures of, 589
Polarimeter
components of, 518
Polarity
bond, 116–120
molecular, 119–121
solubility of solutes and, 184–185
solubility of vitamins and, 186
Polyamide(s) A condensation polymer in which the monomers are linked together by amide linkages, 494
examples of, 494–495
Polyatomic ion(s) Ion formed from a group of atoms, held together by covalent bonds, that has acquired a charge, 94
common, listing of, 94
formula writing conventions for, 95
formulas for compounds containing, 95–97
Lewis structures for, 110
naming compounds containing, 96–97
Polyester(s) A condensation polymer in which the monomers are linked together by ester linkages, 459
examples of, 459–461
Polymer(s) A very large molecule composed of many identical repeating units, 350
addition, 350–352
condensation, 460–461
formaldehyde-based, 425–426
network, 425–426
polyamides, 494–495
polyesters, 459–461
polyurethanes, 495–496
Polymerase chain reaction A method for rapidly producing multiple copies of a DNA nucleotide sequence, 680
Polymerization reaction A reaction in which the repetitious combining of many small molecules produces a very large molecule, 350
Polyprotic acid(s) An acid that can transfer two or more H^+ ions (protons) during an acid-base reaction, 241
writing formulas of, 241–242
Polysaccharide(s) A carbohydrate made up of many monosaccharide units bonded to each other by glycosidic linkages, 508, 539
biologically important, 539–544
Polysome(s) A complex of mRNA and several ribosomes, 676
formation of, protein synthesis and, 675–676
Polyunsaturated fatty acid(s) A fatty acid with a carbon chain in which two or more carbon-carbon double bonds are present, 556
common, listing of, 557
structural formula notation for, 557
Position of equilibrium Specifies, in a qualitative way, the relative amounts of reactants and products present at equilibrium for a chemical reaction, 226
terminology associated with, 226–227
Potential energy Stored energy that matter possesses as a result of its position, condition, and/or composition, 151
cohesive forces and, 152–155
kinetic molecular theory of matter and, 151–152
types of, 151
Pressure(s) The force applied per unit area, that is, the total force on a surface divided by the area of that surface, 155
boiling point magnitude and, 167–168
measurement of, using barometer, 155
partial, 161–163
units for, 155
Primary alcohol An alcohol in which the hydroxyl-bearing carbon atom is attached to only one other carbon atom, 383–384
Primary amide(s), 486
Primary amine(s), 472–473
Primary carbon atom A carbon atom bonded to only one other carbon atom, 311
Primary nucleic acid structure The sequence of nucleotides in the nucleic acid, 657
general characteristics of, 657–658
Primary structure of a protein The sequence of amino acids present in its peptide chain or chains, 598
Primary transcript RNA (ptRNA) The material from which messenger RNA is made, 666
exons and introns and, 668–669
Products
in a chemical reaction, 136
Proenzyme(s) An inactive precursor of an enzyme, 633
Progestins
biological functions of, 575–576
synthesis of, from cholesterol, 762
Propanoic acid
derivatives, pain reliever use for, 442
Property(ies) The distinguishing characteristics of a substance that are used in its identification and description, 3
chemical, 3–4
colligative, 195–196
physical, 3
types of, 3–4
Propionate
derivatives of, transamination and, 773
Propylene glycol
properties of, 379
uses of, 379
Prostaglandin(s) 20-carbon fatty acid derivative that contains a cyclopentane ring and oxygen-containing functional groups, 577
biological functions of, 578
structure of, 577
Prosthetic group The non-amino acid portion of a conjugated protein, 608
Protein(s) A polymer in which the monomer units are amino acids; polypeptides that contain more than 50 amino acid units, 588, 598
amino acid building blocks of, 588–590
complete dietary, 590
components of plasma membranes, 579
conjugated, examples of, 607–608
conjugated, prosthetic groups and, 608
denaturation, examples of, 609
denaturation, human hair and, 610
digestion of, 769–770
fibrous, 605–607
general characteristics of, 587–588
globular, 605
hydrolysis of, 608
levels of structure for, 598
oligomeric, 604
primary structure of, 598–600
secondary structure, alpha helix, 600–601
secondary structure, beta-pleated sheet, 601
secondary structure, triple helix, 601–602
simple, 607–608
structure-summary for, 606
tertiary structure, disulfide bonds, 603
tertiary structure, electrostatic interactions, 603–604
tertiary structure, hydrogen bonding, 604
tertiary structure, hydrophobic attractions, 604
Protein denaturation The partial or complete disorganization of a protein's characteristic three-dimensional shape as a result of disruption of its secondary, tertiary, and quaternary structural interactions, 609
Protein metabolism
relationships between carbohydrate metabolism and, 787–789
relationships between lipid metabolism and, 787–789
Protein synthesis
inhibition of, antibiotics and, 676
overview of, 665–666
site for, ribosomal RNA and, 672
summary-diagram of, 677
transcription phase of, 667–669
translation phase of, 672–676

Protein turnover The repetitive process in which body proteins are degraded and resynthesized, 771
causes of, 771
Protium, 54
Proton(s) Subatomic particles that possess a positive (+) electrical charge, 49
canal rays and, 51
discharge experiments and, 51
discovery of, 51
location of, within atom, 49
nuclear charge and, 49
properties of, 49
Pure substance(s) A single kind of matter that cannot be separated into other kinds of matter by any physical means, 6
characteristics of, 6, 9
compounds as, 7–9
elements as, 7–9
heterogeneous samples of, 8
types of, 7–9
Purine
derivatives of, nucleotides and, 653–654
Pyridoxal phosphate
transamination and, 774–775
Pyrimidine
derivatives of, nucleotides and, 653–654
Pyruvate
amino acid degradation and, 782
Cori cycle and, 732–733
fates of, 724–726
gluconeogenesis and, 731–732
glycolysis and, 722
oxidation to acetyl CoA, 725
reduction to ethanol, 725–726
reduction to lactate, 725

Quaternary ammonium salt(s) An ammonium salt in which all four groups attached to the ammonium nitrogen atom are organic groups, 480
properties of, 480–481
Quaternary carbon atom A carbon atom bonded to four other carbon atoms, 311
Quaternary structure of a protein The associations among the separate chains in an oligomeric protein, 605
hydrophobic interactions and, 604–605
Quinine, 485

Radiation
biological effects of, 283–285
detection of, 285–286
exposure to, sources of, 286–287
free radical formation and, 282
ion-pair formation and, 282
ionizing effects of, 282
penetrating ability of, 283–284
use of, in medicine, 287–290
Radioactive decay The process whereby a radionuclide is transformed into a nuclide of another element as a result of the emission of radiation, 273
alpha particle emission, 273–274
beta particle emission, 274–276
equations for, 273–276
gamma ray emission, 275
modes of, 273–276
rate of, 276–279
Radioactive decay series A sequence of nuclear reactions beginning with a very long-lived radionuclide and ending with a stable nuclide of lower atomic number, 281
example of, 281
Radioactive nuclide(s) [radionuclide(s)] An atom with an unstable nucleus that spontaneously emits energy (radiation), 272
characteristics of, 272
modes of decay for, 273–276
parent and daughter, 273–276
use of, in medicine, 287–290
Radioactivity The radiation spontaneously emitted from the nucleus of an unstable nuclide, 272
discovery of, 272–273
nature of emissions, 272–273
Rancidity
antioxidants and, 566–567
fats and oils and, 566
hydrolytic, 566
oxidative, 566
Rate of a chemical reaction The rate at which reactants are consumed or products are produced in a given time period, 218
factors affecting, 218–221
Reactants
in a chemical reaction, 136
Reaction(s); *see* Chemical reactions
Reaction rate(s)
catalysts and, 220–221
concentration change and, 219–221
factors affecting, 218–221
physical nature of reactants and, 219–221
temperature change and, 220–221
Recombinant DNA DNA molecules that have been synthesized by splicing a segment of DNA (usually a gene) from one organism into the DNA of another organism, 678
clones and, 682
E coli bacteria use and, 678–679
plasmids and, 678–681
steps in formation of, 679
transformation process and, 682
Reducing agent(s) A reactant that causes reduction by providing electrons for the other reactant to accept, 215
identification of, 215–216
Reducing sugar(s) A carbohydrate that gives a positive test with Tollens, Fehling's and Benedict's solutions, 529
characteristics of, 529
Reduction The process whereby a substance in a chemical reaction gains one or more electrons, 215
gain of electrons and, 215
oxidation number decrease and, 215
Representative elements All of the elements of the s and p areas of the periodic table, 72
electronic characteristics of, 72
periodic table positions of, 71–72
Resonance structures Two or more Lewis structures for a molecule that differ only in the arrangement of bonding electrons, 355
Restriction enzyme(s) Enzymes that recognize specific base sequences in DNA and cleave the DNA in a predictable manner at these sequences, 680
recombinant DNA production and, 678–681
Reversible reaction A chemical reaction in which the products formed can react to yield the original reactants, 223–224
Riboflavin
biological functions of, 640
presence of, in FAD, 692
structure of, 640
Ribonucleic acid; *see* RNA
Ribose 5-phosphate
pentose phosphate pathway and, 735
Ribose
as nucleotide subunit, 653
occurrence of, 525
presence of, in NAD^+, 693
structure of, 525
Ribosomal RNA (rRNA) Combines with a series of proteins to form complex structures, called ribosomes, that serve as the physical sites for protein synthesis, 666
protein synthesis site and, 672
Ribosome(s)
sites for protein synthesis, 672
Ribulose 5-phosphate
pentose phosphate pathway and, 734–735
RNA (ribonucleic acid)
backbone of, structure for, 656
differences between DNA and, 666
formation of, transcription and, 667
types of molecules, 666
Rounding off The process of deleting unwanted (nonsignificant) digits from calculated numbers, 26
of numbers, rules for, 26
Rubbing alcohol, 378
Rutherford, Ernest, 51–52, 272, 279

Saccharin, 537
Salt(s) An ionic compound containing a metal or polyatomic ion as the positive ion and a nonmetal or polyatomic ion (except hydroxide) as the negative ion, 245
acid-base neutralization, formation of and, 245
dissociation of, 245
hydrolysis of, 252–256
Saponifiable lipid(s) A lipid that can be hydrolyzed under alkaline conditions to yield salts of fatty acids, 555
classification-summary for, 573
fatty acid building blocks for, 555–558
phosphoacylglycerols as, 567
sphingolipids as, 570
triacylglycerols as, 560–561
waxes as, 569
Saponification reaction(s)
amides, 492–493
esters, 457–458
triacylglycerols, 564–565
Saturated fatty acid(s) Fatty acids that have a carbon chain in which all carbon-carbon bonds are single bonds, 555
common, listing of, 557
structural formula notation for, 555
Saturated hydrocarbon(s) A hydrocarbon in which all carbon-carbon bonds are single bonds, 300
alkanes, 300–312
cycloalkanes, 312–315
Saturated solution(s) A solution that contains the maximum amount of solute that

can be dissolved under the conditions at which the solution exists, 181
characteristics of, 181–183
Scientific notation A system in which an ordinary decimal number is expressed as the product of a number between 1 and 10 times 10 raised to a power, 29
converting from, to decimal notation, 30
division in, 30–32
multiplication in, 30–32
significant figures and, 30
writing numbers in, 29–30
Secondary alcohol An alcohol in which the hydroxyl-bearing carbon atom is attached to two other carbon atoms, 384
Secondary amide(s), 486
Secondary amine(s), 472–473
Secondary carbon atom A carbon atom bonded to two other carbon atoms, 311
Secondary structure of a protein The arrangement in space of the atoms in the backbone of a protein, 600
interactions responsible for, 600–602
types of, 600–602
Selenium
biological function of, 646
Semipermeable membrane A thin-layered material that allows certain types of molecules to pass through but prohibits the pass of others, 196
Shell(s); *see* Electron shell(s)
Shuttle system(s)
carnitine/acyl carnitine, 747–748
dihydroxyacetone/glycerol 3-phosphate, 727
urea cycle, 777
Sigma bond(s) A covalent bond in which atomic orbital overlap occurs along the axis joining the two bonded atoms, 337
orbital overlap and, 338
Significant figure(s) The digits in any measurement that are known with certainty plus one digit that is uncertain, 25
exact numbers and, 28
guidelines for determining number of, 25–26, 29
in addition and subtraction, 27–28
in multiplication and division, 27–28
logarithms and, 248
mathematical operations and, 26–28
rounding off, to specified number of, 26
scientific notation and, 30–32
Simple enzyme(s) Enzymes composed only of protein (amino acid chains), 622
Simple protein(s) Proteins made up solely of amino acid residues, 607
Simple triacylglycerol(s) A triacylglycerol in which all three fatty acid residues are the same, 560
structural formula for, 562
Simplesse, 563
Single covalent bond(s) A bond in which two atoms share one pair of electrons, 104
molecules containing, examples of, 103–104
strength of, 104
Single-replacement reaction(s) A reaction in which an atom or molecule replaces an atom or group of atoms from a compound, 209
examples of, 209–210
Smog
formation of, equations for, 210
health effects of, 210
Soap
cleansing action of, 565
making of, 565
Sodium chloride
food additive use of, 89
health effects and, 89
Sodium cyclamate, 537
Solid(s) Matter that has a definite shape and a definite volume, 2
distinguishing characteristics of, 2–3, 150–151
kinetic molecular theory of matter applied to, 152–153
melting of, 164
sublimation of, 164
Solid state A state characterized by a dominance of potential energy (cohesive forces) over kinetic energy (disruptive forces), 152
Solubility(ies) The amount of solute that will dissolve in a given amount of solvent, 181
controlled-release drugs and, 192
gases, factors affecting, 182
rules for, 184–185
temperature and, for selected solutes, 181
terminology associated with, 182–184
values of, for selected solutes, 181
Solubility rules
for solutes, 184–185
Solute(s) A solution component that is present in a small amount relative to that of the solvent, 180
ionic, dissolving process for, 183–184
solubility rules for, 184–185
Solution(s) A homogeneous combination of two or more substances in which each substance retains its own chemical identity, 179
acidic, 247–248
aqueous, 183
basic, 247–248
colligative properties of, 195–201
concentrated, 183
concentration units for, 186–193
dilute, 183
dilution of, 193–194
formation of, ionic solutes and, 183–184
general characteristics of, 180
hypertonic, 198–201
hypotonic, 198–201
isotonic, 198–201
neutral, 247–248
pH of, 248–251
saturated, 181–183
solubility rules for solutes, 184–185
types of, 180–181
unsaturated, 183
Solvent(s) The component of a solution that is present in the greatest amount, 180
general characteristics of, 180–181
Specific heat The quantity of heat energy, in calories, that is necessary to raise the temperature of 1 gram of the substance by 1 degree Celsius, 41
temperature change and, 41–42
use of, in calculations, 42–43
values for, table of, 41
Sphingolipid(s) A saponifiable lipid derived from the amino dialcohol sphingosine, 570
biological functions of, 571
cerebrosides, 571
gangliosides, 571
general structure for, 570
sphingomyelins, 571
Sphingomyelins, 571
Sphingosine
structure of, 570
Squalene
cholesterol biosynthesis and, 760
Stable nuclide Has a nucleus that does not easily undergo change, 272
Standard amino acid(s) One of the 20 alpha-amino acids normally found in proteins, 588
name abbreviations for, 589
structures of, 589
Starch
amylopectin form of, 540–542
amylose form of, 540–542
animal, 542
iodine test and, 540–541
properties of, 540–542
structure of, 540–541
State of equilibrium A situation in which two opposite processes take place at equal rates, 166
characteristics of, 166
States of matter
compressibility and, 150–151
gaseous, 3
kinetic molecular theory and, 152–155
liquid, 2–3
plasma, 292
property differences among, 150–151
solid, 2
thermal expansion and, 150–151
Statins
cholesterol plasma levels and, 761
Stereoisomer(s) Isomers whose atoms are connected in the same way but differ in their arrangement in space, 511
conditions necessary for, 511–512
diastereomers as, 511–512
enantiomers as, 511–512
types of, 511–512
Steroid(s) Lipids with structures based on a fused-ring system involving three six-membered rings and one five-membered ring, 573
athletes and, 576
bile salts as, 574
cholesterol as, 573–574
common structural feature of, 573
hormones as, 574–577
synthetic, 575–576
Strained bond(s)
energy considerations and, 695–696
Strong acid(s) A substance that transfers 100%, or very nearly 100%, of its protons to water, 242
commonly encountered, 242
Strong base(s)
commonly encountered, 243
Strong electrolyte(s) A substance that completely (or almost completely) dissociates into ions in solution, 262
characteristics of, 262

Structural formulas
condensed, 301–302
expanded, 301
generation of, from compound names, 310
line-angle drawings for, 313–314
Structural isomers Compounds with the same molecular formula but different structural formulas, that is, different connections between atoms, 303
alkanes, number possible, 304
Subatomic particle(s) Very small particles that are the building blocks from which atoms are made, 48
arrangement of, within atom, 49
evidence for existence of, 50–52
properties, table of, 49
types of, 48–49
Sublimation, 164
Subshell(s); *see* Electron subshell(s)
Substance, 6
Substituent(s) An atom or group of atoms attached to a chain (or ring) of carbon atoms, 307
alkyl, 307
Substituted ammonium ion(s) An ammonium ion in which one or more alkyl, cycloalkyl, or aryl groups have been substituted for hydrogen atoms, 476
generalizations concerning, 476–477
nomenclature for, 476–478
Substitution reaction(s)
for benzene, 360–361
Substrate(s) The reactant in an enzyme catalyzed reaction, 621
Substrate-level phosphorylation The direct transfer of a high-energy phosphate group from an intermediate compound (substrate) to an ADP molecule to produce ATP, 721
glycolysis and, 721
Succinate
citric acid cycle and, 700–701
derivatives of, transamination and, 773
Succinyl CoA
amino acid degradation and, 782
citric acid cycle and, 700–701
Sucrose
derivatives of, and flatulence, 538
hydrolysis of, 536–537
occurrence of, 536
structure of, 536
Sugar(s) A general designation for either a monosaccharide or disaccharide, 521
amino, 532
artificial sweeteners and, 537
blood, 524
brain, 524
fruit, 524
invert, 536–537
malt, 533
milk, 536
reducing, 529
sweetness scale for, 537
table, 536
Sulfa drugs
mode of action, 634–635
Sulfhydryl group
in thiols, 396
Sulfide(s)
nomenclature of, 397
properties of, 397
Sulfonation
of aromatic hydrocarbons, 361
Sulfoxides
synthesis of, from sulfides, 427
Sulfuric acid
properties of, 141
uses of, 141
Symmetrical addition reaction(s) Identical atoms (or groups of atoms) are added to each carbon of a multiple bond, 346
halogenation of alkenes, 346
hydrogenation of alkenes, 346
Synthetic elements, 279–280

Temperature
absolute zero, Charles's law and, 158
enzyme activity and, 626
lowest possible value of, 39
normal, for human body, 42
optimum, for enzymes, 626
scales for measuring, 39–40
solubility of solutes and, 181–183
vapor pressure magnitude and, 166–167
Temperature Scales
Celsius, characteristics of, 39–40
conversions between, 40–41
Fahrenheit, characteristics of, 39–40
Kelvin, characteristics of, 39–40
Terpene(s) Compounds whose carbon skeleton can be divided into two or more units that are identical to the carbon skeleton of isoprene, 343
alkene, 343
isoprene units and, 343–344
selected examples of, 343–344
Tertiary alcohol An alcohol in which the hydroxyl-bearing carbon atom is attached to three other carbon atoms, 384
Tertiary amide(s), 486
Tertiary amine(s), 472–473
Tertiary carbon atom A carbon atom bonded to three other carbon atoms, 311
Tertiary structure of a protein The overall three-dimensional shape that results from the attractive forces between amino acid side chains (R groups) that are widely separated from each other within the chain, 602
interactions responsible for, 602–604
types of, 602–604
Thermal expansion
states of matter and, 150–151
Thiamin
biological functions of, 640
structure of, 640
Thiocarbonyl group, 427
Thioester(s) A sulfur-containing analog of an ester in which an —SR group has replaced the —OR group, 458
examples of, 458–459
Thioether(s)
nomenclature of, 397
properties of, 397
Thiol(s)
nomenclature of, 396
oxidation of, 397
properties of, 396–397
synthesis of, from disulfides, 397
Thomson, John Joseph, 50
Thromboxane(s) 20-carbon fatty acid derivatives that contain a cyclic ether and oxygen-containing functional groups, 578
biological functions of, 578
structure of, 577–578
Thymine
as nucleotide subunit, 654
Titration(s)
acid-base, 264–265
Tobacco
radioactivity associated with, 281
Tollens test
aldehyde oxidation and, 419
polysaccharides and, 539
reducing sugars and, 529
Torr
pressure unit of, 155
Torricelli, Evangelista, 155
Trace elements
importance of, in human body, 62
Trace mineral(s) Dietary minerals needed by the body in amounts less than 20 mg/day, 643
categories of, 643–644
recommended dietary amounts of, 644
***Trans* isomer** An isomer in which two atoms or groups are on different sides of a restricted rotation "barrier" in a molecule, 315
trans-Enoyl CoA
fatty acid spiral and, 748
Transamination The interchange of the amino group of an amino acid with the keto group of an alpha-keto acid, 774
examples of, 774
Transcription The process by which DNA directs the synthesis of RNA molecules that carry the coded information needed for protein synthesis, 666
base-pairing associated with, 667–668
steps in, 667
Transfer RNA (tRNA) Delivers specific individual amino acids to the ribosomes, the sites of protein synthesis, 666
activation of, protein synthesis and, 672–673
amino acids and, 671–672
anticodons and, 671–672
general shape of, 671
Transferases
examples of, 623
Transformation The process of incorporating foreign DNA into a host cell, 682
recombinant DNA production and, 682
Transition elements All of the elements of the d area of the periodic table, 73
electronic characteristics of, 73
periodic table positions of, 71–72
Translation The process by which the codes within RNA molecules are deciphered and a particular protein molecule is formed, 666
elongation phase of, 675
general steps of, 672–676
initiation of, 673
post-translation processing and, 675
summary-diagram for, 674
termination of, 675
Transmutation process A nuclear reaction in which a nuclide of one element is changed into a nuclide of another element, 279
types of, 279
Triacylglycerol(s) A compound formed by esterification of three fatty acid molecules to glycerol, 560
chemical reactions of, 563–567

dietary considerations and, 563
digestion of, 743–745
fatty acid composition of, 562
hydrogenation of, 565–566
hydrolysis of, 564
mixed, 561
mobilization of, 746
occurrence of, 561
rancidity of, 566
saponification of, 564–565
simple, 560–561
storage of, adipose cells and, 745–746
structure of, 561–562
types of, 560–561

Triacylglycerol mobilization The hydrolysis of triacylglycerols stored in adipose tissue, followed by release of the fatty acids and glycerol so produced into the blood stream, 746

Triple covalent bond(s) A bond in which two atoms share three pairs of electrons, 104
molecules containing, examples of, 104–105
strength of, 104

Triprotic acid(s) An acid that can transfer three H^+ ions (three protons) per molecule during an acid-base reaction, 241
examples of, 241

Tritium, 54

Turnover number The number of substrate molecules transformed per second by one molecule of enzyme under optimum conditions of temperature, pH, and saturation, 628

UDP-glucose
glycogenesis and, 729

Units
conversion factors between, table of, 33
English and metric compared, 23
heat energy, 41
mathematical operations and, 34
metric, of length, 23
metric, of mass, 23–24
metric, of volume, 24

Universe
elemental composition of, 10

Unsaturated fatty acids
fatty acid spiral and, 750–751
lipogenesis and, 758–759

Unsaturated hydrocarbon(s) A hydrocarbon that contains one or more carbon-carbon multiple bonds: double bonds, triple bonds, or both, 300
alkenes, 333–352
alkynes, 352–354
aromatic hydrocarbons, 354–362

Unsaturated solution(s) A solution where less solute than the maximum amount possible is dissolved in the solution, 183

Unstable nuclide Has a nucleus that spontaneously undergoes change, 272

Unsymmetrical addition reaction(s) Different atoms (or groups of atoms) are added to the carbon atoms of a multiple bond, 346
hydration of alkenes, 347
hydrohalogenation of alkenes, 347

Uracil
as nucleotide subunit, 654

Urea cycle A cyclic biochemical pathway that produces urea from ammonium ions and carbon dioxide, 776
linkage to citric acid cycle, 779–780
steps in, 777–779

Urea
physical properties of, 776
production of, urea cycle and, 777–779
properties of, 488

Urine
chemical composition of, 780

Vaccine(s) Substance that contains inactive or slightly altered forms of viruses or bacteria, 678
relationship to viruses, 678

Valence electron(s) Electrons in the outermost electron shell, which is the shell with the highest shell number, 80
determining number of, in atoms, 81
generalizations concerning, 82
Lewis structures and, 80–82
number of covalent bonds formed and, 105–106
octet rule and, 82–83

Vapor Gaseous molecules of a substance at a temperature and pressure at which it would ordinarily be thought of as a liquid or solid, 165

Vapor pressure(s) Pressure exerted by a vapor above a liquid when the liquid and vapor are in equilibrium, 166
factors affecting magnitude of, 166
lowering of, 195–196
table of, for water, 167

Vasopressin, 598

Virus(es) Tiny disease-causing agents that are composed of an outer protein coat and an inner nucleic acid core, 678
mode of operation of, 678
nucleic acid content of, 678
vaccines and, 678

Vision
cis-trans isomers and, 342

Vitamin(s) Organic compounds that must be obtained from dietary sources and that are essential in trace amounts for the proper functioning of the human body, 636
fat-soluble, 639–642
general characteristics of, 636–637
solubility-polarity relationships for, 186
water-soluble, 638–640

Vitamin A
β-carotene and, 639
biological functions of, 639
structural forms of, 639

Vitamin B_6
biological functions of, 640
derivative of, transamination and, 774–775
structural forms of, 640

Vitamin B_{12}
biological functions of, 640
structure of, 640

Vitamin C
biological functions of, 638
structural forms of, 638

Vitamin D
biological functions of, 641–642
structural forms of, 641

Vitamin E
biological functions of, 642
structure of, 642

Vitamin K
biological functions of, 642
structural forms of, 642

VLDL (Very low density lipoprotein)
biological functions of, 614

Volatile substance A substance that readily evaporates at room temperature because of a high vapor pressure, 166

Volume
calculation of, using density, 38–39
conversion factors involving, 33
metric units for, 24

VSEPR theory Valence shell electron pair repulsion theory; a set of procedures for predicting the three-dimensional shape of a molecule from information contained in the molecule's Lewis structure, 111
angular atomic arrangements in, 112–113
linear atomic arrangements in, 112
steps involved in applying, 112
tetrahedral atomic arrangements in, 113–114
trigonal planar atomic arrangements in, 112
trigonal pyramidal atomic arrangements in, 113–114
use of, in determining molecular shape, 111–116

Water
fresh, 85
hard, 85
ion-product for, 246
sea, 85
self-ionization of, 246
soft, 85

Watson, James, 658

Wax(es) A monoester formed from the reaction of a long-chain monohydroxy alcohol with a fatty acid molecule, 569
biological functions of, 569
general structure of, 569

Weak acid(s) A substance that transfers only a small percentage of its protons to water, 242
extent of proton transfer for, 242

Weak base(s)
ammonia as a, 243

Weak electrolyte(s) A substance that only partially ionizes into ions in solution, 262
characteristics of, 262

Weight A measure of the force exerted on an object by the pull of gravity, 24
distinction between mass and, 24

Wood alcohol, 376

Zaitsev, Alexander, 385

Zaitsev's rule The major product in an intramolecular alcohol dehydration reaction is the alkene that has the greater number of alkyl groups attached to the carbon atoms of the double bond, 385
use of, 385–386

Zinc
biological function of, 646

Zwitterion(s) A molecule that has a positive charge on one atom and a negative charge on another atom, 591
amino acids, guidelines for determining structural form, 593
amino acids, structure change with pH, 591–593

Zymogen(s) An inactive precursor of an enzyme, 633

Photo Credits

Photographs by Sean Brady: Pages i (top), v, xiii, xv, 2 (bottom), 7 (left), 9 (silver, gold alloy, and granite), 10, 22 (right), 23 (center and right), 26, 37 (two photos), 39, 59 (two photos), 62, 91, 111, 113, 128, 129 (two photos), 131, 138 (two photos), 152 (two photos), 159 (three photos), 167, 180 (top left and right), 182 (bottom), 186, 188 (two photos), 189, 193 (four photos), 194, 195 (two photos), 212 (four photos), 217, 220 (two photos), 229, 238, 245, 251 (two photos), 263 (three photos), 264 (three photos), 346 (two photos), 350, 377 (three photos), 419 (four photos), 426 (three photos), 445 (top), 495 (top), 509, 520, 529, 541 (two photos)

Photographs from the Edgar Fahs Smith Collection, Van Pelt-Dietrich Library, University of Pennsylvania: Pages 57, 82, 130, 156, 158, 227, 347, 355, 513

Computer-generated models by Frank Cox: Pages iii, 94 (two models), 105 (three models), 107, 108, 109 (two models), 114 (three models), 121, 132

Computer-generated models by James P. Birk and Kara M. Birk: Pages 301 (six models), 303 (three models), 322 (four models), 333 (four models), 341 (four models), 352 (three models), 353 (two models), 357 (two models), 359, 372 (three models), 376, 379, 380, 381 (two models), 394 (four models), 397 (two models), 408 (two models), 410, 423 (three models), 439 (four models), 441, 450 (two models), 455, 460, 472 (four models), 477, 478, 480 (two models), 482, 483, 488 (three models), 489, 527 (six models), 536, 560 (four models), 562 (two models), 573, 654

Credits for all other photographs are listed below, with page numbers in boldface.

i (middle and bottom) © 2000 PhotoDisc Inc.; **vii** © 2000 PhotoDisc Inc.; **x** © 2000 PhotoDisc Inc.; **xvi** © 2000 Photo Disc Inc.; **1** Lou Jacobs Jr./Grant Heilman Photography; **2 (top)** Phil Degginger/Color-Pic; **3 (top)** Stone/Ed Simpson; **(bottom)** Andy Levin/Photo Researchers; **4** Phil Degginger/Color-Pic; **5** Stone/Andy Sacks; **6 (two photos)** James Scherer; **7 (middle and right)** Ken O'Donoghue; **9 (emerald)** Courtesy of Smithsonian Institution; **13** Digital Instruments; **21** Dan McCoy/Rainbow; **22 (left)** Stone/Don and Pat Valenti; **23 (left)** E.R. Degginger; **38** Jeffrey Gage/Michael Pollack/Center for Exercise Science, University of Florida; **41 (left)** Stone/David Olsen; **(right)** Grant Heilman/Grant Heilman Photography; **43** AP/Wide World Photos; **48** Novastock/PhotoEdit; **61** Stone/Joseph Nettis; **67** Stone/William S. Helsel; **79** M.S. Davidson/Photo Researchers; **86** Blair Seitz/Photo Researchers; **89** E.R. Degginger; **90 (four photos)** E.R. Degginger; **102** NASA; **114 (bottom)** Archive Photos; **127** Stone/Ralph H. Wetmore II; **142** Georg Gerster/Photo Researchers; **145** Courtesy of Chrysler Corporation; **150** Stone/Bruce Forster; **151** Betty Weiser/Photo Researchers; **154** Phil Degginger, Color Pic; **161** Courtesy Manchester Literary and Philosophical Society; **165 (two photos)** James Scherer; **168** Brian Bailey/Photographers Aspen; **179** Steve Allen/Peter Arnold, Inc.; **180 (bottom)** Coco McCoy/Rainbow; **182 (top)** David Pilosof/Phototake NYC; **184** Stone/David Woodfall; **198** John Mead/Science Photo Library/Photo Researchers; **200 (left and center)** David M. Phillips/Visuals Unlimited; **(right)** Stanley Flegler/Visuals Unlimited; **208** Science Photo Library/Photo Researchers; **209** James Scherer; **211** James Scherer; **213** James Scherer; **219 (top left)** Stone/Vince Streano; **(top right)** Cecile Brunswick/Peter Arnold, Inc.; **(bottom left)** S.C. Fried/Photo Researchers; **(bottom right)** Myrleen Ferguson/PhotoEdit; **222** Stone/Greg Mancuso; **223** Tom McHugh/Photo Researchers; **226** NASA; **237** Dr. Paul A. Zahl/Photo Researchers; **239** Ken O'Donoghue; **241** Ken O'Donoghue; **259 (two photos)** Ken O'Donoghue; **271** © 2000 PhotoDisc Inc.; **272** © Bettmann/CORBIS; **279 (top)** © Bettmann/CORBIS; **(bottom)** Myrleen Ferguson/PhotoEdit; **285** Doug Plummer/Photo Researchers; **286** Charles D. Winters/Photo Researchers; **288** Simon Fraser/Science Photo Library/Custom Medical Stock Photo; **289 (top)** Yoav Levy/Phototake; **(bottom)** Dr. Mony De Leon/Peter Arnold, Inc.; **290** Albert J. Copley/Visuals Unlimited; **291** © Bettmann/CORBIS; **292** NASA; **298** Stone/Arnuls Husmo; **302** Stone/Natalie Fobes; **317** Richard Megna/Fundamental Photos; **318** Daryl Solomon/Envision; **319** Michael Newman/PhotoEdit; **320** Phil Degginger/Color-Pic; **332** Junebug Clark/Photo Researchers; **337** Alan & Linda Detrick/Photo Researchers; **343 (top)** Donald C. Booth/Color-Pic; **(bottom)** Martha Cooper/Peter Arnold, Inc.; **345** © 2000 PhotoDisc Inc.; **351** James Scherer; **352** Bill Stanton/Rainbow; **371** Ed Reschke/Peter Arnold, Inc.; **376** Gary Gold/Liaison; **379** Hank Morgan/Rainbow; **380** Stone/James Cotier; **387** D. & I. MacDonald/Envision; **390** Henryk T. Kaiser/Envision; **391** Michael Viard/Peter Arnold, Inc.; **393** SIU/Peter Arnold, Inc.; **397** Jeff Lapore/Photo Researchers; **406** © 2000 PhotoDisc Inc.; **412** Harvey Lloyd/Peter Arnold, Inc.; **413** Norbert Wu; **414** Steven Needham/Envision; **415** S. McCutcheon/Visuals Unlimited; **427** Coco McCoy/Rainbow; **436** Hans Pfletschinger/Peter Arnold, Inc.; **441** George Mattei/Envision; **445 (bottom)** F. Stewart Westmoreland/Photo Researchers; **446** AP/Wide World Photos; **449** Sherman Thomson/Visuals Unlimited; **454** Fred Whitehead/Animals Animals; **471** Stone/J.F. Causse; **485** Nardin/Jacana/Photo Researchers; **486** Scott Camazine/Photo Researchers; **495 (bottom)** Peter Skinner/Photo Researchers; **496** Dan McCoy/Rainbow; **506** Norris Blake/Visuals Unlimited; **507** Bobby Long/The Stock Shop; **524** Tom Raymond/Medichrome; **527** © Hulton-Deutsch Collection/CORBIS; **531** Larry Mulvehill/Rainbow; **535** Erica Stone/Peter Arnold, Inc.; **538** Paul Skelcher/Rainbow; **540** Steven Needham/Envision; **542** Dr. Dennis Kunkel/Phototake NYC; **543** Don Kreuter/Rainbow; **544** J.M. Barey/Vandystadt/Photo Researchers; **554** Dan Guravich/Photo Researchers; **556** IFA/Peter Arnold, Inc.; **563** Manfred Kage/Peter Arnold, Inc.; **569** Kevin Schaefer/Peter Arnold, Inc.; **572 (far left)** Phil Degginger/Color-Pic; **(second from left)** Don Fawcett/Photo Researchers; **(second from right)** Hans Pfletschinger/Peter Arnold, Inc.; **(far right)** Quest/Science Photo Library/Photo Researchers; **573** C. James Webb/Phototake NYC; **574** Howard Sochurek Medichrome; **577** Stu Forster/Allsport; **578** Dr. Andrew Staehelin/Peter Arnold, Inc.; **587** F.H. Kolwicz/Visuals Unlimited; **603** Professor P.M. Motta & E. Vizza/Science Photo Library/Photo Researchers; **607** © 2000 PhotoDisc Inc.; **609** E.R. Degginger; **610** Rosenfield Images, Ltd./Science Photo Library/Photo Researchers; **620** Steven Needham/Envision; **621** Curzon Studio/Envision; **627 (two photos)** Leonard Lessin/Peter Arnold, Inc.; **636** Saturn Stills/Science Photo Library/Photo Researchers; **638** Jeff Greenberg/Peter Arnold, Inc.; **641** Melissa Grimes/Guy/Photo Researchers; **652** Dr. Nikas/Jason Burns/Phototake NYC; **663** Erika Stone/Peter Arnold, Inc.; **678** NIBSC/Science Photo Library/Photo Researchers; **680** Hank Morgan/Rainbow; **688** Luiz C. Marigo/Peter Arnold, Inc.; **690** R. Bhatnagar/Visuals Unlimited; **709** Lynn Rogers/Peter Arnold, Inc.; **715** Skip Moody/Rainbow; **716** Ed Reschke/Peter Arnold, Inc.; **726** Dan McCoy/Rainbow; **737** Saturn Stills/Science Photo Library/Photo Researchers; **743** Howard Sochurek/Medichrome; **744** Peter Johansky/Envision; **752** Adam G. Sylvester/Photo Researchers; **769** Professor P. Motta/Science Photo Library/Photo Researchers; **784** Leonard Lessin/Peter Arnold, Inc.

Common Functional Groups

Name of class	Structural feature
Alkane	$-\overset{\vert}{\underset{\vert}{C}}-$
Alkene	$>C=C<$
Alkyne	$-C \equiv C-$
Aromatic hydrocarbon	(benzene ring, Kekulé structure) or (benzene ring with inscribed circle)
Alcohol	$-\overset{\vert}{\underset{\vert}{C}}-OH$
Phenol	(benzene ring)–OH
Ether	$-\overset{\vert}{\underset{\vert}{C}}-O-\overset{\vert}{\underset{\vert}{C}}-$
Thiol	$-\overset{\vert}{\underset{\vert}{C}}-SH$
Aldehyde	$-\overset{O}{\overset{\Vert}{C}}-H$ (—CHO)
Ketone	$-\overset{\vert}{\underset{\vert}{C}}-\overset{O}{\overset{\Vert}{C}}-\overset{\vert}{\underset{\vert}{C}}-$
Carboxylic acid	$-\overset{O}{\overset{\Vert}{C}}-OH$ (—COOH or $-CO_2H$)
Ester	$-\overset{O}{\overset{\Vert}{C}}-O-\overset{\vert}{\underset{\vert}{C}}-$ (—COOR or $-CO_2R$)
Amine	$-\overset{\vert}{\underset{\vert}{C}}-NH_2$
Amide	$-\overset{O}{\overset{\Vert}{C}}-NH_2$